AF598459

ENCYCLOPEDIA OF

MOLECULAR BIOLOGY

VOLUME 1

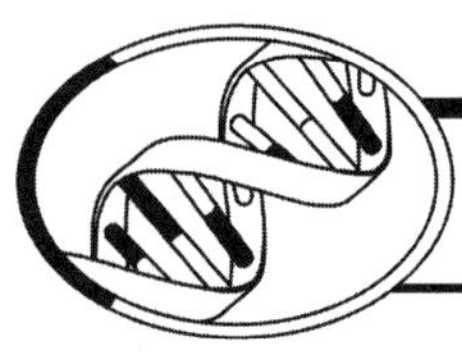

WILEY BIOTECHNOLOGY ENCYCLOPEDIAS

Encyclopedia of Bioprocess Technology: Fermentation, Biocatalysis, and Bioseparation
Edited by Michael C. Flickinger and Stephen W. Drew

Encyclopedia of Molecular Biology
Edited by Thomas E. Creighton

Encyclopedia of Cell Technology
Edited by Raymond E. Spier

Encyclopedia of Ethical, Legal, and Policy Issues in Biotechnology
Edited by Thomas J. Murray and Maxwell J. Mehlman

ENCYCLOPEDIA OF MOLECULAR BIOLOGY

ENCYCLOPEDIA OF

MOLECULAR BIOLOGY

VOLUME 1

Thomas E. Creighton

European Molecular Biology Laboratory
London, England

A Wiley-Interscience Publication

John Wiley & Sons, Inc.

New York / Chichester / Weinheim / Brisbane / Singapore / Toronto

This book is printed on acid-free paper. ♾

For ordering and customer service, call 1-800-CALL-WILEY.

Library of Congress Cataloging-in-Publication Data:

Creighton, Thomas E., 1940–
The encyclopedia of molecular biology / Thomas E. Creighton.
p. cm.
Includes index.
ISBN 0-471-15302-8 (alk. paper)
1. Molecular biology–Encyclopedias. I. Title.
QH506.C74 1999 99-11575
572.8′03—dc21 CIP

Printed in the United States of America.

10 9 8 7 6 5 4 3 2 1

PREFACE

The Wiley Biotechnology Encyclopedias, composed of the *Encyclopedia of Molecular Biology*, the *Encyclopedia of Bioprocess Technology: Fermentation, Biocatalysis, and Bioseparation*, the *Encyclopedia of Cell Technology*, and the *Encyclopedia of Ethical, Legal, and Policy Issues in Biotechnology*, cover very broadly four major contemporary themes in biotechnology. The series comes at a fascinating time in that as we move into the twenty-first century, the discipline of biotechnology is undergoing striking paradigm changes.

Biotechnology is now beginning to be viewed as an informational science. In a simplistic sense, there are three types of biological information. First, there is the digital or linear information of our chromosomes and genes, with the four-letter alphabet composed of G, C, A, and T (the bases Guanine, Cytosine, Adenine, and Thymine). Variation in the order of these letters in the digital strings of our chromosomes or our expressed genes (or mRNAs) generates information of several distinct types: genes, regulatory machinery, and information that enables chromosomes to carry out their tasks as informational organelles (eg, centromeric and telomeric sequences).

Second, there is the three-dimensional information of proteins, the molecular machines of life. Proteins are strings of amino acids employing a 20-letter alphabet. Proteins pose four technical challenges: (i) Proteins are synthesized as linear strings and fold into precise three-dimensional structures as dictated by the order of amino acid residues in the string. Can we formulate the rules for protein folding to predict three-dimensional structure from primary amino acid sequence? The identification and comparative analysis of all human and model organism (bacteria, yeast, nematode, fly, mouse, etc.) genes and proteins will eventually lead to a lexicon of motifs that are the building block components of genes and proteins. These motifs will greatly constrain the shape space computational algorithms must search to successfully correlate primary amino acid sequence with the correct three-dimensional shapes. The protein-folding problem will probably be solved within the next 10 to 15 years. (ii) Can we predict protein function from knowledge of the three-dimensional structure? Once again the lexicon of motifs with their functional as well as structural correlations will play a critical role in solving this problem. (iii) How do the myriad of chemical modifications of proteins (eg, phosphorylation, acetylation) alter their structures and modify their functions? The mass spectrometer will play a key role in identifying secondary modifications. (iv) How do proteins interact with one another and/or with other macromolecules to form complex molecular machines (eg, the ribosomal subunits)? If these functional complexes can be isolated, the mass spectrometer, coupled with a knowledge of all protein sequences that can be derived from the complete genomic sequence of the organism, will serve as a powerful tool for identifying all the components of complex molecular machines.

The third type of biological information arises from complex biological systems and networks. Systems information is four-dimensional because it varies with time. For example, the human brain has 1012 neurons making approximately 1015 connections. From this network arises systems properties such a memory, consciousness, and the ability to learn. The important point is that systems properties cannot be understood from studying the network elements (eg, neurons) one at a time; rather, the collective behavior of the elements needs to be studied together. To study most biological systems, three issues need to be stressed. First, most biological systems are too complex to study directly; therefore they must be divided into tractable subsystems whose properties in part reflect those of the system. These subsystems must be sufficiently small to analyze all their elements and connections. Second, high-throughput analytic or global tools are required for studying many systems elements at one time (see below). Finally, the systems information needs to be modeled mathematically before systems properties can be predicted and ultimately understood. This will require recruiting computer scientists and applied mathematics into biology—just as the attempts to decipher the information of complete genomes and the protein folding and structure/function problems have required the recruitment of computational scientists.

I would be remiss not to point out that there are many other molecules that generate biological information—amino acids, carbohydrates, lipids, etc. These too must be studied in the context of their specific structures and specific functions.

The deciphering and manipulation of these various types of biological information represent an enormous technical challenge for biotechnology. Yet major new and powerful tools for doing so are emerging.

One class of tools for deciphering biological information is termed high-throughput analytic or global tools. These tools can study many genes or chromosome features (genomics), many proteins (proteomics), or many cells rapidly: large-scale DNA sequencing; genome-wide genetic mapping; cDNA or oligonucleotide arrays; two-dimensional gel electrophoresis and other global protein separation technologies; mass spectrometric analysis of proteins and protein fragments; multiparameter, high-throughput cell and chromosome sorting; and high-throughput phenotypic assays.

A second approach to the deciphering and manipulation of biological information centers around combinatorial strategies. The basic idea is to synthesize an informational string (DNA fragments, RNA fragments, protein fragments, antibody combining sites, etc.) using all combinations of the basic letters of the corresponding alphabet—thus creating many different shapes that can be used to activate, inhibit, or complement the biological functions of designated three-dimensional shapes (eg, a molecule in a signal transduction pathway). The power of combinational chemistry is just beginning to be appreciated.

A critical approach to deciphering biological information will ultimately be the ability to visualize the functioning of genes, proteins, cells, and other informational elements within living organisms (*in vivo* informational imaging).

Finally, there are the computational tools required to collect, store, analyze, model, and ultimately distribute the various types of biological information. The creation presents a challenge comparable to that of developing of new instrumentation and new chemistries. Once again, this means recruiting computer scientists and applied mathematicians to biology. The biggest challenge in this regard is the language barriers that separate different scientific disciplines. Teaching biology as an informational science has been a very effective means for breeching these language barriers.

The challenge is, of course, to decipher these various types of biological information and then be able to use this information to manipulate genes, proteins, cells, and informational pathways in living organisms to eliminate or prevent disease, produce higher yield crops, or increase the productivity of animal products and meat.

Biotechnology and its applications raise a host of social, ethical, and legal questions; for example, genetic privacy, germline genetic engineering, cloning animals, genes that influence behavior, cost of therapeutic drugs generated by biotechnology, animal rights, and the nature and control of intellectual property.

The challenge clearly is to educate society so that each citizen can thoughtfully and rationally deal with these issues, for ultimately society dictates the resources and regulations that circumscribe the development and practice of biotechnology. Ultimately, I feel enormous responsibility rests with scientists to inform and educate society about the challenges as well as the opportunities arising from biotechnology. These are critical issues for biotechnology that are developed in detail in the *Encyclopedia of Ethical, Legal, and Policy Issues in Biotechnology*.

The view that biotechnology is an informational science pervades virtually every aspect of this science—including discovery, reduction to practice, and societal concerns. These Encyclopedias of Biotechnology reinforce the emerging informational paradigm change that is powerfully positioning science as we move into the twenty-first century to more effectively decipher and manipulate for humankind's benefit the biological information of relevant living organisms.

LEROY HOOD
University of Washington

CONTRIBUTORS

Hanna E. Abboud, *University of Texas, Health Science Center, San Antonio, TX*
Sankar Adhya, *National Cancer Institute, National Institutes of Health, Bethesda, MD*
Hiroji Aiba, *Nagoya University, Chikusa, Nagoya, Japan*
Philip Aisen, *Albert Einstein College of Medicine, Bronx, NY*
Rudolf K. Allemann, *University of Birmingham, Birmingham, United Kingdom*
Nicholas Allen, *The Babraham Institute, Cambridge, United Kingdom*
Suresh Ambudkar, *National Cancer Institute, National Institutes of Health, Bethesda, MD*
Vernon E. Anderson, *Case Western Reserve University, Cleveland, OH*
Bertil Andersson, *Stockholm University, Stockholm, Sweden*
Ruth Hogue Angeletti, *Albert Einstein College of Medicine, Bronx, NY*
Rodolfo Aramayo, *Schering-Plough Research Institute, Kenilworth, NJ*
Yari Argon, *University of Chicago, Chicago, IL*
K. Arora, *University of California, Irvine, CA*
Leonie K. Ashman, *Hanson Center for Cancer Research, Adeleide, Australia*
John F. Atkins, *University of Utah, Salt Lake City, UT*
William M. Atkins, *University of Washington, Seattle, WA*
Daniel Atkinson, *UCLA, Los Angeles, CA*
David Auld, *Harvard Medical School, Boston, MA*
Paul Babitzke, *Pennsylvania State University, University Park, PA*
Andrew Baird, *Prizm Pharmaceuticals, San Diego, CA*
Stacey J. Baker, *Temple University School of Medicine, Philadelphia, PA*
Tom Baldwin, *Texas A&M University, College Station, TX*
Michael Bamshad, *University of Utah Health Sciences Center, Salt Lake City, UT*
Probal Banerjee, *College of Staten Island, New York, NY*
Ruma Banerjee, *University of Nebraska, Lincoln, NE*
William R. Bauer, *University of California, San Francisco, CA*
Christopher Baum, *Heinrich-Pette Institut, Hamburg, Germany*
Edward A. Bayer, *Weizmann Institute of Science, Rehovot, Israel*
Miguel Beato, *Philipps–Universitaet Marburg, Institut fuer Molekularbiologie, Marburg, Germany*
Dorothy Beckett, *University of Maryland, Baltimore, MD*
Samuel Benchimol, *University of Toronto, Toronto, Ontario, Canada*
Steven J. Benkovic, *Pennsylvania State University, University Park, PA*
Tomas Bergman, *Karolinska Institutet, Stockholm, Sweden*
Gerald Bergtrom, *University of Wisconsin, Milwaukee, WI*
Alan Bernstein, *Mount Sinai Hospital, Toronto, Ontario, Canada*
Harris D. Bernstein, *NIDDKD, National Institutes of Health, Bethesda, MD*
Sophie Bertrand, *Universiteit Gent, Gent, Belgium*
D. M. Bethea, *Thomas Jefferson University, Philadelphia, PA*
Dieter Beyer, *Rhone Poulenc Rorer Recherche Developpement, Vitry-sur-Seine, France*
Timothy R. Billiar, *University of Pittsburgh Medical Center, Pittsburgh, PA*
Asgeir Bjornsson, *University of Aarhus, Aarhus, Denmark*
Colin C. F. Blake, *Norfolk, United Kingdom*
F. Bonomi, *Universita Degli Studi Di Milano, Milano, Italy*
Ralph A. Bradshaw, *University of California, Irvine, CA*
Bertram Brenig, *Georg-August University, Göttingen, Germany*
Kenneth J. Breslauer, *Rutgers University, Piscataway, NJ*
Roy J. Britten, *California Institute of Technology, Corona del Mar, CA*
Maurizio Brunori, *University of Rome, "La Sapienza," Rome, Italy*
Bernd Bukau, *Albert-Ludwigs Universitat Freiburg, Freiburg, Germany*
Jens R. Bundgaard, *University of Copenhagen, Copenhagen, Denmark*
Arsene Burny, *Universite Libre de Bruxelles, Rhode Saint Genese, Belgium*
Kenneth Burtis, *University of California, Davis, CA*
Ana Busturia, *Universidad Aut noma de Madrid, Madrid, Spain*
Giulio L. Cantoni, *NIMH, National Institutes of Health, Bethesda, MD*
M. Stella Carlomango, *Università Degli Studi Di Napoli Federico II, Napoli, Italy*
Gerald M. Carlson, *University of Missouri, Kansas City, MO*
Graham Carpenter, *Vanderbilt University, Nashville, TN*
Robin W. Carrell, *University of Cambridge, Cambridge, United Kingdom*
James Castelli-Gair, *University of Cambridge, Cambridge, United Kingdom*
Enrique Cerda-Olmedo, *Universidad de Sevilla, Sevilla, Spain*
Jonathan Chaires, *University of Mississippi Medical Center, Jackson, MS*
Kung-Yao Chang, *MRC Laboratory of Molecular Biology, Cambridge, United Kingdom*
Lee Chao, *Medical University of South Carolina, Charleston, SC*
Ansuman Chattopadhyay, *Vanderbilt University School of Medicine, Nashville, TN*
Walter Chazin, *Scripps Research Institute, La Jolla, CA*
Elizabeth H. Chen, *University of Texas, Southwestern Medical Center, Dallas, TX*
Donald P. Cheney, *Northeastern University, Boston, MA*
Andreas Chrambach, *National Institutes of Health, Bethesda, MD*
Jason Christiansen, *Virginia Polytechnic Institute, Blacksburg, VA*
Jon A. Christopher, *Texas A&M University, College Station, TX*
Brian F. C. Clark, *Aarhus University, Aarhus, Denmark*
Dennis Clegg, *University of California, Santa Barbara, CA*
F. Cliften, *University of Colorado Health Sciences Center, Denver, CO*
Georges N. Cohen, *Institut Pasteur, Paris, France*
Roberta Colman, *University of Delaware, Newark, DE*
Orla M. Conneely, *Baylor College of Medicine, Houston, TX*
Barry S. Cooperman, *University of Pennsylvania, Philadelphia, PA*
Pamela Correll, *The Pennsylvania State University, University Park, PA*
Pascale Cossart, *Institut Pasteur, Paris, France*
Nancy Craig, *Johns Hopkins University, Baltimore, MD*
Elliott Crooke, *Georgetown University, Washington, DC*
Stanley T. Crooke, *ISIS Pharmaceuticals, Inc., Carlsbad, CA*
Richard D. Cummings, *University of Oklahoma, Oklahoma City, OK*
James Curran, *Wake Forest University, Winston-Salem, NC*
Michael A. Cusanovich, *University of Arizona, Tucson, AZ*
Giuseppe D'Alessio, *Universita di Napoli Federico II, Napoli, Italy*
Antoine Danchin, *Institut Pasteur, Paris, France*

P. L. Davies, *Queen's University, Kingston, Ontario, Canada*
Dennis R. Dean, *Virginia Polytechnic Institute & State University, Blacksburg, VA*
Pieter DeHaseth, *Case Western Reserve University, Cleveland, OH*
Andrew J. DeMello, *Imperial College, London, United Kingdom*
David T. Denhardt, *Rutgers University, Piscataway, NJ*
W. A. Denny, *University of Auckland, Auckland, New Zealand*
Zygmunt Derewenda, *University of Virginia, Charlottesville, VA*
Claude Desplan, *Rockefeller University, New York, NY*
Raymond Devoret, *Centre Universitaire, Orsay, France*
Beth Didomenico, *Schering-Plough Research Institute, Kenilworth, NJ*
Patrick DiMario, *Louisiana State University, Baton Rouge, LA*
Hal B. F. Dixon, *Cambridge University, Cambridge, United Kingdom*
Mark Dodson, *University of Arizona, Tucson, AZ*
Jan Drenth, *Laboratory of Biophysical Chemistry, Groningen, The Netherlands*
Gerhart Drews, *Institut fur Biologie, Freiburg, Germany*
Bernard Dujon, *Institut Pasteur, Paris, France*
Edward Egelman, *University of Minnesota, Minneapolis, MN*
Jean Marc Egly, *Parc d'Innovation, Illkirch, France*
John Ellis, *University of Warwick, Coventry, United Kingdom*
Vincent Ellis, *Thrombosis Research Institute, London, United Kingdom*
Walther Ellis, *Utah State University, Logan, UT*
Volker Erdmann, *Freie Universitat Berlin, Berlin, Germany*
Gerard Evan, *ICRF, London, United Kingdom*
Kenneth Ewan, *Max-Planck Institute of Biophysical Chemistry, Göttingen, Germany*
Hong Fang, *Vanderbilt University, Nashville, TN*
Lynette R. Ferguson, *University of Auckland Medical School, Auckland, New Zealand*
Carol A. Fierke, *Duke University, Durham, NC*
J. R. S. Fincham, *University of Edinburgh, Edinburgh, Scotland*
Anthony L. Fink, *University of California, Santa Cruz, CA*
Gunter Fischer, *MPG Arbeitsgruppe, Halle, Germany*
Darrell R. Fisher, *Pacific Northwest Laboratory, Richland, WA*
Catherine Florentz, *Institut de Biologic Moleculaire et Cellulaire, Strasbourg, France*
Ricardo Flores, *Instituto de Biologia Molecular y Cellularde Plantas, Valencia, Spain*
P. J. Foley, *Thomas Jefferson University, Philadelphia, PA*
Josiane Fontaine-Perus, *CNRS, Nantes Cedex, France*
Jefferson Foote, *Fred Hutchinson Cancer Research Center, Seattle, WA*
Patricia Foster, *Boston University, Boston, MA*
Michel Fougereau, *Centre d'Immunologie Marseille Luminy, Marseille, France*
Francois Franceschi, *Max-Planck Institute for Molecular Genetics, Berlin, Germany*
Murray J. Fraser, *Sydney Children's Hospital, Randwick, Australia*
Ian Freshney, *University of Glasgow, Glasgow, United Kingdom*
Terrence G. Frey, *San Diego State University, San Diego, CA*
Laura Frontali, *University of Rome, "La Sapienza," Rome, Italy*
Yasuo Fukami, *Kobe University, Nada, Kobe, Japan*
Carl W. Fuller, *Amersham Pharmacia Biotech, Cleveland, OH*
Betty Gaffney, *Florida State University, Tallahassee, FL*
Fernando Garcia-Arenal, *Ciudad Universitaria, Madrid, Spain*
Jean-Renaud Garel, *Laboratoire de Biologie Structurale, CNRS, Gif-sur-Yvette, France*
D. R. Garrod, *University of Manchester, Manchester, United Kingdom*
Frank R. Gasparro, *Jefferson University, Philadelphia, PA*
C. Gaspin, *Centre National de la Researche Scientifique, Strasbourg, France*
Anthony A. Gatenby, *E. I. duPont de Nemours & Co., Wilmington, DE*
Kunihiko Gekko, *Hiroshima University, Higashi-Hiroshima, Japan*
Herbert Geller, *UMDNJ—Robert Wood Johnson Medical School, Piscataway, NJ*
John Gerig, *University of California, Santa Barbara, CA*
Norman L. Gershfeld, *National Institutes of Health, Bethesda, MD*
Michael Gershon, *Columbia University, New York, NY*
Jonathan M. Gershoni, *Tel Aviv University, Tel Aviv, Israel*
Jean-Marie Ghuysen, *Universite de Liege, Liege, Belgium*
Toby Gibson, *European Molecular Biology Laboratory, Heidelberg, Germany*
J. A. Girault, *Chaire de Neuropharmacologie, College de France, Paris, France*
Adam Godzik, *Scripps Research Institute, La Jolla, CA*
Jefferey Godzik, *Scripps Research Institute, La Jolla, CA*
J. Peter Gogarten, *University of Connecticut, Storrs Mansfield, CT*
Takashi Gojobori, *National Institute of Genetics, Shizuoka, Japan*
Alfred L. Goldberg, *Harvard Medical School, Boston, MA*
Paul Gollnick, *State University of New York at Buffalo, Buffalo, NY*
Pierre Goldstein, *Centre d'Immunologie INSERM-CNRS de Marseille, Marseille, France*
H. Maurice Goodman, *University of Massachusetts Medical School, Worcester, MA*
Horace B. Gray, *University of Houston, Houston, TX*
Neil Green, *Vanderbilt University School of Medicine, Nashville, TN*
William Griffiths, *Karolinska Institutet, Stockholm, Sweden*
Frank Grosse, *Institut fur Molekulare Biotechnologie, Jena, Germany*
Marianne Grunberg-Manago, *Institut de Biologie Physico-Chemique, Paris, France*
Peter Gruss, *Max-Planck Institute for Biophysical Chemistry, Göttingen, Germany*
Richard Gumport, *University of Illinois, Urbana, IL*
Andras Guttman, *Genetich Biosystems, San Diego, CA*
William Hagan, *College of St. Rose, Albany, NY*
Klaus Hahlbrock, *Max-Planck Institut fur Biochemie, Martinsried, Germany*
Joshua W. Hamilton, *Dartmouth Medical School, Hanover, NH*
Susan Hamilton, *Baylor College of Medicine, Houston, TX*
Xianlin Han, *Washington University School of Medicine, St. Louis, MO*
Ryo Hanai, *Rikkyo University, Toshima, Tokyo, Japan*
Yusuf Hannun, *Duke University Medical Center, Durham, NC*
Stephen E. Harding, *University of Nottingham, Sutton Bonington, United Kingdom*
Peter Harland, *University of California, Berkeley, CA*
Ulrich Hartl, *Max-Planck Institut fur Biochemie, Martinsried, Germany*
R. W. Hartley, *National Institutes of Health, Bethesda, MD*
Enno Hartmann, *Georg-August Universitaet, Göttingen, Germany*
David L. Haviland, *The University of Texas, Houston Health Science Center, Houston, TX*
Toshihiko Hayashi, *University of Tokyo, Tokyo, Japan*
J. K. Heath, *University of Birmingham, Birmingham, United Kingdom*
Robert Helling, *University of Michigan, Ann Arbor, MI*
Ali Hemmati-Brivanlou, *Rockefeller University, New York, NY*
M. A. Hemminga, *Agricultural University of Wageningen, Daagenlaa, Wageningen, The Netherlands*
Beric Henderson, *University of Sydney, Westmead, Australia*
Roger W. Hendrix, *University of Pittsburgh, Pittsburgh, PA*
Leonard Herzenberg, *Stanford University, Stanford, CA*
George P. Hess, *Cornell University, Ithaca, NY*
Christa Heyting, *Agricultural University, Wageningen, The Netherlands*
D. J. Hill, *St. Joseph's Health Center, London, Ontario, Canada*
James A. Hoch, *Scripps Research Institute, La Jolla, CA*
Denis Hochstrasser, *Geneva University Hospital, Geneva, Switzerland*
Nicholas A. Hoenich, *The University of Newcastle upon Tyne, Newcastle, United Kingdom*
Wayne L. Hoffman, *University of Texas, Southwestern Medical Center, Dallas, TX*

J. John Holbrook, *University of Bristol, Bristol, United Kingdom*
Arne Holmgren, *Karolinska Institute, Stockholm, Sweden*
Thomas F. Holzman, *Abbot Laboratories, Abbot Park, IL*
H. Homareda, *Kyorin University, Tokyo, Japan*
Roger Hull, *John Innes Centre, Colney, Norwich, United Kingdom*
Martin Hum, *University of Texas, Southwestern Medical Center, Dallas, TX*
Jennifer A. Hunt, *Duke University Medical Center, Durham, NC*
Atsushi Ikai, *Tokyo Institute of Technology, Yokohama, Japan*
Kiyohiro Imai, *Osaka University, Osaka, Japan*
Taiji Imoto, *Kyushu University, Fukuoka, Japan*
Antonio Incardona, *University of Tennessee, Memphis, TN*
Edward E. Ishiguro, *University of Victoria, Victoria, B.C., Canada*
Akira Ishihama, *National Institute of Genetics, Shizuoka-ken, Japan*
Judith A. Jaehning, *University of Colorado, Denver, CO*
Yuh-Nung Jan, *University of California, San Francisco, CA*
Joel Janin, *Laboratoire de Biologie Structurale, Paris, France*
S. Jansen, *Georg-August University Göttingen, Göttingen, Germany*
Patricia A. Jennings, *University of California, San Diego, La Jolla, CA*
Randy L. Jirtle, *Duke University, Durham, NC*
Jonathan Jones, *John Innes Institute, Norwich, United Kingdom*
Rick Jones, *Southern Methodist University, Dallas, TX*
Hans Jornvall, *Karolinska Institutet, Stockholm, Sweden*
Akia Kaji, *University of Pennsylvania, Philadelphia, PA*
Barton A. Kamen, *University of Texas, Southwestern Medical Center, Dallas, TX*
Heather Kaminski, *Schering-Plough Research Institute, Kenilworth, NJ*
Minoru Kanehisa, *Kyoto University, Kyoto, Japan*
Jack Kaplan, *Oregon Health Sciences University, Portland, OR*
Francois Karch, *University of Geneva, Geneva, Switzerland*
Thom Kaufmann, *Indiana University, Bloomington, IN*
Jack D. Keene, *Duke University, Durham, NC*
Regis B. Kelly, *University of California, San Francisco, CA*
Zvi Kelman, *Memorial Sloan-Kettering Cancer Center, New York, NY*
Byron Kemper, *University of Illinois, Urbana, IL*
James Kennison, *National Institute of Child Health and Human Development, National Institutes of Health, Bethesda, MD*
Kathleen M. Kerr, *National Institutes of Health, Bethesda, MD*
John R. Kirby, *University of California, Berkeley, CA*
Jennifer Kitchen, *Georgetown University, Washington, DC*
Horst Kleinkauf, *Technische Universitat, Berlin, Germany*
Jorg Klug, *Phillips–Universitat Marburg, Marburg, Germany*
Peter Knight, *University of Leeds, Leeds, United Kingdom*
Charlotte Knudsen, *University of Aarhus, Aarhus, Denmark*
Yuji Kobayashi, *Osaka University, Osaka, Japan*
Ralf Koebnik, *Biozentrum Basel, Basel, Switzerland*
Andrzej Kolinski, *Scripps Research Institute, La Jolla, CA*
Gerald B. Koudelka, *State University of New York at Buffalo, Buffalo, NY*
Joseph Kraut, *University of California, San Diego, La Jolla, CA*
Toshio Kuroki, *Showa University, Tokyo, Japan*
Sidney R. Kushner, *University of Georgia, Athens, GA*
Kunihiro Kuwajima, *University of Tokyo, Tokyo, Japan*
Harold Lahm, *Institute of Molecular Animal Breeding, Munich, Germany*
Ratnesh Lal, *University of California, Santa Barbara, CA*
Marc Lalande, *Children's Hospital, Boston, MA*
David Lane, *Centre National de la Researche Scientifique, Strasbourg, France*
Janos Lanyi, *University of California, Irvine, CA*
Paul Lasko, *McGill University, Montreal, Quebec, Canada*
Michael Laskowski Jr., *Purdue University, West Lafayette, IN*
David S. Latchman, *University College London Medical School, London, United Kingdom*
Peter Lawrence, *Medical Research Council, Cambridge, United Kingdom*
Judith E. Layton, *Ludwig Institute for Cancer Research, Victoria, Australia*
Claude J. Lazdunski, *Laboratoire d'Ingeniere des Systèmes Macromoléculaires, Marseille, France*
Robin Leake, *University of Glasgow, Glasgow, United Kingdom*
Stewart H. Lecker, *Harvard Medical School, Boston, MA*
Stuart LeGrice, *Case Western Reserve School of Medicine, Cleveland, OH*
Ruth Lehmann, *New York University, New York, NY*
Arthur M. Lesk, *University of Cambridge, Cambridge, United Kingdom*
Michael Levine, *University of California, Berkeley, Berkeley, CA*
Joe D. Lewis, *Institute of Cell & Molecular Biology, Edinburgh, United Kingdom*
Uno Lindberg, *Stockholm University, Stockholm, Sweden*
Halina Lis, *Weizmann Institute of Science, Rehovot, Israel*
Uriel Littauer, *Weizmann Institute of Science, Rehovot, Israel*
John Little, *University of Arizona, Tucson, AZ*
Lawrence A. Loeb, *University of Washington, Seattle, WA*
Peter C. Loewen, *University of Manitoba, Winnipeg, Manitoba, Canada*
Douglas J. Loftus, *National Cancer Institute, National Institutes of Health, Bethesda, MD*
Kenton Longenecker, *University of Virginia Health Sciences Center, Charlottesville, VA*
J. Michael Lord, *University of Warwick, Conventry, United Kingdom*
Martin G. Low, *Columbia University, New York, NY*
Reinhard Luhrmann, *Institut fur Molekularbiologie und Tumorforschung, Marburg, Germany*
Paul MacDonald, *Stanford University, Stanford, CA*
Robert M. Macnab, *Yale University, New Haven, CT*
Neil B. Madsen, *University of Alberta, Edmonton, Alberta, Canada*
Hasaji Maki, *Ikoma City, Japan*
Gregory S. Makowski, *University of Connecticut, Farmington, CT*
Jack Maniloff, *University of Rochester, Rochester, NY*
Jeffrey R. Mann, *Beckman Research Institute of the City of Hope, Duarte, CA*
Bengt Mannervik, *Uppsala University, Uppsala, Sweden*
Armen Manoukian, *Ontario Cancer Institute, Toronto, Ontario, Canada*
Ahmed Mansouri, *Max-Planck Institute for Biophysical Chemistry, Göttingen, Germany*
Mohamed A. Marahiel, *Philipps–Universitat Marburg, Marburg, Germany*
Guglielmo Marin, *Universita Degli Studi di Padova, Padova, Italy*
Susan Marqusee, *University of California, Berkeley, CA*
Mike Marsh, *University College London, London, United Kingdom*
Garland Marshall, *Washington University, St. Louis, MO*
Jennifer Martin, *University of Queensland, Brisbane, Australia*
Nancy C. Martin, *University of Louisville, Louisville, KY*
Florence Maschat, *Institut de Genetique Humaine, Montpellier, France*
Christopher Mathews, *Oregon State University, Corvallis, OR*
Abdul Matin, *Stanford University, Stanford, CA*
Mark P. Mattson, *University of Kentucky, Lexington, KY*
Russell A. Maurer, *Case Western Reserve University, Cleveland, OH*
Nicola McCarthy, *Imperial Cancer Research Fund, London, United Kingdom*
Richard McCarty, *Johns Hopkins University, Baltimore, MD*
David R. McClay, *Duke University, Durham, NC*
John R. Menninger, *University of Iowa, Iowa City, IA*
Carl R. Merril, *National Institute of Mental Health, National Institutes of Health, Bethesda, MD*
David Metzler, *Iowa State University, Ames, IA*
Robert H. Michell, *The University of Birmingham, Birmingham, United Kingdom*
Marek Mlodzik, *European Molecular Biology Laboratory, Heidelberg, Germany*
Evita Mohr, *Universitat Hamburg, Hamburg, Germany*
Ian J. Molineux, *University of Texas, Austin, TX*

Cesare Montecucco, *Università degli Studi di Padova, Padova, Italy*
Carol Moore, *City University of New York Medical School, NY*
Henning D. Mootz, *Phillips–Universitat Marburg, Marburg, Germany*
Gines Morata, *University Autonoma De Madrid, Madrid, Spain*
Richard I. Morimoto, *Northwestern University, Evanston, IL*
John F. Morrison, *The Australian National University, Canberra, Australia*
Gisela Mosig, *Vanderbilt University, Nashville, TN*
Jan Mous, *F. Hoffmann-LaRoche Ltd., Basel, Switzerland*
Benno Müller-Hill, *Institut fur Genetik / Universitat Koln, Koln, Germany*
J. K. Myers, *Texas A&M University, College Station, TX*
Kyoshi Nagai, *MRC Laboratory of Molecular Biology, Cambridge, United Kingdom*
Haruki Nakamura, *Biomolecular Engineering Research Institute, Osaka, Japan*
Yoshikazu Nakamura, *University of Tokyo, Tokyo, Japan*
Aiguo Ni, *Medical University of South Carolina, Charleston, SC*
Allen W. Nicholson, *Wayne State University, Detroit, MI*
Christof Niehrs, *Deutsches Krebsforschungszentrum, Heidelberg, Germany*
Koich Nishigaki, *Saitama University, Urawa, Japan*
Ken Nishikawa, *National Institute of Genetics, Shizuoka, Japan*
Ralph A. Nixon, *Harvard Medical School, Belmont, MA*
Akio Nomoto, *University of Tokyo, Tokyo, Japan*
Bengt Norden, *Chalmers University of Technology, Gothenburg, Switzerland*
Shiao Li Oei, *Institute fur Biochemie, Berlin, Germany*
Naotake Ogasawara, *Ikoma City, Japan*
Yoshio Okada, *Osaka University, Osaka, Japan*
Charles Ordahl, *University of California, San Francisco, CA*
George Ordal, *University of Illinois, Urbana, IL*
George Oster, *University of California, Berkeley, CA*
Malcolm G. P. Page, *F. Hoffmann-LaRoche Ltd., Basel, Switzerland*
Renato Paro, *University of Heidelberg, Heidelberg, Germany*
D. A. Parry, *Massey University, Palmerston North, New Zealand*
Premal H. Patel, *University of Washington School of Medicine, Seattle, WA*
Barbara Pearse, *MRC Laboratory of Molecular Biology, Cambridge, United Kingdom*
Israel Pecht, *Weizmann Institute of Science, Rehovot, Israel*
Iain K. Pemberton, *Institut Pasteur, Paris, France*
Bernard Perbal, *Centre Universitaire, Orsay, France*
Richard N. Perham, *University of Cambridge, Cambridge, United Kingdom*
Francine Perler, *New England Biolabs, Beverly, MA*
Charles L. Perrin, *University of California, San Diego, La Jolla, CA*
Steven Perrin, *ARIAD Pharmaceuticals, Cambridge, MA*
Bengt Persson, *Karolinska Institute, Stockholm, Sweden*
Donald W. Pettigrew, *Texas A&M University, College Station, TX*
Larry Pfeffer, *University of Tennessee, Memphis, TN*
K. Kevin Pfister, *University of Virginia, Charlottesville, VA*
Margaret A. Phillips, *University of Texas, Dallas, TX*
Don Phillips, *LaTrobe University, Victoria, Australia*
Thomas L. Poulos, *University of California, Irvine, CA*
Linda S. Powers, *Utah State University, Logan, UT*
Peter E. Prevelige, *University of Alabama, Birmingham, AL*
Peter L. Privalov, *Johns Hopkins University, Baltimore, MD*
Gudrun Rappold, *Institut fur Humangenetik Universitatsklinikum, Heidelberg, Germany*
Stephen W. Raso, *Texas A&M University, College Station, TX*
P. D. Rathjen, *University of Adelaide, Adelaide, Australia*
E. Prekumar Reddy, *The Fels Institute for Cancer Research and Molecular Biology, Philadelphia, PA*
Colin Reese, *King's College London, University of London, United Kingdom*
Peter Reichard, *Medical Nobel Institute, Karolinska Institutet, Stockholm, Sweden*
Steffen Reinbothe, *Universite Joseph Fourier, Grenoble, France*
Dietmar Richter, *Institut fur Zellbiochemie und klinische Neurobiologie, Hamburg, Germany*
P. G. Righetti, *University of Milan, Milan, Italy*
Richard J. Riopelle, *Queen's University, Kingston, Ontario, Canada*
James Riordan, *Harvard University, Boston, MA*
Anthony J. Robertson, *National Institutes of Health, Research Triangle Park, NC*
Jean-David Rochaix, *University of Geneva, Geneva, Switzerland*
Mario Roederer, *Stanford University, Stanford, CA*
William A. Rosche, *Boston University School of Medicine, Boston, MA*
Elliott M. Ross, *University of Texas, Southwestern Medical Center, Dallas, TX*
Gregory M. Ross, *Queen's University, Kingston, Ontario, Canada*
Lawrence Rothfield, *University of Connecticut, Farmington, CT*
Harry Rubin, *University of California, Berkeley, CA*
Thomas S. Rush, *Princeton University, Princeton, NJ*
M. S. Saedi, *Hybritech Co., San Diego, CA*
Alan Saltiel, *Parke-Davis Pharmaceutical Research, Ann Arbor, MI*
Aziz Sancar, *University of North Carolina, Chapel Hill, NC*
Matti Saraste, *European Molecular Biology Laboratory, Heidelberg, Germany*
Zuben E. Sauna, *National Institutes of Health, Bethesda, MD*
Lindsay Sawyer, *University of Edinburgh, Edinburgh, United Kingdom*
Walter Schaffner, *Universitat Zurich, Zurich, Switzerland*
Paul Schimmel, *Scripps Research Institute, La Jolla, CA*
Franz Schmid, *Universitat Bayreuth, Bayreuth, Germany*
Manfred Schnarr, *Institut de Biologie Moleculaire et Cellulaire, CNRS, Strasbourg, France*
J. Martin Scholtz, *Texas A&M University, College Station, TX*
Trudi Schupbach, *Princeton University, Princeton, NJ*
James Scott, *Royal Postgraduate Medical School, London, United Kingdom*
Barbara A. Seaton, *Boston University, Boston, MA*
Constantin E. Sekeris, *University of Wurzburg, Wurzburg, Germany*
Igor L. Shamovsky, *Queen's University, Kingston, Ontario, Canada*
Shmuel Shaltiel, *Weizmann Institute of Science, Rehovot, Israel*
Nathan Sharon, *Weizmann Institute of Science, Rehovot, Israel*
G. Shaw, *University of Florida, Gainesville, FL*
William V. Shaw, *Leicester University, Leicester, United Kingdom*
David Sherman, *University of Minnesota, St. Paul, MN*
Ben-Zion Shilo, *Weizmann Institute of Science, Rehovot, Israel*
Makoto Shimoyama, *Shimane Medical University, Izumo, Japan*
Israel Silman, *Weizmann Institute of Science, Rehovot, Israel*
Richard J. Simpson, *Ludwig Institute for Cancer Research, Victoria, Australia*
Gary Siuzdak, *Scripps Research Institute, La Jolla, CA*
James Skeath, *Washington University, St. Louis, MO*
Stephen J. Small, *New York University, New York, NY*
Colleen M. Smith, *Mercer University, Macon, GA*
Gerald R. Smith, *Fred Hutchinson Cancer Research Center, Seattle, WA*
Sarah M. Smolik, *Oregon Health Sciences University, Portland, OR*
Suzanne G. Sobel, *Yale University, New Haven, CT*
Kunitsugu Soda, *Nagaoka University of Technology, Niigata, Japan*
Lila Solnica-Krezel, *Vanderbilt University, Nashville, TN*
Mark Solomon, *Yale University, New Haven, CT*
Joseph Sperling, *The Hebrew University of Jerusalem, Jerusalem, Israel*
Ruth Sperling, *The Hebrew University of Jerusalem, Jerusalem, Israel*
Martin Spiess, *Biozentrum, Basel, Switzerland*
Tom Spiro, *Princeton University, Princeton, NJ*
Mathias Sprinzl, *Universitaet Bayreuth, Bayreuth, Germany*
R. B. Spruijt, *Wageningen Agricultural University, Wageningen, The Netherlands*
F. W. Stahl, *University of Oregon, Eugene, OR*
Martin F. Steiger, *F. Hoffmann-LaRoche Ltd., Basel, Switzerland*

Scott Strobel, *Yale University, New Haven, CT*
Shintaro Sugai, *Soka University, Tokyo, Japan*
Roy Sundick, *Wayne State University, Detroit, MI*
H. Eldon Sutton, *University of Texas, Austin, TX*
Sandra S. Szegedi, *University of Illinois at Urbana-Champaign, Urbana, IL*
Sevec Szmelcman, *Institut Pasteur, Paris, France*
Hatsumi Taniguchi, *University of Occupational and Environmental Health, Fukuota, Japan*
Joyce Taylor-Papadimitriou, *Imperial Cancer Research Fund Laboratories, London, United Kingdom*
Mariella Tegoni, *CNRS, Marseille, France*
Jean Thomas, *Cambridge University, Cambridge, United Kingdom*
Carl Thummel, *University of Utah, Salt Lake City, UT*
Serge Timasheff, *Brandeis University, Waltham, MA*
Thea D. Tlsty, *University of California, San Francisco, CA*
Andrew Travers, *MRC Laboratory of Molecular Biology, Cambridge, United Kingdom*
Jill Trewhella, *Los Alamos National Laboratory, Los Alamos, NM*
Edward Trifonov, *National Institute of Genetics, Shizuoka-ken, Japan*
Toshiki Tsurimoto, *Osaka, Japan*
Agnes Ullmann, *Institut Pasteur, Paris, France*
Dominiqe van der Straeten, *Universiteit Gent, Gent, Belgium*
M. van Lijsebettens, *Universiteit Gent, Gent, Belgium*
Marc Van Montagu, *Universiteit Gent, Gent, Belgium*
Klaas van Wijk, *Stockholm University, Stockholm, Sweden*
G. Varani, *MRC Laboratory of Molecular Biology, Cambridge, United Kingdom*
Serge N. Vinogradov, *Wayne State University, Detroit, MI*
Hans von Dohren, *Technische Universitat Berlin, Berlin, Germany*
Edward Voss, *University of Illinois, Urbana, IL*
Richard Walden, *Horticulture Research International, Kent, United Kingdom*
Melinda Wales, *Texas A&M University, College Station, TX*
Valeria A. Wallace, *University College London, London, United Kingdom*
Frederick G. Walz, *Kent State University, Kent, OH*
Andrew H. J. Wang, *University of Illinois, Urbana, IL*
Hongyun Wang, *University of California, Berkeley, CA*
James C. Wang, *Harvard University, Cambridge, MA*
Michael J. Waring, *Cambridge University, Cambridge, United Kingdom*
Amy Warren, *Dartmouth College, Hanover, NH*
Graham Warren, *Imperial College Research Fund, London, United Kingdom*
Hans Warrick, *Stanford University, Stanford, CA*
Arieh Warshel, *University of Southern California, Los Angeles, CA*
Steven A. Wasserman, *University of Texas, Southwestern Medical Center, Dallas, TX*
Hans Weber, *Institut fur Molekularbiologie, Zurich, Switzerland*
Patricia C. Weber, *Schering-Plough Research Institute, Kenilworth, NJ*
Eric Westhof, *Institut de Biologie Moleculaire et Cellulaire, du CNRS, Strasbourg, France*
Richard A. Wetsel, *University of Texas, Houston, TX*
Paul Whiting, *Merck Sharp and Dohme Research Laboratories, Essex, United Kingdom*
Peter A. Whittaker, *Saint Patrick's College, Maynooth, Kildare, Ireland*
Pernilla Whittung-Stapshede, *California Institute of Technology, Pasadena, CA*
Meir Wilchek, *Weizmann Institute of Science, Rehovot, Israel*
Jim R. Wild, *Texas A&M University, College Station, TX*
Allison K. Wilson, *University of Wisconsin, Madison, WI*
David B. Wilson, *Cornell University, Ithaca, NY*
Leslie Wilson, *University of California, Santa Barbara, CA*
Sam H. Wilson, *NIEHS, National Institutes of Health, Research Triangle Park, NC*
R. Wolfenden, *University of North Carolina, Chapel Hill, NC*
Alan Paul Wolffe, *NICDH, National Institutes of Health, Bethesda, MD*
C. J. A. M. Wolfs, *Wageningen Agricultural University, Wageningen, The Netherlands*
William B. Wood, *University of Colorado, Boulder, CO*
Robert Woody, *Colorado State University, Fort Collins, CO*
Kenneth Yamada, *NIDR, National Institutes of Health, Bethesda, MD*
Charles Yanofsky, *Stanford University, Stanford, CA*
Ada Yonath, *Weizmann Institute of Science, Rehovot, Israel*
Hiroshi Yoshikawa, *Nara Institute of Science & Technology, Ikoma, Japan*
Jennifer Zamanian, *University of California, San Francisco, CA*
Guiliana Zanetti, *University of Milan, Milan, Italy*
Amanda Zeffman, *MRC Laboratory of Molecular Biology, Cambridge, United Kingdom*
Suisheng Zhang, *Institute of Molecular Biotechnology, Jena, Germany*
Mark Zoller, *Ariad Pharmaceuticals, Cambridge, MA*

COMMONLY USED ACRONYMS AND ABBREVIATIONS

Å	angstrom (10^{-10} m)
Ab	antibody
Ac	acetyl group
ADP	adenosine diphosphate
Ala	alanine residue (also A)
AMP	adenosine monophosphate
Arg	arginine residue (also R)
Asn	asparagine residue (also N)
Asp	aspartic acid residue (also D)
ATP	adenosine triphosphate
ATPase	adenosine triphosphatase
Bq	becquerel
bp	base-pair
BSA	bovine serum albumin
C	cytosine base
cal	calorie (4.18 J)
cAMP	3′,5′-cyclic AMP
CD	circular dichroism
CoA	coenzyme A
cDNA	complementary DNA
cGMP	3′,5′-cyclic GMP
Cmc	critical micelle concentration
Cys	cysteine residue (also C)
Da	dalton
DMSO	dimethyl sulfoxide
DNA	deoxyribonucleic acid
DNase	deoxyribonuclease
DTT	dithiothreitol
EDTA	ethylenediamine tetraacetic acid
EG	[ethylenebis(oxonitrilo)] tetraacetic acid
EGF	epidermal growth factor
ELISA	enzyme-linked immunosorbent assay
EM	electron microscopy
EPR, ESR	electron paramagnetic (or sign) resonance
ER	endoplasmic reticulum
EXAFS	extended X-ray absorption fine structure
FAD	flavin-adenine dinucleotide
fMet	*N*-formyl methionine
FMN	flavin mononucleotide
FPLC	fast protein liquid chromatography
G	Gibbs free energy of a system
G	guanine base
G-protein	guanine-nucleotide binding regulatory protein
GABA	g-aminobutyric acid
GalNAc	*N*-acetylgalactosamine residue
GdmCl	guanidinium chloride (guanidine hydrochloride)
GDP	guanosine diphosphate
Glc	glucose residue
GlcNAc	*N*-acetylglucosamine residue
Gln	glutamine residue (also Q)
Glu	glutamic acid residue (also E)
Gly	glycine residue (also G)
GMP	guanoine monophosphate
GSH	glutathione, thiol form
GSSG	glutathione, disulfide form
GTP	guanosine triphosphate
GTPase	guanosine triphosphatase
H	enthalpy of a system
h	hour
His	histidine residue (also H)
HLA	histocompatibility locus antigen
HPLC	high pressure liquid chromatography
Hz	hertz, frequency
IEF	isoelectric focusing IgG, IgA, IgM, etc. immunoglobuin G, A, M. etc.
Ile	isoleucine residue (also I)
Ins	inositol residue
IR	infrared
J	joule
K	degrees Kelvin, absolute temperature
k	rate constant for a specified reaction
K_a	association constant
K_d	dissociation constant
K_{eq}	equilibrium constant for a specified reaction
K_m	Michaelis constant
kb	kilobases
kcal	kilocalorie (4.18 kJ)
kDa	kilodalton
LDL	low density lipoprotein
Leu	leucine residue (also L)
Lys	lysine residue (also K)

m meter
mAb monoclonal antibody
Met methionine residue (also M)
MHC major histocompability locus
min minute
mol mole
M_r relative molecular mass
mRNA messenger RNA
MS mass spectroscopy
mtDNA mitochondrial DNA
Mur muramic acid residue
MurNAc *N*-acetylmuramic acid residue

NAD^+ nicotinamide adenine dinucleotide (oxidized)
NADH nicotinamide adenine dinucleotide (reduced)
$NADP^+$ nicotinamide adenine dinucleotide phosphate (oxidized)
NADPH nicotinamide adenine dinucleotide phosphate (reduced)
NMR nuclear magnetic resonance

PAGE polyacrylamide gel electrophoresis
PBS phosphate-buffered saline
PCR polymerase chain reaction
PEG polyethylene glycol
pH negative logarithm of hydrogen ion concentration
Phe phenylalanine residue (also F)
P_i inorganic phosphate
PP_i inorganic pyrophosphate
p.p.m. parts per million
Pro proline residue (also P)

R gas constant (8.21451 J/(K^1 mol^{-1}))
RIA radioimmunoassay
RNA ribonucleic acid
RNase ribonuclease
RNP ribonucleoprotein
rRNA ribosomal RNA

S entropy of a system
S Svedberg sedimentation unit (10^{-13} s)
s second
s sedimentation coefficient
SDS sodium dodecyl sulfate
snRNA small nuclear RNA
SRP signal recognition particle

T temperature
T thymine residue
TF transcription factor
Thr threonine residue (also T)
T_m midpoint of temperature induced transition
Tri tris(hydroxymethyl) aminomethane
tRNA transfer RNA
Trp tryptophan residue (also W)
TTP thymidine triphosphate
Tyr tyrosine residue (also Y)

U uracil residue
UDP uridine diphosphate
UMP uridine monophosphate
UTP uridine triphosphate
UV ultraviolet

V volume
V_{max} maximum enzyme catalyzed velocity
Val valine residue (also V)

WT wild-type

Xyl xylose residue

CONVERSION FACTORS

SI Units (Adopted 1960)

The International System of Units (abbreviated SI), is being implemented throughout the world. This measurement system is a modernized version of the MKSA (meter, kilogram, second, ampere) system, and its details are published and controlled by an international treaty organization (The International Bureau of Weights and Measures) (1).

SI units are divided into three classes:

BASE UNITS

length	meter[†] (m)
mass	kilogram (kg)
time	second (s)
electric current	ampere (A)
thermodynamic temperature[‡]	kelvin (K)
amount of substance	mole (mol)
luminous intensity	candela (cd)

SUPPLEMENTARY UNITS

plane angle	radian (rad)
solid angle	steradian (sr)

DERIVED UNITS AND OTHER ACCEPTABLE UNITS

These units are formed by combining base units, supplementary units, and other derived units (2–4). Those derived units having special names and symbols are marked with an asterisk in the list below.

Quantity	Unit	Symbol	Acceptable equivalent
*absorbed dose	gray	Gy	J/kg
acceleration	meter per second squared	m/s^2	
*activity (of a radionuclide)	becquerel	Bq	1/s
area	square kilometer	km^2	
	square hectometer	hm^2	ha (hectare)
	square meter	m^2	
concentration (of amount of substance)	mole per cubic meter	mol/m^3	
current density	ampere per square meter	A/m^2	
density, mass density	kilogram per cubic meter	kg/m^3	g/L; mg/cm^3
dipole moment (quantity)	coulomb meter	C·m	

[†]The spellings "metre" and "litre" are preferred by ASTM; however, "-er" is used in the *Encyclopedia.*

[‡]Wide use is made of Celsius temperature (t) defined by

$$t = T - T_0$$

where T is the thermodynamic temperature, expressed in kelvin, and $T_0 = 273.15$ K by definition. A temperature interval may be expressed in degrees Celsius as well as in kelvin.

Quantity	Unit	Symbol	Acceptable equivalent
*dose equivalent	sievert	Sv	J/kg
*electric capacitance	farad	F	C/V
*electric charge, quantity of electricity	coulomb	C	A·s
electric charge density	coulomb per cubic meter	C/m^3	
*electric conductance	siemens	S	A/V
electric field strength	volt per meter	V/m	
electric flux density	coulomb per square meter	C/m^2	
*electric potential, potential difference, electromotive force	volt	V	W/A
*electric resistance	ohm	Ω	V/A
*energy, work, quantity of heat	megajoule	MJ	
	kilojoule	kJ	
	joule	J	N·m
	electronvolt†	eV†	
	kilowatt-hour†	kW·h†	
energy density	joule per cubic meter	J/m^3	
*force	kilonewton	kN	
	newton	N	$kg{\cdot}m/s^2$
*frequency	megahertz	MHz	
	hertz	Hz	1/s
heat capacity, entropy	joule per kelvin	J/K	
heat capacity (specific), specific entropy	joule per kilogram kelvin	J/(kg·K)	
heat-transfer coefficient	watt per square meter kelvin	$W/(m^2{\cdot}K)$	
*illuminance	lux	lx	lm/m^2
*inductance	henry	H	Wb/A
linear density	kilogram per meter	kg/m	
luminance	candela per square meter	cd/m^2	
*luminous flux	lumen	lm	cd·sr
magnetic field strength	ampere per meter	A/m	
*magnetic flux	weber	Wb	V·s
*magnetic flux density	tesla	T	Wb/m^2
molar energy	joule per mole	J/mol	
molar entropy, molar heat capacity	joule per mole kelvin	J/(mol·K)	
moment of force, torque	newton meter	N·m	
momentum	kilogram meter per second	kg·m/s	
permeability	henry per meter	H/m	
permittivity	farad per meter	F/m	
*power, heat flow rate, radiant flux	kilowatt	kW	
	watt	W	J/s
power density, heat flux density, irradiance	watt per square meter	W/m^2	
*pressure, stress	megapascal	MPa	
	kilopascal	kPa	
	pascal	Pa	N/m^2
sound level	decibel	dB	
specific energy	joule per kilogram	J/kg	
specific volume	cubic meter per kilogram	m^3/kg	
surface tension	newton per meter	N/m	
thermal conductivity	watt per meter kelvin	W/(m·K)	
velocity	meter per second	m/s	
	kilometer per hour	km/h	
viscosity, dynamic	pascal second	Pa·s	
	millipascal second	mPa·s	
viscosity, kinematic	square meter per second	m^2/s	
	square millimeter per second	mm^2/s	
volume	cubic meter	m^3	
	cubic diameter	dm^3	L (liter) (5)
	cubic centimeter	cm^3	mL
wave number	1 per meter	m^{-1}	
	1 per centimeter	cm^{-1}	

†This non-SI unit is recognized by the CIPM as having to be retained because of practical importance or use in specialized fields (1).

In addition, there are 16 prefixes used to indicate order of magnitude, as follows:

Multiplication factor	Prefix	Symbol	Note
10^{18}	exa	E	
10^{15}	peta	P	
10^{12}	tera	T	
10^{9}	giga	G	
10^{6}	mega	M	
10^{3}	kilo	k	
10^{2}	hecto	h[a]	[a]Although hecto, deka, deci, and centi are SI prefixes, their use should be avoided except for SI unit-multiples for area and volume and nontechnical use of centimeter, as for body and clothing measurement.
10	deka	da[a]	
10^{-1}	deci	d[a]	
10^{-2}	centi	c[a]	
10^{-3}	milli	m	
10^{-6}	micro	μ	
10^{-9}	nano	n	
10^{-12}	pico	p	
10^{-15}	femto	f	
10^{-18}	atto	a	

For a complete description of SI and its use the reader is referred to ASTM E380 (4).

A representative list of conversion factors from non-SI to SI units is presented herewith. Factors are given to four significant figures. Exact relationships are followed by a dagger. A more complete list is given in the latest editions of ASTM E380 (4) and ANSI Z210.1 (6).

Conversion Factors to SI Units

To convert from	To	Multiply by
acre	square meter (m^2)	4.047×10^{3}
angstrom	meter (m)	$1.0 \times 10^{-10\dagger}$
are	square meter (m^2)	$1.0 \times 10^{2\dagger}$
astronomical unit	meter (m)	1.496×10^{11}
atmosphere, standard	pascal (Pa)	1.013×10^{5}
bar	pascal (Pa)	$1.0 \times 10^{5\dagger}$
barn	square meter (m^2)	$1.0 \times 10^{-28\dagger}$
barrel (42 U.S. liquid gallons)	cubic meter (m^3)	0.1590
Bohr magneton (μ_B)	J/T	9.274×10^{-24}
Btu (International Table)	joule (J)	1.055×10^{3}
Btu (mean)	joule (J)	1.056×10^{3}
Btu (thermochemical)	joule (J)	1.054×10^{3}
bushel	cubic meter (m^3)	3.524×10^{-2}
calorie (International Table)	joule (J)	4.187
calorie (mean)	joule (J)	4.190
calorie (thermochemical)	joule (J)	$4.184^{\dagger}$
centipoise	pascal second (Pa·s)	$1.0 \times 10^{-3\dagger}$
centistokes	square millimeter per second (mm^2/s)	$1.0^{\dagger}$
cfm (cubic foot per minute)	cubic meter per second (m^3/s)	4.72×10^{-4}
cubic inch	cubic meter (m^3)	1.639×10^{-5}
cubic foot	cubic meter (m^3)	2.832×10^{-2}
cubic yard	cubic meter (m^3)	0.7646
curie	becquerel (Bq)	$3.70 \times 10^{10\dagger}$
debye	coulomb meter (C·m)	3.336×10^{-30}
degree (angle)	radian (rad)	1.745×10^{-2}
denier (international)	kilogram per meter (kg/m)	1.111×10^{-7}
	tex[‡]	0.1111
dram (apothecaries')	kilogram (kg)	3.888×10^{-3}
dram (avoirdupois)	kilogram (kg)	1.772×10^{-3}
dram (U.S. fluid)	cubic meter (m^3)	3.697×10^{-6}
dyne	newton (N)	$1.0 \times 10^{-5\dagger}$
dyne/cm	newton per meter (N/m)	$1.0 \times 10^{-3\dagger}$
electronvolt	joule (J)	1.602×10^{-19}
erg	joule (J)	$1.0 \times 10^{-7\dagger}$

†Exact.

‡See footnote on p. xvi.

To convert from	To	Multiply by
fathom	meter (m)	1.829
fluid ounce (U.S.)	cubic meter (m^3)	2.957×10^{-5}
foot	meter (m)	0.3048†
footcandle	lux (lx)	10.76
furlong	meter (m)	2.012×10^{-2}
gal	meter per second squared (m/s^2)	1.0×10^{-2}†
gallon (U.S. dry)	cubic meter (m^3)	4.405×10^{-3}
gallon (U.S. liquid)	cubic meter (m^3)	3.785×10^{-3}
gallon per minute (gpm)	cubic meter per second (m^3/s)	6.309×10^{-5}
	cubic meter per hour (m^3/h)	0.2271
gauss	tesla (T)	1.0×10^{-4}
gilbert	ampere (A)	0.7958
gill (U.S.)	cubic meter (m^3)	1.183×10^{-4}
grade	radian	1.571×10^{-2}
grain	kilogram (kg)	6.480×10^{-5}
gram force per denier	newton per tex (N/tex)	8.826×10^{-2}
hectare	square meter (m^2)	1.0×10^{4}†
horsepower (550 ft·lbf/s)	watt (W)	7.457×10^{2}
horsepower (boiler)	watt (W)	9.810×10^{3}
horsepower (electric)	watt (W)	7.46×10^{2}†
hundredweight (long)	kilogram (kg)	50.80
hundredweight (short)	kilogram (kg)	45.36
inch	meter (m)	2.54×10^{-2}†
inch of mercury (32°F)	pascal (Pa)	3.386×10^{3}
inch of water (39.2°F)	pascal (Pa)	2.491×10^{2}
kilogram-force	newton (N)	9.807
kilowatt hour	megajoule (MJ)	3.6†
kip	newton (N)	4.448×10^{3}
knot (international)	meter per second (m/S)	0.5144
lambert	candela per square meter (cd/m^3)	3.183×10^{3}
league (British nautical)	meter (m)	5.559×10^{3}
league (statute)	meter (m)	4.828×10^{3}
light year	meter (m)	9.461×10^{15}
liter (for fluids only)	cubic meter (m^3)	1.0×10^{-3}†
maxwell	weber (Wb)	1.0×10^{-8}†
micron	meter (m)	1.0×10^{-6}†
mil	meter (m)	2.54×10^{-5}†
mile (statute)	meter (m)	1.609×10^{3}
mile (U.S. nautical)	meter (m)	1.852×10^{3}†
mile per hour	meter per second (m/s)	0.4470
millibar	pascal (Pa)	1.0×10^{2}
millimeter of mercury (0°C)	pascal (Pa)	1.333×10^{2}†
minute (angular)	radian	2.909×10^{-4}
myriagram	kilogram (kg)	10
myriameter	kilometer (km)	10
oersted	ampere per meter (A/m)	79.58
ounce (avoirdupois)	kilogram (kg)	2.835×10^{-2}
ounce (troy)	kilogram (kg)	3.110×10^{-2}
ounce (U.S. fluid)	cubic meter (m^3)	2.957×10^{-5}
ounce-force	newton (N)	0.2780
peck (U.S.)	cubic meter (m^3)	8.810×10^{-3}
pennyweight	kilogram (kg)	1.555×10^{-3}
pint (U.S. dry)	cubic meter (m^3)	5.506×10^{-4}
pint (U.S. liquid)	cubic meter (m^3)	4.732×10^{-4}
poise (absolute viscosity)	pascal second (Pa·s)	0.10†
pound (avoirdupois)	kilogram (kg)	0.4536
pound (troy)	kilogram (kg)	0.3732
poundal	newton (N)	0.1383
pound-force	newton (N)	4.448
pound force per square inch (psi)	pascal (Pa)	6.895×10^{3}
quart (U.S. dry)	cubic meter (m^3)	1.101×10^{-3}
quart (U.S. liquid)	cubic meter (m^3)	9.464×10^{-4}

To convert from	To	Multiply by
quintal	kilogram (kg)	$1.0 \times 10^{2\dagger}$
rad	gray (Gy)	$1.0 \times 10^{-2\dagger}$
rod	meter (m)	5.029
roentgen	coulomb per kilogram (C/kg)	2.58×10^{-4}
second (angle)	radian (rad)	$4.848 \times 10^{-6\dagger}$
section	square meter (m^2)	2.590×10^{6}
slug	kilogram (kg)	14.59
spherical candle power	lumen (lm)	12.57
square inch	square meter (m^2)	6.452×10^{-4}
square foot	square meter (m^2)	9.290×10^{-2}
square mile	square meter (m^2)	2.590×10^{6}
square yard	square meter (m^2)	0.8361
stere	cubic meter (m^3)	$1.0^{\dagger}$
stokes (kinematic viscosity)	square meter per second (m^2/s)	$1.0 \times 10^{-4\dagger}$
tex	kilogram per meter (kg/m)	$1.0 \times 10^{-6\dagger}$
ton (long, 2240 pounds)	kilogram (kg)	1.016×10^{3}
ton (metric) (tonne)	kilogram (kg)	$1.0 \times 10^{3\dagger}$
ton (short, 2000 pounds)	kilogram (kg)	9.072×10^{2}
torr	pascal (Pa)	1.333×10^{2}
unit pole	weber (Wb)	1.257×10^{-7}
yard	meter (m)	$0.9144^{\dagger}$

†Exact.

BIBLIOGRAPHY

1. The International Bureau of Weights and Measures, BIPM (Parc de Saint-Cloud, France) is described in Appendix X2 of Ref. 4. This bureau operates under the exclusive supervision of the International Committee for Weights and Measures (CIPM).
2. *Metric Editorial Guide* (*ANMC-78-1*), latest ed., American National Metric Council, 5410 Grosvenor Lane, Bethesda, Md. 20814, 1981.
3. *SI Units and Recommendations for the Use of Their Multiples and of Certain Other Units* (*ISO 1000-1981*), American National Standards Institute, 1430 Broadway, New York, 10018, 1981.
4. Based on *ASTM E380-89a* (*Standard Practice for Use of the International System of Units* (*SI*)), American Society for Testing and Materials, 1916 Race Street, Philadelphia, Pa. 19103, 1989.
5. *Fed. Reg.*, Dec. 10, 1976 (41 FR 36414).
6. For ANSI address, see Ref. 3.

R. P. LUKENS
ASTM Committee E-43 on SI Practice

A-DNA

A. H.-J. Wang
H. Robinson

A-DNA was first recognized as a **DNA structure** using fiber X-ray diffraction analysis. **B-DNA** can be converted to A-DNA under conditions of low hydration, and the process is reversible. The A-DNA double helix is short and fat, with the base pairs and backbone wrapped farther away from the helix axis (see Fig. 2 of **DNA structure**). The **base pairs** are significantly tilted (~19°) with respect to the helix axis. The major groove has a very narrow width of ~3 Å and a depth of ~13 Å, whereas the minor groove has a broad width of 11 Å and a shallow depth of 3 Å. The base pairs also display a minor propeller twist.

There is increasing evidence that A-DNA may play an important role in biological processes such as protein recognition and **transcription** regulation, as in the **TATA-box** sequence bound with the TATA-box binding proteins (1,2). There is a sequence-dependent propensity to form A-DNA. Guanine-rich regions readily form A-DNA, whereas stretches of adenine resist it. The crystal structures of DNA oligonucleotides having guanine-rich sequences showed a characteristic intrastrand guanine–guanine stacking interaction in the A-DNA double helix, which may explain the propensity of these sequences to adopt the A-DNA conformation. Moreover, the packing of the helices in the crystal lattice revealed a characteristic pattern of the terminal base pairs from one helix abutting the minor groove surface of the neighboring helix, thus minimizing the accessibility of solvent to the wide minor groove. Since a low humidity environment favors formation of A-type helix, the displacement of surface solvent molecules by **hydrophobic** base pairs provides a driving force in stabilizing short oligonucleotides in the A-DNA conformation. Recently, it has been demonstrated that complex ions such as spermine, cobalt(III)hexamine, and **neomycin** can facilitate the B-DNA to A-DNA transition for DNA containing stretches of $(dG)_n \cdot (dC)_n$ sequences.

Finally, the crystal structures of a number of DNA:RNA hybrids, such as the self-complementary r(GCG)d(TATACGC), showed that DNA–RNA hybrid helices are of the A-DNA type. All the ribose and 2′-deoxyribose sugars are in the C3′-*endo* conformation. The 2′-hydroxyl groups of the ribose are involved in different types of hydrogen bonding to the adjacent nucleotides in the chain.

BIBLIOGRAPHY

1. J. L. Kim, D. B. Nikolov, and S. K. Burley (1993) *Nature* **365**, 520–527.
2. Y. Kim, J. H. Geiger, S. Hahn, and P. B. Sigler (1993) *Nature* **365**, 512–520.

ABL ONCOGENES

Stacey J. Baker
E. Premkumar Reddy

The *abl* gene was first identified as the transforming element of Abelson murine leukemia virus (A-MuLV), a replication-defective **retrovirus** that was isolated after inoculating Moloney murine leukemia virus (M-MuLV) into prednisolone-treated BALB/c mice (1). A-MuLV induces **B-cell** lymphomas *in vivo* and transforms both lymphoid and fibroblastic cells *in vitro*. The proviral genome of A-MuLV encodes a single polypeptide chain that is a fusion product of the virally-derived gag and cell-derived *abl* sequences (2). Sequence analysis of c-*abl* sequences revealed that, like the ***src* gene**, the *abl* gene codes for a **tyrosine kinase** that also contains the unique, SH3, SH2, and tyrosine kinase domains (3). Unlike the *src* gene, however, the *abl* gene product contains an additional C-terminal domain whose function is not entirely clear. In addition, the gene product of c-*abl* does not contain the negative-regulatory **tyrosine** residue at its C-terminus (see **Oncogenes**). The tyrosine kinase activity of the c-Abl protein is negatively regulated by its SH3 domain, which is deleted from the v-*abl* gene product. Interestingly, it was also found that the v-*abl* gene product contains a point mutation in its C-terminal sequence, which enhances its tyrosine kinase and transforming activities (4). Thus, both the v-*src* and v-*abl* genes have alterations in their regulatory sequences that result in the constitutive activation of their tyrosine kinase activities, which correlates with their transforming function.

Gene mapping studies have established that the c-*abl* oncogene is located on human chromosome 9q34, the location where the break point occurs in the Philadelphia chromosome. The Philadelphia chromosome is generated when a portion of the c-*abl* gene is **translocated** to chromosome 22 and is fused to a portion of the gene called *bcr*, which itself is disrupted during the translocation process (Fig. 1). This process results in generating a new gene, called *bcr-abl*, that has enhanced oncogenic activity and whose expression leads to the development of leukemia (5,6). It is interesting to note that the BCR-ABL gene is structurally similar to the *gag-abl* gene encoded by the Abelson murine leukemia virus. Both the *gag-abl* and *bcr-abl* genes exhibit high levels of tyrosine kinase activity, which is essential for their transforming activity.

The function of c-abl in normal cell growth is not fully established. The c-abl gene codes for two 145-kDa proteins as a result of **alternative splicing** of the two first exons (see **RNA splicing** and **Introns, exons**). This results in synthesizing two proteins that differ in their amino-terminal sequences. Both forms of c-Abl are found in the cytoplasm and in the nucleus. The c-Abl protein can bind to DNA and to **cell-cycle** regulators like the **retinoblastoma** protein. Recent studies show that c-abl gene expression is induced during cellular stress caused by agents, such as ionizing radiation and certain other genotoxic agents. These agents induce the formation of a complex involving c-Abl, DNA-dependent protein kinase, and Ku antigen (7). The DNA-dependent protein kinase in this

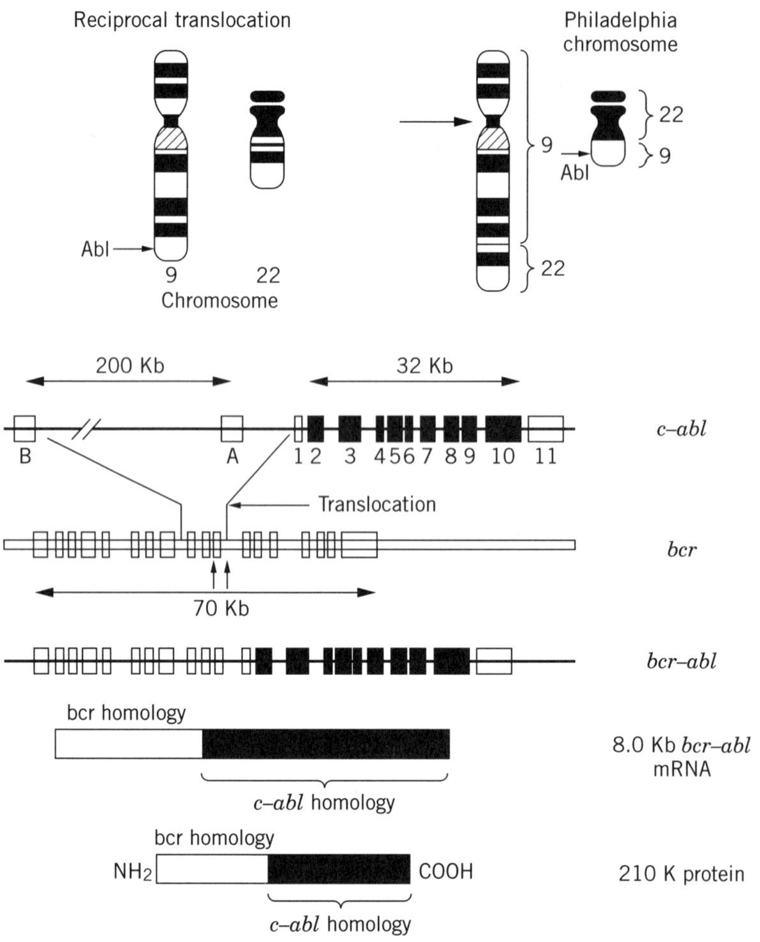

Figure 1. Generation of the bcr-abl oncoprotein by chromosomal translocation. The abl gene in a normal cell, is located on chromosome 9 and encodes a tyrosine kinase. During malignant transformation of myeloid cells, a portion of chromosome 9 that contains the abl locus translocates to chromosome 22 at the breakpoint cluster region (bcr) locus and generates the chimeric bcr-abl oncoprotein. Because the translocation results in deleting the sequences that negatively regulate abl tryosine kinase activity, the fusion protein has constitutive and increased levels of enzymatic activity.

complex is activated by **DNA damage** and in turn phosphorylates and activates c-Abl. Recent studies have also shown that c-Abl associates with the product of the ATM gene, which in turn activates c-Abl in response to ionizing radiation. DNA damage also induces binding of c-Abl to **p53** and contributes to cell-cycle arrest at the G1-phase, which is mediated by p53. Recent studies also show that the c-Abl protein binds to **protein kinase C-**δ and phosphorylates the latter, resulting in its activation and translocation to the nucleus, where it participates in inducing **apoptosis** (8). Thus, a substantial amount of evidence gathered in the past few years indicates that c-Abl protein has a pivotal role in mediating cellular growth arrest and the apoptotic effects that occur during exposure to ionizing radiation and genotoxic stress.

BIBLIOGRAPHY

1. H. T. Abelson and L. S. Rabson (1970) *Cancer Res.* **30**, 2213–2222.
2. E. P. Reddy, M. J. Smith, and A. Srinivasan (1983) *Proc. Natl. Acad. Sci. USA* **80**, 3623–3627.
3. C. Oppi, S. K. Shore, and E. P. Reddy (1987) *Proc. Natl. Acad. Sci. USA* **84**, 8200–8204.
4. S. K. Shore, S. L. Bogart, and E. P. Reddy (1990) *Proc. Natl. Acad. Sci. USA* **87**, 6502–6506.
5. J. Groffen, J. R. Stephenson, N. Heisterkamp, A. De Klein, C. B. Bartam, and G. Grosveld (1984) *Cell* **36**, 93–99.
6. E. Shtivelman, R. P. Lifshitz, R. P. Gale, and E. Canaani (1985) *Nature* **315**, 550–553.
7. R. Baskaran, L. D. Wood, L. L. Whittaker, Y. Xu, C. Barlow, C. E. Canman, S. E. Morgan, D. Baltimore, A. Wynshaw-Boris, M. B. Kastan, and J. Y. J. Wang (1997) *Nature* **387**, 516–519.
8. Z-M. Yuan, T. Utsugisawa, T. Ishiko, S. Nakada, Y. Huang, S. Kharbanda, R. Weichselbaum, and D. Kufe (1998) *Oncogene* **16**, 1643–1648.

ABSCISIC ACID

DOMINIQUE VAN DER STRAETEN
ANTJE ROHDE
MARC VAN MONTAGU

HISTORY

The first evidence for the existence of an acidic inhibitor of coleoptile growth that promoted abscission and seed maturation dates back to the early 1950s, but it was not until 1963 that abscisic acid (ABA) was identified by Frederick Addicott and coworkers (1,2). Addicott's team was studying compounds that stimulated abscission in cotton fruits and had named the active substances *abscisin I* and *abscisin II*; the latter proved to be ABA. Two other independent efforts also culminated in the discovery of ABA. A British group headed by Wareing (3) was investigating bud dormancy of woody plants and called the most active molecule *dormin*. In New Zealand, van Stevenick (4) studied compounds that accelerated abscission of flowers and fruits of *Lupinus luteus*. In 1964, it became evident that the three groups had discovered the same **plant hormone**, which was renamed *abscisic acid* 3 years later.

BIOSYNTHESIS AND METABOLISM

ABA is a universal compound in vascular plants; it is not found in bacteria, but has been reported in green algae, certain fungi, and mosses (5). ABA is a sesquiterpenoid, with mevalonic acid as a precursor. In certain fungi, ABA is produced by a direct, C15 pathway from farnesyl pyrophosphate. In higher plants, ABA is derived from xanthophylls, via an indirect C40 pathway (Fig. 1). Substantial progress in understanding the ABA biosynthetic pathway has been achieved by a combination of molecular and genetic techniques, primarily on mutants in *Arabidopsis thaliana, Zea mays, Nicotiana plumbaginifolia*, and tomato. **Arabidopsis** mutants impaired in ABA biosynthesis were isolated on the basis of their lack of seed dormancy due to ABA deficiency and their ability to overcome a **gibberellin** requirement for germination, which allowed them to germinate in the presence of inhibitors of gibberellin biosynthesis or in a gibberellin-deficient background (6,7). ABA-deficient mutants show a wilty **phenotype** when subjected to water stress.

Mevalonic acid is converted to farnesyl pyrophosphate, a C15 compound, via several intermediates (Fig. 1). The subsequent conversion of farnesyl pyrophosphate to zeaxanthin, a C40 carotenoid, again involves multiple steps (8). A number of *viviparous* (*vp*) mutants of maize identify loci corresponding to these early conversions. The transformation of zeaxanthin to all-*trans*-violaxanthin consists of two epoxidations at the double bonds in both cyclohexenyl rings, with antheraxanthin as an intermediate. The *aba1* mutant impaired in zeaxanthin epoxidase activity has been characterized biochemically in *Arabidopsis* (6,9,10). The *N. plumbaginifolia ABA2* **gene** encoding zeaxanthin epoxidase was recently cloned by **transposon** tagging (11). The gene encodes a **chloroplast**-imported **polypeptide chain** with similarity to bacterial monooxygenases and oxidases. When the Nicotiana *ABA2* gene was expressed in *Escherichia coli* heterologously, the protein was shown to catalyze both epoxidations *in vitro*. Moreover, the **complementary DNA** (cDNA) complemented both the *N. plumbaginifolia aba2* mutant and the *Arabidopsis aba1* mutant. The ortholog in *Arabidopsis* (*ABA1*) was cloned by **homology** with *Nicotiana ABA2* (11). Although presumably a null **allele**, the *Nicotiana aba2* mutant retains up to half the ABA content of wild type in the absence of water stress. This characteristic might indicate the existence of a secondary biosynthetic pathway for ABA. Downstream of all-*trans*-violaxanthin, two isomerizations yield 9′-*cis*-neoxanthin.

The subsequent oxidative cleavage with xanthoxin as a product is thought to be the rate-limiting step in ABA biosynthesis (12,13). A gene that most probably encodes the **enzyme** performing this reaction was cloned from maize, and by homology also from *Arabidopsis* (*VP14*) (13). Its overproduction in *E. coli* indicated that VP14 is probably a member of a novel class of dioxygenases.

Xanthoxin is subsequently converted to ABA-aldehyde by oxidation and isomerization. Biochemical analysis indicates that the *Arabidopsis aba2* mutant is blocked at this step in the pathway (7,14). A final oxidation, catalyzed by ABA-aldehyde oxidase, produces ABA. Mutants in this step have been identified in several species (8). The *Arabidopsis aba3* mutant lacks several aldehyde oxidase activities that require a molybdenum cofactor (14). Yet the mutation blocks modification of the mobybdenum cofactor, rather than impairing the activity of the apoprotein. Cloning of the *ABA2* and *ABA3* genes should provide further clues as to the regulation of ABA biosynthesis in these final steps.

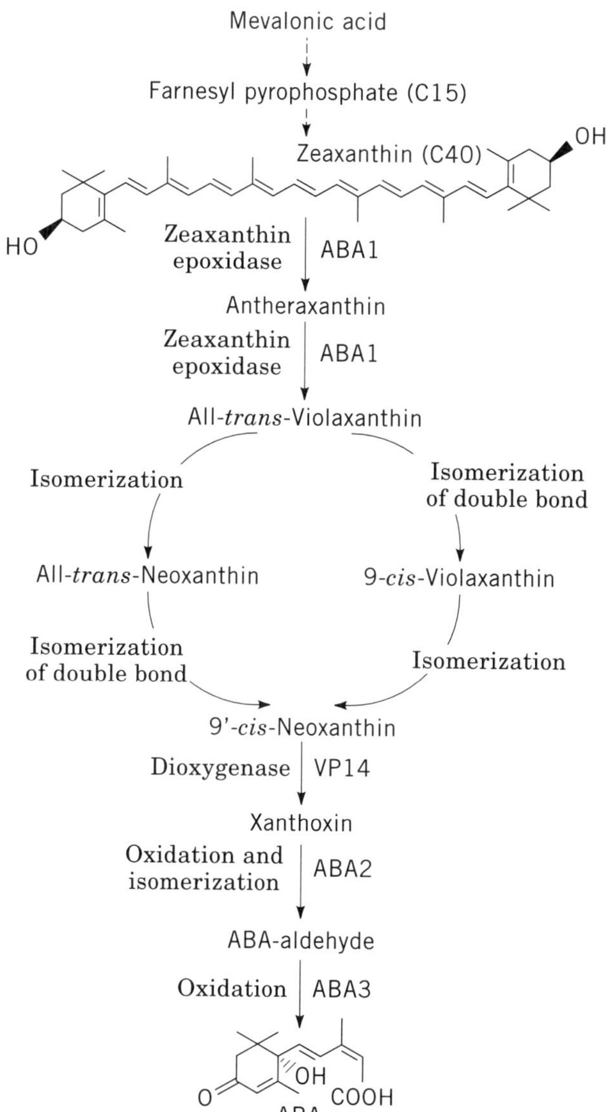

Figure 1. ABA biosynthesis in higher plants. Steps mediated by ABA1, ABA2, and ABA3 in *Arabidopsis* and by VP14 in *Zea mays* are indicated. (Adapted from Ref. 19.)

Metabolism of ABA is mainly to phaseic acid, dihydrophaseic acid, and their respective conjugates (15,16). Direct conjugation of ABA to an ABA glucose ester, or an ABA-β-glucopyranoside, can also occur. Conjugation of ABA does not lead to its storage, probably because it is irreversible and yields unstable compounds. With the exception of phaseic acid, none of the metabolites carries biological activity (16).

SIGNAL PERCEPTION AND TRANSDUCTION

Insight into the molecular basis of ABA responses was gathered by a combination of molecular-genetic, biochemical, and electrophysiological studies (12,17–20). Currently, little is known about the perception of ABA, despite numerous efforts to isolate ABA receptors using different approaches (20). Two groups of mutants, ABA-insensitive and ABA-hypersensitive, have largely contributed to the understanding of ABA **signal transduction**. Whereas the five *abscisic acid-insensitive* (*abi*) mutants of *Arabidopsis* were identified by their ability to germinate on inhibiting concentrations of ABA (21,22), the three *enhanced response to ABA* (*era*) mutants showed an increased dormancy in the presence of low ABA concentrations, compared to wild type (23). Besides their effects on seed development and dormancy, these mutants show altered responses to drought adaptation. The *abi* mutants differ from the above-mentioned *aba* mutants, because they do not have reduced endogenous ABA content and their phenotype cannot be reverted by exogenously supplied ABA. The characterization of *abi* mutants revealed that at least two ABA-response pathways exist: one primarily active in vegetative tissues, involving ABI1 and ABI2, and a second one operating predominantly during seed development and involving ABI3, ABI4, and ABI5 (19). Genes corresponding to four loci have been **cloned**. *ABI1* and *ABI2* encode highly homologous **serine/threonine protein phosphatases** 2C (24–26). These phosphatases might act in a phosphorylation/dephosphorylation cascade, as the protein phosphatase activity of recombinant ABI1 has been demonstrated *in vitro* (27). The functions of ABI1 and ABI2 are partially redundant (25). *ABI3* encodes a putative **transcription factor** that acts mainly during seed development and is the ortholog of *VP1* in maize (28–30). The suggested function of ABI3 consists of either activating the maturation program or preventing germination (30,31). Finally, the predicted amino acid sequence of ERA1 shares similarity with the β-subunit of farnesyl transferases. In yeast and mammalian systems, this enzyme is known to modify signal transduction proteins for **membrane anchoring** (23). In wild-type *Arabidopsis*, ERA1 is proposed to modify a negative regulator of ABA signaling.

The current model of ABA signal transduction integrates ABI1/ABI2-dependent cascades in both seeds and vegetative tissues (Fig. 2). These cascades interact with the ABI3 protein in seeds only, because ABI3 is expressed exclusively in seeds (32). The role of ABI3 is not confined to ABA signaling (19); it is also thought to mediate developmental signals. This role is supported by the seed phenotype of *abi3*, which is more complex than that of *abi1*, *abi2*, or *aba1* mutants (19). The evidence that VP1 can also regulate **transcription** independently of ABA further supports the possibility of a less strict limitation of ABI3 to ABA signaling (33,34). The same holds true for ABI4 and ABI5 (22).

Intermediates in the ABA transduction chain were revealed by microinjection of putative **second messengers** into hypocotyl cells of the *aurea* mutant of tomato (35). Following injection of ABA-inducible *KIN2* and *RD29A* **promoters** linked to β-glucuronidase, coinjection of either Ca^{2+}, cyclic ADP-ribose (cADP), ADP-ribosyl cyclase, or inositol (1,4,5)-triphosphate (IP_3) (see **Calcium signaling** and **Inositol phosphates**) resulted in expression of the ABA-inducible genes in the absence of ABA, implying that cADP ribose,

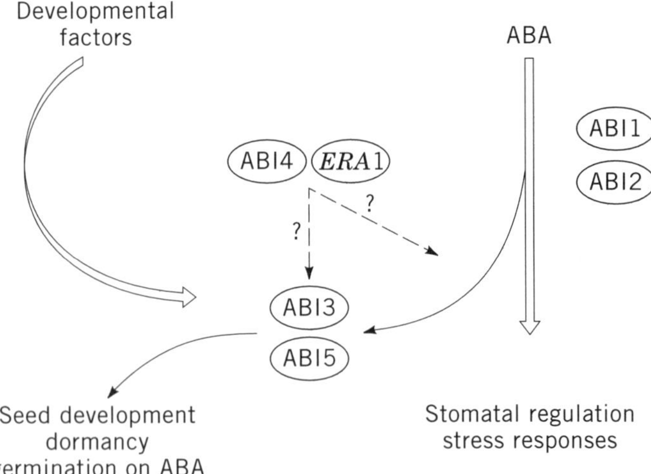

Figure 2. A model for ABA signal transduction.

a Ca^{2+}-mobilizing second messenger, is involved in ABA responses. At present, it is difficult to integrate these data into the current model for the ABA signal transduction pathway (19). Electrophysiological studies (voltage clamp or patch clamp) on the ABA-insensitive mutants *abi1* and *abi2* have clearly shown an altered **ion channel** behavior in their stomatal guard cells and provided further evidence that phosphorylation/dephosphorylation cascades are involved in ABA signaling (36,37).

Given the power of mutational approaches to identify signal transduction components, several groups are currently aiming at the identification of additional ABA signal transduction factors, exploiting novel screening procedures in *Arabidopsis* (12). At least eight *growth control via ABA* (*gca*) loci have been identified on the basis of their insensitivity to ABA inhibition of root growth (38). Furthermore, a *freezing sensitive* (*frs1*) mutant with a wilty phenotype appeared to be nonallelic to other ABA-deficient or -insensitive mutants of *Arabidopsis* (39). Other screens employ **transgenic** lines carrying chimeric promoter-**luciferase** constructs with promoters derived from ABA-responsive genes. Putative mutants are identified by aberrant expression of these genes in the mutagenized progeny (40). The identification of intragenic or extragenic **enhancers** and **suppressors** of known signaling components is yet another alternative.

Whereas part of the ABA signaling pathway is well established in seeds and stomatal guard cells, the implications of ABA signaling in the function of the vegetative meristem remain to be clarified. Furthermore, it will be most interesting to characterize the regulatory influence of light on ABA signaling, as well as the crosstalk with gibberellins, which are known to counteract ABA in many physiological processes (5).

DOWNSTREAM TARGETS

Of all the plant hormones, ABA is probably best known from the point of view of responsiveness at the target gene level. ABA-responsive genes encode proteins with diverse functions and are often induced under water stress (drought, salinity, low temperature). It is important to mention, however, that certain water stress-related genes are independent of ABA

(17,18). ABA-responsive genes can be classified into two groups. Primary ABA response genes are rapidly induced and are independent of **protein biosynthesis**, implying that *trans*-acting factors controlling these genes are under post-translational control. A second class of ABA response genes consists of those that require the expression of other genes.

***Cis*-acting** and ***trans*-acting** elements involved in ABA induction of **gene expression** have been studied extensively, often in relation to water stress (17,18). A bipartite model for regulation of transcription of the primary response genes is proposed, as for the barley *HVA1* and *HVA22* genes (41,42). The specificity of ABA responsiveness relies on the combination of an ABA **response element** (ABRE) and a coupling element. The ABRE has a **consensus sequence** (PyACGTGGC) that contains a G-box core sequence and has been shown to interact with bZIP factors (eg, wheat *Em* gene-binding protein EmBP-1) (43). A set of unique coupling elements would provide specificity in ABA response at the level of individual genes. In addition, in the case of *Em* induction, it has been demonstrated that a conserved domain of 18 amino acid residues of the VP1 protein enhances DNA-binding activity of the EmBP-1 bZIP factor, thereby acting as a DNA chaperone (44). A yeast **two-hybrid system** screen was used to identify proteins that interact specifically with VP1 on the promoter of the *Em* gene. One such protein was related to 14–3–3 proteins (45) and thus may help to stabilize and/or activate the regulatory complex at the *Em* promoter (20).

ABREs are not the only *cis* elements that confer ABA responsiveness. Both in the case of seed desiccation and in water stress conditions, other promoter elements can be involved. For example, ABA responses can be mediated in conjunction with the Sph box in germination-specific promoters. A direct interaction between the Sph box and a 140-residue conserved domain in the *C*-terminal region of VP1 has been demonstrated (46).

The second pathway for ABA-dependent gene expression requires protein biosynthesis, exemplified by the *Arabidopsis* ABA-inducible *RD22* gene. A 67-bp region in the promoter of this gene contains conserved motifs for MYC and MYB factor binding (47) (see **Oncogenes, oncoproteins**). Factors binding to these promoter elements (RD22BP-1 and ATMYB2) were shown to activate transcription of *RD22* in transient transactivation assays (48). Cooperative binding of these factors has been proposed to control ABA-dependent expression of *RD22*.

One of the challenges for the future consists in making the link between the ABA-induced proteins and the tissue- and development-specific physiological responses, summarized below.

EFFECTS

Although abscission was originally considered to be a process regulated by ABA, and the hormone was named accordingly, it is now clear that **ethylene** is the primary factor controlling abscission. Nevertheless, ABA also plays a key role in such diverse plant growth and developmental processes as seed maturation, germination, root growth, and stomatal closure (5). Stomatal movements involve changes in ion fluxes that occur very rapidly in response to ABA and do not require gene expression. Furthermore, ABA is a major stress signal orchestrating responses to dehydration stress, including water stress induced by drought and high-salt concentration as well as by low-temperature conditions.

BIBLIOGRAPHY

1. T. A. Bennet-Clark and N. P. Kefford (1953) *Nature* **171**, 645–647.
2. K. Ohkuma, O. E. Smith, J. L. Lyon, and F. T. Addicott (1963) *Science* **142**, 1592–1593.
3. P. F. Wareing (1956) *Annu. Rev. Plant Physiol.* **7**, 191–214.
4. R. F. M. van Stevenick (1957) *J. Exp. Bot.* **8**, 373–381.
5. P. J. Davies (1995) *Plant Hormones: Physiology, Biochemistry and Molecular Biology*, Kluwer, Dordrecht, The Netherlands.
6. M. Koornneef, M. L. Jorna, D. L. C. Brinkhorst-van der Swan, and C. M. Karssen (1982) *Theor. Appl. Genet.* **61**, 385–393.
7. K. M. Léon-Kloosterziel, M. Alvarez Gil, G. J. Ruijs, S. E. Jacobsen, N. E. Olszewski, S. H. Schwartz, J. A. D. Zeevaart, and M. Koornneef (1996) *Plant J.* **10**, 655–661.
8. I. B. Taylor (1991) In *Abscisic Acid, Physiology and Biochemistry* (W. J. Davies and H. G. Jones, eds.), Bios Scientific, Oxford, UK, pp. 23–37.
9. S. C. Duckham, R. S. T. Linforth, and I. B. Taylor (1991) *Plant Cell Environ.* **14**, 631–636.
10. C. D. Rock and J. A. D. Zeevaart (1991) *Proc. Natl. Acad. Sci. USA* **88**, 7496–7499.
11. E. Marin, L. Nussaume, A. Quesada, M. Gonneau, B. Sotta, P. Hugueney, A. Frey, and A. Marion-Poll (1996) *EMBO J.* **15**, 2331–2342.
12. M. Koornneef, K. M. Léon-Kloosterziel, S. H. Schwartz, and J. A. D. Zeevaart (1998) *Plant Physiol. Biochem.* **36**, 83–89.
13. S. H. Schwartz, K. M. Léon-Kloosterziel, M. Koornneef, and J. A. D. Zeevaart (1997) *Plant Physiol.* **114**, 161–166.
14. S. H. Schwartz, B. C. Cai, D. A. Gage, J. A. D. Zeevaart, and D. R. McCarty (1997) *Science* **276**, 1872–1874.
15. J. A. D. Zeevaart and R. A. Creelman (1988) *Annu. Rev. Plant Physiol. Plant Mol. Biol.* **39**, 439–473.
16. G. Sembdner, R. Atzorn, and G. Schneider (1994) *Plant Mol. Biol.* **26**, 1459–1481.
17. J. Ingram and D. Bartels (1996) *Annu. Rev. Plant Physiol. Plant Mol. Biol.* **47**, 377–403.
18. K. Shinozaki and K. Yamaguchi-Shinozaki (1996) *Curr. Opin. Biotechnol.* **7**, 161–167.
19. S. Merlot and J. Giraudat (1997) *Plant Physiol.* **114**, 751–757.
20. R. S. Quatrano, D. Bartels, T. H. D. Ho, and M. Pagés (1997) *Plant Cell* **9**, 470–475.
21. M. Koornneef, G. Reuling, and C. M. Karssen (1984) *Physiol. Plant.* **61**, 377–383.
22. R. R. Finkelstein (1994) *Plant J.* **5**, 765–771.
23. S. Cutler, M. Ghassemian, D. Bonetta, S. Cooney, and P. McCourt (1996) *Science* **273**, 1239–1241.
24. J. Leung, M. Bouvier-Durand, P.-C. Morris, D. Guerrier, F. Chefdor, and J. Giraudat (1994) *Science* **264**, 1448–1452.
25. J. Leung, S. Merlot, and J. Giraudat (1997) *Plant Cell* **9**, 759–771.
26. K. Meyer, M. P. Leube, and E. Grill (1994) *Science* **264**, 1452–1455.
27. N. Bertauche, J. Leung, and J. Giraudat (1996) *Eur. J. Biochem.* **241**, 193–200.
28. D. R. McCarty, T. Hattori, C. B. Carson, V. Vasil, M. Lazar, and I. K. Vasil (1991) *Cell* **66**, 895–905.
29. J. Giraudat, B. M. Hauge, C. Valon, J. Smalle, F. Parcy, and H. M. Goodman (1992) *Plant Cell* **4**, 1251–1261.
30. F. Parcy, C. Valon, M. Raynal, P. Gaubier-Comella, M. Delseny, and J. Giraudat (1994) *Plant Cell* **6**, 1567–1582.

31. E. Nambara, K. Keith, P. McCourt, and S. Naito (1995) *Development* **121**, 629–636.
32. F. Parcy and J. Giraudat (1997) *Plant J.* **11**, 693–702.
33. V. Vasil, W. R. Marcotte Jr., L. Rosenkrans, S. M. Cocciolone, I. K. Vasil, R. S. Quatrano, and D. R. McCarty (1995) *Plant Cell* **7**, 1511–1518.
34. C.-Y. Kao, S. M. Cocciolone, I. K. Vasil, and D. R. McCarty (1996) *Plant Cell* **8**, 1171–1179.
35. Y. Wu, J. Kuzma, E. Maréchal, R. Graeff, H. C. Lee, R. Foster, and C.-H. Hua (1997) *Science* **278**, 2126–2130.
36. F. Armstrong, J. Leung, A. Grabov, J. Brearley, J. Giraudat, and M. R. Blatt (1995) *Proc. Natl. Acad. Sci. USA* **92**, 9520–9524.
37. Z.-M. Pei, K. Kuchitsu, J. M. Ward, M. Schwarz, and J. I. Schroeder (1997) *Plant Cell* **9**, 409–423.
38. G. Benning, T. Ehrler, K. Meyer, M. Leube, P. Rodriguez, and E. Grill (1996) In *Abscisic Acid Signal Transduction in Plants.* Centro de Reuniones Internacionales sobre Biología Workshop 60 (R. S. Quatrano and M. Pagès, eds.), Instituto Juan March de Estudios e Investigaciones, Madrid, p. 34.
39. J. Salinas, F. Llorente, and J. M. Martinez-Zapater (1996) In *Abscisic Acid Signal Transduction in Plants.* Centro de Reuniones Internacionales sobre Biología Workshop 60 (R. S. Quatrano and M. Pagès, eds.), Instituto Juan March de Estudios e Investigaciones, Madrid, p. 35.
40. N.-H. Chua, Y. Wu, and R. Foster (1996) In *Abscisic Acid Signal Transduction in Plants.* Centro de Reuniones Internacionales sobre Biología Workshop 60 (R. S. Quatrano and M. Pagès, eds.), Instituto Juan March de Estudios e Investigaciones, Madrid, p. 55.
41. Q. Shen and T.-H. D. Ho (1995) *Plant Cell* **7**, 295–307.
42. Q. Shen, P. Zhang, and T.-H. D. Ho (1996) *Plant Cell* **8**, 1107–1119.
43. M. J. Guiltinan, W. R. Marcotte Jr., and R. S. Quatrano (1990) *Science* **250**, 267–271.
44. A. Hill, A. Nantel, C. D. Rock, and R. S. Quatrano (1996) *J. Biol. Chem.* **271**, 3366–3374.
45. T. F. Schultz, J. Medina, A. Hill, and R. S. Quatrano (1998) *Plant Cell* **10**, 837–847.
46. M. Suzuki, C. Y. Kao, and D. R. McCarty (1997) *Plant Cell* **9**, 799–807.
47. T. Iwasaki, K. Yamaguchi-Shinozaki, and K. Shinozaki (1995) *Mol. Gen. Genet.* **247**, 391–398.
48. H. Abe, K. Yamaguchi-Shinozaki, T. Urao, T. Iwasaki, D. Hosokawa, and K. Shinozaki (1997) *Plant Cell* **9**, 1859–1868.

ABSORPTION SPECTROSCOPY

FRANZ X. SCHMID

Light in the ultraviolet (UV) and visible (vis) range of the electromagnetic spectrum shows an energy that is equivalent to about 150 to 400 kJ/mol. Light with the appropriate energy is used to promote electrons from the ground state to an excited state. The absorption of energy from the incident light as a function of its wavelength is measured in absorption spectroscopy. Molecules with electrons that participate in delocalized aromatic systems often absorb light in the near-UV or visible region.

Absorption spectroscopy is usually performed on solutions of molecules in a transparent solvent. The absorbance of a solute depends linearly on its concentration, and therefore absorption spectroscopy is ideally suited for quantitative measurements. The spectral properties of a molecule depend on the molecular environment and the mobility of its chromophores. Absorption spectroscopy and difference spectroscopy (see **Difference spectroscopy**) are therefore well-suited to follow **enzyme**-catalyzed reactions, **ligand binding**, and conformational transitions in proteins and nucleic acids. Spectroscopic measurements are very sensitive, nondestructive, and require only small amounts of material for analysis. Spectrophotometers are standard laboratory equipment, and the measurement of absorbance is technically simple.

ABSORBANCE OF PROTEINS

In proteins, the **peptide bond** absorbs light in the range of 180 to 230 nm (which is called the "far-UV" range). The aromatic residues, **tyrosine** (Tyr), **tryptophan** (Trp), and **phenylalanine** (Phe), also absorb light in this region and, in addition, show bands near 260 to 280 nm (in the "near-UV"). **Disulfide bonds** absorb weakly near 260 nm. Some protein cofactors, such as the heme group, show absorbance in the visible range. When the peptide groups and aromatic residues are part of an asymmetric structure, or when they are immobilized within an asymmetric environment (as in folded proteins), left- and right-handed circularly polarized light is absorbed to different extents. This phenomenon is called **circular dichroism** (CD).

Spectra of the aromatic amino acids and model proteins are found in (1). In the near-UV, the molar absorbance of a phenylalanine residue is much smaller than that of a tyrosine or a tryptophan, and the spectrum of a protein between 240 and 300 nm is therefore dominated by the contributions from the Tyr and Trp side-chains. Phe residues contribute fine structure ("wiggles") to the spectrum between 250 and 260 nm. The aromatic amino acids do not absorb above 310 nm, and therefore protein absorbance should be zero at wavelengths greater than 310 nm. Proteins without Trp residues do not absorb above 300 nm (see Fig. **1c**).

A fraction of the aromatic residues are buried in the **hydrophobic** core of a native, folded protein molecule (see **Protein structure**). When these residues become exposed to the aqueous solvent upon unfolding, their absorption is shifted slightly to shorter wavelengths. This is clearly seen for two forms of **ribonuclease T_1** (Figs. **1a** and **1c**). The difference spectra in Figures **1b** and **d** show that the maximal differences in absorbance occur in the 285 to 295 nm region. The **difference spectrum** for the form without a Trp residue (Fig. **1d**) shows a prominent maximum at 287 nm. It is typical for proteins that contain Tyr residues only. The form with a single Trp residue (Fig. **1b**) shows additional shoulders between 290 and 300 nm in the difference spectrum. They originate from the single Trp residue of this protein. The differences in the absorption spectra between the native and unfolded state of a protein are generally small, but they can be determined with good accuracy and are extremely useful for monitoring conformational changes of a protein.

Spectroscopic methods are, in general, the methods of choice (i) to investigate changes in the behavior of a protein under different solvent conditions, and (ii) to compare the properties of related molecules, such as homologous or mutated forms of a protein. In addition, they are widely used to measure **protein stability** and to follow structural transitions such as unfolding and refolding under a variety of conditions. Absorbance

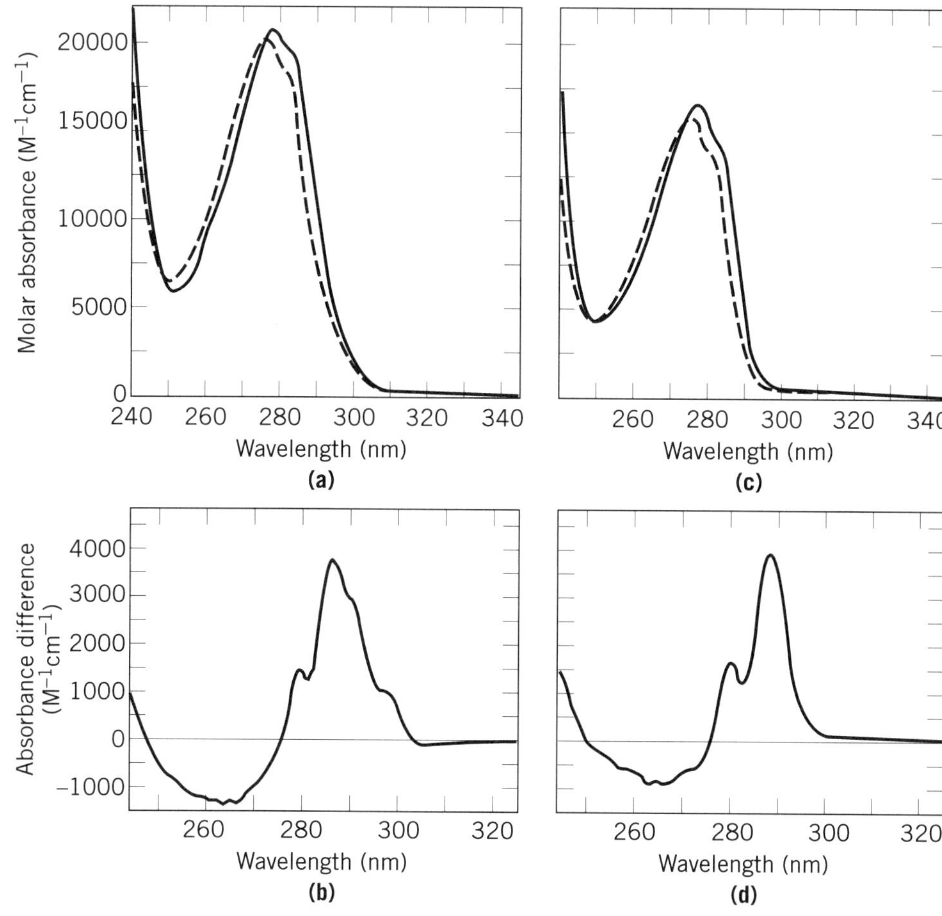

Figure 1. Ultraviolet absorption spectra of (**a**) the wild-type form and (**c**) the Trp59Tyr variant of RNase T_1. The spectra of the native proteins (in 0.1 *M* sodium acetate, pH 5.0) are shown by the continuous lines. The spectra of the unfolded proteins (in 6.0 *M* GdmCl in the same buffer) are shown by broken lines. The difference spectra between the native and unfolded forms are shown in (**b**) and (**d**). Spectra of 15 μM protein were measured at 25°C in 1-cm cuvettes in a double-beam instrument with a band width of 1 nm at 25°C. The spectra of the native and unfolded proteins were recorded successively, stored, and subtracted. From Ref. 1.

changes during fast reactions that occur in the milliseconds range can be followed by using rapid mixing techniques, such as in stopped-flow spectrometry (see **Kinetics**).

ABSORBANCE OF NUCLEIC ACIDS

Nucleic acids show a strong absorbance in the region of 240 to 275 nm. It originates from the $\pi \rightarrow \pi*$ transitions of the **pyrimidine** and **purine** ring systems of the nucleobases. The bases can be protonated, and therefore the spectra of DNA and RNA are sensitive to pH. At neutral pH, the absorption maxima range from 253 nm (for guanosine) to 271 nm (for cytidine), and therefore polymeric DNA and RNA show a broad and strong absorbance near 260 nm.

In native DNA, the bases are stacked in the hydrophobic core of the double helix (see **Base stacking**), and therefore their absorbance is considerably decreased relative to the absorbance of single-stranded DNA, and even more so relative to oligonucleotides. This phenomenon is called hypochromism. It is widely used to follow the melting of DNA double helices.

ABSORBANCE TO DETERMINE CONCENTRATIONS

Absorbance measurements are the methods of choice to determine the concentration of proteins or nucleic acids in solution. Spectrophotometers are standard laboratory equipment, and absorbance can be measured quickly and accurately. The absorbance A is related with the intensity of the light before I_0 and after I passage through the protein solution by Equation 1, and the absorbance depends linearly on concentration, according to the Lambert–Beer relationship (Eq. 2):

$$A = -\log_{10}(I/I_0) \quad (1)$$

$$A = \varepsilon \times c \times l \quad (2)$$

where c is the molar concentration, l the pathlength in cm, and ε the molar absorption coefficient. The concentration of a substance in solution can be determined very rapidly and accurately from its absorbance by using Equation 2. The measurement of absorbance values greater than 2 should be avoided, because only 1% of the incident light is transmitted through a solution with an absorbance of 2 (and is quantified by the photomultiplier). The value of ϵ can be determined by a number of experimental techniques (2) or can be calculated for proteins by adding up the contributions of the constituent aromatic amino acids of a protein (2,3).

The absorbance of nucleic acids does not vary much. The concentrations of nucleic acids in solution are routinely determined from the absorbance at 260 nm. In fact, amounts of nucleic acid are often given as "A_{260} units." For double-stranded DNA, one A_{260} unit is equivalent to 50 μg DNA; for single-stranded DNA, it is equivalent to 33 μg DNA; for single-stranded RNA, it is equivalent to 40 μg RNA. All these amounts would produce an A_{260} of 1 when dissolved in 1 mL and measured in a 1-cm cuvette. Proteins absorb much more weakly than nucleic acids. In a 1:1 mixture of nucleic acids and proteins, the proteins contribute only about 2% to the total absorbance at 260 nm. Consequently, these quantities of contaminating protein hardly affect the concentrations of nucleic acids measured by A_{260}.

ABSORBANCE SPECTROPHOTOMETERS

Absorbance is measured by a spectrophotometer. Spectrophotometers consist usually of two light sources: a deuterium lamp, which emits light in the UV region, and a tungsten/halogen lamp for the visible region. After passing through a monochromator (or through optical filters), the light is focused into the cuvette, and the amount of light that passed through the sample is detected by a photomultiplier or photodiode. In diode array spectrophotometers, the sample is illuminated by the full lamp light; after passage through the cuvette, the transmitted light is spectrally decomposed by a prism into the individual components and quantified by an array of diodes, often in intervals of 2 nm. In diode array spectrometers, the entire spectrum is recorded at the same time and not by a time-dependent scan as in conventional instruments. This is of advantage for measuring time-dependent changes at several wavelengths simultaneously (see **Kinetics**).

BUFFERS FOR ABSORBANCE SPECTROSCOPY

Good **buffers** for measuring difference spectra ideally should not absorb light in the wavelength range of the experiment. For work in the near-UV, buffer absorbance should be small above 230 nm, and indeed most of the solvents commonly used in biochemical experiments do not absorb in this spectral region (1). Buffer absorbance is a major problem, however, in the far-UV region below 220 nm, because buffers that contain carboxyl and/or amino groups absorb light in this wavelength range. Buffers with negligible absorbance in the far-UV include phosphate, cacodylate, and borate. Detailed procedures for the measurement of difference spectra are found in Ref. 1.

BIBLIOGRAPHY

1. F. X. Schmid (1997) Optical spectroscopy to characterize protein conformation and conformational changes. In *Protein Structure: A Practical Approach*, 2nd ed. (T. E. Creighton, ed.), IRL Press, Oxford, UK, pp. 261–297.
2. S. C. Gill and P. H. von Hippel (1989) *Anal. Biochem.* **182**, 319–326.
3. C. N. Pace and F. X. Schmid (1997) How to determine the molar absorption coefficient of a protein. In *Protein Structure: A Practical Approach*, 2nd ed. (T. E. Creighton, ed.), IRL Press, Oxford, UK, pp. 253–259.

Suggestions for Further Reading

S. B. Brown (1980) Ultraviolet and visible spectroscopy. In *An Introduction to Spectroscopy for Biochemists* (S. B. Brown, ed.), Academic Press, London, pp. 14–69.

D. B. Gordon (1994) Spectroscopic techniques. In *Principles and Techniques of Practical Biochemistry* (K. Wilson and J. Walker, eds.), Cambridge University Press, Cambridge, UK, Chap. 7, pp. 324–380.

D. A. Harris and C. L. Bashford (1987) *Spectrophotometry and Spectrofluorimetry: A Practical Approach*, IRL Press, Oxford, UK.

ACCEPTOR STEM

C. Florentz
N. C. Martin

The acceptor stem is the site of attachment of **amino acids** to **transfer RNA** (tRNA). It is formed by 7 base pairs and has 4 single-stranded nucleotides. Nucleotides 1 to 7 from the 5′ end of the tRNA base pair with nucleotides 72–66, respectively, from the 3′ end of the molecule. Whereas the 5′ end of the RNA has a monophosphate group, the 3′ end contains a 3′-hydroxyl group, which is the site of esterification to the amino acid. The four 3′-end single-stranded nucleotides include residue 73, the discriminator base and the well-conserved nucleotides C74, C75, A76, the 3′-CCA sequence. In the three-dimensional structures of tRNA, the acceptor stem is stacked on the T arm, forming the acceptor "branch" of the L-shaped RNA fold (see Fig. 1 of **Transfer RNA**). The acceptor branch usually forms a regular A-type double helix, with the four single-stranded nucleotides extending in a regular helical continuity. In tRNA-like domains of some **viruses**, the acceptor stem is formed by a single RNA chain that folds into a helix due to the presence of a **pseudoknot** (1).

Specific aminoacylation of tRNA by their cognate aminoacyl-tRNA synthetases is dependent on the presence of a series of identity elements. Limited in number, these elements are preferentially located in the **anticodon** loop and in the acceptor stem (2–4). Residue 73, next to the CCA end, is called the "discriminator" base. The hypothesis that it contributes to discrimination of tRNA by cognate **aminoacyl-tRNA synthetases** (5) has been largely confirmed. Residue 73 contributes strongly to specific aminoacylation of 17 different *Escherichia coli* tRNAs (6). It is the element making the major thermodynamic contribution toward aminoacylation of *E. coli* $tRNA^{Cys}$, *E. coli* $tRNA^{His}$, yeast $tRNA^{His}$ and *E. coli* $tRNA^{Leu}$. Replacement of A73 to G73 in human $tRNA^{Ser}$ converts it to a tRNA for isoleucine (7). The crystallographic structures of two tRNA-cognate synthetase complexes suggests two possible mechanisms by which the discriminator base contributes to aminoacylation specificity. One is a direct mechanism, involving direct hydrogen interactions with the synthetase. The other is an indirect mechanism that confers a conformational change to the acceptor end of the tRNA to facilitate aminoacylation. In the complex of yeast $tRNA^{Asp}$/aspartyl-tRNA synthetase, nucleotide G73 fits into the **active site**, forming **hydrogen bond** interactions with side chains of the synthetase. It does not cause a conformational change in the acceptor stem (8). Alternatively, the 2-amino group of G73 of *E. coli* $tRNA^{Gln}$ hydrogen bonds with the phosphate oxygen of the previous nucleotide, folding the backbone of G73 back toward the 3′ end of the tRNA. The formation of this fold-back hairpin enables the synthetase to open the first base pair of the acceptor stem and reach the second and third base pairs for specific interactions (9). In both complexes, additional contacts exist between the enzyme and the acceptor stems of the tRNA. The aspartyl-tRNA synthetase interacts with the $tRNA^{Asp}$ acceptor stem via the major groove of the RNA helix, whereas glutaminyl-tRNA synthetase interacts via the minor groove.

Acceptor stem identity-element nucleotides, as well as additional structural features, are important for synthetase recognition. Alanyl-tRNA synthetase is sensitive to both the exocyclic NH_2 group of G3 and local conformation of the helix due to structural characteristics of the G-U **wobble** pair (10–13). Alanine acceptance by alanyl-tRNA is modulated by additional signals within the acceptor stem, namely, A73, G1-C72, G2-C71 and G4-C69. The 5′ end of histidine-specific tRNA has an additional nucleotide, residue number −1, which is a guanosine. This nucleotide, opposite the discriminator base, does form a base pair with the discriminator in many, but not all, histidyl-tRNA. This −1 guanosine is the major histidine identity element. Its influence is complemented by base pairs U2-A71 and G3-C70. Base pairs 1–72, 2–71, 3–70, and/or 4–69 contribute to aminoacylation in several cases. Glutamine identity requires a weak 1–72 base pair, G2-C71, G3-C70, in addition to signals elsewhere in the tRNA. Glycine identity is dependent on C2-G71, G3-C70 sequences, and serine identity involves G2-C71, among other signals. Methionine identity is based on the presence of A73, G2-C71, C3-G70, in addition to the CAU anticodon.

At the three-dimensional level, tRNA molecules fold into a two-domain L-shaped structure with the amino acid acceptor terminus and the anticodon at opposite ends. Minihelices (small RNA molecules containing only part of a tRNA) that mimic the amino acid acceptor domain, specifically, the acceptor stem stacked on top of the TY stem, have been shown to be useful tools for determining the contribution of the acceptor stem to tRNA function (14,15). Minihelices containing the identity elements from a tRNA are efficient substrates for its cognate aminoacyl-tRNA synthetases. Thus, alanyl-tRNA synthetase, glycyl-tRNA synthetase, and histidyl-tRNA synthetase aminoacylate minihelices efficiently. Minihelices containing partial identity sets of six additional tRNAs are specific substrates for their cognate synthetases. Since identity elements are located very close to the CCA end, minihelices may be reduced in size to microhelices, consisting only of the amino acceptor stem closed by a loop and remain active. The smallest substrate for an aminoacyl-tRNA synthetase is derived from yeast $tRNA^{Asp}$ and consists of only 14 nucleotides. There are three base pairs closed by a tetraloop, plus the discriminator base and the CCA sequence (16). Double-stranded RNA duplexes and RNA/DNA heteroduplexes may also be aminoacylated (15).

The aminoacylation efficiency of minihelices is generally reduced significantly compared to that of the corresponding full-length tRNA. The amino acid charging is, however, very specific and depends on just a few nucleotides. These minimal RNA substrates are devoid of the **anticodons** that read the **genetic code** and provide the link between amino acid and **codon**. Since the relationship between the sequences and structures of the minisubstrates and the specific amino acids is maintained, there appears to be an *operational RNA code* for primitive aminoacylation. This operational code may be the primitive code from which the contemporary genetic code evolved (17–20)

Acceptor stem properties are important for tRNA recognition by proteins other than synthetases. Aminoacylated initiator tRNA in **prokaryotes** is a substrate for a methionyl-tRNA transformylase that converts methionyl-$tRNA^{Met}$ to formylmethionine $tRNA^{Met}$. Recognition of initiator tRNA by the formylase depends on the presence of methionine. Sequence and/or structural elements in the tRNA that are important for formylation by methionyl-tRNA transformylase are clustered at the end of the acceptor stem (21). The key determinants appear to be a mismatch or a weak base pair between nucleotides 1 and 72, a G-C base pair between nucleotides 2 and 71, and a C-G or, less preferably, a G-C base-pair between nucleotides 3 and 70. Mutations at G4-C69 also affect formylation kinetics slightly. In addition to the positive elements A73, G2-C71, C3-G70, and G4-C69, the occurrence of a G-C or a C-G base pair between positions 1 and 72 acts as a major negative determinant for the formylase. Formylation is a prerequisite for interaction with initiation factor IF2, which delivers the initiator tRNA to the P site of the **ribosome**. Special structural features within the acceptor stem of eubacterial initiator tRNA contribute to their discrimination from elongator tRNA. They lack a Watson–Crick base pair between nucleotides 1 and 72 at the end of the acceptor stem. Eukaryotic initiator tRNA almost always has an A1-U72 pair, a feature not found in eukaryotic elongator tRNA.

A high-resolution **X-ray crystallography** structure of the ternary complex, formed by aminoacylated yeast $tRNA^{Phe}$, **elongation factor** Tu from *Thermus aquaticus*, and an analog of GTP, revealed numerous contacts between the protein and the acceptor stem of the tRNA (22). These contacts include specific interactions with residue A76, with phosphates 74, 75, 67, 64, 3, and with riboses 2, 3, 63, 64, 65 along the acceptor stem and the T stem.

The stability of the 1–72 base pair also governs the degree of sensitivity of a peptidyl-tRNA to peptidyl-tRNA hydrolase. This enzyme converts the peptidyl-tRNA of *N*-acetylaminoacyl-tRNA into free tRNA plus peptides or *N*-acetyl amino acids. It is believed to play a role in the translational apparatus through the recycling of free tRNA from immature peptidyl-tRNA created by abortive protein synthesis (23).

The acceptor stem also contains specific information required during synthesis of the tRNA. **Ribonuclease P**, an enzyme that removes extra nucleotides from the 5′ end of tRNA during **tRNA biosynthesis**, recognizes the -74CCA76- sequence in addition to nucleotides within the TY loop. Moreover, this enzyme locates its cleavage site by "measuring" the length of the helix formed by the amino acid acceptor stem fused to the TY stem. Processing at the 3′ end of the tRNA, as well as the addition and repair of the conserved CCA sequence by ATP(CTP)-tRNA nucleotidyltransferase, requires information within the acceptor stem. This enzyme interacts with tRNA at the corner of the structure, where the T and D loops interact, and extends that interaction across the aminoacyl stem to the 3′ end (24,25).

BIBLIOGRAPHY

1. K. Rietveld, C. W. A. Pleij, and L. Bosch (1983) *EMBO J.* **2**, 1079–1085.
2. R. Giegé, J. D. Puglisi, and C. Florentz (1993) *Prog. Nucleic Acid Res. Mol. Biol.* **45**, 129–206.
3. W. H. McClain (1993) *J. Mol. Biol.* **234**, 257–280.
4. M. E. Saks and J. R. Sampson (1995) *J. Mol. Evol.* **40**, 509–518.
5. D. M. Crothers, T. Seno, and D. G. Söll (1972) *Proc. Natl. Acad. Sci. USA* **69**, 3063–3067.
6. Y. M. Hou (1997) *Chem. Biol.* **4**, 93–96.
7. K. Breitshopf and H. J. Gross (1994) *EMBO J.* **13**, 3166–3169.
8. J. Cavarelli, B. Rees, M. Ruff, J.-C. Thierry, and D. Moras (1993) *Nature* **362**, 181–184.
9. M. A. Rould, J. J. Perona, D. Sâll, and T. A. Steitz (1989) *Science* **246**, 1135–1142.

10. Y. M. Hou and P. Schimmel (1988) *Nature* **333**, 140–145.
11. Y. M. Hou and P. Schimmel (1989) *Biochemistry* **28**(17), 6800–6804.
12. W. H. McClain and K. Foss (1988) *Science* **240**, 793–796.
13. K. Musier-Forsyth and P. Schimmel (1992) *Nature* **357**, 513–515.
14. C. Francklyn, K. Musier-Forsyth, and P. Schimmel (1992) *Eur. J. Biochem.* **206**, 315–321.
15. S. A. Martinis, and P. Schimmel (1995) In *tRNA: Structure, Biosynthesis, and Function* (D. Söll, and U. L. RajBhandary, eds.), American Society for Microbiology Press, Washington, DC, pp. 349–370.
16. M. Frugier, C. Florentz, and R. Giegé (1994) *EMBO J.* **13**, 2218–2226.
17. P. Schimmel, R. Giegé, D. Moras, and S. Yokoyama (1993) *Proc. Natl. Acad. Sci. USA* **90**, 8763–8768.
18. P. Schimmel and L. Ribas de Pouplana (1995) *Cell* **81**, 983–986.
19. P. Schimmel (1995) *J. Mol. Evol.* **40**, 531–536.
20. P. Schimmel (1996) *Proc. Natl. Acad. Sci. USA* **93**, 4521–4522.
21. J. M. Guillon, T. Meinel, Y. Mechulam, C. Lazennec, S. Blanquet, and G. Fayat (1992) *J. Mol. Biol.* **224**, 359–367.
22. P. Nissen, M. Kjeldgaard, S. Thirup, G. Polekhina, L. Reshetnikova, B. F. C. Clark, and J. Nyborg (1995) *Science* **270**, 1464–1472.
23. S. Dutka, T. Meinnel, C. Lazennec, Y. Mechulam, and S. Blanquet (1993) *Nucleic Acids Res.* **21**, 4025–4030.
24. P. Spacciapoli, L. Doviken, J. J. Mulero, and D. L. Thurlow (1989) *J. Biol. Chem.* **264**, 3799–3805.
25. L. A. Hegg, and D. L. Thurlow (1990) *Nucleic Acids Res.* **18**, 5975–5979.

ACCESSIBLE SURFACE

K. GEKKO

The surfaces of folded macromolecules, especially proteins, and the internal packing of their atoms have generally been analyzed using the procedure of Lee and Richards (1). The cross section of part of the surface of a native protein is depicted in Figure 1, which demonstrates a number of different surfaces and volumes (see **Protein structure**). The **van der Waals surface** is defined by the spherical atoms that comprise the structure, but it is not very relevant to a folded macromolecule like a protein, where internal atoms and cavities are normally inaccessible to the solvent.

The most relevant surface is the accessible surface, which is defined by its contact with molecules of the solvent. This surface is defined by rolling a spherical probe of appropriate radius R_1 on the outside of the molecule, while maintaining contact with the van der Waals surface. For most purposes, the appropriate probe is a **water** molecule. The accessible surface is that depicted by the center of the probe as it moves over the surface of the protein. In Figure 1 the probe does not contact atoms 3, 9, or 11, and they have no accessible surface area. Such atoms are considered to be interior atoms, not part of the surface of the molecule. Those parts of the van der Waals surface in contact with the surface of the probe are designated the contact surface; they comprise a series of disconnected patches. When the probe is simultaneously in contact with more than one protein atom, its interior surface defines the reentrant surface. The contact surface and reentrant surface together make a continuous surface, which is defined as the **molecular surface**.

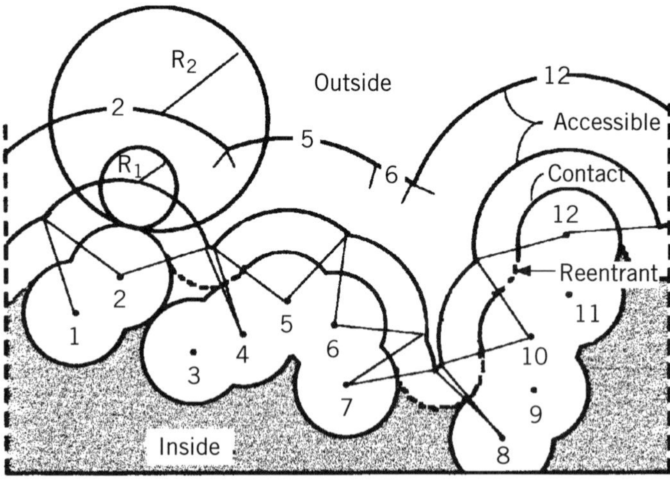

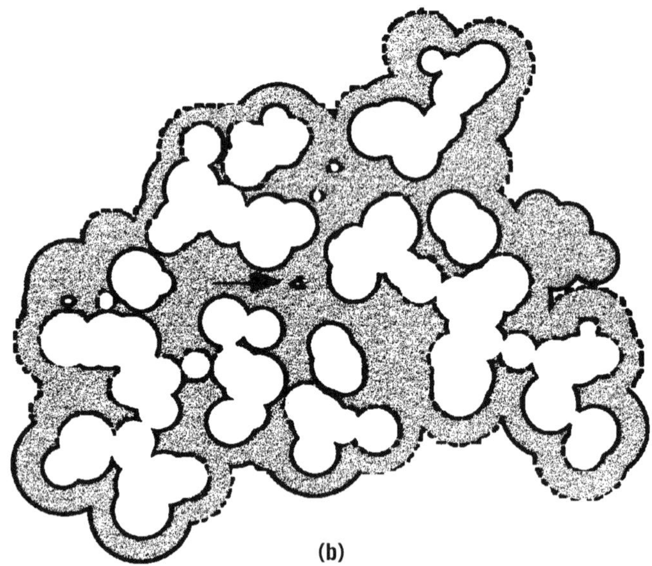

Figure 1. Schematic representation of possible molecular surface definitions. (**a**) A two-dimensional section through part of the van der Waals envelope of a hypothetical protein including 12 atoms with the centers numbered 1–12. R_1 and R_2 show the radius of the probes. (**b**) Superposition of sections through the van der Waals and accessible surfaces of ribonucrease S. In places, the accessible surface is controlled by atoms above or below the section shown. The solid outline is the surface of carbon and sulfur atoms; the dashed outline nitrogen and oxygen. The arrow indicates a cavity inside the molecule large enough to accommodate a solvent molecule with a radius of 1.4 Å. [Taken from F. M. Richards (1977) *Ann. Rev. Biophys. Bioeng.* **6**, 151–176; B. Lee and F. M. Richards (1971) *J. Mol. Biol.* **55**, 279–400.]

The accessible surface area depends on the size of the probe. When the radius of the probe increases from R_1 to R_2, the number of noncontact or interior atoms in Figure 1 increases from three to eight, and the accessible surface is much smoother. Thus, the smaller the probe, the larger the number of features

Table 1. Accessible Surface Areas of Amino Acid Residues

Residue	Accessible surface area[a] (Å^2)
Ala	113
Arg	241
Asn	158
Asp	151
Cys	140
Gln	189
Glu	183
Gly	85
His	194
Ile	182
Leu	180
Lys	211
Met	204
Phe	218
Pro	143
Ser	122
Thr	146
Trp	259
Tyr	229
Val	160

[a] Values estimated for a Gly-X-Gly tripeptide in an extended conformation (3).

that will be revealed. The probe is frequently taken to be a water molecule and approximated as a sphere with a radius of 1.4 Å. The accessible surface areas of individual amino acid residues are given in Table 1.

The structures of protein determined by **X-ray crystallography** indicate that the total accessible surface area A_s of a protein is approximately proportional to the two-third power of its molecular weight. The A_s (in Å^2) of a typical small monomeric protein is usually related to its molecular weight M_w by the approximate relationship (2)

$$A_s = 6.3M_w^{0.73} \quad (1)$$

For oligomeric proteins

$$A_s = 5.3M_w^{0.76} \quad (2)$$

These equations are usually accurate to within ±4% on average for monomers and 5% for oligomers. Equations 1 and 2 imply that an **oligomeric protein** has a larger accessible surface area (by 7 to 13%) than a monomeric protein of the same molecular weight in the molecular weight range up to 35,000.

For an unfolded polypeptide chain, the total accessible surface area A_t in an extended conformation is directly proportional to its molecular weight, within ±3%:

$$A_t = 1.45M_w \quad (3)$$

Thus, for those proteins whose accessible surface area is given by Equation 1, the potential surface area buried by the protein's folding A_b will be approximately given by

$$A_b = 1.45M_w - 6.3M_w^{0.73} \quad (4)$$

This indicates that 55 to 75% of the surface area of the unfolded polypeptide chains is buried in the folding process.

The methodology of Lee and Richards (1) is also applicable to the calculation of the accessible surface area of **nucleic acids** (4). Two-thirds of the water-accessible surface area become buried on **double-helix** formation of **DNA** and **RNA**. When a probe corresponding to a single water molecule is used, the total accessible surface area is similar for **A-DNA** and **B-DNA**, although marked differences appear in the major and minor groove exposures. For the larger probes, there exist considerable differences in accessible surface area between the two conformations.

BIBLIOGRAPHY

1. B. Lee and F. M. Richards (1971) *J. Mol. Biol.* **55**, 379–400.
2. J. Janin, S. Miller, and C. Chothia (1988) *J. Mol. Biol.* **204**, 155–164.
3. S. Miller et al. (1987) *J. Mol. Biol.* **196**, 641–656.
4. C. J. Alden and S. Kim (1979) *J. Mol. Biol.* **132**, 411–434.

ACENTRIC FRAGMENT

ALAN P. WOLFFE

Fragments of **chromosomes** that lack **centromeres** are described as acentric fragments (Fig. 1). They are formed as a result of chromosomal damage. Lacking a centromere, there is no means of attachment to the **mitotic spindle**, which leads to a failure to segregate replicated **chromatids** in the acentric fragments to the daughter cells during **cell division**. Therefore the acentric fragments are normally lost from most cells with progressive cell cycles. The breakage of the chromosome that separates the acentric fragment from the remainder of the chromosome containing the centromere can be a result of either inherent chromosomal fragility, radiation damage or defective **DNA repair** mechanisms.

Chromosome fragile sites lead to acentric fragments in man. The majority of recurrent cancer chromosome breakpoints are found at fragile sites (1). These breakpoints may promote **oncogenic** rearrangements and **translocations**. Chromosomal fragile sites can be induced by growing cells under abnormal conditions, by the addition of drugs or chemicals, and by infection with **viruses**. Fragility seems to be a consequence of incomplete compaction of the **chromatin**. A

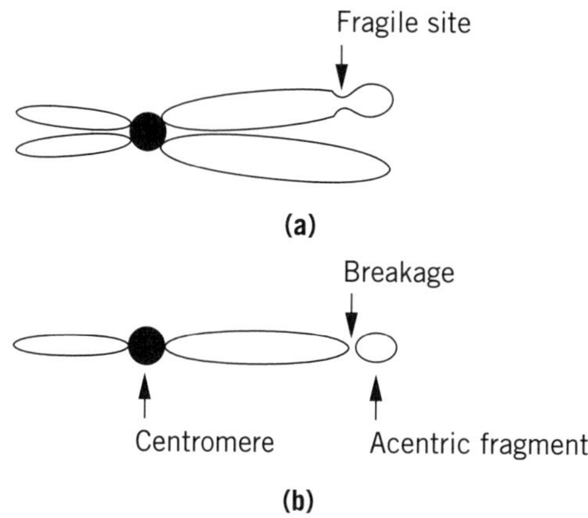

Figure 1. (**a**) A mitotic chromosome is shown with a fragile site indicated. (**b**) Mitosis can create forces that break the chromosome at the fragile site creating a residual chromosome that contains the centromere and an acentric fragment as indicated.

more open and accessible chromatin structure might explain the tendency of these sites to break, to recombine with other chromosomal regions, and to be sites of viral integration (2). A fragile site associated with a form of inherited mental retardation known as fragile X mental retardation has been characterized in some detail. A gene called FMR-1 (fragile X mental retardation gene 1) is present in the fragile site. This gene contains within it a **trinucleotide repeat** of CGG. A normal individual has between 6 and 50 of these trinucleotide repeats. Individuals with fragile X syndrome contain expansions of these repeat sequences such that they have several hundred CGG triplets. This expansion leads to chromosome fragility, to inactivation of FMR-1 **gene expression**, and to **methylation** of regulatory DNA within the gene. The expansion of the CGG trinucleotide sequences is associated with the assembly of an aberrant chromatin structure (3).

Defective chromosomal repair mechanisms lead to a much higher number of chromosomal breaks and translocations than found in normal cells. This high degree of chromosomal damage can eventually lead to malignancy. For example in Bloom's syndrome there is a very high rate of sister chromatid exchange, and in Falconi's anemia and ataxia telangiectasia there is a high incidence of chromosomal breakage and rejoining, leading to multicentric chromosomes and acentric fragments. Individuals with these syndromes are especially sensitive to environmental insults, such as chemical **mutagens** and ionizing radiation that further increase chromosomal damage (4). Detection of acentric fragments is diagnostic of serious abnormalities in chromosomal metabolism.

BIBLIOGRAPHY

1. J. J. Yunis and A. L. Soreng (1984) *Science* **226**, 1199–1204.
2. A. D. Bailey, Z. Li, T. Pavelitz, and A. M. Weiner (1995) Mol. Cell. Biol. **15**, 6246–6255.
3. R. S. Hansen et al. (1993) Cell **73**, 1403–1409.
4. M. S. Meyn (1993) *Science* **260**, 1327–1330.

Suggestion for Further Reading

M. S. Clark and W. J. Wall (1996) *Chromosomes. The Complex Code,* Chapman and Hall, London.

ACETYL COENZYME A

C. M. SMITH

Acetyl coenzyme A (acetylCoA) consists of a two-carbon activated acetyl unit attached to coenzyme A in thioester linkage. AcetylCoA is central to energy generation from the degradative pathways of oxidative fuel metabolism and to a number of biosynthetic pathways that utilize the activated two-carbon acetyl unit. In aerobic cells, it is the product of all the major catabolic pathways of fuel metabolism, including β- oxidation of fatty acids, ketone body degradation, glycolysis and pyruvate oxidation, ethanol oxidation, and the oxidative degradation of many amino acids. The two-carbon acetyl unit of acetylCoA formed from these pathways can be completely oxidized to CO_2 in the tricarboxylic acid cycle (TCA cycle), thus providing aerobic cells with energy from the complete oxidation of fuels. The acetyl unit of acetylCoA is also the basic building block of fatty acids, cholesterol, and other compounds, and it can be transferred to other molecules in acetylation reactions (eg, synthesis of N-acetylated sugars).

STRUCTURE

The ability of acetylCoA to participate in these diverse metabolic pathways is derived from the thioester bond formed between the acyl carbon of the acetyl unit and the sulfhydryl group of coenzyme A (CoASH) (Fig. 1). Because sulfur does not share its electrons, the carbonyl carbon of the thioester bond carries a more positive partial charge than that of an oxygen ester, and electrons are pulled away from the C-H bonds of the terminal methyl group. This distribution of charge facilitates nucleophilic attack at the carbonyl carbon and enhances the ability of the terminal methyl carbon to act as an electrophilic agent in condensation reactions. The acetylCoA thioester bond is a high energy bond, with a $\Delta G^{0'}$ for hydrolysis of -32.2 kJ/mol. Transfer of the acetyl unit to other molecules, therefore, usually occurs with the release of energy.

FORMATION OF ACETYLCoA IN OXIDATIVE PATHWAYS

Three basic types of reactions exist in the oxidative pathways of fuel metabolism that generate acetylCoA: the activation of acetate, the thiolytic cleavage of β-ketoacyl CoAs and β-hydroxy acids, and the oxidative decarboxylation of pyruvate

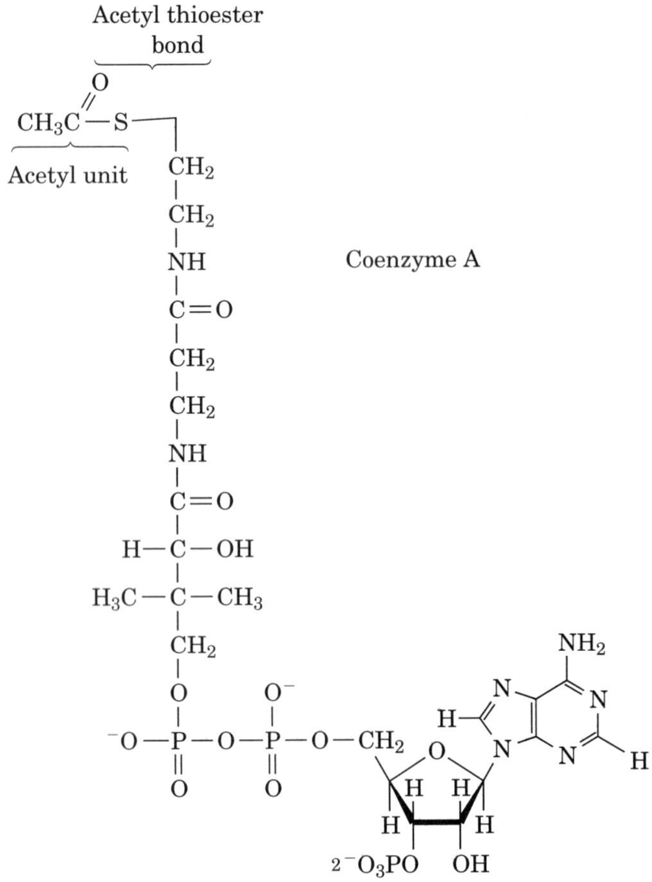

Acetyl thioester bond

Acetyl unit

Coenzyme A

Figure 1. Structure of acetyl coenzyme A (acetyl CoA).

Table 1. Enzymes that Form Acetyl CoA

AcetylCoA synthetase
(1) acetate + ATP + CoASH ⟶ acetyl CoA + AMP + PP_i
β-Ketothiolase
(2) $CH_3(CH_2)_nCH_2C(=O)—CH_2C(=O)—SCoA \longrightarrow CH_3(CH_2)_nCH_2C(=O)—SCoA + CH_3C(=O)—SCoA$
Pyruvate dehydrogenase complex (thiamine-PP, lipoate, FAD)
(3) pyruvate + CoASH + NAD^+ ⟶ Acetyl CoA + NADH + H^+ + CO_2
Citrate lyase
(4) citrate + CoASH + ATP ⟶ acetyl CoA + oxaloacetate + ADP + P_i

(Figure 2). In mammalian cells, acetate is the end product of ethanol metabolism and of threonine degradation. It is activated to acetylCoA in a single ATP- requiring step by the enzyme acetylCoA synthetase (Table 1). In an alternate route, bacterial cells can synthesize acetyl phosphate from acetate and then transfer the activated acetyl group to CoASH. Thiolytic cleavage of β-ketoacyl or β-hydroxy acylCoA derivatives to acetylCoA occurs in the pathways for oxidation of fatty acids, synthesis of the ketone bodies acetoacetate and β-hydroxybutyrate, and oxidative degradation of the amino acids isoleucine, leucine, lysine, tryptophan, phenylalanine, and tyrosine. (Most biochemistry textbooks contain general outlines of these pathways.) This type of reaction is illustrated by the β-oxidation spiral for fatty acids, in which one molecule of the C_{16}-fatty acid palmitate is converted to eight molecules of acetylCoA by enzymes that sequentially oxidize the molecule to a β-ketoacyl compound and then cleave acetylCoA from the carboxylic acid end (Table 1). The third type of reaction, the oxidative decarboxylation of pyruvate by the pyruvate dehydrogenase complex, provides the connecting link between pathways that produce pyruvate and the TCA cycle. The inhibition of the pyruvate dehydrogenase complex by acetylCoA has a regulatory role in controlling the flow of carbon into the various pathways of intermediary metabolism.

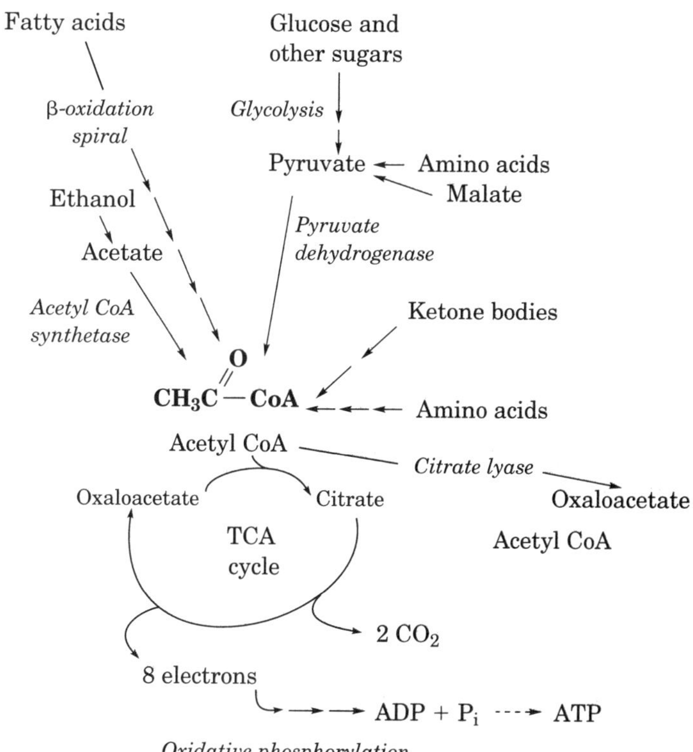

Figure 2. The role of acetyl CoA in oxidative fuel metabolism.

OXIDATION OF THE ACETYL UNIT OF ACETYLCoA IN THE TCA CYCLE

It is estimated that about two thirds of the energy requirements of aerobic cells is met by the complete oxidation of the acetyl group of acetylCoA to CO_2 in the TCA cycle (also called the citric acid cycle or the Krebs cycle.) In this cyclical sequence of reactions, acetylCoA condenses with oxaloacetate to form citrate in a reaction catalyzed by the enzyme citrate synthase. Subsequent reactions of the cycle donate electrons to the **coenzymes NAD^+** and **FAD** for ATP generation from oxidative phosphorylation, regenerate oxaloacetate, and release two carbons as CO_2. The net reaction of the TCA cycle is:

$$\begin{aligned}&\text{acetylCoA} + 3\ \text{NAD}^+ + \text{FAD} + \text{GDP} + P_i\\&\quad + 2\ H_2O \rightarrow \text{CoASH} + 2\ CO_2\\&\quad + 3\ \text{NADH} + \text{FAD(2 H)} + 3\ H^+ + \text{GTP}\end{aligned}$$

Most of the pathways that produce acetylCoA in aerobic cells of **eukaryotic** organisms are located in the **mitochondrial** matrix, and most of the biosynthetic pathways that utilize acetylCoA are outside of the mitochondrion. AcetylCoA is not directly transported through the inner mitochondrial membrane but is "transferred" from the mitochondrial matrix to the **cytosol** as citrate. It is then regenerated in the cytosol by the enzyme citrate lyase.

ACETYLCoA AS A PRECURSOR IN BIOSYNTHETIC REACTIONS

The two-carbon acetyl unit of acetylCoA is the precursor of a number of compounds synthesized in cells. It is the basic building block of fatty acids, cholesterol, and other compounds derived from the five-carbon isoprenoid unit (Figure 3). In the synthesis of fatty acids, acetylCoA is carboxylated to malonyl

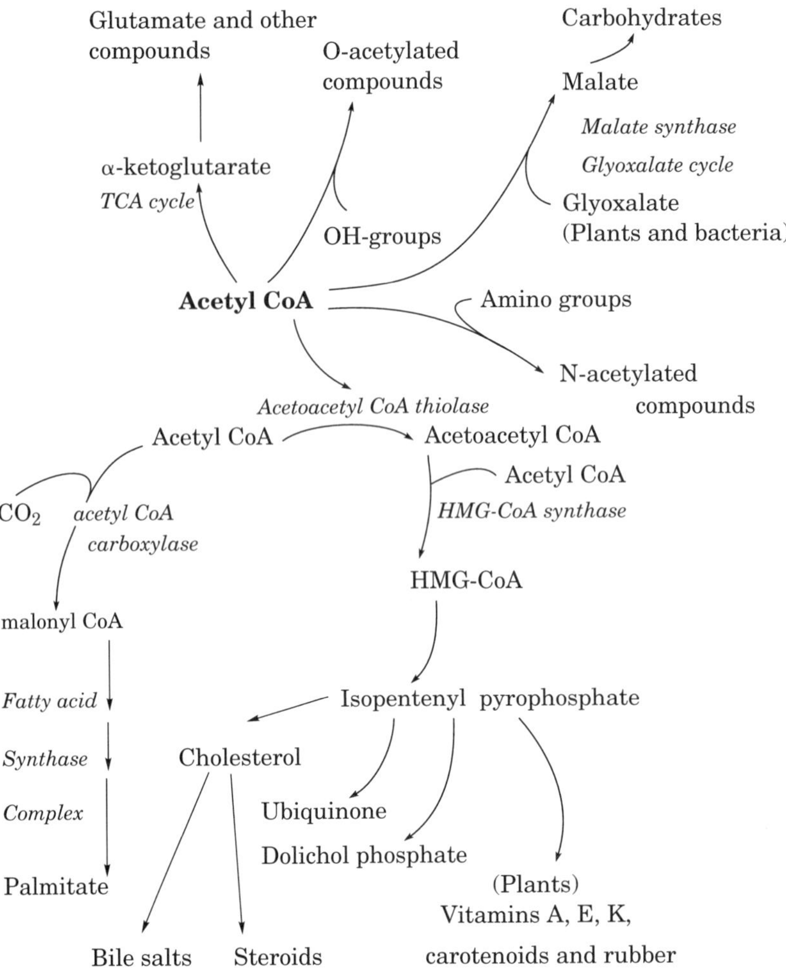

Figure 3. The central role of acetyl CoA in biosynthetic pathways.

CoA by the **biotin**-requiring enzyme, **acetylCoA carboxylase** (Table 2). Subsequent reactions build the C_{16} fatty acid palmitate and other fatty acids from successive additions of the portion of malonyl CoA derived from acetylCoA, thereby providing the cell with the diverse fatty acids required for membrane lipids. In the synthesis of cholesterol, acetyl units from three acetylCoA molecules condense to form 3-hydroxy 3-methylglutaryl CoA (HMG CoA), which is subsequently decarboxylated and converted to the five-carbon isoprenoid unit isopentenyl pyrophosphate. This isoprenoid unit is one of the most common structural units of a number of compounds in mammalian cells, bacteria, and plants. AcetylCoA also contributes one carbon to compounds synthesized from the five-carbon intermediate α- ketoglutarate in the TCA cycle. The acetyl unit of acetylCoA can also be transferred either to hydroxyl groups of compounds to form an acetyl oxygen ester (eg, the neurotransmitter acetylcholine; see **Acetylcholine receptor** and **Acetylcholine esterase**) or to an amino group to form an amide in an N-acetylation reaction (eg, acetylated amino sugars such as N-acetylglucosamine).

In mammalian cells, acetylCoA cannot provide a net source of carbon for the synthesis of glucose or other sugars. However, plants, yeast, and bacteria contain the glyoxylate cycle, which serves as a bypass of the TCA cycle. The net result of the glyoxylate cycle is the conversion of two molecules of acetylCoA to succinate, utilizing TCA cycle enzymes plus isocitrate lyase and malate synthase. As a result, *Escherichia coli* and many other bacteria are able to convert acetate to carbohydrates and amino acids and can thus utilize and grow on acetate as their sole carbon source.

Table 2. Enzymes That Utilize AcetylCoA

Citrate synthase
(5) acetyl CoA + oxaloacetate ⟶ citrate + CoASH
AcetylCoA carboxylase(biotin)
(6) acetyl CoA + CO_2 + ATP ⟶ malonyl CoA + ADP + P_i
AcetoacetylCoA thiolase
(7) 2 acetyl CoA ⟷ acetoacetyl CoA + CoASH
HMGCoA synthase
(8) acetyl CoA + acetoacetyl CoA ⟶ HMG CoA + CoASH
Choline acetyltransferase
(9) acetyl CoA + choline ⟶ acetylcholine + CoASH
Malate synthase
(10) acetyl CoA + glyoxalate ⟶ malate + CoASH

Suggestions for Further Reading

A. L. Lehninger, D. L. Nelson, and M. M. Cox (1990) *Principles of Biochemistry*, 2nd ed., Worth Publishers, New York. Although the pathways mentioned above can be found in more detail in most textbooks, this book is particularly good for outlines of major pathways found in mammals, bacteria, and plants.

D. B. Marks, A. D. Marks, and C. M. Smith (1996) *Basic Medical Biochemistry: A Clinical Approach*, Williams & Wilkins, Baltimore, MD. Provides a general outline of pathways found in humans and their relationship to human physiologic and pathologic conditions.

R. H. Bethal, D. B. Buxtion, J. G. Robertson, and M. S. Olson (1993) Regulation of the pyruvate dehydrogenase multienzyme complex. *Ann. Rev. Nutr.* **13**, 497-520. This article and the following one provide further information about the pyruvate dehydrogenase complex and acetylCoA carboxylase, two of the key enzymes in acetylCoA metabolism, which are highly regulated and the subject of current research.

G. M. Mabrouk, I. M. Helmy, K. G. Thampy, and S. J. Wakil (1990) Acute hormonal control of acetylCoA carboxylase: the roles of insulin, glucagon, and epinephrine. *J. Biol. Chem.* **265**, 6330–6338.

J. L. Goldstein, H. H. Hobbs, and M. S. Brown (1995) Familial hypercholesterolemia. In *The Metabolic and Molecular Bases of Inherited Disease*, 7th ed. (C. R. Scriver, A. L. Beaudet, W. S. Sly, and D. Valle, eds.), McGraw Hill, New York: pp. 1981–2030. Covers certain aspects of pathway regulation for the synthesis of cholesterol from acetylCoA, which is of great interest for protection from heart disease.

ACETYLCHOLINE RECEPTOR

ISRAEL SILMAN

Acetylcholine (ACh) is a widely distributed neurotransmitter in both the peripheral and central nervous systems (1). It is synthesized by the **enzyme** choline acetyltransferase within the cholinergic nerve terminal, where it is packaged into cholinergic vesicles. Arrival of an action potential at a cholinergic nerve terminal triggers quantal release of ACh by fusion of these vesicles with the presynaptic membrane and concomitant **exocytosis** of their contents into the synaptic cleft. The ACh released **diffuses** across the synapse to the postsynaptic membrane, where it activates its target, the acetylcholine receptor (AChR). Termination of transmission at cholinergic synapses involves rapid hydrolysis of ACh, by the enzyme acetylcholinesterase (AChE), which hydrolyzes it to choline and acetic acid (2).

The early observations of Dale (3) showed in various pharmacological preparations that ACh evoked responses similar to those evoked by either nicotine or muscarine. This provided the basis for the grouping of all AChRs into two families, nicotinic (nAChRs) and muscarinic (mAChRs). Dale's pioneering classification, based on the action of these two plant alkaloids, is still valid and used, although subtypes have been recognized and defined within each of the two families, both on the basis of specificity with respect to binding of agonists and antagonists and, in recent years, on the basis of gene **cloning**. Two additional plant alkaloids, *d*-tubocurarine and atropine, also provide useful tools by serving as specific antagonists of nAChRs and mAChRs, respectively (4).

nAChRs are ligand-gated **ion channels**, and their activation, by ACh or other agonists, causes a rapid change in ion permeability of the membrane in which they are embedded. mAChRs are members of the family of **G-protein-coupled receptors** and produce much slower responses, either excitatory or inhibitory, via their corresponding **second messengers** (5). Initial molecular cloning studies provided the **primary structures** of both nAChRs (6) and mAChRs (7), revealing that they belong to distinct families of proteins that share neither sequence identity nor a similar fold. As will be discussed in detail below, nAChRs are composed of pentamers of one or more subunits that display substantial sequence **homology** and, most likely, similar overall folds and transmembrane topologies. Muscarinic receptors are glycoproteins of molecular weight ~80,000 which, as already mentioned, belong to the family of G-protein-coupled receptors (8). Just as for other members of this family, **hydrophobicity** plots predict seven **transmembrane** sequences (8). The nAChR belongs to a larger family of ligand-gated ion channels, with which it too appears to share an overall fold, a common pattern of transmembrane sequences and other structural similarities (9,10; see text below).

The recognition that the large electric organs of electric fish such as the electric eel, *Electrophorus electricus*, and the electric ray, *Torpedo* sp., provide a highly enriched and homogeneous preparation of cholinergic synapses (11) led to their use as an experimental model and also as the source of choice for purification and characterization of the nAChR, of AChE, and of homogeneous cholinergic synaptosomes and synaptic vesicles (2). The fact that the cholinergic synapse at the neuromuscular junction was the parallel system of choice for studying the electrophysiological characteristics of synaptic transmission (12), as well as many ultrastructural and developmental parameters (for recent reviews see Refs. 13–15), meant that many of the structural data garnered by use of the electric organ model system could be directly correlated with the functional data obtained with the neuromuscular preparation (2). Thus, the nAChR of electric organ closely resembles that of muscle (16), and the molecular forms of AChE present in electric organ tissue are homologous to those found at muscle endplates (17,18).

The finding that injection into rabbits of purified nAChR elicited an **autoimmune** condition closely resembling the human muscle disease myasthenia gravis (19) was followed by detection of circulating **antibodies** to the nAChR in patients suffering from the disease. This opened up a fertile area of research in which clinical aspects were closely coupled to advances in basic research on the nAChR (20).

The cholinergic synapse has thus served as the prototypic synapse for understanding many fundamental aspects of synaptic transmission at chemical synapses. Even now, when improved electrophysiological techniques, and approaches such as Ca^{2+} imaging, allow better access to the CNS, and when genetic engineering allows expression of the receptor of choice in a system like the *Xenopus* oocyte that is amenable to user-friendly patch-clamp techniques, as well as in sufficient amounts to carry out structural characterization, the neuromuscular junction and electric organ tissue still occupy a position at the center of the stage.

THE NICOTINIC ACETYLCHOLINE RECEPTOR

Identification and Characterization

As early as 1955, it was suggested by Nachmansohn (21) that the nAChR might be a protein in which a conformational change elicited by the neurotransmitter ACh could induce a change in permeability that would trigger the postsynaptic response, and it was in his laboratory, in the 1960s, that the first evidence was presented that the nAChR is indeed a protein. Thus, Karlin and Bartels (22), using the electric eel electroplaque preparation, showed that the response of the nAChR to carbamylcholine could be modified reversibly by reagents that either modified **thiol groups** or reduced **disulfide bonds**. Subsequently, Karlin and co-workers (23,24) and Changeux et al. (25), also in the Nachmansohn laboratory, demonstrated *in situ* **affinity labeling** of the nAChR in the intact electroplaque. O'Brien and co-workers (26,27), using specific binding of the reversible ligand muscarone and homogenates of electric organ tissue of *Torpedo*, were able to obtain a good estimate (~2 nmol/g tissue) of the number of nicotinic binding sites.

A major breakthrough in the identification of the nAChR, its localization, and its molecular characterization came about by use of (a) the neurotoxin α**-bungarotoxin** (αBgt), a 74-residue polypeptide chain purified by Lee and co-workers from the venom of the banded krait, *Bungarus multicinctus* (28), and (b) the homologous α**-neurotoxins** purified from cobra venom (29). Lee's group presented electrophysiological and pharmacological evidence that these neurotoxins acted as very high-affinity (dissociation constant in the pico-to-femtomolar range) antagonists of the nAChR (30). ^{125}I-αBgt was developed as a powerful tool for localization and quantification of the nAChR by high-resolution **autoradiography** (31). It was shown to be present as a dense quasi-crystalline array at the top of the folds of the postsynaptic membrane of skeletal muscle, opposite the putative ACh-release sites on the presynaptic membrane, with a density very similar to that in purified preparations of postsynaptic membranes obtained from *Torpedo* electroplax. A further step forward was taken when it was shown that binding of ^{125}I-αBgt was retained after solubilization in nonionic **detergents**, such as Triton X-100 and cholate, and that a single toxin-binding component, migrating at ~9 S, could be identified on sucrose gradient **sedimentation velocity centrifugation** (32,33). Furthermore, the receptor so solubilized could be purified by **affinity chromatography** using resins to which either the α-neurotoxin (34) or a quaternary nitrogen ligand (35) had been attached. This permitted characterization of the nAChR as a multisubunit protein. Thus it was shown that it was a pentamer, of molecular weight ~250,000 (36), that contains four polypeptide subunits, α, β, γ, and δ, with apparent molecular weights of 39,000, 48,000, 58,000, and 64,000 and a stoichiometry of $\alpha_2\beta\gamma\delta$. In the **electron microscope**, such purified preparations displayed a rosette-like appearance (37), closely resembling similar, highly organized arrays of rosettes observed in purified postsynaptic membrane preparations (38). This led to the working hypothesis, still currently accepted, that the five subunits lie in the plane of the plasma membrane, surrounding the ion channel running through the center. A fifth polypeptide, the 43-kDa polypeptide (now known as rapsyn), which was found to be weakly associated with the pentamer (39), is believed to be involved in nAChR-clustering and anchoring to the **cytoskeleton** at the synapse (13). In skeletal muscle the γ subunit is present in the nAChR of embryonic muscle, but is replaced by the ε subunit in adult muscle (40).

Studies on model systems were important in establishing the role of the nAChR in **signal transduction** at the cholinergic synapse. Thus the microsac preparation (41) was used to show that ACh could induce a permeability change in such nAChR-enriched vesicles. Furthermore, prolonged exposure to high concentrations of ACh was shown to produce a closed state, with high affinity for ACh (42), equivalent to the desensitized state first described in skeletal muscle by Katz and Thesleff (43). The availability of purified native receptor permitted reconstitution studies into liposomes and into lipid bilayers, permitting recording of single channels induced in the presence of ACh (44,45). The purified pentamer thus contained not only the ACh-binding site, but also the ion channel and the transducing elements involved in activation and desensitization (16).

Affinity labeling studies by Karlin and co-workers (46), using a radioactive reagent, established that the disulfide bond that was susceptible to reducing reagents, as originally demonstrated by Karlin and Bartels (22), was located on the α subunit, in proximity to the binding site for ACh and various quaternary ligands.

An important step forward was the finding of Raftery et al. (47) that the NH_2-terminal sequences of all four subunits display substantial **sequence homology**. When, not long after, the subunits were cloned (6), it was found that such homology extends throughout the whole polypeptide chain, and the receptor pentamer can be viewed as displaying pseudo-fivefold symmetry. One issue that still remains open is the arrangement of the subunits around the lumen. This has been approached primarily by electron microscopy using subunit-specific **antibodies**, with supplementary information coming from use of affinity labels directed toward the ACh-binding site, which label both the α subunit and an adjacent subunit (see text below). As many as 12 permutations are possible, but it is generally accepted that the two α-subunits are not adjacent to each other (48), and discussion focused on which of the β-, γ-, and δ-subunits is flanked on both sides by the α-subunit. Karlin et al. (49) proposed that it is the γ-subunit that lies between the two α-subunits. Although Kubalek et al. (50) proposed that the β-subunit might occupy this position, the assignment of Karlin and co-workers is generally favored (for detailed discussions see Refs. 10 and 51).

Advances in cloning and expression, taken together with development of the patch-clamp technique, had a dramatic influence on research on the nAChR, just as they did on research on other receptors and on ion channels. In the case of the nAChR, however, the availability of large amounts of highly purified receptor from *Torpedo* electric organ resulted in fruitful synergy between these new techniques and the techniques of protein chemistry and structural biology.

In general, research has proceeded on two fronts, one concerned with the structure of the ACh-binding site and the other with that of the ion channel, with the eventual objective of understanding the physical mechanism by which ligand-binding causes channel opening. In the following, the overall topology of the receptor and of the individual subunits will first be reviewed briefly. The current status of our knowledge of the ACh-binding site, and then of the ion channel, will be summarized, followed by a discussion of the recent structural work from the laboratory of Unwin, which is beginning to give us a first glimpse of how the receptor may be functioning. For a number of recent reviews that cover these issues in more detail than is possible here, see Refs. 9,10,16, and 52, as well as the recent paper of Unwin (53), which summarizes the current status of the structural work.

Conformation and Topology

Analysis of the sequences of the four subunits, obtained on the basis of their **complementary DNA** sequences, showed that they shared a very similar topology: A long extracellular NH_2-terminal domain is followed by three putative transmembrane α-**helices**, identified on the basis of hydrophobicity, M1–M3, by a cytoplasmic loop containing ~100–150 residues, and finally, by a fourth transmembrane helix, M4, so that the COOH-terminus is believed to be on the extracellular side of the postsynaptic membrane (54). Other ligand-gated ion channels that belong to the same superfamily, including receptors for γ-aminobutyric acid, glycine, and serotonin, display a similar topology (9,10). A number of consensus sites for *N*-**glycosylation** are present in the extracellular domains, giving rise to a sugar content of ~7% (for literature see Ref. 55); and a number of consensus sites for **phosphorylation**, present on the cytoplasmic loops, presumably fulfill regulatory functions (56).

Photoaffinity labeling, using lipid-soluble probes, has been used to assign the arrangement of the transmembrane sequences in relation to the lipid bilayer and the putative central ion channel (57–59). From such studies, it has been concluded that the M4 sequence, which is also the least conserved, is the most exposed to the lipid environment, with the M1 and M3 sequences displaying more limited exposure to the lipid environment, and M2 facing the lumen of the ion channel. This assignment of M2 as making the principal contribution to the lining of the ion channel is also supported by labeling by channel blockers and by use of the SCAM technique (see text below).

Various spectroscopic techniques show that the nAChR contains substantial amounts of both α-helices and β-sheets (for literature see Ref. 10). In particular, Naumann et al. (60), using Fourier transform infrared (FTIR) spectroscopy on native receptor-rich membranes, have shown a predominance of β structure, 36 to 43%, and an α-helical content of 32 to 33%. Of particular interest are the putative conformations of the membrane-spanning sequences. Görne-Tschelnokow et al. (61) attacked this problem directly, by performing FTIR spectroscopy on membrane preparations from which the extramembrane domains had been shaved by **proteolysis**. Their data suggested **a beta-sheet** content of ~40% for the transmembrane sequences. Modeling studies also suggested a substantial β-sheet component (62). Various labeling studies, which attempt to correlate the degree of labeling of residues with orientation toward either the ion-channel lumen or the lipid bilayer, and thus with either an α-helical or a β-strand pitch, yield complex results (see, for example, Refs. 9,10,59, and 63); and precise assignments will, most likely, have to await a high-resolution 3D protein structure.

ACh-Binding Site

Identification of the residues comprising the ACh-binding site of the nAChR was approached by affinity labeling. Karlin and co-workers used the same radioactive affinity label that they had used to label the α subunit selectively, a quaternary derivative of maleimide, to identify the residues labeled. Because they had demonstrated that labeling occurred subsequent to reduction of the receptor, it was not surprising that they found that the residues labeled were the adjacent cysteines, Cys192 and Cys193, which apparently form an intrachain disulfide bond in the native nAChR, and which are present only in the α-subunit (64). As predicted, these residues are localized in the sequence that had been assigned to the extracellular domain, and would thus be the natural candidate for the ACh-binding site. The importance of this region for agonist and antagonist binding was confirmed by the observation that short synthetic peptides containing Cys192 and Cys193 (eg, from residue 185 to 196) displayed specific binding to αBgt, albeit with much lower affinity than the native receptor (65). Moreover, the α-subunits of the nAChR of both snakes and of the mongoose, which are resistant to αBgt, display mutations in this region that can provide a structural basis for resistance (66).

Changeux and co-workers conducted an extensive study of the residues labeled by the tertiary photoaffinity label, DDF (67). In addition to Cys192 and Cys193, this probe labeled primarily aromatic residues, including Tyr93, Trp149, Tyr190 and Tyr198, all, again, in the putative cytoplasmic domain. Use of other labeling agents, by other laboratories, broadly confirmed these assignments and identified additional aromatic residues (for literature see Ref. 68). Thus the ACh-binding pocket of the nAChR can be viewed as an "aromatic basket" (16) in which the quaternary group of ACh interacts with these aromatics via the π electron–cation interactions (69) which X-ray **crystallographic** studies have shown to play a prominent role within the "aromatic gorge" of AChE (70,71).

Although the studies discussed above indicate a prominent role for the α-subunit in binding of both agonists and antagonists, numerous labeling studies have revealed contributions of adjacent subunits to ligand-binding. These studies have been critically reviewed by Hucho et al. (10). Taken together, the data clearly indicate that the two ACh-binding sites are at the interfaces between the two α-subunits and the γ- and δ-subunits. This, in turn, implies structural difference between the two sites that can serve to explain nonequivalence of the two binding sites in terms of affinity for both α-neurotoxins (72) and antagonists (73).

Ion Channel

Pharmacological studies on the nAChR revealed a number of noncompetitive antagonists, both natural and synthetic, that blocked the change in permeability elicited by ACh without inhibiting the binding of ACh itself. It was, therefore, suggested that at least some of these compounds act by entering the ion channel and sterically blocking ion movement through it (for literature see Refs. 9,10, and 16). Based on the currently held view that the ion channel is the hole in the rosette seen in the electron microscope, such "channel blockers" are believed to exert their action by plugging this hole. Some channel blockers, such as chlorpromazine (74) and trimethylphosphonium (75), bind irreversibly upon UV irradiation; and their labeled derivatives can, accordingly, be used to identify putative elements of the channel lining. Rapid-mixing studies, using nAChR-enriched microsacs, showed that exposure to ACh leads, initially, to greatly increased labeling, followed by diminished incorporation which is consistent with desensitization as measured by ion flux measurements (76). Such labeling studies showed dominant labeling of the residues in the M2 transmembrane sequences of all four subunits, suggesting that they make the principal contribution to the lining of the lumen of the ion channel, in the open state of the channel. One such noncompetitive channel blocker, quinacrine azide, was, however, shown to label amino acid residues in M1 (77).

In parallel to such labeling studies on the purified receptor, use of **site-directed mutagenesis**, combined with expression in *Xenopus* oocytes (78) and use of the patch-clamp technique, permitted assessment of the contributions of individual amino acid residues to the conductance properties of the ion channel, an approach pioneered by the joint efforts of the Numa and Sakmann laboratories (see, for example, Refs. 79 and 80).

A third approach, developed more recently, is the substituted-cysteine-accessibility method, abbreviated as SCAM (81), which combines site-directed mutagenesis with chemical modification, so as to identify the amino acid residues lining the ion channel and their involvement in function. Thus, residues believed to line the ion channel (primarily from the M2 transmembrane sequence, but also from M1) are replaced one-by-one by cysteine residues, using site-directed mutagenesis, and expressed in *Xenopus* oocytes. The reactivity of their thiol groups with thiol reagents bearing negative or positive charges is examined *in situ*, both in the presence and absence of ACh, thus permitting assessment of their accessibility in both the closed and open states of the channel. Furthermore, these sulfhydryl reagents can be added both intra- and extracellularly, yielding information concerning accessibility of channel-lining residues from both surfaces (63).

These various experimental approaches have led to the identification of residues within the M2 sequences of all five subunits that determine the conductance and selectivity of the nAChR channel. In particular, using nAChR mutants expressed in *Xenopus* oocytes, three rings of negatively charged residues which may be referred to as the extracellular (or outer), the intermediate, and the cytoplasmic (or inner) ring, have been shown to play important roles in determining channel conductance (80). A ring of polar (serine and threonine) residues, named the central ring, is located above the intermediate ring, forming a constriction that may serve as part of the selectivity filter (82,83). The bulk of the transmembrane sequence of M2 lies between this central ring and the outer ring.

A more detailed topographical picture has been obtained by the SCAM technique, as summarized by Karlin and Akabas (9). Thus, broadly speaking, the residues in M2 of the α-subunit can be fitted to an **amphipathic** helix so that, with one exception, all the residues on one face of the helix are labeled in the SCAM protocol; some are equally accessible in the presence and absence of ACh, some are more accessible in its presence, and some are more accessible in its absence. Furthermore, certain residues at the extracellular end of the M1 helix of the α-subunit are also labeled. The pattern of labeling of M2 in the absence of ACh is consistent with an α-helical conformation, except for a short stretch in the middle (αLeu250 to αSer252), while in the presence of ACh it is consistent with an uninterrupted α-helix. The pattern of labeling of M1 cannot be ascribed to a specific conformational motif, although it too is changed in the presence of ACh. Karlin and Akabas (9) hypothesize that a movement of M1 and M2 relative to each other, with a concomitant change in **secondary structure**, may flip open the gate of the channel. More recently, the SCAM technique has also located the gate, using comparative studies in which labeling was carried out from either the extracellular or cytoplasmic sides (63). In the absence of ACh—that is, in the closed state of the channel—there is a barrier to the sulfhydryl agents, when added to either side, reacting with residues αGly240 and to αThr244. ACh binding removes this barrier, which serves as an activation gate. Residues αGly240, αGlu241, αLys242, and αThr244 line a narrow part of the channel in which this gate is located. It should be noted that the second of these residues, αGlu241, belongs to the inner anionic ring, and αThr244 to the central ring, as defined above (80,82).

Structural Studies

Despite the extensive studies performed on the characterization of the ACh-binding site and of the ion channel, the question with which all structural and functional studies on the nAChR are ultimately concerned is how binding of ACh is transduced into a permeability change in the ion channel. This is ultimately a problem best addressed by a structural biology approach. In spite of attempts from many laboratories to obtain nAChR crystals that would diffract X-rays (for a published example, see Ref. 84), this has not yet been achieved. Our current structural knowledge is, therefore, derived from the work of Unwin (for a recent summary see Ref. 53). As mentioned above, the postsynaptic membrane of *Torpedo* electric organ contains densely packed, partially crystalline arrays of nAChR (38). When isolated, such membranes have a natural propensity to recrystallize in tubular form (85,86). Since the early 1980s, Unwin has been using electron microscopy to probe the structure of the nAChR from *Torpedo marmorata* and is now able to compare the receptor in its open and closed states at 9 Å resolution (53). The receptor is revealed as an elongated structure, ~125 Å long; the extracellular portion of each subunit extends ~60 Å above the membrane surface, and the intracellular portions extend ~20 Å from the cytoplasmic surface. A density underlying each receptor most likely represents the cytoskeletal protein rapsyn (see text above), normally present in 1:1 stoichiometry with the receptor (87). A view from above shows the five subunits arranged around a fivefold axis of pseudosymmetry, as predicted. The opening of the putative ion channel in the center is about 20 Å at its entrance, narrowing sharply at the membrane surface. At the current resolution, it is possible to begin to glimpse some elements of secondary structure. Thus in each subunit, in a region about 30 Å away from the extracellular surface, a group of three short rods can be detected, presumably α-helices. The two subunits identified as α-subunits (50) contain cavities shaped by these rods, which may correspond to the ACh-binding pockets. The narrowest portion, corresponding to the pore, appears to be shaped by five bent α-helical rods, each contributed by one of the subunits (Fig. 1, left). The bends in the rods, near the middle of the membrane, are the parts closest to the axis of the pore and may represent the gate. In order to visualize the open state of the nAChR, Berriman and Unwin (88) devised a rapid-freezing device, designed to trap the open state, after application of ACh, by preventing the transition to the desensitized state. Comparison of the ACh-activated receptor with the nonactivated form revealed only slight changes (Fig. 1, right). Concerted twisting motions appeared to occur within the rods lining the putative ACh-binding pockets, which were consistent with the requirement that two ACh molecules, binding simultaneously to both α-subunits, are required to open the channel. These localized disturbances were associated with small rotations extending along the axis of all the subunits, and with a switching of the helices within the pore from the bent shape seen in the nonactivated receptor

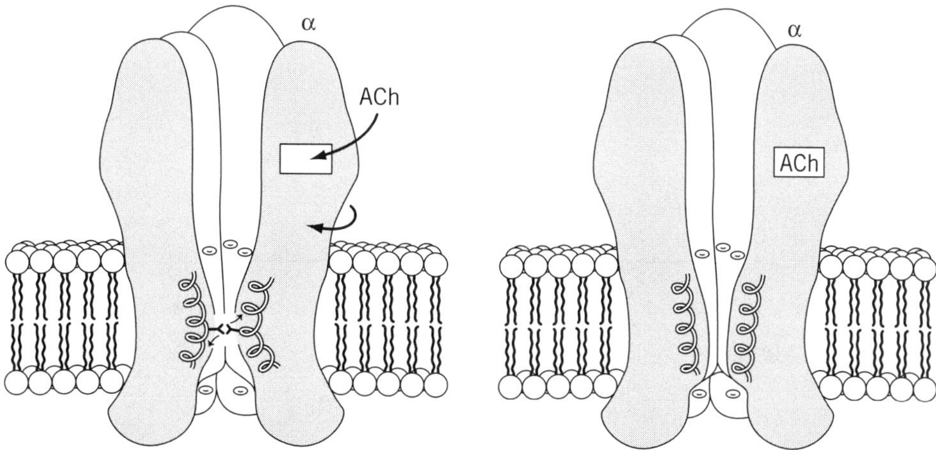

Figure 1. Simplified diagram of the nAChR in its closed (left) and open (right) states, as suggested by the structural results. An ACh molecule first enters a binding site in one of the α-subunits (rectangle), but significant displacements are blocked by the neighboring subunit, lying between the two α-subunits. Interaction of a second ACh with the other site then attempts to draw the neighboring subunit out of the way. A concerted localized displacement thereby takes place, initiating small rotations of the subunits along the shaft to the membrane. The rotations disrupt the gate by disrupting the association of α-helices around the pore in the closed state (left), switching over to an alternate configuration in which a widened polar pathway for ion movement is created. (From Ref. 53, with permission.)

to a tapered, more open configuration. If it is assumed that the threonine side chains of the central ring, which are believed to serve as the pore (see text above, along with Ref. 83) face into the pore, a minimum diameter of 10 Å can be estimated, comparable to the diameter obtained by use of various organic cations (89). Such a concerted rotation of the transmembrane sequences surrounding the lumen of the putative ion channel is consistent with the data accumulated by the various studies using site-directed mutagenesis and labeling, some of which were discussed above (see, in particular, Refs. 59 and 63).

Obviously, when higher-resolution images become available, a more detailed mechanistic description will be feasible, but the conceptual stage appears to have been set.

Neuronal Nicotinic Receptors

Already in the late 1970s, it was proposed, on the basis of ^{125}I-αBgt-binding studies, that nAChRs were present not only in skeletal muscle, but also in the central nervous system (90,91). These studies were not the subject of widespread attention, primarily due to the fact that functional correlates were lacking. When, in the mid-1980s, cloning techniques revealed the presence in brain of an nAChR **gene family** homologous to, but clearly distinct from, those of electric organ and muscle, and widely distributed in different brain areas (92,93), it became obvious that nAChRs, like muscarinic receptors, must have important functions in brain. For many years, however, the lack of specific agonists and antagonists for a given neuronal nAChR subtype severely hampered the identification of functional nAChRs in various brain areas. These difficulties were further aggravated by the unusually fast kinetics of inactivation of some nAChR subtypes, compared to their peripheral counterpart (for literature, see Ref. 94). Development of techniques permitting rapid application and removal of agonists was necessary to overcome this limitation (see, for example, Refs. 95 and 96).

The cloning approach revealed the existence of a large number of AChR receptor subunits in the brain. In contrast to muscle and electric organ, however, these appear to fall into only two categories, α and β. By now nine α-subunits (97) and nine β-subunits have been described in vertebrate brain (98), and multiple nAChR subunits have also been described in invertebrates such as *Caenorhabditis elegans* (99) and *Drosophila* (100).

As discussed by Sargent (98), the various subtypes of subunits are expressed differentially in one brain region or another. The most recently discovered α9-subunit, for example, has a pattern of expression restricted to cochlear hair cells (97). The pharmacology of neuronal nicotinic receptors in relation to such topics as nicotine addiction (101) and Alzheimer's disease (102) are currently topics of intensive investigation; so too is their involvement in brain plasticity (103) and in behavior, learning, and memory (104,105). A recent development is the observation that choline serves as a selective agonist for α7 nAChRs in rat hippocampal neurons (106), in line with earlier reports for α7 ectopically expressed in oocytes (107,108). Because α7 belongs to what appears to be the evolutionarily oldest group of nAChrs (109), it is feasible that choline, rather than ACh, was the primeval transmitter for cholinergic receptors.

Neuronal nAChRs have not yet been purified or expressed in large amounts. Thus, essentially all our knowledge of their functional properties comes from studies in which subunits were expressed singly or in combination, most frequently by injection into *Xenopus* oocytes. Such expression studies were augmented, as for *Torpedo* and muscle nAChRs, by site-directed mutagenesis. These studies have been reviewed in detail by Sargent (98), and only a few points will be discussed here briefly.

Presence of an α-subunit is necessary to obtain a functional receptor; and in the case of the $\alpha 7$-subunit, injection of it alone into oocytes is sufficient for it to form functional oligomers (110). Changeux, Bertrand, and co-workers took advantage of this experimental system to perform expression, combined with site-directed mutagenesis, which showed that the ligand-gated ion channel produced by the $\alpha 7$-subunit closely resembles that of the pentameric muscle and *Torpedo* receptors (see, for example, Ref. 111 and the review in Ref. 16). Direct experimental evidence for a pentameric stoichiometry was provided for the chick nAChR produced in oocytes by expression of the $\alpha 4$ and $\beta 2$ subunits with radioactive label incorporated into them. The stoichiometry of the subunits in the purified receptor was consistent with an $\alpha 4_2 \beta 2_3$ pentamer (112). This does not mean, however, that expression of the mRNAs for any mixture of α-subunits, with or without a β-subunit, will produce a functional oligomer. Thus far, only $\alpha 7$ has been shown to produce functional oligomers containing only α subunits. Furthermore, a recent study of Yu and Role (113) has shown that $\alpha 5$ can participate in the function of an ACh-gated ion channel only if it is coexpressed with another α- and a β-subunit. The structural basis for such restrictions with respect to assembly and/or function are, at this stage, unknown. Patrick and co-workers (114) observed, however, that a **cyclophilin** is required for expression of functional homo-oligomeric, but not hetero-oligomeric, nAChRs. They raise the possibility that the α-subunits in the homo-pentamer may not assume identical folds. Cyclophilins may thus play a critical role in the maturation of such homo-oligomers, acting directly or indirectly as prolyl *cis/trans* **isomerases** or as **molecular chaperones** (115).

BIBLIOGRAPHY

1. P. Taylor (1990) In *The Pharmacological Basis of Therapeutics* (A. G. Gilman, T. W. Rall, A. S. Nies, and P. Taylor, eds.), 8th Edition, Pergamon Press, New York, pp. 122–130.
2. V. P. Whittaker (1992) *The Cholinergic Neuron and its Target: The Electromotor Innervation of the Electric Ray Torpedo as a Model*, Birkhäuser, Boston.
3. H. H. Dale (1914) *J. Pharmacol. Exp. Ther.* **6**, 146–190.
4. R. J. Lefkowitz, B. B. Hoffman, and P. Taylor (1990) In *The Pharmacological Basis of Therapeutics* (A. G. Gilman, T. W. Rall, A. S. Nies, and P. Taylor, eds.), 8th ed., Pergamon Press, New York, pp. 84–121.
5. D. A. Brown et al. (1997) *Life Sci.* **60**, 1137–1144.
6. S. Numa et al. (1983) *Cold Spring Harbor Symp. Quant. Biol.* **48**, 57–69.
7. T. Kubo et al. (1986) *Nature* **323**, 411–416.
8. J. Wess et al. (1997) *Life Sci.* **60**, 1007–1014.
9. A. Karlin and M. H. Akabas (1995) *Neuron* **15**, 1231–1244.
10. F. Hucho, V. I. Tsetlin, and J. Machold (1996) *Eur. J. Biochem.* **239**, 539–557.
11. D. Nachmansohn (1959) *Chemical and Molecular Basis of Nerve Activity*, Academic Press, New York.
12. B. Katz (1966) *Nerve, Muscle and Synapse*, McGraw-Hill, New York.
13. S. C. Froehner (1993) *Annu. Rev. Neurosci.* **16**, 347–368.
14. Z.-Z. Wang et al. (1996) *Cold Spring Harbor Symp. Quant. Biol.* **61**, 363–371.
15. J. R. Sanes (1997) *Curr. Opin. Neurobiol.* **7**, 93–100.
16. J.-P. Changeux (1995) *Biochem. Soc. Trans.* **23**, 195–205.
17. I. Silman and A. H. Futerman (1987) *Eur. J. Biochem.* **170**, 11–22.
18. J. Massoulié et al. (1993) *Prog. Neurobiol.* **41**, 31–91.
19. J. Patrick and J. Lindstrom (1973) *Science* **180**, 871–872.
20. A. Vincent et al. (1995) *J. Physiol. Paris* **89**, 129–136.
21. D. Nachmansohn (1955) *Harvey Lect.* **39**, 57–99.
22. A. Karlin and E. Bartels (1966) *Biochim. Biophys. Acta* **126**, 525–535.
23. A. Karlin and M. Winnik (1968) *Proc. Natl. Acad. Sci. USA* **60**, 668–674.
24. I. Silman and A. Karlin (1968) *Science* **164**, 1420–1421.
25. J.-P. Changeux, T. Podleski, and L. Wofsy (1967) *Proc. Natl. Acad. Sci. USA* **58**, 2063–2070.
26. R. D. O'Brien and L. P. Gilmour (1969) *Proc. Natl. Acad. Sci. USA* **63**, 496–503.
27. R. D. O'Brien, L. P. Gilmour, and M. E. Eldefrawi (1970) *Proc. Natl. Acad. Sci. USA* **65**, 438–445.
28. C. C. Chang and C. Y. Lee (1963) *Arch. Int. Pharmacodyn.* **144**, 241–257.
29. C. C. Chang and C. Y. Lee (1966) *Brit. J. Pharmacol.* **28**, 172–181.
30. C. Y. Lee (1972) *Annu. Rev. Pharmacol.* **12**, 265–286.
31. H. C. Fertuck and M. M. Salpeter (1976) *J. Cell Biol.* **69**, 144–158.
32. J.-P. Changeux, M. Kasai, and C. Y. Lee (1970) *Proc. Natl. Acad. Sci. USA* **67**, 1241–1247.
33. R. Miledi, P. Molinoff, and L. T. Potter (1971) *Nature* **229**, 554–557.
34. E. Karlsson, E. Heilbronn, and L. Widlund (1972) *FEBS Lett.* 28, 107–111.
35. A. Karlin and D. A. Cowburn (1973) *Proc. Natl. Acad. Sci. USA* **70**, 3636–3640.
36. J. A. Reynolds and A. Karlin (1978) *Biochemistry* **17**, 2035–2038.
37. J. Cartaud et al. (1973) *FEBS Lett.* **33**, 109–113.
38. J. E. Heuser and S. R. Salpeter (1979) *J. Cell Biol.* **82**, 150–173.
39. R. R. Neubig, E. K. Krodel, N. D. Boyd, and J. B. Cohen (1979) *Proc. Natl. Acad. Sci. USA* **76**, 690–694.
40. M. Mishina et al. (1986) *Nature* **321**, 406–411.
41. M. Kasai and J. P. Changeux (1971) *J. Membr. Biol.* **6**, 1–80.
42. M. Weber, T. David-Pfeuty, and J. P. Changeux (1975) *Proc. Natl. Acad. Sci. USA* **72**, 3443–3447.
43. B. Katz and S. Thesleff (1957) *J. Physiol.* **138**, 63–80.
44. N. Nelson, R. Anholt, J. Lindstrom, and M. Montal (1980) *Proc. Natl. Acad. Sci. USA* **77**, 3057–3061.
45. G. Boheim et al. (1981) *Proc. Natl. Acad. Sci. USA* **78**, 3586–3590.
46. C. L. Weill, M. G. McNamee, and A. Karlin (1974) *Biochem. Biophys. Res. Commun.* **61**, 997–1003.
47. M. A. Raftery, M. W. Hunkapiller, C. D. Strader, and L. E. Hood (1980) *Science* **208**, 1454–1456.
48. D. S. Wise, J. Wall, and A. Karlin (1981) *J. Biol. Chem.* **256**, 12624–12627.
49. A. Karlin et al. (1983) *J. Biol. Chem.* **258**, 6678–6681.
50. E. Kubalek, S. Ralston, J. Lindstrom, and N. Unwin (1987) *J. Cell Biol.* **105**, 9–18.
51. P. C. Kearney et al. (1996) *Neuron* **17**, 1221–1229.
52. H. A. Lester (1997) *Harvey Lect.*, **91**, 79–98.
53. N. Unwin (1998) *J. Struct. Biol.* **121**, 181–190.

54. J.-L. Popot and J.-P. Changeux (1984) *Physiol. Rev.* **64**, 1162–1239.

55. H. Shoji et al. (1992) *Eur. J. Biochem.* **207**, 631–641.

56. R. L. Huganir and P. Greengard (1987) *Trends Pharmacol. Sci.* **8**, 472–477.

57. J. Giraudat, C. Montecucco, R. Bisson, and J.-P. Changeux (1985) *Biochemistry* **24**, 3121–3127.

58. M. P. Blanton and J. B. Cohen (1994) *Biochemistry* **33**, 2859–2872.

59. M. P. Blanton et al. (1998) *J. Biol. Chem.* **273**, 8659–8668.

60. D. Naumann, C. Schultz, U. Görne-Tschelnokow, and F. Hucho (1993) *Biochemistry* **32**, 3162–3168.

61. U. Görne-Tschelnokow et al. (1994) *EMBO J.* **13**, 338–341.

62. M. O. Ortells and G. G. Lunt (1996) *Prot. Eng.* **9**, 51–59.

63. G. G. Wilson and A. Karlin (1998) *Neuron* **20**, 1269–1281.

64. P. N. Kao et al. (1984) *J. Biol. Chem.* **259**, 11662–11665.

65. D. Neumann, D. Barchan, M. Fridkin, and S. Fuchs (1986) *Proc. Natl. Acad. Sci. USA* **83**, 9250–9253.

66. D. Barchan, M. Ovadia, E. Kochva, and S. Fuchs (1995) *Biochemistry* **34**, 9172–9176.

67. J. Langenbuch-Cachat et al. (1988) *Biochemistry* **27**, 2337–2345.

68. J.-P. Changeux, J.-L. Galzi, A. Devillers-Thiéry, and D. Bertrand (1992) *Q. Rev. Biophys.* **25**, 395–432.

69. D. A. Dougherty and D. A. Stauffer (1990) *Science* **250**, 1558–1560.

70. M. Harel et al. (1993) *Proc. Natl. Acad. Sci. USA* **90**, 9031–9035.

71. M. Harel et al. (1996) *J. Am. Chem. Soc.* **118**, 2340–2346.

72. A. Maelicke and E. Reich (1976) *Cold Spring Harbor Symp. Quant. Biol.* **40**, 231–235.

73. S. E. Pedersen and J. B. Cohen (1990) *Proc. Natl. Acad. Sci. USA* **87**, 2785–2789.

74. T. Heidmann, R. E. Oswald, and J. P. Changeux (1983) *Biochemistry* **22**, 3112–3127.

75. F. Hucho, W. Oberthür, and F. Lottspeich (1986) *FEBS Lett.* **205**, 137–142.

76. T. Heidmann, J. Bernhardt, E. Neumann, and J. P. Changeux (1983b) *Biochemistry* **22**, 5452–5459.

77. M. DiPaola, P. N. Kao, and A. Karlin (1990) *J. Biol. Chem.* **265**, 11017–11029.

78. E. A. Barnard, R. Miledi, and K. Sumikawa (1982) *Proc. R. Soc. Lond. B* **215**, 241–246.

79. K. Imoto et al. (1986) *Nature* **324**, 670–674.

80. K. Imoto et al. (1988) *Nature* **335**, 645–648.

81. M. H. Akabas, D. A. Stauffer, M. Xu, and A. Karlin (1992) *Science* **258**, 307–310.

82. K. Imoto et al. (1991) *FEBS Lett.* **289**, 193–200.

83. A. Villarroel, S. Herlitze, M. Koenen, and B. Sakmann (1991) *Proc. R. Soc. B* **243**, 69–74.

84. S. Hertling-Jaweed et al. (1988) *FEBS Lett.* **241**, 29–32.

85. J. Kistler and R. M. Stroud (1981) *Proc. Natl. Acad. Sci. USA* **78**, 3678–3682.

86. A. Brisson and P. N. T. Unwin (1984) *J. Cell Biol.* **99**, 1202–1211.

87. W. J. LaRochelle and S. C. Froehner (1986) *J. Biol. Chem.* **261**, 5270–5274.

88. J. A. Berriman and N. Unwin (1994) *Ultramicroscopy* **56**, 241–252.

89. B. N. Cohen, C. Labarca, N. Davidson, and H. A. Lester (1992) *J. Gen. Physiol.* **100**, 373–400.

90. Z. Vogel and M. Nirenberg (1976) *Proc. Natl. Acad. Sci. USA* **73**, 1806–1810.

91. J. Patrick and W. B. Stallcup (1977) *Proc. Natl. Acad. Sci. USA* **74**, 4689–4692.

92. D. Goldman et al. (1987) *Cell* **48**, 965–973.

93. S. Heinemann et al. (1990) *Prog. Brain Res.* **86**, 195–203.

94. E. X. Albuquerque et al. (1997) *J. Pharmacol. Exp. Ther.* **280**, 1117–1136.

95. N. G. Castro and E. X. Albuquerque (1993) *Neurosci. Lett.* **164**, 137–140.

96. Z.-W. Zhang, S. Vijarayaghavan, and D. K. Berg (1994) *Neuron* **12**, 167–177.

97. A. B. Elgoyhen et al. (1994) *Cell* **79**, 705–715.

98. P. B. Sargent (1993) *Annu. Rev. Neurosci.* **16**, 403–443.

99. J. T. Fleming et al. (1997) *J. Neurosci.* **17**, 5843–5857.

100. E. D. Gundelfinger (1992) *Trends Neurosci.* **15**, 206–211.

101. J. A. Dani and S. Heinemann (1996) *Neuron* **16**, 905–908.

102. H. Schröder et al. (1991) *Neurobiol. Aging* **12**, 259–262.

103. J.-P. Changeux et al. (1996) *Cold Spring Harbor Symp. Quant. Biol.* **61**, 343–362.

104. N. Le Novère, M. Zoli, and J. P. Changeux (1996) *Eur. J. Neurosci.* **8**, 2428–2439.

105. E. D. Levin et al. (1997) *Brain Res. Bull.* **43**, 295–304.

106. M. Alkondon et al. (1997) *Eur. J. Neurosci.* **9**, 2734–2742.

107. A. Mandelzys, P. De Koninck, and E. Cooper (1995) *J. Neurophysiol.* **74**, 1212–1221.

108. R. L. Papke, M. Bencherif, and P. Lippiello (1996) *Neurosci. Lett.* **213**, 201–204.

109. N. Le Novère and J. P. Changeux (1995) *J. Mol. Evol.* **40**, 155–172.

110. S. Couturier et al. (1990) *Neuron* **5**, 847–856.

111. D. Bertrand et al. (1993) *Proc. Natl. Acad. Sci. USA* **90**, 6971–6975.

112. R. Anand et al. (1991) *J. Biol. Chem.* **266**, 11192–11198.

113. C. R. Yu and L. W. Role (1998) *J. Physiol.* **509**, 667–681.

114. S. A. Helekar, D. Char, S. Neff, and J. Patrick (1994) *Neuron* **12**, 179–189.

115. S. A. Helekar and J. Patrick (1997) *Proc. Natl. Acad. Sci. USA* **94**, 5432–5437.

Suggestions for Further Reading

E. X. Albuquerque et al. (1997) Properties of neuronal acetylcholine receptors: pharmacological characterization and modulation of synaptic function. *J. Pharmacol Exp. Ther.* **280**, 1117–1136.

J.-P. Changeux (1995) The acetylcholine receptor: a model for allosteric membrane proteins. *Biochem. Soc. Trans.* **123**, 195–205.

F. Hucho, V. I. Tsetlin, and J. Machold (1996) The emerging three-dimensional structure of a receptor: the nicotinic acetylcholine receptor. *Eur. J. Biochem.* **239**, 539–557.

A. Karlin and M. H. Akabas (1995) Towards a structural basis for the function of nicotinic acetylcholine receptors and their cousins. *Neuron* **15**, 1231–1244.

P. B. Sargent (1993) The diversity of neuronal nicotinic acetylcholine receptors. *Ann. Rev. Neurosci.* **16**, 403–443.

N. Unwin (1998) The nicotinic acetylcholine receptor of the *Torpedo* electric ray. *J. Struct. Biol.* **121**, 181–190.

V. P. Whittaker (1992) The cholinergic neuron and its target: the electromotor innervation of the electric ray *Torpedo* as a model, Birkhäuser, Boston.

ACHAETE–SCUTE COMPLEX

J. B. SKEATH

A central theme in developmental biology is to understand how the mechanisms that specify particular cell types integrate with those that govern where and when cell type-specific structures form during animal **development**. Over the better part of the past century, the development of the *Drosophila* nervous system has served as a model with which to unravel the interconnected processes of cell fate specification and pattern formation. The **genes** of the *achaete–scute* complex (AS-C) bridge these two processes. The AS-C is composed of four related genes found within ~90 kbp of DNA in the distal tip of the **X-chromosome**. The four genes—*achaete* (*ac*), *scute* (*sc*), *lethal of scute* (*l'sc*) and *asense* (*ase*)—all encode for basic **helix-loop-helix** transcriptional activator proteins and were initially identified by mutational analysis. Loss-of-function mutations in the AS-C remove sensory organs in the peripheral nervous system (PNS) and neural structures in the central nervous system (CNS), whereas gain-of-function mutations induce the formation of ectopic neural structures. The "proneural" *ac*, *sc*, and *l'sc* genes are expressed prior to neural precursor formation in precise patterns of cell clusters that forecast where neural precursors will form. During both embryonic and adult development, the primary patterning genes in *Drosophila* act through a large array of ***cis-acting*** regulatory regions found within the AS-C to create the stereotyped pattern of AS-C-expressing proneural cell clusters. Within each cluster (equivalence group), a cell communication process mediated by the ***Notch* signaling** pathway (lateral inhibition) restricts AS-C gene expression to a single cell. Within this cell, *ac*, *sc* and *l'sc*, either alone or in combination, are thought to activate a cassette of genes that directs this cell to acquire the neural precursor fate. All other cells within the cluster extinguish AS-C gene expression and are directed towards epidermal development. One target of the AS-C proneural genes in neural precursors is *ase*, which promotes the differentiation of neural structures. The AS-C thus links the process of cell fate specification to that of pattern formation: the AS-C genes receive and interpret global positional information through their regulatory regions and translate this information through their function into the formation (and differentiation) of a two-dimensional array of neural structures.

GENETIC AND MOLECULAR CHARACTERIZATION OF THE *ACHAETE–SCUTE* COMPLEX

The embryonic and adult *Drosophila* nervous systems are composed of a variety of different neural structures, each organized in reproducible but distinct patterns. For example, the neural precursors of the embryonic CNS form in orthogonal rows, whereas in the adult PNS large innervated bristles arise in an irregular, yet stereotyped pattern. Each bristle, or sensory organ, forms from a single neural precursor cell that occupies a fixed location in the developing fly. Cell migration is limited in the *Drosophila* PNS and CNS. Thus, the initial neural precursor pattern largely determines the later pattern of neural structures.

Genetic analyses identified a set of four genetic activities, designated the AS-C, located near the distal tip of the X-chromosome critical for the formation, patterning, and differentiation of neural structures (1). Loss-of-function mutations in any one of these genes remove neural structures, although the precise set of structures removed differs for each gene. For example, loss of *l'sc* activity primarily affects the CNS, whereas loss of *ac* or *sc* activity primarily results in PNS defects. Loss of *ase* function removes sensory organs along the wing blade but also causes differentiation defects in other sensory organs. Conversely, mis-expression of any AS-C gene promotes the formation of ectopic neural structures. These analyses suggested that the AS-C promotes the initial commitment of a cell to become a neural precursor and that their gene products are at least partially redundant. Detailed genetic mosaic analyses in the *Drosophila* wing **imaginal disc** (the larval structure that generates the wing and notum of the adult fly) showed that the activities of *ac* and *sc* define zones of neural competency from which neural precursors arise (2). These zones of neural competency are known as proneural cell clusters and are "equivalence groups" within which all cells can (although only one or a few will) become a neural precursor.

The genes *ac*, *sc*, and *l'sc* are each expressed in a complex and dynamic pattern of cell clusters, each of which quickly resolves to a single cell, the putative neural precursor (Fig. 1). The gene expression patterns of *ac* and *sc* are identical throughout neurogenesis and partially overlap with the expression pattern of *l'sc*. On the other hand, *asense* is expressed exclusively in neural precursors and their progeny. Based on mutant phenotypes and expression patterns, *ac, sc*, and *l'sc* are called proneural genes whereas *ase* is called a neural precursor gene.

The molecular **cloning** of the AS-C in the 1980s identified all four genes as encoding putative transcriptional activator proteins that belong to the class B type of basic helix-loop-helix (bHLH) proteins. These four genes are contained within ~90 kbp of DNA, separated from one another by an average of ~30 kbp of DNA replete with *cis*-regulatory regions required to activate different members of the AS-C in complex and partially overlapping patterns (3). Transcriptional activation requires the formation of a heterodimer between A and B class bHLH proteins (4); these heterodimers recognize the core DNA sequence CANNTG, where N is not specified (an 'E' box). An A-class bHLH protein with which AS-C gene products heterodimerize and activate transcription is the product of the *daughterless* (*da*) gene. Heterodimers of one AS-C gene product and Da mediate transcriptional activation of genes expressed throughout proneural clusters and in neural precursors. The *da* gene is expressed ubiquitously throughout *Drosophila* development, and AS-C genes are expressed in complex yet invariant patterns of cell clusters that predict where neural precursors form. Thus, the spatiotemporal specificity of where and when neural precursors develop is a function of the pattern of AS-C gene expression.

CIS- AND TRANS-REGULATION OF PRONEURAL GENE EXPRESSION

Activation of the proneural genes of the AS-C in precise and reproducible patterns of proneural clusters is the first step in the

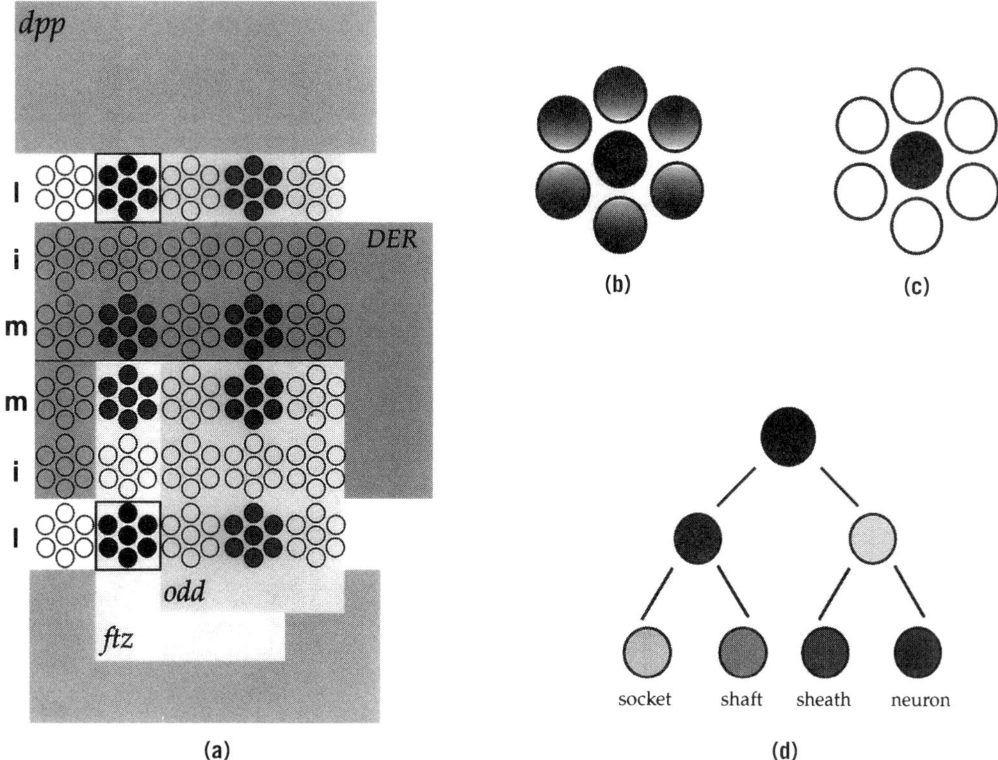

Figure 1. Patterning and specification of neural precursors. (**a**) Patterning of AS-C gene expression. Schematic representation of the ectoderm of a *Drosophila* embryo. The expression of the *ac* and *sc* genes in four proneural clusters per hemisegment is shown in red and blue. The activity of the early patterning genes *ftz*, *odd*, *DER*, and *dpp* subdivide the embryo and lead to the activation of *ac/sc* gene expression in reproducible quadrants. For example, the combined action of *ftz* activation and *odd* repression leads to *ac/sc* gene expression in domains that express *ftz* but not *odd* (yellow). Along the D/V axis, *dpp* (light blue) and *DER* (green) restrict *ac/sc* expression to the lateral column. Together, the action of these four genes delimit *ac/sc* gene expression within one proneural cluster, shown in blue. Note that other factors can overcome *DER* mediated repression in the medial column. (**b**) Singling out of a neural precursor. Within each proneural cluster, a single cell (in this example, the central cell) comes to express the AS-C proneural genes to the highest level, and this cell becomes the presumptive neural precursor (dark blue cell). (**c**) Resolution of cell fate within a proneural cluster. The presumptive neural precursor retains proneural gene expression and acts through the *Notch* signaling pathway to remove proneural gene expression in the other cells of the cluster (white). These cells are directed toward epidermal development. (**d**) Neural precursor lineage. Each neural precursor goes through an invariant cell lineage to produce a particular neural structure. The lineage of a sensory bristle of the PNS is shown. The neural precursor divides to yield two differently fated daughter cells, which in turn divide to produce a total of four cells. Each cell acquires a different fate, and together the four cells make up a functional sensory organ—an innervated bristle.

formation of the stereotyped two-dimensional pattern of neural precursors in the *Drosophila* nervous system. The genetic regulatory mechanisms that dictate where and when AS-C expressing proneural clusters form are best understood for the development of the embryonic CNS (5). The anterior/posterior (A/P) and dorsal/ventral (D/V) register of AS-C-expressing cell clusters in the early embryo suggests that the segmentation genes, which segment the embryo along the A/P axis, and the dorsoventral genes, which specify pattern along the D/V axis, function to lay down directly the initial pattern of AS-C proneural clusters. A similar process mediated by ***wingless*** and ***decapentaplegic*** (*dpp*), as well as the genes of the *iroquois* complex (6), probably acts directly on the AS-C to create the pattern of AS-C-expressing cell clusters in adult structures.

During development of the embryonic *Drosophila* CNS, the AS-C proneural genes are expressed in an invariant orthogonal pattern of proneural cell clusters. For example, *ac* and *sc* are co-expressed in two rows of cell clusters in the medial and lateral—but not intermediate—columns of each segment. Prior to AS-C gene activation, the activities of different A/P and D/V patterning genes have subdivided the early embryo into an orthogonal pattern of squares, reminiscent of a checkerboard. Each square expresses a unique combination of these patterning genes, and the borders of the AS-C proneural clusters match precisely the limits of these squares (Fig. 1). This suggests a model whereby the combined activities of particular combinations of patterning genes within a square either does or does not activate a member of the AS-C. The composite pattern of AS-C-positive cell clusters would then result from the integration of the activities of each square.

Support for the checkerboard model comes from genetic analyses that assayed the expression of AS-C genes in embryos mutant for various known patterning genes. For example, the anterior border of every fourth transverse row of *ac/sc*-positive cell clusters coincides with the anterior edge of the expression of the *fushi-tarazu* (*ftz*) **pair-rule gene**, and the posterior border of these proneural clusters abuts the anterior border of the *odd-skipped* (*odd*) pair-rule gene (Fig. 1). In embryos mutant for *ftz*, this row of *ac/sc*-expressing proneural clusters disappears, whereas in embryos mutant for *odd* it expands posteriorly to fill the entire *ftz* domain. Thus, the combined action of *ftz* activation and *odd* repression defines the anterior and posterior boundaries respectively, of proneural clusters in this row. The D/V patterning genes control the mediolateral extent of AS-C proneural clusters. For example, the medial boundary of the lateral column of AS-C proneural clusters abuts the lateral border of the activity of the *Drosophila* **epidermal growth factor** (EGF) receptor (*DER*), and the lateral boundary of these clusters abuts the medial boundary of *dpp* activity (Fig. 1). In embryos mutant for *DER*, the lateral column of AS-C-expressing proneural clusters expands medially, whereas in *dpp* mutant embryos these clusters expand laterally. Thus, *DER* and *dpp* together restrict *ac/sc* gene expression to the lateral column in the developing CNS. The concerted action of *ftz* and *odd* along the A/P axis, and *DER* and *dpp* along the D/V axis, can account for the activation of *ac/sc* in one eighth of its proneural clusters. Combinations of other patterning genes probably act similarly to initiate AS-C proneural gene expression in the other regions of the developing CNS and PNS.

Scattered throughout the AS-C are regulatory regions that decode the positional information contained within the checkerboard pattern of spatial regulators and translate it into the transcriptional activation of *ac*, *sc*, and *l'sc*. The locations of many of these regulatory regions have been identified through the use of DNA rearrangements within the AS-C and of **reporter gene** constructs that contain specific genomic regions from the AS-C region (7). For example, *ac* and *sc* share a number of regulatory regions located between the two genes that function to activate *ac* and *sc*, but not *l'sc*, in identical patterns throughout development. It is thought that the gene products of many of the patterning genes act directly through these regulatory regions to create the precise and invariant pattern of AS-C-expressing proneural clusters.

RESOLUTION OF PRONEURAL CELL CLUSTERS: THE SINGLING OUT OF THE NEURAL PRECURSOR

Within each proneural cluster, AS-C expression quickly becomes restricted to one (or a few) cells (Fig. 1). This cell initiates *ase* expression, enlarges, and segregates as a neural precursor to a position below the proneural cluster from which it arose. All clusters exhibit identical dynamics of proneural gene expression, which directly reflect the cell-fate decisions made by the cells of a cluster. Initially, all cells express one or more of the proneural genes; proneural gene expression confers on all cells the ability to become a neural precursor. Then, via a cell-cell communication pathway mediated by the *Notch* signaling pathway, one cell comes to express the proneural genes to the highest level. This cell is the presumptive neural precursor; it then acts, again through the *Notch* pathway, to inhibit and eventually to extinguish proneural gene expression from all other cells of the cluster (8). The cells that lose proneural gene expression are directed toward the epidermal fate. Thus the fate of cells within a proneural cluster correlates—not with the initial expression of the proneural genes—but rather with the fate of proneural gene expression in that cell: cells that retain proneural gene expression become neural precursors, whereas cells that lose proneural gene expression are directed toward the epidermal fate.

The analysis of the AS-C provides a paradigm for how specific cell fates (and cellular structures) arise in reproducible and invariant patterns, but many questions still remain with respect to the AS-C. For example, roughly half of all CNS neural precursors still form in embryos devoid of AS-C function. Thus other "proneural genes" must exist that promote neural precursor formation in the CNS. In addition, AS-C function is not restricted to the nervous system. The gene *l'sc* mediates the initial selection of muscle progenitor cells in the mesoderm in much the same way that it (as well as *ac* and *sc*) single out neural precursors in the ectoderm. Furthermore, Galant et al (9) recently demonstrated that a butterfly AS-C **homologue** is expressed in, and thus may promote the formation of, the precursors to the pigment-producing scale cells that cover the butterfly wing. Thus, AS-C genes may mediate additional developmental events in *Drosophila* and other animals. Finally, in *Drosophila* the AS-C is an integral unit composed of four structurally related genes. Has the overall genomic structure of the AS-C and the relative position and expression of each AS-C gene been strongly conserved throughout evolution, as is observed for the genes of the **homeobox** clusters? Or does the AS-C and its constituent genes display significant plasticity over evolutionary time? Research on the AS-C has led to an understanding of some of the molecular mechanisms that specify particular cell types and those that determine where and when these cell types form. It is likely that further insights will derive from continued analysis of AS-C expression, regulation, and function.

Acknowledgments

I thank Michelle Hresko, Grace Panganiban, Lisa Nagy, and Ellen Ward for their comments. J.B.S is supported by a grant from NINDS (RO1-NS36570), the Cancer Research Fund of the Damon Runyon-Walter Winchell Foundation Award, DRS-9 and by a HHMI Research Resources Program for Medical Schools Junior Faculty Award #76296-538202.

BIBLIOGRAPHY

1. A. Garcia-Bellido (1979) *Genetics* **91**, 491–520.
2. C. Stern (1954) *American Scientist* **42**, 213–247.
3. S. Campuzano, L. Carramolino, C. V. Cabrera, M. Ruiz-Gomez, R. Villares, A. Boronat, and J. Modolell (1985) *Cell* **40**, 327–338.

4. C. Murre, P. Schonleber-McCaw, H. Vaessin, M. Caudy, L. Y. Jan, Y. N. Jan, C. V. Cabrera, J. N. Buskin, S. D. Hauschka, A. B. Lassar, H. Weintraub, and D. Baltimore (1989) *Cell* **58**, 537–544.
5. J. B. Skeath, G. Panganiban, J. Selegue, and S. B. Carroll (1992) *Genes Dev.* **6**, 2606–2619.
6. J. L. Gomez-Skarmeta, R. Diez del Corral, E. De La Calle-Mustienes, D. Ferres-Marco, and J. Modolell (1996) *Cell*, **85**, 95–105.
7. J. L. Gomez-Skarmeta, I. Rodriguez, C. Martinez, J. Culi, D. Ferres-Marco, D. Beamonte, and J. Modolell (1995) *Genes Dev.* **9**, 1869–1882.
8. J. A. Campos-Ortega (1993) In *The Development of* Drosophila melanogaster (M. Bate and A. Martinez-Arias, eds.), Cold Spring Harbor Press, Cold Spring Harbor, NY, pp. 1091–1130.
9. R. Galant, J. B. Skeath, S. Paddock, D. L. Lewis, and S. B. Carroll (1998) *Current Biology* **8**, 807–813.

ACHILLES' CLEAVAGE

ALLISON K. WILSON

Achilles' cleavage (AC) is a procedure developed to permit cutting double-stranded **DNA** at a single site (the AC site) in a complex **genome** (Fig. 1). The AC site is first protected from **methylation** by a sequence-specific **DNA-binding protein**, and the methylated genome is subsequently cut by a methylation-sensitive **restriction enzyme**. Other sites in the genome containing the same restriction/methylation site as the AC site are methylated and thus are not cut. The two key components of any AC system are (i) a restriction site that is recognized by both a methylation-sensitive restriction enzyme and its corresponding **methyltransferase** and (ii) the ability to block methylation of the chosen AC site by sequence-specific binding of a protein to that site. The overlap of the AC site and the recognition sequence for the DNA binding protein can occur naturally, or it can be engineered using **recombination** or **site-directed mutagenesis**. These requirements somewhat limit the applications of the AC technique.

Koob et al. (1) originally developed this procedure using two different **repressor/operator** systems. In one instance, **plasmids** carrying the operator of the ***lac* operon** were incubated with the **Lac repressor** and then methylated by the *Hha* I methyltransferase. Repressor binding prevented methylation of the *Hha* I site located within the *lac* operator. The methyltransferase and the operator protein were subsequently removed, leaving only the unmethylated AC site free to be cut by *Hha* I (1). Variations of the original AC method were also used to demonstrate the ability to make only one or a few cuts in the genomes of **yeast** and *Escherichia coli* (2), to turn frequent-cutting restriction enzymes into rare cutters (3), and to analyze the strength of protein-DNA interactions (4).

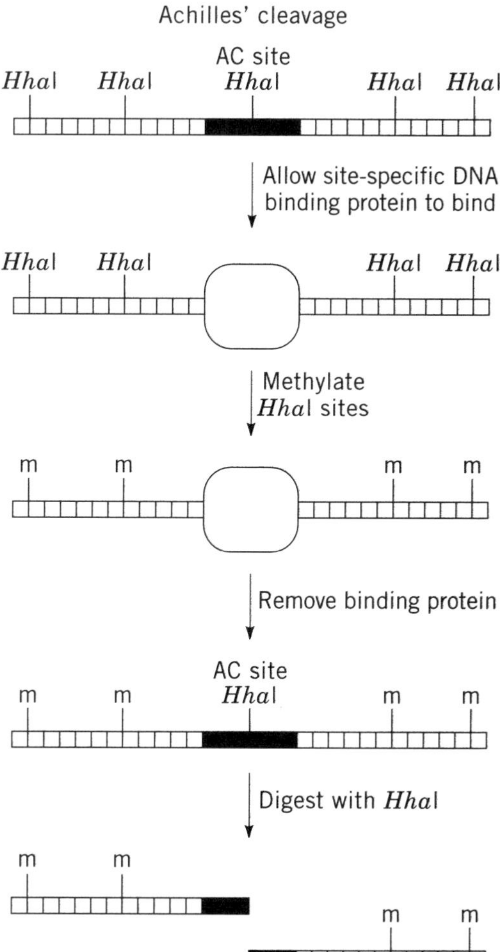

Figure 1. Achilles' Cleavage. Diagram of the method used to prevent restriction enzyme digestion at all sites except for the Achilles' cleavage (AC) site. Horizontal lines connected by short vertical lines represent double-stranded DNA. *Hha*I restriction sites are marked by bold vertical lines above the double-stranded DNA. The binding site for the DNA-binding protein is represented by a solid rectangle. The DNA-binding protein is represented by an empty oval. Sites of methylation are indicated by a vertical line and the letter 'm.'

RECA-AC AND RARE

The original AC method was made more universally applicable by using **RecA** protein and a sequence-specific **oligonucleotide** to form the protective complex, obviating the need for a sequence-specific binding protein. Two groups simultaneously developed this procedure, and their protocols were designated RecA-AC (5) and RARE (6). The RecA-AC/RARE technique still requires selection of a cleavage site that is recognized by both a restriction enzyme and its cognate methyltransferase. An oligonucleotide of 40 to 60 nucleotides in length is designed that matches the DNA sequence containing the cleavage site. In the presence of Mg^{2+} and a non-hydrolyzable ATP analogue [eg, ATP(γ-S)], RecA protein will bind to the oligonucleotide, forming a complex. When double-stranded DNA is added, the complex will bind to the matching genomic sequence to form a so-called 'triple-stranded' synaptic complex. This complex protects the cleavage site from methylation. As in the AC procedure, the methyltransferase and the RecA protein are removed after methylation, before the methylation-protected site is cleaved with the appropriate restriction enzyme.

As an alternative, the RecA complex can be used to protect specific restriction sites from restriction enzyme digestion, rather than from methyltransferase. This technique can be useful in some **cloning** procedures.

The RecA-AC/RARE technique has been used in various genome mapping experiments, for example to map gaps in human contiguous (**contig**) DNA and to map **telomeres** (7). The ability to cut genomic DNA at a few specific sites allows mapping large pieces of DNA by using **pulsed-field gel electrophoresis** and **Southern blotting** techniques to size DNA between two cleavage sites, or between a marker and the end of a telomere (7).

The RecA-AC/RARE technique is a useful technique for mapping and manipulating DNA and should continue to prove helpful in large-scale genome analysis studies of the future. One limitation to the technique is the number of sequences recognized by both a restriction enzyme and a methyltransferase. As more restriction enzyme/methyltransferase pairs are isolated or developed, the technique will be come even more useful.

BIBLIOGRAPHY

1. M. Koob, E. Grimes, and W. Szybalski (1988) *Science* **241**, 1084–1086.
2. M. Koob and W. Szybalski (1990) *Science* **250**, 271–273.
3. J. Kur, M. Koob, A. Burkiewicz, and W. Szybalski (1992) *Gene* **110**, 1–7.
4. P. M. Skowron, R. Harasimowicz, and S. M. Rutkowska (1996) *Gene* **170**, 1–8.
5. M. Koob, A. Burkiewicz, J. Kur, and W. Szybalski (1992) *NAR* **20**, 5831–5856.
6. L. J. Ferrin and R. D. Camerini-Otero (1991) *Science* **254**, 1494–1497.
7. L. J. Ferrin and R. D. Camerini-Otero (1994) *Nature Genet.* **6**, 379–383.

Suggestions for Further Reading

W. Szybalski (1997) RecA-mediated Achilles' heel cleavage. *Curr. Opinion Biotechnol.* **8**, 75–81.

L. J. Ferrin (1995) Manipulating and mapping DNA with RecA-assisted restriction endonuclease (RARE) cleavage, In *Genetic Engineering, Principles and Methods* (J. K. Setlow, ed.), Plenum Press, New York, Vol. 17, pp. 21–30.

ACRIDINE DYES

LYNNETTE FERGUSON
WILLIAM DENNY

Acridines are fused linear tricyclic aromatic molecules of planar geometry. Aminoacridine (Fig. 1) was originally developed as a topical antibacterial agent and is one of the most widely used and studied acridines. It is a strong base (pK_a 9.9) due to the resonant **amino group** and is largely ionized at physiological pH. Acridines are typical DNA-**intercalating** ligands, binding tightly but reversibly to double-stranded DNA by insertion of the aromatic chromophore between adjacent base pairs, with the long axis of the acridine parallel to the base-pair axis, ensuring maximum overlap in the binding site (1,2). It was originally hypothesized by Brenner et al. (3) that acridine-induced mutations arise through the addition or deletion of a single base pair, an event described as **frameshift mutagenesis**.

NH_2 N

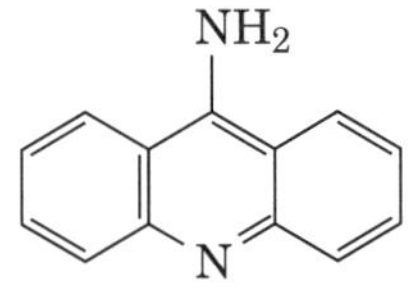

Figure 1. Structure of the mutagen 9-aminoacridine.

9-Aminoacridine is a strong mutagen in the rII region of **bacteriophage** T4 (4), in the CI gene of **lambda phage** (5), at various loci in *Escherichia coli* (6), and at the hisC3076 locus in *Salmonella typhimurium* (7). This and related simple acridines cause frameshift mutagenesis, especially in monotonous runs of repeated sequences. In mammalian cells their activity is somewhat weak and variable. They appear not to be causing frameshifts (at least in the most commonly used mammalian assays), but are mutagenic through weak ability to break chromosomes (8).

Substitutions on the acridine ring profoundly affect the mutagenic characteristics of these compounds, which have been divided into seven subgroups according to their distinctive mutagenic characteristics (2). 9-Aminoacridine and related simple compounds form the first group, while proflavine and other 3,6-diaminoacridines form the second group. The 3,6-diamino groups increase DNA binding affinity, and these compounds resemble 9-aminoacridine in having frameshift mutagenic activity, but are also substrates for metabolic activation (9). They are also readily activated by visible light into other products that are mutagenic in the *Salmonella* mutagenicity test (**Ames test**) (10). Some of these compounds and also the third class of quaternized acridines have considerable affinity for mitochondrial DNA. The fourth group of acridines retain DNA intercalation activity, but also act as topoisomerase II (topo II) poisons (see **DNA topology**).

Type II topoisomerases produce double-strand breaks in DNA during replication and **transcription**, permitting strand passage through the transient break, which is then resealed by the enzyme. The potent antitumor activity of 9-anilinoacridines of the fourth group, such as amsacrine (Fig. 2), is thought to relate to their ability to stabilize the cleavable complex formed between topo II enzymes and DNA, thereby leading to an accumulation of double-strand breaks (11). These DNA breaks also result in these compounds showing strong clastogenic and recombinogenic properties. A clastogen is a physical, chemical or viral agent that produces chromosome breaks and other chromosomal mutagenesis. The fifth group of benzacridines are characterized by a higher susceptibility to metabolic activation, primarily through oxidative mechanisms (12). Benzacridines are primarily base-pair

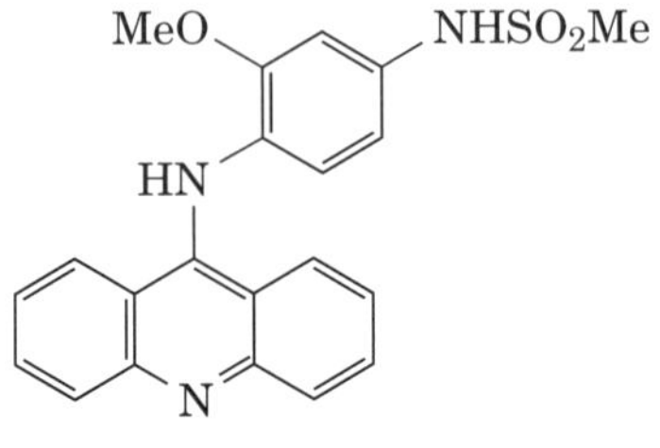

MeO $NHSO_2Me$ HN N

Figure 2. Structure of the 9-anilinoacridine amsacrine.

substitution mutagens, producing a spectrum of mutagenic events that is fundamentally different from that produced by other acridines and in which the acridine ring plays only a minor role. The sixth group of acridines, epitomized by those originally developed in the Institute of Cancer Research, Philadelphia, and widely known as ICR compounds (13), all carry an alkylating moiety, usually a nitrogen mustard. These compounds are not only potent frameshift mutagens in various organisms, but also cause base-pair substitution events in mammalian cells (2,14). Although the nitroacridines can be distinguished as a seventh group, they may alternatively be considered as a subgroup of the previous class, because they can act as either simple acridines or as alkylators, depending upon the position of the nitro substitution and the nitroreductase capability of the organism concerned.

BIBLIOGRAPHY

1. L. R. Ferguson and W. A. Denny (1990) *Mutagenesis* **5**, 529–540.
2. L. R. Ferguson and W. A. Denny (1991) *Mutat. Res.* **258**, 123–160.
3. S. Brenner, L. Barnett, F. H. C. Crick, and A. Orgel (1961) *J. Mol. Biol.* **3**, 121–124.
4. A. Orgel and S. Brenner (1961) *J. Mol. Biol.* **3**, 762–768.
5. T. A. Skopek and F. Hutchinson (1984) *Mol. Gen. Genet.* **195**, 418–423.
6. S. M. Thomas and D. G. MacPhee (1985) *Mutat. Res.* **151**, 49–56.
7. L. R. Ferguson and D. G. MacPhee (1983) *Mutat. Res.* **116**, 289–296.
8. W. R. Wilson, N. M. Harris, and L. R. Ferguson (1994) *Cancer Res.* **44**, 4420–4431.
9. M. Fukenaga, Y. Mitzuguchi, and L. W. Yielding (1987) *Chem. Pharm. Bull. Japan* **35**, 792–797.
10. Y. Iwamoto, L. R. Ferguson, A. Pearson, and B. C. Baguley (1992) *Mutat. Res.* **268**, 35–41.
11. L. R. Ferguson and B. C. Baguley (1996) *Mutat. Res.* **355**, 91–101.
12. R. E. Lehr, A. W. Wood, W. Levin, A. H. Conney, and D. M. Jerina (1988) In *Polycyclic Aromatic Hydrocarbon Carcinogenesis: Structure–Activity Relationships* (S. K. Yang and B. D. Silverman, eds.), CRC Press, Boca Raton, FL, pp. 31–58.
13. H. J. Creech, R. K. Preston, R. M. Peck, and A. P. O'Connell (1972) *J. Med. Chem.* **15**, 739–746.
14. L. F. Stankowski, K. R. Tindall, and A. W. Hsie (1986) *Mutat. Res.* **160**, 133–147.

ACROCENTRIC CHROMOSOME

ALAN P. WOLFFE

An acrocentric **chromosome** has a single **centromere** that is localized at or near one end of the chromosome (acro = extremity). One of the most common forms of chromosomal **translocation** occurs when two acrocentric chromosomes fuse at the ends containing the centromeres (1). This is called a Robertsonian event (Fig. 1). This generally leads to the formation of a **metacentric** chromosome, where the new centromeric region appears at or near the middle of the new chromosome. Robertsonian events lead to a decrease in the apparent number of chromosomes. Much of the variation in the number of chromosomes in the members of a family or genus is related to this type of acrocentric chromosomal fusion. Whereas the number of chromosomal arms remain relatively constant, counting two for a metacentric chromosome and one for an acrocentric, the number of chromosomes can vary widely. For example, most dogs have either 32 to 38 or 18 to 21 **autosomes** (any chromosome other than the **X-** or **Y-chromosomes**), but 34–38 chromosomal arms. Those with the greater number of autosomes have larger numbers of acrocentric chromosomes (2,3).

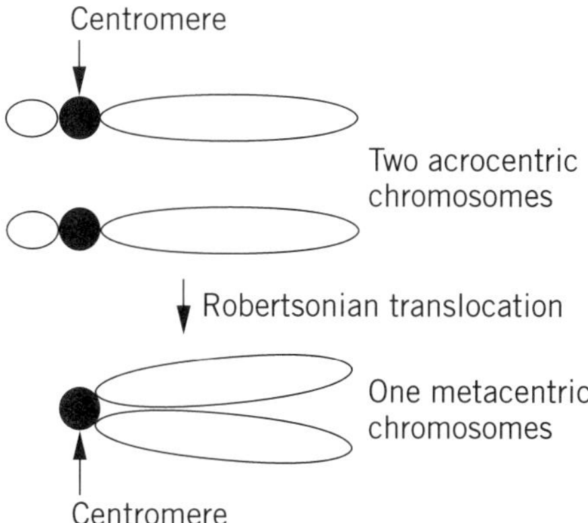

Figure 1. Robertsonian translocation occurs when two acrocentric chromosomes fuse to create a single metacentric chromosome.

BIBLIOGRAPHY

1. W. Robertson (1916) *J. Morph.* **27**, 179–197.
2. R. K. Wayne, W. G. Nash, and S. J. O'Brian (1987) *Cytogenet. Cell Genet.* **44**, 123–133.
3. R. K. Wayne, W. G. Nash, and S. J. O'Brian (1987) *Cytogenet. Cell Genet.* **44**, 134–141.

Suggestion for Further Reading

R. P. Wagner, M. P. Maguire, and R. L. Stallings (1993) *Chromosomes. A Synthesis*. Wiley-Liss, New York.

ACROSOME

STEPHAN JANSEN
BERTRAM BRENIG

The acrosome is an organelle of the spermatozoon that covers the apical part of the **sperm** head. This organelle comprises an outer acrosomal **membrane**, an inner acrosomal membrane, and plasma membranes. The acrosome plays an important role during the early stages of **fertilization**, in that the acrosome and its associated molecules mediate recognition of the egg to be fertilized. Although acrosome morphology varies with the species, the function of the acrosome and its contents as regulating elements during the early stages of fertilization are fairly similar among species. The acrosome contains **proteolytic** enzymes (see **Proteinases**) as well as proteins for binding to the zona pellucida and plasma membrane of the **egg** (1). Therefore, the associated molecules of the acrosome have been proposed as targets for nonhormonal antifertilization agents.

PRIMARY LIGANDS

The primary ligands are proteins that are located on or close to the acrosome; their function is to ensure that recognition between the gametes is species-specific. This is especially necessary in animals with extracorporal fertilization (eg, frog and sea urchin). On the surface of the acrosomal cap, there is a plethora of adhesion molecules (2), but their origin is not in all cases clear. Sperm adhesins become associated only upon ejaculation (3). Many of these primary ligands, such as galactosyltransferase (4), are **lectins** and bind to **O-linked oligosaccharides** and glycans of the ZP3 receptor of the oocyte's **extracellular matrix**, the zona pellucida (5,6). Other proteins were discovered that induce metabolic events; a 95-kDa protein was detected that induces **phosphorylation** events (7).

One of the most fascinating events during the early stages of fertilization is the acrosome reaction. After the initial contact of the sperm with the zona pellucida of the oocyte, and under the influence of progesterone (8), a unique event called the acrosome reaction takes place, which is calcium-dependent (9). Although in mammals the acrosome reaction can be induced without zona pellucida glycoproteins, they act in a catalytic manner (10). The acrosome reaction involves **exocytosis** of the outer membrane, causes the formation of hybrid vesicles, and leads to exposure of the inner contents of the acrosome. In many invertebrates, such as starfish, sea cucumbers, or sea urchin, a polymerization event takes place that results in the formation of **actin** filaments. In mollusks, the actin filaments are already present in the unreacted spermatozoon and exposed on activation.

Recent investigations report the presence of the **inositol phosphate** system in mammalian sperm. It involves the activation of a sperm **receptor** that stimulates a **G-protein**. This activates **phospholipase C**, which cleaves subsequently phosphatidyl inositol diphosphate (PIP_2) into **diacylglycerol** and inositol triphosphate (IP_3). IP_3 triggers the release of calcium from intercellular stores, and diacylglycerol activates protein kinase**kinase** C, which is calcium-dependent. This causes proteins to be phosphorylated and leads subsequently to the acrosome reaction (11). Upon their exposure, the molecules within the acrosome become important for the fertilization process.

SECONDARY LIGANDS: ACROSIN/PROACROSIN

A major component of the inner acrosome is the protein proacrosin. This is the best-characterized molecule of those involved in the early stages of fertilization. It is important for local lysis of the zona pellucida and therefore plays a central role at the stage of sperm penetration. Proacrosin is stored as its inactive form, a **zymogen**. On raising the pH, or on contact with zona pellucida glycoproteins, the zymogen converts into its active proteolytic form, acrosin. This is accomplished by several intramolecular reorganizations (12). Proacrosin is a two-polypeptide chain molecule and, typical of **serine proteinases**, the heavy chain starts at residue Val24. Major alterations of the molecule involve proteolytic cleavage of the proline-rich C-terminal tail, cleavage of the light chain, and linkage to the heavy chain via **disulfide bonds**.

Proacrosin is a multifunctional protein; it does not act solely as a proteinase, but is also a fucose-binding protein (13). The molecule shows high binding affinity for fucoidan, heparin, and zona pellucida glycoprotein preparations (14). The primary mechanism of proacrosin binding to the zona pellucida is not only an interaction with carbohydrates, but involves positively charged amino acid residues of the proacrosin and negatively charged sulfate groups of the zona pellucida glycoproteins. **Homology modeling** of proacrosin revealed that the groove for binding to the zona pellucida contains loops in which the positively charged amino acids are located. The active site for proteolytic activity is located in the center, surrounded by the positively charged amino acids. The binding groove and proteolytic center appear to form a single reactive unit, even though they are located on different regions on the **polypeptide chain**. The proteolytic activity could be blocked without affecting proacrosin's binding characteristics.

The arrangement of the binding site and proteolytic center suggests two major tasks of the enzyme during fertilization. The molecule recognizes the zona pellucida specifically, binds to it, and digests it locally to promote sperm entry through the extracellular matrix. However, there must be further binding and proteolytic proteins involved in fertilization, as mouse **knockout strains** lacking the proacrosin gene did not produce infertile males (15); other molecules must be able to take over the role of proacrosin. On the other hand, male mice with a functionally active proacrosin gene demonstrated a significantly greater fitness in terms of reproductive success in a competition-mating experiment with the proacrosin knockout mice (16). Although proacrosin is the major compound of the acrosomal vesicle, there are further adhesion proteins. Fertilin, formerly designated PH-30, is a molecule that is involved in interactions between the sperm and egg (17,18).

BIBLIOGRAPHY

1. A. P. Aguas and P. Pinto da Silva (1985) *J. Cell. Biol.* **100**, 528–534.
2. R. Jones (1990) *J. Reprod. Fert. Suppl.* **42**, 89–105.
3. J. J. Calvete, D. Solis, L. Sanz, T. Diaz-Maurino, W. Schäfer, K. Mann, and E. Töpfer-Petersen (1993) *Eur. J. Biochem.* **218**, 719–725.
4. X. Gong, D. H. Dubois, D. J. Miller, and B. D. Shur (1995) *Science* **269**, 1718–1721.
5. S. Kitazume-Kawaguchi, S. Inoue, Y. Inoue, and W. J. Lennarz. (1997) *Proc. Natl. Acad. Sci. USA* **94**, 3650–3655.
6. R. A. Kinloch, Y. Sakai, and P. M. Wassarman (1995) *Proc. Natl. Acad. Sci. USA* **92**, 263–267.
7. L. Leyton, P. LeGuen, D. Bunch, and P. M. Saling (1992) *Proc. Natl. Acad. Sci. USA* **89**, 11692–11695.
8. E. R. Roldan, T. Murase, and Q. X. Shi (1994) *Science* **266**, 1578–1581.
9. C. Arnoult, R. A. Cardullo, J. R. Lemos, and H. M. Florman (1996) *Proc. Natl. Acad. Sci. USA* **93**, 13004–13009.
10. C. N. Tomes, C. R. McMaster, and P. M. Saling (1996) *Mol. Reprod. Dev.* **43**, 196–204.
11. R. Yanagimachi (1994) Mammalian fertilization. *In The Physiology of Reproduction*, 2nd ed. (E. Knobil and J. D. Neill, eds.), Raven Press, New York, pp. 189–317.
12. T. Baba, W. Kashiwabara, K. Watanabe, H. Itoh, Y. Michikawa, K. Kimura, M. Takada, A. Fukamizu, and Y. Arai (1989) *J. Biol. Chem.* **264**, 11920–11927.
13. E. Töpfer-Petersen and A. Henschen (1987) *FEBS Lett.* **226**, 68–42.
14. R. Jones and C. R. Brown (1987) *Exp. Cell Res.* **171**, 505–508.

15. T. Baba, S. Azuma, S. Kashiwabara, and Y. Toyoda (1994) *J. Biol. Chem.* **269**, 31845–31849.

16. I. M. Adham, K. Nayernia, and W. Engel (1997) *Mol. Reprod. Dev.* **46**, 370–376.

17. E. A. Almeida, A. P. Huovila, A. E. Sutherland, L. E. Stephens, P. G. Calarco, L. M. Shaw, A. M. Mercurio, A. Sonnenberg, P. Primakoff, and D. G. Myles (1995) *Cell* **81**, 1095–1104.

18. J. P. Evans, R. M. Schultz, and G. S. Kopf (1997) *Dev. Biol.* **187**, 94–106.

Suggestions for Further Reading

B. T. Storey (1995) Interactions between gametes leading to fertilization: The sperm's eye view. *Reprod. Fertil. Dev.* **7**, 927–942. Describes the interactions between the gametes leading to fertilization.

E. Töpfer-Petersen and J. J. Calvete (1995) Molecular mechanisms of the interaction between sperm and the zona pellucida in mammals: Studies on the pig. *Int. J. Androl. Suppl. 2* **18**, 20–26. Summarizes established similarities and differences between sperm oocyte interactions of mammalian species.

ACTIN

EDWARD H. EGELMAN

Actin is one of the most ubiquitous and conserved eukaryotic proteins. While actin was originally isolated and studied as part of the contractile apparatus of muscle (see **Thin filament**), we now know that actin is found in virtually all eukaryotic cells and can be the most abundant **protein** present. Actin forms part of the **cytoskeleton** in nonmuscle cells and is involved in the maintenance of cell shape as well as cell dynamics and motility. The active form of actin is a helical polymer called F-actin (F for filamentous), assembled from monomeric subunits of G-actin (G for globular). Actin exists in a number of very similar isoforms (88% **amino acid** identity between yeast cytoplasmic actin and human muscle actin), and these isoforms display tissue, rather than species, specificity (1,2). Thus, human cytoplasmic actin is much closer in sequence to yeast cytoplasmic actin than it is to human muscle actin. The actin monomer contains 374–376 residues, varying with the isoform, and is about 42k MW.

Actin interacts specifically with over 40 other proteins, and many of these interactions fall into separate classes. There are proteins that control the polymerization state of actin by nucleating, capping, or severing filaments [such as **gelsolin**, **profilin**, cofilin, and the β-thymosins (3–7)], proteins that crosslink actin filaments together into gels and bundles [such as α-**actinin**, villin, and fimbrin (8–10]), and proteins that regulate the interactions of **myosin** with actin [such as **tropomyosin**, **troponin**, calponin, and caldesmon (11), (see **Thin filament**)]. In addition, many glycolytic enzymes bind actin, and it has been suggested that this provides spatial organization to metabolic processes via an association with the cytoskeleton (12). Other high-affinity interactions, such as between G-actin and the endonuclease DNase I (see **DNase I sensitivity**) (13), do not lend themselves to simple explanations at this time.

Extensive biochemical, structural, and genetic studies have explored the properties of both G- and F-actin. The most detailed structural information exists for G-actin, since X-ray crystal structures have been obtained for complexes of G-actin with DNase I (13), gelsolin segment 1 (14), and profilin (15). The G-actin structure consists of two domains, each with two subdomains. These domains were originally called the "large" and "small" domains, but this terminology is not appropriate, as we now know that the "large" domain only contains 51% of the mass of the subunit. The N- and C-terminii are both located in the largest subdomain, subdomain-1. Between the two domains is a nucleotide-binding cleft, normally occupied by ATP in G-actin and ADP in F-actin, as well as the high-affinity metal-binding site. The subunit structure of G-actin is homologous to that of yeast hexokinase and the ATP-binding domain of HSC-70, a **chaperonin**, suggesting a common evolutionary origin for all of these proteins (16). It has been proposed that the bacterial protein FtsA has the same fold (17), which would indicate that FtsA is a prokaryotic ancestor of the eukaryotic actins. It has already been established that the bacterial FtsZ protein has a common structure with eukaryotic **tubulins**. Since neither hexokinase nor HSC-70 polymerize, it is evident that actin's ability to polymerize does not reside in any unique features of the secondary or tertiary structure, but rather evolved later. Thus, establishing that FtsA has a common subunit structure with actin will not, in and of itself, reveal any information about possible quaternary structures for FtsA.

Electron microscopic observations of F-actin (18–20) have been very useful in defining the orientation of the subunit in the filament, as well as revealing aspects of structural dynamics. X-ray fiber diffraction from oriented gels of F-actin has led to an atomic model for F-actin (21,22), which provides a starting point for understanding the molecular details of the interactions of actin with many other proteins and ligands. However, many of the atomic details are still underdetermined by the method of fitting the monomeric structure into the filament, since the X-ray fiber diffraction pattern does not introduce all of the constraints needed. We thus have a general picture of the orientation of the subunit in the filament, but still lack information about many of the atomic interactions.

The subunit is oriented in the filament so that the two domains are nearly perpendicular to a plane containing the helix axis (Fig. 1). The resulting filament is about 100 Å in diameter, and subunits are spaced axially by a 27.3 Å rise along the filament axis. The change in the axial rise per subunit is less than .08 Å per subunit (<0.3%) when actin filaments in muscle are placed under full tension (23,24), establishing that these filaments are relatively inextensible. Nevertheless, this amount of extensibility is enough to complicate the interpretation of muscle mechanics. Subunits are related to their axial neighbors by an ~167° rotation about the filament axis, giving rise to a 59 Å pitch left-handed helix (sometimes called the genetic helix). An interesting property of the actin filament is the ability of subunits to rotate within the helix (25), and it has recently been shown that an actin-binding protein, cofilin, can change the twist of the actin filament by 5° per subunit, without significantly changing the axial rise per subunit (7). The helical geometry of actin also gives rise to two right-handed long-pitch strands containing actin subunits related by a 54.6 Å axial separation. These strands typically have a mean pitch of ~700–760 Å (with the differences in the mean being due to the isoform, preparative conditions, associated proteins and ligands, etc.). The double-stranded character of these helices give rise to "crossovers" in projection

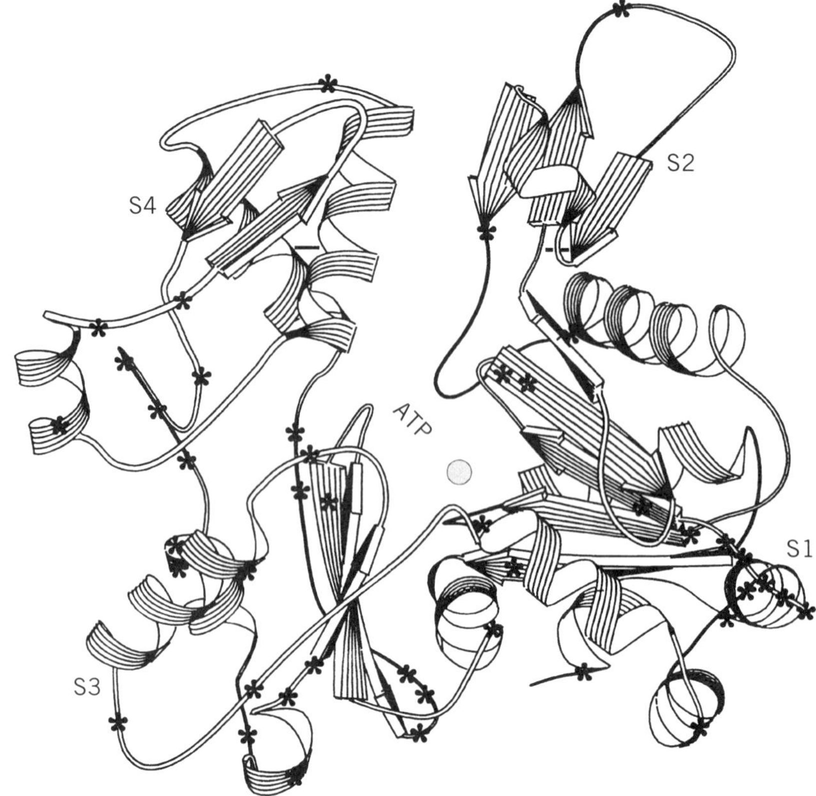

Figure 1. The G-actin subunit is shown by a ribbon representation (39) based upon the structure determined by X-ray crystallography (13). The degree of sequence conservation in actin is indicated by the fact that the only residues (45 out of 375) that differ between vertebrate striated skeletal muscle and yeast cytoplasmic actin are indicated by the asterisks. The four subdomains of actin are labeled S1–S4, and the most conserved part of the molecule is subdomain 2, with only 2 substitutions out of 38 residues between these two isoforms. The cleft between the two domains is occupied by ATP and a bound metal (indicated by the sphere). Adapted from Orlova et al. (40).

at half of the pitch, or about 350–380 Å, but these crossover spacings can be variable due to the angular freedom within the actin filament (25,26).

Every subunit in the actin filament is oriented with the same polarity, which gives rise to the overall polarity of the filament. This polarity has been described as the filament having a "barbed" end and a "pointed" end, based upon the morphological polarity that appears in electron micrographs when actin filaments are extensively decorated by the binding of proteolytic fragments of myosin, such as S1 or HMM (heavy meromyosin). This decoration generates a chevron appearance caused by all myosin molecules binding to the actin filament with an angle of about 45°. The morphological polarity has been related to a kinetic polarity apparent during *in vitro* polymerization, since the barbed end is the fast-growing end and the pointed end is the slow-growing end.

The helical symmetry of the actin filament is often described as 13/6 or 28/13, meaning that it has a helical repeat of 13 subunits in 6 turns of the 59 Å helix, or 28 subunits in 13 turns. The helical repeat is the distance that a subunit must be translated axially to bring it into register with another subunit. An ideal 13/6 helix would have crossovers every 355 Å, with each crossover containing a helical repeat, while an ideal 28/13 helix would have crossovers every 382 Å, with two crossovers per helical repeat. However, there is no reason to believe that the helical symmetry needs to be described as a ratio of small integers, as there is an infinitesimally small change in structure from a symmetry of 13/6 to one of 1301/600 (a rotation of ~0.1° per subunit), but the helical repeat changes from 355 Å to 35,517 Å! The actual symmetry of the actin filament is entirely defined by the interactions between one actin subunit and four nearest neighbors (−2 and 2 along the same long-pitch strand, and −1 and +1 on the opposite strand). In the absence of accessory proteins [such as **tropomyosin**, which binds to seven actin subunits along the same long-pitch strand (see **Thin filament**)], all changes of state must be propagated by such local interactions. Nevertheless, many *in vitro* experimental observations have shown that pure actin (in the absence of accessory proteins) can have long-range cooperative interactions within a filament (20,27–29).

The flexibility of the actin filament has been studied by many methods, including **spectroscopy** (30), light microscopy (31), and **electron microscopy** (32,33). Most studies have suggested a persistence, or correlation, length of about 6–7 μm. This indicates that an actin filament much shorter than 6–7 μm can be treated as a rigid rod, while a filament much longer than 6–7 μm can be treated as a random coil. The persistence length does not suggest that a 6–7 μm filament is approximated well by a rigid rod, since an actin filament only 0.6 μm in length (one-tenth of the persistence length) will have a characteristic fluctuation of tangents at the two ends of about 25°.

Actin polymerization has been studied *in vitro* in great detail and is induced by raising the salt concentration. As for almost all other protein polymers in the cell, the assembly process involves the noncovalent binding of subunits to each other, such that every subunit (with the exception of subunits at the filament ends) is in the same environment. Probably more is known about the *in vitro* polymerization of actin than for any other protein polymer, but new findings suggest that the *in vivo* regulation of actin assembly involves the interactions with many other proteins and may be greatly different from what occurs *in vitro* (34). Polymerization can occur *in vitro* when the subunit concentration is above the critical concentration, which is the concentration of the monomer that would be at equilibrium with a population of polymers (see **Treadmilling**). Polymerization *in vitro* also requires that a monovalent salt, such as NaCl, be present at a concentration of ~100 mM, or a divalent salt such as $MgCl_2$ be present at ~1 mM.

Actin has a single high-affinity metal binding site, which has a greater affinity for Ca^{2+} than for Mg^{2+}. However, due to the much higher concentrations of Mg^{2+} than Ca^{2+} present *in vitro*, it is expected that this site will be occupied by Mg^{2+}. This has been shown experimentally to be true in muscle (35), where the Mg^{2+} concentration is three orders of magnitude higher than the Ca^{2+} concentration. Actin also has 5–10 lower affinity metal binding sites (36). The spontaneous nucleation of actin filaments is often the rate-limiting step in actin polymerization *in vitro*, since the subsequent growth of existing filaments can occur rapidly. Within the cell, however, there is good reason to believe that actin polymerization occurs under the control of other proteins which serve to nucleate and cap growing filaments, and it is possible that spontaneous self-nucleation of actin filaments may never occur.

Actin polymerization is also loosely coupled to the hydrolysis of the bound nucleotide. While almost all attention has been focused on the primary nucleotide-binding site, actin also has a second nucleotide-binding site with mM affinity (37). Since this site is likely to be occupied in cells such as muscle where there exists a millimolar concentration of ATP, it remains to be seen whether this second site plays a physiological role. It is known that the nucleotide hydrolysis is not required for polymerization, since G-ADP actin can polymerize, as can G-actin subunits containing nonhydrolyzable analogs of ATP. The role of ATP hydrolysis is to actually destabilize the filament (38), allowing for depolymerization to occur more readily. ATP hydrolysis lags behind the initial polymerization event and occurs once a subunit is part of the polymer. The cell therefore uses the energy of ATP hydrolysis to allow for the dynamic assembly and disassembly of actin filaments, in contrast to less dynamic components of the cytoskeleton, such as **intermediate filaments**. The energy of ATP hydrolysis can also be used to drive **treadmilling**, where there is a unidirectional flux of subunits through a filament.

BIBLIOGRAPHY

1. P. A. Rubenstein (1990) *BioEssays* **12**, 309–315.
2. E. S. Hennessey, D. R. Drummond, and J. C. Sparrow (1993) *Biochem. J.* **291**, 657–671.
3. E. Ballweber, E. Hannappel, T. Huff, and H. G. Mannherz (1997) *Biochem. J.* **327**, 787–793.
4. M. F. Carlier and D. Pantaloni (1997) *J. Mol. Biol.* **269**, 459–467.
5. D. A. Schafer and J. A. Cooper (1995) *Ann. Rev. Cell Dev. Biol.* **11**, 497–518.
6. L. D. Burtnick, E. K. Koepf, J. Grimes, E. Y. Jones, D. I. Stuart, P. J. McLaughlin, and R. C. Robinson (1997) *Cell* **90**, 661–670.
7. A. McGough, B. Pope, W. Chiu, and A. Weeds (1997) *J. Cell Biol.* **138**, 771–781.
8. P. Matsudaira (1991) *Trends Biochem. Sci.* **16**, 87–92.
9. D. Hanein et al. (1998) *Nature Struct. Biol.* **5**, 787–792.
10. K. A. Taylor and D. W. Taylor (1994) *Biophys. J.* **67**, 1976–1983.
11. L. S. Tobacman (1996) *Ann. Rev. Physiol.* **58**, 447–481.
12. H. R. Knull and J. L. Walsh (1992) *Curr. Top. Cell. Reg.* **33**, 15–30.
13. W. Kabsch, H. G. Mannherz, D. Suck, E. F. Pai, and H. C. Holmes (1990) *Nature* **347**, 37–44.
14. P. J. McLaughlin, J. T. Gooch, H. G. Mannherz, and A. G. Weeds (1993) *Nature* **364**, 685–692.
15. C. E. Schutt, J. C. Myslik, M. D. Rozycki, N. C. W. Goonesekere, and U. Lindberg (1993) *Nature* **365**, 810–816.
16. K. M. Flaherty, D. B. McKay, W. Kabsch, and K. C. Holmes (1991) *Proc. Natl. Acad. Sci.* **88**, 5041–5045.
17. P. Bork, C. Sander, and A. Valencia (1992) *Proc. Natl. Acad. Sci.* **89**, 7290–7294.
18. R. A. Milligan, M. Whittaker, and D. Safer (1990) *Nature* **348**, 217–221.
19. A. Orlova and E. H. Egelman (1995) *J. Mol. Biol.* **245**, 582–597.
20. A. Orlova, E. Prochniewicz, and E. H. Egelman (1995) *J. Mol. Biol.* **245**, 598–607.
21. K. C. Holmes, D. Popp, W. Gebhard, and W. Kabsch (1990) *Nature* **347**, 44–49.
22. M. Lorenz, D. Popp, and K. C. Holmes (1993) *J. Mol. Biol.* **234**, 826–836.
23. K. Wakabayashi, Y. Sugimoto, H. Tanaka, Y. Ueno, Y. Takezawa, and Y. Amemiya (1994) *Biophys. J.* **67**, 2422–2435.
24. H. E. Huxley, A. Stewart, H. Sosa, and T. Irving (1994) *Biophys. J.* **67**, 2411–2421.
25. E. H. Egelman, N. Francis, and D. J. DeRosier (1982) *Nature* **298**, 131–135.
26. J. Hanson (1967) *Nature* **213**, 353–356.
27. F. Oosawa (1983) In *Muscle and Non-Muscle Motility* (A. Stracher, ed.) Academic Press, New York, p. 151.
28. G. Drewes and H. Faulstich (1993) *Eur. J. Biochem.* **212**, 247–253.
29. A. Muhlrad, P. Cheung, B. Phan, C. Miller, and E. Reisler (1994) *J. Biol. Chem.* **269**, 11852–11858.
30. T. Yanagida and F. Oosawa (1978) *J. Mol. Biol.* **126**, 507–524.
31. H. Isambert, P. Venier, A. C. Maggs, A. Fattoum, R. Kassab, D. Pantaloni, and M. F. Carlier (1995) *J. Biol. Chem.* **270**, 11437–11444.
32. T. Takebayashi, Y. Morita, and F. Oosawa (1977) *Biochim. Biophys. Acta* **492**, 357–363.
33. A. Orlova and E. H. Egelman (1993) *J. Mol. Biol.* **232**, 334–341.
34. M. F. Carlier (1998) *Curr. Opin. Cell Biol.* **10**, 45–51.
35. T. Kitazawa, H. Shuman, and A. P. Somlyo (1982) *J. Musc. Res. Cell Motil.* **3**, 437–454.
36. M. F. Carlier, D. Pantaloni, and E. D. Korn (1986) *J. Biol. Chem.* **261**, 10778–10784.
37. P. Kiessling, B. Polzar, and H. G. Mannherz (1993) *Biol. Chem. Hoppe-Seyler* **374**, 183–192.
38. C. Combeau and M. F. Carlier (1988) *J. Biol. Chem.* **263**, 17429–17436.
39. P. J. Kraulis (1991) *J. Appl. Crystallogr.* **24**, 946–950.
40. A. Orlova, X. Chen, P. A. Rubenstein, and E. H. Egelman (1997) *J. Mol. Biol.* **271**, 235–243.

Suggestions for Further Reading

P. Sheterline, J. Clayton, and J. Sparrow (1995) Actin, *Protein Profile* **2**, 1–103.

E. Reisler (1993) Actin molecular structure and function, *Curr. Opin. Cell Biol.* **5**, 41–47.

K. R. Ayscough and D. G. Drubin (1996) Actin: General principles from studies in yeast, *Ann. Rev. Cell Dev. Biol.* **12**, 129–160.

ACTIN-BINDING PROTEINS

UNO LINDBERG
CLARENCE E. SCHUTT

THE ACTIN MICROFILAMENT SYSTEM

Actin was identified more than 50 years ago as an essential element of the force generating apparatus of muscle cells (1,2). Together with myosin, tropomyosin, and a large number of other proteins, actin filaments constitute sarcomeres, which are ordered serially along contractile myofibrils (3,4). In the mid 1960s, actin (5) and myosin (6,7) were recognized as major structural components of the eukaryotic cytoplasm (8–10); for an early review of the field see (11). The unexpected observation that actin was the inhibitor of DNase I (12), and that the DNase inhibitor could be crystallized (13), led to the identification of the actin monomer-binding protein, **profilin** (14). The properties of profilin seemed to explain how unpolymerized actin could exist at high concentrations in cells without spontaneously polymerizing (15). It is now known that there are several other proteins controlling actin polymerization, and today the family of actin-binding proteins contains over 100 members including proteins that sequester actin monomers (profilins, thymosins), sever (**gelsolins**), depolymerize (cofilins), and cross-link (villin, fimbrin) actin filaments. Further characterization of the nonmuscle myosins (16), and an increased appreciation of the role of polymerization of actin suggests how cells can extend pseudopods and drag themselves along extracellular matrices (17,18), and exhibit the high degree of motility seen on the surface of lymphocytes and in malignant cancer cells. This entry will describe general characteristics of actin-binding proteins, the analysis of which in many cases has reached the atomic level.

Actin is an obligatory component of all eukaryotes, and is one of the most highly conserved proteins during **evolution**. Its sequence has varied less than 10% during the last 1200 million years (19). In mammalian cells, six different isoforms are known, of which the muscle specific α-isoform and the nonmuscle β-isoform are the most studied [for review, see (20)]. On the basis of structural **homology** an actin-related protein, FtsA, which is involved in cell division, has been recognized in prokaryotes (21). Also, in eukaryotes, there are several actin-related proteins, ARPs (22).

Actin can be kept in a monomeric form only under low salt conditions or in the presence of actin **sequestering** proteins (review 20). Under physiological salt conditions, actin polymerizes into long filaments. In this form, actin participates in force generation and motile activity. The actin filament is polar, with a rapidly growing (+)-end (barbed end) and a slow growing (−)-end (pointed end). In the cell, actin filaments grow by addition of monomers at the (+)-end, which is located at the outer edge of cell surface protrusions: lamellipodia and filopodia. In the submembraneous zone all round the cell, actin filaments are organized into sheets and bundles held together along their lengths by a variety of **crosslinking** proteins, which differ in length and flexural rigidity.

The myosins vary in structure; some are globular in shape, like myosin I, and do not polymerize, whereas others have a long α-helical, supercoiled, rod domain and can form filamentous superstructures in the cytoplasm (16). Together, actin and myosin transduce chemical free energy of adenosine trisphosphate (ATP) to mechanical work, not just in muscle cells, but in all eukaryotic cells. In different states of organization, the actomyosin system, is responsible for muscle contraction, cell locomotion, and cell division, and for many different local cellular transport processes (4). In nonmuscle cells, the actomyosin system is referred to as the **microfilament** system. It is based on the organization of actin filaments, whose formation, organization and activity depend on the activity of a large number of actin-binding proteins, including myosin.

Microfilaments are present in particularly high concentrations in close apposition to the inner leaflet of the lipid bilayer (23–27), where they are directly or indirectly linked to cell surface proteins (receptors and adhesion proteins) continuously probing the extracellular environment. The lipid bilayer of the plasma membrane has been likened with "light machine oil" (28). It is an excellent electrochemical barrier, but it is the dense cortical weave of microfilaments which confers mechanical stability and deformability to the cell surface. Thus, the microfilament system can be viewed as an integral part of the physical barrier that separates the cellular contents from the outside world.

One of the first submembraneous actin-containing superstructures to be characterized was the red blood cell cortical network consisting of actin and the actin-binding proteins **tropomyosin**, **spectrin** and **band 4.1** (29). Linkage to the plasma membrane is provided by the spectrin-binding protein ankyrin, which binds to the transmembrane anion channel band III, and band 4.1 (an ERM protein, see below), which interacts directly with the transmembrane protein glycophorin. The adaptability of the shape of the red blood cell to the compressive forces present in the low-bore blood capillaries reflects the elastic properties of the spectrin:actin network. In the intestinal epithelium (30, 31, and refs. therein), microvilli on the apical surface of the cell are formed by tightly packed actin filament bundles stabilized by the actin crosslinking proteins villin and fimbrin (Fig. 1). The presence of nonmuscle myosin in these structures suggests that active movement may be involved in the resorptive process.

Specializations of the membrane-associated actomyosin system are found in all sensory organs, ie chemomechanical transduction in the actomyosin system is essential to the ability of the organism to perceive perturbations in the environment. For instance, the cellular basis for the sense of hearing lies in the microfilament-rich stereocilia in the hair cells of the inner ear, which convert subtle changes in the ambient air pressure to electrical signals transmitted to the brain via nerve impulses (32). Similarly, the mechanosensory cells in the skin confer the sense of touch (33).

Macrophages, lymphocytes, and neutrophils, which are highly motile cells, exploit the full potential of the cortical microfilament system as they forcibly move through and along extracellular matrices. When macrophages contact foreign invaders, the versatility of the cell cortex allows them to adapt their own surface structures as they engulf the pathogen.

The process of cell migration depends on the coordination of four major events involving microfilament reorganization: extension at the leading edge of membrane lamellae and filopodia, attachment to extracellular structures via transmembrane adhesion molecules, development of intracellular

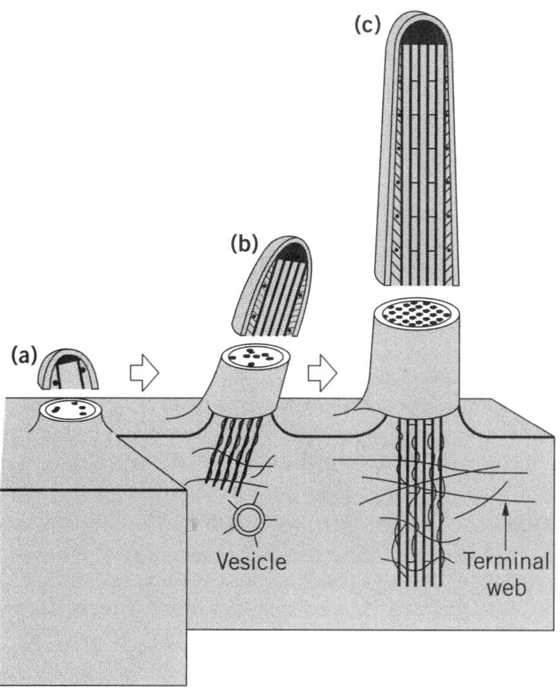

Figure 1. Structural organization of a microvillus. Among the simplest and best studied cell surface structures are the microvilli on polarized epithelial cells, especially those on cells of the intestinal epithelium. The majority of the proteins involved in the architecture of the microvillus have now been identified and characterized. Most of them are essential for the survival of the organism. In some cases their three-dimensional structure has been determined. It is still unclear how the assembly of the microvillus takes place and how it is controlled, how dynamic microvilli are, and what their contribution to uptake of nutrients is. The construction of a microvillus on apical surface of an intestinal epithelial cell is shown here. Many of the microvillar proteins also occur in other cell surface structures. An exception is villin, which is confined to the microvilli of a few cell types. In addition to villin, the major actin-binding proteins of the microvillar core are ezrin, fimbrin, and myosin-I. Drawing from Fath and Burgess 1995, *Curr. Biol.* **5**, 591–593.

tension, and release of trailing end attachments with retraction of trailing ends (18,27,34,35). The protrusion of lamellae at the advancing edge is driven by polymerization of actin filaments at focal complexes containing their fast growing ends at the plasma membrane, and crosslinking of formed filaments (24,25,36,37). Clustering of transmembrane adhesion proteins, integrins, gives rise to focal adhesion sites (Fig. 2) linking **extracellular matrix** proteins to bundled actin filaments on the inside of the plasma membrane (38). The importance of the focal adhesions in transmembrane signalling is emphasized by their association with kinases regulating signal transducing protein:protein interactions and enzymes generating second messengers controlling not only the microfilament system, but also gene transcription (39). Several actin-binding proteins are targetted to these sites, contributing to formation of the integrin:actin filament linkage. The association of linkage proteins like talin, tensin, and vinculin appears to be controlled by phosphorylation. Myosin is activated through Ca^{2+}-dependent processes (40,41), and the regulated deployment of myosin molecules leads to the generation of the forces needed for cell motility and other translocation processes. Release of adhesion at the trailing end of a migrating cell is controlled by Ca^{2+} and is followed by **endocytosis** of the integrins and their recycling to advancing cell edges (42). Used filaments are depolymerized, a process which involves the actin monomer- and filament-binding protein ADF (or cofilin) (17), and sequestered forms of unpolymerized actin are reformed (see Fig. 3).

Thus, the combined action of protein factors controlling polymerization, crosslinking and tension development in the cell cortex is initiated by transmembrane signalling. Controlling the timing and position of sites of polymerization at the cell surface is a potent means of regulating shape change. An astonishing variety of well-ordered, yet dynamic, structures are possible, reflecting the mechanical properties and specificities of the different kinds of actin bundling proteins and junctional connectors. Their timed and coordinated formation prevents the appearance of random rigid networks, which would be incompatible with the well-orchestrated movements seen.

SIGNAL TRANSDUCTION AND RESTRUCTURING OF THE ACTIN MICROFILAMENT SYSTEM

Cell surface receptors continuously monitor the extracellular milieu and pass signals to the cytoplasm by generating second messengers of various kinds which either stimulate or inhibit the activity in the microfilament system. Addition of **growth factors** to serum-starved cultured cells causes the immediate outgrowth of membrane lamellae and filopodia, events that depend on the polymerization and organization of actin. About 60 seconds later, there is a generalized increase in motile activity of the cells, including restructuring of the actin-containing stress fiber system and translocation of the cell. Observations of this kind have been made both with epidermal growth factor (EGF) (43) and platelet-derived growth factor (26). In the case of EGF, a significant fraction of the receptors was found in direct association with microfilaments, and a binding site for actin on the receptor has been identified (44,45).

The phosphatidylinositol-cycle (PI cycle) is coupled to activation of the microfilament system (46). Ligand-induced, receptor-mediated activation of kinases generate phosphatidyl inositol 4,5-bisphosphate (PtdIns 4,5-P2) and phosphatidyl inositol 3,4,5-trisphosphate (PtdIns 3,4,5-P3) (47,48), both of which affect the activity of actin-binding proteins. **Profilin**, **gelsolin**, α-actinin, vinculin, ERM family of proteins, and myosin all bind PtdIns 4,5-P2, and/or PtdIns 4,5-P3. Stimulation of platelets with low concentrations of thrombin causes a transient increase in polyphosphoinositides during the first 10 sec (49). This is accompanied by an equally transient transformation of unpolymerized actin into filaments. This initial polymerization is over in about 10–20 sec (49). The immediate precursor for this polymerization appears to be profilin:actin, since increased amounts of free profilin appears in cell extracts after stimulation. Within 2 min, these filaments become organized into supramolecular structures by the action of crosslinking proteins stabilizing the protrusions seen on the platelet surface (50–52). Simultaneous activation of phospholipase $C\gamma$-1 results in the hydrolysis of PtdIns 4,5-P2, releasing the second messenger inositol trisphosphate, which causes release of Ca^{2+} ions from intracellular Ca^{2+} stores

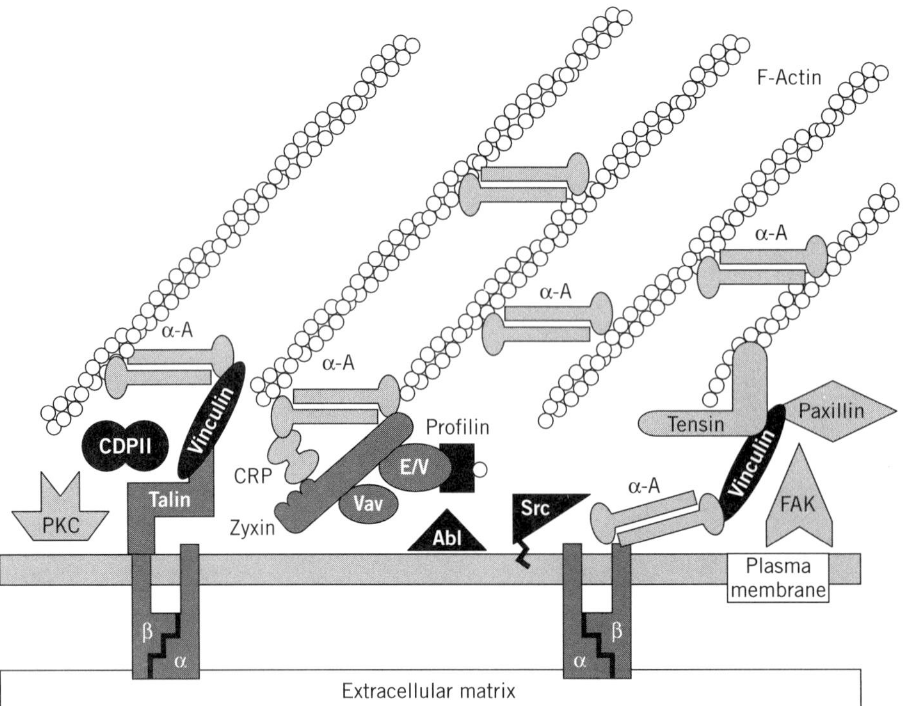

Figure 2. Proteins of focal adhesion sites. Adherens junctions comprise both cell:extracellular matrix contacts, exemplified by focal adherens in tissue cultured cells, and intercellular junctions such as the zonula adherens between epithelial cells. The regulated assembly of adherens junctions is essential to a diversity of cellular functions including wound healing, anchorage-dependent cell proliferation, adhesion, and migration. Assembly of an adherens junction may be initiated when an integrin of the $\beta 1$ subfamily binds to its extracellular matrix ligand, when a cadherin recognizes an identical molecule on a neighboring cell, or when a growth factor activates its receptor. These ligand–receptor interactions result in establishment of a physical connection between the extracellular ligand and the actin cytoskeleton. Efforts to elucidate these linkages have implicated vinculin as one of the connecting molecules in a chain that involves binding of the $\beta 1$ subunit of integrin to talin, talin to vinculin, vinculin to α-actinin, and α-actinin to filamentous actin. It is now known that vinculin also can bind directly to actin filaments. How the initial ligand–receptor interactions trigger recruitment and organization of the multiple proteins found at an adherens junction is unknown, but recent evidence suggest that the activities of tyrosine kinases, tyrosine phosphatases, protein kinases and members of the Rho family of small GTPases are involved. (Drawing from M. C. Beckerle (1997) *Bioassays* **18**, 949–957)

(53). Calcium ions are involved in activation of the actomyosin interaction and thereby force generation.

In addition to connections to the phosphatidyl inositol signal transduction cycle, the microfilament system is linked to the cAMP-dependent signalling system. Agonists causing increasing levels of cAMP often inhibit actin polymerization (54). A connection between adenylate kinase pathways and the actin monomer-binding protein profilin was suggested by genetic studies showing that increased levels of profilin could compensate for the deletion of the C-terminal domain CAP of the cyclase (55,56). In platelets, the actin-binding protein, vasodilator-stimulated protein VASP is phosphorylated by cAMP- and cGMP-dependent kinases (57), which then correlates with the inhibition of microfilament reorganization.

The proteins involved in controlling the organization and function of the microfilament system are part of multiprotein signal transduction complexes. Several of these proteins have been identified as products of protooncogenes, ie proteins which in mutated form may cause cancer (58,59). In these signal transduction complexes the proteins are linked together by adaptor modules in linker proteins such as SH2, SH3, and WW domains which recognize specific features of short polypeptide stretches, especially phosphotyrosine-containing and polyproline sequences (60). An example is the oncogene product **Vav** that has several such binding motifs suggesting how it links signal transduction at the cell surface to restructure the cortical cytoplasm (61). Vav possesses many of the homology domains commonly found in signal transduction complexes, comprising a phosphotyrosine-binding SH2 domain, two proline-rich peptide recognizing SH3 domains, a phosphatidyl inositol 4,5-bisphosphate recognizing PH domain, a guanine nucleotide exchange factor DH-domain, two cysteine-

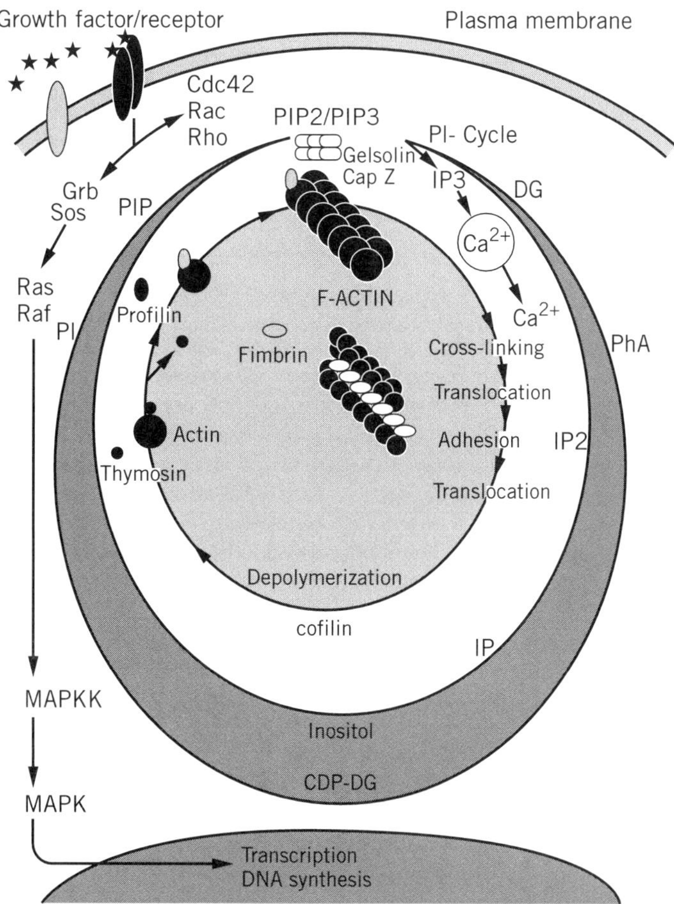

Figure 3. Links between the phosphatidylinositol and the cell motility cycles. Binding of growth factors like platelet-derived growth factor and epidermal growth factor to their cognate receptors activate the kinases activities of the receptors resulting in phosphorylation of a number of tyrosines on the cytoplasmic part of the receptor. This leads to activation of the signal transduction elements Cde42, Rac and Rho which control outgrowth of filopodia, membrane lamellae and the formation of stress fibers, respectively (64). There is only little information so far regarding the mechanism of activation of these small GTPases and how they relay the signal to the phosphatidylinositol (PI) cycle. Receptor activation is followed by an immediate increase in the activity of kinases that form PtdIns 4,5-bisphosphate (PIP2) and PtdIns 3,4,5-trisphosphate (PIP3). Profilin, gelsolin, CapZ, α-actinin, vinculin, the ERM family of proteins all bind these phosphoinositides, which influence their activity in the microfilament-based cell motility cycle (CM-cycle). Phosphoinositides binding to CapZ or gelsolin unblock the (+)-end of actin filaments allowing incorporation of profilin:actin at that site. Profilin:actin is then dissociated and the actin monomer becomes stably incorported into the growing filament. Cross-linking of filaments by proteins like fimbrin gives rise to the ensembles of filaments seen in membrane lamellae and filopodia which function together with myosin in the contractile phase of cell motility. Calcium ions play important roles in regulating both the actomyosin interaction and the depolymerization of actin filaments. The depolymerization is facilitated by the dephosphorylated form of cofilin and the actin monomer sequestering proteins thymosin and profilin (rev. 163). Another branch of signal transduction from the growth factor receptor activates the gene program that leads to DNA synthesis and cell division. Grb and Sos are small proteins acting as signal transduction elements activating the small GTPases Ras, which in turn activates the serine/threonine kinase Raf. The further signalling pathway is complex invlving mitogen activated protein kinases, MAPKK and MAPK (for review, see *Cell* 1998, **95**, 447–450). Other abbreviations: IP3, inositol trisphosphate; IP2, inositol bisphosphate; IP, inositol monophosphate; DG, diacylglycerol; Pha, phosphatidic acid; CDP-DG, cytidyldiphosphate-diacylglycerol; PI, phosphatidylinositol.

rich nuclear transcriptionally important LIM domains, and an actin-binding calponin homology domain. The clearest demonstration that these functional linkages actually lead to nuclear transcription is that the production of Interleukin-2 (IL2) requires Vav-dependent actin cap formation in activated T-cells (62,63).

Outgrowth of filopodia and membrane lamellae, and the formation of stress fibers, which are manifestations of the activation of the microfilament system by different kinds of ligand:receptor interactions, all depend on the activation of members of the Rho subfamily of small GTPases for their control. The small GTPase Cdc42 appears to control formation of filopodia, Rac controls membrane lamellae and Rho stress fibers (64). There is only little information so far as to how the receptor-mediated transmembrane signal is relayed to the GTPases and what their downstream targets are. It has been shown with permeabilized platelets that ligation of the thrombin receptor, and activated Rac, uncap actin filament (+)-ends through phosphoinositide synthesis (65). Some of the microfilament-controlling, signal transduction elements mentioned below interact with small GTPases. Interestingly, moesin (see below), an actin-binding protein linking actin filaments to transmembrane receptors, is required for Rho and Rac effects on the actin filament system (66).

VASP belongs to a growing collection of related proteins that are involved in stimulation of assembly of actin filaments *in vivo* (67–69). In moving cells, it is found in highly dynamic membrane regions and in integrin-rich adhesion plaques. Molecular cloning of VASP has established that it has both a central proline-rich core, which interacts directly with profilin (68,70), and an actin-binding site. In cells, the primary site of growth of actin filaments is thought to be at advancing cell edges, and profilin:actin heterodimers are believed to be guided into this site by VASP (71). High concentrations of VASP appear to be targetted to focal adhesion sites by zyxin (72). Zyxin also binds α-actinin and Vav suggesting that orderly assembly of crosslinked actin networks can proceed at these locales. In addition to binding profilin and zyxin, VASP binds to the SH3 domains of the src-family of protein kinases (69), thus integrating signal transduction via phosphorylation with the ordered assembly of cortical structures in locally specialized sites (63). VASP and vinculin complexes can be immunoprecipitated out of cell extracts with either antivinculin or antiVASP antibodies, and recent experiments have demonstrated that the interaction between the two proteins is enhanced by PtdIns 4,5-bisphosphate binding to vinculin (73).

VASP has come to prominence owing to its relevance to the understanding of the movement of the bacterium *Listeria* through the host cell cytoplasm (74). *Listeria* expresses a proline-rich protein on its surface called ActA (zyxin-like), which binds to the modular proline-binding SH3-domain of VASP. Apparently, VASP can attract profilin:actin complexes to sites of polymerization on the surface of the bacterium, generating a propulsive force. There are many examples of pathogenic microorganisms that have acquired host cell genes that code for proteins that modulate the behavior of the microfilament system for the purpose of invasion, locomotion in the cytoplasm, and cell-to-cell spreading.

WASP is a multidomain, actin-binding protein, which is defective in patients afflicted with Wiskott-Aldrich syndrome (75,76), a disease characterized by severe thromifocytopenia and immunodeficiency. WASP is specific to hematopoietic cells, which in case of WAS have a paucity of surface filopodia and other shape abnormalities (77). The small GTPases Cdc42 and Rac bind directly to WASP (78–80). WASP is also a binding partner for Src family of protein tyrosine kinases (81).

A protein homologous to WASP, N-WASP, is more widely distributed. N-WASP is concentrated at nerve terminal regions in the post synaptic density, and coexpression of Cdc42 and N-WASP induces extremely long actin-based filopodia in cells (82,83). Its involvement in the formation of membrane lamellae and filopodia may be controlled by phosphoinositides in the plasma membrane. It contains a pleckstrin homology PH domain and a cofilin homology domain. Furthermore, N-WASP has been shown to be essential for the actin polymerization-dependent movement of the bacterium *Shigella flexneri* (84).

WIP is a recently discovered WASP interacting protein that is important for cortical actin assembly (85). WIP contains both profilin- and actin-binding regions, which like in VASP, are near each other in the protein sequence suggesting that they constitute the binding site for a profilin:actin heterodimer to be added to a growing actin filament. The small signal transduction GTPase Cdc42 binds to WASP (78,79), but not to WIP. Thus, WIP function may be regulated by Cdc42 via WASP. Stimuli that activate Cdc42 may target the WASP-WIP complex to the actin filament system via interactions between the WH1 domain of WASP and the proline-rich ABM1 motifs of structural proteins such as zyxin and vinculin. The presence of SH3 binding motifs in both WIP and WASP, and the capacity to bind to the adaptor protein Nck (86), suggest that the WASP-WIP complex couples additional signaling pathways to the actin filament system.

Formins are profilin and actin-binding proteins from *S. cerevisiae*. They direct the determination of cell polarization and cytokinesis (87 for refs.). In response to α-mating factor, the formin Bni1p forms complexes with the active form of Cdc42, actin, profilin, and actin associated protein Aip3p (88–90). These proteins localize to the tips of mating projections suggesting that the formin is a target for Cdc42, linking the pheromone response pathway to activation of the actin system.

The formin family of proteins is also important in limb and kidney development in vertebrates (91). Mutations in formins lead to defects in the contractile ring formation during cytokinesis, causing the appearance of multinucleated cells. Formins are localized both in the cytoplasm and the nucleus of the cells. FH-proteins, defined by the presence of 'formin homology' regions, are important for a number of actin-dependent processes, including polarized cell growth and cytokinesis. They are large, probably multi-domain, proteins and their function may in part be mediated by an interaction with profilin.

IQGAP is yet another actin-binding, multidomain, regulatory protein (92). It was originally described as a 190-kD protein with extensive sequence similarity to the catalytic domain of RasGaps (91). These proteins control the activity of small regulatory GTPases by stimulating their GTPase activity. Subsequently an IQGAP2 was described as a liver-specific protein with a 62% homology to IQGAP1 (93). However, so far no GAP activity has been found in relation to the GTPases examined. Both IQGAPs have several copies of a ca 50 amino acid long internal repeat (IR), a single WW (SH3-like) domain presumably binding to proline-rich sequences, a number of calmodulin-binding IQ motifs, and a putative actin-interacting calponin homology CH domain. In transient transfection of COS-7 cells with epitope-tagged Cdc42, it was demonstrated that IQGAP is present in the advancing lamellae of motile cells, and a major *in vivo* target of activated Cdc42 (94). The IQGAP1 binds to filamentous actin *in vitro* via its N-terminal domain and cross-links actin filaments in a Cdc42-dependent manner (95). Recently IQGAP1 was implicated in the regulation of E-cadherin-mediated cell-cell adhesion (96).

In fission yeast, *Schizosaccharomyces pombe*, the rng2 gene codes for an IQGAP-related protein, rng2p. The rng2p is located in the actomyosin ring and the spindle pole body and is required for the assembly of the actomyosin ring at cytokinesis (97).

Paxillin is not known as an actin-binding protein, but it is involved in the control of the microfilament system at the adhesion contacts. It was detected as a 68-kD phosphotyrosine-containing protein in RSV-transformed cells, and was later purified from smooth muscle cells (98,99). Many features of paxillin suggest that it functions in signalling events at the adhesive membrane. It is concentrated at focal adhesions, and it is tyrosine phosphorylated in response to extracellular matrix binding and cell transformation. It binds *in vitro* to the elongated tail domain of vinculin, and thereby links vinculin to the SH3 domain of the pp60^{csrc}, a component of focal contacts. Integrins, which are the principal transmembrane proteins at these sites, do not seem to bind directly to paxillin, and paxillin does not seem to bind either α-actinin or actin, but there is evidence for direct association of paxillin to growth factor receptors. Adhesion of cells to an extracellular matrix like fibronectin or laminin via transmembrane integrins results in increased tyrosine phosphorylation of paxillin, and a similar increase in phosphorylation takes place in connection with clustering of integrins in the plane of the membrane with antibodies. This phosphorylation appears to be caused by activation of the focal adhesion-associated tyrosine kinase, pp125^{FAK}, and tyrosine phosphorylation of paxillin and pp125^{FAK} initiated by cell adhesion is accompanied by the appearance of organized actin-containing stress fibres. Mutagenesis of paxillin has identified one of the paxillin LIM domains as being responsible for targetting this protein to focal contacts (100)

Zyxin is a low-abundance phosphoprotein localized at sites of cell-substratum adhesion in fibroblasts (72). It has an architecture of an intracellular signal transducer with a proline-rich domain, a nuclear export signal, and three copies of the LIM motif. LIM is a double **zinc finger** domain found in many proteins that play central roles in regulation of cell differentiation. Zyxin interacts with α-actinin, members of the cysteine rich protein CRP family, proteins that display Src homology 3 (SH3 domains) and Vasp/Ena family members, which in turn bind to both actin and profilin. Zyxin and its partners have been implicated in the spatial control of actin filament assembly as well as in pathways important for cell differentiation. Based on its repertoire of binding partners and its behaviour, zyxin is thought to serve an organizing centre for the assembly of multimeric protein machines that function both at sites of cell adhesion and in the nucleus.

In summary, the major signal transduction pathways and mechanisms of activation have been shown to be linked to the changes in the dynamic restructuring of the actin cortex: the phosphatidyl inositol cycle and the cAMP second messenger systems and the small GTPase coupled signal transduction mechanisms. All of this points to the growing realization that cytoplasmic signal transduction and transcriptional events in the nucleus are so interwoven that discussions of sequential control by one system or the other are oversimplifications. The idea that actin cortical movements are "downstream" events of "upstream" effectors is similarly unjustified, since actively motile cells are required for the generation of the signal itself.

There are perplexing aspects of polymerization-based propulsion of bacterial pathogens, and by implication filopodial extension. WASP and N-WASP have actin-binding sequences. It is not clear what the role of these proteins is in the regulation of actin polymerization at focal growth sites. It is known that the polar actin filaments are oriented such that their fast growing ends are "pushing" against the moving bacterial wall, or against the membranes of the advancing edges of lamellipodia and filopodia. Actin-binding proteins such as WASP are anchored in these nascent growth complexes (to ActA in the case of *Listeria*). The problem is to explain what happens when an incoming actin monomer is presented as a profilin:actin precursor to the end of the growing filament, since it would appear to be blocked by WASP. Another problem is to account for the free energy of ATP hydrolysis that is made available by the actin polymerization reaction.

FURTHER LINKS OF ACTIN FILAMENTS TO THE EXTRACELLULAR MATRIX

Cortactin is the name of two related proteins p80 and p85, which cross-link actin filaments close to the plasma membrane. Cortactin was first recognized as a substrate for tyrosine-phoshorylation in pp60^{src}-transformed cells, but its presence has also been demonstrated in a variety of non-transformed cells (101). Cortactins have internal repeats of 37 amino acid residues in the N-terminal end of the molecule, which are necessary for their binding to actin filaments. In the C-terminus there is an SH3-domain, and preceding that in the sequence there are proline and serine/threonine rich regions. These latter parts of the molecule are not needed for actin-binding, and their functions remain to be elucidated.

In non-transformed cells, cortactin is normally phosphorylated on serine and threonine, but becomes transiently phosphorylated on tyrosine in response to growth factor stimulation, and on transformation by activated cSrc. In platelets, cortactin becomes transiently tyrosine-phosphorylated in reponse to stimulation with thrombin. However, tyrosine phosphorylation of cortactin does not appear to affect its ability to bind to actin filaments, although it does influence its localization to the cortical actin filament system.

Stimulation of cultured cells with various growth factors and platelets with thrombin results in translocation of cortactin into the actin filament system of advancing lamellae and filopodia (102,103). Activation of fibroblast growth factor receptor-1 results in an association of the activated receptor with cSrc, and a correlated association of cortactin with SH2 domain of cSrc. The redistribution from the cytoplasm into advancing lamellae (ruffles and lamellipodia) appears to depend on the Rac-1-induced activation of the serine/threonine kinase PAK1, a downstream effector of Rac1 and Cdc42 (104). These GTPases are involved in the control of kinases in the phosphatidyl inositol signalling pathway, and there is evidence that PtdIns 4,5-bisphosphate could be involved in the association of cortactin with the activated microfilament system (105).

Eps8 is a putative actin-binding protein which resembles cortactin in its distribution in the cell (106). It has an unusual

Src homology-3 domain in the carboxy terminus, formed by an intertwined dimer of Esp8 molecules which may affect its peptide specificity. Eps8 is mostly localized in the perinuclear region. However, on stimulation of cells with growth factors, it accumulates in the peripheral cell extensions, where it is found colocalized with cortactin and filamentous actin. The Eps8 pool associated with newly formed lamellipodia and membrane lamellae appears to be mostly detergent-insoluble, as is the tyrosine phosphorylated population of Eps8 molecules generated by activation of vSrc kinase, suggesting Eps8 recruitment to specific sites during cell remodeling.

Ezrin, radixin, moesin (ERM) belong to the band 4.1 superfamily on the basis of their homology to the erythrocyte band 4.1, a protein that connects the actin/spectrin network to the erythrocyte membrane protein glycophorin C (31,107,108). Like cortactin, the ERM proteins appear to play structural and regulatory roles in stabilizing specialized organizations of actin filaments in connection with the plasma membrane both during development and in adult tissues. They are generally present in microvilli, filopodia, and membrane ruffles, sometimes together, but more often differentially distributed between different types of cells. They link transmembrane proteins with the cortical actin filaments, a process governed by signal transduction involving the Rho-GTPase (109). Ezrin has been shown to bind specifically to nonmuscle β-actin with high affinity (110).

The N-terminal domain of ezrin associates with plasma membrane components, and the C-terminal half connects these structures with the submembranous actin filaments. Purified ezrin exists in an inactive conformation that requires activation to expose sites that allow it to associate with the membrane components and with actin filament (107). It has been shown that binding of ezrin to the transmembrane hyaluronan receptor CD44 (111) requires activation of ezrin by PtdIns 4,5-bisphosphate binding to the N-terminal domain of ezrin (112–114). These are not the only transmembrane proteins binding ERM, since ERM proteins have a membrane location also in cells that do not express CD44. It is now known that ERM proteins, in addition to CD44, also bind to intercellular adhesion molecules, ICAM-1 and ICAM-2. A positively charged amino acid cluster of CD44, CD43, and ICAM2 positioned close to the lipid bilayer has been implicated in the binding (115).

A group of phosphorylated, 50–55 kDa, ezrin-binding protein has been purified from human placenta and bovine brain by affinity chromatography on immobilized N-terminal domains of ezrin or moesin (31). Isolated EBP50 has two peptide-binding PDZ domains in the N-teminal half of the molecule with capacity to bind to cytoplasmic tails of transmembrane proteins. The EBP50 can be coprecipitated with ERM and may serve to bind ERM members to one or more integral membrane proteins.

Cells opened up by the use of the detergent digitonin are still amenable to activation with non-hydrolyzable analogues of GTP. Addition of guanosine 5′-O-(3-thiotriphosphate), GTPγS, to such permeabilized cell models induces assembly of both actin filaments and focal adhesion complexes through activation of endogenous Rho and Rac (66). The sensitivity to GTPγS is lost after a few minutes of incubation, but can be restored by the addition of the actin-binding protein moesin indicating that moesin is required for the effect of Rho and Rac on actin filament system. This is an important step towards the resolution of the signal relay which initiates the reorganization of the microfilament system.

Talin is an elongated, flexible actin-binding protein first recognized in smooth muscle (116,117) and platelets (118,119). It is present in focal contacts in tissue cultured cells, in adhesion plaques formed by activated platelets, in myotendinous junctions, and in dense plaques of smooth muscle, but is absent from cell:cell interaction sites (for refs, see 38). In motile cells talin is present in membrane ruffles. It is a major protein in platelets constituting more than 3% of the total protein. In response to activation, it is redistributed to the actin-rich platelet cortex (120). Thus talin is clearly of central importance for the association of actin filaments to membrane components to sites where cells make close contact with extracellular matrix proteins, and it has been shown to interact with the cytoplasmic domains of transmembrane β1-integrin (116,121).

Talin has a molecular mass of of 270 kD as estimated from the gene sequence (122). It is about 60 nm in length, and appears to be composed of a series of globular domains as judged by electron microscopy (123). In solution, it forms homodimers by antiparallel association of the monomers. It binds to integrin, vinculin, and actin *in vitro*. There are reports that talin can nucleate, cap and cross-link actin filaments, and that its activities may be controlled by phosphorylation by both serine/threonine and tyrosine kinases (124,125). An N-terminal domain of talin is homologous to ezrin (122), and may be the part of the molecule which associates with membrane lipids. The larger C-terminal domain contains the binding site for vinculin (see 38 and 125 for further refs.). Disruption of the talin gene in embryonic stem cells leads to a disturbance in the formation of vinculin and paxillin-containing focal contacts and cell spreading (126). The talin (−/−) cells do form embryoid bodies, and notably two morphologically distinct cell types have been observed that were able to spread and form focal adhesion-like structures with vinculin and paxillin on fibronectin. Thus, although talin was essential for the expression of β1-integrin, a subset of differentiated cells managed without it. What substitutes for talin in these cells is unclear.

Vinculin is a 117-kD, actin-binding protein found concentrated at all sites of attachment of filamentous actin to transmembrane components of adherens junctions (38,127–131). At these sites, it enters into multiple interactions with other proteins at the cytoplasmic face of the contacts. It is critically required as a structural component in the formation of the junctions. On ligand-induced clustering of integrins, vinculin is recruited from the cytoplasm to sites of integrin-ligand interaction, where it is a substrate for tyrosine phosphorylation by pp60^{src}.

There is electron microscopic evidence that the vinculin molecule can attain different conformational states. It is found either as a compact, globular molecule, or it is unfolded into a globular head with an extended tail (132–134). In the compact globular state, its actin-binding capacity is masked by an intramolecular association of the 95-kD head and 30-kD tail domains. The actin-binding site can be exposed by binding of acidic phospholipids to the protein suggesting that the affinity of vinculin for target molecules is regulated by modulation of the head-to-tail interaction, and that this may be used in the control of the assembly of adherens junctions (73,135). The phospholipid PtdIns 4,5-bisphosphate binds to two descrete regions of the vinculin tail, disrupts the intramolecular head-tail interaction and induces vinculin oligomerization. The binding of PtdIns 4,5-bisphosphate to vinculin also enhances the association of vinculin to VASP the actin- and profilin-binding protein present in focal adhesion sites. The C-terminal tail region contains a binding site for the focal adhesion protein paxillin (136). Furthermore, vinculin has been identified as part of the cadherin-catenin junctional complex (137–139).

Tensin is a dimer of two 200 kD polypeptide chains. It is found in different types of cell contacts, including focal adhesions, zonula adherens, intercalated discs, and myotendinous and neuromuscluar junctions (140–143). It contains three distinct, actin-binding sites per monomer and appears to cap as well as cross-link actin filaments. Tensin also binds vinculin and possibly tyrosine kinase receptors through a Src homology 2 (SH2) domain (144). It has been suggested that a tensin dimer with its 6 actin-binding sites might wrap around the end of an actin filament, and bind it to a tyrosine kinase receptor through its SH2 domain. Induction of the formation of a focal contact, by the binding of extracellular ligands to integrins, causes tensin to be phosphorylated on tyrosine. Insertin, an actin-binding protein suggested to be involved in the actual incorporation of monomers into filaments, has been recognized as a proteolytic breakdown product of tensin (145).

PROTEINS CONTROLLING POLYMERIZATION OF ACTIN

DNase I (mw 31 000) was the first actin monomer-binding protein found (12). However it is doubtful whether DNase I plays a role as an actin monomer-binding protein in the cell. Since the enzyme is produced by acinar cells in the pancreas and secreted into the duodenum, it has been thought of as a digestive enzyme. However, intracellular forms of the enzyme have been observed, and there is evidence that it might be involved in programmed cell death, apoptosis. (146).

The crystallization of DNase I was reported in 1948 by Kunitz (147). The same year Laskowsky and collaborators described the presence of a proteinaceous inhibitor in tissues from vertebrates (148,149). Two DNase I inhibitors (I and II) were isolated and crystallized in 1966 (13,150). They were shown to form gelifying high molecular weight aggregates. Searching for the identity of the inhibitors eventually led to the finding that actin was the ubiquitous inhibitor of DNase I (12). Analysis of the crystals of the DNase I inhibitor II revealed that they contained actin together with an equimolar amount of a small protein, now known as profilin (14,15). Later, the inhibitor I and II were identified as profilin in complex with γ-actin and β-actin, respectively (151).

DNase I is a glycoprotein. It forms a stable 1:1 complex with monomeric actin, and when mixed with equimolar amounts filamentous actin, it causes depolymerization of the filaments and formation of the 1:1 complex with actin (152–154). Kabsch and coworkers provided the first insights into the structure of actin by solving the structure of the DNase I:actin complex (155; see special entry on actin). The high resolution structure of DNase I alone and complexed to an oligonucleotide has been reported as well (156,157). The current model of the actin filament was arrived at using the structure of the actin monomer, as it occurs in the DNase I:actin crystal, together with diffraction data obtained with oriented gels of F-actin (158). This model is claimed to fit well into reconstructions of actin filaments from electron microscopic images (159,160), and reconstructions of actin filaments with bound actin-binding proteins are interpreted in terms of this model (see below). However, doubts about this model being a relevant representation of the actin filament have been expressed, and an alternative way of deriving a model of F-actin has been suggested from the analysis of the organization of actin monomers in crystals of profilin and actin (161,162). To ascertain the validity of the current model, independent information about the orientation of the actin monomer in the actin filament needs to be acquired.

Profilin (mw 12 000–15 000) is an abundant actin monomer-binding protein in eukaryotic cells (163, review). (See separate article **Profilin**.) Particularly high concentrations of profilin are found in lymphoid cells and brain cells. In lymphoid tissues the concentration of unpolymerized actin is about 100 μM and in platelets it may be as a high as 200 μM. In both cases about 50% of the unpolymerized actin can be accounted for by profilin:β-actin and profilin:γ-actin isoforms. The remaining unpolymerized actin appears to be sequestered by β-thymosins (see below). For references on specific aspects of profilin, see (164–189).

Observations point at **profilin** being an important factor in the control of the release of inositol trisphosphate and thus of the generation of Ca^{2+} pulses in the cell. The primary targets for Ca^{2+} regulation in cells is the actomyosin system with global changes in structure and activity as the result. It is now known that many of the microfilament-associated proteins interact with components formed in the phosphatidylinositol-cycle as a result of receptor-mediated activation of the cell, and that these interactions modulate the activity of these proteins viś-a-viś actin. Although the exact physiological roles of these interactions remain to be elucidated, it suggests that the phosphatidylinositol-cycle is directly involved in controlling the microfilament-based motility cycle (see Fig. 3).

Beta thymosins constitute a conserved family of polypeptides (mw 5 000), the members of which were originally isolated from mammalian thymus and thought to be important for the immunoregulatory activities of this organ (190). Homologues of Tβ4 have been found also in distantly related organisms (191). The thymosin β4 isoform is abundant in mammalian tissues, and has attracted much attention after the discovery of its association with unpolymerized actin in platelet extracts. It has a poorly defined structure as determined by NMR in solution (192). Binding studies suggested that it binds to the (+)-end of the actin monomer in competition with profilin. However, crosslinking studies imply a more complex interaction involving sites at both ends of the actin molecule (193).

Binding of thymosin β4 to actin monomers inhibits incorporation of actin monomers at both ends of the actin filament. Thymosin β4 has a rather low affinity for actin (K_d = 5 μM), but since it is present in high concentrations in cells (in platelets ca 500 μM) it is considered to be the most important actin sequestering protein. In the presence of polymerization-inhibiting concentrations of thymosine β4, addition of profilin overcomes the thymosin effect, due to the higher affinity of profilin for actin (K_d = 0.25μM), and the fact that profilin:actin can add actin monomers onto the fast growing ends of actin filaments (see **Profilin**).

In vitro, thymosin can be shown to bind also to actin filaments and decrease the critical concentration for actin polymerization (194), and experiments on cultured cells indicate that β-thymosins may not be just actin sequestering factors (195). In the NIH3T3 cell line thymosin β10 is the predominant beta thymosin isoform, being present in about equimolar concentration to proflin and ADF/cofilin. Overexpression of thymosin $\beta 10$ in these cells led to an increase in the cellular content of polymerized actin without any change in the total actin content of the cells. An intimate relationship with the regulation of the dynamic functioning of the microfilament system is also indicated by the increase in motility caused by overexpressing thymosin $\beta 10$. Consonant with this an isoform of beta thymosine, thymosin $\beta 15$, is upregulated in some forms of malignant cells (196).

The ADF/cofilin family of proteins include actin depolymerizing factor, cofilin, destrin, depactin, actophorin (197–199). These proteins are found in all kinds of eukaryotes, and when tested genetically they have proved to be essential. *In vitro*, they can form 1:1 complexes with actin monomers as well as with actin in filamentous form. The structure of the prototype ADF/cofilin, actophorin from *Acanthamoeba* have been solved by crystallography (200) and there are NMR structures of yeast cofilin and destrin (201,202). It has recently been recognized that an ADF homology domain is present in each member of a newly identified protein family consisting of the ADF/cofilins, the twinfilins, and the drebrin/Abp1s (203).

In the cell, most of the ADF/cofilin is found in the perinuclear area, but there are also significant amounts in highly motile advancing lamella and filopodia, where the reorganization of actin through polymerization/depolymerization is most intensive. The interaction between ADF/cofilin and actin is sensitive to the pH of the medium, and as in the case of profilin, polyphosphoinositides interfer with the ADF:actin interaction. Physiologically, however, the most important mechanism controlling the activity of ADF/cofilin depends on phosphorylation/dephosphorylation of the protein. Phosphorylation of ADF/cofilin at a site near the N-terminus (S3) blocks its binding to actin. In cell extracts about 60% of the protein is phosphorylated, and in connection to receptor-mediated stimulation of cells dephosphorylation of ADF/cofilin takes place. The phosphatase involved in this reaction has not been identified, but a specific protein kinase, LIM kinase, has been found responsible for the *in vivo* phosphorylation of ADF/cofilin (204,205).

The turnover of actin filaments inside living cells appears to be up to 100-fold faster than in *in vitro* experiments with actin alone. It has now been demonstrated in *in vitro* as well as in *in vivo* experiments that ADF/cofilin can accelerate the turnover of actin filaments (17,199,206–208). ADF/cofilin binds with higher affinity to ADP-containing actin, monomeric as well as filamentous, and facilitation of filament growth by active ADF/cofilin is likely to be brought about by increased depolymerisation of the actin filaments at their proximal ADP-containing ends increasing treadmilling activity. Nucleotide exchange, possibly enhanced by profilin, would in this model precede the reallocation of the ATP-containing precursor to the fast growing, distal (+)-end of the population of growing filaments. The presence of ADF in lamellipodia, where filaments are long, argues against a severing function *in vivo*, and the end-specific effects of ADF on filament assembly seen *in vitro* also indicate that the severing activity of ADF/cofilin is less important *in vivo*.

Twinfilin is a newly discovered protein composed of two cofilin-like regions (209). It was identified and characterized as an actin monomer-binding protein in budding yeast. Genes coding for homologous proteins have been recognized also in *Caenorhabditis elegans*, humans and mice. The two halves of the molecule twinfilin I and II resemble the corresponding domain from other species more than they resemble each other. Thus, twinfilins have evolved from a common ancestor and the twinfilins represent a single protein family (199). Twinfilin does not appear to act as an actin filament depolymerizing factor, but there are *in vivo* observations that suggest its involvement in the control of the dynamics of the microfilament system.

Arp

The first reports describing a family of actin-related proteins, the Arps, appeared a few years ago (2,209). These are highly conserved proteins that share a 30–60% homology with actin, but are functionally distinct from actin. Two of these actin-related proteins, Arp2 (44 kD) and Arp 3 (47 kD) have attracted a great deal of attenion, since they were discovered to exist in a large complex together with five other proteins (40 kD, 35 kD, 19 kD, 18 kD, and 15 kD) that could be isolated by affinity chromatography on immobilized profilin (210). The Arp2/Arp3 complex has been reported to nucleate actin filament formation, to bind along the sides of actin filaments and to express filament crosslinking activity. A similar complex was identified in searching for factors that initiate actin assembly at the surface of Listeria (211). The Arp2/Arp3 complex is highly conserved among eukaryotes. Null mutations are lethal. It is localized in the cortex of amoebae and yeast and in the lamellipodia of higher eukaryotes (212,213). In yeast, the Arp2/Arp3 complex is required for the integrity and motility of actin patches and for endocytosis (214). Thus the complex appears to play a major role in actin-based motility.

It has now been demonstrated that the Arp2/Arp3 complex isolated from *Acanthamoeba* binds profilin (215) and to the (−)-ends of actin filaments (168). It is not an efficient nucleator of actin polymerization. Capped filaments grow by the addition of monomers to the (+)-end. The complex can be seen by electron microscopy to attach the (−)-end of one filament to the side of another filament. This suggests that the Arp2/Arp3 complex might control the assembly of a branching network of actin filaments in the advancing lamellae of motile cells. Similar branching of actin filaments have been recognized earlier in electron microscopic pictures of the weave of microfilaments in advancing lamellae in moving keratinocytes (36). The cellular concentrations of the Arp2/Arp3 complex has been estimated to be high enough to cap the (−)-ends of all filaments in a cell. These observations imply that models assuming that actin filaments *in vivo* undergo an ADF/cofilin-aided depolymerization from the (−)-end during motile activity may be too simplistic, since there must also be mechanisms for the controlled uncapping of the (−)-ends of the filaments.

Further studies of the *Listeria* system have shown that the bacterial protein ActA, which is essential for the actin polymerization-dependent propulsion of the bacterium through the cytoplasm, acts synergistically with the Arp2/Arp3 complex *in vitro*. Working together, they eliminate the lag phase in the polymerization of actin and increase the initial rate of assembly 50-fold (216). This activity has been shown to reside in the N-terminal domain of the ActA protein. A cellular protein that corresponds to ActA has not yet been identified, but zyxin has been pointed out as a plausible candidate.

Capping protein Z (CapZ) is a heterodimeric protein, which was first detected in skeletal muscle sarcomers, where it is associated with actin filaments at the Z-line end of sarcomers, i.e. at the (+)-end of the filament (217). *In vitro*, it binds with high affinity ($K_d \approx$ nM) to the (+)-end of actin filaments. A homologous protein that exists in non-muscle cells, capping protein β_2 (CPβ_2), is enriched at the edge of the advancing lamellum of spreading platelets. By binding to the (+)-end of actin filaments, it blocks the addition of actin monomers to that end, and since actin filaments *in vivo* appear to grow by assembly primarily at that end, these capping proteins may be important regulators of actin polymerization. The fact that CPβ_2 is necessary for the proper organization of actin bundles in the morphogenesis of bristles in *Drosophila*, the view is strengthened that CPβ_2, gelsolin in some cases, and Cap G may play active roles in actin-based motility in response to signaling. There is evidence also in this case that polyphosphoinositides are involved in the regulation of the protein.

Adducin

In erythrocytes there is, in addition to CapZ, a second actin filament (+)-end capping protein, adducin (218,225). This protein is associated with the short erythrocyte actin filaments (one α,β heterodimer per filament). Adducin has been shown to bind to spectrin:actin complexes in the cell cortex, and promote the binding between these two proteins. It also appears to be able to bind directly to actin filaments and to bundle them. Adducin has a relatively low affinity for the filament (+)-end, but even so it appears to be the capping protein bound to the erythrocyte actin filaments, and not CapZ. Its binding to actin filaments is downregulated by calcium and calmodulin, an effect possibly modulated by reversible phosphorylation of the protein. The complex relationship between adducin and the spectrin:actin network indicates that adducin represents a new type of regulated actin filament-binding and barbed end capping protein. The entire molecule is required for capping activity, and the association of adducin with actin filaments seems to be regulated by Rho-associated kinase and myosin phosphatase.

Gelsolin is an abundant and ubiquitously expressed actin-binding protein, which confers calcium sensitivity to the regulation of the dynamics of the microfilament system (219). (See separate article **Gelsolin**.) For references on specific aspects of gelsolin, see (219–223).

Tropomodulin is a 43 kD protein that caps the (−)-ends of actin filaments in erythrocyte cortical network and in striated muscle sarcomeres (225). The interaction is particularly tight when tropomyosin is bound simultaneously. It is thought that tropomodulin is important in controlling the length of the actin filaments in these situations.

ARCHITECTURAL ELEMENTS OF THE ACTIN CORTEX

The dynamic changes in cell surface morphology and motile behaviour, seen in response to growth hormones and other cell stimulating agents, reflect changes in the formation, organization, and activity of the microfilament system. Actin-binding proteins that cross-link actin filaments into tightly packed bundles and more loosely organized networks are necessary for the formation of the weave of microfilaments seen in membrane lamellae and filopodia and in the rest of the submembraneous zone around the cell. Actin-based surface projections on cells take many forms, such as the elaborate stereocilia of cochlear hair cells, microvilli of intestinal epithelium, and the membrane ruffles and filopodia on moving fibroblasts. The formation of these structures depends on the activity of actin-crosslinking proteins. Important members of this class of proteins belong to the spectrin superfamily which are reviewed in ref. 237.

Villin is closely related to gelsolin, an actin severing protein (described above), but distinguished from gelsolin by the presence of an additional domain comprised of 76-amino acid residues, called the "headpiece" (224–229). The headpiece contains an additional actin-binding site, which confers actin filament bundling activity to the protein. Villin is found in the microvilli of certain types of epithelia. Crosslinking of actin filaments occurs at low Ca^{2+} concentrations. Like gelsolin, villin severs actin filaments in the presence of micromolar concentrations of Ca^{2+}, and after breakage of the actin filament, villin remains bound to the (+)-end of one of the fragments. The actin-binding properties, tissue distribution and expression during cell differentiation suggest that villin is important in organizing the actin filament core of microvilli in epithelial cells forming brush borders. In support of this, villin expressed transiently in transfected fibroblasts, results in loss of stress fibers and appearance of large numbers of microvillar protrusions on the dorsal surface of the cells. Microinjection of high concentrations of villin into cells that normally lack this protein, lose their stress fibres, and incorporate the villin into cell surface microspikes and large microvilli, an activity which was shown to depend on the presence of the C-terminal head piece.

The N-terminal domain of villin (14T) consists of 126 amino acid residues. It binds calcium and actin. The structure of this domain has been determined by NMR (230). The 76-amino acid residue long, C-terminal domain of villin is an actin-binding module that is also used in another actin-bundling protein **dematin** (band 4.9), but whose core domain is unrelated to villin. The 3-dimensional structure of the stably-folded 35 residue long subdomain of the villin headpiece has been determined by NMR (231). It is the smallest autonomously folding protein unit found.

Fimbrin belongs to the plastin family of actin-binding proteins (31,232,237). Members of this family are present in all types of eukaryotic cells. Fimbrin is the smallest of the actin crosslinking proteins. It consists merely of an N-terminal EF-domain and two actin-binding domains, each composed of two CH domains. Fimbrin is present together with villin in the microvilli of the intestinal brush border, and in the membrane lamellae and microspikes of most other eukaryotic cell types. A fimbrin-binding site on actin has been suggested from analysis of actin mutations suppressing fimbrin mutations (233). The structure of the N-terminal actin-crosslinking domain of fimbrin has been solved by crystallography (234), and electron cryomicroscopic images of fimbrin-decorated actin filaments has been used to localize the fimbrin-binding site on the surface of the actin filament (235).

In a suggested scenario for the formation of a microvillus (236), the first phase consists of the "streaming" of actin filaments from electron-dense plaques on the apical plasma membrane. These filaments then gathered into loose bundles that extend into rudimentary microvilli. Villin may contribute to an initial bundling of the filaments. Ezrin (see section III), a protein controlled by tyrosine phosphorylation, links the microvillar actin filaments to the membrane components. During a second phase, the microvilli elongate and become highly organized into a full core of hexagonally-packed actin filaments. This phase correlates in time with the localization of fimbrin to the core of the microvillus. In this way a stable surface protrusion is thought to take form. Similar schemes may apply to the development of surface protrusions on the surface of other types of cells.

ABP-280, filamin, and ABP-120, are all rod-like, dimeric, actin filament crosslinking proteins, where each monomer has an N-terminal actin-binding CH-domain and a long repeat region consisting of tandemly repeated, 100 amino acid residue long, homologous segments (237). The length of the repeat region is different in the different ABPs. The ABP120 was found in *Dictyostelium discoideum*. The vertebrate variants, ABP-280 and filamin, are major cellular proteins, and products of different genes. Filamin is located specifically in the Z-lines structure of the sarcomeres in muscles. ABP-280 is a major cellular protein in non-muscle cells, where it is present in the cell cortex. The function of ABP-280 appears to be to cross-link actin filaments into orthogonal networks. The 3D-structure of the repeat region of ABP-120 (gelation factor, 6 repeats) has been determined by NMR spectroscopy, which shows it to have an immunoglobulin-like fold (226). Both filamin and ABP-280 (24 repeats) share conserved residues that form the core of the gelation factor repetitive segment structure as determined by NMR. This distinguishes these actin crosslinking proteins from the spectrins and α-actinins, which have tandem repeats of an α-helical domain.

On the basis of their actin filament crosslinking activity, depending on the presence of a calponin homology CH domain, and their elongated shape, as seen by electron microscopy, these proteins have been grouped together with the spectrin and fimbrin families into a superfamily. It is true that the actin binding domains of all these proteins are homologous. However, the presence or absence of repeat domains, and the widely different structures of the repeat domains of the spectrin and filamin families, rather suggests that the three families should be considered as separate protein families employing a common actin-binding domain, since it is possible that the different families will turn out to have quite different functions in relation to the actin filament system.

Spectrin (fodrin) was first isolated as part of the matrix proteins of the erythrocyte plasma membrane (237–240). Underneath the lipid bilayer of these cells, and linked to transmembrane proteins glycophorin and the ion channel band 3, there is a regular organization of units consisting of five or six spectrin molecules attached to a short actin filament. The spectrin molecules are attached to spectrins of adjacent units to form a sheet of five- and six-sided polygons. The membrane skeletons, and isolated junctional complexes, reproducibly contain four proteins in the molar ratios of 1 spectrin dimer:2-3 actin monomers:1 band 4.1 protein:0.1–0.5 protein 4.9 (protein designations refer to mobility on gel electrophoresis). In addition tropomyosin, tropomyosin-binding protein and adducin are thought to be part of the junctional complex.

Spectrin consists of two subunits, an α and a β subunit, which form heterodimers, which in turn associate to form tetramers in the membrane lattice. The β-chain has an N-terminal actin binding CH domain and 12 α-helical repeat domains, whose structure has been determined by NMR spectroscopy (241). The α-chain associates in an antiparallel fashion with the β-chain. It has a C-terminal EF-hand domain, which is close to the actin-binding site of the β-chain followed towards the N-terminus by a binding site for protein 4.1, and a Src-like sequence in the middle of the somewhat longer repeat region. Between repeat 11 and 12, there is a calmodulin-binding sequence inserted. The structures of the spectrin repeat, the phosphoinositide-binding pleckstrin (PH) domain, the SH3 domain, and the CH domain have all been determined (241–246).

The spectrin-actin based network of proteins endows the erythrocyte with an exquisite elasticity which makes it possible for it to go through quite drastic shape changes. However, it is clear that the structural role played by the spectrin:actin system is not its sole function. The close relationship between the spectrin-actin system and transmembrane ion channels implies that the range of controlling functions is wider. The spectrin-actin system is not unique to the erythrocytes, but exists also in other tissues and in cells of as widely separated organisms as *Drosophila*, *Acanthamoeba*, *Dictyostelium*, echinoderms, and possibly also in higher plants.

Alfa-actinin is a classical skeletal muscle protein identified a long time ago (237,247). It is highly enriched at the Z-disc-end of the actin filaments in the sarcomeres of the myofibrils. It is now known also to be an integral part of the microfilament system in nonmuscle cells, where an interaction between transmembrane $\beta1$ integrins and α-actinin has been demonstrated (see Fig. 2). There are several isoforms of α-actinin, which are products of three different genes. *In vitro*, it crosslinks actin filaments into regular ladder-like structures. It is therefore thought to contribute to the organization of actin in cells. Alfa-actinin is an elongated homodimer (200-215 kD) with antiparallel arrangement of the two subunits. Each subunit consists of a CH-domain, 4 α-helical domains ("spectrin repeats"), and an EF-hand domain. The binding of nonmuscle α-actinin to actin filaments is modulated by Ca^{2+} binding to the EF-domain, whereas the muscle isoform of α-actinin is insensitive to Ca^{2+}. The 3D-structure of the CH-domain is known by analogy to the structure of the corresponding domain from calponin solved by x-ray crystallography, and the structure of the repeated domain is known by analogy to the known α-helical repeat domain in spectrin, and the EF-hand can be compared with calmodulin.

Alfa-actinin can be extracted from myofibrils without major destruction of the sarcomeric organization of the actin filaments in the myofibril, and when added back to the myofibril preparation it rebinds at the Z-disc. This implies that α-actinin may serve some function, other than a structural, in relation to the actin filaments. In nonmuscle cells, α-actinin is found enriched in spots along stress fibres and in cell adhesion plaques.

Dystrophin is the product of the gene responsible for the X-linked myopathies Duchenne and Becker muscular dystrophy. It is an elongated protein of high molecular weight (430 kD) present in low amounts in muscle and nerve cells. A closely related protein, utrophin, is more widely distributed, and there are many dystrophin/utrophin isoforms produced, due to the operation of alternative promotors and alternative splicing. The structure of these proteins place them in the spectrin/α-actinin family of proteins. They are multidomain proteins; the largest isoform is comprised of 3 N-terminal and 2 internal actin-binding sites, a peptide-binding WW domain, a calcium-binding EF domain, a zinc finger domain. The

N-terminal actin binding region has sequence homology with the calponin homology CH domains of the spectrin/α-actinin family of proteins.

Dystrophin/utrophin connect cortical actin filaments with transmembrane proteins, dystroglycans and sarcoglycans, which in turn associate with extracellular matrix proteins, laminin and merosin, respectively. Skeletal muscle dystrophin can be purified from muscle cells as a large multiprotein dystrophin-glycoprotein complex, which stabilizes actin filaments *in vitro* through lateral associations. Both dystrophin and utrophin bind with higher affinity to β- than to α-actin. Comprehensive reviews on the structure and function of the dystrophin/utrophins are found in ref (237,248).

ION CHANNEL-ASSOCIATED ACTIN-BINDING PROTEINS

The spectrin-actin network in erythrocytes is involved in anchoring ion exchange proteins in the membrane via the linker protein **ankyrin**, a critical link in maintaining the characteristic biconcave shape of the cell (249). In axons, an interaction between actin, **fodrin**, and ankyrin appears to be responsible for concentrating sodium channels at the nodes of Ranvier (249). At the neuromuscular junction, the clustering of acetycholine receptors appears to be mediated by spectrin, actin, and the **rapsyn/43 kD actin-binding protein** (250). Various lines of evidence suggest that intact actin filaments, and perhaps actin polymerization itself, are required for calcium regulation of the postsynaptic response of NMDA receptors in the central nervous system (251). Postsynaptic excitatory synapses are most often found on the small actin-rich budlike structures, called spines, that protrude from the dendrites of highly arborized neurons such as cerebellar Purkinje cells (252). Dendritic spines have pronounced electron-dense (as seen by electron microscopy) post-synaptic densities, called PSD's, that are the likely sites of proteins linking receptors to actin. Candidate proteins include PSD-95/SAP90, chapsyn/PSD-93, SAP102, and **alpha-actinin-2**. Only alpha-actinin has a demonstrated actin binding affinity, and it has been suggested that clustering for some classes of receptors might be mediated by a looser, more "corral-like", entrapment mechanism (253). The PSD contains brain **dystrophin**, an isoform of the muscle protein originally identified as the mutated gene product involved in muscular dystrophy that is known to link transmembrane ion channels to actin filaments (254).

The precise role for actin in the functioning of ion channels is unclear. The provision of clustered anchorage sites may confer cooperativity in the opening of channels, or support the formation of multiprotein complexes capable of integrating different signalling or environmental factors. It is not unlikely that the actin network might passively or actively transmit forces to the channels providing the impetus behind stretch-activated gating phenomena.

MYOSIN AS AN ACTIN-BINDING PROTEIN

The myosins belong to a large superfamily of proteins, which together with actin transduce chemical energy to force generating tension and movements in the eukaryotic cells. Myosin II is the prime partner of actin in the generation of force in muscle, as well as in nonmuscle cells. Generally myosins are thought of as the force generators, and therefore referred to as motor molecules. However, the actual mechanism of force generation is still unknown. Although, the ATP-binding head domain is largely conserved, there are many different molecular forms of myosin serving in as diverse functions as muscle contraction, cell motility, membrane traffic, and sensory perception. Myosin hydrolysis ATP to ADP.Pi still bound to the head domain. Interaction with actin filaments brings about product release, and in the presence of actin filaments, there is a rapid ATP-dependent cycling of heads on and off the actin filaments as ATP is hydrolysed. A current review of the myosin superfamily and their functional involvements is found in ref. 255. For a more detailed account of the structure and function of myosin see the entry on myosin in this encyclopedia.

REGULATORS OF THE ACTOMYOSIN INTERACTION

The N-terminus of actin is involved in the binding of a large number of actin binding proteins, including myosin (S1), tropomyosin, troponin I, α-actinin, caldesmon, gelsolin, cofilin, actobindin. This actin interface is important for muscle contraction, since it is implicated not only in the activation of the myosin ATPase, but also in the regulation of actomyosin interaction by interacting with troponin I.

Tropomyosin is an elongated, dimeric, coiled-coil α-helical protein that binds along the actin filament (see special entry). The tropomyosin molecule has a sevenfold repeat of nonpolar and polar amino acid residues that bind seven actin monomers in the filament. Multiple genes for tropomyosin and alternative splicing can generate a number of tropomyosins that are differentially expressed during development and in different cell types. Muscle cells express 1–2 and most vertebrate nonmuscle cells 3–8 tropomyosin isoforms. Tropomyosin in complex with the heterotrimeric protein troponin makes muscle contraction Ca^{2+} dependent, but the function of tropomyosin in nonmuscle cells has not been clarified. However, the involvement of tropomyosin in the control of chemomechanical transduction also in nonmuscle cells is indicated by the occurrence of tropomyosin in tightly bundled and contractile organizations of actin filaments (stress fibres). For further information see special entry on tropomyosin.

Troponin is a heterotrimeric protein complex that interacts directly with actin through one of its subunits. Together with tropomyosin it confers calcium sensitivity to muscle contraction in striated muscles (see special entry).

Caldesmon is involved in the regulation of the actomyosin interaction in smooth muscle (256,257). It inhibits actomyosin ATPase and filament sliding *in vitro* has been shown to bind to subdomain 2 of actin. Three-dimensional image reconstruction of reconstituted thin filaments consisting of actin and smooth muscle tropomyosin, both with and without a caldesmon derivative added has been reported. In filaments containing the caldesmon derivative, tropomyosin was found in a different position as compared to the situation in the absence of the caldesmon. The observations suggest that caldesmon causes changes in the relationship between tropomyosin and the actin filament, which are different from those seen with troponin.

The giant modular protein nebulin (mw 600- 900 kD) spans the whole length of the thin filament of the striated muscle sarcomeres in vertebrates, and has been proposed to function as a "ruler" for control of the length of actin filaments in sarcomeres (258–260). It is thought to bind and stabilize F-actin. It comprises 2-3 % of the myofibrillar protein mass of skeletal muscles. There are tissue and development-specific isoforms. The C-terminal part of human nebulin is anchored in the sarcomere Z-disk and contains an SH3 domain. The nebulin SH3 sequence from several species has been determined and found strikingly conserved. Its 3D-structure has been determined in solution by NMR spectroscopy, and its interaction with poly(L-proline) has been modelled. Nebulin consists of nearly 200 tandem repeats of homologous modules about 35 residues long. These are organized into about 20 tandem superrepeats consisting of 7 different modules, and this superrepeat segment is flanked near the N- and C-termini by single-repeat regions containing 8 modules of the same type. It has been proposed that the 35 residue module is the basic actin-binding domain and that the superrepeats reflect tropomyosin/troponin binding sites along the nebulin molecule. Exactly how the nebulin molecules are linked to actin filaments in the sarcomere of muscle is not yet known. Nebulin promotes actin nucleation and stabilizes actin filaments. Crosslinking experiments have identified the first two residues in actin to be involved in binding nebulin.

FUTURE DIRECTIONS

One of the most remarkable aspects of the contractile apparatus in muscle cells is the capacity of the system to ramp up its power output in response to increased demands for mechanical work. This phenomenon is known as the Fenn Effect, and it suggests that the acto-myosin system can increase its ATPase activity in response to higher imposed loads. It is apparently the lattice organization of filaments into sarcomeres that somehow enables the effect of external loads to be transmitted directly to the proteins producing the biochemical reactions that produce work from ATP hydrolysis. It remains to be seen if the bundles and meshworks found in non-muscle cells display similar, or even more unusual, chemomechanical feedback mechanisms. Muscle cells also perform work at nearly 100% thermodynamic efficiency, and it is interesting to ask whether the free energy transduction pathways in the cytoplasm are equally optimized for their tasks.

The newly discovered connections between the microfilament system and signal transduction raise a host of questions that will be the subject of future research. What roles do mechanical forces play in transmitting signals from the surfaces of cells to their nuclei? What regulates the assembly of these focal sites of signalling potential and what leads to their breakup after a signalling cascade has outlived its usefulness?

The large number of actin-binding proteins obscures the possible branching points that must have occurred during evolution to establish this incredibly diverse dynamical system. The analysis of evolutionary relationships in actin-binding proteins from amoeba, plants, and mammals will be of extraordinary interest for understanding how the immune system and the mammalian nervous system developed as specializations of more primitive motile mechanisms.

BIBLIOGRAPHY

1. F. B. Straub (1942) Studies from the Institute of Medical Chemistry, Univ. Szeged **2**, 3–15.
2. F. B. Straub (1942) Studies from the Institute of Medical Chemistry, Univ. Szeged **3**, 23–37.
3. A. G. Engel and C. Franzini-Armstrong (1994) *Myology*, vol. 1, 2nd edition. McGraw-Hill, Inc. New York.
4. B. Alberts, D. Bray, A. Johnson, J. Lewis, M. Raff, K. Roberts, and P. Walter (1998) *Essential Cell Biology*, Garland Publishing Inc., New York & London.
5. S. Hatano, T. Totsuka, and F. Oosawa (1966) *Biochim. Biophys. Acta* **127**, 488–498.
6. S. Hatano and M. Tazawa (1968) *Biochim. Biophys. Acta* **154**, 507–519.
7. M. R. Adelman and E. W. Taylor (1969) *Biochemistry* **8**, 4964–4975.
8. H. Ishikawa, R. Bishoff, and H. Holzter (1969) *J. Cell Biol.* **43**, 312–328.
9. E. Lazarides and K. Weber (1974) *Proc. Natl. Acad. Sci* **71**, 2268–2272.
10. K. Weber and U. Groeschel-Stewart (1974) *Proc. Natl. Acad. Sci* **71**, 4561–4564.
11. T. D. Pollard and R. R. Weihing (1974) *Critical Reviews in Biochemistry* **2**, 1–65.
12. E. Lazarides and U. Lindberg (1974) *Proc. Natl. Acad. Sci* **71**, 4742–4746.
13. U. Lindberg (1966) *J. Biol. Chem.* **241**, 1246–1248.
14. L. Carlsson, L.-E. Nyström, U. Lindberg, K. K. Kannan, H. Cid-Dresdner, S. Lövgren, and H. Jörnvall (1976) *J. Mol. Biol.* **105**, 353–366.
15. L. Carlsson, L.-E. Nyström, I. Sundkvist, F. Markey, and U. Lindberg (1977) *J. Mol. Biol.* **115**, 465–483.
16. V. Mermall, P. L. Post, and M. S. Mooseker (1998) *Science* **279**, 527–533.
17. M.-F. Carlier and D. Pantaloni (1997) *J. Mol. Biol.* **269**, 459–467.
18. M. D. Welch, A. Mallavarapu, J. Rosenblatt, and T. J. Mitchison (1997) *Curr. Opin. Cell Biol.* **9**, 54–61.
19. R. F. Doolittle (1995) *Philos. Trans. R. Soc. Lond. B. Biol. Sci.* **349**, 235–240.
20. P. Sheterline, J. Clayton, and J. Sparrow (1996) *Protein profile; Actin*, 3rd edition. Acad. Press, London.
21. M. Sanchez, A. Valencia, M. J. Ferrandiz, C. Sander, and M. Vicente (1994) *EMBO J* **13**, 4919–4925.
22. S. Frankel and M. S. Mooseker (1996) *Curr. Opin. Cell Biol.* **8**, 30–37.
23. J. V. Small and J. E. Celis (1978) *Cytobiologie* **16**, 308–325.
24. A.-S. Höglund, R. Karlsson, E. Arro, B.-A. Fredriksson, and U. Lindberg (1980) *J. Muscle Res. Cell Motil.* **1**, 127–146.
25. J. V. Small, G. Rinnerthaler, and H. Hinsen (1981) *Cold Spring Harb. Symp. Quant. Biol.* **46**, 599–611.
26. K. Mellström, A.-S. Höglund, M. Nistér, C.-H. Heldin, B. Westermark, and U. Lindberg (1983) *J. Muscle Res. Cell Motil.* **4**, 589–609.
27. J. V. Small, M. Herzog, and K. Anderson (1995) *J. Cell Biol.* **129**, 1275–1286.
28. G. L. Nicolson, G. Poste, and T-H. Ji (1977) Cell Surface Reviews, vol. 3., Dynamic aspects of cell surface organization (G. Poste and G. L. Nicolson, eds.), North-Holland, Amsterdam, p. 148.
29. A. Viel and D. Branton (1996) *Curr. Opin. Cell Biol.* **8**, 49–55
30. M. Arpin, M. Algrain, and D. Louvard (1994) *Curr. Opin. Cell Biol.* **6**, 136–141.
31. A. Bretscher, D. Reczek, and M. Berryman (1997) *J. Cell Sci.* **110**, 3011–3018.
32. A. J. Hudspeth (1997) *Curr. Opin. Neurobiol.* **7**, 480–486.
33. M. Kernan and C. Zuker (1995) *Curr. Opin. Neurobiol.* **5**, 443–448.

34. M. Abercrombie, J. E. Heysman, and S. M. Pegrum (1970) *Exp. Cell Res.* **59**, 393–398.
35. G. Albrecht-Buehler and R. M. Lancaster (1976) *J. Cell Biol.* **71**, 370–382.
36. T. M. Svitkina, A. B. Verkhovsky, K. M. McQuade, and G. G. Borisy (1997) *J. Cell Biol.* **139**, 397–415.
37. A. Y. Chan, S. Raft, M. Bailly, J. B. Wyckoff, J. E. Segall, and J. S. Condeelis (1998) *J. Cell Sci.* **111** 199–211.
38. B. M. Jockusch, P. Bubeck, K. Giehl, M. Kroemker, J. Moschner, M. Rothkegel, M. Rüdiger, K. Schlüter, G. Stanke, and J. Winkler (1995) *Ann. Rev. Cell Dev. Biol.* **11**, 379–416. Review.
39. A. Howe, A. E. Aplin, S. K. Alahari, and R. L. Juliano (1998) *Curr. Opin. Cell Biol.* **10**, 220–231.
40. J. Kendrick-Jones, R. C. Smith, R. Craig, and S. Citi (1987) *J. Mol. Biol.* **198**, 241–252.
41. K. M. Trybus (1991) *Curr. Opin. Cell Biol.* **3**, 105–111.
42. M. A. Lawson and F. R. Maxfield (1995) *Nature* **377**, 75–79.
43. M. Chinkers, J. A. McKanna, and S. Cohen (1979) *J. Cell Biol.* **83**, 260–265
44. J. C. den Hartigh, P. M. van Bergen en Henegouwen, A. J. Verkleij, and J. Boonstra (1992) *J. Cell Biol.* **119**, 349–355
45. M. A. Van der Heyden, P. A. Oude Weernink, B. A. Van Oirschot, P. M. Van Bergen en Henegouwen, J. Boonstra, and G. Rijksen (1997) *Biochim. Biophys. Acta* **1359** (3), 211–221.
46. I. Lassing and U. Lindberg (1985) *Nature* **314**, 472–474.
47. X. D. Ren and M. A. Schwartz (1998) *Curr. Opin. Gent. Dev.* **8**, 63–67.
48. C. L. Carpenter, K. F. Tolias, A. C. Couvillon, and J. H. Hartwig (1997) *Adv. Enzyme Regul.* **37**, 377–390.
49. I. Lassing and U. Lindberg (1990) *FEBS Lett.* **262**, 231–233.
50. L. Carlsson, F. Markey, I. Blikstad, T. Persson, and U. Lindberg (1979) *Proc. Natl. Acad. Sci. USA* **76**, 6376–6380.
51. F. Markey, T. Persson, and U. Lindberg (1981) *Cell* **23**, 145–153.
52. J-H. Hartwig and K. Barkalow (1997) *Curr. Opin. Hematol.* **4**, 351–356.
53. M. J. Berridge (1993) *Nature* **361**, 315–325.
54. M. Halbrügge and U. Walter (1993) In *Protein Kinases in Blood Cell Function* (C.-K. Huang and R. I. Sha'afi, eds.), CRC Press, Boca Raton, FL, pp. 245–298.
55. J. Field, A. Vojtek, R. Ballester, G. Bolger, J. Colicelli, K. Ferguson, J. Gerst, T. Kataoka, T. Michaeli, and S. Powers (1990) *Cell* **61**, 319–327.
56. M. Fedor-Chaiken, R. J. Deschenes, and J. R. Broach (1990) *Cell* **61**, 329–340.
57. C. Haffner, T. Jarchau, M. Reinhard, J. Hoppe, S. M. Lohman, and U. Walter (1995) *EMBO J.* **14**, 19–27.
58. A. Ben-Ze'ev (1997) *Curr. Opin. Cell Biol.* **9**, 99–108.
59. E. Weisberg, M Sattler, D. S. Ewaniuk, and R Salgia (1997) *Crit. Rev. Oncog.* **8**, 343–358.
60. G. B. Cohen, R. Ren, and D. Baltimore (1995) *Cell* **80**, 237–248.
61. R. L. Collins, M Deckert, and A. Altman (1997) *Immunology Today* **18**, 221–224.
62. K. D. Fischer, Y. Y. Kong, H. Nishina, K. Tedford, L. E. Marengere, I. Kozieradzki, T. Sasaki, M. Starr, G. Chan, S. Gardener, M. P. Nghiem, D. Bouchard, M. Barbacid, A. Bernstein, and J. M. Penninger (1998) *Curr. Biol.* **8**, 554–562
63. L. J. Holsinger, I. A. Graef, W. Swat, T. Chi, D. M. Bautista, L. Davidson, R. S. Lewis, F. W. Alt, and G. R. Crabtree (1998) *Curr. Biol.* **8**, 563–572.
64. A. Hall (1998) *Science* **279**, 509–514.
65. J. H. Hartwig, G. M. Bokoch, C. L. Carpenter, P. A. Janmey, L. A. Taylor, A. Toker, and T. Stossel (1995) *Cell*, **82**, 643–653.
66. D. J. G. Mackay, F. Esch, H. Furthmayr, and A. Hall (1997) *J. Cell Biol.* **138**, 927–938.
67. M. Reinhard, M. Halbrügge, U. Scheer, C. Wiegand, B. M. Jockusch, and U. Walter (1992) *EMBO J.* **11**, 2063–2070.
68. C. Haffner, T. Jarchau, M. Reinhard, J. Hoppe, S. M. Lohmann, and U. Walter (1995) *EMBO J.* **14**, 19–27.
69. F. B. Gertler, K. Niebuhr, M. Reinhard, J. Wehland, and P. Soriano (1996) *Cell* **87**, 227–239.
70. M. Reinhard, K. Giehl, K. Abel, C. Haffner, T. Jarchau, V. Hoppe, B. M. Jockusch, and U. Walter (1995) *EMBO J.* **14**, 1583–1589.
71. L. G. Cao, G. G. Babcock, P. A. Rubenstein, and Y. L. Wang (1992) *J. Cell Biol.* **117**, 1023–1029
72. M. C. Beckerle (1997) *BioEssays* **19**, 949–956.
73. S. Hüttelmaier, O. Mayboroda, B. Harbeck, T. Jarchau, B. M. Jockusch, and M. Rüdiger (1998) *Curr. Biol.* **8**, 479–488.
74. I. Lasa and P. Cossart (1996) *Trends Cell Biol.* **6**, 109–114. Review
75. J. M. J. Derry, H. E. Ochs, and U. Francke (1994) *Cell* **78**, 635–644.
76. T. Kirchhausen and F. S. Rosen (1996) *Curr. Biol.* **6**, 676–678.
77. U. Molina, D. M. Kenney, F. D. Rosen, and E. Remold-O'Donnell (1993) *J. Exp. Med.* **151**, 4383–4390.
78. P. Aspenström, U. Lindberg, and A. Hall (1997) *Curr. Biol.* **6**, 70–75.
79. M. Symons, J. M. J. Derry, B. Kartak, S. Jiang, V. Lemahieu, F. McCormick, U. Francke, and A. Abo (1996) *Cell* **84**, 723–734.
80. R. Kolluri, K. F. Tolias, C. L. Carpenter, F. S. Rosen, and T. Kirchhausen (1996) *Proc. Natl. Acad. Sci. USA* **93**, 5615–5618.
81. S. Banin, O. Truong, D. R. Katz, M Waterfield, P. M. Brickell, and I. Gout (1996) *Curr. Biol.* **6**, 981–988.
82. H. Miki, K. Miura, and T. Takenawa (1996) *EMBO J.* **14**, 5326–5335.
83. H. Miki, T. Sasaki, Y. Takai, and T. Takenawa (1998) *Nature* **391**, 93–96.
84. S. T. Suzuki, H. Miki, T Takenawa, and C. Sasakawa (1998) *EMBO J.* **17**, 2767–2776.
85. N. Ramesh, I. M. Anton, J. H. Hartwig, and R. S. Geha (1997) *Proc. Natl. Acad. Sci. USA* **94**, 14671–14676.
86. I. M. Anton, W. Lu, B. J. Mayer, N. Ramesh, and R. S. Geha (1998) *J. Biol. Chem.* **273**, 20992–20995.
87. J. A. Frazier and C. M. Field (1997) *Curr. Biol.* **7**, 414–417.
88. M. Evangelista, K. Blundell, M. S. Longtine, C. J. Chow, N. Adames, J. R. Pringle, M. Peter, and C. Boone (1997) *Science* **276**, 118–122.
89. J. Petersen, O. Nielsen, R. Egel, and I. M. Hagan (1998) *J. Cell Biol.* **141**, 1217–1228.
90. N. Watanabe, P. Madaule, T. Reid, T. Ishizaki, G. Watanabe, A. Kakizuka, Y. Saito, K. Nakao, B. M. Jockusch, and S. Narumiya (1997) *EMBO J.* **16**, 3044–3056.
91. L. M. Machesky (1998) *Curr. Biol.* **8**, 202–205.
92. L. Weissbach, J. Settleman, M. F. Kalady, A. J. Snijders, A. E. Murthy, Y.-X. Yan, and A. Bernards (1994) *J. Biol. Chem.* **269**, 20517–20521.
93. S. Brill, S. Li, C. W. Lyman, D. M. Church, J. J. Wasmuth, L. Weissbach, A. Bernards, and A. J. Snijders (1996) *Mol. Cell Biol.* **16**, 4869–4878.
94. J. W. Erickson, R. A. Cerione, and M. J. Hart (1997) *J. Biol.Chem.* **272**, 24443–24447.
95. M. Fukata, S. Kuroda, K. Fujii, T. Nakamura, I. Shoji, Y. Matsuura, K. Kawa, A. Iwamatsu, A. Kikuchi, and K. Kaibuchi (1997) *J. Biol. Chem.* **272**, 29579–29583.

96. S. Kuroda, M. Fukata, M. Nakagawa, K. Fujii, T. Nakamura, T. Ookubo, I. Izawa, T. Nagase, N. Nomura, H. Tani, I. Shoji, Y. Matsuura, S. Yonehara, and K. Kaibuchi (1998) *Science* **281**, 832–835.
97. K. Eng, N. I. Naqvi, K. C. Wong, and M. K. Balasubramanian (1998) *Curr. Biol.* **8**, 611–621
98. C. E. Turner, J. R. Glenney Jr., and K. Burridge (1990) *J. Cell Biol.* **111**, 1059–1068.
99. C. E. Turner (1994) *BioEssays* **16**, 47–52.
100. M. C. Brown, J. A. Perrotta, and C. E. Turner (1996) *J. Cell Biol.* **135**, 1109–1123.
101. H. Wu and J. T. Parsons (1993) *J. Cell Biol.* **120**, 1417–1426.
102. X. Zhan, C. Plourde, X. Hu, R. Friesel, and T. Maciag (1994) *J. Biol. Chem.* **269**, 20221–20224.
103. K. Ozawa, K. Kashiwada, M. Takahashi, and K Sobue (1995) *Exp. Cell Res.* **221**, 197–204.
104. S. A. Weed, Y. Du, and J. T. Parsons (1998) *J. Cell Sci.* **111**, 2433–2443.
105. H. He, T Watanabe, X. Ahan, C. Huang, E. Schuuring, K. Fukami, T. Takenawa, C. C. Kumar, R. J. Simpson, and H. Maruta (1998) *Mol. Cell Biol.* **18**, 3829–3837.
106. C. Provenzano, R. Gallo, R. Carbone, P. P. Di Fiore, G. Falcone, L. Castellani, and S. Alema (1998) *Exp. Cell Res.* **242**, 186–200.
107. K. V. Kishan, G. Scita, W. T. Wong, P. P. Di Fiore, and M. E. Newcomer (1997) *Nat. Struct. Biol.* **4**, 739–743.
108. A. Vaheri, O. Carpén, L. Heiska, T. S. Helander, J. Jääskeläinen, P. Majander-Nordenswan, M. Sainio, T. Timonen, and O. Turunen (1997) *Curr. Opin. Cell Biol.* **9**, 659–666.
109. R. J. Shaw, M. Henry, F. Solomon, and T. Jacks (1998) *Mol. Biol. Cell* **9**, 403–419.
110. C. B. Shuster and I. M. Herman (1995) *J. Cell Biol.* **128**, 837–848.
111. J. W. Legg and C. M. Isacke (1998) *Curr. Biol.* **8**, 705–708.
112. V. Niggli, C. Andreoli, C. Roy, and P. Mangeat (1995) *FEBS Lett.* **376**, 172–176.
113. S. Tsukita, S. Yonemura, and S. Tsukita (1997) *Trends Biochem. Sci.* **22**, 53–58.
114. M. Hirao, N. Sato, T. Kondo, S. Yonemura, M. Monden, T. Sasaki, Y. Takai, S. Tsukita, and S. Tsukita (1996) *J. Cell Biol.* **135**, 37–51.
115. S. Yonemura, M. Hirao, Y. Doi, N. Takahashi, T. Kondo, and S. Tsukita (1998) *J. Cell Biol.* **140**, 885–895.
116. K. Burridge and L. Connell (1983) *Cell Motil.* **3**, 405–417.
117. L. Molony, D. McCaslin, J. Abernethy, B. Paschal, and K. Burridge (1987) *J. Biol. Chem.* **262**, 7790–7795.
118. N. C. Collier and K. Wang (1982) *FEBS Lett.* **143**, 205–210.
119. T. O'Halloran, M. C. Beckerle, and K. Burridge (1985) *Nature* **317**, 449–451.
120. M. C. Beckerle, K. Burridge, G. N. De Martino, and D. E. Croall (1987) *Cell* **51**, 569–577.
121. A. Horwitz, K. Duggan, C. Buck, M. C. Beckerle, and K. Burridge (1986) *Nature* **320**, 531–533.
122. D. J. G. Rees, S. E. Ades, S. J. Singer, and R. O. Hynes (1990) *Nature* **347**, 685–689.
123. J. Winkler, H. Lunsdorf, and B. M. Jockusch (1997) *Eur. J. Biochem.* **243**, 430–436.
124. V. Niggli, S. Kaufmann, W. H. Goldmann, T. Weber, and G. Isenberg (1994) *Eur. J. Biochem.* **226**, 951–957.
125. W. H. Goldmann, R. M. Ezzel, E. D. Adamson, V. Niggli, and G. Isenberg (1996) *J. Muscle Res. Cell Motil.* **17**, 1–5.
126. H. Priddle, L. Hemmings, S. Monkley, A. Woods, B. Patel, D. Sutton, G. A. Dunn, D. Zicha, and D. R. Critchley (1998) *J. Cell Biol.* **142**, 1121–1133.
127. B. Geiger (1979) *Cell* **18**, 193–205.
128. B. Geiger, K. T. Tokuyasu, A. H. Dutton, and S. J. Singer (1980) *Proc. Natl. Acad. Sci. USA* **77**, 4127–4131.
129. B. Geiger (1982) *J. Mol. Biol.* **159**, 685–701.
130. K. M. Yamada and B. Geiger (1997) *Curr. Opin. Cell Biol.* **9**, 76–85.
131. S. W. Craig and R. P. Johnson (1996) *Curr. Opin. Cell Biol.* **8**, 74–85.
132. R. P. Johnson and S. W. Craig (1994) *J. Biol. Chem.* **269**, 12611–12619.
133. L. Molony and K. Burridge (1985) *J. Cell. Biochem.* **29**, 31–36.
134. L. M. Milam (1985) *J. Mol. Biol.* **184**, 543–545.
135. J. Week, S. T. Barry, and D. R. Critchley (1996) *Biochem. J.* **314**, 827–832.
136. C. K. Wood, C. E. Turner, P. Jackson, and D. R. Critchley (1994) *J. Cell Sci.* **107**, 709–717.
137. R. B. Hazan, L. Kang, S. Roe, P. I. Borgen, and D. L. Rimm (1997) *J. Biol Chem.* **272**, 32448–32453.
138. E. E. Weiss, M. Kroemker, A.-H. Rüdiger, B. M. Jockusch, and M. Rüdiger (1998) *J. Cell Biol.* **141**, 755–764.
139. M. Watabe-Uchida, N. Uchida, Y. Imamura, A. Nagafuchi, K. Fujimoto, T. Uemura, S. Vermeulen, F. van Roy, E. D. Adamson, and M. Takeichi (1998) *J. Cell Biol.* **142**, 847–857.
140. J. A. Wilkins, M. A. Risinger, and S. Lin (1986) *J. Cell Biol.* **103**, 1483–1494.
141. S. M. Bockholt, C. A. Otey, J. Glenney Jr., and K. Burridge (1992) *Exp. Cell Res.* **203**, 39–46.
142. S. H. Lo, P. A. Janmey, J. J. Hartwig, and L. B. Chen (1994a) *J. Cell Biol.* **125**, 1067–1075.
143. S. H. Lo, E. Weisberg, and L. B. Chen (1994b) *BioEssays* **16**, 817–823.
144. S. Davis, M. L. Lu, S. H. Lo, S. Lin, J. A. Butler, B. J. Druker, T. M. Roberts, Q. An, and L. B. Chen (1991) *Science* **252**, 712–715.
145. C. Weight, A. Gaertner, A. Wegner, H. Korte, and H. E. Meyer (1992) *J. Mol. Biol.* **227**, 593–595.
146. H. G. Mannherz, M. C. Peitsch, S. Zanotti, R. Paddenberg, and B. Polzar (1995) *Curr. Top. Microbiol. Immunol.* **198**, 161–174.
147. M. Kunitz (1948) *Science* **108**, 19–20.
148. W. Dabrowska, E. J. Cooper, and M. Laskowski (1945) *J. Biol. Chem.* **177**, 991–992.
149. E. J. Cooper, M. J. Trautman, and M. Laskowski (1950) *Proc. Soc. Exptl. Biol. Med.* **73**, 219–222.
150. U. Lindberg (1967) *Biochemistry* **6**, 323–335.
151. M. Segura and U. Lindberg (1984) *J. Biol. Chem.* **259**, 3949–3954.
152. S. E. Hitchcock, L. Carlsson, and U. Lindberg (1975) Cell Motility Cold Spring Harbor Symp. pp. 545–559.
153. S. E. Hitchcock, L. Carlsson, and U. Lindberg (1976) *Cell* **7**, 531–542.
154. H. G. Mannherz, H. Brehme, and U. Lamp (1975) *Eur. J. Biochem.* **60**, 109–116.
155. W. Kabsch, H. G. Mannherz, D. Suck, E. F. Pai, and K. C. Holmes (1990) *Nature* **347**, 37–44.
156. C. Oefner and D. Suck (1986) *J. Mol. Biol.* **192**, 605–632.
157. S. A. Weston, A. Lahm, and D. Suck (1992) *J. Mol. Biol.* **226**, 1237–1256.
158. K. C. Holmes, D. Popp, W. Gebhard, and W. Kabsch (1990) *Nature* **347**, 44–49.
159. K. C. Holmes, M. Tirion, D. Popp, M. Lorenz, W. Kabsch, and R. A. Milligan (1993) *Adv Exp. Med. Biol.* **332**, 15–22.

160. I. Rayment, H. M. Holden, M. Whittaker, C. B. Yohn, M. Lorenz, K. C. Holmes, and R. A. Milligan (1993) *Science* **261**, 58–65.
161. C. E. Schutt, M. D. Rozycki, J. C. Myslik, and U. Lindberg (1995) *J. Struct. Biol.* **115**, 186–198.
162. R. Page, U. Lindberg, and C. E. Schutt (1998) *J. Mol. Biol.* **280**, 463–474.
163. K. Schlüter, B. M. Jockusch, and M. Rothkegel (1997) *Biochim. Biophys. Acta* **1359**, 97–109.
164. F. Buss, C. Temm-Grove, S. Henning, and B. M. Jockusch (1992) *Cell Motil. Cytoskeleton* **22**, 51–61.
165. M. Evangelista, K. Blundell, M. S. Longtine, C. J. Chow, N. Adames, J. R. Pringle, M. Peter, and C. Boone (1997) *Science* **276**, 118–122.
166. N. Watanabe, P. Madaule, T. Reid, T. Ishizaki, G. Watanabe, A. Kakizuka, Y. Saito, K. Nakao, B. M. Jockusch, and S. Narumiya (1997) *EMBO J.* **16**, 3044–3056.
167. J. Petersen, O. Nielsen, R. Egel, and I. M. Hagan (1998) *J. Cell Sci.* **111**, 867–876.
168. R. D. Mullins, J. F. Kelleher, J. Xu, and T. D. Pollard (1998) *Mol. Biol. Cell* **9**, 841–852.
169. V. Magdolen, U. Oechsner, G. Muller, and W. Bandlow (1988) *Mol. Cell Biol.* **8**, 5108–5115.
170. M. Haugwitz, A. A. Noegel, J. Karakesisoglou, and M. Schleicher (1994) *Cell* **79**, 303–314.
171. D. A. Kaiser, M. Sato, R. F. Ebert, and T. D. Pollard (1986) *J. Cell Biol.* **102**, 221–226.
172. A. Lambrechts, J. van Damme, M. Goethals, J. Vandekerckhove, and C. Ampe (1995) *Eur. J. Biochem.* **230**, 281–286.
173. T. Fujiwara, K. Tanaka, A. Mino, M. Kikyo, K. Takahashi, K. Shimizu, and Y. Takai (1998) *Mol. Biol. Cell* **9**, 1221–1233.
174. 15. W. Witke, A. H. Sharpe, and D. J. Kwiatkowski (1993) *Mol. Biol. Cell* **4** 149a.
175. C. Björkegren, M. Rozycki, C. E. Schutt, U. Lindberg, and R. Karlsson (1993) *FEBS Lett.* **333**, 123–126.
176. C. E. Schutt, J. C. Myslik, M. D. Rozycki, N. C. Goonesekere, and U. Lindberg (1993) *Nature* **365**, 810–816.
177. E. S. Cedergren-Zeppezauer, N. C. W. Goonesekere, M. D. Rozycki, J. C. Myslik, Z. Dauter, U. Lindberg, and C. E. Schutt (1994) *J. Mol. Biol.* **240**, 459–475.
178. K. S. Thorn, H. E. Christensen, R. Shigeta, D. Huddler, L. Shalaby, U. Lindberg, N. H. Chua, and C. E. Schutt (1997) *Structure* **5**, 19–32.
179. N. M. Mahoney, P. A. Janmey, and S. C. Almo (1997) *Nature Struct. Biol.* **4**, 953–960.
180. W. J. Metzler, A. J. Bell, E. Ernst, T. B. Lavoie, and L. Mueller (1994) *J. Biol. Chem.* **269**, 4620–4625.
181. H. Larsson and U. Lindberg (1988) *Biochim. Biophys. Acta* **953**, 95–105.
182. E. Korenbaum, P. Nordberg, C. Björkegren-Sjögren, C. E. Schutt, U. Lindberg, and R. Karlsson (1998) *Biochemistry* **37**, 9274–9283.
183. V. K. Vinson, E. M. De La Cruz, H. N. Higgs, and T. D. Pollard (1998) *Biochemistry* **37**, 10871–10880.
184. D. Pantaloni and M.-F. Carlier (1993) *Cell* **75**, 1007–1014.
185. M.-F. Carlier and D. Pantaloni (1993) *Sem. Cell Biol.* **5**, 183–191.
186. F. Markey, H. Larsson, K. Weber, and U. Lindberg (1982) Biochim. Biophys. Acta **704**, 43–51.
187. S. C. Mockrin and E. D. Korn (1980) *Biochemistry* **19**, 5359–5362.
188. P. J. Goldschmidt-Clermont, L. M. Machesky, S. K. Doberstein, and T. D. Pollard (1991) *J. Cell Biol.* **113**, 1081–1089.
189. M. Tanaka and H. Shibata (1985) *Eur. J. Biochem.* **151**, 291–297.
190. U. Lindberg, C. E. Schutt, E. Hellsten, A.-C. Tjäder, and T. Hult (1988) *Biochim. Biophys. Acta* **967**, 391–400.
191. D. Safer and V. T. Nachmias (1994) *BioEssays* **16**, 473–479.
192. V. T. Nachmias (1993) *Curr. Opin. Cell Biol.* **5**, 56–62.
193. M. Czisch, M. Schleicher, S. Horger, W. Voelter, and T. A. Holak (1993) *Eur. J. Biochem.* **218**, 335–344.
194. D. Safer, T. E. Sosnick, and M. Elzinga (1997) *Biochemistry* **36**, 5806–5816.
195. M. F. Carlier, D. Didry, I. Erk, J. Lepault, M. L. Van Troys, J. Vandekerckhove, I. Perelroizen, H. Yin, Y. Doi, and D. Pantaloni (1996) *J. Biol. Chem.* **271**, 9231–9239.
196. H. Q. Sun, K. Kwiatkowska, and H. L. Yin (1996) *J. Biol. Chem.* **271**, 9223–9230.
197. L. Bao, M. Loda, P. A. Janmey, R. Stewart, P. Anand-Apte, and B. R. Zetter (1996) *Nat. Med.* **12**, 1322–1328.
198. M. F. Carlier and D. Pantaloni (1997) *J. Mol. Biol.* **269**, 459–467.
199. J. A. Theriot (1997) *J. Cell. Biol.* **136**, 1165–1168.
200. A. Moon and D. G. Drubin (1995) *Mol. Biol. Cell* **6**, 1423–1431.
201. S. A. Leonard, A. G. Gittis, E. C. Petrella, T. D. Pollard, and E. E. Lattman (1997) *Nat. Struct. Biol.* **4**, 369–373.
202. A. A. Fedorov, P. Lappalainen, E. V. Fedorov, D. G. Drubin, and S. C. Almo (1997) *Nat. Struct. Biol.* **4**, 366–369.
203. H. Hatanaka, K. Ogura, K. Moriyama, S. Ichikawa, I. Yahara, and F. Inagaki (1996) *Cell* **85**, 1047–1055.
204. P. Lappalainen, M. M. Kessels, J. M. T. V. Cope, and D. G. Drubin (1998) *Mol. Cell Biol.* **9**, 1951–1959.
205. S. Arber, F. A. Barbayannis, H. Hanser, C. Schneider, C. A. Stanyon, O. Bernard, and P. Caroni (1998) *Nature* **393**, 805–808.
206. N. Yang, O. Higuchi, K. Ohashi, K. Nagata, A. Wada, K. Kangawa, E. Nishida, and K. Mizuno (1998) *Nature* **393**, 809–812.
207. M.-F. Carlier, V. Laurent, J. Santolini, R. Melki, D. Didry, G. X. Xia, Y. Hong, N.-H. Chua, and D. Pantaloni (1997) *J. Cell Biol.* **136**, 1307–1323.
208. J. Rosenblatt, B. J. Agnew, H. Abe, J. R. Bamburg, and T. J. Mitchison (1997) *J. Cell Biol.* **136**, 1323–1332.
209. B. L. Goode, D. G. Drubin, and P. Lappalainen (1998) *J. Cell. Biol.* **142**, 723–733.
210. T. A. Schroer (1994) *J. Cell Biol.* **127**, 1–4.
211. L. M. Machesky (1997) *Curr. Biol.* 164–167.
212. M. D. Welch, A. Iwamatsu, and T. J. Mitchison (1997) *Nature* **385**, 265–269.
213. L. M. Machesky, E. Reeves, F. Wientjes, F. J. Mattheyse, A. Grogan, N. F. Totty, A. L. Burlingame, J. J. Hsuan, and A. W. Segal (1997) *Biochem. J.* **328**, 105–112.
214. M. D. Welch, A. H. DePace, S. Verma, A. Iwamatsy, and T. J. Mitchison (1997) *J. Cell Biol.* **138**, 375–384.
215. D. Winter, A. V. Podtelejnikov, M. Mann, and R. Li (1997) *Curr. Biol.* **7**, 519–529.
216. R. D. Mullins, J. A. Heuser, and T. D. Pollard (1998) *Proc. Natl. Acad. Sci. USA* **95**, 6181–6186.
217. M. D. Welch, J. Rosenblatt, J. Skoble, D. A. Portnoy, and T. J. Mitchison (1998) *Science* **281**, 105–108.
218. J. F. Casella, S. W. Craig, D. J. Maack, and A. E. Brown (1987) *J. Cell Biol.* **105**, 371–379.
219. P. A. Kuhlman, C. A. Hughes, V. Bennet, and V. M. Fowler (1996) *J. Biol. Chem.* **271**, 7986–7991.
220. H. L. Yin (1987) *Bioessays* **7**, 176–179.
221. T. Azuma, W. Witke, T. P. Stossel, J. H. Hartwig, and D. J. Kwiatkowski (1998) *EMBO J.* **17**, 1362–1370.

222. P. J. McLaughlin, J. T. Gooch, H.-G. Mannherz, and A. G. Weeds (1993) *Nature* **364**, 685–692.
223. L. D. Burtnick, E. K. Koepf, J. Grimes, E. Y. Jones, D. I. Stuart, P. J. McLaughlin, and R. C. Robinson (1997) *Cell* **90**, 661–670.
224. A. McGough, W. Chiu, and M. Way (1998) *Biophys. J.* **74**, 764–772.
225. V. M. Fowler (1996) *Curr. Opin. Cell Biol.* **8**, 86–96.
226. J. Hartwig (1994) Actin-binding proteins 1. Spectrin Superfamily. *Protein Profile.* **1**, 711–778.
227. A. Bretscher and K. Weber (1979) *Proc. Natl. Acad. Sci. USA* **76**, 2321–2325.
228. M. S. Mooseker (1985) *Annu. Rev. Cell Biol.* **1**, 209–241.
229. D. Louvard (1989) *Curr. Opin. Cell Biol.* **1**, 51–57.
230. J. J. Otto (1994) *Curr. Opin. Cell Biol.* **6**, 105–109.
231. M. A. Markus, P. Matsudaira, and G. Wagner (1997) *Protein Sci.* **6**, 1197–1209.
232. C. J. McKnight, P. T. Matsudaira, and P. S. Kim (1997) *Nat. Struct. Biol.* **4**, 180–184.
233. A. Bretscher and K. Weber (1980) *J. Cell Biol.* **86**, 335–340.
234. J. E. Honts, T. S. Sandrock, S. M. Brower, J. L. O'Dell, and A. E. M. Adams (1994) *J. Cell Biol.* **126**, 413–422.
235. S. C. Goldsmith, N. Pokala, W. Shen, A. A. Fedorov, P. Matsudaira, and S. C. Almo (1997) *Nat. Struct. Biol.* **4**, 708–712.
236. D. Hanein, P. Matsudaira, and D. J. DeRosier (1997) *J. Cell Biol.* **139**, 387–396.
237. K. R. Fath and D. R. Burgess 1995, *Curr. Biol.* **5**, 591–593.
238. P. Fucini, C. Renner, C. Herberhold, A. A. Noegel, and T. A. Holak (1997) *Nat Struct Biol* **4**, 223–230.
239. T. W. Tillack, S. L. Marchesi, V. T. Marchesi, and E. Steers Jr. (1970) *Biochim. Biophys. Acta* **200**, 125–131.
240. G. L. Nicolson, V. T. Marchesi, and S. J. Singer (1971) *J. Cell Biol.* **51**, 265–272.
241. V. Bennett (1990) *Curr. Opin. Cell Biol.* **2**, 51–56.
242. J. Pascual, M. Pfuhl, G. Rivas, A. Pastore, and M. Saraste (1996) *FEBS Lett* **383**, 201–207.
243. P. Fucini, C. Renner, C. Herberhold, A. A. Noegel, and T. A. Holak (1997) *Nat. Struct. Biol.* **4**, 223–230.
244. J. Pascual, M. Pfuhl, D. Walther, M. Saraste, and M. Nilges (1997) *J. Mol. Biol.* **273**, 740–751.
245. M. Nilges, M. J. Macias, S. I. O'Donoghue, and H. Oschkinat (1997) *J. Mol. Biol.* **269**, 408–422.
246. F. J. Blanco, A. R. Ortiz, and L. Serrano (1997) *J. Biomol. NMR* **9**, 347–357.
247. K. D. Carugo, S. Banuelos, and M. Saraste (1997) *Nat. Struct. Biol.* **4**, 175–179.
248. S. Ebashi and F. Ebashi (1965) *J. Biochem.* **58**, 7–13.
249. S. J. Winder (1997) *J. Muscle Res. Cell Motil.* **18**, 617–629.
250. V. Bennett (1990) *Physiol. Rev.* **70**, 1029–1065.
251. Y. Srinivasan, L. Elmer, J. Davis, V. Bennett, and K. Angelides (1988) *Nature* **333**, 177–180.
252. M. Colledge and S. C. Froehner (1998) *Curr. Opin. Neurobiol.* **8**, 357–363.
253. C. Rosenmund and G. L. Westbrook (1993) *Neuron* **10**, 805–814.
254. K. M. Harris and S. B. Kater (1994) *Annu. Rev. Neurosci.* **17**, 341–371.
255. D. W. Allison, V. I. Gelfand, I. Spector, and A. M. Craig (1889) *J. Neurosci.* 18, 2423–2436.
256. T. Kim, K. Wu, J. Xu, and I. B. Black (1992) *Proc. Natl. Acad. Sci. USA* **89**, 11642–11644.
257. P. A. Huber (1997) *Int. J. Biochem. Cell Biol.* **29**, 1047–1051.
258. J. M. Squire (1997) *Curr. Opin. Struct. Biol.* **7**, 247–257.
259. J. Trinick (1992) *FEBS Lett.* **307**, 44–48.
260. R. Littlefield and V. M. Fowler (1998) *Annu. Rev. Cell Dev. Biol.* **14**, 487–525.

ACTIN POLYMERIZATION TOXINS

C. MONTECUCCO

Several bacteria release **toxins** that modify **actin** or proteins controlling its polymerization (1). A group of *Clostridia spp.* secrete binary toxins, whose most studied member is the C2 toxin released by *Cl. botulinum*. Botulinum C2 toxin is not neurospecific, as are botulinum **neurotoxins**, but it affects many nonneuronal cells (2). In the intestinal loop model, it induces an acute inflammatory reaction, characterized by the alteration of endothelia and a large increase in vascular permeability. At the same time, epithelial degeneration, exfoliation, and necrosis are also observed. In cultured cells, C2 causes rounding up, with formation of blebs, followed by cell death.

Two polypeptide chains, I (55 kDa) and II (100 kDa), are needed for cell intoxication. Chain I catalyzes the **ADP-ribosylation** of Arg177 of soluble G-actin, a residue located in an area involved in protein-protein contact within the polymerized form, F-actin. The ADP-ribosylated G-actin binds to the barbed end of F-actin and prevents polymerization, whereas depolymerization at the opposite end is unaffected (3). This leads *in vitro* and *in vivo* to the disassembly of actin microfilaments, with cell rounding and release of focal adhesion plaques.

Single-chain enzymes that catalyze the ADP-ribosylation of small G proteins are released by several bacteria (4). They cannot intoxicate cells because they lack the second polypeptide chain, but they are active after cell permeabilization or injection. C3 toxin is released by some strains of *Cl. botulinum*; it catalyzes the specific ADP-ribosylation of Asn-41 of Rho, a small **GTP-binding protein** involved in the regulation of actin polymerization. C3 induces the depolarization of F-actin, with rounding up and binucleation of injected cells. A different type of Rho modification is caused by cytotoxic necrotizing factors released by *E. coli* strains associated with gastroenteritis, urogenital infections, and septicemia (5). These factors cause cell ruffling, stress fiber formation, and multinucleation, by a covalent modification of Rho protein to lock it in its GTP-bound active form (6).

Clostridia spp. involved in the induction of diarrhea, associated with pseudomembraneous colitis, release in the intestine enterotoxins of very large dimensions (7). They are termed large clostridial toxins (LCTs) because of their size, which is in the range of 250 to 308 kDa. They are organized as A-B toxins, which are cleaved proteolytically into two polypeptide chains, A and B (see **Toxins**). The carboxyl-terminal part includes segments of 20 to 50 residues repeated 14 to 30 times, which are believed to mediate LCT binding to cell surface **receptors.** Such an organization in the absence of an oligomer of type B gives rise to a multivalent type of cell binding, as in the case of cholera toxin. Binding is followed by internalization of LCTs into coated **vesicles** and then into **endosomes**. By an unknown mechanism, the amino-terminal catalytic domain of the LCTs is released from the rest of the molecule and translocates into the cytosol, where it catalyzes the transfer of a sugar residue from the corresponding UDP derivative to an actin polymerization controlling protein (7,8). All LCTs induce rounding of different cells in culture, but the effects of the various LCT toxins can be differentiated by staining actin filaments: *Cl. difficile* and *Cl. novyi* LCTs cause a breakdown of the F-actin microfilaments, whereas *Cl. sordellii* LCT induces the formation of filopodialike structures on the cell surface, with some membrane blebbing 7. This is related to the different protein(s) targeted by the LCTs. All of them modify a threonine residue of small **GTP-binding protein**(s) of the **Ras** superfamily, in such a way that its GTPase activity is unaltered but it can no longer interact with its effector molecule. This provides a further example of the ability of bacterial toxins to "choose" essential cell targets and to modify essential cell functions.

BIBLIOGRAPHY

1. K. Aktories (1997) In *Guidebook to Protein Toxins and Their Use in Cell Biology* (R. Rappuoli and C. Montecucco, eds.), Sambrook and Tooze, Oxford University Press, Oxford.
2. L. L. Simpson (1989) *J. Pharmacol. Exp. Ther.* **251**, 1223–1228.
3. M. Wille, I. Just, A. Wegner, and K. Aktories (1992) *J. Biol. Chem.* **267**, 50–55.
4. K. Aktories and A. Wegner (1992) *Curr. Top. Microbiol. Immunol.* **175**, 115–131.
5. A. Caprioli et al. (1984) *Biochem. Biophys. Res. Commun.* **118**, 587–593.
6. E. Oswald et al. (1994) *Proc. Natl. Acad. Sci. USA* **91**, 3814–3818.
7. C. von Eichel-Streiber, P. Boquet, M. Sauerborn, and M. Thelestam (1996) *Trends Microbiol.* **4**, 375–382.
8. I. Just et al. (1995) *Nature* **375**, 500–503.

ACTINOMYCIN D

M. WARING

Actinomycins are chromopeptide antibiotics produced by various species of streptomycetes (1). They differ only in their content of certain **amino acids** in their peptide rings (Fig. 1). Actinomycin D, which is active against several forms of cancer, has the distinction of being the first antibiotic discovered (in 1943) to possess useful antitumor activity. It is still used in the clinic for that purpose. It occupies a unique place in the origins of molecular biology for quite a different reason, however, namely its astounding specificity as an inhibitor of DNA-dependent RNA synthesis, which enabled it to play a vital part in the discovery of **messenger RNA** and helped uncover the early evidence that **transcriptional** regulation of gene expression is fundamental to cell growth and development (2). The result of its near-absolute specificity for binding tightly to double helical B-form DNA is a total blockade of transcription by **RNA polymerase**, leading to cessation of synthesis of all forms of RNA, stable as well as unstable. Elongation of growing RNA chains is promptly terminated; there is no particular effect on the process of initiation of transcription. Thus actinomycin was used in the 1960s to measure the kinetics of breakdown of pulse-labeled RNA, thought to reflect at least partly the turnover of mRNA, and to establish the requirement for new RNA transcription in such processes as induced enzyme synthesis, mechanisms of steroid **hormone** and peptide hormone action, and early embryonic **development**. Replication of DNA **viruses** is usually strongly inhibited by actinomycin, but many RNA viruses are completely insensitive to the antibiotic; this is clear evidence that their life cycles do not require the participation of DNA at any point. Because actinomycin acts essentially identically to block transcription in all cells, once it has gained access to the nucleus, it is equally applicable to the study of gene activity in eukaryotes as well as prokaryotes (it is generally ineffective against **gram-negative bacteria** simply because of a permeability problem, for if the walls of such cells are digested away, or otherwise weakened, the **protoplast** is revealed as fully sensitive to the antibiotic).

Actinomycin does not bind to RNA or to single-stranded DNA, and its affinity for double-helical DNA is related to the latter's content of guanine nucleotides; synthetic polynucleotides composed entirely of A · T base pairs do not interact with the antibiotic at all. It took some time to establish the nature of the actinomycin–DNA complex, and for many years the matter was controversial. Eventually the issue was resolved in favor of intercalation as a result of careful structure–activity comparisons (1), binding measurements with a wide variety of synthetic as well as naturally occurring DNAs (3,4), hydrodynamic experiments with circular DNA (5), **X-ray crystallography** (6), and high-resolution nuclear magnetic resonance (**NMR**) (7). A landmark was the determination of the structure of a crystalline 1:2 actinomycin:deoxyguanosine complex, which revealed for the first time that the antibiotic has an axis of near-perfect twofold rotational symmetry, which allows it to react with two guanine nucleosides in a symmetry-related fashion (6). This observation immediately suggested plausible intercalation of the phenoxazinone chromophore between two G · C base pairs of DNA at a rotationally symmetrical site centered around a 5′-GpC-3′ step. The strong preference of actinomycin for such sites in DNA was elegantly confirmed by **footprinting** experiments a decade later (8,9). With its tricyclic aromatic chromophore firmly embedded between the base pairs, the cyclic pentapeptide rings of the antibiotic are left neatly filling the minor groove of the distorted B-form helix, where they form numerous additional **van der Waals** contacts that help to stabilize the complex; positioned this way, their intrinsic right-hand twisted disposition with respect to the intercalated chromophore makes perfect sense, and the whole antibiotic molecule occupies a site covering about six base pairs in the minor groove.

Detailed **kinetic** studies have revealed that both the association and dissociation reactions are complicated and require several rate constants to fit the data (3,10). The slowest processes are characterized by time constants in the range of minutes, conspicuously slower than those measured for most other DNA-binding drugs, and there is a correlation between the slowest rate constant for the dissociation reaction and the efficiency of inhibition of chain elongation by RNA polymerase. This has led to the notion that reversibly bound actinomycin molecules serve as relatively long-lived blocks to the progression of the transcribing enzyme along its template, providing a partial explanation for the extreme effectiveness of actinomycin as an inhibitor of RNA synthesis. What is not so obvious is why the functioning of DNA as a template in replication should remain unaffected even at much higher concentrations of actinomycin (2).

The critical determinant that enables the antibiotic to recognize its preferred binding sites in DNA seems to be the

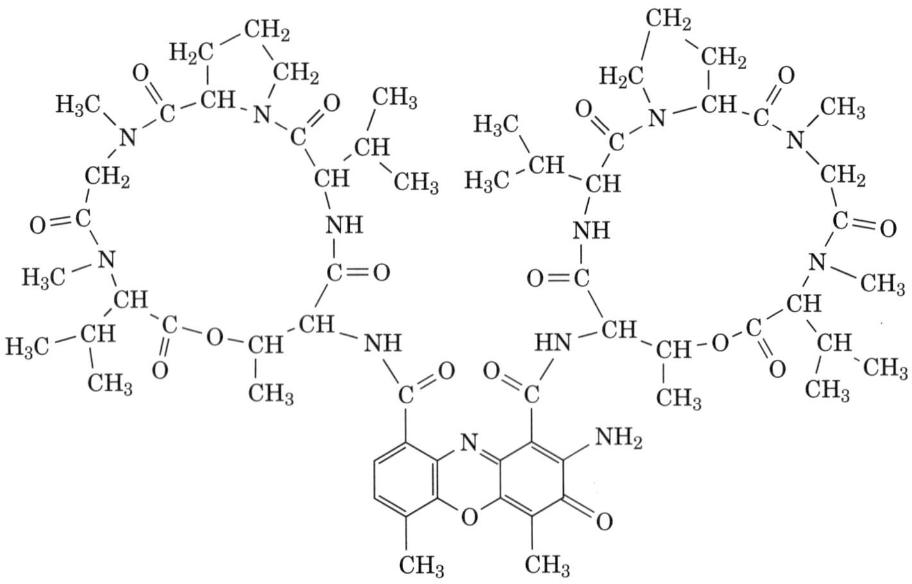

Figure 1. The structures of the actinomycin antibiotics.

actinomycin D

2-amino group of the guanine nucleotide, as is also true for numerous other ligands (11). If the 2-amino group is removed from guanines (leaving inosine–cytosine base pairs) and transferred to adenines (forming 2,6-diaminopurine-thymine base pairs), the sites to which actinomycin binds are relocated accordingly. The molecular recognition process includes the formation of **hydrogen bonds** between the purine 2-amino groups and the carbonyl substituents of the L-**threonine** residues in the peptide rings of the antibiotic; there are also hydrogen bonds from the same threonine residues to the N(3) atoms of the purine nucleotides at the binding site. Some of the kinetic complexity of the association/dissociation reactions no doubt reflects a "shuffling" process, whereby actinomycin initially interacts with a variety of potential binding sites on the DNA lattice, then migrates one-dimensionally to locate its preferred binding sites marked by a pair of purine 2-amino groups suitably disposed in the minor groove of the double helix (12).

BIBLIOGRAPHY

1. J. Meienhofer and E. Atherton (1977) In *Structure–Activity Relationships Among the Semisynthetic Antibiotics* (D. Perlman, ed.), Academic Press, New York, pp. 427–529.
2. E. Reich and I. H. Goldberg (1964) *Prog. Nucleic Acid Res. Mol. Biol.* **3**, 183–234.
3. W. Müller and D. M. Crothers (1968) *J. Mol. Biol.* **35**, 251–290.
4. R. D. Wells and J. E. Larson (1970) *J. Mol. Biol.* **49**, 319–342.
5. M. J. Waring (1970) *J. Mol. Biol.* **54**, 247–279.
6. H. M. Sobell, S. C. Jain, T. D. Sakore, and C. E. Nordman (1971) *Nature New Biol.* **231**, 200–205.
7. D. J. Patel (1974) *Biochemistry* **13**, 2396–2402.
8. M. J. Lane, J. C. Dabrowiak, and J. N. Vournakis (1983) *Proc. Natl. Acad. Sci. USA* **80**, 3260–3264.
9. M. W. Van Dyke, R. P. Hertzberg, and P. B. Dervan (1983) *Proc. Natl. Acad. Sci. USA* **79**, 5470–5474.
10. R. Bittman and L. Blau (1975) *Biochemistry* **14**, 2138–2145.
11. C. Bailly and M. J. Waring (1995) *Nucleic Acids Res.* **23**, 885–892.
12. K. R. Fox and M. J. Waring (1985) *Nucleic Acids Res.* **14**, 2001–2014.

Suggestions for Further Reading

E. F. Gale, E. Cundliffe, P. E. Reynolds, M. H. Richmond, and M. J. Waring (1981) *The Molecular Basis of Antibiotic Action*, 2nd ed., Wiley, London, pp. 314–333.

H. M. Sobell (1973) The stereochemistry of actinomycin binding to DNA and its implications in molecular biology. *Prog. Nucleic Acid Res. Mol. Biol.* **13**, 153–190.

ACTIVATION ENERGY

D. S. AULD

The **kinetic** rates of chemical and enzyme-catalyzed reactions depend on temperature. The relationship for this dependence is known as the Arrhenius law:

$$k = Ae^{E/RT} \quad (1)$$

where A and E are constants, R is the gas constant, and T is the absolute temperature in Kelvin units. The activation energy E is the height of the energy barrier that the reaction must exceed to pass from reactants to products. It is usually expressed in kilocalories per mole or in joules per mole. Equation 1 can be expressed as:

$$\log k = \log A - E/(2.303RT) \quad (2)$$

Therefore the magnitude of the activation energy can be obtained from the slope of a plot of the log of a rate constant for a reaction as a function of $1/T$. Many chemical and enzyme-catalyzed reactions increase the rate of reaction by two- to threefold for each 10°C increase in temperature. Although this relationship is useful in explaining the temperature dependence of reactions, it does not explain the rate in the thermodynamic terms of **enthalpy** H **entropy** S, or **free**

energy G. This analysis comes from **transition state** theory. The Arrhenius activation energy does correspond to the standard enthalpy of the reaction in the van't Hoff equation (1).

BIBLIOGRAPHY

1. A. Cornish-Bowden (1979) *Fundamentals of Enzyme Kinetics*, Butterworths, London, Boston, pp. 9–12.

Suggestion for Further Reading

D. Piszkiewicz (1977) *Kinetics of Chemical and Enzyme-Catalyzed Reactions*, Oxford University Press, New York, pp. 27–52.

ACTIVE SITE

JOHN F. MORRISON

The folding of a **polypeptide chain** that produces the final **protein structure** of an **enzyme** also leads to formation of the active site. From **X-ray crystallography** studies, it is apparent that the active site of an enzyme is a groove, cleft, or pocket that has access to the solvent and forms only a small part of the total solvent-**accessible surface** of the protein. The relatively large sizes of enzymes are undoubtedly due to the need to obtain, at the active site, the correct spatial relationships of the amino acid residues that are involved in binding of substrates, **catalysis**, and the release of products, as well as for any conformational changes associated with these steps. The binding of substrates or inhibitors at the active site pocket of an enzyme involves matching up of the **nonpolar** groups of the substrate with the nonpolar side-chains of amino acid residues, **hydrogen bonding** between the **polar** appropriate groups on the substrate with the backbone NH and CO groups within the active site, and even **salt bridge** formation. For substrates, these initial interactions are followed by the conformational changes that lead to the formation of the transition-state complex (see **Transition state analogue**) and the chemistry for catalyzing the reaction brought about by reactive groups with the correct alignments. These may be the acidic, basic, and nucleophilic groups of the protein component of the enzyme, or the electrophilic groups of a prosthetic group (see **Coenzyme, cofactor**).

ACTIVE SITE-DIRECTED IRREVERSIBLE INHIBITORS

JOHN F. MORRISON

Active site-directed irreversible inhibitors of **enzymes** are also known as *active site-direcetd inactivating reagents*, *affinity labels*, and *photoaffinity labels*. They combine the features of a substrate, or substrate analogue, with those of a group-specific reagent, as in **affinity labeling**, and have been used to determine the amino acid residues that are present in the **active site** and involved in enzymic **catalysis**. They are capable of binding specifically and reversibly at the active site of an enzyme and then causing inactivation through time-dependent covalent modification of an adjacent amino acid residue. The functional group of the inhibitor is usually an electrophile that can interact with an appropriately positioned nucleophile of the enzyme, to generate a covalent bond between them. The electrophilic groups include epoxides and α-haloketones. Since these compounds are reactive in solution, they could also cause some nonspecific enzyme modifications.

The action of an active-site directed irreversible inhibitor I can be illustrated by

$$E + I \rightleftharpoons EI \rightarrow E - I \qquad (1)$$

where EI represents a Michaelis complex (see **Michaelis–Menten kinetics**) and $E - I$ denotes the covalently cross-linked form of the complex. On the basis of this formulation, it would be expected that at the early stages of the interaction, I would behave as a **competitive inhibitor** with respect to the substrate. Examples of the action of an active-site directed irreversible inhibitors are the acetylation of amino acid residues at the active site of prostaglandin synthase by aspirin (acetyl salicylate) (1) and the alkylation by L-**TPCK** (N-tosylphenylalanine chloromethyl ketone) of a **histidine** residue at the active site of α-**chymotrypsin** (2).

Photoaffinity labels, such as diazoketones and aryl azides, introduce a greater degree of specificity to the modification of amino acid residues, as they are not reactive in solution. It is only after the reversible interaction of the affinity label at the active site of an enzyme, and exposure of the resulting complex to light of the correct wavelength, that a highly reactive group is formed. Such treatment with diazoketones and aryl azides leads to the formation of carbenes and azines that are extremely reactive and can add across O—H bonds of unionized carboxyl groups or unsaturated carbon–hydrogen bonds (3).

BIBLIOGRAPHY

1. G. J. Roth, N. Stanford, J. W. Jacobs, and P. W. Majerus (1977) *Biochemistry* **16**, 4244–4248.
2. C. Walsh (1979) *Enzymatic Reaction Mechanisms*, W. H. Freeman and Company, San Francisco, Calif., p. 86.
3. V. Chowdry and F. H. Westheimer (1979) *Ann. Rev. Biochem.* **48**, 293–325.

ACTIVE-SITE TITRANTS

JOHN F. MORRISON

For determination of values for the **kinetic** rate constants and **ligand-binding** stoichiometries associated with an **enzyme**-catalyzed reaction, it is necessary to know the concentration of functional **active sites** of the enzyme. The latter value cannot be calculated simply from the total concentration of protein and molecular weight of the enzyme; even when the enzyme preparation has been shown to be homogeneous by a variety of techniques, it is possible that inactive enzyme is present. The inactivity could be due to the inability of the enzyme to bind substrate or perform the chemistry of the reaction, or both. Therefore, before undertaking detailed kinetic investigations on any enzyme, it is important to determine the concentrations of the active sites that can both bind the ligands and are active.

The concentration of binding sites can be obtained from data for the reversible inhibition of an enzyme by a tight-binding substrate analogue that gives rise to **competitive inhibition**. Tight-binding inhibition occurs under conditions where the total inhibitor concentration I_t is comparable to the total enzyme concentration E_t, and a substantial fraction of the inhibitor is bound, so that allowance has to be made for the reduction of free inhibitor concentration as a result of formation of the enzyme-inhibitor complex (1). The variation of the steady-state velocity as a function of the concentrations of I_t and E_t is described by Equation 1:

$$v = \frac{kA}{2(K_a + A)}[\{(K_i' + I_t - E_t)^2 + 4K_i'E_t\}^{1/2} - (K_i' + I_t - E_t)] \tag{1}$$

where k denotes the maximum rate of product formation in terms of moles per mole of enzyme per second, K_a is the Michaelis constant for the substrate present at concentration A, and K_i' represents an apparent inhibition constant whose relationship to the true inhibition constant K_i for the interaction of I with E is given by

$$K_i' = K_i(1 + A/K_a)$$

Equation 1 also applies to multisubstrate reactions, provided that the nonvaried substrates are present at saturating concentrations. When no assumptions are made about the purity of the enzyme, E_t of Equation 1 is replaced by αEx, where α represents the degree of purity of the enzyme and Ex denotes the total protein concentration. Fitting to the modified form of Equation 1 of steady-state velocity data obtained at varying concentrations of I_t and E_t would yield values for K'_i as well as α, which would yield a measure of the concentration of binding sites (2,3).

The technique of active-site titration is used to determine the concentration of catalytically active enzyme (4,5). It requires that on mixing enzyme and the titrating substrate, there is an initial rapid burst of product formation because of the accumulation of an enzyme-bound intermediate whose rate of breakdown is much slower than its rate of formation. The procedure can be illustrated by reference to the Uni–Bi reaction:

$$\begin{array}{ccccccc} E & \rightleftharpoons & EA & \rightarrow & EQ & \rightarrow & E \\ + & & & & + & & + \\ A & & EPQ & & P & & Q \end{array}$$

for which P is released before Q, and at a very much faster rate. Specific measurement of the release of P as a function of time would yield a plot (Fig. 1) that consists of curved (burst) and linear sections. The intersection of the extrapolated linear curve with the vertical ordinate would give the concentration of P that is equal to the concentration of functional active sites.

Active-site titrations may be performed with chromogenic or radioactive substrates, and rapid reaction techniques are often required. This is not the case, however, for determination of the active-site concentration of *hexokinase* (6). This can be done by incubating the enzyme with ^{14}C-glucose and ATP complexed with Cr^{3+}, rather than the usual Mg^{2+}, and separating the enzyme-CrADP-glucose-6-phosphate complex on a Sepharose column. In this case, the enzyme undergoes only a single turnover, and the product complex has a half-life of over 10 min, so the amount of ^{14}C associated with the enzyme gives a measure of the active-site concentration (6).

Comparison of the results obtained from tight-binding inhibition and active-site titration studies could yield information about the presence in an enzyme preparation of substantial proportion of sites that can bind substrate, but cannot catalyze the reaction.

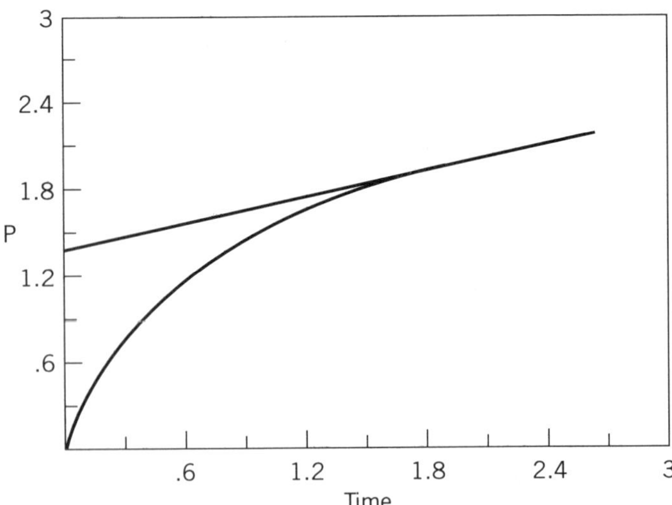

Figure 1. Active-site titration curve showing burst formation of a covalent enzyme-intermediate complex.

BIBLIOGRAPHY

1. J. W. Williams and J. F. Morrison (1979) *Meth. Enzymol.* **63**, 437–467.
2. J. F. Morrison and S. R. Stone (1985) *Comments Mol. Cell. Biophys.* **2**, 347–368.
3. M. J. Sculley and J. F. Morrison (1986) *Biochem. Biophys. Acta* **874**, 44–53.
4. A. Fersht (1977) *Enzyme Structure and Mechanism*, W. H. Freeman and Company, San Francisco, Calif., pp. 122–126.
5. F. J. Kezdy and E. T. Kaiser (1970) *Meth. Enzymol.* **19**, 3–20.
6. K. D. Danenberg and W. W. Cleland (1975) *Biochemistry* **14**, 28–39.

ADENOVIRUS

KEI FUJINAGA

The adenoviruses constitute a large group of DNA **viruses**, the *Adenoviridae* family. Genera include *Mastadenovirus* (human, simian, bovine, equine, porcine, ovine, canine, and opossum) and *Aviadenovirus* (birds). Adenoviruses are nonenveloped icosahedral particles of 70 to 100 nm in diameter, with 252 capsomeres, of which 240 are hexons and 12 are penton bases. A projecting fiber attaches to each penton base to form a penton.

Human adenoviruses (Ads) include 47 serotypes, which can be classified into six subgroups, A to F, based on their ability for red-blood-cell agglutination and oncogenicity in rodents, plus their DNA **homology**. Ads cause respiratory tract infections, conjunctivitis, hemorrhagic cystitis, and gastroenteritis. Highly oncogenic group A (eg, Ad12 and Ad18), weakly oncogenic group B (eg, Ad3 and Ad7), and Ad9 of group D can induce tumors in rodents, but no conclusive evidence has been reported linking adenoviruses with malignant diseases in the human. All Ads that have been tested can transform cultured rodent cells. No infectious virus is present, but viral transforming genes (E1A and E1B) are present and expressed in tumors and in transformed cells induced by adenoviruses.

The viral genome consists of a single linear molecule of double-stranded DNA [MW ~23×10^6, size ~36 kilobase pairs (kbp)], varying somewhat with the type. Human adenovirus type 2 (Ad2), the first to be sequenced completely, has a total of 35,937 bp. The Ad genome has **inverted terminal repeats** of 103 bp to 163 bp, depending on the type, and two identical **replication origins**, one in each terminal repeat. A 55-kDa terminal protein (TP) covalently bound to each 5′ end of the viral DNA molecule serves as a primer for protein-primed viral **DNA replication**. The viral genome carries five early **transcription** units (E1A, E1B, E2, E3, E4), two delayed early transcription units (IX and IVa2), and one late transcription unit (major late), all of which are transcribed by **RNA polymerase II**. The viral genome also carries one or two (depending on the type) VA genes transcribed by RNA polymerase III. The *E1A*, *E1B*, *IX*, *major late*, *VA*, and *E3* are on the rightward reading DNA strands (r strand), and others are on the leftward reading DNA strand (l strand).

The E1A is the first transcription unit to be expressed shortly after infection, using cellular **transcription factors**, encoding two major mRNAs of 12S and 13S. The encoded E1A proteins are required for productive viral replication and play a key role in cell transformation. E1A proteins transactivate early and late viral genes and cellular transcription units, and they bind cellular proteins, including p300, a family member of CBP (CREB binding protein)/p300, plus the RB family members, RB (**retinoblastoma** susceptibility gene product, tumor suppressor), p107, and p130. Binding to these cellular proteins is required for cell transformation. Binding of E1A to RB disrupts the RB · E2F complex to activate a cellular transcription factor, E2F, resulting in the activation of E2F-dependent **cell cycle**-related genes.

Two major proteins of 19 kDa and 55 kDa, encoded by 13S and 22S **messenger RNAs**, respectively, are generated from the E1B region. The E1B proteins are required for efficient viral growth and cooperate with E1A products to transform rodent cells, preventing E1A-induced **apoptosis**. The E1B 19-kDa protein has a functional similarity to the cellular anti-apoptotic Bcl-2, and the E1B 55-kDa protein interacts with **tumor suppressor** and apoptosis-related **p53**.

Region E2 encodes proteins involved in viral replication, including the viral terminal protein precursor, pTP (coded by *E2B*), viral DNA polymerase (coded by *E2B*), and the DNA-binding protein (coded by *E2A*). None of the *E3* proteins are required for productive infections of adenoviruses in cultured cells, but are important for *in vivo* infections in humans, suppressing host defense mechanisms by **cytotoxic T lymphocytes** or **tumor necrosis factor**-α. The E4 proteins are involved in transcriptional regulation, preferential transport of viral mRNA, and efficient viral DNA replication. Other functions of the E4 gene, such as transformation with E1A, tumor-suppressor p53 binding, and apoptosis-inhibiting activities, have been recently reported. The Ad9 E4 gene can oncogenically transform rat cells in the absence of the E1A and E1B genes and is required for mammary tumorigenesis.

The transcription of adenovirus late genes is triggered by the onset of viral DNA replication, yielding at least 18 distinct mRNAs by differential **poly(A)** site utilization and **alternative splicing**. Viral DNA replication activates a major late **promoter** (MLP) located at 16.4 map units of the viral genome, and one large primary transcript is generated, terminating at 99 map units at the right end of the genome. This transcript is processed to five families of late mRNAs, L1 to L5, based on the use of common poly(A) addition sites. The late mRNAs encode structural polypeptides of the viral particle and proteins involved in polypeptide processing, capsomere assembly, and packaging of the viral genomic DNA.

Adenoviruses are being used as **vectors** for delivery of therapeutic genes to target organs *in vivo*. Recombinant adenoviruses can be constructed by replacing the E1A and E1B genes with foreign ones. The E3 region also can be deleted without significant changes in virus growth. Recombinant adenoviruses can propagate efficiently in 293 cells, which complement the defect. The advantages are that the growth of recombinant viruses does not require host cell division, and a high titer of recombinant viruses can be easily obtained.

Suggestions for Further Reading

T. Shenk (1996) *Adenoviridae*: The Viruses and Their Replication. In *Fields Virology*, 3rd ed. (B. N. Fields et al., eds.), Lippincott-Raven, Philadelphia, pp. 2111–2148.

M. S. Horwitz (1996) Adenoviruses. In *Fields Virology*, 3rd ed. (B. N. Fields et al., eds.), Lippincott-Raven, Philadelphia, pp. 2149–2171.

J. Tooze (1981) *DNA Tumor Viruses*, 2nd ed., Cold Spring Harbor Laboratory Press, Cold Spring Harbor, NY, pp. 943–1054.

ADENYLATE CHARGE

D. E. ATKINSON

The adenylate charge, or adenylate energy charge, is a linear measure of the energy stored in the adenylate system ([ATP] + [ADP] + [AMP]) in a living cell. It is analogous to the charge of a storage battery. Its value is defined by the expression:

$$\text{Adenylate charge} = \frac{[\text{ATP}] + 0.5[\text{ADP}]}{[\text{ATP}] + [\text{ADP}] + [\text{AMP}]}$$

The status of the adenylate system is best described by the adenylate charge and appears to be the most ubiquitous regulatory signal in metabolism; it affects the rates of nearly all metabolic conversions and the partitioning of metabolically available substances between oxidation, synthesis of cell substance, and production of storage compounds. In living organisms, the same controls that regulate metabolic partitioning also maintain the value of the adenylate charge near 0.9.

BACKGROUND

Organisms are exquisitely regulated material- and energy-transducing systems. Material from the environment must be converted to the many compounds of which the organism is made, and energy, usually obtained either by absorption of sunlight or from oxidation of foodstuffs, must be converted for use in biosynthesis, movement, and membrane activities (see **Energy transduction**). Partitioning of material and energy between those functions must be adjusted continuously to meet the changing needs of the cell or organism. Transduction of energy through the adenylate system is an integral part of each of those functions.

Any system that is not at chemical and physical equilibrium can, in principle, supply energy. The two types of nonequilibrium situations that are mainly used in energy storage and transduction by organisms are (1) difference in electric charge or chemical potential across a membrane (see **Membrane potential**) and (2) a ratio of [ATP] to [ADP] that is far from equilibrium. All cells contain enzyme systems that catalyze transfer of energy between the two.

The energy status of a chemical storage system may be defined by either of two parameters: the molar Gibbs **free energy** change of the relevant chemical reaction or the mole fraction of the higher-energy state of the system. The two are computationally interconvertible but are not linearly related. The molar free energy change is a function of the ratio of activities of the products and reactants of the reaction; the mole fraction is a linear measure of the extent of reaction. In the familiar case of a lead storage battery, the free energy is measured by the voltage and the mole fraction by the charge (measured by a hydrometer in this case).

If a generic energy-transducing reaction is indicated by the type of reaction $A \leftrightharpoons B$, the molar free energy change is given by

$$\Delta G = \Delta G^\circ + RT \ln\left(\frac{[B]}{[A]}\right) \quad (1)$$

and the fractional charge (mole fraction of the more energetic state) by

$$\text{Fractional charge} = \frac{[B]}{[A] + [B]} \quad (2)$$

In the case of a lead storage battery, A is ($2Pb(SO_4) + 2H_2O$) and B is ($Pb + PbO_2 + 2H_2SO_4$). For the metabolic adenylate system, A is ($ADP + P_i$) and B is (ATP).

A system is fully discharged when only A is present and fully charged when it consists of B alone. For intermediate states, the stoichiometrically linear charge function indicates how much work is available as a fraction of that of a totally charged system. [This metabolic use of the term "work" is slightly nonstandard. In thermodynamic terms, the maximum work available is the integrated product of the charge multiplied by the changing value of the Gibbs free energy as the system goes to equilibrium. But the energy of the adenylate system is used stoichiometrically—for example, to affect chemical change or mechanical movement—and charge is therefore proportional to available work, defined in terms of biological effect.] It is because of its stoichiometric nature that the charge function is more relevant in most contexts than the molar free energy change for both storage batteries and the metabolic adenylate system.

If the adenylate system merely alternated between ADP and ATP, the charge function would be the simple ATP mole fraction, [ATP]/([ATP] + [ADP]). Some enzymes, however, couple the use of ATP to synthetic reactions by converting ATP to AMP and pyrophospate, and so the concentration of AMP must also be taken into account.

$$ATP + H_2O \leftrightharpoons AMP + PP_i \quad (3)$$

The pyrophosphate is rapidly hydrolyzed enzymically by pyrophosphatase, and this removal of product pulls the reaction forward (increases the numerical value of the negative free energy change).

Accumulation of AMP as a consequence of such reactions would deplete the cell's stores of ATP and ADP and so would be rapidly lethal. That outcome is prevented by the action of adenylate kinase, which catalyzes the phosphorylation of AMP:

$$ATP + AMP \leftrightharpoons 2ADP \quad (4)$$

The sum of reactions (3) and (4) is conversion of two molecules of ATP to two of ADP. In such cases, a single molecule of ATP, in effect, supplies twice as much metabolic energy to a metabolic reaction as is obtained from action of an ordinary kinase that converts ATP to ADP, but a second molecule of ATP is required to make good the energy balance.

Because of the participation of AMP in metabolic energy transduction, the simple mole fraction of ATP is not an adequate measure of the energy status of the adenylate system. Reaction (4) shows that two molecules of ADP are energetically equivalent to one of ATP. Thus, the linear charge function for the adenylate system (the effective mole fraction of ATP) is seen to be

$$\text{Adenylate charge} = \frac{[ATP] + 0.5[ADP]}{[ATP] + [ADP] + [AMP]} \quad (5)$$

The adenylate charge is the linear measure of the metabolic work available. A system containing only ATP is fully charged, with an adenylate charge value of 1.0, and one containing only AMP is fully discharged, with an adenylate charge of 0. The charge value would be 0.5 if only ADP were present. If reaction (4) catalyzed by adenylate kinase is at equilibrium, the concentrations of ATP, ADP, and AMP are fixed for any particular value of the adenylate charge (Figure 1).

REGENERATION AND UTILIZATION OF ATP

In aerobic organisms, ATP is regenerated mostly by oxidation of foods or storage compounds to carbon dioxide (see **ATP synthase**). In anaerobic organisms, substrates undergo other energy-yielding conversions—for example, fermentation of glucose to ethanol and carbon dioxide—rather than oxidation. Carbohydrates are converted to pyruvate by way of the glycolytic pathway. The pyruvate is oxidized to a derivative of acetic acid, acetyl coenzyme A, which is oxidized to carbon dioxide in the reactions of the citrate cycle, or Krebs cycle.

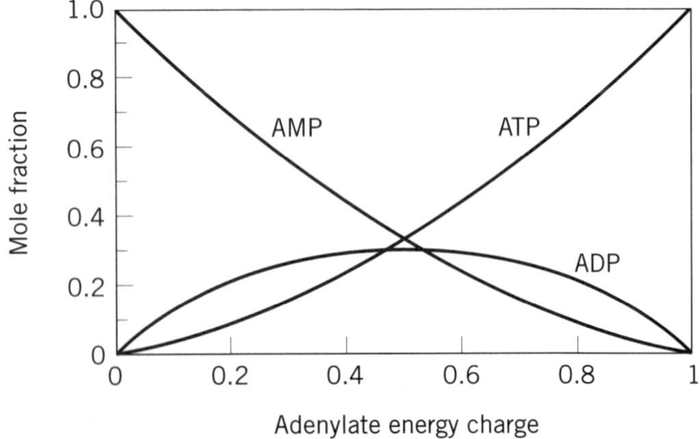

Figure 1. Mole fractions of the components of the adenylate nucleotide system as a function of the adenylate charge. The reaction catalyzed by adenylate kinase is assumed to be near equilibrium, with an apparent equililbrium constant irrespective of differential ionization and magnesium binding) of 0.8.

The electrons lost in those oxidations are passed on to oxygen by mediation of a series of membrane-associated enzymes. Those electron transfers are coupled to the conversion of ADP to ATP, thus supplying metabolically available energy to the adenylate system. The routes by which other classes of foods are metabolized feed into this central pathway. Fats are broken down to produce **acetyl coenzyme A**, which joins the carbohydrate pathway at that point. **Amino acids** derived from **protein degradation** are metabolized by individual pathways to produce various intermediates of glycolysis or the citrate cycle. Thus, the same central pathways are taken in the utilization of all foods.

Glycolysis and the citrate cycle are also centrally involved in biosynthesis. The biosynthetic pathways leading to the many components of a cell all begin with intermediates of glycolysis or the citrate cycle. That is, 10 or 12 intermediates of these degradative pathways are also the starting points for all synthetic sequences. Each such intermediate occupying a metabolic branchpoint must be partitioned between two competing pathways, one leading to oxidation to carbon dioxide and the other to synthesis of one or more components of the cell. The adenylate system provides energy for the chemical activities of the cell, including biosynthesis, for **active transport** of nutrients and ions across **membranes** against chemical potential gradients, and for most other biological requirements, including mechanical movement. All those functions are catalyzed or affected by **proteins**. Proteins involved in those functions have evolved affinities for ATP and ADP (or AMP) that maximize their functional usefulness to the organism. Thus, it is essential that the ratio of ATP to ADP remain virtually constant. The problem would be equivalent to the regulation of voltage by the power supply of a complex electronic device, if it were not so much more complicated.

REGULATION OF THE REGENERATION AND UTILIZATION OF ATP

The rate of regeneration of ATP from ADP is regulated in large part by controlling the rate at which substrate is made available to the electron transport phosphorylation system. At least five enzymes that catalyze reactions in glycolysis or the citrate cycle respond sensitively to the status of the adenylate pool and adjust the properties of the catalytic site accordingly. An increase in the energy charge causes a decrease in the rate of the reaction catalyzed by the enzyme, and a decrease in charge causes an increase in rate. This **feedback inhibition** system acts to adjust the rate of regeneration of ATP to meet momentary requirements and thus to stabilize the value of the charge. The multiplicity of control sites may be surprising; a single throttle point would seem to be sufficient to regulate the rate of supply of substrate to the electron transport system. The regulatory requirements are, however, much more complex because the central pathways also supply starting materials for synthesis.

Core metabolism consists of the central pathways by which foodstuffs or storage materials of all types are prepared for oxidation and of branches from those pathways that lead to synthesis of one or more products. The primary function of oxidative pathways is regeneration of ATP from ADP, which puts energy into the adenylate system. Biosynthesis is powered by conversion of ATP to ADP, which removes energy from the adenylate system. Adjustment of the partitioning of resources between these oppositely directed pathways to meet the momentary metabolic needs of the cell or organism is the most central of the regulatory requirements that underlie cellular function and survival. At each branchpoint, the partitioning of resources must respond at least to the energy status of the cell as reflected in the ATP/ADP system and to the momentary need for the end products of the synthetic branch.

At such branchpoints, the next enzyme in the degradative pathway and the first enzyme in the biosynthetic branch compete for the branchpoint metabolite, their common substrate. Maintenance of an appropriate balance between oxidation and biosynthesis requires that both enzymes must be regulated. As well as decreasing the rate of the degradative reaction, an increase in the value of the charge leads to an increase in the rate of the competing reaction that channels the branchpoint metabolite into biosynthesis. Thus, enough substrate is oxidized to maintain the normal status of the adenylate system, and biosynthesis is allowed to the extent that the products are needed and the supply of resources allows.

This partitioning is not affected by turning enzymes on and off but by changing the affinities of the competing enzymes for the substrate, which allows for much more sensitive control. Affinity is usually expressed in terms of the Michaelis constant, $\boldsymbol{K_m}$, the concentration of substrate at which half the catalytic sites bind substrate, leading to a reaction velocity half the maximal rate. Thus, competition between the enzymes is regulated in the most direct and effective way—by modulating their relative abilities to capture the common substrate. The competition is shown generically in Figure 2. Reactions in pathways that lead to degradation of substrate and the regeneration of ATP respond to variation in the energy status of the adenylate pool as indicated by curve R, and reactions that direct substrate into biosynthetic pathways that utilize ATP re-

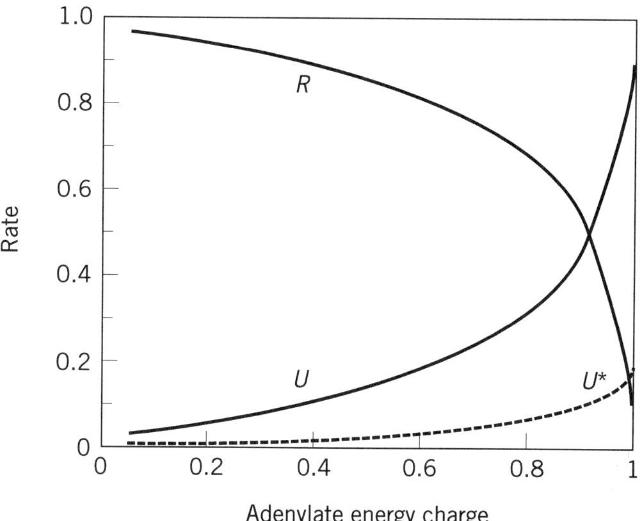

Figure 2. Generalized illustration of the effects of the adenylate charge on the rates of reactions in which ATP is regenerated (R) and in which ATP is utilized (U). Curve U was calculated for an enzyme for which the K_m for ATP at the catalytic site is six times that of ADP. Curve U^* represents 80% depression of rate as a consequence of feedback inhibition of the enzyme by the end product of the biosynthetic sequence.

spond as shown by curve U. The result is that the value of the energy charge is maintained within a narrow range of values near the intersection of the curves. If the charge drifts upward slightly, the rate of use of ATP in biosynthesis tends to increase and the rate of regeneration of ATP decreases, counteracting the drift. A downward drift has the opposite effect.

The regulatory interactions illustrated by curves R and U are necessary but not adequate. By themselves, they would adjust the rate of all biosynthetic sequences similarly and only on the basis of the availability of energy, without regard to the need for the individual products. Those controls are supplemented by product feedback inhibition. The first enzyme in nearly every biosynthetic sequence that has been studied bears a regulatory site that binds the end product of the sequence. When that site is occupied, the conformation of the enzyme changes so as to decrease the affinity for substrate at the catalytic site (increase the K_m) (see **Allostery**). Thus, as the concentration of end product rises, the fraction of enzyme molecules that bind it increases, the affinity of enzyme for substrate decreases, and the enzyme competes less vigorously for substrate. When the concentration of end product falls—for example, when an amino acid is being used more rapidly for protein synthesis—a smaller fraction of the regulatory sites of the first enzyme involved in its synthesis is occupied by end product. The resulting decrease in K_m causes the substrate to be more successful in its competition with the enzyme that catalyzes the next step in the degradative pathway, and the rate of synthesis increases. Such interactions adjust synthetic rates to meet changing metabolic needs. Curve U^* in Figure 2 illustrates the response of an enzyme when the end product of its sequence is available in adequate amount from external sources. Any response between curve U^* and somewhat above curve U is possible. Under ordinary conditions, the response will fluctuate in the vicinity of curve U to adjust production to meet metabolic demand. These interactions stabilize the pools of amino acids, assuring that they will neither be depleted when demand for protein is high nor build up to unnecessary or injurious levels when demand is small. Pool levels of other metabolites are regulated similarly.

The curves of Figure 2 thus provide a general overview of the regulation of the central pathways in their roles of regenerating ATP and providing starting materials for biosynthesis (1). The interaction of curves R and U adjust degradative metabolism and overall biosynthetic rates so as to maintain a nearly constant value of energy charge and, in the process, necessarily cause an appropriate partitioning of resources between degradation and synthesis. Superposition of product feedback inhibition, illustrated by curve U^*, adjusts the rate of each individual synthetic sequence to meet the momentary needs of the organism or cell. Even when the overall rate of biosynthesis is high, the synthetic pathway leading to any given metabolite will be suppressed when that product is in good supply. Thus, the components of this control system act together to assure that the adenylate system, the immediate energy source for nearly all cell activities, is maintained at a high and constant charge, near 0.9 (2), and that the rate of production of each biosynthetic product is determined by interaction between the overall availability of resources and the cell's need for the individual product. Regulation of biosynthetic rates is roughly equivalent to the decisions made by a human consumer who must balance how strongly an item is desired and how readily it can be afforded.

REGULATORY PROPERTIES OF ENZYMES

Enzymes that show R-type responses (see **Allostery**) catalyze reactions in the pathways that supply substrates for electron transport–linked regeneration of ATP but usually do not involve ATP or ADP directly. Such enzymes have, therefore, evolved regulatory, allosteric sites, distinct from the catalytic site, where ADP or AMP can bind, causing conformational changes that increase the binding affinity for the substrate at the catalytic site. Some such enzymes decrease substrate affinity when ATP binds at a regulatory site. Typically, the nucleotide binds cooperatively to two or more sites, and the effect on K_m is related to the square or higher power of the nucleotide concentration. Such interactions underlie the sensitive response of R-type enzymes to variation in the value of the charge (3).

In contrast, ATP and ADP are directly involved in most of the reactions at branchpoints that direct substrate into synthetic sequences. Thus, separate nucleotide-binding regulatory sites are not required. The U-type response is a consequence of higher affinity at the catalytic site for ADP, a product of the reaction, than for ATP, a reactant. This reversal of the usual pattern of higher affinity for reactants than for products results in pronounced inhibition by ADP across most of the energy charge range. At a charge of 0.5, for example, when the concentrations of ATP and ADP are approximately equal, ATP would be excluded from most of the catalytic sites because of competition by ADP. The result would be that most of the kinetic response to variation in the ATP/ADP ratio would occur in a rather narrow range near the high end of the energy charge scale. Curve U of Figure 2 is calculated for an enzyme for which the value of K_m for ATP is six times that for ADP.

The regulatory stability illustrated by Figure 2 depends on (1) precise evolutionary adjustment of the affinities for adenylates at regulatory sites of R-type enzymes and at catalytic sites of U-type enzymes and of the conformational links by which binding at those sites causes conformational changes that modulate affinity for substrate at catalytic sites and (2) the relative affinity at catalytic sites of U-type enzymes for ATP and ADP. As a consequence of such interactions, the ATP/ADP ratio, or the energy charge, affects reaction rates more strongly than do the absolute concentrations of the adenylate nucleotides. A mutant strain of *Escherichia coli* unable to synthesize adenylate nucleotides grew on adenine-limiting media at essentially normal rates when the intracellular concentrations of ATP, ADP, and AMP were half the normal levels. The energy charge and the ATP/ADP ratio retained their normal values. That mutant, like other organisms that have been studied, did not grow if the energy charge fell slightly below its usual value of about 0.9 (4).

The controls illustrated in Figure 2 should not be confused with thermodynamic or mass-action effects; they are strictly kinetic. The value of the Gibbs free energy change is high and negative for U-type reactions under all physiological conditions, as is also true for R-type reactions. Thus, degradation of fuels and synthesis of products are both thermodynamically favorable at all times, and evolved kinetic control mechanisms determine which conversions actually occur. The adenylate energy transduction system, which links metabolic sequences and nearly all other cell activities energetically, is ideally placed for the additional role of mediating the regulatory interactions that underlie the integrated activities of cells and organisms.

BIBLIOGRAPHY

1. D. E. Atkinson (1972) In *Horizons of Biochemistry* (A. San Pietro and H. Gest, eds.), Academic Press, New York, pp. 83–96.
2. A. G. Chapman and D. E. Atkinson (1977) In *Advances in Microbiological Physiology*, vol. 15 (A. H. Rose and D. W. Tempest, eds.), Academic Press, New York, pp. 253–308.
3. D. E. Atkinson (1970) In *The Enzymes*, 3rd ed., vol. 1 (P. D. Boyer, ed.), Academic Press, New York, pp. 461–489.
4. J. S. Swedes, R. J. Sedo, and D. E. Atkinson (1975) *J. Biol. Chem.* **250**, 6930–6938.

Suggestions for Further Reading

D. E. Atkinson (1977) *Cellular Energy Metabolism and Its Regulation*, Academic Press, New York. A general discussion of metabolic interactions and regulation. Chapters 4 (pp. 85–107) and 7 (pp. 201–224) are most relevant to the subject of this article.

R. H. Garrett and C. M. Grisham (1995) *Biochemistry*, Saunders/Harcourt Brace, New York, Chapters 17 and 25. An excellent general biochemistry textbook with strong coverage of metabolism.

ADENYLATE CYCLASES

ANTOINE DANCHIN

The ubiquity of **cyclic AMP** (cAMP) in regulating **enzymatic** activity and/or **genetic expression** in all kingdoms of life, except for the **archaea** (but see later), accounts for the interest displayed in its mode of synthesis and the vast amount of literature devoted to the enzymes that produce it, the adenylate cyclases. These enzymes, which catalyze synthesis of cAMP from ATP and yield pyrophosphate as a by-product, can be classified into four different classes according to their common features: (1) cyclases related to enterobacterial adenylate cyclases; (2) toxic adenylate cyclases isolated from bacterial pathogens; (3) a large and probably ancient class that comprises cyclases from both eukaryotes and prokaryotes and is strongly related to **guanylate cyclases**; and (4) one example, presently known only from the eubacteria *Aeromonas hydrophila* and *yersinia pestis*, that differs entirely from all other classes.

Class I: The Enterobacterial Type

The first complete adenylate cyclase gene, *cya*, was cloned and sequenced from *Escherichia coli*. Work on other enterobacteria, such as *Erwinia chrysanthemi*, *Proteus mirabilis*, *Salmonella typhimurium*, *Yersinia intermedia*, and *Yersinia pestis* demonstrates that both the environment of the genes and the proteins specified are similar in size and overall organization to those of *E. coli* at the corresponding locus (1). Analysis of the *cya* gene from other bacterial species, related to enterobacteria but distinct from them, using genetic **complementation** of appropriate *cya* defective strains of *E. coli* (and more recently by direct sequencing of whole **genomes**) reveals that the genes from many other bacteria, in particular *Aeromonas caviae*, *Aeromonas hydrophila*, *Haemophilus influenzae*, *Pasteurella multocida*, and *Vibrio cholerae*, directly synthesize a protein structurally and phylogenetically related to the *E. coli* cyclase (1,2) (see also the database at http://tigr.org).

No long stretch of **hydrophobic** amino acid residues is present to explain the membrane-bound localization of the adenylate cyclases. In all cases, the proteins are very rich in **cysteine** residues, an uncommon feature for proteins located in the cytoplasm or at the cytoplasmic border of the membrane. This might account for the extreme difficulty in purifying the enzymes. In addition, they are also rich in **histidine** residues, which could indicate that metal ions take part in the folding and/or activity of the polypeptide chain, but no experimental data support this hypothesis. Finally, the protein is made of two functionally well-defined **domains**. The catalytic domain is NH_2-terminal, whereas the glucose-sensitive regulatory domain is COOH-terminal. Comparison of the polypeptide sequence of the catalytic domain of the *E. coli* enzyme with sequences in the protein data libraries do not reveal significant identities with other known proteins. The catalytic domain sequence has been experimentally identified to be about 420 residues. Differences in the amino acid sequence of the *E. chrysanthemi* enzyme often result from the presence of complementary charged residues in place of neutral ones. This suggests that there are more **electrostatic interactions** (including **salt bridges**) that stabilize the protein at the lower growth temperature of this bacterium. This observation might be helpful when trying to understand the tertiary structure of the protein, which is still not known.

The carboxy-terminal domain of the protein is involved in regulating of the enzymatic activity, in particular its inhibition by glucose. A component of the phosphorylation cascade that mediates import of glucose in the cell, enzyme IIAGlc, is involved in this regulation, but in a way not yet understood. An aspartate residue (Asp414 in the *E. coli* enzyme) is involved in the process in an unknown way. Tonic inhibition of the catalytic domain by the regulatory domain could be relieved by phosphorylation of this residue, although such phosphorylation has never been demonstrated (3).

Class II: The Calmodulin-Activated Toxic Class

Whooping cough is caused by the **gram-negative** bacterium *Bordetella pertussis*, which secretes many toxic proteins into the medium, including an adenylate cyclase. In 1980 it was discovered that this enzyme is activated by a host protein, **calmodulin**, which does not occur in bacteria (4). Two years later, Leppla (5) demonstrated that another toxic adenylate cyclase, secreted by a gram-positive bacterium, *Bacillus anthracis*, the etiological agent of anthrax, is also activated by host calmodulin. These observations stimulated intense efforts, but several years were required before the *cya* genes from either organisms could be cloned. However, in 1988 a simple idea, predating its generalization under the name "**two-hybrid system**," *in vivo* complementation by a **plasmid** encoding an activator of the function (in this case, calmodulin), permitted the cloning of adenylate cyclase genes coding for the calmodulin-dependent cyclases (6).

Bordetella pertussis adenylate cyclase is synthesized as a large bifunctional polypeptide chain of 1706 amino acid residues. This contrasts with the various low values reported for the molecular weight of the purified protein (from 43 to 70 kDa). The explanation became apparent when it was demonstrated that the N-terminal segment of the protein (400 residues) alone displays calmodulin-activated adenylate cyclase activity, whereas the rest of the molecule is responsible for hemolytic activity and for transporting the toxin. Sequence, molecular genetic, and physiological studies indicate that the adenylate cyclase domain is fused to a polypeptide chain similar to that of the *E. coli* hemolysin toxin. Therefore the name cyclolysin was coined for the toxic adenylate cyclase from *B. pertussis*.

The adenylate cyclase of *B. anthracis* has been named after the symptom it triggers in the infected host, edema factor. It is encoded in a plasmid, together with another toxin, the lethal factor and a carrier protein, the protective antigen, necessary to internalize both the edema factor and the lethal factor into host target cells. The adenylate cyclase (edema factor) protein, 800 amino acid residues long, comprises four regions of different function. The first region is a **signal peptide**, permits secretion of the protein. The second region corresponds to the domain that binds with the protective antigen. The third region encodes the adenylate cyclase function. It is followed by the fourth region of unknown function. These toxic adenylate cyclases have been subjected to a most thorough biochemical analysis, but they have not yet been crystallized.

In spite of several attempts to isolate other members of this class, until 1998, we knew only three examples of such proteins, isolated from extremely distant bacteria, one gram-positive and two gram-negative (the adenylate cyclase from *B. bronchiseptica* is very similar to the *B. pertussis* enzyme) (7). Several examples of similar proteins have now been discovered in *Pseudomonas aeruginosa*, and in *Yersinia* species. Comparison of the catalytic regions of the *B. pertussis* and *B. anthracis* adenylate cyclases identified four conserved regions that are involved in catalysis, calmodulin binding and activation. The first region comprises a sequence, Gly-XXXX-Gly(Ala)-Lys-Ser, similar to the nucleotide-binding motif found in many ATP- or **GTP-binding proteins**. Therefore it was proposed as part of the catalytic site, and *in vitro* mutagenesis substantiated this interpretation. A second region, with the sequence Pro-Leu-Thr-Ala-Asp-Ile-Asp having some similarity in 6-**phosphofructokinase**, is also involved in catalysis, and it was proposed that the aspartate residues present in this region are involved in binding ribose and magnesium-phosphate. Although it is strongly conserved, however, the first proline residue does not seem very important because it could be replaced by a leucine residue without any measurable influence on the activity or calmodulin activation of the wild-type enzyme.

Although many calmodulin-dependent enzymes have been identified, the mechanism of activation by calmodulin is still poorly understood (see **Calmodulin**). In several cases, limited **proteolysis** releases active, calmodulin-independent forms of the enzymes. Accordingly, it was proposed that the calmodulin-binding domain of these enzymes blocks access of substrates to the **active site** and that activation results because an inhibitory domain is removed upon binding calmodulin. The most original feature of the *B. pertussis* protein is that it can be split into two separate domains which recover most of the initial activity when combined. This observation, together the with analysis of mutants in the region conserved between the *B. anthracis* and *B. pertussis* enzymes, indicates that these proteins may form a catalytic center from the cooperation of two halves. The function of calmodulin may be to trigger the appropriate conformational movement necessary to form an active catalytic center (3).

Class III: The "Universal" Class

Adenylate cyclases from multicellular eukaryotes have long remained elusive because purifying the corresponding catalytic subunit is extremely difficult. Following intense work all over the world, however, they have been the first adenylate cyclases to be crystallized and analyzed by **X-ray crystallography** (see Fig. 1). The activity of these enzymes is subject to a complex regulatory pattern, in particular by **GTP-binding proteins**. Class III enzymes were first discovered in yeast, but, for convenience, we start with the eubacterial enzymes.

Eubacteria. Class III adenylate cyclases form a very diverse collection in eubacteria, both in length and in regulation. Gram-positive bacteria such as *Corynebacterium liquefaciens* secrete large amounts of cAMP because they have a very active class III adenylate cyclase, which is activated by pyruvate. *S. coelicolor* synthesizes a much less active adenylate cyclase involved in aeromycelium formation. Gram-negative bacteria, such as *Rhizobium meliloti*, synthesize at least two different adenylate cyclases. Disruption of both genes simultaneously does not alter cAMP production and suggests the presence of further enzymes. The gram-negative sliding myxobacterium *Stigmatella aurantiaca*, that exhibits an elaborated differentiation pattern, harbors at least two adenylate cyclase genes, each of them corresponding to class III enzymes. They have been partially purified and are inhibited by adenosine, as are the mammalian enzymes (see later). They comprise two domains. The catalytic domain is carboxy-terminal and the regulatory domain is a likely ion transporter in one case and the phosphorylated moiety of a two-component regulatory system in the other. Many other bacteria possess class III cyclases, in particular **cyanobacteria** (8,9). These enzymes generally comprise two domains. The catalytic domain is carboxy-terminal. There are no indications that they must oligomerize to be active.

Lower Eukaryotes. *Saccharomyces cerevisiae* was the first organism from which class III adenylate cyclase genes were cloned and sequenced. The enzyme is activated by the *RAS* gene product (10,11). Two forms of the enzyme may exist. A long form contains repetitions of a leucine-rich motif that plays a regulatory role and whose significance was recently substantiated and extended. Then it became clear that this eukaryotic adenylate cyclase is completely different from the enterobacterial class because of sequence differences in the catalytic center and also because the organization of the gene is different. The catalytic domain is located at the COOH-terminus in *S. cerevisiae* cyclase, whereas it is found at the NH_2-terminus in *E. coli*. The yeast enzyme remained the only example of its class until Garbers, Goeddel and co-workers (12) recognized that the genes encoding **guanylate cyclase**s that had been cloned from several metazoans were derived from an ancestor common to the yeast adenylate cyclase. Another member of class III was subsequently discovered by Young *et al.* (13), who cloned the adenylate cyclase gene from *Schizosaccharomyces pombe* by hybridization using the catalytic domain gene sequence from *S. cerevisiae* as a probe. Finally, the first higher eukaryote adenylate cyclase gene isolated in Gilman's laboratory (14) from bovine brain displayed features clearly reminiscent of this class. Since then, many other genes or **cDNA** for adenylate cyclases belonging to this

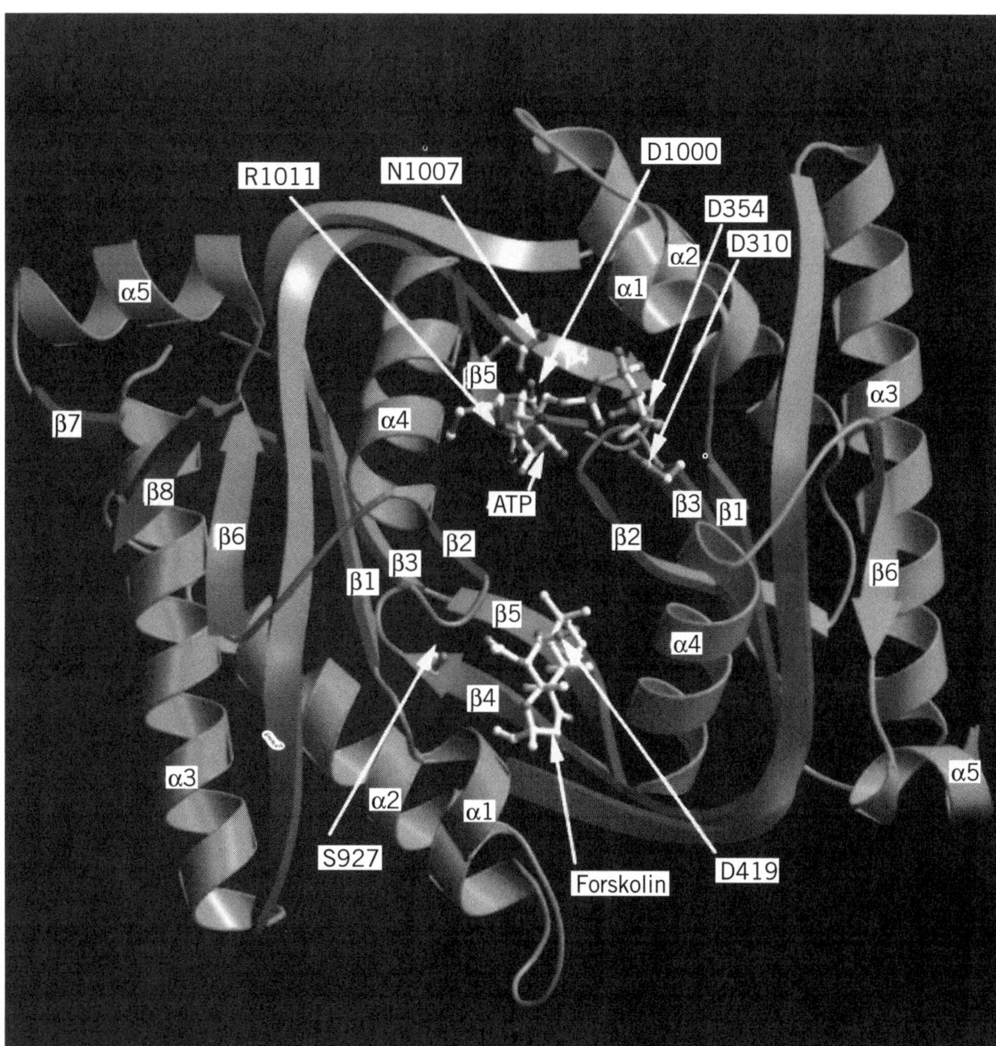

Figure 1. Three-dimensional model of the C1 and C2 domains of the catalytic core of type I adenylate cyclase derived from the X-ray crystallographic structure of the C2 homodimer of the type II enzyme (19). The polypeptide backbone of the C1 domain is green and that of C2 is red. The C-terminal nonhomologous segment of C1 is truncated after the α-helix 5 (lower right). Only a few side chains in the active site are shown, designated by one-letter abbreviations. When they differed from the residues in the crystal structure, they were placed according to Liu *et al.* [*Proc. Natl. Acad. Sci.* (1997) **94**, 13414–13419.] ATP was placed as described by Liu *et al.* and modified according to the interactions reported by Tesmer *et al.* [*Science* (1997) **278**, 1907–1916.] The ligand atoms are indicated by C, white; N, blue; O, red; P, green, and Mg, purple. The figure is kindly provided by James Hurley. See color insert.

class have been isolated and sequenced from lower eukaryotes: *Saccharomyces kluyveri*, *Trypanosoma brucei* and *T. equiperdum*, *Plasmodium falciparum*, *Neurospora crassa*, and *Dictyostelium discoideum* (3).

Higher Eukaryotes: Nine Types. Many class III adenylate cyclases have been identified in higher eukaryotes, in particular in vertebrates, but the most thorough study is in mammals, where several types differing in their regulatory properties have been identified (15). All are regulated in more or less complex ways by **G-proteins** (16). Mammalian adenylate cyclases are informally grouped into nine types according to their tissue location and activity regulation. All but type 9 are activated by the diterpene forskolin, and some are activated by protein kinase C and/or other regulators. Type 1 enzymes were described as calmodulin-activated enzymes from brain. Type 2 proteins are found in brain, lung and other tissues. Type 3 are abundant in olfactory tissue, and the smaller type 4 enzymes are present in testicular tissue. Adenylate cyclase 1, 2, and 8 are positively regulated by calcium/calmodulin, whereas types 5 and 6 are directly inhibited by calcium. Adenylate cyclase 2 and 4 are sensitive to multiple regulatory effects from diverse receptors (3,15). Adenylate cyclase 9 mRNA, found in rat brain, is particularly abundant in the hippocampus, cerebellum, and neocortex (17). However, the classification into types is somewhat arbitrary (for example, type 4 can also be calmodulin-activated). They all have overall similar structures. Two phylogenetically related cytoplasmic domains are required for catalysis. All types are connected by an integral membrane domain and have variable integral membrane domains at the NH_2 terminus of the protein. Among their many functions, their role in synaptic plasticity and memory is particularly interesting (18) and will certainly make adenylate cyclases extremely fashionable again.

Three-Dimensional Structure. The diverse origins of class III adenylate cyclases are reflected in the wide variation in their general organizations and molecular weights. The smallest protein is the *R. meliloti* enzyme, which contains only a catalytic domain, although some data suggest that an upstream sequence may yield a much longer protein that has a complex regulatory pattern. The next shortest protein, also bacterial, is the enzyme from *C. liquifaciens*. The yeasts produce long proteins, as do higher eukaryotes (except in the case of the testicular enzyme). In all cases, the catalytic domain is located at the COOH-terminus. The mammalian enzymes consist of twelve hydrophobic membrane-spanning regions that form two distinct domains, two cytoplasmic regions that contain both variable and conserved regions, and, in particular, two well-conserved domains that are responsible for catalysis.

Comparison of the catalytic domain sequences of the class III proteins shows that four amino acid stretches are strongly conserved: (1) (Met/Leu/Ile/Val)-(Met/Leu/Ile/Val)-Phe-(Ala/Thr)-(Asp/Ser)-(Leu/Ile)-X-(Asn/Asp)-(Phe/Ser); (2) (Ile/Val)-Lys-Thr-X-Gly-(Ser/Asp)-(Ala/Ser/Thr)-(Tyr/Phe)-Met; (3) (Met/Leu/Ile/Val)-(Arg/Lys)-(Met/Leu/Ile/Val)-Gly-(Met/Leu/Ile/Val)-(His/Asn)-X-Gly-X-(Val/Ala)-(Val/Leu)-(Ala/Ser)-Gly; and (4) (Trp/Tyr/Phe)-Gly-(Asn/Asp/Pro)-Thr-Val-Asn-X-Ala-Ser-Arg-(Met/Leu/Ile/Val) (see Fig. 2). Crystallization of the catalytic core of one protein and determination of its structure by X-ray diffraction (19) showed that these regions are part of the organization of the structure (Fig. 1). The crystal structure contains the forskolin-binding site but unfortunately not the

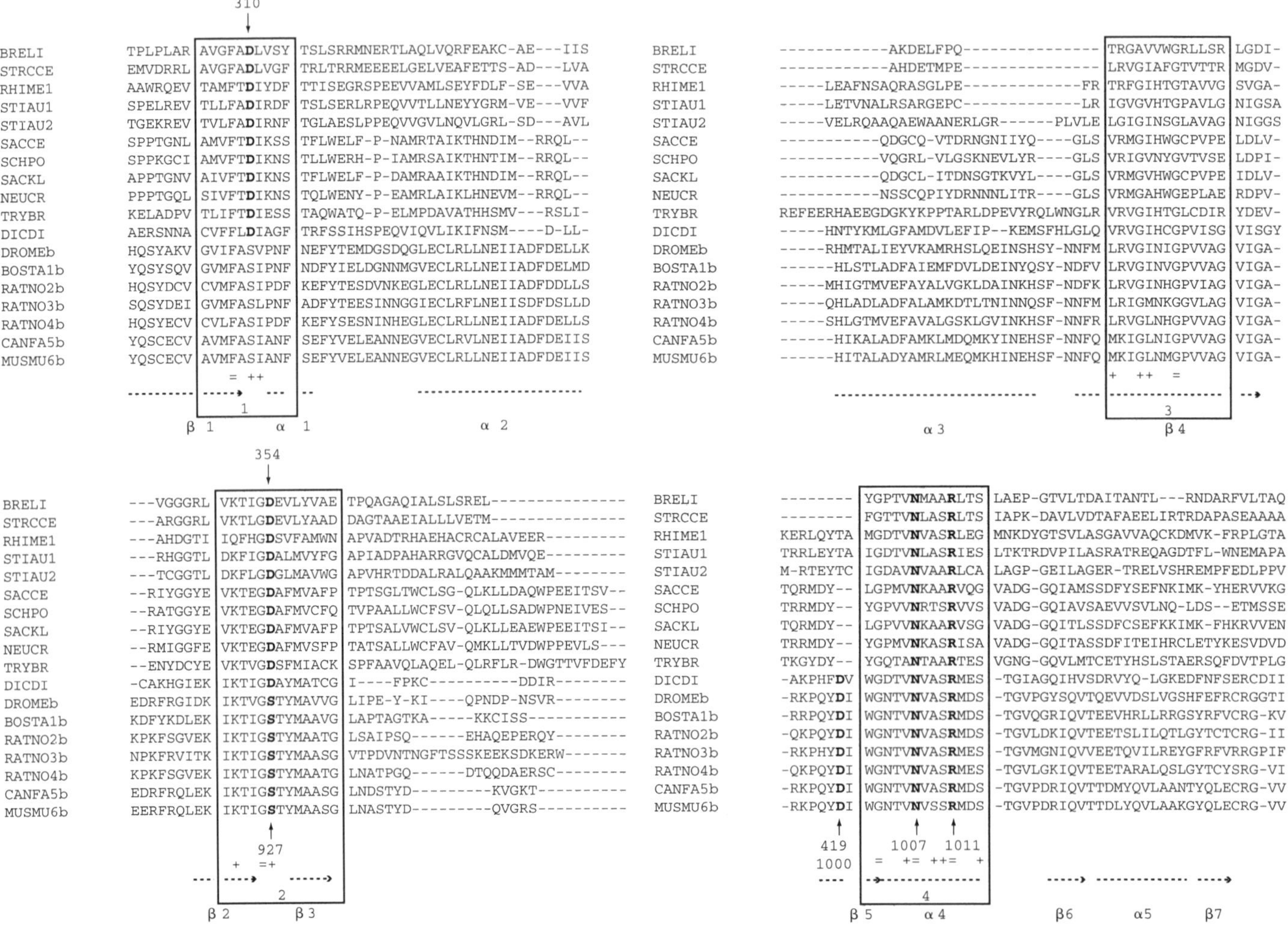

Figure 2. Alignment of the amino acid sequences of Class III adenylate cyclases from various organisms. The four regions of conserved sequence described in the text are enclosed in the four boxes. Secondary structural elements found in the crystal structure of Fig. 1 are displayed underneath the alignments, and the corresponding residues of the active site depicted in Fig. 1 are indicated.

nucleotide binding site. As a step toward understanding the evolution and function of class III cyclases, enzymes displaying significant guanylyl cyclase activity that have evolved from an adenylyl cyclase ancestor were isolated. A single amino acid residue change (Gly-Asp-Thr-Val-Asn to Gly-Asp-Thr-Ile-Asn in the region of the fourth α-helix of the catalytic core) alters the nucleotide specificity of the enzyme (20). This corresponds to a pocket situated in a region of the protein that might accommodate the heterocyclic base (19).

Class IV Adenylate Cyclases

The preceding three classes of structurally unrelated adenylate cyclases already pose a challenging problem. So it was a surprise that *Aeromonas hydrophila* synthesizes another enzyme, a very small cyclase of 193 residues, which has an optimal temperature for activity of 65°C and is at least ten times more active than the class I adenylate cyclase in the same organism (21). No function has yet been discovered for this protein. As yet, it has been found only in various isolates of *A. hydrophila* and in *Y. pestis* (unpublished). There was one report of the presence of cAMP in Archaea, but this was later proven to be the result of an artifact of the growth culture. Therefore it was interesting to see that the sequence of adenylate cyclase from *A. hydrophila* is significantly similar to a gene product of the archaebacterium *Methanococcus jannaschii*. The gene is expressed in *E. coli*, where it is toxic, but it does not restore cAMP synthesis. Therefore, nothing is yet known about the nature of adenylate cyclases, if they exist, in Archaea, but we may expect that, at some point, they might be discovered and that they would belong to this new class.

BIBLIOGRAPHY

1. A. Danchin (1993) *Adv. Sec. Mess. Phosphoprot. Res.*, **27**, 109–161.
2. P. Trotot, O. Sismeiro, C. Vivarès, P. Glaser, A. Bresson-Roy, and A. Danchin (1996) *Biochimie* **78**, 277–287.
3. O. Barzu and A. Danchin (1994) *Prog. Nucleic Acid Res. Mol. Biol.* **49**, 241–283.
4. J. Wolff, G. H. Cook, A. R. Goldhammer, and S. A. Berkowitz (1980) *Proc. Natl Acad. Sci. USA* **77**, 3840–3844.
5. S. H. Leppla (1982) *Proc. Natl Acad. Sci. USA* **79**, 3162–3166.
6. P. Glaser, D. Ladant, O. Sezer, F. Pichot, A. Ullmann, and A. Danchin (1988) *Mol. Microbiol.* **2**, 19–30.
7. F. Betsou, O. Sismeiro, A. Danchin, and N. Guiso (1995) *Gene* **162**, 165–166.
8. M. Kasahara, K. Yashiro, T. Sakamoto, and M. Ohmori (1997) *Plant Cell. Physiol.* **38**, 828–836.
9. M. Katayama and M. Ohmori (1997) *J. Bacteriol.* **179**, 3588–3593.
10. T. Kataoka, D. Broek, and M. Wigler (1985) *Cell* **43**, 493–505.
11. P. Masson, G. Lenzen, J. M. Jacquemin, and A. Danchin (1986) *Curr. Genet.* **10**, 343–352.
12. M. Chinkers, D. L. Garbers, M. S. Chang, D. G. Lowe, H. Chin, D. V. Goeddel, and S. Schulz (1989) *Nature* **338**, 78–83.
13. D. Young, M. Riggs, J. Field, A. Vojtek, D. Broek, and M. Wigler (1989) *Proc. Natl. Acad. Sci. USA* **86**, 7989–7993.
14. J. Krupinski, F. Coussen, H. A. Bakalyar, W.-J. Tang, P. G. Feinstein, K. Orth, C. Slaughter, R. R. Reed, and A. G. Gilman (1989) *Science* **244**, 1558–1564.
15. J. Hanoune, Y. Pouille, E. Tzavara, T. Shen, L. Lipskaya, N. Miyamoto, Y. Suzuki, and N. Defer (1997) *Mol. Cell. Endocrinol.* **128**, 179–194.
16. A. Marjamaki, M. Sato, R. Bouet-Alard, Q. Yang, I. Limon-Boulez, C. Legrand, and S. M. Lanier (1997) *J. Biol. Chem.* **272**, 16466–16473.
17. R. T. Premont, I. Matsuoka, M. G. Mattei, Y. Pouille, N. Defer, and J. Hanoune (1996) *J. Biol. Chem.* **271**, 13900–13907.
18. M. D. Nielsen, G. C. K. Chan, S. W. Poser, and D. R. Storm (1996) *J. Biol. Chem.* **271**, 33308–33316.
19. G. Zhang, Y. Liu, A. E. Ruoho, and J. H. Hurley (1997) *Nature* **386**, 247–253.
20. A. Beuve, E. Krin, and A. Danchin (1993) *Compt. Rend. Acad. Sci. Paris* **316**, 553–559.
21. O. Sismeiro, P. Trotot, F. Biville, C. Vivarès, and A. Danchin (1998) *J. Bacteriol.* **180**, 3339–3344.

ADENYLYLATION

WILLIAM ATKINS

Adenylylation is the process in which adenosine-5′-monophosphate (AMP) is covalently attached to a **protein**, nucleic acid, or small molecule via a phosphodiester or phosphoramidate linkage. Most often, the AMP is derived from ATP, but in some bacterial adenylylation reactions $NADP^+$ is the source. Similarly, deadenylylation is the process in which AMP is removed from the adenylylated molecule. The adenylylation/deadenylylation processes may provide regulatory control of **enzyme** activity, contribute to intermediate steps in individual enzymatic reaction mechanisms, or occur as intermediate steps along the biosynthetic pathway of cofactors. In this sense, adenylylation is analogous to **phosphorylation, sulfation, methylation**, and other intracellular covalent modification reactions for which multiple functions exist. Adenylylation occurs in a wide range of organisms, including bacteria, yeast, and mammals, although it is less common than many other **post-translational modification** reactions as a source of enzyme regulation. The general chemical reaction for ATP-dependent adenylylation is shown in Figure 1.

It is important to distinguish adenylylation from the related processes of phosphorylation, adenylation, and **ADP-ribosylation**:

1. Phosphorylation is readily distinguished from each of the other processes by the lack of incorporation of sugar or adenine in the acceptor.
2. Adenylation results in covalent attachment of ADP via the β-phosphoryl group. Relatively few examples of adenylation are known, and they appear to be limited to adenylation of carbohydrates to yield, for example, glucose-1-ADP.
3. ADP-ribosylation results in a covalent bond between the ribose moiety of NADPH and an acceptor.

These covalent modification reactions also contribute to multiple biological functions, including regulation of enzymatic activity. Analytically, adenylation may be distinguished from adenylylation with the appropriate radiolabeled substrates. Specifically, α-^{32}P-ATP, but not β-^{32}P-ATP or

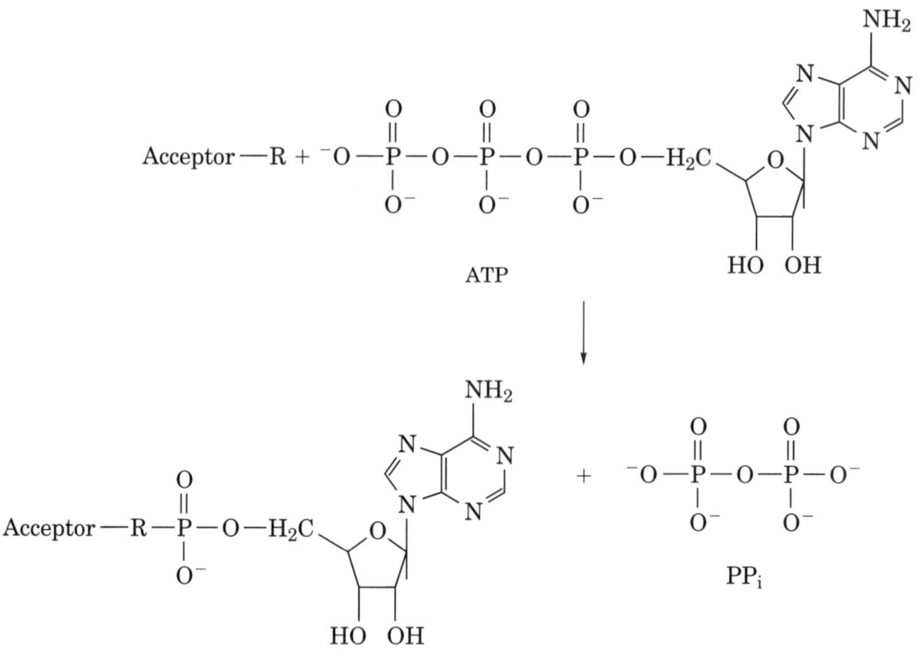

Figure 1. General reaction scheme for ATP-dependent adenylylation. R = O or N. The α, β, and γ phosphorous atoms are labeled.

γ-^{32}P-ATP, will yield radiolabeled acceptor if adenylylation occurs. Typically, ATP that is radiolabeled with ^{3}H or ^{13}C in the adenine moiety is also used, in separate experiments, to confirm that label incorporated into the acceptor is not a result of phosphorylation, without adenylylation. An additional criterion that is often applied to distinguish adenylylation from phosphorylation is its sensitivity to phosphodiesterases. Phosphodiesterases will cleave the AMP from an adenylylated substrate, but they will not hydrolyze phosphate from phosphorylated protein. Specific examples of molecules that are adenylylated are discussed separately below.

GLUTAMINE SYNTHETASE

In all organisms, glutamine synthetase (GS) plays a critical role in intermediary metabolism by catalyzing the ATP-dependent condensation of ammonia with glutamate, to yield glutamine. It was discovered by Stadtman and co-workers (1,2) that *Escherichia coli* glutamine synthetase exhibited dramatically different kinetic properties when isolated from cultures grown in nitrogen-rich vs. nitrogen-starved media. They demonstrated that the differences in enzymatic activity of GS corresponded to the presence or absence of covalently attached adenylyl groups. Since their initial findings, bacterial GS and associated regulatory enzymes have provided a paradigm for understanding biological regulation of nitrogen assimilation in some prokaryotic organisms. Comparison of the molecular details of adenylylation-dependent regulation of prokaryotic GS remains an active research area. Mammalian and plant GS, in contrast, are not regulated by adenylylation.

Regulation of glutamine biosynthesis is characterized most thoroughly for *E. coli* and includes a complex bicyclic cascade that controls the adenylylation of Tyr397 on the surface of GS. Bacterial GS are dodecameric oligomers with two face-to-face hexameric rings. In the subset of bacterial strains that regulate GS activity via adenylylation, the adenylylation state of GS may vary from 0 to 12 AMPs/GS dodecamer, and the

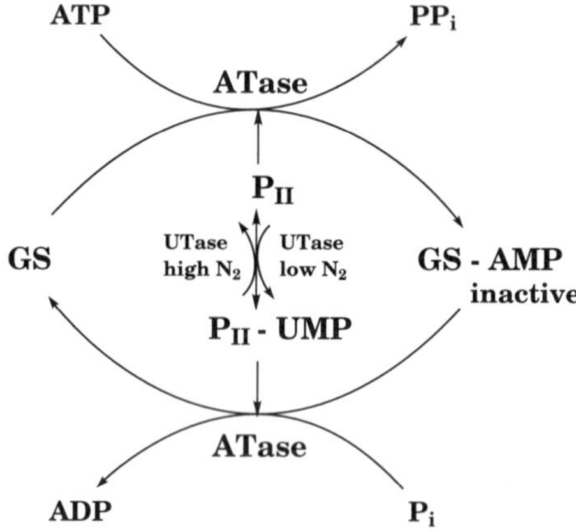

Figure 2. Bicyclic cascade for regulation of *E. coli* glutamine synthetase (GS). ATase catalyzes the adenylylation/deadenylylation of GS; P$_{II}$ stimulates the adenylylation activity, and uridylylated P$_{II}$ (P$_{II}$-UMP) stimulates the deadenylylation activity. Uridylylation of P$_{II}$ is catalyzed by UTase. UTase uridylylates P$_{II}$ when nitrogen levels are low, and it hydrolyzes the uridylyl group from P$_{II}$-UMP when nitrogen levels are high.

enzymatic activity decreases with increasing extent of adenylylation. The regulatory cascade is summarized in Figure 2. It includes the adenylyl transferase (ATase), a **signal transduction** enzyme (P$_{II}$), and a uridylyl transferase (UTase) enzyme that responds directly to nitrogen levels. When the nitrogen levels are low, UTase uridylylates P$_{II}$ at Tyr51 to form P$_{II}$-UMP. P$_{II}$-UMP stimulates the deadenylylation activity of ATase, to decrease the proportion of GS in the adenylylated form and thus to increase the rate at which nitrogen is "fixed" in the amino acid glutamine, derived from glutamate by GS. When nitrogen levels are high, the UTase cleaves the UMP from P$_{II}$-UMP to generate P$_{II}$. Unmodified P$_{II}$ stimulates the adenylylation activity of ATase, to increase the adenylylation state of GS and reduce its efficiency in converting glutamate to glutamine. The complexity of these regulatory cascades underscores the importance of adenylylation as a mechanism for controlling nitrogen metabolism in some prokaryotes.

The structural basis for the decrease in enzymatic activity of *E. coli* GS upon adenylylation is not known. Although **X-ray crystallography** structures are available for the unadenylylated GS from *Salmonella typhimurium* (3,4), no structure for an adenylylated GS has been determined. **Spectroscopic** and **kinetic** analyses indicate that adenylylation causes an increase in the K_m **(Michaelis constant)** for substrates and for the metal cofactors (Mg^{2+} or Mn^{2+}) that are required at each GS **active site**, although no specific interactions between the adenylyl group and protein residues have been identified (5–7). On the basis of many biophysical criteria and **hydrodynamic volume** properties, the adenylylated and unadenylylated forms of GS do not differ greatly in conformation (8,9). Presumably, adenylyl groups attached at Tyr397 of each subunit in dodecameric GS induce subtle changes in the local conformation of the active sites to which they are adjacent. Spectroscopic methods have suggested that the adenine moiety of AMP attached to each subunit within the hexameric ring structure is highly dynamic, and it collides by **diffusion** with the subunit adjacent to it (10,11). Thus, specific **hydrogen bonds** or **electrostatic interactions** between the adenylyl groups and protein residues may be limited to the phosphate and ribose moieties of AMP. Alternatively, AMP may simply block the enzyme active sites sterically.

Based on sequence comparisons of GS from numerous prokaryotes, and on mutational analysis of the *E. coli* and *Anabaena 7120* GS, a minimal **consensus sequence** including the Tyr 397 to be adenylylated and a properly positioned proline residue are required for efficient adenylylation by ATase. These two residues are indicated in bold in the local sequence of *E. coli* GS: Met-Asp-Lys-Asn-Leu-**Tyr**-Asp-Leu-**Pro**-Pro-Glu-Glu-Ser-Lys. Individual mutation of several residues in this sequence has negligible or modest effects on the rate of adenylylation by ATase. In contrast, replacement of the bold proline nearly abolishes the ATase-catalyzed adenylylation, and substitution of serine by proline at the analogous position in the *Anabaena 7120* is sufficient to make this GS a substrate for ATase from *E. coli* (12,13).

The *E. coli* Atase has been characterized genetically and biochemically; it is a constitutively expressed 945 amino acid residue protein (~115 kDa) that contains two highly homologous **domains** (8,14). Genetically engineered N- and 3C-terminal domains that have been expressed separately and purified are both catalytically active. However, the two nonoverlapping constructs exhibit different catalytic activities. The N-terminal domain (residues 1 to 423) catalyzes deadenylylation of GS, requires P_{II}-UMP, and is inhibited by P_{II}. In contrast, the C-terminal domain (residues 425 to 945) catalyzes the adenylylation of GS and is not regulated by P_{II} or P_{II}-UMP. It is striking that P_{II} is required for the adenylylation activity of the full-length ATase, whereas the C-terminal domain by itself is unresponsive to P_{II} or P_{II}-UMP. Presumably, P_{II} and P_{II}-UMP regulate the adenylylation reaction of the C-terminal domain by interacting with the N-terminal domain, or the full-length construct is required to maintain the structural integrity of this regulatory site. In addition, glutamine activates the adenylylation reaction of the C-terminal domain. It also is noteworthy that the adenylylation and deadenylylation reactions catalyzed by ATase are not the reverse of each other. Deadenylylation requires inorganic phosphate (P_i) rather than PP_i, and the reaction product is ADP rather than ATP. In effect, this deadenylylation reaction may be considered formally as transfer of an adenylyl group from GS to P_i, or the adenylylation of phosphate anion. Although the two functional domains of ATase exhibit significant homology at the amino acid level, they have clearly distinct functional properties. The structural basis for these differences remains unknown.

DNA LIGASES, RNA LIGASES, AND RELATED ENZYMES

DNA ligases and **RNA ligases** catalyze the formation of phosphodiester bonds at single strand breaks with adjacent 3′-hydroxyl and 5′-phosphate termini in DNA or RNA, respectively. For bacterial, mammalian, and virus ligases, the first step in the catalytic cycle is the adenylylation of an active-site ϵ-amino group of a **lysine** side chain (Fig. 3). Mammalian and virally encoded ligases utilize ATP to adenylylate the lysine, whereas bacterial ligases exploit $NADP^+$. A consensus sequence found in ligases from each of these sources that includes the adenylylated Lys, in bold, is -**Lys**-X-(Asp/Asn)-Gly-(15,16). The phosporamidate bond formed upon adenylylation of the lysine residue is more stable toward hydrolysis than is the phosphoester bond formed upon adenylylation of tyrosine residues or hydroxyl groups of aminoglycosides. Interestingly, a similar consensus sequence is found in **messenger RNA**-capping enzymes that transfer guanylylate to the 5′-terminus of mRNA, followed by methylation to generate the mature RNA message.

Functionally related RNA 3′-phospho-cyclases have been identified, and they appear to be present in mammals, yeast, enteric bacteria, and archaebacteria. This wide distribution suggests that these enzymes have a critical role, but no physiological function has been determined. As part of the catalytic cycle, the 3′-phosphoryl group on the terminal nucleotide of **transfer RNA** or **small nuclear RNA** (snRNA) undergoes ATP-dependent adenylylation, to generate a phosphodiester bond. This intermediate is attacked subsequently by the 2′-hydroxyl group to release AMP and produce a terminal cyclic phosphate (17). As with the ligase-catalyzed reactions, this cyclization proceeds via an initial adenylylation reaction.

OTHER MAMMALIAN PROTEINS

On the basis of radiolabeling experiments, it has been suggested that plasma **membrane** proteins from liver or parotid glands of rats are adenylylated (18,19). No function has been ascribed to any of these proteins. The hepatic proteins are glycosylated, and their adenylylation is inhibited by **lectins** (20). An interesting possibility is that the adenylyl group is attached to these proteins via their carbohydrates. The same proteins are phosphorylated, and it has been suggested that some protein kinases may be capable of catalyzing adenylylation, in addition to phosphorylation. Further studies that demonstrate covalent attachment of adenylyl groups unambiguously are required, and the physiological function for adenylylation of these proteins remains unclear.

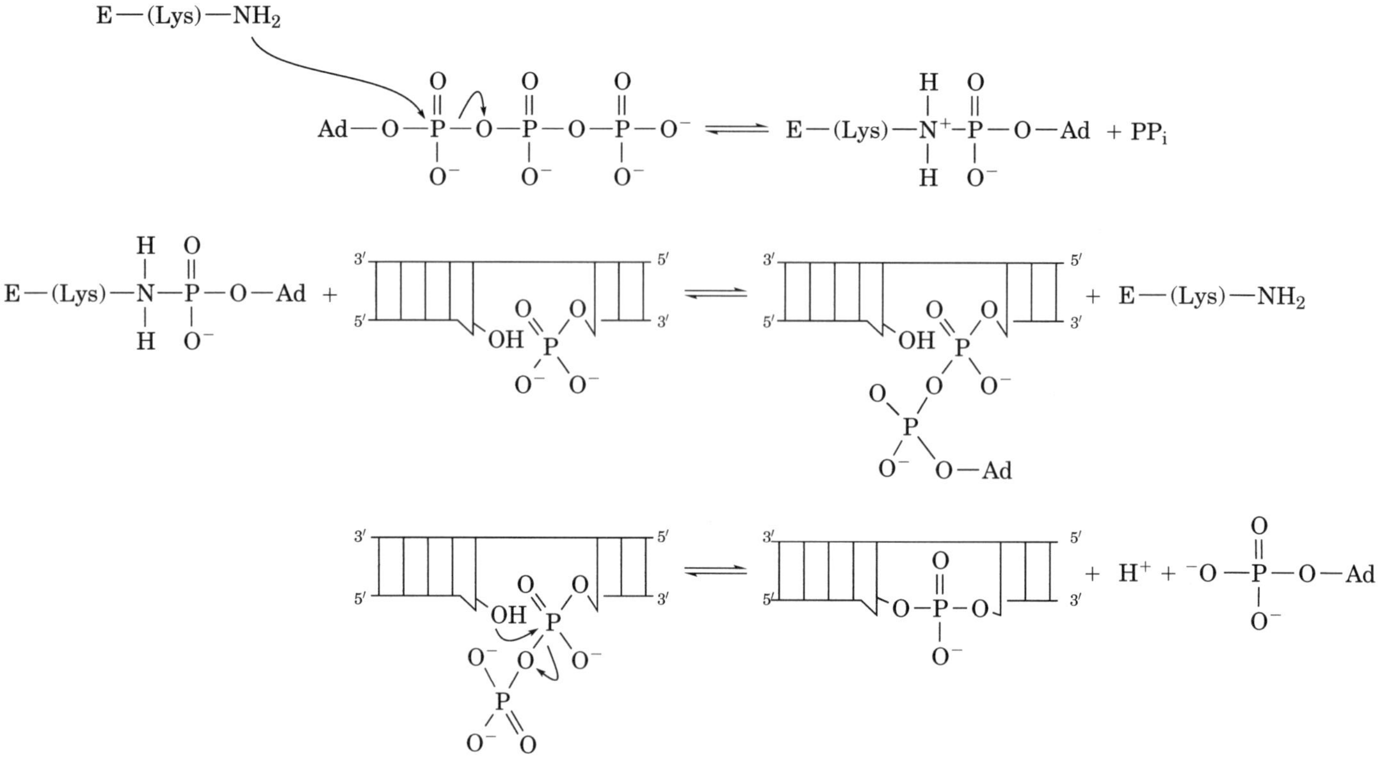

Figure 3. Mechanism for DNA and RNA ligases. The catalytic reaction begins with ATP-dependent adenylylation of the 3′-OH of the nicked DNA or RNA. Ad, adenine; E, a ligase.

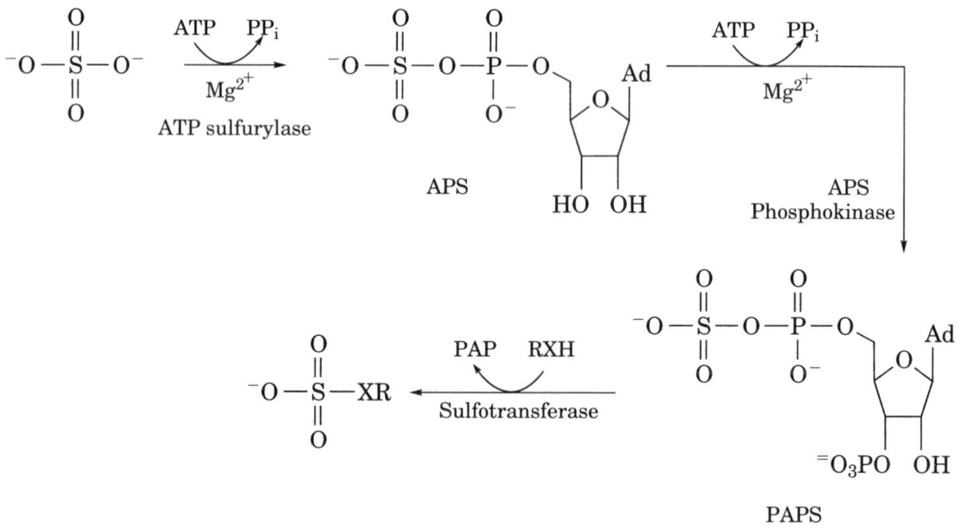

Figure 4. Biosynthesis of 3′-phospho-adenosine-5′-phosphosulfate (PAPS). The first step in the biosynthetic pathway is adenylylation of sulfate to generate adenosine-5′-phosphosulfate (APS).

AMINOGLYCOSIDE ANTIBIOTICS

Several aminoglycoside antibiotics, including tobramycin, gentamycin, and **kanamycin**, are adenylylated by bacterial nucleotidyl transferases, and this contributes to antibiotic resistance in some strains. The enzymes responsible are not specific for ATP, and they also utilize GTP or UTP, thus resulting in guanylylation or uridylylation as well. These nucleotidyl transferases are **plasmid**-encoded. The adenylylation (nucleotidylylation) site on various aminoglycoside antibiotics differs, but it is most frequently either (a) the 2′- or the 3′-hydroxyl of the 3-aminoglucose ring or (b) the 4′-hydroxyl of the 6-aminoglucose ring. Based on the X-ray crystallography structure of kanamycin nucleotidyl transferase (21), the catalytic mechanism has been proposed to be an in-line displacement of PP_i by the hydroxyl group of the aminoglycoside, with an active-site glutamate residue acting as a general base. The presence of only a few contacts between the adenine ring of ATP and the nucleotidyl transferase active site is consistent with the lack of specificity for the nucleotide substrate.

ADENOSINE-5-PHOSPHOSULFATE

An additional example of adenylylation is provided by the biosynthetic pathway of the cofactor 3′-phospho-adenosine-5′-phosphosulfate (PAPS), which is summarized in Figure 4. This cofactor is used by several sulfotransferases in the sulfation of catechols, phenols, and other alcohols. An intermediate formed en route to PAPS is adenosine-5′-phosphosulfate (APS). APS is formed by ATP sulfurylase, in an adenylylation reaction that requires ATP and sulfate ion, SO_4^{2-}, to yield APS and PP_i. The formation of APS is partially rate-limiting in PAPS biosynthesis, and deficiency of ATP-sulfurylase may cause impaired sulfation of proteoglycans required for maintenance of extracellular matrix, cartilage, and connective tissues. Thus, the adenylylation reaction catalyzed by this enzyme may have direct clinical impact, manifested as chondrodysplasias, if it operates inadequately (22).

BIBLIOGRAPHY

1. B. M. Shapiro, H. S. Kingdon, and E. R. Stadtman (1967) *Proc. Natl. Acad. Sci. USA* **58**, 642–649.
2. B. M. Shapiro and E. R. Stadtman (1968) *J. Biol. Chem.* **243**, 3769–3771.
3. R. J. Almassy, C. A. Janson, R. Hamlin, N. H. Xuong, and D. Eisenberg (1986) *Nature* **323**, 304–309.
4. M. M. Yamashita, R. J. Almassy, C. A. Janson, D. Lasek, and D. Eisenberg (1989) *J. Biol. Chem.* **264**, 17681–17689.
5. A. Ginsburg, J. Yeh, S. B. Hennig, and M. D. Denton (1970) *Biochemistry* **9**, 633–648.
6. L. M. Abell and J. J. Villafranca (1991) *Biochemistry* **30**, 1413–1418.
7. L. P. Reynaldo, J. J. Villafranca, and W. DeW. Horrocks Jr. (1996) *Prot. Sci.* **5**, 2532–2544.
8. S. B. Hennig and A. Ginsburg (1971) *Arch. Biochem. Biophys.* **144**, 611–627.
9. W. M. Atkins (1994) *Biochemistry* **33**, 14965–14973.
10. J. J. Villafranca, S. G. Rhee, and P. B. Chock (1978) *Proc. Natl. Acad. Sci. USA* **75**, 1255–1259.
11. W. M. Atkins, B. M. Cader, J. Hemmingsen, and J. J. Villafranca (1993) *Protein Sci.* **2**, 800–813.
12. E. D. Longton, B. M. Cader, J. Hemmingsen, and J. J. Villafranca (1992) In *Biosynthesis and Molecular Regulation of Amino Acids in Plants* (B. K. Singh, H. E. Flores, and J. C. Shannon, eds.), American Society of Plant Physiologists, Rockville, MD, pp. 59–68.
13. J. Hemmingsen (1991) Doctoral dissertaion, The Pennsylvania State University, University Park, PA.
14. R. Jaggi, W. C. van Heeswijk, H. V. Westerhoff, D. Ollis, and S. G. Vasudevan (1997) *EMBO J.* **16**, 5562–5571.
15. T. Lindahl and D. E. Barnes (1992) *Annu. Rev. Biochem.* **61**, 251–281.
16. A. E. Tomkinson, N. F. Totty, M. Ginsburg, and T. Lindahl (1991) *Proc. Natl. Acad. Sci. USA* **88**, 400–404.
17. P. Genschik, E. Billy, M. Swianiewicz, and W. Filipowicz (1997) *EMBO J.* **16**, 2955–2967.
18. E. San Jose, A. Benguria, and A. Villalobo (1990) *J. Biol. Chem.* **265**, 20653–20661.
19. M. Hara-Yokoyama, H. Sugiya, and S. Furuyama. (1994) *Int. J. Biochem.* **26**, 1103–1109.
20. E. San Jose, H. J. E. Villalobo, H-J. Gabius, and A. Villalobo (1993) *Biol. Chem. Hoppe-Seyler* **374**, 133–141.
21. L. C. Pedersen, M. M. Benning, and H. M. Holden (1995) *Biochemistry* **34**, 13305–13311.
22. A. Superti-Furga (1994) *Am. J. Hum. Genet.* **55**, 1137–1145.

Suggestions for Further Reading

Reviews of Nitrogen Regulation of GS in Bacteria and GS Evolution

J. R. Brown, F. T. Robb, and W. F. Doolittle (1994) *J. Mol. Evolution* **38**, 566–576.

B. Magasanik (1993) *J. Cell. Biochem.* **51**, 34–40.

Genetic Relationships of Aminoglycoside-Modifying Enzymes, Including Adenylyl Transferases

K. J. Shaw, P. N. Rather, R. S. Hare, and G. H. Miller (1993) *Microbiol. Rev.* **57**, 138–163.

Overview of PAPS Biosynthesis

C. D. Klaasen and J. W. Boles (1997) *FASEB J.* **11**, 404–418.

ADJUVANTS

MICHEL FOUGEREAU

The term *adjuvant* designates substances that enhance the **immune response** without affecting the specificity of recognition. "Adjuvanticity" was first described in the 1920 by Ramon, who made the observation that mineral substances (such as metal salts and aluminum) or crude materials (such as tapioca) considerably augment the immune response to various vaccines. He then invented the name of "substances stimulantes et adjuvantes de l'immunité." When an immune response is monitored by the kinetics of occurrence of circulating antibodies, it can easily be shown that, in the presence of adjuvants, the antibody titer is considerably higher and is maintained for a much longer period of time. Although first discovered over 70 years ago, very little progress has been made in this area, and the mode of action of adjuvants remains somewhat elusive. It is generally accepted that they act essentially in two ways: (*1*) retain the antigen in emulsion or aggregates, depending on the nature of the adjuvant, ensuring a slow but relatively constant release of antigen that may continuously restimulate the immune system; and (*2*) behave as a nonspecific activator of some partners of the immune response, like mobilizing **macrophages** that are acting as **antigen-presenting** cells or by exerting a polyclonal activation of lymphocytes, which is the case for the bacterial lipopolysaccharide (LPS), a potent polyclonal activator of **B cells**.

To date, the best and universally used substance is the so-called Freund's complete adjuvant, which is an emulsion prepared with a suspension of killed mycobacteria in mineral oil. Unfortunately, it cannot be used for human purposes and must be strictly confined to laboratory animals. Attempts have been made to isolate active molecules from mycobacteria (and also from many other microorganisms). This resulted in a long list of molecules, from which the muramyl dipeptide (*N*-acetylmuramyl-L-alanyl-D-isoglutamine, or MDP) was the most extensively studied. Endowed with good adjuvant properties, it is still too toxic for human use; and many attempts have been made, and are still being made, to define nontoxic homologues. For vaccination purposes in humans thus far, the old recipes, among which are aluminum hydroxide, aluminum phosphate, or alum precipitate, are still the most widely used.

An interesting advance was made with ISCOM (which stand for immuno-stimulating complex), proposed by Morein et al. (1) ISCOM is a cage-like matrix made up of cholesterol and Quil A, which is a substance extracted from the bark of a tree, *Quillaja saponaria*, that contains, after partial purification, five main components with triterpenoid structures. The matrix spontaneously organizes in the presence of **antigen**, most often protein isolated from a viral coat envelope. The construction ensures high immunogenicity and is used in a number of animal vaccines.

It is likely that one key event played by adjuvants is at the steps of antigen uptake and processing by antigen-presenting cells. Supporting this idea is the fact that bacterial, or more generally particulate antigens, are far better immunogens than the purified molecules isolated from cells. This constitutes an obvious difficulty for designing sophisticated "pure" vaccines, which would at first sight appear advantageous over using complete bacteria that usually contain toxic components, but that also serve as carrier with an adjuvant effect for the desired antigen. To date, an ideal vaccine would have to reconcile the purity of a **recombinant protein** with a well-characterized potent adjuvant. We know much about making recombinant proteins, but have at present no convincing pure adjuvant. One interesting approach is to use a living vector, such as **vaccinia virus**, that has been genetically modified to express a given antigen at the surface of the microorganism. One may also couple a gene expressing the antigen with a gene encoding a **cytokine** that would specifically favor the amplification of a helper **T cell** compartment. Another possibility is to embed the antigen in liposomes that could be eventually targeted to well-defined cells of the immune system.

All these approaches have been attempted with variable success. To date, one must unfortunately admit that, despite the fantastic progress made in recent years in the understanding of the basic mechanisms that operate in the immune system, no really significant advances have been made in defining more efficient and more rational vaccines.

See also the entries **Immune response**, **Immunization**, and **Immunogen**.

BIBLIOGRAPHY

1. B. Morein, B. Sundquist, S. Höglund, K. Dalsgaard, and A. Osterhaus (1984) Iscom, a novel structure for antigen presentation of membrane proteins from envelopped viruses. *Nature*, **308**, 457–460.

Suggestion for Further Reading

F. R. Vogel (1995) Immunologic adjuvants for modern vaccine formulations. *Ann. N.Y. Acad. Sci.*, **754**, 153–160.

ADP-RIBOSYLATION, MONO

MAKOTO SHIMOYAMA

MonoADP-ribosylation is a **post-translational modification**, catalyzed by ADP-ribosyltransferase (ADPRT), that transfers the ADP-ribose moiety from NAD^+ to a specific **amino acid** residue of target **protein** and releases the nicotinamide moiety. To date, four amino acid-specific ADPRT have been reported, specific for **arginine**, **cysteine**, diphthamide (a modified form of **histidine**), and **asparagine** residues. These transferases were originally discovered as the **cholera**, **pertussis**, and **diphtheria toxins** and the *Clostridium botulinum* C3 enzyme (1,2), in the order of amino acid-specificity described above. The native form of such ADP-ribosylating toxins is a heteromultimer, and one of the subunits has ADP-ribosyltransferase activity (1). The catalytic subunit penetrates the host cell with the assistance of the other subunits. The prokaryotic ADPRT modify target proteins, including a variety of **GTP-binding proteins**, in eukaryotic cells (Fig. 1). This modification leads to alterations in the target protein, and consequently in cell functions (1). The bacterial toxins and C3 enzyme have proven extremely useful in studies on **signal transduction** pathways in various cell types of eukaryotes. MonoADP-ribosylation must be distinguished from **poly-ADP ribosylation**.

The reversible Arg-specific ADP-ribosylation (Fig. 2), like protein **phosphorylation**, was initially documented as a regulatory mechanism for control of **nitrogen fixation** in the **photosynthetic** bacteria *Rhodospirillum rubrum* and *Azospirillum braziliense* (3). Endogenous dinitrogenase reductase-ADPRT modifies Arg101 on the target protein, dinitrogenase reductase, resulting in inactivation of the reductase. Recovery from the inactive form is achieved through cleavage of the Arg-ADP-ribose linkage by dinitrogenase reductase-activating glycohydrolase. This is the only example of metabolic regulation through endogenous ADP-ribosylation. Purification, characterization, and molecular **cloning** of eukaryotic Arg-specific ADP-ribosyltransferase (4–7) and of ADP-ribosyl-Arg hydrolase (8) were reported. By analogy to prokaryotes, metabolic regulation through reversible ADP-ribosylation in eukaryotes has been postulated.

Among the eukaryotic ADPRTs detected, the best-defined one is Arg-specific. Purification and characterization of the ADPRT from turkey erythrocytes (4), rabbit skeletal muscle (9), and chicken peripheral heterophils (polymorphonuclear leukocytes) (5) revealed that these enzymes exhibit different physical and regulatory properties, kinetics, and intracellular localization. Guanidino compounds, such as arginine and agmatine, function *in vitro* as acceptors for Arg-specific ADPRT (4). With β-NAD^+ and arginine as substrate, α-ADP-ribosylated arginine is formed by Arg-specific ADP-ribosyltransferase; and the α anomer, but not the β anomer, is utilized as substrate by ADP-ribosyl-Arg hydrolase (10).

Some of the ADPRTs catalyze NAD glycohydrolysis or auto-Arg-specific ADP-ribosylation, when either water or the enzyme itself serve as acceptor for ADP-ribose (1,2). Upon incubation of intact cells or cell lysates with $[^{32}P]NAD^+$, it would appear that the ADP-ribose released from NAD^+ by cellular NAD glycohydrolase is attached to some proteins nonenzymatically (11,12). Thus, enzymatic and non-enzymatic ADP-ribosylations should be differentiated.

Molecular cloning and expression studies of **complementary DNA** for Arg-specific ADPRT revealed that the rabbit and human skeletal muscle transferases are glycosylphosphatidylinositol-anchored (**GPI-anchored**) (6)

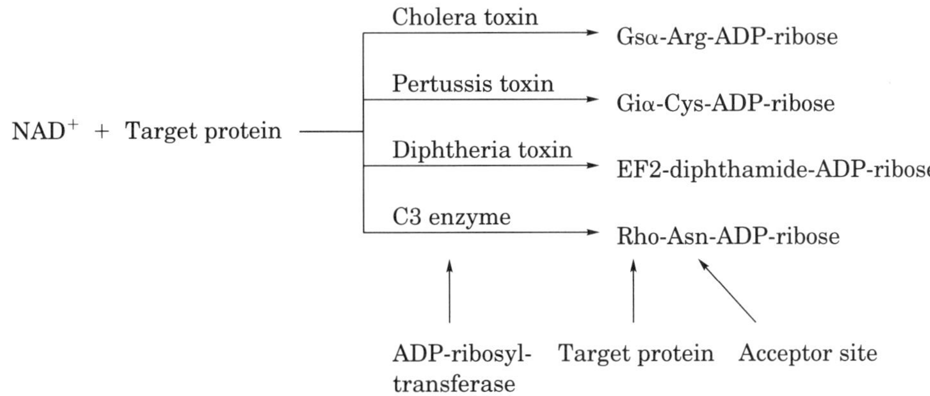

Figure 1. Target protein and acceptor site for ADP-ribosylation in eukaryotic cells by prokaryotic toxins and C3 enzyme.

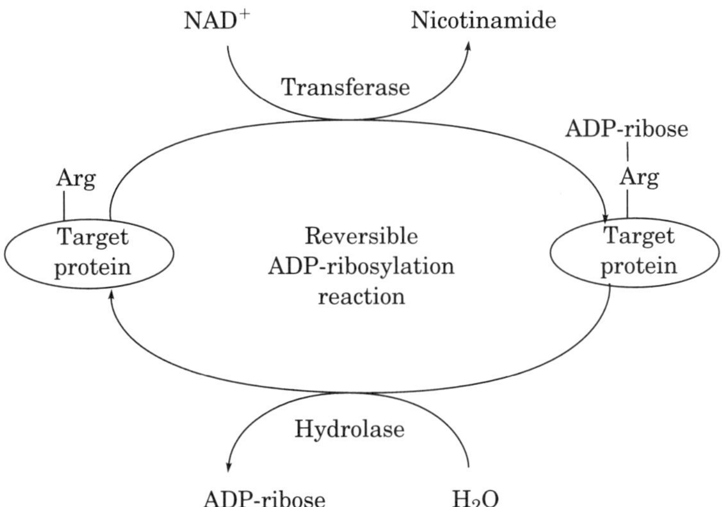

Figure 2. Arg-specific ADP-ribosylation and de-ADP-ribosylation reactions.

RT6.1	202:	SFYPDQEEVLIPG (* * over Q and E)	(20)
RT6.2	202:	SFRPDQEEVLIPG	(13)
Mouse homologue of the rat RT6, Rt6.1	202:	SYYTHEEEVLIPG	(21)
Rt6.2	202:	SSFPREEEVLIPG	(21)
Chicken bone marrow ADPRT1	217:	SFFPSEDEVLIPP	(7)
ADPRT2	217:	SFYPSEDEVLIPP	(7)
Chicken erythroblast ADPRT	217:	STFPGEDEVLIPP	(22)
Rabbit skeletal muscle ADPRT	233:	SFFPGEEEVLIPP	(6)
Human skeletal muscle ADPRT	233:	SFFPGEEEVLIPP	(23)
Cholera toxin	105:	SPHPDEQEVSALG	(24)
E. coli heat-labile enterotoxin	105:	SPHPYEQEVSALG	(25)
C. perifringens iota toxin	414:	PGYAGEYEVLLNH	(26)
Bacteriophage T2	583:	LGIATEAEVILPR	(27)

Figure 3. Comparison of the amino acid sequences of a Glu-rich motif in eukaryotic, prokaryotic, and viral Arg-specific ADP-ribosyltransferases and T-cell antigen RT6. References are cited by the numbers in parentheses. ADPRT: arginine-specific ADP-ribosyltransferase.

and that two forms of transferases from chicken bone marrow cells are secreted (7).

The previously known rat and mouse **T-cell** marker RT6 have significant sequence **homology** to the ADPRT. It is now accepted that these proteins are GPI-anchored and possess transferase and/or NAD glycohydrolase activities (13–15).

A common feature of the vertebrate Arg-specific ADP-ribosyltransferase is their resemblance to the transferase from cholera toxin and to *Escherichia coli* heat-labile enterotoxin. Strictly conserved regions around the **active site** containing two Glu residues (E207 and E209, indicated with an asterisk in Fig. 3) are seen in all Arg-specific ADPRT detected in vertebrates, bacteria, and viruses. On the other hand, RT6.1 and RT6.2 (rat), which have NAD glycohydrolase activity, but not Arg-specific ADPRT, contain Gln in place of Glu207. The **recombinant protein** of a mutant Gln207Glu-RT6.1 possesses Arg-specific ADPRT activity (16,17), but alterations in the kinetic parameters of the NAD glycohydrolase reaction are slight (17). Furthermore, the mouse homologues of rat RT6, Rt6.1, and Rt6.2 have Glu207 and ADPRT activity.

When the Glu is replaced with Gln, Rt6.1 loses the activity (17).

It has been reported that defects in RT6 expression are associated in various animal models with the pathogenesis of autoimmune insulin-dependent diabetes and systemic lupus erythematosus (18,19), although a possible mechanism for RT6-mediated protection from these autoimmune diseases is unknown. Definite physiological functions of monoADP-ribosylation in eukaryotes must be established by further investigation.

BIBLIOGRAPHY

1. S. F. Carroll and R. J. Collier (1994) *Methods Enzymol.* **235**, 631–639.
2. F. R. Althaus and C. Richiter (1987) *ADP-Ribosylation of Proteins*, Springer-Verlag, Berlin, pp. 131–194.
3. P. W. Ludden (1994) *Mol. Cell. Biochem.* **138**, 123–129.
4. J. Moss, S. J. Stanley, and P. A. Watkins (1980) *J. Biol. Chem.* **255**, 5838–5840.
5. K. Mishima, M. Terashima, S. Obara, K. Yamada, K. Imai, and M. Shimoyama (1991) *J. Biochem.* **110**, 388–394.
6. A. Zolkiewska, M. S. Nightingale, and J. Moss (1992) *Proc. Natl. Acad. Sci. USA* **89**, 11352–11356.
7. M. Tsuchiya, N. Hara, K. Yamada, H. Osago, and M. Shimoyama (1994) *J. Biol. Chem.* **269**, 27451–27457.
8. J. Moss, S. J. Stanley, M. S. Nightingale, J. J. Murtagh Jr., L. Monaco, K. Mishima, H.-C. Chen, K. C. Williamson, and S.-C. Tsai (1992) *J. Biol. Chem.* **267**, 10481–10488.
9. J. E. Peterson, S.-A. L. Jacquiline, and D. J. Graves (1990) *J. Biol. Chem.* **265**, 17602–17609.
10. J. Moss, N. J. Oppenheimer, R. E. West Jr., and S. J. Stanley (1986) *Biochemistry* **25**, 5408–5414.
11. H. Hilz, R. Bredehost, P. Adamietz, and Wielckens (1982) *ADP-Ribosylation Reactions. Biology and Medicine,* Academic Press, San Diego, CA, 207–219.
12. D. Cervantes-Laurean, D. E. Mintaer, E. L. Jacobson, and M. K. Jakobson (1993) *Biochemistry* **32**, 1528–1534.
13. F. Koch, F. Haag, A. Kashan, and H.-G. Thiele (1990) *Proc. Natl. Acad. Sci. USA* **87**, 964–967.
14. T. Takada, K. Iida, and J. Moss (1994) *J. Biol. Chem.* **269**, 9420–9423.
15. F. Koch-Nolte, D. Petersen, S. Balasubramanian, F. Haag, D. Kahlke, T. Willer, R. Kastelein, F. Bazan, and H. G. Thiele (1996) *J. Biol. Chem.* **271**, 7686–7693.
16. T. Maehama, S. Hoshino, and T. Katada (1996) *FEBS Lett.* **388**, 189–191.
17. N. Hara, M. Tsuchiya, and M. Shimoyama (1996) *J. Biol. Chem.* **271**, 29552–29555.
18. D. L. Greiner, E. S. Handler, K. Nakano, J. P. Mordes, and A. A. Rossini (1986) *J. Immunol.* **136**, 148–151.
19. F. Koch-Nolte, J. Klein, C. Hollmann, M. Kuhl, F. Haag, H. R. Gaskins, E. Leiter, and H.-G. Thiele (1995) *Int. Immunol.* **7**, 883–890.
20. F. Haag and H.-G. Thiele (1990) *Nucl. Acids Res.* **18**, 1047.
21. C. Hollmann, F. Haag, M. Schlott, A. Damaske, H. Bertuleit, M. Matthes, M. Kuhl, H. G. Thiele, and F. Koch-Nolte (1996) *Mol. Immunol.* **33**, 807–817.
22. T. Davis and S. Shall (1995) *Gene* **164**, 371–372.
23. I. J. Okazaki, A. Zolkiewska, M. S. Nightingale, and J. Moss (1994) *Biochemistry* **33**, 12828–12836.
24. J. J. Mekalanos, D. J. Swartz, G. D. Pearson, N. Harford, F. Groyne, and M. de Wilde (1983) *Nature* **306**, 551–557.
25. T. Yamamoto, T. Gojobori, and T. Yokota (1987) *J. Bacteriol.* **169**, 1352–1357.
26. S. Perelle, M. Gibert, P. Boquet, and M. R. Popoff (1995) *Infect. Immun.* **63**, 4967.
27. T. Koch and W. Ruger (1994) *Virology* **203**, 294–298.

Suggestions for Further Reading

F. Haag and F. Koch-Nolte (1997) *ADP-Ribosylation in Animal Tissues; Structure, Function, and Biology of Mono(ADP-ribosyl)transferases and Related Enzymes. Advances in Experimental Medicine and Biology*, Vol. 419, Plenum Press, New York.

J. Moss and P. Zahradka (1994) *ADP-Ribosylation: Metabolic Effects and Regulatory Functions. Molecular and Cellular Biochemistry*, Vol. 138, Kluwer, Dordrecht, The Netherlands.

ADRIAMYCIN

DON PHILLIPS
CARLEEN CULLINANE

Adriamycin is the trade name for an anthraquinone-containing antibiotic (Fig. 1) that also has the international nonproprietary name (INN) of doxorubicin (1) and the NCI internal identification number NSC 123127. It is a member of the anthracycline group of compounds that contain an anthraquinone chromophore and a polycyclic ring system (1). It has good anticancer activity against a wide spectrum of tumors (and still remains the anti-tumor agent with the widest spectrum of anticancer activities) and is one of the most widely used of the 50 or so chemotherapeutic compounds currently in clinical use (2–4). Its full potential as an anticancer agent has not been reached because of a dose-limiting cardiotoxicity (1–4). Despite enormous effort over the past 30 years to develop more effective and less cardiotoxic derivatives, it remains one of the best proven anticancer drugs. It also exhibits good antibacterial activity.

	R_1	R_2	R_3	R_4
Adriamycin	OCH_3	OH	OH	H
Daunomycin	OCH_3	H	OH	H
Epirubicin	OCH_3	OH	H	OH
Idarubicin	H	H	OH	H

Figure 1. Structure of Adriamycin and clinically relevant derivatives.

STRUCTURE AND CHEMISTRY

Adriamycin consists of a planar tetracyclic ring system linked by a glycosidic bond to daunosamine, an amino sugar (Fig. 1). It is a bright red compound that is normally isolated as the hydrochloride salt, with a chemical composition of $C_{27}H_{29}O_{11} \cdot HCl$ and a molecular weight of 579.98. It is soluble in **water**, physiological saline, and methanol (5,6), although it self-associates in neutral aqueous buffers (7). The pK_a of the **amino group** is 8.2 (5,6). It has **absorbance** maxima at 233, 480, and 530 nm; at concentrations less than 25 μM, it has an extinction coefficient at 480 nm of 11,500 $M^{-1}cm^{-1}$ (7). It is **fluorescent**, and it can be quantified in aqueous solutions using excitation and emission wavelengths of 480 nm and 595 nm, respectively. The solid compound is quite stable if kept in the dark at 4°C and free of moisture. It is photosensitive, and solutions must therefore also be kept in the dark at 4°C; even under these conditions, some decomposition will occur in a week or so. It chelates Fe^{3+} with exceedingly high affinity and readily forms the $Fe(Adriamycin)_3$ complex (8).

HISTORY

In 1958 a new species of *Streptomyces* was identified in a soil sample taken from southern Italy. A red antibiotic, **daunomycin**, was isolated from this fungus and exhibited good activity against a range of murine tumors (1). In a search for other, potentially more active derivatives of daunomycin, the soil fungus was subjected to the **mutagen** *N*-nitroso-*N*-methyl urethane, which generated a mutant strain that produced a modified form of the antibiotic. This new antibiotic was named Adriamycin because of the closeness of the original soil sample to the Adriatic Sea. It exhibited both a wider spectrum of activity than daunomycin and an improved anticancer response against animal tumors (1); it was also remarkably successful in treating human tumors. Adriamycin was rapidly introduced into clinical trials in Italy and the United States, and it was approved for clinical use in the United States in 1974 (1).

CLINICAL USE

Adriamycin is particularly useful for the treatment of solid tumors such as breast, lung, ovarian, and thyroid carcinomas, as well as soft-tissue sarcomas (2–4); it is also active against lymphoid and myelogenous leukemia (2–4). The most significant problem is a cumulative, dose-dependent cardiomyopathy that can lead to heart failure in up to 10% of patients who receive the maximum recommended dose of 550 mg/m^2 (2,3). The cardiotoxicity appears to be due to the **redox** activity of the drug (9): Adriamycin can undergo a one-electron reduction that is catalyzed by a range of enzymes, such as microsomal cytochrome P450 reductase, mitochondrial NADH dehydrogenase, cytochrome b5 reductase, and xanthine reductase (10). This leads to production of the semiquinone that, in the presence of molecular oxygen, generates superoxide, then hydrogen peroxide, and ultimately the extremely reactive and highly toxic hydroxyl radical. These hydroxyl radicals react with any nearby molecule, hence leading to oxidative damage of critical targets, such as DNA and membrane lipids. In most tissues, this lethal process is minimized by superoxide dismutase (which converts superoxide to hydrogen peroxide) and **catalase** (which subsequently converts hydrogen peroxide to water), as well as glutathione peroxidase (which also converts hydrogen peroxide to water). Heart tissues are particularly sensitive to Adriamycin because of their greatly compromised capacity for protection against hydroxyl radicals, due to absence of the bulk hydrogen peroxide-detoxifying enzyme catalase, and also because Adriamycin inhibits glutathione peroxidase activity (10).

The pharmacokinetics of Adriamycin have been extensively documented (11–14). The major metabolite of Adriamycin involves metabolism of the C9 side-chain alcohol to adriamycinol, with the conversion being catalyzed by the ubiquitous cytoplasmic NADPH-dependent aldo-keto reductase (15). Subsequently, microsomal glycosidases that are present in most tissues convert adriamycinol into the inactive deoxyadriamycinol aglycone and daunosamine. The aglycone is then demethylated and conjugated to polar groups to yield more hydrophilic metabolites, which are excreted mainly in the bile (15).

Adriamycin is taken up into cells by passive diffusion (16) and localizes in the nucleus (17), and this has been confirmed recently in single squamous carcinoma cells by quantitative **confocal** laser scanning microscopy (18). The nuclear localization is readily attributed to the known high affinity of Adriamycin for DNA (see text below). The nature of the cytoplasmic localization is less clear, but the distribution as a multitude of small patches of fluorescence throughout the cytoplasm (19) is consistent with some degree of mitochondrial localization (with the drug bound to mitochondrial membranes, and to a lesser extent, also to the mitochondrial DNA), because it is known that there are up to a thousand mitochondria per cell (20).

Resistance

When cells in culture are exposed to Adriamycin for an extended period of time, the cells gradually become resistant to the drug. This type of resistance is due primarily to amplification of the multidrug **transporter** gene *mdr1*, leading to overexpression of a 170-kDa glycoprotein now known as P-glycoprotein (21). This protein forms a pore in the cell membrane and actively pumps Adriamycin (and many other drugs) out of the cell, using ATP as an energy source. In patients treated with Adriamycin, much of the resistance that develops over several months (and diminishes the antitumor effect of the drug) appears to derive from overexpression of P-glycoprotein. In order to enhance the usefulness of Adriamycin, there have been some attempts to develop specific inhibitors of this protein. This approach is no longer considered viable because of the multitude of other drug efflux pumps that have been identified in recent times (22).

In addition to the phenomenon of multidrug resistance, other mechanisms of resistance that have been identified for Adriamycin include elevated levels of **glutathione** and decreased levels of topoisomerase II (see **DNA topology**) in cells in culture (although not yet demonstrated in tumors) (21).

INTERACTIONS WITH DNA

Numerous studies have shown that Adriamycin binds to DNA with high affinity, by intercalating between adjacent base pairs of DNA. In this process, the DNA unwinds to accommodate the drug, and this is reflected by either (a) an increased length of small linear fragments of DNA (hence an increase of **viscosity** of the solution) or (b) loss of the number of negative

supercoils for supercoiled DNA, such as a plasmid (resulting in an initial increased viscosity and a decreased **sedimentation coefficient**); these processes have been summarized previously (3,23). The structure of the intercalated species has been fully characterized by **X-ray crystallography** studies of the drug-oligonucleotide complex. The dominant features are that the drug chromophore lies virtually at right angles to the adjoining base pairs and that the amino sugar fits snugly into the minor groove of DNA (24,25).

Binding studies prior to about 1980 misinterpreted curved **Scatchard plots** as indicating more than one class of DNA binding site. It is now recognized that this curvature results from an extreme form of **negative cooperativity** displayed when the drug binds to DNA. It is fully described by the neighbor-exclusion principle, in which binding at one site within a DNA molecules precludes binding at nearby sites, with values for the intrinsic **association constant** and the number of occluded base pairs being approximately $2 \times 10^6\ M^{-1}$ and 3.0 bp, respectively, at near physiological ionic strengths (26,27). Although this interpretation assumes that the drug intercalates randomly between all possible base pair combinations, this is not strictly correct, as there is a slight sequence selectivity for some sites. Theoretical quantum mechanical calculations indicated the requirement for a three-base pair site, with ACG as the preferred site (28,29). *In vitro* transcription **footprinting** studies confirmed the requirement for a triplet site, but showed that the preferred consensus sequence was TCA (with the drug intercalating between the C and A) (30); there were, however, clearly only small energetic differences between this and other triplet sites.

Adriamycin free and bound to DNA are in rapid equilibrium. The rate of binding is **diffusion-controlled**, and the off-rate is also fast; the latter is usually measured by stopped-flow **detergent** sequestration, which has revealed that Adriamycin has an overall half-life on DNA of approximately 0.5 sec at room temperature (31). By analogy with the structurally similar drug daunomycin, however, the kinetic processes are likely to be much more complex, especially after longer periods of dissociation (32).

MECHANISM OF ACTION

The exact molecular events involved in the mechanism of action of Adriamycin have not yet been resolved. Although many possible mechanisms have been proposed, and the problems in identifying the critical factors have been well summarized (33), three potential mechanisms have been identified: (i) impairment of topoisomerase II activity; (ii) bioreductive action of the drug; and (iii) membrane-related effects (33,34). There is a convincing body of evidence to show that the primary target is DNA (33,35): (i) Almost all of the drug in the nucleus (>99.8%) of single, living cells has been shown to be associated with DNA (36); (ii) over 80% of all the drug present in human tumor biopsies is associated with DNA (37); (iii) increasing drug activity (over three orders of magnitude) correlates with increasing DNA binding, damage, or impairment of DNA template activity (38). DNA binding therefore appears to be related in some way to the anticancer properties. At present there are two types of interactions with DNA that appear to relate to the drug action: (i) impairment of topoisomerase II activity and (ii) formation of DNA adducts and interstrand cross-links.

IMPAIRMENT OF TOPOISOMERASE II

When Adriamycin is added to tumor cells in culture, this results in protein-associated double- and single-strand DNA breaks. These breaks have been shown to occur close to nuclear **matrix attachment regions**, where topoisomerase II is localized. This enzyme regulates the topological state of DNA (see **DNA topology**), and the DNA breaks appear to arise from intercalation of the drug at the same site, resulting in structural distortions that interfere with the re-ligation step of topoisomerase II (2,39,40). There are several reasons why this appears to contribute to the mechanism of action of Adriamycin: (i) tumor cells that are resistant to Adriamycin have a reduced activity of the enzyme (39,40); (ii) a correlation has been shown between the induction of double-strand DNA breaks and the cytotoxicity of Adriamycin in P388 leukemia cells (41); (iii) this type of DNA damage occurs with clinical levels of the drug (39,40). There are also, however, factors that question the role of topoisomerase II in the mechanism of action of Adriamycin: (i) Some studies failed to detect DNA strand breaks in cell cultures at Adriamycin concentrations where cytotoxic responses were observed (42); (ii) topoisomerse II-mediated damage is rapidly reversed following removal of the drug (39), whereas DNA double-strand breaks are known to increase long after removal of the drug (43); (iii) some derivatives that induce a high level of double-strand breaks exhibit a low level of cytotoxicity (44); (iv) there is little evidence of the involvement of topoisomerase II in some tumors (10).

DNA INTERSTRAND CROSS-LINKS

There have been many reports of Adriamycin–DNA adducts formed by enzymatic, microsomal, or cellular activation of the drug (33,45,46), but it was not until 1990 that it became clear from *in vitro* transcription footprinting studies that these adducts formed predominantly at 5′-GC-3′ sequences (47). It was shown subsequently that the adduct at this site behaved as an interstrand cross-link (48,49), although with limited stability [half-life of 5 to 40 h depending on the DNA sequence and fragment length (49,50)]. It is now clear that adducts at GC sequences and interstrand cross-links at GC sequences are one and the same lesion, which act as virtual interstrand cross-links (51,52). Most surprisingly, the cross-link has been shown to be mediated by formaldehyde and to yield a characteristically unstable aminal linkage on one end (51–53) (Fig. 2). This unusual adduct functions as an interstrand cross-link because of the additional stability arising from the intercalated chromophore and from additional **hydrogen bonds** that are formed to the second DNA strand (53). The cross-link is unstable to heat and to alkali, and this explains why it has proven to be so difficult to find this lesion in cells in the past (50). Conditions have recently been established to isolate these lesions from tumor cells (53–56), and this has led to a resurgence in cellular studies of this drug. There is now good evidence that these interstrand cross-links are involved in the mechanism of action of Adriamycin: The capacity of different derivatives of Adriamycin correlates well with their

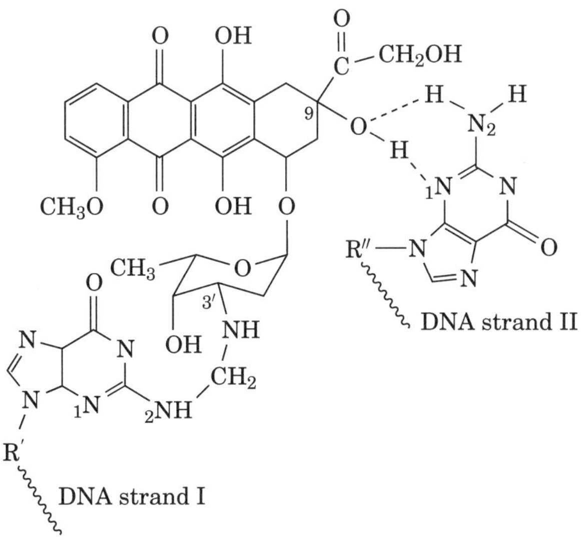

Figure 2. Proposed structure of the Adriamycin–DNA virtual interstrand cross-link (52).

cytotoxicity (55), and gene-specific cross-linking assays have detected sufficient lesions in both the nuclear and mitochondrial genomes for this lesion to be cytotoxic at clinical levels of the drug (56).

Other Possible DNA-Related Effects

Two additional DNA-related effects also occur at low drug levels *in vitro*: (i) impairment of DNA ligase activity (56), and (ii) impairment of helicase activity (58). At this stage, however, their contribution to the mechanism of action of Adriamycin under clinical conditions is unknown.

Current Status

The evidence at present indicates that there is no single mechanism of action of Adriamycin. Both impairment of topoisomerse II activity (resulting in protein-associated double-strand breaks) and formation of interstrand cross-links occur at clinical levels of the drug. Adriamycin is known to activate the general cellular response to **DNA damage** in which the tumor suppressor protein **p53** is induced, resulting in arrest of the **cell cycle** and induction of **apoptosis** (59). Although apoptosis is known to occur at submicromolar concentrations of the drug (60), this response is yet to be fully understood in the context of drug-induced cytotoxicity (61). The development of new derivatives of Adriamycin in the future is likely to rely heavily on an even more detailed understanding of the molecular consequences of the DNA damage induced by this drug.

THE SEARCH FOR NEW DERIVATIVES

Because of the dose-limiting toxicities of Adriamycin, there has been an intense international effort over three decades to find new derivatives with improved anticancer activity and/or reduced cardiotoxicity; this endeavor has been extremely well summarized by Weiss (1). Unfortunately, the outcome from this search has been disappointing. From over 2000 derivatives tested by 1992 (and perhaps as many as 2500 tested to date), none is substantially superior to Adriamycin for proven anticancer activity, although several offer advantages for clinical use: Idarubicin (lacking the methoxy group at the C4 position, Fig. 1) is more amenable to oral administration than Adriamycin; epirubicin (Fig. 1) is less cardiotoxic and results in less nausea and vomiting than Adriamycin at equimolar doses; its similar anticancer activity to Adriamycin, but reduced side effects, has facilitated its wide use as an alternative to Adriamycin. High-dose clinical trials are still in progress with this derivative.

BIBLIOGRAPHY

1. R. B. Weiss (1992) *Semin. Oncol.* **19**, 670–686.
2. B. A. Chabner and C. E. Myers (1993) In *Cancer: Principles and Practice of Oncology* (V. T. DeVita, S. Hellman, and S. A. Rosenberg, eds.), Lippincott, Philadelphia, PA, pp. 376–381.
3. W. B. Pratt, R. W. Ruddon, W. D. Ensminger, and J. Maybaum (1994) *The Anticancer Drugs*, 2nd ed., Oxford University Press, New York, pp. 155–165.
4. T. W. Sweatman and M. Israel (1997) In *Cancer Therapeutics, Experimental and Clinical Agents* (B. A. Teicher, ed.), Humana Press, Totowa, NJ, pp. 113–135.
5. F. Arcamone (1982) *Doxorubicin: Anticancer Antibiotics*, Academic Press, New York, pp. 17–20.
6. R. T. Dorr and D. S. Alberts (1982) In *Current Concepts in the Use of Doxorubicin Chemotherapy* (S. E. Jones, ed.), Farmitalia Carla Erba S.p.A., Milan, Italy, pp. 1–20.
7. J. B. Chaires, N. Dattagupta, and D. M. Crothers (1982) *Biochemistry* **21**, 3927–3932.
8. A. Garnier-Suillerot (1988) In *Anthracycline and Anthracenedione-Based Anticancer Agents* (J. W. Lown, ed.), Elsevier, Amsterdam, pp. 129–161.
9. J. H. Doroshow (1995) In *Anthracycline Antibiotics: New Analogues, Methods of Delivery, and Mechanisms of Action* (W. Priebe, ed.), American Chemical Society, Washington, DC, pp. 259–267.
10. Ref. 3, pp. 160–164.
11. Ref. 2, p. 378.
12. Y. N. Lee, K. K. Chan, P. A. Harris and J. L. Cohen (1980) *Cancer* **45**, 2231–2239.
13. P. A. J. Speth, P. C. M. Linssen, J. B. M. Boezeman, J. M. C. Wessels, and C. Haanen (1986) **377**, 414–422.
14. C. E. Myers, E. G. Mimnaugh, G. C. Yeh, and B. K. Sinha (1988) In *Anthracycline and Anthracenedione-Based Anticancer Agents* (J. W. Lown, ed.), Elsevier, Amsterdam, p. 529.
15. Ref. 3, pp. 158–160.
16. M. Dalmark and H. H. Storm (1981) *Gen. Physiol.* **78**, 349–364.
17. A. DiMarco (1975) *Cancer Chemother. Rep.* **6**, 91–106.
18. K. Kawai, Y. Minamiya, M. Kitamura, I. Matsuzaki, M. Hashimoto, H. Suzuki, and S. Abo (1997) *Cancer* **79**, 214–219.
19. V. Pillay, R. D. Martinus, J. S. Hill, and D. R. Phillips (1998) *J. Cell. Biochem.* **69**, 1–7.
20. J. W. Shay and H. Werbin (1987) *Mutat. Res.* **186**, 149–160.
21. Ref. 3, pp. 50–66.
22. J. S. Lee, S. Scala, Y. Matsumoto, B. Dickstein, R. Robey, Z. Zhan, G. Altenberg, and S. E. Bates (1997) *J. Cell. Biochem.* **68** 513–526.
23. S. Neidle and M. R. Sanderson (1983) In *Molecular Aspects of Anti-cancer Drug Action* (S. Neidle and M. J. Waring, eds.) Macmillan Press, London, pp. 35–56.
24. A. H. Wang, G. Ughetto, G. J. Quigley, and A. Rich (1987) *Biochemistry* **26**, 1152–1163.

25. G. Ughetto (1988) In *Anthracycline and Anthracenedione-Based Anticancer Agents* (J. W. Lown, ed.), Elsevier, Amsterdam, pp. 295–334.
26. F. Barcelo, J. Martorell, F. Gavilanes, and J. M. Gonzalez-Ros (1988) *Biochem. Pharmacol.* **37**, 2133–2138.
27. E. Stutter, H. Schuetz, and H. Berg (1988) In *Anthracycline and Anthracenedione-Based Anticancer Agents* (J. W. Lown, ed.), Elsevier, Amsterdam, pp. 245–293.
28. K. Chen, N. Gresh, and B. Pullman (1986) *Nucleic Acids Res.* **14**, 2251–2267.
29. B. Pullman (1991) *Anti-Cancer Drug Design* **7**, 95–105.
30. H. Trist and D. R. Phillips (1989) *Nucleic Acids Res.* **17**, 3673–3688.
31. B. Gandecha and J. R. Brown (1985) *Biochem. Pharmacol.* **34**, 733–736.
32. D. R. Phillips, P. Greif, and R. C. Boston (1988) *Molec. Pharmacol.* **33**, 225–230.
33. Ref. 14, pp. 528–569.
34. C. E. Myers (1992) *Cancer Chemother. Biol. Resp. Modifiers Ann.* **13**, 45–52.
35. Ref. 14, pp. 546–551.
36. M. Gigli, S. M. Doglia, J. M. Millot, L. Valantini, and M. Manfait (1988) *Biochim. Biophys. Acta* **950**, 13–20.
37. J. Cummings and C. S. McArdle (1986) *Br. J. Cancer* **53**, 835–838.
38. L. Valentini, V. Nicolella, E. Vannini, M. Menozzi, S. Penco, and F. Arcamone (1985) *Il. Farmaco Ed. Sci.* **40**, 377–389.
39. C. Holm, J. Covey, D. Kerrigan, K. W. Kohn, and Y. Pommier (1991) In *DNA Topoisomerases in Cancer* (M. Pomesil and K. Kohn, eds.), Oxford University Press, New York, pp. 161–171.
40. Y. Pommier (1995) In *Anthracycline Antibiotics: New Analogues, Methods of Delivery and Mechanisms of Action*, ACS Symposium Series No 574, pp. 183–203.
41. G. J. Goldenberg, H. Wang, and G. W. Blair (1986) *Cancer Res.* **46**, 2978–2983.
42. F. A. Fornari, W. D. Jarvis. M. S. Orr, J. K. Randolph, S. Grant, and D. A. Gerwitz (1996) *Biochem. Pharmacol.* **51**, 931–940.
43. M. Binaschi, G. Capranico, P. De Isabella, M. Marini, R. Supino, and S. Tinelli (1990) *Int. J. Cancer* **45**, 347–352.
44. M. Binaschi, G. Capranico, L. Dal Bo, and F. Zunino (1997) *Mol. Pharmacol.* **51**, 1053–1059.
45. D. R. Phillips (1990) In *Molecular Basis of Specificity in Nucleic Acid–Drug Interactions* (B. Pullman and J. Jortner, eds.), Kluwer Academic, Dordrecht, The Netherlands, pp. 137–155.
46. J. Cummings, L. Anderson, N. Willmott, and J. F. Smyth (1991) *Eur. J. Cancer* **27**, 532–535.
47. C. Cullinane and D. R. Phillips (1990) *Biochemistry* **29**, 5638–5646.
48. C. Cullinane, A. van Rosmalen, and D. R. Phillips (1994) *Biochemistry* **33**, 4632–4638.
49. S. M. Cutts and D. R. Phillips (1995) *Nucleic Acids Res.* **23**, 2450–2456.
50. A. van Rosmalen, C. Cullinane, S. M. Cutts, and D. R. Phillips (1995) *Nucleic Acids Res.* **23**, 42–50.
51. D. J. Taatjes, G. Guadiano, K. Resing, and T. H. Koch (1996) *J. Med. Chem.* **39**, 4135–4138.
52. D. J. Taatjes, G. Guadiano, K. Resing, and T. H. Koch (1997) *J. Med. Chem.* **40**, 1276–1286.
53. S. M. Zeman, D. R. Phillips, and C. M. Crothers (1998) *Proc. Natl. Acad. Sci. USA* **35**, 11561–11565.
54. A. Skladanowski and J. Konopa (1994) *Biochem. Pharmacol.* **47**, 2269–2278.
55. A. Skladanowski and J. Konopa (1994) *Biochem. Pharmacol.* **47**, 2279–2287.
56. C. Cullinane, S. M. Cutts, C. Panousis, and D. R. Phillips (1998) *Proc. Am. Assoc. Cancer Res.* **39**, 424.
57. G. Ciarrochi, M. Lestingi, M. Fontana, S. Spadari, and A. Montecucco (1992) *Biochem J.* **279**, 141–146.
58. N. R. Bachur, R. Johnson, F. Yu, R. Hickey, N. Appelgren, and L. Malkas (1993) *Mol. Pharmacol.* 44, 1064–1069.
59. G. Zaleskis, E. Berleth, S. Verstovek, M. J. Ehrke, and E. Mihich (1994) *Mol. Pharmacol.* **46**, 901–908.
60. A. Skladanowski and J. Konopa (1993) *Biochem. Pharmacol.* **46**, 375–382.
61. R. B. Lock and L. Stribinskiene (1996) *Cancer Res.* **56**, 4006.

Suggestions for Further Reading

C. E. Myers, E. G. Mimnaugh, G. C. Yeh, and B. K. Sinha (1988) Biochemical Mechanisms of Tumor Cell Kill by the Anthracyclines. In *Anthracycline and Anthracenedione-Based Anticancer Agents* (J. W. Lown, ed.), Elsevier, Amsterdam, pp. 527–569. (Contains an excellent discussion of the criteria for proof of the mechanism of action of Adriamycin and why it is difficult to prove how any anticancer drug kills cells.)

R. B. Weiss (1992) The anthracyclines: Will we ever find a better doxorubicin? *Semin. Oncol.* **19**, 670–686. (A most comprehensive review of the history of Adriamycin, of the search for new derivatives, and of the clinical status of those derivatives.)

W. B. Pratt, R. W. Ruddon, W. D. Ensminger, and J. Maybaum (1994) *The Anticancer Drugs*, 2nd ed., Oxford University Press, New York.

B. A. Chabner and C. E. Myers (1993) In *Cancer: Principles and Practice of Oncology* (V. T. DeVita, S. Hellman, and S. A. Rosenberg, eds.), Lippincott, pp. 376–381. (An excellent concise review of Adriamycin, with an emphasis on cellular and clinical aspects.)

AFFINITY CHROMATOGRAPHY

SHMUEL SHALTIEL

THE BASIC PRINCIPLE

Affinity chromatography (AC) is a general **chromatographic** method for the selective extraction and purification of biological macromolecules on the basis of their biorecognition (1–3). The method makes use of the specific physiological affinity between a desired macromolecule (M) and one of its physiological ligands (L). The ligand, or its analogue (L′), actually acts as a "bait" and is used to extract or "fish out" a desired macromolecule (M_1) (Fig. 1) from a mixture of macromolecules (M_1; M_2; M_3; M_4; M_5; M_6...). The other macromolecules have a very low (if any) affinity for L, presumably because they are designed to refrain from interfering *in vivo* with the physiological recognition of L by M.

GENERAL PROCEDURE FOR AN AC PURIFICATION: KEY STEPS

1. *Immobilization (anchoring) of the ligand on an inert carrier*: L is anchored on a carrier to yield an insoluble material, usually in a beaded form. This carrier should be as inert as possible (eg, beaded **agarose**) to achieve true **active-site**-mediated AC. Also, the attachment point of the ligand should not involve

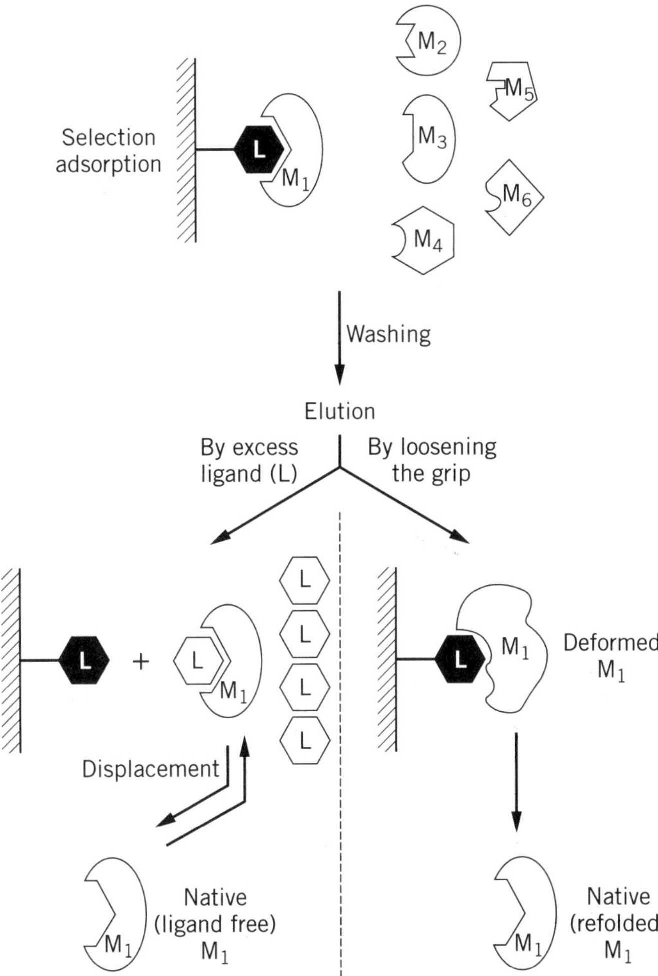

Figure 1. Schematic representation of the general procedure for an AC purification. For further details, see text.

groups that are involved in binding to the macromolecule. Over the years, different carriers and various methods for ligand immobilization were developed. These were reviewed and evaluated by Wilchek et al. (3). In general, the anchoring of L onto an inert carrier involves (i) the introduction of chemically reactive groups to the inert carrier, (ii) the covalent attachment of L to the activated carrier, and (iii) inactivation of the excess of reactive groups (if any) that may remain on the activated carrier after completion of the ligand anchoring step. The immobilized ligand can be used either batchwise or as a column. It may also find other uses—for example, to detect or demonstrate specific **protein–protein interactions** by the binding of a specific protein. Furthermore, it may use the resulting column material to bind and fish out another protein that interacts with it. Such immobilized ligands have been used also for labeling of cells, for the localization of proteins on cell surfaces, for the demonstration of leakage of enzymes or specific proteins from damaged tissues, and so on.

Historically, the pioneering work of Axén et al. (4) on the CNBr activation of beaded agarose had a great influence on the development of AC and the conversion of this methodology into a most widely used tool in separation science. To this day, beaded agarose continues to be the inert carrier of choice, and its activation with CNBr for ligand binding is still an activation method of choice. A thorough analytical study of the mechanism of activation of agarose by CNBr (5) showed that three major products are formed: a carbamate (chemically inert), a linear or a cyclic imidocarbonate (slightly reactive), and a cyanate ester (chemically very reactive). Analysis of freshly activated agarose showed that 60% to 85% of the total coupling capacity of the agarose is due to the formation of the cyanate esters. They are the ones that actually react and immobilize the ligand (Fig. 2). On the basis of this mechanism of activation by CNBr, it became possible to develop more efficient activation procedures, which are reviewed in Refs. 3 and 6.

2. *Selective adsorption.* The key selective step in AC is obviously the extraction of the desired macromolecule M, which is singled out and removed by the immobilized L, from the mixture in which it is present. The macromolecule—be it an **enzyme**, an **antibody**, a receptor, a **hormone**, a **growth factor**, or the like—is selectively bound by the biospecific ligand L, which can be another protein, a peptide, a polynucleotide or a nucleotide, a polysacharide or a carbohydrate, a lipid, a vitamin, or just a metal ion. Functionally, L may be a substrate, a substrate analogue, an inhibitor, an **antigen**, a coenzyme, a cofactor, or a regulatory metabolite. In many cases, the biospecific ligand used for the immobilization is a structural analogue of the physiological ligand (L′). It is imperative, however, to ensure that it still retains the property of selective binding to M, and ideally to M only. In choosing the ligand for an affinity chromatography column, it is often possible to adjust the grip of M onto the anchored L, and thus to optimize both the adsorption and the elution steps. It should be noted that the adsorption conditions used (buffer, pH, ionic strength, temperature) should also be carefully chosen to secure an optimal and selective adsorption.

3. *Washing out nonspecifically bound impurities.* This is usually carried out with an excess of the buffer used for selective adsorption.

4. *Elution of the desired macromolecule.* The detachment of M from the column (elution) is one of the most important steps in purification by AC. Obviously, the ideal elution is by a specific displacement of M with an excess of its biospecific ligand (Fig. 1). This procedure preserves the native structure of M by forming the more stable complex of M with its biospecific ligand L. When such elution is achieved, it strongly suggests that true **active-site**-mediated AC is involved. However, very often biospecific ligands fail to elute the desired protein, and nonspecific means have to be applied. These usually include a change in solvent or buffer composition, a change in pH or in ionic strength, the addition of a **chaotropic** or a "deforming" buffer, a change in temperature, or a change in the electric field (**electrophoretic** desorption) (3). All these bring about a deformation of the protein (7,8), a concomitant loosening of the grip of M for L, and consequently elution. In some cases, the binding of M to the L column is so tight that it is not possible to recover M in a fully active form. If M is an enzyme, this may yield a less active preparation (part of the M molecules may be totally inactive, or all molecules may have a lower affinity for the substrate or a lower **turnover number**). In some instances, the purified enzyme is fully active, but it may lose its ability to be regulated—for example, if the regulatory domain of M loses its affinity for a regulatory metabolite. Under such circumstances, immobilized ligands with lower affinity for M must be tried. Among the remedies that can be used to improve

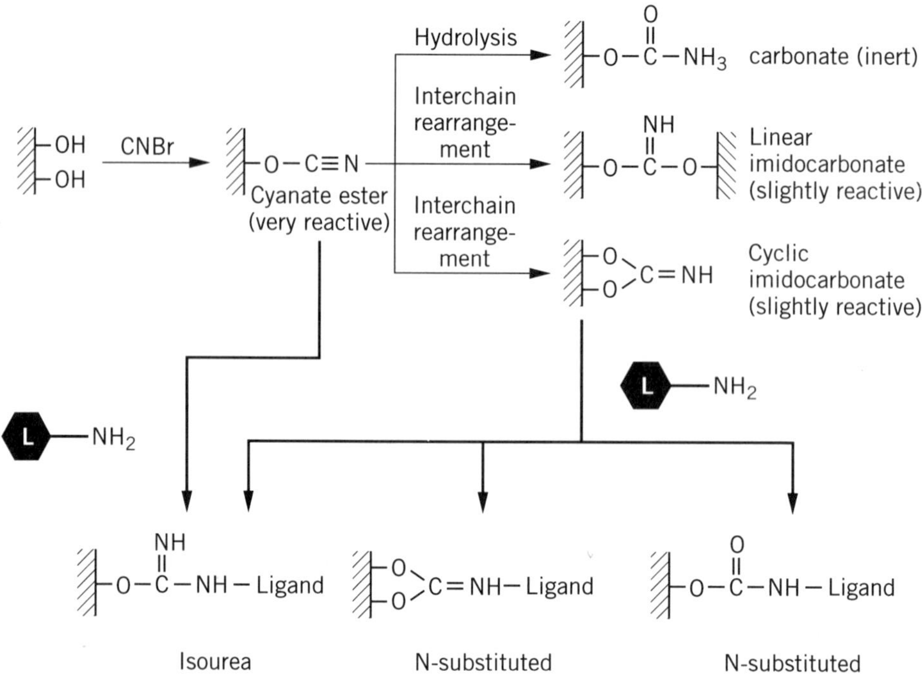

Figure 2. The mechanism of the CNBr activation of agarose. (Modified from Ref. 3.)

the elution step, one should note the possibility of binding the ligand to the matrix by means of an easily cleavable form—for example, through an ester bond (9,10), which can be readily hydrolyzed with a mild base; through a link that includes vicinal hydroxyl groups, which can be readily cleaved with periodate; or through diazo bonds, which can be readily reduced with dithionates (11). It should be remembered, however, that such columns are of limited value, because they can be used only once. Electrophoresis has also been used for elution (12). Because proteins are charged, they will detach from the column and migrate toward the appropriate electrode, if the column with the adsorbed M is exposed to a strong enough electric field. This mild method of elution was successfully applied with high yields in immunoaffinity chromatography and in some AC systems.

INTERPOSING AN "ARM" BETWEEN THE LIGAND AND THE MATRIX BACKBONE

While developing the basic principles of AC, it was observed that the purification of M is often improved by interposing a hydrocarbon chain (an "arm" or a "spacer") between L and the matrix backbone (1). It was presumed that such an arm relieves the steric restrictions imposed by the backbone on the ligand, thereby increasing its flexibility and its availability to the protein (13). Such arms were found to improve significantly the extraction of proteins and the efficacy of the purification by AC. Initially, it was assumed that such hydrocarbon arms do not alter the inert nature of the matrix, a condition that obviously has to be ensured to preserve an active-site mediated adsorption of the extracted protein. This assumption seemed reasonable at the time because it had just been shown that at least some water-soluble proteins are quite well described as "an oil drop with a polar coat" (14), implying that the surface of water-soluble proteins is polar and thus not attracted to lipophilic "baits." We now know that such arms, in and of themselves, may bind proteins. In fact, this observation led to the discovery of **hydrophobic chromatography**.

THE LIMITATIONS OF BIOSPECIFICITY—INTERACTIONS THAT ARE NOT ACTIVE SITE-MEDIATED

Proteins and their physiological ligands are multifunctional molecules whose functions involve a variety of physical interactions: hydrophobic, **electrostatic**, ion-dipole, and so on. Therefore, it is reasonable to assume that a protein might interact with a column coated with a ligand (very often anchored to the beads at a local concentration much higher than its concentration *in vivo*) not only by means of its active site. While it is sometimes possible to minimize these nonspecific interactions, it is not always possible to avoid such interfering effects, because they may be an intrinsic property of the system. For example, if ATP is linked to a matrix through its amino group or its ribose moiety; the column thus obtained may retain an enzyme having a biospecific site for ATP; but at the same time, this very column would be negatively charged due to its triphosphate groups, and it would have hydrophobic loci due to its adenine residues. Other proteins, in addition to the desired one, may therefore "regard" the column material as an **ion exchanger** by virtue of its triphosphate groups, or as a hydrophobic column by virtue of its adenine moieties. The efficiency of resolution will then depend on the magnitude of the affinity produced by charge–charge or hydrophobic interactions, as compared to the affinity between the active site of the desired macromolecule and its immobilized substrate or effector analogue. With columns of macromolecular ligands (eg, enzyme subunits, antibodies, lectins), the probability of encountering such built-in interfering effects is considerably higher, because their immobilization usually involves different anchoring points. This leads to a heterogeneous presentation of the various regions of the ligand macromolecule. In some of these presentations, the biospecific active site is available for interaction, while in other presentations the active site itself is inaccessible or sterically hindered. Hydrophobic patches in such ligands may be available for interaction not only in the biospecifically functional presentation, but also in other presentations. In fact, the tendency of a **lectin** such as concanavalin A to adsorb onto hydrophobic substances, in addition to its binding to sugars of the mannosyl configuration, was observed in several laboratories.

THE RELATIVITY OF BIOLOGICAL RECOGNITION: DIFFERENT PROTEINS MAY SHARE A TASTE FOR A BIORECOGNITION ELEMENTS

The occurrence of common biorecognition sites in different enzymes is obvious when they are functionally similar, acting on the same substrate (eg, ATP), or utilizing the same cofactor (eg, NAD). This actually forms the basis for general ligand-affinity chromatography (15). However, common biorecognition elements may also be found with proteins having no apparent functional similarity. For example, the free catalytic subunit of cAMP-dependent protein kinase (**protein kinase A**) is preferentially retarded on immobilized soybean **trypsin inhibitor** (16). Though initially unexpected, this is actually not surprising; in spite of the fact that **trypsin** and this **kinase** catalyze two different chemical reactions (hydrolysis of **peptide bonds** versus a phosphotransferase reaction), these two enzymes do have similar biorecognition elements (or subsites) at their active site: trypsin cleaves peptide bonds adjacent to positively charged amino acid residues (**arginine** and **lysine**), while cAMP-dependent protein kinase phosphorylates **serine** residues that are vicinal (in the sequence of amino acids) to the same positively charged arginine and lysine residues (17–20). Similarly, it was shown (21) that **TLCK** (α-*N*-tosyl-L-lysine chloromethyl ketone), an **affinity labeling** reagent originally designed for labeling the active site of trypsin, specifically attacks a **thiol group** at the active site of the catalytic subunit of cAMP-dependent protein kinase. It seems, therefore, that the retardation of the free catalytic subunit on the immobilized inhibitor is due (at least in part) to an affinity between the inhibitor and recognition subsites at the active site of the enzyme.

BIBLIOGRAPHY

1. P. Cuatrecasas, M. Wilchek, and C. B. Anfinsen (1968) *Proc. Natl. Acad. Sci. USA* **61**, 636.
2. P. Cuatrecasas and C. B. Anfinsen (1971) *Annu. Rev. Biochem.* **40**, 259.
3. M. Wilchek, T. Miron, and J. Kohn (1984) *Methods Enzymol.* **104**, 3.
4. R. Axén, J. Porath, and S. Ernbäck (1967) *Nature (London)* **214**, 1302.
5. J. Kohn and M. Wilchek (1981) *Anal. Biochem.* **115**, 375.
6. S. B. Mohan and A. Lyddiatt (1997) In *Affinity Separations* (P. Matejtschuk, ed.), IRL Press, Oxford University Press, New York, p. 1.
7. S. Shaltiel, J. L. Hedrick, and E. H. Fischer (1966) *Biochemistry* **5**, 2108.
8. J. L. Hedrick, S. Shaltiel, and E. H. Fischer (1969) *Biochemistry* **8**, 2422.
9. R. J. Brown, N. E. Swaisgood, and H. R. Horton (1979) *Biochemistry* **18**, 4901.
10. P. Singh, S. D. Lewis, and J. A. Shafer (1979) *Arch. Biochem. Biophys.* **193**, 284.
11. P. Singh, S. D. Lewis, and J. A. Shafer (1980) *Arch. Biochem. Biophys.* **203**, 776.
12. M. R. Morgan, P. J. Brown, M. J. Lieiand, and P. D. Ocan (1978) *FEBS Lett.* **87**, 239.
13. P. Cuatrecasas (1970) *J. Biol. Chem.* **245**, 3059.
14. D. C. Phillips (1966) *Sci. Am.* **November**, 78.
15. K. Mosbach (1978) *Adv. Enzymol.* **46**, 205.
16. E. Alhanaty, N. Bashan, S. Moses, and S. Shaltiel (1979) *Eur. J. Biochem.* **101**, 283.
17. H. G. Nimmo and P. Cohen (1977) *Adv. Cyclic Nucl. Res.* **8**, 145.
18. O. Zetterquist et al. (1976) *Biochem. Biophys. Res. Comun.* **70**, 696.
19. B. E. Kemp, E. Benjamin, and E. G. Krebs (1976) *Proc. Natl. Acad. Sci. USA* **73**, 1038.
20. P. Daile, P. R. Carnegie, and J. D. Young (1975) *Nature* **257**, 416.
21. A. Kupfer, V. Gani, J. S. Jimenez, and S. Shaltiel (1979) *Proc. Natl. Acad. Sci. USA* **76**, 3073.

AFFINITY ELECTROPHORESIS

A. CHRAMBACH

By analogy to **affinity chromatography**, it is possible to introduce specific **ligands** for a **macromolecule** into the gels of **gel electrophoresis** and to measure the specific retardation of the macromolecule due to its interaction with such a reagent. The advantage of such affinity methods lies in the augmented resolving power conferred by the specificity of the binding interaction.

The procedures used to introduce affinity reagents into gels have varied. In cross electrophoresis, a ligand with a net charge opposite to the species of interest migrates electrophoretically into the gel in the opposite direction. Alternatively, uncharged ligands can simply be added to the gelation mixture. Macromolecular substrates within a gel may serve as immobilized affinity reagents, either by themselves or as carriers of covalently attached affinity groups. The magnitude of the electrophoretic retardation depends on the concentration of the affinity reagent in the gel; quantitative determination of this relationship makes it possible to estimate the apparent association constant for binding of the ligand to the sample. Further information concerning the interaction can be gained from affinity electrophoresis by variation of the buffer composition (eg, the addition of metal ions to the buffer), the pH, or the temperature.

Suggestions for Further Reading

T. C. Bog-Hansen and J. J. Hau (1981) Glycoproteins and glycopeptides (affinity electrophoresis). In *Electrophoresis: A Survey of Techniques and Applications*, Vol. 18B (Z. Deyl, A. Chrambach, F. M. Everaerts, and Z. Prusik, eds.), Elsevier, Amsterdam, pp. 219–252.

T. C. Bog-Hansen and K. Takeo, eds. (1989) Symposium on affinity electrophoresis. *Electrophoresis* **10**, 811–870.

K. Takeo (1987) Affinity electrophoresis. *Adv. Electrophoresis* **1**, 229–280.

AFFINITY LABELING

R. F. COLMAN

Affinity labeling is a strategy to modify chemically an **amino acid** residue within a specific **ligand-binding** site of an **enzyme**, either at the **active site** or at a regulatory, **allosteric** site. In this approach, a reagent is designed that resembles structurally the natural ligand of the enzyme but features in addition a functional group capable of reacting covalently and indiscriminately with many different amino acid residues. The designed reagent is intended to mimic the natural ligand in forming a reversible enzyme-reagent complex analogous to the enzyme-substrate complex and, once directed to that specific site, to react irreversibly with an amino acid residue accessible from that site. In the case of a purified enzyme, such a reagent allows identification of a particular ligand binding site or **domain**, which can be an experimental evaluation of a predicted binding site location based on recognition of a **protein motif** from its amino acid sequence. Affinity labeling constitutes a valuable starting point for selecting appropriate target sites for subsequent **site-directed mutagenesis** experiments and is an important tool in probing structure–function relationships in enzymes. If the synthesized reagent has a characteristic absorbance or fluorescence spectrum, affinity labeling offers a means of introducing a chromophore at a specific substrate site to report on the enzyme conformation or on distances between designated sites in the enzyme (see **Energy transfer**). For medicinal chemistry, affinity labeling permits mapping of the substrate binding site and establishment of its size, so that inhibitory drugs directed toward that site can be designed more rationally. In the case of a heterogeneous cell preparation or of a complex protein mixture, a specific affinity label can be used to identify a **receptor** protein or to tag a class of macromolecules, such as **nucleotide binding** proteins.

One characteristic of affinity labeling is the initial formation of a reversible enzyme–reagent complex (ER), as indicated below.

$$E + R \underset{k_{-1}}{\overset{k_1}{\rightleftharpoons}} ER \xrightarrow{k_{\max}} ER' \tag{1}$$

where E and R represent the free enzyme and reagent, respectively, and ER' is the covalently modified enzyme. The formation of a reversible ER complex is often indicated by a "rate saturation effect" in which the rate constant for modification (k_{obs}) increases as the reagent concentration is elevated until the enzyme site is saturated with reagent; subsequently, the rate constant ($k_{\max}$) is not changed by further increases in reagent concentration (see **Kinetics**). This kinetic pattern contrasts with the linear dependence on reagent concentration of the rate of a direct bimolecular chemical modification. For an affinity label, k_{obs} can be described by the equation

$$k_{\text{obs}} = \frac{k_{\max}}{1 + \frac{K_R}{[R]}} \tag{2}$$

where the apparent dissociation constant for the enzyme–reagent complex K_R, is given by $(k_{-1} + k_{\max})/k_1$ and where $k_{\max}$ is the maximum rate of modification at saturating concentrations of reagent. This type of kinetic behavior is often presented as a double reciprocal plot of $1/k_{\text{obs}}$ versus $1/[R]$, based on the equation

$$\frac{1}{k_{\text{obs}}} = \frac{1}{k_{\max}} + \left(\frac{K_R}{k_{\max}}\right)\left(\frac{1}{[R]}\right) \tag{3}$$

Examples of **purine** nucleotide-based affinity labels are shown in Figures 1 and 2. Experiences with their use illustrate the possibilities of affinity labeling.

BDB-TAMP

The compound 6-(4-bromo-2,3-dioxobutyl)thioadenosine 5′-monophosphate (6-BDB-TAMP) shown in Figure 1 was synthesized as a reactive nucleotide analogue to target nucleotide binding sites in enzymes (1). The bromoketo group, adjacent to the 6-position of the purine ring, can potentially react with several nucleophiles found in proteins, including those of the side chains of **cysteine**, **histidine**, **tyrosine**, **lysine**, **methionine**, **glutamate**, and **aspartate** residues. In addition, the dioxo group provides the possibility of reaction with arginine residues.

As shown in Figure 1, the structure of 6-BDB-TAMP also exhibits a remarkable resemblance to that of adenylosuccinate, a key metabolic intermediate in the conversion of inosine monophosphate to adenosine monophosphate. The last step in that pathway is catalyzed by adenylosuccinate lyase, an enzyme proposed to initiate the cleavage reaction to AMP by attack of an enzymic general base on the β hydrogen of adenylosuccinate. Elimination of the amino group is then facilitated by protonation of the leaving group by an enzymic general

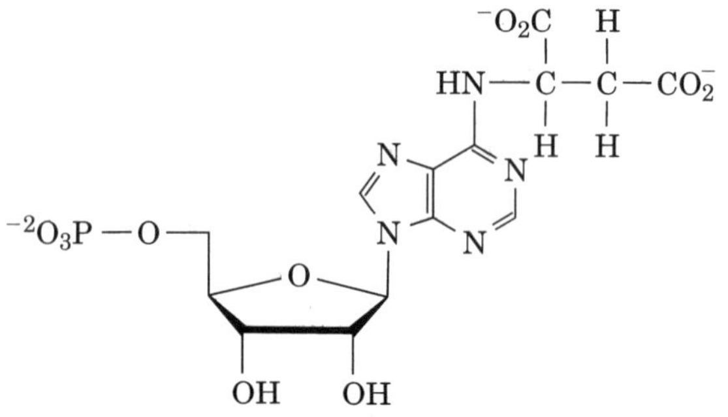

Adenylosuccinate

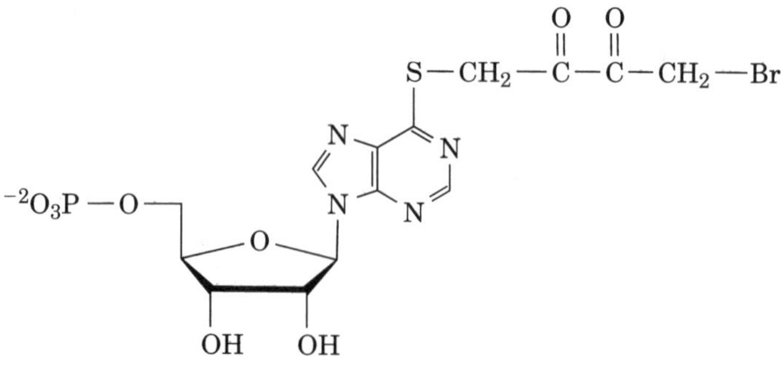

6-BDB-TAMP

Figure 1. Comparison of the structures of the natural ligand adenylosuccinate and of the affinity label 6-(4-bromo-2,3-dioxobutyl)thioadenosine 5′-monophosphate (6-BDB-TAMP).

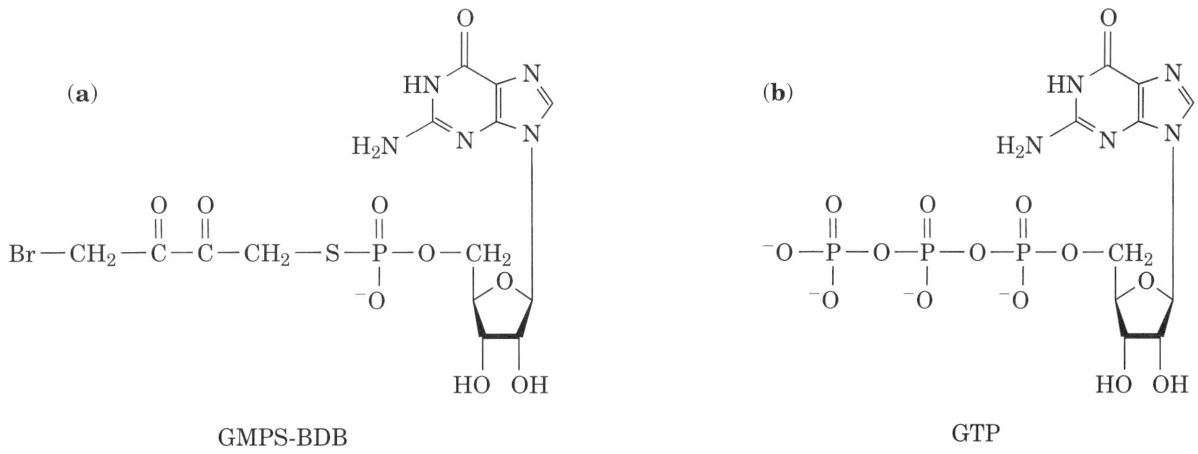

Figure 2. Comparison of the structures of the affinity label (**a**) guanosine 5′-0-[S-(4-bromo-2,3- dioxobutyl)thio]phosphate (GMPS-BDB) and (**b**) the natural ligand GTP.

acid. Despite this proposal, the key general base and acid of the enzyme had not been identified. The similarity between the structures of adenylosuccinate and 6-BDB-TAMP suggested that the latter would bind irreversibly to the adenylosuccinate site of adenylosuccinate lyase, where the bromodioxobutyl group would be likely to occupy the succinyl subsite at which the critical catalytic steps should occur.

6-BDB-TAMP has recently been confirmed to function as an affinity label of *Bacillus subtilis* adenylosuccinate lyase (2). The initial inactivation rate constant exhibits nonlinear dependence on the concentration of 6-BDB-TAMP, with an apparent reversible $K_R = 30\ \mu$M prior to irreversible inactivation at pH 7.0 and 25°C. The tetrameric enzyme incorporates about 1 mol of 6-BDB-[^{32}P]TAMP per mol of enzyme subunit on complete inactivation. The substrate adenylosuccinate or the products AMP plus fumarate protect against inactivation and incorporation of radioactive reagent, indicating that 6-BDB-TAMP targets the adenylosuccinate binding site. Purification of the only radioactive peptide labeled by 6-BDB-TAMP led to the identification of His^{141} as the modified amino acid (2). These results indicated that 6-BDB-TAMP is an affinity label of His^{141} in the substrate binding site of adenylosuccinate lyase, where it may serve as a general base accepting a proton from the succinyl group during catalysis (2) (see **Histidine** residues). This study illustrates many of the desirable characteristics of affinity labeling: binding of the reagent by the enzyme prior to modification of a single site under mild conditions, competition between the reagent and the natural ligand (adenylosuccinate), and identification of the modified residue within a protein of known amino acid sequence.

GMPS-BDB

A reactive guanine derivative is shown in Figure 2, guanosine 5′-0-[S-4-bromo-2,3-dioxobutylthio]phosphate (GMPS-BDB), which represents a novel class of compounds containing a bromodioxobutyl group linked to the sulfur of a purine nucleotide thiophosphate (3,4). The reactive moiety of GMPS-BDB is located at a position equivalent to that of the pyrophosphate region of GTP, as illustrated in Figure 2. Thus, it might be expected that GMPS-BDB would act as an affinity label for GTP sites in proteins (see **GTP-binding proteins**).

Adenylosuccinate synthetase catalyzes the first of the two enzymatic reactions in the conversion of IMP: the condensation of IMP and aspartate to form adenylosuccinate, as GTP is hydrolyzed to GDP and inorganic phosphate. The phosphoryl group has been proposed to be transferred to IMP from the terminal phosphate of GTP to form a 6-phosphoryl-IMP intermediate.

GMPS-BDB has been used as an affinity label of adenylosuccinate synthetase from *Escherichia coli*, and it has been demonstrated that inactivation occurs concomitantly with the modification of Arg^{143} of each subunit of the dimeric enzyme (5). The modification of Arg^{143} and the inactivation by GMPS-BDB is prevented by adenylosuccinate or by IMP plus GTP, implying that the reaction target is in the region of the active site. This result pointed to Arg^{143} as a logical target for **site-directed mutagenesis**; when the positively charged arginine was replaced by the neutral leucine, the expressed mutant enzyme exhibited a significant decrease in its affinity for nucleotides (5).

The crystal structure of *E. coli* adenylosuccinate synthetase has been determined, allowing a comparison to be made between the substrate sites identified by affinity labeling of the enzyme in solution and those assigned within the crystalline form by **X-ray crystallography**. Arg^{143} from one subunit projects into the putative active site of the second subunit, indicating that both subunits of dimeric adenylosuccinate synthetase contribute to each active site and that Arg^{143} plays an important role in nucleotide binding.

AMPS-BDB

The adenosine analogue of GMPS-BDB has also been synthesized: adenosine 5′-0-[S-(4-bromo-2,3-dioxobutyl)-thiophosphate] (AMPS-BDB) (4) and can be considered as a reasonable mimic of ADP or ATP. Bovine liver glutamate dehydrogenase is an allosteric enzyme that is reversibly activated by ADP. On incubation of AMPS-BDB with glutamate dehydrogenase, the enzyme reacts covalently, resulting in an irreversibly *activated* enzyme that is no longer responsive to externally added ADP (6). AMPS-BDB appears to function as an ADP substitute that is covalently bound to Arg^{459} within the activator site of the allosteric bovine liver glutamate dehydrogenase (6).

The above are representative examples of enzymes that have been studied using the strategy of affinity labeling. They illustrate how affinity labeling, X-ray crystallography and site-directed mutagenesis can be used as complementary approaches in evaluating the functional role of particular amino acids in a protein.

BIBLIOGRAPHY

1. R. F. Colman, Y.-C. Huang, M. M. King, and M. Erb (1984) *Biochemistry* **23**, 3281–3286.
2. T. T. Lee, C. Worby, J. E. Dixon, and R. F. Colman (1997) *J. Biol. Chem.* **272**, 458–465.
3. D. H. Ozturk, I. Park, and R. F. Colman (1992) *Biochemistry* **31**, 10544–10555.
4. S. H. Vollmer, M. B. Walner, K. V. Tarbell, and R. F. Colman (1994) *J. Biol. Chem.* **269**, 8082–8090.
5. O. A. Moe et al. (1996) *Biochemistry* **35**, 9024–9033.
6. K. O. Wrzeszczynski and R. F. Colman (1994) *Biochemistry* **33**, 11544–11553.

Suggestions for Further Reading

R. F. Colman (1983) Affinity labeling of purine nucleotide sites in proteins. *Ann. Rev. Biochem.* **52**, 67–91.

R. F. Colman (1990) Site-specific modification of enzyme sites. In *The Enzymes*, 3rd ed., Vol. 19 (D. S. Sigman and P. D. Boyer, eds.), Academic Press, San Diego, Calif., pp. 285–321.

R. F. Colman (1997) Affinity labels for NAD(P)-specific sites. *Methods Enzymol.* **280**, 186–203.

R. F. Colman (1997) Affinity labelling. In *Protein Function: A Practical Approach*, 2nd ed. (T. E. Creighton, ed.) Oxford University Press, Oxford UK, pp. 155–183.

W. B. Jakoby and M. Wilchek, eds. (1977) Affinity labeling. *Methods Enzymol.* **46**.

AFFINITY MATURATION

MICHEL FOUGEREAU

It was observed long ago that the affinities of **antibodies** increase with time during **immunization** with a T-dependent **antigen**. When such an antigen is injected for the first time, it induces the occurrence of low-affinity **IgM** antibodies, which are rapidly replaced by **IgG**. This takes place in the first two weeks of the primary response. At the same time, an increase of the average affinity is usually observed. If the same antigen is given a second time, IgG antibodies are subsequently produced at a higher yield, and with an average affinity that is still increasing. This is characteristic of the secondary response, a phenomenon that may be repeated and amplified by multiple administrations of the antigen. This is why long-term immunization schedules are used whenever both high titers and high affinity are wanted. Affinity maturation, which is a unique property of **B cells** stimulated by T-dependent antigens, is the consequence of (*1*) the clonal organization of lymphocytes, (*2*) the expansion of clones that have been stimulated specifically by the antigen, and (*3*) the presence of a very peculiar mechanism that generates **somatic mutations** at each cellular division of stimulated B cells.

1. Antibody specificities are spread randomly in the huge collection of clones that represent at any given time the **repertoire** expressed by B lymphocytes that have rearranged their immunoglobulin genes and thus have **IgM** at their surface. Clones that are potentially reactive to the antigen proliferate, as the result of the cooperation between **T cells** and **B cells** and the phenomenon of **antigen processing** and presentation. This ensures that the first wave of antibodies are generally of low or moderate affinity.

2. As the immunization proceeds, expansion of the B-reactive cell population takes place, giving rapidly growing colonies, termed *germinal centers*, in secondary lymphoid organs (lymph nodes, tonsils, spleen). Germinal centers are highly organized structures that favor cell–cell interactions that will amplify local B-cell division and expansion.

3. At the same time, somatic mutations will be triggered by a mechanism that is **probably** enzyme-driven, but still not completely elucidated. The rate of the mutations that are introduced is of the order of 1×10^3 per base pair and per generation, which is excessively high. As a result of these mutations, some clones will gain affinity, whereas others will lose. One therefore needs mechanisms to select clones of high affinity and eliminate those of lower affinity; otherwise the system would expand indefinitely and explode. Stimulation by antigen will favor the saving and expansion of clones with the highest affinity, whereas **apoptosis** will operate on cells that did not receive stimuli as they lost recognition ability.

PCR and **DNA sequencing** at the level of single cells allowed the group of Rajewsky to trace the successive mutations that occurred at each cell division, so that it was possible to reconstruct a genealogical tree of these mutations within the germinal center. Interestingly, hypermutated cells can ultimately follow two types of **differentiation**: One leads to the terminal plasma cells, which will secrete circulating antibodies, and the other leads to the so-called **memory cells** that may persist for long periods of time and may be restimulated with great efficiency in the course of another run of immunization.

The extraordinary plasticity and adaptability of the immune system are thus quite remarkable and allow it to respond efficiently to pathogens. Raising memory cells endowed with the ability to produce rapidly high-affinity antibodies is an obvious goal of vaccination.

See also entries **Immune response**, **Class switching**, and **Somatic hypermutation**.

Suggestions for Further Reading

D. M. Tarlinton, A. Light, G. J. Nossal, and K. G. Smith (1998) Affinity maturation of the primary response by V gene diversification. *Curr. Top. Microbiol Immunol.* **229**, 71–83.

C. J. Jolly, S. D. Wagner, C. Rada, N. Klix, C. Milstein, and M. S. Neuberger (1996) The targeting of somatic hypermutation. *Semin. Immunol.*, **8**, 159–168.

A. Ehlich, V. Martin, W. Muller, and K. Rajewsky (1994) Analysis of the B-cell compartment at the level of single cells. *Curr. Biol.* **4**, 573–583.

AFFINITY SELECTION

BARRY HENDERSON
JACK KEENE
DANIEL KENAN

A key requirement of **combinatorial library** approaches is that desirable molecules must be segregated from the remaining library population. Segregation is most often accomplished by affinity partitioning, in which an immobilized target is used to capture interacting molecules from a solution-phase library. Target proteins can be immobilized either using **antibodies** bound to *Staphylococcus* **protein A**, using absorption of targets onto wells of plastic microtiter plates, or covalent attachment to a variety of polymeric supports. Alternatively, combinatorial libraries attached to solid supports—for example, one-bead, one-compound (OBOC) libraries—can be used to bind soluble targets. Solid-phase libraries have also been used to screen highly complex targets, such as living cells on which a variety of receptors are expressed (1). In all cases, unbound material is removed by washing, and the bound material is recovered for amplification or identification, depending on the nature of the library.

Plastic pins were used to develop the first peptide-based combinatorial libraries. The pins are arranged such that they fit neatly into a single well of a microtiter plate. Thus, the pin system provides a convenient format for both synthesis and screening. The power of this method derives largely in the ease of handling large numbers of discrete syntheses in parallel. Sequential steps of the synthesis can be carried out by transferring the pin arrays through various microtiter reaction chambers. The peptides are then directly available for affinity screening, either on the solid phase or following cleavage from the solid support.

Lam et al. (2) developed another approach for peptide affinity selection involving split synthesis of peptides on solid support beads (OBOC libraries: see **Combinatorial Synthesis**). From a pool of millions of beads, a binding reaction is performed using a target molecule such as an **antibody**, receptor, **enzyme**, or even whole cells. Beads displaying affinity for the target are isolated, and the peptide on the bead can be microsequenced.

Fodor et al. (3) have developed immobilized combinatorial libraries on silicon microchips. Chip-based addressable libraries of peptides, oligonucleotides, and small organic molecules can be readily prepared. These methods use photolithography to control regions accessible for subsequent chemical modification. This method enables a miniaturized, fully addressable library to be generated on the surface of a silicon chip. The resulting arrays can be screened using standard affinity methods.

The affinity selection methods described above are extremely broad in their applications. Although classical nucleic acid libraries have been screened by **hybridization**, which is a highly specialized type of affinity partitioning, modern DNA and RNA aptamer libraries are screened in identical fashion to other combinatorial libraries by affinity selection over immobilized targets. Although selection strategies have been introduced that are based on properties other than affinity, such methods are highly specialized and are not likely to displace the current reliance on affinity methods for combinatorial screening.

See also **Combinatorial libraries**, **Libraries**, **Combinatorial synthesis**, **DNA libraries**, **Genomic libraries**, **cDNA libraries**, **Expression libraries**, **Peptide libraries**, and **Phage display libraries**.

BIBLIOGRAPHY

1. M. E. Pennington, K. S. Lam, and A. E. Cress (1996) *Mol. Diversity* **2**, 19–28.
2. K. S. Lam, S. E. Salmon, E. M. Hersh, V. J. Hruby, W. M. Kazmierski, and R. J. Knapp (1991) *Nature* **354**, 82–84.
3. S. P. Fodor, J. L. Read, M. C. Pirrung, L. Stryer, A. T. Lu, and D. Solas (1991) *Science* **251**, 767–773.

AGAROSE

A. CHRAMBACH

Agarose is one of the two most popular materials used to prepare gels for use in **gel electrophoresis**, the other being **polyacrylamide**. Agarose is primarily a polymer of molecular weight approximately 120 kDa of agarobiose (an anhydrogalactose-galactose disaccharide). At sufficiently high concentrations and low temperatures, the polymer forms β-helical strands, which interact by **hydrogen bonding** to form the agarose gel; they dissociate at higher temperatures to form agarose solutions. Such thermally controlled gelation is the unique feature of agarose, which provides both a high reproducibility and simplicity of preparing the gel. Since gelation requires noncovalent interactions between the copolymer strands, denaturing agents, such as solutions of **urea**, prevent gelation.

Agarose is derived from a natural product of oceanic algae, agar. It is processed to reduce the content of acidic groups, primarily sulfate, so as to reduce the degree of **electroendosmosis** that occurs during electrophoresis. Further industrial processing produces a large number of commercially available agarose fractions distinguished by (1) the degree of electroendosmosis in an electric field, determined by the number of acidic groups; (2) the degree of covalent substitution of the acidic groups with chemical groupings such as hydroxyethyl or vinyl (allyl) groups, which tends to lower the melting point in proportion to the degree of substitution; (3) the addition of linear carbohydrate polymers, such as clarified locust bean gum, to increase the viscosity of the gel and thereby eliminate measurable electroendosmosis.

Most agaroses at concentrations greater than about 0.4% (w/v) form gels at room temperature. Agaroses with unusually high gel strengths can gel at concentrations as low as about 0.05%. The pore sizes of such agarose gels make them ideal for separating **DNA** molecules and **oligonucleotides**; their mobility is determined primarily by their size, as nucleic acids of the same class have the same charge density per nucleotide. The least concentrated gels exhibit mean pore sizes of up to 500 to 1,000 nm, sufficiently large for the penetration of large particles or small viruses (see **Particle electrophoresis**). Highly soluble agarose species substituted to a maximum extent by hydroxyethyl groups can be used to prepare gels with concentrations of up to 9 to 10%. Such gels have pores of similar size to those of 3 to 6% polyacrylamide gels, and with greater resolving power for native proteins. Agarose can also be used as a copolymer with polyacrylamide, which provides larger pore sizes than polyacrylamide by itself. The copolymer is produced by adding agarose to the mixture of acrylamide monomers before its polymerization. For the separation by size of DNA fragments, or of other large particles, by **capillary zone electrophoresis**, solutions of agarose can be used.

The adherence of agarose gels to glass or plastic apparatus walls is very weak, and only horizontally oriented gel electrophoresis apparatus can normally be used. The adherence of agarose to vertical glass surfaces may, however, be strengthened by drying a thin film of agarose onto glass walls; this can permit electrophoresis in vertical glass slabs or tubes. Horizontal thin-layer agarose gels with high Joule heat dissipation capacity, which are therefore amenable to electrophoresis at high field strength without melting, can be formed on hydrophilic surfaces of thin plastic sheets (eg, "Gel-Bond").

For preparative purposes, bands of a macromolecular sample in agarose gels may be recovered by solubilizing the agarose at increased temperature or adding the enzyme agarase. The macromolecule of interest is subsequently separated from the low-molecular-weight agarose fragments by filtration or precipitation methods.

Suggestions for Further Reading

I. C. M. Dea, A. A. McKinnon, and D. A. Rees (1972) Tertiary and quaternary structure in aqueous polyacrylamide systems which model cell wall cohesion: Reversible changes in conformation and association of agarose, carrageenan and galactromannans. *J. Mol. Biol.* **68**, 153–172.

T. G. L. Hickson and A. Polson (1968) Some physical characteristics of the agarose molecule. *Biochim. Biophys. Acta* **165**, 43–58.

FMC Corporation (1982) *The Agarose Monograph*, Bioproducts Dept., Rockland, ME.

J. O. Jeppson, C. B. Laurell, and B. Franzen (1979) Agarose gel electrophoresis. *Clin. Chem.* **25**, 629–638.

P. Serwer (1983) Agarose gels: Properties and use for electrophoresis. *Electrophoresis* **4**, 375–382.

D. Tietz (1987) Gel electrophoresis of intact subcellular particles. *J. Chromatogr.* **418**, 305–344.

AGGLUTINATION

J. FOOTE

"Agglutinin" is an obsolete term for an **antibody** capable of causing cells to come together in macroscopic clumps. Agglutination (clumping) of bacteria by serum was perhaps the earliest observed manifestation of *in vitro* immune reactions. The phenomenon is due to crosslinking of cells through the reaction of multivalent antibody molecules with cell-surface **antigens**. The quantity of antibody necessary for crosslinking is generally very small, yet the cell masses involved in agglutination generate a response visible to the naked eye. Because of these characteristics, this oldest of techniques remains one of the most sensitive and widely used types of **immunoassay**.

Agglutination immunoassays can employ cells or synthetic particles. The two broad categories of cell-based agglutination assay are termed "direct" and "passive". In the direct assay, antibodies cause clumping of microbes or blood cells by reaction with surface antigens endogenous to those cells. Typically, the immune status of serum in a test sample will be assessed, based on the highest dilution that will still agglutinate target cells. Direct agglutination assays are widely used to diagnose infectious disease by the presence of specific antibodies in acute or convalescent serum. The "monospot" assay for infectious mononucleosis is an example of this application (1). In passive agglutination immunoassays, the target antigen is exogenous, and is coated covalently or by adsorbtion onto the cells used for the assay. Test samples are then assayed as for the direct agglutination method. Many variants of direct and passive agglutination are in use for research or diagnostic procedures. For example, the test for Rh blood type is a two-step method in which antibodies in the test sample react with endogenous Rh antigen on erythrocytes, then addition of anti-human **IgG** induces agglutination (2).

Availability of uniformly sized latex particles allowed development of a chemically defined substrate on which to attach antigen or antibody (3). The role of the particles was precisely analogous to that of cells, in that their aggregation in response to antibody–antigen crosslinking generated an optically detectable signal. The advantage of inert particles was in their stability, compared to cells, and in interassay reproducibility. Measurement of agglutination was at first by turbidometry. Development of laser **light-scattering** techniques have greatly improved the sensitivity of detection (4) and have led to development of automated instruments for quantitation of agglutination reactions.

BIBLIOGRAPHY

1. C. L. Lee, I. Davidsohn, and R. Slaby (1968) *Am. J. Clin. Pathol.* **49**, 3–11.
2. R. R. A. Coombs, A. E. Mourant, and R. R. Race (1945) *Br. J. Exp. Pathol.* **26**, 255–266.
3. J. M. Singer and C. M. Plotz (1956) *Am. J. Med.* **21**, 888–892.
4. V. Schulthess, R. J. Cohen, N. Sakato, and G. B. Benedek (1974) *Immunochemistry* **13**, 955–962.

AGROBACTERIUM

RICHARD WALDEN

Agrobacterium tumefaciens and *Agrobacterium rhizogenes* are members of the *Agrobacterium* genus of soil **bacteria** responsible, respectively, for tumor formation and hairy root disease in dicotyledenous **plants**. The molecular basis of inducting neoplastic growth in plants has been one of the most intriguing areas of research in plant pathology for more than 50 years. Advances in understanding the molecular basis of tumor formation has led to the development of **vector** systems based on *Agrobacterium* to carry out **plant genetic engineering**.

Agrobacterium belongs to the *Rhizobiacae* family. Classification of the bacterium has been based on physiological and biochemical criteria, although increasingly the taxonomic structure of the genus has been based on **DNA** analysis of **chromosomal** groups of the differing biovars (1). The best studied members of the genus are *A. tumefaciens* and *A. rhizogenes*. *A. tumefaciens* is the causative agent of crown gall disease. It has a wide, obvious host range, including the majority of dicotyledonous plants and some monocotyledonous plants (2). Practically, crown gall can be a serious disease of fruit crops, including grape vines. The physical basis of crown gall disease is the transfer of a defined region of DNA, transferred or T-DNA (see **T-complex**), from a **plasmid** maintained in the bacteria, known as the Ti or tumor-inducing plasmid (see **Ti plasmid**) to the **genome** of the infected plant cell. *A. rhizogenes* induces hairy root disease analogously. A T-DNA is transferred from the Ri, or root-inducing, plasmid to the genome of the infected plant cell (see **Ri plasmid**). This natural form of plant cell **transformation** has been adapted and used extensively in creating **transgenic** plants (see **Plant Genetic Engineering**).

BIBLIOGRAPHY

1. H. Bouzar, J. B. Jones, and A. Bishop (1995) In *Agrobacterium Protocols* (K. M. A. Gartland and M. R. Davey, eds.), Humana Press, Totowa, NJ, pp. 9–13.
2. M. DeCleene and J. DeLey (1976) *Bot. Rev.* **113**, 81–89.

Suggestions for Further Reading

K. Kersters and J. De Ley (1984) Genus III Agrobacterium Conn 1943, In *Bersey's Manual of Systematic Bacteriology*, Vol 1 (N. R. Krieg and J. G. Holt, eds.), Williams and Wilkins, Baltimore, pp. 244–254.

G. Kahl and J. Schell (1982) *Molecular Biology of Plant Tumors*, Academic Press, London.

ALANINE (ALA, A)

T. E. CREIGHTON

The **amino acid** alanine is incorporated into the nascent **polypeptide chain** during **protein biosynthesis** in response to four **codons**—GCU, GCC, GCA, and GCG—and represents approximately 8.3% of the residues of the proteins that have been characterized. The alanyl residue incorporated has a mass of 71.09 Da, a **van der Waals volume** of 67 Å^3, and an **accessible surface** area of 113 Å^2. Ala residues are usually relatively variable during **divergent evolution**, as they are frequently interchanged in **homologous** proteins with **serine**, **threonine**, **valine**, **glutamic acid** and **proline** residues.

The Ala side chain is simply a methyl group:

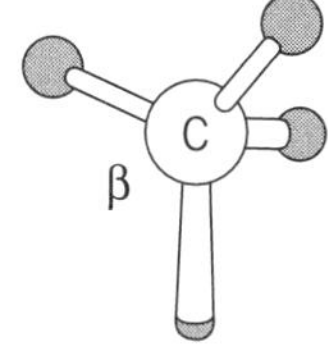

This **nonpolar** side chain makes Ala residues unreactive chemically, relatively **hydrophobic**, and not very **hydrophilic**; consequently, 38% of the Ala residues are fully buried in the folded conformations of proteins. There, the methyl side chain undergoes rapid rotations about the $C_\alpha-C_\beta$ single bond. Ala has the greatest tendency of all the normal amino acid residues to adopt the **alpha-helical** conformation in model **peptides**. It occurs frequently in that conformation in folded **protein structures**, but at about only half that frequency in **beta-sheets** and in reverse **turns**.

Suggestion for Further Reading

T. E. Creighton (1993) *Proteins: Structures and Molecular Properties*, 2nd ed., W. H. Freeman, New York.

ALBUMINS

ATSUSHI IKAI

Albumins are operationally defined as **proteins** that (1) remain soluble in pure **water** after dialysis of protein samples such as egg white or blood serum against distilled water and (2) are not precipitated in 50% saturated ammonium **sulfate**. This contrasts with another group of proteins, called **globulins**, which are precipitated in both distilled water and 50% saturated ammonium sulfate. This operational classification is rather obsolete today, but names like **serum albumin**, **ovalbumin**, and lactalbumin have remained. Other albumins include muscle albumin and plant albumins (1).

BLOOD ALBUMINS

Serum albumin and other plasma proteins have been most closely studied, mainly for their medical interest. After the cells have been removed from blood by light **centrifugation** in the presence of chelating agents such as **EDTA**, a yellowish fluid called plasma remains. By the addition of Ca^{++}, **fibrinogen** is converted to fibrin and the clotting process produces a soft gel. Removal of this gel by centrifugation or other methods will leave a clear fluid called serum. Serum thus obtained contains 70 to 80 g/L of total protein, consisting of more than 150 different kinds of proteins. Dialysis of serum against distilled water will precipitate the globulins, while albumins stay in solution. The albumin fraction contains serum albumin as the major protein (2,3). (see **Serum albumin**).

EGG ALBUMINS

Egg proteins are first classified into yolk proteins and egg white proteins. Yolk contains proteins called vitellogenin, phosvitin, and lipovitellin, all in association with yolk lipid. Egg white is a reservoir of several kinds of proteins as a protective and nutritious environment for the fetus. Egg white proteins are further divided into albumins and globulins, as for serum proteins. The following is a list of egg white proteins that are not globulins. Descriptions of egg white globulins are found under **Globulins**.

Ovalbumin

This is the major **glycoprotein** in the white of eggs, comprising 65% of the total egg white protein, with a molecular weight of 43,000 and **isoelectric point** (pI) of 4.7. The N-terminal glycine residue is acetylated. Two species containing one or two sites of **phosphorylation** can be separated by **electrophoresis**. Oligosaccharide containing three moles of N-acetylglucosamine and five moles of mannose is linked to an **asparagine** residue through an **N-glycosidic** linkage.

Treatment of ovalbumin with subtilisin yields a readily crystallizable plakalbumin, named after the platelike appearance of the resulting crystals. Ovalbumin is often used as an effective antigen in immunological studies. The protein has recently been found to have a similar amino acid sequence and three-dimensional structure to **α_1-antitrypsin** of the **serpin** family (4) (see **Ovalbumin**).

Ovotransferrin (or *conalbumin*)

This protein binds ferric ions and is the same as apo**transferrin**, but differing in the carbohydrate moiety. In egg white, the protein is present in an almost iron-free form. This protein constitutes about 10% of the egg protein.

Ovomucoid

Ovomucoid is a glycoprotein (carbohydrate content about 25% by weight) with a molecular weight of 28,000 and constitutes 1.5% of the total egg white protein. It inhibits **trypsin** and **chymotrypsin** but not **plasmin**, **thrombin**, **elastase**, or collagenase. Its pI falls in the range 3.9 to 4.5. It remains active in the supernatant after egg white has been coagulated by heat. It has three **homologous** domains connected by linking peptides, each domain being homologous to **bovine pancreatic trypsin inhibitor** (BPTI). Thus, it is speculated that ovomucoid has evolved by two tandem repeats of the BPTI gene. Carbohydrates are linked to asparagine residues at residue numbers 10, 53, 69, 75, and 175. Similar inhibitors can be purified from the egg white of the quail, goose, and turkey.

Ovoinhibitor

This is a 48-kDa multiheaded **proteinase inhibitor** that can simultaneously inhibit trypsin, chymotrypsin, and elastase. Unlike ovomucoid, it also inhibits the proteinases of bacterial origins. It is a glycoprotein with 5 to 10% content of carbohydrate.

Avidin

This protein occupies a special position in biochemistry in that it is widely used as a probe based on its strong affinity for **biotin**. Their binding constant reaches $10^{14}\ M^{-1}$ under optimal conditions. **Avidin-biotin systems** are widely used for labeling macromolecular interacting systems, not only **protein–protein interactions**, but also systems based on DNA and other macromolecules. **Avidin** can be substituted by **streptavidin**, which is of bacterial origin but with a similar activity. Avidin in the egg white is largely free of biotin, and feeding rats with purified avidin as the sole protein source will cause various symptoms known to accompany biotin (vitamin H) deficiency.

Ovomucin

This is a glycoprotein of high molecular weight responsible for the mucous character of egg white. The treatment of ovomucin with reagents that react with **thiol groups** reduces the viscous nature of egg white.

PLANT ALBUMIN

Ricin is an example of plant albumins (from *Ricinus communis*) and has an interesting toxic function. It consists of an A subunit, of 32 kDa, pI of 7.5, and 2.4% carbohydrate, and a B subunit, of 34 kDa, pI of 4.8, and 6.5% carbohydrate. It inactivates the 60S subunit of **ribosomes** after being internalized into the cell and thus inhibits a peptide elongation step (see **Translation**). The A subunit carries out the inactivation reaction, while the B subunit, which binds to cell-surface galactose residues, helps the A subunit to be internalized into the cell. Plant albumins are more readily precipitable in 50% ammonium sulfate than animal albumins.

MILK ALBUMINS

α-Lactalbumin

This is a single-chain 14-kDa protein that is the same as the B chain of lactose synthetase in lactating granules. The A chain of lactose synthetase alone catalyzes the synthesis of N-acetylgalactosamine from N-acetylglucosamine and UDP-galactose, but when in association with α-lactalbumin, it uses glucose as a substrate and synthesizes lactose. The amino acid sequence of α-lactalbumin with 143 amino acids (human, bovine, guinea pig) is homologous to that of **lysozyme**, and its three-dimensional structure is also similar. A recent review of the relationship between lysozyme and lactalbumin is recommended (5) (see also **alpha-Lactalbumin** and **Lysozyme**).

Lactoferrin

This is an iron-binding protein of 88,000 in humans and 86,000 in cows. It is similar in functional properties to serum transferrin, but immunologically and structurally distinguishable (6).

BIBLIOGRAPHY

1. B. Blombäck and L. A. Hanson (1979) *Plasma Proteins*, John Wiley & Sons, New York.
2. M. Perutz (1992) *Protein Structure: New Approach to Disease and Therapy*, W. H. Freeman and Co., New York.
3. T. Peters Jr. (1985) *Adv. Prot. Chem.* **37**, 161–245.
4. P. E. Stein, G. W. Leslie, J. T. Finch, and R. W. Carrell (1991) Crystal structure of uncleaved ovalbumin at 1.95 Å resolution, *J. Mol. Biol.* **221**, 942–959.
5. M. A. McKenzie and F. H. White Jr. (1991) Lysozyme and α-Lactalbumin: Structure, function, and interrelationships, *Adv. Prot. Chem.* **41**, 173–315.
6. B. F. Anderson, H. M. Baker, E. J. Dodson, G. E. Harris, S. V. Rumball, J. M. Waters, and E. M. Baker (1987) Structure of lactoferrin at 3.2 Å resolution, *Proc. Nat. Acad. Sci. USA* **84**, 1769–1773.

ALCOHOL DEHYDROGENASE (ADH)

H. JÖRNVALL
B. PERSSON

ADH is a type of **enzyme** that catalyzes the reversible interconversion of an alcohol to an aldehyde/ketone [EC 1.1.1.1]. The substrate specificity is often wide, but it typically includes at least some activity with ethanol. The **coenzyme** is often **NAD^+**. ADH was initially purified from yeast (YADH) and from horse liver (LADH), sources from which it is commercially available. It is, however, now known to be widespread, with at least some forms present in **eukaryotes**, **prokaryotes** and **archaeabacteria** in general. It has been well studied at the protein and **DNA** levels from many sources. The "classical" forms of ADH are Zn-containing **metalloproteins** with subunits of ~40 kDa (~350 to 390 residues) (1). They are now considered part of a larger protein family (see **Protein evolution**), known as medium-chain dehydrogenases/reductases (**MDR**) (2).

Although ADH is frequently regarded as well understood, much is still unknown, and novel findings of general interest are constantly obtained:

1. ADH is no longer just an enzyme, but a whole system of great complexity, involving other protein families and a multiplicity of MDR forms.
2. another such family is (short-chain dehydrogenases/reductases) **SDR** (3). The major ADH of insects is this type, and SDR enzymes are numerous, including **hormone**-converting and regulatory enzymes in humans and all living forms.
3. MDR-ADH occur as different classes (4) and **isozymes**. They exhibit distinct evolutionary patterns, repeated evolutionary origins, and varying properties. At least six classes and eight genes have been discerned in mammals, but these numbers are likely to grow; they differ among species because of recent **gene duplications**. Expressions differ with organ, tissue, and age, including special fetal forms.
4. Functionally, the ADH system has long been an enigma, but recent results suggest that it has important roles. Metabolically, ADH is associated with aldehyde dehydrogenase, which is part of another large family. Medically, both activities are important in alcohol metabolism and in several disease states but also in vision and in cellular regulation and differentiation (5,6).

MDR-ADH

Classes in Vertebrates

Mammals and most vertebrates contain different classes of MDR Zn-containing ADH enzymes. These classes differ substantially in substrate specificity and in their immunological, chromatographic, and electrophoretic properties (4). They constitute separate forms that are intermediate in their properties between isozymes and distinctly different enzymes, and they differ in primary structures by about 30% but have closely related conformations (Fig. 1). The classes reflect a series of gene duplications, which occurred largely at early vertebrate times (8). Later duplications in individual lines have given rise to intraclass isozymes. The subunits of isozymes can hybridize into mixed dimers that reflect the subunit mixture, but those of different classes do not.

Class I contains the classical ADH enzyme, which is present in large amounts in liver but in lower amounts in many other organs. It is involved in ethanol metabolism and in liver function. It has good enzymatic activity with primary alcohols, cyclohexanol, retinols, and many other alcohols, plus their corresponding aldehydes. Isozymes occur and are largely species specific. The horse ADH subunits E and S derive from two separate genes and the human subunits α, β, and γ from three genes. The α subunit is expressed in the fetus and β and γ in the adult. The β and γ types also exhibit further variability from alleles that have different population distributions in Caucasians and Orientals. This population variation, together with other alleles of the next enzyme of alcohol metabolism (aldehyde dehydrogenase), explains the different sensitivity to ethanol consumption of Caucasians and Orientals.

Class II ADH is highly variable. It is also expressed in the liver but in lower amounts. Thus far it is without a recognized functional role, but in humans it has a higher K_m for ethanol, and hence is of little importance in ethanol metabolism.

Class III ADH is expressed universally in cells and organs, and is identical to **glutathione** (GSH)-dependent formaldehyde dehydrogenase (EC 1.2.1.1). In the absence of GSH, it is a dehydrogenase toward long-chain alcohols and is largely inactive with ethanol. The K_m for ethanol is $>3\ M$, so there is activity only at high ethanol concentrations. In the presence of GSH, class III ADH is active with the GSH/formaldehyde adduct, hydroxymethyl-GSH, when it functions as a formaldehyde dehydrogenase, producing formic acid.

Class IV ADH is present in the stomach and in skin (epithelia). It is the most ethanol-active form and has been discussed in relation to first-pass ethanol metabolism and to retinol dehydrogenase function. Both roles, however, are far from established.

Little is known about the remaining enzyme classes in vertebrates.

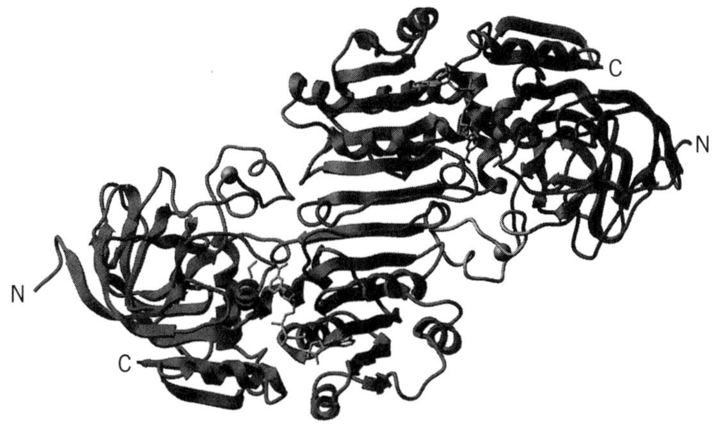

Figure 1. Structure of human ADH class I (7). The diagram and the modeling of ethanol were made with the program ICM (Molsoft, Metuchen, NJ) from coordinates in the data bank.

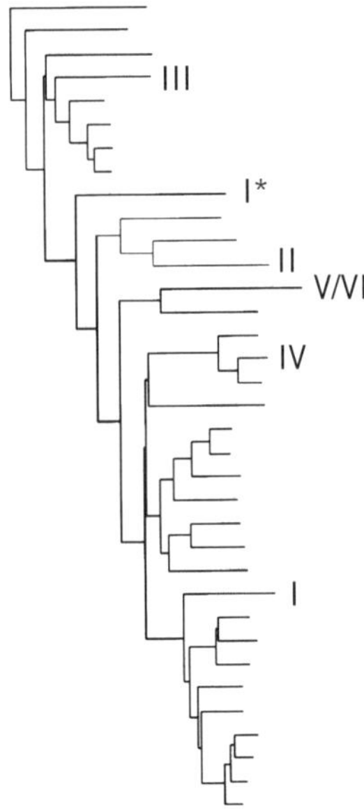

Figure 2. Evolutionary tree of animal ADH. The deviating class I form of bony fish is indicated by a star. The tree was calculated using ClustalW (2) with corrections for multiple substitutions.

Class III is the evolutionary ancestor, which is apparent from **phylogenetic trees** (Fig. 2), the rates of change in present forms, and in apparently being the only MDR-ADH in invertebrates. However, final judgment of the phylogeny must await characterization of further forms. Additional gene duplications probably remain to be discovered, and the true evolutionary relationships may be complex, involving both extant and extinct forms.

In molecular terms, the classes exhibit different evolutionary patterns. Class III is like an enzyme of basic metabolism, and has relatively constant functional properties among species and variation primarily in the nonfunctional parts of the protein (Fig. 3). Class I is the result of an early gene duplication from class III, whereas class IV in turn diverged from the class I line. Special hybrid forms of class I/III (in fish), II/I (in birds) and IV/I (in amphibians) have been traced and may reflect either the phylogeny or simply additional gene duplications. In short, the ADH system has many interesting evolutionary features (8).

Yeast, Plants, Other Nonvertebrate Forms

Nonvertebrate MDR-ADH are also frequent and have at least two classes. One is the class III enzyme, which has GSH-dependent formaldehyde dehydrogenase activity, as expected from the ancestral properties noted previously. The other is ethanol-active MDR-ADH: yeast ADH (YADH, from at least three genes) and plant ADH (PADH). YADH has greater ADH activity than LADH because of weaker **coenzyme** binding and consequently a greater turnover rate, because coenzyme dissociation is the rate-limiting step. Although YADH and PADH are related to LADH structurally and functionally, they are products of gene duplications separate from that leading to class I ADH. Combined, all of these ethanol-active forms suggest that ADHs active with alcohol have arisen several times during evolution, probably by functional **convergence**.

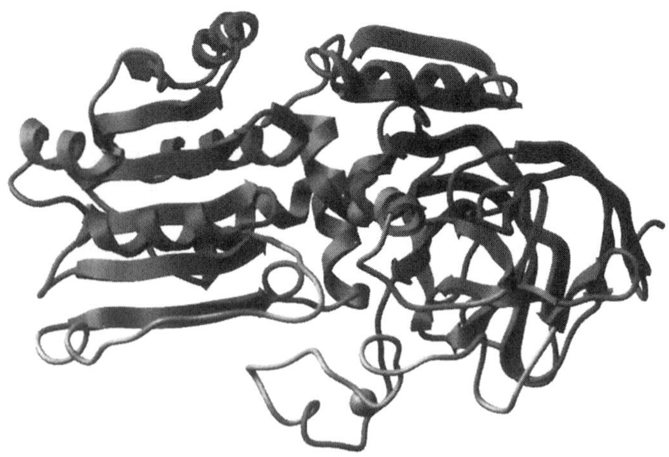

Figure 3. Variable segments in subunits of ADH. Those subunits that vary between class I enzymes affect functional segments, typical of a functionally variable protein. Those subunits variable between class III enzymes affect superficial segments, typical of a functionally constant protein (3).

Presumably all or many of these ethanol-active forms are derived from the ancestral class III line (Fig. 4), but they can differ in **quaternary structure**. Ethanol-active YADH is a tetramer, whereas yeast class III ADH is a dimer.

Function

Liver ADH (class I) is the major metabolic enzyme for ethanol, either ingested or produced in the intestine, and for other alcohols. Overall, the formaldehyde specificity of the ancestral form (class III), the multiplicity of all ADH systems, and the gradual changes in substrate specificity suggest that the whole ADH system is a basic enzyme, together with aldehyde dehydrogenase, in cellular defense reactions against alcohols and aldehydes, converting them to acidic end products. The validity of this general role of ADH is supported by the fact that the only animals apparently lacking ADH (except for formaldehyde-active class III) are marine invertebrates (10), whose environment is low in alcohol and aldehyde levels. The formaldehyde dehydrogenase activity of class III ADH is universal and constant throughout the living world, although the K_m and enzymatic efficiencies are increased in microorganisms. This permits life at higher aldehyde concentrations (11) and suggests a general defense function for the ADH system.

Additional functions of individual classes are likely. Class IV has been discussed relative to its possible role in retinal formation important for **rhodopsin** and vision (6) and in **retinoic acid** formation, which important for vertebrate growth and **differentiation** (5). Similarly, special forms of ADH and aldehyde dehydrogenase are involved in the fatty acid cycle in dermal wax and plasmalogen formation in the central nervous system and are associated with an inborn error of metabolism (Sjögren-Larsson syndrome) affecting the corresponding genes (12).

Other MDR Enzymes

Many enzymes active on other alcohols or polyols are related to MDR-ADH in structure and origin. Among these are sorbitol

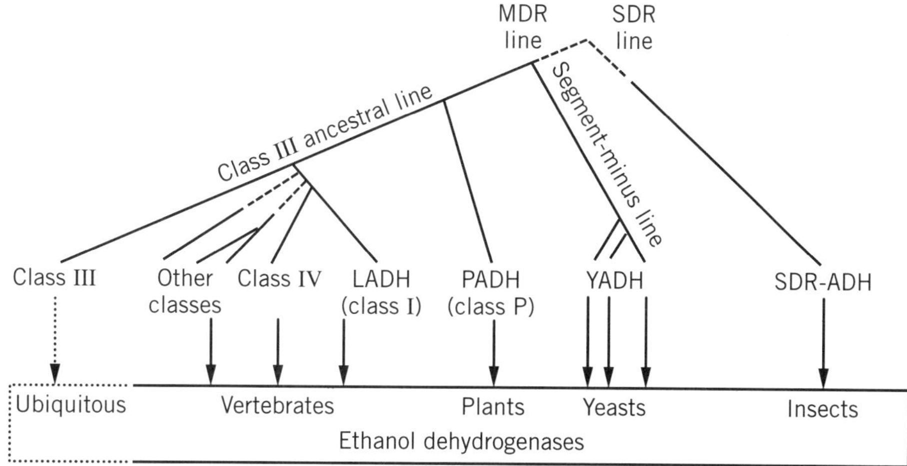

Figure 4. Functional convergence toward ethanol dehydrogenases in different lines of living organisms, emphasizing the repeated appearance of ethanol-activity, many separate gene duplications, and an apparent MDR-ancestral nature of class III. Dashed lines indicate unknown relationships, and dotted lines low activity with ethanol. Segment-minus forms lack internal segments (Fig. 1) and are tetrameric. The remaining forms are dimers. Apart from the gene duplications shown, which lead to ethanol-active dehydrogenases, additional branchings lead to cinnamyl ADH (in plants), sorbitol dehydrogenase (in animals), threonine dehydrogenase (in prokaryotes), several reductases (in prokaryotes and eukaryotes), and additional enzymes.

dehydrogenase, threonine dehydrogenase, quinone oxidoreductase, the enoyl reductase component of the fatty acid synthesis machinery, and enzymes with undefined functions, like VAT-1 of synaptic membranes (2). These enzymes are similar conformationally and in their overall active-site relationships. But they also differ markedly, including functionally, in that all do not have the active-site zinc typical of MDR-ADH. This suggests that they have different enzymatic mechanisms. Zinc is also absent in quinone oxidoreductase, which is a crystallin in ocular lenses of some animals.

SDR-ADH

Insect ADH is the only well known SDR-ADH, but hundreds of other SDR enzymes are known. These enzymes work on hydroxyl groups of a multitude of sugar, steroid, prostaglandin, and other compounds. They also include reductases, with other activities (like double-bond saturation), and enzymes of no less than three of the enzyme classes in general (oxidoreductases, lyases, and isomerases (3)).

OTHER ADH FORMS

ADH activity is also known in enzymes other than the MDR and SDR forms, including iron-activated forms (13). Prokaryotes have several different forms, including long-chain alcohol dehydrogenases (14) and methanol dehydrogenases. They may even have MDR relationships, and this definitely applies to other special prokaryotic forms, such as factor-dependent formaldehyde DH (15), which is apparently the equivalent of class III MDR-ADH in **Gram-positive** bacteria.

BIBLIOGRAPHY

1. C.-I. Brändén, H. Jörnvall, H. Eklund, and B. Furugren (1975) *The Enzymes*, 3rd ed., chap. 11, pp. 103–190.
2.. B. Persson, J. S. Zigler Jr., and H. Jörnvall (1994) *Eur. J. Biochem.* **226**, 15–22.
3. H. Jörnvall et al. (1995) *Biochemistry* **34**, 6003–6013.
4. B. L. Vallee and T. J. Bazzone (1983) *Curr. Top. Biol. Med. Res.* **8**, 219–244.
5. G. Duester (1996) *Biochemistry* **35**, 12221–12227.
6. A. Simon et al. (1995) *J. Biol. Chem.* **270**, 1107–1112.
7. T. D. Hurley et al. (1994) *J. Mol. Biol.* **239**, 415–429.
8. O. Danielsson et al. (1994) *Proc. Natl. Acad. Sci. USA* **91**, 4980–4984.
9. J. D. Thompson, D. G. Higgins, and T. J. Gibson (1994) *Nucleic Acids Res.* **22**, 4673–4680.
10. M. R. Fernández et al. (1993) *FEBS Lett.* **328**, 235–238.
11. M. R. Fernández et al. (1995) *FEBS Lett.* **370**, 23–26.
12. V. De Laurenzi et al. (1996) *Nature Genet.*, **12**, 52–57.
13.. R. K. Scopes (1983) *FEBS Lett.* **156**, 303–306.
14. T. Inoue et al. (1989) *J. Bacteriol.* **171**, 3115–3122.
15. P. W. van Ophem, J. Van Beeumen, and J. A. Duine (1992) *Eur. J. Biochem.* **206**, 511–518.

Suggestions for Further Reading

H. Weiner et al., eds. (1987 to 1997) *Enzymology and Molecular Biology of Carbonyl Metabolism*, Vols. 1–6, Plenum, New York. Up-to-date summaries on ADH and the metabolically-related enzymes. Published biannually from the conference series with the same name, Vol. 6 (1997) now in press.

H. Eklund and C.-I. Brändén (1987) Alcohol dehydrogenase. *Biol. Macromol. Assembl.* **3**, 74–142.

W. F. Bosron, T. Ehrig, and T.-K. Li (1993) Genetic factors in alcohol metabolism and alcoholism. *Seminars Liver Dis.* **13**, 126–135.

J. Shafqat et al. (1996) Pea formaldehyde-active class III alcohol dehydrogenase: Common derivation of the plant and animal forms but not of the corresponding ethanol-active forms (classes I and P). *Proc. Natl. Acad. Sci. USA* **93**, 5595–5599.

ALIGNING SEQUENCES

TOBY GIBSON
PEER BORK

The most basic activity in **sequence analysis** involves aligning protein or nucleotide sequences together. This need arises due to the processes of molecular **evolution**: **gene duplication** followed by continual **divergence** of the sequences through the accumulation of **mutations** over time. Comparative biological analysis, which has long been such a powerful tool for biologists (as exemplified by Linnaeus and Darwin), is arguably even more applicable in sequence analysis than in any other branch of biology, because it can be applied to an enormous number of character states at the level of individual residues in nucleic acid or protein sequences. First, however, related sequences must be correctly aligned before the power of comparative analysis can be brought to bear. Because of the difficulty of aligning highly diverged sequences, and the many applications of sequence alignment, this is one of the most active areas for method development in computational biology.

Alignment tasks generally divide into pairwise sequence alignment and multiple sequence alignment, although the underlying algorithms may share many details. The most sensitive methods for aligning sequences belong to the class of algorithms known as dynamic programming (or minimum string edit) that were initially developed for applications in text comparison. Two of the dynamic programming algorithms most used in biology are usually known as Needleman–Wunsch (1) and Smith–Waterman (2), after the researchers who first applied them to biological sequences. Because these algorithms allow gaps to be inserted at any position in the sequences, they are computationally slow. By contrast, word comparison algorithms, which do not allow gaps, are much faster, but at the expense of accuracy and sensitivity in aligning sequences.

PAIRWISE SEQUENCE ALIGNMENT ALGORITHMS

Dynamic Programming

The basic algorithm works through a two-dimensional matrix in which every residue in one sequence is scored against every residue in the other sequence (1). The algorithm begins in one corner (eg, top left) of the matrix and ends in the opposite corner (bottom right). At each point in the matrix, the algorithm iterates the same set of choices. Typically, it chooses which of three existing paths scores best when extended into the current point: (i) match the residues and continue aligning from the previously aligned residue pair, or (ii) pay the penalty and insert a one-residue gap into sequence X, or (iii) pay the penalty and insert a one-residue gap into sequence Y (see **Gap penalty**). The algorithm is guaranteed to find the best path through the matrix, allowing for gaps at any position in either sequence. Scores for matching the residues are taken from residue exchange, or mutation, matrices. For nucleotide sequences, these are usually quite simple: for example, +1 for an *identity*, 0 for a *transition*, and −1 for a *transversion*. For proteins, typical exchange matrices are the more complex 20 * 20 *PAM* (**point accepted mutation**) matrices introduced by Margaret Dayhoff (3), or subsequently derived PAM series derived from larger alignment datasets (4,5). A PAM 250 matrix is shown in Figure 1. Gap penalties are used to control the frequency and length of gaps inserted in the sequences. Where appropriate, varying gap penalties in a position-specific manner can improve the alignment.

Dynamic programming algorithms work through a two-dimensional matrix of area M*N in aligning sequences of lengths M and N. Therefore the computational requirement [usually symbolized as O(MN)] has a constant factor for the calculation, multiplied by the two sequence lengths. To obtain an alignment, the algorithm makes a first pass to determine the end of the highest scoring matched segment and then a second pass working back to obtain the alignment. If only the highest score, but not the actual path, is needed (as in a database search), then only the first pass need be done. Naive implementations that first plot the whole matrix to an array will also use O(MN) memory. Because the algorithm works through the array systematically, however, it is unnecessary to store the whole array in memory. Memory-efficient implementations of the first pass are straightforward, because the alignment is not being kept (6). The second pass, to obtain the alignment, is more complicated, but memory-efficient recursive methods have been developed that have allowed large alignment tasks to be ported to small personal computers, with a small but acceptable loss in calculation speed (7–9).

Global Alignment

The standard Needleman–Wunsch algorithm (1) finds the optimal full-length alignment for a pair of sequences. Global alignment is appropriate where sequences are known to be both **homologous** and collinear and is therefore often used for multiple alignment of sequence families.

Best Local Alignment

Variants of the Smith–Waterman algorithm (2) find the optimal alignment that has a positive value for the path for a given pair of sequences. Except for highly related sequences, the best local alignment is a partial match between the sequences. Residue exchanging mutation matrices, such as *PAM*250, provide log-odds scores for the likelihood that a pair of residues will exchange as a result of mutation: Similar residues that exchange easily have positive log scores, while dissimilar residues have negative log scores. The best local alignment is taken as the highest-scoring continuously positive path. This algorithm is appropriate under conditions where sequences are not known to be both fully homologous and collinear—for example, multi*domain* proteins, or DNA regions containing rearrangements. Smith–Waterman type algorithms underlie the most sensitive methods for database searching by sequence homology yet to be devised. Because of the computational cost, they are not applied as often as search methods using ungapped alignment algorithms.

The Waterman–Eggert (10) extension of the algorithm will return sets of suboptimal paths that do not intersect with the optimal path, and in this way it can find repeats in sequences.

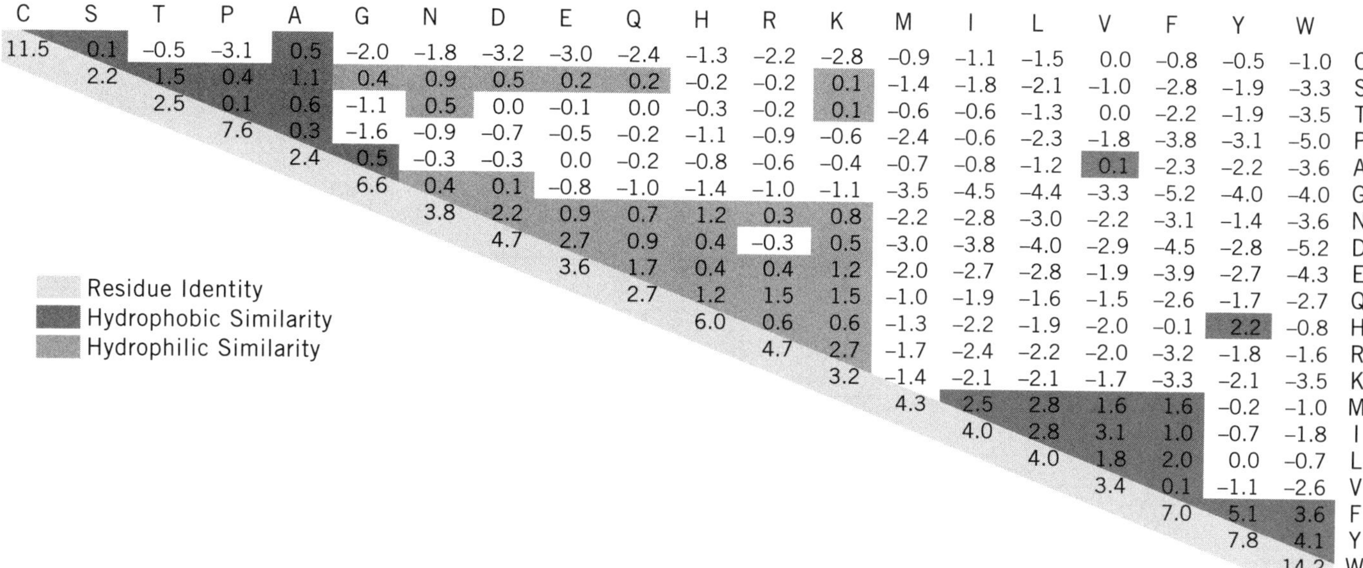

	C	S	T	P	A	G	N	D	E	Q	H	R	K	M	I	L	V	F	Y	W
C	11.5	0.1	−0.5	−3.1	0.5	−2.0	−1.8	−3.2	−3.0	−2.4	−1.3	−2.2	−2.8	−0.9	−1.1	−1.5	0.0	−0.8	−0.5	−1.0
S		2.2	1.5	0.4	1.1	0.4	0.9	0.5	0.2	0.2	−0.2	−0.2	0.1	−1.4	−1.8	−2.1	−1.0	−2.8	−1.9	−3.3
T			2.5	0.1	0.6	−1.1	0.5	0.0	−0.1	0.0	−0.3	−0.2	0.1	−0.6	−0.6	−1.3	0.0	−2.2	−1.9	−3.5
P				7.6	0.3	−1.6	−0.9	−0.7	−0.5	−0.2	−1.1	−0.9	−0.6	−2.4	−0.6	−2.3	−1.8	−3.8	−3.1	−5.0
A					2.4	0.5	−0.3	−0.3	0.0	−0.2	−0.8	−0.6	−0.4	−0.7	−0.8	−1.2	0.1	−2.3	−2.2	−3.6
G						6.6	0.4	0.1	−0.8	−1.0	−1.4	−1.0	−1.1	−3.5	−4.5	−4.4	−3.3	−5.2	−4.0	−4.0
N							3.8	2.2	0.9	0.7	1.2	0.3	0.8	−2.2	−2.8	−3.0	−2.2	−3.1	−1.4	−3.6
D								4.7	2.7	0.9	0.4	−0.3	0.5	−3.0	−3.8	−4.0	−2.9	−4.5	−2.8	−5.2
E									3.6	1.7	0.4	0.4	1.2	−2.0	−2.7	−2.8	−1.9	−3.9	−2.7	−4.3
Q										2.7	1.2	1.5	1.5	−1.0	−1.9	−1.6	−1.5	−2.6	−1.7	−2.7
H											6.0	0.6	0.6	−1.3	−2.2	−1.9	−2.0	−0.1	2.2	−0.8
R												4.7	2.7	−1.7	−2.4	−2.2	−2.0	−3.2	−1.8	−1.6
K													3.2	−1.4	−2.1	−2.1	−1.7	−3.3	−2.1	−3.5
M														4.3	2.5	2.8	1.6	1.6	−0.2	−1.0
I															4.0	2.8	3.1	1.0	−0.7	−1.8
L																4.0	1.8	2.0	0.0	−0.7
V																	3.4	0.1	−1.1	−2.6
F																		7.0	5.1	3.6
Y																			7.8	4.1
W																				14.2

Figure 1. PAM250 amino acid exchange matrix developed by Gonnet and colleagues (4) and superseding the original Dayhoff matrix (3). Similar pairs of amino acids have positive log-odds exchange values while dissimilar pairs have negative values. All positive scores are colored: Red shows scores for exact matches, purple highlights similar pairs of hydrophobic residues, and green indicates similar pairs of hydrophilic residues. The highest residue exchange scores are for bulky aromatic residues (Phe, Tyr, Trp) and are stronger than exact matches of highly mutable residues such as Ser. The strongest mismatches are between bulky hydrophobic residues and small or negatively charged residues.

Best Local Ungapped Alignment

Widely used database search tools, such as BLAST (11) and FASTA (12), search for the highest scoring matched regions without allowing for gaps. Word searches and other ungapped alignment methods are much faster than dynamic programming approaches, but at the expense of sensitivity. Thus these methods are likely to miss homologous but divergent matches. To improve the results, FASTA does a second dynamic pass on the set of top hits. For a small reduction in search speed, BLAST2 examines the gap cost between the set of ungapped positive matches between two sequences and returns composite best locally aligned regions, including gaps whenever the score is still positive. The latter algorithm is likely to approach Smith–Waterman sensitivity except for the most unusual alignment circumstances.

MULTIPLE-SEQUENCE ALIGNMENT

This is a set of homologous protein or nucleotide sequences that have been correctly aligned, allowing for the presence of **indels**. Figure 2 shows an aligned region for some **elongation factor** TU sequences.

Uses of Multiple Alignments

Multiple alignments are indispensable in computational biology. They are the basic dataset used to construct **phylogenetic trees**, which are themselves important computational tools (eg, for weighting sequences by divergence) as well as providing insight into past evolution. They reveal conserved residues that are likely to be structurally or functionally (eg, in catalysis) important and unconserved positions that either are unimportant or have acquired a change of function (Fig. 2). They improve the accuracy of many sequence analysis functions as compared to single sequences, such as **secondary structure prediction** (13), **coiled-coil** prediction (14), and **transmembrane helix** prediction (15). They are useful as the query input for the most sensitive homology searches (alignment profile (16) and hidden Markov model searches (17)) and can be used to detect divergent homologues that single-sequence queries cannot pick up. They are useful in the detection of **domains** and in defining their boundaries in modular, **mosaic proteins** (18). Multiple alignments of folded RNAs are a prime resource used in determining their secondary and tertiary structures, by mapping residue conservation and long-distance-coupled mutations (19,20). DNA multiple alignments are used for identifying conserved signals, such as **promoter** elements and **RNA splice** sites (21).

Multiple-Alignment Algorithms

So far it has been necessary to adopt heuristic strategies to generate multiple alignments, because formally correct methods have been computationally impractical to implement. The ideal method to align N sequences would be N-dimensional dynamic programming, as this would be guaranteed to find the optimal path (ie, the optimal multiple alignment) in an N-dimensional matrix. Unfortunately the computer time required to align N sequences of length l is $O(l^N)$ and is impractical for more than three or four sequences, although by limiting the search space to likely regions, the MSA program can align

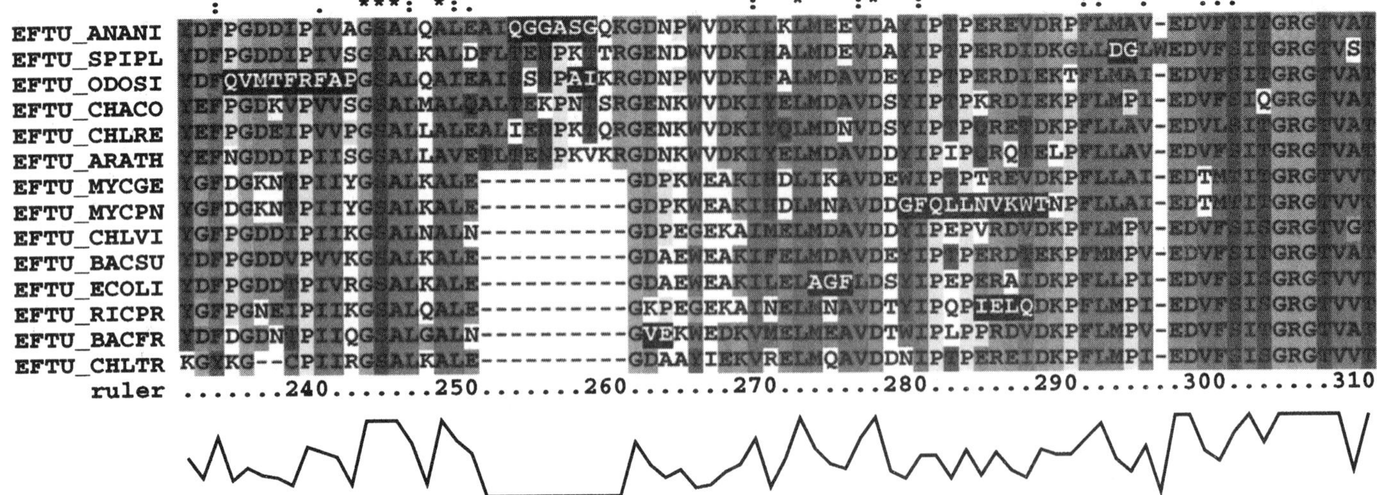

Figure 2. Part of a multiple alignment of 14 prokaryotic EFTU sequences using the one-letter code for amino acids. Gaps are indicated by dashes. Columns marked by asterisks are completely conserved, columns marked by colons are strongly conserved, and columns marked by periods are weakly conserved. The graph at the bottom shows the conservation in the columns. Color is an essential aid to sequence analysis and is used here to highlight conservation according to the amino acid properties. Inverted characters indicate poorly matching regions of sequence. Some of these are due to natural sequence divergence, but there are four errors in sequence determination causing frameshifted regions: EFTUSPIPL 203–206, EFTUODOSI 234–243, EFTUMYCPN 279–289, and EFTURICPR 284–287. These errors may lead to false inferences: If R-285 were completely conserved, it would be a candidate functional residue, while the false gapped column at 206 incorrectly suggests a surface loop. See color insert.

up to eight sequences (22). Another broad class of methods, those that iterate toward an optimized score for the alignment, are still computationally intensive but are becoming practical with increasing computer power. These include a number of approaches, some of which may be used in combination, such as global minimization, genetic algorithms, and trained **neural networks** (23,24). The goal is to harness a good model description of a multiple-sequence alignment with an effective iteration strategy, so as to get high-quality alignments in a practical timescale. In the meantime, widely used alignment programs such as Clustal W (25) follow the heuristical clustered alignment strategy.

Progressive Clustered Alignment

This two-step approach was introduced by Feng and Doolittle (26) in an attempt to minimize errors in the final multiple alignment, by aligning the most similar sequences first and the most divergent ones last. The set of unaligned sequences are first aligned in pairwise fashion to each other, so that a matrix of the approximate pairwise similarities may be obtained. A matrix-based tree construction method, such as Neighbor-Joining (27), is used to construct a dendrogram linking the sequences according to their observed similarities. Guided by the dendrogram branching order, the sequences are then sequentially aligned together using dynamic programming, beginning with the most closely related sequences and ending by merging the most divergent groups by profile alignment. This procedure minimizes alignment errors, which become more likely with increasing sequence divergence, becoming a problem for proteins less than 25% to 30% identical in sequence. The final alignment should always be examined for misalignment, especially of the more divergent sequences, because such errors are likely to be present in any but the most straightforward multiple alignment task.

The sensitivity of the basic clustered alignment strategy can be improved by several modifications, such as weighting the sequences by divergence and position-specific **gap penalties** (25). Where **tertiary structure** information is available, gap penalty masks can be employed to guide the gaps into regions of sequence that are expected to be tolerant of **indels** (28).

Profile Alignment

A set of aligned sequences can be aligned to one or more new sequences by treating the group much as a single sequence. The score for one alignment column can be obtained by summing the log-odds residue exchange scores for the observed set of amino acids (16), correcting for sequence relatedness by downweighting similar sequences (25). Conserved alignment columns score more highly than unconserved columns, where the log-odds scores tend to cancel each other. Gap penalties can be lowered at existing indels, because gaps in new sequences are more likely at these positions than at ungapped positions. The improvement in signal to noise provided by the extra information in the alignment means that a profile alignment is more accurate than independent pairwise alignment of the same set of sequences. As well as being used to merge aligned groups in the clustered alignment strategy, profiles provide a sensitive search strategy to find highly divergent homologues and are one of the main sequence analysis tools for identifying protein domain families (18).

Hidden Markov Model (HMM) Alignment

HMMs are a class of probabilistic models, applicable when the components of a complex linked system behave independently (a so-called Markov chain). Up to a point, this is valid for residue mutations in globular protein sequences, and HMMs are being applied increasingly in multiple-alignment algorithms and profile-type database searches (17,29). The models are more complex than the widely used PAM model for protein evolution introduced by Margaret Dayhoff (3), providing both advantages and disadvantages. On the plus side, the models provide direct probabilities for evaluating database search matches, which can include multiple matches in a repeated sequence and can formally treat biological complexities such as splice junctions in genomic sequence. On the debit side, the extra parameters lead to more complex optimization problems at several levels. Thus, inexperienced users may set up poorly optimized HMMs while, for program developers, there is a problem as to whether the most appropriate HMMs are being applied to sequence evolution. It is important for the user not to be seduced by technical jargon, but to justify the use of newer methods such as these on the basis of convincing results.

ERROR IN SEQUENCE ALIGNMENTS

Errors are very common and invalidate any conclusions obtained by methods based on sequence alignments (see Fig. 2). The three main sources of error are: the input sequences, mistakes by the user, or alignment algorithm failure. Causes and effects of error are manifold; some major ones are discussed below.

Errors in the Input Sequences

Experimental errors in determining sequences include double-insert cloning errors, truncated cDNA clones, base insertion or deletion causing translation **frameshifts**, and translation with inappropriate **genetic code** (eg, for plastid-encoded proteins). Figure 3 shows an example of a frameshift error in a database entry. Algorithm failure during alignment may be induced by sequence error, such as frameshifted regions, in which the sequences are no longer similar, and may induce incorrect gap placement (Fig. 2). Errors in sequences are very common and likely to be present in any sequence family (30,31).

Further errors can arise in preparing sequence database entries. These can affect any part of the entry and can have particularly unpredictable consequences. However, the most common annotation errors are undoubtedly in predicted translation products and are a consequence of the limited accuracy of current gene prediction algorithms; that is, despite high-quality DNA sequence, the predicted protein sequence might contain artificially translated **introns**, missed exons, or terminal truncations or might be an artificial fusion product of two independent genes. Protein sequence alignments can often help in identifying **translation** problems.

Mistakes by the User

Generally these are due to inadequate attention to detail. Erroneous inclusion of nonhomologous sequences in the input set is quite common, particularly if keyword searches are used to extract a set of sequences; there are many examples of unrelated proteins performing equivalent functions, as well as sequences that have functions incorrectly ascribed to them. Another source of error occurs during the alignment process, when parameters may be set poorly. Trial and error is usually needed to find the parameters that best suit a particular group of sequences. For example, insertion of too few or too many gaps may suggest that the gap penalties are not optimal. It is important to take the time to get a basic understanding of how a given program works, or it is unlikely to do the best job.

Algorithm Failure

Clustered alignment is a heuristic strategy that is not guaranteed to find the optimal multiple alignment. The alignment process is likely to compound errors introduced by the user or in the input sequences. Alignment mistakes will also be made in difficult alignment cases, even when there are no input errors. There are many instances of homologous proteins

```
                            ******************************************  ***!!*
M.genitalium EFTA 151       AEEVRDLLTSYGFDGKNTPIIYGSALKALEGDPKWEAKIHDLIKAVDE^W
Translation                 AEEVRDLLTSYGFDGKNTPIIYGSALKALEGDPKWEAKIHDLMNAVD^^W
M.pneumoniae DNA  619       ggggcgttattgtggaaacaatgtgcagcgggcatggaacgtaagggGAt
                            caatgattccagtagaaccttagcctactagacagacataattacta  g
                            aagatcaattccttcgcctttttatagtatttggatgcttagtatt  g

                            **** ****!   ********************************* ***
M.genitalium EFTU 200       IPTPTREVDG+++PFLLAIEDTMTITGRGTVVTGRVERGELKVGQEVEIV
Translation                 IPTPEREVD^^^^PFLLAIEDTMTITGRGTVVTGRVERGELKVGQEIEIV
M.pneumoniae DNA  765       acacgcgggAAACctttgaggaaaaagcgaggagcggcggtaggcgagag
                            tcccagata    ctttctaactctcgggcttcggtaggatatgaatatt
                            tattatagc    gcggacaccggttcttcgtctgtattagaataatact
```

Figure 3. Frameshifted segment in the *Mycoplasma pneumoniae* EFTU sequence revealed by comparing the DNA to the closely related *Mycoplasma genitalium* EFTU using a dynamic programming three-frame comparison allowing shifts between translation frames (30). Exclamation points mark the frameshift sites. The first frameshift is caused by a base being dropped, and the second frameshift is caused by a base being added, thereby returning to the original translation frame. See Figure 2 for a multiple alignment spanning this region.

having diverged over long periods of time, so that they have apparently very little sequence similarity. The "twilight zone" where sequence similarity merges with sequence dissimilarity is in the range 20% to 25% identity. (Five percent identity would be a random match for protein sequence, neglecting residue biases. In practice, given residue biases and gap insertions to maximize the similarity, the expectation for random sequence matches approaches ~15% identity, but higher for short or biased sequences. There is, however, no *a priori* reason why correctly aligned but extremely divergent proteins should not be found with 0% pairwise identity.) Divergent nucleotide sequences are even harder to align, because random similarity is reached at ~25% identity before gaps are added. Wherever possible, alignments need to be checked against additional data available for a sequence family, such as known **tertiary structures** and whether invariant catalytic residues, or any other known conserved motifs, are correctly aligned.

Consequences of Errors in Aligned Sequences

Errors in multiple alignments can have disastrous consequences for **phylogenetic** inference on the basis of sequence trees. Sequences with misaligned segments or translation frameshifts are apparently more divergent than they should be from the other sequences. The branch leading to that sequence will then have a longer length (which erroneously implies a more rapid **molecular clock**) and the branch point may migrate toward the centre of the tree, giving a false order of divergence. Such incorrect phylogenies may be quite exciting when they seem to refute established viewpoints. Two rules of thumb are useful: (i) Infinitely more wrong tree topologies can be generated than right ones, and (ii) wrong phylogenies are more interesting than right phylogenies. Incautious advocacy of a wrong phylogeny can waste many peoples' time.

Errors also disrupt evaluation of conserved sites in sequences (Fig. 2). Catalytic residues are often absolutely conserved, so a single misaligned sequence may lead to rejection of the correct site. Most conserved residues have a structural role: Structure prediction for protein uses conserved **hydrophobic** residues and, for RNA, conserved base-pairing residues. Misalignments disrupt the conservation periodicities, leading to rejection of the correct structures. Terminally truncated sequences of multidomain proteins can lead to false inferences for the domain boundaries, which are very usefully defined by coincidence with the protein *N*- or *C*-termini.

BIBLIOGRAPHY

1. S. B. Needleman and C. D. Wunsch (1970) *J. Mol. Biol.* **48**, 443–453.
2. T. F. Smith and M. S. Waterman (1981) *Adv. Appl. Math.* **2**, 482–489.
3. M. O. Dayhoff, R. M. Schwartz, and B. C. Orcutt (1978) In *Atlas of Protein Sequence and Structure*, Vol. 5, Suppl. 3 (M. O. Dayhoff, ed.), NBRF, Washington, pp. 345–352.
4. S. A. Benner, M. A. Cohen, and G. H. Gonnet (1994) *Protein Eng.* **7**, 1323–1332.
5. G. Vogt, T. Etzold, and P. Argos (1995) *J. Mol. Biol.* **249**, 816–831.
6. O. Gotoh (1982) *J. Mol. Biol.* **162**, 705–708.
7. E. W. Myers and W. Miller (1988) *CABIOS* **4**, 11–17.
8. J. D. Thompson (1995) *CABIOS* **11**, 181–186.
9. J. A. Grice, R. Hughey, and D. Speck (1997) *CABIOS*, **13**, 45–53.
10. M. S. Waterman and M. Eggert (1987) *J. Mol. Biol.* **197**, 723–728.
11. S. F. Altschul, W. Gish, W. Miller, E. W. Myers, and D. J. Lipman (1990) *J. Mol. Biol.* **215**, 403–410.
12. W. R. Pearson and D. J. Lipman (1988) *Proc. Natl. Acad. Sci. USA* **85**, 2444–2448.
13. B. Rost and C. Sander (1993) *J. Mol. Biol.* **232**, 584–599.
14. A. Lupas, M. Van Dyke, and J. Stock (1991) *Science*, **252**, 1162–1164.
15. B. Persson and P. Argos (1994) *J. Mol. Biol.* **237**, 182–192.
16. M. Gribskov, A. D. McLachlan, and D. Eisenberg (1987) *Proc. Natl. Acad. Sci. USA*, **84**, 4355–4358.
17. S. Eddy (1996) *Curr. Opin. Struct. Biol.* **6**, 361–365.
18. P. Bork and T. J. Gibson (1996) *Methods Enzymol.* **266**, 162–184.
19. C. R. Woese, R. R. Gutell, R. Gupta, and H. F. Noller (1983) *Microbiol. Rev.* **47**, 621–669.
20. F. Michel and E. Westhof (1990) *J. Mol. Biol.* **216**, 585–610.
21. R. F. Doolittle (1990) *Molecular Evolution: Computer Analysis of Protein and Nucleic Acid Sequences, Methods in Enzymology*, Vol. 183, Academic Press, San Diego, CA.
22. D. J. Lipman, S. F. Altschul, and J. D. Kececioglu (1989) *Proc. Natl. Acad Sci USA* **86**, 4412–4415.
23. O. Gotoh (1996) *J. Mol. Biol.* **264**, 823–838.
24. C. Notredame and D. G. Higgins (1996) *Nucleic Acids Res.* **24**, 1515–1524.
25. J. D. Thompson, D. G. Higgins, and T. J. Gibson (1994) *Nucleic Acids Res.* **22**, 4673–4680.
26. D.-F. Feng and R. F. Doolittle (1987) *J. Mol. Evol.* **25**, 351–360.
27. N. Saitou and M. Nei (1987) *Mol. Biol. Evol.* **4**, 406–425.
28. A. M. Lesk, M. Levitt, and C. Chothia (1986) *Protein Eng.* **1**, 77–78.
29. A. Krogh, M. Brown, S. Mian, K. Sjölander, and D. Haussler (1994) *J. Mol. Biol.* **235**, 1501–1531.
30. E. Birney, J. D. Thompson, and T. J. Gibson (1996) *Nucleic Acids Res.* **24**, 2730–2739.
31. J. M. Claverie (1993) *J. Mol. Biol.* **234**, 1140–1157.

Suggestions for Further Reading

R. F. Doolittle (1990) *Molecular Evolution: Computer Analysis of Protein and Nucleic Acid Sequences, Methods in Enzymology*, Vol. 183, Academic Press, San Diego, CA.

R. F. Doolittle (1996) *Computer Methods for Macromolecular Sequence Analysis, Methods in Enzymology*, Vol. 266, Academic Press, San Diego, CA.

D. Sankoff and J. B. Kruskal (1984) *Time Warps, String Edits and Macromolecules: The Theory and Practice of Sequence Comparison*, Addison-Wesley, Reading, MA.

M. S. Waterman (1989) *Mathematical Methods for DNA Sequences*, CRC Press, Boca Raton, FL.

ALKALINE PHOSPHATASE

NICHOLAS ALLEN

Alkaline phosphatases (E.C. 3.1.3.1) belong to a family of orthophosphoric monoester phosphohydrolases that have an alkaline pH optimum. Their genes are very frequently used as a **reporter** gene. Alkaline phosphatase activity is most commonly detected by the hydrolysis of 5-bromo-4-chloro-3-indolyl phosphate (BCIP), which, when coupled to the reduction of nitro blue tetrazolium (NBT), forms a formazan and an indigo dye that together form a strong black/purple precipitate (1). In addition, a number of fluorogenic substrates are also available (1). 4-Methylumbelliferyl phosphate (MUP) gives a blue **fluorescent** product upon hydrolysis, and 2-hydroxy-3-naphthoic acid-2-phenylanilide phosphate (HNPP/Fast Red TR) fluoresces with a broad emission peak between 540 and 590 nm and can be observed using either fluorescein or rhodamine filter sets. Molecular Probes Inc. have also developed a proprietary substrate called ELF-97 (Enzyme Linked Fluorescence-97), in which cleavage of the molecule converts it from a soluble phosphate to an insoluble alcohol, with an accompanying shift from weak blue fluorescence to a bright yellow fluorescence (2).

Human placental alkaline phosphatase (hpAP) is most commonly used as a reporter enzyme in nonradioactive detection systems, where the enzyme is linked to other molecular probes (eg, specific antibodies)—for example, for the detection of proteins and nucleic acids by **Western blot**, **Southern blot**, and **Northern blot** analysis, and most commonly in ***in situ* hybridization**. Elegant studies have also been performed in which hpAP is fused to soluble extracellular domains of receptor molecules, which are then used as probes to detect the sites of ligand production *in vivo* and facilitate the subsequent **cloning** of the ligand genes (3).

Although animals express a number of alkaline phosphatase genes, the human placental isoform has been developed as a reporter gene, because it can be distinguished from other endogenous isoforms through its high thermostability (4). Thus all background endogenous alkaline phosphatase activities, from embryonic, intestinal, and nonspecific genes, can be minimized by preheating tissue preparations up to 80°C for prolonged periods. hpAP also retains its activity following histological processing for wax imbedding and sectioning tissues. In addition, background from endogenous alkaline phosphatases can be further inhibited by the amino acids L-phenylalanine or L-homoarginine.

hpAP has been used in a wide range of applications, including *in vitro* transfection studies and transgenic studies *in vivo*. The sensitivity of hpAP in transient expression assays is equivalent to that of chloramphenicol acetyltransferase (CAT) (see **Reporter genes**). A particularly useful variant of hpAP is a **cDNA** encoding a secreted form of the protein (5,6) that allows hpAP activity to be assayed by sampling tissue culture medium, giving the benefit of monitoring changes in gene expression with time. In this system, background activities are further eliminated, because the endogenous isoforms are anchored to the cell membranes and do not contribute to the activity in the culture supernatants.

hpAP is also an effective reporter gene to analyze gene expression *in situ* in tissue preparations. hpAP was first used in retroviral vectors to infect small numbers of cells in developing embryos as a tool to study cell fate and lineage analysis; subsequently, hpAP has been used as a robust reporter gene in transgenic mice (7). Indeed, mice that express high levels of hpAP from a ubiquitously expressed **promoter** thrive with no adverse effects. hpAP is particularly useful to use in combination with a second reporter gene, such as *lacZ* (**beta-galactosidase**), in dual labeling studies, as the common substrates are quite distinct and give different colored products (8).

BIBLIOGRAPHY

1. D. A. Knecht and R. L. Dimond (1984) *Anal. Biochem.* **136**, 180–184.
2. K. D. Larison et al. (1995) *J. Histochem. Cytochem.* **43**, 77–83.
3. J. G. Flanagan and P. Leder (1990) *Cell* **63**, 185–194.
4. P. Henthorn (1988) *Proc. Natl. Acad. Sci.* **85**, 6342–6346.
5. T. T. Yang, P. Sinai P. A. Kitts, and S. R. Kain (1997) *Biotechniques*, **23**, 1110–1114.
6. B. R. Cullen and M. H. Malim (1992) *Methods Enzymol.* **216**, 362–368.
7. S. E. DePrimo, P. J. Stambrook, and J. R. Stringer (1996) *Transgenic Res.* **5**, 459–466.
8. X. Li, W. Wang, and T. Lufkin (1997) *Biotechniques* **23**, 874–878.

ALKYLATION

T. IMOTO

Alkyl groups have the general formula $C_nH_{2n+1}-$. Alkylation is a reaction that introduces an alkyl group into a compound. Commonly used alkylating agents are olefins, alkyl halides, and alcohols. Biologically important alkylation occurs on nitrogen and oxygen atoms in polynucleotides. Alkylating agents that cause these reactions are **mutagens** and **carcinogens** and include methyl methanesulfonate, dimethyl sulfate, *N*-methyl-*N*-nitrosourea, dimethylnitrosourea, 1-methyl-3-nitro-1-nitrosoguanidine, etc. Some alkylating agents are used as anticancer drugs, including cyclophosphamide, nitrogen mustard and its oxide, and triethylenephosphoramide.

Alkylating agents easily alkylate nucleophiles in proteins such as those on **cysteine** (1), **histidine** (2), and **methionine** (3) residues, **amino groups** (4), and sometimes **tyrosine** residues and **carboxyl groups** (5). The most well-known alkylation is that for blocking free **thiol groups** of cysteine residues to prevent their oxidation, especially in the course of determining the **primary structure** of a protein (see **Protein sequencing**).

REDUCTION AND ALKYLATION OF PROTEINS

The protein at a concentration of 2% (w/v) is dissolved in 0.5 M **Tris** buffer, pH 8.1, containing 6 M **guanidinium chloride** and 0.002 M **EDTA** and reduced with **dithiothreitol** (50 mol/mol of **disulfide bond**) at 30° to 40°C for 4 h under nitrogen (6). Then the solution is cooled to room temperature, and iodoacetamide (100 mol/mol of the original disulfide bond) is added, with the addition of NH_4OH as necessary to maintain a constant pH. Other alkylating agents that can be employed instead are iodoacetic acid, 4-vinylpyridine, N-ethylmaleimide, and ethyleneimine.

BIBLIOGRAPHY

1. J. Bridgen (1972) *Biochem. J.* **126**, 21–25.
2. A. M. Crestfield, W. H. Stein, and S. Moore (1963) *J. Biol. Chem.* **238**, 2413–2420.
3. J. M. Gleisner and R. L. Blaklay (1975) *J. Biol. Chem.* **250**, 1580–1587.
4. H. J. Goren and E. A. Barnard (1970) *Biochemistry* **9**, 974–983.

5. K. Takahashi, W. H. Stein, and S. Moore (1967) *J. Biol. Chem.* **242**, 4682–4690.
6. M. J. Waxdal et al. (1968) *Biochemistry* **7**, 1959–1966.

ALLELIC EXCLUSION

MICHEL FOUGEREAU

The **clonal selection theory** put forward by Burnet (1) and Jerne (2) proposes that each single lymphocyte expresses only one single type of **antibody**. This theory received its first structural confirmation when it was shown that patients with multiple myeloma, a proliferative disorder of plasma cells, had in their serum a highly elevated level of only one molecular species of **immunoglobulin**, originating from the corresponding malignant clone. Single-cell experiments performed by the Nossal group in Australia (3) indicated that each cell isolated from mouse **immunized** with *Salmonella* **antigens** produced only one antibody of one given specificity. These observations were in agreement with the clonal theory, but did not explain by which mechanism this was achieved. An elegant approach to that problem was made with the analysis of allotypes in heterozygous animals. Allotypes, as defined by Oudin (4), are antigenic specificities that are shared by a group of animals within a given species. These genetic markers are **allelic** variants of immunoglobulins that are found on both the heavy and light chains. They have been studied extensively in the rabbits, mice, and humans. In most cases, they are the result of a few amino acid substitutions that confer specific **epitopes** (or allotopes) that can be recognized by specific antibodies. In a rabbit heterozygous for H- and L-chain allotypes, Oudin showed that any given immunoglobulin molecule always had two identical heavy chains and two identical light chains with respect to the corresponding allotypes. This ensured that the Ig molecule was symmetrical and therefore had two identical antibody combining sites, an obvious prerequisite of the clonal theory. The next step was made at the intracellular level, when Pernis et al. (5) showed with specific anti-allotype antibodies that, in a heterozygous rabbit, any given cell always expressed only one allotype of each heavy or light chain. This suggested that a **B cell**, although **diploid**, had only one of its two alleles functional, and the phenomenon was termed "allelic exclusion."

The molecular mechanisms that account for this phenomenon could be understood only when the complex mosaic structures of immunoglobulin (Ig) genes were deciphered. Each Ig heavy or light chain contains an NH_2-terminal variable and a COOH-terminal constant region, each encoded by a multiplicity of gene segments. Diversity of the variable regions, which is directly linked to the necessity to produce a very large number of different antibody molecules with a limited number of genes, is generated by specific mechanisms of **gene rearrangements** that take place exclusively in lymphocytes. The heavy-chain **variable region** is encoded after the rearrangement of three discrete sets of germline gene segments (V_H, D, J_H), and two for the light chain (V_L, J_L). Gene rearrangements take place sequentially during B-cell differentiation in the bone marrow, in a strictly regulated manner, as initially proposed by Alt and Baltimore. The heavy-chain gene segments rearrange D to J_H and V_H to D–J_H. The gene-segment joinings are random, so only one-third of the rearranged genes have an open reading frame. As soon as one rearrangement is in frame, the corresponding μ chain is expressed. Whenever the first rearrangement is successful, the second allele remains in the germline configuration, which indicates that a regulatory mechanism has taken place. It was shown by transfection experiments and by making μ transgenic mice that the negative feedback inhibition of the gene rearrangement of the second allele was controlled by the μ chain itself, when expressed at the cell surface. It was also shown that, at this "preB stage," the μ chain was associated with a surrogate light chain (ΨL), which is monomorphic and composed of two polypeptide chains encoded by the $\lambda 5$ and VpreB genes. Expression of $\mu-\Psi$L has been shown to down-regulate the heavy-chain gene rearrangement, whereas rearrangement of the light chain locus is switched on, in the same random reading frames as above. The final result of this complex sequence of events is that, at each locus, only one allele is functional; the other one either remains in the germline configuration or has an out-of-frame rearranged gene, thereby accounting for the allelic exclusion phenomenon.

See also entries **Clonal selection theory**, **Gene rearrangement**, and **Antibodies**.

BIBLIOGRAPHY

1. M. F. Burnet (1959) *The Clonal Selection Theory of Acquired Immunity*, Vanderbilt University Press, Nashville, TN.
2. N. K. Jerne (1955) The natural selection theory of antibody formation. *Proc. Natl. Acad. Sci. USA* **41**, 849–857.
3. G. J. V. Nossal, A Szenberg, G. L. Ada, and C. M. Austin (1964) Single cell studies on 19 S antibody production. *J. Exp. Med.* **119**, 485–502.
4. J. Oudin (1960) Allotypy of rabbit serum proteins. I. Immunochemical analysis leading to the inoliviolualization of seven main allotypes. *J. Exp. Med.* **112**, 107–124.
5. B. Pernis, G. Chiappino, A. S. Kelus, and P. G. H. Gell (1965) Cellular localisation of immunoglobulins with different allotypic specificities in rabbit lymphoid tissues. *J. Exp. Med.* **122**, 853–876.
6. F. Alt (1984) Exclusive immunoglobulin genes. *Nature* **312**, 502–503.

Suggestions for Further Reading

E. ten Boeckel, F. Melchers, and A. G. Rolink (1998) Precursor B cellsshowing H chain allelic inclusion display allelic exclusion at the level of pre-B cell receptor surface expression. *Immunity* **8**, 199–207.

D. Löffert, A. Ehlich, W. Müller, and K. Rajewsky (1996) Surrogate light chain expression is required to establish immunoglobulin heavy chain allelic exclusion during early B cell development. *Immunity* **4**, 133–144.

ALLOANTIBODY, ALLOANTIGEN

MICHEL FOUGEREAU

Alloantigens are **allelic** variants that can induce, when injected into animals of the same species but having a distinct genetic background, the production of the corresponding alloantibodies. The most commonly known example of alloantigens are the blood group substances, which define the basic

ABO blood groups in humans. Blood group substances have long been identified as being derived from a polysaccharide core that is found in pneumococcus group XIV, onto which three genes encoding for glycosyl transferases will serially add units that will confer a specific antigenicity to the newly derived molecule. Three genes operate this system in humans: A, B, and H. The H gene is present in all individuals and will add one residue of fucose, leading to the so-called H substance. In individuals who possess the A gene, an additional GalNAC (*N*-acetylgalactosamine) will be added, providing the A substance, which is expressed at the red blood cell surface and confers the A group. In those who have the B gene, a galactose residue is added to H, making the B substance, which is also expressed at the cell surface of red cells in individuals of group B. The A and B genes are codominant, so an individual will express the A, the B, or both the A and B molecules at the red blood cell surface, depending upon what A and B genes are present. If neither the A nor B allele is present, the group will be O. There are many other blood groups in humans, with special reference to the Lewis groups, which are also derived from the same polysaccharide backbone by addition of different units. What is peculiar regarding the ABO blood group system is the fact that individuals who lack the A or the B specificity spontaneously produce alloantibodies directed against the missing substance. So an individual of group A will make anti-B alloantibodies, an individual of group B will produce anti-A, and one of group O will have both. Very severe accidents in blood transfusion would result from the agglutination of the donor red blood cells by the alloantibodies of the recipient. The reverse situation, agglutination of the host red blood cells by antibodies of an incompatible donor, should also be avoided, although accidents are less severe. Antigen compatibility between donor and recipient is therefore a must. ABO compatibility should also be observed in allografts.

Even if the genetic determinism of these alloantigens is straightforward, it still is not clear why the alloantibodies corresponding to the nonexpressed substance(s) are produced. The prevalent explanation is that blood groups cross-react with bacteria normally present in the intestinal flora, which would stimulate the immune system to produce the corresponding crossreacting antibodies. It should be stressed that these "natural" antibodies are of the **IgM** type, which may be directly related to the fact that they are the result of the stimulation by a T-independent polysaccharide antigen. The titer of alloantibodies varies greatly between individuals, but may be considerably elevated upon an incompatible transfusion.

Another famous case of alloantigen in the blood groups is the Rhesus factor, which was responsible for the dramatic hemolytic disease of the newborn, due to the immunization of an Rh mother against red blood cells of an Rh^+ fetus. This occurs during the delivery because some fetal red cells may enter the maternal circulation and induce the formation of **IgG** antibodies that will actively cross the placental barrier in a subsequent pregnancy and then provoke lysis of Rh^+ fetus red cells. Treatment involves complete transfusion of the newborn with Rh blood. It has now been generalized to prevent the immunization by injecting the Rh mother with anti-D (anti-Rhesus) antibodies immediately after delivery, to trap the red blood cells from the newborn that would have penetrated the maternal circulation at birth.

Many other alloantigens are known, but alloantibodies are generally not produced unless the antigen is given. This is the case of the major histocompatibility complex (MHC) molecules, which constitute a major problem in transplantation. The name of **major histocompatibility complex** indicates by itself how these molecules were first discovered as a major target for graft rejection, as the result of a very severe alloimmune response, characterized primarily by the production of **cytotoxic T lymphocytes**. Many other systems may behave as potential alloantigens, which simply reflects the existence of allelic variants. The case of allotypes of **immunoglobulins** has been studied particularly by immunologists and has provided remarkable genetic markers for the study of **immunoglobulin biosynthesis** and diversity at the time.

Suggestion for Further Reading

P. L. Mollison, C. P. Engelfriet, and M. Contreras (1987) *Blood Transfusion in Clinical Medicine*, 8th ed., Blackwell, Oxford, UK.

ALLOPHENIC

JAMES KENNISON

Allophenic individuals are composed of cells of two different genotypes (often called mosaics or **chimeras**). Allophenic mice are the basis of the revolution in **mouse** genetics, allowing mice with specific gene replacements to be generated and studied. Allophenic mice are made by mixing cells from two **embryos** of different genotypes (1). These new composite embryos are implanted into foster mothers and allowed to develop. The early embryos are able to compensate and form a single individual from the mixture of embryonic cells. The mosaic progeny that result from these manipulations have tissues that contain cells of the two different genotypes. The allophenic mice have also been called tetraparental mice. When the germ cells are also of mixed origin, progeny can be recovered from both genotypes.

The ability to make allophenic mice in the laboratory was first used to study the contributions of cells to individual tissues. The number of cells that give rise to a particular tissue can be estimated from the proportions of the two genotypes in a large number of allophenic mice. The cellular autonomy of mutant phenotypes can be determined by generating allophenic mice between mutant and wild-type mouse embryos. The two cell types usually differ both in the mutant of interest and in some marker genotype, such as an **enzyme** polymorphism.

Two advances in the technology of allophenic mice have contributed to a revolution in mouse genetics in the last few years. The first advance was the ability to use embryonic teratocarcinoma cells from cell culture as one of the two cell types used to make the allophenic mice (2). The cultured cells are injected into the interior of a normal mouse embryo and are incorporated into the embryo, to form an allophenic mouse. The teratocarcinoma cells are totipotent and can even form germ cells that give rise to the **gametes** for the next generation. The second major advance was the development of techniques for **site-directed mutagenesis** and gene replacement in mammalian cells in culture (3) (see **Gene targeting**). A gene of interest is altered in a specific way in teratocarcinoma cells in culture. A clone of cells with the altered genotype is produced, and cells from the clone are injected into early mouse embryos to produce allophenic mice. These allophenic mice can be studied as mosaic individuals or can be allowed to develop to fertile adults. If some of the germ cells in the adult allophenic mice are derived from the genetically altered teratocarcinoma cells, progeny can be recovered from the gametes produced by the teratocarcinoma cells and used to found a mutant strain of the specific genotype desired. This has led to the production of hundreds of new mouse strains mutant for developmentally important genes. It has also made the construction of mouse models for human disease far easier than in the past. These technologies have been extended from mice to a variety of other mammals; and they could be used to alter the human **genome** genetically, with far-reaching consequences.

BIBLIOGRAPHY

1. B. Mintz (1962) *Science* **138**, 594.
2. B. Mintz and K. Illmensee (1975) *Proc. Natl. Acad. Sci. USA* **72**, 3585–3589.
3. M. R. Capecchi (1989) *Science* **244**, 1288–1292.

Suggestions for Further Reading

B. Mintz (1974) *Annu. Rev. Genetics* **8**, 411–470.
R. Jaenisch (1988) *Science* **240**, 1468–1474.

ALLOSTERY

DONALD PETTIGREW

Allostery is used to describe regulatory phenomena in biological systems. The term "allosteric" (other shape) was coined by Monod and coworkers to describe a particular type of regulatory behavior and marks an important synthesis in the development of biochemistry and molecular biology. Since its inception, the term has evolved to describe several related concepts and is used today to describe a variety of phenomena. To some, it is associated with a particular type of regulatory behavior, while to many it is used to describe several different aspects of regulation. The development of the concept, the meanings it has come to have, its application in describing regulatory properties of **proteins** and **enzymes**, and recent developments that indicate the need to use it in its original sense are described here.

The allosteric concept was conceived in the early 1960s by Monod and coworkers to describe the properties of regulatory enzymes. In the mid-1950s, several investigators described enzymes whose catalytic properties were dependent on effector molecules other than the substrate. The first well-characterized example of this behavior was the finding by Cori and colleagues that 5′-AMP is required for the *in vitro* catalytic activity of **glycogen phosphorylase** b from resting skeletal muscle (1). Studies of the regulation of biosynthetic pathways in bacteria showed that the catalytic properties of the first enzyme in a pathway are often modulated by the end product of the pathway, which has a structure different from that of the substrate—for example, isoleucine inhibition of threonine deaminase (see **Threonine operon**) (2) and CTP inhibition of **aspartate transcarbamoylase** (3). The differences in the structures of the substrates and effectors were first noted by Monod and Jacob, who postulated in a seminal paper in 1961 that the effectors bind to separate and distinct sites (4). They coined the term *allosteric site* for the effector binding site to distinguish it from the **active site** and postulated that the effectors act indirectly by causing changes in the conformation of the enzyme that alter its catalytic site and kinetic properties. In 1963, Monod and coworkers compiled the available data on regulatory enzymes (5). These enzymes all contained more than one subunit per enzyme molecule–that is, dimers, tetramers, and dodecamers. For several of these enzymes, changes in **quaternary structure** (association–dissociation of the subunits) had been shown to occur upon addition of the effectors. Monod was also aware of unpublished crystallographic results showing that the oxygen-binding sites on **hemoglobin** are located far from one another and that the distances between the amino acid residues labeled by heavy atoms change upon binding of oxygen (Ref. 6, pp. 577–578). On this basis, the key elements of a general model for the functional structures of regulatory enzymes were formulated:

1. Each molecule of the native enzyme contains multiple subunits with binding sites for substrates and ligands, and the regulatory behavior depends on the relationships between the subunits.
2. No direct interactions between substrate(s) or effectors are required, that is, the effectors act indirectly on the catalytic site.
3. The actions of the effectors are due entirely to a reversible conformational change in the protein that is induced by effector binding.

Monod et al. postulated that protein conformational change is the basis for control and coordination of chemical events in living cells, and Monod considered this to be the "second secret of life" (Ref. 6, p. 576). Structural studies using **X-ray crystallography** methods support the concepts of this general model (see **Enzyme regulation**).

These concepts were used to develop the concerted model, a specific molecular model for describing regulatory behavior (7). This model was developed as a plausible explanation for the positive cooperativity in the binding of oxygen to hemoglobins. Because the apparent affinity for binding of oxygen changes as its concentration is varied, these are called *homotropic interactions*. The concerted model retains the aspects of the general model with respect to oligomeric structure of the protein, multiple binding sites located far from one another, and conformational changes upon binding of one ligand that are transmitted by protein conformational changes to the other ligand binding sites, thereby altering the binding affinities. This model defines the relations between the conformational changes and ligand binding. The word *concerted* refers to the all-or-none nature of the conformational transition. That is, only two conformations are allowed for the subunits in an oligomer and all of the subunits in a given oligomer have the same conformation. The conformations are denoted by the terms *R-state* and *T-state*, and they differ in affinity for binding ligands. The R-state has the higher affinity for ligands. Positive cooperativity in oxygen binding is explained by a large predominance of the unliganded protein in the T-state with a shift to the R-state upon binding of oxygen. The model was extended to enzymes by assuming that the interactions between the substrate and enzyme are in rapid equilibrium and that all forms of the enzyme have the same V_{max}. At this juncture, the term *allosteric* takes a different meaning. The conformational change between the two states is called an *allosteric transition*. So, in addition to referring to sites for molecules other than the substrate, it refers to interactions between substrate binding sites on different subunits. The concept of the "other shape" is extended to the enzymes, which are termed "allosteric enzymes."

This initial formulation of the concerted model describes homotropic interactions. The original considerations of the allosteric concept were developed to treat cases in which the behavior of one molecule, the substrate, depends on the concentration of another molecule. These are called *heterotropic interactions*. The regulatory enzymes whose properties were catalogued by Monod et al. showed both homotropic and heterotropic effects (5). The phenomenon of desensitization—that is, treatments that result in loss of substrate homotropic interactions and heterotropic interactions, but do not affect catalytic activity, had been described for some of these enzymes. Monod et al. postulated that desensitization means that the molecular basis might be the same for both homotropic and heterotropic interactions. They stated that if this were the case, the concerted model could also be used to describe heterotropic effects—that is, the original allostery. In the model, activating effectors have higher affinity for the R-state, while inhibitory effectors have higher affinity for the T-state. This issue has resulted in great confusion in the use of the model. All subsequent treatments presume that a single allosteric transition is the basis for both homotropic and heterotropic effects. This presumption is particularly evident in structural studies of regulatory proteins, where the terms R-state and T-state dominate discussions.

In the three cases of regulatory proteins that have been examined with care, it is clear that the concerted model fails. More than two states are involved in oxygen binding to hemoglobin (8,9). Homotropic and heterotropic interactions are not due to the same transition in aspartate transcarbamoylase (10,11). Although the same T-state is predicted for **phosphofructokinase** with different inhibitors, experimental results show that the properties of the inhibited forms are different (12). These results argue strongly that the term *allosteric* should be used in a stricter sense to refer to effects of one ligand on the binding of another, independently of assumptions about the mechanism of the interaction.

BIBLIOGRAPHY

1. E. H. Fischer, A. Pocker, and J. C. Saari (1971) *Essays Biochem.* **6**, 23–68.
2. H. E. Umbarger (1956) *Science* **123**, 848.
3. R. A. Yates and A. B. Pardee (1956) *J. Biol. Chem.* **221**, 757–780.
4. J. Monod and F. Jacob (1961) *CSHSQB* **26**, 389–401.
5. J. Monod, J.-P. Changeaux, and F. Jacob (1963) *J. Mol. Biol.* **6**, 306–329.
6. H. F. Judson (1979) *The Eighth Day of Creation: The Makers of the Revolution in Biology*, Simon & Shuster, New York.
7. J. Monod, J. Wyman, and J.-P. Changeaux (1965) *J. Mol. Biol.* **12**, 88–118.
8. G. K. Ackers and J. H. Hazzard (1993) *Trends Biochem. Sci.* **18**, 385–390.
9. J. M. Holt and G. K. Ackers (1995) *FASEB J.* **9**, 210–218.
10. W. N. Lipscomb (1994) *Adv. Enzymol. Rel. Areas Mol. Biol.* **68**, 67–151.
11. R. C. Stevens, Y. M. Chook, C. Y. W. N. Cho, Lipscomb, and E. R. Kantrowitz (1991) *Protein. Eng.* **4**, 391–408.
12. V. L. Tlapak-Simmons and G. D. Reinhart (1994) *Arch. Biochem. Biophys.* **308**, 226–230.

ALPHA-BUNGAROTOXIN AND CURARE-MIMETIC TOXINS

C. MONTECUCCO

The venom of the banded krait snake (*Bungarus multicinctus*) contains a variety of protein **toxins** that are collectively called bungarotoxins. As is often the case, an animal venom is, in fact, a mixture of different toxins specific for different target molecules. The α-neurotoxins, or curare-mimetic toxins, bind specifically to the **acetylcholine receptor** with high affinities ($\boldsymbol{K}_d$ in the 10^{-9} to 10^{-12}M range for neuronal and muscular receptors of different species) and prevent the opening of the ion **channel** caused by acetylcholine binding (1,2). Such inhibition of the transmission of the nerve-muscle impulse results in a flaccid paralysis, which may end in respiratory failure and death.

α-bungarotoxin is the prototype of a large group of monomeric toxins, produced by many snakes, including cobra and sea snakes, and composed of 66 to 74 amino acid residues with four or five **disulfide bonds**. Their three-dimensional structure is rather flat, with three adjacent β-sheet loops (3) that expose residues essential for receptor binding: Lys-27, Trp-29, Arg-39, and Lys-55 (numbering of erabutoxin) (Fig. 1, see top of next page) (4).

Many species of snakes, including *B. multicinctus*, also produce α-neurotoxins of smaller size. Their polypeptide chains are 60 to 62 residues long but adopt the same three-finger β-sheet fold, stabilized by four disulfide bonds. At variance from the longer α-neurotoxins, these toxins are dimers, but the same three positively charged residues are essential for receptor binding. The corresponding area of the acetylcholine receptor involved in α-neurotoxin binding has not been mapped in detail, but the segment of residues 185 to 199 plays a major role, together with residues 128 to 142.

The various α-neurotoxins bind with a range of affinities and specificities to the multitude of muscular and neuronal acetylcholine receptors; fluorescent and **radiolabeled** toxin derivatives are invaluable tools for the study of the structure and function of such receptors (5).

BIBLIOGRAPHY

1. A. L. Harvey, ed. (1991) *Snake Toxins*, Pergamon Press, New York.
2. R. Rappuoli and C. Montecucco (1997) *Guidebook to Protein Toxins and Their Use in Cell Biology*, Oxford University Press, Oxford, UK.
3. M. D. Walkinshaw, W. Saenger, and A. Maelicke (1980) *Proc. Natl. Acad. Sci. USA* **77**, 2400–2404.
4. O. Tremeau et al. (1995) *J. Biol. Chem.* **270**, 9362–9369.
5. A. Devillers-Thiery et al. (1993) *J. Membr. Biol.* **136**, 97–112.

ALPHA-HELIX (3_{10}-HELIX AND PI-HELIX)

JENNIFER MARTIN

The right-handed α-helix is one of two regular types of protein **secondary structure**, the other being the **β-strand** that

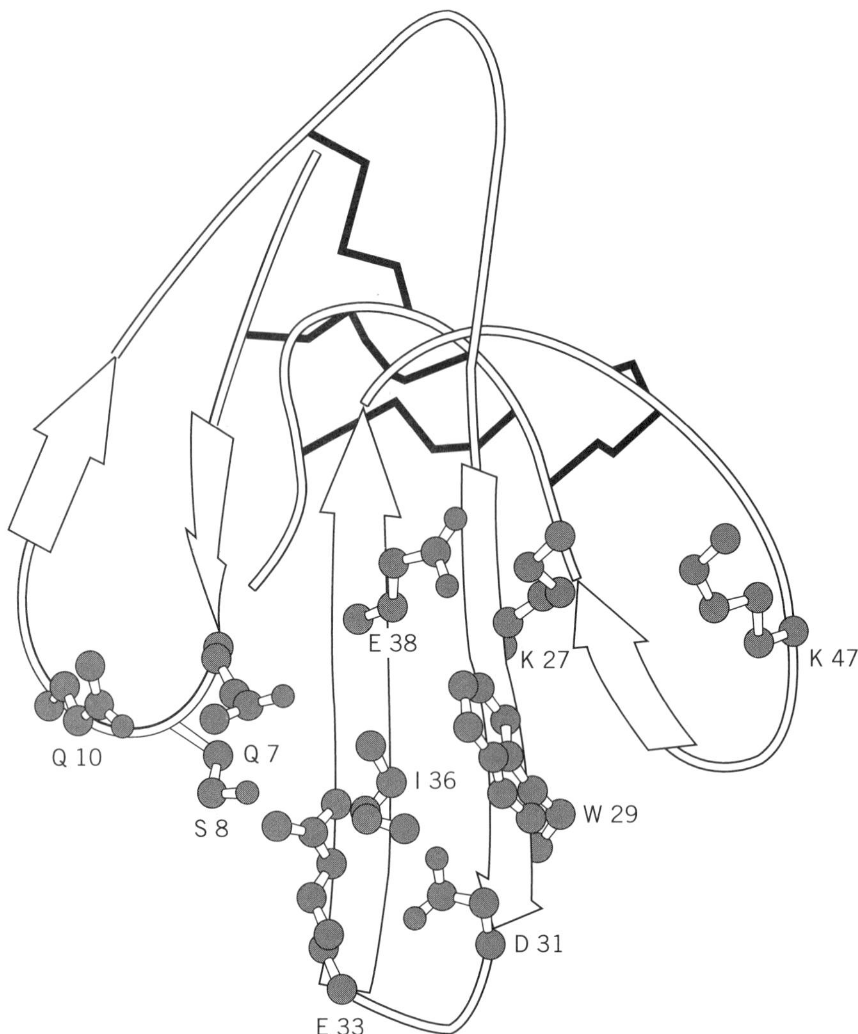

Figure 1. Erabutoxin, a snake toxin that binds to the acetylcholine receptor. Folding of the polypeptide chain shows the typical three-finger folding of curare-mimetic snake toxins. Several residues determine the specific binding and inhibition of the acetylcholine receptor with Lys-27, Trp-29, Arg-33, and Lys 47 playing a major role (4). Reproduced from (2) with permission.

forms **β-sheets**. Helices are formed from consecutive stretches of **residues** of the **polypeptide chain** and are characterized by a right-handed coiled **backbone** and a regular repeating pattern of backbone **hydrogen bonds**. α-Helices are usually depicted as coils or cylinders in protein structure diagrams (Fig. 1). There are 3.6 residues in every turn of α-helix, and for each turn the backbone is translated by 5.4 Å (or 1.5 Å per residue). In **protein structures**, the average length of an α-helix is 10 residues, although much shorter and longer examples have been observed. The backbone angles are approximately $-60°$ and $-50°$ for ϕ and Ψ, respectively, corresponding to the allowed region in the lower left of the **Ramachandran plot**. The atoms of the polypeptide chain pack closely together in the α-helical conformation, making favorable **van der Waals interactions**. The **side chains** of each residue are oriented outward from the axis of the helix, with the $C\alpha-C\beta$ bond pointing toward the N-terminal end of the helix.

A regular hydrogen bond pattern is formed in the α-helix, corresponding to bonds formed between the backbone carbonyl oxygen of residue (i) and the backbone amide of residue ($i + 4$) in the polypeptide chain, so that there are 13 atoms between each acceptor and donor pair. Thus, apart from the first few amide groups and last few carbonyl oxygen atoms, all the peptide bonds in an α-helix form hydrogen bond interactions. The hydrogen bonds are all oriented in the same direction, as are the dipoles of each of the peptide bonds. Therefore, the helix itself also has a significant dipole moment (a partial positive charge at the N-terminus and a partial negative charge at the C-terminus), and charged residues that interact with the dipole are often found in the sequence at the appropriate end of a helix (1). For example, the negatively charged Asp residue is highly favored at the N-terminus. The helix dipole has also been implicated in protein functions, such as substrate binding and **catalysis**.

Different amino acids have different tendencies for forming α-helices (2). The amide nitrogen of **proline** is cyclized with its backbone and thus cannot act as a hydrogen bond donor. Proline residues therefore are not found frequently in α-helices; when they are, they cause irregularities such as kinks and bends in the middle of helices. On the other hand, proline is a preferred residue at the N-terminus of the helix where other residues would have an unpaired backbone amide.

The staggered arrangement of side chains around the helix can be visualized in two dimensions through the use of a **helical wheel** (a projection down the axis of the helix). Helices in **membrane proteins** are often **amphipathic**, having **polar**

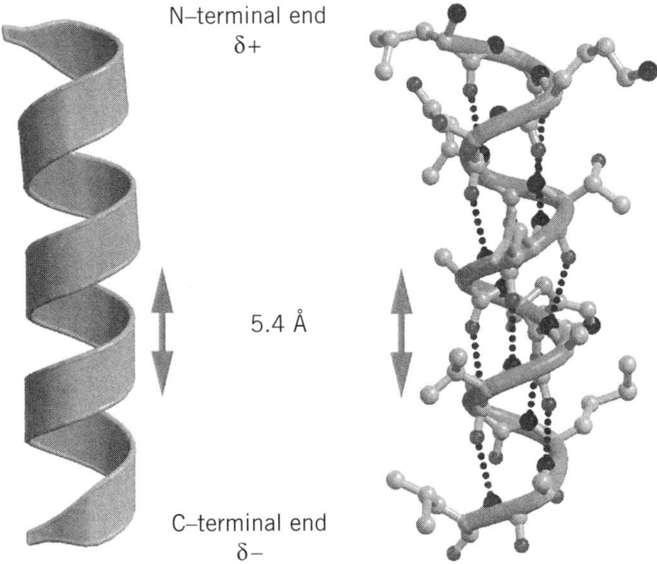

Figure 1. A typical protein α-helix. (**Left**) The helix is depicted schematically as a coil, and the *N*- and *C*-terminal ends, with their respective partial positive and negative charges, are indicated. (**Right**) The atomic detail of the α-helical structure is shown, with hydrogen bonds between backbone carbonyl oxygens (*i*) and backbone amide nitrogens of residue (*i* + 4) indicated by dotted lines. Nitrogen atoms are shown in blue and oxygen atoms are shown in red. This figure was generated using Molscript (3) and Raster3D (4,5). See color insert.

or charged side chains arranged on one side and **hydrophobic** side chains on the opposite side of the helix.

There are several variations to the α-helix conformation of proteins. The 3_{10}-helix is more tightly wound than the α-helix with backbone hydrogen bonds formed between residues (*i*) and (*i* + 3). Its name is based on its structure, with three residues per turn and 10 atoms between hydrogen bond donor and acceptor (although in real protein structures this geometry may be somewhat distorted). The 3_{10}-helix is usually only observed at the ends of α-helices or in short stretches of 4 to 5 residues. The Π-helix is more loosely wound than the α-helix with backbone hydrogen bonds formed between the (*i*) and (*i* + 5) residues and is very rarely observed in protein structures. The left-handed helix, having the same hydrogen bonding pattern as α-helices but backbone ϕ and Ψ angles of +60° and +50° (rather than the −60° and −50° angles of the right-handed α-helix), is not often observed, because this conformation results in steric clashes between backbone and side chain atoms. A **glycine** residue, however, having only a hydrogen atom as its side chain, can adopt this conformation. The poly(Pro)II helix, with no backbone hydrogen bonds, is also left-handed, with three residues per turn and a translation of 3.12 Å per residue. This helical structure is observed in **polyproline** and in proline-rich regions of proteins.

[See also **Protein structure** and **Secondary structure, protein**.]

BIBLIOGRAPHY

1. L. Serrano and A. R. Fersht (1989) *Nature* **342**, 296–299.
2. J. S. Richardson and D. C. Richardson (1988) *Science* **240**, 1648–1652.
3. P. J. Kraulis (1991) *J. Appl. Crystallogr.* **24**, 946–950.
4. E. A. Merritt and M. E. P. Murphy (1994) *Acta Crystallogr.* **D50**, 869–873.
5. D. J. Bacon and W. F. Anderson (1988) *J. Mol. Graphics* **6**, 219–222.

Suggestions for Further Reading

D. J. Barlow and J. M. Thornton (1988) Helix geometry in proteins. *J. Mol. Biol.* **201**, 601–619.

C. Branden and J. Tooze (1991) *Introduction to Protein Structure*, Garland, New York.

W. G. J. Hol, P. T. Van Duijnen, and H. J. C. Berendsen (1978) The α-helix dipole and the properties of proteins. *Nature* **273**, 443–446.

J. S. Richardson (1981) The anatomy and taxonomy of protein structure. *Adv. Protein Chem.* **34**, 167–339.

C. Toniolo and E. Bennedetti (1991) The polypeptide 3_{10} helix. *TIBS* **16**, 350–353.

ALPHA-HELIX FORMATION

J. MYERS
J. SCHOLTZ

α-Helices are the most common type of **secondary structure** in globular proteins (see **α-Helix**). As such, there is intense interest in understanding the factors that contribute to α-helix formation in peptides and proteins (1–4).

All peptide helices unfold with increasing temperature. This indicates that helix formation is an enthalpically favorable process (it proceeds with the release of heat). The main feature of helical structure is the repeating pattern of intramolecular **hydrogen bonds** formed between atoms in the peptide backbone (> C═O · H–N <). These hydrogen bonds probably provide the main enthalpic stabilization to the helical conformation (5). Opposing this **enthalpy** is the conformational **entropy** of freezing the backbone conformation, because the process of change from a disordered conformation (unfolded peptide or protein) to an ordered one (helix) is unfavorable entropically. This unfavorable entropy is about the same magnitude as the favorable enthalpy; that is, in isolation the helices are at best marginally stable.

Because the α-helix has an (*i*, *i* + 4) backbone hydrogen bonding pattern, the first four >N—H and the last four >C═O groups of the helix will not be hydrogen-bonded to a backbone partner. Polar groups that do not form hydrogen bonds to **water** or other protein groups are destabilizing (see **Protein stability**). Therefore it is important that these groups either be hydrated or have other intramolecular partners. The side chain of the first residue of the helix (the "*N*-cap") often fills this role, if it is able to hydrogen-bond to one or two otherwise unsatisfied backbone >N–H groups. These "capping" hydrogen bonds have been proposed to be important factors in determining where helices begin and end in proteins (6,7).

The helical conformation has the peptide groups and their associated dipoles aligned in the same direction, whereas in nonhelical forms the dipoles are distributed randomly. As the helix forms, the dipoles on adjacent peptide groups align in an unfavorable fashion (see **Electrostatic interactions**).

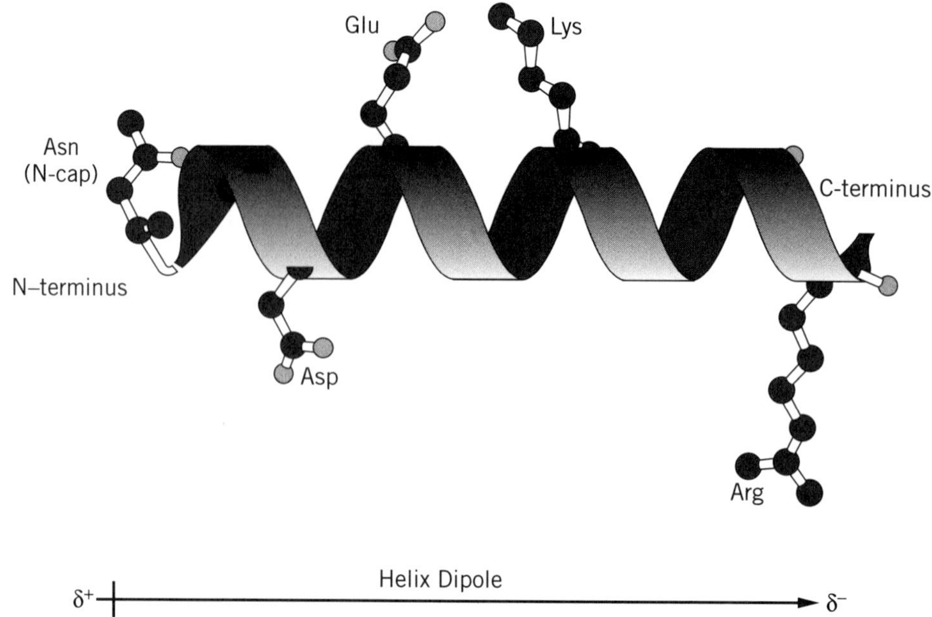

Figure 1. Interactions involving amino acid side chains that affect the stability of isolated α-helices. (**Top**) Ribbon model of an α-helix, with stabilizing interactions involving (i) hydrogen bonding of the N-cap side-chain at the N-terminus, indicated here as Asn, (ii) favorable electrostatic interactions of ionized side chains (Asp and Arg shown here) with the helix dipole, and (iii) electrostatic interactions between oppositely charged side chains on adjacent turns of the helix. (**Bottom**) Quantitative estimates of the net stabilizing interactions in an isolated α-helix and in folded globular proteins.

However, when a full turn of the helix is completed, the first hydrogen bond is formed and the peptide groups involved in the hydrogen bond align favorably. This adds to the **cooperativity** of helix formation: Once the first hydrogen bond is formed, it is easier to continue forming the helix. The alignment of the individual peptide group dipoles in a single direction gives rise to a helix *macrodipole* (simply the additive effects of the individual dipoles). The N-terminus of the helix has a partial positive charge, while the C-terminus carries a partial negative charge. Residues with charged side chains at the ends of a helix can interact favorably or unfavorably with the helix macrodipole, giving rise to an effect known as the *charge–dipole interaction*.

Different amino acid residues have different tendencies to form helices (see **Helix–coil theory**). For example, alanine is found in protein helices much more often than is glycine (8). The intrinsic preference to be in a helical conformation has been termed *helix propensity*. Since this is an intrinsic property, it should be characteristic of a given amino acid and independent of context. However, quantitative measurements of helix propensity in different systems have provided markedly different results (1,2,4). This reflects the difficulty of measuring intrinsic properties in complex molecules like proteins and peptides (however, see Ref. 9).

Interactions between side chains can also affect the stability of helices. Side chains three and four residues away from each other in the amino acid sequence will be close in space in a helical conformation. Therefore, electrostatic interactions, hydrogen bonds, or **hydrophobic** contacts between these side chains can contribute to helix stability. Likewise, if interactions are possible in the nonhelical form of the peptide or protein, this can alter the amount of α-helix observed in a peptide or protein. Figure 1 shows examples of various types of side-chain interactions possible in the helix, an estimate of the magnitudes of the contributions of the main factors contributing to helix stability, and comparison of these estimates with the **free energy** provided by typical interactions in proteins.

In addition to the stability of helices, their rates of folding and unfolding are of interest as well. The transition from helix to random coil occurs very rapidly, on the order of nanoseconds, two to three orders of magnitude faster than the fastest folding proteins (10).

BIBLIOGRAPHY

1. J. M. Scholtz and R. L. Baldwin (1992) *Annu. Rev. Biophys. Biomol. Struct.* **21**, 95–118.
2. A. Chakrabartty and R. L. Baldwin (1995) *Adv. Protein Chem.* **46**, 141–176.
3. J. W. Bryson, S. F. Betz, H. S. Lu, D. J. Suich, H. X. Zhou, K. T. O'Neil, and W. F. DeGrado (1995) *Science* **270**, 935–941.
4. N. R. Kallenbach, P. Lyu, and H. Zhou (1996) In *Circular Dichroism and the Conformational Analysis of Biopolymers* (G. D. Fasman, ed.), Plenum Press, New York.
5. J. M. Scholtz, S. Marqusee, R. L. Baldwin, E. J. York, J. M. Stewart, M. Santoro, and D. W. Bolen (1991) *Proc. Natl. Acad. Sci. USA* **88**, 2854–2858.
6. L. G. Presta and G. D. Rose (1988) *Science* **240**, 1632–1641.
7. E. T. Harper and G. D. Rose (1993) *Biochemistry* **32**, 7605–7609.
8. P. Y. Chou and G. D. Fasman (1978) *Adv. Enzymol.* **47**, 45–148.
9. J. K. Myers, C. N. Pace, and J. M. Scholtz (1997) *Proc. Natl. Acad. Sci. USA* **94**, 2833–2837.
10. S. Williams, T. P. Causgrove, R. Gilmanshin, K. S. Fang, R. H. Callender, W. H. Woodruff, and R. B. Dyer (1996) *Biochemistry* **35**, 691–697.

Suggestion for Further Reading

The first four references cited above provide excellent reviews of helix formation in peptides and the relationship between helix formation in peptides and proteins.

ALPHA-LACTALBUMIN

S. SUGAI

It requires patience to achieve quiet research. The investigation of properties of α-lactalbumin (α-LA) was not worthy of much notice until about 1965, especially from the standpoint of molecular biology. It is only a protein component of the complex enzyme *lactose synthase* and elucidation of its function had made little progress. However, the hidden connection to molecular biology was ferreted out in about 1967 and initiated the great explosion of activity in studying the **molten globule** (MG) conformation of proteins. These studies of the MG proteins opened a new area of structural biology connected to various physiological functions. Modern molecular biology sometimes requires the MG protein model to understand such physiological processes. The best-characterized MG is that formed by α-LA, and it can easily be prepared. α-LA is now one of the star proteins in molecular biology. The development of research into α-LA has been deeply affected by changing times.

One (α) of three peaks (α, β, γ) in the **sedimentation velocity** patterns of proteins in the noncasein fraction of skim milk was found responsible for a lactalbumin isolated from milk, which has subsequently been called α-lactalbumin. In the middle 1960s, it was confirmed that lactose is synthesized in the mammary gland from UDP-Galactose (UDP-Gal) and glucose (Glc) by an enzyme "lactose synthase", which can be resolved into two fractions. One of them is α-LA, and the other is *galactosyltransferase* (GT). Alone in the **Golgi apparatus**, but with metal ions such as Mn^{2+}, the latter catalyzes the transfer of Gal to GlcNAc on **glycoproteins**:

$$\text{UDP-Gal} + \text{GlcNAc} \rightarrow \text{Lactosamine-NAc} + \text{UDP} \qquad (1)$$

where NAc is the *N*-acetyl group. α-LA is synthesized in the mammary gland during lactation. It passes through the Golgi apparatus, where it combines with GT and alters its substrate specificity to inhibit reaction 1. A decrease of three orders of magnitude in the $\boldsymbol{K}_m$ for Glc facilitates the synthesis of lactose instead:

$$\text{UDP-Gal} + \text{Glc} \rightarrow \text{lactose} + \text{UDP} \qquad (2)$$

STRUCTURES

α-LA is normally a single polypeptide chain of 123 residues, four **disulfide bonds**, and a molecular weight of about 14,000, except for rat α-LA, which has 140 residues. About 1970, similarities were noted in the molecular weight, chemical composition, and amino acid sequence of α-LA and hen egg white **lysozyme** (HEWL), a lytic enzyme. Subsequently, it was found that the exon-intron organization of the genes for both proteins are the same (1). These suggested **divergent** evolution of the two proteins from a common ancestral gene, and α-LA was assumed to have a three-dimensional structure similar to that of HEWL, in spite of marked differences in their functions. Model-building studies showed that α-LA could adopt a molecular conformation similar to that of HEWL. **X-ray crystallographic** determination of the α-LA structure was not successful for a long time, but the conformations of both proteins in aqueous media were compared. Until 1980, significant differences in their physical behavior were observed: (1) an intermediate conformation appears in the unfolding equilibrium of α-LA, but not of HEWL, at intermediate concentrations of **guanidinium chloride** (GdmCl) at a neutral pH, which is also stable at acid pH and is designated the "A-state"; (2) one Ca^{2+} ion is tightly bound to α-LA, but not to HEWL, and stabilizes the native (N) conformation. Removal of Ca^{2+} from α-LA induces the A-state near room temperature. Figure 1 shows a phase diagram of bovine α-LA for the N- A- and U-(unfolded) states in solutions of GdmCl at 25 °C. Subsequently, it was shown that the kinetic intermediate of bovine α-LA captured at the early stage of refolding from U has properties similar to those of the equilibrium A-state. Later, it was reported that a number of proteins including HEWL assume the equilibrium and kinetic intermediate(s) similar to the A-state of α-LA in the MG form, and the MG is now considered a general physical state of proteins (see **Molten globule**).

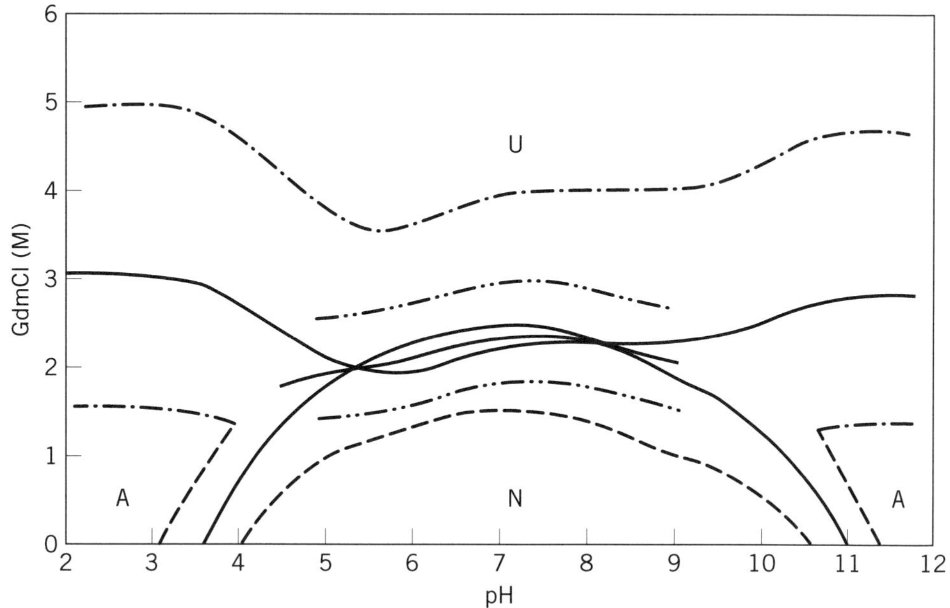

Figure 1. Phase diagram for the three states (N, A, and U) of bovine α-lactalbumin in a solution of GdmCl at any concentration and pH at 25°C.* The solid curves represent the pH and [GdmCl] where $K_{eq}{}^{NA}$, $K_{eq}{}^{AU}$, or $K_{eq}{}^{NU}$ is unity. $K_{eq}{}^{ij}$ is for the conformational transition between *i* and *j*. The other curves represent the pH and [GdmCl] where K_{eq}^{NA} = 10 or 0.1 (- - -), K_{eq}^{AU} = 10 or 0.1 (— · —), and K_{eq}^{NU} = 10 or 0.1 (— · ·—).
*K. Kuwajima, Y. Ogawa, and S. Sugai (1983) *J. Biochemistry* (Tokyo) **89**, 759–770.

Since 1987, the crystal structures of baboon, human, goat, guinea pig, and recombinant bovine α-LA have been determined at resolutions of 1.7 to 2.3 Å (2). The overall features of the α-LA conformation are similar to those of HEWL. Both structures are divided into two (α and β) sub**domains** by a deep cleft: the α-domain comprises residues 1 to 34 and 86 to 123, and the β-domain includes residues 35 to 85. In baboon α-LA, the α-domain contains four **α-helices**, A(residues 5 to 11), B(23 to 34), C(86 to 99), and D(105 to 109), plus single turns of 3_{10}-helices (12 to 16, 101 to 104 and 115 to 119). The β-domain consists of a three-stranded antiparallel **β-sheet** (41 to 56), a 3_{10}-helix (76 to 82), and loop regions (Fig. 2). In α-LA from some other species, however, the D-helix is replaced by a loop. Variations in the backbone structure occur in the flexible C-terminal region of residues 100 to 123. The conformation and the ligand coordination of a Ca^{2+} – binding site (K_a of 10^9 to $10^{10} M^{-1}$ for α-LA in the N-state) are conserved at the interface of both the domains in all of the α-LA. The Ca^{2+} binding site is composed of Asp residues at 82, 87, and 88 and peptide carbonyl oxygens at residues 79 and 84 (the other two ligands are water molecules) and is in a slightly distorted, rigid pentagonal bipyramidal form (designated the "α-LA elbow"). The Asp residues are not found in the corresponding region of most lysozymes, but some have the Ca^{2+}-binding site in an α-LA elbow (Table 1). The latter lysozymes bind a Ca^{2+} ion in the N-state and are important for studying the **evolution** of the α-LA family. A Zn^{2+}-binding site was also found in the crystal structure of human α-LA, located at the entrance of the cleft between the two sub-domains and coordinated with Glu49, Glu116, and two water molecules. α-LA binds numerous metal ions: $Na^+, K^+, Ba^{2+}, Ca^{2+}, Zn^{2+}, Mg^{2+}, Mn^{2+}, Co^{2+}, Cu^{2+}, Pb^{2+}, Hg^{2+}, Cd^{2+}, Sr^{2+}, Al^{3+}, Tb^{3+}, Eu^{3+}, Sc^{3+}$, and Y^{3+}. Elucidation of their roles in lactose synthesis is awaited. Only the binding sites for Ca^{2+} and Zn^{2+} have been observed in X-ray structures. The structures of α-LA determined by both X-ray crystallography and by NMR in solution exhibit two **hydrophobic** clusters of aromatic residues: cluster I of Phe31, His32, Tyr36, and Trp118, and cluster II of Phe53, Trp60, Tyr103, Trp104, and Trp26 (3). These make up the hydrophobic core in α-LA by combining with other hydrophobic parts in the α-domain.

MOLTEN GLOBULE INTERMEDIATE

The **molten globule** of α-LA is compact (a Stokes radius only 10 to 20% larger than that of the N-form) and has N-like **secondary structure**, but without the specific **tertiary structural** packing interactions. α-LA assumes the following stable MG forms in aqueous media: (1) an equilibrium unfolding state at intermediate concentrations of GdmCl or **urea**; (2) the acid state; (3) a partially unfolded state obtained by removing of the bound Ca^{2+} ion at neutral pH and low salt concentrations; and (4) after partial reduction of the disulfide

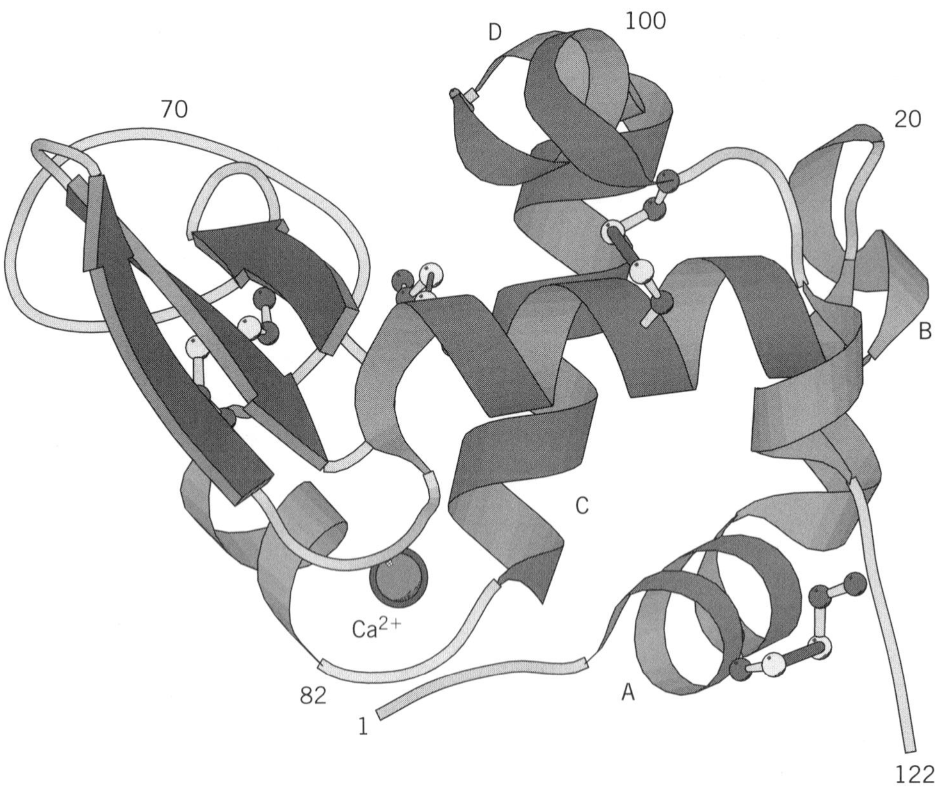

Figure 2. Schematic drawing of the native structure of baboon α-lactalbumin produced with the program MolScript.* The A, B, C, and D α-helices are labeled, β-sheets are arrows, the bound Ca^{2+} is labeled, and sulfur atoms in the disulfide bonds are in light gray. The C-terminal residue was not observed.
*P. J. Kraulis (1991) *J. Appl. Cryst.* **24**, 946–950.

Table 1. Comparisons of "Equivalent Amino Acid Residues" in Some α-Lactalbumins and Lysozymes in Positions at the Calcium-Binding Sites in α-Lactalbumins

Protein	Source	Residue[a]				
		79	82	84	87	88
α-Lactalbumin	Baboon	Lys	Asp	Asp	Asp	Asp
	Human	Lys	Asp	Asp	Asp	Asp
	Cow	Lys	Asp	Asp	Asp	Asp
	Horse	Lys	Asp	Asp	Asp	Asp
	Rabbit	Asn	Asp	Asp	Asp	Asp
	Wallaby	Lys	Asp	Asp	Asp	Asp
	Rat	Lys	Asp	Gly	Asp	Asp
	Guinea pig	Lys	Asp	Asp	Asp	Asp
	Goat	Lys	Asp	Asp	Asp	Asp
	Camel	Lys	Asp	Asp	Asp	Asp
Lysozyme[a]	Human	Ala	Gln	Asn	Asp	Ala
	Horse	Lys	Asp	Asn	Asp	Asp
	Echidna	Lys	Asp	Asp	Asp	Asp
	Dog	Lys	Asp	Asn	Asp	Asp
	Donkey	Lys	Asp	Asn	Asp	Asp
	Hen egg	Ala	Ser	Asp	Ala	Ser
	Pigeon	Lys	Asp	Asn	Asp	Asp
	Cow	Glu	Glu	Asp	Lys	Ala
	Mouse	Ala	Gln	Asp	Ala	Ala
	Baboon	Ala	Gln	Asn	Asp	Ala
	Peking duck	Val	Arg	Asp	Glu	Ala
	Turkey	Ala	Ser	Asp	Ala	Ser
	Rat	Ala	Gln	Asp	Gln	Ala

[a] Residue numbers for lysozymes are the equivalent numbers for (human) α-lactalbumin.

bonds. NMR and amide **hydrogen exchange** show heterogeneous structures of the MG with the four native disulfide bonds intact, an α-domain relatively structured by hydrophobic interactions, and a more unfolded β-domain. Hydrophobic clusters I and II are not found, but a rearranged local cluster is present at some residues in the region of residues 101 to 110. Some form of the hydrophobic core seems to persist in the MG form. The disulfide bond between Cys 6 and Cys 120 in native α-LA can be reduced extremely quickly by thiol reagents, although the corresponding bond in HEWL is not. The superreactivity of the 6–120 bond to reducing agents is one of the characteristics of α-LA. The free **thiol groups** in the three-disulfide intermediate (3SS) of α-LA are easily modified by **iodoacetic acid** or iodoacetamide and generate carboxymethylated or carboxyamidomethylated 3SS α-LA, respectively. These trapped derivatives of bovine α-LA retain the N-conformation, with only slight changes in the local conformation very near Cys6 and Cys120, but they assume the MG form in the absence of Ca^{2+} or at acid pH (4). A two-disulfide species and its trapped derivatives can also be obtained by subsequently reducing the 28–111 disulfide bond. The 2SS derivatives have partial MG characteristics. These disulfide intermediates and their derivatives are frequently used as models of partly unfolded proteins in the molecular biology because they are well characterized and easily prepared (4).

FUNCTIONS

Lactose synthase (reaction 2) has long been studied to determine the binding sites of the donor and acceptor saccharides and of the metal ions, plus the interaction site between GT and α-LA. The structure of the α-LA-GT complex has not been determined, but it has been shown by indirect methods, such as **chemical modification**, that Ca^{2+}, a stabilizing factor of the N-state of α-LA, and some residues of α-LA adjacent to the cleft, including His 32, which interacts with GT, are crucial for the function of lactose synthase. Moreover, a part of the α-LA cleft binds glucose by analogy with HEWL. A recent study by **site-directed mutagenesis** of α-LA pointed out the importance for function of Leu110, His32, and Phe31, which are in the region corresponding to subsite F of the **active site** of HEWL. Also, mutation of Ala106 perturbs the activity (5). The flexible C-terminal region, part of the hydrophobic clusters, and the cleft are essential for the lactose synthase function. The binding sites on GT for α-LA and UDP-Gal have also been tentatively identified.

It has been confirmed that α-LA in the MG is involved in various physiological processes. **Molecular chaperones** bind the protein and assist its folding in the cell. Physicochemical studies *in vitro* indicate that the **chaperonin** GroEL binds apo- or disulfide-reduced α-LA in the MG state, although it scarcely interacts with the N-state of α-LA (6). GroEL recognizes the hydrophobic surface exposed on the MG-conformation. α-LA bound to GroEL is also in the MG-state. α-LA and its disulfide-reduced forms in the MG-conformation interact with other chaperones and are frequently used as model substances for studies of protein folding in the cell. The insertion of soluble proteins into **membranes** and the conformations of the membrane-bound protein have also been topics of considerable interest (7). The insertion of α-LA into model membranes occurs under conditions that favor formation of the MG with its hydrophobic surface. The inserted α-LA is also in the MG-conformation, and its association with a lipid bilayer affects the chain mobility of the lipids. Recently it has been shown that α-LA has some biological functions in addition to lactose biosynthesis.

α-LA is a protein of interesting structure, properties, and functions, and it will be important in molecular biology in the future.

BIBLIOGRAPHY

1. D. C. Phillips et al. (1986) *Biochem. Soc. Trans.* **15**, 737–744.
2. R. Acharya et al. (1989) *J. Mol. Biol.* **208**, 99–127; (1991) **221**, 571–581; (1996) *Structure* **4**, 691–703.
3. A. T. Alexandrescu et al. (1992) *Eur. J. Biochem.* **210**, 699–709; (1993) *Biochemistry* **32**, 1707–1718.
4. J. J. Ewbank and T. E. Creighton (1991) *Nature* **350**, 518–520; (1993) *Biochemistry* **32**, 3677–3707; (1994) *Biochemistry* **33**, 1534–1538; P. S. Kim et al. (1995) *Trans. Roy. Soc. London* (B) **348**, 43–47; (1996) *Biochemistry* **35**, 859–863.
5. K. Brew et al. (1991) *J. Biol. Chem.* **266**, 698–703; (1996) *Biochemistry* **35**, 9710–9715.
6. M. K. Hayer et al. (1994) *EMBO J.* **13**, 3192–3202; C. V. Robinson et al. (1994) *Nature* **372**, 646–651.
7. S. Banuelos and A. Muga (1995) *J. Biol. Chem.* **270**, 29910–29915; (1996) *FEBS Lett.* **386**, 21–25; *Biochemistry* **35**, 3892–3898; K. M. Cauthern et al. (1996) *Protein Sci.* **5**, 1349–1405.

Suggestions for Further Reading

L. J. Berliner and J. D. Johnson (1988) α-Lactalbumin and calmodulin In *Calcium Binding Proteins* Vol. II (M. P. Thompson, ed.), CRC Press, Boca Raton, FL, pp. 79–116.

K. Brew and J. A. Grobler (1992) *α-Lactalbumin Advanced Dairy Chemistry Proteins* Vol. 1 (P. F. Fox, ed.), Elsevier Applied Science, London and New York, pp. 191–229.

M. J. Kronman (1989) Metal-ion binding and the molecular conformational properties of α-lactalbumin *Crit. Rev. Biochem. Mol. Biol.* **24**, 565–667.

K. Kuwajima (1989) The molten globule state as a clue for understanding the folding and cooperativity of globular-protein structure *Proteins: Struct. Funct. Genet.* **6**, 87–103; (1996) The molten globule state of α-lactalbumin *FASEB J.* **10**, 102–109.

H. A. McKenzie and F. H. White Jr. (1991) Lysozyme and α-lactalbumin: Structure, function and interrelationship *Adv. Protein Chem.* **41**, 171–315.

S. Sugai and M. Ikeguchi (1994) "Conformational comparison between α-lactalbumin and lysozyme" *Adv. Biophys.* **30**, 37–84.

ALTERNATIVE SPLICING

CINDY L. WILL
REINHARD LÜHRMANN

Most eukaryotic **genes** that code for proteins contain noncoding sequences (**introns**) that are interspersed among the coding regions (exons). **Transcription** of these genes generates so-called pre-mRNA molecules, which are converted to mature mRNAs by a process termed RNA **splicing**. During this process, the introns are precisely excised and the exons are ligated together (see **RNA splicing**). The majority of nuclear pre-mRNAs are spliced constitutively; that is, only one mature mRNA species is generated from a single pre-mRNA in all tissues. In some cases however, alternative 5′ and/or 3′ splice sites are used during splicing, resulting in the production of more than one mRNA species from a single pre-mRNA. Alternative splicing has been documented for many eukaryotic genes, and a variety of alternative splicing patterns have been observed, as depicted schematically in Figure 1. The utilization of alternative 5′ and/or 3′ **splice sites** (also referred to as donor and acceptor sites, respectively) can result in structurally distinct mRNAs by either excluding potential exon sequences or incorporating otherwise noncoding intron sequences. For some pre-mRNAs, alternative splicing is a nonregulated event such that two or more alternatively spliced mRNAs are produced at a given ratio to one another in all cell types. For others, the choice of alternative splice sites is regulated in a tissue-specific or developmental manner. This type of regulation is mediated by ***trans*-acting** factors that are differentially expressed in a particular tissue or at a specific time during development (see text below). These *trans*-acting factors may be positive or negative regulators that activate or repress the use of an alternative splice site either directly or indirectly, for example by modulating the affinity of general splicing factors.

Alternative splicing can lead to both quantitative and qualitative changes in gene expression. Quantitative changes can arise if the alternatively spliced mRNA contains a prematurely truncated open reading frame (ie, due to the presence of a **stop codon**) or exhibits an altered stability or **translation** efficiency. In many cases, alternative splicing leads to the production of so-called protein isoforms that are structurally identical in all but a specific region or domain (1). Such structural variants of a given protein often exhibit significant functional differences. Thus, through the generation of multiple protein isoforms, alternative splicing can enhance the phenotypic variability of a single gene.

A central question in both constitutive and alternative splicing that has yet to be clearly resolved is how the correct pairs of 5′ and 3′ splice sites are selected for cleavage and subsequent ligation (see also **Splice sites**). Pre-mRNAs contain multiple authentic 5′ and 3′ splice sites, which must be properly paired in order to prevent the random skipping of one or more exon. Furthermore, these sites must be distinguished from other nonauthentic sites that are also repeatedly present in most pre-mRNAs. In the case of alternative splicing, several different factors appear to be responsible for the preferential selection of one splice site over another. Firstly, features of the pre-mRNA itself (so-called ***cis*-acting** elements) can influence splice-site utilization. For example, the conserved sequences found at the 5′ and 3′ splice sites and the branch site contribute to splice site selection (see **Splice sites**). In recent years it has become clear that the relative strengths of competing splice sites is often a deciding factor in determining which site is used preferentially. The strength of a particular 5′ or 3′ splice site is generally a measure of how efficiently it binds spliceosomal components, such as the U1 and U2 snRNPs or the splicing factor U2AF, which play important roles during the early stages of **spliceosome** assembly. This in turn is often, but not always, a function of how closely its sequences match the 5′ and 3′ splice site consensus sequences, or a function of the length and uridine content of the polypyrimidine tract, which determines its affinity for U2AF. The selection of weak 5′ and 3′ splice sites, on the other hand, can be enhanced by splicing factors that promote U1 or U2 snRNP binding (eg, SR proteins; see text below). Some 5′ splice sites also activate usage of an upstream 3′ splice site (eg, in the preprotachykinin pre-mRNA); and, vice versa, some 3′ splice sites can promote the use of a downstream 5′ splice site (eg, in the B-tropomyosin pre-mRNA) (2,3). In the former case, factors bound at the 5′ splice site (the U1 snRNP) enhance the interaction of U2AF with the upstream 3′ splice site via a network of molecular interactions across the exon (4,5)

In some cases, regions of the pre-mRNA other than the 5′ and 3′ splice sites also contribute to alternative splice-site selection. Purine-rich sequences (so-called splicing enhancers), which are typically located in exons and often bind SR proteins, are known to enhance the use of adjacent 5′ or 3′ splice sites by stabilizing the interactions of spliceosomal components with them (reviewed in Ref. 6). In some instances, intron sequences or sequences in the 3′-untranslated region that regulate **polyadenylation** have also been shown to influence alternative splicing. Finally, pre-mRNA secondary structure (eg, stem-loop structures) can affect splice-site selection, for example by blocking the interaction of spliceosomal components with a particular splice site (7,8).

Splice-site selection during alternative splicing is also regulated by *trans*-acting factors. Many of the currently identified *trans*-acting factors also function in constitutive pre-mRNA splicing. In this case, variations in their concentration, or the concentration of factors that compete with them, are thought to lead to tissue-specific modulation of splice-site usage. Foremost in this catagory of factors are members of the evolutionarily conserved SR protein family, which are characterized by an amino-terminal **RNA-binding domain** and a *C*-terminal domain rich in arginine–serine (RS) dipeptides (reviewed in Ref. 9). SR proteins (eg, SF2/ASF and SC35) play

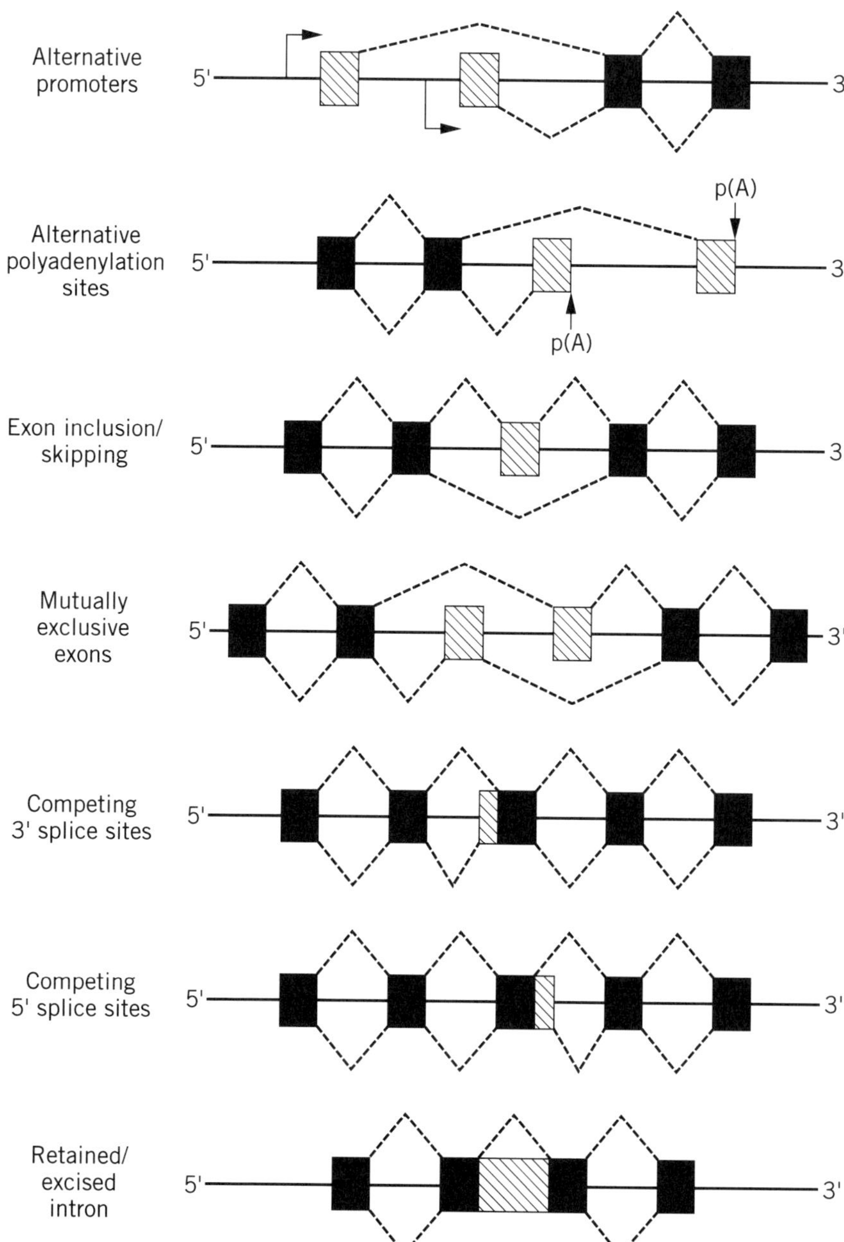

Figure 1. Patterns of alternative splicing. Constitutively spliced exons are shown as black boxes, alternatively spliced exons are shown as shaded boxes, and introns are shown as a solid lines between the exons. Transcription start sites are indicated by an arrow, and polyadenylation sites are indicated by p(A). Splicing events are depicted by a dashed line.

essential roles in constitutive nuclear pre-mRNA splicing, particularly at the earliest stages of spliceosome assembly (see **Spliceosome**). Moreover, in pre-mRNAs that contain multiple 5′ or 3′ splice sites (see Fig. 1), high concentrations of SR proteins generally enhance the use of the more proximal (downstream) 5′ splice site (so-called switching activity) or more proximal 3′ splice site (9,10). However, apparently due to differences in the affinities of individual SR proteins for different pre-mRNAs, the effect of a particular SR protein on 5′ splice site selection can vary from one pre-mRNA to the next (11,12). SR proteins appear to act by facilitating the interaction of the U1 snRNP with the 5′ splice site, an initial step for 5′ splice-site recognition, or by promoting the association of U2AF with the 3′ splice site (see **Spliceosome**). High concentrations of SR proteins can also inhibit exon skipping (Fig. 1), which likewise involves promoting the use of a proximal 5′ splice site. Interestingly, the activity of SR proteins such as SF2/ASF or SC35 in 5′ splice-site selection (but not in constitutive splicing) can be antagonized by the hnRNP proteins A/B (13,14). High concentrations of these proteins generally favor the use of more distal 5′ splice sites, and the ratio of 5′ splice site usage is determined by the relative amounts of hnRNP A/B and SR protein. Thus, the amount of some alternatively spliced pre-mRNAs can be modulated by varying the cellular concentrations of these proteins (15). SR proteins are also involved in activating weak 5′ and 3′ splice sites that are located adjacent to purine-rich splicing enhancers (reviewed in Refs. 9 and 16). The binding of specific SR proteins to these exon enhancer sequences promotes the interaction of U2AF, and thus the U2 snRNP, with the 3′ splice site or the U1 snRNP with the 5′ splice site.

Trans-acting alternative splicing factors that are expressed in a tissue-, sex- or developmental-specific manner have also been identified. In mammals, concrete, well-understood examples of cell-specific regulators of alternative splicing are

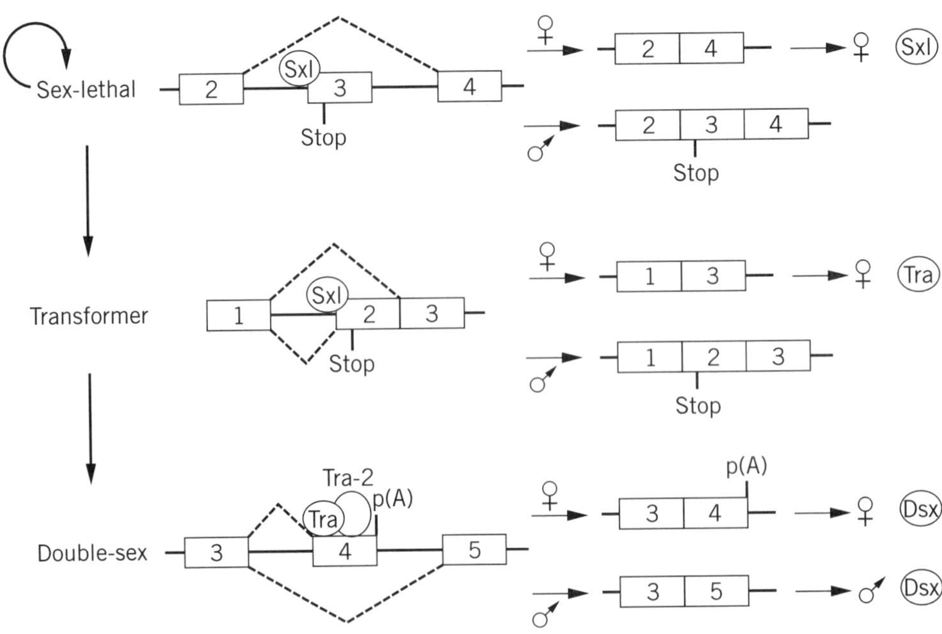

Figure 2. Cascade of alternative splicing events in the sex determination pathway of *Drosophila*. For simplicity, only those exons and introns involved in alternative splicing events are shown. In female flies, the sex lethal protein regulates its own synthesis and that of the tra protein by blocking the use of the 3′ splice site of the third and second exons of the sex-lethal and tra pre-mRNAs, respectively. Both of these exons contain a stop codon, as indicated. The tra protein, together with tra-2, interacts with a splicing enhancer in the fourth exon of the double-sex pre-mRNA and activates use of the upstream 3′ splice site. Exons are depicted as numbered boxes, and introns are depicted as solid lines. Dashed lines above the introns represent splicing events in females, and those below indicate splicing events in males. The sex-lethal (sxl), transformer (tra), and tra-2 proteins are indicated by circles. Polyadenylation is indicated by p(A). (Adapted from Ref. 21).

currently limited. The best-characterized factors are those responsible for alternative splicing events in the fruit fly *Drosophila melanogaster*. For example, sex determination in fruit flies involves a cascade of alternative splicing events that are regulated by sex-specific proteins (17) (see Fig. 2). In males, splicing of the Sex-lethal (Sxl), transformer (tra) and doublesex (dsx) pre-mRNAs are nonregulated events that appear to require only the general splicing machinery (ie, they represent the so-called default splicing patterns). In females, alternative splice sites are activated by female-specific factors, either directly or indirectly, through the inactivation of the male-specific site. The first of these factors known to act in this cascade is the female-specific Sxl protein, which regulates its own synthesis by an alternative splicing event. In females, Sxl is thought to block (by a currently unknown mechanism) use of the 3′ splice site of the third exon of the Sxl pre-mRNA, which leads to exon 3 skipping (18,19). Because exon 3 contains a stop codon, functional Sxl protein is produced only from the female-specific Sxl mRNA (Fig. 2). In the next step of this cascade, the Sxl protein inhibits the use of the upstream 3′ splice site of exon 2 of the *tra* pre-mRNA, which in turn activates splicing at a weaker downstream 3′ splice site. Specifically, Sxl has been shown to bind to the stronger polypyrimidine tract of the upstream 3′ splice site and thereby to inhibit the binding of U2AF (20). This results in U2AF binding to the downstream site, for which it has a lower affinity. The resulting exclusion of exon 2, which also contains a stop codon, leads to the production of functional *tra* protein in females. In the last step, the tra protein, in conjunction with tra-2, directly activates use of the weak 3′ splice site of exon 4 of the doublesex pre-mRNA. Tra and Tra-2, both of which contain RS domains, interact with a purine-rich splicing enhancer present in exon 4 and recruit SR proteins, as well as U2AF, to the upstream 3′ splice site (reviewed in Ref. 6). As a result, exon 4 is included in female-specific dsx mRNA, and **polyadenylation** occurs at the end of this exon. The resulting female dsx protein represses male differentiation, whereas the male protein represses female differentiation. Although in this particular case much has been learned about the molecular mechanisms responsible for splice-site selection, in most cases a clear understanding of the complex processes of alternative splicing awaits further investigation.

BIBLIOGRAPHY

1. R. E. Breitbart, A. Andreadis, and B. Nadal-Ginard (1987) *Annu. Rev. Biochem.* **56**, 467–495.
2. F. H. Nasim, P. A. Spears, H. M. Hoffmann, H.-C. Kuo, and P. J. Grabowski (1990) *Genes Dev.* **4**, 1172–1184.

3. T. Tsukahara, C. Casciato, and D. M. Helfman (1994) *Nucleic Acids Res.* **22**, 2318–2325.
4. B. E. Hoffman and P. J. Grabowski (1992) *Genes Dev.* **6**, 2554–2568.
5. S. M. Berget (1995) *J. Biol. Chem.* **267**, 14902–14908.
6. K. J. Hertel, K. W. Lynch, and T. Maniatis (1997) *Curr. Opin. Cell Biol.* **9**, 350–357.
7. P. A. Estes, N. E. Cooke, and S. A. Liebhaber (1992) *J. Biol. Chem.* **267**, 14902–14908.
8. P. Sirand-Pugnet, P. Durosay, B. Clouet d' Orval, E. Brody, and J. Marie (1995) *J. Mol. Biol.* **251**, 591–602.
9. X.-D. Fu (1995) *RNA* **1**, 663–680.
10. D. S. Horowitz and A. R. Krainer (1994) *Trends Gen.* **10**, 100–106.
11. A. M. Zahler, K. M. Neugebauer, W. S. Lane, and M. B. Roth (1993) *Science* **260**, 219–222.
12. A. M. Zahler and M. B. Roth (1995) *Proc. Natl. Acad. Sci. USA* **92**, 2642–2646.
13. A. Mayeda and A. R. Krainer (1992) *Cell* **68**, 365–375.
14. X. Yang, M. R. Bani, S. J. Lu, S. J. Rowan, Y. Ben-David, and B. Chabot (1994) *Proc. Natl. Acad. Sci. USA* **91**, 6924–6928.
15. J. F. Caceres, S. Stamm, D. M. Helfman, and A. R. Krainer (1994) *Science* **265**, 1706–1709.
16. J. L. Manley and R. Tacke (1996) *Genes Dev.* **10**, 1569–1579.
17. B. S. Baker (1989) *Nature* **340**, 521–524.
18. J. I. Horabin and P. Schedl (1993) *Mol. Cell. Biol.* **13**, 1408–1414.
19. B. Granadino, L. O. F. Penalva, M. R. Green, J. Valcárcel, and L. Sánchez (1998) *Proc. Natl. Acad. Sci. USA* **94**, 7343–7348.
20. J. Valcárcel, R. Singh, P. D. Zamore, and M. R. Green (1993) *Nature* **362**, 171–175.
21. M. J. Moore, C. C. Query, and P. A. Sharp (1993) In *The RNA World* (R. F. Gesteland and J. F. Atkins eds.), Cold Spring Harbor Laboratory Press, Cold Spring Harbor, NY, pp. 303–358.

Suggestions for Further Reading

Y.-C. Wang, M. Selvakumar, and D. M. Helfman (1997) Alternative Pre-mRNA Splicing. In *Eukayotic mRNA Processing* (A. R. Krainer, ed.), IRL Press, Oxford, UK, pp. 242–279.

B. Chabot (1996) Directing alternative splicing: cast and scenarios. *Trends Gen.* **12**, 472–478.

J. L. Manley and R. Tacke (1996) SR proteins and splicing control. *Genes Dev.* **10**, 1569–1579.

D. C. Rio (1993) Splicing of pre-mRNA: mechanism, regulation and role in development. *Curr. Opin. Genet. Dev.* **3**, 574–584.

ALU SEQUENCES

GEORGES N. COHEN

The **genomes** of almost all higher **eukaryotes** contain highly **repetitive DNA** sequences that are not clustered together. They are distributed throughout the genome, interspersed with longer stretches of DNA with unique (or moderately repetitive) sequences. In the human genome, the majority of such sequences belong to a single family of **SINEs** called the **Alu family**. Each sequence is about 300 base pairs long. Although the many copies present are recognizably related, they are not precisely conserved in sequence. Their name derives from the fact that most contain a single site of cleavage for the **restriction enzyme AluI** near their middle. More than 500,000 Alu sequences are present in the human genome, accounting for 3 to 6% of the total DNA. Any particular segment of DNA of 5000 bp or longer has a high probability of containing at least one Alu sequence. Most Alu sequences are flanked by tandem **direct repeats** of DNA and move like **transposable elements** creating target-site duplications when they insert.

On average Alu DNA sequences contain about 80% identity between members of the family, but certain internal regions are more conserved: an internal 40-bp region and two sets of sequences, one near the 5′ end and another one farther down in the transcriptional direction, that are homologous to sequences found in the **promoter** for **RNA polymerase** III.

One end of the Alu DNA segment is defined precisely by comparing several Alu sequences. The other end occurs at, or is adjacent to, a run of A bases of variable length that may or may not be interrupted occasionally by other bases. The internal structure of an Alu sequence is dimeric and may consist of an ancestral duplication of a segment of approximately 150 bp. In some rodents, a major SINE is 130 bp long and has sequence similarities with half of the primate Alu sequences. As in Alu, it is bound on one side by a poly(dA) sequence.

ORIGIN

The Alu sequence derives from an internally deleted host cell 7SL RNA gene that encodes the RNA component of the **signal-recognition particle** (SRP) that functions in **protein biosynthesis** (1,2). Consequently, an Alu sequence can be considered to be a transposable element or an unusually mobile **pseudogene**. Alu sequences are transcribed from the 7SL RNA promoter, a polymerase III promoter internal to the transcript, so that it carries the information necessary for its own **transcription** wherever it moves. However, it needs to borrow a reverse transcriptase to transpose.

EVOLUTION

The Alu sequences may be grouped into discrete subfamilies on the basis of their sequences. Distinct families have amplified within the human genome in recent evolutionary history (3). The Human Specific or Predicted Variant subfamily, one of the most recently formed group of Alu sequences, amplified to 500 copies within the human genome sometime after the human/great ape divergence, which is thought to have occurred 4 to 6 million years ago. Comparisons of the sequence and locations of the Alu sequences in different mammals suggest that they have multiplied only recently.

Polymorphism of the Alu family member differs from other types of **polymorphism**, such as Variable Number of Tandem Repeat (VNTR, or **minisatellite DNA**) or Restriction Fragment Length Polymorphism (**RFLP**), because individuals share Alu insertions based upon identity by descent from a common ancestor as a result of a single event that occurred one time within the human population (4). In contrast the VNTR and RFLP polymorphisms have arisen multiple times within a population. Alu sequences represent a unique source of human genetic variation and a molecular fossil record of genomic evolutionary history. These sequences are natural landmarks for physical gene mapping and for reconstructing the evolutionary history/expansion of tandemly arrayed **gene families** (4).

POSSIBLE FUNCTIONS

The physiological role of Alu elements is unknown, although it has been proposed that they are involved in **DNA replication**, regulation of transcription, and transport of signal recognition particle RNA to the nucleus. For example, Alu RNA and proteins that bind to Alu elements have been identified in human cells. In particular, it has been demonstrated that some Alu sequences in human gene regions have been altered in sequence so that they are now important in controlling and enhancing transcription (5). The consensus sequence of one of the major Alu families contains a functional **retinoic acid** binding element (see **Response elements**). The random insertion throughout the primate genome of thousands of Alu repeats containing a retinoic acid response element might have altered the expression of numerous genes, thereby contributing to evolutionary potential (6).

BIBLIOGRAPHY

1. A. M. Weiner (1980) *Cell* **22**, 209–218.
2. E. Ullu, S. Murphy, and P. M. Melli (1982) *Cell* **29**, 195–202.
3. M. A. Batzer et al. (1996) *J. Mol. Evol.* **42**, 22–29.
4. R. J. Britten (1994) *Proc. Natl. Acad. Sci. USA* **91**, 6148–6150.
5. R. J. Britten (1996) *Proc. Natl. Acad. Sci. USA* **93**, 9374–9377.
6. G. Vansant and W. F. Reynolds (1995) *Proc. Natl. Acad. Sci. USA* **92**, 8229–8233.

Suggestion for Further Reading

P. Jagadeeswaran, B.G. Forgeti, and S. M. Weissman (1981) "Short interspersed repetitive DNA elements in eucaryotes: Transposable DNA elements generated by reverse transcription of RNA pol III transcripts?" *Cell* **26**, 141–142.

AMBER MUTATION

WILLIAM A. ROSCHE
PATRICIA L. FOSTER

An amber **mutation** is a **nonsense mutation** that changes a sense **codon** (one specifying an amino acid) into the translational **stop codon** UAG, causing premature termination of the **polypeptide chain** during **translation**. The mutation, the codon, and the mutant are all called amber. Amber mutations arise by single base changes in the codons for eight amino acids (and in the UAA stop codon, although this is not a nonsense mutation). Mutations in the anticodons of the **transfer RNAs** that read those eight codons, in principle, could give rise to **amber suppressors**, but suppressors are recovered only if another tRNA exists that reads the codon. In *Escherichia coli*, five amber suppressors that arise by a single base change have been identified. In addition, amber mutations are suppressed by **ochre suppressors** because of **wobble pairing** in the third position (5′) of the **anticodon**. Amber mutations in *E. coli* and its **bacteriophages** are easily identified by their pattern of suppression by known suppressors. In **bacteria**, amber suppressors have relatively mild effects. Many laboratory strains and even natural isolates of *E. coli* carry amber suppressors. This might be surprising, because amber suppressors are expected to prevent the proper termination of many proteins, but amber codons are used relatively infrequently in *E. coli* and related bacteria.

The name amber was originally given to mutants of bacteriophage T4 that grow on *E. coli* strain K12 (λ) but not on *E. coli* strain B (1). It turned out that the K12 strain used has an amber suppressor, whereas the B strain does not. The word amber was inspired by Harris Bernstein who participated in the original experiment (Bernstein means amber in German), although published versions of the story disagree on whether the mutants were named after Harris Bernstein himself or his mother (2,3). It also could be significant that at nearly the same time that amber mutants were being discovered, Seymour Benzer was also analyzing nonsense mutations in the *rII* genes of phage T4 and calling them "ambivalent" (4).

BIBLIOGRAPHY

1. R. H. Epstein, A. Bolle, C. Steinberg, E. Kellenberger, E. Boy de la Tour, R. Chevalley, R. Edgar, M. Susman, C. Denhardt, and I. Lielausis (1964) *Cold Spring Harbor Symp. Quant. Biol.* **28**, 375–392.
2. R. S. Edgar (1966) In *Phage and the Origins of Molecular Biology* (J. Cairns, G. S. Stent, and J. D. Watson, eds.), Cold Spring Harbor Laboratory Press, Cold Spring Harbor, NY, pp. 166–170.
3. F. W. Stahl (1995) *Genetics* **141**, 439–442.
4. S. Benzer and S. P. Champe (1962) *Proc. Natl. Acad. Sci. USA* **48**, 1114–1121.

AMBER SUPPRESSOR

J. F. CURRAN

Amber suppressors are mutant **transfer RNAs** that translate the UAG (amber) **stop codon** as a sense codon. **Amber mutations** cause **translation** of the **messenger RNA** during **protein biosynthesis** to terminate prematurely, resulting in truncated and usually inactive **polypeptide chains**. Amber suppressors allow for protein synthesis to continue beyond the mutational block in translation, usually resulting in active protein. These suppressors have been used extensively in prokaryotic **genetic** studies, and in studies of the apparatus and mechanisms of translation. For complete discussions of these and other suppressors, see **Nonsense suppression, Suppressor tRNA**, and **Genetic suppression**.

AMES TEST

LYNNETTE FERGUSON
WILLIAM DENNY

The *Salmonella*/mammalian **microsome** test for **mutagens** was originally developed in the laboratory of Bruce Ames (1) and has become sufficiently used and well-recognized to be familiarly described by his name. The assay utilizes several specially constructed strains of *Salmonella typhimurium* that normally require histidine for growth and can be reverted to **prototrophy** by a wide range of different mutagens. The assay requires that test chemicals and bacteria be plated onto a minimal agar petri dish, incorporating trace amounts of histidine and **biotin**, which are required for growth, to allow all the

Table 1. Genotype and Reversion Characteristics of Some *Salmonella typhimurium* Strains Commonly Used for Mutagenicity Testing

Histidine Mutation	Strain Number	Additional Mutations: Permeability	Repair	*R* Factor	Nature of Mutation	
hisC3076	TA1537	*rfa*	Δ*uvrB*	-	WT sequence unknown. Mutant thought to be +1 near CCC	Frameshifts
hisD3052	TA1538	*rfa*	Δ*uvrB*	-	WT:GAC-ACC-GCC-CGG-CAG...	Frameshifts
	TA98			pKM101	Mutant:GAC-ACC-GCC-GGC-AGG...	
HisD6610	TA97	*rfa*	Δ*uvrB*	PKM101	WT:GTC-ACC-CCT-GAA-GAG-A*TC-GCC Mutant:GTC-ACA-CCC-CCC-TGA (opal)	Frameshifts
hisG46	TA1535	*rfa*	Δ*uvrB*	-	WT:.......-CTC-...	Some base-pair substitution events
	TA100			pKM101	Mutant:...-CCC-...	Extragenic suppressors
HisG428	TA102	*rfa*		PAQ1	WT:CAG-AGC-AAG-CAA-GAG...	Transitions and transversions
	TA104				Mutant:CAG-AGC-AAG-TAA (ochre)	Extragenic suppressors Small deletions (−3, −6)

bacteria to grow through a small number of generations. In the absence of mutagen, a small number of colonies will grow on these plates, whereas mutagenic chemicals may increase this number very considerably. Mutations are scored as the number of revertant colonies per dish, usually as a function of applied dose. The test protocol incorporates homogenates of (usually) rat liver directly into the petri dish, thereby permitting mammalian metabolism of many compounds that require activation before they will interact with cellular DNA.

The DNA sequence around the original mutation has been determined in those strains most commonly used for mutagenicity testing (Table 1). The bacteria have been made more sensitive to mutagens by the introduction of several additional characteristics. Many of the strains carry a deletion in the *uvrB* gene and are defective in the ability for **DNA repair**. The bacterial cell wall has increased permeability to bulky chemicals because of the *rfa* mutation, and certain introduced **plasmids** may increase the sensitivity of the bacteria to mutation by some types of chemicals. The Ames test was originally developed as a screen for chemical carcinogens (1), but this has only proved appropriate to certain chemical classes (eg, Ref. 2). Nevertheless, because of the enormous number of chemicals tested in this assay, it must still occupy a premier position in testing for mutagenic properties of chemicals.

BIBLIOGRAPHY

1. B. N. Ames, J. McCann, and E. Yamasaki (1975) *Mutat. Res.* **31**, 347–364.
2. J. Ashby and R. W. Tennant (1994) *Mutagenesis* **9**, 7–16.

AMIDINATION

T. IMOTO

Amidination takes place when an **amino group** is treated with an imide ester.

The reaction proceeds with reasonable yield when conducted under alkaline conditions (pH around 10). Side reactions, such as **cross-linking** take place at lower pH. Both α- and ε-amino groups can be amidinated, and the basicity of the modified residue increases. The amidinyl groups incorporated are stable in acidic media. The role of amino groups in protein function can be investigated by amidination. Proteins can be readily **radiolabeled** by amidinating with ^{13}C- or ^{3}H-labeled imide ester. Amidination of the ε-amino group on **lysine** residues is particularly useful in **peptide mapping** and in determining the **primary structure** of a protein (see **Protein sequencing**). The proteolytic enzyme **trypsin** does not cleave at the modified lysine residues, thereby limiting cleavage to **arginine** residues. Moreover, amidinyl groups are removed by aminolysis, and the resulting deprotected **peptides** are cleaved further with trypsin.

ACETAMIDINATION OF PROTEINS

After reduction and **alkylation** of the protein (100–700 nmol), it is dissolved in a few mL of 0.2 M triethylamine-HCl **buffer**, pH 10.3, containing 5.0 M **guanidinium chloride** (GdmCl) (1). Ethyl (or methyl) acetamide hydrochloride is dissolved in an equivalent amount of NaOH solution to maintain a final acetamide concentration of 0.1 to 0.15 M (100-fold molar excess of acetamide over amino groups). The reaction mixture is incubated for 1 hr at 25°C, **dialyzed** against 0.05 M NH_4HCO_3 containing 2.5 M GdmCl, and then against 0.05 M NH_4HCO_3. The protein is finally lyophilized.

DEAMIDINATION BY METHYLAMINOLYSIS

Acetamidinated protein or peptide (2.7 mg) is dissolved in 1.6 mL of 6 to 9 M **urea**. Then 0.9 mL of methylamine-formic acid buffer (9.6 M methylamine adjusted with HCOOH to pH 11.5) is added, and the reaction mixture is held for 4 h at 25°C. The final concentration of methylamine is 3.5 M. The reaction mixture is exhaustively dialyzed against deionized water at 4°C or, for peptides, isolated by gel filtration on Sephadex G-10, equilibrated and eluted with 0.1 M NH_4HCO_3.

BIBLIOGRAPHY

1. G. C. DuBois et al. (1981) *Biochem. J.* **199**, 335–340.
2. J. K. Inman et al. (1983) *Methods Enzymol.* **91**, 559–569.

AMINO ACID ANALYSIS

T. E. CREIGHTON

The **amino acid** compositions of **proteins** are routinely determined by completely hydrolyzing the **peptide bonds** of the **polypeptide chain**, and then determining quantitatively the constituent amino acids that were released.

PEPTIDE BOND HYDROLYSIS

The traditional method of hydrolyzing polypeptide chains has been to incubate them anaerobically in 6 *M* HCl at approximately 110°C for 24–72 h (1). More modern methods use other acids, higher temperatures, and shorter periods of time. Most peptide bonds hydrolyze at similar rates, but those between the large **nonpolar** amino acid residues, particularly **Val**, **Leu**, and **Ile**, are hydrolyzed more slowly and require longer hydrolysis times or the addition of organic acids such as trifluoroacetic acid. Hydrolysis is presumably hindered sterically by the bulky side chains.

Any chemical procedure that hydrolyzes the peptide bonds of the backbone will also hydrolyze the chemically similar amide side chains of **Asn** and **Gln** residues, to produce the amino acids **aspartic acid** and **glutamic acid**, respectively. It is feasible to measure the total number of Asn and Gln residues by measuring the amount of ammonia released during the hydrolysis, but otherwise it is not possible to distinguish between Asp and Asn and between Glu and Gln after hydrolysis of the polypeptide chain. In this case, it is common practice to designate such uncertain residues by the three-letter abbreviations Asx and Glx, and by the one-letter abbreviations B and Z, respectively.

Trp residues are usually destroyed completely by acid hydrolysis, probably as a result of reaction with chlorine produced by oxidation of the HCl. They can be protected by the addition of thiol or sulfonic acid compounds or of phenol to scavenge the chlorine (2). **Tyr** residues are also susceptible to chlorination, but they are usually lost only partially. The **thiol groups** of **Cys** residues are oxidized and the amino acid partially destroyed by acid hydrolysis; this residue is best analyzed after performic acid oxidation of the protein to convert all the Cys residues to cysteic acid.

Some of the problems with acid hydrolysis can be overcome by using other procedures, such as hydrolysis by alkali or by **proteinases**. Other amino acids, notably Ser and Thr, are destroyed by alkaline hydrolysis, however, and total proteinase digestion to amino acids is not straightforward. Consequently, acid hydrolysis remains in common use.

QUANTIFYING THE AMINO ACIDS

The identities and quantities of the various amino acids present in a protein hydrolyzates are normally determined by automated amino acid analyzers. The amino acids are separated chromatographically and quantified as they emerge from the column. Traditional methods used ion-exchange chromatography of the free amino acids, followed by detection with **ninhydrin** or fluorescent reagents such as **fluorescamine**. **Proline** does not react in the usual manner with such reagents, due to absence of an amino group, so special procedures are required to measure it. More rapid and sensitive methods now predominate, in which the amino acids are reacted with suitable reagents prior to the chromatographic separation, rather than after. The favored method at present is to react the amino acids with phenylisothiocyate (see **Edman degradation**) and then to separate the colored derivatives by **reverse-phase chromatography**. With this procedure, a complete quantitative amino acid analysis can be carried out in just a few minutes with only picomole quantities of amino acids (3).

The relative numbers of aromatic residues (**Phe**, **Tyr**, and **Trp**) in intact proteins and peptides can usually be determined from the UV absorbance spectrum under conditions in which the polypeptide chain is fully unfolded so that its spectrum is the sum of its constituent residues (4).

Amino acid analysis does not give directly the number of residues of each amino acid per polypeptide chain. The most accurate result is the molar ratios of the various amino acids. The true molecular weight of the polypeptide chain, in the absence of any non-amino acid moieties, must be known for the amino acid analysis results to be converted to the number of residues of each amino acid per chain. Only with very accurate results, or with very small proteins, are such values usually close to the actual integer values. An alternative procedure is to use progressive chemical modification of one type of amino acid side chain for **counting residues**.

BIBLIOGRAPHY

1. R. L. Hill (1965) *Adv. Protein Chem.* **20**, 37–107.
2. L. T. Ng et al. (1987) *Anal. Biochem.* **167**, 47–52.
3. S. A. Cohen and D. J. Strydom (1988) *Anal. Biochem.* **174**, 1–16.
4.. H. Edelhoch (1967) *Biochemistry* **6**, 1948–1954.

AMINO ACIDS

T. E. CREIGHTON

Twenty amino acids are the building blocks of **proteins**. They are linked together in a linear **polypeptide chain** by forming **peptide bonds** between them, in an order ordained by the nucleotide sequence of the corresponding **gene** for the protein, **translated** from the corresponding **messenger RNA**.

Nineteen of the amino acids have the general structure

$$\begin{array}{c} \mathbf{R} \\ | \\ H_2N-CH-CO_2H \end{array}$$

and differ only in the chemical structures of the side chain R. The **amino group** and the **carboxyl group** give this class of compounds its name. At physiological pH values, both groups are ionized, and the zwitterion is the common form of the amino acid. The exceptional amino acid, **proline**, differs in that its side chain is bonded to the nitrogen atom of the amino

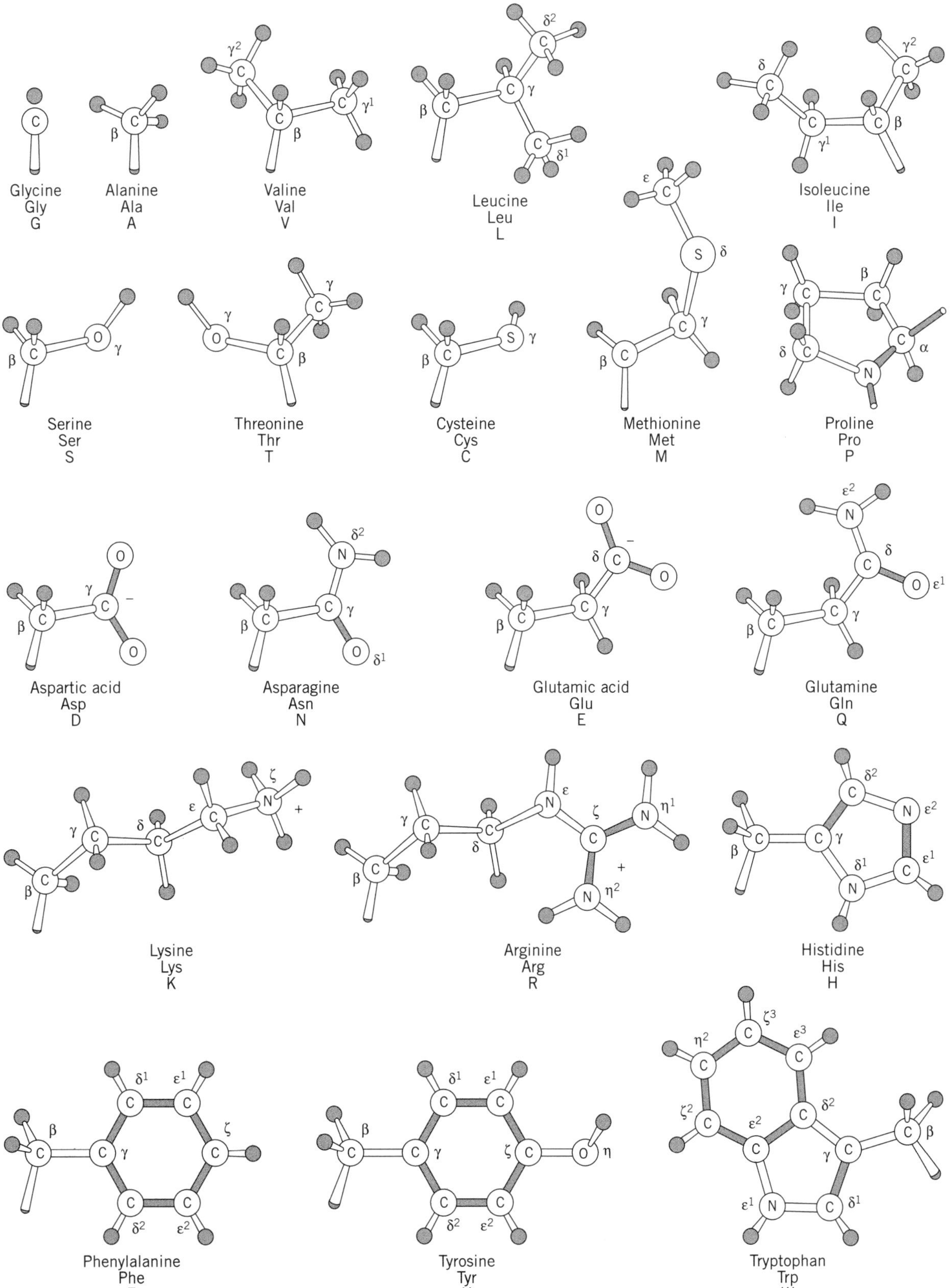

Figure 1. The side chains of the 20 amino acids that occur naturally in proteins. Small unlabeled spheres are hydrogen atoms; other atoms are labeled. Double bonds are black, and partial double bonds are shaded. In the case of proline, the bonds of the polypeptide backbone are included and are black. The three- and one-letter abbreviations commonly used appear below the name of the amino acid. Note that isoleucine and threonine have asymmetric centers in their side chains, and only the isomer illustrated is used biologically.

group, which is then a secondary amine, and proline is an imino acid:

```
      H2
      C
 H2C     CH2
   \     /
   HN—CH—CO2H
```

The central α-carbon atom is asymmetric in 19 of the amino acids and is always the L isomer:

The exception is **glycine**, in which the side chain is simply another hydrogen atom, so the C_α atom is no longer asymmetric.

The structures of the side chains of the 20 normal amino acids used in protein biosynthesis are described in Figure 1, and their properties are described in individual entries. The central, asymmetric carbon atom is designated as α, and the atoms of the side chains are commonly designated β, γ, δ, ε and ζ in order away from the C_α atom. Chemical groups are, however, usually designated by the carbon atom to which they are bonded; hence, the N_ζ atom of a **lysine** residue is part of the ε-amino group. A 21st amino acid that is used in protein biosynthesis in only a few instances is **selenocysteine**. Many variations of these 21 amino acids can be found in proteins as a result of **post-translational modifications** after synthesis of the polypeptide chain.

Linking the amino acids into a polypeptide chain involves the condensation of the α-carboxyl group of one with the α-amino group of the next, with the elimination of one water molecule. The remaining amino acid within the polypeptide chain is then known as a *residue*. Each type of residue is frequently designated with either three- or one-letter abbreviations, which are given in Figure 1. The three-letter abbreviations are obvious, but the one-letter are preferred with long sequences, as they save space and are less likely to be confused; for example, Gln, Glu, and Gly can easily be confused, but not Q, E, and G. The sequences of amino acids in proteins are usually written with either abbreviation, starting at the left with the *N*-terminal residue, with the free α-amino group, which is considered the first residue of the polypeptide chain. Amino acid residues in polypeptide chains are properly referred to by changing the ending of their amino acid names, adding or replacing the frequent "-ine" ending with "-yl" (eg, glycyl or alanyl residues), but the amino acid names are commonly used for the residues also. This complication can be minimized by using the one- or three-letter abbreviations. Note that these abbreviations should be used for residues in proteins only, not for the free amino acids.

Depending on the organism and its circumstances, the amino acids are derived from breaking down proteins ingested in feeding and recycling them, and also in some cases by synthesis de novo. In the well-known case of humans, the following amino acids cannot be synthesized and must be obtained from the diet: **histidine**, **isoleucine**, **leucine**, **lysine**, **threonine**, **tryptophan**, and **valine**. **Arginine** is synthesized, but not at a rate sufficient during growth. **Methionine** is required in large amounts to produce **cysteine** if that amino acid is not supplied adequately; similarly, **phenylalanine** is required in the absence of **tyrosine**.

Suggestions for Further Reading

E. J. Cohn and J. T. Edsall (1943) *Proteins, Amino Acids and Peptides*, Van Nostrand-Reinhold, Princeton, NJ.

R. E. Marsh and J. Donohue (1967) Crystal structures of amino acids and peptides, *Adv. Protein Chem.* **22**, 235–256.

AMINO GROUPS

T. IMOTO

Amino groups are widely distributed in biological substances such as proteins, **polynucleotides**, polysaccharides, and **lipids**. They play important roles in **electrostatic interactions** because of their nucleophilicity and positive charge. The amino groups in the **nucleotides** of nucleic acids are involved in pairing bases. Modification of these groups causes serious errors in nucleic acid replication, **translation**, and **gene expression**. Amino groups in proteins are important for maintaining their structure and solubility and, sometimes, for manifestating their biological function. In particular, an amino group plays a pivotal role in the enzymes that utilize **pyridoxal phosphate** derivatives as **coenzymes**.

There are two kinds of amino groups in proteins. One is the N-terminal α-amino group, that has a pK_a of 6 to 8, and the other is the ε-amino group of **lysine** residues that have pK_a values generally between 8 and 10.5. The N-terminal amino group is very important for elucidating the **primary structure** of a protein because a free N-terminal amino group is indispensable for **Edman degradation**. N-terminal amino groups in proteins are often protected by acylation as a **post-translational modification**. Attaching **ubiquitin** molecules to the amino groups of proteins induces their degradation (see **Protein degradation**). Amino groups are reactive nucleophiles and are widely utilized to immobilize proteins on solid supports for **affinity chromatography**.

CHEMICAL MODIFICATION OF AMINO GROUPS IN PROTEINS

Many modification methods for amino groups have been invented based on their excellent nucleophilicities (1). An amino group is a strong nucleophile only in its nonprotonated form, so it is most reactive at high pH. Because of the differences in pK_a values between α- and ε-amino groups, the former may be selectively modified by controlling the pH of the reaction medium. Because many amino groups are exposed and not involved in the protein function, modification of amino groups is suitable for introducing **reporter groups**, such as chromophores, into proteins and for **radiolabeling** them. The number of amino groups present and subsequently the extent of their modification with some modifying reagents, is determined by trinitrophenylation with 2,4,6-**trinitrobenzene sulfonic acid**. The integral number of amino groups present

Table 1. Chemical Modification of Amino Groups of Proteins

Reaction	Reagent	pH
Acylation	Acetic anhydride	5.5–8
	Acetylimidazole	>5
	N-Acetylsuccinimide	>4
	Citraconic anhydride	8.2
	N-Hydroxysuccinimide acetate	6.9–8.5
	Maleic anhydride	6–10
	Succinic anhydride	7–10
Alkylation and arylation	1-Fluoro-2,4-dinitrobenzene	7–11
	Iodoacetic acid	7.5–9
	2,4,6-Trinitrobenzene sulfonic acid	9.5
Amidination	Methyl acetamidate	7–10.5
Carbamylation	Potassium cyanate	> 7
Guanidination	1-Guanyl-3,5-dimethylpyrazole nitrate	9.5
	O-Methylisourea	10–11
Reductive alkylation	Formaldehyde + sodium borohydride	8–10

in a protein can be counted. Representative modification methods for amino groups are shown in Table 1.

BIBLIOGRAPHY

1. T. Imoto and H. Yamada (1989) In *Protein Structure: A Practical Approach* (T. E. Creighton, ed.), IRL Press, Oxford, UK, pp. 247–277.

AMINOACYL tRNA SYNTHETASES

P. SCHIMMEL

The aminoacyl tRNA synthetases catalyze reactions that establish the rules of the **genetic code**. For this reason, there is great interest in these **enzymes** and their **evolutionary** development, which is thought to be closely connected to the establishment of the code. Research on the synthetases has led to the concept of an operational RNA code for amino acids that is imbedded in the **acceptor stems** of **transfer RNA (tRNA)** (1). The operational RNA code may have played an important role in the assembly of the genetic code and in the overall design of tRNA synthetases.

In the flow of genetic information, **messenger RNA** (mRNA) is **transcribed** from DNA, and the mRNA, in turn, is the template for protein synthesis (Fig. 1). The triplet **codons** of mRNA interact with the **anticodons** of tRNA through **complementary base pairing**. Amino acids joined to tRNA are incorporated into the growing **polypeptide chain**. The algorithm of the genetic code relates each amino acid to a specific trinucleotide codon. The triplet associated with a particular amino acid is determined in the aminoacylation reaction, where a given amino acid is linked to a tRNA bearing the anticodon trinucleotide that corresponds to that amino acid. These aminoacylation reactions are catalyzed by aminoacyl tRNA synthetases.

AMINOACYLATION REACTION AND THE GENETIC CODE

For most tRNA synthetases, aminoacylation is carried out in a two-step reaction:

$$\text{E} + \text{AA} + \text{ATP} \leftrightarrow \text{E(AA-AMP)} + \text{PP}_\text{i} \quad (1)$$

$$\text{E(AA-AMP)} + \text{tRNA} \leftrightarrow \text{AA-tRNA} + \text{AMP} + \text{E} \quad (2)$$

In the first reaction, the enzyme, E, uses ATP to activate an amino acid, AA, to yield the firmly bound *aminoacyladenylate* (AA-AMP). In the second step, the activated amino acid is transferred to the 3′-end of the tRNA, where it is connected by an ester linkage to the 2′- or 3′-hydroxyl group (after initial attachment, the amino acid can migrate back and forth between the 2′ and the 3′ positions). While most synthetases can carry out amino acid activation in the absence of tRNA, there are a few exceptions (such as glutaminyl-, glutamyl-, and argininyl-tRNA synthetases) that require the presence of the cognate tRNA for amino acid activation.

Because each of the 20 natural amino acids used in protein synthesis has a corresponding, or **cognate**, aminoacyl tRNA synthetase, there are 20 of these enzymes in each cellular compartment where proteins are synthesized. Each of them must distinguish its amino acid from all others and, at the same time, recognize the cognate tRNA that bears the anticodon corresponding to that amino acid. In **prokaryotes** and in the cytoplasm of **eukaryotes**, there is typically one tRNA synthetase for each amino acid (eukaryotic **mitochondria** have an additional set of synthetases that are essential for mitochondrial protein synthesis). However, the degeneracy of the genetic code means that there are 61 trinucleotides coding for the 20 amino acids. Reading these 61 triplets requires more than just 20 tRNAs. As a consequence, there are multiple tRNA isoacceptors for many of the synthetases. The synthetases for a particular amino acid must, therefore, recognize and aminoacylate all tRNA isoacceptors for that amino acid. This consideration in itself has important implications.

For example, the codons for serine are sixfold degenerate. In order to read these six codons, the serine tRNA isoacceptors must collectively permute all three anticodon nucleotides. Thus, for a single seryl-tRNA synthetase to aminoacylate all these tRNA^{Ser} isoacceptors, the anticodon is not suitable for discrimination. Direct experiments *in vitro* and the **X-ray crystallography** structure of the seryl-tRNA synthetase-tRNA^{Ser} complex have demonstrated that, in fact, seryl-tRNA synthetase does not contact the anticodon trinucleotide (2). This observation, and others described below, showed that, for at least some amino acids, the relationship between an amino acid and the triplet of the genetic code is not direct.

CLASSES OF AMINOACYL tRNA SYNTHETASES

The synthetases are heterogeneous in **quaternary structures** and subunit sizes, and this heterogeneity obscured more fundamental relationships between these enzymes. For example, in *Escherichia coli*, the quaternary structures of synthetases include α, α_2, α_4, and $\alpha_2\beta_2$ (3). Subunit sizes vary from 303 to 951 amino acid residues (4). The 20 aminoacyl tRNA synthetases are now known to be divided into two classes of 10 enzymes each (Table 1) (8,9). These classes are based on conserved sequence motifs and the structural architecture of the catalytic domains (8,9). The classification is also based on the fact that the site of initial amino acid attachment on the tRNA differs between the two classes (8). The classes appear to be fixed in evolution, because there is no example of an enzyme switching classes depending on the organism to which

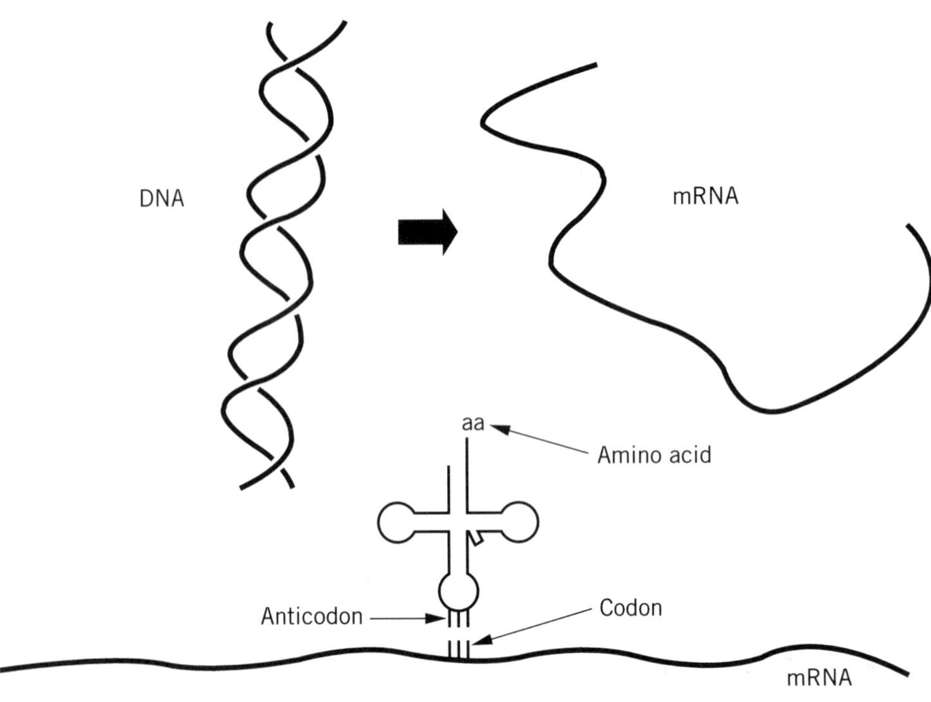

Figure 1. Flow of genetic information. Messenger RNA is synthesized from DNA. The mRNA has a string of trinucleotide codons that are translated into a polypeptide whose amino acid sequence is determined by the codons, according to the rules of the genetic code. The amino acid that is inserted into the polypeptide is determined by the codon–anticodon interaction with the tRNA that bears the amino acid corresponding to the particular anticodon. Therefore, the genetic code is determined by the linking of a particular amino acid with a particular anticodon triplet within a tRNA. The joining of amino acids to tRNA is catalyzed by aminoacyl tRNA synthetases. Thus, the genetic code is determined at the biochemical level in the aminoacylation reaction. (This figure was provided by Arturo Morales.)

Table 1. Classes of Aminoacyl tRNA Synthetases

Class I	Class II
Arginine	Alanine
Cysteine	Asparagine[a]
Glutamic	Aspartate
Glutamine1	Glycine
Isoleucine	Histidine
Leucine	Lysine
Methionine	Phenylalanine
Tryptophan	Proline
Tyrosine	Serine
Valine	Threonine

[a] Gram-positive bacteria, plant chloroplasts, and animal mitochondria have been shown to have less than 20 tRNA synthetases. Instead, glutamyl-tRNA synthetase catalyzes attachment of glutamic acid to both $tRNA^{Glu}$ and $tRNA^{Gln}$ and, similarly, aspartyl-tRNA synthetases catalyzes attachment of aspartate to both $tRNA^{Asp}$ and $tRNA^{Asn}$. An amidotransferase then catalyzes the amidation of Glu-$tRNA^{Gln}$ to give Gln-$tRNA^{Gln}$, and, likewise, amidation of Asp-$tRNA^{Asn}$ gives Asn-$tRNA^{Asn}$ (5–7).

it belongs. Thus, the two classes may have developed early in evolution.

Class I

These enzymes are usually monomers and are characterized by an architecture that is similar to that seen in dehydrogenases and other **nucleotide-binding** proteins. This structural motif is a **Rossmann** nucleotide-binding fold, which consists of alternating **β-strands** and **α-helices** (Fig. 2) (10–12). In the case of class I tRNA synthetases, the fold is divided into two $\beta_3\alpha_2$ halves to give an overall $\beta_6\alpha_4$ structure. In this structure, the β-strands are arranged in parallel. A polypeptide of variable length, designated as *connective polypeptide 1* (CP1), links together the two halves of the active site (13). In some class I enzymes, this insertion plays a role in **translational editing**. It also contains some of the residues for binding the synthetase to the tRNA acceptor helix (14).

The nucleotide-binding fold contains the site for adenylate synthesis. This catalytic **domain** may be identified by two characteristic sequence motifs, without any knowledge of three-dimensional structure. One motif is the 11–amino acid element known as the signature sequence, which ends in the sequence–His–Ile–Gly–His, or HIGH in one-letter code (10,11). This element is located in the first half of the nucleotide-binding fold at the end of the first β-strand and the beginning of the first α-helix. It was designated as a signature sequence because it served as a clear signature for a subgroup of related synthetases, before many crystal structures were determined. The second element is the KMSKS motif, located in the second half of the nucleotide-binding fold (15). These elements are critical parts of the **active site**.

Class II

The class II enzymes are mostly α_2 dimers. The active sites of class II enzymes have a completely different architecture that harbors three characteristic sequence motifs. The structure consists of a seven-stranded antiparallel β-sheet with three α-helices (9,16–18) (Fig. 3) (8,9,19). The three characteristic sequence motifs are known as motifs 1, 2, and 3. The sequences of these motifs are highly degenerate (8,9). They consist of a helix–loop–strand, strand–loop–strand, and strand–helix, respectively. All three of these motifs form part of the active site.

Amino Acid Attachment

The site of initial amino acid attachment for class I enzymes is the 2′-hydroxyl, whereas the 3′-hydroxyl is used by class II enzymes (20). This distinction is now understood to result from a difference in the ways that the two enzymes approach the end of the tRNA. In particular, class I enzymes approach the end

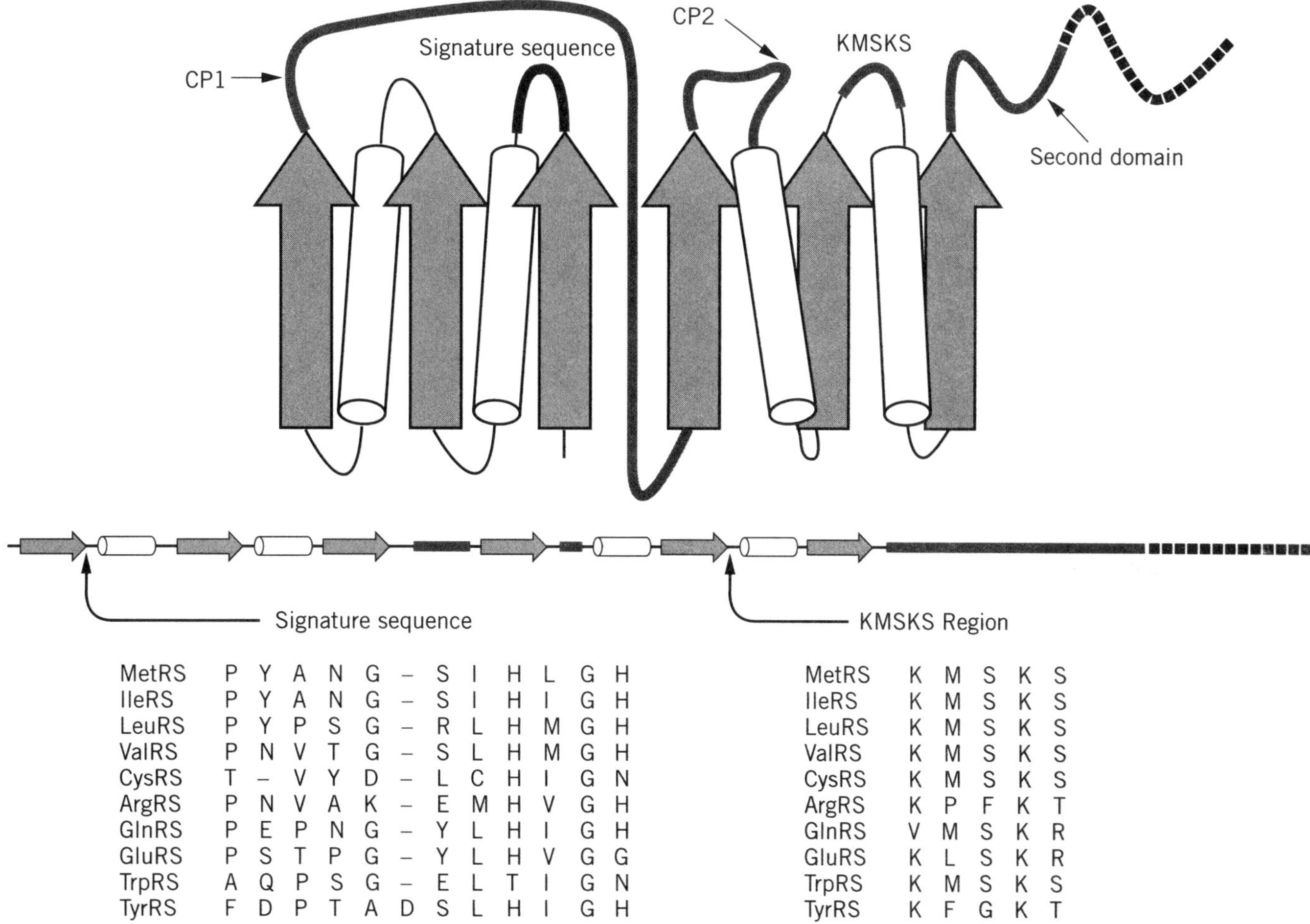

MetRS	P	Y	A	N	G	–	S	I	H	L	G	H
IleRS	P	Y	A	N	G	–	S	I	H	I	G	H
LeuRS	P	Y	P	S	G	–	R	L	H	M	G	H
ValRS	P	N	V	T	G	–	S	L	H	M	G	H
CysRS	T	–	V	Y	D	–	L	C	H	I	G	N
ArgRS	P	N	V	A	K	–	E	M	H	V	G	H
GlnRS	P	E	P	N	G	–	Y	L	H	I	G	H
GluRS	P	S	T	P	G	–	Y	L	H	V	G	G
TrpRS	A	Q	P	S	G	–	E	L	T	I	G	N
TyrRS	F	D	P	T	A	D	S	L	H	I	G	H

MetRS	K	M	S	K	S
IleRS	K	M	S	K	S
LeuRS	K	M	S	K	S
ValRS	K	M	S	K	S
CysRS	K	M	S	K	S
ArgRS	K	P	F	K	T
GlnRS	V	M	S	K	R
GluRS	K	L	S	K	R
TrpRS	K	M	S	K	S
TyrRS	K	F	G	K	T

Figure 2. Design of a class I tRNA synthetase. The nucleotide-binding fold of class I tRNA synthetases consists of alternating β-strands (arrows) and α-helices (cylinders) that form a $\beta_6\alpha_4$ structure. A two-dimensional spatial arrangement of these elements is shown at the top, and a linear arrangement is shown directly below. A second domain of variable size occurs after the nucleotide-binding fold. The fold is split into two $\beta_3\alpha_2$ halves by an insertion known as *connective polypeptide 1* (CPI). A second, smaller insertion (CP2) splits the second half of the fold. Two sequence elements were used to identify and classify the class I enzymes. These are known as the *12-residue signature sequence*, which ends in the HIGH tetrapeptides (10,11) and as the *KMSKS* sequence (12). Their locations in the schematic structure are shown near the label signature sequence and KMSKS. By way of example, an alignment of these regions in the sequences of the 10 class I *E. coli* enzymes is shown beneath the schematic figures. Similar alignments can be made for class I enzymes from other species throughout evolution. (This figure was provided by Arturo Morales.)

of the tRNA acceptor helix from the minor groove side, while class II enzymes approach from the major groove side (21).

OVERALL STRUCTURAL DESIGN AND THE SYNTHETASE-tRNA COMPLEX

The class-defining active-site domain is only a part of the tRNA synthetase structure. Inserted into the active-site domain are sequence motifs that enable the tRNA acceptor helix to bind with its 3′-end near the aminoacyl adenylate. These insertions are typically idiosyncratic to the synthetase. Two examples are the CP1 insertions of class I enzymes and the variable loops of motif 2 of class II enzymes, both of which have a role in docking the acceptor helices to the enzymes. However, in addition to insertions into the active-site domain, all synthetases have a second major domain. This domain, also typically idiosyncratic to the enzyme, provides for contacts with parts of the tRNA that are distal to the amino acid attachment site. For many (but not all) synthetases, this includes contacts with the anticodon. For the class I methionyl- and glutaminyl-tRNA synthetases, this second, anticodon-binding, domain is largely α-helical and largely β-structure, respectively (22,23). This difference demonstrates that, even for enzymes in the

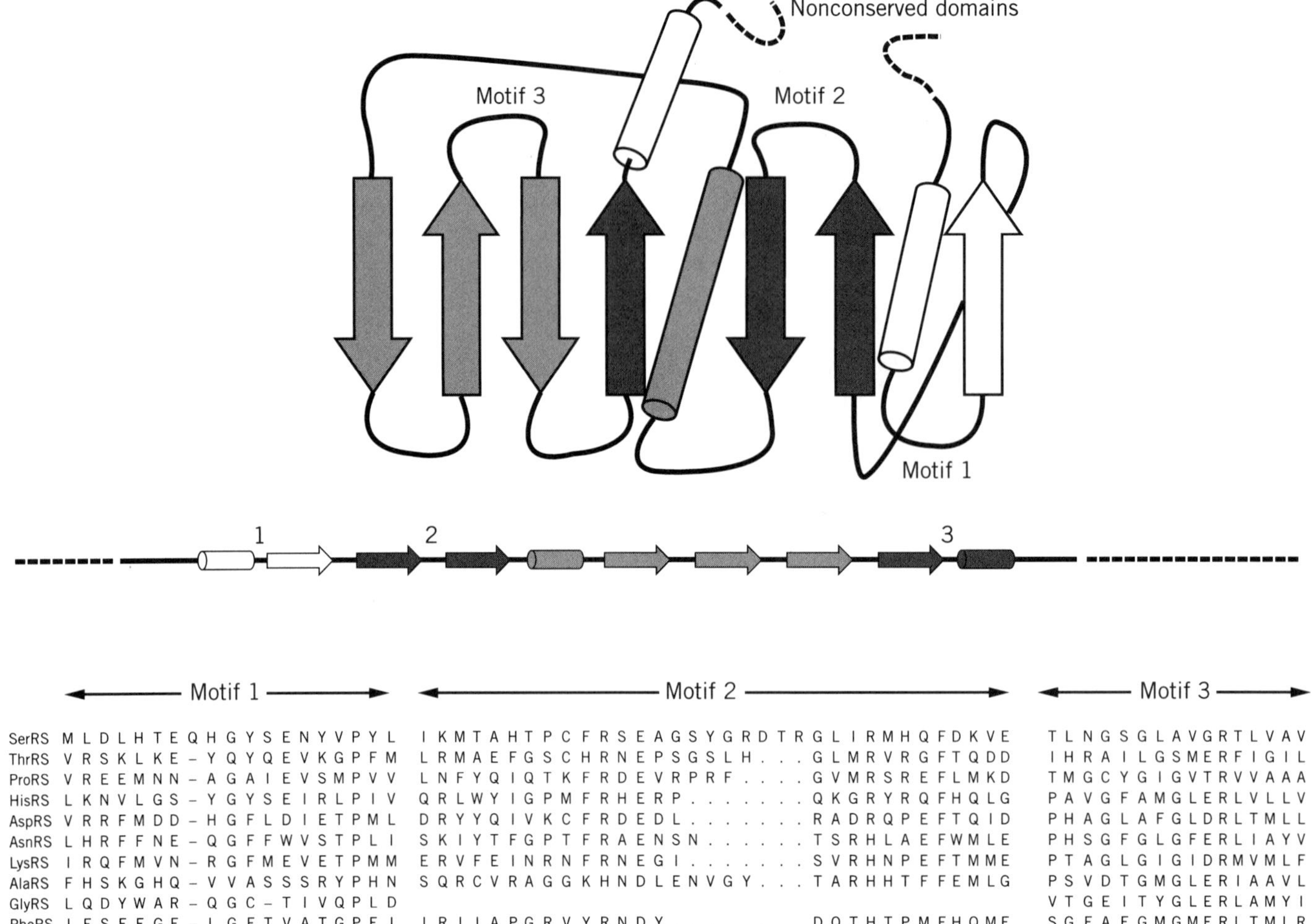

Figure 3. Design of a class II tRNA synthetase. The seven-stranded β-sheet with three α-helices is shown. The variable-sized second domain is shown as a dashed line that may occur on either the *N*- or the *C*-terminal side of the class-defining domain. Three characteristic sequence motifs are present in class II enzymes (8,9) and are distinguished in this illustration by their different shadings. These motifs are highly degenerate in sequence and consist of a helix–loop–strand (motif 1), strand–loop–strand (motif 2), and strand–helix. The locations of these motifs in the class-defining domain are shown. An example of the sequences of these motifs for *E. coli* tRNA synthetases is also shown (8,19). Note the high degeneracy of these sequence motifs, especially when compared with the conserved sequence elements in class I enzymes (Fig. 2). (This figure was provided by Dr. Arturo Morales.)

same class, their second domains are completely unrelated. In the case of the class II seryl-tRNA synthetase, an unusual **coiled-coil** protrudes from the N-terminus of the enzyme. This structure, which is not found in many other class II enzymes, provides for contacts with the variable loop of $tRNA^{Ser}$ (2,24).

Thus, to a rough approximation, the synthetases comprise two major domains (9,14,16,23,25–31). The tRNA molecule also comprises two major domains, which consist of the four arms of the cloverleaf secondary structure (Fig. 4). One domain is the acceptor-TψC minihelix, where the amino acid acceptor end and the TψC stem stack together to make a helix of 12 base pairs. The second domain is formed by stacking of the dihydrouridine stem with the anticodon stem. The result is an L-shaped three-dimensional structure where the amino acid acceptor end and the anticodon-containing template-reading-head are segregated into separate structural units (see **Transfer RNA**).

The two major domains of the synthetase make contact with the two domains of the L-shaped tRNA molecule (Fig. 5). The catalytic domain, with insertions containing RNA-binding determinants, interacts with the acceptor-TψC minihelix. The second major domain interacts with the second domain of the tRNA, where interactions may extend as far as the anticodon trinucleotide or may involve contacts with special structures, such as the large variable loop of $tRNA^{Ser}$. The details of the

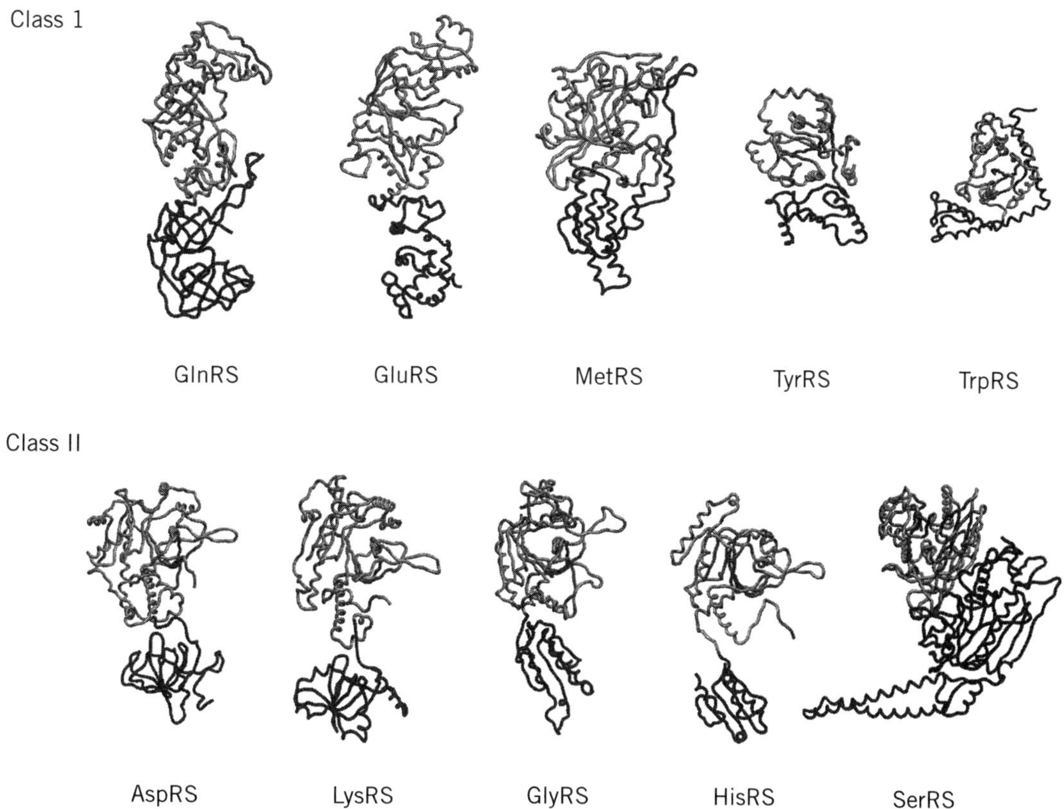

Figure 4. Examples of crystal structures of class I and class II tRNA synthetases. Regardless of the class to which an enzyme is assigned, its structure can be approximated as comprising two major domains. One is the class-defining active site domain, which is shared by all members of the same class. This domain (gray) is thought to be the historical tRNA synthetase. The second domain (dark gray) is typically idiosyncratic to the synthetase and is not shared by all members of the same class. This domain was probably added later to the synthetase structure. In these illustrations, the domains have been defined by the obvious visual divisions in the structures and, for that reason, the catalytic domain may extend somewhat beyond the region that contains the class-defining motifs. Some of these enzymes are homodimers; in all cases, only a single subunit is shown. In the case of the tyrosyl tRNA synthetase, TyrRS, the structure of the second domain is not complete and therefore is truncated in the structural representation. Not shown is PheRS (25), which has an $\alpha_2\beta_2$ quaternary structure. These structures were determined for GlnRS (14), GluRS (26), MetRS (23), TyrRS (27), TrpRS (28), AspRS (16), LysRS (29), GlyRS (30), HisRS (31), and SerRS (9). (This figure was provided by Arturo Morales.)

interactions between tRNA and synthetase are described in **RNA-binding proteins**.

OPERATIONAL RNA CODE FOR AMINO ACIDS

In addition to seryl-tRNA synthetase, alanyl-tRNA synthetase is an example of a synthetase that makes no contact with the anticodon (34). Instead, an acceptor helix G3:U70 **wobble** base pair is a major determinant of the identity of an alanine tRNA (35,36). Alteration of this base pair to G:C, A:U, I:U, or U:G abolishes aminoacylation with alanine (37–39). Transfer of this base pair into other, nonalanine tRNA confers alanine acceptance on them. Thus, the G3:U70 pair marks a tRNA for charging with alanine.

Because the G3:U70 base pair is located in the acceptor helix, further experiments tested whether the 12-bp acceptor-TψC minihelix by itself would be a substrate for aminoacylation (Fig. 6) (40–42). Not only the minihelix, but also a 7-base-pair(bp) microhelix consisting of just the acceptor stem, is efficiently charged with alanine, provided that it contains the G3:U70 base pair (40,43). Transfer of this base pair into other microhelices confers alanine acceptance on them. Thus, the charging behavior of the mini- and microhelices reproduces that seen with the full tRNA.

The charging of specific RNA helices has now been demonstrated with at least 11 different tRNA synthetases, even for cases where the anticodon is known to play a significant role in the recognition of the related tRNA (41,44–47). In the case of

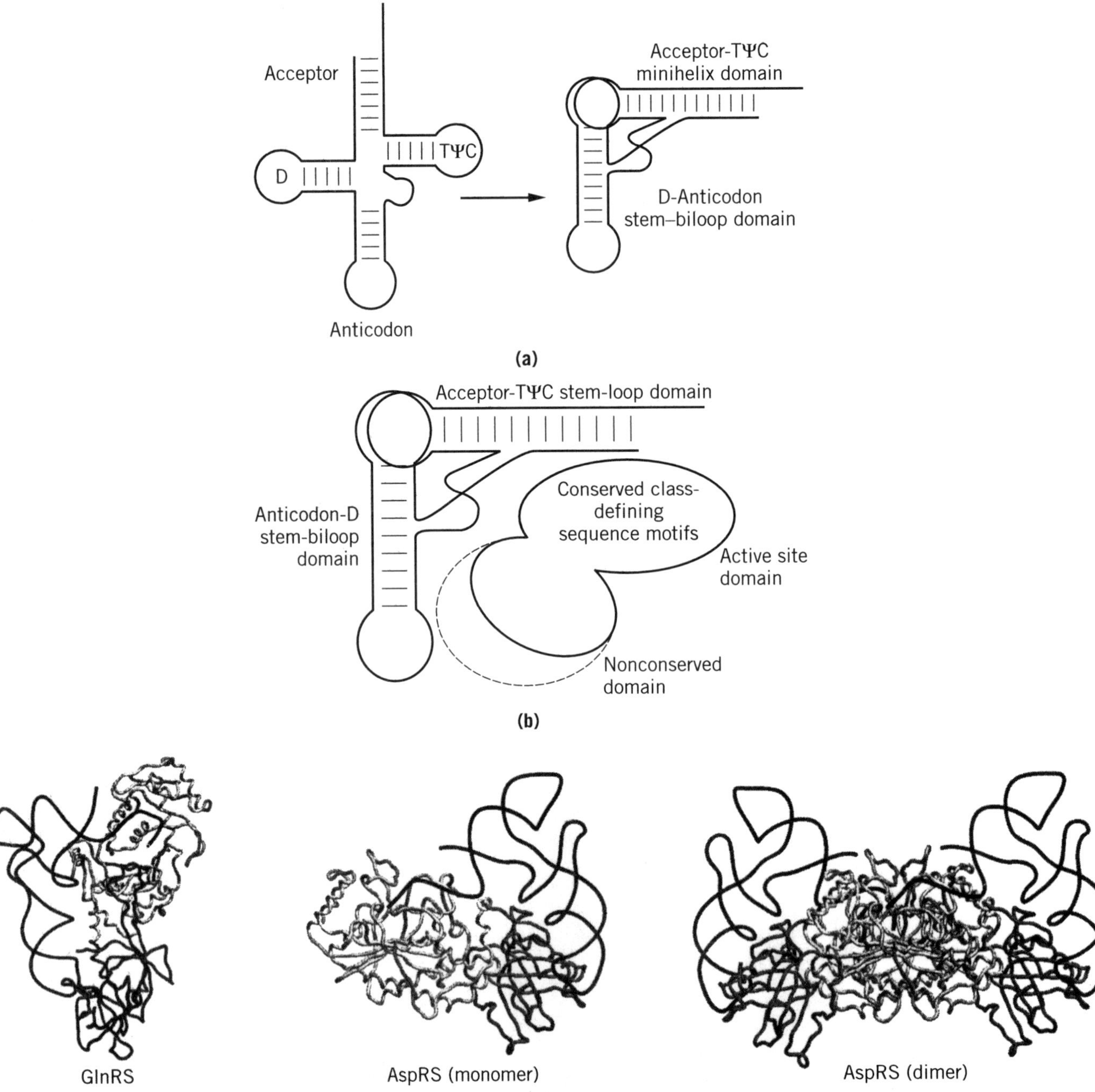

Figure 5. Two major domains of a synthetase interacting with two domains of a tRNA. (**a**) Schematic representation of the folding of the tRNA structure into two domains that segregate the amino acid attachment site into a 12-bp minihelix stem–loop and the anticodon triplet into a second domain. The dihydrouridine (D) and TψC loops are indicated as common landmarks found in most all tRNA. (Adapted from Ref. 32.) (**b**) Domains of a synthetase (see Fig. 4) interacting with the domains of a tRNA. The second domain of the synthetase is shown with a dotted line to indicate that it varies in size and often extends to the anticodon. (Adapted from Ref. 33.) (**c**) Examples of synthetase-tRNA complexes, with the active-site class-defining domain interacting with the acceptor-TψC minihelix portion of the tRNA, and the second, idiosyncratic domain of the synthetase interacting with the second domain of the tRNA (14,16). Separate shades of gray have been used to show the separate domains. In the case of aspartyl tRNA synthetase, the protein is a dimer. Binding of a single tRNA to the monomer is shown to more clearly illustrate the domain–domain interactions. The dimeric complex with two bound tRNA is also shown at the right. (This figure was provided by Arturo Morales.)

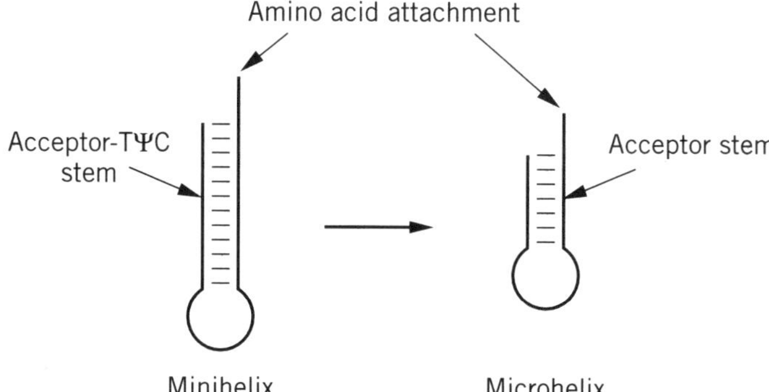

Figure 6. Minihelix and microhelix substrates for aminoacylation. The minihelix is derived from the 12-bp acceptor-TψC domain of the tRNA, while the microhelix is a hairpin helix whose base pairs consist of the 7-bp acceptor stem portion (40). These substrates for aminoacylation are devoid of the anticodon trinucleotides of the genetic code. About 11 examples of aminoacylation of minihelix or microhelix structures have been demonstrated. These substrates have been used to delineate the operational RNA code for amino acids (41). (Adapted from Ref. 42.)

histidine, for example, an extra nucleotide at the 5′ end of the acceptor helix is characteristic of and unique to histidine tRNA throughout evolution. RNA microhelices that contain the extra base are charged with histidine (48). Thus, the 5′-appended nucleotide marks a molecule for charging with histidine. The smallest substrates seen to be charged with specific amino acids are stem–loop hairpins with as few as four base pairs stabilized by an RNA **tetraloop** motif (Fig. 7) (49).

The charging of microhelix substrates is sometimes considerably less efficient than that for the corresponding tRNA. In each case, however, charging is sequence-specific and depends on two to four nucleotides near the amino acid attachment site. In a fine-structure mapping of the efficiently charged alanine microhelix, a constellation of atoms was identified as

A	A	A	A
C	C	C	C
C	C	C	C
A	U	$^{-1}$G-C	A
^{1}G-C	^{1}G-C	^{1}G-C	^{1}C-A
G-C	C-G	U-A	G-C
G-U^{70}	G-C^{70}	G-C^{70}	C-G^{70}
C-G	C-G	C-G	G-C
U G	U G	U G	U G
U C	U C	U C	U C
Ala	Gly	His	Met

Figure 7. RNA tetraloop substrates for aminoacylation. The amino acid that can be charged onto the designated structures is indicated (49). These short helices are stabilized by an RNA tetraloop motif that confers unusual stability to short RNA helices. (Adapted from Ref. 49.)

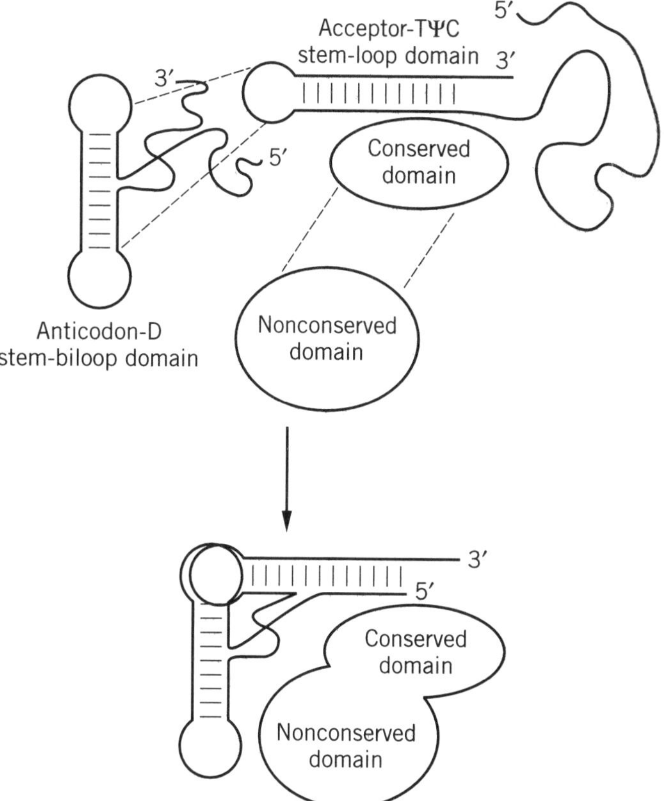

Figure 8. Assembly of a tRNA synthetase in evolution. A primordial tRNA synthetase is envisioned as interacting with a minihelix-like structure. As the tRNA structure developed, a template reading head (anticodon domain) was added, along with the second domain for the synthetase. (Adapted from Ref. 1.)

critical for the aminoacylation signal. Prominent among these atoms was the exocyclic 2-amino group of G of the G3:U70 base pair. Also important were specific 2′-hydroxyl groups that fell within a 5-Å (Å $= 10^{-10}$ m) radius of the critical 2-amino group (39).

These anticodon-independent aminoacylations of oligonucleotide substrates demonstrate that specific RNA sequences and structures per se, rather than the trinucleotides of the genetic code, correspond to specific amino acids. The relationship between these RNA sequences and structures and particular amino acids constitutes an operational RNA code for amino acids that is distinct from, but related to, the genetic code. This operational RNA code may have predated the genetic code (1).

ASSEMBLY OF THE SYNTHETASE-tRNA COMPLEX AND RELATIONSHIP OF THE OPERATIONAL RNA CODE TO THE GENETIC CODE

Several considerations led to the proposal that the minihelix and anticodon-containing second domain of tRNA had distinct origins (1,50–52). The minihelix, with its amino acid attachment site, is viewed as the historical, or earliest, part of the tRNA. Similarly, the class-defining catalytic domain of tRNA synthetases is thought to be the historical enzyme, with the idiosyncratic second domain added later. Experiments with alanyl-tRNA synthetase have demonstrated that a relatively small piece of the enzyme (containing the active site) can by itself charge an RNA microhelix. This result directly demonstrated a domain–domain interaction between discrete units of the tRNA and the synthetase that may somewhat resemble an evolutionarily earlier system (33,54).

Thus, the early synthetase may have consisted solely of a domain for adenylate synthesis. Insertions into this domain allowed the docking of RNA substrates near the activated amino acid so that aminoacylation could occur (Fig. 8). Addition of the second domain of the tRNA, with its anticodon-containing template reading head, and of the second domain of the synthetases was a second, later event. This event led to the joining of the operational RNA code to the genetic code. This scheme also suggests that the relationship between a particular amino acid and the triplet of the code is random, and simply depends on which anticodon-containing domain happened to be fused to the minihelix domain of the tRNA.

CONCLUSIONS

The tRNA synthetases may be among the earliest proteins, arising in evolution contemporaneously with the development of the genetic code. The first synthetases may have been **ribozymes** that catalyzed aminoacylation reactions with a specificity that depended on the sequences and structures of the RNA substrates (54,55). The proteins that replaced these ribozymes were probably small. As a result, they could not extend much beyond the amino acid attachment site and, for that reason, gave rise to a system of interactions that based specificity of aminoacylation on interactions near the end of the acceptor helix. How these charged RNA substrates were used to synthesize specific proteins is a question of great interest.

BIBLIOGRAPHY

1. P. Schimmel, R. Giegé, D. Moras, and S. Yokoyama (1993) *Proc. Natl. Acad. Sci. USA* **90**, 8763–8768.
2. V. Biou, A. Yaremchuk, M. Tukalo, and S. Cusack (1994) *Science* **263**, 1404–1410.
3. P. Schimmel (1987) *Ann. Rev. Biochem.* **56**, 125–158.
4. S. A. Martinis and P. Schimmel (1996) in *Escherichia coli and Salmonella* (F. C. Neidhardt Jr., ed.), ASM Press, Washington, DC, pp. 887–901.
5. A. Schon, C. G. Kannangara, S. Gough, and D. Söll (1988) *Nature* **331**, 187–190.
6. A. W. Curnow, M. Ibba, and D. Söll (1996) *Nature* **382**, 589–590.
7. Y. Gagnon, L. Lacoste, N. Champagne, and J. Lapointe (1996) *J. Biol. Chem.* **271**, 14856–14863.
8. G. Eriani, M. Delarue, O. Poch, J. Gangloff, and D. Moras (1990) *Nature* **347**, 203–206.
9. S. Cusack, C. Berthet-Colominas, M. Hartlein, N. Nassar, and R. Leberman (1990) *Nature* **347**, 249–255.
10. T. A. Webster, H. Tsai, M. Kula, G. A. Mackie, and P. Schimmel (1984) *Science* **226**, 1315–1317.
11. S. W. Ludmerer and P. Schimmel (1987) *J. Biol. Chem.* **262**, 10801–10806.
12. C. Hountondji, F. Lederer, P. Dessen, and S. Blanquet (1986) *Biochemistry* **25**, 16–21.
13. R. M. Starzyk, T. A. Webster, and P. Schimmel (1987) *Science* **237**, 1614–1618.
14. M. A. Rould, J. J. Perona, D. Söll, and T. A. Steitz (1989) *Science* **246**, 1135–1142.
15. C. Hountondji, P. Dessen, and S. Blanquet (1986) *Biochimie* **68**, 1071.
16. M. Ruff, S. Krishnaswamy, M. Boeglin, A. Poterszman, A. Mitschler, A. Podjarny, B. Rees, J. C. Thierry, and D. Moras (1991) *Science* **252**, 1682–1689.
17. S. Cusack, M. Hartlein, and R. Leberman (1991) *Nucl. Acid Res.* **19**, 3489–3498.
18. D. Moras (1992) *Trends Biochem. Sci.* **17**, 159–164.
19. S. Cusack (1993) *Biochimie* **75**, 1077–1081.
20. P. R. Schimmel and D. Söll (1979) *Annu. Rev. Biochem.* **48**, 601–648.
21. J. Cavarelli and D. Moras (1993) *FASEB J.* **7**, 79–86.
22. M. A. Rould, J. J. Perona, D. Söll, and T. A. Steitz (1991) *Nature* **352**, 213–218.
23. S. Brunie, C. Zelwer, and J. Risler (1990) *J. Mol. Biol.* **216**, 411–424.
24. S. Cusack, A. Yaremchuk, and M. Tukalo (1996) *EMBO J.* **15**, 2834–2842.
25. L. Mosyak, L. Reshetnikova, Y. Goldgur, M. Delarue, and M. G. Safro (1995) *Nature Struct. Biol.* **2**, 537–547.
26. O. Nureki, D. G. Vassylyev, K. Katayanagi, T. Shimizu, S. Sekine, T. Kigawa, T. Miyazawa, S. Yokoyama, and K. Morikawa (1995) *Science* **267**, 1958–1965.
27. P. Brick, T. N. Bhat, and D. M. Blow (1988) *J. Mol. Biol.* **208**, 83–98.
28. S. Doublie, G. Bricogne, C. Gilmore, and C. W. Carter Jr. (1995) *Structure* **3**, 17–31.
29. S. Onesti, A. D. Miller, and P. Brick (1995) *Structure* **3**, 163–176.
30. D. T. Logan, M.-H. Mazauric, D. Kern, and D. Moras (1995) *EMBO J.* **14**, 4156–4167.
31. J. G. Arnez, D. C. Harris, A. Mitschler, B. Rees, C. S. Francklyn, and D. Moras (1995) *EMBO J.* **14**, 4143–4155.
32. J. J. Burbaum and P. Schimmel (1991) *J. Biol. Chem.* **266**, 16965–16968.
33. D. D. Buechter and P. Schimmel (1993) *Crit. Rev. Biochem. Mol. Biol.* **28**, 309–322.
34. S. J. Park and P. Schimmel (1988) *J. Biol. Chem.* **263**, 16527–16530.
35. Y.-M. Hou and P. Schimmel (1988) *Nature* **333**, 140–145.
36. W. H. McClain and K. Foss (1988) *Science* **240**, 793–796.
37. S. J. Park, Y.-M. Hou, and P. Schimmel (1989) *Biochemistry* **28**, 2740–2746.
38. K. Musier-Forsyth, N. Usman, S. Scaringe, J. Doudna, R. Green, and P. Schimmel (1991) *Science* **253**, 784–786.
39. K. Musier-Forsyth and P. Schimmel (1992) *Nature* **357**, 513–515.
40. C. Francklyn and P. Schimmel (1989) *Nature* **337**, 478–481.
41. S. A. Martinis and P. Schimmel (1995) in *tRNA: Structure, Biosynthesis and Function* (D. Söll and U. L. RajBhandary, eds.), American Society for Microbiology, Washington, DC, pp. 349–370.
42. P. Schimmel (1993) in *The Translation Apparatus* (K. H. Nierhaus, F. Franceschi, A. R. Subramanian, V. A. Erdmann, and B. Wittmann-Liebold, eds.), Plenum Press, New York, pp. 13–21.
43. C. Francklyn, J.-P. Shi, and P. Schimmel (1992) *Science* **255**, 1121–1125.
44. M. Frugier, C. Florentz, and R. Giegé (1994) *EMBO J.* **13**, 2218–2226.
45. Y.-M. Hou, T. Sterner, and R. Bhalla (1995) *RNA* **1**, 707–713.
46. C. L. Quinn, N. Tao, and P. Schimmel (1995) *Biochemistry* **34**, 12489–12495.

47. M. E. Saks and J. R. Sampson (1996) *EMBO J.* **15**, 2843–2849.
48. C. Francklyn and P. Schimmel (1990) *Proc. Natl. Acad. Sci. USA* **87**, 8655–8659.
49. J.-P. Shi, S. A. Martinis, and P. Schimmel (1992) *Biochemistry* **31**, 4931–4936.
50. A. M. Weiner and N. Maizels (1987) *Proc. Natl. Acad. Sci. USA* **84**, 7383–7387.
51. H. F. Noller (1993) in *The RNA World* (R. F. Gesteland and J. F. Atkins, eds.), Cold Spring Harbor Laboratory Press, Cold Spring Harbor, NY, pp. 137–156.
52. N. Maizels and A. Weiner (1994) *Proc. Natl. Acad. Sci. USA* **91**, 6729–6734.
53. D. D. Buechter and P. Schimmel (1995) *Biochemistry* **34**, 6014–6019; Correction, p. 16352.
54. J. A. Piccirilli, T. S. McConnell, A. J. Zaug, H. F. Noller, and T. R. Cech (1992) *Science* **256**, 1420–1424.
55. M. Illangasekare, G. Sanchez, T. Nickles, and M. Yarus (1995) *Science* **267**, 643–647.

AMINOPEPTIDASES

J. RIORDAN

Enzymes that catalyze the hydrolytic cleavage of the **peptide bond** that connects the N-terminal residue to the rest of a **peptide**, **polypeptide**, or **protein** are referred to as aminopeptidases (E.C. 3.4.11) (see **Peptidase**). The products of hydrolysis are, therefore, the released N-terminal **amino acid** and the remainder of the peptide chain. The latter can, in turn, also be an aminopeptidase substrate; hence these enzymes can release amino acids sequentially, ultimately resulting in complete hydrolysis of the polypeptide (though in practice this is not always observed). The principal determinant of specificity appears to be the free **α-amino group** of the N-terminal residue; hence these enzymes can release most, though not all, of the known amino acids, albeit at different rates. There are some aminopeptidases (eg, methionine aminopeptidase) that are quite limited in specificity. Usually, aminopeptidases can also act on amino acid amides and esters, which provide convenient substrates for routine assays. Some aminopeptidases sequentially remove dipeptides from the N-terminus of substrates, and these are referred to as dipeptidyl-peptidases. Others remove tripeptides and, hence, are tripeptidyl-peptidases.

Many aminopeptidases are zinc **metalloenzymes**, but some are **serine proteinases** or **thiol proteases**. Of those that require zinc, some have an **active site** containing a single ion (see **Thermolysin**), whereas others have a co-catalytic site that involves two closely spaced zinc ions.

Aminopeptidases are widely distributed in various tissues and cells. They can be monomeric (a single polypeptide chain) or have up to 12 subunits. Many are integral components of cell **membranes**, but they are also found in the **cytosol**. They have a broad range of biological functions, including regulation of **hormone** concentration, control of the **cell cycle**, and recovery of amino acids from dietary peptides and proteins (1). All proteins synthesized by **eukaryotic** cells begin at their N-terminus with **methionine**, and its removal by methionine aminopeptidase is often crucial, not only for biological function of the protein but even for cell survival. Aminopeptidases also play important roles in the food industry—for example, ripening of cheese (2) and production of soy sauce (3).

Aminopeptidases are inhibited by the antitumor antibiotic bestatin [(2*S*,3*R*)-3-amino-2-hydroxy-4-phenylbutanol]-L-leucine, isolated from culture filtrates of *Streptomyces olivoreticuli*. It is a potent inhibitor of bovine lens aminopeptidases, with a K_i of 1.3 nM, and has numerous biological activities when administered to laboratory animals.

Membrane-bound aminopeptidases have often been identified on the basis of some other property, and then, once their amino acid sequence has been established, they are recognized to be aminopeptidases. Thus, the B-lymphocyte differentiation factor BP-1/6C3, whose expression correlates with proliferation and transformation of immature **B cells** (antibody-producing lymphocytes), has been shown to be identical to glutamyl aminopeptidase, also known as aminopeptidase A (4). Also the myeloid leukemia antigen CD-13 has been identified as aminopeptidase N (5), and the amino acid sequence of leukotriene A4 hydrolase revealed an aminopeptidase-like structure that led to the recognition of its aminopeptidase activity (6). Aminopeptidase N has also been shown to have a function unrelated to its enzymatic activity; that is, it serves as a cell-surface **receptor** for certain coronaviruses that cause upper respiratory infections (7).

The amino acid sequence and three-dimensional structure of leucine aminopeptidase from bovine lens have been determined (8). This is a broad-specificity cytosolic enzyme found in tissues of all organisms. It has a high degree of sequence similarity to several other aminopeptidases, which suggests that they all share similar structures and catalytic mechanisms.

BIBLIOGRAPHY

1. A. Taylor (1993) *FASEB J.* **7**, 290–298.
2. J. Meyer, D. Howald, R. Jordi, and M. Fuerst (1989) *Milchwissenschaft* **44**, 678–681.
3. T. Nakadai (1988) *Nippon Shoyu Kenkyusho Zasshi* **14**, 50–56.
4. Q. Wu et al. (1990) *Proc. Natl. Acad. Sci. USA* **87**, 993–997.
5. T. Inoue et al. (1994) *J. Clin. Endocrinol. Metab.* **79**, 171–175.
6. J. Z. Haeggstrom, A. Wetterholm, B. L. Vallee, and B. Samuelsson (1990) *Biochem. Biophys. Res. Commun.* **173**, 431–437.
7. B. Delmas et al. (1992) *Nature* **357**, 417–420; Yeager et al. *ibid.* 420–422.
8. S. K. Burley, P. R. David, A. Taylor, and W. N. Lipscomb (1990) *Proc. Natl. Acad. Sci. USA* **87**, 6878–6882.

Suggestions for Further Reading

A. Taylor (1993) Aminopeptidases: towards a mechanism of action. *Trends Biochem. Sci.* **18**, 167–172.

H. Kim and W. N. Lipscomb (1994) Aspartate transcarbamylase from *Escherichia coli*: activity and regulation. *Adv. Enzymol.* **68**, 153–213.

2-AMINOPURINE (AP)

LYNNETTE FERGUSON
WILLIAM DENNY

AP is a base analogue that is a **mutagen** in a wide range of systems and species (1,2) (Fig. 1). Freese (3,4) originally

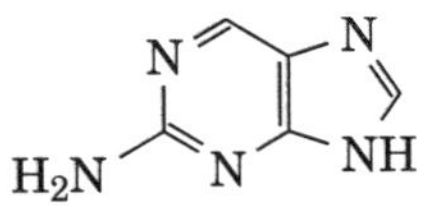

Figure 1. Structure of the base analogue 2-aminopurine.

described A.T → G.C and G.C → A.T transition **mutations** following AP treatment of **bacteria** and **bacteriophage**. The results were interpreted as resulting from **tautomeric** shifts and changes in base pairing behavior following the incorporation of AP into DNA. AP is readily metabolised to form deoxy-2-aminopurine triphosphate (dAPTP) (5), which may be incorporated opposite thymine during **DNA replication**, to form an AP.T base pair. During subsequent rounds of replication, incorporation of dCMP opposite the AP would lead directly to an A.T → G.C transition mutation. Law et al. (6) found that a DNA duplex containing AP.C is thermodynamically more stable than a DNA duplex containing A.C. This is in agreement with the suggestion that the rate of insertion of an improper base by a **DNA polymerase** is determined by differences in stability between the newly formed mismatch site and the corresponding normal **Watson–Crick base pair** (7). The probability that transition mutations will occur results from a combination of the likelihood of tautomeric shifting, the availability of the analogues to the DNA replicating machinery, and the discriminatory behavior and fidelity of the DNA polymerases.

Although best known as a base-pair substitution mutagen, the earliest reports of AP activity focused on the production of unequal, multipolar **mitosis** in mouse **tissue culture** cells, producing **aneuploid** daughter cells also bearing chromosome mutations (1). AP also causes **frameshift mutations**—for example, in a set of lacZ mutant variants of *Escherichia coli* (2). *E. coli dam* mutants have reduced ability for methylation of adenine. Many of these mutants also have enhanced sensitivity to AP and increased mutability by this base analogue (8). It appears that AP is partially able to saturate or inactivate the methylation-directed **mismatch repair** system, allowing the escape from repair of replication errors that lead to frameshift mutation. This results in indirect mutations that can be detected at certain sites (2).

BIBLIOGRAPHY

1. A. Ronen (1979) *Mutat. Res.* **75**, 1–47.
2. C. G. Cupples, M. Cabrera, C. Cruz, and J. H. Miller (1990) *Genetics* **125**, 275–280.
3. E. Freese (1959a) *Proc. Natl. Acad. Sci. USA* **45**, 622–633.
4. E. Freese (1959b) *J. Mol. Biol.* **1**, 87–105.
5. E. G. Rogan and M. J. Bessman (1970) *J. Bacteriol.* **103**, 622–633.
6. S. M. Law, R. Eritja, M. F. Goodman, and K. J. Breslauer (1996) *Biochemistry* **35**, 12329–12337.
7. M. F. Goodman, R. L. Hoskins, R. Lasker, and D. N. Mhaskar (1993) *Basic Life Sci.* **31**, 409–423.
8. B. W. Glickman, P. van den Elsen, and M. Radman (1978) *Mol. Gen. Genet.* **163**, 307–312.

AMINOPTERIN, METHOTREXATE, TRIMETHOPRIM, AND FOLIC ACID

MARTINA HUM
BARTON A. KAMEN

Aminopterin, methotrexate, and trimethoprim are all analogs of *folic acid* (Fig. 1) that antagonize folate-dependent metabolic pathways (1,2). Folate is a water-soluble vitamin. Although some bacteria synthesize folate, mammalian cells cannot, and consequently it is an absolute dietary requirement. Reduced folates function as **cofactors** in many metabolic processes common to nearly all cells, such as **thymidylate** and **purine** synthesis and donation of methyl groups (Fig. 2) (1) (see **Methylation**). Folates gain entry into mammalian cells by either the reduced folate carrier (RFC) and/or a hydrophobic membrane-associated folate receptor (FR) found in placental, choroid plexus and kidney cells (3). The FR is unidirectional and, once inside, cellular retention of folate is enhanced by its polyglutamation, catalyzed by the **enzyme** folylpolyglutamyl synthetase (FPGS) (4). FPGS adds up to six or seven **glutamic acid** residues via an unusual **peptide bond** through their γ-carboxyl group, rather than the normal α. The number of glutamic acid residues added may play a role in regulating and distributing reduced folates, and in guaranteeing their availability as cofactors. Polyglutamation also increases the affinity of folate for folate-dependent enzymes, such as **thymidylate synthase** (TS), aminoimidazole carboxamide ribonucleotide and glycinamide ribonucleotide transformylases. The latter two are involved in purine synthesis. In cells synthesizing DNA, 5,10-methylenetetrahydrofolate serves in thymidylate synthesis as a methyl group donor for converting dUMP to dTMP. This is the only reaction in which tetrahydrofolate is partially oxidized to dihydrofolate (1,4). **Dihydrofolate reductase** (DHFR) is the critical enzyme involved in converting dihydrofolate back to tetrahydrofolate, thus maintaining reduced folate pools to serve as one-carbon group carriers (see Fig. 2).

Folate homeostasis is recognized as important. The structure of DHFR has been determined by **X-ray crystallography**. One proposed mechanism of resistance to methotrexate involves DHFR **gene amplification** (5–7). The FR is overexpressed in some carcinomas, and its gene has been localized to the 11q13 region (3). Human FPGS has been cloned and mapped to chromosome 9q (4). The **complementary DNA** for RFC has been **cloned** and the RFC gene localized to the long arm of chromosome 21 (8–11).

METHOTREXATE (MTX)

Folate analogs entered cancer therapy in the 1940s, when aminopterin was successfully used to induce temporary remissions in children with acute lymphoblastic leukemia (ALL) (12). Because of a better therapeutic index, MTX eventually emerged as the antifolate used clinically in treating cancers, such as leukemias, lymphomas, osteosarcoma, breast cancer, choriocarcinoma, head and neck cancers, and nonmalignant disorders, such as arthritis and asthma (4,13).

Like folates, MTX enters mammalian cells via the RFC. MTX has an apparent **dissociation constant** in the micromolar range, and via the FR has higher affinity in the

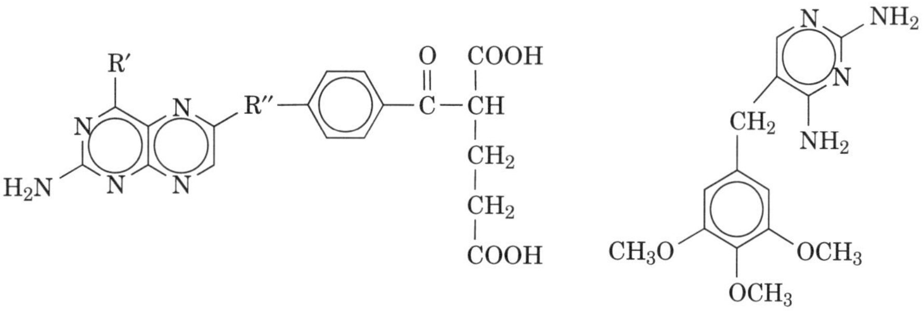

	R′	R″
Folic acid	-OH	$-CH_2NH-$
Methotrexate	$-NH_2$	$-CH_2N(CH_3)-$
Aminopterin	$-NH_2$	$-CH_2NH-$

(a)

(b)

Figure 1. (**a**) The structures of folic acid and folate analogs: folic acid and analogs; (**b**) the structure of trimethoprim.

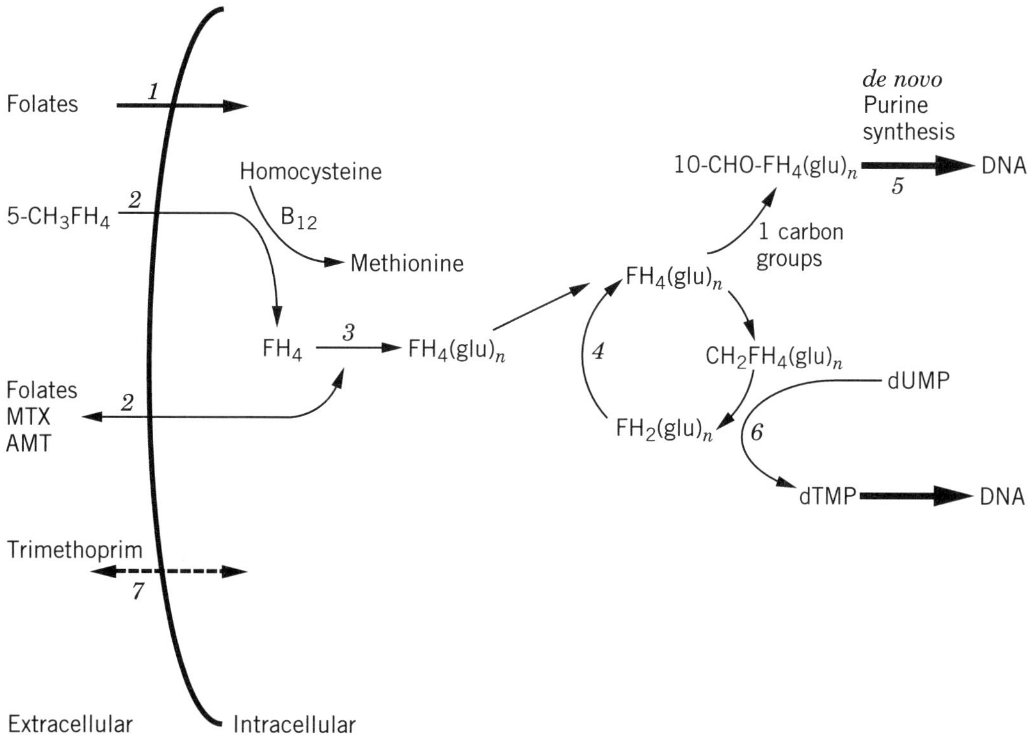

Figure 2. Folate transport, accumulation, and target enzymes. (*1*) folate receptor (FR); (*2*) reduced folate carrier (RFC), which is bidirectional; (*3*) folylpolyglutamyl synthetase (FPGS); (*4*) dihydrofolate reductase (DHFR); (*5*) aminoimidazole carboxamide ribonucleotide and glycinamide ribonucleotide tranformylases; (*6*) thymidylate synthase (TS); (*7*) passive diffusion of lipid-soluble compounds. FH_4 and FH_4 $(glu)_n$, tetrahydrofolate with and without the added glutamyl residues; FH_2, dihydrofolate; CH_2FH_4, 5,10 methylene-FH_4; 10-CHO-FH_4, 10-formyl FH_4; B_{12}, vitamin $\mathbf{B_{12}}$ [Adapted from Hum and Kamen (1996) *Investigational New Drugs* **14**, 110–111.]

nanomolar range (3). Cellular retention of MTX is enhanced by polyglutamation, which also enhances the affinity of the drug for enzymes. Once inside the cell, MTX acts as a tight-binding **competitive inhibitor** of DHFR. This leads to an accumulation of dihydrofolate and depletion of the reduced folate pools in cells actively making dTMP via the *de novo* pathway. Accumulated dihydrofolate polyglutamates are also inhibitors of TS and the enzymes involved in the *de novo* synthesis of purines (1). The resulting imbalance of nucleotides causes base substitutions, which lead to errors in DNA synthesis and ultimately to cell death.

AMINOPTERIN (AMT)

Aminopterin (AMT) was the first antifolate drug used clinically in childhood ALL (12) (see previous). Although more potent than MTX, in preclinical studies the toxicity of AMT was more severe and more unpredictable (13–15). Transport and metabolic studies *in vitro* have shown that AMT is the preferred substrate (16, 17). This results in greater AMT accumulation at lower concentrations and more complete polyglutamation, leading to improved cellular retention for the cytotoxic effect.

The greater potency of AMT compared to MTX has led to renewed clinical interest. AMT may find a role in treating patients with resistant or refractory malignancies or those patients in whom *in vitro* studies indicate AMT is the better choice on the basis of metabolism and accumulation (16).

TRIMETHOPRIM

Trimethoprim is an antibacterial agent developed in the 1950s. Studies by Hitchings (2) showed that its mechanism of action is the competitive inhibition of DHFR. Unique to trimethoprim is its much greater affinity (50,000- to 100,000-fold) for bacterial DHFR than for mammalian DHFR (18,19). Trimethoprim is lipid-soluble and enters cells rapidly without requiring specific transport mechanisms. Its selective toxicity is further enhanced by the ability of folinic acid to reverse even the slight effects of trimethoprim on mammalian cells, whereas bacterial cells, unable to transport folinic acid, are not rescued by folinic acid administration (19,20). In addition to its antibacterial activity, the drug has activity against *Pneumocystis carinii*, an opportunistic infection of the lungs encountered in severely immunocompromised patients (especially **HIV** patients).

BIBLIOGRAPHY

1. M. C. Hum and B. A. Kamen (1996) *Investigational New Drugs* **14**, 110–111.
2. G. H. Hitchings (1973) *J. Infect. Dis.* **128**(suppl), S433–S436.
3. S. Weitman, R. G. W. Anderson, and B. A. Kamen (1994) In *Vitamin Receptors: Vitamins as Ligands in Cell Communities* (K. Dakshinamurti, ed.), Cambridge University Press, Cambridge, UK, pp. 106–136.
4. T. A. Garrow, A. Admon, and B. Shane (1992) *Proc. Natl. Acad. Sci. USA* **89**, 9151–9155.
5. F. W. Alt et al. (1978) *J. Biol. Chem.* **253**, 1357–1370.
6. J. R. Bertino (1993) Ode to Methotrexate, *J. Clin. Oncology* **11**(1), 5–14.
7. E. Chu and C. H. Takimota (1993) In *Principles & Practice of Oncology* (V. T. De Vita, S. Hellman, and S. A. Rosenburg, eds.), Lippincott, Philadelphia, pp. 358–374.
8. J. A. Moscow et al. (1995) *Cancer Res.* **55**, 3790–3794.
9. P. D. Prasad et al. (1995) *Biochem. Biophys. Res. Commun.* **206**, 681–687.
10. S. C. Wong et al. (1995) *J. Biol. Chem.* **270**, 17468–17475.
11. T. L. Yang-Feng et al. (1995) *Biochem. Biophys. Res. Commun.* **210**(3), 874–879.
12. S. Farber et al. (1948) *N. Engl. J. Med.* **238**, 787–793.
13. A. Goldin et al. (1955) *J. Natl. Cancer Inst.* **15**, 1657–1664.
14. F. M. Sirotnak and R. C. Donsbach (1975) *Biochem. Pharmacol.* **24**, 156–158.
15. F. S. Philips et al. (1973) *Cancer Res.* **33**, 153–158.
16. A. Smith et al. (1996) *Clin. Cancer Res.* **2**(1), 69– 73.
17. B. G. Rumberger, J. R. Barrueco, and F. M. Sirotnak (1990) *Cancer Res.* **50**, 4639–4643.
18. Anonymous (1971) Trimethoprim-sulfamethoxazole. *Drugs*, **1**(1), 8–53.
19. J. J. Burchall (1973) *J. Infect. Dis.* **128**(suppl), S437–S441.
20. W. Brumfitt and J. M. T. Hamilton-Miller (1980) *Brit. J. Hosp. Med.* **23**(3), 281–288.

Suggestions for Further Reading

J. R. Bertino, B. A. Kamen, and A. Romanini (1997) Folate antagonists. In *Cancer Medicine* (J. F. Holland, R. C. Bast, D. L. Morton, E. Frei, D. W. Kufe, and R. R. Weicheslbaum, eds.), Williams and Wilkins, Baltimore, pp. 907–922.

R. L. Blakley and S. J. Benkovic (1984) *Folates and Pterins: Chemistry and Biochemistry of Folates*, Wiley, New York, Vol. 1.

C. H. Takimoto and C. J. Allegra (1995) New antifolates in clinical development, *Oncology* **9**, 649–659.

AMPHIPATHIC

N. GERSHFELD

Compounds that contain both highly **polar** and **nonpolar**, **hydrophobic** moieties are *amphipathic*. In the presence of **water**, these compounds tend to aggregate, forming structures that are strongly influenced by the relative extents of the polar and hydrophobic characters of the molecule, with the polar groups acting to shield the hydrophobic moieties from the solvent. In **lipids**, the hydrophobic group is the hydrocarbon domain of the aliphatic chain (see **Lipids** and **Fatty acids**). Typical examples are (a) the sodium salts of fatty acids, which form micelles in water, and (b) phospholipids, which form bilayers in membranes. The presence of significant amounts of nonpolar **amino acids**, such as **alanine**, **valine**, **leucine**, **isoleucine**, **proline**, **methionine**, **phenylalanine**, and **tryptophan**, make certain **proteins** amphipathic; these amino acids are believed to increase the accessibility of the proteins to the hydrophobic domain of the membrane bilayer (see **Fluid mosaic model**).

AMPHOTERIC

HARUKI NAKAMURA

An amphoteric molecule is one that has both acidic and basic chemical features. For example, proteins are amphoteric electrolytes because they generally have both acidic and basic groups on the side chains of their amino acid residues. Extreme examples of amphoteric molecules are the ampholytes used to establish pH gradients in **isoelectric focusing**.

AMPICILLIN

E. E. Ishiguro

Ampicillin (D[−]-α-aminobenzylpenicillin; Fig. 1) is a member of a growing family of antimicrobial agents known as the semisynthetic **penicillins**. The semisynthetic penicillins are derivatives of the natural product, 6-aminopencillanic acid, that have been deliberately modified chemically (1). The chemical modifications are introduced to create new compounds with specific desirable properties. In the case of ampicillin, the addition of the aminobenzyl side chain results in a product with an increased resistance to acidic pH. Thus, ampicillin is medically significant because it was the first penicillin to be administered orally in chemotherapeutic practice. Furthermore, ampicillin exhibits a broader antibacterial spectrum than do the naturally occurring penicillins and is effective against many **Gram-negative bacterial** species. The outer membrane component of the Gram-negative bacterial cell wall represents a permeability barrier to many antibiotics, including the natural penicillins (2). The broad activity spectrum and the relative low cost of ampicillin have made it an invaluable tool in molecular biology and genetics for studying Gram-negative model organisms, such as *Escherichia coli* and *Haemophilus influenzae*. Applications that employ ampicillin are summarized below.

The first application of penicillin in genetics was for the selection of auxotrophic mutants of *E. coli* (3,4). This technique, penicillin selection, was based on the fact that penicillin kills only actively growing bacteria, whereas nongrowing bacteria are penicillin-tolerant (see **Penicillin**). Although benzylpenicillin (penicillin G) was employed in the original studies describing this technique, ampicillin would be far more effective for this purpose because of its broad spectrum.

Molecular biologists have extensively exploited genetic elements encoding **β-lactamase** (see **penicillin-binding proteins**), especially the enzyme designated TEM-1. TEM-1 is a broad spectrum β-lactamase that confers effective high level resistance to penicillins (5). The gene encoding TEM-1 was incorporated into the first plasmid **cloning** vectors (6) and has been widely used in cloning vectors since. Ampicillin has been used routinely for the selection and maintenance of bacteria carrying recombinant plasmids encoding β-lactamase, and this is undoubtedly its most common application in molecular biology. For this purpose, ampicillin is incorporated into bacteriological media at final concentrations ranging from about 50–100 μg/mL. For *E. coli*, these levels are about 10–20 times higher than the minimum inhibitory concentration (MIC) of ampicillin. The MIC is defined as the minimum concentration of the antibiotic that is necessary to inhibit bacterial growth.

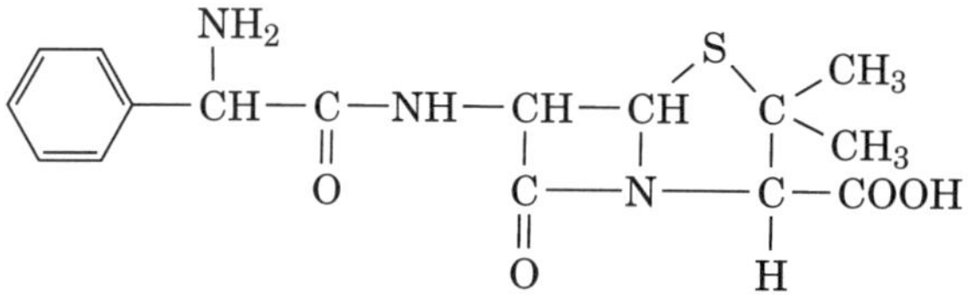

```
              NH2
              |                          S
   (C6H5)—— CH—C—NH—CH——CH      /  \   CH3
                 ||      |     |      C /
                 O       |     |      | \ CH3
                         C——————N——————C—COOH
                         ||            |
                         O             H
```

Figure 1. Structure of ampicillin.

When a mixture of ampicillin-sensitive and ampicillin-resistant bacteria are plated on solid media containing ampicillin, such as for selection of transformants carrying a β-lactamase-encoded plasmid, it is not uncommon to find large colonies formed by ampicillin-resistant bacteria surrounded by a zone of smaller colonies. The ampicillin-resistant bacteria in the large central colonies produce and secrete β-lactamase. The activity of the secreted β-lactamase creates a zone of reduced ampicillin concentration around the resistant colonies, and this permits the ampicillin-sensitive bacteria in the vicinity to grow. The subsequent growth of the ampicillin-sensitive bacteria results in the formation of the smaller so-called satellite colonies. The satellite colonies normally do not represent a major hindrance in these procedures, because the desired ampicillin-resistant bacteria can be readily purified by streaking on an ampicillin-containing medium. However, the problem of satellite colony formation can be minimized by substituting carbenicillin, another semisynthetic penicillin (see Fig. 1 in **Penicillin**) for ampicillin in the selection medium at a concentration of 50–100 μg/mL (7). Carbenicillin is less susceptible to hydrolysis by the β-lactamase and is therefore less likely to promote satellite colony formation. It is therefore often used in place of ampicillin.

The β-lactamase gene has also been introduced into plaque-forming and defective derivatives of the *E. coli* bacteriophage Mu (8). These ampicillin-selectable phages have been used for mutagenesis and for the construction of *lac* fusions in applications that take advantage of the ability of Mu to **transpose** randomly.

The concept of **transposon** mutagenesis has been applied to transposable genetic elements encoding β-lactamases for the generation of random **gene fusions** that are directly selectable with ampicillin (or carbenicillin). For example, a derivative of Tn*3* designated Tn*3*-HoHo1 (9) is a *lacZ*-containing transposon capable of producing both transcriptional and translational **beta-galactosidase** fusions. Although it was originally developed for studies on ***Agrobacterium*** *tumefaciens*, it has been adapted for use in other bacterial genera.

β-Lactamase is a periplasmic enzyme. It has served as an important model for studying protein export to the bacterial periplasm. Urbain et al. (10) have recently developed a technique for quantifying β-lactamase activity in cultures of *E. coli* carrying recombinant plasmids that confer ampicillin resistance. Their assay involved determining the conversion of ampicillin to aminobenzylpenicilloic acid in periplasmic extracts of cells by quantitative high performance liquid chromatography (**HPLC**). The procedure may be useful for investigating the mechanism of β-lactamase translocation. It may also prove useful in studies on protein expression. For example, since β-lactamase is expressed constitutively from ampicillin-selectable recombinant plasmids, its activity could serve as a useful internal standard in protein coexpression studies.

β-lactamase has also been used as a genetic tool for studying **membrane proteins**, and these applications are based on the fact that β-lactamase is an exported protein (11,12). A plasmid vector that permits the *in vitro* construction of translational fusions between a gene of interest, encoding either a membrane protein or an exported protein, and the mature form (ie, the exported form) of the TEM β-lactamase has been described (13). Transformants carrying recombinant plasmids with in-frame fusions are ampicillin-selectable. This technique may be used for the analysis of protein export signals or for determining the topological organization of membrane proteins. A strategy for maximizing yields of membrane and exported proteins based on this vector has also been described (14). It is notable that alkaline phosphatase has been used widely for studying protein export signals and for topological mapping of membrane proteins (15). The β-lactamase system is an attractive alternative to **alkaline phosphatase** for these purposes (13,14) (see **Reporter genes**). For example, β-lactamase fusions are directly selectable (with ampicillin), whereas the identification of alkaline phosphatase fusions is based on phenotypic screening. Moreover, only the periplasmic form of alkaline phosphatase is enzymatically active, whereas both the cytoplasmic and periplasmic forms of β-lactamase are active. Consequently, the cellular location of the β-lactamase fusions can be determined on the basis of the levels of ampicillin resistance; cytoplasmic β-lactamase confers ampicillin resistance only at high cell density, whereas periplasmic β-lactamase confers ampicillin resistance at low cell density. As an extension of these studies, Broome-Smith et al. (16) have constructed a transposable β-lactamase element, designated *TnblaM*, that is equivalent to the transposable alkaline phosphatase element, *TnphoA*. The β-lactamase fusions constructed with *TnblaM* are directly selectable with ampicillin, and this is a major advantage over alkaline phosphatase system, which as already noted is based on phenotypic screening.

BIBLIOGRAPHY

1. J. H. C. Nayler (1991) in *50 Years of Penicillin Application. History and Trends*, Public Ltd., Czech Republic, 64–74.
2. H. Nikaido and M. Vaara (1985) *Microbiol. Rev.* **49**, 1–32.
3. J. Lederberg and N. Zinder (1948) *J. Am. Chem Soc.* **70**, 467–468.
4. B. D. Davis (1949) *Proc. Natl. Acad. Sci. USA* **35**, 1–10.
5. N. Datta and M. D. Richmond (1966) *Biochem. J.* **98**, 204–210.
6. F. Bolivar, R. L. Rodriguez, M. C. Betlach, and H. W. Boyer (1977) *Gene* **2**, 75–93.
7. F. M. Ausubel, R. Brent, R. E. Kingston, D. D. Moore, S. D. Seidman, J. A. Smith, and K. Struhl (1994) *Current Protocols in Molecular Biology*, Wiley, New York, p. 1.8.7.
8. E. A. Groisman (1991) *Meth. Enzymol.* **204**, 180–212.
9. S. E. Stachel, G. An, C. Flores, and E. W. Nester (1985) *EMBO J.* **4**, 891–898.
10. J. L. Urbain, C. M. Wittich, and S. R. Campion (1998) *Anal. Biochem.* **260**, 160–165.
11. J. K. Broome-Smith, M. Tadayyon, and Y. Zhang (1990) **4**, 1637–1644.
12. M. Tadayyon, Y. Zhang, S. Gnaneshan, L. Hunt, F. Mehraein-Ghomi, and J. K. Broome-Smith (1992) *Biochem. Soc. Trans.* **20**, 598–601.
13. J. K. Broome-Smith and B. G. Spratt (1986) *Gene* **49**, 341–349.
14. J. K. Broome-Smith, L. D. Bowler, and B. G. Spratt (1989) *Mol. Microbiol.* **3**, 1813–1817.
15. C. Manoil, J. J. Mekalanos, and J. Beckwith (1989) *J. Bacteriol.* **172**, 515–518.
16. M. Tadayyon and J. K. Broome-Smith (1992) *Gene* **111**, 21–26.

AMYLOID

C. C. F. BLAKE

Amyloid is an insoluble, proteinaceous, fibrous material associated with a number of prominent disease states (1), which can also form spontaneously *in vitro* from **oligopeptides** and **denatured** proteins. The disease states in which amyloid has been implicated as an important, or even causal, factor include Alzheimer's disease (see **Amyloid precursor protein**), the transmissible spongiform encephalopathies (see **Scrapie**), non-insulin-dependent (type II) diabetes, and a number of polyneuropathies. It is thought that the extreme stability of amyloid fibrils permits their progressive accumulation in the extracellular spaces of vital organs whose functioning is thereby inhibited, leading to organ failure and death. The *in vivo* development of amyloid deposits from globular precursor proteins is linked to either genetic **mutation**, incorrect processing, or the abnormal accumulation of wild-type proteins. Probably the most remarkable feature of amyloid is that its molecular structure appears to be constant and independent of the protein precursors. For example, all amyloids, regardless of the disease involved or the source of the fibrils, share similar morphological (2), tinctorial (3), and structural (4,5) properties. This implies that amyloid formation is not a simple aggregation process but a structural conversion of globular proteins to a state that can be incorporated into a particular type of **fibrous protein** structure. The amyloid diseases may therefore be defined as diseases of protein misfolding.

At least 15 different proteins can form amyloid fibrils *in vivo* (1), and recently the involvement of amyloid fibrils in Huntingdon's disease has also been demonstrated (6). In the **electron microscope**, amyloid fibers are about 100 Å (10 nm) in diameter and usually straight or only slightly curved. Closer examination shows the fibrils to be composed of a number of smaller diameter protofilaments, arranged in a more or less parallel array. In those fibrils so far examined, the protofilaments give the fibrils the appearance of hollow cylinders or ribbons when observed in cross section. Although the number, size, and arrangement of these constituent protofilaments appear to differ somewhat, the ribbons may be merely unrolled or unformed cylinders, and hence the variation may not be so great as appears. The molecular structure of the amyloid fibrils has been established by X-ray **fiber diffraction**. The first diffraction patterns (7,8) showed the intense 4.7 Å (0.47 nm) meridional and 10 Å (1 nm) equatorial reflections that have been subsequently shown to characterize all amyloid X-ray patterns. These reflections indicate that the molecular structure of amyloid is composed of **beta-sheets** arranged parallel to the fiber axis, with their constituent **beta-strands** at right angles to the axis of the fibril (4,9). This so-called "cross-β" structure is quite different from the more common insect silk **chorions**, whose fibers are formed from β-sheets having their β-strands parallel to the fiber axis. The first use of intense synchrotron X-ray sources on amyloid (10) extended the observable X-ray pattern to 2 Å (0.2 nm) and revealed how this "classic" amyloid model could be reconciled with low-energy twisted β-sheets (11). These new data revealed a previously unobserved repeat distance of 115 Å (11.5 nm) along the fiber axis of the transthyretin amyloid of familial amyloidotic polyneuropathy (FAP), which was proposed (10) to correspond to the repeat distance of a complete helical turn of a twisted β-sheet, whose helix axis was parallel to the fiber axis. The basic helical unit therefore consists of a segment of β-sheet whose 24 β-strands complete one helical turn. This **beta-helix** model of the molecular structure of the amyloid protofilament is shown in Figure 1. Given that all amyloid fibrils share common morphological, tinctorial, and structural properties, it is reasonable to see the β-helix model as characteristic of all amyloid protofilaments. The β-helix model is distinct from other fibrous proteins, showing that the amyloid structure is unique. Its features enable low energy twisted β-sheets to be incorporated in a linear fibril in such a way that continuous β-type **hydrogen bonding** can be extended over the total length of the fibril. This, together with the opportunity for a continuous **hydrophobic** core to stabilize the β-sheet interactions along the length of the fibril, can be reasonably held to account for the known extreme stability of amyloid fibrils, which is central to their role in disease.

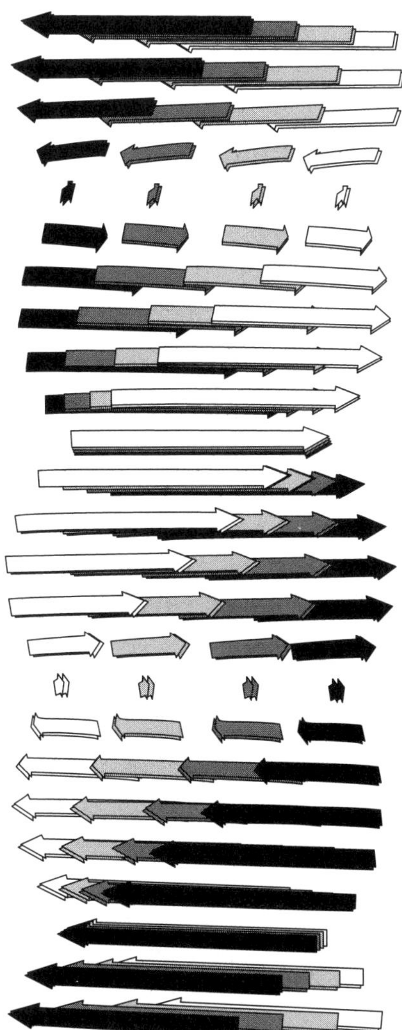

Figure 1. Drawing of one complete turn of the proposed β-helix structure of the amyloid fibril. The arrows represent β-strands and are shaded to represent the different β-sheets. The direction of the arrows has no significance, and no β-strand connections are shown.

A remarkable property of amyloid fibrils is that they appear to be very similar whatever their source. This has recently been critically tested (5) by showing that the high resolution synchrotron X-ray patterns from eight different amyloids, of both disease and synthetic origins, have the same structural characteristics. This observation suggests that amyloid is a highly stable generic structure capable of accommodating proteins and peptides with a wide range of chain lengths and structures within a common fibrillar form. The repetition of β-strands in the fibril model allows polypeptide chains of different lengths to occupy the same structural framework. For example, a 10-residue peptide could form a single β-strand of the amyloid structure, consistent with the known lower limit on the length of peptides capable of forming amyloid, while longer polypeptide chains could fold back and forward on themselves, thereby forming a number of consecutive β-strands of the amyloid β-sheets. The need for loops of chain of different lengths or conformation between successive β-strands to maximize the β-strand propensity, and to generate a stable hydrophobic core between the β-sheets, is likely to give each fibril a characteristic structure within the core amyloid framework. Thus it is probably realistic to consider amyloids as a closely related family of fibrils, each with its own detailed structural differences, while maintaining a similar overall core structure.

The nature of the processes that convert the globular, soluble protein precursors into amyloid fibrils, known as either *amyloidogenesis* or *fibrillogenesis*, are important, as they underpin the development of amyloid diseases and represent a novel field of scientific investigation. It is evident that some of the proteins forming the predominantly β-structured amyloid fibrils are themselves largely α-helical in their normal soluble states, for example, **lysozyme**, the **prion** protein, and the Aβ peptide in Alzheimer's disease (see **Amyloid precursor protein**). The nature of this structural rearrangement has been particularly studied in two amyloidogenic proteins: transthyretin (TTR) and lysozyme. Transthyretin amyloidosis is associated with familial amyloidotic polyneuropathy (FAP) and senile systemic amyloidosis (SAA). Both the wild-type protein (SAA) and more than 50 different genetic variants (FAP) (13) give rise to amyloid. That so many variants of transthyretin are amyloidogenic (nearly 40% of all its residues), suggests that the structure of this protein allows a particularly facile transformation into the fibrillar form. As transthyretin is a 55-kDa homotetramer, with the four monomers associating in such a way as to generate a stack of four 8-stranded β-sheets (14), there is a superficial structural similarity with amyloid fibril (10). Nevertheless there is evidence that the tetrameric form of TTR is not a building block for the fibril (15), and the discovery of a molecular hotspot in the pattern of amyloidogenic variants (16) suggests that structural change to the monomer is also required prior to its incorporation into the fibril. It is therefore evident that even all-β proteins, such as TTR, must undergo a significant degree of structural rearrangement to form amyloid.

The human lysozyme molecule (see **Lysozymes**) has about 35% α-helix and only 10% β-sheet structure, segregated into an α- and a β-domain. Hence the structural changes that must accompany its incorporation into the predominantly β-structured amyloid fibril will need to be extensive. Two

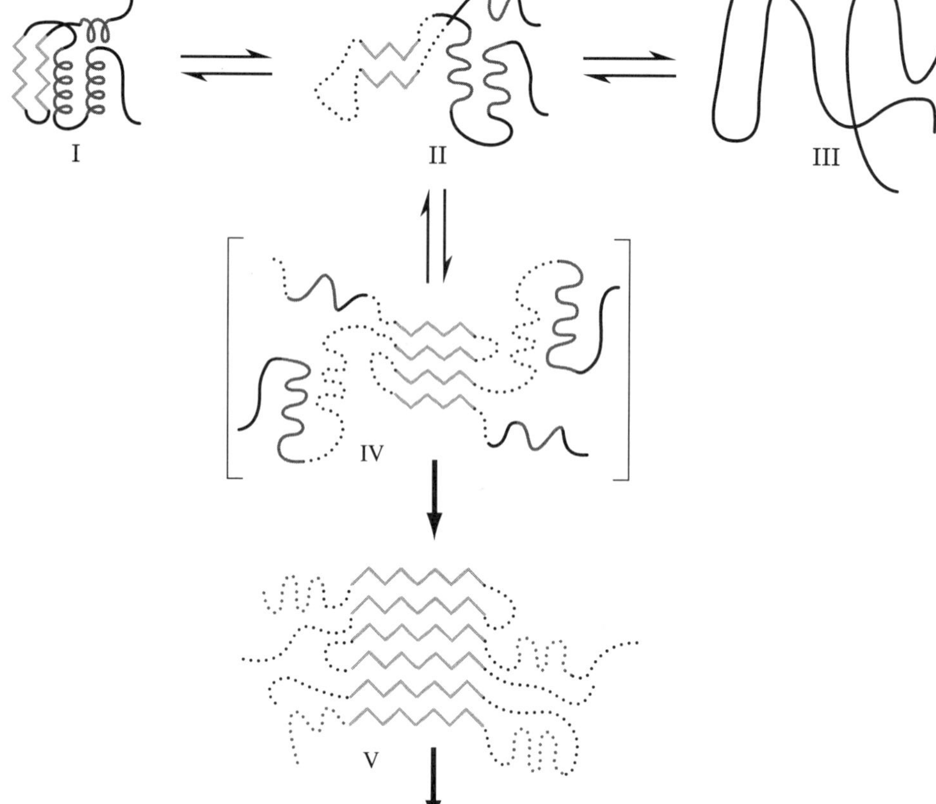

Figure 2. Schematic drawing of the possible mechanism of fibril formation in human lysozyme. (—) β-sheet structure; (**—**) α-helical structure. A partly-folded, molten-globule form (II), distinct from the native (I), and denatured (III) forms, self-associates through the β-domain (IV) to initiate fibril formation. This provides a template for further deposition and the development of a β-sheet core structure of the amyloid fibril (V). Undefined structures are shown as broken lines.

genetic variants of human lysozyme are known to form amyloid *in vivo*, Ile56Thr and Asp67His (17). A thorough biophysical study of these variants has highlighted major differences in their stability and folding behavior as compared with the wild type (18). It has generated a possible mechanism of the α- to β-structural transformation of fibril formation of lysozyme, shown in Figure 2, which may be applicable *mutatis mutandis* to other amyloidogenic proteins. Thermal denaturation studies show that the variants unfold at temperatures at least 10°C lower than wild-type lysozyme and, unlike the wild type, do not regain activity when returned to physiological conditions. Fourier transform infrared spectroscopy (see **Vibrational spectroscopy**) of Asp67His lysozyme shows a significant gain in β-structure and loss of α-structure in the soluble material, demonstrating that an α-to-β structural interconversion is associated with fibril formation. Detection of an unfolding intermediate displaying the characteristic binding to the hydrophobic dye 1-anilinonaphthalene sulfonic acid (**ANS**) is consistent with a **molten globule**-like intermediate, as shown in Figure 2. **Hydrogen exchange** shows that the flexibilities of the native folds of the variants have been dramatically increased by the mutations. Inspection of the **X-ray crystallography** structures of the two variants (18) suggests that the key to both amyloidogenic mutations in human lysozyme lies in the effect they produce at the interface between the α- and β-domains, with the result that domain adhesion may be weakened. Both variant lysozymes unfold dramatically faster than the wild-type protein, because the docking of the two domains, required for achieving the final rigid protein fold (19), is compromised by the presence of a **threonine** side-chain in the place of the wild-type **hydrophobic** anchoring residue, **isoleucine**.

Detailed study of the structure and behavior of variants of lysozyme has led to the proposal that transient populations of amyloidogenic proteins in a molten-globule state that lacks global cooperativity are an important feature of the conversion from a soluble to the fibrillar state (18). These observations lead to a model of the mechanism for amyloid formation for amyloidogenic lysozyme shown in Figure 2. Such a mechanism may also operate for other amyloidogenic proteins. For example, the structure of the 121–231 domain of the prion protein (20) also demonstrates that mutations associated with prion disease are involved in the maintenance of the hydrophobic core, which may be related to the earlier suggestion of a molten globule intermediate in the formation of aggregates in scrapie (21) (see **Prion**). Recent evidence supports the hypothesis that conformational plasticity is a key feature in prion fibril formation. A similar observation has been made for the Aβ protein forming the amyloid in Alzheimer's disease (22) (see **Amyloid precursor protein**). The mechanism for helix-to-sheet conversion in lysozyme, proceeding from soluble forms of the amyloidogenic precursor proteins through transient populations of intermediates with the characteristics of molten globules, and on to intermolecular β-sheet association, seems to parallel the processes in the aggregation of prions and the Aβ amyloid peptides, and it may occur generally in the amyloidoses. Studies of these processes is certain to extend our knowledge of **protein folding** behavior into new areas and coincidentally address some of the more intractable diseases of our times.

BIBLIOGRAPHY

1. M. B. Pepys (1994) in *Santer's Immunologic Diseases* (M. M. Frank, K. F. Austen, H. N. Claman, and E. R. Unanue, eds.), Little, Brown & Co., Boston, pp. 637–655.
2. A. S. Cohen, T. Shirahama, and M. Skinner (1981) in *Electron Microscopy of Protein*, Vol. 3 (I. Harriss, ed.), Academic Press, London & New York, pp. 165–205.
3. G. G. Glenner, E. D. Eanes, and D. L. Page (1972) *J. Histochem. Cytochem.* **20**, 821–826.
4. G. G. Glenner (1980) *N. Eng. J. Med.* **303**, 1283–1292.
5. M. Sunde, L. C. Serpell, M. Bartlam, P. E. Fraser, M. B. Pepys, and C. C. F. Blake (1997) *J. Mol. Biol.* **273**, 729–739.
6. E. Scherzinger et al. (1997) *Cell* **90**, 549–558
7. E. D. Eanes and G. G. Glenner (1968) *J. Histochem. Cytochem.* **16**, 673–677.
8. L. Bonar, A. S. Cohen, and M. Skinner (1967) *Proc. Soc. Exp. Biol. Med.* **131**, 1373–1375.
9. J. H. Cooper (1976) in *Amyoidosis* (O. Wehelius and A. Paternak, eds.), Academic Press, London & New York, pp. 61–68.
10. C. C. F. Blake and L. C. Serpell (1996) *Structure* **4**, 989–998.
11. C. Chothia (1973) *J. Mol. Biol.* **75**, 295–302.
12. L. C. Serpell, M. Sunde, P. E. Fraser, P. Luther, E. Morris, E. Sandgren, E. Lundgren, and C. C. F. Blake (1995) *J. Mol. Biol.* **254**, 113–118.
13. M. J. M. Saraiva (1995) *Hum. Mutat.* **5**, 191–196.
14. C. C. F. Blake, M. J. Geisow, S. J. Oatley, B. Rerat, and C. Rerat (1978) *J. Mol. Biol.* **121**, 339–356.
15. W. Colon and J. W. Kelly (1992) *Biochemistry* **31**, 8654–8660.
16. L. C. Serpell, G. Goldsteins, I. Dacklin, E. Lundgren, and C. C. F. Blake (1996) *Amyloid: Int. J. Exp. Clin. Invest.* **3**, 75–85.
17. M. B. Pepys et al. (1993) *Nature* **362**, 553–557.
18. D. R. Booth et al. (1997) *Nature* **385**, 187–793.
19. S. D. Hooke, S. E. Radford, and C. M. Dobson (1994) *Biochemistry* **33**, 5867–5876.
20. R. Riek, S. Hornemann, G. Wider, M. Billeter, R. Glockshuber, and K. Wüthrich (1996) *Nature* **382**, 180–182.
21. J. Safar, P. Roller, D. Gajdusek, and C. Gibbs (1994) *Biochemistry* **33**, 8375–8383
22. C. Soto, E. Castano, B. Frangione, and N. Inestrosa (1995) *J. Biol. Chem.* **270**, 3063–3067.

Suggestions for Further Reading

G. R. Bock and J. A. Goode, eds. (1996) *The Nature and Origin of Amyloid Fibrils*, Ciba Foundation, John Wiley and Sons Ltd., Chichester, UK.

L. C. Serpell, M. Sunde, and C. C. F. Blake (1997) The molecular basis of amyloidosis. *Cell. Mol. Life Sci.* **53**, 871–887.

AMYLOID PRECURSOR PROTEIN

C. C. F. BLAKE

At the molecular level, the necessary, and probably sufficient, cause for the brain dysfunction in Alzheimer's disease is the deposition of aggregated forms of the **amyloid** β-protein (Aβ or βA4), a **proteolytic** fragment of the amyloid precursor protein (APP), in the brain parenchyma and vascular system (1). In the Western world, dementia is the most common neurological diagnosis and is the third leading cause of natural death, and Alzheimer's disease is by far the dominant cause of dementia (2). It has been estimated that there are 3 million patients in the United States alone with Alzheimer's disease (3). In addition, older patients with Down's syndrome (**trisomy** 21) tend to develop a neuropathology that is very similar to that seen in Alzheimer's patients (4) The disease is chronic, progressive, and at present untreatable. The progression of the disease begins with loss of short-term memory and disorientation, followed by complete impairment of memory, judgment, and reasoning. Firm diagnosis of the disease can be given only after postmortem examination of the brain, which is characterized by extensive loss of neurons and particular microscopic lesions, which include what are now called "neurofibrillar tangles" and "senile plaques." The neurofibrillar tangles are composed of **paired helical filaments** consisting of phosphorylated **tau protein**, which is normally associated with **microtubules**. The senile plaques are composed mainly of Aβ deposited in the brain parenchyma. Also present in the brain is congophilic amyloid angiopathy from the accumulation of the Aβ peptide in the walls of blood vessels. The Aβ protein in the senile plaques and in the vascular deposits is present in the form of **amyloid** (5,6), long, stable fibrils resulting from the accumulation of precursor peptides that adopt a continuous **beta-sheet** structure. In this regard, Alzheimer's disease is related to Creutzfeldt–Jakob disease (CJD) and other spongiform encephalopathies, which are also characterized by the presence of amyloid deposits, in that case composed of the **prion** protein. All known risk factors for Alzheimer's disease appear to influence one or more of the following: (1) the concentration of Aβ (7); (2) the amount of the longer, more amyloidogenic Aβ chains (8,9); or (3) the initiation of amyloid formation (10). Increases in any of these factors appear to increase the risk of Alzheimer's disease. Study of the origins of the β-amyloid protein from APP, its aggregation into amyloid fibrils, and the identification of genes involved in its inherited susceptibility are therefore central to understanding of, and devising therapies against, the scourge of Alzheimer's disease.

The amyloid precursor protein (APP), is a 110–130-kDa protein with the features of a transmembrane cell-surface **glycoprotein**. It is encoded by a gene localized on **chromosome** 21 encoded by 18 exons (see **Introns, exons**). A family of eight transmembrane glycoproteins is generated by **alternative splicing** of some of these exons (11). The amino acid sequence predicts that the single transmembrane domain of APP is near the *C*-terminus (Fig. 1) (12). The Aβ sequence consists of a maximum 43-residue segment straddling the transmembrane and extracellular portions of the APP chain, and encoded by parts of exons 16 and 17. APP is processed in two distinct pathways : a major nonamyloidogenic route involving **proteolytic** cleavage within the region carrying the Aβ segment to separate the extracellular and membrane-bound domains (13): and an amyloidogenic route, shown in Figure 1, leading to the release of the Aβ segment (7). The cleaving of the Aβ sequence from an internal site in the APP chain implies that two distinct proteolytic events are required to generate the *N*-terminus and the *C*-terminus of the Aβ peptide. The proteolytic enzymes involved in processing APP are termed *secretases*: α-secretase for cleavage of APP within the Aβ

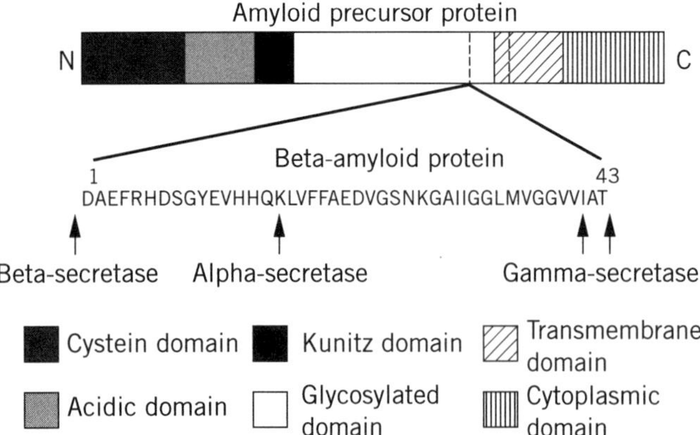

Figure 1. Schematic diagram of the derivation of the 4-kDa β-amyloid protein (Aβ) from the 110–130-kDa amyloid precursor protein (APP) by the action of the α-, β-, and γ-secretase enzymes. The probable domain structure of APP has been simplified.

segment, and β- and γ-secretases for cleavage of APP at the *N*- and *C*-terminal sides of the Aβ segment, respectively. Soluble, *C*-truncated forms of APP are generated by two pathways: (1) by α-secretases cutting within the Aβ sequence, thereby precluding the release of Aβ (13); and (2) by β-secretases cleaving near the *N*-terminus of Aβ producing *C*-terminal fragments containing the complete Aβ (14). The generation of intact Aβ chains suggests that γ-secretases generate the *C*-terminus of β-amyloid from these *C*-terminal fragments of APP after release of the transmembrane domain from the lipid bilayer (7). The identity and location of the secretases have proved elusive, but clearly they are of prime importance in Alzheimer's disease, and targets for therapies.

Electron microscope and X-ray **fiber diffraction** analyses, often using synthetic analogs or fragments of the Aβ peptide, have revealed the molecular characteristics of the Aβ amyloid fibrils. In the electron microscope, different forms of Aβ amyloids are seen at different pH values, but the physiological form appears to be represented by a fibril about 90 Å (9 nm) diameter, composed of five or six parallel protofilaments 25–30 Å (2.5–3 nm) in diameter arranged around a hollow core (15). The X-ray fiber diffraction patterns of Aβ amyloid (16,17) show the usual 4.7 Å (0.47 nm) meridional reflection and 10 Å (1 nm) equatorial reflection that demonstrate that the molecular structure of amyloid consists of paired β-sheets running parallel to the fiber axis, with their constituent β-strands perpendicular to the fiber axis. This structure is very similar to that observed in other amyloids produced from a variety of proteins and peptides (see **Amyloid**). Some shorter Aβ chains, such as residues 11–28, particularly when aligned in a strong magnetic field, form pseudocrystals (18), which give more detailed X-ray diffraction patterns that may be capable of yielding greater structural detail and hence begin to define the features that dispose them to form amyloid.

The discovery that Aβ is produced and secreted by cells continuously under normal metabolic conditions, and is present in a soluble form in biological fluids (14,19), suggests that other factors are involved in Aβ amyloidosis. It has been proposed that the isolated Aβ protein exists in two forms: a soluble form that may represent a normal host protein and a modified amyloidogenic form (20). These two forms have identical sequences and hence are likely to represent conformation isomers. The soluble form is easily degraded and appears to have an **alpha-helical/random coil** structure: **NMR** analysis of synthetic Aβ 1–40 (21) has shown that residues 15–23 and 31–35 form α-helices, while the rest of the peptide is in a random-coil conformation, and no stable **tertiary structure** is present. The amyloidogenic form is more resistant to degradation, has a high β-sheet content, and forms the fibrillar aggregates found in brains with Alzheimer's disease. This behavior is very reminiscent of the normal and scrapie forms of the prion protein, which is also an agent for amyloid brain diseases (see **Prion**, **Scrapie**, and **Amyloid**). The existence of two structural forms for the Aβ protein implies the possibility that other proteins may be involved in regulating their interconversion. It is known, for example, that in the rare early-onset (30–60 years of age) form of Alzheimer's disease, the disease has been linked to mutations on chromosomes 1 and 14, in addition to chromosome 21, which carries the APP gene. The genes on chromosomes 14 and 1 have been identified with presenilin 1 (22) and presenilin 2 (23,24), respectively. The two presenilins have 65% amino acid sequence identity, and **hydrophobicity** plots predict seven transmembrane regions resembling a structural **membrane protein**. The genetic variants of presenilin 1 allow the secretion of Aβ peptides of longer length, and hence more amyloidogenic (25), possibly through a mechanism involving protein sorting and trafficking of APP.

The more common sporadic and familial late onset (>65 years) form of the disease has been linked with mutations on chromosome 19, subsequently narrowed to the $\varepsilon 4$ allele of the **apolipoprotein** E (apoE) gene (26,27). These studies have shown that carriers of the $\varepsilon 4$ allele have an increased risk of developing Alzheimer's disease and that the inheritance of the e4 allele correlates with an increased deposition of Aβ amyloid in blood vessels and plaques. The apoE4 **isoform** differs from apoE3, the most common isoform, by an Arg/Cys substitution at position 112. The apoE protein is known to be one of the proteins associated with amyloid deposits (28). It has been found that apoE can form complexes with synthetic Aβ analogues and to enhance amyloid fibril formation by Aβ *in vitro* (29), possibly by a direct physical interaction. Frangione and his colleagues (30) have made the intriguing proposal that apoE can itself possibly form an amyloid-like structure (there is experimental evidence for this in its *C*-terminal domain), which may be able to induce other proteins such as Aβ to misfold into a β-sheet structure that could allow them to be incorporated into a growing amyloid fibril. They term this process "conformational mimicry," and certainly the most recent analyses of *in vivo* amyloid formation characterize it as a disease of protein misfolding, which is possibly **cooperative** (see **Amyloid**).

BIBLIOGRAPHY

1. A. I. Bush, K. Beyreuther, and C. L. Masters (1992) *Pharmacol. Ther.* **56**, 97–117.
2. A. Ott, M. M. B. Bretaler, and F. van Harskamp (1995) *Br. Med. J.* **310**, 970–973.
3. R. E. Scully et al. (1993) *New Engl. J. Med.* **324**, 1255–1263.

4. K. E. Wisniewski, A. J. Dalton, D. R. Crapper-McLachlan, G. Y. Wen, and H. M. Wisniewski (1985) *Neurology* **35**, 957–961.
5. G. G. Glenner and C. W. Wong (1984) *Biochem. Biophys. Res. Comm.* **120**, 885–890.
6. C. L. Masters, G. Simms, N. A. Weinman, G. Multhaup, B. L. McDonald, and K. Beyreuther (1985) *Proc. Natl. Acad. Sci. USA* **82**, 4245–4249.
7. C. Haas and D. J. Selkoe (1993) *Cell* **75**, 1039–1042.
8. C. Hilbich, B. Kisters-Woike, J. Reed, C. L. Mastrers, and K. Beyreuther (1992) *J. Mol. Biol.* **228**, 460–473.
9. N. Suzuki et al. (1994) *Science* **264**, 1336–1340.
10. B. Hymen et al. (1995) *Proc. Natl. Acad. Sci. USA* **92**, 3586–3590.
11. R. Sandbrink, C. L. Masters, and K. Beyreuther (1994) *J. Biol. Chem.* **269**, 1510–1517.
12. J. Kang et al. (1987) *Nature* **325**, 733–736.
13. F. S. Esch et al. (1990) *Science* **248**, 1122–1124.
14. P. Seubert et al. (1993) *Nature* **361**, 260–263.
15. P. E. Fraser, J. Nguyen, W. Surewicz, and D. A. Kirschner (1991) *Biophys. J.* **60**, 1190–1201.
16. D. A. Kirschner, C. Abraham, and D. A. Selkoe (1986) *Proc. Natl. Acad. Sci. USA* **83**, 503–507.
17. P. E. Fraser et al. (1992) *Biochemistry* **31**, 10716–10723.
18. H. Inouye, P. E. Fraser, and D. A. Kirschner (1993) *Biophys. J.* **64**, 502–519.
19. M. Shoji et al. (1992) *Science* **258**, 126–129.
20. C. Soto, E. Castano, B. Frangione, and N. Inestrosa (1995) *J. Biol. Chem.* **270**, 3063–3067.
21. H. Sticht, P. Bayer, D. Willbold, S. Dames, C. Hilbich, K. Beyreuther, R. W. Frank, and P. Rosch (1995) *Eur. J. Biochem.* **233**, 293–298.
22. R. Sherrington et al. (1995) *Nature* **375**, 754–760.
23. E. Levy-Lahad et al. (1995) *Science* **269**, 973–977.
24. E. Rogaev et al. (1995) *Nature* **376**, 775–778.
25. M. Barinaga (1995) *Science* **268**, 1845–1846.
26. E. H. Corder et al. (1993) *Science* **261**, 921–923.
27. W. J. Strittmatter (1993) *Proc. Natl. Acad. Sci. USA* **90**, 1977–1980.
28. T. Wisniewski and B. Frangione (1992) *Neurosci. Lett.* **135**, 235–238.
29. J. Ma, A. Yee, H. B. Brewer, S. Das, and H. Potter (1994) *Nature* **372**, 92–94.
30. B. Frangione, E. M. Castaño, T. Wisniewski, J. Ghiso, F. Prelli, and R. Vidal (1996) in *The Nature and Origin of Amyloid Fibrils*, Ciba Foundation Symposium 199 (G. R. Bock and J. A. Goode, eds.), John Wiley and Sons Ltd., Chichester, UK, pp. 132–145.

Suggestions for Further Reading

D. J. Selkoe (1994) Normal and abnormal biology of the beta-amyloid precursor protein, *Annu. Rev. Neurosci.* **17**, 489–517.

G. Evin, K. Beyreuther, and C. L. Masters (1994) Alzheimer's disease amyloid precursor protein, *Amyloid: Int. J. Exp. Clin. Invest.* **1**, 263–280.

ANALOGY

T. GOJOBORI

Analogy is a similarity caused by factors other than a common genetic ancestry. In general, analogy is a similarity arising by **convergent evolution**. Therefore, in contrast to **homology**, analogy between two traits of interest is not based on the sharing of a common ancestor. Originally, the term "analogy" was used mostly for morphological characters. For example, the wings of bats and insects are said to be analogous organs, as they perform the same function in different species, even though they do not show a common underlying plan of structure. It is obvious that the wings of bats and insects were not derived evolutionarily from the same organ in a common ancestor.

When the similarity between two different molecules is found at the level of **nucleotide sequences**, amino acid sequences, and the **tertiary structure** of **proteins**, it is important to know whether the similarity is homologous or analogous. In particular, characteristic patterns of amino acids ("amino acid sequence motifs") and the tertiary structures of protein molecules may exhibit analogy rather than homology. In contrast, any sequence similarities of nucleotides and amino acids are explained mostly by homology, because the probability that the two sequences of interest are similar by chance can be computed easily and is usually extremely small. For more details, see **Convergence**.

ANALYTICAL ULTRACENTRIFUGATION

T. HOLZMAN

Analytical ultracentrifugation experiments fall into two different categories, termed **sedimentation velocity** and **sedimentation equilibrium**. However, they are often employed sequentially on a given macromolecule under analysis (Fig. 1), and they yield related, complementary information about the solution behavior of a macromolecule that is of great interest for a variety reasons. The need may be simple, for example, to determine whether a macromolecule aggregates when prepared or stored in various ways, e.g., refrigerated, frozen, or lyophilized. On a much more complex level and much more important in the context of modern protein biochemistry and molecular biology is to detect and measure the interactions of macromolecules with themselves or with other molecules and to relate this to their functional roles. For these studies, the precise analysis of their behavior in solution is necessary. Finally, analytical ultracentrifugation plays an important role in medicine and industry. It is the reference method for analyzing the concentration-dependent solution behavior of **recombinant proteins** and other macromolecules employed in pharmaceutical/pharmacological research and in structure-based drug design (1).

Analytical ultracentrifugation differs from preparative **centrifugation** in several important respects. First and foremost as the name implies, it is an *analytical* technique. If carefully applied, it can give hydrodynamic information that is precise, accurate, and unobtainable by any other single method. The hydrodynamic parameters (**molecular weights** and **sedimentation coefficients**) obtained from analytical ultracentrifugation are absolute measurements because the technique is an absolute physical method in the strictest sense. It measures either the intensive or extensive physical properties of solutes directly and requires only knowledge of the rotational rate of the centrifuge rotor, the sample temperature, and the solution density and viscosity (which are extensive properties of the solute–solvent system).

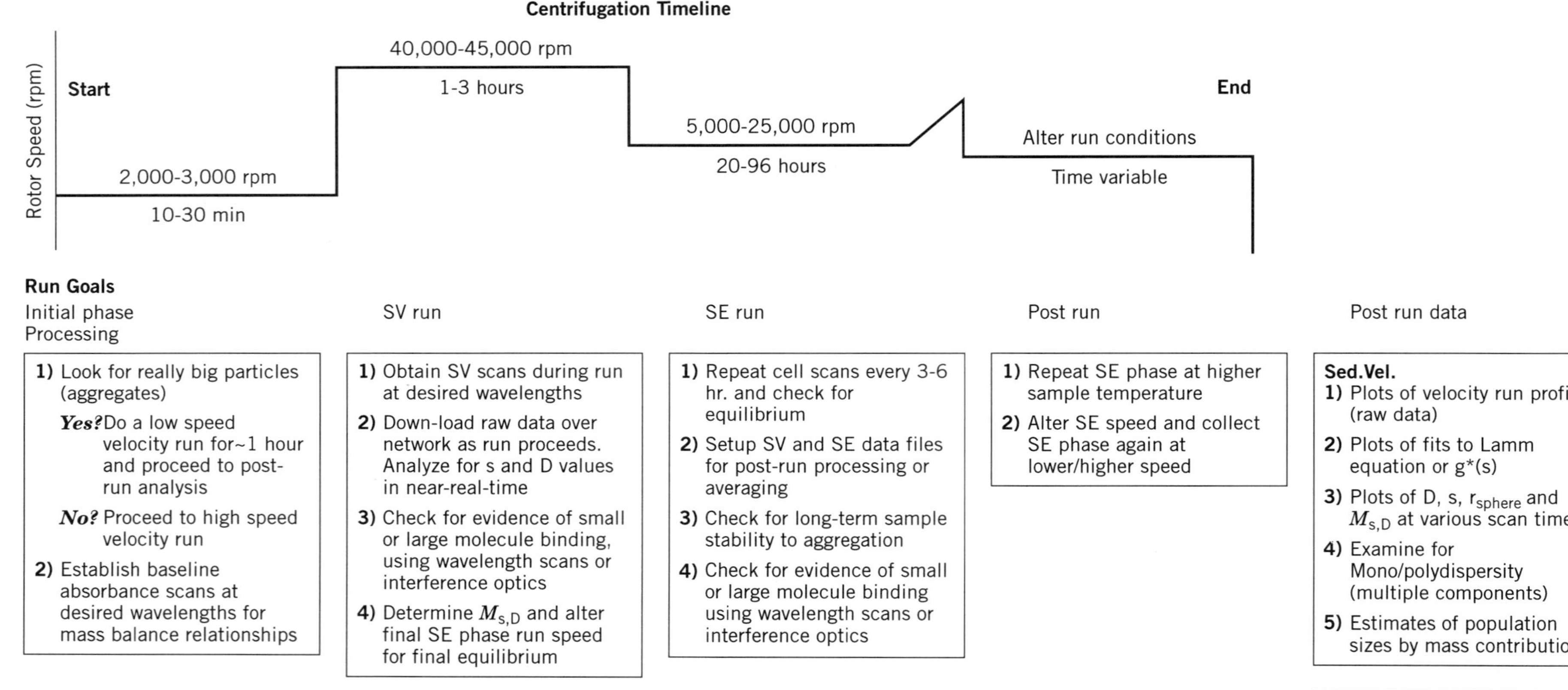

Figure 1. Typical execution scheme for sedimentation velocity (SV) and sedimentation equilibrium (SE) runs.

Similarly, calculations of molecular properties from ultracentrifugation data are based on direct, first-principle analyses of solute behavior in solution. Unlike indirect methods, such as static **light-scattering**, the calculations in analyzing data from the analytical ultracentrifuge do not rely on comparison to a known standard. For example, in static light scattering the scattering behavior of a low molecular weight organic molecule (toluene, for example) is required to calibrate the instrument. **Dynamic light scattering** provides absolute **diffusion** coefficients, but the molecular weights estimated from the diffusion coefficients are indirect and depend upon the solution density, viscosity, etc. and on assumptions about molecular dimensions.

Another major difference between preparative centrifugation and analytical ultracentrifugation is that the latter is usually applied to highly purified solute samples. Although samples must be pure, analytical analyses usually require only small amounts of sample material, in the range of 10 to 1000 μg per sample. Finally, although analytical ultracentrifugation is inherently a nondestructive technique, it is not routine to recover or reuse samples that have been analyzed because analytical ultracentrifuge cells and rotors (Figs. 2 and 3), especially those that are most useful for analyzing numerous samples, are not designed for facile sample recovery. Although analytical ultracentrifugation was at one time a useful tool for evaluating the purity of certain biological macromolecules, its use as a simple tool for purity analysis has been superseded by a variety of other bioanalytical techniques. For example, inexpensive **SDS-PAGE** is routinely the technique of choice for analyzing protein purity during a protein separation protocol and for estimating the molecular weights of polypeptide chains present.

The present (and likely future) value of analytical ultracentrifugation lies in its unmatched ability to provide direct information in two primary areas: (1) the extent to which an otherwise chemically pure sample of a macromolecular solute exhibits monodispersity or polydispersity and (2) the extent to which different chemically pure molecules interact. Because these two questions are fundamental to all areas of modern cellular and molecular biology and biochemistry, analytical ultracentrifugation remains a unique and extremely powerful biophysical tool.

THE MODERN ANALYTICAL ULTRACENTRIFUGE

The technique requires only small sample volumes, ~20 μl to ~500 μl depending upon the type of centerpiece employed (Figs. 2 and 3). A computer-controlled optical system detects the exact extent and position of the radial displacement of solute molecules at various times during the centrifugation run. Optical detection is accomplished by directly measuring the concentration of the solute as a function of radial position in the sample cell, either by its absorbance or by the change in refractive index, using Rayleigh interference optics. Like dual-beam optical **spectroscopy**, the technique is also differential in that it employs a sample solution to be measured and a reference solution used to "blank out" or "subtract" a signal background. The only modern, commercially available analytical ultracentrifuges are the XLA/XLI models manufactured by Beckman Instruments, which operate at speeds ranging from ~2,000 to ~50,000 rpm. The precise speed or speed range of operation for an experiment is determined by

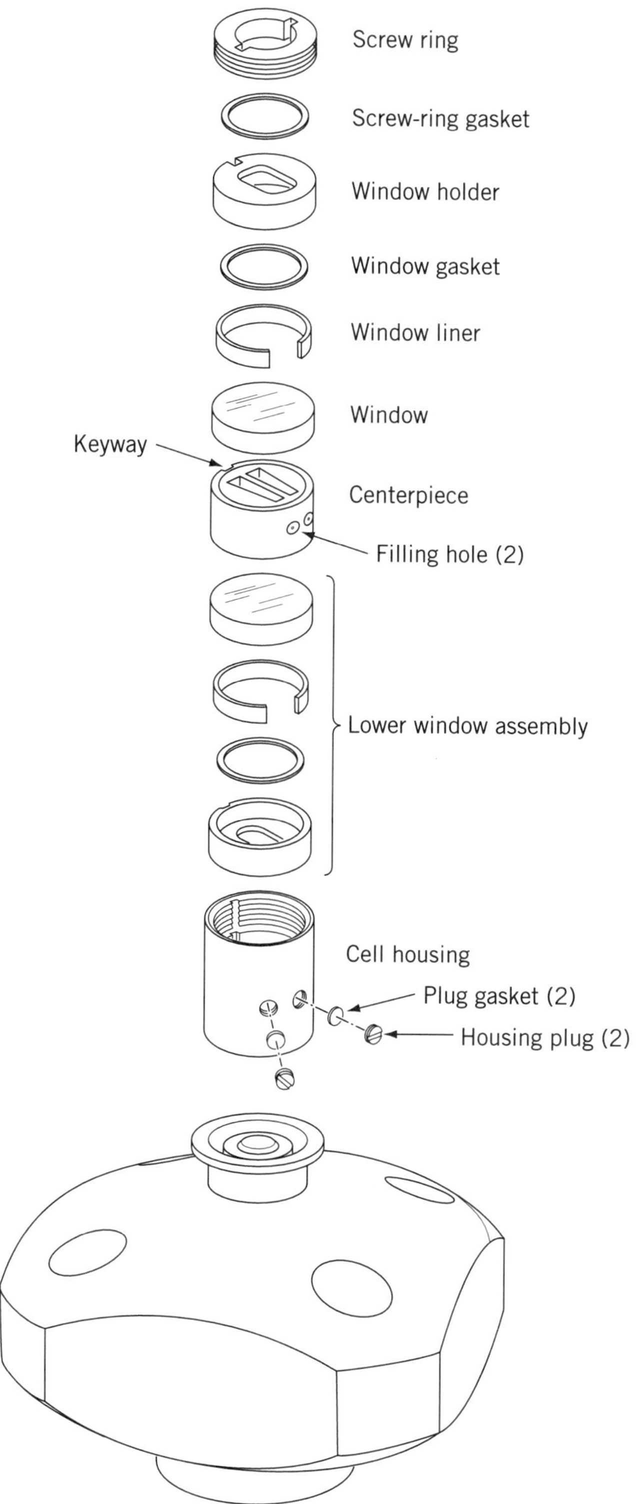

Figure 2. A four-hole analytical ultracentrifuge rotor and a vertically exploded view of the cell assembly with a double-sector, long-channel centerpiece. Sample and reference solutions for analysis (~400 to 500 μl) are introduced into the sealed centerpiece with an appropriate syringe system. Then the cell is placed in the rotor and subjected to centrifugation. Depending upon the analyses performed, it is possible after centrifugation to recover a portion of the sample using the same syringe system (figure courtesy of Beckman Instruments, by permission).

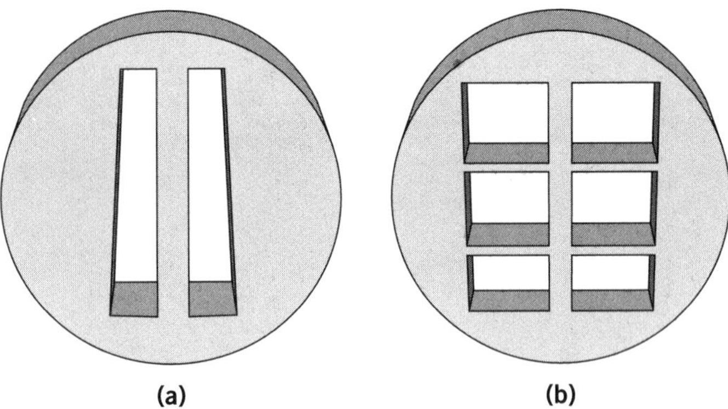

Figure 3. Illustrations of centerpieces used in analytical ultracentrifugation. Left, a long-column, double-sector centerpiece used for sedimentation velocity and equilibrium studies that require large data sets. Right, the multichannel centerpiece (or "Yphantis" centerpiece after its inventor, Dr. David Yphantis) used primarily for sedimentation equilibrium studies where analysis of a larger number of samples is desired.

the type of measurement(s) to be made and depends on the size of the molecules being studied and on the type of sample cell centerpieces, centrifuge rotor, and the data collection system used.

The XLA/XLI analytical ultracentrifuges offer either absorbance-based optical detection (XLA, Fig. 4) or both absorbance and interference-based optical detection (XLI, Fig. 5). In addition to measuring the molecular weights and sedimentation coefficients of individual molecules, their distributions in samples can also be measured. The analytical ultracentrifuge can also be also used for direct determination of **ligand-binding** constants and stoichiometries. Further, when combined with other physical measurements, these instruments are useful for determining the **diffusion** coefficients and molecular shapes of macromolecules. Thus, changes in shape related to ligand binding or solvent perturbations can also be evaluated.

Choosing an Optical System for Measurements

Although it is often useful to employ both absorbance and optical interference measurements when examining a sample, some considerations make one type of measurement preferable to the other. The possible choices for various types of samples are outlined in Table 1. In addition to the general preferences for detecting various analytes, other important factors influence the choice of detection methods. The relative sensitivities and dynamic ranges of absorbance versus interference detection can be a consideration. The typical useful range for absorbance measurements in a **sedimentation equilibrium centrifugation** experiment for an "average" protein with an absorbance at 280 nm of 1.0 $(mg/ml)^{-1}(cm)^{-1}$ is in the range of ~10 μg/ml to ~4 mg/ml, when the measurements employ centerpieces with both standard (12 mm) and reduced (3 mm) path lengths. In **sedimentation velocity experiments** using absorbance detection at 280 nm, the practical low end of the concentration range is about 50 to 100 μg/ml. Sensitivity for proteins can be increased by using light with wavelengths in the range of 215 to 240 nm, where the protein peptide backbone absorbs and the absorbance of a protein is typically 3- to 10-fold greater, although this is technically more demanding. Alternatively, the protein may have a natural cofactor chromophore with a large extinction coefficient, or **chemical modification** can be used to introduce such a chromophore. Optical interference measurements have roughly the same absolute dynamic range as absorbance measurements, but they are routinely more useful in the mid-to-high concentration range, i.e., ~0.1 to ~10 mg/ml.

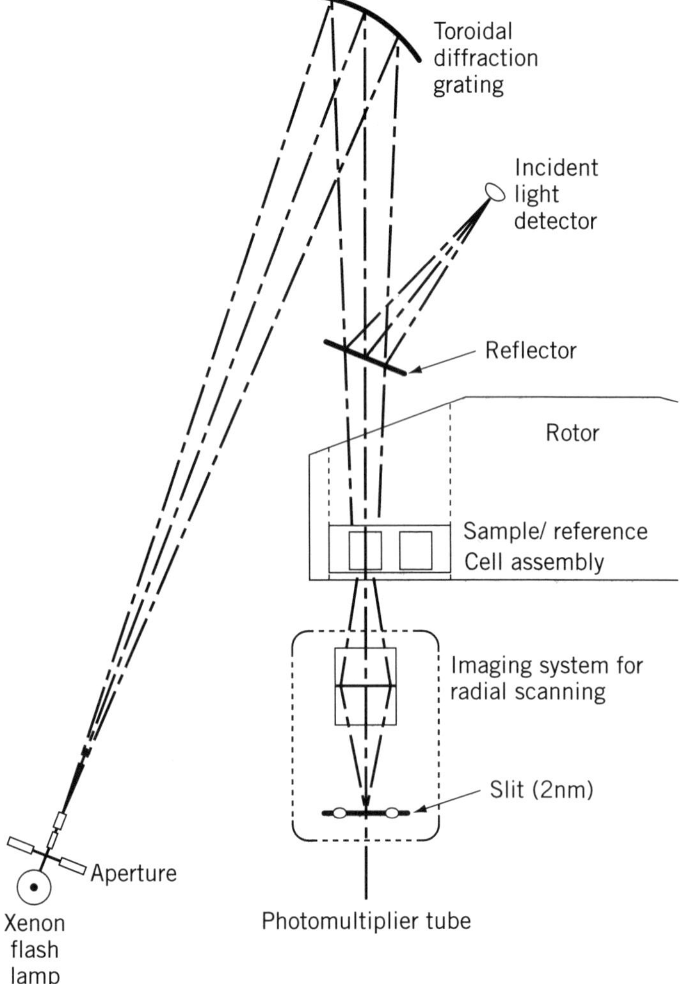

Figure 4. Absorbance-based optical system of the XLA analytical ultracentrifuge from Beckman Instruments. Light from a high intensity flash lamp passes along an optical tube to a computer-controlled toroidal diffraction grating, which selects the wavelength(s) of observation. Light at the selected wavelength, within the range ~200–800 nm, is focused on the sample cell in the rotor. Light passing through the sample is recorded by the radially tracking photomultiplier. Then the data are analyzed using specialized software on an associated computer (figure courtesy of Beckman Instruments, by permission).

The selectivity of the detection technique can also be a factor in performing an experiment. The use of this approach is illustrated in Fig. 6 for examining the potential for interactions between the Alzheimer's disease beta-amyloid peptide and the serum C1Q component of the **complement system**,

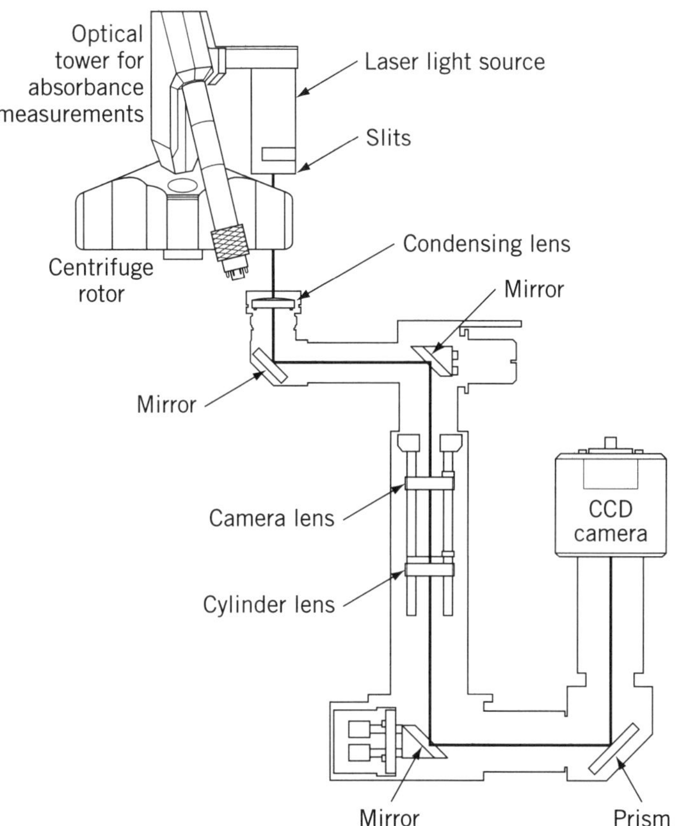

Figure 5. Interference and/or absorbance-based optical system of the XLI ultracentrifuge from Beckman Instruments. The absorbance-based optical system is outlined in Fig. 4. For interference measurements, a laser light source is attached to the absorbance optical tower, and an additional optical path and detector system is employed (lower right quadrant of figure). The interference system uses Rayleigh interference optics and a computer-controlled CCD camera detection system. Data are analyzed using specialized software on an associated computer. Using an eight-hole centrifuge rotor and centerpieces that hold three sample and reference solutions each, it is possible to analyze 21 samples in a single ultracentrifuge run (figure courtesy of Beckman Instruments, by permission).

which other studies had suggested is a trigger for activating the complement cascade and for potential neuronal cell loss. In this work, a 5-kDa peptide was labeled with the chromophore *fluorescein*. Binding of the labeled peptide to a ~150 kDa antifluorescein **antibody** was used as a positive control to examine the potential for β-amyloid peptide to bind to C1q. The labeled peptide binds to the antibody as expected, but it does not bind to C1q.

In analytical ultracentrifugation, another important consideration is sample and reference solution buffer matching. If the solutions are poorly matched, the resulting data are usually poor, even with the extensive signal averaging employed with advanced detectors and software.

Interference detection measurements are based on the change in the refractive index of the solution caused by a solute, which is a function of its concentration. For these measurements, the system records the differential change in the refractive index of the sample versus the reference cell, as a function of radial position. The refractive index of a solution is a function of the mass of all solutes, not of their absorbance or molarity, and it is not possible to distinguish between different solutes. In fact, buffers and salts routinely contribute most of the change in refractive index because they are present in much higher concentrations than the protein. Therefore, interference measurements are extremely sensitive to small mismatches in refractive index between the buffers of the sample and reference solutions. As a result, samples analyzed by interference measurements routinely require exhaustive **dialysis**. With appropriate sample/reference solution preparation, it is possible to achieve protein concentrations of >100 mg/ml routinely. The only effective upper limit is the intrinsic solubility of the protein itself under the conditions of analysis (rotor velocity, temperature, pH, etc., which can be easily changed). For both absorbance and interference measurements, the lower limit on size is a molecular weight of ~200 Da, and the upper limit is effectively about 30×10^6 Da.

Table 1. General Preferences in Analytical Ultracentrifugation for Detecting Different Analytes

Analyte	Absorbance Detection	Interference Detection
Carbohydrates	Usually not useful unless solute has appreciable UV or visible absorbance properties	Usually the method of choice
Nucleic acids	Follow absorbance at ~260 nm	Possible, can be used at high analyte concentrations
Organic polymers	Usually not useful unless solute has appreciable UV or visible absorbance properties	Usually the method of choice
Polysaccharides	Usually not useful unless solute has appreciable UV or visible absorbance properties	Usually the method of choice
Proteins	Useful if protein contains Trp and/or Tyr residues, so that it absorbs at 280 nm	Useful if protein lacks Trp and Tyr residues

Absorbance optics permit the differential absorbance of two molecules to be employed to distinguish their sedimentation properties (as in Fig. 6). This is not possible with interference optics, however, so it is not possible to distinguish between two different solute molecules with similar masses.

Another feature of interference measurements is their use in a differential mode. Because the interference optical system is double-beam, differential centrifugation experiments can be performed where the protein is present in both cells and a small ligand only in one. Consequently, the signal from the protein is blanked out, leaving only the signal from the ligand. A ligand that binds to the protein (even weakly) is displaced along the radial axis of centrifugal force in the centrifuge cell. Because the protein signal is blanked out, the only change in refractive index observed along the radial (force) axis is caused by the change in radial distribution of the ligand (2,3).

Absorbance measurements generally have less stringent referencing requirements. In practice it is necessary only that the reference buffer has an absorbance at the

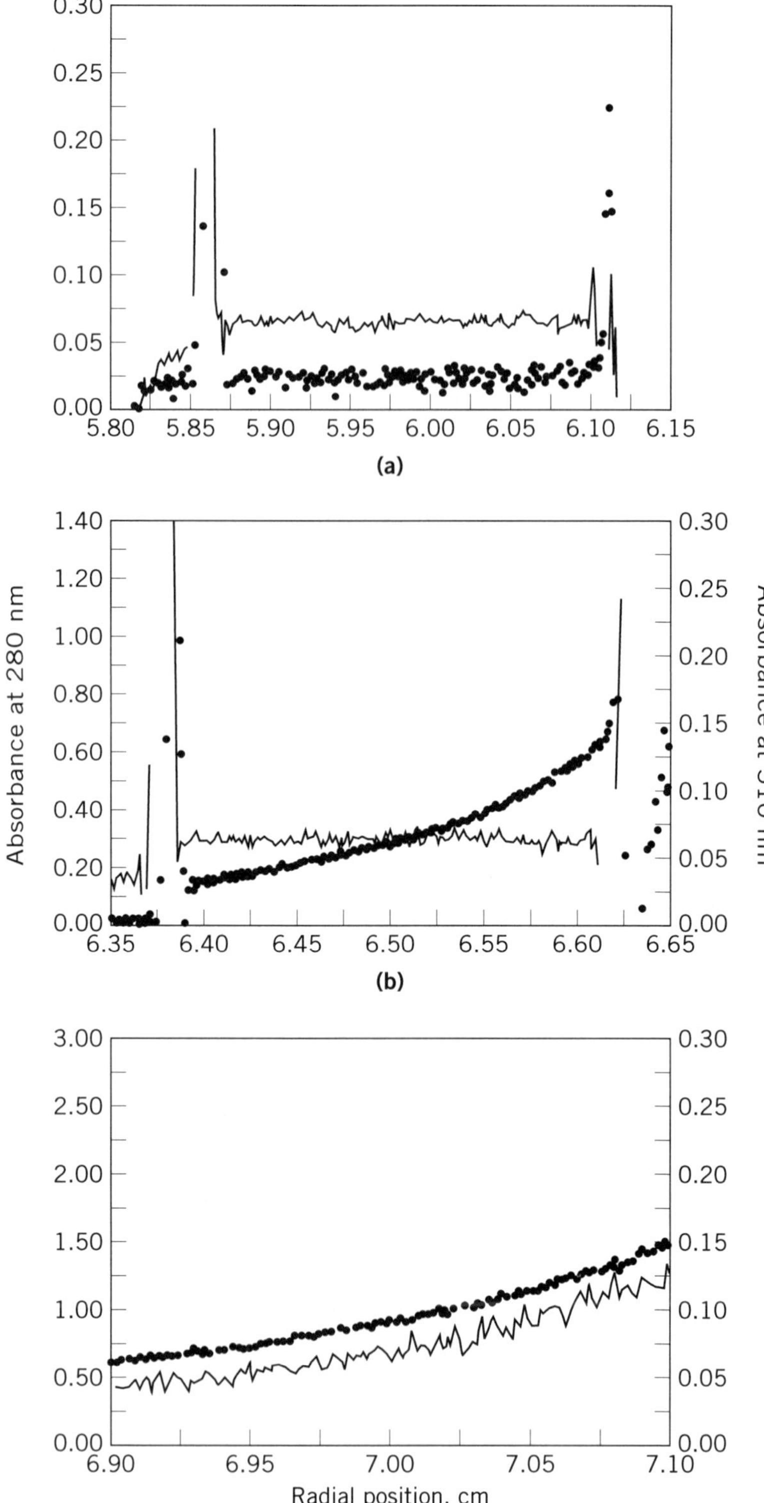

Figure 6. Binding of a xanthene-labeled Alzheimer's β-amyloid peptide to an antifluorescein antibody but not to serum complement C1q, as analyzed at sedimentation equilibrium at low speed (5000 rpm). The fluorescein analog, xanthene, absorbs light at 510 nm (solid line), and the proteins absorb light at 280 nm (dotted line). Panel (**a**) shows the radial distributions of labeled peptide alone at 280 and 510 nm (no sedimentation). Panel (**b**) shows the labeled peptide sedimented in the presence of C1q, C1q tended to sediment, but the peptide did not. Panel (**c**) shows the control experiment where the labeled peptide was sedimented in the presence of an anti fluorescein antibody. The peptide binds to the antibody and sediments with a molecular weight equivalent to that of the antibody. These results demonstrate that the monomeric form of the amyloid peptide implicated in Alzheimer's disease is unlikely to bind to complement C1q and thus induce complement-mediated neural cell death (from Ref. 8, with permission).

wavelength of analysis at which it is electronically possible for the instrument to perform the desired subtraction. Therefore, absorbance measurements are relatively more forgiving for buffer matching in sample/reference preparation. They are not without other intrinsic problems, however, in particular those related to the absorbance of buffers and solute extinction coefficients. Most biologically useful buffers and common solution additives have some characteristic absorbance in the wavelengths used to analyze proteins in solution. Because of the high total sample/reference solution absorbance, it may be difficult or impossible to make the requisite difference absorbance measurements. However, under these conditions

Table 2. Summary of Types of Solute Hydrodynamic Behavior

Hydrodynamic Characteristic	Example	Sedimentation Velocity Behavior	Sedimentation Equilibrium Behavior	Consequences
Monodisperse	Solute is pure, a single spherelike chemical entity that does not undergo self-interaction	Velocity profiles fit a single-term Lamm equation (see **Sedimentation velocity centrifugation**), and the single sedimenting boundary gives unique values of s and D during run	Data fit a single exponential term at all rotor speeds and solute concentrations	Molecular weight, s, and D have minimal or no dependencies on concentration; solute solution behavior is "ideal" or nearly so
Paucidisperse	Sample comprises a small number of chemically distinct solutes, which may or may not interact	Velocity profiles fit a multiple-term Lamm equation, and sedimenting boundaries give multiple values of s and D during run	Data do not fit a single exponential term at all rotor speeds and solute concentrations	Molecular weight, s, and D depend on concentration
		It may be possible to resolve separate sedimenting g*(s) distributions	Chances for extracting binding constants for a macromolecular assembly are usually severely diminished	
Polydisperse	Sample comprises multiple solutes, which may or may not interact	Velocity profiles fit a multiple-term Lamm equation, and sedimenting boundaries give multiple values of s and D during run	Data do not fit a single exponential term at all rotor speeds and solute concentrations	Molecular weight, s, and D depend on concentration, solute solution behavior is usually termed "nonideal"
	Sample is pure, a single spherelike chemical entity that aggregates	It may be possible to resolve separate sedimenting $g^*(s)$ distributions	Depending upon nature of self-association, it may be possible to extract binding constants for a macromolecular assembly process	
Nonideal	Sample comprises a single, highly asymmetric chemical entity that undergoes self-interaction	Velocity profiles fit a multiple-term Lamm equation, and sedimenting boundaries give multiple values of s and D during run	Data exhibit skewed behavior and cannot be fit by conventional analyses	s and D depend on concentration, solute solution behavior is "nonideal"
		It may be possible to resolve separate sedimenting $g^*(s)$ distributions	Usually not possible to extract binding constants for a macromolecular assembly process	

protein samples that have been carefully prepared by dialysis but are optically opaque at UV wavelengths, may be routinely analyzed by interference rather than absorbance detection (as above). Finally, one interesting drawback to absorbance measurements, which is routinely neglected in sedimentation equilibrium analyses (in particular), is that the analyte solution extinction coefficient(s) may vary with the composition of the solution, so that its absorbance can change during a centrifugation run. Homo- or heteroassociation processes may alter the protein extinction coefficients, which contributes nonlinear perturbations to the total absorbance. Thus careful analyses require that the radial distributions of molecules undergoing such associations be measured at several wavelengths. Interference measurements, in comparison, are insensitive to these alterations in solute extinction coefficient.

Monodispersity, Polydispersity, Paucidispersity, and Nonideality

A solute is said to be monodisperse if it exhibits behavior characteristic of a single species in solution. If the solute undergoes indefinite self-association under the solution conditions examined, it is said to exhibit *polydispersity*. If a sample of solute molecules is composed of a mixture of a limited number of species with different chemical compositions, its solution behavior is *poly*disperse, for example, with respect to its molecular weight distribution, but the sample is termed *pauci*disperse. The differences between various types of solute behavior in solution are outlined in Table 2. It is often useful to define molecular weight distributions in terms of the molecular weight averages M_n, M_w, and M_z, which are the number-, weight-, and z-averages, respectively (Table 3). They are valuable for estimating sample homogeneity. A pure solute gives identical values of M_n, M_w, and M_z. Different methods of measurement give different averages.

Nonideality is typically observed at high concentrations of a solute and is caused by the effects of **excluded volume** (versus actual molecular volume) on solute hydrodynamic behavior. The effects of nonideality are small for spherical molecules, because their excluded and actual volumes are essentially equivalent. With such molecules, their hydrody-

Table 3. Molecular Weight Averages

Type	Equation[a]	Methods to Determine
M_n, number-average	$\frac{\sum_i c_i}{\sum_i (c_i/M_i)}$	Osmotic pressure, sedimentation equilibrium
M_w, weight-average	$\frac{\sum_i c_i M_i}{\sum_i c_i}$	Sedimentation equilibrium; light scattering
M_z, z-average	$\frac{\sum_i c_i M_i^2}{\sum_i c_i M_i}$	Sedimentation equilibrium

[a] C_i is the concentration, and M_i is the molecular weight of the ith species in weight per unit volume.

namic behavior varies in direct proportion to the size of the molecular sphere. When molecules are rodlike, nonideality becomes more pronounced. As the concentration increases, the effect is nonlinear and it is significantly worse at greater ratios of molecular length to width. Intermolecular interactions are another cause of nonideality. For example, electrostatic charge–charge repulsion can be significant at high solute concentrations and low solution ionic strengths. Experimentally, such charge–charge effects are usually reduced by increasing the solution ionic strength. The role of nonideality in sample behavior is identified by performing a series of studies at different solute concentrations and extrapolating the measured hydrodynamic parameters to infinite dilution.

Other Physical Measurements Necessary to Analyze Ultracentrifuge Data

Calculating results from sedimentation equilibrium and velocity experiments requires several additional physical measurements of the solute and of the solutions employed. To calculate the anhydrous weight-average molecular weight, the density of the solvent and the **partial specific volume** of the solute must be known.

There are typically two ways to obtain these values. They can be measured with an ultrasonic density meter or pycnometer or estimated from various tables if the specific buffer and solute chemical compositions are known. For a protein, the approximate partial specific volume is calculated from its amino acid composition (4,5). The density of a buffer of exact composition can be estimated from an appropriate reference table (for example, from the *CRC Handbook of Chemistry and Physics*). Of the two direct methods for determining solvent densities and solute partial specific volumes, ultrasonic densitometry is preferred because it requires less solute for measurements. It is good experimental practice to measure both the solvent density and the partial specific volume of a solute sample of interest. In the case of proteins, this is particularly true if the studies analyze the tendencies of the proteins to undergo association or conformational changes when bound to a ligand. Such interactions are often accompanied by significant changes in the volume of the protein that are unaccounted for if the specific volumes of the individual amino acids comprising the protein are used in the calculation (6,7). Additionally, if the protein contains carbohydrate (a **glycoprotein**, for example) or bound lipid (a **lipoprotein**), estimates from reference tables can, at best, be charitably characterized as a guess.

The final physical measurement necessary is the measurement of the solution viscosity . This value can be estimated from tables (5) (as for buffer density), but again it is good experimental practice to measure the solution viscosity directly with a simple, commercially available viscometer.

BIBLIOGRAPHY

1. S. W. Snyder, R. P. Edalji, F. G. Lindh, K. A. Walter, L. Solomon, S. Pratt, K. Steffy, and T. F. Holzman (1996) *J. Protein Chem.* **15**, 763–774.
2. I. Z. Steinberg and H. K. Schachman (1966) *Biochemistry* **12**, 3728–3747.
3. T. M. Lohman, C. G. Wensley, J. Cina, J. R. R. Burgess, and M. T. Record (1980) *Biochemistry* **19**, 3516–3522.
4. E. J. Cohn and J. T. Edsall (1943) In *Proteins, Amino Acids, and Peptides as Ions and Dipolar Ions* (ACS Monograph series), Reinhold Publishing, New York, pp. 155–176.
5. H. Durchschlag (1986) In *Thermodynamic Data for Biochemistry and Biotechnology* (H. J. Hinz, ed.), Springer-Verlag, New York, pp. 45–128.
6. H. Durchschlag and R. Jaenicke (1982) *Int. J. Biol. Macromol.* **5**, 143–148.
7. H. Durchschlag and R. Jaenicke (1982) *Biochem. Biophys. Res. Commun.* **108**, 1074–1079.
8. S. W. Snyder, G. T. Wang, L. Barrett, U. S. Ladror, D. Casuto, C. M. Lee, G. A. Krafft, R. B. Holzman, and T. F. Holzman (1994) *Exp. Neurol.* **128**, 136–142.

Suggestions for Further Reading

K. E. Van Holde (1971) *Physical Biochemistry*, Prentice–Hall, Inc., Englewood Cliffs, NJ, pp. 70–121. A short, highly readable and concisely illustrated treatment of the mechanical model of centrifugation for the biologist.

D. Eisenberg and D. Crothers (1979) Physical Chemistry with Applications to the Life Sciences, Benjamin–Cummings, Menlo Park, CA, pp. 701–745. An excellent introductory physical chemistry text with a good review of fundamental flow equations and a number of useful examples.

C. Tanford (1961) *Physical Chemistry of Macromolecules*, Wiley, New York, pp. 317–456. A classic and still relevant text on the nature of transport processes as applied to biological systems.

T. M. Schuster and T. M. Laue (1994) Modern Analytical Ultracentrifugation: Acquisition and Interpretation of Data for Biological and Synthetic Polymer Systems, Birkhäuser, Boston. The most recent compendium of modern data analytical methods and applications pertaining to the new XLA analytical ultracentrifuge.

ANDROGENESIS

GEORGES N. COHEN

Androgenesis is the mode of **reproduction** of **eukaryotes** where both sets of **chromosomes** of the offspring are of paternal origin. In higher **plants**, androgenesis can be obtained either *in vitro* with anther cultures or with isolated pollen grains. In animals, androgenesis can be obtained only experimentally. A technique similar to that described in the entry **Digynism** can be used, that of pronuclear transplantation: the male **pronucleus** is removed from a fertilized egg and injected into another egg, from which the female pronucleus is then removed, so that now the zygote contains two paternally derived sets of chromosomes.

Rarely, diandric embryos arise spontaneously when a fertilized egg loses its female pronucleus. These embryos are not viable. In humans, their implantation results in a uterine tumor called a hydatiform mole. In spontaneous abortions, diandric embryos are a minority among triploid conceptuses (1).

Diandric embryos are useful for studies of genetic **imprinting**.

BIBLIOGRAPHY

1. D. E. McFadden and J. T. Pantzar (1996) *Hum. Pathol.* **27**, 1018–1020.

ANEUPLOIDY

GEORGES N. COHEN

An aneuploid cell has one or more complete **chromosomes** either in excess or less than the normal **haploid**, **diploid**, or **polyploid** number characteristic of the species from which the cell derives. In other words, the chromosome number is not a multiple of the haploid cell number of chromosome.

Fluorescence **in situ hybridization** (FISH) with chromosome-specific probes has been successfully applied to rapidly detect numerical aberrations in metaphase and interphase amniotic cells (1). Aneuploidy arises primarily by the process of **nondisjunction**, probably caused by nonrandom premature **centromere** division in the first meiotic division of maternal **meiosis**. This varies among chromosomes, however, and a significant proportion of paternal and/or meiosis II errors have been described.

The presence of **kinetochores** or kinetochore proteins (detected immunochemically) in the **micronuclei** of binucleated cells indicates a cell with a high probability for aneuploidy following **cytokinesis** (2,3). The relationship between **Y-chromosome** aneuploidy in male humans, the number of micronuclei, the status of kinetochore proteins, and aging has been reviewed (4).

BIBLIOGRAPHY

1. W. L. Kuo et al. (1991) *Am. J. Hum. Genet.* **49**, 112–119.
2. D. A. Eastmond and J. D. Tucker (1989) *Mutat. Res.* **224**, 517–525.
3. B. K. Vig, H. J. Yoo, and D. Schiffmann (1991) *Mutagenesis* **5**, 361–367.
4. J. Nath, J. D. Tucker, and J. C. Hando (1995) *Chromosoma* **103**, 725–731.

Suggestion for Further Reading

D. K. Griffin (1996) The incidence, origin, and etiology of aneuploidy. *Int. Rev. Cytol.* **167**,263–296. An exhaustive review.

ANGIOGENIN

J. RIORDAN

Angiogenin (Ang) is one of a group of proteins that are potent inducers of angiogenesis, the process by which new blood vessels are formed (see **Epidermal growth factor**, **Fibroblast growth factor**, **Transforming growth factor**, **Tumor necrosis factor**). It was first isolated from medium conditioned by human adenocarcinoma (HT-29) cells based on the premise that tumor cells must secrete angiogenic factors in order to attract blood vessels so that they can grow. Ang was identified by its ability to stimulate angiogenesis on the chorioallantois, the outer membrane that surrounds the embryo in fertilized chicken eggs. Subsequently, it was shown to be present in normal human plasma, bovine milk, and bovine and mouse serum. It is a 14-kDa basic protein that has 33% sequence identity to bovine pancreatic ribonuclease A and is one of several members of the RNase A superfamily of proteins that have unusual biological activities (see **Ribonuclease A superfamily**). It is the only angiogenic protein that is a ribonuclease, and the only ribonuclease that is angiogenic. It is an important protein that may be involved in wound healing, is critical for the growth of solid tumors, and has become a target for the treatment of metastatic cancer and other angiogenesis-related diseases. For recent reviews on Ang see Refs. 1 and 2.

PROTEIN CHEMISTRY

Human Ang is a single **polypeptide chain** of 123 **amino-acid** residues and three **disulfide bonds** (Fig. 1). All of the main components of the catalytic site (see **Active site**) of RNase A are present in Ang, and thus it is not surprising that it has enzymatic activity toward RNA and dinucleotides and which, like that of RNase A, is limited to cleavage after pyrimidines. What is surprising is that this activity is four to six orders of magnitude less than that of RNase A. Nevertheless, it is essential for angiogenic activity. Part of the reason for such low activity was revealed by **X-ray crystallography** (Fig. 2). The structure of Ang is closely similar to that of RNase A, but with two major differences. On the one hand, the pyrimidine-binding (B_1) component of the catalytic active site was found to be occluded by the side chain of Gln117. On the other, the second (B_2) component of this site is structurally quite different from that in RNase A. In the latter, it is part of an eight-residue disulfide loop that is found in all of the 55 known mammalian members of the ribonuclease superfamily but missing in all of the nine known Ang. Instead, it is replaced by a sequence of residues that interacts with a cell-surface binding protein (see text below). Despite these likely impediments to substrate binding, Ang does catalyze RNA hydrolysis, albeit very weakly.

It has been suggested that through **evolution** Ang was endowed with a unique structure that attenuates its catalytic potency toward RNA in general, but that undergoes a conformational change when Ang is brought together with its spe-

<QNDSRYTHFL TQ**H**YDAKPQG RDDRYCESIM RRRGLTSPC**K**

DINTFIHGNK RSIKAICENK NGNPHRENLR ISKSSFQVTT

CKLHGGSPWP PCQYRATAGF RNVVVACENG LPV**H**LDQSIF

RRP

Figure 1. Amino acid sequence of human angiogenin. The active site histidine (H) and lysine (K) residues are in bold, and the disulfide bridges between cysteine (C) residues are indicated by dotted lines.

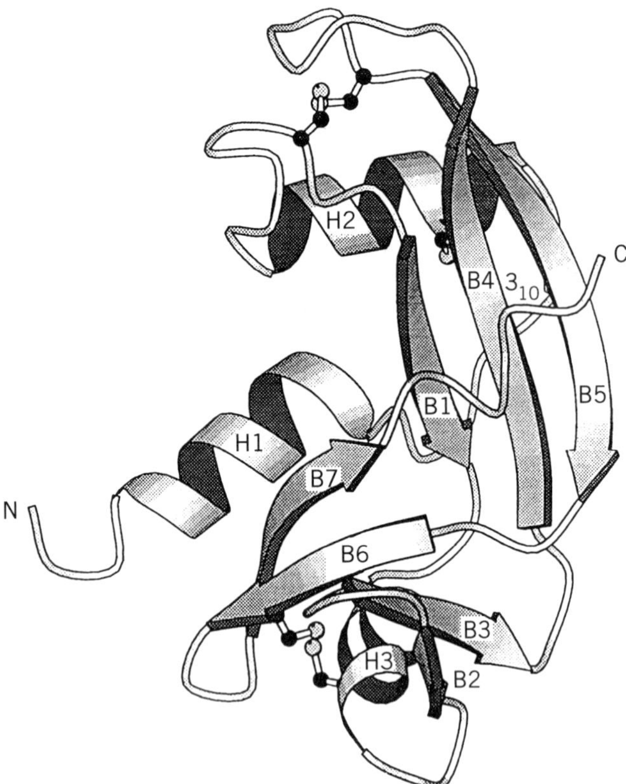

Figure 2. The polypeptide fold for human angiogenin drawn with the program MOLSCRIPT. (From Ref. 9, with permission.)

cific, yet unidentified, substrate (3). Whether this structural rearrangement activates Ang or merely ensures its specificity is unknown. It may be that the very low activity of Ang is perfectly adequate for carrying out its unique biological function. It should be noted, however, that Ang is an effective inhibitor of cell-free protein synthesis by virtue of its ability to cleave 18S rRNA when this substrate is present in the intact **ribosome**, and in fact is an even more effective inhibitor than RNase A. When isolated 18S RNA is used as substrate, the activity of Ang is markedly decreased relative to RNase A. This suggests that the activity of Ang depends on the environment of its substrate.

MOLECULAR BIOLOGY

The human *Ang* gene, localized to **chromosome** band 14q11, is present as a single copy per haploid **genome,** with no **introns** in the protein-coding region (1,2). The gene has been cloned and expressed in both *Escherichia coli* and BHK cells, and well over 40 Ang variants have been prepared in order to explore various aspects of its structure–function relationships. Thus, **site-directed mutagenesis** of His13, His114, or Lys40 (the equivalents of the catalytic residues His12, His119, and Lys41 in RNase A) abolishes *both* the ribonucleolytic and the angiogenic activity of Ang. Although angiogenically inactive, these variants block the angiogenic activity of native angiogenin. By contrast, an Ang variant in which residues 58 to 70 were replaced by residues 59 to 73 of RNase A had 300- to 600-fold more activity toward RNA substrates but was angiogenically inactive. Thus, ribonuclease activity is essential but not sufficient for angiogenic activity. Moreover, this variant, unlike the active site variants, was unable to block the angiogenic activity of native Ang, suggesting that the mutated region of the protein is involved in cell binding, probably interacting with a cell-surface binding protein. Recall that in RNase A this region of the molecule constitutes the B_2 component of the substrate binding site. In Ang it apparently has evolved into a cell binding site, with loss of catalytic activity but concomitant acquisition of a new biological potential.

CELL BIOLOGY

Blood vessels are composed of specialized cells known as endothelial cells. It is not surprising, given its function, that ^{125}I-labeled Ang binds to endothelial cells, and it does so in a manner that is characteristic of receptor binding—that is, time- and concentration-dependent, reversible, saturable, and competitive with unlabeled Ang. Yet Ang is a plasma protein (200 to 400 mg/L), and obviously it does not continuously stimulate new blood vessel formation as it circulates through the vasculature. This is explained by the fact that the angiogenin **receptor** is only expressed in sparsely cultured endothelial cells, not in confluent cells (as exist in blood vessels) (4). The receptor, not yet fully characterized, is a 170-kDa transmembrane protein whose presence on the cell surface correlates with the mitogenic activity of Ang. When the receptor is present, the cells respond to Ang by both increased thymidine uptake and cell proliferation. Confluent cells lack the receptor and, hence, do not respond this way to Ang.

Experiments with protein **cross-linking** reagents have demonstrated that Ang also interacts with a 42-kDa endothelial cell-surface protein that is a member of the **actin** family (1,2). On binding Ang, it dissociates from the cell surface as an Ang–actin complex. Remarkably, this complex activates tissue plasminogen activator (tPA) and thus generates plasmin (see **Plasmin** and **Plasminogen**). This, in turn, stimulates cell-associated proteolytic activity to degrade the **extracellular matrix** and thereby facilitates cell migration, an essential feature of the angiogenesis process. Ang also acts as a **cell adhesion molecule** and, when coated on a plastic surface, mediates the binding of endothelial and tumor cells. *In vivo* it may help direct migrating blood vessels toward Ang-secreting tumor cells.

Addition of Ang to cultured endothelial cells stimulates a transient increase in cellular **diacylglycerol**, seemingly the result of an Ang-induced activation of **phospholipase C**, as well as an increase in prostacyclin secretion owing to activation of phospholipase A_2. Interpretation of these **second messenger** responses has not been possible, thus far, largely because the systems are complex and may only be stimulated indirectly by Ang.

NUCLEAR TRANSLOCATION

Angiogenin undergoes endocytosis by sparsely cultured endothelial cells and is rapidly translocated from the cell surface to the **nucleus**, where it accumulates in the **nucleolus** (1,2). This process is receptor-mediated and regulated by a nuclear localization signal (see **Nuclear import**, **export**) that involves three basic residues, Arg31-Arg32-Arg33, of Ang. This signal is essential for angiogenesis and suggests that the substrate for the ribonucleolytic activity of Ang is located in the nucleolus, a highly specialized region where biogenesis of ribosomes takes place (see **Nucleolus** and **Ribosome**). One possibility is that the ribonucleolytic activity of Ang enhances the **transcription** and processing of rRNA. Two enzymatically inactive variants of angiogenin whose cell-binding site is intact also accumulate in the nucleolus when added to endothelial cells, but they are angiogenically inactive. This suggests that not only is nuclear localization essential for the biological activity of Ang but, in addition, the translocated protein must be enzymatically active. This has been confirmed by recent studies in which a DNA aptamer (an oligonucleotide that binds fairly tightly to Ang) that inhibits both the enzymatic and angiogenic activities of Ang is also translocated to nucleus, but only when added to endothelial cells together with Ang. The inhibitor and the protein accumulate in the nucleus in a 1:1 stoichiometric ratio (5).

MECHANISM OF ACTION

Although Ang may well be involved in a variety of angiogenesis-related situations, its mechanism of action is likely to be quite similar for each of them; hence it is only summarized here in the context of tumor-induced angiogenesis. All tumor cells that have been examined have been found to secrete Ang, and in at least one case, pancreatic cancer, the aggressiveness of the tumor is related to the amount of angiogenin produced (6). Ang secreted by tumor cells migrates through the extracellular matrix until it reaches an endothelial cell. There it combines with cell-surface actin and dissociates as a complex that stimulates the activity of tPA and the formation of plasmin. Degradation of the extracellular matrix—for example, by plasmin-activated matrix **metalloproteinases**—may stimulate a few endothelial cells to migrate, and this could trigger expression of the Ang receptor. Binding of Ang to the receptor would then, on the one hand, initiate a second-messenger response, perhaps through activation of the above-mentioned phospholipases, and, on the other, promote endocytosis and nuclear localization of Ang. The ribonucleolytic action of Ang within the nucleolus would, together with signals generated via the second messengers, activate processes leading to cell proliferation. The proliferating endothelial cells would migrate through the now degraded extracellular matrix toward the tumor cell from which the Ang was released. The cell-adhesion properties of Ang may be important for ensuring cell migration in the proper direction. At present, this view of the mode of action of Ang is largely speculative, particularly because nothing is known about the events that occur within the nucleus and lead to cell division. Nevertheless, it summarizes current thinking and is a useful basis for further investigation.

TUMOR BIOLOGY

The levels of Ang and its **messenger RNA** are increased in the tissues and cells of patients with various types of cancers, indicative of an *in vivo* role for Ang in the process of tumor angiogenesis. This being the case, anti-angiogenin agents could have potential importance in the treatment of cancer and other angiogenesis-related diseases. The most potent inhibitor of Ang is a 50-kDa protein originally isolated from placenta as an inhibitor of RNase A, hence known as placental ribonuclease inhibitor (RI) (see **Ribonuclease inhibitor**) (1,2). It inhibits Ang with a K_i of 0.7 fM, one of the strongest **protein–protein interactions** known, and is 60-fold stronger than its inhibition of RNase A. Unfortunately, RI is evidently unstable in extracellular fluids and has not been found to be effective in treating tumors in laboratory animals.

Monoclonal antibodies raised against Ang have been used to treat athymic mice injected with HT-29 colon carcinoma cells, and they reduced the incidence of tumors by up to 65%. Similar results have been observed with other types of tumors, and when used in conjunction with conventional therapeutic agents, a synergistic effect was seen (2). Actin was also tested for its anti-tumor activity, and it too was capable of preventing tumor growth in more than 60% of treated animals. It should be noted that anti-angiogenin agents do not affect the growth of already-established tumors. They are only effective when administered to animals at the same time as the tumor cells. They are thought to act by specific extracellular inactivation of tumor-secreted Ang and the consequent inhibition of tumor angiogenesis. These experiments are important in that they provide clear evidence of a crucial role for angiogenin in the early stages of development of these tumors.

OTHER BIOLOGICAL PROPERTIES

One of the major complications associated with regular hemodialysis is the increased morbidity and mortality arising from infections. This has been attributed to dysfunction of polymorphonuclear leukocytes. Indeed, a number of compounds have been isolated from uremic serum and shown to inhibit the biological activity of these white cells. One of these is an inhibitor of leukocyte degranulation that turns out to be Ang (7). Nanomolar concentrations of Ang inhibit both spontaneous and peptide-stimulated degranulation by 60% and 30%, respectively. However, Ang has no other effect on the cellular responses of polymorphonuclear leukocytes, such as **chemotaxis**, **phagocytosis** or their peptide-stimulated oxidative respiratory burst.

Ang has also been reported to suppress significantly the proliferation of human lymphocytes stimulated by a mixed-lymphocyte culture or by phytohemagglutinin or concanavalin A (8). A maximal immunosuppressive effect was seen at an Ang dose of 50 to 100 mg/mL. It was thought that this effect might synergize with the effect of Ang on neovascularization of tumors and thereby contribute to tumor development.

CONCLUSIONS

Ang is now recognized as a pleiotropic molecule capable of inducing several intra- and extracellular activities. It induces most of the individual events in the process of angiogenesis including binding to endothelial cells, stimulating second messengers, mediating cell adhesion, activating cell-associated proteinases, inducing cell invasion, stimulating DNA synthesis and cell proliferation, and organizing the formation of tubular structures from cultured endothelial cells. Characterization of its cellular receptors, elucidation of its mechanism of nuclear translocation, identification of its intranucleolar target substrate, and understanding how these events result in cellular proliferation and blood vessel formation will provide important and novel information that can be utilized for either promoting or inhibiting angiogenesis for therapeutic purposes.

BIBLIOGRAPHY

1. J. F. Riordan (1997) Structure and Function of Angiogenin In *Ribonucleases: Structures and Functions* (G. D'Alessio and J. F. Riordan, eds.), Academic Press, New York, pp. 445–489.
2. B. L. Vallee and J. F. Riordan (1997) *Cell. Mol. Life Sci.* **53**, 803–815.
3. R. Shapiro (1998) *Biochemistry* **37**, 6847–6856.
4. G.-f. Hu, J. F. Riordan, and B. L. Vallee (1997) *Proc. Natl. Acad. Sci. USA* **94**, 2204–2209.
5. V. Nobile, N. Russo, G.-f. Hu, and J. F. Riordan (1998) *Biochemistry* **37**, 6857–6863.
6. S. Shimoyama et al. (1996) *Cancer Res.* **56**, 2703–2706.
7. H. Tschesche, C. Kopp, W. H. Horl, and U. Hempelmann (1994) *J. Biol. Chem.* **269**, 30274–30280.
8. J. Matousek et al. (1995) *Comp. Biochem. Physiol.* **112B**, 235–241.
9. P. J. Kraulis (1991) *J. Appl. Crystallogr. A* **24**, 946–950.

Suggestions for Further Reading

F. Bussolino, A. Mantovani, and G. Persico (1997) Molecular mechanisms of blood vessel formation. *Trends Biochem. Sci.* **22**, 251–256.

J. Folkman and P. A. D'Amore (1996) Blood vessel formation: What is its molecular basis? *Cell* **87**, 1153–1155.

ANHYDRIDES

T. IMOTO

"Anhydride" means "without water," and many anhydrides are present in nature. Anhydrides are easily hydrated in the presence of water to acids or bases. Inorganic anhydrides include SO_3, P_2O_5, CaO, Na_2O, etc. Acid anhydrides are the most important anhydrides in biochemistry. An acid anhydride [I] is formed by eliminating a molecule of water from two acids (Scheme 1):

$$R^1\text{—}\underset{\underset{\text{O}}{\|}}{\text{C}}\text{—OH} + R^2\text{—}\underset{\underset{\text{O}}{\|}}{\text{C}}\text{—OH} \longrightarrow R^1\text{—}\underset{\underset{\text{O}}{\|}}{\text{C}}\text{—O—}\underset{\underset{\text{O}}{\|}}{\text{C}}\text{—}R^2 + H_2O$$

[I] Carboxylic acid anhydride

[II] Acetic anhydride

[III] Maleic anhydride

[IV] Succinic anhydride

[V] Phthalic anhydride

Adding a molecule of water reverses the reaction. Practically, a carboxylic acid anhydride is prepared by the nucleophilic displacement of chloride ion from acyl chlorides by carboxylate ion.

Acid anhydrides are reactive and are good acylating agents. They react with water to yield acids, with **amino groups** to yield amides, and with alcohols to yield esters. Acid anhydrides are utilized as reactive intermediates in organic syntheses, such as **peptide synthesis**, and sometimes as **enzyme**-substrate intermediates. Acid anhydrides are unstable and therefore have large standard **free energies** of hydrolysis. The two terminal phosphate linkages in ATP are anhydride linkages, and acid anhydrides play important roles in bioenergetics (see **Adenylate charge**). Acid anhydrides are widely employed for **chemical modification** of proteins, particularly of amino groups. The acetylation reaction by an acid anhydride with amino groups in a protein proceeds so easily that this reaction is suitable for the **radiolabeling** of a protein with a radioisotope-labeled acid anhydride (1). Acetic anhydride [II], maleic anhydride [III], and succinic anhydride [IV] are frequently employed in chemical reactions and chemical modifications of proteins. Phthalic anhydride [V] is an aromatic anhydride and is a good leaving group.

ACETIC ANHYDRIDE

Acetic anhydride is the most important acid anhydride. Its boiling point is 139.6°C, and it has a characteristic penetrating odor. It is widely employed in organic syntheses. It is used to introduce the acetate group into organic compounds and is called an acetylating agent. To acetyle amino groups in a protein (2), prepare a protein solution (2 to 10%, w/v) in half-saturated sodium acetate. Over a period of 1 h at 0°C, add, in five equal portions, a weight of acetic anhydride equal to that of the protein, and continue stirring for an additional hour.

BIBLIOGRAPHY

1. T. N. M. Schumacher and T. J. Tsomides (1995) In *Current Protocols in Protein Science* (J. E. Coligan et al., eds.), Wiley, pp. 3.3.11–3.3.12.
2. T. Imoto and H. Yamada (1989) In *Protein Structure: A Practical Approach* (T. E. Creighton, ed.), IRL Press, Oxford, UK, pp. 247–277.

8-ANILINONAPHTHALENE-1-SULFONIC ACID

ANTHONY L. FINK

8-Anilinonaphthalene-1-sulfonic acid (ANS) (Fig. 1) and its dimer, 1,1′-bis(4-anilino-5-naphthalenesulfonic acid) (bis ANS) are two widely used "**hydrophobic**" probes in **fluorescence** studies of **proteins** and **membranes** (1). These dyes are minimally fluorescent in **polar** environments, such as aqueous solutions, but their fluorescent emission is dramatically increased in **nonpolar** environments. An increase in quantum yield (intensity) and a blue shift in the wavelength

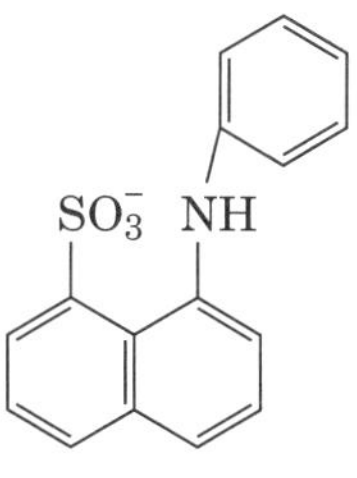

Figure 1. The structure of 8-anilinonaphthalene-1-sulfonic acid (ANS).

of maximum emission (λ_{max}) is observed when binding to macromolecules.

The most common uses of ANS (and bis-ANS) are (1) to monitor conformational changes in proteins (2,3); (2) to follow the kinetics of **protein folding** and unfolding (4–6); (3) to detect and characterize partially folded intermediates of proteins (both transient and equilibrium) (7,8); (4) to measure changes in membrane properties (9,10); (5) to detect the presence of exposed hydrophobic **accessible surfaces** (11); and (6) to study **ligand binding** through displacement assays (12,13). Photoincorporation of bis-ANS has been used to locate solvent-exposed hydrophobic regions in proteins (14) and more specifically (1) to investigate **virus** capsid structure and assembly (15,16); (2) to monitor **tubulin** assembly and interaction with drugs (17) because bis-ANS specifically inhibits assembly of tubulin (18); (3) to measure distances via fluorescent energy transfer from **tryptophan** residues to bound ANS (17,19); (4) to study **enzyme**–substrate interactions (20), including **nucleotide** binding (21); and (5) as a sensitive staining method for proteins in **SDS-PAGE** (22). ANS has been used to detect conformational changes in a wide variety of functional proteins during ligand binding (ranging from small molecules to DNA) (13,23).

The advantages of ANS as a probe reflect the simplicity of fluorescence experiments and the sensitivity of ANS to its immediate environment. Although most commonly used in the fluorescent steady-state mode, it has also been used in more complex experiments (6,25). The **absorbance** spectrum of ANS has maxima at ~270 nm and in the 345 to 380 nm range. The fluorescent emission spectrum is very sensitive to the environment and has a maximum of ~520 nm in **water**, which becomes progressively blue-shifted with decreasing polarity and reaches 464 nm in very nonpolar environments. A corresponding change occurs in the fluorescent lifetime from ~0.25 ns in water to ~15 ns in hydrophobic environments. The overlap between ANS excitation and tryptophan emission in the vicinity of 350 nm provides a useful donor/acceptor energy transfer system. Although the properties of ANS and bis-ANS are similar, the dimer has substantially higher affinity for many protein sites (26).

It is commonly assumed that ANS and bis-ANS bind to proteins and membranes by hydrophobic interactions mediated by the two aromatic rings, although there is little direct data (such as **X-ray crystallography** or **NMR** structures) to support this hypothesis. These dyes are anions over most of the commonly used pH range and **electrostatic interactions** are also important (possibly even critical) in ANS binding to proteins and membranes (27).

The applications of ANS to biomembrane studies have been summarized (24) and are not presented in any depth here. ANS binds to membranes in the vicinity of the phospholipid polar headgroups. The nonpolar part of the ANS molecule is directed toward the region of the fatty acid side chains, but the ANS does not penetrate very far into the nonpolar region of the bilayer. The binding is very sensitive to the nature of the phospholipids and to the presence of other membrane components, such as cholesterol.

For proteins, it is likely that only those ANS molecules that bind in hydrophobic regions exhibit the characteristic strong fluorescence. The remaining ANS molecules are bound to cationic sites exposed to aqueous solvent and hence are quenched (28). The thermodynamics of ANS binding to a protein have been determined: the binding is enthalpically driven at all temperatures and entropically opposed at temperatures greater than 14°C (27). The self-association of ANS and bis-ANS in water correlates with their tendency to form complexes with proteins (29). Thus care must be taken, especially when working at low pH and high ANS concentrations, to avoid ANS aggregation.

It has been recognized for some time that ANS and bis-ANS are excellent probes for partially folded intermediate states of proteins, such as the **molten globule** (7,8–30). Because the dyes do not normally bind significantly to the native states of most proteins nor to the fully **unfolded** states, they are widely used as a diagnostic for the presence of partially folded intermediates. However, ANS and bis-ANS do bind to some native proteins, usually to a specific site, e.g., the heme-binding site in apo-**myoglobin** (31). The predilection of ANS for nonnative conformations has been useful in investigating designed proteins to determine whether they have a rigid, tightly packed core, or are more like a molten globule (32), and in studies to determine the conformational state of substrate proteins interacting with molecular chaperones (33).

Fluorescent changes that correspond to binding and release of ANS during protein folding are frequently used to characterize the kinetics of folding, especially the buildup and decay of transient intermediates (34–36). Differences in partially folded intermediate conformations of proteins are distinguished by differences in the properties of bound ANS (5,37).

The interpretation of results from ANS probe experiments is not without hazards, in part because the dyes perturb the system under investigation. For example, in experiments to monitor the kinetics of protein folding, the presence of ANS or bis-ANS and their strong affinity for transient intermediates significantly perturb the kinetics of folding (35). Similarly, it is customarily assumed that if ANS or bis-ANS is added to the native protein and binding is observed, it is to the native state. However, when the native state is only nominally predominant, preferential binding of the dye to a partially folded intermediate effectively shifts the equilibrium to favor the intermediate conformation (19). The presence of tight-binding ligands (eg, substrates) to the native state shifts the equilibrium back to the native conformation (19). Another source of potential complications in interpreting results arises from the fact that the dye concentration used in many studies drastically exceeds the protein concentration. It is important to use purified ANS. Most commercial preparations contain some bis-ANS, which produces misleading results because bis-ANS often binds much more tightly than ANS. The bis-ANS is removed by Sephadex LH-20 chromatography (38).

BIBLIOGRAPHY

1. L. Stryer (1965) *J. Mol. Biol.* **13**, 482–495.
2. T. Korte and A. Herrmann (1994) *Eur. Biophys. J.* **23**, 105–113.
3. H. Tsuruta and T. Sano (1990) *Biophys. Chem.* **35**, 75–84.
4. G. de Prat Gay et al. (1995) *J. Mol. Biol.* **254**, 968–979.
5. M. C. Shastry and J. B. Udgaonkar (1995) *J. Mol. Biol.* **247**, 1013–1027.
6. B. E. Jones, J. M. Beechem, and C. R. Matthews (1995) *Biochemistry* **34**, 1867–1877.
7. G. V. Semisotnov et al. (1991) *Biopolymers* **31**, 119–128.
8. Y. Goto and A. L. Fink (1989) *Biochemistry* **28**, 945–952.
9. A. Azzi (1975) *Q Rev. Biophys.* **8**, 237–316.
10. M. G. Tozzi-Ciancarelli, C. Di Massimo, and A. Mascioli (1992) *Cell. Mol. Biol.* **38**, 303–310.
11. M. Cardamone and N. K. Puri (1992) *Biochem. J.* **282**, 589–593.
12. C. D. Kane and D. A. Bernlohr (1996) *Anal. Biochem.* **233**, 197–204.
13. I. Taylor and G. G. Kneale (1994) *Methods Mol. Biol.* **30**, 327–337.
14. J. W. Seale, J. L. Martinez, and P. M. Horowitz (1995) *Biochemistry* **34**, 7443–7449.
15. A. T. Da Poian, J. E. Johnson, and J. L. Silva (1994) *Biochemistry* **33**, 8339–8346.
16. J. Secnik et al. (1990) *Biochemistry* **29**, 7991–7997.
17. A. Bhattacharya, B. Bhattacharyya, and S. Roy (1996) *Protein Sci.* **5**, 2029–2036.
18. P. Horowitz, V. Prasad, and R. F. Luduena (1984) *J. Biol. Chem.* **259**, 14647–14650.
19. L. Shi, D. R. Palleros, and A. L. Fink (1994) *Biochemistry* **33**, 7536–7546.
20. P. M. Horowitz and N. L. Criscimagna (1985) *Biochemistry* **24**, 2587–2593.
21. R. Takashi, Y. Tonomura, and M. F. Morales (1977) *Proc. Natl. Acad. Sci. USA* **74**, 2334–2338.
22. P. M. Horowitz and S. Bowman (1987) *Anal. Biochem.* **165**, 430–434.
23. A. R. Walmsley, G. E. Martin, and P. J. Henderson (1994) *J. Biol. Chem.* **269**, 17009–17019.
24. J. Slavik (1982) *Biochim. Biophys. Acta* **694**, 1–25.
25. E. E. Gussakovsky and E. Haas (1995) *Protein Sci.* **4**, 2319–2326.
26. C. G. Rosen and G. Weber (1969) *Biochemistry* **8**, 3915–3920.
27. W. R. Kirk, E. Kurian, and F. G. Prendergast (1996) *Biophys. J.* **70**, 69–83.
28. D. Matulis, R. Lovrien, and T. I. Richardson (1996) *J. Mol. Recognition* **9**, 433–443.
29. B. Stopa et al. (1997) *Biochimie.* **79**, 23–26.
30. Y. Goto, T. Azuma, and K. Hamaguchi (1979) *J. Biochem.* **85**, 1427–1438.
31. E. Bismuto et al. (1996) *Protein Sci.* **5**, 121–126.
32. S. F. Betz et al. (1996) *Fold Des* **1**, 57–64.
33. M. Gross et al. (1996) *Protein Sci.* **5**, 2506–2513.
34. A. F. Chaffotte et al. (1992) *Biochemistry* **31**, 4303–4308.
35. M. Engelhard and P. A. Evans (1995) *Protein Sci.* **4**, 1553–1562.
36. O. B. Ptitsyn et al. (1990) *FEBS Lett.* **262**, 20–24.
37. E. Zerovnik et al. (1997) *Eur. J. Biochem.* **245**, 364–372.
38. S. S. York, R. C. Lawson Jr., and D. M. Worah (1978) *Biochemistry* **17**, 4480–4486.

Suggestions for Further Reading

C. A. Royer (1995) Fluorescence spectroscopy. *Methods Mol. Biol.* **40**, 65–89.

J. Slavik (1982) Anilinonaphthalene sulfonate as a probe of membrane composition and function. *Biochim. Biophys. Acta* **694**, 1–25.

ANIMAL POLE, VEGETAL POLE

STEPHAN JANSEN
BERTRAM BRENIG

The cell mass of an **egg** is not uniformly distributed, but it exhibits significant differences in terms of morphology and at the molecular level. In order to describe this polarity, the terms animal pole and *vegetal pole* were invented to describe the two opposite poles of the egg.

Unlike the eggs of insects, which are elliptically shaped, most **oocytes** of amphibians and mammalians exhibit a less pronounced asymmetry. One element of asymmetry is the location of the cell **nucleus**, which is normally not right in the center of the oocyte, but is more peripheral, sometimes even adjacent to the egg membrane. Due to gravitational forces, the yolk within settles to the bottom of the egg and forms the vegetal pole; consequently, the polarity becomes more apparent when more yolk is present in the egg. Another component causing asymmetry is that **polar bodies** extruded during **meiosis** are frequently located in the region of the animal pole. In ascidians, **sperm** enters the egg somewhere in the animal hemisphere (1), causing cytoplasmic movements and rotation of the egg cortex. As a consequence, a further distinct region, the *gray crescent*, becomes visible in the fertilized egg. The gray crescent is also visible in amphibians, but is not very apparent in sea urchins. In the egg of *Unio elongatulus*, sperm entry occurs only at the vegetal pole. This is attributed to the presence of a 220-kDa binding protein that is concentrated in a restricted region of the crater region within the vegetal pole (2).

The naming of the two poles, animal and vegetal, is not based on a precise function; rather, the names have arisen from the idea that the "higher" organs evolve in the animal polar region, whereas the vegetal pole was assumed to be destined to form the "lower" organs necessary for reproduction and providing nutrition. The two poles form one of three possible coordinates, and further developmental changes in a number of amphibians correlate with the subsequent dorsal-ventral body axis of the animal. In mammals, the mechanism by which the inner cell mass settles in certain places is not fully understood. However, it was shown in the mouse that the bilateral symmetry of the early blastocyst is normally aligned with the animal-vegetal axis of the zygote. The embryonic-abembryonic axis is oriented orthogonally to the animal-vegetal axis (3).

After a sperm activates the egg, **karyogamy** of the male and female **pronuclei** occurs, and the egg starts to divide by **mitosis**. Most important, the cell mass does not increase during the first cell divisions; starting from the one cell, two cells are formed, after another round of divisions four cells, then eight cells, 16 cells, and so forth, until a great number of smaller cells are formed at the **morula** stage. These cells do not all have the same size; the smaller ones are called micromeres, whereas the larger ones are named *macromeres*. During these cleavage steps, there is relatively little **gene expression** from the nucleus of the new individual cells. In other words, the **genome** of the new cell does not determine its own development after **fertilization** and karyogamy at these very early stages of development. With regard to the new cells, the regulating elements are external, maternally derived gene products. These substances, mostly **RNA** molecules, are already present in the unfertilized egg, and they are the important factors that determine the fate of the divided cells. This has been shown by a number of experiments. Chemical inactivation or enucleation in embryos has shown that the nucleus is not necessary for the initial rounds of cleavage. Even enucleated egg fragments are able to perform developmental changes, and cross-fertilization experiments revealed that cells follow the maternal pattern of development. These maternally derived gene products are not evenly distributed in the egg, indicating that the animal and vegetal pole are not only a matter of morphological appearance, but are also related to the presence of a concentration gradient of different gene products. This was demonstrated in experiments in which sections of the animal or vegetal pole were excised and recultivated. When micromeres of the vegetal pole at the 16-cell stage were implanted into the animal pole of a donor embryo, a complete second gut developed; the micromeres are capable of changing the fate of neighboring cells (4). Cutting sea urchin eggs at the eight-cell stage into two halves, to produce two embryos each with an animal and a vegetal cell, generates two pluteus capable of normal **development**. In contrast, cutting the cell into an animal and an vegetal hemisphere causes considerable aberrations from the normal development.

One of the most quoted experiments to show that animal and vegetal pole cells differ significantly involved excision in which the fate of sections at the 64-cell stage were investigated. The individual cells alone were capable of forming only a **blastula**. However, adding at least four micromeres from the vegetal pole could compensate for this defect and lead to a pluteus. Fewer numbers of micromeres resulted in forms intermediate between blastula and normal pluteus. Interestingly, the vegetalization occurs not only after the addition of micromeres, but also when Li^+ ions were provided. At the animal pole, **messenger RNA** were found that code for cell-surface proteins. These mRNA were found only in the macromeres and mesomeres, not in the micromeres (5). In the vegetal pole region are localized dorsal determinants that are necessary for dorsal axis development in **Xenopus**. The dorsal determinants move from the vegetal pole to a subequatorial region, where they are incorporated into gastrulating cells (6). However, dorsal development may be also activated by the contact between the cortical dorsal determinant and the equatorial core cytoplasm that are brought together by cortical rotation upon fertilization (7).

Gastrulation is the next morphogenic event, and it starts in the region of the vegetal pole or near to the gray crescent, depending on the species. At least five different epigenic movements are observable in the further process of differentiation: invagination of cells, immigration, delamination, proliferation, and epiboly. This means that, although the polarity of the egg determines the fate of the cells at different positions, further movements and reorganization take place that are necessary for organogenesis in higher mammals. One predominantly epigenic movement is invagination during gastrulation. It was shown that vegetal egg cytoplasm is responsible for the specification of vegetal **blastomeres** and promotes gastrulation. Vegetal-deficient embryos of *Halocynthia roretzi* fail to enter gastrulation. They are animalized and arrested at the blastula stage. Reimplantation of the vegetal pole cytoplasm into vegetal-deficient embryos caused gastrulation at the site where implantation occurred (8).

BIBLIOGRAPHY

1. J. E. Speksnijder, L. F. Jaffe, and C. Sardet (1989) *Dev. Biol.* **133**, 180–184.
2. R. Focarelli and F. Rosati (1995) *Dev. Biol.* **171**, 606–614.
3. R. L. Gardner (1997) *Development* **124**, 289–301.
4. A. Ransick and E. H. Davidson (1993) *Science* **259**, 1134–1138.
5. M. Di Carlo, D. P. Romancino, G. Montana, and G. Ghersi (1994) *Proc. Natl. Acad. Sci. USA* **91**, 5622–5626.
6. M. Sakai (1996) *Development* **122**, 2207–2214.
7. H. Kageura (1997) *Development* **124**, 1543–1551.
8. H. Nishida (1996) *Development* **122**, 1271–1279.

Suggestion for Further Reading

H. Eyal-Giladi (1997) Establishment of the axis in chordates: Facts and speculations. *Development* **124**, 2285–2296. A review of the establishment of the axis in chordates.

ANKYRINS

DAVID PARRY

Ankyrins are a family of conserved proteins whose prime function is to act as a link between integral **membrane proteins** and the **spectrin**- or fodrin-based framework lying on the cytoplasmic side of the **plasmalemma**. The ankyrins were first characterized as those proteins that displayed a strong affinity for spectrin in **erythrocyte** membranes. They are expressed in particularly high levels in vertebrate brain but are also found in tissues such as skeletal muscle, lymphocytes, neutrophils and many epithelial tissues (1); they are also found in such primitive organisms as worms (**nematodes**) and fruit flies (***Drosophila***). Associations between ankyrins and a diverse selection of membrane proteins such as **cell adhesion molecules** and **ion channels** (the **anion exchanger**, or band 3 protein, Na^+/K^+ **ATPase** and the voltage-sensitive **sodium channel**) have been characterized and confirm that a key role of the ankyrins is to mediate interactions between spectrin in erythrocytes (known as fodrin in other cells) and proteins constituting the cell membrane. Deficiencies in ankyrin (as well as spectrin and some other membrane-associated proteins) have been correlated with structurally weak erythrocytes.

Ankyrins from erythrocytes (Ank1) and brain (Ank2) are monomeric proteins with a simple **domain** substructure. Each protein has an *N*-terminal domain (about 89 to 95 kDa) with membrane-binding abilities that can interact with the anion exchanger and **tubulin** (amongst others). Most of this globular region comprises 24 tandem quasi-repeats of a 33-residue motif. Because six such repeats appear to represent the smallest structural entity, it is possible that there are four subdomains within the N-terminal domain. It has also been suggested that the 33-residue structures, while folding up in a common manner, could position nonconserved residues on surface sites that would enable them to interact with a wide range of different protein ligands, rather than with a single ligand multiple times. The sequence of the 33-residue motif is very similar to those found in **transcription factors**, in **cell-cycle** control proteins, and in proteins regulating tissue **differentiation**. The second domain of ankyrin (about 62 kDa) can be subdivided into two parts; the first is acidic, proline-rich, and about 80 residues in length. In erythrocytes it links the cytoplasmic domain of the transmembrane anion exchanger to β-spectrin at about the midpoint of the tetrameric structure that β-spectrin forms with α-spectrin (see **Spectrin**). The second portion is basic and much larger (almost 500 residues). It contains the consensus sequence (Arg–Arg–Arg–Lys–Phe–His–Lys/Arg) that is also required for spectrin- or fodrin-binding. It is also much more highly conserved in sequence than the acidic 80-residue subdomain. The C-terminal domain is about 70 kDa in both Ank1 and one variant of Ank2, but about 290 kDa in an **isoform** of Ank2 that is generated by **alternative splicing**. The smaller isoform of Ank2 (molecular weight about 220 kDa) is common in adult rats, while the larger one (about 440 kDa) is expressed highly in neonatal animals. Furthermore, the 220-kDa protein is found widely in the brain (neuron cell bodies, dendrites and glia), whereas the 440-kDa protein seems to be found specifically in unmyelinated axons and dendrites (2). In both Ank1 and Ank2, the C-terminal domain has a regulatory role arising from its ability to modulate the binding affinities of both the membrane-binding and spectrin-binding domains. The sequences of the C-terminal domains differ in size, and the overall homology is not high, although there are some regions of similarity.

Ank3, which is the major ankyrin in kidney, is found in most epithelial cells, axons and muscle cells and has many structural similarities to Ank1 and Ank2. It too has an N-terminal domain (89 kDa) containing a 33-residue repeat with the **consensus sequence** (Asp/Asn–Gly–apolar–Thr–Pro/Ala–Leu–His–apolar–Ala–Ala–X–X–Gly–His/Asn–apolar–X–Val/Ile–Val/Ala–X–apolar–Leu–Leu–X–X–Gly–Ala–X–Apolar/Pro–Asn/Asp–Ala–X–Thr–Basic), a 65-kDa domain with spectrin-binding ability, and a C-terminal domain (56 kDa) with regulatory attributes (2). Interestingly, however, multiple transcripts are expressed, some of which lack the repeat domain at the N-terminal end of the molecule. It is possible that these ankyrins are involved in intracellular vesicle stabilization, sorting, or targeting (2).

BIBLIOGRAPHY

1. V. Bennett and D. M. Gilligan (1993) The spectrin-based membrane skeleton and micron-scale organization of the plasma membrane. *Annu. Rev. Cell Biol.* **9**, 27–66.
2. L. L. Peters, K. M. John, F. M. Lu, E. M. Eicher, A. Higgins, M. Yialamas, L. C. Turtzo, A. J. Otsuka, and S. E. Lux (1995) *Ank3* (epithelial ankyrin), a widely distributed new member of the ankyrin gene family and the major ankyrin in kidney, is expressed in alternatively spliced forms, including forms that lack the repeat domain. *J. Cell Biol.* **130**, 313–330.

ANNEALING, NUCLEIC ACIDS

G. E. PLUM
K. J. BRESLAUER

When two complementary nucleic acid strands are mixed under conditions favoring complex formation, the fully associated duplex is the lowest free-energy state of the system. The annealing process involves formation and disruption of partially **base-paired** regions and expansion of those regions as the system samples progressively more extensively base-paired intermediate states on the path to the lowest energy fully base-paired state. This lowest energy state may not be achieved due to kinetically trapped structures.

Denaturation of short oligonucleotide duplexes is a thermodynamically and kinetically reversible process. When the solution is cooled after melting of the duplex, the optical and hydrodynamic properties of oligonucleotide duplexes return to the initial values. The denaturation of synthetic polymers of repeating sequence are optically reversible, which indicates that most bases are in intact base pairs. However, the hydrodynamic properties display hysteresis. The change in hydrodynamic properties is due to the formation of branched structures comprised of hairpin loops and concatenated aggregates. Natural DNAs display hysteresis in both optical and hydrodynamic properties, indicating failure of a significant fraction of the bases to find a partner base and the presence of a nonlinear structure.

When two nucleic acid polymers encounter each other during hybridization or renaturation, the probability of an exact match of base pairs in the initial complex is very small. Instead, small stretches of duplex are nucleated. These complexes may grow if the sequences of the two strands permit. If there is sufficient thermal energy, ie, sufficiently high temperature, the nucleated regions will form and disassociate as the system seeks the lowest free-energy state. If there is not sufficient thermal energy, unproductive complexes will dominate, and the equilibrium configuration will not be achieved. Conversely, if there is too much thermal energy, formation of the duplex regions will be disfavored, and the strands will separate.

This leads to a common strategy of duplex formation. The annealing process is optimized by incubation at the "melting" temperature, T_m, followed by slow cooling to permit the facile reorganization of the duplex regions, to optimize sequence alignment.

Intramolecular hairpins and internal loops in single-stranded nucleic acid polymers present a significant impediment to annealing (Fig. 1). These structures form rapidly

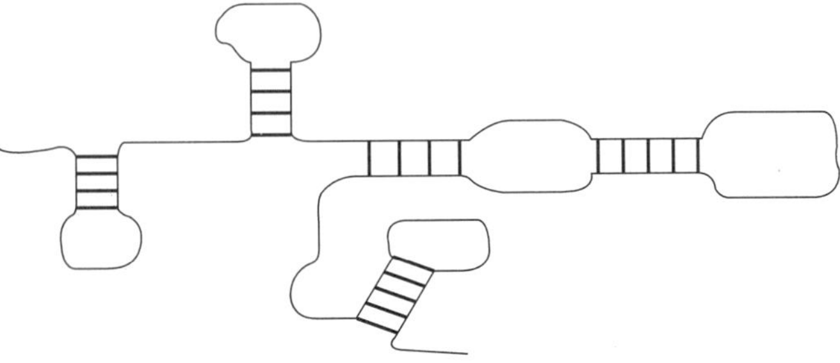

Figure 1. Intramolecular structures formed by nucleic acid single strands.

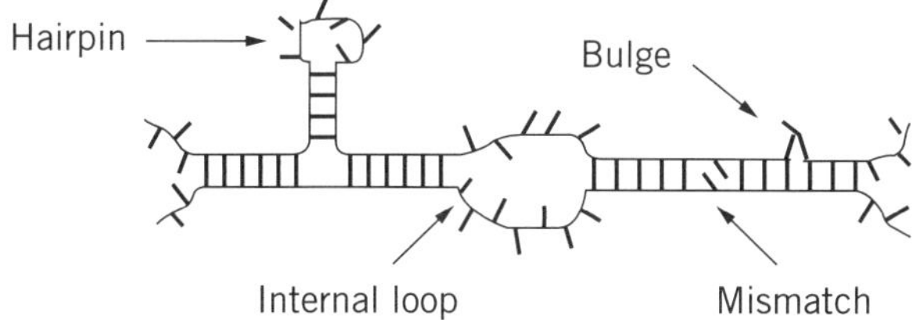

Figure 2. Misalignment structures in nucleic acid duplexes.

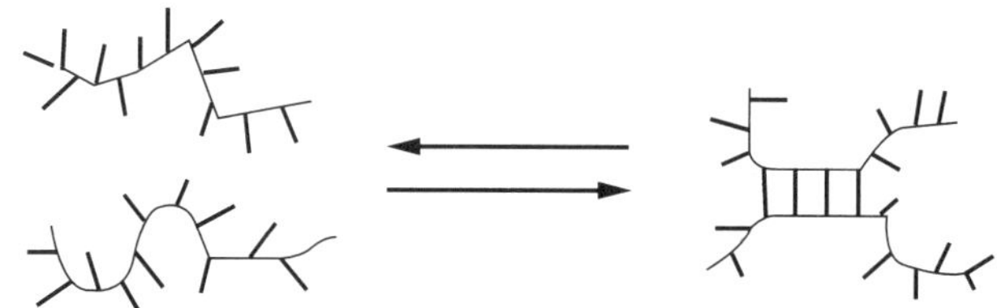

Figure 3. The nucleation event. The number of strands that are interacting changes. The process is bimolecular and depends on strand concentration.

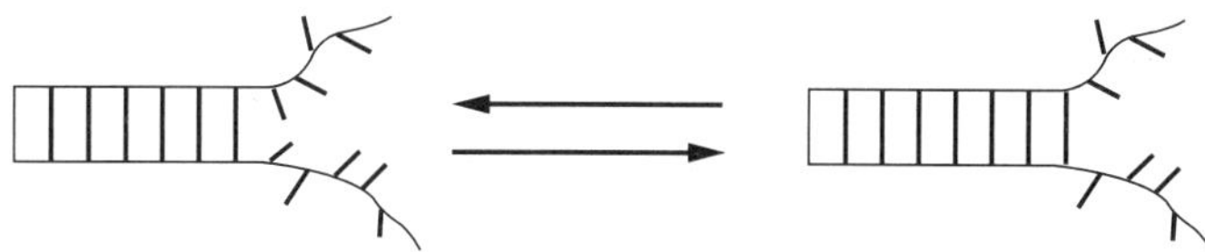

Figure 4. The propagation (zipping) event. The number of strands in the complex does not change; therefore, the process is pseudomonomolecular.

compared to duplex structures. The rate constant for hairpin formation is $10^5–10^7\ s^{-1}$, meaning the structures form on the microsecond timescale. The second-order rate constant for oligonucleotide duplex formation is $10^5–10^7\ M^{-1}\ s^{-1}$. Therefore, at the low concentration of annealing experiments, duplexes form on a time scale of seconds or longer. Intramolecular structures are at a lower energy than is the fully unfolded single-stranded structure. These competing structures must be unfolded for the annealing process to proceed toward the formation of the complementary duplex. The **activation energy** associated with this unfolding process is large, so that at temperatures well below the T_m of these single-stranded structures, the rate of duplex formation is very slow.

Defects in the alignment of base sequences (Fig. 2), including hairpins and internal loops, as well as mismatches and bulges, can occur in duplexes. Here the activation energy, and thus the time required for conversion, is even greater than for single-stranded structures. Because the alphabet of nucleic acids contains only four letters, there are regions of nucleotide sequence complementarity even in random sequences. These regions may form nucleation complexes that cannot be extended, resulting in large numbers of nonproductive intermediate states. In highly repetative sequences, large numbers of such structures may be formed. Under permissive conditions, they are readily interconvertable, and the proper sequence alignment can be achieved.

In contrast to polymers where sequence alignment dominates the kinetics of annealing, formation of short oligonucleotide duplexes is dominated by the frequency of encounter of the complementary strands. Once nucleation occurs, the pairing of the remaining bases is rapid.

Short oligonucleotide duplexes can form without appreciable competing structures. However, competition between the duplex and single-stranded hairpin structures can occur with oligonucleotides of appropriate sequence. The *unfolded single-strand-to-hairpin* and *unfolded single-strand-to-duplex* equilibria are in competition. The two equilibria differ in molecularity, so they can be distinguished experimentally by the concentration dependence. The hairpin structure will be favored at low concentration; and the duplex, at high strand concentration. The salt concentration also affects these equilibria. Because of its lower charge density, the hairpin is favored at low salt concentrations. Therefore, a combination of salt and oligonucleotide concentration can be used to isolate one of the structures.

Because the formation of a complex involves more than one nucleic acid strand, the concentration and the length of the constituent strands also influence the stability of the nucleic acid complex. There is a complex interplay between the length and concentration in determining complex stability. Two events contribute to the formation and stabilization of the complex. The nucleation event (Fig. 3), in which the strands make initial productive contacts, is a multimolecular (bimolecular for duplex formation) process that depends on strand concentration. The propagation process (Fig. 4), also called zipping, in which the prenucleated base-paired chain is extended, is a pseudomonomolecular process. Formally more than one strand is involved, but the formation of a closed base pair adjacent to a preformed base pair is effectively a monomolecular process analogous to extension of a hairpin.

For short oligonucleotides, the nucleation event dominates complex formation. The concentration dependence of the stability is simple and described adequately by the conventional mass-action model. For polymeric complexes, nucleation accounts for a relatively small part of the total stabilizing free energy. The concentration dependence of complex formation disappears for polymeric nucleic acids. For complexes of intermediate length, both nucleation and propagation make nontrivial contributions to complex formation and stability. The concentration dependence of intermediate length complex stability is reduced relative to that for short complexes of short oligonucleotides, but not vanishing as for polymers.

Suggestion for Further Reading

V. A. Bloomfield, D. M. Crothers, and I. Tinoco (1974) *Physical Chemistry of Nucleic Acids*, Harper & Row, New York.

ANNEXINS

BARBARA A. SEATON

Annexins are a family of structurally **homologous, calcium-** and **membrane**-binding **proteins** that exist in all **eukaryotes,** except **fungi.** To date, twenty distinct annexins from protozoans and higher eukaryotes and eleven plant annexins have been identified. Many of these have no mammalian counterparts, but at least ten are found in humans. Individual annexins may be intracellular or extracellular, and they vary in their organ, tissue, cell and subcellular distribution and localization. The mechanism of release of annexins into the extracellular milieu is unclear because annexins lack **signal sequences.** Depending on the source, some annexins constitute a significant proportion (>1%) of total cell protein. Though their *in vivo* functions are not yet established, annexins have been implicated in many processes, including cell proliferation and **differentiation, signal transduction**, membrane trafficking and secretion, **vesicle** aggregation, cell adhesion, interactions with **cytoskeletal** elements, **apoptosis,** anticoagulation, **phospholipase** A_2 inhibition, and **ion channel** activity or regulation. **Phosphorylation** may regulate some annexin-mediated processes. At least two annexins are major substrates for the tyrosine kinases, epidermal growth factor receptor, and retrovirus-encoded protein tyrosine kinase pp60v-src. Phosphorylations through other **tyrosine kinases** or **serine/threonine kinases,** for example, protein kinase C and cAMP-dependent protein kinase A, also have been observed. Some annexins strongly inhibit both types of protein kinases.

Before annexins were recognized as a family, the proteins were identified in many laboratories by their calcium-dependent binding to particulate fractions or **hydrophobic chromatographic** column matrices. Initially these proteins were given diverse names, which reflected either their source or putative function, including lipocortins, calpactins, endonexins, placental anticoagulant proteins (PAPs), chromobindins, calcimedins, and calelectrins. Subsequent sequence analysis revealed that many of these proteins have a common identity. The general term *annexin* subsequently was adopted in reference to their common ability to *annex to* membranes. In common nomenclature, individual annexins are designated by Roman numerals for example, annexin VII.

MACROMOLECULAR INTERACTIONS

The binding of calcium ions to annexins is relatively weak, with **dissociation constants** K_d in the millimolar-to-micromolar range. Complexation with membranes or other proteins increases their calcium-binding affinities. *In vitro*, annexins bind lanthanides and divalent cations, such as strontium, barium, and zinc. Calcium-dependent binding to membranes is high-affinity, and K_d is in the nanomolar range, whereas phospholipid monomers are bound poorly. In many systems, the annexin-membrane association is reversible upon addition of calcium chelators. In other systems, annexins behave more like integral **membrane proteins.** Typically, acidic phospholipids, such as phosphatidylserine, are strongly preferred by annexins, whereas binding to pure phosphatidylcholine membranes has not been observed. Such phospholipid binding preferences may target annexins to specific locations. The strong response of annexin V to extracellular exposure of phosphatidylserine is widely used as the basis of flow cytometric assays to detect membrane phospholipid asymmetry in apoptosis and other cellular processes.

In addition to their membrane lipid-binding properties, some annexins interact with other proteins. Some annexins interact with members of the **EF-hand** family, which suggests structural complementarity between the two protein families.

Annexins II and XI each form heterocomplexes in which the heavy chain is an annexin and the light chain resembles S-100 protein. Extracellular annexins have been described as cell surface **receptors** for some **viruses** (1,2) and matrix proteins, such as **collagen** (reviewed in Ref. 3) and tenascin C (4). Annexin II is identified as a co-receptor for **plasminogen/** tissue plasminogen activator (reviewed in Ref. 5). Calcium-dependent **lectin** activity has been identified in some annexins that bind to specific sialoglycoproteins and glycosoaminoglycans, such as heparin (6,7). The biological importance of this interaction is not yet understood, but it may be related to cell surface function.

PRIMARY STRUCTURES AND MOLECULAR EVOLUTION

Annexin sequences are comprised of two regions, a variable N-terminal region and a conserved C-terminal core region. The core region consists of four or eight canonical repeats of approximately 70 amino acid residues. Sequence homology exists between repeats within an individual annexin and between annexins. The sequence data suggest that **gene duplication** and subsequent **gene fusion** produced a four-**domain** ancestral precursor. Further gene fusion of two four-domain annexins produced annexin VI, which is unique in its possession of eight repeats. The core region is associated with the calcium-dependent phospholipid-binding properties of annexins. Each repeat contains a highly conserved stretch of amino acid residues with a **consensus** calcium-binding sequence (Lys/Arg)-Gly-X-Gly-Thr-X_{38}-(Asp/Glu).

The variable N-terminal regions of annexins exhibit little homology and confer the distinct functional properties of individual annexins. Here the **intron**-exon structures are not highly conserved, whereas they are in the core region. The N-terminal region in annexins may consist of only a few amino acid residues or hundreds. Annexins VII (synexin) and XI have the largest and most hydrophobic N-terminal regions. These may considerably modify the properties of these proteins that arise from the common annexin core. In many annexins, phosphorylation sites occur within the N-terminal regions, as do some **protein–protein interaction** sites.

CRYSTAL STRUCTURES OF ANNEXINS

Annexin crystal structures reveal a highly conserved **alpha-helical** fold in the C-terminal core region characteristic of this family. Each basic repeating unit is made up of five α-helices, four of which run antiparallel, in a four-helix bundle and the fifth helix is perpendicular. The four-domain annexin structure is nearly planar, and the domains are arranged as a symmetrical array. The eight-domain annexin VI resembles two four-domain annexin structures that are approximately perpendicular to each other and are linked by a long, α-helical segment (8,9). The variable N-terminal regions of annexins remain poorly characterized structurally, as they are truncated by gene deletion or by **proteolysis.**

In annexin crystal structures, numerous bound calcium ions are observed in loops along one surface of the protein molecule. These loops have structural motifs that are distinct from those in EF-hand proteins or C2-domain proteins. There are several structural motifs for these calcium-binding sites. The site corresponding to the consensus calcium-binding sequence has the highest affinity for calcium and structurally resembles a type of site observed in phospholipase A_2. Other

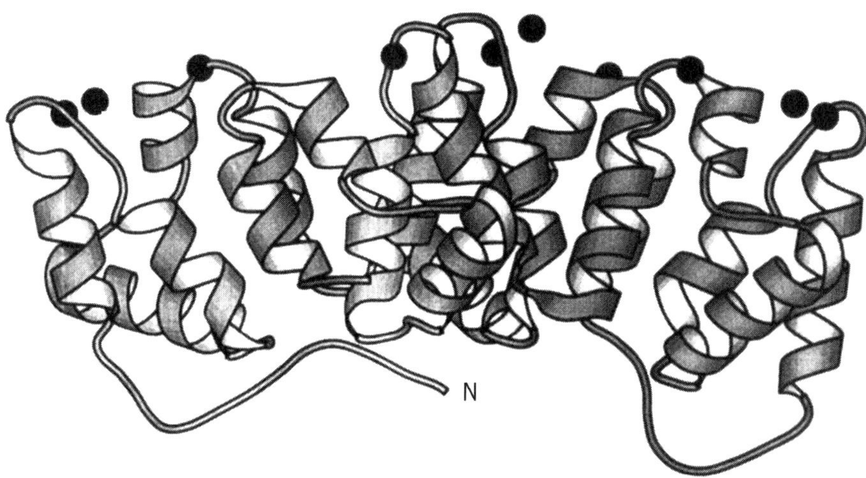

Figure 1. Ribbon diagram of annexin V (10). N-terminus as labeled, calcium ions as filled spheres.

sites bind lanthanides more strongly than calcium. Annexin V (Fig. 1) adopts two distinct molecular conformations, depending on whether a **tryptophan** side chain in the third domain extends outward from the protein surface or lies buried within the protein core. **Spectroscopic** and biochemical evidence indicates that the two states are interrelated and may describe a calcium-dependent conformational change that influences membrane binding. The crystal structures of annexin V complexes with calcium and phospholipid head group analogs indicate that calcium ions are involved directly in the attachment to membranes (10).

MEMBRANE-BOUND ANNEXINS

The membrane-bound structures of two annexins have been investigated by **electron microscopy,** and there is little evidence that the protein penetrates into the lipid bilayer. The molecular structures of membrane-bound annexins are essentially the same as in their crystals, except that the molecules reorient themselves so that their calcium-binding sites are in contact with the membrane (8,11,12). Several annexins exhibit calcium-dependent self-association and/or form two-dimensional arrays on membrane surfaces, a property that may underlie their biological functions (13,14).

Structure-based mechanisms for annexin function have been proposed, but are not fully established. Two hypotheses have been presented to explain the annexin-induced calcium channel activity observed *in vitro.* In the "microscopic electroporation" model described for annexin V, the peripheral binding of the protein changes the electrostatic properties of the membrane (15). The calcium ion is translocated to the **cytosol** through a putative central pore in the annexin molecule. In an alternative model of calcium channel activity, based on hydra annexin XII, the protein hexamer inserts into the bilayer and creates a transmembrane structure resembling an inverted micelle (16,17). Experimental data have been offered to support both hypotheses, but the mechanism and physiological relevance of the channel activity remains controversial. Mechanistic models in which annexins restrict access to membrane phospholipids and/or surfaces have been proposed to explain their inhibition of phospholipase A_2 (18) and thrombin (13), and various models have been suggested for processes, such as vesicle aggregation and membrane fusion. The *in vivo* functions of annexins remain under intensive investigation.

BIBLIOGRAPHY

1. K. Hertogs et al. (1993) *Virology* **197,** 549–557.
2. J. F. Wright, A. Kurosky, and S. Wasi (1994) *Biochem. Biophys. Res. Commun.* **198,** 983–989.
3. K. von der Mark and J. Mollenhauer (1997) *Cell. Mol. Life Sci.* **53,** 539–545.
4. C. Y. Chung and H. P. Erickson (1994) *J. Cell. Biol.* **126,** 539–548.
5. K. A. Hajjar and J. S. Menell (1997) *Ann. N.Y. Acad. Sci.* **811,** 337–349.
6. K. Kojima et al. (1996) *J. Biol. Chem.* **271,** 7679–7685.
7. G. Kassam et al. (1997) *J. Biol. Chem.* **272,** 15093–15100.
8. J. Benz et al. (1996) *J. Mol. Biol.* **260,** 638–643.
9. H. Kawasaki, A. Avila-Sakar, C. E. Creutz, and R. H. Kretsinger (1996) *Biochim. Biophys. Acta* **1313,** 277–282.
10. M. A. Swairjo et al. (1995) *Nat. Struct. Biol.* **2,** 968–974.
11. A. Oloffson, V. Mallouh, and A. Brisson (1994) *J. Struct. Biol.* **113,** 199–205.
12. D. Voges et al. (1994) *J. Mol. Biol.* **238,** 199–213.
13. H. A. M. Andree et al. (1992) *J. Biol. Chem.* **267**, 17907–17912.
14. C. Pigault et al. (1994) *J. Mol. Biol.* **236**, 199–208.
15. P. Demange et al. (1994) *Trends Biochem. Sci.* **19**, 272–276.
16. H. Luecke et al. (1995) *Nature* **378**, 512–515.
17. S. E. Moss (1995) *Nature* **378**, 446–447.
18. F. F. Davidson and E. A. Dennis (1989) *Biochem. Pharmacol.* **38**, 3645–3651.

Suggestions for Further Reading

J. Mollenhauer and others (1997) Annexins, *Cell. Mol. Life Sci.* **53**, 506–555. A multiauthor review with nine concise articles on selected topics, including some material not reviewed elsewhere (eg, plant annexins). Up-to-date reviews on annexin genetics and molecular structure; phosphorylation; collagen-binding; and the roles of annexins in apoptosis, secretion, cancer, and autoimmune diseases.

B. A. Seaton, ed. (1996) *Annexins: Molecular Structure to Cellular Function*, R. G. Landes, Austin TX. Sixteen chapters covering a wide range of topics, including comprehensive reviews on annexin gene and molecular structure; biology of annexins I and XI; annexins in phagocytic leukocytes; nematode annexins; annexin binding to phospholipids and lipid assemblies; roles in membrane trafficking; calcium-independent annexin functions, including annexin/protein interactions; annexin functions involving phospholipid membrane asymmetry and clinical applications; and current experimental approaches to understanding annexin function.

P. Raynal and H. B. Pollard (1994) Annexins: A novel family of calcium- and membrane-binding proteins in search of a function, *Biochim. Biophys. Acta* **1197**, 63–93. A comprehensive survey covering most aspects of putative annexin function.

M. A. Swairjo and B. A. Seaton (1994) Annexin structure and membrane interactions: *a* molecular perspective, *Ann. Rev. Biophys. Biomol. Struct.* 23, 193–213. Review with emphasis on structural and biophysical aspects of annexin-membrane associations and effects on lipid bilayer.

ANOMALOUS DISPERSION

JAN DRENTH

If an atom is hit by an X-ray beam, as in **X-ray crystallography**, it scatters the beam in all directions. The scattered radiation can have the same wavelength as the primary beam (Rayleigh or coherent or elastic scattering) or a longer wavelength (Compton or incoherent or inelastic scattering). For diffraction, only the coherent part of the scattering is of interest. Incoherent scattering simply increases the background.

Electrons in an atom are bound by the nucleus and are, in principle, not free electrons. However, they can be regarded as such if the frequency of the incident radiation ω is large compared with the natural absorption frequencies ω_n of the scattering atom, or if the wavelength of the incident radiation is short compared with the absorption edge wavelength. This is normally true for the light atoms but not for the heavy atoms (Table 1).

If the electrons in an atom are regarded as free electrons, the atomic scattering amplitude (atomic scattering factor in units of electron scattering) is a real quantity f because the electron cloud is centrosymmetric. If they are not free electrons, the atomic scattering factor becomes an imaginary quantity, the scattering amplitude per electron:

$$f(\text{per electron}) = \frac{E_0 e^2}{mc^2} \times \frac{\omega^2}{\omega^2 - \omega_n^2 - i\kappa_n \omega} \quad (1)$$

where E_0 is the amplitude of the electric vector of the incident beam, and κ_n is a damping factor for the nth orbit (K or L or ...). For $\omega > \omega_n$, Eq. (1) approaches $f = E_0 e^2/mc^2$, the scattering amplitude of a free electron (1–3).

In practice, the complex atomic scattering factor, called $f_{\text{anomalous}}$, is separated into three parts: $f_{\text{anomalous}} = f + f' + if''$. f is the contribution to the scattering if the electrons were free electrons, f' is the real part of the correction to be applied for non-free electrons, and f'' is the imaginary correction. $f + f'$ is the total real part of the atomic scattering factor. Values for f, f', and f'' are always given in units equal to the scattering by one free electron and are listed in Ref. 3. Because the anomalous contribution to the atomic scattering factor is mainly due to the electrons close to the nucleus, the value of the correction factors diminishes slowly as a function of the scattering angle, slower than for f.

Anomalous scattering causes a violation of Friedel's law: $I(h\,k\,\ell)$ is no longer equal to $I(\bar{h}\,\bar{k}\,\bar{\ell})$. This fact can be used profitably for determining the absolute configuration (4). Moreover, it can assist in the structural determination of proteins (see **MAD**).

Table 1. The Position of the Kα-Edge for Some Elements

Atomic number	6	16	26	34	78
Element	C	S	Fe	Se	Pt
Kα edge (Å)	43.68	5.018	1.743	0.980	0.158

BIBLIOGRAPHY

1. R. W. James (1965) *The Optical Principles of the Diffraction of X-rays*, G. Bell and Sons, London, p. 135.
2. H. Hönl (1933) *Ann. der Physik, 5. Folge* **18**, 625–655.
3. International Union of Crystallography (1995) *International Tables for Crystallography*, Vol. C (A. J. C. Wilson, ed.), Kluwer Academic Dordrecht, Boston, London.
4. J. M. Bijvoet, A. F. Peerdeman, and A. J. van Bommel (1951) *Nature* **168**, 271–271.

Suggestions for Further Reading

J. Drenth (1999) *Principles of Protein X-ray Crystallography*, Springer, New York.

ANTENNAPEDIA COMPLEX

T. C. KAUFMANN

The antennapedia complex (ANT-C) is a group of tightly linked **genes** that is found in the proximal portion of the right arm of the third chromosome of *Drosophila melanogaster*. The most prominent members of the complex produce striking **homeotic** transformations in adult flies carrying **mutations** in these genes. Thus mutations in the *Antennapedia* (*Antp*) locus cause a transformation of the antenna of the adult fly into a leg, while lesions in the *proboscipedia* (*pb*) gene cause the adult mouth parts to develop into legs rather than the normal palps used in feeding. The existence of the homeotic ANT-C was originally proposed based on the tight linkage of the *proboscipedia* (*pb*), *Sex combs reduced* (*Scr*), and *Antennapedia* (*Antp*) loci. Subsequent genetic analyses have shown that two other homeotic loci, *labial* (*lab*) and *Deformed* (*Dfd*), are also members of the complex. The homeotic loci of the ANT-C are involved in the specification of segmental identity in the posterior head (gnathocephalic) and anterior thoracic regions of the embryo and adult. Moreover, the linear order of the homeotic loci in the complex, *lab*, *pb*, *Dfd*, *Scr*, and *Antp*, corresponds to the anterior–posterior order of altered segments (intercalary, mandibular, maxillary, labial, and thoracic) found in animals bearing mutations in each of the resident loci. Taken together, the results of mutational analyses indicate that members of the complex are necessary to repress head development in the thorax (*Antp*) and elicit normal segmental identity in the anterior thorax (*Scr*) and posterior head (*Scr*, *Dfd*, *pb*, and *lab*).

A similar group of homeotic genes called the **Bithorax Complex** (BX-C) is found more distally on the third chromosome. This set of three homeotic genes (*Ultrabithorax*, *abdominal-A*, and *Abdominal-B*) acts in a similar fashion to the ANT-C, but in the posterior of the thorax and in the abdomen. The ANT-C is distinguished from the BX-C not only by virtue of the domain of action of its homeotic loci (anterior versus posterior) but also by the presence of loci that are not overtly homeotic in character. Two of these, *fushi tarazu* (*ftz*) and *zerknullt* (*zen*), have been shown to affect segment enumeration (*ftz*) and the formation of dorsal structures (*zen*) in the early embryo. A third nonhomeotic gene is ***bicoid*** (*bcd*). Mutations in this locus result in female sterility and maternal effect lethality. Eggs laid by *bcd* females fail to develop normal anterior ends and instead produce mirror-image duplications of structures normally produced at the posterior terminus of the embryo.

In addition to these genetically defined loci, several other genes have been found in the ANT-C by molecular mapping. The first of these is a cluster of cuticle-protein-related genes that map between the *lab* and *pb* loci. Eight small (about 1 kbp) **transcription** units make up the cluster, and all have sequence similarities to known cuticle protein genes. These genes (*cc1* through *cc8*) are also apparently regulated by **ecdysone** in **imaginal discs**. Deletion of the entire cluster has no apparent effect on the development or cuticle morphology of embryos, larvae, or adults. The second molecularly identified gene is the *Amalgam* (*Ama*) transcription unit. The encoded protein places the gene in the **immunoglobulin** superfamily and, like the cuticle cluster, the locus can be deleted from the **genome** with no discernible effect on the organism. Finally, there is the *zen2* or *z2* transcription unit that resides immediately adjacent to the *zen* gene. This locus is similar in structure and sequence to *zen*, but like *Ama* and the *cc* genes has no discernible function.

The entire complex has been **cloned** and sequenced and shown to cover 335 kbp of genomic DNA. The most distal transcription unit is *Antp*, which covers the distalmost 100 kbp of the complex and is made up of eight exons. Proximally, the next 75 kbp contain the *Scr* and *ftz* loci. The distal 50 kbp of this interval contain sequences necessary for *Scr* expression, as well as the two exons of the *ftz* locus and its associated regulatory elements. The proximal 25 kbp contain the three identified exons of the *Scr* transcription unit. The five exons of the *Dfd* gene are found in the central portion of the next-most-proximal 55-kbp interval. The *Dfd* transcription unit covers only 11 kbp of this region, and the 20-kbp interval flanking the gene proximally is the location of ***cis*-acting** regulatory elements for the locus. The next 25-kbp interval contains four of the nonhomeotic transcription units that help distinguish the ANT-C and BX-C. The distalmost is *Ama*, next *bcd*, and finally *zen* and *z2*. The *z2*, *zen*, and *Ama* transcription units are all relatively small (1 to 2 kbp) and comprise two exons each. The *bcd* gene is somewhat larger (3.6 kbp) and is made up of four exons. Immediately proximal to the *z2* transcription unit (about 1 kbp from its 3′ end) is the 5′ end of *pb*, which extends over the next 35 kbp of genomic DNA and contains nine exons. The next 25 kbp of the complex contain the cuticle cluster and its eight identified transcription units. The final 25 kbp are the sites of the *lab* gene, which is made up of three exons. Despite the nonhomeotic nature of three of the smaller transcription units (*zen*, *bcd*, and *ftz*) resident in the complex, these loci are tied to the larger homeotic genes of the region by the nature of their protein products. All five of the large homeotics (*Antp*, *Scr*, *Dfd*, *pb*, and *lab*), and the three small genes, have a **homeobox** motif, and their protein products are found in the **nuclei** of the cells in which they are expressed. Thus eight of the genes in the ANT-C encode regulatory proteins that act as **transcription factors**. The *z2* gene also contains a homeobox; however, the biological significance of the gene is not known, as deletions of this transcription unit have no discernible effect. The cuticle-like genes and *Ama* do not contain a homeobox.

The reasons for the clustering of these developmentally significant loci of similar function are not known. The existence of common or overlapping regulatory elements, the need to insulate regulatory sequences from chromosomal position effect, and the possibility of higher-order **chromatin** structures for proper expression have all been proposed. Whatever the reason, the homeotic complex structure has a long evolutionary standing. Similar clusters are found in vertebrates, an observation consistent with a very early origin of these genes, probably predating the separation of protostomes and deuterostomes.

Suggestions for Further Reading

M. Affolter, A. Schier, and W. J. Gehring (1990) Homeodomain proteins and the regulation of gene expression, *Curr. Opin. Cell Biol.* **2**, 485–495.

M. D. Biggin and McGinnis W. (1997) Regulation of segmentation and segmental identity by Drosophila homeoproteins: the role of DNA binding in functional activity and specificity, *Development* **124**, 4425–4433.

A. Dorn, M. Affolter, W. J. Gehring, and W. Leupin (1994) Homeodomain proteins in development and therapy, *Pharmacol. & Therap.* **61**, 155–183.

D. Duboule and G. Morata (1994) Colinearity and functional hierarchy among genes of the homeotic complexes, *Trends Genet.* **10**, 358–364.

R. Finkelstein and N. Perrimon (1991) The molecular genetics of head development in *Drosophila melanogaster*, *Development* **112**, 899–912.

W. J. Gehring, M. Affolter, and T. Burglin (1994) Homeodomain proteins, *Ann. Rev. Biochem.* **63**, 487–526.

G. Gellon and W. McGinnis (1998) Shaping animal body plans in development and evolution by modulation of Hox expression patterns, *BioEssays* **20**, 116–125.

G. Jurgens and V. Hartenstein (1993) The terminal regions of the body pattern. In *The Development of Drosophila melanogaster* (M. Bate and A. Martinez Arias, eds.), Cold Spring Harbor Laboratory Press, pp. 687–746.

T. C. Kaufman, M. A. Seeger, and G. Olsen (1990) Molecular and genetic organization of the Antennapedia gene complex of *Drosophila melanogaster*, *Adv. Genet.* **27**, 309–362.

P. A. Lawrence and G. Morata (1994) Homeobox genes: their function in Drosophila segmentation and pattern formation, *Cell* **78**, 181–189.

R. S. Mann (1997) Why are Hox genes clustered? *BioEssays* **19**, 661–664.

W. McGinnis and R. Krumlauf (1992) Homeobox genes and axial patterning, *Cell* **68**, 283–302.

G. Morata (1993) Homeotic genes of *Drosophila*, *Curr. Opin. Genet. Dev.* **3**, 606–614.

A. Popadic, A. Abzhanov, D. Rusch, and T. C. Kaufman (1998) Understanding the genetic basis of morphological evolution: the role of homeotic genes in the diversification of the arthropod bauplan, *Int. J. Dev. Biol.* **42**, 453–461.

B. T. Rogers and T. C. Kaufman (1997) Structure of the insect head in ontogeny and phylogeny: a view from *Drosophila, Int. Rev. Cytol.* **174**, 1–84.

ANTIBIOTIC RESISTANCE

KATHLEEN M. KERR
ZUBEN E. SAUNA
SURESH V. AMBUDKAR

In the 1940s, the clinical use of antibiotics first curbed the widespread threat of deadly bacterial infection. These drugs effectively inhibited bacterial growth that had gone unchecked for decades. Antibiotics did not, however, eradicate the threat of bacterial infection. In fact, the widespread use of antibiotics gave a selective advantage to bacteria that had antibiotic resistance. Many strains of bacteria have developed antibiotic resistance, or insensitivity to antibiotic drugs, in response to antibiotic selection pressures. Now, bacteria employ a myriad of resistance mechanisms to circumvent the best efforts of antibiotic researchers and clinicians. Only with the development of new and potent antibiotics and the appropriate use of existing antibiotics will researchers regain control over this resilient lifeform, bacteria.

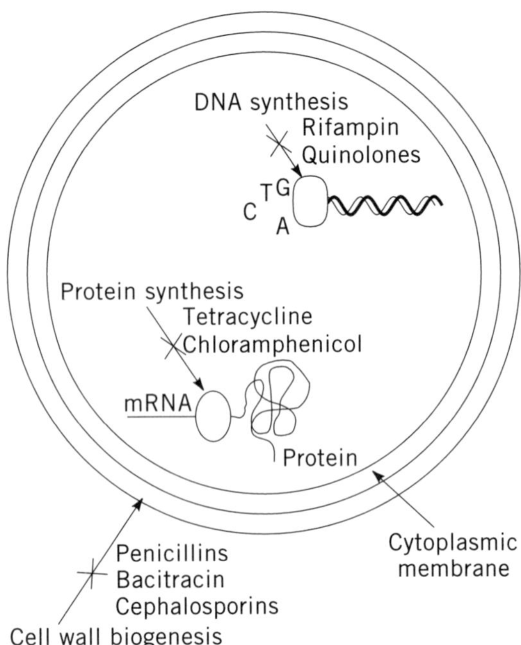

Figure 1. Mechanisms of action of some common antibiotics. A schematic of some basic functions of bacterial cells to which antibiotics are targeted. The X indicates an inhibition of that function by the antibiotics noted (see text for details).

A HISTORICAL PERSPECTIVE

The development of antibiotics as therapeutic agents began in the late 1930s to combat the most common cause of death, infectious disease. In the preantibiotic era, any infection could prove mortal. Subsequently, over 150 different antibiotics have been synthesized or discovered, and these drugs are used to treat bacterial infections, such as pneumonia, malaria, and tuberculosis (1,2).

Antibiotics are a collection of natural products and synthetic compounds that kill bacteria. Naturally occurring antibiotics are isolated from molds, yeasts, and bacteria. These organisms use antibiotics as defense mechanisms to kill other bacteria. Alternatively, synthetic antibiotics are developed by understanding the architecture and function of bacteria (see Fig. 1). Some bacteria have cell walls, and many effective antibiotics, such as **penicillin**, bacitracin, and cephalosporin, inhibit the synthesis of this cell wall. The bacterial machinery for **protein biosynthesis** differs from that of many host organisms and, therefore, is a good target for antibiotics, such as tetracycline and **chloramphenicol**. Additionally, antibiotics, such as **rifampin** and quinolones, specifically inhibit **DNA replication** in bacteria (1).

Antibiotics were considered the "wonder drugs" of their time, and in retrospect, this highly favored opinion resulted in their overuse. Antibiotics were commonly prescribed by physicians to cure and to appease their patients. Some patients would have genuine bacterial infections, for which antibiotic treatment is appropriate, whereas others would request antibiotics for viral infections that are not susceptible to these drugs. In addition, individuals who have suppressed immune systems, such as AIDS patients or organ transplants patients, would harbor bacteria that acquire resistance more easily. Antibiotics were also used prophylactically in agriculture and aquaculture industries to keep livestock healthy (3).

By the late 1960s, infectious disease appeared to be under control by the use of a variety of antibiotics. However, antibiotics simply depressed the propagation of bacteria. They did not eradicate it. Nonetheless, research in human health changed its focus from infectious diseases to chronic diseases, and new antibiotics were no longer being developed (2,4). Many microbiologists warned the human health community that bacteria were potent, infectious pathogens that should not be underestimated. Bacteria have survived for more than three billion years despite numerous environmental changes on earth, and industrial wastes, insecticides, and herbicides. Obviously, the mechanisms bacteria use to survive and adapt to these adversities were very effective.

THE ORIGINS OF ANTIBIOTIC RESISTANCE

Even before the first clinical application of antibiotics, antibiotic resistance, the ability of bacteria to evade the deleterious effects of antibiotics, was postulated. In 1940, Abraham and Chain identified a bacterial enzyme that inactivates one of the first antibiotics, penicillin. Then, any bacterium that produces this enzyme would be resistant to penicillin (5). Moreover, microorganisms that use an antibiotic as defense mechanisms would require immunity to that antibiotic. This inherent resistance to a particular antibiotic was defined as a naturally-occurring trait called intrinsic resistance. A few strains of bacteria with intrinsic resistance to a particular antibiotic would not constitute a clinical threat because many diverse antibiotics are available. However, the ability of bacteria to propagate this antibiotic resistance to other strains of bacteria had been underestimated.

The exchange of genetic information between bacteria of the same strain is a common, yet typically slow process. Mechanisms of exchange include (1) **conjugation**—a single DNA strand from one bacterium enters another bacterium and is replicated as a part of that genome, (2) **transduction**—foreign DNA is introduced into bacteria by transducing bacteriophages, and (3) transformation—autonomously replicating circular DNA plasmids are obtained by bacteria (1). Originally, it was believed that these genetic exchange mechanisms were restricted to bacteria of the same strain. However, a new method of gene exchange, using integrons, was recently identified (6,7).

Integrons are independent, **mobile elements** that encode **genes** for protein functions, and encode additional DNA to guarantee the integron's expression and integration into the bacterial **genome**. Integrons effectively generate widespread antibiotic resistance by donating antibiotic resistance genes to any strain of bacteria. Ironically, it is widely believed that integrons evolved only recently in response to antibiotic selection pressure. In other words, the use of antibiotics advanced the widespread occurrence of antibiotic resistance.

In addition to the acquisition of genes by the exchange of genetic information, bacteria also have a high mutation rate that allows them to respond to the selective pressure of antibiotics by using their own genome. For example, if bacteria were subjected to tetracycline, a random mutation in the 30S **ribosome** to weaken tetracycline binding would be advantageous and, therefore, would be perpetuated by the survival of the tetracycline-resistant bacteria (1). It has also been postulated that **housekeeping genes**, like acyltransferases, may have mutated to gain the ability to modify and inactivate aminoglycoside antibiotics (8,9,10).

The widespread phenomenon of antibiotic resistance has developed from the promiscuity of bacteria and their genomes. Initially, intrinsic antibiotic resistances were isolated incidents, but the threat of antibiotics has been readily circumvented using the acquisition of antibiotic resistance genes and the high mutational frequency of individual bacteria.

MECHANISMS OF ANTIBIOTIC RESISTANCE

Using both newly acquired genes and their own mutated genes, bacteria utilize three basic mechanisms to support antibiotic resistance. Enzymes that degrade antibiotics inside the cell are key players in antibiotic resistance. Bacteria can also alter their permeability barriers to keep antibiotic concentrations below toxic levels inside the cell. Furthermore, the cellular targets of antibiotics can be modified to evade the effects of the antibiotics. These mechanisms of antibiotic resistance are distinct, but all are obtained through the acquisition or mutation of genes.

Enzymatic Inactivation of Antibiotics. Degradative enzymes are a common mechanism by which bacteria become resistant to antibiotics. Such enzymes chemically modify antibiotics so that they no longer function. The genes for these degradative enzymes are obtained by acquisition of exogenous genes or mutation of endogenous genes. The expression of these genes also governs the level of antibiotic resistance in bacteria.

β-Lactamases are common examples of degradative enzymes that generate antibiotic resistance. β-Lactamases inactivate β-lactam antibiotics, a structurally similar group of penicillin-like antibiotics, all of which have a β-lactam ring structure. β-Lactamases are typically grouped into two major classes, penicillinases and cephalosporinases, based on their substrate affinity (11,12). In addition to β-lactamases, other enzymes also degrade different antibiotics, such as the aminoglycosides, gentamycin, tobramycin, and amikacin.

All β-lactam antibiotics function similarly. Their β-lactam ring structure inhibits the final step of bacterial cell wall synthesis. Bacterial cell walls are constructed of alternating *N*-acetylglucosamine and *N*-acetyl-muramic acid residues that form long peptidoglycan chains, and the final step in cell wall synthesis involves the enzymatic crosslinking of these peptidoglycan chains by a transpeptidase. Because the β-lactam bond resembles a portion of the peptidoglycan chains, this transpeptidase can mistake a β-lactam antibiotic for its natural substrate and hydrolyze the β-lactam bond. This hydrolysis covalently links the β-lactam drug to the transpeptidase and renders it nonfunctional (13) (see also **Penicillin-binding proteins**).

Although β-lactamases are effective at degrading some antibiotics, their mere presence is not sufficient to cause clinically relevant antibiotic resistance. In fact, these enzymes are found ubiquitously in almost all bacteria, and in some blue-green algae and mammalian tissues. β-lactamases must be present in sufficient quantities to degrade the β-lactam antibiotics effectively before they inhibit cell wall synthesis. The cellular concentration of a β-lactamase depends on its gene expression, and β-lactam antibiotics are inducers of β-lactamase expression. Furthermore, particular β-lactamases have variable affinities for β-lactam antibiotics. Therefore, the degree of antibiotic resistance due to β-lactamases is based on a combination of the ability of the β-lactam antibiotic to induce the expression of β-lactamase, and its ability to be a substrate for β-lactamase (14).

As researchers began to understand this mechanism of antibiotic resistance, more effective β-lactam antibiotics were developed. The early cephalosporins, like penicillin and amoxicillin, are extremely sensitive to β-lactamases, because these β-lactam antibiotics are potent inducers of β-lactamase expression and good substrates for the β-lactamase. In contrast, the more recently developed antibiotic, imipenem, is a strong inducer of β-lactamase expression but maintains its antibiotic activity because it is a poor substrate for most β-lactamases (15). In addition to these new antibiotics, combination therapies are also being implemented to combat antibiotic resistance. Such therapies are comprised of a β-lactam antibiotic together with β-lactamase inhibitors, like clavulanic acid, sulbactam, and tazobactum (16).

Altered Permeability Barriers: Pore Proteins and Efflux Systems. The bacterial cell **membrane** is the major permeable barrier separating the outside of the cell from the inside. The fluidity of the membrane is generally balanced to include most nutrients, while excluding many **toxins**. Adjusting this fluidity impedes the function of the membrane. Therefore, bacteria cannot protect themselves by changing the fluidity of their membrane. Instead, bacteria have additional structures that surround the cytoplasmic membrane or form pores through it. The alteration of these structures to exclude antibiotics is another mechanism of antibiotic resistance.

Gram-positive and **Gram-negative bacteria** have distinct structures that surround their cytoplasmic membranes. Most Gram-positive bacteria have thick cell walls that are mechanically quite strong, although very porous. Although the cell wall helps Gram-positive bacteria retain their shape, it does not exclude most antibiotics and, therefore, is not a good barrier. Thus, Gram-positive bacteria are relatively susceptible to the influx of antibiotics. Alternatively, a more effective barrier, a second lipid bilayer or membrane, surrounds Gram-negative bacteria. This outer membrane is partially composed of a lipid, lipopolysaccharide (LPS), that is not commonly found in cytoplasmic membranes. The distinguishing feature of LPS is its decreased fluidity, which makes the LPS bilayer an efficient barrier that prevents the permeation of most **hydrophobic** antibiotics into Gram-negative bacteria (17).

Enveloped by effective barriers, bacteria use pore-forming proteins, called **porins**, to obtain nutrients from outside the cell. Porins are transmembrane proteins that function as nonspecific, aqueous channels, and allows nutrients to diffuse across the membrane. Porins generally exclude antibiotics because they are narrow and restrictive. Most antibiotics are large, uncharged molecules that cannot easily traverse the narrow porin channels that are lined with charged amino acid residues. However, some antibiotics enter the bacteria through porins, and the deletion or alteration of these porins to exclude particular antibiotics is linked to antibiotic resistance.

Because bacteria cannot develop barriers that are impermeable to all molecules, some toxins do diffuse into bacteria along with nutrients. Therefore, bacterial cell membranes also contain transport proteins that cross the membranes and use energy to remove toxins. They are called active efflux systems, and some are directly identified as another significant cause of antibiotic resistance.

Many active efflux systems resemble other transport proteins that catalyze the efflux of common, small molecules, like glucose or cations, and it is likely that mutation has modified them to transport antibiotics. Based on their overall structure, mechanism, and sequence homologies, these transport proteins are classified into four families: (1) the major facilitator family; (2) the resistance nodulation division family; (3) the staphylococcal multidrug resistance family; and (4) the ATP-binding cassette (ABC) transporters. Of these four families, only the ABC transporters use the chemical energy generated from the hydrolysis of ATP to drive molecules across the membrane. Members of the three other families use an electrochemical proton gradient, or **proton-motive force**, as the source of energy (18,19).

Some active efflux systems exclude a variety of unrelated toxins from the cell. These multidrug resistance (MDR) efflux systems in bacteria are comparable to those found in mammalian cells (see **Drug Resistance**). An example of a bacterial MDR efflux system is the Bmr transporter that transports drugs which have diverse chemical structures and physical properties and include cationic dyes, rhodamine-6G, **ethidium bromide**, and the antibiotics netropsin, puromycin, and fluoroquinone (20). Other MDR efflux systems characterized in bacteria include NorA in *Staphylococcus aureus*, MexB in *Pseudomonas aeruginosa*, and EmrB in *Escherichia coli*. If MDR efflux systems occur extensively in bacteria as the source of many antibiotic resistances, they pose a far more formidable challenge than more specific mechanisms of resistance.

Modification of the Antibiotic's Target. Antibiotics inhibit bacterial growth by inactivating different key proteins that are essential for bacterial survival (see Fig. 1). However, the antibiotic sensitivity of these target proteins can be altered. Typically, antibiotic targets are altered by reducing their affinity for the antibiotic. Bacteria accomplish this change in affinity several ways. Bacteria acquire exogenous DNA for a mutated target protein that no longer interacts with the antibiotic, yet retains the original target protein's function. Alternatively, bacteria's endogenous genes can be mutated to achieve the same end. In contrast, DNA for novel modifying enzymes can be acquired to alter the antibiotic target post-translationally, reducing its affinity for the antibiotic.

Altering an antibiotic's target protein directly at the DNA level is a common mechanism of target modification. An example of this modification is the mutation of genes for **penicillin-binding proteins** (PBPs). PBPs are transpeptidases, previously discussed, that catalyze the final step in bacterial cell wall synthesis. These PBPs have high affinity for penicillin and its derivatives, and the binding of penicillin permanently inactivates PBPs. Originating from both endogenous and exogenous DNA sources, mutated PBPs can have a lower affinity for penicillin. Therefore, PBPs are resistant to the antibiotic, yet still provide a crucial function in bacterial cell wall synthesis (21,22). Another example of target modification via mutated DNA is a single amino acid mutation in the quinolone resistance-determining region of the **DNA gyrase** gene, gyrA, that can provide up to a 20-fold increase in quinolone resistance (23).

In addition to using mutated antibiotic targets, bacteria can acquire new genes that produce proteins that, in turn, alter antibiotic targets. A well-studied example is the resistance of Staphylococci to **erythromycin**. It is known that Staphylococci have acquired genes to produce a protein that methylates a residue on the 23S ribosome. The 23S ribosome is the target of erythromycin, but methylated 23S ribosome has a low affinity for erythromycin. This exogenous gene is expressed and prevents the binding of erythromycin to the ribosomes, making the bacteria erythromycin-resistant (24).

BIBLIOGRAPHY

1. H. C. Neu (1992) *Science* **257**, 1064–1073.
2. J. Travis (1994) *Science* **264**, 360–362.
3. M. Castiglia and R. A. J. Smego (1997) *J. Am. Pharm. Assoc.* **NS37**, 383–387.
4. G. H. Cassell (1997) *FEMS Immunol. Med. Microbiol.* **18**, 271–274.
5. E. P. Abraham and E. Chain (1940) *Nature* **146**, 837.
6. H. W. Stokes and R. M. Hall (1989) *Mol. Microbiol.* **3**, 1669–1683.
7. C. M. Collis, G. Grammaticopoulos, J. Briton, H. W. Stokes, and R. M. Hall (1993) *Mol. Microbiol.* **9**, 41–52.
8. T. Udou, Y. Mizuguchi, and R. J. J. Wallace (1989) *FEMS Microbiol. Lett.* **48**, 227–230.
9. K. J. Shaw et al. (1992) *Antimicrob. Agents Chemother.* **36**, 1447–1455.
10. P. N. Rather, E. Orosz, K. J. Shaw, R. Hare, and G. Miller (1993) *J. Bacteriol.* **175**, 6492–6498.
11. M. H. Richmond and R. B. Sykes (1973) *Adv. Microb. Physiol.* **9**, 31–88.
12. K. Bush (1989) *Antimicrob. Agents Chemother.* **33**, 259–276.

13. A. Tomasz (1979) *Annu. Rev. Microbiol.* **33**, 113–137.
14. D. M. Livermore (1993) *J. Antimicrob. Chemother.* **31** (suppl. A), 9–21.
15. J. Y. Jacobs, D. M. Livermore, and K. W. M. Davy (1984) *J. Antimicrob. Chemother.* **14**, 221–229.
16. K. Coleman et al. (1994) *J. Antimicrob. Chemother.* **33**, 1091–1116.
17. P. R. Cullis and M. J. Hope (1985) In *Biochemistry of Lipids and Membranes* (D. E. Vance and J. E. Vance, eds.), Benjamin and Cummings, New York, Chap. 2.
18. S. B. Levy (1992) *Antimicrob. Agents Chemother.* **36**, 695–703.
19. K. Lewis, D. C. Hooper, and M. Ouellette (1997) *ASM News* **63**, 605–610.
20. A. A. Neyfakh, V. E. Bidnenko, and L. B. Chen (1991) *Proc. Natl. Acad. Sci. USA* **88**, 4781–4785.
21. B. G. Spratt and K. D. Cromie (1988) *Rev. Infect. Dis.* **10**, 699–711.
22. J. M. Ghuysen (1991) *Annu. Rev. Microbiol.* **45**, 37–67.
23. G. A. Jacoby and A. A. Medeiros (1991) *Antimicrob. Agents Chemother.* **35**, 1697–1704.
24. R. Leclercq and P. Courvalin (1991) *Antimicrob. Agents Chemother.* **35**, 1267–1272.

Suggestions for Further Reading

A short news article: J. Davies (1996) Bacteria on the rampage, *Nature* **383**, 219–220.

Two well written reviews in an excellent anthology: 1. A. Bauernfeind and N. H. Georgopapadalou (1995) In *Drug Transport in Antimicrobial and Anticancer Chemotherapy* (N. H. Georgopapadalou, ed.), Dekker, New York, Vol. 17, pp. 1–19. 2. R. E. W. Hancock (1995) In *Drug Transport in Antimicrobial and Anticancer Chemotherapy* (N. H. Georgopapadalou, ed.), Dekker, New York, Vol. 17, pp. 289–306.

Three exhaustive articles in a single issue of *Science* devoted to antibiotic resistance: 1. J. Davies (1994) Inactivation of antibiotics and the dissemination of resistance Genes, *Science* **264**, 375–382. 2. H. Nikaido (1994) Prevention of drug access to bacterial targets: Permeability barriers and active efflux systems, *Science* **264**, 382–388. 3. B. G. Spratt (1994) Resistance to antibiotics mediated by target alterations, *Science* **264**, 388–393.

ANTIBODY

MICHEL FOUGEREAU

Antibodies are specific **proteins**, termed **immunoglobulins**, that are produced by **B cells** upon stimulation with **antigens**, which may be proteins, polysaccharides, nucleic acids, and so on, either in soluble form or as part of complex cellular organisms, such as bacteria, parasites, **viruses**, eukaryotic cells of animal or plant tissues, and pollens. In fact, the immune system is so built as to have the potential to make antibodies against any **macromolecule** of the living world. It was reported at the end of the nineteenth century by von Behring and Kitasato that a group of guinea pigs immunized against a sublethal dose of **diphtheria toxin** became resistant to diphtheria, whereas another group that received tetanus toxin resisted further challenge with a normally lethal dose of tetanus toxin. This key experiment showed that the protection was specific toward the immunizing agent. Thus, one master word in immunology is specificity, and this immediately raises the problem of what immunologists call the **repertoire**. How is it possible to make the billions of different molecules that are potentially necessary to assume specific recognition of an obviously astronomical number of potential antigens? What structures are recognized on an antigen? What are the structural bases for antibody specificity? All these are basic questions that have been solved progressively during the past 40 years.

The protein nature of antibodies was established by M. Heidelberger in 1928, by showing that protein nitrogen was present in a specific antigen–antibody precipitate in which the antigen was a polysaccharide, which must have originated from the antibody part of the precipitate. Progress in the knowledge of the structure of antibodies paralleled the emergence of new technological tools in biochemistry and in molecular biology. **Analytical ultracentrifugation** performed by Kabat and Tiselius at the end of the 1930 indicated that antibodies were found in several classes of sizes, mostly with **sedimentation coefficients** of between 19S and 7S. **Electrophoretic** characterization of serum, also studied by Kabat in the same period, revealed that antibodies were located in the **globulin** part of the spectrum, mostly in the slowest moving γ-globulin fraction. This name remained in use for years until it was replaced by a more generic one, that of **immunoglobulins**. Immunoglobulins are thus defined as the immune proteins synthesized specifically by B cells. They are expressed in two forms: (*1*) as integral **membrane proteins** at the surface of B lymphocytes, where they represent the B-cell receptor (BCR), by analogy with the **T-cell receptor** (TCR) expressed at the surface of **T cells**; and (*2*) as a soluble form, secreted in the bloodstream by the plasma cells, which represent the terminal stage of differentiation of the B lineage. This soluble form is what immunologists call circulating antibodies, or simply antibodies.

A main difficulty encountered in studying antibody structure was due to their extraordinary heterogeneity, which deserves some comments. As a rule, when an antigen such as **hemoglobin** or **serum albumin**, or in fact any protein, is used as an **immunogen**, a large array of antibodies with discrete fine specificities is synthesized. At the surface of a protein, many discrete regions are recognized by as many distinct antibodies. These regions are called *antigenic determinants*, or **epitopes**. For a protein of molecular weight 50 kDa, there are 10, 20, or more epitopes that could be identified by distinct antibodies. But antibody heterogeneity is not limited to the mosaic of epitopes. It may be shown that in fact one given epitope may be recognized by slightly different antibodies, and this contributes an extraordinary multiplication of the heterogeneity. This results from the clonal organization of lymphocytes, as originally proposed by Burnet (1). B lymphocytes (and T cells as well) are organized as discrete clones, each of which makes one and only one type of antibody molecule. In humans, the average population of lymphocytes at any given time is of the order of 10^{12}, of which approximately one-fifth are B cells and four-fifths are T cells. These B cells make about 10^{20} molecules of circulating immunoglobulins, a number that can be calculated from the Ig concentration in serum (10 mg/mL). The size of one clone is largely dependent on the immunization state and can vary from one single resting cell to several thousand or more. What happens when an animal is being immunized is that all clones that happen to interact somewhat with any potential epitope of the administered antigen will proliferate, expand, and synthesize as many different antibodies, thereby contributing to a very great heterogeneity. This of course made it very difficult, if not impossible, to study the fine structure of antibody molecules, especially when considering their amino acid sequences. A bias was brought to this by the existence of multiple myeloma in humans, or its equivalent as experimentally induced plasmacytoma in the mouse. These are lymphoproliferative disorders that affect plasma cells. As a result of the malignancy, one clone will expand specifically so that, very rapidly, it will become the major B-cell clone expressed in the body. The corresponding immunoglobulin will be secreted in excessively large amounts, and this provides a homogeneous material available for structural studies. An obvious drawback with this material is that the antibody specificity is generally unknown, which prevents its use for the study of **antigen–antibody interactions**. This difficulty was turned around when Kohler and Milstein (2) succeeded in making somatic hybrids between myeloma cells and normal B lymphocytes from hyperimmunized mice. These **hybridomas** combined the properties of myeloma cells, which can grow indefinitely in culture or upon transplantation in syngeneic strains of mice, with the clones of immunized lymphocytes that could be selected for a known antigenic specificity, leading to the fantastic expansion of "**monoclonal antibodies**" (mAbs). Monoclonal antibodies were not only suitable for detailed structural analysis of antibodies, including determination of the three-dimensional **protein structure** of antibody–antigen complexes. They also allowed access, through the hybridoma cells, to isolation of **messenger RNA** encoding Ig polypeptide chains and the corresponding **complementary DNA**, and thus to the complete gene organization of the three main Ig gene loci.

The basic structure of antibodies, elucidated from the pioneer work of Porter in England (3) and Edelman in the United States (4), is given by that of the **IgG** molecule—known initially as 7S-γ-globulin. It is a symmetrical molecule, containing two identical heavy chains (H, 52 kDa) and two identical light chains (L, 23 kDa), which are either **kappa** (κ) **or lambda** (λ) **chains**. Each H–L pair contains one combining site for the antigen, so the conventional H_2 L_2 IgG molecule is bivalent. The symmetry of the molecule is in agreement with the **clonal selection theory**, which postulated that each B cell expressed one and only one antibody specificity. This basic model could be extrapolated to other antibody classes, which differ from each other in the nature of their heavy chains and the degree of polymerization of the basic H_2L_2 unit. Five classes have thus been described in higher vertebrates: **IgG**, **IgM**, **IgA**, **IgD**, and **IgE**, having the corresponding γ, μ, α, δ, and ϵ heavy chains, respectively. IgM are expressed as H_2L_2 monomers at the cell surface of B cells, but are pentamers in the serum. IgA is mostly a dimer, whether other classes remain as monomers. The diversification of classes allows the antibodies to assume two types of function: (*1*) antigen recognition, common to all classes, and (*2*) biological or effector functions, such as **complement** fixation, active transplacental transfer, or fixation to various cell types, which amplify the action of antibodies and generally favor the elimination of a pathogen.

The functional duality of antibody molecules was clearly established in the late 1950 by Porter (5), who succeeded in cleaving the IgG antibody by **papain**, a proteinase, which led to the isolation of two identical fragments that were antigen binding, or Fab, and one Fc fragment that did not recognize the antigen but fixed the first component of the complement cascade. Structural support for this organization was clearly demonstrated by Hilschmann in 1965 (6), who reported the first amino acid sequence of two monoclonal human light chains. This is another example of the importance of monoclonal materials in this saga of the determination of the antibody structure. The work was performed on **Bence-Jones proteins**, which are free light chains isolated from the urine of patients with multiple myeloma. They are the result of a

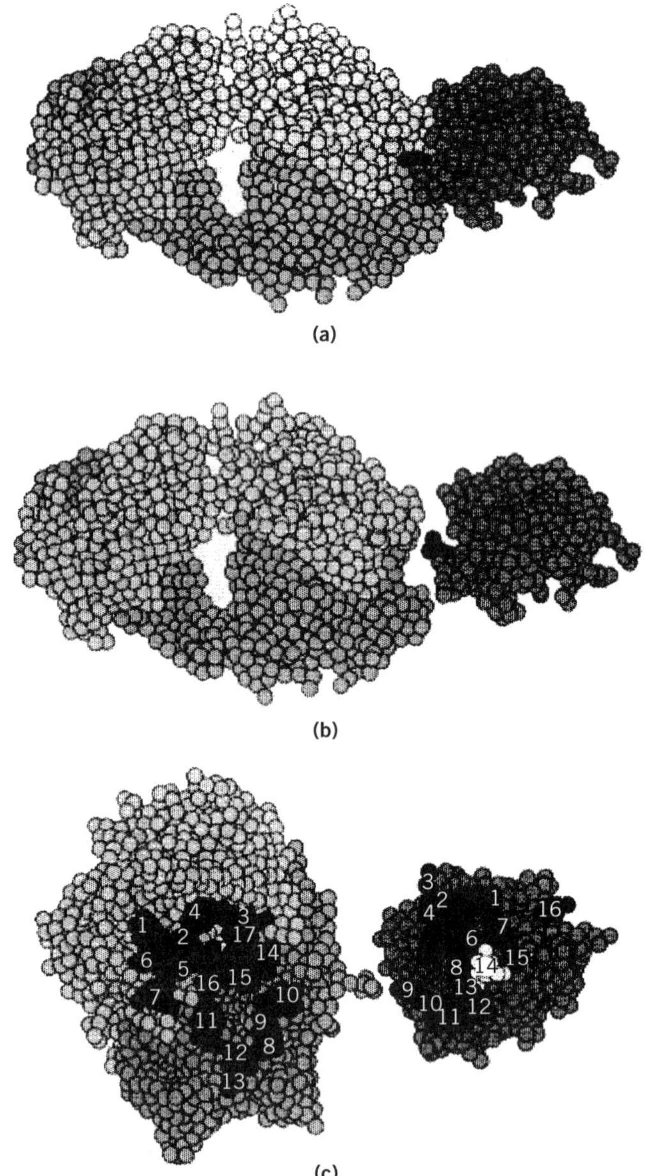

Figure 1. Three-dimensional structure of a lysozyme–Fab antilysozyme, showing the amino acid contributions from the heavy chain (upper part of Fab) and from the light chain (lower part). The antigen is on the right, and the Fab is on the left. (**a**, **b**) Side views. (**c**) Front view of the interacting residues of the antibody combining site (left) and epitope (right). (From Ref. 1, with permission from R. Poljak.)

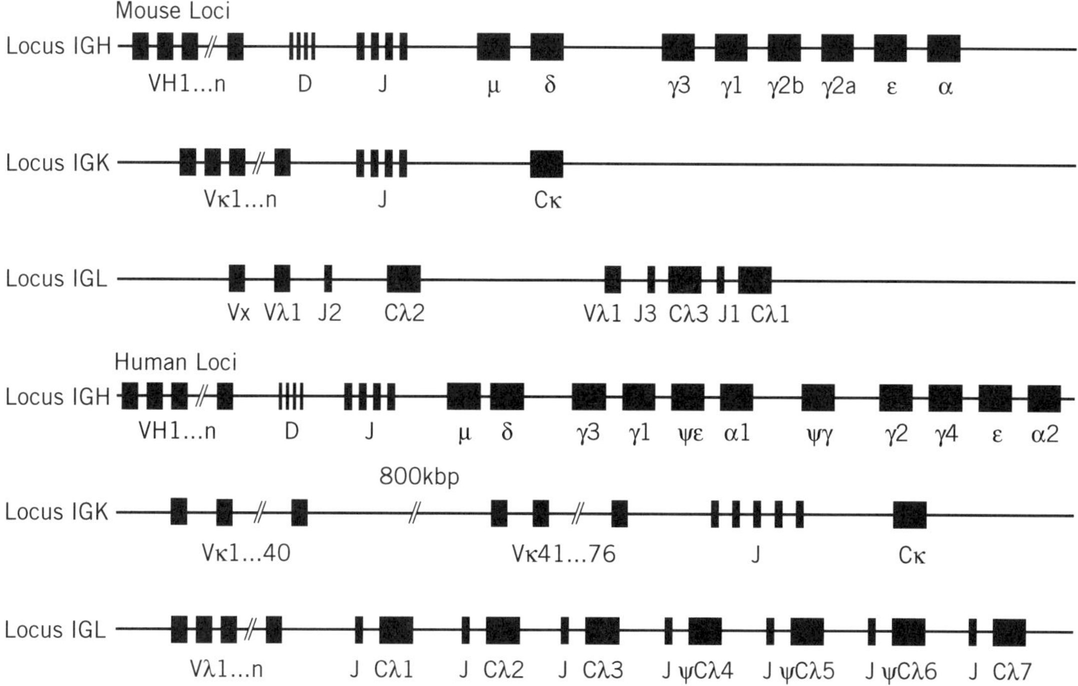

Figure 2. Schematic organization of the three Ig gene loci in (**a** mice and (**b**) humans. The IGK, IGL, and IGH loci, coding for the κ light chain, the λ light chain, and the heavy chain, respectively, are located on mouse chromosomes 6, 16, and 12 and on human chromosomes 2, 22, and 14, respectively.

dysbalanced synthesis between the monoclonal heavy and light chains by the malignant plasma cells. Because of their smaller size, they pass the renal barrier and are thus easily isolated in large amounts from patient's urine. The major discovery was that the Bence-Jones light chains, or Ig light chains, were composed of one NH_2-terminal half of 110 amino acid residues that profoundly differed between two patients, whereas the COOH-terminal half, also of 110 residues, was identical. This gave clear evidence for the existence of a huge structural diversity in antibodies, providing a unique basis for antibody specificity. Further extensive work indicated that heavy chains had also a variable region of similar size and a much longer constant region. The antibody combining site could thus be visualized as resulting from a contribution of both the V_H and the V_L regions. A more detailed analysis, computed by Kabat (7), indicated that within the V **variable regions**, subregions of hypervariability could be identified around positions 30, 50, and 100, which were later proven to participate directly in making up the antibody combining site. Three hypervariable regions, also called *complementarity determining regions* (CDR), were identified on each heavy and light chain, and their fine analysis revealed an extraordinary diversity that certainly could account for the expected huge repertoire of antibodies necessary to accommodate the potential repertoire of epitopes of the living world. Direct proof that the CDRs were implicated in antigen recognition came first from **affinity labeling** experiments, until a final confirmation was provide by **X-ray crystallography** analysis of crystals of antigen–antibody complexes. In fact, crystals were prepared from the Fab fragment of an antilysozyme antibody, combined with **lysozyme**, because the complete IgG antibody molecule contains a floppy "hinge" region that prevents crystallization. This structure, obtained by the group of Poljak (8), presented in Figure 1, indicates that about 20 amino acid residues of the antibody participate in binding the antigen, plus a similar number of residues from the antigen. On average, the area of interaction of the two partners is of the order of 600 Å^2. Care should be taken in attempts to generalize this model to all other antigen-antibody pairs, because the size and shape of the antibody combining site varies greatly from one system to another. This variation is reflected in the wide variation of the association constants K_A, between 10^5 and 10^{12} M^{-1}. The interactions between antigen and antibody are exclusively non covalent and involve primarily **salt bridges**, **van der Waals interactions**, and **hydrogen bonds**. The contributions of **enthalpy** and **entropy** vary immensely from one system to another, stressing again, if needed, the fantastic diversity potential of antibody molecules.

Ultimately, the main problem that the immune system has to face is how to generate such a huge diversity with a number of genes that must necessarily be limited. At the time of the first data concerning the amino acid sequences of the light chains, many hypotheses were put forward to account for this diversity. There were two extremes: One considered that the diversity was exclusively the result of **somatic hypermutations**, whereas the other claimed that everything was encoded at the germline level, arguing that, provided that any light chain might pair with any heavy chain, 10,000 L-chain genes and 10,000 H-chain genes might generate 10^8 antibodies, a number that was already considered reasonable. Besides the fact that 20,000 genes would represent a large portion of the **genome**, this theory did not account for the conservation of the constant regions, a very serious objection that led Dreyer and Bennet (9) to propose that genes encoding the V and the C regions were separate in the germline. This was shown to be the case when the basic principles of the Ig gene organization were elucidated by the elegant experiments of Tonegawa in 1978 (10). In brief, V and C regions are encoded by separate regions within each of the three H, K, and L Ig loci, with a small number of **C genes** and a large number of **V genes** (Fig. 2). In addition, the V regions are in fact encoded by a mosaic of two gene segment clusters for the light chains (V_L and J_L) and three for the heavy chains (V_H, D, and J_H). Random combination of these elements takes place exclusively during B-cell differentiation and leads to a large collection of clones expressing various combinations from the basic gene mosaic. Diversity is further amplified greatly by other mechanisms, including somatic hypermutation. As a result, the number of distinct B-cell clones present at any time certainly far exceeds that necessary, especially in view of antigen–antibody recognition being partly degenerate.

See also entries **B Cell**, **Immunoglobulin**, **Clonal selection theory**, **Gene rearrangement**, and **Repertoire**.

BIBLIOGRAPHY

1. M. F. Burnet (1959) *The Clonal Selection Theory of Acquired Immunity*. Vanderbilt University Press, Nashville, TN.
2. B. Köhler and C. Milstein (1975) Continuous culture of fused cells secreting antibody of predefined specificity. *Nature* **256**, 495–499.
3. J. B. Fleischman, J. B. Pain, and R. R. Porter (1962) Reduction of gammaglobulins. *Arch. Biochem. Biophys. Suppl.* **1**, 174–180.
4. G. M. Edelman and M. D. Poulik (1961) Studies on structural units of the γ-globulins. *J. Exp. Med.* **113**, 861–884.
5. R. R. Porter (1959) The hydrolysis of rabbit gammaglobulin and antibodies by cristalline papain. *Biochem. J.* **73**, 119–126.
6. N. Hilschmann and L. Craig (1965) Amino acid sequence studies with Bence-Jones proteins. *Proc. Natl. Acad. Sci.* **53**, 1403–1409.
7. T. T. Wu and E. A. Kabat (1970) An analysis of the sequences of the variable regions of the Bence-Jones proteins and mycloma light chains and their implications for antibody complementarity. *J. Exp. Med.* **132**, 211–250.
8. A. G. Amit, R. A. Mariuzza, S. E. V. Phillips, and R. J. Poljak (1986) Three-dimensional structure of an antigen–antibody complex at 2.8 A resolution. *Science* **233**, 747–753.
9. W. J. Dreyer and J. C. Bennett (1965) The molecular basis of antibody formation: a paradox. *Proc. Natl. Acad. Sci. USA* **54**, 864–869.
10. S. Tonegawa (1983) Somatic generation of antibody diversity. *Nature*, **302**, 575–581.

Suggestions for Further Reading

G. M. Edelman, B. A. Cunningham, W. E. Gall, P. D. Gottlieb, U. Rutishauser, and M. J. Waxdal (1969) The covalent structure of an entire gamma G immunoglobulin molecule. *Proc. Natl. Acad. Sci. USA* **63**, 78–85.

F. Alt, T. K. Blackwell, and G. D. Yancopoulos (1987) Development of the primary antibody repertoire. *Science* **238**, 1079–1087.

ANTIBODY–ANTIGEN INTERACTIONS

J. FOOTE

Binding of **antigens** has long been a central paradigm for molecular recognition. In addition, the biological nuances of antigen recognition have such profound ramifications for medicine that the study of antibody–antigen interactions remains a key branch of molecular immunology. In this entry we first describe the structural chemistry of antigen binding, then follow with aspects of antibody–antigen interaction that lead to unique biological phenomena.

CHEMICAL ASPECTS OF ANTIBODY–ANTIGEN INTERACTION

General Properties

The chemical interactions between **antibody** and antigen do not differ substantially from other protein–ligand interactions. **Hydrogen bonds**, **van der Waals interactions**, and sometimes **salt bridges** are used to form the antibody-antigen contact. Extremely close steric complementarity between antibody and antigen surfaces seems to be a common characteristic of interfaces (1). Gaps between the opposing antibody and antigen surfaces are sometimes filled by **water** molecules (2). The association constant for antibody–antigen interactions ranges from 10^5 to 10^{12} M^{-1}, with typical values for protein antigens around 10^8–10^9 M^{-1} (3). Rate constants for binding low molecular weight **haptens** can be as high as 10^8 $M^{-1}s^{-1}$. Reactions with macromolecular antigens are slower, $\leq 10^6$ $M^{-1}s^{-1}$, except for highly charged antigens, which sometimes enhance the rate through **electrostatic interactions** (4). In most cases, antigen binding does not cause easily observed changes in the binding site of an antibody. In some cases, conformational changes in the antibody are intrinsic to the binding mechanism. These changes can be caused subsequent to antigen binding, termed an "**induced fit**" mechanism (5); alternatively, antigen can bind selectively to one of several preexisting antibody conformations (6). Generalizations about the nature of the antibody–antigen interaction are hazardous, as is clear from the observation that the same antibody can bind one ligand through a few strong contacts and another ligand through many weak interactions (7).

Combining Site Structure and Function

Comparative studies showed that antibody variability was concentrated in three short stretches of sequence within each variable domain (8). These regions were postulated to confer antigenic specificity on an antibody molecule, and were termed "complementarity-determining residues" (CDRs). X-ray **crystallographic** studies confirmed that the CDRs, although noncontiguous in the antibody **primary structure**, formed loops that were adjacent to each other in three-dimensional space (9). The position of CDRs within an IgG molecule is shown in Figure 1. Structural studies on antibody–antigen complexes have confirmed that the CDRs form the vast majority of intermolecular contacts. Non-CDR residues can also form contacts, however, and in some cases not all CDRs participate in the interaction with antigen (10). Of the six CDRs, heavy chain CDR3, formed at the genetic level by joining of a D segment to V_H and J_H, is the most structurally diverse, and usually the most energetically significant in binding antigen. (See **Immunoglobulin structure**.)

Antigenicity

Antibodies bind to structurally precise surfaces on protein antigens (11). These surface areas can be composed of a contiguous length of polypeptide chain or from several parts of a chain that are separate in primary structure but adjacent

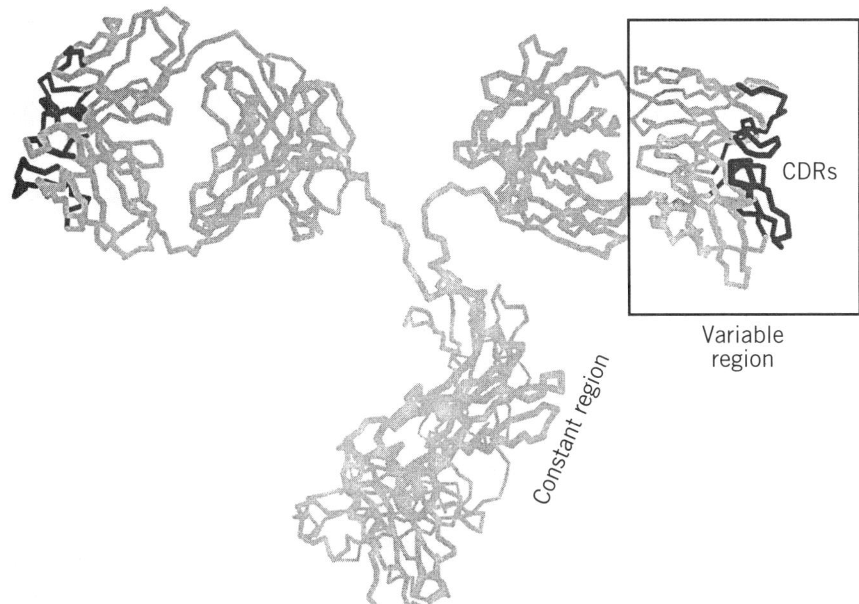

Figure 1. Position of CDRs (dark lines) within an IgG molecule.

in three-dimensional space. The only true prerequisite for a portion of a protein to be recognized as an antigen is surface **accessibility**, although other factors, such as **hydrophilicity** and mobility, are often considered in predicting antigenicity from protein sequence (12). In the case of peptide–antibody complexes, about 7–10 peptide residues fit in the binding site of the antibody and form an ordered structure, even if the free peptide itself is not strongly ordered in solution (13). Sometimes a bound peptide structure resembles the conformation of the same sequence in an intact protein, a phenomenon that underlies the utility of peptide vaccines (14).

BIOLOGICAL ASPECTS OF ANTIBODY–ANTIGEN INTERACTION

T-Dependent and T-Independent Immune Responses

"Normal" immune responses depend on the participation of **T cells**. In outline, antibody-producing B cells capture antigen (protein or protein–hapten conjugate) on their surface, internalize and fragment the antigen, and represent short peptide fragments on the cell surface, embedded in the binding groove of **major histocompatibility complex (MHC)** molecules (see **Antigen presentation, processing**). T-cells recognize the peptide–MHC combination and produce signals that cause **B cells** to enter pathways of proliferation and differentiation (15). Molecular genetic processes activated at the immunoglobulin loci include heavy-chain **class switching** and **somatic hypermutation**, leading to production of soluble **IgG**, **IgE**, and **IgA** of high affinity. T-independent antigens include virtually all nonprotein macromolecules. B-cell proliferation and differentiation also occur in a T-independent response, but chain switching and somatic mutation are difficult to detect. Structural attributes of antibodies in a T-independent response are that they are of the **IgM** isotype, contain germline (unmutated) **variable region** sequences, and show low antigen affinity.

Maturation of the Immune Response

Antibodies isolated soon after an initial antigen exposure, termed "primary response antibodies", differ from those obtained later in the response or after a second administration of antigen. This transformation of the antibody **repertoire**, which leads to progressive increases in the affinity for antigen (16), is termed "maturation of the immune response." The structural basis of this phenomenon has been determined from studies of immune responses to haptens, and can be outlined as follows. Contact with antigen induces a process within lymphocytes that introduces point mutations in the variable regions of antibody genes (17). Most mutations have a neutral or deleterious effect on antigen affinity (18). However, some mutations improve the interaction with antigen. Higher affinity confers a selective advantage on lymphocytes that express this mutation, which may be competing with nearby lymphocytes for a limited amount of antigen. Selected lymphocytes proliferate and can undergo further rounds of mutation and selection. Maturation of the immune response leads to improved affinity universally in anti-hapten responses. The same progressive affinity increase is presumed to occur with protein antigens, but unequivocal evidence for this is lacking.

The diversity of the initial repertoire is determined by germline variable gene diversity and processes involved in **gene rearrangement**. At a structural level, antibody diversity is concentrated around the center of the antigen combining region. Somatic mutation acts to form the mature repertoire, which for protein antigens will show association constants in excess of $10^8\ M^{-1}$. Somatic point mutations can occur anywhere in the variable region, but those selected during maturation are in general located in a band immediately peripheral to the area of diversity expressed in the early repertoire (19).

Anti-idiotype Antibodies

"**Idiotype**" is an immunological word that corresponds in structural terms to the unique CDRs of an individual antibody. If an antibody (Ab1) is used to **immunize** an animal, its CDRs can be recognized as a foreign structure, even in an animal of the same species. An antibody that reacts with the CDRs of Ab1 will be made, termed an **anti-idiotype** antibody (Ab2). The anti-idiotype can be used in the same way to generate an anti-anti-idiotype (Ab3). In a theory of immune regulation (20), if Ab1 is complementary to an antigen and Ab2 is complementary to Ab1, then the binding site of Ab2 should resemble the initiating epitope on the antigen. ***In vivo***, Ab2 is then said to carry an "internal image" of the antigen. This chain of alternating complementary recognition properties is diagrammed in Figure 2).

Structural studies have investigated the extent to which the internal image model is accurate. No gross structural similarity between an antigen and Ab2 is necessary for both molecules to bind to Ab1 with high affinity. However, the structure of an antilysozyme (Ab1) complexed with an Ab2 showed that many of the Ab1 residues used for binding lysozyme were also used to bind Ab2 (21). Furthermore, many Ab2 atoms contacting Ab1 occupied positions analogous to contact atoms in lysozyme, and intermolecular hydrogen bonds and even water molecules at the interface were conserved between the Ab1:lysozyme and Ab1:Ab2 complexes. Thus Ab2 can mimic an antigen through close chemical complementarity with Ab1, even in the absence of sequence or structural homology. In the case of Ab3 molecules, structural similarity to Ab1 is unequivocal, although not surprising. CDR sequences are often nearly identical between Ab1 and Ab3, and the structure of an antigen:Ab3 complex shows chemical complementarity as stringent as would be expected at the interface between antigen and an antibody raised directly (22).

Heterophile Antibodies and Molecular Mimicry

An antibody that reacts with an antigen to which the host has not been exposed is termed a "heterophile antibody" (23). The origin of heterophile antibodies is obscure; no antigen has been unambiguously identified that in a natural situation causes heterophile antibodies to appear. Structurally, heterophile antibodies tend to be low affinity IgM that react with carbohydrate antigens. Some heterophile antibodies, termed "natural antibodies," appear to arise spontaneously and persist for the lifetime of the organism. The heterophile antibodies that determine ABO blood group compatibility are of this type. In other cases, appearance of heterophile antibodies clearly results from infection. For example, infectious mononucleosis in humans reproducibly induces antibodies specific for an antigen on sheep red blood cells (24). The supposition in such cases is of a process of "molecular mimicry": a pathogen-specified product is made during infection that possesses sufficient structural similarity to the heterophile antigen that heterophile antibodies are induced (25). When the reactivity induced is to a self-antigen normally subject to immune **tolerance** mechanisms, **autoimmune disease** may result. For example, streptococcal respiratory infection is the immediate antecedent of rheumatic fever, an autoimmune attack on the heart and vascular tissue. A carbohydrate structure on the bacterial surface is thought to mimic structural glycoproteins in heart valves and vessels (26). Experimental induction of autoimmune diseases by immunization supports this type of antibody–antigen interaction as an initiating mechanism for autoimmunity (27).

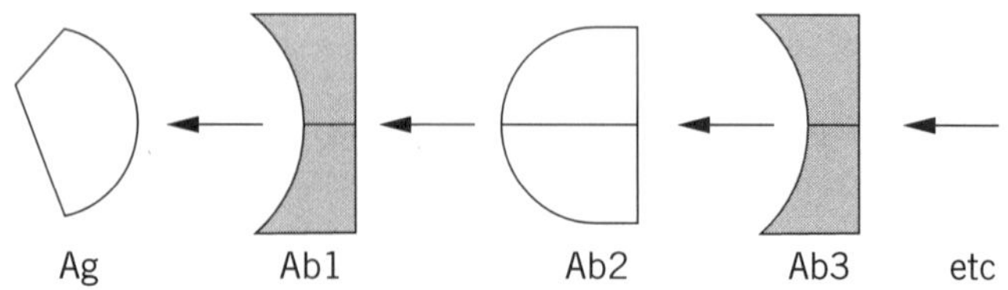

Figure 2. Chain of complementary binding activities within an idiotype network.

Polyreactive or Polyspecific Antibodies

Polyreactive antibodies do not fit the standard paradigm of one antibody recognizing a single antigen. Instead, a single polyreactive antibody can recognize a panel of many antigens that bear no obvious chemical similarity to each other. Many polyreactive antibodies are IgM, and others are IgG and other isotypes and can bind antigens with high affinity. For example, a set of **hybridomas** screened for simultaneous reactivity with **tubulin**, **actin**, myosin, and single- and double-stranded DNA showed a median antibody affinity for all five antigens of $10^7\ M^{-1}$ (28). The molecular basis of polyreactivity is unknown. Numerous **V genes** have appeared in polyreactive antibodies, and there is as yet no characteristic sequence motif that predicts polyreactivity. Nevertheless, sequence studies implicated heavy-chain CDR3 as a major determinant of polyreactivity (29). Structure–function studies comparing a polyreactive and monoreactive anti-insulin with similar sequences confirmed that polyreactivity was associated with heavy-chain CDR3 (30). This CDR loop is presumed to adopt multiple conformations to accommodate different antigens, but crystallographic data on the three-dimensional structure of polyreactive antibodies are lacking.

BIBLIOGRAPHY

1. A. G. Amit, R. A. Mariuzza, S. E. V. Phillips, and R. J. Poljak (1986) *Science* **233**, 747–753.
2. T. N. Bhat, G. A. Bentley, G. Boulot, M. I. Green, D. Tello, W. Dall'Acqua, H. Souchon, F. P. Schwarz, R. A. Mariuzza, and R. J. Poljak (1994) *Proc. Natl. Acad. Sci. USA* **91**, 1089–1093.
3. I. Pecht (1982) Dynamic aspects of antibody function, in *The Antigens* (M. Sela, ed.), Academic Press, New York, Vol. 6, pp. 1–68.
4. C. S. Raman, R. Jemmerson, B. T. Nall, and M. J. Allen (1992) *Biochemistry* **31**, 10370–10379.
5. J. M. Rini, U. Schulze-Gahmen, and I. A. Wilson (1992) *Science* **255**, 959–965.
6. D. Lancet and I. Pecht (1976) *Proc. Natl. Acad. Sci. USA* **73**, 3548–3553.
7. W. Dall'Acqua, E. R. Goldman, E. Eisenstein, and R. A. Mariuzza (1996) *Biochemistry* **35**, 9667–9676.
8. T. T. Wu and E. A. Kabat (1970) *J. Exp. Med.* **132**, 211–250.
9. R. J. Poljak, L. M. Amzel, H. P. Avey, B. L. Chen, R. P. Phizackerly, and F. Saul (1973) *Proc. Natl. Acad. Sci. USA* **70**, 3305–3310.
10. D. R. Davies, E. A. Padlan, and S. Sheriff (1990) *Annu. Rev. Biochem.* **59**, 439–473.
11. D. C. Benjamin, J. A. Berzofsky, I. J. East, F. R. N. Gurd, C. Hannum, S. J. Leach, E. Margoliash, J. G. Michael, A. Miller, E. M. Prager, M. Reichlin, E. E. Sercarz, S. J. Smith-Gill, P. E. Todd, and A. C. Wilson (1984) *Annu. Rev. Immunol.* **2**, 67–101.
12. J. A. Berzofsky (1985) *Science* **229**, 932–940.

13. R. Stanfield and I. A. Wilson (1995) *Curr. Opin. Struct. Biol.* **5**, 103–113.
14. R. A. Lerner (1982) *Nature* **299**, 592–596.
15. A. Lanzavecchia (1985) *Nature* **314**, 537–539.
16. H. N. Eisen and G. W. Siskind (1964) *Biochemistry* **3**, 996–1008.
17. M. G. Weigert, I. M. Cesari, S. J. Yonkovich, and M. Cohn (1970) *Nature* **228**, 1045–1047.
18. C. Chen, V. A. Roberts, and M. B. Rittenberg (1992) *J. Exp. Med.* **176**, 855–866.
19. I. M. Tomlinson, G. Walter, P. T. Jones, P. H. Dear, E. L. L. Sonnhammer, and G. Winter (1996) *J. Mol. Biol.* **256**, 813–817.
20. N. Jerne (1974) *Ann. Inst. Pasteur (Immunol)* **125C**, 373–389.
21. B. A. Fields, F. A. Goldbaum, X. Ysern, R. J. Poljak, and R. A. Mariuzza (1995) *Nature* **374**, 739–742.
22. K. C. Garcia, S. V. Desiderio, P. M. Ronco, P. J. Verroust, and L. M. Amzel (1992) *Science* **257**, 528–531.
23. J. Forssmann (1911) *Biochem. Z.* **37**, 78–115.
24. J. R. Paul and W. W. Bunnell (1932) *Am. J. Med. Sci.* **183**, 90–104.
25. R. T. Damian (1964) *Am. Naturalist* **98**, 129–149.
26. J. B. Zabriskie, K. C. Hsu, and B. C. Seegal (1970) *Clin. Exp. Immunol.* **7**, 147–159.
27. R. S. Fujinami and M. B. A. Oldstone (1985) *Science* **230**, 1043–1045.
28. L. Diaw, C. Magnac, O. Pritsch, M. Buckle, P. M. Alzari, and G. Dighero (1997) *J. Immunol.* **158**, 968–976.
29. C. Chen, M. P. Stenzel-Poore, and M. B. Rittenberg (1991) *J. Immunol.* **147**, 2359–2367.
30. Y. Ichiyoshi and P. Casali (1994) *J. Exp. Med.* **180**, 885–895.

Suggestions for Further Reading

E. A. Padlan (1994) Anatomy of the antibody molecule, *Mol. Immunol.* **31**, 169–217. (A superb, comprehensive review of antibody structure and function.)

FASEB J. **9**, 1–147 (1994). An entire issue devoted to structural immunology, including reviews and research papers.

Immunol. Rev. **163**, "Molecular anatomy of the immune response" (1998). (An entire volume of reviews of structural immunology.)

ANTIBODY-CONJUGATED TOXINS

C. Montecucco

Soon after the discovery of the high specificity of **antibody** binding, Paul Ehrlich proposed to exploit such specificity to redirect and restrict the activity of **toxins** to pathological tissues such as tumors (1–3). Such "magic bullets" could be developed only long after his proposal because of the need of antibodies of well-defined specificity, such as **monoclonal antibodies**, and to improve the understanding of the biochemistry and mechanism of the action of toxins. Such conjugates have been prepared by linking one of the many available cell-killing plant or bacterial toxins via a **disulfide** or thioether bond to an antibody specific for a cell surface antigen of cancer and noncancer cells. Chimeric toxins have also been prepared by fusing a toxin gene to a gene encoding a **hormone**, **growth factor**, or **cytokine** with the intention of depleting receptor-rich cell populations (1–3). In general, it is important that the linkage between the two immunotoxin parts be easily cleaved on or inside cells. Moreover, to restrict immunotoxin binding only to the cells to be intoxicated, the portion of the toxin determining its intrinsic specificity (the B subunit) must be either deleted or inactivated mutationally, and the F_c portion of the antibody must be removed. In this respect, **ribosome**-inactivating plant toxins lacking the B subunit have been frequently used. To avoid unspecific binding to F_c-receptor bearing cells, the toxin can be linked to the $F(\mathrm{ab})_2$ portion of a monoclonal antibody. An additional problem may be presented by the presence of circulating antitoxin antibodies elicited in previous vaccination protocols (eg, anti**diphtheria**) or during the treatment with the immunotoxin. In the latter case, one can use another immunotoxin made with the same antibody and an immunologically unrelated toxin.

Several antibody-conjugated toxins are very effective in killing target cells in culture or in isolated dispersed tissues, ie, bone marrow, but are less effective *in vivo*, because of the lack of accessibility of the cancer cells in solid tumors. However, protocols are being developed to overcome these problems and exploit the potentials of immunotoxins to reach small cancer populations that are undetected by physical and surgical methods (3).

BIBLIOGRAPHY

1. S. Olsnes, K. Sandvig, O. W. Petersen, and B. van Deurs (1989) *Immunol. Today* **10**, 291–295.
2. D. A. Vallera (1994) *Blood* **83**, 309–317.
3. G. R. Trush, L. R. Lark, and E. S. Vitetta (1996) *Ann. Rev. Immunol.* **14**, 49–71.

ANTICODON

C. Florentz
N. C. Martin

The anticodon of **transfer RNA** (tRNA) is in the central part of the linear RNA sequence, at positions 34, 35, and 36. This triplet of nucleotides forms transitory base pairs with three nucleotide codons of the **messenger** RNA (mRNA) during **protein biosynthesis**. These codon/anticodon interactions enable the mRNA to direct the order of incorporation of **amino acids** into the **polypeptide chain**. These interactions occur on the small subunit of the ribosome. The presence of the anticodon in the seven-membered central loop of tRNA led to the designation of this subdomain as the anticodon loop. The anticodon stem plus the loop constitute the anticodon arm (stem and loop), which contributes to one branch of the L-shaped 3D structure of tRNA. In this structure, the anticodon is about 80 Å from the amino acid acceptor end of the tRNA (see Fig. 1 of **Transfer RNA**). Besides its fundamental contribution to translation of the genetic information through complementary pairing to the codon, the anticodon of tRNA may contain important identity elements, recognized by cognate **aminoacyl-tRNA synthetases**, which lead to specific aminoacylation of the tRNA. Besides the anticodon, other members of the loop and of the nearby stem, including modified nucleotides, also contribute both to the efficiency and specificity of aminoacylation.

Codon–anticodon interaction occurs basically through classic **Watson–Crick base pairs**. However, additional types of interactions allow **wobble** to occur between the third base of the codon and the first base of the anticodon (1). The original wobble rules suggested that the first nucleoside of the anticodon can pair with more than one nucleoside at the third position of the codon. Thus, anticodons with a U at the first position could interact with codons having either A or G at the third position. Those presenting a G at position 34 could interact with codons terminating with U or C. More interestingly, tRNA presenting an inosine (deaminated adenosine) at position 34 could recognize codons terminating with either C, U, or A. For example, yeast $tRNA^{Ala}$, anticodon 5′-IGC-3′, interacts with three codons: 5′-GCC-3′, 5′-GCU-3′, and 5′-GCA-3′.

Numerous data accumulated over the years led to revised wobble rules (2,3), which reflect new possible interactions between classic bases and take into account the vast occurrence of modified nucleosides in the anticodons of tRNA, especially those at the wobble position (residue 34, first position of the anticodon). One minor tRNA, *Escherichia coli*, $tRNA^{Ile}2$, has a methionine anticodon CAU. However, in the mature tRNA, C34 is modified by covalent attachment of lysine on position 2 of the pyrimidine ring, converting it to lysidine or a "k2C" base. Lysidine at the wobble position leads to recognition of the isoleucine-specific codon 5′-AUA-3′ instead of the methionine codon AUG. Interestingly, this modification is also required for specific recognition of the tRNA by isoleucyl-tRNA synthetase (see discussion below). Other examples of modification at position 34 that contribute to the specificity of codon–anticodon recognition have been reported. This is also true for nucleotide 37, at the 3′ side of the anticodon triplet. Many different modified purine nucleosides have been identified at this position, and these modifications contribute to the fidelity of protein synthesis. For example, hypermodification of A to i6A (*N*-6-isopentenyladenosine), t6A (*N*-6-threonylcarbamoyladenosine), and their derivatives, stabilizes the relatively weak A-U and U-A base pairing between the third position of the anticodon and the first position of the codon. The presence of the hypermodified nucleoside ms2i6A (*N*-6-(D-2-isopentenyl)-2-methylthioadenosine), in contrast, prevents codon misreading by *E. coli* $tRNA^{Phe}$. Modifications introduce conformational flexibility or rigidity that restrict or enlarge the number of potential base pairs. Thus the molecular mechanism by which modified bases alter codon recognition are largely structural in nature.

UAG (**amber**), UAA (**ocher**), and UGA (opal) codons do not code for an amino acid, because there are no tRNA with corresponding anticodons. Known as **nonsense codons**, they normally signal termination of **translation**, but such codons can be created by **mutation**, when they cause premature termination of protein synthesis. Interestingly, nonsense **suppressor** tRNA, with mutations in the anticodon, can recognize and "suppress" nonsense mutations by inserting a specific amino acid during translation (4). The first suppressor tRNA identified was a tyrosine-specific tRNA in which a single base substitution converted the anticodon from GUA (Tyr) to CUA (amber). Conversion of a classic tRNA to an efficient suppressor RNA may also involve mutations outside the anticodon. Suppressor tRNA have been used as a tool in the search of tRNA identity elements *in vivo* and as a way of inserting desired amino acids at specific sites in proteins.

The anticodon triplet specifies which amino acid the tRNA will insert in response to a codon and is, of course, directly correlated with the amino acid bound to the -CCA end. A simple analysis of the **genetic code**, however, reveals that the anticodon is not a common signal for synthetase recognition. For example, there are six different serine codons and, consequently, the potential for six different anticodon sequences in serine-specific tRNA. This great variability in isoacceptor tRNA precludes a common signal in the anticodon for recognition by seryl-tRNA synthetase, but it is clear that anticodon nucleotides, as well as other nucleotides within the anticodon loop, are critical for which amino acids are charged by aminoacylation (5–10).

Single-point mutations at any anticodon nucleotide of yeast $tRNA^{Phe}$ or yeast $tRNA^{Asp}$ lead to severe losses in aminoacylation efficiency (11,12). A simple anticodon switch changes the aminoacylation specificity of methionine and valine tRNA (13). Depending on the tRNA, one, two, or three anticodon nucleotides may be involved in tRNA specificity. Position 35 is required for charging of arginine, both 35 and 36 are important for valine or threonine, and 34, 35, and 36 are involved in specificity for Asn, Asp, Cys, Gln, Ile, Lys, Met, Phe, Trp, and Tyr. In *E. coli* $tRNA^{Gln}$, residue 35 is of greater importance than residues 34 and 36. In yeast $tRNA^{Asp}$, nucleotides 34 and 35 contribute more than do nucleotide 36. Transplantation of anticodons into noncognate tRNA results in increases of four to six orders of magnitude in aminoacylation of the chimeric tRNA with the amino acid corresponding to the transplanted anticodon. Since it does not preclude aminoacylation of the chimeric tRNA with the original amino acid, however, it is clear that additional identity elements elsewhere in the tRNA structure dictate amino acid specificity (see **Transfer RNA**). The extent and the relative contribution of the different nucleotides of the anticodon to this specificity varies. The losses in specificity for yeast $tRNA^{Asp}$ are about 500-fold, while those for *E. coli* $tRNA^{Val}$ are 100,000-fold.

Other nucleotides within the anticodon domain often contribute to amino acid specificity. Residue 38 is involved in yeast $tRNA^{Asp}$, and both residues A37 and U38 are involved in *E. coli* $tRNA^{Gln}$. An original approach to anticodon domain function uses the *in vitro* synthesis of minihelices (shortened forms of tRNA) mimicking isolated anticodon stem and loop domains. These minihelices possess identity elements and were tested for stimulation of aminoacylation rates of the acceptor stem as substrate and/or for inhibition of full-length tRNA charging (14). Yeast valyl-tRNA (15) and *E. coli* isoleucine-tRNA (16) anticodon minihelices stimulated, up to threefold, the aminoacylation of an acceptor arm minihelix by cognate aminoacyl-tRNA synthetase. The isolated anticodon domain of $tRNA^{fMet}$ binds to *E. coli* methionyl-tRNA synthetase (17).

The anticodon domain is further implicated in discrimination between initiator and elongator forms of tRNA (18) and in alternate functions of tRNA, such as initiation of **reverse transcription** in **retroviruses** (19) (see **Transfer RNA**).

BIBLIOGRAPHY

1. F. H. C. Crick (1966) *J. Mol. Biol.* **19**, 548–555.
2. S. Yokoyama and S. Nishimura (1995) in *tRNA: Structure, Biosynthesis, and Function* (D. Söll and U. L. RajBhandary, eds.), American Society for Microbiology, Washington, DC, pp. 207–223.
3. K. Watanabe and S. Osawa (1995) in *tRNA: Structure, Biosynthesis, and Function* (D. Söll and U. L. RajBhandary, eds.), American Society for Microbiology, Washington, DC, pp. 225–250.

4. H. Ozeki, H. Inokuchi, F. K. M. Yamao, H. Sakano, T. Ikemura, and Y. Shimura (1980) in *Transfer RNA: Biological Aspects* (D. Söll, J. Abelson, and P. Schimmel, eds.), Cold Spring Harbor Laboratory, New York, pp. 341–362.
5. L. Kisselev (1985) *Prog. Nucleic Acid Res. Mol. Biol.* **32**, 237–266.
6. L. Pallanck, M. Pak, and L.-H. Schulma (1995) in *tRNA: Structure, Biosynthesis, and Function* (D. Söll and U. L. RajBhandary, eds.), American Society for Microbiology, Washington DC, pp. 371–394.
7. R. Giegé, J. D. Puglisi, and C. Florentz (1993) *Prog. Nucleic Acid Res. Mol. Biol.* **45**, 129–206.
8. W. H. McClain (1993) *J. Mol. Biol.* **234**, 257–280.
9. M. E. Saks, J. R. Sampson, and J. N. Abelson (1994) *Science* **263**, 191–197.
10. W. H. McClain (1995) in *tRNA: Structure, Biosynthesis, and Function* (D. Söll and U. L. RajBhandary, eds.), American Society for Microbiology Press, Washington, DC, pp. 335–347.
11. J. R. Sampson, A. B. DiRenzo, L. S. Behlen, and O. C. Uhlenbeck (1989) *Science* **243**, 1363–1366.
12. J. Putz, J. D. Puglisi, C. Florentz, and R. Giegé (1991) *Science* **252**, 1696–1699.
13. L. H. Schulman and H. Pelka (1988) *Science* **242**, 765–768.
14. S. A. Martinis and P. Schimmel (1995) in *tRNA: Structure, Biosynthesis, and Function* (D. Söll and U. L. RajBhandary, eds.), American Society for Microbiology Press, Washington, DC, pp. 349–370.
15. M. Frugier, C. Florentz, and R. Giegé (1992) *Proc. Natl. Acad. Sci. USA* **89**(9), 3990–3994.
16. O. Nureki, T. Niimi, T. Muramatsu, H. Kanno, T. Kohno, C. Florentz, R. Giegé, and S. Yokoyama (1994) *J. Mol. Biol.* **236**, 710–724.
17. T. Meinnel, Y. Mechulam, S. Blanquet, and G. Fayat (1991) *J. Mol. Biol.* **220**, 205–208.
18. N. Mandal, D. Mangroo, J. J. Dalluge, J. A. McCloskey, and U. L. RajBhandary (1996) *RNA* **2**, 473–482.
19. C. Isel, J.-M. Lanchy, S. F. J. Le Grice, C. Ehresmann, B. Ehresmann, and R. Marquet (1996) *EMBO J.* **15**, 917–924.

ANTI CONFORMATION

VERNON ANDERSON

Anti is used as a prefix to describe the relative orientation of two substituents on adjacent atoms of a molecule when their **torsion angle** is about 180° (1) (see **Conformation**). In addition to its formal usage, *anti* has become broadly used to describe the orientation of the **nucleic acid** bases relative to the ribose ring when describing the torsion angle about the glycosidic bond (2). The glycosidic bond torsion is defined by the relative positions of the four atoms surrounding this bond,

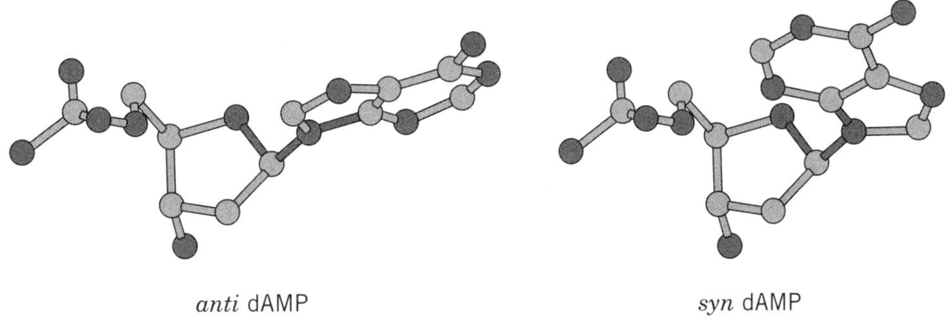

Figure 1. The *anti* and *syn* conformations of dAMP are shown. The bonds defining the glycosidic torsion angle are darkened for emphasis in both conformers. In the *anti* conformer, the torsion is 110°, and in the *syn* conformer the torsion angle is +60.

O4′—C1—N9(1*—C4*(2)) for the purine (pyrimidine) nucleosides (3). When C4 is not over the ribose ring, the conformation is *anti*, corresponding to torsion angles of −80 to −180 and 100 to 180°. Conformations where the purine ring is over the sugar are *syn* and correspond to torsion angles of −80 to 100°. These conformations for adenine are shown in Figure 1.

BIBLIOGRAPHY

1. W. Klyne and V. Prelog (1960) *Experientia* **16**, 521.
2. A. E. V. Haschemeyer and A. Rich (1967) *J. Mol. Biol.* **27**, 369–384.
3. IUPAC-IUB Commission on Biochemical Nomenclature (1983) *Eur. J. Biochem.* **17**, 193–201.

Suggestions for Further Reading

C. R. Cantor and P. R. Schimmel (1980) *Biophysical Chemistry, Part I: The Conformation of Biological Macromolecules*, W. H. Freeman, San Francisco, CA.

E. L. Eliel et al. (1967) *Conformational Analysis*, Wiley-Interscience, New York.

W. Saenger (1984) *Principles of Nucleic Acid Structure*. Springer Advanced Texts in Chemistry (C. R. Cantor, ed.), Springer-Verlag, New York.

ANTIFREEZE PROTEINS

PETER L. DAVIES

Antifreeze proteins (AFP) adsorb specifically to the surface of ice. By doing so, they inhibit its growth and cause a local, nonequilibrium depression of the freezing point below the colligative melting/freezing point (1). This depression, which is called thermal hysteresis (°C), helps organisms to avoid or resist freezing when they encounter ice at temperatures below their colligative freezing points. For example, in polar and cold temperate marine environments bony fishes (teleosts) can encounter icy seawater at its freezing point of ~ −1.9°C, over

1°C colder than their colligative freezing point of ~ −0.7°C. Those fishes that have adapted to this environment produce high concentrations of AFP in their blood (10 to 25 mg/ml) that depress their freezing point by the crucial 1°C needed for survival. For practical reasons, AFP (sometimes known as thermal hysteresis proteins) have been best characterized in fish, but have also been found in insects, plants, and bacteria that encounter subzero conditions. Not all of these organisms resist freezing. In those that do freeze, AFP may enhance their freeze tolerance by keeping ice crystals from growing (recrystallizing) in the frozen state (2). Recrystallization of ice occurs as the melting temperature is approached and ice becomes more fluid. It is effectively inhibited by very low AFP concentrations (as low as 10 μg/ml). Although there are reports that AFP neutralizes ice nucleators (3) and has protective effects on cells above 0°C (4), the two well-established functions of macromolecular antifreezes, thermal hysteresis and inhibition of ice recrystallization, are related activities that depend on AFP interaction with ice by absorption-inhibition.

INHIBITION OF ICE GROWTH

In ice, each **water** molecule makes four tetrahedrally oriented **hydrogen bonds** to neighboring molecules, and the resulting ice lattice has an underlying hexagonal symmetry. When an aqueous solution is cooled in the presence of an ice crystal, water adds preferably to the atomically rough prismatic surfaces of the crystal much faster than to the two smoother basal planes. At a microscopic level, the prismatic surfaces are curved, and the ice grows radially as a disk (Fig. 1). Most impurities are swept aside by the expanding ice front. AFP have a specific affinity, however, for an ice surface plane and become adsorbed, inhibiting growth at that spot and typically shaping the ice crystal into a hexagonal bipyramid. The reason for growth inhibition at the surface of ice is that water is forced to add to the lattice between the bound AFP (1,5). These submicroscopic ice fronts are constrained to grow with a surface curvature that makes it harder for water to join the lattice, which in turn leads to a local depression of the freezing point below that of the bulk solvent. Rather than completely covering the surface, AFP have been likened to buttons on a mattress.

The relationship between thermal hysteresis and AFP concentration is non linear. Thermal hysteresis increases hyperbolically with increasing AFP concentration and approaches a plateau value of between 1 to 1.5°C for fish AFP. If the temperature falls below the nonequilibrium freezing point at moderate to high AFP concentrations, the ice crystal grows uncontrollably. At low AFP concentrations and moderate undercooling, the ice front overgrows bound AFP. This phenomenon has been used to deduce the ice surface to which AFP bind by growing ice in a dilute AFP solution and then allowing it to sublime (ice etching) (5). The orientation of the AFP residue on the ice indicates its binding plane, which is expressed in crystallographic terms as a Miller index (see below).

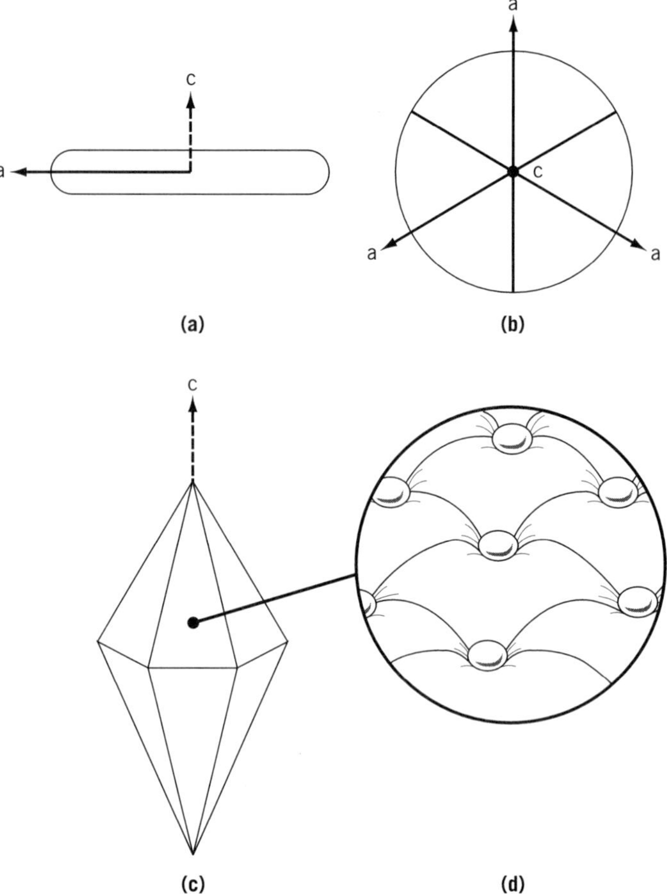

Figure 1. Structural models for the inhibition of ice growth by fish AFP. a) In the absence of AFP, a single ice crystal from the melt grows fastest along the crystallographic a axes to form a disk with curved prismatic surfaces. b) View of disk perpendicular to a) with c axis perpendicular to the page. c) AFP adsorption to any aspect of the prismatic surface forces the crystal to become a hexagonal bipyramid. d) The inset shows a section of the surface illustrating the submicroscopic ice front curvature between bound AFP.

ADSORPTION OF AFP TO ICE

One of the most intriguing features of AFP is their structural diversity. In fish, there are at least four unrelated types (Fig. 2). A fifth one has been recently described (6), and it is likely that additional types will be discovered. They are so different that it has not been possible to identify a shared structural motif to explain their common ice-binding activity. An added complication is that they bind to different ice surface planes (7). One principle that they have in common is the ability to make a specific hydrogen-bonding match between the protein surface and water in the ice lattice. The **alpha-helical** type I AFP is an example where **threonine** residues repeated every 11 amino acids are aligned along the same face of the helix, spaced 16.5 Å apart. This spacing matches the distance between exposed oxygen atoms on the 20−21 pyramidal plane in the direction ⟨01−12⟩ that has been established by ice etching as the orientation of type I AFP binding to ice (5). A different, but equally nonrandom hydrogen-bonding match is made between the small globular type III AFP and the prism plane of ice 10−10 (8). It seems unlikely that these few hydrogen bonds are sufficient to bind AFP irreversibly to ice. A second binding principle might be a closely contoured fit to the ice surface that allows **van der Waals attractions** to strengthen the binding. Solvation effects may also contribute to the energetics of binding (9)

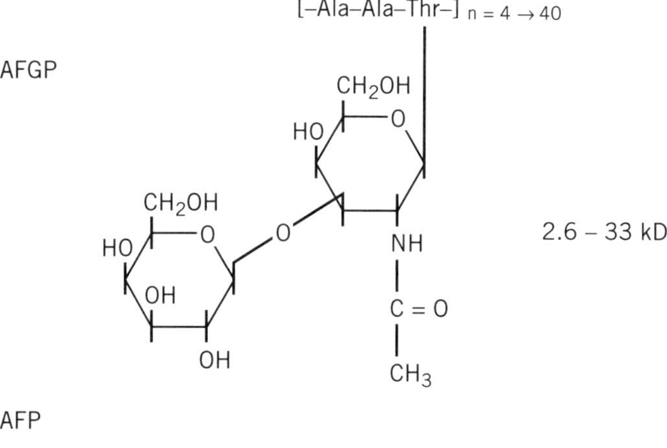

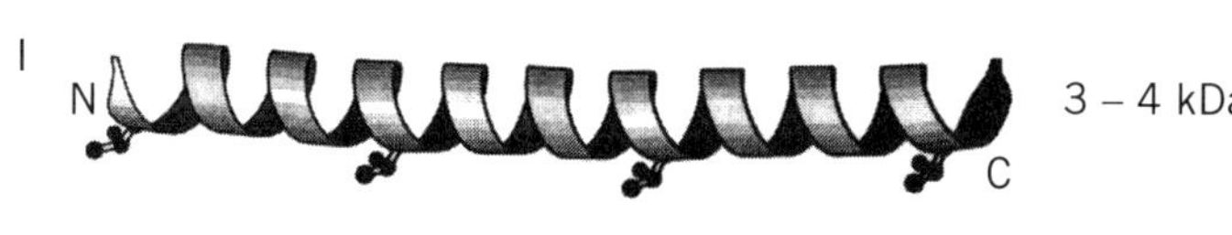

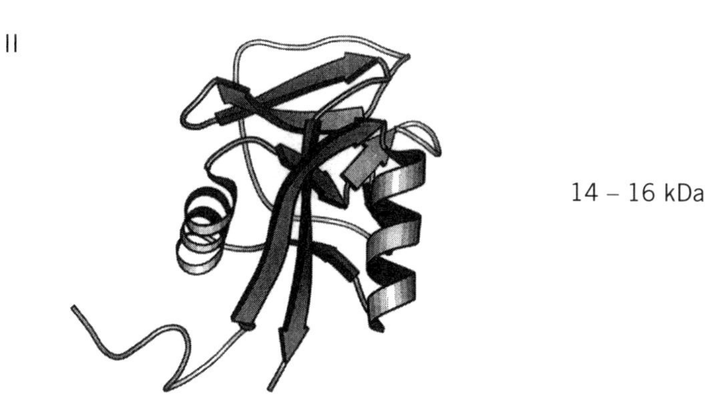

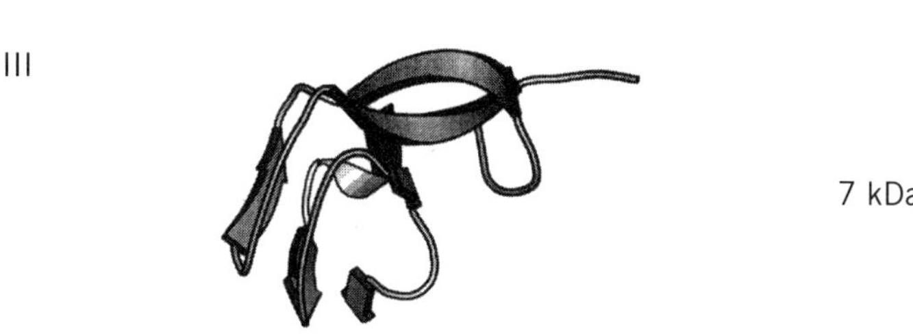

Figure 2. Representative structures of fish AFP. The antifreeze glycoprotein (AFGP) is a glycopeptide repeat of 4 to ~40 units. The 3-D structures of the AFP are presented in MOLSCRIPT (8). Type I AFP is from winter flounder, and the Thr side chains are shown to illustrate the 11-residue repeats. Type II is a model structure for sea raven AFP, and the Type III AFP structure is that of isoform from ocean pout. The approximate masses of the proteins are indicated on the right.

EVOLUTION OF AFP

The classification of fish AFP is based on clear-cut similarities and differences at the amino acid sequence level that have now been substantiated by 3-D structural analysis (Fig. 2). The four types described are the antifreeze **glycoproteins**, which are polymers of the tripeptide repeat Ala-Ala-Thr with a disaccharide (β-D-galactopyranosyl-(1→3)-2-acetamido-2-deoxy-α-D-galactopyranose) attached to the Thr, and three AFP types. Type I AFP is an alanine-rich single α-helix. The subtype isolated from flounder serum has a characteristic 11-residue sequence repeat beginning with Thr. The subtype from flounder skin (10) is more like the sculpin AFP, which binds to a different plane of ice (7) and does not have a prominent repeat. Type II AFP is a globular AFP with sequence homology and structural similarity to the carbohydrate recognition domain of calcium-dependent **lectins** and to lithostathin (a protein that binds to and prevents growth of calcium carbonate crystals in pancreatic secretions). Type III AFP is a smaller protein with β-like secondary structure and a rather angular 3-D fold (8,9).

It is clear that the different AFP types have distinct evolutionary origins, which suggests that they have been independently co-opted as antifreezes relatively recently in the 175 million year radiation of the bony fish (11). Indeed, now there are several examples of very closely related fish that produce different AFP types and at least one clear example of distantly related fish that produce the same type through **convergent** evolution (12). This pattern of recent evolution is reinforced by the observation that AFP **gene families** show evidence of extensive **gene amplification**, often with the AFP genes organized as tandem repeats and with substantial variation in gene dosage (13). The diversity of AFP types is also reflected in their expression and processing. AFP genes are either constitutively expressed or are regulated by a variety of cues, including temperature, photoperiod, and hormonal cycles (14). AFP are typically produced in the liver as preproteins for secretion into the serum with or without a prosequence (see **Protein secretion**). The small antifreeze glycoproteins are produced from a large **polyprotein** by both **proteolysis** and glycosylation (12). In winter flounder, the external tissues, skin, scales, fins, and gills, produce **isoforms** representing a subclass of Type I AFP that lacks a signal sequence (10).

Not unexpectedly, plant and insect AFP also differ. AFP from winter rye are homologues of three types of extracellular plant pathogenesis-related sequences, namely, endochitinases, endoglucanases, and thaumatin-like proteins (15). AFP from moths and beetles are distinct (16,17). Both are hydrophilic **disulfide-bond**-rich proteins of the hemolymph that produce 5°C of thermal hysteresis and are thus significantly more active than fish AFP. In some terrestrial insects that resist freezing at temperatures down to −30°C, AFP may be part of a suite of antifreeze mechanisms that include elevated polyol levels and the elimination of ice nucleators that interfere with supercooling.

CONCLUSION

Adsorption-inhibition is the generally accepted mechanism of AFP action, but the energetics of tight binding to ice are not yet understood. Other areas of current research include the basis for the hyperactivity of insect AFP, activities of AFP other than thermal hysteresis, and applications for AFP in conferring freeze tolerance/resistance to unprotected organisms.

BIBLIOGRAPHY

1. P. W. Wilson (1993) *Cryo-Lett.* **14**, 31–36.

2. C. A. Knight, A. L. DeVries, and L. D. Oolman (1984) *Nature* **308**, 295–296.
3. A. Parody-Morreale et al. (1988) *Nature* **333**, 782–783.
4. B. Rubinsky, A. Arav, and G. L. Fletcher (1991) *Biochem. Biophys. Res. Commun.* **180**, 566–571.
5. C. A. Knight, C. C. Cheng, and A. L. DeVries (1991) *Biophys. J.* **59**, 409–418.
6. G. Deng, D. W. Andrews, and R. A. Laursen (1997) *FEBS Lett.* **402**, 17–20.
7. C. C. Cheng and A. L. DeVries (1991) *Life Under Extreme Conditions* (G. di Prisco, ed.), Springer-Verlag, Berlin, pp. **1–14**.
8. Z. Jia et al. (1996) *Nature* **384**, 285–288.
9. F. D. Sönnichsen et al. (1996) *Structure* **4**, 1325–1337.
10. Z. Gong et al. (1996) *J. Biol. Chem.* **271**, 4106–4112.
11. G. K. Scott, G. L. Fletcher, and P. L. Davies (1986) *Can. J. Fisheries and Aquatic Sci.* **43**, 1028–1034.
12. L. Chen, A. L. DeVries, and C. H. C. Cheng (1997) *Proc. Natl. Acad. Sci. USA* **94**, 3817–3822.
13. C. L. Hew et al. (1988) *J. Biol. Chem.* **263**, 12049–12055.
14. P. L. Davies, C. L. Hew, and G. L. Fletcher (1988) *Can. J. Zool.* **66**, 2611–2617.
15. W. C. Hon et al. (1995) *Plant Physiol.* **109**, 879–989.
16. L. A. Graham et al. (1997) *Nature* **388**, 727–728.
17. M. G. Tyshenko et al. (1997) *Nat. Biotechnol.* **15**, 887–890.

Suggestions for Further Reading

K. E. Zachariassen (1985) Physiology of cold tolerance in insects, *Physiol. Rev.* **65**, 799–832.

P. L. Davies, K. V. Ewart, and G. L. Fletcher (1993) The diversity and distribution of fish antifreeze proteins: New insights into their origins, In *Fish Biochemistry and Molecular Biology*, Vol. 2 (T. P. Mommsen and P. W. Hochachka, eds.), Elsevier, Amsterdam, pp. 279–291.

M. Griffith and K. V. Ewart (1995) Antifreeze proteins and their potential use in frozen foods, *Biotechnol. Adv.* **13**, 375–402.

Y. Yeh and R. E. Feeney (1996) Antifreeze proteins: Structures and mechanisms of function, *Chem. Rev.* **96**, 601–617.

P. L. Davies and B. D. Sykes (1997) Antifreeze proteins, *Current Opinions in Structural Biology* (in press).

ANTIGEN

MICHEL FOUGEREAU

The historical definition of an antigen, given for many years in immunology textbooks, is a rather circular one, because it describes the antigen as a foreign substance that induces upon penetration in an animal (or a human being) the production of an **antibody** with which it will combine specifically, *in vivo* or *in vitro*. This classical antibody response is a characteristic of vertebrates. Although not basically wrong, this definition necessitates some comments addressing three points: (*1*) antibody, (*2*) induction, and (*3*) foreignness.

ANTIBODY

When this definition was proposed, the **immune response** was characterized solely by the production of antibodies. When it was realized, in the 1960, that the specific functions of the immune system relied on two distinct types of lymphocytes, **B cells** and **T cells**, it became clear that the immune response could no longer be limited to the production of antibodies. B and T cells "recognize" the antigen, but in very different ways. B cells express **immunoglobulins** at their surface, whereas T cells express the so-called **T-cell receptor** (TCR). By extension, surface immunoglobulins are also given the name of B-cell receptor (BCR), by analogy with the TCR. Once stimulated by an antigen, in combination with other signals (**cytokines**), the B cells differentiate into plasma cells that will express a soluble form of immunoglobulins, also known as *antibodies* or *circulating antibodies*, because they are free in the bloodstream. Immunoglobulins and TCR are constructed on the same general pattern, but are encoded by discrete sets of **genes**. They interact with the antigen in very different manners. The immunoglobulins—or antibodies—interact directly with a nativeform of the antigen, through a small portion of this antigen called the *antigenic determinant* or **epitope**. The TCR does not recognize the antigen as a whole, but interacts witha fragment of it, which results from **antigen processing** and presentation. This has been well studied for **protein** antigens, which are cleaved into **peptides** (the processing step) in the so-called antigen-presenting cells (**macrophages**, dendritic cells, etc.), which will then reexpress these peptides at their surface, in close association with molecules encoded by genes of the **major histocompatibility complex** (MHC) (presentation step). Finally, the TCR will bind to both the peptide and the MHC-presenting molecule. The peptide derived from the antigen and that interacts with the corresponding TCR is describedas the "T-epitope."

INDUCTION

For a long period of time, in fact until it became possible to work on the chemistry of antibodies in the beginning of the 1960, immunochemists centered their interest on the most accessible partner of the antigen–antibody complex—that is, the antigen, for which simple models of well-defined structure could be worked out. Although somewhat frustrating for a physiologist (this would be like an enzymologist exclusively studying substrates), this approach shed light on the exquisite specificity of recognition by the immune system, thanks mostly to the pioneering work of Landsteiner (1) and the discovery of haptens. Haptens are small molecules that are unable to stimulate the immune system, but may bind to an antibody. To produce this antibody, the hapten had to be conjugated to a protein that served as a carrier, conferring the property of stimulating the immune system. This observation clearly indicated that induction and recognition were two separate properties of an antigen. To clarify this, it was proposed to distinguish antigenicity from immunogenicity and, consequently, an antigen from an **immunogen**. Antigenicity describes the chemical structures that condition interaction with an antibody or with a TCR, whereas immunogenicity defines the properties of a molecule to induce an immune response. This distinction is far from pure semantics. A protein, considered as a typical "natural" immunogen, has the dual characteristics of a carrier, which is involved in processing and presentation, and of a hapten (in fact a mosaic of haptens) represented by the B epitope. This duality is of central importance for the immune system, in that it implies a close cooperation between

B, T, and antigen-presenting cells. In most cases, an antibody response requires cooperation with the T-cell compartment, leading to the distinction of T-dependent antigens, as opposed to T-independent ones, which can stimulate B cells directly without this T-cell "help."

FOREIGNNESS

At the end of the nineteenth century, the first antigens that were identified were pathogens: bacteria, **viruses**, parasites, or **toxins** of various origins. It was soon realized that pathogenicity and immunogenicity were not linked and that a huge variety of cells or molecules could induce an immune response, providing that they were foreign to the animal it was injected into. Furthermore, this requirement of foreignness was strengthened by the famous aphorism "horror autotoxicus" put forward in 1901 by Ehrlich, to indicate that the immune system could not be stimulated by self-components, leading to the concept that it was a master of self–nonself discrimination. Due to independent observations of Grubb, Oudin, and Kunkel (2–4), it became apparent that this distinction between self and nonself was not as clear as initially thought. Oudin described three types of antigenic specificities, termed *isotypy*, *allotypy*, and *idiotypy*. Isotypy refers to those specificities that are characteristic of one molecule of one given animal species, for instance the mouse serum albumin. An antibody raised against albumin of one given mouse will react with the albumin of anymouse. Allotypy defines antigenic specificities that are shared by a group of individuals within a given species, as initially shown by Oudin for the rabbit immunoglobulins. This is the case of the human blood groups, for which people of, say, group A share this specificity. Allotypy simply reflects the existence of epitopes (or allotopes) that are encoded by allelic variants (see **Alloantibody**, **alloantigen**). The last type of specificity, idiotypy, is more subtle, and is inherent to the molecules of the immune system itself. It was initially described by Oudin as the characteristics of one antibody molecule, synthesized by one given animal and specific for one given antigen. So, if a rabbit is given antigen X, it will produce an anti-X antibody (or Ab1). If a second rabbit, expressing the same immunoglobulin allotypes, is immunized against the anti-X antibody, it will produce an anti-anti-X antibody (or Ab2). Ab1 is an **idiotype**, Ab2 an **anti-idiotype**, and the antigenic specificities recognized on Ab1 by Ab2 are called the *idiotopes*. Because it was later shown that Ab1 and Ab2 could be produced as discrete waves within the same animal, it was clearly realized that the self–nonself distinction was not so obvious. The horror autotoxicus dogma had also to be revisited when it was shown by Avrameas, and extended by Coutinho (5), that a low level of natural autoantibodies was consistently present in every individual, as well as autoreactive T-cell clones. Such natural antibodies have no pathogenic effect,as opposed to those identified in certain **autoimmune diseases**, for reasons that are still not completely understood.

The term *antigen* covers a huge number of structures, from cells and macromolecular complexes, to relatively small molecules, the lower limit in molecular weight being of the order of a few thousand. Ultimately, and whatever the source of the materials considered, proteins and, to a lesser extent, polysaccharides are the main antigens. Proteins are representative of the T-dependent antigens, which implies processing and presentation. Depending upon their size, the potential number of discrete epitopes of one given antigen may vary from a very small number to several tens, in roughly linear proportion with the exposed area of the molecule. Polysaccharides behave most frequently as T-independent antigens. They are frequently constituents of the bacterial surface and are thus important for the preparation of vaccines against these microorganisms. Another example of polysaccharide antigens is given by the major blood groups in humans. Some macromolecular complexes, such as bacterial lipopolysaccharides, have been studied extensively because they also behave as polyclonal potent mitogenic activators of B cells. Nucleic acids are considered poor immunogens, although they clearly constitute a target for autoantibodies in systemic lupus erythematosus, a severe autoimmune disease. More recently, DNA has been tentatively used as a vaccine, in the form of expression vectors encoding a protein that acted secondarily as an immunizing antigen.

Detailed analysis of the antigenic determinants or epitopes of natural antigens is a difficult task and has necessitated the use of models, among which the most extensively studied in the 1960s and 1970s were polysaccharides by the group of Kabat at Columbia University (6). Based on the binding inhibition of dextran (a polymer of glucose) to anti-dextran antibodies by oligosaccharides of various lengths, the range of size of an epitope, and hence that of the corresponding antibody combining site, was estimated to be between 3 and 6 monosaccharide units. A similar approach was taken by the group of Sela at the Weizmann Institute, with synthetic peptides that mimicked the structure of natural proteins. Direct analysis of natural peptide epitopes with the corresponding antibody was made much later by **X-ray crystallography** and gave a clear picture of the organization of the antibody combining site. The central message is that there is a huge diversity in size and form of natural epitopes. At the surface of a protein, the number of amino acid residues that interact with the antibody combining site varies from a few to about 20. The variation in both size and number of epitopes at the antigen surface contribute a large part of the heterogeneity of the immune response.

See also entries **Antigen processing and presentation**, **Epitope**, **Immunogen**, **Idiotypes**, **Anti-idiotypes**, and **Superantigen**, **Xenogeneic**.

BIBLIOGRAPHY

1. K. Landsteiner and J. van der Scheer (1936) On cross-reactions of immune sera to azoproteins. *J. Exp. Med.* **63**, 325–339.
2. R. Grubb (1956) Agglutination of erythrocytes coated with incomplete anti-Rh by certain rheumatoid arthritic sera and some other sera. *Acta Path. Microbiol. Scand.* **39**, 195–197.
3. J. Oudin (1960) Allotypy of rabbit serum proteins. I. Immunochemical analysis leading to the individualization of seven main allotypes. *J. Exp. Med.* **112**, 107–124.
4. M. Harboe, C. K. Osterland, and H. G. Kunkel (1962) Genetic characters of human γ-globulins in myeloma proteins. *J. Exp. Med.* **116**, 719–738.
5. A. Coutinho, M. D. Kazatchkine, and S. Avrameas (1995) Natural autoantibodies. *Curr. Opin. Immunol.* **7**, 812–818.
6. E. A. Kabat (1956) Heterogeneity in extent of the combining regions of human antidextran. *J. Immunol.* **77**, 377–380.

Suggestions for Further Reading

Y. Paterson (1991) The structural basis of antigenicity. *Intern. Rev. Immunol.* **7**, 121–218.

M. van Regenmortel (1989) Structural and functional approaches to the study of protein antigenicity. *Immunol. Today* **10**, 266–271.

ANTIGEN PROCESSING, PRESENTATION

MICHEL FOUGEREAU

Antigen processing and presentation are crucial steps for recognition of protein **antigens** by **T cells**. The **T-cell receptor** (TCR) will not bind a protein directly, as do **immunoglobulins**, but only as peptides derived from the original protein (processing step) and bound to class I or class II molecules of the **major histocompatibility complex**, or MHC (presentation step). This applies to helper as well as to effector **cytotoxic T lymphocytes** and therefore condition severy aspect of the T-dependent **immune response**. These functions strictly condition immunogenicity and correlate well with the dichotomy between carrier and hapten that had indicated long ago that the hapten part of the complex was recognized as such by the B cell, but was unable to initiate an immune response unless conjugated to a carrier protein (see **Antigen**). Later it was shown that carrier recognition involved the T helper cell but was also strictly dependent on MHC molecules (MHC restriction). Antigen processing and presentation take place in specialized cells and involve a number of molecules, most of which are encoded by genes of the MHC.

The antigen-presenting cells that are most efficient and most extensively studied are the dendritic cells and the epidermal Langerhans cells. They originate from $CD34^+$ bone marrow precursors and migrate to different tissues. They express FcRϵI and FcRII, as well as class I and class II molecules of MHC (in humans, HLA-DR). They internalize the antigen by the **endosomal** pathway and, after processing, reexpress at the cell surface antigen-derived peptides associated with class I molecules. Alternatively, they may present peptides associated with class I MHC after macropinocytosis. They may become interdigitated cells and migrate to secondary lymphoid organs, where they present peptides specifically to CD4 T cells. B cells also behave as antigen-presenting cells, and this is of major importance for the antibody response induced by a T-dependent antigen, because it conditions the necessary direct interaction between T and B cells. Finally, **macrophages** and monocytes also can present antigens, in addition to their other functions.

There are two main pathways for antigen processing and presentation. One involves association with MHC class I molecules and presentation to CD8 T lymphocytes, while the other involves MHC class II molecules and interaction with CD4 lymphocytes.

CLASS I PATHWAY

MHC Class I molecules are composed of one α **polypeptide chain** of 43 kDa that forms a heterodimer with the β2 microglobulin. A peptide binding site, which accommodates nonapeptides, is shared by the first two **domains** of the α chain. There is a huge degeneracy in peptide recognition by one such site, as expected from the low number of different MHC molecules expressed in any individual. Class I molecules are specialized in the presentation of endogenous proteins, synthesized *in situ* by a virus or any intracellular bacteria or parasite. Processing starts as a fraction of these cytosolic proteins are hydrolyzed in the **proteasome**, a **proteinase**-rich complex involved in **protein degradation**. The resulting peptides will be transported by the specialized TAP gene products to class I molecules in the **endoplasmic reticulum**. The class I peptide complex will pass through the **Golgi apparatus**, where the α chain becomes glycosylated before being included in a **secretory vesicle** and finally expressed at the plasma membrane. Interaction with the CD8 T cell involves the TCR, which makes contact with amino acid residues of both the peptide and the class I molecule, whereas CD8 binds solely to the MHC molecule. In addition to this specific interaction, many other **cell adhesion molecules** contribute to this "immunological synapse," as immunologists sometimes call it.

CLASS II PATHWAY

MHC class II molecules are composed of two chains, α and β, each containing two domains. The peptide binding site is shared by the α1 and β1 domains. The $\alpha\beta$ heterodimer associates with the Ii (CD74) invariant chain that prevents any association of an endogenous peptide with the class II molecule. After glycosylation during passage through the Golgi, the complex is packed in the class II vesicles that fuse with **endosomes**, where the invariant Ii chain is hydrolyzed, leaving only the CLIP peptide still associated with the $\alpha\beta$ heterodimer. The last step, controlled by HLA-DM molecules, involves an exchange between CLIP and a peptide derived from an exogenous protein antigen. The complex of peptide and class II molecules may now be expressed at the cell surface. A fraction of the class II molecules is reinternalized and may bind new peptides, before being reexpressed at the cell surface. The size of peptides that bind to the class II molecule may vary from between 10 and 30 amino acid residues. Interaction with T cells is somewhat similar to the case previously described, except that it involves essentially CD4 T cells.

CD4 T cells are helper cells that will either (a) interact with the T effector compartment and promote the emergence of cytotoxic T cells or (b) interact with B cells that will ultimately produce antibodies. The latter interaction is of particular interest, because the B cell will act as both a presenting cell and an antibody-producing cell. The T-cell–B-cell interaction is initiated because the B cell presents peptides derived from the specific antigen that had been internalized after binding to the surface immunoglobulin. The antigenic peptide–class II complex will trigger the interacting TCR and activate the T-cell CD3 signaling module. Critical molecules are then produced by the T cell, amongst which are soluble cytokines and one membrane ligand, CD40L, which will bind to its receptor, the CD40 molecule, which is constitutively expressed on the B cell. This is the major signal that will ultimately activate the final phase of differentiation of the B cell and provide the key to clonal expansion and antibody production.

See also entries **B cell**, **Antibody**, **T cell**, **T-cell receptor**, **Antigen**, **Immunogen**, and **Immune response**.

Suggestions for Further Reading

J. L. Whitton (1998) An overview of antigen presentation and its central role in the immune response. *Curr. Top. Microbiol. Immunol.* **232**, 1–13.

R. M. Steinman and J. Banchereau (1998) Dendritic cells and the control of immunity. *Nature* **392**, 245–252.

ANTIGENIC VARIATION

MICHEL FOUGEREAU

Antigenic variation is a sophisticated molecular mechanism that allows parasites, **viruses**, and some bacteria to escape the **immune response** of the invaded host, by changing the nature of their expressed surface **antigens**. Classical examples are (a) **trypanosomes** amongst parasites and (b) **influenza virus**.

Trypanosomes are infectious agents transmitted by glossinas and are responsible for severe tropical diseases, such as sleeping sickness in humans. Trypanosomes express surface **glycoproteins** that induce an immune response in the host. Although at any given time one trypanosome expresses one surface glycoprotein, it may switch to the expression of another one, having a slightly different structure. There are large numbers of these different forms (up to 1000), called **variable surface glycoproteins** (VSG), each being encoded by a distinct **gene**. There are two known mechanisms that account for the successive expression of different VSG genes. One is **transposition** that brings a gene that was present initially as part of one of several gene clusters, scattered on different chromosomes, to one **telomere** end. As a consequence of the transposition, that gene is expressed, until another one becomes functional. An alternative mechanism is a change in the control of **transcription** of expression sites. As the structure varies from one VSG to another, there is a permanent change in antigenicity of the parasite population, thereby ensuring escape from the immune mechanisms of the host. Although a large number of genes are potentially active, only a limited fraction is used, so that the **repertoire** of different VSGs expressed in one host remains limited; this may explain the appearance of chronic immunity in tropical populations.

Viruses use a different mechanism to escape the immune system. One example is that of influenza A, which expresses two surface antigens, hemagglutinin and neuraminidase. It is well known that new variants always arise, sometimes being highly dangerous as they spread rapidly. New variants are the result of **mutations**, which induce *antigenic drift*, and exchange of genetic material between different virus strains, leading to *antigenic shift*, which has a more drastic impact on spreading of the virus. Antigenic variation of viruses is a major drawback in preventing infection (see **Virus infection, animal**). One approach is to prepare modified strains by **site-directed mutagenesis** and thus produce vaccines that might anticipate their natural variation.

A more recent problem of antigenic variation is that by **HIV**, which mutates rapidly, leading to the rapid accumulation of numerous variants within the same patient and providing a particularly efficient way for the virus to escape the immune system. Added to the fact that lymphocytes themselves area major target for this virus, this stresses the unusual danger of this virus.

Suggestions for Further Reading

G. A. Cross, L. E. Wirtz, and M. Navarro (1998) Regulation of vsg expression site transcription and swtiching in Trypanosoma brucei. *Mol. Biochem. Parasitol.* **91**, 77–91.

S. Subbarao and G. Schochetman (1996) Genetic variability of HIV-1. *AIDS* **10** (Suppl. A), S13–S23.

ANTI-IDIOTYPE IMMUNOGLOBULINS

MICHEL FOUGEREAU

Every biologist is familiar with the fact that injection of an **antigen** into an animal induces an **immune response** that results, among other things, in the production of specific **antibodies**. So, if a mouse is given antigen X, it will produce the corresponding anti-X antibody. If we now purify this anti-X antibody and inject it into a mouse of the same strain, (ie, expressing the same **allotypes** as the first mouse), the second mouse will produce an antibody that will react with the anti-X antibody and that represents an anti-anti-X antibody. This defines the idiotypic specificities, which were first described independently in the 1960s by Oudin (1) in the rabbit and by Kunkel (2) in humans. These specificities are located on the **variable regions** of both the heavy and the light chains and are the result of epitope(s) [in this case we use the term of idiotope(s)], related at least in part to the antibody combining site of anti-X. It was shown subsequently that this game could be played one step further: If we now inject a third mouse with the anti-anti-X molecule, the mouse will make anti-anti-anti-X antibodies. Immunologists proposed a simplified nomenclature that reads

$$X \neg Ab1 \neg Ab2 \neg Ab3 \cdots$$

where Ab1 is the idiotype, Ab2 is the anti-idiotype, and Ab3 is the anti-anti-idiotype. This constitutes the so-called idiotypic cascade. One observation of particular interest, independently made by the groups of Oudin in Paris and Urbain (3) in Bruxelles, was that some antibodies in the Ab3 population behaved like Ab1, in that they could interact with antigen X, which the animal had never "seen." This led them to consider that Ab2 had structures resembling X, because it could induce the formation of antibodies (Ab3) that behaved as anti-X. Ab2 was thus considered an "internal image" of the antigen. The observation was generalized by Jerne (4), once it was shown that this cascade could take place spontaneously within the same animal and that the level of expression of each antibody could be modulated (up or down, depending upon the experimental conditions). Jerne proposed that the entire population of antibodies of a given individual were organized as a huge network of interactions, leading to a dynamic equilibrium of B lymphocytes. This was called the *idiotypic network*.

Modulation of the **repertoire** of **B cells** expressed at any given time by injection of either partner of the cascade was thoroughly investigated during a number of years. Particular attention was paid to the diverse roles exerted by the Ab2 antibodies, because it was demonstrated that two subpopulations, termed Ab2β and Ab2α, had completely antagonistic regulatory effects, resulting either in the production of Ab3 antibodies that were Ab1-like, or in the down regulation of Ab1. Ab2β are anti-idiotype antibodies that contain a typical internal image and are therefore able to stimulate the production of anti-X antibodies, without disposing of the original antigen. This led to the development of numerous attempts to make "idiotypic vaccines." The second type of population, Ab2α, recognizes idiotopes that are outside the antibody combining site of Ab1, and they appeared quite suitable to repress undesirable antibodies (ie, **autoantibodies** that have a pathogenic effect in **autoimmune diseases**).

The theoretical interest of idiotypic vaccines is to function as surrogate for antigens that are difficult to use for **immunization**, either because they are difficult to isolate, are poorly immunogenic, or present a potential hazard. Many attempts have been made, especially for **viruses**, including **HIV**, and parasites such as schistosomiasis. Antibodies were generally produced, but it turned out that in most (if not all) cases they were poorly neutralizing, and the protective effect *in vivo* was too low relative to the high efficiency that is expected from an acceptable vaccine.

Down-regulation of pathogenic autoantibodies encountered in autoimmune disease has also been extensively investigated, and it is still currently used in therapeutics. In fact, one uses immunoglobulin preparations assembled from very large pools of donors that are administered intravenously (IVIg). Although experimental arguments are compatible with the presence of anti-idiotypes, many other factors may account for the down regulation of the patient antibodies.

Finally, an interesting application of the internal image is the isolation of cellular receptors by Ab2β that mimic the corresponding ligand. Numerous examples have been reported and successfully used; the first was the Ab2β of the **insulin** cascade, which was shown to activate the insulin receptor of the β islet of pancreas. Adrenergic receptors or T3 and T4 **thyroid hormone** receptors are other examples. This could facilitate receptor isolation by the use of immunoadsorbents prepared with these anti-idiotypic internal image antibodies.

See also entries **Autoantibodies**, **Autoimmunity**, **Idiotypes**, **Tolerance**.

BIBLIOGRAPHY

1. J. Oudin and M. Michel (1969) Idiotypy of rabbit antibodies. Comparison of idiotypy of various kinds of antibodies formed in the same rabbit against *S. typhi*. *J. Exp. Med.* **130**, 619–629.
2. H. G. Kunkel, M. Mannick, and R. C. Williams (1963) Individual antigenic specificity of isolated antibodies. *Science* **140**, 1218–1220.
3. J. Urbain, M. Wilker, J. D. Frannsen, and C. Collignon (1977) Idiotypic regulation of the immune system by the induction of antibodies by anti-idiotypic antibodies. *Proc. Natl. Acad. Sci. USA* **74**, 5126–5129.
4. N. K. Jerne (1974) Towards a network theory of the immune system. *Ann. Immunol.* **125 C**, 373–389.

Suggestions for Further Reading

C. A. Bona (1996) Internal image concept revisited. *Proc. Soc. Exp. Biol. Med.* **213**, 32–42.

R. J. Poljak (1994) An idiotope-anti-idiotope complex and the structural basis of molecular mimicking. *Proc. Natl. Acad. Sci. USA* **91**, 1599–1600.

ANTIPARALLEL BETA-BARREL MOTIFS

JENNIFER MARTIN

Antiparallel β-barrel motif is a general term that describes several **protein motifs** observed in **protein structures** that have in common a barrel-shaped **β-sheet** structure formed from antiparallel **beta-strands**. These include the **up-and-down β-barrel**, the **Greek key** barrel (formed from two Greek key motifs), and the **jelly roll** barrel (Fig. 1). The different motifs have different connections between the strands in the barrel. Other types of antiparallel β-barrels include the

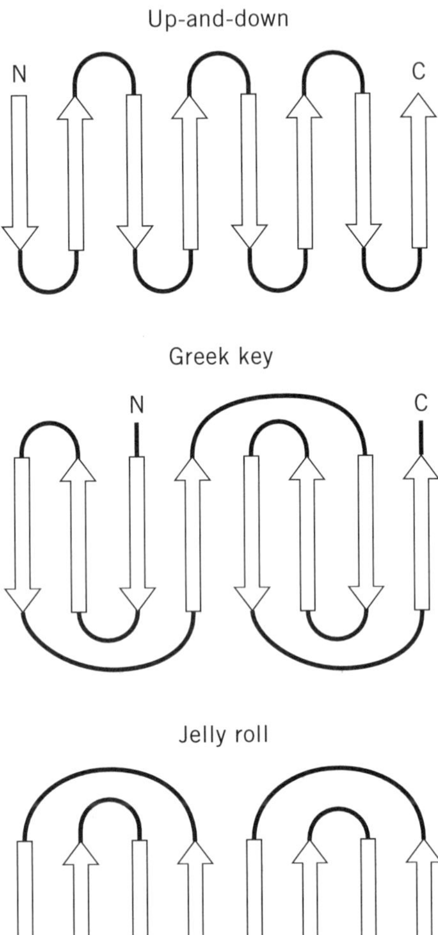

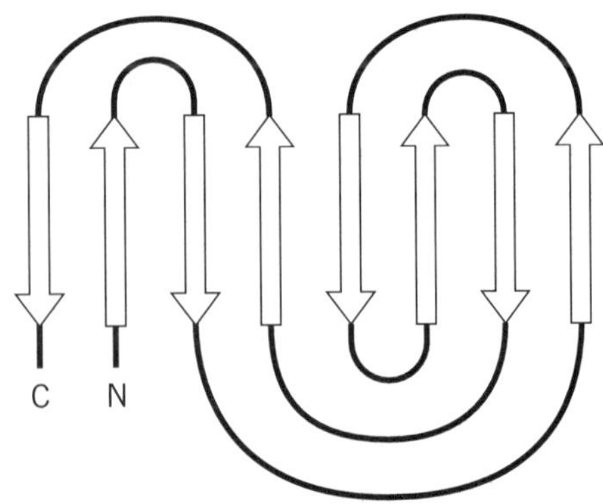

Figure 1. Schematic representation of the up and down (or β-meander), Greek key, and jelly roll topologies. Antiparallel β-barrels can be formed from each of these three protein motifs, which represent different ways of linking consecutive β-strands together into a β-sheet.

interleukin-1 motif and the **neuraminidase** superbarrel. All these protein barrels comprise only antiparallel β-strands and are therefore classified as all-β structural **domains**. They therefore differ significantly from the **TIM barrel**, which is classified as an α/β domain and is formed from a parallel β-barrel surrounded on the exterior by α-helices.

[See also **Up and down beta-barrel**, **Greek key motif**, **Jelly roll motif**, and **TIM barrel**.]

ANTISENSE OLIGONUCLEOTIDES

S. T. CROOKE

Antisense oligonucleotides are used to **hybridize** to a specific **RNA** molecule *in vivo* and thereby to inhibit its subsequent use. In most cases, the target RNA is a **messenger RNA** (mRNA), which cannot then be translated into a **protein** (see **Gene expression**). All that is required is knowledge of the gene's nucleotide sequence, so that an antisense oligonucleotide with the complementary sequence can be synthesized. In this way, the expression of one **gene** can be blocked, and the consequent observed effects help to elucidate the physiological role of the gene and its product.

Selection of the sites in a RNA molecule at which optimal antisense activity may be induced is complex, dependent on the terminating mechanism and influenced by the chemical class of the oligonucleotide. Each RNA displays a unique pattern of sites of sensitivity. Within the phosphorothioate oligodeoxynucleotide class, studies have shown that antisense activity can vary from undetectable to 100% by shifting an oligonucleotide by just a few bases in the RNA target (1,2). Significant progress has been made in developing general rules that help to define potentially optimal sites in RNA species, but to a large extent this must be determined empirically for each RNA target and every new chemical class of oligonucleotides.

THERAPEUTIC USES

Besides being useful for characterizing the roles of genes, antisense oligonucleotides have the potential for therapeutic use to regulate the expression of certain genes or to block an infection by a pathogenic **bacterium** or **virus**. The therapeutic use of oligonucleotides represents a new paradigm for drug discovery, because oligonucleotides have not been studied before as potential drugs and they are being used to intervene in processes that have not been considered sites at which drugs might act. The affinity and the specificity of binding derive from the hybridization of two oligonucleotides and therefore are theoretically much greater than can be achieved with small molecules. Furthermore, the rational design of the nucleotide sequence of antisense oligonucleotide is much more straightforward than the design of small molecules interacting with proteins (see **Ligand binding**). Finally, it is possible to consider the design of antisense drugs to treat a very broad range of disorders, including those not amenable to other types of treatment.

This use of antisense oligonucleotides is still in its infancy, and the initial enthusiasm must be tempered by appropriate reservations concerning practical aspects. To be useful as a drug, an antisense oligonucleotide must be much more stable than ordinary nucleic acids, and it must be able to reach its desired site of action, the interior of a cell. Questions about the technology reduce to, can oligonucleotide analogs be created that have appropriate properties to be drugs? Specifically, what are the pharmacokinetic, pharmacological, and toxicological properties of these compounds and what are the scope and potential of the medicinal chemistry of oligonucleotides? Answers to many of these questions are available now.

PHOSPHOROTHIOATE OLIGODEOXYNUCLEOTIDES

To address the problem of stability, nonconventional oligonucleotides not susceptible to degradation or to hydrolysis by nucleases have been designed. Of the first generation of oligonucleotide analogs, the phosphorothioate class is best understood and has produced the broadest range of activities (3). Phosphorothioate oligonucleotides differ from normal in that one of the nonbridging oxygens in the phosphate group is replaced by a sulfur. The resulting compound is negatively charged, chiral at each phosphorothioate, and much more resistant to nucleases than a phosphodiester bridge.

Hybridization

The hybridization of phosphorothioate oligonucleotides to DNA and RNA has been thoroughly characterized. The melting temperature (T_m) of a phosphorothioate oligodeoxynucleotide bound to RNA is lower than for a corresponding phosphodiester oligodeoxynucleotide by approximately 0.5°C per nucleotide. Compared to RNA duplex formation, a phosphorothioate oligodeoxynucleotide has a T_m approximately 2.2°C lower per nucleotide. This means that, to be effective *in vitro*, phosphorothioate oligodeoxynucleotides must be relatively long, typically at least 17 to 20 nucleotides, and invasion of double-stranded regions of the target RNA is difficult (4–7).

Interactions with Proteins. Phosphorothioate oligonucleotides bind to proteins. These interactions can be (3) nonspecific, sequence-specific, or structure-specific, each of which may have different characteristics and effects. Nonspecific binding to a wide variety of proteins has been demonstrated, most thoroughly with **serum albumin**. The affinity of such interactions is low. The **dissociation constant** (K_d) for albumin is approximately 200 μM, about the same as for its binding of aspirin or penicillin (8–10). Phosphorothioates also interact with **nucleases** and **DNA polymerases**; they are slowly metabolized by both endo and **exonucleases** and are **competitive inhibitors** of these **enzymes** (11). In an RNA–DNA duplex, phosphorothioates are substrates for **ribonuclease H** (RNaseH) (12). At higher concentrations, phosphorothioates inhibit the enzyme, presumably by binding as a single strand (11). Again, the oligonucleotides are competitive antagonists for the DNA–RNA substrate.

Phosphorothioates are competitive inhibitors of DNA polymerase α and β with respect to the DNA template and **noncompetitive inhibitors** of DNA polymerases γ and δ (12). They are also competitive inhibitors for the **reverse transcriptase** of **HIV** and inhibit its associated RNase H activity (13,14). They bind to the cell surface protein CD4 and to protein kinase C (15). Phosphorothioates inhibit various viral polymerases (16), and they also cause potent, nonsequence-specific inhibition of **RNA splicing** (17).

In Vivo Pharmacokinetics. Binding of phosphorothioate oligonucleotides to serum albumin and alpha-2 **macroglobulin** provides a repository for these drugs in the serum and prevents their rapid renal excretion. Because serum protein binding is saturable, however, intact oligomer may be found in urine with high doses (18,19), eg, 15 to 20 mg kg^{-1} administered intravenously to rats. Phosphorothioate oligonucleotides are rapidly and extensively absorbed after parenteral administration, as much as 70% within 4 h (20,21). Distribution of phosphorothioate oligonucleotides from blood after absorption or intravenous administration is extremely rapid. Distribution **half-lives** are less than one hour (18–20,22). Clearance from the blood and plasma exhibits complex kinetics, with a terminal elimination half-life of 40 to 60 h in all species except man, where it may be somewhat longer (21). Phosphorothioates distribute broadly to all peripheral tissues, although no evidence for significant penetration of the blood brain barrier has been reported. Liver, kidney, bone marrow, skeletal muscle, and skin accumulate the highest amounts (20,22). Liver accumulates the drug most rapidly (20% of a dose within 1 to 2 hours) and also eliminates it most rapidly (eg, the terminal half-life from liver is 62 h and from renal medulla 156 h). Within the kidney (23), oligonucleotides are probably filtered by the glomerulus, then reabsorbed by the proximal convoluted tubule epithelial cells, perhaps mediated by interactions with specific proteins in the brush border membranes. At relatively low doses, clearance of phosphorothioate oligonucleotides is caused primarily by metabolism (19,20,22), mediated by exo- and endonucleases.

Pharmacological Activities. Phosphorothioates also have effects inconsistent with the antisense mechanism for which they were designed. Some of these effects are sequence- or structure-specific. Others result from nonspecific interactions with proteins. These effects are particularly prominent in *in vitro* tests for antiviral activity, when high concentrations of cells, viruses, and oligonucleotides are often incubated together (24,25). Human immune deficiency virus (HIV) is particularly problematic, because many oligonucleotides bind to the gp120 protein (26). Moreover, uncertainty as to the mode of action of antisense oligonucleotides is certainly not limited to antiviral or just *in vitro* tests (27–29). These observations indicate that, before drawing conclusions, careful analysis of dose-response curves, direct analysis of target protein or RNA, and inclusion of appropriate controls are required. In addition to interactions with proteins, other factors can contribute to unexpected results, such as overrepresented sequences of RNA and unusual structures that may be adopted by oligonucleotides (26).

A relatively large number of reports of *in vivo* activities of phosphorothioate oligonucleotides have now appeared documenting activities after both local and systemic administration (30). However, for only a few of these reports have sufficient studies been performed to draw relatively firm conclusions concerning the mechanism of action by directly examining target RNA levels, target protein levels, and pharmacological effects, using a wide range of control oligonucleotides and examining the effects on closely related isotypes (31–34). Thus, there is a growing body of evidence that phosphorothioate oligonucleotides induce potent systemic and local effects *in vivo*, suggesting highly specific effects difficult to explain via any mechanism other than antisense.

Toxicological Properties. Phosphorothioate oligonucleotides are not toxic in themselves, but their use is limited by side effects. In rodents, this is immune stimulation (35,36). In monkeys, it is sporadic reductions in blood pressure associated with bradycardia, which is often associated with activation of C-5 **complement** involving activation of the alternative complement pathway (37). All phosphorothioate oligonucleotides tested to date induce these effects, although there are slight variations in potency depending on their sequence and/or

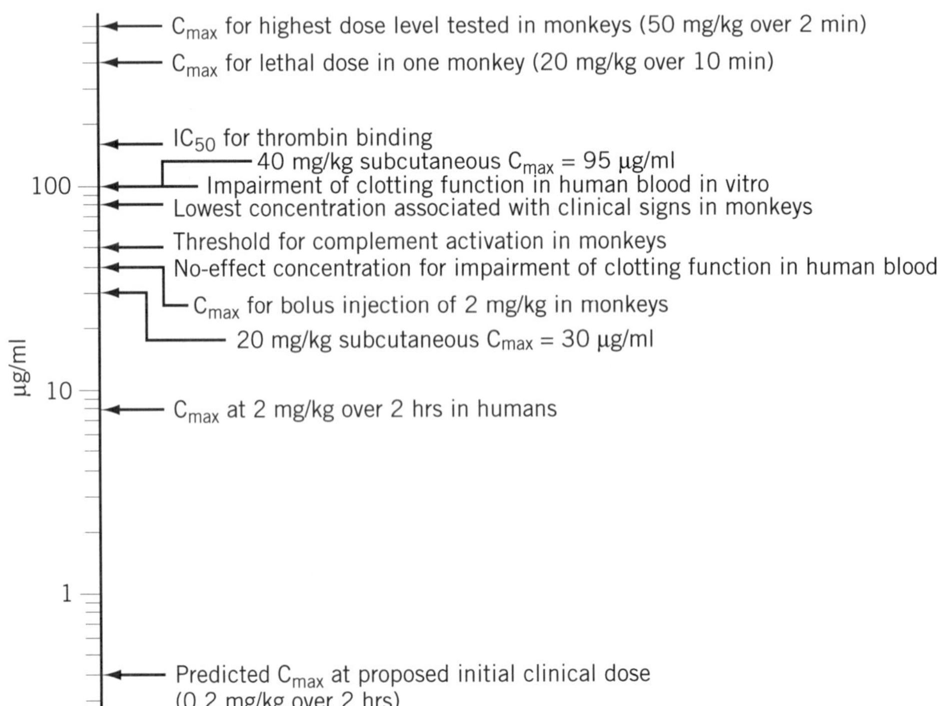

Figure 1. Plasma concentrations of ISIS 2302 at which various activities are observed. These concentrations are those of intact ISIS 23902 and were determined by extracting and analyzing plasma by **capillary zone electrophoresis**.

length (36,38,39). A second prominent toxicological effect in the monkey is on **blood clotting**. The mechanisms responsible for these effects are probably very complex, but preliminary data suggest that direct interactions with **thrombin** are at least partially responsible (40).

In man, the toxicological profile differs. When ISIS 2922 is administered intravitreally to patients with **cytomegalovirus** retinitis, the most common adverse event is anterior chamber inflammation, which is easily managed with steroids. A relatively rare and dose-related adverse event is morphological changes in the retina associated with loss in peripheral vision (41). ISIS 2105, a 20-mer phosphorothioate designed to inhibit the replication of human papilloma viruses that cause genital warts, has been administered intradermally at doses as high as 3 mg/wart weekly for three weeks. Essentially no toxicities have been observed (42).

Therapeutic Index. An attempt to put the toxicities and their dose-response relationships into a therapeutic context is shown in Fig. 1. This is particularly important because considerable confusion has arisen concerning the potential utility of phosphorothioate oligonucleotides for selected therapeutic purposes as a result of unsophisticated interpretation of toxicological data. As can be readily seen, the immune stimulation induced by these compounds is particularly prominent in rodents and unlikely to be dose-limiting in humans. Nor have hypotensive events in humans been observed to date. This toxicity occurs at lower doses in monkeys than in man and certainly is not dose-limiting in man. On the basis of present experience, the dose-limiting toxicity in man is likely to result from blood clotting abnormalities, associated with peak plasma concentrations well in excess of 10 μg/ml. Thus, it phosphorothioate oligonucleotides have a therapeutic index that supports their evaluation for a number of therapies.

Clinical Activities. Significant therapeutic benefit has been reported in patients with cytomegalovirus retinitis treated locally with fomivirsen (43). ISIS 2302 administered every other day for one month also resulted in statistically significant improvement for five to six months in patients with steroid-dependent Crohn's disease in a randomized, double-blind placebo, controlled trial (44).

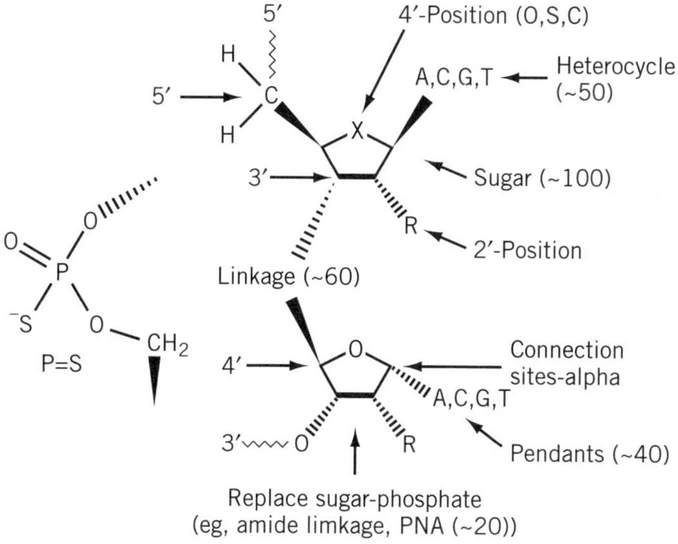

Figure 2. Isis oligonucleotide modifications.

Medicinal Chemistry of Oligonucleotides

The core of any rational drug discovery program is medicinal chemistry. Although the synthesis of modified **nucleic acids** has been a subject of interest for some time (see **DNA synthesis**), the intense focus on the medicinal chemistry of oligonucleotides dates perhaps no more than the past five years. Modifications have been made to the base, sugar, and phosphate moieties of oligonucleotides (Fig. 2). The subjects of medicinal chemical programs include approaches to (3) create enhanced and more selective affinities for RNA or duplex structures, the ability to cleave nucleic acid targets, enhanced nuclease stability, (4) cellular uptake and distribution, and *in vivo* tissue distribution, metabolism, and clearance. Arguably, the most interesting modifications to date are those that alter the sugar moiety (45) and the backbon. Modifications such as 2′ methoxyethoxy enhance affinity for RNA, potency *in vivo*, provide a dramatic increase in stability, and reduce the potency for blood clotting and inflammatory effects. Also of interest are a number of modifications that replace the phosphate or the entire phosphate sugar backbone. Several novel chemical classes are being evaluated in animals and will shortly be studied in man, so it seems likely that a variety of chemical classes with differing properties will be available in the near future.

BIBLIOGRAPHY

1. M. Y. Chiang et al. (1991) *J. Biol. Chem.* **266**, 18162–18171.
2. C. F. Bennett and S. T. Crooke (1996) Oligonucleotide-based inhibitors of cytokine expression and function. In *Therapeutic Modulation of Cytokines* (B. Henderson, and M. W. Bodmer eds.), CRC Press, Boca Raton, pp. 171–193.
3. E. De Clercq, F. Eckstein, and T. C. Merigan (1969) *Science* **165**, 1137–1140.
4. W. F. Lima et al. (1992) *Biochemistry* **31**, 12055–12061.
5. T. Vickers et al. (1991) *Nucleic Acids Res.* **19**, 3359–3368.
6. B. P. Monia et al. (1993) *J. Biol. Chem.* **268**, 14514–14522.
7. B. P. Monia et al. (1992) *J. Biol. Chem.* **267**, 19954–19962.
8. S. T. Crooke et al. (1996) *J. Pharmacol. Exp. Ther.* **277**(2), 923–937.
9. R. W. Joos and W. H. Hall (1969) *J. Pharmacol. Exp. Ther.* **166**, 113–118.
10. S. K. Srinivasan, H. K. Tewary, and P. L. Iversen (1995) *Antisense Res. Dev.* **5**(2), 131–139.
11. S. T. Crooke et al. (1995) *Biochem. J.* **312**(2), 599–608.
12. W. Y. Gao et al. (1992) *Mol. Pharmacol.* **41**, 223–229.
13. C. Majumdar et al. (1989) *Biochemistry* **28**, 1340–1346.
14. Y. Cheng, W. Gao, and F. Han (1991) *Nucleosides Nucleotides* **10**, 155–166.
15. C. A. Stein et al. (1991) *Acquired Immune Defic. Syndr.* **4**, 686–693.
16. C. A. Stein and Y. C. Cheng (1993) *Science*, **261**, 1004–1012.
17. D. Hodges and S. T. Crooke (1995) *Mol. Pharmacol.*, **48**, 905–918.
18. S. Agrawal, J. Temsamani, and J. Y. Tang (1991) *Proc. Natl. Acad. Sci.* USA **88**, 7595–7599.
19. P. Iversen (1991) *Anticancer Drug Des.* **6**(6), 531–8.
20. P. A. Cossum et al. (1994) *J. Pharmacol. Exp. Ther.* **269**, 89–94.

21. S. T. Crooke et al. (1994) *Clin. Pharm. Ther.* **56**, 641–646.
22. P. A. Cossum et al. (1993) *J. Pharmacol. Exp. Ther.* **267**, 1181–1190.
23. J. Rappaport et al. (1995) *Kidney Int.* **47**, 1462–1469.
24. R. F. Azad et al. (1993) *Antimicrob. Agents Chemother.* **37**(9), 1945–1954.
25. R. W. Wagner et al. (1993) *Science* **260**, 1510–1513.
26. J. R. Wyatt et al. (1994) *Proc. Natl. Acad. Sci.* USA **91**, 1356–1360.
27. C. M. Barton and N. R. Lemoine (1995) *Br. J. Cancer* **71**, 429–437.
28. T. L. Burgess et al. (1995) *Proc. Natl. Acad Sci* USA **92**, 4051–4055.
29. M. Hertl, L. M. Neckers, and S. I. Katz (1995) *J. Invest. Dermatol.* **104**, 813–818.
30. S. T. Crooke (1995) *Therapeutic Applications of Oligonucleotides*, R. G. Landes, Austin, TX.
31. N. M. Dean and R. McKay (1994) *Proc. Natl. Acad. Sci.* USA **91**, 11762–11766.
32. N. M. Dean et al. (1996) *Cancer Res.* **56**(15), 3499–3507.
33. B. P. Monia et al. (1995) *Nature Med.* **2**(6), 668–675.
34. B. Monia et al. (1996) *J. Biol. Chem.* **24**(14), 14533–14540.
35. S. P. Henry et al. (1997) *Toxicology* **116**(1–3), p. 77–88.
36. S. P. Henry et al. *Antisense Nucleic Acid Drug Dev.*, In Press.
37. S. Henry et al. *J. Pharmacol. Exp. Ther.*, In Press.
38. K. G. Cornish et al. (1993) *Pharmacol. Commun.*, **3**, 239–247.
39. W. M. Galbraith et al. (1994) *Antisense Res. Dev.* **4**(3), 201–206.
40. S. Henry, W. Novotny, and J. Leeds. Submitted.
41. S. L. Hutcherson et al. (1995) Abstracts of the SFO, CA, *35th ICAAC*, p. 204.
42. J. M. Glover et al. (1996) *Pharmacol. Exp. Ther.*, In Press.
43. D. S. Boyer et al. (1997) In *AAO Annual Meeting*, San Francisco.
44. B. R. Yacyshyn et al. *N. Engl. J. Med.*, Submitted.
45. Y. S. Sanghvi, and P. D. Cook (1994) ACS Symposium Series No. 580 American Chemical Society, Washington, DC, p. 232.

Suggestions for Further Reading

S. T. Crooke and B. Lebleu (1993) *Antisense Research and Applications*, CRC Press, Boca Raton.

S. T. Crooke (1992) *Ann. Rev. Pharmacol. Toxicol.* **32**, 329–376.

S. T. Crooke (1993) *FASEB J.* **7**, 533–539.

S. M. Freier (1993) Hybridization considerations affecting antisense drugs, In *Antisense Research and Applications* (S. T. Crooke and B. Lebleu, eds.), CRC Press, Boca Raton, pp. 67–82.

S. T. Crooke (1995) *Therapeutic Applications of Oligonucleotides*, R. G. Landes, Austin.

ANTISERA

MICHEL FOUGEREAU

Antisera have long been the major reagent in immunology, and this field was for years known as serology. An antiserum is the serum of an animal that has been **immunized** against an **antigen**, or more commonly hyperimmunized, to get the highest possible titer of the desired **antibodies**. Very clearly, vaccination and serotherapy were the first great miracle in fighting against pathogens, long before the discovery of antibiotics. From a medical standpoint, antisera were used to fight quickly against a pathogen or a **toxin**, because vaccination takes a long time to generate a host **immune response**. There are several concerns regarding the use of antisera for therapy in humans or in animals. First, one has to be certain that protection is indeed ensured by antibodies and not exclusively by cell-mediated immunity. This is not a trivial matter, because there are only a limited number of instances in which protection is due primarily to antibodies. Second, the use of antisera from a foreign animal species (heterologous antisera) will induce a strong immunization of the host against all the injected antigens. This results in the appearance, within a week or two, of a local and generalized syndrome, called *serum sickness*, that may include urticaria, local edema, rashes, arthralgia, fever, lymphadenopathy, and, ultimately, severe glomerulonephritis resulting from the formation of immune complexes. This was observed when serotherapy with horse serum was extensively used in humans, especially against **diphtheria** or tetanus. Diphtheria, a bacterial disease, can now be cured with antibiotics, and children undergo a regular schedule of vaccination that actively prevents the disease. Tetanus remains a very severe disease, with an elevated rate of mortality. Serotherapy is still used because antibodies are extremely effective against the toxin; nevertheless, everyone should be revaccinated every 10 years, because this provides the best and safest protection.

Heterologous antisera should be used only in very exceptional occasions—for example, after bites from highly dangerous snakes. Heterologous antisera are being systematically replaced, when appropriate, by purified **immunoglobulins**, preferentially of human origin. Intravenous immunoglobulins (IVIgs) are used to correct severe immunodeficiencies of the **B-cell** compartment. For example, this is the case with X-linked agammaglobulinemia (Bruton disease), which is transmitted genetically through the mother and occurs in young boys. This disease is due to a variety of mutations of the BTK gene (for Bruton tyrosine kinase) that result in an early blockage of B-cell differentiation. IVIgs are also used to help patients with chronic or transient hypoglobulinemia. Whenever the deficiency is severe and would necessitate a lifelong treatment, an allogenic bone marrow graft is performed, which restores the immune system of the patient. IVIgs are also used in therapy of **autoimmune diseases**, although the mechanisms are not absolutely understood, but may be due in part to a reequilibration of the **idiotype** network. Specific purified antibody of human origin can be prepared in some selected cases. The most popular preparation is the anti-D (Rhesus) immunoglobulins that are used worldwide for prevention of the hemolytic disease of the newborn (see **Alloantibody, alloantigen**).

With the start of organ transplantations, many attempts were made to prevent rejection by blocking the immune system of the recipient. Antilymphocyte serum, prepared in rabbits or in horses, has been used but had the usual limitation linked to the induction of serum sickness mentioned above. It could, however, help for a transient difficult acute period of rejection. It has been replaced by a **monoclonal antibody**, prepared in the mouse and directed against the CD3 (signaling module) of the **T-cell receptor**. Used for a short period of time, it proved efficient in down-modulating the immune response of the host, with limited risk of immunization against the heterologous immunoglobulin. Genetic engineering has also been proposed to minimize immunization against heterologous determinants. One classical approach is to insert the six specific hypervariable regions of a murine monoclonal antibody in place of the corresponding regions of a human antibody framework, using **protein engineering**. This engineered antibody is certainly less immunogenic, but it still is, because of the **idiotype** determinants that cannot be avoided. Fully engineered human monoclonal antibodies would certainly be ideal for human use. This goal is being worked on actively by pharmaceutical companies. Its is not yet practical, and no one has yet succeeded in making stable "natural" human monoclonal antibodies.

Besides the roles in human therapy mentioned above, antisera remain of course reagents of choice for experimental purposes. They may be used to identify new molecules and to purify them with techniques like **immunoaffinity chromatography** or **immunoprecipitation**, which are powerful tools to isolate rare components, such as **membrane proteins**, receptors, **hormones**, or diverse ligands. Monoclonal antibodies have tended to replace antisera, which is partly unfortunate and sometimes a mistake, because conventional antisera can be exquisitely specific and are very efficient, potent tools, with a mosaic of specificities that may of great help to the molecular biologist.

See also entries **Immunization** and **Immunogen**.

Suggestion for Further Reading

E. Harlow and D. Lane, eds. (1988) *Antibodies: A Laboratory Manual*, Cold Spring Harbor Laboratory Press, Cold Spring Harbor, NY.

ANTITERMINATION CONTROL OF GENE EXPRESSION

P. GOLLNICK
P. BABITZKE

Bacteria have evolved many different complex mechanisms to control both **transcription** and **translation** of **genes** in response to environmental changes. In many cases, transcription is controlled at the level of initiation by **DNA-binding proteins** that either inhibit (**repressors**) or stimulate (activators) initiation. In addition, transcription can be regulated at the level of elongation. In some cases, transcription of a gene or **operon** will terminate prematurely in the absence of the action of a positive regulatory molecule. In these cases, antitermination factors allow transcription to read through termination signals and to generate full-length transcripts.

Two fundamentally different mechanisms for antitermination have been described. In one case, RNA polymerase is modified so as to allow it to read through transcription terminators. This type of mechanism controls phage development (1) and expression of rRNA operons (2). The second mechanism, covered in this chapter, involves trans-**acting factors** that interact with **RNA** and prevent formation of the terminator structure. This mechanism is very similar to **attenuation**, but antitermination can be distinguished from attenuation in that the action of the regulatory molecule results in transcription readthrough, with the default pathway being premature termination. In attenuation, the action of the regulatory molecule induces transcription termination, and the default pathway is readthrough.

Three distinct mechanisms that regulate gene expression by antitermination will be reviewed here. These mechanisms differ primarily in the type of biomolecule used as the regulator. The first mechanism uses antiterminator proteins that are activated to bind RNA targets in response to environmental stimuli. In the second mechanism, **transfer RNA** is used as the regulator. In this case, the degree of aminoacylation (or charging) of the tRNA is used to sense the availability of the cognate amino acid within the cell, to induce expression of genes involved in metabolism of this amino acid. Finally, in the case of the **Escherichia coli** tryptophanase operon, it appears that **ribosomes** are used as the regulatory molecule. Thus, bacteria have evolved a large number of mechanisms to use different biomolecules to all perform the same task—that is, to alter the conformation of the nascent mRNA to signal RNA polymerase whether it should terminate prematurely or continue transcription of the particular structural gene(s).

RNA-BINDING PROTEIN-MEDIATED ANTITERMINATION: THE SAC/BGL FAMILY OF ANTITERMINATOR PROTEINS

Expression of several catabolic operons in bacteria is regulated by antitermination involving **RNA-binding proteins**. These proteins prevent formation of Rho-independent transcription terminators in the nascent mRNA upstream of the regulated gene(s) (3). One such system in *E. coli* and several in *Bacillus subtilis*, appear to be highly related based on similarities of their antiterminator proteins as well as their RNA targets. In addition, several other systems function similarly but appear to have arisen independently. A general model for this mechanism is shown in Figure 1.

The *E. coli bgl* Operon

The *E. coli bglGFB* operon encodes all the functions necessary for the regulated uptake and utilization of aromatic β-glucosides. The operon is cryptic in wild-type strains but

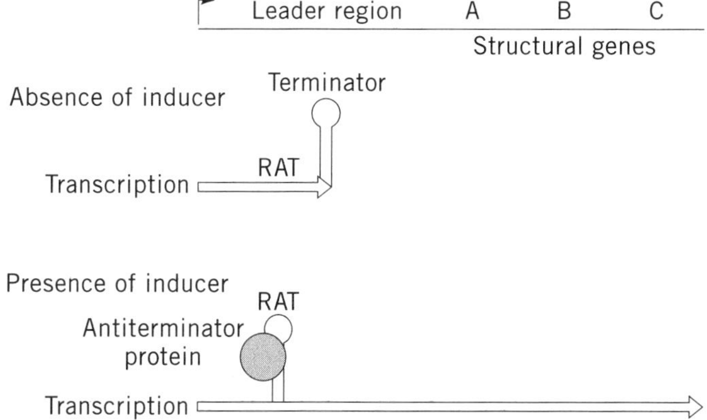

Figure 1. A general model for antitermination control by the Sac/Bgl family of antiterminator proteins. Under noninducing conditions, transcription starts at the promoter (designated by the arrow) and terminates prematurely, often in a leader region prior to the structural genes. In the presence of inducer, the antiterminator protein is activated to bind to the RAT (ribonucleic antiterminator) RNA. This binding stabilizes an RNA secondary structure involving the RAT, which prevents formation of the overlapping terminator, and transcription continues into the structural genes.

can become functional through spontaneous mutations. When functional, expression of this operon is regulated by antitermination mediated by the BglG protein in response to the levels of β-glucosides (4). In the absence of inducer, BglG does not bind RNA, and most transcripts terminate at one of two Rho-independent transcription terminators present in the leader region upstream of *bglG* and between *bglG* and *bglF*. When β-glucoside levels are high, BglG binds to an RNA target, named RAT for ribonucleic antiterminator, just upstream of the terminators. This binding stabilizes an alternative antiterminator RNA structure, which prevents formation of the terminator, thus allowing transcription to continue and the operon to be expressed.

The RNA-binding activity of BglG is regulated by **phosphorylation** mediated by BglF. In the absence of β-glucosides, BglF phosphorylates BglG, which prevents it from dimerizing and binding to the RAT (5). In the presence of β-glucosides, BglF dephosphorylates BglG, which now dimerizes and binds to the RAT. Phosphorylation of both β-glucosides and BglG is accomplished by transfer of the phosphate group from the same phosphorylated residue, Cys24, in BglF (6). These results suggest that, under conditions in which β-glucoside levels are high, the phosphate group can be transferred from BglG back to Cys24 in BglF. A model has been proposed in which unliganded BglF phosphorylates BglG, and β-glucoside binding induces BglF to undergo a conformational change that activates it to dephosphorylate BglG.

A similar system for β-glucoside utilization exists in the related **Gram-negative** enteric bacterium *Erwinia chrystanthemi*, although in this case the *arb* operon is not cryptic (7). ArbG shows high sequence similarity to BglG, suggesting that it functions analogously in antitermination control of the *E. chrystanthemi arb* genes. Antitermination also appears to control β-glucoside operons in several **Gram-positive bacteria** as well. A putative β-glucoside (*bgl*) operon has also been identified in *B. subtilis* and may be regulated by a similar antitermination mechanism (8). In addition, a protein, BglR, with homology to BglG also controls β-glucoside usage in *Lactococcus lactis* (9).

The *B. subtilis sac* Genes

Expression of two sucrose utilization operons in *B. subtilis*, *sacPA* (10) and *sacB* (11), is induced by sucrose via transcription antitermination mediated by the RNA-binding proteins SacT and SacY, respectively. SacT and SacY show extensive sequence similarity to each other, as well as to BglG from *E. coli*. The antitermination mechanisms that control these genes also appear to be quite similar to that described above for the *E. coli bgl* operon. Rho-independent transcription terminators exist in leader regions upstream of both *sacPA* and *sacB* and prevent transcription of the structural genes in the absence of the inducer, which is sucrose. In the presence of sucrose, SacT and SacY are activated to bind RAT sequences in the *sacPA* and *sacB* leader transcripts, respectively, and allow transcription to read through into the structural genes (12). Like BglG in *E. coli*, both of these antiterminator proteins are phosphorylated. In the case of SacY, phosphorylation negatively regulates RNA-binding activity and appears to be mediated by SacX (13). SacT is phosphorylated by HPr, a component of the phosphoenolpyruvate **phosphotransferase system**, but the role of this phosphorylation in sucrose-mediated antitermination is less clear (12).

Recently, the **protein structure** of the RNA-binding domain of SacY has been determined by both **NMR** (14) and **X-ray crystallography** (15). The domain exists as a dimer, with each monomer consisting of a four-stranded antiparallel **beta-sheet**. Several amino acid residues have been identified through genetic, biochemical, and preliminary NMR studies as being important for RNA binding. These residues are clustered on the surface of one side of the protein structure (15).

Other Examples of Bgl/Sac Type Antiterminators

In addition to the *bgl* and *sac* systems described above, several other operons are regulated by RNA-binding antiterminator proteins with homology to BglG, SacY, and SacT. LicT regulates the *licS* gene, which is involved in β-glucan utilization in *B. subtilis* (16). There is also a RAT sequence overlapping a potential Rho-independent terminator upstream of *licS*.

In *Lactobacillus casei*, the lactose (*lac*) operon is regulated in response to lactose levels by LacT, which shows sequence homology to the other members of the Bgl/Sac family of antiterminators (17). The 5′-leader region of the *lac* mRNA contains a region with sequence similarity to the RAT sequence, as well as a potential stem-loop structure resembling a Rho-independent terminator.

Antiterminators with No Similarity to the Bgl/Sac Family

Several other systems are regulated by RNA-binding antiterminator proteins that are unrelated to those of the Bgl/Sac family; furthermore, these proteins do not appear to be related to each other. These regulatory systems thus appear to have arisen independently.

In *B. subtilis*, both the *glp* regulon, which is involved in usage of glycerol-3-phosphate, and a histidine-utilization (*hut*) operon are regulated by RNA-binding antiterminator proteins; GlpP (18) and HutP (19), respectively. The amino acid sequences of these antiterminator proteins are not similar to any other antiterminator proteins. Further, the mechanisms by which these antiterminator proteins function appear to be different from those described above, because there are no clear antiterminator RNA secondary structures near the terminators in these operons.

The amidase (*ami*) operon of *Pseudomonas aeruginosa* is regulated by antitermination in response to short-chain aliphatic amides, such as acetamide. The *amiR* gene encodes an antiterminator protein (AmiR), which is negatively regulated by AmiC, apparently through formation of an AmiC-AmiR complex (20). Acetamide destabilizes the AmiC-AmiR complex, leading to antitermination and expression of the operon. AmiR interacts with an RNA target in the 5-leader region of the *ami* mRNA that contains a Rho-independent terminator. However, no clear antiterminator RNA secondary structure is predicted. AmiR binding has been suggested to function in antitermination by interfering directly with formation of the terminator stem-loop structure (20).

In addition to all the catabolic operons described above, one anabolic operon has been shown to be regulated by antitermination. Expression of the *nas* operon of *Klebsiella pneumoniae*, which encodes enzymes required for nitrate assimilation in this bacterium, is induced by nitrate or nitrite. The NasR protein mediates transcription antitermination through a terminator in the leader region of the operon (21). This protein shows weak homology with AmiR in the carboxyl-terminal region.

TRANSFER RNA-MEDIATED ANTITERMINATION

An interesting variation on the antitermination mechanism involves the use of tRNA as the regulatory molecule. This mechanism regulates a large number of **aminoacyl-tRNA synthetase** genes in **Gram-positive bacteria** (22,23) and several amino acid biosynthetic operons, including the *ilv-leu*, in *B. subtilis* (24,25), and the *his* and *trp* operons in *Lactococcus lactis* (26). Expression of these genes is induced specifically by starvation for the corresponding amino acid. In the case of the amino acid operons, insufficient levels of the amino acid leads to increased expression of the corresponding biosynthetic operon. For the aminoacyl-tRNA synthetase genes, increasing the level of the synthetase is thought to allow more efficient charging of the cognate tRNA when the corresponding amino acid pool is low.

A long (approx. 300-nucleotide) untranslated leader region exists upstream of the structural gene(s) of these operons that contains several conserved features, including three stem-loop structures preceding a Rho-independent transcription terminator. Hence, in the absence of the inducing signal, transcription terminates prematurely in the leader region prior to the coding sequences. In addition to the conserved secondary structures, there is an important conserved 14-nucleotide sequence known as the T-box present in each leader region; hence these genes are known as the T-box family. An alternate arrangement of the leader region involving base-pairing between a portion of the T-box and a conserved sequence in the 5′ side of the terminator stem has been proposed to form an antiterminator structure that allows transcription to read through into the structural genes (Fig. 2) (27).

Another important conserved feature of the leader region of these genes is the presence of a triplet sequence corresponding to a codon for the appropriate amino acid for each operon. For example, in *tyrS*, which encodes tyrosyl-tRNA synthetase, the leader contains a UAC tyrosine codon, while the ilv-leu operon leader contains a CUC leucine codon. This triplet is always present in a bulged sequence in Stem-loop I (Fig. 2) and has been shown to be critical for induction in several systems. It was the presence of these triplets that led to the hypothesis that tRNAs play a role in this regulatory mechanism. This triplet was designated the "specifier sequence" because, in the case of the *B. subtilis tyrS* gene, altering the sequence to correspond to a codon for another amino acid switched induction to respond to starvation for the new amino acid (27). Other experiments demonstrated that translation of this codon was not involved in induction and that uncharged tRNA was the inducer (27). In addition, a second interaction between the CCA sequence at the 3′ end of the uncharged tRNA and the complementary UGG sequence in the T-box have been shown to be important (28).

A model for tRNA-regulated antitermination regulation is presented in Fig. 2. Under starvation conditions for the corresponding amino acid, the cognate uncharged tRNA interacts with two sites in the leader region, to induce formation of the antiterminator structure and allow transcription to read through into the coding region. Aminoacylation of this tRNA is predicted to interfere with the interaction at the CCA end and prevent the charged tRNA from binding; the leader transcript then folds into the conformation with the terminator, halting transcription. It is not known if factors in addition to tRNA are required for antitermination. To date, however, it has not been possible to reconstitute tRNA-mediated antitermination in vitro, and several other lines of evidence also suggest that other factors may be involved in this mechanism (23).

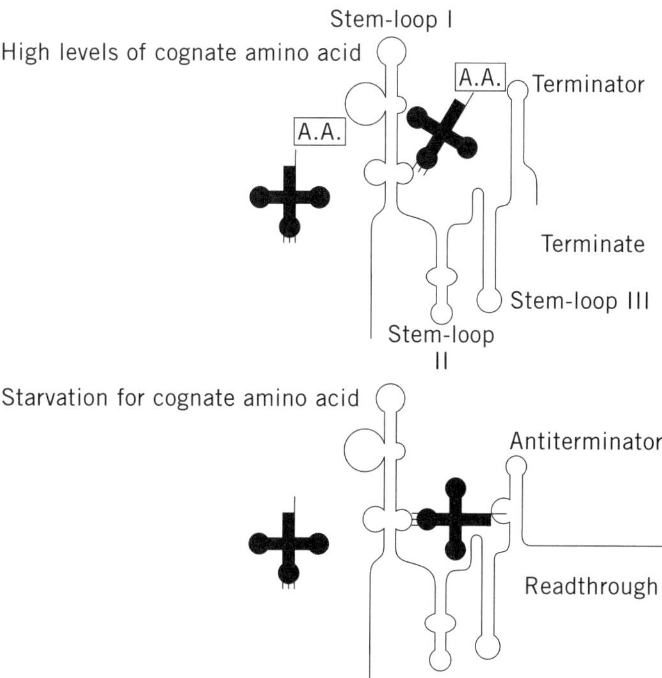

Figure 2. Model for antitermination control by tRNA. Under conditions with adequate levels of the cognate amino acid (aa), the charged tRNA does not interact with the leader region, and the terminator forms. Under conditions of starvation for the appropriate amino acid, the uncharged tRNA interacts with the leader region via base-pairing between the anticodon and the specifier sequence, and by base-pairing between the CCA sequence at the acceptor end of the tRNA with the side bulge of the antiterminator in the leader. These interactions stabilize formation of the antiterminator conformation of the leader transcript, resulting in induction of expression of the gene. The tRNA is shown as the shaded cloverleaf structure, and a boxed "A.A." attached to the tRNA indicates it is aminoacylated. Adapted from T. M. Henkin (1996) *Annu. Rev. Gen.* **30**, 35–57.

In addition to the antitermination mechanism described above, processing of the leader RNA has been shown to play a role in regulating expression of the *B. subtilis thrS* gene (29). Cleavage occurs in the loop of the antiterminator near the T-box sequence and is more efficient under threonine starvation conditions, suggesting that bound tRNA induces both antitermination and RNA processing. This processing increases the stability of the mRNA, which would allow for increased translation and production of the threonyl-tRNA synthetase. Thus induction of expression of this gene in response to threonine starvation occurs at both the level of transcription antitermination and mRNA stability.

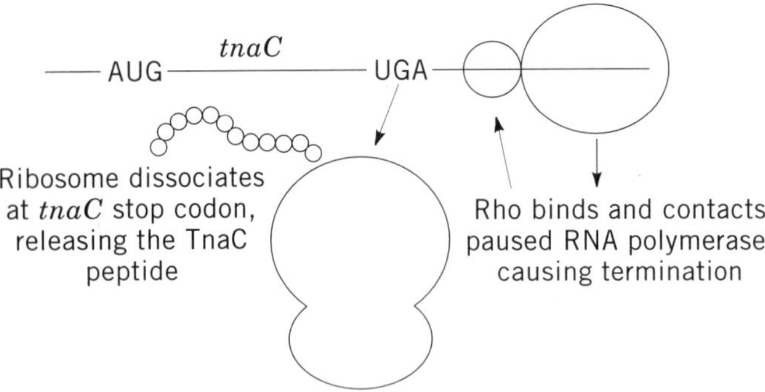

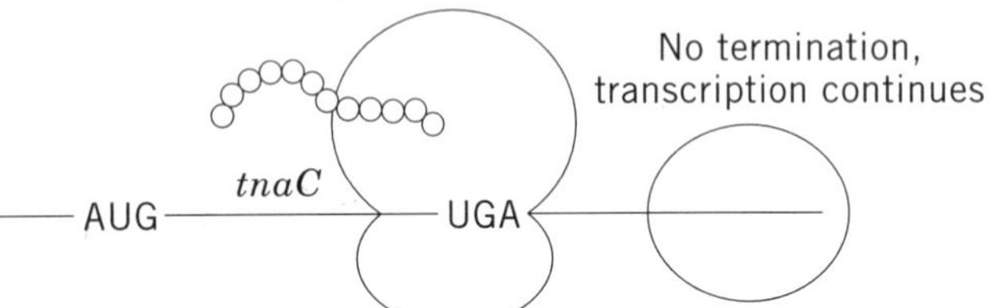

Figure 3. Model of *E. coli tna* operon regulation. Under noninducing conditions (no extracellular tryptophan), ribosome dissociation at the *tnaC* stop codon exposes a *rut* site, allowing Rho binding. Rho translocates to the paused RNA polymerase, leading to transcription termination. Under inducing conditions (extracellular tryptophan), ribosome stalling at the *tnaC* stop codon prevents Rho association, leading to transcription readthrough. See text for details. Adapted from C. Yanofsky, K. V. Konan and J. P. Sarsero (30).

THE *E. COLI* TRYPTOPHANASE OPERON

E. coli and several other microorganisms have the capacity to degrade tryptophan as a source of carbon, nitrogen, and/or energy (30). The degradative tryptophanase operon (*tnaCAB*) of *E. coli* is regulated by **catabolite repression** (31) and by an antitermination mechanism. Antitermination involves translation of a ***cis**-acting* 24-residue leader peptide (*tnaC*) containing a critical Trp codon (32,33), one or more RNA polymerase pause sites between *tnaC* and *tnaA* (34), and Rho termination factor (34). While the precise antitermination mechanism responsible for controlling the *tna* operon is not firmly established, all of the data are consistent with the following model (Fig. 3) (35). During growth in a medium lacking both tryptophan and a catabolite-repressing carbon source, transcription initiation is efficient. As transcription proceeds, translation of the leader peptide occurs as soon as the coding sequence becomes available. Once the translating ribosome reaches the UGA stop codon, ribosome release exposes a *rut* (Rho utilization) site that immediately follows the stop codon. Rho then binds to the *rut* site and begins to translocate in the 3′-direction, until it encounters paused RNA polymerase, ultimately leading to transcription termination upstream of *tnaA*. When cells are growing with inducing levels of tryptophan, TnaC, or a complex of TnaC with an unidentified protein, prevents ribosome release at the *tnaC* stop codon, thereby masking the *rut* site and, hence, blocking Rho interaction with the transcript. Eventually RNA polymerase would overcome the pause signal and transcribe the structural genes encoding tryptophanase (*tnaA*) and a tryptophan **permease** (*tnaB*). This model assumes that there is a fundamental difference between the TnaC peptide, or the TnaC peptide-protein complex, in cells growing with or without tryptophan. It was proposed that such a complex under inducing conditions would prevent ribosome release (35), reminiscent of characterized translation attenuation mechanisms (36). The tryptophanase operon of *Proteus vulgaris* is thought to be regulated by a mechanism essentially identical to that of *E. coli* (37).

BIBLIOGRAPHY

1. D. I. Freidman and D. L. Court (1995) *Mol. Microbiol.* **18**, 191–200.
2. C. Condon, C. Squires, and C. L. Squires (1995) *Microbiol. Rev.* **59**, 623–645.
3. B. Rutberg (1997) *Mol. Microbiol.* **23**, 413–421.
4. S. Mahadevan and A. Wright (1987) *Cell* **50**, 485–494.
5. O. Amster-Choder and A. Wright (1990) *Science* **249**, 540–542.
6. Q. Chen, J. C. Arents, R. Bader, P. W. Pestma, and O. Amster-Choder (1997) *EMBO J.* **16**, 4617–4627.
7. M. El Hassount, B. Henrissat, M. Chippaux, and F. Barras (1992) *J. Bacteriol.* **174**, 765–777.
8. D. Le Coq, C. Lindner, S. Krüger, M. Steinmetz, and J. Stulke (1995) *J. Bacteriol* **177**, 1527–1535.
9. J. Bardowski, D. S. Ehrlich, and A. Chopin (1994) *J. Bacteriol.* **176**, 5681–5685.
10. M. Debarbouille, M. Arnaud, A. Fouet, A. Klier, and G. Rapoport (1990) *J. Bacteriol.* **172**, 3966–3973.
11. A. M. Crutz, M. Steinmetz, S. Aymerich, R. Richter, and D. Le Coq (1990) *J. Bacteriol.* **172**, 1043–1050.
12. M. Arnaud, M. Debarbouille, G. Rapoport, M. H. Saier Jr., and J. Reizer (1996) *J. Biol. Chem.* **271**, 18966–18972.
13. M. Idelson and O. Amster-Choder (1998) *J. Bacteriol.* **180**, 1043–1050.
14. X. Manival, Y. Yang, M. P. Strub, M. Kochoyan, M. Steinmetz, and S. Aymerich (1997) *EMBO J.* **16**, 5019–5029.
15. H. van Tilbeurgh, S. Manival, S. Aymerich, J. M. Lhoste, C. Dumas, and M. Kochoyan (1997) *EMBO J.* **16**, 5030–5036.
16. K. Schnetz, J. Stulke, S. Gertz, S. Krüger, M. Krieg, M. Hecker, and B. Rak (1996) *J. Bacteriol.* **178**, 1971–1979.
17. C. A. Alpert and U. Siebers (1997) *J. Bacteriol.* **179**, 1555–1562.
18. E. Glatz, R. P. Rutberg, and B. Rutberg (1996) *Mol. Microbiol.* **19**, 319–328.
19. L. Wray Jr. and S. Fisher (1994) *J. Bacteriol.* **176**, 5466–5473.
20. S. A. Wilson, S. J. M. Wachira, R. A. Norman, L. H. Pearl, and R. E. Drew (1996) *EMBO J.* **15**, 5907–5916.
21. J. T. Lin and V. Stewart (1996) *J. Mol. Biol.* **256**, 423–435.
22. T. M. Henkin (1994) *Mol. Microbiol.* **13**, 381–387.
23. C. Condon, M. Grunberg-Manago, and H. Putzer (1996) *Biochimie* **78**, 381–389.
24. J. A. Grandoni, S. A. Zahler, and J. M. Calvo (1992) *J. Bacteriol.* **174**, 3212–3219.
25. J. A. Grandoni, S. B. Fulmer, V. Brizio, S. A. Zahler, and J. M. Calvo (1993) *J. Bacteriol.* **175**, 7581–7593.
26. P. Renault, J. J. Godon, C. Delorme, G. Corthier, and S. D. Erlich (1995) *Dev. Biol. Stand.* **85**, 431–441.

27. F. J. Grundy and T. M. Henkin (1993) *Cell* **74**, 475–482.
28. F. J. Grundy S. M. Rollins, and T. M. Henkin (1994) *J. Bacteriol.* **176**, 4518–4526.
29. C. Condon, H. Putzer, and M. Grunberg-Manago (1996) *Proc. Natl. Acad. Sci. USA* **93**, 6992–6997.
30. C. Yanofsky, K. V. Konan, and J. P. Sarsero (1996) *Biochimie* **78**, 1017–1024.
31. J. L. Botsford and R. D. DeMoss (1971) *J. Bacteriol.* **105**, 303–312.
32. P. Gollnick and C. Yanofsky (1990) *J. Bacteriol.* **172**, 3100–3107.
33. K. Gish and C. Yanofsky (1995) *J. Bacteriol.* **177**, 7245–7254.
34. V. Stewart, R. Landick, and C. Yanofsky (1986) *J. Bacteriol.* **166**, 217–223.
35. K. V. Konan and C. Yanofsky (1997) *J. Bacteriol.* **179**, 1774–1779.
36. P. S. Lovett and E. J. Rogers (1996) *Microbiol. Rev.* **60**, 366–385.
37. A. V. Kamath and C. Yanofsky (1997) *J. Bacteriol.* **179**, 1780–1786.

Suggestions for Further Reading

T. M. Henkin (1996) Control of transcription termination in prokaryotes. *Ann. Rev. Gen.* **30**, 35–57. A comprehensive review of the mechanisms of gene regulation by controlling transcription termination in bacteria.

T. Platt (1998) RNA structure transcription elongation, termination and antitermination. In *RNA Structure and Function* (R. W. Simons and M. Grunberg-Manago, eds.), Cold Spring Harbor Laboratory Press, Cold Spring Harbor, NY, pp. 541–574. A more in-depth look at the processes involved in transcription and regulation of transcription termination.

M. Steinmetz (1993) Carbohydrate catabolism: pathways, enzymes, genetic regulation, and evolution. In Bacillus subtilis and Other Gram-positive Bacteria: biochemistry, physiology, and molecular genetics. (A. L. Sonenshein, J. A. Hoch, and R. Losick, eds.), American Society for Microbiology, Washington, DC, pp. 157–170. A more in-depth look at the Sac systems in *B. subtilis*.

ANTITHROMBIN

R. W. CARRELL

Antithrombin is a **proteinase inhibitor** of the **serpin** family that is the principal anticoagulant in human plasma; it is also known as antithrombin-III. It is composed of a single **polypeptide chain** of 432 amino acid residues, with four carbohydrate side chains (see ***N*-Glycosylation**). It has a reactive-center **peptide bond** between Arg393 and Ser394 that provides a specific target for **thrombin**, factor Xa, and other **proteinases** of the **blood clotting** system. Antithrombin is present in plasma at a concentration of nearly 100 mg/L and circulates in a relatively inactive form, being activated conformationally by the binding of heparin. Genetic deficiency or dysfunction of antithrombin is a significant cause of familial thromboembolic disease.

Suggestions for Further Reading

S. T. Olson and I. Björk (1992) Regulation of thrombin by antithrombin and heparin cofactor II, in *Thrombin, Structure and Function* (L. J. Berliner, ed.), Plenum Press, New York, pp. 159–217.

L. Jin, J. P. Abrahams, R. Skinner, M. Petitou, R. N. Pike, and R. W. Carrell (1997) The anticoagulation activation of antithrombin by heparin, *Proc. Natl. Acad. Sci. USA* **94**, 14683–14688.

α_1-ANTITRYPSIN

R. W. CARRELL

α_1-Antitrypsin is a **proteinase inhibitor** of the **serpin family** that consists of a single **polypeptide chain** of 294 amino acid residues and four carbohydrate side chains (see ***N*-Glycosylation**). It also known as α_1-proteinase inhibitor. Present at a concentration of 1 g/L, it is the predominant proteinase inhibitor in human plasma in terms of concentration. Its reactive-center **peptide bond** between Met358 and Ser359 specifically targets the **elastases** that are involved in connective tissue remodeling. Genetic deficiency of α_1-antitrypsin is relatively common and predisposes to the development of the lung disease emphysema, with an associated risk of liver cirrhosis (see **Serpins**).

Suggestions for Further Reading

R. W. Carrell, J.-O. Jeppsson, C.-B. Laurell, S. O. Brennan, M. C. Owen, L. Vaughan, and D. R. Boswell (1982) Structure and variation of human α_1-antitrypsin, *Nature* **298**, 329–334.

R. Mahadeva and D. A. Lomas (1998) α_1-Antitrypsin deficiency, cirrhosis and emphysema, *Thorax* **53**, 501–505.

AP ENDONUCLEASES

AZIZ SANCAR
CAROL THOMPSON

AP endonucleases are enzymes that cleave the phosphodiester bond on either side of an **AP site** in DNA and are involved in **base excision repair**. Enzymes that cleave the phosphodiester bond to the 3′ side are referred to as type I AP endonucleases, and those that cleave the phosphodiester bond 5′ to the AP site are type II AP endonucleases. Whether or not a bona fide type I AP endonuclease exists that is not associated with **DNA glycosylase** activity is not known. However, type II AP endonucleases, which cleave AP sites by hydrolysis, have been found in many organisms and have been characterized.

AP LYASES (CLASS I AP ENDONUCLEASES)

Type I AP endonucleases are also DNA glycosylases, and it appears that the same **active site** is involved in both the glycosylase and AP endonuclease activities. However, it has also been demonstrated that cleavage of AP site occurs by β-elimination, which involves abstraction of a hydrogen from the C2′ position of the deoxyribose. Because this is a lyase reaction, and not a hydrolyase reaction, the type I AP endonucleases are referred to as AP lyases (1).

Endonuclease III and 8-oxoguanine glycosylase are examples of this class of enzymes (1,2). The α-NH_2 group of the *N*-terminal amino acid residue forms a **Schiff base** with the ring-opened form of the deoxyribose, eventually leading to cleavage of the deoxyribose–phosphate bond without water addition. Following elimination of the base by glycosylase action, the enzyme remains attached to DNA through a protonated Schiff base intermediate where the C1′ of the sugar is covalently bound to the nitrogen of the *N*-terminal α-amino group of the protein (3). β-Elimination of this intermediate, followed by hydrolysis of the Schiff base, results in cleavage of the phosphodiester bond and generation of a *trans* $\alpha-\beta$-unsaturated aldehyde on the 5′ side, and a 5′-phosphate on the 3′ side, of the cleavage, which is concomitant with the release of the enzyme. When the AP lyase activity is associated with a glycosylase activity, the two reactions are, for the most part, coupled (2). In these instances, it is natural to assign a physiological role to the AP lyase. However, cleavage of an AP site on the 3′ side occurs readily even in the absence of enzymes and can be accelerated by alkaline pH and by basic proteins such as **histones** and **cytochrome** *c*. Hence the assignment of AP lyase activity to a protein in the absence of an associated glycosylase activity is virtually impossible.

AP ENDONUCLEASES (CLASS II AP ENDONUCLEASES)

These enzymes hydrolyze the phosphodiester bond 5′ to an AP site. AP endonuclease IV in *E. coli*, which has a **homologue** in yeast and humans, is the only example of a "pure" AP endonuclease—that is, an endonuclease with only hydrolysis activity 5′ to an AP site and with no significant activity on any other substrate. However, the best-characterized AP endonuclease is the *E. coli* exonuclease III and its homologues in eukaryotes, including yeast, Drosophila, and mammals. In contrast to the pure AP nuclease endonuclease IV, the exonuclease III, as the name implies, has exonuclease activity as well.

Exonuclease III/Rrp1/APE

The prototype of this group of enzymes, exonuclease III, has several activities, including **ribonuclease** H, 3′ to 5′ exonuclease, and type II AP endonuclease. The enzyme is specific for double-stranded DNA. It is a simple polypeptide chain of 30 kDa with no cofactor. The **X-ray crystallography** structure of an enzyme–substrate complex demonstrates that the base is flipped out of the double helix, into the active site of the enzyme (4). The *E. coli* enzyme is a potent 3′ to 5′ exonuclease. The exonuclease activity has a preference for blunt double-stranded termini, but it can initiate exonuclease action from a nick as well. An activity related to the 3′ to 5′ exonuclease function is the ability of the enzyme to remove deoxyribose fragments, including 3′-phosphoglycolate esters and 3′-phosphate, from the 3′ termini of DNA strand breaks generated by attack of reactive oxygen species on DNA. Exonuclease III-defective mutants are extremely sensitive to H_2O_2 and ionizing radiation, which most probably kills cells through the strand breaks introduced by reactive oxygen species. As a rule, these are not clean breaks with 3′-OH and 5′-P termini. Instead, they most often contain either 3′-phosphate or fragmented deoxyribose, which must be processed further before they can be utilized by **DNA polymerases** to fill in the gap created by the DNA damage. These structures are removed efficiently by exonuclease III, but not by endonuclease IV.

The *Drosophila* exonuclease III homologue is a monomer of 75 kDa. The 250-amino-acid-residue *C*-terminal region of the protein is homologous to the exonuclease III family of AP endonucleases. However, the *N*-terminal 400 residues are not related to any known sequence, and the function of this putative **domain** is unknown at present. The *Drosophila* enzyme was initially isolated as an activity that catalyzes strand transfer and was thought to perform a **RecA**-like function in this organism, hence the name Rrp1 (Recombination related protein 1). However, later work has not confirmed a role for this protein in **recombination**. The enzyme has a potent type II AP endonuclease activity, but its 3′ to 5′ exonuclease activity is rather modest compared to the *E. coli* enzyme and is only 10^{-2} to 10^{-3} of the AP endonuclease activity. Whether the *N*-terminal 400 residues play any role in recombination or in DNA repair is not known at present.

The human AP endonuclease (APE/HAP1/APEX) is highly homologous (90% to 95% sequence identity) to the enzymes from bovine and rodent sources. These enzymes are 35-kDa monomers and are clearly related to *E. coli* exonuclease III. The mammalian enzymes have potent type II AP endonuclease and 3′-phosphoglycol aldehyde esterase activities necessary for denuding a "jagged" 3′ terminus caused by direct attack of reactive oxygen species or by direct hit by ionizing radiation. In contrast to exonuclease III, however, the 3′ to 5′ exonuclease activity of mammalian type II AP endonucleases is rather weak and undetectable with certain 3′ sequences. The structure of human AP endonuclease indicates that it flips out the deoxyribose moiety into the active site (5), as is observed also in the *E. coli* exonuclease III.

A unique property of mammalian APE proteins is their capacity to reduce critical **cysteine** residues in certain **transcription factors** and by doing so activate these transcription factors. In fact, the human APE was also purified as a factor that activates the AP-1(Jun/Fos) transcription factor and was named Ref (for reducing factor 1) before the realization that it is identical to APE. The APE cysteine residue involved in this reduction reaction is located in the NH_2-terminal region and plays no role in the AP endonuclease activity of this protein. In addition to Jun and Fos, other transcription factors, including USF, NF-κB, c-Myb, and v-Rel, are activated by reduction of an oxidized cysteine by APE *in vitro* (5). However, the physiological role of these *in vitro* activities is currently not known.

Mouse APE **knockout** mutants are lethal at the embryonic stage. This could be due to either (a) the importance of base excision repair for survival of the organism or (b) disruption of an activating mechanism for important transcription factors.

Endonuclease IV

The *E. coli* endonuclease IV, Nfo, is a monomer of 30 kDa that cleaves 5′ to AP sites in a reaction that is independent of divalent metal ions such as Mg^{2+}. It also has potent 3′-PGA diesterase and 3′-phosphatase activities but does not have exonuclease function. Although Nfo^- mutants are not especially sensitive to H_2O_2, they are hypersensitive to killing by the oxidative agents bleomycin and *t*-butylhydroperoxide, indicating that endonuclease IV acts on a subclass of oxidative DNA lesions that are not processed efficiently by exonuclease III. The importance of Nfo in cellular defense against oxidative stress is underscored by the fact that this is the only DNA repair enzyme that is induced by oxidative stress as part of the SoxRS regulon.

The Nfo homologue in yeast, Apn1, is the major AP endonuclease in this organism. Apn1 is a monomer of 40 kDa with 31% sequence identity with the *E. coli* Nfo endonuclease. The biochemical properties of Apn1 are very similar to those of Nfo, and indeed the expression of Apn1 in *E. coli* corrects the oxidant-sensitive phenotype of *nfo* mutants, but not the H_2O_2 sensitivity.

BIBLIOGRAPHY

1. V. Bailly and W. G. Verly (1987) *Biochem. J.* **242**, 565–572.
2. M. L. Dodson, R. D. Schrock, and R. S. Lloyd (1993) *Biochemistry* **32**, 8284–8290.
3. R. P. Cunningham, H. Asahara, J. F. Bank, C. P. Scholes, J. C. Salerno, K. Surerus, E. Munck, J. McCracken, J. Peisach, and M. H. Emptage (1988) *Biochemistry* **28**, 4450–4455.
4. C. D. Mol, C. F. Kuo, M. M. Thayer, R. P. Cunningham, and J. A. Tainer (1995) *Nature* **374**, 381–386.
5. M. A. Gorman, S. Morera, D. G. Rothwell, E. de la Fortelle, C. D. Mol, J. A. Tainer, I. D. Hickson, and P. S. Freemont (1997) *EMBO J.* **16**, 6548–6558.

Suggestions for Further Reading

B. Demple and L. Harrison (1994) Repair of oxidative damage to DNA: enzymology and biology. *Annu. Rev. Biochem.* **63**, 915–948.

P. W. Doetsch and R. P. Cunningham (1990) The enzymology of apurinic/apyrimidinic endonucleases. *Mutation Res.* **236**, 173–201.

E. Seeberg, L. Eide, and M. Bjoras (1995) The base excision repair pathway. *Trends Biochem. Sci.* **20**, 391–397.

R. J. Roberts (1995) On base flipping. *Cell* **82**, 9–12.

AP SITES (APURINIC/APYRIMIDINIC SITES)

Aziz Sancar
Carol Thompson

AP (apurinic/apyrimidinic) sites are deoxyribose residues in the DNA that have lost the purine or pyrimidine bases.

OCCURRENCE

AP sites are generated by a variety of mechanisms:

Spontaneous Base Loss

The glycosylic bond joining the base to the deoxyribose in the DNA backbone is relatively unstable and is cleaved, even under physiological conditions, at a rate incompatible with life. It is estimated that at normal pH and temperature there are base losses at rates of about 10^{-12} s^{-1} for double-stranded DNA and 10^{-10} s^{-1} for single-stranded DNA. This means that humans (with about 1.2×10^{10} bp per diploid cell) lose about 1 base per minute per cell, or, taking into account that there are about 10^{13} cells in the human body, lose 1.5×10^{16} bases from their DNA daily. The rate of base loss is greatly accelerated by heat and low pH. Because of higher susceptibility of their N9 atom to nucleophilic attack, the purine bases are hydrolyzed at a rate about 100-fold faster than are pyrimidines (1). In fact, because of this unique property of purines, one of the procedures for chemical **DNA sequencing** reactions consists of heating the DNA at acidic pH, followed by cleavage of the resulting AP site by alkali treatment.

Ionizing Radiation

Ionizing radiation most often causes base reduction, oxidation, or fragmentation. In particular, a **urea** residue attached to the deoxyribose is a relatively common product of ionizing radiation and, for all practical purposes, may be considered an AP site. Such radiation causes base loss by either (a) generating reactive oxygen species that attack and destroy bases or (b) direct hits by ionizing radiation.

Glycosylases

There are about a dozen **DNA glycosylases** with varying degrees of specificity for bases with minor modifications: uracil glycosylase, 3-methyladenine DNA glycosylase, 8-oxoguanine DNA glycosylase, and glycosylases that act on mismatches, such as T-G glycosylase, specific for the T residue, and A-G glycosylase, specific for the A residue. These enzymes release the modified, abnormal, or mismatched bases by hydrolyzing the glycosylic bond linking the base to the deoxyribose. Most of the glycosylases have dual enzymatic activities in that, after cleaving the glycosylic bond, they also cleave the phosphodiester bond by β-elimination. However, some, such as uracil glycosylase, are "pure" glycosylases with no such associated AP lyase activity.

CONSEQUENCES

AP sites in DNA are, in fact, very unstable. Even at physiological pH and temperature, the abasic sugar can be eliminated by $\beta-\delta$ elimination. The presence of the divalent metal ions that would be expected in biological fluids greatly accelerates the cleavage reaction. The reaction may be further accelerated by basic proteins, such as **cytochrome** *c*. Consequently, AP sites as such do not constitute an important lesion interfering with cell survival. In contrast to β-elimination, however, which occurs readily under a variety of conditions, the δ-elimination reaction, which in combination with β-elimination would release the abasic deoxyribose from DNA, does not occur at physiologically relevant rates. **AP endonucleases** hydrolyze the phosphodiester bond 5 to the AP site, and *Escherichia coli* cells that lack this enzyme are sensitive to agents that generate AP sites or fragmented deoxyribose residues.

BIBLIOGRAPHY

1. T. Lindahl (1976) *Nature* **259**, 64–66.

Suggestions for Further Reading

B. Demple and L. Harrison (1994) Repair of oxidative damage to DNA: enzymology and biology. *Annu. Rev. Biochem.* **63**, 915–948.

E. Seeberg, L. Eide, and M. Bjoras (1995) The base excision repair pathway. *Trends Biochem. Sci.* **20**, 391–397.

APOPTOSIS

Nicola McCarthy
Gerard Evan

Wyllie et al. initially used the term "apoptosis" in a paper in 1972 to describe a newly observed form of **cell death** (1). The discovery of apoptosis has revolutionized many areas of modern biology, especially those areas concerned with disease development (2). This is primarily because apoptosis is an active form of cell death that, unlike **necrosis**, is critical for the maintenance of tissue homeostasis (3,4). Thus, cell populations are numerically controlled through not only **differentiation** and proliferation, but also the physiological loss of cells through apoptosis. Critically, this implies that a cell which loses the capacity to apoptose, or a cell which becomes inappropriately sensitive to death stimuli, can lead to perturbations in cell number regulation and, ultimately, disease development.

Cells undergoing apoptosis are evident in all tissues in response to diverse stimuli. In healthy tissues, apoptosis accounts for all the cell deaths occurring in response to normal physiological signals. Apoptosis (often referred to as **programmed cell death** in a developmental context) occurs at many stages of embryonic **development**, eg, during the formation of digits from a solid limb paddle, where interdigital cells between the digits die (5). Single apoptotic cells are evident in healthy proliferating tissues such as small gut crypts and the dermis of the skin (4). Moreover, apoptosis is also seen during endocrine-induced atrophy of tissues, in cell-mediated **immunity**, in development of the nervous system, and in the "necrotic" oxygen-starved cores of solid tumors (6).

Morphologically, an apoptotic cell is very different from a necrotic cell (1,6,7); see Figure 1. One of the most prominent and easily identifiable stages of apoptosis involves the **nucleus**. **Chromatin** condenses and forms aggregates near the nuclear **membrane**, which becomes convoluted, whereas the **nucleolus** becomes enlarged and appears abnormally granular. The chromatin is also subject to the actions of an activated **endonuclease** that cleaves the DNA into 300- to 50-kbp fragments initially, and 180-bp fragments subsequently (8). Changes in the cytoplasm are also evident at this time; the cell shrinks visibly, adherent cells round up, and distinct protuberances or membrane blebs are discernible. **Organelles** within the shrunken cytoplasm still look normal, except for dilation of the **endoplasmic reticulum**. These "blebbing" cells exclude vital dyes, indicating no structural failure in the cell membrane. It is at this point in the apoptotic process, which may only take 10 to 30 minutes, that apoptotic cells *in vivo* are **phagocytosed**, by either their nearest neighbors or professional **macrophages** (9). Thus, apoptosis occurring in single cells *in vivo* is very easily overlooked due to its rapid progression. Cells in the later stages of apoptosis, especially those *in vitro*, form apoptotic bodies as a result of pinching off of the highly convoluted blebbing areas of the cell. These apoptotic bodies are phagocytosed or, in situations where there is much cell death, the unphagocytosed apoptotic cells undergo secondary necrosis, characteristically swelling and loosing membrane integrity.

REMOVAL OF APOPTOTIC CELLS

A critical part of the apoptotic pathway is the efficient recognition and phagocytosis of apoptotic cells. The rapid disposal of apoptotic cells does not elicit an **immune response**, con-

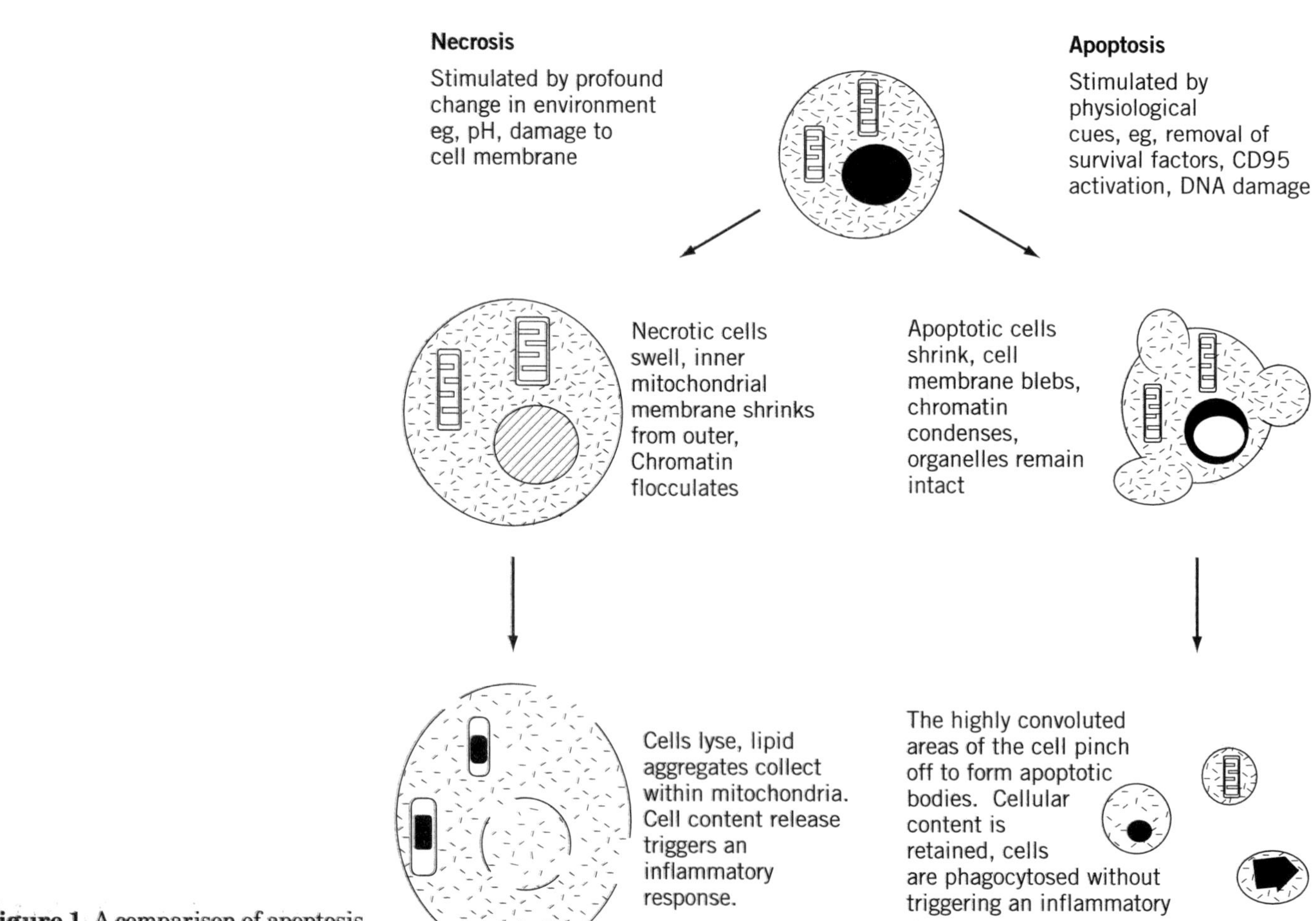

Figure 1. A comparison of apoptosis and necrosis.

sistent with apoptosis being the "no-nonsense" pathway for the disposal of unwanted cells. A breakdown in either the communication between macrophage and apoptotic cell or in the pathway of apoptosis itself may be two of the mechanisms through which chronic inflammatory disorders occur. Recognition of apoptotic cells by professional macrophages is mediated partly by $\alpha_v\beta_3$ **integrin** (the *vitronectin* receptor) and the CD36 ligand. Between these two molecules on the external surface of the cell sits a **glycoprotein** containing the sequence -R-G-D- (-Arg-Gly-Asp-), *thrombospondin*, which acts as a molecular bridge to bind to an unknown anionic site on the apoptotic cell (10–12). Apoptotic cells also express surface phosphatidylserine, which binds to an as yet uncharacterized macrophage **receptor** (13,14). The recognition of apoptotic cells mediated by the vitronectin receptor, CD36 ligand, and thrombospondin is confined to monocyte-derived macrophages. Recognition of apoptotic cells through the expression of phosphatidylserine, which is flipped onto the cellular surface due to the inactivation of an ATP-dependent flipase, is utilized primarily by inflammatory macrophages (9,15). Four other molecules have been implicated in apoptotic cell recognition systems: (1) the 61D3-antigen found on human macrophages, (2) ICAM-3, found on the surface of human **B cells** (16), (3) the ABC1 transporter (17), and (4) the macrophage scavenger receptor (18). As mentioned previously, the interaction between apoptotic cells and macrophages does not elicit an inflammatory response, unless the apoptotic cells start to undergo secondary necrosis. Under these circumstances, other macrophage receptors are employed, and inflammatory **cytokines** are released (19). This change from "silent" death to one that activates the immune system may be a method of recruiting more phagocytes to the site of cell death to cope with the increasing number of corpses.

REGULATION OF APOPTOSIS AND CELL VIABILITY

The discovery that apoptosis is an active form of cell death which is transiently suppressable by both inhibitors of protein and RNA synthesis suggested that the cell needed to synthesize new proteins prior to its death [reviewed in (20)]. This sparked an enthusiastic hunt for a critical cell death gene that had to be translated prior to death. No such gene has been found, but many genes that regulate apoptosis have been identified as a result.

A cell can be triggered to undergo apoptosis in response to diverse stimuli, with each stimulus activating different pathways involving **gene expression**, as well as **post-translational modification**. The discovery that apoptosis can be suppressed by the presence of specific survival factors, either cytokines or proteins expressed within the cell, suggested that regulation of cell viability was paramount to the function of the cell (21). Indeed, it appears that apoptosis represents a default pathway in all cells (22). If a cell does not receive the correct survival stimulus, it will die; consequently, cells within multicellular organisms are maintained in a viable state by a constant supply of survival factors.

Once a cell is triggered to apoptose, activation of a common set of destruction proteins, irrespective of the stimulus involved, precipitates the morphological changes that we call apoptosis. Overall, the apoptotic pathway can be divided into three phases:

1. The *decision phase*, in which the cell receives a stimulus and, depending on both its internal and external environments, may or may not be triggered to die.
2. The *commitment phase*, in which the cell is committed to death and cannot recover.
3. The *execution phase*, in which the decision to die has been made and the activated apoptotic machinery leads to the morphological changes that define an apoptotic cell.

THE DECISION PHASE OF APOPTOSIS

The decision phase can be compared to that of a judge hearing the evidence during a trial. The evidence presented comes from many different sources: genes, cytokines, toxic chemicals, DNA damage, and the presence of viral and bacterial infection. Only once all the evidence is heard will the sentence be passed. The genes that modulate the decision phase are many and varied. Some are more commonly associated with the regulation of the cell cycle, such as *c*-**myc**, *cdc25*, *c-fos*, **p53** and *rb* (**retinoblastoma**), whereas others are members of recently identified gene families, such as the *bcl-2* family. **Cytokines** can either trigger or delay apoptosis. **Tumor necrosis factor** α (TNFα) and CD95 ligand (CD95L, FasL, Apo-1L) are two examples of cytokines that trigger apoptosis upon binding to their receptors (23), whereas the addition of factors such as **interleukin**-3 (IL-3) (24) or ***insulin****-like growth factor 1* (IGF-1) suppresses death (25). Conversely, removal of anti-apoptotic cytokines also induces cell death (26,27). Both viruses and bacteria trigger apoptosis upon infection of a cell, suggesting that the cell commits suicide to prevent the invading pathogen from spreading. However, both viruses and bacteria have evolved mechanisms that manipulate this response. **Adenovirus**, eg, has three genes required for productive infection: E1A, E1Bp19, and p55. E1A stimulates the infected cell to proliferate; however, this stimulus concomitantly triggers apoptosis. E1Bp19 and p55 delay the host cell from triggering cell death, thereby facilitating virus production (28,29). Alternatively, some bacteria such as *Shigella flexneri* actively promote the death of the cell by triggering the execution machinery (30–32). This allows the early release of the bacteria and triggers an immune response, which damages the surrounding tissue and aids the passage of the bacteria into its target cells.

Oncogenes, Tumor Suppressor Genes, and Apoptosis

In terms of suppressing tumorigenesis, long-lived organisms like humans need to limit the chances of their component cells acquiring mutations that lead to increased proliferative capacity and eventual clonogenic outgrowth. This is, in part, limited by restricting the types of cells with proliferative capacity. Many adult cell lineages exist in a senescent or postreplicative state. Moreover, for those cells that must retain the capacity to divide, apoptosis provides a "safety net." Genes that are involved in proliferation are also involved in cell death; in fact, the two processes appear inexorably linked. This paradoxical coupling of two opposing biological processes was first realized through work on the pervasive oncogene *c-myc* (33) but has since been confirmed for several mitotic genes (34).

c-myc and Apoptosis

The protein product of the *c*-**myc** gene, c-Myc, is both necessary and sufficient to ensure fibroblast proliferation; in the presence of **mitogen**, all cells in cycle express c-Myc. Fibroblasts remain in cycle as long as mitogens and c-Myc are present, but only remain viable in the presence of additional survival factors. In the absence of such survival factors, cycling cells expressing deregulated Myc undergo apoptosis. This response to oncogene deregulation is thought to protect the organism against cells that acquire mutations in c-Myc. Mutant cells that continuously express c-Myc will proliferate, but since the survival factor *in vivo* is thought to be limiting, the mutant cells will die once the supply of factor is exhausted (25,35). Thus, small proto-tumors may well arise *in vivo*, but they should regress due to lack of survival signals. If, however, another mutation has occurred producing a cooperating anti-apoptotic lesion, the mutated cells will survive. Several genes that cooperate with *myc* during tumorigenesis are anti-apoptotic. For example, v-Abl, which is a very good suppressor of apoptosis in response to many stimuli, will render haemopoietic cells cytokine-independent for growth and survival (36,37). *bcl-2*, a gene initially identified through its translocation in the t(14:18) mutation found in follicular lymphoma (38), will cooperate with *myc*, suppressing the apoptotic signal induced through Myc expression (39,40). The outcome of this cooperation *in vivo* is lymphoma, as seen in both double *bcl-2*/*Eμ-myc* **transgenic** mice (41,42) and in human patients with the t(14:18) translocation (43,44).

Bcl-2, an Anti-apoptotic Oncogene

The potent ability of Bcl-2 to suppress apoptosis triggered in response to a number of diverse stimuli prompted much research into its biochemical function. This function is still relatively obscure, but several members of the Bcl-2 family of proteins have been identified as a result (45–47). Bcl-2's form and functionality have been conserved throughout multicellular evolution, as exemplified by Ced-9, its counterpart in the nematode *C. elegans*, and the Adenovirus E1Bp19 protein (48). Ten or so members of this family now exist, and all share defined regions of **homology**, although not all act to suppress apoptosis (Table 1). The pro-apoptotic members of the family (Bax, Bak, and Bcl-x_S (49–53)) share three regions of homology, BH-1, -2, -3, whereas the anti-apoptotic members (Bcl-2 and Bcl-x_L) have a fourth region of homology, known as BH-4 (54,55). Removal of the BH-4 **domain** from either Bcl-2 or Bcl-x_L results in a loss of protective function. Family members can dimerize with themselves and with one another (see **Isoenzymes**), and these interactions are important for their function (56,57). For example, Bcl-2 preferentially binds to the pro-apoptotic family member Bax, whereas Bcl-x_L binds to Bak. If the number of Bax homodimers in the cell exceeds the number of Bcl-2 homodimers or Bax/Bcl-2 heterodimers, then the cell is more likely to undergo apoptosis (47). However, it is not clear which forms of these proteins are dominant for determining life or death. A second set of Bcl-2-associated proteins has been identified that influences cell survival through interaction with Bcl-2 family members, but whose members do not have all, if any, of the BH regions (Table 2) (45,46). Bag-1, eg, has no homology to Bcl-2, but binds to it and augments the protective function of Bcl-2 (58). Bad, on the other hand, has homology to the family and binds Bcl-2, an interaction that disrupts Bcl-2's binding with Bax, leading to cell death (59,60).

Both Bcl-x_L and Bcl-2 have effects on cell-cycle transition (61). Cells expressing either gene exhibit a longer-than-average time to reenter the cell cycle after arresting in G_0, suggesting that Bcl-x_L and Bcl-2 interact with components of the cell-cycle machinery.

Table 1. Pro-apoptotic and Anti-apoptotic Members of the Bcl-2 Family[a]

Bcl-2 Family Inhibitors	Bcl-2 Family Promotors
Bcl-2 (mammalian)	Bax (mammalian)
Bcl-xL (mammalian)	Bak (mammalian)
Bcl-w (mammalian)	Bcl-xS (mammalian)
Mcl-1 (mammalian)	
A1 (mammalian)	
NR-13 (mammalian)	
E1B p19 (viral)	
BHRF1 (viral)	

[a] Anti-apoptotic members contain up to 4 Bcl-2 homology (BH) domains and the transmembrane (TM) domain, whereas pro-apoptotic members contain only BH 1-3 and the TM domain, except for Bcl-x_s that contains only BH4, BH3, and the TM domain. Members of this gene family have been conserved throughout evolution with homologues existing in nematodes, mammals, and viruses (46,47, and references therein].

Table 2. Proteins That Bind to Bcl-2 Family Members and Regulate Their Function[a]

Bcl-2-like Inhibitors of Apoptosis	Bcl-2-like Promotors of Apoptosis
Bag-1 (no BH domains)	Hrk (BH 3, TM domain)
Bra (BH 2 and BH 3)	Bik (BH 3, TM domain)
Raf-1 (no BH domains)	Bid (BH 3)
	Bim (BH 3, TM domain
	Bad (BH 1, 2, 3, no hydrophobic tail)

[a] A variety of proteins interact with Bcl-2 proteins. Some contain the BH domains that influence their function, such as Bid, Bim, Bik, whereas others have no homology with Bcl-2, but still bind and influence its function (46,47, and references therein].

Tumor Suppressor Genes and Apoptosis

Tumor suppressor gene products suppress unrestrained cell proliferation through their specific inhibitory effects on the cell cycle and are implicated in the development of neoplasia following their loss or functional inactivation. One tumor suppressor gene product, **p53**, induces apoptosis in some tumor cell lines. p53 is a short-lived protein that is stabilized in the presence of DNA damage and triggers cell-cycle arrest, presumably to facilitate **DNA repair** (62). If the DNA damage is too great to repair, apoptosis is triggered, earning p53 the title "guardian of the genome" (63). At present, it is unclear if p53's induction of apoptosis is effected through its upregulation of $p21^{Waf\text{-}1/Cip\text{-}1}$, leading to cell-cycle arrest (64), through some other p53-regulated gene, or is independent of p53 transcriptional activity (65). Bax, the pro-apoptotic Bcl-2 homologue, is transcriptionally regulated by p53, but not all apoptotic cell deaths induced by p53 stabilization require Bax expression (66). The role of p53 as a sensor of DNA damage was neatly illustrated by examining apoptosis in thymocytes from p53-knockout mice (67,68). p53 null thymocytes undergo death by apoptosis as normal, except when treated with agents that damage DNA, such as etoposide or irradiation. Therefore, in the absence of p53, the cell is neither instructed to leave the cell cycle and repair the damaged DNA, nor to die. Cells therefore progress through subsequent cycles with damaged and mutated DNA. The role of p53 in inducing apoptosis in circumstances in which DNA damage is not evident is at present unclear, although in p53 knockout mice there is no evidence of a defect in tissue homeostasis due to ineffective cell death when cytokines become limiting (67,68).

A second tumor suppressor gene implicated in control of apoptosis is the **retinoblastoma** (*rb*) gene (69–71). The absence of the functional gene product, Rb, results in massive apoptosis in the haemopoietic and nervous system possibly due to the cells being constantly in cycle and unable to arrest or terminally differentiate. This makes it impossible for cells to establish appropriate survival signals and cell:cell contacts. Loss of both p53 and Rb function results in rapid tumor progression (72), suggesting that cells which have overcome restraints upon cell-cycle progression can also escape apoptosis, underlining the link between cell proliferation and cell death.

Cytokine Signaling Pathways and Apoptosis

Although a wealth of information is available about the **signal transduction** pathways activated by cytokines and **growth factors** in many cell types, surprisingly little is known about how apoptosis (and conversely cell viability) is modulated. Initially, many of the factors now referred to as *survival factors* were considered solely to stimulate cells to proliferate. Thus, many of the identified signal transduction pathways are involved in proliferation but not necessarily survival. For survival signals, two pathways can exist. Cells can be stimulated to survive by one specific pathway. Loss of this signal upon removal of the cytokine may simply trigger cell death due to negation of the first signal. However, removal of the cytokine may trigger another independent signaling pathway. Only now are these different possibilities being investigated.

Survival Signals. Cytokines with very different actions activate substantially overlapping signaling circuits, many of which appear to be required, but are not solely responsible, for regulating cell survival. However, even with these complex signaling networks, it is possible to gather some information about survival signals. For example, apoptosis stimulated by the withdrawal of **nerve growth factor** (NGF) in phaeochromocytoma (PC12) cells involves the induction of the AP-1 **transcription factor** c-Jun proximal to the time when neurons become committed to apoptosis. Activation of c-Jun requires phosphorylation by the protein kinase JNK (MKK4) that is, in turn, activated by MAP kinase (73) (see **Phosphorylation, protein**). In contrast, the suppression of apoptosis by NGF in the same cells may involve a discrete survival signaling pathway routed through **Ras** (74), Raf, and MAP kinase (73). Hence, in this case, the induction of apoptosis and promotion of survival may be independent informational processes, rather than mere negation of one other.

In some documented cases, the ability of anti-apoptotic cytokines to suppress apoptosis does not depend on the synthesis of new genes or proteins, indicating that survival is mediated by posttranslational mechanisms. IGF-1 (25), a potent inhibitor of apoptosis in many cell types, and **epidermal growth factor** (EGF) (75) are reported to suppress apoptosis effectively in cells treated with inhibitors of RNA and protein synthesis. The suppression of apoptosis in such instances must be effected by means of preexisting molecular machinery. One possible candidate signal is the activation of PI-3 kinase (PI3-K) through Ras, whose inhibition blocks the ability of NGF to mitigate apoptosis (76). However, specificity is absent, since PI3-K is activated in many signaling pathways in response to signals that do not suppress apoptosis. IGF-1 also signals via PI3-K, and the pro- and anti-apoptotic signaling pathways for this cytokine have been identified. PI3-K can be activated by its upstream effector Ras, and it can activate several downstream pathways, including the ribosomal protein $p70^{S6K}$, the **Rho** family polypeptide Rac, and the serine/threonine protein kinase PKB/Akt. Of these downstream effectors, PKB/Akt is anti-apoptotic in fibroblasts expressing c-Myc and in other cell types in the absence of survival factors, whereas the Ras-mediated Raf signaling pathway is proapoptotic (77–79). These findings underscore the pleiotropic nature of intracellular signaling. Signals emanating from GTP-Ras trigger a plethora of potential biological outcomes, some of which promote apoptosis and some of which suppress it. The net outcome for Ras activation is presumably dictated by downstream interactions that potentiate or mitigate other signals.

Killer Cytokines. Apoptosis can also be triggered by certain **cytokines**, such as TNFα and CD95 ligand. These two pathways represent the best understood triggers of apoptosis, since their activation pathways are essentially mapped. CD95 ligand and TNFα bind and activate their specific receptors on the surface of the cell (80). The TNF type 1 and CD95 (Fas, Apo-1) receptors are functionally similar; both are transmembrane receptors that contain a region within their cytoplasmic tail which is necessary for the initiation of the apoptotic response (81–83). This region of homology is called the *death domain* (DD). Several adaptor proteins bind to this domain and can trigger apoptosis by directly linking to the downstream effector **caspase** machinery (23). Alternatively, in the case of TNFR1, the adaptor proteins can also activate NFκB signaling pathways (84). CD95-induced cell death, eg, results in the association of the ***death-induced signaling c** omplex* (DISC) (80). Upon binding of its ligand, the CD95 receptor trimerizes and binds to its death domain the protein *Fas-associated death domain* (FADD), also known as MORT1 (85,86). FADD contains a C-terminal DD and, at its N-terminus, a *death effector domain* (DED), so called because the expression of this domain is required to trigger CD-95-induced apoptosis. The DED domain of FADD binds to the DED in caspase-8, a protein also called FLICE (FADD-like ICE, where ICE is "interleukin converting enzyme") (87,88). The **thiol proteinase** domain of FLICE, located at its C-terminus, is activated upon binding FADD, and FLICE then activates the *caspase cascade* by cleaving downstream caspases (89) (see **Caspases**). CD95-triggered apoptosis can be suppressed by expression of the anti-apoptotic proteins Bcl-2 and Bcl-x_L (90) or by proteins that inhibit the interaction of FLICE with FADD (91). Thus, the binding of CD95 ligand to the receptor and recruitment of FADD occur within the decision phase of apoptosis. The commitment of cells to CD95-induced apoptosis does not appear to occur until FLICE is recruited to the DISC, cleaved, and therefore activated. TNFR1 exhibits a similar pathway involving an overlapping set of adaptor proteins (23).

THE COMMITMENT PHASE OF APOPTOSIS

As yet, it is unclear what commits a cell to undergo apoptosis. To a certain degree, the final trigger must be cell autonomous, because cells that have received the same stimulus do not all enter apoptosis at the same time; cells may die within 10 minutes of the signal or within 3 days. Initially, the **caspases** were proposed to act at this point, with their activation leading to certain death (92–94) Although exogenous expression of pro-caspases leads to cell death, this death is again stochastic and can be suppressed by the addition of IGF-1 (95) or thiol proteinase inhibitors such as the poxvirus protein CrmA and **baculovirus** inhibitor p35 (96,97). Surprisingly though, not all caspase inhibitors suppress cell death. In the case of c-Myc-induced or Bax-induced apoptosis, the addition of tetrapeptide caspase inhibitors based on the pro-IL-1βcleavage site does not prevent cells from becoming committed to die (98,99). Thus, even when caspases are inhibited, the cell still undergoes commitment, but it takes much longer to progress through the morphological changes of apoptosis (99). This result suggests caspases are required for the rapid execution phase of apoptosis, but not necessarily for commitment.

The unknown entity in mammalian cells is the putative homologue of the *cell death gene 4* (Ced-4) (see **Programmed cell death**). Ced-4 is required for cell death in *C. elegans* and appears to function upstream of Ced-3, suggesting that Ced-4 may be involved in commitment (100). Recent data have shown that Ced-9, Ced-3, and Ced-4 bind one another, with Ced-4 being the critical intermediate (101–104). Without Ced-4 expression, Ced-9 does not interact with Ced-3. The expression of Ced-4 in mammalian cells produces a similar reaction between Bcl-x_L and the Ced-3 homologue, FLICE (caspase-8) (101). Whether Ced-4 is a killer of mammalian cells is also not clear, but Ced-4 expression is toxic to yeast (104). In the fission **yeast** *Schizosaccharomyces pombe*, Ced-4 localizes to condensed **chromatin** and induces caspase-independent death in this apoptotically naïve cell model. Cell death is suppressed upon the expression of Ced-9, which causes relocalization of Ced-4 to the membrane of the nucleus and to the **endoplasmic reticulum**. These data strongly suggest that Ced-9 suppresses Ced-4-induced death by binding to it. This, in turn, implies that the disruption of the Bcl-x_l/ Ced-4-"like protein" and FLICE interaction in mammalian cells may commit the cell to death.

THE EXECUTION PHASE OF APOPTOSIS

The execution phase describes the stage where the cells morphologically change and exhibit all the characteristics of apoptosis. Not all the agents responsible for these changes have been identified, but activation of the endonuclease occurs at this stage, as does activation of the *ced-3* homologues, the **caspases**. Overall, the caspase pathway can be thought of as a packaging process that carries out the rapid dismantling of the cell. Cells still die in the absence of caspase activity, but death is significantly prolonged (99). The cleavage of caspase substrates such as Poly-(ADP ribose) polymerase (PARP), the catalytic subunit of DNA protein kinase, and nuclear lamins [reviewed in (105)], in concert with an endonuclease, allow the nucleus and chromatin to be degraded efficiently. This saves the dying cell energy and facilitates the maintenance of membrane integrity, which in turn prevents release of the cell contents, which would stimulate an immune response.

Other effectors of the execution phase are not so well documented. Both ceramide and reactive oxygen species are found in apoptotic cells, but their exact roles are not clear (110–114): They may well be a consequence of death, rather than an effector. Changes in **mitochondria** also occur within apoptotic cells (115–117). Selective release of ***cytochrome c*** from the mitochondria appears to trigger apoptosis through activation of the caspases. A membrane permeability transition also occurs where the pores that regulate traffic between the mitochondrial outer and inner membranes become disrupted (117). However, it is unclear exactly when this change occurs within an individual cell undergoing apoptosis, making it difficult to assign these changes to a particular phase within the apoptotic pathway.

Overall, the execution phase is one where caspases and endonucleases are active and the cell is dismantled, progressing through the series of morphological changes that originally defined the process of apoptosis. Execution ends with the efficient and rapid phagocytosis of the corpse.

SUMMARY

Apoptosis is an active form of cell death indistinguishable from programmed cell death that occurs during invertebrate and vertebrate development. Apoptosis is thought to be the default state of all cells; hence, cells must receive a survival signal in order to remain viable. Apoptosis is triggered in response to a diverse set of stimuli and plays a critical role in the maintenance of tissue homeostasis. Mutations that disrupt the apoptotic pathway are pivotal in the development of many diseases, including degenerative brain disorders and cancer.

BIBLIOGRAPHY

1. A. H. Wyllie, J. F. Kerr, and A. R. Currie (1972) Cellular events in the adrenal cortex following ACTH deprivation. *J. Pathol.* **106**, 1.
2. C. B. Thompson (1995) Apoptosis in the pathogenesis and treatment of disease. *Science* **267**, 1456–1462.
3. G. Evan, E. Harrington, A. Fanidi, H. Land, B. Amati, and M. Bennett (1994) Integrated control of cell proliferation and cell death by the *c-myc* oncogene. *Phil. Trans. R Soc. Lond. B* **345**, 269–275.
4. M. D. Jacobson, M. Weil, and M. C. Raff (1997) Programmed cell death in animal development. *Cell* **88**, 347–354.
5. J. R. Hinchliffe and D. A. Ede (1973) Cell death and the development of limb form and skeletal pattern in normal and wingless (ws) chick embryos. *J. Embryol. Exp. Morphol.* **30**, 753–772.
6. A. H. Wyllie, J. F. Kerr, and A. R. Currie (1980) Cell death: The significance of apoptosis. *Int. Rev. Cytol.* **68**, 251–306.
7. J. F. Kerr, A. H. Wyllie, and A. R. Currie (1972) Apoptosis: A basic biological phenomenon with wide–ranging implications in tissue kinetics. *Br. J. Cancer* **26**, 239–257.
8. A. H. Wyllie (1980) Glucocorticoid-induced thymocyte apoptosis is associated with endogenous endonuclease activation. *Nature* **284**, 555–556.

9. J. Savill, V. Fadok, P. Henson, and C. Haslett (1993) Phagocyte recognition of cells undergoing apoptosis. *Immunol. Today* **14**, 131–136.
10. J. Savil, I. Dransfield, N. Hogg, and C. Haslett (1990) Vitronectin receptor-mediated phagocytosis of cells undergoing apoptosis. *Nature* **343**, 170–173.
11. J. Savill, N. Hogg, Y. Ren, and C. Haslett (1992) Thrombospondin cooperates with CD36 and the vitronectin receptor in macrophage recognition of neutrophils undergoing apoptosis. *J. Clin. Invest.* **90**, 1513–1522.
12. Y. Ren, R. L. Silverstein, J. Allen, and J. Savill (1995) CD36 gene transfer confers capacity for phagocytosis of cells undergoing apoptosis. *J. Exp. Med.* **181**, 1857–1862.
13. V. A. Fadok, D. R. Voelker, P. A. Campbell, D. L. Bratton, J. J. Cohen, P. W. Noble, D. W. Riches, and P. M. Henson (1993) The ability to recognize phosphatidylserine on apoptotic cells is an inducible function in murine bone marrow-derived macrophages. *Chest* **103**, 102s.
14. V. A. Fadok, D. R. Voelker, P. A. Campbell J. J. Cohen, D. L. Bratton, and P. M. Henson (1992) Exposure of phosphatidylserine on the surface of apoptotic lymphocytes triggers specific recognition and removal by macrophages. *J. Immunol.* **148**, 2207–2216.
15. V. A. Fadok et al. (1992) Different populations of macrophages use either the vitronectin receptor or the phosphatidylserine receptor to recognize and remove apoptotic cells. *J. Immunol.* **149**, 4029–4035.
16. P. K. Flora and C. D.Gregory (1994) Recognition of apoptotic cells by human macrophages: Inhibition by a monocyte/macrophage-specific monoclonal antibody. *Eur. J. Immunol.* **24**, 2625–2632.
17. M. F. Luciani and G. Chimini (1996) The ATP binding cassette transporter ABC1 is required for the engulfment of corpses generated by apoptotic cell death. *Embo. J.* **15**, 226–235.
18. N. Platt and S. Gordon (1995) Role of the murine scavenger receptor in the recognition of apoptotic thymocytes by macrophages. *J. Cell Biochem. Suppl.* **19B**, 300.
19. M. Stern, J. Savill, and C. Haslett (1996) Human monocyte-derived macrophage phagocytosis of senescent eosinophils undergoing apoptosis. Mediation by alpha v beta 3/CD36/thrombospondin recognition mechanism and lack of phlogistic response. *Am. J. Pathol.* **149**, 911–921.
20. N. J. McCarthy, C. A. Smith, and G. T. Williams (1992) Apoptosis in the development of the immune system: Growth factors, clonal selection and bcl-2. *Canc. Metas. Rev.* **11**, 157–178.
21. G. I. Evan, L. Brown, M. Whyte and E. Harrington (1995) Apoptosis and the cell-cycle. *Curr. Opin. Cell Biol.* **7**, 825–834.
22. M. Raff, B. Barres, J. Burne, H. Coles, Y. Ishizaki, and M. Jacobson (1993) Programmed cell death and the control of cell survival: Lessons from the nervous system. *Science* **262**, 695–700.
23. S. Nagata (1997) Apoptosis by death factor. *Cell* **88**, 355–365.
24. M. K. Collins, J. Marvel, P. Malde, and A. Lopez-Rivas (1992) Interleukin 3 protects murine bone marrow cells from apoptosis induced by DNA damaging agents. *J. Exp. Med.* **176**, 1043–1051.
25. E. Harrington, A. Fanidi, M. Bennett, and G. Evan (1994) Modulation of Myc-induced apoptosis by specific cytokines. *Embo. J.* **13**, 3286–3295.
26. G. T. Williams, C. A. Smith, E., Spooncer, T. M. Dexter, and D. R. Taylor (1990) Haemopoietic colony stimulating factors promote cell survival by suppressing apoptosis. *Nature* **343**, 76–79.
27. R. C. Duke and J. J. Cohen (1986) IL-2 addiction: Withdrawal of growth factor activates a suicide program in dependent T cells. *Lymphokine Res.* **5**, 289–299.
28. E. White, P. Sabbatini, M. Debbas, W. Wold, D. Kusher, and L. Gooding (1992) The 19-kilodalton Adenovirus E1B transforming protein inhibits programmed cell death and prevents cytolysis by tumour necrosis factor α. *Mol. Cell. Biol.* **12**, 2570–2580.
29. E. White, R. Cipriani, P. Sabbatini, and A. Denton (1991) Adenovirus E1B 19-kilodalton protein overcomes the cytotoxicity of E1A proteins. *J. Virol.* **65**, 2968–2678.
30. A. Zychlinsky, B. Kenny, R. Menard, M. C. Prevost, I. B. Holland, and P. J. Sansonetti (1994) IpaB mediates macrophage apoptosis induced by *Shigella flexneri*. *Mol. Microbiol* **11**, 619–627.
31. K. Thirumalai, K. S. Kim, and A. Zychlinsky (1997) IpaB, a *Shigella flexneri* invasin, colocalizes with interleukin-1 beta-converting enzyme in the cytoplasm of macrophages. *Infect. Immun.* **65**, 787–793.
32. A. Guichon and A. Zychlinsky (1996) Apoptosis as a trigger of inflammation in *Shigella*-induced cell death. *Biochem. Soc. Trans.* **24**, 1051–1054.
33. G. Evan et al. (1992) Induction of apoptosis in fibroblasts by *c-myc* protein.*Cell* **63**, 119–125.
34. M. Zonig and G. I. Evan (1996) Cell cycle: On target with Myc. *Curr. Biol.* **6**, 1553–1556.
35. E. Harrington, A. Fanidi, and G. Evan (1994) Oncogenes and cell death. *Curr. Opin. Genet. Dev.* **4**, 120–129.
36. C. A. Evans, L. P. Owen, A. D. Whetton, and C. Dive (1993) Activation of the Abelson tyrosine kinase activity is associated with suppression of apoptosis in hemopoietic cells. *Cancer Res.* **53**, 1735–1738.
37. J. L. Cleveland, M. Dean, N. Rosenberg, J. Y. Wang, and U. R. Rapp (1989) Tyrosine kinase oncogenes abrogate interleukin-3 dependence of murine myeloid cells through signaling pathways involving c-myc: Conditional regulation of c-myc transcription by temperature-sensitive v-abl. *Mol. Cell. Biol.* **9**, 5685–5695.
38. Y. Tsujimoto, L. R. Finger, J. Yunis, P. C. Nowell, and C. M. Croce (1984) Cloning of the chromosome breakpoint of neoplastic B cells with the t(14;18) chromosome translocation. *Science* **226**, 1097–1099.
39. A. Fanidi, E. Harrington, and G. Evan (1992) Cooperative interaction between *c-myc* and *bcl-2* proto-oncogenes. *Nature* **359**, 554–556.
40. R. Bissonnette, F. Echeverri, A. Mahboubi, and D. Green (1992) Apoptotic cell death induced by *c-myc* is inhibited by *bcl-2*. *Nature* **359**, 552–554.
41. A. Strasser, A. W. Harris, M. L. Bath, and S. Cory (1990) Novel primitive lymphoid tumours induced in transgenic mice by cooperation between *myc* and *bcl-2*. *Nature* **348**, 331–333.
42. T. J. McDonnell and S. J. Korsmeyer (1991) Progression from lymphoid hyperplasia to high-grade malignant lymphoma in mice transgenic for the t(14;18). *Nature* **349**, 254–256.
43. S. J. Korsmeyer, T. J. McDonnell, G. Nunez, D. Hockenbery, and R. Young (1990) Bcl-2: B cell life, death and neoplasia. *Curr. Top. Microbiol. Immunol.* **166**, 203–207.
44. Y. Tsujimoto, J. Cossman, E. Jaffe, and C. M. Croce (1985) Involvement of the bcl-2 gene in human follicular lymphoma. *Science* **228**, 1440–1443.
45. J. C. Reed (1997) Double identity for proteins of the Bcl-2 family. *Nature* **387**, 773–776.
46. M. D. Jacobson (1997) Apoptosis: Bcl-2-related proteins get connected. *Curr. Biol.* **7**, R227–R281.
47. E. Yang and S. J. Korsmeyer (1996) Molecular thanatopsis: A discourse on the BCL2 family and cell death. *Blood* **88**, 386–401.
48. G. Williams and C. Smith (1993) Molecular regulation of apoptosis—genetic-controls on cell-death. *Cell* **74**, 777–779.

49. Z. Oltvai, C. Milliman, and S. Korsmeyer (1993) Bcl-2 heterodimerizes in vivo with a conserved homolog, Bax, that accelerates programed cell death. *Cell* **74**, 609–619.

50. L. Boise et al. (1993) *bcl*-x, a *bcl-2*-related gene that functions as a dominant regulator of apoptotic cell death. *Cell* **74**, 597–608.

51. T. Chittenden, E. Harrington, R. O'Connor, G. Evan, and B. Guild (1995) Induction of apoptosis by the Bcl-2 homologue Bak. *Nature* **374**, 733–736.

52. M. Kiefer, M. Brauer, V. C. Powers, J. Wu, S. Umansky, L. Tomei, and P. Barr (1995) Modulation of apoptosis by the widely distributed Bcl-2 homologue Bak. *Nature* **374**, 736–739.

53. S. Farrow et al. (1995) Cloning of a novel *bcl-2* homologue by interaction with adnovirus E1B 19K. *Nature* **374**, 731–733.

54. B. S. Chang, A. J. Minn, S. E. Muchmore, S. W. Fesik, and C. B. Thompson (1997) Identification of a novel regulatory domain in Bcl-xL and Bcl-2. *Embo J.* **16**, 968–977.

55. S. W. Muchmore, M. Sattler, H. Liang, R. P. Meadows, J. E. Harlan, H. S. Yoon, D. Nettesheim, B. S. Chang, C. B. Thompson, S. L. Wong, S. L. Ng, and S. W. Fesik (1996) X-ray and NMR structure of human Bcl-xL, an inhibitor of programmed cell death. *Nature* **381**, 335–341.

56. X.-M. Yin, Z. Oltvai, and S. Korsemeyer (1994) Bh1 and bh2 domains of Bcl-2 are required for inhibition of apoptosis and heterodimerization with Bax. *Nature* **369**, 321–323.

57. T. Sato, S. Irie, S. Krajewski, and J. C. Reed (1994) Cloning and sequencing of a cDNA encoding the rat Bcl-2 protein. *Gene* **140**, 291–292.

58. S. Takayama et al. (1995) Cloning and functional-analysis of *bag-1*—a novel *bcl-2*-binding. *Cell* **80**, 279–284.

59. E. Yang, J. P. Zha, J. Jockel, L. H. Boise, C. B. Thompson, and S. J. Korsmeyer (1995) Bad, a heterodimeric partner for Bcl-xL and Bcl-2, displaces Bax and promotes cell death. *Cell* **80**, 285–291.

60. J. Zha, H. Harada, E. Yang, J. Jockel, and S. J. Korsmeyer (1996) Serine phosphorylation of death agonist BAD in response to survival factor results in binding to 14-3-3 not BCL-X(L) *Cell* **87**, 619–628 (see comments).

61. L. O'Reilly, D. Huang, and A. Strasser (1996) The cell death inhibitor Bcl-2 and its homologues influence control of cell cycle entry. *Embo. J.* **15**, 6979–6990.

62. M. B. Kastan, O. Onyekwere, D. Sidransky, B. Vogelstein, and R. W. Craig (1991) Participation of p53 protein in the cellular response to DNA damage. *Cancer Res.* **51**, 6304–6311.

63. D. P. Lane (1992) Cancer. p53, guardian of the genome. *Nature* **358**, 15–16 (news; comment).

64. W. El-Deiry et al. (1993) *WAF1*, a potential mediator of p53 tumor suppression. *Cell* **76**, 817–825.

65. C. Caelles, A. Helmberg, and M. Karin (1994) p53-dependent apoptosis in the absence of transcriptional activation of p53-target genes. *Nature* **370**, 220–223 (see comments).

66. C. Yin, C. M. Knudson, S. J. Korsmeyer, and T. Van Dyke (1997) Bax suppresses tumorigenesis and stimulates apoptosis in vivo. *Nature* **385**, 637–640.

67. S. W. Lowe, E. M. Schmitt, S. W. Smith, B. A. Osborne, and T. Jacks (1993) p53 is required for radiation-induced apoptosis in mouse thymocytes. *Nature* **362**, 847–849 (see comments).

68. A. R. Clarke et al. (1993) Thymocyte apoptosis induced by p53-dependent and independent pathways. *Nature* **362**, 849–852 (see comments).

69. A. Clarke et al. (1992) Requirement for a functional *Rb-1* gene in murine development. *Nature* **359**, 328–330.

70. E.-H. Lee et al. (1992) Mice deficient for Rb are nonviable and show defects in neurogenesis and haematopoiesis. *Nature* **359**, 288–294.

71. T. Jacks, A. Fazeli, E. Schmitt, R. Bronson, M. Goodell, and R. Weinberg (1992) Effects of an *Rb* mutation in the mouse. *Nature* **359**, 295–300.

72. H. Symonds et al. (1994) p53-dependent apoptosis suppresses tumour growth and progression *in vivo*. *Cell* **76**, 703–711.

73. Z. Xia, M. Dickens, J. Raingeaud, R. Davis, and M. Greenberg (1995) Opposing effects of ERK and JNK-p38 MAP kinases on apoptosis. *Science* **270**, 1326–1331.

74. C. D. Nobes and A. M. Tolkovsky (1995) Neutralizing anti-p21ras Fabs suppress rat sympathetic neuron survival induced by NGF, LIF, CNTF and cAMP. *Eur. J. Neurosci.* **7**, 344–350.

75. A. Geier, R. Beery, M. Haimsohn, R. Hemi, Z. Malik, and A. Karasik (1994) Epidermal growth factor, phorbol esters, and aurintricarboxylic acid are survival factors for MDA-231 cells exposed to adriamycin. *In Vitro Cell Dev. Biol. Anim.* **30a**, 867–874.

76. R. Yao and G. M. Cooper (1995) Requirement for phosphatidylinositol-3 kinase in the prevention of apoptosis by nerve growth factor. *Science* **267**, 2003–2006.

77. S. Kennedy, A. Wagner, S. Conzen, J. Jordan, A. Bellacosa, P. Tsichlis, and N. Hay (1997) The PI 3-kinase/akt signaling pathway delivers an anti-apoptotic signal. *Genes & Devel.* **11**, 701–713.

78. H. Dudek et al. (1997) Regulation of neuronal survival by the serine-threonine protein-kinase akt. *Science* **275**, 661–665.

79. A. Kauffmann-Zeh, P. Rodriguez-Viciana, E. Ulrich, C. Gilbert, P. Coffer, and G. Evan (1997) Suppression of c-Myc-induced apoptosis by Ras signalling through PI 3-kinase and PKB. *Nature* **385**, 544–548.

80. P. Krammer, I. Behrmann, P. Daniel, J. Dhein, and K.-M. Debatin (1994) Regulation of apoptosis in the immune sustem. *Curr. Opin. Immunol.* **6**, 279–289.

81. S. Nagata and P. Golstein (1995) The Fas death factor. *Science* **267**, 1449–1456.

82. L. Tartaglia, T. Ayres, G. Wong, and D. Goeddel (1993) A novel domain within the 55 kd TNF receptor signals cell death. *Cell* **74**, 845–853.

83. N. Itoh and S. Nagata (1993) A novel protein domain required for apoptosis. Mutational analysis of human Fas antigen. *J. Biol. Chem.* **268**, 10932–10937.

84. M. Rothe, V. Sarma, V. W. Dixit, and D. V. Goeddel (1995) Traf2-mediated activation of NF-kappa-b by TNF receptor-2 and CD40. *Science* **269**, 1424–1427.

85. M. Boldin, E. Varfolomeev, Z. Pancer, I. Mett, J. Camonis, and D. Wallach (1995) A novel protein that interacts with the death domain of Fas/Apo1 contains a sequence related tothea death domain. *J. Biol. Chem.* **270**, 7795–7798.

86. A. M. Chinnaiyan, K. O'Rourke, M. Tewari, and V. M. Dixit (1995) FADD, a novel death domain-containing protein, interacts with the death domain of Fas and initiates apoptosis. *Cell* **81**, 505–512.

87. M. Muzio et al. (1996) FLICE, a novel FADD homologous ICE/CED-3–like protease, is recruited to the CD95 (Fas/Apo-1) death-inducing signaling complex. *Cell* **85**, 817–827.

88. M. Boldin, T. Goncharov, Y. Goltsev, and D. Wallach (1996) Involvement of MACH, a novel MORT1/FADD-interacting protease, in Fas/APO-1- and TNF receptor-induced cell death. *Cell* **85**, 803–815.

89. A. Chinnaiyan and V. Dixit (1996) The cell-death machine. *Curr. Biol.* **6**, 555–562.

90. X. Zhang et al. (1996) Up-regulation of Bcl-xL expression protects CD40-activated human B cells from Fas-mediated apoptosis. *Cell Immunol* **173**, 149–154.

91. M. Thome et al. (1997) Viral FLICE inhibitory proteins (FLIPs) prevent apoptosis induced by death receptors. *Nature* **386**, 517–521.
92. E. Alnemri, D. Livingston, D. Nicholson, G. Salvesan, N. Thornberry, W. Wong, and J. Yuan (1996) Human ICE/CED-3 protease nomeclature. *Cell* **87**, 171.
93. Y. Lazebnik, A. Takahashi, G. Poirier, S. H. Kaufmann, and W. Earnshaw (1995) Characterization of the execution phase of apoptosis *in vitro* using extracts from condemned-phase cells. *J. Cell Sci.* **19**, 41–49.
94. A. Takahashi and W. Earnshaw (1996) Ice-related proteases in apoptosis. *Curr. Opin. Gen. & Dev.* **6**, 50–55.
95. Y. Jung, M. Miura, and J. Yuan (1996) Suppression of interleukin-1 beta-converting enzyme-mediated cell death by insulin-like growth factor. *J. Biol. Chem.* **271**, 5112–5117.
96. V. Gagliardini, P. A. Fernandez, R. K. Lee, H. C. Drexler, R. J. Rotello, M. C. Fishman, and J. Yuan (1994) Prevention of vertebrate neuronal death by the crmA gene. *Science* **263**, 826–828.
97. N. Bump et al. (1995) Inhibition of ICE family proteases by baculovirus antiapoptotic protein p35. *Science* **269**, 1885–1888.
98. J. Xiang, D. Chao, and S. Korsmeyer (1996) Bax-induced cell death may not require interleukin-1ß-converting enzyme-like proteases. *Proc. Natl. Acad. Sci. USA* **93**, 14559–14563.
99. N. McCarthy, M. Whyte, C. Gilbert, and G. Evan (1997) Inhibition of Ced-3/ICE-related proteases does not prevent cell death induced by oncogenes, DNA damage, or the Bcl-2 homologue Bak. *J. Cell. Biol.* **136**, 215–227.
100. J. Yuan and H. R. Horvitz (1992) The *Caenorhabditis elegans* cell death gene ced-4 encodes a novel protein and is expressed during the period of extensive programmed cell death. *Development* **116**, 309–320.
101. A. M. Chinnaiyan, K. O'Rourke, B. R. Lane, and V. M. Dixit (1997) Interaction of CED-4 with CED-3 and CED-9: A molecular framework for cell death. *Science* **275**, 1122–1126 (see comments).
102. M. S. Spector, S. Desnoyers, D. J. Hoeppner, and M. O. Hengartner (1997) Interaction between the *C. elegans* cell-death regulators CED-9 and CED-4. *Nature* **385**, 653–656.
103. D. Wu, H. D. Wallen, and G. Nunez (1997) Interaction and regulation of subcellular localization of CED-4 by CED-9. *Science* **275**, 1126–1129 (see comments).
104. C. James, S. Gschmeissner, A. Fraser, and G. Evan (1997) Ced-4 induces chromatin condensation in *S. pombe* and is inhibited by direct physical association with Ced-9. *Curr. Biol.* **7**, 246–252.
105. M. Whyte (1996) ICE/Ced-3 proteases in apoptosis. *Trends Cell Biol.* **6**, 245–248.
106. A. Fraser, N. McCarthy, and G. I. Evan (1996) Biochemistry of cell-death. *Curr. Opin. Neurobiol.* **6**, 71–80.
107. T. Fernandes Alnemri et al. (1996) In-vitro activation of CPP32 and Mch3 by Mch4, a novel human apoptotic cysteine protease containing 2 FADD-like domains. *Proc. Natl. Acad. Sci. USA* **93**, 7464–7469.
108. Y. A. Lazebnik, A. Takahashi, R. D. Moir, R. D. Goldman, G. G. Poirier, S. H. Kaufmann, and W. C. Earnshaw (1995) Studies of the lamin proteinase reveal multiple parallel biochemical pathways during apoptotic execution. *Proc. Natl. Acad. Sci. USA* **92**, 9042–9046.
109. A. Fraser and G. Evans (1996) A license to kill. *Cell* **85**, 781–784.
110. P. J. Hartfield, G. C. Mayne, and A. W. Murray (1997) Ceramide induces apoptosis in PC12 cells. *FEBS Lett.* **401**, 148–152.
111. S. C. Wright, H. Zheng, and J. Zhong (1996) Tumor cell resistance to apoptosis due to a defect in the activation of sphingomyelinase and the 24 kDa apoptotic protease (AP24). *Faseb. J.* **10**, 325–332.
112. P. Santana et al. (1996) Acid sphingomyelinase-deficient human lymphoblasts and mice are defective in radiation-induced apoptosis. *Cell* **86**, 189–199.
113. G. J. Pronk, K. Ramer, P. Amiri, and L. T. Williams (1996) Requirement of an ICE-like protease for induction of apoptosis and ceramide generation by REAPER. *Science* **271**, 808–810.
114. M. Jacobson and M. Raff (1995) Programmed cell-death and *bcl-2* protection in very-low oxygen. *Nature* **374**, 814–816.
115. R. M. Kluck, E. Bossy Wetzel, D. R. Green, and D. D. Newmeyer (1997) The release of cytochrome c from mitochondria: A primary site for Bcl-2 regulation of apoptosis. *Science* **275**, 1132–1136 (see comments).
116. J. Yang, X. Liu, K. Bhalla, C. N. Kim, A. M. Ibrado, J. Cai, T. I. Peng, D. P. Jones, and X. Wang (1997) Prevention of apoptosis by Bcl-2: Release of cytochrome c from mitochondria blocked. *Science* **275**, 1129–1132 (see comments).
117. G. Kroemer, N. Zamzami, and S. A. Susin (1997) Mitochondrial control of apoptosis. *Immunol. Today* **18**, 44–51.

Suggestions for Further Reading

M. D. Jacobson (1997) Apoptosis: Bcl-2-related protein get connected. *Curr. Biol.* **7**(5), R277–R281. An up-to-date review on the pathway of apoptosis.

M. D. Jacobson, M. Weil, and M. C. Raff (1997) Programmed cell death in animal development. *Cell* **88**, 347–354.

M. Sluyser, ed. (1996) *Apoptosis in Normal Development and Cancer*, Taylor and Francis, London.

E. Yang and S. J. Korsmeyer (1996) Molecular thanatopsis: A discourse on the Bcl-2 family and cell death. *Blood* **88**, 386–401.

ARA OPERON

ROBERT HELLING

The pentose sugar L-arabinose is distributed widely as a component of **plant** polysaccharides in pectins, gums, and cell walls. It is also present in smaller amounts in the cell walls of some **bacteria**. Thus it is likely that, in its natural environment in the vertebrate intestine, the bacterium *Escherichia coli* periodically encounters arabinose released through degradation of these polymers (1). Not surprisingly, *E. coli*, like many other bacteria and **fungi**, has *ara* **genes** that encode the **enzymes** and **transport** proteins necessary to utilize L-arabinose as a source of carbon and energy.

The *ara* genes of *E. coli* have been of exceptional importance to biology for several reasons. If the ***lac* operon** is the exemplar of regulation of **gene expression** by **repression of expression**, the *ara* operon is the counterpart as the model for positive control of gene expression, even though more complete analysis has revealed that each has both positive and negative features. An **operon** is one gene or a linear sequence of related genes that are transcribed as one unit from a common initiation point (see **Transcription**). The transcript is then **translated** into protein products corresponding to the separate genes. The first clear example of DNA looping as a feature of transcription control was the *ara* operon. The hypothesis of positive control of the *ara* operon was a challenge to establishment scientists, and its validation was the end of a fascinating story in the history of science. Moreover, the details of the operon are complex and interesting in themselves.

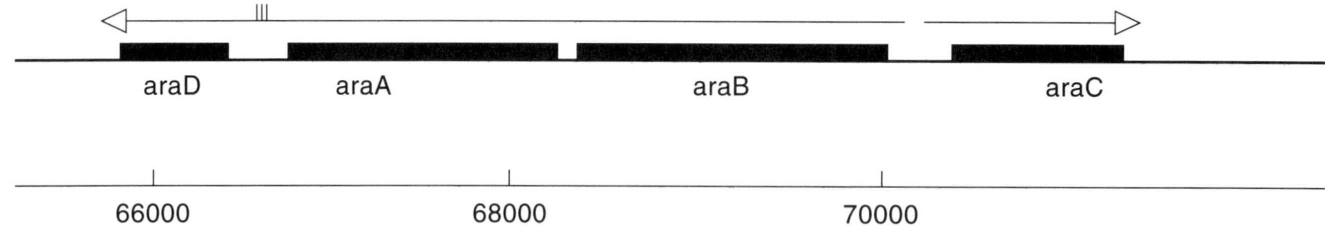

Figure 1. The positions of the *araBAD* operon and the *araC* regulatory gene on the *E. coli* chromosome. Transcription of each is initiated in the 338-nucleotide base-pair region between *araB* and *araC*. Nucleotide location on the *E. coli* chromosome is shown beneath the gene map. Vertical bars correspond to the three **inverted repeat** REP sequence pairs that are assumed to produce three self-paired hairpin structures in the mRNA between *araA* and *araD*. The entire chromosome contains approximately 4,639,221 nucleotide pairs. The primary sequence information is in Refs. 21 and 22.

Table 1. ***ara*** **Genes and Gene Products**

Gene	Location on the *E. coli* Chromosome (minutes)[a]	Size of Gene (Amino Acid Codons)	Activity of Product
araA	1.4	500	L-Arabinose isomerase
araB	1.5	566	L-Ribulokinase
araC	1.5	292	AraC regulatory protein
araD	1.4	231	L-Ribulose-5-phosphate-4-epimerase
araE	64.2	472	L-Arabinose transport, low affinity
araF	42.8	329	L-Arabinose transport, high affinity
araG	42.7	504	L-Arabinose transport, high affinity
araH	42.6	329	L-Arabinose transport, high affinity
araJ	8.5	394	Transport or processing of polymer?

[a] Location is based on a 100-min circular chromosome.

ARA OPERON OF *ESCHERICHIA COLI*

The classical *ara* operon of *Escherichia coli* comprises three genes, *araBAD* (Fig. 1). This and four other operons, *araC*, *araE*, *araFGH*, and *araJ*, are uniquely associated with metabolism of L-arabinose (Table 1). Each set of genes is under control of the activator protein AraC, the product of the *araC* gene. The function of each gene is known, except for *araJ*, which may encode an enzyme that processes or transports an arabinose-containing polymer.

Uptake And Utilization of Arabinose

Two independent systems deliver arabinose from the environment across the cell **membrane** into the cell. *araE* encodes a **membrane protein** that mediates arabinose uptake via proton **symport** and is the lower-affinity transporter ($K_M = 50\ \mu$M) (2). The *araFGH* operon encodes a **periplasmic** arabinose-binding protein (*araF*), a probable **ATPase** subunit (*araG*), and a membrane protein (*araH*), which together mediate ATP-driven arabinose transport. This transporter shows higher affinity for arabinose ($K_M = 1\ \mu$M) than AraE, but lower capacity (2,3).

Internal arabinose is converted in three steps to D-xylulose-5-phosphate, a metabolite in the pentose-phosphate shunt pathway, and one that is not unique to arabinose metabolism. The enzyme mediating the first step, L-arabinose isomerase, has a low affinity for arabinose, with a K_M (**Michaelis constant**) of 60 mM (4); this suggests that cells growing on arabinose have a very high internal arabinose concentration (5). The product of isomerase activity, L-ribulose, is converted to L-ribulose-5-phosphate by L-ribulokinase, and the phosphorylated compound is converted in turn to xylulose phosphate by the epimerase encoded by *araD*. Arabinose inhibits growth of *araD* mutants on other nutrients, presumably because accumulation of high concentrations of ribulose phosphate is toxic. Thus secondary mutants lacking isomerase or kinase activity as the result of *araA*, *araB*, or *araC* mutation, and therefore not forming ribulose phosphate, can be selected by plating an *araD* population on broth plates containing arabinose (6).

Regulation of ara Operon Expression

In the absence of arabinose, the *ara* genes are essentially not expressed, except for the regulatory gene *araC*. On exposure to L-arabinose, all of the *ara* genes are activated, transcription of *araBAD* begins within five seconds, and the Ara proteins appear within several minutes, allowing growth on the sugar (5). The true inducer has been shown by **X-ray crystallography** of the inducer-AraC protein complex to be α-L-arabinose (7). However, *araC* mutants have been obtained that are inducible by D-fucose or by β-methyl-L-arabinoside, normally inhibitors of induction (1,8). *In vivo* studies using cells lacking both arabinose transport systems showed the K_M for induction by arabinose to be unexpectedly high, nearly 10 mM (9), although, as noted above, the internal arabinose concentration is likely to be higher than this during growth on arabinose because of the low affinity of the isomerase for the pentose. The high concentration-dependence of induction suggests that there might be **natural selection** against a mutant isomerase with a higher affinity for arabinose.

Complementation studies showed that the *presence* of AraC protein is necessary for *araBAD* gene expression (10,11), a discovery unanticipated by those who believed all gene activation resulted from *removal* of repressors. Later, AraC was shown to be required for expression of the other *ara* operons as well (except for *araC*). *araC* is the prototype for a large family of regulatory genes that share sequence similarity in the DNA-binding region and are **homologous** and presumably related through **evolution** (12).

Cyclic AMP bound to **cyclic AMP receptor protein (CRP)** is also a positive regulator for all the *ara* operons. Expression of many genes is controlled by availability of CRP-cAMP; lack of expression due to low CRP-cAMP is referred to as **catabolite repression**. *In vitro* transcription of *araBAD* mimics that *in vivo* in that it requires both the AraC and CRP regulatory proteins with their bound ligands. Analysis of transcription *in vitro*, and further *in vivo* studies, have given a broad understanding at the molecular level of control of *araBAD* transcription, although details remain to be determined (5).

The AraC **protein structure** has two **domains** connected by a flexible polypeptide linker. The *N*-terminal domain binds arabinose and is responsible for formation of the active dimeric form of the protein. The *C*-terminal domain binds DNA at specific sites, with similar sequences, upstream of each *ara* operon. In the regulatory region between the divergently transcribed *araBAD* and *araC* operons (Fig. 1), there are five sites at which AraC can bind (Fig. 2).

In the absence of arabinose, the two DNA-binding domains of the AraC dimer are oriented so that they cannot readily bind both of the adjacent I sites at the same time. Instead, the AraC dimer contacts I_1 and O_2, thereby forming a DNA loop within the region between *araB* and *araC* (Fig. 2) (13). On addition of arabinose, the dimer undergoes a conformational shift such that the two DNA-binding domains preferentially bind I_2I_1 (Fig. 2) (14). The presence of AraC at I_2 stimulates addition of **RNA polymerase** and open complex formation, and transcription of *araBAD* commences, if CRP-cAMP is present (5). Although this model was proposed and refined before detailed structural information was available, X-ray crystallographic

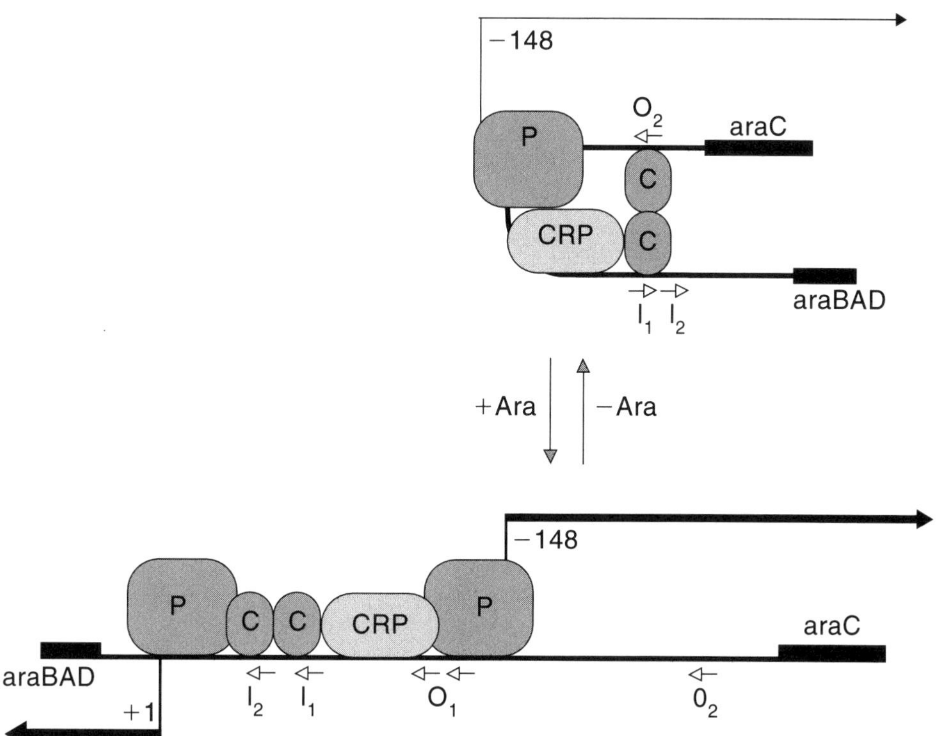

Figure 2. Activation of transcription of the *araBAD* operon by arabinose. CRP refers to CRP-cAMP, and P refers to RNA polymerase. AraC protein is shown as a dimer bound to I_1 and O_2 in the absence of arabinose and to I_1 and I_2 in the presence of arabinose. I_1, I_2, O_1 (a double site), and O_2 are all potential binding sites for the AraC protein and conform to the consensus sequence 5′-TAGCN_7TCCATA-3′ (reading in the direction of the arrows) (14), although there is considerable variation among sites. Note how the mode of pairing of the dimers differs depending on the presence of arabinose (7). On addition of arabinose, transcription of *araBAD* is initiated and expression of *araC* is increased (bottom). After a few minutes, AraC-arabinose dimers bind to O_1 and O_2 so as to form a new DNA loop and reduce *araC* transcription to that characteristic of cells in the absence of arabinose (not shown).

studies of the AraC *N*-terminal domain and linker support the model (7).

Addition of arabinose also affects expression of *araC*. When the I_1–O_2 loop is opened, transcription of *araC* is accelerated (Fig. 2). After a few minutes, AraC-arabinose dimers re-form DNA loops, this time by bridging O_1 and O_2 (not shown). O_1–O_2 looping does not regulate *araBAD* transcription, but it interferes with RNA polymerase binding and initiation at the *araC* promoter; this reduces the rate of *araC* transcription to that characteristic of cells in the absence of arabinose (5).

araC is controlled by CRP-cAMP as well. CRP-cAMP binding increases, but is not essential for, *araC* transcription; it is necessary for substantial expression of the other *ara* operons.

The mechanism by which AraC-arabinose, CRP-cAMP, and RNA polymerase interact to trigger transcription is not clear. Bound CRP-cAMP helps to open the I_1I_2 repression loop on addition of arabinose, but it probably activates RNA polymerase binding or initiation as well, either by direct contact or indirectly through contact with AraC (15). (CRP-cAMP does not aid AraC binding.) Several kinds of CRP-cAMP-independent mutants have been obtained. One class (*rpoD*) has an altered RNA polymerase **sigma factor**-70 subunit (16), while another (*araC*) is altered in AraC itself (17). Other mutants (*araI*) are transcribed by RNA polymerase independently of the AraC or CRP proteins, and they result from changes at the RNA polymerase-binding site in the *araBAD* promoter (18).

OTHER MECHANISMS FOR ARABINOSE DEGRADATION

The pathway for arabinose degradation that is common to *E. coli* and many other bacteria is not the sole means for utilizing L-arabinose in the biological world, nor is the regulatory model described above the only means for control of *ara* enzyme synthesis. At least five different pathways exist (1), and different regulatory systems are found even among organisms using the same pathway. Although the work on other organisms is still at an early stage, it is known that the **gram-positive** bacterium ***Bacillus** subtilus* utilizes the same degradatory pathway for arabinose utilization as *E. coli* but appears to use a simple repressor as the means of control of *ara* gene expression (19). The fungus ***Aspergillus*** uses a completely different pathway for arabinose utilization and, not surprisingly, appears to have different control systems as well (20). It will be remarkable if the *ara* genes in these other organisms are regulated by the simple repressors once hypothesized for the *ara* genes in *E. coli*.

BIBLIOGRAPHY

1. M. E. Doyle (1974) Ph.D thesis, The University of Michigan, University Microfilms, Ann Arbor, MI, p. 71.
2. R. Schleif (1969) *J. Mol. Biol.* **46**, 185–196.
3. R. W. Hogg (1977) *J. Supramol. Struct.* **6**, 411–417.
4. J. W. Patrick and N. Lee (1968) *J. Biol. Chem.* **243**, 4312–4318.
5. R. Schleif (1996) In *Escherichia coli and Salmonella typhimurium* (F. C. Neidhardt et al., eds.), American Society for Microbiology, Washington, DC, pp. 1300–1309.
6. H. Boyer, E. Englesberg, and R. Weinberg (1962) *Genetics* **47**, 417–425.
7. S. M. Soisson et al. (1997) *Science* **276**, 421–425.
8. S. Beverin, D. E. Sheppard, and S. S. Park (1971) *J. Bacteriol.* **106**, 107–112.
9. M. E. Doyle et al. (1972) *J. Bacteriol.* **110**, 56–65.
10. R. B. Helling and R. Weinberg (1963) *Genetics* **48**, 1397–1410.
11. E. Englesberg et al. (1965) *J. Bacteriol.* **90**, 946–957.
12. M.-T. Gallegos et al. (1997) *Microbiol. Mol. Biol. Rev.* **61**, 393–410.
13. K. Martin, L. Huo, and R. Schleif (1986) *Proc. Nat. Acad. Sci. USA* **83**, 3654–3658.
14. N. Lee, C. Francklyn, and E. P. Hamilton (1987) *Proc. Nat. Acad. Sci. USA* **84**, 8814–8818.
15. X. Zhang and R. Schleif (1998) *J. Bacteriol.* **180**, 195–200.
16. J. C. Hu and C. A. Gross (1985) *Mol. Gen. Genet.* **199**, 7–13.
17. L. Heffernan, R. Bass, and E. Englesberg (1976) *J. Bacteriol.* **126**, 1111–1131.
18. A. H. Horwitz, C. Morandi, and G. Wilcox (1980) *J. Bacteriol.* **142**, 659–667.
19. I. Sa-Nogueira and S. S. Ramos (1997) *J. Bacteriol.* **179**, 7705–7711.
20. G. J. G. Ruijter et al. (1997) *Microbiol.* **143**, 2991–2998.
21. R. G. Wallace, N. Lee, and A. V. Fowler (1980) *Gene* **12**, 179–190.
22. N. W. Lee et al. (1986) *Gene* **47**, 231–244.

Suggestions for Further Reading

J. Beckwith (1987) The operon: an historical account. In *Escherichia coli and Salmonella typhimurium. Cellular and Molecular Biology* (F. C. Neidhardt et al., eds.), American Society for Microbiology, Washington, DC, pp. 1439–1443. (Describes the difficulty in achieving acceptance of the model of positive control.)

J. Gross and E. Englesberg (1959) Determination of the order of mutational sites governing L-arabinose utilization in *Escherichia coli* B/r by transduction with phage P1bt. *Virology* **9**, 314–331. (This is the research paper in which the initial studies with the *ara* operon are described. It has served as a model guiding analysis of many other genes.)

R. Schleif (1993) Induction, repression and the *araBAD* operon. In *Genetics and Molecular Biology*, 2nd ed., Johns Hopkins University Press, pp. 359–383. [Systematic description of the arabinose genes and enzymes, and how it has been studied. Thoughtful and stimulating, with problems (and answers at the end of the book).]

R. Schleif (1996) Two positively regulated systems, *ara* and *mal*. In *Escherichia coli and Salmonella typhimurium* (F. C. Neidhardt et al., eds.), American Society for Microbiology, Washington, DC, pp. 1300–1309. (Recent review of control of the *ara* genes, but before detailed structural studies of the AraC regulatory protein had been done.)

ARABIDOPSIS

MIEKE VAN LIJSEBETTENS
NANCY TERRYN
MARC VAN MONTAGU

Arabidopsis thaliana (L.) Heynh is a dicotyledonous **plant** that is ideally suited for molecular-genetic studies. It is generally accepted as a model for unraveling the molecular mechanisms involved in plant growth and **development**, biochemical pathways, cell biology, physiology, and pathogenic interactions. Minimal **genomic** DNA content, few **repetitive DNA** sequences, and small **gene families** account for its technical and biological simplicity. *Arabidopsis* is a natural diploid amenable to laboratory-scale genetic experiments. It is small (15 cm to 30 cm) and is grown at high density (100 plants per 0.5 m^2) without seed contamination. It has a short life cycle (6 weeks to 3 months), high seed production by self-fertilization (up to 10,000 seeds per plant), and is easily **mutagenized**. International coordination of *Arabidopsis* research has resulted in fast exchange of material, methodology, and information, and the creation of large research programs, such as the **genome-sequencing** and **expressed-sequence tag** (EST) projects. As a consequence, *Arabidopsis* research is leading in the field of plant science.

HISTORY

The first botanical description of *Arabidopsis* by Johannes Thal goes back to 1577. In 1907, Laibach studied the continuity of the **chromosomes** (5 per haploid genome) by using the plant, and he was the first to emphasize the advantages of the species for genetic analyses (1). Bennett and Smith (2) showed that *Arabidopsis* contains the smallest nuclear DNA content of the angiosperms analyzed. In this period, extensive chemical and irradiation mutagenesis of the plant was performed (3). Koornneef *et al.* (4) published the first **genetic map** containing 76 morphological markers. In 1984, Meyerowitz and co-workers demonstrated the small size and low complexity of the genome, which provoked general interest in using the plant as an experimental model (5). In 1988, efficient **transformation** methods opened the potential of **transgenic** research in the species (6). Saturation mutagenesis of the genome by insertion of heterologous DNA was initiated (7). In 1989, the US National Science Foundation launched the "Multinational Coordinated Long-Range Plan for Arabidopsis Genome Research," steered by an international board of scientists (8), with the aim of promoting *Arabidopsis* as a model system for plants, in analogy to other models such as *Drosophila melanogaster* and *Escherichia coli*. A seed and DNA stock center in Ohio (ABSCR, USA) and in Nottingham (NASC, UK), a **database** (AtDB), and joint efforts for physically mapping and sequencing the genome and for gene identification are major achievements of this initiative.

CLASSIFICATION, GEOGRAPHICAL DISTRIBUTION, AND ECOTYPES

Arabidopsis thaliana is an annual herb of the mustard family (Brassicaceae, previously named Cruciferae), has bisexual flowers, and is typified by a cross-shaped corolla, tetradynamous stamen (four long and two short ones), and capsular fruit (siliques). The genus *Arabidopsis* comprises 27 species and has been classified under a new tribe, the Arabidae, based on classical morphological and molecular **phylogenetic** studies. *Arabidopsis thaliana* is a facultative long-day plant, meaning that long days accelerate the initiation of flowering. It originates from Eurasia and North Africa and is now a common weed in the temperate regions of the Northern Hemisphere. Its broad geographical distribution resulted in natural variation. Approximately 200 ecotypes (wild populations) have been registered. These ecotypes represent a natural source of heritable variation. They differ, for instance, in flowering time, response to cold, fresh weight production, and pathogenic resistance. Frequently used ecotypes are Columbia, Niederzenz, Wassilewskija, and the laboratory strain Landsberg *erecta*. Among ecotypes, a DNA sequence **polymorphism** of up to 1.4% in low-copy DNA has been measured. Deletions and substitutions have occurred. For example, the Landsberg *erecta* genome contains three copies of the transposon Tag1, which is absent from the Columbia genome.

GENOME STRUCTURE AND ORGANIZATION

The size of the nuclear genome of *Arabidopsis thaliana* was estimated to be between 50 to 150 Mb by reassociation kinetics (see **C-value**), **flow cytometry**, and **electron microscopy**. From recent physical mapping data, this number has now been refined to 100 to 140 Mb. This is 3- to 400-fold smaller than that of other members of the angiosperms. The genome of *Arabidopsis* consists of 80% single- and low-copy DNA (9). Four classes of highly repeated DNA have been identified. They represent only 10% of the genome and are located mainly at the **telomeres** and around the **centromeres**. **Ribosomal** DNA accounts for 6% of the genome and is localized at the top of chromosomes 2 and 4.

From the sequence data in public databases, some conclusions can be drawn about **gene** organization and DNA composition: the gene density is one gene every 4 to 5 kb; the average gene size is 1.24 kb; there are 20,000–25,000 predicted **transcripts**; genes have between 0 and 30 **introns**, which are between 59 and 1500 bp long; small gene families occur either dispersed or in tandem arrays; long stretches (approximately 120 kb) of unique or low-copy DNA are interspersed with short stretches of a few kb of moderately repeated DNA; the GC content of the DNA is 41.4%; and **methylation** occurs on 6% of the cytosine bases.

The *Arabidopsis* genome sequencing project was initiated in Europe at the end of 1993, followed by an American and a Japanese initiative, whose aim is sequencing one-third of the genome by 1999 and the entire genome by the year 2002. Over 30,000 redundant *Arabidopsis* ESTs and genes have already been sequenced and submitted to public databases. Two approaches have been taken: an American team used whole-plant cDNA collections (10) and sequenced them from one side, whereas a French consortium (11) used tissue-specific libraries and sequenced them from both sides. The first analyses in 1998 of the genome sequencing program indicate that approximately half of the genes are represented by an EST and that almost 40% of the genes have no **homology** with available database entries (12). Based on sequence information, **PCR** screens are performed on the **transgenic** collection to identify insertions in these newly identified genes and to study their function.

Information on the *Arabidopsis thaliana* genome project, on *Arabidopsis thaliana* research in general is very well centralized in the so-called *Arabidopsis thaliana* database (AtDB). This project has a home-page on the Internet where much information on ongoing *Arabidopsis* research can be found (genetic and physical maps, links to public databases, seed and DNA stock centers, Weeds World, public information, etc.). The WWW address is: http://genome-www.stanford.edu/arabidopsis/.

GENETICS AND PHYSICAL MAPPING

Based on **metaphase** staining and genetic **linkage group** analyses, the haploid chromosome number is 5, ranging in size from 13.4 Mb to 25.4 Mb. The genetic map comprises more than 460 loci. The ratio of physical to genetic distance between markers, on average, is 200 kb per cM. However, from the physical map construction of chromosome 4, it was noticed that **recombination** hot spots (30 to 50 kb/cM) and low spots ($\geq$ 550 kb/cM) do occur (13). Many tools are available to map a mutation, for example, recessive visible markers, codominant embryo-lethal markers, dominant selectable markers on the located T-DNA and *Ac/Ds* insertions, **restriction fragment length polymorphism** (RFLP)-derived and PCR-based molecular markers, such as **microsatellites**. Several RFLP, rapid-amplified polymorphic DNA, or amplified fragment length polymorphism (AFLP) molecular marker maps have been constructed, based on different mapping populations. A combined map, made by statistical integration, gives an approximate position and order of the markers. Recombinant inbred (RI) lines, derived from a cross between Columbia and Landsberg *erecta* (14), have been used to locate more than 750 molecular markers unambiguously. Genes are mapped by RFLP segregation analysis by using RI lines or by matrix-based PCR analysis of pooled **yeast artificial chromosome** clones (YACs). The physical map consists of **contigs** of DNA clones that are correlated with the mapped markers. Currently, YAC, **bacterial artificial chromosome** (BAC), and **phage P1 artificial chromosome** (PAC) contig. maps are available that cover almost the entire genome. Genetic and physical maps are updated through AtDB.

SCIENTIFIC ADVANCES AND APPLICATIONS

The molecular-genetic approach in *Arabidopsis* research has led to major breakthroughs in plant developmental biology. Tremendous progress has been made in understanding the molecular control of meristem identity during flower initiation and flower organ formation, embryo development and pattern formation during embryogenesis, root development, epidermal cell fate specification in root hair and trichome formation, and cell determination in the vegetative meristem. Genes have been identified that are involved in hormone perception, biosynthesis, and **signal transduction**. The first **hormone receptor** for plants has been characterized in *Arabidopsis* (15). Much of the molecular insights into light perception and signal transduction, **cell cycle** regulation, and disease resistance in higher plants comes from studies in *Arabidopsis* (16–18).

The *Arabidopsis* genes and mutants are resources exploited either to isolate orthologs from other species and to test their functional conservation (19) or to be used straight away for the genetic modification of even distantly related crop plants (20). The molecular markers within contigs in *Arabidopsis* have been used for comparative mapping with *Brassica* spp. Colinearity in 5- to 10-cM regions has been demonstrated between the *Arabidopsis* genome and that of *Brassica nigra* (21). This implies that information and markers obtained from the physical mapping in *Arabidopsis* can be applied to **syntenic** genomic regions in mustard crops to analyze important traits in breeding programs.

BIBLIOGRAPHY

1. F. Laibach (1943) *Bot. Archiv* **44**, 439–455.
2. M. D. Bennett and J. B. Smith (1976) *Philos. Trans. R. Soc. Lond. B Biol. Sci.* **274**, 227–274.
3. G. P. Rédei (1975) *Ann. Rev. Genet.* **9**, 111–127.
4. M. Koornneef et al. (1983) *J. Hered.* **74**, 265–272.
5. L. S. Leutwiler, B. R. Hough-Evans, and E. M. Meyerowitz (1984) *Mol. Gen. Genet.* **194**, 15–23.
6. D. Valvekens, M. Van Montagu, and M. Van Lijsebettens (1988) *Proc. Natl. Acad. Sci. USA* **85**, 5536–5540.
7. K. A. Feldmann (1991) *Plant J.* **1**, 71–82.
8. National Science Foundation (1990) *A long-range plan for the multinational coordinated* Arabidopsis thaliana *genome research project* (NSF 90-80), National Science Foundation, Washington, DC. (published annually).
9. R. E. Pruitt and E. M. Meyerowitz (1986) *J. Mol. Biol.* **187**, 169–183.
10. T. Newman et al. (1994) *Plant Physiol.* **106**, 1241–1255.
11. R. Cooke et al. (1996) *Plant J.* **9**, 101–124.
12. M. Bevan et al. (1998) *Nature (London)* **391**, 485–488.
13. R. Schmidt et al. (1995) *Science* **270**, 480–483.
14. C. Lister and C. Dean (1993) *Plant J.* **4**, 745–750.
15. G. E. Schaller and A. B. Bleecker (1995) *Science* **270**, 1809–1811.
16. J. L. Dangl (1995) *Cell* **80**, 363–366.
17. C. Lin et al. (1995) *Science* **269**, 968–970.
18. A. Hemerly et al. (1992) *Proc. Natl. Acad. Sci. USA* **89**, 3295–3299.
19. V. F. Irish and Y. T. Yamamoto (1995) *Plant Cell* **7**, 1635–1644.
20. D. Weigel and O. Nilsson (1995) *Nature* **377**, 495–500.
21. U. Lagercrantz, J. Putterill, G. Coupland, and D. Lydiate (1996) *Plant J.* **9**, 13–20.

Suggestions for Further Reading

M. Anderson and J. Roberts (1998) *Arabidopsis* (Annual Plant Reviews, Vol. 1), Sheffield Academic Press, Sheffield, UK.

J. Bowman (1994) *Arabidopsis: an Atlas of Morphology and Development*, Springer-Verlag, New York.

C. Koncz, N.-H. Chua and J. Schell (1992) *Methods in Arabidopsis Research*, World Scientific, Singapore.

J. M. Martínez-Zapater and J. Salinas (1998) *Arabidopsis Protocols* (Methods in Molecular Biology, Vol. 82) The Humana Press, Totowa, NJ.

E. M. Meyerowitz and C. R. Somerville (1994) *Arabidopsis* (Cold Spring Harbor Monograph Series, Vol. 27), Cold Spring Harbor Laboratory Press, Cold Spring Harbor, NY.

ARGININE (ARG, R)

T. E. CREIGHTON

The **amino acid** arginine is incorporated into the nascent **polypeptide chain** during **protein biosynthesis** in response to six **codons**—CGU, CGC, CGA, CGG, AGA, and AGG—and represents approximately 5.7% of the residues of the proteins that have been characterized. The arginyl residue incorporated has a mass of 156.19 Da, a **van der Waals volume** of 148 Å^3, and an **accessible surface** area of 241 Å^2. Arg residues have average conservation during **divergent evolution**; they are interchanged most frequently in **homologous** proteins with **lysine**, the other basic residue.

The Arg side chain consists of three nonpolar methylene groups and the strongly basic δ-guanido group:

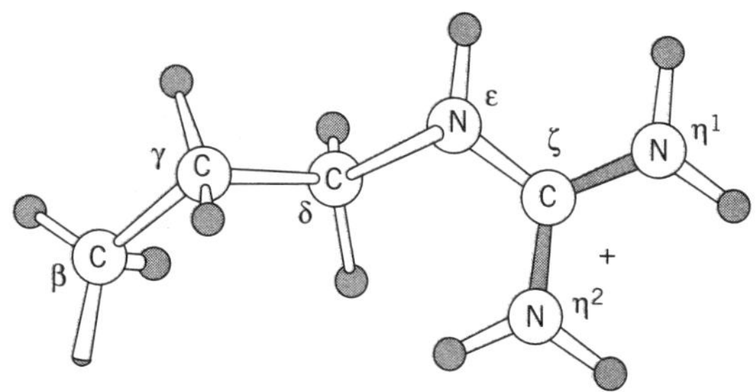

With a pK_a value usually of about 12, the guanido group is ionized over the entire pH range in which proteins exist naturally. The ionized guanido group is planar as a result of resonance:

$$-(CH_2)_3-NH-C\begin{cases}=\overset{+}{N}H_2\\ -NH_2\end{cases}$$

$$\rightleftharpoons$$

$$-(CH_2)_3-NH-C\begin{cases}-NH_2\\ =\underset{+}{N}H_2\end{cases}$$

$$\rightleftharpoons$$

$$-(CH_2)_3-\overset{+}{N}H=C\begin{cases}-NH_2\\ -NH_2\end{cases}$$

$$\rightleftharpoons$$

$$-(CH_2)_3-NH-C^{+}\begin{cases}-NH_2\\ -NH_2\end{cases}$$

and the positive charge is effectively distributed over the entire group. In the protonated form, the guanido group is unreactive, and only very small fractions of the nonionized form are present at physiological pH values. The guanido groups of Arg residues are almost invariably at the surfaces of native protein structures, and virtually no Arg residues are fully buried, but the nonpolar part of the side chain, and the adjoining polypeptide backbone, are frequently buried within the interior. Arg residues favor the ***alpha*-helical** conformation in model peptides and also occur most frequently in that **secondary structure** in folded **protein structures**.

Proteinases frequently cleave polypeptide chains adjacent to Arg residues, as in the processing of pro-hormones, such as pro-**insulin**, at pairs of basic residues.

The guanido group can form heterocyclic condensation products with 1,2- and 1,3-dicarbonyl compounds, such as phenylglyoxal, 2,3-butanedione, and 1,2-cyclohexanedione:

$$\underset{\text{Arg}}{-(CH_2)_3-NH-C\begin{cases}-NH_2\\ =\underset{+}{N}H_2\end{cases}} + \underset{\text{cyclohexanedione}}{\text{(1,2-cyclohexanedione)}}$$

$$\rightleftharpoons$$

$$-(CH_2)_3-NH-C\begin{cases}-NH-\\ =\underset{+}{N}H-\end{cases}\text{(cyclohexane ring bearing two OH)}$$

These reactions occur readily because the distance between the two carbonyl groups of the reagents closely matches that between the two unsubstituted nitrogen atoms of the guanido group. The adduct formed can be stabilized further by the presence of borate, which complexes with the adjacent hydroxyl groups.

The guanido group can be cleaved by hydrazine (H_2NNH_2) to produce the side chain of ornithine:

$$\underset{\text{Arg}}{-CH_2-CH_2-CH_2-NH-C\begin{cases}=NH\\ -NH_2\end{cases}}$$

$$\downarrow H_2N-NH_2$$

$$\underset{\text{Orn}}{-CH_2-CH_2-CH_2-NH_2}$$

This reaction is, however, often accompanied by cleavage of the polypeptide backbone.

Suggestions for Further Reading

E. L. Smith (1977) Reversible blocking at arginine by cyclohexanedione, *Meth. Enzymol.* **47**, 156–161.

A. Honegger et al. (1981) Chemical modification of peptides by hydrazine, *Biochem. J.* **199**, 53–59.

R. B. Yamasaki et al. (1980) Modification of available arginine residues in proteins by *p*-hydroxyphenylglyoxal, *Anal. Biochem.* **109**, 32–40.

ASCARIS

W. WOOD

Ascaris is a genus of large animal parasitic **nematodes**. Common species include *Ascaris equorum*, a horse parasite, *Ascaris suum*, a pig parasite, and *Ascaris lumbricoides*, which parasitizes humans. Adult ascarid nematodes, which reside in the host intestine, range from several inches to several feet in length, and females can produce more than 2 million eggs per day. Fertilized eggs are released into the environment by defecation, where they remain dormant until ingested by another host. The acidic environment of the stomach and the increase in temperature trigger resumption of embryonic development, resulting in first-stage (L1) juveniles (often termed *larvae*) inside the egg shell. After molting once, the larvae hatch from the egg as second-stage (L2) juveniles. Like all nematodes (*q.v.*), Ascaris molts through three more juvenile stages (L3–L4) and then to a sexually mature adult. Stages in the parasitic life cycle are tissue-specific: The L2 burrows through the wall of the small intestine and is carried via the bloodstream or lymphatic system to the lungs, bronchi, and trachea. It molts to L3 and migrates up the trachea and down the esophagus, returning to the small intestine where it molts to L4 and then, about two to three months after the initial infection, to a mature adult that begins producing eggs. About 25% of the human population is infected by *Ascaris lumbricoides*, primarily in developing countries. Ascaris infections are generally debilitating but not life-threatening, although heavy infections can cause severe complications resulting from intestinal obstruction, peritonitis, or allergic reaction.

As an experimental organism, Ascaris was important to early embryologists and cytologists such as Theodor Boveri and Otto zur Strassen in the 1890s, who took advantage of several favorable properties of the early **embryos** that are still exploited by present-day researchers. The embryos are obtainable in large numbers from gravid females; they are transparent, facilitating observation by light **microscopy**; they can be stored in their original dormant state and their development initiated synchronously by high temperature or acid shock; and the pattern and timing of their embryonic cleavages, like those of most nematodes (*q.v.*), are invariant from embryo to embryo, with many cell fates apparently determined as the cells are born. The species studied most extensively by Boveri, *Ascaris megalocephala*, has the additional advantage of having only one large **chromosome**, which was convenient for cytological studies. Ascaris also has the peculiarity that there is chromosome fragmentation and diminution in all of the **somatic** precursor cells as they separate from the **germ line** in the early embryo. Although later work has shown that most of the discarded DNA does not include coding sequences, this feature allowed Boveri to demonstrate that cytoplasmic components specific to the germ line were responsible for maintaining chromosomal integrity in these cells, which supported the view that nonrandomly segregating cytoplasmic determinants were responsible for dictating early cell fates.

In the modern era, Ascaris has also become an important system for research in neurobiology. Part of the rationale behind Sydney Brenner's choice of the small free-living soil nematode *Caenorhabditis elegans* (see ***Caenorhabditis***) for genetic and ultrastructural analysis of a simple nervous system and its development was the hope that Ascaris, only distantly related but much more suitable for electrophysiology because of its larger size, would allow functional analysis of a homologous nervous system. This hope has been largely realized: The two nervous systems appear to be very similar in both structure and function, and information from each has been valuable in understanding the other.

Suggestion for Further Reading

G. O. Poinar (1983) *The Natural History of Nematodes*, Prentice-Hall, Englewood Cliffs, NJ, p. 323.

ASEXUAL

ENRIQUE CERDÁ-OLMEDO
JAVIER AVALOS

The term "asexual" means without **sex**; it is an adjective applied to multicellular organisms with missing or atrophied sexual organs and to forms of reproduction that do not involve the fusion of male and female gametes (see **Asexual reproduction**).

The initial concept of sex referred to diploid, multicellular species in which new individuals are produced by the fusion of two haploid gametes, a sperm and an ovum (sexual reproduction). Individuals of many of these species may be classified into two sexes, *male* and *female*, depending on whether they are able to produce **sperm** or ova (**eggs**), respectively. This is the situation in most animals and in dioecious plants. The sex of an individual may be determined genetically or by the environment (see **Sex determination**). Individuals are *asexual* if the organs that produce gametes are missing or atrophied; *intersexual* if their sexual organs are intermediate or ambiguous; or ***hermaphrodite*** if they have both kinds of sexual organs, whether they are functional or not. Hermaphroditism is prevalent in many species of plants, whether both pollen and ova are produced in the same *hermaphrodite* flowers, or in separate flowers of the same plant (*monoecious* plants).

These concepts cannot be transferred to unicellular organisms without some modifications. The **haploid** cells that predominate in the life cycles of many unicellular eukaryotes can often be classified into different **mating types**, sometimes more than two, with the criterion that only those of different mating type are able to fuse and form diploids. This situation is called *heterothallism*. The mating types of the lower eukaryotes are sometimes called sexes, even though cells of different mating types are not homologous to sperm and ova of the animals.

In the situation called *homothallism*, diploid cells may be formed by the fusion of two genetically identical haploid cells. This would happen if there were a single mating type or if haploid cells change their mating type when they find no partner of the opposite one, as occurs in the homothallic strains of the yeast *Saccharomyces*.

Asexual bacteria are unable to transfer or receive DNA via conjugation (see **Conjugation**; **Hfrs and F-primes**; **F plasmid**).

ASEXUAL REPRODUCTION

ENRIQUE CERDÁ-OLMEDO
JAVIER AVALOS

Asexual reproduction includes all forms of reproduction that do not involve the fusion of male and female gametes. The asexual reproduction of a cell consists in the formation of genetically identical daughter cells; in the eukaryotes, this implies nuclear division via mitosis. Multicellular organisms are formed through repeated cycles of asexual reproduction from an initial zygote (see **Development** and **Embryology**). The daughter cells need not be completely identical, because cells with the same genotypes may differ in the quantitative distribution of **transcription factors** (see **Differentiation**). The genotypes of the daughter cells may differ because of mutation, **mitotic**

recombination, chromosomal nondisjunction, and other infrequent events.

There are two kinds of *asexual* reproduction of multicellular organisms, namely, **parthenogenesis** and vegetative reproduction. Parthenogenesis imitates sexual reproduction, in that embryogenesis is apparently normal, but without participation of a male gamete. **Haploid** individuals are produced through the parthenogenetic development of a normal female gamete. **Diploid** individuals are produced through the fusion of the haploid nuclei of a female gamete and a neighboring haploid cell, such as the synergids of plants and the polar bodies of animals. Other diploid individuals are produced through the parthenogenetic development of a diploid cell after failure of meiosis. In vegetative reproduction, individuals are derived from one or many somatic cells. It is a common natural process in many plants and lower animals, and it is the only multiplication mechanism in some of them.

Artificial interventions permit new forms of asexual reproduction. These include (a) the production of haploid plants from the culture of anthers and (b) the production of diploid animals through the replacement of the nucleus of the ovum by a somatic nucleus. Vegetative reproduction and ameiotic parthenogenesis lead to the formation of *clones*, which are sets of individuals with identical genotype. Clonal propagation is usually carried out through vegetative reproduction from a single cell or a group of identical cells.

The **parasexual cycle** may be seen as a form of asexual reproduction in the lower eukaryotes and in cultured cells of multicellular organisms. Haploid cells, not necessarily of different **mating type**, fuse to form **heterokaryons**, and the nuclei of the heterokaryon fuse to produce diploid nuclei; these revert to haploidy, not by meiosis, but by random chromosome losses during mitosis, accompanied by **mitotic recombination**.

The sexual processes of bacteria imply the conjugation of two compatible cells and the transfer of part of the genetic information of one to the other (see **Conjugation**; **Hfrs and F primes**; **F plasmid**). Formally similar are asexual processes in which some genetic information from a cell is transferred to another via a **virus** or via naked DNA (see **Transformation**). These processes are not exclusive to the prokaryotes.

In evolutionary terms and in comparison with the various sexual processes, asexual reproduction has advantages of speed, economy, and the conservation of certain genotypes that may be lost or made less frequent after meiosis (see **Heterosis**, **Polyploidy**, and **Aneuploidy**). The main drawback of asexual reproduction is the limited creation of new genetic variation (see **Genetic diversity**).

ASPARAGINE (ASN, N)

T. E. CREIGHTON

The **amino acid** asparagine is incorporated into the nascent **polypeptide chain** during **protein biosynthesis** in response to six **codons**—CGU, CGC, CGA, CGG, AGA, and AGG—and represents approximately 4.4% of the residues of the proteins that have been characterized. The asparaginyl residue incorporated has a mass of 114.11 Da, a **van der Waals volume** of 96 Å^3, and an **accessible surface** area of 158 Å^2. Asn residues are those most variable during **divergent evolution**; they are interchanged frequently in **homologous** proteins with **serine**, **aspartic acid**, **lysine**, and **histidine** residues.

The side-chain of Asn residues is the same as that of **aspartic acid**, except that the **carboxyl group** has been converted to the amide:

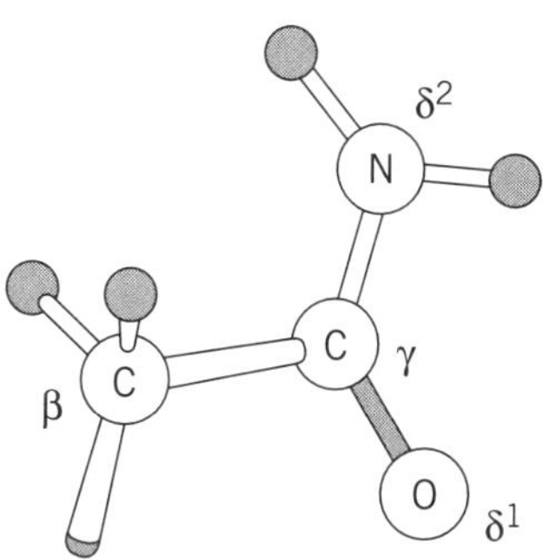

Both amino acids occur naturally and are incorporated directly into proteins during their biosynthesis; Asn residues do not arise from amidation of Asp in proteins. The amide side chain does not ionize and is not very reactive chemically. It is **polar**, however, because it is both a **hydrogen bond** donor and acceptor. The amide group is labile at extremes of pH and at high temperatures, and Asn residues can **deamidate** to Asp. The Asn residue is especially labile at alkaline pH because its side chain is sterically suited to interact with the –NH– group of the following residue in the polypeptide chain, to form transiently a cyclic succinimidyl derivative (see **Deamidation**). This derivative can undergo racemization and hydrolysis to cleave the polypeptide chain or to produce a mixture of D and L isomers of Asp and isoAsp residues. In an isoAsp residue, the normal backbone peptide bond is through the side-chain carboxyl group rather than the usual α-carboxyl. The deamidation reaction of Asn residues occurs 30–50 times more rapidly if the following residue is **Gly**, because the absence of a side chain there favors succinimide formation, and –Asn–Gly– sequences are especially prone to deamidation. The rate depends on the polypeptide conformation, however, because only some conformations permit succinimide formation. If the succinimide ring reacts with hydroxylamine instead of water, peptide cleavage results. Therefore, Asn–Gly peptide bonds are readily cleaved by incubation with hydroxylamine.

Asn residues are the site of **N-glycosylation** of proteins, one of the most prevalent **post-translational modifications**. The Asn residue that is so glycosylated always occurs in a characteristic sequence: –Asn–Xaa–Ser–, –Asn–Xaa–, Thr–, or –Asn–Xaa–Cys–, where Xaa can be any residue except **Pro**, which also cannot immediately follow the tripeptide sequence.

Suggestions for Further Reading

S. J. Wearne and T. E. Creighton (1989) Effect of protein conformation on rate of deamidation: ribonuclease A, *Proteins Struct. Funct. Genet.* **5**, 8–12.

W. J. Chazin et al. (1989) Identification of an isoaspartyl linkage formed upon deamidation of bovine calbindin D_{9k} and structural characterization by 2D ^{1}H NMR, *Biochemistry* **28**, 8646–8653.

P. Bornstein and G. Balian (1977) Cleavage at Asn-Gly bonds with hydroxylamine, *Meth. Enzymol.* **47**, 132–144.

F. Wold (1985) Reactions of the amide side-chains of glutamine and asparagine *in vivo*, Trends Biochem. Sci. **10**, 4–6.

ASPARTATE TRANSCARBAMOYLASE

MELINDA E. WALES
JAMES R. WILDA

Aspartate transcarbamoylase (carbamoylphosphate:L-aspartate carbamoyl-transferase, ATCase, EC 2.1.3.2) is a ubiquitous **enzyme** of **pyrimidine** biosynthesis in which it catalyzes the first unique step in the *de novo* synthetic pathway:

$$\text{L-aspartate} + \text{Carbamoyl phosphate} \longrightarrow \text{N-carbamoyl aspartate} + P_i$$

THE *ESCHERICHIA COLI* PARADIGM

Although the enzyme exists in a variety of oligomeric and multifunctional complexes in various organisms, the most studied ATCase is that of *Escherichia coli*. It is a structurally complex, **allosterically** regulated enzyme containing two catalytic trimers associated with three regulatory dimers, $2(c_3){:}3(r_2)$ (Fig. 1). The catalytic polypeptide chain is encoded by the *pyrB* **gene**, whereas *pyrI* encodes the regulatory polypeptide. These genes are preceded by two tandem **promoters**, designated P_1 and P_2, and an open reading frame encoding a 44-residue leader polypeptide, *pyrL*. Promoter P_2 has been identified as the physiologically significant promoter (see Fig. 2).

The *pyrLBI* operon is located on the linkage map of *E. coli* at 97 minutes and unlinked to any of the genes or small operons of the other six enzymes involved in pyrimidine biosynthesis. The expression of these genes is noncoordinately regulated by the intracellular levels of uridine or cytidine nucleotides, with the expression of the *pyrBI* operon negatively regulated over 300-fold. Most of this regulation (50-fold) occurs through a UTP-sensitive **attenuation** control mechanism, whereas attenuator-independent mechanisms are responsible for approximately a 6.5-fold range of regulation, including a pyrimidine-sensitive **transcriptional** initiation mechanism, and **stringent** control by ppGpp (1). According to the current model for attenuation control, transcriptional termination at the *pyrBI* attenuator (a **rho**-independent transcriptional terminator) is regulated by the relative rates of transcription and **translation** within the *pyrBI* leader region. Low intracellular levels of UTP cause **RNA polymerase** to pause at the uridine-rich region in the leader transcript (this pausing is enhanced by NusA, a general **transcription factor** that increases the efficiency of termination). This allows time for a **ribosome** to initiate translation and catch up to the stalled RNA polymerase before it transcribes the attenuator region. As RNA polymerase eventually makes its way through the attenuator, the adjacent translating ribosome blocks formation of the *terminator hairpin*. This permits RNA polymerase to read through into the *pyrBI* structural genes, with translation terminating before the *pyrB* initiation codon. When intracellular levels of UTP are high, RNA polymerase does not pause during the transcription of *pyrL*. Without this pausing, the ribosomes cannot catch up to RNA polymerase before it transcribes the attenuator. This allows formation of the RNA hairpin, thus terminating transcription prior to the structural genes.

In addition to the genetic regulation of *pyrLBI*, the gene product ATCase also exerts significant control over the rate of pyrimidine biosynthesis (2). This is achieved by allosteric modification of the enzymatic activity in response to substrate concentration (**homotropic** cooperativity) and the nucleotide end products of both the purine and pyrimidine pathways (**heterotropic** regulation; see Tables 1 and 2). In the oligomeric complete enzyme, catalysis proceeds by a preferred order mechanism, with carbamoyl phosphate binding before aspartate and N-carbamoyl-L-aspartate leaving before inorganic phosphate. Homotropic cooperativity is induced by aspartate in the presence of a saturating concentration of carbamoyl phosphate and involves a structural transition of the enzyme consistent with a two-state, **concerted model**. As aspartate is bound, the enzyme shifts from the **T-state**, characterized by low activity and low affinity for substrates, to the **R-state** with high activity and high affinity for substrates. The kinetic consequence of this positive cooperativity is a sigmoidal substrate saturation curve (Fig. 3). Heterotropic activation is caused by the purine ATP, whereas the pyrimidine end products CTP and UTP **feedback inhibit** the enzyme. The pattern of inhibition exhibited by pyrimidine nucleotides is synergistic: CTP inhibits ATCase activity approximately 60% and UTP has minimal effect, while CTP in combination with UTP inhibits ATCase activity >95% (3). CTP, ATP,

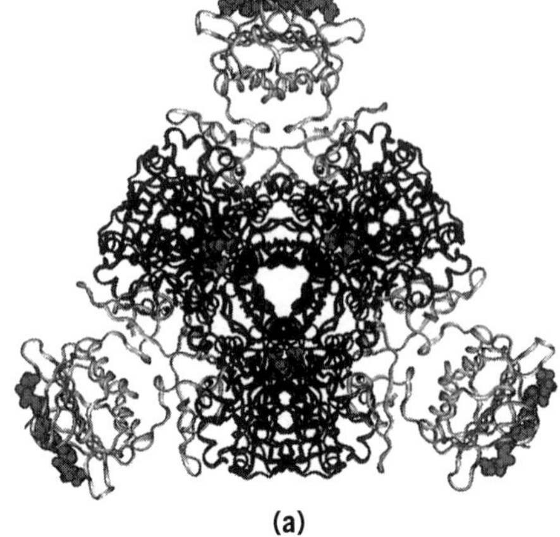

(a)

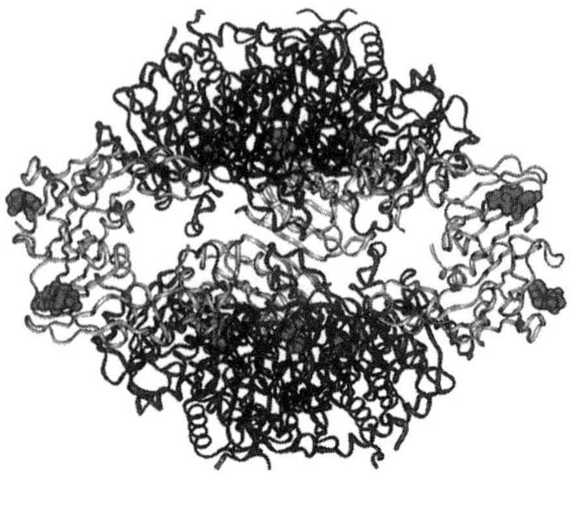

(b)

Figure 1. Structure of the intact *E. coli* ATCase viewed along the 3-fold axis (**a**) and along the approximate 2-fold axis (**b**). The active sites are located within the catalytic trimer. The allosteric sites are located within the regulatory dimers. The PDB file 8at1 was used to generate this figure.

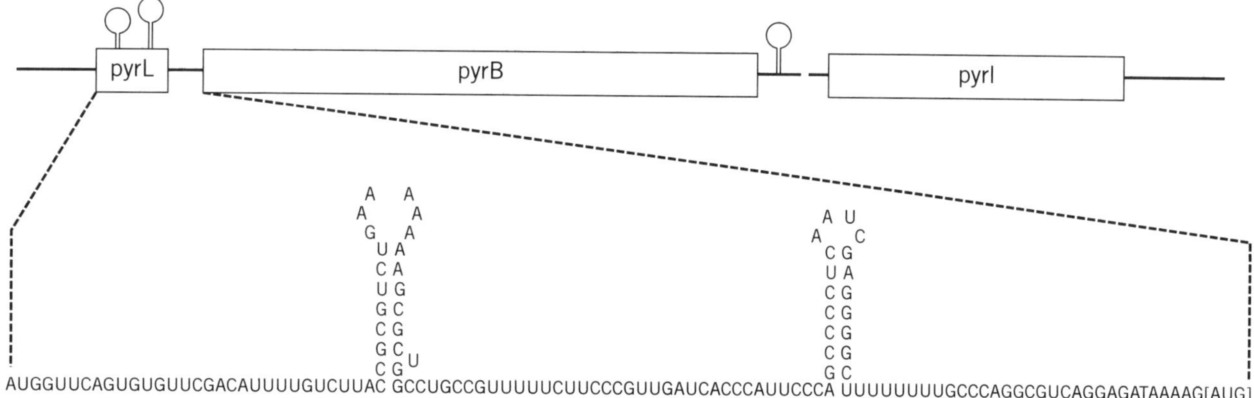

Figure 2. Genetic organization of the *E. coli pyrLBI* operon. The position of the three cistrons on the chromosome is shown at the top, along with the positions of three hairpin structures adopted by the messenger RNA. The sequence around the first two is shown below. The first is the transcriptional *pause hairpin*, flanked on each side by uridine-rich sequences. The second is the *terminator hairpin*, encoded by the attenuator sequence.

Table 1. Physiological Significance of Catalytic Characteristics of the *E. coli* Aspartate Transcarbamoylase

Catalytic Characteristic	Functional Consequence
• Ordered substrate binding (CP→Asp)	• Places catalytic emphasis on aspartate
• Cooperative aspartate binding	• Allows large changes in catalytic activity with only small changes of physiological substrate concentrations
• Distinct regulatory and catalytic subunits	• Increased sophistication in the allosteric modulation of enzymatic activity
• Presence of shared active sites and catalytic trimers	• Facilitates modulation of catalytic activity with subtle structural movements
• T→R structural transition	• Provides for dramatic differences in aspartate binding and modulation of catalytic capacity at physiological levels of substrate

Table 2. Physiological Significance of Allosteric Characteristics of the *E. coli* Aspartate Transcarbamoylase

Allosteric Characteristic	Functional Consequence
• Allosteric inhibition by CTP	• 50–70% inhibition of activity by nucleotide end-products to balance new synthesis with utilization
• Synergistic inhibition by UTP and CTP	• UTP and CTP are both end-products of the biosynthetic pathway and the synergism provides 90–95% inhibition of activity
• Competitive allosteric activation by ATP	• ATP competes with CTP and UTP for the same binding sites, thus balancing intracellular purine and pyrimidine nucleotide pools
• Competitive allosteric inhibition by CTP and CTP + UTP	• Allows for displacement of ATP (activation) by inhibitors (CTP or CTP + UTP)
• Negative cooperativity in binding CTP	• Permits modulation of enzymatic activity over a large range of inhibitor concentration
• Communication of allosteric signals across protein–protein interfaces	• Facilitates modulation of allosteric signal
• Asymmetry of the holoenzyme	• Provides two distinct classes of allosteric binding sites within a single enzyme

and UTP bind competitively, with different affinities, to a common allosteric site. CTP binds the most tightly with a pattern consistent with two classes of three sites each, whose **dissociation constants** differ by a factor of 20 ($K_{d\text{-CTP}}$: 5 to 20 μM). The binding of ATP follows a pattern similar to that of CTP with two classes of affinity sites, except that ATP binding is an order of magnitude weaker than that of CTP ($K_{d\text{-ATP}}$: 60 to 100 μM). The binding of UTP appears to be limited to three sites ($K_{d\text{-UTP}}$: 800 μM), although there may be a second class of sites too weak to be measured. Interestingly, the binding of CTP to three sites appears to enhance the binding of UTP to the remaining three sites almost 100-fold, resulting in a predicted K_d for UTP in the presence of CTP of 10 μM (4). These biochemical binding characteristics are well suited to the physiological requirements of the pathway: Although the intracellular concentration of CTP (500 μM) and UTP (900 μM) is 3- to 6-fold lower than that of ATP (3 to 5 *mM*), their stronger binding can effectively displace ATP at the allosteric sites of the enzyme.

Comparison of the **X-ray crystallography** structures of the enzyme has revealed that each of the catalytic chains is composed of two independently folding structural **domains**: the carbamoyl phosphate (CP) domain and aspartate (Asp) domain (Fig. 4). The **active sites** are located at the interface between the CP and Asp domains of one catalytic chain and the CP domain of an adjacent catalytic chain. Likewise, the regulatory chains are composed of two structural domains: the zinc-binding (Zn) domain and allosteric (Allo) domain. Differences between the two states of the enzyme have been

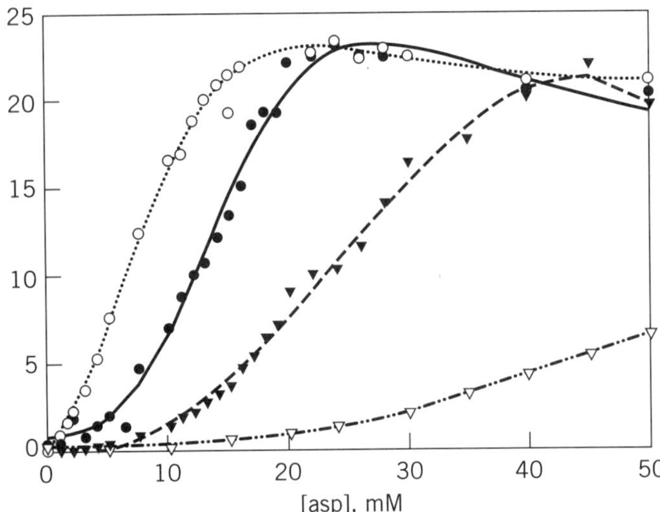

Figure 3. Substrate saturation profile of the *E. coli* ATCase. The velocity of the enzyme at varying concentrations of aspartate [asp] was determined in the presence of no allosteric effectors •, the activator ATP °, the inhibitor CTP ▼, and both pyrimidine end-products CTP and UTP ▽.

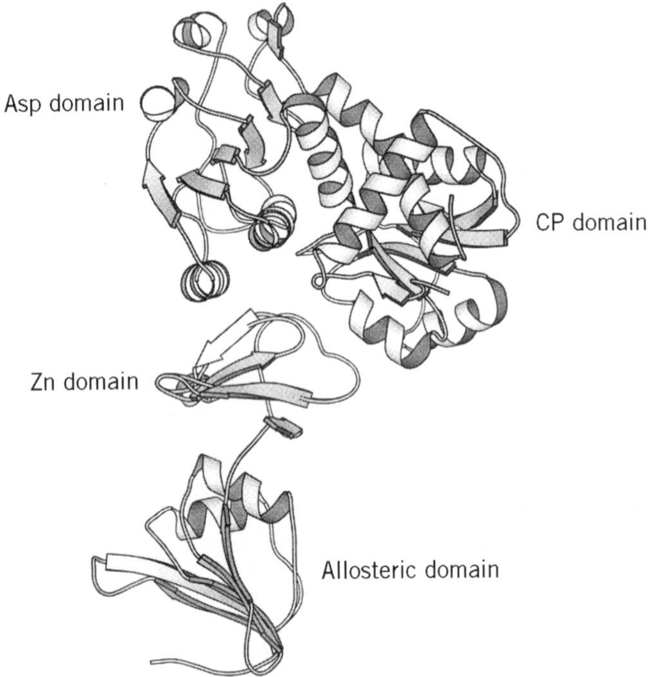

Figure 4. Domain organization of the *E. coli* ATCase illustrated with one catalytic chain (C1) and its associated regulatory chain (R1). The catalytic chain is composed of two independently folding domains: the aspartate-binding (Asp) domain and carbamoyl phosphate-binding (CP) domain. Likewise, the regulatory chain is composed of the zinc-binding (Zn) domain and nucleotide-binding (allosteric) domain. This R1:C1 unit was drawn with Rasmol and the MOLSCRIPT programs.

identified by comparison of the T-state structure with bound inhibitor (CTP) and the R-state structure binding a **bisubstrate analogue** [N-(phosphonacetyl)-L-aspartate; PALA] (5,6). The T → R change in **quaternary structure** involves substantial conformational rearrangements as the catalytic trimers separate by 11 Å and mutually reorient 10° around the 3-fold axis, while each regulatory dimer rotates 15° around the 2-fold axis. During substrate binding, the two domains of each catalytic chain (Asp and CP) undergo domain closure, whereas the two domains of the regulatory chain (Allo and Zn) undergo domain separation. **Site-directed mutagenesis** studies have shown that closure of the CP and Asp domains is important for the formation of the high-affinity, high-activity R-state, which is required to attain the proper active-site conformation needed for catalysis, and for homotropic cooperativity.

In addition to these studies, well over 100 site- and region-specific mutations have been created in the *E. coli* ATCase. These studies, along with the many structural and biochemical analyses, have contributed to our understanding of ATCase by verifying active-site and allosteric site locations, identifying interactions important to the stabilization of the T- and R-states, and confirming catalytically significant residues. More recently, site-directed mutagenesis studies have been directed toward the allosteric mechanism. Mutations have been created that can effectively separate the homotropic from the heterotropic effect, ATP activation from inhibition, and CTP inhibition from CTP + UTP synergism. Although the mechanism of allosteric regulation remains obscure, current mutagenesis and structural studies are attempting to identify and differentiate between possible mechanisms, such as (1) discreet pathways for each nucleotide signal, (2) complex and multiple interlocking pathways, or (3) modulation of general global energy changes.

DIVERSITY OF ASPARTATE TRANSCARBAMOYLASE

Prokaryotes

With the exception of some anaerobic protozoan parasites, all examined organisms are capable of *de novo* pyrimidine biosynthesis and have been found to possess the enzyme aspartate transcarbamoylase. The earliest classification of bacterial ATCases by Bethell and Jones in 1969 (16), based on enzyme size and response to nucleotides, partitioned the enzyme into three classes: (1) Class A ATCases are the largest ATCases found in bacteria and were initially described for the pseudomonads; (2) class B includes the ATCases of *E. coli* and other members of the *Enterobacteriaceae*; whereas (3) class C is the class with the smallest ATCase and until recently was restricted to **Gram-positive** bacteria such as *Bacillus* (Fig. 5). The class A ATCases are approximately 450 kDa in size, and the intact enzyme exists as two catalytic trimers of 35 kDa in dodecameric association with three dimers of 45 kDa chains ($2A_3{:}3D_2$). In *P. putida*, this 45-kDa chain has been shown to be a nonfunctional homologue of *dihydroorotase* (DHOase*), the third enzyme in *de novo* pyrimidine biosynthesis immediately following ATCase in the biosynthetic pathway.

The *Bacillus subtilis* enzyme at 100 kDa is a typical class C enzyme and corresponds in size, architecture, and function to the free catalytic trimer of the *E. coli* enzyme. Class C enzymes are composed of three identical polypeptide chains (c_3) of approximately 34 kDa each and have no associations with other enzymes in the pathway. Furthermore, these enzymes have no associated regulatory subunit nor the attendant allosteric regulation. As a consequence of the active site being shared

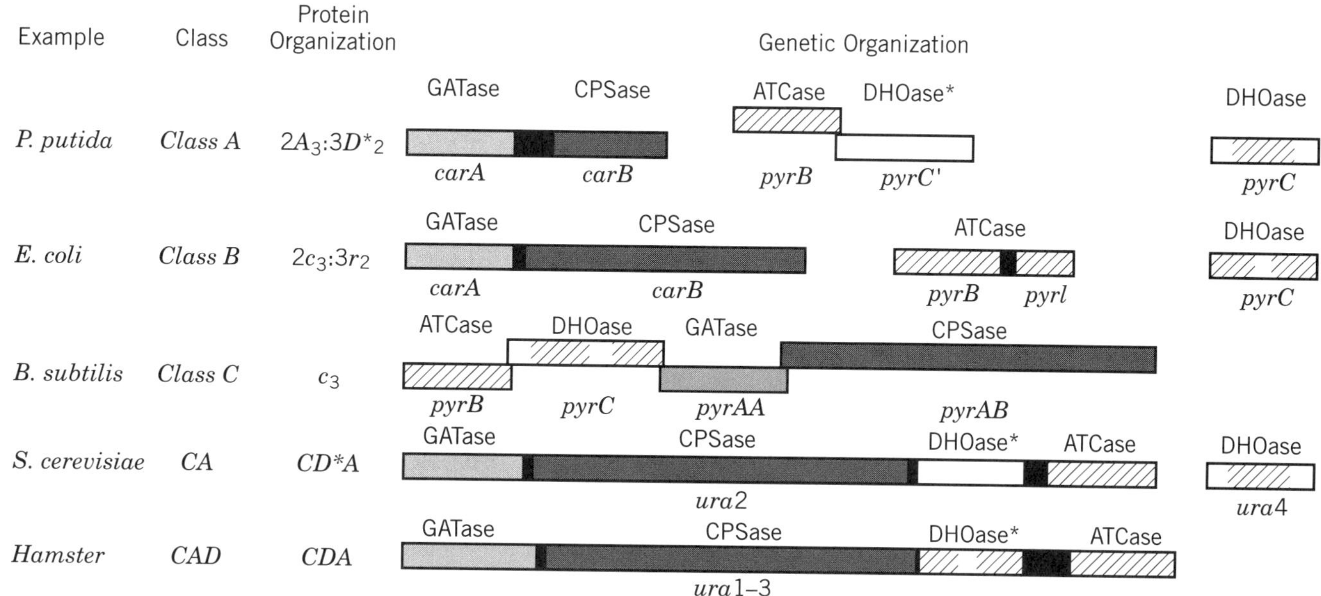

Figure 5. Organization of the first three enzymes and genes of pyrimidine biosynthesis: glutamine amidotransferase (GATase)/CPSase, ATCase, and DHOase. Although multiple examples are known for each category, the examples selected to represent each class are from *P. putida* (10) *E. coli* (11,12), *B. subtilis* (13), *S. cerevisiae* (9), and hamster (14). The genetic organization is represented according to the following: Boxes represent coding sequences with enzyme designations above and gene designations below. Noncoding DNA that joins genes transcribed into polycistronic messages is represented by black segments. Overlapping genes are represented by overlapping boxes. Although all seven cistrons of *pyr* metabolism are clustered in *B. subtilis*, the organization of only the first three is shown here.

between adjacent catalytic chains, the class C homotrimer may be considered the catalytic core of the bacterial ATCases. Unlike in *E. coli* where the *pyr* genes are scattered around the chromosome and are not coordinately regulated, all seven *B. subtilis* genes appear to be part of the same **operon** and transcribed on a single **messenger RNA**. The 3′ ends of the reading frames overlap with the 5′ ends of the downstream open reading frames for all **cistrons** in the cluster except *pyrB* and the preceding ORF1. Expression of the *pyr* genes is repressed by pyrimidines, and expression of aspartate transcarbamoylase has been shown to decline rapidly as *B. subtilis* cells enter **sporulation**. The mechanisms for the nutritional and developmental regulation of *pyr* gene expression in *B. subtilis* have not yet been elucidated.

Although the class B enzymes appeared to be restricted to the *Enterobacteriaceae*, recent studies suggest some euryarchaebacterium (7) may also contain very similar c_6:r_6 enzymes. The *E. coli* ATCase provides the textbook example of allosteric regulation with inhibition by pathway end-products CTP and UTP, and activation by the end-product of the parallel purine biosynthetic pathway, ATP. However, the various ATCases from different tribes of the *Enterobacteriaceae* display divergent patterns of feedback inhibition or activation. For example, while CTP serves as an allosteric feedback inhibitor in the *E. coli* enzyme, CTP can have no effect or serve as an allosteric activator in other *Enterobacteriaceae* (Table 3). Some of these diverged allosteric patterns contradict the basic regulatory paradigm of inhibition by pathway end-products as a mechanism to conserve energy and prevent the synthesis of unneeded products. Nonetheless, the discovery of CTP + UTP synergistic inhibition provided a common logic that applies to all the allosterically regulated enzymes: Since the concentration of ATP in actively growing cells is 2 to 5 *mM*, ATCase would be continually activated by ATP unless CTP or CTP + UTP are present to reduce the activated enzyme level. Even in those organisms in which CTP is an activator, the combination of CTP and UTP always serves as an effective antagonist to activation by ATP.

Due to the oligomeric nature of the class B enzymes, it has been possible to form hybrids by combining the catalytic subunits of one enzyme with the regulatory subunits of different enzymes that have diverged allosteric patterns. These hybrid protein complexes have demonstrated that the regulatory dimer determines the nature of the allosteric control of ATCase (Table 4). In addition, chimeric proteins have been constructed by intragenic fusion of the CP domain with the Asp domain (or the Zn domain with the Allo domain) forming several novel protein structures. In one instance, this type of **protein engineering** has led to the generation of a chimeric enzyme by fusion of the domains of the hamster ATCase cDNA and the bacterial *pyrB* genes. This chimera formed weakly active ATCase trimers that, although unstable, could marginally satisfy physiological requirements.

Eukaryotes

The ability to form chimeric proteins opens the possibility that intragenic fusions could provide a mechanism for the evolu-

Table 3. Classification and Allosteric Characteristics of Bacterial Aspartate Transcarbamoylases

ATCase Class Characteristics	Bacterial Species[a]	Allosteric
ATCase A	*Pseudomonas fluorescens* *Acintobacter calcoaceticus* *Azomonas agilis* *Azotobacter vinelandii*	No activation UMP Inhibition
ATCase B1 (I)[b]	*Escherichia coli* *Salmonella typhimurium*	ATP activation CTP, CTP + UTP inhibition
ATCase B2 (IV)	*Yersinia intermedia*	ATP activation CTP, UTP inhibition
ATCase B3 (V)	*Erwinia carnegiana* *Erwinia herbicola*	No activation sCTP[c], CTP + UTP inhibition
ATCase B4 (IV)	*Yersinia entercolitica*	ATP activation No inhibition
ATCase B5 (IV)	*Yersinia kristensenii* *Yersinia frederiksenii*	No activation No inhibition
ATCase B6 (II)	*Aeromonas hydrophila* *Serratia marcescens*	CTP, ATP activation CTP + UTP antagonism
ATCase B7 (III)	*Proteus vulgaris*	CTP, ATP activation CTP + UTP Inhibition
ATCase C	*Bacillus subtilis* *Streptococcus faecalis* *Staphylococcus epidermidis*	None

[a] Class A and C examples are from (15).
[b] Tribal classifications (given in parentheses) are according to Bergey's *Manual of Determinative Bacteriology*. B1–B7 indicate subgroups of the ATCase class B enzymes.
[c] sCTP indicates only slight inhibition by CTP (>20%).

Table 4. The Regulatory Chain of Class B ATCases Dictates the Allosteric Response

Source of subunit		Response to Effectors[a]	
Catalytic Subunit	Regulatory Subunit	ATP	CTP
E. coli	*E. coli*	+	−
	S. marcescens	+	+
	P. vulgaris	+	+
S. marcescens	*E. coli*	+	−
	S. marcescens	+	+
	P. vulgaris	+	+
P. vulgaris	*E. coli*	+	−
	S. marcescens	+	+
	P. vulgaris	+	+

[a] + = activation; − = inhibition.

tionary development of new proteins by the fusion of domain modules. Among the biosynthetic pathways of **eukaryotes**, there are several examples of single polypeptides that carry multiple enzymatic activities. Eukaryotic ATCases provide one of the best-studied examples of a multienzymatic protein. In lower eukaryotes, such as **yeast**, the first enzyme in the pathway, *carbamoylphosphate synthetase* (CPSase), and ATCase are physically linked, forming the CD*A protein fusion complex. For both *Saccharomyces cerevisiae* and *S. pombe*, the enzyme architecture includes four domains; three functional domains corresponding to *glutamine amidotransferase* (GLNase), CPSase, ATCase activities, and one dihydrorotase-like (D*; DHOase) domain (8,9). Dihydroorotase is the third enzymatic activity of the pyrimidine biosynthetic pathway just following ATCase. This structural organization is similar to the architecture previously discussed for the class A prokaryotic enzymes, which possess a domain with significant homology to DHOase. The regulation of this complex includes the feedback inhibition of both CPSase and ATCase by UTP. This occurs even though the carbamoylphosphate produced by GLNase/CPSase is tightly channeled to ATCase and a single feedback inhibition of the GLNase/CPSase alone should be sufficient to regulate pyrimidine metabolism.

In higher eukaryotes, the CA complex is associated with a functional DHOase domain producing a multienzyme *CAD complex* (physically arranged in a CDA sequence). This complex provides the central metabolic control for pyrimidine biosynthesis, with CPSase subjected to allosteric inhibition by UTP and activation by 5-phosphoribosyl 1-pyrophosphate (PRPP). Since all these regulatory and catalytic functions involve a single polypeptide chain, the multifunctional complex must have a well-defined domain structure. A series of **proteolytic** and genetic truncation studies over the last 15 years have provided evidence that the domain structure of CAD is well-defined and simple: Each enzymatic domain is separated from the others by a proteinase-sensitive polypeptide linker, and each domain can function in the absence of the other activities.

There are a number of proposals regarding the evolutionary role of the multienzymatic architecture. In yeast, significant substrate channeling occurs between CPSase and ATCase of the bifunctional CD*A complex. Channeling of substrates can occur when successive enzymatic activities are carried out on the same protein complex and could conceivably provide an evolutionary advantage by limiting the loss of intermediate products. However, this is not always the case, as the trifunctional CAD complex freely releases the products of CPSase and ATCase, whereas ATCase and DHOase can readily utilize substrates provided from outside the enzymatic complex. In the case of CAD and similar enzymatic complexes, coordinate **gene expression** provides an alternative regulatory advantage for the evolution and maintenance of large enzymatic complexes. The evolution of multienzyme complexes would provide for a smaller number of independent genes to regulate and simplify the coordination, production, and subcellular localization of multiple enzymatic activities.

In summary, the *de novo* biosynthetic pathway involves the set of reactions that supplies UMP from metabolites of the central metabolic pathways: aspartate, glutamine, ATP, PRPP and carbon dioxide. The *de novo* pyrimidine pathway is universal in its enzymatic steps from carbamoyl phosphate synthetase (CPSase) to orotidylate decarboxylase (OMP decarboxylase); however, evolution has developed a variety of regulatory controls and genetic organizations. Independent of its architectural variability, ATCase provides an important regulatory component of *de novo* pyrimidine biosynthesis in all free-living organisms. It is clear that maintaining homeostasis in intracellular nucleotide pools is an essential consideration for metabolism, and the various genetic and enzymological regulation mechanisms are critical for balancing the nucleotide precursors of DNA/RNA synthesis.

BIBLIOGRAPHY

1. C. Liu, J. P. Donahue, L. S. Heath, and C. L. Turnbough Jr. (1993) *J. Bacteriol.* **175**, 2363–2369.
2. J. C. Gerhart and A. B. Pardee (1962) *J. Biol. Chem.* **237**, 891–896.
3. J. R. Wild, S. J. Loughrey, and T. C. Corder (1989) *Proc. Natl. Acad. Sci. USA* **86**, 52–56.
4. P. England and G. Hervé (1993) *Biochemistry* **31**, 9725–9732.
5. R. P. Kosman, J. E. Gouaux, and W. N. Lipscomb (1993) *Proteins* **15**, 147–176.
6. H. Ke, W. N. Lipscomb, Y. Cho, and R. B. Honzatko (1988) *J. Mol. Biol.* **204**, 725–747.
7. C. Purcarea, G. Hervé, M. M. Ladjimi, and R. Cunin (1997) *J. Bacteriol.* **179**, 4143–4157.
8. M. Lollier et al. (1995) *Curr. Genet.* **28**, 138–149.
9. L. Jaquet et al. (1995) *J. Mol. Biol.* **248**, 639–652.
10. M. J. Schurr et al. (1995) *J. Bacteriol.* 177, 1751–1759.
11. T. A. Hoover et al. (1983) *Proc. Natl. Acad. Sci. USA* **80**, 2462–2466.
12. C. D. Pauza, M. J. Karels, M. Navre, and H. K. Schachman (1987) *Proc. Natl. Acad. Sci. USA* **79**, 4020–4024.
13. C. L. Quinn, B. T. Stephenson, and R. L. Switzer (1991) *J. Biol. Chem.* **266**, 9113–9127.
14. J. P. Simmer et al. (1989) *Proc. Natl. Acad. Sci. USA* **86**, 4382–4386.
15. M. J. Kenny, D. McPhail, and M. Shepherdson (1996) *Microbiology* **142**, 1873–1879.
16. M. R. Bethell and M. E. Jones (1969) *Arch. Biochem. Biophys.* **134**, 352–365.

Suggestions for Further Reading

J. N. Davidson et al. (1993) The evolutionary history of the first three enzymes in pyrimidine biosynthesis. *Biol. Essays* **15**, 157–164.

E. R. Kantrowitz and W. N. Lipscomb (1990) *Escherichia coli* aspartate transcarbamoylase: The molecular basis for a concerted allosteric transition. *Trends Biochem. Sci.* **15**, 53–59.

W. N. Lipscomb (1996) Aspartate transcarbamoylase from *Escherichia coli*: Activity and regulation. *Adv. Enzymol.* **68**, 67–151.

H. K. Schachman (1993) Aspartate transcarbamoylase. *Curr. Opin. Struct. Biol.* **3**, 960–967.

J. R. Wild and M. E. Wales (1990) Molecular evolution and genetic engineering of protein domains involving aspartate transcarbamoylase. *Ann. Rev. Microbiol.* **44**, 93–118.

ASPARTIC ACID (ASP, D)

T. E. CREIGHTON

The **amino acid** aspartic acid is incorporated into the nascent **polypeptide chain** during **protein biosynthesis** in response to only two **codons**—GAU and GAG—and represents approximately 5.3% of the residues of the proteins that have been characterized. The aspartyl residue incorporated has a mass of 115.09 Da, a **van der Waals volume** of 91 Å^3, and an **accessible surface** area of 151 Å^2. Asp residues are frequently changed during **divergent evolution**; they are interchanged in **homologous** proteins most frequently with **asparagine** and **glutamic acid** residues.

The side chain of Asp residues is dominated by its **carboxyl group**:

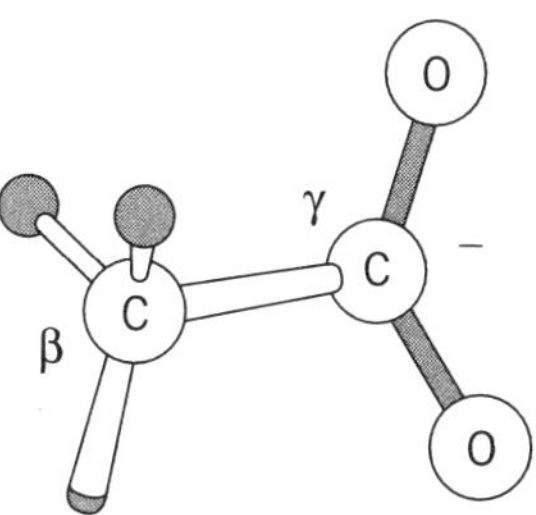

This carboxyl group is normally no more reactive than are those of corresponding organic molecules, such as acetic acid. Its intrinsic pK_a value is close to 3.9, so Asp residues are ionized and very polar under physiological conditions; consequently, very few Asp residues are buried in folded protein structures, and nearly all have at least the carboxyl group on the surface. The pK_a can be shifted in folded proteins, however, and either the ionized or nonionized form can be used in the protein's function. For example, **carboxyl proteinases** have one active-site carboxyl group function in the ionized form, and another when nonionized. Asp carboxyl groups have a weak intrinsic affinity for Ca^{2+} ions, and they are used in many **calcium-binding proteins**.

Asp residues differ from Glu only in having one methylene group, rather than two, so it might be thought that they would be very similar chemically and functionally in proteins, but this is not so. The slight difference in length of the side chains causes them to have different tendencies in their chemical interactions with the peptide backbone, so they have markedly different effects on the conformation and chemical reactivity of the peptide backbone. For example, Asp residues favor the ***alpha*-helical** conformation much less than Glu residues. In folded **protein structures**, Asp residues occur most frequently in reverse **turns**, whereas Glu residues are most frequently found in α-helices. The polypeptide chain can be cleaved relatively easily at Asp residues, because the side-chain carboxyl group participates in the reaction. –Asp–Pro– peptide bonds are especially labile in acid, because the carboxyl group interacts with the unique tertiary N atom of the Pro residue.

Suggestions for Further Reading

M. Landon (1977) Cleavage at aspartyl-prolyl bonds, *Meth. Enzymol.* **47**, 145–149.

T. E. Creighton (1993) *Proteins: Structures and Molecular Properties*, 2nd ed., W. H. Freeman, New York.

ASPARTYL PROTEINASE INHIBITORS, PROTEIN

M. LASKOWSKI JR.

The synthetic peptide aspartyl proteinase inhibitors [see **Proteinase inhibitors, protein**] are among the most intensively designed molecules. The principal recent reason for this is that HIV proteinase inhibitors turned out to be successful drugs for fighting AIDS. In contrast, the protein inhibitors of aspartyl proteinases are very little studied. It is difficult to believe that this neglect is a result of the rare occurrence of such inhibitors in nature. The activation of zymogens of aspartyl proteinases (pepsinogen) has been a problem of long-standing interest. The propeptide is clearly a pepsin inhibitor. Much more potent and typical is the pepsin inhibitor from the roundworm (*Ascaris lumbricoides*). The determination of the three-dimensional structure of its complex with pepsin is now in progress and may awaken this long-dormant field. A large number of reports deal with cathepsin D inhibitors classified as members of the soybean trypsin inhibitor (Kunitz), or STI family, [see **Soybean trypsin inhibitor**, (Kunitz)]. Many of the cathepsin D inhibitors are also said to inhibit trypsin, but the details of their interaction either with cathepsin D or with trypsin have been little studied.

ASYMMETRIC UNIT

JAN DRENTH

The asymmetric unit is the fundamental structure that is repeated in crystals of macromolecules (see **Crystallography; X-ray Crystallography**). Depending on the symmetry of the crystallographic lattice, each particle is present in multiple copies in the **unit cell** of the crystal. Only if symmetry is absent (**space group** P1) is a single copy present in the unit cell. In this case, the entire unit cell is an asymmetric unit. But if the unit cell has symmetry elements, it can be divided into identical parts related by this symmetry. Such a part, called the asymmetric unit, does not contain any crystallographic symmetry element. In Fig. 1, a projection of the unit cell in space group P2 is drawn. The twofold axes are perpendicular to the plane of the drawing. This unit cell has two asymmetric units. If the structure of the asymmetric unit is known, the structure of the entire crystal can be reproduced by applying the crystallographic symmetry.

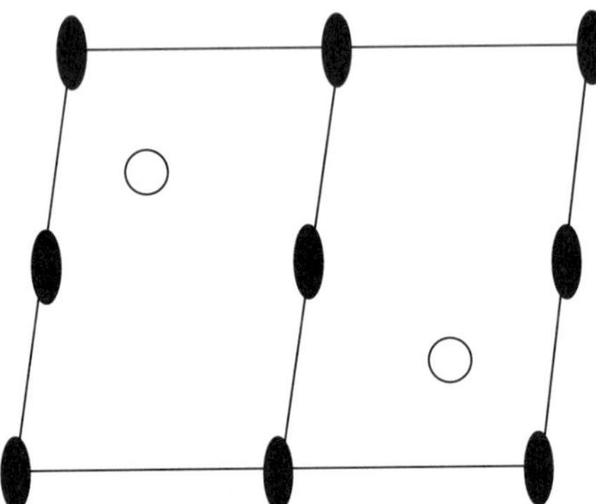

Figure 1. Projection of a unit cell in space group P2 that has twofold axes (black ellipses) perpendicular to the plane of the drawing. The circles are particles related by the symmetry. The symmetry divides the unit cell into two exactly equal parts. Therefore, this cell has two asymmetric units.

BIBLIOGRAPHY

1. International Union of Crystallography (1992) *International Tables for Crystallography*, Vol. A (T. Hahn, ed.), Kluwer Academic, Dordrecht, Boston, London.

Suggestions for Further Reading

J. P. Glusker, M. Lewis, and M. Rossi (1994) *Crystal Structure Analysis for Chemists and Biologists*, VCH, New York, Weinheim, Cambridge.

J. Drenth (1999) *Principles of Protein X-ray Crystallography*, Springer, New York.

ATOMIC FORCE MICROSCOPY

RATNESH LAL

Binnig, Quate, and Gerber developed a **scanning probe technique** known as atomic force microscopy (AFM) and as *scanning force microscopy*. Unlike its predecessor, scanning tunneling microscope (STM), AFM also images nonconducting samples, such as biological specimens, in a liquid environment at molecular and even atomic resolution. Currently, structural information about biological materials at the molecular level is obtained from other **microscopic** techniques, including **electron microscopy**, **electron crystallography**, **X-ray crystallography**, nuclear magnetic resonance (**NMR**) spectroscopy, and **vibrational spectroscopy**. These techniques require extensive sample preparation and unfavorable operating environments, and they are unsuitable for providing real-time functional information. Molecular function is studied by various molecular biological, biochemical, and electrophysiological techniques, but it is difficult to combine both structural and functional studies in one technique. Moreover, these techniques provide incomplete information about the surfaces of biological macromolecules, the very sites of their interactions with other molecules. In contrast, AFM images the surfaces of biological specimens, where most of the regulatory biochemical and other signals are directed. Other microscopic techniques also image surfaces, for example, the **scanning electron microscope** (SEM), but AFM images living cells and molecules in a liquid environment at comparable and often greater resolution.

PRINCIPLE OF OPERATION

AFM is based on the general physical principle that the interactive force between two bodies is inversely proportional to some power of the distance separating them and on the physicochemical natures of the interacting bodies. A tip that is sharp on the molecular scale is attached to a cantilevered spring: As it is moved across the surface of the specimen, it is deflected by the interactive forces between the atoms of the tip and those of the specimen (see Fig. 1 of **Scanning probe techniques**). Because the spring constant of the commonly used cantilevers (10^{-1} to 10^{-2} N/m) is much smaller than the intermolecular vibrational spring constant of the atoms in the specimen (10 N/m), the cantilever senses exquisitely small forces exerted by the individual sample atoms. The tip's deflection is a measure of the forces sensed by the cantilever, which are transduced to generate molecular images.

In practice, a microfabricated cantilevered tip is pressed against a sample surface by a small tracking (loading) force. The tip is raster-scanned in the x-y plane over the specimen by moving the sample beneath the tip or by moving the tip over the sample. The sample's vertical position (z) is also

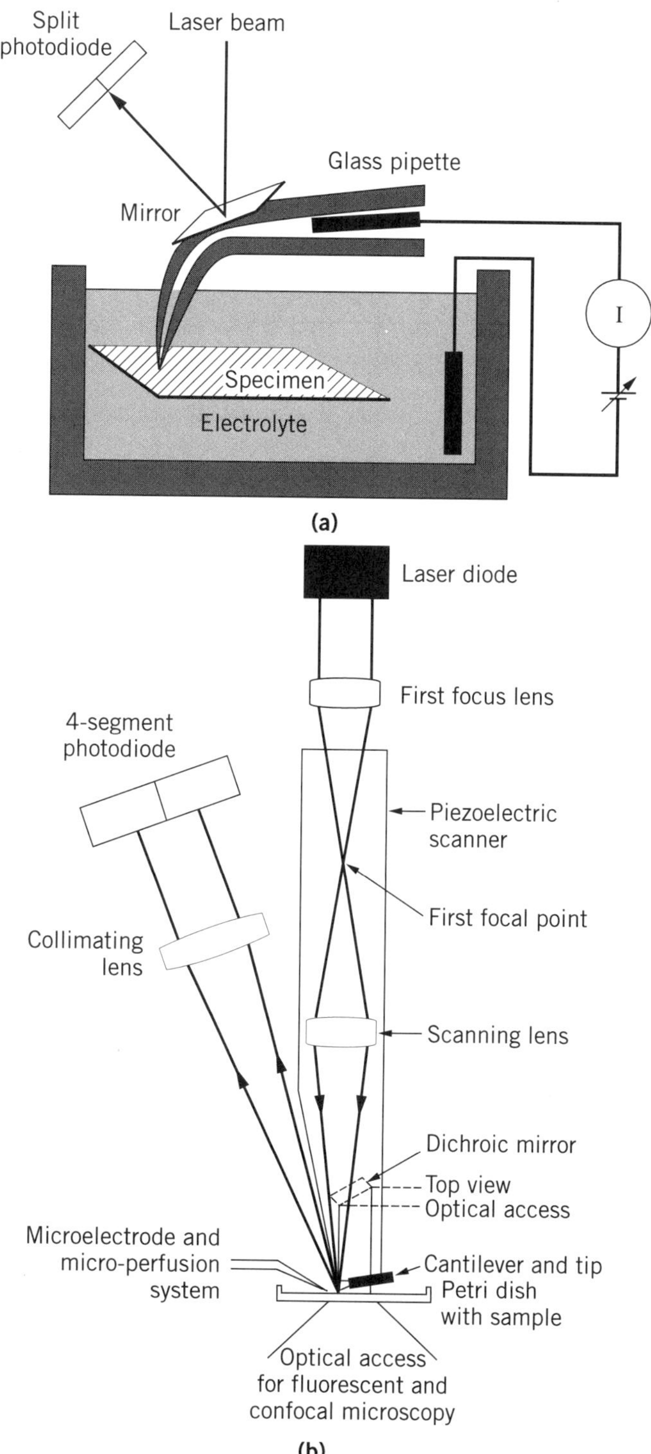

Figure 1. Schematic illustration of the operating principle of multimodal atomic force microscopes. (**a**) Schematic of the combined scanning ion conductance and atomic force microscope. A pipette serves as the probe: A laser beam reflecting from a mirror glued to the back of the pipette provides the deflection signal for the topographic image as the pipette is moved across the surface. The electrode has a nanometer-sized hole. Electrodes within the pipette and in the bath measure electrical currents (R. A. Proksch, R. Lal, P. K. Hansma, G. Morse, and G. Stucky (1996) *Biophys. J.* **71**, 2155–2157). (**b**) A combined light and atomic force microscope. The first focal point is located inside the upper portion of the piezoelectric scanner. After the positions of the lenses are adjusted, the scanning focused spot accurately tracks the cantilever, and the zero-deflection signal from the four-segment photodiode is independent of the position within the scan area. One of the key advantages of this new AFM is that there is optical access to the sample from above and below. Thus, the new AFM can be combined with an optical microscope of high numerical aperture. The other key advantage is that the sample is stationary during scanning and can be large, so that techniques for on-line perturbations and recordings are easily incorporated. For details, see P. K. Hansma, B. Drake, D. Grigg, C. B. Prater, F. Yasher, G. Gurley, V. Elings, S. Feinstein, and R. Lal (1994) *J. Appl. Phys.* **76**, 796–799; R. Lal and S. A. John (1994) *Am. J. Physiol.* **256**, C1–C21.

monitored. The three planes of movement are controlled by a piezoelectric xyz scanner, and the information about the three coordinates is used to create the image (Fig. 1). The cantilevered tip is brought sufficiently close to touch the sample (known as "contact" mode), or it oscillates at a finite distance (>a few nanometers) from the surface ("noncontact" mode). A noncontact mode microscope has the advantage that it does not perturb the sample, but the lateral resolution in these microscopes is poor, and hence they are not commonly used for biological imaging.

The deflective force is translated into a detectable signal in several ways. The most common is by an optical lever system (Fig. 1). A monochromatic laser beam reflects from the upper face of the cantilever, the angular direction of which changes as the cantilevered tip undergoes deflections. The reflected beams are captured and converted into electrical signals by position-sensitive multisegmented photodetectors. Such an optical lever amplifies the cantilever's deflection as much as a thousandfold, so deflections even less than a nanometer are measured.

Modes of Operation

Using an appropriate feedback system, the cantilever's deflection is kept constant or left to respond freely to the sensed forces. In the constant deflection mode (also called "constant force mode"), the feedback loop changes the height of the sample (to maintain the constant deflection) by adjusting the voltage applied to the z portion of the xyz piezoelectric scanner. The amount of z change corresponds to the sample's topological height at each point in the x-y raster. Combining the information from the three coordinates generates the 3-D image.

In the variable deflection mode ("constant height mode"), the feedback loop is open so that the cantilever deflection is proportional to the change in the tip–sample interaction, that is the force sensed by the cantilever. The surface image is constructed from the deflection information. It is called the constant height mode because the z component of the piezoelectric scanner does not change appreciably. This is usually unsuitable for a sample with large surface corrugation (e.g., cells), because the force fluctuations, and thus the cantilever's deflections, are enormous and often result in disengagement of the tip from the sample.

"Error mode" imaging relies on the imperfection in the feedback loop to operate in the constant-deflection mode. The error signal is amplified to yield contour information in the z plane. In the error mode, the feedback loop gathers high-frequency information that is normally not acquired in the constant deflection mode. This high-frequency information provides details of sharp contour changes (edges) in the sample. Measurements of actual height in error mode imaging are not accurate, however, in contrast to the other modes of operation. The main advantage of error mode operation is that imaging occurs without exerting high forces on the sample.

In "tapping mode" imaging, the cantilever is oscillated at very high frequency, normally near its resonance frequency, as it scans the sample. As the tip approaches the sample surface, its oscillatory amplitude decreases because of energy loss when the tip "taps" the surface. The amplitude of the cantilever oscillation is detected and used by the feedback system to adjust the tip–sample distance for constant amplitude. This ensures a much shorter tip–sample contact time, and smaller lateral forces are exerted on the cantilever. In this way, this mode has been successfully used for imaging delicate and individual macromolecules. The disadvantage of this mode of operation is that the vertical imaging force can be large, thus increasing the possibility of sample damage.

Sample Preparation

AFM is used to image specimens in aqueous, semiaqueous, or dry conditions. The imaging condition is normally chosen to maintain the specimen in as near a lifelike condition as possible. Where resolution takes priority over the physiological condition, however, the investigator is not as constrained. Imaging conditions also influence the choice of substrate, the stability of the specimen with respect to the tip interaction, and the preservation of the specimen with respect to its physiological or biochemical functions. At present, investigators rely on an empirical approach to find a suitable method, and probably will for some time to come. When it "works", the search stops for the "ideal support" or buffer.

The physicochemical characteristics of the sample determine or suggest ways under which it can be imaged. Problems encountered are as simple as getting the sample to attach to the support. Techniques for sample support include drying down of samples and adsorption to specially prepared surfaces. For example, imaging of **plasmid** DNA is vastly improved in to both resolution and consistency, under propanol, which increases the humidity and produces a more hydrated condition. This allows reducing the tip-tracking force exerted on the sample to <1 nN, thus decreasing sample deformation.

The interaction of the sample with the support determines the magnitude of the imaging force. If the sample does not adhere tightly to the support, low tracking forces must be used for imaging, or the tip literally sweeps the sample from the support. Originally, graphite (**hydrophobic** and uncharged), mica (**hydrophilic** and negatively charged), and glass (usually negatively-charged) were the supports most routinely used. These supports can also be modified chemically to adjust their **hydropathy**, charge density, and polarity. Today the repertoire has expanded greatly, and examples include gold treated with a variety of agents for DNA imaging. Gold supports maintained under potential (voltage) control have been used for DNA imaging with the **scanning transmission microscope**, and they may also prove useful for AFM.

The specimen support can also be modified or coated chemically so that it acts as a **ligand** for the sample and thus orients the specimen in a defined way. It is also possible to use artificial systems to generate constraints where there were none before. Examples include imaging isolated **cholera toxin** molecules incorporated into synthetic phospholipid bilayers, followed by covalent **cross-linking** or imaging the **vaccinia virus** protrusion out of living cells held by a suction pipette.

Forces in AFM

Interactive forces that deflect the cantilevered-tip are attractive or repulsive, and they vary depending on the mode of operation and the conditions used for imaging.

In contact mode imaging, the tip is deflected mainly by the repulsive forces from the overlapping electron orbitals of the atoms of the tip and sample. The dominant attractive force is a **van der Waals interaction** due primarily to the nonlocalized dipole–dipole interactions among atoms of the tip and specimen. When imaging in air, (attractive) surface tension is also present because of adsorbed **water** layers. For imaging in fluids, **electrostatic interactions** between charges on the specimen and the tip (occurring either naturally or induced by to polarization), osmotic pressure due to charge movements and rearrangements, and structural forces due to **hydration**, solvation, or adhesion enable a reduction in the net imaging force although both the meniscus and surface tension forces are abolished.

The interplay of local forces determines the stability of the specimen and the resolution. Theoretically the force should be $\leq 10^{-10}$ N for nonperturbed biological imaging. The sensitivity of AFM is sufficient to record small interactive forces, including the breaking of **hydrogen bonds**. Imaging in contact mode under liquid, but with a net attractive rather than repulsive force, increases the resolution significantly and has produced true atomic resolution, even with an imaging force of 10^{-11} N. As explained below, however, successful imaging of cells, membranes, and isolated proteins has been obtained with forces as large as 10^{-7} N.

In principle, any movement of the tip caused by its interaction with the sample in the x, y, or z directions provides information about the specimen's topography. To date, most information has been obtained from z deflections. Improvements in hardware and software have, however, allowed recording movements in the zy or zx planes and measurement of lateral forces for image generation. The contribution of lateral forces to image contrast generation can be substantial.

Resolution in AFM

Spatial Resolution. The limit of spatial resolution for AFM is not well defined because, unlike conventional microscopies, the images are formed by reconstructing the contours of interacting forces between the specimen and tip. The operating resolution in AFM is defined as the minimum size of two adjacent features that can be distinguished clearly. By selecting a small scan size and suitable operating conditions, one can distinguish two structures that are less than a nanometer apart. Image processing tools used to define resolution in **X-ray crystallography** and **electron microscopy** studies may not give correct results for AFM. The operating resolution can be divided into three categories:

1. Instrumental resolution: The lateral resolution is about 1 Å and is determined by the limitations of the hardware. The vertical resolution is 0.1 Å, and hence molecular perturbations on a sample surface can be imaged.
2. Target resolution: The lateral resolution achieved depends on the characteristics of the tip, the operating environment, and the nature of the specimen. For crystalline solid specimens and many inorganic materials, atomic resolution of 1 to 2 Å has been achieved.
3. Resolution in biological specimens: The nature of the biological samples and their preparation play a key role in determining the resolution limits. For the surface of a living cell, the resolution is relatively poor (~10 nm) but greater than that by light **microscopy** and comparable to that by **scanning electron microscopy**. In a biological specimen whose the density of particles is high and mobility is limited (e.g., proteins in a membrane), the resolution is comparable to that of a crystalline specimen.

Temporal Resolution. Temporal resolution is limited by the maximum speed at which a specimen can be scanned and still have the tip accurately track surface features. Preliminary studies suggest that the scan speed should be ≤ 2.2 μm/s for 1 nm spatial resolution on soft and deformable biological materials imaged in aqueous solution. Thus, membrane macromolecules whose dimensions are 10 nm × 10 nm (such as **channels** and **receptors**) divided into 10 × 10 pixels (with pixel size ~1 nm) require about 45 to 50 ms to image. However, if only a single line is scanned, the image can be repeated every 4 to 5 ms. Measuring at a single point, rather than scanning, increases the temporal resolution significantly, and hence it is possible to obtain spatial information at very short time intervals. The temporal resolution also depends on the mode of operation (constant deflection or constant height mode), operating environment (solvents, pH, viscosity, elasticity), and the nature of the interactions between tip and sample. In the constant-height mode, the scan speed is limited by the speed with which the deflection of the cantilever changes in reacting to surface features. In the constant deflection mode, the scan speed is limited by how the speed with which the piezo scanner changes its z component. There is ultimately a limit to the temporal resolution imposed by the low-pass filter used to eliminate sampling noise. These filters typically have a cutoff frequency of ~15 kHz, corresponding to a time resolution of 67 μs.

Temporal resolution also depends on the material being imaged. Individual molecules at molecular resolution require faster scan speeds than cells at lower resolution. Molecular movements of biological macromolecules can be correlated with their lateral **diffusion** constant. Lipids in a bilayer have a typical diffusion constant of 10^{-8} cm^2/s, corresponding to a mean velocity of ~2 μm/s. The proteins embedded in natural biomembranes have diffusion constants many orders of magnitude lower (e.g., the **acetylcholine receptor** in myoblast patches has a diffusion constant $<3.0 \times 10^{-12}$ cm^2/s. Thus it is quite possible to obtain images at molecular resolution of proteins and other macromolecules that are properly anchored in a lipid bilayer or immobilized on a substrate.

Identity of Imaged Structure

Although AFM provides molecular-resolution surface information for crystalline and amorphous materials, it is often difficult to define the nature of individual components, especially if the specimen contains a heterogeneous population of structures. This is the case with most biological systems, except in favorable systems like membranes that contain a crystalline patch of similar protein molecules. For mixed macromolecules, it is essential to compare the information obtained from AFM with that from alternative or complementary techniques, such as structural probes of electron microscopy and X-ray crystallography, biochemical and immunological binding assays, pharmacological labeling, and electrophysiological measurements.

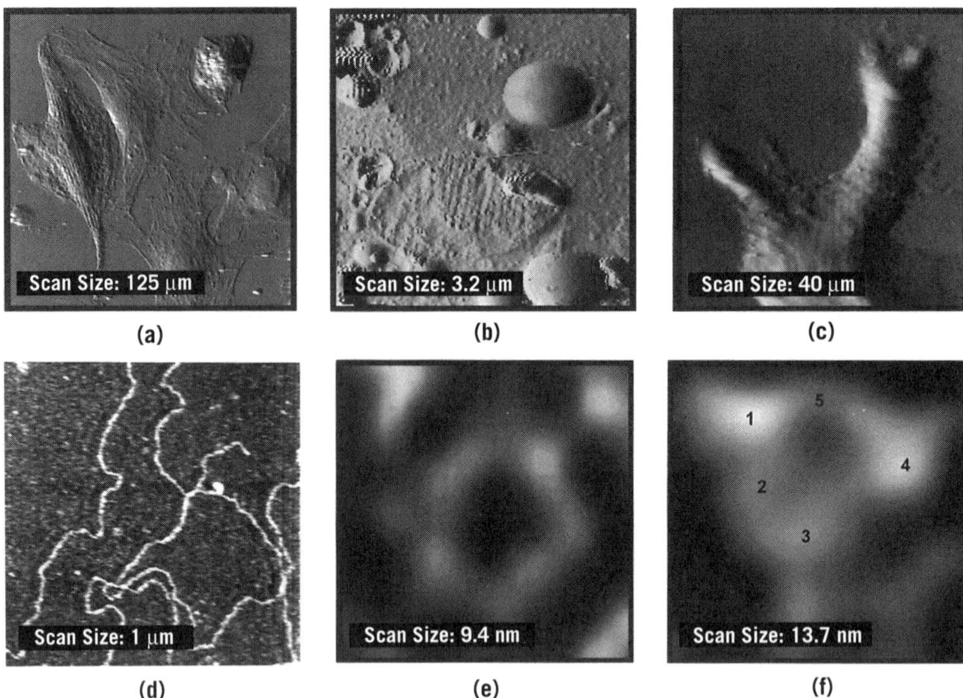

Figure 2. Images obtained with AFM. (**a**) Error-mode AFM image of a fixed atrial cell. The cytoskeletal network and cell nucleus are visible. [S. Shroff, D. Saner, and R. Lal (1995) *Am. J. Physiol.* **269**, C286–292.] (**b**) AFM error mode image of a **freeze-fracture** replica of atrial tissue. Details can be identified in the mitochondrion and in the atrial granules and vesicles. Scale Bar: 1 mm. [L. Kordylewsky, D. Saner, and R. Lal (1994) *J. Microsc.* **173**, 173–181.] (**c**) Error-mode AFM image of a neurite outgrowth in PC-12 cells treated with **nerve growth factor** and dibutyryl **cyclic AMP**. For details, see R. Lal, B. Drake, D. Blumberg, D. Saner, P. K. Hansma, S. Feinstein (1995) *Am. J. Physiol.* **269**, C275–C285. (**d**) Lambda DNA under propanol using a regular silicon nitride tip. Note the sharp bends in the strands and a fairly regular series of lumps along the strands, 6 to 8 nm apart. The strand width is 7 to 9 nm, greater than expected, which may result from the relatively large size of the tip. Scan size is 1000 nm × 1000 nm (courtesy H. Hansma, H. G. Hansma, R. L. Sinsheimer, M. Q. Li, and P. K. Hansma (1992) *Nucleic Acids Res.* **20**, 3585–3590). (**e**) AFM height mode image of a single connexon (a gap junction hemichannel) imaged on its cytoplasmic face. The subunit structure, a central pore, and the spacing between connections are apparent [S. A. John, D. Saner, J. Pitts, M. Finbow, and R. Lal (1997) *J. Struct. Biol.* **120**, 22–31).] (**f**) High-resolution AFM imaging of a single acetylcholine receptor expressed in *Xenopus* oocytes. The channel has a diameter of ~10.5 nm, and the five subunits (~1 to 1.5 nm in diameter) that have a central pore-like structure are shown. The protrusion of one unit is not apparent in the image [R. Lal (1998) *Scanning Microsc.* **10**, 81–96].

Examples of AFM Imaging

Cells and Cellular Processes. AFM images cellular and subcellular structures in physiological conditions at a resolution far exceeding that of optical microscopes. Living cells have been imaged in aqueous conditions with a resolution as small as 10 nm. By applying a larger imaging force, intracellular **organelles** and **cytoskeleton** networks have also been examined (Fig. 2). The ability to view structures beneath the plasma membrane is puzzling. Two possible mechanisms are (1) the tip penetrates the bilayer and images the substructure or (2) the plasma membrane drapes around the cytoskeletal fibers and the tip images the contour of the plasma membrane. If the "drape" model is correct, these observations give wonderful demonstrations of the flexibility of biological membranes and a potential tool for measuring the "drape characteristics" of natural and synthetic membranes.

In AFM imaging, the scan area can be varied from the micron to nanometer range, and hence it is possible to image features ranging from whole cells to individual macromolecules, such as **ion channels** and receptors (Fig. 2). In air-dried, hydrated ***Xenopus*** oocytes in which **acetylcholine receptor** proteins are expressed, the characteristic pentameric subunit structure of the expressed receptor has been observed after removing of the follicle layer. This technique of imaging expressed proteins on or in the surface of the plasma membrane of oocytes will shortly enable the characterization of a myriad of ion channels and receptors expressed in an appropriate expression system.

The major factor limiting resolution in imaging a cell surface is the mobility of the upper plasma membrane, plus the mobility of the macromolecules within the plasma membrane: the lower membrane is anchored to the substrate. Improvements in resolution may be made by (1) increasing the surface rigidity (e.g., suction of cells onto patch pipettes and thus reducing the lateral mobility of proteins); (2) by imaging with low forces (e.g., attractive force mode imaging or magnetic tapping imaging).

Membranes and Membrane Proteins. Imaging membranes, both native and reconstituted, has received wide attention because of their flattened 2-D sheet-like structure and the ease in preparing them. Purified membrane proteins, such as **bacteriorhodopsin**, **gap junctions**, **acetylcholine receptor**, and the hexagonally packed intermediate (HPI) layer from *Drosophila radiodurans*, have been imaged in aqueous conditions without fixation (Fig. 2). These membrane proteins have been characterized extensively by alternative techniques, such as electron microscopy and electron and X-ray crystallography. The results from AFM studies agree remarkably with those from other techniques. In addition, AFM images provide a direct observation of membrane polarity. For example, extracellular and cytoplasmic surfaces of gap junctions are distinguished unambiguously. The thickness measurement in AFM study is also direct and often very precise.

Synthetic membranes (Langmuir–Blodgett) and reconstituted vesicles have been imaged at molecular resolution. Images of Langmuir–Blodgett films provide direct measurement of lipid membrane thickness, obtained previously by indirect methods and theoretical extrapolations, because the height resolution in AFM is subnanometer. AFM images of Langmuir–Blodgett films show individual polar head groups and their molecular arrangement, including their long-range packing. An advantage of studying these membranes with AFM is that one can change the lipid composition on-line and study lipid–lipid interactions, lipid fluidity, and lipid–protein interactions. AFM images of proteins that are naturally embedded within membranes and form 2-D crystalline arrays and images of LB films provide some of the best evidence that the imaging of biological specimens generally agrees with that by electron microscopy, with the advantage, of course, that AFM imaging occurs in nearly physiological environments. It is also worth noting that such a correlation adds weight to the interpretation of images gathered by electron microscopy.

Proteins immobilized by synthetic membranes have also been imaged. Bacterial **porins** are one of the best-studied channel-forming **membrane proteins**. Porins reconstituted as 2-D crystals in lipid vesicles have been imaged in a liquid environment. AFM images at molecular resolution show the trimeric structure of porins illustrated by X-ray crystallography and electron microscopic **single-particle reconstruction**. In addition, recent studies show that molecular resolution can be obtained on noncrystalline specimens in liquid media. This opens a new avenue for studying the molecular structure of biological macromolecules (e.g., ion channels, receptors) that are easily expressed in an appropriate expression systems, such as the *Xenopus* oocyte, or simply isolated and anchored properly on suitable substrates. For example, purified **cholera toxin** molecules incorporated into synthetic phospholipid bilayers by covalent **cross-linking**, observed at molecular resolution, have the expected pentameric structure. Other membrane proteins imaged by AFM include the hexagonally packed intermediate (HPI) layer of *Drosophila radiodurans*, Na^+, K^+-**ATPase**, vacuolar **proton pumps** (V-H^+-ATPase), and Ca^{2+}-ATPase.

As mentioned before, AFM imaging of whole cell membranes has also been achieved. Acetylcholine receptor expressed in *Xenopus* oocytes has been observed. **Calcium channels** have been localized on the calyx-type nerve terminal of fixed chick ciliary ganglion in culture by imaging **avidin**-coated 30-nm gold particles incubated with ω-**conotoxin** GVIA linked to **biotin**, although the molecular structure of individual calcium channels was not reported. The interchannel spacing of 40 nm was noticed, which may reflect the spatial limitation due to tagging with 30-nm gold particles. Individual calcium channels have much smaller diameters. Other membrane channels, such as **sodium channels**, **potassium channels**, and gap junctions, are 6 to 10 nm in diameter. Isolated cellular organelles have been imaged with AFM, including **nuclear pore complexes**, ~134 nm in outer diameter, with a central pore-like trough.

Isolated Macromolecules, such as DNA, Amino Acids, and Proteins. Imaging isolated macromolecules is challenging because it is difficult to find suitable surfaces to which to anchor the molecules for repeatable and reliable imaging. The recent development of cryo-AFM shows good promise for obtaining high-resolution images of isolated macromolecules. Images at molecular resolution have been obtained of DNA at the **plasmid** and **chromosomal** levels, **polyamino acids**, isolated proteins, and ligand–receptor complexes. Large protein fibers, such as **actin** and **microtubules**, have also been imaged at molecular resolution, and it was possible to discern individual actin molecules. Isolated protein molecules show dynamic changes while imaging with the AFM. For example, when **glycogen phosphorylase** *b* binds to **phosphorylase kinase**, the dimensions and shapes of the proteins change noticeably.

Imaging DNA and nucleic acids has been appealing for many reasons. Given their well-known geometry and easy availability, they are readily identified and hence used for calibration and for studies of the interaction between tip and sample. Intriguingly, this may also open a door for structure-based sequencing and mapping of DNA AFM. However, **DNA sequencing** AFM will require an order-of-magnitude improvement in resolution (to ~2 to 3 Å). This increase may come from improvements in hardware and software, but methods to prepare DNA in extended conformations will probably be just as important in improving resolution.

Images of double-stranded DNA at molecular resolution (2 to 3 nm), in which the helical pitch and turns could be deciphered, have been obtained in air and liquid. Occasional images at higher resolution showing individual base pairs have also been obtained. Single-stranded DNA, though, has proven more intractable to image at any molecular resolution.

Images of complexes of DNA and **protein A** deposited onto mica show single proteins bound to the end of the DNA strands. In addition some single protein molecules bind to up to four

DNA strands per protein molecule. When **RNA polymerase** binds to DNA, AFM images show that the modified DNA is bent at marked angles where the polymerase binds. One appeal of these approaches is in searching for **DNA-binding proteins** and, intriguingly, perhaps to image the effects of **topoisomerase** on DNA.

Imaging Dynamic Processes. AFM, unlike other molecular level imaging systems, allows imaging in an aqueous environment. In an elegant set of studies, AFM was used to visualize real-time surface processes on **vaccinia virus** *pox viridae*-infected monkey kidney cells. Real-time changes in surface morphology and the **exocytosis** of enveloped virus and proteins were observed over a period of 19 hours. In contrast, cells not infected with virus showed no appreciable change in surface morphology. The real-time contractile activity of cultured atrial myocytes was also imaged. As the concentration of calcium increased, cells underwent rapid contraction, and a corresponding shortening in cytoplasmic fibers (perhaps cross-bridges) was observed.

Dynamic studies have been conducted on isolated proteins, such as formation of glycogen phosphorylase-phosphorylase kinase complexes, **antibody–antigen interactions**, dynamics of **immunoglobulin** adsorption, and binding of **streptavidin** to a biotinylated lipid bilayer. Real-time polymerization of **fibrin**, a protein important in blood clotting, shows that the polymer chain grows by the fusion of many short chains, rather than by successive addition of monomers to a few long chains. A change in Langmuir–Blodgett film morphology has been observed as trace amounts of a fluorescent dye are added, suggesting that the perturbation of molecular conformation by the tracer molecules may not be as insignificant as is commonly believed.

AFM can be used for *in situ* studies of the growth of protein 3-D crystals in their native solution environment and of the role of nucleation centers, lattice defects, and saturation level. These studies may provide clues for growing the 3-D crystals essential for high-resolution **X-ray crystallography**.

Structure-Function Studies

AFM can be combined with other techniques, which opens the possibility, as various biochemical, pharmacological, and other perturbations are introduced on-line, of real-time dynamic studies for direct structure–function correlations at the molecular level. For example, "single cell" experiments have been reported where electrical activity and AFM images were obtained from *Xenopus* oocytes expressing acetylcholine receptor. Electrical recording of acetylcholine-sensitive current and labeling by specific binding of α-**bungarotoxin** were also conducted in parallel. The receptor density calculated from the AFM studies correlates well with that from electrical measurements and toxin binding, but the clustering of acetylcholine receptor differs from the uniform distribution of the expressed receptors that is commonly assumed in electrophysiological studies. A correlation between patch-clamp electrical recording and AFM of **transcription factor** IID (TFIID) interactions with the **nuclear pore complex** has also been reported, showing unplugging of the nuclear pore complex accompanied by prolonged electrical current through the channel, perhaps reflecting the reopening of the channels. A combined AFM and patch-clamp study measured the electrical current in the membrane patches excised from *Xenopus* oocytes and attached to the pipette tip, while imaging the surface topology with AFM. Although the resolution of such a study is limited, it nevertheless shows the promise of direct structure-function studies of membrane macromolecules. Another study simultaneously measured images of **bacteriorhodopsin** in purple membranes adsorbed onto a lipid monolayer, ion transport through the membrane, and the electrical properties of the membrane. A recent study simultaneously measured the surface structure of nuclear pore filters and electrical current passing across the filter through pores of different diameters. For such a study, a scanning ion-conductance microscope was developed that records electrical activity, while imaging the 3D-structures of various membranes (Fig. 1).

Imaging force can be varied considerably during experiments. Such a feature has been used to nanomanipulate protein and membrane structures, and two different conformations of membrane proteins have been reported.

As already mentioned, action is at biological surfaces. The potential for rapidly characterizing and cataloguing the structures of synthetic peptides opens vistas for synthesizing drugs that can interact with "similar surfaces" within the body. Because AFM measures the force of an interaction between substrate and cells, interactions of a ligand or agonist should be measurable if well-defined molecules are placed on the tip of the AFM. Then it may be possible to use this ligand as a probe to determine the presence or absence of a receptor in impure, natural preparations. Perhaps more importantly, it may be possible to measure the interactive forces between agonist and receptor. This information will prove useful in designing of drug inhibitors or mimics.

Analysis of Micro-mechanical Properties

Contrast mechanism and image formation in AFM reflect a sum of many local forces and the micromechanical properties of the specimen. As the choice of specimen is shifted from rigid and hard materials (such as mica, graphite, tungsten) to soft and deformable biological materials, the dominating micro-mechanical properties shift from pure frictional to viscoelastic. Frictional forces on the atomic scale have been measured between two silica surfaces, two thin films, and between tungsten and graphite. By using local frictional force contours, fluorinated and hydrocarbon regions were distinguished in a Langmuir–Blodgett film. The hydrocarbons and fluorocarbons had separate domain structures: hydrocarbons as circular domains, fluorocarbons as the surrounding flat films. The frictional force was higher in the fluorinated region than in the hydrocarbon region. Such compositional studies provide a mechanism to identify individual components in a multicomponent sample like a cell membrane.

The viscoelastic properties of several soft, deformable biological materials, including cells, has also been obtained by a technique called *force mapping*. Such a study can be undertaken during on-line pharmacological and biochemical perturbations. In each imaging pixel, the tip is brought into proximity with the surface until a preset deflection is reached. The tip then retracts to its original position, and this process is repeated in every pixel. A height image is obtained from the amount of vertical piezo movement necessary at each point to obtain the preset deflection of the cantilever. For each pixel, a deflection versus distance curve is stored, which can be fitted to different models to obtain properties, such as the

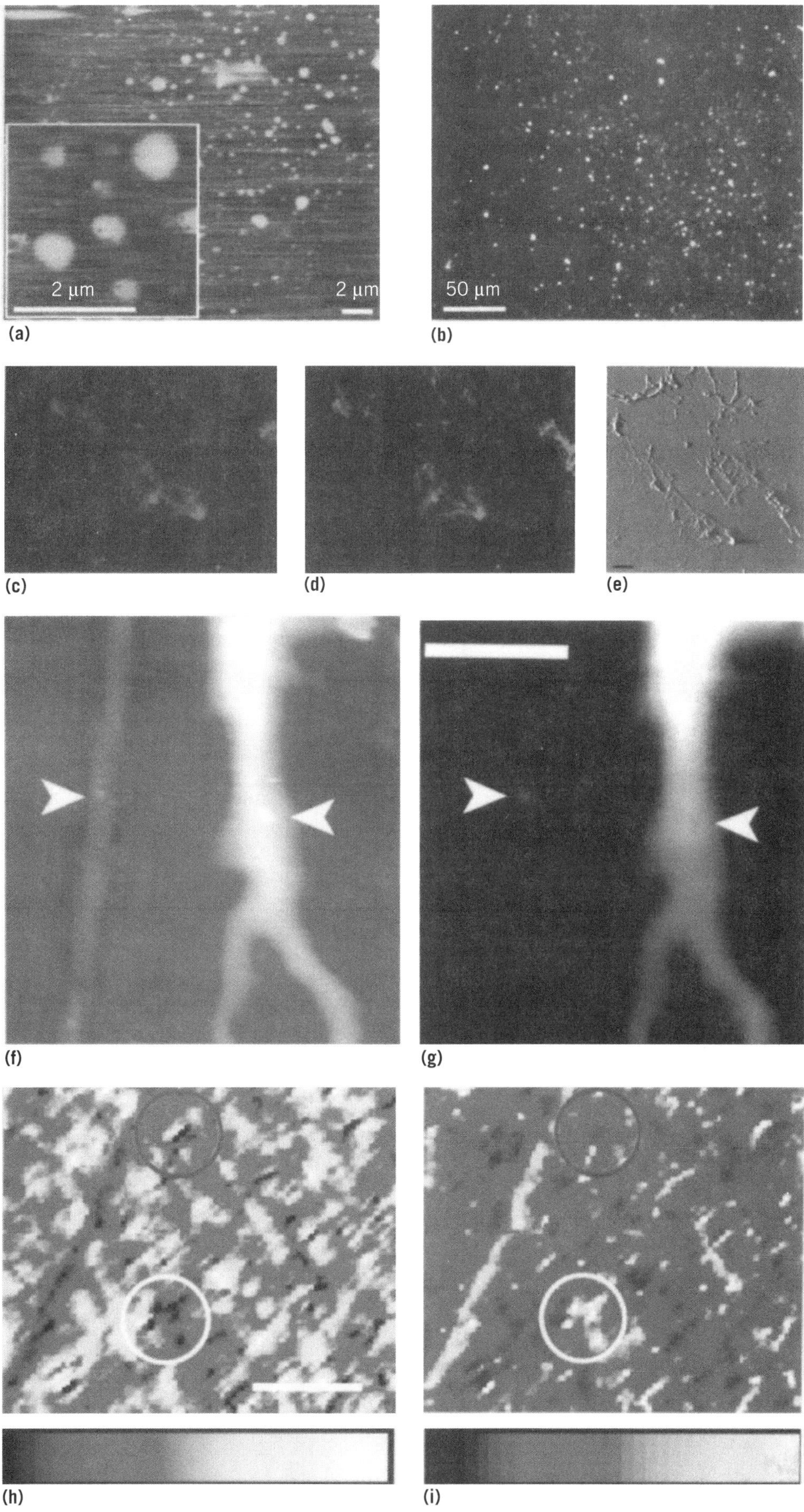
2 μm
2 μm
50 μm
(a)
(b)
(c)
(d)
(e)
(f)
(g)
(h)
(i)

Figure 3. Examples of AFM multimodal imaging. (**a**) and (**b**) Simultaneous immunofluorescence and atomic force microscopy of amyloid β peptide (AβP) reconstituted into liposomes. All liposomes, with or without AβP, were imaged with AFM (**a**), a few are shown at higher magnification in the inset. The liposomes were treated with anti-AβP antibody and subsequently identified with fluorescein-conjugated second antibody. The AβP-carrying liposomes showed strong fluorescence signals (**b**). [S. K. Rhee, A. P. Quist, and R. Lal (1998) *J. Biol. Chem.* **273**, 13379–13382]. (**c**), (**d**) and (**e**) Adhesion sites between a *Xenopus* retinal glial cell (XR1 cell line) and **extracellular matrix** material in a cell culture. The fluorescent images show the location of β-**integrin** (**c**) and f-**actin** (**d**) fibers detected by immunofluorescence, and the tapping mode AFM image (**e**) reveals the 3-D architecture of the focal point after removing of the cell body [R. Lal and R. Proksch (1997) *Int. J. Imaging Syst. Technol.* **8**, 293–300]. (**f**) and (**g**): Simultaneously combined AFM and fluorescence-confocal microscopic images. The sample was a suspension of fluorescently labeled latex beads that were dried into a gel on a plastic diffraction grating. The lines of the grating are visible in the topographic AFM image (**f**) but not in the confocal fluorescent image (**g**). The two images allow distinguishing a nonfluorescent particle (left arrow) from a fluorescent particle (right arrow), although both appear as raised bumps in the AFM image. [P. E. Hillner, D. A. Walters, R. Lal, H. G. Hansma, and P. K. Hansma (1995) *J. Micro. Soc. Am.* **1**, 123–126]. (**h**) and (**i**): Simultaneously combined AFM and scanning ion-conductance microscope (SICM) electrophysiology. (**h**) shows a tapping mode AFM image of a nucleopore membrane, and (**i**) shows the associated ionic conductivity image obtained by tapping mode SICM. Note that there are some differences in the pores detected by the two procedures. For example, the area circled in white shows a groups of pores that appear to be deep in the AFM image and highly conductive in the SICM image. The area circled in grey contains a large pore that appears deep in the AFM image but is nonconductive in the SICM image. The scale bars at the bottom are intensity-coded. Brighter is a greater height in the AFM image and a greater conductance in the SICM image. [R. A. Proksch, R. Lal, P. K. Hansma, G. Morse, and G. Stucky (1996) *Biophys. J.* **71**, 2155–2157.]

elastic modulus of the sample surface. Usually, the tip apex is approximated by a semisphere or a cone and the specimen by a spherical or planar model, depending on the shape of the features on the surface.

Thus, AFM is an imaging tool and also a system for analyzing micromechanical properties of cells, subcellular organelles, and macromolecules. It may be possible to study localized viscoelastic properties of molecular motor units, the distribution and propagation of contraction waves in a muscle cell, and the correlation between the calcium concentration wave and electrical propagation. One can also induce local shearing (frictional) force or pressure to assess the effects on the vascular system (mimicking the role of blood flow-induced shearing on vasorelaxation) or to distinguish pressure- or shear-sensitive ion channels in plasma membranes.

Simultaneous Multimodal Imaging

The simple design of AFM allows integrating it with other techniques, such as light **fluorescence microscopy**, laser **confocal microscopy**, and **near-field scanning optical microscopies**. Such integrated systems permit simultaneous multimodal imaging and provide independent verification with appropriately labeled markers. For example, using appropriately labeled **fluorescent** signals, one can identify specific areas and then use AFM to obtain high-resolution details.

Combined AFM and Light Fluorescence Microscope. Although conventional AFMs are ideal for high-resolution imaging, they could not be combined with large-aperture optical microscopes. In a few AFMs, however, the cantilever moves and the sample is stationary, permitting the addition of optical microscopes that have high numerical apertures. The most promising of these AFMs has the scanned-cantilever mode in which the cantilever position is accurately tracked by a scanned focused spot (Fig. 1) and is incorporated into an inverted fluorescence microscope. This combined fluorescence and force microscope has been used to image **immunolabeled** membranes and whole cells (Fig. 2). Fluorescent labels show remarkable correspondence among AFM images and the specificity of the molecules: such correspondence that one can obtain structural information at molecular resolution on biological macromolecules present individually or in small clusters, long as they have detectable fluorescent signals.

Combined Atomic Force Microscope and Confocal Microscope. Early combined AFM and laser-scanning confocal microscopes (LSCM) included features like a stationary sample stage, an AFM with an optical tracking system for the scanned cantilever, and either a scanned-beam or tandem design confocal microscope. The limitations of such systems include a limited scan range of both the AFM and confocal images. The scan range of the independently scanned confocal spot is limited to the size of the field of view of the objective and off-axis optical aberrations, and the scanned-cantilever AFMs with optical level detection require optical tracking of the cantilever for a large scan range. The latter constraint was overcome by the scanned-cantilever (tip) design with an optical tracking feature (the features explained previously in the combined AFM-fluorescence microscope), where the sample is scanned by a piezo system and the AFM tip and objective remain stationary. The AFM registers the topography of the sample surface, and the LSCM laser scans the surface to obtain fluorescent data on the same scan area. Although the AFM scan size is increased in this improved design, the confocal scan size is still limited by the objective. Moreover, although the AFM images are of the sample surface, the confocal image plane may not be the sample surface, but anywhere within the confocal slice, which could be no more than 100 nm thick.

A new combined AFM and LSCM allows simultaneous imaging of the sample surface in both modes, in addition to the conventional confocal imaging through the sample thickness. The salient features of such a combined microscope include a scanned-sample approach wherein the specimen is scanned above an inverted microscope objective with a fixed optical path for fluorescent LSCM imaging. An AFM positioned directly above the sample simultaneously measures the surface topography. Therefore, in this design the confocal spot and AFM cantilever remain stationary. Optical cantilever tracking is not required, and the confocal spot can be centered in the microscope's objective. The AFM feedback system ensures that the focal point is on or near the surface of the specimen, so that when the cantilever is positioned at the confocal spot, the LSCM and AFM images are acquired in direct registration, allowing image features to be easily correlated.

In this combined multimodal system, the confocal plane can be selected from the topmost region on a specimen surface or anywhere through its depth, so it is quite possible to follow, for example, cytoplasmic **signal transduction** processes leading to changes in the cellular plasma membrane surface conformations.

Combined Atomic Force Microscopy and Electrophysiological Recording. A combined tapping-mode AFM and a scanning ion-conductance microscope have been developed recently (Fig. 1). One of the salient features of this combined microscope is a bent glass pipette used as both the force sensor and the conductance probe. The force-sensing capability allows measuring of the pipette deflection, which then is used to create surface images in both regular contact mode and tapping mode. The conductance-measuring capability allows recording the electrical current flow across pores in a suitable specimen. Using such a microscope, it is possible to image the structures of channels and receptors and to measure their functional states (conducting vs. non-conducting) (Fig. 3).

Another approach is combining AFM with the patch-clamp technique in the same experiment. Such a combined technique records electrical current in the excised membrane patches from *Xenopus* oocytes that are attached to the patch pipette tip, while simultaneously observing the surface topology with AFM. Also, The membrane surface is also deformed by applying pressure through the patch pipette and observing the lateral displacement of features. However, the resolution is limited to about 10 nm.

Suggestions for Further Reading

G. Binnig, C. F. Quate, and C. Gerber (1986) Atomic force microscope, *Phys. Rev. Lett.* **56**, 930–933.

B. Drake, S. A. C. Gould, A. L. Weisenhorn, H. G. Hansma, P. K. Hansma, C. F. Quate, C. B. Prater, T. R. Albrecht, and D. S. Cannell (1989) Imaging crystals, polymers, and processes in water with the atomic force microscope, *Science* **243**, 1586–1589.

P. K. Hansma, V. B. Elings, C. E. Bracker, and O. Marti (1988) Scanning tunneling microscopy and atomic force microscopy—application to biology and technology, *Science* **242**, 209–216.

E. Henderson, P. G. Haydon, and D. S. Sakaguchi (1992) Actin filament dynamics in living glial cells imaged by atomic force microscopy, *Science* **257**, 1944–1946.

P. E. Hillner, D. A. Walters, R. Lal, H. G. Hansma, and P. K. Hansma (1995) Combined atomic force and confocal laser scanning microscope, *J. Micro. Soc. Am.* **1**, 123–126.

J. H. Hoh, R. Lal, S. A. John, J. P. Revel, and M. F. Arnsdorf (1991) Atomic force microscopy and dissection of gap junctions, *Science* **253**, 1405–1408.

R. Lal (1998) Imaging molecular structure of channels and receptors with an atomic force microscope, *Scanning Microsc.* **10**, 81–96.

R. Lal and S. A. John (1994) Biological applications of atomic force microscopy, *Am. J. Physiol. Cell Physiol.* **266**, C1–C21.

R. Lal and R. Proksch (1997) Multimodal imaging with atomic force microscopy: Combined atomic force, light fluorescence, and laser confocal microscopy and electrophysiological recordings of biological membranes, *Int. J. Imaging Syst. Technol.* **8**, 293–300.

F. Ohnesorge and G. Binnig (1993) True atomic resolution by atomic force microscopy through repulsive and attractive forces, *Science* **260**, 1451–1456.

B. N. J. Persson (1987) The atomic force microscope—can it be used to study biological molecules, *Chem. Phys. Lett.* **141**, 366–368.

R. A. Proksch, R. Lal, P. K. Hansma, D. Morse, and G. Stucky (1986) Imaging the internal and external pore structures of membrane in fluid: Tapping mode scanning ion conductance microscopy, *Biophys. J.* **71**, 2155–2157.

D. Sarid (1991) *Scanning Force Microscopy: With Applications to Electric, Magnetic and Atomic Forces*, Oxford University Press, New York.

Y. Z. Zhang, S. Sheng, and Z. Shao (1996) Imaging biological structures with the cryoatomic force microscope, *Biophys. J.* **71**, 2168–76.

ATP SYNTHASE

GEORGE OSTER
HONGYUN WANG

ATP synthase, also called F_0F_1 **ATPase**, or simply F-ATPase, is the universal **protein** that terminates oxidative phosphorylation by synthesizing ATP from ADP and phosphate. Nearly identical proteins are found in eukaryotic **mitochondria** and bacteria, and they all operate on the same principle. Electron-driven ion **pumps** set up concentration and electrical gradients across a membrane (see **Chemiosmotic coupling** and **Proton motive force**). ATP synthase utilizes the energy stored in this electrochemical gradient to drive nucleotide synthesis. It does this in a surprising way by converting the electromotive force into a rotary torque that promotes phosphate binding and liberates ATP from the catalytic site where it was formed. Remarkably, this process can be reversed in certain circumstances. ATP hydrolysis can drive the engine in reverse, so that F-ATPase functions as a proton pump. Indeed, the vacuolar V-ATPases, the most ubiquitous intracellular proton pumps, are structurally similar to ATP synthase and operate according to the same principles.

ATP synthase is composed of at least eight subunit types, whose stoichiometries are denoted by the subscripts (α_3, β_3, γ, δ, ε, a, b_2, c_{12}), that combine into two distinct regions. The geometric arrangement of the subunits is shown schematically in

Fig. 1(a). The soluble F_1 portion consists of a hexamer, $\alpha_3\beta_3$. This hexamer is arranged in an annulus about a central shaft consisting of the **coiled-coil** γ subunit. Subunits δ and ε are also generally isolated with F_1. The F_0 portion consists of three transmembrane subunits, a, b_2, and c_{12}. The 12 copies of the c-subunit form a disk into which the γ and ε subunits insert. The remainder of F_0 consists of the transmembrane subunits a and b_2. The latter is attached by the δ subunit to an α subunit, so that it anchors the a subunit to F_1. Thus there are two "stalks" connecting F_0 to F_1, $\gamma\varepsilon$ and $b_2\delta$.

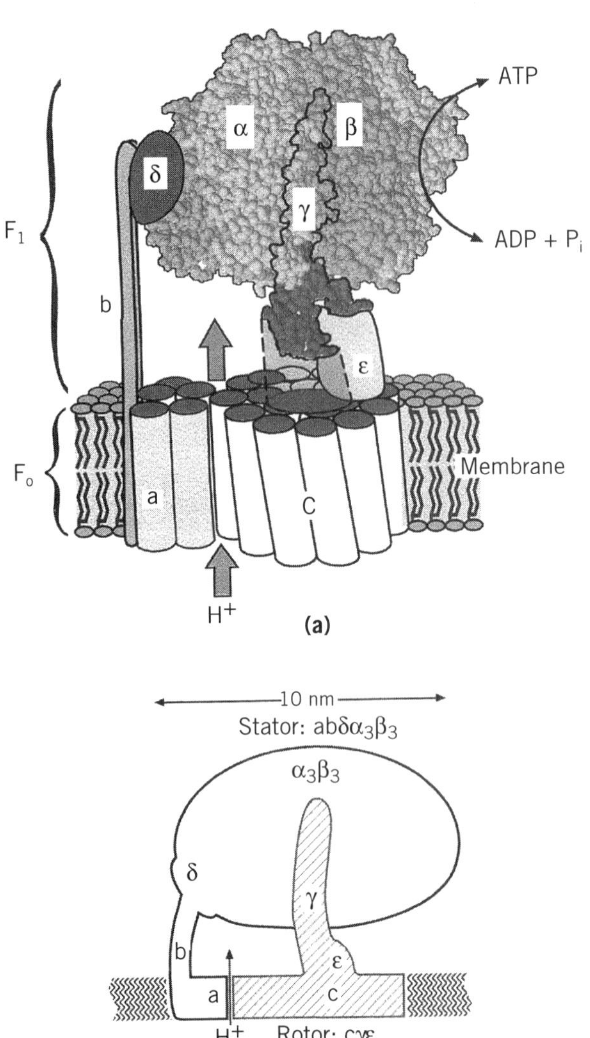

Figure 1. Schematic diagram of the subunit organization of ATP synthase (1). (**a**) The $\alpha_3\beta_3$ hexamer and a portion of the γ shaft. The lower part of γ has not been resolved. The c-subunit consists of 12 pairs of transmembrane α-helices, and the a subunit consists of six transmembrane α-helices. The ε subunit abuts c and γ and interacts with the DELSEED region of β. The a subunit is attached to an α subunit via the b and δ subunits. (**b**) The proton motive force at the a–c interface leads to the functional subdivision into two counterrotating assemblies, usually denoted as the "rotor" and "stator." The rotor consists of subunits c_{12}-γ-ε, and the stator consists of subunits a-b_2-δ-$\alpha_3\beta_3$.

The key to understanding how ATP synthase carries out its catalytic and synthetic roles lies in this geometric organization. The entire protein can be divided into two operational regions denoted suggestively as the "rotor" and the "stator" for reasons that derive from the rotary mechanism by which the protein operates (Fig. 1b). Indeed, it turns out that ATP synthase has two rotary engines in one. The F_0 motor converts transmembrane electrochemical energy into a rotary torque on the γ shaft, and F_1 uses ATP hydrolysis to turn the γ shaft in the opposite direction. Because they are connected, one drives the other in reverse. When F_0 dominates, the rotor turns clockwise (looking upward in Fig. 1), so that F_1 synthesizes ATP. When F_1 dominates, so that F_0 is driven counterclockwise, it can pump protons against an electrochemical gradient. Deciphering how this remarkable dual energy transduction works is one of the great triumphs of modern chemistry.

ATP SYNTHESIS IN F_1 IS DRIVEN BY ROTATION OF THE γ-SHAFT

We begin with the F_1 motor, because we now know precisely what it looks like. This is due to John Walker and the **X-ray crystallography** group at Cambridge, who worked out the exact structure of the $\alpha_3\beta_3$ hexamer and most of the γ shaft (1). A stereo view of the structure is shown in Fig. 2. Walker was awarded the Nobel Prize in 1997 because his structure revealed essential asymmetries in the molecule's structure that were the key to understanding how it worked. In the early 1980's, Paul Boyer at UCLA proposed the surprising theory that, in the catalytic sites of F_1, ATP is in chemical equilibrium with its reactants, ADP and phosphate (2). So the formation of ATP is essentially without cost energetically. However, because each ATP hydrolyzed under cellular conditions liberates about 12 kcal/mol, this energetic price must be paid at some point. Boyer proposed that F_1 pays this price in the mechanical work necessary to liberate the nucleotide from the catalytic site. Further, release of product (ATP) proceeds sequentially and cyclically around the $\alpha_3\beta_3$ hexamer because the synthetic reactions are synchronized in a fixed-phase relationship by rotation of the γ shaft and cooperative coupling between the three catalytic sites. Boyer's "binding change" mechanism neatly fits the Walker structure, and Boyer shared the 1997 Nobel Prize. A schematic diagram of the binding change mechanism is shown in Fig. 3.

There are actually six nucleotide-binding sites on the $\alpha_3\beta_3$ hexamer. All lie at the interfaces between the α and β subunits. The catalytic sites lie mostly in the β subunit, whereas the noncatalytic sites lie mostly in the α subunit. The role of the noncatalytic sites is uncertain, but they may help to hold the hexamer together. Each catalytic site traverses the synthetic cycle sequentially:

$$\begin{array}{ccc} \text{Empty} & \overset{④}{\rightleftharpoons} & \text{ATP} \\ \updownarrows ① & & ③ \updownarrows \\ \text{ADP} & \underset{②}{\rightleftharpoons} & \text{ADP}\cdot\text{P}_i \end{array}$$

In steps 1 and 2, a site binds ADP and phosphate (not necessarily in that order). While trapped in the catalytic site in

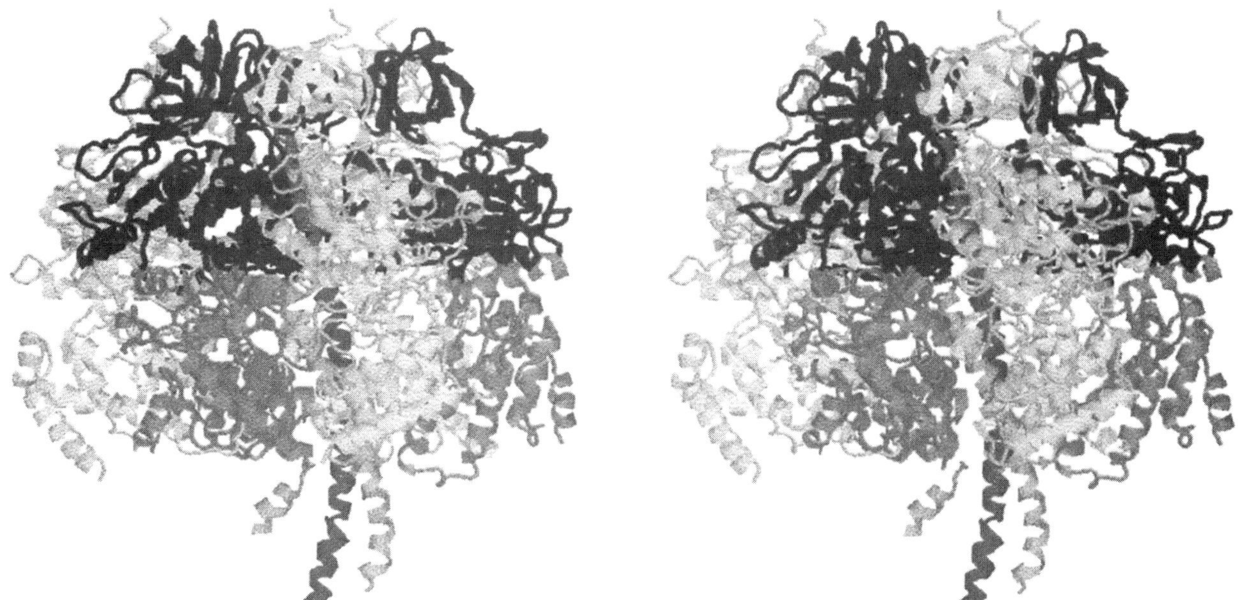

Figure 2. Stereo pair showing the molecular structure of the F_1 subunit run through center, coils extending out from middle [from (1)]. The α subunit is in light gray, and the two coils of the γ subunit medium gray. Its asymmetrical structure is evident. In the β subunit: the stationary upper barrel segment is in dark gray, and the lower hinge segment is medium gray. The structure of the ε subunit (not shown) is known, but it is not clear how it attaches to the γ and c subunits.

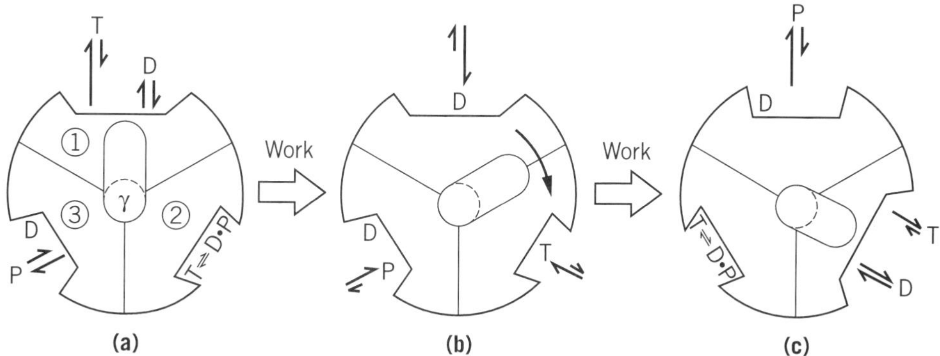

Figure 3. The binding change mechanism. Notation for site occupancies: T = ATP bound, DP = ADP • P_i bound, D = ADP bound. The β subunits are numbered clockwise. The lengths of the arrows indicate the relative binding affinities. (**a**) The system starts with either $(\beta_1, \beta_2, \beta_3) = (E, T \Leftrightarrow D \bullet P, D)$ or $(\beta_1, \beta_2, \beta_3) = (D, T \Leftrightarrow D \bullet P, D)$. (**b**) Clockwise rotation of γ increases the binding affinity of ADP in β_1, traps ATP in β_2, and promotes P_i binding on β_3. (**c**) Further rotation of γ traps ADP and allows P_i binding in β_1, releases the tightly bound ATP and allows ADP binding in β_2, and traps P_i in β_3.

step 3, the reactants (ADP and P_i) and product (ATP) are in chemical equilibrium. Step 4 requires the input of mechanical torque from F_0 on γ to trap the reactants in the ATP state and to pry open the site, releasing the tightly bound ATP. Most of the 12 kcal/mol price of synthesis is paid in step 4. The way in which this works is found in the shape of the $\alpha_3\beta_3$ hexamer and the γ shaft.

The γ subunit is asymmetric and bowed. It fits into a central annulus in $\alpha_3\beta_3$, which is itself asymmetric (Fig. 4). At the top of the $\alpha_3\beta_3$ hexamer is a **hydrophobic** "sleeve" in which the γ shaft rotates. Further down, however, the annulus is offset from the center, so that as γ rotates clockwise, it sequentially pushes outward on each catalytic site. In addition, the ε subunit is located eccentrically and attached to the γ and c subunits, so that as γ rotates, it comes into contact sequentially with each β subunit in a conserved region called the DELSEED sequence (named for the single-letter abbreviation of its constituent amino acid residues). Together, this asymmetrical

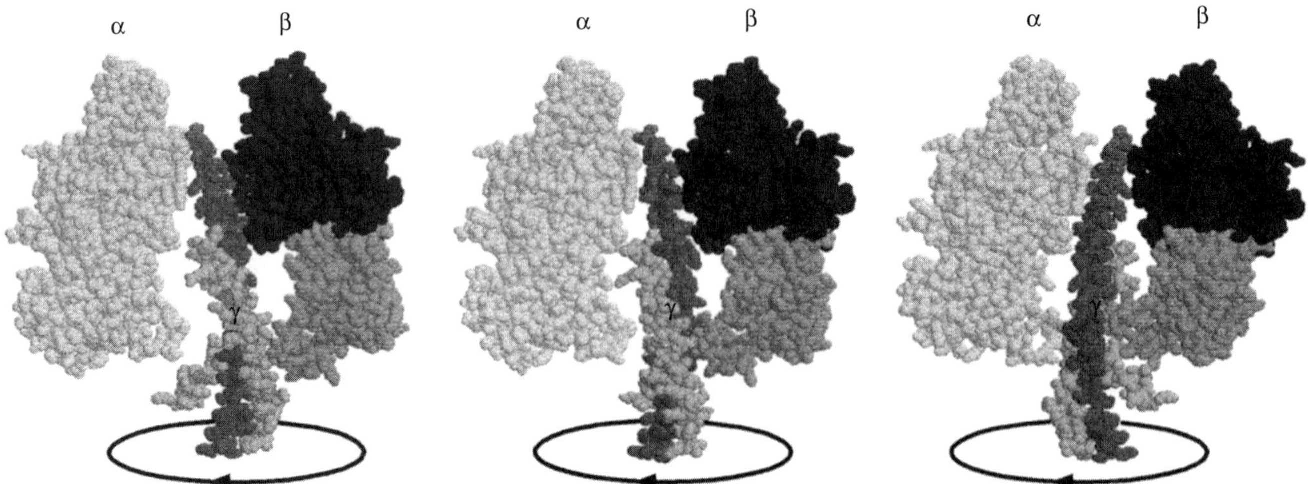

Figure 4. Cross section of F_1 showing the conformational changes in the β subunits that drive rotation of γ. The α subunit is on the left and the stationary barrel region of β is on the right. During the hydrolytic cycle, the lower segment of β undergoes a hinge-bending motion that rotates it about 30° inward. This motion pushes on the eccentric γ coiled coil, causing it to rotate within the barrel bearing. During synthesis, the rotation of γ pushes on each catalytic site. The panels show three snapshots of the motion during a 180° rotation. Movies of the rotational sequence can be downloaded from the authors' World Wide Web site: http://www.cnr.berkeley.edu/~goster/ATP_movies.html.

rotation exerts stress on the catalytic site, loosening its grip on ATP, so that thermal fluctuations can free it into solution.

The catalytic sites do not act independently. Instead, they are synchronized, so that each site traverses the synthetic cycle in a more or less fixed phase with respect to the others. This synchronization is orchestrated in several ways. As the γ shaft rotates, it stresses each catalytic site, and it also interacts electrostatically with the β subunits at two locations (3). These interactions may mediate phosphate- and nucleotide-binding, the necessary precursors to synthesis. In addition, the catalytic sites are elastically coupled, so that the occupancy of one site affects the affinity of the other two sites. The consequence of this coupling is that when ATP concentrations are low enough so that only one site is occupied, hydrolysis proceeds much more slowly than when more than one site is occupied.

Together with the F_1 molecular structure, the binding change model strongly supports the idea that catalysis involves rotation of the γ subunit. However, dramatic visual confirmation was provided by *in vitro* experiments in which the $\alpha_3\beta_3\gamma$ subunits were isolated and attached to a bead. A fluorescently tagged **actin** filament was attached to the γ shaft, and when ATP was supplied, the filament could be clearly seen rotating. In fact, a complete revolution takes place in three steps and consumes a single ATP per step (4).

The viscous drag on the actin filament was estimated, which allowed computing the torque developed by the F_1 motor and comparing it with the **free energy** available from ATP hydrolysis. The startling result was that the motor generates an average torque of more than 40 piconewton nanometers (40×10^{-12} N $\times$ 10^{-9} m), more than six times the maximum force developed by **kinesin** or myosin. More impressively, the motor operates near 100% mechanical efficiency, this precludes any sort of heat engine, which would be limited by the Carnot efficiency (4). Several models have been proposed that address the issue of torque generation and efficiency (4–6).

The energy to drive this motion derives from the hydrolytic cycle of ATP at the catalytic site. Moreover, the conformational change that drives the hydrolysis motor must be nearly the reverse of the motion that frees ATP from the catalytic site during synthesis. Examination of the structure reveals that the major conformational change is a hinge-bending motion in the β subunits. The bottom portion of each β subunit below the nucleotide-binding site rotates inward approximately 30°, during which it pushes on the bowed γ subunit, turning it much like one cranks an automobile jack (Fig. 4).

F_0 CONVERTS PROTON MOTIVE FORCE INTO ROTARY TORQUE

Currently, there have been no direct observations of rotation in the F_0 portion of ATP synthase (7). However, current thinking is that the F_0 assembly converts the energy contained in the transmembrane proton motive force into a rotary torque at the interface of the a and c subunits (Fig. 1). This torque turns the rotor (the c, γ, and ε subunits), which couples to the F_1 synthetic machine.

The c assembly consists of 12 subunits, each consisting of two transmembrane **alpha-helices** (8). There is one essential acidic amino acid (Asp61 in the *Escherichia coli* ATP synthase) which binds protons. Because variants of ATP synthase can operate on sodium rather than protons, the interaction between the c subunit and the translocated ion has the property of an electrostatic carrier mechanism (9).

The a subunit consists of six transmembrane α-helices that contain at least one essential basic residue (Arg210 in *E. coli*)

(8,10). The **electrostatic interaction** of these rotor and stator charges is essential for torque generation, and several proposals have been put forward for the way this could work (11–14). Whatever the mechanism, the F_0 motor must generate a torque sufficient to liberate three ATP's from the three catalytic sites in F_1 for each revolution. If the proton flux through the stator is tightly coupled to the rotation of the c subunit, then a rotation of $2\pi/3$ carries four protons down the electromotive potential of 230 mV typical of the mitochondrial inner membrane (14). This is sufficient to account for the mechanical energy required for synthesis of one ATP.

Under anaerobic conditions, the ATP synthase of the bacteria *E. coli* can reverse its operation, hydrolyzing ATP and turning the c subunit backward so that it functions as a proton pump. This is not surprising because the F-ATPases are structurally similar to the most common proton pumps, the vacuolar, or V-ATPases (7). These pumps may have been the **evolutionary** precursors of ATP synthase (15). A striking difference between the two is that the F-ATPases have 12 acidic rotor charges, whereas the V-ATPases have six. It can be shown that this enables the V-ATPases to function more efficiently as ion pumps at the expense of relinquishing their capability to synthesize ATP.

SUMMARY

Both the F_1 and F_0 motors can operate in both directions. F_1 is a hydrolytically driven three-piston engine that can be driven in reverse to synthesize ATP from ADP and phosphate. F_0 is an ion-driven rotary engine that can be driven in reverse to function as an ion pump. The F-ATPases are structurally similar to and presumably evolutionarily related to the V-ATPase ion pumps (15). It is thought that most ion pumps function by an "alternating access" mechanism, whereby an ion is first bound strongly on the dilute side, then energy is supplied to move the ion so that it communicates with the concentrated side and weakens its binding affinity (16). In contrast with other ion pumps, however, the F and V-ATPases accomplish this by a rotary mechanism that is driven indirectly by nucleotide hydrolysis, rather than by direct phosphorylation (17). It is thought that the F_0 motor is also related to the bacterial **flagellar** motor. Both can operate on sodium, although the flagellar motor has eight or more "stators" and develops far more torque than F_0 (18,19).

The mechanism driving the F_1 hydrolytic motor may carry hints for other nucleotide hydrolytically fueled motors, such as kinesin, myosin, and **dynein**. However, important structural differences may make the comparison difficult (4). For example, the motors previously mentioned all "walk" along a polymer track to which they bind tightly during a portion of their mechanochemical cycle. The power stroke of the F_1 motor is driven by the β subunit, which pushes on the γ shaft, but does not bind tightly to it, that is, it does not "walk" around the γ shaft. Moreover, no other motor operates with nearly the efficiency of the F_1 motor, implying that there are important entropic steps in other motors that are absent in the F_1 motor.

BIBLIOGRAPHY

1. J. Abrahams, A. Leslie, R. Lutter, and J. Walker (1994) *Nature* **370**, 621–628.
2. P. Boyer (1993) *Biochim. Biophys. Acta* **1140**, 215–250.
3. M. Al-Shawi, C. Ketchum, and R. Nakamoto (1997) *J. Biol. Chem.* **272**, 2300–2306.
4. K. Kinosita, R. Yasuda, H. Noji, S. Ishiwata, and M. Yoshida (1998) *Cell* **93**, 21–24.
5. F. Oosawa and S. Hayashi (1986) *Adv. Biophys.* **22**, 151–183.
6. H. Wang and G. Oster (1998) *Nature* **396**, 279–282.
7. M. Finbow and M. Harrison (1997) *Biochem. J.* **324**, 697–712.
8. R. H. Fillingame (1997) *J. Exp. Biol.* **200**, 217–224.
9. P. Dimroth (1997) *Biochim. Biophys. Acta* **1318**, 11–51.
10. R. H. Fillingame (1996) *Curr. Opinion Struct. Biol.* **6**, 491–498.
11. S. B. Vik and B. J. Antonio (1994) *J. Biol. Chem.* **269**, 30364–30369.
12. W. Junge, H. Lill, and S. Engelbrecht (1997) *Trends Biochem. Sci.* **22**, 420–423.
13. G. Kaim, U. Matthey, and P. Dimroth (1998) *EMBO J.* **17**, 688–695.
14. T. Elston, H. Wang, and G. Oster (1998) *Nature* **391**, 510–514.
15. R. Cross and L. Taiz (1990) *FEBS Lett.* **259**, 227–229.
16. B. Alberts, D. Bray, J. Lewis, M. Raff, K. Roberts, and J. Watson (1994) *Molecular Biology of the Cell*, Garland, New York.
17. S. Khan (1997) *Biochim. Biophys. Acta* **1322**, 86–105.
18. H. Berg (1995) *Biophys. J.* **68**, 163s–166s.
19. K. Muramoto, I. Kawagishi, S. Kudo, Y. Magariyama, Y. Imae, and M. Homma (1995) *J. Mol. Biol.* **251**, 50–58.

Suggestions for Further Reading

P. Boyer (1993) The binding change mechanism for ATP synthase—some probabilities and possibilities. *Biochim. Biophys. Acta* **1140**, 215–250.

J. Weber and A. E. Senior (1997) Catalytic mechanism of F1-ATPase. *Biochim. Biophys. Acta* **1319** (1), 19–58.

ATP-BINDING MOTIF

JENNIFER MARTIN

The ATP-binding motif, also called the *Walker motif*, is a structural **protein motif** that is frequently found in proteins that bind ATP or GTP. It can usually be identified from just the **primary structure** of a protein using a fingerprint sequence that characterizes proteins as ATP or GTP-binding (see **GTP-binding proteins**). Walker et al. (1) identified two regions of sequence conservation: the A region has a stretch of small **hydrophobic** residues followed by [Gly/Ala]–X–X–Gly–X–Gly–Lys–Thr/Ser, where X is any residue. The second region of sequence conservation also has a stretch of small hydrophobic residues, this time ending in a conserved **aspartate** residue. In the three-dimensional structures of proteins having the ATP-binding motif, such as adenylate kinase, the hydrophobic residues of the A-region form a buried β-strand, and the glycine-rich region forms a loop, called the **P-loop**, that interacts with the phosphate of the bound nucleotide. The second conserved region codes for another hydrophobic β-strand, and the conserved aspartate is required for binding the magnesium ion that usually accompanies nucleotides bound to proteins.

The ATP-binding motif represents a common, but not the only, mode of interaction between proteins and ATP/GTP. For example, it is structurally distinct from the actin fold of proteins such as **actin**, hsp70 (see **Hsc, hsp proteins**), and hexokinase, which bind and hydrolyze ATP or GTP and where a conformational change is implicated in nucleotide binding (2).

[See also **Nucleotide-binding motif** and **P-loop**.]

BIBLIOGRAPHY

1. J. E. Walker, M. Saraste, M. Runswick, and N. Gay (1982) *EMBO J.* **1**, 945–951.
2. W. Kabsch and K. C. Holmes (1995) *FASEB J.* **9**, 167–174.

Suggestions for Further Reading

G. E. Schulz (1992) Binding of nucleotides by proteins. *Curr. Opin. Struct. Biol.* **2**, 61–67 (Excellent review of nucleotide binding motifs.)

ATPASE

R. E. McCarty

Any **enzyme** that catalyzes the hydrolysis of ATP to ADP and inorganic phosphate (P_i) is classified as an ATPase. In some cases, the phosphate is transiently transferred to the protein before its release as a product. The hydrolysis of ATP liberates much energy, so this reaction is usually coupled to another, energetically unfavorable reaction (see **Coupled reactions**). Three major types of ATPase are associated with membranes that couple ATP hydrolysis to the translocation of specific ions across a membrane: P-, V-, and F-ATPases (see **Active transport**).

P-ATPases have a relatively simple polypeptide composition (one or two subunits) and are phosphorylated as part of their catalytic cycle. Examples of P-ATPases are (*1*) the Na^+,K^+ – ATPase of the plasma membrane of animal cells; (*2*) the H^+ – ATPase of the plasma membranes of **yeast**, **fungi**, and plants; and (*3*) the Ca^{2+} – ATPase of the sarcoplasmic reticulum (the **endoplasmic reticulum** of muscle). The plasma membrane Na^+,K^+ – ATPase and H^+ – ATPase function to generate and maintain the plasma membrane electrical potential difference (inside negative), as well as the ionic disequilibria across that membrane. Ca^{2+} – ATPases function in Ca^{2+} homeostasis (see **Calcium signaling**).

V-ATPases have a much more complicated polypeptide composition than do P-ATPases and are not phosphorylated during catalysis. V-ATPases are found on the membranes within the interiors of **eukaryotic** cells (endomembranes), including membranes from vacuoles (from which the "V" is derived), **lysosomes**, **Golgi**, **secretory vesicles**, **clathrin**-coated vesicles, and, in some instances, plasma membranes. V-ATPases are H^+ – ATPases that show some similarity to the proton-linked **ATP synthases**. The function of V-ATPases is to catalyze proton transport into the endomembrane interior compartments at the expense of ATP hydrolysis. Acidification of the interior of endomembranes is required for some of their functions. In **archaebacteria**, an enzyme with similarity to V-ATPases functions as an ATP synthase.

F-ATPases (also known as "F_1–F_0") have a complex polypeptide composition and, as in V-ATPases, there is no phosphorylated intermediate in the reaction mechanism. In non**photosynthetic** eukaryotes, the F-ATPase is found exclusively on the inner membrane of **mitochondria**, whereas in green plants and algae there are two distinct F-ATPases; one in mitochondria and the other on the thylakoid membrane of **chloroplasts**. In bacteria, F-ATPase is present in the plasma membrane. ATPases couple the flow of protons (or Na^+ in some cases) to ATP hydrolysis and synthesis. The activity of the F-ATPases is very tightly regulated. Several mechanisms combine to prevent wasteful ATP hydrolysis by F-ATPase, but to allow rapid ATP synthesis at the expense of electrochemical proton (or Na^+) potentials generated by proton translocation linked to electron transport (see **Chemiosmotic coupling**). F-ATPases operate *in vivo* as ATP synthases; consequently, the term *ATP synthase* is preferred to *F-ATPase* when referring to the entire enzyme. The catalytic portion of the ATP synthase may be readily removed from coupling membranes and purified as a large 400-kDa water-soluble protein. Such preparations may have high ATPase activity and are called "F_1" or "F_1-ATPase" (see **ATP synthase**).

ATTENUATION OF TRANSCRIPTION

P. Babitzke
P. Gollnick

The ability to modulate **gene** expression in response to changing environmental signals is crucial for the survival of all organisms. Virtually every stage involved in the synthesis, function, and degradation of **macromolecules** is a potential target for one or more regulatory events (1). Regulatory mechanisms have been identified for all three stages of **transcription** (initiation, elongation, and termination). Several postinitiation regulatory mechanisms have been categorized as transcription attenuation mechanisms. Transcription attenuation can be defined as any mechanism that utilizes transcription pausing or transcription termination to modulate expression of downstream genes. For the purpose of this article, however, the definition will be restricted to situations in which the action of the regulatory molecule promotes transcription termination, with the default situation being transcriptional readthrough. There are also related **antitermination** mechanisms in which the action of the regulatory molecule promotes transcriptional readthrough.

Once transcription of a gene is initiated, the transcription elongation complex and its nascent transcript are potential targets for regulation. As transcription proceeds, the nascent transcript may fold into specific secondary and tertiary structures that signal the transcribing **RNA polymerase** to pause or terminate transcription before reaching the structural genes (1). Transcription attenuation mechanisms allow the organism to modulate the extent of transcriptional readthrough past the terminator structure in response to changing environmental signals, thereby regulating expression of the downstream genes. As will be seen, several different transcription attenuation mechanisms have been identified.

TRANSCRIPTION ATTENUATION OF BIOSYNTHETIC OPERONS OF ENTERIC BACTERIA

Transcription attenuation was the first demonstration that organisms can exploit **RNA structure** to modulate gene expression. The first attenuation mechanism was elucidated by Charles Yanofsky and his coworkers for the *Escherichia coli* tryptophan biosynthetic operon (*trpEDCBA*; see ***trp* operon**) (2). In addition, many other amino-acid biosynthetic **operons** in enteric bacteria are regulated by transcription attenuation (eg, *his*, *leu*, *ilv*, *pheA*). In each case, the genetic information required for transcription attenuation is encoded within a 150–300-bp leader region located between the **promoter** and the first structural gene of the operon (1). Because the salient features of transcription attenuation are conserved in each system, the *E. coli trp* operon will be discussed, and the key differences with respect to other operons will be pointed out where appropriate.

Transcription initiation of the *E. coli trp* operon is regulated by TrpR, a **DNA-binding** repressor protein. Once transcription starts, the elongating transcription complex is subject to control by transcription attenuation (1). The combined actions of repression (80-fold) and transcription attenuation (eight-fold) result in approximately 600-fold regulation in response to changing concentrations of intracellular tryptophan (3). A simplified model of the *E. coli trp* operon transcription attenuation mechanism is depicted in Figure 1 (4). The 141-nucleotide leader transcript can form three overlapping RNA secondary structures, referred to as the pause structure, the antiterminator, and a Rho-independent terminator. In addition, the nascent *trp* leader transcript contains a small open reading frame that encodes a 14-amino acid residue leader peptide. Soon after transcription of the *trp* operon is initiated, a secondary structure (structure1:2) forms in the nascent transcript that signals RNA polymerase to pause. The paused RNA polymerase complex allows sufficient time for a **ribosome** to initiate **translation** of the leader peptide. The translating ribosome then disrupts the paused RNA polymerase complex, and transcription resumes, with the ribosome closely following the molecule of RNA polymerase, thereby coupling transcription and translation. At this point, two different outcomes can occur, depending on the level of tryptophan in the cell. Under conditions of limiting tryptophan, the level of charged **transfer RNA** $tRNA^{Trp}$ is low. As a result of the low tryptophanyl-$tRNA^{Trp}$ concentration, the translating ribosome stalls at one of two tandem Trp **codons** strategically placed within the leader peptide coding sequence. Ribosome stalling at the Trp codons effectively uncouples transcription and translation. As transcription proceeds, therefore, the antiterminator structure (structure 2:3) forms and prevents formation of the overlapping Rho-independent terminator (structure 3:4), resulting in transcriptional readthrough into the *trp* structural genes. Under conditions of tryptophan excess, the level of charged $tRNA^{Trp}$ is sufficiently high to allow efficient translation of the tandem Trp codons, and the ribosome continues to the end of the leader peptide. When the ribosome reaches the leader peptide **stop codon**, it physically blocks formation of the antiterminator structure, thereby promoting terminator formation and, hence, termination of transcription, before RNA polymerase reaches the *trp* structural genes. Thus, expression of the *trp* operon is decreased when the cell has an adequate supply of tryptophan. As can be seen by the above example, the regulatory signal is charged $tRNA^{Trp}$, and the sensory event is the capacity to translate a short peptide-coding sequence (1). The transcription attenuation mechanisms for several other amino acid biosynthetic operons, such as the *his*, *phe*, and *leu* operons, are essentially identical to that for the *trp* operon, except that the leader peptides contain seven His (5), seven Phe, and four Leu codons (6), respectively.

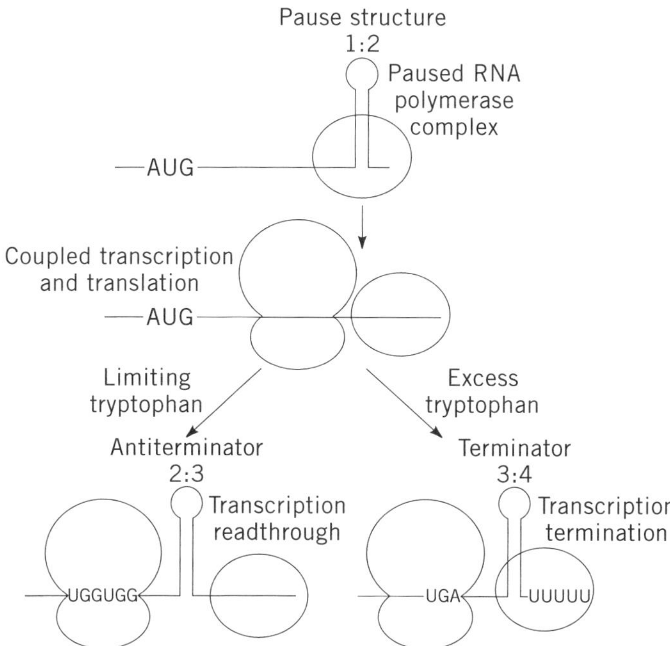

Figure 1. Model of transcription attenuation for the *E. coli trp* operon. RNA polymerase pauses following formation of the pause structure provides time for a ribosome to initiate translation of the leader peptide. Under tryptophan-limiting conditions, the ribosome stalls at the tandem Trp codons, resulting in transcription readthrough. Under conditions of tryptophan excess, the ribosome reaches the leader peptide stop codon. This ribosome position blocks formation of the antiterminator, leading to terminator formation and transcription termination. (See text for details.) Adapted from Landick and Yanofsky (4).

Expression of the *E. coli* pyrimidine biosynthetic operon, *pyrBI*, is also regulated by transcription attenuation (7,8). In this case, the concentration of UTP serves as the regulatory signal, in conjunction with a UTP-dependent pause signal consisting of an RNA hairpin and several U residues just after the hairpin. A transcription terminator exists approximately 60 nucleotides downstream of the pause structure, but the leader transcript does not have the potential to form an antiterminator structure. Finally, there is a 44-residue leader peptide encoded by an open reading frame beginning prior to the pause signal and extending past the terminator. The model for this attenuation mechanism is as follows: When there is a deficiency of UTP, the transcribing RNA polymerase pauses at the leader pause site. This provides time for a ribosome to initiate translation of the leader peptide, which results in coupled transcription and translation. As transcription proceeds, the translating ribosome prevents formation of the transcription terminator, allowing transcription of the structural genes. However, pausing is inefficient when the cell contains an adequate supply of UTP. In this case, RNA polymerase transcribes and recognizes the terminator before the ribosome reaches this segment of the leader transcript, thus halting transcription in the leader region prior to the structural genes.

TRANSCRIPTION ATTENUATION OF GRAM-POSITIVE BIOSYNTHETIC OPERONS

The transcription attenuation mechanisms that have been identified for the *trp* and *pyr* operons in the **Gram-positive bacterium** *Bacillus subtilis* differ dramatically from those described for the enteric bacteria. Most notably, ribosomes and tRNA molecules are not involved. Instead, sequence-specific **RNA-binding proteins** are responsible both for sensing the level of tryptophan or UMP in the cell and, ultimately, for the decision to terminate transcription or to readthrough into the structural genes.

Expression of the *trpEDCFBA* operon of *B. subtilis* is regulated by TRAP, the *trp* RNA-binding attenuation protein (9,10). TRAP is composed of 11 identical subunits arranged in a single ring (11). Tryptophan binding between each adjacent subunit in a cooperative manner activates TRAP to bind RNA (11). Transcription initiation of the *trp* operon appears to be constitutive, occurring 203 nucleotides upstream of the first structural gene. A transcription attenuation model for the *B. subtilis trp* operon is depicted in Figure 2. The *B. subtilis trp* leader transcript contains inverted repeats that can form mutually exclusive antiterminator and Rho-independent terminator structures (12), although there is no apparent transcription pause signal in this case. When cells are growing in excess tryptophan, tryptophan-activated TRAP binds to 11 closely spaced (G/U)AG repeats present in the *trp* leader transcript (11,13). Recent studies have shown that TRAP binds to these repeats by wrapping the RNA around the protein ring, with the bases of the (G/U)AG repeats interacting with several amino acid residues on adjacent subunits in the protein (14). TRAP binding blocks formation of the antiterminator, which allows formation of the overlapping terminator structure. Thus, transcription halts in the leader region prior to the *trp* structural genes. Under conditions of limited tryptophan, TRAP is not activated and does not bind to the *trp* leader transcript. Thus, as transcription proceeds, the antiterminator forms, allowing transcription readthrough into the *trp* structural genes (9,12).

Expression of the *B. subtilis pyr* operon is also controlled by transcription attenuation mediated by an RNA-binding protein. This operon encodes 10 polypeptide chains involved in de novo synthesis of pyrimidines (15,16). Transcription attenuation occurs in response to UMP levels through the action of PyrR, which is encoded by the first gene of the operon (17). There are several novel features of the attenuation mechanism that controls this operon. In contrast to the systems described previously, the *pyr* operon contains three attenuators: one located in the 5′-untranslated leader region, the second between the first and second genes, and the third between the second and third genes of the operon. Each attenuator can form three alternative RNA secondary structures. In addition to terminator and antiterminator structures similar to those described previously, a structure called the anti-antiterminator can form upstream of, and overlapping, the antiterminator. In the presence of UMP, PyrR is activated to act at each of the three attenuators to promote transcription termination and down-regulate expression of the operon. In contrast to the mechanism by which TRAP controls attenuation in the *trp* operon, PyrR binding does not interfere directly with formation of the antiterminator structure, but rather functions by stabilizing the anti-antiterminator, which thereby indirectly stabilizes the terminator (18). The nature of the binding sites for PyrR and TRAP reflects the different effects these proteins have on RNA structure when they bind. TRAP binds to an entirely single-stranded site (19, 20), whereas PyrR binds to a stem-loop structure (21).

Another novel feature of this system is that PyrR is not only an RNA-binding regulatory protein, it is also an **enzyme**, uracil phosphoribosyltransferase (UPRTase), which catalyzes formation of UMP from uracil and 5-phosphoribosyl-1-pyrophosphate. The physiological role of this enzyme is not clear, as *B. subtilis* has an additional UPRTase that has been shown to be more important for UMP synthesis. The structure

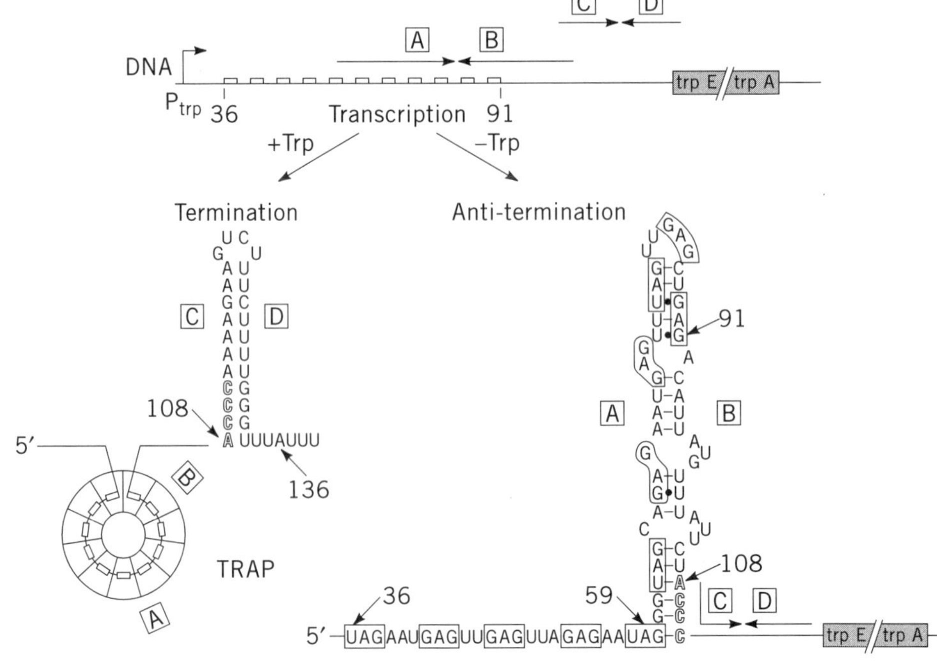

Figure 2. Model of transcription attenuation of the *B. subtilis trp* operon. The large boxed letters designate the complementary strands of the terminator and antiterminator RNA structures. Small rectangles represent the GAG and UAG repeats involved in TRAP binding; these triplet repeats are also outlined in the sequence of the antiterminator structure. Numbers indicate the residue positions relative to the start of transcription. Nucleotides 108 to 111 overlap between the antiterminator and terminator structures and are shown as outlined letters. The TRAP protein is represented as an 11-subunit ring, and the bound RNA is shown forming a matching circle on binding to TRAP, with each triplet repeat interacting with one subunit. From Antson et al. (11).

of PyrR in the absence of UMP shows two oligomeric forms of the protein, one as a dimer and the other as a hexamer (21). Both forms appear to exist in solution, but PyrR is thought to bind RNA as a dimer. Neither the amino acid sequence nor the structure of PyrR show significant similarity to TRAP. Thus it appears that these two similar attenuation mechanisms evolved independently.

ANTISENSE RNA-MEDIATED TRANSCRIPTION ATTENUATION

Antisense RNA control of gene expression has been documented for many prokaryotic genes, some of which involve an interesting form of transcription attenuation (22–24). For example, the copy number of the *Streptococcus agalactiae* plasmid pIP501 is regulated by antisense RNA-mediated transcription attenuation (23,24). The antisense RNA (RNA III) inhibits expression of *repR*, the gene encoding the essential RepR initiator protein, by binding to the nascent *repR* leader transcript (RNA II). This interaction promotes formation of a Rho-independent terminator upstream of the *repR* coding sequence. Interaction of RNA III with RNA II is initiated by the formation of a "kissing complex" between single-stranded loops of both molecules, followed by propagation of the RNA helix. In the absence of RNA III interaction, formation of the transcriptional terminator is prevented, and expression of *repR* can proceed normally. A model illustrating this mechanism is depicted in Figure 3.

TRANSCRIPTION ATTENUATION OF EUKARYOTIC GENES

Expression of many eukaryotic genes is also regulated at the level of elongation of transcription by processes similar to attenuation (for reviews see 25,26). In many cases, the action of an inducer protein is required to release the stalled RNA polymerase II elongation complex. In several systems, including *c-myc* (27–31), *N-myc* (32), *c-fos* (33), and adenosine deaminase (34), potential stem-loop structures, similar to prokaryotic transcription terminators, have been identified near the sites of attenuation. However, the precise transcription attenuation mechanisms of these systems are not known.

The best-characterized eukaryotic system involving control of transcription elongation is transcriptional activation of **HIV**-1 by the transactivator protein Tat (35). In this system, transcription initiates at the HIV-1 promoter in the viral **long terminal repeat** (LTR). In the absence of Tat, transcription terminates prematurely prior to the structural genes. Tat recognizes an RNA target called TAR, located near the 5′-end of the viral transcript (36,37). After binding to TAR, Tat interacts with several host cell proteins to enhance the processivity of RNA polymerase II and allow expression of the HIV-1 genes. Transcriptional pausing of RNA polymerase II just downstream of the TAR RNA structure was recently demonstrated. This could allow time for Tat to interact with TAR before RNA polymerase terminates prematurely (38). Interestingly, the pause signal is an RNA stem-loop similar to those seen in attenuation control in enteric bacterial systems.

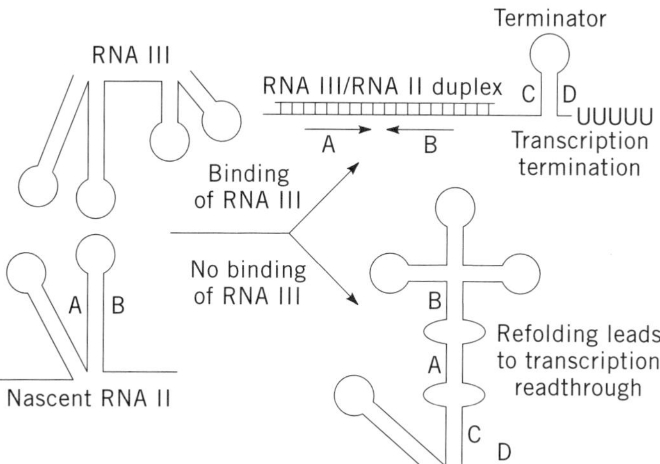

Figure 3. Model of plasmid-encoded *repR* transcription attenuation. Interaction of RNA III with the nascent RNA II (*repR*) transcript promotes terminator formation and transcription termination. In the absence of RNA III interaction, refolding of the nascent transcript blocks formation of the terminator, leading to transcription readthrough. (See text for details.) Adapted from Brantl and Wagner (24).

BIBLIOGRAPHY

1. C. Yanofsky (1988) *J. Biol. Chem.* **263**, 609–612.
2. C. Yanofsky (1981) *Nature* **289**, 751–758.
3. C. Yanofsky, R. L. Kelley, and V. Horn (1984) *J. Bacteriol.* **158**, 1018–1024.
4. R. Landick and C. Yanofsky (1987) In Escherichia coli *and* Salmonella typhimurium: *cellular and molecular biology* (F. C. Neidhardt, J. L. Ingraham, K. B. Low, B. Magasanik, M. Schaechter, and H. E. Umbarger, eds.), American Society for Microbiology, Washington, DC, pp. 1276–1301.
5. S. W. Artz and D. Holzschu (1983) In *Amino Acids: Biosynthesis and Genetic Regulation* (K. M. Herrmann and R. L. Summerville, eds.), Addison-Wesley Publishing Co., Reading, Mass, pp. 379–404.
6. P. W. Carter, J. M. Bartkus, and J. M. Calvo (1986) *Proc. Natl. Acad. Sci USA* **83**, 8127–8131.
7. K. L. Roland, F. E. Powell, and C. E. Turnbough (1985) *J. Bacteriol.* **163**, 991–999.
8. S. P. Lynn et al. (1987) *J. Mol. Biol.* **194**, 59–69.
9. P. Babitzke and C. Yanofsky (1993) *Proc. Natl. Acad. Sci. USA* **90**, 133–137.
10. J. Otridge and P. Gollnick (1993) *Proc. Natl. Acad. Sci. USA* **90**, 128–132.
11. A. A. Antson et al. (1995) *Nature* **374**, 693–700.
12. H. Shimotsu, M. I. Kuroda, C. Yanofsky, and D. J. Henner (1986) *J. Bacteriol.* **166**, 461–471.
13. P. Babitzke, J. T. Stults, S. J. Shire and C. Yanofsky (1994) *J. Biol. Chem.* **269**, 16597–16604.
14. M. Yang et al. (1997) *J. Mol. Biol.* **270**, 696–710.
15. C. G. Lerner, B. T. Stephenson, and R. L. Switzer (1987) *J. Bacteriol.* **169**, 2202–2206.
16. C. L. Quinn, B. T. Stephenson, and R. L. Switzer (1991) *J. Biol. Chem.* **266**, 9113–9127.
17. R. J. Turner, Y. Lu, and R. L. Switzer (1994) *J. Bacteriol.* **176**, 3708–3722.
18. Y. Lu, R. J. Turner, and R. L. Switzer (1996) *Proc. Natl. Acad. Sci. USA* **93**, 14462–14467.
19. P. Babitzke, J. Yealy, and D. Campanelli (1996) *J. Bacteriol.* **178**, 5159–5163.

20. S. Xirasagar, M. B. Elliott, W. Bartolini, P. Gollnick, and P. Gottlieb (1998) *J. Biol. Chem.* **272**, 19863–19869.
21. D. R. Tomchick, R. J. Turner, R. L. Switzer, and J. L. Smith (1998) *Structure* **6**, 337–350.
22. R. P. Novick, S. Iordanescu, S. J. Projan, J. Kornblum, and I. Edelman (1989) *Cell* **59**, 395–404.
23. S. Brantl, E. Birch-Hirschfeld, and D. Behnke (1993) *J. Bacteriol.* **175**, 4052–4061.
24. S. Brantl and E. G. H. Wagner (1994) *EMBO J.* **13**, 3599–3607.
25. D. L. Bently (1995) *Curr. Opin. Genet. Dev.* **5**, 210–216.
26. S. Wright (1993) *Mol. Biol. Cell.* **4**, 661–668.
27. D. Eick and G. W. Bornkamm (1986) *Nucl. Acids. Res.* **14**, 8331–8346.
28. T. K. Kerpola and C. M. Kane (1988) *Mol. Cell Biol.* **8**, 4389–4394.
29. L. London, R. G. Keene, and R. Landick (1991) *Mol. Cell Biol.* **11**, 4599–4615.
30. N. Mechti et al. (1986) *Nucl. Acids. Res.* **14**, 9653–9666.
31. S. Wright, L. F. Mirels, M. Clara, B. Calayag, and J. M. Bishop (1991) *Proc. Natl. Acad. Sci. USA* **88**, 11383–11387.
32. L. Xu, Y. Meng, R. Wallen, and R. A. DePhino (1995) *Oncogene* **11**, 1865–1872.
33. R. Treisman (1986) *Cell* **46**, 567–574.
34. V. Ramamurthy et al. (1990) *Mol. Cell. Biol.* **10**, 1484–1491.
35. K. A. Jones (1997) *Genes and Dev.* **11**, 2593–2599.
36. B. Berkhout, R. H. Silverman, and K. T. Jeang (1989) *Cell* **59**, 273–282.
37. C. Dingwall et al. (1989) *Proc. Natl. Acad. Sci.* **86**, 6925–6929.
38. M. Palangat, T. I. Meier, R. G. Keene, and R. Landick (1998) *Mol. Cell.* **1**, 1033–1042.

Suggestions for Further Reading

P. Babitzke (1997) Regulation of tryptophan biosynthesis: Trp-ing the TRAP or how *Bacillus subtilis* reinvented the wheel. *Mol. Microbiol.* **26**, 1–9. (A recent review covering TRAP-mediated regulation of the *B. subtilis trp* genes.)

R. Landick and C. L. Turnbough (1992) Transcriptional attenuation. In *Transcriptional Regulation*, (S. L. McKnight and K. R. Yamamoto, eds.) Cold Spring Harbor Laboratory Press, Cold Spring Harbor, NY, pp. 407–446. (An in-depth review of attenuation, particularly amino acid operons in enteric bacteria.)

T. M. Henkin (1996) Control of transcription termination in prokaryotes. *Ann. Rev. Gen.* **30**, 35–57. (A review covering control of transcription termination in prokaryotes.)

AUTOANTIBODY

MICHEL FOUGEREAU

Autoantibody is a very misleading term for certain **antibodies**, because it covers very different situations. This probably is largely due to Ehrlich's initial proposals, formulated at the beginning of the twentieth century, of the *horror autotoxicus* dogma, leading later to the discrimination between self and nonself (1). If one sticks to this view, one must consider that autoantibodies escape from the dogma and produce a pathological situation, as is the case for **autoimmune diseases**. Two main observations have led us to revisit this view: one is the existence of **idiotypes** and **anti-idiotypes**, and the other is the presence of the so-called natural autoantibodies.

It was proposed by Jerne (2) that **immunoglobulins** are organized as a large idiotypic network in which Ig molecules regulate each other by their interactions. Another, more pragmatic approach would be to consider that the millions of different Ig molecules present at any given time in one individual provide a sufficient number of different specificities to recognize any potential epitope, including those expressed by the immunoglobulins themselves, and more generally by any self-**epitope**, unless subject to strict negative selection during differentiation, leading to an absolute tolerance state, which would be in agreement with Ehrlich's concept. Is this the case? During B-cell differentiation, before leaving bone marrow as immature B lymphocytes, newly formed cells are screened against the local environment, and self-aggressive cells are eliminated, providing the major basis for **B-cell** tolerance. But this is not an all-or-none phenomenon. It is believed to be relatively flexible, based on the average affinity of **IgM** molecules expressed at the surface of immature B cells. This view leaves open the possibility that immature B cells circulating in the periphery have retained some autoreactive potential against a possible self antigen.

The next question is, Are there natural immunoglobulins endowed with properties of recognition of self molecules? The answer is undoubtedly yes and is supported by many observations, especially those from the groups of Avrameas and Coutinho (3). Natural antibodies in whole serum were first described as having the essential characteristic of being polyreactive, a conclusion that was not *a priori* surprising because of their polyclonality. When analysis at the **monoclonal** level was made possible with the **hybridoma** technology, it was realized that most natural antibodies had a wide recognition spectrum, as observed from their patterns of binding to many different molecules used for systematic screening. Most natural antibodies were IgM, and **complementary DNA** sequencing of their V_H and V_L regions revealed that they contained very few mutations, indicating that they reflected **transcription** of germline **genes**. Affinity measurements revealed moderate but significant association constants. Natural monoclonal autoantibodies thus seem best defined as primarily germline-encoded and polyreactive. Polyreactivity contains in itself the possibility of having both anti-self and anti-nonself specificities. Presumably due to their relatively modest affinities, low concentration, and, of course, the fact that they passed the screen for potentially aggressive anti-self specificities, these antibodies qualify as harmless and may be regarded as the pool of basic circulating immunoglobulins of wide spectrum, endowed with a high connectivity, and constituting the primary germline idiotypic network. By contrast, aggressive autoantibodies found in the serum of patients suffering autoimmune diseases generally contain mutated sequences and are often of isotypes other than IgM.

One may thus consider active immunization as bringing a transient perturbation of the background of the natural equilibrium, leading to the emergence of clones that acquire a greater affinity and an increased specificity through the occurrence of **somatic hypermutations**.

See also entries **Autoimmunity, Autoimmune diseases, Idiotypes, Anti-idiotypes**, and **Tolerance**.

BIBLIOGRAPHY

1. P. Ehrlich (1900) The Croonian lecture: On immunity. *Proc. Roy. Soc. Lond. (Biol.)* **66**, 424.
2. N. K. Jerne (1974) Towards a network theory of the immune system. *Ann. Immunol.* **125C**, 373–389.
3. A. Coutinho, M. D. Kazatchkine, and S. Avrameas (1995) Natural autoantibodies. *Curr. Opin. Immunol.* **7**, 812–818.

Suggestion for Further Reading

A. Coutinho, M. D. Kazatchkine, and S. Avrameas (1995) Natural autoantibodies. *Curr. Opin. Immunol.* **7**, 812–818.

AUTOIMMUNE DISEASES

MICHEL FOUGEREAU

Autoimmunity is not pathologic *per se* because natural **autoantibodies**, natural autoreactive **T-cell** clones, and the connected **idiotypes** and **anti-idiotypes** are part of the basic organization of the immune system. Autoimmune diseases (AIDS) are the consequence of a major failure in the regulation of the immune system, as the emergence of self-reactivity becomes aggressive and harmful to the organism. They represent a major cause of morbidity, and their incidence increases with aging.

The primary causes of autoimmune diseases (AIDS) are still very poorly understood, and the precise target of the autoimmune reactions is not always known. In some cases, one organ, and eventually one target molecule, can be identified. Examples of these are myasthenia gravis, in which antibodies directed against the **acetylcholine receptor** have been identified, or autoimmune thyroiditis, where the presence of autoantibodies directed against receptors for thyroid stimulating hormone have been documented. Such antibodies may have very different impacts upon binding. Some may act as a mimic of the hormone and therefore activate thyrocytes, thereby inducing hyperthyroidism, which is characteristic of Graves' disease. Conversely, others will block receptor activation, leading to a hypothyroidism state, as seen in Hashimoto's disease. Another example of an organ-specific AID is insulin-dependent diabetes mellitus, in which the cellular target is the β islets of the endocrine pancreas. Some AIDs appear to be not organ-specific, such as systemic lupus erythematosous, when a number of autoantibodies are directed against a large variety of antigens, including DNA and nucleoproteins.

Several characteristics generally underline the occurrence of AIDs: (a) In most cases there is a genetic predisposition, with a frequent linkage to certain **major histocompatibility complex** (MHC) haplotypes; (b) they are chronic diseases that persist for years; and (c) although the molecular target of autoantibodies or autoreactive T-cell clones may be identified, this does not clarify the nature of the initial antigen, if any.

Linkage to an MHC haplotype is indicated by the increase of the relative risk for a given AID to occur. The most spectacular example is that of ankylosing spondylitis, a very severe arthritis that most frequently affects hips, for which 95% of patients are HLA-B27. This indicates a relative risk of about 70 (ie a $B27^+$ individual has 70 times greater chance to have the disease than the average population). This AID is also particularly interesting because it has been observed that microorganisms such as *Klebsiella* share an identical sequence of six amino acid residues, Gln–Thr–Asp–Arg–Glu–Asp (QTDRED), with an **epitope** of the HLA27 molecule. It was postulated that this AID might be the consequence of an epitope mimicry, an infection with *Klebsiella* inducing an immune response that subsequently would cross-react with the HLA epitope of the host. A similar hypothesis was proposed for rheumatoid arthritis, a very common rheumatological disorder that may develop for years, because it was found to be associated with certain DR4 alleles expressing the Gln–Lys–Arg–Ala–Ala (QKRAA) sequence. A similar sequence was found in Hsp 70, which might provide another example of epitope mimicry as the starter for the initial immunization.

The V_H and V_L regions of autoantibodies isolated from AID patients have been sequenced extensively, with the hope of detecting significant differences from natural, nonaggressive autoantibodies. Frequently, dominant expression of some V_H and/or V_L is observed, as in the case of anti-DNA autoantibodies isolated from lupus patients. In most occasions, but not always, these antibodies have mutated **variable regions**, as opposed to the germline sequences generally found in natural antibodies. The significance of this remains unclear, however, primarily because the autoantibodies developed by a rather large number of AID patients seem to be only passive witnesses of an autoimmune response, rather than being involved in pathogenesis, which is frequently attributed to **cytotoxic T lymphocytes**.

Numerous animal models have been defined and extensively studied. Many have stressed the importance of the genetic background, such as the obese strain of chickens that develop a hereditary spontaneous autoimmune thyroiditis that is quite similar to Hashimoto's disease. Another example is the nonobese diabetic (NOD) mice that develop an autoimmune diabetes, which spontaneously develops an insulitis that becomes full insulin-dependent diabetes mellitus at 7 months of age. New Zealand black (NZB) mice and other strains that spontaneously develop a systemic lupus erythematosous-like syndrome are another example. These models are interesting because they provide a possible key to understanding the genetic basis for the equivalent human diseases. Other approaches are centered more on attempts to isolate **antigens** that might induce an AID-like syndrome. An example of this is experimental autoimmune encephalomyelitis, which is a possible model for multiple sclerosis; it can be induced upon injection of myelin basic protein, from which encephalitogenic peptides and precise epitopes have been described. This disease can also be transferred with lymphocytes but not by serum, pointing to the role of cell-mediated immunity in this AID. Interestingly, **T cells** with a suppressive activity have also been identified, offering the possibility of modulating the immune system negatively by this approach. Such a treatment is eagerly awaited for most AIDs, particularly multiple sclerosis. Thus far, primarily nonspecific immunosuppressive agents are used, with the obvious problem of generating severe secondary effects.

See also entries **Autoantibody**, **Idiotype**, **Anti-idiotype**, and **Tolerance**.

Suggestions for Further Reading

G. Del Prete (1998) The concept of type-1 and type-2 helper T cells and their cytokines in humans. *Int. Rev. Immunol.* **16**, 427–455.

C. C. Goodnow (1997) Balancing immunity, autoimmunity and self-tolerance. *Ann. NY. Acad. Sci.* **815**, 55–66.

E. G. Spack (1997) Treatment of autoimmune diseases through manipulation of antigen presentation. *Crit. Rev. Immunol.* **17**, 529–536.

AUTOIMMUNITY

MICHEL FOUGEREAU

The immune system is able to discriminate between self and nonself, but there are often ambiguities with this formulation. First, it certainly is intended to stress the fact that the immune system is designed for the important role of fighting against external pathogens, whereas it should be harmless for the components of the organism. There are two ways to be harmless. One is simply not to "show off," the second is to be present, but silent. Immunologists long lived with the idea originally put forward by Ehrlich in 1901 of the *horror autotoxicus*, thus considering that **autoantibodies**, and more generally **autoimmunity**, could not exist. This was even strengthened by Burnet, when he proposed the **clonal selection theory**, based on the elimination of autoreactive lymphocytes (the so-called "forbidden clones"), leading to immune **tolerance**. On these bases, autoimmunity may only be considered to be a pathological deviation, with the emergence of **autoimmune diseases**. Discovery of the interactions between **idiotype** and **anti-idiotype** and of the existence of natural antibodies endowed with auto reactivity made it necessary to reconsider the problem.

A long debate has taken place for years between immunologists, first on the existence, and then on the nature, of natural antibodies and their possible autoimmune properties. Identification of "natural" monoclonal antibodies that were polyspecific clarified the situation, because it was demonstrated that these antibodies contained **variable (V) regions** of both the heavy and the light chains that were mostly not mutated, and like the germline gene from which they were derived. As a result, these antibodies could be considered the first circle of antibody defense, with some polyspecificity that is the best witness of a certain degeneracy of recognition by the immune system. Deliberate immunization will stimulate clones that have the best affinity for the antigen, and this affinity will be amplified by the mechanism of **somatic hypermutation**, as the specificity becomes more precise. To be polyspecific immediately implies that both auto- and xenospecificities may be found, as was shown using a large panel of random antigens for screening. Autoimmunity has also been described in the **T-cell** population and is subject to the same comments, with the exception that T cells do not mutate upon antigenic stimulation.

The entire **repertoire** expressed at a given time by the immune system is probably best visualized as a large network of interacting molecules that regulate each other. Deliberate immunization would result in transient perturbation of this equilibrium. Deviation to the production of aggressive B- and T-cell clones would thus be a central alteration of this equilibrium, ensuring some leakiness for the negative selection of potentially aggressive clones.

See also entries **Autoantibody**, **Autoimmune diseases**, **Idiotype**, **Anti-idiotype**, and **Tolerance**.

Suggestions for Further Reading

C. Goodnow (1996) Balancing immunity and tolerance: deleting and tuning lymphocyte repertoires. *Proc. Natl Acad. Sci. USA* **93**, 2264–2271.

S. Lacroix-Desmazes, L. Mouthon, S. H. Spalter, S. Kaveri, and M. Kazatchkine (1996) Immunoglobulins and the regulation of autoimmunity through the immune network. *Clin. Exp. Rheumatol.* **14** (Suppl. 15), S9–S15.

R. J. Cornall, C. C. Goodnow, and J. G. Cyster (1995) The regulation of self-reactive B cells. *Curr. Opin. Immunol.* **7**, 804–811.

AUTONOMOUS CONTROLLING ELEMENTS

N. CRAIG

Transposable elements are discrete pieces of DNA that can move between nonhomologous positions in the **genome**. When Barbara McClintock discovered transposable elements in maize, she called them **controlling elements**, emphasizing that they could affect gene expression, in addition to being able to move from place to place within the genome (1). She discovered two classes of elements; one she termed *activator* (Ac), which could itself translocate within the genome. She also determined that Ac encoded a product that could promote the movement of a second class of elements called *dissociation* (Ds). In the presence of Ac, Ds could move and cause **chromosome** breaks that altered the expression of nearby genes, but was unable to move in the absence of Ac. Ds elements are so-named because their translocation could be associated with chromosome breakage.

We now know that Ac is an intact transposable element encoding a **recombinase**, a **transposase**, that promotes element translocation and has special sequences at the termini of the element upon which the transposase acts to promote translocation. Such intact elements are said to be "autonomous" because they can translocate without the assistance of another element. Ds, by contrast, is a nonautonomous element. It doesn't encode a transposase itself and is thus dependent on the presence of other Ac elements to supply a transposase to promote its transposition. Ds is actually an Ac element that has undergone internal deletions that have removed the transposase gene but have left intact the special terminal sequences on which transposase acts intact (Fig. 1). Thus the terminal sequences of Ds are the same as the terminal sequences of Ac, so that the Ac transposase can promote the translocation of Ds (2); there are several different Ds elements that contain different extents of the internal, nonterminal sequences. The deletions are thought to result from abortive repair of the gapped donor site after element excision (2–4).

Virtually all bacterial elements that have been observed are autonomous; that is, encode a transposase and have cognate terminal sequences. Nonautonomous and autonomous elements are frequently observed in other organisms such as *Drosophila*, ***Caenorhabditis*** *elegans*, and plants (5).

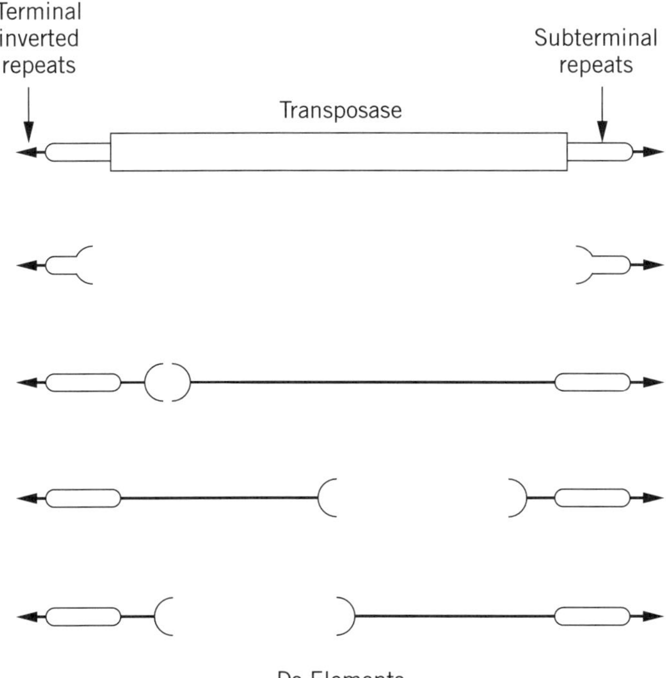

Figure 1. Autonomous and nonautonomous elements. A schematic representation on an intact (autonomous) Ac element is shown (top line). It encodes a transposase formed from several exons (not shown) and special recombination sequences at the ends of the element. The arrows represent short perfect inverted repeats and the oval other repeated sequences; these are not shown to scale. Only some of the subterminal repeats are necessary for transposition. Various deletion derivatives of Ac are shown (lines 2–5); these internal deletions make the element dependent on an intact Ac for transposase and hence for translocation.

BIBLIOGRAPHY

1. B. McClintock (1956) *Cold Spring Harbor Symp. Quant. Biol.* **21**, 197–216.
2. E. Rubin and A. A. Levy (1997) *Mol. Cell. Biol.* **17**, 6294–6302.
3. E. S. Coen, T. P. Robbins, J. Almeida, A. Hudson, and R. Carpenter (1989) In *Mobile DNA* (D. A. Berg and M. M. Howe, ed.), American Society for Microbiology, Washington, DC, pp. 413–436.
4. W. R. Engels, D. M. Johnson-Schlitz, W. B. Eggleston, and J. Sved (1990) *Cell* **62**, 515–525.
5. H. Saedler and A. Gierl (1996) *Curr. Top. Microbiol. Immunol.* **204**, 27–48.

AUTONOMOUSLY REPLICATING SEQUENCES

GEORGES N. COHEN

The **chromosomes** of **yeast** contain many genomic segments at which **DNA replication** is ordinarily initiated. On average, one is found every 40 kb in the *Saccharomyces cerevisiae* **genome**. These regions can act as the **origin of replication** for a circular **plasmid**. When such a segment is linked to a nonreplicative DNA fragment, it can confer on that fragment the ability to replicate autonomously as a plasmid in a yeast cell in the S-phase of the cell cycle. This construction is called an *autonomously replicating fragment* (ARS).

An ARS-containing DNA fragment can **transform** yeast cells at much higher frequencies than a fragment without an ARS, because it can replicate autonomously without undergoing a rare homologous **recombination** before its **genes** can be expressed. In contrast to the latter transformants, however, those formed with an ARS-containing DNA are extremely unstable and lose the transforming DNA even under selective conditions. For unknown reasons, they tend to remain associated with the mother cell without being evenly distributed to the daughter cell upon division.

ARS have been found in *Saccharomyces cerevisiae* and but also in other yeasts, such as *Candida maltosa*, *Hansenula polymorpha*, *Kluyveromyces lactis*, and *Schizosaccharomyces pombe*. They are not restricted to yeasts and have also been described in other lower **eukaryotes**, such as *Aspergillus nidulans* and *Entamoeba histolytica*, and even in at or near potential replicative origins from *Drosophila*, Chinese hamster, and humans.

AUTORADIOGRAPHY

D. R. FISHER

Autoradiography is the technique of recording an image of a preparation containing beta-particle emitting **radioactivity**, using photographic film, X-ray-sensitive film, an emulsion, or other radiation-sensitive medium. The samples are placed directly against the film for a period of time to allow radioactive emissions from the sample to interact with the film emulsion and create an image. A photographic emulsion is a suspension of crystals of silver bromide embedded in gelatin. When crystals of silver bromide are struck by charged-particle or photon radiation, the silver atoms are ionized and form an invisible latent image. After exposure to the sample, the photographic grains in the emulsion are fixed using standard photographic developing, which removes silver bromide that has not been ionized. After the emulsion is developed, each small aggregate of reduced silver atoms becomes a visible dark spot on the emulsion; collectively, they make up the photographic image.

Production of a visible silver grain requires a number of ionization events, so the photographic response is not exactly linear to the amount of radiation present. Preflashing the film with a uniform low intensity of light "primes" each grain of silver to become reduced and visible after absorbing just one or a very few additional beta particles from the sample. This increases substantially the sensitivity of the film and also makes the photographic response more directly proportional to the amount of radiation in the sample. The signal-to-noise ratio is often increased by exposing the film to low temperatures. The sensitivity can also be enhanced by using scintillation screens that emit visible light on encountering a beta particle; the light is recorded by the film (see **Fluorography**).

Autoradiography has a large number of practical applications in the biological, chemical, and physical sciences, because it provides both qualitative and quantitative information (eg., images and amounts present). It may be used to image large, small, and microscopic specimens, including sectioned whole organisms, organs, tissues, cellular structures, and nucleic acids that contain some radiolabeled compound. An example of a whole-animal autoradiograph is shown in Figure 1. Cells may be autoradiographed either in culture as a monolayer, on a glass slide, or on thinly sectioned living tissues from an animal organ or tumor. Microautoradioagraphy involves coating the sample directly with a radiation-sensitive emulsion; cellular constituents that have incorporated the radiolabel can be clearly identified. Autoradiography is used with **electrophoresis** or **chromatography** to image radiolabeled macromolecules and other separated chemicals for quantitative analysis. For example, autoradiography is useful for indicating the position of hybridized nucleic acids on **Southern blots** and **Northern blots**, and of proteins on **Western blots.**

HISTORICAL DEVELOPMENT

The first autoradiograph, or **autoradiogram**, was made in 1859 by Crookes, who placed uranium rocks on photographic plates (1). Crookes would not understand the process by which the images were created until 37 years later when he visited the laboratory of Henri Becquerel. Bequerel discovered gamma rays from naturally radioactive uranium salts in 1896 when he placed uranium together with photographic plates that had been placed inside black paper to protect them from sunlight (1). This discovery closely followed the discovery of X rays by Wilhelm Roentgen in 1895 using a Crookes electron tube.

The French photographer Nicéphore Niepce had observed the darkening of silver grains on photographic plates by uranium in 1867, but he did not recognize any practical applications of this phenomenon (2). Although George de Hevesy, the pioneer of radioactive tracers, had used bismuth isotopes in animals in 1923, the first autoradiographs of radioactively labeled animal tissues were made in 1924 by Lacassagne in France (1–3). Lacassagne fed polonium-210 to rabbits and later placed thin sections of rabbit kidney tissues in paraffin blocks against photographic plates. Gettler made the first autoradiographs of human tissues from deceased persons who had ingested radium chloride in 1933 (1).

The earliest images were of poor quality. Autoradiographic techniques were improved by Leblond, who, in 1943, showed the microdistribution of iodine-131 in cells (1,2). In 1946, Bélanger and Leblond (4) developed a method for locating radioactive elements in tissues by covering histological sections with a photographic emulsion. In 1954, Gabriel introduced techniques for identifying radioiodine using human **serum albumin** zone paper **electrophoresis** (1). In 1953, techniques were developed for labeling nucleic acids and studying kinetics of cell mitosis and division (5). High resolution techniques have more recently been employed for study of other metabolic and pharmacokinetic processes in molecular biology.

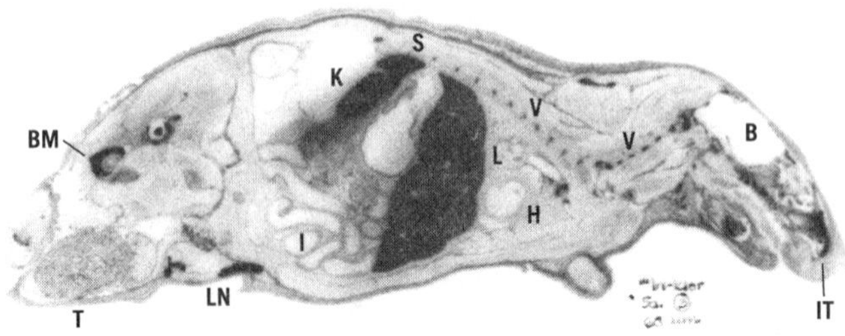

Figure 1. Whole-body autoradiograph of a rat that had been injected with indium-111 chloride. Courtesy of B. Anders Jönsson, Lund University, Sweden.

FILM-LESS AUTORADIOGRAPHY

Modern trends in autoradiography involve replacing high speed X-ray film with radiation detector systems, laser scanners, and computer-based imaging systems. A variety of radiation-detecting crystals and phosphors have been developed for this purpose. Storage phosphor screens are more sensitive, by a factor of about 20–100, for beta-emitting radionuclides (2), and they are reusable. The exposure time is also much less, by a factor of about 10, over conventional X-ray film, and samples may be processed at room temperature and without a darkroom or chemicals for film developing. Applications of filmless autoradiography include **two-dimensional gels**, **Southern blots**, **Northern blots**, immunoblots, and quantitative polymerase chain reaction (**PCR**) (2).

Microchannel array detectors have been introduced to replace both X-ray films and phosphor screens (8). The new instruments are faster (by a factor of ~ 10) and have greater image resolution than do phosphor screens for detecting latent images from hybridization studies using macromolecules labeled with carbon-14, sulfur-35, phosphorus-32, and iodine-125 from flat gels, blots, membranes, tissue slices, and other flat specimens.

BIBLIOGRAPHY

1. M. Brucer (1990) *A Chronology of Nuclear Medicine*, Heritage Publications Inc., St. Louis, MO.
2. R. Wegmann, N. Balmain, S. Ricard-Blum, and S. Guha (1995) *Cell. Mol. Biol.* **41**, 1–20.
3. A. Lacassagne, J. Lattes, and J. Lavedan (1925) *J. Radiol. Electr.* **9**, 1–14.
4. L.-F. Bélanger and C. P. Leblond (1946) *Endocrinology* **39**, 386–400.
5. A. Howard and S. R. Pelc (1970) *Heredity* **6** (suppl.), 261–273.
6. J. G. Gall and M. Pardue (1969) *Proc. Natl. Acad. Sci. USA* **63**, 378–383.
7. E. M. Southern (1975) *J. Mol. Biol.* **98**, 503–517.
8. Packard Instrument Company (1993) *Enter the New Era of Instant Autoradiography: The InstantImager™*, promotional literature from Packard Instrument Company, Meriden, Conn.

Suggestions for Further Reading

G. A. Boyd (1955) *Autoradiography in Biology and Medicine*, Academic Press, New York.

E. J. Hall (1994) *Radiobiology for the Radiologist*, 4th ed., J. B. Lippincott Company, Philadelphia, pp. 92–95, 397–402.

J. C. Kaplan and M. Delpech (1989) "Biologie Moléculaire et Médecine", in *Médecine et Sciences Flammarion*, Paris, p. 610.

D. LeGuellec and J. -Y. Sire (1991) in *Microscopie électronique. Cryométhodes, Immunocytologie, Autoradiographie, Hybridation in situ* (G. Morel, ed.), INSERM, pp. 299–304.

R. Wegmann, N. Balmain, S. Ricard-Blum, and S. Guha (1995) *Cell. Mol. Biol.* **41**, 1–20.

AUTOSOME

ALAN P. WOLFFE

An autosome is any **chromosome** other than the **sex** chromosomes. For example, the human **genome** has 24 chromosomes, including 22 autosomes plus the **X**- and the **Y-chromosomes**. This is described as the **haploid** set. If a cell contains a complete haploid set of chromosomes, it is described as **euploid**. If the chromosome set is altered by duplication or deletion then the cell is **aneuploid**. If a cell contains two sets of genomes per nucleus, it is **diploid**, and if it contains more than two, it is described as **polyploid**.

Organisms that have a largely vegetative existence, such as some **fungi** and **algae**, have mainly haploid cells. In metazoans the **gametes**, such as the **spermatozoa** and **oocytes** in humans, are haploid. Sexual **reproduction** involves the fusion of two haploid gametes to form a diploid **egg** in an animal. The vast majority of the cells in an animal that develop from an egg are diploid. Thus there are two active copies of each autosome in most animal cells. In **plants**, a polyploid state is common where the number of genomic copies in somatic cells may range from 3 to 10 (1). The number of genomic copies can actually be **polymorphic** within a particular species (2). Thus plants are relatively insensitive to the number of copies of a particular **gene** in a given cell. This is not the case with animals, where somatic cells are almost always diploid. The only exceptions are those animals with **parthenogenetic** life styles in which either males or females develop as a default state from an unfertilized egg. In general, animal cells cannot accommodate more than two copies of a particular chromosomal region.

Aneuploidy occurs when parts of chromosomes or whole chromosomes are absent from the genome. In humans the nondisjunction of homologous chromosomes during **meiosis** or of sister **chromatids** during **mitosis** can lead to aneuploidy. When a chromosome is in excess of the normal euploid number, the condition is called hyperploidy. For example, a hyperploid state in humans occurs when there are three copies of a particular chromosome, a condition known as **trisomy**. All human autosomes except chromosome 1 have been found as trisomies (3). Most trisomic embryos are spontaneously aborted, but live births occur with trisomies of chromosomes 13, 18, and 21 (4). The most common human trisomy to reach term is that of chromosome 21, resulting in Down's syndrome. These children are mentally retarded, predisposed to cancer and have a shortened life span. Presumably the deleterious effects of three copies of chromosome 21 follow from an imbalance in the expression of one or several sets of genes on chromosome 21 relative to the other genes on the other chromosomes in the cell.

Experiments in *Drosophila melanogaster* indicate that changes in the dosage of a single gene might be tolerated, but a deletion that includes several genes or a large section of a chromosome is likely to cause sterility (5). Surprisingly in trisomies of *D. melanogaster* involving whole chromosome arms, many of the duplicated autosomal genes express only diploid levels of gene product. The expression of genes outside the duplicated region is also affected, however, indicating the pleiotropic effects of altering gene copy number. An interesting aspect of sexual reproduction requiring the use of two distinct haploid gametes is that there are two copies of each autosome in a diploid cell, whereas there are be one or two copies of each sex chromosome. **Dosage compensation** is the phenomenon whereby the **transcriptional** activity of genes on the sex chromosome is adjusted to similar levels independent of the number of copies, so that it is constant relative to the activity of a gene on an autosome (see **X-chromosome inactivation**).

Humans, again have two X-chromosomes in one sex (female) and one X-chromosome in the other (male). To achieve dosage compensation, an entire X-chromosome is silenced in female diploid cells. This process is called X-chromosome inactivation and results in the formation of the **Barr body**. The capacity to monitor the activity of genes on autosomes relative to the sex chromosomes has facilitated the discovery of novel mechanisms of regulating genes at the level of whole chromosomes.

BIBLIOGRAPHY

1. J. E. Averett (1980) Polyploidy in plant taxa: summary. In *Polyploidy* (W. H. Lewis, ed.), Plenum, New York.
2. K. Irifune (1990) *J. Sci. Hiroshima Univ.* **23**, 163–181.
3. T. J. Hassold (1986) *Trends Genet.* **2**, 105–109.
4. E. B. Hook, B. B. Topol, and P. K. Cross (1989) *Am. J. Hum. Genet.* **45**, 855–861.
5. D. L. Lindsley et al. (1972) *Genetics* **71**, 157–184.

Suggestion for Further Reading

M. S. Clark and W. J. Wall (1996) *Chromosomes. The Complex Code*, Chapman and Hall, London.

AUXINS

DOMINIQUE VAN DER STRAETEN
MARC VAN MONTAGU

HISTORY

Auxins were the first class of **plant hormones** to be discovered. Initial observations by Charles Darwin and his son Francis on the phenomenon of bending of canary grass coleoptiles toward a unilateral source of light (known as phototropism) led the foundation for a series of experiments that culminated in the discovery of the first phytohormone (1). The active substance was isolated from oat coleoptiles by Went in 1926 (2) and later identified and named auxin (from the Greek "to increase"). By analogy with the animal hormone concept from Bayliss and Starling (3), plant physiologists interpreted auxin as a phytohormone.

BIOSYNTHESIS AND METABOLISM

The most abundant natural auxin is indole-3-acetic acid (IAA). It is mainly synthesized in the shoot apex and in young leaves. Other substances with auxin activity that are found in plants are indole-3-butyric acid, 4-chloro-indole-acetic acid, and phenyl acetic acid (4,5); these compounds are active only at higher concentrations, and their roles in development remain largely unknown. Multiple pathways exist for auxin biosynthesis; some of these are dependent on tryptophan, while others are independent (Fig. 1). In certain species, different

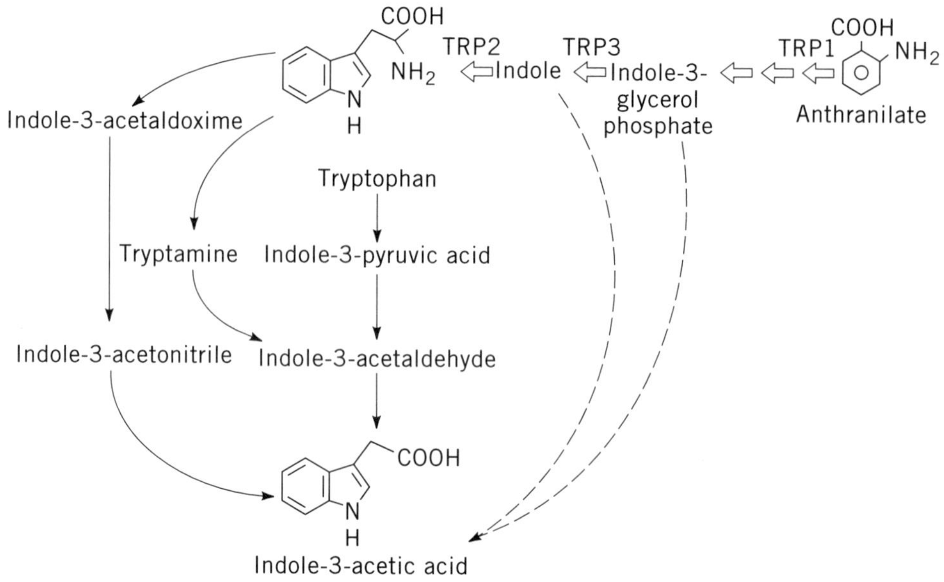

Figure 1. Auxin biosynthesis in higher plants. (Based on Ref. 4.)

routes are operational (6), whereas in others, essentially all of the IAA is derived from tryptophan (7). Three possible routes for tryptophan-dependent IAA synthesis have been proposed (4,5), which include indole-3-acetonitrile, tryptamine, and indole-3-pyruvic acid as the respective intermediates. Genes encoding tryptophan decarboxylase (which catalyzes the conversion of tryptophan to tryptamine) and nitrilase (forming IAA from indole-3-acetonitrile) have been cloned (8–10). The route via indole-3-acetonitrile has been found primarily in Brassicaceae. Likewise, the tryptamine pathway is not supposed to be of general importance, because tryptamine is not universally present in plants. A unique tryptophan-dependent pathway via indole-3-acetamide is used by bacteria and by plant cells that are transformed with ***Agrobacterium*** *tumefaciens*. The corresponding genes have been used to alter the IAA levels in **transgenic** plants (11).

Tryptophan-auxotroph mutants have revealed the existence of a tryptophan-independent pathway for IAA biosynthesis. Both maize (12) and **Arabidopsis** (13) tryptophan mutants overproduce IAA or IAA conjugates, respectively. On the other hand, mutants that block tryptophan biosynthesis before indole-3-glycerol phosphate have a **phenotype** suggestive of auxin deficiency (14). Based on these observations, IAA was proposed to be synthesized through a tryptophan precursor, either indole or indole-glycerol phosphate, thereby bypassing tryptophan (4). An auxin-overproducing mutant that accumulates both free and conjugated auxins has been isolated (15,16). The wild-type gene product might inhibit auxin biosynthesis and shows similarity to aminotransferases (17).

Of all the auxin present in plant tissues, 95% resides in conjugated form, coupled to **amino acids**, **peptides**, or **proteins** via amide bonds, or to sugars via ester linkages (18). The role of these conjugates is diverse: the storage of free IAA (released by hydrolysis), its transport, and eventually its catabolism. A genetic approach was used to characterize a hydrolase that catalyzes the formation of free IAA from IAA-Leu (10). The *ILR1* (*IAA-leucine resistant 1*) **gene** encodes a protein with similarity to a bacterial aminoacylase that catalyzes amide cleavage.

TRANSPORT

Another important aspect determining auxin levels is transport of the hormone. Auxin transport can be either (a) polar, basipetally directed, and active or (b) nonpolar and inactive (19). The latter occurs through phloem tissue. Polar transport is probably involved specifically in processes of apical dominance, vascular differentiation, and shoot–root communication. Evidence exists for the presence of influx and efflux carriers, although protonated auxin can enter cells by simple **diffusion**. Proton–auxin **symport**, energized by the plasma membrane **proton pump**, allows a more rapid uptake than by diffusion. The AUX1 protein has been suggested to represent an auxin uptake carrier (20), based on its similarity to amino acid uptake carriers, which also function as symporters. Although several mutants have been characterized that show growth defects similar to those upon treatment with polar transport inhibitors (21), only one of them appears to be a candidate for the auxin efflux carrier. The PIN1 (*pin formed 1*) gene product shows the characteristics of a transmembrane protein (22). The corresponding mutant displays abnormal cotyledon development, together with twisted, narrow leaves with branching midveins. Auxin transport is reduced at all stages of development (23). Recently, factors that may modulate auxin transport have been characterized (24,25). Both the loci *RCN1* (*ROOT CURLING on NPA1*) (coding for a protein phosphatase A regulatory subunit) and *MP1* (*MONOPTEROS1*) (coding for a homologue of the auxin response factor ARF1) (26) (see text below) have been isolated based on their altered response to auxin transport inhibitors. Mutations in the *MP1* gene interfere with vascular strand formation and with initiation of the body axis in the embryo (27).

SIGNAL PERCEPTION AND TRANSDUCTION

The perception and transduction of the auxin signal remains largely enigmatic (Fig. 2). Two different approaches

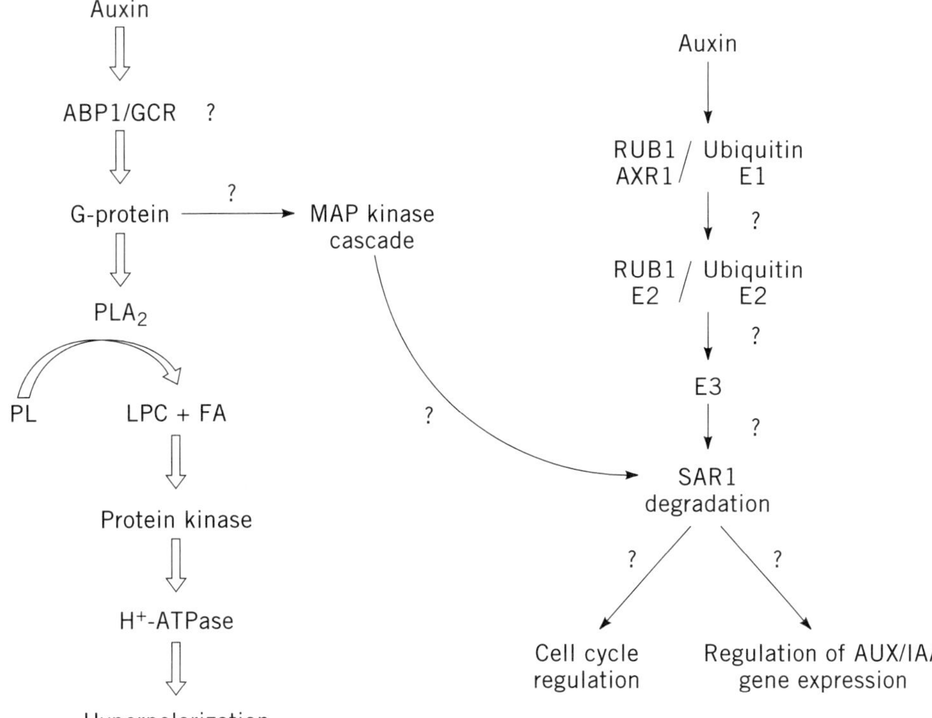

Figure 2. A model for auxin signal transduction. PL, phospholipids; PLA_2, phospholipase A_2; LPC, lysophospholipids; FA, fatty acids. (Modified from Refs. 29 and 39.)

have been followed toward the identification of putative auxin receptors: (a) the characterization of auxin-resistant mutants and (b) the isolation of auxin-binding proteins (28,29). Because auxin is present both intra- and extracellularly, a site of perception may be needed on both sides of the membrane. Although many auxin-binding proteins have been identified, none has been proven to function as an internal receptor at the molecular level. In contrast, an external receptor has been characterized. There is evidence that auxin causes changes in the plasma membrane potential via stimulation of the proton **ATPase** and modulation of **ion channels**. These rapid responses result from auxin binding to a protein named ABP1 (auxin-binding protein 1). Thus far, it is unknown whether ABP1 mediates cell growth. Although ABP1 possesses a *C*-terminal signal for retention in the endoplasmic reticulum (ER) and largely resides in the ER, it does not seem to bind auxin in this cellular compartment. Instead, ABP1 can reach the cell surface by a yet unknown mechanism, as demonstrated by immunolocalization experiments (30). It is hypothesized that ABP1 is retained at the membrane periphery by interaction with a docking protein, because ABP1 does not contain a potential membrane-spanning domain. The nature of the docking protein is undefined, but biochemical evidence supports the concept of a **G-protein**-coupled receptor (GCR) (31). These receptors contain seven transmembrane domains and are directly coupled to heterotrimeric G proteins. A possible effector could be phospholipase A_2, which produces lysophospholipids and free **fatty acids**. The former have been proposed to act as second messengers that can activate the plasma membrane proton ATPase (32,33). However, G-protein-coupled receptors may activate different G proteins in different cell types, as has been shown in animal cells (34), and therefore may control a downstream branched transduction pathway. At least one report suggests the involvement of a MAP **kinase** cascade in the initiation of cell division by auxin in cell cultures (35).

The second approach to molecular characterization of auxin signal transduction involves the isolation and characterization of auxin-resistant mutants. At least seven loci were identified by screening for resistance to auxin-induced root growth inhibition (21). AXR1 (auxin resistant 1) probably plays a role at an early stage in the signaling chain, because the recessive *axr1* mutant cannot respond to auxin in all the assays tested. The *AXR1* gene encodes a protein with homology to the amino-terminal half of the **ubiquitin**-activating enzyme E1 (36). Thus, a central part of the mechanism of auxin action appears to play a regulatory role in protein degradation. Recently, two genes of *Saccharomyces cerevisiae* were found to be similar to *AXR1* (37). One of these regulates the stability of key components of **cell cycle** regulation, such as the **cyclin**-dependent kinase inhibitor Sic1p, through conjugation of the ubiquitin-like protein RUB1 to CDC53p, a component of a E3 ubiquitin—protein ligase complex (37). The speculation that *AXR1* plays a role in cell cycle control is considerably strengthened by the characterization of the *tir1* (*transport inhibitor response 1*) locus, a semidominant mutation that also confers auxin resistance (38). Sequence analysis revealed similarity between TIR1 and F-box proteins found in E3 ubiquitin-protein ligase complexes in yeast. The product of the *SAR1* (*suppressor of auxin resistance*) gene has been proposed as a candidate target for AXR1-dependent ubiquitin degradation (39). The *sar1* mutant was identified as a suppressor of the Axr1 phenotype (40). SAR1 may function as a repressor of *AUX/IAA* genes, or it may be involved in cell cycle regulation (40). The mostly root-specific *axr4* mutation probably defines an AXR1-independent path. Double-mutant analysis indicates that at least two interacting pathways exist, in which AXR4 and AUX1 define a first route while AXR1 defines a second one (41).

DOWNSTREAM TARGETS

Ultimately, the auxin signal leads to induction or repression of a series of target genes, some of which are responsive in a few minutes (42,43). The products of these primary response genes are likely candidates to play a role in auxin-stimulated growth. The induction of the early genes is independent of *de novo* **protein biosynthesis**, implying that **post-translational processes** control the activity of factors that affect their expression. The transcriptional activation of early genes is based on the presence of auxin-responsive, ***cis*-acting**, **response elements** (AuxREs). Two auxin response-element domains were defined in the *Pisum sativum IAA4* and *IAA5* genes, one of which acts as an auxin switch while the other acts as an **enhancer** element. The former AuxRE is present in the promoter of *AUX/IAA*, *SAUR* (*small auxin-upregulated mRNAs*, and auxin-inducible 1-aminocyclopropane-1-carboxylate synthase genes. An alternative AuxRE is found in the promoter of the *GH2* and *GH4* genes. Essentially, this element consists of a G-box motif, binds bZIP factors, and is generally responsive to stress signals. The biological role of most of the auxin early-response gene products remains to be elucidated. Nevertheless, based on similarities with proteins of known functions, it is proposed that *GH2* and *GH4* genes are **glutathione-*S*-transferases**. *AUX/IAA* polypeptide chains are short-lived nuclear proteins that probably function as an auxin switch in a secondary phase of signal transduction (42), implying that their target gene products probably mediate auxin responses. Thus, the *AUX/IAA* polypeptides might be involved in homo- and heterodimerization with auxin response factors, such as ARF1, that bind to AuxREs (26,44). The *IAA17* and *AXR3* genes are identical (44). *Axr3* is a dominant auxin-resistant mutant with extremely resistant roots, although the general phenotype of the mutant resembles auxin hypersensitivity (45).

EFFECTS

Depending on temporal and spatial factors, a particular subset of downstream target genes is activated and results in one of many described effects. Auxin has been implicated in a variety of growth and developmental processes, such as cell division, stem and root elongation, leaf expansion, formation of adventitious and lateral roots, induction of vascular tissue, apical dominance, tropisms, flowering, and fertility (46).

BIBLIOGRAPHY

1. C. Darwin and F. Darwin (1881) *The Power of Movement in Plants*, Appleton-Century-Crofts, New York.
2. F. W. Went (1927) *Proc. Kon. Nederl. Akad. Wetensch.* **30**, 10–19.
3. W. M. Bayliss and E. H. Starling (1902) *Proc. R. Soc.* **69**, 352–353.
4. J. Normanly, J. P. Slovin, and J. D. Cohen (1995) *Plant Physiol.* **107**, 323–329.
5. B. Bartel (1997) *Annu. Rev. Plant Physiol. Plant Mol. Biol.* **48**, 51–66.
6. L. Michalczuk, T. J. Cooke, and J. D. Cohen (1992) *Phytochemistry* **31**, 1097–1103.
7. K. Bialek, L. Michalczuk, and J. D. Cohen (1992) *Plant Physiol.* **100,** 509–517.
8. V. De Luca, C. Marineau, and N. Brisson (1989) *Proc. Natl. Acad. Sci. USA* **86**, 2582–2586.
9. D. Bartling, M. Seedorf, A. Mithöferr, and E. W. Weiler (1992) *Eur. J. Biochem.* **205**, 417–424.
10. B. Bartel and G. R. Fink (1995) *Science* **268**, 1745–1748.
11. H. J. Klee and C. P. Romano (1994) *Crit. Rev. Plant Sci.* **13**, 311–324.
12. A. D. Wright, M. B. Sampson, M. G. Neuffer, L. Michalczuk, J. P. Slovin, and J. D. Cohen (1991) *Science* **254**, 998–1000.
13. R. L. Last, P. H. Bissinger, D. J. Mahoney, E. R. Radwanski, and G. R. Fink (1991) *Plant Cell* **3**, 345–358.
14. R. L. Last and G. R. Fink (1988) *Science* **240**, 305–310.
15. W. Boerjan, M.-T. Cervera, M. Delarue, T. Beeckman, W. Dewitte, C. Bellini, M. Caboche, H. Van Onckelen, M. Van Montagu, and D. Inzé (1995) *Plant Cell* **7**, 1405–1419.
16. J. J. King, D. P. Stimart, R. H. Fisher, and A. B. Bleecker (1995) *Plant Cell* **7**, 2023–2037.
17. M. Gopalraj, T.-S. Tseng, and N. Olszewski (1996) *Plant Physiol.* **111** (Suppl.), 114 (♯469).
18. J. Normanly (1997) *Physiol. Plant.* **100**, 431–442.
19. T. L. Lomax, G. K. Muday, and P. H. Rubery (1995) In *Plant Hormones* (P. J. Davies, ed.), Kluwer Dordrecht, pp. 509–530.
20. M. J. Bennett, A. Marchant, S. T. May, H. G. Green, S. P. Ward, P. A. Millner, A. R. Walker, B. Schulz, and K. A. Feldmann (1996) *Science* **273**, 948–950.
21. O. Leyser (1997) *Physiol. Plant.* **100**, 407–414.
22. L. Galweiler, E. Wisman, A. Yephremov, and K. Palme (1996) http://nasc.nott.ac.uk/Norwich/contents/Norwich.html P307
23. K. Okada, J. Ueda, M. K. Komaki, C. J. Bell, and Y. Shimura (1991) *Plant Cell* **3**, 677–684.
24. C. Garbers, A. DeLong, J. Deruére, P. Bernasconi, and D. Söll (1996) *EMBO J.* **15**, 2115–2124.
25. C. S. Hardtke and T. Berleth (1998) *EMBO J.* **17**, 1405–1411.
26. T. Ulmasov, G. Hagen, and T. J. Guilfoyle (1997) *Science* **276**, 1865–1868.
27. G. K. H. Przemeck, J. Mattsson, C. S. Hardtke, Z. R. Sung, and T. Berleth (1996) *Planta* **200**, 229–237.
28. H. Barbier-Brygoo (1995) *Crit. Rev. Plant Sci.* **14**, 1–25.
29. H. Macdonald (1997) *Physiol. Plant.* **100**, 423–430.
30. J. Deikman and M. Ulrich (1995) *Planta* **195**, 440–449.
31. P. A. Millner, D. A. Groarke, and I. R. White (1996) *Plant Growth Regul.* **18**, 143–147.
32. G. F. E. Scherer and B. André (1993) *Planta* **191**, 515–523.
33. H. Yi, D. Park, and Y. Lee (1996) *Physiol. Plant.* **96**, 359–368.
34. E. J. Neer (1995) *Cell* **80**, 249–257.
35. T. Mizoguchi, Y. Gotoh, E. Nishida, K. Yamaguchi-Shinozaki, N. Hayashida, T. Iwasaki, H. Kamada, and K. Shinozaki (1994) *Plant J.* **5**, 111–122.
36. H. M. O. Leyser, C. A. Lincoln, C. Timpte, D. Lammer, J. Turner, and M. Estelle (1993) *Nature* **364**, 161–164.
37. D. Lammer, N. Mathias, J. M. Laplaza, W. Jiang, Y. Liu, J. Callis, M. Goebl, and M. Estelle (1998) *Genes Dev.* **12**, 914–926.
38. M. Ruegger, E. Dewey, W. M. Gray, L. Hobbie, J. Turner, and M. Estelle (1998) *Genes Dev.* **12**, 198–207.
39. W. M. Gray and M. Estelle (1998) *Curr. Opin. Biotechnol.* **9**, 196–201.
40. A. Cernac, C. Lincoln, D. Lammer, and M. Estelle (1997) *Development* **124**, 1583–1591.

41. C. Timpte, C. Lincoln, F. B. Pickett, J. Turner, and M. Estelle (1995) *Plant J.* **8**, 561–569.
42. S. Abel and A. Theologis (1996) *Plant Physiol.* **111**, 9–17.
43. Y. Takahashi, S. Ishida, and T. Nagata (1995) *Plant Cell Physiol.* **36**, 383–390.
44. D. Rouse, P. Mackay, P. Stirnberg, M. Estelle, and O. Leyser (1998) *Science* **279**, 1371–1373.
45. H. M. O. Leyser, F. B. Pickett, S. Dharmasiri, and M. Estelle (1996) *Plant J.* **10**, 403–413.
46. P. J. Davies (1995) *Plant Hormones: Physiology, Biochemistry and Molecular Biology*, Kluwer, Dordrecht, The Netherlands.

AUXOTROPH

WILLIAM A. ROSCHE
PATRICIA L. FOSTER

An auxotroph is a **mutant** that requires a factor for growth that is not required by the **wild-type**. The opposite of an auxotroph is a **prototroph**, an organism that grows in minimal medium containing only inorganic salts and a carbon source (plus an additional energy source if the organism is an **autotroph**). Typical auxotrophic mutations are in genes encoding biosynthetic **enzymes** for **amino acids**, vitamins, and **nucleotides**. Some organisms are natural auxotrophs. For example, humans require that all vitamins and several amino acids be supplied in their food. In contrast many, but not all, free-living **microorganisms**, such as *Escherichia coli*, are prototrophs. A specific auxotrophic **phenotype** is called an auxotrophy, eg, an arginine auxotrophy, and the mutant is called an arginine auxotroph.

AVIDIN

MEIR WILCHEK
EDWARD A. BAYER

Avidin is a minor **glycoprotein** component of egg white that binds the vitamin **biotin** with the largest **association constant** known ($\mathbf{K_a} \sim 10^{15}\ M^{-1}$) for a **ligand**–protein interaction (1). Its function in egg white is not known precisely, although protective or antibiotic activities have been suggested. It gained importance not because of its natural functionality, but because of its utility as a tool in molecular biology [see **Avidin-biotin system**]. Together with its bacterial relative **streptavidin** and in complex with biotin, avidin has become one of the most important tools in molecular biology because of its myriad of applications in many fields of biology, biotechnology, and even chemistry (2). Originally, the avidin-biotin complex was introduced to localize, detect, and isolate biologically active molecules (3,4), but over the years it has became a standard system for replacing radioactivity in **immunoassays**, for DNA probes, and for various clinical and medical applications (5,6).

The avidin molecule forms a very stable tetramer that consists of four identical monomers. The monomer is glycosylated at asparagine-17 and is strongly basic (**isoelectric point** of about 10.5) due to a surplus of **arginine** and **lysine** residues. The relatively high molar **absorption** ($A^{1\%}_{[280\ nm]} = 1.54$) reflects the four **tryptophan** and one **tyrosine** residue per monomer. The molecular mass of the glycosylated tetramer is about 62,400 Da. When the oligosaccharide residue is removed, the estimated mass is 57,120 Da.

Each avidin monomer binds one molecule of biotin. The binding is considered to be kinetically noncooperative among the subunits (7), although one of the tryptophan residues from a neighboring monomer is inserted physically into the binding site and thus contributes to binding biotin and to the enhanced stability of the tetrameric structure (8). Avidin also binds to the dye HABA (4′-hydroxyazobenzene-2-carboxylic acid), but the association constant (ca. $10^6\ M^{-1}$) is considerably lower than that of the avidin-biotin complex.

The unusually strong binding and the many uses of the system caused scientists to try to understand the molecular basis for this interaction. Moreover, since its inception, many groups have sought to improve various components of the system or the system as a whole. Concurrently, structure-function studies have been implemented, in which early results show the importance of aromatic amino acids in the biotin-binding sites of both avidin and streptavidin. The final structural proof is the determination of the crystal structure of both proteins (Fig. 1), which shows that their overall fold, organization into the tetramer, and content and position of amino acids in the binding sites are all similar (8–10).

The avidin monomer forms a β-barrel, composed of eight antiparallel **β-strands**, connected sequentially by loops. The biotin-binding site is a **hydrophobic** pocket that, in the absence of biotin, contains five molecules of structured **water**. The combined structure of these water molecules emulates that of biotin itself, and as a consequence, they prevent the binding pocket from collapsing. The water molecules are expelled upon binding of biotin. Because no conformational change occurs during complex formation, the binding has a gain of **free energy**. In fact, the avidin-biotin system can be considered a natural host–guest complex. It is clear that any change of this "ideal system" leads to reduced affinity. Indeed, **site-directed mutagenesis** or **chemical modification** of residues in the binding site have reduced the affinity constant. An example is the **nitration** of the essential binding-site tyrosine (11), which decreases the association constant to $\sim 10^{10}\ M^{-1}$, and renders the binding reversible at alkaline pH. Another example is the site-directed modification of tryptophan 120 in streptavidin to phenylalanine, which is also accompanied by a loss in affinity (12).

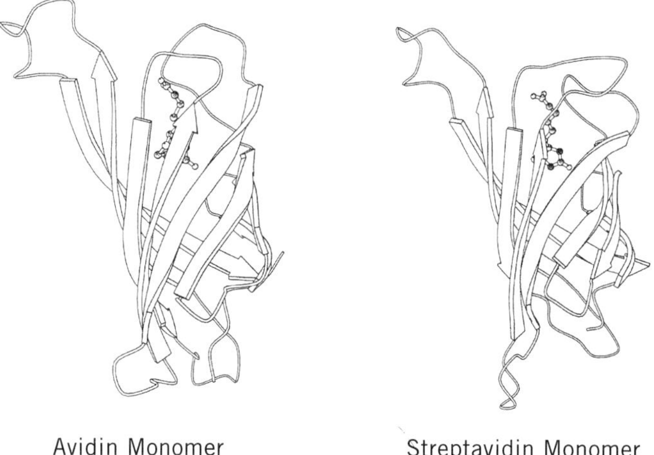

Figure 1. Ribbon diagrams of the avidin- and streptavidin–biotin monomers, showing the eight strands of the β-barrel. The biotin molecule is shown in a ball-and-stick model.

Different modifications of avidin were introduced over the years, including Extravidin, Neutralite avidin, and Lite avidin (13). Extravidin and Neutralite avidin are avidins in which the arginine residues were modified, whereas the oligosaccharide moieties were removed from Lite and Neutralite avidins. These derivatives are useful in many applications, because avidin, being a basic glycoprotein, is subject to a variety of extraneous nonspecific or undesired binding, particularly to DNA. Similar qualities of genetically engineered forms of avidin are possible because avidin was successfully expressed in a eukaryotic host (14,15). This should enable preparation of different derivatives of avidin with high affinities, provided that the binding site residues are preserved. It seems that the nonbinding-site residues can be altered extensively.

BIBLIOGRAPHY

1. N. M. Green (1975) *Adv. Protein Chem.* **29**, 85–133.
2. M. Wilchek and E. A. Bayer (1989) *Trends Biochem. Sci.* **14**, 408–412.
3. E. A. Bayer and M. Wilchek (1978) *Trends Biochem. Sci.* **3**, N237–N239.
4. E. A. Bayer and M. Wilchek (1980) *Methods Biochem. Anal.* **26**, 1–45.
5. M. Wilchek and E. A. Bayer (1984) *Immunol. Today* **5**, 39–43.
6. M. Wilchek and E. A. Bayer (1988) *Anal. Biochem.* **171**, 1–32.
7. M. L. Jones and G. P. Kurzban (1995) *Biochemistry* **34**, 11750–11756.
8. O. Livnah, E. A. Bayer, M. Wilchek, and J. L. Sussman (1993) *Proc. Natl. Acad. Sci. USA* **90**, 5076–5080.
9. W. A. Hendrickson, A. Pähler, J. L. Smith, Y. Satow, E. A. Merritt, and R. P. Phizackerley (1989) *Proc. Natl. Acad. Sci. USA* **86**, 2190–2194.
10. P. C. Weber, D. H. Ohlendorf, J. L. Wendoloski, and F. R. Salemme (1989) *Science* **243**, 85–88.
11. E. Morag, E. A. Bayer, and M. Wilchek (1996) *Biochem. J.* **316**, 193–199.
12. T. Sano and C. R. Cantor (1995) *Proc. Natl. Acad. Sci. USA* **92**, 3180–3184.
13. E. A. Bayer and M. Wilchek (1994) In *Egg Uses and Processing Technologies* (J. S. Sim, and S. Nakai, eds.), CAB International, Wallingford, UK, pp. 158–176.
14. K. J. Airenne, P. Sarkkinen, E.-L. Punnonen, and M. S. Kulomaa (1994) *Gene* **144**, 75–80.
15. K. J. Airenne, C. Oker-Blom, V. S. Marjomäki, E. A. Bayer, M. Wilchek, and M. S. Kulomaa (1997) *Protein Express. Purif.* **9**, 100–108.

Suggestions for Further Reading

M. Wilchek and E. A. Bayer, eds. (1990): *Avidin-Biotin Technology, Methods Enzymol.*, Vol. 184, Academic Press, San Diego, p. 746.

N. M. Green (1990) Avidin and streptavidin, *Methods Enzymol.* **184**, 51–67.

AVIDIN-BIOTIN SYSTEM

MEIR WILCHEK
EDWARD A. BAYER

The avidin-biotin system has become one of the major mainstays in biochemical analysis and has far-reaching application in biotechnology, industry, and clinical medicine (1). The general idea of the avidin-biotin system is that **biotin**, a low molecular weight vitamin, can be chemically coupled to other low or high molecular weight molecules (e.g., **proteins**, **hormones**, **DNA** molecules, etc.). The biotin moiety is still recognized by **avidin** or **streptavidin**, either as the native protein or in derivatized form containing any one of a number of **reporter groups** or probes. The principle of this system is illustrated in Fig. 1.

The avidin-biotin system has become a "universal" tool in most of the fields of the biological sciences, because of studies that commenced in the mid 1970s and the constant development that has continued until today (2–6). The system has been applied for a wide variety of purposes (Fig. 1) and has recently been adapted for clinical use for the localization, imaging, and therapy of cancer (7). The binder and the target can be any of the components listed in Figure 1. What is required for this system is the capacity to biotinylate a binding entity so that the specificity and activity of the binding function is retained. For different approaches to biotinylation and the reagents used for binding to different functional groups, see **Biotin**. As can be seen, most of the functional groups on biological molecules can be modified with biotin.

Because of its charge neutrality and lack of **glycosylation**, streptavidin is generally preferred over egg-white avidin in many applications, although new derivatives of avidin (e.g., Neutravidin) may prove advantageous and less expensive [see **Avidin** and **Streptavidin**). The final component added to the system is the probe. The various probes and their potential uses are shown in Fig. 1. The probes are prepared in two ways. They are chemically conjugated directly to avidin or streptavidin (Fig. 1, Approach A) and are **fluorescent**, **radioactive**, or other types of macromolecules (proteins, polysaccharides, etc.). A second approach is to biotinylate the probes and to interact them with streptavidin under subsaturating ratios, thus leaving extra binding sites vacant (Fig. 1, Approach B). More recently, **fusion proteins** have been prepared of streptavidin with different **enzymes** and native fluorescent proteins.

In many cases, such as **affinity chromatographic** applications, more reversible binding of biotin to avidin would be a distinct advantage. In this context, streptavidin has a critical disadvantage in that the interaction between its subunits is very strong, and it cannot be used when reversibility of the interaction is desired. This is in contrast to avidin, from which an immobilized monovalent form can be produced (8), and the immobilized avidin monomer binds biotinylated compounds reversibly (9).

Site-directed mutagenesis and **chemical modification** studies are currently being performed on both avidin and streptavidin to understand better their interaction with biotin, the interaction between their subunits, and, perhaps eventually, for better application in the previously-mentioned systems (10,11). In this regard, the single, critical **tyrosine** residue of the binding sites of avidin and streptavidin was **nitrated**, and the tetrameric structures of the resultant "nitro avidin" and "nitro streptavidin" were retained (12). The biotin-binding property also had sufficiently high affinity for a variety of applications, including affinity chromatography, enzyme immobilization, and **phage-display** technology (13,14). The major difference between the nitro avidins and the native molecules is that biotinylated compounds are released by competition with free biotin, and then the latter is liberated

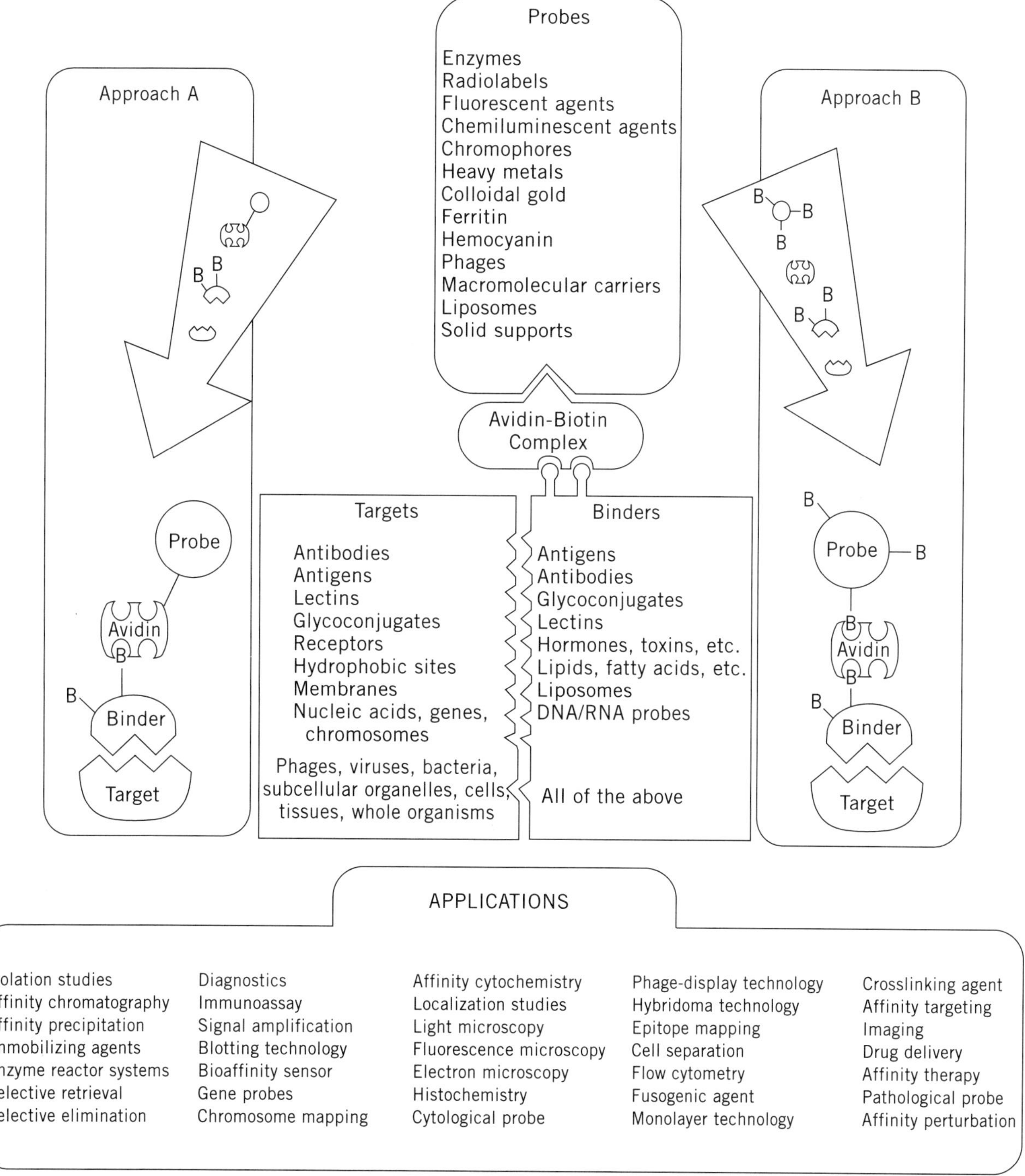

Figure 1. Overview of the avidin-biotin system and the two major strategies for the various applications. In both approaches, a target molecule in a desired experimental system is combined with a biotinylated binder molecule. Approach A involves direct interaction with an avidin-conjugated probe. In approach B, avidin is a sandwich between the biotinylated binder and the biotinylated probe. Various targets, binders, probes, and many of the applications are listed.

by treating the column with basic solutions (pH 10), thereby regenerating the original biotin-binding capacity of the nitro avidin affinity column. Reduced or altered binding characteristics are also conferred on the binding sites of avidin or streptavidin by site-directed mutagenesis of selected binding site residues, such as **tryptophans**.

Today, the avidin/streptavidin-biotin system is a real necessity in most fields of biological study.

BIBLIOGRAPHY

1. M. Wilchek and E. A. Bayer, eds. (1990) *Avidin-Biotin Technology Methods in Enzymology*, Vol. 184, Academic Press, San Diego.
2. E. A. Bayer and M. Wilchek (1978) *Trends Biochem. Sci.* **3**, N237–N239.
3. E. A. Bayer and M. Wilchek (1980) *Methods Biochem. Anal.* **26**, 1–45.
4. M. Wilchek and E. A. Bayer (1984) *Immunol. Today* **5**, 39–43.
5. M. Wilchek and E. A. Bayer (1989) In *Protein Recognition of Immobilized Ligands* (T. W. Hutchens, ed.), Alan R. Liss, New York, pp. 83–90.
6. E. A. Bayer and M. Wilchek (1996) In *Immunoassay* (E. P. Diamandis and T. K. Christopoulos, eds.), Academic Press, San Diego, pp. 237–267.
7. G. Paganelli, P. Magnani, A. G. Siccardi, and F. Fazio (1995) In *Cancer Therapy with Radiolabeled Antibodies* (D. M. Goldenberg, ed.), CRC Press, Boca Raton, FL, pp. 239–254.
8. N. M. Green and E. J. Toms (1973) *Biochem. J.* **133**, 687–700.
9. K. P. Henrikson, S. H. G. Allen, and W. L. Maloy (1979) *Anal. Biochem.* **94**, 366–370.
10. A. Chilkoti, P. H. Tan, and P. S. Stayton (1995) *Proc. Natl. Acad. Sci. USA* **92**, 1754–1758.
11. A. Chilkoti, B. L. Schwartz, R. D. Smith, C. J. Long, and P. S. Stayton (1995) *Bio/Technology* **13**, 1198–1204.
12. E. Morag, E. A. Bayer, and M. Wilchek (1996) *Biochem. J.* **316**, 193–199.
13. E. Morag, E. A. Bayer, and M. Wilchek (1996) *Anal. Biochem.* **243**, 257–263.
14. M. E. M. Balass, E. A. Bayer, S. Fuchs, M. Wilchek, and E. Katchalski-Katzir (1996) *Anal. Biochem.* **243**, 264–269.

Suggestion for Further Reading

M. D. Savage, G. Mattson, S. Desai, G. Nielander, S. Morgensen, and E. J. Conklin (1992) *Avidin-Biotin Chemistry: A Handbook* Pierce Chemical Co., Rockford, Ill.

AZURIN

ISRAEL PECHT
OLE FARVER

Azurins are **proteins** of bacterial origin that belong to the family of cupredoxins (ie, blue single-copper proteins) as is **plastocyanin** (1). Azurins consist of a single **polypeptide chain** that contains ca. 128 amino acid residues (~14 kDa molecular weight) (2). Several **X-ray crystallography** studies of azurins isolated from various bacteria, as well as of single-site mutated forms, have yielded high-resolution, three-dimensional **protein structures** (Fig. 1) (1,3–6). The protein consists of eight β-**strands** that form a β-sandwich (barrel) structure. A segment between β-strands 4 and 5 lies outside this barrel and contains a short α-**helix** (residues 55 to 67). The copper ion is bound about 7 Å below the protein's surface and is coordinated by five ligands: three equatorial ones [a **cysteine**(112) thiolate and two imidazoles (His46 and His117)] and two weak ligands in axial positions (the carbonyl O atoms of Gly45 and Met121). This configuration (the carbonyl O atom in Gly45 excepted), is conserved in all known structures of "blue" type 1 (T1) copper sites determined thus far (1).

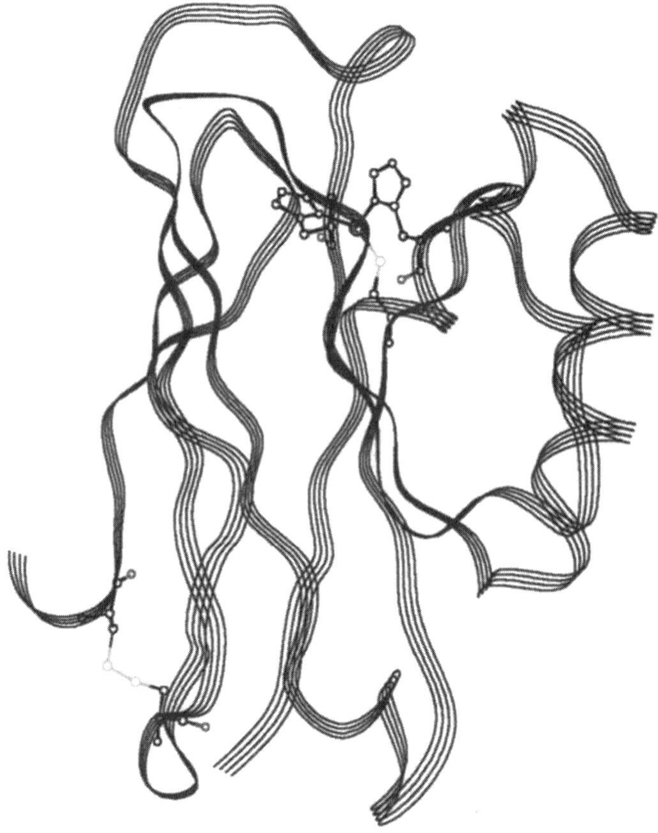

Figure 1. The three-dimensional structure of azurin. The copper center with its ligands is seen near the top of the molecule. The disulfide bond is found at the opposite end of the molecule.

These protein–metal interactions yield a copper site that has a distorted, trigonal bipyramidal geometry, which may explain the stabilization of the Cu(I) state relative to Cu(II) (7) (see **Redox enzymes**). Structure-imposed Cu → L π-backbonding has been proposed to account further for stabilization of the cuprous state, because strong π-interaction with the d_π orbitals results in an increase of the ligand field strength. The **oxidation/reduction potentials** of the Cu(II)/Cu(I) in azurins (~310 mV) are higher than is generally observed for copper complexes. Studies of several mutant forms in which Met121 is replaced, demonstrated that large **hydrophobic** residues raise the Cu(II)/(I) potential, whereas negatively charged ones lower it (8): The span in reduction potential going from Glu121 to Leu121 azurin is 300 mV, or about half the range of the potentials for the naturally occurring T1 sites. The large increase in reduction potentials caused by substitutions involving bulky hydrophobic side chains at position 121 is rationalized by exclusion by such side chains of **water** from the metal site or simply by providing a low dielectric environment for the copper ion. In contrast, negatively charged hydrophilic residues or water are expected to stabilize the copper ion in the more positively charged +2 state and thus lower the potential, as is indeed observed. Smaller effects on the reduction potential were observed when mutations were made in other positions. It should be emphasized, however, that a thiolate copper ligand *per se* is not a prerequisite for high reduction potentials (8).

As with all other cupredoxins, azurins are also spectroscopically characterized by an intense blue color and a uniquely narrow hyperfine coupling in the electron paramagnetic resonance (EPR) spectrum (see **Electron magnetic resonance**). The intense blue color is due to a $\pi S \rightarrow Cu(d_{x^2-y^2})$ ligand-to-metal charge transfer involving the thiolate ligand as the electron donor; it has a molar extinction coefficient of 4000 to 6000 $M^{-1}cm^{-1}$ (9). This is more than 100 times larger than what is found for simple Cu(II) complexes. A second characteristic property associated with the blue (T1) Cu (II) is an EPR spectrum displaying a hyperfine splitting in the $g_{\|}$ region, due to interaction of the Cu(II) nuclear and electron spins, which is unusually narrow ($\sim$0.008 cm^{-1}), or approximately 50% smaller than those of inorganic copper complexes. This is attributed to delocalization of the unpaired $Cu(d_{x^2-y^2})$ electron onto the Cys(S) $p\pi$ orbital, thus reducing the nuclear–electron interaction (10). The unique spectroscopic properties of the T1 site are most likely related to the electron-mediating function of cupredoxins, and this relationship is a central issue in studies of biological **electron transfer proteins** (11).

Recent experimental investigations of the electron transfer mechanism by azurin have in part been directed toward the intramolecular process in single-site mutated forms of the proteins (12–14). This was motivated by the finding that azurin provides a useful model system for studies of long-range, intraprotein electron transfer (LRET). One way by which this reaction can be induced is from the half-reduced **disulfide bond** to the Cu(II) ion (12,14). Another one is by attaching an extraneous redox center to the protein (11,13). Results of these studies yielded insights into the mechanisms of intramolecular electron transfer in proteins and the parameters that determine the degree of electronic coupling between electron donor and acceptor. Effective electron transfer in proteins requires minimal reorganization of the redox centers as a result of the process. As described above, the metal ion in the blue copper proteins is coordinated to a site that is intermediate between the preferred geometries for Cu(II) (tetragonal planar) and Cu(I) (tetrahedral) (6). Furthermore, the relatively high reduction potentials provides an increased driving force. Finally, the tightly knit, antiparallel **β-sheet** structure of azurin probably improves the electronic coupling between the reaction partners through the protein matrix.

BIBLIOGRAPHY

1. E. T. Adman (1991) *Adv. Protein Chem.* **42**, 145–198.
2. L. Ryden and J.-O. Lundgren (1976) *Nature* **261**, 344–346.
3. E. T. Adman and L. H. Jensen (1981) *Isr. J. Chem.* **21**, 8–12.
4. H. Nar, A. Messerschmidt, and R. Huber (1991) *J. Mol. Biol.* **218**, 427–447.
5. W. E. B. Shephard, B. F. Anderson, D. A. Lewandoski, G. E. Norris, and E. N. Baker (1990) *J. Am. Chem. Soc.* **112**, 7817–7819.
6. A. Messerschmidt, L. Prade, S. J. Kroes, J. Sanders-Loehr, R. Huber, and G. W. Canters (1998) *Proc. Natl. Acad. Sci. USA* **95**, 3443–3448.
7. C. S. St. Clair, W. R. Ellis Jr., and H. B. Gray (1992) *Inorg. Chim. Acta* **191**, 149–155.
8. T. Pascher, B. G. Karlsson, M. Nordling, B. G. Malmström, and T. Vanngård (1993) *Eur. J. Biochem.* **212**, 289–296.
9. H. B. Gray and E. I. Solomon (1981) In *Copper Proteins*, Vol. **3** (T. G. Spiro, ed.), Wiley, New York, pp. 1–39.
10. W. E. Antholine, P. M. Hanna, and D. R. McMillin (1993) *Biophys. J.* **64**, 267–272.
11. A. J. Di Bilio et al. (1997) *J. Am. Chem. Soc.* **119**, 9921–9922.
12. O. Farver, L. K. Skov, G. Gilardi, G. van Pouderoyen, G. W. Canters, S. Wherland, and I. Pecht (1996) *Chem. Phys.* **204**, 271–277.
13. J. J. Regan et al. (1995) *Chem. Biol.* **2**, 489–496.
14. O. Farver and I. Pecht (1997) *J. Biol. Inorg. Chem.* **2**, 387–392.
15. T. J. Mizoguchi, A. J. Di Bilio, H. B. Gray, and J. H. Richards (1992) *J. Am. Chem. Soc.* **114**, 10076–10078.

B

B CELL

MICHEL FOUGEREAU

Specific recognition by the immune system is mediated fundamentally by B and T lymphocytes. These cells derive from the hematopoietic **stem cells** through discrete differentiation pathways, although both originate in the bone marrow (and in the liver during embryonic and fetal life) (see **Hematopoiesis**). **T cells** mature in the thymus, whereas B cells continue to differentiate in the bone marrow itself. B cells express **immunoglobulins** that directly interact with native **epitopes**—that is, subregions of an **antigen** with a well-defined three-dimensional structure. Surface immunoglobulins are associated with a signaling module made of a Igα-Igβ heterodimer to form the B-cell receptor (BCR). In contrast to the BCR, **T-cell receptors** (TCRs) do not interact directly with native epitopes, but they identify peptides derived from the original antigen as presented by molecules encoded by genes of the **major histocompatibility complex** (MHC).

Both the B and the T cells have to face the **repertoire** problem—that is, how to generate extremely large number of different immunoglobulins and TCRs to meet the requirements of recognizing an extraordinary large number of discrete epitopes. A theoretical approach suggests that this number might be as large as 10^{17}. Because the total number of lymphocytes in a human adult averages 10^{12} (roughly divided into one-fifth B cells and four-fifths T cells), and taking into account that they are clonally organized, the number of BCRs and/or TCRs expressed at one given time appears much less than 10^{17}, which implies some basic degeneracy in the immune recognition system. Nevertheless, the number of different structures must still be quite large, which therefore raises the problem of how to generate such a large repertoire with, necessarily, a limited number of genes. This is the main challenge for B cells (and T cells as well), to which the way they differentiate brings the answer.

During fetal life, B lymphocytes differentiate in the liver and then in the bone marrow, which remains active throughout life. The first steps of B-cell differentiation take place in the bone marrow, which is a primary lymphoid organ, and drives precursors derived from the hematopoietic stem cells to the immature B lymphocytes. This period of differentiation is antigen-independent and is essentially devoted to generation of the basic **immunoglobulin** repertoire, which is the result of a complex sequence of events that involve multiple gene rearrangements. Immature B lymphocytes migrate to secondary lymphoid organs, namely spleen, lymph nodes, tonsils, and gut-associated lymphoid tissues, including Peyer's patches. Further differentiation necessitates antigen encounter and a number of complex cellular interactions, among which cooperation between the B and T cells plays a major role. A second round of diversity is then generated, which results mostly from somatic mutations, and the final steps of differentiation, including isotype switching, lead to the emergence of plasma cells and memory B cells. Two discrete lineages of B cells have been well-characterized in the mouse and are designated B1 and B2. B1 cells express the CD5 marker, have a low level of surface **IgD**, and are mostly encountered in peritoneal and pleural cavities. They seem preferentially responsible for autoantibody synthesis. B2 cells represent the major and conventional B-cell population.

FROM THE HEMATOPOIETIC STEM CELL TO THE IMMATURE B CELL

All differentiation events that drive the emergence of the various lineages (granulocytes, erythrocytes, monocytes, megakaryocytes, **lymphocytes**) derive from a common hematopoietic stem cell that originates in the bone marrow. The B lineage presumably initiates from a precursor common to the B and the T lymphocytes. The main features of the molecular and cellular events that take place in the bone marrow and lead to B cell commitment are indicated in Figure 1. Three major subpopulations define the B lineage: proB, preB, and immature B cells. The successive steps of differentiation proceed from the periphery toward the center of the bone marrow and may be followed in several ways: (a) acquisition and/or loss of surface antigens (especially CD markers) identified by **monoclonal antibodies**, (b) identification of cytoplasmic proteins and/or **messenger RNA** transcripts, and (c) analysis of the gene rearrangement status for each of the three Ig gene loci: H, L, and K.

The first steps of differentiation involve a sequence of direct interactions of the precursors with the stromal cells, ensured by various sets of **cellular adhesion molecule**s (CAMs). The VLA4 **integrin** expressed at the cell surface of the precursors interacts with stromal VCAM1, and the resulting early proB cells now express Kit, a receptor that binds the stem cell factor of stromal cells. This interaction triggers proliferation of the proB cells, which will continue to differentiate through other stimuli. Late proB cells are stimulated by a soluble **growth factor** also of stromal origin, **interleukin**-7 (IL-7), which drives proliferation of preB cells. As these steps of differentiation proceed, the cells are dividing, and Ig gene rearrangements that generate the basic repertoire take place in a sequential fashion. CD marker characterization indicates that CD34, which is already expressed in the hematopoietic stem cells, remains present at the surface of precursors and proB cells. Late proB cells can be characterized by the coexpression of CD34 with CD19, which is a very specific marker of the B lineage because it is found on all subsequent stages, with the exception of the plasma cells. CD10, also known as the common acute lymphocytic leukemia antigen (CALLA), is expressed up to the immature B-cell stage a seems expressed slightly before CD19. Other markers are expressed as differentiation proceeds, such as CD20, CD21, CD22, CD23, and CD24.

THE BASIC IG REPERTOIRE DEVELOPS IN THE BONE MARROW AS B CELLS DIFFERENTIATE

As already stated, the main feature of B-cell differentiation is the acquisition of the basic repertoire of immunoglobulins

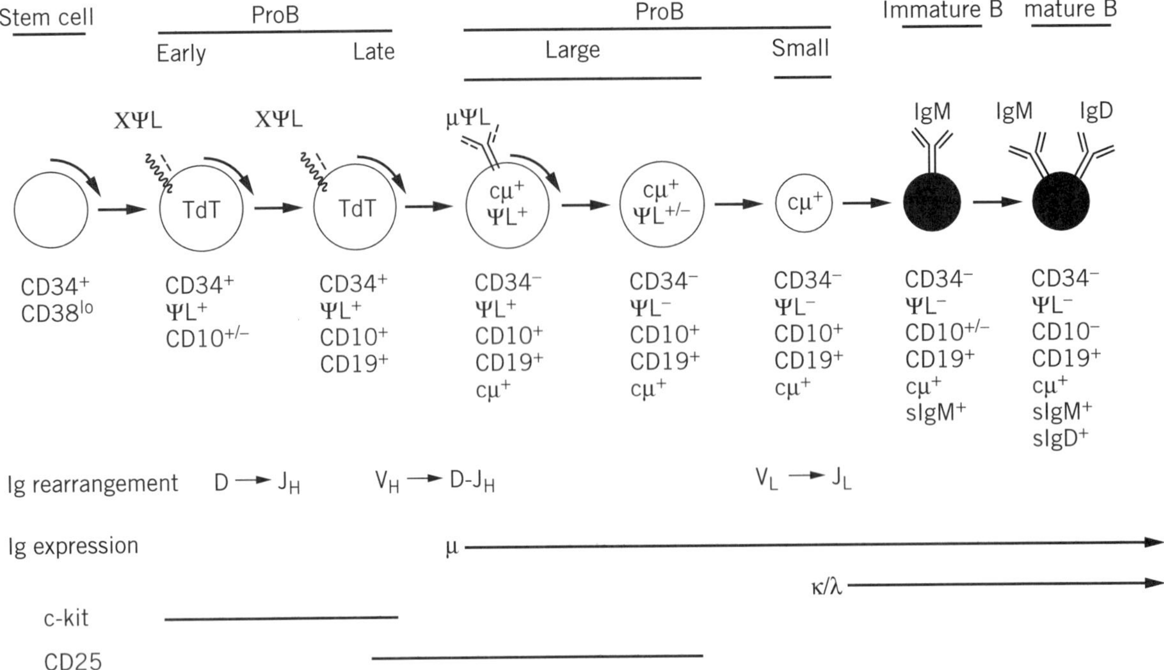

Figure 1. B-cell differentiation in humans. Early steps of B-cell differentiation take place in the bone marrow and are antigen independent. The major discrete steps that lead from the hematopoietic stem cell (HSC) to the immature B lymphocyte are identified by the successive Ig gene rearrangements, by the presence of cytoplasmic markers, and by the expression of characteristic markers at the cell surface. The very early stages that go from HSC to the early proB cell encompass several precursors that are not yet completely defined. TdT, terminal deoxynucleotidyl transferase, responsible for generating N diversity; ΨL, surrogate light chain that combines with the Igμ chain to form the preB receptor. Late events of B-cell differentiation, which take place in the periphery, are antigen-dependent. The germinal center organization plays a key role in mounting a T-dependent B-cell response, during which affinity is tuned upon the acquisition of somatic mutations and the biological functions of the antibodies are adapted by isotype switching. Ultimately, this differentiation results in plasma cells that secrete antibodies and in memory B cells that are involved in secondary responses. Each step of differentiation may be followed by various CD surface markers, as indicated.

that becomes expressed as IgM on immature B cells. IgM has the classical H_2L_2 organization, and both the H and L chains have variable and constant regions. The variable regions of the heavy and light chains interact in the antibody combining site, which is responsible for specific antigen recognition. Heavy and light chains are encoded by genes that are localized on three discrete gene clusters, H, κ and λ, located on chromosomes 14, 2, and 22, respectively. The unique feature of Ig gene organization is that they must be rearranged by the specific **recombinases** RAG1 and RAG2 (for recombinase activating genes) before becoming functional. The first recombination event, DH to JH, takes place in the early proB cells (see Fig. 1) and is rapidly followed by the rearrangement of one of the V_H segments to D-J. Another enzyme, terminal deoxynucleotidyl-transferase, is also active in proB cells, for which it represents an additional cytoplasmic marker. This enzyme adds nucleotides in a random fashion during the joining of D to J and of V to D-J rearrangements. ProB cells that have rearranged the IgV_H locus must "make" two decisions. One is to remain monoclonal with respect to H-chain production, the second is to activate rearrangement of the light chains. Both events are regulated by the μ chain itself, which must be in its membrane form. Once the first allele of the IgH locus has completed the rearrangement process, the resulting gene may be functional or not, depending upon whether the recombination has generated a sequence of nucleotides with an open reading frame. Because of the triplet organization of the **genetic code**, this happens only once every three rearrangements. If the first rearrangement is out of frame, the second allele will recombine, with the same probability of success. A second failure will lead to **cell death**. Conversely, once a functional gene has been generated, the resulting heavy chain will exert a negative feedback on a further rearrangement of the IgH locus, ensuring monoclonal expression of the μ chain. Much evidence suggests that the heavy chain must be expressed at the cell surface to regulate these events. Expression of the μ chain at the cell surface requires that it associate with another partner, which resembles the light chain, and for this reason it is named ΨL or surrogate light chain. First described in the mouse and then identified in humans, the surrogate light chain is composed of two polypeptide chains, encoded by the λ-like (or $\lambda 5$ in the mouse) and the VpreB genes, part of the regular IgL locus. The μ-ΨL complex becomes expressed at the surface of what is now a large preB cell, which also expresses CD10 and CD19, but no longer CD34. The cell is able to undertake light-chain gene rearrangements, which occur in the order $\kappa \rightarrow \lambda$. The recombination process is regulated in the same manner as that of the heavy chains, so there is only one light chain that is expressed in any given cell, either κ or λ (see **Kappa and lambda light** chains). The negative feedback on further rearrangements of the light chain genes is exerted by the complete IgM, which is now expressed at the surface of the immature B cell and has replaced the $\mu\Psi$L complex. At this point, the Igα–Igβ module is associated with the preB cell complex, strongly suggesting that the $\mu\Psi$L complex might be considered a "preB" receptor.

IMMATURE B CELLS ARE SELECTED BEFORE LEAVING THE BONE MARROW

Before leaving the bone marrow, immature B cells that express surface IgM are confronted with the local "self" antigenic environment. They are particularly sensitive to triggering by multivalent antigens, resulting in their clonal deletion by **apoptosis**. Alternatively, soluble self antigens do not cause the immature B cells to die, but instead induce an anergic state that does not seem to prevent the cells migrating to the periphery. It should be noted, however, that this negative selection has a threshold that leaves a fraction of autoreactive cells going to the periphery. These cells are responsible for the presence in the bloodstream of natural **autoantibodies** that must be of some physiological relevance. Once validated by this "quality control," immature B cells circulate to the periphery through blood vessels and lymphatic system, colonize the secondary lymphoid organs, and actively recirculate. They now have become mature B cells, expressing both IgM and IgD isotypes at their surface.

FINAL STEPS OF B-CELL DIFFERENTIATION TAKE PLACE IN THE PERIPHERY AND ARE ANTIGEN-DEPENDENT

The B lymphocytes that are now circulating in the periphery are nondividing, short-lived cells. To achieve their ultimate differentiation, they must be triggered by an antigen, most often T-cell-dependent (Fig. 1). Within the first days after antigenic stimulation and T-cell help, activated B cells may mature to plasma cells that have an abundant **endoplasmic reticulum** and secrete immunoglobulins. Activated B cells may also evolve to colonize the primary follicle of a lymph node and generate a germinal center, where they interact with follicular dendritic cells, divide rapidly, switch to another **isotype**, most frequently **IgG**, and start to accumulate **somatic hypermutations** that considerably amplify diversity and affinity. Clones of high affinity will be positively selected by antigen and can now progress toward plasma cell differentiation or become long-lived B **memory cells**.

CONCLUSIONS

Differentiation of B cells is a highly sophisticated process that takes place continuously in the bone marrow and results in the constant emergence of a very large repertoire of immunoglobulins, expressed both as membrane B-cell receptors (BCRs) and soluble antibodies. The numerous molecular events that lead from the hematopoietic stem cell to the mature B cell are under constant selective pressure, implying a high level of cell death. Therefore, the available repertoire appears, at any given time, to be a compromise between the necessary economy in the gene number, compensated by the recombination processes, and an unavoidable wastage due to the stochastic aspects of gene rearrangements and to the negative selection of clones having a high affinity for self components.

See also entries **Antibody**, **Clonal selection theory**, **Gene rearrangement**, **Immunoglobulin**, and **repertoire**.

Suggestions for Further Reading

P. D. Burrows and M. D. Cooper (1997) B cell development and differentiation. *Curr. Opin. Immunol.*, **9**, 239–244.

Y. J. Liu and C. Arpin (1997) Germinal center development. *Immunol. Rev.*, **156**, 111–126.

H. Karasuyama, A. Rolink, and F. Melchers (1996) Surrogate light chain on B cell development. *Adv. Immunol.*, **63**, 1–41.

A. Galy, M. Travis, D. Cen, and B. Chen (1995) Human T, B, natural killer and dendritic cells arise from a common bone marrow progenitor cell subset. *Immunity* **3**, 459–473.

J. Banchereau and F. Rousset (1992) Human B lymphocytes: phenotype, proliferation, and differentiation. *Adv. Immunol.* **52**, 125–262.

L. A. Herzenberg and A. B. Kantor (1993) B-cell lineages exist in the mouse. *Immunol. Today* **14**, 79–83.

B_{12} (COBALAMIN)

R. BANERJEE

Cobalamins are complex organometallic cofactors containing a central cobalt atom that is coordinated equatorially to four nitrogen atoms provided by the corrin ring. The two cofactor forms of this vitamin are *adenosylcobalamin* (AdoCbl) and *methylcobalamin* (MeCbl). A variety of other forms have been described in which the upper ligand is a water, *glutathione*, or cyano group, but the physiological significance of these forms is unknown. In nature, this cofactor is found in association with two **enzyme** subfamilies, the AdoCbl-dependent *isomerases* that catalyze 1,2 rearrangement reactions and the MeCbl-dependent *methyltransferases* that catalyze transmethylation reactions. Members of both subfamilies are fairly prevalent in the bacterial world, where the isomerases are involved in fermentative pathways, and the methyltransferases are involved in pathways leading to methionine, acetate, or methane. In mammals, only two B_{12}-dependent enzymes are known, the cytoplasmic *methionine synthase* and the mitochondrial *methylmalonyl-CoA mutase*. In this review, the molecular biological aspects of B_{12} biosynthesis, enzymology and diseases will be discussed briefly.

B_{12} BIOSYNTHESIS

The structural cousins, vitamin B_{12}, heme, and chlorophyll, are constructed from a common template that is derived from the precursor, 5-aminolevulinic acid. Methylation at C2 converts the common intermediate, uroporphyrinogen III, to precorrin-1 and commits it to B_{12} biosynthesis. Thereafter, a biosynthetic assembly line, which includes a series of methylations, ring contraction (between rings A and D), cobalt insertion, alkylation, amidations, and nucleotide loop assembly reactions, takes the cofactor to its final form. Two separate pathways lead to corrin ring biosynthesis in the aerobic and anaerobic worlds (for reviews, see Refs. 1–4). They run parallel at some steps and diverge at others. A unique structural feature of B_{12} is the presence of a nucleotide loop that terminates in dimethylbenzimidazole, which can serve as the lower axial ligand to cobalt. The pathways leading to the assembly of dimethylbenzimidaole in aerobic and anaerobic organisms are distinct, but their details remain to be elucidated (5). The **genes** encoding corrin biosynthesis functions have been **cloned** from *Pseudomonas denitrificans*, where they are organized in clusters scattered along the **genome** (reviewed in Ref. 3) and from *Salmonella typhimurium*, where they are clustered at 41 min (reviewed in Ref. 2). The **transcription factor** PocR regulates the transcription of the cobalamin biosynthetic and propanediol utilization (dependent on a B_{12} enzyme) genes in *S. typhimurium* (6,7).

B_{12} ENZYMES

The cofactor role of AdoCbl was first described by Barker and co-workers in glutamate mutase (8), followed a few years later by the discovery of MeCbl in methionine synthase (9). B_{12}-dependent enzymes catalyze chemically difficult reactions and manipulate the reactive cobalt–carbon bond in radically different ways (reviewed in Ref. 10). In the methyltransferase subfamily, cobalamin is the initial acceptor of the methyl group donated from substrates such as methyltetrahydrofolate and methanol, and alkyltransfer to and from the cobalamin occurs *via* heterolytic cleavage of the cobalt–carbon bond. In the isomerases, the cobalt–carbon bond is broken homolytically, as the enzymes catalyze 1,2 rearrangement reactions. The migration of a diversity of groups, ranging from carbon (in methylmalonyl-CoA mutase, glutamate mutase, and methylene glutarate mutase) to nitrogen (in ethanolamine ammonia lyase and β-lysine mutase) and oxygen (in diol dehydrase), is catalyzed by the isomerases. The genes encoding a number of methyltransferases and isomerases have been cloned and the structures of the B_{12}-binding domain of the *Escherichia coli* methionine synthase (11) and of the *Propionibacterium shermanii* methylmalonyl-CoA mutase (12) have been determined. In both enzymes, the cofactor is bound in an extended conformation in which the intramolecular base, dimethylbenzimidazole, is removed from the cobalt and replaced by a **histidine** residue donated by the protein. Another important member of the B_{12} family of enzymes is **ribonucleotide reductase**, which converts ribonucleotides to deoxyribonucleotides (13). The reaction mechanisms of B_{12}-dependent enzymes are discussed in a number of reviews (10,14–18).

B_{12}-RELATED DISEASES

Functional B_{12} deficiency can result from either nutritional insufficiency or from genetic defects (19,20). It is manifested clinically by a combination of symptoms including hematological and neurological abnormalities, methylmalonic aciduria and homocystinuria, depending on whether one or both B_{12}-dependent enzymes is affected. The genetic defects or inborn errors of cobalamin metabolism can result from impairments in uptake, transport, or enzymatic function and are inherited as autosomal recessive traits. Cobalamin absorption and transport abnormalities resulting from impairments in intrinsic factor, its receptor, or in transcobalamin (TC) II have been described. The cDNA encoding human intrinsic factor (21) and TC II (22) have been cloned and mapped, and genetic defects in TC-II in patient cell lines have been identified (23).

Intracellular cobalamin metabolism is complex, compartmentalized, and dependent on several enzymes. Defects in the early steps following internalization of TC II/cobalamin affect both MeCbl and AdoCbl syntheses and fall into the *cbl* C, D, and F genetic complementation groups. The identities of the proteins encoded by these loci and their precise functions are unknown. MeCbl synthesis is specifically compromised in *cbl*G and *cbl*E patients. Cloning of the cDNA encoding methionine synthase (24–26) and identification of mutations correlated with the *cbl*G phenotype (25,27) indicate that this locus represents methionine synthase. AdoCbl synthesis is specifically impaired in *cbl*A, B, and *mut* patients. The defect in *cbl*B and *mut* patients is in cobalamin adenosyltransferase and methylmalonyl-CoA mutase, respectively. The cDNA encoding methylmalonyl-CoA mutase has been cloned (28), and a number of pathogenic mutations have been described (29).

BIBLIOGRAPHY

1. A. Battersby (1994) *Science* **264**, 1551–1557.
2. A. I. Scott (1994) *Tetrahedron* **50**, 13315–13333.
3. F. Blanche, B. Cameron, J. Crouzet, L. Debussche, D. Thibaut, M. Vuilhorgne, F. J. Leeper, and A. R. Battersby (1995) *Angew. Chem. Int. Ed. Engl.* **34**, 383–411.
4. A. I. Scott (1996) *Pure Appl. Chem.* **68**, 2057–2063.
5. P. Renz (1998) In *Vitamin B_{12} and B_{12} Proteins* (B. Kraeutter, D. Arigoni, and B. T. Golding, eds.), Wiley-VCH, Weinheim, Germany.
6. T. A. Bobik, M. Ailion, and J. R. Roth (1992) *J. Bacteriol.* **174**, 2253–2266.
7. M. R. Rondon and J. C. Escalante-Semerena (1992) *J. Bacteriol.* **174**, 2267–2272.
8. H. A. Barker, H. Weissbach, and R. D. Smyth (1958) *Proc. Natl. Acad. Sci. USA* **44**, 1093–1097.
9. J. R. Guest, S. Friedman, D. D. Woods, and E. L. Smith (1962) *Nature* **195**, 340–342.
10. R. Banerjee (1997) *Chem. Biol.* **4**, 175–186.
11. C. L. Drennan, S. Huang, J. T. Drummond, R. Matthews, and M. L. Ludwig (1994) *Science* **266**, 1669–1674.
12. F. Mancia, N. H. Keep, A. Nakagawa, P. F. Leadlay, S. McSweeney, B. Rasmussen, P. Bösecke, O. Diat, and P. R. Evans (1996) *Structure* **4**, 339–350.
13. S. Licht, G. J. Gerfen, and J. Stubbe (1996) *Science* **271**, 477–481.
14. C. L. Drennan, R. G. Matthews, and M. L. Ludwig (1994) *Curr. Opin. Struct. Biol.* **4**, 919–929.
15. B. T. Golding and W. Buckel (1997) In *Comprehensive Biological Catalysis* (M. L. Sinnott, ed.), Vol. 4, Academic Press, London.
16. B. Babior (1988) *Biofactors* **1**, 21–26.
17. J. Stubbe (1988) *Biochemistry* **27**, 3893–3900.
18. R. G. Matthews, R. V. Banerjee, and S. W. Ragsdale (1990) *Biofactors* **2**, 147–152.

19. W. A. Fenton and L. E. Rosenberg (1995) *Inherited Disorders of Cobalamin Transport and Metabolism*, McGraw-Hill, New York, pp. 3111–3128.
20. M. I. Shevell and D. S. Rosenblatt (1992) *Can. J. Neurol. Sci.* **19**, 472–486.
21. B. K. Dieckgraefe, B. Seetharam, L. Banaszak, J. F. Leykam, and D. H. Alpers (1988) *Proc. Natl. Acad. Sci. USA* **85**, 46–50.
22. O. Platica, R. Janeczko, E. V. Quadros, A. Regec, R. Romain, and S. P. Rothenberg (1991) *J. Biol. Chem.* **266**, 7860–7863.
23. L. Qian, E. V. Qaudros, and S. P. Rothenberg (1997) *Meth. Enzymol.* **281**, 269–281.
24. Y. N. Li, S. Gulati, P. J. Baker, L. C. Brody, R. Banerjee, and W. D. Kruger (1996) *Human Mol. Genet.* **5**, 1851–1858.
25. D. Leclerc, E. Campeau, P. Goyette, C. E. Adjalla, B. Christensen, M. Ross, P. Eydoux, D. S. Rosenblatt, R. Rozen, and R. A. Gravel (1996) *Human. Mol. Genet.* **5**, 1867–1874.
26. L. H. Chen, M.-L. Liu, H.-Y. Hwang, L.-S. Chen, J. Korenberg, and B. Shane (1997) *J. Biol. Chem.* **272**, 3628–3634.
27. S. G. Gulati, P. Baker, B. Fowler, Y. Li, W. Kruger, L. C. Brody, and R. Banerjee (1996) *Human Mol. Genet.* **5**, 1859–1866.
28. R. Jansen, F. Kalousek, W. A. Fenton, L. E. Rosenberg, and F. D. Ledley (1989) *Genomics* **4**, 198–205.
29. F. D. Ledlay and D. S. Rosenblatt (1997) *Human Mutation* **9**, 1–6.

Suggestion for Further Reading

R. Banerjee, ed. (1999) *Chemistry and Biochemistry of* B_{12}, Wiley, New York.

B-CELL RECEPTOR (BCR)

MICHEL FOUGEREAU

Immunoglobulins can exist as **membrane proteins**, expressed at the surface of **B cells**. When B cells differentiate in the bone marrow from the initial hematopoietic **stem cells**, they reach a stage, called *immature B lymphocytes*, that is characterized by the presence of **IgM** at the cell surface (Fig. 1). This IgM, which is a basic $\mu 2\kappa 2$ or $\mu 2\lambda 2$ monomer, is anchored into the membrane bilayer by an extra portion at the COOH-terminus of the heavy chains that is not present in the soluble form. Differences in length result simply from the **alternative splicing** of membrane exons (see **Introns/exons**) present in the germline. The **transmembrane** region is terminated by only three amino acid residues located in the cytoplasm of the B cell.

Because the surface Ig clearly acts as a B-cell receptor (BCR) upon triggering by the antigen, the necessity of a signaling module, similar to CD3 in **T cells**, is apparent. This was indeed the case. The BCR-activating module is a heterodimer composed of one Igα and one Igβ chain (also designated as CD79 a and b), each of which has an external Ig-like domain, a transmembrane portion, and an cytoplasmic region that contains several signaling motifs, called *immunoreceptor tyrosine-based activation motif* (ITAM). ITAM were first described on polypeptide chains of CD3 and consist of two Tyr–X–X–Leu/Ile sequences separated by a stretch of seven amino acid residues. They are found in a number of **signal transduction** modules associated with most crucial receptors of the immune system, such as the **T-cell receptor** (TCR), BCR, and FcR. Upon stimulation of BCR with cross-linking antigens, ITAM interact with protein **tyrosine kinases** of the **src** family (p59^{fyn}, p53^{lyn}, p56^{lyn}, p55^{lck}), leading to the translocation of p72^{syk} to the membrane and activation of the **ras** pathway and **phospholipase C**γ**2**. As a consequence, the BCR is internalized, which is the first step of processing of antigens by B cells. Full activation of B cells require other signals, involving molecules other than the BCR.

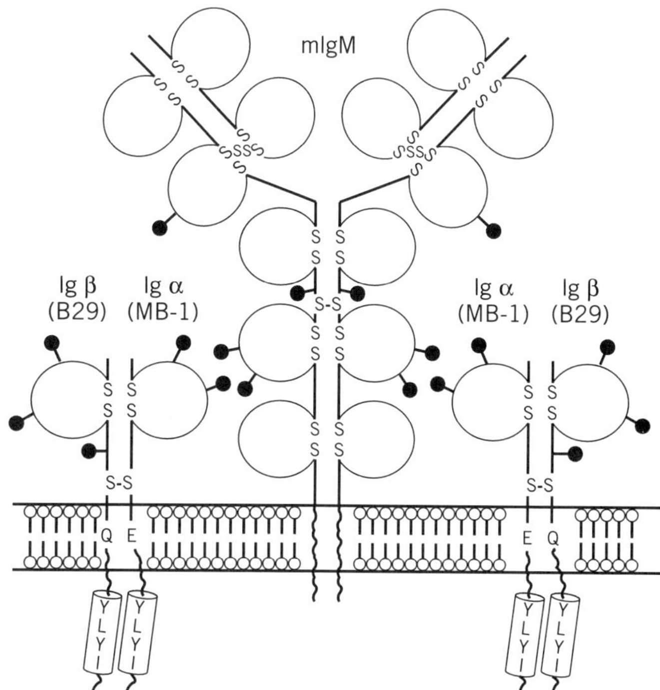

Figure 1. Schematic representation of the B-cell receptor (BCR). The BCR is composed of one antigen binding module, the surface IgM, connected to the signaling module, a heterodimer containing the Igα and Igβ, encoded by the MB-1 and B-29 genes, respectively. Note the Ig domain organization of the ectodomains of the Igα and Igβ chains that contain in their cytoplasmic part the ITAM (immunoreceptor tyrosine-based activation motif) motifs that play a central role in signaling by contacting intracellular protein tyrosine kinases.

Suggestion for Further Reading

M. Reth (1995) The B cell receptor complex and co receptors. *Immunol. Today* **16**, 310–313.

B-DNA

A. H.-J. WANG
H. ROBINSON

B-DNA is the predominant type of **DNA structure** under physiological conditions. The B-DNA double helix is right-handed, with its **base pairs** perpendicular to the helix axis that passes through the center of the base pairs. The major groove and minor groove are roughly equivalent in depth, 8.5 and 7.5 Å, respectively, whereas the width of the major groove is ~12 Å and that of the minor groove is ~6 Å.

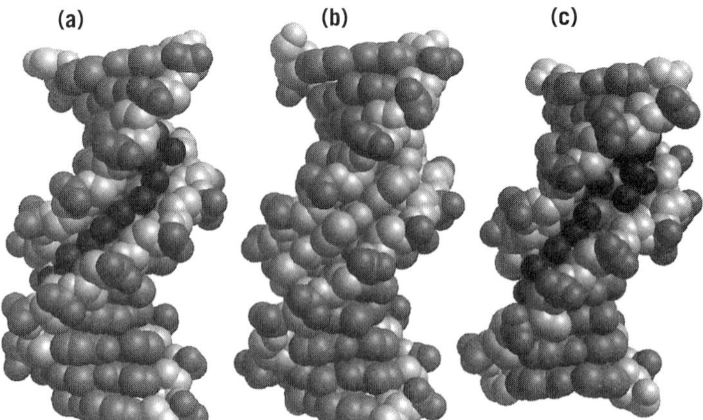

Figure 1. Examples of the structure of B-form DNA. (**a**) Crystal structure of the double-stranded dodecamer d(CGCGAATTCGCG) (Protein Data Base BDL084), with the water molecules in the minor groove shown as spheres. (**b**) Crystal structure of the double-stranded d(CGCAAATTTGCG)-distamycin A complex (Protein Database GDL003). The drug displaces the waters in the minor groove. (**c**) Crystal structure of the double-stranded decamer d(CCAGGCCTGG) (Protein Database BDJS30), with the waters bound to the minor groove shown as spheres. See color insert.

The first **X-ray crystallography** structure of B-DNA (Fig. 1A) was that of a dodecamer sequence $d(CGCGAATTCGCG)_2$ that contains the EcoRI **restriction** sequence GAATTC (1). There are a number of interesting structural features associated with this B-DNA structure. A distinctive feature is that the minor groove is narrower at the AATT region of the double helix than the CGCG ends. The narrow minor groove at the AATT region is filled by a spine of **water** molecules that form **hydrogen bonds** to both the O2 of thymines and the N3 of adenines, although this spine of waters has been reinterpreted recently using a higher resolution structural analysis as a spine of waters on sodium ions (2). Additionally the base pairs in the central AATT region of the helix have high propeller twist angles. The propeller twisting enhances the stacking of the bases along each strand of the double helix, and the AT base pairs are in general less restrictive to propeller twisting because of fewer hydrogen bonds than the GC base pairs (3). Lastly, the sugar pucker of the deoxyribose ring favors the C2′-*endo* conformation, although a range of conformations from C1′-exo to O4′-*endo* is also present. This may reflect the greater flexibility associated with B-DNA structures.

The narrow minor groove region associated with the A-T sequences in B-DNA affords an excellent binding site for the minor groove-binding drugs, such as distamycin A, netropsin, Hoechst 33258, Hoechst 33342, and DAPI. The crystal structure and solution structures of the complexes of several DNA oligonucleotides with those minor groove-binding drugs, most of them in 1–1 drug-to-duplex complexes (Fig. 1B), have been determined (4,5). Those structures revealed that the drug replaces the spine of hydration in the narrow minor groove and stabilizes the DNA structure, without perturbing the overall conformation significantly. The narrow minor groove associated with the central AATT base pairs is essential for the 1–1 binding mode of drugs. The binding energy derives in part from the gain in **entropy** associated with the displacement of the water molecules. The sequence preference for A-T regions by minor groove-binding drugs is due to the greater negative electrostatic potential at the bottom of the minor groove at A-T regions. Finally the presence of N2 amino groups of guanines provides both a charge and a steric hindrance to drug binding.

NMR study of the binding of distamycin A (a pyrrole-containing anti-tumor antibiotic) to DNA containing A-T sequences revealed a more complex pattern. Distamycin A can bind DNA not only in a 1–1 drug/duplex complex but also in a 2–1 mode, with two distamycins bound to the minor groove in an antiparallel, side-by-side manner (6). The latter binding mode requires that the minor groove width at the binding site be expanded so as to accommodate two drug molecules, reflecting the flexibility of B-DNA. This new finding has stimulated an active study in the design of new minor groove-binding compounds that can bind to all four base pairs, (A-T, T-A, C-G, and G-C), with high specificity. It was found that compounds with imidazole-containing units do not have a great discriminating power regarding the recognition toward G/C versus A/T base pairs. However, compounds that combine the imidazole units with the pyrrole units can be designed to possess excellent sequence-specific binding properties. Rules for such designs have been proposed recently (7).

Additional work on the high-resolution crystal structures of several DNA decamer oligonucleotides with "mixed" sequences, including d(CCAGGCCTGG) (Fig. 1C), showed that the minor groove width of those structures in general is wider, but with some variations. Those DNA decamers have a slightly narrow or wide minor groove, depending on the sequence. In general, the A-T regions have a narrower minor groove than the G-C regions. The **hydration** structure in the groove is dependent on the groove width. It was noted that a single spine of water molecules along the floor of the minor groove is associated with a narrow minor groove, whereas a ribbon of double water molecules, bridging the base edge N or O atom to the O4′ atoms of the sugar ring, is associated with the wide minor groove.

BIBLIOGRAPHY

1. R. M. Wing, H. R. Drew, T. Takano, C. Broka, S. Tanaka, K. Itakura, and R. E. Dickerson, (1980) *Nature* **287**, 755–758.
2. X. Shui, L. McFail-Isom, G. G. Hu, and L. D. Williams (1998) *Biochemistry* **37**, 834–8355.
3. C. R. Calladine (1982) *J. Mol. Biol.* **161**, 343–352.
4. M. Coll, C. A. Frederick, A. H.-J. Wang, and A. Rich (1987) *Proc. Natl. Acad. Sci. USA* **84**, 8385–8389.
5. H. Robinson, Y.-G. Gao, C. Bauer, C. Roberts, C. Switzer, and A. H.-J. Wang (in press) *Biochemistry*, and references cited therein.
6. J. G. Pelton and D. E. Wemmer (1989) *Proc. Natl. Acad. Sci. USA* **86**, 5723–5727.
7. S. White, J. W. Szewczyk, J. M. Turner, E. E. Baird, and P. E. Dervan (1998) *Nature* **391**, 468–471.

BACKBONE

JENNIFER MARTIN

Backbone, or main chain, is the general term used to describe the connecting chain in **polymers**. Different kinds of polymers have different chemical backbones. For example, in **proteins** the backbone is a **polypeptide chain**, but **nucleic acids** have a sugar phosphate backbone. The backbone of a polymer, which may adopt a regular structural **conformation**, is the constant or repeating part of the polymer, as opposed to the attached **side chains**, which can be variable.

[See also **Polymer** and **Side chain**.]

Suggestions for Further Reading

L. Mandelkern (1983) *An Introduction to Macromolecules*, Springer-Verlag, New York.

P. Munk (1989) *Introduction to Macromolecular Science*, Wiley-Interscience, New York.

BACTERIOCINS

C. J. LAZDUNSKI
V. GELI

Broadly defined, bacteriocins are substances produced by one **bacterium** that adversely affect another. Most of them are **peptide antibiotics** that are synthesized on **ribosomes**, as in normal **protein biosynthesis**. Others, such as bacitracin and **gramicidin**, are synthesized in bacteria by multienzyme complexes or sequential enzyme reactions (1). In a strict sense, bacteriocin is an abbreviation of bacterial toxin (with antibacterial activity). Many bacterial toxins act against **eukaryotic** cells and are not called bacteriocins. Some of them, however, like **diphtheria**, tetanus or **cholera** toxins, resemble bacteriocins, such as **colicins** or pyocins, in having domain structures that convey the binding and toxic activities specific to each molecule (2,3). The current basis for allocating a name to the agent responsible for bacteriocinlike activity produced by **Gram-positive** or -negative bacteria is to adopt some derivation of either the genus or species name of the producer strain, together with an alphabetical and/or numerical code designation specifying that strain. To avoid confusion, bacteriocins that have the same amino acid sequence, irrespective of the species of origin, should have the same name, that first published. There is a wide variety of locations of the genetic determinants of bacteriocins. They can be encoded by conjugative or nonconjugative **plasmids**, by conjugative **transposons**, or be encoded on the **chromosome** (4).

Because of the different cell-wall structures of gram-positive or -negative bacteria, their bacteriocins have evolved differently in size and specificity. In gram-negative bacteria, the outer membrane necessitates receptor-mediated antagonistic activities, and very specific proteins are produced with domains for receptor binding, translocation, and activity (see **Colicin**). In contrast, gram-positive bacteria possess a multilayered peptidoglycan wall without an outer membrane. This favors peptides of small size that penetrate the murein network without receptor binding and specific translocation. Consequently, bacteriocins produced by gram-positive bacteria have a broad host range and are sometimes active on taxonomically unrelated genera. Thus, many of the bacteriocinlike agents produced by gram-positive bacteria kill species other than those likely to have the same ecological niche. Production of antibiotic peptides is the rule in gram-positive bacteria. Proteins the size of colicins are hardly ever encountered. In general the structures of these peptides must be stabilized by **posttranslational modifications**.

Presently there is a burgeoning interest in bacteriocins because of the possible applications in the food industry (particularly lactic acid bacteria) and as a strategy for preventing certain infectious diseases. This review gives the general features of the main classes of bacteriocins and emphasizes the common features that are beginning to emerge.

An organism that produces a peptide antibiotic must be able to (1) synthesize the antibiotic, (2) export the antibiotic into the extracellular medium, (3) protect itself from the action of the antibiotic, and (4) interact with the sensitive cell and interfere with its growth or survival. This interaction may or may not require entry of the antibiotic into the cell. Such entry is characteristic of some of the better-characterized ribosomally synthesized low molecular weight bacteriocins. For a detailed description of these bacteriocins, the reader is referred to the Suggestions for further reading.

CLASSES OF BACTERIOCINS

A list of representative examples of lantibiotic and non-lantibiotic bacteriocins, together with their organisms of origin and mode of action is presented in Table 1. Four distinct classes of bacteriocins from lactic acid bacteria have been defined (5).

Class I. The lantibiotics (lanthionine-containing peptides) are small antibiotic peptides that are distinguished from other bacteriocins by their content of dehydro and thioether amino acids (lanthionine and 3-methyllanthionine) (6). Two subgroups have been defined on the basis of their distinctive ring structure. Type A comprises screw-shaped, amphipathic molecules that have molecular masses of 2.2 to 3.5 kDa and two to seven net positive charges. Type B consists of more globular molecules that have molecular masses of 2 kDa and either no net charge or a net negative charge. The lantibiotics include a growing list of modified peptides that

Table 1. Representative Examples of Lanthionine-Containing Bacteriocins

Bacteriocin	Type	Mass	Organism of Origin	Mode of Action (References)
Nisin A	L-A	3353	*Lactococcus lactis*	Pore formation (8)
Pep5	L-A	3488	*Staphylococcus epidermis*	Pore formation (44)
Subtilin	L-A	3317	*Bacillus subtilis* ATCC 6633	Pore formation (45)
Epilancin K7	L-A	3032	*Staphylococcus epidermis*	Pore formation (46)
Epidermin	L-A	2164	*Staphylococcus epdermis*	Pore formation (47)
Gallidermin	L-A	2164	*Staphylococcus gallinarum*	Pore formation (48)
Lacticin 481	L-A	2901	*Lactococcus lactis*	Pore formation (49)
Streptococcin A-FF22	L-A	2795	*Streptococcus pyognes*	Pore formation (50)
Salivaricin A	L-A	2315	*Streptococcus salivarus*	Pore formation (51)
Mutacin	L-A	3245	*Streptococcus mutans*	Pore formation (52)
Lactocin S	L-A	3764	*Lactobacillus sake*	Pore formation (36)
Carnocin U149	L-A	4635	*Carnobacterium piscicola*	Pore formation (53)
Cytolysin L1	L-A	4164	*Enterococcus faecalis*	Pore formation (54)
Cinnamycin	L-B	2042	*Streptomyces cinnamoneus*	Membrane disorganization (55)
Duramycin	L-B	2014	*Streptomyces cinnamoneus*	Membrane disorganization; pore formation (55)
Mersacidin	L-B	1825	*Bacillus* sp.	Cell-wall synthesis (56)
Actagardine	L-B	1890	*Actinoplanes* sp.	Cell-wall synthesis (57) inhibition

have unique structures produced by bacilli, lactococci, lactobacilli, staphylococci, streptococci, and streptomyces (Fig. 1). The functions of the modified amino acids are twofold. First, these small peptide molecules are less likely to achieve stable conformations by the same means as larger proteins. Yet, they are generally extremely heat stable. Unusual cross-links like the thioether bridge from lanthionine allow the molecules to achieve and retain their proper folded structure despite their small size (7). Secondly, unusual sidechains, such as dehydroalanine, provide a wider repertoire of chemical reactivity that is be important for the biological activity of these peptides (8).

Class II. These bacteriocins are small (<10 kDa), relatively heat stable, non-lanthionine-containing peptides that are membrane-active. They have been subdivided into Class IIa, *Listeria*-active peptides with the N-terminal consensus sequence Tyr-Gly-Asn-Gly-Val-Xaa-Cys-; Class IIb, poration complexes requiring two different peptides for activity; Class IIc, thiol-activated peptides requiring reduced **cysteine** residues for activity. Unmodified peptide antibiotics (not containing lanthionine) share several characteristic features. They are produced in a prepeptide that has an N-terminal extension (**leader peptide**). Leader peptides are amphiphilic and lack the distinctive **signal peptide** hydrophobic region of the **sec**-dependent exported proteins (see **Protein targeting**). The leader peptides feature some sequence homology, particularly at the proteolytic processing site, where two glycine residues are found at position -1 and -2 with respect to the cleavage site (4,9). The processed, active bacteriocins contain a substantial amount of **hydrophobic** amino acids and can form **amphiphilic** helices. Accordingly, the plasma membrane is the target of most of these bacteriocins.Flanking regions upstream and downstream of the structural **gene** of non-lantibiotic peptides contain additional **open reading frames** for which a function in bacteriocin production was either demonstrated or postulated on the basis of sequence homologies. Such genes code for immunity peptides, for two-component regulatory proteins consisting of a histidine kinase and the respective regulator, and for proteins with significant homology to the ABC transporter family. Some of these genes have an operon-like organization.

Class III. These bacteriocins are large (>30 kDa) heat-labile proteins that include many bacteriolytic extracellular enzymes (hemolysin and muramidases) that mimic the physiological action of bacteriocins.

Class IV. These are complex bacteriocins that contain essential lipid or carbohydrate moieties in addition to protein. Chemical analysis of the purified antibacterial components is still necessary to confirm that the presence of the additional moieties is indeed essential to the biological activity of these molecules, thereby justifying the establishment of this class IV.

Microcins

An additional class of bacteriocins comprises a family of antibiotic substances called "microcins" produced by diverse members of the *Enterobacteriacae* (10). They are distinguished from the majority of **colicins** by much lower molecular weight (<10 kDa) and because their synthesis is not induced by conditions that lead to induction of the **SOS repair** pathway. Instead, they are synthesized during the stationary phase, similar to that observed with most conventional antibiotics (11). The microcins were operationally defined as substances produced by gram-negative bacteria that pass through a cellophane membrane and inhibit the growth of an indicator *Escherichia coli* strain (12). Several substances known to be ribosomally synthesized fall within this group. The best

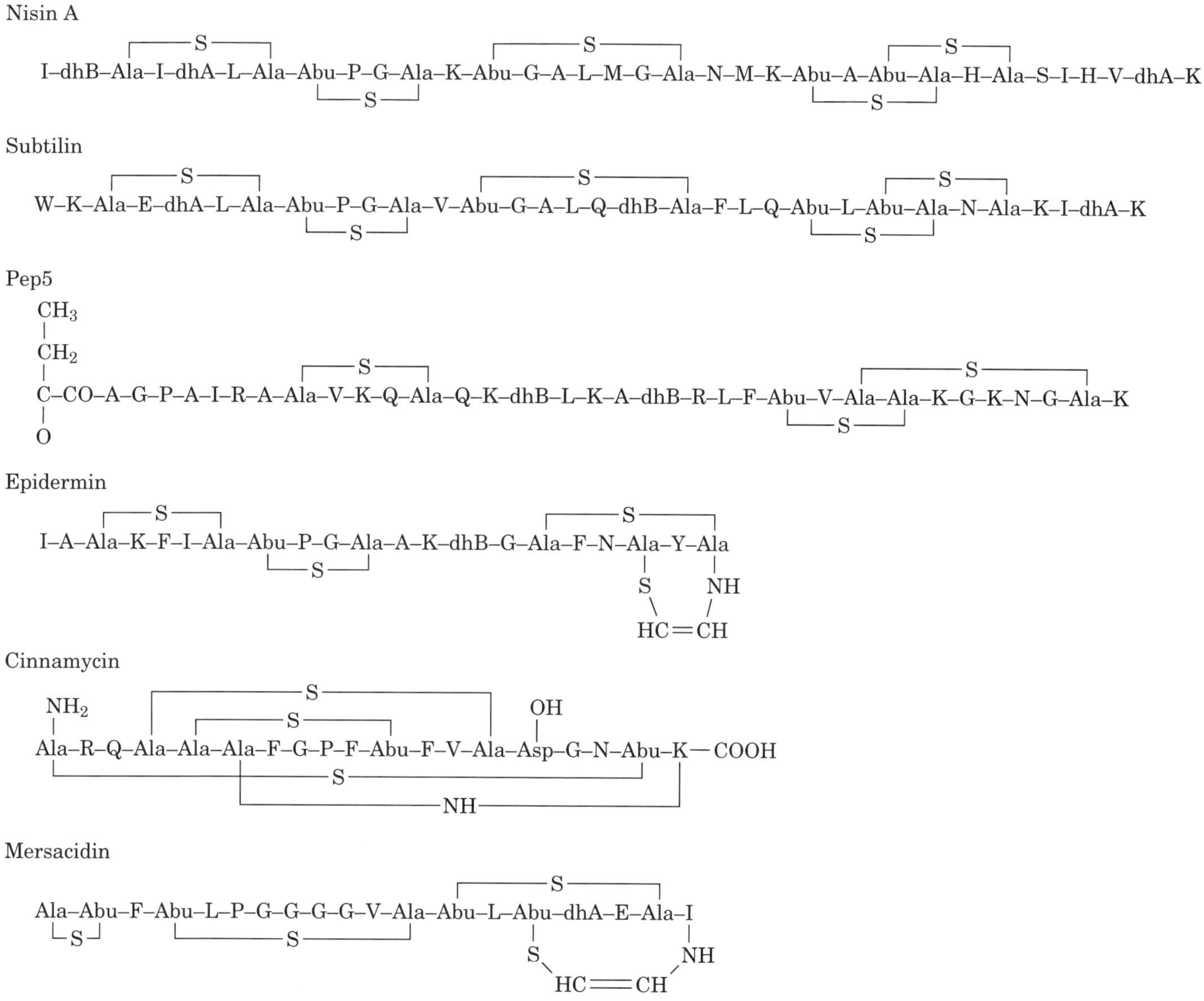

Nisin A

I–dhB–Ala–I–dhA–L–Ala–Abu–P–G–Ala–K–Abu–G–A–L–M–G–Ala–N–M–K–Abu–A–Abu–Ala–H–Ala–S–I–H–V–dhA–K

Subtilin

W–K–Ala–E–dhA–L–Ala–Abu–P–G–Ala–V–Abu–G–A–L–Q–dhB–Ala–F–L–Q–Abu–L–Abu–Ala–N–Ala–K–I–dhA–K

Pep5

CH_3–CH_2–C(=O)–CO–A–G–P–A–I–R–A–Ala–V–K–Q–Ala–Q–K–dhB–L–K–A–dhB–R–L–F–Abu–V–Ala–Ala–K–G–K–N–G–Ala–K

Epidermin

I–A–Ala–K–F–I–Ala–Abu–P–G–Ala–A–K–dhB–G–Ala–F–N–Ala–Y–Ala (S–HC=CH–NH)

Cinnamycin

NH_2–Ala–R–Q–Ala–Ala–Ala–F–G–P–F–Abu–F–V–Ala–Asp(OH)–G–N–Abu–K—COOH (S, S, S, NH bridges)

Mersacidin

Ala–Abu–F–Abu–L–P–G–G–G–G–V–Ala–Abu–L–Abu–dhA–E–Ala–I (S–HC=CH–NH)

Figure 1. Primary structure of representative lantibiotics. Abbreviations used: dhA, didehydroalanine; dhB, didehydrobutyrine; Ala-S-Ala, lanthionine: Abu-S-Ala, ß-methyl lanthionine; Ala-NH-Lys, lysinoalanine; Asp-OH, hydroxyaspartic acid.

known of these antibiotics peptides are microcin C7, microcin B17, and colicin V. Microcin C7 is a heptapeptide containing modifications at both the N- and C-termini that block protein synthesis *in vivo* and *in vitro* (13). Microcin B17 (Mcc B17) is a 43-residue peptide containing posttranslational modifications at the peptide backbone including serine, cysteine, and glycine residues, that result in thiazole and oxazole rings (14). Mcc B17 has much in common with lantibiotics because it is derived from a precursor peptide that is posttranslationally modified; 26 out of the 43 residues are glycine. Mcc B17 inhibits DNA replication and induces the SOS response (15). The target of this antibiotic is a DNA gyrase (16) (see **DNA topology**). Colicin V (ColV) was first described in 1925 in the first report of an antibiotic substance produced by *E. coli* (17). It has a molecular weight of only 6000. The structure of mature colicin V is not totally known, even regarding side-chain modifications or N-terminal processing. Colicin V kills sensitive cells by disrupting their membrane potential (18).

BIOSYNTHESIS AND POSTTRANSLATIONAL MODIFICATIONS

Generally the low molecular weight bacteriocins of gram-positive bacteria are first formed in an inactive precursor form that has a leader peptide. Following a variety of posttranslational modification reactions, the C-terminal propeptide domain is cleaved from the N-terminal leader sequence to yield the mature antimicrobial molecule. These prepeptides range from 18 to 30 amino acid residues and are not highly homologous except near the cleavage site. The prepeptides of all non-lanthionine-containing bacteriocins characterized thus far have two Gly residues at positions -2 and -1 relative to the processing site and *β*-**turn** promoting residues near the cleavage site. Possible functions of the prepeptide include (1) stabilizing the propeptide during **translation**; (2) keeping the bacteriocin inactive, particularly after completion of modifications and thus helping to protect the producing strain; (3) allowing recognition of the ABC transporter system; (4)

Table 2. Non-Lanthionine-Containing Bacteriocins

Bacteriocin	Type	Mass	Organism of Origin	Mode of Action (References)
Pediocin PA-1	NLC	4600	*Pedicoccus acido lantici* PAC 1.0	Pore formation (58)
Mesentericin Y-105	NLC	3666	*Leuconostoc mesenterides* Y105	Pore formation (59)
Carnobacteriocin A	NLC	5100	Carnobacterium piscicola LV17A	? (60)
Sakacin A	NLC	4308	*Lactobacillus sake* LB706	? (61)
Lactacin F	NLC	5600	*Lactobacillus* sp.	Pore formation (62)
Lactococcin A	NLC	5800	*Lactococcus lactis* (cremoris)	Pore formation (63)

directing the precursor through a specific recognition motif toward biosynthetic enzymes (modification enzymes); (5) interacting with the propeptide region to stabilize a conformation that is essential for correct modification and thioether formation. There is evidence that the modifications of lantibiotics are made at the prepeptide stage. Cross-similarities between leaders of the modified and unmodified bacteriocins indicate that the overall features of leader peptides are important during regulation, synthesis, and generation of the immunity of peptide bacteriocins, regardless of whether or not they are modified.

The primary amino acid sequences of the propeptide components of many of the low molecular weight bacteriocins produced by gram-positive bacteria are known now. One important feature of the bacteriocins of gram-positive bacteria is their cysteine content. Those in which one or more cysteine residues are linked to dehydrated serine and threonine residues to form the thioether-linked amino acids lanthionine and methyl-lanthionine are called lantibiotics as previously mentioned (Fig. 1). Alternatively, bacteriocins in which pairs of cysteine residues undergo modification to form **disulfide bonds** are called cystibiotics (cysteine-containing antibiotics). A third subgroup of the bacteriocins, of which lactococin B is an example, are designated thiolbiotics, because they contain only a single cysteine residue that must be present in the reduced **thiol** form to be active (19). At neutral pH, many of the low molecular weight bacteriocins are cationic. This is a unifying feature of lantibiotic and non-lanthionine containing bacteriocins and may have some significance for their activity.

Little is known about the posttranslational modification reactions and the biosynthetic enzymes involved in the maturation of lantibiotics. The lantibiotic B and C genes (designated *Lan*B and *Lan*C, as stipulated by the 1st and 2nd International Workshops on Lantibiotics) are probably involved in the maturation pathways, because mutation studies indicate that the gene products are essential for producing functional lantibiotics. The products of these genes, designated LanB and LanC, but also known as NisB and NisC, respectively, may be involved in the dehydration and thioether bond formation. Limited similarity was reported between LanB and *E. coli* IlvA, a threonine dehydratase, thus suggesting a similar function (20). This leads to the hypothesis that LanC and other similar proteins are involved in the enzyme-catalyzed formation of thioether bonds from the dehydrated residues and cysteine. Another modification enzyme that has been identified is EpiD, which is encoded by the epidermin gene cluster located on the 54-kb plasmid pTu32 (21). EpiD has no homologue in other lantibiotic gene clusters, and it catalyzes the biosynthesis of the C-terminal aminovinylcysteine residue of epidermin (22).

EXTRACELLULAR AND MATURATIONAL RELEASE

After biosynthesis, class I and II bacteriocins are released to the extracellular medium. For each bacteriocin there is a relatively specific membrane protein system whose function is to translocate a precursor form of the antibiotic across the cytoplasmic membrane to the outside of the cell. The secretion of several peptide bacteriocins is mediated by dedicated transmembrane translocators belonging to the ATP-binding cassette (ABC) transporter superfamily (23–26). The transporters are encoded in the same operons as the bacteriocin structural gene or on a neighboring operon. The C-terminal ATP-binding domain and the N-terminal hydrophobic integral membrane domain are expressed as a single polypeptide or as separate polypeptides. Recently, strong evidence has been reported that, very likely, all precursor peptides of the lantibiotic and non-lantibiotic bacteriocins that have leader peptides of the double glycine type (see above) are processed by their dedicated ABC-transporters, concomitant with export (27). It has also been proposed that the N-terminal domains of bacteriocin transporters are essential for initial recognition and subsequent proteolytic processing. These new proteolytic enzymes are **thiol proteinases**. In fact, genes for potential ABC transporters have been found in all the known lantibiotic gene clusters. The lantibiotic transporters range in size from 535 to 714 amino acid residues and generally have the ATP-binding **domain** and the membrane-spanning domain in the same single protein. Sometimes a second ABC transporter system has been identified, which is not essential for production but participates in producer self-protection, thus providing some type of additional immunity (28). Although the leader sequence of the double-glycine type may be cleaved concomitant with export by the ABC-transporter itself, this is not the general case.

The ABC transporters have thiol proteinase activity, and putative proteinase genes (designated P) have been described for the leader peptides of other lantibiotics. These gene products share similarities with **subtilisin**-like **serine proteinases**. For example, the *Nis*P gene encodes an extracellular serine proteinase with a C-terminal extension anchoring it on the outer side of the cytoplasmic membrane (29). With nisin, it has been shown that cleavage of the leader peptide is not a prerequisite for export by the ABC-transporter, and this conclusion may be extended to other peptides with antimicrobial activity (30).

A detailed maturational pathway has been proposed for nisin (31). First the inducing signal (which could be nisin itself) activates the transcription at the *nis*A promoter (which has features also found in other positively regulated **promoter** sequences) via a two-component response-regulator system

(32). This results in the production of pre-nisin containing free cysteines and no dehydrated residues. The pre-nisin is directed, presumably by virtue of the leader peptide, to a membrane-located complex containing the modifying enzymes LanB (probably involved in dehydration) and LanC (probably involved in thioether bonds, as previously mentioned). At this stage, the leader helps to maintain the peptide in an inactive form. Subsequently, the precursor nisin is translocated via the ABC exporter NisT at the expense of ATP hydrolysis. Finally, precursor nisin is activated by proteolytic cleavage by the extracellular protease NisP attached to the outside of the cell envelope. The role of the leader is still not fully understood, apart from its function in producing an inactive conformation. In particular, it is not known how it functions in targeting the pre-lantibiotic to the maturation and export proteins, what its fate is after cleavage, and how it contributes to processes, such as self protection, after it has been cleaved off.

This proposed sequence of events may also apply to other lantibiotics. However, additional intracellular conversions are required in specific modification reactions, such as those involving the N- and C-termini of epidermin (33), lantibiotic Pep5, epilamin K7 (34,35), and lactocin S (36). In these cases it is likely that the leader peptide cleavage occurs intracellularly.

MODES OF ACTION

The various structural properties of the subgroups of bacteriocins are reflected in their three different modes of action. The primary activity of Class I type A lantibiotics is based on forming voltage-dependent, short-lived pores in the cytoplasmic membrane (37,38). The peptides rapidly induce leakage of ions and small metabolites from bacterial cells and a collapse of the electrochemical proton gradient, leading to cessation of biosynthetic processes and eventually to cell death. Type A lantibiotics require a membrane potential of between 50 and 100 mV for pore formation, depending on the individual peptide. It is assumed that pores are formed by a transiently associated peptide oligomer in a transmembrane orientation (barrel stave model), as suggested for alamethicin. The susceptibility toward a particular peptide of different bacterial species, or even of strains within one species, varies much more than one would expect on the basis of the pore-formation model. *In vivo* pore formation or pore stability may be positively or negatively influenced by such factors as phospholipid composition of the membrane, interactions of the peptides with integral membrane components, or the presence of surface layers. The models may need to be refined, and pore formation may depend on local perturbation of the bilayer (39). In addition, the peptides could exert secondary effects that contribute to bactericidal activity, such as autolysis of cells by activating cell-wall hydrolyzing enzymes (40).

The antibacterial effects of class I type B lantibiotics are rather weak. They bind to the head groups of phospholipids, preferentially to phosphoethanolamine, which may cause membrane permeabilization and allow the release of cations and other small solutes. Duramycin also induces the formation of complex pores (41). Mersacidin and actagardine are distinguished from other type B lantibiotics by their mode of action. Both interfere with cell-wall biosynthesis in gram-positive bacteria, which may eventually offer new possibilities in antimicrobial therapy (42).

Generally, the bactericidal action of class II bacteriocin against sensitive cells is produced principally by destabilizing membrane function, such as energy transduction, rather than disrupting the structural integrity of the membrane. This effect results from the energy-independent dissipation of the **proton motive force** and loss of the permeability barrier of the cytoplasmic membrane. It contrasts with the energy-dependent bactericidal action of the lantibiotics. In addition, before pores are formed, all of the non-lanthionine containing bacteriocins interact with membrane-associated receptor proteins, in contrast to class I lantibiotic bacteriocins.

IMMUNITY OF BACTERIOCIN-PRODUCING CELLS

One of the definitive features of bacteriocin-producing cells is their ability to resist the action of their own inhibitory substances through a specific immunity mechanism. Such a mechanism is based on dedicated peptides or proteins called immunity proteins, which specifically antagonize the bacteriocin. For nisin and subtilisin, their antagonist proteins of 245 and 165 amino acid residues, respectively, display the features of bacterial lipoproteins and are similar in amino acid sequence. Despite the high degree of similarity of these lantibiotics, there is no cross-immunity between producing cells. In other cases, as with pepS, the immunity is much more closely related to the immunity systems of unmodified bacteriocins, which are short, have only 69 residues, and have a hydrophobic N-terminal segment and a strongly hydrophilic C-terminal part.

Currently, there are no clues to the mechanism of the immunity phenomenon. The proposed location of the peptides outside the cell and the observation that the cytoplasmic membrane of immune clones is not depolarized by externally added bacteriocin suggest that the bacteriocins are directly antagonized by the immunity peptide, like colicin A, for example (43).

BIBLIOGRAPHY

1. H. Kleinhauf and H. von Döhren (1990) *Eur. J. Biochem.* **192**, 1–15.
2. M. Kageyama, M. Kobayashi, Y. Sano, and H. Masaki (1996) *J. Bacteriol.* **178**, 103–110.
3. H. Bénédetti and V. Géli (1996) In *Handbook Biological Physics* (W. N. Konings, H. R. Kaback, and J. S. Lolkema eds.), Elsevier, Amsterdam, The Netherlands, Vol. 2, pp. 665–691.
4. H. G. Sahl (1994) In *Antimicrobial Peptides* (Ciba Foundation Symposium 186), Wiley, Chichester, pp. 27–53.
5. T. R. Klaenhammer (1988) *Biochimie* **70**, 337–349.
6. G. Jung (1991) *Angew. Chem. Int. Ed. Engl.* **30**, 1051–1068.
7. E. Gross and J. L. Morell (1967) *J. Amer. Chem. Soc.* **53**, 2791–2792.
8. E. Gross and J. L. Morell (1971) *J. Amer. Chem. Soc.* **93**, 4634–4635.
9. W. M. de Vos, O. P. Knipers, J. R. van der Meer, and R. J. Siezen (1995) *Mol. Microbiol.* **17**, 427–437.
10. F. Baquero and F. Moreno (1984) *FEMS Microbiol. Lett.* **23**, 117–124.
11. J. F. Martin and A. L. Demain (1980) *Microbiol. Rev.* **44**, 230–251.
12. C. Asensio, C. Perez-Diaz, M. C. Martinez, and F. Baquero (1976) *Biochem. Biophys. Res. Commun.* **69**, 7–14.

13. R. Kolter and F. Moreno (1992) *Ann. Rev. Microbiol.* **46**, 141–163.
14. A. Bayer, S. Freund, C. Nicholson, and C. Jung (1993) *Angew. Chem. Int. Ed. Engl.* **32**, 1336–1339.
15. M. Herrero and F. Moreno (1986) *J. Gen. Microbiol.* **132**, 393–402.
16. J. L. Vizan, C. Hernandez-Chico, I. del Castillo, and F. Moreno (1991) *EMBO J.* **10**, 467–476.
17. A. Gratia (1925) *C.R. Soc. Biol.* **93**, 1040–1041.
18. C. Yang and J. Konisky (1984) *J. Bacteriol.* **158**, 757–759.
19. G. Bierbaum and H. Sahl (1991) In *Nisin and Novel Antibiotics* (G. Jung and H. G. Sahl, eds.), Escom Publishers, Leiden, The Netherlands, pp. 386–396.
20. Z. Gutowski-Eckel, C. Klein, K. Siegers, K. Bohm, M. Hammelmann, and K. D. Entian (1994) *Appl. Environ. Microbiol.* **60**, 1–11.
21. N. Schnell, G. Engelke, R. Augustin, F. Rosenstein, F. Götz, and K. D. Entian (1991) In *Nisin and Novel Antibiotics* (G. Jung and H. G. Sahl, eds.), Escom Publishers, Leiden, The Netherlands, pp. 269–276.
22. T. Kupke, S. Stefanovic, H. G. Sahl, and F. Götz (1992) *J. Bacteriol.* **174**, 5354–5361.
23. L. Gilson, H. K. Mahanty, and R. Kolter (1990) *EMBO J.* **9**, 3875–3884.
24. M. S. Gilmore, R. A. Segarra, and M. C. Booth (1990) *Infect. Immunol.* **58**, 3914–3923.
25. J. D. Marugg, C. F. Gonzalez, B. S. Kunka, A. M. Ledeboer, M. J. Pucci, M. Y. Toonen, S. A. Walker, L. C. Zoetmulder, and P. A. Vandebergh (1992) *Appl. Environ. Microbiol.* **58**, 2360–2367.
26. G. W. Stoddard, J. P. Petzel, M. J. van Belkum, J. Kok, and L. L. McKay (1992) *Appl. Environ. Microbiol.* **58**, 1952–1961.
27. L. S. Havarstein, H. Holo, and I. F. Nes (1994) *Microbiol.* **140**, 2393–2389.
28. K. Venema, G. Venema, and J. Kok (1995) *Trends Biochem. Sci.* **3**, 299–304.
29. J. R. van der Meer, J. Polman, M. M. Beerthuyzen, R. J. Siezen, O. P. Kuipers, and W. M. de Vos (1993) *J. Bacteriol.* **175**, 2578–2588.
30. J. R. van der Meer, H. S. Rollema, R. J. Siezen, M. M. Bethuyzen, O. P. Kuipers, and W. M. de Vos (1994) *J. Biol. Chem.* **269**, 3555–3562.
31. W. M. de Vos, O. P. Kuipers, J. R. van der Meeer, and R. Siezen (1995) *Mol. MIcrobiol.* **17**, 427–437.
32. W. M. de Vos and G. F. Simons (1994) In *Genetics and Biotechnology of Lactic Acid Bacteria* (M. J. Gasson and W. M. de Vos, eds.), Backie Academic , Glasgow, pp. 52–105.
33. T. Kupke, C. Kempter, V. Gnan, G. Jung, and F. Götz (1994) *J. Biol. Chem.* **269**, 5653–5659.
34. M. Reis, M. Eschbach-Bludau, M. Iglesias-Wind, T. Kupke, and H. G. Sahl (1994) *Appl. Environ. Microbiol.* **60**, 2876–2883.
35. M. van de Kamp, H. W. van den Hooven, R. N. Konings, C. W. Hilbers, C. W. van de Ven, G. Bierbaum, H. G. Sahl, O. P. Kuipers, R. J. Seizen, and W. M. de Vos (1995) *Eur. J. Biochem.* **230**, 587–600.
36. M. Skangen, J. Nissen-Meyer, G. Jung, S. Stefanovic, K. Sletten, C. I. Mortreveldt-Abilgaard, and I. F. Nes (1994) *J. Biol. Chem.* **269**, 27183–27185.
37. H. G. Sahl (1991) In Nisin and novel lantibiotics *Proceedings of the First International Workshop on Lantibiotics* (G. Jung and H. G. Sahl, eds.), Leiden, Escom Publishers, pp.347–359.
38. R. Benz, G. Jung, and H. G. Sahl (1991) In Nisin and novel lantibiotics *Proceeding of the First International Workshop on Lantibiotics* (G. Jung and H. G. Sahl, eds.), Leiden, Escom Publishers, pp. 359–372.
39. A. J. Driessen, H. W. van den Hooven, W. Kuiper, M. van de Kamp, H. G. Sahl, R. N. Konings, and W. Konings (1995) *Biochemistry* **34**, 1606–1614.
40. G. Bierbaum and H. G. Sahl (1987) *J. Bacteriol.* **169**, 5452–5458.
41. T. Sheth, R. M. Henderson, S. B. Hlady, and W. A. Cuthbert (1992) *Biochim. Biophys. Acta* **1107**, 179–185.
42. S. Chatterjee, D. Chatterjee, K. Jani, H. Blumbach, B. Ganguli, N. Klesel, M. Limbert, and G. Seibert (1992) J. *Antibiot.* **45**, 839–845.
43. D. Espesset, D. Duché, D. Baty, and V. Géli (1996) *EMBO J.* **15**, 2356–2364.
44. R. Kellner, G. Jung, and H. G. Sahl (1991) In *Nisin and Novel Lantibiotics* (G. Jung and H. G. Sahl, eds.), Escom Publishers, Leiden, The Netherlands, pp. 141–158.
45. E. Gross, H. H. Kitz, and E. Nebelin (1973) *Hoppe-Seyler's Z. Physiol. Chem.* **354**, 810–812.
46. M. van de Kamp, L. Horstink, H. W. van den Hooven, R. N. Konings, C. W. Hilbers, A. Frey, H. G. Sahl, J. W. Metzger, and F. J. van de Ven (1995) *Eur. J. Biochem.* **227**, 757–771.
47. H. Allgaier, G. Jung, R. Werner, U. Schneider, and H. Zähner (1986) *Eur. J. Biochem.* **160**, 9–22.
48. R. Kellner, G. Jung, T. Hörner, H. Zähner, N. Schnell, K. Entian, and F. Götz (1988) *Eur. J. Biochem.* **177**, 53–59.
49. J. C. Piard, C. Delorme, M. Novel, M. Desmageaud, and G. Novel (1993) *FEMS Microbiol. Lett.* **112**, 313–318.
50. R. W. Jack, A. Carne, J. Metzger, S. Stefanovic, H. G. Sahl, G. Jung, and J. R. Tagg (1994) *Eur. J. Biochem.* **220**, 455–462.
51. K. F. Ross, C. Ronson, and J. R. Tagg (1993) *Appl. Environ. Microbiol.* **59**, 2014–2021.
52. J. Novak, P. W. Canfield, and E. J. Miller (1994) *J. Bacteriol.* **176**, 4316–4320.
53. G. Stoffels, J. Niessen-Meyer, A. Gudmundsdottir, K. Sletten, H. Halo, and I. F. Nes (1992) *Appl. Environ. Microbiol.* **58**, 1417–1422.
54. M. S. Gilmore, R. A. Serraga, M. C. Booth, C. P. Bogie, L. R. Hall, and D. B. Clewell (1994) *J. Bacteriol.* **176**, 7335–7344.
55. A. Fredenhagen, F. Märki, G. Fendich, W. Märki, J. Gruner, J. van Oostrum, F. Raschdorf, and H. H. Peter (1991) In *Nisin and Novel Lantibiotics* (C. Jung and H. G. Sahl eds), Escom Publishers, Leiden, The Netherlands, pp. 131–140.
56. H. Kogler, H. Bauch, H. W. Fehlhaber, C. Griesinger, W. Schubert, and V. Teetz (1991) In *Nisin and Novel Lantibiotics* (G. Jung and H. G. Sahl, eds.), Escom Publishers, Leiden, The Netherlands, pp. 159–170.
57. N. Zimmermann, N. S. Freud, A. Fredenhagen, and G. Jung (1993) *Eur. J. Biochem.* **216**, 419–428.
58. M. L. Chikindas, M. J. Garcia-Garcera, A. J. Driessen, A. M. Ledeboer, J. Niessen-Meyer, I. F. Nes, T. Abee, W. N. Konings, and G. Venema (1993) *Microbiology* **59**, 3577–3584.
59. A. Maftah, T. Renault, C. Viguoles, Y. Hechard, P. Bressolier, M. Ratineaud, Y. Cenatiempo, and R. Julian (1993) *J. Bacteriol.* **175**, 3232–3235.
60. R. W. Woboro, T. Henkel, M. Sailer, K. L. Roy, J. C. Vederas, and M. E. Skiles (1994) *Microbiology* **140**, 517–526.
61. A. Holck, L. Axelson, S. S. E. Birkeland, T. Aukrust, and H. Bloom (1992) *J. Gen. Microbiol.* **138**, 2715–2720.
62. P. Muriana and T. R. Klaenhammer (1991) *J. Bacteriol.* **173**, 1779–1788.
63. H. Holo, O. Niessen, and I. F. Ness (1991) *J. Bacteriol.* **173**, 3879–3887.

Suggestions for Further Reading

R. James, C. Lazdunski, and F. Pattus, eds. (1992) *Bacteriocins, Microcins and Lantibiotics*, NATO ASI Series, Springer-Verlag, Berlin, Vol. H65.

R. W. Jack, J. R. Tagg, and B. Ray (1995) Bacteriocins of gram-positive bacteria, *Microbiol. Rev.* **59**, 171–200.

T. R. Klaenhammer (1993) Genetics of bacteriocins produced by lactic acid bacteria, *FEMS Microbiol. Rev.* **12**, 39–86.

F. Moreno, J. L. San Millan, C. Hernandez-Chico, and R. Kolter (1995) Microcins, *Biotechnology 28*, 307–321.

H-G. Sahl, R. W. Jack, and G. Bierbaum (1995) Biosynthesis and biological activities of lantibiotics with unique post-translational modifications, *Eur. J. Biochem.* **230**, 827–853.

BACTERIORHODOPSIN

JANOS K. LANYI

Bacteriorhodopsin is a small (26 kDa) integral **membrane protein**, the prototype of seven-helical **G-protein**-linked **receptors**, that upon illumination transports protons across the membrane. It forms extended two-dimensional hexagonal arrays in the cytoplasmic membrane of **halobacteria**. Its transmembrane **α-helices** surround the **prosthetic group** retinal, which is linked via a **Schiff base** to Lys216 near the middle of helix G and lies at a small angle to the membrane plane. Photoisomerization of the retinal from all-*trans* to 13-*cis* sets off a sequence of thermal reactions (the "photocycle") in which the interaction of the retinal and the protein causes proton transfers between various donor and acceptor groups. Together, these transfers result in the complete translocation of a proton from the cytoplasmic to the extracellular surface, thus generating a transmembrane electrochemical gradient for protons. The **proton gradient** is utilized in the way usual in **bacteria** for the synthesis of ATP, the uptake of nutrients (amino acids) and K^+, and the transport of Na^+ out of the cells (see **Proton motive force**).

Bacteriorhodopsin forms trimers that assemble in the two-dimensional hexagonal array that constitutes the patches termed "**purple membrane**." The purple membrane contains only bacteriorhodopsin and **lipids**, and the regular crystalline lattice made it possible to determine its structure by cryo-**electron crystallography** at 7 Å (1) and then 3 Å (2) resolution. The protein was also crystallized from a lipid cubic phase, and its structure has now been determined by **X-ray crystallography** at 2.5 Å resolution (3). The protein consists of seven transmembranous α-helices, with short interhelical loops and short N and C termini. Three of the helices, B, C, and D, are normal to the plane of the membrane and the other four, A, E, F, and G, are inclined at various small angles to the perpendicular (Fig. 1). The retinal is bound to the **ε-amino group** of Lys216, forming a protonated Schiff base near the middle of helix G, and its polyene chain lies at about 23° from the membrane plane. Thus the Schiff base divides the protein into extracellular and cytoplasmic halves.

The trajectory of the transported proton from one surface to the other is through proton-conductive "half-channels" through each of the two halves of the molecule. Identification of the residues that participate in the two half-channels has been the objective of much research in the past few years. The extracellular half-channel is quite complex and contains **polar** and **hydrogen-bonding** residues. Many of them play roles in the release of protons to the extracellular surface. Asp85 and Asp212 are anionic residues located near the Schiff base. A possible pathway for protons leads from Asp85, via Arg82, Glu204, and Glu194, to the extracellular surface and is consistent with functional studies. Another plausible pathway is through Glu9 or Glu74, but the results of **mutagenesis** indicate that these two residues are dispensable. The cytoplasmic half-channel is simpler. It contains mostly **hydrophobic** residues between the retinal and Asp96, which is the proton donor in the reprotonation of the Schiff base. Thr46 is close enough to interact with Asp96, and mutagenesis confirms its significant role in the proton donor and acceptor function of Asp96. Acidic residues at the cytoplasmic aqueous interface include residues 36, 38, 102, and 104. It has been suggested that they form a "funnel" at the surface that directs protons into the cytoplasmic half-channel (2). Although immobilization of the protein in the purple membrane is responsible for the well-known, extraordinary thermal and photostability of bacteriorhodopsin, neither the trimeric arrangement of the monomers nor the rigid lattice structure have functional roles in the transport activity (4). Thus, availability of the structure of bacteriorhodopsin, even at the current limited resolution, has generated considerable insights into the proton transport mechanism and has provided the framework for mechanistically interpreting a wealth of spectroscopic information.

The chromophore with all-*trans*, 15-*anti* retinal has a broad **absorption** band with a maximum at 568 nm. This is considerably red-shifted from that of retinal in isolation (380 nm) or a retinal analog with a protonated Schiff base (440 nm). This red-shift results primarily from the diffuse counterion to the charged Schiff base, which comprises principally Asp85 with a contribution from Asp212, which is several angstroms removed and providing only partial compensation of the charge (5). The all-*trans* chromophore exists in thermal equilibrium with the 13-*cis*, 15-*syn* configuration, which absorbs at 555 nm. Sustained illumination converts the retinal to 100% all-*trans*, 15-*anti*, known as "light-adaptation" (6). The photoreaction of this isomer is normally active in transport.

The photocycle of the all-*trans* chromophore is described by the intermediate states J, K, L, M, N, and O, their substates, and the sequence of their interconversions (Fig. 2). Each intermediate is characterized by a distinct absorption maximum in the visible wavelength region, numerous vibrational bands of the retinal in the infrared and Raman region and of the protein in the infrared region (see **Vibrational spectroscopy**). The kinetics of the photocycle describe the sequence and the energetics of the chemical reactions that translocate protons across the membrane. In the K intermediate, which arises on the nanosecond timescale after decay of the excited state, the retinal assumes a twisted 13-*cis*, 15-*anti* configuration, as indicated by high-amplitude hydrogen-out-of-plane vibrations. Its absorption maximum is red-shifted by 30 to 40 nm from the initial state. It is converted in about one microsecond into the L state, which absorbs at 540 to 550 nm. In L, the polyene of the retinal is more relaxed, but other changes begin to appear in the protein in both the extracellular and cytoplasmic regions (7). The Schiff base forms stronger hydrogen bonds that are correlated with structural changes of bound **water** detected in the infrared wavelength region. One of these water molecules is bound to the Schiff base, Asp85, and Asp212 and may play a role in the pK_a shifts that result in proton transfer from the

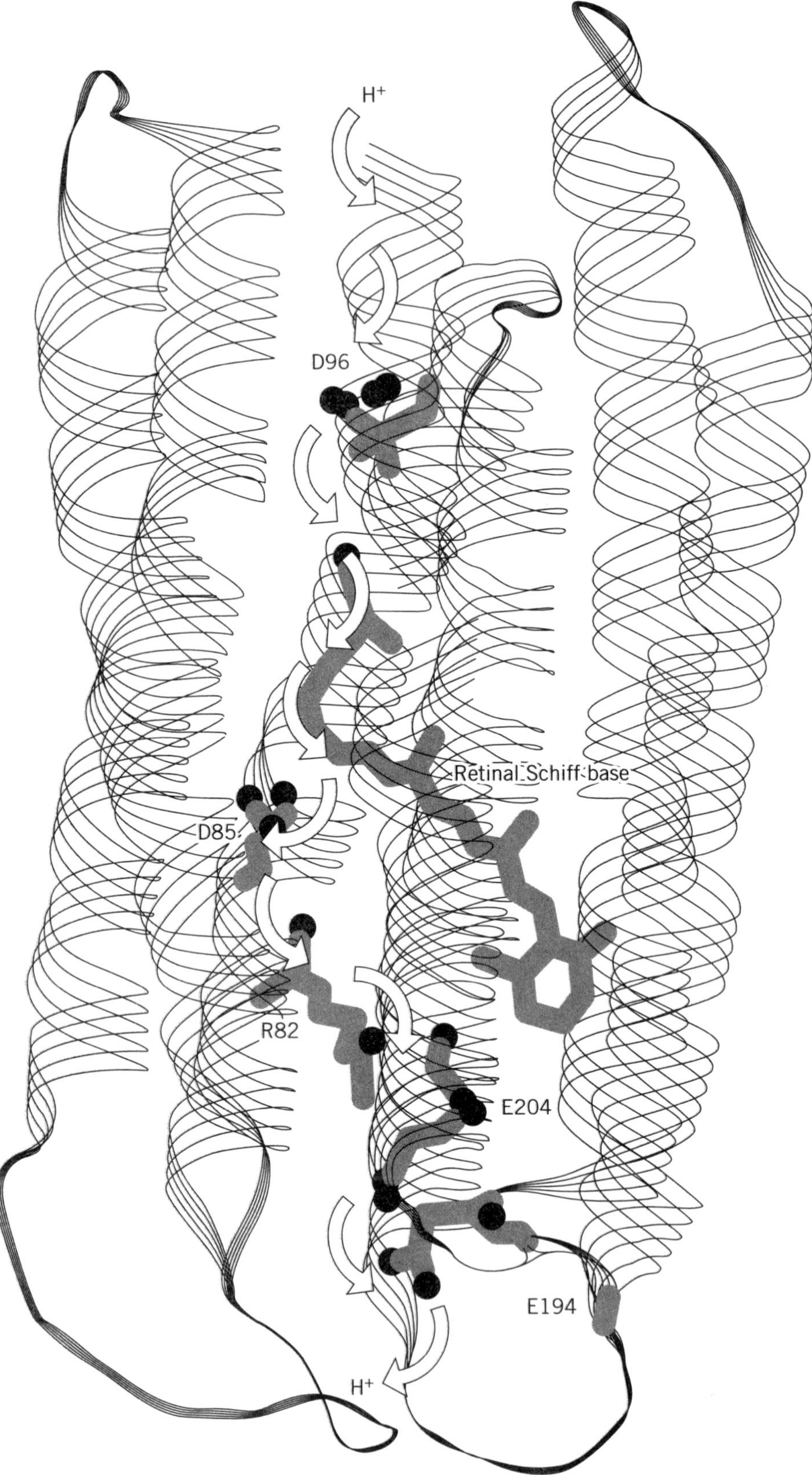

Figure 1. Structure of bacteriorhodopsin (1), and the pathway of proton transport. The seven transmembranous α-helices are shown, along with only the all-*trans* retinal and the most important residues. The curved arrows identify the proton transfers that occur at different times in the photocycle (see text) and add up to the complete transport of a proton from the cytoplasmic (upper) to the extracellular surface of the membrane.

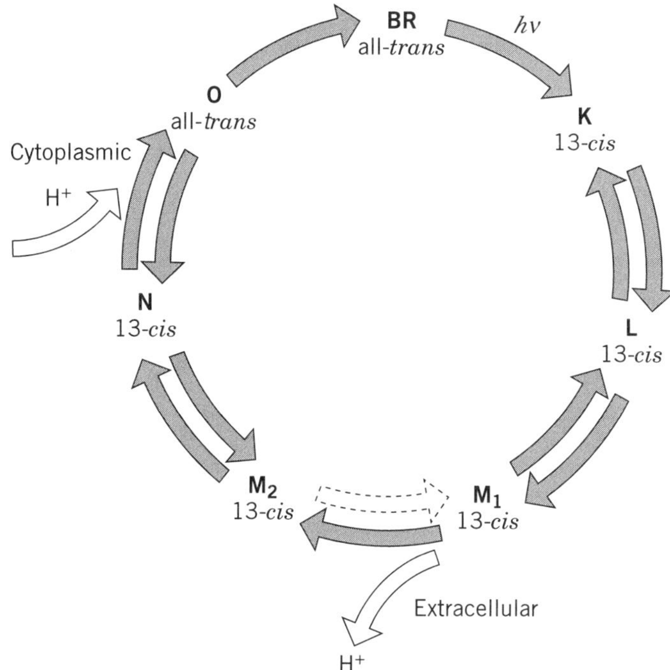

Figure 2. The photocycle of bacteriorhodopsin. The intermediate states are shown, and the isomeric configuration of the retinal is indicated. The Schiff base is protonated in all but the M states.

Schiff base to Asp85. The hydrogen bonding of two other water molecules near Asp96 are also affected in L, suggesting that the structural changes near the Schiff base are transmitted all the way across the protein to the cytoplasmic region.

The M state is formed by transfer of a proton from the retinal Schiff base to Asp85 (8). It has a strongly blue-shifted absorption maximum at 410 nm. The kinetics of this conversion, measured at visible and infrared wavelengths, suggest that L is in equilibrium with an early M state, M_1, and that the mixture of L and M_1 decay together to form the late M state, M_2 (9–11). At pH > 6 this occurs in a unidirectional reaction, and thus L disappears as M_2 is formed. At pH < 6, however, the $M_1 \rightarrow M_2$ reaction is not unidirectional, and both L and M_1 remain present and coexist with M_2 because at the higher pH a proton dissociates from a site in the extracellular region that interacts with Asp85. This site is either Glu204 or depends on the carboxyl group of Glu204. Its proton is passed to Glu194 and is released from there to the extracellular surface. The anomalous titration properties of Asp85 indicate that the nature of the interaction between the proton affinities of this aspartate residue and the proton release site is such that either may be protonated but not both (12,13). When Asp85 becomes protonated by the Schiff base, the pK_a of the proton release group is lowered, and the proton dissociates. When this proton is released to the bulk solution at a pH greater than the pK_a for the release, the pK_a of Asp85, in turn, is driven higher, and deprotonation of the Schiff base becomes complete. This prevents reprotonation of the Schiff base from the extracellular direction.

Reprotonation of the Schiff base is from the cytoplasmic direction by Asp96 (8,14), which produces the N intermediate that absorbs near 560 nm but with a lower extinction than the initial state. Large-scale protein conformational changes in N are evident from a pair of negative and positive difference features in the infrared spectrum that originate from a shift of the amide I band (15). Electron and X-ray diffraction studies of the M and the N states, measured either at various times after flash illumination and freezing or in a photostationary state at ambient temperature, indicate considerable changes of conformation at the cytoplasmic surface. The most conspicuous of these is an outward tilt of the cytoplasmic end of helix F (16). Its occurrence in the M intermediate may be transitory during the M → N conversion of the unperturbed wild-type photocycle, but this feature is clearly observable when M is stabilized in the wild-type or in mutant forms of the protein. The movement of helix F as a rigid body is confirmed by distance measurements using pairs of spin labels (17). The effects of osmotic agents, humidity, and in-plane cooperativity in the purple membrane lattice on the M → N reaction and on the protein conformation change suggest that the rationale of the helical tilt is to increase the **hydration** of the cytoplasmic region and thereby to decrease the pK_a of Asp96. Thus, Asp96 becomes a proton donor to the Schiff base. The tilt of helix F is recovered during decay of the N state, presumably recovering the initial high pK_a of Asp96 and causing its reprotonation from the cytoplasmic surface.

Reisomerization of the retinal to all-*trans* occurs in the N → O transition (18). This is made possible by the lowered barrier-to-bond rotation in the polyene chain upon protonation of the Schiff base. Residues that contact the chain near the 9-methyl and 13-methyl groups, such as Trp182 (19) and Leu93 (20), facilitate the reisomerization, probably through steric interactions that transmit residue displacements in the protein to the retinal and vice versa. The O state has a strongly red-shifted maximum at visible wavelengths, at least partly because Asp85 is still protonated. Consequently, the main component of the counterion to the protonated Schiff base is lacking. Large-amplitude hydrogen-out-of-plane vibrations indicate that, as in the K state, the retinal chain is twisted. These features disappear in the final O → BR reaction, which appears to be limited by the rate of proton transfer from Asp85 to the still unprotonated proton release site (21). As expected from the recovery of the low initial pK_a of Asp85, this reaction is unidirectional under all conditions, and it ensures the full repopulation of the initial state and also the functioning of the **proton pump** against large transmembranous proton gradients.

Bacteriorhodopsin is one of three types of similar retinal proteins in halobacterial membranes. Their functions are all based on the photoisomerization of all-*trans* retinal to 13-*cis*, 15-*anti* and the protein reactions that accompany the thermal reisomerization. Halorhodopsin is an inwardly-directed, light-driven chloride ion pump. It lacks Asp85 and Asp96, and the retinal Schiff base does not deprotonate during the photocycle (22). Sensory **rhodopsins** I and II are receptors for phototactic behavior (23). A profound similarity in the mechanisms of these proteins with different functions is indicated by the fact that their activities are interconvertible with minimal perturbations. Thus, the Asp85Thr mutant of bacteriorhodopsin binds chloride, exhibits a photocycle similar to that of halorhodopsin, and transports chloride from the extracellular to the cytoplasmic direction (24). Halorhodopsin, in turn, transports protons when the weak acid, azide, is added, by binding near the Schiff base and functioning as a proton acceptor (25). Sensory rhodopsin I transports protons like bacteriorhodopsin when the transducing protein that is normally tightly bound to it is genetically deleted (26,27).

BIBLIOGRAPHY

1. N. Grigorieff et al. (1996) *J. Mol. Biol.* **259**, 393–421.
2. Y. Kimura et al. (1997) *Nature* **389**, 206–211.
3. E. Pebay-Peyroula, G. Rummel, J. P. Rosenbusch, and E. M. Landau (1997) *Science* **277**, 1676–1681.
4. N. A. Dencher and M. P. Heyn (1979) *FEBS. Lett.* **108**, 307–310.
5. K. Nakanishi et al. (1980) *J. Am. Chem. Soc.* **102**, 7945–7947.
6. G. S. Harbison et al. (1984) *Proc. Natl. Acad. Sci. USA* **81**, 1706–1709.
7. A. Maeda et al. (1997) *J. Biochem. (Tokyo)* **121**, 399–406.
8. M. S. Braiman et al. (1988) *Biochemistry* **27**, 8516–8520.
9. L. Zimányi et al. (1992) *Biochemistry* **31**, 8535–8543.
10. S. Dickopf and M. P. Heyn (1997) *Biophys. J.* **73**, 3171–3181.
11. B. Hessling, J. Herbst, R. Rammelsberg, and K. Gerwert (1997) *Biophys. J.* **73**, 2071–2080.
12. S. P. Balashov, E. S. Imasheva, R. Govindjee, and T. G. Ebrey (1996) *Biophys. J.* **70**, 473–481.
13. H. T. Richter, L. S. Brown, R. Needleman, and J. K. Lanyi (1996) *Biochemistry* **35**, 4054–4062.
14. K. Gerwert, B. Hess, J. Soppa, and D. Oesterhelt (1989) *Proc. Natl. Acad. Sci. USA* **86**, 4943–4947.
15. M. S. Braiman, O. Bousché, and K. J. Rothschild (1991) *Proc. Natl. Acad. Sci. USA* **88**, 2388–2392.
16. S. Subramaniam, M. Gerstein, D. Oesterhelt, and R. Henderson (1993) *EMBO J.* **12**, 1–8.
17. T. E. Thorgeirsson et al. (1997) *J. Mol. Biol.* **273**, 951–957.
18. S. O. Smith et al. (1983) *Biochemistry* **22**, 6141–6148.
19. O. Weidlich et al. (1996) *Biochemistry* **35**, 10807–10814.
20. J. K. Delaney, G. Yahalom, M. Sheves, and S. Subramaniam (1997) *Proc. Natl. Acad. Sci. USA* **94**, 5028–5033.
21. H. T. Richter et al. (1996) *Biochemistry* **35**, 15461–15466.
22. D. Oesterhelt (1995) *Israel J. Chem.* **35**, 475–494.
23. W. D. Hoff, K. H. Jung, and J. L. Spudich (1997) *Annu. Rev. Biophys. Biomol. Struct.* **26**, 223–258.
24. J. Sasaki et al. (1995) *Science* **269**, 73–75.
25. G. Váró, L. S. Brown, R. Needleman, and J. K. Lanyi (1996) *Biochemistry* **35**, 6604–6611.
26. R. A. Bogomolni et al. (1994) *Proc. Natl. Acad. Sci. USA* **91**, 10188–10192.
27. U. Haupts, C. Haupts, and D. Oesterhelt (1995) *Proc. Natl. Acad. Sci. USA* **92**, 3834–3838.

Suggestions for Further Reading

T. G. Ebrey (1993) Light energy transduction in bacteriorhodopsin. In *Thermodynamics of Membranes, Receptors and Channels*, (M. Jackson, ed.), CRC Press, Boca Raton, FL, pp. 353–387.

J. K. Lanyi (1993) Proton translocation mechanism and energetics in the light-driven pump bacteriorhodopsin. *Biochim. Biophys. Acta* **1183**, 241–261.

J. K. Lanyi and G. Váró (1995) The photocycles of bacteriorhodopsin. *Israel J. Chem.* **35**, 365–386.

R. A. Mathies, S. W. Lin, J. B. Ames, and W. T. Pollard (1991) From femtoseconds to biology: Mechanism of bacteriorhodopsin's light-driven proton pump. *Ann. Rev. Biophys. Biophys. Chem.* **20**, 491–518.

D. Oesterhelt, J. Tittor, and E. Bamberg (1992) A unifying concept for ion translocation by retinal proteins. *J. Bioenerg. Biomembr.* **24**, 181–191.

BAL 31 NUCLEASE

H. B. GRAY, JR.

The **enzymes** commonly referred to as BAL 31 nuclease have been widely used for manipulation of nucleic acids, most frequently through their ability to shorten both strands of double-stranded DNA from the ends in a controllable manner. Other activities include endonucleolytic attack on double-stranded DNA in response to the presence of a variety of covalent and noncovalent alterations in DNA structure (eg, DNA-carcinogen adducts and junctions between right- and left-handed helical regions) and highly processive unidirectional exonucleolytic degradation of single-stranded DNA.

SOURCE AND SOME PHYSICAL PROPERTIES

BAL 31 is the strain designation given the original isolates of the marine bacterium that produces the nucleases, which was originally determined to be a species of *Pseudomonas* (1) but was reclassified as *Alteromonas espejiana* (2). The **nuclease** activity was discovered adventitiously as a contaminant in preparations of the *Alteromonas* bacteriophage PM2, but these enzymes proved to be bacterial products secreted into the culture medium (3,4). The common name also appears as BAL31 and BAL-31; all should be used when searching electronic databases.

The bulk of the nuclease activity in culture supernatants is owing to two molecularly and kinetically distinct forms, both of which are active as single **polypeptide chains** (5). The smaller of these, designated the "slow" (S) form, is produced by **proteolysis** of the larger, "fast" (F) form, which in turn derives from an even larger precursor that has not been characterized (6). *Alteromonas espejiana* copiously produces extracellular **proteinase** activity (6), which accounts for the progressive conversion of larger to smaller species in the culture fluid as growth proceeds. At least the S species is not a fully homogenous protein, because its N-terminal amino acid is not unique; either the presumed endoproteolytic event that generates this species from the F form does not occur at a unique site or there is exoproteolytic activity as well. Conversion of the F form to a species indistinguishable in molecular size and catalytic properties from the S form can be done by proteolysis *in vitro* (6).

The American Type Culture Collection strain of this organism (ATCC 29659) is suitable for production of the nucleases. No overproducing strains, either of *Alteromonas* itself or of heterologous hosts containing **cloned** nuclease **genes**, have been reported. Purification procedures that effect separation of the S and F forms have been described (5) and modified (6). BAL 31 nuclease is available commercially from several sources, but these products are mixtures of the S and F species. A partial amino acid sequence of an internal fragment produced by cleavage with cyanogen bromide has been obtained (7). This new sequence did not have significant **homology** with any reported sequences. Some physical properties of the S and F forms are presented in Table 1 (5).

Table 1. Some Physical Properties of F and S Forms of BAL 31 Nuclease

	Molecular Weight (kDa)	Isoelectric pH	Molar Absorption Coefficient (10^5 liter/mol-cm)	Weight Absorption Coefficient (dl/g-cm)
F	109.0	4.2	1.10 ± 0.09	10.2 ± 0.8
S	85.0	4.2	1.00 ± 0.03	11.7 ± 0.3

REACTIONS CATALYZED

Five activities appear to account for the degradation—ultimately to mononucleotides—of DNA: (1) a 5′ → 3′ exonuclease activity that acts on single-stranded DNA in a highly processive manner (ie, many nucleotides are removed in a single productive enzyme-substrate encounter) (8); (2) a 3′ → 5′ directed exonuclease activity that acts only on duplex structures, leaving a 5′-terminated single-stranded "tail" (9); (3) an endonuclease activity against single-stranded DNA, (much slower than the exonuclease activity on single-stranded DNA [8]), that is also elicited by a variety of covalent or noncovalent distortions or lesions in duplex DNA (3–5, 10–14); (4) an exonuclease activity that can excise a small number of nucleotides starting from the site of a strand break (nick) in duplex DNA (15), which is inferred to be 5′ → 3′-directed; and (5) an activity that can remove short 3′-terminated tails from otherwise duplex DNA. The F and S forms differ greatly in the rates of removal of nucleotides, at given molar concentrations of DNA ends and enzyme, via the 3′ → 5′ exonuclease action on double-stranded DNA, which led to their "fast" and "slow" designations (5). The two forms have comparable kinetic behavior toward single-stranded DNA (5,8).

The BAL 31 nucleases also catalyze the terminally directed hydrolysis of double-stranded RNA and degrade RNA that contains nonduplex structure (16). Whether an analog of the endonuclease activity on altered duplex DNA exists has not been determined, except that cleavage probably occurs in response to a strand break in duplex RNA.

The 5′ → 3′ exonuclease activity on single-stranded DNA is highly processive, as the nucleases are able to degrade DNA polymers of nearly 500 nucleotides in length completely without dissociation of enzyme from substrate (8). The much larger (5,400-nucleotide) *ϕ***X174** DNA did not appear to be degraded processively, however, so there may be limits to the number of nucleotides removed in a single productive enzyme-substrate encounter (5). The values of V_{max} and the turnover number (k_{cat}), on the basis of the rates of internucleotide bond hydrolysis, are similar for the 500-nucleotide polymer described above and the minimal DNA substrate, a dinucleotide diphosphate (8). The kinetic data appear to require a **facilitated diffusion** mechanism for the activity on macromolecular substrates, in which nuclease molecules bind randomly and diffuse along the contour of the chain until a 5′ end is located, at which time catalysis can begin (8). This apparently was the first example of a requirement for facilitated diffusion in an enzymatic reaction.

The 3′ → 5′ exonuclease activity on duplex DNA and the 5′ → 3′ exonuclease activity on single-stranded DNA can account for the progressive reduction in the length of duplex DNA. Above certain nuclease concentrations, short 5′-terminated single-stranded tails from the 3′ → 5′ exonuclease activity are evident in partial digests (9). If 3′-terminated tails are present, they cannot be more than a few nucleotides in length. The 5′ tails have a limiting length, at the higher enzyme concentrations examined, of about 7 nucleotide residues when about 100 nucleotides are removed by the 3′ → 5′ exonuclease activity. Up to 50% of the ends could be joined in **DNA ligase** reactions under conditions favoring the joining of fully base-paired ends. This indicates that the 5′ → 3′ exonuclease activity on single-stranded DNA can terminate at the junction between single- and double-stranded regions, to leave fully based-paired ends on a significant fraction of molecules in a partially digested population. **DNA polymerase**-mediated repair, to render ends with 5′-terminated tails fully base-paired, markedly increased the fraction of ends joinable by DNA ligase, as expected (9).

As the nuclease concentrations were decreased below those mentioned above, the average length of the single-stranded tails per 100 nucleotides removed increased dramatically, and the ligase-mediated joining of ends in the absence of polymerase-mediated repair became undetectable. This apparent dependence on the concentration of the enzyme of the relative velocities of two exonuclease reactions that it catalyzes has not been explained. The velocity of the overall reaction to shorten DNA, as measured by the hyperchromicity associated with the hydrolysis of internucleotide bonds, is proportional to enzyme concentration, as expected (17). Repair of ends generated at the low nuclease concentrations noted above with DNA polymerase rendered a high percentage of the molecules ligatable (9).

There must be an activity that can remove short 3′-terminated protruding single-stranded tails from duplex DNA, because the nuclease readily degrades such starting substrates (9). No evidence exists, however, for 3′ terminally directed hydrolysis of single-stranded DNAs (8).

The value of the K_m (**Michaelis constant**), in terms of the molar concentration of DNA ends, for the overall exonucleolytic shortening of duplex DNA by the S form is so large that it is not practical for the substrate concentration to approach it in actual reactions. Consequently, the enzyme velocity is directly proportional to substrate concentration over the accessible range (18). A value for this K_m was reported in earlier work (5), but it appears to have been determined under conditions for which the substrate concentration did not exceed that of the enzyme, violating a premise of Michaelis-Menton-based kinetics. Kinetic parameters (V_{max} per unit concentration of nuclease and K_m) for the length reduction of duplex DNA by the F form can be measured accurately, because the K_m value is much lower than for the S form and lies in the range of experimentally accessible concentrations of DNA ends. More recent values (18) are in reasonable agreement with the earlier data (5).

These kinetic parameters for the F nuclease depend on the length of the DNA substrate (7) (not determinable for the S form as noted). This has been shown (7) to be consistent with a model in which nonspecific binding of the nuclease away from the ends is followed by a "search" process to form a productive enzyme-substrate complex, with the enzyme bound to a terminus, as for the degradation of single-stranded DNA. However, such a mechanism is not required by the kinetics as in the case of single-stranded DNA (noted in the text above).

The kinetics of length reduction of duplex DNA are also dependent on its guanine + cytosine(G + C) content (7,18). This is significant in the case of the S form of the nuclease, for which the rate of nucleotide removal from DNA ends decreases over fourfold over the range of 37–66 mole% G + C residues. This dependence is somewhat less for the F enzyme. Data are available to predict these effects of DNA base composition on the kinetics (18).

In contrast to the high processivity for the $5' \rightarrow 3'$ exonuclease action on single-stranded DNA noted earlier, the $3' \rightarrow 5'$ exonuclease is quasi-processive, removing only about 18 and 28 nucleotide residues per productive binding event for the S and F enzymes, respectively (9).

The length reduction of duplex RNA (16) presumably proceeds by a similar mechanism but has not been characterized. The BAL 31 nucleases are the most efficacious enzymes known for the controlled length reduction of duplex DNA, and they are apparently the only enzymes that can catalyze this reaction for duplex RNA.

Duplex DNA is not attacked endonucleolytically (away from an end) at a significant rate unless there is some alteration of the duplex structure. Hence, nonsupercoiled, closed circular duplex DNA (form I° DNA) is extremely resistant to attack by the BAL 31 nucleases (11,13,14). The very limited attack on form I° DNA at high enzyme concentrations and long incubations (14), plus the kinetic parameters for exonucleolytic degradation of linear duplex DNA, lead to estimates of the relative rates of introduction of endonucleolytic breaks to exonucleolytic scissions of 8×10^{-11} and 7×10^{-12} for the S and F enzymes, respectively.

Negative supercoiling in closed circular DNA can elicit the endonuclease activity (12). In the majority of molecules of such supercoiled DNA, an endonucleolytic event (cleavage in one strand) is followed by the removal, in a processive manner, of several nucleotides (6.5 and 2.8 nucleotides for the F and S forms, respectively) from the initially nicked strand, to yield a gapped circular DNA intermediate (15). A fraction contain only a strand break, and no nucleotides are excised. The percentage of molecules with no nucleotides removed is significantly lower for the F nuclease than for the S species. The nicks and gaps are bounded by 5′-phosphoryl and 3′-hydroxyl termini. It is assumed that other alterations that result in endonucleolytic attack give rise to such gapped circular intermediates.

The removal of a small number of nucleotides could be accomplished exonucleolytically, starting from the site of the initial nick. Or, there could be a second endonucleolytic cut a few nucleotides away from the first one; the short oligonucleotide between endonucleolytic breaks would dissociate at room temperature to leave a gap. Available evidence supports the exonucleolytic mechanism. It was reasoned that the presumed exonucleolytic activity producing the gaps should be $3' \rightarrow 5'$–directed, as the exonucleolytic activity attacking base-paired ends has this characteristic. However, this proved to be inconsistent with the results of further experiments carried out assuming the $3' \rightarrow 5'$ mode of attack; the $5' \rightarrow 3'$ attack is thus inferred. Because this activity operates on nominally duplex DNA, it might be expected that it could remove nucleotides in a $5' \rightarrow 3'$ direction from fully based-paired ends, to leave 3′-terminated single-stranded ends. It was noted that, if these are present, they could not be more than a few nucleotides in length, but this does not rule out very limited activity comparable to that producing the gaps.

The nicked and gapped circular DNA intermediates are then converted to linear duplex DNA by a second endonucleolytic event in the other strand, which requires a second encounter with a nuclease molecule (15). The resulting linear duplex DNA is further degraded by the mechanism noted in the text above. As expected, circular duplex DNA containing a nick introduced by means other than BAL 31 nuclease action is converted to linear duplex DNA, with the occurrence of gaps of the average sizes noted above in most of the circular molecules before their linearization (15).

EFFECTS OF SOLVENT VARIABLES AND DENATURING AGENTS

Both Ca^{2+} and Mg^{2+} at concentrations above 10–12 mM are required for maximum activity on both single- and double-stranded substrates (9). All activities are optimal near neutral pH (4,9). The nucleases are remarkably resistant to inactivation at elevated salt concentrations (4,10). The reaction buffer in which most of the work described above was done contained 0.6 M NaCl). This salt tolerance is not surprising, considering that sea water is the natural milieu of these extracellular enzymes.

High concentrations of agents that normally **denature** proteins fail to eliminate the nuclease activities. Single-stranded DNA is degraded at 40% of the maximum rate in 6.5 M **urea** (4), and substantial activity on this substrate was found in the presence of 6 M **guanidinium** chloride. Crude preparations maintained activity against both single- and double-stranded substrates in the presence of 5% (w/v) SDS (3). At least a portion of the structure of the S nuclease does not become disrupted under the stringent denaturation conditions used for denaturing SDS-PAGE (6).

The nucleases are not remarkably resistant to thermal inactivation, as the half-life for disappearance of the activity on single-stranded DNA is only 3–5 minutes at 50°C (7). However, tests on preparations stored at 4°C imply that most of the activity should be retained for years in a buffer containing 5 mM Mg^{2+} and Ca^{2+} (4). Commercial preparations are often supplied in 50% (v/v) glycerol for storage at −20°C, under which conditions they should retain full activity indefinitely.

APPLICATIONS

By far the most extensive use of the BAL 31 nucleases, represented by hundreds of literature citations, has been the controlled length reduction of linear duplex DNA. The bulk of these reports describe the production of deletion mutants and/or the elucidation of sequences required for a particular biological activity of a (usually) cloned DNA segment. Cloned sequences can be deleted unidirectionally by cleavage of the vector containing the cloned insert at a unique restriction site on one side of the insert (see **Restriction enzymes**), carrying out a partial BAL 31 degradation, releasing the shortened insert from the shortened vector by use of a restriction site on the other side of the insert, and ligation to an intact linearized vector DNA (19).

Numerous reports have appeared of BAL 31 nuclease-mediated identification of **telomeric** sequences, which occur at the ends of linear eukaryotic chromosomes and hence are degraded first by the exonuclease in intact DNA; Yao and Yao (20) and De Lange and Borst (21) apparently represent the earliest work. Where these DNAs are too large to isolate as intact molecules in solution, in situ lysis in **agarose** has been used (22) so that the initial substrate DNA is largely intact.

The progressive removal of sequences from duplex ends allows the determination of the **restriction map** of a DNA, either naturally linear or linearized at a unique site, by noting the order in which fragments from subsequent digestion with the restriction enzyme in question disappear from **gel electrophoresis** patterns of progressively shortened aliquots (19, 23). This technique is greatly enhanced if unidirectional deletions can be done, as noted above (19), as ambiguity arising from loss of sequences from both ends of the fragment is eliminated. The sites of bound proteins, including **nucleosomes**, and interstrand crosslinks that block the exonuclease can be elucidated because intact duplex sequences between such sites will not be attacked (24,25).

The endonuclease activity that cleaves in response to alterations in duplex structure has been used in several laboratories to detect such alterations. This attack will be followed by cleavage of the other strand and exonucleolytic attack from the ends thus generated, as noted in the text above. Where chemical modification is done, form I° DNA is used (this requires that the modification does not introduce nicks), and the rate of its loss on incubation with nuclease is monitored. Some of the chemical lesions that give rise to endonucleolytic attack are pyrimidine dimers and possibly other photoproducts of ultraviolet irradiation, adducts with **carcinogens** such as *N*-acetoxy-*N*-2-acetylaminofluorene and *N*-methyl-*N*-nitrosourea, interstrand cross-links produced by reaction with nitrous acid, apurinic sites, and adducts with Hg^{2+} and Ag^{+} ions (11,13,14). Strand breaks were noted in the text above as eliciting cleavage of the opposite strand. Noncovalent alterations eliciting endonucleolytic attack include moderate degrees of negative supercoiling and very high degrees of positive supercoiling (12), junctions between right-handed B-DNA and left-handed Z-DNA regions (10), the presence of unpaired nucleotides in one strand of an otherwise duplex DNA (26), and **cruciform** structures resulting from the extrusion of inverted repeated sequences under supercoiling stress (27,28). **Matrix attachment sites** for **nucleolar** DNA are apparently sensitive to the nuclease (29). Finally, the lack of sensitivity of form I° DNAs has been used to help identify such species in DNA populations (30).

BIBLIOGRAPHY

1. R. T. Espejo and E. S. Canelo (1968) *J. Bacteriol.* **95**, 1887–1891.
2. K. Y. Chan, L. Baumann, M. M. Garza, and P. Baumann (1978) *J. Syst. Bacteriol.* **28**, 217–222.
3. H. B. Gray, Jr., D. A. Ostrander, J. L. Hodnett, R. J. Legerski, and D. L. Robberson (1975) *Nucleic Acids Res.* **2**, 1459–1492.
4. H. B. Gray, Jr., T. P. Winston, J. L. Hodnett, R. J. Legerski, D. W. Nees, C.-F. Wei, and D. L. Robberson (1981) In *Gene Amplification and Analysis* (J. G. Chirikjian and T. S. Papas, eds.), Elsevier North-Holland, New York, pp. 169–203.
5. C.-F. Wei, G. A. Alianell, G. H. Bencen, and H. B. Gray, Jr. (1983) *J. Biol. Chem.* **258**, 13506–13512.
6. C. R. Hauser and H. B. Gray, Jr. (1990) *Arch. Biochem. Biophys.* **276**, 451–459.
7. T. Lu (1992) Some physical and catalytic properties of BAL 31 nuclease, Ph.D. dissertation, University of Houston, Houston, Texas.
8. T. Lu and H. B. Gray, Jr. (1995) *Biochim. Biophys. Acta* **1251**, 125–138.
9. X.-G. Zhou and H. B. Gray, Jr. (1990) *Biochim. Biophys. Acta* **1049**, 83–91.
10. M. W. Kilpatrick, C.-F. Wei, H. B. Gray, Jr., and R. D. Wells (1983) *Nucleic Acids Res.* **11**, 3811–3822.
11. R. J. Legerski, H. B. Gray, Jr., and D. L. Robberson (1977) *J. Biol. Chem.* **252**, 8740–8746.
12. P. P. Lau and H. B. Gray, Jr. (1979) *Nucleic Acids Res.* **6**, 331–357.
13. C.-F. Wei, G. A. Alianell, H. B. Gray, Jr., R. J. Legerski, and D. L. Robberson (1983) In *DNA Repair: A Laboratory Manual of Research Procedures* (E. C. Friedberg and P. C. Hanawalt, eds.), vol. 2, pp. 13–40.
14. C.-F. Wei, R. J. Legerski, G. A. Alianell, D. L. Robberson, and H. B. Gray, Jr. (1984) *Biochim. Biophys. Acta* **782**, 404–414.
15. A. Przykorska, C. R. Hauser, and H. B. Gray, Jr. (1988) *Biochim. Biophys. Acta* **949**, 16–26.
16. G. H. Bencen, C.-F. Wei, D. L. Robberson, and H. B. Gray, Jr. (1984) *J. Biol. Chem.* **259**, 13584–13589.
17. X.-G. Zhou (1989) Ethidium bromide-mediated renaturation of denatured closed circular DNAs in alkaline solution: mechanistic aspects and fractionation of closed circular DNAs on a molecular weight basis. Some catalytic properties and mechanism of exonuclease action of BAL 31 nuclease, Ph.D. dissertation, University of Houston, Houston, Texas.
18. H. B. Gray, Jr. and T. Lu (1993) In *Enzymes of Molecular Biology* (M. M. Burrell, ed.), Humana Press, Totowa, New Jersey, pp. 231–251.
19. C. R. Hauser and H. B. Gray, Jr. (1991) *Gene Anal. Tech. Appl.* **8**, 139–147.
20. M.-C. Yao and C. H. Yao (1981) *Proc. Natl. Acad. Sci U.S.A.* **78**, 7436–7439.
21. T. De Lange and P. Borst (1982) *Nature* **299**, 451–453.
22. R. F. Wintle, T. G. Nygaard, J. A. Herbrick, K. Kvaloy, and D. W. Cox (1997) *Genomics* **40**, 409–414.
23. R. J. Legerski, J. L. Hodnett, and H. B. Gray, Jr. (1978) *Nucleic Acids Res.* **5**, 1445–1464.
24. W. A. Scott, C. F. Walter, and B. L. Cryer (1984) *Mol. Cell. Biol.* **4**, 604–610.
25. W.-P. Zhen, O. Buchardt, H. Nielsen, and P. E. Nielsen (1986) *Biochemistry* **25**, 6598–6603.
26. D. H. Evans and A. R. Morgan (1982) *J. Mol. Biol.* **160**, 117–122.
27. L. H. Naylor, H. A. Yee, and J. H. van de Sande (1988) *J. Biomol. Struct. Dyn.* **5**, 895–912.
28. N. M. Morales, S. D. Coburn, and U. R. Muller (1990) *Nucleic Acids Res.* **18**, 2777–2782.
29. O. V. Iarovaia, M. A. Lagarkova, and S. V. Razin (1995) *Biochemistry* **34**, 4133–4138.
30. E. Cuzzoni, L. Ferretti, C. Giordani, S. Castiglione, and F. Sala (1990) *Mol. Gen. Genet.* **222**, 58–64.

Suggestions for Further Reading

H. B. Gray, Jr., T. P. Winston, J. L. Hodnett, R. J. Legerski, D. W. Nees, C.-F. Wei, and D. L. Robberson (1981) The extracellular nuclease from *Alteromonas espejiana*: an enzyme highly specific for nonduplex structure in nominally duplex DNA, In *Gene Amplification and Analysis* (J. G. Chirikjian and T. S. Papas, eds.), Elsevier North- Holland, New York, pp. 169–203.

H. B. Gray, Jr. and T. Lu (1993) The BAL 31 nucleases (EC 3.1.11), In *Enzymes of Molecular Biology* (M. M. Burrell, ed.), Humana Press, Totowa, New Jersey, pp. 231–251.

BALBIANI RING

ALAN P. WOLFFE

The **polytene chromosomes** of insects have extraordinary utility in the investigation of **chromosomal** structure and function. Polytene chromosomes display the phenomenon of puffing in which the **chromatin** associated with chromosomal bands is decondensed (see **Puff**). In the larval salivary glands of the midge *Chironomus*, several puffs become very large and are described as *Balbiani rings*. Balbiani is the cytologist who discovered polytene chromosomes in 1881. The visible aspect of a puff is essentially the complex of **RNA** and protein that accumulates as a result of vigorous **transcriptional** activity. Among the gene products encoded by the Balbiani rings are the glue proteins required to attach the midge pupa to its substrate. All of the Balbiani ring genes are expressed coordinately and maximally during the second larval instar. Then they show differential expression in prepupae (1). The most detailed ultrastructural analysis has been performed on Balbiani ring 2 in chromosome IV of *Chironomus tentans*. Although the polytene chromosome band itself contains more than 470 kbp of DNA, the major gene is 37 kbp long and contains two very well-defined transcription units that are differentially expressed at the prepupal stage. **Electron microscopy** detects the first engaged **RNA polymerase** molecule at the approximate start site of transcription and the last RNA polymerase at the site of termination (2,3). Within each puff, the Balbiani ring gene forms a loop of transcriptionally active chromatin with attached **ribonucleoprotein** complexes. As synthesis proceeds, the nascent transcripts fold up into compact ribonucleoprotein complexes that head toward the **nuclear pores** when they are exported to the cytoplasm. Each 37-kb pre-mRNA transcript is processed in the nucleus by staged association with specific proteins (4).

Contour measurements of chromosomal structure in the puff indicate that the chromatin is fully extended into an array of **nucleosomes** 10 nm in diameter at the site of transcription, whereas once transcription is completed, chromatin coils back up into a 30-nm diameter fiber, which is then finally packaged into a supercoiled loop. Upstream of the start site of transcription is a region free of nucleosomes, presumably corresponding to the **promoter**, and compacted chromatin fibers are further upstream and downstream of the gene. Immunologic analysis allows defining the structural components of chromatin on active and inactive segments of the Balbiani ring. Surprisingly, proteins, such as **histone** H1 and the core histones, remain on chromatin even while it is actively transcribed (5). Thus keen observation allows a rather complete ultrastructural picture of the transcription process.

BIBLIOGRAPHY

1. U. Lendahl and L. Weislander (1987) *Develop. Biol.* **121**, 130–138.
2. B. Bjorkroth, C. Ericsson, M. M. Lamb, and B. Daneholt (1988) *Chromosoma* **96**, 333–340.
3. C. Ericsson et al. (1989) *Cell* **56**, 631–639.
4. H. Mehlin, B. Daneholt, and U. Skoglund (1992) *Cell* **69**, 605–613.
5. C. Ericsson, U. Grossbach, B. Bjorkroth, and B. Daneholt (1990) *Cell* **60**, 73–83.

Suggestion for Further Reading

B. Daneholt (1997) A look at messenger RNP moving through the nuclear pore. *Cell* **88**, 585–588.

BARNASE AND BARSTAR

R. W. HARTLEY

Barnase is a **ribonuclease** secreted by the bacterium ***Bacillus** amyloliquefaciens*. It is a small protein of 110 **amino acid** residues with no **disulfide bonds** or nonpeptide components (see **Protein structure**). As such, it was recognized in the early 1960s to be an ideal subject for the then-emerging study of **protein folding**. Its specific inhibitor, barstar, is produced intracellularly by the same organism and is an equally simple and even smaller protein of only 89 residues. Together they form a two-subunit complex in which the **active site** of barnase is buried, providing a model system for the study of **protein–protein interactions**. The system has contributed substantially to our understanding of how proteins fold and interact and has potential for further uses.

The three-dimensional structures of both proteins and their complex have been determined (Fig. 1) (1). The genes for both proteins have been **cloned** into **plasmid vectors** in ***Escherichia** coli* and the expressed products obtained in high yield. Expression of barnase alone is lethal, so simultaneous expression of barstar is necessary for production of barnase.

The catalytic mechanism of barnase and its relatives is essentially the same as that of the pancreatic ribonuclease family (see **Ribonuclease A**), with a histidine and a glutamic acid residue acting respectively as proton donor and acceptor, in place of the two histidines of the latter. The barnase **active site** lies in a broad groove on the side of the **β-sheet** opposite the major **α-helix** (see Fig. 1). In the complex, which has a **dissociation constant** on the order of 10^{-14} M, barstar completely blocks the active site, with most of its contacts involving its second helix and an adjacent loop. An aspartic acid carbonyl group of barstar occupies the position of the attacked phosphate group of substrate RNA. The rate at which barnase and barstar associate is controlled largely by charged groups, positive on barnase, negative on barstar. Mutation to alanine of any of these charged residues reduces binding but increases the stability of each protein, indicating the evolutionary balance between function and stability.

Both barnase and barstar may be unfolded reversibly by heat or by **denaturants** such as **urea** or **guanidinium chloride** (see **Protein unfolding**). Both fold and unfold in a highly **cooperative** two-state manner; that is, only the native and unfolded states are significantly occupied at **equilibrium** under most conditions. During the kinetics of folding, however, barnase passes through an observable intermediate state. In this it exemplifies the behavior of most larger proteins or protein **domains** (see **Protein folding**). Barstar, on the other

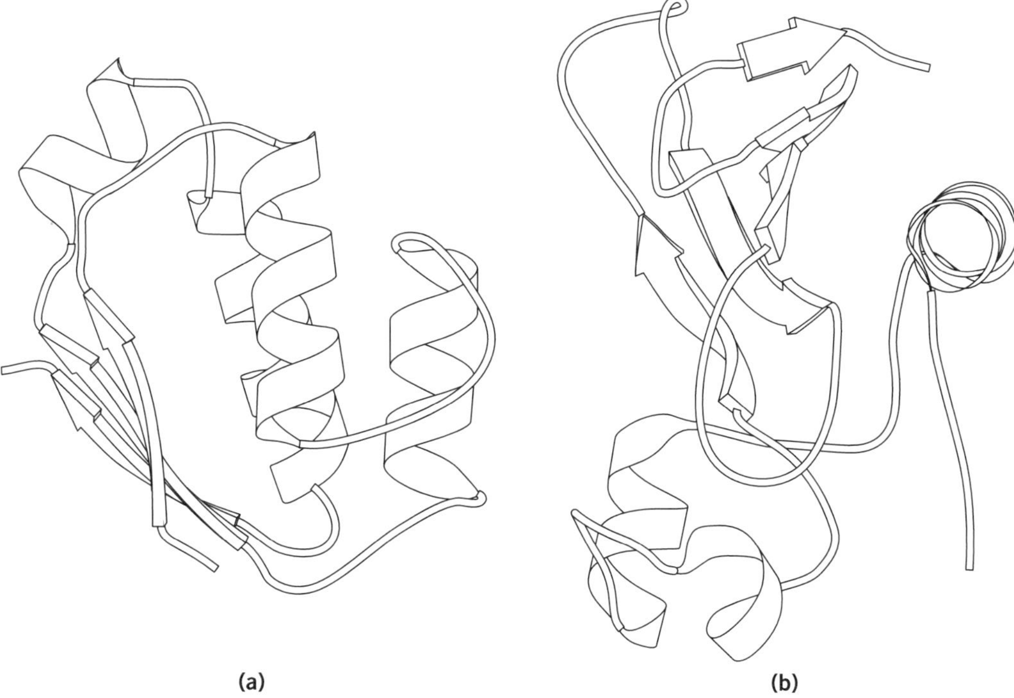

Figure 1. Molscript (1) representation of the structures (**a**) barstar and (**b**) barnase in their complex.

hand, appears to collapse more directly to the folded state, as do most other smaller proteins. Studies of barnase folding, using **kinetic** and equilibrium techniques and **site-directed mutagenesis**, have traced the order in which its parts come together to form the native fold and have estimated the free energies of various interactions in several states that occur during folding.

Barnase is a member of a large family of microbial ribonucleases that share the same basic **tertiary structure** but have quite divergent sequences; this provides a wider system for approaching the central problem of how the fold is determined by the sequence (see **Protein structure prediction**). The homologous ribonucleases of the *Streptomyces* share less than 25% sequence identity with barnase, but the structure of the active-site region is so well conserved that these enzymes are also inhibited by barstar, with dissociation constants as low as 2×10^{-12} M; consequently, coexpression of barstar permits the high level expression of the cloned genes for these enzymes as well. As the *Streptomyces* also produce barstar homologues, it seems clear that the extreme structural conservatism of the active site of the ribonucleases is based on the strict requirement of strong binding to the inhibitor.

Separate from the continuing application of barnase and barstar as small model proteins to general problems of protein structure and chemistry are more recent uses of their genes that take advantage of the toxic effect of barnase expression in tissues of heterologous organisms. Under control of a specific **promoter**, expression of the barnase gene can ablate cells under conditions or in tissues where the promoter is turned on. This property is being applied, for example, in studies of **development**, strategies against **viruses**, design of a conditionally lethal selective **cloning vector**, and, most spectacularly, in the development of male-sterile **plants**. In the last instance, male sterility can be reversed by inclusion of the barstar gene.

There have been several efforts at **computer simulation** of aspects of barnase folding and unfolding, and more can be expected. The continued application of directed mutagenesis, physical chemistry, and computer modeling to the folding mechanisms of barnase and barstar and their homologues will provide insight into the manner in which the sequences of these two interdependent families of proteins determine their native folds. *In vivo* use of the barnase gene, with its lethal effect limited to specific conditions or tissues, will be useful to developmental biologists and should have many practical applications in agriculture and possibly in medicine as well.

BIBLIOGRAPHY

1. P. Kraulis (1991) *J. Appl. Crystallog.* **24**, 946–950.

Suggestions for Further Reading

R. W. Hartley (1997) Barnase and Barstar, in *Ribonucleases, Structure and Functions* (G. D'Alessio and J. F. Riordan, eds.), Academic Press, New York, pp. 51–100. (Contains an exhaustive bibliography through 1995.)

A. Matouschek, L. Serrano, and A. R. Fersht (1994) Analysis of protein folding by protein engineering, in *Mechanisms of Protein Folding* (R. H. Pain, ed.), IRL Press, Oxford, pp. 137–159.

BARR BODY

ALAN F. WOLFFE

In eutherian (placental) mammals, **dosage compensation** mechanisms operate in female cells to silence one of the two **X-chromosomes**, so that female cells have as many X-chromosome-derived transcripts as male cells containing a single X-chromosome. This **X-chromosome inactivation** process was first proposed by Mary Lyon (1961) and is known as the **Lyon hypothesis** (1). Female mammalian embryos begin development with two active X-chromosomes, but very early in embryogenesis almost all of the genes on one of the two X-chromosomes become inactivated (see **Random X inactivation**). Although the initial choice between inactivation of the maternal or paternal X-chromosome is random, once established in a repressed state the same X-chromosome are inactivated after every cell division. The inactivation process occurs over the entire chromosome, and practically all of the genes on the chromosome are silenced. This transcriptional inactivation is concomitant with the chromosome taking on the appearance of **heterochromatin** and also becoming late replicating during **S phase** (see **Facultative heterochromatin**). Only a small fraction of the inactive X-chromosome, including the genes located in the pseudoautosomal region at Xp22.3, escapes the global silencing process (2). The inactive chromosome remains in the nucleus and can be detected cytologically as a *Barr body* in somatic cells (3). The Barr body is found only in cells containing more than one X-chromosome, if cells are **trisomic** for the X-chromosome, two Barr bodies will be detected. The staining procedures used to detect Barr bodies argue for a global difference in chromatin condensation (4). Differences in staining of heterochromatin compared to transcriptionally competent **euchromatin** may be caused by differences in compaction or differences in the association of many more accessory proteins in heterochromatin (5).

The use of advanced microscopy and molecular **cytogenetics**, in which fluorescent ***in situ*** **hybridization** is used to "paint" chromosomal territories, has recently allowed detailed dissection of chromosomal organization (see **Denaturation mapping**). Light microscopic optical serial sectioning of the active and inactive X-chromosome territories reveal that they occupy similar volumes. However, reconstructed active X-chromosomes have a flatter shape and a more extended, folded surface area than the inactive X-chromosome (6). The conclusion is that the differential staining properties of the Barr body are caused mainly by the association of a distinct group of accessory proteins and RNA with this inactive chromosome (see **X-chromosome inactivation**).

Even on the active X-chromosome, most of the chromatin is not transcriptionally active. Thus on both the inactive and the active X-chromosomes the great majority of the chromatin is maintained in a folded state typical of a transcriptionally repressed state. There are, however, global differences in chromatin organization between active and inactive X-chromosomes. The Barr body contains **methylated** DNA, hypoacetylated histones, a specialized structural RNA (Xist), and is late replicating during the S phase. Association with Xist RNA is an important causal factor in X-inactivation. The three-dimensional distribution in the nucleus of Xist RNA coincides with the chromosomal territory occupied by the inactive X-chromosome (7). Even after removing bulk chromatin in the preparation of a nuclear matrix (which is responsible for the overall morphology of the nucleus), the Xist RNA remains in the matrix. This is consistent with the hypothesis that Xist RNA has a structural role in establishing the inactive X-chromosome territory. The Barr body still has a great deal to teach scientists about the molecular mechanisms that establish and maintain nuclear compartments.

BIBLIOGRAPHY

1. M. F. Lyon (1961) *Nature* (London) **190**, 372–373.
2. C. M. Disteche (1995) *Trends Genet.* **11**, 17–22.
3. M. L. Barr and E. G. Bertram (1949) *Nature* (London) **163**, 676–677.
4. S. M. Gartler and A. D. Riggs (1983) *Annu. Rev. Genet.* **17**, 155–190.
5. S. W. Brown (1966) *Science* **151**, 417–425.
6. R. Eils et al. (1996) *J. Cell Biol.* **135**, 1427–1440.
7. C. M. Clemson, J. A. McNeil, H. F. Willard, and J. B. Lawrence (1996) *J. Cell Biol.* **132**, 259–275.

Suggestion for Further Reading

T. Cremer et al. (1993) Role of chromosome territories in the functional compartmentalization of the cell nucleus. *Cold Spring Harbor Symp. Quant. Biol.* **58**, 777–792.

BASE EXCISION REPAIR

AZIZ SANCAR
CAROL THOMPSON

Base excision repair is the **DNA repair** system for abnormal bases in DNA, such as uracil, hydroxymethyluracil, and hypoxanthine, and for nonbulky base modifications, such as thymine glycol and hydrates, 8-hydroxyguanine, O^4-methylthymine, and 3-methyladenine (Fig. 1). The abnormal bases and simple base lesions are removed from DNA by the combined actions of **DNA glycosylases** (1–3) and AP (apurinic/apyrimidinic) endonucleases. The basic reaction carried out by glycosylases is the cleavage of the glycosylic bond joining the base to deoxyribose. Some of the glycosylases cleave both the glycosylic bond and the phosphodiester bond 3′ to the resulting AP site in a more or less concerted reaction. Hence, DNA glycosylases have been classified either as "pure" glycosylases or as glycosylase/AP lyases. AP endonucleases are also classified as either 3′- or 5′-AP endonucleases. All 3′-AP endonucleases are lyases that cleave the phosphodiester bond by β-elimination. The 5′-AP endonucleases carry out true hydrolysis.

The combined actions of glycosylase, AP lyase, and AP endonuclease/exonuclease creates a one- to two-nucleotide gap in the damaged strand. This gap is filled in by **DNA polymerase** I in *Escherichia coli*, and mostly by DNA polymerase β (4) and to a lesser degree by DNA polymerases δ and ε in yeast and humans (5). The resulting nick is ligated by **DNA ligase** in *E. coli*, and in eukaryotes by the ligase I– (6) or ligase III–XRCC1 complex (7). Depending on the polymerase/ligase combination involved in filling in and sealing the gap, a repair patch of 1 to 10 nucleotides may be produced.

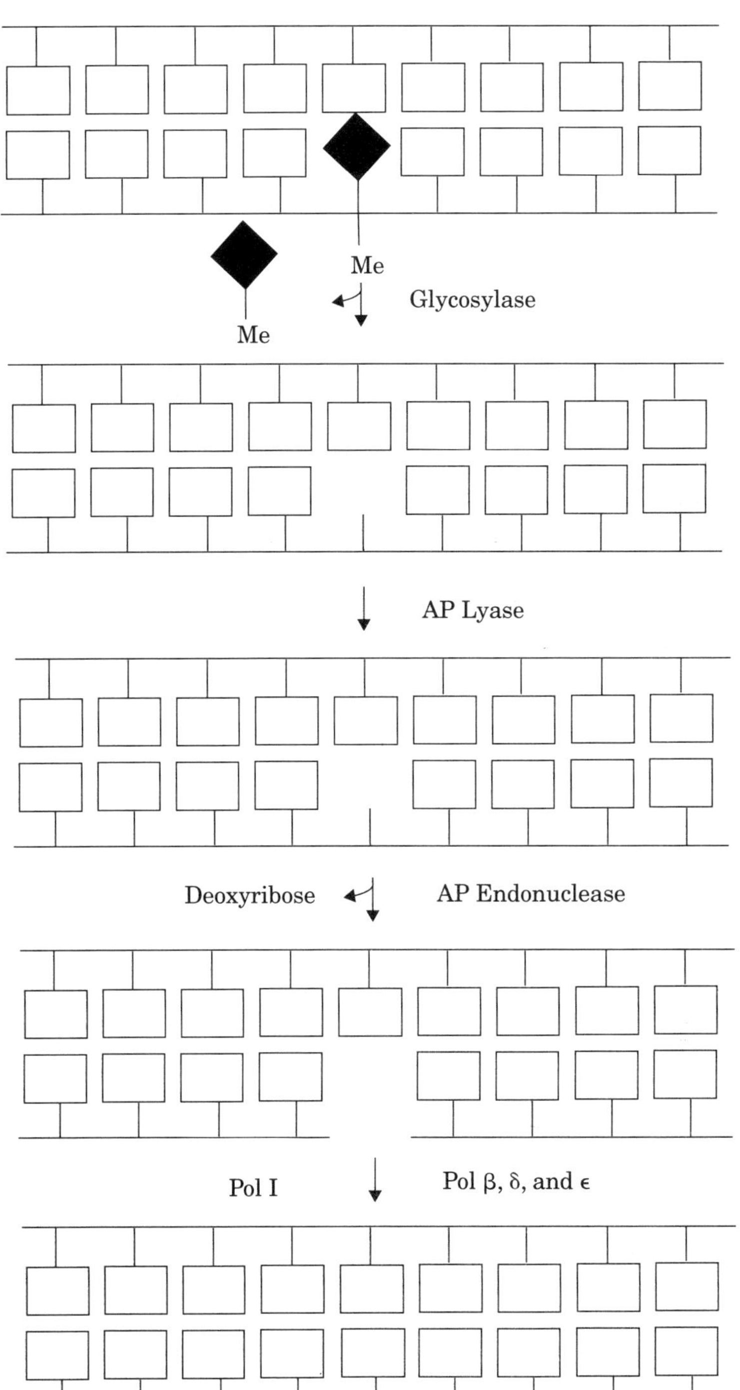

Figure 1. Repair of 3-methyl adenine by base excision repair. Methyladenine DNA glycosylase cleaves the glycosylic bond of the alkylated base; this releases the alkylated base and leaves an AP site. An AP lyase and an AP endonuclease cleave 3′ and 5′, respectively, of the AP site, releasing the abasic sugar. The resulting gap is filled in by Pol I in *E. coli* and by Pol β or Po lδ and ϵ in humans, to produce repair patches of 1 to 10 nucleotides.

BIBLIOGRAPHY

1. T. Lindahl (1974) *Proc. Natl. Acad. Sci. USA* **71**, 3649–3653.
2. T. Lindahl (1976) *Nature* **259**, 64–66.
3. J. Laval (1977) *Nature* **269**, 828–832.
4. R. W. Sobol, J. K. Horton, R. Kohn, H. Gu, R. K. Singhal, R. Prasad, K. Rajewsky, and S. H. Wilson (1996) *Nature* **379**, 183–186.
5. Z. Wang, X. Wu, and E. C. Friedberg (1993) *Mol. Cell. Biol.* **13**, 1051–1058.
6. R. Prasad, R. K. Singhal, D. K. Srivastava, J. T. Molina, A. E. Tomkinson, and S. H. Wilson (1996) *J. Biol. Chem.* **271**, 16000–16007.
7. K. W. Caldecott, C. K. McKeown, J. D. Tucker, S. Ljungquist, and L. H. Thompson (1994) *Mol. Cell. Biol.* **14**, 68–76.

Suggestions for Further Reading

B. Demple and L. Harrison (1994) Repair of oxidative damage to DNA: enzymology and biology. *Annu. Rev. Biochem.* **63**, 915–948.

E. Seeberg, L. Eide, and M. Bjoras (1995) The base excision repair pathway. *Trends Biochem. Sci.* **20**, 391–397.

BASE PAIRS

A. H.-J. WANG
H. ROBINSON

DNA exists predominantly as an antiparallel double-stranded helix. The two strands are helically coiled, which maximizes the exposure of the negatively charged sugar–phosphate backbone to water and shields the **hydrophobic** aromatic bases in the middle from **water**. Figure 1 shows the Watson–Crick base pairs in which the specificity of base pairing is provided by **hydrogen bonds**. The complementary arrangements of hydrogen bond donors and acceptors allow the Watson–Crick base pairing of guanine with cytosine (G:C) and adenine with thymine (A:T). The two C1′ atoms within a G:C base pair and an A:T base pair are equidistant (~10.5 Å). It should be noted that other types of base pairs are also known to exist, and they often are involved in unusual DNA structures.

Bases in a DNA double helix are stacked on top of each other. The stacking interaction is stabilized by the electronic $\pi-\pi$ interaction of the aromatic rings, and it also minimizes the exposure of the bases to the solvent. The base stacking has preferred interactions, depending on the characteristic dipole–dipole interactions between adjacent base pairs. Thus, the stacking patterns vary with the helix type and the particular base–base step. In B DNA, the 5′-purine-*p*-pyrimidine-3′ (Pu-Py) base-pair step has a good stacking overlap, and is thus

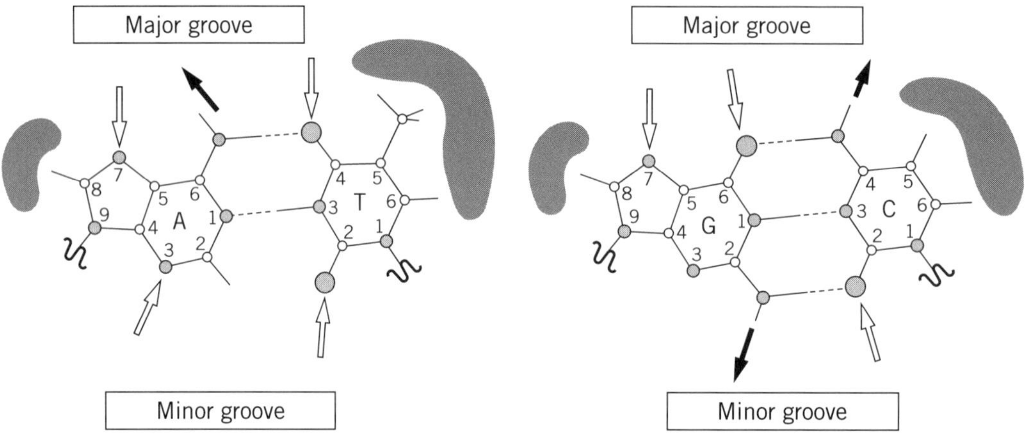

Figure 1. Schematic illustration of the A:T and G:C Watson–Crick base pairs. Hydrogen bonds are shown as dashed lines. Hydrogen bond donors (gray arrow) and acceptors (white arrow) of the bases are shown. Note that the bases can pair in more than one way. Hydrophobic edges of the base pairs are emphasized by gray shade. The major and minor groove edges of the base pair are shown.

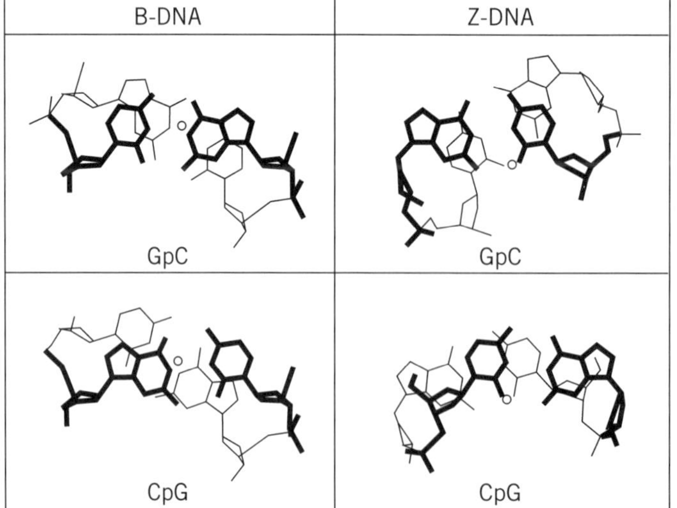

Figure 2. Stacking interactions in B-DNA (left) and Z-DNA (right) showing Pu-Py step stacking (top) and Py-Pu step stacking (bottom). The circle shows the position of the helix axis. Note that the Py-Pu steps (including CpG, TpA, and CpA (= TpG)) in B-DNA have very small stacking interactions, causing those steps to be easily deformed.

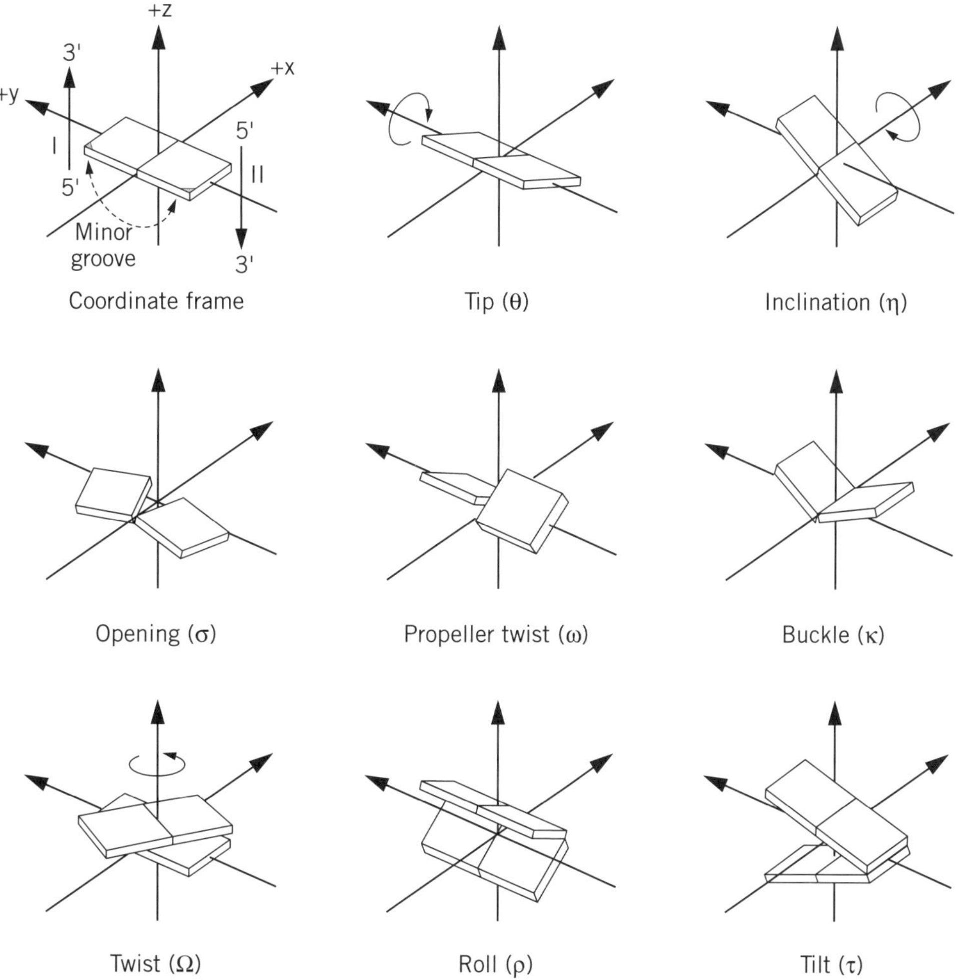

Figure 3. Definitions of various base pair conformations. (upper two rows) or two successive base pairs (bottom row). In the top row the motions of the bases are coordinated, while in the middle row their motions are opposed. The left, center, and right columns describe rotations about the z, y, and x axes, respectively. The standard coordinate frame is defined at the upper left.

more stable, whereas the 5′-pyrimidine-*p*-purine-3′ (Py-Pu) base-pair step has a poor stacking overlap, and is thus less stable (Figure 2). Such differences in stacking energy have strong implications in the bendability of DNA associated with protein binding. It has been noted that DNA sequences with Py-Pu steps in B-DNA are deformed more easily; therefore steps such as CpG, TpA, or TpG (= CpA) are often associated with kinked DNA structures.

Through the high resolution X-ray structures of DNA oligonucleotides, it has been found that base pairs within a DNA duplex (or other structures) have significant departures from a strict coplanar geometry. The conformational parameters associated with the DNA base pairs are defined in Figure 3. For a more complete definition, the reader is referred to the Cambridge Convention (1).

BIBLIOGRAPHY

1. R. E. Dickerson et al. (1989) *EMBO J.* **8**, 1–4.

BASE-PAIR SUBSTITUTION

LYNNETTE FERGUSON
WILLIAM DENNY

These **mutations** of DNA affect only one base pair at a time, possibly producing either amino acid substitutions or chain termination codons in a gene (see **Genetic code**). Amber (UAG), ochre (UAA), and UGA codons will all terminate polypeptide synthesis (see **Stop codons**). Amino acid substitutions often produce leaky and/or **temperature-sensitive** mutations, whereas chain-terminating mutants are usually nonleaky. It has been suggested that there is an important difference between the consequences of frameshift and base-pair substitution mutations in diploid organisms. Frameshifts are often **recessive**, whereas base-pair substitution mutations may exert partial **dominance** (1).

Freese (2) divided base-pair substitutions into two classes, *transitions* and *transversions*. Transitions are mutations in which the purine–pyrimidine orientation is conserved; that is, a purine is replaced by another purine, a pyrimidine by another pyrimidine. The eight possible transversion mutations in double-stranded DNA are: A.T ↔ T.A; T.A ↔ G.C; G.C ↔ C.G; C.G ↔ A.T. Transition mutations have the purine–pyrimidine orientation conserved, as in A.T ↔ G.C or C.G ↔ T.A.

Transitional mutations have been traditionally considered to be due to **tautomerism** of nucleotides. All of the four common DNA bases can exist in tautomeric forms, of which the biologically most interesting involve keto–enol pairs for thymine and guanine and amino–imino pairs for cytosine and adenine. The tautomer, although obeying the usual purine–pyrimidine pairing rules, is capable of hydrogen bonding with a base that is normally noncomplementary. If this occurs and is not repaired, it will lead directly to base-pair substitution mutagenesis in subsequent cell generations. Some base analogues are highly mutagenic, presumably through direct incorporation into DNA through mispairing with a normal base in a tautomeric form (see 2-**Aminopurine**, 5-**Bromouracil**). Some DNA alkylating chemicals may be mutagenic through adding an alkyl group onto a normal nucleotide base, thereby affecting its base-pairing properties in subsequent rounds of replication (see **Dimethyl sulfate**, **Ethyl methane sulfate**). Such premutagenic events will be fixed at a low level during subsequent rounds of replication, typically leading to transition mutations of various types.

BIBLIOGRAPHY

1. C. Wills (1968) *Proc. Natl. Acad. Sci. USA* **61**, 937–944.
2. E. Freese (1959) *Brookhaven Symp. Biol.* **12**, 63–73.

BENCE-JONES PROTEINS

MICHEL FOUGEREAU

By the middle of the nineteenth century, a British physician made a curious observation on the urine of a patient suffering from a bone disease (in fact, a multiple myeloma). Perhaps because his name was Dr. Watson, this practitioner was curious. So, once back home he heated the urine and observed the formation of a precipitate at 60°C, which redissolved upon boiling. Because he was also clever, he realized that this was not normal, but could not provide any explanation. He thus sent a sample to a member of the faculty in London, Dr. Henry Bence-Jones, who not only confirmed the fact but also gave the answer. The mysterious substance was tagged as "hydrated albumin deutoxide." Dr. Watson was certainly pleased with the answer, but was forgotten. Dr. Bence-Jones became immortal, although the nature of the Bence-Jones protein (BJP) remained a puzzle until it was solved by Edelman and Gally (1), who showed, more than a century later, that BJP were the free form of the monoclonal **immunoglobulin** light chain that is associated with a heavy chain in the corresponding myeloma protein.

BJPs have been of great interest for quite a while because they are present in large amounts in urine of most patients with multiple myeloma, thus providing a simple source of monoclonal materials, perfectly suitable for structural analysis of light chains, including amino acid sequencing. The first two sequences were obtained by Hilschmann and Craig in (2) in 1965 for two BJPs of the κ **isotype**, giving the first evidence that immunoglobulin chains contained **variable** and **constant** regions. In normal physiological situations, the synthesis rates of the heavy and light chains are similar, whereas they are most often imbalanced in malignant conditions. In most cases of multiple myeloma, an excess of light chain is produced and secreted, quite often as **disulfide-bonded** dimers, causing an associated kidney disease. The presence of BJP is still a valid sign for the diagnosis and prognosis of multiple myeloma.

In the mouse, an equivalent disease may be induced, at a frequency dependent upon the strain, with a high response in BALB/c, as shown by Potter (3). In the mouse too, BJPs are present in the urine of animals that develop a plasmacytoma. Because the tumor may be transplanted for generations and produce unchanged myeloma proteins and BJP, this provided most of the materials that allowed the immunologists to solve the basic features of **immunoglobulin structure** and to open the way to **monoclonal antibodies**, as well as the key to Ig genes and thus to the elucidation of the genetic basis for antibody diversity.

See also entries **Antibody**, **Immunoglobulin**, and **κ and λ light chain**.

BIBLIOGRAPHY

1. G. M. Edelman and J. A. Gally (1962) The nature of Bence-Jones proteins. Chemical similarities to chains of myeloma globulins and normal γ-globulins. *J. Exp. Med.* **116**, 207–227.
2. N. Hilschmann and L. Craig (1965) Amino acid sequence studies with Bence-Jones proteins. *Proc. Natl. Acad. Sci. USA* **53**, 1403–1409.
3. J. L. Fahey and M. Potter (1959). Bence-Jones proteinemia associated with a transplantable mouse plasma cell neoplasm. *Nature* **184**, 654–655.

Suggestion for Further Reading

R. A. Kyle (1994) The monoclonal gammopathies. *Clin. Chem.* **40**, 2154–2161.

BETA-BULGE

JENNIFER MARTIN

The β-bulge is a term used to define a specific **secondary structure** feature of **protein structures**, where the regular **hydrogen bond** interactions and **backbone** conformation of the **β-sheet** are disrupted by the presence of an extra residue. The additional residue is usually located on a **β-strand** at one edge of the β-sheet, where the bulge is more easily accommodated within the structure of the protein (Fig. 1). β-Bulges occur frequently, and often there are two or more per protein (1). In some cases, β-bulges have been found to be conserved in the structures of related proteins, where they may play a functional role. The more general role of the β-bulge is to alter the direction of the β-strand, so that it may be considered a type of **turn**. However, the change in direction of the polypeptide chain is not as pronounced as that caused by other types of turns.

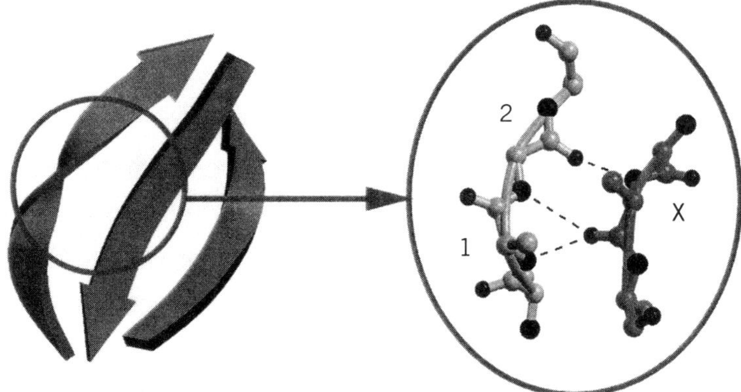

Figure 1. Example of a classic β-bulge in an antiparallel β-sheet. Three β-strands of the antiparallel β-sheet are shown. The normal twist of the β-sheet is clearly accentuated by the β-bulge, caused by an additional residue in the middle β-strand at the edge of the β-sheet. The close-up view of the β-bulge **(right)** shows residue X from the β-strand forming backbone hydrogen bonds (shown as dashed lines) with two residues (1 and 2) from the outer β-strand. For clarity, side-chain atoms have been removed, backbone nitrogen atoms and backbone oxygen atoms are shown as dark spheres. This figure was generated using Molscript (3) and Raster3D (4,5).

A β-bulge is that region between two consecutive β-type hydrogen bonds that includes two residues (called positions 1 and 2) on one β-strand opposite a single residue (position X) on the other β-strand (2). There are several different classes of β-bulge, most (90%) occurring between antiparallel β-strands. The two most common are the classic and the G1, accounting for about 80% of β-bulges. In the classic β-bulge, the residue at position 1 has backbone dihedral angles ($\phi = -100, \psi = -25$) (see **Ramachandran plot**), closer to those of an α-helical than to a β-strand conformation, but residue 2 has angles closer to the β-strand conformation ($\phi = -180$ and $\psi = 160$). In the G1 β-bulge, the residue at position 1 has a positive ϕ value ($\phi = 85$ and $\psi = 0$) and is therefore almost always glycine (thus the name G1). The residue at position 2 of the G1 class has dihedral angles corresponding to β-strand ($\phi = -90$ and $\psi = 150$). The G1 β-bulge often occurs in combination with a type II β-turn (which requires a glycine at position 3). Compared to the usual β-sheet structure, a β-bulge disrupts the alternating side chain placement on one of the β-strands and increases the right-handed twist of the β-strand from the usual 10° to 35° to 45°.

[See also **Beta-sheet**.]

BIBLIOGRAPHY

1. A. W. E. Chan, E. G. Hutchinson, D. Harris, and J. M. Thornton (1993) *Protein Sci.* **2**, 1575–1590.
2. J. S. Richardson, E. D. Getzoff, and D. C. Richardson (1978) *Proc. Natl. Acad. Sci. USA* **75**, 2574–2578.
3. P. J. Kraulis (1991) *J. Appl. Crystallogr.* **24**, 946–950.
4. E. A. Merritt and M. E. P. Murphy (1994) *Acta Crystallogr.* **D50**, 869–873.
5. D. J. Bacon and W. F. Anderson (1988) *J. Mol. Graphics* **6**, 219–222.

BETA-GALACTOSIDASE OF *ESCHERICHIA COLI*

AGNES ULLMANN

Exploration of the lactose (*lac*) system of *Escherichia coli* was started in the late 40s at the Pasteur Institute in Paris by Jacques Monod and his collaborators. As a result, most of the major concepts of **gene expression** and regulation were established with this system (see ***lac* operon**). The major gene product of the *lac* operon is β-galactosidase, the enzyme that splits lactose into glucose and galactose and the product of the *lacZ* gene. In addition to its importance for studies of enzyme induction, this protein has become very useful as a **reporter gene** in very many studies of **gene expression**.

CHARACTERIZATION

Because of its importance in genetic studies of *lac* operon, *E. coli* β-galactosidase was purified and characterized by several groups in the early 50s (9,10). It could be easily measured quantitatively because of a **chromogenic substrate**, *o*-nitrophenyl-β-D-galactoside (ONPG). The enzyme is a hydrolytic transglucosidase and accounts for up to 5% of the total protein in haploid, fully induced or constitutive strains of *E. coli*. The purified enzyme was crystallized, although its three-dimensional structure had to wait 40 years to be solved (see later). Several procedures for producing large amounts of enzyme have been described (reviewed in Ref. 6). β-galactosidase is a particularly stable enzyme. It can be stored in 40% ammonium sulfate for several years at 4 °C without significant loss of activity. In buffered solution, the enzyme is stable at 37 °C for several weeks. At 57 °C the half-life of heat inactivation is 10 min (11). Active enzyme is recovered with 100% yield after **denaturation** with 8 M **urea**, and fairly good recovery is possible after treatment by 6 M **guanidinium chloride** (12).

β-Galactosidase is specific for the β-D-galactopiranoside configuration and cleaves the bond between the anomeric carbon (C1) and the glycosyl oxygen (10). The **enzyme** acts as a hydrolase, but also as a transferase. The galactosyl moiety can be transferred to monosaccharides, oligosaccharides, alkyl alcohols, and phenols and also to **mercaptoethanol**. β-Galactosidase also transfers the galactosyl moiety from the C4 position to the C6 position of glucose to form *allolactose*, which is the natural **inducer** of the *lac* operon (13). β-Galactosidase requires Mg^{2+} and Na^{+} ions for maximal activity (9), but the reason for needing these cations is still unclear.

That β-galactosidase is a tetramer of four identical subunits, possessing one **active site** per subunit, is well documented, and the tetramer is the only active form of the enzyme. Because of its high turnover number (about 6000 moles of ONPG hydrolyzed/sec/mole of enzyme, one molecule of β-galactosidase per bacterial cell can be accurately measured. Several residues (Glu 461, Met 502, Tyr 503, Glu 537) have been identified as important for catalytic function or to be near the active site (14,15). These residues are found in the three-dimensional structure in close proximity to one another and form a pocket likely to be the substrate-binding site (16).

The amino acid sequence of β-galactosidase, determined before DNA **sequencing** techniques had been developed, was the longest protein sequence ever determined (17). According to this sequence, the protein contains 1021 amino acid residues and a subunit molecular weight of 116,248. Subsequently, the nucleotide sequence of the *lacZ* gene was also determined (18), from which it was predicted that β-galactosidase consists of 1023 residues, with a molecular weight of 116,353 per subunit. It is quite remarkable that the amino acid sequence deduced from the DNA sequence differs only in ten amino acid residues from that obtained by amino acid sequencing of such a large protein.

There is also a **homologue** of β-galactosidase, evolved β-galactosidase (EBG) . Normally *E. coli* does not grow on lactose when the *lacZ* gene is deleted. However, cells with *lacZ* deletions were selected that grow on lactose and produce a new β-galactosidase, called EBG (19). The *ebgA* gene maps at 66 min on the *E. coli* chromosome, whereas that for *lacZ* maps at 8 min. The active EBG protein is a hexamer containing 964 residues per monomer. The sequence identity between the *ebgA* and *lacZ* genes at the DNA nucleotide level (50%) exceeds that at the amino acid level (30%). This led to the conclusion that the two genes descended by **divergence** from a common ancestral gene. β-galactosidase and EBG do not cross-react with **antisera** prepared against each type of protein.

β-GALACTOSIDASE COMPLEMENTATION

The term **complementation** has been generally used in genetics for the phenomenon by which a biological function lost or altered by a mutation is restored through mutual compensation by two differently altered mutant genes. In the case of *intracistronic complementation*, where the two mutations occur within different copies of the same **cistron**, it is generally accepted that the repair of function occurs at the level of the protein product of the gene and involves the interaction of differently altered polypeptide chains (see **Interallelic complementation**). Such complementation involves noncovalent interaction of polypeptide chains and may occur by reassociation of differently altered subunits of an oligomeric protein or by interaction between fragments of a single polypeptide chain. Studies of β-galactosidase complementation contributed to elucidating the structure of the enzyme and to understanding the mechanism of the specific recognition, reassociation, and folding of proteins (for reviews, see Refs. 20,21).

Complementation Between Point Mutants

Jacob and Monod (8) showed that heterogenotes carrying different $lacZ^-$ point mutants may become $lacZ^+$ by complementation. *In vitro* studies carried out by Perrin (22) showed that many of the inactive proteins containing these point mutations exhibit monomeric structures, but cross-react immunologically with β-galactosidase. In contrast, the active, complemented, β-galactosidase had a sedimentation coefficient like tetrameric wild-type 16 S, although it was more heat-labile. The restoration of enzyme activity in this type of complementation can be accounted for by a mechanism suggested by Crick and Orgel (23), repair of lesions by reassociation of differently altered subunits.

Some inactive β-galactosidase proteins with point mutations exist as dimers. A specific class of these mutants can be activated up to 1000 times by specific antibodies raised against wild-type β-galactosidase (24,25). The activating antibody directs the formation of active tetramers. In a strict sense this is not a complementation reaction, but the kinetics of activation are similar to those of complementation between point mutant proteins *in vitro*: In both cases, the appearance of β-galactosidase activity is a relatively slow process, requiring about 300 min at 37 °C.

Complementation Between Deletion Mutants

The isolation of deletion mutants of the *lacZ* gene involving quite large genetic segments (26) led to the discovery of a new type of intracistronic complementation (for a review, see Ref. 27). Inactive *lacZ* deletion mutants, lacking part of the gene corresponding to either the amino- or the carboxy-terminal region of the β-galactosidase polypeptide chain, would complement another inactive deletion mutant containing the region missing in the first. The N- and C-terminal regions involved in the two distinct classes of complementation are known as the α and ω regions, respectively (Fig. 1). Both α- and ω-complementation represent noncovalent reassociation of complementary fragments of the β-galactosidase subunit polypeptide chain, which then can reassemble into an enzymatically active tetrameric structure. α- and ω-complementation takes place *in vivo* in partial diploids and *in vitro*, with extracts of relevant mutant strains.

α-Complementation

All mutants that have their promoter-proximal segment (α-segment) intact are α-donors. All genes that contain partial deletions of the N-terminus not extending beyond a barrier (which is located by the dashed line in Figure 1 between mutants 274 and X64) are α-acceptors when the proteins they produce are mixed with that of an α-donor (30). The time course of *in vitro* α-complementation is relatively slow and reaches a plateau in about 3 h. The α-complemented enzyme, like wild-type β-galactosidase, is a tetramer but is less stable to heat or urea treatment. The equilibrium constant for the complementation reaction was estimated to be approximately 1 to

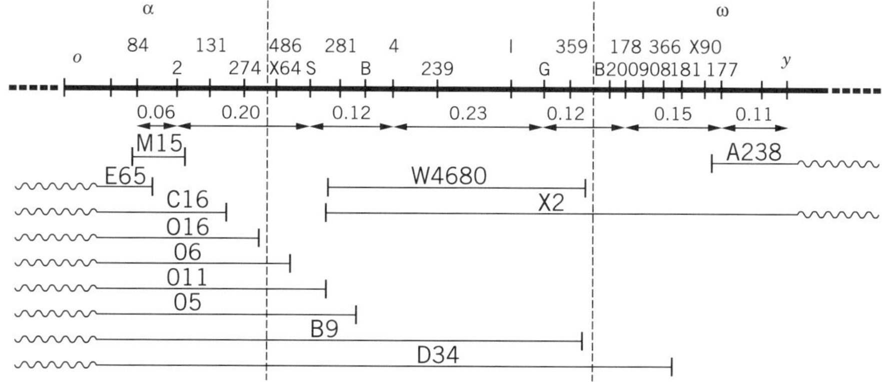

Figure 1. Diagrammatic representation of the *lacZ* gene. The figures and letters above the solid line indicate the position of point mutations in the *lacZ* gene: *o*, operator; *y*, structural gene of Lac permease. The figures below the line represent recombination frequencies, as determined by crosses between two point mutants. The lines below indicate the extent of various deletions: M 15 isolated by Beckwith (28); A 238 and W 4680, isolated by Cook and Lederberg (29); all others by Jacob, Ullmann and Monod (26). Reprinted from (20) by permission of Cold Spring Harbor Laboratory.

$2 \times 10^9 M^{-1}$ (21). Given the slow association rate constant, one can predict a very low dissociation rate, indicating that the complementation process is virtually irreversible under physiological conditions.

The most unusual property of the α-peptide is its high temperature resistance in 6 M guanidinium chloride. Bacterial extracts of *lacZ* strains containing α-donors liberate soluble α-peptides after boiling in this denaturant (30) or after autoclaving in its absence (31). The smallest α-peptide still capable of complementation has a size of about 7400 Da. Upon cleavage of β-galactosidase with **cyanogen bromide**, Langley et al. (32) isolated and sequenced a peptide that has α-donor activity corresponding to residues 3 to 92 of the polypeptide chain.

The best studied α-acceptor is the M15 protein, produced by the *lacZ* mutant ΔM15 isolated by Beckwith (28) and shown genetically to be a small deletion in the early part of the gene (Fig. 1). The M15 protein lacks residues 11 through 41 of β-galactosidase and is a dimer (33). It has a trace level of enzyme activity and can be significantly activated by addition of anti-β-galactosidase antibodies (34), up to 15% of that obtained by α-complementation.

The nature of the α-complementation process and its probable pathway have been studied extensively (21). It is believed that the α-peptide modifies the **quaternary structure** of the α-acceptor by causing its tetramerization and consequently the recovery of enzymatic activity. The three-dimensional structure of β-galactosidase (16) fully confirms this prediction by showing that the α-peptide segment of the polypeptide chain participates in forming specific subunit contacts (see later).

ω-Complementation

All mutants that have their promoter-distal region (ω-region) intact are ω-donors. They complement ω-acceptors that are products of deletion or point mutants in the ω-region (27) (Fig. 1), which represents about one-third of the total genetic length of *lacZ*. The purified ω-peptide has a molecular weight of 40,000 Da (35). The kinetics of ω-complementation are faster than those of α-complementation and reach a maximum extent in about 1 h. Although ω-peptide is reversibly **renatured** from 8 M urea or 6 M guanidinium chloride, it is highly heat-labile.

ω-Acceptor proteins active in complementation have not been obtained in pure form. In crude bacterial extracts (36,37), ω-acceptor exists in different forms of aggregation, depending on the ionic strength. At low ionic strength it is probably a tetramer that dissociates at high ionic strength, most probably into monomers. Both forms are active in complementation. The ω-complemented enzyme is a tetramer, like the wild type, and exhibits very similar catalytic properties. The main difference appears after denaturation in 8 M urea. Recovery of its enzyme activity after **renaturation** does not exceed 5%, whereas it is practically 100% with wild-type enzyme. Moreover, the ω-peptide is recovered quantitatively from the complemented enzyme after urea treatment (11).

One of the most remarkable features of ω-complementation is the high affinity and specificity that the peptide fragments exhibit. They reassociate even at high dilution and in the presence of a large excess of foreign proteins. The yield of complementation is considerably increased in the presence of a β-galactosidase substrate analog or in the presence of anti-β-galactosidase antibodies (36,37). Comparing wild-type and ω-complemented β-galactosidase, Goldberg (38) proposed a model for the tertiary structure of the two monomers. After having shown that the active, complemented enzyme involves reassociation into a tetrameric structure of four ω-acceptor polypeptides with four ω-peptides, he showed that both ω and acceptor are folded to form compact globular structures, close to those of the corresponding segments of the complete polypeptide within the native enzyme. This "globule model" involves the notion that large polypeptide chains contain several distinct **domains** that assemble specifically to give the polypeptide chain its final tertiary structure. The well-defined compact ω-domain is apparent within the native monomer in the three-dimensional structure of β-galactosidase (16), which also shows an unstructured 10-residue exposed segment that joins the ω-domain to the rest of the chain. This explains, *a posteriori*, why mild proteolytic treatment of the wild-type β-galactosidase liberates significant amounts of ω-peptide (11).

THREE-DIMENSIONAL STRUCTURE

The structure of tetrameric β-galactosidase was solved by **X-ray crystallography** at 2.5 Å resolution by B.W. Matthews et al. (16). It is the longest polypeptide chain for which an atomic structure has been determined. The crystal structure shows that β-galactosidase is a tetramer with dimensions of roughly 175 × 135 × 90 Å along the respective twofold axes. The constituent monomers interact through two different monomer–monomer contacts. The β-galactosidase monomer consists of five compact domains and a relatively extended 50-residue N-terminal segment, corresponding to the α-peptide, which contributes to the activating interface (Fig. 2). The monomer is relatively flat and elongated and has approximate dimensions of 50 × 100 × 40 Å. The first domain consists of residues 52 to 217, the second of residues 220 to 334, the third of residues 334 to 627, the fourth of residues 627 to 736, and the final fifth domain, corresponding to the ω peptide, of residues 737 to 1023. Topologically, the second domain is identical to the fourth domain. The modular structure of the monomer corroborates earlier findings (37,38), which suggested that the folding of this rather long polypeptide chain is facilitated by each domain acting as an independent unit.

At present, no structure of a complex of β-galactosidase with a substrate or analog is available, so the specific interaction sites between enzyme and substrate are not known. Nevertheless, earlier biochemical studies (14,15) concluded that residues important for catalytic function or near the **active site** (Glu461, Met502, Tyr503, Glu537) are found close to one another and located around a deep pit within the third domain, which is postulated to be the substrate-binding site. Residues from distant regions of the sequence (Trp568, Trp999) also contribute to the active site.

The X-ray structure of β-galactosidase provides a structural rationale for both α- and ω-complementation. The N-terminal segment, representing the α peptide, participates in forming a subunit interface, which in turn allows formation of the active site, made up of elements from two different subunits. The ω-fragment, corresponding to the fifth domain of the subunit, exhibits a specific topology. Rotation of this domain by more than 90° positions Trp 999 at the active site in the vicinity of the third domain.

β-GALACTOSIDASE AS A TOOL

The β-galactosidase of *E. coli* is one of the most commonly used enzymes in molecular biology for several reasons: (*1*) numerous genetic and biochemical aspects of the *lac* system are known; (*2*) several indicator media are available for detecting lactose metabolism and monitoring Lac^+ and Lac^- bacteria; (*3*) the assay procedure based on o-nitrophenyl-β-D-galactoside (ONPG) hydrolysis is exquisitely sensitive; (*4*) the protein is highly stable and easy to purify; and (*5*) the availability of substituted galactoside derivatives permits selecting and screening *lac* mutants exhibiting a variety of phenotypes. For example, the chromogenic noninducing and colorless substrate, 5-bromo-4-chloro-3-indolyl-β-D-galactoside (X-Gal), forms an insoluble blue dye upon hydrolysis, thus providing a sensitive test for β-galactosidase in solid media. X-Gal also discriminates between strains that produce high and low levels of β-galactosidase, allowing detection of plaque-forming **bacteriophages** that carry the *lacZ* gene.

β-GALACTOSIDASE FUSIONS

The observation that the N-terminal 23 residues of β-galactosidase can be replaced with other amino acids without affecting enzymatic activity enabled Müller-Hill and Kania (39) to obtain fusions between *lacI* and *lacZ* genes that produce a hybrid protein that has both **Lac repressor** and β-galactosidase activities. Several genetic methods have been developed to construct *lac* fusions that produce hybrid genes which specify hybrid proteins. Subsequently, *in vitro* construction of hybrid genes facilitated obtaining a variety of fusions (40). A number of commercially available vectors exist for constructing *lacZ* fusions *in vitro*. They contain a *lacZ* gene truncated at the 5′-end and preceded by a synthetic oligonucleotide containing multiple **restriction enzyme** cleavage sites, a *polycloning* site. Thus, if a 5′-coding sequence is cloned into one of these sites so that **transcription** and **translation** are restored across *lacZ*, a hybrid protein with β-galactosidase activity is produced. The sensitive detection of this activity with X-Gal has made β-galactosidase one of the most commonly used *in vivo* **reporter enzymes**. Using the *lacZ* gene as a reporter provides a sensitive method to detect genes subject to specific regulatory signals or to study the mechanism localizing a protein to a given cellular compartment. The regulation, identification, and localization of many proteins of various origins have been uncovered by using the β-galactosidase activity of fusion proteins as a marker.

During the last few years, a number of new eukaryotic expression vectors that produce β-galactosidase fusion proteins have been designed. They are currently being used to analyze developmentally regulated genes, transgenic expression, tissue-specific expression, and a number of other aspects involved in embryogenesis or developmental biology. This approach has been facilitated by using replication-defective retroviral vectors that encode the *lacZ* gene (41), which has become a widely used cell lineage marking system. The labeled cells are easily identified by histochemical staining, using X-Gal or **immunofluorescence** with antibodies raised against β-galactosidase.

β-galactosidase fusions provide a unique and versatile experimental tool. The hybrid proteins are easy to purify (42) and can be used as **antigen** to raise antibodies against the target gene product. In addition, because there is no restriction on the size of the target gene DNA fused to *lacZ* and because many hybrid proteins retain all or a portion of the activity of the target gene product, these fusions provide a means to identify functional domains within the target gene product.

α-Complementation

Perhaps the most widely exploited property of β-galactosidase is α-complementation, described previously. Whereas the region of the foreign protein that can replace the N-terminus of β-galactosidase is more or less random in hybrid fusion proteins, replacements in the α-peptide for α-complementation are more restrictive. That insertion of a few amino acid

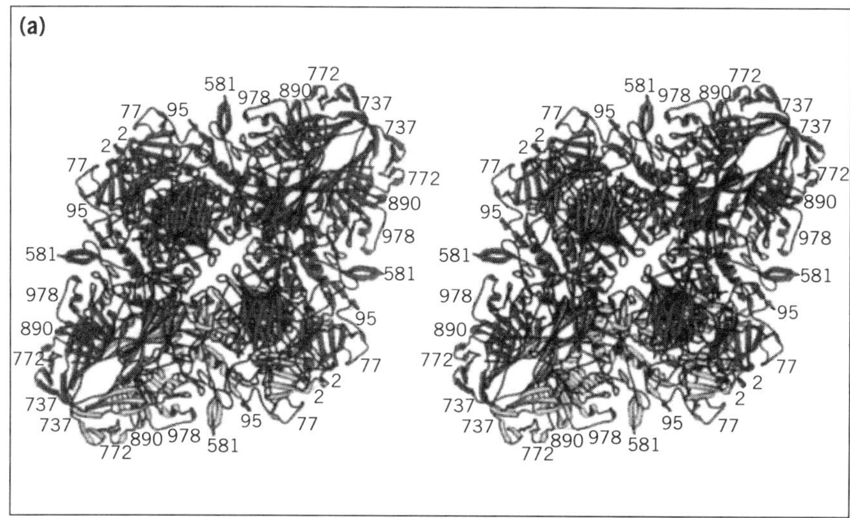

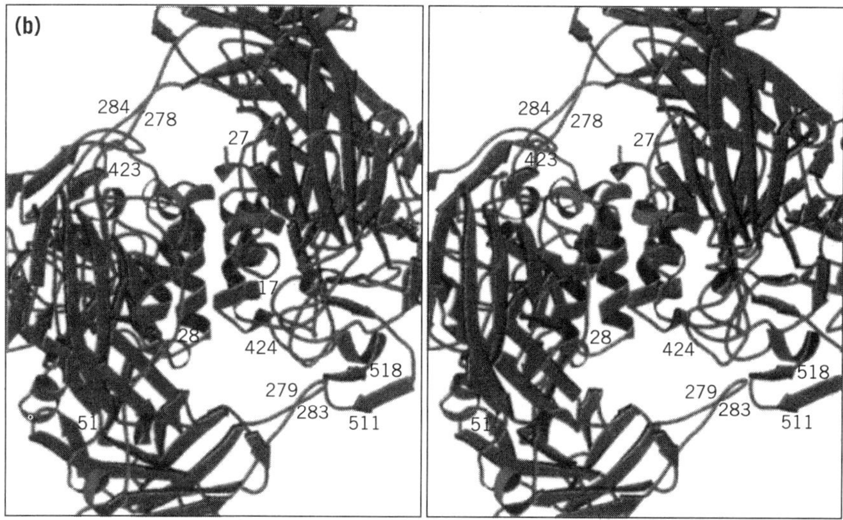

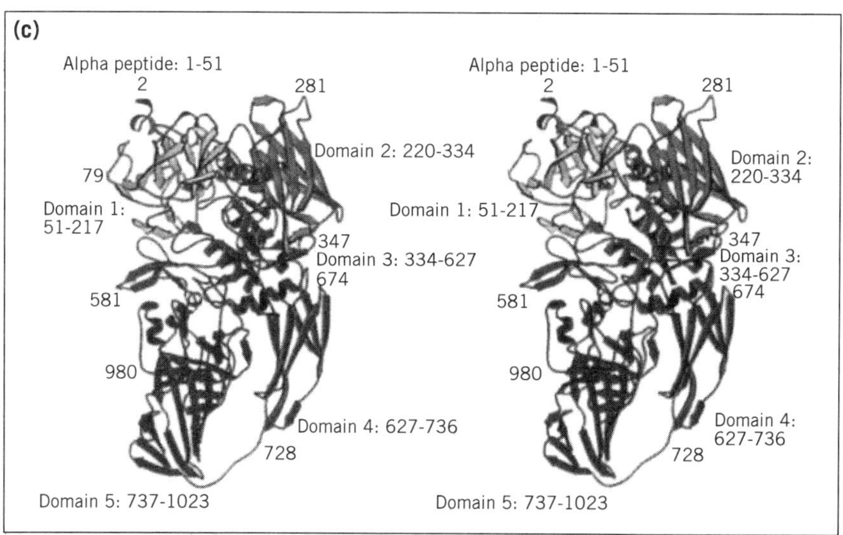

Figure 2. Three-dimensional structure of β-galactosidase. (**a**) Ribbon representation of the β-galactosidase tetramer showing the largest face of the molecule. Contacts between red/green and blue/yellow dimers form the long interface. Contacts between the red/yellow and blue/green dimers form the activating interface. Formation of the tetrameric particle results in two deep clefts that run across opposite faces of the molecule. Each contains two active sites. (**b**) Ribbon diagram of the blue/green dimer viewed down the molecular two-fold axis, showing the composition of the activating interface. Residues 1 to 50 from each chain, which form the α-complementation region (see text), are shown in red. The interface includes contacts between the respective complementation peptides, between two α-**helices** from the respective monomers that pack together to form a four-helix bundle and between an extended loop (residues 272 to 288) from each monomer that reaches across the interface and extends into the active-site region of the neighboring monomer, stabilizing the active site structure. (**c**) Stereo ribbon diagram of the β-galactosidase monomer showing the domain organization of the chain. Residues corresponding to successive domains are colored in successive spectral colors. See color insert.

residues does not always interfere with complementation capacity (43) was widely used during the late 70s to develop new cloning vectors, now commercially available. Indeed, α-complementation has become one of the most commonly used methods for identifying bacterial colonies containing recombinant DNA.

The first vector based on α-complementation was developed by Messing et al. (44). Within the major intergenic region of the filamentous bacteriophage M13, they inserted the *lac* regulatory region and the DNA sequence that codes for the first 146 amino acids of β-galactosidase that have α-complementation capacity. When infected with these phages, bacteria that produce the M15 α-acceptor protein produce active β-galactosidase by complementation and form blue plaques in the presence of the chromogenic substrate X-Gal. Insertion of foreign DNA into the α region of M13 interferes with α-complementation and gives rise to recombinants that form pale blue or colorless plaques. This simple test has made **cloning** in filamentous bacteriophages a routine procedure. The insertion of a series of synthetic restriction enzyme sites into the α-region provides a variety of targets for cloning foreign DNA fragments. Insertion of these polycloning sites, representing up to 18 extra amino acid residues, into the α-region has practically no effect on α-complementation capacity. Insertion of additional DNA into a cloning site, however, generally abolishes α-complementation and creates recombinant bacteriophages that produce colorless plaques when grown on X-Gal plates (*blue/white screen*). The great advantage of using vectors generated from single-stranded DNA phage, like M13, is that one obtains both the double-stranded **replicative form** from the infected bacteria and the single-stranded DNA, from the progeny phage. The latter is frequently used as a **template** for sequencing with the dideoxy-mediated chain-terminating **Sanger method** (45) and also for *in vitro* **mutagenesis**. Many other vectors that permit identification of recombinant clones based on α-complementation have been designed subsequently. The most commonly used are the pUC vectors (46).

Recently, Mohler and Blau (47) adapted intracistronic complementation of the *lacZ* gene for use in eukaryotic cells. By constructing replication-defective **retroviral** vectors that express α-donors and α-acceptors and ω-donors and ω-acceptors, complementation of the relevant *lacZ* mutants in mammalian cells permits analyzing cell fusion and detecting colocalized interacting proteins within single intact cells. α- and ω-complementation can be exploited for a wide range of studies, including **transgenic** animals that express complementing *lacZ* mutants from two **promoters** of interest, which then could reveal cell lineages in which the products of both genes coincide spatially and temporally. In line with these results, Rossi et al. (48) applied β-galactosidase complementation to monitor protein–protein interactions in intact eukaryotic cells.

Given the wide range of applications for β-galactosidase during the last decades, one can anticipate that further development of the β-galactosidase fusion and complementation systems will contribute to our understanding of many features of cellular regulation and organization.

BIBLIOGRAPHY

1. M. Cohn, J. Monod, H. R. Pollock, S. Spiegelman, and R. Stanier (1953) *Nature* **172**, 1096–1097.
2. J. Monod, G. Cohen-Bazire, and M. Cohn (1951) *Biochem. Biophys. Acta* **7**, 585–599.
3. J. Lederberg and E. L. Tatum (1946) *Cold Spring Harbor Symp. Quant. Biol.* **11**, 113–114.
4. E. L. Wollman, F. Jacob, and W. Hayes (1956) *Cold Spring Harbor Symp. Quant. Biol.* **21**, 141–162.
5. H. W. Rickenberg, G. N. Cohen, G. Buttin, and J. Monod (1956) *Ann. Inst. Pasteur* **91**, 829–857.
6. I. Zabin and A. V. Fowler (1978) In *The Operon* (J. H. Miller and W. S. Reznikoff, eds.), Cold Spring Harbor Laboratory, Cold Spring Harbor, New York, pp. 89–121.
7. A. B. Pardee, F. Jacob, and J. Monod (1959) *J. Mol. Biol.* **1**, 165–178.
8. F. Jacob and J. Monod (1961) *J. Mol. Biol.* **3**, 318–356.
9. M. Cohn (1957) *Bact. Rev.* **21**, 140–168.
10. K. Wallenfels and O. P. Malhotra (1961) *Adv. Carbohydrate Chem.* **16**, 239–298.
11. A. Ullmann, F. Jacob, and J. Monod (1968) *J. Mol. Biol.* **32**, 1–13.
12. A. Ullmann and J. Monod (1969) *Biochem. Biophys. Res. Commun.* **35**, 35–42.
13. C. Burstein, M. Cohn, A. Kepes, and J. Monod (1965) *Biochem. Biophys. Acta* **95**, 634–639.
14. M. Ring and R. E. Huber (1990) *Arch. Biochem. Biophys.* **283**, 342–350.
15. C. G. Cupples, J. H. Miller, and R. E. Huber (1990) *J. Biol. Chem.* **265**, 5512–5518.
16. R. H. Jacobson, X-J. Zhang, R. F. DuBose, and B. W. Matthews (1994) *Nature* **369**, 761–766.
17. A. V. Fowler and I. Zabin (1977) *Proc. Natl. Acad. Sci. USA* **74**, 1507–1510.
18. A. Kalnins, K. Otto, U. Rüther, and B. Müller-Hill (1983) *EMBO J.* **2**, 593–597.
19. H. W. Stokes, P. W. Betts, and B. G. Hall (1985) *Mol. Biol. Evol.* **2**, 469–477.
20. A. Ullmann and D. Perrin (1970) In *The Lactose Operon* (J. R. Beckwith and D. Zipser, eds.), Cold Spring Harbor Laboratory, Cold Spring Harbor, New York, pp. 143–172.
21. I. Zabin (1982) *Mol. Cell. Biochem.* **49**, 87–96.
22. D. Perrin (1963) *Cold Spring Harbor Symp. Quant. Biol.* **28**, 529–532.
23. F. H. C. Crick and L. E. Orgel (1964) *J. Mol. Biol.* **8**, 161–165.
24. B. Rotman and F. Celada (1968) *Proc. Natl. Acad. Sci. USA* **60**, 660–667.
25. E. Conway de Macario, J. Ellis, R. Guzman, and B. Rotman (1978) *Proc. Natl. Acad. Sci. USA* **75**, 720–724.
26. F. Jacob, A. Ullmann, and J. Monod (1965) *J. Mol. Biol.* **31**, 704–719.
27. A. Ullmann, D. Perrin, F. Jacob, and J. Monod (1965) *J. Mol. Biol.* **12**, 918–923.
28. J. R. Beckwith (1964) *J. Mol. Biol.* **8**, 427–430.
29. A. Cook and J. Lederberg (1962) *Genetics* **47**, 1335–1353.
30. A. Ullmann, F. Jacob, and J. Monod (1968) *J. Mol. Biol.* **24**, 339–343.
31. S. L. Morrison and D. Zipser (1970) *J. Mol. Biol.* **50**, 359–371.
32. K. E. Langley, A. V. Fowler, and I. Zabin (1975) *J. Biol. Chem.* **250**, 2587–2592.
33. K. E. Langley, M. R. Villarejo, A. V. Fowler, P. J. Zamenhof, and I. Zabin (1975) *Proc. Natl. Acad. Sci. USA* **72**, 1254–1257.
34. R. S. Accolla and F. Celada (1976) *FEBS Lett.* **67**, 299–302.
35. M. E. Goldberg and S. J. Edelstein (1969) *J. Mol. Biol.* **46**, 431–440.
36. A. Ullmann and J. Monod (1970) in *The Lactose Operon* (J. R. Beckwith and D. Zipser, eds.), Cold Spring Harbor Laboratory, Cold Spring Harbor, New York, pp. 265–272.
37. F. Celada, A. Ullmann, and J. Monod (1974) *Biochemistry* **13**, 5543–5547.

38. M. E. Goldberg (1969) *J. Mol. Biol.* **46**, 441–446.
39. B. Müller-Hill and J. Kania (1974) *Nature* **249**, 561–563.
40. T. J. Silhavy and J. Beckwith (1985) *Microbiol. Rev.* **49**, 398–418.
41. J. R. Sanes, J. L. R. Rubenstein, and J-F. Nicolas (1986) *EMBO J.* **5**, 3133–3142.
42. A. Ullmann (1984) *Gene* **29**, 27–31.
43. U. Rüther, M. Koenen, K. Otto, and B. Müller-Hill (1981) *Nucleic Acids Res.* **9**, 4087–4098.
44. J. Messing, B. Gronenborn, B. Müller-Hill, and P.H. Hofschneider (1977) *Proc. Natl. Acad. Sci. USA* **74**, 3642–3646.
45. F. Sanger, S. Nicklen, and A. R. Coulson (1977) *Proc. Natl. Acad. Sci. USA* **74**, 5463–5467.
46. J. Vieira and J. Messing (1982) *Gene* **19**, 259–268.
47. W. A. Mohler and H. M. Blau (1996) *Proc. Natl. Acad. Sci. USA* **93**, 12423–12427.
48. F. Rossi, C. A. Charlton, and H. M. Blau (1997) *Proc. Natl. Acad. Sci. USA* **94**, 8405–8410.

Suggestions for Further Reading

The Lactose Operon (1970). (J. R. Beckwith and D. Zipser, eds.), Cold Spring Harbor Laboratory, Cold Spring Harbor, New York, 437 pages.

A. Ullmann (1992) Complementation in β-galactosidase: From protein structure to genetic engineering, *BioEssays* **14**, 201–205.

B. Müller-Hill (1996) *The lac Operon. A Short History of a Genetic Paradigm*, Walter de Gruyter. Berlin, New York.

BETA-GLUCURONIDASE

NICHOLAS ALLEN

β-Glucuronidase (E.C. 3.2.1.31; GUS) is the most widely used **reporter gene** in plant molecular biology (1). The *gusA* gene cloned from *Escherichia coli* encodes a 68-kDa β-glucuronidase that forms a stable homotetramer and catalyzes the hydrolysis of a large number of glucuronides, in which D-glucuronic acid is conjugated through a β-*O*-glycosidic linkage to any aglycone. β-Glucuronidase has emerged as the reporter gene of choice to be used in plants, because there is very little or no endogenous glucuronidase activity across the phyla. This minimizes the detection of background activity and allows very small quantities of GUS to be detected. GUS activity may be followed in cell lysates or *in situ*. The major use of GUS has been to study gene expression patterns in transgenic plants by expressing GUS under the control of regulatory sequences of interest. A second use has been to monitor the intracellular fate of chimeric proteins by generating GUS fusion genes. For example, GUS has been fused to gene leader sequences that target the fusion gene to different organelles and allow intracellular trafficking of proteins to be studied (2).

GUS activity may be detected using a variety of substrates. As in the case of **beta-galactosidase** and X-gal, GUS expression is most commonly detected using histochemical substrates, such as X-GlcU (5-bromo-4-chloro-3-indolyl β-D-glucuronide), which gives a dark blue precipitate upon hydrolysis. Alternative substrates include 5-bromo-6-chloro-3-indolyl β-D-glucuronide, 6-chloro-3-indolyl β-D-glucuronide, and indoxyl β-D-glucuronide that give magenta, pink, and blue precipitates, respectively (3). The variety of substrates also makes GUS a valuable component of dual (or multiple) reporter gene systems; for example, GUS and *lacZ* expression could be detected in the same tissues by using substrates that give different colored histochemical precipitates. In addition to histochemical substrates, a number of fluorescent and chemiluminescent substrates have been developed that are analogous to the substrates for β-galactosidase, such as CFDG-GlcU (Molecular Probes Inc.) and 4-methylumbelliferyl β-D-glucuronide (MUGlcU) (4).

In the case of *lacZ*, problems may be encountered in loading substrates into cells; to load FDG, for example, cells must undergo a moderate osmotic shock that can affect cell viability. In the case of GUS, however, a second gene, *gusB*, may be expressed that encodes a **permease** that actively takes up and transports glucuronide substrates into the cell. In addition to substrates that allow reporter gene visualization, a number of other bioactive molecules can be conjugated to glucuronides, which could then be released by hydrolysis in GUS-expressing cells. Thus, combined use of gusA and gusB increases the use of the reporter gene, from merely indicating gene expression to controlling a specific cell manipulation (1).

BIBLIOGRAPHY

1. R. A. Jefferson (1989) *Nature* **342**, 837–838.
2. R. A. Jefferson, T. A. Kavanagh, and M. W. Bevan (1987) *EMBO J.* **6**, 3901–3907.
3. G. A. Hull and M. Devic (1995) *Methods. Mol. Biol.* **49**, 125–141.
4. C. E. Olesen, J. J. Fortin, J. C. Voyta, and I. Browstein (1997) *Methods Mol. Biol.* **63**, 61–70.

BETA-HELIX

JENNIFER MARTIN

A β-helix is a type of **protein motif** or **domain** found in a very few **protein structures**. It is characterized by an unusual parallel **β-sheet** topology formed from three parallel β-sheets wound together into a right-handed helical structure (Fig. 1). The **β-strands** in each β-sheet are short, having only 2 to 5 residues. Each coil of the β-helix has the same three-dimensional arrangement of a group of **secondary structure**

Figure 1. Orthogonal views of the β-helix of pectate lyase E (1), showing the three parallel β-sheets. For clarity, only the β-strand secondary structure (as purple arrows) and loops connecting the strands (yellow) are shown. Short helical regions and the *N*- and *C*-terminal loops have been removed. This figure was generated using Molscript (2) and Raster3D (3,4). See color insert.

elements and is thus similar to other coiled repeating structures, such as the β-roll, composed of two parallel β-sheets forming a β-helix, and the **leucine-rich repeat** of **ribonuclease inhibitor** that is composed of repeating α**-helices** and β-strands forming a horseshoe-shaped structure. In the β-helix, the side chains of repeating residues are packed into the center of the helix and interact with one another to form, for example, "asparagine ladders," "serine stacks," and/or "aromatic stacks."

[See also **Beta-sheet**, **protein motif**, **domain**, **protein**, and **protein structure**.]

BIBLIOGRAPHY

1. M. D. Yoder, S. E. Lietzke, and F. Jurnak (1993) *Structure* **1**, 241–251.
2. P. J. Kraulis (1991) *J. Appl. Crystallogr.* **24**, 946–950.
3. E. A. Merritt and M. E. P. Murphy (1994) *Acta Crystallogr.* **D50**, 869–873.
4. D. J. Bacon and W. F. Anderson (1988) *J. Mol. Graphics* **6**, 219–222.

Suggestions for Further Reading

M. D. Yoder and F. Jurnak (1995) The parallel β-helix and other coiled folds. *FASEB J.* **9**, 335–342.

C. Kisker et al. (1996) A left-handed β-helix revealed by the crystal structure of carbonic anhydrase from the archaeon bacteria *Methanosarcina thermophila*. *EMBO J.* **15**, 2323–2330.

BETA-LACTAMASES

M. G. P. PAGE

β-Lactamases constitute one of the oldest known mechanisms of bacterial **antibiotic resistance** and one of the most widely distributed. The **enzymes** hydrolyze β**-lactam** antibiotics, such as **penicillin**, rendering them inactive (Fig. 1).Most of these enzymes also react with other cyclic structures, such as γ-lactams and isatoic anhydride derivatives (1), and some have low reactivity with simple esters, **peptides**, and depsipeptides. Richmond and Sykes (2) and Bush (3) classified β-lactamases according to their catalytic activity and apparent substrate specificity, which remains the most useful classification for the clinical microbiologist. Today, the ready availability of β-lactamase amino acid sequences obtained from their **genes** provides the basis for a structural classification first proposed by Ambler (4). There are two types of β-lactamase:

1. those with a **serine** residue at their **active site** that is transiently acylated by the β-lactam substrate during hydrolysis (Richmond and Sykes groups I and II; Ambler class A, plus the more recently distinguished classes C and D); for an example, see **Serine proteinases**.
2. those with metal ion cofactors, which are Zn(II) in the physiological state, and with apparently no acyl-enzyme intermediate (Richmond and Sykes group III; Ambler class B) (see **Metalloproteinases**).

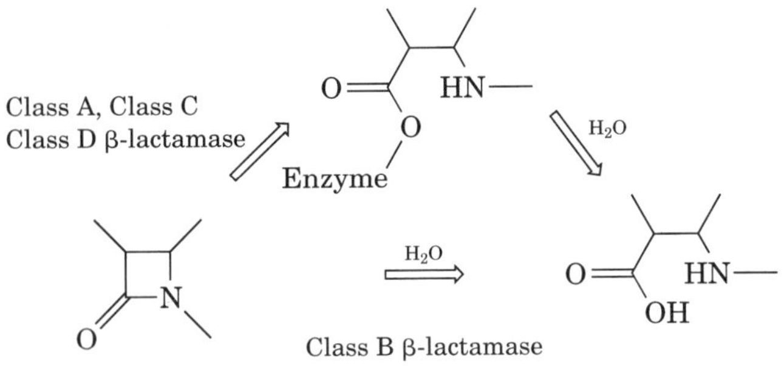

Figure 1. The reaction catalyzed by β-lactamases.

The serine β-lactamases include enzymes that have both narrow and broad substrate specificities and are sensitive, to varying degrees, to β-lactam **enzyme-activated inhibitors**. This group is the more abundant and currently poses the greater clinical problem. The metallo-β-lactamases are active on a broad spectrum of β-lactam substrates and are much less sensitive to the mechanism-based inhibitors, although they can be inhibited by metal-ion chelating agents. This group has been relatively scarce in nature, but its prevalence is rising because of the increased use of β-lactam antibiotics that resist hydrolysis by serine β-lactamases.

SERINE β-LACTAMASES

Sequence comparisons of the active serine enzymes indicate six major groups:

1. The class A β-lactamases: Richmond and Sykes Group II enzymes; Bush groups 2a (penicillinases), 2b (broad spectrum β-lactamases), 2b′ (extended broad spectrum β-lactamases), 2c (carbenicillinases), 2e (cephalasporinases). This group also includes a few proteins identified as D-Ala-D-Ala carboxypeptidases (DAC).
2. The class C β-lactamases: Richmond and Sykes and Bush Group 1, together with three proteins from *Streptomyces*, *Nocardia*, and *Bacillus subtilis*, identified as carboxypeptidase/transpeptidases.
3. The Class D β-lactamases from **gram-negative bacteria** (Bush Group 2d oxacillinases), together with β-lactam receptor proteins from gram-positive bacteria.
4. The bifunctional transglycosylase/transpeptidase, class A penicillin-binding proteins (PBP).
5. The monofunctional transpeptidase, class B PBP.
6. D-Ala-D-Ala carboxypeptidases (DAC).

The last three groups are biosynthetic enzymes of the bacterial cell wall that react with D-Ala-D-Ala.

All six groups have three sets of conserved residues in common, which comprise the active site of the enzyme (see later). The first comprises the sequence Ser-X-X-Lys, where X is any residue and the serine residue is the one that is transiently acylated by substrates and the only residue that is absolutely conserved in all of the proteins. The second has

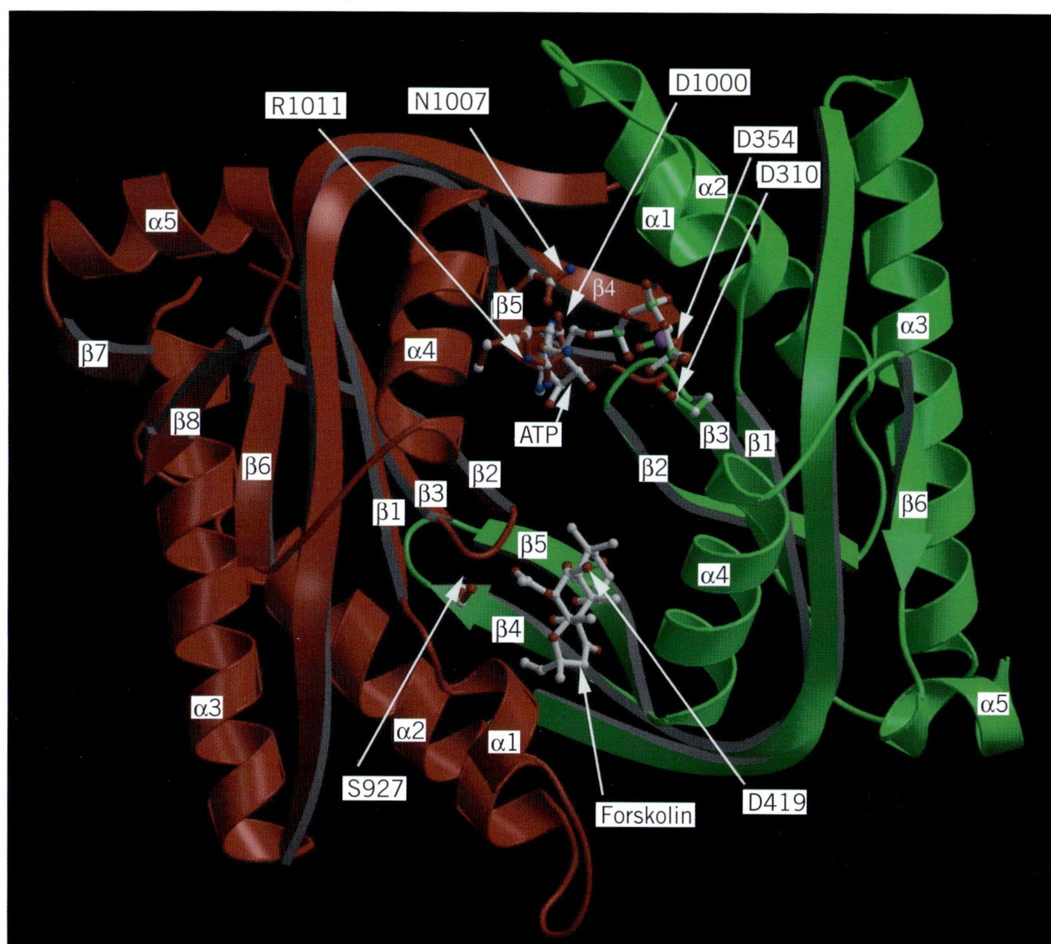

Adenylyte Cyclases, Figure 1, see page 59. Three-dimensional model of the C1 and C2 domains of the catalytic core of type I adenylate cyclase derived from the X-ray crystallographic structure of the C2 homodimer of the type II enzyme (19). The polypeptide backbone of the C1 domain is green and that of C2 is red. The C-terminal nonhomologous segment of C1 is truncated after the α-helix 5 (lower right). Only a few side chains in the active site are shown, designated by one-letter abbreviations. When they differed from the residues in the crystal structure, they were placed according to Liu *et al.* [*Proc. Natl. Acad. Sci.* (1997) **94,** 13414–13419.] ATP was placed as described by Liu *et al.* and modified according to the interactions reported by Tesmer *et al.* [*Science* (1997) **278,** 1907–1916.] The ligand atoms are indicated by C, white; N, blue; O, red; P, green, and Mg, purple. The figure is kindly provided by James Hurley.

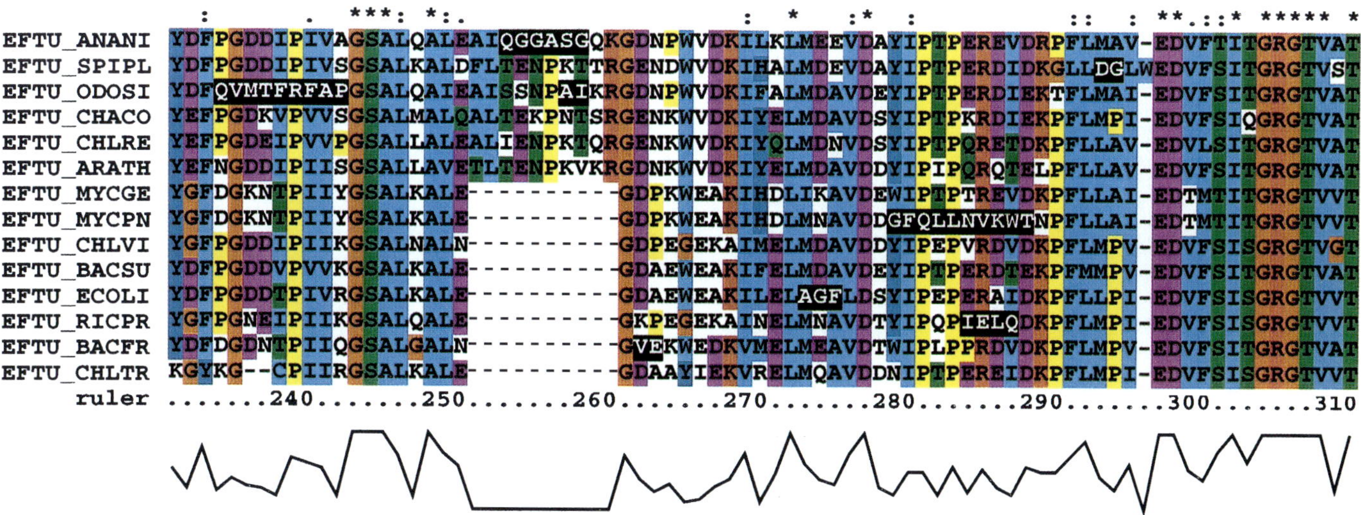

Aligning Sequences, Figure 2, see page 88. Part of a multiple alignment of 14 prokaryotic EFTU sequences using the one-letter code for amino acids. Gaps are indicated by dashes. Columns marked by asterisks are completely conserved, columns marked by colons are strongly conserved and columns marked by periods are weakly conserved. The graph at the bottom shows the conservation in the columns. Color is an essential aid to sequence analysis and is used here to highlight conservation according to the amino acid properties. Inverted characters indicate poorly matching regions of sequence. Some of these are due to natural sequence divergence, but there are four errors in sequence determination causing frameshifted regions: EFTUSPIPL 203–206, EFTUODOSI234–243, EFTUMYCPN 279–289 and EFTURICPR 284–287. These errors may lead to false inferences: If R-285 were completely conserved, it would be a candidate functional residue, while the false gapped column at 206 incorrectly suggests a surface loop.

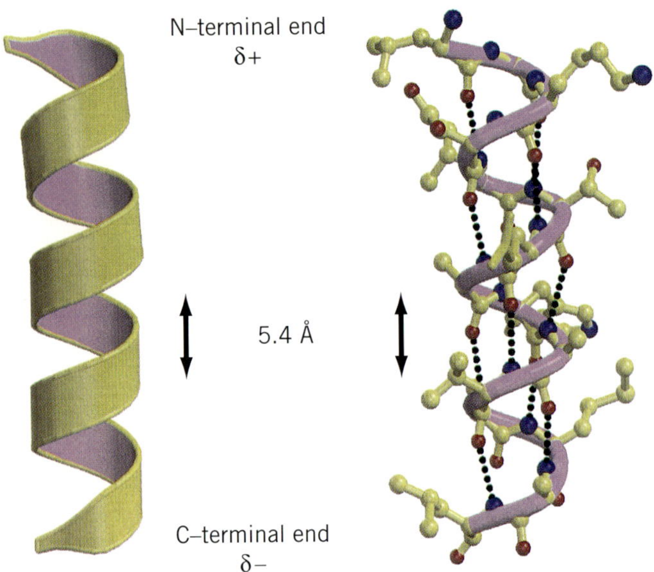

Alpha-helix, Figure 1, see page 97. A typical protein α-helix. (**Left**) The helix is depicted schematically as a coil, and the N- and C-terminal ends, with their respective partial positive and negative charges, are indicated. (**Right**) the atomic detail of the α-helical structure is shown, with hydrogen bonds between backbone carbonyl oxygens (i) and backbone amide nitrogens of residue (i + 4) indicated by dotted lines. Nitrogen atoms are shown in blue and oxygen atoms in red. Figure generated using Molscript (3) and Raster3D (4,5).

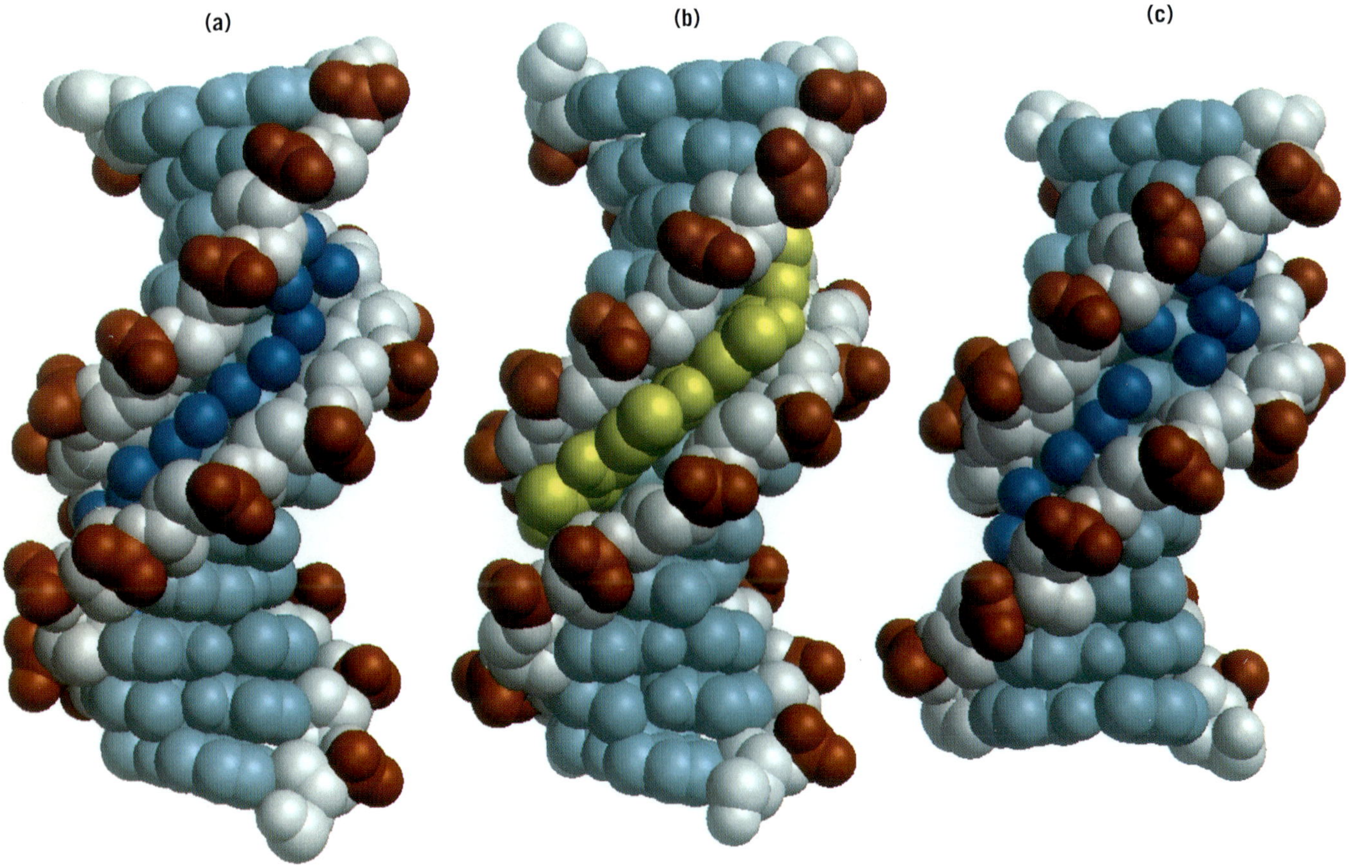

(a) (b) (c)

B-DNA, Figure 1, see page 240. Examples of the structure of B-form DNA. (**a**) Crystal structure of the double-stranded dodecamer d(CGCGAATTCGCG) (Protein Data Base BDL084), with the water molecules in the minor groove shown as spheres. (**b**) Crystal structure of the double-stranded d(CGCAAATTTGCG)-distamycin A complex (Protein Data Base GDL003). The drug displaces the waters in the minor groove. (**c**) The crystal structure of the double-stranded decamer d(CCAGGCCTGG) (Protein Data Base BDJS30), with the waters bound to the minor groove shown as spheres.

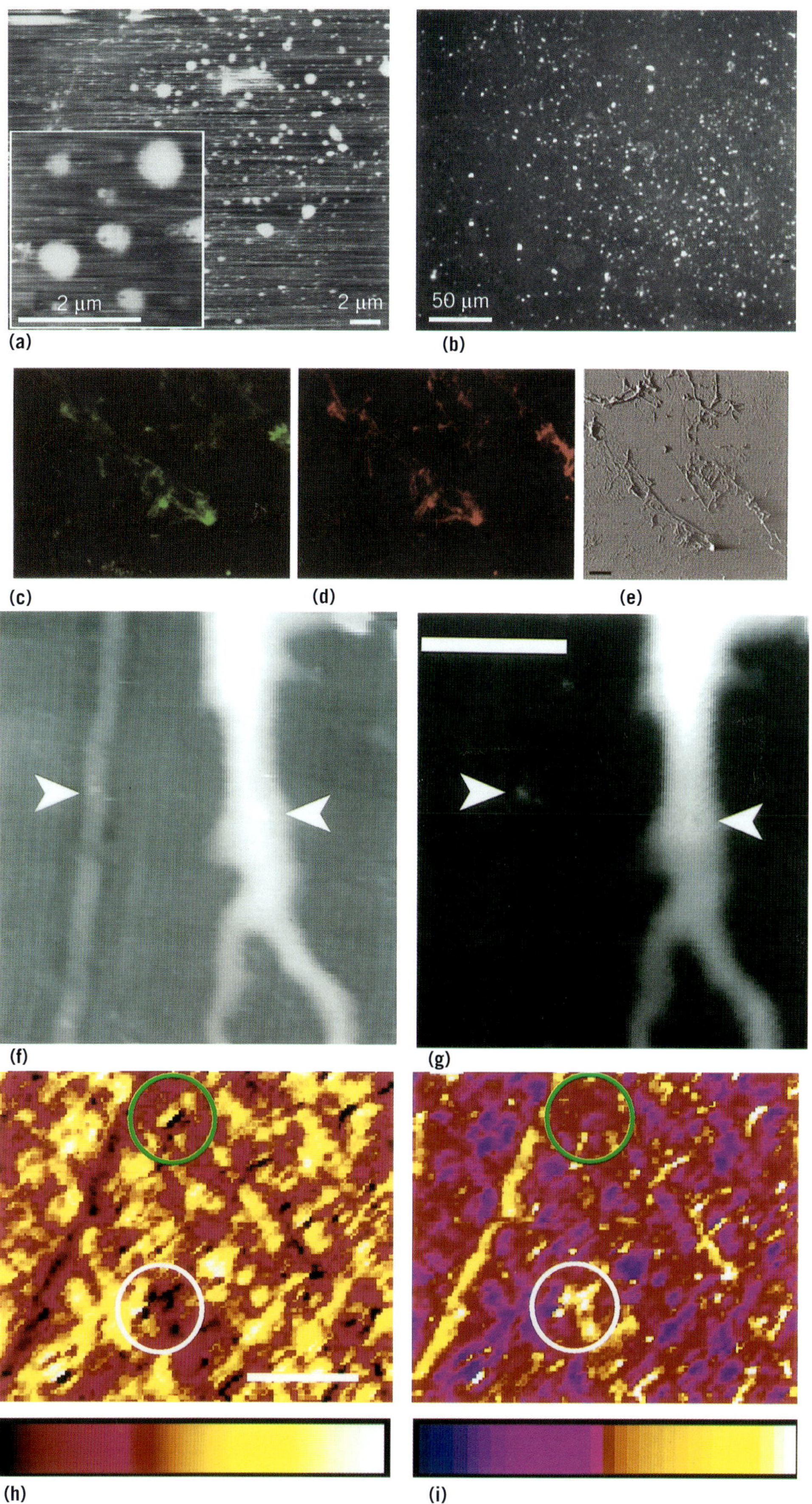

Atomic Force Microscopy, Figure 3, **see page 209.** Examples of AFM multimodal imaging. (**a**) and (**b**): Simultaneous immunofluorescence and atomic force microscopy of amyloid (β peptide (AβP) reconstituted into liposomes. All liposomes, with or without AβP, were imaged with AFM (**a**), a few are shown at higher magnification in the inset. The liposomes were treated with anti-AβP antibody and subsequently identified with fluorescein-conjugated second antibody. The AβP-carrying liposomes showed strong fluorescence signals (**b**). [S. K. Rhee, A. P. Quist, and R. Lal (1998) *J. Biol. Chem.* **273**, 13379–13382]. (**c**), (**d**), and (**e**): Adhesion sites between a Xenopus retinal glial cell (XR1 cell line) and **extracellular matrix** material in cell culture. The fluorescent images show the location of β-**integrin** (**c**) and f-**actin** (**d**) fibers detected by immunofluorescence, and the tapping mode AFM image (**e**) reveals the 3-D architecture of the focal point after removing of the cell body [R. Lal and R. Proksch (1997) *Intl. J. Imaging Syst. Technol.* **8**, 293–300]. (**f**) and (**g**): Simultaneous combined AFM and fluorescence-confocal microscopic images. The sample was a suspension of fluorescently labeled latex beads that were dried into a gel on a plastic diffraction grating. The lines of the grating are visible in the topographic AFM image (**f**) but not in the confocal fluorescent image (**g**). The two images allow distinguishing of a nonfluorescent particle (left arrow) from a fluorescent particle (right arrow), although both appear as raised bumps in the AFM image [P. E. Hillner, D. A. Waiters, R. Lal, H. G.Hansma, and P. K. Hansma (1995) *J. Micro. Soc. Am.* **1,** 123– 126]. (**h**) and (**i**): Simultaneously combined AFM and scanning ion-conductance microscope (SICM) electrophysiology. (**h**) shows a tapping mode AFM image of a nucleopore membrane, and (**i**) shows the associated ionic conductivity image obtained by tapping mode SICM. Note that there are some differences in the pores detected by the two procedures. For example, the area circled in white shows a group of pores that appear to be deep in the AFM image and highly conductive in the SICM image. The area circled in grey contains a large pore that appears deep in the AFM image but is nonconductive in the SICM image. The scale bars at the bottom are intensity-coded. Brighter is a greater height in the AFM image and a greater conductance in the SICM image [R. A. Proksch, R. Lal, P. K. Hansma, G. Morse, and G. Stucky (1996) *Biophys J.* **71**, 2155–2157].

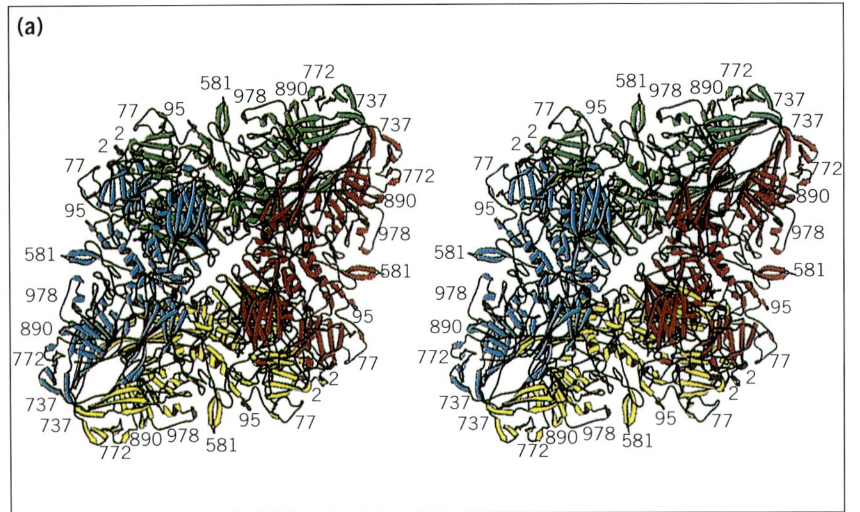

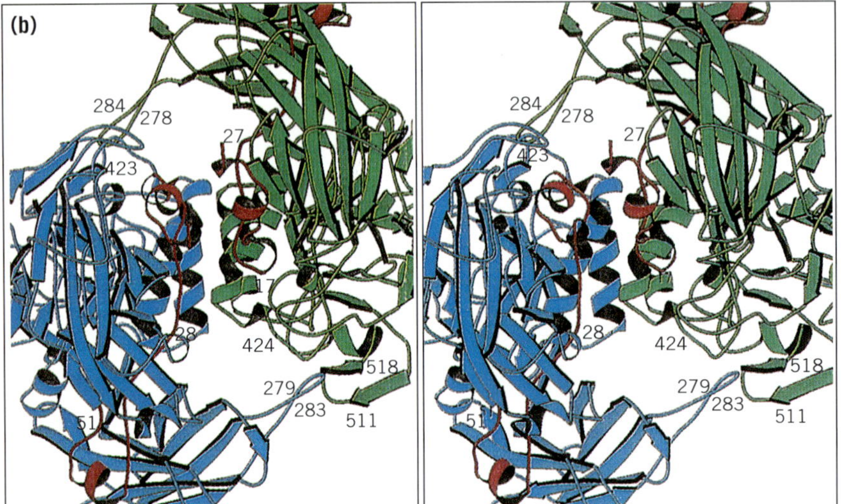

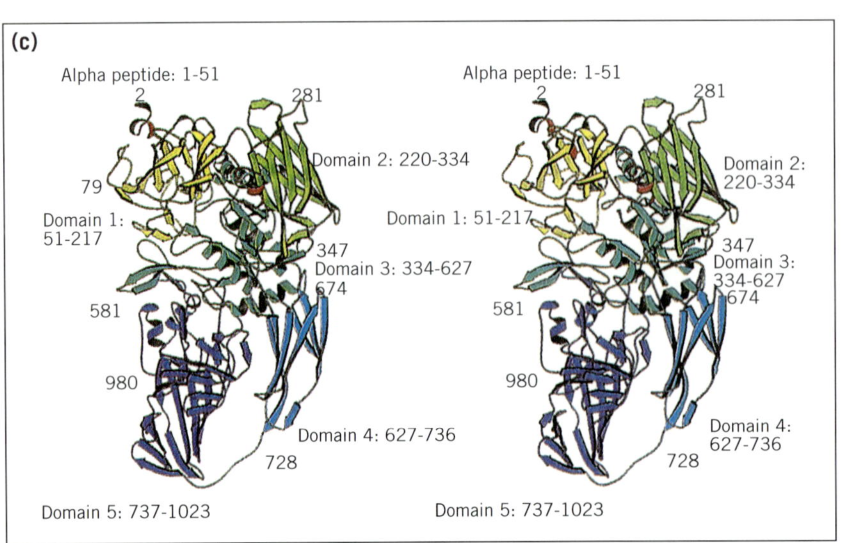

Beta-galactosidase of *Escherichia coli*, Figure 2, see page 265. Three-dimensional structure of β-galactosidase. (**a**) Ribbon representation of the β-galactosidase tetramer showing the largest face of the molecule. Contacts between red/green and blue/yellow dimers form the long interface. Contacts between the red/yellow and blue/green dimers form the activating interface. Formation of the tetrameric particle results in two deep clefts that run across opposite faces of the molecule. Each contains two active sites. (**b**) Ribbon diagram of the blue/green dimer viewed down the molecular two-fold axis, showing the composition of the activating interface. Residues 1 to 50 from each chain, which form the α-complementation region (see text), are shown in red. The interface includes contacts between the respective complementation peptides, between two α-**helices** from the respective monomers that pack together to form a four-helix bundle and between an extended loop (residues 272 to 288) from each monomer that reaches across the interface and extends into the active-site region of the neighboring monomer, stabilizing the active site structure. (**c**) Stereo ribbon diagram of the β-galactosidase monomer showing the domain organization of the chain. Residues corresponding to successive domains are colored in successive spectral colors.

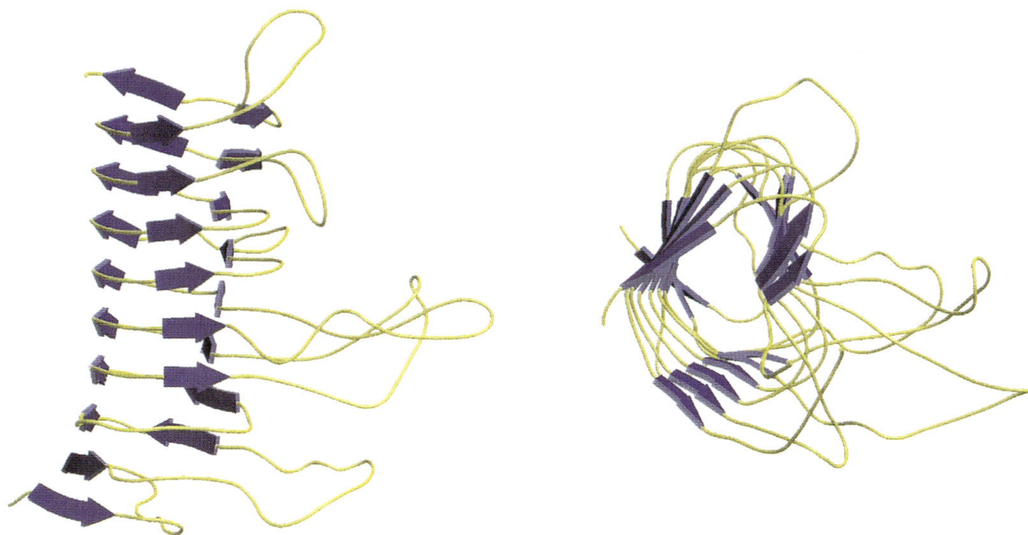

Beta-helix, Figure 1, see page 267. Orthogonal views of the β-helix of pectate lyase E (1), showing the three parallel β-sheets. For clarity, only the β-strand secondary structure (as purple arrows) and loops connecting the strands (yellow) are shown. Short helical regions and the *N*- and *C*-terminal loops have been removed. This figure was generated using Molscript (2) and Raster3D (3,4).

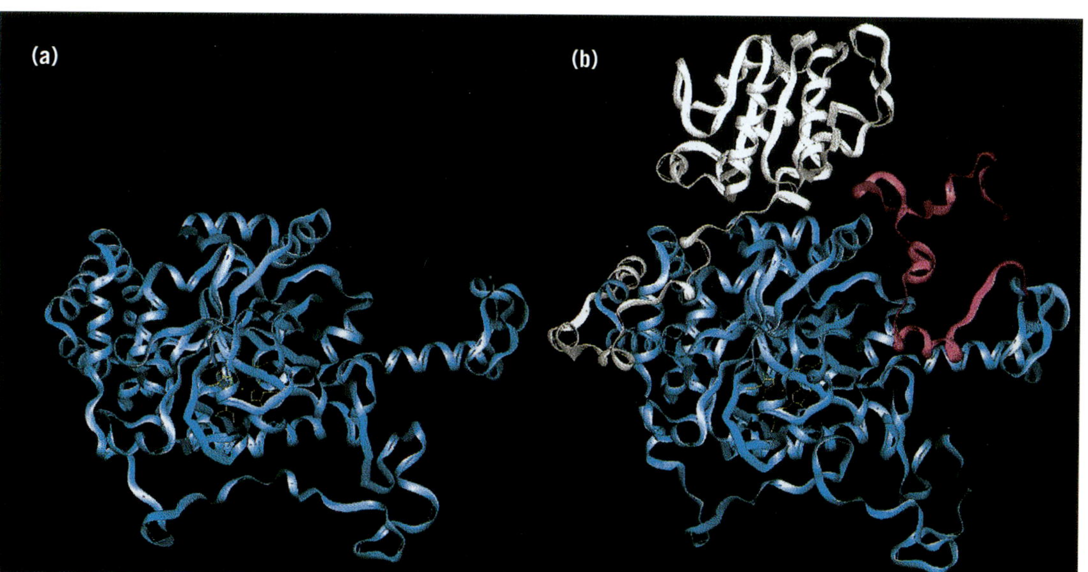

Catalase, Figure 1, see page 352. Comparison of the structures of the small subunit of *Proteus mirabilis* catalase (**a**) (8) and the large subunit of *E. coli* HPII catalase (**b**) (9). The conserved β-barrel core structure is colored blue in both subunits. The 78 additional residues at the *N* terminus of HPII in (**b**) are colored red, and the 195 additional residues at the *C* terminus of HPII in (**b**) are colored white.

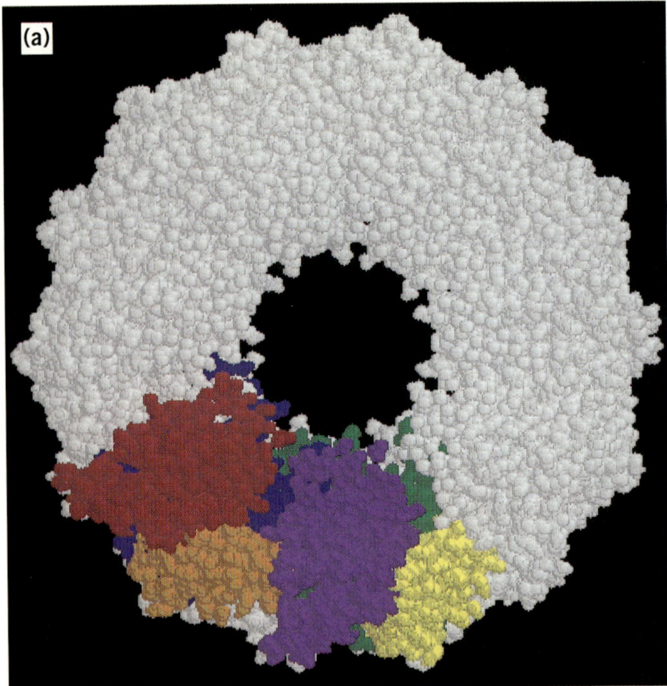

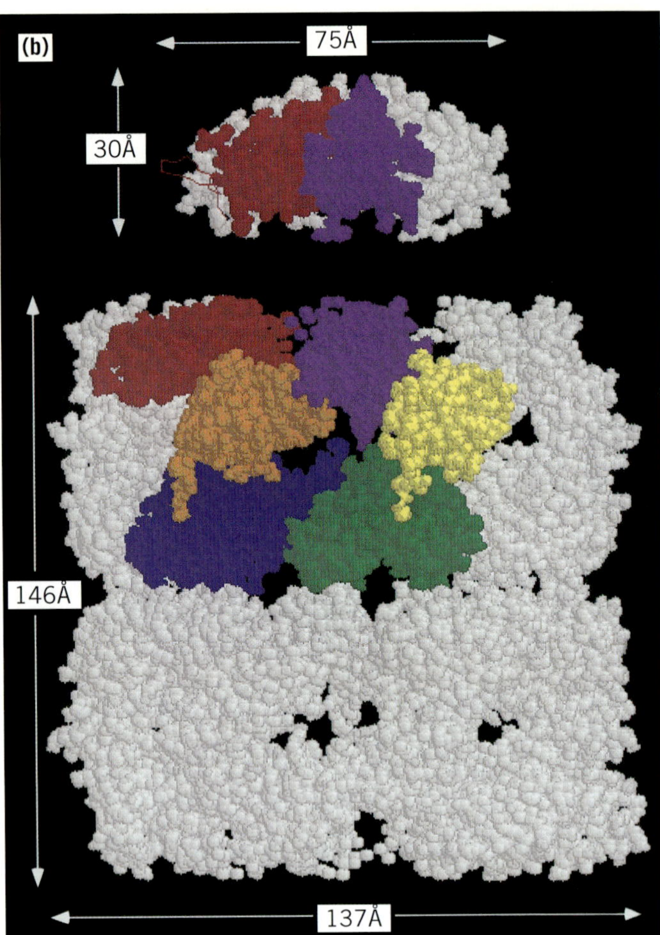

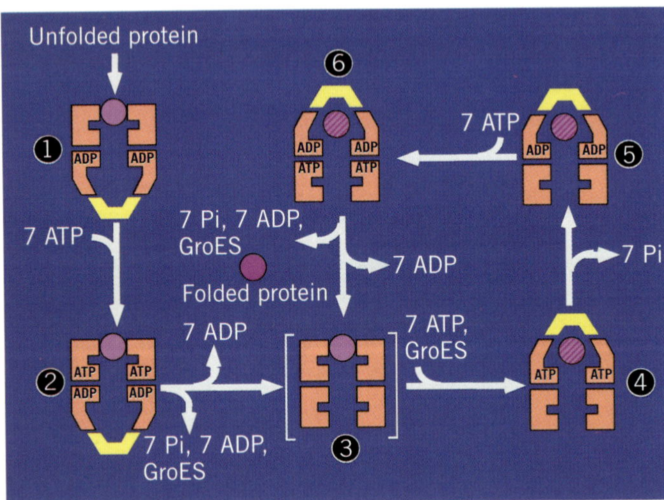

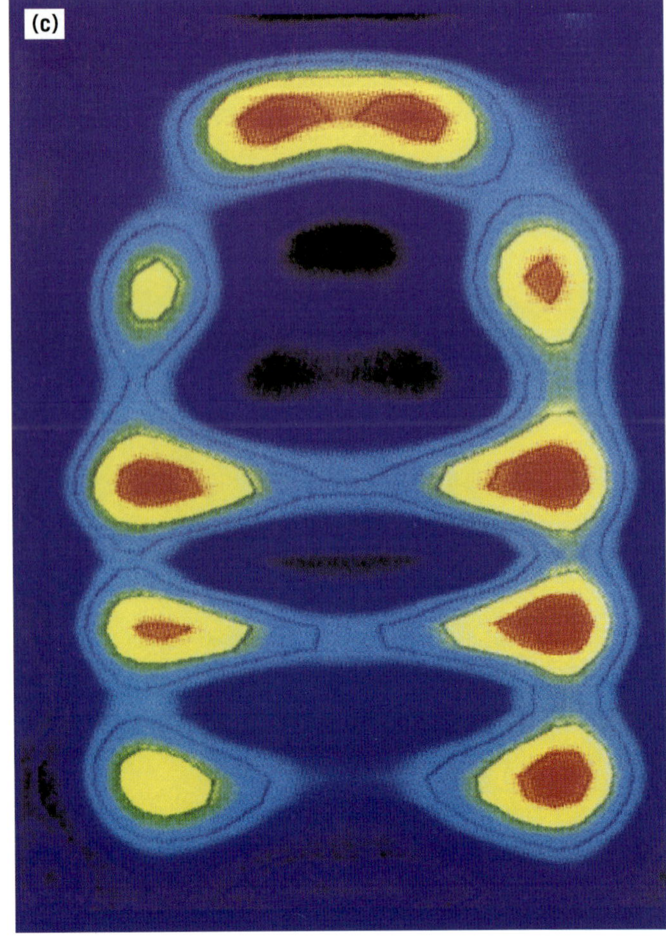

Chaperonin, Figure 1, see pages 396–397. Structure and function of the chaperonin system. (**a,b**) Space-filling representations showing a top and side view, respectively, of the crystal structure of a double mutant form of GroEL (26). Two adjacent subunits are colored with the apical domains in red and purple, the intermediate domains in orange and yellow, and the equatorial domains in blue and green, respectively. Free passage between the GroEL rings is obstructed by *N*- and *C*-terminal residues not resolved in the crystal structure. (**b**) Side view of GroES (28) at the top; two adjacent domains and a single mobile loop that is structured in the GroES crystal due to crystal packing are colored. The loop regions normally protrude from the base of GroES downward toward the GroEL. (**c**)Asymmetrical GroEL/GroES complexes as revealed by cryoelectron microscopy (45). Note the upward and outward movement of the apical GroEL domains interacting with GroES. (**d**) Model of the GroEL/GroES reaction cycle in assisting protein folding (30); see text for details. The term *unfolded protein* refers to a partially folded intermediate that is represented by the light pink spheres; folded protein is represented by the dark pink sphere, while the hatched spheres represent a mixture of folded and partially folded proteins expected in a population of GroEL molecules. At step 4, GroES may associate with either the protein-containing ring of GroEL or with the opposite empty ring; the latter possibility is not shown. Reprinted from *Nature* **381**, 571–580 (1996) with permission; copyright (1996) Macmillan Magazines Ltd.

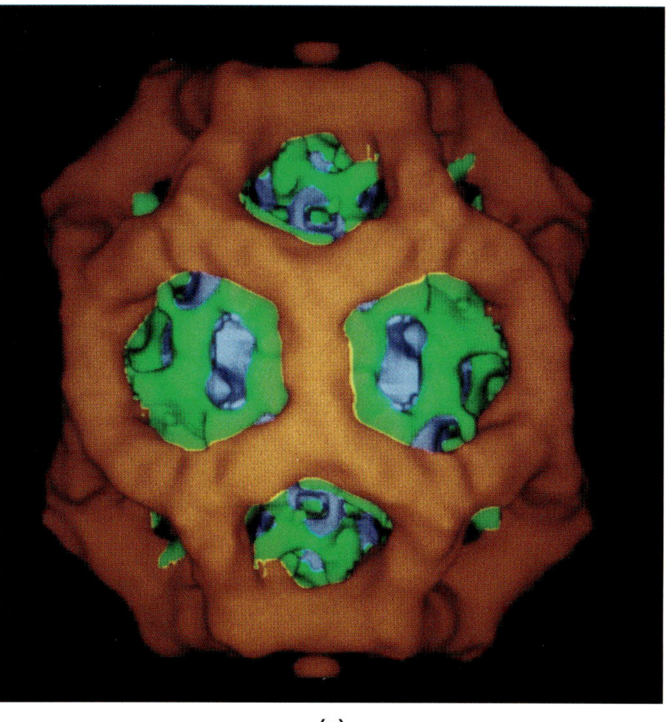

(a)

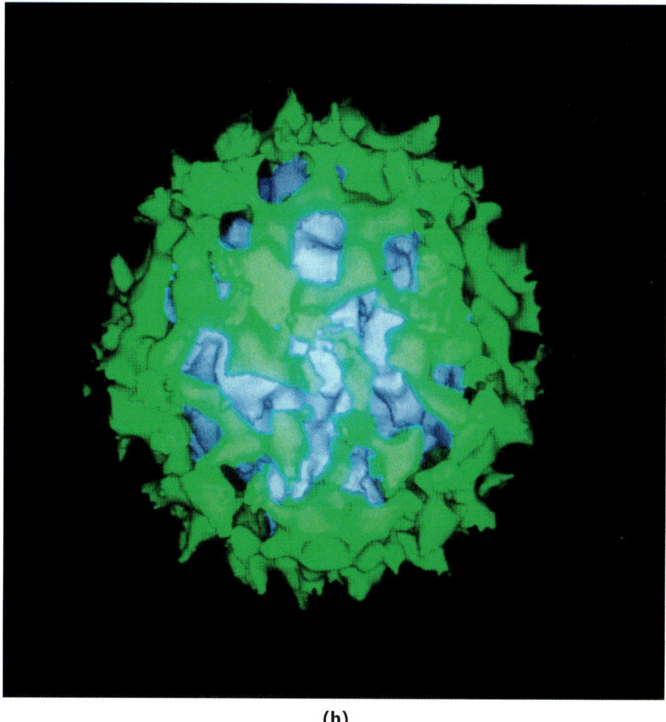

(b)

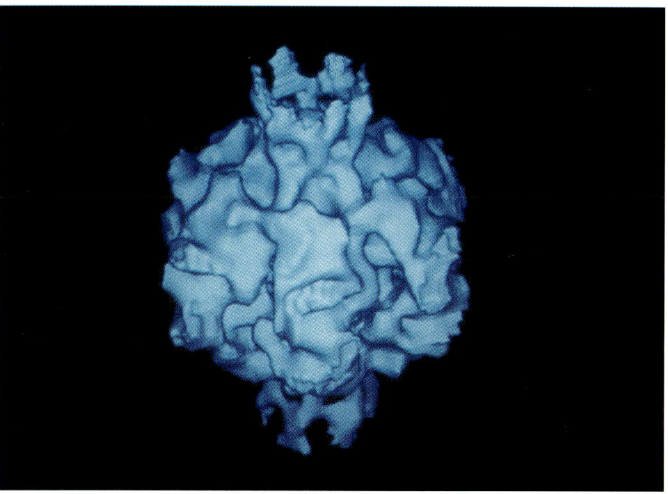

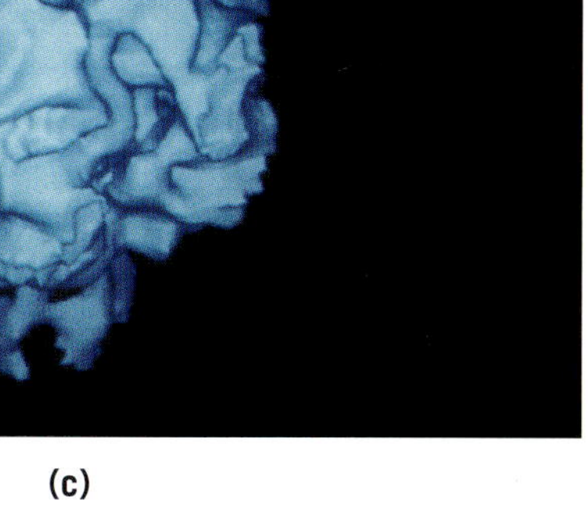

(c)

Clathrin, Figure 6, see page 479. Clathrin encapsulates coated vesicles, the organelles responsible for uptake of essential nutrients into living cells (eg, especially across the placenta in mammals and into oocytes in chickens). (**a**) A three-dimensional map of a clathrin coat generated from electron micrographs of specimens reassembled from their constituent protein complexes. Colors highlight the outer polyhedral clathrin lattice (red), the shell of clathrin terminal domains (green) revealed in (**b**) and an inner shell of adaptors (blue) exposed in (**c**). These three maps led to the description of this structure as a cell biologist's LEGO model. [Courtesy of G. P .A. Vigers, R. A. Crowther and B. M. F. Pearse from data presented in *EMBO J.* (1986) **5,** 2079–2085.]

Coiled-coils, Figure 2, see page 490. The five-stranded coiled-coil oligomerization domain of cartilage oligomeric matrix protein (COMP). Each of the α-helices forms about one-third of a complete turn over the length of the structure. The *N*-terminal end is at the bottom of the figure. (Courtesy of R. A. Kammerer.)

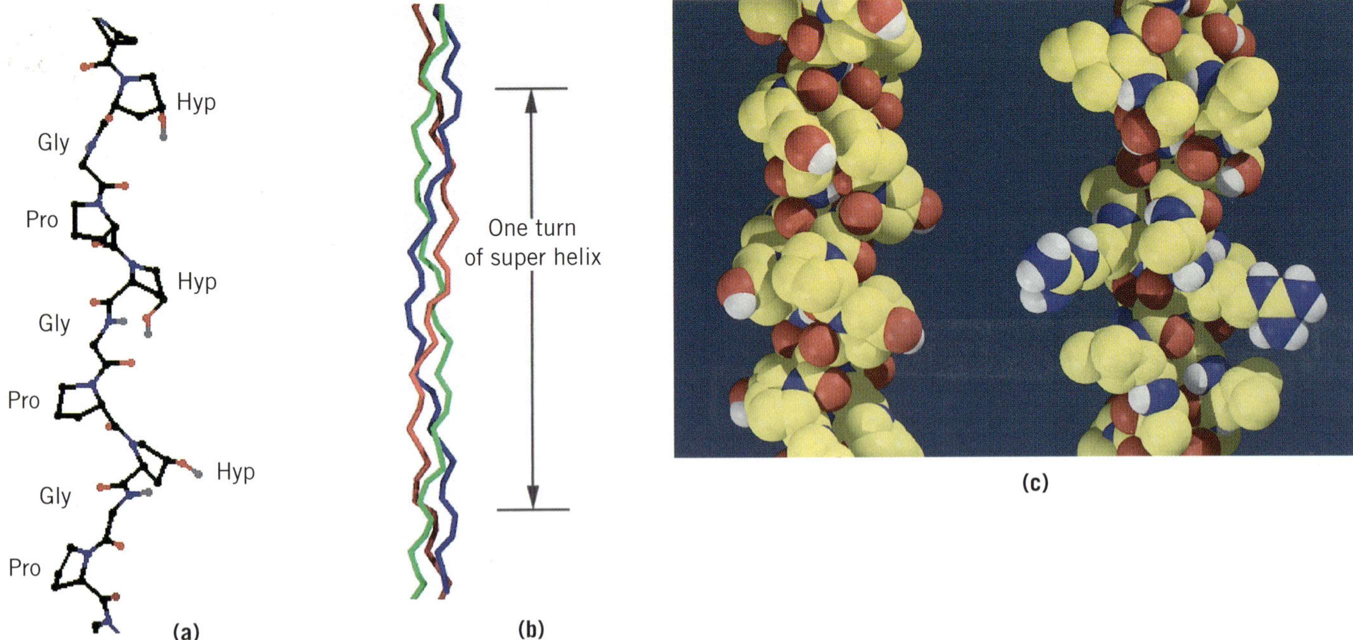

Collagen, Figure 2, see page 502. Collagenous triple helix.(**a**) Gly–Pro–Hyp repeated sequence of α chain in a model of part of one α chain in the collagenous triple helix. Both NH and CO groups project perpendicular to the fibrillar axis (C, black; N, blue; O, red; H, gray). (**b**) Backbone of the collagenous triple helix. (**c**) Gly–Pro–Hyp trimer (*left*). Note that there is a groove on the surface of the helix. A part of human type III collagen molecule, $[\alpha 1(\text{III})]_3$ in the sequence of GITGARGLAGP (*right*). Note that the all residues except Gly project to the outer surface of the molecule (C, yellow; N, blue; O, red; H, white).

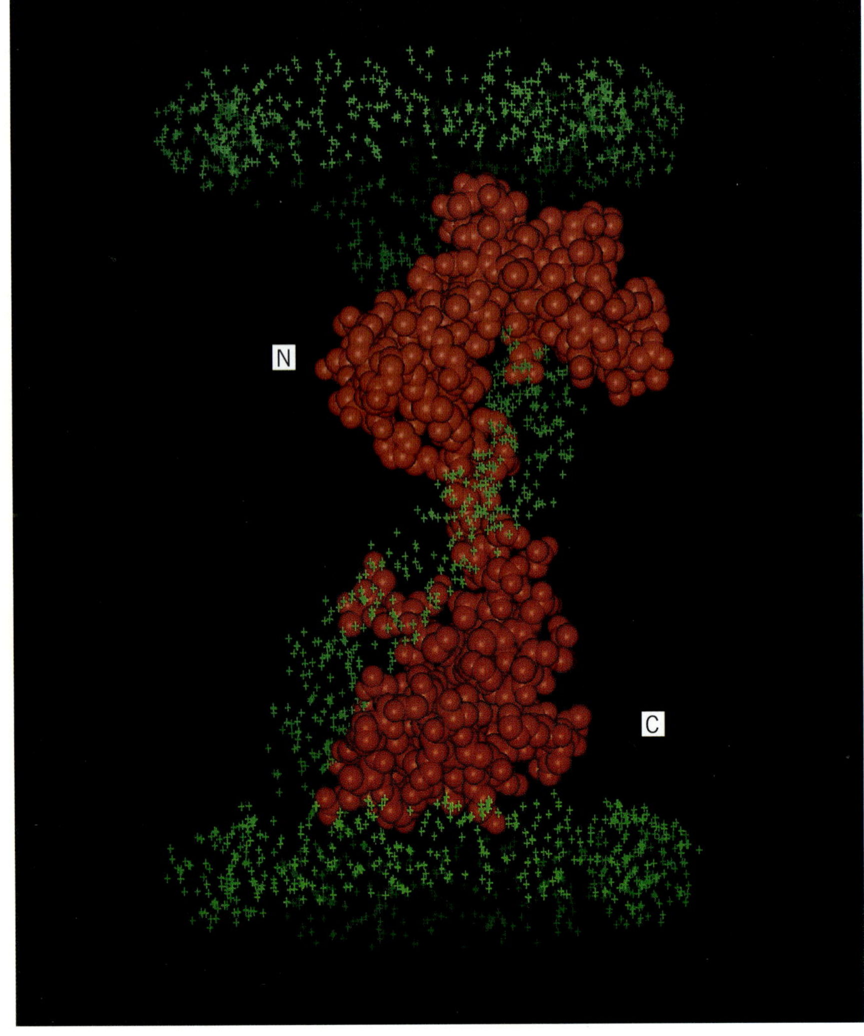

Contrast Variation, Figure 2, page 574. Model structure for the muscle protein complex troponin C/troponin I derived from small-angle neutron scattering and contrast variation (14). Troponin C is represented as a space filling model based on its crystal structure, and the green crosses represent the volume occupied by the troponin I. This figure is based on work done at Los Alamos National Laboratory.

either a serine (class A) or a **tyrosine** residue (in class C and class D β-lactamase groups), followed after one residue by an **asparagine** (except in one class D β-lactamase). The third conserved segment is usually Lys-Thr-Gly, but only the **glycine** residue is absolutely conserved. Sequence analyses have suggested that β-lactamase activity may have arisen several times by a process of evolutionary **convergence**.

Three-dimensional structures of class A enzymes from *Staphylococcal aureus* (5), *Bacillus cereus* (6), *B. licheniformis* (7), *Streptomyces albus* (8), and *Escherichia coli* TEM-1 (9,10) have been published, which include only two of the five activity groups of class A enzymes. Structures of the class C β-lactamases from *Citrobacter freundii* (11) and *Enterobacter cloacae* (12) have been published, and this group is extended by the homologous structure of the transpeptidase domain of the *Streptomyces* carboxypeptidase/transpeptidase (13). The overall structures of all of these enzymes and a DAC are similar. The core of each molecule is a five-stranded, antiparallel β-**sheet** flanked on one side by three α-**helices** and on the other by a larger, more diverse α/β domain (see **Protein structure**). The active site lies between the two domains and is bound by one edge of the β-sheet. As shown in Figure 2, the conserved residues listed previously make up the active site regions and occupy very similar positions in all of the structures (Figure 2). Even the alternative serine and tyrosine residues in the second conserved segment of class A (DAC) and class C (DAC), respectively, have their hydroxyl groups in the same positions; see Ser 130 and Tyr150 in Fig. 2 a and b, respectively. Major differences between the two classes of enzymes that are believed to have functional consequences occur in the area that serves as a recognition pocket for the side chain of the substrate. The class A enzymes have a loop (the Ω loop) that forms the base of the acyl amino side-chain binding pocket. This loop includes residues Glu166 and Asn170, which localize and activate a water molecule for attack on the ester bond of the substrate (14,15). The projection of this loop into the active site also limits the scope for binding bulky acyl side chains, which was exploited in developing β-lactamase-resistant antibiotics, such as methicillin.

Evolutionary accumulation of peripheral mutations that modify the conformation of the Ω loop and of neighboring surface loops that define the edges of the side-chain recognition pocket has led to the appearance of enzymes that belong to groups 2b′, 2c, and 2e that have extended substrate recognition profiles (16). In the class C enzymes, the water molecule that attacks the ester of the acyl intermediate is located on the opposite side of the plane of the ester and is activated by direct interaction with components of a hydrogen-bonded relay formed by the conserved residues. Identification of the individual residues involved in activating of the water has not yet been achieved, although Tyr150 has been implicated (11). The class C enzymes have a deeper, more **hydrophobic** side-chain binding pocket, lined by Tyr122, that enables them to bind substrates with large 7-acyl amino side chains.

Conformational changes during the β-lactamase reaction have been proposed for the class A, C, and D β-lactamases on a variety of grounds. With simple substrates, **NMR** and **circular dichroism** spectroscopy have suggested changes in the structure accompanying acyl-enzyme formation (17,18). A number of substrates show nonstoichiometric bursts of hydrolysis (19), and a conformation change leading to a substrate-induced inactivation has been invoked (1,20). These observations in solution contrast with the finding of very few differences in the structures of the free enzyme and the acyl-enzyme complex determined by **X-ray crystallography** (9,11). An unambiguous description of the reaction mechanism must wait until these observations can be reconciled.

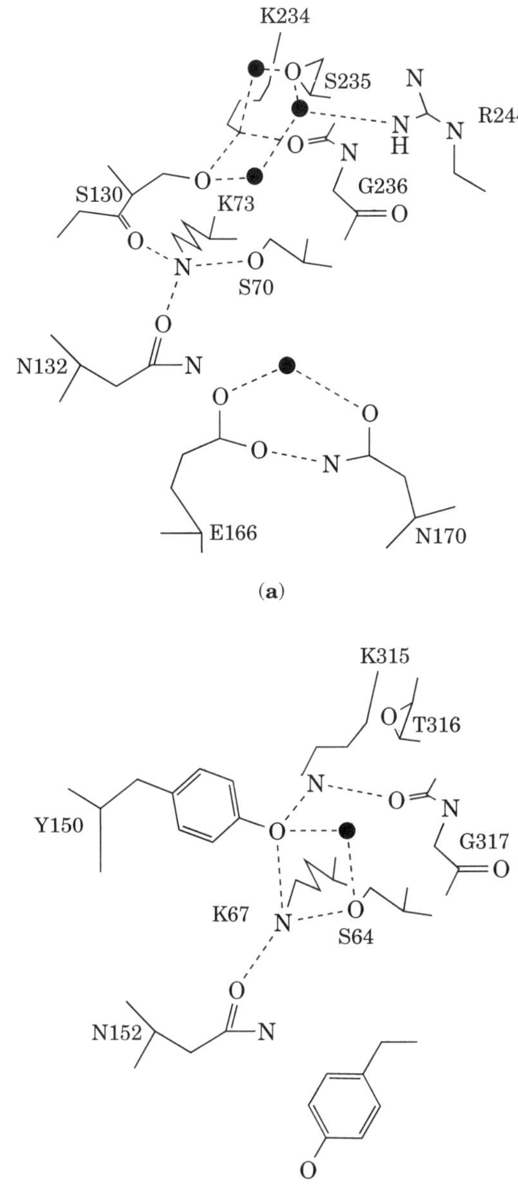

Figure 2. Comparison of the similar active-site regions of β-lactamases of class A (a) and class C (b). The residues are identified with the one-letter code. The lack spheres are water molecules observed crystallographically that are probably hydrogen-bonded (dashed lines) to various groups of the protein. The active-site serine residue that is acylated in the catalytic reaction is S70 in (a) and S64 in (b).

METALLO-β-LACTAMASES

Sequence comparisons suggest that there are four groups of this type of β-lactamase that are not closely related, but share

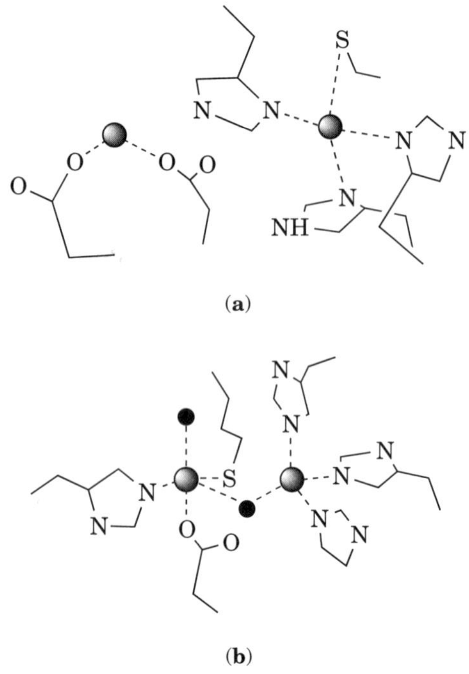

(a)

(b)

Figure 3. Comparison of the similar active-site regions of metallo-β-lactamases from *Bacillus cereus* (a) and *Bacteriodes fragilis* (b). The shaded spheres are the two metal ions, cadmium (II) in (a) and zinc (II) in (b), and the smaller solid spheres are water molecules observed crystallographically.

similarities with a protein from the actinorhodin biosynthetic gene cluster of *Streptomyces*. A few residues involved in binding the metal ion cofactors are absolutely conserved within this diverse group. Three structures are available, that representing two of the major subgroups of metallo-β-lactamases (Fig. 3). All three structures are similar overall to each other, but not to the serine β-lactamases, and the core of each molecule is formed from two antiparallel β-sheets, each with additional parallel strands and flanked by α-helices. The structure of *Bacillus cereus* β-lactamase II was solved with Cd(II) in place of the natural Zn(II) metal ion (21). One Cd(II) ion is bound tightly, with a dissociation constant of about 1 μM, by the side chains of three **histidine** and one **cysteine** residue (22–24). In the homologous enzyme from *Bacteroides fragilis*, one Zn(II) ion is similarly coordinated by three histidine side chains, as in the *B. cereus* enzyme, but the fourth ligand is a water molecule that is shared with a second Zn(II) ion (25). The second zinc ion is further coordinated by three other side chains, those of His, Cys, and Asp residues. Two more Cd(II) ions are located in the *B. cereus* structure, but the coordination of one of them is much weaker (23), so it may not be physiological. Early stoichiometries determined for the binding of Zn(II) and Cd(II) suggested that two ions per molecule are bound (22). The second Cd(II) ion is bound by two Asp residues and lies only 10 Å from the tightly bound ion (Fig. 3), where it might modulate the catalytic activity. It is believed that the water molecule complexed with the two Zn(II) ions in the *B. fragilis* enzyme is activated to form hydroxide for attack on the β-lactam ring, and a corresponding water molecule is probably activated by the tightly bound ion in the *B. cereus* enzyme.

INHIBITION

The important role of β-lactamases in antibiotic resistance has led to the development of specific β-lactamase inhibitors that can be used to protect β-lactamase-sensitive antibiotics. For the serine β-lactamases, several classes of mechanism-based

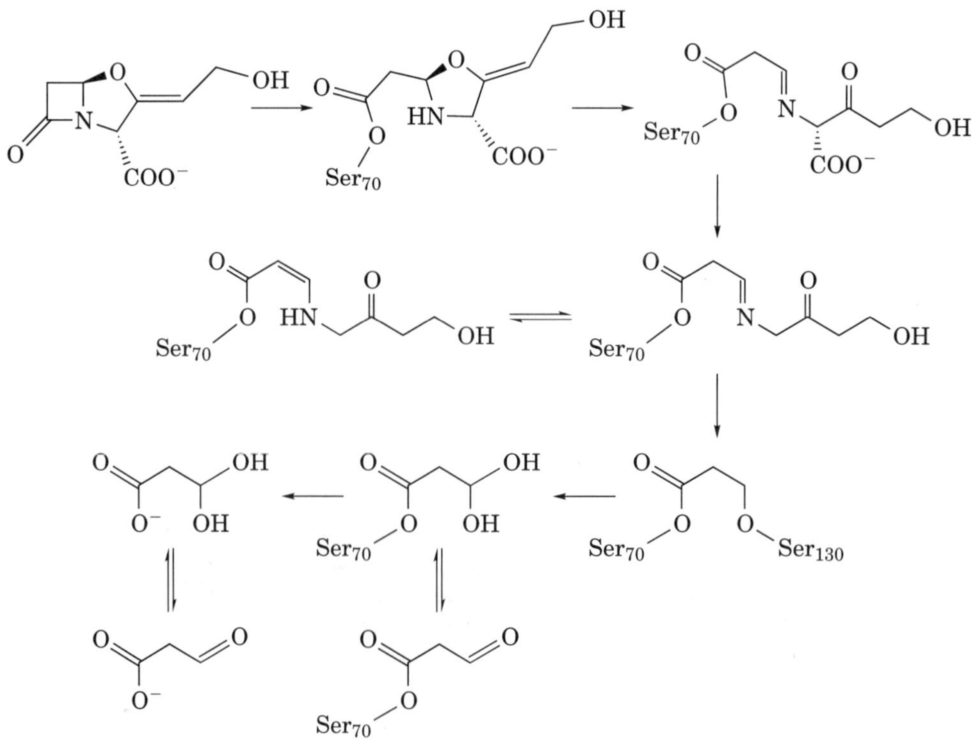

Figure 4. Suggested mechanism for the reaction of TEM-2 class A β-lactamases with clavulanic acid (27).

inhibitors have been discovered, and several inhibitor proteins are known (26). In many cases, the mechanism of action of the small-molecule inhibitors is not clear, and only two classes have been used in combination with antibiotics.

Clavulanic acid has been widely used clinical application. It is selective for the class A and class D β-lactamases. Opening of the β-lactam ring of clavulanic acid by the initial attack of the β-lactamase triggers a series of chemical rearrangements in the inhibitor moiety (Fig. 4) that results in an acyl-enzyme that is resistant to attack by water and may even be cross-linked by the inhibitor moiety (27).

The penam sulfone acids *sulbactam* and *tazobactam* are also used clinically as β-lactamase inhibitors. As with clavulanic acid, a series of chemical rearrangements are provoked by the reaction with β-lactamase. Although sulbactam and tazobactam are relatively selective for class A and class D β-lactamases, analogs that have increased activity against class C β-lactamases have been reported (28).

REGULATION OF BIOSYNTHESIS

Expression of chromosomal β-lactamase in *Bacillus* and of **plasmid**-encoded β-lactamases in staphylococci is under the control of a typical **repressor** protein (BlaI and MecI, respectively) and a second regulatory protein (BlaR and MecR, respectively). The BlaR and MecR regulatory proteins are similar. Both are integral **membrane proteins** that have a β-lactam-binding **domain** homologous to the class D serine β-lactamases (29). The predicted topology of the polypeptide in the membrane is three to five membrane-spanning α-**helices**, and the N-terminus in the cytoplasm and the β-lactam-binding domain is on the outer surface of the membrane, where it could act as a **receptor** for β-lactams.

Expression of chromosomal class C β-lactamases in Enterobacteriaceae from the *ampC* gene is under the control of the *ampD*, *ampE*, *ampG*, and *ampR* gene products. AmpR protein is a typical **transcription** activator that is lost in organisms, such as *E. coli*, that have constitutive AmpC production. AmpR responds to the binding of 1,6-anhydro-*N*-acetylmuramyl-[L]-alanyl-[D]-glutamyl-*meso*-diaminopimelic acid (MurNAc tripeptide) produced by breakdown of the peptidoglycan of the cell wall (30). The three other proteins are involved in transmembrane signaling for induction of the β-lactamase. Inactivation of AmpD results in massive overproduction of β-lactamase, whereas loss of either AmpE or AmpG activity results in a total block of induction. AmpD is an amidase that cleaves the tripeptide from the MurNAc tripeptide, thus inactivating it as an inducer. It is thought that AmpG is an integral membrane protein that acts as a transporter for the MurNAc tripeptide, whereas it has been suggested that AmpE, also an integral membrane protein, provides energy for its uptake (31).

BIBLIOGRAPHY

1. M. G. P. Page (1993) *Biochem. J.* **295**, 295–304.
2. M. H. Richmond and R. B. Sykes (1973) *Adv. Microb. Physiol.* **9**, 31–88.
3. K. Bush (1989) *Antimicrob. Agents Chemother.* **33**, 264–270, 271–276.
4. R. P. Ambler (1980) *Phil. Trans. R. Soc. Lond.* **B239**, 321–331.
5. O. Herzberg and J. Moult (1987) *Science* **236**, 694–701.
6. B. Samaraoui, B. Sutton, R. Todd, P. Artimyuk, S. G. Waley, and D. Phillips (1986) *Nature* **320**, 378–380.
7. P. C. Moews, J. R. Knox, O. Dideberg, P. Charlier, and J.-M. Frère (1990) *Proteins: Struct. Function Genet.* **7**, 156–171.
8. O. Dideberg, P. Charlier, J. P. Wery, P. Dehottay, J. Dusart, T. Erpicum, J.-M. Frère, and J.-M. Ghuysen (1987) *Biochem. J.* **245**, 911–913.
9. N. C. J. Strynadka, H. Adachi, S. E. Jensen, K. Johns, A. Sielecki, C. Betzel. K. Sutoh, and M. N. G. James (1992) *Nature* **359**, 700–705.
10. C. Jelsch, L. Mourey, J. M. Masson, and J.-P. Samama (1992) *Proteins: Struct. Function Genet.* **16**, 364–383.
11. C. Oefner, A. D'Arcy, J. J. Daly, K. Gubernator, R. L. Charnas, I. Heinze, C. Hubschwerlen, and F. K. Winkler (1990) *Nature* **343**, 284–288.
12. E. Lobkovsky, P. C. Moews, J. Liu, H. Zhao, J.-M. Frère, and J. R. Knox (1993) *Proc. Natl. Acad. Sci. USA* **90**, 11257–11261.
13. J. A. Kelly and A. P. Kuzin (1995) *J. Mol. Biol.* **254**, 223–236.
14. R. M. Gibson, H. Christensen, and S. G. Waley (1990) *Biochem. J.* **272**, 613–619.
15. W. A. Escobar, A. K. Tan, and A. L. Fink (1991) *Biochemistry* **30**, 10783–10787.
16. G. A. Jacoby and A. A. Medeiros (1991) *Antimicrob. Agents Chemother.* **35**, 1697–1704.
17. M. Jamin, C. Damblon, A. M. Baudin-Misselyn, F. Durant, G. C. K. Roberts, P. Charlier, G. Llabres, J.-M. Frère. (1994) *Biochem. J.* **301**, 199–203.
18. P. Taibi-Tonche, I. Massova, S. B. Vakulenko, S. A. Lerner, and S. Mobashery (1996) *J. Am. Chem. Soc.* **118**, 7441–7448.
19. A. Samuni and N. Citri (1975) *Biochem. Biophys. Res. Commun.* **62**, 7–11.
20. S. G. Waley (1991) *Biochem. J.* **279**, 87–94.
21. B. J. Sutton, P. J. Artymiuk, A. E. Cordero-Borboa, C. Little, D. C. Phillips, and S. G. Waley (1987) *Biochem. J.* **248**, 181–188.
22. R. B. Davies and E. P. Abraham (1974) *Biochem. J.* **143**, 129–135.
23. G. S. Baldwin, A. Galdes, A. O. Hill, B. E. Smith, S. G. Waley, and E. P. Abraham (1978) *Biochem. J.* **175**, 441–447.
24. A. Carfi and O. Dideberg (1995) *EMBO J.* **14**, 4919–4921.
25. N. O. Concha, B. A. Rasmussen, K. Bush, and O. Herzberg (1996) *Structure* **4**, 823–836.
26. N. C. J. Strynadka, S. E. Jensen, K. Johns, H. Blanchard, M. Page, A. Matagne, J.-M. Frère, and M. N. G. James (1994) *Nature* **368**, 657–660.
27. R. P. A. Brown, R. T. Aplin, and C. J. Schofield (1996) *Biochemistry* **35**, 12421–12432.
28. H. G. F. Richter, P. Angehrn, C. Hubschwerlen, M. Kania, M. G. P. Page, J.-L. Specklin, and F. K. Winkler (1996) *J. Med. Chem.* **39**, 3712–3722.
29. T. Kobayashi, Y. F. Zhu, N. J. Nicholls, and J. O. Lampen (1987) *J. Bacteriol.* **169**, 3873–3878.
30. J. T. Park (1995) *Mol. Microbiol.* **17**, 4521–426.
31. S. Lindquist, M. Galleni, F. Lindberg, and S. Normark (1989) *Mol. Microbiol.* **3**, 1091–1102.

Suggestion for Further Reading

M. I. Page, ed. (1992) *The Chemistry of β-Lactams* Blackie Academic & Professional, Glasgow. An excellent coverage of all aspects of the interaction of β-lactamases with substrates and inhibitors. Particularly useful chapters are β-Lactamase: mechanism of action by S. G. Waley (pp. 198–228) and β-Lactamase: Inhibition by R.F. Pratt (pp. 229–271).

BETA-LACTOGLOBULIN

L. SAWYER

β-Lactoglobulin is the major whey **protein** in the milk of ruminants, and it has also been reported in milk from many other species, although not human, lagomorph, or rodent. Where found, genetic variants have also been reported in most cases. Isolation of whey protein from bovine milk is straightforward (1), and its concentration is 2 to 3g/L. Consequently, the protein has been used since its first isolation in 1934 as a convenient, small, soluble protein with which to develop and calibrate new techniques. For example, it is used as a component of standard mixtures for **isoelectric focusing** and **SDS-PAGE** and a calibration sequence for automatic sequenators (see **Protein sequencing**). The extensive literature on β-lactoglobulin, largely that from the domestic cow, can be broadly divided into studies related to the molecule as a protein and those related to its importance to the dairy industry. Reviews of both abound (2–6).

The polypeptide chain of the protein contains about 160 residues. That of ruminant species is dimeric, and in other species it is monomeric. Figure 1 shows some of the known sequences from the SWISSPROT database (7). **Disulfide bonds** link cysteines 66 to 160 and 106 to 119. The asterisks show structurally conserved regions that were recognized following the structural determination first of the homologous plasma *retinol-binding protein* and then of bovine β-lactoglobulin (8,9). These motifs define a widely distributed family called the *lipocalins* (10,11), whose structure is an eight-stranded

```
            ******************
BOVINE A    LIVTQTMKGL DIQKVAGTWY SLAMAASDIS LLDAQSAPLR VYVEELKPTP EGDLEILLQK 60
PIG I       VEVTPIMTEL DTQKVAGTWH TVAMAVSDVS LLDAKSSPLK AYVEGLKPTP EGDLEILLQK 60
HORSE II    TDIPQTMQDL DLQEVAGRWH SVAMVASDIS LLDSESVPLR VYVEELRPTP EGNLEIILRE 60
DONKEY      TNIPQTMQDL DLQEVAGKWH SVAMAASDIS LLDSEEAPLR VYIEKLRPTP EDNLEIILRE 60
DOG I       IVVPRTMEDL DLQKVAGTWH SMAMAASDIS LLDSETAPLR VYIQELRPTP QDNLEIVLRK 60
CAT I       ATVPLTMDGL DLQKVAGMWH SMAMAASDIS LLDSETAPLR VYVQELRPTP RDNLEIILRK 60
DOLPHIN     VSVIRTMEDL DIQRVAGTWH SVAMAASDIS LLDTEEAPLR VNVEELRPTP QGDLEIFLQK 60
KANGAROO    VENIRSKNDL GVEKFVGSWY LREAAKT--- -MEF-SIPLFDMDIKEVNLTP EGNLELVLLE 56
MONKEY      IDSPQTMQDV DLPKLAGTWH SMAMAA....                                 26

               G                                           *******************
BOVINE A    WENDECAQKK IIAEKTKIPA VFKIDALNEN --KVLVLDTD YKKYLLFCME NSAEPEQS-L 117
PIG I       RENDKCAQEV LLAKKTDIPA VFKINALDEN --QLFLLDTD YDSHLLLCME NSASPEHS-L 117
HORSE II    GANHACVERN IVAQKTEDPA VFTVNYQGER --KISVLDTD YAHYMFFCVG PPLPSAEHGM 118
DONKEY      GENKGCAEKK IFAEKTESPA EFKINYLDED --TVFALDSD YKNYLFLCMK NAATPGQS-L 117
DOG I       WEDGRCAEQK VLAEKTEVPA EFKINYVEEN --QIFLLDTD YDNYLFFCEM NADAPQQS-L 117
CAT I       WEDNRCVEKK VLAEKTECAA KFNINYLDEN --ELIVLDTD YENYLFFCLE NADAPDQN-L 117
DOLPHIN     RDKNGCVKEK IIAEKTEIPA VFKINFLNEN --KIFVLDSD YTNYLLFCME NTADPERS-L 117
KANGAROO    -KTDRCVEKK LLLKKTKKPT EFEIYISSES SYTFCVMETD YDSYFLFCLY NISDREK--M 113

            A    ***************
BOVINE A    VCQCLVRTPE VDDEALEKFD KALK-ALPMH IRLSFNPTQL EEQCHI 162
PIG I       VCQSLARTLE VDDQIREKFE DALK-TLSVP MRI--LPAQL EEQCRV 160
HORSE II    VCQYLARTQK VDEEVMEKFS RALQ-PLPGR VQIVQDPSGG QERCGF 163
DONKEY      VCQYLARTQM VDEEIMEKFR RALQ-PLPGR VQIVPDLTRM AERCRI 162
DOG I       MCQCLARTLE VDENVMEKFN RALK-TLPVH MQLLN-PTQA EEQCLI 161
CAT I       VCQCLTRTLK ADNEVMEKFD RALQ-TLPVH VRLFFDPTQV AEQCRI 162
DOLPHIN     TCAYLARTLQ VDDGVMEKFN KAIKPALPMH IRL-FSPTQL EEQCHV 162
KANGAROO    ACAHYVRRIE -ENKGMNEFK KILR--T-LA MPYTVIEVRT RDMCHV 155
```

Figure 1. Sequences of mature β-lactoglobulins with the totally conserved residues in bold. The sites of possible genetic variation within the ruminant species are shown in the bovine sequence in italics, and changes in the bovine B variant are shown above. The N-terminal sequence of the macaque protein is also shown (34). The asterisks show the structurally conserved regions indicative of the lipocalin family.

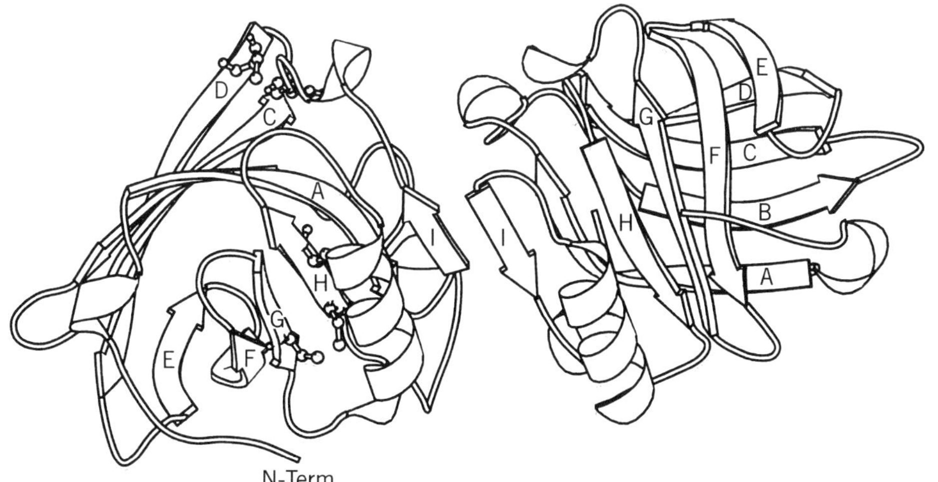

Figure 2. A MOLSCRIPT (35) diagram of the dimer of β-lactoglobulin showing the fold of the main chain.

antiparallel β-barrel with $(+1)_8$ topology (see **β-sheet**) and with an **α-helix** on the outer surface. The structure of β-lactoglobulin is shown in Fig. 2 (12). The family relationship has also been observed in the gene sequences, which show similar arrangements of **introns** and **exons** (13). The 400 bp upstream of the **open reading frame** controls the high level of expression and its restriction to the mammary gland (14). This has led to successful use of the β-lactoglobulin operon in **transgenic technology** (15).

No definite function has yet been ascribed to β-lactoglobulin, but its presence in a family containing mostly small, secreted proteins that transport **hydrophobic** molecules perhaps implies that its function, other than the obvious nutritional one, is associated with the transport or uptake of fatty acids or retinol, both of which it binds (16,17). Indeed, over 20 different **ligands** have been reported that have **association constants** varying between 5×10^7 and 4×10^2 M (3). By analogy with other lipocalins, the binding site should be in the central pocket formed by the β-strands, but there is no definitive evidence that this is the case. Recently it has been shown that palmitate and retinol can bind independently and simultaneously (18).

Some physical parameters of the protein are summarized in Table 1. The bovine protein is stable down to pH 2, and conformational studies have shown that there are several pH-dependent changes between pH 2 and pH 9 (19,20). At the extremes of pH, the dimer dissociates, and it aggregates above pH 8.5 to 9.0. The conformational change between pH 6 and 8 leads to increased reactivity of the free **cysteine** residue, perhaps triggered by the titration of a **carboxyl group** with a pK_a of 7.3. An association of four dimers forms an octamer at 5°C and pH 4.5 mainly in the A variant. This probably results from a carboxyl–carboxylate interaction involving the Gly64Asp change between the most common B and A genetic variants of β-lactoglobulin (21).

The complex behavior during refolding from **urea** between 2° and 25°C is ascribed to **disulfide interchange**. Cys106 pairs with either Cys119 or Cys121, and the former is the native (22,23). It is not yet clear whether this also explains the partially folded form observed at pH 2, where the protein is predominantly a monomer (24). Refolding from **guanidinium chloride** at low pH shows a transient increase in helical content (25), an effect that is stabilized in ethanolic solution (26).

Table 1. Selected Fundamental Parameters of Bovine β-Lactoglobulin

Number of amino acids	162
A variant	Asp64, Val118
B variant	Gly64, Ala118
Relative molecular mass	18,400
Sedimentation coefficient (S°_{20w})	2.83×10^{13} s^{-1}
Diffusion coefficient	7.70×10^{-7} cm^2s^{-1}
Stokes' radius	2.68 nm
Radius of gyration	2.17 nm
Axial ratio	2:1
Dipole moment	730 Debye
Absorbance at 280 nm	0.96 (1.0 g/l solution, 1 cm cell)

β-Lactoglobulin has been cloned and expressed both in *Escherichia coli* and in **yeast**, allowing creation of **site-directed mutations** (27) for probing the solution properties (28) or modifying the thermal denaturing behavior (29). Thermal denaturation is of particular interest to the dairy industry because heating milk is the basis of much processing and also the genetic variants exhibit different behavior (30). In addition, protein concentration, pH, temperature, ionic strength, dielectric constant, and buffer type all contribute to the denaturing behavior. Differential scanning **calorimetry** and thermal aggregation studies show that the A variant is less stable than the B variant (31), but studies at lower protein concentration show that the reverse is true (32). Thermal aggregation also involves intermolecular disulfide interchange, presumably initiated by the free **thiol group** (33).

BIBLIOGRAPHY

1. R. Aschaffenburg and R. Drewry (1957) *Biochem. J.* **65**, 273–277.
2. J. M. A. Tilley (1960) *Dairy Sci. Abstr.* **22**, 111–125.
3. R. L. J. Lyster (1972) *J. Dairy Res.* **39**, 279–318.
4. S. G. Hambling, A. S. McAlpine, and L. Sawyer (1992) In *Advanced Dairy Chemistry I.* (P. F. Fox, ed.), Elsevier, Amsterdam, pp. 141–190.

5. M. McSwiney, H. Singh, O. Campanella, and L. K. Creamer (1994) *J. Dairy Res.* **61**, 221–232.
6. J. E. Kinsella and D. M. Whitehead (1987) *Adv. Food Nutr. Res.* **33**, 343–438.
7. A. Bairoch and B. Boeckmann (1994) *Nucleic Acid Res.* **22**, 3578-
8. M. Newcomer et al. (1984) *EMBO J.* **3**, 1451–1454.
9. M. Z. Papiz et al. (1986) *Nature* **324**, 383–385.
10. S. Pervaiz and K. Brew (1987) *FASEB J.* **1**, 209–214.
11. D. R. Flower (1996) *Biochem. J.* **318**, 1–14.
12. S. Brownlow et al. (1997) *Structure*, **5**, 481–495.
13. S. Ali and A. J. Clark (1988) *J. Mol. Biol.* **189**, 415–426.
14. J. P. Simons, M. McClenaghan, and A. J. Clark (1987) *Nature* **328**, 530–532.
15. A. L. Archibald et al. (1990) *Proc. Natl. Acad. Sci. USA* **87**, 5178–5182.
16. A. A. Spector and J. E. Fletcher (1970) *Lipids* **5**, 403–411.
17. F. D. Fugate and P. S. Song (1980) *Biochim. Biophys. Acta* **652**, 28–42.
18. M. Narayan and L. J. Berliner (1997) *Protein Sci.* **7**, 150–157.
19. C. Tanford, L. G. Bunville, and Y. Nozaki (1959) *J. Am. Chem. Soc.* **81**, 4032–4036.
20. S. N. Timasheff, L. Mescanti, J. J. Basch, and R. Townend (1966) *J. Biol. Chem.* **241**, 2496–2501.
21. J. Witz, S. N. Timasheff, and V. Luzzati (1964) *J. Am. Chem. Soc.* **86**, 168–173.
22. H. A. McKenzie, G. B. Ralston, and D. C. Shaw (1972) *Biochemistry* **11**, 4539–4547.
23. T. E. Creighton (1980) *J. Mol. Biol.* **137**, 61–80.
24. L. Ragona et al. (1997) *Folding and Design* **2**, 281–290.
25. D. Hamada, S. Segawa, and Y. Goto (1996) *Nature Struct. Biol.* **3**, 868–873.
26. E. Dufour, H. C. Bertrand, and T. Haertlé (1993) *Biopolymers* **33**, 589–598.
27. C. A. Batt, L. D. Rabson, D. W. S. Wong, and J. E. Kinsella (1990) *Agric. Biol. Chem.* **54**, 949–955.
28. Y. Katakura, M. Totsuka, A. Ametani, and S. Kaminogawa (1994) *Biochim. Biophys. Acta* **1207**, 58–67.
29. C. A. Batt, J. Brady, and L. Sawyer (1994) *Tr. Food Sci.* **5**, 261–265.
30. E. Jakob and Z. Puhan (1992) *Int. Dairy J.* **2**, 157–178.
31. X. L. Huang, G. L. Catignani, and H. E. Swaisgood (1994) *J. Agric. Food Chem.* **42**, 1276–1280.
32. G. I. Imafadon, K. F. Ng-Kwai-Hang, V. R. Harwalkar, and C.-Y. Ma (1991) *J. Dairy Res.* **74**, 2416–2422.
33. S. P. F. M. Roefs and K. G. Dekruif (1994) *Eur. J. Biochem.* **226**, 883–889.
34. N. Azuma and K. Yamauchi (1991) *Comp. Biochem. Physiol.* **99B**, 917–921.
35. P. J. Kraulis (1991) *J. Appl. Cryst.* **24**, 946–950.

Suggestions for Further Reading

L. Banaszak, N. Winter, Zh. Xu, D. A. Bernlohr, S. Cowan, and T. A. Jones (1994) Lipid-binding proteins—a family of fatty-acid and retinol transport proteins, *Adv. Protein Chem.* **45**, 89–151.

L. K. Creamer (1995) Effect of sodium dodecyl sulphate and palmitic acid on the equilibrium unfolding of bovine β-lactoglobulin *Biochemistry* **34**, 7170–7176.

P. F. Fox (1995) *Heat Induced Changes in Milk*, 2nd ed., International Dairy Federation, Brussels.

T.-R. Kim et al. (1997) High level expression of bovine β-lactoglobulin in *P. pastoris* and characterization of its physical properties, *Prot. Eng.* **10**, 133.

H. A. McKenzie (1971) β-Lactoglobulins. In *Milk Proteins—II*. Academic Press, New York. pp. 257–330.

B. Y. Qin et al. (1998) Structural basis of the Tanford transition of bovine β-lactoglobulin, *Biochemistry*. **37**, in press.

R. Townend, T. T. Herskovits, S. N. Timasheff, and M. J. Gorbunoff (1969) The state of amino acid residues in β-lactoglobulin *Arch. Biochem. Biophys.* **129**, 567–580.

BETA-MEANDER

JENNIFER MARTIN

The β-meander is a **protein motif** frequently observed in **protein structures**. It has a particular type of antiparallel **β-sheet** structure, with a very simple topology in which two or more **β-strands** that are consecutive in sequence are also adjacent to one another in the three-dimensional structure (Fig. 1). β-Meanders are equivalent to multiply linked **hairpins**.

[See also **Beta-sheet**.]

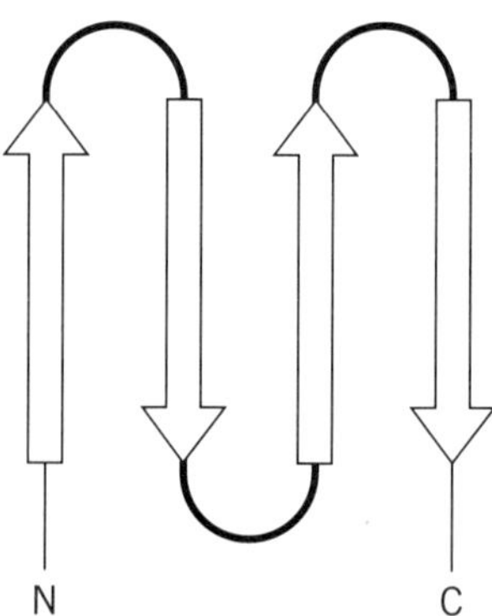

Figure 1. Schematic representation of a β-meander consisting of four β-strands (shown as arrows).

BETA-PLEATED SHEET

JENNIFER MARTIN

β-Pleated sheet is an alternative name for **β-sheet**, a type of **secondary structure** occurring in **protein structures**. The name arises from the pleated nature of the sheet and the individual **β-strands** when viewed from the side.

[See also **Beta-sheet**.]

BETA-SHEET

JENNIFER MARTIN

The β-sheet is a **secondary structure** component of **protein structure**, comprising two or more adjacent **β-strands** linked by **hydrogen bonds**. In contrast to the coiled **backbone** of the **α-helix**, the structure of a β-strand is characterized by an extended backbone **conformation**, corresponding to the favorable upper left region of the **Ramachandran plot**. β-Sheets are formed by a side-by-side arrangement of β-strands, with backbone hydrogen bonds between adjacent strands linking the sheet together. The side chains of consecutive residues on a β-strand do not interact with each other, but lie alternately above and below the sheet, giving a pleated appearance to the β-sheet backbone (the β-sheet is sometimes referred to as a **β-pleated sheet**).

β-Strands are often depicted as arrows in protein structure diagrams, with the direction of the arrow running from the ***N*-terminal** end to the ***C*-terminal** end of the β-strand. When packed together in a β-sheet, the β-strands may all be oriented in the same direction (parallel β-sheet) or in alternating directions (antiparallel β-sheet). In some cases, the β-strands will be a mixture of parallel and antiparallel (mixed β-sheet). The backbone hydrogen bonding patterns of parallel and antiparallel β-sheet structures differ (Fig. 1). In the parallel β-sheet, hydrogen bonds are evenly spaced along the β-strand and point at an angle to the other strand. The antiparallel β-sheet has alternately wide and narrow spacings between adjacent hydrogen bonds. The average backbone angles for parallel and antiparallel β-sheets also vary, with ϕ of $-119°$ and Ψ of $113°$ for parallel structures and ϕ of $-139°$ and Ψ of $135°$ for antiparallel structures (see **Ramachandran plot**). Most β-sheets have a right-handed twist of about 10°. The regular hydrogen bonding pattern and backbone angles of β-sheets are sometimes disrupted at the edges by the formation of a **β-bulge**, accentuating the twist to 35° to 45°.

Because a β-sheet can be formed from sections of polypeptide chain that are distant from each other in the primary structure, many different types of β-sheet have been identified in protein structures. The **β-meander** is a simple antiparallel β-sheet topology where β-strands that are consecutive in sequence are also located adjacent to one another in the three-dimensional structure. The adjacent β-strands in a β-meander are often connected by **hairpin** turns. Other types of antiparallel β-sheet structure include the **jelly roll motif**, the **Greek key motif**, and the **up and down β-barrel**. The **β-helix** is formed from parallel β-sheets.

[See also **Protein structure** and **Secondary structure, protein**.]

BIBLIOGRAPHY

1. P. J. Kraulis (1991) *J. Appl. Crystallogr.* **24**, 946–950.
2. E. A. Merritt and M. E. P. Murphy (1994) *Acta Crystallogr.* **D50**, 869–873.
3. D. J. Bacon and W. F. Anderson (1988) *J. Mol. Graphics* **6**, 219–222.

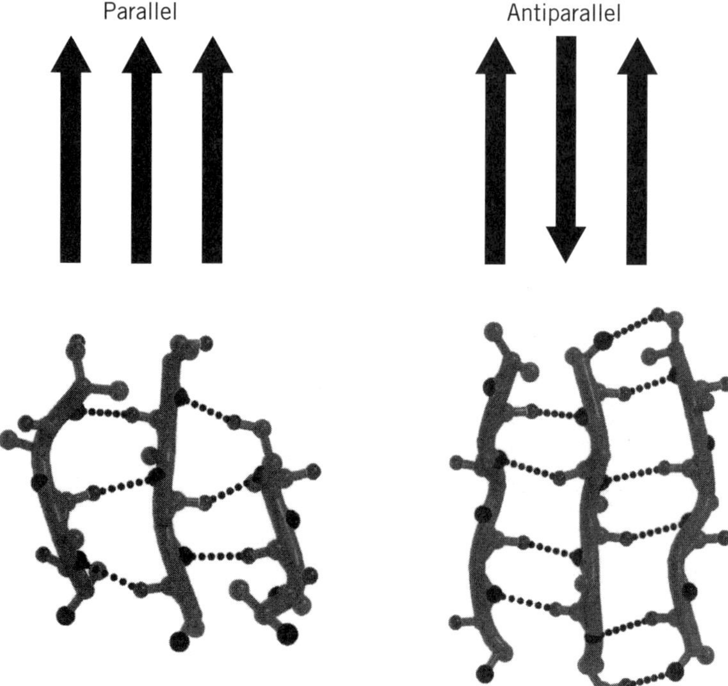

Figure 1. β-Sheet structure in proteins. Schematic representation of parallel and antiparallel β-sheets, with each strand of the sheet depicted as an arrow having direction from the *N*-terminal to the *C*-terminal end. In the lower half of the figure, the atomic detail of the β-sheet structure is shown, with hydrogen bonds between backbone carbonyl oxygen atoms and amide nitrogen atoms indicated by dotted lines. Note the approximately equal spacing of hydrogen bonds in the parallel β-sheet compared with the alternating wide and narrow spacing in the antiparallel β-sheet. For clarity, only $C\beta$ atoms of side chains are shown. Nitrogen atoms and oxygen atoms are shown as dark spheres. This figure was generated using Molscript (1) and Raster3D (2,3).

Suggestions for Further Reading

C. Branden and J. Tooze (1991) *Introduction to protein structure*, Garland, New York.

T. E. Creighton (1993) *Proteins: Structures and Molecular Properties*, W. H. Freeman, New York.

J. S. Richardson (1981) The anatomy and taxonomy of protein structure. *Adv. Protein Chem.* **34**, 167–339.

BETA-STRAND

JENNIFER MARTIN

A β-strand is one of the two regular types of **secondary structure** (the other being the α-**helix**) that is commonly observed in **protein structures**. The β-strand is characterized by an extended **backbone** conformation. The β-strand is not stable as a discrete structural unit by itself, but is stable only when it associates through **hydrogen bond** interactions with adjacent β-strands to form a β-**sheet**.

[See also **Beta-sheet** and **Secondary structure, protein**.]

BETA-TURNS

JENNIFER MARTIN

A β-turn is a type of nonregular **secondary structure** in **proteins** that causes a change in direction of the **polypeptide chain**. In the β-turn, a **hydrogen bond** is formed between the **backbone** carbonyl oxygen of one residue (i) and the backbone amide NH of the residue three positions further along the chain $(i + 3)$ (Fig. 1) (1). This (i) to $(i + 3)$ interaction distinguishes β-turns from γ-**turns**, which have an (i) to $(i + 2)$ hydrogen bond. There are several different types of β-turns, classified according to the backbone dihedral angles of the two intervening residues, $(i + 1)$ and $(i + 2)$. The type I and II turns are the most common, together accounting for about two-thirds of all β-turns, but they have very different amino acid preferences (2). Type I turns prefer Asn, Asp, or Ser in the (i) position; Asp, Ser, Thr or Pro in $(i + 1)$; Asp, Ser, Asn, or Arg in $(i + 2)$; and Gly, Trp, or Met in $(i + 3)$. In contrast, type II turns have a preference for Pro at position $(i + 1)$, Asn or Gly at position $(i + 2)$ and Gln or Arg at $(i + 3)$ (2). Other β-turn types include type I′ (related to type I by inversion of the sign of the ϕ and ψ angles of the **Ramachandran plot**), II′ (related to type II in the same way), IV, VIa, VIb, and VIII. Types I′ and II′ are almost exclusively found in **hairpins**; that is, they usually connect two adjacent antiparallel β-**strands**. The type IV β-turn is a miscellaneous type that includes any (i) to $(i + 3)$ hydrogen-bonded turn in a protein structure where the ϕ and ψ angles are different (by >40°) from the values that define the other β-turn types (Table 1). Types VIa and VIb have a proline *cis*-peptide bond at the $(i + 2)$ position; in type VIa the proline adopts a backbone conformation close to that of α-**helical** residues, whereas in type VIb the proline has a conformation similar to that of residues in a β-**strand**.

[See also **Secondary structure, protein**, **Turns**, **Gamma-turns**, and **Omega loops**.]

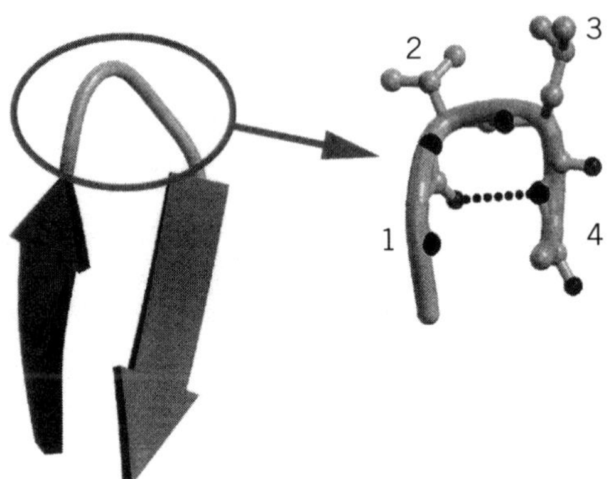

Figure 1. Schematic representations of a β-turn in a protein structure. (**Left**) The β-turn is shown connecting two β-strands in a hairpin motif. (**Right**) The detailed atomic structure of the β-turn is shown. Residues are numbered 1 to 4 for (i) to $(i + 3)$. The hydrogen bond between the backbone carbonyl oxygen of residue 1 and the backbone amide nitrogen of residue 4 is shown as a dotted line. This example is a type I β-turn, because the backbone angles for the $(i + 1)$ residue are $\phi = -63°$ and $\psi = -33°$; and for the $(i + 2)$ residue, ϕ is $-86°$ and $\psi = -3°$. Oxygen atoms and nitrogen atoms are shown as dark spheres. This figure was generated by Molscript (3) and Raster3D (4,5).

Table 1. Classification of Beta-Turn Types

Beta-Turn Type	Backbone Dihedral Angles of Central Residues $\phi i + 1$	$\psi i + 1$	$\phi i + 2$	$\psi i + 2$
I	−60°	−30°	−90°	0°
I′	60°	30°	90°	0°
II	−60°	120°	80°	0°
II′	60°	−120°	−80°	0°
IV	Any (i) to $(i + 3)$ hydrogen-bonded turn having angles that differ by more than 40° from those of other β-turn types			
VIa	−60°	120°	−90°	0°
VIb	−120°	120°	−60°	0°
VIII	−60°	−30°	−120°	120°

BIBLIOGRAPHY

1. J. S. Richardson (1981) *Adv. Protein Chem.* **34**, 167–339.
2. C. M. Wilmot and J. M. Thornton (1988) *J. Mol. Biol.* **203**, 221–232.
3. P. J. Kraulis (1991) *J. Appl. Crystallogr.* **24**, 946–950.
4. E. A. Merritt and M. E. P. Murphy (1994) *Acta Crystallogr.* **D50**, 869–873.
5. D. J. Bacon and W. F. Anderson (1988) *J. Mol. Graphics* **6**, 219–222.

BICOID

ERNST A. WIMMER
CLAUDE DESPLAN

Mutational analysis in the fruitfly *Drosophila melanogaster* has demonstrated that many aspects of anteroposterior patterning are already programmed in the oocyte by maternal gene activity (1,2). During embryonic development, cells within an embryonic field receive positional information and are instructed by the concentration gradient of **morphogens** (3). Historically, the graded distribution of the Bicoid protein (Bcd) has provided the first evidence for the existence of such a morphogenetic gradient. Maternally derived *bcd* **messenger RNA** is localized to the anterior pole of the oocyte, where it serves as the source for the Bcd concentration gradient. Then, Bcd acts as a concentration-dependent activator of **transcription** and as a **translational repressor**. However, Bcd is only part of a wider regulatory network of maternal determinants, and its presence might actually be restricted to dipteran flies.

Anterior Localization of the Maternal Determinant *Bicoid*

Evidence for an anterior organizing center has been obtained by experimental embryology, for example, the removal and transplantation of polar cytoplasm (4) (see **Organizer**). The nature of the patterning agents has been identified as RNA-containing particles (5). Genetic experiments have shown that the anterior morphogenetic function requires the activity of the *bicoid* maternal gene (*bcd*). Mothers homozygous mutant for *bcd* are normal, but all of their progeny lack head, thorax, and some abdominal structures. Instead posterior terminal structures are duplicated at the anterior (Fig. 1; 6). **Cloning** of the *bcd* **gene** revealed that *bcd* is expressed during oogenesis and that its mRNA is localized to the anterior tip of the oocyte and early embryo (Fig. 2) (7). This localization depends on the presence of the *bcd* 3′-UTR, which contains a specific structure recognized by factors that carry it to the anterior pole along the polarized **microtubule** array of the oocyte (8).

The Morphogenetic Gradient of the Bicoid Transcription Factor

After egg deposition, the anteriorly localized *bcd* mRNA starts to be **translated**, producing a source of the Bcd protein. Then, Bcd **diffuses** passively in the syncytial environment of the blastoderm embryo and generates an anteroposterior concentration gradient whose high point is at the anterior pole of the embryo and whose low point is near the posterior pole (Fig. 2; 9,10). Decreasing or increasing the copy number of the *bcd* gene changes the slope of the Bcd gradient and, consequently, changes the **fate map** of the early embryo. This suggests that anterior positional values are specified by Bcd in a concentration-dependent manner (9,10).

The Bcd protein contains a homeodomain (7) that does not belong to any specific class (see **Homeobox genes**). It is characterized by a **lysine** residue at residue 50 (Lys50) of the homeodomain that determines its **DNA-binding** specificity (14). Bcd functions as a transcriptional activator in cell culture and in yeast (11–13). The morphogenetic gradient of Bcd differentially activates the first zygotically active segmentation genes, the gap genes. At high concentrations of Bcd, the head gap genes *orthodenticle* (*otd*), *empty spiracles* (*ems*), and *buttonhead* (*btd*) are activated (15–18), whereas lower levels of Bcd are required to activate *hunchback* (hb), which is expressed in a broad anterior domain (Fig. 3; 13). A detailed study of the proximal *hb* **promoter** suggested that the affinity of Bcd binding sites in the promoters of the different target genes determines at which threshold level, and therefore at which position along the anterior-posterior axis, these genes are activated (19,20). The ability of Bcd to activate target genes differentially provided the first molecular explanation of how a morphogen might act.

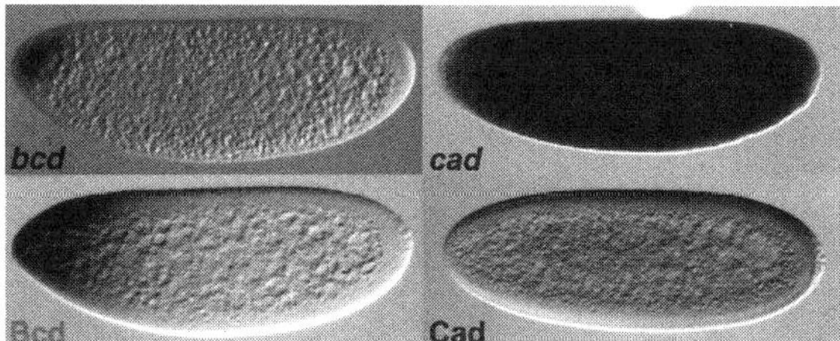

Figure 2. mRNA and protein expression of bcd and cad. The *bcd* mRNA is localized at the most anterior tip of the embryo. Its translation and diffusion of its protein product lead to the formation of the Bcd morphogenetic gradient. Maternal *cad* mRNA is distributed throughout the embryo, but its translation is blocked by the Bcd protein. This leads to the formation of a posterior to anterior gradient of the Cad protein.

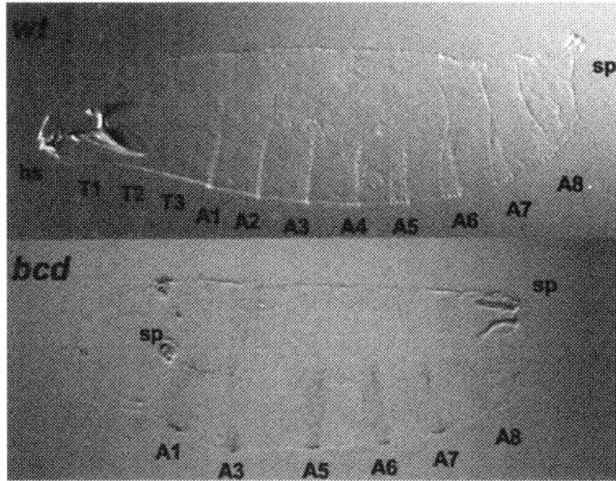

Figure 1. Phenotype of *bcd* embryos. Mothers homozygous mutant for *bcd* lay embryos that develop cuticles lacking all head structures (hs), thorax (T1, T2, T3), and two abdominal segments (A2, A4). Furthermore, the posterior structures like the spiracles (sp) are duplicated at the anterior.

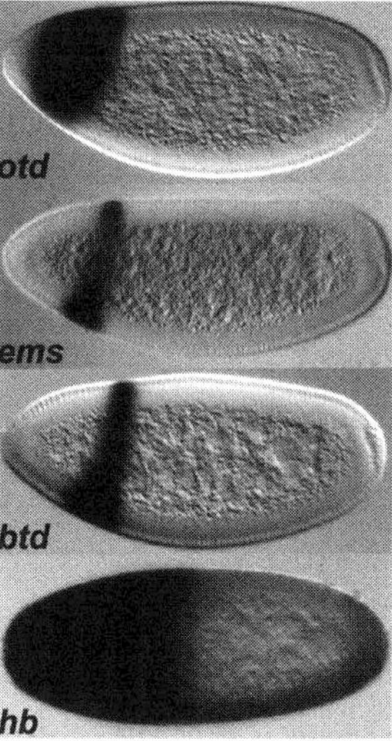

Figure 3. Expression pattern of *bcd* target genes. *bcd* is a maternal gene that controls the expression of the zygotic gap target genes. The morphogenetic gradient of Bcd protein differentially activates head (*otd, ems*, and *btd*) or thoracic (*hb*) gap genes.

Bicoid Acts as a Translational Regulator

Although Bcd is a DNA-binding protein, it also controls translation of the *caudal* gene (*cad*). *cad* is a gene required for posterior development, and its function is evolutionarily conserved in diverse organisms, including vertebrates (21). Maternal *cad* mRNA is ubiquitously distributed (22–24). Its translation is blocked at the anterior of the embryo by *bcd*, leading to a posterior-to-anterior gradient of the Cad protein (Fig. 3). *cad* is expressed ectopically in *bcd* mutant embryos throughout the embryo. Bcd binds to a sequence in the 3′-untranslated region of the *cad* mRNA (25–27). Like DNA binding, Bcd's RNA binding is conferred by the homeodomain and depends on the presence of Lys50 in the homeodomain. The Bcd homeodomain is the only one that binds RNA (25,27).

Bicoid is Integrated into a Network of Maternal Determinants

Bcd is not the only maternal morphogen involved in anteroposterior patterning. A morphogenetic gradient of the Hunchback protein (Hb) creates the correct polarity in the embryo in the absence of *bcd*. The anteroposterior Hb gradient is established by translational repression of ubiquitous maternal mRNA by the *nanos* posterior determinant (28–30). Bcd and Hb are redundant for abdominal segmentation because high levels of maternal Hb rescue the abdominal phenotype of *bcd* mutant embryos (28,29) and the activation of the central gap gene *Krüppel*. At the anterior of the embryo, Bcd and Hb synergize to activate the anterior gap genes (31). Because each morphogen has instructive functions on its own, both must be removed from the embryo to disrupt the polarity of the embryo completely. In more posterior regions, Bcd is also redundant with *cad* for activating the gap gene *knirps* (32). This shows that *bcd* function is partially redundant with other maternal functions. It should be noted that maternal *hb* and *cad* functions are not essential for viability in the presence of *bcd*. It is likely that they represent remains of ancestral systems that patterned the embryo before the appearance of *bcd*.

At the anterior pole of the embryo, Bcd function is influenced by the activity of the terminal system, which involves locally activating the Torso (Tor) **receptor tyrosine kinase** (33). Tor activity leads to two opposite effects on Bcd function. Tor activity increases the activator function of Bcd by lowering the threshold level of Bcd where target genes are activated (18,34,35). At the anterior pole, however, Bcd function is blocked in a Tor-dependent manner, and most *bcd* target genes retract from the anterior pole where Bcd concentration is the highest (36). The effect of *tor* activity on Bcd function might be mediated by *tor*-dependent phosphorylation of Bcd (36), or it may be indirect and may involve other factors (34).

Evolutionary Considerations on Bicoid

There are several reasons to believe that the critical functions of *bcd* in patterning the anteroposterior axis of the *Drosophila* embryo (Fig. 4) may not be conserved throughout evolution. *bcd* is unlikely to represent an ancestral morphogen. No *bcd* homologous genes have been identified outside higher dipterans (38), despite its homeobox and its specific location in the Hox cluster (7). Moreover, *bcd* shows an unusually high rate of **divergence** for a homeobox gene (39), and its function is not even conserved within higher dipterans (38,40). Therefore, it is reasonable to assume that *bcd* is the result of recent **gene duplication** in the Hox cluster of dipterans and that it evolves relatively freely. Therefore, although the role of Bcd in *Drosophila* represents a striking example of a powerful morphogenetic **transcriptional factor**, it is likely that this function is not conserved in other insects and may exist only to allow the *Drosophila* embryo to develop quickly.

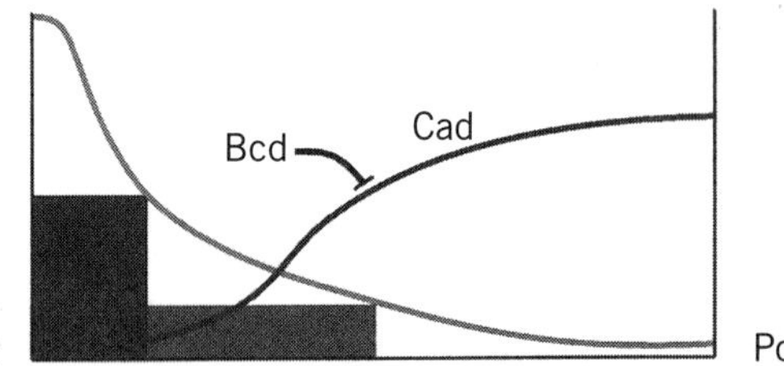

Figure 4. Morphogenetic functions of the Bcd gradient. The Bcd gradient differentially activates zygotic target genes (dark boxes) in head and thorax anlagen. The Bcd gradient also leads to the formation of the Cad protein gradient by blocking translation of its mRNA.

BIBLIOGRAPHY

1. C. Nüsslein-Volhard and E. Wieschaus (1980) *Nature* **287**, 795–801.
2. C. Nüsslein-Volhard, H. G. Frohnhofer, and R. Lehmann (1987) *Science* **238**, 1675–1681.
3. L. Wolpert (1969) *J. Theor. Biol.* **25**, 1–47.
4. K. Sander (1969) *Advances in Insect Physiology* (J. E. Treherne, M. J. Berridge, and V. B. Wigglesworth, eds.), Academic Press. Vol. 12, pp. 125–235.
5. K. Kalthoff (1979) *Determinants of Spatial Organization* (S. Subtelney and I. R. Konisberg, eds.), Academic Press, N.Y., pp. 97–126.
6. H. G. Fröhnhofer and C. Nüsslein-Volhard (1986) *Nature* **324**, 120–125.
7. T. Berleth, M. Burri, G. Thoma, D. Bopp, S. Richstein, G. Frigerio, M. Noll, and C. Nüsslein-Volhard (1988) *EMBO J.* **7**, 1749–1756.
8. D. Ferrandon, L. Elphick, C. Nüsslein-Volhard, and D. St Johnston (1994) *Cell* **79**, 1221–1232.
9. W. Driever and C. Nüsslein-Volhard (1988) *Cell* **54**, 95–104.
10. W. Driever and C. Nüsslein-Volhard (1988) *Cell* **54**, 83–93.
11. S. D. Hanes and R. Brent (1989) *Cell* **57**, 1275–1283.
12. W. Driever, J. Ma, C. Nüsslein-Volhard, and M. Ptashne (1989) *Nature* **342**, 149–154.
13. G. Struhl, K. Struhl, and P. M. Macdonald (1989) *Cell* **57**, 1259–1273.
14. J. Treisman and C. Desplan (1989) *Nature* **341**, 335–337.
15. D. Dalton, R. Chadwick, and W. McGinnis (1989) *Genes and Devel.* **3**, 1940–1956.
16. R. Finkelstein and N. Perrimon (1990) *Nature* **346**, 485–488.
17. U. Walldorf and W. J. Gehring (1992) *EMBO J.* **11**, 2247–2259.
18. E. A. Wimmer, M. Simpson-Brose, S. M. Cohen, C. Desplan, and H. Jäckle (1995) *Mechanisms of Development* **53**, 235–245.
19. W. Driever, G. Thoma, and C. Nüsslein-Volhard (1989) *Nature* **340**, 363–367.
20. G. Struhl (1989) *Nature* **338**, 741–744.
21. V. Subramanian, B. I. Meyer, and P. Gruss (1995) *Cell* **83**, 641–653.
22. M. Mlodzik and W. J. Gehring (1987) *Cell* **48**, 465–478.
23. M. Mlodzik, A. Fjose, and W. J. Gehring (1985) *EMBO J.* **4**, 2961–2969.

24. P. M. MacDonald and G. Struhl (1986) *Nature* **324**, 537–545.
25. R. Rivera-Pomar, D. Niessing, U. Schmidt-Ott, W. J. Gehring, and H. Jäckle (1996) *Nature* **379**, 746–749.
26. J. Dubnau and G. Struhl (1996) *Nature* **379**, 694–699.
27. S. K. Chan and G. Struhl (1997) *Nature* **388**, 634.
28. G. Struhl, P. Johnston, and P. A. Lawrence (1992) *Cell* **69**, 237–249.
29. M. Hülskamp, C. Pfeifle, and D. Tautz (1990) *Nature* **346**, 577–580.
30. C. Schülz and D. Tautz (1995) *Development* **121**, 1023–1028.
31. M. Simpson-Brose, J. Treisman, and C. Desplan (1994) *Cell* **78**, 855–865.
32. R. Rivera-Pomar, X. Lu, N. Perrimon, H. Taubert, and H. Jackle (1995) *Nature* **376**, 253–256.
33. N. Perrimon and C. Desplan (1994) *Trends in Biochemical Sciences* **19**, 509-513.
34. Q. Gao, Y. Wang, and R. Finkelstein (1996) *Mechanisms of Development* **56**, 3–15.
35. U. Grossniklaus, K. M. Cadigan, and W. J. Gehring (1994) *Development* **120**, 3155–3171.
36. E. Ronchi, J. Treisman, N. Dostatni, G. Struhl, and C. Desplan (1993) *Cell* **74**, 347–355.
37. Y. Bellaïche, R. Bandyopadhyay, C. Desplan, and N. Dostatni (1996) *Development* **122**, 3499–3508.
38. R. Schöder and K. Sander (1993) *Roux's Arch. Dev. Biol.* **203**, 34–43.
39. R. Sommer and D. Tautz (1991) *Development* **113**, 419–430.
40. F. Bonneton, P. J. Shaw, C. Fazakerley, M. Shi, and G. A. Dover (1997) *Mechanisms of Development* **66**, 143–156.

Suggestions for Further Reading

P. A. Lawrence (1992) *The Making of a Fly, The Genetics of Animal Design*, Blackwell Scientific Publications.

P. A. Lawrence and G. Struhl (1996) *Cell* **85**, 951–961.

D. St Johnston (1995) *Cell* **81**, 161–170.

D. St Johnston and C. Nüsslein-Volhard (1992) *Cell* **68**, 201–219.

BIFUNCTIONAL CROSSLINKING REAGENTS

G. M. CARLSON
O. W. NADEAU

TERMINOLOGY

If a bifunctional chemical **crosslinking** agent is designed so that its two reactive groups are identical, it is referred to as a *homobifunctional* reagent; if its two reactive groups are different, it is a *heterobifunctional* reagent. If one or both reactive groups of the crosslinker become so only as the result of a photochemical reaction caused by exposing the reagent to light of an appropriate wavelength, then the crosslinking reagent is photoactivatable, photoreactive, photosensitive, or **light-activated (caged)**. The portion of the crosslinking reagent other than, and usually between, its two reactive groups is referred to as the *spacer arm*, or *crossbridge*. If the spacer arm contains a covalent bond that can be easily broken (by an oxidant, reductant, base, etc), then the crosslinker is termed *cleavable*; otherwise, it is classified as *noncleavable*. With most chemical crosslinkers, a portion of the reagent is incorporated into the final crosslinked complex; however, there is a special class of reagents that facilitate crosslinking without being incorporated into the final covalent complex. These reagents are referred to as *zero-length* crosslinkers, because the groups they crosslink are not separated by a spacer arm.

VARIABLES IN THE DESIGN OF CROSSLINKING REAGENTS

By varying the chemical and physical properties of the reactive groups and spacer arm of a crosslinking agent, one can obtain crosslinkers with different properties and uses. A heterobifunctional reagent, with its two different types of reactive groups, allows for different selectivities in the functional groups that become crosslinked, compared to a homobifunctional reagent containing a pair of the same reactive groups. The selectivity shown by the reactive groups of crosslinkers for the side chains of particular amino acids mirrors the selectivity of those reactive groups when used as monofunctional reagents in the chemical modification of proteins, from which crosslinking is an outgrowth.

A number of variables pertain to spacer arms: (*1*) the length of the crossbridge, as in the case of bisimidoesters (Table 1) with different numbers of methylene groups in the spacer arm; (*2*) its geometry, which in the case of the phenylenebismaleimides (Table 1) also causes an increase in length, progressing from *ortho* to *para*; (*3*) its chemical nature, especially the number of charged or polar groups on the spacer arm, which influences the crosslinker's solubility properties and uses. For example, **membrane proteins** would more likely be crosslinked by bifunctional reagents that are **hydrophobic**, rather than **hydrophilic**. **Reporter groups** (chromophores, radioactive tracers, etc) can also be incorporated into spacer arms to facilitate quantification of the extent of protein modification, studies on the crosslinked complex, or identification of the crosslinked components. Incorporation into the spacer arm of cleavable bonds, such as **disulfide bonds**, is also useful in the identification of crosslinked components, especially by **diagonal methods** (1).

EXAMPLES OF BIFUNCTIONAL CROSSLINKERS

The compounds listed in Table 1 are representative of some of the most frequently used chemical classes of crosslinking reagents, but they represent only a small fraction of the literally hundreds of bifunctional crosslinking reagents that have been described in the scientific literature. The widely used *N*-substituted bismaleimides are selective for **thiol groups**, especially at pH values near neutrality, where **amino groups** are predominantly protonated. This class of crosslinkers is available with many different spacer arms, having large variations in length and solubility, or having chromophores or cleavable bonds incorporated. The alkylating bishaloacetyl crosslinkers are also selective at neutral pH for sulfhydryl groups, but they will also react with imidazole side chains of some **histidine** residues under this condition. It should be remembered that few reagents used to modify proteins chemically are absolutely specific for any given amino acid side chain; on the other hand, high selectivity can often be achieved empirically through varying the pH, reaction time, and concentrations of reactants.

Table 1. Examples of Bifunctional Crosslinkers

Compound	Structure	Class	Selectivity
Homobifunctional			
1. *N,N′-p*-Phenylenebismalemide	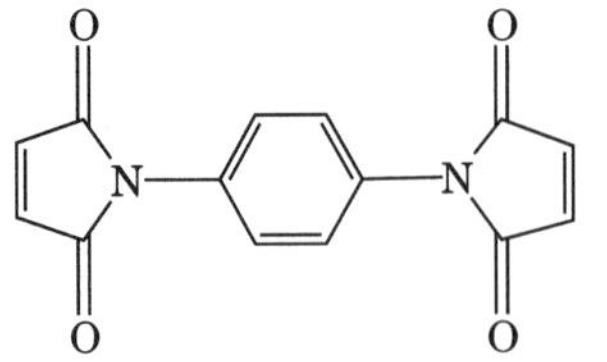 (maleimide)N—C_6H_4—N(maleimide)	Bismalemide	Sulfhydryl
2. *N,N′*-Ethylene-bis(iodoacetamide)	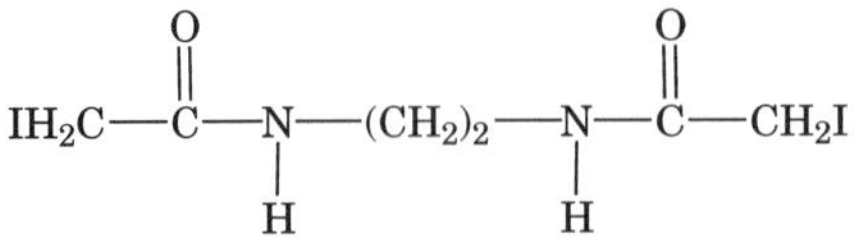 $IH_2C-C(=O)-N(H)-(CH_2)_2-N(H)-C(=O)-CH_2I$	Bishaloacetylamide	Sulfhydryl
3. Dimethylsubermidate	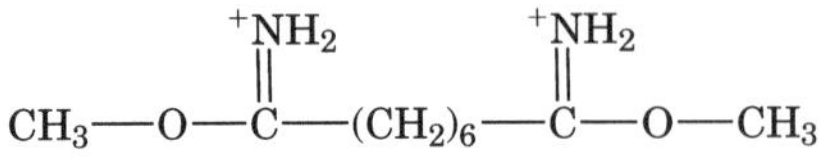 $CH_3-O-C(={}^{+}NH_2)-(CH_2)_6-C(={}^{+}NH_2)-O-CH_3$	Bisimidoester	Amine
4. *N*-Hydroxysuccinimidylsuberate	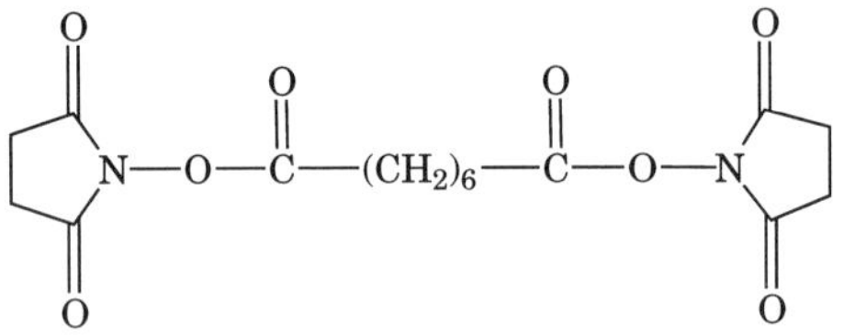 (succinimide)N—O—C(=O)—$(CH_2)_6$—C(=O)—O—N(succinimide)	*N*-Hydroxysuccinimide ester	Amine
5. Glutaraldehyde	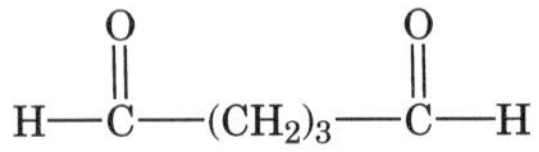 $H-C(=O)-(CH_2)_3-C(=O)-H$	Dialdehyde	Amine
Heterobifunctional			
6. *N*-Succinimidyl-3-maleimidopropionate	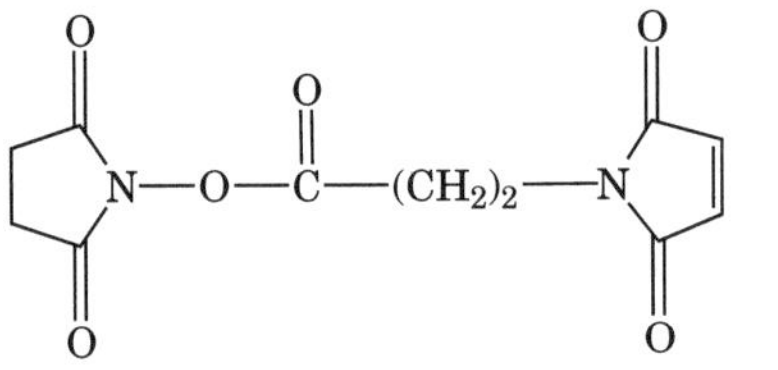 (succinimide)N—O—C(=O)—$(CH_2)_2$—N(maleimide)	*N*-Hydroxysuccinimide ester/ malemide	Amine/ sulfhydryl
7. *N*-5-Azido-2-nitrobenzoyloxysuccinimide	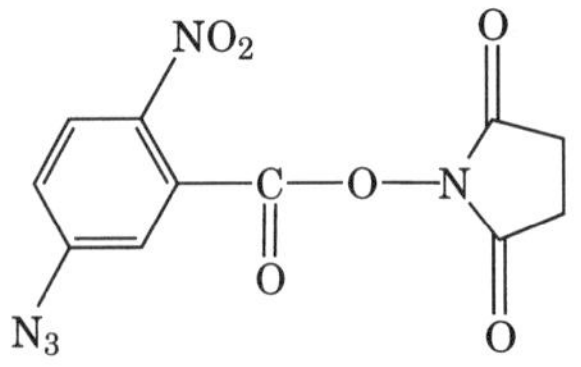 (NO_2, N_3)C_6H_3—C(=O)—O—N(succinimide)	Arylazide/N-hydroxysuccinimide ester	Broad/amine

The bisimidoesters were among the earliest used crosslinkers and are also available with a variety of spacer arms. These compounds preferentially react at alkaline pH with amino groups, with which they form amidine bonds, retaining the positive charge of the original amine. A more frequently used class of amine-selective reagents is the ***N*-hydroxysuccinimide** esters, which produce an amide-crosslinked product that is relatively stable in aqueous solution. The *N*-hydroxysuccinimide ester group is frequently combined with maleimide, such as in compound 6 of Table 1, to produce heterobifunctional crosslinkers with selectivity toward crosslinking amino and sulfhydryl groups. Another common class of heterobifunctional crosslinkers contains at one end a photoactivatable group, which is generally assumed to be nonselective for the functional groups it modifies. For this class of reagents, the two-step reaction protocol is particularly useful, with the chemical reaction being carried out first in the absence of activating radiation, followed by irradiation of the photosensitive group to bring about crosslinking. The arylazide shown in Table 1 (compound 7) is an example of a photosensitive crosslinker that is widely used. The bis-homobifunctional aldehyde **glutaraldehyde** (Table 1, compound 5) is perhaps the most frequently used chemical crosslinker.

BIBLIOGRAPHY

1. J. A. Cover, J. M. Lambert, C. M. Norman, and R. R. Traut (1981) *Biochemistry* **20**, 2843–2852.

Suggestions for Further Reading

S. S. Wong (1993) *Chemistry of Protein Conjugation and Cross-Linking*, CRC Press, Boca Raton, FL. (An outstanding book describing all aspects of the chemistry and uses of bifunctional reagents in crosslinking and conjugation, with an extensive bibliography for each chapter.)

Annual products catalog of Pierce Chemical Co., Rockford, IL. [Contains structures, helpful information, and specific uses (with references) for a large number of bifunctional reagents.]

BINDING

SERGE N. TIMASHEFF

Total binding B_L is the number of ligand molecules that at any moment are in contact with (occupy sites on) the **accessible surface** of a **protein** molecule (see **Ligand binding**. It is related to **preferential binding** (as measured by **equilibrium dialysis**) by

$$B_L = \left(\frac{\partial m_L}{\partial m_{pr}}\right)_{T,\mu_w,\mu_L} + \frac{m_L}{m_w} B_w \tag{1}$$

where B_L and B_w are the numbers of ligand and water molecules, respectively, that are in contact with the protein surface.

When the interactions are weak, the ligand must be used at high concentration (eg, 1 *M* sucrose, 3 *M* glycerol, 8 *M* **urea**). As a consequence, the total binding is *not* given by equilibrium dialysis, as the last term of Eq. (1) (binding of water) becomes significant. Total binding can be measured by nonthermodynamic techniques that respond to contacts between ligand molecules and protein [1]. Such techniques include **calorimetric** titration, which detects the heat of protein–ligand contact, or spectroscopic techniques (eg, **fluorescence** or **uv absorbance**) that detect spectral perturbations each time a ligand–protein contact is made. If the total **hydration** is known from other techniques, then total binding can be derived from dialysis equilibrium results by the application of Eq. (1).

BIBLIOGRAPHY

1. S. N. Timasheff (1995) In *Protein-Solvent Interactions* (R. B. Gregory, ed.), Marcel Dekker, New York, Chap. 11.

BIOINFORMATICS

MINORU KANEHISA

Bioinformatics is a new branch of molecular biology, also kown as computational biology. Computational physics and computational chemistry have emerged as third branches in their respective disciplines, after the experimental and theoretical branches, due largely to the advance of computational capabilities in modern computers. In contrast, bioinformatics has emerged in molecular biology as the result of advances in experimental technology, especially in the high-throughput **DNA sequencing**, which is generating a vast amount of **gene** and **protein** sequence data. Since initiation of the Human Genome Project in the late 1980s, bioinformatics has become an integral part of the coordinated efforts to sequence the entire **genomes** of a number of organisms from **bacteria** to human. Consequently, bioinformatics is a data-driven discipline requiring large-scale **databases** and associated technologies for data management and interpretation. This contrasts with the model-driven, theory-based approaches in computational physics and computational chemistry.

Bioinformatics covers a diverse range of topics, which broadly have two roles. First, bioinformatics is widely used in experimental projects, such as in determining the optimal **genetic mapping** strategy and how best to assemble raw sequence data, and in the storing and handling information and materials. Secondly and more important, the major task of bioinformatics is to develop new databases and new computational technologies that will help to understand the biological meaning encoded in the sequence data. Interpretation of sequence data cannot be achieved by numerical calculations based on first-principle equations, which are virtually nonexistent in biology, but it requires using and processing empirical knowledge acquired from experimental data. This is what human experts would do if the amount of data and knowledge were manageable in size. Thus, bioinformatics was born from the marriage of molecular biology, in the age of massive data production, with artificial intelligence, which is a mature branch of computer science, to automate knowledge processing.

Table 1 summarizes topics in bioinformatics for interpretating of sequence data. The methods of computer science have been used traditionally in artificial intelligence applications, such as speech recognition and natural language processing, but they are also quite effective in solving problems in molecular biology. There is a rough distinction between how data are organized and consequently what kinds of computational methods are used in the three categories of Table 1. In the first category of similarity searches, the database is a collection of all the known primary data, such as the **sequence database** and the **structure database**, which are the repositories of all reported sequences and three-dimensional structures, respectively, of **nucleic acids** and **proteins**. A basic operation in this category is comparing of individual sequences or structures to detect any similarity. For example, to understand the functional implications of a newly determined gene or protein sequence, it is customary to perform a sequence similarity, or **homology**, search, comparing the query sequence with each of the sequences in the database. If any similar sequence is found in the database, the query sequence is assumed to have a similar or related function. This reasoning is based on the empirical observation that homologous sequences generally share similar functions and similar 3-D structures, because they arose from a common evolutionary ancestor and because of functional constraints.

The procedure of comparing two sequences is a problem of sequence alignment, which is a major topic in **sequence analysis**. Given a similarity measure of amino acids or nucleotides, the problem of obtaining the best sequence alignment is equivalent to optimizing a given score function that represents the overall similarity. There are variations in how the optimal alignment is made, either globally for the entire sequences or locally to detect localized regions, how many sequences are aligned at a time, and which similarity measure is to be used. By allowing insertions and deletions (**indels**), the number of possible alignments grows exponentially as the number and length of the sequences to be compared increases.

Table 1. Bioinformatics for Interpretation of Sequence Data

Problems in Biological Science		Methods in Computer Science
Similarity search	Pairwise sequence alignment Sequence similarity search Multiple sequence alignment RNA secondary structure prediction 3-D protein structure alignment	Dynamic programming Simulated annealing Genetic algorithms Other optimization algorithms
Structure/function prediction	Motif extraction Functional site prediction Cellular localization prediction Coding region prediction Transmembrane segment prediction Protein secondary structure prediction 3-D protein structure prediction 3-D RNA structure prediction	Discriminant analysis Neural network Hidden Markov model Formal grammar Other pattern recognition and learning algorithms
Molecular classification	Superfamily classification Fold classification Ortholog/paralog grouping of genes	Clustering algorithms

Thus, the sequence alignment problem is a typical combinatorial optimization problem in computer science. Although the alignment of two or three sequences can be solved rigorously by using dynamic programming algorithms (1), which effectively evaluate all possibilities, multiple alignment of many sequences requires heuristic algorithms to obtain approximate solutions. Furthermore, to search large databases effectively for similar sequences, heuristic algorithms such as FASTA (2) and BLAST (3) have been developed for rapidly identifying regions of two sequences that are similar locally. Other extensions of the sequence alignment algorithms include analysis of a single RNA sequence to predict any **secondary structure** (see **RNA structure prediction**) and compare two 3-D **protein structures**.

When humans try to interpret a sentence written in a foreign language, they use dictionaries and a knowledge of grammar. However, the process of using a homology search to interpret a sentence written in the DNA language is like comparing it with all the sentences written in the past and checking precedents as to how they were interpreted. Sequence interpretation would become more efficient if our knowledge of sequence-functional relationships were more advanced. The second and third categories in Table 1 involve such reorganizations of the primary data. In the second category of structure/function prediction, the primary data are classified into a group based on a given criterion, such as functional identity, and higher level knowledge is abstracted from this group. This is based on our empirical observation that the function of a protein molecule is usually exerted at a functional site in its three-dimensional structure, which is usually a **ligand-binding** site for other molecules, and that the site is formed by one or more linear polypeptide segments that have conserved sequence patterns. These conserved patterns are called sequence motifs, and they can be extracted, for example, by multiple alignment of a set of sequences that have the same function. A motif search against a **library** of known motifs is widely used as an alternative method for interpreting sequence data, when no similar sequences have been found by a homology search. However, the bioinformatics problem here is not how to search the known motifs, but how best to extract and define new motifs effectively. Thus, the problem is related to machine learning in computer science, where a higher level concept or a generalized pattern is abstracted from a set of given examples.

Although the second category in Table 1 is targeted at a specific function and a specific group of sequences, the third category of molecular classification represents a more global analysis, where the grouping is attempted with all the available data, and the relationship among groups, which is often hierarchical, is examined. For example, all known protein sequences can be hierarchically classified into families and **superfamilies** according to their level of sequence similarity. Because complete genome sequences are available, it has become possible to establish similarity relationships of all **genes** within species (**paralogous genes**) and across species (**orthologous genes**). These analyses are carried out by clustering algorithms, and the similarity score is used as a measure of the closeness of two sequences. In practice, however, there are biological difficulties, notably due to the abundance of multi**domain**, **multifunctional proteins**, which make automatic clustering a complicated task.

Molecular biology has been dominated by a reductionist approach where, starting from a selected functional aspect of life, such as metabolism, **signal transduction**, or **development**, molecular components and their interactions are identified experimentally. In contrast, sequencing an entire genome identifies a complete set of genes and proteins in an organism, but it does not tell how they should be interrelated to form a functioning, living system. This is where a synthetic approach based on bioinformatics will play an increasingly more dominant role in the basic research into understanding life and in the applied research into biomedical relevance. Table 2 summarizes the role of bioinformatics relative to the level of abstraction in nature. At the molecular level, an amino acid residue is represented by a symbol, for example, C for **cysteine**, which is a compound consisting of carbon, hydrogen, nitrogen, oxygen, and sulfur atoms. When comparing two amino acid sequences, only the one-dimensional connection of symbols is considered biologically relevant, neglecting any atomic details. At a higher level of abstraction, which may be called a network level, a gene

Table 2. Level of Abstraction

Level	Constituents	Entity	Informatics
Subatomic level	Elementary particles	Atom	Computational physics
Atomic level	Atoms	Amino acid, etc.	Computational chemistry
Molecular level	Amino acids, nucleotides, etc.	Protein, gene, etc.	Bioinformatics
Network level	Proteins, genes, etc.	Molecular pathway, molecular assembly	

or a protein is represented by a symbol, for example, Ras for an **oncogene** product. A primary concern here is the Ras signal transduction pathway, which is generated by a network of interacting molecules. Bioinformatics is relatively well established at the molecular level, such as in the sequence and structure databases and in the computational methods to analyze them. To step up to a higher level, however, it is necessary to develop new databases and new computational technologies, such as a network database containing all the wiring-diagram information about genes and molecules, plus the algorithms to analyze them. This direction would eventually lead to *in silico* reconstruction of a biological organism.

BIBLIOGRAPHY

1. S. B. Needleman and C. D. Wunsch (1970) *J. Mol. Biol.* **48**, 443–453.
2. W. J. Wilbur and D. J. Lipman (1983) *Proc. Natl. Acad. Sci. USA* **80**, 726–730.
3. S. F. Altschul et al. (1990) *J. Mol. Biol.* **215**, 403–410.

BIOPOLYMER

JENNIFER MARTIN

Biopolymers, sometimes referred to as biological **macromolecules**, are **polymers** of biological importance such as **proteins**, **nucleic acids**, and carbohydrates. Like all polymers, biopolymers are formed by joining together many copies of repeating units to form long chains that have a repeating or constant **backbone** and variable **side chains**. The repeating units of proteins are the 20 naturally occurring **amino acids** that form a **polypeptide chain**. Nucleic acids are formed from **nucleotides** and have a sugar-phosphate backbone with **purine** or **pyrimidine** side chains.

[See also **Polymer** and **Macromolecule**.]

Suggestions for Further Reading

A. G. Walton and J. Blackwell (1973) *Biopolymers*, Academic Press, New York.

P. Munk (1989) *Introduction to Macromolecular Science*, Wiley-Interscience, New York.

BIOTIN

MEIR WILCHEK
EDWARD A. BAYER

Biotin is a vitamin (vitamin H) and a **growth factor**, first isolated by Kögl in 1935 (1). Its structure (Fig. 1) was established by du Vigneaud in 1942 (2,3).

Biotin is a prosthetic group for a family of "biotin-requiring" **enzymes** (carboxylases, decarboxylases and transcarboxylases) (4). These multisubunit enzymes contain two types of **active sites**, one wherein a molecule containing a carboxylic acid group serves as a donor of the carboxyl group and the second wherein another molecule acts as an acceptor. The biotin moiety is attached through a **lysine** residue in a particular sequence (Ala-Met-Lys-Met) that is exposed on the surface of a special type of biotin-binding subunit. The biotin group participates as a transient carrier to which the carboxyl group is bound covalently in the process of being transferred from the donor to the acceptor molecule. The biotin moiety of the enzyme is readily available for binding to **avidin**, which inactivates the enzyme. After serving their respective function, the biotin-requiring enzymes undergo degradation, whereby ϵ-N-biotinyl lysine (termed biocytin) is released into the circulation. There the enzyme biotinidase hydrolyzes biocytin to regenerate biotin and lysine. A rare mutation in the gene in humans causes a deficiency in the level of this enzyme in the blood stream, which, in turn, causes a disease due to a deficiency of biotin (5).

Biotin attained true fame when it was recognized that the high-affinity interaction between avidin and biotin can be applied for a variety of biotechnological and biomedical purposes (as described in **Avidin-biotin system**)(6,7). The varied application of this system is usually achieved upon coupling biotin covalently to different biologically active binder molecules, **antibodies**, **hormones**, **DNA**, etc. The binding of biotin to either avidin or **streptavidin** greatly enhances the stability of these proteins, such that the avidin–biotin or streptavidin–biotin complex is essentially irreversible. Because both proteins primarily recognize the ureido ring of the biotin molecule, the carboxylic acid side-chain can be modified almost at will, and the resultant derivatives can be used to incorporate biotin covalently into virtually any other molecule (Fig. 2).

Both polyclonal and **monoclonal antibodies** have been produced against biotin. Comparing the sequence of the V_H region of a monoclonal anti-biotin preparation (8) with the sequences of avidin and streptavidin revealed an astonishing similarity in their binding sites, indicating that such functional elements are conserved in nature for a given purpose, irrespective of the protein type.

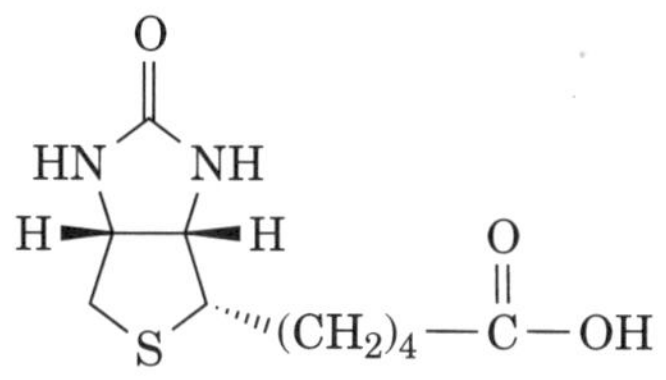

O
HN NH
H H O
S $(CH_2)_4$—C—OH

Figure 1. The structure of biotin.

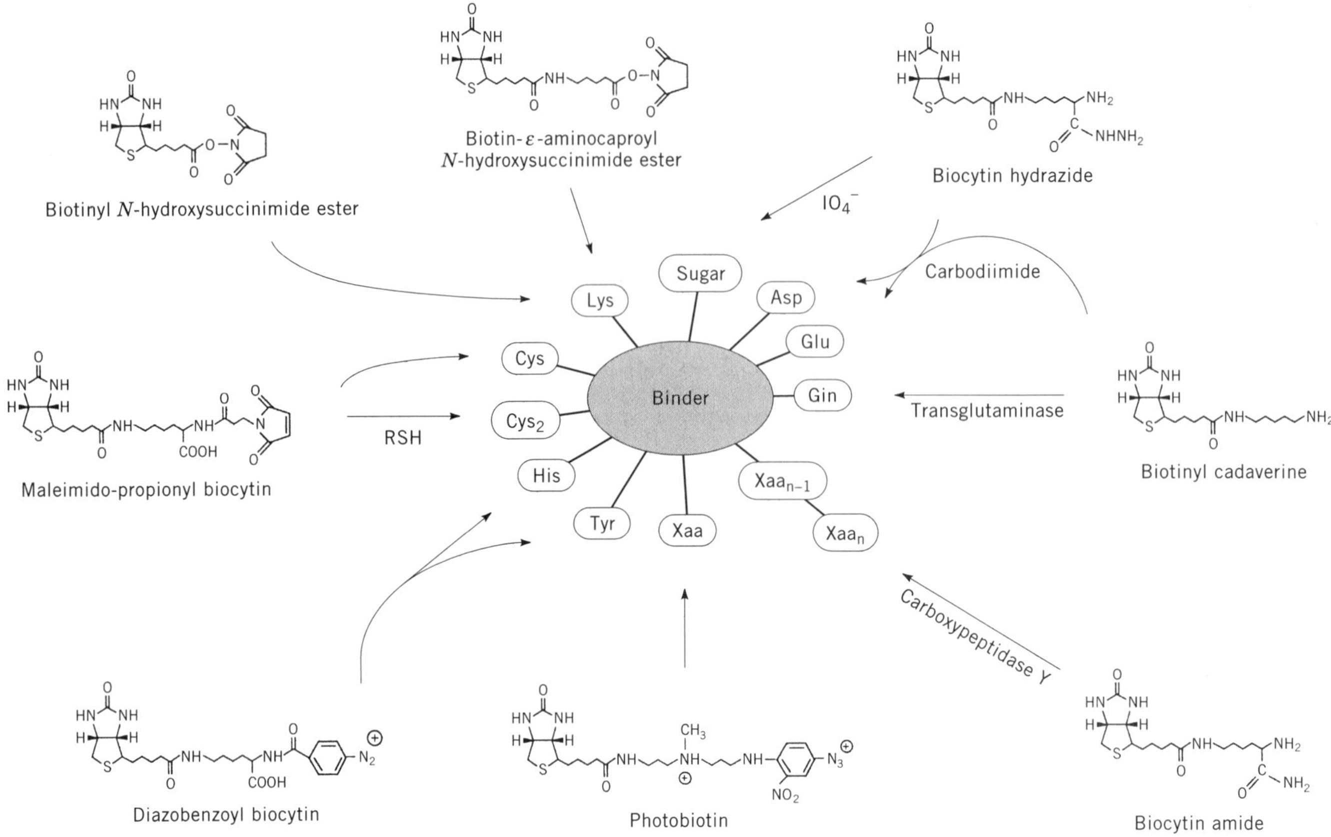

Figure 2. Selected approaches using different biotin-containing reagents for biotinylating functional groups on proteins.

BIBLIOGRAPHY

1. F. Kögl and B. Z. Tönnis (1936) *Physiol. Chem.* **242,** 43–73.
2. V. du Vigneaud, D. B. Melville, K. Folkers, D. E. Wolf, R. Mozingo, J. C. Kereszteolf, and S. A. Harris (1942) *J. Biol. Chem.* **146,** 475–485.
3. V. du Vigneaud, K. Hoffman, and D. B. Melville (1942) *J. Amer. Chem. Soc.* **64,** 188–189.
4. J. Moss and M. D. Lane (1971) *Adv. Enzymol.* **35,** 321–372.
5. B. Wolf, G. S. Heard, J. R. McVoy, and R. E. Grier (1985) *Ann. N.Y. Acad. Sci.* **447,** 252–62.
6. E. A. Bayer and M. Wilchek (1978) *Trends Biochem. Sci.* **3,** N237–N239.
7. E. A. Bayer and M. Wilchek (1980) *Methods Biochem. Anal.* **26,** 1–45.
8. H. Bagci, F. Kohen, U. Kuscuoglu, E. A. Bayer, and M. Wilchek (1993) *FEBS Lett.* **322,** 47–50.

Suggestion for Further Reading

M. Wilchek and E. A. Bayer, eds. (1990) *Avidin-Biotin Technology, Methods Enzymol.* Vol. 184, Academic Press, San Diego.

BIOTIN REPRESSOR

GEORGES N. COHEN

All genes for the biosynthesis of **biotin** of *Escherichia coli* are grouped in a single **operon** located at 17 min on the **chromosome**, except for except for *bioH*, which is located at 75 min. The genes of the operon *bioA,BFCD* are transcribed divergently from a single regulatory region located between the *bioA* and the *bioB* genes. **Transcription** in both directions is corepressed by biotinyl-5′-adenylate and the biotin **repressor**. The following table shows the respective function of these genes:

bioC and bioH	Unidentified steps in pimeloyl CoA biosynthesis
bioF	7-Keto-8-aminopelargonic acid synthetase
bioA	7,8-Diaminopelargonic acid synthetase
bioD	Dethiobiotin synthetase
bioB	Biotin synthase (introduction of the S atom)

No precise information is available on the function of the *bioC* and *bioH* genes, other than that a mutation in either gene results in no excretion of a known intermediate in the pathway. Therefore the products of these genes have been assigned to some early steps before 7-keto-8-aminopelargonic acid synthesis. All biotin genes are coordinately repressed when biotin is added to the growth medium in excess of 1 ng/Ml.

The biotin repressor is a very interesting bifunctional protein of 321 amino acid residues, which acts at two different levels. In addition to its repressor function, it is endowed with acetyl CoA carboxylase biotin holoenzyme synthetase activity, which activates biotin to biotinyl-5′-adenylate and transfers the biotin to acceptor proteins. As soon as these proteins are totally biotinylated, biotinyl-5′-adenylate accumulates and serves as the corepressor of the biotin operon. This is a case of an enzyme synthesizing its own repressor, a unique property thus far among **DNA-binding proteins**. Mutations in the corresponding gene (*birA*) inactivate the repressor function partially or totally and also alter the enzymatic function. This repressor binds to a 40-bp symmetrical biotin **operator** site to prevent **transcription** of the biotin biosynthetic genes. The structure of the repressor is highly asymmetrical and consists of three **domains**. The N-terminal domain is mostly **alpha-helical**, contains a **helix–turn–helix motif** and is loosely connected to the remainder of the molecule. The central domain consists of a seven-stranded mixed **beta-sheet**, with α-helices covering one face. The other side of the sheet is largely exposed to the solvent and contains the enzyme's **active site**. The C-terminal domain comprises a six-stranded antiparallel β-sheet sandwich. The location of biotin is consistent with mutations affecting enzymatic activity.

Two molecules of the monomeric repressor bind cooperatively to one molecule of operator in the nanomolar concentration range. The data suggest that one molecule of repressor monomer binds to each of the two operator half sites and that they form a dimer only after they bind. Because the complex between repressor and DNA has not yet been crystallized, details of its structure remain an open question.

BIP (HSP70)

R. J. ELLIS

The acronym BiP stands for binding protein, and it is the only member of the hsc,hsp70 family (see **Hsc,hsp proteins**) known to occur in the lumen of the **endoplasmic reticulum** (ER) of mammalian cells. It was identified originally as an ER protein that increases in amount when cells are starved of glucose (1) and subsequently as a protein that binds noncovalently to the heavy (H) chains of **immunoglobulins** as they enter the ER lumen (2). The alternative term of "Grp78" stands for **glucose-regulated protein** with an apparent subunit mass of about 78 kDa. BiP is now regarded as having a general **molecular chaperone** role for transport in the ER lumen and for **protein folding *in vivo***. BiP proteins are highly conserved in mammals, and there is 67% identity between mouse BiP and a homologue in the ER lumen of *Saccharomyces cerevisiae* called "Kar2p" (3). Like all hsc,hsp70 proteins, BiP exhibits a weak **ATPase** activity.

STRUCTURE

BiP is encoded in a nuclear gene located on human **chromosome** 9q34, and cDNA from mammalian species indicate a primary translation product of 635 amino acid residues, including an 18-residue *N*-terminal signal sequence for ER targeting and the *C*-terminal tetrapeptide **KDEL sequence** for ER retention (GenBank accession numbers M19645 and M17169). The *N* and *C* termini contain clusters of acidic residues thought to be involved in Ca^{2+} binding (4). No crystal structure is available for mammalian BiP, but a structure is known for the *N*-terminal 45-kDa fragment of bovine hsc70, which retains the ATPase activity (5), and for the *C*-terminal polypeptide-binding domain of the *Escherichia coli* hsp70 homologue called **DnaK** (6). Purification protocols for BiP utilize **affinity chromatography** on ATP columns (7). Recombinant hamster BiP has been purified from *E. coli* cells (8) and is marketed by StressGen Biotechnologies Corporation, as are polyclonal **antibodies** to rodent BiP. *In vivo* BiP exists in interconvertible **oligomeric** and monomeric forms and is subject to **phosphorylation** on serine and threonine residues, as well as to **ADP-ribosylation**. However, only monomeric, unmodified species of BiP are found in complexes with unfolded or unassembled polypeptides (9).

FUNCTION

BiP binds transiently to a range of newly synthesized secretory proteins, as they traverse the ER membrane and enter the ER lumen, and more permanently to misfolded, underglycosylated, or unassembled proteins whose transport from the ER is blocked (10); it does not bind to native, folded proteins. The binding is reversed by the addition of ATP and is believed to exert a molecular chaperone function by preventing premature folding and/or aggregation; this function is achieved by the shielding of potentially interactive **hydrophobic** surfaces during the time when BiP is bound. There is also evidence that the BiP homologue in yeast functions as a molecular **motor protein** to promote the transport of proteins across the ER membrane (11). This transport function requires the binding of the BiP homologue to the *J* domain of the yeast ER **membrane protein** Sec63p (12) (see also text later in this article). Studies of the binding of synthetic **peptides** and bacteriophage **peptide libraries** show that the optimum peptide length for binding is seven to eight residues, with extensive sequence diversity but a marked preference for hydrophobic residues (13,14). These observations support the idea that BiP binds to a wide range of sequences that normally occur inside fully folded proteins. A computer program is available that scores potential BiP binding sites in protein sequences (14); it has been used to map such sequences in immunoglobulin heavy chains to the regions that interact with light chains (15).

INTERACTION WITH OTHER CHAPERONES

Cytosolic hsc,hsp70 proteins in the bacterial and eukaryotic cytosol interact with other chaperones of the DnaJ (or hsp40) family that contain *J* domains (see **DnaK/DnaJ proteins**). The yeast BiP homologue, Kar2p, interacts with two other ER proteins that contain *J* domains: Sec63p, a membrane protein involved in protein translocation across the ER membrane (16);

and Scj1p, a lumenal protein (17). BiP also binds either sequentially or simultaneously with chaperones such as **calnexin** and Grp94 during the folding in the ER of proteins such as immunoglobulin light (L) chains (18), thyroglobulin (19), vesicular stomatitis virus G protein (20), and **major histocompatibility complex** class II chains (21). BiP is thus one component in a complex set of interactions in the ER lumen between different chaperones and polypeptide chains that are folding.

INDUCTION OF BIP

BiP is an abundant protein under normal growth conditions, constituting about 5% of the ER lumenal proteins, but its amount increases greatly under conditions that result in the accumulation of proteins within the ER lumen that are unable to fold correctly. These conditions include the biosynthesis of mutant chains, glucose starvation, and treatment with amino acid analogs, drugs that inhibit glycosylation, and calcium **ionophores** (22). The **promoters** of BiP genes in mammals contain several ***cis-acting*** regulatory elements required for high basal-level expression and for inducibility (22,23).

BIBLIOGRAPHY

1. J. Pouyssegur, R. P. C. Shiu, and I. Pastan (1977) *Cell* **11**, 941–947.
2. I. G. Haas and M. Wabl (1983) *Nature* **306,** 387–389.
3. K. Normington, K. Kohno, Y. Kozutsumi, M. J. Gething, and J. Sambrook (1989) *Cell* **57**, 1223–1236.
4. D. R. J. Macer and G. L. E. Koch (1988) *J. Cell. Sci.* **91**, 61–70.
5. K. M. Flaherty, C. DeLuca-Flaherty, and D. B. McKay (1990) *Nature* **346**, 623–628.
6. X. Zhu, X. Zhao, W. F. Burkholder, A. Gragerov, C. M. Ogata, M. E. Gottesman, and W. A. Hendrickson (1996) *Science* **272**, 1606–1614.
7. P. J. Rowling, S. H. McLaughlin, G. S. Pollock, and R. B. Freedman (1994) *Protein Exp. Purif.* **5**, 331–336.
8. J. Wei and L. M. Hendershot (1995) *J. Biol.Chem.* **270**, 26670–26676.
9. P. J. Freiden, J. R. Gaut, and L. M. Hendershot (1992) *EMBO J.* **11**, 63–70.
10. M. J. Gething, S. Blond-Elguindi, K. Mori, and J. F. Sambrook (1994) in *The Biology of Heat Shock Proteins and Molecular Chaperones* (R. I. Morimoto, A. Tissieres and C. Georgopoulos, eds.), Cold Spring Harbor Laboratory Press, Cold Spring Harbor, pp. 111–135.
11. S. Panzner et al. (1994) *Cell* **81**, 561–570.
12. J. L. Brodsky and R. Schekman (1993) *J. Cell Biol.* **123**, 1355–1363.
13. G. C. Flynn, J. Pohl, M. T. Flocco, and J. E. Rothman (1991) *Nature* **353**, 726–730.
14. S. Blond-Elguindi, S. E. Cwirla, W. J. Dower, R. J. Lipshutz, S. R. Sprang, J. F. Sambrook, and M. J. Gething (1993) *Cell* **75**, 717–728.
15. G. Knarr, M. J. Gething, S. Modrow, and J. Buchner (1995) *J. Biol. Chem.* **270**, 27589–27594.
16. D. Feldheim, J. Rothblatt, and R. Schekman (1992) *Mol. Cell. Biol.* **12**, 3288–3296.
17. G. Schlenstedt, S. Harris, B. Risse, R. Lill, and P. A. Silver (1995) *J. Cell Biol.* **129**, 979–988.
18. J. Melnick, J. L. Dul, and Y. Argon (1994) *Nature* **370**, 373–375.
19. P. S. Kim and P. Arvan (1995) *J. Cell Biol.* **128**, 29–38.
20. C. Hammond and A. Helenius (1994) *Science* **266**, 456–458.
21. M. S. Marks, R. N. Germain, and J. S. Bonifacino (1995) *J. Biol. Chem.* **270**, 10475–10481.
22. A. S. Lee (1992) *Curr. Opin. Cell Biol.* **4**, 267–273.
23. W. W. Li, L. Sistonen, R. I. Morimoto, and A. S. Lee (1994) *Mol. Cell. Biol.* **14**, 5533–5546.

Suggestion for Further Reading

D. N. Herbert, J. F. Simons, J. R. Peterson, and A. Helenius (1995) Calnexin, calreticulin and BiP/Kar2p in protein folding, *Cold Spring Harbor Symp. Quant. Biol.* **60**, 405–415.

F.-U. Hartl (1996) Molecular chaperones in cellular protein folding, *Nature* **381**, 571–580.

J. L. Brodsky (1996) Post-translational protein translocation: not all hsc70s are equal, *Trends Biochem. Sci.* **21**, 122–126.

M.-J. Gething, ed. (1997) *Molecular Chaperones and Protein Folding Catalysts*, Oxford University Press, Oxford (this volume contains much detailed information about BiP).

BISUBSTRATE ANALOGUE

JOHN F. MORRISON

Bisubstrate analogues were developed originally for mechanistic studies on **enzymes** that catalyze reactions with two substrates or products. However, they have also proved to be of value in studies on enzymes by **X-ray crystallography**. Bisubstrate analogues are characterized by the fact that they embody in a single molecule the structural features of each of the two substrates (or products). Hence, it was expected that they would bind simultaneously at the binding sites for the two substrates, and therefore more tightly than the individual substrate molecules, and act as potent enzyme inhibitors. They have also been referred to as **transition state analogues**, but such a classification seems inappropriate.

Two early, and now classical, examples of bisubstrate enzyme inhibitors are P^1, P^5-di-(adenosine-5′)pentaphosphate (AP_5A) and *N*-phosphonoacetyl-L-aspartate (PALA), whose structures are given in Figure 1. AP_5A was developed as an inhibitor of adenylate kinase, which is a monomeric enzyme that catalyzes the reaction:

$$MgATP^{2-} + AMP^{2-} \rightleftharpoons MgADP^{-} + ADP^{3-}$$

The reaction conforms to a rapid equilibrium, random **kinetic mechanism**, which implies that the enzyme possesses two distinct nucleotide-binding sites within its active site. One is for either $MgATP^{2-}$ or $MgADP^{-}$ and the other is for ADP^{3-} or AMP^{2-}. Irrespective of whether they are substrates or products, the moieties bound by the enzyme include two adenosine moieties and four phosphate moieties, whereas AP_5A differs only in having five phosphate groups linked covalently. AP_5A is a potent inhibitor of the adenylate kinase reaction, with the apparent inhibition constant K_i value being in the nanomolar region; the inhibition is competitive with respect to both MgATP and AMP, as would be expected (1,2). Binding studies indicate that the stoichiometry of binding is 1:1 and the **dissociation constant** is 15 nM (3). The binding affinity is reduced seven-fold in the absence of Mg^{2+}. It is of interest that the binding to adenylate kinase of AP_4A, which is the equivalent of

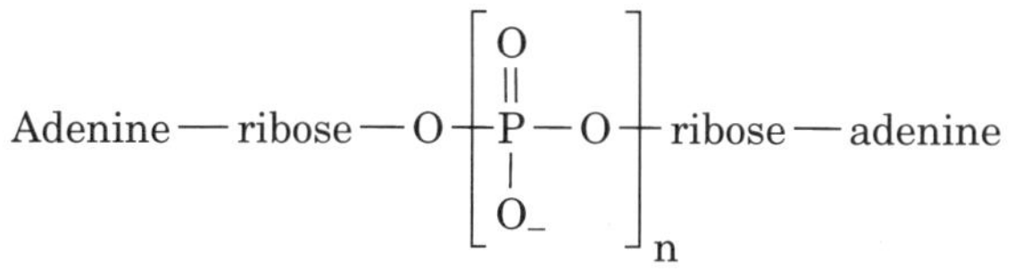

$$\text{Adenine}-\text{ribose}-\text{O}\left[\begin{array}{c}\text{O}\\ \|\\ \text{P}-\text{O}\\ |\\ \text{O}^-\end{array}\right]_n\text{ribose}-\text{adenine}$$

(a)

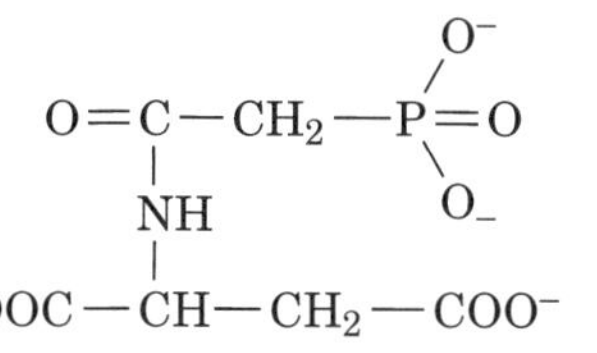

$$\begin{array}{l}\text{O}=\text{C}-\text{CH}_2-\text{P}(\text{O}^-)(=\text{O})\text{O}^-\\ \quad\;|\\ \quad\text{NH}\\ \quad\;|\\ {}^-\text{OOC}-\text{CH}-\text{CH}_2-\text{COO}^-\end{array}$$

(b)

Figure 1. Examples of bisubstrate analogues. (**a**), P^1, P^n-di(adenosine-5′) *n*-phosphate, where n = 4 (AP_4A), 5 (AP_5A), or 6 (AP_6A). (**b**), *N*-phosphonoacetyl-L-aspartate (PALA).

covalently linking ATP to AMP, or ADP to ADP, is almost 3,000-fold weaker. An increase in the number of phosphoryl groups to six also reduces the binding affinity by 400-fold. The crystal structures of three adenylate kinases and enzyme–AP_5A complexes were solved well ahead of the spatial assignment of the substrate binding sites. A review of the problems associated with the determination of the structural relationship between the two nucleotide binding sites of adenylate kinase has been presented (4).

AP_5A also acts as a strong inhibitor of ATP:NMP phosphotransferase from *Dictyostelium discoideum*, which utilizes either UMP or CMP as the acceptor of a phosphoryl group from ATP (5). Binding studies indicate that the dissociation constant of the enzyme-AP_5A complex is 160 μM. However, the corresponding bisubstrate analogue with uridine replacing one adenine nucleotide (UP_5A) is a more potent inhibitor and binds with a dissociation constant of 3 nM. The enzyme has been cocrystallized with UP_5A.

PALA has been used extensively for kinetic and structural studies on **aspartate transcarbamoylase (ATCase)**, which catalyzes the reaction:

$$\text{Carbamoyl phosphate} + \text{aspartate} \rightleftharpoons \text{carbamoylaspartate} + P_i$$

PALA is considered an analogue of the two substrates linked covalently. The intact ATCase enzyme consists of both catalytic and regulatory units, does not show Michaelis–Menten kinetics, and is subject to **allosteric** control by nucleoside triphosphates. However, it is possible to prepare an active trimer of just the catalytic subunits that shows **Michaelis–Menten kinetics** and is not subject to allosteric control. This form of enzyme has been very useful for elucidating fundamental information about the catalytic sites and mechanism of action of ATCase. Investigations with the catalytic trimer have shown that one molecule of PALA binds to each subunit of the trimer (6) and the inhibition is linear competitive with respect to carbamoyl phosphate and linear noncompetitive relative to aspartate (7). Additional kinetic data have indicated that the kinetic mechanism for the aspartate transcarbamoylase reaction is essentially ordered, with carbamoyl phosphate being the first substrate to add; this explains the inhibition pattern. The pH-independent K_i value for the enzyme-PALA complex is 7.2 nM, which is three orders of magnitude lower than the dissociation constant for the corresponding enzyme-carbamoyl phosphate complex (7).

PALA has also played an important role in the demonstration that the co-operativity observed with intact ATCase can be explained in terms of a two-state allosteric **concerted model** (8). Thus, the binding of up to three molecules of PALA to the intact enzyme, which has six **active sites**, leads to activation, even though PALA is occupying active sites, as a result of displacing the equilibrium between the T (inactive) and R (active) forms of the enzyme toward R. Only with greater occupancy is PALA inhibitory. Structural studies support the idea that PALA is a true bisubstrate analogue of the substrates for the enzyme (9).

BIBLIOGRAPHY

1. G. E. Lienhard and I. I. Secemski (1973) *J. Biol. Chem.* **248**, 1121–1123.
2. P. Feldhaus, T. Frohlich, R. S. Goody, M. Isakov, and R. H. Schirmer (1975) *Eur. J. Biochem.* **57**, 197–204.
3. J. Reinstein, I. R. Vetter, I. Schlichting, P. Rosch, A. Wittinghofer, and R. S. Goody (1990) *Biochemistry* **29**, 7440–7450.
4. M.-D. Tsai and H. Yan (1991) *Biochemistry* **30**, 6806–6818.
5. L. Wiesmuller, K. Scheffzek, W. Kliche, R. S. Goody, A. Wittinghofer, and J. Reinstein (1995) *FEBS Lett.* **363**, 22–24.
6. G. R. Jacobson and G. R. Stark (1973) *The Enzymes* **9**, 225–308.
7. J. L. Turnbull, G. L. Waldrop, and H. K. Schachman (1992) *Biochemistry* **31**, 6562–6569.
8. L. E. Parmentier, M. H. O'Leary, H. K. Schachman, and W. W. Cleland (1992) *Biochemistry* **31**, 6598–6602.
9. W. N. Lipscomb (1994) *Adv. Enzymol.* **68**, 67–151.

BITHORAX COMPLEX

F. KARCH

The **homeotic genes** of the bithorax complex (BX-C) have been the subject of many landmark discoveries in the field of **developmental** biology. Homeotic genes were first identified in *Drosophila* by mutations that affect their expression. These mutations lead to spectacular effects on the morphology of the fly; they cause the development of a specific body structure (ie, segment, antenna, leg, wing) at the place where another structure normally develops. The first homeotic mutations, *bithorax* (*bx*) and *bithoraxoid* (*bxd*), were described in 1923 by Bridges and Morgan (1). For example, in flies homozygous for *bxd* mutations, their first abdominal segment (A1) develops like a copy of the segment immediately adjacent anteriorly—namely, the third thoracic segment (T3). Thus, the role of the bxd^+ function is to assign the identity of A1. In the late 1940s, Ed Lewis pioneered the field of developmental genetics by discovering the existence of additional homeotic mutations near *bx* and *bxd*. Although these mutations seem to affect different body segments of the thorax and abdomen, their **complementation** patterns turned out to be rather complex. Because they all mapped at the same genetic and cytological position, Lewis decided to name the locus the bithorax complex (BX-C). The work of Lewis was compiled in a milestone article in 1978 (2) in which he describes a series of homeotic mutations of BX-C that affect the identities of the third thoracic segment and all of the abdominal segments. Mutations in any of them transform the considered segment into a copy of the segment immediately anterior. Remarkably, genetic mapping revealed that these mutations are arranged on the **chromosome** in the same order along the anteroposterior axis as the segments they affect. By combining three mutations, Lewis produced flies with four wings instead of two (transformation of T3 into T2). Because such animals look like more ancestor forms of insect, Lewis proposed that homeotic genes played an instrumental role in **evolution**. The correspondence between the order of the genes on the chromosome and the order of the segments on the body of the fly has now attained almost mystical status and has turned out to be true also of the vertebrate homologues of BX-C. The Nobel Prize Committee has recognized this pioneering works by awarding its 1995 Nobel Prize of Medicine to Ed Lewis and two other *Drosophila* geneticists (Y. Nusslein-Volhard and E. Wieschaus).

In 1978, BX-C was the first *Drosophila* gene **cloned** from the chromosome without any prior knowledge of the products. To clone what turned out to be a large complex, Bender, Spierer, and Hogness developed the method of **chromosome walking** (3). Finally, in 1983, the molecular characterization of the BX-C led to the discovery of the homeobox in the laboratories of W. Gehring and of M. Scott (4,5).

MOLECULAR GENETICS OF BX-C

Figure 1 summarizes the molecular genetics of the BX-C. The complex covers 300 kbp of DNA, which are represented by the thin horizontal line marked off in kb (6,7). Above the DNA line are represented the sites of the homeotic mutations that affect the identities of each of the segments under the control of the BX-C. The vertical arrows represent the sites of chromosomal rearrangement breaks, the triangles the sites of insertion of **transposons**, and the horizontal lines the extent of deletions. Expression studies and analysis of the homeotic **phenotypes** in embryos have revealed that the unit transformed in each of these nine classes of mutations does not correspond to body segments; instead, it is composed of the posterior part of one segment and the anterior compartment of the next segment. These units are named parasegments (PSs) (8). For example, *bxd* mutations cause the transformation of the posterior part of T3 (pT3) and the anterior part of A1 (aA1) into p(T2) and aT3. This corresponds to the transformation of parasegment 6 (PS6) into PS5. The mutations affecting parasegment identity form nine discrete entities that, as predicted by Lewis, are aligned on the chromosome in the same order as the parasegments in which they act on the body of the fly (*abx*/*bx*, *bxd*/*pbx*, *iab-2* through *iab-8*). These mutations define nine PS-specific functions, and the arrows point toward the parasegments of the adult fly that are most affected in each class of mutations (Fig. 1). For reasons that will become clear later, it is worthwhile noting that all the mutations affecting the PS-specific functions are due to chromosomal rearrangements (more than 100 have been mapped). Thus, it seems impossible to affect these functions by point mutations, and it is unlikely that they correspond to individual **genes** coding for distinct **proteins**.

Northern blots and the isolation and sequencing of **complementary DNA** revealed the existence of only three **transcription** units *Ubx*, *abd-A*, and *Abd-B*. These three transcriptions units are all transcribed from right to left (of Fig. 1) and contain a **homeobox** sequence at their 3′ end. The *Ubx* transcription unit (see **Ultrabithorax**) covers 70 kb of DNA and can generate 12 different transcripts by **alternative splicing** and **polyadenylation**. **Translation** of these **messenger RNAs** yields a family of six proteins characterized by constant amino- and carboxy-proximal regions of 247 and 99 amino acid residues, respectively. The latter homeobox sequence is encoded by the 3′-terminal common exon. The members of this family are distinguishable by a short variable region that links the constant regions and consists of different combinations of three optional elements of 9, 17, and 17 residues (9–11). The spectrum of RNA products changes with time and tissue. It has been demonstrated that the *Ubx* protein containing the portion coded by the second microexon is not expressed in the nervous system (12). However, the different **isoforms** of the *Ubx* product are not all essential, since a mutation that eliminates the second microexon (and thus four isoforms) has no effect on fly viability and development (13).

The *abd-A* transcription unit is spread over a 20-kb region of DNA, and the mature *abd-A* transcript is composed of at least eight exons (14). As in *Ubx*, the *abd-A* transcription unit contains a microexon, but no alternative splicing generating multiple forms of the *abd-A* product has been detected thus far. The homeobox homology is found near the middle of the 330-residue protein. Analysis of the sequence of the whole BX-C (15) indicates that the open reading frame extends further upstream from the published ATG **initiation codon**, raising the possibility that *abd-A* may be slightly larger than originally thought. Inasmuch as this upstream open reading frame is conserved in *Tribolium*, it is likely that *abd-A* consists of 590 residues (16).

Finally, the *Abd-B* transcription unit consists of three classes of transcripts that are generated by the use of three alternate **promoters** and differential spicing (Fig. 1). While the alpha class of transcripts produces a protein of 55 kDa, the beta and gamma forms produce a product of 30 kDa that is truncated at the *N*-terminus (17,18).

Mutations that affect the *Ubx*, *abd-A*, and *Abd-B* transcription units (shown below the DNA line in Fig. 1) are all lethal at the embryo stage of development, and cuticle analysis of the dead embryos detects homeotic transformations. In *Ubx* mutants, PS5 and PS6 are transformed into PS4. If such an embryo could survive, it would give rise to a fly with T3 and A1 transformed into T2 (ie, a fly with three pairs of wings). Homozygous *abd-A* embryos have PS7, 8, and 9 transformed into PS6 (in the adult this would correspond to a transformation of A2, A3, and A4 into A1). Finally, *Abd-B* mutations have PS10, 11, 12, and 13 transformed into PS9 [A5–A8 transformed into A4 (2,7,19,20)].

EXPRESSION

The genetic and molecular data that have been described thus far appear to conflict. On the one hand, genetic analysis reveals the existence of nine parasegment-specific functions that are responsible for the identity of PS5 to 13, which will form the posterior thorax and the abdominal segments of an adult fly. On the other hand, molecular studies indicate that BX-C encodes only three protein products. This apparent discrepancy was solved when **antibodies** directed against the *Ubx*, *abd-A*, and *Abd-B* products became available, allowing

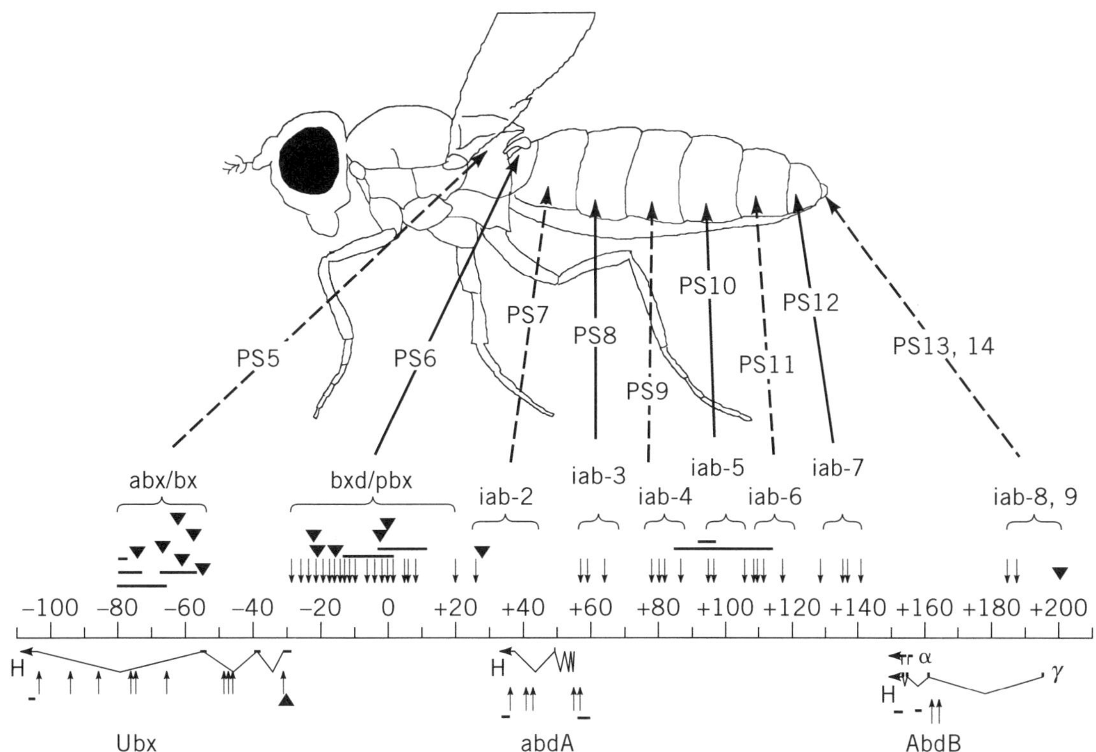

Figure 1. Genetic map of the bithorax complex (BX-C). The 300 kilobases of DNA that compise the complex are indicated by the bottom line, which is marked off in 10^4 base pairs. Above the DNA line are represented the sites of the homeotic mutations that affect the identities of each of the segements under control of BX-C. The vertical arrows represent the sites of chromosomal rearrangement breaks, the triangles the sites of insertion of transposons, and the horizontal lines the extent of deletions. The nine parasegments (PS5 to PS13, 14) that each part of the complex controls are indicated above, in the adult fly. The three transcription units (*Ubx*, *abdA*, and *AbdB*) are indicated below the DNA; in each case, transcription occurs from right to left. The exons are indicated by the thick horizontal lines, the introns by the thin V's connecting them. H indicates the homeobox domain. The alternative promoters α and γ are indicated for *AbdB*. Mutations that affect the transcription units are indicated below them.

determination of which part of the embryo these genes are expressed. A first observation derived from these studies is that each of the *Ubx*, *abd-A*, and *Abd-B* genes are expressed in domains composed of several parasegments. *Ubx* is expressed from PS5 to 13 (9,21), *abd-A* from PS7 to 12 (14,22), and *Abd-b* from PS10 to 14 (23,24). Second, the expression patterns are complex, intricate, and dynamic. Comparisons of the expression patterns between wild-type embryos and those carrying mutations in the PS-specific functions revealed that the latter correspond to large *cis*-regulatory regions that are responsible for construction of the complex expression patterns of *Ubx*, *abd-A*, and *Abd-B*. The *abx/bx* and *bxd/pbx cis*-regulatory regions are responsible for UBX expression in PS5 and 6, respectively (21,25,26). The *iab-2*, *iab-4*, and *iab-4 cis*-regulatory regions control expression of *Abd-A* in PS7, 8, and 9, respectively, while *iab-5*, *iab-6*, *iab-7*, and *iab-8* are responsible for the pattern of ABD-B expression (initiated from the α promoter) in PS10 to 13 (14,24,27). PS14 expresses a truncated form of ABD-B, resulting from transcription initiated from the β and γ promoters.

BX-C REGULATION

BX-C gene regulation can be divided into two phases, initiation and maintenance. During the early phases of embryogenesis, when parasegment identity is initially selected, the PS-specific *cis*-regulatory regions are the targets of the **gap gene** and **pair-rule gene** products (28–31). These gap and pair-rule proteins activate the *cis*-regulatory regions in successively more posterior parasegments.

The gap and pair-rule gene products are present only transiently during early development. The fact that homeotic genes are expressed throughout development implies the existence of a mechanism that maintains the activity state of each of the *cis*-regulatory regions. This maintenance system requires the **Polycomb** *group* (*Pc-G*) and the **trithorax** *group* (*trx-G*) genes (32–34). While the products of the *Pc-G* function as negative regulators, the products of the *trx-G* act as positive regulators. The products of the *Pc-G* exert their regulatory effects by interacting with specific elements in each of the *cis*-regulatory domains called *polycomb*-**response elements** (35–39). There may be equivalent or overlapping *trx* response elements for the *trx-G* proteins (36,40). Though their precise mode of action is unknown, the products of the *Pc-g* and *trx-g* are thought to stabilize the expression patterns in each parasegment by **imprinting** an inactive or active **chromatin** conformation of the PS-specific *cis*-regulatory subregions (33,41,42).

REGULATORY ELEMENTS OF BX-C

Molecular studies using **reporter gene** constructs have revealed the existence of elements within the PS-*cis*-regulatory units that seem to be responsible for the initiation and maintenance phases of BX-C regulation. Some DNA fragments are able to initiate expression of a *Ubx-lacZ* reporter gene in the proper parasegments during early embryonic development (28,30,43,44). In most cases, however, these patterns are not maintained, and expression expands into more anterior parasegments around the time when BX-C regulation would switch to the maintenance mode. Other BX-C DNA fragments are capable of retaining the appropriate parasegmental restrictions in lacZ expression after the gap and pair-rule gene products disappear. These fragments contain "maintenance elements," also known as *Pc-g* response elements because their activity depends on *Pc-g* gene products (35–39,43,44). Finally, a third type of regulatory elements that has been identified in experiments with *Ubx-lacZ* reporter constructs are tissue- or cell-type-specific enhancers. They induce *lacZ* expression in specific tissue or cell types, with no restriction along the anteroposterior axis.

Many observations suggest that the PS-specific *cis*-regulatory units are organized into functionally independent domains. This is best illustrated by the expression patterns of "enhancer trap" transposons integrated in different domains of the complex (45,46). These enhancer traps are subject to regulatory elements located within the same domain, but they are insensitive to regulatory elements in adjacent domains. The autonomy of each domain is ensured by elements that are believed to function as boundaries. Two such regulatory elements, *Mcp* and *Fab-7*, have been identified. *Mcp* is located between the *iab-4* and *iab-5 cis*-regulatory units or domains, while *Fab-7* is located between *iab-6* and *iab-7* (46–50).

CONCLUDING REMARKS

Molecular analysis of the BX-C has confirmed most of the predictions that Lewis had foreseen in his 1978 model. He had initially envisioned activation of a new gene product in each parasegment. It is now clearly established that there are only three major groups of related protein products encoded by BX-C (UBX, ABD-A, and ABD-B). Discrete genetic units exist, however, that are sequentially activated in each parasegment. These units function as transcription regulatory regions (PS-specific *cis*-regulatory regions). The complex *cis*-regulation that they mediate results in a very intricate pattern of expression, both between and within parasegments. Each parasegment is a mosaic of cells expressing different homeotic products. Under the direction of these proteins, different cells adopt different fates, yielding the complex array of pattern elements that characterizes a given parasegment (or segment). The PS-specific *cis*-regulatory regions are large (*bxd* is spread over more than 30 kb; see Fig. 1) and can act from remote distances on their target promoters (*iab-5* is localized 50 kb away from its target *Abd-B* promoter). These properties suggest that the structure of the chromatin plays an important role to allow such long-distance interactions. Chromatin structure is also evoked by the properties of the *Pc-G* gene. The products of these *Pc-G* genes function as cellular memory to maintain the repressed state of the homeotic genes in body regions where they have not been activated during early development. There are analogies between the *Pc-G* repression and **mating-type** silencing in yeast or heterochromatic **position-effect** variegation in *Drosophila*. Though little is known at the molecular level, these analogies suggest that *Pc-G* repression involves the formation of a complex of Pc-G proteins leading to a chromatin structure that is refractory to transcription. The finding of boundary elements insulating adjacent PS-specific *cis*-regulatory regions has led to a model in which the sequential activation the *cis*-regulatory regions would be due to the stepwise opening of chromosomal domains (45,46,49,51). Although no molecular clues exist to support such a model, it provides a rationale for the remarkable correspondence between the genomic organization of the BX-C and the anteroposterior axis of the fly. A similar model has been discussed recently in the case of the clusters of homeotic genes in mice, the Hox clusters (52).

BIBLIOGRAPHY

1. C. B. Bridges and T. Morgan (1923) *Carnegie Inst. Washington Publ.* **327**, 137.
2. E. B. Lewis (1978) *Nature* **276**, 565–570.
3. W. Bender, P. Spierer, and D. S. Hogness (1983) *J. Mol. Biol.* **168**, 17–33.
4. W. McGinnis, M. S. Levine, E. Hafen, A. Kuroiwa, and W. J. Gehring (1984) *Nature* **308**, 428–433.
5. M. P. Scott and A. J. Weiner (1984) *Proc. Natl. Acad. Sci. USA* **81**, 4115–4119.
6. W. Bender, M. Akam, F. Karch, P. A. Beachy, M. Peifer, P. Spierer, E. B. Lewis, and D. S. Hogness (1983) *Science* **221**, 23–29.
7. F. Karch, B. Weiffenbach, M. Peifer, W. Bender, I. Duncan, S. Celniker, M. Crosby, and E. B. Lewis (1985) *Cell* **43**, 81–96.
8. A. Martinez-Arias and P. Lawrence (1985) *Nature* **313**, 639–642.
9. P. A. Beachy, S. L. Helfand, and D. S. Hogness (1985) *Nature* **313**, 545–551.
10. M. B. O'Connors, R. Binari, L. A. Perkins, and W. Bender (1988) *EMBO J.* **7**, 435–445.
11. K. Kornfeld, R. B. Saint, P. A. Beachy, P. J. Harte, D. A. Peattie, and D. S. Hogness (1989) *Genes & Dev.* **3**, 243–258.
12. R. Weinzierl, J. M. Axton, A. Ghysen, and M. Akam (1987) *Genes & Dev.* **1**, 386–397.
13. A. Busturia, I. Vernos, J. Casanova, and G. Morata (1990) *EMBO J.* **9**, 3551–3555.
14. F. Karch, W. Bender, and B. Weiffenbach (1990) *Genes Dev.* **4**, 1573–1587.
15. C. H. Martin, C. A. Mayeda, C. A. Davis, C. L. Ericsson, J. D. Knafels, D. R. Mathog, S. E. Celniker, E. B. Lewis, and M. J. Palazzolo (1995) *Proc. Natl. Acad. Sci. USA* **92**, 8398–8402.
16. T. D. Shippy, S. J. Brown, and R. E. Denell (1998) *Dev. Genes Evol.* **207**, 446–452.

17. M. Zavortink and S. Sakonju (1989) *Genes Dev.* **3**, 1969–1981.
18. S. E. Celniker, D. J. Keelan, and E. B. Lewis (1989) *Genes Dev.* **3**, 1424–1436.
19. E. Sanchez-Herrero, I. Vernos, R. Marco, and G. Morata (1985) *Nature* **313**, 108–113.
20. J. Casanova, E. Sanchez-Herrero, A. Busturia, and G. Morata (1987) *EMBO J.* **6**, 3103–3109.
21. R. A. H. White and M. Wilcox (1985) *Nature* **318**, 563–567.
22. A. Macias, J. Casanova, and G. Morata (1990) *Development* **110**, 1197–1207.
23. M. DeLorenzi, N. Ali, G. Saari, C. Henry, M. Wilcox, and M. Bienz (1988) *EMBO J.* **7**, 3223–3231.
24. S. E. Celniker, S. Sharma, D. J. Keelan, and E. B. Lewis (1990) *EMBO J.* **9**, 4277–4286.
25. C. Cabrera, J. Botas, and A. Garcia-Bellido (1985) *Nature* **318**, 569–571.
26. R. A. H. White and M. E. Akam (1985) *Nature* **318**, 567–569.
27. E. Sanchez-Herrero (1991) *Development* **111**, 437–449.
28. S. Qian, M. Capovilla, and V. Pirrotta (1991) *EMBO J.* **10**, 1415–1425.
29. J. Muller and M. Bienz (1992) *EMBO J.* **11**, 3653–3661.
30. M. J. Shimell, J. Simon, W. Bender, and M. B. O'Connor (1994) *Science* **264**, 968–971.
31. F. Casares and E. Sanchez Herrero (1995) *Development* **121**, 1855–1866.
32. A. Shearn (1989) *Genetics* **121**, 517–525.
33. R. Paro (1990) *Trends Genet.* **6**, 416–421.
34. J. Simon, A. Chiang, and W. Bender (1992) *Development* **114**, 493–505.
35. J. Simon, A. Chiang, W. Bender, M. J. Shimell, and M. O'Connor (1993) *Dev. Biol.* **158**, 131–144.
36. C. S. Chan, L. Rastelli, and V. Pirrotta (1994) *EMBO J.* **13**, 2553–2564.
37. B. Christen and M. Bienz (1994) *Mech. Dev.* **48**, 255–266.
38. A. Chiang, M. B. O'Connor, R. Paro, J. Simon, and W. Bender (1995) *Development* **121**, 1681–1689.
39. S. Poux, C. Kostic, and V. Pirrotta (1996) *EMBO J.* **15**, 4713–4722.
40. H. Strutt, G. Cavalli, and R. Paro (1997) *EMBO J.* **16**, 3621–3632.
41. V. Pirrotta and L. Rastelli (1994) *Bioessays* **16**, 549–556.
42. J. Simon (1995) *Curr. Opin. Cell Biol.* **7**, 376–385.
43. J. Simon, M. Peifer, W. Bender, and M. O'Connor (1990) *EMBO J.* **9**, 3945–3956.
44. J. Muller and M. Bienz (1991) *EMBO J.* **10**, 3147–3155.
45. K. McCall, M. B. O'Connor, and W. Bender (1994) *Genetics* **138**, 387–399.
46. M. Galloni, H. Gyurkovics, P. Schedl, and F. Karch (1993) *EMBO J.* **12**, 1087–1097.
47. H. Gyurkovics, J. Gausz, J. Kummer, and F. Karch (1990) *EMBO J.* **9**, 2579–2585.
48. F. Karch, M. Galloni, L. Sipos, J. Gausz, H. Gyurkovics, and P. Schedl (1994) *Nucleic Acids Res.* **22**, 3138–3131.
49. J. Mihaly, I. Hogga, J. Gausz, H. Gyurkovics, and F. Karch (1997) *Development* **124**, 1809–1820.
50. J. Mihaly, I. Hogga, S. Barges, M. Galloni, R. Mishra, K. Hagstrom, M. Müller, P. Schedl, L. Sipos, J. Gausz, H. Gyurkovics, and F. Karch. (1998) *CMLS*, **54**, 60–70.
51. M. Peifer, F. Karch, and W. Bender (1987) *Genes Dev.* **1**, 891–898.
52. T. Kondo, J. Zákány, and D. Duboule (1998) *Molec. Cell* **1**, 289–300.

BIURET REACTION

T. IMOTO

Proteins form biuret-like compounds (structure I) in alkaline solution by losing a proton from the nitrogen atoms of the peptide bonds. In strong alkaline solution and Cu^{2+}, this compound forms a blue colored complex (structure II). This color reaction is called the biuret reaction. The sensitivity of the reaction is not very great, but the color yield is similar from protein to protein. Thus, this reaction has been employed to quantify proteins in widely varying samples.

According to the method of Gornall et al. (1), 1 to 10 mg of protein is determined by **absorbance** at 540 nm after incubation with biuret reagent ($CuSO_4$, potassium sodium tartrate, NaOH) for 30 min at 20° to 25°C. An improved biuret method by Westley and Lambeth (2), which removes the free Cu^{2+} ion with an ion-exchange resin and adds Na-diethyldithiocarbonate as a coloring reagent, detects 0.05 to 1 mg of protein. A simple and sensitive micro-biuret method was developed by Itzhaki and Gill (3), where UV absorbance at 310 nm is employed to quantify protein.

$$H_2H{-}\underset{\underset{O}{\|}}{C}{-}\underset{H}{N}{-}\underset{\underset{O}{\|}}{C}{-}NH_2$$

[I] Biuret; carbamylurea

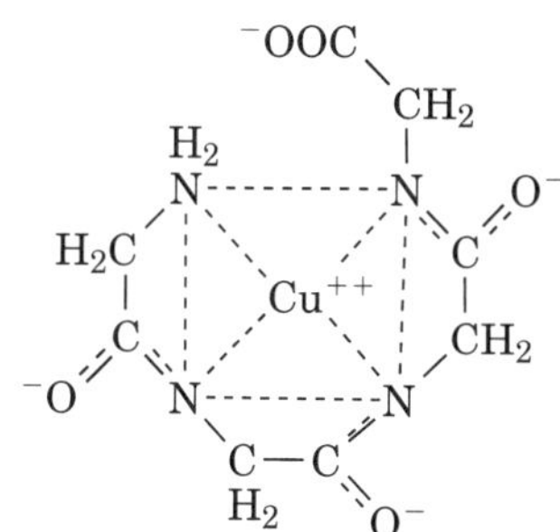

[II] A kind of biuret compound; copper glycine peptide complex under alkaline conditions

BIBLIOGRAPHY

1. A. G. Gornall, C. S. Bardawill, and M. M. David (1949) *J. Biol. Chem.* **177**, 751–766.
2. J. Westley and J. Lambeth (1960) *Biochim. Biophys. Acta* **40**, 364–366.
3. R. F. Itzhaki and D. M. Gill (1964) *Anal. Biochem,* **9**, 401–410.

BLASTODERM

JAMES KENNISON

The blastoderm is an **embryo** at the stage that consists of a single cell layer. This cell layer is formed by the cleavage divisions of the fertilized **zygote**, largely under the control of maternally deposited gene products [see **Maternal control**].

Suggestions for Further Reading

B. I. Balinsky (1975) *An Introduction to Embryology*, 4th ed., W. B. Saunders, Philadelphia, p. 5.

S. Shostak (1991) *Embryology: An Introduction to Developmental Biology*, Harper Collins, New York.

S. F. Gilbert (1997) *Developmental Biology*, Sinauer Associates, Sunderland, MA.

L. Wolpert et al. (1998) Principles of Development, Oxford University Press, Oxford, U.K.

J. M. W. Slack (1991) *From Egg to Embryo: Regional Specification in Early Development*, 2nd ed., Cambridge University Press, Cambridge, U.K.

BLASTOMERE

JAMES KENNISON

After **fertilization**, the **zygote** begins the cleavage divisions. During the cleavage divisions, the cells of the embryo are called *blastomeres*.

Suggestions for Further Reading

B. I. Balinsky (1975) *An Introduction to Embryology*, 4th ed., W. B. Saunders, Philadelphia, p. 5.

S. Shostak (1991) *Embryology: An Introduction to Developmental Biology*, Harper Collins, New York.

S. F. Gilbert (1997) *Developmental Biology*, Sinauer Associates, Sunderland, MA.

L. Wolpert et al. (1998) *Principles of Development*, Oxford University Press, Oxford, U.K.

J. A. Moore (1972) *Heredity and Development*, Oxford University Press, New York.

BLASTOPORE

JAMES KENNISON

The blastopore is a line of invagination on amphibian **embryos** for gastrulation (see **Gastrula**) (1). After fertilization, the amphibian **zygote** divides to form a mass of cells called blastomeres. The upper surface is called the *animal pole*, and the lower surface is called the *vegetal pole*. During gastrulation, the cells at the vegetal pole move inside the embryo, and the cells of the animal pole are left to form the outer surface of the embryo. The animal cells left at the surface will form the ectoderm, including the nervous system. The cells that move inside during gastrulation form the mesoderm and endoderm.

The process of gastrulation begins with the formation of a small indentation between the animal and vegetal poles on one side of the embryo. This small indentation is the beginning of formation of the blastopore. As gastrulation proceeds, the blastopore expands laterally to form a crescent, and eventually it encircles the entire embryo as gastrulation proceeds. The vegetal cells and the marginal cells between the vegetal and animal cells invaginate through the blastopore and form a second layer inside the embryo. As gastrulation continues, the blastopore shrinks to cover the remaining vegetal cells, called the *yolk plug*. When the blastopore almost completely covers the yolk plug, organogenesis begins. Interactions between various groups of cells during gastrulation cause cells to become determined to form various organs and structures. The determination of cells caused by interactions with their neighboring cells is called *developmental induction*. One of the most well-known examples is the **organizer** region of the amphibian embryo that specifies the determination of the nervous system. This is the dorsal lip of the blastopore and is often called the *Spemann organizer*, because it was first characterized by Hans Spemann (2).

BIBLIOGRAPHY

1. B. I. Balinsky (1975) *An Introduction to Embryology*, 4th ed., W. B. Saunders, Philadelphia, pp. 161–164.
2. H. Spemann (1938) *Embryonic Development and Induction*, Yale University Press, New Haven.

Suggestions for Further Reading

S. Shostak (1991) *Embryology: An Introduction to Developmental Biology*, Harper Collins, New York.

S. F. Gilbert (1997) *Developmental Biology*, Sinauer Associates, Sunderland, MA.

L. Wolpert et al. (1998) *Principles of Development*, Oxford University Press, Oxford, U.K.

J. A. Moore (1972) *Heredity and Development*, 2nd ed., Oxford University Press, New York.

BLEOMYCIN

CAROL WOOD MOORE

Bleomycin (BLM), discovered in 1962, is a family of low-molecular-weight metalloglycopeptides (M_r approximately 1500) produced by *Streptomyces verticillus* and isolated from the culture media of *S. verticillus* as copper complexes (1–3). It is widely used as an anticancer antibiotic in patients. BLM acts as a limited endonuclease and produces a variety of lesions in DNAs by a mechanism involving free radical attack on deoxyribose in both DNA strands. BLM damages DNAs in ways that mimic ionizing radiation. The detailed knowledge of the complicated chemistry of BLM and its interactions with DNA *in vivo* and *in vitro* drives much of its use in applications to molecular biology. BLM and structurally related analogues are considered radiomimetic and oxidative DNA-cleaving reagents, and as such they are utilized as chemical tools to study and understand the activities of this class of agents. Tools of molecular biology are also employed to understand the chemical, biological, and clinical aspects of the mechanism of action of BLMs. In addition, broadly used **cloning** strategies in several organisms use a gene conferring resistance to BLM and structurally related analogues as a selectable marker. Phleomycin (PLM) is used as the selective agent and is available in commercial formulations. BLM causes multiple changes to cells and is cytocidal, and its cytotoxicity is high where there is no or only a limited barrier to BLM reaching cellular targets. BLM effectively kills all types of cells tested. Cellular resistance is conferred in several ways, including protection by **nucleosomes** in **chromatin**, **DNA repair**, metabolic inactivation of BLM, BLM-resistance proteins, and restricted entry of BLM. An overview of resistance mechanisms was recently published (4).

STRUCTURE AND ACTIVATED COMPLEX

The structurally complex BLMs contain a metal-binding domain and a **DNA-binding** domain (Fig. 1). BLMs require metals and oxygen species for their activity (5–9). The metal-binding domain is also the site of oxygen activation and is attached to a disaccharide group; it binds redox-active transition metals, such as Fe(II), Co(II), Cu(II), Ni(II), and Zn(II). The most stable BLM–O_2–metal complex is formed with cobalt (8). When a BLM-Fe(II)-O_2 complex binds to DNA and the Fe(II) oxidizes to Fe(III), the complex attacks the C4′ position of DNA deoxyribose (10–13). The complex thereby behaves as a limited endonuclease.

The coplanar bithiazole moiety partially intercalates into the minor groove between bases of DNA. The cationic *C*-terminal amines are also involved in interactions with nucleic acids. The terminal amines in BLM A_2 and BLM B_2 are dimethylsulfonium propylamine and agmatine, respectively, and are similar in length and bear one positive charge (Fig. 1). BLM B_2 produces considerably more DNA breaks and killing than BLM A_2 over a wide range of chemical concentrations (14). Without the terminal amine, the BLM molecule no longer cleaves DNA or possesses antitumor activity.

BLM and structurally related PLM (15) differ in the oxidation state of their sulfur heterocycles (Fig. 1). One of the two conjugated thiazole rings of the BLM bithiazole is modified by hydrogenation to 4,5-dihydrothiazole (thiazoline) in PLM (16,17). In addition, the *C*-terminal amines in clinical preparations of BLMs differ chemically and quantitatively from prepared mixtures of PLMs. Tallysomycin is closely related to BLM (18,19).

ANTICANCER USE

BLM is an important therapeutic agent that is useful as a single agent in treating several human cancers and is widely used in combination chemotherapy and radiotherapy. The water-soluble product used in cancer treatment, Blenoxane, is a family of 11 metal-free congeners differing in their terminal amines. The clinical mixture (20) is comprised mainly of BLM A_2 [approximately 55% to 70% (usually 68% to 69%)] and BLM B_2 (approximately 25% to 32%). The effectiveness of BLM as an anticancer agent is associated with its ability to produce lesions in DNA (21,22).

BLM is principally used in patients with lymphoma or a variety of solid tumors. It has been included in regimens for treating malignant and peripheral **T-cell** lymphomas, as well as in combination BEP chemotherapy (BLM, etoposide,

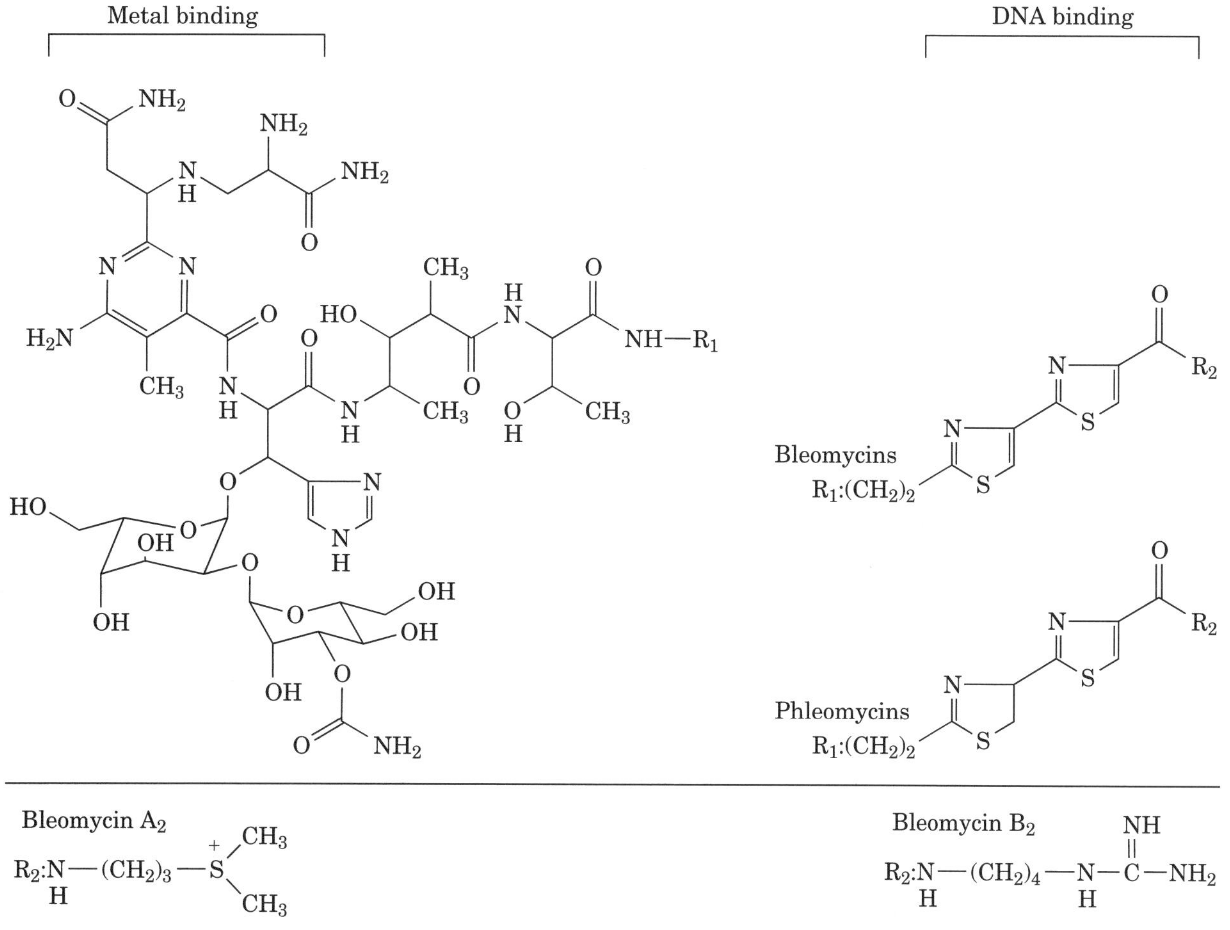

Metal binding

DNA binding

Bleomycins R_1:$(CH_2)_2$

Phleomycins R_1:$(CH_2)_2$

Bleomycin A_2

R_2:$\mathrm{NH-(CH_2)_3-S^+(CH_3)_2}$

Bleomycin B_2

R_2:$\mathrm{NH-(CH_2)_4-NH-C(=NH)-NH_2}$

Figure 1. Structures of bleomycins and phleomycins (16,117–119).

cisplatin) for metastatic testicular teratoma and Hodgkins and non-Hodgkins lymphoma. BLM is also used in the chemotherapy and management of Kaposi's sarcoma (eg, pulmonary, gastrointestinal, epidemic, disseminated), as well as with vincristine in combination chemotherapy for epidemic Kaposi's sarcoma. BLM does not cause bone-marrow, hepatic, or renal toxicities, nor does it cause cardiosuppression. Pulmonary fibrosis is a side effect of BLM treatment that limits its use. The molecular mechanism of lung fibrosis is being investigated, but is not fully known. Patients with genetic susceptibility are particularly susceptible to lung fibrosis (23,24). In animal models, taurine (25) or taurine and niacin (26) counter BLM-induced lung fibrosis. The newer BLM derivatives, peplomycin and liblomycin, were developed because of their lower pulmonary toxicity and broader antitumor spectrum in animal studies (27).

MECHANISM OF ACTION ON NUCLEIC ACIDS

Chemical Action on DNA: DNA Damage

The unique chemical action of BLMs in the presence of oxygen and Fe(II) catalytically cleaves double-stranded DNA *in vivo* and *in vitro*. Single- and double-stranded breaks are produced, leaving 5′-phosphate- and 3′-phosphoglycolate-termini (11,28–32). The most genotoxic and lethal lesions for cells are double-stranded breaks. BLM recognizes 5′-phosphoguanylyl(3′,5′)thymidine or 5′-phosphoguanylyl (3′,5′)cytosine sequences most frequently, releasing the pyrimidines when they are located to the 3′ side of guanosine (28–30,33–36) and leaving DNA alkali-labile (32,37). While cleavage at the first site is G-Py-specific, the second nucleophilic attack by a BLM–Fe(II)–O_2 complex on the opposite DNA strand is not a sequence-specific cleavage and instead is probably targeted by the structural perturbation of the DNA at the first cleavage site (38,39). Deoxyribose degradation also produces 3-(pyrimidin-1′-yl)-2-propenal and 3-(purin-9′-yl)-2-propenal (11,31,40) and oligonucleotide 3′-(phosphoro-2-*O*-glycolic acid) derivatives (11,31,41). DNA strand breaks are stoichiometric with the production of base propenals (42).

Preferential Cleavage Between Nucleosomes

BLM preferentially cleaves **linker regions** of **chromatin** between **nucleosomes** (43–47). The enhanced resistance of DNA bound to nucleosomes seems to be related to the necessity of a conformational change for BLM binding and intercalation. In yeast, internucleosomal cleavage and DNA degradation (46) and killing (48) are less pronounced in logarithmic phase cells than in cells that are in stationary phase. The cellular and molecular basis for this may relate to the highly effective use of BLM on particular solid tumors.

Bleomycin and Phleomycin

PLM was discovered before bleomycin, but was too toxic in patients to be used as an anticancer drug. PLM is also substantially more effective than BLM on a per mole basis in yeast in producing cell killing (48), DNA breaks in intracellular DNAs (49), release of nucleosomes from chromatin (45), and genetic changes (50). Thus, the DNA lesions produced by the BLM and PLM could differ in their nature or frequency, or they could be processed differently by the cells. The mechanisms of BLM and PLM interaction with DNA *in vitro* also appear to differ (51–54). PLM exhibits a higher requirement than BLM for ferrous ions (49). Bithiazole intercalation in DNA is thought to be necessary for producing double-stranded, but not single-stranded, breaks *in vitro* (52). Accordingly, BLM produces more double-strand breaks than PLM in PM2 phage DNA (52,54). Cu(II)BLM, but not Cu(II) PLM, intercalates (51). BLM, but not PLM, degrades relaxed DNA to a greater extent than either positively or negatively **superhelical** DNA (54). On the other hand, BLM and PLM cleave DNA *in vitro* at similar preferred sites at similar frequencies (55,56) and produce comparable numbers of DNA breaks under some conditions (57,58).

RNA

BLM cleaves a variety of RNAs. The most studied have been **transfer RNA** and tRNA precursors. In contrast to DNA cleavage, BLM-induced RNA cleavage is usually wholly or predominately at a single site, often at a junction between single- and double-stranded RNA regions (59–62).

DNA REPAIR

Biochemical and Genetic Evidence for Radiomimetic Properties

The cellular processes that repair BLM-induced DNA lesions are not entirely known. The most important mechanisms for the repair of BLM-induced DNA damage are **recombinational repair**, **base-excision repair**, and post-replication repair. BLM and ionizing radiation produce similar lesions in DNA, although BLM produces a narrower spectrum of products than ionizing radiation. DNA breaks are introduced approximately linearly with increasing concentrations of BLM (14) and ionizing radiation (63). Some of the chromosomal lesions produced after either BLM treatments or ionizing irradiation can be ligated immediately and lead to a quick component of DNA rejoining, but other lesions require more time for processing before the termini of the DNA molecules become substrates for ligation (14,63–65). For example, the unusual phosphoglycolate must be removed by the DNA 3′-repair diesterase activity of apurinic and apyrimidinic endonuclease (66,67). After extension of the remaining DNA strand by one nucleotide, DNA ligase resynthesizes intact DNA molecules by forming a phosphodiester bond between adjacent 3′-hydroxyl and 5′-phosphoryl termini (14). Rapid and slower components of DNA rejoining have been identified in several laboratories. An ultrarapid phase of cellular recovery accompanies rapid repair in human cells (64,68), suggesting that some of the cytotoxic treatment effects of BLM could be counteracted in the clinic and reduce chemotherapeutic effectiveness.

The importance of DNA repair is shown by the many studies in various organisms of mutants hypersusceptible to killing by BLM and defective in repair of DNA lesions. All *rad* mutations of *Saccharomyces cerevisiae* (69–72) that confer hypersensitivity to killing by BLM analogues also confer cross-sensitivity to ionizing radiation (73–78), so pathways are shared for the repair of chromosomal damage by bleomycins and ionizing radiation. Moreover, mutant strains with altered resistance to lethal effects of BLM have been isolated and characterized in several laboratories, and direct selection for mutations conferring hypersensitivities to lethal effects of BLM resulted in mutants exhibiting cross-hypersensitivities to ionizing radiation.

Genetic Recombination and Mutation

Mechanisms of recombination are important for DNA repair and rejoining **recombinant DNA**. BLM is recombinogenic and mutagenic (38,50,79–85). The amount of recombination or mutation caused by BLM depends upon the assay system and treatment conditions. BLM was found to be weakly mutagenic to mitochondrial DNA (86).

ADDITIONAL CELLULAR TARGETS

Membrane

The plasma membrane restricts BLM internalization in some mammalian cells (87,88). This could be due to the polar and charged groups on the BLM molecule. Membrane damage by BLM (68,89) or cell electropermealization (88) circumvents this restriction.

Cell Wall

BLM molecules are readily taken up into yeast cells, but are not equally distributed from cell to cell (89). BLM initially localizes to cell walls, causes cell wall and membrane damage, and aids the enzymatic conversion of cells to spheroplasts (75,89–91). BLM alters the anchorage of several mannoproteins in the cell wall matrix of intact cells or isolated cell walls, and it disrupts essential cell wall polymers. These activities facilitate the entry of BLM into yeast cells.

BLM Hydrolase

BLM hydrolase hydrolyzes and inactivates BLM. The enzyme is a **thiol proteinase** that has DNA-binding and peptide-cleavage domains. It binds DNA and RNA. BLM hydrolase activity protects cells from BLM toxicity, but limits the use of BLM in cancer chemotherapy. Although its normal cellular function is unknown, the enzyme is present in diverse organisms, including humans (92–96) and yeast (97–100). Expression of the yeast BLM hydrolase in mammalian cells results in resistance to BLM (101). A member of a galactose regulatory system (102,103), yeast BLM hydrolase binds to nicked double-stranded DNA, single-stranded DNA, and RNA, without sequence specificity (104). This enzyme associates with plasma membranes and is in the cytosol (105). Human BLM hydrolase was recently shown to exhibit endopeptidase activity (106). This enzyme is thought to play a role in the development of resistance to BLM during chemotherapy (95,107–109).

BLM Resistance Proteins

Several proteins in microorganisms that produce BLM or structural analogues actually confer resistance to these products. Vectors bearing genes encoding these proteins confer high levels of resistance to BLM and related antibiotics and are used as cloning vehicles. The proteins bind and form stable complexes with BLM with high specificity, thereby preventing BLM from complexing with DNA. One of these proteins is encoded by the *Streptoalloteichus hindustanus* (*Sh*) *ble* gene (110–112) found on the Tn5 bacterial **transposon**, along with additional genes encoding resistance to other antibiotics. Expression of the *Sh ble* gene in transgenic mice reduced BLM toxicity in the mice and protected against lung fibrosis (113,114). High levels of the protein were detected in lungs, kidney, and spleen.

Multiple Targets and Cytotoxicity

Multiple cellular targets of BLM are expected because of the chemical mechanism of action of the molecule and because some of the BLM-hypersensitive mutants isolated in different organisms do not exhibit hypersensitivity to lethal effects of radiation. Multiple cellular enzymes are important for surviving the toxicities of BLM, and deficiencies in their function could reduce chances of survival. The relationship and significance of each cellular target of BLM to cellular toxicity are not known. Why BLM causes **apoptosis** in some types of cells and not in others is also unknown. BLM is not considered apoptotic at low concentrations. Far less is known about targets of BLM in cells than about the chemical mechanism of action of BLM *in vitro* on defined substrates.

Additional factors modulate BLM activity. Generally, BLM preferentially affects mitotic cells in the G2-M phase of the **cell cycle**. BLM is unlikely to be a useful agent to study meiosis because of the abundant lesions the molecule produces in DNA. Intracellular metal ion concentrations and pH also modulate BLM activities (48,115,116). Studies elucidating the roles of the multiple targets of BLM and factors that modulate BLM activities will improve our understanding and efficacious uses of the widely studied BLM family of related compounds.

BIBLIOGRAPHY

1. H. Umezawa, K. Maeda, T. Takeuchi, and Y. Okami (1966) *J. Antibiotics* **19**, 200–209.
2. H. Umezawa, Y. Suhara, T. Takita, and K. Maeda (1966) *J. Antibiotics* **19**, 210–215.
3. H. Umezawa (1976) *GANN* **19**, 3–36.
4. W. S. El-Deiry (1997) *Curr. Opin. Oncol.* **9**, 79–87.
5. E. A. Sausville, J. Peisach, and S. B. Horwitz (1976) *Biochem. Biophys. Res. Commun.* **91**, 871–877.
6. R. M. Burger, J. Peisach, and S. B. Horwitz (1981) *J. Biol. Chem.* **256**, 11636–11639.
7. R. M. Burger, S. J. Projan, S. B. Horwitz, and J. Peisach (1986) *J. Biol. Chem.* **261**, 15955–15959.
8. J. Stubbe and J. Kozarich (1987) *Chem. Rev.* **87**, 1107–1136.
9. P. C. Dedon and I. H. Goldberg (1992) *Chem. Res. Toxicol.* **5**, 311–332.
10. R. M. Burger, J. Peisach, and S. B. Horwitz (1981) *Life Sci.* **28**, 715–727.
11. L. Giloni et al. (1981) *J. Biol. Chem.* **256**, 8608–8615.
12. J. W. Sam and J. Peisach (1993) *Biochemistry* **43**, 1488–1491.
13. L. F. Povirk, Y. H. Han, and R. J. Steighner (1989) *Biochemistry* **28**, 5808–5814.
14. C. W. Moore (1990) *Biochemistry* **29**, 1342–1347.
15. K. Maeda, H. Kosaka, K. Yagishita, and H. Umezawa (1956) *J. Antibiotics* **9**, 82–85.
16. T. Takita et al. (1972) *J. Antibiotics* **25**, 197–199.
17. D. A. McGowan, U. Jordis, D. K. Minster, and S. M. Hecht (1977) *J. Am. Chem. Soc.* **99**, 8078–8079.
18. H. Kawaguchi et al. (1977) *J. Antibiotics* **30**, 779–788.
19. M. Konishi et al. (1977) *J. Antibiotics* **30**, 789–805.
20. C. Crooke and W. Bradner (1976) *J. Med.* **7**, 333–425.
21. S. K. Carter, S. T. Crooke, and H. Umezawa (1978) *Bleomycin, Current Status and New Developments*, Academic Press, New York.
22. S. M. Hecht, ed. (1979) *Bleomycin: Chemical, Biochemical and Biological Aspects*, Springer-Verlag, New York.

23. C. K. Haston, C. I. Amos, T. M. King, and E. L. Travis (1996) *Cancer Res.* **56**, 2596–2601.
24. R. P. Marshall, R. J. McAnulty, and G. J. Laurent (1997) *Int. J. Biochem. Cell. Biol.* **29**, 107–120.
25. R. E. Gordon, R. F. Heller, and R. F. Heller (1992) *Adv. Exp. Med. Biol.* **315**, 319–328.
26. G. Gurujeyalakshmi, M. A. Hollinger, and S. N. Giri (1998) *Am. J. Respir. Cell. Mol. Biol.* **18**, 334–342.
27. T. Takita and T. Ogino (1987) *Biomed. Pharmacother.* **41**, 219–226.
28. A. D. D'Andrea and W. A. Haseltine (1978) *Proc. Natl. Acad. Sci. USA* **75**, 3608–3612.
29. M. Takeshita, P. Grollman, E. Ohtsubo, and H. Ohtsubo (1978) *Proc. Natl. Acad. Sci. USA* **75**, 5983–5987.
30. A. P. Grollman and M. Takeshita (1980) *Adv. Enzymol. Regul.* **18**, 67–83.
31. N. Murugesan et al. (1985) *Biochemistry* **24**, 5735–5744.
32. H. Sugiyama, C. Xu, N. Murugesan, and S. M. Hecht (1985) *J. Am. Chem. Soc.* **107**, 4101–4105.
33. M. Takeshita and P. Grollman (1979) In *Bleomycin: Chemical, Biochemical and Biological Aspects* (S. Hecht, ed.), Springer-Verlag, New York, pp. 207–221.
34. C. K. Mirabelli, C.-H. Huang, A. W. Prestayko, and S. T. Crooke (1982) *Cancer Chemother. Pharmacol.* **8**, 57–65.
35. C. K. Mirabelli et al. (1982) *Cancer Res.* **42**, 2779–2785.
36. V. Murray and R. F. Martin (1985) *J. Biol. Chem.* **260**, 10389–10391.
37. J. C. Wu, J. W. Kozarich, and J. Stubbe (1983) *J. Biol. Chem.* **258**, 4694–4697.
38. R. J. Steighner and L. F. Povirk (1990) *Proc. Natl. Acad. Sci. USA* **87**, 8350–8354.
39. L. F. Povirk, Y.-H. Han, and R. J. Steighner (1989) *Biochem.* **28**, 8508–8514.
40. R. M. Burger, A. R. Berkowitz, J. Peisach, and S. B. Horwitz (1980) *J. Biol. Chem.* **255**, 11832–11838.
41. S. Uesugi et al. (1984) *Nucleic Acids Res.* **12**, 1581–1592.
42. R. M. Burger, J. Peisach, and S. B. Horwitz (1982) *J. Biol. Chem.* **257**, 8612–8614.
43. M. T. Kuo and T. C. Hsu (1978) *Nature* **271**, 83–84.
44. M. T. Kuo and T. C. Hsu (1978) *Chromosoma* **68**, 229–240.
45. C. W. Moore, (1988) *Cancer Res.* **48**, 6837–6843.
46. C. W. Moore, C. S. Jones, and L. A. Wall (1989) *Antimicrob. Agents Chemother.* **33**, 1592–1599.
47. K. Sidik and M. J. Smerdon (1990) *Biochem.* **29**, 7501–7511.
48. C. W. Moore (1982) *Cancer Res.* **42**, 929–933.
49. C. W. Moore (1989) *Cancer Res.* **49**, 6935–6940.
50. J. F. Mc.Koy et al. (1995) *Mutat. Res. DNA Repair* **336**, 19–27.
51. L. F. Povirk, M. Hogan, N. Dattagupta, and M. Buechner (1981) *Biochem.* **20**, 665–671.
52. C.-H. Huang, C. K. Mirabelli, Y. Jan, and S. T. Crooke (1981) *Biochem.* **20**, 233–238.
53. C.-H. Huang, A. W. Prestayko, and S. T. Crooke (1982) *Biochem.* **21**, 3704–3710.
54. C.-H. Huang, C. K. Mirabelli, S. Mong, and S. T. Crooke (1983) *Cancer Res.* **43**, 2849–2856.
55. M. Takeshita et al. (1981) *Biochem.* **20**, 7599–7606.
56. J. Kross, W. D. Henner, S. M. Hecht, and W. A. Haseltine (1982) *Biochem.* **21**, 4310–4318.
57. H. Suzuki et al. (1969) *J. Antibiot.* **22**, 446–448.
58. R. Stern, J. A. Rose, and R. M. Friedman (1974) *Biochemistry* **13**, 307–312.
59. B. J. Carter et al. (1990) *Proc. Natl. Acad. Sci. USA* **87**, 9373–9377.
60. A. Huttenhofer, S. Hudson, H. F. Noller, and P. K. Mascharak (1992) *J. Biol. Chem.* **267**, 24471–24475.
61. L. L. Guan et al. (1993) *Biochem. Biophys. Res. Commun.* **191**, 1338–1346.
62. M. V. Keck and S. M. Hecht (1995) *Biochemistry* **34**, 12029–12037.
63. C. W. Moore (1982) *J. Bacteriol.* **150**, 1227–1233.
64. C. W. Moore and J. B. Little (1985) *Cancer Res.* **45**, 1982–1986.
65. C. W. Moore (1988) *J. Bacteriol.* **170**, 4991–4994.
66. A. W. Johnson and B. Demple (1988) *J. Biol. Chem.* **263**, 18017–18022.
67. J. Laval (1996) *Pathol. Biol.* **44**, 14–24.
68. C. W. Moore, A. W. Malcolm, K. N. Tomkinson, and J. B. Little (1985) *Cancer Res.* **45**, 1978–1981.
69. J. Game (1993) *Semin. Cancer Biol.* **4**, 73–83.
70. S. Prakash, P. Sung, and L. Prakash (1993) *Annu. Rev. Genet.* **27**, 33–70.
71. E. C. Friedberg (1995) *DNA Genes and Mutagenesis*. ASM Press, Washington, D.C.
72. D. Ramotar and J.-Y. Masson (1996) *Mol. Cell. Biochem.* **158**, 65–75.
73. C. W. Moore (1978) *Mutat. Res.* **51**, 165–180.
74. C. W. Moore (1980) *J. Antibiot.* **33**, 1369–1375.
75. C. W. Moore (1982) *Antimicrob. Agents Chemother.* **21**, 595–600.
76. C. W. Moore (1991) *J. Bacteriol.* **173**, 3605–3608.
77. D. J. Keszenman, V. A. Salvo, and E. Nunes (1992) *J. Bacteriol.* **174**, 3125–3132.
78. C. H. He, J.-Y. Masson, and D. Ramotar (1996) *Can. J. Microbiol.* **42**, 1263–1266.
79. M. A. Hannan and A. Nasim (1978) *Mutat. Res.* **53**, 309–316.
80. M. A. Hannan, A. Nasim, and T. Brychcy (1978) *Mutat. Res.* **58**, 107–110.
81. B. K. Vig and R. Lewis (1978) *Mutat. Res.* **55**, 121–145.
82. C. W. Moore (1978) *Mutat. Res.* **58**, 41–49.
83. A. Severgnini, O. Lillo, and E. Nunes (1991) *Environ. Mol. Mutagen.* **18**, 102–106.
84. L. F. Povirk and M. J. F. Austin (1991) *Mutat. Res.* **257**, 127–143.
85. L. F. Povirk et al. (1994) *J. Mol. Biol.* **243**, 216–226.
86. L. R. Ferguson and P. M. Turner (1988) *Eur. J. Cancer Clin. Oncol.* **24**, 591–596.
87. G. Pron, J. Belehradek, Jr., S. Orlowski, and L. M. Mir (1994) *Biochem. Pharmacol.* **48**, 301–310.
88. L. M. Mir, O. Tounekti, and S. Orlowski (1996) *Gen. Pharmacol.* **27**, 745–748.
89. C. W. Moore, R. Del Valle, J. F. Mc.Koy, A. Pramanik, and R. E. Gordon (1992) *Antimicrob. Agents Chemother.* **36**, 2497–2505.
90. R. Beaudouin et al. (1993) *Antimicrob. Agents Chemother.* **37**, 1264–1269.
91. S. T. Lim, C. K. Jue, C. W. Moore, and P. N. Lipke (1995) *J. Bacteriol.* **177**, 3534–3539.
92. S. Akiyama et al. (1981) *Biochem. Biophys. Res. Commun.* **101**, 55–60.
93. J. S. Lazo, S. M. Sebti, and A. E. Filderman (1987) In *Metabolism and Mechanism of Action of Anti-cancer Drugs* (G. Powis and R. A. Prough, eds.), Taylor and Francis, London, pp. 194–210.
94. C. Nishimura, H. Suzuki, N. Tanaka, and H. Yamaguchi (1989) *Biochim. Biophys. Acta* **1012**, 29–35.
95. S. M. Sebti et al. (1989) *Biochemistry* **28**, 6544–6548.

96. S. E. Montoya, R. E. Ferrell, and J. S. Lazo (1997) *Cancer Res.* **57**, 4191–4195.
97. N. G. Kambouris, D. J. Burke, and C. E. Creutz (1992) *J. Biol. Chem.* **267**, 21570–21576.
98. C. Enenkel and D. H. Wolf (1993) *J. Biol. Chem.* **268**, 7036–7043.
99. U. Magdolen, G. Müller, V. Magdolen, and W. Bandlow (1993) *Biochim. Biophys. Acta* **1171**, 299–303.
100. H. E. Xu and S. A. Johnston (1994) *J. Biol. Chem.* **269**, 21177–21183.
101. A. Pei, T. P. Calmels, C. E. Creutz, and S. M. Sebti (1995) *Mol. Pharmacol.* **48**, 676–681.
102. L. Joshua-Tor, H. E. Xu, S. A. Johnson, and D. C. Rees (1995) *Science* **269**, 945–950.
103. W. Zheng, H. E. Xu, and S. A. Johnston (1997) *J. Biol. Chem.* **272**, 30350–30355.
104. W. Zheng and S. A. Johnston (1998) *Mol. Cell. Biol.* **18**, 3580–3585.
105. I. Niemer, G. Müller, and G. Strobel (1997) *Curr. Genet.* **32**, 41–51.
106. R. P. Koldamova, I. M. Lefterov, V. G. Gadjeva, and J. S. Lazo (1998) *Biochemistry* **37**, 2282–2290.
107. D. Drocourt, T. Calmels, J. P. Reynes, M. Baron, and G. Tiraby (1990) *Nucleic Acids Res.* **18**, 4009.
108. D. Bromme et al. (1996) *Biochemistry* **35**, 6706–6714.
109. A. A. Ferrando, A. Velasco, E. Campo, and C. Lopez-Otin (1996) *Cancer Res.* **56**, 1746–1750.
110. A. Gatignol, M. Baron, and G. Tiraby (1987) *Mol. Gen. Genet.* **207**, 342–348.
111. A. Gatignol, M. Dassain, and G. Tiraby (1990) *Gene* **91**, 35–41.
112. D. Drocourt et al. (1990) *Nucl. Acids Res.* **18**, 4009.
113. P. L. Tran et al. (1997) *J. Clin. Invest.* **99**, 608–617.
114. J. Weinbach et al. (1996) *Cancer Res.* **56**, 5659–5665.
115. C. W. Moore and D. A. Vossler (1980) *Biochim. Biophys. Acta* **610**, 425–429.
116. C. W. Moore (1994) *Antimicrob. Agents Chemother.* **38**, 1615–1619.
117. T. Takita et al. (1972) *J. Antibiot.* **25**, 755–758.
118. H. Naganawa, Y. Muraoka, T. Takita, and H. Umezawa (1977) *J. Antbiot.* **30**, 388–396.
119. T. Takita et al. (1978) *J. Antibiot.* **31**, 801–804.

Suggestions for Further Reading

R. M. Burger (1998) Cleavage of nucleic acids by bleomycin. *Chem. Rev.* **98**, 1153–1170. (Reviews the chemical mechanism of bleomycin action.)

S. M. Hecht (1994) RNA degradation by bleomycin, a naturally occurring bioconjugate. *Bioconjugate Chem.* **5**, 513–526. (Reviews and compares mechanisms of cleavage of DNAs and RNAs.)

J. S. Lazo and S. M. Sebti (1997) Bleomycin. *Cancer Chemother. Biol. Response Modif.* **17**, 40–45. (Together with earlier reviews in this series by the same authors, this article provides overview of clinical uses of bleomycin and mechanisms of tumor resistance to bleomycin cytotoxicity.)

D. H. Petering, Q. Mao, W. Li, E. DeRose, and W. E. Antholine (1996) Metallobleomycin–DNA interactions: structures and reactions related to bleomycin-induced DNA damage. *Met. Ions Biol. Syst.* **33**, 619–648. (Reviews mechanisms of interactions of metallobleomycin and DNA in cells and *in vitro*.)

L. F. Povirk (1996) DNA damage and mutagenesis by radiomimetic DNA-cleaving agents: bleomycin, neocarzinostatin and other enediynes *Mutat. Res.* **355**, 71–89. (Summarizes the chemical action and relationship of DNA lesions to types of mutations produced by radiomimetic antibiotics.)

BLOOD CLOTTING

V. ELLIS

The clotting or coagulation of blood is necessary for the maintenance of vascular integrity, and is therefore essential to survival. Cessation of bleeding can be achieved within minutes, involving both cellular and soluble protein components present in blood. The overall function of the blood coagulation system is the rapid amplification of a small initial stimulus via a series of sequential reactions that activate various **zymogens** and pro-cofactors of **serine proteinases**, culminating in the generation of **thrombin**, which proteolytically converts **fibrinogen** to fibrin. Fibrin monomers subsequently polymerize to form the insoluble fibrin clot. The major enzymatic components of the coagulation pathways are a series of **homologous** serine proteinases whose function is determined primarily by differences in the organization of their *N* terminal, noncatalytic **domains**. Most of the reactions are catalyzed by complexes between these serine proteinases, nonenzymatic protein cofactors, and serine proteinase zymogens that are assembled on phospholipid surfaces in the presence of calcium ions. These complexes are able to catalyze activation of the proteinase zymogen by limited **proteolysis** at rates up to 10^6-fold greater than those in the absence of cofactor/phospholipid. The proteinase product of one reaction then participates in another analogous complex, thus providing the amplification, as well as the multiple levels at which both positive and negative regulation can be exerted.

Figure 1 shows the two classical pathways of blood coagulation, termed the *intrinsic* and *extrinsic* pathways. These were originally envisaged as elegant, essentially linear sequences of reactions, but **feedback inhibition** reactions and cross-reactions between the two pathways are now known to be important and to provide further amplification. The terminology used to describe the two pathways of blood coagulation is somewhat counterintuitive, tending to imply the opposite to what is now known to be the physiological clotting mechanism. Genetic deficiencies in the extrinsic pathway lead to severe bleeding tendencies, but initiation of this pathway *in vitro* requires components in the form of tissue extracts (containing tissue factor and phospholipid) that are not present in blood. By contrast, the physiological role of the "intrinsic" pathway is unclear, as genetic deficiencies in its components do not give rise to bleeding disorders, but all the components are present in blood, and it can be initiated by simply removing blood from the circulation. This is due to contact with a negatively charged surface, such as glass, giving the pathway its alternative name of "contact activation," and involves proteinases and cofactors quite distinct from those of the extrinsic pathway.

A complete description of the coagulation system and the many proteolytic and inhibitory reactions that lead to its exquisite functional regulation is beyond the scope of this article. We will therefore focus on two key proteolytic steps that are representative of the two types of multicomponent complexes of the extrinsic pathway and that demonstrate the fundamental interactions and mechanisms underlying catalysis of the proteolytic reactions of blood coagulation; one

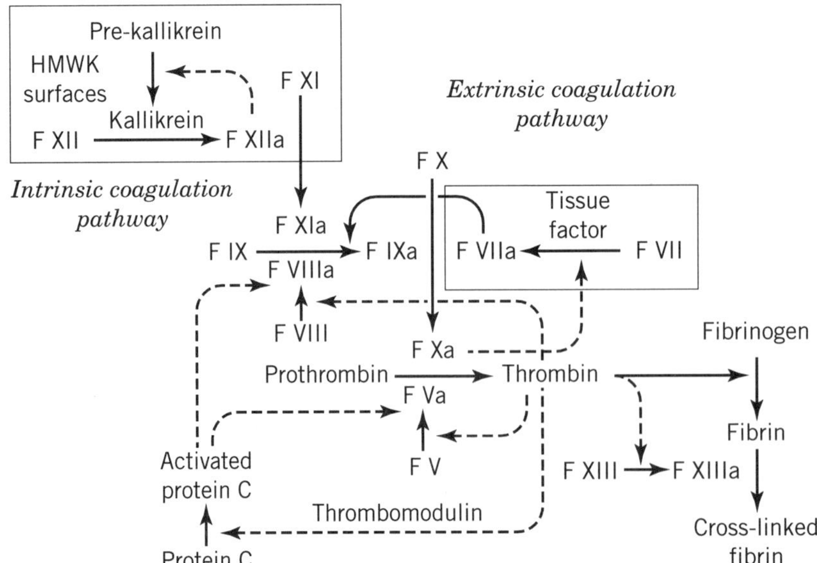

Figure 1. The proteolytic reactions of the blood clotting pathways. The zymogen/proteinase pairs are shown in black with the "forward" reactions as solid lines, the cofactors are shown in blue with the positive feedback reactions as dashed blue lines. The proteolytic negative feedback reactions are shown in red. The complexes initiating the two pathways of coagulation are boxed: the factor VII–tissue factor complex and the contact activation complex that assembles on high molecular-weight kininogen (HMWK). The Roman numeral system used for the coagulation proteinases and cofactors is based loosely on the order in which they were discovered, with the exceptions of the first four components fibrinogen, prothrombin, tissue factor–phospholipid complex, and Ca^{2+}. The proteolytically activated forms of these factors are denoted by a subscript "a," for example, factor X_a. Additional nonproteolytic inhibitory mechanisms exist, the principal of which are the **serpin** antithrombin (an inhibitor of most of the Gla-proteinases) and tissue factor pathway inhibitor (TFPI), a Kunitz-type **proteinase inhibitor** of factors VII_a and X_a.

requiring proteolytically activated soluble cofactors, the other requiring integral **membrane protein** cofactors that do not require proteolytic activation.

REGULATION OF BLOOD CLOTTING BY COMPLEXES INVOLVING SOLUBLE PROTEIN COFACTORS

The only proteolytic reaction common to both pathways is the activation of prothrombin, the final serine proteinase zymogen of the coagulation cascade. Factor X_a (i.e., activated factor X) is the proteinase component of the "prothrombinase complex" that catalyzes the generation of thrombin. Both zymogen and active proteinase have an *N*-terminal "Gla domain," which contains **γ-carboxyglutamic acid** residues. The Gla domain is also present in prothrombin (as well as factors VII and IX, and protein C, the so-called vitamin K–dependent or Gla proteinases); it binds Ca^{2+} and participates in binding to negatively charged phospholipid surfaces, which *in vivo* are provided principally by activated platelets. Therefore, both **enzyme** and substrate bind to phospholipid surfaces, with **dissociation constants** of 0.1–1 μM (1). The cofactor of the prothrombinase complex is factor V_a, which is generated by thrombin-catalyzed limited proteolysis of factor V, providing one of the major feedback activation reactions. Factor V_a also binds to negatively-charged phospholipid surfaces, but with a much lower dissociation constant of ~ 5 nM (2). It does not contain a Gla domain, and binds through a different mechanism involving partial insertion of the protein into the lipid bilayer; it is often viewed as a membrane binding site for factor X_a. The latter thus incorporates into the complex by both protein–phospholipid and **protein–protein interactions**. Kinetic experiments suggest that the two phospholipid-bound components associate by lateral **diffusion**, but the interactions involved are complex and to some extent **cooperative**. Although prothrombin also binds to phospholipid, it is not clear whether the relevant substrate is phospholipid-bound prothrombin or prothrombin in the solution phase but at a high local concentration in the vicinity of the membrane surface, which can be several orders of magnitude above that in bulk solution. The kinetic effect of prothrombin binding to phospholipid is a reduction in the apparent **K_m (Michaelis constant)** for prothrombin activation ("apparent" as the K_m does not reflect the true concentration at the phospholipid–surface interface). This is of physiological significance, since the K_m falls from ~ 100 μM to < 1 μM (3), compared to a concentration of 1.4 μM in blood plasma. In addition to this effect, the catalytic rate constant, k_{cat}, for prothrombin activation is greatly increased, primarily as a consequence of the X_a–V_a interaction (Table 1). This suggests that the active-site

Table 1. Effect of Various Components of Prothrombinase Complex on Kinetic Parameters for Prothrombin Activation[a]

Components	K_m, μM	k_{cat}, s^{-1}	Relative Rates
X_a Ca^{2+} PL V_a	0.2	30	100
X_a Ca^{2+} PL	0.06	0.04	0.42
X_a Ca^{2+} V_a	34	6.2	0.12
X_a Ca^{2+}	84	0.01	0.00009
X_a	131	0.01	0.00005

[a] *Source*: Data taken from Rosing et al. (3).

environment of the proteinase is significantly affected by complex formation, although the molecular basis of this effect is not known. The prothrombinase complex also helps to achieve maximum catalytic advantage by ordering the sequence of the two proteolytic cleavages required for prothrombin activation (see **Thrombin**).

The phospholipid-bound complex of factor IX_a (proteinase) and factor VIII_a (cofactor) that activates factor X has molecular and functional characteristics very similar to those of the prothrombinase complex. Factor VIII shares homology with factor V and is also activated by thrombin; factor IX has the same domain structure as factor X and is also activated by factor VII_a. There is evidence that the phospholipid-binding properties of these proteinase–zymogen components enable the proteinase product of one complex to be transferred to the subsequent complex as its enzyme without dissociation from the membrane, thereby giving a further catalytic advantage (4).

REGULATION OF BLOOD CLOTTING BY COMPLEXES INVOLVING INTEGRAL MEMBRANE PROTEINS

Two integral membrane proteins with quite distinct and contrasting functions are involved in the regulation of blood coagulation: tissue factor and thrombomodulin, which are involved, respectively, in the initiation and termination of coagulation. Blood coagulation subsequent to injury of the vasculature is initiated by the exposure of tissue factor, a nonenzymatic cofactor widely expressed by cells of the subendothelium of blood vessels (although its expression can also be induced in endothelial cells and monocytes). Tissue factor binds factor VII_a, a Gla-proteinase, with high affinity (K_d = 1–5 nM) (5). The structure of this complex has been determined by **X-ray crystallography** (6); the extracellular part of tissue factor is composed of two homologous C2-type **immunoglobulin**-like modules, and the protein is most closely related to the IFNγ receptor, a member of the class 2 **cytokine** receptors (see **Interferon**). Factor VII_a is engaged in multiple contacts with both of the tissue factor modules.

The classical substrate for factor VII_a in this complex is factor X, but it is now known that factor IX is also a physiologically relevant substrate (7) (see **Hemophilia**). Activation of these substrates, and the subsequent incorporation of their activated forms into their cognate complexes, thus propagates the initial stimulus proteolytically. In contrast to factors V and VIII, tissue factor is constitutively active as a cofactor and does not require proteolytic processing. However, factor VII does need proteolytic activation for the complex to be active, and what initially catalyzes this cleavage and thus constitutes the initial proteolytic event in the coagulation cascade has not been a trivial problem to address, because of the extraordinary sensitivity of the system to proteolytic activation, and it remains a topic of controversy. Proposals have included (*1*) that factor VII is not a true zymogen and possesses a significant degree of intrinsic proteolytic activity, (*2*) induction of activity in factor VII by tissue factor, (*3*) autoactivation of factor VII, (*4*) and activation of factor VII by trace amounts of activated proteinases. Although not demonstrated unambiguously, mechanism 4 appears to be favored, as factor VII can be activated by factors VII_a, IX_a, X_a, and thrombin *in vitro*. These may be generated by the contact activation system, alternative mechanisms, or "leakage" of the system, but in such low amounts that they are insignificant in the absence of tissue factor. The principal activator of factor VII once the system is initiated is considered to be factor X_a (8). Tissue factor has little or no effect on factor VII activation, but assembly of the tissue factor complex increases factors IX and X activation by up to 10^4-fold (9). Both the kinetic mechanisms and the molecular interactions responsible for this effect are analogous to those described for the prothrombinase complex, which involves the optimal presentation of the zymogen substrate to the catalytically efficient proteinase:cofactor complex for maximum catalytic advantage.

The other integral membrane protein of the coagulation system, thrombomodulin, is expressed by endothelial cells and acts as a cofactor for the proteolytic activation of the negative regulator of the coagulation system, protein C (10). Activated protein C (a Gla-proteinase) proteolytically inactivates the procoagulant cofactors factors V_a and VIII_a. In common with the principal feedback activation reactions, thrombin is again involved. But in this case, rather than proteolytically activating the cofactor, it is the enzymatic component of the thrombomodulin complex; thus thrombin can fully modulate its own generation. The interaction with thrombomodulin dramatically alters the substrate specificity of thrombin, such that protein C activation is increased 10^4-fold, while its activity toward factor V and fibrinogen is diminished (1). Although protein C can bind phospholipid through its Gla domain, this interaction plays only a minor part in these effects and, because thrombin lacks a Gla domain, protein–protein interactions are the major determinants in thrombomodulin complex anticoagulant function, in contrast to the procoagulant complexes described above.

BIBLIOGRAPHY

1. C. T. Esmon, N. L. Esmon, and K. W. Harris (1982) *J. Biol. Chem.* **257**, 7944–7947.
2. S. Krishnaswamy and K. G. Mann (1988) *J. Biol. Chem.* **263**, 5714–5720.
3. J. Rosing, G. Tans, J. W. Govers-Riemslag, R. F. Zwaal, and H. C. Hemker (1980) *J. Biol. Chem.* **255**, 274–283.
4. K. G. Mann, M. E. Nesheim, W. R. Church, P. Haley, and S. Krishnaswamy (1990) *Blood* **76**, 1–16.
5. R. F. Kelley, K. E. Costas, M. P. O'Connell, and R. A. Lazarus (1995) *Biochemistry* **34**, 10383–10392.
6. D. W. Banner, A. D'Arcy, C. Chene, F. K. Winkler, A. Guha, W. H. Konigsberg, Y. Nemerson, and D. Kirchhofer (1996) *Nature*, **380**, 41–46.
7. K. A. Bauer, P. M. Mannucci, A. Gringeri, F. Tradati, S. Barzegar, B. L. Kass, H. ten Cate, A. S. Kestin, D. B. Brettler, and R. D. Rosenberg (1992) *Blood* **79**, 2039–2047.
8. S. Butenas and K. G. Mann, (1996) *Biochemistry* **35**, 1904–1910.
9. M. Zur and Y. Nemerson (1980) *J. Biol. Chem.* **255**, 5703–5711.

10. C. T. Esmon (1995) *FASEB J.* **9**, 946–955.

Suggestion for Further Reading

K. G. Mann, R. J. Jenny, and S. Krishnaswamy (1988) Cofactor proteins in the assembly and expression of blood clotting enzyme complexes. *Annu. Rev. Biochem.* **57**, 915–956.

BLOT OVERLAYS

J. M. GERSHONI

Regardless of whether a blot contains immobilized **DNA**, **RNA** or **protein** (see **Blotting**), it is ultimately reacted with a specific probe that defines the type of **ligand-binding** interaction being studied (see **Southern blots**, **RNA blots**, and **Protein blots**). **Radiolabeled** probes are detected by **autoradiography**, and the radioactive complexes are quantified by computer-digitized imaging or simply by excising out the bands and "counting" the radioactivity. Often the primary probe itself is not easily detectable, so a secondary probe is used, such as an **antibody** against the first probe conjugated to an **enzyme**, or **avidin** or **streptavidin** when the primary probe is biotinylated (see **Avidin-biotin system**). When using an enzyme-conjugated probe, one must ensure that the quencher or blocking reagent used on the blot has no interfering enzymatic activity. For example, **hemoglobin** should not be used when horseradish **peroxidase** is employed as the detection system.

DNA PROBES

Traditionally, Southern and RNA blots of DNA and RNA, respectively, are probed with single-stranded DNA or double-strand DNA that has been **denatured**. The probes are usually radioactive, so that any duplex formed upon **hybridization** of the probe to a nucleic acid on the blot is detected by autoradiography. The rationale of such experiments is to identify the position of the immobilized nucleic acid in the electrophoretogram that is complementary to the sequence of the probe, so that they hybridize by **Watson-Crick base pairing**. The conditions for hybridization, the *stringency*, are regulated according to the degree of homology and complementarity between the probe and the target (1,2). DNA is also used to probe protein blots, a procedure that has been named "Southwestern blotting." As expected, this type of experiment is designed to reveal specific interactions between defined DNA sequences and the **DNA-binding proteins** that bind them, such as **transcription factors** (3). DNA probing of *colony* or *plaque* blots is a routine approach in recombinant DNA **cloning** (4).

IMMUNOBLOTTING

The original application of protein blots was for identifying a particular **antigen** in a protein gel pattern by probing the blot with its corresponding **antibody** (5). Detecting the immunocomplex is often accomplished by using a secondary probe, for example, a goat antimouse IgG conjugated to an enzyme, such as horseradish peroxidase, when a murine **monoclonal antibody** is used for the primary probe. The sensitivity of these assays is increased by combining **immunoprecipitation** prior to **gel electrophoresis**. Thus, a crude sample of proteins is first immunoprecipitated with relevant antibodies, and the precipitated proteins are subsequently resolved by electrophoresis, blotted, and probed with a monoclonal antibody of interest. The use of antibodies to probe colony or plaque blots is very effective in screening **expression libraries** (6).

LECTIN BLOTTING

To identify **glycoproteins**, **lectins** can be used to probe protein blots (7). A radioactive or enzyme-conjugated lectin is employed as the probe. A particular case of interest is the fact that the enzyme horseradish peroxidase is itself a glycoprotein. Thus, for example, a blot can be probed with the mannose-specific lectin concanavalin *A*, washed and further incubated with horseradish peroxidase directly (8). The multivalent concanavalin A binds to blotted glycoproteins containing mannose and subsequently also binds to the horseradish peroxidase without the need for chemical conjugation. When using lectins, ensure that the quencher used does not itself contain sugar.

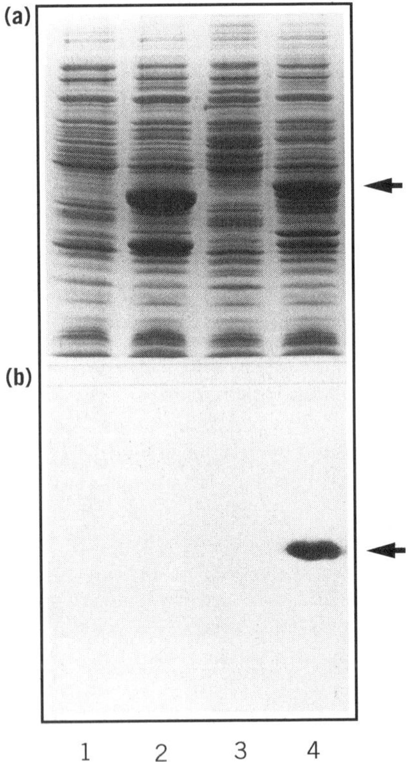

Figure 1. Neurotoxin overlays of protein blots. *Escherichia coli* transformed with pATH2 expression **vectors** containing DNA inserts corresponding to the neurotoxin binding site (lanes 3 and 4) or not (lanes 1 and 2) were induced for gene expression (lanes 2 and 4) or not (lanes 1 and 3). Then samples of all four treatments were resolved by SDS-PAGE and stained with **Coomassie brilliant blue** (**a**) or blotted onto nitrocellulose membrane filters and probed using radioactive neurotoxin. Then the washed filters were autoradiographed (**b**). The arrows indicate the position of the fusion protein containing the toxin binding site.

LIGAND BLOTTING

Blotting is usually considered for detecting **antigens** or nucleic acid hybrids, but it is also a very powerful way to identify all sorts of protein–protein interactions, even interactions involving nonpeptide ligands. Thus when studying any **receptor**, one should consider probing protein blots with any corresponding ligands that are detectable (9). In such experiments, it is usually advisable not to boil the protein sample or subject it to **disulfide bond** reduction before electrophoresis. Furthermore, the blot can be incubated in solutions that promote **renaturation** of the immobilized proteins. Omission of methanol from the transfer buffers used in electroblotting is also helpful for retaining functional conformations of the blotted protein. Ligand blotting has been used successfully to identify peptides that bind to **hormones**, **cytoskeletal** components, **neurotoxins**, nucleotides, **calmodulin**, and even ions such as Ca^{2+} (9–11) (Fig. 1).

CELL BLOTTING

Blots have even been probed with intact cells, which has proven useful for identifying interactions between proteins and cells (12). Furthermore, bacteria can be used to probe a blot, and the interaction with a protein is detected by subsequently allowing the bacteria to grow on the surface of the blot, so that colonies are observed at the site of the immobilized protein (13). **Viruses** have also been used to probe blots to detect their corresponding receptor proteins.

In summary, one should consider probing blots with selected ligands any time bimolecular interactions are to be analyzed. Surveying literature **databases** using specific conjugations, such as "ligand-blot", "calmodulin-blot" or "cell-blot," as key words usually produces results that provide the imaginative investigator with starting points from which to proceed.

BIBLIOGRAPHY

1. J. Meinkoth and G. Wahl (1984) *Anal. Biochem.* **138**, 267–284.
2. L. G. Davis, M. D. Dibner, and J. F. Battey (1986) *Basic Methods in Molecular Biology,* Elsevier, New York, pp. 62–65.
3. C. N. Flytzanis (1994) In *Protein Blotting: A Practical Approach* (B. S. Dunbar, ed.), IRL Press, Oxford, UK, pp. 163–168.
4. Ref. 2 pp. 227–229.
5. H. Towbin and J. Gordon (1984) *J. Immunol. Methods* **72**, 313–340.
6. T. V. Huynh, R. A. Young, and R. W. Davis (1986) In *DNA Cloning, A Practical Approach*, Vol. 1 (D. M. Glover, ed.), IRL Press, Oxford, UK, pp. 49–78.
7. S. Bar-Nun and J. M. Gershoni (1994) In *Cell Biology; a laboratory handbook*, Vol 3 (J. E. Celis, ed.), Academic Press, San Diego, pp. 323–331.
8. J. C. S. Clegg (1982) *Anal. Biochem.* **127**, 389–394.
9. J. M. Gershoni (1988) *Methods of Biochemical Analysis* **33**, 1–58.
10. P. Hossenlopp and M. Binoux (1994) In *Protein Blotting: A Practical Approach* (B. S. Dunbar, ed.), IRL Press, Oxford, UK, pp. 169–188.
11. A. Vieira, R. G. Elkin, and K. Kuchler (1994) In *Cell Biology: A laboratory handbook*, Vol. 2 (J. E. Celis, ed.), Academic Press, San Diego, pp. 314–321.
12. E. G. Hayman, E. Engvall, E. A'Hearn, D. Barnes, M. Pierschbacher, and E. Rouslahti (1982) *J. Cell Biol.* **95**, 20–23.
13. J. M. Gershoni (1987) Protein blotting: a tool for the analytical biochemist, *Adv. Electrophoresis* **1**, 141–176.

BLOTTING

J. M. GERSHONI

Blotting is a method in which a **macromolecule** is immobilized on a blotting matrix and subsequently probed with a detectable ligand to determine whether the macromolecule binds that specific **ligand**. The immobilized macromolecule can be **DNA**, **RNA** or protein, in which case one generates DNA blots (**Southern blots**), **RNA blots** (**Northern blots**) (1), or **protein blots** (**Western blots**) (2,3). Blots of lipids have also been produced (4). The macromolecule can be applied to the blotting matrix directly (dot blot), or it can be derived and eluted from an electrophoretic gel (gel blot) or even from a **bacterial** colony or **bacteriophage** plaque (colony blot).

TYPICAL BLOT ANALYSIS

The most common application of blotting involves a complex mixture of DNA, RNA, or protein to be resolved by using a standard **gel electrophoresis** procedure, such as **agarose** gel separation of DNA fragments or RNA or **SDS-PAGE** of protein samples. After electrophoresis, the gel is dismantled from its cassette or glass plates, etc., and a sheet of an appropriate blotting matrix (eg, a **nitrocellulose** membrane filter; see **Blotting matrices**) is cut to size and applied to the surface of the gel. The transfer of the resolved polynucleotides or peptides is accomplished via a procedure called "blotting", and the blotted macromolecules adsorb to the surface of the matrix while retaining their relative positions, thus creating a faithful replica of the original electrophoretic pattern. The "blot" thus produced is subsequently incubated with a ligand probe, which might be **radioactive** for detection via **autoradiography** or conjugated to an **enzyme** whose activity is detectable (see **Blot overlays**). Extensive washing of the blot removes the excess probe from that which is specifically associated with the immobilized macromolecule and remains bound. Subsequent detection of the retained ligand in the complex formed identifies the relevant bimolecular interaction.

METHODS FOR BLOTTING

Dot Blotting

This is the simplest method of applying a sample to be tested (5,6). The sample can be an unfractionated polynucleotide or protein mixture in solution. A small volume (typically 2 to 5 μl) is applied directly to the surface of a dry blotting matrix by micropipetting the sample onto the matrix or by using commercial vacuum manifolds that enable the application of larger sample volumes (e.g., 100 μl to 1 ml) by filtration. Such manifolds often create focused and uniform dots or thin slots of sample, thus leading to the terms "dot blots" or "slot blots", respectively. Typically, an 8 cm × 12 cm piece of blotting matrix contains as many as 96 different dots that correspond to the geometry of a standard 96-well **ELISA** plate. Once created, the dot blot is processed and probed as any other blot (see **Southern blots**, **RNA blots**, **Protein blots**, and **Blot overlays**). The advantages of the dot blot procedures are that they do not require any separation process and thus do not subject the sample to undue chemical modifications that could, for example, **denature** a protein sample (although some denaturation of protein occurs upon adsorption to the matrix). Moreover, dot blots are simple, cheap and quick. Where quantification is intended, direct dot blotting ensures maximal yields of sample recovery (see **Filter hybridization assays**). Obviously, however, chromatography or electrophoresis is necessary to resolve a complex sample to ascribe the signal to a specific component.

Gel Blots

Gel electrophoresis is routinely employed to resolve macromolecules of DNA, RNA, or protein. Agarose gels and **polyacrylamide gels** can be blotted, and the common goal is to elute the "bands" efficiently from the gel to be immobilized on the surface of the blotting matrix, so as to generate a faithful replica of the electrophoretic pattern. This can be accomplished in a number of ways:

1. Diffusion blotting simply relies on the fact that the macromolecules in the gel spontaneously **diffuse** out of the gel (7). Consequently, blots are produced when a blotting matrix is simply applied to one or both sides of the gel. This approach is usually time-consuming (24 to 72 h) and of low efficiency, but it produces two equal copies simultaneously.
2. Convection blotting (also called capillary transfer) is the process of eluting the resolved macromolecules by mass flow of buffer through the gel. This is the original procedure introduced by Edwin Southern for transferring DNA restriction fragments from agarose gels onto nitrocellulose membrane filters (thus the term Southern blotting) (8). In this method, the gel is placed on top of a paper wick, which draws buffer from a reservoir. The gel is covered with a piece of blotting matrix that, in turn, is covered by a stack of paper towels or absorbent paper and a weight that ensures uniform pressure over the surface of the gel. The stack of paper towels draws a continuous flow of buffer vectorially through the gel, and with it the polynucleotides or peptides, which are eluted and deposited on the surface of the blotting matrix (9).
3. Vacuum blotting is an elaboration of blotting by convection in which the flow is accelerated by employing negative pressure (i.e., suction). Positive pressure has also been used to enhance blotting.
4. Electroblotting is achieved by applying an electric field so as to elute the proteins or polynucleotides from their corresponding gels by **electrophoresis**. This technique has been used primarily for proteins because binding DNA and RNA to nitrocellulose requires high salt conditions that are incompatible with electroblotting. Electroblotting of nucleic acids is possible, however, with alternative blotting matrices, such as nylon membranes. A variety of commercial apparatus equipment and home-made systems are available, including those that generate gradient electric fields and provide different elution efficiencies to compensate for differences in the molecular mass of the polymers to be eluted (10,11). Electroblotting is performed by using tank systems, which require 2 to 4 liters of transfer buffer, or semidry blotting systems, which conserve buffer and are flexible because different buffers can be employed for the anode and cathode (12).
5. Direct blotting is an elaboration of electroblotting in which a conveyor belt passes the blotting matrix by the exposed bottom of the polyacrylamide gel during electrophoresis. As the proteins or DNA fragments reach the bottom of the gel, they are deposited directly onto the slowly moving blotting matrix and generate a blot. The resolution of the bands is modulated by regulating the conveyor belt speed (13).

Colony Blots

At times it is necessary to identify a specific bacterial colony or phage that contains DNA of interest or expresses a particular protein. **Libraries** of **recombinant DNA**-containing bacteria or **lambda phage expression libraries** are used to produce colony or plaque blots to be processed and screened like any other blot. The library of bacteria or phage is plated on agar after suitable incubation, and replicas are produced by simply placing a sheet of blotting matrix, such as nitrocellulose membrane or nylon membrane, onto the colony- or plaque-containing surface. The matrices pick up sufficient material from each colony or plaque at precisely the same relative position corresponding to that on the original agar plate. Then the colony blots are processed, for example, by standard DNA filter hybridization, immunoblotting, or ligand blotting (14,15) (Fig. 1).

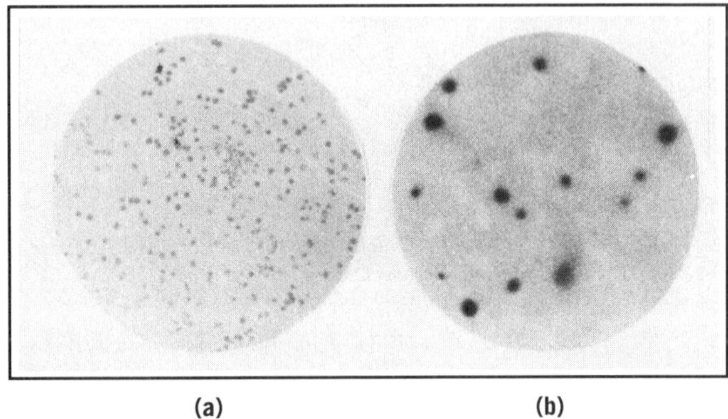

Figure 1. Bungarotoxin overlay of bacterial colonies. *Escherichia coli* were transformed with a pATH2 expression **vector** containing a DNA insert coding for a fragment of the α subunit of the nicotinic **acetylcholine receptor**. This fragment is responsible for the receptor's binding of the neurotoxin α**-bungarotoxin**. The bacteria were plated to a density of approximately 200 colonies per agar plate (**a**) and expression was induced. A replica of the plate was produced by using a nitrocellulose filter disc, which was processed for overlay with radio-**iodinated** α-bungarotoxin. The filter was subsequently washed and autoradiographed, illustrating those colonies that contain the relevant DNA fragment (**b**) (for details see Ref 15).

BIBLIOGRAPHY

1. J. Meinkoth and G. Wahl (1984) *Anal. Biochem.* **138**, 267–284.
2. J. M. Gershoni and G. E. Palade (1983) *Anal. Biochem.* **131**, 1–15.
3. H. Towbin and J. Gordon (1984) *J. Immunol. Methods* **72**, 313–340.
4. T. Taki, S. Handa, and D. Ishikawa (1994) *Anal. Biochem.* **221**, 312–316.
5. J. M. Gershoni (1988) *Methods Biochem. Anal.* **33**, 1–58.
6. L. G. Davis, M. D. Dibner, and J. F. Battey (1986) *Basic Methods in Molecular Biology*, Elsevier, New York, pp. 147–149.
7. B. Bowen, J. Steinberg, U. K. Laemmli, and H. Weintraub (1980) *Nucleic Acids Res.* **8**, 1–20.
8. E. M. Southern (1975) *J. Mol. Biol.* **98**, 503–517.
9. Ref. 6 pp. 62–65.
10. M. Bittner, P. Kupferer, and C. F. Morris (1980) *Anal. Biochem.* **102**, 459–471.
11. J. M. Gershoni, F. E. Davis, and G. E. Palade (1985) *Anal. Biochem.* **144**, 32–40.
12. G. Jacobson (1994) In *Protein Blotting: A Practical Approach* (B. S. Dunbar, ed.), IRL Press, Oxford, UK, pp. 53–72.
13. S. Beck (1993) *Methods Mol. Biol.* **23**, 219–223.
14. Ref. 6 pp. 185–189.
15. J. M. Gershoni (1987) *Proc. Natl. Acad. Sci USA* **84**, 4318–4321.

BLOTTING MATRICES

J. M. GERSHONI

Blotting is the process of transferring macromolecules from **electrophoretic** gels to immobilizing matrices, called blotting matrices. The variety of matrices available for blotting is quite diverse, and the characteristics of each directly affects the ultimate result and quality in different blot analyses [see **Blotting**, **Southern blots**, **Protein blots (Western blots)**, **RNA blots (Northern blots)**, and **Blot overlays**]. Generally, two kinds of blot matrices are used: (1) chemically modified paper filters and (2) microporous membrane filters. Although filters are used, blotting depends on chemical adsorption of transferred molecules to the filter material itself, rather than filtration per se, where separation is achieved by size exclusion. Thus porosity is less important than the chemical composition, density, and thickness of the material. After binding of the desired ligand, normally the remaining binding sites on the matrix are blocked with a neutral compound, known as a quencher or blocking reagent.

PAPER FILTERS

Initially, it was thought that transferred molecules would be best suited for blotting if they were covalently bound to the blotting matrix. This led to chemically modified paper filters that contained active moieties to bind DNA, RNA, and protein covalently. For example, **cyanogen bromide** (CNBr)-activated paper was produced (1), as was diazobenzyloxymethyl (DBM) paper, which was the most popular filter of this type (2). These filters covalently immobilize the blotted macromolecules, but they are cumbersome to handle, have the fibrous texture characteristic of blotting paper, and therefore are rarely used today.

MEMBRANE FILTERS

Microporous membrane filters have become the matrices of choice for blotting. These materials are thin films of synthetic polymers with very fine, uniform surfaces. Filters with average porosities of 0.2 to 0.45 μm are extremely suitable for all types of blotting. A variety of materials exist, and each offers unique advantages.

Nitrocellulose

Nitrocellulose is by far the most commonly used blotting matrix. It binds DNA, RNA, and protein reasonably well, although the mechanism is not clear (3,4). **Hydrophobic** interactions definitely play a role, and the salt conditions are critical, particularly for the adsorption of RNA. Nitrocellulose is often produced as a mixed ester with cellulose acetate, which reduces to some extent its binding capacity for proteins. The advantages of nitrocellulose are that it binds protein well and affords very good signal-to-background ratios in Western blotting assays. It allows staining of the immobilized protein patterns with such dyes as **Ponceau S** and Amido black. It becomes brittle, however, after baking at 80°C (typical for Southern and Northern blotting), which reduces the repeated use of these filters. Furthermore, nitrocellulose is less resistant to various organic solvents, such as methanol (5).

Nylon Membranes

Nylon membranes were introduced initially as an alternative to nitrocellulose for protein blotting (5), and subsequently they were applied in RNA and DNA transfers (4). These membranes are usually derivatives of nylon 66 and often are modified with positive charges. They have proven exceptional binders of DNA, RNA, and protein, but at times this presents some difficulty by producing high backgrounds, particularly in protein blotting. These filters are mechanically stable and thus can be reprobed numerous times without lossing band definition or sensitivity.

Polyvinyl Difluoride

Polyvinyl difluoride (PVDF) membrane filters have a special application in blotting where subsequent chemical manipulation of the immobilized macromolecule is desired. A case in point is the use of these membranes in protein blots, where the individual bands are excised out of the blot and subjected to **Edman degradation** for N-terminal amino acid sequencing of the resolved and blotted polypeptide chains (6,7).

BIBLIOGRAPHY

1. L. Clarke, R. Hitzman, and J. Carbon (1979) *Methods Enzymol.* **68**, 436–442.
2. J. C. Alwine, D. J. Kemp, and G. R. Stark (1977) *Proc. Natl. Acad. Sci. USA* **74**, 5350–5354.
3. A. De Maio (1994) In *Protein Blotting: A Practical Approach* (B. S. Dunbar, ed.), IRL Press, Oxford, UK, pp. 11–32.
4. J. Meinkoth and G. Wahl (1984) *Anal. Biochem.* **138**, 267–284.
5. J. M. Gershoni and G. E. Palade (1982) *Anal. Biochem.* **124**, 396–405.
6. M. A. Mansfield (1994) In *Protein Blotting: A Practical Approach* (B. S. Dunbar, ed.), IRL Press, Oxford, UK, pp. 33–52.

7. C. Eckerskorn and F. Lottspeich (1993) *Electrophoresis* **14**, 831–838.

BLUNT-END LIGATION

DAVID B. WILSON

Blunt-end ligation is covalently joining two or more double-stranded **DNA** fragments with flush ends by the enzyme **DNA ligase** (1). Many **restriction enzymes** produce blunt ends, when they cleave both polynucleotide chains at the same site, and some **cohesive ends** are converted to blunt ends by the **fill-in reaction.** Blunt end ligation is a much less efficient reaction than ligation of DNA fragments with cohesive ends. One technique for overcoming the low efficiency of blunt-end ligation is ligating a short oligonucleotide **linker fragment** to the DNA ends that are to be joined (2). Because a large molar concentration of the linker can be added to the reaction, its ligation to the ends of the DNA occurs more readily than joining of the ends themselves. Linkers are designed to contain a unique restriction site so that the ligation mixture is digested with the corresponding enzyme to create cohesive ends, which can be joined by conventional ligation.

The efficiency of blunt-end ligation is greatly increased by carrying out the reaction in the presence of 15% polyethylene glycol, 6,000 MW (**PEG**), although only linear molecules are produced in this reaction. Circular **plasmids** are required for efficient **transformation** of *Escherichia coli*. An efficient way of producing circular molecules from the linear product of blunt-end ligation is including a bacteriophage P1 lox **recombination** site in one of the molecules being ligated (3). After ligation, the DNA is treated with Cre recombinase, which by recombination between the lox sites produces a circular molecule from a linear molecule that contains two lox sites. This type of molecule is created when the molecule without a lox site is ligated to two molecules with a lox site, one at each end.

It is possible to amplify the products of a low-efficiency ligation containing the desired insert by **PCR**, using primers that bind to vector DNA on each side of the cloning site. The PCR product of the expected size is purified and recloned. This results in transformants from ligations that give no positive colonies on direct transformation (4).

BIBLIOGRAPHY

1. T. Maniatis, E. F. Fritsch, and J. Sambrook (1982) *Molecular Cloning; a Laboratory Manual*, 2nd ed., Cold Spring Harbor Press, Cold Spring Harbor, NY.
2. G. J. Bhat, M. J. Lodeg, P. J. Myler, and K. D. Stuart (1991) *Nucleic Acids Res.* **19**, 398.
3. A. C. Boyd (1993) *Nucleic Acids Res.* **21**, 817–821.
4. H. W. Son and E. Lolis (1995) *BioTechniques* **18**, 644–650.

Suggestion for Further Reading

A. E. T. Konson and D. S. Levin (1997) Mammalian DNA ligases, *Bioessays* **19,** 893–901; a review of DNA ligases.

BOHR EFFECT

K. IMAI

Bohr et al. (1) discovered that the oxygen dissociation curve of blood shifts to higher partial pressures of oxygen in the presence of increasing concentrations of carbon dioxide by lowering the oxygen affinity of **hemoglobin**. This effect, it was concluded, results from the concomitant decrease in pH, and the effect of pH on the oxygen affinity of hemoglobin has subsequently been known as the Bohr effect. It was subsequently discovered that CO_2 also lowers the oxygen affinity of hemoglobin (Hb) directly by binding to its α-amino groups, forming carbamino compounds (2,3):

$$O_2 - Hb - NH_2 + CO_2 \leftrightarrow Hb - NHCOOH + O_2 \quad (1)$$

The original observations of Bohr et al. (1) are now known as the "classical Bohr effect," which is a composite of the specific effects of H^+ and CO_2.

The CO_2 content of blood is reduced upon oxygenation of hemoglobin at a constant CO_2 pressure (4). This phenomenon, known as the "classical Haldane effect," is a combination of the oxygen-linked dissociation of the carbamino groups (Eq. 1) and release of protons

$$Hb(H^+)_\delta + O_2 \leftrightarrow HbO_2 + \delta\ H^+ \quad (2)$$

followed by dehydration of bicarbonate:

$$H^+ + HCO_3^- \leftrightarrow H_2CO_3 \leftrightarrow CO_2 + H_2O \quad (3)$$

The release of protons upon oxygenation or their uptake upon deoxygenation of hemoglobin is known as the "Haldane effect."

The Bohr effect and the Haldane effect are thermodynamically equivalent (5) because they are linked functions (6). Whatever effect oxygen binding has on the affinity of hemoglobin for protons, changes in the pH must have the same effect on the affinity of hemoglobin for oxygen:

$$\left(\frac{\partial H^+}{\partial Y}\right)_{pH} = \left(\frac{\partial \log P_{O_2}}{\partial pH}\right)_Y \quad (4)$$

where H^+ is the number of protons bound per heme group, Y is the fraction of hemoglobin binding sites occupied by O_2, and P_{O2} is the partial pressure of O_2. The left-hand side of Eq. 4 is known as the "Haldane coefficient," and the right-hand side is the "Bohr coefficient." They give the magnitude of the Haldane and Bohr effects, which must be the same.

These effects are physiologically important for efficient and regulated transport of O_2 and CO_2 in opposite directions (see **Hemoglobin**). When the shape of the oxygen dissociation curve expressed by Y versus log P_{O2} is independent of changes in pH, at least within the range $0.1 < Y < 0.9$, Eq. 4 is simplified to

$$\delta = \frac{\Delta \log P_{50}}{\Delta pH} \quad (5)$$

where P_{50} is the P_{O2} at $Y = 0.5$ and δ is the Bohr coefficient, which is independent of the degree of O_2 binding. The Bohr coefficient expresses the number of protons (per heme group) bound to hemoglobin upon full oxygenation and is the same as δ of Eq. 2.

δ is negative above pH 6.3, and protons are released upon oxygenation of hemoglobin, whereas δ is positive at lower pH values, and protons are taken up upon oxygenation. These two phenomena are known respectively as the "alkaline" and "acid" Bohr effects. At physiological pH 7.4, human adult hemoglobin

A (HbA) has $\delta = -0.6$ in the presence of 0.1 *M* NaCl. This value is halved in the absence of Cl^-.

The Bohr effect arises from changes in the pK_a values of particular ionizable groups of hemoglobin as a result of changes in their environment upon binding oxygen. The group primarily responsible for the alkaline Bohr effect of human HbA is the imidazole side chain of the C-terminal His146 of the β chain. Its pK_a value changes from 8.0 to 7.1 upon oxygenation, contributing the greater part of the alkaline Bohr effect in the absence of Cl^-. In the presence of Cl^-, binding of these ions to several ionizable groups that line the central cavity of the hemoglobin molecule, including the α-amino groups of the α chains and the ε-amino groups of Lys82 of the β chains, affects the protonation of those groups. Consequently, the release of Cl^- ions upon binding of oxygen produces an additional, Cl^--dependent Bohr effect (7). It is thought that His143 of the β chains is one of the groups responsible for the acid Bohr effect (8), but the others are still unidentified.

The term "Bohr effect" is sometimes used for pH-dependent binding of other ligands, such as CO_2 and alkylisocyanides, to hemoglobin. It is also used for pH-dependent ligand binding to **oxygen-binding proteins** other than hemoglobin.

BIBLIOGRAPHY

1. C. Bohr, K. A. Hasselbalch, and A. Krogh (1904) *Skand. Arch. Physiol.* **16**, 402–412.
2. J. K. W. Ferguson and F. J. W. Roughton (1934) *J. Physiol.* (London) **83**, 87–102.
3. J. V. Kilmartin and L. Rossi-Bernardi (1969) *Nature* **222**, 1243–1246.
4. J. Christiansen, C. C. Douglas, and J. S. Haldane (1914) *J. Physiol.* (London) **48**, 244–277.
5. I. Tyuma and Y. Ueda (1975) *Biochem. Biophys. Res. Commun.* **65**, 1278–1283.
6. J. Wyman (1964) *Adv. Protein Chem.* **19**, 223–286.
7. M. F. Perutz, G. Fermi, C. Poyart, J. Pagnier, and J. Kister (1993) *J. Mol. Biol.* **233**, 536–545.
8. M. F. Perutz, J. V. Kilmartin, K. Nishikura, J. H. Fogg, P. J. G. Butler, and H. S. Rollema (1980) *J. Mol. Biol.* **138**, 649–670.

Suggestions for Further Reading

J. V. Kilmartin and L. Rossi-Bernardi (1973) Interaction of hemoglobin with hydrogen ions,carbon dioxide, and organic phosphates, *Physiol. Rev.* **53**, 836–890.

I. Tyuma (1984) The Bohr effect and the Haldane effect in human hemoglobin, *Jpn. J. Physiol.* **34**, 205–216.

BOWMAN–BIRK INHIBITORS

M. LASKOWSKI JR.

Bowman–Birk proteinase inhibitor is a member of a widely studied family of serine proteinase inhibitors, protein [see **Serine proteinase inhibitors**]. Most such inhibitors consists of a single polypeptide chain of 65–80 amino acid residues containing two homologous regions, each with a reactive site. In some Bowman–Birk inhibitors, both sites are specific for trypsin; in others, one is specific for trypsin and one for chymotrypsin; and, in still others, one is for trypsin and one for elastase. Ternary complexes of one inhibitor and two enzyme molecules are commonly found. The polypeptide chain is crosslinked by seven disulfide bridges, some within the homology regions, others between them, endowing the inhibitor with great stability.

Suggestion for Further Reading

T. Ikenaka and S. Norioka (1986) Bowman–Birk family serine proteinase inhibitors. In *Proteinase Inhibitors* (A. Barrett and G. Salveson, eds.), Elsevier New York, 361–374.

BPTI (BOVINE PANCREATIC TRYPSIN INHIBITOR)

M. LAKOWSKI JR.

Bovine pancreatic trypsin inhibitor, BPTI (Kunitz), is also called trypsin kallikrein inactivator, TKI, aprotinin, and Trasylol™. It is the first protein inhibitor of proteinases [see **Proteinase inhibitors, protein**] to be isolated and characterized by Moses Kunitz as an inhibitor of bovine trypsin. It was independently discovered as a kallikrein inactivator. BPTI was the first protein proteinase inhibitor to be sequenced. Shortly afterward, sequencing the kallikrein inactivator revealed the identity of the two inhibitors. They consist of 58 amino acid residues, crosslinked by three disulfides. Lys^{15} is at the P_1 position of the reactive site. The K_I for BPTI-bovine β trypsin interaction, 5×10^{-14} M, is often—probably incorrectly—referred to as the strongest of the proteinase–protein proteinase inhibitor interactions. BPTI is also extremely stable. Its T_m at neutral pH is 103°C. BPTI was the first protein proteinase inhibitor to have its three-dimensional structure determined both in free form and in complex with bovine β trypsin. It has been—and still is—widely used in Europe as a drug to avoid postsurgical complications. For this use, it is isolated from bovine lungs. BPTI has been the first superbly characterized small protein that was widely available. Therefore, it became the favorite substance of many protein researchers. It was the first protein to have its three-dimensional structure determined by nuclear magnetic resonance (NMR). It served as the object of numerous molecular dynamics and structure prediction calculations. It was the first protein for which the pattern of closure of disulfide bridges was studied in detail. BPTI (Kunitz) gave its name to a widely studied family of standard-mechanism canonical protein inhibitors of serine proteinases [see **Serine proteinase inhibitors**].

Suggestion for Further Reading

W. Gebhard, H. Tschesche, and H. Fritz (1986) Biochemistry of aprotinin and aprotinin-like inhibitors. In *Proteinase Inhibitors* (A. Barrett and G. Salvesen, eds.), Elsevier New York, 375–388.

BRAGG ANGLE

JAN DRENTH

The Bragg angle is important for analyzing the diffraction from crystals in **X-ray crystallography**. W.L. Bragg noted that X-ray diffraction by a crystal can be understood without going into the details of diffraction theory (1). The process is similar to ordinary reflection of light from a mirror. In the crystal, the mirrors are the lattice planes. They are the planes constructed through the lattice points, which are the corners of the **unit cells**. Within a set, the planes are parallel and equidistant with perpendicular distance d (Fig. 1). θ, the reflective angle, is called the Bragg angle. The beams reflected from the upper and the lower lattice plane are in phase and reinforce each other if the difference in path lengths of the two beams, which is given by the path A–B–C, is an integral number of the wavelength λ. The path difference is also equal to $2d \sin \theta$. This results in Bragg's law: $2d \sin \theta = n\lambda$, where n is an integer, 1, 2 3, etc.

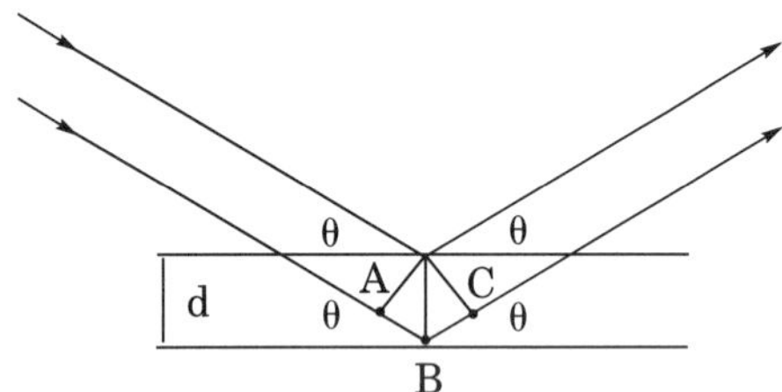

Figure 1. Bragg's view on X-ray diffraction. The incident beam is reflected by lattice planes. d is their distance, and θ is the reflective angle.

BIBLIOGRAPHY

1. W. L. Bragg (1913) *Proc. Cambridge Phil.Soc.* **17**, 43–57.

Suggestions for Further Reading

J. P. Glusker, M. Lewis, and M. Rossi (1994) *Crystal Structure Analysis for Chemists and Biologists*, VCH Publishers, New York, Weinheim, Cambridge.

J. Drenth (1999) *Principles of Protein X-ray Crystallography*, Springer, New York.

BRASSINOSTEROIDS

DOMINIQUE VAN DER STRAETEN
SOPHIE BERTRAND
MARC VAN MONTAGU

HISTORY

Unlike the five other "classical" **plant hormones**, which were all discovered more than 35 years ago, brassinosteroids (BRs) have been identified only recently. In 1979, Grove et al. (1) reported on the first growth-promoting steroid from plants. The substance was isolated from *Brassica napus* (rapeseed) pollen and was called *brassinolide*. Brassinolide was found in minute quantities in pollen grains; about 250 kg of pollen were needed to obtain only 10 mg of the pure compound. BRs are widely present in angiosperm families, but are also found in gymnosperms, ferns, and green algae (2).

Recently, BRs have gained recognition as true plant hormones (3). Genetic experiments provided evidence that they are indispensable for normal plant growth. A second breakthrough in our knowledge of BRs owes to the recent elucidation of the entire biosynthetic pathway (2,4) (Fig. 1).

BIOSYNTHESIS AND METABOLISM

Brassinosteroids are unique as plant hormones because they are structurally related to the insect molting hormone **ecdysone** and to mammalian **steroid hormones**. More than 40 BRs occur naturally, all of which share a common 5α-cholestane skeleton. The relative activities of the BRs increase according to their position in the biosynthetic pathway: brassinolide is most active, whereas early precursors up to 6-oxocampestanol are virtually inactive (5). Hydroxylation at position C22 is critical for biological activity.

Brassinolide is synthesized from campesterol via two pathways (2) (Fig. 1). These pathways were established by classical labeled-precursor feeding experiments. Campesterol is derived essentially from the mevalonate pathway. Initially, the double bond in the B ring of campesterol is reduced to campestanol (6) over 3-hydro-Δ4,5-campesterol and 3-dehydro-campestanol. From campestanol, two possible routes lead to brassinolide. In this branch, campestanol can be oxidized on C6 to yield 6-oxocampestanol, with 6α-hydroxycampestanol as an intermediate.

The first branch is called the early C6-oxidation pathway and seems ubiquitous in higher plants. 6-Oxocampestanol is subsequently hydroxylated at position 22 on the side chain (7), resulting in cathasterone, and further hydroxylated at position 23, to yield teasterone (4). This step is followed by epimerization of the hydroxyl function on C3, yielding typhasterol, with 3-dehydroteasterone as an intermediate (8,9). Typhasterol is further hydroxylated on C2, resulting in castasterone that is finally converted to brassinolide by lactamization of the B ring (10). Two steps in the early C6-oxidation pathway appear to be rate-limiting: (a) the formation of cathasterone from 6-oxocampestanol and (b) the final conversion of castasterone to brassinolide (2).

A second possible route from campestanol to castasterone is the so-called late C6-oxidation path, with 6-deoxo brassinosteroids as intermediates. The latter compounds are the least active BRs and were therefore considered to be dead-end products. The second biosynthetic path has been discovered only recently; it was demonstrated in cultured cells and seedlings of *Catharanthus roseus*, as well as in seedlings of rice and tobacco (11,12). Recent evidence indicates that 6-deoxoteasterone is formed from campestanol via 6-deoxocathasterone, by subsequent hydroxylations (7). This step is followed by an epimerization to yield 6-deoxotyphasterol, comparable to the conversion from teasterone to typhasterol in the early oxidation pathway. Finally, 6-deoxotyphasterol is hydroxylated at position 2, forming castasterone.

Compelling evidence for an essential role for BRs in plant growth and development came from recent genetic studies

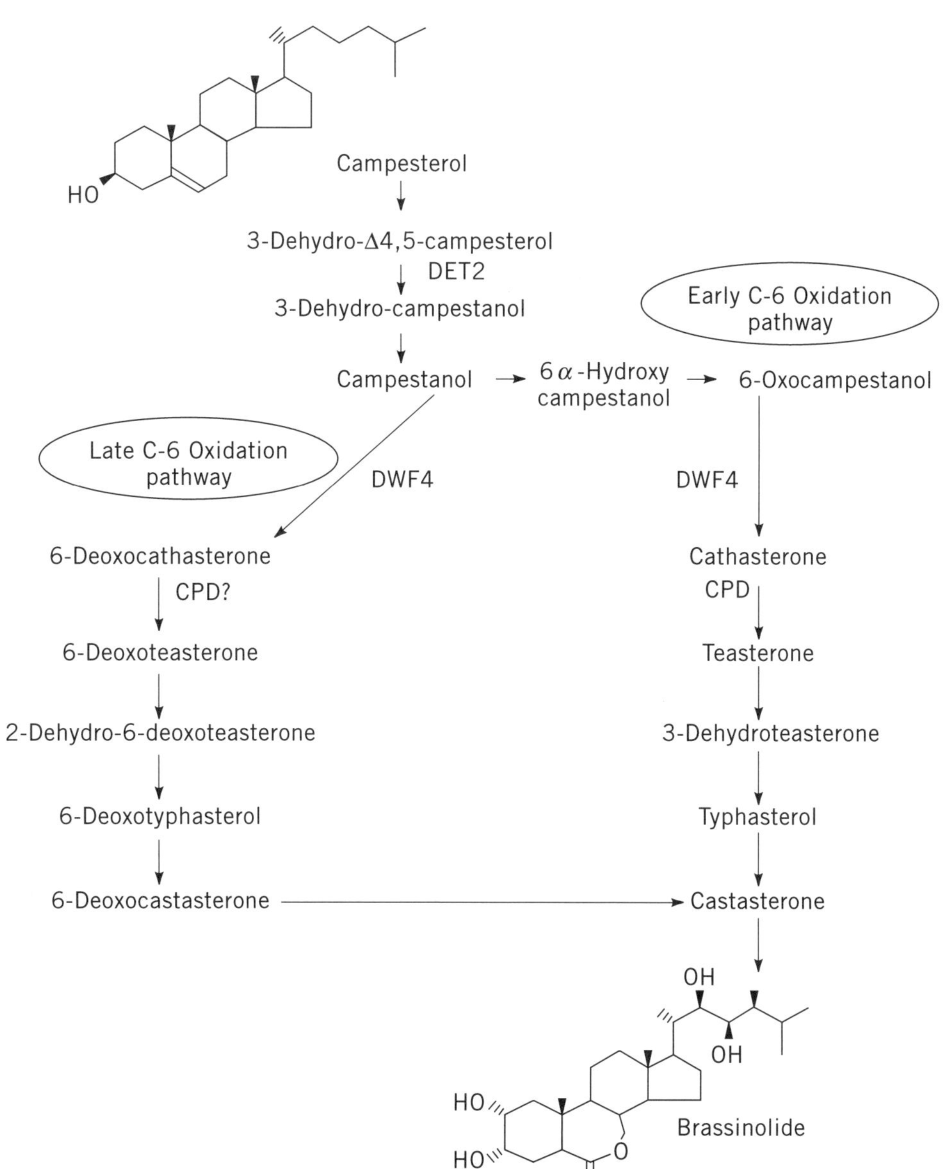

Figure 1. Brassinosteroid biosynthesis in higher plants. Steps mediated by DET2, DWF4, and CPD in *Arabidopsis* are indicated. (Adapted from Ref. 7.)

in ***Arabidopsis*** *thaliana* (3,13). The *det2* (*de-etiolated 2*), *dwf4* (*dwarf4*), *cpd* (*constitutive photomorphogenesis and dwarfism*), and *dim* (*diminuto*) mutants identify different steps in BR biosynthesis (7,14–17). All of these mutants display drastic **phenotypic** effects that can be reverted by application of brassinolide. In the light, the plants are smaller than wild type because of reduced cell size, the leaves are darker green, apical dominance is reduced, and the plants are less fertile. In the dark, a de-etiolation phenotype is observed, characterized by short and thick hypocotyls, open and expanded cotyledons, and formation of primary leaves. This phenotype is accompanied by derepression of light-regulated genes, with the exception of the *dim* mutant, where a normal repression of light-regulated genes is observed in the dark (18). Collectively, these observations support the existence of cross-talk between the light and BR signaling pathways. The four **genes** corresponding to the above-mentioned loci have been **cloned** recently. The *DET2* gene product shares similarity with mammalian steroid 5α-reductases. Definite proof for its function was given by two experiments that support functional conservation between mammalian and plant steroid 5α-reductases (19). When expressed in human embryonic kidney cells, DET2 catalyzed 5α-reduction of several animal steroid substrates. In addition, *det2* mutants could be rescued by expression of human steroid 5α-reductases. The *CPD* and *DWF4* genes were cloned by T-DNA tagging (see **T complex**), and both encode **cytochrome P450** monooxygenases that share **homology** with steroid hydroxylases (7,15). Feeding studies indicated that DWF4 acts as a 22α-hydroxylase (7), whereas CPD functions as a 23α-hydroxylase (15). It is reasonable to assume that each enzyme would play a role in both branches of BR biosynthesis, although there is yet no evidence that CPD uses 6-deoxocathasterone as a substrate. Finally, the *DIM* gene encodes a protein with a putative FAD-binding domain and a nuclear localization motif (see **Nuclear import, export**) (18,20). Feeding experiments indicate that *dim* mutations affect a step before typhasterol formation, so it might be involved in epimerization of teasterone (13).

As is the case for most other plant hormones, glucosylation plays a role in deactivation of BRs. Some of these conjugates may serve as a storage form, as for instance the C23-glucosylated brassinolide and esters at position C3 (21,22). Permanent inactivation of BRs is achieved by C25 and C26 hydroxylation, as well as by cleavage of the side chain (13). It should be noted that transport of BR storage forms is facilitated after glucosylation, as a result of their increased **hydrophilicity**. Transport of labeled BRs has been shown to occur from root to shoot, probably via xylem (23). It is not yet clear, however, whether this also plays a role in determining the endogenous levels of BRs.

SIGNAL PERCEPTION AND TRANSDUCTION

The conservation of steroid-like compounds as signaling molecules across several phyla including fungi, invertebrates, and vertebrates made it tempting to speculate on the existence of soluble nuclear BR receptors in plants (24,25). Two BR-insensitive mutants, *bri1* (*brassinosteroid insensitive 1*) and *cbb2* (*cabbage 2*), have been isolated in *Arabidopsis* (16,26). In both mutants, hormone insensitivity is specific to BRs, and neither one can be rescued by external BR application. Expression of BR-regulated genes was abolished in BR-insensitive mutants (16). *Bri1* and *cbb2* mutations result in dramatic effects on development, including an exacerbated dwarfism, bushy phenotype, dark green and thickened leaves, and male sterility. In addition, de-etiolation of dark-grown seedlings was observed (26). *Bri1* and *cbb2* are alleles of a single locus, and the corresponding gene has been cloned using a map-based approach (27). Surprisingly, BRI1 does not share similarity with **steroid receptors**, but belongs to the class of receptor-like transmembrane **kinases** (28). Members of this family have an *N*-terminal extracellular **leucine-rich repeat** (LRR); there are 25 such repeats in BRI1, with a unique stretch of 70 amino acid residues between the 21st and 22nd LRR. These are followed by a transmembrane domain and an internal **serine/threonine kinase** domain that relays the signal. Other LRR kinases include proteins that influence meristem size and organ formation, such as clavata1, erecta, and Xa21, which confers resistance against *Xanthomonas oryzae pv. Oryzae* in rice (29–31). The *BRI1* gene is ubiquitously expressed in plant tissues in light and dark, consistent with a presumed role as a BR receptor in different cell types (27). The BRI1 protein could bind BR either directly or indirectly, via a protein intermediary. Although none of the previously identified LRR kinases can bind nonproteinaceous ligands, the island of 70 unique amino acid residues in the LRR of BRI1 could serve the purpose of directly binding BR. Mutation of this region severely affects BRI1 activity. In animal systems, LRR kinases are known to form homo- or heterodimers upon ligand binding, thereby activating the intrinsic kinase activity (28). Heterodimerization of BRI1 and XA21, or other LRR domains of disease resistance proteins, could control cross-talk between BR and pathogenic signaling (27).

DOWNSTREAM TARGETS

To date, genes involved in the primary response to BR have not been identified. Likewise, ***cis*-acting** sequences or ***trans*-acting** factors involved in the BR response remain to be characterized. In contrast, the expression of at least 50 genes was altered upon treatment of soybean hypocotyls and epicotyls with BR for 2 to 17 h (32,33). One of these genes was cloned, and the predicted polypeptide chain shared similarity with xyloglucan endo-*trans*-glycosylases (XETs) (34). The gene was termed *BRU1* (*BR up-regulated*) and is not responsive to either **auxin** or **gibberellin**. Its recombinant product catalyzed endo-*trans*-glycosylation *in vitro* (35). The *in vivo* function of *BRU1* is thus to control expansion growth, either by cell-wall loosening or by integration of newly synthesized xyloglucans in the wall (36).

In *Arabidopsis*, the *TCH4* (*touch 4*) gene encodes an active XET, and its transcriptional activation is maximal within 2 h of BR treatment (37). XET is encoded by a **multigene family** in the *Arabidopsis* **genome**. Some family members, such as *TCH4*, are responsive to BR and auxin, but not to gibberellins, whereas others, such as *meri5* (*meristem 5*), are regulated by gibberellins, less generally by BR, and not by auxin (16). Certain *XET* genes do not respond to BR. The *XET* gene family thus appears to be differentially regulated by a combination of hormonal and environmental cues (such as light, heat, cold, and touch) (38). ***In situ* hybridization** of soybean epicotyl cross sections with a *BRU1* probe revealed highest gene expression in vascular tissues (phloem cells and xylem parenchyma), as well as in the cortical starch sheath layer (35). In contrast to the rather restricted action of auxin on cell elongation, BRs have been shown to have an effect both on epidermal and internal tissues (39). Nevertheless, the action is more profound on internal tissues, correlating well with the spatial expression pattern of *BRU1*. Comparative kinetic analysis of auxin and BR action indicates that auxin possibly initiates cell wall elongation at the epidermis, whereas BR continues to stimulate elongation of both outer and inner layers (39–41). BRs thus seem to have a prolonged effect on cell elongation. Brassinolide application results in a predominantly transverse orientation of cortical microtubules, allowing expansion in the longitudinal direction (42).

EFFECTS

The final effects of BRs result from expression of an array of secondary response genes, mentioned above. A multitude of BR functions in plant development are known, including leaf bending and unrolling, stem elongation, root inhibition, xylogenesis, and pollen tube growth (5). It is clear that several of these activities may result from cross-talk with light and other hormonal signaling cascades.

BIBLIOGRAPHY

1. M. D. Grove, G. F. Spencer, W. K. Rohwedder, N. Mandava, J. F. Worley, J. D. Warthen, G. I. Steffens, J. L. Flippen-Anderson, and J. C. Cook (1979) *Nature* **281**, 216–217.
2. S. Fujioka and A. Sakurai (1997) *Physiol. Plant.* **100**, 710–715.
3. S. D. Clouse (1996) *Curr. Biol.* **6**, 658–661.
4. S. Fujioka, T. Inoue, S. Takatsuto, T. Yanagisawa, T. Yokota, and A. Sakurai (1995) *Biosci. Biotech. Biochem.* **59**, 1543–1547.
5. T. Yokota (1997) *Trends Plant Sci.* **2**, 137–143.
6. H. Suzuki, T. Inoue, S. Fujioka, T. Saito, S. Takatsuto, T. Yokoto, N. Murofushi, T. Yanagisawa, and A. Sakurai (1995) *Phytochemistry* **40**, 1391–1397.

7. S. Choe, B. P. Dilkes, S. Fujioka, S. Takatsuto, A. Sakurai, and K. A. Feldmann (1998) *Plant Cell* **10**, 231–243.
8. H. Suzuki, S. Fujioka, S. Takatsuto, T. Yokota, N. Murofushi, and A. Sakurai (1994) *J. Plant Growth Regul.* **13**, 21–26.
9. H. Suzuki, T. Inoue, S. Fujioka, S. Takatsuto, T. Yanagisawa, T. Yokota, N. Murofushi, and A. Sakurai (1994) *Biosci. Biotech. Biochem.* **58**, 1186–1188.
10. T. Yokota, Y. Ogino, N. Takahashi, H. Saimoto, S. Fujioka, and A. Sakurai (1990) *Agric. Biol. Chem.* **54**, 1107–1108.
11. Y.-H. Choi, S. Fujioka, A. Harada, T. Yokota, S. Takatsuto, and A. Sakurai (1996) *Phytochemistry* **43**, 593–596.
12. Y.-H. Choi, S. Fujioka, T. Nomura, A. Harada, T. Yokota, S. Takatsuto, and A. Sakurai (1997) *Phytochemistry* **44**, 609–613.
13. M. Szekeres and C. Koncz (1998) *Plant Physiol. Biochem.* **36**, 145–155.
14. J. Li, P. Nagpal, V. Vitart, T. C. McMorris, and J. Chory (1996) *Science* **272**, 398–401.
15. M. Szekeres, K. Németh, Z. Koncz-Kálmán, J. Mathur, A. Kauschmann, T. Altmann, G. P. Rédei, F. Nagy, J. Schell, and C. Koncz (1996) *Cell* **85**, 171–182.
16. A. Kauschmann, A. Jessop, C. Koncz, M. Szekeres, L. Willmitzer, and T. Altmann (1996) *Plant J.* **9**, 701–713.
17. R. Azpiroz, Y. Wu, J. C. LoCascio, and K. A. Feldmann (1998) *Plant Cell* **10**, 219–230.
18. T. Takahashi, A. Gasch, N. Nishizawa, and N.-H. Chua (1995) *Genes Dev.* **9**, 97–107.
19. J. Li, M. G. Biswas, A. Chao, D. W. Russell, and J. Chory (1997) *Proc. Natl. Acad. Sci. USA* **94**, 3554–3559.
20. A. R. Mushegian and E. V. Koonin (1995) *Protein Sci.* **4**, 1243–1244.
21. H. Suzuki, S. K. Kim, N. Takahashi, and T. Yokota (1993) *Phytochemistry* **33**, 1361–1367.
22. S. Asakawa, H. Abe, N. Nishikawa, M. Natsume, and M. Koshioka (1996) *Biosci. Biotech. Biochem.* **60**, 1416–1420.
23. N. Nishikawa, S. Toyama, A. Shida, and F. Futatsuya (1994) *J. Plant Res.* **107**, 125–130.
24. M. Beato, P. Herrlich, and G. Schutz (1995) *Cell* **83**, 851–857.
25. S. D. Clouse (1996) *Plant J.* **10**, 1–8.
26. S. D. Clouse, M. Langford, and T. C. McMorris (1996) *Plant Physiol.* **111**, 671–678.
27. J. Li and J. Chory (1997) *Cell* **90**, 929–938.
28. D. M. Braun and J. C. Walker (1996) *Trends Biochem. Sci.* **21**, 70–73.
29. S. E. Clark, R. W. Williams, and E. M. Meyerowitz (1997) *Cell* **89**, 575–585.
30. W.-Y. Song, G.-L. Wang, L.-L. Chen, H.-S. Kim, L.-Y. Pi, T. Holsten, J. Gardner, B. Wang, W.-X. Zhao, L.-H. Zhu, C. Fauquet, and P. Ronald (1995) *Science* **270**, 1804–1806.
31. K. U. Torii, N. Mitsukawa, T. Oosumi, Y. Matsuura, R. Yokoyama, R. F. Whitier, and Y. Komeda (1996) *Plant Cell* **8**, 735–746.
32. S. D. Clouse and D. Zurek (1991) In *Brassinosteroids: Chemistry, Bioactivity and Applications*, ACS Symposium Series, Vol. 474 (H. G. Cutler, T. Yokota, and G. Adam, eds.), American Chemical Society, Washington, D.C., pp. 122–140.
33. S. D. Clouse, D. M. Zurek, T. C. McMorris, and M. E. Baker (1992) *Plant Physiol.* **100**, 1377–1383.
34. D. M. Zurek and S. D. Clouse (1994) *Plant Physiol.* **104**, 161–170.
35. S. D. Clouse (1997) *Physiol. Plant.* **100**, 702–709.
36. D. J. Cosgrove (1997) *Plant Cell* **9**, 1031–1041.
37. W. Xu, M. M. Purugganan, D. H. Plisensky, D. M. Antosiewicz, S. C. Fry, and J. Braam (1995) *Plant Cell* **7**, 1555–1567.
38. W. Xu, P. Campbell, A. K. Vargheese, and J. Braam (1996) *Plant J.* **9**, 879–889.
39. R. Tominaga, N. Sakurai, and S. Kuraishi (1994) *Plant Cell Physiol.* **35**, 1103–1106.
40. M. Katsumi (1985) *Plant Cell Physiol.* **26**, 615–625.
41. D. M. Zurek, D. L. Rayle, T. C. McMorris, and S. D. Clouse (1994) *Plant Physiol.* **104**, 505–513.
42. K. Mayumi and H. Shibaoka (1995) *Plant Cell Physiol.* **36**, 173–181.

BREFELDIN A

YAIR ARGON

Brefeldin A (BFA) is a fungal metabolite that is a powerful tool for dissecting membrane traffic and organelle dynamics. It is named after *Penicilium brefeldianum*, which produces it. BFA is a macrolide antibiotic with a molecular weight of 280.36 Da and the structure

H_3C — O — O — OH — OH

BFA is normally used on cells in culture at concentrations of 1 to 10 μM. As first discovered by Misumi et al. (1), BFA primarily inhibits secretion, without significantly affecting **endocytosis**. It does not affect protein biosynthesis significantly, does not deplete cellular ATP levels, and does not affect **cytoskeletal** elements. The unique feature of BFA, which distinguishes it from other inhibitors of protein traffic, is that it causes coalescence of organelles. Within minutes of BFA application, the **Golgi apparatus** fragments and fuses with the **endoplasmic reticulum** (ER), which does not fragment (2), leading to mixing of the two compartments.

The effects of BFA are completely reversible. Retrograde movement of Golgi proteins into the ER occurs via long, tubulovesicular processes extending out of the Golgi along **microtubules**. This retrograde traffic pathway can be separated into two distinct phases: movement of Golgi proteins into the intermediate compartment between the ER and Golgi complex, followed by cycling between the modified ER and the intermediate compartment (3). Remarkably, many Golgi **enzymes** are still active in this environment and can modify proteins in the ER (4).

One of the important uses of BFA-induced redistribution has been the functional demarcation of compartments along the secretory pathway (5). It is clear that BFA affects the early Golgi compartments, the *cis* and medial cisternae. Whether BFA also affects the *trans*-most cisternae of the Golgi stack and the *trans*-Golgi network seems to be dependent on tissue and cell type. Thus, in pancreatic acinar cells, exit from the Golgi complex and formation of **secretory granules** are not BFA-sensitive (6), while in other cells the formation of secretory granules is inhibited (7). Secretion of preformed secretory granules is not inhibited by BFA. Thus, the effects of BFA point to a common mechanism in the traffic of proteins that follow either the constitutive or the regulated pathways of **exocytosis**.

BFA's effects are not limited to the Golgi apparatus and are reiterated throughout the **endosome/lysosome** system.

BFA treatment induces tubulation of these compartments (8). Similar to the mixing of the Golgi with the ER, the *trans*-Golgi network mixes with the recycling endosomal system. Remarkably, this mixed system remains partly functional, with normal cycling between plasma membrane and endosomes, but with impaired traffic between endosomes and lysosomes (8). This suggests that the vesicle-budding mechanisms (see below) in the endocytic pathway are similar to, yet distinct from, those in the exocytic pathway. Furthermore, these observations reinforce the plasticity of the endocytic pathway, which is largely functional even when some traffic steps are inhibited.

The major effects of BFA on protein traffic are explained by its ability to prevent binding of cytosolic coat proteins onto membranes (9). The target of BFA's action is a **nucleotide exchange factor** for **ADP-ribosylation** factor (ARF), a small **GTP-binding protein**. BFA inhibits the ability of Golgi membranes to catalyze the exchange of GTP onto ARF specifically, thereby preventing ARF from interacting with the membrane. As a result, the association of the coat protein β-COP with the Golgi membrane is inhibited (10,11). β-COP is a subunit of a cytosolic protein complex, the coatomer, that reversibly associates with Golgi membranes and controls vesicular transport.

One possible mode of ARF action is via its ability to stimulate phospholipase D in Golgi membranes (12). Phospholipase D converts phospholipids into phosphatidic acid and thus can alter the **lipid** content of **membranes**. Stimulation of the Golgi-associated phospholipase D activity is BFA-sensitive, suggesting a possible link between transport events and the underlying architecture of the lipid bilayer.

The pathogenic bacterium *Staphylococus aureus* mimics the action of BFA and disassembles the Golgi apparatus (13,14). This effect is mediated by the secretion of EDIN (epidermal-cell differentiation inhibitor), an extracellular enzyme that ADP-ribosylates **Rho GTPase**. As a result, all the manifestations of BFA treatment are reproduced. Thus, the regulatory circuit of coat-membrane assembly may involve more than one small GTP-binding protein. In addition to the effects of the small GTP-binding proteins, **heterotrimeric G proteins** of the Gi/Go subfamily also contribute to the regulation of the cycle of coatomer binding (15,16). Activation of heterotrimeric G proteins promotes binding of β-COP to Golgi membranes and antagonizes the effect of BFA.

Several BFA-resistant **cell lines** exist, including the PtK1 rat kangaroo cell line, a derivative of monkey kidney Vero cells, and two derivatives of the human epidermoid carcinoma KB cell line (17). The BFA resistance is dominant and is due to a Golgi-associated factor that is **homologous** to the target of BFA in cells that are sensitive to the drug. In Golgi membranes from BFA-resistant PtK1 cells, the basal phospholipase D activity is high and insensitive to BFA (12), suggesting a mechanism for the drug resistance. Another mechanism may involve a member of the ATP-binding cassette superfamily of transport proteins, as suggested by BFA-resistant mutants in yeast (18).

The ability of BFA to affect budding of **clathrin**-coated vesicles from the *trans*-Golgi network (observed in many, but not all cells) is mediated through the rapid and reversible redistribution of γ-adaptin (19). This component of the clathrin coat is specific to Golgi-derived coated vesicles and is absent from plasma membrane-derived coated vesicles. The kinetic and pharmacological similarities between BFA's effects on γ- adaptin and β-COP underscore the biochemical similarities between membrane budding mechanisms that are mediated by various coat complexes.

The effects of BFA are readily reversible, and this reversibility is due partly to its detoxification. The mechanism of BFA detoxification was studied in CHO cells and is mediated by the **glutathione *S*-transferase** system via conversion of the antibiotic to its **glutathionyl** and **cysteinyl** derivatives, followed by secretion (20).

BFA affects cellular compartmentalization not only via protein traffic, but also through lipid metabolism (21). Its main effect on lipids is enhanced hydrolysis of sphingomyelin, a key regulator of cell proliferation and differentiation. Sphingomyelin is produced from ER-derived ceramide and is delivered to other membranes, presumably via the same trafficking system used for proteins. The effect of BFA on the level of sphingomyelin seems to be a result of the mixing of cellular membranes. BFA treatment also increases the rate of sphingomyelin biosynthesis from phosphatidylcholine, and this effect may be related to the above-mentioned action on phospholipase D. In this context, it is noteworthy that C6 ceramide, a cell-permeable ceramide analogue, partially restores BFA sensitivity in a BFA-resistant cells (22).

The various effects of BFA on a number of cellular membranous organelles indicate both the common and distinct mechanisms that operate to sort membrane components. BFA seems to inhibit a common key step in the mechanism of vesicle budding, namely, regulation of the nucleotide status of different small G proteins. The BFA-sensitive G proteins are each endowed with organelle specificity, coupled perhaps to tissue specificity, which explains the variable sensitivity of membranes to BFA. On the other hand, the use of BFA has also revealed a physiological connection between cellular **signal transduction** pathways and the traffic of membrane lipids and proteins through the cell.

BIBLIOGRAPHY

1. Y. Misumi, Y. Misumi, K. Miki, A. Takatsuki, G. Tamura, and Y. Ikehara (1986) *J. Biol. Chem.*, **261**, 11398–11403.
2. J. Lippincott-Schwartz, L. C. Yuan, J. S. Bonifacino, and R. D. Klausner (1989) *Cell* **56**, 801–813.
3. J. Lippincott-Schwartz, J. G. Donaldson, A. Schweizer, E. G. Berger, H. P. Hauri, L. C. Yuan, and R. D. Klausner (1990) *Cell* **60**, 821–836.
4. N. E. Ivessa, C. De Lemos-Chiarandini, Y. S. Tsao, A. Takatsuki, M. Adesnik, D. D. Sabatini, and G. Kreibich (1992) *J. Cell Biol.* **117**, 949–958.
5. R. D. Klausner, J. G. Donaldson, and J. Lippincott-Schwartz (1992) *Cell Biol.* **116**, 1071–1080.
6. L. C. Hendricks, S. L. McClanahan, G. E. Palade, and M. G. Farquhar (1992) *Proc. Nat. Acad. Sci. USA* **89**, 7242–7246.
7. S. G. Miller, L. Carnell, and H. H. Moore (1992) *J. Cell Biol.* **118**, 267–283.
8. J. Lippincott-Schwartz, L. Yuan, C. Tipper, M. Amherdt, L. Orci, and R. D. Klausner (1991) *Cell* **67**, 601–616.
9. L. Orci, M. Tagaya, M. Amherdt, A. Perrelet, J. G. Donaldson, J. Lippincott-Schwartz, R. D. Klausner, and J. E. Rothman (1991) *Cell* **64**, 1183–1195.
10. J. G. Donaldson, J. Lippincott-Schwartz, G. S. Bloom, T. E. Kreis, and R. D. Klausner (1990) *J. Cell Biol.* **111**, 2295–2306.
11. J. G. Donaldson, D. Finazzi, and R. D. Klausner (1992) *Nature* **360**, 350–352.

12. N. T. Ktistakis, H. A. Brown, P. C. Sternweis, and M. G. Roth (1995) *Proc. Nat. Acad. Sci. USA* **92**, 4952–4956.
13. M. Sugai, C. H. Chen, and H. C. Wu (1992) *J. Biol. Chem.* **267**, 21297–21299.
14. M. Sugai, C. H. Chen, and H. C. Wu (1992) *Proc. Nat. Acad. Sci. USA* **89**, 8903–8907.
15. J. G. Donaldson, R. A. Kahn, J. Lippincott-Schwartz, and R. D. Klausner (1991) *Science* **254**, 1197–1199.
16. N. T. Ktistakis, M. E. Linder, and M. G. Roth (1992) *Nature* **356**, 344–6.
17. N. T. Ktistakis, M. G. Roth, and G. S. Bloom (1991) *J. Cell Biol.* **113**, 1009–1023.
18. T. G. Turi and J. K. Rose (1995) *Biochem. Biophys. Res. Commun.* **213**, 410–418.
19. M. S. Robinson and T. E. Kreis (1992) *Cell* **69**, 129–138.
20. A. Bruning, T. Ishikawa, R. E. Kneusel, U. Matern, F. Lottspeich, and F. T. Wieland (1992) *J. Biol. Chem.* **267**, 7726–32.
21. C. M. Linardic, S. Jayadev, and Y. A. Hannun (1992) *J. Biol. Chem.* **267**, 14909–14911.
22. T. Oda, C. H. Chen, and H. C. Wu (1995) *J. Biol. Chem.* **270**, 4088–4092.

Suggestions for Further Reading

R. D. Klausner, J. G. Donaldson, and J. Lippincott-Schwartz (1992) Brefeldin A: insights into the control of membrane traffic and organelle structure. *J. Cell Biol.* **116**, 1071–1080.

J. Lippincott-Schwartz, L. C. Yuan, J. S. Bonifacino, and R. D. Klausner (1989) Rapid redistribution of Golgi proteins into the ER in cells treated with brefeldin A: evidence for membrane cycling from Golgi to ER. *Cell* **56**, 801–813.

N. T. Ktistakis, H. A. Brown, P. C. Sternweis, and M. G. Roth (1995). Phospholipase D is present on Golgi-enriched membranes and its activation by ADP ribosylation factor is sensitive to brefeldin A. *Proc. Natl. Acad. Sci. USA* **92**, 4952–4956.

5-BROMOURACIL

LYNNETTE FERGUSON
WILLIAM DENNY

Litman and Pardee (1) found that three **thymine** analogues (5-bromouracil, 5-chlorouracil and 5-iodouracil) were all incorporated fairly efficiently into the **DNA** of **bacteriophage** T4, and all were powerful **mutagens** in this organism. Further studies on 5-bromouracil (BU) mutagenesis by Freese (2,3) were important in defining transition and transversion mutations and distinguishing these events from **frameshift mutation**. All the mutants induced by BU could be reverted by either BU or 2-**aminopurine**, although spontaneous and **acridine**-induced mutants could not. The BU-induced mutations were suggested to be transitions, because it was considered that the most likely effect of incorporating this base analogue into DNA would be to induce mispairing with guanine, by a mechanism that was unknown at that time (3). Mispairings with guanine do occur and can lead to BU mutagenesis (for example, as shown in Fig. 1).

There is some evidence that ionization of bromouracil leads to a high probability of mispairing with guanine and that ionized structures are more likely to be involved in this type of mutagenesis than a neutral **wobble-pairing** structure containing the favored keto structures of BU (4). Goodman et al. (5) suggested that BU may exhibit substantially different base-pairing behavior depending upon whether it is present as a template base or as a deoxyribonucleoside triphosphate substrate. They interpreted differences in mutagenesis by this base analogue in different organisms and different systems as due partly to the relative size of and imbalance of the deoxynucleoside triphosphate pools. Classic interpretations of BU mutagenesis invoke direct base mispairing, and this is likely to be involved in the mutagenesis seen in bacterial and bacteriophage systems. However, studies in **yeast** (6) suggest that the majority of BU-induced mutants do not occur at the site where BU is incorporated. Szyszko et al. (7) have identified uracil as a major lesion after BU incorporation into DNA of *E. coli*, presumably because of dehalogenation of incorporated BU. They suggested that this would be followed by formation of apyrimidinic sites by the enzyme uracil-DNA glycosylase, followed by single-stranded nicks. Efficient BU mutagenesis is dependent upon various **DNA repair** gene functions, suggesting a major role for **SOS response** repair in many of the BU-induced mutations (8).

BIBLIOGRAPHY

1. R. M. Litman and A. B. Pardee (1956) *Nature* **178**, 529–531.
2. E. Freese (1959) *Proc. Natl. Acad. Sci. USA* **45**, 622–633.
3. E. Freese (1959) *Brookhaven Symp. Biol.* **12**, 63–73.
4. H. Yu, R. Eritja, L. B. Bloom, and M. F. Goodman (1993) *J. Biol. Chem.* **268**, 15935–15943.
5. M. F. Goodman, R. L. Hopkins, R. Lasken, and D. N. Mhaskar (1985) *Basic Life Sci.* **31**, 409–423.

1st generation	2nd generation	3rd generation	4th generation	
A.T →	A.BU → ↘	A.BU →	A.BU	(Mutations in later generations)
		G.BU → ↘	G.BU	(Mutations in later generations)
			G.C	(Transition mutation)

Figure 1. Mechanism of generation of transition mutation by 5-bromouracil (BU) after DNA replication. The presence of BU in one DNA strand in place of the original thymine (T), causes guanine (G) to be inserted in the complementary position upon DNA replication, instead of the original adenine (A).

6. V. Noskov, K. Negishi, A. Ono, A. Matsuda, B. Ono, and H. Hayatsu (1994) *Mutat. Res.* **308**, 43–51.
7. J. Szyszko, I. Pietrzykowska, T. Twardowski, and D. Shugar (1983) *Mutat. Res.* **108**, 13–27.
8. I. Pietrzykowska, M. Krych, and D. Shugar (1985) *Mutat. Res.* **149**, 287–296.

BUFFERS

H. B. F. DIXON

BASIC PRINCIPLES

The term *buffer* solution usually refers to a solution that minimizes changes in pH when hydrogen ions (hydrons, H^+) are added to the solution or removed from it. (As discussed in reference 1, IUPAC has approved *hydron* to designate a hydrogen cation when a term independent of hydrogen isotope is desired.) Such a solution is therefore said to buffer the pH. Solutions can also be designed to buffer other species, particularly metal ions, as will be described below.

Because the concentration of hydrogen ions in biological media is low, usually about 0.1 μM, it could be greatly changed by the production or consumption of very small quantities of hydrogen ions in chemical reactions. This is what makes buffering commonly necessary in experiments in molecular biology. The principle of a buffer is simple. Relative to the low concentration of free hydrogen ions, the buffer consists of much higher concentrations of both a base, A^- (ie, a substance that can combine reversibly with the hydrons), and its conjugate acid, HA, which is formed when the base combines with a hydron. Hence the equilibrium

$$HA \rightleftharpoons H^+ + A^- \tag{1}$$

is established. Subsequently, any addition of H^+ causes a much smaller increase in its free concentration, $[H^+]$, because some of the hydrons are used up in making more HA to displace this equilibrium to the left. Likewise, any fall in $[H^+]$ is diminished because it leads to the dissociation of HA.

SIMPLE THEORY

Users of buffers need to be aware of many features of their action if they are not to make mistakes that can easily ruin their experiments. For this, a little theory of buffer solutions is required.

Dissociation of Water

Water spontaneously dissociates into one hydrogen ion and one hydroxide ion:

$$H_2O \rightleftharpoons H^+ + OH^- \tag{2}$$

Because of the equilibrium, the product $[H^+][OH^-]$ is constant; its value, about 10^{-14} M^2, changes only slightly with temperature. Both the hydrons and the hydroxide ions will be largely hydrated, reversibly combined with one or more water molecules. In this article, the term *acid* is used in the sense of a Brønsted acid, a substance that can donate a hydron, and *base* is used to mean a Brønsted base, a substance that can combine with one. If no such acid or base is added to pure water, the values of $[H^+]$ and $[OH^-]$ are equal, namely, 10^{-7} M, ie, 0.1 μM. Given that the pH is defined as

$$pH = -\log_{10}[H^+] \tag{3}$$

the pH is 7. The pH scale is useful in water over about the range 0 to 14, over which $[H^+]$ varies from 1 M to 10^{-14} M, and $[OH^-]$ from 10^{-14} M to 1 M.

The pK_a and the pH

The acid HA will have a dissociation constant, K_a, the equilibrium constant for reaction (1), which is defined as

$$K_a = \frac{[H^+][A^-]}{[HA]} \tag{4}$$

Rearranging this to find $[H^+]$ gives

$$[H^+] = \frac{K_a[A^-]}{[HA]} \tag{5}$$

and taking the negative logarithm of each side to find the pH [eq. (3)] gives

$$pH = pK_a - \log\frac{[A^-]}{[HA]} \tag{6}$$

if we define pK_a as $-\log K_a$. This is known as the Henderson–Hasselbalch equation, and it is fundamental for understanding buffer action and how single groups titrate.

Figure 1 shows the fractional concentration of the base, ie, $[A^-]/([HA] + [A^-])$, as a function of pH. The first point to note is that, when the pH = pK_a, $\log([A^-]/[HA] = 0$, and so $[A^-]/[HA] = 1$, and $[A^-]$ is 0.5 of its maximal value. Hence, the pK_a is the pH at which the acid is half-dissociated. When the pH is one unit above the pK_a, $\log([A^-]/[HA]) = 1$, and so $[A^-]/[HA] = 10$, and hence $[A^-]$ is 10/11, ie, 0.91, of its maximal value as shown; when the pH is two units above the pK_a, $\log([A^-]/[HA]) = 2$, and so $[A^-]/[HA] = 100$ and, hence, $[A^-]$ is 100/101, ie, 0.99 of its maximal value. Similarly, when the pH is one unit below the pK_a, $\log([A^-]/[HA]) = -1$, and

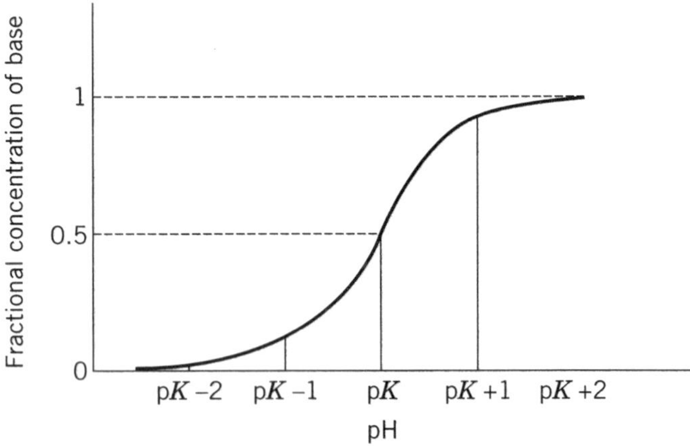

Figure 1. The dependence of the fractional concentration of the basic form of a monobasic acid on the pH. The vertical axis shows the concentration of A^- expressed as a fraction of the sum of the concentrations of A^- and HA. Vertical lines mark: (1) pH = pK_a − 1, when $[A^-] = 0.1[HA]$, so the fractional concentration of A^- is 1/11; (2) pH = pK_a, when $[A^-] = [HA]$, so the fractional concentration of A^- is 1/2; (3) pH = pK_a + 1, when $[A^-] = 10[HA]$, so the fractional concentration of A^- is 10/11.

so $[A^-]/[HA] = 0.1$, and $[A^-]$ is 0.091 of its maximal value; when the pH is two units below the pK_a, $[A^-]$ is 0.01 of its maximal value. These figures illustrate that comparing the pH with the pK of a substance instantly indicates how much of the substance or group is in its hydronated and unhydronated forms. All simple acids follow the curve of Figure 1; they differ only in their pK_a values.

Theory of Buffering

With properly buffered solutions, both $[A^-]$ and [HA] will be much greater than $[H^+]$ and $[OH^-]$. In this case, any H^+ produced in a reaction converts A^- into HA, and any H^+ taken up dissociates from HA to convert it into A^-. To see the consequences of this, it is convenient to modify the Henderson–Hasselbalch equation as follows:

$$\mathrm{pH} = \mathrm{p}K_a - \log\frac{[\mathrm{A}^-]}{[\mathrm{HA}]} = \mathrm{p}K_a - \log[\mathrm{A}^-] + \log[\mathrm{HA}] \quad (7)$$

For good buffering, it is necessary for *both* $[A^-]$ and $[HA^-]$ to be large enough so the logarithm of neither is changed enough to produce an unacceptable change in pH when some of one is converted into the other.

This means that there are two basic requirements for a buffer: (1) It must be sufficiently concentrated (in terms of the total of $[A^-] + [HA^-]$), and (2) the pH must not be too far from the pK_a of the buffer so that the concentrations of *both* A^- and HA will be large. It is the second of these that is most often forgotten. If the pH and pK_a differ by one unit, only 1/11 of the total buffer concentration will be in the minor form.

This point may be illustrated by the example of a worker who wondered why mercaptoethanol and **iodoacetate** inhibited his enzyme when neither one by itself did so. He was using more than 10 mM of each reagent in a 10-mM Tris buffer of pH 7. Hence, the 10-mM H^+ released when the mercaptoethanol and iodoacetate reacted exceeded by tenfold the concentration of free Tris, which was less than 1 mM in a solution far below its pK_a of 8.1.

To a first approximation, dilution does not affect the pH of a buffer solution because it does not affect the ratio $[A^-]/[HA^-]$. There are, however, two qualifications to this conclusion. First, it is assumed that the dissociation constant of the buffering acid is unaltered whereas, in fact, it is likely to change with ionic strength (see below). It is also assumed that dilution is not so extreme that either $[A^-]$ or $[HA^-]$ become comparable to $[H^+]$ or $[OH^-]$. In any case, dilution decreases the concentrations of both A^- and HA; hence dilution makes the solution a worse buffer in that a given amount of conversion of one form into the other will produce a larger change in the ratio.

PRACTICAL POINTS

pH Measurement

Methods of measuring pH are outside the scope of this article but the process is important because reproducibility of buffer preparation often depends on it (see below). The most common method uses a glass electrode. This method depends on dipping the electrode into the solution to be tested in such a way that an electrical cell is created and that the potential given by this cell depends on the pH of the solution. Normally, a layer of thin glass separates the solution under test from one containing one of the electrodes of the cell. This glass is selectively permeable to hydrons, so that the potential across it depends on the pH of the solution. Provided that the other junction potential between this solution and the other electrode is negligible—a condition approached by a bridge containing concentrated KCl—the potential should reliably indicate the pH.

Such pH meters require calibration, and standard buffers are used for this. The makers normally give reliable instructions for calibrating the electrode, but some of these are easily overlooked. The first concerns the temperature. The pH standards have pH values that may themselves vary with temperature. Standardization is valid at only one particular temperature, and it cannot be assumed that an electrode standardized at one temperature gives a meaningful reading at a different one. The pH range specified by the makers of the apparatus should also be noted. Glass membranes may be slightly permeable to Na^+, so that when $[H^+]$ is low, as in alkaline solution, even a slight sensitivity to Na^+ may cause overestimation of $[H^+]$ and hence too low an estimate of a high pH.

Specifying a Buffer Solution

Buffer solutions are often specified in a conventional way, such as 0.2-M sodium acetate buffer, pH 4.8. The concentration given should refer to the sum of the two forms of the buffering species, designated here as $[AcO^-]$ and [HOAc] in the case of acetate buffer, but such a specification does not make fully clear how this solution has been obtained. It might have been prepared by (1) mixing the appropriate amounts of sodium acetate and acetic acid, as calculated from the pK and the Henderson–Hasselbalch equation, (2) adding 0.2-M sodium acetate to 0.2-M acetic acid until a pH of 4.9 was reached, or (3) adding strong NaOH to 0.2-M acetic acid to reach pH 4.9, on the assumption that the volume added diluted the acetic acid negligibly. It would not have been appropriate, however, to have added strong HCl to 0.2-M sodium acetate, as this would have produced NaCl in addition to the buffering species.

Measuring pH is not straightforward, especially in view of the effects of temperature on the measurements (see above); for reproducibility, especially between different laboratories, it is much safer to specify the concentration of *each* form of the buffering substance, given that weighing material out with an accuracy of 1% to 2% is easy. Even an error of 2% in each would change the pH by only log(1.04), ie, by 0.017 of a pH unit. Hence, it would be better to describe the buffer just mentioned as 0.12-M sodium acetate, 0.08-M acetic acid (pH 4.8), thus adding the pH for information rather than as part of the definition.

Effects of Temperature

Users should remember that the pH of a buffer may change with temperature. This is because the dissociation constant of the buffering acid, like any other equilibrium constant, depends on temperature if the dissociation has an appreciable **enthalpy** change, ΔH. Dissociations of carboxylic acids and phosphoric acid have low values of ΔH, and so this effect is small with buffers based on them, but the dissociation of ammonium to ammonia and a hydron is accompanied by a large heat intake (ie, ΔH is positive), and increasing temperature promotes dissociation, ie, lowers the pK_a. Hence, the pH of a buffer that uses an amino group decreases on heating. The effect is quite large, a fall of 0.028 per degree Celsius.

The importance of this needs to be considered in each experiment. If a user wished the glutamate residues of an enzyme to be hydronated to the same extent in an experiment at 0 °C as they would be in a buffer of pH 7 at 20 °C, it would be wrong to use an amino buffer with a pH of 7 at 20 °C and simply cool it. Cooling would raise the pH of the buffer, but not the pK_a of the residues, so these residues would dissociate more on cooling. But this would be a proper procedure if the experimenter was concerned with the state of the **lysine** residues because their pK_a would rise with that of the buffering amine so that their degree of hydronation would remain the same.

Effects of Concentration

The dissociation constants referred to above are those actually exhibited by the buffering species in the solution being used; ie, they are not idealized values extrapolated to standard conditions of ionic strength and temperature. They therefore change with ionic strength and the addition of other components to the solution, eg, organic solvents or deuterium oxide. Many experiments may require a solution with a particular ionic strength or particular concentrations of specified ions. Remember that these may affect the apparent value of the pK of the buffering species.

An acid and its conjugate base differ by one H^+ and therefore by one charge. Typically, they will have charges 0 and −1, or +1 and 0. If, however, one of the forms has more than unit charge, positive or negative, the pK will have a greater dependence on ionic strength. The most commonly used buffer to which this applies is **phosphate buffer**, where the acid is $H_2PO_4^-$ and its conjugate base is the doubly charged HPO_4^{2-}. This makes its pK_a particularly sensitive to ionic strength. For example, a solution of 0.2-M phosphate may increase its pH by over 0.2 of a unit on tenfold dilution.

Many chemical reactions are catalyzed by acids or bases, so that an increased buffer concentration can accelerate them even at a fixed pH. This should be borne in mind when altering buffer concentrations. It is not very common for enzyme-catalyzed reactions because natural selection often ensures that if such catalysis accelerates the reaction, a corresponding acidic or basic group will be present in the **enzyme** to provide it.

Volatile Buffers

Separation methods such as **chromatography** and **electrophoresis** often require a buffered solution. This solution needs appreciable ionic strength (in **ion-exchange chromatography** to give a constant competition for the sites of the exchanger, in electrophoresis to give a constant conductivity and therefore electric field) that is negligibly affected by the substances being separated. Because it is convenient to be able to remove the buffer at the end of the procedure, a volatile buffer that can be removed by evaporation is desirable. The two forms of the buffering substance must differ in charge, however, so that one at least will have a net charge and hence be nonvolatile. Nevertheless, volatile buffers can be prepared by mixing a volatile acid with a volatile base, provided that their pK_a values are not greatly separated. An example is a mixture of pyridine (pK_a 5.2) and acetic acid (pK_a 4.8). On mixing, some of the acetic acid dissociates to acetate and hydronates some of the pyridine to pyridinium. Hence, the pH will be given by

$$\text{pH} = 5.2 + \log\frac{[\text{pyridine}]}{[\text{pyridinium}]} = 4.8 + \log\frac{[\text{acetate}]}{[\text{acetic acid}]} \quad (8)$$

The concentrations of pyridinium ($C_5H_5N^+$-H) and acetate (AcO^-) ions will be equal, because each acetic acid molecule that dissociates to acetate forms one pyridinium ion. As the solution is evaporated off, water, pyridine (C_5H_5N) and acetic acid (AcOH) evaporate, and so the equilibrium

$$AcO^- + C_5H_5N^-\text{-}H \rightleftarrows AcOH + C_5H_5N \quad (9)$$

is displaced to the right to replace the un-ionized forms as they are removed.

Ammonium bicarbonate forms a useful buffer of this type. Although it is highly volatile in solution, it is not volatile when dry, as it has a stable crystal lattice. Hence to allow ammonium and bicarbonate to react to form ammonia and carbon dioxide, it may be necessary after one drying to add a little water and then dry it again. Alternatively, triethylammonium bicarbonate, prepared from triethylamine and carbon dioxide, may be used, as it does not form nonvolatile crystals. Ammonium acetate and ammonium formate have been used in this way (2), but the procedures for drying them are complicated because their pK_a values are so far apart that there is very little of the un-ionized forms at equilibrium.

Appropriate pK_a Values

Biological material usually has a pH near 7, and it is not always easy to find a buffering species with the desired pK_a. Carboxylic acids possess values that are too low, whereas amines have values that are too high. A large number of the commonly used buffers are amines with electron-attracting substituents, which make the lone pair of electrons on the nitrogen atom less available and so lower the atom's pK_a. Examples include:

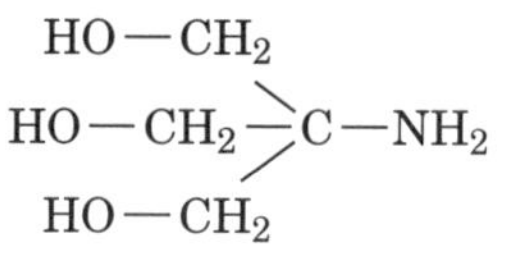

```
HO—CH2                     CH2—CH2
      \                   /       \
HO—CH2—C—NH2             O         N—CH2
      /                   \       /    \
HO—CH2                     CH2—CH2      CH2—SO3⁻
```

Tris

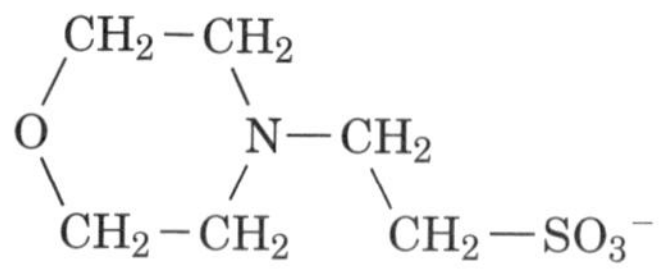

2-Morpholinoethanesulfonic acid

with pK_a values of 8.1 and 6.2, respectively (see **Tris buffer**). **Phosphate buffers** are commonly used because they have a convenient pK_a of about 7 between $H_2PO_4^-$ and HPO_4^{2-}. Their main disadvantages are their propensity to support fungal and algal growth and the sensitivity of their pK_a to ionic strength (see above).

The Need for Buffers

It should be remembered that it is pointless to add a buffer if the components of the reaction mixture are already highly buffering. Glycolytic intermediates, for example, are phosphate esters, and adjusting them to a pH near 7 creates a buffered solution.

Chemical Reactivity

It is often important to use buffering species that will not react chemically under the conditions of an experiment. Hence an experiment to acylate **amino groups** of proteins should use a buffer whose basic component has as low a nucleophilic reactivity as possible. Exclusion of primary amines, such as Tris, may be all that is needed. But extreme conditions may demand introduction of steric hindrance; eg, 2,6-dimethylpyridine has a

pK_a close to that of pyridine but vastly less nucleophilic reactivity because the methyl groups have little steric effect on protonation of the nitrogen atom but greatly slow its reaction with larger molecules.

Another form of reactivity is the possibility of precipitating desired components in the solution, eg, cations such as Ca^{2+}, by phosphate. Occasionally, reactivity may even be of advantage. **Urea** is often used as a protein **denaturant** but, if kept for long in concentrated solution, especially at raised temperatures, its equilibration

$$NH_2—CO—NH_2 \rightleftarrows NH_4^+ + CNO^- \quad (11)$$

can produce about 20-mM cyanate (CNO^-) from 8 M urea (3). The harm in this is the reverse reaction, because the cyanate can carbamoylate **lysine** residues in proteins, forming *uncharged* ureas and destabilizing the folded protein. This process is minimized if ammonium ions are present in the buffer as they compete with the protein for the cyanate.

"Good" Buffers

A number of buffers were recommended by Good and colleagues (4) for a variety of reasons. The most important was that both acidic and basic forms were charged and, therefore they could not rapidly permeate biological **membranes**. If **organelles** are suspended in a buffer without this feature, acetate, for example, molecules of acetic acid, but not acetate ions, may penetrate them and lower the internal pH below that of the external buffer. Many of the Good buffers have strongly acid groups, such as sulfonate, $-SO_3^-$. An example is 2-morpholinoethanesulfonate (eq. 10); its sulfonate group renders it impermeant even when its nitrilo group is un-ionized. Other desirable characteristics of buffering species are also listed (4). For many purposes, transparency at UV wavelengths is necessary.

Under other circumstances, the possession of a further charge can be a disadvantage. Ideally, buffers for ion-exchange chromatography possess only one species with a charge opposite in sign to that of the exchanger so that the equilibration of the exchanger can be followed by pH changes; for example, a buffer cation cannot adsorb to, or desorb from, the exchanger except by exchange with H^+, which changes the pH.

METAL-ION BUFFERS

The free concentration of many metal ions in biological media is very low, and it may be necessary to buffer it. The principles are exactly the same as those that apply to buffering $[H^+]$. Some compound that binds the metal ion is introduced in roughly equal concentrations of the free and metal-bound forms. A substance that buffers in the correct concentration range can be chosen from a tabulation of binding constants (5).

A complication is that many of the compounds that ligate metal ions with suitable affinity also bind hydrons, resulting in competition between the cations for the ligand. This means that the affinity of the ligand for the metal ion varies with pH. The complex of the buffering species with the metal ion is often a chelate, so that two or more ligating groups are involved. Hence, the competition is not a simple 1:1 replacement, and the pH dependence of the net affinity may be complex.

BIBLIOGRAPHY

1. J. F. Bunnett and R. A. Y. Jones (1988) *Pure Appl. Chem.* **60**, 1115–1116.
2. C. H. W. Hirs, S. Moore, and W. H. Stein (1952) *J. Biol. Chem.* **195**, 669–683.
3. G. R. Stark, W. H. Stein, and S. Moore (1960) *J. Biol. Chem.* **235**, 3177–3181.
4. N. E. Good and S. Izawa (1972) *Methods Enzymol.* **24**, 53–68.
5. R. M. Smith and A. E. Martell (1974-89) *Critical Stability Constants*, vols. 1–6, Plenum Press, New York.

Suggestion for Further Reading

R. J. Beynon and J. S. Easterby (1996) *Buffer Solutions: The Basics*, IRL Press, Oxford, U.K. This gives not only an excellent account but also WWW addresses for programs to calculate buffer compositions (http://www.bi.umist.ac.uk/buffers.html) and other help on the Internet.

BZIP DOMAIN

ANDREW TRAVERS

The bZip DNA binding domain and the closely related **helix-loop-helix** (HLH) domain are found in many eukaryotic **transcription** factors. Both bind in the major groove of DNA and have the capability of forming homo- or heterodimers with **proteins** of the same class. The simplest of these domains is the bZip domain which, when bound to DNA, forms a continuous α-helix. The lower basic part of this helix (the b of bZip) forms sequence specific contacts with DNA whereas the upper part can dimerize with an appropriate partner through the formation of a leucine zipper. The dimer thus has the appearance of a pair of scissors binding to a palindromic DNA sequence in which each half-site is exposed on the opposite face of DNA. The helix-loop-helix domain binds to DNA in a similar manner with the difference that the α-helix involved in dimerization is separated from the DNA binding α-helix by a short loop.

BUOYANT DENSITY

T. HOLZMAN

The buoyant density of a macromolecule is its effective density in solution. It is defined relative to the fluid medium in which it resides and/or through which it is sedimented (see **Centrifugation**). The dimensions of buoyant density are in cm^3/g. If a macromolecule is sedimented through a fluid column in which the density of the column is adjusted so that it spans a gradient of density *both* less than *and* greater than the macromolecule, the macromolecule moves to a fixed position in the fluid column (see **Density gradient centrifugation**). At this position the density of the macromolecule is equal to that of the solvent. This is an equilibrium position that is independent of time. This process or technique is termed density gradient sedimentation-equilibrium, equilibrium banding or isopycnic gradient centrifugation. These techniques have uses that range from the analysis of macromolecule solvation to the separation of both living cells and subcellular **organelle** systems (1). Recently the evaluation of macromolecular values of buoyant density have been been employed in a variety of research situations including (1) the analysis of viral infectivity during therapy (2,3) (see **Virus infection**); (2) the nature of circulating immune complexes of **viruses** (4); (3) as a preselection method for **PCR** methods (5).

BIBLIOGRAPHY

1. S. Daenen, W. Huiges, E. Modderman, and M. R. Halie (1993) *Leukemia Res.* **17**, 37–41.
2. A. Nagasaka, S. Hige, T. Matsushima, J. Yoshida, Y. Sasaki, I. Tsunematsu, and M. Asaka (1997) *J. Med. Virol.* **52**, 190–194.
3. T. Kanto, N. Hayashi, T. Takehara, H. Hagiwara, E. Mita, M. Naito, A. Kasahara, H. Fusamoto, and T. Kamada (1995) *J. Hepatol.* **22**, 440–448.
4. M. Hijikata, Y. K. Shimizu, H. Kato, A. Iwamoto, J. W. Shih, H. J. Alter, R. H. Purcell, and H. Yoshikura (1993) *J. Virol.* **67**, 1953–1958.
5. R. J. Carrick, G. G. Schlauder, D. A. Peterson, and I. K. Mushahwar (1992) *J. Virol. Meth.* **39**, 279–290.

Suggestions for Further Reading

C. H. Chervenka (1973) A Manual of Methods for the Analytical ultracentrifuge, Spinco Division, Beckman Instruments, Inc. Palo Alto, CA. *The* classic manual of methods for the original model E analytical ultracentrifuge. Although the mechanical components related to model E have been superseded by the modern XLA and XLI analytical ultracentrifuges, the methods themselves are still applicable.

R. Hinton and M. Dobrota (1980) *Laboratory Techniques in Biochemistry and Molecular Biology: Density Gradient Centrifugation* (T. S. Work and E. Work, eds.), North-Holland, New York. A somewhat dated, but useful, compendium of techniques for preparative density gradient centrifugation.

C

C GENES OF IMMUNOGLOBULINS

MICHEL FOUGEREAU

C genes encode for the **constant (C) regions** of **immunoglobulin** (Ig) heavy and light chains and therefore define the **isotypes** of H (γ, μ, α, δ and λ) and L (κ, λ) chains. The simplest structure is that of the Cκ gene, which is unique in both the human and mouse species. It encodes for the entire constant region of the κ chain. In humans, the Cκ gene has three **allelic** variants based on single amino acid substitutions at two positions, which constitute the Km allotypes. The Cκ gene is separated from the Jκ region by a noncoding region of 1.2 kbp that contains a κ **enhancer**. A second enhancer region is present in the 3′ flanking region. The organization of the Cλ genes is somewhat more complicated, and there are some minor differences between humans and mouse. In both cases, they are present as tandem duplications of a JλCλ unit, in which each J gene is separated from the C gene by a noncoding sequence of about 1 kbp. There are four discrete tandems in the mouse, Jλ1Cλ1 to Jλ4Cλ4, of which only three are functional. In humans, four functional tandem JC pairs are present, in addition to three pairs of **pseudogenes**. An enhancer region is present 3′ of the coding region in both species. In humans, the IgK and IgL loci are located at **chromosomal** positions 2p12 and 22q11, respectively.

The heavy-chain constant gene locus, located at 14q32 in humans, spans over 200 kbp. The detailed organization of the individual human CH genes is given in Figure 1. Some characteristic features may be identified. First, each isotype is encoded by a mosaic of exons (see **Introns/exons**) that precisely reflects the **domain** structure of the Ig. The constant region of Cμ is encoded by four exons, Cμ1 to Cμ4, corresponding to the four constant domains of the μ chain, whereas in the case of the Cγ isotypes one of the exons encodes the hinge region. Second, each isotype encoding C gene has additional membrane exons (one or two, depending on the isotype) that encode the transmembrane section and the short COOH-terminal cytoplasmic tail. Coding units that make secreted or membrane heavy chains result from **alternative splicing** and terminate at two distinct **polyadenylation** sites (see Fig. 1). The overall CH region also contains two pseudogenes. It will be noticed that a switch region is located at 5′ of each set of C genes, with the exception of the δ locus, which accounts for the coexpression of **IgM** and **IgD** at the surface of mature B cells.

A final remark should be made regarding the general organization of the regulating elements that control IgH gene expression. **Promoter** regions are present at 5′ of each V gene and

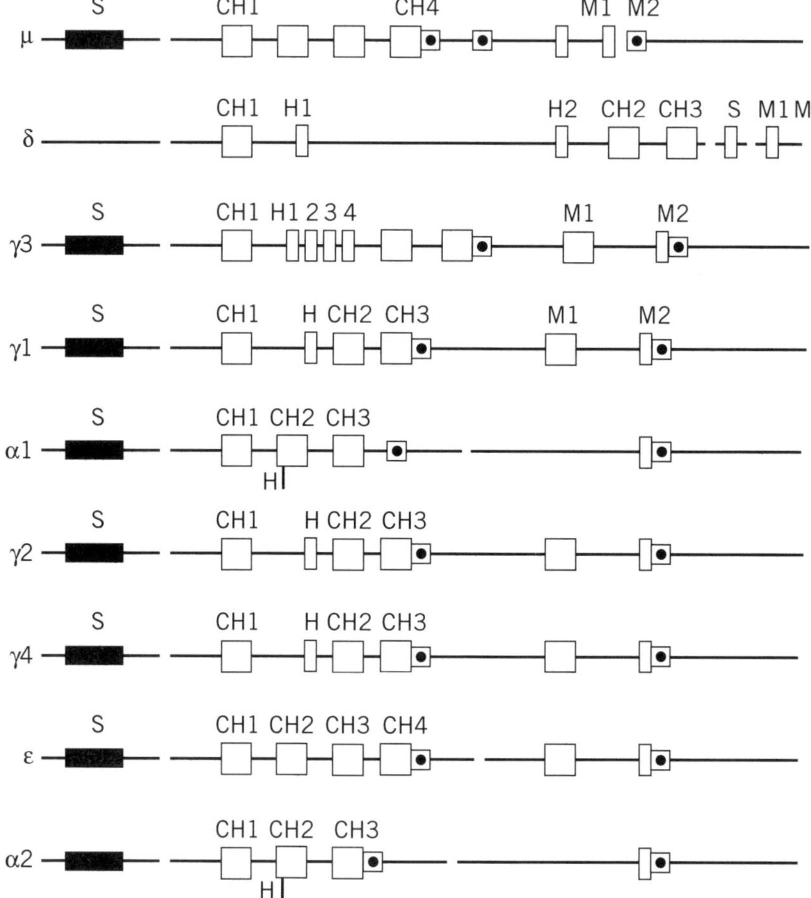

Figure 1. Organization of human heavy-chain constant region genes. Exons encoding discrete constant domains (CH), hinge regions (H), or transmembrane portion of surface heavy chains (M) are represented by open squares or rectangles. Polyadenylation sites are indicated by a dot. (From Ref. 1, with permission; adapted from Ref. 2.)

will be discussed elsewhere. Enhancers are split into two regions; one is located at 5′ of the Sμ switch region, so that with each switching event it will control the newly adjacent isotype gene; the other is located at 3′ of the entire IgCH locus and may thus control whatever gene is to be turned on.

See also entries **Antibody**, **Class switching**, **Isotype**, and **Switch region**.

BIBLIOGRAPHY

1. J.-P. Revillard (1998) In *Immunologie*, De Boeck Université, Paris-Bruxelles, p. 57.
2. J. D. Capra and J. D. Duker (1989) *J. Biol. Chem.* **264**, 12745.

Suggestions for Further Reading

V. Giudicelli et al. (1997) IMGT, the international ImMunoGeneTics database. *Nucleic Acids Res.* **25**, 206–2011.

T. Honjo, F. W. Alt, and T. H. Rabbitts (1989) *Immunoglobulin Genes*, Academic Press, New York.

F. Matsuda and T. Honjo (1996) Organization of the human immunoglobulin heavy chain locus. *Adv. Immunol.* **62**, 1–29.

C-BANDING

ALAN P. WOLFFE

The procedure of C-banding is used to highlight regions of constitutive **heterochromatin**. In man, sites of C-banding include the **centromeres** of all of the **chromosomes** and large segments of heterochromatin in chromosomes 1, 3, 9, 16 and the **Y-chromosome**. The DNA sequences in constitutive heterochromatin are highly **repetitive** and include many **satellite DNA** sequences that are thought to be genetically inert. The distribution of satellite DNA sequences in chromosomes varies considerably from species to species and can even provide information that is specific to a particular individual.

C-banding involves the depurination of DNA in the chromosomes by treatment with hydrochloric acid. The sugar-phosphate backbone of DNA remains intact. Treatment with hydrochloric acid (0.2 M) is halted before depurination is complete (a treatment of several minutes). Then the chromosomes are incubated for several minutes with either sodium hydroxide (0.07 M) or barium hydroxide (0.07 M). This **denatures** the DNA, which aids solubilization. Then the chromosomes are washed overnight with 2 × SSC (0.3 M NaCl plus 0.03 M trisodium citrate) at 66°C. This warm salt wash breaks the sugar-phosphate backbone and allows DNA fragments to dissolve. Then the chromosomes are stained with **Giemsa**. Staining relies on the retention of intact nucleoprotein complexes in the highly compacted heterochromatin.

C-banding of metaphase chromosomes is greatly enhanced in resolution by two methodological improvements: first, by using chromosomes isolated early in **metaphase**, when the contraction of chromatin is incomplete; second, by the use of a sophisticated microscopic technique known as epi-illumination, in which the chromosome is viewed from the same side as it is illuminated. This provides enhanced contrast between stained and unstained bands.

Suggestion for Further Reading

R. P. Wagner, M. P. Maguire, and R. L. Stallings (1993) *Chromosomes. A Synthesis.* Wiley-Liss, New York.

C-TERMINUS

JENNIFER MARTIN

The *C*-terminus is the term for one end of the **polypeptide chain** of a **protein**. Proteins are **polymers** formed by condensation of the **amino groups** and **carboxyl groups** of **amino acids**. That end of the polypeptide chain having an uncondensed or "free" carboxyl group is called the carboxyl terminus, or ***C*-terminus** (Fig. 1). By convention, the *C*-terminus is the last residue in the protein sequence or **primary structure**. Similarly, the other end of the polypeptide chain, having an uncondensed or free amino group, is termed the amino terminus, or ***N*-terminus**, and by convention it is the first residue in the protein sequence. The *C*-terminal carboxyl group is usually negatively charged at physiological pH.

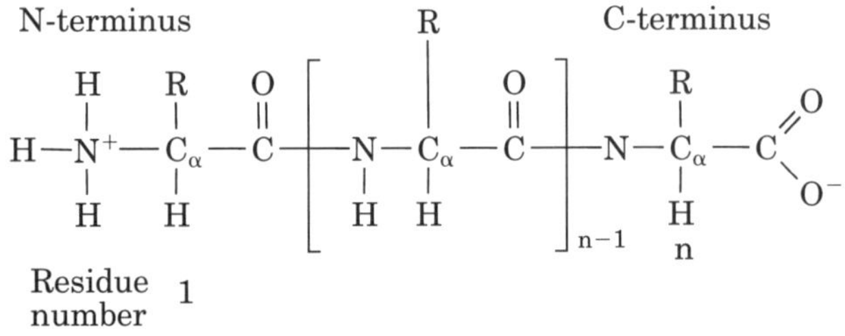

N-terminus　　　　　　C-terminus

$$\mathrm{H_3N^+{-}C_\alpha HR{-}CO}\left[\mathrm{{-}NH{-}C_\alpha HR{-}CO{-}}\right]_{n-1}\mathrm{NH{-}C_\alpha HR{-}COO^-}$$

Residue number　1　　　n

Figure 1. Schematic representation of a protein, showing the *C*-terminus as the last residue in the polypeptide chain. The square brackets indicate the repeating part of the chain (the backbone), and the R group denotes the variable side chain of each amino acid residue. The *N*-terminus is the first residue in the chain.

The *C*-terminal carboxyl group can be changed by **post-translational modification** such as **amidation**. An amidated *C*-terminus is more common in smaller proteins, or **peptides**, including **hormones** and **toxins**, and may be important for biological function or stability. Another important covalent modification that occurs only at the *C*-terminal residue is the attachment of **membrane anchors** to secure an otherwise water-soluble protein to a hydrophobic **membrane**. Proteins can be degraded by **carboxypeptidases** that specifically hydrolyze the *C*-terminal peptide bond.

[See also ***N*-terminus** and **Polypeptide chain**.]

C-VALUE

R. J. BRITTEN

The *C*-value is the long-used term for the mystery, or paradox, of the wide range of **genome** sizes present among **eukaryotes**. They range from 13 Mbp (megabase pairs, 1.3×10^7)

for **yeast** to 1.7×10^{11} bp for *Amphiuma*, or the mudpuppy, a urodele amphibian. The units used above, the number of nucleotide pairs in one **haploid** set of DNA, should be used consistently. In much of the literature, the amount of DNA per cell is expressed in picograms (1 pg is 9.8×10^8 nucleotides), and the classic tables are often in error due to the limited accuracy of the methods used. The most accurate genome sizes are those for which complete nucleotide sequences are available; for example, the yeast genome has 13,105,020 bp (base pairs), give or take a small uncertainty because not all copies of repeated sequences (see **Repetitive DNA**) have been sequenced. The smallest bacterial genome accurately known from sequence is that of ***Mycoplasma*** *genitalium*, with 580,073 bp. Viruses have even smaller genomes, but they are not free-living and depend on their hosts for much of the required information. A part of the range in DNA content is apparently due to the different requirements of different species for genetic information, but that is not all there is to it, by any means.

The familiar fruit fly, **Drosophila**, is a complete animal, with all of the necessary structures, and it is not surprising that it has more DNA than yeast, which has an apparently simpler structure. The genome of *Drosophila melanogaster* is about 170 Mbp. It is easy to imagine that more genes and associated regulatory systems are needed to program the development and specify the complex structure of an animal. That is clearly true, but life is not so simple. The yeast genome has little space between its **genes** and almost no **introns**, and it is this compactness of organization that accounts for the relatively small size of its genome. The Drosophila genome, and plant and animal genomes in general, do not share this very compact organization, although some are more compact than others. They typically have large intergenic spaces, and most of the genome is not made up of genes and regulatory systems and has no known function. Most mammals, including humans, have about 3×10^9 bp, and it is not obvious why we require 20 times more DNA than does *Drosophila*. Some believe it is required for the brain, but humans have no more DNA than do mice. It is not reasonable that the mudpuppy requires almost 50 times more genetic information than we do, and its large genome must have some other explanation. One might expect that the very large urodele genomes are burdened with repeated sequences, and they do have a larger fraction of repeated sequences (90%) than *Drosophila* (30%). That is not the only source of the large genome size, however, since the so-called single-copy DNA is also very large in absolute quantity in the large urodele genomes. The genome sizes of single-celled protozoa range from about 1.0×10^9 to 1.0×10^{12}, but that information does nothing to reduce the mystery.

The "*C*-value paradox" remains as paradoxical as when it was first observed. It is known that a more compact genome is typically more compact in many features, such as smaller introns and smaller amounts of repeated sequences. **Natural selection** forces controlling genome size must exist, but remain unknown. How genomes have evolved is not well known, but it seems that there have been events of growth, some due to doubling of the entire set of chromosomes, others due to increases in the interspersed repeated sequences. Extraordinary reductions in DNA have also occurred in some species, while closely related species have retained their large genomes. There is a correlation between genome size and cell size, but it is not known how the two are connected, although cell size does affect metabolic rates. All that can be said is that genome size is not a parameter that is understood, although it is nevertheless intimately connected with gene structure and organization. That mammals have a quite uniform genome size is evidence for selective forces that control the *C*-value. Some geneticists consider the extra DNA in large genomes as a possible limitation in searching for genomic and regulatory function and have favored organisms with relatively small genomes, such as the puffer fish (*Fugu rubripes*) and a weed (***Arabidopsis*** *thaliana*). Of course, the small genomes do reduce the amount of work required to determine the complete genome sequence.

Suggestions for Further Reading

T. Cavalier-Smith, ed. (1985) *The Evolution of Genome Size*, John Wiley & Sons, Chichester, U.K.

B. John and G. L. G. Miklos (1988) *The Eukaryote Genome in Development and Evolution*, Allen & Unwin, London.

CAAT BOX

PIETER L. DEHASETH
DAVID R. SETZER

A CAAT box is a *cis*-acting regulatory element in **transcription** (see **Promoter**, Eukaryotic promoters) that contains the sequence 5′-CCAAT-3′ at its core and is bound by one or more different **transcription factors**. These include but are not limited to C/EBP and various members of the CTF/NF1 family of proteins. Because of the multiplicity of transcription factors that recognize a CAAT box, it is usually not possible to determine the identity of the particular factor that regulates expression of a gene by simple sequence inspection. CAAT boxes are found upstream of a number of genes transcribed by **RNA polymerase** II and increase the transcriptional rate of the gene above the level that could be achieved by the transcriptional core machinery in the absence of a CAAT box. These *cis*-acting sequence elements are generally found upstream of the **TATA box**, but their exact position relative to it is variable, and they may be found in either orientation relative to the direction of transcription from the downstream initiation site.

CADHERINS

DAVID PARRY

The cadherins are a family of **transmembrane** proteins that regulate cell–cell adhesion in a Ca^{2+}-dependent manner during **embryogenesis** and as the animal matures (1). Three general groups of cadherins exist within the superfamily: (i) the "classical" cadherins, (ii) the desmosomal cadherins, and (iii) the protocadherins (1). The classical cadherins, of which there are more than 40 examples, display much conservation in their structures. The sequences of their cytoplasmic domains may be as much as 90% identical for proteins within the same species and up to 60% between widely divergent species. In contrast,

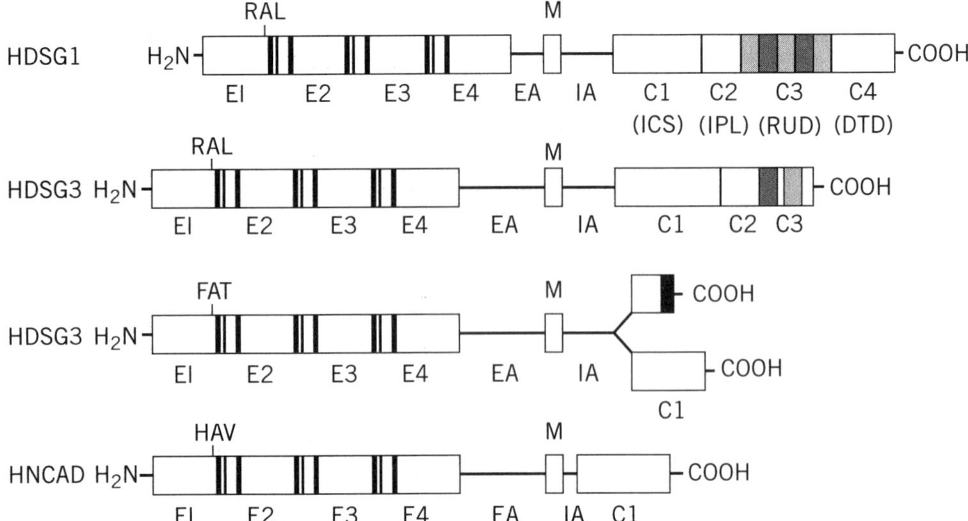

Figure 1. Comparison of the domain structures of human desmoglein 1 (HDSG1), desmoglein 3 (HDSG3), desmocollin 3 (HDSC3), and *N*-cadherin (NCAD). This illustrates the structural similarities between the classical cadherins and those in the desmosomes. The transmembrane α-helical segment (M) separates the extracellular domains E1 to E4 and the extracellular anchor EA from the intracellular anchor IA and the intracellular domains C1 to C4. Putative calcium-binding domains in the E domains are shown as shaded rectangles. The two desmocollin 3 forms result from alternative splicing. The black rectangle represents 11 amino acids at the *C*-terminal end that are unique to the shorter structure. (From Ref. 5, with permission.)

there is more variation in sequence within the extracellular domains, and 30% to 60% identity is common. These cadherins are located at the adherens junction and at other points of contact between cells that lack a highly organized structure. The desmosomal cadherins, on the other hand, are the Ca^{2+}-dependent **cell adhesion molecules** found in **desmosomes**. Within the desmosomal cadherins there are two subclasses: the desmogleins and the desmocollins. **Isoforms** of both have been characterized. The desmogleins and desmocollins are structurally distinct, implying different roles *in vivo*. The third group of cadherins, the protocadherins, are more variable in their structures, and their claim for inclusion in the cadherin family lies with the repeating structures in their extracellular domains that are similar to those displayed by the classical and desmosomal cadherins. The protocadherins also have cell adhesion properties, which again link them to the cadherin family.

Cadherins are classified as Type I integral **membrane glycoproteins** (ie, their *N*-termini lie on the extracellular side of the cell membrane) and, with one exception (glycosylphosphatidylinositol-anchored T-cadherin), each molecule contains a single α-**helical** transmembrane-spanning region (M) separating the cytoplasmic from the extracellular domain (1). The extracellular domain generally contains four quasi-repeating motifs (E1, E2, E3, and E4), each about 112 residues long and with one or two potential Ca^{2+}-binding sites per repeat (Fig. 1). A (partial) fifth domain of quite variable length has been designated the *extracellular anchor* (EA). Some at least of the first four repeats participate in an antiparallel overlap with similar domains in cadherin molecules emanating from other cells, thus facilitating homotypic assembly under Ca^{2+} control. Other proteins are also believed to be involved in this interaction. Details of the conformation of a cadherin repeat have recently been obtained using nuclear magnetic resonance (NMR) and X-ray diffraction methods (2, 3). These repeats are also important in determining the specificity of adhesion between the different cadherins. The cytoplasmic domain, which is able to bind a variety of proteins, contains a substructure that shows considerable variations in size between different members of the cadherin family. Consider, for example, human desmoglein and human *N*-cadherin (4). Both display a proline-rich region (albeit of different lengths) termed the *intracellular anchor* (IA), followed by a highly charged region (C1). This represents the full extent of the human *N*-cadherin sequence, in contrast to human desmoglein, which contains a further 320 residues. The C1 domain, being highly conserved between *N*-cadherin and desmoglein, is likely to be involved in interactions with cytoskeletal **microfilaments** and other key proteins in the cytoplasm, such as plakoglobin. The remaining part of human desmoglein can be subdivided into a 59-residue proline-rich domain (C2), a region with novel 29-residue repeats (C3) that contain a high density of putative **phosphorylation** sites, and a *C*-terminal domain (C4) that is quite divergent in sequence between human and cow, for example. The first half of domain C4 (known as C4a) is very **nonpolar** and contains two partial repeats of length 28 residues that are rich in **glycine** residues. They are not related to the 29-residue repeat seen in domain C3. Subdomain C4b is clearly basic in character, in contrast to all of the other cytoplasmic domains, which are acidic. A model of the spatial arrangement of desmoglein in the desmosome has been proposed (4).

BIBLIOGRAPHY

1. J. A. Marrs and W. J. Nelson (1996) Cadherin cell adhesion molecules in differentiation and embryogenesis. *Int. Rev. Cytol.* **165**, 159–205.
2. M. Overduin, T. S. Harvey, S. Bagby, K. I. Tong, P. Yau, M. Takeichi, and M. Ikura (1995) Solution structure of the epithelial cadherin domain responsible for selective cell adhesion. *Science* **267**, 386–389.
3. L. Shapiro, A. M. Fannon, P. D. Kwong, A. Thompson, M. S. Lehmann, G. Grubel, J.-F. Legrand, J. Als-Nielsen, D. R. Colman, and W. A. Hendrickson, (1995) Structural basis of cell-cell adhesion by cadherins. *Nature (London)* **374**, 327–337.
4. L. A. Nilles, D. A. D. Parry, E. E. Powers, B. D. Angst, R. M. Wagner, and K. J. Green (1991) Structural analysis and expression of human desmoglein: a cadherin-like component of the desmosome. *J. Cell Sci.* **99**, 809–821.
5. A. P. Kowalczyk, T. S. Stappenbeck, D. A. D. Parry, H. L. Palka, M. L. A. Virata, E. A. Bornslaeger, L. A. Nilles, and K. J. Green (1994) Structure and function of desmosomal transmembrane core and plaque molecules. *Biophys. Chem.* **50**, 97–112.

Suggestions for Further Reading

K. J. Green and J. C. R. Jones (1996) Desmosomes and hemidesmosomes: structure and function of molecular components. *FASEB J.* **10**, 871–881.

M. Takeichi (1995) Morphogenetic roles of classic cadherins. *Curr. Opin. Cell Biol.* **7**, 619–627.

O. Huber, C. Bierkamp, and A. Kemier (1996) Cadherins and catenins in development. *Curr. Opin. Cell Biol.* **8**, 685–691.

CAENORHABDITIS

W. WOOD

Caenorhabditis is a genus of primarily free-living soil **nematodes**. The free-living species *Caenorhabditis elegans* has become well known after being chosen by Sydney Brenner in the 1960s for a concerted genetic, ultrastructural, and molecular analysis of animal development (1). A few other Caenorhabditis species have been characterized in less detail, largely for comparison, but the findings summarized in this article are for *C. elegans* unless otherwise indicated.

C. elegans was chosen for a variety of features, including its short life cycle and **hermaphroditic** mode of **development** (facilitating genetic analysis), its anatomical and cellular simplicity (<1000 somatic cells in the adult hermaphrodite), and its transparency and small size (facilitating light **microscopy** of developing animals and ultrastructural analysis, respectively). Its genome size has turned out to be relatively small as well (100 Mb), making molecular biology convenient. An entire field of *C. elegans* biology has grown out of Brenner's original investigations. As a result of this effort and the relative simplicity of "the worm," as it is sometimes referred to, *C. elegans* is now the most completely understood metazoan in terms of genetics, ultrastructure, development, and physiology, and it has taken its place with the fruit fly ***Drosophila*** and the laboratory **mouse** as one of the most important model organisms for research biology (2,3). Its entire cell lineage from zygote to adult has been determined; its ultrastructure including the complete connectivity of its nervous system has been described in detail; an extensive genetic map has been assembled; and DNA sequencing of its entire **genome** will be completed in 1998. It is beautifully suited for the genetic approach: identifying **genes** required for any biological process of interest by mutations that affect it, molecularly characterizing these genes after isolation by positional **cloning**, and eventually studying the biochemical functions of the corresponding gene products. As in all organisms, the genetic control of development and physiology is still only beginning to be understood in *C. elegans*. The rapidly accumulating detailed knowledge of the worm's biology, as well as the recent realization that many developmental and physiological mechanisms are remarkably conserved throughout the animal kingdom, make *C. elegans* a powerful experimental system for investigating a variety of the most important questions in current biological research.

GENERAL DESCRIPTION

C. elegans inhabits soil in most parts of the world and feeds on soil **bacteria**. The adults are about 1 mm in length, and the two sexes, hermaphrodites and males, are morphologically distinguishable (Fig. 1). The hermaphrodites produce both eggs and sperm and can self-fertilize. Males can mate with hermaphrodites to give cross progeny; hermaphrodites cannot fertilize each other. A hermaphrodite can produce about 300 self progeny, in addition to cross progeny if mated. **Embryos** begin development in the hermaphrodite uterus and are laid at the gastrulation stage. They hatch as first-stage juveniles (L1 larvae), which as in all nematodes grow through three subsequent larval stages punctuated by molts, before a final molt to the sexually mature adult. The adult reproductive period lasts for about 4 days, after which the animals live for an additional 2 weeks.

C. elegans is one of the simplest metazoans and has fixed numbers of **somatic cells**: 959 in the adult hermaphrodite and 1031 in the adult male. In the laboratory it can be cultured conveniently on agar plates spread with *Escherichia coli* bacteria, or in liquid culture when larger quantities are desired for biochemical work. Individual animals can be observed and transferred from plate to plate with a platinum wire "worm pick" under a dissecting microscope, which is sufficient for scoring many mutant phenotypes. Taking advantage of the animal's transparency throughout the life cycle, it is possible to follow individual cells during development at higher magnification using a compound microscope, preferably equipped with Nomarski differential interference-contrast optics (see **Microscopy**).

GENETICS AND THE GENOME

C. elegans is **diploid**, with five **autosomes** (I–V) and a sex **chromosome** (X). Hermaphrodites are XX and males XO; sex is determined by the X chromosome to autosome ratio. Males arise spontaneously in self-fertilizing hermaphrodite populations at a frequency of about 0.2%, as the result of X chromosome **nondisjunction**.

The **haploid** genome size of *C. elegans* is 100 Mb, about 10 times that of **yeast**, one-half that of *Drosophila*, and one-thirtieth that of mammals. Based on frequency of predicted coding units in the genomic DNA sequence, which is now nearly complete, the estimated total number of functional **genes** is about 13,000. Over 1000 genetic loci have been identified by

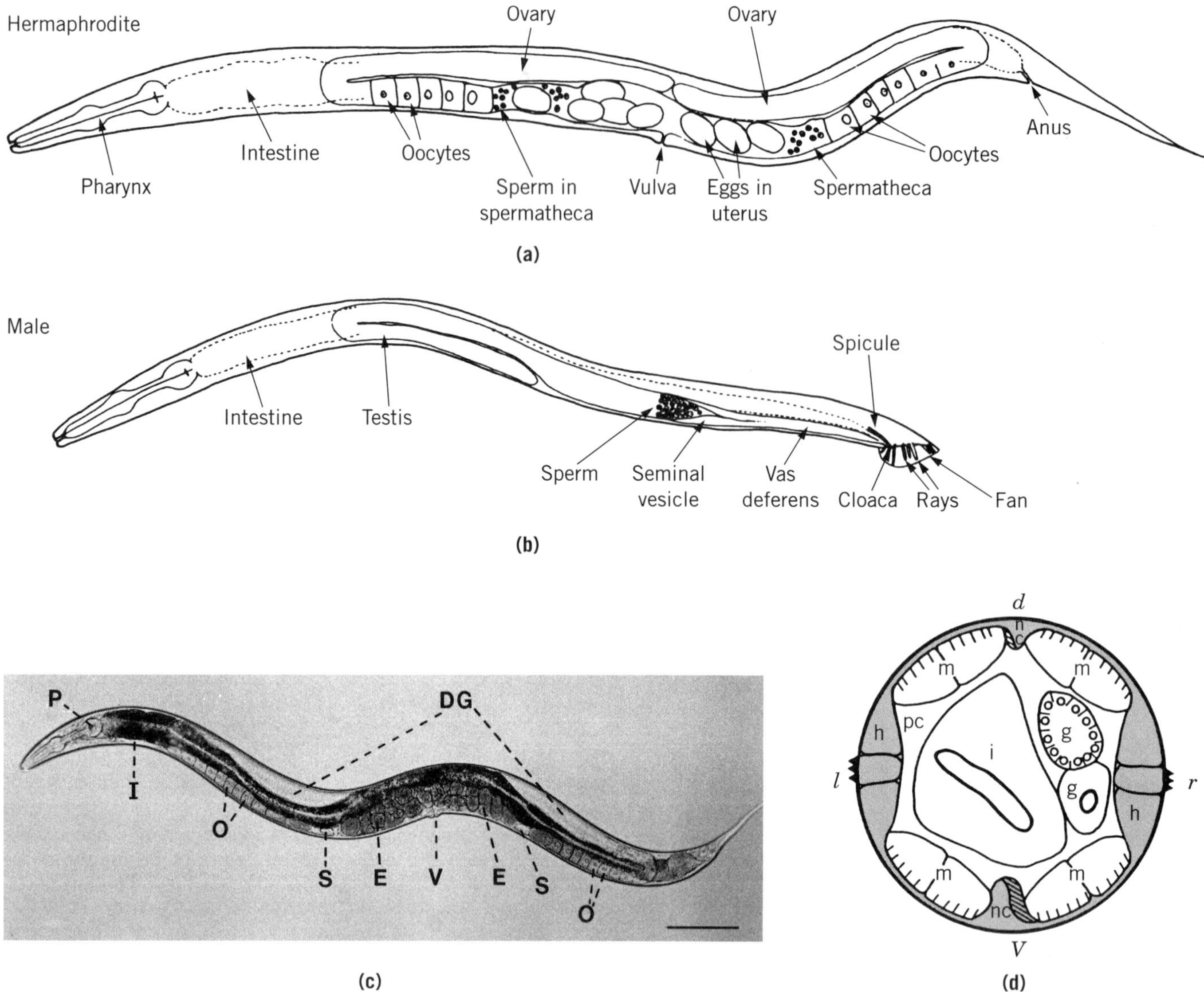

Figure 1. Anatomy of hermaphrodite and male *C. elegans* adults. Major structural features mentioned in the text are labeled in the diagrams and the photograph. (**a**) Hermaphrodite. (**b**) Male. (**c**) Hermaphrodite, bright-field photomicrograph. Scale bar 0.1 mm. P, pharynx; I, intestine; O, oocytes; S, spermatheca; E, eggs; V, vulva; DG, distal gonad. (**d**) Hermaphrodite, cross section through the anterior, viewed toward the anterior. *d*, *v*, *l*, *r*, dorsal, ventral, left, right, respectively; m, muscle; h, hypodermis; I, intestine; g, gonad; pc, pseudocoelom; nc, nerve cord. Note the treads (alae) on the left and right sides of the body.

mutation following chemical **mutagenesis** and mapped to the six linkage groups. Positional cloning of mutationally identified genes is facilitated by an extensive physical map of overlapping clones, which is anchored to the genetic map at many locations. The DNA sequence of the entire genome will soon be completely known.

ANATOMY

C. elegans has the typical nematode body plan (see **Nematodes**): an outer tube of hypodermis with attached musculature and neurons, surrounding an interior space (the *pseudocoelom*), which contains the gut and, in adults, the gonad (see Fig. 1). The hypodermis secretes a three-layered, **collagenous** cuticle that is shed and replaced at each larval molt (see text below). Four bands of body-wall muscles run the length of the animal. The neuromuscular system bends the body only in the dorsal–ventral direction, so that the animal always lies on one side or the other on a solid surface. Specialized hypodermal cells on each side secrete "treads," known as *alae*, on which the worm moves. The digestive system consists of a muscular pharynx in the head, which crushes bacteria and pumps them into the intestine, which empties at the anus near the tail. An excretory system controls the hydrostatic pressure on which the worm depends to maintain its body shape.

The entire nervous system of the hermaphrodite consists of only 302 neurons and 56 supporting and glial cells. The major ganglion is the nerve ring at the base of the pharynx, which sends processes down ventral and dorsal nerve cords to a secondary ganglion in the tail. Motor neurons from the cords enervate the body-wall muscles. Sensory neurons extend into the nerve ring from chemoreceptors and touch receptors in the head. Hermaphrodites have specialized neurons that control egg-laying (see text below); males lack these but have additional neurons that provide input from sensory structures in the tail and control male mating behavior.

The hermaphrodite reproductive system is comprised of a bilobed gonad, with one lobe extending anteriorly and one posteriorly from the uterus near the middle of the animal. The gonad consists of a somatic sheath surrounding the germ cells and has several specialized regions. The distal arm of each lobe contains **mitotically** dividing germ-cell **nuclei** in a common **cytoplasm**. The nuclei enter **meiosis** as they move away from the distal tip. The earliest nuclei to mature differentiate into **sperm** during the fourth larval stage; at the molt to adulthood the germ line switches sex, and subsequent meiotic nuclei are recruited to form **oocytes** as they round the bend into the proximal arm, which contains the oviduct. At the proximal end of the oviduct is the spermatheca, containing stored sperm, through which the oocytes pass and become fertilized on their way to the uterus. A muscular vulva connects the uterus to the outside and serves as the egg-laying apparatus. In the male, the gonad is single lobed and produces only **sperm**, which are stored in the vas deferens and released through the cloaca in the tail. The fan-shaped male tail is specialized for mating, which is accomplished by deposition of sperm through the hermaphrodite vulva into the uterus, where they move to the spermatheca and compete for oocytes with the resident hermaphrodite sperm.

FERTILIZATION AND EMBRYONIC DEVELOPMENT

Following fertilization, a tough chitinous egg shell forms around the zygote, and embryonic cleavage begins. The spheroidal embryos, about 70 μm in length, are viable if dissected out of the hermaphrodite at this time, but normally they remain in the uterus for about 2 h until the onset of gastrulation, when they are laid through the vulva. Gastrulation is simple; only 53 cells move from the exterior to the interior, but the result is a triploblastic embryo with outer ectodermal precursor cells that will form the hypodermis and nervous system, inner endodermal precursors that will form the gut, and in between a layer of mesodermal precursors that will give rise to muscles and the somatic gonad. By 6 h after fertilization, midway through embryogenesis, the embryo consists of about 600 cells. At this point, cell proliferation essentially ceases and a process of **morphogenesis** begins which literally squeezes the spheroidal embryo into the shape of a worm as organogenesis proceeds internally and cuticle is formed externally. During this time about 40 cells undergo **programmed cell death** and are engulfed by neighboring cells. At the end of embryogenesis the worm, about 3.5 times the length of the original embryo, digests the shell from the inside and hatches out of the egg.

The process of **embryogenesis** is essentially invariant at the cellular level. It proceeds by a stereotyped series of **cell divisions** (Fig. 2), whose timing and relative spatial orientations are the same in every embryo. This feature and the transparency of the embryo made it possible for John Sulston and his colleagues to trace out the entire embryonic cell lineage from fertilization to hatching, so that the ancestry of each of

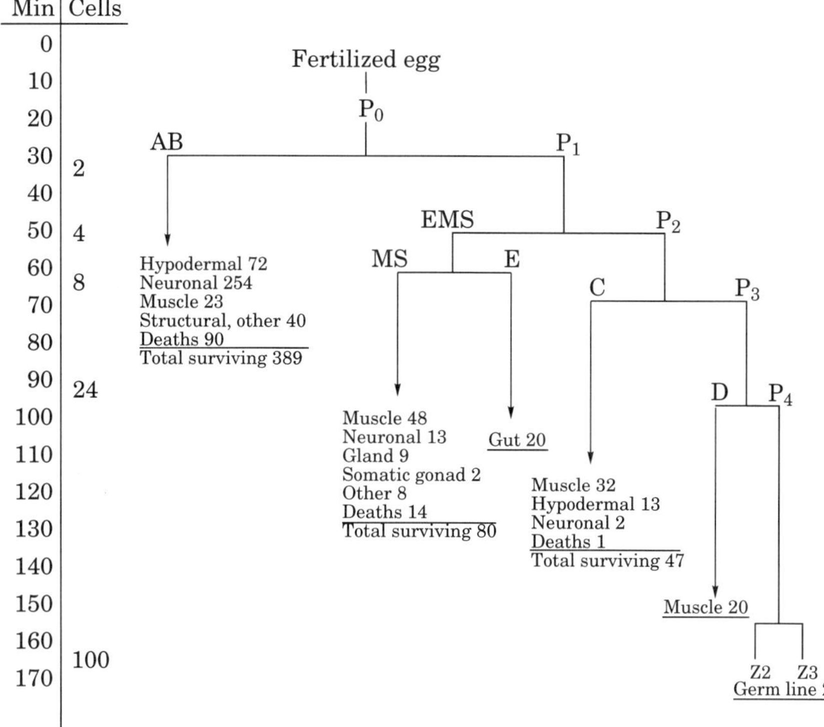

Figure 2. Early divisions in the *C. elegans* embryonic cell lineage, showing cells and tissues derived from each of the major branches. The vertical axis shows time after fertilization and total number of cells in the embryo. Horizontal lines indicate times of cell divisions. P_0 through P_4 are germ line cells; the cells named AB, MS, E, C, and D are somatic founder cells for the various branches of the lineage. The number of cells of each different type produced in each branch is indicated. Note that many cells are programmed to die during embryonic development, especially in the AB and MS branches.

the 558 cells in the L1 is known (4). Most cells are born close to their final locations; only about 12 cells undergo long-range migrations during embryogenesis.

Genetic and molecular analysis of the cleavage stage has revealed that cell fates in the early embryo (Fig. 2) are determined by a combination of maternally derived factors, which segregate asymmetrically to specific cells during cleavage, and cell-signaling mechanisms by which certain cells induce new cell fates in their neighbors. This patterning process appears to proceed by evolutionarily conserved mechanisms that are common to all embryos; for example, the signaling pathways so far known to be employed include several that are also important for insect and vertebrate **development**, such as the *wingless* (Wnt), and ***Notch*** pathways (5).

LARVAL DEVELOPMENT

Larval Stages

After hatching as an L1, the worm grows through three more larval stages, L2–L4 (Fig. 3), separated by molts at which a new cuticle is secreted by the underlying hypodermis and the old one shed. The newly hatched L1, 250 μm long with 558 cells in the hermaphrodite and 560 in the male, includes functional digestive, neuromuscular, and neurosensory systems and a gonad primordium consisting of two somatic and two germline cells. Most of the cells in the L1 divide no further during larval development. However, 55 are *blast* cells that undergo additional divisions during larval development to produce, primarily, the adult reproductive structures and the neurons that control them: the hermaphrodite and male gonads, the vulva and egg-laying muscles in the hermaphrodite, and the sex muscles and specialized tail mating structures in the male. Additional motor neurons are also produced in the nerve cords for finer control of the body-wall muscles. As in the embryo, the invariance of these processes, the transparency of all stages, and the small number of cells involved made it possible to map the larval cell lineages in both sexes from L1 to adulthood (6). Consequently, the entire cell ancestry of *C. elegans* is known, from fertilized egg to adult.

Cell Interactions during Larval Development

Elaboration of larval structures involves cell interactions, again occurring by evolutionarily conserved signaling pathways such as those involving homologues of Notch, Wnt, **epidermal growth factor** (EGF), and **transforming growth factor** (TGF)-β-superfamily ligands as the signaling molecules. Application of the genetic approach to the process of vulval development, for example, has revealed details of a ***ras***-based pathway by which the developing gonad signals underlying hypodermal cells to differentiate using an EGF-like ligand. The molecular components are very similar to those of *ras* pathways found in mammalian cells, in which conversion of *ras* to an **oncogene** causes a variety of cancers, and the *C. elegans* pathway has become an important model system for cancer research.

Cell interactions during larval development also include guidance cues by which several cells, including the growing axonal processes of motorneurons and the distal tip cells of the enlarging hermaphrodite gonad, find their way along the basement membrane of the pseudocoelom en route to their final destinations. Again, application of the genetic approach in *C. elegans* has shown that this process is accomplished by a conserved mechanism: The so-called *netrin* ligands and their **receptors** that guide these cells in the worm are remarkably similar to the molecules that guide axonal growth from the neural floor plate to the spinal cord in developing avian and mammalian embryos, making *C. elegans* a potentially useful model system for study of nerve regeneration.

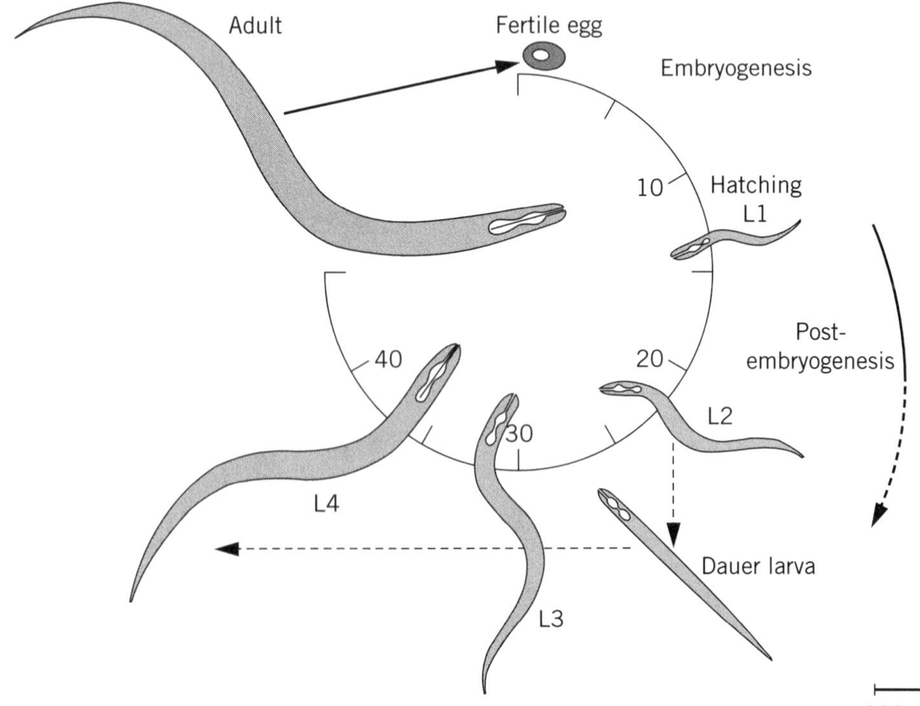

Figure 3. The *C. elegans* life cycle. Clock circle indicates hours of development after fertilization at 25°C. Eggs containing developing embryos are laid at about 2 h. The L1 larvae, hatching at about 14 h, undergo four more molts at the times indicated as they grow to adulthood. See text for further explanation.

The cell divisions and molts in larval development occur in a precisely timed sequence. Again, application of the genetic approach has identified many of the genes in the timing mechanism and is providing new information on how developmental clocks function.

The Dauer Larva

Like most free-living nematodes, *C. elegans* can molt to an alternative form of the L3, called a *dauer larva* (German for "enduring larva") when conditions are unsuitable for reproduction. Dauer larvae, often simply called dauers, do not feed, are relatively resistant to drying, and can live for up to a year if desiccation is prevented. If conditions improve, they can molt to the L4 stage and resume the normal developmental pathway, with no change in subsequent lifespan. Dauer development is triggered by lack of food and overcrowding, signaled by a pheromone of unknown nature that stimulates chemosensory structures in the head. This process is of interest for at least two reasons. First, the signaling mechanism involves homologues of ligands, receptors, and downstream components of the evolutionarily conserved TGF-β pathway, which is of widespread importance but not fully understood in mammalian development. Second, entry into the pathway of dauer development appears to turn off genes that normally limit the *C. elegans* lifespan to less than 3 weeks, allowing the dauer to live much longer. This finding, and the experimentally convenient short normal lifespan, make *C. elegans* an exciting model system for studying genetic control of aging, which is under active investigation (7).

BEHAVIOR AND THE NERVOUS SYSTEM

C. elegans can move forward or backward and change direction, by coordinated flexing of its body-wall muscles. It is touch-sensitive, moving forward in response to a touch on the tail and backward in response to a touch on the head. Chemosensors in the head, connected by sensory neurons to the ring ganglion, can detect a variety of ions as well as volatile odorants and elicit either an attractive or repulsive response (see **Chemotaxis**). For example, *C. elegans* is attracted to Na^+, K^+, Cl^-, and several alcohols and ketones; it is repelled by Cu^{2+}, acid pH, D-tryptophan, and benzaldehyde. Males are attracted by a pheromone of unknown nature that is produced by hermaphrodites. Although some of its relatives have photoreceptors, *C. elegans* does not appear to respond to light or use it as a sensory cue. It does, however, sense and respond to temperature and will move to a preferred point in a temperature gradient.

Is an animal with only 302 neurons capable of learning? Simple conditioning experiments suggest that *C. elegans* not only can habituate to several stimuli, but also is capable of associative learning, that is, learning to use a normally neutral stimulus to predict the arrival of a second more significant stimulus. For example, if presence of food is paired with one of two equally attractive ions and absence of food with the other during conditioning, the conditioned animals will preferentially move toward a source of the paired ion when tested subsequently in the absence of food, and this preference lasts up to 7 h after training. In addition, *C. elegans* not only can sense a temperature gradient, but also can "remember" the temperature at which it has previously fed and move to the same temperature when placed in a new gradient without food. The genetic approach should allow the genes involved in learning and memory in *C. elegans* to be identified and the mechanisms of the proteins they encode to be elucidated.

C. elegans does not show obvious circadian rhythms, but it exhibits much higher frequency (ultradian) rhythms, such as a regular defecation cycle that is repeated about once every 45 s when the animal is feeding, regardless of the temperature. As in other organisms, the mechanisms of the temperature-compensated molecular clocks that control such cycles (like the human heartbeat) are just beginning to be understood. Application of the genetic approach to cyclical behaviors in *C. elegans* is identifying the genes and proteins that control ultradian rhythms.

CURRENT INVESTIGATION AND ONLINE INFORMATION ABOUT *C. ELEGANS*

Knowledge about *C. elegans* genetics, development, and behavior is accumulating rapidly in several areas of currently exciting biological research, some of which are mentioned above. These include mechanisms of pattern formation in development, origins of left–right asymmetry in animal body plans, cell fate determination, programmed cell death, cell migration and guidance, developmental timing mechanisms, physiological sensory mechanisms, organization of animal genomes, and animal **evolution**.

As a supplement to the references listed below, much current knowledge of *C. elegans* is accessible electronically on the World Wide Web. A convenient access site can be found at *http://eatworms.swmed.edu*.

BIBLIOGRAPHY

1. S. Brenner (1974) *Genetics* **77**, 71–94.
2. W. B. Wood et al. (eds.) (1988) *The Nematode Caenorhabditis elegans*, Cold Spring Harbor Laboratory Press, Cold Spring Harbor, NY.
3. D. L. Riddle, T. Blumenthal, B. J. Meyer, and J. R. Priess (eds.) (1997) *C. elegans II*, Cold Spring Harbor Laboratory Press, Cold Spring Harbor, NY.
4. J. Sulston, E. Schierenberg, J. White, and J. Thomson (1983) *Dev. Biol.* **100**, 64–119.
5. R. Schnabel and J. R. Priess (1997) In *C. elegans II* (D. L. Riddle, T. Blumenthal, B. J. Meyer, and J. R. Priess, eds.), Cold Spring Harbor Laboratory Press, Cold Spring Harbor, NY, pp. 361–382.
6. J. Sulston and H. Horvitz (1977) *Dev. Biol.* **56**, 110–156.
7. C. Kenyon (1997) In *C. elegans II* (D. L. Riddle, T. Blumenthal, B. J. Meyer, and J. R. Priess, eds.), Cold Spring Harbor Laboratory Press, Cold Spring Harbor, NY, pp. 791–813.

Suggestions for Further Reading

S. Brenner (1974) The genetics of *Caenorhabditis elegans*. *Genetics* **77**, 71–94. (The classic paper in which Brenner first presented his rationale, descriptions of morphological and behavioral mutants, and preliminary genetic analysis of *C. elegans*.)

D. L. Riddle, T. Blumenthal, B. J. Meyer, and J. R. Priess (eds.) (1997) *C. elegans II*, Cold Spring Harbor Laboratory Press, Cold Spring Harbor, NY. (The sequel to the previous reference book. A good up-to-date summary of current knowledge about *C. elegans*.)

J. Sulston and H. Horvitz (1977) Post-embryonic cell lineages of the nematode *Caenorhabditis elegans*. *Dev. Biol.* **56**, 110–156. (The first extensive cell lineaging of *C. elegans*, describing all the cell lineages during larval development.)

J. Sulston, E. Schierenberg, J. White, and J. Thomson (1983) The embryonic cell lineage of the nematode *Caenorhabditis elegans*. *Dev. Biol.* **100**, 64–119. (The monumental paper in which Sulston and his co-workers published the complete embryonic cell lineage of *C. elegans*.)

W. B. Wood (1988) In *The Nematode Caenorhabditis elegans* (W. B. Wood and the community of *C. elegans* researchers, eds.), Cold Spring Harbor Laboratory Press, Cold Spring Harbor, NY, pp. 1–16. (The introductory chapter to a useful general reference on *C. elegans*.)

CAGED ATP

JACK KAPLAN

The synthesis of caged ATP (caged ADP and caged phosphate) and many of the properties that make these reagents useful in a variety of biological applications were first published in 1978 (1). The driving force for the development of such a reagent was the desire to be able to introduce ATP rapidly (and synchronously) at sites of biological interest at a desired time. A pulse of ATP would be released by **light activation** and would then initiate the processes being studied (Fig. 1). In contrast to **affinity labeling**, here the ATP analogue should not bind to its site prior to activation and ATP is released freely in solution. The strategy employed was an approach that had previously been used in synthetic organic chemistry (ie, photodeprotection). In the chemical applications, a desired functionality in a molecule was modified and protected from a variety of reagents and conditions during a multistep synthesis, to be deprotected at a later step by a light-activated process. In biological applications, the caged substrates are "protected" from their receptor, **enzyme**, or binding site by the chromophoric moiety, and following the pulse of light they are released to their biological target.

The properties of molecules appropriate for the study of biological systems are subject to more constraints than those used in synthetic organic chemistry. Much of the subsequent work on various caged ATP molecules and other caged compounds has been aimed at refining or improving their basic properties to expand the areas of application or overcome a particular limitation. The design requirements for a caged ATP are illustrative of the needs of any caged biological ligand or substrate [see **Light-activated (caged) ligands**] (2).

The basic requirements for a biologically useful caged ATP (and for most other caged compounds) are as follows: (i) The activating light must be at wavelengths long enough to avoid damage to biological material (usually greater than 300 nm). (ii) The photorelease process must be as efficient as possible; that is, the quantum yield (ratio of molecules of product obtained to molecules of caged compound excited) must be as high as possible. (iii) The photorelease process should be as rapid as possible, relative to the rate of the process of interest; this has been achieved in the tens of microseconds to millisecond range, at normal temperatures and pH values. (iv) The photoreleased fragment, which usually is the modified chromophore, should not be harmful to biological materials (see text below). (v) The **absorbance** of the caged compound should be reasonably high at the exciting wavelength or wavelength range. Initially it was thought that too high an absorbance might lead to nonuniform release of substrate across the depth of an illuminated sample. Interestingly, for the recent application of multiphoton excitation, very high absorbances are an advantage. (vi) Prior to photolysis, the caged substrate should neither bind to nor interact with the biological material of interest. In many respects, the first caged ATP reported and subsequently used (but the second synthesized) fulfilled many of these requirements. In a variety of systems, one or other of these properties has been less than ideal, and this has led to various variations on the original structure in attempts to provide an improved caged ATP.

$$L{-}X \xrightarrow{h\nu} L + X'$$

Figure 1. Basis of photorelease of a caged substrate. The biological ligand (L) or substrate is rendered inactive by the attachment of a photocleavable chromophore (x). Following photoactivation the ligand is released accompanied by the modified chromophore, or photofragment.

SYNTHESIS AND PHOTOCHEMICAL PROPERTIES OF CAGED ATP

The initial synthesis of caged ATP was based on a strategy of first preparing caged phosphate and then coupling this to ADP. This had the advantage that the photochemical lability, quantum yield, and so on, could be first characterized, using caged phosphate and the readily assayed free substrate, inorganic phosphate (1). This route suffered from the following disadvantages: (i) It was less direct than was desirable to achieve a gamma ^{32}P-labeled caged ATP for a variety of **phosphorylation** studies, and (ii) it was not generally applicable to a wide range of phosphate-bearing biological ligands that might be desirable to cage. It would be better to have a more universal caging moiety that could be readily attached to the nucleotide or other organophosphates. These disadvantages were overcome by the development of a synthesis by Trentham and co-workers, who were able to attach the protecting 2-nitrobenzyl-based chromophore directly to the terminal phosphate of ATP, using a diazoprecursor (3). This was based on an approach that had been used by others to prepare a photosensitive **cyclic AMP** phosphotriester analogue (4). This method was used to prepare a wide array of caged phosphorylated biological ligands, including GTP, ADP, inositol trisphosphate, and so on (3).

The quantum yield for the first successful caged ATP was 0.54 and has not been bettered by subsequent analogues. The first caged ATP that was synthesized, the primary 2-nitrobenzyl analogue, yielded on photolysis some ATP, but only very low levels, even though photolysis was complete. It was hypothesized that this was due to a reaction between ATP and the released photofragment (2-nitrosobenzaldehyde), so that a chemically modified ATP resulted (1). The 2-nitrophenylethyl analogue (from the secondary benzyl alcohol) then became the molecule of choice, because photolysis yielded the less reactive 2-nitrosoacetophenone and free ATP in high yield (Fig. 2). This is the caged ATP that has been used most frequently in biological studies. The absorbance properties of caged ATP are simply the sum of the spectra of the 2-nitrobenzyl moiety and of ATP. This results in a long tail of

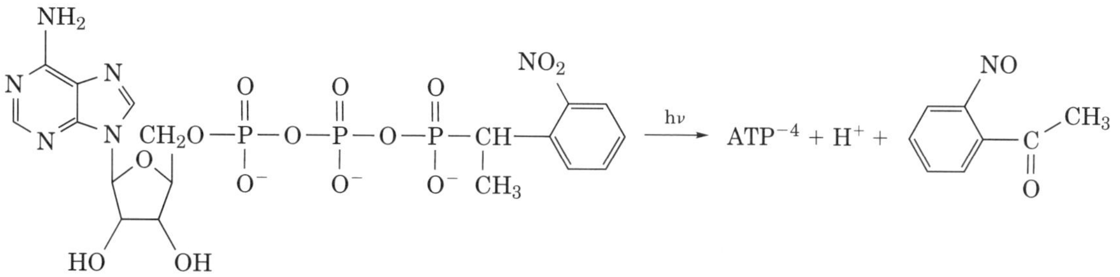

$$\text{Caged ATP} \xrightarrow{h\nu} \text{ATP}^{-4} + \text{H}^+ + \text{2-nitrosoacetophenone}$$

Figure 2. Photolysis of caged ATP. Photolysis of caged ATP produces a proton, free ATP, and the photofragment, 2-nitrosoacetophenone.

absorbance that extends to above 350 nm. This has enabled illumination above 300 nm to be employed (using lasers or flash-lamps) so that photodamage due to absorption by most biological samples is avoided.

MECHANISM AND KINETICS OF PHOTORELEASE

Although the earliest studies showed that ATP release could be achieved in a subsecond time frame, ideas about the kinetics of the release process and the likely mechanism were initiated by the report of McCray et al. in 1980 (5). From this work it was apparent that the rate of release of ATP from caged ATP was pH-dependent and in the millisecond time range.

Importantly, it was also pointed out in this work that there was a chromophoric intermediate on the reaction pathway, identified tentatively as an aci-nitro intermediate. The intermediacy of this type of chromophore has proven useful in later mechanistic studies of the photorelease of caged substrates, because it is central to almost all photorelease processes in which the protecting moiety is a 2-nitrobenzyl residue. Another tool that has been useful in mechanistic studies, and profitably used by Trentham and colleagues, is the reaction of the photoreleased nitrosoketone with **thiol groups**. This reaction had been initially employed by Kaplan et al. (1) to protect biological samples against any damaging effects of the photofragment, because it was known that nitrosoketones react readily with thiols. It was shown that the inhibition of the Na,K-ATPase enzyme due to the actions of the photofragment could be prevented by the simultaneous presence of **dithiothreitol** (DTT) or similar reagents. Furthermore, the protecting thiol had to be present in amounts at least stoichiometric with the released photofragment. In many subsequent applications, it has become routine to include reduced **glutathione** or DTT in the reaction media to mop up the released photofragment. Trentham and colleagues (6) used the rapid rate of this reaction to examine the release rates of the photofragment. This proved valuable because with caged ATP, as with many photoreleased biological substrates, it is often difficult to identify a biological process that is sufficiently fast to be used unambiguously as a bioassay to determine the photorelease rate of substrate. It should be emphasized, however, that for all new caged substrates it is essential to measure the release rate of the photoproduct of interest and not merely the rate of decay of an intermediate on the release pathway. There have been several reports of biphasic decay of putative aci-nitro intermediates when their breakdown was followed using transient ultraviolet spectroscopy (7,8).

The most recent estimates (obtained from time-resolved infrared spectroscopy) of the formation of free ATP following the photolysis of caged ATP are about 220 s^{-1} at pH 7 and 22°C, a rate that is the same as the aci-nitro anion intermediate decay rate (9). An outline of the steps involved in the photolysis pathway are shown in Figure 3.

OTHER CAGED ATP STRUCTURES

The most familiar caged ATP is P^3-(1-(2-nitrophenyl)ethyl) adenosine 5′-triphosphate. Variations on the basic theme have been made in the hope of increasing the long wavelength absorbance of the molecule or of speeding the rate of release. Unfortunately, no molecule yet reported has achieved this aim while simultaneously maintaining the relatively high

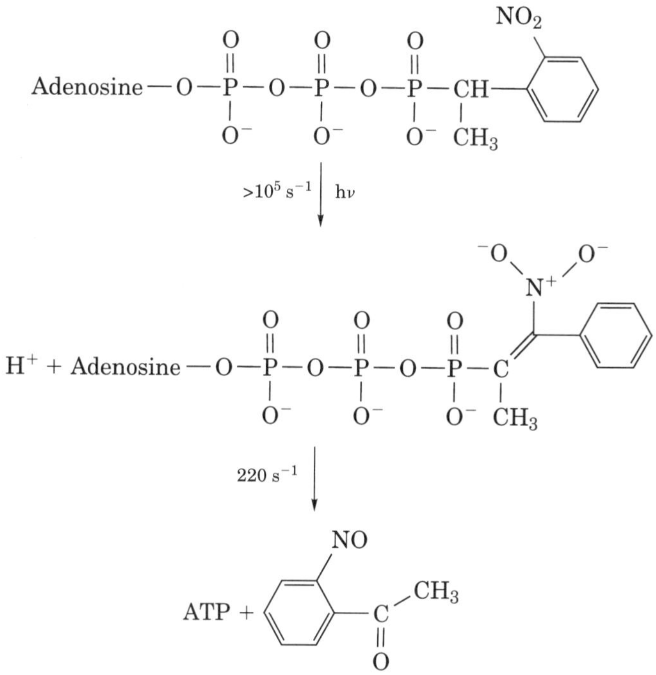

$$\text{Adenosine}-O-P(O)(O^-)-O-P(O)(O^-)-O-P(O)(O^-)-CH(CH_3)-C_6H_4NO_2 \xrightarrow[>10^5\ s^{-1}]{h\nu} H^+ + \text{Adenosine}-O-P(O)(O^-)-O-P(O)(O^-)-O-P(O)(O^-)-C(CH_3)=C_6H_4=N^+(O^-)_2 \xrightarrow{220\ s^{-1}} \text{ATP} + \text{2-nitrosoacetophenone}$$

Figure 3. Scheme of breakdown of caged-ATP following excitation. The release rate of ATP is controlled by dark reactions that follow photoexcitation and relaxation to ground states. A detailed study of the products and intermediates has been made using time-resolved infrared spectroscopy and isotopomers of caged ATP (6).

quantum yield. Most attempts have involved variations in the chromophore, while maintaining the same nitrobenzyl photochemical cleavage mechanism (10). Recently potentially useful alternative photochemistries have begun to be exploited, but as yet no generally useful molecule has emerged (11–13).

BIOLOGICAL APPLICATIONS OF CAGED ATP

The two areas where caged ATP has been used most extensively are in studying the mechanism of **active transport** by **ion pumps** and in the mechanism of muscle contraction and its regulation. In single turnover studies of the Na,K-ATPase (see text below), Forbush showed that caged ATP did in fact bind with low affinity to the enzyme prior to photolysis (14). This leads to extra complications in kinetic modeling of events following photolysis. Subsequently, similar effects have been noted in muscle fibers (15). Such observations suggest that a caged ATP with a bulkier chromophore might reduce the prephotolysis binding.

STUDIES ON ION PUMPS

The ion pumps are P-type **ATPases** that couple the transport of ions against their electrochemical potential gradient to the hydrolysis of ATP. This family of membrane proteins includes the Na pump or Na,K-ATPase, the Ca pump [from intracellular membranes (SERCA) or plasma membranes (PMCA)], **proton pumps** from a variety of organisms such as yeast, bacteria, and so on, and a range of heavy metal pumps that transport Cu, Cd, Ni, and so on, across cellular membranes throughout the animal and plant kingdom (16). The studies on ion pumps that have used caged ATP illustrate well several of the different advantages of this experimental strategy. These proteins carry out a series of biochemical transformations that are thought to accompany the transport of cations across the membrane. These transformations arise from the intermediate involvement of a phosphoenzyme, which is formed by transferring the terminal phosphate of ATP to the protein at an early step in the cycle, and from the protein to water in a later step. These **kinase** and **phosphatase** activities are linked to ion binding, translocation, and ion release steps (17). In order to probe the cation activation of some of these processes, it was necessary in the case of the Na pump to be able to initiate the pump cycle in a sealed system. This was required because the activating and inhibitory cations have very asymmetric effects, depending on whether they act at the extracellular or intracellular surface. The strategy employed was to trap caged ATP within resealed human erythrocyte ghosts. The extracellular medium could then be altered at will, and the Na pump process could be initiated by illumination of the ghost suspension and release of intracellular ATP from caged ATP. This enabled characterization of the side effects of activating and inhibiting cations on the transphosphorylation reactions under genuine initial rate conditions (18,19). The essential properties of the photorelease strategy here were the stability of caged ATP during experimental procedures, until ultraviolet irradiation, when ATP could be released in good yield synchronously inside the cells in suspension. The possibility of releasing ATP from caged ATP in a rapid synchronous fashion within an ordered structure was also exploited in structural studies of the Ca pump in oriented multilayers where diffraction before and after the release of ATP from caged ATP showed movements of the protein mass relative to the membrane structure during the reaction cycle (20). Such studies would not have been possible without the photorelease approach.

Ion pumps transfer charge during their reaction cycle across cell membranes; and electrophysiological measurements of such movements, along with their analysis, have been a central area of study. In recent years, Bamberg and co-workers (21) and Apell and co-workers (22) have initiated the use of caged ATP in a novel experimental system to analyze the mechanistic consequences of these phenomena in a variety of ion pumps. These workers have used black lipid membranes and have either attached or fused to them biomembrane fragments or vesicles containing ion pumps. Following the rapid, synchronous release of ATP from caged ATP in the medium bathing the membranes, the current-carrying and charge translocating steps can be analyzed with a conventional electrophysiological measuring system (23,24). Such studies have probed the basis of the electrogenic nature of the Na pump and of the electrically neutral basis of the gastric proton pump, for example (25). Recent studies have dissected out the electrical properties of the bacterial active K transporter, the Kdp-ATPase of *Escherichia coli* (26).

Regulation and Control of Muscle Contraction

An adequate understanding of the mechanism of muscle contraction can only be achieved from experiments in a fairly intact and complex tissue. This is because the structure of the muscle fiber and its organization is an inherent feature of its function. The sliding filament model for the molecular basis of muscle contraction is the prevailing paradigm in this field, and its central feature is this relationship (27,28). Thus it is necessary to be able to perform high-resolution kinetic studies in an ordered array of macromolecules. Recent efforts have been aimed at understanding the precise steps in the ATP hydrolysis cycle carried out by myosin that lead to the generation of force during contraction. Caged ATP has found application in detailed studies on the kinetics of ATP hydrolysis, structural changes in the fiber organization associated with contraction and relaxation (at the macroscopic and molecular levels), spectroscopic measurements on muscle fibers, and the process of relaxation associated with Ca pumping by the sarcoplasmic reticulum. Studies in all these areas have received considerable impetus by the introduction of the photorelease strategies. It has been shown that it is possible to photorelease ATP (or other substrates) in chemically skinned muscle fibers in the millisecond time range and to cause the synchronized initiation of biochemical processes that produce contraction (or relaxation) in the bundle of fibers (29,30). Prior to this technology, such experiments were limited by the diffusional delays that were inherent in mechanically adding substrate and allowing it to diffuse into the fiber to its site of action. Initial studies that employed this approach to analyze the kinetics of the mechanical processes have now been greatly extended; by using covalently attached chromophoric reporters or paramagnetic reporters, the kinetics of conformational changes in the myosin protein can be monitored (31,32). As well as such structural studies, it has also been possible to carry out time-resolved diffraction studies of fibers in a synchnotron beam prior to and following the release of caged ATP within the fiber (33,34). These applications use the highly ordered skeletal muscle system to understand the basis of muscle contraction and relaxation. In smooth muscle, the contractile system has evolved to a highly complex level of cellular regulation, and many **signal transducing** systems and effectors play a role in regulating smooth muscle activity (35). In this area, the photorelease of caged ATP (and a variety of **second messengers**) continues to be a highly fruitful strategy. Recent studies by the Somlyos and their group provide a wide range of applications of this technology to the smooth muscle system (36,37).

Other Systems

Since 1980 to the present time (early 1999) there have been more than 200 articles published using caged ATP, and many more using related substrates. These have extended from **difference Fourier** structural determinations of proteins with or without bound substrates (38,39), to the cellular effects of occupancy of receptors (40). Structural biology of macromolecules, especially proteins, continues to provide enormous insights into molecular biological processes, and it is in this area that several interesting applications have great potential. The possibility of obtaining not merely a static picture of a protein at atomic resolution, but also one that contains dynamic information, is enticing. The accessibility of high-intensity radiation sources and a revival of interest in the **Laue diffraction** method have made fast **X-ray crystallography** studies an attainable goal. The idea that such measurements could be made with protein crystals before and after laser-induced release of caged substrates within a crystal has received recent attention; and although many of the technical difficulties (uniformity of release, fate of photofragment, local heating effects) have not been overcome, this approach offers great promise for the future (41,42). The number of physiological systems that have begun to be probed using caged ATP continues to increase; recent additions to this group include L-type Ca channel modulation (43), single **kinesin** molecules in optical traps (44), ATP-sensitive K channels in cholinergic interneurons (45), sea-urchin sperm **flagella** motility (46), and limulus photoreceptors (47).

SUMMARY

The introduction of the photorelease strategy that followed the description of caged ATP has now been extended to a wide array of biological substrates, including nucleoside phosphates, cyclic AMP, cyclic GMP, protons, Ca^{2+}, Mg^{2+}, peptides, inositol phosphates, sugars, amino acids, neurotransmitters, toxins, and so on. In almost all cases, the photochemical process has the same basis as the original caged ATP strategy; along with a wide array of biological substrates, the approach can now be employed to probe systems that range in complexity from single protein crystals to brain slices.

BIBLIOGRAPHY

1. J. H. Kaplan, B. Forbush III, and J. F. Hoffman (1978) *Biochemistry* **17**, 1929–1935.
2. G. P. Hess (1999) *Encyclopedia of Molecular Biology* (this encyclopedia).
3. J. W. Walker, G. P. Reid, J. A. McGray, and D. R. Trentham (1988) *J. Am. Chem. Soc.* **110**, 7170–7177.
4. J. Engels and E.-J. Schlaeger (1977) *J. Med. Chem.* **20**, 907–911.
5. J. A. McCray, L. Herbette, T. Kihura, and D. R. Trentham (1980) *Proc. Natl. Acad. Sci. USA* **77**, 7237–7241.
6. A. Barth et al. (1997) *J. Am. Chem. Soc.* **119**, 4149–4159.
7. J. E. T. Corrie (1993) *J. Chem. Soc., Perkins Trans* **1993**, 2161–2166.
8. G. C. R. Ellis-Davies, J. H. Kaplan, and R. J. Barsotti (1996) *Biophys. J.* **70**, 1006–1016.
9. A. Barth et al. (1995) *J. Am. Chem. Soc.* **117**, 10311–10316.
10. J. F. Wootton and D. R. Trentham (1989) *NATO ASI Series C* **272**, 277–296.
11. J. E. Baldwin et al. (1990) *Tetrahedron* **46**, 6879–6884.
12. R. S. Givens et al. (1993) *J. Am. Chem. Soc.* **115**, 6001–6012.
13. J. E. T. Corrie and D. R. Trentham (1992) *J. Chem. Soc., Perkins Trans I*, 2409–2417.
14. B. Forbush III (1984) *Proc. Natl. Acad. Sci. USA* **81**, 5310–5314.
15. J. Sleep, C. Hermann, T. Barman, and F. Travers (1994) *Biochemistry*, **33**, 6038–6042.
16. S. Lutsenko and J. H. Kaplan (1995) *Biochemistry* **34**, 15607–15613.
17. J. H. Kaplan (1985) *Annu. Rev. Physiol.* **47**, 534–544.
18. J. H. Kaplan and R. J. Hollis (1980) *Nature* **288**, 587–589.
19. J. H. Kaplan (1982) *J. Gen. Physiol.* **80**, 915–937.
20. D. Pascolini et al. (1988) *Biophys. J.* **54**, 679–688.
21. K. Fendler, E. Grell, M. Haubs, and E. Bamberg (1985) *EMBO J.* **4**, 3079–3085.
22. R. Borlinghaus, H.-J. Apell, and P. Lauger (1987) *J. Membr. Biol.* **97**, 161–178.
23. E. Bamberg, H.-J. Butt, A. Eisenrauch, and K. Fendler (1993) *Q. Rev. Biophys.*, **26**, 1–25.
24. I. Wuddel and H.-J. Apell (1995) *Biophys. J.* **69**, 909–921.
25. M. Stengelin, K. Fendler, and E. Bamberg (1993) *J. Membr. Biol.* **132**, 211–227.
26. K. Fendler, S. Drose, K. Altendorf, and E. Bamberg (1996) *Biochemistry*, **35**, 8009–8017.
27. A. F. Huxley and R. Niedergerke (1954) *Nature* **173**, 971–973.
28. H. E. Huxley and J. Hanson (1954) *Nature* **173**, 973–976.
29. Y. E. Goldman, M. G. Hibberd, J. A. McCray, and D. R. Trentham (1982) *Nature* **300**, 701–705.
30. E. Homsher and N. C. Millar (1990) *Annu. Rev. Physiol.* **52**, 875–896.
31. J. W. Tanner, D. D. Thomas, and Y. E. Goldman (1992) *J. Mol. Biol.* **223**, 185–203.
32. C. L. Berger, F. C. Svensson, and D. D. Thomas (1989) *Proc. Natl. Acad. Sci. USA* **86**, 8753–8757.
33. K. J. V. Poole, G. Rapp, Y. Maeda, and R. S. Goody (1988) *Adv. Exp. Med. Biol.* **226**, 391–404.
34. K. Horiuti et al. (1994) *J. Biochem.* **115**, 953–957.
35. A. P. Somlyo and A. V. Somlyo (1994) *Nature* **372**, 231–236.
36. A. P. Somlyo and A. V. Somlyo (1990) *Annu. Rev. Physiol.* **52**, 857–874.
37. B. Zimmerman et al. (1995) *J. Biol. Chem.* **270**, 23966–23974.
38. A. J. Scheidig et al. (1995) *J. Mol. Biol.* **253**, 132–150.
39. C. Raimbault et al. (1997) *Eur. J. Biochem.* **250**, 773–782.
40. G. D. Housley, N. P. Raybould, and P. R. Thorne (1998) *Hear. Res.* **119**, 1–13.
41. J. Hadju and L. N. Johnson (1990) *Biochemistry* **29**, 1669–1678.
42. I. Schlichting et al. (1990) *Nature* **345**, 309–314.
43. B. O'Rourke, P. H. Backx, and E. Marban (1992) *Science* **257**, 245–248.

44. M. Higuchi, E. Muto, Y. Inoue, and T. Yanagida (1997) *Proc. Natl. Acad. Sci. USA*, **94**, 4395–4400.
45. K. Lee, A. K. Dixon, T. C. Freeman, and P. J. Richardson (1998) *J. Physiol.* **510**, 441–453.
46. T. Tani and S. Kamimura (1998) *J. Exp. Biol.* **201**, 1493–1503.
47. M. N. Faddis and J. E. Brown (1992) *J. Gen. Physiol.* **100**, 547–570.

CALCIUM SIGNALING

ALAN SALTIEL

Calcium ion is an important **second messenger** involved in cell signaling and **signal transduction**. Most calcium in cells is sequestered in intracellular vesicles, the **endoplasmic reticulum** (ER) or **mitochondria**, where it is stored for release when needed. Small, localized increases in calcium result from its regulated release from the ER, which is produced by inositol trisphosphate, IP_3. IP_3 is produced by the **hormone**-dependent hydrolysis of phosphoinositides, to produce **inositol phosphates**, including IP_3. The IP_3 binds to a specific ER protein, a specialized **calcium channel** with four identical subunits, each with a single membrane-spanning segment and a single IP_3-binding site (1). The binding of IP_3 to this receptor results in a stereospecific release of calcium from the ER.

IP_3-sensitive calcium channels are expressed ubiquitously in tissues. However, the process of regulated calcium release is most tightly regulated in endocrine and neuronal cells, and it is often involved in coupling stimulus to secretion, which is usually a calcium-dependent process. Calcium release from the ER is oscillatory in nature. Increases in intracellular calcium are modulated not by increases in the amount of the ion released, but instead by the frequency of the oscillations. This may explain why the release of some hormones is pulsatile.

One of the critical targets in calcium signaling is CAM kinase, a multigene family of protein **kinases** that are sensitive to regulation by calcium/**calmodulin**. There are at least four members of this protein kinase family, with differences in distribution and substrate specificity. CAM kinases appear to be especially critical to modulation of signaling in the central nervous system (2).

BIBLIOGRAPHY

1. M. J. Berridge (1993) *Nature* **361**, 315–325.
2. P. I. Hanson and H. Shulman (1992) *Annu. Rev. Biochem.* **61**, 229–601.

CALCIUM-BINDING PROTEINS

M. R. NELSON
C. WEBER
W. J. CHAZIN

Calcium-binding proteins (CaBPs) are a key component linking the inorganic and organic systems that produce biological activities. Calcium is one of the most abundant inorganic elements in nature and is ubiquitous throughout biology, playing roles at the organismal, cellular, and molecular levels. Calcium is an essential component of shells and bones, where it is essentially deposited in crystals to create macroscopic support structures. At the cellular level, Ca^{2+} is one of the crucial currencies of most living organisms, acting as a **second messenger** in a wide range of key intracellular and extracellular systems. Calcium-binding proteins play important roles in mediating each of these effects. These proteins can be grouped into four primary categories: (*1*) intracellular proteins involved in Ca^{2+}-mediated **signal transduction**, (*2*) **enzymes**, (*3*) extracellular cell surface and **extracellular matrix proteins**, and (*4*) proteins containing **γ-carboxyglutamic acid** (Gla) residues. Although a variety of structural motifs bind Ca^{2+}, all involve oxygen atoms of the protein backbone or side chains, reflecting the intrinsic affinity of Ca^{2+} for oxygen atoms.

INTRACELLULAR CALCIUM-SIGNALING PROTEINS

EF-Hand Calcium-Binding Proteins

This family is the most extensively studied class of CaBPs. These proteins are characterized by a highly conserved helix–loop–helix motif termed the **EF-hand motif**, which consists of a 12-residue Ca^{2+}-binding loop flanked by two **alpha-helices**. Almost all EF-hand CaBPs are composed of pairs of EF-hands. This pairing of Ca^{2+}-binding sites is presumed to stabilize the protein conformation, increase the Ca^{2+} affinity of each site over that of isolated sites, and provide a means for cooperativity in Ca^{2+} binding. The cooperativity between sites allows for an "all or nothing" response to Ca^{2+}-binding, which is crucial for the function of these CaBPs as intracellular Ca^{2+} sensors.

Calmodulin and troponin C are the best known members of this family of CaBPs. They each have two largely independent domains connected by a flexible linker. Each domain contains two EF-hands, so these proteins each bind four Ca^{2+} ions. In the resting cell, they exist in an inactive state, with either Mg^{2+} or no ion bound. When the intracellular Ca^{2+} concentration rises in response to a signal, the proteins bind Ca^{2+}. This induces a dramatic conformational change, exposing a large **hydrophobic** surface within each domain that can interact with target proteins. There are many other CaBPs thought to function in a similar manner including caltractin and the calmodulin-like domain of plant Ca^{2+}-dependent protein kinase.

Less is known about the S100 proteins, another large and important subfamily of EF-hand CaBPs. The proteins in this subfamily are composed of two EF-hands each, the first of which is a variant version with a 14-residue binding loop, termed a pseudo-EF-hand. The ligands in the pseudo EF-hand are mainly carbonyl oxygens of the peptide backbone, whereas the canonical 12-residue loop coordinates Ca^{2+} primarily with oxygen atoms of side chains. Many S100 proteins are found as hetero- or homodimers, and they exhibit tissue-specific expression patterns. While these proteins are thought to be involved in signal transduction, the molecular mechanism of this function is not known, nor have any target proteins been positively identified. However, the expression of S100 proteins is deregulated in some diseases, including cancer, rheumatoid arthritis, and Down's syndrome, and **antibodies** against these proteins are commonly used as markers for screening for these diseases.

Most EF-hand CaBPs are directly involved in intracellular calcium signal transduction. However, at least three fulfill other cellular requirements. **Parvalbumin** and calbindin D_{9k}

(an S100 protein) are thought to play roles in Ca^{2+} buffering and in Ca^{2+} uptake and transport, respectively. These and other members of the EF-hand protein family are active in various aspects of intracellular Ca^{2+} homeostasis. The diversity of the roles of EF-hand CaBPs is further illustrated by the recent structure of BM-40, a glycoprotein found in the extracellular matrix (1), which was shown to contain two EF-hands. Prior to this discovery, the EF-hand motif was thought to be unique to intracellular CaBPs. This diversity in function of the EF-hand CaBPs provides evidence of very extensive evolutionary optimization of the fit between the EF-hand CaBP fold and the calcium ion.

Annexins

The **annexin** family is a second important class of intracellular CaBPs. The exact function of the proteins in this family is unknown, but they do exhibit an intriguing Ca^{2+}-dependent high-affinity binding to phospholipids (see **Membranes**). Annexins bind a large number of Ca^{2+} ions with high cooperativity. They assist transport of Ca^{2+} ions across membranes *in vitro*, although the physiological relevance has not yet been established. The mechanism of membrane binding is not known. The current model involves the Ca^{2+} ions acting as "glue" by simultaneously interacting with the protein and the membrane (2).

Ca^{2+}-BINDING ENZYMES

Two well-known enzymes exhibit Ca^{2+}-dependent translocation to the membrane fraction, presumably through a mechanism similar to that used by the annexins. Some **isoforms** of protein kinase C bind to phospholipids with high affinity in the presence of Ca^{2+} (2). The intracellular group IV **phospholipase** A_2 also requires Ca^{2+} for membrane association (3).

Many other enzymes require Ca^{2+} ions for stability. This includes many **serine proteases**: members of both the **trypsin** and the **subtilisin** families have been found to bind Ca^{2+}. These proteases each bind one to three Ca^{2+} ions, which are required for structural stability but do not directly participate in catalysis. In most cases, the Ca^{2+} ions are coordinated by ligands dispersed throughout the structure. Trypsin is an exception, however, and binds Ca^{2+} using a single 12-residue surface loop. The use of Ca^{2+} ions to stabilize the protein fold is not unique to the serine proteases. Another example of an enzyme that uses Ca^{2+} in this manner is **thermolysin**, a Zn^{2+}-dependent protease.

In other enzymes, the Ca^{2+} ions are directly involved in catalysis. This includes the secreted forms of phospholipase A_2, in which the required Ca^{2+} ion is thought to be involved in **transition state** stabilization (4). **Staphylococcal nuclease**, a secreted protein that catalyzes DNA and RNA hydrolysis (see **Nuclease**), also relies on Ca^{2+} for catalysis. The structure of this protein shows Ca^{2+} bound in the active site, where it is thought to be used to polarize the phosphate at the scissile phosphoester bond (5,6).

EXTRACELLULAR CELL SURFACE AND EXTRACELLULAR MATRIX Ca^{2+}-BINDING PROTEINS

Cadherins

These are tissue-specific **cell adhesion molecules** that are strongly Ca^{2+}-dependent (see **Cadherin**). At the levels of Ca^{2+} found in the extracellular milieu, cadherins bind several Ca^{2+} ions. A major conformational change is seen when Ca^{2+} is removed from these proteins. The resulting change to the apo state conformation is thought to prevent cadherins from interacting with each other.

C-Type Lectins

The C-type **lectins** are a Ca^{2+}-dependent class of lectins, which mediate many cell surface carbohydrate recognition events. Selectins, concanavalin A, and mammalian mannose-binding protein are examples of this type of CaBP. The structure of mannose-binding protein shows that Ca^{2+} is directly involved in binding the carbohydrate to this protein. In concanavalin A, on the other hand, the Ca^{2+}-binding site is $1.0-1.4 \times 10^{-9}$ m from the carbohydrate binding site; Ca^{2+} does not participate directly in binding the carbohydrate, but instead stabilizes the protein structure. It is not known whether selectins use Ca^{2+} directly in carbohydrate binding or to stabilize the fold required for carbohydrate recognition.

EGF Modules

A subset of **EGF motifs** present as **domains** in a number of proteins contain β-hydroxy–aspartic acid residues and have Ca^{2+}-binding activity. This type of EGF motif is found in some proteins involved in the **blood clotting** cascade and also in the extracellular matrix protein fibrillin. Fibrillin has 54 EGF modules, 43 of which are thought to bind Ca^{2+}. It is thought that Ca^{2+} is important in maintaining the structure of the fibrillin monomers and also in allowing the monomers to aggregate into microfibrils. A sheath of these microfibrils covers and stabilizes the **elastin** fibers that give tissues such as skin, blood vessels, and lungs their needed elasticity.

GLA-CONTAINING PROTEINS

The Gla-containing proteins contain **γ-carboxyglutamic acid** residues. This is a glutamic acid residue that has been carboxylated in a vitamin K–dependent process. These proteins are found in the bones and teeth, in the kidney, and in the blood. Osteocalcin (bone Gla-protein) and matrix Gla-protein are found in bone. They are believed to be involved in the mineralization of this tissue (7). Similar proteins are thought to be involved in the mineralization of teeth. In the kidney, the Gla-containing protein nephrocalcin inhibits the nucleation, aggregation, and growth of calcium oxalate crystals. An abnormal form of this protein is found in the urine of people suffering from kidney stones. However, it is not entirely clear whether this defect is responsible for the growth of kidney stones (8).

By far the best studied of the Gla-containing proteins are those found in blood. The majority of Gla-containing proteins are **zymogen** forms of serine proteases involved in the **blood clotting** (coagulation) cascade. These proteins are: pro**thrombin**, factor VII, factor IX, factor X, protein C, protein S, and protein Z. The protease domains of these proteins are very similar to the pancreatic digestive proteases **trypsin**, **chymotrypsin**, and **elastase**. The Gla-containing coagulants require Ca^{2+} binding in order to bind to phospholipids. This Ca^{2+}-dependent phospholipid binding is responsible for the membrane association properties required for the function of the proteins. Ca^{2+} binding also appears to be necessary for the formation of the native conformation of the Gla domain (9).

The Gla-containing coagulants have two classes of metal ion binding sites. There are approximately three higher affinity sites that are not metal ion–specific, and three to four lower affinity sites, which are specific for Ca^{2+}. Only Sr^{2+} can substitute for Ca^{2+} in both types of binding site.

The high resolution three-dimensional structure of the Gla domain of prothrombin, with seven Ca^{2+} ions bound (10), reveals the Gla domain to involve nine to 10 turns of **alpha-helix**, in three separate helices. Seven Ca^{2+} ions interact with 24 oxygen atoms from 16 of the 18 carboxylate groups of the nine Gla residues that are ordered in the structure. A 10th Gla residue is disordered and does not participate in Ca^{2+} binding. The coordination geometries of the Ca^{2+} ions do not correspond to any idealized polyhedra. Five of the Ca^{2+} ions are involved in a polymeric array with 18 of the liganding oxygen atoms. Four of these Ca^{2+} ions are completely buried in the protein. This complex structure is thought to nucleate the folding of the Gla domain, and is essentially electrically neutral. The complexity and irregularity of this structure explains the selectivity for Ca^{2+} ions. Ca^{2+} is able to adopt different and distorted coordination geometries. Mg^{2+}, on the other hand, is fairly rigid in its requirement for six ligands, and cannot accommodate the unusual network of ligands in the Gla domain. The remaining two metal ion sites in the Gla domain of prothrombin are solvent accessible, and carry a net charge of about +0.5 each. Because of the positive charge, these sites are thought to be involved in neutralizing the negatively-charged phospholipids, allowing the protein to associate with membranes.

BIBLIOGRAPHY

1. E. Hohenester, P. Maurer, C. Hohenadl, R. Timpl, J. N. Jansonius, and J. Engel (1996) *Nature Struct. Biol.* **3**, 67–73.
2. M. D. Bazzi and G. L. Nelsestuen (1993) *Cell. Signal.* **5**, 357–365.
3. J. Y. Channon and C. C. Leslie (1990) *J. Biol. Chem.* **265**, 5409–5413.
4. E. A. Dennis (1994) *J. Biol. Chem.* **269**, 13057–13060.
5. P. J. Loll and E. E. Lattmam (1989) *Prot. Struct. Funct. Genet.* **5**, 183–201.
6. F. A. Cotton, E. E. Hazen, and M. J. Legg (1979) *Proc. Natl. Acad. Sci. USA* **76**, 2551–2555.
7. M. F. Young, J. M. Kerr, K. Ibaraki, A.-M. Heegaard, and P. G. Robey (1992) *Clin. Orthoped. Rel. Res.* **281**, 275–294.
8. F. L. Coe, Y. Nakagawa, J. Asplin, and J. H. Parks (1994) *Miner. Electrolyte Metab.* **20**, 378–384.
9. J. W. Suttie (1993) *FASEB J.* **7**, 445–452.
10. M. Soriano-Garcia, K. Padmanabhan, A. M. de Vos, and A. Tulinsky (1992) *Biochemistry* **31**, 2554–2566.

Suggestions for Further Reading

M. Celio, ed. (1996) *Guidebook to the Calcium-Binding Proteins*, Oxford University Press, Oxford.

H. Kawasaki and R. H. Kretsinger (1995) *Protein Profile; Calcium-Binding Proteins 1: EF-hands*, Vol. 2, Academic Press, London (a good general source for information about EF-hand proteins).

S. Liemann and A. Lewit-Bentley (1995) Annexins: a novel family of calcium- and membrane-binding proteins in search of a function, *Structure* **3**, 233–237.

P. Maurer, E. Hohenester, and J. Engel (1996) Extracellular calcium-binding proteins, *Curr. Opin. Cell Biol.* **8**, 609–617.

C. A. McPhalen, N. C. J. Strynadka, and M. N. G. James (1991) Calcium-binding sites in proteins: a structural perspective, *Adv. Protein Chem.* **42**, 77–144.

L. J. Van Eldik, J. G. Zendegui, D. R. Marshak, and D. M. Watterson (1982) Calcium-binding proteins and the molecular basis of calcium action, *Intern. Rev. Cytol.* **77**, 1–61. (A review of Gla-containing proteins, Ca^{2+}-binding enzymes, and EF-hand proteins.)

CALMODULIN

M. R. Nelson
C. Weber
W. J. Chazin

Calmodulin is the quintessential member of the **EF-hand** family of **calcium-binding proteins** and functions as a key mediator in numerous **signal transduction** pathways. It is an acidic **protein** (**isoelectric point** of 4.2) of molecular weight 16.8 kDa that is found in most eukaryotic cells, from yeast to humans. It is composed of two largely independent globular **domains** connected by a flexible central α-**helix**. The affinity of calmodulin for Ca^{2+} is fine-tuned to respond to intracellular calcium signals. Conformational changes within each of the domains induced by the binding of Ca^{2+} leads to the transduction of the Ca^{2+} signal.

BIOLOGICAL FUNCTION

Calmodulin is a signal transduction protein. It has been implicated in the control of a wide range of cellular functions, including cell proliferation, smooth muscle contraction, the regulation of **ion channels**, long-term potentiation and memory, and **exocytosis**. Furthermore, it has recently been found inside the **nucleus**, where it is thought to be involved in the regulation of **DNA replication**, **gene expression**, and **DNA repair**. In a resting cell with basal levels of Ca^{2+}, calmodulin exists in the inactive apo state. When a calcium signal is initiated and the intracellular levels of Ca^{2+} increase, calmodulin binds Ca^{2+} ions. This causes the protein to undergo a dramatic conformational change that exposes a large **hydrophobic** patch on the protein. This newly exposed hydrophobic surface then interacts with various cellular proteins, modulating their activity and thereby transducing the calcium signal.

Protein kinases and **phosphatases** are among the best studied of calmodulin's targets. These enzymes are activated by the release of an autoinhibitory **domain** that is bound to the **active site** in the resting state. Ca^{2+}-loaded calmodulin activates these proteins by binding to a site near to or overlapping with the autoinhibitory domain, causing the auto-inhibitory domain to dissociate from the active site. Myosin light chain kinase and calcium-calmodulin-dependent protein kinase II are two well-known examples of this class of calmodulin-regulated proteins. Calmodulin also regulates proteins involved in the generation of other second messengers, such as calmodulin-dependent cyclic nucleotide phosphodiesterase and nitric oxide synthase. The mechanism of regulation of these enzymes is thought to be very similar to that used to regulate the kinases and phosphatases. Calmodulin has also been shown to interact with proteins of the cytoskeleton and their regulatory proteins, such as **spectrin** and brush border myosin. However, the biological significance of these interactions is still unclear.

Recently, interactions between apocalmodulin and targets such as neuromodulin and unconventional myosins have been described. These interactions are thought to be mediated by an "IQ motif" (with the consensus sequence IQXXXRGXXXR, where X is any amino acid) in the target (see **Protein motifs**). This motif and its interaction with a Ca^{2+}-free EF-hand calcium-binding protein was first described in myosin (1,2). The functional significance of the interaction of proteins containing the IQ motif with apocalmodulin is not clear. It has been suggested, however, that perhaps neuromodulin, which is associated with the plasma membrane, functions as a trap for calmodulin, sequestering it near the membrane in the absence of the activating calcium signal (3).

ION-BINDING PROPERTIES

Calmodulin binds four calcium ions, with a **dissociation constant** of about 10^{-6} *M*. Binding is selective for Ca^{2+}; calmodulin does not bind Mg^{2+} or monovalent ions with appreciable affinity. The binding of Ca^{2+} is also highly **cooperative**, which is very important for calmodulin's biological function as a calcium sensor, because it allows for a tightly controlled "all or nothing" response to the calcium signal: the four binding events all occur within a very narrow range of Ca^{2+} concentration. If these binding events were spread over a large range of Ca^{2+} concentrations, calmodulin would be at least partially activated over the entire range. The system would therefore lack the required sharp separation between the activate and inactive states.

STRUCTURE

Calmodulin is a largely helical protein composed of four EF-hand motifs organized into two independent globular domains, each of which contain two EF-hands. The two domains are connected by a long, flexible central α-helix, giving the protein a dumbbell-like appearance (Figure 1a). This **quaternary structure** is relatively similar in the apo- and Ca^{2+}-loaded states of the protein. However, significant conformational changes occur within the individual domains on Ca^{2+} binding (Fig. 1b). In the apo state, each domain occupies the "closed" conformation (4–6). In this conformation, the four helices in the domain are nearly antiparallel, with interhelical angles near 180°. In Ca^{2+}-loaded calmodulin, each domain occupies an "open" conformation in which the four α-helices are nearly perpendicular to each other (7,8). Each domain exposes a hydrophobic surface of about 1.25×10^{-8} m^2 in this open conformation (5).

INTERACTIONS WITH TARGET PEPTIDES

The biophysical characterization of the interaction between calmodulin and its targets is often studied using small **peptides** (10–20 amino acid residues) derived from the calmodulin-binding domain of target enzymes. Calmodulin binds to these target peptides extremely tightly, with dissociation constants ranging from 10^{-7} to 10^{-11} M. The peptides adopt an amphiphilic helical conformation, and tend to have

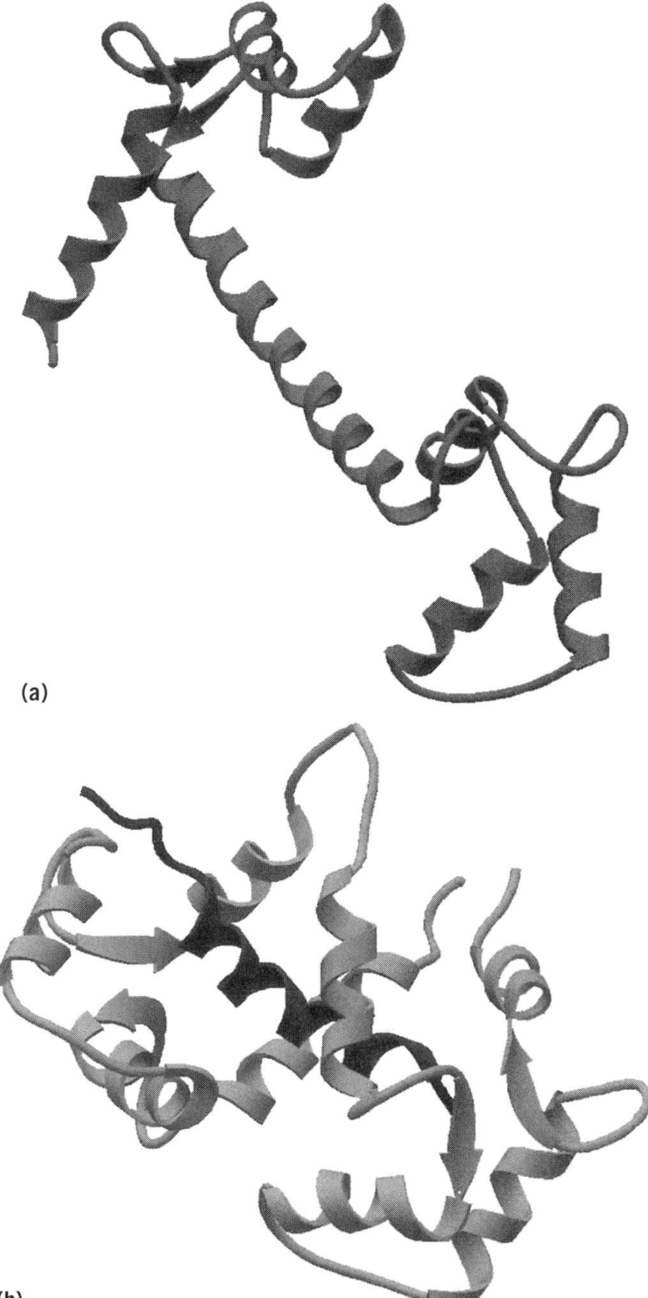

Figure 1. Ribbon representations of the three-dimensional conformations of $(Ca^{2+})_4$-calmodulin in the (**a**) absence and (**b**) presence of a target peptide. The peptide-free diagram was constructed using the coordinates of 1CLL (8). The two domains at the ends of the central α-helix consist of two EF-hand motifs. The diagram of $(Ca^{2+})_4$-calmodulin and the peptide analog of the myosin light chain kinase was constructed using the coordinates of 2BBM (10). The bound peptide is depicted darker than calmodulin.

a bulky hydrophobic residue at either end, often (but not always) spaced 12 residues apart. However, the sequences of the target peptides are not highly similar. Calmodulin is able to bind so tightly to such a wide array of targets because of its own plasticity. The flexible central α-helix connecting its two domains can function as an "expansion joint," allowing calmodulin to bind to peptides with different numbers of residues between the two bulky hydrophobic anchors. Furthermore, **van der Waals** interactions, which can be rather nonspecific, tend to dominate as the critical components stabilizing the interaction between calmodulin and the peptides. **Hydrogen bonds**, which are more structurally specific, are not as important in these interactions. The adaptability of the peptide binding surface of calmodulin may be further aided by its high proportion of **methionine** residues. Methionine is an unusually flexible and polarizable amino acid, which may allow calmodulin to mold its peptide-binding surface to meet the requirements of many different peptide sequences (9,10).

High resolution three-dimensional structures of three complexes of calmodulin and peptides derived from target enzymes have been reported (11–13). These structures show that the relative disposition of the two domains of calmodulin is altered by the binding of a target peptide, but there is little change within the Ca^{2+}-activated domains themselves. The calmodulin–peptide complex forms a well-packed ellipsoid, which contrasts sharply with the dumbbell shape of calmodulin observed in the absence of target (Fig. 1). The two domains of calmodulin essentially wrap around the target peptide, forming a hydrophobic tunnel in which the peptide binds. As of yet, there are no structures of calmodulin bound to an entire target protein. It is thought, however, that the mode of binding to the target sequence on the intact enzyme will be very similar to that seen in the calmodulin–peptide complexes.

BIBLIOGRAPHY

1. Xie, D. H. Harrison, I. Schlichting, R. M. Sweet, V. N. Kalabokis, and A. G. Szent-Gyorgyi (1994) *Nature* **368**, 306–312.
2. A. Houdusse and C. Cohen (1996) *Structure* **4**, 21–32.
3. Y. Liu and D. R. Storm (1990) *Trends Pharmacol. Sci.* **11**, 107–111.
4. H. Kuboniwa, N. Tjandra, S. Grzesiek, H. Ren, C. B. Klee, and A. Bax (1995) *Nature Struct. Biol.* **2**, 768–776.
5. M. Zhang, T. Tanaka, and M. Ikura (1995) *Nature Struct. Biol.* **2**, 758–767.
6. B. E. Finn, J. Evenäs, T. Drakenberg, J. P. Waltho, E. Thulin, and S. Forsén (1995) *Nature Struct. Biol.* **2**, 777–783.
7. Y. S. Babu, C. E. Bugg, and W. J. Cook (1988) *J. Mol. Biol.* **204**, 191–204.
8. R. Chattopadhyaya, W. Meador, A. Means, and F. Quiocho (1992) *J. Mol. Biol.* **228**, 1177–1192.
9. H. J. Vogel and M. Zhang (1995) *Mol. Cell. Biochem.* **149/150**, 3–15.
10. K. T. O'Neil and W. F. DeGrado (1990) *Trends Biochem. Sci.* **15**, 59–64.
11. M. Ikura, G. M. Clore, A. M. Gronenborn, G. Zhu, C. B. Klee, and A. Bax (1992) *Science* **256,** 632–638.
12. W. E. Meador, A. R. Means, and F. A. Quiocho (1992) *Science* **257**, 1251–1255.
13. W. E. Meador, A. R. Means, and F. A. Quiocho (1993) *Science* **262**, 1718–1721.

Suggestions for Further Reading

O. Bachs, N. Agell, and E. Carafoli (1994) Calmodulin and calmodulin-binding proteins in the nucleus, *Cell Calcium* **16**, 289–296 (a review of the current knowledge about calmodulin's nuclear functions).

A. Crivici and M. Ikura (1995) Molecular and structural basis of target recognition by calmodulin, *Annu. Rev. Biophys. Biomol. Struct.* **24**, 85–116 (an exhaustive review of interactions between calmodulin and its targets).

R. D. Hinrichsen (1993) Calcium and calmodulin in the control of cellular behavior and motility, *Biochim. Biophys. Acta* **1155**, 277–293.

L. J. Van Eldik and D. M. Watterson, eds. (1997) *Calmodulin and Signal Transduction*, Academic Press, San Diego (a book containing articles about many aspects of calmodulin's function and structure).

C. B. Klee and E. Carafoli, eds. (1997) *Calcium as a Cellular Regulator*, Oxford University Press, Oxford.

CALNEXIN/CALRETICULIN

R. J. Ellis

Calnexin (also known as *p88*, *IP90*, and *CATCHER*) is a membrane-bound **molecular chaperone** present in the **endoplasmic reticulum** (ER) that binds selectively to many monoglucosylated **glycoproteins** that fold in this compartment during synthesis by ribosomes bound to the cytosolic face of the ER membrane. Calreticulin is a soluble homologue of calnexin, with the same glycan specificity, but occurs in the ER lumen. Together with other proteins, calnexin and calreticulin provide a means for improving the efficiency of protein folding and assembly in the ER compartment (see **Molecular chaperone**) and form part of the quality-control system that ensures that only correctly folded and assembled proteins are transported from the ER compartment along the pathway followed by secretory proteins (1). Unlike some other molecular chaperones, calnexin and calreticulin are not **stress response** proteins, do not exhibit **ATPase** activity, and appear to recognize glucose residues in core glycans rather than **hydrophobic** surfaces in partially folded polypeptides (1). In some cell types, calreticulin also occurs in the **nucleus** and cytosol, where it binds to **steroid hormone** receptors and **integrin** molecules, respectively; these interactions may indicate roles for calreticulin in **gene expression** and cell signaling.

DISCOVERY

An 88-kDa protein termed *p88* was found in transient association with class I **major histocompatibility** molecules synthesized by several murine lymphoma cell lines (2). Newly synthesized-class I heavy chains bound rapidly to p88 before they associated with β_2-microglobulin chains. In mutant cells that lack microglobulin, the incompletely assembled class I molecules exhibited prolonged interaction with p88 and were correspondingly impaired in their transport to the **Golgi** stacks (3). Moreover, microglobulin and peptides derived from the cytosol need to be assembled with the heavy chain before the latter can be released from p88. These findings led to the proposal that p88 is a molecular chaperone that promotes assembly of class I molecules by retaining intermediates in the ER until complete complexes are formed.

Independent **pulse-chase** experiments on a human ER protein termed *IP90* showed that it transiently associates with many newly synthesized proteins, including the T-cell receptor (TCR), the B-cell antigen receptor, and class I molecules (4,5). The binding of IP90 to TCR complexes lacking α chains was prolonged, consistent with a retention role for this protein in the ER for incompletely assembled complexes (6). Subsequently, p88 and IP90 were found to be identical with the Ca^{2+}-binding phosphoprotein calnexin (4–7), a protein originally identified as one of several proteins labeled by **phosphorylation** of canine pancreatic **microsomes** with [γ-^{32}P]-GTP (8).

STRUCTURE

Calnexin is a type I nonglycosylated integral **membrane protein** of 573 amino acid residues, with a predicted molecular weight of 65,400 and with its substrate-binding domain in the lumen of the ER. It has a cytosolic tail of 89 residues that is phosphorylated and contains a *C*-terminal RKPRRE (–Arg–Lys–Pro–Arg–Arg–Glu) sequence that acts as an ER-retention signal (9) (see **KDEL sequence**). It also possesses four **domains** with high sequence similarity to calreticulin, the major **calcium-binding protein** of the ER lumen. Calreticulin contains 400 amino acid residues (46,000 molecular weight), which include both a KDEL ER-retrieval sequence and a strongly negatively charged region involved in Ca^{2+} binding at its *C* terminus (10). Both calnexin and calreticulin appear to function as monomers.

FUNCTION

Calnexin and calreticulin are both **lectins** that specifically recognize glycoproteins that contain monoglucosylated core glycans; they do not recognize glycans containing two, three or no glucose residues (11,12). When nascent chains of glycoproteins enter the ER lumen, a core glycan containing three terminal glucose residues is added en bloc to specific **asparagine** sidechains (see ***N*-glycosylation**). This addition of glycan may have evolved initially to make folding intermediates more soluble under the conditions of high protein concentration that characterize the ER lumen. These glucose residues are then rapidly removed one at a time by glucosidases I and II in the ER lumen. Another lumenal enzyme called UDP-Glc:glycoprotein glucosyltransferase then adds back one glucose residue. Thus glycans containing single terminal glucose residues can arise either as intermediates in the glucose-trimming process or after regeneration by the transferase. Such monoglucosylated glycans then bind to calnexin and calreticulin. In the case of pancreatic ribonuclease B, this binding occurs solely through the glycan and does not involve additional recognition of the protein moiety by the lectins (13,14), but such additional recognition may occur in the case of other proteins (11,15).

The function of the binding of calnexin and calreticulin to monoglucosylated glycoproteins appears to be to assist their correct folding by preventing aggregation of partially folded or misfolded chains and ensuring retention of the latter until they are degraded (see **Protein degradation *in vivo***). Evidence for this conclusion comes from experiments in which proteins unable to fold correctly because of mutation remain bound to these chaperones for much longer than normal proteins and are eventually degraded (16–18). Additional possible roles include the suppression of formation of nonnative **disulfide bonds** (19) and the premature assembly of **oligomers** (18).

The distinction between correctly folded and partially folded chains is not made by these lectins but by the glucosyltransferase. This enzyme has the unusual property of distinguishing, by an unknown mechanism, between subtly different protein conformations; it adds glucose to the core glycan only if the protein attached to this glycan is not correctly folded (20,21). Such reglucosylated glycoproteins will then bind back to calnexin and calreticulin. The removal and addition of the terminal glucose residue then continues until the protein has folded sufficiently to be no longer recognized by the transferase. The binding of calnexin to ribonuclease B *in vitro* is dynamic, indicating that the binding is readily reversible; glucosidase II cannot remove terminal glucose residues while calnexin is bound to the glycoprotein (14).

Calnexin and calreticulin associate with a wide range of proteins in the ER, including soluble secretory proteins, **extracellular matrix** molecules, **ion channels**, membrane **receptors**, and various glycoproteins of **viruses**. They also retain misfolded proteins that accumulate in some ER-storage diseases such as cystic fibrosis and **α_1-antitrypsin** deficiency (22). Addition of glucosidase inhibitors to cells results in reduced rates of secretion of many glycoproteins, presumably due to the inhibition of their binding to the chaperones (23). All these observations support the view that the functions of calnexin and calreticulin are important for cell function. Nevertheless, viable mammalian cell lines are known that lack either glucosidase I or II or calnexin, while mutants of *Saccharomyces cerevisiae* that lack genes for any one of these three proteins show no growth defect. However, mutants of *Saccharomyces pombe* that lack calnexin are not viable. Both species of *Saccharomyces* appear to lack calreticulin. These discrepancies may indicate some redundancy and overlap of function amongst the chaperones in the ER compartment, which include **BiP**, hsp90, and **protein disulfide isomerase**, as well as calnexin and calreticulin (1).

BIBLIOGRAPHY

1. A. Helenius, E. S. Trombetta, D. N. Hebert, and J. S. Simons (1997) *Trends Cell Biol.* **7**, 193–200.
2. E. Degen and D. B. Williams (1991) *J. Cell Biol.* **112**, 1099–1115.
3. E. Degen, M. F. Cohen-Doyle, and D. B. Williams (1992) *J. Exp. Med.* **175**, 1653–1661.
4. F. Hochenstenbach, V. David, S. Watkins, and M. B. Brenner (1992) *Proc. Natl. Acad. Sci. USA* **89**, 4734–4738.
5. K. Galvin et al. (1992) *Proc. Natl. Acad. Sci USA* **89**, 8452–8456.
6. V. David, F. Hochstenbach, S. Rajagopalan, and M. B. Brenner (1993) *J. Biol. Chem.* **268**, 9585–9592.
7. N. Ahluwalia, J. J. M. Bergeron, I. Wada, E. Degen, and D. B. Williams (1992) *J. Biol. Chem.* **267**, 10914–10918.
8. I. Wada, D. Rindress, P. H. Cameron, W.-J. Ou, J. J. Dohery, D. Louvard, A. W. Bell, D. Dignard, D. Y. Thomas, and J. J. M. Bergeron (1991) *J. Biol. Chem.* **266**, 19599–19610.
9. M. Michalak, R. E. Milner, K. Burns, and M. Opas (1992) *Biochem. J.* **285**, 681–692.
10. B. Sonnichsen, J. Fullefrug, P. Nguyen, W. Diekmann, D. G. Robinson, and G. Mieskes (1994) *J. Cell Sci.* **107**, 2705–2717.
11. F. E. Ware, A. Vassilakos, P. A. Petersen, M. R. Jackson, M. A. Lehrmann, and D. B. Williams (1995) *J. Biol. Chem.* **270**, 4697–4704.

12. R. G. Spiro, Q. Zhu, V. Bhoyroo, and H.-D. Soling (1996) *J. Biol. Chem.* **271**, 11588–11594.
13. A. R. Rodan, J. F. Simons, E. S. Trombetta, and A. Helenius (1996) *EMBO J.* **15**, 6921–6930.
14. A. Zapun, S. M. Petrescu, P. M. Rudd, A. R. Dwek, D. Y. Thomas, and J. J. M. Bergeron (1997) *Cell* **88**, 29–38.
15. D. B. Williams (1995) *Biochem. Cell Biol.* **73**, 123–132.
16. C. Hammond and A. Helenius (1994) *Science* **266**, 456–458.
17. D. F. Qu, J. H. Teckman, S. Omura, and D. H. Perlmutter (1996) *J. Biol. Chem.* **271**, 22791–22795.
18. D. N. Hebert, B. Foellmer, and A. Helenius (1996) *EMBO J.* **15**, 2961–2968.
19. W. Chen, J. Helenius, I. Braakman, and A. Helenius (1995) *Proc. Natl. Acad. Sci. USA* **92**, 6229–6233.
20. M. C. Sousa, M. A. Ferrero-Garcia, and A. J. Parodi (1992) *Biochemistry* **31**, 97–105.
21. S. E. Trombetta and A. J. Parodi (1992) *J. Biol. Chem.* **267**, 9236–9240.
22. P. J. Thomas, B. Qu, and P. L. Pederson (1995) *Trends Biochem. Sci.* **20**, 456–459.
23. V. Gross, T. Andus, T. A. Tranhl, R. T. Schwartz, K. Decker, and P. C. Heinrich (1983) *J. Biol. Chem.* **258**, 12203–12209.

Suggestions for Further Reading

C. Hammond and A. Helenius (1993) A chaperone with a sweet tooth, *Curr. Biol.* **3**, 884–886.

J. J. M. Bergeron, M. B. Brenner, D. Y. Thomas, and D. B. Williams (1994) Calnexin: a membrane-bound chaperone of the endoplasmic reticulum, *Trends Biochem. Sci.* **19**, 124–128.

C. Hammond and A. Helenius (1995) Quality control in the secretory pathway, *Curr. Opinion Cell Biol.* **7**, 523–529.

D. B. Williams and T. A. Watts (1995) Molecular chaperones in antigen presentation, *Curr. Opin. Immunol.* **7**, 77–84.

M.-J. Gething, ed. (1997) *Molecular Chaperones and Folding Catalysts*, Oxford University Press, Oxford (contains detailed articles on calnexin and calreticulin).

S. Dedhar (1994) Novel functions for calreticulin: interactions with integrins and modulation of gene expression? *Trends Biochem. Sci.* **19**, 269–271.

M. G. Coppolino, M. J. Woodside, N. Demaurex, S. Grinstein, R. St-Arnaud, and S. Dedhar (1997) Calreticulin is essential for integrin-mediated calcium signalling and cell adhesion, *Nature* **386**, 843–847.

5′ CAP

J. D. LEWIS

The 5′cap is characteristic of all **RNAs** transcribed by **RNA polymerase** II, and it is added to the nascent transcript soon after transcriptional initiation. Many lines of evidence suggest that the cap contributes to many aspects of RNA metabolism, including RNA stability (1), pre-mRNA **splicing** (2,3), **polyadenylation** (4), U **small nuclear RNA** (snRNA) export (5), and **translation** of **messenger RNA** (6). The influence of the cap on RNA metabolism, in both the **nucleus** and the cytoplasm, is mediated by **cap-binding proteins**.

The cap consists of an inverted 7-methyl guanosine nucleotide linked by a 5′-5′-triphosphate linkage to the first template-encoded nucleotide of the RNA, to give the following structure: m^{7}G(5′)ppp(5′)N. The cap modification is one of the earliest covalent modifications detected on a nascent RNA transcript, and it occurs cotranscriptionally (7). The actual capping reaction requires the sequential action of several enzymatic activities and is well understood (see Fig. 1). The capping enzyme possesses an RNA 5′-triphosphatase activity, which converts the 5′ terminal triphosphate of the RNA to a diphosphate. RNA guanylyltransferase then catalyzes the transfer of GMP, derived from GTP, to the diphosphate, resulting in the formation of the G(5′)ppp(5′)N cap. This cap structure is the substrate for methylation by RNA (guanine-7)-methyltransferase, which converts it to the monomethylated cap structure. In the absence of cap methylation, the reaction is reversible. The genes encoding the capping enzyme and methylase are essential in the **yeasts** *Saccharomyces cerevisiae* and *Schizosaccharomyces pombe*, underlining the importance an intact cap for viability (8–10). The cause of death in these mutants is probably due to cumulative defects in the various processes in which the cap is involved, such as RNA stability and **splicing** (11).

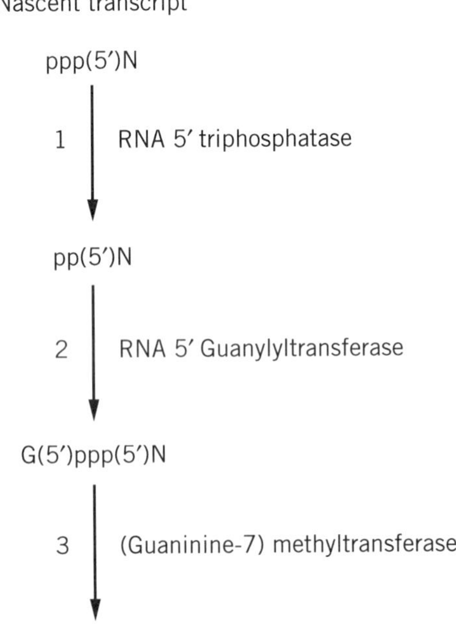

Figure 1. The capping reaction at the 5′ nucleotide (N) of an RNA transcript: (*1*) the 5′-triphosphate of the nascent transcript is hydrolyzed by RNA 5′-triphosphatase to a diphosphate; (*2*) mRNA guanylyltransferase adds the cap from GMP (which is derived from GTP). The cap is then methylated on the N-7 (position 7 nitrogen) of guanosine by mRNA (guanine-7) methylase.

BIBLIOGRAPHY

1. Y. Furuichi, A. LaFiandra, and A. J. Shatkin (1977) *Nature* **266**, 235–239.
2. M. Konarska, R. Padgett, and P. Sharp (1984) *Cell* **38**, 731–736.
3. A. Krainer, T. Maniatis, B. Ruskin, and M. Green (1984) *Cell* **36**, 993–1005.
4. C. Cooke and J. C. Alwine (1996) *Mol. Cell. Biol.* **16**, 2579–2584.

5. J. Hamm and I. W. Mattaj (1990) *Cell* **63**, 109–118.
6. A. Shatkin (1985) *Cell* **40**, 223–224.
7. M. Salditt-Georgieff, M. Harpold, S. Chen-Kiang, and J. E. Darnell Jr. (1980) *Cell* **19**, 69–78.
8. S. Shuman and B. Schwer (1995) *Mol. Microbiol.* **17**, 405–410.
9. Y. Shibagaki, N. Itoh, H. Yamada, S. Nagata, and K. Mizumoto (1992) *J. Biol. Chem.* **267**, 9521–9528.
10. B. Schwer and S. Shuman (1994) *Proc. Natl. Acad. Sci. USA* **91**, 4328–4332.
11. B. Schwer and S. Shuman (1996) *RNA* **2**, 574–583.

Suggestions for Further Reading

J. D. Lewis, S. Gunderson, and I. W. Mattaj (1995) The influence of 5′ and 3′ end structures on pre-mRNA metabolism, *J. Cell Sci. Suppl.* **19**, 13–19.

J. D. Lewis and E. Izaurralde (1997) The role of the cap structure in RNA processing and nuclear export, *Eur. J. Biochem.* **247**, 461–469.

CAP-BINDING PROTEINS

J. D. LEWIS

Most of the functions in the **nucleus** and cytoplasm of the **5′ cap** of **messenger RNA** (mRNA) are mediated by cap-binding proteins (CBP), which show highly specific binding to the cap structure, $m^7G(5')ppp(5')N$. Two major cap-binding activities have been characterized in **eukaryotes**. The first is the nuclear cap-binding complex (CBC), which plays a role in pre-mRNA **splicing** (1–3), **polyadenylation** (4), and U **small nuclear RNA** (snRNA) export (5). The second, eIF4E (6), is part of a trimeric complex, eIF4F, which is required for cap-dependent **translation** of mRNA.

CBC was initially identified in nuclear extracts from HeLa cells as an activity that bound specifically to a normal 7-methyl guanosine-5′-capped RNA. Purification of the activity showed that it was a heterodimeric complex comprising two subunits; CBP80 and CBP20 (1). Neither protein alone can bind specifically to the 5′cap—only as a heterodimer. Current evidence has shown that CBC is required for efficient (*1*) pre-mRNA splicing, (*2*) polyadenylation, and (*3*) the export of RNA polymerase II–transcribed U snRNA (see below).

CBC is required for efficient splicing of the cap-proximal **intron**, where its major function is to facilitate the binding of U1 snRNP to the cap-proximal 5′ splice site (3) (see **Gene splicing**). Experiments in the yeast *Saccharomyces cerevisiae* and in HeLa splicing extracts have demonstrated that this function of CBC is conserved in **evolution** (2,7).

A role for CBC in polyadenylation has also been demonstrated *in vitro* (4): (see **Polyadenylation**). Depletion of CBC from HeLa cell polyadenylation extracts results in a reduction in the efficiency of polyadenylation. Further analysis of this defect demonstrated that CBC was required for efficient cleavage of the pre-mRNA, but not for the polyadenylation reaction itself.

Nucleocytoplasmic transport of some classes of RNA, specifically the 5′capped uracil-rich (U) snRNAs, is facilitated by CBC. **Antibodies** raised against one of the subunits, CBP20, can specifically inhibit the interaction of CBC with the cap and, as a consequence, inhibit the export of these U snRNAs from the nucleus to the cytoplasm (5).

The other CBP, eIF4E, is required for cap-dependent translation of mRNA in the cytoplasm, and it can bind directly to the cap as a monomer (6). In order to function in translation, eIF4E has to assemble with two other **polypeptide chains** in a heterotrimeric complex of eIF4A, eIF4G, and eIF4E, to form a complex known as eIF4F. This trimeric complex binds to the cap of mRNAs and complexes with a second translational initiation complex, eIF3, which has an **RNA helicase** activity. This helicase activity is thought to unwind **secondary structure** in the mRNA and to allow the cap-dependent association of the 40S **ribosomal** subunit with the mRNA. The 40S subunit then scans for the **initiation codon** and complexes with the 60S ribosomal subunit to initiate translation.

BIBLIOGRAPHY

1. E. Izaurralde, J. Lewis, C. McGuigan, M. Jankowska, E. Darzynkiewicz, and I. Mattaj (1994) *Cell* **78**, 657–668.
2. J. Lewis, D. Görlich, and I. Mattaj (1996) *Nucl. Acids Res*, **24**, 3332–3336.
3. J. D. Lewis, E. Izaurralde, A. Jarmolowski, C. McGuigan, and I. W. Mattaj (1996) *Genes Devel.* **10**, 1683–1698.
4. S. M. Flaherty, P. Fortes, E. Izaurralde, I. W. Mattaj, and G. M. Gilmartin (1997) *Proc. Nat. Acad. Sci. USA* **94**, 11893–11898.
5. E. Izaurralde, J. Lewis, C. Gamberi, A. Jarmolowski, C. McGuigan, and I. W. Mattaj (1995) *Nature* **376**, 709–712.
6. N. Sonnenberg, M. Rupprecht, W. Merrick, and A. Shatkin (1979) *Proc. Natl. Acad. Sci. USA* **75**, 4345–4349.
7. H. Colot, F. Stutz, and M. Rosbash (1996) *Genes Devel.* **10**, 1699–1708.

Suggestions for Further Reading

J. D. Lewis and E. Izaurralde (1997) the role of the cap structure in RNA processing and nuclear export, *Eur. J. Biochem.* **247**, 461–469.

W. Merrick and J. Hershey (1996) in J. Hershey, M. Mathews, and N. Sonnenberg, eds., *Origins and Targets of Translational Control*, Cold Spring Harbor Laboratory Press, Cold Spring Harbor, New York, pp. 1–29.

CAPILLARY ZONE ELECTROPHORESIS

A. CHRAMBACH

Capillary zone electrophoresis (CZE) is **electrophoresis** in very thin capillaries. It is an analytical separation method that is compatible with high field strength (200 to 800 V/cm) and small samples (micrograms to nanograms). The high field strength makes the separation rapid and gives high resolving power [1). The very small capillary diameter, less than 0.2 mm, counteracts dispersion of the zone, presumably through interaction of the analyte with the inner wall of the capillary. The sample is detected by its absorbance or fluorescence as it proceeds past a stationary detector at the end of the migration path.

The inner walls of the silicate capillary carry negatively charged groups, so **electroendosmosis** is significant; this can be used for the purpose of separation, but it is usually suppressed by coating the inner wall with a polymer such as

polyacrylamide gel or dextran. Separations based predominantly on size and shape differences are achieved in CZE in the presence of soluble polymers to produce a **molecular sieve** effect. A variety of uncharged hydrophilic polymers in a wide range of molecular weights are available for that purpose [2–4]. Although providing high resolving power (see **Gel electrophoresis**), gels are less applicable to CZE because the gelation process may introduce air bubbles and inhomogeneities within the capillary; also, a gel-filled capillary can only be used once or a few times, while polymer solutions can be replaced easily. One drawback of CZE is that the analyte bands are enclosed within the tube and therefore are unavailable for detection by immunological, hybridization, and staining techniques. However, the eluate of CZE can be subjected to **mass spectrometry** to provide molecular weights of the species detected, in addition to their mobilities [5].

BIBLIOGRAPHY

1. F. Foret and P. Bocek (1989) *Adv. Electrophoresis* **3**, 273–347.
2. K. Ganzler, K. S. Greve, A. S. Cohen, B. L. Karger, A. Guttman, and N. C. Cooke (1992) *Anal. Chem.* **64**, 2665–2671.
3. M. C. Ruiz-Martinez, J. Berka, A. Belenkii, F. Foret, A. W. Miller, and B. L. Karger (1993) *Anal. Chem.* **65**, 2851–2858
4. D. Tietz, A. Aldroubi, H. Pulyaeva, T. Guszczynski, M. M. Garner, and A. Chrambach (1992) *Electrophoresis* **13**, 614–615.
5. D. Figeys, I. van Oostveen, A. Ducret, and R. Aebersold (1996) *Anal. Chem.* **68**, 1822–1828.

Suggestions for Further Reading

P. Gebauer and P. Bocek (eds.) (1995) Symposium on capillary electrophoresis theory. *Electrophoresis* **16**, 1985–2174.

P. D. Grossman and J. C. Colburn (1992) *Capillary Electrophoresis*, Academic Press, New York, pp. 1–352.

B. L. Karger (ed.) (1993) Symposium on capillary electrophoresis. *Electrophoresis* **14**, 371–560.

CAPSIDS, VIRAL

ROGER HENDRIX

The capsids of **viruses**—the protein part of the virus particle (virion) that surrounds and protects the nucleic acid genome—have been the subject of intense study by biochemists and structural biologists for over 50 years (see **Virus structure**). Capsids are also noteworthy in that they provide one of the few examples in which the detailed properties of a biological system have been predicted successfully from "first principles." This entry describes the principles on which we understand capsid structure—that is, what we expect capsids to be like and why we expect that, and then describes the ways in which real viruses do or do not follow those expectations.

A **gene** of double-stranded DNA can encode a **protein** of only about 1/20 the mass of the gene itself. As a consequence, if a virus is to encode its own capsid (and in almost all cases, viruses do so), it will only be able to make enough protein to produce a useful sized capsid if it can use multiple copies of the protein(s) encoded in its genes. Crick and Watson (1), who made the first concrete proposal for how proteins might be arranged in virus capsids, assumed that the capsids would be made of multiple identical virus-encoded protein subunits. They also made another assumption, based on a prominent property of proteins, namely that proteins are very specific in the interactions they make with other molecules, presumably including other proteins in the structure of a virus capsid. They assumed that the identical protein subunits would be packed into the capsid structure in such a way that they all made the identical set of contacts with their neighbors—so-called **equivalent packing**. Mathematically, there are only two general ways to satisfy these assumptions when packing asymmetric objects like proteins; these are to arrange the protein subunits with helical symmetry or to arrange them with one of the cubic symmetries. Helical symmetry is easily visualized: The subunits are arranged in a helical array as if they were on successive steps of a spiral staircase. Formally, each subunit is related to the preceding one by a characteristic rotation and a translation in the direction of the helix axis. "Cubic" symmetry refers to a small group of symmetries characterized by having multiple axes of rotational symmetry. These correlate with the five classical Platonic solids, which have these symmetry axes. Besides the cube, from which the entire group of symmetries takes its name, the Platonic solids include the tetrahedron, the octahedron, the dodecahedron, and the icosahedron. Of these, the icosahedron allows (or more properly, arranging protein subunits according to the five-, three-, and twofold rotational symmetry axes of an icosahedron) allows the largest structure to be made with a given size of subunit for any of the group. More to the point, of this group of symmetries, it is icosahedral symmetry that is used by real viruses.

HELICAL SYMMETRY

Many viruses can be shown to have helical symmetry. The best studied of these is the well-known **tobacco mosaic virus** (TMV), which has a virion with only one type of protein subunit, present in something over 2000 copies and arranged in a simple helix with 16.33 subunits per turn. There is in principle no limit on how long a helical structure such as this can be, but in TMV it is limited during assembly by the length of the genomic RNA. In the mature virion, the single-stranded RNA genome follows the helix of the proteins and lies in the groove between successive layers of the helix on the inside of the helical tube, three nucleotides per protein subunit. A domain of each subunit closes over this groove after the RNA has entered it during assembly and effectively seals off the RNA from the solvent (2).

Structurally somewhat more complex examples of helically symmetric viruses are the filamentous bacteriophages, such as phage M13. The major protein subunit of these viruses is arranged in what might best be described as a small number of helices running in parallel up a cylindrical tube. The circular single-stranded DNA genome of these viruses is stretched the length of the helical tube. The filamentous phages have small numbers of copies of additional virus-encoded proteins present on the ends of the virions. These have roles in virion assembly and in **virus infection** (3).

Many enveloped viruses, for example, **influenza virus**, have their genome—single-stranded RNA in the case of Influenza—wrapped in a helical array with a virus-encoded protein. The resulting flexible nucleoprotein rods are enclosed in the viral envelope, one such structure for each of the eight different genome segments in the case of influenza.

ICOSAHEDRAL SYMMETRY AND QUASI-EQUIVALENT PACKING

An icosahedron (Fig. 1) has 20 faces, each an equilateral triangle. Its symmetry is defined by axes of rotational symmetry, each passing through the center of the icosahedron: five-fold axes passing through the corners, three-fold axes passing through the centers of the faces, and twofold axes passing through the centers of the edges. Strictly speaking, an icosahedral virus capsid is not an icosahedron; instead, its protein subunits are related to each other by the symmetry axes of an icosahedron. However, thinking of the protein subunits as lying on the surface of an icosahedron provides a convenient way to visualize and discuss the structure of the capsid, as well as a reasonable approximation to the truth. To make an icosahedrally symmetric capsid, we place three identical protein subunits on each face of the icosahedron, symmetrically arrayed about the center. These subunits will necessarily also find themselves in symmetrical relations with their fellows around all the five-, three-, and twofold axes of the icosahedron, and inspection of a model of such a structure shows that all of the subunits lie in equivalent relationships to their neighbors.

Current knowledge of virus structure allows us to see that some small virus capsids—for example, the Parvoviruses—can be adequately described as such 60-subunit icosahedrally symmetric structures. However, it was already clear by about 1960 that the majority of "spherical" viruses are too large and have too many subunits to be so described. Because there is mathematically no way to make a larger structure while preserving strict equivalence in subunit packing, it was not clear that this way of describing capsid structure would be applicable to most viruses. However, Caspar and Klug (4) proposed a clever extension of Crick and Watson's ideas that has provided a framework for understanding and describing the great majority of spherical virus capsids. Caspar and Klug introduced the idea of "quasi-equivalence" in protein subunit packing. In a nutshell, they suggested that if the requirement for equivalence were relaxed—but not abandoned entirely—it would be possible to make much larger capsids with identical protein subunits, containing multiples of 60 subunits.

Figure 1. An icosahedron.

A simple way to imagine the process of making a larger capsid is to start with a 60-subunit capsid as described above, group the subunits together into the pentamers that surround each corner (fivefold axis) of the icosahedron, and move those pentamers away from the center to a larger radius. Caspar and Klug's insight was that there are certain ways to fill the resulting spaces between the pentamers with *hexamers* of the same subunit that preserve, at least approximately, the geometrical packing arrangements among all of the subunits in the structure.

A more systematic way to describe this process, shown in Figure 2, is to consider constructing an icosahedron from a piece of paper that has a hexagonal lattice of "protein subunits" printed on it. (Experience has shown, in fact, that the most effective way to understand the concepts described here is to actually take scissors in hand and to construct and examine some capsid models.) The simplest way to make an icosahedron from such a sheet of paper is to cut out 20 of the small triangles in the lattice that each contain three subunits, and join them together into an icosahedron. This icosahedron will have 60 subunits, equivalently packed, and will be identical to the 60-subunit structure described above. It is possible, however, to cut larger triangles out of the paper and use them as the faces of the icosahedron; as long as the corners of these larger triangles fall on the intersection points of the lattice, the resulting icosahedron will be biologically sensible and geometrically acceptable. The figure shows an example of such a larger triangle encompassing four of the small unit triangles. If 20 of these are cut out and assembled into an icosahedron, the structure will have $4 \times 60 = 240$ protein subunits. The subunits in this case are not all equivalently packed—some are in pentamers and some in hexamers, and there are other subtle differences in their geometrical relationships to their neighbors—but the basic relationships between subunits are nonetheless preserved: They are quasi-equivalent. The faces of this icosahedron are said to be "triangulated" into the four unit triangles, and the structure as a whole is said to have a "triangulation number" (T) of 4. The figure also shows a somewhat more complex way of defining the icosahedral face, in which the edges of the icosahedral face are not parallel to the lattice lines. In the example shown, the area of the face is equal to the area of seven small triangles, and the corresponding icosahedron is given the triangulation number 7. More generally, the triangulation number of any icosahedron can be calculated from the formula $T = h^2 + hk + k^2$, where h and k are the number of steps along the lattice in the **h** and **k** directions required to go from one corner of the icosahedral face to another. The "allowed" triangulation numbers derived in this way are a subset of the integers and form an infinite series that starts 1, 3, 4, 7, 9, 12, 13,....

ICOSAHEDRAL VIRUS CAPSIDS IN NATURE

As the structures of actual virus capsids have been determined in increasing numbers and in increasing detail over the past 30 years, the general predictions of the Caspar and Klug theory have largely been confirmed. At the same time,

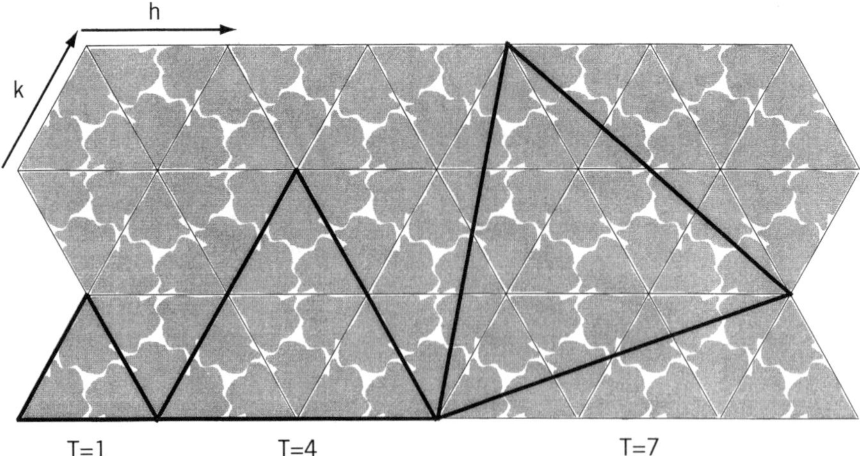

Figure 2. A hexagonal lattice of protein subunits with indications of how they can be fit into a strictly equivalent virus capsid ($T = 1$) or into two different kinds of quasi-equivalent capsids ($T = 4$, $T = 7$). The irregular wedge-shaped objects, of which three are packed into each small triangle, represent protein subunits. The bold triangles show the faces of the capsid icosahedron for the three indicated triangulation numbers. The **h** and **k** unit vectors are used to calculate the triangulation number, as described in the text.

there are now many examples of "variations on the theme"—features of the structures that are not explicitly predicted by the theory, but are not at odds with its basic concepts. In addition, there are at least two examples of structures that should not have occurred. The definitive test of how the capsid structure is organized is a high-resolution structure by **X-ray crystallography**; however, it is often possible to get useful information from lower resolution structural measurements. For example, the hexamers and pentamers of the capsid subunit often cluster in such a way that they can be seen by **electron microscopy** as a distinct morphological unit, called a **capsomere**. Examining the geometrical relationships among the capsomeres on the surface of a capsid can lead to a good idea of the triangulation number of the capsid.

Parvoviruses, as mentioned above, have 60-subunit $T = 1$ structures (5). The same is true for the small **ϕX174, bacteriophage**, but in this case there are two proteins present in 60 copies each, rather than only one (6). Here we can think of the heterodimer as the basic building block, with 60 copies of that heterodimer arranged with $T = 1$ icosahedral symmetry.

$T = 3$ capsids, the first group that requires that quasi-equivalence be invoked, are especially well-populated, with examples from plant, animal, and bacterial viruses. **Tomato bushy stunt virus** (TBSV), the first of this group to be solved to atomic resolution, conforms quite well to the expectations of the theory, having 180 chemically identical subunits, each of which occupies one of three similar but nonidentical positions in the $T = 3$ lattice (7). One set of subunits in TBSV, those occupying the "C" position, send an end of their polypeptide chain toward the threefold symmetry axes, where they intertwine with the corresponding parts of the two symmetrically related subunits. This was the first of what are now several examples of capsid proteins interdigitating and intertwining with their neighbors; this is presumed to provide mechanical strength to the capsid. **Poliovirus** (a Picornavirus) has a structure very similar to that of TBSV, including the positioning of subunits and even the fold of the polypeptide chains within those subunits, but in this case the three quasi-equivalent positions are occupied by three similar but chemically distinct proteins (8). We might imagine that two of the three coding regions encoding these three proteins were the result of **gene duplication** in an ancestral virus that, like TBSV, had only one capsid protein gene. A slightly different variation on this theme is found in **Cauliflower Mosaic virus** (a Comovirus), in which the three quasi-equivalent positions are occupied by two proteins, one of which consists of two very similar **domains**, each occupying one of the three quasi-equivalent positions. These last two examples can be regarded either as slightly noncanonical $T = 3$ structures or as $T = 1$ structures made of 60 copies of the heterotrimer (Picornavirus) or heterodimer (Comovirus).

Many capsid structures with larger triangulation numbers have been identified and characterized. Among these are **lambda phage**, **P22 phage**, bacteriophage HK97, and others ($T = 7$), **T4 phage** and **Reovirus** ($T = 13$), **Herpesvirus** ($T = 16$), and **Adenovirus** ($T = 25$). The largest triangulation number that has been definitively determined is that of the algal virus PBCV1 ($T = 169$). A particularly instructive example is provided by bacteriophage P2 and its satellite phage P4. The P2 capsid protein assembles into a $T = 7$ shell; but in the presence of P4, which does not encode its own capsid protein, the P2 capsid protein assembles into a $T = 4$ structure, just big enough to enclose the P4 genome. This is accomplished through the agency of a P4-encoded protein that associates transiently with the P2 capsid protein during assembly and directs it into the $T = 4$ geometry (9).

Although these large viruses all fit into the general picture envisioned by Caspar and Klug, they are replete with exceptions and extensions to the original picture, and these variations all expand our view of how capsids can be constructed. For bacteriophage T4 and Adenovirus, the proteins that make the pentamers at the five-fold symmetric corners of the capsid are encoded by a different gene from the proteins at the six-fold positions in the remainder of the capsid. This is analogous to the Picornavirus example above and presumably reduces the amount of conformational or bonding versatility demanded of any one protein, at the relatively minor expense of encoding an additional protein. In T4, the major component of the capsid, which occupies the hexamer positions, is, as expected, a hexamer, but in Adenovirus these positions are occupied by *trimers* of the "hexon protein" (10). However, the hexon protein is organized into two similarly folded domains, and the structure of the trimeric hexon is very close to a sixfold symmetric arrangement of those domains.

Some viruses—particularly but not exclusively some of the double-stranded DNA phages—have prolate capsids, which have a standard icosahedral arrangement of subunits, except that the shell is elongated along one of the fivefold axes of symmetry and an extra band of hexamers is inserted around the equator. Thus bacteriophage T4 has an elongated $T = 13$ capsid, and bacteriophage $\phi 29$ has an elongated $T = 4$ capsid. These structures pose interesting questions with regard to how their length is specified and accurately achieved, but they cause no serious problems for the idea of quasi-equivalent packing.

Most viruses have other capsid subunits in addition to the main, icosahedrally packed, subunit protein. In some cases, these are present in equimolar amounts with the main subunit and packed with the same symmetry. Thus phages λ and T4 have well-studied examples of such "decoration proteins"; Herpesvirus has a protein clustered as trimers that fits this description, but is systematically absent from positions immediately surrounding the pentamers. Proteins in this category are known in some cases to provide additional strength and stability to the capsid. Structurally, they can be regarded simply as additional domains of the main capsid protein, albeit ones that are encoded by separate genes and generally join the structure at a different time.

Many capsids also have "minor" proteins that are not arrayed with icosahedral symmetry. The portal protein of the double-stranded DNA phages, for example, forms a grommet-like 12-subunit oligomer that replaces a pentamer at one of the five-fold corners of the icosahedral shell and provides an attachment site for the six-fold symmetric helical tail (11). Adenovirus has a trimeric spike extending out from the shell along each of its five-fold symmetry axes.

The first radical deviation from the expectations of the Caspar–Klug ideas was found in the virion of Papovavirus **SV40**. The capsomeres of SV40 visible by electron microscopy are arranged as expected for a $T = 7$ structure. However, all of the capsomeres are pentamers of the VP1 subunit (12), including the 60 capsomeres situated at positions of sixfold local symmetry, which would be expected to be hexamers in order to interact quasi-equivalently with their environment. The subunits adapt to this extraordinary state of affairs by having the part of the polypeptide chain that contacts the neighboring capsomere located on a flexible arm corresponding to the *C*-terminus of the subunit. The *C*-terminal arm leaves its home subunit at dramatically different angles in different cases, allowing it to interact with its neighbor in essentially the same way in each case. This it does by invading the structure of the neighbor and forming one strand of a **β-sheet** structure in that subunit. This might be regarded as a form of quasi-equivalent interaction, but of a sort requiring a much more radical subunit flexibility than envisioned by Caspar and Klug.

Another surprising arrangement of subunits is found in fungal virus L-A and bacteriophage $\phi 6$. These capsids have 120 identical protein subunits, not one of the "allowed" numbers. These are $T = 1$ structures, built of 60 asymmetric homodimers. The unexpected feature of this arrangement is that the two chemically identical subunits occupy nonequivalent positions. Nonetheless, these capsids, as for those of SV40, evidently function well enough to have survived **natural selection** (ie, very well indeed), and an understanding of their structures enlarges our view of the capabilities of proteins.

BIBLIOGRAPHY

1. F. Crick and J. D. Watson (1956) *Nature* **177**, 473–475.
2. J. N. Champness et al. (1976) *Nature* **259**, 20–24.
3. L. Makowski and M. Russel (1997) In *Structural Biology of Viruses* (W. Chiu, R. M. Burnett, and R. L. Garcea, eds.), Oxford University Press, New York, pp. 352–380.
4. D. Caspar and A. Klug (1962) *Cold Spring Harbor Symp. Quant. Biol.* **27**, 1–24.
5. J. Tsao et al. (1991) *Science* **251**, 1456–1464.
6. R. McKenna, L. Ilag, and M. Rossmann (1994) *J. Mol. Biol.* **237**, 517–543.
7. A. J. Olson, G. Bricogne, and S. C. Harrison (1983) *J. Mol. Biol.* **171**, 61–93.
8. J. Hogle, M. Chow, and D. Filman (1985) *Science* **229**, 1358–1365.
9. O. Marvik et al. (1995) *J. Mol. Biol.* **245**, 59–75.
10. R. M. Burnett (1997) In *Structural Biology of Viruses* (W. Chiu, R. M. Burnett, and R. L. Garcea, eds.), Oxford University Press, New York, pp. 209–238.
11. C. Basinet and J. King (1985) *Annu. Rev. Microbiol.* **39**, 109–129.
12. R. L. Garcea and R. C. Liddington (1997) In *Structural Biology of Viruses* (W. Chiu, R. M. Burnett, and R. L. Garcea, eds.), Oxford University Press, New York, pp. 187–208.

Suggestions for Further Reading

W. Chiu, R. M. Burnett, and R. L. Garcea (eds.) (1997) *Structural Biology of Viruses* Oxford University Press, New York.

B. N. Fields, D. M. Knipe, and P. M. Howley (eds.) *Virology* (1996) Lippincott-Raven, Philadelphia.

J. E. Johnson and J. A. Speir (1997) *J. Mol. Biol.* **269**, 665–675.

J. E. Johnson and A. J. Fisher (1994) In *Encyclopedia of Virology*, Vol. 1 (R. G. Webster and A. Granoff, eds.), Academic Press, London, pp. 1573–1586.

CARBON ISOTOPES

D. R. FISHER

Carbon is element number 6 in the periodic table and has valence states of 2, 3, or 4 (1). Thirteen isotopes of carbon have been identified (2), ranging in atomic mass number from ^{8}C (half-life = 3×10^{-22} s) to ^{20}C (half-life = 0.01 s) (see **Radioactivity** and **Radioisotopes**). Two stable isotopes of carbon are found in nature: ^{12}C at 98.9%, and ^{13}C at 1.1% abundance.

The most important radioactive isotope of carbon is ^{14}C (half-life = 5730 years). Carbon-14 decays by beta-minus emission to nitrogen-14, which is stable. Carbon-14 decay yields one beta particle, with an energy of 0.156 MeV maximum, 0.0495 MeV on average. Carbon-14 is produced naturally in the earth's atmosphere by cosmic ray interactions with stable nitrogen-14 according to the reaction $^{14}N(n,p)^{14}C$. Atmospheric $^{14}CO_2$ is inhaled by animals and respired by plants in photosynthesis, incorporating small amounts of carbon-14 in all living organisms. Consequently, carbon-14 dating is useful for determining the age of organic matter such as wood and archeological specimens. Hundreds of different organic molecules have been labeled with carbon-14 as a tracer for studying biochemical processes and are available from commercial suppliers.

Another radioactive isotope, carbon-11 (half-life = 20.38 min), is used in nuclear medicine diagnostics with positron-emission tomography. Carbon-11 decays by beta-plus (positron) emission to boron-11, which is stable. Carbon-11 decay yields 0.98 beta particles, with an energy of 0.960 MeV maximum, 0.386 MeV on average. Positron emission is characterized by twin 0.511-MeV photons that result from the annihilation of a positron and an electron and allow it to be detected externally. Carbon-11 is produced by proton accelerators according to the reaction $^{11}B(p,n)\,^{11}C$. Carbon-11 positrons are useful for ^{11}C-acetate (3) and ^{11}C-palmitate (3) metabolism kinetic imaging in the assessment of heart disease (4).

Carbon-13 is very useful in studies involving nuclear magnetic resonance (**NMR**), because it gives an NMR signal, whereas the normal isotope, carbon-12, does not.

Carbon-14 was first produced artificially for metabolism studies in 1939. At present, it is produced commercially by neutron irradiation of beryllium nitride or aluminum nitride according to the reaction $^{14}N(n,p)^{14}C$. The low cross section of this reaction and the long half-life of carbon-14 results in low specific activities, which can make carbon-14 labeling of organic materials quite difficult.

Carbon-14 labeling of organic molecules has numerous applications in the biomedical sciences. Examples include studies of glucose metabolism and conversion to $^{14}CO_2$, galactose oxidation, carbohydrate metabolism, the Krebs cycle, **nucleic acid** synthesis and **hybridization**, **autoradiography**, or **fluorography**, and for studying many other biochemical pathways. Among the most important carbon-14 compounds used in biochemistry and metabolism studies are the carbon dioxides and monoxide, carboxyl-labeled acids, methanols, cyanides, carbides, formic acids, and cyanamides (5). Carbon-14 is detected by beta-particle liquid scintillation counting, **autoradiography**, and **fluorography**.

BIBLIOGRAPHY

1. D. R. Lide and H. Pr. Frederikse, eds. (1995) *CRC Handbook of Chemistry and Physics*, CRC Press, Boca Raton, Fla.
2. Knolls Atomic Power Laboratory (1966) *Chart of the nuclides*, 15th ed., available from General Electric Company, San Jose, Calif.
3. H. R. Schön, H. R. Schelbert, A. Najafi et al. (1982) *Am. Heart J.* **103**, 532–547.
4. T. L. Rosamond, D. R. Abendschein, B. E. Sobel et al. (1987) *J. Nucl. Med.* **28**, 1322–1329.
5. *The Radiochemical Manual*, 2nd ed. (1966) Amersham, Bucks, England.

Suggestions for Further Reading

J. R. Catch (1961) *Carbon-14 Compounds*, Butterworths, London.

J. C. Harbert, W. C. Eckelman, and R. D. Neumann, eds. (1996) *Nuclear Medicine: Diagnosis and Therapy*, Thieme Medical Publishers, Inc., New York.

CARBONIC ANHYDRASE

J. A. HUNT
C. A. FIERKE

The enzyme carbonic anhydrase (CA) catalyses the reversible hydration of carbon dioxide to bicarbonate and a proton, using a catalytic zinc ion bound at the active site (1,2). Carbonic anhydrases are of interest because of their varied metabolic roles, efficient catalytic mechanisms, and high metal ion specificity. The enzyme is found in animals, plants, and bacteria, where it plays roles in respiration, **photosynthesis**, and CO_2 fixation. Three genetically distinct families of CA proteins have been identified: αCA isozymes, found in all animals; βCA, commonly found in plants and bacteria; and γCA, found in a variety of bacteria (3). Of these classes of CA proteins, the α-CA family is the best characterized. This article provides a brief overview of the metabolic roles and genetic structure of the α-CA human isozymes and describes the catalytic reaction mechanism and metal-binding properties of the high activity, well-characterized human α-CA isozyme II (CAII). Recent work on the β- and γ-CA enzymes, which catalyze CO_2 hydration using radically different protein structures, is summarized.

METABOLIC FUNCTION AND GENETIC STRUCTURE OF HUMAN α-CA ISOZYMES

Carbonic anhydrase II (CAII) was first discovered in 1933 in human erythrocytes (4), where it facilitates CO_2 transport in respiration by converting CO_2, released as a metabolic by-product from tissues, to bicarbonate. Bicarbonate is carried by the bloodstream to pulmonary capillaries, where its conversion back to CO_2 is catalyzed by CA for release by the lung. CAII is also expressed in many other tissues, including ocular epithelium, where it plays a role in maintaining intraocular pressure (5). A human genetic disease in which CAII is nonfunctional indicates that this isozyme is also crucial for bone resorption and kidney function (6). Furthermore, the inhibition of CAII by aromatic sulfonamides (RSO_2NH_2) is used clinically to treat glaucoma and altitude sickness (5). Additionally, at least six more CA human **isozymes** with varied activities and locations have been discovered (5). The cytosolic isozymes CAI, CAIII, and CAVI are expressed in high concentrations in red blood cells, **muscle** cells, and salivary glands, respectively. CAIV is a membrane-bound isozyme important for regulating bicarbonate levels in the kidney, whereas CAV is a **mitochondrial** isozyme. CAVI is secreted by the salivary glands. The exact physiological role of several of these isozymes is still under investigation.

The genomic structure of the human CA isozymes lends insight into their **evolution** and regulation of their **transcription**. CAI, CAII, and CAIII have been mapped to **chromosome** 8q22, CAV and VII to chromosome 16, CAIV to chromosome 17q, and CAVI to chromosome 1 (3). This chromosomal mapping, along with comparisons of DNA sequence homologies, indicates that **gene duplication** events over 450 million years ago led to distribution of the CA isozymes on four different chromosomes; further gene duplications occurred later, resulting in closely linked isozymes CAI, CAII, and CAIII on one chromosome, and CAIV and VI on another (3). The structures of the various human CA genes currently are being studied to determine the mechanism of their tissue-specific expression. CAI has an unusually long 5′-**untranslated region** containing two **promoters** that direct expression of CAI in erythroid cells and in the colon (7). Other CA genes are also likely to have complex structures to direct their expression in different locations and developmental stages.

CATALYTIC MECHANISM AND ZINC-BINDING PROPERTIES OF HUMAN CAII

Carbonic anhydrase II is the most thoroughly characterized and most highly active of the CA isozymes, catalyzing the hydration of CO_2 with a rate constant of $\sim 10^6\ s^1$ (1). The crystal structure of CAII reveals a tightly bound zinc ion coordinated by the nitrogen atoms of three protein **histidine** residues, His94, His96, and His119, located at the bottom of a deep active site cleft (Fig. 1) (8). At physiological pH, a hydroxide ion is also bound by zinc, completing the tetrahedral geometry of the metal site. Residue Thr199 forms a **hydrogen bond** with the zinc hydroxide, orienting it for efficient catalysis. Catalysis occurs in two main steps (1), initiated by nucleophilic attack of the hydroxide ion on the carbonyl carbon of CO_2 to form bicarbonate.

$$\text{E-ZnOH}^- + CO_2 + H_2O \leftrightarrow \text{E-ZnHCO}_3^- + H_2O$$
$$\leftrightarrow \text{E-ZnOH}_2 + HCO_3^- \quad (1)$$

Water displaces bicarbonate, and the product is released. In the second, rate-limiting step in catalysis (Eq. 2), the catalytically active enzyme species is regenerated by transfer of the product proton from the zinc-water to protein residue His64. This residue is exposed to solvent and transfers the proton to buffer B:

$$\text{E-ZnOH}_2 + B \leftrightarrow {}^+\text{H-E-ZnOH}^- + B$$
$$\leftrightarrow \text{E-ZnOH}^- + BH^+ \quad (2)$$

This proton shuttle is essential for rapid catalysis, as transfer of the proton to buffer is much faster than transfer to water. Carbonic anhydrase is one of the few biological systems in which proton transfer can be studied (9).

Just as the active site of CA has evolved for rapid catalysis, the protein structure is optimized for tight, specific binding of the zinc ion, with a K_d of ~2 pM (2,10) (see **Zinc-binding proteins**). CAII binds only Cu^{2+}, and Hg^{2+} with comparable affinity, and neither of these metals confers catalytic activity to the enzyme (10). CAII has a much lower affinity, with a K_d of nanomolar, for metals such as Co^{2+}, Mg^{2+}, Cd^{2+}, and Ni^{2+}, and only the Co^{2+}-substituted enzyme is active. As the metal binding properties of this protein have been studied extensively, CAII is often used as a model for creating novel metal sites in proteins.

Studies in which **site-directed mutagenesis** has been used to alter the nature of conserved residues in the zinc site (2) reveal the importance of several factors for high metal affinity. First, in almost all CAII variants studied, the zinc ion retains tetrahedral geometry even if the surrounding protein structure must be rearranged to accommodate new side-chain positions. Second, the distance between the metal and ligand is crucial. Third, conserved residues that form hydrogen bonds with the histidine zinc ligands each contribute modestly to zinc affinity but play a large role in controlling the rates of metal equilibration (2,11). The histidine ligands, the residues that form hydrogen bonds to these ligands, and the surrounding protein structure contribute to the reactivity and binding properties of the zinc ion.

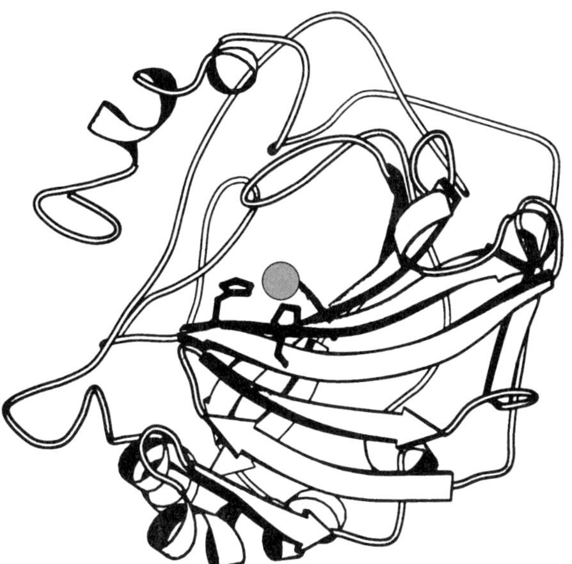

Figure 1. The crystal structure of human CAII determined by X-ray crystallography (8). It reveals a twisted β-sheet structure in which three histidine residues coordinate the catalytic zinc ion. Figure generated using MOLSCRIPT (13).

GENETICALLY DISTINCT CA FAMILIES AROSE THROUGH FUNCTIONAL CONVERGENCE

Examples of the β-CA and γ-CA carbonic anhydrase families are only now being discovered and characterized. In an amazing example of evolutionary functional **convergence**, these protein families are completely unrelated genetically to one another and to the α-CA family, yet these enzymes catalyze the same reaction using a catalytic zinc ion (3). The structure determined by **X-ray crystallography** of a γ-CA isolated from **archaebacteria** (12) reveals a trimeric protein containing three active sites; in each active site, a zinc ion is bound at the monomer interface by three histidine residues from different polypeptide chains. As in α-CA proteins, the zinc is bound in tetrahedral geometry, the fourth ligand being a solvent molecule. In this case, a **glutamic acid** residue located near the zinc ion in γ-CA may be the functional analog of Thr199 in human CAII. No structure of β-CA has been solved yet, but phylogenetic and spectroscopic data suggest that the zinc ion may be coordinated by glutamate or **cysteine** residues rather than histidine (3).

SUMMARY

In recent years, the techniques of site-directed mutagenesis and X-ray crystallography have greatly expanded our knowledge of the relationships between protein structure and function in human CAII (2). The structure of CAII appears to have evolved for maximum catalytic activity and metal affinity. Applying these methods to the carbonic anhydrase enzymes that have evolved independently to catalyze the same reaction will give further insight into how protein structure can dictate the reactivity and affinity of metals.

BIBLIOGRAPHY

1. D. N. Silverman and S. Lindskog (1988) *Acc. Chem. Res.* **21**, 30–36.
2. D. W. Christianson and C. A. Fierke (1996) *Acc. Chem. Res.* **29**, 331–339.
3. D. Hewett-Emmett and R. E. Tashian (1996) *Mol. Phylogenet. Evol.* **5**, 50–77.
4. W. C. Stadie and H. O'Brien (1933) *J. Biochem.* **103**, 521–529.
5. S. J. Dodgson, R. E. Tashian, G. Gros, and N. D. Carter (1991) *The Carbonic Anhydrases: Cellular Physiology and Molecular Genetics*, Plenum Press, New York.
6. W. S. Sly et al. (1983) *Proc. Natl. Acad. Sci. USA* **80**, 2752–2756.
7. H. J. M. Brady et al. (1991) *Biochem. J.* **277**, 903–905.
8. K. Hakansson, M. Carlsson, L. A. Svensson, and A. Liljas (1992) *J. Mol. Biol.* **227**, 1192–1204.
9. D. N. Silverman (1995) *Meth. Enzymol.*. **249**, 479–503.
10. S. Lindskog and P. O. Nyman (1964) *Biochim. Biophys. Acta* **85**, 462–474.
11. C.-C. Huang et al. (1996) *Biochemistry* **35**, 3439–3446.
12. C. Kisker et al. (1996) *EMBO J.* **15**, 2323–2330.
13. P. Kraulis (1991) *J. Appl. Crystallogr.* **24**, 946–950.

Suggestions for Further Reading

D. W. Christianson (1991) Structural biology of zinc. *Adv. Prot. Chem.* **42**, 218–355.

S. J. Dodgson, R. E. Tashian, G. Gros, and N. D. Carter (1991) *The Carbonic Anhydrases: Cellular Physiology and Molecular Genetics*, Plenum Press, New York.

W. S. Sly and P. Y. Hu (1995) Human carbonic anhydrases and carbonic anhydrase deficiencies. *Ann. Rev. Biochem.* **64**, 375–401.

R. E. Tashian (1992) Genetics of the mammalian carbonic anhydrases. *Adv. Gen.* **30**, 321–356.

CARBOXYL GROUPS

T. IMOTO

$$C_2H_5-N{=}C{=}N-\underset{H_2}{C}-\underset{H_2}{C}-\underset{H_2}{C}-N(CH_3)_2$$

[I] EDC

$$RCOOH + R^1N{=}C{=}NR^2 \xrightarrow{H^+} R^1N^+H{=}\overset{RCOO}{\overset{|}{C}}-NHR^2 \xrightarrow{H_2O} RCOOH + R^1NHCONHR^2$$

$$R^1N^+H{=}C(OOCR)-NHR^2 \xrightarrow{R^3NH_2} RCONHR^3 + R^1NHCONHR^2$$

A carboxyl group has the formula —COOH. Among compounds that have a carboxyl group are carboxylic acids. A carboxyl group shows relatively weak acidity. Many carboxylic acids are found in nature as free acids or in the form of esters, amides, or salts. Biologically important acids are those found in the metabolic pathways of glycolysis, fatty acids, **amino acids**, and so on. A carboxylic acid reacts with a base to form a salt and can be reduced to an aldehyde and an alcohol. A carboxylic acid forms an ester with an alcohol, an amide with an amine, and an acid anhydride with an carboxylic acid [see **Anhydrides**]. Carboxyl groups are not very reactive, and some catalyst or activation processes are required for their reactions.

There are three kinds of carboxyl groups in proteins, the C-terminal α-carboxyl, the β-carboxyl of **aspartic acid**, and the γ-carboxyl of **glutamic acid**. They have intrinsic pK_a values of 3.8, 4.0, and 4.4, respectively. Carboxyl groups are often employed as catalytic groups in **enzymes**, and they often play critical roles in **ligand binding**. Thus, carboxyl groups are important in protein chemistry.

CHEMICAL MODIFICATION OF CARBOXYL GROUPS IN PROTEINS

Esterification and amidation are the usual chemical modifications of carboxyl groups (1). Carboxyl groups in proteins are esterified by treating the dry protein with methanol-HCl. However, this method is drastic and not suitable for selective modification of carboxyl groups. Trialkyloxonium salts, such as triethyloxonium tetrafluoroborate, are commonly used for mild esterification of carboxyl groups in proteins (2). Diazoacetyl compounds, such as diazoacetamide, methyl diazoacetate, and *N*-diazoacetylglycinamide, react with carboxyl groups at acidic pH to form esters. Other reagents, such as acid halogen compounds and ethyleneimine have also been used to esterify the carboxyl groups in proteins.

Carboxyl groups are amidated with simple amines after activating the carboxyl groups with water soluble carbodiimides, such as 1-ethyl-3-(3-dimethylaminopropyl) carbodiimide (EDC) [I]. At pH 4.75 to 5, the reaction proceeds as in Scheme 1. Exhaustive amidation is achieved by using high concentrations of both amine (1 M) and EDC (0.1 M) at pH 4.75 under **denaturing** conditions (3). Limited amidation is possible using low concentrations of amine (0.1 M) and EDC (2.6 mM) under mild conditions at pH 5.0 (4). Isoxazolium salts, including Woodward's Reagent K (2-ethyl-5-phenylisoxazolium 3′-sulfonate), can be used in the place of EDC at pH 3 to 5. Amidation is also achieved by ammonolysis of the esters in liquid ammonia at −55°C. In this case, selective amidation is possible starting with a selectively esterified protein (5).

BIBLIOGRAPHY

1. T. Imoto and H. Yamada (1989) In *Protein Structure: A Practical Approach* (T. E. Creighton, ed.), IRL Press, Oxford, U.K., pp. 247–277.
2. S.M. Parsons et al. (1969) *Biochemistry* **8**, 700–712.
3. T.-Y. Lin and D.E. Koshland Jr. (1969) *J. Biol. Chem.* **244**, 505–508.
4. H. Yamada et al. (1981) *Biochemistry* **20**, 4836–4842.
5. R. Kuroki et al. (1986) *J. Biol. Chem.* **261**, 13571–13574.

CARBOXYL PROTEINASE

J. RIORDAN

The carboxyl proteinases (E.C. 3.4.23) are also known as *aspartyl proteinases* or *acid proteinases*, so called because the mechanism by which they catalyze the hydrolysis of **peptide bonds** in **proteins** (see **Proteinases**) involves the participation of two **carboxyl groups**, usually provided by the side chains of two **aspartic acid** residues (1). The prototypical member of this large family of structurally-related enzymes is the gastric proteinase, **pepsin**, which is optimally active at acid pH. Other members include *rennin* (*chymosin*) from the fourth stomach of the calf; *renin*, a kidney enzyme found in blood plasma; **cathepsin** D, found in **lysosomes**; numerous fungal enzymes, such as penicillopepsin; and a number of retroviral proteinases from **retroviruses**, including the human immunodeficiency virus **HIV** proteinase that is the target of the proteinase inhibitors used in the treatment of acquired immune deficiency syndrome (AIDS). Pepsin and renin illustrate the range of specificity exhibited by proteinases in general and carboxyl proteinases in particular. Pepsin cleaves multiple bonds in virtually all proteins found in the diet, whereas renin cleaves a single bond in its only known biological substrate, *angiotensinogen*—the precursor of *angiotensin*.

Carboxyl proteinases occur in all **eukaryotic** organisms and are important for protein processing and degradation.

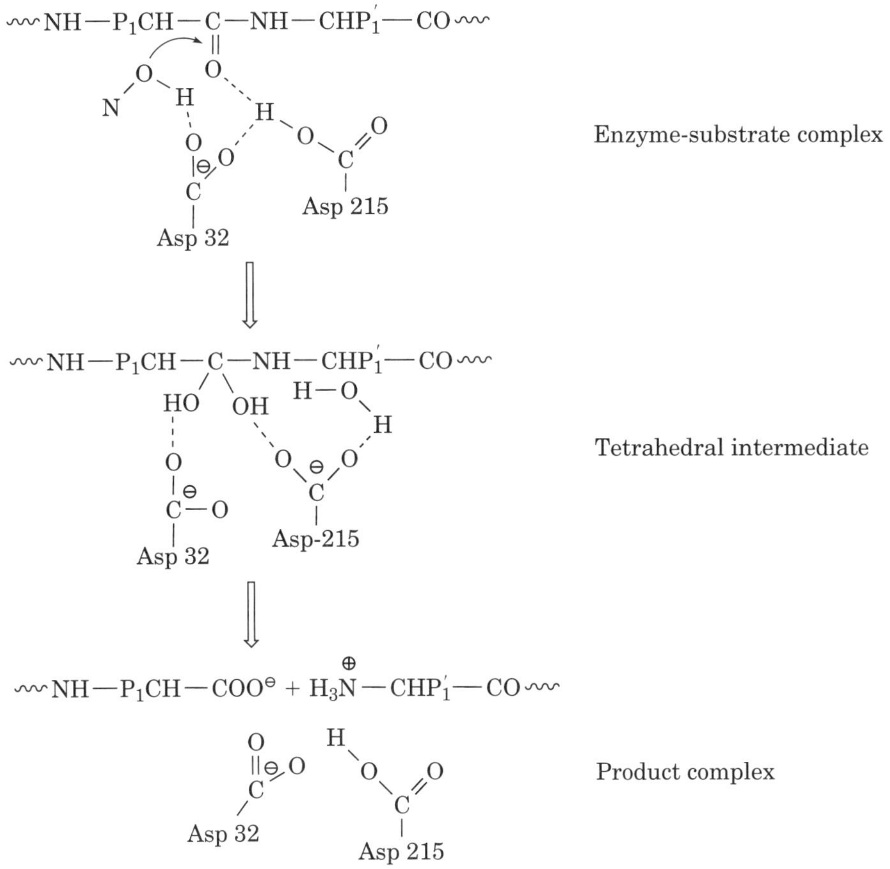

~NH—P$_1$CH—C(=O)—NH—CHP$'_1$—CO~ (Asp 32, Asp 215)

Enzyme-substrate complex

⇓

~NH—P$_1$CH—C(OH)(OH)—NH—CHP$'_1$—CO~ (Asp 32, Asp-215)

Tetrahedral intermediate

⇓

~NH—P$_1$CH—COO$^\ominus$ + H$_3$N$^\oplus$—CHP$'_1$—CO~ (Asp 32, Asp 215)

Product complex

Figure 1. Schematic representation of the mechanism of action of carboxyl proteinases. The substrate (top line) binds to the active site of the enzyme [represented by the carboxyl groups of Asp32 and Asp215 (porcine pepsin numbering)] to form an enzyme-substrate complex. P_1 and P'_1 are the sidechains of the main specificity-determining amino acid residues that contribute the -CO- and -NH- groups to the peptide bond that will be hydrolyzed. The carboxylate group of Asp 32 promotes the attack of a water molecule on the carbonyl carbon of the peptide bond, to form a tetrahedral intermediate that is stabilized by transfer of a proton from Asp 215. The intermediate undergoes rearrangement to form a product complex, which subsequently dissociates, thereby regenerating the original enzyme.

They are generally synthesized as inactive precursors that spontaneously activate under acidic conditions. Protein and peptide substrates bind to the **active site** of these enzymes, and one of the carboxylate groups (in the case of pepsin, it is from Asp32) activates a water molecule to attack the carboxyl group of the peptide bond that is to be cleaved (Figure 1). The other carboxyl group, which is protonated (in pepsin, it is Asp215), donates a proton to the peptide nitrogen atom to form an activated intermediate that dissociates into the products. It is believed that this general mechanism pertains to all of the carboxyl proteinases.

It is interesting that in carboxyl proteinases such as pepsin and renin, both catalytic carboxyl groups are present in the same molecule. In the retroviral proteinase from HIV-1, which only contains 99 amino acids, two molecules form a dimer with a single active site containing one catalytic aspartic acid from each subunit. The larger mammalian and microbial carboxyl proteinases contain approximately 325 residues and appear to have **homologous** amino- and carboxyl-terminal **domains** that likely evolved from an early **gene duplication** and **fusion** event. Each domain contributes a catalytically active aspartic acid residue.

A characteristic feature of carboxyl proteinases is their susceptibility to inhibition by **pepstatin**, an acylated pentapeptide analogue produced by **Streptomyces**. Recently a number of carboxyl proteinase inhibitors have been found useful for the treatment of AIDS, particularly when used in conjunction with other viral enzyme inhibitors. Carboxyl proteinase inhibitors have also been proposed for the treatment of malaria. Two such proteinases are believed to be essential for degradation of **hemoglobin** in the red blood cell phase of the life cycle of the parasite.

BIBLIOGRAPHY

1. D. R. Davies (1990) *Annu. Rev. Biophys. Biophys. Chem.* **19**, 189–215.

CARBOXYPEPTIDASES

J. RIORDAN

Enzymes that catalyze the hydrolytic cleavage of the carboxyl-terminal **amino acid** residue from an oligo- or **polypeptide chain** are called carboxypeptidases.

One group of these enzymes uses a catalytic mechanism analogous to that of the **serine proteinases** and, in fact, are inhibited by the prototypical serine proteinase inhibitor, diisopropylfluorophosphate (**DIFP**). These enzymes are found in the **vacuoles** of higher **plants** and **fungi** and in the **lysosomes** of animal cells, and they presumably participate in intracellular **protein degradation**. Many fungi also secrete serine carboxypeptidases. *Carboxypeptidase Y* (E.C. 3.4.16.1) is a serine carboxypeptidase present in yeast vacuoles (1). It is a very useful enzyme for determining the amino acid sequence of peptides and proteins (see **Protein sequencing**), because it can release all C-terminal amino acids, including **proline**. Moreover, it is active in the presence of **denaturants**, such as **urea** or sodium dodecyl sulfate (**SDS**), and in the pH range from 4 to 7.

The other group contains metallocarboxypeptidases, which have a zinc ion at the catalytic site (2). The best known of these are the pancreatic enzymes, carboxypeptidases A and B, which are synthesized as largely inactive **zymogens**, procarboxypeptidases A and B, and stored in the zymogen granules of the pancreatic acinar cells. On ingestion of a meal, they are released into the duodenum through the pancreatic duct, become activated by the action of trypsin, and help digest dietary proteins and peptides. The specificity of carboxypeptidase A (E.C. 3.4.17.1) is toward aromatic and bulky **hydrophobic** amino acids, nicely complementary to that of **chymotrypsin**, which generates peptides with just such amino acids at their C-termini. Similarly, the specificity of carboxypeptidase B (E.C. 3.4.17.2) is toward peptides with the C-terminal basic amino acids **arginine** and **lysine**, which can be generated by the action of **trypsin**. The combined action of these endo- and exopeptidases ensures optimum formation of essential amino acids. Carboxypeptidase N circulates in plasma and is a peptidyl-L-arginine hydrolase that is thought to be responsible for the degradation of bradykinin and other **hormones**.

An important dipeptidyl carboxypeptidase, *angiotensin converting enzyme* (ACE), cleaves C-terminal dipeptides from a variety of substrates, most notably angiotensin I and bradykinin. The former reaction generates angiotensin II, a potent vasopressor, and the latter inactivates a vasodilator. This dual effect on blood pressure has led to the widespread use of ACE inhibitors as antihypertensive agents.

BIBLIOGRAPHY

1. K. Breddam (1986) *Carlsberg Res. Commun.* **51**, 83–128.
2. F. X. Aviles et al. (1993) *Eur. J. Biochem.* **211**, 381–389.

CARCINOGEN

LYNNETTE FERGUSON
WILLIAM DENNY

Translated literally, the term "carcinogen" means giving rise to carcinomas, or epithelial malignancies. In practice, "carcinogen" is used to describe any physical or chemical agent that increases the incidence of tumors in animal models or in human populations. Chemical carcinogens are a diverse group of agents with a range of properties. Much of the early work on carcinogenesis in animal models identified two stages: tumor initiation, thought to involve a mutational event, and tumor promotion, which probably did not. Although this classification of events is still useful in animal studies, it is now recognized as an oversimplification. Cancer develops through a series of stages, which have been exceptionally well-characterized in certain cancers such as those of the colon (1). (An historic perspective on the development of understanding the nature of carcinogenesis is given by Harris (2)). Many of the steps in carcinogenesis involve either **mutations** or **epigenetic** changes, with cells gaining a selective advantage and undergoing clonal expansion as a result of activation of **protooncogenes** and/or inactivation of **tumor suppressor genes**. "Epigenetic" is defined as "all processes relating to the expression (transcription and translation) and the interaction of the genetic material. Epigenetic mechanisms may act at three levels of cell organization: (1) Turn genes off or modulate **protein biosynthesis**, (2) regulate the **translation** of **messenger RNA** into proteins, and (3) regulate the topographic

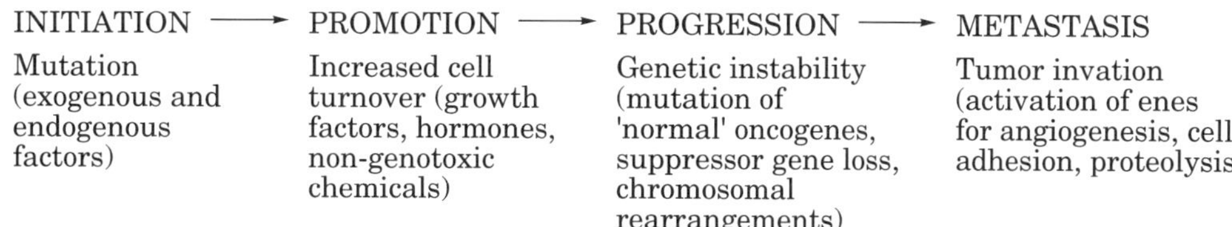

Figure 1. Factors involved in the various stages of carcinogenesis.

distribution and function of proteins." (3). Metastatic spread of the cancer involves the activation of further oncogenes controlling various aspects of cell adhesion and movement. A scheme showing some of the events involved in various stages of cancer development is shown in Figure 1.

Mutagenicity is the major mechanism for the activation of proto-oncogenes, and most carcinogens are also **mutagens**. Miller and Miller (4) claimed that the property in common to all the diverse carcinogens is that they either interact directly with DNA or can be metabolically activated to nucleophilic intermediates that are reactive with DNA. While this is certainly true for most carcinogens, there is increasing recognition that not all carcinogens are DNA reactive. For example, inhibitors of topoisomerase II enzymes have been shown to induce human cancers (5), and these act to cause DNA breaks indirectly by poisoning the enzyme (see **DNA topology**). A number of carcinogens may act through epigenetic mechanisms. There is also a class of carcinogens that have no action at all on the genetic material (eg, **peroxisome** proliferators, which are thought to work through **receptor** binding and modification of fatty acid metabolism (6)).

While there is no doubt that mutagenic mechanisms are involved in carcinogenesis, there is considerable doubt as to the relative roles of exogenous and endogenous mutagens in the development of human cancers. Exogenous mutagens include alkylating chemicals such as nitrogen mustards and **nitrosamines**, usually encountered in cancer therapy, and industrial chemicals such as vinyl chloride, **dimethyl sulfate**, and bis(chloromethyl) ether. Prominent endogenous mutagenic processes include oxy-radical DNA damage, DNA depurination, **DNA polymerase** infidelity, and deamination of **5-methylcytosine**.

BIBLIOGRAPHY

1. B. Vogelstein, E. R. Fearon, S. R. Hamilton, S. E. Kern et al. (1988) *N. Engl. J. Med.* **319**, 525–532.
2. C. C. Harris (1991) *Cancer Res.* **51** (Suppl.), 5023s–5044s.
3. R. Reiger, A. Michaelis, and M. M. Green (1991) *Glossary of Genetics: Classical and Molecular*. Springer-Verlag, New York.
4. E. C. Miller and J. A. Miller (1966) *Pharmacol. Rev.* **18**, 805.
5. L. R. Ferguson and B. C. Baguley (1994) *Environ. Mol. Mutagen.* **24**, 245–261.
6. S. Green (1995) *Mutat. Res.* **333**, 101–109.

CASEIN

ATSUSHI IKAI

Several kinds of phosphoproteins collectively called casein constitute 80% of the **protein** of cow's milk and 40% of that of humans. Casein is made up of all 20 kinds of **amino acids**, which is an important characteristic from a nutritional point of view. Skim milk, which is the milk left after the cream is removed, is brought to pH 4.7 to form a precipitate that contains the casein. The supernatant is called whey and contains 20% of the total protein. **Electrophoresis** at alkaline pH separates casein into the α, β, and γ fractions. The α fraction is further separated into α_{s1} and α_{s2}, the common subscript s signifying sensitivity to precipitation by the addition of calcium and κ-casein. Both α_s-casein and β-casein are precipitated with Ca^{++}, whereas κ_s-casein is not. In milk, these caseins constitute micelles (calcium caseinate-phosphate complex) of a diameter in the range of 30 to 300 nm. The α- and β- caseins are rich in phosphate, mainly in the form of O-phosphoserine residues, and they form a calcium complex that is not soluble, but precipitated. When surrounded by the **hydrophilic** parts of κ-casein in milk, however, it is soluble as large complexes. The γ-casein fraction is a heterogeneous mixture primarily of proteolytic fragments of β-casein.

Bovine β-casein contains 209 amino acid residues (with a molecular weight of 23,600) and is very hydrophobic. Five phosphoserine residues are found in the N-terminal region. In milk, the phosphoserine residues are associated with Ca^{++} ions, forming calcium caseinate. β-casein and its partial hydrolysis products present in milk do not have definite three-dimensional structures. They are considered to have a **random coil** structure that is easily digestible, an advantage of casein as a nutritional protein. The relative casein composition of bulk milk is given in Table 1.

The fourth stomach of ruminating animals (abomasum) contains a proteolytic enzyme called *chymosin*, which liberates a glycopeptide from κ-casein at neutral pH. The remaining molecule, called para-casein, reacts with Ca^{++} and then forms an insoluble curd, the source of cheese. The enzyme preferentially cleaves the peptide bond between Phe105-Met106 of κ-casein to produce para-κ-casein. Humans do not have chymosin, but **pepsin** in the stomach converts κ-casein to para-casein to start the digestion of milk in the infant stomach. Chymosin and pepsin are structurally related.

The casein micelle has a complex structure, and the relative disposition of the four subspecies of casein, and especially

Table 1. Average Casein Composition of Bovine Bulk Milk Samples

Casein Type	Weight %
α_{s1}	38.1
α_{s2}	10.2
β	35.7
κ	12.8
γ	3.2
Total	100.0

Figure 1. Structural model of casein micelles. Reproduced from [1] (Fig. 13, p. 108) with permission.

that of κ-casein, has long been the focus of discussion. Since chymosin preferentially digests κ-casein, this casein is thought to be on the periphery of the micelle. Recent **electron micrographs** of variously treated micelles show granular structures covering the surface, prompting a suggestion that a casein micelle is composed of smaller-sized submicelles. The idea of a submicellar structure has actually been a long standing one from reconstitution experiments, and by incorporating various observations, a currently accepted model of casein micelles is presented by Holt [1], as reproduced in Figure 1.

Secretory vesicles of casein micelles are found abundantly in mammary secretory cells. After the casein proteins are synthesized, they undergo **O-glycosylation** and **phosphorylation** before being secreted. These changes occur in concert with lactose synthesis by the enzyme known as lactose synthetase. All the enzymes involved in glycosylation, phosphorylation, and lactose synthesis can tolerate Ca^{++} concentrations in the range of 1 *mM*. Accumulation of Ca^{++} in the mammary gland cell, together with phosphorylation of some serine residues, is necessary for the formation of casein micelles.

BIBLIOGRAPHY

1. C. Holt (1992) Structure and stability of bovine casein micelles, *Adv. Prot. Chem.* **43**, 63–151.

CASPASES

NICOLA MCCARTHY
GERARD EVAN

Caspases are **proteinases** involved in **apoptosis** and **programmed cell death**. They are mammalian **homologues** of

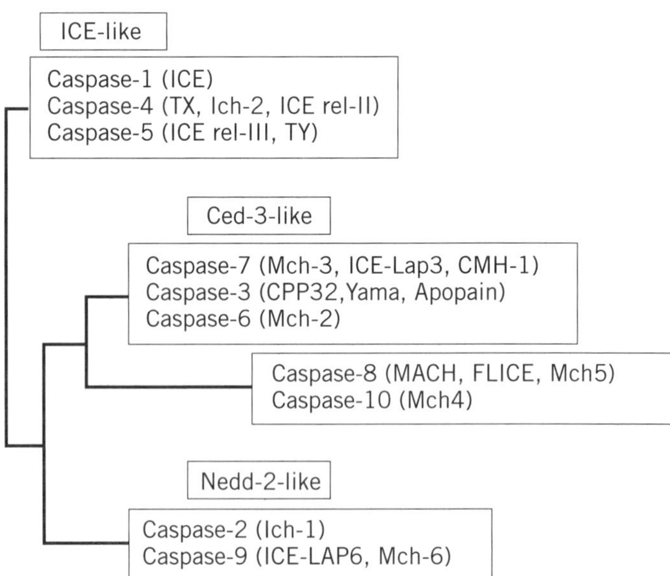

Figure 1. The mammalian caspase family. The various members have been arranged in a phylogeny on the basis of their relative amino acid sequence homologies. The members of the Ced-3-like family of caspases are thought to be the downstream effectors of apoptosis and to be activated in a caspase cascade by the apical members represented by caspases 2, 8, 9, and 10 (1).

the Ced-3/interleuking-1β converting enzyme (ICE) products of **cell death** genes (1). The caspase family in mammalian cells currently numbers 14, as shown in Figure 1. They are **thiol proteinases** that are synthesized as inactive **pro-proteins** that are processed to a pro-domain and two subunits of approximately 10 and 20 kDa in size. The pro-caspases are inert until proteolytic cleavage of the pro-domain and/or the p10 subunit from the p20 subunit. Caspases share a conserved **active-site** sequence, -Gln-Ala-Cys-Arg-Gly- (QACRG) or -Gln-Ala-Cys-Gln-Gly- (QACQG), and cleave their substrates at the peptide bond after an aspartate residue; hence, their name- **c**ysteine **asp**artate-specific protein**ases** (1,2). Caspases cleave specific substrates during **apoptosis** and are required for the rapid degradation of the cell (2–6); however, none is thought to be critical for apoptosis to occur.

Caspases can be divided into subgroups, with each sharing a specific substrate specificity beyond the overall caspase requirement for Asp at the P1 position of the substrate (see **Proteinases**). Caspases 1, 4, and 5, for example, preferentially cleave after -Tyr-Val-Ala-Asp- (YVAD) sequences and have not been shown to be critical for apoptosis to occur. Caspases 3 and 6, however, are active during cell death and cleave primarily substrates with the sequence -Asp-Xaa-Xaa-Asp- (DxxD). Caspases 2, 8, and 10 are different again and are thought to activate primarily the downstream caspases, such as caspase 3 and 6 (7). This suggests that caspases act in a hierarchy, with activated upstream caspases cleaving downstream caspases and activating them in turn, to generate a *caspase cascade* (4,8). The apical caspases are proposed to be those with large pro-domains such as caspases 2, 8, and 10. The pro-domains contain a protein structure that facilitates their interaction with other proteins leading to caspase activation, as seen in CD95 (Fas/Apo-1)-induced apoptosis (9).

BIBLIOGRAPHY

1. E. Alnemri, D. Livingston, D. Nicholson, G. Salvesan, N. Thornberry, W. Wong, and Yuan, J. (1996) Human ICE/CED-3 protease nomenclature. *Cell* **87**, 171.
2. A. Takahashi and W. Earnshaw (1996) Ice-related proteases in apoptosis. *Curr. Opin. Gen. Dev.* **6**, 50–55.
3. A. Chinnaiyan and V. Dixit (1996) The cell-death machine. *Curr. Biol.* **6**, 555–562.
4. Y. Lazebnik, A. Takahashi, G. Poirier, S. H. Kaufmann, and W. Earnshaw (1995) Characterization of the execution phase of apoptosis *in vitro* using extracts from condemned phase cells. *J. Cell. Sci.* **19**, 41–49.
5. M. Whyte (1996) ICE/Ced-3 proteases in apoptosis. *Trends Cell Biol.* **6**, 245–248.
6. A. Fraser, N. McCarthy, and G. I. Evan (1996) Biochemistry of cell death. *Curr. Opin. Neurobiol.* **6**, 71–80.
7. T. Fernandes Alnemri, R. C. Armstrong, J. Krebs, S. M. Srinivasula, L. Wang, F. Bullrich, L. C. Fritz, J. A. Trapani, K. J. Tomaselli, G. Litwack, and E. S. Alnemri (1996) *In vitro* activation of CPP32 and Mch-3 by Mch-4, a novel human apoptotic cysteine protease containing 2 FADD-like domains. *Proc. Natl. Acad. Sci. USA* **93**, 7464–7469.
8. Y. A. Lazebnik, A. Takahashi, R. D. Moir, R. D. Goldman, G. G. Poirier, S. H. Kaufmann, and W. C. Earnshaw (1995) Studies of the lamin proteinase reveal multiple parallel biochemical pathways during apoptotic execution. *Proc. Natl. Acad. Sci. USA* **92**, 9042–9046.
9. A. Fraser and G. Evan (1996) A license to kill. *Cell* **85**, 781–784.

CASSETTE MUTAGENESIS

LYNNETTE FERGUSON
WILLIAM DENNY

Cassette mutagenesis involves replacing a wild-type **DNA** sequence with synthetic double-stranded oligonucleotides in order to introduce one or more **mutations** (1–6). The technique allows saturation of a target amino acid codon with **mutations**, and can be used to probe the role of that particular residue in a protein and the effect of specific point mutations in it. In their description of the method, Wells et al. (4) synthesized single-stranded oligonucleotides containing different codons over the target in separate pools. Oligonucleotide-**site-directed mutagenesis** procedures are used to generate **restriction sites** that closely flank the target codon in the **plasmid** containing the gene of interest. Use of the appropriate restriction endonucleases permits insertion of the synthetic duplex cassettes. These are designed to restore fully the wild-type coding sequence except over the target codon, and also to eliminate one of the restriction sites. This latter point is important for selection of the clones containing the mutant cassette.

Several variations on the original technique have been developed for specific purposes. For example, Reidhaar-Olsen and Sauer (5) described a combinatorial method, in which they randomized two or three positions by oligonucleotide cassette mutagenesis, selected for functional protein and then sequenced to determine the spectrum of allowable substitutions at each position. They applied the method repeatedly in order to examine the role of different substitutions on the DNA-binding domain of **lambda repressor**. Kegler-Ebo et al. (6) described codon cassette mutagenesis as a way of depositing single codons at specific sites in double-stranded DNA. Using this method, a series of 11 cassettes is sufficient to insert all possible amino acids at any constructed target site.

BIBLIOGRAPHY

1. S. Green (1955) Mutat. Res. **333**, 101–109.
2. M. D. Matteuci and H. L. Heyneker (1983) *Nucleic Acids Res.* **11**, 3113–3121.
3. J. B. McNeil and M. Smith (1985) *Mol. Cell. Biol.* **5**, 3545–3551.
4. J. A. Wells, M. Vasser, and D. B. Powers (1985) *Gene* **34**, 315–323.
5. J. F. Reidhaar-Olsen and R. T. Sauer (1988) *Science* **241**, 53–57.
6. D. M. Kegler-Ebo, G. W. Polack, and D. DiMaio (1994) *Methods Mol. Biol.* **57**, 297–310.

CATABOLITE REPRESSION

H. AIBA

The ability of glucose to inhibit the synthesis of certain **enzymes**, referred to as the *glucose effect*, was recognized in **bacteria** at an early date by Monod (1). He observed that when *Escherichia coli* encounters both glucose and lactose, for example, it metabolizes the glucose first and represses the use of lactose, resulting in biphasic growth (diauxie). This phenomenon is due to the repressive effect of glucose on the synthesis of enzymes required for the metabolism of other sugars (see **Gene expression**). Later, the term catabolite repression was introduced as a general name for the glucose effect because compounds closely related to glucose elicited varying degrees of repression of glucose-sensitive enzymes, and the catabolites derived from the repressing compound were assumed to cause the glucose effect (2). Studies on catabolite repression up to the early 1970s brought about the discovery of a positive control system of **transcription** by **cyclic AMP** (cAMP) and the **cyclic AMP receptor protein** (CRP) (also called the catabolite gene activator protein, (CAP)) and led to the concept that catabolite repression is caused by reduction of the level of intracellular cAMP (3). We now know that multiple and different mechanisms are operating, depending on the growth conditions and the target **operons**, and that the cAMP-dependent mechanism is just one aspect of catabolite repression. Ironically, the preferential utilization of glucose over lactose, the prototype of catabolite repression, is not due to the reduction of the cAMP-CRP complex (4,5).

Sometimes the term catabolite repression has been used to describe the glucose effect that is independent of the operon-specific regulator, such as the **Lac repressor**. Here, we shall use it as a general name for the glucose effect as originally defined. According to this view, which is apparently generally accepted, catabolite repression includes all forms of the glucose effect, regardless of their mechanisms. We also concentrate on catabolite repression in *E. coli*. Although several mechanisms of glucose catabolite repression are now understood in *E. coli*, a common feature is that glucose causes catabolite repression ultimately by modulating the activity of **transcription factors** involved in the regulation of target operons.

CAMP–CRP COMPLEX

The best characterized mechanism of catabolite repression involves the regulation of the intracellular concentration of the cAMP–CRP complex. It is well known that the presence of glucose in the growth medium lowers the intracellular cAMP level under certain conditions (3,6). Cyclic AMP is synthesized from ATP by the enzyme **adenylate cyclase**. Although the mechanism of regulation of the cAMP level remains elusive, glucose is thought to decrease cAMP by decreasing the level of the phosphorylated form of enzyme IIA^{Glc}, which is involved in the activation of adenylate cyclase. IIA^{Glc} is one of the enzymes of the phosphoenolpyruvate-dependent carbohydrate **phosphotransferase system** (PTS) and is directly responsible for the **active transport** and phosphorylation of glucose (7). Recently, it was discovered that the concentration of CRP is also lowered by the presence of glucose and that this is an additional factor contributing to catabolite repression (8). The decreased CRP is a consequence of the complex autoregulation of expression of the *crp* gene (9). It should be noted that the reduction in the cAMP-CRP level by glucose is usually rather moderate (in the range of several-fold).

INDUCER EXCLUSION

The second mechanism of catabolite repression is inducer exclusion, by which glucose lowers the intracellular concentration of inducers necessary for the induction of catabolic operons (7). The target of glucose signaling in inducer exclusion is operon-specific regulators, such as the Lac repressor. The dephosphorylated enzyme IIA^{Glc}, which accumulates in the presence of glucose, binds to and inactivates (for example) the Lac permease, resulting in an increase of the active unliganded Lac repressor (see ***lac* operon**). Inducer exclusion is a mechanism by which glucose inhibits more strictly the expression of target operons.

CATABOLITE REPRESSOR/ACTIVATOR PROTEIN

The third mechanism of catabolite repression is mediated by the catabolite repressor/activator (Cra) protein, which acts as a global regulator of genes encoding enzymes of central carbohydrate metabolism (10). The unliganded form of Cra binds to the operator regions of target operons, causing either activation or inhibition of transcription. The presence of glucose or other PTS sugars produces glycolytic catabolites, such as fructose-1-phosphate, which bind to the Cra protein and cause it to dissociate from the target DNA, resulting in either catabolite repression or catabolite activation.

RELATIONSHIPS BETWEEN THE VARIOUS MECHANISMS

While multiple mechanisms of catabolite repression have been identified in *E. coli*, their signaling pathways appear to be interrelated to each other. For example, the PTS plays a pivotal role in the regulation of the intracellular concentrations of cAMP, inducer, and glycolytic catabolites. In addition, it is particularly important to realize that the contribution of each mechanism varies, depending on growth conditions and the target genes. For example, the Cra-mediated mechanism may play no role in catabolite repression of the *lac* operon, because this operon is not under the control of Cra. An unexpected finding is that the presence of glucose in the lactose medium does not affect the intracellular cAMP level (4). This means that catabolite repression mediated by the reduction in cAMP never happens in glucose–lactose diauxie. The presence of unliganded Lac repressor through inducer exclusion is the principal mechanism for this historical phenomenon (4,5).

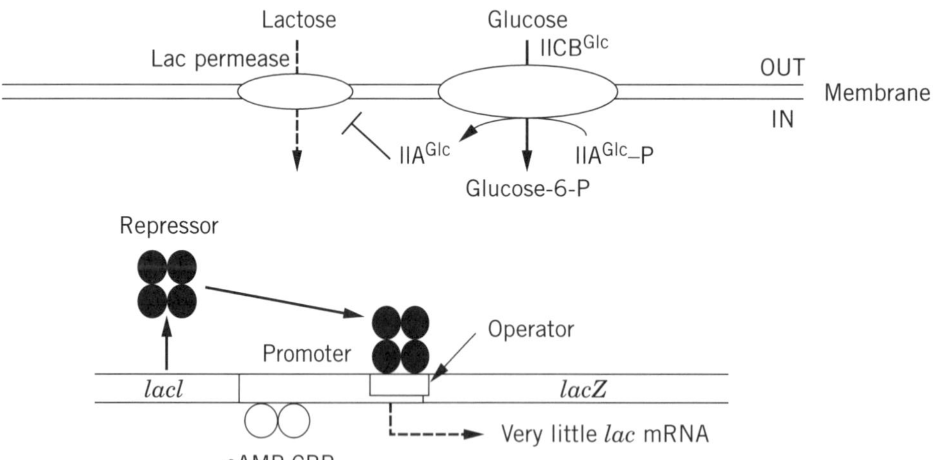

Figure 1. Mechanism of catabolite repression in the glucose-lactose system (4). When both lactose and glucose are present, glucose is transported and phosphorylated by the glucose PTS (IIA^{Glc} + $IICB^{Glc}$), increasing the concentration of the nonphosphorylated form of IIA^{Glc}, which prevents the uptake of lactose by inhibiting the Lac permease activity. Thus, the concentration of *lac* inducer is very low in the presence of glucose, so the Lac repressor is active and represses transcription of the *lac* operon. It should be noted that glucose does not affect the binding of cAMP–CRP to the promoter, because the levels of cAMP and CRP are not reduced by the presence of glucose.

Figure 1 illustrates the present understanding of catabolite repression in the glucose–lactose system.

There are more stories yet to be elucidated before catabolite repression in *E. coli* is fully understood. Further diversity in the mechanism of catabolite repression is known in **Gram-positive** bacteria (11).

BIBLIOGRAPHY

1. J. Monod (1947) *Growth* **11**, 223–289.
2. B. Magasanik (1961) *Cold Spring Harbor Symp. Quant. Biol.* **26**, 249–256.
3. I. Pastan and R. Perlman (1970) *Science* **169**, 339–344.
4. T. Inada, K. Kimata, and H. Aiba (1996) *Genes to Cells* **1**, 293–301.
5. K. Kimata, H. Takahashi, T. Inada, P. Postma, and H. Aiba (1997) *Proc. Natl. Acad. Sci. USA* **94**, 12914–12919.
6. R. S. Makman and E. W. Sutherland (1965) *J. Biol. Chem.* **240**, 1309–1314.
7. P. W. Postma, J. W. Lengeler, and G. R. Jacobson (1993) *Microbiol. Rev.* **57**, 543–594.
8. H. Ishizuka, A. Hanamura, T. Kunimura, and H. Aiba (1993) *Mol. Microbiol.* **10**, 341–350.
9. H. Ishizuka, A. Hanamura, T. Inada, and H. Aiba (1994) *EMBO J.* **13**, 3077–3082.
10. M. H. Saier and T. M. Ramseier (1996) *J. Bacteriol.* **178**, 3411–3417.
11. C. J. Hueck and W. Hillen (1995) *Mol. Microbiol.* **15**, 395–401.

Suggestions for Further Reading

B. Magasanik (1970) Glucose effects: inducer exclusion and repression, in *The Lactose Operon*, J. Beckwith and D. Zipser, eds., Cold Spring Harbor Laboratory Press, Cold Spring Harbor, New York, pp. 189–220.

A. Ullmann and A. Danchin (1983) Role of cAMP in bacteria, *Adv. Cyclic Nucleotide Res.* **15**, 1–53.

N. D. Meadow, D. K. Fox, and S. Roseman (1990) The bacterial phosphoenolpyruvate:glycose phosphotransferase system. *Annu. Rev. Biochem.* **59**, 497–542.

M. H. Saier Jr., T. M. Ramseier, and J. Reizer (1996) Regulation of carbon utilization, in *Escherichia coli and Salmonella typhimurium: Cellular and Molecular Biology*, F. C. Neidhardt, R. Curtiss III, J. L. Ingraham, E. C. C. Lin, J. K. B. Low, B. Magasanik, W. S. Reznikoff, M. Riley, M. Schaechter, and H. E. Umbarger, eds., American Society for Microbiology, Washington, DC, pp. 1325–1343.

CATALASE

P. C. LOEWEN

Catalase, also called *hydroperoxidase*, is a protective **enzyme** that has been studied for over a century; the concept of a specific protein catalyzing (hence the name *catalase*) the degradation of hydrogen peroxide to oxygen and water first appeared in 1900 (1). Removal of H_2O_2, as shown in the following reaction, avoids unwanted side reactions and prevents the formation of the even more reactive hydroxyl radical:

$$2\,H_2O_2 \rightarrow 2\,H_2O + O_2 \qquad (1)$$

Although the removal of these reactive oxygen species is not essential for growth, catalases do enhance long-term survival in an aerobic environment (2).

A number of developments have spawned considerable interest in the enzyme and spurred extensive studies of both the enzyme and its genes. The first was the simplicity of the "drop test" assay, which involves the visual monitoring of oxygen evolution after the application of a drop of 30% H_2O_2 to the edge of a bacterial colony. Such ease of **phenotypic** scoring has resulted in extensive use of the enzyme as a diagnostic tool for microbiological strain identification. Catalases have been found to be synthesized as part of a variety of **stress response** systems, resulting in their frequent use as indicators of stress response activation. More recently, the catalase–peroxidase family has gained notoriety from its role as the *in vivo* activator of isoniazid into an effective antibiotic in *Mycobacterium tuberculosis* (3). Loss of the enzyme through **mutation** results in isoniazid resistance, one of the reasons for the increasing spread of tuberculosis.

ASSAY

Two quantitative assay methods are commonly used for catalase activity. One involves the measurement of oxygen evolution using an oxygraph equipped with a Clark electrode (4); the second is a spectrophotometric assay of H_2O_2 by its **absorbance** at 240 nm (5). The two assay procedures produce comparable values in reasonably clear protein solutions, but the oxygraph protocol has the advantage that it can be used for catalase determinations in whole-cell suspensions and in quite turbid extracts. Catalase activity can be visualized on **polyacrylamide** gels following **electrophoresis** under nondenaturing conditions (6), producing a clear band in a brown background. This same visualization procedure can be easily modified to visualize **peroxidase** activities as brown bands on a clear background.

CLASSIFICATION

Because of the protection afforded against active oxygen species by catalases, most aerobic organisms produce at least one catalase from among the three main classes of the enzyme. The largest class includes the "typical," heme-containing, monofunctional catalases with either small (~60 kDa) or large (>80 kDa) subunits. The next largest class includes the bifunctional catalase–peroxidases, which are also heme-containing but with sequence similarity to plant and fungal peroxidases. The third and, to date, smallest class includes the non-heme- or Mn-containing catalases. Small-subunit, monofunctional catalases have been found in prokaryotes and eukaryotes but not in an archael species, and the large-subunit catalases are restricted to fungi and bacteria. The catalase-peroxidases are found in **prokaryotes** and **archaebacteria**, and the non-heme-containing enzymes have been found only in bacteria (7).

REACTION

Catalases and peroxidases both degrade H_2O_2 but, whereas catalase employs H_2O_2 as both electron donor and acceptor (reaction 1), a peroxidase uses organic substrates as the electron donor to reduce H_2O_2:

$$H_2O_2 + RH_2 \rightarrow 2\,H_2O + R \qquad (2)$$

The catalase-peroxidases catalyze both reactions 1 and 2 at significant rates.

The well-characterized catalytic reaction pathway is a two-step process. First, H_2O_2 is converted to water and an oxyferryl species, compound I, with the iron in the +5 oxidation state but with part of the charge delocalized in a heme cation radical:

$$\text{Heme—Fe}^{\text{III}} + H_2O_2 \rightarrow \text{heme—Fe}^{\text{V}}\text{=O(cpd I)} + H_2O \quad (3)$$

Then compound I reacts with a second molecule of H_2O_2 to produce water and molecular oxygen:

$$\text{Heme—Fe}^{\text{V}}\text{=O} + H_2O_2 \rightarrow \text{heme—Fe}^{\text{III}} + O_2 + H_2O \quad (4)$$

In the absence of a suitable substrate for the reduction of compound I, other intermediates can be formed such as the catalytically inactive compound II.

STRUCTURE AND PROPERTIES

Detailed structural information is now available for all three types of catalases. Small- and large-subunit catalases share a common highly conserved core sequence of about 350 residues organized in a β-barrel structure (see **Beta-sheet**). This is illustrated in Figure 1 (8,9), which compares the small subunit of *Proteus mirabilis* catalase (Fig. 1**a**) and the large subunit of *Escherichia coli* HPII (Fig. 1**b**). An *N*-terminal extension of 10–90 residues (colored red in Fig. 1B) and a *C*-terminal extension of 150–170 residues (colored white in Fig. 1**b**), the latter organized in a flavodoxin-like structure, are added to the core structure in the large-subunit enzyme. The tetrameric organization of small-subunit enzymes is sufficiently stable to resist denaturation by organic reagents and to retain activity over a broad pH range from 4 to 11. In large-subunit enzymes, the extended sequences allow the amino terminus of one subunit to be overlapped or trapped by the carboxyl terminus of an adjacent subunit (10), and this interweaving further enhances the resistance of the tetramer to denaturation by heat, **SDS**, and **urea**.

Besides size, the small- and large-subunit catalases have two other characteristic differences. The first is that small-subunit enzymes contain heme *b*, which in large-subunit enzymes is converted to a *cis*-hydroxy γ-spirolactone heme *d* and flipped 180°. The second is that small-subunit, but not large-subunit, catalases contain NADPH (11), which is postulated to have a role in the reduction of inactive compound II. A number of unusual modifications have been identified in catalases, including

1. A **methionine** sulfone in the active site of the *P. mirabilis* enzyme (12)
2. A blocked **cysteine** residue in HPII from *E. coli* (13)
3. A **histidine–tyrosine** bond in HPII

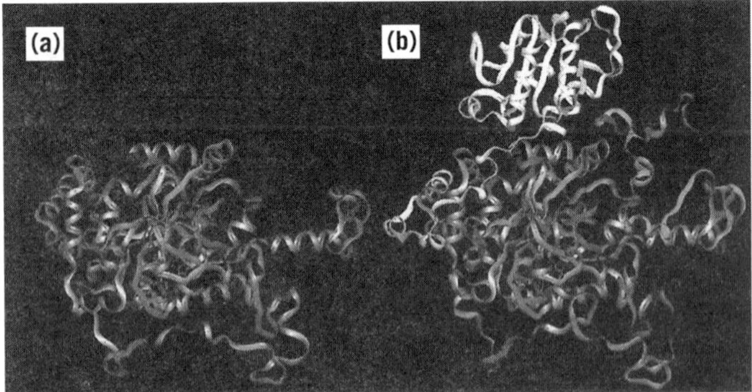

Figure 1. Comparison of the structures of the small subunit of *Proteus mirabilis* catalase (**a**) (8) and the large subunit of *E. coli* HPII catalase (**b**) (9). The conserved β-barrel core structure is colored blue in both subunits. The 78 additional residues at the *N* terminus of HPII in (**b**) are colored red, and the 195 additional residues at the *C* terminus of HPII in (**b**) are colored white. See color insert.

The narrow channel leading to the deeply buried **active site** of catalases limits access to small substrates and inhibitors, and results in a 10-fold higher apparent K_m for H_2O_2 compared to the non-heme-containing catalases and catalase–peroxidases. Within the catalase active site, a histidine residue immediately above the heme, and an asparagine residue situated to one side, orient the H_2O_2 over the heme iron (12) to facilitate a very fast reaction in both small- and large-subunit enzymes [$k_{cat} \cong 3.5 \times 10^5\ s^{-1}$ (15) and $1.2 \times 10^5\ s^{-1}$, respectively] as compared to the catalase–peroxidases and non-heme-containing catalases [$k_{cat} \cong 1.6 \times 10^4\ s^{-1}$ (16) and $3.9 \times 10^5\ s^{-1}$ (17), respectively].

CLONING AND EXPRESSION

The number of catalase enzymes in a given organism varies: one in animals, one or two in **fungi**, one to three in plants, and zero to three in bacteria. The availability of catalase-deficient hosts, the ease of assay, and the development of protocols using oligonucleotide and **PCR** probes have facilitated the **cloning** and characterization of over 90 catalase genes. Most of the isolated genes are from plants and bacteria, with a small number from fungi and animals. A **phylogenetic** comparison of the core protein sequences has revealed separate branches for the small-subunit enzymes from bacterial, plant, animal, and fungal sources, and one additional branch containing the large-subunit enzymes from both fungal and bacterial origins.

The most detailed studies of catalase gene expression have been carried out in plants and bacteria. In maize, catalase expression is tied to developmental changes during kernel development and seed germination and to environmental factors such as light and growth regulators (18). A multitude of species-specific mechanisms control catalase and catalase–peroxidase expression in bacteria, but two common threads are evident among these expression patterns involving induction either by an active oxygen species such as H_2O_2 or in stationary phase. In most cases, the control mechanisms can be rationalized as either protective responses to oxidative stress or as a means of enhancing long-term survival during metabolic limitation (7,19).

BIBLIOGRAPHY

1. O. Loew (1900) U.S. Dept. Agriculture Rpt. 65, Govt. Printing Office, Washington, DC.
2. M. R. Mulvey, J. Switala, A. Borys, and P. C. Loewen (1990) *J. Bacteriol.* **172**, 6713–6720.
3. J. S. Blanchard (1996) *Annu. Rev. Biochem.* **65**, 215–239.
4. M. Rorth and P. K. Jensen. (1967) *Biochim. Biophys. Acta* **139**, 171–173.

5. A. G. Hildebrandt and I. Roots (1975) *Arch. Biochem. Biophys.* **171**, 385–397.
6. E. M. Gregory and I. Fridovich. (1974) *Anal. Biochem.* **58**, 57–62.
7. P. C. Loewen (1997) In J. G. Scandalios, ed., *Oxidative Stress and the Molecular Biology of Antioxidant Defenses*, Cold Spring Harbor Laboratory Press, Cold Spring Harbor, NY, pp. 273–308.
8. P. Gouet, H. M. Jouve, and O. Dideberg (1995) *J. Mol. Biol.* **249**, 933–954.
9. J. Bravo, N. Verdaguer, J. Tormo, C. Betzel, J. Switala, P. C. Loewen, and I Fita (1995) *Structure* **3**, 491–502.
10. W. R. Melik-Adamyan, V. V. Barynin, A. A. Vagin, V. V. Borisov, B. K. Vainshtein, I. Fita, M. R. N. Murthy, and M. G. Rossmann (1986) *J. Mol. Biol.* **188**, 63–72.
11. H. N. Kirkman and G. F. Gaetani (1984) *Proc. Natl. Acad. Sci. USA* **81**, 4343–4347.
12. A. Buzy, V. Bracchi, R. Sterjiades, J. Chroboczek, P. Thibault, J. Gagnon, H.-M. Jouve, and G. Hudry-Clergeon (1995) *J. Protein Chem.* **14**, 59–72.
13. S. Sevinc, W. Ens, and P. C. Loewen (1995) *Eur. J. Biochem.* **230**, 127–132.
14. I. Fita and M. G. Rossmann (1985) *J. Mol. Biol.* **185**, 21–37.
15. A. Deisseroth and A. L. Dounce (1970) *Physiol. Rev.* **50**, 319–375.
16. A. Claiborne and I. Fridovich (1979) *J. Biol. Chem.* **254**, 4245–4252.
17. Y. Kono and I. Fridovich (1983) *J. Biol. Chem.* **258**, 6015–6019.
18. J. G. Scandalios (1992) in J. G. Scandalios, ed., *Molecular Biology of Free Radical Scavenging Systems*, Cold Spring Harbor Laboratory Press, Cold Spring Harbor, NY, pp. 117–152.
19. H. Schellhorn (1994) *FEMS Microbiol. Lett.* **131**, 113–119.

CATALYSIS

JOHN F. MORRISON

Catalysis by **enzymes** starts with the collision between the enzyme E and its substrate A to form an enzyme-substrate complex EA:

$$\mathrm{E + A \rightleftharpoons EA \rightleftharpoons EA^* \rightleftharpoons EP^* \rightleftharpoons EP \rightleftharpoons E + P}$$

[collision clamping] conversion [unclamping dissociation]

binding catalysis release

This is a second-order reaction, as it involves two reactants, and the collision rate is equal to $k[E_t]\ [A_t]$, where k is a second-order rate constant. There is a distinct limit to the value of k, as it cannot exceed the **diffusion** rate (see **Diffusion-controlled reaction** and **Turnover number**). Nonproductive complexes can also form as a result of the interaction of a substrate with an enzyme. If the structure of a substrate is complementary to that of the **active site** of the enzyme, the interaction rate is high and the binding tight. If, however, this is not the case, some of the binding energy has to be utilized to facilitate the interaction through a conformational change or clamping reaction on the enzyme. Binding is the product of the initial interaction and the subsequent conformational change.

After being bound to the enzyme, the substrate must undergo activation. That is, a proportion of the total number of substrate molecules in the ground state must be raised to the **transition state**. This is a short-lived, unstable species with a structure intermediate between that of the substrate and the product. These activated molecules can fall back to the ground state or be converted to EP^* at a rate that is independent of the structure of the reactants. The **activation energy** barrier for the conversion of ground-state molecules to transition-state molecules is very much lower for an enzyme-catalyzed reaction than for either a noncatalyzed reaction or chemically catalyzed reaction. In the activated state, the substrate is bound more tightly to the enzyme than it is in the ground state (see **Transition state analogue**). If such tighter binding did not occur, there would be no catalysis. In fact, the increase in rate is proportional to the increase in binding. It is for this reason that transition state analogues are potent inhibitors of enzymes. After formation of the EP^* complex, the enzyme undergoes another conformational or unclamping reaction to release the product into the bulk medium and regenerate the free form of enzyme. Such regeneration is an essential part of enzymic catalysis.

BIBLIOGRAPHY

1. D. D. Hackney (1990) *The Enzymes* **19**, 1–36.

CATALYTIC ANTIBODIES

I. LEE
S. J. BENKOVIC

The concept of *catalytic antibodies* owes its origin to Pauling and Jencks, who proposed that an **antibody** normally differs from an enzyme by its inability to bind selectively and to stabilize the **transition state** of a chemical reaction. An antibody that did happen to be specific for a transition state should therefore function like an enzyme and promote chemical catalysis of the corresponding reaction (1). Advances in chemical synthesis and in **hybridoma** technology to produce **monoclonal antibodies** have enabled the exploitation of the diversity and affinity of the immune repertoire to develop catalytic antibodies that are elicited against **transition state analogues**. The binding sites of these antibodies are anticipated to bind substrates structurally related to the transition state and to process them to products through a pathway lower in free energy, and therefore more rapid, than the normal one that occurs in the absence of antibody. By clever design of appropriate transition state analogues, "tailor-made catalysts" should be created to catalyze reactions with no enzymic counterparts. Challenge of the immune system with a transition state analogue to induce a catalytic antibody is a necessary, but not a sufficient, condition to generate an effective catalyst; factors other than high affinity for the transition state, such as precise orientation of catalytic residues and the effective release of reaction products at the active site, also contribute to the overall catalytic efficiency of an antibody. Strategies have evolved to design transition state analogue **immunogens** that would solicit catalytic functions within the antibody combining site for efficient catalysis.

REACTIONS CATALYZED BY ANTIBODIES

There are now approximately 100 reactions that have been catalyzed by antibodies (1–3). These reactions include (*1*) pericyclic processes (oxy-Cope rearrangement, Diels–Alder condensation, Claisen rearrangement), (*2*) elimination reactions (decarboxylation, dehydration, *syn* elimination of HF from fluoroketones, E2 (biomolecular) elimination of benisoxazole), (*3*) hydrolyses (carbonate esters, esters, amide, lactones, enol ethers), (*4*) bond-forming reactions (lactonization, **peptide synthesis**, cationic cyclization, aldol condensation), and (*5*) redox reactions (ketone reduction, epoxidation, sulfoxide oxidation). Antibodies can catalyze, with high stereo- and regiospecificity, reactions that may not generally be catalyzed by enzymes. This property has appeal in the potential application of antibody catalysis to commercial synthetic and medical uses, such as pharmaceutical synthesis and prodrug activation.

THE NATURE OF THE CATALYTIC SITE

The **active sites** of catalytic antibodies elicited by a given immunogen exhibit high sequence homologies and are often structurally convergent (4). These sites harbor shallow clefts complementing the structural and electronic features of the immunogen. The positioning of active site residues in the binding pocket is accomplished by somatic mutation of the germline ancestral antibody arising from an immunological response (5). The germline antibody undergoes a substantial amount of **induced-fit** conformational change on binding the immunogen. By the process of **affinity maturation**, the active site residues in the mature antibody become preoriented such that a rigid binding pocket is generated for optimal binding (lock-and-key fit).

NATURE OF CATALYSIS

Catalysis by antibodies is like that by enzymes, in that it exhibits **Michaelis–Menten** *kinetics* in which substrate binding precedes a chemical transformation, followed by dissociation of the product. Kinetic characterization experiments (6) indicate that the antibody-catalyzed reaction recapitulates a number of characteristics expected of an enzyme, but the chemical transformation step is often rate-limiting. The rate enhancement is generally less than that observed for enzymes catalyzing similar reactions. Transition state analogues are only approximations of the true transition state, and they also do not demonstrate the extremely tight binding to enzymes that is predicted theoretically (see **Transition state analogues**). Furthermore, antibodies catalyze reactions primarily through restrictions on the translational and rotational movement of substrates, and only to a lesser extent through the active-site acid/base or nucleophilic catalysis that occurs in enzymes. The difference in catalytic efficiency between antibody and enzyme can also be attributed in part to the latter's ability to provide more extensive **electrostatic interactions** and **hydrogen bonding** during catalysis through conformational mobility (7).

GENERATION OF CATALYTIC ANTIBODIES

The recovery of catalytic antibodies relies on an efficient means of sampling the immune repertoire and screening for effective catalysts. The murine immune repertoire is estimated to have a diversity of 10^8, and it can be further expanded by immunization. The majority of catalytic antibodies have been derived from hybridomas that generally capture only 0.01% of the immune repertoire. The development of recombinant antibody fragments as combinatorial **libraries** consisting of $\geq 10^6$ members in **lambda** and in **M13 phage** has increased the access to the immune repertoire and possibly the recovery of catalytic antibodies. The diversity of an antibody combinatorial library, and potentially the yield of efficient catalytic antibodies, can be further expanded by polypeptide chain-shuffling experiments in which the heavy-chain fragment of a catalytic antibody is allowed to cross with a library of light-chain fragments elicited by the same immunoglobulin, and vice versa (8).

SCREENING FOR CATALYTIC ANTIBODIES

Catalytic antibodies have been identified by their high affinities for the appropriate haptens, as well as their catalytic activities for the desirable reactions. Given the often low activity of catalytic antibodies, sensitive but specific methods have been developed for identifying the catalyst. CatELISA is based on screening for the presence of antibody-catalyzed reaction products using product-specific antibodies as primary antibody indicators in **enzyme-linked immunosorbent assay** (ELISA) (9). An alternative method depends on the catalytic release of a reactive species that can be trapped via covalent modification of the phage-displayed antibody, which then permits the recovery of DNA encoding catalytic antibody clones from the phage particles (10). Another method involves the recovery of antibody clones that sustain growth of **microorganisms**, such as **yeast** or bacterial **auxotrophs**, through catalysis of essential biological pathways (11). The host organisms, however, may also impose an additional selection on what catalytic antibodies can be recovered from the screen, since the host will not survive expression of antibodies that are toxic.

BIBLIOGRAPHY

1. R. A. Lerner, S. J. Benkovic, and P. G. Schultz (1991) *Science* 252, 659–667.
2. S. J. Benkovic (1992) *Annu. Rev. Biochemistry* **61**, 29–54.
3. J. R. Jacobsen and P. G. Schultz (1995) *Curr. Opin. Struct. Biol.* **5**, 818–824.
4. J. B. Charbonnier et al. (1997) *Science* **275**, 1140–1142.
5. G. J. Wedemayer et al. (1997) *Science* **276**, 1665–1669.
6. J. D. Stewart et al. (1994) *Proc. Natl. Acad. Sci. USA* **91**, 7404–7409.
7. M. R. Haynes et al. (1994) *Science* **263**, 646–652.
8. B. Posner et al. (1994) *Trends Biochem. Sci.* **19**, 145–150.
9. D. S. Tawfik et al. (1993) *Proc. Natl. Acad. Sci. USA* **90**, 373–377.
10. K. D. Janda et al. (1997) *Science* **275**, 945–948.
11. Y. Tang, J. B. Hicks, and D. Hilvert (1991) *Proc. Natl. Acad. Sci. USA* **88**, 8784–8786.
12. J. A. Smiley and S. J. Benkovic (1994) *Proc. Natl. Acad. Sci. USA* **91**, 8319–8323.

CATALYTIC TRIAD

J. RIORDAN

The term *catalytic triad* is used to describe the arrangement of amino acid residues in the **active sites** of **serine proteinases** that underlies their mechanism of **enzyme** action. It was first

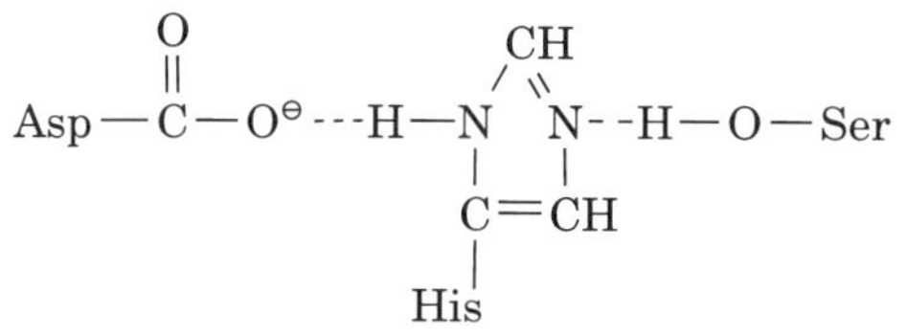

$$\text{Asp}-\overset{\overset{\displaystyle O}{\|}}{C}-O^{\ominus}\cdots H-N\diagup\!\!\overset{CH}{}\diagdown N\cdots H-O-\text{Ser}$$

$$\begin{array}{c} | \quad | \\ C{=}CH \\ | \\ \text{His} \end{array}$$

Figure 1. The catalytic triad, the hydrolytic apparatus of serine proteinases. It consists of the side chains of an aspartic acid, a histidine, and a serine residue (eg, residues 102, 57 and 195, respectively, in bovine chymotrypsin). In the absence of substrate, the imidazole group of the histidine is unprotonated, but it potentiates the nucleophilic properties of the hydroxyl side chain of the serine. When substrate binds to the active site, the serine proton is transferred to the imidazole group and the oxygen attacks the carbonyl group of the substrate. The resulting positively charged imidazole is stabilized by interaction with the negative charge of the aspartate carboxyl group. The activated serine is also highly reactive toward inhibitors such as PMSF and DIFP.

applied to the active site of **chymotrypsin** when the structure of that enzyme was determined by **X-ray crystallography** (1). Earlier studies had shown that chymotrypsin could be inactivated by chemical modification of one of its serine residues with the nerve gas, diisopropylfluorophosphate (**DIFP**). Examination of the enzyme structure revealed that the hydroxyl group of this **serine**, residue 195, was "activated" by the imidazole group of **histidine** 57, which in turn was "activated" by the carboxyl group of **aspartate** 102 (Fig. 1). Originally described as a *charge relay system*, these three residues are more commonly called a *catalytic triad*.

Similar triads have been observed in both the chymotrypsin and **subtilisin** evolutionary families of serine proteinases (2), as well as in other hydrolytic enzymes (3). It is remarkable that the bacterial serine proteinase **subtilisin** has the same geometrical arrangement of aspartate, histidine, and serine residues as chymotrypsin, but all other structural aspects of the two proteins are quite different. This is a classic example of **convergent** evolution. The activated serine interacts with the carbonyl carbon atom of the peptide bond to be hydrolyzed, forming an oxyanion intermediate, which is converted to an acylated serine with release of the amine component of the peptide. Subsequent hydrolysis of the acylenzyme occurs by the reverse reaction in which **water** substitutes for the activated serine.

BIBLIOGRAPHY

1. D. M. Blow (1976) *Acc. Chem. Res.* **9**, 145–152.
2. T. A. Steitz and R. G. Schulman (1982) *Annu. Rev. Biochem. Biophys.* **11**, 419–444.
3. D. M. Blow (1990) *Nature* **343**, 694–695.

CATHEPSINS

J. RIORDAN

The term *cathepsin* is derived from the Greek word meaning "to digest" and is used to describe a broad range of intracellular **proteinases** that serve important biological functions. Two major pathways have been identified that control the degradation of cellular proteins (see **Protein degradation**). One operates within the **cytoplasm** and is called the **ubiquitin**-mediated pathway. The other functions within specific cellular compartments, especially the **lysosomes**, but also in **endosomes** and the **endoplasmic reticulum**. Lysosomes are the major site of intracellular protein degradation (1). They are amply endowed for this function because they contain as many as 20 different proteolytic enzymes and as many other hydrolytic enzymes (eg, **phosphatases**, **lipases**, **nucleases**, and sulfatases) (1).

The lysosomal proteinases are called *cathepsins*. They are single-chain proteins that range in molecular mass from 20 kDa to about 40 kDa. Typically they have a broad substrate specificity, are optimally active at somewhat acidic pH, and belong to either the **thiol proteinase** or **carboxyl proteinase** classes. For no particular reason, they have been designated by letters as cathepsins B, L, H, M, N, S, and T, all of which are thiol proteinases or, more commonly, *cysteine proteinases*; and cathepsins D and E, which are carboxyl (or *aspartyl*) proteinases (2). All of these are **endopeptidases**, enzymes that cleave **peptide bonds** in the internal part of a protein. In addition, there are several lysosomal *exopeptidases*, including *cathepsin C*, which removes dipeptides from the amino-terminus of polypeptide chains and is therefore a dipeptidyl aminopeptidase, and at least two carboxycathepsins that are **carboxypeptidases**. Cathepsin G is a chymotrypsin-like serine proteinase found in neutrophils, and cathepsin R is another serine proteinase found in the **endoplasmic reticulum**.

The lysosomal cathepsins have been considered to be the most active proteinases in the body, because they degrade four times as much protein as the pancreatic and gastric proteinases combined (3). Much of this is essentially turnover of cytosolic proteins to balance **protein synthesis** and maintain homeostasis. However, cathepsins are also important components of the **immune system**: In order for a foreign protein to generate an **antibody** response, it must be taken up by a specialized **antigen-presenting** cell and degraded within endosomes. The resultant peptide fragments are then "presented" on the cell surface, where they trigger antibody production (4).

BIBLIOGRAPHY

1. F. Authier, B. I. Posner, and J. J. M. Burgeron (1994) In *Cellular Proteolytic Systems* (A. J. Ciechanover and A. L. Schwartz, eds.), Wiley-Liss, New York, pp. 89–113.
2. J. S. Bond and P. E. Butler (1987) *Annu. Rev. Biochem.* **56**, 333–364.
3. A. J. Barrett and H. Kirschke (1981) *Methods Enzymol.* **80**, 535–561.
4. C. V. Harding (1994) In *Cellular Proteolytic Systems* (A. J. Ciechanover and A. L. Schwartz, eds.), Wiley-Liss, New York, pp. 163–180.

CAUDAL PROTEIN

MAREK MLODZIK

Caudal protein (Cad) was identified originally as a **homeodomain** protein in *Drosophila*, but **hom-**

ologues have subsequently been found in vertebrates as well. It belongs to the superfamily of **DNA-binding** homeodomain **transcription factors**. Caudal **messenger RNA**, which is supplied by the mother in the unfertilized **egg**; is required for proper segmentation of posterior segments in insects and for the development of the posterior lineage in vertebrates. In *Drosophila*, Caudal has been shown to bind to **promoter** elements and to activate directly the **transcription** of several segmentation genes. In this process, it cooperates with other transcriptional activators of segmentation genes, eg, **Bicoid** (another homeodomain protein). At the preblastoderm stage, Caudal protein forms a gradient, which is generated by **translational repression** caused by Bicoid binding to the *cad* mRNA. Although this early function appears to be specific to *Drosophila* (and other long-germ-band insects), the zygotic function of Caudal, specification of the most posterior segments in insects or posterior cell lineage in vertebrates, appears to be conserved throughout the animal kingdom.

The *caudal* (*cad*) gene of *Drosophila* was first isolated as a **homeobox**-containing gene by cross-hybridization to the homeobox of ***Ultrabithorax*** (*Ubx*) and others (1,2). The *cad* homeodomain shares 58% identity with that of *ftz*, and 53% and 52%, respectively, with those of *Antp* and *Ubx*. During *Drosophila* development, *cad* is expressed both maternally and zygotically. The maternal expression is detected in the **germ line** in nurse cells during **oogenesis**, and its mRNA accumulates in the developing **oocyte**. At the end of oogenesis, the *cad* mRNA is distributed evenly throughout the oocyte. No Cad protein is detected during oogenesis. Cad protein is first detected beginning with nuclear division cycles 6 or 7, just prior to the formation of the syncitial blastoderm. From this stage on to the 13 nuclear cycle, the distribution of Cad protein forms a concentration gradient throughout the embryo and along the anterior-posterior axis, with its maximum at the posterior end. The *cad* mRNA remains uniformly distributed throughout the cytoplasm to the 12 nuclear cleavage; subsequently, it follows the protein gradient, when it is then distributed as a concentration gradient itself. Nevertheless, the protein gradient precedes the mRNA gradient by several nuclear cycles, so that the mRNA gradient is not relevant for the generation of its protein gradient (see below). Following the final nuclear cleavage during cellularization of the blastoderm, the maternally-derived *cad* mRNA disappears, and with it the protein translated from it (1–3).

Formation of the protein gradient from the initially uniformly distributed mRNA is generated by translational repression in the anterior regions of the embryo. Another maternally provided homeodomain protein, Bicoid [see "***Bicoid* gene**";], binds the *cad* mRNA and prevents its translation (4,5). This is the only reported example of a homeodomain protein controlling the expression of another homeodomain factor at the level of its mRNA by translational repression. Caudal protein also accumulates in the pole cells during their formation and is present there for several hours of embryogenesis. There is no *de novo* expression in the pole cells, however, and all the protein present was synthesized from the maternally provided mRNA. Formation of the Caudal protein gradient does not depend on zygotic gene expression. In older unfertilized eggs, an anterior-posterior gradient is detected that is similar to that normally present in embryos of the same stage (2,3). Maternally expressed Caudal has also been reported in other insects, such as *Bombyx mori* (6).

Zygotic expression of Caudal protein in *Drosophila* begins during cellularization of the blastoderm as a stripe in the posterior region of the embryo corresponding to the anlagen of the most posterior segments (the presumptive abdominal segments A9/A10, parts of the telson) and the hindgut. At the extended-germ-band stage, and throughout the rest of embryogenesis, *cad* is expressed in posterior midgut cells, the Malpighian tubules, the anal plate and, at weaker levels, during germ-band extension in pair-rule-like stripes (1–3). In third instar larvae, *cad* is expressed in the Malpighian tubules, the posterior midgut, the gonads of both sexes, and in the anlagen of the anal plates and the hindgut, within the genital (3). Most aspects of the zygotic expression are conserved in those vertebrates that have been analyzed (7–10).

In *Drosophila*, the function of the Caudal transcription factor is required for segment specification in the thoracic and abdominal regions by regulating the transcription of several segmentation genes (11–14). Embryos lacking both maternal and zygotic Cad protein are severely shortened, with variable deletions of many of the thoracic and abdominal segments. Zygotically provided protein can, however, largely rescue these defects. Ectopic expression of Caudal protein throughout the embryo at the blastoderm stage leads to **phenotypes** that are somewhat opposite to the phenotypes observed in *cad*-defective embryos (15). Analysis of the ***cis*-acting** elements of several segmentation genes has demonstrated that Cad is directly activating transcription of *fushi tarazu* (11) and other segmentation genes (12–14). Caudal acts in concert with Bicoid, and is partially redundant, in controlling the expression of the downstream targets, the segmentation genes (12,13).

Embryos from wild-type mothers that lack zygotic function display an absence of parts of the most posterior segments, the terminalia; in particular, the anal tuft, parts of the anal pads, and the terminal sense organs, all structures that are derived from a cryptic tenth abdominal segment, are deleted (2). The isolation and developmental expression analysis of Caudal homologues in vertebrates (7–10) has suggested that this aspect of Caudal function, specification of most posterior segments or posterior lineage in vertebrates, has been conserved throughout **evolution**. Based on the expression patterns of some of the vertebrate Caudal homologues, additional roles for this subfamily of homeodomain transcription factors have been proposed; eg, participation in rostrocaudal axial patterning or in the development and regeneration of the liver (16,17). However, all the assumptions of Caudal function(s) in vertebrate development are based purely on the respective expression patterns, and no loss-of-function mutants have been reported to date. Thus, all these proposed roles of vertebrate Caudal homologues await confirmation by the analysis of specific mutants, eg, **knockout** mice, or the potential identification of mutants in other model systems, such as the **zebrafish**.

BIBLIOGRAPHY

1. M. Mlodzik, A. Fjose, A. Gehring, and W. J. Gehring (1985) *EMBO J.* **4**, 2961–2969.
2. P. M. MacDonald and G. Struhl (1986) *Nature* **324**, 537–545.
3. M. Mlodzik and W. J. Gehring (1987) *Cell* **48**, 465–478.
4. J. Dubnau and G. Struhl (1996) *Nature* **379**, 694–699.
5. R. Rivera-Pomar, D. Niessing, U. Schmidt-Ott, W. J. Gehring, and H. Jäckle (1996) *Nature* **379**, 746–749.

6. X. Xu, P. X. Xu, and Y. Suzuki (1994) *Development* **120**, 277–285.
7. A. Frumkin, Z. Rangini, A. Ben-Yehuda, Y. Gruenbaum, and A. Fainsod (1991) *Development* **112**, 207–219.
8. A. Frumkin et al. (1993) *Development* **118**, 553–562.
9. J. S. Joly et al. (1992) *Differentiation* **50**, 75–87.
10. F. Beck, T. Erler, A. Russell, and R. James (1995) *Dev. Dyn.* **204**, 219–227.
11. C. Dearolf, J. Topol, and C. Parker (1989) *Nature* **341**, 3430–3433.
12. R. Rivera-Pomar, X. Lu, N. Perrimon, H. Taubert, and H. Jäckle (1995) *Nature* **376**, 253–256.
13. C. Schulz and D. Tautz (1995) *Development* **121**, 1023–1028.
14. T. Hader et al. (1998) *Mech. Dev.* **71**, 177–186.
15. M. Mlodzik, G. Gibson, and W. J. Gehring (1990) *Development* **109**, 271–277.
16. A. V. Morales, E. J. de la Rosa, and F. Pablo (1996) *Dev. Dyn.* **206**, 343–353.
17. U. Doll and J. Niessing (1996) *Eur. J. Cell Biol.* **70**, 260–268.

Suggestions for Further Reading

R. Rivera-Pomar and H. Jäckle (1996) From gradients to stripes in *Drosophila* embryogenesis: filling in the gaps. *Trends Genet.* **12**, 478–483.

D. Duboule (1994) *Guidebook to Homeobox Genes*. Sambrook and Tooze Publication, Oxford University Press, Oxford, U.K.

CAULIFLOWER MOSAIC VIRUS

R. HULL

Cauliflower mosaic virus (CaMV) was the first plant **virus** shown to have a DNA **genome** and the first shown to **replicate** by **reverse transcription**. The virus is found worldwide but only causes significant losses locally. It is transmitted by aphids in the externally borne or stylet-borne manner (see **Virus infection, plant**) and encodes a helper factor that is required for transmission. It has a narrow host range, being restricted primarily to the Cruciferae. CaMV is the type member of the Caulimovirus genus, which contains 11 species and 6 possible members.

This virus has had a significant impact on plant virology and plant molecular biology. The unravelling of the reverse transcription replication mechanism led to the concept of *pararetroviruses* (1,2). The virus is an important source of gene regulatory elements, which have been used extensively in the genetic manipulation of plants.

VIRUS STRUCTURE

CaMV particles are isometric, about 50 nm in diameter (Fig. 1). Although the virus particles have been crystallized, **X-ray crystallography** did not yield detailed information, most probably because of the heterogeneity of the coat protein (3) The structure has been shown by **cryoelectron microscopy** and three-dimensional **image reconstruction** to comprise three concentric layers of solvent-excluded density (4). The outermost layer is made up of a total of 420 coat protein subunits arranged in $T = 7$ **icosahedral symmetry**; it is the first example of a $T = 7$ virus that obeys the stoichometric rules of icosahedral symmetry. The DNA genome is distributed in layers II and III, together with some of the viral coat protein. The viral coat protein is expressed from **open reading frame** IV (see below) as a 58-kDa precursor molecule that is processed **proteolytically** to several molecular sizes, the major ones being 42 and 37 kDa. The protein in virions is **glycosylated** and **phosphorylated**, and the virions also contain minor amounts of the products of open reading frames III and V.

CaMV particles are very stable and are only dissociated at high pH (5) or by digestion of the coat protein with proteinases.

VIRAL GENOME

Each virus particle contains a single molecule of circular double-stranded DNA of 8 kbp. The DNA of most isolates of CaMV has three discontinuities (D1, D2, and D3 in Fig. 2), one in one strand (the α strand) and two in the other (yielding the β and γ strands); one strain has only two discontinuities, one in each strand. The discontinuities have an unusual structure,

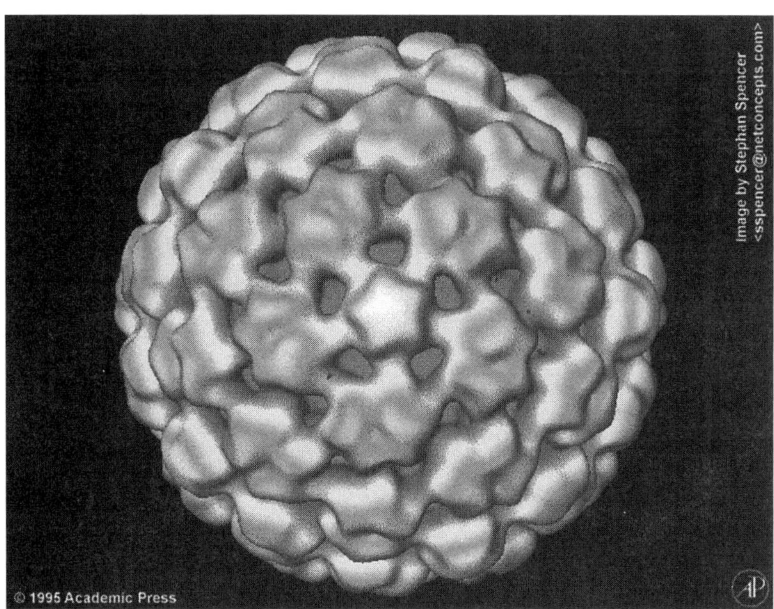

Figure 1. External structure of cauliflower mosaic virus, viewed down a local five-fold axis.

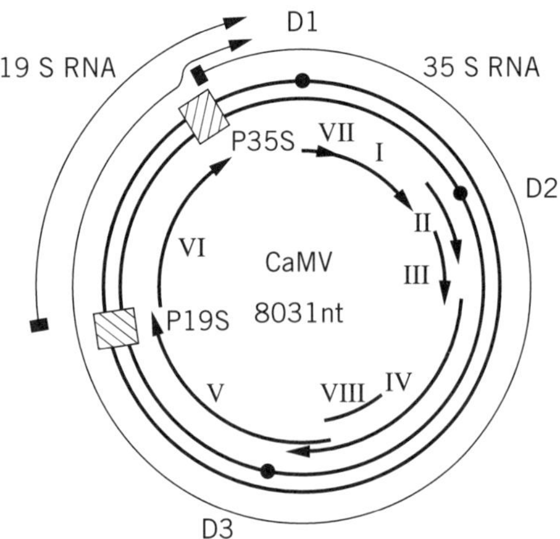

Figure 2. Genome organization of CaMV. The double-stranded DNA genome is represented by the thick double circle, with the discontinuities shown as ●. The promoters for the 35S and 19S transcripts are indicated by ▨, and the positions of the transcripts are shown by the outer arcs. The inner arcs are the open reading frames I to VIII.

with a fixed 5′ DNA nucleotide (with sometimes one or two ribonucleotides attached) and variable 3′ end, overlapping the 5′ end by 10 to 30 nucleotides. They are the sites of the priming of (+)- and (−)-strand **DNA replication**, and this unusual structure results from viral replication (see below). The DNA has a twisted conformation that, because of the discontinuities, cannot be **supercoiled**, the constraining forces being unknown.

CaMV DNA has six, or possibly eight, open reading frames (ORFs) on the complement to the α strand (Fig. 2). ORF I product (37 kDa) is involved in the cell-to-cell movement of the virus, and ORF II product (18 kDa) is the aphid transmission protein (see **Virus infection, plant**). The product of ORF III is a protein of 15 kDa that is processed to 11 kDa and thought to be involved in packaging the DNA genome (6). ORF IV encodes the viral coat protein, which contains the "cys" motif C^{ys}–X_2C^{ys}–X_4H^{is}–X_4C^{ys}– characteristic of **retrovirus** *gag* or nucleoproteins and a highly basic lysine-rich region. The viral **polymerase** (reverse transcriptase +ribonucleaseH) is encoded by ORF V (79 kDa). The product of ORF VI forms the matrix of the **inclusion bodies** in which virus particles are found embedded in the **cytoplasm** of infected cells. It has several functions, including being the site of the reverse transcription phase of viral replication (7), involved in virus assembly (8), a transactivator during viral expression (see below), and a symptom determinant. No products have been found for ORFs VII and VIII, and it is uncertain whether they are expressed.

TRANSCRIPTION AND TRANSLATION

CaMV DNA is transcribed asymmetrically from the α strand to give a more-than-genome length transcript (the 35S RNA), with a terminal repeat of 180 nucleotides and subgenomic transcript (the 19S RNA) that is the **messenger RNA** for ORF VI (Fig. 2); the transcripts are **polyadenylated**. There are suggestions of other transcripts, such as one for ORF V (9), but these have not been characterized. **Splicing** events between the leader sequence and an acceptor site in ORF II (10) are essential for virus infectivity. It is possible that these events might form the mRNA for expression of ORF III and, possibly, ORFs IV and V.

The **promoter** for the 35S RNA is well characterized and widely used in the expression of transgenes in plants. It directs constitutive, high-level expression in most tissues of most plant species. The promoter has a modular structure, with regions that control the tissue specificity of expression [reviewed in 11). Nuclear factors have been identified that bind to specific regions of the promoter. The 19S promoter is much weaker than the 35S promoter and is less well characterized.

The CaMV genome also contains the signal sequence for the polyadenylation of both the 35S and 19S transcripts. Due to the terminal repeat, the transcription of the 35S RNA has to pass this signal the first time it is encountered and recognize it the second time. It is thought that the bypass is due to the proximity of the promoter (12).

Translation of the 35S RNA is effected by several unusual mechanisms [reviewed in 11 and further refined in 10). The leader sequence upstream of ORFI is long (>600 nucleotides) and contains several **AUG** start codons, which could inhibit downstream translation. However, this leader sequence is folded into a complex stem-loop structure, which enables small ribosomal subunits to "shunt" from the 5′ to 3′ end of the leader sequence without scanning the sequence in between; this then opens up ORFI. Translation of downstream ORFs on this polycistronic message is dependent on the presence of a **transactivator**, which is the product of ORF VI, and possibly on the splicing mechanism described above. This process of transactivation seems to enable **ribosomes** that have translated one ORF to remain competent to translate the next ORF downstream.

VIRAL REPLICATION

The replication cycle of CaMV (Fig. 3) has been studied in much detail, but there are are still several aspects that are not fully understood. Virus particles enter the uninfected cell and are unencapsidated by an unknown mechanism. The viral DNA genome enters the **nucleus**, where the discontinuities are sealed and the molecule associates with **histones** forming a **minichromosome** (Fig. 3, step 2). This is the template for transcription by the host DNA-dependent **RNA polymerase**, giving the 35S and 19S RNAs, which pass to the cytoplasm (Fig. 3, step 3). The 19S RNA is translated to yield the inclusion body protein. The 35S RNA has two functions, being the mRNA for at least the products of ORFs I and II and the template for the reverse transcription phase of replication. There is some evidence that this phase of replication takes place in viruslike particles in the inclusion bodies (7). The first event in reverse transcription is the annealing of the 3′ end of $tRNA^{met}{}_{init}$ to a site, that of discontinuity 1 (D1), about 600 nucleotides from the 5′ end of the 35S RNA (Fig. 3 step 4). This acts as a primer for the synthesis of α-strand DNA toward the 5′ end of the template. RNaseH activity removes the RNA moiety of the DNA–RNA duplex thus formed and leaves DNA (Fig. 3, step 5) that, because of the terminal repeat of the 35S RNA, is complementary to the 3′ end of the template. Thus, a strand switch is effected, and synthesis of the α strand DNA

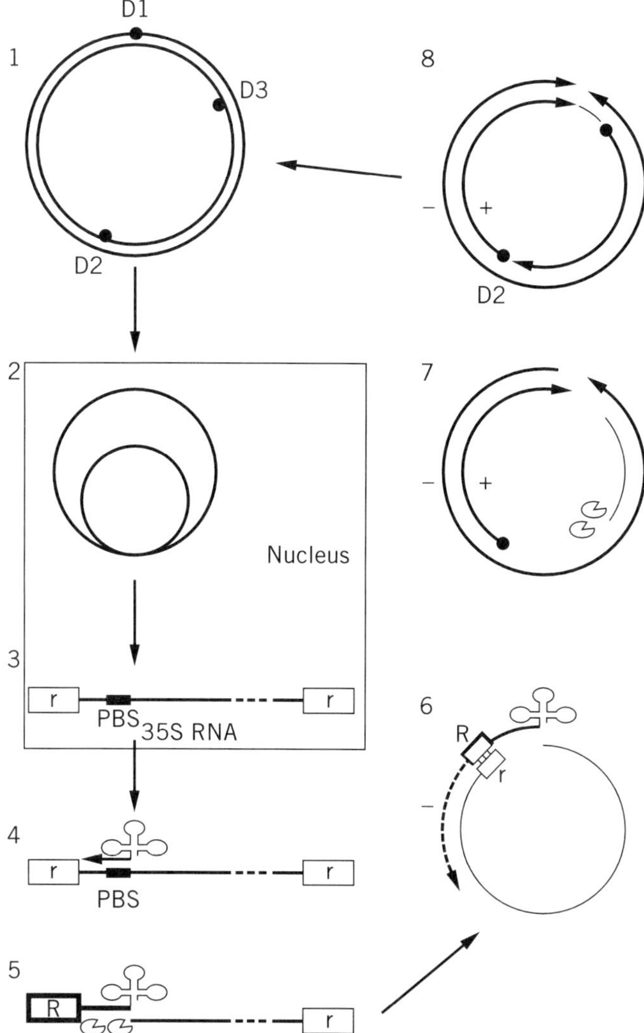

Figure 3. Replication cycle of CaMV. Step 1: The input DNA containing discontinuities 1–3 (D1–D3) passes to the nucleus where it forms minichromosomes (step 2). These are the template for the transcription of the 35S RNA (step 3) with terminal repeats [r], which moves to the cytoplasm. Priming of the 35S RNA occurs by the annealing of the 3′ end of $\text{tRNA}^{\text{met}}{}_{\text{init}}$ to the primer binding site (PBS) on the 35S RNA and leads to the synthesis of (−)-strand DNA (step 4). RNaseH removes the RNA of the RNA DNA duplex, leaving the complement to the terminal repeat sequence [R] (step 5). This anneals to the 3′ end of the 35S RNA, and DNA synthesis continues (step 6). RNAseH activity leaves purine-rich regions ●, which act as primers for (+)-strand DNA synthesis (steps 7 and 8). The oncoming strand displaces the primer sequence, thus giving the characteristic discontinuities, and a second strand-switch is effected at D1 to complete the molecule.

continues (Fig. 3, step 6). The digestion of the RNA by RNaseH leaves purine-rich regions at the positions of discontinuities 2 and 3 (D2 and D3), which act as primers for the synthesis of the β and γ strands (Fig. 3 steps 7 and 8). When the oncoming strand of (+)-strand DNA reaches the next priming site, it displaces the primer and some newly synthesized DNA, resulting in the triple-strand discontinuity described above. A second strand switch occurs at the tRNA priming site between the oncoming α and γ strands, allowing the synthesis of the γ strand to continue and making discontinuity 1.

BIBLIOGRAPHY

1. R. Hull and H. Will (1989) *Trends Genet.* **5**, 357–359.
2. H. Temin (1989) *Nature* **339**, 254–255.
3. Z. X. Gong, H. Wu, R. H. Cheng, R. Hull, and M. G. Rossmann (1990) *Virology* **179**, 941–945.
4. R. H. Cheng, N. H. Olson, and T. S. Baker (1992) *Virology* **186**, 655–668.
5. R. Al Ani, P. Pfeiffer, G. Lebeurier, and L. Hirth (1979) *Virology* **93**, 175–187.
6. J. L. Mougeot, T. Guidasci, T. Wurch, G. Lebeurier, and J. M. Mesnard (1993) *Proc. Natl. Acad. Sci. USA* **90**, 1470–1473.
7. C. M. Thomas, R. Hull, J. A. Bryant, and A. J. Maule (1985) *Nucl. Acids Res.* **13**, 4557–4576.
8. A. Himmelbach, Y. Chapdelaine, and T. Hohn (1996) *Virology* **217**, 147–157.
9. A. L. Plant, S. N. Covey, and D. Grierson (1985) *Nucl. Acids Res.* **13**, 8305–8321.
10. Z. Kiss-László, S. Blanc, and T. Hohn (1995) *EMBO J.* **14**, 3552–3562.
11. H. M. Rothnie, Y. Chapdelaine, and T. Hohn (1994) *Adv. Virus Res.* **44**, 1–67.
12. H. Sanfaçon and T. Hohn (1990) *Nature (Lond.)* **346**, 81–84.

Suggestion for Further Reading

R. J. Shepherd (1994) Caulimoviruses. In *Encyclopedia of Virology* (R. G. Webster and A. Granoff, eds.), Academic Press, London, pp. 223–226.

CDNA LIBRARIES

DANIEL KENAN
BARRY HENDERSON
JACK KEENE

In contrast to **genomic libraries**, which contain raw DNA sequences harvested from an organism's **chromosomes**, cDNA libraries are composed of processed nucleic acid sequences harvested from an organism's **RNA** pools (see **Complementary DNA**). cDNA can be prepared from any RNA source, including in RNA and noncoding RNA, isolated from single cell organisms, cultured metazoan cells, and isolated tissues. cDNA libraries provide a powerful means of examining cell- and tissue-specific gene expression. For example, mammalion cDNA libraries contain only the fraction of sequences that are expressed in the subject tissues at the time of harvest, typically in the range of 10% of the **genes** carried by the **genome**. Also, cDNA is typically prepared enzymatically from **polyadenylated** mRNA, a population of RNA that has previously undergone post-transcriptional editing and removal of any intervening sequences (**introns**). Whereas genes in genomic DNA are often interrupted by numerous introns, the copy of a gene in a cDNA or mRNA commonly contains an uninterrupted sequence encoding the gene product. Thus, cDNA inserts are directly expressible as both RNA and protein gene products. Vectors used in cDNA libraries often contain expression elements that mediate transcription and translation of the cloned cDNAs.

In constructing cDNA libraries, the goal is usually to represent full-length protein coding sequences for all of the mRNAs expressed in the subject cells or tissue. Thus, it is essential to preserve the integrity of cellular mRNA during extraction and purification protocols. Libraries made from fragmented RNA will not preserve the full open reading frame, rendering the resultant partial clones less useful. Unfortunately, RNA tends to be a transient species by design, as cells have developed complex post-transcriptional mechanisms to regulate gene expression through mRNA decay. The problem is worse in certain tissues, such as pancreas, that secrete large amounts of degradative enzymes as part of their normal function. Though troublesome, the RNA decay problem can be overcome through specialized procedures that seek to inactivate **ribonuclease** enzymes concomitantly with RNA extraction from intact tissue.

Once total RNA is extracted from the subject tissue, the mRNA fraction is purified by annealing to oligodeoxythymidine (oligo-(dT)) cellulose. Oligo-(dT) is a short polymer that hybridizes specifically to polyadenylate. Because the majority of cellular mRNAs are **polyadenylated**, affinity chromatography over oligo-(dT) cellulose provides a simple means to remove contaminating RNA species, such as ribosomal and transfer RNAs, which do not contain **poly A** tails. The purified mRNA is then annealed to oligo-(dT), which serves as a primer that can be extended by the enzyme **reverse transcriptase** (RT). RT extension thus generates **antisense** DNA copies of the purified mRNA species. Under appropriate conditions, RT can be induced to complete a second-strand DNA synthesis, using the first DNA strand as a template. The result is a collection of double-stranded DNA copies of the parental mRNA population (hence the "c" in "cDNA"). The double-stranded DNA population can then be cloned into an appropriate plasmid or bacteriophage vector to constitute a cDNA library.

cDNA libraries may be screened either by **hybridization** to the cloned inserts or by detection of RNA or protein expression products encoded by the cDNA inserts. Hybridization screening is similar in principle for both genomic and cDNA libraries and uses synthetic oligonucleotide probes or gene probes derived from existing clones. The probes are labeled with radioactivity or a fluorescent tag and then annealed to the total nucleic acids contained within a population of bacterial colonies or viral plaques. Colonies or plaques that contain nucleic acids complementary to the probe can then be detected and purified. Protein expression libraries can be screened based on affinity to **antibodies**, proteins, nucleic acids, or small-molecule ligands. In addition, functional properties can be screened for, such as **catalysis** or induction of **transcription** of a **reporter gene**.

Useful information regarding tissue-specific patterns of gene expression can be ascertained through screening cDNA libraries. By definition, a cDNA population is derived from the fraction of genes expressed in a given tissue through sampling the mRNA pool. These expressed sequences can be sampled and characterized in brute force fashion by high-throughput random sequencing of cDNA clones isolated from a library. It is not necessary to determine the full-length cDNA sequence in order to derive useful information from this approach. The partial sequences obtained are entered into an **Expressed Sequence Tag** (EST) **database** where the frequency of each partial sequence (known as a "tag") can be scored (1). In this manner, it is possible to determine the expression level of individual genes for any tissue. Furthermore, investigators that identify a gene product of interest in the EST database are already one step ahead on their sequencing efforts, because much of the open reading frame may already be in the public database. Finally, it is possible to request many EST clones from the I.M.A.G.E. Consortium (Integrated Molecular Analysis of Genomes and their Expression; http://www-bio.llnl.gov/bbrp/image/image.html), providing an easy source of genetic material for subcloning projects.

The concept of ESTs as quantitative indicators of gene expression levels has been further refined through a new method termed Serial Analysis of Gene Expression (SAGE) (2). SAGE provides a means of tagging every mRNA in the cell with a small, intrinsic nine-nucleotide sequence that is essentially unique to each transcript. In addition, SAGE provides a means to eliminate artifacts arising through processing and polymerase chain reaction (PCR) amplification of the original cDNA, thereby rendering the method highly quantitative. Finally, SAGE enables high throughput sequencing of the nine-nucleotide tags on a scale sufficient to quantify several hundred thousand cDNA species for a given tissue (3). Although the sequence information obtained from a single SAGE tag is minimal, it is usually sufficient to identify the transcript via hybridization or database searching methods. SAGE is being conducted on a large scale through the Cancer Genome Anatomy Project (CGAP; http://inhouse.ncbi.nlm.nih.gov/ncicgap/index.html), with the goal of rendering in detail the molecular differences between various normal tissues and tumors, in order to understand better the molecular basis of cancer. Ultimately the CGAP database, among others, will provide the capability of performing "virtual" **Northern blots**, in which any sequence segment can be screened through WWW-based databases to ascertain all known sequence, genetic, and biological information associated with mRNAs containing the sequence segment.

Tissue- or cell-specific gene expression can also be studied through specialized cDNA libraries derived from the subset of genes that are differentially expressed between two preparations of mRNA (4). For example, a cell culture line can be treated with a **hormone**, such as **dexamethasone**, and RNA can be prepared from samples of the cell culture before and after treatment. cDNAs prepared from the pretreatment population can then be used to remove selectively any complementary mRNAs from the post treatment preparation through a process known as **subtractive hybridization**. The majority of the mRNA species remaining in the post treatment pool should be those transcripts that were synthesized as a direct result of hormone treatment. These specific mRNAs can then be converted into cDNA and cloned into a suitable library vector, to constitute a differential cDNA library. Screening or sequence tagging a subtractive library greatly increases the efficiency of characterizing tissue-specific or stimulus-specific differences in gene expression.

See also **Combinatorial libraries**, **Libraries**, **Combinatorial synthesis**, **Affinity selection**, **DNA libraries**, **Genomic libraries**, **Expression libraries**, **Peptide libraries**, and **Phage display libraries**.

BIBLIOGRAPHY

1. M. D. Adams, M. Dubnick, A. R. Kerlavage, R. Moreno, J. M. Kelley, T. R. Utterback, J. W. Nagle, C. Fields, and J. C. Venter (1992) *Nature* **355**, 632–634.
2. V. E. Velculescu, L. Zhang, B. Vogelstein, and K. W. Kinzler (1995) *Science* **270**, 484–487.
3. L. Zhang, W. Zhou, V. E. Velculescu, S. E. Kern, R. H. Hruban, S. R. Hamilton, B. Vogelstein, and K. W. Kinzler (1997) *Science* **276**, 1268–1272.
4. J. S. Wan, S. J. Sharp, G. M. Poirier, P. C. Wagaman, J. Chambers, J. Pyati, Y. L. Hom, J. E. Galindo, A. Huvar, P. A. Peterson, M. R. Jackson, and M. G. Erlander (1996) *Nat Biotechnol* **14**, 1685–1691.

CELL ADHESION MOLECULES

K. M. YAMADA

Cell adhesion molecules play central roles in embryonic **development**, tissue organization, and overall maintenance of the structure and form of multicellular organisms (1,2) They provide the physical links between cells and other cells, and between cells and the **extracellular matrix**. Adhesion molecules are also involved in dynamic cell interactions, such as during tissue **morphogenesis**, cell migration, **immunological** responses, and possibly memory formation (1,3–7). Adhesive proteins are also intimately involved in intracellular signaling processes that regulate cell growth, **differentiation**, **gene regulation**, and **programmed cell death** (8–11).

Cell adhesion molecules mediate adhesive interactions by forming specific **protein–protein interactions** or between proteins and complex carbohydrates. In addition, cell adhesion molecules frequently link directly to multimolecular protein complexes on the cytoplasmic face of the plasma **membrane**, which, in turn, mediate interactions with the **cytoskeleton** and **signal transduction** pathways (3,11). Consequently, these cell adhesion and signaling complexes not only link cells with other cells and the extracellular matrix but also help to integrate extracellular physical information with the major signal transduction pathways within cells (see **Integrins**). For example, they appear to transduce information from inputs as diverse as transient contacts with other cells, binding of different connective tissue molecules, and tension or torsion at the cell surface. Each of these inputs connects to intracellular signaling pathways, such as **phosphorylation** cascades (9,10).

TYPES OF ADHESION

The physical linkages involved in cell adhesion are provided by adhesion molecules that function as parts of general adhesion systems or in specialized adhesive structures. For example, epithelial cells can adhere to other cells along broad expanses of plasma membrane using general-purpose adhesive molecules such as **cadherins** (12,13). They can also adhere to each other by specialized adhesive structures, such as adherens junctions, **desmosomes**, and **tight junctions** (1). Each type of specialized junctional complex requires specific adhesion molecule components, which can include specialized cadherins and other proteins. Fibroblasts can also form complexes with other cells, but they most characteristically adhere to extracellular matrix. Cell-to-matrix adhesions can be broad and flat, as when epithelial cells adhere via integrins to basement membranes, or they can involve specialized structures that include hemidesmosomes or the **focal contacts** of fibroblastic and endothelial cells. The protein complexes involved in cell-to-matrix adhesion contain different complements of adhesive, cytoskeletal, and signal transduction molecules than those involved in cell-to-cell adhesion (1,11,14). This entry will first provide a general overview of specific types of cell adhesion proteins, then briefly discuss how they are integrated into cytoskeletal and signaling networks.

GENERAL FEATURES OF CELL ADHESION PROTEINS

Cell adhesiveness is generally based on the specific binding of a protein to another molecule at the cell surface. When adhesion results from binding of an adhesion molecule to the same type of protein on a neighboring cell, the interaction is termed '*homophilic*.' The cadherin family is a major mediator of such homophilic interactions (12,13) (see **Cadherins**). Cadherins form complexes in which both intercellular and lateral binding interactions cooperate to create tightly packed adhesion complexes that mediate adhesion with high avidity.

In many systems, however, adhesiveness involves binding of a **receptor** to a specific **ligand**. In cell-to-cell adhesion, the target protein of an adhesion receptor can be a "counter-receptor" or a complex carbohydrate on a protein anchor in the plasma membrane. In cell-to-matrix interactions, a plasma membrane adhesion protein such as an integrin can bind to an extracellular matrix protein that is itself considered to be an adhesive protein. For example, the protein bound could be a **fibronectin** or a **laminin**, which are complex, multifunctional proteins involved in both cell adhesion and anchorage to structural components of the extracellular matrix (2,15).

Consequently, there are two broad classes of adhesion molecules or receptors (Fig. 1). One class is bound to the plasma membrane, often as a **transmembrane protein** (Fig. 1). This type of molecule is generally a receptor, a homophilic adhesion molecule, or a counterreceptor. It often consists of an extracellular **domain** containing one or more cell-interaction domains or sites, as well as a stalk region, a **hydrophobic** transmembrane domain, and (usually) a

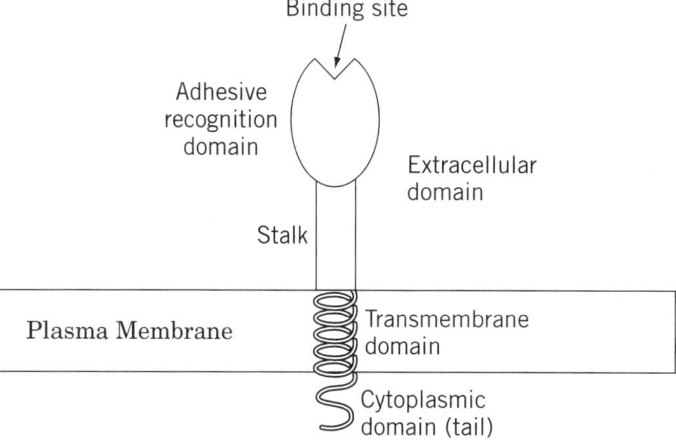

Figure 1. Schematic diagram of a generalized transmembrane adhesion molecule consisting of an extracellular domain, transmembrane segment, and cytoplasmic domain. See text for discussion.

cytoplasmic domain or tail. This type of molecule is often also involved in transmembrane transmission of signals.

The second broad class of adhesion molecules consists of proteins that are often classified as cell surface or extracellular matrix proteins (see **Extracellular matrix**), but that contain domains or sites involved in cellular adhesion (Fig. 2). Molecules in this class include fibronectins, laminins, vitronectin, tenascins, thrombospondins, and the **collagens**. All contain one or more cell-binding domains, which consist of a primary recognition motif consisting of a short peptide sequence (eg, Arg–Gly–Asp), and sometimes a synergy site that provides a substantial increase in receptor-binding specificity and affinity. As discussed below, many of these proteins also contain a variety of other functional domains.

In general, adhesion molecules frequently have the following properties (1–3,6,15–18):

- Composed of multiple repeats of **protein motifs**, such as the immunoglobulin (Ig) motif, the **epidermal growth factor** (EGF) repeat, or the fibronectin motif.
- Specialized functional domains, including a domain for formation of dimers or higher polymers.
- Moderate affinity, for example, with **dissociation constants** (K_d) in the range of 10^{-6}–10^{-7} M for fibronectin, and even as great as 10^{-4} M for leukocyte adhesion molecules that mediate rolling adhesion. In contrast, many well-known protein–protein interactions have affinities with $K_d = 10^{-9}$–10^{-11} M.
- High avidity after clustering. Adhesion molecules often form functional clusters or aggregates in the plane of the plasma membrane, which causes them to develop strong total avidity due to the cooperation of the otherwise weak binding of individual molecules.
- Regulation by activation, such as the "inside-out" signaling that increases the affinity of integrins in platelet activation and leukocyte adhesion, or regulation by phosphorylation of the cadherin system.

SPECIFIC CELL ADHESION MOLECULES

The broad class of adhesion molecules embedded in the plasma membrane contains several large groups of proteins that share common structural motifs, especially the Ig motif (Fig. 3). Cadherins represent a large family of proteins involved in homophilic cell-to-cell adhesive interactions (12) (see **Cadherins**). Cadherins bind to cadherins of the same type on other cells via cell interaction sites that can include the recognition sequence His–Ala–Val. Besides the "classic" cadherins, such as E-cadherin and N-cadherin, there are now known to be a number of other types of cadherins, as well as the highly specialized cadherins termed *desmocollins* and *desmogleins* found exclusively on desmosomes, which link cells together at particularly strong attachment sites connected to **intermediate filaments** such as **keratins** or **vimentin** (see **Desmosome**). Cadherins are quite sensitive to depletion of calcium in the surrounding medium, accounting for the ability of calcium chelators such as **EDTA** to dissociate tissues into component cells (see **Calcium-binding proteins**). Although desmosomal cadherins are present in epithelia of all ages, cadherins appear to be of particular importance during embryonic development, when they mediate cell–cell adhesion and help define tissues by their propensity to bind primarily to cadherins of the same type, rather than to other cadherins on unrelated cell types (12,13).

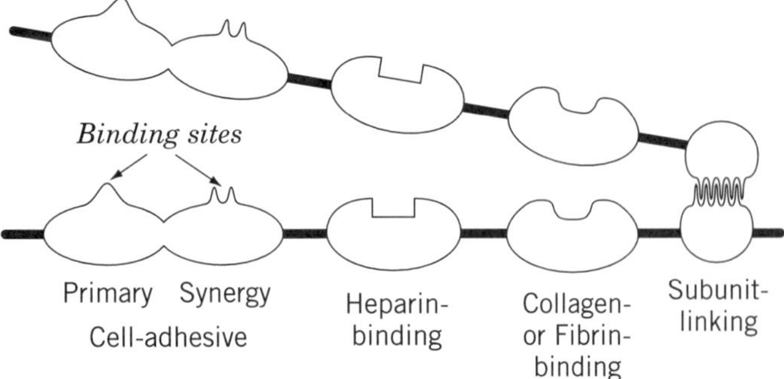

Figure 2. Schematic summary of common features of extracellular matrix adhesion molecules. Linear arrays of structural and functional domains provide a series of specific binding sites. See text for further discussion.

Integrins are major cell-surface receptors for a host of extracellular matrix proteins, as well as "counterreceptors" on other cells (8,19) (see **Integrins**). There are more than 20 types of integrin subunits. Genetic absence of most integrin subunits leads to disease or death, often in the embryo or near the time of birth. Integrins are heterodimers of one α and one β subunit, each of which provides adhesive specificity. They have a highly distinctive appearance, with a bulbous head domain contributed by each subunit, a binding site that also appears to involve part of each subunit, and two spindly legs (Fig. 3). The legs penetrate through the plasma membrane and end in rather short cytoplasmic domains. Although short, these cytoplasmic tails can mediate a remarkable range of signaling events (see text below). Integrin functions are also often inhibited by depletion of divalent cations.

Although cell adhesion molecules (CAMs) and counterreceptors are structurally related by their use of the Ig motif, and they often even share the -CAM designation, they can differ functionally (Fig. 3). Molecules such as NCAM are homophilic adhesive molecules that bind to the same type of molecule on adhering cell surfaces (20). In contrast, counterreceptors such as the ICAMs and VCAM have specialized peptide recognition sites that are bound specifically by integrins such as LFA-1 (CD11a/CD18 or $\alpha_\mu\beta_2$) or VLA-4 ($\alpha_4\beta_1$). The functions and sites of expression of these molecules differ widely. For example, molecules such as NCAM are implicated along with cadherins in embryonic developmental events, such as axonal bundling and guidance, whereas the ICAM counterreceptors are present as targets for binding by cells in the blood circulation. Levels of counterreceptors on the cell surface can often also be regulated rapidly in response to **cytokines**.

A new family of transmembrane proteins termed *ADAMs*, which are membrane proteins with a disintegrin and a **metalloproteinase** domain (21), contain a **proteinase**-like domain, as well as an integrin recognition site in the disintegrin domain. A member of this family is present on **sperm** and is likely to play a role in adhesive interactions, but the functions of other members of this rapidly growing family remain to be characterized. Syndecans are cell-surface heparan sulfate proteoglycans, with a protein core that crosses the plasma

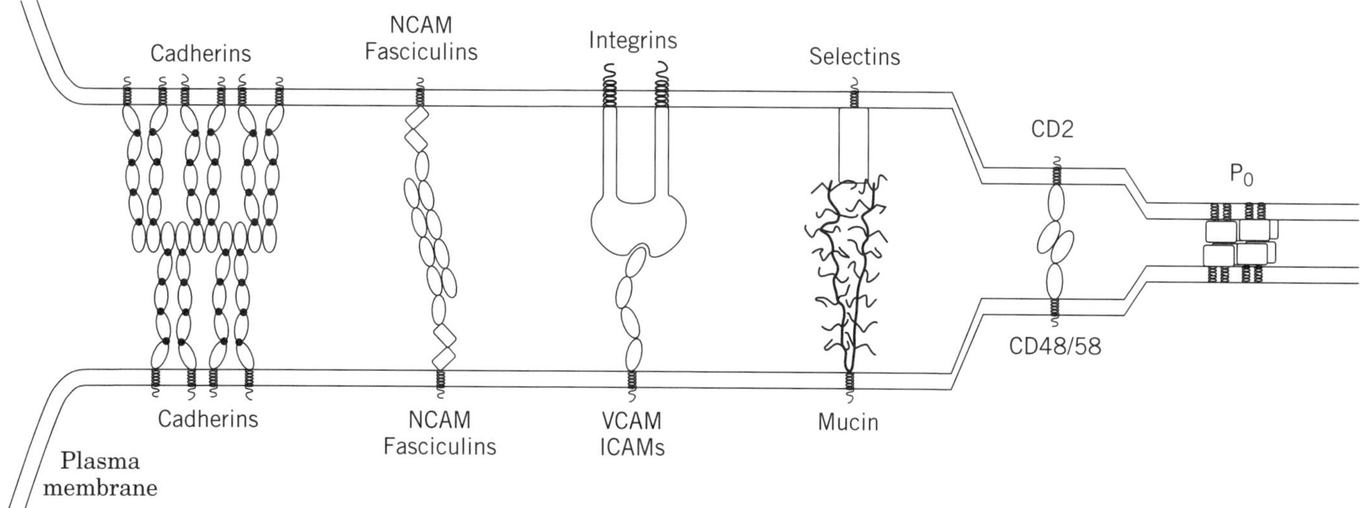

Figure 3. Examples of cell adhesion protein complexes linking plasma membranes of one cell (top) with another (bottom). Several of the major adhesion systems are listed. Adhesion molecules comprised of repeating immunoglobulin motifs include cadherins, NCAM, fasciculins, VCAM, and the ICAMs; all are anchored in the plasma membrane. Cadherins appear to exist as dimers that form lateral and intercellular connections. Selectins bind via lectin domains to complex carbohydrates in certain mucins. CD2, CD48, and CD58 contain only two Ig repeats. Myelin protein zero (P_0) forms homophilic complexes between closely spaced plasma membranes.

membrane and terminates in a cytoplasmic tail. Syndecans appear to function as "coreceptors," mediating signaling in association with a primary adhesion molecule (22). For example, syndecans bind to fibronectin at the same time that cell surface integrins bind to a cell-binding domain of fibronectin, and these binding partners cooperate during formation of the specialized adhesion sites known as **focal contacts**. Syndecans have several other signaling functions as well, and are reported to interact with both cytoskeletal proteins and protein kinase C (23).

Extracellular Adhesion Molecules

The second major class of adhesive molecules exists extracellularly, and its members are generally targets for cell-surface adhesion receptors such as the integrins. A gallery of such molecules is presented in Figure 4, showing functional binding sites for recognition by cell-surface receptors and domains for binding to a variety of ligands. Many contain one or more heparin-binding domains, which are used to bind to heparan sulfate proteoglycans in the extracellular matrix or in cell-surface syndecan molecules. Many also contain other types of binding domains for extracellular molecules, such as for collagen/gelatin, **fibrin**, entactin (nidogen), and fibulin (Figs. 2 and 4). Many of these molecules are both large and multimeric, with a specific domain for covalent crosslinking to other subunits. Finally, most of these proteins are either **alternatively spliced** or encoded by a set of closely related genes, thereby generating a number of **isoforms**. A number of reviews provide details about these complex proteins, including Refs. 2 and 18 (see **Extracellular matrix**).

Peptide Adhesive Recognition Sites

A striking characteristic of a number of cell adhesion molecules is their frequent use of short peptide recognition sites (Refs. 16, 24–26 and references cited therein). Short peptides with these sequences can at least partially mimic the binding function of intact molecules to adhesion receptors. Examples include Arg–Gly–Asp (RGD) and Leu–Asp–Val (LDV), which can bind directly to certain integrin receptors. Nevertheless, binding can be 25- to 200-fold more active in larger peptides or proteins, such as in association with a "synergy" site that functions synergistically with the primary adhesive peptide site (26). Some peptide recognition sites appear to be "cryptic" in the intact molecule, and may be active only in small fragments of the protein, such as after **proteolysis**. Table 1 lists the putative peptide recognition sequences in cell adhesion molecules. It is important to stress that a number of adhesion molecules also function by adhering to other proteins over much broader intermolecular contact areas, via multiple molecular contacts and with no simple peptide motif, in analogy to classic high affinity noncovalent **protein–protein interactions** (27). It is nevertheless striking that so many cell adhesion molecules use simple, specific, peptide adhesive-recognition motifs.

Major Functions of Cell Adhesion Molecules Beyond Adhesiveness

Cell adhesion molecules and receptors can be intuitively understood as mechanisms for attaching cells to other cells, for permitting cell adherence to extracellular matrix molecules, and for mediating traction during cell migration. It has become clear, however, that they also play critical roles in cellular signaling. In fact, their signaling functions may be of equal biological importance. A current view is that "adhesion" proteins are actually "cell-interaction" proteins that have multiple functions involving the bidirectional transfer of information at the cell surface. Thus, some of these proteins, such as thrombospondin and tenascin, can have antiadhesive

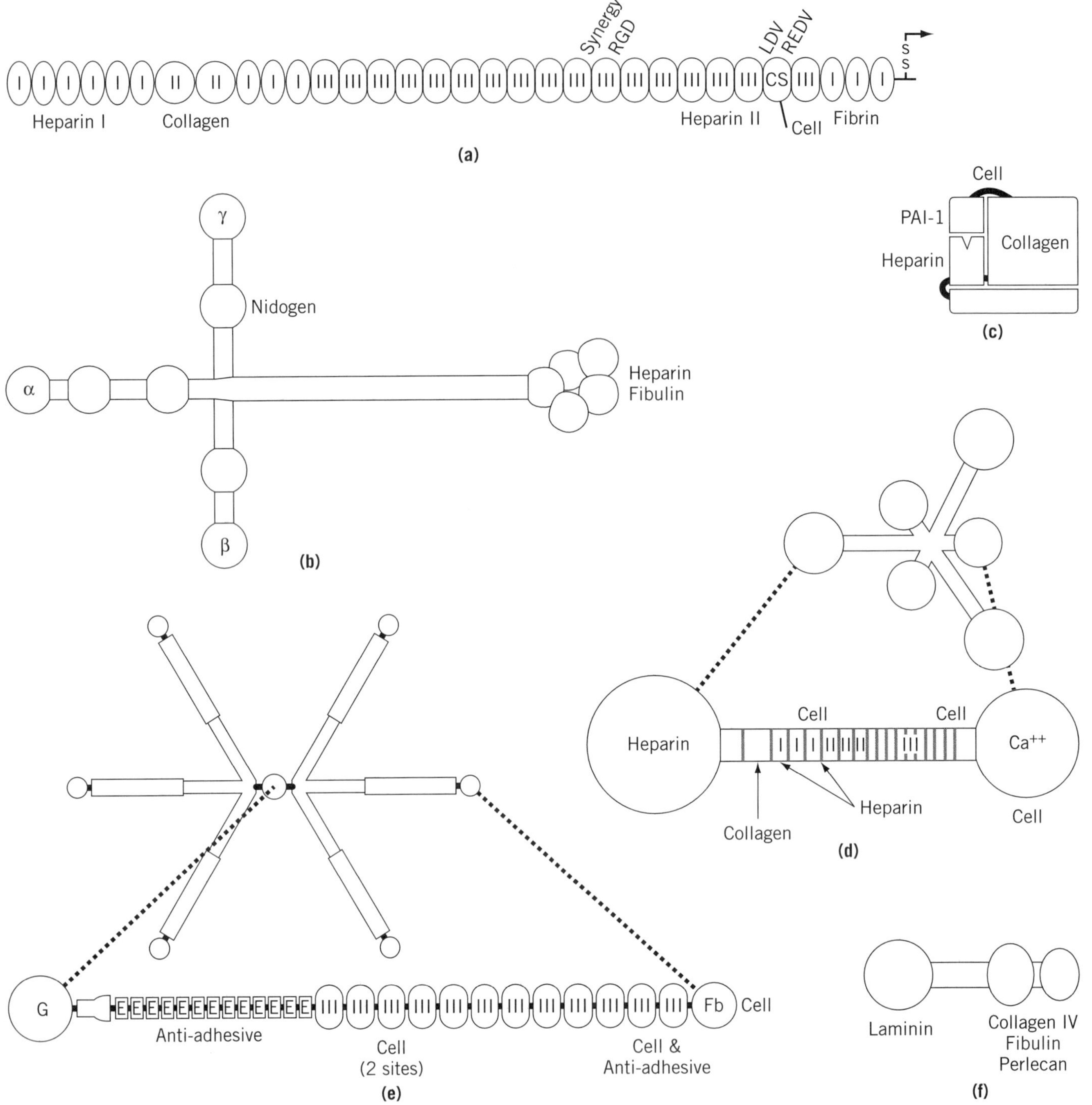

Figure 4. Composite of six examples of cell adhesion molecules of the extracellular matrix: (**a**) Fibronectin, (**b**) Laminin (multiple cell-binding sites), (**c**) Vitronectin, (**d**) Thrombospondin, (**e**) Tenascin, (**f**) Nidogen (entactin). These molecules vary widely in size and shape, and often consist of repeating protein motifs organized into functional binding domains. Specific molecular or cell-binding sites are labeled.

activities in certain situations and can modulate a host of cellular functions. Even classic "adhesive" proteins, such as fibronectin, can have a bewildering range of activities, mediated by integrin binding, that range from activating or modulating nearly every known mammalian signal transduction pathway (tyrosine phosphorylation by **tyrosine kinases**, **mitogen**-activated protein kinases (MAP kinases), Ca^{2+} and H^{+} fluxes, **inositol phosphate** pathways, preventing **apoptosis**, and activating specific gene **transcription** (9–11).

Table 1. Adhesive Recognition Sequences

Protein and Peptide Name	Recognition Sequence
Fibronectin	
Cell-binding determinant	GRGDS
Synergy site	PHSRN
Potential synergistic sites	RNS, SDV
CS1 site of IIICS domain	DELPQLPHPNLHGPEILDVPS
CS5 site of IIICS domain	REDV
FN-C/H-I	YEKPGSPPREVVPRPRPGV
FN-C/H-II	KNNQKSEPLIGRKKT
FN-C/H-III	YRYRYTPKEKTGPMKE
FN-C/H-IV	SPPRRARVT
FN-C/H-V	WQPPRARI
Type III5 site	KLDAPT
E1	TDIDAPS
Laminin α_1 chain	
PA22.2 (IKVAV)	SRARKNAASIKVAVSADR
RGD site	RGDN
GD-1	KATPMLKMRTSFHGCIK
GD-2	KEGYKVRLDLNITLEFRTTSK
GD-3	KNLEISRSTFDLLRNSYGVRK
GD-4	DGKWHTVKTEYIKRKAF
GD-6	KQNCLSSRASFRGCVRNLRLSR
AG-10	NRWHSIYITRFG
AG-32	TWYKIAFQRNRK
AG-73	RKRLQVQLSIRT
Laminin β_1 chain	
YIGSR	YIGSRC
PDSGR	PDSGR
F9 (RYVVLPR)	RYVVLPRPVCFEKGKGMNYVR
LGTIPG	LGTIPG
Integrin $\alpha_2\beta_1$ binding site	YGYYGDALR
Laminin β_2 chain	
LRE site	LRE
Laminin γ_1 chain	
P20	RNIAEIIKDI
C-16	KAFDITYVRLKF
C-28	TDIRVTLNRLNTF
C-64	SETTVKYIFRLHE
C-68	TSIKIRGTYSER
Fibrinogen	
RGD sites	RGDS and RGDF
γ chain peptide	HHLGGAKQAGDV
Integrin $\alpha_M\beta_2$ binding site	KRLDGGS
von Willebrand factor	
RGD site	RGDS
GPIb site	CQEPGGLVVPPTDAP plus LCDLAPEAPPPTLPP
Asp-514–Glu-542	DLVFLLDGSSRLSEAEFEVLKA-FVVDMME
Vitronectin	RGDV
Entactin/nidogen	SIGFRGDTC
Fertilin	TDE
Circumzooite protein	VTCG
Thrombospondin	
(CD36-binding motif)	CSVTXG
Heparin-binding motif	...YSXY
N-Terminal domain site	VDAVRTEKGFLL-LASLRQMKKTRGTLLALERKDHS
GAG-independent site	FQGVLQNVRFVF
4NIs	RFYVVMWK
Collagen type I	
RGD sites	RGDTP and SRGDTG
DGEA site	DGEA
Collagen type IV	
IV-H1	GVKGDKGNPGWPGAP
Hep-I	TAGSCLRKFSTM
Hep III	GEFYFDLRLKGDK
α_3(IV) 185–203	CNYYSNSYSFWLASLNPER
$\alpha_2\beta_1$ binding site	FYFDLR
Amyloid P component	FTLCPR
Amyloid precursor protein	RHDS
Bone sialoprotein	EPRGDNYR (cyclic)
L1	PSITWRGDGRDLQEL
ICAM-1	
JF9	VLYGPRLDERDAPGNWTWPEN-SQQTPMC
ICAM40-51	KELLLPGNNRKV
Cyclic peptide 1	PSKVILPRGGC (cyclic)
$\alpha_L\beta_2$ binding site	LET, IET
ICAM-2	GKSFTIECRVPTVEP
NCAM	KYSFNYDGSE
VCAM-1	
Domain 1 C-D loop	QIDSP (cyclic peptide with terminal Cs)
(binding motifs)	IDSP, GNEH, KLEK
N-Cadherin	LRAHAVDVNG
Myelin protein 0 (P_0)	YSDNGTF
Ninjurin	PPRWGLRNRPIN
Leishmania pg63	...SRYD...
Streptavidin	GRYDS[a]

[a] See Refs. 14, 16, and 22 for additional information.

Two key sites of membrane adhesion molecules that allow them to function as signal transduction receptors are their **ligand-binding** domains and their cytoplasmic domains. The ligand-binding domains are of obvious importance in binding to extracellular molecules, but their roles are more intriguing in at least some cases. There appear to be separable functions for ligand occupancy, specifically, filling the binding site with a ligand or an **antibody**, as opposed to receptor clustering (which can be induced by a multivalent ligand, such as fibrils of fibronectin or collagen). These two inputs can cooperate in promoting accumulation of specific cytoskeletal proteins, such as α-actinin and **actin**, which are thought to be crucial for forming strong adhesions (28). Even though they lack intrinsic enzymatic activities, integrins appear to be able to function as signaling receptors and as regulators of actin cytoskeletal organization by recruiting other molecules to their cytoplasmic domains. Integrins can reportedly bind directly to certain cytoplasmic proteins directly, such as to talin, α-actinin, and focal adhesion kinase, perhaps regulated by ligand occupancy on the outside of the cell. Integrin clustering appears to play a central role in forming large complexes of over 30 molecules that can serve as signaling centers, such as for MAP kinase activation (Fig. 5; Refs. 14,28).

This same theme of complex formation, binding of cytoskeletal proteins such as actin, and accumulation of signaling molecules also appears to occur for cell-to-cell adhesion molecules (11). Cadherin cytoplasmic domains interact

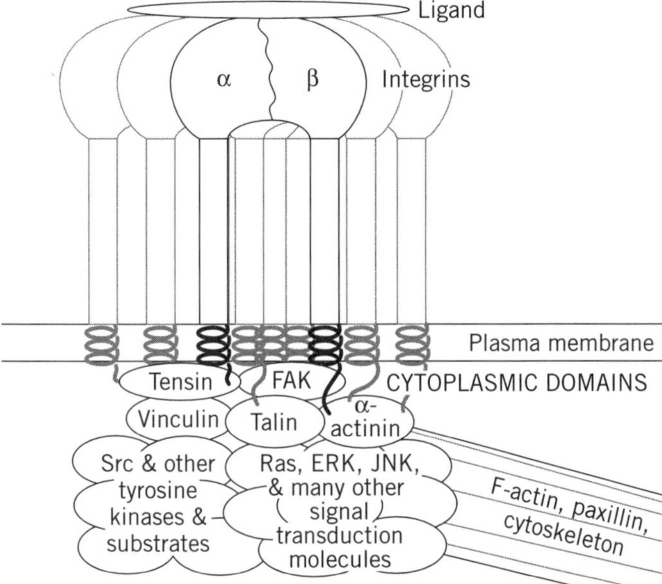

Figure 5. Integrin adhesion and signaling complex formed after cell adhesion molecule interactions. Binding of a ligand and aggregation of adhesion receptors in the plane of the plasma membrane results in the formation of multimolecular complexes that can contain over 30 cytoskeletal and signal transduction molecules, which link to the actin-containing cytoskeleton and mediate intracellular signaling.

with catenins, which then link to the actin cytoskeleton. In addition, however, the binding of β-catenin to cadherins can titrate its cytoplasmic levels; β-catenin molecules that are not bound by cadherins can enter the **nucleus** and regulate **gene expression** (3). A variety of other signaling pathways exist in cell–cell and cell–matrix adhesion systems. Although much more remains to be learned about precise pathways and overall integration of the many cytoplasmic effects of cell adhesion molecules, it is clear that adhesion molecules are crucial components of the basic regulatory mechanisms of cells and provide dynamic links to the external environment.

BIBLIOGRAPHY

1. B. Alberts, D. Bray, J. Lewis, M. Raff, K. Roberts, and J. D. Watson (1994) in *Molecular Biology of the Cell*, 3rd ed., Garland Publishing, Inc., New York, pp. 949–1009.
2. E. D. Hay, ed. (1991) *Cell Biology of Extracellular Matrix*, Plenum Press, New York.
3. B. M. Gumbiner (1996) *Cell* **84**, 345–357.
4. S. F. Gilbert (1994) *Developmental Biology*, Sinauer Associates, Sunderland, MA.
5. R. O. Hynes (1996) *Devel. Biol.* **180**, 402–412.
6. G. M. Edelman and J. P. Thiery (1985) *The Cell in Contact: Adhesions and Junctions as Morphogenetic Determinants*, John Wiley & Sons, New York.
7. M. S. Grotewiel, C. D. Beck, K. H. Wu, X. R. Zhu, and R. L. Davis (1998) *Nature* **391**, 455–460.
8. R. O. Hynes (1992) *Cell* **69**, 11–25.
9. E. A. Clark and J. S. Brugge. (1995) *Science* **268**, 233–239.
10. M. A. Schwartz, M. D. Schaller, and M. H. Ginsberg (1995) *Annu. Rev. Cell Devel. Biol.* **11**, 549–599.
11. K. M. Yamada and B. Geiger (1997) *Curr. Opin. Cell Biol.* **9**, 76–85.
12. M. Takeichi (1990) *Annu. Rev. Biochem.* **59**, 237–252.
13. A. S. Yap, W. M. Brieher, and B. M. Gumbiner (1997) *Annu. Rev. Cell Devel. Biol.* **13**, 119–146.
14. B. M. Jockusch, P. Bubeck, K. Giehl, M. Kroemker, J. Moschner, M. Rothkegel, M. Rudiger, K. Schluter, G. Stanke, and J. Winkler (1995) *Annu. Rev. Cell Devel. Biol.* **11**, 379–416.
15. S. Ayad, R. P. Boot-Handford, M. J. Humphries, K. E. Kadler, and C. A. Shuttleworth (1994) *The Extracellular Matrix FactsBook*, Academic Press, New York.
16. K. M. Yamada (1991) *J. Biol. Chem.* **266**, 12809–12812.
17. P. D. Richardson and M. Steiner (1995) *Principles of Cell Adhesion*, CRC Press, Boca Raton, FL.
18. C. Chothia and E. Y. Jones. (1997) *Annu. Rev. Biochem.* **66**, 823–862.
19. R. M. Lafrenie and K. M. Yamada (1996) *J. Cell Biochem.* **61**, 543–553.
20. G. M. Edelman and K. L. Crossin (1991) *Annu. Rev. Biochem.* **60**, 155–190.
21. T. G. Wolfsberg and J. M. White (1996) *Devel. Biol.* **180**, 389–401.
22. D. J. Carey (1997) *Biochem. J.* **327**, 1–16.
23. E. S. Oh, A. Woods, and J. R. Couchman (1997) *J. Biol. Chem.* **272**, 11805–11811.
24. Y. Yamada and H. K. Kleinman (1992) *Curr. Opin. Cell Biol.* **4**, 819–823.
25. E. Ruoslahti (1996) *Annu. Rev. Cell Devel. Biol.* **12**, 697–715.
26. S. Aota, M. Nomizu, and K. M. Yamada (1994) *J. Biol. Chem.* **269**, 24756–24761.
27. M. Nomizu, Y. Kuratomi, S. Y. Song, M. L. Ponce, M. P. Hoffman, S. K. Powell, K. Miyoshi, A. Otaka, H. K. Kleinman, and Y. Yamada (1997) *J. Biol. Chem.* **272**, 32198–32205.
28. K. M. Yamada and S. Miyamoto (1995) *Curr. Opin. Cell Biol.* **7**, 681–689.

Suggestions for Further Reading

B. Alberts, D. Bray, J. Lewis, M. Raff, K. Roberts, and J. D. Watson (1994) "Cell junctions, cell adhesion, and the extracellular matrix," in *Molecular Biology of the Cell*, 3rd ed., Garland Publishing, Inc., New York, pp. 949–1009.

E. D. Hay, ed. (1991) *Cell Biology of Extracellular Matrix*, Plenum Press, New York.

C. Chothia and E. Y. Jones. (1997) The molecular structure of cell adhesion molecules, *Annu. Rev. Biochem.* **66**, 823–862.

B. M. Gumbiner (1996) Cell adhesion: the molecular basis of tissue architecture and morphogenesis, *Cell* **84**, 345–357.

Another excellent source of current information on adhesion molecules is found each year in issue 5 of *Current Opinion in Cell Biology* (eg, see these issues for the past 6 years). This yearly issue is devoted to "cell-to-cell contact and extracellular matrix," and provides authoritative updates in the field.

CELL CYCLE

MARK SOLOMON

A dramatic change in how we understand the progression of the eukaryotic cell cycle, from phenomenological to biochemical, occurred in the late 1980s. The cell cycle consists of four main phases: G1, S, G2, and M. **DNA replication** occurs in S phase, and the **chromosomes** are segregated to daughter cells during mitosis (M phase). The so-called gap phases (G1 and G2) are defined simply as the periods separating DNA replication and mitosis. The length of the cell cycle ranges from just a few minutes in certain early embryos (which don't need to increase their mass between divisions), to 1.5 h in budding yeast, to approximately one day in typical mammalian **tissue culture** cells. In mammalian cells, the durations of S phase (~7 h) and mitosis (~1 h) are relatively inflexible, whereas the lengths of the gap phases can vary greatly. The length of G1 is particularly variable and is sensitive to the levels of nutrients and growth factors. In general, once a cell progresses past the "Restriction Point" in G1, it becomes committed to complete the cell cycle. A convergence of research in yeast, marine invertebrates, and frogs led to a phenomenal growth in our mechanistic understanding of how all eukaryotic cells regulate these events and the transitions between them. The basic machinery controlling the cell cycle consists of a subfamily of protein **kinases**, termed cyclin-dependent kinases (Cdk's). These enzymes are in turn regulated by a number of mechanisms, including **transcription**, inhibitory and activating **phosphorylations**, binding to inhibitory proteins, and **ubiquitin**-mediated protein degradation. Checkpoints that prevent the execution of one cellular event before a prior step has been completed further modulate cell cycle progression.

HISTORICAL THREADS: YEAST GENETICS, MPF, AND CYCLINS

Genetic studies of the cell cycle in the yeasts *Saccharomyces cerevisiae* and *Schizosaccharomyces pombe* illuminated important principles underlying the logic of cell cycle control and identified many of its key players. Screens were performed for conditional **temperature-sensitive mutations**, causing cells to arrest quickly with uniform morphologies, indicative of defects in executing individual cell cycle steps. Analysis of a large collection of such *cdc* (cell division cycle) mutants from *S. cerevisiae* placed them into series of dependent and independent pathways (1). For example, inactivation of *CDC28* causes cells to arrest at "START" (equivalent to the restriction point of mammalian cells) and blocks pathways leading to formation of the bud, DNA replication, and duplication of the spindle pole body. Inactivation of genes acting after *CDC28* can block one of these pathways, while the other two continue. A similar screen in the distantly related yeast *Schizosaccharomyces pombe* successfully identified the most proximal regulators of the G2 to mitosis transition (Fig. 1) (2). $cdc2^+$ was found to be the key

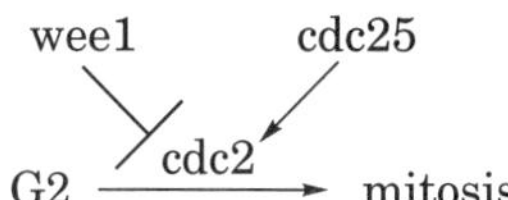

Figure 1. Regulation of the G2-to-mitosis transition in *Schizosaccharomyces pombe*. $cdc2^+$ is required for this transition. In the absence of its activity, cells arrest in G2. $cdc25^+$ is a positive regulator of $cdc2^+$, and its inactivation also gives a G2 arrest. In contrast, $wee1^+$ negatively regulates $cdc2^+$, and its inactivation leads to premature activation of Cdc2 and entry of cells into mitosis at an unusually small size (a "wee" phenotype).

regulator of entry into mitosis and to encode a protein kinase. Inactivation of $cdc2^+$ causes cells to arrest in G2, whereas its premature activation leads to early entry into mitosis. Genetic analysis indicated that the $cdc25^+$ gene product functioned in a pathway leading to activation of Cdc2 and that the product of $wee1^+$ functioned in a pathway leading to inhibition of Cdc2. Sequence analysis of these genes indicated that both $cdc2^+$ and $wee1^+$ encoded protein kinases, but initially shed little light on the function of Cdc25 (see text below).

Two other lines of research provide essential historical threads leading to our current understanding of the cell cycle. The first involves **maturation promoting factor** (MPF), which was originally defined as a developmental factor found in the cytoplasm of metaphase-arrested frog eggs that could induce oocytes to proceed through meiosis and to "mature" into eggs following its injection into their cytoplasm (3). Subsequent research found that MPF activity is present in all eukaryotic cells during M phases (meiosis or mitosis). Thus, MPF is a universal regulator of entry into mitosis. The second thread concerns the **cyclins**. The first cyclins were found during studies of translational control before and after fertilization of sea urchin eggs conducted as part of the Physiology course at the Marine Biological Laboratory in Woods Hole (4). These proteins were synthesized continuously, and they accumulated until their abrupt degradation during mitosis. This pattern of accumulation hinted that cyclins might play an important role during the cell cycle. Later work showed that the injection of cyclin **messenger RNA** led to the maturation of frog oocytes, suggesting that cyclins, like MPF, functioned as positive regulators of the G2-to-mitosis transition (5).

Work on cyclins, MPF, and the yeast *cdc* genes came to an explosive convergence in the late 1980s with the purification of MPF and the identification of its subunits as homologs of $cdc2^+$ and cyclin (6–9). Thus, this key regulator of entry into mitosis was a protein kinase (Cdc2) whose activity was controlled at least in part via the cyclic accumulation and degradation of a regulatory subunit (cyclin). The activity of Cdc2 is typically assayed by its ability to **phosphorylate** a convenient substrate, **histone** H1. Cdc2 is a workhorse master regulator. Rather than sitting at the apex of a large regulatory pathway, Cdc2 generally phosphorylates the downstream-most targets of mitotic regulation. For example, its direct phosphorylation of nuclear lamins causes the disassembly of the lamin-containing filaments of the nuclear lamina (10,11).

REGULATION OF CDC2 ACTIVITY: PHOSPHORYLATION AND CYCLIN DEGRADATION

In addition to cyclin binding, enzymes such as Cdc2 are regulated by multiple phosphorylations (12,13) (see Fig. 2). Cdc2 is negatively regulated via phosphorylation of Thr14 and Tyr15. (The amino acid positions in this section refer to the phosphorylation sites in human Cdc2.) These sites are phosphorylated by Wee1 (the negative regulator of Cdc2 identified genetically in *Schizosaccharomyces pombe* (see Fig. 1)) and by Wee1-like protein kinases (Mik1 in *Schizosaccharomyces pombe* and the membrane-bound Myt1 protein in vertebrates). Dephosphorylation of these inhibitory sites is carried out by Cdc25 proteins, genetically identified activators of Cdc2 that share distant similarity to protein **tyrosine phosphatases**. Finally,

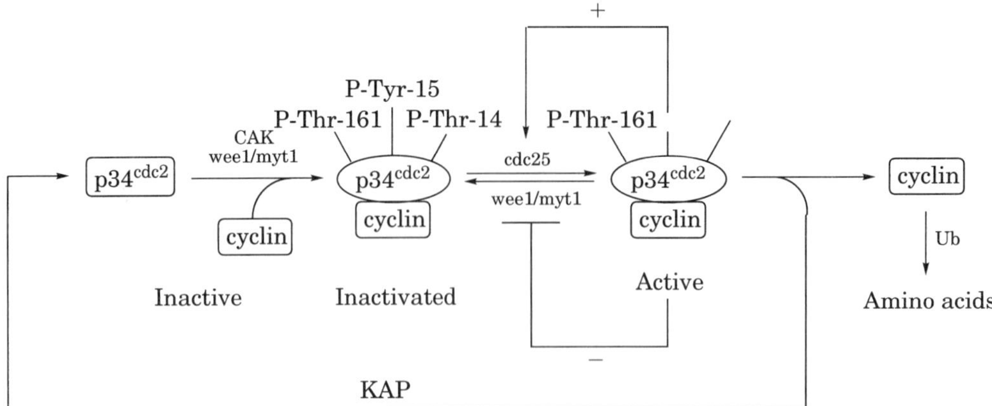

Figure 2. Activation of Cdc2. Monomeric Cdc2 is unphosphorylated and inactive. Binding to cyclin induces its phosphorylation on inhibitory sites (Thr14 and Tyr15) by Wee1-like protein kinases and its phosphorylation on an activating site (Thr161) by CAK. The transition into mitosis is accompanied by two feedback loops involving stimulation of Cdc25 (the phosphatase acting on Thr14 and Tyr15) and inhibition of Wee1-like enzymes, resulting in the abrupt and irreversible activation of Cdc2 and entry into mitosis. The degradation of cyclin by the ubiquitin system toward the end of mitosis leads to the inactivation of Cdc2 and its dephosphorylation, probably by a phosphatase called KAP.

activating phosphorylation of Cdc2 occurs on Thr161 within the so-called T-loop and is carried out by the Cdk-activating kinase (CAK). Dephosphorylation of Thr161 has not been as well studied. It occurs following cyclin degradation, possibly by a phosphatase termed KAP. Two feedback loops operate as cells enter mitosis and lead to the inhibition of Wee1-like enzymes and the stimulation of Cdc25, thus ensuring the abrupt and irreversible transition from G2 to mitosis. The roles of cyclin binding and activating phosphorylation have been studied using **X-ray crystallography** of Cdk2 (14–16), a close relative of Cdc2. Monomeric Cdk2 is inactive for two major reasons. First, residues that interact with the phosphates of ATP are out of position, resulting in the misalignment of the γ-phosphate of ATP for phosphoryl transfer. Second, the T-loop is positioned so that it would physically prevent protein substrates from gaining access to the site of catalysis. Binding to cyclin relieves this steric block and moves residues that interact with ATP into their proper positions. The cyclin A-Cdk2 complex is about 1% active. Full activation requires activating phosphorylation (on Thr160 in Cdk2), which alters the positions of some T-loop residues and creates an acidic patch for the binding of optimal substrates containing a key basic residue.

One of the irreversible ratchet steps in the cell cycle is the degradation of cyclins by the ubiquitin system (17). Ubiquitin is a 76-residue protein whose covalent attachment to proteins can target them for **proteolysis** by the **proteasome**, a huge multiproteinase unwinding and degrading machine. Ubiquitin is activated by its ATP-dependent covalent attachment to an enzyme called E1. E1 then transfers ubiquitin to one of many E2 enzymes. The E2s can ubiquitinate substrates, often with the help of an E3. The E3s may be the most diverse and interesting components of this system. Some E3s receive the covalently bound ubiquitin; others serve as matchmakers that bring together a substrate and the appropriate E2. For the mitotic cyclins, the E3 was first termed the cyclosome, though it is now generally termed the *anaphase promoting complex* (APC), an 8- to 12-subunit complex. The APC is the regulated component of the ubiquitin system for the degradation of the mitotic cyclins and serves as the target for checkpoint signals (see below) that can block cyclin degradation. The ubiquitin-dependent degradation of cyclins that act earlier in the cell cycle generally does not require the APC and has been best studied in *S. cerevisiae*.

RELATIVES: CDKS AND CYCLINS

Both Cdc2 and the mitotic cyclins are members of large families of related proteins, many of which function in cell cycle regulation. The Cdc2 relatives are termed Cdks (for cyclin-dependent kinases) and the cyclin relatives are designated by letter. The different cyclins accumulate at various times during the cell cycle and pair with a limited subset of the Cdks to execute their respective functions. The major cyclin–Cdk complexes and their functions are shown in Table 1. Complexes of D-type cyclins with Cdk4 and Cdk6 function in progression through G1 and are the first kinases to phosphorylate the cell-cycle inhibitor retinoblastoma protein (Rb). The cyclin E–Cdk2 complex stimulates the initiation of DNA synthesis, and the cyclin A–Cdk2 complex is generally thought to promote passage through S phase. Cyclin B partners with Cdc2 to promote the G2-to-mitosis transition. All of these complexes are regulated by phosphorylations on sites equivalent to those used in Cdc2. In addition to these enzymes, a large number of cyclin–Cdk complexes function in processes other than cell cycle regulation, including transcription and neuronal differentiation. In general, the regulatory subunits of these Cdks are stable proteins and are classified as cyclins only on the basis of sequence similarity.

Table 1. Functions of Cyclin–Cdk Complexes Involved in Cell Cycle Control

Catalytic Subunit	Major Cyclin Partner	Function
Cdc2	Cyclin B	G2 to M
Cdk2	Cyclin A	S
Cdk2	Cyclin E	G1 to S
Cdk4	Cyclin D	G1
Cdk6	Cyclin D	G1

CDK INHIBITORS

In addition to cyclin binding and phosphorylation, Cdks are also regulated by the binding of inhibitory proteins (18). These can be divided into two classes: the CIP/KIP family (consisting of $p21^{CIP1}$, $p27^{KIP1}$, and $p27^{KIP2}$) and the INK4 family (consisting of $p15^{INK4b}$, $p16^{INK4a}$, $p18^{INK4c}$, and $p19^{INK4d}$). The CIP/KIP proteins bind preferentially to the cyclin–Cdk complex and cause its inactivation. These proteins bind to numerous Cdks, but show a preference for Cdk2. In contrast, the INK4 proteins bind preferentially to monomeric Cdks (inhibiting the binding of cyclin) and are specific for Cdk4 and Cdk6. Cdk inhibitors play very important roles both in cell cycle progression (particularly in responses to **growth factors**) and during development. One of the best-studied roles for a Cdk inhibitor involves the regulation of the initiation of DNA synthesis in *S. cerevisiae* by Sic1p. Entry into S phase in *S. cerevisiae* requires the activity of complexes containing an S-phase cyclin (Clb5p or Clb6p) and the budding yeast Cdk, Cdc28p. These complexes are inhibited by Sic1p. Degradation of Sic1p by the ubiquitin system is triggered following the phosphorylation of Sic1p by Cdc28p bound to G1 cyclins, thereby freeing the Clb5p and Clb6p complexes to initiation DNA replication (19).

CHECKPOINTS

Among the key mechanisms for ensuring the orderly progression of cell cycle events are checkpoints, which monitor the completion of certain events and prevent the execution of subsequent events if the monitored event has not been completed (20,21). For example, checkpoints can block entry into mitosis if the DNA is damaged or has not been fully replicated, or they can block exit from mitosis if the spindle has not been properly assembled. In *S. cerevisiae* a checkpoint even monitors whether a bud has formed. One of the defining features of a checkpoint sensor is that it is not part of the machinery being sensed and that it is dispensable, though only for the short term. The checkpoint that provides the paradigm for all others involves the sensing of **DNA damage** and unreplicated DNA by Rad9p in *S. cerevisiae*. A number of yeast *cdc* mutants cause G2 arrests that are dependent on this checkpoint. For example, inactivation of *CDC9*, which encodes **DNA ligase**, leads to G2 arrest if *RAD9* is functional. In contrast, inactivation of *CDC9* in a *rad9* mutant does not lead to an immediate arrest. Instead, cells progress into mitosis and continue through a couple of cell cycles, until they arrest as microcolonies. The G2-arrested *RAD9* cells remain viable while they attempt to repair their DNA, whereas the *rad9* cells rapidly lose viability as they proceed through the checkpoint. Unperturbed wild-type cells grow normally in the absence of this checkpoint. The importance of checkpoints to a proper cell cycle and to human health is underscored by the **p53** protein. p53 is the most frequently mutated protein in human cancers and functions as part of a checkpoint responding to DNA damage by transcriptionally inducing the p21 Cdk inhibitor (22), thereby blocking cell cycle progression. The absence of this checkpoint leads to genomic instability (the accumulation of further mutations, loss of chromosomes, etc.) and a greatly increased risk of tumor formation.

BIBLIOGRAPHY

1. J. R. Pringle and L. H. Hartwell (1981) In *The Molecular Biology of the Yeast Saccharomyces* (J. Strathern, E. Jones, and J. Broach, eds.), Cold Spring Harbor Laboratory Press, Cold Spring Harbor, NY, pp. 97–142.
2. P. Nurse (1990) *Nature* **344**, 503–508.
3. Y. Masui and C. L. Markert (1971) *J. Exp. Zool.* **177**, 129–146.
4. T. Evans, E. T. Rosenthal, J. Youngblom, D. Distel, and T. Hunt (1983) *Cell* **33**, 389–396.
5. K. I. Swenson, K. M. Farrell, and J. V. Ruderman (1986) *Cell* **47**, 861–870.
6. W. G. Dunphy, L. Brizuela, D. Beach, and J. Newport (1988) *Cell* **54**, 423–431.
7. J. Gautier, C. Norbury, M. Lohka, P. Nurse, and J. Maller (1988) *Cell* **54**, 433–439.
8. J. C. Labbé, J. P. Capony, D. Caput, J. C. Cavadore, J. Derancourt, M. Kaghad, J. M. Lelias, A. Picard, and M. Dorée (1989) *EMBO J.* **8**, 3053–3058.
9. J. Gautier, J. Minshull, M. Lohka, M. Glotzer, T. Hunt, and J. L. Maller (1990) *Cell* **60**, 487–494.
10. M. Peter, J. Nakagawa, M. Dorée, J. C. Labbé, and E. A. Nigg (1990) *Cell* **61**, 591–602.
11. G. E. Ward and M. W. Kirschner (1990) *Cell* **61**, 561–577.
12. M. J. Solomon (1993) *Curr. Opin. Cell Biol.* **5**, 180–186.
13. M. J. Solomon and P. Kaldis (1998) In *Results and Problems in Cell Differentiation*, Vol. 22: *Cell Cycle Control* (M. Pagano, ed.), Springer, Heidelberg, pp. 79–109.
14. J. L. DeBondt, J. Rosenblatt, J. Jancarik, H. D. Jones, D. O. Morgan, and S.-H. Kim (1993) *Nature* **363**, 595–602.
15. P. D. Jeffrey, A. A. Russo, K. Polyak, E. Gibbs, J. Hurwitz, J. Massagué, and N. P. Pavletich (1995) *Nature* **376**, 313–320.
16. A. A. Russo, P. D. Jeffrey, and N. P. Pavletich (1996) *Nat. Struct. Biol.* **3**, 696–700.
17. R. W. King, R. J. Deshaies, J.-M. Peters, and M. W. Kirschner (1996) *Science* **274**, 1652–1659.
18. T. J. Soos, M. Park, H. Kiyokawa, and A. Koff (1998) In *Results and Problems in Cell Differentiation*, Vol. 22: *Cell Cycle Control* (M. Pagano, ed.), Springer, Heidelberg, pp. 111–131.
19. R. M. Feldman, C. C. Correll, K. B. Kaplan, and R. J. Deshaies (1997) *Cell* **91**, 221–230.
20. L. H. Hartwell and T. A. Weinert (1989) *Science* **246**, 629–634.
21. S. J. Elledge (1996) *Science* **274**, 1664–1672.
22. J. W. Harper, G. R. Adami, N. Wei, K. Keyomarsi, and S. J. Elledge (1993) *Cell* **75**, 805–816.

Suggestions for Further Reading

W. G. Dunphy, ed. (1997) *Methods in Enzymology*, Vol. 283: *Cell Cycle Control*, Academic Press, San Diego, CA.

D. O. Morgan (1997) Cyclin-dependent kinases: engines, clocks, and microprocessors. *Annu. Rev. Cell Dev. Biol.* **13**, 261–291.

A. Murray and T. Hunt (1993) *The Cell Cycle*, W. H. Freeman, New York.

M. Pagano, ed. (1998) *Results and Problems in Cell Differentiation*, Vol. 22: *Cell Cycle Control*, Springer, Heidelberg.

CELL DEATH

NICOLA MCCARTHY
GERARD EVAN

It is not initially obvious why cell death should be considered important. The death of single-celled organisms, such as **amoeba** or **bacteria**, has no apparent advantage for the individual cell. However, in multicellular organisms damage to an individual cell can have repercussions for the whole animal. It therefore makes biological sense to dispose of this cell and replace it with a healthy one. Cells become damaged in a variety of different ways, but for long-lived, multicellular animals, the greatest risk is from a damaged cell that has acquired an **oncogenic** mutation leading to unrestricted clonal growth. To minimize this risk, several pathways have evolved that limit the expansion of somatic cells, one of which is cell death.

Tissue size is restricted by several different methods, not all of which actually require the physical loss of the cell. The problem facing multicellular organisms is how to maintain the capacity to divide and replace damaged cells while minimizing the risk of mutation leading to enhanced growth potential. One way around this problem is to make tissues post-replicative, as occurs in the central nervous system (CNS). All the neurons within the CNS are produced during embryogenesis and are then maintained throughout the life span of the organism. For some tissues, however, such an approach is not practical due to persistent physical damage to the cells, such as occurs in the epithelial lining of the gut or skin. Here, cells are continuously produced by a select number of proliferating **stem cells**. The daughter cells they produce move upward, away from the basement membrane, terminally differentiate, and are eventually shed. Because these cells are continuously replaced, any damaged cell should be automatically eliminated. Tissue size is also restricted by the need for vascularization. If there is no blood supply, only a small proportion of cells can be maintained by the diffusion of solutes, which is one of the reasons why many solid tumors have central necrotic zones.

For tissues that require a significant number of cells in cycle, the risk of mutation is minimized by both **senescence** and cell death. Senescence is a pathway that is invoked once cells have undergone a specific number of doublings, causing them to arrest permanently. Cells in this senescent state are unable to reenter the cell cycle (1). Thus, a cell that does acquire a mutation allowing constant proliferation is limited in its capacity to divide, restricting clonal outgrowth [see **Senescence**]. Finally, cells can be triggered to die when no longer required, or when damaged. This "programmed cell death" is essential for the maintenance of tissue homeostasis and is invoked in disparate situations including DNA damage, absence of survival signals, and oncogene activation (2,3) [see **Programmed cell death** and **Apoptosis**]. Thus, multicellular organisms are protected against somatic mutation by a number of independent mechanisms that act in concert to limit the potential for **neoplastic transformation**.

THE MEANING OF CELL DEATH

The concept that cell death is fundamentally important for restricting cell population expansion took some time to be accepted for several reasons. Dead cells are not obvious in healthy tissues, whereas they are abundant in areas of damaged tissue, such as ischemic heart tissue resulting from myocardial infarction, allowing dead cells to be fundamentally associated with disease. Only during embryogenesis are dying cells seen in abundance, but these were generally regarded as cells deleted as a result of overproduction during development. In addition, death is seen as a bad outcome in human terms, a misconception that has allowed human anthropomorphic confusions to outweigh scientific evidence. More recently, however, several of these objections have been overruled, allowing cell death to become an accepted, essential daily process.

Early investigators of both invertebrate and vertebrate development observed that developmental cell deaths occurred in response to several biological cues and could be suppressed by inhibitors of both protein and RNA synthesis (4,5), suggesting a requirement for macromolecular synthesis (6,7). Moreover, they made the important connection that these cell deaths were an essential part of the developmental program of the organism concerned; hence the term "programmed cell death" (PCD) (8). It is now possible in less complex invertebrate models, such as the **nematode** *Caenorhabditis elegans*, to map both the fate of all individual cells within the developing organism and the genes that dictate these fates (9) [see **Programmed cell death**].

For vertebrates, the importance of cell death was not appreciated until the detailed characterization of the form of cell death termed **apoptosis** (10). The programmed cell deaths observed during **development** are identical to apoptotic cells in mammalian tissues, suggesting that a regulated form of cell death has been conserved throughout **evolution** (11,12). From these early observations, our understanding of apoptosis/PCD has grown to include a complex pathway that is both morphologically and biochemically distinct from classical necrosis or accidental cell death. **Necrosis** is a passive event, in which cells that become irreversibly damaged, and therefore useless, die (13,14). This form of cell death, which requires no input from the dying cell, occurs in cells subject to physical disruption or severe toxic stress. Conversely, apoptosis/PCD describes a pathway of events in which the cell is actively involved. Due to the active nature of apoptosis, which is triggered by many physiological and toxic stimuli, this form of cell death is sometimes referred to as "cell suicide." Overall, the importance of "active" cell death is underlined by the fact that death is the default state for all cells. Hence, all cells are programmed to die, unless signaled to survive (15).

The importance of cell death in the regulation of tissue homeostasis is paramount. The number of proliferating cells determines the cell population number, as does the number of differentiating cells and the number of dying cells. Research over the last 20 years has graphically shown how cells that have mutations which disrupt their capacity to undergo apoptosis in response to physiological stimuli are involved in the etiology of several diseases, including Alzheimer's, AIDS, and cancer (16).

Aside from the three pathways of death described above, the "death" of a cell does not always result in the loss of viability or cellular function. Within a cell population, few cells are actually proliferating, many are in a resting state outside of the **cell cycle**, known as $\mathbf{G_0}$ phase. Certain cells can reenter the cell cycle given the appropriate stimuli, but cells that enter a permanent G_0 state do not respond to these proliferation signals. Cells in the latter state are termed senescent and can complete all normal functions except division (1) [see **Senescence**]. Cells that escape the confines of senescence are able to proliferate continuously and, more important, the mutations that lead to the evasion of senescence are common and perhaps mandatory in tumor cells (17).

Thus, the description "cell death" encapsulates several diverse processes. This can involve the physical loss of the cells through apoptosis/PCD or, in some circumstances, necrosis. Alternatively, cell death can refer to a genetic death, where cells no longer retain the ability to replicate, but continue to survive. Which method of cell death is induced depends on several factors, including the cell's internal environment, its external environment, and its developmental history. It is not yet clear whether the genes that regulate apoptosis/PCD and senescence overlap. For example, genes such as **p53** and **retinoblastoma** appear to be required for the regulation of apoptosis and senescence, but it is not clear whether these genes perform similar or different roles in each pathway. Indeed, cells that have evaded the first signal to senesce enter *crisis*, which is defined as the point at which the culture exhibits both apoptosis and proliferation. This suggests that there must be some overlap in the control of different types of cell death, and this may enable one type of cell death to be employed when another either is not suitable or is not able to be induced.

BIBLIOGRAPHY

1. C. A. Afshari and J. C. Barrett (1996) Molecualr genetics of in vivo cellular senescence. In *Cellular Aging and Cell Death* (N. J. Holbrook, G. R. Martin, and R. A. Lockshin, eds.), Wiley-Liss, New York, pp. 109–122.
2. A. J. Hale, C. A. Smith, L. C. Sutherland, V. E. Stoneman, V. Longthorne, A. C. Culhane, and G. T. Williams (1996) Apoptosis: Molecular regulation of cell death. *Eur. J. Biochem.* **237**, 884.
3. S. J. Martin, and D. R. Green (1995) Apoptosis and cancer: The failure of controls on cell death and cell survival. *Crit. Rev. Oncol. Hematol.* **18**, 137–153.
4. J. R. Tata (1966) Requirement for RNA and protein synthesis for induced regression of the tadpole tail in organ culture. *Dev. Biol.* **13**, 77–94.
5. R. A. Lockshin (1969) Programmed cell death. Activation of lysis by a mechanism involving the synthesis of protein. *J. Insect Physiol.* **15**, 1505–1516.
6. A. Glucksmann (1965) Cell death in normal development. *Arch. Biol. Liege* **76**, 419–437.
7. P. G. Clarke and S. Clarke (1996) Nineteenth century research on naturally occurring cell death and related phenomena. *Anat. Embryol. Berl.* **193**, 81–99.
8. R. A. Lockshin and C. M. Williams (1964) Programmed cell death. II. Endocrine potentiation of the breakdown of the intersegmental muscles of silkmoths. *J. Insect Physiol.*, **10**, 643.
9. R. E. Ellis, J. Y. Yuan, and H. R. Horvitz (1991) Mechanisms and functions of cell death. *Ann. Rev. Cell Biol.* **7**, 663–698.
10. A. H. Wyllie, J. F. Kerr, and A. R. Currie (1972) Cellular events in the adrenal cortex following ACTH deprivation. *J. Pathol.* **106**, ix.
11. M. D. Jacobson, M. Weil, and M. C. Raff (1997) Programmed cell death in animal development. *Cell* **88**, 347–354.
12. A. Fraser, N. McCarthy, and G. I. Evan (1996) Biochemistry of cell-death. *Curr. Opin. Neurobiol.* **6**, 71–80.
13. B. F. Trump, J. M. Valigorsky, J. H. Dees, W. J. Mergner, K. M. Kim, R. T. Jones, R. E. Pendergrass, J. Garbus, and R. A. Cowley (1973) Cellular change in human disease. A new method of pathological analysis. *Hum. Pathol.* **4**, 89–109.
14. A. H. Wyllie, J. F. Kerr, and A. R. Currie (1980) Cell death: The significance of apoptosis. *Int. Rev. Cytol.* **68**, 251–306.
15. M. C. Raff (1992) Social controls on cell survival and cell death. *Nature* **356**, 397–400.
16. C. B. Thompson (1995) Apoptosis in the pathogenesis and treatment of disease. *Science* **267**, 1456–1462.
17. W. E. Wright and J. W. Shay (1996) Mechanisms of escaping senescence in human diploid cells. In *Celluar Aging and Cell Death* (N. J. Holbrook, G. R. Martin, and R. A. Lockshin, eds.), Wiley-Liss, New York, pp. 153–166.

Suggestions for Further Reading

N. J. Holbrook, G. R. Martin, and R. A. Lockshin (1996) *Cellular Aging and Cell Death*. Vol. 16, Modern Cell Biology (J. B. Harford, series ed.), Wiley-Liss, New York. A good introductory text to both senescence and cell death.

M. Sluyser (1996) *Apoptosis in Normal Cancer and Development*, Taylor & Francis, London. This book covers all aspects of apoptosis, giving both a physical discription of the process and covering all the molecular aspects of the pathway.

CELL FUSION, CELL HYBRIDS

Y. OKADA

Fusions of both the external and intracellular **membranes** of cells are important for **differentiation** and **development**. Moreover, enveloped **viruses** infect cells via fusion of their envelopes with cell or **endosome** membranes.

Cell fusion is the process of fusion of the membranes of two or more cells and results in the formation of cells with multiple **nuclei**. It occurs at various stages of the natural development of organisms, such as in the first step of fertilization of an **oocyte** with **sperm** and in myotube formation by fusion of myoblasts during differentiation of skeletal muscles. Artificial cell fusion can be induced by the addition of a high concentration of **Sendai virus**, an enveloped virus of the paramyxovirus group, as was demonstrated in 1957–63 (1–3). This finding coincided with cell biology's beginning focus on the culture of **somatic cells** of birds and mammals. In 1961, Barski et al. (4) reported the appearance of hybrid cells after a few months of mixed culture of two different mouse cancer cell lines. These hybrid cells had a single nucleus containing **chromosomes** from both parent cell lines, and their appearance was considered to be due to the spontaneous fusion of cells of the two cell lines.

Sendai virus proved to be an important development in this field because it has some useful characteristics for the fusion of somatic cells:

1. Its targets are sialoglycoproteins and sialolipids, which are present in the cell membranes of almost all mammalian and fowl cells. Consequently, it can induce the fusion of a wide range of cells (3).

2. Its cell fusion activity is not affected by procedures that inactivate the viral **genome**, such as exposure to ultraviolet (UV) light. Thus, fused cells could be prepared under conditions inhibiting virus growth, using a UV-inactivated virus.
3. There is no species specificity in its fusion of cells, so interspecific **heterokaryons** can be induced easily (5,6), unlike in the case of fertilization.
4. The frequency of virus-induced hybrid formation is at least 1000 times greater than that of spontaneous hybridization.

Littlefield reported in 1964 (7) further progress in methods for the selection of hybrid cell clones, by fusing two different mutants defective in the **salvage pathway** for nucleotide biosynthesis and culturing them in a medium containing **aminopterin**, which inhibits *de novo* nucleotide synthesis. One of the mutant cell lines that were fused lacked **thymidine kinase**, the other **hypoxanthine-guanine phosphoribosyl transferase**. The only cells that could grow from the mixed culture were fused cells that had acquired mutual **complementation** of the two mutant defects. Based on these new techniques, the field of *somatic cell genetics* was established in the 1960s.

That the chemical fusogen **polyethylene glycol** is effective for the fusion of **protoplasts** of **plant** cells was first reported in 1974 (8). With this fusogen, hybrid plants, such as the "pomato," can be prepared in cultures of somatic hybrid cells. Moreover, *electroporation* (see **Tranfection**) was found to be useful as a physical method for cell fusion in the 1980s (10,11). Cell fusion by these two methods does not require viral receptors, so they have made possible the fusion of cells from all kinds of organisms.

In the early stage of somatic cell genetics in the 1960s and 70s, many reports provided important information on some basic phenomena in cell biology, permitting a focus on molecular biology in the next stage of the 1980s and 90s. The main findings were as follows:

1. Immediately after multinucleated cell formation by artificial cell fusion, some nuclear proteins derived from the parents are rapidly transferred and mixed between the various nuclei. This tends to induce synchronization of the stage of **DNA replication** of the nuclei after the fusion of randomly growing cells. The degree of this synchronization is greatest in cells with only two nuclei and decreases sharply with an increasing number of nuclei in polykaryocytes (12,13). This may be why almost all randomly isolated hybrid cells have in their nucleus one set of chromosomes derived from each parent. As a special case, the fusion of cells in **mitosis** with cells in interphase causes the rapid dissolution of the nuclear membrane of the interphase cells, followed by the condensation of its chromosomes, a process named *chromosomal pulverization* (14) or *premature chromosome condensation* (15). Reactivation of dormant nuclei from chick erythrocytes was demonstrated upon their fusion with cultured mammalian cells (16,17). In the case of cell **neoplastic transformation** by tumor viruses, induction of Simian Virus 40 (**SV40**) production was observed after the fusion of SV40-transformed nonproducer hamster cells with monkey cells, which are a permissive host for SV40 (18). Later, a similar observation was reported on the fusion of **Rous sarcoma virus** (RSV)-transformed rat cells with chicken cells (a permissive host of RSV) (19). In the case of cell differentiation, the distinction between luxury and household functions of cells was observed by the formation of hybrids of cells with different phenotypes (20). The findings were important for choosing the correct combination of cells for hybridization of differentiated cells. In 1974, Köhler and Milstein (21) reported that **monoclonal antibodies** could be prepared from hybrid cells (**hybridomas**) of **B cells** and **myelomas** (tumor cells obtained from B cells). This method of preparing monoclonal antibodies is now a major technique in molecular, cell, and developmental biology, and in medicine.
2. Weiss and Green (22) observed that the chromosomal balance is unstable in interspecific man/mouse hybrids and human chromosomes disappear randomly during serial passage in culture. Based on this finding, mapping of the human chromosomes became possible (23) and was followed by the technique of direct ***in situ* hybridization** of the chromosomes with **complementary DNA**.
3. In 1972, Bootsma and colleagues (24) demonstrated that genetic complementation groups in a hereditary disease, *xeroderma pigmentosum*, could be classified by the cell fusion technique. This was the first successful genetic analysis of a human hereditary disease.

MECHANISM OF CELL FUSION

Artificial cell fusion using the various fusogens mentioned above has been established as a routine laboratory method. These fusogens, viral, chemical, and physical, have different modes of action on cell membranes, but induce similar changes in the cell membrane for fusion of the **lipid** bilayers. In general, **glycoproteins** are distributed on cell surfaces, with their **hydrophobic** domains embedded in the lipid bilayer of the cell membrane, their **hydrophilic** domains exposed to the outside, while their intracytoplasmic domains are associated with the **cytoskeletal** system. **Water** molecules are also associated with the outside of cell membranes. These structures on the exteriors of cells inhibit the close contact of lipid bilayers of neighboring cell membranes, which is essential for cell fusion. Close contact of membranes requires on the cell membrane surface the transient appearance of areas from which glycoproteins are excluded. All the fusogens can induce such areas.

Sendai virus has two glycoproteins: HN with **receptor** binding and destroying activities, and F with fusogenic activity that is essential for fusion. The first step in cell fusion is the attachment of the virus to cell surfaces and agglutination of the cells, which is produced by HN activity. The second step is insertion of the fusogenic domain of F into the lipid bilayer. This domain is located at the N-terminus of F and consists of 15 relatively hydrophobic residues (25); it has the unique characteristic of trapping cholesterol molecules in its tertiary structure at 37°C (26). The amino acid sequence of this domain is well conserved in paramyxoviruses.

Simultaneous removal of cholesterol molecules from multiple sites in the cell membrane by the attachment of several hundred virus particles perturbs the cell membrane and causes breakdown of the normal barrier to ions. At this stage, calcium ions from the medium promptly penetrate the cell and induce changes in cell structures, such as separating the connection between the cytoplasmic domains of membrane proteins and the cytoskeletal system (see **Calcium signaling**). Macromolecules can also **diffuse** through the cell membrane, and membrane fluidity increases, so that membrane proteins can move more freely in the lipid bilayer. This results in clustering of intramembrane domains of the proteins (demonstrable as cold-induced clustering) and the appearance of areas with no membrane proteins. The appearance of these areas is followed by close attachment of the lipid bilayers of neighboring cells, due to strong hydrophobic interactions, and cell fusion (27).

Polyethylene glycol is reported to induce clustering of membrane proteins similar to that induced by Sendai virus (28). Electroporation by electric pulses causes pores to be formed in cell membranes that may allow Ca^{2+} ions into the cytoplasm, as is also caused by Sendai virus. Thus, the fundamental mechanisms of these three fusogens appear to be similar.

It is known that calcium ions (29) and an energy (ATP) supply (2) are required for cell fusion. In the absence of either one, cells rapidly degenerate and fusion is greatly decreased. Evidence suggests that calcium ions associate directly with phospholipid molecules to normalize the perturbed membrane structure and promote a connection between the two lipid bilayers, in addition to acting in the cytoplasm as mentioned above. Why does cell fusion require an energy supply? In simple terms, cell fusion would be expected to produce a favorable decrease in **free energy**, by changing from a number of small vesicles to one large one, in which case an energy supply would not be necessary. Energy may be required for the removal of excess calcium ions introduced into the cytoplasm during cell fusion, but not for the membrane fusion itself. On culture of fused cells, the calcium ions are rapidly sequestered in **organelles** of the cells, and their level in the cytoplasm returns to normal. If the supply of energy is delayed, the cells degenerate.

SUPPLEMENT

1. In the case of Sendai virus, cell-to-cell fusion is also possible as a result of viral envelope fusion itself, if one virus envelope fuses with the membranes of two different cells. But envelope fusion is slower than that induced by a high concentration of the virus and is observed to occur after completion of cell-to-cell fusion. Finally, as a result of viral envelope fusion, many viral glycoproteins are integrated into the fused cell surfaces, which are excluded from the cell surface by internalization by coated vesicles in culture (30) (see **Clathrin**).

Polykaryocytes (syncytia) are often observed in pathological tissues infected by an enveloped virus, such as paramyxoviruses, **retroviruses** and the **Herpes virus**. This polykaryocyte formation may occur by cell fusion via viral envelope fusion. Newly synthesized viral glycoproteins are distributed massively on the surfaces of infected cells, which consequently are quite similar to the viral envelope, and they may fuse with neighboring noninfected cells, as in viral envelope fusion.

2. Enveloped viruses cause infection by fusion of their envelopes either (a) with cell membranes at neutral pH or (b) with endosome membranes at acidic pH after their internalization from the cell surface (see **Virus infection**). Infections by paramyxoviruses, retroviruses, and the Herpes virus are of the first type and induce syncytia formation *in vivo* and *in vitro*. **Influenza virus** is of the second type and does not induce syncytia *in vivo*, but the viral envelope can fuse with cell membranes under acidic conditions *in vitro*. This is because the influenza fusogenic glycoprotein (HA, or hemagglutinin) is not functional at neutral pH, but becomes functional under acidic conditions by a conformational change, forming trimers (31).

The fusogenic glycoproteins of various enveloped viruses differ, but all of them contain a hydrophobic domain that can interact with the lipid bilayer of cellular membranes. In some cases, the glycoproteins of the viruses are known to be synthesized in an inactive form in which their hydrophobic domain is hidden and then activated by **proteolytic** cleavage exposing this domain. This was first demonstrated by Homma (32) with the Sendai virus. In this virus, the F glycoprotein is synthesized as an inactive form F_0 and is then cleaved to F_1 and F_2, with the fusogenic domain being exposed at the N-terminus of F_1. The glycoprotein (gp160) of human immunodeficiency virus (**HIV**) a retrovirus, is cleaved to gp120 and gp41, and the fusogenic domain is exposed at the N-terminus of gp41 (33). The HA glycoprotein of influenza virus is also activated by the cleavage of inactive HA_0 to HA_1 and HA_2, with the hydrophobic domain being exposed at the N-terminus of HA_2 (34) and becoming functional by trimer formation under acidic conditions.

APPLICATION OF FUSOGENIC REACTIONS TO CELL ENGINEERING

Various biotechniques for the reconstitution of cells or introduction of macromolecules into cells have been developed using the unique characteristics of the interactions of fusogens with cell membranes. Introduction of macromolecules such as DNA, RNA, and proteins from the medium into the cytoplasm has become possible by perturbation of the cell membrane. Electroporation is usually used for that purpose (35), but treatment of the cells with Sendai virus can also be used (36).

Another technique for the introduction of macromolecules is based on the mechanism of infection of Sendai virus with cell membranes. In 1979, Uchida et al. (37) reported on the reassembly of viral envelopes as artificial **vesicles**, with HN and F glycoproteins embedded in their surface and any macromolecules present trapped inside. These pseudovirus particles containing the macromolecules instead of viral nucleocapsids will introduce those macromolecules into a cell that they "infect." This technique is especially useful *in vivo*. Subsequently, modifications have been made to simplify the preparation procedure, using spontaneous fusion of the UV-inactivated virus with simple, artificial **liposomes** containing the desired macromolecules. Using such a preparation, the human **hepatitis B virus** genome has been introduced into rat liver cells *in vivo* to induce hepatitis (38). This technique is useful for drug delivery *in vivo* and gene therapy (39).

Reconstruction of cells was first reported by Veomett et al. in 1974 (40). On incubation with **cytochalasin**, cells could be separated into *nucleoplasts* (enclosed by cell membranes but lacking cytoplasm) and *cytoplasts* (without a nucleus). Viable cells could be reconstituted by the fusion of nucleoplasts with cytoplasts. Heterologous combinations could also be constituted. This technique has been expanded to the preparation of cloned animals. Introduction of nuclei from somatic cells at the early stages after cleavage of fertilized eggs into enucleated eggs, by Sendai virus-mediated fusion, is reported to result in a high frequency of development of animals (41).

BIBLIOGRAPHY

1. Y. Okada, T. Suzuki, and Y. Hosaka (1957) *Med. J. Osaka Univ.* **7**, 709–717.
2. Y. Okada (1962) *Exp. Cell Res.* **26**, 98–128.
3. Y. Okada and J. Tadokoro (1963) *Exp. Cell Res.* **32**, 417–430.
4. G. Barski, S. Sorieul, and F. Cornfert (1961) *J. Natl. Cancer Inst.* **26**, 1269–1277.
5. H. Harris and J. F. Watkins (1965) *Nature* **205**, 640–646.
6. Y. Okada and F. Murayama (1965) *Exp. Cell Res.* **40**, 154–158.
7. J. Littlefield (1966) *Exp. Cell Res.* **41**, 190–196.
8. K. N. Kao and M. R. Michayluk (1974) *Planta* **115**, 355–367.
9. G. Melchers, M. D. Sacristan, and A. A. Holder (1978) *Carlsberg Res. Commun.* **43**, 203–218.
10. M. Senda, J. Takeda, S. Abe, and T. Nakamura (1979) *Plant Cell Physiol.* **20**, 1441–1443.
11. U. Zimmermann and J. Vienken (1982) *J. Membrane Biol.* **67**, 165–182.
12. T. Yamanaka and Y. Okada (1966) *Biken's J.* **9**, 159–175.
13. T. Yamanaka and Y. Okada (1968) *Exp. Cell Res.* **49**, 461–469.
14. H. Kato and A. A. Sanderberg (1968) *J. Nat. Cancer Inst.* **41**, 1117–1123.
15. P. N. Rao and R. T. Johnson (1972) *J. Cell Sci.* **10**, 495–513.
16. H. Harris (1965) *Nature* **206**, 583–588.
17. N. R. Ringertz (1974) In *Somatic Cell Hybridization* (R. L. Davidson and F. de la Cruz, eds.), Raven Press, New York, pp. 239–264.
18. P. Gerver (1966) *Virology* **28**, 501–509.
19. J. Svoboda, O. Machala, and I. Holzanek (1967) *Acta Virol.* **13**, 155–157.
20. R. L. Davidson and K. Yamamoto (1968) *Proc. Nat. Acad. Sci. USA* **60**, 894–901.
21. G. Köhler and C. Milstein (1975) *Nature* **256**, 495–497.
22. M. C. Weiss and H. Green (1967) *Proc. Nat. Acad. Sci. USA* **58**, 1104–1111.
23. R. J. Klebe, T. Chen, and F. H. Ruddle (1970) *Nat. Acad. Sci. USA* **66**, 1220–1227.
24. E. A. de Weerd-Kastelein, W. Keijzer, and D. Bootsma (1972) *Nature New Biol.* **238**, 80–83.
25. M. J. Gething, J. M. White, and M. D. Waterfield (1978) *Proc. Nat. Acad. Sci USA* **75**, 2737–2740.
26. K. Asano and A. Asano (1985) *Biochem. Int.* **10**, 115–122.
27. J. Kim and Y. Okada (1981) *Exp. Cell Res.* **132**, 125–136.
28. D. S. Roos, J. M. Robinson, and R. L. Davidson (1983) *J. Cell Biol.* **97**, 909–917.
29. Y. Okada and F. Murayama (1966) *Exp. Cell Res.* **44**, 527–551.
30. J. Kim and Y. Okada (1982) *Exp. Cell Res.* **140**, 127–136.
31. I. A. Wilson, J. J. Skehel, and D. C. Wiley (1981) *Nature (Lond.)* **289**, 366–373.
32. M. Homma (1971) *J. Virol.* **8**, 619–629.
33. J. M. McCune et al. (1988) *Cell* **53**, 55–67.
34. S. G. Lazarowitz and P. W. Choppin (1975) *Virology* **68**, 440–454.
35. U. Zimmermann, J. Vienken, and G. Pilwat (1980) *Bioelectrochem. Vioenerg.* **7**, 553–574.
36. K. Tanaka, M. Sekiguchi, and Y. Okada (1975) *Proc. Nat. Acad. Sci. USA* **72**, 4071–4075.
37. T. Uchida et al. (1979) *Biochem. Biophys. Res. Commun.* **87**, 371–379.
38. K. Kato et al. (1991) *J. Biol. Chem.* **266**, 3361–3364.
39. V. J. Dzau, M. J. Mann, R. Morishita, and Y. Kaneda (1996) *Pro. Nat. Acad. Sci. USA* **93**, 11421–11425.
40. G. Veomett, D. M. Prescott, J. Shay, and K. R. Porter (1974) *Proc. Nat. Acad. Sci. USA* **71**, 1999–2002.
41. J. McGrath and D. Solter (1983) *Science* **220**, 1300–1302.

Suggestions for Further Reading

N. Düzgunes, ed. (1993) *Membrane Fusion Technique*, Part B, Methods in Enzymology, Vol. 221, Academic Press.

N. Düzgunes and F. Bronner, eds. (1988) *Membrane Fusion in Fertilization, Cellular Transport, and Viral Infection*, Current Topics in Membranes and Transport, Vol. 32, Academic Press.

J. Wilschut and D. Hoekstra, eds. (1991) *Membrane Fusion*, Marcel Dekker, Inc.

CELL JUNCTIONS

D. GARROD

A cell junction is a cell surface structure observed by **electron microscopy** that mediates cellular interactions. Cell junctions are of two types, those that mediate cell–cell interactions and those that mediate cell–substratum interactions. In the former category, the cells of vertebrates possess four principal types, all of which are represented at the lateral surfaces of simple epithelial cells, such as those that line the intestine and kidney tubules (Fig. 1). The four principal types of cell–cell junction are (i) the **tight junction** or zonula occludens (pl. zonulae occludentes); (ii) the **intermediate junction** or zonula adherens (pl. zonulae adherentes); (iii) the **desmosome** or macula adherens (pl. maculae adherentes); (iv) the **gap junction** or nexus. The principal type of cell–substratum junction is the **hemidesmosome**, which is present in contact with the basement membrane in the basal cells of stratified epithelia, such as epidermis, in some simple epithelial cells, especially amnion, and mammary epithelium. Cells that are well spread in culture exhibit a second type of cell–substratum junction, the focal adhesion or **focal contact** (Fig. 1). This is a region where the lower cell surface is most closely apposed to the culture substratum (separation 10 to 15 nm). It is not clear whether the focal contact has a strict equivalent *in vivo*, but its study has made an extensive and important contribution to the understanding of cell adhesion. The basic structures of the principal types of cell junctions were elucidated by electron microscopy in the 1960s (1–3). Since then, a substantial amount of detailed knowledge of the molecular composition of each junctional type has been revealed.

The intermediate junction, the desmosome, the hemidesmosome, and the focal contact are primarily adhesive junctions that bind cells to each other or to their substratum. Each represents a cell-surface site at which cell adhesion molecules (see **Cell surface adhesion receptors**) are concentrated. In desmosomes and intermediate junctions, the adhesion molecules are specific types of **cadherins** (4,5); in hemidesmosomes and focal contacts, they are **integrins** (6). These

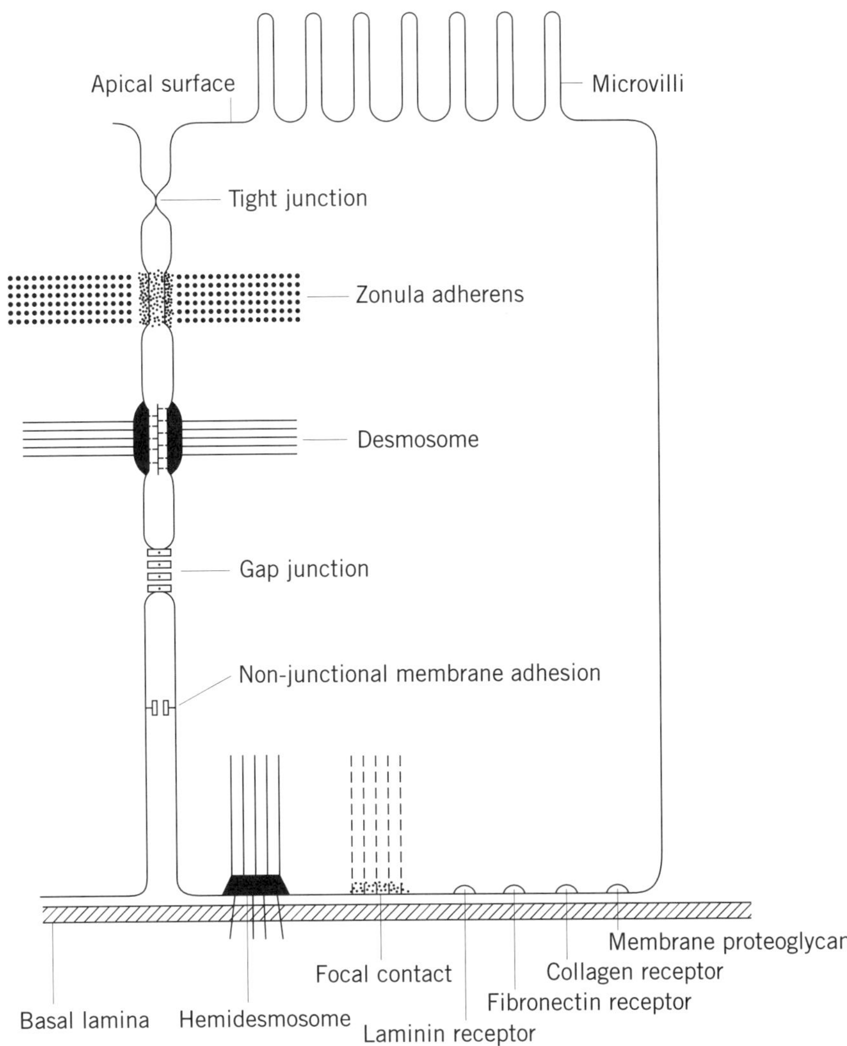

Figure 1. Generalized simple epithelial cell showing cell junctions and other adhesion mechanisms. The lateral surface mediates cell–cell interactions via adhesive junctions, the zonula adherens (intermediate junction) and the desmosome, and communicating gap junctions that allow small molecules to be exchanged between the cytoplasm of adjacent cells. The tight junction, or zonula occludens, restricts and regulates paracellular permeability. The basal surface mediates adhesion to the basement membrane. Hemidesmosomes are present in some simple epithelia but are most numerous in epidermis. Integrin and non-integrin receptors for several matrix components are present here. When cells are cultured, many of these become localized in focal contacts, small areas where adhesion to the substratum is strongest. (From *Journal of Cell Science*, with permission of the Company of Biologists.)

junctions also represent cell-surface attachment points for elements of the **cytoskeleton**. In desmosomes and hemidesmosomes, the associated cytoskeletal elements are **intermediate filaments**, also known as tonofilaments. These are important in providing structural continuity throughout a tissue. This continuity is disrupted in certain **autoimmune** and genetic diseases of desmosomes (eg, pemphigus) (7), hemidesmosomes (eg, bullous pemphigoid, junctional epidermolysis bullosa) (8), or intermediate filaments (eg, epidermolysis bullosa simplex) (9). Such diseases result in structural disruption of the epidermis and/or mucosal tissues, with effects ranging from mild to lethal. In intermediate junctions and focal contacts, the cytoskeletal elements are **actin** filaments, also known as **microfilaments**. The actin filaments associated with focal contacts are organized into large bundles, called stress fibers, of several micrometers in length. During animal development, morphogenetic events such as gastrulation and neural tube rolling involve gross changes in cell shape that appear to be caused by contraction of submembranous microfilament rings associated with zonula adherens-type junctions (10,11). More limited contractile activity of such rings may be involved in regulation of the permeability of tight junctions, with which zonulae adherentes are frequently closely associated in the simple epithelial junctional complex, a region where junctions are concentrated at the extreme apicolateral interface between the cells.

Transduction of signals across the plasma membrane is a vital function in the regulation of normal cell behavior in both developing and adult organisms. Many adhesion receptors have now been shown to transduce signals, either from inside the cell to outside regulating the function of the receptors themselves, or from outside to inside regulating signalling pathways and gene expression in response to adhesive stimuli. Such **signal transduction** is an additional important function of adhesive junctions (12).

The principal function of the tight junctions is to occlude the paracellular channels of epithelia, thereby restricting intercellular leakage of molecules between the distinct biological compartments separated by the epithelium (13). Tight junctions are also important in forming tight seals between some endothelial cells, such as those of the blood–brain barrier. Their sealing properties arise because they are zonular in nature, encircling the entire apicolateral margins of simple epithelial cells, and because there is direct contact between the plasma membranes of adjacent cells. Tight junctions are also important in maintaining the composition of the different membrane domains of epithelial cells, because they prevent **diffusion** of molecules within the outer leaflet of the plasma membrane, thereby restricting them to either the apical or the basolateral domain (14). This property is sometimes referred to as the "fence" function of tight junctions, and their occluding properties are described as a "gate" function.

Gap junctions, so-called because they exhibit a regular plasma membrane separation of 2 nm, are punctate membrane sites that provide **hydrophilic** channels for direct cell-to-cell communication (15). They permit diffusion of molecules of less that 1000 daltons between cells. These ubiquitous junctions are important in both excitable and nonexcitable tissues. In the former they facilitate direct transmission of electrical impulses, while in the latter they mediate a phenomenon called metabolic cooperation, and they may also transmit signaling molecules such as **calcium** and **inositol phosphates** between cells. Their function is essential in embryonic development (16) and in the normal functioning of adult tissues—for example, in coordination of the contraction of cardiac muscle and smooth muscle in the uterus during parturition (17,18).

Gap junctions appear to be widespread in the animal kingdom, being present even in primitive organisms such as *Hydra*. Intermediate junctions have also been clearly demonstrated in invertebrates (eg, in *Drosophila*). The occurrence of desmosomes and tight junctions in invertebrates is much less clear. Desmosome-like structures are clearly present in insects—for example, between the apposed epithelia of the upper and lower surface of the wing blade in *Drosophila*. However, these desmosome-like structures appear to be associated with **microtubules** rather than intermediate filaments, and their adhesion may be mediated by integrins rather than cadherins. Tight junctions are absent from insects, which instead possess septate junctions, structures that have a ladder-like ultrastructural appearance between adjacent plasma membranes and which, like tight junctions, appear to restrict paracellular permeability (19).

BIBLIOGRAPHY

1. M. G. Farquhar and G. E. Palade (1963) *J. Cell Biol.* **17**, 375–412.
2. D. E. Kelly (1966) *J. Cell Biol.* **28**, 51–72.
3. J-P. Revel and M. Karnovsky (1967) *J. Cell Biol.* **33**, 7–12.
4. J. L. Holton et al. (1990) *J. Cell Sci.* **97**, 239–246.
5. P. J. Koch et al. (1990) *Eur. J. Cell Biol.* **53**, 1–12.
6. A. Sonnenberg et al. (1991) *J. Cell Biol.* **113**, 907–917.
7. J. R. Stanley (1995) In "Cell adhesion and human disease." *Ciba Found. Symp.* **189**.
8. R. E. Burgeson and A. M. Christiano (1997) *Curr. Opin. Cell Biol.* **9**, 651–658.
9. W. H. I. McLean and E. B. Lane (1995) *Curr. Opin. Cell Biol.* **7**, 118–125.
10. T. E. Schroeder (1970) *J. Embryol. Exp. Morphol.* **23**, 427–462.
11. P. Karfunkel (1971) *Dev. Biol.* **25**, 30–56.
12. C. Rosales et al. (1995) *Biochem. Biophy. Acta.* **1242**, 77–98.
13. J. L. Madara and K. Darmsathaphorn (1985) *J. Cell Biol.* **101**, 2124–2133.
14. K. Simons (1990) In *Morphoregulatory Molecules* (G. M. Edelman, B. A. Cunningham, and J.-P. Thiery, eds.), Wiley, New York, pp. 341–356.
15. N. B. Gilula, O. R. Reeves, and A. Steinbach (1972) *Nature* **235**, 262–265.
16. A. E. Warner, S. C. Guthrie, and N. B. Gilula (1984) *Nature* **311**, 127–131.
17. W. C. Cole and R. E. Garfield (1985) In *Gap Junctions* (M. V. L. Bennett and D. C. Spray, eds.), Cold Spring Harbor Laboratory Press, Cold Spring Harbor, NY, pp. 215–230.
18. E. C. Beyer, D. L. Paul, and D. A. Goodenough (1987) *J. Cell Biol.* **195**, 2621–2629.
19. P. J. Bryant (1994) In *Molecular Mechanisms of Epithelial Cell Junctions: From Development to Disease* (S. Citi, ed.) R. G. Landes, Austin, TX, pp. 1–21.

Suggestions for Further Reading

D. R. Garrod and J. E. Collins (1992) Intercellular Junctions and Cell Adhesion in Epithelial Cells. In "Epithelial Organization and Development (T. P. Fleming, ed.), Chapman and Hall, London, pp. 1–52. (A source of reference to the early literature.)

D. R. Garrod, M. A. J. Chidgey, and A. J. North (eds.) (1998) *Adhesive Interactions of Cells*, JAI Press, Greenwich, CT. (Reviews on desmosomes by the editors, tight junctions by S. Citi, and adherens junctions and focal contacts by A. Ben-Ze'ev, as well as other reviews in cell adhesion.)

D. A. Goodenough, J. A. Goliger, and D. L. Paul (1996) *Annu. Rev. Biochem.* **65**, 475–502. (A thorough review of gap junctions.)

L. Barradori and A. Sonnenberg (1996). *Curr. Opin. Cell Biol.* **8**, 647–656. (An introduction to the literature on hemidesmosomes.)

CELL LINE

R. I. FRESHNEY

When a primary cell culture (see **Tissue culture**) is subcultured, it becomes a cell line. This implies that multiple lineages of cells, not necessarily distinct from each other, coexist in the culture (1). Serial subculture, while maintaining these parallel lineages, will tend to show convergence towards whatever common **phenotype** is best adapted to the culture conditions being employed, as the cell type with the greatest proliferative ability will predominate. The cell line may be finite and die out after a fixed number of population doublings, or become a continuous cell line (see **Immortalization**) (Fig. 1). Cell lines, particularly continuous cell lines, can become a valuable resource, particularly if preserved in liquid nitrogen after appropriate characterization and validation. Many continuous cell lines are currently available, with wide-ranging properties, including **drug resistance** markers, inducibility for specific **enzymes**, estrogen sensitivity, cornification, blood vessel formation, **hemoglobin** synthesis, myogenesis, and **transfection** susceptibility (Table 1). Finite cell lines with distinct phenotypic properties are also becoming available through the development of serum-free selective media (2–4). (see **Serum dependence**) and can be derived in the laboratory or purchased from commercial suppliers.

ORIGIN OF CELL LINES

Regenerating tissues *in vivo* are made up of a small, self-repopulating, **stem cell** pool, an expandable pool of proliferating **progenitor cells**, and a nonproliferating differentiated cell pool. On demand, cells leave the stem cell pool and enter the progenitor compartment, where expansion is regulated to meet the current demand in the differentiated cell compartment. When cells enter the differentiated cell compartment,

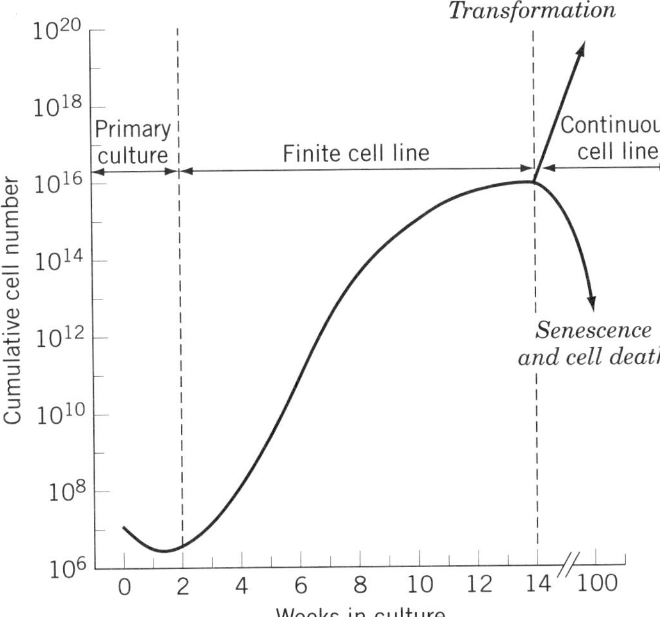

Figure 1. Effect of continued culture passage on cumulative cell number, assuming that no cells are discarded. The curve shows the initial decline due to selection, then exponential growth during the replicative phase, and then growth arrest and eventual deterioration following senescence, in a finite cell line, or continued proliferation, often at an enhanced rate, following transformation. Reproduced from Freshney's *Culture of Animal Cells, A Multi-Media Guide*, 1999, Wiley-Liss, N.Y.

this may be an irreversible process, as seen with erythrocytes, keratinocytes, and neurons; or it may be reversible, as seen with hepatocytes, endothelial cells, and fibrocytes. Cell lines can be derived from any tissue with a proliferating compartment, or from cells that can re-enter a proliferating compartment. It is possible that cell lines will contain stem cells, but, except for hemopoietic cells (see **Hematopoiesis**), markers are unavailable to determine this. As many cell lines express lineage markers and can, under appropriate conditions, differentiate, it seems most likely that they are derived from the progenitor cells of the tissue.

As cell lines are derived from random outgrowth or from disaggregated cells, multiple lineages of cells will be present. The purity of the culture will be determined by how well these lineages are matched, that is, derived from cells with the same phenotypic fate, or are derived from cells with dissimilar fates. In practice, the selective pressure exerted by the medium will tend to limit the cell population to cells of like phenotype with similar survival capacity, and cells that proliferate less rapidly will gradually be overgrown. It is possible that cells within one phenotypic group have different proliferative capacities but interactions between them produces a uniform proliferative rate in the entire population.

FINITE CELL LINES

When a cell line is generated at the first subculture of a primary culture, its lifespan is determined, initially, by the environmental conditions. If the conditions are inadequate, the cell line will die out within one or two subcultures, but if the medium, substrate, and other conditions are satisfactory, then the cell line may progress through several serial subcultures. The ultimate limit to the number of subcultures will be determined by the potential regenerative life-span of the cells (see **Senescence**; **immortalization**). Most cell lines from normal tissues will undergo a fixed number of population doublings (Fig. 1), given by the product of the number of passages and the estimated number of doublings per passage (5). These are known as finite cell lines. Normal human skin fibroblasts will usually achieve around 50–60 population doublings and then enter **senescence**. Senescent cells no longer proliferate, may be partially differentiated, and can remain viable for several months. Cultures of finite cell lines must, therefore, be stored frozen at an early passage level, thawed when required, and used between predetermined passage levels, usually between 10 and 30, to ensure an adequate and consistent supply.

CONTINUOUS CELL LINES

Some cell lines, such as those derived from mice or from tumors of many species, are not limited by a finite lifespan. They progress either by smooth transition to an immortal cell line (see **Immortalization**) or undergo **neoplastic transformation** at some stage and, instead of dying out after a set number of population doublings, continue to proliferate unchecked. Some cultures show evidence of selection and go through a period called *crisis*, when most cells in the population die out by senescence, but a transformed subpopulation survives, usually with an enhanced growth rate, increased cloning efficiency, loss of **contact inhibition**, acquisition of anchorage-independent growth, ability to grow to a higher saturation density, and increased tumorigenicity in animals (see: **Neoplastic transformation**).

There are a large number of continuous cell lines in existence, many of which are banked in repositories such as the American Type Culture Collection (ATCC) or the European Collection of Animal Cell Cultures (ECACC). They form a valuable resource of vigorously growing cultures, capable of unlimited expansion, but are subject to several caveats.

1. Their origin must be validated before use. Acquisition from a reputable cell bank, or the originator, will usually guarantee this, but there are many instances of continuous cell lines being cross-contaminated by other, more vigorously growing cell lines such as **HeLa cells** (6–8). The identity of a cell line needs to be confirmed prior to extensive use. This is done most effectively by performing a **DNA fingerprint**, but it is also possible by immunotyping, **isoenzyme** analysis, or **chromosome** analysis.

2. The cells must be shown to be free from infection. Continuous cell lines form an ideal substrate for **mycoplasma**, which can grow undetected in the culture (9). Again, most reputable cell banks will be able to demonstrate that cell lines being distributed are mycoplasma-free, but stocks still need regular checks every 1 or 2 months to ensure that they remain uninfected.

3. Continuous cell lines are genetically unstable. They are usually **aneuploid** (the chromosomal complement differs in number, and by chromosomal rearrangements, from the donor) and heteroploid (contain subpopulations with differing chromosomal constitutions). This is reflected in their phenotypic

Table 1. Cell Lines in Common Use[a]

Cell line	Morphology	Origin	Age	Status	Ploidy	Characteristics	Reference
MRC5	Fibroblast	Human lung	Embryonic	Normal	Diploid	Susceptible to human viral infection	Jacobs (1970) *Nature* **227**, 168
WI38	Fibroblast	Human lung	Embryonic	Normal	Diploid	Susceptible to human viral infection	Hayflick and Moorhead (1961) *Exp. Cell Res.* **25**, 585
IMR90	Fibroblast	Human lung	Embryonic	Normal	Diploid	Susceptible to human viral infection	Nichols et al. (1977) *Science* **196**, 60
A2780	Epithelial	Human ovary	Adult	Neoplastic	Aneuploid	Chemosensitive with resistant variants	Tsuruo et al. (1986) *Jpn. J. Cancer Res.* **77**, 941
A549	Epithelial	Human lung	Adult	Neoplastic	Aneuploid	Synthesizes pulmonary surfactant	Giard et al. (1972) *J. Natl. Cancer Inst.* **51**, 1417
A9	Fibroblast	Mouse subcutaneous	Adult	Neoplastic	Aneuploid	HGPRT-ve: deriv. L929	Littlefield (1964) *Nature* **203**, 1142
BHK21-C13	Fibroblast	Syrian hamster kidney	NB	Normal	Aneuploid	Transformable by polyoma virus	Macpherson and Stoker (1962) *Virology* **16**, 147
BRL3A	Epithelial	Rat liver	NB	Normal		Produce IGF-II	Coon (1968) *J. Cell Biol.* **39**, 29a
Caco-2	Epithelial	Human colon	Adult	Neoplastic	Aneuploid	Transports ions and amino acids	Fogh (1977) *J. Natl. Cancer Inst.* **58**, 209
CHANG liver	Epithelial	Human liver	Embryonic	Normal?	Aneuploid	HeLa contaminated	Chang (1954) *Proc. Soc. Exp. Biol. Med.* **87**, 440
CHO-K1	Fibroblast	Chinese hamster ovary	Adult	Normal	Diploid	Simple karyotype	Puck et al. (1958) *J. Exp. Med.* **108**, 945
EB-3	Lymphocytic	Human	Juvenile	Neoplastic	Diploid	EB virus + ve	Epstein and Barr (1964) *Lancet* **1**, 252
GH1, GH3	Epithelial	Rat	Adult	Neoplastic	Aneuploid	Produce growth hormone	Yasumura et al. (1966) *Science* **154**, 1186
HeLa	Epithelial	Human cervix	Adult	Neoplastic	Aneuploid	G6PD type A	Gey et al. (1952) *Cancer Res.* **12**, 364
HeLa-S3	Epithelial	Human cervix	Adult	Neoplastic	Aneuploid	High plating efficiency; will grow well in suspension	Puck and Marcus (1955) *Proc. Natl. Acad. Sci. USA* **41**, 432
Hep-2	Epithelial	Human larynx	Adult	Neoplastic	Aneuploid	HeLa contaminated⟨su⟩d	Moore et al. (1955) *Cancer Res.* **15**, 598
HT-29	Epithelial	Human colon	Adult	Neoplastic	Aneuploid	Differentiation inducible with NaBt	Fogh and Trempe (1975) in *Human Tumor Cells in vitro*, J. Fogh, ed., Academic Press, New York, p. 115
KB	Epithelial	Human oral	Adult	Neoplastic	Aneuploid	HeLa contaminated	Eagle (1955) *Proc. Soc. Exp. Biol. (N.Y.)* **89**, 362
L1210	Lymphocytic	Mouse	Adult	Neoplastic	Aneuploid	Rapidly growing; suspension	Law et al. (1949) *J. Natl. Cancer Inst.* **10**, 179
L5178Y	Lymphocytic	Mouse	Adult	Neoplastic	Aneuploid	Rapidly growing suspension	
L929	Fibroblast	Mouse	Adult	Normal	Aneuploid	Clone of L cell	Sanford et al. (1948) *J. Natl. Cancer Inst.* **9**, 229
LS	Fibroblast	Mouse	Adult	Neoplastic	Aneuploid	Grow in suspension: deriv. L929	Paul and Struthers (unpublished)

Table 1. Cell Lines in Common Use[a] *(continued)*

Cell line	Morphology	Origin	Age	Status	Ploidy	Characteristics	Reference
MCF7	Epithelial	Human breast pleural effusion	Adult	Neoplastic	Aneuploid	Estrogen recep +ve	Soule et al. (1973) *J. Natl. Cancer Inst.* **51**, 1409
P388D	Lymphocytic	Mouse	Adult	Neoplastic	Aneuploid	Grow in suspension	Dawe and Potter (1957) *Am. J. Pathol* **33**, 603; Koren et al. (1975) *J. Immunol.* **114**, 894
S180	Fibroblast	Mouse	Adult	Neoplastic	Aneuploid	Cancer chemotherapy screening	Dunham and Stewart (1953) *J. Natl. Cancer Inst.* **13**, 1299
STO	Fibroblast	Mouse	Embryonic	Normal	Aneuploid	Used as feeder layer for embryonal stem cells	Bernstein (1975) *Proc. Natl. Acad. Sci. USA* **72**, 1441
3T3-L1	Fibroblast	Mouse, Swiss	Embryonic	Normal	Aneuploid	Adultipose differentiation	Green and Kehinde (1974) *Cell* **1**, 113
3T3 A31	Fibroblast	Mouse BALB/c	Embryonic	Normal	Aneuploid	Contact inhibited; readily transformed	Aaronson and Todaro (1968) *J. Cell Physiol.* **72**, 141
NRK49F	Fibroblast	Rat kidney	Adult	Normal	Aneuploid	Induction of suspension growth by transforming growth factors	DeLarco and Todaro (1978) *J. Cell Physiol.* **94**, 335
Vero	Fibroblast	Monkey kidney	Adult	Normal	Aneuploid	Viral substrate and assay	Hopps et al. (1963) *J. Immunol.* **91**, 416
ZR-75-1	Epithelial	Human breast, ascites fluid	Adult	Neoplastic	Aneuploid	ER-ve, EGFr + ve	Engel (1978) *Cancer Res.* **38**, 3352

[a] Modified from R. I. Freshney *Culture of Animal Cells*, 1994, 3rd ed. Wiley-Liss, N.Y., p. 151.

diversity, including variations in morphology, enzyme activity, antigenic expression, and growth characteristics. It is normal practice to clone such populations, select a clone with the required properties, and cryopreserve sufficient ampoules (12–100) for future use. Cultures are generally maintained for a limited period, usually about 3 months, and then replaced from frozen stock.

SUBCULTURE

Subculture, or passage, is the transfer of a culture from one vessel to another, usually using **trypsin** to dislodge the cells from the substrate and dissociate them from each other. If the cell line is proliferating, then subculture will also imply diluting the cell concentration achieved at the end of the culture period, to a new seeding concentration to initiate a new culture. Growth from seeding until the next subculture is known as a growth cycle, which is repeated each time the culture is passaged. A series of growth cycles constitute serial propagation, and this should not normally proceed beyond about 3 months without replacing stock from the freezer (see above).

Each growth cycle is composed of a lag phase (Fig. 2), where the cells are recovering from trypsinization, synthesizing new **extracellular matrix**, adhering to the substrate, spreading, repolymerizing the **actin** microfilament **cytoskeleton**, and responding to signaling from extracellular matrix adhesions and cytoskeletal rearrangements, which leads to expression of **cyclins** and, eventually, reentry to the cell cycle. When the cells start to divide, they enter the exponential or "log" phase of the growth cycle and will continue exponential growth until all the available growth surface is occupied or the medium is exhausted. If the medium is limiting, it is customary to replace it half way through the exponential phase of growth. When all the available growth surface is occupied, the culture is said to be *confluent* and will require to be subcultured again. If the culture is allowed to progress beyond confluence, growth will slow down, as a result of density limitation of growth (see **Contact inhibition**), and the culture enters the plateau phase of the growth cycle. If the cells are normal and respond to normal growth control signals, there will be little evidence of cell proliferation or cell loss in plateau, specifically, minimal turnover. If, however, the cells are transformed, they will continue to proliferate for several cell generations after reaching confluence and reach a higher saturation density at plateau, due to deficiencies in contact inhibition of cell motility and density limitation of cell proliferation (10,11). Even in plateau, there will be continued cell proliferation, although at a lower level than in exponential growth, balanced by increased cell loss from **apoptosis**, resulting in a greater turnover than occurs in normal cells in plateau.

The numerical parameters that can be derived from the growth curve are the duration of the lag period, the doubling

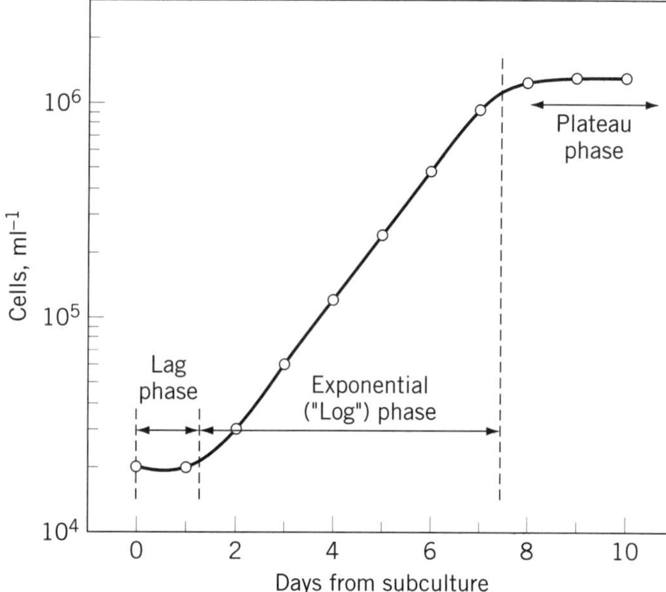

Figure 2. Phases in the growth cycle of cultured cells following subculture. Modified from Freshney's *Culture of Animal Cells, A Multi-Media Guide*, 1999, Wiley-Liss, N.Y.

Table 2. Examples of Selective Media[a]

Cells or Cell Line	Medium	Reference
Fibroblasts	MCDB 202	McKeehan et al. (1977) *In Vitro* **13**, 399.
Fibroblasts	MCDB 110	Bettger et al. (1981) *Proc. Natl. Acad. Sci. USA* **78**, 5588.
Keratinocytes	MCDB 153	Tsao et al. (1982) *J. Cell Physiol.* **110**, 219
Bronchial epithelium	LHC	Lechner and LaVeck (1985) *J. Tissue Cult. Meth.* **9**, 43
Mammary epithelium	MCDB 170	Hammond et al. (1984) *Proc. Natl. Acad. Sci. USA* **81**, 5435
Prostate epithelium (rat)	WAJC 401	McKeehan et al. (1982) *In Vitro* **18**, 87.
Prostate epithelium (human)	WAJC 404	McKeehan et al. (1984) *Cancer Res.* **44**, 1998
Glial cells		Michler-Stuke and Bottenstein (1982) *J. Neurosci. Res.* **7**, 215
Melanocytes		Naeyaert et al. (1991) *Br. J. Dermatol.* **125**, 297
Small cell lung cancer	HITES	Carney et al. (1981) *Proc. Natl. Acad. Sci. USA* **78**, 3185
Adenocarcinoma of lung		Brower et al. (1986) *Cancer Res.* **46**, 798
Colon carcinoma		Van der Bosch et al. (1981) *Cancer Res.* **41**, 611
Endothelium	MCDB 130	Knedler and Ham (1987) *In Vitro* **23**, 481

[a] Modified from R. I. Freshney *Culture of Animal Cells*, 1994, 3rd ed., Wiley-Liss, New York.

time in mid–log phase, and the saturation density in plateau. The last will depend on the feeding regimen and, if cell kinetics are being determined, should be measured under nonlimiting medium conditions. Where these parameters are consistent, they can be used to calculate the split ratio, the degree of dilution required to reinitiate a new growth cycle with a short lag period and to achieve the appropriate density for subculture at a convenient time in the future, usually around one week. The split ratio should be a power of 2 when handling finite cell lines. This allows an approximate calculation of the number of generations that have elapsed since the last subculture; for instance, a split ratio of 8 would imply that the cells had undergone 3 population doublings in each growth cycle. Split ratios are less useful with continuous cell lines. They are not needed to determine the stage in the life cycle of the cell line, as it is immortal, and the population doubling time is so much shorter that split ratios of the order of 1–100 are required to produce a week-long growth cycle. At this level of dilution, it is more advisable to determine the cell concentration at each subculture and to dilute to give the seeding concentration that will allow growth to the next subculture in a suitable interval, say, one week.

SELECTIVE CULTURE

Alterations in the choice of medium and/or substrate will determine which cells survive in primary and early passage cultures. Most selective media are serum-free (see **Serum Dependence**), and recipes are available for the culture of many different cell types, including fibroblasts, epithelial cells (epidermal, mammary), glial cells, melanocytes, endothelial cells, and smooth muscle Table 2. Selection of a purified cell culture, with defined characteristics, by cloning, use of a selective medium, or physical separation, allows the resultant culture to be called a *cell strain*.

CHARACTERIZATION

Cell morphology and growth characteristics are simple criteria for characterization. Spindle-shaped cells are usually considered to be fibroblast-like, without necessarily implying that they are known to be of the fibrocyte lineage. Similarly, cells that grow in patches of polygonal, or pavement-like, cells with distinct boundaries between the cells are said to be epithelial-like, although numerous other cell types can assume a similar morphology. Characterization of a cell strain employs such criteria as (*1*) identification of the type of **intermediate filament** protein (eg, **cytokeratin** subtype in different epithelial cells, desmin in muscle cells, glial fibrillary acidic protein in astrocytes, and **neurofilament** protein in neurons and some neurendocrine cells) (12); (*2*) expression of cell-surface **antigens**, such as epithelial membrane antigen (13); (*3*) enzymes, such as tyrosine aminotransferase and its inducibility by glucocorticoids in hepatocytes (14); (*4*) marker chromosomes (15); (*5*) **isoenzymes** (16); and (*6*) the **DNA fingerprint** (17,18). The same criteria may be used to characterize a cell line, but would give only an average for the population in a cell strain.

CELL BANKS

Most of the cell lines in common use are lodged in cell banks containing records of their identifying characteristics. Cell lines or strains that have been characterized and shown to be free of mycoplasma can be submitted to a cell bank for safe-keeping, with or without authority to distribute to other

Table 3. Cell Banks[a]

United States and Canada	American Type Culture Collection (ATCC), 12301 Parklawn Drive, Rockville, MD 20852
	National Institute of General Medical Sciences (NIGMS), Human Genetic Mutant Cell and National Institute on Ageing Cell Culture Repositories, Coriell Institute for Medical Research, Copewood Street, Camden, NJ 08103
	Repository for Human Cell Strains, and Cell Repository for Neuromuscular Diseases, Children's Hospital Research Instsitute, McGill University, 2300 rue Tupper Street, Montreal, Quebec H3H 1P3, Canada
Europe	European Collection for Animal Cell Cultures (ECACC), PHLS/CAMR, Porton, Salisbury, England
	European Collection for Biomedical Research, Dept. Cell Biology and Genetics, Erasmus University, P.O. Box 1783, Rotterdam, Netherlands
	Collection Nationale des Cultures des Microorganismes, Institute Pasteur, 25 rue du Dr. Roux, F-75724 Paris Cédex 15, France
	Human Genetic Cell Repository, Hospices Civils de Lyon, Hôpital Debrousse, 29 rue Soeur Bourier, F-69322 Lyon Cédex 05, France
	Tumorbank, Institut für Experimentelle Pathologie (dkfz), Deutsche Krebsforschungszentrum, in Neuenheimer Feld 280, Postfach 101949, D-6900 Heidelberg 1, Germany
	Centro Substrati Cellulari, Instituto Zooprofilattico Sperimentale della Lombardia, e dell'Emilia, Via A.Biandri 7, I-25100 Brescia, Italy
	National Bank for Industrial Microorganisms and Cell Cultures (NBIMCC), Blvd. Lenin 125 BL 2, V floor, Sofia, Bulgaria
	National Collection of Agricultural and Industrial Microorganisms (NCAIM), Dept. Microbiology, University of Horticulture, Somloi ut 14-16, H-1118 Budapest, Hungary
Japan	Japanese Cancer Research Resources Bank (JCRB), National Institute of Hygienic Sciences, Kami-Yoga, Setagaya-Ku, Tokyo
	General Cell Bank, Institute of Physical and Chemical Research of RIKEN, 2-1 Hirosawa, Wako, Saitama, 351-01
Australia	Commonwealth Serum Laboratories, 45 Poplar Road, Parkville, Victoria 3052

[a] Modified from R. I. Freshney, *Culture of Animal Cells*, 1994, 3rd ed., Wiley-Liss, New York.

users. Cell lines may be obtained from cell banks for a set charge (Table 3) (19).

Information on cell lines is also available through a number of data banks (20) such as ATCC ⟨www.atcc.org⟩ and ECACC ⟨www.ecacc.org.uk⟩.

BIBLIOGRAPHY

1. W. I. Schaeffer (1990) *In Vitro Cell. Dev. Biol.* **26**, 97–101.
2. D. W. Barnes, D. A. Sirbasku, and G. H. Sato, eds. (1984) *Methods for Serum-Free Culture of Epithelial and Fibroblastic Cells*, Alan R. Liss, New York, pp. 3–24.
3. R. Maurer (1992) in *Animal Cell Culture, a Practical Approach*, 2nd ed. R. I. Freshney, ed., IRL Press at Oxford University Press, Oxford pp. 15–46.
4. D. W. Jayme and D. F. Gruber (1994) in *Cell Biology, a Laboratory Handbook*, J. E. Celis, ed., Academic Press, Nw York, pp. 18–24.
5. L. Hayflick, P. S. Moorhead (1961) *Exp. Cell Res.* **25**, 585–621.
6. C. S. Stulberg, W. D. Peterson Jr., and W. F. Simpson (1976) *Am. J. Hematol.* **1**, 237–42.
7. W. A. Nelson Rees, D. W. Daniels, and R. R. Flandermeyer (1981) *Science* **212**, 446–452.
8. R. J. Hay (1991) *Dev. in Biol. Stand.* **75**, 193–204.
9. G. T. McGarrity, D. G. Murphy, and W. W. Nichols eds., *Mycoplasma Infection of Cell Cultures*, Plenum Press, New York, pp. 87–104.
10. R. Dulbecco and J. Elkington (1973) *Nature* **246**, 197–199.
11. B. Westermark (1974) *Int-J. Cancer* **12**, 438–451.
12. F. C. S. Ramaekers, J. J. G. Puts, A. Kant, O. Moesker, P. H. K. Jap, and G. P. Vooijs (1982) *Cold Spring Harbor Symp. Quant. Biol.* **46**, 331–339.
13. E. Heyderman, K. Steele, and M. G. Ormerod (1979) *J. Clin. Pathol.* **32**, 35–39.
14. T. Ikeda, N. Sawada, M. Satoh, and M. Mori (1998) *J. Cell. Physiol.* **175**, 41–49.
15. T. R. Chen (1988) *Cytogen. Cell Genet.* **48**, 19–24.
16. K. G. Steube, D. Grunicke, and H. G. Drexler (1995) *In Vitro Cell. Dev. Biol.* **31**, 115–119.
17. G. N. Stacey, B. J. Bolton, D. Morgan, S. A. Clark, and A. Doyle (1992) *Cytotechnology* **8**, 13–20.
18. G. Stacey, B. Bolton, A. Doyle, and B. Griffiths (1992) *Cytotechnology* **9**, 211–216.
19. A. Doyle, R. Hay, and B. E. Kirsop, eds. (1990) *Living Resources for Biotechnology*, Cambridge Univ. Press, Cambridge, U.K., pp. 1–15.
20. A. Doyle, R. Hay, and B. E. Kirsop, eds. (1990) *Living Resources for Biotechnology*, Cambridge Univer. Press, Cambridge, England, pp. 17–49.

CELL LINEAGE

MICHAEL GERSHON

The daunting complexity of vertebrate organisms has led investigators to adopt a reductionist approach to their investigation. This approach is based on the assumption that an otherwise impossibly complicated system can be effectively understood through an analysis of its constituent cells and molecules. A reductionist concept of development envisions the process as the assembly of elementary units in many combinations and permutations, unfolding over time in stages that escalate progressively in their complexity. The transition from one stage to the next is influenced by the conditions that prevailed during the earlier period. Developmental biology thus seeks to infer the mechanisms of morphogenesis and histogenesis from the properties, activities, and responses of the precursors that give rise to the various lineages of the cells of the body.

That the concept of lineage should be an important one in developmental biology is something of a paradox. In a simplistic sense, it seems odd that lineage can be a meaningful issue, when all of the cells of an animal descend from a single precursor, the fertilized zygote. Distinct lineages, however, arise very early in embryogenesis. Formation of a body plan in lower vertebrates, including amphibia, is a process that begins in the oocyte, continues after fertilization and cleavage, and is finally completed at gastrulation (1,2). Polarity in the oocyte causes cytoplasmic constituents to be distributed asymmetrically in the egg and newly formed blastula (3–6). This asymmetric distribution establishes different lineages in blastomeres, which do not each inherit identical cytoplasmic determinants from their predecessors. The resulting differential inheritance creates cellular differences between the blastomere-derived lineages and allows cells to interact with one another (7,8). Regional inequalities and cellular interactions may then be expressed as functional differences between cells in different locations. These processes form an ordered cascade that leads to the establishment of a pattern of small groups of like cells that follow a developmental agenda common to themselves, but different from those followed by other small groups of cells located elsewhere in the embryo (4). Functional differences between cells located in particular sites in the developing embryo are necessary for the differential cellular behaviors that underlie morphogenetic movements, like gastrulation and neurulation (9–11). It is also these different cellular loci, and the progeny to which they give rise, that provide developmental meaning to the concept of lineage.

In mammals, the establishment of lineages is delayed relative to that in amphibian embryos. Early cleavage cells of mammalian embryos appear to be identical to one another and totipotent (12). Each of the blastomeres of a two-cell embryo is capable of generating a complete adult, and will do so if the embryo is split (13,14). Destruction of a single blastomere, moreover, may not prevent the normal development of a mature animal. **Chimeric** animals can also be constructed by combining morulae (15). It is thus unlikely that localized determinants play a role in mammalian development prior to or during the eight-cell stage. The pattern of cell diversification must be generated at subsequent stages, through the interactions of cells with one another and with their microenvironment.

LINEAGE AS A PEDIGREE

Lineage, understood as a set of cells for which the ancestry can be traced back to a single progenitor, can possibly be delineated in simple transparent embryos, such as that of ***Caenorhabditis*** *elegans* by direct observation (16,17). Cells can be recognized and followed from egg to adult in a living embryo as they divide, migrate, differentiate, and/or die. Keeping track of the cells is facilitated by the fact that nematode development is virtually invariant. Because the process is so constant, it is possible to predict the fates of descendant cells from their positions on the lineage tree. It is, of course, not possible to use similar direct observation by itself to determine the pedigrees of the cells of an adult mammal. Not only are the numbers of cells in a mammalian embryo too large to permit the progeny of unmarked individual cells to be followed, the details of cell lineage in mammals may also show random variations, even between genetically identical individuals. Lineage can possibly be followed in mammals by marking individual precursor cells in one or another location, for example, with a replication-deficient **retrovirus** that carries the gene encoding *Escherichia coli* **beta-galactosidase** (18–20). The activity of the bacterial enzyme or the protein can then be detected histochemically or immunocytochemically, respectively, thereby permitting all of the cells of the clone that ultimately develops from the single virus-infected progenitor to be identified. This type of analysis has been especially powerful in studies of the development of the central nervous system (CNS).

Studies of *C. elegans* have been particularly important in establishing that the terminally differentiated cells of an adult animal do not necessarily constitute the lineal descendants of a single **founder cell** (16,17,21). Different phenotypic classes of cells (with the exception of intestinal and germ cells) in *C. elegans* thus arise from several founder cells, which themselves are found on different arms of the lineage tree. The descendants of a single primitive embryonic precursor, therefore, may become cells of the hypodermis, musculature, and nervous system of the mature animal (21). Lineage in the straightforward pedigree sense thus does not predict the commitment of multipotent precursor cells to a specific phenotype. That choice, the moment when a cell's fate has become determined, may be made long before the terminally differentiated phenotype becomes evident in the appearance of the cell, and may be shared by cells originating from more than a single progenitor (22). Determination in this sense implies that a cell has undergone a lasting and autoperpetuating change that will forevermore distinguish that cell and its progeny from other kinds of cell and will commit the determined cell and its descendants to a common specialized developmental program. Cells may become determined either before, or when, they overtly differentiate (display their ultimate phenotype in a detectable manner); moreover, all of the cells that are determined to follow a shared developmental program do not necessarily arise from a common precursor, and multipotent precursor cells may divide asymmetrically so that not all of their progeny are identically determined. The observation that cells of a particular terminally differentiated phenotype may not share an origin from a single "founder" has led the concept of lineage to acquire a meaning distinct from its linkage to cellular pedigrees.

LINEAGE AND DETERMINATION

Cells of embryos that share a determined fate, as well as those that descend from a common founder, are often referred to as members of a common lineage. For example, the multipotent **stem cells** that are found in the neuroepithelium of the neural tube are self-sustaining, but they also give rise to progeny determined to be neurons or glia; following determination, the cells are said to be developing in neuronal or glial lineages (23). Lineage is used to denote their common determined phenotype, not their origin from a common precursor. The neuroepithelial stem cells proliferate in response to the actions of defined mitogens, including **epidermal growth factor** (EGF) and **fibroblast growth factor** (FGF), and differentiate as neurons or glia under the influence of a variety of microenvironmental factors. Factors that have been found under at least some circumstances (*in vitro* or *in vivo*) to promote the restriction of subsets of central nervous system stem cells to the glia lineage include **sonic hedgehog** (Shh) (24), ciliary neurotrophic factor (CNTF) (25), leukemia inhibitory factor (LIF) (25), and bone morphogenetic protein 4 (BMP4) (26).

The generation of the phenotypic diversity of cells in various brain regions occurs in stages, progressing from pluripotentiality to lineage restriction (a phenomenon that prevents cells from giving rise to certain types of progeny), to the determination of specific phenotypes, and finally to the manifestation of the features of the terminally differentiated cells (27). Because of their responsiveness to their microenvironment, transplanted stem cells can integrate into a variety of brain regions, and thus potentially may provide a tool for treating the diseased or injured brain (23). The progenitor cells of the vertebrate brain, however, are themselves heterogeneous and thus members of different lineages (27). These differences help to explain how and when neuroepithelial cells begin to respond to the molecular signals provided by their microenvironment. Environmental factors, be they mitogens, trophic **growth factors**, or **morphogens**, can only affect cells that have prepared themselves to be affected. To respond to an extracellular factor, the responding cell must have previously acquired the relevant receptors. Most of the stem cells of the vertebrate nervous system express the **intermediate filament** protein, nestin (23,25). Lineages of progenitors, however, can be defined by the expression in common of molecular properties that distinguish the cells of the lineage from other progenitors in the neuroepithelium; such lineage-defining molecular traits include expression of the ***achaete/scute*** family of basic helix–loop–helix (bHLH) **transcription factors**, **EGF receptors**, ventricular zone gene 1 (*vzg-1*), and the embryonic neural **cell adhesion molecule** (E-NCAM) (27). Progenitors in different locations may respond differently to the gliogenesis-promoting factors noted above.

LINEAGES IN THE DEVELOPMENT OF THE NEURAL CREST

A source of precursor cells in developing vertebrates that has been extremely valuable for studies of the role(s) of cell lineage in development has been the neural crest (28,29). This is a transitory structure that appears during embryogenesis and disappears as its component cells disperse, migrating through the embryo to give rise to a wide variety of terminally differentiated cell types (30,31). The lineages of cells as diverse as melanocytes, fibroblasts, endocrine cells, smooth and skeletal muscle, cartilage, bone, meningeal cells, Schwann cells, satellite cells, enteric glia, autonomic, enteric, and sensory neurons can all be traced back to the neural crest. A great deal of investigation has been devoted to understanding when crest-derived cells become determined, and what factors cause them to become so. Considerable evidence indicates that at least some of the cells of the premigratory crest are pluripotent. Clones of these cells give rise in culture to several different classes of terminally differentiated cells (32–35). Lineage studies that have marked (by microinjection) individual cells of the premigratory crest (36,37), as well as early-migrating crest-derived cells (38), have shown that progeny of the marked precursors reach different destinations and thus give rise to a variety of cell types. Furthermore, cultures of stem cells can be prepared from the neural crest (39,40), just as they can from the neuroepithelium of the neural tube (23,25).

Although it is clear that the neural crest contains multipotent cells, the population of cells in the premigratory crest is actually heterogeneous. Some of the cells, even in the premigratory crest itself, are already determined (41). Cells determined to develop in the melanocytic lineage, for example, migrate selectively along the dorsolateral pathway (42). Among the uncommitted cells, there appears to be a progressive loss of developmental potential that occurs as a function of age (28,29,43). Whether the loss of developmental potential is simply determined by the age of the cells, or by the signals the cells receive as a result of their position in the embryo, has yet to be determined.

Some of the crest-derived cells that arrive in the organs that lie at the ends of their migratory pathways are still multipotential when the crest-derived cells arrive. The bowel has been carefully investigated in this regard and is a good model to illustrate the role of lineages in neural crest development (44). Two techniques have been employed to investigate the developmental potential of the crest-derived cells that colonize the gut. One, back-transplantation, consists of placing a segment of avian gut that has received a complement of émigrés from the neural crest in a migration pathway of a younger host embryo (45,46). When this is done, crest-derived cells leave the grafts and remigrate in their new hosts. Where they go depends on where the grafts are placed. If the grafts are situated so as to replace the host's crest in locations that normally provide crest-derived émigrés for the bowel, the donor cells migrate to the host's gut and participate in forming the enteric nervous system (ENS). On the other hand, if the grafts are placed anywhere else in the embryo, the enteric crest-derived cells of the donor fail to migrate to the host's bowel. Instead, they migrate to peripheral nerves, sensory, and sympathetic ganglia. Donor cells that arrive in sympathetic ganglia become catecholaminergic sympathetic neurons, even though catecholaminergic neurons are not found in the enteric ganglia of the avian gut. Donor cells that find themselves in peripheral nerves develop as Schwann cells, even though the ENS contains glia rather than Schwann cells. Enteric crest-derived cells thus learn nothing from their previous journey to the bowel and evidently remember nothing of it. Crest-derived cells from the donor gut, however, never give rise to muscle, melanocytes, or connective tissue, indicating that they have lost at least some of developmental potential of their ancestors; nevertheless, they remain pluripotent to the extent that they are able to give rise to cell types that they would not have done had they been left *in situ*.

Clonal analysis of crest-derived cells isolated from the embryonic avian (47) or fetal mammalian bowel (48,49) has also suggested that the population of crest-derived émigrés that colonizes the gut is a heterogeneous mixture of multipotent and determined cells. One or more terminally differentiated types of cells may be found in clones arising from individual precursors. None of the clones developing *in vitro* from enteric crest-derived cells contain melanocytes, endocrine cells, or skeletal muscle, although some contain smooth muscle. The heterogeneity of the crest-derived cell population that colonizes the bowel is reflected in the birthdays of the different types of enteric neuron (50). Some of these cells, containing acetylcholine or serotonin, are born very early in embryonic life, beginning in the mouse at E8.5, an age that precedes the arrival in the bowel of even the most precocious of enteric neurons. The crest-derived population that arrives in the gut, therefore, must contain a mixture of already-determined postmitotic neurons, as well as the undetermined pluripotent precursors that were revealed by the experiments involving back-transplantation and clonal analysis described above. On the other hand, other types of enteric neuron, such as those that contain calcitonin gene-related peptide (CGRP), do not even begin to leave the mitotic cycle until after the last serotonergic neurons have been born. The fact that the timing of the birthdays of enteric neurons is a function of their phenotypic class was among the earliest observations to suggest that multiple lineages of crest-derived precursors participate in the formation of the ENS. It is at least plausible that neurons in the same lineage might be born at similar times.

Confirmation of the idea that the ENS is formed by multiple lineages of crest-derived precursors came from a study that combined an *in vitro* investigation of enteric neuronal development with an analysis of the effects of the targeted deletion of the mash-1 transcription factor (51). It had previously been known that cells in the developing mammalian (but not avian) bowel are transiently catecholaminergic (TC) (52–55), and that these cells are derivatives of the neural crest (56–58). The TC cells were found not to be neurons, because they proliferate (56,57,59), although it was postulated that the TC cells might be neural precursors. That hypothesis was confirmed, and the disappearance of TC cells was found to occur not because they die, but because they lose their catecholaminergic properties as they acquire the characteristics of their terminally differentiated phenotype (56).

When it was discovered that TC cells express the same cell-surface differentiation antigens as precursors in the sympathoadrenal lineage, it was proposed that enteric and sympathetic neurons arise from a common lineage of crest-derived progenitors (60,61). This idea was furthered by the demonstration that *mash-1* is expressed during the development of both sympathetic and enteric neurons (62,63). On the other hand, knockout of *mash-1* is accompanied by the loss of almost the entire sympathetic nervous system (the superior cervical ganglion is spared), while the development of the ENS (except for that of the esophagus, which is lost) is merely delayed (63,64).

The shared features expressed by sympathoadrenal and ENS precursors made it possible to test the hypothesis that sympathetic and enteric neurons are derivatives of a common precursor lineage. **Complement**-mediated lysis of all of the cells isolated from the bowel that express **antigens** common to the sympathoadrenal lineage reduces the numbers of enteric neurons developing *in vitro*, but it does not prevent their formation (51). No serotonergic neurons develop *in vitro* following lysis of TC cells, although neurons expressing CGRP appear normally. The cells in the developing bowel that express *mash-1* were found to be TC cells, and the knockout of *mash-1* in mice was demonstrated to result in the complete loss of early-developing serotonergic neurons, while late-born CGRP-containing cells appear on schedule. The apparent delay in timing of the formation of the ENS in *mash-1* knockout mice is thus due to the selective loss of the early-developing lineage of enteric neurons. These observations established for the first time that the ENS is, in fact, derived from at least two lineages of crest-derived precursors. One of these, which may be common to the precursors of sympathoadrenal cells, obligatorily expresses *mash-1*, gives rise to neurons that are born early, and is the origin of all of the serotonergic neurons of the bowel. The other lineage is never catecholaminergic, does not express *mash-1*, gives rise to neurons that are born late, and is the origin of peptidergic enteric neurons, including those that contain CGRP.

Since the discovery of the role of *mash-1* in the development of one lineage of enteric neurons, additional lineages or sublineages of enteric neuronal precursor have been discovered. All of the crest-derived cells that colonize the bowel appear to express $p75^{NTR}$, the common neurotrophin receptor (44,56,65). The cells (from the vagal and sacral regions of the neural crest) evidently also express Ret (66,67), which must be stimulated by its functional ligand, glial cell line-derived neurotrophic factor (GDNF) (68–70). Actually, in order to stimulate Ret, GDNF has to form a complex with an α-component, GFRα-1 (71–73). If Ret (74,75), GDNF (69,76), or GFRα-1 (77) are knocked out, there are no neurons or glia in the entire bowel below the esophagus and in the immediately adjacent region of the stomach. GDNF is a mitogen for early crest-derived enteric neuronal precursors and later supports their development as neurons (78–80). Because the defect in the *mash-1* knockout mice is much more limited than that seen after the elimination of Ret or the components of its ligand complex, it follows that the *mash-1*-dependent lineage is a subset of a larger lineage of GDNF-dependent precursors.

More limited sublineages have been identified by the dependence of subsets of enteric neurons on other growth factor/receptor complexes, such as a still-to-be identified neuropoietic cytokine that activates the alpha component of the receptor for CNTF (65). Other factors, such as NT-3/TrkC (81) and **laminin**-1 (82,83), have been identified that promote the development of enteric neurons, but these factors do not define particular lineages. Endothelin-3 (ET-3) and its preferred receptor, endothelin B (ET_B), also appear to be necessary for a subset of enteric neurons (84,85). This factor/receptor complex, however, does not appear to define a particular lineage of enteric neurons. When either ET-3 (84) or ET_B (85) is knocked out in mice, the terminal colon becomes aganglionic, just as it does in humans with Hirschsprung's disease, some of whom also display ET-3 (86) or ET_B (87) mutations. The lesion that results from deletion of ET-3 or ET_B thus is geographically restricted, but within the affected zone all lineages of crest-derived enteric neurons are affected. ET-3 has been found to inhibit enteric neuronal development, and it has been postulated to act by preventing the premature differentiation of enteric neurons before their crest-derived have completed their colonization of the gut (78,88).

SUMMARY

The concept of lineage is used in two ways in describing development. In the pedigree sense, lineage is used to refer to the descendants of a common progenitor. Sharing a lineage defined as a pedigree, however, does not necessarily mean that the descendants of the common progenitor share the same developmental program or fate. Lineage as a pedigree is thus independent of the determination of its members. In a second developmental sense, the term *lineage* is used to refer to cells that do follow a common developmental program, regardless of whether these cells trace their ancestry back to a single shared precursor. Usually they do not. Identity of determination thus is now the common denominator that defines lineage.

BIBLIOGRAPHY

1. J. B. L. Bard (1990) *Morphogenesis: The Cellular and molecular processes of Developmental Anatomy*, Cambridge University Press, Cambridge, U.K.
2. J. P. Trinkaus (1984) *Cells into Organs: The Forces That Shape the Embryo*, 2nd ed., Prentice-Hall, Engelwood Cliffs, NJ.
3. M. R. Rebagliati et al. (1985) *Cell* **42**, 769–777.
4. J. B. Gurdon (1992) *Cell* **68**, 185–199.

5. J. P. Vincent, G. F. Oster, and J. C. Gerhart (1986) *Dev. Biol.* **113**, 484–500.
6. M. Kirchner, J. Newport, and J. Gerhart (1985) *Trends Genet.* **1**, 41–47.
7. S. C. Guthrie and N. B. Gilula (1989) *Trends Neurosci.* **12**, 12–16.
8. J. D. Hardin and L. Y. Cheng (1986) *Dev. Biol.* **115**, 490–501.
9. H. Spemann (1938) *Embryonic Development and Induction*, Yale University Press, New Haven, CT.
10. W. C. Smith et al. (1993) *Nature* **361**, 547–549.
11. K. W. Y. Cho et al. (1991) *Cell* **66**, 1111–1120.
12. R. L. Gardner (1985) *Philos. Trans. R. Soc. Lond. (Biol.)* **312**, 163–178.
13. S. J. Kelly (1977) *J. Exp. Zool.* **200**, 365–376.
14. A. K. Tarkowski (1959) *Nature* **184**, 1286–1287.
15. A. McLaren (1976) *Mammalian Chimeras*, Cambridge University Press, Cambridge, U.K.
16. C. Kenyon (1985) *Philos. Trans. R. Soc. Lond. (Biol.)* **312**, 21–38.
17. J. E. Sulston and H. R. Horvitz (1977) *Dev. Biol.* **56**, 110–156.
18. D. D. Galileo et al. (1990) *Proc. Natl. Acad. Sci. USA* **87**, 458–462.
19. C. Walsh and C. Reid (1995) *Ciba Found. Symp.* **193**, 21–40.
20. J. G. Parnavelas, M. C. Mione, and A. Lavdas (1995) *Ciba Found Symp.* **193**, 41–58.
21. J. E. Sulston, E. Schierenberg, and J. G. White (1983) *Dev. Biol.* **100**, 64–119.
22. B. Christ, H. J. Jacob, and M. Jacob (1977) *Anat. Embryol.* **150**, 171–186.
23. R. McKay (1997) *Science* **276**, 66–71.
24. D. M. Orentas and R. H. Miller (1996) *Dev. Biol.* **177**, 43–53.
25. K. K. Johe et al. (1997) *Genes Dev.* **10**, 3129–3140.
26. R. E. Gross et al. (1996) *Neuron* **17**, 595–606.
27. L. Lillien (1998) *Curr. Opin. Neurobiol.* **8**(1), 37–44.
28. N. M. Le Douarin and E. Dupin (1993) *J. Neurobiol.* **24**, 146–161.
29. D. J. Anderson (1989) *Neuron* **3**, 1–12.
30. N. M. Le Douarin (1982) *The Neural Crest*, Cambridge University Press, Cambridge, U.K.
31. N. M. Le Douarin (1986) *Science* **231**, 1515–1522.
32. K. Ito, T. Morita, and M. Sieber-Blum (1993) *Dev. Biol.* **157**, 517–525.
33. M. Sieber-Blum and A. M. Cohen (1980) *Dev. Biol.* **80**, 96–106.
34. M. Sieber-Blum et al. (1993) *J. Neurobiol.* **24**(2), 173–184.
35. A. Baroffio, E. Dupin, and N. M. Le Douarin (1988) *Proc. Natl. Acad. Sci. USA* **85**, 5325–5329.
36. M. Bronner-Fraser and S. E. Fraser (1988) *Nature* **335**, 161–164.
37. M. Bronner-Fraser and S. Fraser (1989) *Neuron* **3**, 755–766.
38. S. E. Fraser and M. Bronner-Fraser (1991) *Development* **112**, 913–920.
39. D. L. Stemple and D. J. Anderson (1993) *Dev. Biol.* **159**, 12–23.
40. D. Stemple and D. J. Anderson (1992) *Cell* **71**, 973–985.
41. P. D. Henion and J. A. Weston (1997) *Development* **124**(21), 4351–4359.
42. C. A. Erickson and T. L. Goins (1995) *Development* **121**, 915–924.
43. D. J. Anderson (1993) *Annu. Rev. Neurosci.* **16**, 129–158.
44. M. D. Gershon (1997) *Curr. Opin. Neurobiol.* **7**, 101–109.
45. T. P. Rothman et al. (1990) *Development* **109**, 411–423.
46. T. P. Rothman et al. (1993) *Dev. Dyn.* **196**, 217–233.
47. F. Sextier-Sainte-Claire Deville, C. Ziller, and N. M. Le Douarin (1994) *Dev. Biol.* **163**, 141–151.
48. L. Lo and D. J. Anderson (1995) *Neuron* **15**(3), 527–539.
49. L. Lo, L. Sommer, and D. J. Anderson (1997) *Curr. Biol.* **7**, 440–450.
50. T. D. Pham, M. D. Gershon, and T. P. Rothman (1991) *J. Comp. Neurol.* **314**, 789–798.
51. E. Blaugrund et al. (1996) *Development* **122**, 309–320.
52. P. Cochard, M. Goldstein, and I. B. Black (1978) *Proc. Natl. Acad. Sci. USA* **75**, 2986–2990.
53. G. M. Jonakait et al. (1979) *Proc. Natl. Acad. Sci. USA* **76**, 4683–4686.
54. G. Teitelman, T. H. Joh, and D. J. Reis (1978) *Brain Res.* **158**, 229–234.
55. M. D. Gershon et al. (1984) *J. Neurosci.* **4**, 2269–2280.
56. G. Baetge, J. E. Pintar, and M. D. Gershon (1990) *Dev. Biol.* **141**, 353–380.
57. G. Baetge and M. D. Gershon (1989) *Dev. Biol.* **132**, 189–211.
58. G. Baetge, K. A. Schneider, and M. D. Gershon (1990) *Development* **110**, 689–701.
59. G. Teitelman et al. (1981) *Dev. Biol.* **86**, 348–355.
60. D. J. Anderson et al. (1991) *J. Neurosci.* **11**, 3507–3519.
61. J. F. Carnahan, D. J. Anderson, and P. H. Patterson (1991) *Dev. Biol.* **148**, 552–561.
62. F. Guillemot and A. L. Joyner (1993) *Mech. Dev.* **42**, 171–185.
63. F. Guillemot et al. (1993) *Cell* **75**, 463–476.
64. L. Lo et al. (1994) *Perspect. Dev. Neurobiol.* **2**, 191–201.
65. A. Chalazonitis et al. (1998) *Dev. Biol.* **198**(2), 343–365.
66. V. Pachnis, B. Mankoo, and F. Costantini (1993) *Development* **119**, 1005–1017.
67. P. L. Durbec et al. (1996) *Development* **122**, 349–358.
68. P. Durbec et al. (1996) *Nature* 381–793.
69. S. Jing et al. (1996) *Cell* **85**, 1113–1124.
70. M. Trupp et al. (1996) *Nature* **381**, 785–789.
71. M. Trupp et al. (1997) *J. Neurosci.* **17**, 3554–3567.
72. A. M. Davies et al. (1997) *Neuron* **19**, 485.
73. J. Widenfalk et al. (1997) *J. Neurosci.* **17**, 8506–8519.
74. A. Schuchardt et al. (1994) *Nature* **367**, 380–383.
75. J. J. S. Treanor et al. (1996) *Nature* **382**, 80–83.
76. M. Sénchez et al. (1996) *Nature* **382**, 70–73.
77. G. Cacalano et al. (1998) *Neuron* **21**, 53–62.
78. C. J. Hearn, M. Murphy, and D. Newgreen (1998) *Dev. Biol.* **197**, 93–105.
79. R. O. Heuckeroth et al. (1998) *Dev. Biol.* **200**(1), 116–129.
80. A. Chalazonitis et al. (1998) *Dev. Biol.* **204**, 385–406.
81. A. Chalazonitis et al. (1994) *J. Neurosci.* **14**, 6571–6584.
82. T. P. Rothman et al. (1996) *Dev. Biol.* **178**, 498–513.
83. A. Chalazonitis et al. (1997) *J. Neurobiol.* **33**, 118–138.
84. A. G. Baynash et al. (1994) *Cell* **79**, 1277–1285.
85. K. Hosoda et al. (1994) *Cell* **79**, 1267–1276.
86. P. Edery et al. (1996) *Nat. Genet.* **12**(4), 442–444.
87. E. G. Puffenberger et al. (1994) *Cell* **79**, 1257–1266.
88. J. J. Wu, T. P. Rothman, and M. D. Gershon (1997) *Neurosci. Abst.* **23**, 24.

Suggestions for Further Reading

B. Alberts et al. (1994) *Molecular Biology of the Cell*, 3rd ed., Garland Publishing, New York, pp. 1037–1107.

N. M. Le Douarin and E. Dupin (1993) *J. Neurobiol.* **24**, 146–161.

L. Lillien (1998) *Curr. Opin. Neurobiol.* **8**(1), 37–44.

CELL SURFACE ADHESION RECEPTORS

D. GARROD

Cell surface adhesion receptors are molecules that mediate cell adhesion by binding to other molecules on the surface of an adjacent cell or to a component of the **extracellular matrix**. Many adhesion receptors have been described. Most of them belong to one of a small number of families of related molecules in which individual members share the same basic molecular structure. The principal families are the **cadherins**, the **immunoglobulin** (Ig) superfamily, the **integrins**, and the selectins (Fig. 1) as follows.

CADHERINS

In most tissues, a major contribution to intercellular adhesion is made by calcium-dependent cell adhesion molecules knows as cadherins (1). In general, these are simple transmembrane glycoproteins. The extracellular domain has an adhesion site toward the *N*-terminal region and several **calcium-binding** sites. Adhesive binding is mainly homophilic: a cadherin molecule on one cells binds to another cadherin molecule of the same type on the next cell. Linkage of the cytoplasmic **domain** to the **cytoskeleton** through proteins known as catenins is necessary for cadherin function. The best characterized is epithelial cadherin, E-cadherin. This appears very early in development, when it is involved in compaction of the eight-cell embryo and cell polarization. In adult epithelia—for example, intestinal epithelium—it is present on the lateral cell surfaces but is concentrated in intercellular junctions, known as the zonulae adherentes (see **Cell junction**; **Intermediate junction**), which ring the apicolateral margins of cells. These junctions are characterized by a cortical ring of cytoskeleton, the major component of which is **actin**. Cadherins are also present in the nervous system, where they play an essential role in neural development.

The adhesive glycoproteins of the other major intercellular junctions of epithelia, the **desmosomes** are also members of the cadherin family (2). Their extracellular domains are very like those of cadherin, but their cytoplasmic domains differ, being specialized for forming desmosomal plaques and, thereby, attachment to the keratin **intermediate filament** cytoskeleton, rather than to actin.

IMMUNOGLOBULIN SUPERFAMILY

Another major group of cell-to-cell adhesion molecules are members of the Ig superfamily. Their extracellular portions are characterized by the presence of at least one, and usually multiple, immunoglobulin-like domains (3,4). Included in this group are several nervous system **cell adhesion molecules**, such as the neural cell adhesion molecule (N-CAM), L-1, and TAG, which are involved in neuronal guidance and fasciculation. Several members of the immunoglobulin family are concerned with **antigen** recognition and adhesion in **T cells**. These include the following: the **T-cell receptor** (CD3) and its co-receptors CD4 and CD8, which together recognize the complexes of antigen peptide and **major histocompatibility complexes** on other cells; the major histocompatibility complex molecules themselves; and lymphocyte function related antigen 2 (LFA-2 or CD2), a receptor for another immunoglobulin-like molecule, LFA-3, expressed on other cells. Another group of immunoglobulin-like cell adhesion molecules includes the so-called intercellular adhesion molecules, ICAM-1, -2, and -3, which are more widely expressed, for example, on epithelial and endothelial cells, and V-CAM on endothelial cells. These are involved in the inflammatory response.

The immunoglobulin superfamily is large and diverse, probably because the basic structure of the immunoglobulin domain is versatile and readily adaptable to different binding functions (see **Immunoglobulin structure**). Among these molecules, however, the T cell receptor and the immunoglobulins themselves have somatically-variable domains necessary for antigen recognition. Members of the superfamily are even present in insects, where they are involved in nerve connections; the association of immunoglobulin-like domains in cellular recognition preceded the immune system in evolution.

INTEGRINS

Both cell-to-cell and cell-to-matrix receptors are contained within this large family of adhesion molecules. Integrins are heterodimers consisting of one α chain and one β chain, both of which are necessary for adhesive binding (5). Seventeen different α chains and eight different β chains are now known. Integrins may be classified into subfamilies according to which β subunit is involved in the complex. Thus $\beta 1$ integrin may associate with one of nine different α subunits, to give a

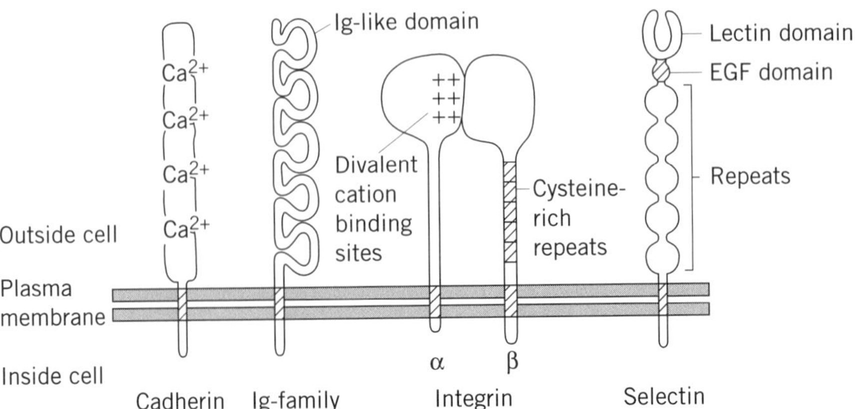

Figure 1. Diagram showing the generalized molecular structures of the four principal families of cell-surface adhesion receptors. (From Ref. 7, with permission of BMJ Publishing Group.)

series of matrix receptors of differing specificity. The $\beta 2$ integrins, on the other hand, are a family of cell-to-cell adhesion molecules of lymphoid cells with three alternative α subunits. The classification is made more complicated because some α subunits can associate with different β subunits (for example, $\alpha 6\beta 1$ and $\alpha 6\beta 4$).

Some integrins are apparently quite specific in their ligand-binding properties—for example, $\alpha 5\beta 1$ for the Arg–Gly–Asp tripeptide sequence of **fibronectin**—whereas others are promiscuous—for example, $\alpha v\beta 3$, once regarded as the vitronectin receptor, also binds fibronectin, **fibrinogen**, von Willebrand factor, thrombospondin, and ostepontin. An interesting example is $\alpha 4\beta 1$, which binds both the IIICS domain of fibronectin and V-CAM on endothelial cells. To complicate matters further, individual cell types usually express multiple integrins. A good example to consider here is the blood platelets that express predominantly αIIb$\beta 3$ (GPIIb/IIIa), which binds fibrinogen, fibronectin, von Willebrand factor, and vitronectin, but also lesser amounts of $\alpha V\beta 3$, $\alpha 5\beta 1$, $\alpha 2\beta 1$ (**collagen**), and $\alpha 6\beta 1$ (**laminin**).

SELECTINS

Most cellular interactions seem to entail homophilic or heterophilic **protein–protein interactions**. However, the selectins constitute a family of cell adhesion proteins that bind to carbohydrate. Selectins have **lectin**-like carbohydrate-binding domains at their extracellular *N*-terminal extremities (6). One of these, L-selectin (L = leucocyte) is a "homing receptor," mediating regionally specific adhesion of lymphocytes to endothelium in peripheral lymph nodes. This molecule is also involved in the adhesion of neutrophils to endothelium during the inflammatory response. Two other members of this family, E-selectin (E = endothelial) and P-selectin (P = platelet), also participate in the inflammatory response. E-selectin is unregulated on endothelial cells over a period of hours after stimulation by inflammatory mediators. P-selectin is contained within Wiebel–Palade bodies of endothelial cells and platelet α granules, from which it is rapidly mobilized on activation, mediating adhesion to neutrophils and monocytes.

BIBLIOGRAPHY

1. M. Takeichi (1990) Cadherins: a molecular family important for selective cell–cell adhesion. *Annu. Rev. Biochem.* **59**, 237–252.
2. D. R. Garrod, M. A. J. Chidgey, and A. J. North (1996) *Curr. Opin. Cell Biol.* **8**, 670–678.
3. T. A. Springer (1990) *Nature* **346**, 425–434.
4. G. M. Edelman and K. L. Crossin (1991) *Annu. Rev. Biochem.* **60**, 155–190.
5. R. O. Hynes (1992). *Cell* **69**, 11–25.
6. M. P. Bevilaqua and R. M. Nelson (1993) *J. Clin. Invest.* **91**, 379–387.
7. D. R. Garrod (1997). Cell to cell and cell to matrix adhesion. In *Basic Molecular and Cell Biology*, 3rd ed., D. S. Latchman (ed.) BMJ Publishing Group, pp. 79–96.

Suggestions for Further Reading

D. R. Garrod, M. A. J. Chidgey, and A. J. North (eds.) (1999) *Adhesive Interactions of Cells*, JAI Press, Geenwich, CT. (A series of up-to-date reviews on cell adhesion receptors and intercellular junctions.)

"Cell adhesion and human disease." *Ciba Found. Symp.* **189**. (A valuable collection of reviews on cell adhesion molecules and their involvement in human disease.)

M. Hortsch and C. S. Goodman (1991) *Annu. Rev. Cell Biol.* **7**, 505–557. (Adhesion receptors in *Drosophila*—the invertebrate situation.)

CEN SEQUENCES

GEORGES N. COHEN

Autonomously replicating sequences (ARS) are not stable *in vivo* during cell division and cannot be used readily as plasmid **vectors**. But, if certain DNA sequences that function as **centromeres** in yeast (CEN SEQUENCES) are grafted onto a **plasmid** already containing an ARS, the resulting vector plasmid is stabilized and segregates accurately during **mitosis** and **meiosis**. The CEN sequences are necessary for attaching **chromosomes** to the **mitotic spindle**, presumably through some specific connector proteins that join the **microtubules** to the centromere. In addition, CEN sequences can be used to construct linear chromosomes by preventing circularization by the addition of **telomeric** sequences that are normally at the end of chromosomes (1,2).

Conserved features of *Saccharomyces cerevisiae* CEN sequences are confined to a region of about 120 bp. The highly conserved 8 bp at the left (PuTCACPuTG, where Pu = purine) constitute the left boundary of a functional CEN sequence. The right boundary lies within or just beyond the 25 bp at the right, with the consensus sequence TGT-T-TG–TTCCGAA——AAA, where - indicates no specific base. A mutant that lacks the left conserved element can still assemble into a centromere that is partially functional in mitosis and well functioning in meiosis. The sequences between the two conserved terminal DNA elements are not essential for centromere function. Their lengths can be increased by 50% or their sequence from 6 to 12% in GC content without measurable changes in mitotic or meiotic segregation of plasmids carrying such CEN mutations (3). The right boundary sequence appears to be a binding site for a protein, as evidenced by various inactive mutations, especially in the central CCG sequence, and by an exonuclease blocking assay (4). The left element, which carries the **palindromic** sequence CACPuTG, binds the **helix–loop–helix motif** protein CPF1 (5).

BIBLIOGRAPHY

1. L. Clarke and J. Carbon (1980) *Nature* **287**, 504–509.
2. C–L Hsiao and J. Carbon (1981) *Gene* **15**, 157–166.
3. L. Panzeri et al. (1985) *EMBO J.* **4**, 1867–1874.
4. R. Ng., J. Ness and J. Carbon (1986) *Basic Life Sci.* **40**, 479–492.
5. R. Niedenthal, R. Stoll, and J. H. Hegemann (1991) *Mol. Cell. Biol.* **11**, 3545–3553.

Suggestion for Further Reading

L. Clarke and J.Carbon (1985) "The structure and function of yeast centromeres," *Ann. Rev. Genet.* **19**, 29–57.

CENTRIFUGATION

T. HOLZMAN

Centrifugation is a technique often employed during isolation or analysis of various cells, **organelles**, and biopolymers, including **proteins**, **nucleic acids**, lipids, and carbohydrates dissolved or dispersed in biologically relevant solvents (ie, typically aqueous buffers). It is also applicable to synthetic macromolecules dispersed or dissolved in nonaqueous, organic solvents. In this technique, the sample (comprising a liquid phase and a solute) is placed in a suitable vessel, and the vessel is spun in a centrifugal rotor. The centrifugal force created by the spinning rotor causes the solute sample to sediment out of solution (typically, though not always, toward the base of the vessel) (Fig. 1). The centrifugal force applied to the sample is akin to gravitational force (acceleration) and is measured in gravities, as in $n \times G$ (the gravitational constant G equals 6.6720×10^{-11} N-m²/Kg²). The extent to which the sample sediments toward the base of the vessel is a function of a series of complex interacting factors related to the properties of the solute alone and of the system as a whole (solvent and solute) (Fig. 1, Tables 1 and 2).

The factors that define the entire sedimenting system include (1) the rate ω at which the rotor spins; (2) the length of time for which the centrifugal force is applied; (3) the solvent density ρ; and (4) the system temperature T. The factors specific to the sedimented solute include (1) the radial distance r of the solute from the central axis of rotation; (2) the solute **partial specific volume** $\bar{v}$; (3) the solute **diffusion** coefficient D; (4) the solute **frictional coefficient** f; and (5) the solute **sedimentation coefficient** s. During centrifugation (Fig. 1), the centrifugal force is opposed by **buoyant density** and frictional forces. Under these conditions it can be shown that the rate of sedimentation of the solute molecules of weight-average molecular weight M_w and moving as a radial boundary r_b is defined by

$$\text{Boundary velocity} = v = \frac{M_w(1 - \bar{v}\rho)}{Nf}\omega^2 r_b. \qquad (1)$$

Operationally, the technique can be divided into preparative and analytical modes, which differ principally in whether the actual recovery of the materials that have been centrifugally separated is desired or practical (Fig. 2).

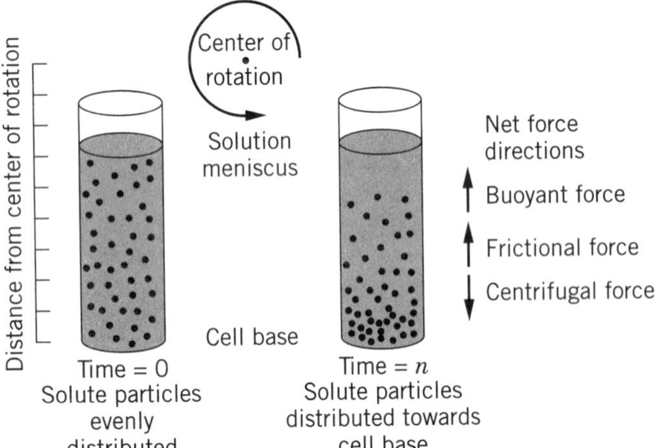

Figure 1. A schematic diagram of a centrifugation experiment.

Table 1. The Forces Involved in Centrifugation

Proportionality Constant or Force Equation	Cause or Definition	Dimensions
Centrifugal force $= \omega^2 rm$	Centrifugation of solute of mass, m	(g-cm)(radians/sec)
Buoyant force $= -\omega^2 rm_s$	Solvent of mass m_s; displacement by sedimenting solute	(g-cm)(radians/sec)
Frictional force $= -fv$	Resistance to movement of solute particle at velocity v through solvent	g-cm/sec²

Table 2. General Proportionality and Descriptive Constants for Centrifugation

Term	Definition	Dimensions
Angular velocity $= \omega = (2\rho/60) \times$ rpm	$2\pi\left(\frac{\text{radians}}{\text{revolution}}\right)\left(\frac{\text{revolutions}}{\text{min}}\right)\left(\frac{1\text{min}}{60\text{sec}}\right)$	radians/sec
Frictional coefficient $= f = 6\pi\eta r_s$	Stokes' law frictional coefficient for a sphere of equivalent radius r_s in a solvent of viscosity η.	g/sec
Diffusion coefficient D	Defined by Fick's first law (see Refs. 1 and 2) for an ideal solution at infinite dilution as $D = RT/Nf$, e.g., 1 Fick $= 10^7$ cm²/sec	cm²/sec
Sedimentation coefficient s	The velocity of sedimentation per unit of centrifugal force	sec
Solution density ρ	Mass per unit volume	g/cm³
Solution viscosity η	The resistance to change within a fluid, a type of internal friction	poise or dyne-sec/cm²
Solute partial specific volume $\bar{v}$	The change in volume per unit mass	cm³/g
Solute buoyant density	Effective density determined by isopycnic centrifugation in a specific gradient medium	g/cm³
Avagadro's number N	6.022×10^{23}	mole⁻¹
Gas constant R	8.314×10^7	ergs/K-mole

PREPARATIVE CENTRIFUGATION AND ULTRACENTRIFUGATION

Preparative centrifugation (Fig. 2) is routinely employed in isolating, separating, and purifying a variety of biological and cellular/subcellular components. Some examples include **chromatin**, coated vesicles (see **clathrin**), cytosolic proteins, DNA, **Golgi**, **inclusion bodies**, **lipoproteins**, **lysosomes**, **Microorganisms**, **Microsomes**, **mitochondria**, myelin, **nuclei**, **glycoproteins**, **RNA**, **Peroxisomes**, plasma **membranes**, **polysomes**, proteoglycans, **ribosomes**, synaptosomes, and **viruses**. Preparative centrifugation has two

Preparative Isolation and Recovery of Solute(s)	Solute Analytical Information
Common Methods	Common Methods
Selective pelleting	Sedimentation velocity
Density gradient separations	Sedimentation equilibrium
Rate zonal	
Flotation	
Isopynic	

Figure 2. Common preparative and analytical methods used in centrifugation.

primary modes of separation. Solutes can be selectively pelleted from solution, as described here, or they can be subjected to separation along a solution density gradient (see **Density gradient centrifugation**).

Selective Pelleting or Differential Centrifugation

Clearing macromolecular components from solution is probably the most commonly employed preparative centrifugation technique. In this method the sample is subjected to centrifugation, usually, though not necessarily, at a constant rotor velocity. Over time, a pellet of sedimented material is deposited along the most radially distant wall of the tube containing the sample (Table 3). At a sufficiently high rotor speed (and therefore high centrifugal force), virtually all macromolecular solute components heavier than the solvent are pelleted out of solution over time. Because it is possible to select a variety of rotor velocities, sample tube dimensions, rotor types, centrifugation run times, etc., it is often possible to accomplish a significant level of purification by pelleting alone because the method of separation relies on the differential sedimentation rates of small and large solute components. For example, large components are typically removed by sedimentation at low speeds for short periods of time. If the supernatant from such a centrifugation is placed in a new container and centrifuged again at a higher rotor velocity for an appropriate period of time, it is possible to collect another distribution of particles from solution, this time of smaller dimensions than the first. Therefore repeating this process yields a series of pelleted samples with distributions of solute molecules, each of which have accumulated under conditions differing only in their centrifugal forces and thus have different sedimentation rates. Most samples of biological interest are complex mixtures of interacting solutes in solution at widely different concentrations. Because the size, shape, and concentration distributions of these components are routinely broad, together they exhibit very broad distributions of sedimentation rates. Thus the technique of solute pelleting, though widely applied, is only rarely sufficient to purify a single component completely. Typically it is used to selectively enrich, often to a high degree, a preparation of cellular or subcellular components in a single component. For small biological solutes, such as soluble cytosolic proteins prepared by a protein isolation procedure, pelleting is routinely employed along with differential **precipitation (salting out)** using ammonium sulfate (see **Sulfate salts**) to

Table 3. Descriptive Constants for Preparative Centrifugation

Term	Definition
Clearing factor, k	time (hours) $= k/s$, $s =$ *svedbergs*; the value of k varies with each rotor type and compares the relative rotor efficiencies for pelleting
Clearing time, t	$t = k/s$ (see clearing factor, above); the time required to pellet a particle to the base of the tube
Flotation Coefficient, f_s	For particles which float in a particular solution in a centrifugal field, it is similar to the sedimentation coefficient, units are in sec.
Relative centrifugal force or (RCF)	RCF $= 1.12\ r\ (\text{rpm}/1000)^2$, where r is in mm; the ratio of the centrifugal force at a specific rotation rate in rpm for a particular rotor at a particular radius r from the center of rotation of that rotor to the gravitational force of the Earth at sea level.
Square-root speed reduction law	Maximum allowed rotor speed $= \dfrac{\text{Desired speed}}{\sqrt{\dfrac{\text{Max allowed solution density}}{\text{actual solution density}}}}$ Used to calculate the permissible rotor speed in rpm when using nonprecipitating gradient materials at densities greater than the manufacturer permits in a rotor

yield highly enriched, sedimented precipitates of the desired protein components.

Pelleting is also of great use when applied to harvesting viruses or cells from various growth media. When cells or viruses are used to produce various recombinant proteins of pharmaceutical interest, centrifugal harvesting is often employed to capture the biological component from the (usually large) fermentation broth. In these applications, a preparative centrifuge and rotor capable of accepting a continuous flow into and through the rotor are employed. The centrifugal force applied to the sample stream as it passes through the rotor is adjusted to permit selective pelleting of cells within the centrifuge rotor. The cell-depleted broth passes out of the rotor/centrifuge and into a receiver vessel, and, then the harvested cells are recovered from the centrifuge rotor.

BIBLIOGRAPHY

1. K. E. Van Holde (1971) *Physical Biochemistry*, Prentice Hall, Englewood, Cliffs, NJ, pp. 70–121.
2. H. Fugita (1962) *Mathematical Theory of Sedimentation Analysis*. Academic Press, New York.

Suggestions for Further Reading

K. E. Van Holde (1971) *Physical Biochemistry*, Prentice–Hall, Englewood Cliffs, NJ, pp. 70–121. A short, highly readable and concisely illustrated treatment of the mechanical model of centrifugation for the biologist.

D. Eisenberg and D. Crothers (1979) *Physical Chemistry with Applications to the Life Sciences*, Benjamin–Cummings, Menlo Park, CA, pp. 701–745. An excellent introductory physical chemistry text with a good review of fundamental flow equations and a number of useful examples.

C. Tanford (1961) *Physical Chemistry of Macromolecules*, Wiley, New York, pp. 317–456. A classic and still relevant text on the nature of transport processes as applied to biological systems.

CENTROMERES

ALAN P. WOLFFE

The centromere (kentron = center; meros = part) is the region of the mitotic **chromosome** that participates in chromosomal movement. The **mitotic spindle** attaches to a specialized structure at the centromere known as the **kinetochore**. The motor responsible for the movement of the chromosomes toward the spindle poles during mitosis is also located at the centromere. Morphologically, centromeres are distinguished by their appearance at **metaphase** and **anaphase** as constrictions in the chromatin and by their **heterochromatin** staining pattern. A chromosome without a centromere is described as **acentric**. It does not segregate properly during **cell division** and is rapidly lost in successive cell cycles.

Centromeres occupy different positions in chromosomes, and they are useful markers. Chromosomes with a single centromere are called **monocentric**. Chromosomes with centromeres near one end are called **acrocentric**, those with the centromere visible at or near the middle are **metacentric**, and those with the centromere truly at the end are **telocentric**. The monocentric chromosomes found in most metazoan plants and animals almost always have centromeres embedded in segments of heterochromatin. Some organisms, such as **plants** in the genus *Luzula*, have **holocentric** chromosomes with diffuse centromeres. In these chromosomes the spindle attaches to centromeric heterochromatin that is distributed along the entire length of the chromosome (1). Mammalian tissue culture cells occasionally develop chromosomes with multiple centromeres (2). A chromosome with two centromeres is called **dicentric**. Dicentric chromosomes are often the result of chromosome breakage followed by fusion. A dicentric chromosome normally breaks at anaphase, when the centromeres are pulled in opposite directions. The holocentric chromosomes found in certain plants have developed as yet unknown mechanisms to avoid this problem.

Centromeres represent highly specialized chromosomal organelles. The simplest types of centromeres are found in the **yeast** *Saccharomyces cerevisiae*. The presence of these centromeric sequences, together with the **autonomously replicating sequences** that serve to direct **DNA** replication within exogenous **plasmid** DNAs allow these small **minichromosomes** to be stably maintained through cell division. The centromeric sequence allows the minichromosomes to be segregated in **mitosis** and **meiosis** with accuracy. Maintenance of minichromosomes through cell division also provides a simple assay for the definition of functional centromeric sequences in yeast.

All of the yeast centromeres are functional as small segments of DNA about 1,000 base pairs (1 kbp) or less in length. **Nucleosomes** are assembled in specific positions on these DNA sequences, and they may contain specialized **histone** variants (3). Within each segment of centromeric DNA are similar nucleotide sequences. In the middle of the centromeric DNA is a 220-bp segment that constitutes the centromeric core, that contains the minimal sequence necessary for centromeric function. Only a single spindle fiber attaches to a yeast centromere (4). If this attachment is to the core, then the nucleoprotein complex assembled on the core also functions as a kinetochore.

Mammalian centromeres are considerably more complex than those in *S. cerevisiae*. Early experiments on the centromere of the human Y-chromosome established that much more DNA is required for the chromosomal segregation function than in yeast. Over 300 kbp of DNA, including 200 kbp of α-satellite DNA, are required to generate a functional centromere. Deletion of the α-satellite DNA makes the centromere inactive (5).

The centromere can be broken up into three distinct structures, the kinetochore and the central and pairing domains (Fig. 1). The protein constituents of the centromere have been identified primarily through the use of autoantibodies from patients with rheumatic disease. These include the inner centromeric proteins (INCENP) and the CENP A, B, and C proteins. Immunological staining reveals that the INCENPs are in the pairing domain. The central domain contains dense chromatin known as constitutive heterochromatin. The kinetochore is anchored to this heterochromatin (Fig. 1). The DNA within this heterochromatin is composed primarily of various families of repetitive DNA (satellite DNA).

The α-satellite family of DNA sequences (which comprises 5% of the human genome) is probably present at the centromeres of all human chromosomes. The basic repeat, 171 bp long, occurs in large arrays up to 3×10^6 bp long. The α-satellite repeats assemble a specialized chromatin structure that provides many insights into the molecular

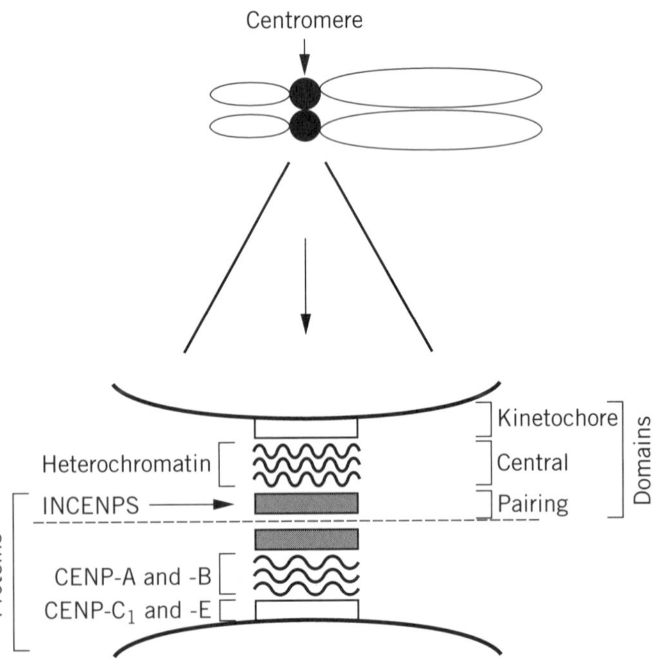

Figure 1. An expanded view of the centromere of a mammalian chromosome. The centromere is a bipartite structure. The central axis between the pairing domains is indicated by the dashed line. The relative positions of heterochromatin, the INCENPS, and CENPS are indicated within the kinetochore and the central and pairing domains.

nature of constitutive heterochromatin. It has been long that nucleosomes are positioned with respect to the nucleotide sequence found in α-satellite DNA so that histone-DNA contacts would begin and end at particular sites (6). An important point is that a single nucleosome is believed to exist on every 171 bp repeat. In modified form, this DNA sequence also provides the foundation for crystallization of the nucleosome core (7). Several interesting specialized chromosomal proteins also associate with α-satellite DNA. A 10-kDa protein, high mobility group protein (**HMG**-I/Y) binds to α-satellite DNA specifically *in vitro* (8). The HMG-I/Y protein binds selectively in the minor groove of the **double helix**, associating with runs of six or more AT base pairs. In addition HMG-I/Y also probably recognizes certain **secondary structural** features of the DNA molecule. CENP-B specifically recognizes a 17-bp DNA sequence (5′ CTTCGTTGGAAA CGGGA 3′) in a subset of α-satellite DNA repeats. The DNA-binding domain of CENP-B is necessary and sufficient for localization to the centromere in vivo (9). CENP-B contains anionic regions rich in **aspartic** and **glutamic acid** residues, which are characteristic of many proteins that interact with chromatin. CENP-B is found throughout the centromeric heterochromatin beneath the kinetochore plates (see **Kinetochore**). Although the exact functions of CENP-B in heterochromatin are not known, it may play a role in the higher order folding of centromeric chromatin through self-assembly mechanisms. Although CENP-B is found at all mammalian centromeres examined, the CENP-B recognition element is not found in all α-satellite DNA (10). It is possible that CENP-B can also be targeted to centromeres through interactions with other protein components.

CENP-A is also specifically associated with centromeric DNA and is especially interesting because it shares homology with core histone H3. Recent sequence analysis has revealed that CENP-A is a very specialized H3 variant (11). Each core histone and CENP-A have two **domains**, an amino-terminal tail domain that lies on the outside of nucleosomal DNA and a carboxyl-terminal, histone-fold domain that is involved in protein–protein interactions and in wrapping the DNA. CENP-A has a highly divergent amino-terminal tail domain and a relatively conserved carboxy-terminal, histone-fold domain. The targeting of CENP-A to centromeric DNA is directed by the histone-fold domain, although the exact mechanism by which this is achieved is still unknown. Characterization of a CENP-A homologue in yeast (CSE4, Ref. 3) establishes that this specialized histone is essential for normal chromosomal segregation during mitosis. Genetic evidence suggests that CSE4 heterodimerizes with histone H4 (12). This establishes that specialized nucleosomal structures are present at the centromere, in addition to the presence of positioned nucleosomes. Clearly, the centromere represents a highly differentiated chromosomal domain even at the most fundamental level of chromatin structure.

CENP-C is a large 107-kDa protein that is highly **hydrophilic** and basic (pI = 9.4) (13). CENP-C is concentrated in a narrow band immediately below the inner kinetochoric plate at the interface between the chromosome and the kinetochore (Figure 1). CENP-C binds DNA directly and is required for normal kinetochoric assembly. CENP-C is found only at the active centromere of a stable dicentric chromosome, suggesting a direct role in centromeric function (14). Other components of the centromere include CENP-E, which resembles **microtubule**-binding motor proteins. CENP-E is a 312-kDa **polypeptide chain** with a tripartite structure consisting of globular domains at the N- and C-termini, separated by a 1,500-residue α-**helical** domain that is predicted to form **coiled-coils**. CENP-E colocalizes to the centromere and kinetochore during metaphase, but it is released from the centromere at the onset of anaphase, when it is degraded. CENP-E functions as a kinetochoric motor during the early part of mitosis (15) (see **Kinetochore**).

The assembly of the centromere and the mechanisms of kinetochoric association with chromatin offer perhaps our best opportunity to understand the construction of a specialized chromosomal domain at the biochemical level.

BIBLIOGRAPHY

1. S. Pimpinelli and C. Goday (1989) *Trends Genet.* **5**, 310–313.
2. B. K. Vig and N. Paweletz (1988) *Chromosoma* **96**, 275–283.
3. S. Stoler, K. C. Keith, K. E. Curnick, and M. Fitzgerald-Hayes (1995) *Genes Dev.* **8**, 573–586.
4. J. B. Peterson and H. Ris (1976) *J. Cell Sci.* **22**, 219–226.
5. C. Tyler-Smith, R. Oakley, and Z. Larin (1993) *Nature Genet.* **5**, 368–375.
6. R. T. Simpson (1991) *Prog. Nucl. Acids Res. Mol. Biol.* **40**, 143–184.
7. K. Luger et al. (1997) *Nature* **389**, 251–260.
8. M. J. Solomon, F. Strauss, and A. Varshavsky (1986) *Proc. Natl. Acad. Sci. USA* **83**, 1276–1280.
9. A. F. Pluta et al. (1995) *Science* **270**, 1591–1594.
10. I. G. Goldberg et al. (1996) *Mol. Cell Biol.* **16**, 5156–5168.
11. K. F. Sullivan, M. Hechenberger, and K. Masri (1994) *J. Cell Biol.* **127**, 581–592.
12. M. M. Smith et al. (1996) *Mol. Cell Biol.* **16**, 1017–1026.
13. H. Saitoh et al. (1992) *Cell* **70**, 115–125.
14. J. Tomkiel et al. (1994) *J. Cell Biol.* 125, 531–545.
15. K. D. Brown, K. W. Wood, and D. W. Cleveland (1996) *J. Cell Sci.* **109**, 961–969.

ČERENKOV RADIATION

D. R. FISHER

Čerenkov radiation, or the Čerenkov effect, is the visible bluish-white light surrounding a radioactive source observed in a pool of water. The phenomenon was first reported by the Russian physicist P. A. Čerenkov in 1934 (1), who was one of several scientists who observed the glow of light in close proximity to intense gamma-radiation sources. This effect is the reason why radiation sources are said to "glow in the dark." Its importance for molecular biology is that it is routinely used to measure the amount of the isotope ^{32}P in a sample (see **Radioactivity** and **Phosphorous isotopes**).

The Čerenkov effect can be explained by classic electromagnetic theory and principles of optic science. Analysis of Čerenkov radiation has shown fundamental relationships between the velocity of charged particles, light, the intensity of the light, and its wavelength spectrum (2). Only a small fraction (<0.1%) of the charged-particle radiation is emitted by the absorbing medium as coherent light. According to Jelley (3), who provided the first scientific explanation for the Čerenkov effect, the observed light results from charged particles traversing a transparent dielectric medium, one that does not conduct electricity. Charged particles are produced in the absorbing medium (water) when gamma radiation from the radioactive source interacts with the absorber. These charged particles produce local polarization along their path in the dielectric. Light in the visible spectrum is emitted when the polarized molecules in the medium return to their rest state soon after passage of the charged particle. If the velocity of the charged particles is less than that of light in the same medium, the light emitted from molecules in the dielectric is overridden and not observed. However, if the velocity of the charged particles is greater than that of light in the dielectric, a wavefront of light is produced from individual molecules in the dielectric, and the emission is reinforced by constructive interference. The Čerenkov effect is analogous to the bow wave from a ship that travels faster than the velocity of the surface waves, or to the shock wave trailing a supersonic aircraft passing though air. In other words, if the velocity v of a charged particle traversing a transparent dielectric material of refractive index n ($= \beta c$) exceeds the velocity of light (c/n) in the medium, or $v > c/n$ and $\beta > 1/n$, then Čerenkov radiation is emitted at an angle ξ relative to the particle direction, where $\xi = \arccos\ (1/\beta c)$. The velocity c of light in a vacuum is 2.997×10^{10} cm/s, and $\beta = v/c$. The Čerenkov photons form a conical wavefront of half-angle ($90° - \xi$) behind the particle.

The polarization effect actually decreases the energy lost by a charged particle that traverses a condensed medium, whereas the production of Čerenkov radiation increases the loss of energy by the particle. The visible spectrum is produced at a frequency interval of about 3×10^{14} Hz. Applications of Čerenkov radiation theory have been developed for detecting single high energy charged particles, measuring the energy of charged particles, and determining angles of incidence.

The light pulses emitted by charged particles traveling in a transparent medium can be collected and counted by modern scintillation counters. The amount of light produced in water is small compared to that produced in the presence of a scintillator, but it can be detected from beta-emitting radionuclides if their energy is greater than the threshold energy of 265 keV. The average energy of phosphorous-32 beta particles is 695 keV, so the majority of those emitted can be detected.

BIBLIOGRAPHY

1. P. A. Čerenkov (1934) *Compt. Rend. Acad. Sci. URSS* **2**, 451.
2. I. Frank and I. Tamm (1937) *Compt. Rend. Acad. Sci. URSS* **3**, 109.
3. J. V. Jelley (1953) *Atomics* **4**, 81.

Suggestions for Further Reading

F. H. Attix (1986) *Introduction to Radiological Physics and Radiation Dosimetry*, Wiley, New York.

W. J. Price (1958) *Nuclear Radiation Detection*, 2nd ed., McGraw-Hill, New York.

CHAIN-TERMINATION (DIDEOXY) DNA SEQUENCING

CARL W. FULLER

The relative length of a DNA chain is determined easily and accurately by **gel electrophoresis**, and the most common types of **DNA sequencing** use methods that map sequence information to DNA chain lengths. For chain-termination DNA sequencing, this is done by synthesizing the complement of the DNA using a **DNA polymerase** under conditions that terminate synthesis at sites where only one of the four bases occurs. Thus, all sequencing experiments are done in three steps. First, a single pure DNA segment is isolated for sequencing. Second, this DNA is used as a template for synthesis catalyzed by a **DNA polymerase** with mixtures of normal and chain-terminating nucleotides. Finally, the products of this synthesis are separated according to size by gel electrophoresis. Numerous variations on each of these steps are commonly used.

ISOLATION OF SPECIFIC SEGMENTS OF DNA

Most of the techniques of molecular biology rely on isolating specific segments of DNA, and sequencing is no exception. In fact, the first practical application of chain-termination sequencing relied on **cloning** specific DNA segments using vectors derived from **M13 bacteriophage**. These bacteriophages contain a single-stranded DNA **chromosome** that accommodates inserts of more than 5000 bases. Isolating the single-stranded DNA in pure form from these phages is simple and inexpensive, so these vectors are still commonly used. Similarly, virtually all plasmid vectors are commonly used for DNA sequencing, and essentially any clone of up to about 200 kb can be sequenced by cycle sequencing techniques, provided sufficient DNA can be purified. Another popular way of isolating DNA for sequencing is by using the polymerase chain reaction (**PCR**). With nested PCR primers, it is now relatively simple to amplify segments of **genomic** DNA for direct sequencing in a matter of hours.

CHAIN TERMINATION REACTIONS

Chain termination reactions require the isolated template DNA, a suitable primer, 2′-deoxynucleoside triphosphates (dNTPs), 2′,3′-dideoxynucleoside triphosphate (ddNTP) chain terminators, and a DNA polymerase. The most critical component is the DNA polymerase. Many DNA polymerases have been used for sequencing, including those from eubacteria, such as the DNA polymerase I (**Klenow fragment**) of *Escherichia coli*, **reverse transcriptases** from retroviruses, such as avian myeloblastosis virus, polymerases from bacteriophage, such as **T7 phage** DNA polymerase, and polymerases from archaea, such as *Thermococcus litoralis*. Virtually all of these have been genetically or chemically modified to eliminate exonuclease activities or to improve reaction rates with the ddNTPs.

The most recent examples of polymerases specifically engineered for DNA sequencing are enzymes derived from *Thermus aquaticus*, **Taq DNA polymerase**. Two regions of this polymerase are modified to produce particularly effective polymerases for DNA sequencing. First, *Taq* DNA polymerase has 5′-3′ exonuclease activity, which degrades sequencing primers. The first 300 amino acid residues at the N-terminus of this enzyme are required for this exonuclease activity. Portions of this **domain** can be deleted, or the activity can be eliminated by point mutation. Secondly, normally *Taq* DNA polymerase is relatively inefficient at using ddNTPs. As discovered by Tabor and Richardson, this can be improved more than 10^4-fold by changing residue Phe667 to Tyr. This modification improves ddNTP usage, and it also greatly improves the quality of the sequence data obtained. Native *Taq* DNA polymerase produces sequencing bands that vary in intensity more than 15-fold, depending on the nearby sequence. In contrast, Tyr667 polymerase produces bands that vary in intensity by less than threefold. This makes interpretating the results of the electrophoretic separation much more accurate. A number of polymerases that have this modification are now commercially available, as is T7 DNA polymerase, which naturally has a tyrosine at the corresponding position.

CYCLE SEQUENCING

Cycle sequencing is the process of using repeated cycles of thermal denaturation and polymerization to produce greater amounts of product in a DNA sequencing reaction. The amount of product DNA increases linearly with the number of cycles. (This distinguishes it from PCR, which uses two primers so that the amount of product increases exponentially with the number of cycles.) During each cycle, the thermostable DNA polymerase extends the annealed primer molecules, typically at 60° to 70°C. The mixture is heated above the melting temperature of DNA (95°C), dissociating the extended primer from the template. Then, the mixture is cooled, allowing another molecule of primer (which is present in excess) to anneal to the limited supply of template. Further cycles of extension and denaturation result in producing much more extended primer than the amount of template used. This improves the sensitivity of the sequencing experiment, and it also allows ready use of double-stranded templates for sequencing. Generally, cycle sequencing works much more reliably over a wider range of template concentrations than noncycled protocols. This accounts for its nearly universal application for large-scale DNA sequencing projects.

METHODS FOR LABELING DNA SEQUENCES

The products of the chain termination reactions must be labeled for all practical DNA sequencing methods. The original label was α-^{32}P dATP that was simply added to the chain-termination reaction. Newly synthesized DNA was labeled with radioactive phosphorous, and detected by simple **autoradiography**. More recently, the lower energy isotopes ^{33}P and ^{35}S (in the form of α-thio-dATP) have been used because they generate autoradiograms with higher resolution. These offer the advantage of using less total radioactivity than other methods. In addition, only specifically terminated, elongated DNA chains are labeled and therefore visualized by autoradiography, which eliminates the background bands and stops normally observed on DNA sequencing autoradiograms and results in extremely clean sequence data.

Automated, **fluorescent** DNA sequencing methods were introduced in 1987 and have become essential tools for large-scale sequencing efforts. The sequence products used by these automated systems are labeled by fluorescent primers (dye primers) or fluorescent dideoxynucleotides (dye terminators). These have been used in single-color detection instruments, and in four-color multiplex instruments in which the four bases are distinguished by color. Recent innovations include fluorescent dye-labeled DNA primers that exploit fluorescent **energy transfer** to optimize the absorption and emission properties of the label. These primers carry a fluorescein derivative at the 5′-end as a common donor and rhodamine derivatives attached to a modified thymidine within the primer sequence as acceptors. Adjustment of the donor–acceptor spacing by placing the modified thymidine in the primer sequence allows generating four primers. All have strong absorption at a common excitation wavelength (488 nm) and efficient fluorescent emission at 525, 555, 580, and 605 nm. These improve the sensitivity and accuracy of the automated sequencing system.

Fluorescent dye-labeled ddNTP terminators have also been used extensively for DNA sequencing, and those that use the energy-transfer principle are also commercially available. Like the radio-labeled terminators, they have the advantage of labeling only specifically terminated, elongated DNA chains, so that background bands are eliminated.

ELECTROPHORESIS AND AUTOMATED SEQUENCING

The high-resolution separation of DNA fragments by size is essential for all sequencing methods. For radioactively labeled DNA sequencing experiments, this is done by using gels cast in glass plates that are 0.2 to 0.4 mm thick, 40 to 80 cm long, and wide enough to accommodate 32 to 96 samples. Typically, the gels are 4 to 8% polyacrylamide cross-linked with *N,N′*-methylene bisacrylamide (see **Polyacrylamide**) and contain tris borate buffer (0.089 M, pH 8.3) and 7 to 8 M urea. After electrophoresis for 2 to 18 hours, the gels must be removed from the glass plates for autoradiography and reading of the sequence of 200 to 400 nucleotides. Because these gels are cumbersome to make and use, considerable effort has been made to improve separation methods. The most commonly used methods involve a sensitive fluorescent detection instrument that continuously monitors the migration of fluorescent-labeled DNA past a fixed position on the gel. The results are collected and evaluated directly by computer, producing finished sequence information. This saves considerable labor in "reading" the sequence from the gels and improves the resolution sufficiently to read 500 or more bases routinely from a single sequence experiment. Noncross-linked "gels" have also been introduced that run in 50 to 100 micron diameter, 40 to 70 cm long capillaries with fluorescent detection. The efficient heat transfer of these electrophoresis media allow faster, high-resolution separations.

Suggestions for Further Reading

C. W. Fuller and M. A. Reeve (1995) Thermo sequenase - A novel thermostable polymerase for DNA sequencing, *Nature* **376**, 796–797.

J. Ju et al. (1995) Design and synthesis of fluorescence energy transfer dye-labeled primers and their application for DNA sequencing and analysis, *Anal. Biochem.* **231**, 131–140.

L. G. Lee et al. (1992) DNA sequencing with dye-labeled terminators and T7 DNA polymerase: Effect of dyes and dNTPs on incorporation of dye-terminators, and probability analysis of termination fragments, *Nucleic Acids Res.* **20**, 2471–2483.

J. Messing (1983) Bacteriophage M13 vectors for DNA sequencing, *Methods Enzymol.* **101**, 33–38.

G. M. Prober et al. (1987) system for rapid DNA sequencing with fluorescent chain-Terminating Dideoxynucleotides, *Science* **238**, 336–341.

B. B. Rosenblum et al. (1997) New dye-labeled terminators for improved DNA sequencing patterns, *Nucleic Acids Res* **25**(22), 4500–4504.

F. Sanger, S. Nicklen, and A. R. Coulson (1977) DNA sequencing with chain terminating inhibitors, *Proc. Natl. Acad. Sci. USA* **74**(12), 5463–5467.

L. M. Smith et al. (1986) Fluorescence detection in automated DNA sequence analysis, *Nature* **321**(6071), 674–679.

S. Tabor and C. C. Richardson (1995) A single hydroxyl moiety in Pol I-T DNA polymerase is responsible for distinguishing between deoxy- and dideoxyribonucleotides, *Proc. Natl. Acad. Sci. USA* **92**, 6339–6343.

P. B. Vander Horn et al. (1997) Thermo sequenase DNA polymerase and *T. acidophilum* pyrophosphatase: New thermostable enzymes for DNA sequencing, *BioTechniques* **22**, 758–765.

CHAOTROPES: KOSMOTROPES

SERGE N. TIMASHEFF

Compounds that increase and decrease, respectively, the aqueous solubility of proteins are classified as chaotropes and kosmotropes (1). Therefore, the **Hofmeister series** of ions may be divided into these two categories. For the series of anions, solubilization in water is promoted in the order

$$SCN^- > CLO_4 > I^- > Br^- > Cl^- > F^- > HPO_4^{2-} > SO_4^{2-}$$

Those on the left-hand side of Cl^- generally increase the solubility; they are called *chaotropes*. Those on the right-hand side of Cl^- act as salting-out agents; they are called *kosmotropes*. Chaotropes are **water**-structure breakers; kosmotropes, to the contrary, are water-structure makers. A similar division may be made about the Na^+ ion in the cationic Hofmeister series. In the salts, the predominant effect is that of anions. An important characteristic is that the effects are additive for all the species in the solution.

Organic molecules also affect water structure and can be classified as chaotropes and kosmotropes. Thus, **urea**, glycine, formamide, and acetamide are chaotropes. Organic kosmotropes are sucrose, polyols, the methylacetamines, methyl formamides, and methyl ureas. Except for sugar and the polyols, caution must be exercized in using organic kosmotropes, since their action may reverse itself at high concentration.

The mechanism of action of these molecules in stabilizing or destabilizing protein structure, and as a corollary in acting as **salting in** or **salting out** agents, is related to their effects on the orientation of water molecules through polar interactions with the water **hydrogen bond** donor and acceptor properties. By perturbing or strengthening the interactions between water molecules (maximization of hydrogen bond formation), these molecules exercise their chaotropic or kosmotropic actions. At interfaces, the water structure is already perturbed by the nonavailability at the surface of other water molecules for hydrogen bond formation. This is true for the water–air interface. Similar perturbations may occur at interfaces with proteins and other biological entities—hence, the effect on their properties. Control of reactions by the chaotropic/kosmotropic effect is very widespread in biological systems. Its principal diagnosis is that the compound which affects a particular process must be present at a high concentration (local concentration for cell compartments and organelles).

BIBLIOGRAPHY

1. K. D. Collins and M. W. Washabaugh (1985) *Quart. Rev. Biophys.* **18**, 323–422.

CHAPERONIN

R. J. ELLIS
F. U. HARTL

The chaperonins are a family of **molecular chaperone** found in all types of cell whose function is to assist the correct folding of certain polypeptide chains that have been either newly synthesized or generated from native proteins by environmental stresses that cause partial unfolding.

NOMENCLATURE

The term *chaperonin* was suggested by Sean Hemmingsen (1) to describe a family of highly sequence-related molecular chaperones found in **chloroplasts**, **mitochondria**, and **eubacteria** such as *Escherichia coli*. This term was proposed to simplify the existing complex nomenclature for different members of this family whose close relationship was only realized when cDNA for the chloroplast chaperonin was sequenced (1). The term was subsequently extended to distant homologues found in **archaebacteria** and the eukaryotic cytosol (2,3). The two subfamilies of the chaperonin family are referred to as follows:

1. The GroE or group I subfamily, found in chloroplasts and other plastids, mitochondria, and all eubacteria
2. The TCP-1 or group II subfamily, found in archaebacteria and the eukaryotic cytosol

The GroE terminology reflects the fact that the eubacterial protein was first identified by genetic studies in four laboratories in 1972/73 as a bacterial protein required for the replication of **bacteriophage** such as lambda in *E. coli* (4). "Gro" refers to phage growth, and the suffix "E" refers to the observation that the phage growth defect is overcome when the phage carries a mutation in the head gene E. The TCP-1 terminology is derived from the identification of a protein called the *t*-complex polypeptide encoded by the mouse T locus (5). The mitochondrial members of the GroE subfamily are sometimes referred to as "hsp60" proteins, but this terminology should not be used to describe the chaperonins as a whole, since the eukaryotic TCP-1 members and the chloroplast GroE members are not **heat shock** proteins.

Table 1 presents a suggested nomenclature and useful abbreviations for members in both subfamilies, and lists other names that are used. Some authors confine the abbreviation

Table 1. Nomenclature of the Chaperonins

Preferred Name	Other Names	Useful Abbreviations
GroE subfamily		
Eubacterial chaperonin 60	GroEL (*E. coli*), 65-kDa antigen	Eu cpn60
Eubacterial chaperonin 10	GroES (*E. coli*), cochaperonin	Eu cpn10
Mitochondrial chaperonin 60	hsp60, mitonin, HuCha60	Mt cpn60
Mitochondrial chaperonin 10	hsp10, cochaperonin	Mt cpn10
Chloroplast chaperonin 60	Rubisco subunit binding protein	Ch cpn60
Chloroplast chaperonin 10	chaperonin 21, cochaperonin	Ch cpn10, Ch cpn21
TCP-1 subfamily		
Cytosolic chaperonin 60	*t*-Complex polypeptide 1, chaperonin-containing TCP-1, TCP-1 ring complex	Cyt cpn60, TCP-1, CCT, TRiC
Archaebacterial chaperonin 60	Thermophilic factor 55, thermosome	Ar cpn60, TF55

"cpn60" to the GroE subfamily, but this restriction is inconsistent with the fact that the TCP-1 members share with the GroE members subunit molecular masses of approximately 60 kDa.

FUNCTION

The main function of the chaperonins is to assist the folding of certain newly synthesized polypeptide chains into their biologically active conformations (see **Protein folding *in vivo***). They achieve this end, not by providing steric information required for each chain to fold correctly, but by sequestering each partially-folded chain in a protected compartment generated by the oligomeric structure of the chaperonin molecule. In this compartment, each chain can continue to fold to a point where aggregation with other partially folded chains is no longer a problem (6–9). This aggregation arises because some proteins fold via intermediate states that expose **hydrophobic** surfaces transiently and are thus subject to intermolecular interactions with similar states. This aggregation effect can sometimes be observed when chemically denatured proteins refold spontaneously in dilute solution *in vitro* (see **Protein folding *in vitro***), but its magnitude is expected to be enhanced in the intact cell because of the phenomenon of macromolecular crowding, or **excluded volume**, thus increasing the effective concentrations of these states by two or three orders of magnitude (8,9). It should be stressed that, although the chaperonins increase the yield of correctly folded proteins by minimizing aggregation, they do not increase the rate of folding above that achieved by the fastest folding fraction of spontaneously refolding chains that manage to avoid aggregation.

The best evidence for this view of chaperonin function comes from genetic and **pulse-chase** labeling studies with living cells. Several proteins imported into the mitochondria of yeast cells form aggregates when the function of the mitochondrial chaperonin is impaired by mutation (10), whereas about 30% of newly synthesized soluble cytoplasmic proteins in *E. coli* cells become either insoluble or inactive when cpn60 function is switched off by means of a **temperature-sensitive mutation** (11). *In vitro* studies of the ameliorating effect of added GroE proteins on the aggregation of various pure denatured proteins during their refolding after removal of denaturant are consistent with this antiaggregation role (12–14). *In vitro* studies also suggest that a side effect of this antiaggregation role is the partial unfolding of nonaggregated chains that have become kinetically trapped in misfolded conformations, thus allowing these chains another chance to fold correctly (15); it is not clear how important such an effect may be *in vivo*. In addition, the GroE chaperonins in eubacteria and mitochondria (16) and the TCP-1 chaperonins in the archaebacteria (17) are stress proteins that prevent and reverse the denaturation of some fully folded proteins by stresses such as high temperature (see **Stress response**).

It must be emphasized that the actual spectrum of proteins that use chaperonin function to fold correctly in intact cells is not known, but preliminary calculations and observations suggest that it is probably a minority in the case of both the GroE chaperonins (9,18) and the TCP-1 chaperonins (7). Genetic evidence derived from the study of yeast mutants suggests that a major function of the eukaryotic TCP-1 is to assist the folding of **tubulin** and **actin** (19), consistent with the observation that newly synthesized chains of actin and tubulin can be isolated from pulse-labeled cells of **CHO** cells in the form of complexes with TCP-1 chaperonin oligomers (20). *In vitro* protein refolding experiments also suggest that the substrate specificities of the GroE chaperonins and the TCP-1 chaperonins are different (6).

STRUCTURE

The chaperonins occur as large oligomeric structures consisting of two stacked rings of subunits, each about 55–60 kDa in size, surrounding a central cavity or cage in which the protein substrate binds; each subunit catalyzes the slow hydrolysis of ATP to ADP. The GroE chaperonins have seven subunits per ring, whereas the TCP-1 chaperonins have eight or nine. The subunit sequence identity between members of the GroE subfamily is in the range 42–76%, whereas that between members of the TCP-1 subfamily is around 30–40%, but the identity between the subfamilies is much less (15–20%) and is confined to regions corresponding to the **ATPase** domain in GroE cpn60. The evolutionary implications of these similarities are under debate (7,21).

A convenient abbreviation for the chaperonin(s) is cpn(s). Thus the subunits of the ring can be called *cpn60 subunits*, and the oligomer can be called *cpn60*. The GroE subfamily also contains another type of oligomeric ring made of seven smaller 10-kDa subunits, called *cpn10*, because there is a slight similarity between the cpn10 subunit sequence and part of the cpn60 subunit sequence (21). GroE cpn60 and cpn10 oligomers bind to each other in 1:1 and 1:2 ratios in the presence of either ATP or ADP, and these binary complexes play an essential role in the protein folding function of these molecules (8,9). The chloroplast cpn10 oligomer is unusual in that each subunit consists of two copies of a cpn10 sequence fused head to tail and thus is often referred to as *cpn21* (22).

The TCP-1 oligomers found in the eukaryotic cytosol are much more variable than those found in either the archaebacteria or the GroE subfamily, since subunits of the latter

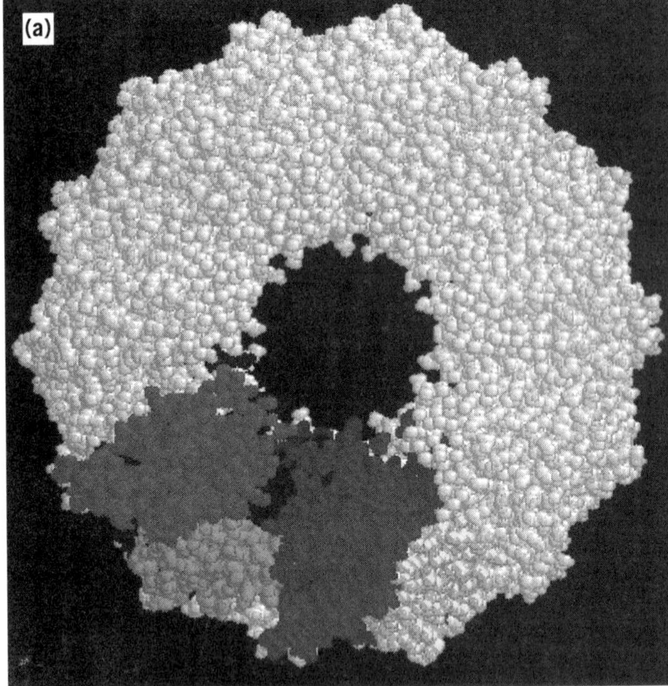

Figure 1. Structure and function of the chaperonin system. (**a**,**b**) Space-filling representations showing a top and side view, respectively, of the crystal structure of a double mutant form of GroEL (26). Two adjacent subunits are colored with the apical domains in red and purple, the intermediate domains in orange and yellow, and the equatorial domains in blue and green, respectively. Free passage between the GroEL rings is obstructed by *N*- and *C*-terminal residues not resolved in the crystal structure. (**b**) side view of GroES (28) at the top; two adjacent domains and a single mobile loop that is structured in the GroES crystal due to crystal packing are colored. The loop regions normally protrude from the base of GroES downward toward the GroEL. (**c**) Asymmetrical GroEL/GroES complexes as revealed by cryoelectron microscopy (45). Note the upward and outward movement of the apical GroEL domains interacting with GroES. (**d**) Model of the GroEL/GroES reaction cycle in assisting protein folding (30); see text for details. The term *unfolded protein* refers to a partially folded intermediate that is represented by the light pink spheres; folded protein is represented by the dark pink sphere, while the hatched spheres represent a mixture of folded and partially- folded proteins expected in a population of GroEL molecules. At step 4, GroES may associate with either the protein-containing ring of GroEL or with the opposite empty ring; the latter possibility is not shown. Reprinted from *Nature* 381, 571–580 (1996), with permission; copyright (1996) Macmillan Magazines Ltd. See color insert.

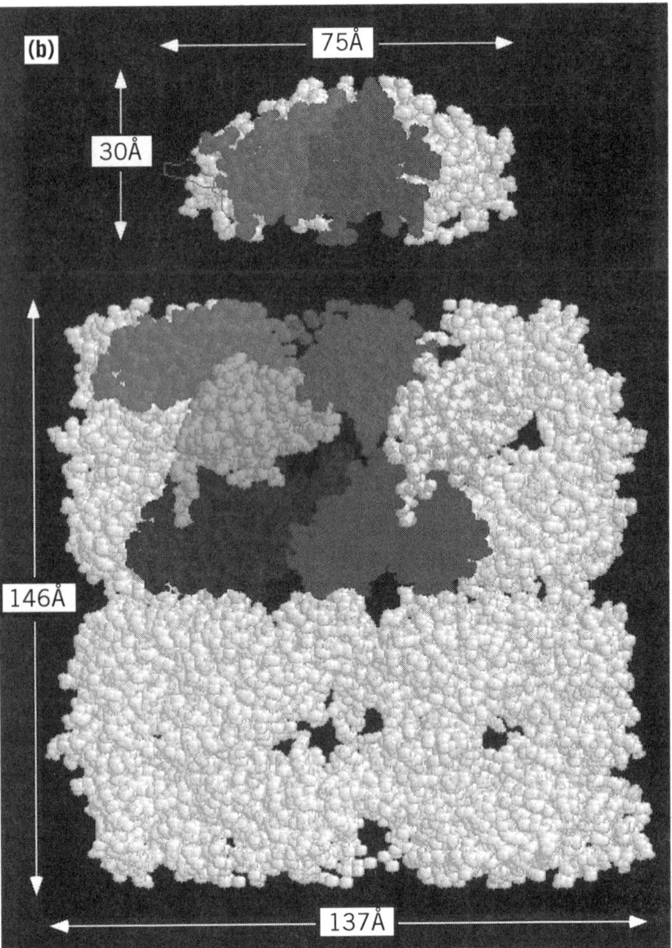

Figure 1. (*Continued*)

consist of only one or two sequences, whereas the former have at least eight different, but related, subunit sequences in each ring (6,7,23). For this reason, the eukaryotic TCP-1 oligomers are also called *CCT complexes*, for chaperonin-containing TCP-1 complexes, or *TriC* for TCP-1 ring complexes (see Table 1). The TCP-1 chaperonins do not appear to contain cpn10-like members.

The best-studied chaperonins are the cpn60 and cpn10 oligomers from *E. coli*, termed GroEL and GroES, respectively. GroEL has been extensively studied by both **negative stain** and **cryoelectron microscopy** (24,25) and by **X-ray crystallography** of a double mutant form (26). GroEL is a cylindrical structure containing a central cavity about 50 Å in diameter (Fig. 1a). Each subunit consists of (*1*) an apical domain to which the protein substrate and GroES bind; (*2*) an equatorial domain that protrudes into the central cavity, which contains the ATPase site and is responsible for most of the intersubunit interactions; and (*3*) an intermediate domain that contains potential hinge sites responsible for the considerable movements of the other two domains revealed by electron microscopy (24,25). These conformational changes result from the binding of nucleotide and GroES to the GroEL (26) cylinder. The crystal structure of GroES shows a dome-shaped structure with a hydrophilic inner surface that fits over one end of the GroEL cylinder (24,27,28) (see also Fig. 1b,c), creating a large enclosed dome-shaped space about 65 × 80 Å in size, inside which the protein substrate continues to fold (29–32) (see also Fig. 1c,d).

No crystal structure for any TCP-1 chaperonin is yet available, but electron-microscopic studies suggest that the general architecture and overall domain structure of the TCP-1 cylinder resembles that of GroEL (33,34). The archaebacterial chaperonin was originally termed thermophilic factor 55, because it is virtually the only protein made by thermophilic

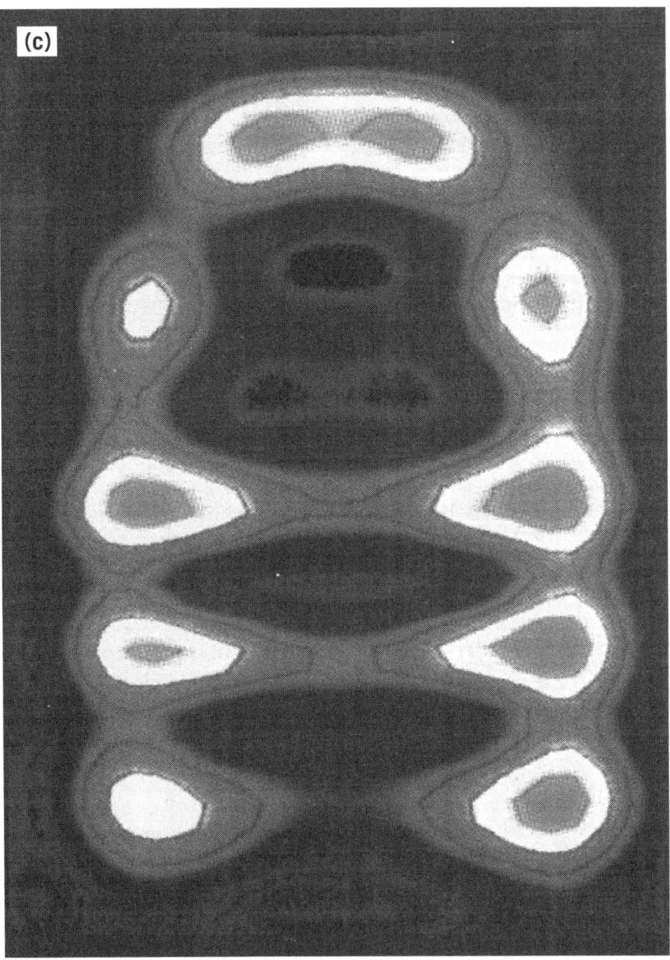

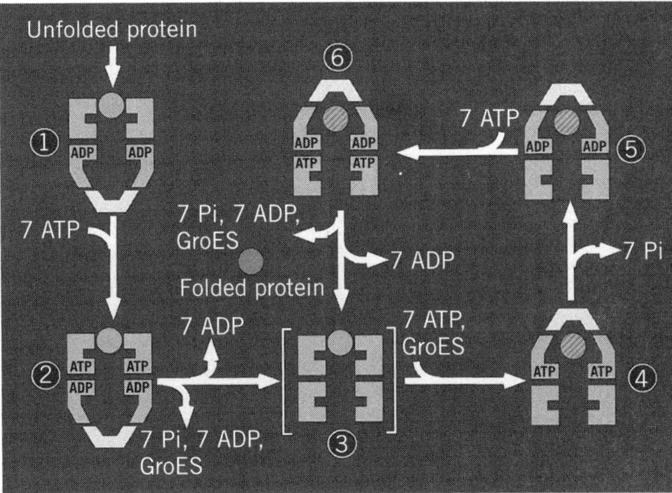

Figure 1. *(Continued)*

archaebacteria under heat-shock conditions (35), but it is now termed the *thermosome* (36). Thermosomes generally contain eight subunits per ring, but there are several reports that a significant subset of particles contain nine subunits per ring, while the chaperonins of some *Sulfolobus* species appear all to contain nine subunits per ring (37). The thermosomes of *Pyrodictium*, *Thermoplasma*, and *Sulfolobus* consist of two different but related subunits of almost identical molecular mass (33). Electron-microscopic image analysis suggests that the two types of subunit in the thermosome of *Thermoplasma acidophilum* alternate within each ring, generating a four-fold symmetry (38).

Averaged electron-microscopic images of the mammalian TCP-1 particle reveal a ringlike structure with eightfold quasi-rotational symmetry with a central channel (33). Primary structures of eight distinct types of related subunit have been determined by cloning and sequencing of mouse cDNA; these types share about 30% amino acid residue identity. A striking observation is that the sequence of each type is more highly conserved among different eukaryotic species than are the sequences between types in any one species; in mammals the sequence identity is over 96% for each type of subunit, and around 60% between yeast and mouse. These observations have prompted the conclusion that each type of subunit diverged early in evolution, has changed only slowly during the evolution of eukaryotes, and thus may have a specific function in binding subsets of sequences in substrate proteins and/or other molecular chaperones (6,7). Consistent with this suggestion is the observation that the different types of subunit in each TCP-1 chaperonin ring deviate most from one another in the sequences of the apical domains, where the protein substrate binds.

MECHANISM OF ACTION

The molecular details of how the chaperonins function have, and continue to be, studied by means of *in vitro* protein refolding experiments using defined components, but such experiments suffer from the disadvantage that the conditions under which they are performed are different in important respects from those operating during protein folding in the intact cell (9,18). Thus it is not surprising that this field is subject to intense debate between people supporting different views (39). Since the chaperonins evolved to fold proteins inside cells,and not inside test tubes, it is essential when evaluating the validity of these views to define those aspects of chaperonin action observed *in vitro* that are mechanistically essential to improve the yield of correctly folded protein, when extrapolated as far as possible to *in vivo* conditions. In our opinion, the model outlined in Figure 1d best meets this criterion for the GroE chaperonins (30,40).

In this model, newly synthesized polypeptide chains are released from **ribosomes** in the form of partially folded intermediates that may contain bound molecular chaperones of the hsp70/DnaJ families (see **Protein folding *in vivo***). The latter chaperones prevent aggregation and premature folding before each chain is complete. Such intermediates are commonly referred to as *unfolded proteins*, but in fact they often are **molten globules**. These partially folded, compact intermediates lose the hsp70/DnaJ chaperones by an uncharacterized mechanism that results in each intermediate molecule binding by hydrophobic interactions to the apical domains at one end of a GroEL cylinder; the other end of the cylinder is occupied by GroES as a result of ADP binding to the GroEL ring proximal to the GroES (see Fig. 1d, step 1). The binding of polypeptide and the hydrolysis of ATP by the ring not occupied by GroES causes the GroES and the ADP to dissociate (step 2). GroES and ATP then rebind with equal probability to either ring (not both rings), resulting in 50% of the bound polypeptide being encapsulated inside the cavity capped by GroES (step 4). This rebinding of GroES results in the displacement of the bound polypeptide into the cavity, where it can continue to fold; it is for this reason that this model has been dubbed the "Anfinsen cage" model to stress the idea that the protein in the cavity folds in a similar, but not necessarily identical, manner to that by which it folds in a protein renaturation experiment of the type pioneered by Anfinsen (41,42). Thus one essential difference between chaperonin-assisted protein folding *in vivo* and spontaneous protein refolding *in vitro* is that the former process segregates each chain into the protected compartment provided by the GroEL/GroES cylinder to avoid the problem of aggregation.

The time available for this protected folding is set by the time it takes each ring of GroEL to hydrolyze seven ATP molecules. This hydrolysis is positively cooperative in each ring, and two rounds take about 10 s at 37°C (steps 4 and 5). The GroES then dissociates, and the protein is free either to diffuse into the cytosol or to rebind to the apical domains of the same GroEL cylinder (step 6). Rebinding occurs only if sufficient hydrophobic residues are still exposed on the surface of the compact intermediate. The cycle of release into the cage then repeats until the protein is folded sufficiently to the point where aggregation with other partially folded proteins is no longer a problem for the cell. The number of reaction cycles will vary, depending on the rate at which a given protein can fold when released into the cavity. Even with a single type of protein substrate, the population of folding chains will not be in synchrony or occupy the same number of cycles, since some chains fold more rapidly than others of the same sequence (see **Protein folding *in vitro***). Usually only a fraction of the polypeptide chains fold to completion in a single cycle of GroES binding and unbinding, depending on the intrinsic rate of folding of a particular polypeptide and its tendency to become trapped in misfolded states. Rebinding of these intermediates to the apical domains of GoEL results in partial unfolding in preparation for a subsequent folding trial (30). This unfolding function distinguishes the GroEL/ES machinery from a simple folding cage,which acts solely to prevent intermolecular aggregation.

Uncertainties about this model include the fate of the 50% of polypeptide bound to the GroEL ring not capped by GroES; such polypeptides would be prone to aggregation and degradation if released into the cytosol. Release of such polypeptides *in vitro* is reduced in the presence of high concentrations of synthetic polymers that mimic the macromolecular crowding, **excluded volume** effect present in the cytosol (40). However, some release of partially folded chains may be essential *in vivo* because of the danger that proteins that are unable to fold correctly as a result of mutation will clog up the GroEL apical domains; an escape route for such molecules may be provided by release from the ring not capped by GroES.

Another problem concerns the significance of GroEL molecules with GroES bound to each end of the cylinder that have been observed *in vitro* (43). The ratio of total GroES subunits to total GroEL subunits in extracts of *E. coli* cells is equimolar, consistent with such GroES/GroEL/GroES complexes occurring *in vivo*, where their formation may be favored by the effect of macromolecular crowding on protein association. Nevertheless, such complexes are not mechanistically essential, at least as deduced from *in vitro* refolding experiments, nor can they be permanent if the compact intermediates are to enter the GroEL cavities.

Another limitation imposed by this model is the size range that can be accommodated inside the cage, which is likely to have an upper limit around 60 kDa. Larger proteins than this, up to about 150 kDa, exist inside bacterial cells and even larger ones in the cytosol of eukaryotic cells. A GroES equivalent to cap the TCP-1 chaperonin oligomer has not been identified, so it is possible that the latter assists protein folding in a manner somewhat different from that for the GroE chaperonins. There is evidence that, unlike GroEL,the mammalian TCP-1 chaperonin binds to chains of actin being synthesized by extracts of reticulocytes before the chains have been released from the ribosomes (44).

There are four known intracellular compartments in which proteins fold that do not appear to contain any type of chaperonin: the **endoplasmic reticulum** lumen, the intermembrane **mitochondrial** space, the intrathylakoidal lumen, and the **periplasmic** space of **Gram-negative** bacteria. Additionally, the eukaryotic cytosol seems to lack a general chaperonin, since the TCP-1 chaperonin appears to be restricted to a small subset of protein substrates, including actin and tubulin. Thus it is certainly not the case that the folding of all proteins in all cells involves the chaperonins. Given that the intracellular environment strongly favors aggregation, it is likely that the chaperonin-independent proteins utilize the assistance of other types of molecular chaperone for their folding. This assistance may also apply to proteins larger than 60 kDa that cannot fit into the chaperonin cage. For these proteins, the cotranslational and sequential folding of protein **domains** may circumvent the requirement for a sequestered folding compartment.

It is also possible that some chaperonins have roles additional to those discussed above with respect to protein folding, since there are sporadic reports of chaperonin-like molecules occurring on the cell surface in both prokaryotic and eukaryotic cells and in the blood serum, as well as stimulating **cytokine** production by animal cell cultures. The bacterial chaperonins are also the major immunogens in all human bacterial infections, and their possible roles in protective immunity and autoimmune disease is an active area of research (45).

NOTE ADDED IN PROOF

The cage mechanism for chaperonin-assisted protein folding is strongly supported by recent structural and functional data. The crystal structure of the asymmetrical GroEL:GroES complex with bound ADP shows that the cage is sufficient to accept proteins up to at least 70 kDa and that the wall of the cage provides a hydrophilic environment permissive for folding (Z. Xu et al. *Nature* **388**, 741–750, 1997). Crystal structures of the archaean thermosome suggest that in this chaperonin the folding cage is closed by loop sequences that emanate from the apical domains of the chaperonin subunits, explaining the lack of a separate GroES-like factor (M. Klumpff et al. *Cell* **91**, 263–270, 1997; L. Ditzel et al. *Cell* **93**, 125–138, 1998). A single ring mitochondrial chaperonin is fully functional in protein folding in a GroES-dependent manner *in vivo* (K. L. Nielsen et al. *Mol. Cell* **2**, 93–99, 1998). The size range of GroEL substrate proteins and their kinetics of interaction *in vivo* are consistent with the cage mechanism (K. L. Ewalt et al. *Cell* **90**, 491–500, 1997). Oligomeric GroEL with an intact central cavity is essential for the maintenence of growth of *E. coli* (F. Weber et al. *Nature Struct. Biol.* **11**, 977–985, 1998).

BIBLIOGRAPHY

1. S. M. Hemmingsen, C. Woolford, S. M. van der Vies, K. Tilly, D. T. Dennis, G. C. Georgopoulos, R. W. Hendrix, and R. J. Ellis (1988) *Nature* **333**, 330–334.
2. R. S. Gupta (1990) *Biochem. Int.* **20**, 833–841.
3. R. J. Ellis (1990) *Science* **250**, 954–959.
4. C. Georgopoulos, D. Ang, K. Liberek, and M. Zylicz (1990) in R. I. Morimoto, A. Tissieres, and C. Georgopoulos, eds., *Stress Proteins in Biology and Medicine*, Cold Spring Harbor Laboratory Press, New York, pp. 191–221.
5. L. Silver, K. Artzt, and D. Bennett (1979) *Cell* **17**, 275–284.
6. H. Kubota, G. Hynes, and K. Willison (1995) *Eur. J. Biochem.* **230**, 3–16.
7. K. R. Willison and A. L. Horwich (1996) in R. J. Ellis, ed., *The Chaperonins*, Academic Press, San Diego, pp. 108–136.
8. F. U. Hartl (1996) *Nature* **381**, 571–580.
9. R. J. Ellis and F. U. Hartl (1996) *FASEB J.* **10**, 20–26.
10. E. M. Hallberg, Y. Shu, and R. L. Hallberg (1993) *Mol. Cell Biol.* **13**, 3050–3057.
11. A. L. Horwich, K. B. Low, F. A. Fenton, I. N. Hirshfield, and K. Furtak (1993) *Cell* **74**, 909–917.
12. P. Goloubinoff, J. P. Christeller, A. A. Gatenby, and G. H. Lorimer (1989) *Nature* **342**, 884–889.
13. J. Buchner, M. Schmidt, M. Fuchs, R. Jaenicke, R. Rudolph, F. X. Schmid, and T. Kiefhaber (1991) *Biochemistry* **30**, 1586–1591.
14. J. Martin, T. Langer, R. Boteva, A. Schramel, A. L. Horwich, and F.-U.Hartl (1991) *Nature* **352**, 36–42.
15. G. S. Jackson, R. A. Staniforth, D. J. Halsall, T. Atkinson, J. J. Holbrook, A. R. Clarke, and S. G. Burston (1993) *Biochemistry* **32**, 2554–2563.
16. J. Martin, A. L. Horwich, and F. U. Hartl (1992) *Science* **258**, 995–998.
17. A. Guagliardi, L. Cerchia, S. Bartolucci, and M. Rossi (1994) *Protein Sci.* **3**, 1436–1443.
18. R. J. Ellis (1996) in R. J. Ellis, ed., *The Chaperonins*, Academic Press, San Diego, pp. 1–25.
19. D. Ursic, J. C. Sedbrook, K. L. Himmel, and M. R. Culbertson (1994) *Mol. Biol. Cell* **5**, 1065–1080.
20. H. Sternlicht, G. W. Farr, M. L. Sternlicht, J. K. Driscoll, K. R. Willison, and M. B. Yaffe (1993) *Proc. Natl. Acad. Sci USA* **90**, 9422–9426.
21. R. S. Gupta (1996) in R. J. Ellis, ed., *The Chaperonins*, Academic Press, San Diego, pp. 27–64.
22. U. Bertsch, J. Soll, R. Seetharam, and P. V. Viitanen (1992) *Proc. Natl. Acad. Sci. USA* **89**, 8696–8700.
23. K. F. Liou and K. R. Willison (1997) *EMBO J.* **16**, 101–106.
24. A. M. Roseman, S. Chen, H. White, K. Braig, and H. R. Saibil (1996) *Cell* **87**, 241–251.
25. O. Llorca, S. Marco, J. L. Carrascosa, and J. M. Valpuesta (1997) *J. Struct. Biol.* **118**, 31–42.
26. K. Braig, Z. Otwinowski, R. Hegde, D. C. Boisvert, A. Joachimiak, A. L. Horwich, and P. B. Sigler (1994) *Nature* **371**, 578–586.
27. S. C. Mande, V. Mehra, B. Bloom, and W. G. J. Hol (1996) *Science* **271**, 203–207.
28. J. F. Hunt, A. J. Weaver, S. J. Landry, L. Gierasch, and J. Diesenhofer (1996) *Nature* **379**, 37–42.
29. J. Martin, M. Mayhew, T. Langer, and F. U. Hartl (1993) *Nature* **366**, 228–233.
30. M. Mayhew, A. C. R. da Silva, J. Martin, H. Erdjument-Bromage, P. Tempst, and F. U. Hartl (1996) *Nature* **379**, 420–426.
31. J. S. Weissman, H. S. Rye, W. A. Fenton, J. M. Beecham, and A. L. Horwich (1996) *Cell* **84**, 481–490.
32. S. Chen, A. M. Roseman, A. S. Hunter, S. P. Wood, S. G. Burston, N. A. Ranson, A. R. Clarke, and H. R. Saibil (1994) *Nature* **371**, 261–264.
33. T. Waldmann, E. Nimmesgern, M. Nitsch, J. Peters, G. Pfeifer, S. Muller, J. Kellermann, A. Engel, F. U. Hartl, and W. Baumeister (1995) *Eur. J. Biochem.* **227**, 848–856.
34. V. A. Lewis, G. M. Hynes, D. Zheng, H. Saibil, and K. Willison (1992) *Nature* **358**, 249–252.
35. B. M. Phipps, A. Hoffmann, K. O. Stetter, and W. Baumeister (1991) *EMBO J.* **10**, 1711–1722.
36. B. M. Phipps, D. Typke, R. Hegerl, S. Volker, A. Hoffmann, K. O. Stetter, and W. Baumeister (1993) *Nature* **361**, 475–477.
37. S. Marco, D. Urena, J. L. Carrascosa, T. Waldmann, J. Peters, R. Hegerl, G. Pfeifer, H. Sack-Kongehl, and W. Baumeister (1994) *FEBS Lett.* **341**, 152–155.
38. M. Nitsch, M. Klumpp, A. Lupas, and W. Baumeister (1997) *J. Mol. Biol.* **267**, 142–149.
39. A. C. Clarke and P. A. Lund (1996) R. J. Ellis, ed., in *The Chaperonins* Academic Press, San Diego, pp. 168–212.
40. J. Martin and F. U. Hartl (1997) *Proc. Natl. Acad. Sci. USA* **94**, 1107–1112.
41. R. J. Ellis (1994) *Curr. Biol.* **4**, 633–635.
42. R. J. Ellis (1996) *Fold. Des.* **1**, R9–R15.
43. Z. Torok, L. Vigh, and P. Goloubinoff (1996) *J. Biol. Chem.* **271**, 16180–16186.
44. J. Frydman and F. U. Hartl (1996) *Science* **272**, 1497–1502.
45. A. R. M. Coates (1996) in R. J. Ellis, ed., *The Chaperonins* Academic Press, San Diego, pp. 268–296.

Suggestions for Further Reading

R. J. Ellis, ed. (1996) *The Chaperonins*, Academic Press, San Diego (this is the first book devoted to the chaperonins and contains 10 chapters by different authors covering, as well as the main topics discussed above, other aspects such as gene regulation, gene accession numbers, primary structures, and immunological and medical implications).

M.-J. Gething, ed. (1997), *Molecular Chaperones and Protein Folding Catalysts*, Oxford University Press, Oxford (this book contains detailed information about the chaperonins, as well as other molecular chaperones).

CHEMICAL MODIFICATION

T. IMOTO

Chemical modification is one of the most useful methods of identifying the functional groups of a **protein**. Chemical modification is also used for labeling proteins with **reporter groups** to monitor their conformations for **radiolabeling** and to increase their stability. If proteins are modified chemically under mild conditions, the modifications occur to the native conformation, and the modified proteins usually retain their native conformations. However, conformational changes in the modified proteins sometimes occur, and one must be careful about checking the conformation of the modified proteins, especially when they lose their functions. One must also be careful about the side reactions that often accompany otherwise specific chemical modifications.

It is possible to modify chemically the residues of **aspartic acid**, **glutamic acid**, **histidine**, **lysine**, **arginine**, **methionine**, **tryptophan**, **tyrosine**, and **cysteine**, and less easily **serine**, **threonine**, **asparagine**, and **glutamine**, but it is not possible to modify specifically **glycine**, **alanine**, **valine**, **leucine**, **isoleucine**, **phenylalanine**, and **proline** residues.

Residues that participate in the function are usually **accessible** to the **solvent** and consequently susceptible to reaction with a chemical reagent. The residues that show different reactivity in the presence and absence of a **ligand** are important for the function. A good method for discriminating the functional residues after modification involves analyzing the responsible residues within the proteins retained by an appropriate affinity column [see **Affinity chromatography**]. **Affinity labeling** is a sophisticated way of modifying active site residues selectively. Modification with a **suicide substrate** is a very good method for modifying catalytic residues specifically.

The stability of a protein can be altered by chemical modification (see **Protein stability**). Proteins are stabilized by introducing additional stabilizing forces such as **hydrophobic** forces, **electrostatic interactions**, and **hydrogen bonds**. Intra- or intermolecular **cross-links** are also effective for stabilizing proteins, as is attaching polymers to their surfaces.

Amino groups are reactive nucleophiles that can be modified by many chemical reactions, including acylation, **amidination**, and **guanidination**. **Carboxyl groups** can be modified after activation with **carbodiimides** or by esterification. Tyrosine residues are modified by **nitration** with **tetranitromethane**, **iodination**, or **acetylation** with acetylimidazole. The **thiol group** is the strongest nucleophile among all of the functional groups of amino acids, so there are many reagents that react specifically with it. Among these reactions are metal binding with *p*-mercurylbenzoate (PCMB); mixed-disulfide formation with disulfide reagents, such as 5,5′-dithiobis(2-nitrobenzoic acid) (DTNB, or **Ellman's reagent**) or dithiodipyridine; **alkylation** with **iodoacetate**, methyl iodide, ethyleneimine or ***N*-ethylmaleimide**, cyanylation with 2-nitro-5-thiocyanobenzoic acid (NTCB), and oxidation with many oxidants. Reduction of the **disulfide bonds** of proteins and alkylation of the resulting thiol groups is an important step in protein chemistry. The imidazole group of histidine residues can be modified with **diethylpyrocarbonate**, or they can be photooxidized in the presence of photosensitizing dyes. The sulfur of methionine residues can be oxidized to the sulfoxide by air or by oxidants, or alkylated with agents like methyl iodide under acidic conditions. The latter reaction can be reversed by thiols, so an isotopic label can be introduced in 50% of the terminal methyl group of methionine residues using labeled methyl iodide. The guanido group of arginine residues forms heterocyclic condensation products with 1,2- and 1,3-dicarbonyl compounds, such as phenylglyoxal, 2,3-butanedione, and 1,2-cyclohexanedione. The indole ring of the tryptophan residue can be modified with various oxidants, such as *N*-bromosuccinimide, iodine, or ozone, or by electrophilic reagents, such as 2-hydroxy-5-nitrobenzylbromide or 2-nitrophenylsulfenylchloride.

Suggestions for Further Reading

C.H.W. Hirs (ed.) (1967) Modification reactions. Specific modification reactions, *Methods Enzymol.* **11**, 481–711.

C.H.W. Hirs (ed.) (1972) Modification reactions. Specific modification reactions, *Methods Enzymol.* **25**, 387–671.

C.H.W. Hirs and S.N. Timasheff (eds.) (1977) Chemical modification, *Methods Enzymol.* **47**, 407–498.

C.H.W. Hirs and S.N. Timasheff (ed.) (1983) Chemical modification. Active-site labeling, *Methods Enzymol.* **91**, 549–642.

T. Imoto and H. Yamada (1989) Chemical modification. In *Protein Structure: A Practical Approach* (T. E. Creighton, ed.), IRL Press, Oxford, U.K., pp. 247–277.

CHEMICAL SHIFT

J. T. GERIG
W. E. PALKE

A nuclear magnetic resonance (NMR) spectrum of the protons (hydrogen atoms), ^{13}C, or other nuclei of a molecule typically features groups of signals displayed over a range of frequencies. The frequencies can ultimately be related to the Larmor (resonance) frequencies of the nuclei under study. The frequencies of absorbed or emitted energy represented in, for example, the proton NMR spectrum are different for protons in structurally or chemically distinct environments. Chemical shift refers to the sensitivity of the Larmor frequency to the covalent structure of which the observed nucleus is part, sensitivity to the intra- and intermolecular interactions present, and sensitivity to other sample variables, such as temperature, concentration, and pressure. The chemical shift is an important aspect of the NMR experiment because it ultimately provides a unique "signature" for each spin of a molecule of interest.

Atomic nuclei such as ^{1}H, ^{13}C, and ^{15}N, which have the property referred to as spin, take up certain quantum-mechanically allowed orientations when they are in a magnetic field. In the magnetic field, these nuclei undergo a precessional motion around the direction of the magnetic field. The frequency of this motion (ν) is characteristic of the type of nucleus and its local chemical environment and can be expressed as

$$\nu = \gamma B_0(1 - \sigma)/2\pi$$

where γ is the gyromagnetic ratio of the nucleus, B_0 is the magnitude of the laboratory magnetic field, and σ is the chemical shielding parameter, or screening constant. (The precessional frequency is also dependent on possible scalar and dipolar coupling interactions between nuclei; in terms of energy these are smaller influences on the precessional frequency and are neglected for the present discussion [see scalar coupling].) Chemical and structural information is implicit in the value of the shielding parameter; it is the purpose of NMR experiments to measure nuclear precessional frequencies and thus define the values of σ for nuclear spins of a sample. The shielding parameter is dimensionless and typically is in the range $1-1000 \times 10^{-6}$. Values for σ are conveniently discussed in terms of parts per million (ppm).

The chemical shift, as represented by the shielding parameter (σ), for a given spin in a molecule depends principally on the chemical bonds that hold the spin to the molecule. The carbon atom of a methyl group has a shielding parameter that is about 180 ppm different from the shielding parameter of the carbon atom of a carbonyl group, primarily because the local electronic structures about the two carbons are so different (see **NMR**). Because local electronic structures for a given chemical group tend to be similar from molecule to molecule, the shielding parameters for the molecules' nuclei tend to fall into narrow, identifiable bands. Thus constructing correlation tables that reliably indicate the approximate expected values of the shielding parameters of each nucleus in a molecule is possible. Organic chemistry textbooks generally contain such tables.

Shielding parameters also subtly reflect the entire electronic structure of a molecule and the electronic structures of solvent molecules or other solutes that are present in a sample. The tertiary structures of biopolymers result in characteristic effects on the shielding parameters of their component nuclei (1,2).

When considered in detail, it becomes apparent that the value of the shielding parameter is dependent on the orientation of the molecule in a magnetic field. That is, σ is anisotropic and must be represented by a tensor. In gases and most liquid samples, rotational and translational motions of the molecules are rapid, and molecules change their positions and orientations rapidly in the magnetic field. Consequently, in these samples the value of the shielding parameter detected experimentally is the average of all possible orientations of the molecule. If the motions of the molecules in a sample are restricted, as would be the case in solids or in liquid crystalline solutions that are partially ordered, a range of σ values will be detected for a given nucleus.

By convention, NMR spectra are displayed so that shielding parameters (or their equivalents) increase from left to right along a chemical shift axis. That is, the NMR absorption or emission signals for nuclei with the largest shielding parameter appear to the right in the spectrum, whereas nuclei with smaller shielding parameters appear progressively to the left. Also, by convention the algebraic signs of the numbers along the shielding parameter axis are the negatives of the actual number. Finally, measuring absolute values of shielding parameters with confidence is impractical. The NMR signal from a convenient material is thus chosen as a standard or reference signal, and shielding parameters are measured relative to this signal. Therefore, a signal appearing in a proton NMR spectrum at the position labeled 7 ppm has a shielding parameter that is 7 ppm smaller than the shielding parameter that is characteristic of the reference signal, arbitrarily assigned 0 ppm.

BIBLIOGRAPHY

1. D. S. Wishart and B. D. Sykes (1994) *Methods Enzymol.* **239**, 363–392.
2. S. S. Wijmenga, M. Kruithof, and C. W. Hilbers (1997) *J. Biomol. NMR* **10**, 337–350.

Suggestions for Further Reading

R. J. Abraham, J. Fisher, and P. Loftus (1988) *Introduction to NMR Spectroscopy*, Wiley, New York.

F. A. Bovey (1988) *Nuclear Magnetic Resonance Spectroscopy*, Academic, San Diego.

R. K. Harris (1983) *Nuclear Magnetic Resonance Spectroscopy: a physicochemical view*, Pitman, Marshfield, Mass.

S. W. Homans (1992) *A Dictionary of Concepts in NMR*, Clarendon, Oxford.

R. Kitamaru (1990) *Nuclear Magnetic Resonance: principles and theory*, Elsevier, New York.

R. S. Macomber (1998) *A Complete Introduction to Modern NMR Spectroscopy*, Wiley, New York.

C. H. Yoder and C. D. Schaeffer Jr. (1987) *Introduction to Multinuclear NMR*, Benjamin/Cummings, Menlo Park.

The NMR literature is replete with theoretical and experimental studies of chemical shielding effects. A database of proton, carbon, and nitrogen chemical shifts in proteins in maintained (http://www.bmrb.wisc.edu).

CHEMILUMINESCENCE

J. A. CHRISTOPHER
S. W. RASO
T. O. BALDWIN

Chemiluminescence is the production of light via a chemical reaction. As in all **luminescence**, the source of the light is the decay of an electron from a higher energy excited state to the ground state. In chemiluminescence, the reactants are generally in their ground-state configuration, but one or more of the products is formed with an electron in a high-energy orbital (the product molecule is formed in an excited state). Thus, chemiluminescence is the conversion of chemical energy (Gibbs **free energy**) to radiant energy (light). A necessary precondition for chemiluminescence is that the change in free energy for the reaction be sufficiently large to produce one of the products in an excited-state configuration.

Perhaps the simplest example of a chemiluminescent reaction is the thermal dissociation of the cyclic peroxide tetramethyl 1,2-dioxetane to two molecules of acetone (Fig. 1). The free energy of the **transition state** of the reaction is about 90 kcal/mol above the ground state for the acetone products. This is sufficient energy for one of the acetone molecules to be produced in an electronically excited state. When this molecule decays to the ground state, a photon of light is emitted.

In many cases, a chemiluminescent reaction is carried out in the presence of a dye. The energy from the excited state

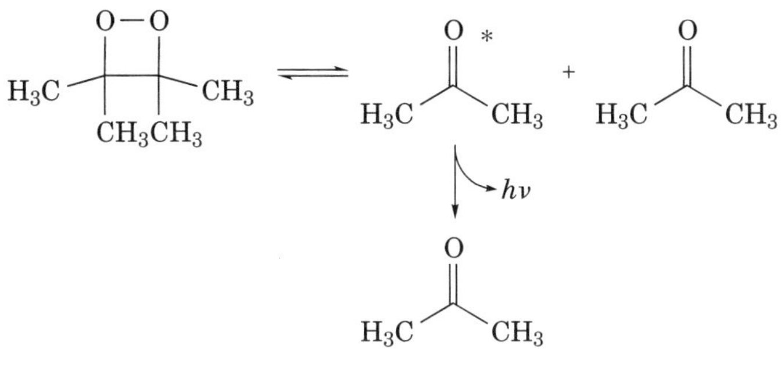

Figure 1. Simple example of a chemiluminescent reaction, the thermal dissociation of the cyclic peroxide tetramethyl 1,2-dioxetane to two molecules of acetone.

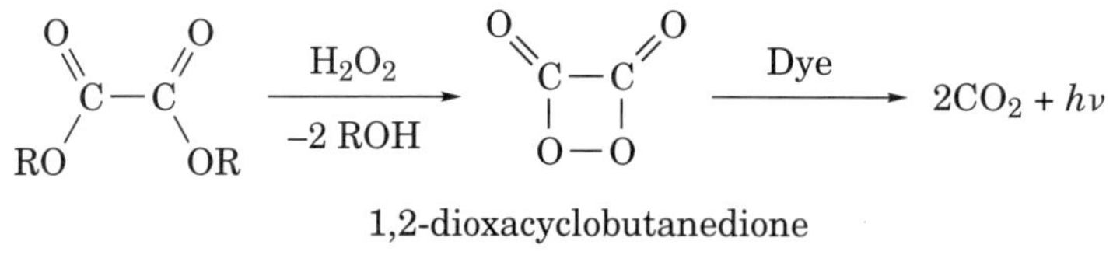

$$\mathrm{RO{-}C({=}O){-}C({=}O){-}OR} \xrightarrow[-2\,\mathrm{ROH}]{\mathrm{H_2O_2}} \underset{\text{1,2-dioxacyclobutanedione}}{\mathrm{C_2O_4}} \xrightarrow{\text{Dye}} 2\mathrm{CO_2} + h\nu$$

Figure 2. Chemiluminescent reaction coupled with a phosphorescent dye. The radiant energy from the chemical reaction is transferred to the dye and is then released at a much slower rate, producing a longer-lasting, but less intense glow. This is the basis of such commercial products as the glow-stick for emergency lighting and luminescent necklaces and other novelty items.

of the reaction product is transferred to the dye, so that the reaction product is in the ground state and the dye is now in an electronically excited state. The dye will then decay to the ground state and emit a photon of a different color, depending on the chemical structure of the dye. Thus, a reaction that produces UV light may be coupled with a dye that emits in the visible region in order to provide useful illumination. A second reason for coupling a chemiluminescent reaction with a dye is to increase the lifetime of the light emission. Many chemiluminescent reactions are over quickly, and the light is released in a brief, intense flash. If, however, the reaction is coupled with a phosphorescent dye (see **Luminescence**), the radiant energy is transferred to the dye and is then released at a much slower rate, producing a longer-lasting, but less intense glow. This is the basis of such commercial products as the glow-stick for emergency lighting, as well as the luminescent novelty necklaces or other items often seen at carnivals. An example of the of reaction used in these devices is shown in Figure 2.

Chemiluminescent reactions are common in living organisms, where they are known as bioluminescent reactions. In some natural systems, energy from the chemical reaction will be passed to "antenna proteins," which serve the same functions as dyes in the strictly chemical reaction (see **Luciferases and luciferins**).

Suggestion for Further Reading

J. D. Roberts and M. C. Caserio (1977) *Basic Principles of Organic Chemistry*, 2nd ed., W. A. Benjamin, Menlo Park, CA, pp. 1371–1418.

CHEMIOSMOTIC COUPLING

R. E. McCarty

Many forms of energy are interconverted by living systems. Chemical energy may be used to drive mechanical processes, to generate osmotic and electrical gradients, and even to emit light (see **Energy transduction**). In **photosynthesis**, light energy is converted to chemical energy and is used indirectly to drive the thermodynamically unfavorable synthesis of ATP from ADP and P_i. In oxidative metabolism in **mitochondria** and some bacteria, ATP synthesis is driven by the energy released by oxidation of metabolites. During the 1950s, it was clearly established that both oxidative phosphorylation and photosynthetic phosphorylation are dependent on electron transport. How electron transport could be linked to the formation of a phosphoanhydride bond in ATP was a major question in bioenergetics for the next two decades.

Oxidation–reduction reactions are clearly quite different in character from the removal of the elements of water from ADP and P_i to form ATP. Yet, there is not doubt that these two processes are coupled to one another. In biochemical parlance, two reactions are said to be coupled when they share a common intermediate (see **Coupled reactions**). The nature of the intermediate common to electron transfer and to ATP synthesis was elusive. In 1961 in a short, entirely theoretical article, Peter Mitchell proposed a radically different idea for coupling (1). Rather than a chemical intermediate linking electron transport and phosphorylation, as was then generally thought, Mitchell suggested that transmembrane electrochemical proton potentials $\Delta\tilde{\mu}_{H^+}$ could couple the two reactions. In addition, Mitchell realized that specific **active transport** systems could be linked to $\Delta\tilde{\mu}_{H^+}$.

THE CHEMIOSMOTIC HYPOTHESIS

The postulates of the chemiosmotic hypothesis (1–3) in brief are as follows:

1. Membranes that catalyze oxidative phosphorylation (the inner mitochondrial membrane and some bacterial plasma membranes) and photosynthetic phosphorylation (**chloroplast** thylakoids and plasma membranes of some bacteria) are poorly permeable to protons.
2. Electron transport generates $\Delta\tilde{\mu}_{H^+}$ by vectoral transport of electrons and protons.
3. An ATPase is driven in reverse (ATP synthesis) by the energetically favorable flow of protons down their electrochemical potential.
4. Coupling membranes contain specific exchange metabolite transport systems that may be linked to the $\Delta\tilde{\mu}_{H^+}$.

Before considering these postulates in further detail, we will define $\Delta\tilde{\mu}_{H^+}$. The chemical potential of a substance, μ, is the free energy of a system per mole. The energetics of the movement of an ion across a membrane has two components: chemical (osmotic) and electrical. Chemical work must be done to generate a concentration (actually activity) gradient and electrical work, to generate the charge imbalance (the membrane potential). The combined electrochemical potential is $\tilde{\mu}$. Just as is the case for Gibbs free energy, G, it is the change in electrochemical potential ($\Delta\tilde{\mu}$) that is of interest. In general, at constant temperature and pressure

$$\Delta\tilde{\mu} = RT\ \ln[x^{z+}]_a/[x^{z+}]_b + zF\Delta\psi \quad (1)$$

where R is the gas constant (8.3 kJ K^{-1} mol^{-1}); T, the absolute temperature; ln, $\log_e$; x^{z+}, a cation of z positive charges; a and b, two compartments separated by a membrane; z, the charge on the cation; F, Faraday's constant (96.5 kJ $mol^{-1}V^{-1}$); and $\Delta\psi$, the membrane potential. For protons, $z = 1$, and Equation 1 becomes

$$\Delta\tilde{\mu}_{H^+} = RT\ \ln[H^+]_a/[H^+]_b + F\Delta\psi \quad (2)$$

or

$$\Delta\tilde{\mu}_{H^+} = 2.30RT\ \log[H^+]_a/[H_+]_b + F\Delta\psi \quad (3)$$

or, since pH = $-\log[H^+]$,

$$\Delta\tilde{\mu}_{H^+} = 2.30RT(pH_b - pH_a) + F\Delta\psi \quad (4)$$

or

$$\Delta\tilde{\mu}_{H^+} = 2.30RT(\Delta pH) + F\Delta\psi \quad (5)$$

Mitchell preferred to express Equation 5 in electrical units and coined the term "proton motive force" (pmf or Δp), which is simply $\Delta\tilde{\mu}_{H^+}$ divided by F:

$$\Delta p = \Delta\tilde{\mu}_{H^+}/F = 2.30\ RT/F(\Delta pH) + \Delta\psi \quad (6)$$

As forms of energy may be interconverted, so may their units. In some ways it makes sense to consider the energetics of proton gradients in electrical units. Proton fluxes across energy-coupling membranes are analogous to electric circuits. Some people prefer to use $\Delta\tilde{\mu}_{H^+}$, others Δp. Since the two terms are readily interconvertible, this usage, although potentially confusing, is not a problem. At 25°C, Equation 5 may be written

$$\Delta\tilde{\mu}_{H^+} = 5.69\ \Delta pH + 96.5\Delta\psi \quad (7)$$

where $\Delta\tilde{\mu}_{H^+}$ is also given in kilojoules per mole and $\Delta\psi$ is expressed in volts. The proton motive force (in millivolts) may be expressed as

$$\Delta p = 59\ \Delta pH + \Delta\psi \quad (8)$$

where $\Delta\psi$ is also given in millivolts.

As in any thermodynamic analysis, sign conventions are very important to keep straight for $\Delta\tilde{\mu}_{H^+}$. Consider a topologically closed phospholipid bilayer vesicle (Fig. 1). This vesicle has an inside aqueous compartment, a, and an external compartment, b, the suspending medium. ΔpH is defined as $pH_{out} - pH_{in}$, which is 2.0 units for the case shown in Figure 1. A charge difference ($\Delta\psi$) where the inside of the membrane is more positive than the outside is defined as having a positive sign. Suppose $\Delta\psi$ is +100 mV. From the data given in Figure 1. the $\Delta\tilde{\mu}_{H^+}$ or Δp may be calculated as $\Delta\tilde{\mu}_{H^+} = 5.69(2) + 96.5(0.1) = 21.0$ kJ mol^{-1} or $\Delta p = 59(2) + 100 = 218$ mV. Notice the sign of the answers, which are positive numbers. These values give the energy cost per mole ($+\Delta\tilde{\mu}_{H^+}$ or $+\Delta p$) to generate the electrochemical protein potential or Δp. The flow of protons down their electrochemical potential is exergonic, and the absolute magnitudes of $\Delta\tilde{\mu}_{H^+}$ and Δp are the same, but the sign changes. This statement is equivalent to saying that if a chemical reaction has a positive ΔG value in the forward reaction, the reverse reaction must have a negative ΔG of the same absolute magnitude.

The essence of chemiosmotic coupling is that exergonic electron transport by the respiratory and photosynthetic electron chains is obligatorily linked to transmembrane proton flux, resulting in the generation of $\Delta\tilde{\mu}_{H^+}$. The flow of protons down their electrochemical potential provides the driving force for ATP synthesis. There is compelling evidence in support of these proposals.

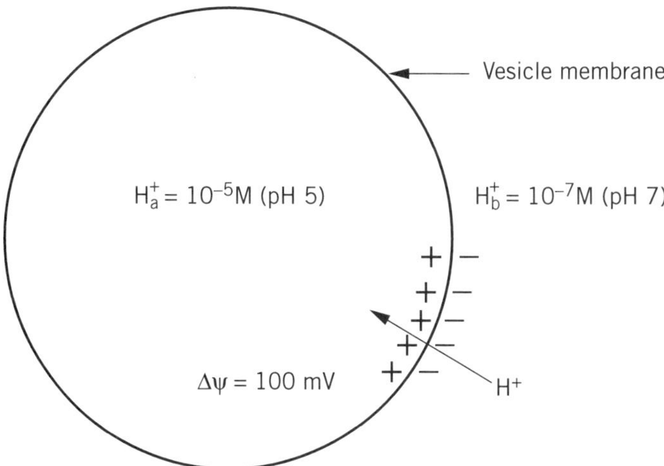

Figure 1. $\Delta\tilde{\mu}_{H^+}$ Across a vesicular membrane. ΔpH is defined as pH_{out} (compartment b) $-pH_{in}$ (compartment a), and $\Delta\psi$ as positive since the inside of the membrane is more positive than the outside.

EVIDENCE IN SUPPORT OF THE CHEMIOSMOTIC COUPLING HYPOTHESIS

As Mitchell predicted (1–3), the mitochondrial, bacterial, and chloroplast membranes that couple ATP synthesis to electron transport are poorly permeable to protons, except when proton-linked processes, such as ATP formation, occur at high rates. Proton transport was shown to be linked to electron transport in mitochondria (4), chloroplasts (5), and bacteria (6). The measurements of the magnitudes of $\Delta\tilde{\mu}_{H^+}$ across these membranes turned out to be difficult, but in most instances, $\Delta\tilde{\mu}_{H^+}$ values approaching 20 kJ mol^{-1} ($\Delta p \sim 200$ mV) have been measured during steady state, rapid electron transport. As predicted by Mitchell, lipophilic weak acids (eg, 2,4-dinitrophenol) could collapse the $\Delta\tilde{\mu}_{H^+}$ by shuttling protons across the membrane. ATP synthesis is also inhibited by these reagents, which are termed "uncouplers" because they uncouple electron flow and ATP synthesis.

When Mitchell proposed the chemiosmotic hypothesis, the understanding of biological membrane structure, especially how proteins may be integrated into membranes, was rudimentary. We take for granted today that membrane proteins may be plugged into the lipid bilayer in an asymmetric manner (see **Membranes**). But in the 1960s Mitchell's proposals for membrane protein asymmetry, transmembrane electron transport, and membrane sidedness were novel. Although not all of Mitchell's predictions turned out to be correct (cytochrome oxidase, for example, translocates protons via a mechanism that Mitchell did not foresee), many were. His model for proton and electron transport in thylakoid membranes is compared to a much more recent view in Figure 2. There is a remarkable similarity between the two schemes.

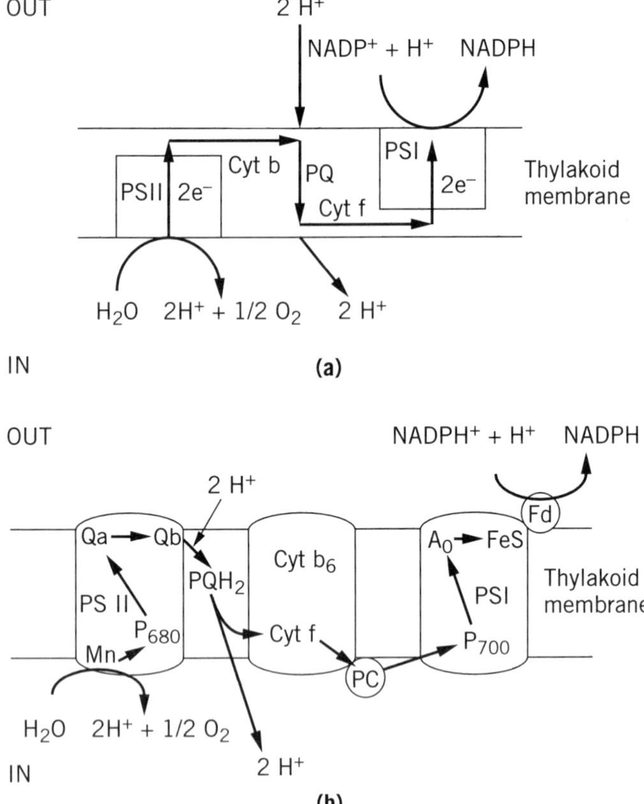

Figure 2. Schemes for electron and proton transport in chloroplast thylakoid membranes. In (**a**) Mitchell's early scheme (2) is illustrated with the membrane in the horizontal position more commonly used by others. In (**b**) a modern interpretation of electron and proton transport in thylakoids is given. The similarity between the two schemes is remarkable. Note in particular the transmembrane electron transport by photosystem II (PS II) and photosystem I (PS I), the oxidation of water to $2H^+$ and 1/2 O_2 inside (the thylakoid lumen) and proton translocation coupled to plastoquinone (PQ) oxidation/reduction. These elements are essentially the same in the two schemes. Cytochrome *b* (Cyt b) was not properly placed by Mitchell. In (**b**) Mn stands for the manganese of the oxygen evolving complex, Qa and Qb, for the bound plastoquinone electron acceptors of PS II; P_{680} and P_{700}, for the reaction center chlorophylls; Cyt f, for cytochrome f; PC, for plastocyanin; A_0, for the primary electron acceptor in PSI; FeS, for iron–sulfur electron transport proteins; Fd, for **ferredoxin**. Ferredoxin–$NADP^+$ oxidoreductase and the FeS protein of the cytochrome b6f complex were omitted for clarity.

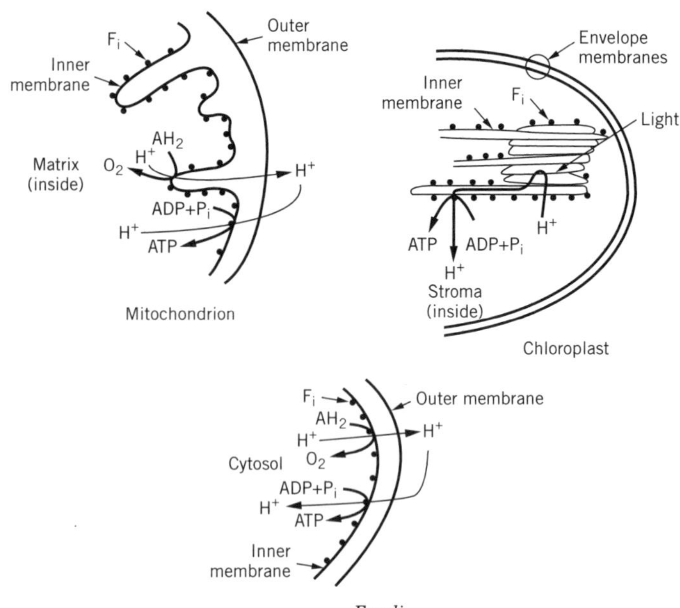

Figure 3. Sidedness of coupling membranes. Mitochondrial and bacterial (*E. coli*) electron transport is coupled to the ejection of protons, whereas in thylakoids, protons are accumulated. Note that, in bacteria and chloroplasts, ATP is generated within the same compartment in which it is used. In contrast, most of the ATP synthesized within mitochondria is exported to the cytoplasm. From Ref. 47, with permission.

As illustrated in Figure 3 (7), the sidedness of the **chloroplast** thylakoid membrane is opposite that of mitochondria and bacteria (8). As a result of photoelectron transport, thylakoids accumulate protons from the stroma. In contrast, oxidative electron transport in mitochondria and bacteria cause protons to be ejected from the mitochondria or bacteria. The steady state $\Delta\psi$ across the thylakoid membrane is very low, and the ΔpH is high (as much as 3.5 units). The ΔpH component of the $\Delta\tilde{\mu}_{H^+}$ in mitochondria and bacteria is, however, usually small, and $\Delta\psi$ is high. These observations make sense when it is realized that the mitochondrial matrix and bacterial cytoplasm are the internal compartments bounded by the coupling membranes. The pH of these compartments must be controlled to allow the numerous enzymes they contain to function. In contrast, the thylakoid lumen contains no metabolic enzymes. Electrical neutrality in thylakoids is maintained by Cl^- flux. Essentially, thylakoids accumulate HCl in the light.

Dramatic evidence in favor of the chemiosmotic hypothesis was obtained in André T. Jagendorf's laboratory during the mid-1960s (9,10). Simply by adjusting the pH of thylakoid suspensions to pH 4 and rapidly raising the pH to 8.0 in the presence of ADP + P_i, relatively large amounts of ATP are formed *in the dark.* ATP synthesis induced by acid to base transitions is abolished by uncouplers and reagents that block ATP synthesis. Later, it was shown (11) that artificially generated electric fields could also generate ATP. In mitochondrial vesicles, Thayer and Hinkle (12) showed that the rate of Δp-driven ATP synthesis is as fast or faster than that coupled to electron transfer.

The membrane of the halophilic bacterium, *Halobacter halobium* (*salinarium*) elaborates purple patches composed entirely of **bacteriorhodopsin**. Bacteriorhodpsin is a light-dependent proton pump that, when incorporated into lipid membranes, can generate substantial $\Delta\tilde{\mu}_{H^+}$ (12). When the enzyme from bovine heart mitochondria that catalyzes ATP synthesis was coreconstituted with bacteriorhodopsin, vesicles were obtained that catalyzed light-dependent, uncoupler-sensitive ATP formation (14).

Mitchell's proposal that ATP synthesis is accomplished by driving a proton-linked ATPase in reverse was also novel. By the time Mitchell was developing the chemiosmotic hypothesis, it had been clearly established that the energy of ATP hydrolysis could be used to generate ion gradients. For example, the Na^+ and K^+ concentration gradients and $\Delta\psi$ across mammalian plasma membranes are generated by an ATPase, the Na^+,K^+-ATPase (see **Ion pumps**). If, Mitchell reasoned, an ATPase acts as an ion pump, might an ATPase operate in reverse to utilize the $\Delta\tilde{\mu}_{H^+}$ generated by electron transport to drive ATP synthesis? Work on "coupling factors" during the 1960s established that mitochondria and chloroplasts contain an enzyme that is required for ATP synthesis that could under appropriate conditions also hydrolyze ATP (15). The catalytic part of what is now known as "ATP synthase" is called "F"$_1$ and is an extrinsic membrane protein. F_1 may be removed from the membrane and is very water-soluble. (See article on **ATP synthase**.)

Mitochondrial ATP synthesis and hydrolysis are strongly inhibited by the antibiotic, **oligomycin**. The ATPase activity of F_1, after its removal from the inner mitochondrial membrane, is insensitive to oligomycin—an observation that seemed incompatible with a role of F_1 in oxidative phosphorylation. Racker then showed that a crude detergent fraction of mitochondrial inner membranes contain a factor that confers sensitivity to oligomycin to F_1. This factor was called "F_0"; F_0 is not a coupling factor, and it is improper to denote F_0 as "F zero" or "F naught." Kagawa and Racker (16) pioneered the isolation and purification of the entire ATPase complex, F_1–F_0, which could be incorporated into phospholipid vesicles that catalyzed energy-linked activities. Similar studies were carried out with the chloroplast (CF_1–CF_0) and bacterial enzymes.

Mitchell predicted that F_0 contains a mechanism for transmembrane proton movement. Removal of CF_1 was shown (17) to greatly enhance the proton permeability of the thylakoid membrane. *N,N'*-Dicyclohexylcarbodiimide (DCCD) at low concentrations inhibits ATP synthesis and ATP hydrolysis by F_1–F_0 (17). DCCD acts by blocking proton conductance by F_0 and, in thylakoids, restores net light-dependent proton uptake and high $\Delta\tilde{\mu}_{H^+}$ to membranes from which CF_1 had been removed. The DCCD reacts with an Asp or Glu residue in a very hydrophobic, 8-kDa polypeptide of F_0 (18).

The evidence that F_1–F_0 is a proton-translocating ATPase/ATP synthase is overwhelming. Phosphorylation driven by imposed ΔpH or $\Delta\psi$ is abolished by specific inhibitors of either F_1 or F_0 function. As described above, F_0 conducts protons rapidly, and blocking F_0 proton transport also inhibits ATP synthesis. ATP hydrolysis by F_1–F_0 is linked to proton translocation. In thylakoids ATP hydrolysis in the dark can generate $\Delta\tilde{\mu}_{H^+}$ values of significant magnitude to drive some ATP synthesis. This ATP-dependent ATP synthesis is very sensitive to inhibition by uncouplers. The $\Delta\tilde{\mu}_{H^+}$ generated by ATP hydrolysis can drive electron transport in reverse (19), a process that is blocked by ATP synthase inhibitors and uncouplers. Other evidence that the electron-transport chain and ATP synthase are linked via $\Delta\tilde{\mu}_{H^+}$ includes the facts that phosphorylation decreases the magnitude of $\Delta\tilde{\mu}_{H^+}$ and stimulates the rate of electron transport. Specific inhibition of ATP synthesis increases the $\Delta\tilde{\mu}_{H^+}$ and slows electron transport. Uncouplers strongly decrease $\Delta\tilde{\mu}_{H^+}$ and inhibit ATP synthesis, but stimulate electron transport. An extensive analysis of the coupling between photosynthetic electron transport and ATP synthesis in the steady state gave results strongly indicating that $\Delta\tilde{\mu}_{H^+}$ is the intermediate linking the two processes.

METABOLITE TRANSPORT AND MOTILITY

Mitchell's fourth postulate, that coupling membranes contain specific exchange transport systems, has also received broad experimental support, especially for mitochondrial and bacterial coupling membranes. The chloroplast thylakoid membrane does not appear to contain metabolite transporters. ATP is synthesized in chloroplasts within the same compartment (the stroma) in which it is utilized in photosynthesis (see Fig. 3). The inner membrane of the envelope, not the thylakoid membrane, is the permeability barrier of the chloroplast. In contrast, the inner membrane of the mitochondrion and the plasma membrane of bacteria are both coupling membranes and permeability barriers. In these membranes, transport must coexist with ATP synthesis.

Most of the ATP generated by oxidative phosphorylation in mitochondria is exported to the cytoplasm and is hydrolyzed to drive cellular processes. Mitochondria must contain mechanisms to export ATP and import ADP and P_i. In addition, mitochondria will take up pyruvate, di- and tricarboxylic acids, and some amino acids. Cation exchangers are also present. Many of these transporters utilize the $\Delta\tilde{\mu}_{H^+}$ established by the electron transport chain (Fig. 4). A few transporters will be briefly considered in this article, with emphasis on ADP/ATP and P_i transporters in mitochondria. (see **Mitochondria** and **Membrane transport**).

At physiological pH, ADP exists as ADP^{3-}, and ATP as ATP^{4-}. The inner membrane contains a well-characterized transporter, known as the *adenine nucleotide translocator* (20). This transporter catalyzes the counterexchange translocation of ADP^{3-} and ATP^{4-}. The one-for-one exchange transport

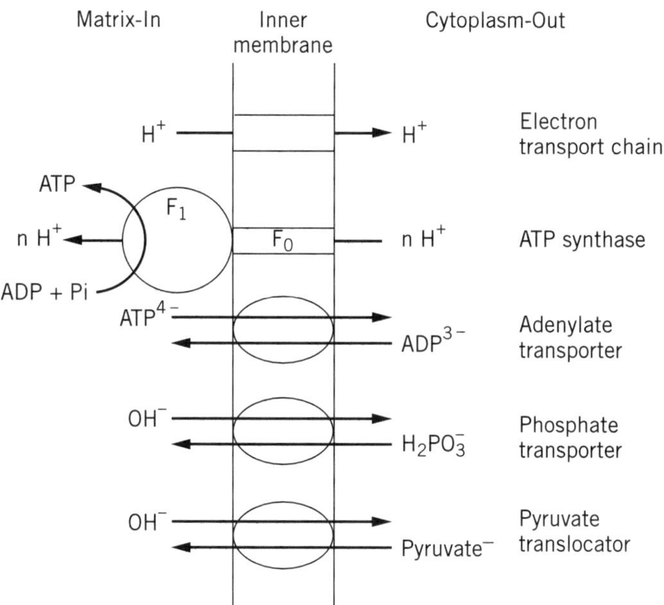

Figure 4. Proton circuits in the mitochondrial inner membrane. Electron transport ejects protons from the mitochondrion, resulting in the formation of $\Delta\tilde{\mu}_{H^+}$. A number of processes including ATP synthesis, ADP/ATP exchange transport, P_i–OH^- exchange transport, and the uptake of certain metabolites (illustrated by pyruvate uptake in exchange for OH^-) are linked to the $\Delta\tilde{\mu}_{H^+}$.

(**antiport**), with ATP^{4-} exported from the mitochondrion and ADP^{3-} imported, is electrogenic; there would be the net transfer of one negative charge out of the mitochondrion per transport event. The $\Delta\psi$ across the mitochondrial inner membrane is outside positive. Thus, ATP^{4-}/ADP^{-3} exchange transport would be promoted in the direction of ATP export and ADP import by the $\Delta\psi$ generated by electron flow.

P_i accumulation by mitochondria is also catalyzed by a transporter. P_i is either cotransported with H^+ or counterexchanged for OH^-. Since the form of P_i that is preferentially transported is $H_2PO_4^-$, one-for-one transport of P_i with either H^+ (**symport**) or OH^- (antiport) is electrically neutral and is not affected by $\Delta\psi$. P_i transport is, however, linked to ΔpH. The pH of the matrix of respiring mitochondria is about 0.5 unit more alkaline than that of the cytoplasm. Either OH^- antiport or H^+ symport would favor P_i uptake into mitochondria. This mechanism and the rapid utilization of P_i by oxidative phosphorylation assure that the P_i is delivered to the mitochondrial matrix from the cytoplasm. The transport of ATP/ADP and P_i by mitochondria costs the equivalent of one proton/ATP synthesized. Pyruvate, the end product of aerobic glycolysis, is transported into mitochondria via an electrically neutral counterexchange with $OH^{-\prime}$, and its transport, like that of P_i, is linked to ΔpH.

Bacteria have the potential to carry out the active transport of a large number of compounds, including sugars and amino acids (21). Transport may be linked to the $\Delta\tilde{\mu}_{H^+}$ generated by electron transport, or, under anaerobic conditions, by ATP hydrolysis by F_1–F_0. The best-studied example of $\Delta\tilde{\mu}_{H^+}$-dependent active transport in bacteria is the lactose transporter (lac **permease**) of *Escherichia coli* (22). Lactose uptake is obligatorily linked to H^+ uptake into the cell, at a probable stoichiometry of one lactose to one proton. The bacterium maintains, by either electron transport or ATP hydrolysis, a negative $\Delta\psi$ (outside positive), and, depending upon the conditions, a negative ΔpH (inside alkaline). The flow of protons down the $\Delta\tilde{\mu}_{H^+}$ provides the driving force for the accumulation of lactose to concentrations in excess of 1,000 times that of the medium. In addition to proton-linked transport systems, bacteria contain ABC (ATP binding cassette) transporters (23) that utilize ATP for transport to metabolites or ions directly, as well as the phosphotransferase system (24) that utilizes phosphoenolpyruvate as both an energy source and a source of phosphate for the transport and concurrent phosphorylation of some sugars.

The concepts developed by Mitchell for metabolite and ion transport were subsequently found to apply to membranes other than coupling membranes. The plasma membranes of higher plant cells, yeast, and fungi contain an enzyme that is structurally, functionally, and mechanistically related to the Na^+, K^+-ATPase of animal cell plasma membranes. This enzyme, the H^+-ATPase (25), is the primary ion pump in plant, yeast, and fungal plasma membranes. ATP hydrolysis pumps protons out of the cells, generating the transplasma membrane $\Delta\tilde{\mu}_{H^+}$. The ΔpH component in plant cells is 1.0–2.0 pH units, whereas $\Delta\psi$ may be in excess of -200 mV. The active transport across the plasma membrane is driven by the energetically favorable flow of protons down the $\Delta\tilde{\mu}_{H^+}$. As is the case for lactose transport in *E. coli*, specific transporters that mediate the translocation of metabolites across the membrane are obligatorily coupled to H^+ transport. An important example of H^+: metabolite cotransport (symport) in plants is the active loading of sucrose (26) into the phloem. As a consequence of sucrose: H^+ cotransport, the sucrose concentration within the phloem may be as high as 0.5 *M*.

An entirely new class of H^+-ATPases was discovered during the 1980s and characterized in the 1990s (27). The members of this class are called "V-ATPases," in which the V stands for *vacuolar*. All eukaryotic cells contain V-ATPases. In plant and yeast cells, the V-ATPase is present on the vacuolar membrane and in the **Golgi apparatus**. In animal cells, V-ATPases are present in **clathrin**-coated vesicles, chromaffin granules (adrenal medulla), Golgi membranes, **lysosomes**, synaptic vesicles, and, in some specialized cells, the plasma membrane. The V-ATPases are structurally much more complex than plasma membrane H^+-ATPases, and, although the V-ATPases large subunits have some sequence similarity to the α and β subunits of F_1, V-ATPases and F_1–F_0 ATPases are distinct.

The function of V-ATPases is to hydrolyze ATP and to pump protons from the cytoplasm into the internal aqueous compartment of endomembranes. The interiors of vacuoles and lysosomes are more acidic than that of the cytoplasm. Metabolite and ion fluxes are linked to the $\Delta\tilde{\mu}_{H^+}$ generated by the V-ATPase. A Ca^{2+}/H^+ exchange transporter has, for example, been found in the vacuolar membrane of higher plants (28). Acidification of the vesicles that are formed by receptor-mediated endocytosis is required for dissociation of the ligand from its receptor (29).

The V-ATPase of eukaryotes never functions as an ATP synthase. Even in the best of conditions *in vitro*, imposed proton gradients will not cause the pump to reverse. The H^+-ATPase of **archaebacteria** (eg, *Halobacter salinarium* [*halobium*], (30)) more closely resembles V-ATPases than F_1–F_0. Yet, the V-ATPase (sometimes called "A-ATPase" for Archeon), of the archeon can make ATP at the expense of the $\Delta\tilde{\mu}_{H^+}$ generated by the light-dependent bacteriorhodopsin proton pump. Eukaryotes tailored the V-ATPase to fit their needs.

In addition to its utilization in ATP synthesis and membrane transport, $\Delta\tilde{\mu}_{H^+}$ can be converted to mechanical energy in the form of motility of bacteria with **flagella** such as *E. coli* and *Bacillus subtilis*. Flagellar rotation is driven by $\Delta\tilde{\mu}_{H^+}$, rather than directly by ATP. Starved *B. subtilis* is immotile, but motility may be restored in transient manner by the artificial imposition of $\Delta\psi$. ΔpH (inside alkaline) is also capable of driving flagellar motion (31).

SODIUM ELECTROCHEMICAL POTENTIAL

The electrochemical sodium ion potential, $\Delta\tilde{\mu}_{Na^+}$, is given by

$$\Delta\tilde{\mu}_{Na^+} = RT\ \ln[Na^+]a/[Na+]b + F\ \Delta\psi \qquad (9)$$

where R, T, ln, $\Delta\psi$, and F are as defined in Equation 1. In some bacteria, $\Delta\tilde{\mu}_{Na^+}$ is exploited for ATP synthesis, active transport, and motility. In addition, Na^+/metabolite cotransport systems are present in animal cells. Transporters within the plasma membrane of these cells link the translocation of the substrates (eg, amino acids) to the flux of Na^+ down the $\Delta\tilde{\mu}_{Na^+}$ generated by the Na^+, K^+-ATPase (32). In bacteria, part of the $\Delta\tilde{\mu}_{H^+}$ generated by respiration may be converted to $\Delta\tilde{\mu}_{Na^+}$ by counterexchange transport of Na^+ for H^+. In some bacteria, Na^+ translocation may be linked to metabolism or electron transport without the involvement of $\Delta\tilde{\mu}_{H^+}$. For example, in

Propionigenium modestum, the anaerobic decarboxylation of succinate to form propionate is coupled to the transport of Na^+ out of bacterium, generating a $\Delta\tilde{\mu}_{Na^+}$ (33). In **halophiles**, Na^+ transport has been shown to be driven by electron transport through the NADH dehydrogenase region of the electron transport chain.

Vesicles of the *P. modestum* plasma membrane catalyze ATP-dependent Na^+ transport, and an imposed $\Delta\tilde{\mu}_{Na^+}$ drives ATP synthesis (34). The membrane contains a Na^+-ATPase (ATP synthase) that is analogous to the F_1–F_0 ATP synthase of coupling membranes. It is interesting to note that F_0 subunits of the *P. modestum* Na^+-ATPase confer to F_1 from *E. coli* the ability to translocate Na^+ (35). One archeon, *Methanosarina mazei* Gö 1, contains an F_1–F_0 that uses $\Delta\tilde{\mu}_{Na^+}$ and a V-ATPase (V_1–V_0) that uses $\Delta\tilde{\mu}_{H^+}$ (36). It is clear, therefore, that proton transport per se is not required for ATP synthesis by F_1–F_0. This observation makes mechanisms of ATP synthesis in which translocated protons are directly involved in catalysis very unlikely. $\Delta\tilde{\mu}_{Na^+}$ may also be the driving force for motility in some marine bacteria.

PYROPHOSPHATE AND ENERGY COUPLING

Pyrophosphate (PP_i) is merely two P_i molecules joined by an anhydride bond:

$$2\ HPO_4^{2-} + 2H^+ \leftrightarrow H_2P_2O_7^{2-} + H_2O$$

The equilibrium for this reaction lies far to the left; at pH 7 and 25°C, the standard free-energy change for (PP_i) synthesis from 2 P_i is about +16.7 kJ mol^{-1}. PP_i is a product of several biosynthetic reactions, including aminoacyl tRNA synthesis, fatty-acid activation, and DNA and RNA synthesis. These reactions occur with equilibrium constants close to 1 and are freely reversible. By coupling these reactions to PP_i hydrolysis, the reactions are pulled in the synthetic direction. Cells contain soluble **pyrophosphatases** that hydrolyze PP_i to 2 P_i.

Other roles for PP_i in plants have recently been elucidated. The vacuolar membrane contains a pyrophosphatase (37) that couples pyrophosphate hydrolysis to inward proton translocation. Both the V-ATPase and the pyrophosphatase contribute significantly to the $\Delta\tilde{\mu}_{H^+}$ across the vacuolar membrane. Thus, some of the energy of PP_i hydrolysis is conserved by generation of the $\Delta\tilde{\mu}_{H^+}$. In addition, PP_i may be utilized as a phosphoryl donor, as an alternative to ATP, in glycolysis.

In the photosynthetic bacterium, *Rhodospirillum rubrum*, a pyrophosphatase, and F_1–F_0 ATPase (ATP synthase) exist within the same membrane (38). Net PP_i synthesis driven by the photoelectron transport chain of the bacterium has been observed. PP_i synthesis is inhibited by proton ionophores (uncouplers), but not by reagents that block ATP synthesis by interacting with F_1. Membrane vesicles (chromatophores) from *R. rubrum* are oriented so that the catalytic sites of both the ATP synthase and the pyrophosphatase face the outside. ATP hydrolysis by these vesicles in the dark is coupled to inward H^+ translocation, resulting in the generation of substantial $\Delta\tilde{\mu}_{H^+}$ values. It is even possible to detect ATP-dependent pyrophosphate synthesis that is also uncoupler-sensitive. ATP hydrolysis and pyrophosphate synthesis are linked by the $\Delta\tilde{\mu}_{H^+}$.

CONCLUDING REMARKS

The four postulates of the chemiosmotic coupling hypothesis have received broad experimental support. Coupling membranes are poorly permeable to H^+. Electron transport results in the generation of $\Delta\tilde{\mu}_{H^+}$ at values that are consistent with the energetic demands of ATP synthesis. An ATPase is present that is reversible and can make ATP, provided the magnitude of $\Delta\tilde{\mu}_{H^+}$ is sufficient, or hydrolyze ATP if the $\Delta\tilde{\mu}_{H^+}$ is low. Artificially imposed $\Delta\tilde{\mu}_{H^+}$ can drive ATP synthesis at rates as least as fast as in oxidative or photosynthetic phosphorylation. $\Delta\tilde{\mu}_{H^+}$ is energetically and kinetically competent to be the intermediate common between electron transport and ATP synthesis. The fourth postulate, the existence of exchange transporters and their possible coupling to the $\Delta\tilde{\mu}_{H^+}$, has had a tremendous impact on the field of metabolite transport, not only in mitochondria but also in many other membranes. In part, the chemiosmotic hypothesis had its origins in Mitchell's long-standing interest in active transport in bacteria. The chemiosmotic hypothesis provided a new theoretical basis for the interpretation of transport experiments. Very early on, Mitchell proposed that membrane proteins may be organized in specific orientations within membranes and that electron transfer may in part occur across membranes. For his remarkable insights and experimental contributions, Peter Mitchell was awarded the Nobel Prize in Chemistry in 1978.

That H^+ (or Na^+) translocation is coupled to ATP synthesis and hydrolysis is generally accepted. Some evidence suggests that under some circumstances, the protons translocated during electron transport need not equilibrate with the bulk aqueous phase prior to their passage through F_1–F_0 to make ATP. Because of the high buffering capacity of the interior of thylakoids, 50 ms or more of illumination may be required to generate ΔpH values of magnitudes sufficient to drive ATP synthesis. When the generation of $\Delta\psi$ by proton transport is prevented, permeant buffers should delay the onset of ATP synthesis by increasing the internal buffering capacity. The delay is not as long as it should be, if it is assumed that the added buffer is uniformly distributed throughout the internal volume of the thylakoids (39, 40). These and other observations kept alive the localized proton theory of R. J. P. Williams (41). According to this theory, dubbed "microchemiosmosis" by Mitchell, protons are trapped in localized regions of the membrane that are not necessarily in equilibrium with each other. The transmembrane bulk phase $\Delta\tilde{\mu}_{H^+}$ was proposed to be used only for energy storage.

It is difficult to design experiments that provide unequivocal support for the localized proton hypothesis. If protons may be delivered to the ATP synthase without equilibrating with the bulk aqueous phase, some barrier must exist to prevent equilibration. Disruption of thylakoid structure has been suggested to convert the coupling mechanism from localized to delocalized (ie, chemiosmotic). It would thus be very difficult to establish the nature of the barrier. Results of a number of other reported experiments at first glance seem at odds with the chemiosmotic hypothesis. In most cases, flaws in the methods used or in the interpretation of the results can explain the apparent discrepancy.

The chemiosmotic hypothesis has stood the test of time. During the past few years, much has been learned about the structure of some of the proton and electron translocators of coupling membranes. Yet, we know little about how cytochrome oxidase pumps protons or how the ATP synthase exploits the $\Delta\tilde{\mu}_{H^+}$ to drive ATP synthesis. The emphasis of bioenergetics has shifted away from *what* couples electron transport to ATP synthesis to *how* the proton translocating elements work.

BIBLIOGRAPHY

1. P. Mitchell (1961) *Nature* **191**, 144–148.
2. P. Mitchell (1966) *Chemiosmotic Coupling in Oxidative and Photosynthetic Phosphorylation,* Glynn Research, Bodmin, Cornwell, England.
3. P. Mitchell (1979) Les Prix Nobel en 1978, Nobel Foundation, Stockholm, pp. 142–143.
4. P. Mitchell and J. Moyle (1965) *Nature 298*, 147–151.
5. J. S. Neuman and A. T. Jagendorf (1964) *Arch Biochem Biophys.* **107**, 109–119.
6. P. Mitchell (1962) *J. Gen. Microbiol.* **29**, 144–148.
7. R. E. McCarty (1985) *Bioscience* **35**, 27.
8. P. C. Hinkle and R. E. McCarty (1978) *Sci. Am.* **238**, 104–123.
9. A. T. Jagendorf and E. Uribe (1966) *Proc. Natl. Acad. Sci. USA 55*, 170–177.
10. A. T. Jagendorf (1967) *Fed. Proc.* **26**, 1361–1369.
11. H. T. Witt (1979) *Biochim. Biophys. Acta 505*, 355–427.
12. W. S. Thayer and P. C. Hinkle (1975) *J. Biol. Chem.* **250**, 5330–5335.
13. W. Stoekenius (1985) *Trends Biochem Sci.* **10**, 483–485.
14. E. Racker and W. Stoeckenius (1974) *J. Biol. Chem.* **249** 662–663.
15. E. Racker (1976) *A New Look at Mechanisms in Bioenergetics*, Academic Press, New York.
16. Y. Kagawa, A. Kandrach, and E. Racker (1973) *J. Biol. Chem.* **248**, 676–684.
17. R. E. McCarty and E. Racker (1967) *Brookhaven Symp. Biol.* **19**, 202–212.
18. K. J. Cattell, C. R. Lindop, I. G. Knight, and R. B. Beechey (1971) *Biochem. J.* **125**, 169–177.
19. B. Chance and G. Hollunger (1961) *J. Biol. Chem.* **258**, 1474–1486.
20. M. Klingenberg (1989) *Arch. Biochem Biophys* **270**, 1–14.
21. R. D. Simoni and P. W. Postman (1975) *Annu. Rev. Biochem.* **44**, 523–554.
22. H. R. Kaback (1986) *Annu. Rev. Biophys. Chem.* **15**, 279–319.
23. C. F. Higgins (1992) *Annu. Rev. Cell Biol.* **8**, 67–113.
24. N. D. Meadow, D. K. Fox, and S. Roseman (1990) *Ann. Rev. Biochem.* **59**, 497–542.
25. R. Serrano (1990) in C. Larsson and I. M. Moller, eds., *The Plant Plasma Membrane: Structure, Function and Molecular Biology*, Springer-Verlag, New York, pp. 127–153.
26. N. Sauer (1992) in D. T. Cooke and D. T. Clarkson, eds., *Transport and Receptor Proteins of Plant Membranes: Molecular Structure and Function*, Plenum Press, New York, pp. 67–75.
27. N. Nelson (1989) *J. Bioenerg. Biomembr.* **21**, 533–571.
28. K. S. Schumaker and H. Sze (1986) *J. Biol. Chem.* **261**, 12172–12178.
29. I. Melman (1992) *J. Exp. Biol.* **172**, 39–45.
30. K. Ihara, T. Abe, K.-I. Sugimura, and Y. Mukohata (1992) *J. Exp. Biol.* **172**, 475–485.
31. S. Matsuura, J. Shioi, and Y. Imae (1977) *FEBS Lett.* **82**, 187–190.
32. E. J. Collarini and D. L. Oxender (1987) *Annu. Rev. Nutr.* **7**, 75–90.
33. P. Dimroth (1991) *Bioessays 13*, 463–468.
34. W. Hilpert, B. Schink, and P. Dimroth (1984) *EMBO J.* **3**, 1665–1670.
35. W. Laubinger, G. Dekers-Hebestreit, N. Altendorf, and P. Dimroth (1990) *Biochemistry* **25**, 5458–5463.
36. R. Wilms, C. Frieberg, E. Wegerk, I. Meier, F. Mayer, and V. Müller (1996) J. *Biol. Chem* **271**, 18843–18852.
37. P. A. Rea, Y. Kim, V. Sarafian, R. J. Poole, J. M. Davies, and D. Sanders (1992) *Trends Biochem. Sci.* **17**, 348–353.
38. P. Nyring, B. F. Nore, and A. Strid (1991) *Biochemistry 30*, 2883–2887.
39. R. D. Horner and E. N. Moudrianakis (1983) *J. Biol. Chem.* **258**, 11643–11647.
40. R. A. Dilley, S. M. Theg, and W. A. Beard (1987) *Annu. Rev. Plant Physiol.* **38**, 347–389.
41. R. J. P. Williams (1969) *Curr. Top. Bioenerg.* **3**, 79–156.

Suggestions for Further Reading

W. A. Cramer and D. B. Kraff (1990) *Energy Transduction in Biological Membranes. A Text of Bioenergetics*, Springer-Verlag, New York.

G. D. Greville (1969) A scrutiny of Mitchell's chemiosmotic hypothesis of respiratory chain and photosynthetic phosphorylation, *Curr. Top. Bioenerg.* **3**, 1–78 (a very clear and useful review of the state of chemiosmotic coupling in the early days).

D. G. Nichols and S. J. Ferguson (1992) *Bioenergetics*, Academic Press, London.

R. N. Robertson (1968) *Protons, Electrons, Phosphorylation and Active Transport*, Cambridge University Press, London (in this slender volume, the author describes thoughts about charge separations and ion fluxes that predated the chemiosmotic hypothesis).

V. P. Skulachev and P. C. Hinkle, eds. (1982) *Chemiosmotic Proton Circuits in Biological Membranes*, Addison-Wesley, London.

CHEMOKINES

David T. Denhardt
Herbert M. Geller
Heather R. Kaminski

Chemokines are small secreted **proteins**, generally 8 to 15 kDa in mass, that were originally identified by their ability to stimulate **chemotaxis** and/or activate leukocytes (lymphocytes, neutrophils, eosinophils, mast cells, monocytes, and macrophages). They are now known to be multifunctional and to be secreted by many different cell types in response to various stimuli (for reviews see Refs. 1–5). In inflamed tissues, they may induce effector functions such as the generation of an oxidative burst or the secretion, from storage granules, of **proteinases** (by neutrophils and monocytes), histamine (by basophils), and cytotoxic proteins (by eosinophils). Chemokines activate **cell-surface receptors** to promote adhesion of circulating leukocytes to the vascular endothelium and subsequently to facilitate extravasation (diapedesis) of the adherent leukocyte into adjacent tissue, the source of the chemokine. They have proinflammatory effects and can give rise to acute or chronic inflammatory responses; some have been shown to be involved in movement and activation of cells in both angiogenesis/vasculogenesis and neural pathfinding, thus implicating them in aspects of **development** not directly involved in the **immune response**.

CHEMOKINE STRUCTURE AND CHROMOSOME LOCATION

The chemokine family (at least 40 members have been identified in humans) consists of four classes distinguished by the positioning of four conserved **cysteine** residues near the N-terminus of the protein. These four cysteine residues define a distinct structural motif based on the two **disulfide bonds** (C16C3, C26C4) that they form. All the chemokines possess a central **β-pleated sheet** region consisting of three antiparallel **beta-strands** (a **Greek key motif**) followed in most cases by a C-terminal **alpha-helix**. There is evidence that the chemokines can form dimers and higher oligomers, but whether this is important for their function is controversial (2). The N-terminal sequence of the chemokines is of particular importance in determining their specificity. Alteration or removal of one or a few **amino acid** residues can dramatically affect their activity or target cell specificity. This provides the opportunity for the specificity of a chemokine to be altered by local factors subsequent to its secretion.

The CXC or α-class chemokines possess a single residue between the first and second cysteines. This family can be further subdivided by the presence or absence of the amino acid sequence Glu-Leu-Arg (ELR) preceding the first **cysteine**. This sequence is recognized by chemokine receptors on specific cell types, and thus, has functional implications. The CC or β-class chemokines have no residues between the first and second cysteines. Some members of this group have two additional cysteine residues that form a third disulfide bond (6).

Two additional chemokines have been identified that do not fit into either the CC or CXC classes. The γ class is missing both the first and third cysteines. This C chemokine has lymphotactin as its prototype member. The δ class possesses a CXXXC motif, and thus far is represented only by fractalkine/neurotactin. This atypical membrane-bound CXXXC chemokine has, on the C-terminal side of the β-sheet sequences, a long glycosylated mucin structure, followed by a transmembrane segment and a short C-terminal cytoplasmic **domain**. The chemokines, with their alternative names, are listed in Table 1 by class.

Members within a class are linked both by their structural similarities and by their location in the human genome. Although there are exceptions in each class, members of the CXC group generally map to the chromosomal locus 4q13. CC chemokines are found primarily on chromosomes 7, 8, 9, and 17. Chromosomal location may provide a useful tool in identifying new members of the γ and δ classes.

Table 1. Human Chemokines and Their Receptors

Chemokine (and Alternative Names)	Receptors
ELR + CXC	
IL-8	CXCR1, R2
GROα (MGSA-α)	CXCR2, R1
GROβ (MGSA-β,MIP-2α)	CXCR2
GROγ (MIP-2β)	CXCR2
ENA-78	CXCR2
LDGF-PBP	CXCR2
GCP-2	CXCR2
Non-ELR CXC	
PF4	Unknown, GAGs[a]
Mig	CXCR3
IP-10	CXCR3, GAGs[a]
ITAC	CXCR3
SDF-1α/β	CXCR4
BCA-1/BLC	CXCR5
CC	
MIP-1α (LD78α)	CCR1, CCR5, CCR9
MIP-1β (Act-2, HC21)	CCR1, CCR5, CCR9
MDC (STCP-1)	CCR4
TECK	Unknown
TARC	CCR4
RANTES	CCR1, CCR3, CCR4, CCR5
HCC-1	CCR9
HCC-4 (NCC-4, LEC)	Unknown
DC-CK1 (PARC, MIP-4, AMAC-1)	Unknown
MIP-3α (LARC, Exodus)	CCR6
MIP-3β (ELC)	CCR7
MCP-1	CCR2, CCR9
MCP-2	CCR2, CCR9
MCP-3	CCR2, CCR9
MCP-4	CCR2, CCR3, CCR9
Eotaxin	CCR3, CCR9
Eotaxin-2/MPIF-2	CCR3
Six Cysteine CC	
I-309	CCR8
MIP-5/HCC-2(Lkn-1)	CCR1, CCR3
MPIF-1 (CKβ-8)	Unknown
6Ckine (SLC,Exodus-2, TCA-4)	CCR7, CXCR3
Others	
Lymphotactin (SCM-1)	XCR1
CX3C (Fraktalkine, Neurotactin)	CX3CR1
Multiple Chemokines	DARC (Duffy antigen)

[a] GAG = Glycosyl amino glycan.

CHEMOKINE EXPRESSION

Chemokines are expressed by virtually all cell types. Regulation of chemokine expression is complex and highly dependent on specific signaling pathways. Cells can express chemokines constitutively or in response to stimulation by, for example, **interleukin**-1 (IL-1), interleukin-4, **tumor necrosis factor** α, **interferon**-γ (IFN-γ), or pathogenic agents such as **endotoxin** (lipopolysaccharide). Low molecular-weight fragments (200 kDa) of the glycosaminoglycan hyaluronan (resulting from tissue destruction, for example) can enhance expression of some chemokines (eg, MIP-1, RANTES, or MCP-1) in specific macrophage types (eg, bone marrow-derived macrophages or elicited peritoneal macrophages) (7). Depending upon the situation, the expression of these same chemokines can be either suppressed or enhanced by inflammatory mediators, such as IL-10 or IFN-γ (8).

One well-characterized example of induction of chemokines is provided by the **macrophage**. Macrophages can be stimulated to produce-chemokines during T cell-directed delayed type hypersensitive reactions in tissues by contact with activated T cells displaying CD40L. Kornbluth et al. (9) found that, although LPS was also a potent inducer of chemokines, the macrophage-stimulating **lymphokines** IFN-γ and GM-CSF (granulocyte macrophage colony stimulating factor) produced by activated T cells were not good inducers.

CHEMOKINE RECEPTORS

Chemokines bind to seven-transmembrane-spanning G-protein-coupled receptors (see **Heterotrimeric G proteins**) whose sequence **homology** is distinct from that of other such receptors. The interaction of a chemokine with its receptor can activate diverse signaling pathways, depending upon the specific signaling elements expressed by the target cell. Chemokine receptors are named on the basis of the cysteine motif that they recognize, and most will bind more than one chemokine ligand in that group. Likewise, a particular chemokine can interact with more than one receptor. Table 1 lists the chemokines and their respective receptors. Some receptors are expressed constitutively by particular cell types, whereas the expression of others may require an inducing signal. Thus CCR1 and CCR2 are constitutively expressed by monocytes, but are expressed by lymphocytes only after stimulation by IL-2 (10). Some chemokine receptors are expressed by nonhematopoietic cells, eg, endothelial cells, neurons, astrocytes, and epithelial cells. This emphasizes recent findings that chemokine receptors may have other important functions besides that of leukocyte chemotaxis. The Duffy antigen receptor, the determinant of the Duffy blood group, is found on a variety of cells, including erythrocytes and endothelial cells, as well as cells of the nervous system. It is unusual in its ability to bind both CC and CXC chemokines without eliciting any known biologic response. Possibly it serves to store or scavenge chemokines (11).

Specific chemokines generally act on specific subsets of leukocytes as a function of the receptors expressed by the cell (Table 2). Most CXC chemokines containing the ELR sequence are chemotactic for neutrophils, whereas non-ELR chemokines generally attract lymphocytes. Neutrophils express the CXCR1 receptor, which is stimulated by IL-8 and granulocyte chemotactic protein-2 (GCP-2), and the CXCR2 receptor, which responds to IL-8, GCP-2, the LPS-induced CXC chemokine (LIX), the neutrophil-activating peptide-2 (NAP-2), the epithelial cell-derived neutrophil activating peptide 78 (ENA-78), and the growth-regulated oncogenes, GRO-α, GRO-β and GRO-γ. IL-2-stimulated lymphocytes express CXCR3, the receptor for MIG, IP-10, ITAC, and 6Ckine.

Eosinophils, basophils, monocytes, NK cells, and T cells respond to members of the family of CC chemokines. These include monocyte chemoattractant proteins (MCP-1, -2, -3, -4, and -5), macrophage inflammatory protein-1α and -1β (MIP-1α, MIP-1β), RANTES (regulated on activation, normal T expressed and secreted), eotaxin-1 and -2, and I-309. Receptors for these chemokines expressed by eosinophils are CCR1 (recognized by MCP-3 and -4; MIP-1; RANTES) and CCR3 (recognized by MCP-3 and -4; eotaxin-1 and -2; RANTES). Basophils express CCR3 and CCR2, which is activated by all the MCPs. Receptors expressed on monocytes are CCR1, 2, 5, and 8. Ligands for CCR5 include MIP-1α, MIP-1β, and RANTES, whereas the ligand for CCR8 is I-309. Activated T cells express CCR1, CCR2, CCR4 (recognized by TARC, the thymus and activation-regulated chemokine), CCR5, and CCR7, whose ligand is MIP-3 and 6Ckine (3,4).

Table 2. Primary Expression of Chemokine Receptors[a]

Receptor (Alternative Names)	Primary Expression
CCR1	Monocytes, eosinophils, activated T cells
CCR2	Monocytes, eosinophils, activated T cells
CCR3	Basophils, eosinophils, subset of Th2, monocytes
CCR4	Activated T lymphocytes
CCR5	Monocytes, T lymphocytes
CCR6 (GPR-CY4, DRY6, CKR-L3)	T and B lymphocytes, bone marrow-derived dendritic cells
CCR7 (EBI1)	Activated T and B lymphocytes
CCR8 (Ter1, CemR1)	Monocytes, T lymphocytes
CCR9 (D6)	Bone marrow, T lymphocytes, monocytes
CXCR1 (IL-8RA)	Polymorphonuclear cells (PMNs, neutrophils)
CXCR2 (IL-8RB)	Polymorphonuclear cells (PMNs, neutrophils)
CXCR3	Activated T lymphocytes
CXCR4 (LESTR,HMSTR,Fusin)	Monocytes, T lymphocytes, PMNs
CXCR5 (BLR-1)	B cells
XCR1 (GPR5)	Unknown
CX3CR1 (V28)	Neurons, endothelial cells
DARC (Duffy antigen)	Red blood cells, postcapillary endothelial cells

[a] This table lists cell types that have been widely accepted as expressors of chemokine receptors. It should be noted that many other cell types express these receptors as well.

CHEMOKINE SIGNALING

The ability of **pertussis toxin** to suppress many of the chemokine-induced signals suggests that the heterotrimeric $G_{\alpha\beta\gamma}$ proteins mediating chemokine signaling belong to the G_i group, which is defined by their ability to inhibit **adenylate cyclase**. Activation of the G_i protein leads to replacement of GDP with GTP and dissociation into α_i•GTP and $\beta\gamma$ subunits, both membrane-associated. Downstream mediators of general $G_{\alpha\beta\gamma}$ signaling may include the ***src*** of nonreceptor protein **tyrosine kinases**, phosphatidylinositol 3-kinase (PI3-K), **phospholipase C**, protein kinase C species, and the Ras-Raf-MEK-ERK kinase cascade (12,13). Phospholipase C hydrolyzes phosphatidylinositol 4,5-bisphosphate to produce inositol trisphosphate, a mobilizer of intracellular Ca^{2+}, and **diacylglycerol**, which, together with Ca^{2+}, activates protein kinase C (see **Second messengers** and **Calcium signaling**). Activation of phospholipase D, which generates phosphatidic acid, and phospholipase A2, which releases **arachidonic acid**, may be direct or indirect downstream effects of chemokine stimulation. Immediate effects (within seconds) on cell behavior are generally mediated by reversible changes in protein **phosphorylation** of **cytoskeletal** proteins, directed in part by the small **GTPases** Rho, Rac and Cdc42, thereby affecting the polymerization and depolymerization of **actin**, which underlies the extension and retraction of lamellipodia, implementors of leukocyte migration (3). Longer term changes are effected by **transcriptional** or **posttranscriptional** changes in gene expression.

Evidence for specific signaling intermediates includes the following: The response of neutrophils to chemoattractants can be suppressed by inhibitors of **tyrosine** phosphorylation, possibly acting on members of the *src* of nonreceptor protein tyrosine kinases (12). In human T lymphocytes, RANTES induces PI3-K; also, wortmannin, which is a potent PI3-K inhibitor, can inhibit RANTES-induced T-lymphocyte responses (14). In a T-cell hybrid, MCP-1 mobilized intracellular

Ca^{2+} concentrations, stimulated Ca^{2+} import, and transiently increased tyrosine phosphorylation of the ERKs by a pertussis toxin-sensitive process (15). In human monocytes, Yen et al. (16) obtained evidence for a putative G_q-mediated pathway, distinct from the pertussis toxin-sensitive G_i-dependent pathway, which activated ERK2 via a protein kinase C-dependent process that probably did not require Ras. Both the G_i and the proposed G_q pathways were required for a chemotactic response to MCP-1. MIP-1 activation of the CCR5 receptor expressed in a transfected murine pre-B lymphoma cell line led to phosphorylation and activation of a protein **tyrosine kinase**, known variously as RAFTK, Pyk2, or CAK-β, that is related to focal adhesion kinase (17). The cytoskeletal protein paxillin became phosphorylated and associated with RAFTK, and the downstream signaling kinases JNK/SAPK and p38 also appeared to be stimulated. Thus, chemokines have the ability to activate a variety of **signal transduction** mechanisms that connect the extracellular environment with the cytoskeleton.

CHEMOKINE FUNCTIONS

As part of the immune defense system, lymphocytes patrol the body, continually surveying it for pathogens or cells expressing abnormal surface **antigens** (11,18). They move from the blood stream into tissues and then back into the blood stream via the lymphatic circulation. Granulocytes and monocytes exit the blood stream, but cannot recirculate. Extravasation is a coordinated multistep process that requires a series of complex changes in the activity of molecules on the cell's surface as it undergoes the transition from a freely floating cell to a weakly adherent rolling cell, and then to a strongly adherent stationary cell able to insinuate itself between the endothelial cells and penetrate the underlying stroma. Chemokines help drive this process and, as noted in Table 3, determine what types of cells infiltrate into the tissues in various inflammatory diseases. Imai et al. (19) report that the MDC and TARC chemokines, which are produced by dendritic cells in the thymus, act via the CCR4 receptor to recruit and activate T lymphocytes in the thymus. DC-CK1, another C-C chemokine derived from dendritic cells, is a chemoattractant for naive T cells (20). In allergic diseases such as asthma, expression of eotaxin and monocyte chemoattractant proteins appear particularly important in stimulating the accumulation and activation of eosinophils and mast cells, leading to histamine release and the allergic response (21). Similar events are thought to occur in rhinitis and atopic dermatitis.

Subsets of lymphocytes express different surface molecules that allow them to target different tissues in the body. Extravasation of the lymphocyte from the blood stream is initiated when a free flowing lymphocyte becomes tethered to the endothelium such that it can still roll in the direction of flow. Rolling of leukocytes is controlled in part by the interaction of selectins with specific sialylated carbohydrate determinants. L-selectin, expressed by lymphocytes, monocytes, neutrophils, and eosinophils, interacts with receptors such as CD34 on the activated endothelium. P-selectin and E-selectin molecules are expressed by activated endothelial cells and bind structures related to the tetrasaccharide sialyl Lewis antigens on the leukocyte surface. Thrombin and histamine can mobilize P-selectin from intracellular stores, whereas up-regulation of E-selectin requires induction by inflammatory agents such as

Table 3. Chemokine Involvement in Inflammatory Diseases

Inflammatory Disease	Infiltrate	Chemokines Involved
Acute respiratory distress syndrome	Neutrophils	IL-8, GRO-α,β,γ, ENA-78
Asthma	Eosinophils, T-cells, monocytes, basophils	MCP-1,4, MIP-1α, Eotaxin, Rantes
Bacterial pneumonia	Neutrophils	IL-8, ENA-78, GRO-α,β,γ
Sarcoidosis	T-cells, monocytes	IP-10, MIP-1α, MCP-1, Rantes
Glomerulonephritis	Monocytes, T-cells, neutrophils	MCP-1, GRO
Rheumatoid arthritis	Monocytes, neutrophils	IL-8, MCP-1, IL-8, ENA-78
Osteoarthritis	Monocytes, neutrophils	MIP-1b, IL-8
Atherosclerosis	T-cells, monocytes	IL-8, IP-10, Rantes, MCP-1
Inflammatory bowel disease	Monocytes, neutrophils, T-cells, eosinophils	MCP-1, IL-8, ENA-78, Eotaxin
Psoriasis	T-cells, neutrophils	MCP-1, MIG, IL-8, Rantes, MIP-1
Bacterial meningitis	Neutrophils, monocytes	IL-8, GRO-α, MCP-1, MIP-1α,β
Viral meningitis	T-cells, monocytes	MCP-1, IP-10
Multiple sclerosis	T-cells	Rantes, IP-10, MIP-1α
Seasonal allergic rhinitis	Eosinophils	Eotaxin, Rantes, IL-8, MIP-1, MCP-1
Periodontal disease	Monocytes, PMNs	IL-8, MCP-1

IL-1, tumor necrosis factor, or LPS. This interaction appears to result in activation of leukocyte 1 and 2 integrins, enabling them to bind to ligands of the **immunoglobulin** superfamily (eg, VCAM-1, the ICAMs; see **Cell adhesion molecules**) either expressed constitutively by the endothelial cells or induced by inflammatory agents. Various subsets of lymphocytes rolling on an ICAM-1 monolayer can be induced to arrest within one second by SDF-1, 6Ckine, MIP-3, or MIP-3; adhesion, which can be prevented by pertussis toxin, is transient, lasting some 5 to 8 minutes (22).

In inflammatory diseases, it is likely that sentinel cells at the site of inflammation produce chemokines that induce the strong interactions required to elicit leukocyte invasion. Chemokine expression has been demonstrated in glomerulonephritis, asthma, inflammatory bowel disease, and allogenic transplant rejection (2,4). Because multiple inflammatory mediators are generally elaborated at sites of inflammation, the task of determining which is responsible for any particular pathogenic response is difficult. Although there is considerable redundancy built into the system, some chemokines predominate in some forms of injury. For example, IL-8 is a major causative factor associated with reperfusion injury and acid-induced pneumonitis, which is the consequence of neutrophil invasion and the release of inflammatory mediators (23).

Angiogenesis (see **Angiogenin**), the induction of synthesis of new blood vessels, typically with reference to the vascularization of a tumor, is a complex response affected by many factors, including chemokines. Although the specific role of chemokines in angiogenesis is unclear, evidence has suggested that CXC chemokines possessing the ELR motif generally promote angiogenesis, whereas those lacking the ELR motif tend to inhibit it (24). Mice lacking either SDF-1 or the counterpart receptor CXCR4 are defective in formation (vasculogenesis) of the large vessels supplying the intestinal tract (25). Vessels supplying the small intestine were missing major branches, and this led to hemorrhaging. Likewise, the vascular supply to the stomach in these mice was bereft of all large arteries and veins.

Many functions regulated by chemokines in the immune system, including cell migration and adhesion, as well as cytoskeletal reorganization, are important in the development and function of the nervous system. Defects in brain development in the CXCR4-deficient mouse (and its counterpart, the SDF-1 deficient mouse) are likely to be harbingers of other developmental and brain-specific functions of chemokines (26). The CXCR4-deficient mouse is embryonic-lethal, although with some animals surviving to term. The most prominent defects are found in the development of the cerebellum, where the granule neurons migrate prematurely. Since the radial glial cells are preserved in these animals, the evidence is consistent with an inhibitory role of SDF-1 on granule cell migration. This inhibition could occur by blocking the action of a migratory-promoting substance, or by altering adhesion molecules on the cell surface of the radial glial cell or the granule neuron. Since some of the guidance cues for this migratory process have been identified, it is of obvious importance to investigate if SDF-1 has any action to modulate the expression or conformation of these proteins.

Additional evidence for central nervous system actions of chemokines has been provided by the **transgenic** (see **Transgenic Technology**) expression of chemokines under the influence of the myelin basic protein **promoter**. Directed expression of the murine chemokine KC stimulated neutrophil infiltration of the brain at various sites (27). Likewise, overexpression of MCP-1 in both thymus and brain led to infiltration of either MAC-1, F4/80, or macrophages and monocytes, respectively (28). These studies demonstrate that each organ is capable of recruiting mononuclear cells under specific conditions.

VIRAL ADAPTATION TO CHEMOKINES

Viruses often exploit endogenous cell surface receptors, including chemokine receptors, in the initial stages of infection. For example, mice deficient in MIP-1 as the result of a targeted gene disruption (**knockout** mice) exhibit attenuated responses to certain viral infections (Coxsackie virus, **influenza virus**). Several of the chemokine receptors serve as entry mechanisms for various strains of **HIV**. Roger (29) has reviewed the role of chemokine receptors and the HLA **haplotype** in the development of AIDS. The CCR5 receptor appears to be the preferred coreceptor for macrophage-tropic, nonsyncytium-forming HIV-1 strains, whereas the CXCR4 receptor (fusin) primarily services the syncytium-inducing, T-cell-line-tropic HIV-1 strains (30). Some HIV isolates can utilize both receptors, while a few other variants appear to have acquired the ability to co-opt other chemokine receptors as coreceptors for virus infection. Individuals with mutations in chemokine receptors may exhibit considerable resistance to infection or to progression of the disease. One mutation in particular should be noted. People homozygous for a deletion mutation in the CCR5 gene (CCR5 $\delta 32$) are resistant to infection by the virus; heterozygotes for the same mutation had a significantly slower onset of disease (31). Chemokines can also affect the progression of an HIV infection. SDF-1, a ligand for CXCR4, inhibits HIV entry into cells (32,33), and the CC chemokines MIP-1α, MIP-1β, and RANTES, all ligands for CCR5, likewise seem to retard HIV-1 infection (34).

CONCLUSION

The chemokine field is evolving rapidly. Although the initial focus has been on the role of chemokines in the immune system, future directions for chemokine research lie in elucidating chemokine expression and function in other organs and during **development**. Importantly, the involvement of chemokines and their receptors in many disease processes makes them primary targets for drug development.

ACKNOWLEDGMENTS

Research in the authors' laboratories has been supported by the National Institutes of Health.

BIBLIOGRAPHY

1. N. W. Schluger and W. N. Rom (1997) *Current Opin. Immunol.* **9**, 504–508.
2. B. J. Rollins (1997) *Blood* **90**, 909–928.
3. M. Baggiolini (1998) *Nature* **392**, 565–568.
4. A. D. Luster (1998) *New England J. Med.* **338**, 436–445.
5. A. Zlotnick, J. Morales, and J. A. Hedrick (1998) *Adv. Immunol.* in press.
6. A. Pardigol et al. (1998) *Proc. Natl. Acad. Sci. USA* **95**, 6308–6313.
7. C. McKee et al. (1996) *J. Clin. Invest.* **98**, 2403–2413.
8. M. R. Horton et al. (1998) *J. Immunol.* **160**, 3023–3030.
9. R. S. Kornbluth, K. Kee, and D. D. Richman (1998) *Proc. Natl. Acad. Sci. USA* **95**, 5205–5210.
10. P. Loetscher, M. Seitz, M. Baggiolini, and B. Moser (1996) *J. Exp. Med.* **184**, 569–577.
11. M. B. Furie and G. J. Randolph (1995) *Amer. J. Pathol.* **146**, 1287–1301.
12. G. M. Bokoch (1995) *Blood* **86**, 1649–1660.
13. J. S. Gutkind (1998) *J. Biol. Chem.* **273**, 1839–1842.
14. L. Turner, S. G. Ward, and J. Westwick (1995) *J. Immunol.* **155**, 2437–2444.
15. P. M. Dubois et al. (1996) *J. Immunol.* **156**, 1356–1361.
16. H.-h. Yen, Y. Zhang, S. Penfold, and B. J. Rollins (1997) *J. Leukoc. Biol.* **61**, 529–532.
17. R. K. Ganju et al. (1998) *Blood* **91**, 791–797.
18. T. A. Springer (1994) *Cell* **76**, 301–314.
19. T. Imai et al. (1998) *J. Biol. Chem.* **273**, 1764–1768.
20. G. J. Adema et al. (1997) *Nature* **387**, 713–717.
21. B. Lamkhioued et al. (1997) *J. Immunol.* **159**, 4593–4601.
22. J. J. Campbell et al. (1998) *Science* **279**, 381–384.

23. H. G. Folkesson, et al. (1995) *J. Clin. Invest.* **96**, 107–116.
24. R. M. Strieter et al. (1995) *J. Leukocyte Biol.* **57**, 752–762.
25. K. Tachinaba et al. (1998) *Nature* **393**, 591–594.
26. Y-R. Zou et al. (1998) *Nature* **393**, 595–599.
27. M. E. Fuentes et al. (1995) *J. Immunol.* **155**, 5769–5776.
28. M. Tani et al. (1996) *J. Clin. Invest.* **98**, 529–539.
29. M. Roger (1998) *FASEB J.* **12**, 625–632.
30. Y. Feng, C. C. Broder, P. E. Kennedy, and E. A. Berger (1996) *Science* **272**, 872–877.
31. M. W. Smith et al. (1997) *Science* **277**, 959–965.
32. C. C. Bleul (1996) *Nature* **382**, 829–833.
33. E. Oberlin et al. (1996) *Nature* **382**, 833–835.
34. D. Zagury et al. (1998) *Proc. Natl. Acad. Sci. USA* **95**, 3857–3861.

CHEMOTAXIS

G. W. ORDAL
J. R. KIRBY

Chemotaxis is the movement of a motile cell in response to chemical changes in the environment and implies a directional sense. In most cases where **bacteria** migrate up concentration gradients of attractants, direction is not sensed directly. The small size of bacteria would make spatial comparisons of concentration unreliable. Instead, bacteria usually sense temporal changes in concentration by using chemoreceptors, **transducer proteins** that bind or release attractants and repellents. This temporal measurement allows the cells to alter motion of their **flagella** briefly so that a "biased random walk" brings the bacteria to the right destination (1,2) (Fig. 1). Eukaryotic cells carry out chemotaxis by an unrelated mechanism, and they are much larger and probably do sense direction.

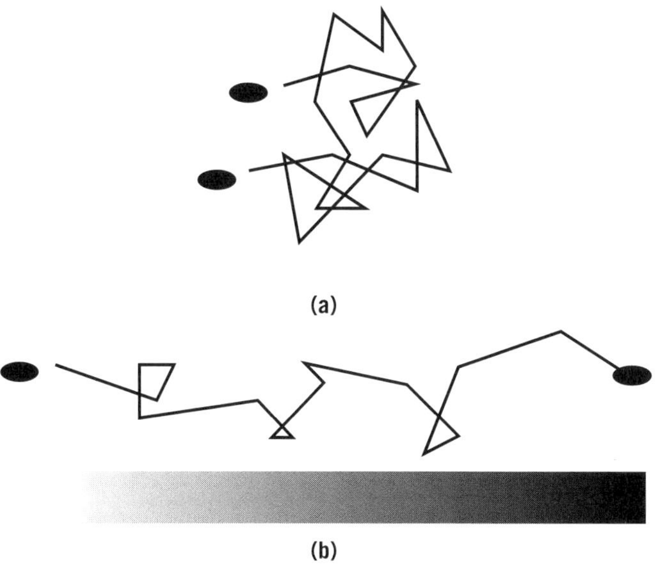

Figure 1. Motion of peritrichous bacteria. (**a**) Random walk in an isotropic medium. (**b**) Biased random walk in a gradient of an attractant or repellent.

BACTERIA

There are three main types of motile bacteria:

1. the peritrichous (flagella over the surface),
2. the polar (flagella at one or both poles),
3. the gliding bacteria, which lack flagella.

The peritrichous bacteria, such as ***Escherichia** coli* and ***Bacillus** subtilis*, normally swim smoothly by rotating their flagella counterclockwise (looking along the flagella toward the cell body) and tumble by rotating their flagella clockwise, which discoordinates the bundle of flagella. Chemotaxis in *E. coli* occurs when the cell happens to move toward higher attractant concentrations. It decreases the probability of tumbling and thus increases the tendency to travel up the attractant gradient (1). This decreased probability of tumbling results (Fig. 2) because the attractant molecules bind to methylated chemoreceptors to cause a brief inactivation of the cytoplasmic CheA **kinase** (see **Transducer proteins**). Consequently, there is less phosphorylated CheA (CheA-P) and less CheY-P (2) (see **Phosphorylation**). CheY-P dissociates from the switch that controls flagellar rotation, so that smooth swimming ensues, and CheZ hydrolyzes the phosphoryl group from CheY-P. The enhanced smooth swimming is short-lived because CheZ hydrolyzes the phosphoryl group from CheY and the CheR **methyltransferase** increases the degree of methylation of the chemoreceptor and restores the prestimulus activity of CheA (3) (see also **Methylation, protein**). Thus, there is an inverse relationship between the levels of CheY-P and termination of a smooth swimming event.

Logically, it seems plausible that bacteria moving by chance toward *lower* attractant concentrations would show an increased tendency to tumble and hence to begin heading in another direction. In the case of *E. coli*, however, there is no demonstrable increased tendency for these bacteria to tumble compared to bacteria in an isotropic medium (1). Interestingly, CheY-P cannot affect rotational direction of the flagella by itself. Fumarate needs to be released or synthesized for switching to occur (4).

The role of CheY-P is less clear in the case of polar bacteria, such as *Caulobacter crescentus* and ***Pseudomonas** aeruginosa*. When these bacteria move by chance toward lower concentrations of attractant, they reverse motion by rotating their flagella in the opposite direction, to head back up the gradient. If repellent is added, however, they show a series of reversals of motion. This result implies that CheY-P does not govern the *direction* of rotation of the flagella but instead causes repeated reversals of behavior. Such an event might ensue if binding of CheY-P to the switch reduces the energy of activation of switching the flagellar rotational direction. In the absence of binding, the energy of activation is presumed to be so high that reversals are very infrequent. In the case of *Halobacterium salinarium*, negative stimuli cause significant increases in the concentrations of free fumarate in the cytoplasm (4). The connection between CheY-P and fumarate in controlling direction of flagellar rotation has not been elucidated.

Other bacteria, like *Rhizoboium meliloti*, a **Gram-negative** bacterium phylogenetically rather distant from *E. coli*, rotate their flagella unidirectionally. In this case, the CheA/CheY

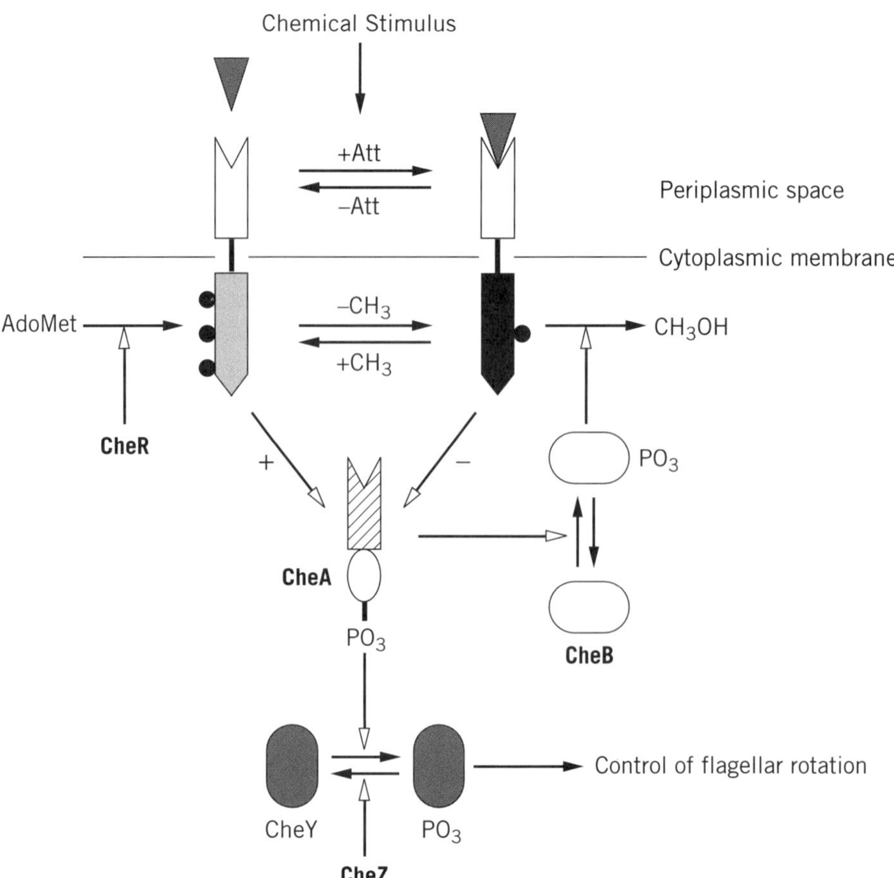

Figure 2. Diagram of the mechanism of chemotaxis in *Escherichia coli*. Open arrows indicate control of the designated reaction or signaling step. At the *top*, the attractant (Att) binds to its transmembrane chemoreceptor, which is methylated on Glu residues. The methyl groups are depicted as closed black circles. This binding decreases the phosphorylated form of the protein CheA (*center*), which decreases phosphorylation of the proteins CheY and CheB. Dephosphorylated CheB is much less active in removing the methyl groups from the chemoreceptor, so CheR restore them. Phosphorylated CheY dissociates from the switch that controls flagellar rotation, and smooth swimming ensues. CheZ hydrolyzes the phosphoryl group from CheY, ending the response to the attractant.

system also exists with the complication that there are two species of CheY. CheY2-P is the major species that affects flagellar movement, and it acts by slowing flagellar rotation (5). Thus it appears that the basic system involving the two-component couple CheA and CheY is universal, but how it functions varies for different groups of organisms.

The potential role of CheY-P is still less clear in the case of the gliding bacteria. There have been many efforts to understand the basis of motility in these bacteria, but no structure resembling flagella or the corresponding **basal bodies** (protein complexes that rotate within the membrane to cause flagellar rotation) have ever been identified by **electron microscopy**. However, it is clear that at least some gliding bacteria, including *Myxococcus xanthus*, carry out chemotaxis and have proteins homologous to the chemotactic proteins of *E. coli*. In *M. xanthus* CheA and CheY are fused into one protein (6). It is hard to imagine that there would not also be a motor apparatus, homologous to that in *E. coli*, with which the chemotactic proteins would interact to bring about chemotaxis.

One issue that has existed for many years has been whether the rotation of multiple flagella are coordinated or independent. The finding that bundles of flagella located 50 μm apart on *Spirillum volutans* were observed switching synchronously within 10 ms implies electrical coordination of direction of the flagellar rotation (1). In contrast, certain mutant strains of ***Salmonella*** *typhimurium* show random and uncorrelated switching of their flagellar direction when observed under partially deenergized conditions (7). *E. coli* filamentous cells switch their flagella asynchronously although biases of nearby, but not distant, motors were correlated (8). In *H. salinarium*, an **archaebacterium**, the rotational direction of flagellar bundles on each is end also uncoordinated, and cells were frequently observed to be immobilized because the bundles at both ends worked against each other. Thus, there is no general rule governing flagellar coordination. Indeed, the mechanism by which an electrical connection could work, as in *S. volutans*, is obscure.

EUKARYOTES

Chemotaxis in eukaryotes occurs by an unrelated mechanism. The process is best known from studies in the aggregating slime mold ***Dictyostelium*** *discoidium* (9–12) and in leucocytes (13), it is probably quite similar in both. In the slime mold, chemotaxis plays a role in feeding and in the aggregation that precedes **differentiation** and **sporulation** (9,10). During aggregation, **cyclic AMP** (cAMP) is produced and binds to a "seven-helix, membrane-spanning" **receptor**, which activates a heterotrimeric **G-protein** (9). **Guanylate cyclase** is activated by operation of this G-protein, thereby producing cGMP (10,11). Consequently, **myosin** I and II light chains are phosphorylated, myosin II is translocated to the **cytoskeleton**, and its heavy chain is phosphorylated (10). After a brief lag period, **calcium channels** in the plasma membrane open (10–12). In both prokaryotic and eukaryotic cells, calcium ion is maintained at very low concentrations, usually around 0.1 μM (14). Thus, when the calcium channel opens, calcium ion enters the cell and binds to **calmodulin** (12). These events lead to **actin** polymerization (10) and growth of filopodia in the direction of the signal's arrival. Near the point of stimulation, the cytoskeleton is formed (10), and dissolution occurs at the distal end of the organism. The cells relay the signal through the timing of activation of **adenylate cyclase**, production and secretion of cAMP, and hydrolysis of cAMP by **phosphodiesterase** (10). These events are accompanied by release of calcium ion from internal stores, opening of calcium ion channels independent of G-proteins and cGMP, activation of **phospholipase C**, and activation of a **mitogen**-activated protein (MAP) kinase (10,11). A G-protein α-subunit is phosphorylated and binds to the cAMP receptor, so that the latter becomes phosphorylated and desensitized (10). Binding of calcium ion to calmodulin activates the calmodulin-dependent protein **phosphatase**, calcineurin A, and also increases **protein synthesis** by binding to the L19 protein of the 60S **ribosome** subunit (12).

Finally, one point of similarity for the prokaryotic and eukaryotic mechanisms is the use of calcium ions (see **Calcium signaling**). In *E. coli*, calcium ion concentration increases briefly when bacteria encounter repellents and decreases when they encounter attractants (14). Release by light of calcium ion from a "caged" calcium ion compound causes tumbling: conversely, binding of cytoplasmic calcium ion by a chelator that takes up calcium ion upon illumination causes smooth swimming (4). Potentially, calcium ion may influence the stability of CheY-P, but its role is still obscure. Thus, *both* systems, albeit differently, take advantage of the potential **second messenger** role of calcium ion, whereby, because of its very low initial concentration, a modest influx causes manyfold changes in its concentration.

BIBLIOGRAPHY

1. H. C. Berg (1975) *Ann. Rev. Biophys. Bioeng.* **4**, 119–136.
2. J. B. Stock and M. G. Surette (1996) In *Escherichia coli and Salmonella, Cellular and Molecular Biology*, 2nd ed. (F.C. Neidhardt, ed.), ASM Press, Washington, DC, Vol. I, pp. 1103–1129.
3. J. S. Parkinson (1993) *Cell* **73**, 857–871.
4. R. Barak and M. Eisenbach (1996) *Curr. Top. Cell. Reg.* **34**, 137–158.
5. V. Sourjik and R. Schmitt (1996) *Mol. Microbiol.* **22**, 427–436.
6. W Shi and D.R. Żusman (1994) *Res. Microbiol.* **45**, 431–435.
7. R. M. Macnab and D. P. Han (1983) *Cell* **32**, 109–117.
8. A. Ishihara, J. E. Segall, S. M. Block, and H.C. Berg (1983) *J. Bacteriol.* **155**, 228–237.
9. M.Y. Chen, R. H. Insall, and P. N. Devreotes (1996) *Trends Genet.* **12**, 52–57.
10. C. A. Parent and P. N. Devreotes (1996) *Ann. Rev. Biochem.* **65**, 411–440.
11. P. J. Van Haastert (1995) *Experientia* **51**, 1144–1154.
12. D. Malchow, R. Mutzel, and C. Schlatterer (1996) *Int. J. Dev. Biol.* **40**, 135–139.
13. G. P. Downey (1994) *Curr. Opinion Immunol.* **6**, 113–124.
14. L. S. Tisa and J. Adler (1995) *Proc. Natl. Acad. Sci. USA* **92**, 10777–10781.

Suggestions for Further Reading

M. Eisenbach (1996) Control of bacterial chemotaxis. *Mol. Microbiol.* **20**, 903–910.

W. D. Hoff, K. -H. Jung, and J. L. Spudich (1997) Molecular mechanism of photosignaling by archaeal sensory rhodopsins, *Ann. Rev. Biophys. Biomol. Struct.* **26**, 221–256.

CHIMERA

J. FONTAINE-PERÚS

Spemann (1,2) was the first to employ the term "chimera" and to consider the great potential for surgically created chimeric embryos in the analysis of **developmental** mechanisms. The chimera method has frequently involved imaginative experimental procedures by which cells of one species are grafted into another. Any animal thus composed of different cell populations that derive from more than one fertilized egg should be considered as a chimera. This type of animal can currently be constructed in amphibians, birds, and mammals.

Many fundamental concepts of embryology have been at least partly formulated on the basis of results of cell or tissue transplants between two different embryos, usually separate species of amphibians. The most spectacular transplantation experiments, published by Spemann and Mangold in 1924 (3), demonstrated the organizing power of the dorsal lip of the blastopore during gastrulation by interspecific transplantations of this area. More recently, a model of lens induction was developed by using chimeric eyes (4). The technical advantages of producing amphibian chimeras are straightforward, owing to the independence of the embryos from their parents. They are easily accessible and receptive to foreign tissue, even across species barriers. All these qualities are not shared by higher vertebrates, such as birds and mammals. Nonetheless, bird embryos have several advantages over other vertebrate embryos, making certain interesting approaches feasible. The greatest advantage is continual accessibility within the egg throughout the developmental period. Another is the ease with which the various rudiments can be delineated, and thus removed and replaced, with extreme precision. An avian chimera obtained by combining quail and chick cells has been the most successful method, having provided a continual source of new data about developmental mechanisms for almost 30 years (5,6). With the advent of the quail-chick nuclear marker, which is particularly simple to employ, easy to identify, and endowed with great resolving power, avian chimeras have been used to study the ontogeny of the nervous system, the development of the hematopoietic and immune systems, and the formation of muscles and skeleton. Quail **heterochromatin** in the nucleus is concentrated around the **nucleolus**. This creates a large, deeply staining mass that is easily distinguishable from the diffuse heterochromatin of chick cells. Moreover, there are some **antigens** that are quail-specific and not detectable in chick cells. These phenomena allow individual quail cells to be readily distinguished, even when most of the cell population is chick.

Although the avian **embryo** is a practical model, perfectly suitable for tissue graft experiments after the incubated egg is opened, it is difficult to undertake this type of investigation in the mammalian fetus *in utero*. Nonetheless, it has become routine to remove postimplanted mammalian embryos from the uterus, manipulate them, and return them to a foster mother for further development. Thus, chimeric mice are the result of two or more early-cleavage (usually 4- or 8-cell) embryos that have been artificially aggregated to form a composite embryo. Since each cell is able to produce any component of the body, the construction of the chimeric mouse has very important consequences for the study of mammalian ontogeny. A very powerful application of this technique is the transfer of genes into every cell of the mouse embryo. During mouse development, there is a stage when only two cell types are present: outer cells, which will form the fetal portion of the placenta, and inner cells, which will give rise to the embryo itself. These inner cells are known as embryonic **stem cells** because each in isolation can generate all the cells of the embryo (7,8). These cells can be grown in culture, where they are treated to incorporate new DNA. The new embryonic stem cells can then be injected into another early-stage mouse embryo, resulting in a chimeric mouse. Mice that derive from these animals are **transgenic** mice. Our understanding of regulatory mechanisms in mammalian development is improving increasingly rapidly as a result of the construction of these genetically modified mice. A combination of the tools of developmental genetics with those of embryology should lead to real advances in the study of such mechanisms. In this field, our group has pioneered the grafting of embryonic tissues from transgenic mice into the chick embryo (9,10). Owing to these interspecies grafting experiments, it is possible to monitor factors that regulate the expression of a particular gene *in vivo*. The value of this technique is greatly increased when a **reporter gene** is used to follow the changes in gene expression of the grafted cells. The possibility of conducting grafts until late stages of *in vivo* development allows the behavior of wild and mutant mouse cells to be observed at any developmental stage and location.

AMPHIBIAN CHIMERA

Using amphibians, Spemann and Mangold (3) improved our understanding of the specification of the nervous system by transplanting dorsal blastopore lip tissue from an early gastrula into the ventral ectoderm of another gastrula. They used differently pigmented embryos from two species of newt: darkly pigmented *Triturus taeniatus* and nonpigmented *Triturus cristatus*. On the basis of color, it was easy to distinguish host and donor tissues. The dorsal blastopore lip tissue from early *Triturus taeniatus* gastrula, once transplanted into an early *Triturus cristatus* gastrula in the region, would normally become ventral epidermis. In fact, the donor tissue did not become belly skin but invaginated and formed a secondary embryo, face to face with its host. The more recent use of nuclear markers has allowed Spemann's results to be confirmed (11). Such chimeras elegantly demonstrate the organizing power of the dorsal lip of the blastopore in amphibian gastrula, since whole secondary embryos formed under the influence of the transplanted tissue.

Considerable advances have also taken place in the field of differentiation and organogenesis through the use of amphibian chimeras. The more common examples are those involving the interaction of epithelia with adjacent mesenchyme. After being separated, embryonic epithelium and mesenchyme can be recombined in different ways (12). In a classic experiment, Spemann and Schotté (13) transplanted flank ectoderm from an early frog gastrula into the region of newt gastrula destined to become part of the mouth. Similarly, the presumptive flank ectodermal tissue of newt gastrula was placed into the presumptive oral regions of frog embryos. The structures of the mouth region differ greatly between salamander and frog larvae. The *Triturus* salamander larva has club-shaped balancers beneath its mouth, whereas frog tadpoles produce mucus-secreting glands and suckers. Frog tadpoles also have a horny jaw without teeth, whereas the salamander has a set of calcareous teeth in its jaw. The larvae resulting from the transplants were chimeras. The salamander larvae had frog-like mouths, and the frog tadpoles had salamander teeth and balancers. In other words, the mesodermal cells instructed the ectoderm to make a mouth, but the ectoderm responded by making the only mouth it "knew" how to make. Thus, instructions sent by mesenchymal tissue can cross species barriers, although the response of the epithelium is species-specific. Thus, organ-type specificity is usually controlled by the mesenchyme within a species, but species specificity is usually controlled by the responding epithelium.

The cells that form the lens are derived from a region of head ectoderm in contact with optic vesicles of the anterior neural plate. Servetnick and Grainger (14) removed animal cap ectoderm from various gastrula stages and then transplanted them into the presumptive lens region of neural-plate-stage embryos. Ectoderm from early gastrula showed little or no competence to form lenses, but ectoderm from slightly later stages was able to. By the end of gastrulation, this ability to respond to the neural plate signal had been lost. This competence was found to be inherent within the ectoderm itself and not induced by other surrounding tissues. These observations showed that only mid- to late-gastrula ectoderm is able to respond to signals from the anterior neural plate. It has recently been demonstrated that the transcription factor Pax6 may play a role in the determination processes of eye tissue (see ***Pax* genes**).

AVIAN CHIMERA

When portions of quail embryo are grafted into a similar region of chick embryo, the cells become integrated into the host and participate in the construction of the appropriate organs. This grafting is done while the embryo is still inside the egg, and the chick that hatches is a "chimera," having a portion of its body composed of quail cells (15). The quail is usually chosen as a donor because it is easier to identify quail cells among chick cells than the reverse.

The ontogeny of the peripheral nervous system is one of the fields in which the use of avian chimeras is the most fruitful. This system arises almost entirely from the **neural crest**, a transient structure that develops at the neural tube apex. To follow the fate of chick neural tube-bearing neural crests, a segment is replaced by a homologous quail fragment. This method has permitted the diverse range of neural crest potentialities to be well-described, and the list of neural crest derivatives well-defined (6).

In relation to its origin, neural crest not only participates in the formation of spinal and cranial sensory ganglia, sympathetic and parasympathetic ganglia and plexuses, and Schwann cells of the peripheral nerves, but it also gives rise to endocrine and paraendocrine cells (calcitonin-producing cells, carotid body type-I cells, adrenomedullary cells) and pigment cells. The construction of an avian chimera has demonstrated the contribution of the cephalic neural crest to mesectodermal derivatives. For example, nasal and maxillary processes are built up partly by crest cells of mesencephalic origin. The mesencephalic crest cells also form the cephalic skeleton, the upper and lower jaws, the palate, and the tongue. The rhombencephalic crest participates in formation of the pre-optic region and the hyoid arch skeleton. In addition to cartilage and bone, the cephalic neural crest takes part in other derivatives of the head and neck, such as dermis and connective tissue.

Brain development has also been characterized by the use of the quail-chick chimera technique (16). Quail-chick brain chimeras can hatch and survive without showing impaired movement or locomotion, which indicates that functional synapses have been established between host and donor neurones, as well as between donor neurones and host muscles. Moreover, the normal behavior of the chimeras demonstrates that proper neuronal connections develop in the brain, which means that quail axons recognize local signals for growth and directionality in the chick environment, as they do in normal development. An interesting application of the chimeras is illustrated by chicks with transplanted quail mesencephalic-diencephalic brain areas, which then exhibit quail vocalization (17). Another important issue is the elucidation of when and where groups of cells in the brain make commitments to particular development pathways. Genes with spatially restricted expression are of particular interest, since they may indicate the existence of committed groups of cells that are important in pattern formation, but are not discernible on the basis of morphology. An example is provided by the **zinc-finger** gene *Krox-20*, which has restricted domains of expression in the early neural plate. *Krox-20* is expressed in the early neural epithelium, first in one stripe, and then in two stripes in the hindbrain. Subsequently, *Krox-20* is expressed in two alternating segments (rhombomeres) in the hindbrain. The generation of stripes of *Krox-20* expression in the early neural plate suggests that rhombomeric precursor cells are committed prior to the morphological appearance of the rhombomere.

Hox **homeotic gene** expression is seen along the dorsal axis, from the anterior boundary through to the tail. Direct evidence for hindbrain plasticity comes from quail-chick chimera experiments showing that anterior-to-posterior transpositions can reprogram Hox expression and induce a transformation in cell fate (18). The quail-chick system has also shown that environmental cues play a significant role in maintaining the Hox code in the neuroepithelium.

Chimeric experiments involving the muscles of the body, limbs, and skeleton have also been carried out in birds. For development, somites are the primitive metameric structures of the vertebrate body from which arise the vertebrae that surround the spinal cord, the muscles and connective tissue holding the vertebrae, the dermal layer of the skin of the back, and the limb musculature. All these data were obtained by constructing a chimeric embryo after interspecific grafts of somites between quail and chick embryos (19,20). The somites appear as pairs of epithelial spheres that bud off from the unsegmented paraxial mesoderm in a craniocaudal direction. They become polarized into a ventral mesenchymal compartment, the sclerotome, which yields the dorsal skeleton, and a dorsal epithelial component, the dermomyotome, from which striated muscle and dermis arise. By constructing a quail-chick chimera after interspecific exchanges of medial or lateral halves of newly-formed somites, Ordhal and Le Douarin (21) determined that, regardless of the somites, cells farthest from the neural tube (ie, lateral) migrate to form the body wall and limb musculature.

The interspecies graft method in birds has contributed considerably to understanding the control mechanisms of somite patterning. This method demonstrates that specification of the somite is accomplished by the interaction of different tissues that form its environment. In fact, the newly formed somite is composed mostly of unspecified cells, and the determination of somite compartments toward the different lineages is regulated by environmental cues. The ventral-medial portion of the somite is induced to become the sclerotome by factors, especially *Sonic hedgehog* protein, that are secreted from the notochord and neural tube floor plate. If portions of the notochord, the source of Sonic **hedgehog**, are transplanted next to other regions of the somite, those regions also become sclerotome cells. These sclerotome cells express a new transcription factor, Pax1, which activates cartilage-specific genes and is necessary for the formation of vertebrae (22). In similar ways, the myotome is induced by two distinct signals. The muscles surrounding the body axis, which arise from the medial portion of the somite, are induced by factors from the dorsal neural tube, probably members of the *Wnt* family (23,24). The muscles derived from the lateral portion of the somite, which form the musculature of the limbs and body wall, are probably induced through the combination of Wnt proteins from the epidermis with bone morphogenetic protein-4 (BMP4) protein from the lateral plate mesoderm (25). These factors cause the myotome cells to express particular transcription factors, MyoD and Myf5, that activate muscle-specific genes. The dermatome differentiates in response to another factor secreted by the neural tube, neurotrophin 3 (NT-3) (26).

MAMMALIAN CHIMERA

The laboratory mouse is by far the most popular mammal used to construct chimeras for developmental studies. Mice offer a large variety of genetically well-characterized strains that provide scope for using genetic markers. Moreover, they breed throughout the year and thus continually supply embryos. The most convenient time for manipulation of a mammalian embryo is before its implantation. During the first three to four days of gestation, mouse embryos have not yet formed an attachment in the mother's reproductive tract and can thus be explanted, manipulated, and then retransplanted into another mouse to continue their development. Chimeric mice are generally formed by the fusion of two early embryos that organize to produce a single mouse with two distinct cell populations.

A skeletal muscle cell (myotube) is an elongated cell containing many nuclei. It has been widely debated whether this cell is derived from the fusion of several mononucleated precursor cells (myoblasts) or from a single cell that undergoes nuclear division without cell division. Evidence for the fusion process has been provided by mouse chimeric constructions. Mintz and Baker (27) fused mouse embryos from two strains that produce different types of the dimeric **enzyme** isocitrate dehydrogenase: one strain makes the A subunit and the other B. If myotubes are formed from one cell whose nuclei divide without cytokinesis, the dimeric enzyme will be purely AA or BB. If, however, myotubes are formed by fusion between cells, some might code for the B subunit and others for A, in which case the molecules of the enzyme will be hybrid (AB). **Electrophoresis** can separate these three types (see **Isozyme, isoenzyme**). The presence of the hybrid AB enzyme in extracts of skeletal muscle tissue confirms the fusion model.

The study of mutations that impair development or function has long been recognized as a valuable means of elucidating the normal role of genes in such processes. Genetically chimeric animals consisting of mixtures of mutant and wild type cells can be of considerable value for studying the genes that are primarily implicated in developmental mechanisms. Methods of manipulating mouse embryos and transferring genes into every embryonic cell are now standard procedures. In recent years, an extraordinary increase in the use of these methods has led to numerous exciting prospects.

The most widely used method to produce transgenic mice is to inject **cloned** DNA into a pronucleus of a fertilized egg. Although this method has been used primarily for studies of **gene expression**, an unexpected benefit is the frequent integration of the injected DNA into genes, causing insertional mutations. Experimental infection of embryos by **retroviruses** has also been developed as a method for gene transfer. As in the case of spontaneous infection, this approach has resulted in insertional mutations. An additional approach to producing transgenic mice involves genetic modification of tissue culture lines of embryonic stem cells, followed by their re-incorporation into growing embryos. By applying somatic cell genetic techniques to these cells while in culture, both random and selected insertional mutations have been produced. Another method for assaying the function of a gene is to eliminate it from the genome of the whole organism. Very recent studies, using strategies to inactivate genes by homologous **recombination** in embryonic stem cells, with the subsequent generation of germ-line chimeras, have made this scenario possible and thereby initiated a new era in mammalian developmental biology.

The power of this technology can be illustrated by the *int-1* gene, which is expressed during central nervous system development in a temporally and spatially restricted fashion. Following inactivation of the gene by homologous recombination in embryonic stem cells, germ-line chimeras were obtained and bred to derive homozygous *int-1*-deficient mice. These homozygous mice were not viable, because specific regions of the brain were missing. This experiment indicates that *int-1* is an essential gene for early development and suggests that its expression might determine the fate of a cell. The *Hox a-3* gene has been found to control segment-specific gene expression in Drosophila, but what is its role in mammals? Chisaka and Capecchi (28) used **gene targeting** to determine the function of *Hox a-3* in the development of the mouse. Homozygous mutants of *Hox a-3* were found to have severe anomalies in development of the hindbrain and in mesectodermal neural crest derivatives and to lack thyroid, parathyroid, and thymus glands. These studies clearly open up new avenues for the study of mammalian development at the molecular level and will certainly be instrumental in unraveling the molecular networks responsible for the functioning of a multicellular organism.

MAMMALIAN AVIAN CHIMERA

In chimeras, the analysis of cell lineages depends on the ability to distinguish between cells of different origins. Markers are required to trace the developmental fate of cells and to recognize them at all ages in chimeric embryos. In mammals, genetic manipulations appear to be valuable means of elucidating the normal role of genes and identifying altered cells. Furthermore, the environment of the chimera allows these cells to survive and to display their phenotype. In chimeras, identification of all the descendants of genetically modified cells has thus far provided a reliable and sensitive means of measuring alterations in embryogenesis. In conjunction with an *in situ* marker, this may help in determining the cell type and the nature of the functions affected by the particular mutant. Extremely sensitive labeling is possible with transgenic mouse lines since cells that integrate foreign DNA, such as a reporter gene encoding for *Escherichia coli* β-galactosidase, are detectable by simple histochemical revelation, which can be used to differentiate grafted cells from host cells. Currently, the major means of exploring this genetic tool is by *in vitro* experiments in which mutant-induced tissue is cultured with wild-type tissue. The culture system is efficient in tissue development for only limited periods, however, and the media used may affect the tissue outcome. *In vivo* micromanipulation remains the most powerful means of studying the fates of cells, their origins, and the cell-cell interactions that govern development. Since implantation of mouse embryo is particularly unsuitable for *in vivo* manipulation, chick embryo is used as the site for developing mouse embryonic cells. Studies in such chimeras provide considerable information on the lineage and the differentiation of mouse embryonic cells, since *in ovo* grafted mouse cells are accessible throughout chick host embryogenesis. Moreover, these results are indicative of the information to be obtained through genetic manipulations in conjunction with embryonic tissue transplantation. It has been clearly demonstrated that *in ovo* transplantation provides a suitable environment for the development of mammalian cells and that the information supplied by this environment is capable of promoting mouse cell differentiation.

Chimeras were prepared by transplanting somites from nine-day postcoitum mouse embryos into two-day-old chick embryos at different axial levels. Mouse somitic cells then differentiated *in ovo* into dermis, cartilage, and skeletal muscle, as they normally do in the course of development, and were able to migrate into chick host limb. To trace the behavior of somitic myogenic stem cells more closely, somites arising from mice bearing a transgene of the desmin gene linked to a reporter gene coding for *E. coli* β-galactosidase were grafted *in ovo*. Interestingly, the transgene was rapidly expressed in myotomal muscles derived from implants. In the limb muscle mass, positive cells were found several days after implantation. This method facilitates investigation of the mechanisms of mammalian development, allowing the normal fate of implanted mouse cells to be studied and providing suitable conditions for identification of descendants of genetically modified cells.

Chimeras were also prepared by transplanting fragments of neural primordium mouse embryos into chick embryos at different axial levels. Mouse neuroepithelial cells differentiated and organized to form the different cellular compartments normally constituting the central nervous system. The graft entered into the development of the peripheral nervous system through migration of neural crest cells associated with mouse neuroepithelium. Depending on the graft level, mouse crest cells participated in the formation of various derivatives, such as head components, sensory ganglia, orthosympathetic ganglionic chain, nerves, and neuroendocrine glands. Knock-out mice with the tenascin genes inactivated, which express LacZ instead of tenascin and show no tenascin production (29), were used specifically to follow Schwann cells lining nerves derived from the implant. Together with the previous results on somite development, this study shows that the chick embryo constitutes a privileged environment, facilitating access to the developmental potentials of normal or defective mammalian cells. It allows study of the histogenesis and precise timing of formation of a known structure, as well as the implications of a given gene at all equivalent mammalian embryonic stages.

BIBLIOGRAPHY

1. J. Spemann (1901) *Arch. Entwicklungsmech. Org.* **12**, 224–264.
2. J. Spemann (1921) *Arch. Entwicklungsmech. Org.* **48**, 533–570.
3. J. Spemann and H. Mangold (1924). *Arch. Mikrost. Anat. Arch. Entwicklungsmech.* **100**, 599–638.
4. J. J. Henry and R. M. Grainger (1987) *Dev. Biol.* **124**, 200–214.
5. N. M. Le Douarin (1969) *Bull. Biol. Fr. Belg.* **103**, 435–452.
6. N. M. Le Douarin (1982) *The Neural Crest*, Cambridge University Press, London and New York.
7. R. L. Gardner (1968) *Nature* **220**, 596–597.
8. L. A. Moustafa and R. L. Brinster (1972) *J. Exp. Zool.* **181**, 193–202.
9. J. Fontaine-Pérus, V. Jarno, C. Fournier Le Ray, Z. L. Li, and D. Paulin (1995) *Development* **121**, 1705–1718.
10. J. Fontaine-Pérus, P. Halgand, Y. Chéraud, T. Rouaud, M. E. Velasco, C. Cifuentes-Diaz, and F. Rieger (1997) *Development* **124**, (16), 3025–3036.
11. R. L. Gimlick and J. Cook (1983) *Nature* **306**, 471–473.
12. J. W. Saunders Jr, J. M. Cairns, and M. T. Gasseling (1957) *J. Morphol.* **101**, 57–88.
13. H. Spemann and O. Schotté (1932) *Naturwissenschaften* **20**, 463–467.
14. M. Servetnick and R. M. Grainger (1991) *Development* **112**, 177–188.
15. N. M. Le Douarin and M. A. Teillet (1973) *J. Embryol. Exp. Morphol.* **30**, 31–48.
16. N. M. Le Douarin (1993) *TINS* **16**, 64–72.
17. E. Balaban (1988) *Proc. Natl. Acad. Sci. USA* **85**, 3657–3660.
18. A. Grappin-Botton, M. A. Bonnin, L. A. McNaughton, R. Krumlauf, and N. M. Le Douarin (1995) *Development* **121**, 2707–2721.
19. A. Chevallier, M. Kieny, A. Mauger, and P. Sengel (1977) *Vertebrate Limb and Somite Morphogenesis*, Cambridge University Press, Cambridge, pp. 421–432.
20. B. Christ, H. J. Jacob, and M. Jacob (1977) *Anat. Embryol.* **150**, 171–186.
21. C. P. Ordhal and N. Le Douarin (1992) *Development* **114**, 339–353.
22. C. A. Smith and R. S. Tuan (1996) *Teratology* **52**, 333–345.
23. A. E. Münsterberg, J. Kitajewski, D. A. Bumcroft, A. P. McMahon, and A. Lassar (1995) *Genes Dev.* **9**, 2911–2922.
24. H. M. Stern, A. M. C. Brown, and S. D. Hauschka (1995) *Development* **121**, 3675–3686.
25. O. Pourquié et al. (1996) *Cell* **86**, 461–471.
26. G. Brill, N. Kahane, C. Carmeli, D. Von Schack, Y. A. Barde, and C. Kalcheim (1995) *Development* **121**, 2583–2594.
27. B. Mintz and W. W. Baker (1967) *Proc. Natl. Acad. Sci. USA* **58**, 592–598.
28. O. Chisaka and M. R. Capecchi (1991) *Nature* **350**, 473–479.
29. Y. Saga, J. Yagi, Y. Ikawa, T. Sakakura, and S. Aizawa (1992) *Genes Dev.* **6**, 1821–1838.

Suggestions for Further Reading

E. Dupin, C. Ziller, and N. M. Le Douarin (1998) "The Avian Embryo as a Model in Development Studies: Chimeras and *in vitro* Clonal Analysis," *Curr. Top. Dev. Biol.* **36**, 1-35.15

N. Le Douarin, F. Dieterlen Lièvre, and M. A. Teillet (1996) "Quail-Chick Transplantations" *Methods Cell Biol.* **51**, 23–59.

CHIRAL AND CHIRAL CENTER

VERNON ANDERSON

The word chiral is derived from the Greek word "cheir" for "hand." It was first used to describe any molecule that could not be superimposed on its mirror image, like left and right hands, by Lord Kelvin (1). This supplanted the less precise term dissymetrical used by Pasteur (2). The nonsuperimposable mirror image isomers are **enantiomers**. The physical property that differentiates enantiomers is the direction that they rotate plane polarized light. Thus, the two enantiomers are differentiated as either dextro(+) or levo(−) rotatory, depending on whether the rotation of the polarized light is clockwise or counterclockwise, respectively. Solutions containing an excess of one of the enantiomers are said to be optically active. A solution containing exactly equal concentrations of both enantiomers is called racemic (see **Racemizic and racemization**).

The most common structural element that generates chiral molecules is the presence of a tetrahedral atom with four different substituents, by definition a chiral center. The α-carbon of the α-amino acids (Fig. 1), except for **glycine**, and the hydroxymethylene carbons of carbohydrates are all chiral centers. The mirror images of chiral molecules have different absolute **configurations**, which are uniquely identified by the rules proposed by Cahn et al. (3) (see **Configuration**).

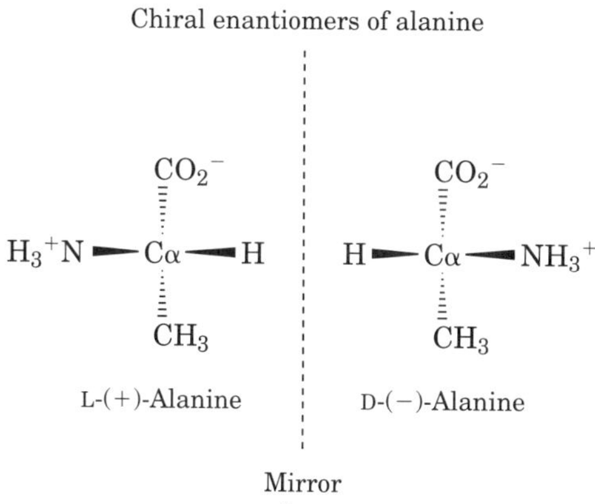

Figure 1. The enantiomers of alanine, which are mirror images. Note that the α-carbon is chiral because of the four different substituents. The (L) and (D) designate the absolute configuration of the chiral center, whereas the (+) and (−) indicate which direction an aqueous solution of the enantiomer will rotate the plane of polarized light.

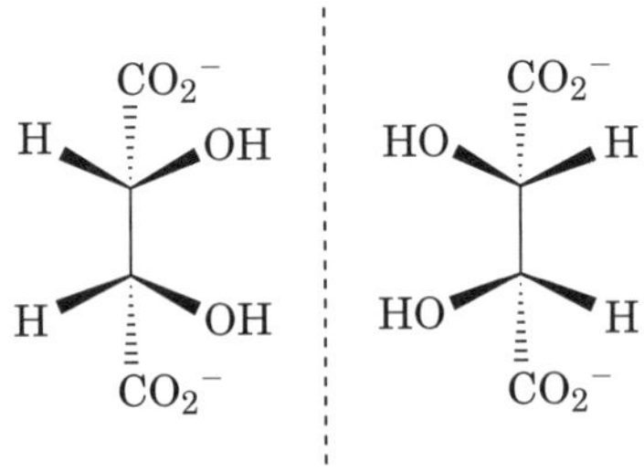

Meso form of tartrate

Figure 2. Meso tartrate, which contains two chiral centers C2 and C3, but is not chiral because there is an internal plane of symmetry. Note that, unlike the enantiomers of alanine, these mirror images can be superimposed by a 180° rotation.

Possessing a tetrahedral chiral center is not required for chirality, nor is the presence of a chiral center proof that the molecule will be chiral. Other arrangements of atoms can generate nonsuperimposable mirror images. One example of biochemical importance is the hexacoordinate complexes of metal ions, such as the Mg ATP complex (4). Other chiral molecular structures have been surveyed by Testa (5). A molecule with an internal mirror plane may have chiral centers but not be chiral. These are meso compounds. Because of their internal symmetry, these molecules can be superimposed on their mirror image. The **diastereomeric** form of tartaric acid, which has two chiral centers, but a mirror plane between C2 and C3, provides an example of a meso compound (Fig. 2).

BIBLIOGRAPHY

1. L. Kelvin (1894) *Boyle Lect. J. Oxford Univ. Junior Sci. Club* (May 25), 25.
2. L. Pasteur (1853) *Pharmacol. J.* **XIII**. 111.
3. R. S. Cahn, C. K. Ingold, and V. Prelog (1966) *Angew. Chem. Int. Ed.* **5**, 385–415.
4. E. A. Merritt et al. (1978) *Biochemistry* **17**(16), 3274–3278.
5. B. Testa (1982) In *Stereochemistry* (C. Tamm, ed.), Vol. 3, Elsevier, Amsterdam, pp. 1–48.

Suggestions for Further Reading

J. March (1985) *Advanced Organic Chemistry,* Wiley-Interscience, New York, pp. 82–100.

K. Mislow (1966) *Introduction to Stereochemistry,* W. A. Benjamin, New York, pp. 50–60.

CHIRONOMUS

G. BERGTROM

The subfamily chironomini includes Chironomus and related genera that by some estimates contain over 5000 species. Because of this diversity, many studies of chironomids focus on environmental biology and systematics. However, it was the giant **polytene chromosomes** of salivary gland cells and the bright red color of the larvae of these ubiquitous flies that first drew the attention of experimental biologists more than 100 years ago.

The alternating dark bands (**chromomeres**) and light bands of the polytene chromosomes are punctuated by **puffed** regions, the largest of which were named **Balbiani rings** to honor their discovery by E. G. Balbiani in 1881. The possible relationship between chromomeres and genetic loci (see **Genetics**) on the one hand, and between puffs and gene activity on the other, excited the interest of cytologists and geneticists. One of these, W. Beermann, examined the morphology of chromosomes in four cells in the "special lobe" of the salivary glands of larvae that were derived from a cross between *C. tentans* and *C. pallidivitatus*. He was able to correlate the production of prominent **secretion granules** by these cells with the presence of a specific chromosome puff in both the hybrid and in *C. pallidivitatus*, and he showed that the puff was absent from the special cell chromosomes in *C. tentans*, which do not contain the secretion granules. C. Pelling later demonstrated that polytene chromosome puffs selectively incorporated radioactive uridine, firmly establishing that they are sites of active gene **transcription**. The sequential appearance of some, and the disappearance of other, chromosome puffs during **development** could be mimicked by treating cultured late-stage larval salivary glands with the **steroid hormone**, 20-hydroxyecdysone. Thus, the same hormone that signals molting and metamorphosis in insects also controls the differential transcription of genes encoded by responsive puffs (see **Ecdysone**).

In a prodigious feat of molecular isolation that predates molecular **cloning** by almost a decade, J. E. Edstrom, B. Daneholt, and their colleagues used micromanipulation to dissect individual Balbiani rings. A very large (75S) mRNA (**messenger RNA**) encoding a **polypeptide** of about 850 kDa was extracted from pooled Balbiani ring 2 isolates. This and other salivary gland secretory proteins encoded by the major Balbiani rings are extruded through the larval mouth and spun together with benthic detritus to make tubular houses. The stickiness of *Chironomus* "silk" results from the aggregation properties of interacting secretory polypeptides from the major Balbiani rings, and several studies focus on those aspects of structure that could explain the adhesive properties of these unusual biopolymers. In *C. tentans*, each major Balbiani ring encodes polypeptides with internal repetitive **domains**. The core repeat of one of these domains has four **cysteine residues** in invariant positions; a subrepeat region contains tandem *basic residue–Pro-acidic residue* motifs. This arrangement might encourage protein strand alignment and polymerization of the "silk." The expression, evolution, and structure of polytene chromosomes themselves remain active areas of study (notably of **centromere** and **telomere** structure and replication). Other studies seek to define sex-determining genes and proteins and to characterize diverse **transposable elements** in chironomids.

Because much attention has centered on *Chironomus* **hemoglobins**, the rest of this article deals with their structure, function and expression during development, and with the structure, organization and evolution of the many **globin** genes in this large **multigene family**.

CHIRONOMUS HEMOGLOBIN STRUCTURE AND FUNCTION

The red pigment so visible through the larval integument was identified as hemoglobin by E. R. Lankester in 1872.

In 1932, T. Svedberg et al. used their then-new technique of **sedimentation velocity ultracentrifugation** to estimate the molecular weight of hemoglobin from *Chironomus* (and other invertebrates); chironomid hemoglobins "weighed in" at 32 kDa. Physiological studies focused on the unusually high oxygen affinity of the hemoglobin (compared to its vertebrate counterparts) and on how it facilitated oxygen supply to larval tissues in a relatively hypoxic benthic habitat.

Chironomus hemoglobin is actually a mixture of many hemoglobin proteins synthesized and secreted by larval fat body. Long before the first chironomid globin gene was cloned, much was known about the structures of these proteins. In contrast to the tetrameric vertebrate hemoglobins (see **Hemoglobin**), chironomid hemoglobins do not show **cooperative** binding of oxygen and exist as either monomers or homodimers under physiological conditions (Svedberg's 32 kDa "hemoglobin" was presumably a mixture of homodimers). Over the course of a decade, the amino acid sequences of 12 *C. thummi* globins were determined in the laboratory of G. Braunitzer. Since none are truly similar in sequence to their vertebrate counterparts, alignments of chironomid and vertebrate globin sequences were possible only because they share several amino acids at crucial positions known to be involved in binding the heme group. On the other hand, comparison of the three-dimensional structure of a monomeric chironomid hemoglobin (determined by **X-ray crystallography**) with that of vertebrate globins reveals many shared structural features, including similarly organized ***alpha*-helical** domains and a characteristic **hydrophobic** heme pocket. These observations clearly indicate that similar molecular functions and three-dimensional structures can be attained in different proteins despite considerable differences in **primary structure**. The differences in primary structure between globins of different organisms reflects adaptation to species-specific oxygen requirements over long evolutionary periods.

Although at least two *C. thummi* globin genes encode identical polypeptides (1), pairwise comparisons of amino acid sequences reveal remarkably dissimilar globins. These differences have also been ascribed to evolutionary specialization of different hemoglobins that now serve different respiratory functions (2). However, the physiological roles of individual hemoglobins are at best only poorly understood. *In vivo* determinations of larval O_2 binding and storage abilities under different conditions of oxygen tension necessarily represent a function of the aggregate set of hemoglobins present at the time of measurement. As expected, the O_2 affinities determined for purified hemoglobins that are present in different proportions in the hemolymph are not easily related to the aggregate *in vivo* data. What does emerge from such experiments is a loose correlation of the generally high oxygen affinities of the hemoglobins and the benthic, often oxygen-depleted habitat of larvae.

Larvae live in the mud at the bottom of freshwater ponds and lakes, developing in burrows constructed of detritus cemented by the sticky secreted salivary gland proteins discussed above. Moreover, chironomids can withstand the oxygen stress of polluted or eutrophic waters, often in such a situation comprising more than 80% of all animal individuals. This tolerance is due in part to the high concentrations of hemolymph hemoglobin. Zebe (3) suggested that the hemoglobins of *C. thummi* may not be required for oxygen delivery at higher partial pressures (oxygen needs under these conditions are presumably satisfied by diffusion). Rather, because they can bind O_2 at low partial pressures (50–10 torr), the hemoglobins allow continued respiration in times of severe oxygen deficit, when they become physiologically significant. From this, one might predict that oxygen stress will favor larval production of hemoglobins with particularly high O_2 binding affinities. Homodimeric hemoglobins have higher O_2 affinities than the monomeric hemoglobins and do indeed increase in concentration in response to circadian and seasonal declines in O_2 availability and water quality, where they may be more involved in O_2 transport under these conditions than monomeric hemoglobins. With a more pronounced **Bohr effect** than monomeric hemoglobins, dimeric hemoglobins may also allow response to more immediate changes in oxygenation conditions due to local fluctuations in pH.

Overall hemoglobin levels reach their highest concentrations in the hemolymph of *C. thummi* in the fourth, or last, larval stage (**instar**). Although it is likely that newly synthesized hemoglobins function mainly in respiration, hemolymph hemoglobin is also taken up by developing **oocytes** and deposited in yolk granules during **oogenesis** in pupae. This hemoglobin undoubtedly serves a nutrient rather than a respiratory function during embryogenesis. At the same time, developing eggs may contain hemoglobins of their own making. Cross-reactive antihemoglobin **antibodies** tagged with colloidal gold particles are localized over rough **endoplasmic reticulum** in newly fertilized eggs, suggesting that hemoglobin synthesis may begin during embryogenesis. Hemoglobins first accumulate sufficiently to become visible as a red pigment in the blood of second instar larvae. Thereafter, hemoglobin synthesis is cyclic, as it is high during the instar and declines at each molt. Hemoglobin synthesis occurs for the last time just after pupariation, ceasing sometime in the pupal stage, by which time most existing hemoglobins have been degraded. Hemoglobins have disappeared and are virtually undetectable several hours after adult emergence, at which time the adult flies have a fully developed tracheal respiratory system. Once again, in spite of these observations, it is not possible to state with certainty the physiological role played by any given hemoglobin at any given moment. As we will see, this issue may be moot for many of the hemoglobins, since there is reason to believe that much of the structural variation between hemoglobins has not been selected to perform a unique physiological function, but instead is the result of **neutral evolution**.

DEVELOPMENTAL REGULATION OF *CHIRONOMUS* HEMOGLOBIN SYNTHESIS

While cyclic hemoglobin production is tied to molting, it should come as no surprise that the synthesis of individual members of such a large family of hemoglobins is also regulated differentially during development. For example, polyacrylamide gel electrophoresis (**PAGE**) of *C. thummi* fourth instar hemolymph reveals two hemoglobins that are absent in electrophoretic separations of third instar larval hemolymph. Another hemoglobin, detectable in embryo homogenates by electrophoresis and immunoblotting, is undetectable in last instar larval hemolymph. In **S1 nuclease** protection assays performed to follow specific globin gene expression during development, transcripts of the ct-1, ct-4, ct-7B4, and ct-7B5 globin genes (ct-, cp-, ctn-, and kc- are prefixes for *C. thummi*, *C. piger*, *C. tentans*, and *K. cornishi* globins and globin genes) are detected at high levels in 3rd instar *C. thummi* larvae. Transcripts of each of these globin genes decline to undetectable levels immediately after molting to the fourth instar, reappearing 24 h later, along with two new fourth-instar-specific globin transcripts (ct-3, ct-6). The expression of each of these globin mRNAs was tracked further during fourth instar development. The results of these experiments (4) indicate that globin transcript steady-state levels follow one of two kinetic patterns. Transcripts of the ct-3 and ct-6 genes rise from very low or undetectable levels post-molt, and then remain elevated through 8 days of the instar. Globin mRNA expressed in both instars also rise from low or undetectable levels just after the molt. But unlike the 4th instar-specific globin mRNAs, they reach maximum levels after 2–3 days and then decline to near basal levels. The absence of detectable globin transcripts just after the third to fourth instar molt is consistent with the observation that the rate of synthesis of hemoglobins of cultured larval tissues exposed to the molting hormone 20-hydroxyecdysone (20HE) is very much reduced. Like other steroid hormones, 20HE binds to an intracellular **receptor**, which then binds to a **hormone response element**(s), directly controlling the transcription of adjacent genes. Such direct action on globin genes is unlikely, however, since a search of DNA lying between or flanking any *C. thummi* globin gene sequences published or recorded in international DNA sequence databases failed to detect 20HE response elements. As yet unidentified factors must mediate the hormone-induced repression of hemoglobin synthesis during molting. There is no evidence at all for a role for 20HE in the selective repression of "non-stage-specific" globin mRNA levels during the fourth instar, indicating the action of still other factors in the selective silencing of some but not all globin genes.

CHIRONOMUS GLOBIN GENE STRUCTURE AND ORGANIZATION

At the same time as the differential expression of hemoglobins or globin transcripts was being described, many more globin genes were being characterized. Known *C. thummi* genes exceed the number needed to account for the 12 hemoglobins that were first isolated. The loci of individual and clustered globin genes were mapped by ***in situ* hybridization** of cloned DNA to polytene salivary gland chromosomes, and as noted, the expression of many globin genes, including several for which no protein product has yet been isolated, has been demonstrated. Table 1 summarizes some key properties of the globin genes characterized thus far. Genes that are clearly inactive include ct-13RT (which suffered insertion of a **retroposon** in coding DNA), ct-V (with an inappropriately located **stop codon**) and ctn-ORFB (with its coding region truncated at the 5′ and 3′ ends). With these exceptions, all of the known chironomid globin genes possess the structural prerequisites of active genes, including appropriately located TATA (**promoter**) and AATAAA (**polyadenylation**) signals, an open reading frame and a **stop codon**. When compared to vertebrate (eg, human) globin genes, a striking difference is that the latter are separated by thousands of nucleotides, whereas chironomid globin genes are often separated by fewer than 500 nucleotides. Another difference: chironomid globin genes lack the hallmark structure of three **exons** separated at conserved positions by two **introns** that characterize vertebrate (and even plant) globin genes. The growing database of chironomid globin genes from several species, their unusual structure, and their organization have sparked considerable interest in the evolution of this unique multigene family.

EVOLUTION OF *CHIRONOMUS* GLOBINS

Phylogenetic analyses support an early origin of monomeric hemoglobins about 500 million years ago and the appearance of homodimeric hemoglobins about 250–300 million years ago. The chironomid globin **gene family** has experienced many of the same evolutionary events as other multigene families, such as point-mutational inactivation, **translocations**, gene loss, *gene* **conversion** or correction, and **gene fusion**. They also support two somewhat unorthodox hypotheses. One is that eukaryotic genes can acquire introns, complete with correctly placed intron/exon junctions and splicing signals. The other hypothesis is that much of the diversity of hemoglobin structure in *Chironomus* has arisen by neutral evolution rather than by functional adaptations of the proteins. There is resistance to ascribe to neutral evolution those mutations causing even minor differences in amino acid sequence between the products of **homologous** genes in different species or **paralogous** *genes* in the same organism. Such differences between globin genes in *Chironomus*, and between the products of multigene family members in general, are widely thought to be the result of adaptive selection. However, evidence from the different chironomid globins and globin genes suggests that many were not selected for their adaptive value, but exist instead as a neutral consequence of positive selection of the expression of high levels of hemoglobin from multiple genes.

Origin of a Central Intron

In the 1980s, the sequencing of vertebrate globin genes became a cottage industry! Each new report presented a globin gene with the same mosaic structure: three exons and two introns, the latter always interrupting the coding DNA at the same location. In a prophetic paper (5), Gō analyzed X-ray crystallography structure data for a vertebrate globin and concluded that the ancestral globin gene had a third intron in the center of the gene, in addition to the outer introns, such that the entire gene was separated into four exons of similar length, each coding for a separate structural domain. Gō's prediction was fulfilled shortly thereafter with the report of the first plant (legume) globin gene, which had a centrally located intron in addition to two introns at the same position as those in the vertebrate globin genes. Apparently the structure of plant globin genes represents the ancestral state. Many of the first chironomid globin genes to be reported were intronless, a state presumably due to the **reverse transcription** of a completely processed transcript (see **Splicing**) and reinsertion of the **complementary DNA** into the genome. The discovery of the *C. thummi* ct-2ß and ct-9.1 genes (6), and of several other invertebrate globin genes with a central intron, however, raised questions about the origin of chironomid globin gene introns. Unlike the conservation of location of the outer vertebrate globin gene introns, the location of central introns is not conserved across kingdoms or phyla, and none are at the position predicted by Gō. Could they, like the outer introns in globin genes, still be descendants of an ancestral intron? Gilbert suggested that introns at nonconserved locations have moved by "intron sliding." Others have argued that the discordant location of introns is more easily (and parsmoniously) explained if introns can be acquired by previously uninterrupted genes, and several plausible mechanisms for intron acquisition have been suggested. In considering these, Kao et al. (6) offered a mechanism whereby a chironomid globin gene could have acquired a central intron by a simple **gene duplication** accompanied (or followed) by deletion of all but the DNA presently found in the intron.

Evolution of a Globin Multigene Family

The DNA sequences of orthologous globin gene clusters in different chironomid species support the adaptive selection of a high globin gene copy number and the essentially neutral evolution of individual globin genes. The cp-Y, cp-W, cp-V, and cp-Z genes and five cp-7B genes in *C. piger* (7) and the ct-Y, ct-W, ct-V, and ct-Z genes and seven ct-7B genes of sibling species *C. thummi* (1) are a case in point. The phylogenetic study summarized in Figure 1 traces the evolutionary history of the two clusters, showing the seven genes that are clearly homologous, and a series of gene duplications, losses, and fusions necessary to explain the current sequence and organization of the clusters. Pairwise comparisons show that at the nucleotide level, homologous globin genes or logical gene pairs (ie, genes of indeterminate orthology, but sharing immediate common ancestry) in the two species share from 96.7–99.2% identity. In three cases, homologous genes in each species code for identical polypeptides, despite several millions of years of species divergence. Within *C. thummi*, the ct-7B5, ct-7B6, ct-7B9, and ct-7B10 paralogous genes are also very similar, as shown by the amino acid differences in the following pairwise comparisons:

	7B6	7B10	7B5	7B9
7B8	6	4	6	6
7B6		2	1	1
7B10			2	2
7B5				0

In addition to identical paralogues (ct-7B5, ct-7B9), the coding region of a gene sequenced from an independent genomic clone is identical to that of ct-7B5 (8). This gene, which is neither a second copy of ct-7B5 nor an allele of ct-7B9, is called ct-7B9/5 because the 5′ and 3′ DNA flanking its coding region share similarities with the 5′ and 3′ noncoding regions of ct-7B9 and ct-7B5, respectively. The ct-7B9/5 gene is a chimeric gene that may be either an independent gene locus or an allelic **haplotype**. Clearly the 7B globin gene subfamily predates the *thummi/piger* divergence. The ability of *C. thummi* and *C. piger* to lose different descendants of an ancestral cluster by

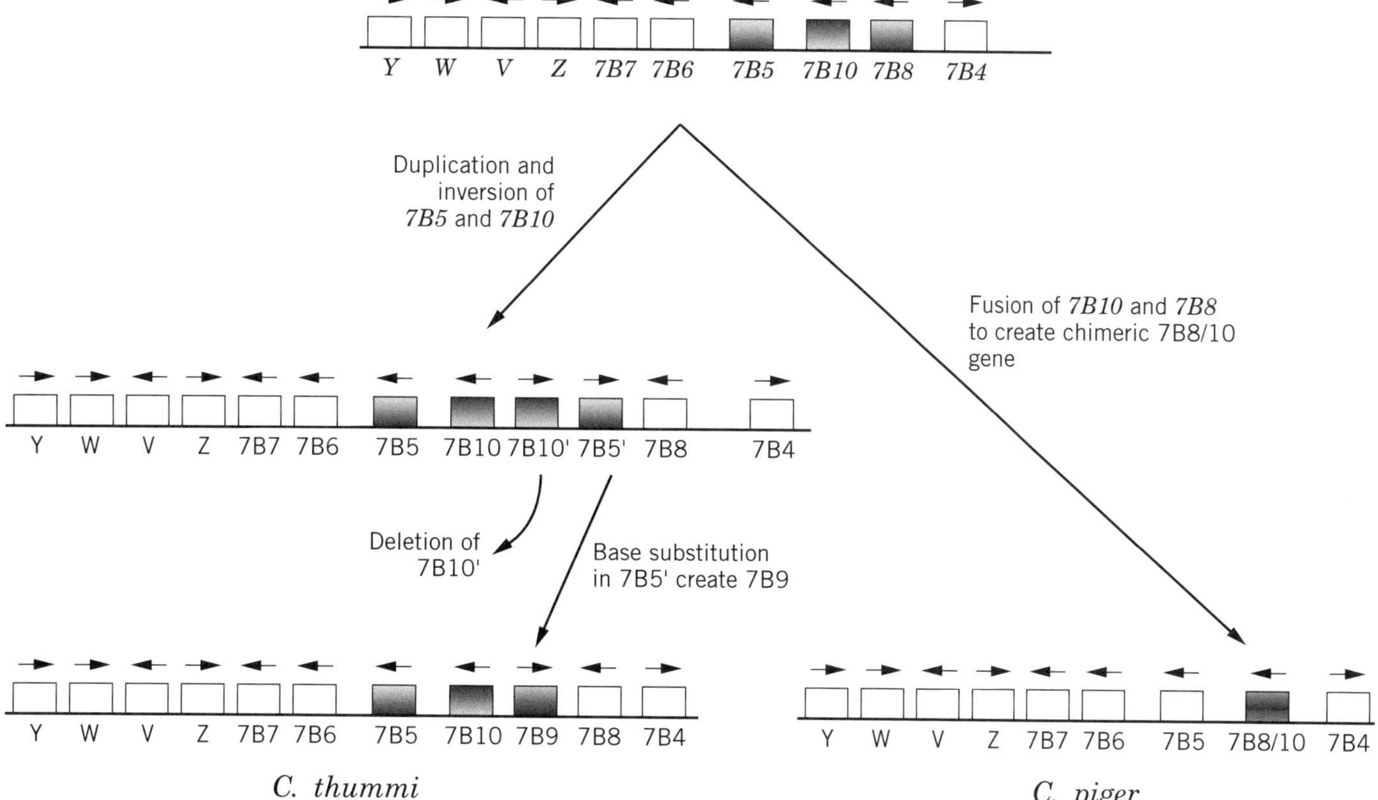

Figure 1. Evolutionary history of orthologous gene clusters in *C. thummi* and its sibling subspecies, *C. piger*. ct- and cp-designations are deleted to save space. Arrows above the genes indicate direction of transcription. Ancestral genes are designated in italics. Shaded genes trace the history of recent duplications, losses and fusions.

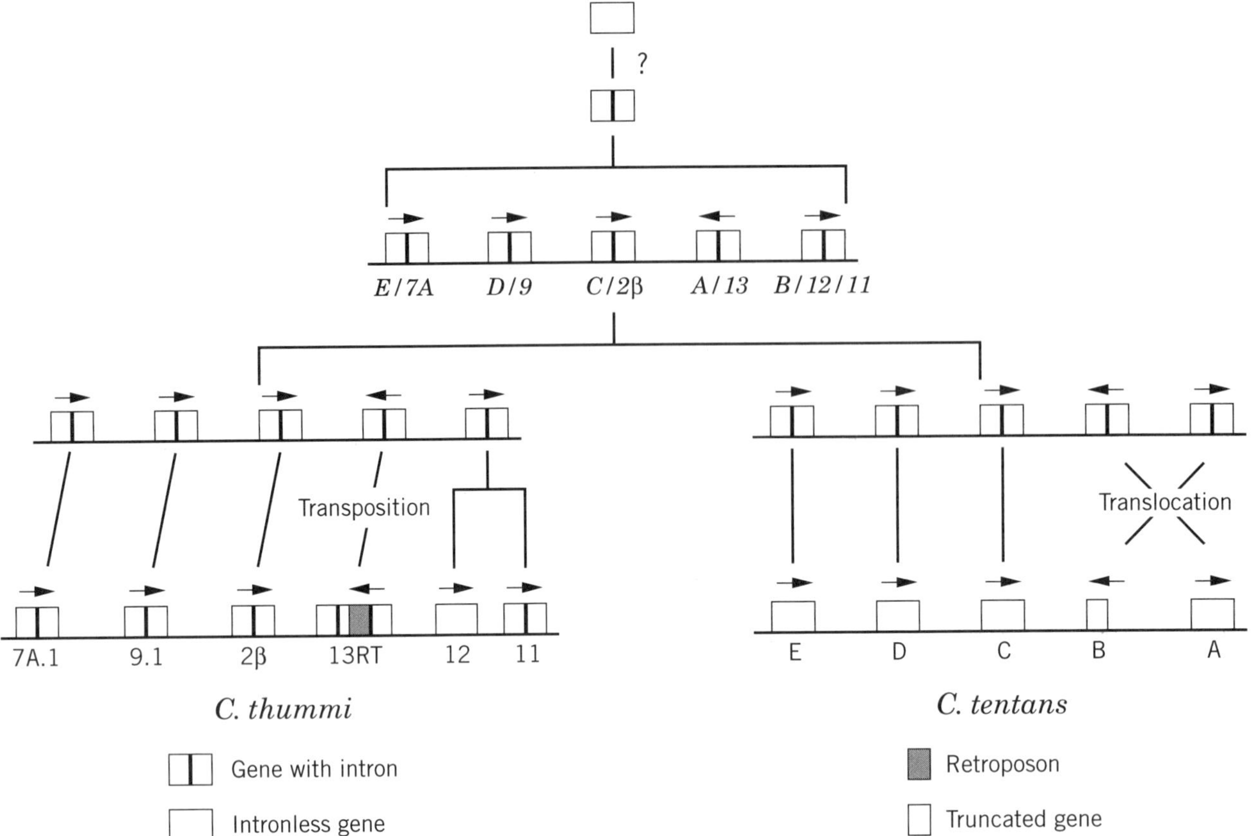

Figure 2. Evolutionary history of orthologous gene clusters in *C. thummi* and its distant relative, *C.tentans*. ct-, ctn- and ORF ("open reading frame") designations are deleted to save space. Arrows above the genes indicate direction of transcription. Ancestral genes are in italics. Details are given in the text.

gene fusion, and the overall similarity of paralogous genes, implies that any small differences that accumulate in the genes before or after fusions and corrections are evolutionarily inconsequential. The 7B gene cluster seems to have undergone periodic expansion (**gene duplication**) and contraction (**gene fusion**), while maintaining the ability to encode identical or nearly identical globins, for millions of years. The best explanation for this phenomenon is that the few differences that accumulate between 7B genes are the result of **random drift** (neutral evolution), but that the cluster itself has experienced positive selection of a high number of copies of globin genes as a means to ensure the synthesis of large amounts of hemoglobin.

Whereas the orthologous globin gene clusters in *C. thummi* and *C. piger* illustrate early evolutionary events in the life of a multigene family, globin gene clusters in *C. thummi* and *C. tentans* are separated by 60 million years of evolution (9,10). Unlike the case of the 7B genes, there are no recent gene duplications, fusions, or losses in the *thummi* or *tentans* clusters. The extant paralogous and homologous genes sequences are much less similar to each other (except for a region of the ct-2ß and ct-9A genes that underwent partial gene correction, homogenizing intronic and exonic DNA in the middle of the two genes). The divergence of the *C. thummi* and *C. tentans* clusters offers a wider window through which to test the concept of gene copy number selection. Figure 2 summarizes the evolution of these two clusters. The cluster in *C. thummi* includes ct-2ß, ct-9.1, three more intron-bearing globin genes, and one that lost an intron (ct-12). The homologs in *C. tentans* are intronless, indicating multiple intron losses after species separation. The possibility that the central intron was originally acquired by the progenitor to the ancestral cluster is suggested in Figure 2. Phylogenetic analysis using branch-and-bound and bootstrapping algorithms supports the polytomous (ie, nearly simultaneous) origin of five linked ancestral, intron-bearing genes (*E/7A, D/9, C/2ß, A/13RT*, and *B/11/12*). Thus, an ancestral gene cluster existed prior to species separation, as did the 7B gene ancestors in *thummi* and *piger*. The translocation shown between ctn-A and ctn-B is one explanation for their present organization in *C. tentans*. Clearly, the same cluster has followed different evolutionary paths in the two species. It is, however, the very recent inactivation of a different (non-homologous) gene in each cluster that suggests that not all globin genes in a cluster are indispensable, even after 60 million years of evolution. In *C. thummi*, ct-13RT suffered the insertion of a retroposon into coding DNA in exon I. This occurred very recently, since the rest of the coding region, the intron, and the flanking DNA containing promoter and polyadenylation motifs are all still intact. In *C. tentans*, the ctn-ORFB gene was truncated at both the 5′ and 3′ ends. Again, this event must have occurred

Table 1. Chironomid Globin Genes

Organism	Gene	Protein Product	Status[a]	Homologs	Locus[b]	Comments
C. th. thummi	ct-1	I	Active	—	F2b3	Allele of ct-1A; intronless; aa (amino acid) sequence available
	ct-1A	IA	Active	—	F2b3	Allele of ct-1; intronless; aa sequence available
	ct-3-1	III		—	AIB2	Intronless; * 3-1 and/or 3-2 = available aa sequence
	ct-3-2	III	*	—	A1B2	Intronless, * 3-1 and/or 3-2 = available aa sequence
	ct-4-1	IV	*	—	AIB2	Intronless, * 4-1 and/or 4-2 = available aa sequence
	Ct-4-2	IV	*	—	AIB2	Intronless, * 4-1 and/or 4-2 = available aa sequence
	ct-E	Putative	Putative	—	AIB2	Intronless
	ct-6	VI	Active	—	F2b3	Intronless; aa sequence available
	Not found	VIIB-1	Active	—	—	aa sequence available
	Not found	VIIB-2	Active	—	—	aa sequence available
	ct-7B3	VIIB-3	Active	—	N.D.	Intronless; protein product assumed
	ct-7B4	VIIB-4	Active	—	F2b3	Intronless; protein product assumed
	ct-7B5	VIIB-5	Active	—	F2b3	Intronless; protein product assumed
	ct-7B6	VIIB-6	Active	—	F2b3	Intronless; protein product assumed
	ct-7B7	Putative	Putative	—	F2b3	Intronless; protein product assumed
	ct-7B8	Putative	Putative	—	F2b3	Intronless; protein product assumed
	ct-7B9	Putative	Putative	—	F2b3	Intronless; protein product assumed
	ct-7B10	Putative	Putative	—	F2b3	Intronless; protein product assumed
	ct-7B9/5	Putative	Putative	—	N.D.	Intronless chimera; protein product assumed
	ct-Y	Putative	Active	—	F2b3	Partial DNA sequence, RT-PCR product
	ct-W	Putative	Active	—	F2b3	Intronless; protein product assumed
	ct-V	None	Inactive	—	F2b3	Stop codon in middle of coding region
	ct-Z	Putative	Active	—	F2b3	Intronless; protein product assumed
	ct-7A.1	VIIA	Active	—	F2b3	Central intron; aa sequence avail.
	ct-9.1	IX	Active	—	F2b3	Central intron; aa sequence of allele avail.
	ct-2ß	II2ß	Active	—	F2b3	Central intron; aa sequence avail.
	ct-13RT	None	Inactive	—	F2b3	Central intron; retrposon insertion in coding DNA; may be transcribed
	ct-12	Putative	Putative	—	F2b3	Intronless
	ct-11	Putative	Putative	—	F2b3	Central intron
C. piger	cp-7B4	Putative	Putative	ct-7B4	F2b3	Intronless; protein product assumed
	cp-7B5	Putative	Putative	ct-7B5	F2b3	Intronless; protein product assumed
	cp-7B6	Putative	Putative	ct-7B6	F2b3	Intronless; protein product assumed
	cp-7B7	Putative	Putative	ct-7B7	F2b3	Intronless; protein product assumed
	cp-7B8	Putative	Putative	*	F2b3	Intronless; protein product assumed; * chimera of ct-like 7B8 and 7B10 genes;
	cp-Y	Putative	Putative	ct-Y	F2b3	Intronless; protein product assumed
	cp-W	Putative	Putative	ct-W	F2b3	Intronless; protein product assumed
	cp-V	Putative	Putative	ct-V	F2b3	Intronless; protein product assumed
	cp-Z	Putative	Putative	ct-Z	F2b3	Intronless; protein product assumed
C. tentans	ctn-ORFE	Putative	Putative	ct-7A.1	—	Intronless; protein product assumed
	ctn-ORFD	Putative	Putative	ct-9.1	—	Intronless; protein product assumed
	ctn-ORFC	Putative	Putative	ct-2	—	Intronless; protein product assumed
	ctn-ORFB	None	Inactive	*ct-11/12	—	Intronless; may be transcribed; * ancestor to ct-11 and ct-12
	ctn-ORFA	Putative	Putative	ct-13RT	—	Intronless; protein product assumed
C. Pallidivittatus	cpa-3-1	Putative	Putative	ct-3-1	—	Intronless; protein product assumed
	cpa-3-2	Putative	Putative	ct-3-2	—	Intronless; protein product assumed
	cpa-4-1	Putative	Putative	ct-4-1	—	Intronless; protein product assumed
	cpa-4-2	Putative	Putative	ct-4-2	—	Intronless; protein product assumed
	cpa-E	Putative	Putative	ct-E	—	Intronless; protein product assumed
	cpa-F	Putative	Putative	?	—	Intronless; protein product assumed
K. cornishi	kc-7BA	Putative	Putative	?	—	Intronless; protein product assumed
	kc-7BB	Putative	Putative	?	—	Intronless; protein product assumed
	kc-7BC	Putative	Putative	?	—	Intronless; protein product assumed
	kc-7BD	Putative	Putative	?	—	Intronless; protein product assumed
	kc-7BE	Putative	Putative	?	—	Intronless; protein product assumed
	kc-7BF	Putative	Putative	?	—	Intronless; protein product assumed
	kc-7BG	Putative	Putative	?	—	Intronless; protein product assumed

[a] Active genes have been shown to be transcribed and/or translated, or to correspond to an identified globin polypeptide. [b] Positions mapped to polytene salivary gland chromosomes of *C. thummi*.

*Asterisked items are explained in "Comments" column.

very recently, since the gene retains all other attributes of a viable, transcribable gene. The phylogenetic analyses of the clusters supports the recent inactivation of the ctt-13RT and ctn-ORFB genes, and the likelihood that accumulated amino acid differences between many (if not all) homologous and paralogous genes are the result of neutral evolution. If the greater diversity of the *tentans* and *thummi* genes cannot be completely explained as the positive adaptation of structurally diverse hemoglobins, then their maintenance, like that of the large number of 7B genes in *thummi* and *piger*, must be the result of positive gene copy number selection favoring high levels of hemoglobin synthesis.

In sum, the evolution of *Chironomus* globin genes spans more than 250 million years, in which time individual sequences within and across species have accumulated many amino acid substitutions. Some especially diverged hemoglobins may have evolved to serve unique functions, for example in environments that undergo cyclic changes in oxygenation. The specific expression of some globins during development might reflect habitat (and therefore, oxygenation) differences as larvae first descend from the waters' surface to the benthos, and later ascend to the surface at the time of eclosion. On the other hand, differences between many of the globin genes may be neutral, selection favoring the proliferation of a large family of genes encoding proteins of physiologically similar function. Kimura (11) suggested that duplicated genes accumulating neutral change are preadapted, later becoming substrates for positive adaptation after speciation. A consequence of high gene copy number selection is that some of the many functionally redundant globin genes can serve as raw material for Darwinian selection, which could explain the evolution and spread of more than 5000 species of Chironomus to diverse habitats.

Note added in proof

Different globin genes from at least 12 more chironomid species have been characterized since the assembly of Table 1 [Hankelu et al. (1997) *Gene* **205**, 151–160]. The divergent location of introns in these genes supports the independent, recent acquisition of the introns.

BIBLIOGRAPHY

1. P. M. Trewitt, R. A. Luhm, F. Samad, S. Ramakrishnan, W.-Y. Kao, and G. Bergtrom (1995) *J. Mol. Evol.* **41**, 313–328.
2. M. Goodman, J. Pedwaydon, J. Czelusniak, T. Suzuki, T. Gotoh, L. Moens, F. Shishikura, D. Walz, and S. Vinogradov (1988) *J. Mol. Evol.* **27**, 236–249.
3. E. Zebe (1991) *Comp. Biochem. Physiol.* **99A**, 525–529.
4. D. A. Saffarini, P. M. Trewitt, R. A. Luhm, and G. Bergtrom (1991) *Gene* **101**, 215–222.
5. M. Go (1981) *Nature* **291**, 90–92.
6. W.-Y. Kao, P. M. Trewitt, and G. Bergtrom (1994) *J. Mol. Evol.* **38**, 241–249.
7. T. Hankeln, C. Luther, P. Rozynek, and E. R. Schmidt (1991) in *Structure and Function of Invertebrate Oxygen Carriers*, S. Vinogradov and O. H. Kapp, eds., Springer-Verlag, New York, pp. 287–296.
8. W.-Y. Kao and G. Bergtrom (1995) *Gene* **153**, 209–213.
9. P. Rozynek, M. Broecker, T. Hankeln, and E. R. Schmidt (1991) in *Structure and Function of Invertebrate Oxygen Carriers*, S. Vinogradov and O. H. Kapp, eds., Springer-Verlag, New York, pp. 287–296.
10. M. C. Gruhl, W.-Y. Kao, and G. Bergtrom (1997) *J. Mol. Evol.* **45**, 499–508.
11. M. Kimura (1991) *Jpn. J. Genet.* **66**, 367–386.

Suggestions for further reading

W. Beermann (1963) Cytological aspects of information transfer in cellular differentiation. *Am. Zool.* **3**, 23–32.

S. T. Case and L. Wieslander (1992) Secretory proteins of Chironomus salivary glands: structural motifsand assembly characteristics of a novel biopolymer. *Results Probl. Cell Differ.* **19**, 187–226.

R. Lewin (1984) Surprise finding with insect globin genes, *Science* **226**, 328.

P. A. Osmulski and W. Leyko (1991) The structure and function of Vhironomus hemoglobins, in *Structure and Function of Invertebrate Oxygen Carriers*, O. H. Kapp and S. N. Vinogradov, eds., Springer-Verlag, New York, pp. 305–312.

J. H. Rogers (1989) How were introns inserted into nuclear genes? *Trends Genet.* **5**, 340–343.

A. Stoltzfus and W. F. Doolittle (1993) Slippery introns and globin gene evolution, *Curr. Biol.* **3**, 215–217.

A. Stoltzfus, D. F. Spencer, M. Zucker, J. M. Logsdon Jr., and W. F. Doolittle (1995) Introns and the origin of protein-coding genes, *Science* **268**, 1366–1367.

CHLORAMPHENICOL

DIETER BEYER

Chloramphenicol is a bacteriostatic agent that inhibits the growth of many species of **Gram-positive** and **Gram-negative bacteria**; it was the first broad-spectrum antibiotic to be used clinically. Originally obtained from cultures of the soil bacterium *Streptomyces venezuelae*, chloramphenicol inhibits bacterial **protein biosynthesis** by interfering with the intrinsic catalytic activity of the peptidyl-transferase of the **ribosome** during the elongation phase of **translation**.

Eukaryotic cells are generally not affected by chloramphenicol. Nevertheless, the clinical application of this drug for the treatment of systemic infections has been severely curtailed, primarily because of the haematotoxicity associated with its use. Chloramphenicol continues to be used topically, particularly in the treatment of eye infections. The emergence of chloramphenicol-resistant bacteria, apparently in response to the selective pressure exerted by the drug (see **Antibiotic resistance**), has also restricted the use of chloramphenicol in the treatment of bacterial infections.

Chloramphenicol resistance is frequently encountered in many genera of bacteria. Several mechanisms of resistance have been described: (i) modification of the target, the bacterial ribosome; (ii) alteration in the permeability of the bacterial cell, leading to low intracellular antibiotic concentrations; and (iii) enzymatic modification of the antibiotic. Resistance most commonly occurs as the result of inactivation of chloramphenicol by **chloramphenicol acetyltransferase** (CAT). This enzyme is of value to molecular biologists, because it can be assayed with specificity and sensitivity.

Chloramphenicol is used today in molecular biology primarily as a component of selective media for genetic studies, to distinguish chloramphenicol-resistant from chloramphenicol-susceptible bacteria. Chloramphenicol is also an essential component of CAT assays, which are used to study gene regulation, generally in eukaryotic cells. Finally, chloramphenicol provides a tool with which to investigate the peptidyl-transferase center of the bacterial ribosome.

CHEMISTRY

Chloramphenicol [D(−)-*threo*-2-dichloroacetamido-1-*p*-nitrophenyl-1,3-propanediol; Fig. 1] has two asymmetric carbon atoms. Of the four stereoisomers, only one, the D-threo isomer, has antibacterial activity. This compound can be produced on an industrial scale by chemical synthesis or by fermentation. Chloramphenicol is poorly soluble in water, but dissolves readily in organic solvents (eg, methanol).

Derivatives having a wide range of para substituents are active; therefore, this part of the molecule is not involved in specific drug–target interactions. The propanediol moiety, the site of the D-threo configuration, is essential for activity. Acetylation of one or both hydroxyl groups inactivates the drug. Removal of the dichloroacetamido side chain virtually eliminates biological activity, although substitution by a number of groups does not render the drug inactive. Some substituents reduce activity against intact bacteria, but enhance the inhibition of cell-free protein synthesis (1). Apparently, these derivatives are not taken up by the bacterial cell as efficiently as the parent compound.

ANTIBACTERIAL SPECTRUM

Chloramphenicol is a broad-spectrum antibiotic, active against archaebacteria and both gram-positive and gram-negative bacteria. Organisms that fall within its spectrum include *Bacillus anthracis*, *Brucella abortus*, *Corynebacterium diphtheriae*, *Escherichia coli*, *Hemophilus influenzae*, *Klebsiella pneumoniae*, *Neisseria meningitidis*, *Salmonella* spp., *Serratia marcescens*, *Staphylococcus aureus*, *Streptococcus* spp., and so on (2).

Chloramphenicol inhibits translation in bacterial cells, intact **chloroplasts**, and intact **mitochondria**. Some of the toxic side effects associated with chloramphenicol may arise from the effect of the drug on mitochondrial metabolism. Translation by mammalian cells, yeast, and protozoa is not inhibited by chloramphenicol.

MODE OF ACTION

Chloramphenicol inhibits bacterial peptidyl-transferase, the enzymatic reaction by which one more amino acid is added to the growing **polypeptide chain**. During the elongation phase of protein biosynthesis in bacteria, chloramphenicol "freezes" **polysomes** on the **messenger RNA**, thereby fixing the peptidyl-tRNA to the ribosomes (3). Chloramphenicol competes for binding to the 50S subunit of the bacterial ribosome with molecules that specifically bind to the A site of the peptidyl-transferase center (aminoacyl-tRNA and its analogues, eg, CACCA-Leu) and is, therefore, thought to bind next to, or within, the A site of the peptidyl-transferase center (4). In contrast, no competition is observed with substrates that bind at the P site (peptidyl-tRNA analogues, eg, CACCA-LeuAc or AcPhe-tRNA). Furthermore, neither the dinucleotide ApC nor amino acids block the binding of chloramphenicol, whereas both **puromycin** and the dinucleotide CpA do (5). Thus, the drug presumably binds in the same region of the A site as the universally conserved CCA part of aminoacyl-tRNA.

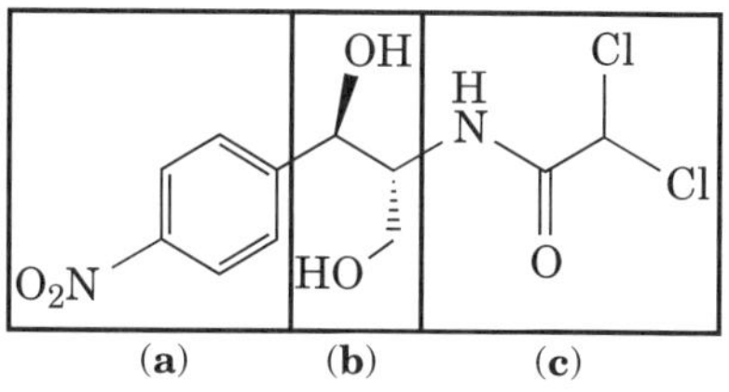

The peptidyl-transferase assay in which **peptide bonds** are formed between a peptidyl-tRNA and puromycin on the 50S subunit (the "puromycin reaction") has been used to unravel the mechanism of action of chloramphenicol (6) (Fig. 2). Chloramphenicol blocks this reaction, provided that the donor peptidyl-tRNA, bound at the P site, is short (eg, AcPhe-tRNA). In this case, chloramphenicol appears to disturb the correct positioning of the nascent peptide during its synthesis, thereby causing the premature release of the peptidyl-tRNA from the ribosome. Chloramphenicol does not, however, affect the puromycin reaction when a larger peptidyl-tRNA—for example, Ac(Phe)$_2$-tRNA—is used as donor (7). It is likely, therefore, that chloramphenicol interferes with the binding of the terminal CCA of the aminoacyl-tRNA at the A site and also interferes with the entry of the nascent polypeptide into the "tunnel" that normally guides it away from the peptidyl-transferase center.

Chloramphenicol inhibits protein synthesis in bacterial extracts with varying potencies depending on the template employed. In particular, synthesis promoted by poly(U) is markedly more resistant to chloramphenicol than that promoted by poly(C) or poly (A) (8). Synthesis occurring with a natural mRNA (eg, MS2 phage RNA) is usually very sensitive to chloramphenicol. The precise reason for the template-dependence is not known. Chloramphenicol may inhibit poly(U)-dependent poly(Phe) synthesis less than it does the synthesis of other polypeptides because the structures of D-threo chloramphenicol and L-phenylalanine are similar (Fig. 3). Competition between these molecules during poly(Phe) synthesis may involve steric hindrance between their phenyl groups and, consequently, chloramphenicol may be chased from the ribosome. Nevertheless, this does not explain why the same template dependence is also observed with several other peptidyl-transferase inhibitors that differ chemically from chloramphenicol and, in particular, that do not have a phenyl group (e.g, the group B streptogramins and the macrolides). Further experiments are needed to elucidate this phenomenon in greater detail.

OH Cl H N Cl O_2N HO O

(a) (b) (c)

Figure 1. Chemical structure of chloramphenicol [D(−)-*threo*-2-dichloroacetamido-1-*p*-nitrophenyl-1,3-propanediol].

RIBOSOMAL BINDING SITE

Chloramphenicol has a single high-affinity binding site located on the 50S ribosomal subunit ($K_d = 2 \times 10^{-6}$ M), as well as one or more low-affinity sites, at least one of which is located on the 30S subunit ($K_d = 2 \times 10^{-4}$ M) (9). A number of 50S

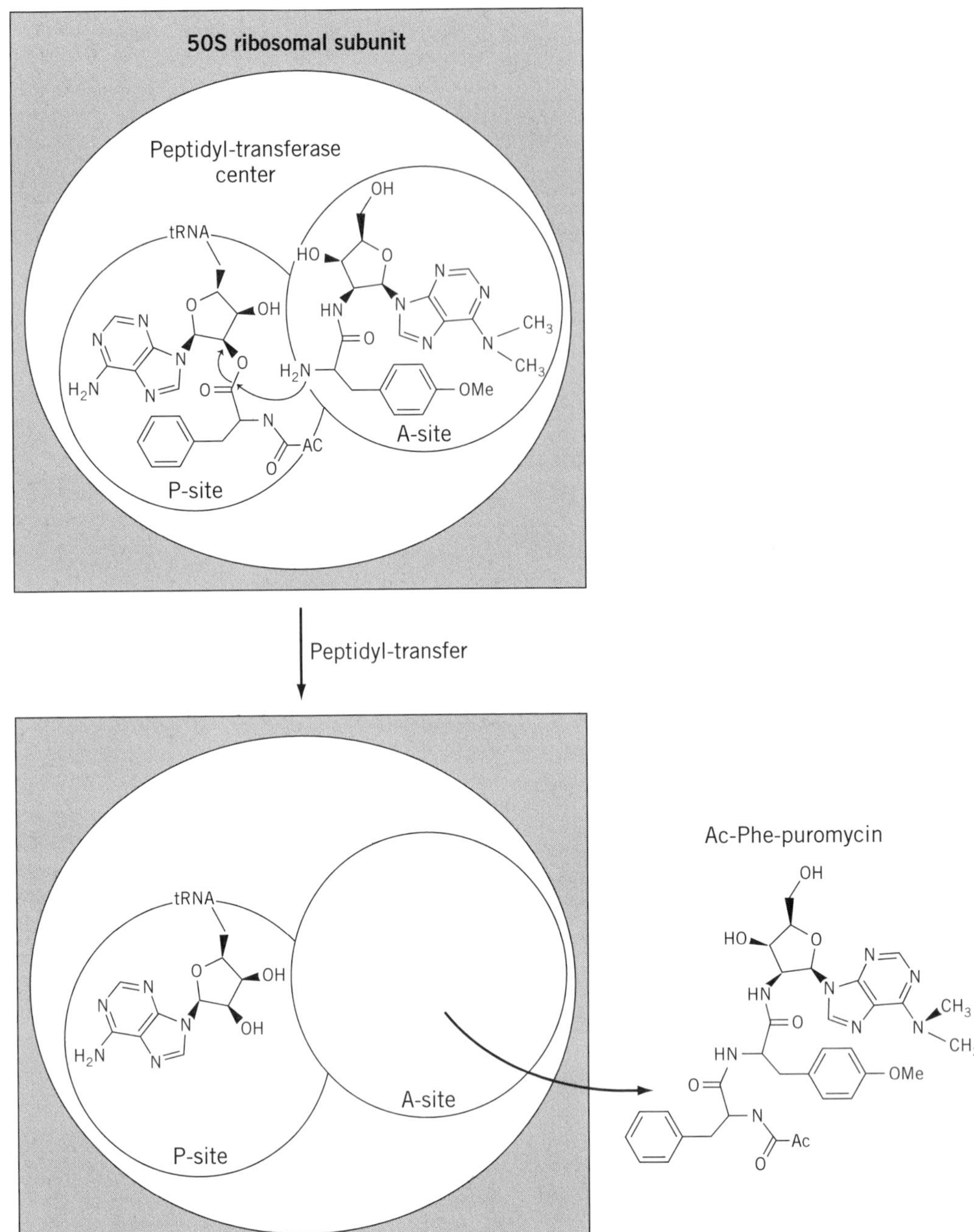

Figure 2. The puromycin reaction. A peptidyl-tRNA analogue [eg, Ac[^{14}C]Phe-tRNA) bound to the ribosomal P site of 50S ribosomal subunit reacts with puromycin (an aminoacyl-tRNA analogue) bound to the A site of the peptidyl-transferase center. The resulting peptidyl-puromycin (eg, Ac[^{14}C]Phe-puromycin) is falling of the 50S subunit and can be separated from the substrate (Ac[^{14}C]Phe-tRNA) by extraction.

proteins were labeled in **affinity-labeling** experiments, including L1, L2, L11, L16, L19, and L27 (10). Most of these proteins are located in or next to the peptidyl-transferase center of the ribosome, and several can be affinity-labeled with other antibiotics that interact with this center (**erythromycin**, puromycin, lincosamides, streptogramins).

That the peptidyl-transferase domain is the primary binding site of chloramphenicol was confirmed by RNA **footprinting** experiments, in which bases of 23S rRNA were compared in antibiotic-free or chloramphenicol-bound ribosomes for their reactivity to chemical probes. Chloramphenicol protected bases A2059, A2062, A2451, and G2505 and enhanced the reactivity of A2058 (11). Protection of A2070 and U2506 has also been observed (12). All these nucleotides are located in the central loop of domain V of 23S rRNA (Fig. 4). Additionally, several mutations of RNA that confer resistance to chloramphenicol have

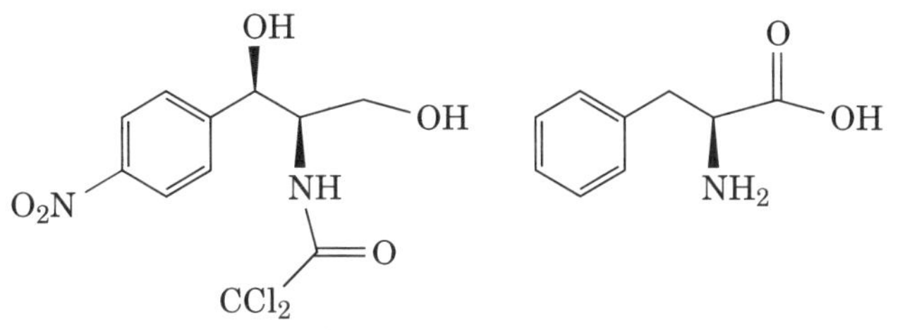

(a) (b)

Figure 3. Structural similarity of D(−)-*threo* chloramphenicol and L-phenylalanine. Competition between these molecules during poly(Phe) synthesis may involve steric hindrance between their phenyl groups.

been identified in the same domain (Fig. 4; see references in Ref. 12).

COLD-SHOCK INDUCTION

Shifting growing *E. coli* from 37°C to 20°C inhibits cellular growth and shuts down protein biosynthesis. To compensate for the stress produced by the temperature shift, the levels of (p)ppGpp drop and the synthesis of a subset of proteins is induced; this induction is known as **cold-shock**. Certain inhibitors of protein synthesis, including chloramphenicol, specifically induce the synthesis of the cold-shock proteins in the absence of a temperature shift (13). Ribosomal inhibitors have, in fact, been classified according to their ability to elicit a cold-shock- or a heat-shock-like response. Inhibitors in whose presence the A site of the peptidyl-transferase center is occupied, either by the antibiotic or by an aminoacyl-tRNA (chloramphenicol, erythromycin, fusidic acid, spiramycin, **tetracycline**), produce a cold-shock-like response. In contrast, inhibitors in whose presence the A site of the peptidyl-transferase remains empty (**kanamycin**, **streptomycin**) produce a heat-shock-like response.

CHLORAMPHENICOL RESISTANCE

Modification of the Bacterial Ribosome

Several mutations in 23S rRNA confer resistance to chloramphenicol. The residues at which these mutations occur (residues G2057, G2061, G2447, A2451, C2452, A2503, U2504; *E. coli* numbering) are located in the peptidyl-transferase loop of 23S rRNA (Fig. 4), which is in accordance with their proposed binding site.

Alteration in Permeability to Chloramphenicol

Protein synthesis in *S. venezuelae* cells that have begun to produce chloramphenicol becomes insensitive to the antibiotic. There is no modification of the ribosome in these cells; instead, the antibiotic is excluded from the cytoplasm during biosynthesis. The amino acid sequence deduced from the gene conferring resistance (*cml*) is markedly similar to those of chloramphenicol-resistance genes from *Streptomyces lividans*, *Rhodococcus fascias*, and other bacteria. These polypeptides contain 12 **hydrophobic** regions characteristic of membrane-associated **transport** proteins and are related to a known family of efflux proteins (14).

Enzymatic Modification of Chloramphenicol

The most common mechanism of chloramphenicol resistance is inactivation by the enzyme chloramphenicol 3′-*O*-acetyltransferase (CAT), encoded by the *cat* gene. The biological and immunological properties of three types of CAT (I, II, III) have been described. The structure of CAT III has been determined by **X-ray crystallography**, as well as its tertiary complex with chloramphenicol and CoA (15). The enzyme

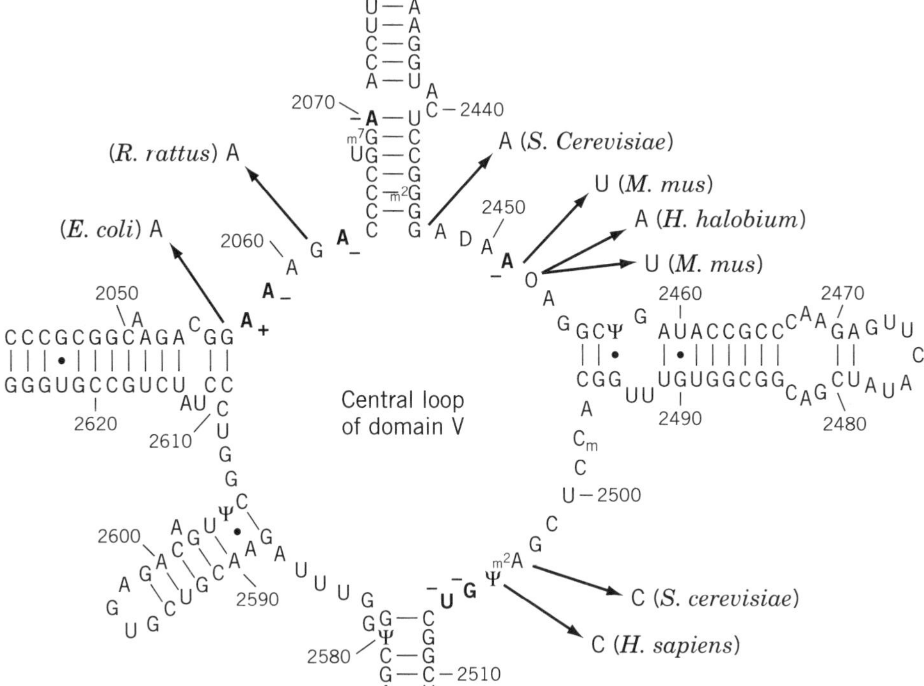

Figure 4. Interaction of chloramphenicol with the central loop of domain V of 23S rRNA. The secondary structure of the central loop of domain V in *E. coli* 23S rRNA is derived from phylogenetic sequence comparisons. Differences in the accessibility of bases to chemical reagents in the presence and absence of chloramphenicol (footprints) are shown in red (−, protection of base; +, enhanced activity). Mutations leading to chloramphenicol resistance are indicated by arrows, and the corresponding organisms are indicated in brackets. All of the eukaryotic mutations were characterized for mitochondrial rRNA. (Figure derived from Refs. 11 and 12.)

is a trimer; its **active sites** lie at each of the interfaces between adjacent subunits. CAT catalyzes the transfer of an acetyl moiety from acetyl-CoA to the 3'-hydroxyl of chloramphenicol. The initial product of the reaction, 3-acetyl chloramphenicol, can undergo nonenzymatic intramolecular rearrangement to 1-acetyl chloramphenicol. The latter compound is a substrate for a second round of enzymatic acetylation at the 3-hydroxyl, which yields 1,3-diacetyl chloramphenicol as the final product (Fig. 5). Neither the monoacetylated nor the diacetylated derivative binds to the bacterial ribosome; both, therefore, are devoid of antibiotic activity.

Recently, a second chloramphenicol 3′-*O*-modifying gene was discovered (16). This enzyme, chloramphenicol 3'-*O*-phosphotransferase (CPT) phosphorylates chloramphenicol, yielding 3'-*O*-phosphochloramphenicol, which binds only very weakly to ribosomes and thus does not inhibit translation.

Chloramphenicol is also inactivated by chloramphenicol hydrolase, which hydrolyzes the amide bond (17) (Fig. 5). This enzyme, encoded by the *cml* gene, was first isolated from the mycelium of a chloramphenicol-producing strain of *Streptomyces*. A similar enzyme has been obtained from several other bacteria, including *E. coli*.

Induction of Chloramphenicol Resistance by Translational Attenuation

Chloramphenicol resistance by enzymatic modification in both Gram-positive and Gram-negative bacteria can be either constitutive or inducible. What is unusual about inducible chloramphenicol resistance is that it occurs by translational attenuation, rather than by transcriptional **attenuation**. The ribosome-binding site for the resistance determinant is sequestered in a secondary "hairpin" structure situated within the encoding mRNA; a short, translated open reading frame, termed the leader, lies upstream of this structure. In the absence of the antibiotic, the secondary structure is maintained, the ribosome does not translate the mRNA that constitutes

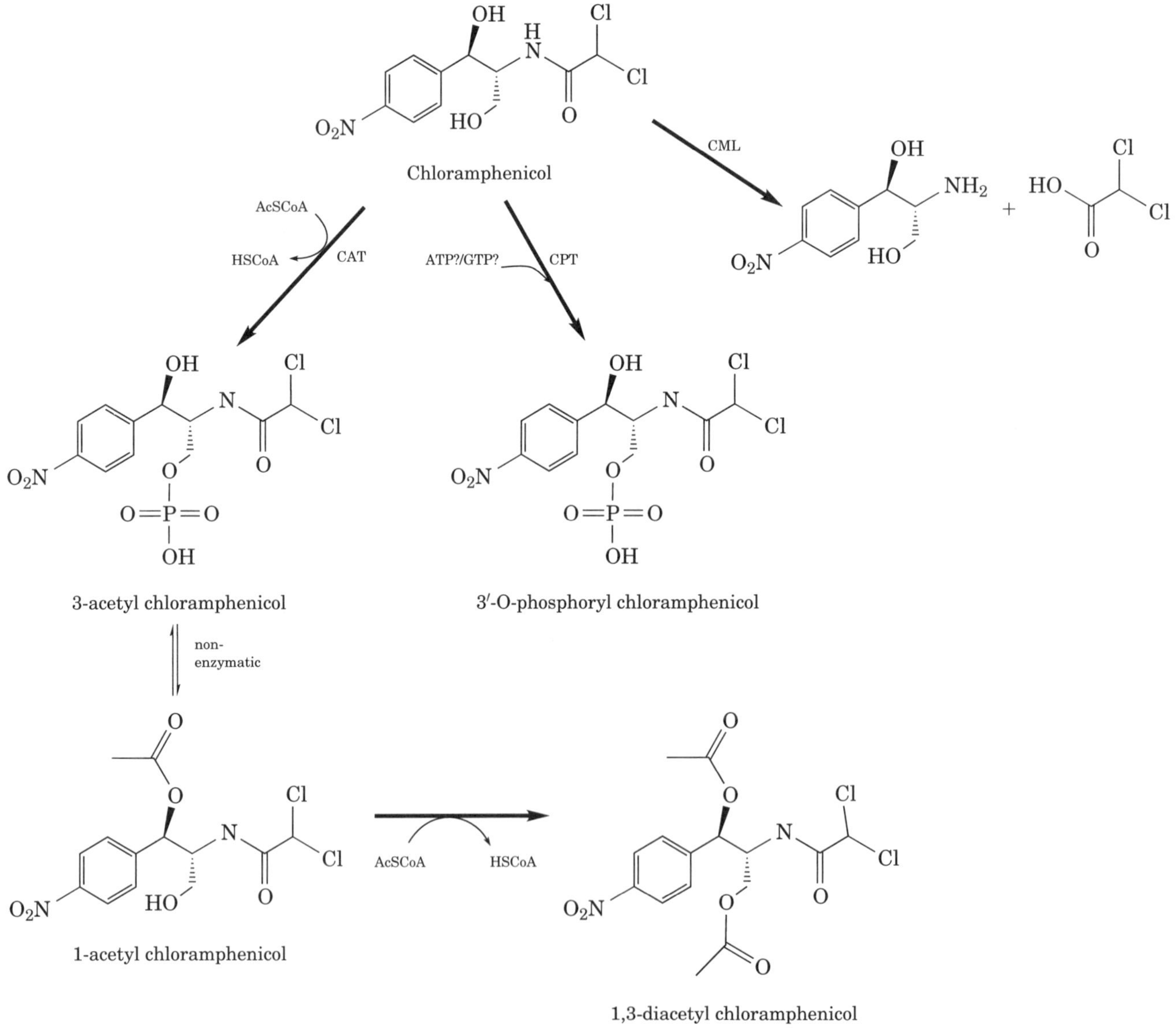

Chloramphenicol

CML

AcSCoA

HSCoA

CAT

ATP?/GTP?

CPT

3-acetyl chloramphenicol

3′-O-phosphoryl chloramphenicol

non-enzymatic

AcSCoA

HSCoA

1-acetyl chloramphenicol

1,3-diacetyl chloramphenicol

Figure 5. Enzymatic inactivation of chloramphenicol. CAT, chloramphenicol 3′-*O*-acetyltransferase; CPT, chloramphenicol 3′-*O*-phosphotransferase; CML, chloramphenicol hydrolase.

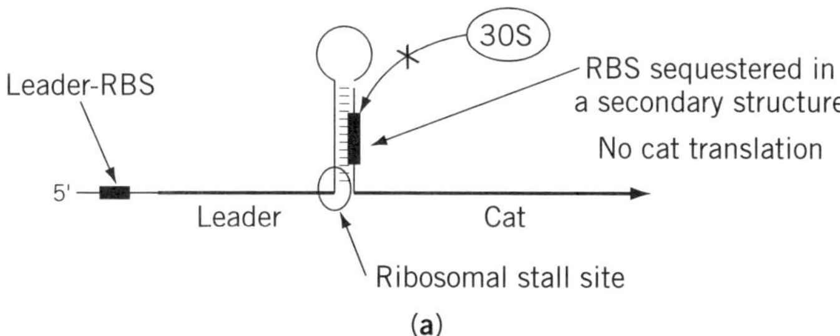

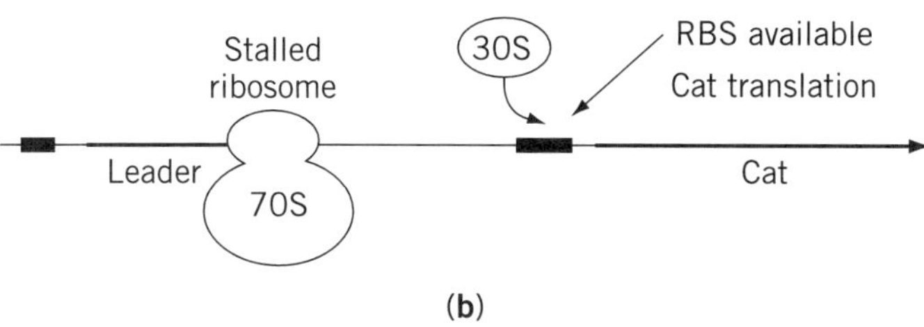

Figure 6. Chloramphenicol resistance by translational attenuation. RBS, ribosomal binding site; 30S, small ribosomal subunit; 70S, entire bacterial ribosome.

the hairpin, and the resistance gene is not translated. In the presence of the antibiotic, the ribosome stalls at the leader sequence and the secondary structure relaxes, allowing the *cat* (or *cml*) determinant to be translated (18) (Fig. 6).

THE USE OF CAT IN MOLECULAR BIOLOGY

CAT is a gene product that can be assayed with specificity and sensitivity, and consequently it is of considerable value to molecular biologists. When the *cat* gene is fused to heterologous regulatory signals, gene expression can be determined by measuring the CAT activity. Eukaryotes lack endogenous CAT activity, and the amount of CAT activity correlates well with the amount of mRNA transcribed, so the CAT assay has proved to be a powerful tool for investigating and quantifying gene expression in eukaryotic cells. Conventional CAT assays involve the incubation of cell extracts with radioactively labeled chloramphenicol, chromatographic separation of the reaction products, and **autoradiography**. The recent use of **fluorescent**, rather than radioactive, substrates has dramatically speeded up and simplified the CAT assay (19).

BIBLIOGRAPHY

1. O. Pongs (1979) In *Antibiotics* Vol. 1 (F. H. Hahn ed.), Springer-Verlag, Heidelberg, Germany, pp. 26–56.
2. A. A. Yunis (1988) *Annu. Rev. Pharmacol. Toxicol.* **28**, 83–100.
3. E. Cundliffe and K. McQuillen (1967) *J. Mol. Biol.* **30**, 137–143.
4. D. Nierhaus and K. H. Nierhaus (1973) *Proc. Natl. Acad. Sci. USA* **70**, 2224–2228.
5. M. L. Celma, R. E. Monro, and D. Vazquez (1971) *FEBS Lett.* **13**, 247–251.
6. R. R. Traut and R. E. Monro (1964) *J. Mol. Biol.* **10**, 63–72.
7. H.-J. Rheinberger and K. H. Nierhaus (1990) *Eur. J. Biochem.* **193**, 643–650.
8. Z. Kucan and F. Lipmann (1964) *J. Biol. Chem.* **239**, 516–520.
9. J. L. Lessard and S. Pestka (1972) *J. Biol. Chem.* **247**, 6909.
10. B. S. Cooperman, C. J. Weitzmann, and C. L. Fernandez (1990) In *The Ribosome: Structure, Function, & Evolution* (W. E. Hill et al., eds.), American Society for Microbiology, Washington, D.C., pp. 491–501.
11. D. Moazed and H. F. Noller (1987) *Biochimie* **69**, 879–884.
12. C. Rodriguez et al. (1995) *J. Mol. Biol.* **247**, 224–235.
13. P. A. Lagosky and F. N. Chang (1981) *J. Biol. Chem.* **256**, 11651–11656.
14. H. Nikaido and M. H. Sailer Jr. (1992) *Science* **258**, 936–942.
15. A. G. W. Leslie (1990) *J. Mol. Biol.* **213**, 167.
16. R. H. Mosher et al. (1995) *J. Biol. Chem.* **270**, 27000–27006.
17. V. S. Malik and L. C. Viking (1971) *Can. J. Microbiol.* **17**, 1287–1290.
18. P. S. Lovett (1996) *Gene* **179**, 157–162.
19. C. K. Lefevre et al (1995) *Biotechniques* **19**, 488–493.

Suggestions for Further Reading

E. F. Gale, E. Cundliffe, P. E. Reynolds, M. H. Richmond, and M. J. Waring (1981) Chloramphenicol. In *The Molecular Basis of Antibiotic Action*, 2nd ed. Wiley, New York, pp. 460–468.

F. E. Hahn (1983) Chloramphenicol. In *Antibiotics*, Vol. VI (F. H. Hahn, ed.), Springer-Verlag, Berlin, pp. 34–45.

L. C. Vining and C. Stuttard (1995) Chloramphenicol. *Biotechnology* **28**, 505–530.

CHLORAMPHENICOL ACETYLTRANSFERASE

W. SHAW

The **enzymic** mechanisms by which bacteria modify biologically active molecules are limited by two constraints: (i) the structure–activity correlations for each class of chemical agent and (ii) the metabolic repertoire available to the bacteria. In the case of **chloramphenicol** (Cml), which has a number of vulnerable functional groups, there are several possibilities. Each functional group of Cml (Fig. 1) contributes to its effectiveness as an inhibitor of ribosomal peptidyltransferase activity (1), and there are examples (reviewed in Ref. 2) of enzyme-mediated resistance to Cml due to dehalogenation, nitro group reduction, hydrolysis of the amide bond, and modification of the hydroxyl groups by phosphorylation (3) or acetylation (2,4,5). Nonetheless, after more than four decades of medical and veterinary use, the preponderant enzymic modification mechanism for Cml resistance in bacteria of clinical importance is that of *O*-acetylation of the 3-hydroxyl group, catalyzed by chloramphenicol acetyltransferase (CAT) (Fig. 1).

Genes for CAT are widespread among **gram-positive** and **gram- negative bacteria**; the *cat* gene in each case is either chromosomal or carried by a mobile genetic element, such as a plasmid or **transposon**. More important biochemically are the properties of representative variants within the CAT "family" and the pattern of conservation of the **primary structure**, deduced from nucleotide sequences of its genes, which in turn is related to its three-dimensional **protein structure**, substrate specificity, and catalysis.

All CAT **polypeptide chains** are in the range 24 to 26 kDa and normally exist in solution as compact and very stable *homo*-trimers. However, some CAT variants associate *in vivo* and *in vitro* to give hybrids, functional $\alpha_2\beta$ and $\alpha\beta_2$ *hetero*-trimers with physical and catalytic characteristics reflecting

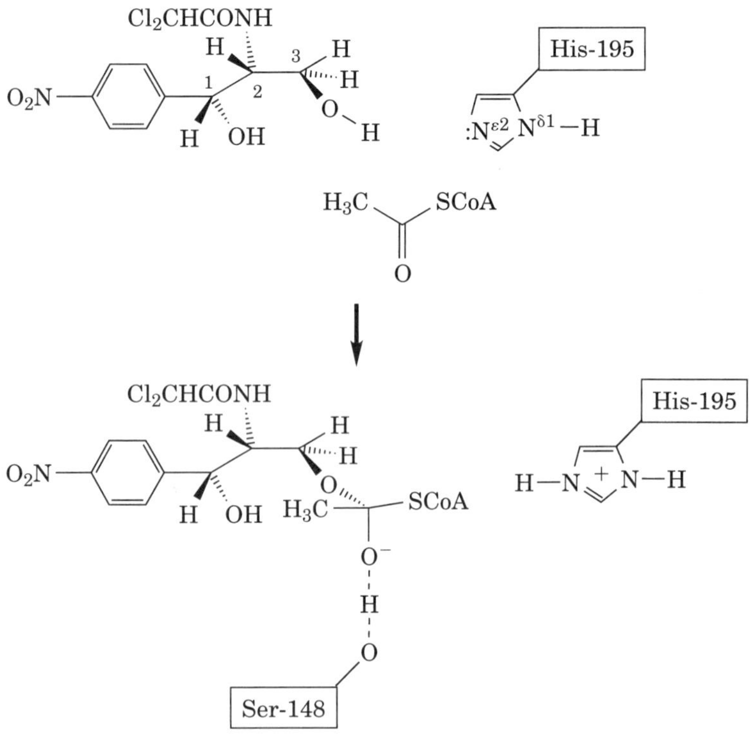

Figure 1. The mechanism of acetylation of the 3-hydroxyl group of chloramphenicol by acetyl-CoA as catalyzed by CAT. The two substrates and the His195 residue of CAT are shown (**top**). Within the transition state or tetrahedral intermediate (**bottom**), His195 has abstracted a proton from the 3-hydroxyl group, to generate an "oxyanion" intermediate that has attacked the carbonyl of acetyl-CoA. The intermediate and transition state are stabilized by hydrogen bonding with the side-chain hydroxyl group of Ser148 of CAT.

the properties of the parental trimers (6). Because natural isolates of Cml-resistant bacteria may occasionally harbor more than one *cat* gene, it is possible in such instances that the intracellular CAT pool will include both parental and hybrid trimers.

The extent of structural variation within the CAT family can be appreciated from a comparison of the deduced primary structures for the products of known *cat* determinants, which yields a lower limit of 28% identity for the most divergent pair of known sequences. Only ~11% of the amino acid residues appear to be identical in all CAT variants, comprising not only residues with side chains that are involved in catalysis and substrate binding, but also those that contribute to critical structural elements necessary for the precise folding and stable packing of the polypeptide chains. A reference point for understanding the variety in primary structures of CAT variants is the type III enzyme (CAT_{III}), for which a wealth of information is available, including the **tertiary structure** at high resolution (7,8) and the structural determinants for the binding of each substrate (2,4,5,7,8). However, the type I enzyme (CAT_I) may be the most widely distributed variant, because it is specified by many "F-like" R plasmids of gram-negative bacteria, by transposon *Tn9* (9), and by promoter-less *cat* "cassettes," constructed in the laboratory for insertion "downstream" of the noncoding sequences of other genes, to study the regulation of their expression (see **Reporter genes**). Type II CAT (CAT_{II}), notable among "enteric" CATs for its particular sensitivity to inhibition by reagents that react with **thiol groups** and by its association with *Haemophilus influenzae* (10), is less commonly encountered. CAT_I has the remarkable properties of a high affinity for triphenylmethane dyes, such as crystal violet (reviewed in Ref. 2), and the ability to bind a steroidal antibiotic (fusidic acid) both tightly and specifically (11). The latter property is sufficient to confer resistance to fusidate in mutant strains of *Escherichia coli* that were selected for their sensitivity to the antibiotic prior to the introduction of the gene for CAT_I. A plausible mechanism is the effective sequestration of fusidate by high levels of CAT_I, thereby impeding access of the antibiotic to its cellular target, ribosomal **elongation factor** G. The structural basis for the curious binding of fusidate, which is competitive with respect to Cml, has been deduced by **protein engineering** and **X-ray crystallography** (11). Of the eight residues in the Cml binding pocket of CAT_{III} that differ from those of CAT_I, only four appear to be responsible for the former's low affinity ($K_i = 279\ \mu M$) for fusidate. Replacement of each of the four with their counterparts in CAT_I is sufficient to confer upon CAT_{III} an affinity for fusidate ($K_i = 5.4\ \mu M$) that approaches that of wild-type CAT_I ($K_i = 1.5\ \mu M$).

The catalytic machinery of CAT_{III} comprises a number of amino acid side chains (as well as **backbone** atoms and ordered **water** molecules) that make precise contacts with one another or with a substrate molecule. The homo-trimers have rotational symmetry at the molecular threefold axis, and consequently have three identical active sites, so only atoms in a single monomer need be addressed. Nonetheless, each of the structurally and functionally equivalent **active sites** lies deep in the interfacial clefts between subunits. Central to catalysis (12) is His195, which arises from one face of each cleft to supply the general base ($N^{\varepsilon 2}$ in Fig. 1) to deprotonate the C3 hydroxyl of Cml, producing an "oxyanion" intermediate that in turn attacks the carbonyl (C2) carbon of acetyl CoA to yield a tetrahedral intermediate (Fig. 1). Essential for the stabilization of the latter, *en route* to the **transition state** for the reaction, and confirmed by **site-directed mutagenesis** (13), is a negatively charged **hydrogen bond** (Fig. 1) between the oxyanion and the hydroxyl of Ser148, another residue conserved in all CATs. A neighboring participant in catalysis is Thr174, also conserved, which is hydrogen-bonded to a water molecule that in turn probably makes two hydrogen bonds with the putative tetrahedral intermediate (Fig. 1), one to the 1-hydroxyl of Cml and the other to the 3-oxygen of the intermediate. Two additional conserved residues (Arg18 and Asp199) facilitate catalysis via a network of hydrogen bonds with His195, anchoring the side chain of the latter in a novel conformation that allows it to fulfill its general base role (7,8,13–15).

Chloramphenicol becomes acetylated on its 1-hydroxyl group also, albeit at a rate much slower than the 3-acetyl derivative is generated (16). The equations below, all of which are reversible, indicate the transformations involved in the two reactions, indicating that in both cases CAT acetylates only the 3-hydroxyl:

$$\text{chloramphenicol} + \text{acetyl-CoA} \rightleftharpoons \text{3-acetyl chloramphenicol} + \text{CoA} \quad (1)$$

$$\text{3-acetyl chloramphenicol} \rightleftharpoons \text{1-acetyl chloramphenicol} \quad (2)$$

$$\text{1-acetyl chloramphenicol} + \text{acetyl-CoA} \rightleftharpoons \text{1,3-diacetyl chloramphenicol} + \text{CoA} \quad (3)$$

Reaction 2 is a nonenzymic acetyl migration, slow and reversible, which yields at equilibrium a mixture of mono-acetyl products. The 1-acetyl Cml so formed is available for a second round of enzymic acetylation at the C3 position, yielding 1,3-diacetyl Cml. Reaction 3 is ~150-fold less efficient than reaction 1, almost certainly due to an unfavorable "fit" of the substrate at the active site because of the bulky 1-acetyl substituent (17). In any case, the sluggish final step (reaction 3) is of little microbiological significance, because both mono-acetyl derivatives of Cml are already devoid of significant antimicrobial activity, making the rate of reaction 1 the prime determinant of the Cml-resistance phenotype.

In summary, the precise geometry and chemical properties of both substrate binding sites (for Cml and acetyl CoA) and of the catalytic center of CAT_{III} each contribute to its extraordinary efficiency, with a **turnover number** of 600 s^{-1} (25°C) with a K_m for Cml of 12 μM (~4μg/mL), reassuringly close to the concentrations at which it inhibits most bacteria of clinical importance. A derived kinetic parameter, the so-called specificity constant (k_{cat}/K_m), which combines a measure of substrate affinity with one for catalytic competence, is actually the second-order rate constant for productive collisions of an enzyme with its substrate(s). The value for CAT_{III} and Cml ($5 \times 10^7\ s^{-1}M^{-1}$) approaches that of well-characterized enzyme reactions that are limited by **diffusion** of the reactants (typically 10^8 to $10^9\ s^{-1}M^{-1}$) and hence at the limit of evolutionary development. By such criteria, CAT_{III} has evolved to a state approaching "perfection" in biological catalysis (18), wherein a fine balance has been struck between rate acceleration (k_{cat}) and specificity (and affinity) for substrate (K_m). Although it is not clear how the specificity and catalytic efficiency of CAT_{III} (and related variants) have evolved, the acetyltransferase (E2p) of the pyruvate dehydrogenase complex, which generates acetyl-CoA for central metabolism, has a three-dimensional structure that is virtually identical to that of CAT, as well as the same mechanism of catalysis, but the two proteins have very few identities in primary structure—only those involved with the active site (19).

It is of interest that there is a large family of "xenobiotic" *O*-acetyltransferases (XATs) with a range of specificities for natural products; several of these enzymes have a low affinity for Cml and hence were first detected as effectors of low-level resistance to the antibiotic (20). All appear to be trimeric but have no sequence homologies with members of the *bona fide* CAT family described above. One such XAT (with CAT activity) has been studied by X-ray crystallography (21) and shown to have a tertiary structure quite different from that of CAT, but the catalytic mechanism may well involve general base catalysis involving a conserved histidine residue.

BIBLIOGRAPHY

1. E. F. Gale et al. (1981) *The Molecular Basis of Antibiotic Action*, 2nd ed., Wiley, London, pp. 462–468.
2. W. V. Shaw (1983) *CRC Crit. Rev. Biochem.* **14**, 1–46.
3. R. H. Mosher et al. (1995) *J. Biol. Chem.* **27**, 27000–27006.
4. W. V. Shaw (1992) *Sci. Progress (Oxford)* **76**, 565–580.
5. W. V. Shaw and A. G. W. Leslie (1991) *Annu. Rev. Biophys. Biophys. Chem.* **20**, 363–386.
6. P. J. Day, I. A. Murray, and W. V. Shaw (1995) *Biochemistry* **34**, 6416–6422.
7. A. G. W. Leslie (1990) *J. Mol. Biol.* **213**, 167–186.
8. A. G. W. Leslie, P. C. E. Moody, and W. V. Shaw (1988) *Proc. Natl. Acad. Sci. USA* **85**, 4133–4137.
9. N. E. Alton and D. Vapnek (1979) *Nature* **282**, 864–869.
10. I. A. Murray, J. V. Martinez-Suarez, T. J. Close, and W. V. Shaw (1990) *Biochem. J.* **272**, 505–510.
11. I. A. Murray et al. (1995) *J. Mol. Biol.* **254**, 993–1005.
12. A. Lewendon et al. (1994) *Biochemistry* **33**, 1944–1950.
13. A. Lewendon, I. A. Murray, W. V. Shaw, M. R. Gibbs, and A. G. W. Leslie (1990) *Biochemistry* **9**, 2075–2080.
14. A. Lewendon et al. (1988) *Biochemistry* **27**, 7385–7390.
15. A. Lewendon and W. V. Shaw (1993) *J. Biol. Chem.* **268**, 20997–21001.
16. J. Ellis, C. R. Bagshaw, and W. V. Shaw (1995) *Biochemistry* **34**, 16852–16859.
17. I. A. Murray et al. (1991) *Biochemistry* **30**, 3763–3770.
18. W. J. Albery and J. R. Knowles (1976) *Biochemistry* **15**, 5631–5640.
19. A. Mattevi et al. (1993) *Biochemistry* **32**, 3887–3901.
20. I. A. Murray and W. V. Shaw (1997) *Antimicrob. Agents Chemother.* **41**, 1–6.
21. T. W. Beaman, M. Sugantino, and S. L. Roderick (1998) *Biochemistry* **37**, 6689–6696.

CHLOROPLAST

JEAN-DAVID ROCHAIX

A distinctive feature of chloroplasts of **plants** and **algae** is their extensive, internal, green, chlorophyll-containing **membrane** system, called thylakoid membranes, where the primary reactions of **photosynthesis** occur. This system of **photosynthetic reaction centers** converts light energy into chemical energy, which is used to drive cellular metabolism. Besides their important role in photosynthesis, chloroplasts are also involved in several biochemical pathways, such as the biosynthesis of **amino acids**, fatty acids, tetrapyrroles including chlorophyll and heme, carotenoids, isoprenoids and pyrimidines. Chloroplasts are also involved in carbon

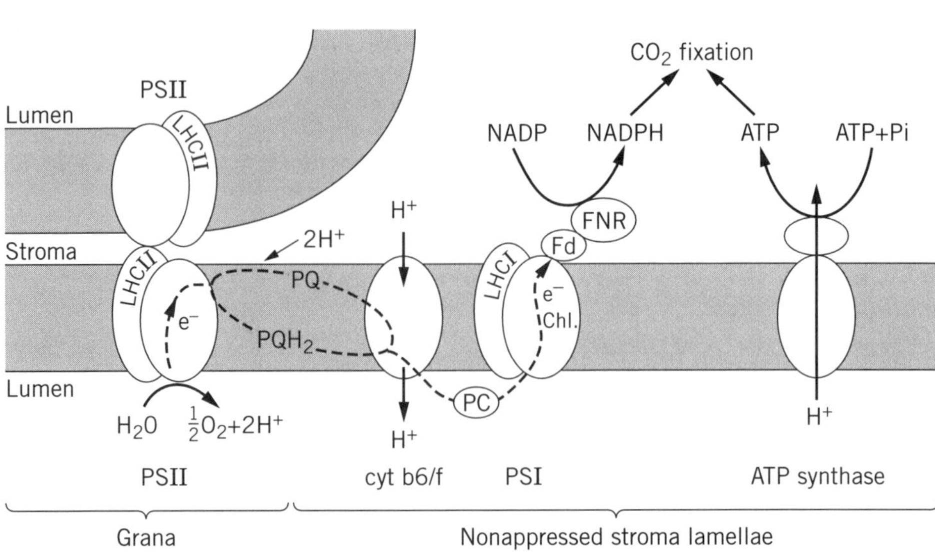

Figure 1. Photosynthetic complexes in the thylakoid membrane of chloroplasts. PSII (photosystem II) is located within the appressed grana region, whereas PSI (photosystem I) is located within the nonappressed stroma lamellae. The photosynthetic electron transfer chain is shown starting with water as electron donor to PSII, to plastoquinone (PQ), to the cytochrome b6/f complex (cytb6/f), to the soluble electron transfer protein plastocyanin (PC), to PSI, to ferredoxin (Fd), to ferredoxin-NADP oxidoreductase (FNR), and to NADP as final electron acceptor. Electron flow is coupled to proton translocation into the lumen. The resulting pH gradient across the thylakoid membrane drives ATP synthesis. Both ATP and NADPH are used for CO_2 fixation.

metabolism and in nitrogen and sulfur assimilation (1). Like **mitochondria**, chloroplasts possess their own genetic system, which cooperates closely with the **nucleus** in biosynthesizing numerous organellar components. Chloroplasts represent one type of plastid derived from colorless proplastids in the meristematic cells of plant leaves and shoots, which have only a rudimentary internal membrane system (1). Light profoundly affects the development of proplastids. They differentiate into chloroplasts in the presence of light, whereas in its absence they differentiate into etioplasts, which lack chlorophyll and contain a prolamellar body. Upon subsequent illumination, the prolamellar body gives rise to lamellae of the thylakoid membrane. Depending on the plant tissues, the developmental stage, and the environmental conditions, proplastids also differentiate into chromoplasts in petals or fruits, into leucoplasts in roots, or into amyloplasts in tubers in which starch is accumulated. Proplastids also develop into elaioplasts in glands, certain fruits and seeds, where they are involved in synthesizing lipids, terpenoids, carotenoids, and carbohydrates. Although these various plastid forms have rather distinct morphologies, plastid differentiation is reversible to a large extent, because chloroplasts develop from leucoplasts or amyloplasts, and viceversa. During transitions from chloroplasts to the other plastid forms, the expression of most organellar genes is reduced, whereas specific nuclear genes encoding plastid proteins are activated (1,2). An important point is that all plastid types contain an internal membrane system that is crucial for their interconversion.

THYLAKOID MEMBRANES AND THE PHOTOSYNTHETIC APPARATUS

The internal thylakoid membrane system consists of appressed and non-appressed flattened membrane vesicles, called grana and stroma lamellae, respectively (Fig. 1). The primary reactions of photosynthesis are catalyzed by four major protein-pigment complexes of the thylakoid membrane: (i) photosystem II and (ii) photosystem I, and their associated chlorophyll antennae, (iii) the **cytochrome** b6/f complex, and (iv) the **ATP synthase** (Fig. 1; see also **Photosynthesis**). Briefly, light energy is captured by the antennae and channeled to the reaction centers of photosystem II and photosystem I. The energy is used to energize an electron in chlorophyll and to create a stable charge separation across the membrane. This triggers a series of oxido-reductions along the photosynthetic **electron-transfer** chain. At one end of this chain, **water** is oxidized by photosystem II with concomitant evolution of oxygen and release of protons into the lumen. Then electrons are transferred to plastoquinone, to the cytochrome b6/f complex, which acts as a **proton pump**, and to the soluble electron carrier plastocyanin in the thylakoid lumen. At the other end of the chain, photosystem I oxidizes plastocyanin upon light absorption and transfers electrons to **ferredoxin** and then to NADP to form NADPH. The resulting pH gradient is used by the fourth complex, ATP synthase, to produce ATP on the stromal side. This enzyme also functions in the opposite direction by hydrolyzing ATP to pump protons into the thylakoid lumen and thus generate a pH gradient. Because the abundance of the thylakoid membrane complexes facilitates their biochemical analysis and because the state of the redox cofactors is monitored readily by **spectroscopic** techniques, the thylakoid membrane has been studied intensively and represents one of the best-studied membrane systems.

Each of the four photosynthetic complexes contains numerous protein subunits, some of which are encoded by the chloroplast **genome**, whereas others are encoded by the nuclear genome (Table 1; Fig. 2). The two principal reaction center polypeptides of photosystems I and II are highly **hydrophobic** and contain 11 and 5 transmembrane α-**helices**, respectively, to which most of the redox cofactors and several chlorophylls are bound with an asymmetrical distribution across the thylakoid membrane. This asymmetry is crucial for the vectorial electron transport in the membrane. The distribution of these complexes is unequal between the appressed (grana) and nonappressed thylakoid membrane regions (3). Photosystem II is localized predominantly in the

Table 1. Informational Content of Chloroplast DNA from Land Plants and Green Algae

Genes Involved in	Number in Chloroplast (Nucleus)[a]
Photosynthesis	
Photosystem II	14 (5)
Photosystem I	5 (8)
Cytochrome b6/f complex	5 (2)
ATP synthase	6 (3)
Ribulose biphosphate carboxlase/oxygenase	1 (1)
NADH dehydrogenase [b]	11
Light-independent chlorophyll synthesis [c]	3
CO_2 uptake[d]	1
Protein Biosynthesis	
Ribosomal RNAs	4
Ribosomal proteins	20
tRNAs	30–31
RNA polymerase subunits	4
Other Functions	
ClpP subunit of ATP-dependent protease	1
ORFs of unknown function	10–20

[a] The number of chloroplast-encoded subunits of photosynthetic complexes. The numbers in parenthesis refer to the number of nucleus-encoded subunits.
[b] These genes are found only in the chloroplast genomes of land plants.
[c] The number of chloroplast genes involved in light-independent chlorophyll synthesis and CO_2 uptake. These genes are found only in the chloroplast genomes of gymnosperms, liverwort, and Chlamydomonas.
[d] This gene is involved in CO_2 uptake.

grana regions, whereas photosystem I and the ATP synthase complex are found exclusively in the nonappresssed regions. The cytochrome b6/f complex is present in both the grana and nonappressed regions. Destacking and restacking of thylakoid membranes are induced experimentally by decreasing and then increasing again the cation concentration. Remarkably, the lateral segregation of the photosynthetic complexes between grana and stromal membranes is lost upon destacking because of random mixing, but it is restored upon restacking the membranes (4).

Other dynamic changes in thylakoid membrane organization occur when plants and algae are subjected to light of different wavelengths that is preferentially absorbed by either photosystems II or I. Under these conditions, part of the chlorophyll antenna is displaced from one photosystem to the other, so as to achieve balanced light absorption and hence optimal functioning of the two photosystems (5). Thus when photosystem II is preferentially activated, the plastoquinone pool is reduced. This leads to the activation of a protein kinase associated with the cytochrome b6/f complex, to the phosphorylation of the **light-harvesting complex** (LHCII), and to the concomitant movement of part of the photosystem II antenna to photosystem I. If photosystem I is preferentially activated, LHCII is dephosphorylated, and it returns to photosystem II in the grana regions.

Stroma and Carbon-Fixation Cycle

The ATP and NADPH produced by the primary light reactions of photosynthesis are used as sources of energy and reducing power to drive the reactions of the carbon fixation cycle, which convert CO_2 into glyceraldehyde 3-phosphate, a precursor to sugars, amino acids, and fatty acids. Although these reactions are also called the "dark reactions," the enzymes involved are inactivated in the dark and need to be reactivated by light through the reducing power generated by photosynthesis. The key reaction, which involves converting one atom of inorganic carbon, as CO_2, into organic carbon, is catalyzed by the enzyme **ribulose 1,5 bisphosphate carboxylase** (Rubisco), a large stromal enzyme that works only sluggishly (6). Therefore, it is required in large amounts and is thought to be the most abundant protein on earth. This enzyme also has an oxygenase activity, which predominates if the concentration of CO_2 is low. Under these conditions it catalyzes the first step of a pathway called photorespiration, which ultimately liberates CO_2 and thereby reverses the photosynthetic reaction. In addition, the stroma includes a large number of proteins involved in several important metabolic pathways (amino acid and fatty acid synthesis, sulfur and nitrogen assimilation). The chloroplast transcription and translation systems are also contained in this compartment.

Chloroplast DNA and Its Informational Content

Chloroplasts together with mitochondria, are the only cellular organelles containing their own apparatus for **protein biosynthesis**. It consists of chloroplast **DNA**, **RNA polymerase**, **enzymes** involved in RNA metabolism, **ribosomes**, **transfer RNA**, and several **translation** factors. Chloroplast ribosomes resemble those of **bacteria** and have similar ribosomal RNA and proteins and sensitivity to a similar spectrum of **antibiotics**. The circular chloroplast DNA molecules range in size between 70 kb and 400 kb (7) and are present in about 100 copies per chloroplast. A typical mesophyll cell contains close to 100 plastids and thus about 10,000 chloroplast DNA circles. A dozen chloroplast genomes from several vascular plants and algae have been sequenced. These sequences have revealed the existence of about 120 chloroplast genes in plants and green algae. They include about 50 genes that encode components of the transcriptional apparatus (subunits of RNA polymerase) and of the translational apparatus (ribosomal RNA, ribosomal proteins, transfer RNA, and translation factors). About 40 genes are involved in photosynthesis, and they encode some of the subunits of photosystems I and II, the cytochrome b6/f complex, ATP synthase and Rubisco (see Table 1). The other subunits of these complexes are encoded by the nuclear genome, translated on cytosolic ribosomes, and imported posttranslationally into the chloroplast. The genes involved in the plastid protein synthesizing system and in photosynthesis have been conserved during evolutionary **divergence** of the chloroplast genomes of plants and green algae. The remaining chloroplast genes, however, have not been universally conserved. Eleven genes encoding subunits of NADH dehydrogenase are present in the chloroplasts of plants, but not in algae. Whereas the role of the mitochondrial NADH dehydrogenase in respiration is well understood, the function of the chloroplast enzyme has not yet been elucidated. It could be involved in a chlororespiratory pathway by reducing the plastoquinone pool in the dark, which is ultimately oxidized by molecular oxygen via unknown redox components (8).

In most plants, one of the last steps of the chlorophyll synthesis pathway, the conversion of protochlorophyllide

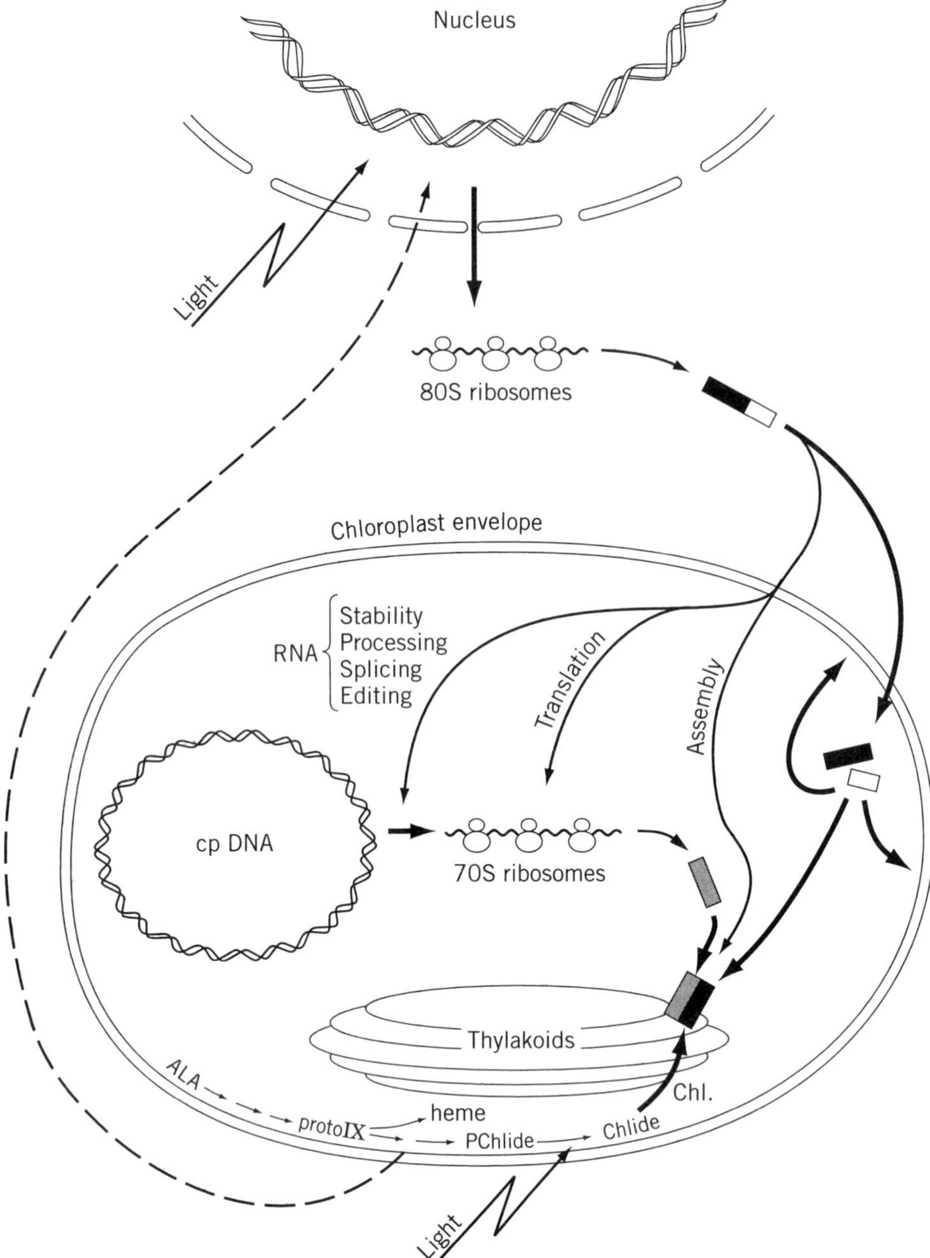

Figure 2. Biosynthesis of the photosynthetic apparatus and protein traffic in the chloroplast. Photosynthetic complexes consist of nucleus- and chloroplast-encoded subunits. The former are synthesized as precursors on cytosolic 80S ribosomes and targeted to the chloroplast. Upon import into the organelle, the N-terminal stromal transit peptide domain is cleaved, and the protein is directed to the stroma, to the envelope, or to the thylakoids. In the latter case, the protein contains an additional cleavable thylakoid targeting domain. Several posttranscriptional steps in the chloroplast, such as RNA stability, processing, splicing, editing, and translation, plus the assembly of protein complexes, require the action of numerous nucleus-encoded factors. Chlorophyll, the major pigment of the thylakoid membrane is synthesized entirely in the chloroplast. Synthesis starts from δ-aminolevulinic acid (ALA) and involves several steps in common with heme biosynthesis until protoporphyrin IX (proto IX). One of the last steps of chlorophyll synthesis, conversion of protochlorophyllide (Pchlide) to chlorophyllide (Chlide), requires light in land plants. Chlorophyll synthesis is tightly coordinated with the synthesis of its apoproteins. Expression of nuclear genes of photosynthetic proteins is strongly stimulated by light. Some of the chlorophyll precursors influence, directly or indirectly, expression of nuclear genes involved in photosynthesis.

into chlorophyllide, is light-dependent. In green algae and **gymnosperms**, an alternative light-independent pathway for chlorophyll synthesis is mediated by three chloroplast genes that are absent in **angiosperms**. The sequences of chloroplast genomes have revealed additional genes whose functions are still unknown.

The chloroplast genomes of nongreen algae contain twice as many genes as those of higher plants. Additional genes include those required for photosynthesis that are nucleus-encoded in plants and green algae, genes involved in the synthesis of fatty acids, amino acids, and pigments, genes required for **protein folding** and transport, and additional genes of unknown function (9). The smallest plastid genome identified, that of the white parasitic plant *Epifagus virginiana*, is only 70 kbp in size. It has lost all the genes involved in photosynthesis, and the remaining genes encode mostly components of the plastid protein synthesizing system.

It is generally admitted that plastids originated as the result of an endosymbiotic event in which a **prokaryotic** photosynthetic organism, probably similar to a **cyanobacterium**, invaded a primitive **eukaryotic** cell. Strong support for this endosymbiotic hypothesis arises from the considerable similarity between the transcriptional and translational systems of prokaryotes and plastids. It is thought that during evolution genetic information from the intruder was gradually lost and transferred to the nucleus of the host. The question thus arises why chloroplast DNA has been maintained. One possibility is that this evolutionary plastid genome size reduction is still in progress and has not yet reached its final stage. Another possibility is that the plastid protein synthesizing apparatus is essential for synthesizing the large hydrophobic polypeptides of the photosynthetic reaction centers, which cannot be translocated across the plastid envelope membrane. A third recently advanced hypothesis is that the presence of the plastid protein synthesizing system is essential to allow a rapid response of plastid gene expression to environmental changes (10).

Chloroplast Gene Expression

Two distinct RNA polymerases are present in the chloroplasts of higher plants. One is similar to its bacterial homologue, and its subunits are encoded by chloroplast genes. This enzyme transcribes primarily genes involved in photosynthesis, which are expressed at a high level. The second plastid RNA polymerase is nucleus-encoded and is required for expressing the nonphotosynthetic plastid functions necessary for plant growth (11). Many chloroplast genes are organized in large transcription units. These units are transcribed into large precursor transcripts, which then are processed into individual **messenger RNA** (mRNA) molecules. Chloroplasts contain RNA **splicing** systems, because several plastid genes contain **introns**, mostly group II and group I, which have a characteristic secondary structure (12). These introns have also been found in mitochondrial genes, and some of them are **self-splicing**. Splicing in the chloroplast is rather complex, as in the case of the *psaA* gene encoding one of the reaction center polypeptides of photosystem I in the green alga *Chlamydomonas*. This gene consists of three coding regions (**exons**) that are widely separated on the chloroplast genome and are flanked by group II intron sequences (13). They are transcribed individually, and maturation of the *psaA* mRNA depends on two trans-splicing reactions in which the separate transcripts of the three exons are spliced together. A particularly intriguing feature is that one of the introns is split into three parts (14). This has interesting evolutionary implications because it is thought that group II introns represent the precursors of nuclear introns and their associated splicing factors. In this view, the split chloroplast intron may represent an intermediate between group II and nuclear introns. The chloroplast genetic system has evolved at a rather slow rate and could have therefore maintained some ancient gene organization.

Another unusual feature of chloroplast RNA metabolism is RNA **editing** in vascular plants (15). Editing in chloroplasts is a posttranscriptional process in which specific C residues of a primary transcript are changed to U. Editing has important implications for interpreting DNA genomic sequence data. As an example, an ACG triplet may be edited to AUG, thereby creating a new initiation codon, which could not be identified in the DNA sequence. Alternatively, an editing event may change an internal codon and thus change the corresponding amino acid predicted by the DNA sequence. Therefore, sequencing of chloroplast genomes may not allow identifying of all of the plastid genes.

Because the subunits, redox cofactors, and pigments of photosynthetic complexes are synthesized by two distinct genetic systems, the process has to occur in a coordinated way (Fig. 2). Genetic studies with ***Chlamydomonas*** and **maize** have indeed revealed the existence of highly complex interactions between nucleus and chloroplast (16). A large number of nuclear genes are involved in chloroplast gene expression. They encode factors targeted to the chloroplast that act at different posttranscriptional steps, such as RNA processing, RNA stability, RNA splicing, translation, and the assembly of photosynthetic complexes. Light strongly enhances some of these steps, especially translation. Several translational activators have been identified, which act at the level of initiating translation. Translation in the chloroplast occurs on chloroplast ribosomes, which are often closely associated with the thylakoid membrane. Cotranslational insertion into the thylakoid membrane has been proposed for the hydrophobic reaction center polypeptides. In addition, synthesis of chlorophyll and its apoproteins needs to be strictly coordinated, because free chlorophyll is highly photoreactive and causes serious damage to the cell.

Chloroplast-Nuclear Cross talk

Chloroplast function and development depend to a large extent on the nucleus. A large number of nuclear genes encode chloroplast structural components and enzymes and are involved in regulating chloroplast gene expression. Reciprocally, chloroplasts also influence nuclear gene activity. This is apparent in mutant plants with defective chloroplasts, where nuclear genes of proteins involved in photosynthesis are no longer expressed. As an example, when carotenoid synthesis is inhibited, chloroplasts rapidly bleach in strong light because chlorophyll is photooxidized in the absence of carotenoids (17). Under these conditions, expression of nuclear genes that code for several abundant chloroplast proteins involved in photosynthesis is specifically repressed. A block in chloroplast protein synthesis has a similar effect (18). These observations imply the existence of a plastid-derived factor that directly or indirectly influences nuclear gene activity. There are mutants of ***Ara-***

bidopsis in which the transduction of this plastid-derived signal to the nucleus is affected (19). The nature of the plastid factor is still unknown in plants, although studies with *Chlamydomonas* suggest that some porphyrin compounds, which act as intermediates in the chlorophyll biosynthetic pathway, are involved in this response (Fig. 2, 20).

Protein Sorting in the Chloroplast

Chloroplasts are bounded by an envelope that consists of the outer and inner membranes. The outer membrane is freely permeable to ions and small molecules, whereas the inner membrane is highly selective and contains specific translocators and **permeases** that allow regulated metabolic transport between cytosol and stroma. The envelope also contains the protein import system.

From just the modest size of the plastid genome, it is clear that the majority of the chloroplast proteins are encoded by nuclear genes and imported into the chloroplast. Six chloroplast compartments can be distinguished: (1) the outer envelope membrane, (2) the intermembrane space, (3) the inner envelope membrane, (4) the stroma, (5) the thylakoid membrane, and (6) the lumen. Nucleus-encoded proteins destined to the chloroplast are synthesized as precursor proteins containing, in most cases, a transient N-terminal **transit peptide** (21). Transit peptides are both necessary and sufficient to import a polypeptide into the chloroplast. Transit peptides of stromal proteins consist of 30 to 120 residues in only a poorly conserved sequence. The only distinguishing feature is that they are rich in hydroxylated amino acids and deficient in acidic residues. Recognition of the protein import **receptor** by the transit peptide is followed by translocation of the precursor protein in an extended conformation across the two envelope membranes. ATP and GTP are the sole energy sources for this process, which also requires the participation of several factors to unfold protein on the outside and to refold protein on the inside of the organelle. Several **molecular chaperones** play an important role in the proper folding of the polypeptides that enter the chloroplast (21). Translocation of the precursor of protochlorophyllide oxidoreductase, an enzyme involved in the last step of chlorophyll synthesis, also requires the presence of its substrate, protochlorophyllide, inside the plastid (22). This raises the possibility that the substrate drives the translocation by inducing or stabilizing folding of the enzyme on the stromal side of the envelope.

Thylakoid precursor proteins contain a bipartite transit peptide. The first domain targets the protein to the stroma, and the second hydrophobic domain, which resembles the **signal sequences** of **secretory proteins**, acts as the thylakoid targeting domain. Surprisingly, there are four pathways for protein translocation into or across the thylakoid membrane (21). The first corresponds to the bacterial protein secretion system and uses **Sec proteins** homologous to the bacterial SecA and SecY proteins. The second uses a system involving a **signal recognition particle**. The third pathway is rather unique because it uses only the trans-thylakoid pH gradient as an energy source (see **Chemiosmotic coupling**). The fourth pathway involves spontaneous insertion of certain proteins into the thylakoid membrane.

Insertion of proteins into the chloroplast envelope occurs by several routes. Some nucleus-encoded polypeptide chains lack a cleavable transit peptide and are inserted directly into the outer and inner membranes. Other envelope membrane proteins containing a cleavable transit peptide use the general import pathway. At least one inner membrane envelope protein is encoded by the chloroplast genome, so it must contain an appropriate targeting signal.

Chloroplast Engineering

A major breakthrough in chloroplast research in 1988 was the development of an efficient method for genetically **transforming** chloroplasts of the green alga *Chlamydomonas* (23), which was subsequently adapted to higher plants (24). In this method, tungsten or gold particles are coated with DNA and bombarded into cells with a particle gun (see **Transfection**). Upon entry of the particles into chloroplasts, the DNA is released and integrated into the chloroplast chromosome by homologous **recombination**. The existence of an efficient chloroplast homologous recombination system and the development of selectable markers for chloroplast transformation have opened the door to manipulating the chloroplast genome. In particular, this new technology allows directed chloroplast gene disruption, a powerful tool for elucidating the role of genes of unknown function (see **Knockout strains**). It has also permitted **site-directed mutagenesis** of specific residues of photosynthetic reaction center polypeptides so as to gain new insights into their structure-function relationship, and it has been very useful for studying chloroplast gene expression. Chloroplast transformation has important applications for plant biotechnology and **protein engineering**. Because the chloroplast genome is present in multiple copies, up to 10,000 per cell, new genetic information introduced into plastids is amplified. In principle, this opens the possibility of expressing foreign proteins of commercial interest in large quantities. The expression of foreign genes in the chloroplast compartment offers the additional advantage of considerably reducing the risk of transfer of new genetic material to the environment because the chloroplasts from the male parent are not transmitted to the progeny in most crop plants.

BIBLIOGRAPHY

1. N. W. Gillham (1994) *Organelle Genes and Genomes*. Oxford University Press, New York.
2. J. K. Hoober (1984) *Chloroplasts*. Plenum Press, New York.
3. J. Olive and O. Vallon (1991) *J. Electron. Microsc. Technol.* **18**, 360–374.
4. G. Ojakian and P. Satir (1974) *Proc. Natl. Acad. Sci. USA* **21**, 2052–2056.
5. J. F. Allen (1992) *Biochim. Biophys. Acta* **1098**, 275–335.
6. R. J. Spreitzer (1993) *Ann. Rev. Plant Physiol. Plant Mol. Biol.* **44**, 411–434.
7. M. Sugiura (1996) In *Molecular Genetics of Photosynthesis.* (B. Andersson, A. H. Salter, and J. Barber, eds.), Oxford University Press, Oxford, New York, pp 58–74.
8. P. Bennoun (1982) *Proc. Natl. Acad. Sci. USA* **79**, 4352–4356.
9. M. Reith and J. Munholland (1995) *Plant Mol. Biol. Rep.* **13**, 333–342
10. J. Allen (1995) *J. Theor. Biol.* **165**, 609–631.
11. L. A. Allison, L. D. Simon, and P. Maliga (1996) *EMBO J.* **15**, 2802–2809.
12. M. Sugita and M. Sugira (1996) *Plant Mol. Biol.* **32**, 315–326.

13. U. Kück, Y. Choquet, M. Schneider, M. Dron, and P. Bennoun (1987) *EMBO J.* **6**, 2185–2195.
14. M. Goldschmidt-Clermont et al. (1991) *Cell* **65**, 135–143.
15. H. Kössel et al. (1993) In *Plant Mitochondria*, A. Brennicke and U. Kück (eds.), VCH, Weinheim, Germany, pp. 93–102.
16. J.-D. Rochaix (1992) *Ann. Rev. Cell Biol.* **8**, 1–28.
17. W. Taylor (1989) *Ann. Rev. Plant Physiol. Plant Mol. Biol.* **40**, 211–233.
18. J. Gray (1996) In *Membranes: Specialized Functions in Plants*, M. Smallwood, J. P. Knox, and D. J. Bowles, eds., Bios Scientific Oxford, pp. 441–455.
19. R. E. Susek, F. M. Ausubel, and J. Chory (1993) *Cell* **74**, 787–799.
20. U. Johanningmeier and S. H. Howell (1984) *J. Biol. Chem.* **259**, 13541–13549.
21. K. Cline and R. Henry (1996) *Annu. Rev. Cell Dev. Biol.* **12**, 1–26.
22. S. Reinbothe, S. Runge, B. Reinbothe, B. von Cleve, and K. C. Apel (1995) *Plant Cell* **7**, 161–172.
23. J. E. Boynton et al. (1988) *Science* **240**, 1534–1538.
24. P. Maliga (1993) *Trends Biotechnol.* **11**, 101–107.

Suggestions for Further Reading

N. W. Gillham (1994) *Organelle Genes and Genomes*, Oxford University Press, New York.

J. K. Hoober (1984) *Chloroplasts*, Plenum Press, New York.

CHO CELLS

R. I. FRESHNEY

ORIGIN

CHO cells were originated as a **cell line** by Theodore Puck in 1958 (1) from the Chinese hamster, *Cricetulus griseus*. CHO/Pro cells requiring proline for growth were subsequently derived by nutritional selection. The parental cells were treated with bromodeoxyuridine (see **5-Bromouracil**) in proline-deficient medium and then exposed to the near-visible UV from a fluorescent light. This killed the growing cells, but not those that required proline. Proline-requiring clones were then grown up by feeding the surviving cells with proline-rich medium. CHO/Pro$^-$ was subcloned by dilution cloning, and cell line CHO/Pro-K1 was isolated. The current designation of this cell line in common use is CHO-K1, and it still has a requirement for proline, which is present in Ham's F12, the medium usually recommended for its propagation. It is listed in the American Type Culture Collection (ATCC) catalogue as CCL-61 (Fig. 1).

PROPERTIES

CHO-K1 (Fig. 1) is a continuous cell line and near-**diploid** with 20 **chromosomes** ($2C = 22$). It has a very short doubling time of around 15 h, making it popular as a host for **transfection** and biotechnology. The plating efficiency is also very high, and can be 100% under optimal conditions, so these cells have always been popular for clonogenic survival studies of nutritional mutants and radiation survival. Chinese hamster cells originally became popular for genetic studies because of their relatively small number of readily distinguishable chromosomes, but the advent of chromosome banding techniques and chromosome painting by fluorescence ***in situ* hybridization** makes the distinction of individual chromosomes easier in many other species.

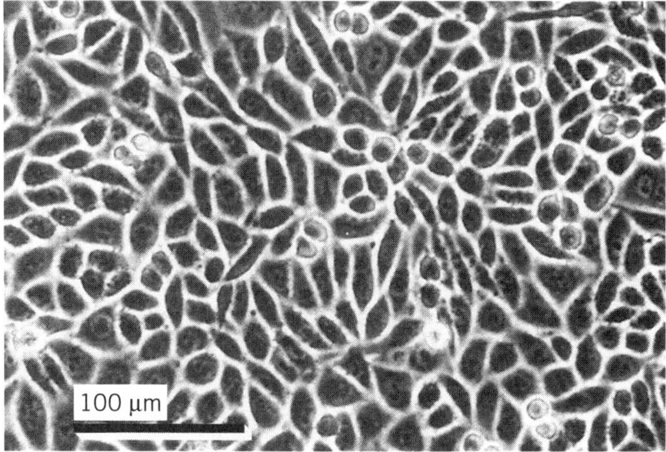

Figure 1. Confluent culture of CHO-K1 cells. Phase contrast, Olympus CK microscope, 20× objective.

CHO-K1 cells, although transformed, still retain nutrient-dependent G_1 **cell-cycle** blockade. Isoleucine deprivation blocks the cells in G_1, and its restoration generates a synchronous population (2).

USAGE

CHO-K1 cells have been used in cell-cycle control and signaling (3) and are used extensively in biotechnology (4). They are used frequently in DNA **transfection** (5) and virally mediated DNA transfer (6). They can be maintained in suspension culture to generate large numbers of cells or product. Using methotrexate-induced co-amplification with cotransfected **dihydrofolate reductase**, they have been used for production of **interferon-**γ (7) and pro**thrombin**-2 (8). They have been adapted to grow in serum-free medium (9) (see **Serum dependence**). Together with V79 cells, another Chinese hamster cell line, they have been used extensively in genotoxicity studies (10).

BIBLIOGRAPHY

1. T. T. Puck, S. J. Cieciura, and A. Robinson (1958) *J. Exp. Med.* **108**, 945–956.
2. K. D. Ley and R. A. Tobey (1970) *J. Cell. Biol.* **47**, 453–459.
3. G. Tortora, S. Pepe, C. Bianco, V. Damiano, A. Ruggiero, G. Baldassarre, C. Corbo, Y. S. Chochung, A. R. Bianco, and F. Ciardiello (1994) *Intl. J. Cancer* **59**, 712–716.
4. K. B. Konstantinov (1996) *Biotechnol. Bioeng.* **52**, 271–289.
5. S. Subramanian and F. Srienc (1996) *J. Biotechnol.* **49**, 137–151.
6. T. Schoneberg, V. Sandig, J. Wess, T. Gudermann, and G. Schultz (1997) *J. Clin. Invest.* **100**, 1547–1556.
7. V. Leelavatcharamas, A. N. Emery, and M. Al-Rubeai (1994) *Cytotechnology* **15**, 65–71.
8. G. Russo, A. Gast, E. J. Schlaeger, A. Angiolillo, and C. Pietropaolo (1997) *Protein Express. Purif.* **10**, 214–225.
9. M. J. Keen and N. T. Rapson (1995) *Cytotechnology* **17**, 153–163.
10. H. F. L. Mark, R. Naram, T. Pham, K. Shah, L. P. Cousens, C. Wiersch, E. Airall, M. Samy, K. Zolnierz, R. Mark, K. Santoro, L. Beauregard, and P. H. Lamarche (1994) *Ann. Clin. Lab. Sci.* **24**, 387–395

CHOLERA TOXIN AND ENTEROTOXINS

C. MONTECUCCO

Cholera toxin (CLT) and the closely similar heat-labile enterotoxins (LTs) are released by toxigenic strains of *Vibrio cholerae* and *Escherichia coli*, respectively, the causative agents of epidemic diarrheas (1). These bacteria attach firmly to the apical portion of the intestinal epithelium and produce several **toxins**, including CLT or LT, encoded by genes contained in a transferable virulence cassette (2). Five B subunits (103 residues) are secreted into the bacterial **periplasm**, where they assemble into a pentamer that binds the A chain. A is then proteolytically cleaved at a single point to generate the catalytic A1 chain (192 residues) linked via a **disulfide bond** to the A2 peptide (3,4) (Fig. 1). The structural organization of these enterotoxins is shared by *Shiga* toxins (other enterotoxins, which cause bloody diarrheas and necrosis of intestinal epithelium and are produced by *Shiga spp.* and *E. coli spp.*) and **pertussis toxin** (3–6). The B pentamer has the shape of an irregular cylinder with a flat surface and convoluted bottom surface. The five B monomers are arranged around a 5-fold axis in the oligomer-binding fold, recently identified in a set of oligonucleotide- and oligosaccharide-binding proteins (7). Each monomer contains five β-strands in an antiparallel **beta-sheet** and one **alpha-helix**. The five helices form a central pore hosting the carboxyl-terminal segment of A2, which emerges on the other side with a Lys/Arg-Asp-Glu-Leu motif that may be important in cell penetration (8). The amino-terminal half of A2 forms a long α-helix, rising from the flat surface, involved in interaction with A1. Thus, very few protein–protein contacts exist between A1 and B. The convoluted oligomer B surface, distal from A1, forms five sugar-binding sites on the outside edge of the cylinder (1,3,4).

The structure of A1 shows a cleft where NAD binds and Arg-7 and Glu-112, two residues essential for activity, are located (9). Each B subunit of CLT or LTs possesses a single binding site specific for the oligosaccharide portion of a ganglioside (10), and one molecule of toxin binds five glycolipid molecules. Oligosaccharide binding affinity is rather low, but a high affinity of the toxin for the cell (K_d of the order of 10^{-10} M) is obtained because of the pentavalent binding. This strategy of multivalent binding to obtain a strong cell association is displayed by other bacterial toxins and viruses (11).

The target of CLT and LTs is localized on the basolateral membrane; therefore, these toxins have to transcytose through the cell to display their activity. Available evidence indicates that CLT and LTs are internalized inside apical **endosomes** that may recycle the glycolipid–toxin complex to the surface or deliver it to later endosomes. As a result of the binding of the **KDEL sequence** of the A subunit to a KDEL **receptor**, the toxin moves retrogradely through the **Golgi** cisternae, and some toxin molecules are then expected to be sorted into basolateral endosomes that fuse with the basolateral membrane (8). At some stage of this intracellular trafficking, these toxins have to be reduced and the A1 subunit has to translocate from the lumenal to the cytosolic side of the membrane to interact with their target on the cytosolic face of the basolateral membrane. Structural and membrane photolabeling data indicate that oligomer B does not penetrate the lipid bilayer (1,10,12), but A1 inserts in the membrane upon reduction of the A1—A2 interchain disulfide bond. Hence, it is likely that as soon as the disulfide bridge is reduced, A1 "rolls over" oligomer B and inserts into the membrane. This is at variance from postulated mechanism of membrane penetration of other toxins, whose protomer B plays an active role in the insertion of the catalytic subunit. Neither the chemical nature of the reducing agent nor the intracellular stage at which this step takes place are known.

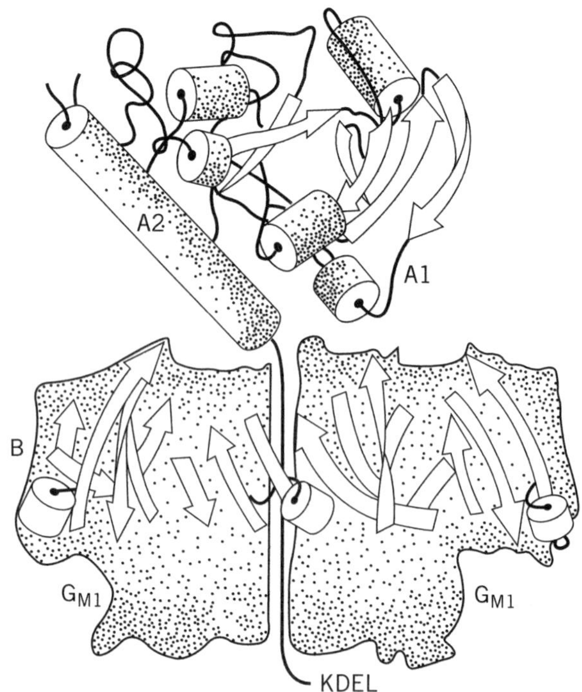

Figure 1. Structural organization of cholera toxin and related toxins, showing a cross section of the molecule of cholera toxin and the *E. coli* heat-labile enterotoxins (1,3,4). A is composed of two polypeptide chains: A1, endowed with ADP-ribosyltransferase activity, and A2, which consists of a long α-helix, involved in interaction with A1, followed by a structureless segment that penetrates the small central hole of the B pentamer. Each of the five B subunits forming the B oligomer has a binding site for the oligosaccharide portion of the ganglioside G_{M1}. Thus, cholera toxin binds to the membrane via multiple interactions, with the catalytic A1 subunit pointing away from the membrane, with little protein–protein contact between A1 and B.

The A1 subunits of CLT and LTs catalyze the transfer of ADP-ribose from NAD (see **ADP ribosylation**) to an Arg residue present in the LRX**R**VXT conserved sequence of the α subunit of the G_S, G_t, and G_{olf} large trimeric **GTP-binding proteins** involved in the coupling of cell surface receptors to the **adenylate cyclase** (13). Such modification results in a permanent activation of this latter enzyme and a large increase in cellular **cyclic AMP** (cAMP) level, which initiates a cascade of **signal transduction** pathways. A1 ADP-ribosylating activity is enhanced by a group of cytosolic or membrane GTP-binding proteins, termed ARF (ADP-ribosylating factors), present in eukaryotic cells (14). ARF are strongly conserved from yeast to humans and are involved in the control of membrane trafficking and protein transport inside cells (15).

The increased cAMP level brought about by CLT or LTs in the enterocyte has a series of consequences, but the inhibition of a **sodium channel** and activation of a **chloride channel** localized on the apical membrane appear to be very relevant to diarrhea. In fact, a decreased sodium reabsorption and increased chloride secretion cause an osmosis-driven loss of water into the intestine. Also important in cholera is the activation of entero-chromaffim cells, which respond to the cAMP increase with release of VIP (vasointestinal peptide), which further lowers intestinal water reabsorption and inhibits muscle cells with an alteration of intestinal peristalsis.

BIBLIOGRAPHY

1. B. D. Spangler (1992) *Microbiol. Rev.* **56**, 622–647.
2. M. K. Waldor and J. J. Mekalanos (1996) *Science* **272**, 1910–1914.
3. T. Sixma et al. (1993) *J. Mol. Biol.* **230**, 890–918.
4. R. G. Zhang et al. (1995) *J. Mol. Biol.* **251**, 563–573.
5. M. E. Fraser, M. M. Chernaia, Y. V. Kozlov, and M. N. G. James (1994) *Nature Struct. Biol.* **1**, 59–64.
6. P. E. Stein et al. (1994) *Structure* **2**, 45–57.
7. A. G. Murzin (1993) *EMBO J.* **12**, 861–867.
8. W. I. Lencer et al. (1995) *J. Cell Biol.* **131**, 951–962.
9. M. Domenighini, C. Magagnoli, M. Pizza, and R. Rappuoli (1994) *Mol. Microbiol.* **14**, 41–50.
10. E. A. Merritt et al. (1994) *Protein Sci.* **3**, 166–175.
11. G. Menestrina, G. Schiavo, and C. Montecucco (1994) *Mol. Aspects Med.* **15**, 81–193.
12. M. Tomasi and C. Montecucco (1981) *J. Biol. Chem.* **256**, 11177–11181.
13. D. M. Gill and M. J. Woolkalis (1991) *Methods Enzymol.* **195**, 267–280.
14. J. Moss and M. Vaugham (1991) *Mol. Microbiol.* **5**, 2621–2627.
15. J. E. Rothman and F. T. Wieland (1996) *Science* **272**, 227–234.

CHORION GENES AND PROTEINS

DAVID PARRY

The outer eggshell of those numerous insects whose **embryos** develop externally is known as the *chorion*, and it must be mechanically robust in order to provide the environment in which the **egg** can develop successfully. The chorion must also have a structure that minimizes water loss, while still permitting the gas exchange vital for embryonic respiration. In ***Drosophila*** the chorion consists of an outer exochorion, an endochorion, a thin inner chorionic layer, a wax layer and the vitelline membrane. The endochorion has the most important structural and gas-transporting roles *in vivo* (1). The wax and inner chorionic layers act as the waterproofing agent for the egg. Proteins in the chorion undergo extensive **post-translational modifications** and consequent peroxidase-catalyzed cross-linking of two or three **tyrosine** residues, which results in a significant increase in the mechanical stability of the entire structure. In contrast to the chorion in the *Drosophila*, however, the gypsy moth and other Lepidoptera have a chorion with an outer "sieve" layer (30 nm thick) overlying about 50 to 60 lamellae of variable thickness (but each about 0.2 μm thick on average). The innermost region of the chorion (ie, that beneath the lamellar region) is called the *trabecular layer* and has a thickness of about 0.5 μm (2). Lamellar substructures of this general type are not found outside Ditrysia. An individual lamella consists of a fibrous layer in which fibrillar elements reported to be about 3 to 4 nm in diameter lie parallel to the surface. The orientation of the lamellae change in a relatively systematic manner from one lamella to the next, thus generating a liquid crystal-like, cholesteric phase ultrastructure. These and other observations reveal that the *Drosophila* and silk moths show remarkable similarities, but equally remarkable differences. For example, they share regulatory elements that direct chorion **gene expression** to the follicular epithelium at the end of **oogenesis** (2). In contrast, they differ in their structural **gene** sequences, their **chromosomal** organization, their chorionic ultrastructures, and their modes of morphogenesis.

The chorion in general is a very complex structure and, in some cases at least, contains a hundred or more different proteins. The proteins in silk moths, however, all have similarities in sequence and fall within either the α or β branches of the chorion superfamily. These proteins in turn are encoded by numerous duplicated genes, all of which contain a single **intron**. Fortunately, the chorion genes too fall into one of a small number of families - A, B, and C. Each family contains multiple genes that occurred during evolution by **gene duplication** and sequence **divergence**. The families themselves are related and constitute a **superfamily** with A genes in one branch and B and C in the other. The A and B genes, which exist in pairs in divergent orientation in the chromosome, are coordinately expressed and **transcribed** in opposite directions under the direction of a bidirectional **promoter**. The chorion genes in *Drosophila* are quite different from those in silkmoths with regard to sequence and organization. In particular, the *Drosophila* genes are tandemly orientated, and each has its own unidirectional promoter that is temporally unique.

Proteins in the mature silkmoth chorion are usually small (10 to 20 kDa) and have *N*- and *C*-terminal **domains** that are variable in sequence and structure. These regions endow the protein with specific functional and structural attributes. Quasi-repeats are not uncommon (3): In *Drosophila* the dipeptide (Gly–His) is repeated five times towards the *N*-terminal end of s36 and nine times in the *C*-terminal region of s38; a hexapeptide repeat based on tyrosine is found in the *C*-terminal region of s36. Peptide repeats of the form (Cys–Gly), (Cys–Gly–Gly), and (Gly–Tyr–Gly–Gly–Leu) are found in the case of silkmoths. Unlike the chorion proteins in *Drosophila*, a central domain is largely conserved in the α and β families of the silkmoth proteins. This is characterized by tandem repeats. In the case of the α family, the central domain is 52 residues in length and displays a six-residue quasi- repeat with a **consensus sequence** (Gly–X–Val/Ile–Y–Val/Ile–Z), where X is generally a charged or large polar residue, Y is variable, and Z is commonly Ala, Gly or Cys. The β family also displays a similar but different six-residue quasi-repeat. The predicted **secondary structure** of both hexapeptide repeats is that of an antiparallel **β-strand** terminated by a **β-turn**. Interestingly, the occurrence of valine and isoleucine residues two apart in the α repeat places them on the same side of the β-sheet in a manner analogous to that in feather **keratin**. Depending on the precise form of the β-turns, one face of the twisted **β-sheet** would be almost entirely apolar and could give rise to a β-barrel structure in which the apolar residues would be located internally. In the case of the gypsy moth Ld15 protein (and the 292a protein of *A. polyphemus*), both penta- and hexapeptide quasi-repeats are found with consensus sequences of (Gly–Leu–X–Pro/Gly–Tyr), (Tyr–Gly–X–X–Gly/Ala), and (Gly–X–Val–X–apolar–Ala/Gly) (2). *Drosophila* chorion proteins s36 and s38, on the other hand, contain three consecutive large apolar residues in a quasi-repeat, and these are likely to form a β-strand. They are followed by three to seven residues rich in proline and basic amino acids that are likely to form a connecting β-turn or loop. These few examples illustrate that although there are significant variations in the detail of the sequence repeats, the antiparallel β-sheet conformation that is predicted to occur in all cases would seem to be a constant feature of all the chorion proteins. The paired organization of the moth chorion genes may imply that the α and β members of the chorion superfamily interact with one another to form the core of the structure, thus leaving the *N*- and *C*-terminal regions in an external location where they are able to provide other important functional properties.

BIBLIOGRAPHY

1. A. C. Spradling (1993) Developmental Genetics of Oogenesis. In *The Development of Drosophila melanogaster*, Vol. 1, (M. Bate and A. M. Arias, eds.), Cold Spring Harbor Laboratory Press, Cold Spring Harbor, NY, pp. 1–70.
2. R. F. Leclerc and J. C. Regier (1993) Choriogenesis in the Lepidoptera: morphogenesis, protein synthesis, specific mRNA accumulation, and the primary structure of a chorion cDNA from the gypsy moth. *Dev. Biol.* **160**, 28–38.
3. S. J. Hamodrakas, A. Batrinou, and T. Christophoratou, Structural and functional features of *Drosophila* chorion proteins s36 and s38 from analysis of primary structure and infrared spectroscopy. (1989) *Int. J. Biol. Macromol.* **11**, 307–313.

Suggestions for Further Reading

F. C. Kafatos, G Tzertzinis, N. A. Spoerel, and H. T. Nguyen (1995) Chorion Genes: An Overview of Their Structure, Function and Transcriptional Regulation. In *Molecular Model Systems in the Lepidoptera* (M. R. Goldsmith and A. Wilkins, eds.) Cambridge University Press, Cambridge, England, pp. 181–215.

T. L. Orr-Weaver (1991) *Drosophila* genes: cracking the eggshell's secrets. *Bioessays* **13**, 97–105.

CHROMATID

ALAN P. WOLFFE

A chromatid is one of the two adjacent strands of a **chromosome** that are generated as a result of chromosomal duplication. The two chromatids are held together at the **centromere** and become visible during **mitosis** (Fig. 1, see top of next page). Chromosomal dynamics during the **cell cycle** are conveniently discussed in terms of the appearance and behavior of chromatids. It is during **interphase** that DNA is replicated (**S-phase**). Before the S-phase there is a gap known as G1, when the chromosomes are considered to be unineme, meaning that a single duplex DNA molecule runs the entire length of the chromosome (see **Lampbrush chromosome**). The amount of DNA in G1 per haploid genome is called the **C-value**. A **diploid** cell has a 2C DNA content. After the S-phase, in the gap known as G2, the chromosomes have duplicated and there is a 4C DNA content. Now each chromosome consists of two unineme sister chromatids. Mitosis begins at the end of G2. Whereas the chromosomes are too diffuse and decondensed to be visualized when somatic cells are in the interphase, during mitosis the chromosomes and chromatids become visible. Mitosis can be divided into four stages. During **prophase**, the **chromatin** in the chromosomes begins to condense so that they appear as morphologically indistinct, diffuse threads. At the end of the prophase, the chromosomes become well separated. Now each chromosome is distinctly double-stranded, consistent with two sister chromatids joined at the centromere. In the metaphase the chromosomes are maximally contracted and align on a plane, known as the metaphase plate. The metaphase plate is midway between the **spindle poles** and represents an integral component of the **mitotic spindle**. In the **anaphase**, the pairing region within the centromere dissolves (see **Centromere**). This allows the mitotic spindle to pull the two sister chromatids apart. Each pole receives a complete set of chromatids, representing all of the chromosomes. Finally in **telophase** the mitotic spindle breaks down, and the nuclear membrane is reassembled around the chromosomes. The chromosomes progressively decondense and cease to be visible by light **microscopy**.

Sister chromatids also interact during **meiosis**, which represents the cell division, chromosomal and chromatid segregation events associated with **gamete** formation. There are two cell divisions during meiosis. Once again chromosomal dynamics are best described in terms of the segregation of chromatids. Initially, the chromosomes replicate, then the cell enters the first meiotic cell division. Homologous chromosomes are attracted to each other, forming stable pairs held together by the **synaptonemal complex** (see **Homologous chromosomes**). Thus each structure consists of two chromosomes, each containing two sister chromatids. **Recombination** occurs between the maternal- and paternal-derived chromosomes. Sites of such genetic **crossing over** are called **chiasmata**. Chiasmata represent sites of physical breakage and reunion between nonsister chromatids. They generally encompass a few hundred to a few thousand base pairs of DNA. Finally, each pair of chromosomes separates. At the start of the second meiotic cell division, each chromosome consists of two chromatids joined by a centromere. Then the chromatids are aligned on a metaphase plate attached to a spindle. The centromere splits, and the two sister chromatids move to opposite poles. Then nuclei are reformed in each of the four haploid daughter cells. Thus for both mitosis and meiosis the mechanics of chromosome condensation, recombination, and segregation emphasize the fate of individual chromatids.

Suggestions for Further Reading

A. T. C. Carpenter (1987) Gene conversion, recombination nodules, and the initiation of meiotic synapsis. *BioEssays* **6**, 232–236.

G. J. Gorbsky (1992) Chromosome motion in mitosis. *BioEssays* **14**, 73–80.

G. S. Roeder (1990) Chromosome synapsis and genetic recombination: Their roles in meiotic chromosome segregation. *Trends Genet.* **6**, 385–389

CHROMATIN

JEAN THOMAS

Chromatin is the complex of **histones** and DNA, associated with smaller amounts of other proteins, into which the

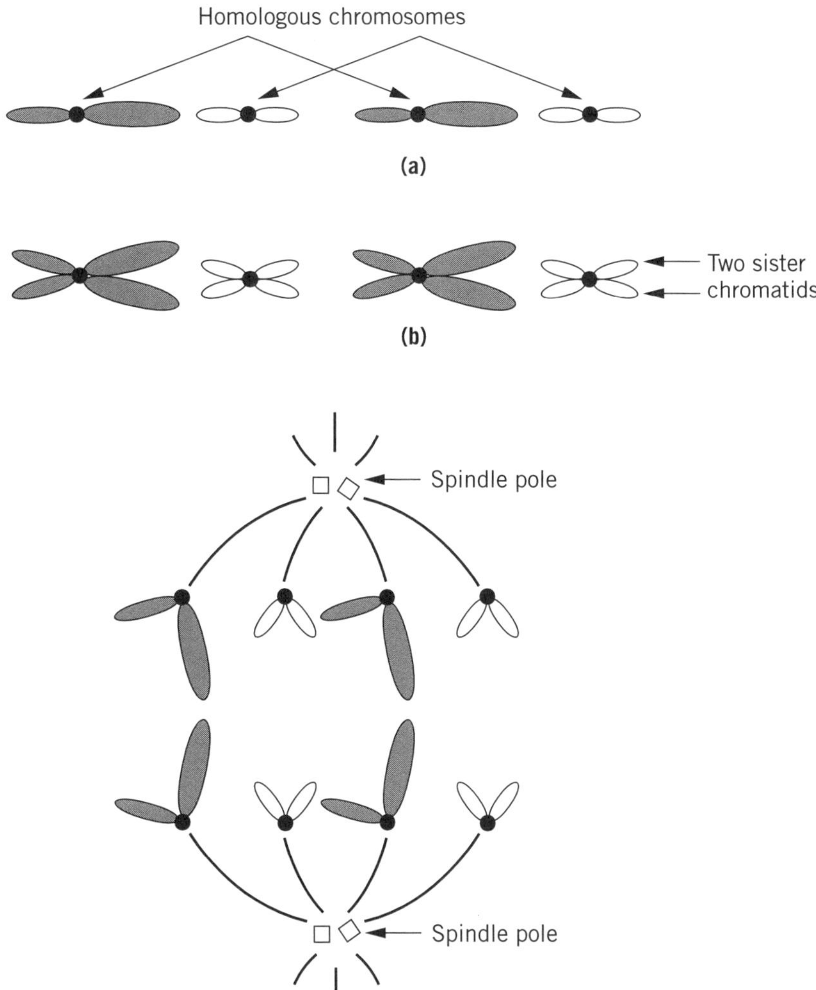

Figure 1. Chromosomal dynamics during the cell cycle. (**a**) At the end of mitosis and in G1, there are two copies of each chromosome, called homologous chromosomes, in a diploid cell. (**b**) At the end of the S-phase and in G2, each chromatid in each chromosome has been duplicated, creating two sister chromatids joined at the centromere. (**c**) During the anaphase of mitosis the two sister chromatids separate and move to opposite spindle poles.

chromosomal DNA of all eukaryotes is organized. As well as being a means of packing DNA in an orderly fashion, chromatin structure plays a crucial role in the regulation of gene expression, both activation and repression. Three books contain much useful information about chromatin structure and function (1–3); for the most up-to-date information in many areas of this rapidly moving field, the reader will be referred to recent articles and reviews.

OVERVIEW

Chromatin has a beaded appearance in the **electron microscope** at low ionic strength, arising from a regularly repeating structure (4). The repeating unit is the **nucleosome**, which contains ~166 to 240 bp of DNA, of which ~166 bp is wound in two left-handed superhelical turns around an octameric complex of the four core **histones** (H3, H4, H2A, and H2B) and stabilized by one molecule of histone H1; the nucleosome also includes a variable length of **linker DNA** (~0 to 76 bp) that connects neighboring nucleosomes. The octamer is organized as a central $H3_2H4_2$ tetramer, which has a rather flat, twisted horseshoe shape, flanked above and below by an H2A–H2B dimer. The details of the organization of histone and DNA in the core of the nucleosome are revealed in the high-resolution structure of the 146-bp **nucleosome** core particle (which contains no linker DNA and no H1) recently determined by **X-ray crystallography** (5). The globular domain of histone H1 binds asymmetrically, bridging a point near the dyad and an entering or exiting DNA double helix (6) (see **Nucleosome** for further details); the basic *C*-terminal domain interacts with the linker DNA, partially neutralizing its charge. The array of nucleosomes is known as a nucleosome filament, or the 10-nm filament (10 nm being the diameter of the nucleosome). The amount of DNA contained in the repeating unit, which reflects the length of linker DNA, is determined from the sizes of the DNA in the chromatin fragments liberated by incomplete digestion with micrococcal nuclease (**Staphylococcal nuclease**), which cuts in the linker. A series of fragments, usually visualized as a "ladder" of bands in **gel electrophoresis**, occurs at multiples of the unit repeat size. This is commonly about 200 bp (when there is a linker length of 200 − 166 = 34 bp), but it may be as long as ~240 bp (in sea urchin sperm), or as short as ~166 bp (essentially no linker DNA) in yeast and in mammalian cerebral cortex neurons (the glial cells in the same tissue have a repeat length of about ~200 bp). The repeat length measured in this way is the average for the tissue or cell population under examination; there will be local variations about this mean (see **Linker DNA**). Packaging of the DNA in nucleosomes achieves about a sixfold length compaction of the DNA, and further packing is achieved by salt-dependent, histone H1-assisted folding of the 10-nm nucleosome filament into a 30-nm filament. Much evidence supports a simple helical coiling of the nucleosome filament into a "solenoid" with six nucleosomes per turn, but there are other models (see text below). Most of the chromatin in the cell is in the form of a 30-nm filament throughout most of the **cell cycle**, and at interphase appears to be looped onto the nuclear matrix; at mitosis, the 30-nm filament is subjected to further levels of folding to give the highly dense metaphase chromosome in which the overall length compaction of the DNA is about 10,000-fold.

The 30-nm filament is not a suitable template for **transcription** by the large eukaryotic **RNA polymerases** and clearly has to be unfolded first; formation of an "open" ("transcriptionally competent") chromatin state is an essential prerequisite for transcription. This involves acetylation of the basic *N*-terminal regions of the core histones, which appear to play a role in chromatin folding, and some depletion of histone H1. It has recently become clear that acetylases are targeted to particular regions of the chromosome through recruitment, directly or indirectly, by gene regulatory proteins bound to specific DNA sequences at **promoters** or upstream activating sequences. Cells appear to contain many acetylase (and deacetylase) activities, which function as components of multiprotein complexes (see **Histone acetylation**). Even in the 10-nm filament form, nucleosomes may act as blocks to the formation of transcription initiation complexes by occluding promoters, which may include strategically positioned nucleosomes. It is now clear that the cell has specialized energy-dependent mechanisms that are required for "remodeling" some promoters, so that they become accessible to transcription factors, as well as mechanisms for facilitating RNA chain elongation. These mechanisms also involve multiprotein complexes, which may have subunits in common with acetylases and deacetylases. Chromatin is thus both an effective means of packing and the target for an integrated network of machines that modify it. There is a wealth of genetic evidence showing that the growing list of factors that regulate transcription includes both *bona fide* chromatin components (eg, histones) and proteins that modify chromatin (eg, acetyltransferases and deacetylases). An unexpected discovery was that a structural motif found in all the core histones is also found in a number of components of the transcription initiation factor TFIID (see **Histone fold**), as well as in some histone acetyltransferase complexes (see **Histone acetylation**), suggesting a structural and evolutionary link between chromatin and the assemblies that act upon it. Another surprising finding has been that histone H1 may have a specific gene regulatory role, in addition to its role in packaging and as a general repressor (see **Histones**).

NUCLEOSOME POSITIONING AND MOBILITY

Histones package the whole of the **genome** and therefore appear to be largely indifferent to DNA sequence. Locally, the position of a nucleosome is determined by the underlying DNA sequence and, in particular (because the DNA, which is normally rigid, is bent around the octamer surface), by the local bendability of DNA, which is sequence-dependent. Analysis of the DNA sequences of bulk nucleosome core particles shows that A and T di- and trinucleotides are likely to occur where the minor groove faces inward, and G and C di- and trinucleotides where the minor groove faces out. This is because the minor groove of AT-rich DNA is naturally narrow and so, by having A and T where the minor groove faces in, the compression of the groove that occurs on wrapping the DNA around the octamer is readily accommodated (7). The dinucleotide periodicity is on average 10.2 bp, exactly the average structural periodicity given by the X-ray crystal structure of the nucleosome core particle (see **Nucleosome**). The helical periodicity in solution is 10.6 bp, so the DNA is overtwisted in the nucleosome, presumably to achieve the match between the structural and sequence periodicities that underlie the rules that govern rotational positioning of the octamer with respect to the DNA sequence (the orientation of a face of the duplex with respect to the histone octamer surface). This is thus determined by the local bendability (flexibility toward curvature) of the DNA, which is determined by the properties of the individual base steps. A rotationally positioned duplex shows a characteristic 10-nucleotide pattern of nicking by DNase I, as each strand rises from the surface about every 10 nucleotides. Translational positioning, on the other hand (the basis of which is not well understood), refers to the choice of a particular stretch of DNA by the histone octamer, rather than other stretches of the same length translated forward or backward along the DNA by about 10 bp, which would allow the same rotational setting. The abundance of AA/TT dinucleotides separated by roughly 10 bp led to the construction of fragments with alternating A/T and C/C sequences [(A or T)$_3$NN(G or C)$_3$NN]$_n$—the so-called TG pentamer—that indeed formed very stable nucleosomes (8). Selection of naturally occurring 146-bp DNA sequences with a high affinity for the histone octamer revealed (in addition to the A/T, G/C periodicity) that sequences with repeated TATAAAACGCC motifs ("phased TATA" sequences) formed nucleosomes that were even more stable, suggesting that high-affinity binding is aided by flexible sequences (9); selection from longer (220 bp) synthetic DNA sequences also revealed a signal (CTAG) that favored high-affinity binding (10). Such sequences, even if they occurred infrequently in the genome, could have important implications for chromatin organization and regulation, as could sequences that are refractory to nucleosome formation (eg, TGGA repeats identified in a negative-selection approach (11). It seems likely that the signals discovered for rotational positioning *in vitro* (7) also function *in vivo*, as revealed by analysis of the complete genome sequences of the yeast *Saccharomyces cerevisiae* and the **nematode** *Caenorhabditis elegans* (12), where dinucleotide periodicities are compatible with nucleosomal constraints. Analysis of 168-bp chicken erythrocyte chromatosome sequences [ie, containing H5(H1] shows that the statistical preferences of the octamers for particular sequences are slightly modulated (13), presumably in order to optimize the stability of the octamer–DNA–H1 ternary complex. This would be reflected in slight differences in translational positions of nucleosomes in the presence of H5(H1), which might have functional consequences (eg, in exposing or revealing short DNA sequences in an H1-dependent manner). An additional feature of chromatosomal DNA is the frequent occurrence of AGGA within half a double helical turn of one terminus, imposing asymmetry on the chromatosome. Whether this is related to the asymmetric binding of H1 to the **nucleosome** remains to be seen. The rules for nucleosome organization on DNA sequences have recently been reviewed (14).

Although histones can, in general, package DNA essentially irrespective of sequence, but capitalize on the local bendability of DNA to promote a better fit on the octamer surface, some DNA sequences have a particularly high affinity for the octamer, leading to a "positioned nucleosome" in which a particular translational position is preferred. Nucleosome reconstitution experiments show that the H3–H4 tetramer alone is sufficient to confer positioning on a defined DNA sequence (15,16). Positioned nucleosomes have been identified *in vivo* by nucleosome mapping techniques and *in vitro* in nucleosome reconstitution experiments on defined DNA sequences. The rules for nucleosome positioning *in vivo* are not well understood: When the TG pentamer (see above) was introduced into yeast, it appeared to exclude nucleosomes rather than provide optimal positioning (17), showing that a strong rotational setting is not sufficient for positioning *in vivo*. Positioned nucleosomes may act as a boundary (18) and determine the positions of several neighboring nucleosomes on either side. More than one positioned nucleosome separated at less than a nucleosome's length of DNA would serve to keep the region clear of histones (eg, at promoters) for the binding of sequence-specific proteins with various gene regulatory or structural roles. At the inducible yeast *PHO5* promoter, events leading to promoter remodeling and transcription of the *PHO5* gene, which encodes a phosphatase, are initiated in response to low phosphate levels by binding of the **transcription factor** PHO4 to a site between two positioned nucleosomes in a set of four over the promoter (19) (see text below). Positioned arrays of nucleosomes, essential for the functioning of the promoter, are also found in many other cases—for example, the mouse mammary tumor virus (MMTV) major late promoter (20) (see also Ref. 2). It has recently become apparent (21) that nucleosome positioning is the probable explanation for the differential regulation of *Xenopus* oocyte and somatic 5S genes *in vivo* (22), where the oocyte genes are repressed and the somatic genes remain active when H1 accumulates after the mid-blastula transition. Sequence differences between the two types of gene mean that the oocyte nucleosome occludes the TFIIIA binding site and binds H1 preferentially (so the gene is turned off), whereas the converse is broadly true for the somatic gene, which remains active.

A nucleosome core can be reconstituted on to DNA *in vitro* by dialysis of an equimolar mixture of the histone octamer (or H3–H4 tetramers and H2A–H2B dimers) and DNA from high ionic strength (eg, 2 M NaCl) to low ionic strength. Even in cases where there are strong nucleosome positioning signals in the DNA, octamers on different fragments will occupy a population of less favorable minor translational positions in addition to a major position, all with the same rotational setting of the DNA. For example, on the *Xenopus borealis* somatic 5S rRNA gene, which is regarded as a strong positioning sequence, octamers can occupy 10 different positions (six being predominant), all related to each other by the 10 bp helical periodicity of the DNA, which must be caused by a strong rotational signal in the DNA sequence (21). The occupancy of less favorable octamer positions on a DNA sequence is higher at low temperatures (eg, 4°C), when the octamers are trapped, than if the temperature is then shifted to 37° (23). The change in positions populated demonstrates that nucleosomes have an intrinsic mobility, which could be useful, for example, in remodeling of promoters. In the presence of H1, nucleosome mobility is greatly reduced (24,25); conversely, loss of H1 might enable nucleosomes to move, for example, to expose binding sites for gene regulatory proteins.

HIGHER-ORDER STRUCTURE: THE 30-NM FILAMENT

The 10-nm nucleosome filament, which represents the first level of chromatin structure, is the form in which chromatin exists at low ionic strength (eg, >20 mM) *in vitro*. As the ionic strength is increased, the 10-nm filament condenses into a 30-nm (diameter) filament, visible by electron microscopy. Most of the chromatin in the nucleus is in this form. The folding is a consequence of ionic-strength induced screening of the residual negative charge on the linker DNA (26), much of the charge having already been neutralized by the basic tail(s) of histone H1 and, probably, the basic *N*-terminal tails of at least some of the core histones, and results in new interactions (eg, nucleosome-nucleosome) within the higher-order structure.

Various models have been proposed for the 30-nm filament (27), some of which are shown in Figure 1. All agree on a radial distribution of nucleosomes with their faces parallel to the filament axis, but they differ fundamentally in the path taken by the linker DNA, whether bent, curved, or straight. The evidence on this point is conflicting (see **Linker DNA**). The evidence for the various models has been extensively reviewed (28–31). Much evidence supports the original "solenoid" model (32), which proposes a simple helical coiling of the nucleosome filament, with about six nucleosomes per turn at physiological ionic strength, and necessarily bent linker DNA (Fig. 1a). [The details are easily modified to accommodate the asymmetric, rather than symmetric, location of the globular domain of H1 near the dyad (6); see text above.] The solenoid places the dyads of the nucleosomes, histone H1, and the linker DNA between nucleosomes, on the inside of the solenoid; H1 has indeed been shown to be located within the 30-nm filament (33). However, other models have been proposed, such as the helical-ribbon model (34,35) (Fig. 1b), which invokes straight linkers and is based on the zigzag appearance of the nucleosome filament in electron micrographs, and these too would be compatible with an internal location of H1. The coiled linker model (36), like that shown in Fig. 1c, which is related to the solenoid, is not necessarily compatible, because the dyads of the nucleosomes, and hence H1, are alternately inside and outside the solenoid for some linker lengths (eg, in chicken erythrocyte chromatin).

Whatever the model that best describes the spatial relationships between nucleosomes in the higher-order structure, the 30-nm filament in the cell nucleus is unlikely to be completely regular, partly because of local heterogeneity in linker length along the filament within the average value, partly due to the presence of different histone variants or subtypes, and so on. The 30-nm filament is also a dynamic structure that undergoes thermal "breathing", and this property is probably central to gene regulation through control of chromatin folding. The stability of the 30-nm filament is relevant to the ease with which it may be unpackaged for transcription and appears to be increased by special variants of H1 (such as H5 and spH1 in the condensed chromatin of nucleated erythrocytes and sea urchin sperm, respectively) (37), and it is probably decreased in

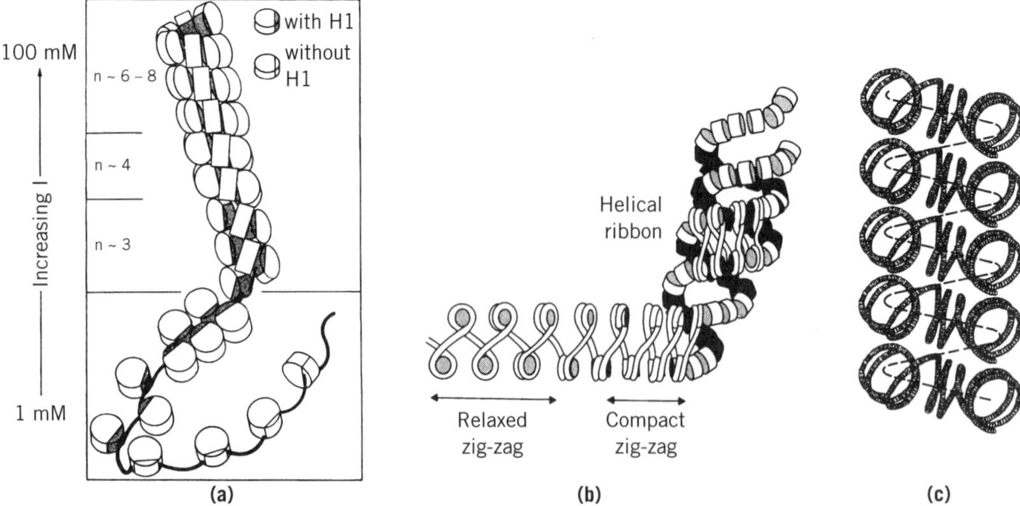

Figure 1. Some models of chromatin higher-order structure. The 30 filament is shown as (**a**) a solenoid (32), (**b**) a helical ribbon (34), and (**c**) as a helical conformation with coiled linkers (36). In (**a**) and (**b**) the most condensed structures are shown at the top, and the extended structures present at low ionic strength at the bottom. (From Refs. 32, 34, and 36, respectively, with permission.)

regions of transcriptionally competent chromatin by (partial) loss of H1 (eg, Ref. 38) and by acetylation of the core histone *N*-terminal tails (see **Histone acetylation**).

CHROMATIN AND TRANSCRIPTION

In eukaryotes, transcription works in a chromatin context (39,40). Some genes are active in all cell types, whereas others are active only in specific cell types. Active genes in higher eukaryotes have a chromatin structure characterized by a greater sensitivity (7- to 10-fold) to the endonuclease DNase I than the bulk of the chromatin in the nucleus (1,2). This state precedes active transcription and marks "transcriptional competence"; it is the first step in control of transcriptional activity. DNase I sensitivity is probably due to (partial) relaxation of the 30-nm filament toward the 10-nm filament state, which is probably facilitated by partial loss of linker histones and acetylation of the core histone *N*-terminal tails (41) (see **Histone acetylation**). DNase I sensitivity associated with a particular gene extends well beyond the coding region and encompasses an entire chromosomal domain (eg, tens of kilobases) and includes all the regulatory elements (eg, **enhancers**) for the genes (2). The boundaries of these chromosomal domains have been characterized in only a few cases, and no common feature is apparent. In at least some instances, the boundaries act as "insulators", preserving the structural and functional autonomy of the domain and also blocking the influence of regulatory elements from the outside (42). Short localized regions, termed DNase I hypersensitive sites (typically a few hundred base pairs), in the genome show a much higher sensitivity to DNase I than described above (43). These are due to the absence of a canonical nucleosome and the presence of gene regulatory (sequence-specific) proteins. They often occur within a domain of general DNase I sensitivity, coinciding with promoters or enhancers, and may also occur at the boundaries of chromosomal domains. Hypersensitive sites at promoters may be constitutive or inducible, depending on the gene, and are often flanked by "positioned nucleosomes".

The presence of a nucleosome at a promoter generally inactivates it (39). Promoters therefore either have to be kept clear of nucleosomes, or one or more nucleosomes has to be perturbed or disrupted to permit assembly of a preinitiation complex and recruitment of RNA polymerase. Promoters may be kept clear either by flanking positioned nucleosomes or by the presence of sequence-specific proteins that bind during replication—for example, the Grf2 protein in yeast, which binds to an extended region of DNA and precludes nucleosome formation (39). The promoters that are packaged in nucleosomes have to be opened up ("remodeled") in order to permit assembly of a preinitiation complex. The nature and extent of the disruption required to open up a promoter is unclear and may be different in different cases, perhaps depending on where the factor-binding sequences are within a nucleosome (eg, whether they can be exposed simply by loss of H1 or by loss of an H2A-H2B dimer, or whether they also require disruption of the central tetramer in the core of the nucleosome; see Ref. 2). It does not appear to result in complete histone loss. Promoter remodeling and disruption *in vivo* appears to require the action of energy-dependent "chromatin remodeling machines" (see text below) which may be recruited to particular promoters in response to the initial signal for activation, and possibly also histone acetyltransferases (see **Histone acetylation**).

In vitro many transcription factors have been shown to be able to can bind to their cognate sites on the surface of a nucleosomes, albeit with lower affinity than to free DNA. The effects can be accounted for by transient "site exposure" as the interactions between the octamer and DNA fluctuate (44); this model predicts that binding of factors to multiple juxtaposed binding sites on the same nucleosome will be cooperative (45), as indeed already observed. Binding occurs without global displacement of the DNA, which will remain anchored at sites not bound by transcription factors. Consistent with this, a recent report shows that the erythroid transcription factor GATA-1 causes extensive disruption of histone–DNA contacts in nucleosome core particles containing GATA-1 sites, except over 50 bp around the dyad that serves to anchor the DNA (46). The complex is stable in solution, and the disruption is reversed on removal of GATA-1. The inherent flexibility of the nucleosome, and the fact that many promoters have clusters of factor-binding sites, suggests that this mechanism of gaining access to binding sites within nucleosomes at promoters could have physiological relevance.

At some inducible promoters, the binding site for the transcription factor that initiates the remodeling events is already exposed, in the linker between two nucleosomes in a short positioned array, rather than within a nucleosome. This is the situation at the *PHO5* promoter in yeast (19). Activation of the promoter in response to low phosphate relies on the accessibility in the uninduced state of a (weak) binding site for a transcription factor (Pho4) in the 70-bp gap between two of the four positioned nucleosomes at the promoter (Fig. 2). Binding of Pho4 leads to disruption of two nucleosomes on either side of the binding site and exposure of a second, stronger Pho4 binding site, as well as binding sites for another transcription factor, Pho2, which binds cooperatively with Pho4. The nature of the disruption is not clear but is reflected in the accessibility to a **restriction enzyme** of a site normally incorporated into a nucleosome and probably involves a "remodeling complex" (see text below). In other cases, a nucleosome at a promoter may be beneficial or even essential for activation of transcription. In the mouse mammary tumor virus (MMTV) **long terminal repeat** (LTR) promoter, six positioned nucleosomes place two **glucocorticoid response elements** in rotational positions (see text above) on one of the nucleosomes, such that they are accessible to glucocorticoid receptor binding in response to hormone; the two sites are also in close proximity on one face of the nucleosome. The binding site for the transcription factor NF1 is occluded until the nucleosomes are disrupted or perturbed in response to receptor binding (20). In a slightly different scenario, wrapping DNA around a nucleosome may bring into proximity two sites outside the nucleosome core, which are separated by about a nucleosome repeat length of DNA. There are several instances of this (2)—for example, the *Drosophila hsp*26 promoter (47), where a critical nucleosome upstream of the start site of transcription performs such a scaffolding role and brings binding sites for **heat shock** transcription factors into proximity with each other and with TFIID, to create a preset promoter, with bound RNA polymerase, waiting only for the binding of heat shock transcription factor to induce it. The positioned nucleosome is flanked by binding sites for the GAGA protein, which appears to play a crucial role in the architecture of the promoter, by helping to determine the position of the nucleosome (2).

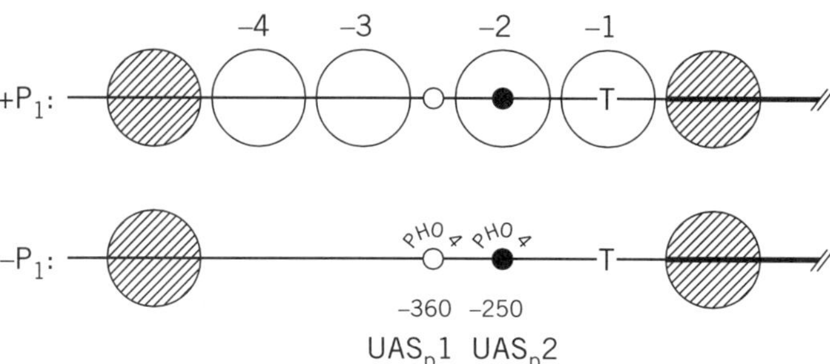

Figure 2. The yeast *PHO5* promoter: chromatin structure in the uninduced (+Pi) and induced (−Pi) states. Large open circles (−1 to −4): the four positioned nucleosomes that are disrupted and become "transparent" upon induction. Small circles: Pho4 binding sites [UASp1 (open) and UASp2 (filled)]. T: TATA box. The thick horizontal line marks the beginning of the coding sequence. From Ref. 75, with permission.

In contrast to initiation, transcriptional elongation can proceed through nucleosomes *in vivo*. In a model system using SP6 polymerase and a single nucleosome, the histone octamer is displaced during elongation (39) and then rebinds upstream of its initial position; the mechanism proposed is transferred from in front of the advancing polymerase to behind it, via a DNA loop that is displaced (48). However, no such studies have yet been carried out with the large eukaryotic RNA polymerases. The indications are that a reconstituted nucleosome array presents a block to elongation by RNA polymerase II, which may be relieved by an energy-dependent mechanism (see next section).

CHROMATIN REMODELING MACHINES

Exciting discoveries in the last couple of years have revealed several multiprotein complexes (molecular masses ranging from 0.5 to 2 MDa) in *Drosophila*, yeast, and human cells that are able to remodel repressive chromatin in an ATP-dependent manner to permit transcription initiation at promoters *in vitro*. These include three complexes derived from a *Drosophila* embryo extract: NURF (*nu*cleosome *r*emodeling *f*actor), CHRAC (*chr*omatin *a*ccessibility *c*omplex) and ACF (*A*TP-dependent *c*hromatin assembly and remodeling *f*actor); the *Saccharomyces cerevisiae* SWI-SNF (*swi*tching-*s*ucrose *n*o*nf*ermenting) and related RSC (*r*emodeling the *s*tructure of *c*hromatin) complexes; and SWI/SNF-related complexes in human cells and *Drosophila*. They have been comprehensively discussed in a number of recent reviews (49–53) that cite the original literature. They all contain an **ATPase** activity; in the case of the three *Drosophila* complexes, this is the same ISWI (*i*mitation *SWI*2) subunit. The various complexes were purified initially using a range of functional assays to follow the purification, and they appear to be functionally and mechanistically distinct; they are likely to differ in their modes of nucleosomal perturbation. NURF (four protein subunits) was purified as a factor required for ATP-dependent remodeling of chromatin templates so that the GAGA protein is able to bind (GAGA binds to GA-rich sites in several *Drosophila* heat-shock promoters, such as the *hsp*26 promoter, already mentioned); and it facilitates transcriptional activation. Interestingly, the 55-kDa subunit of NURF (which contains "WD repeats," which are protein-interaction motifs) is also found in the *Drosophila* chromatin assembly factor dCAF-1 (54), and human homologues are components of histone acetylase and deacetylase complexes (see **Histone acetylation**). CHRAC (five subunits) was purified as a factor that generated ATP-dependent accessibility of sites in a chromatin template, reconstituted *in vitro*, to restriction enzymes. It is also able to generate a regular array from an irregular array of nucleosomes and has been described as an energy-dependent nucleosome spacing factor; it may therefore play a role in chromatin assembly (see text below) (where newly assembled nucleosomes are initially deposited in an irregular array that "matures" to give a regular spacing), although a spacing activity might, of course, also promote transcription factor binding. The third *Drosophila* complex, ACF (four subunits), was also purified as an ATP-dependent nucleosome spacing factor in a fractionated system and plays a role in nucleosome assembly; it can also mediate transcription-factor-mediated chromatin disruption in much the same way as NURF. SWI/SNF (9 to 12 subunits) facilitates transcription factor binding, and both SWI/SNF and RSC (15 subunits) disrupt the repeating 10-nucleotide cleavage pattern given by DNase I on mononucleosomes, showing disruption of the rotational setting of the DNA on the octamer surface. In yeast, the SWI/SNF complex is not abundant and not essential for viability, whereas the RSC complex is abundant and essential. SWI/SNF appears to be necessary for the transcription of only a small subset of genes, and in an *in vitro* system the dependence of transactivation on SWI/SNF activity is much reduced if there are multiple binding sites for the activator on the same nucleosome (55). It is possible that *in vivo* the SWI/SNF complex is needed only for remodeling of weak promoters. Altered forms of mononucleosomes generated by both SWI/SNF (56,57) and RSC (58) complexes have recently been reported. They retain a full complement of histones and have altered physical properties that would be consistent with either a heavily modified mononucleosome or a dinucleosome. The SWI/SNF products have a higher affinity for transcription factors than unmodified nucleosomes. Much remains to be done to (a) elucidate the mechanisms by which the various chromatin remodeling machines act on chromatin, (b) define the roles of the various subunits, and (c) establish which mechanisms, *in vivo*, play a role in promoter disruption for transcription initiation, and whether others are primarily involved in reversing such disruption or in nucleosome assembly, and so on. The relationship between these complexes and the large acetylase, and deacetylase, complexes (see **Histone acetylation**) also remains to be clarified, although the indications are that they cooperate in the remodeling of certain promoters.

The foregoing account has dealt solely with complexes that disrupt chromatin at promoters, allowing transcription initiation complexes to form. The transcriptional machinery also confronts nucleosomes during RNA chain elongation. Experiments on a reconstituted array of nucleosome cores show that the passage of eukaryotic RNA polymerase II is blocked by a nucleosome core and that a heterodimeric protein isolated from human cells will facilitate RNA chain elongation in a chromatin context. The protein has been designated FACT (*fa*cilitates *c*hromatin *t*ranscription) (59) and appears to be a **DNA-binding protein**, one subunit being an HMG-box protein (see **HMG proteins**). Little is yet known about it how it might work, but its action is not ATP-dependent. The *in vitro* assays show a rather modest extension of the transcript (50 to 150 nucleotides) before FACT stalls, and it is not yet clear whether FACT is designed to work only on the promoter-proximal nucleosome or needs cofactors for its normal action, or whether the modest extension is merely due to a technical limitation of the experiment. The chromatin template contains closely packed nucleosome cores (no linker DNA, no H1), and if octamer transfer via a loop of DNA is required, as demonstrated in model systems (48) (see text above), this may not be possible here.

CHROMATIN AND REPRESSION

There are hierarchies of repression, through chromatin structure, as there are of transcriptional activation. The most basic level, promoter occlusion by a nucleosome, was discussed above. Superimposed on this is repression through folding of the nucleosome filament into a 30-nm filament (whether this occurs in yeast is unclear). The higher-order structure is stabilized by a hypoacetylated state of the *N*-terminal tails of the core histones (see **Histone acetylation**). In organisms other than yeast and *Drosophila*, DNA **methylation** is known to contribute to gene repression through chromatin structure, and a **5-methylcytosine**-binding protein (MeCP2) has recently been reported to recruit deacetylases (60,61). This could favor a more stable 30-nm filament, which might be further stabilized by the tighter binding of H1 to methylated DNA (62). The 30-nm filament is not a template for transcription, and in some cases, 30-nm filament stabilization (eg, by specialized linker histone variants) and packing in the nucleus may be at least major players in general repression. This may be the situation in mature nucleated erythrocytes (eg, of birds, fish, and amphibia), where H1 is largely replaced by histone H5 and where there is a global and terminal shutdown of transcription; and in sea-urchin sperm, where highlycondensed chromatin is linked to the presence of sperm-specific H1. Other more complex mechanisms of repression also exist, however, resulting in **heterochromatin**. The essence of these mechanisms—although they differ in many respects in, for example, yeast and *Drosophila,* where they have been extensively studied—is the assembly, over particular chromosomal regions, of repressive multiprotein complexes that interact with chromatin. In the case of *Drosophila* at least, the chromatin is probably in the form of the 30-nm filament, and the mechanisms add an additional layer of repression for genes whose permanent inactivation is crucial and where inappropriate expression cannot be tolerated.

Yeast chromosomes are too small to allow heterochromatin to be seen cytologically, but nonetheless some chromosomal regions show many of the features of heterochromatin in more complex eukaryotes (replication late in S phase, localization at the nuclear envelope, and **position effects** on gene expression that are inherited **epigenetically**). In *Saccharomyces cerevisiae*, such regions occur at the silent mating type loci, *HML* and *HMR*, and at **telomeres** (the ends of the chromosomes; genes placed here are repressed) and have been extensively studied by genetic analysis (63). In each case, a multiprotein complex of Sir2, Sir3, and Sir4 proteins (Sir = *silentinformationregulator* information regulator) interacts with the *N*-terminal regions of histones H3 and H4. At the telomeres, which contain the sequence $[C_{1-3}A]_n$, the **silencers** consist of multiple binding sites for RAP1 (*r*epressor *a*ctivator *p*rotein 1), whereas silencers at *HML* and *HMR* consist of binding sites for ABF1 (*A*RS-*b*inding *f*actor 1), the *o*rigin *r*ecognition *c*omplex (ORC) and RAP1. Residues 16 to 29 of H4 are required for interaction with Sir3 and Sir4 and for silencing, presumably by providing interactions necessary for stabilization of the multiprotein complex. In addition, although the core histones at the silenced loci in yeast are hypoacetylated, there is a specific requirement for Lys12 in acetylated form, possibly for interaction with Sir3 (63). In the current model for telomeric repression, RAP1 binds to the telomeric repeats, initiating polymerization of Sir3/Sir4 over the adjacent chromatin domain, the binding of the Sir proteins being mediated by interaction with the H3 and H4 *N*-termini, possibly stabilized by acetylation of H4 at Lys12. Other repressive mechanisms in yeast also involve the *N*-terminal tails of H3 and H4 (64).

Extensive genetic studies of heterochromatin have also been carried out in *Drosophila*. Genes brought into juxtaposition with heterochromatin (eg, at centromeres) may be silenced in a subset of cells that normally express the gene—an effect known as position effect variegation (PEV) (2). Chromatin structure is altered in variegating genes, as shown by the properties of the *hsp*26 transgene inserted at various chromosomal locations; the heterochromatic *hsp*26 gene has a noninducible promoter and is packaged in an unusually regular nucleosomal array (65). Modifiers of PEV, identified genetically, are either (a) structural proteins that are components of multimeric protein complexes that interact with chromatin or (b) proteins that could play indirect roles in regulating the formation of chromatin—for example, by modifying histones or DNA. One such protein is the histone deacetylase RPD3 (see **Histone acetylation**). However, mutations in this protein that increase acetylation *increase* PEV, rather than decreasing it, as might be expected; this is probably because of the requirement for acetylation at Lys12, as in yeast heterochromatin, against a background of general histone hypoacetylation. The multiprotein complexes that affect PEV, and presumably stabilize chromatin higher-order structure, contain (a) HP1 (heterochromatin protein 1), which binds to chromatin but not to DNA, and (b) the protein termed Su(var)3–7, which interacts with HP1 and chromatin (65). The principles underlying PEV in *Drosophila* seem also to apply to the stable inactivation of **euchromatic** genes needed for maintaining differential expression of developmental regulators (eg, the **homeotic genes** that control the segmental identity of the insect body). Locally silenced heritable chromatin structures are generated by the assembly of multiprotein complexes that contain members of the **Polycomb group** (Pc-G) of proteins, which are homologues of HP1 and contain a "chromodomain" (66). The Pc-G genes are responsible for maintaining long-range repression of homeotic genes over multiple enhancers in the later stages of *Drosophila* development. Like HP1, Pc-G proteins appear not to bind DNA. However, specific DNA sequences that recruit Polycomb group proteins and associated silencing factors [Polycomb response elements; (PREs)] have been identified and appear to be essential for stabilizing determined expression states, which are usually (but not invariably) repressive (see Ref. 66). Various models have been proposed for the mechanism of repression, including propagation of a repressive array of interacting proteins over chromosomal domains (the "spreading model" by analogy with the "spreading" effect of heterochromatin into euchromatin in PEV), an alternative model has been proposed in which clusters of Pc-G proteins interact cooperatively with each other to create tethered loop domains of chromatin which ensure repression (67). Mammalian homologues of HP1-like proteins have been found, as have interacting partners, and it seems likely that the same mechanism of silencing, by assembly into repressive heterochromatin, may be used to regulate some developmentally pivotal mammalian genes whose leaky expression would be disastrous for the cell.

Mammalian **X-chromosome inactivation** is an extreme case of heterochromatin formation over a whole chromosome. The precise molecular basis is not known, but there are some notable features of the chromatin composition: hypoacetylation of histone H4; the presence of an unusual H2A variant, macroH2A (see **Histones**); heavy methylation of the DNA; and the presence of an unusual large untranslated RNA, termed Xist, which is expressed from the inactive X-chromosome, and which is of prime importance in an early event in X-inactivation (68). It would not be too surprising if X-inactivation turned out to have at least some features in common with telomeric and position effect variegation in yeast and *Drosophila*.

REPLICATION-LINKED CHROMATIN ASSEMBLY

During S-phase of the cell cycle in eukaryotes, it is not just the DNA that undergoes replication, but the chromatin as a whole. Chromatin assembly is tightly coupled to **DNA replication** and the passage of the **replication fork**. It involves assembly of nucleosomes on daughter strands, followed by conversion of an initially irregularly spaced array of nucleosomes to a regular array—a process known as maturation. There has been recent progress in understanding the participants in both of these processes. Assembly involves, in equal amounts, both old (parental) histones in preexisting nucleosomes, which are recycled, and newly synthesized histones, which are apparently randomly segregated on to the two nascent duplexes. The temporal order of deposition of newly synthesized histones on to DNA reflects the structural organization of the nucleosome; it begins with the deposition of H3 and H4 to form the central organizing tetrameric core, followed by H2A and H2B as two dimers, and finally addition of H1.

The newly synthesized H4 deposited at the replication fork is diacetylated, at lysines 5 and 12, in all species so far examined (see **Histone acetylation**). Newly synthesized H3 is apparently also modified (acetylated or phosphorylated?), although perhaps not in all cell types. The role, if any, played by acetylation *per se* in the assembly process is not well understood. Acetylation is transient and is erased by deacetylases shortly after deposition of histones on to the DNA; one possibility might be that any tendency for premature formation of internucleosome contacts is avoided by the acetylation (see **Nucleosome**). The deposition of acetylated H3 and H4 involves the participation of chromatin assembly factor 1 (CAF-1), (69), which acts as a core histone **molecular chaperone**. It forms a stable complex with newly synthesized acetylated H3 and H4 and somehow targets them specifically to the replication fork (70); the histone chaperone NAP1 appears to handle H2A and H2B in *Drosophila* (71). Other proteins may also be involved in the assembly process. CAF-1 purified from human cells contains three subunits, indicated by their mass in kDa: p48, p60, and p150; homologues exist in yeast and *Drosophila*. The small subunit (p48) of human CAF-1 is homologous to the regulatory subunit (p46) of the human B-type histone acetyltransferase (responsible for deposition-related acetylation); both subunits bind to histone H4 and, significantly, at a site that is accessible only in the free histone and not in chromatin (as shown by the structure of the nucleosome core particle; see **Nucleosome**). Strikingly, similar subunits also occur in chromatin remodeling machines (eg, the p55 subunit of *Drosophila* NURF) as noted, as well as in histone deacetylases (see **Histone acetylation**), leading to the suggestion that these subunits target the various complexes to their histone substrates in a manner that is regulated by nucleosomal DNA (72). They are members of the highly conserved family of proteins that contain "WD repeats," multiple distinctive sequence motifs that appear to be involved in **protein–protein interactions**. Because the site of interaction of CAF-1 with H4 is accessible only in the free histone, CAF-1 would have to be displaced before nucleosome assembly could occur; the interactions appear to be disrupted by ACF (*A*TP-requiring *c*hromatin assembly and remodeling *f*actor; see text above). ACF was initially identified as a factor capable of regularly spacing an irregularly spaced nucleosome array (51) and would be well-suited to this role during the maturation step of chromatin assembly. It is also potentially able to participate in the remodeling of nucleosomes that accompanies transcriptional activation (see text above: chromatin remodeling machines).

CHROMATIN RECONSTITUTION

Reconstitution *in vitro* of chromatin from its component parts, like any complex biochemical structure, would allow questions to be asked about structure–function relationships. Reconstitution of the nucleosome core is straightforward. It is achieved by gradual dialysis of a 1:1 molar ratio of histone octamer and 146 bp DNA from 2 M NaCl to low salt (eg, 10 mM NaCl). Nucleosome core particles (see **Nucleosome**) for X-ray crystallography were made in this way. The histones can be presented either as the intact octamer or as a 1:1 molar ratio of H2A–H2B dimers and $(H3)_2(H4)_2$ tetramers.

Long "chromatin" reconstituted *in vitro* by mixing histones at high ionic strength (eg, 2M NaCl), and then dialyzing to low ionic strength, shows close packing of octamers, irrespective of the presence of H1, which has no effect because it is the last histone to bind in this procedure. In attempts to determine what cellular factors determine nucleosome spacing (nucleosome repeat length), several cell-free extracts, from *Xenopus* eggs and oocytes and from *Drosophila* embryos, have been developed that will assemble plasmid DNA, irrespective of sequence, into "physiologically" spaced chromatin in an ATP-dependent fashion, independent of replication. Nucleosome spacing (and hence linker length) is increased by H1. The earliest extracts were from *Xenopus* eggs, which contain a large maternal histone pool, and led to the discovery of two acidic histone chaperone proteins, nucleoplasmin and N1/N2, which facilitated chromatin assembly in this system. Work on the *Xenopus* system is summarized in Ref. 2 (Section 3.4.2). The true role of these proteins may be storage of the large histone pool in the egg. Whether similar proteins exist in somatic cells is unclear, although nucleoplasmin-like proteins have been reported. The *Drosophila* embryo extracts are the same ones that led to the discovery of CHRAC (53) and ACF (51) (see text above: Chromatin remodeling) which themselves have nucleosome spacing activity. Spacing may also have an electrostatic component, involving neutralization of charges on linker DNA (73), and a connection between nucleosome spacing and formation of higher-order structure, which is also ionic strength dependent, has been suggested.

A well-defined *in vitro* system starting with pure components, which has recently been further explored (74), will also assemble "properly spaced" chromatin, in an H1-dependent, ATP-independent manner. Irregularly spaced octamers are first deposited on the DNA by dialysis from high to low NaCl concentration, and then a more regularly spaced array is generated by incubation at "physiological ionic strength" (150 mM NaCl) with H1(H5) in the presence of polyglutamate (which probably binds histones reversibly and allows equilibrium to be reached in the histone positions on the DNA). Interestingly, the system is sensitive to the DNA sequence, and regular nucleosome arrays were generated on only about half of the ~2-kbp cloned chicken genomic sequences tested; the sequences tested (by **Southern blotting**) behaved in essentially the same way in chicken liver nuclei. It is argued (74) that genomic chromatin is a mosaic of less well-ordered regions and regular regions that are about 10 nucleosomes long and contain a nucleosome positioning sequence (or sequences), which effectively acts as a boundary (18) against which linker-histone-dependent nucleosome alignment can occur. It is possible that egg, oocyte, and embryo extracts, which appear to be indifferent to DNA sequence as well as ATP-dependent, are designed to assemble chromatin through rapid cycles of cell division, and not to be responsive, for example, to genomic signals that may be necessary to specify arrays of nucleosomes compatible with gene expression and regulation in somatic cells.

BIBLIOGRAPHY

1. K. van Holde (1988) *Chromatin*, Springer-Verlag, New York.
2. A. P. Wolffe (1995) *Chromatin; Structure and Function*, 2nd ed., Academic Press, New York.
3. S. C. R. Elgin (ed.) (1995) *Chromatin Structure and Gene Expression* [*Frontiers in Molecular Biology* series (B. D. Hames and D. M. Glover, eds.)], Oxford University Press, Oxford, U.K.
4. R. D. Kornberg (1977) *Annu. Rev. Biochem.* **46**, 931–954.
5. K. Luger, A. W. Mader, R. K. Richmond, D. F. Sargent, and T. J. Richmond (1997) *Nature* **389**, 251–260.
6. Y.-B. Zhou et al. (1999) *Nature* (in press).
7. A. A. Travers (1987) *Trends Biochem. Sci.* **12**, 108–112.
8. T. E. Shrader and D. M. Crothers (1989) *Proc. Natl Acad. Sci. USA* **86**, 7418–7422.
9. H. R. Widlund et al. (1997) *J. Mol. Biol.* **267**, 807–817.
10. P. T. Lowary and J. Widom (1998) *J. Mol. Biol.* **276**, 19–42.
11. H. Cao, H. R. Widlund, T. Simonsson, and M. Kubista (1998) *J. Mol. Biol.* **281**, 253–260.
12. J. Widom (1996) *J. Mol. Biol.* **259**, 579–588.
13. S. Muyldermans and A. A. Travers (1994) *J. Mol. Biol.* **235**, 855–870.
14. A. Travers and H. Drew (1999) *Biopolymers* (in press).
15. J. J. Hayes, D. J. Clark, and A. P. Wolffe (1991) *Proc. Natl. Acad. Sci. USA* **88**, 6829–6833.
16. C. Spangenberg et al. (1988) *J. Mol. Biol.* **278**, 725–739.
17. S. Tanaka, M. Zatchej, and F. Thoma (1992) *EMBO J.* **11**, 1187.
18. R. D. Kornberg and L. Stryer (1988) *Nucleic Acids Res.* **16**, 6677–6690.
19. J. Svaren and W. Horz (1997) *Trends Biochem. Sci.* **22**, 93–97.
20. T. Archer, M. G. Cordingley, R. G. Wolford, and G. L. Hager (1991) *Mol. Cell. Biol.* **11**, 688–698.
21. G. Panetta et al. (1999) *J. Mol. Biol.* (in press).
22. P. Bouvet, S. Dimitrov, and A. P. Wolffe (1994) *Genes Dev.* **8**, 1147–1159.
23. G. Meersseman, S. Pennings, and E. M. Bradbury (1992) *EMBO J.* **11**, 2951–2959.

24. S. Pennings, G. Meersseman, and E. M. Bradbury (1994) *Proc. Natl. Acad. Sci USA* **91**, 10275–10279.
25. K. Ura, J. J. Hayes, and A. P. Wolffe (1995) *EMBO J.* **14**, 3752–3765.
26. D. J. Clark and T. Kimura (1990) *J. Mol. Biol.* **211**, 883–896.
27. G. Felsenfeld and J. D. McGhee (1986) Structure of the 30 nm fibre. *Cell* **44**, 375–377.
28. J. Widom (1989) *Annu. Rev. Biophys. Biophys. Chem.* **18**, 365–395.
29. K. van Holde and J. Zlatanova (1996) *Prog. Nucleic Acid Res. Mol. Biol.* **52**, 217–259.
30. V. Ramakrishnan (1997) *Crit. Rev. Eukaryot. Gene Expr.* **7**, 215–230.
31. J. Widom (1998) *Annu. Rev. Biophys. Biomol. Struct.* **27**, 285–327.
32. F. Thoma, T. Koller, and A. Klug (1979) *J. Cell Biol.* **83**, 403–427.
33. V. Graziano, S. E. Gerchman, D. K. Schneider, and V. Ramakrishnan (1994) *Nature* **368**, 351–354.
34. C. L. F. Woodcock, L.-L. Frado, and J. B. Rattner (1984) *J. Cell Biol.* **99**, 42–52.
35. C. L. Woodcock and R. Horowitz (1995) *Trends Cell Biol.* **5**, 272–277.
36. J. D. McGhee, J. M. Nickol, G. Felsenfeld, and D. C. Rau (1983) *Cell* **33**, 831–841.
37. J. O. Thomas, C. Rees, and P. J. G. Butler (1986). *Eur. J. Biochem.* **154**, 343–348.
38. R. T. Kamakaka and J. O. Thomas (1990). *EMBO J.* **9**, 3997–4006.
39. R. D. Kornberg and Y. Lorch (1992) *Annu Rev. Cell Biol.* **8**, 563–589.
40. G. Felsenfeld (1992) *Nature* **335**, 219–224.
41. T. R. Hebbes, A. W. Thorne, and C. Crane-Robinson (1988) *EMBO J.* **7**, 1395–1402.
42. P. Geyer (1997) *Curr. Opin. Genet. Dev.* **7**, 242–248.
43. S. Elgin (1988) *J. Biol. Chem.* **263**, 19259–19262.
44. K. J. Polach and J. Widom (1995) *J. Mol. Biol.* **254**, 130–149.
45. K. J. Polach and J. Widom (1996) *J. Mol. Biol.* **258**, 800–812.
46. J. Boyes et al. (1998) *J. Mol. Biol.* **279**, 529–544.
47. G. H. Thomas and S. C. R. Elgin (1988) *EMBO J.* **7**, 2191–2201.
48. V. M. Studitsky, D. J. Clark, and G. Felsenfeld (1994) *Cell* **76**, 371–382.
49. T. Tsukiyama and C. Wu (1997) *Curr. Opin. Genet. Dev.* **7**, 182–191.
50. M. J. Pazin and J. T. Kadonaga (1997) *Cell* **88**, 737–740.
51. T. Ito, J. K. Tyler, and J. T. Kadonaga (1997) *Genes to Cells* **2**, 593–600.
52. B. Cairns (1998) *Trends Biochem. Sci.* **23**, 20–25.
53. P. D. Varga-Weisz and P. B. Becker (1998) *Curr. Opin. Cell Biol.* **10**, 346–353.
54. M. A. Martinez-Balbas, T. Tsukiyama, D. G. Dula, and C. Wu (1998) *Proc. Natl. Acad. Sci. USA* **95**, 132–137.
55. R. T. Utley, J. Cote, T. Owen-Hughes, and J. L. Workman (1997) *J Biol. Chem.* **272**, 12642–12649.
56. J. Coté, C. L. Peterson, and J. L. Workman (1998) *Proc. Natl. Acad. Sci USA* **95**, 4947–4952.
57. G. Schnitzler, S. Sif, and R. E. Kingston (1998) *Cell* **94** 17–27.
58. Y. Lorch, B. R. Cairns, M. Zhang, and R. D. Kornberg (1998) *Cell* **94**, 29–34.
59. G. Orphanides et al. (1998) *Cell* **92**, 105–116.
60. X. Nan et al. (1998) *Nature* **393**, 386–389.
61. P. L. Jones et al. (1998) *Nat. Genet.* **19**, 187–191.
62. M. A. McArthur and J. O. Thomas (1996) *EMBO J.* **15**, 1705–1714.
63. M. Grunstein (1998) *Cell* **93**, 325–328.
64. M. Grunstein (1997) *Nature* **389**, 349–352.
65. L. L. Wallrath (1998) *Curr. Opin. Genet. Dev.* **8**, 147–153.
66. G. Cavalli and R. Paro (1998) *Curr. Opin. Cell Biol.* **10**, 354–360.
67. V. Pirrotta (1998) *Cell* **93**, 333–336.
68. A. M. Keohane, J. S. Lavender, L. P. O'Neill, and B. M. Turner (1998) *Dev. Genet.* **22**, 65–73.
69. S. Smith and B. Stillman (1989) *Cell* **58** 15–25.
70. A. Verreault, P. D. Kaufman, R. Kobayashim, and B. Stillman (1996) *Cell* **87**, 95–104.
71. T. Ito, M. Bulger, R. Kobayashi, and J. T. Kadonaga (1996) *Mol. Cell. Biol.* **16**, 3112–3124.
72. A. Verreault, P. D. Kaufman, R. Kobayashi, and B. Stillman (1997) *Curr. Biol.* **8**, 96–108.
73. T. A. Blank and P. B. Becker (1995) *J. Mol. Biol.* **252**, 305–313.
74. K. Liu and A. Stein (1997) *J. Mol. Biol.* **270**, 559–573.
75. P. D. Gregory and W. Horz (1998) *Eur. J. Biochem.* **251**, 9–18.

Suggestions for Further Reading

S. C. R. Elgin (1996) Heterochromatin and gene regulation in *Drosophila*. *Curr. Opin. Genet. Dev.* **6**, 193–202.

G. Felsenfeld (1996) Chromatin unfolds. *Cell* **86**, 13–19.

E. P. Geiduschek (1998) Chromatin transcription: clearing the gridlock. *Curr. Biol.* **8**, R373–R375.

P. D. Gregory and W. Horz (1998) Chromatin and transcription. *Eur. J. Biochem.* **251**, 9–18.

P. Kaufman (1996) Nucleosome assembly: the CAF and the HAT. *Curr. Opin. Cell Biol.* **8**, 369–373.

R. E. Kingston, C. A. Bunker, and A. N. Imbalzano (1996) Repression and activation by multiprotein complexes that alter chromatin structure. *Genes Dev.* **10**, 905–920.

K. Struhl (1998) Histone acetylation and transcriptional regulatory mechanisms. *Genes Dev.* **12**, 599–606.

CHROMATOFOCUSING

XIANLIN HAN

Biological **macromolecules** differ in their compositions of **buffering** groups, such as **amino acids** and **purine** or **pyrimidine** bases, thus giving each biomolecule its own unique charge properties. Generally, at low or acidic pH values, most biomolecules have a net positive charge, whereas at a higher or basic pH they carry a net negative charge. Of course, at a certain pH, the isoelectric pH or **isoelectric point** (p*I*), these biomolecules do not possess a net electric charge. This unique isoelectric pH provides the basis for the selectivity of chromatofocusing in **chromatographically** separating biomolecules.

Chromatofocusing is a special kind of technique classified within the **ion-exchange chromatography** of biomolecules. A weak ion-exchange matrix and a pH gradient are used in chromatofocusing, instead of a strong ion exchanger and a salt gradient as in normal ion-exchange chromatography. This technique was first described theoretically and experimentally demonstrated by Sluyterman and co-workers (1,2). They proposed that a pH gradient could be produced in an ion-exchange column packed with an appropriate ion-exchange resin with good buffering capacity. Although a pH gradient in a column can be produced in a manner similar to that of a salt gradient by using two different pH buffers in a mixing chamber of a gradient maker, a pH gradient can be created internally in the column by taking advantage of the buffering capacity of the weak ion-exchange resin. In practice, a certain pH buffer is used to equilibrate a column packed with a weak ion exchanger. Then another buffer with a different pH is passed through the column, which generates a pH gradient in the column. If such a pH gradient is used to elute biomolecules bound to the ion-exchange resin, the biomolecules elute in order of their isoelectric pH.

The mechanism of chromatofocusing is based on the buffering action of the charged groups on the ion-exchange resin and on the fact that a biomolecule has a net negative charge at a pH above its isoelectric point. In a descending pH gradient, a single molecular species exists in three charged states—negative, neutral, and positive. When a positively charged column (ie, packed with an anion exchange resin) is equilibrated with a starting buffer of high pH, biomolecules that become negatively charged are initially retained on the column. When an elution buffer of lower pH is passed through the column, a pH gradient develops and the individual molecules continuously change their charged states. The molecules at the rear of the sample zone are the first to be titrated by the low-pH buffer. These molecules become positively charged when the pH is less than their p*I*, so that they are repelled from the column matrix and are carried rapidly to the front of the sample zone, because of the high velocity of the moving buffer. In traveling to the front of the sample zone, the molecules encounter an increase in pH and are titrated from their positive form to neutrality and back to their negative form. Once the molecules become negatively charged again, they readsorb on the matrix and again fall back to the rear of the sample zone. The cycle between the front and rear of the sample zone results in "focusing" (ie, a continuous narrowing of this zone) until the molecules elute from the column. At this point, the pH of the column effluent is approximately the same as p*I* of the components eluting.

Chromatofocusing is an analytical or preparative technique for separating biomolecules according to their p*I*. The details of this technique relating to column packing, sample preparation, and sample application used in this technique are similar to those for **affinity chromatography**. A detailed operational protocol is beyond the scope of this article, so the interested reader is directed to other reviews (3,4).

There are a number of advantages in using chromatofocusing. In this technique, a biomolecule is not subjected to a pH greater than its p*I*, and the resulting focusing effects concentrate the sample into a sharp, highly resolved band. One of the great benefits of chromatofocusing is its ease of operation. No gradient-forming devices or mixers are required. The pH gradient is formed with a single isocratic eluent. Chromatofocusing is used widely in research as the method of choice for resolving **isozymes** and molecular species with very similar charge characteristics, such as **transferrin**, **ferritin**, and **hemoglobins** [see refs. in (3–5)].

BIBLIOGRAPHY

1. L. A. A. Sluyterman and O. Elgersma (1978) *J. Chromatogr.* **150**, 17–30.
2. L. A. A. Sluyterman and J. Wijdenes (1978) *J. Chromatogr.* **150**, 31–44.
3. L. Giri (1990) in *Guide to Protein Purification* (M.P. Deutscher, ed.), *Methods in Enzymology* 182, Academic Press, New York, pp. 380–392.
4. T. W. Hutchens (1989) in *Protein Purification: Principles, High Resolution Methods, and Applications* (J.-C. Janson, and L. Rydén, eds.), VCH, New York, pp. 149–174.
5. Pharmacia (1985) *FPLC Ion Exchange and Chromatofocusing-Principles and Methods*, Offsetcenter AB, Uppsala, Sweden.

CHROMATOGRAPHY

XIANLIN HAN

Chromatography constitutes a family of closely related methods for separating and analyzing a wide variety of chemicals. Today, despite developments in analytical chemistry that link scientists to many modern and extremely sophisticated devices, the classic methodology of chromatography still plays a very important role among analytical techniques. It is almost impossible to imagine a laboratory without chromatographic equipment. Quality control, product purification, and basic research are some of the fields in which chromatography is used.

DEFINITION

The term chromatography, first used by Tsvet (1872–1919), a Russian botanist, derived from the two Greek words *Khromatos* (color) and *graphos* (written). He used the term chromatography to describe his studies on pigment separation using a chalk column (1,2), thus defining chromatography as a method by which the components of a mixture were separated on an adsorbent column in a flowing system (1,2). The International Union of Pure and Applied Chemistry (IUPAC) has further defined chromatography as

> *A method, used primarily for separation of the components of a sample, in which the components are distributed between two phases, one of which is stationary while the other moves. The stationary phase may be a solid, or a liquid supported on a solid, or a gel. The stationary phase may be packed in a column, spread as a layer, or distributed as a film, etc.; in these definitions chromatographic bed is used as a general term to denote any of the different forms in which the stationary phase may be used. The mobile phase may be gaseous or liquid (3).*

HISTORICAL PERSPECTIVE

Although some phenomena that form the basis of chromatographic methods have been known for a long time, Tsvet is generally referred to as the father of chromatography. In 1906, he described the separation of plant pigments by column liquid chromatography using over 100 adsorption media (1,4,5). During the next forty years after Tsvet's work, there were some important developments in the field. For example, Kuhn et al. published two papers in 1931 on separating carotenoids on a calcium carbonate column (6,7). Kuhn, Karrer, and Ruzicka applied the chromatographic technique to their own fields of interest and were awarded the Nobel prize (1937, 1938, and 1939, respectively) for their contributions to chromatography.

Tiselius (8) and Claesson (9) developed the now classical procedures involving the continuous observation of optical properties of solutions flowing out of chromatographic columns. Tiselius was awarded the Nobel prize in 1948 for his research on "Electrophoresis and Adsorption Analysis". The introduction of gradient elution in 1952 (10) was an important contribution to all column chromatographic methods.

Chromatograpic developments greatly accelerated after the famous paper by Martin and Synge appeared in 1941 (11). They presented the invention of liquid-liquid (or partition) chromatography in columns and in planar form (paper chromatography). They also provided a theoretical framework for the basic chromatographic process. Martin and Synge were awarded the Nobel prize in 1952 for this work. When thin layers of supported silica gel were introduced as an alternative for paper in the late 1950s (12,13), the field of **thin-layer chromatography** (TLC) was born and became so popular that it has largely replaced the older technique. Another main development in the progress of chromatography was the introduction of **gas-liquid chromatography** (GLC) by James and Martin (14), which had an unprecedented impact on the analytical chemistry of organic compounds. Porath and Flodin introduced **size-exclusion chromatography** in 1959 (15), allowing easy separation of macromolecules. Modern liquid chromatography (**HPLC**) was introduced in the early 1970s, permitting the efficient separation of a wide range of components.

The theory of chromatography was first studied by Wilson (16), who discussed the quantitative aspects in terms of **diffusion**, adsorption rate, and isothermal nonlinearity. The plate theory was first presented by Martin and Synge (11) and was further explored by Craig (17) and Gluechauf (18). In this theory, chromatography is described in terms closely analogous in its mode of operation to distillation and extraction fractionating columns. Lapidus and Admunson (19), followed by van Deemter and co-workers (20), developed the rate theory, an alternative to the plate theory. In this theory, column efficiency was described as a function of the mobile-phase's flow rate and diffusion properties and the stationary-phase particle size. In 1959, Giddings published another paper on this topic (21), and the rate theory has since become the backbone of chromatographic theory. In 1963, Giddings (22) pointed out that, if the efficiencies of gas chromatography were to be achieved in liquid chromatography, particle sizes of 2 to 20 μm were required. This prediction was found to be correct with the development of HPLC systems. Numerous detailed descriptions of chromatographic theory exist in the literature. Notable examples include a monograph edited by Jönsson (23) and the excellent reviews of Snyder (24) and Horváth and Melander (25).

This is only a small glimpse of the historical development of chromatography. The book *75 Years of Chromatography—A Historical Dialogue* (26), which describes many of the individuals who deserve credit for developing this technique, can be consulted for more complete accounts.

RETENTION MECHANISM CLASSIFICATION

There are three common ways to classify chromatographic methods. The first and most popular classification is based on the mechanisms of retention, the manner in which the analyte interacts with the stationary phase. In this classification, chromatography may be divided into the five following basic types:

Adsorption Chromatography

This technique is based on competition for neutral analytes between the mobile phase (gas or liquid) and a neutral solid adsorbent. Therefore, analytes with **polar groups** are retained longer by a polar adsorbent, and **nonpolar** analytes interact better with a nonpolar stationary phase. In this type of chromatography, the analytes are simultaneously in contact with both the stationary phase and the mobile phase.

Partition Chromatography

This technique is based on competition for neutral analytes between the mobile phase (gas or liquid) and a neutral liquid or liquid-like stationary phase (the latter is usually called a "bonded-phase" when long alkyl chains or their derivatives are bonded to a matrix and behave like a liquid). In partition chromatography, the analyte is transferred from the bulk of one phase into the bulk of the other, so that the analyte molecules are surrounded only by molecules of one phase. Separation in partition chromatography is achieved by differences in the **partition coefficients** of the analytes between the mobile and stationary phases.

Ion-Exchange Chromatography

This technique is based on the **electrostatic interaction** between a charged solute and an oppositely charged solid stationary phase. Separation in **ion-exchange chromatography** is achieved by the differing affinities of ions in solution for oppositely charged ionic groups in the stationary phase. Ion-exchange chromatography is applicable to any solute that acquires a charge in solution. Thus, even carbohydrates, which are largely uncharged below pH 12, are separated by this type of chromatography at sufficiently high pH.

Size-Exclusion Chromatography

This technique is based on the sieving principle and is variously known as gel chromatography, **gel filtration** and gel-permeation chromatography (see **Size-exclusion chromatography**). In this technique, the stationary-phase particles have a wide range of pore sizes, causing the stationary phase to behave like a molecular sieve. Small molecules permeate the pores, and large bulky molecules are excluded. Thus, the solutes are separated on the basis of molecular weight and size, and the larger ones elute first.

Affinity Chromatography

This technique is based on the unique biological specificity of **ligand-binding** interactions (ie, the lock-and-key mechanism) (see **Affinity chromatography**). The ligand is covalently bonded to the matrix that forms packing material for the column. Separation is achieved after the applied **macromolecule** becomes specifically, but not irreversibly, bound to the ligand. The macromolecule is eluted by altering the composition or pH of the eluent so as to weaken its interaction with the ligand, thus promoting dissociation and facilitating elution of the retained compounds.

Many other types of chromatography, such as **hydrophobic-interaction**, **chiral**, **ion-pair**, and **salting-out** chromatography, are also frequently used. Furthermore, in practical chromatography intermediate or mixed types are often used. So, although one dominant mechanism is presented, the chromatographic modes are not mutually exclusive. To know more about the relationships of different chromatographic modes, see a schematic diagram by Saunders (27).

DEVELOPMENT PROCEDURE CLASSIFICATION

The second classification is based on the development procedure, the mechanism by which the sample is removed from the column, and therefore depends on the nature of the mobile phase. This classification was introduced by Tiselius (8) in 1940. There are three chromatographic modes in the classification: (1) elution development, (2) displacement development, and (3) frontal analysis. The principles of each mode are illustrated by Braithwaite and Smith (28). In practice, only elution development and to a lesser extent displacement development are commonly used.

FRACTIONATION PHASE CLASSIFICATION

The third classification is based on the phases between which the fractionation process takes place. In chromatography, one phase is held immobile or stationary, and the other one (the mobile phase) is passed over it. Therefore chromatography is mainly divided into two large groups named according to the state of aggregation of the mobile phase, liquid chromatography and gas chromatography. Further groupings can be made by naming both the mobile and stationary phases, thus liquid-liquid, liquid-solid, gas-liquid, and gas-solid chromatography have been named. More recently, supercritical fluids have been used as mobile phases, and these techniques have been named supercritical fluid chromatography, irrespective of the state of the stationary phase.

BIBLIOGRAPHY

1. M. S. Tsvet (1906) *Ber. Deut. Botan. Ges.* **24**, 316.
2. M. S. Tsvet (1906) *Ber. Deut. Botan. Ges.* **24**, 384.
3. *Recommendations on Nomenclature for chromatography, Rules Approved 1973*, IUPAC Analytical Chemistry Division Commission on Analytical Nomenclature (1974) *Pure Appl. Chem.* **37**, 447.
4. M. S. Tsvet (1910) *Khromofilly v Rastitel'nom i Zhivotnom Mire* Tipogr. Varshavskago Uchebnago Okruga, Warsaw.
5. H. H. Strain and J. Sherma (1967) *J. Chem. Educ.* **44**, 238–242.
6. R. Kuhn and E. Lederer (1931) *Ber.* **64**, 1349.
7. R. Kuhn, A. Winterstein, and E. Lederer (1931) *Hoppe Seyler's Z. Physiol. Chem.* **197**, 141–160.
8. A. Tiselius (1940) *Ark. Kemi. Mineral. Geol.* **14B**, 22.
9. S. Claesson (1946) *Ark. Kemi. Mineral. Geol.* **23A**, 1.
10. R. S. Alm, R. J. P. Williams, and A. Tiselius (1952) *Acta Chem. Scand.* **6**, 826–836.
11. A. J. P. Martin and R. L. M. Synge (1941) *Biochem. J.* **35**, 1358–1368.
12. E. Stahl et al. (1956) *Pharmazie* **11**, 633.
13. E. Stahl (ed.) (1962) *Thin Layer Chromatography*, Academic Press, New York.
14. A. T. James and A. J. P. Martin (1952) *Biochem. J.* **50**, 679–690.
15. J. Porath and P. Flodin (1959) *Nature* **183**, 1657–1659.
16. J. N. Wilson (1940) *J. Am. Chem. Soc.* **62**, 1583–1591.
17. L. C. Craig (1950) *Anal. Chem.* **22**, 1346–1352.
18. E. Glueckauf (1955) *Trans. Faraday Soc.* **51**, 34–44.
19. L. Lapidus and N. R. Amundson (1952) *J. Phys. Chem.* **56**, 984–988.
20. J. J. van Deemter, F. J. Zuiderweg, and A. Klinkenberg (1956) *Chem. Eng. Sci.* **5**, 271.
21. J. C. Giddings (1959) *J. Chem. Phys.* **31**, 1462—1467.
22. J. C. Giddings (1963) *Anal. Chem.* **35**, 2215–2216.
23. J. Å. Jönsson (ed.) (1987) *Chromatographic Theory and Basic Principles (Chromatographic Science Series 38)*, Marcel Dekker, New York.
24. L. R. Snyder (1992) in *Chromatography* (E. Heftmann, ed) (*J. Chromatogr. Library*, Vol. 51A), Elsevier, Amsterdam, pp. A1–A68.
25. C. Horváth and W. R. Melander (1983) in *Chromatography* (E. Heftmann, ed.) (*J. Chromatogr. Library*, Vol. 22A), Elsevier, Amsterdam, pp. A27–A135.
26. L. S. Ettre and A. Zlatkis (1979) *75 Years of Chromatography-A Historical Dialogue* (L.S. Ettre and A. Zlatkis, eds.) (*J. Chromatogr. Library*, Vol. 17), Elsevier, Amsterdam.
27. D. L. Saunders (1975) in *Chromatography*, 3rd ed. (E. Heftmann, ed), Van Nostrand Reinhold, New York, p. 81.
28. A. Braithwaite and F. J. Smith (1985) *Chromatographic Methods*, 4th ed., Chapman and Hall, New York, p. 8.

CHROMOCENTER

ALAN P. WOLFFE

The salivary gland nuclei of *Drosophila melanogaster* contain **polytene chromosomes**. These specialized **chromosomes** essentially contain duplicated repeats of the **DNA** in the entire chromosome. Polytene chromosomes are fused at their **centromeres** to form the chromocenter. This specialized chromosomal domain in *Drosophila* consists of constitutive **heterochromatin**. The DNA sequences within the chromocenter consist of two types of repeats. One type that is assembled into α-heterochromatin is composed of highly repeated simple DNA sequences, whereas the second, β-type is more complex. Structural components of heterochromatin at the chromocenter have been identified (1). The best-characterized protein that accumulates selectively at the chromocenter is known as heterochromatin protein 1 (HP1). **Antibodies** to HP1 colocalize with the type of DNA repeat found in β-heterochromatin called the Dr.D element (2). This is at the edges of the chromocenter in what is termed pericentric heterochromatin.

HP1 contains a protein motif that is found in other chromatin binding proteins and is known as the chromodomain (for chromatin organization modifier). Recent studies have established that the chromodomain family of proteins comprises more than 40 members (3) that can be subdivided into two major groups. Proteins, such as HP1, contain both an amino-terminal chromodomain and a carboxy-terminal "shadow" chromodomain (4). The amino-terminal chromodomain directly binds to heterochromatin, whereas the carboxy-terminal "shadow" chromodomain determines nuclear localization and assists in binding to chromatin (5). The second group of proteins relies on interactions with other proteins to target association with particular chromatin domains (6). The structure of the chromodomain was recently determined using **NMR** (7). The chromodomain has strong homology to two archaebacterial DNA-binding proteins. However, the eukaryotic chromodomain does not interact with DNA and is involved in **protein-protein interactions**. Each chromodomain consists of an amino-terminal, three-stranded, antiparallel **beta sheet** that folds against a carboxy-terminal **alpha-helix**. The presence of both a chromodomain and a shadow chromodomain are thought to allow proteins, such as HP1, to function as adapters in assembling large multicomponent proteins.

The chromocenter is a useful chromosomal domain for identifying the structural components of heterochromatin and potentially for actually understanding how heterochromatin is organized at a molecular level. The formation of the chromocenter indicates how similar **nucleoprotein** complexes that share common structural components can self-associate. It provides a nice example of a specialized nuclear compartment that is assembled so that depends on protein–nucleic acid interactions.

BIBLIOGRAPHY

1. T. C. James and S. C. R. Elgin (1986) *Mol. Cell Biol.* **6**, 3862–3872.
2. G. L. G. Miklos, M.-T. Yamamota, J. Davies, and V. Pirotta (1986) *Proc. Natl. Acad. Sci. USA* **85**, 2051–2055.
3. E. V. Koonin, S. B. Zhou, and J. C. Lucchesi (1995) *Nucl. Acids Res.* **23**, 4229–4233.
4. R. Aasland and A. F. Stewart (1995) *Nucl. Acids Res.* **23**, 3168–3173.
5. J. S. Platero, T. Hartnett, and J. C. Eissenberg (1995) *EMBO J.* **14**, 3977–3986.
6. A. Lorentz, K. Ostermann, O. Fleck, and H. Schmidt (1994) *Gene* **143**, 1–8.
7. L. J. Ball et al. (1997) *EMBO J.* **16**, 2473–2481.

CHROMOGENIC SUBSTRATE

JOHN F. MORRISON

A chromophore is any light-absorbing group (see **Absorption spectroscopy**), and a *chromogenic* substrate is one that is acted on by an **enzyme** so as to increase or decrease the absorption of light at a particular wavelength as substrate is converted to product. An example of a naturally occurring chromogenic substrate is **NADH**, which absorbs light strongly at 340 nm. The absorbance decreases as NADH is oxidized by a pyridine nucleotide-dependent dehydrogenase, because NAD does not absorb substantially at 340 nm. Artificial chromogenic substrates have been used extensively for kinetic studies on a range of enzymes. *p*-Nitrophenylphosphate is hydrolyzed by **phosphatases** to release *p*-nitrophenol, which has a yellow color in alkaline solution. The increase in absorbance with hydrolysis is measured at 400 nm. *p*-Nitrophenylesters are used in a similar manner to measure the activity of *esterases*. Phenazine methosulfate is used as a convenient means of determining the activity of flavoprotein enzymes, as the oxidized form of the electron acceptor is yellow and the reduced form is colorless. An extensive list of artificial, chromogenic substrates has been recorded (1).

BIBLIOGRAPHY

1. R. M. C. Dawson, D. C. Elliott, W. H. Elliott, and K. M. Jones (1986). *Data for Biochemical Research*, Clarendon Press, Oxford, pp. 350–377.

CHROMOMERE

ALAN P. WOLFFE

Chromomeres represent discrete structures in **chromosomes** that are visible under the light microscope in mitotic or meiotic **prophase** (see **Chromatid**). **Electron microscopy** and staining indicate that chromomeres represent regions of move complex, higher order structure than the 30-nm **chromatin** fiber (1). It has been suggested that they represent sites of the preferential chromatin compaction that occurs early in the prophase, coincident with the separation of sister chromatids (2). However, electron microscopic images of native chromatin in the **interphase** nuclei at physiological ionic strengths reveal that chromomeres may be identical to globular clusters of **nucleosomes** [(superbeads)] that probably represent local coiling of the 30-nm fiber. It is also possible to detect bead-like discontinuities in the chromatin fiber in sectioned nuclei. Under certain conditions, these superbeads are remarkably uniform and contain between 8 and 48 nucleosomes. This suggests that chromomeres represent discrete structural units (3). The failure to find a ubiquitous organization of chromatin into chromomeres or superbeads, however, indicates that all chromatin does not adopt this form of higher order structure in interphase nuclei *in vivo*. This irregularity may in fact represent the true state of affairs within the nucleus, because, clearly, eukaryotic DNA is not packaged into structures of a crystalline order and stability.

Chromomeres are useful for discriminating between different chromosomes. They are arranged in a specific and consistent pattern along each chromosome. In **lampbrush chromosomes**, chromomeres occur where there are long regions of inactive chromatin that are consequently compacted into higher order structures.

BIBLIOGRAPHY

1. D. E. Comings (1978) *Ann. Rev. Genet.* **12**, 25–45.
2. G. F. Bahr and P. M. Larsen (1974) *Adv. Cell. Mol. Biol.* **3**, 192–215.
3. H. Zentgraf and W. W. Franke (1984) *J. Cell Biol.* **99**, 272–286.

CHROMOSOMES

ALAN P. WOLFFE

Chromosomes are the **nucleoprotein** complexes that provide the structural framework for the expression of **genes** and mediate the transfer of genetic information from generation to generation. The **nuclei**, **mitochondria**, and **chloroplasts** of all cells, both **eukaryotes** and **prokaryotes**, plus **viruses**, all contain chromosomes. These various chromosomes can vary greatly in size, from very large (see **Polytene chromosomes**, **Lampbrush chromosomes**) to very small (see

Double minute chromosome; **Minichromosomes**). All chromosomes contain **DNA**, which can be packaged by different proteins to assemble a nucleoprotein complex. The DNA in the chromosomes of a eukaryotic cell nucleus is packed with small basic proteins known as **histones**. The most striking property of every chromosome within the eukaryotic cell nucleus is the length of each molecule of DNA incorporated and folded into it. The human **genome** of 3×10^9 bp would extend over a meter if unraveled and straightened, yet it is compacted into a nucleus only 10^{-5} m in diameter. It is an astonishing feat of engineering to organize such a long linear DNA molecule within ordered structures that can reversibly fold and unfold within the chromosome.

The basic architectural matrix of the chromosomes found within eukaryotic cell nuclei is **chromatin**. All of the DNA in the nucleus of a somatic cell is packaged into chromatin by the histone proteins, together with other **DNA-binding proteins**. In specialized **germ cells**, such as **sperm**, **protamines** can replace the histones as the primary means of packaging DNA. There is considerable specialization in the type of chromatin assembled in different regions of a chromosome, depending on its functional requirements. Chromatin containing genes that are being expressed in a cell is called **euchromatin**. A large chromosomal region that contains inactive genes is assembled into **facultative heterochromatin**. Specialized chromosomal structures that contain very few genes but have other essential architectural roles in the chromosome are assembled into constitutive **heterochromatin**. The **centromere** is an essential structure containing constitutive heterochromatin that mediates segregation of the chromosomes. The molecular motors that drive this process are found within the **kinetochore** that is attached to the centromeric heterochromatin. Other heterochromatic domains are at the ends of chromosomes in specialized structures known as **telomeres**. Within the telomere are many reiterated **terminal repeat** sequences resulting from the activity of the enzyme **telomerase**. The telomere protects the end of the chromosome from degradation, fusion, and progressive loss of DNA during chromosomal **duplication**.

The position of the centromere relative to the rest of the chromosome provides an important reference point for describing different types of chromosomes. Chromosomes can be **metacentric**, with a centromere near the middle of the chromosome, or **acrocentric**, when it is near the end of the chromosome, or even potentially **telocentric**, when it is at the very tip. Most eukaryotic chromosomes are **monocentric**, having a single centromere, but some are **holocentric** and have multiple centromeric domains. Holocentric chromosomes have specialized mechanisms for segregating chromosomes during **cell division**that differ from those found in most plant and animal cells.

In animal cells, the presence of multiple centromeres in a single chromosome is often the product of chromosomal rearrangements. Such events usually occur as a result of chromosomal damage and can involve **duplications**, **inversions**, and **translocations**. A **dicentric** chromosome contains two centromeres following one of these rearrangements, whereas an **acentric** chromosome lacks a centromere entirely. **Isochromosomes** might also be formed, where a chromosome contains multiple identical chromosomal arms. All of these rearrangements lead to major problems for segregating chromosomes during the cell division of **meiosis** and **mitosis**. Chromosomal damage can be very useful in mapping the positions of genes relative to each other by methodologies that apply somatic cell genetics (see **Radiation hybrid**).

Each **diploid** cell receives two sex chromosomes, either two **X-chromosomes** in females or one X- and one **Y-chromosome** in males, plus two copies of each of the other chromosomes, known as **autosomes**. Each pair of autosomes consists of two **homologous chromosomes**, one from the father and one from the mother in humans. Immediately after mitotic division, each chromosome consists of a single DNA molecule. This single molecule of DNA is assembled into a **chromatid**, which duplicates during the **S-phase** of the **cell cycle** to form two sister chromatids. Then these sister chromatids are segregated to the two daughter cells at mitosis. In meiosis, **haploid** germ cells are created. This involves segregating of homologous chromosomes at the first meiotic division before the sister chromatids are segregated at the second meiotic division. Because homologous chromosomal regions are aligned during meiosis, chromosomal rearrangements can lead to major problems in segregating homologous chromosomal material. This can lead to **multivalent** chromosomes, whereas a normal meiosis would produce only bivalent chromosomes.

For organisms that undergo sexual reproduction, special **sex** chromosomes exist. In humans, these are known as the X- and Y-chromosomes. Female cells contain two X-chromosomes, whereas male cells contain a single X-chromosome and a Y-chromosome. Mary Lyon proposed that **gene expression** from the two X-chromosomes in a female cell should be the same as that from the single X-chromosome in a male cell (see **Lyon hypothesis**). This is accomplished by silencing one of the two X-chromosomes in a female cell (see **X-chromosome inactivation**; **Random X-inactivation**). The silenced X-chromosome is converted to a **Barr body** made up of facultative heterochromatin. X- and Y-chromosomes share some limited regions of **homology** that mediate their pairing and subsequent segregation during meiosis.

An important aspect of chromosomal biology involves mapping individual genes on chromosomes, which facilitates diagnosing particular diseases. A wealth of mapping procedures exist. These make up the field of **cytogenetics**, which is based on visualizing chromosomes. Mapping procedures involve using particular stains for the DNA and protein components of the chromosome (see **C-banding**; **G-banding**; Giemsa staining). In recent times, these methodologies have been supplemented with high-resolution **denaturation mapping** techniques. Then all of the stained chromosomes of a particular cell that make up the **karyotype** can be displayed formally as an **ideogram**.

At a more refined level, structure at the level of chromosome and chromatin also determines the functions of particular genes. It has long been known from visualizating the large polytene chromosomes that individual chromosomal **domains** exist. Individual domains like the **Balbiani rings** contain chromatin that reversibly changes structure to reflect different functional states. Balbiani rings appear as **puffs** when they are being **transcribed**. Other regions of the polytene chromosome, known as **interbands**, contain the regulatory DNA sequences that control the appearance of puffs. Chromosomal rearrangements can lead to positioning genes next to large domains of constitutive heterochromatin. Placing a gene next to the **chromocenter** in *Drosophila*, where the centromeres fuse together, **represses** expression of the gene.

This **position effect** indicates that structurally specialized domains of chromosomes exist and presents a major problem for biotechnology and gene therapy in expressing foreign DNA in target cells.

Nucleases like **DNase I** have been very useful in mapping regulatory DNA in the chromosome. The **locus control regions**, **enhancers**, and **promoters** that control gene expression exist as sites that are **hypersensitive** to digestion by DNase I. These elements often control the expression of clusters of **contiguous genes** that have related functions in the cell. Sites with extreme **DNase I sensitivity** are also found at the regulatory elements controlling the initiation of **DNA replication** at the **origin of replication**. The **terminally redundant** regulatory elements of **retroviruses** are also assembled into nucleoprotein complexes that retain accessibility to DNase I.

Chromosomes are fascinating structures around which much of modern medicine and biotechnology revolves. They also provide the foundation for considering basic molecular mechanisms that control gene expression and for considering the forces that facilitate **evolution**. Genes require a chromosomal environment to be maintained throughout the generations and within which to realize their full regulatory potential.

CHYMOTRYPSIN, CHYMOTRYPSINOGEN

J. RIORDAN

Chymotrypsin (E.C. 3.4.21.1) is a mammalian digestive **serine proteinase**, of the **trypsin** family, that is synthesized in the acinar cells of the pancreas in the form of an inactive precursor, *chymotrypsinogen*. This **zymogen** is stored, along with other digestive **enzymes** and enzyme precursors, in pancreatic granules, whose contents are released into the duodenum when the pancreas is stimulated by the **hormones** cholecystokinin and acetylcholine, which are secreted in response to eating a meal.

Chymotrypsinogen is a single **polypeptide chain** of 245 amino acid residues that becomes enzymatically active on **proteolytic** cleavage of the peptide bond that connects Arg15 and Ile16. This cleavage is catalyzed by another pancreatic enzyme, **trypsin**, which in turn is generated from its zymogen precursor, trypsinogen, by **enterokinase**, an intestinal serine proteinase. Cleavage of the Arg15–Ile16 bond results in a reorganization of the protein structure, alignment of the **catalytic triad**, and generation of a substrate binding site (1) . Other proteolytic cleavages accompany activation of chymotrypsinogen, but only the one cited is crucial for generating enzymatic activity.

The biological function of chymotrypsin is to digest dietary proteins in the small intestine. It catalyzes the cleavage of **peptide bonds** in which the carbonyl group is supplied primarily by aromatic or bulky **hydrophobic** amino acids (**tyrosine**, **tryptophan**, **phenylalanine**, **leucine**, **isoleucine**, **methionine**). These residues become the C-termini of the product peptides and are subsequently removed by digestion with **carboxypeptidase** A. The catalytic mechanism of chymotrypsin and other serine proteinases is described elsewhere (see **Serine proteinases**).

Numerous chymotrypsin-like enzymes generally referred to as *chymases* have been identified in various animal tissues, particularly in mast cells, neutrophils, and lymphocytes (2). The biological functions of these enzymes are not well understood, but a chymase isolated from human heart is a major angiotensin II -forming enzyme (3). The *prostate-specific antigen*, which is used in the diagnosis of prostate cancer, is a serine proteinase with limited chymotrypsin-like activity (4).

Being a serine proteinase, chymotrypsin is inhibited by diisopropylfluorophosphate (**DIFP**) and also by numerous, naturally occurring **serine proteinase inhibitors** and **serpins** (5), as well as many small synthetic peptide analogues or **active site** reagents (6).

BIBLIOGRAPHY

1. R. M. Stroud, A. A. Kossiakoff, and J. L. Chambers (1977) *Annu. Rev. Biophys. Bioeng.* **6**, 177–193.
2. J. A. Nadel (1991) *Ann. N. Y. Acad. Sci.* **629**, 319–331.
3. Y. Liao and A. Husain (1995) *Can. J. Pharmacol.* **11**(Suppl. F), 13F–19F.
4. A. M. el-Shirbiny (1994) *Adv. Clin. Chem.* **31**, 99–133.
5. J. Potempa, E. Korzus, and J. Travis (1994) *J. Biol. Chem.* **269**, 15957–15960.
6. J. C. Powers et al. (1989) *J. Cell. Biochem.* **39**, 33–46.

Suggestions for Further Reading

R. M. Stroud (1974) A family of protein-cutting enzymes. *Sci. Am.* **231**, 24–88.

T. A. Steitz and R. G. Shulman (1982) Crystallographic and NMR studies of the serine proteases. *Annu. Rev. Biophys. Bioeng.* **11**, 419–444.

CILIA AND EUKARYOTIC FLAGELLA

LESLIE WILSON

Cilia and eukaryotic flagella are long (10 μm to 40 μm), narrow, membrane-bounded structures that contain a highly ordered, stable **microtubule** array, consisting of nine fused outer doublet microtubules surrounding a central pair of singlet microtubules. These microtubules are firmly anchored at the base of the cilium or flagellum in basal bodies. The microtubules of cilia and flagella are extremely stable, and their movement is not complicated by dynamics. In addition to the tubulin backbone, these microtubules contain a large and diverse array of **microtubule-associated proteins** (MAPs), most of which have not been characterized, and several forms of **dynein**, a large multicomponent ATP-transducing mechanochemical **motor protein**. The various MAPs create and maintain the stable axoneme structure, and the dynein is responsible for creating the movement. The movements of cilia and flagella involve bending of these long organelles, which is caused by the sliding of stable double-microtubule pairs past each other. Sliding is effected by the dynein motors; the base of the motor is attached permanently to the A microtubule of one outer doublet pair, and the head, which contains the motor **domain**, transiently attaches and detaches to the B subfiber of an adjacent outer doublet pair. The dynein motor "walks" along the B subfiber toward the minus end, creating the movement. The dynein-mediated sliding of outer double-microtubules occurs in a fashion similar to filament sliding in muscle contraction, in which **actin** filaments and myosin filaments slide past each other through the action of the ATP-transducing myosin motor "heads."

CIRCADIAN RHYTHMS AND CLOCKS IN FUNGI

BETH DIDOMENICO
RODOLFO ARAMAYO

Circadian clocks are ubiquitous and present in organisms as distantly related as **fungi** and mammals. Classically defined as rhythms that persist under constant environmental conditions, circadian clocks control daily rhythms that correlate with changes in the external environment. Most likely, circadian rhythms were selected early in **evolution** for conferring an evolutionary advantage on the organisms containing them.

To compare the properties of circadian rhythms among different organisms, the concept of circadian time (CT) was developed. The circadian day is divided into 24 equal parts, each one circadian hour. By convention, CT0 is subjective dawn, and CT12 represents subjective dusk. The normal circadian cycle in *Neurospora* can be reset in a time-dependent manner by external signals (a property called entrainment) such as light (1) or temperature (2).

Lower eukaryotes are ideal models for clock study because of their genetic and biochemical tractability. *Neurospora crassa* is such an example. Circadian rhythms in *Neurospora* are easily observed with the use of specialized growth tubes called ***race tubes***. When wild-type strains of *Neurospora* are inoculated on one side of a race tube, the organism initiates a rapid vegetative growth toward the opposite side. After growth for a day in constant light, the position of the growth front is marked, and the culture is transferred to constant dark. The light-to-dark transfer synchronizes the culture on CT12 (subjective dusk). Monitoring of the growth front every 24 h reveals a typical pattern of alternate conidial and aconidial bands, whose frequency permits the determination of both period length and phase of the rhythm. The *Neurospora* circadian rhythm has a period of 22 h at 24°C. This period does not change significantly between 18°C and 30°C, as demanded by the circadian definition of a rhythm (a phenomenon called temperature compensation).

Three general questions drive research in fungal chronobiology: What is the biochemical and/or genetic basis for the clock? How does the clock gets its input from its extracellular and extraorganismal environment? How is time, as paced by the clock, transduced within the cell?

THE CLOCK

The central oscillator of the *Neurospora* clock is the *frq* locus. Originally identified by chemical and ultraviolet mutagenesis, mutations at this locus alter the normal circadian clock of the cell but have no severe morphological defects in growth or development (3,4). Cloning of the *frq* locus elucidated a complex pattern of gene **transcription** (3,5). Two overlapping transcripts of 4 and 4.5 kb were identified, both containing long leader sequences, with several upstream open reading frames. These transcripts have been predicted to encode a putative 989-amino-acid-residue protein. Recent data, however, support the existence of two forms of FRQ (6). The long form (FRQ^{1-989}) starts at the first in-frame ATG **start codon**, whereas the short form ($FRQ^{100-989}$) starts at the third in-frame ATG of the longest reading frame in the region (6). Temperature regulates the ratio of FRQ forms by favoring the use of different initiation codons at different temperatures. For example, the long form of FRQ (FRQ^{1-989}) is favored at high temperatures, whereas at lower temperatures the short form of FRQ ($FRQ^{100-989}$) is predominant. These similar but functionally distinct forms of the clock protein FRQ represent a novel adaptive mechanism designed to keep the *Neurospora* clock running over a wide range of temperatures (6). Despite its fine regulation, the biochemical function of FRQ remains unknown (7). However, the presence of a nuclear localization signal [required for FRQ activity (8)], a weak **helix–turn–helix** DNA-binding domain (9), and a conserved acidic region strongly suggest the involvement of FRQ in transcription. As predicted for protein regions important for function, all of these FRQ transcription-factor-like signatures are conserved in FRQ homologues isolated from other distantly related fungal species (10).

The autoregulatory feedback cycle controlled by *frq* takes one day to complete. This loop involves the transcriptional activation and repression of the *frq* locus. Starting with low levels of *frq* transcript and protein (dawn), *frq* transcripts begin to rise, reaching a peak accumulation 4 to 5 h later [just before noon—(8)]. The maximal level of FRQ protein can only be observed 4 to 6 h later (relative to the peak in *frq* mRNA). This time lag is probably due to post-transcriptional regulation of *frq* mRNA or FRQ protein. Synthesized FRQ protein enters the nucleus (8) and acts rapidly, either directly or indirectly, to repress *frq* (11). The repressing activity of FRQ remains until the protein begins to be **phosphorylated**, linking the activity of unknown protein kinases to the clock. Phosphorylation of FRQ increases its turnover (12), until its scarcity allows positive factors like *white collar-1* (*wc-1*) and *white-collar-2* (*wc-2*) (1,7) to restart the cycle (see text below).

INPUT

Light and temperature are among the most important environmental stimuli controlling circadian rhythmicity. The light-dependent **signal transduction** pathway is thought to involve an as yet uncharacterized flavin-dependent blue-light photoreceptor (13,14). When wild-type strains of *Neurospora* grown in the dark receive a pulse of light, the levels of *frq* mRNA increase. Two global regulatory genes, *wc-1* and *wc-2*, have been established to be involved directly or indirectly in the induction and maintenance of *frq* mRNA levels following this light treatment. For example, the light response observed in wild-type strains (ie, the increase in *frq* mRNA levels) cannot be detected in strains mutant in *wc-1*, but it is normal in strains mutant in *wc-2*. Interestingly, the products of both loci (*wc-1* and *wc-2*) are required for the maintenance of high-level expression of *frq* mRNA in the light. Therefore, the pulse of light positively stimulates *frq* mRNA levels in an FRQ-independent, WC-dependent manner (1,15). Moreover, the light dosage positively correlates with the amount of *frq* mRNA, and, depending when this pulse of light is perceived, it can reset the clock, thereby phase advancing or phase delaying the conidiation rhythms of the organism. Complete elucidation of the roles of WC-1 and WC-2 on *frq* gene expression will be essential for a complete understanding of the signal transduction pathway responsible for clock input.

OUTPUT

Genes whose rhythmic regulation equals the period of the strain under study are said to be part of the output component of the oscillator (ie, they execute the commands given by the master oscillator and establish the cellular rhythm). To be defined as such, loss-of-function mutations in these classes of genes must not affect the functioning of the clock. With this rationale in mind, a systematic search for clock-controlled genes (*ccg*) was carried out in *Neurospora* (16). Most, but not all, of the *ccg* genes identified were found to be induced during the asexual **sporulation** pathway. The ones that were not were postulated to be involved in other developmental pathways (eg, sexual sporulation) or other unknown output pathways (7). Although rhythmically regulated, the levels and amplitude of expression differ among different clock-controlled genes. In addition, their expression pattern in both frq^+, and *frq* changes from gene to gene [ie, some genes are repressed, while others are derepressed or unchanged (7)]. To date, a direct link between *frq* (the central master oscillator) and *ccg* (the foot soldiers) has not been established, despite being an area of active investigation. It is not surprising to find that the *ccg* identified have nothing or very little in common. For example, *ccg-1* codes for a small polypeptide of 71 amino acid residues with no known homologues in other organisms (17), despite its elevated level of expression (18), whereas *ccg-12*, encodes copper **metallothionein** (18). See Refs. 18 and 19 for a complete description of *ccg* genes. While the mechanisms of circadian rhythms are incompletely understood at this time, it remains undeniable that the molecular strategies employed by fungi are an excellent model system with which to study similar systems in higher eukaryotes.

BIBLIOGRAPHY

1. S. K. Crosthwaite, J. J. Loros, and J. C. Dunlap (1995) *Cell* **81**, 1003–1012.
2. Y. Liu, M. Merrow, J. J. Loros, and J. C. Dunlap (1998) *Science* **281**, 825–829.
3. J. C. Dunlap (1996) *Annu. Rev. Genet.* **30**, 579–601.
4. J. F. Feldman (1982) *Annu. Rev. Plant Physiol.* **33**, 583–608.
5. C. R. McClung, B. A. Fox, and J. C. Dunlap (1989) *Nature* **339**, 558–562.
6. Y. Liu, Y. Garceau, J. J. Loros, and J. C. Dunlap (1997) *Cell.* **89**, 477–486.
7. D. Bell-Pedersen (1998) *Microbiol.* **144**, 1699–1711.
8. C. Luo, J. J. Loros, and J. C. Dunlap (1998) *EMBO J.* **17**, 1228–1235.
9. M. T. Lewis, L. Morgan, and J. F. Feldman (1996) *Mol. Gen. Genet.* **253**, 401–414.
10. M. Merrow and J. C. Dunlap (1994) *EMBO J.* **13**, 2257–2266.
11. M. W. Merrow, N. Y. Garceau, and J. C. Dunlap (1997) *Proc. Natl. Acad. Sci. USA* **94**, 3877–3882.
12. N. Y. Garceau, Y. Liu, J. J. Loros, and J. C. Dunlap (1997) *Cell* **89**, 469–476.
13. E. Klemm and H. Ninnemann (1979) *Photochem. Photobiol. Rev.* **4**, 207–265.
14. J. Paietta and M. L. Sargent (1981) *Proc. Natl. Acad. Sci. USA* **78**, 5573–5577.
15. S. K. Crosthwaite, J. J. Loros, and J. C. Dunlap (1997) *Science* **276**, 763–769.
16. J. J. Loros, S. A. Denome, and J. C. Dunlap (1989) *Science* **243**, 385–388.
17. J. J. Loros (1995) *Semin. Neurosci.* **7**, 3–13.
18. D. Bell-Pedersen, M. Shinohara, J. J. Loros, and J. C. Dunlap (1996) *Proc. Natl. Acad. Sci. USA* **93**, 13096–13101.
19. D. Bell-Pedersen, N. Garceau, and J. J. Loros (1996) *J. Genet.* **75**, 387–401.

CIRCULAR CHROMOSOME

GEORGES N. COHEN

All **Hfr strains** (for high frequency of **recombination**) of *Escherichia coli* originate from F^+ strains (males, possess the *F* sex factor). When mated with F^- recipients (females), they generate about a thousand times more recombinants than the equivalent F^+ strain under identical conditions (1). The recombinants obtained inherit a particular segment of the donor chromosome exclusively (1). The way in which *Hfr* donors transfer their chromosome to recipients was elucidated by violent agitation of a mixture of mating Hfr and F^- in a high-speed mixer. The mating was interrupted at different time intervals and the mixture was plated for detection of recombinants. It was found that recombinants acquire different selected characters from the donor cell at different times. The order in which these characters appear is the order of their arrangement on the chromosome. The only reasonable interpretation of these results is that the effect of agitation is a rupturing of the chromosome during the transit from the donor to recipient bacteria. Only those genes that have already penetrated the recipient bacteria at the time of treatment can appear in recombinants. Thus, the strain Hfr used comprises a homogeneous population of donor bacteria, all of which transfer their chromosome in the same specifically oriented and linear way, so that a particular extremity, designated O (for origin), is always the first to penetrate the recipient bacteria (2,3).

A number of Hfr stains of independent origin were analyzed with respect to the markers they transferred to recipients and to the orientation of the transfer. The outcome was remarkable and unforeseen. Although all of the bacteria of any one Hfr strain transfer a given set of markers in a particular sequence, different strains transfer different chromosomal segments, parts of which may overlap. Some strains transfer their genes in a direction reverse to others. The relationship of the various genes to one another however, is the same irrespective of the Hfr strain used to map them. This gives an unequivocal general linkage map for *E. coli*, consisting of a single linkage group. Each strain transfers its chromosome as a linear, oriented structure with a specific head and tail, for example,

```
Z Y                CBAO
___________ ----- ________
```

Another strain can always be found which transfers genes A and Z as closely linked markers, for example,

```
Y                   CBAZO                         OBAZY
-------------------------------- ,  or  ____ -------- __________
```

Thus, in spite of this linear transfer, it is impossible to define any extremities on the chromosome. It is concluded that the chromosome of F+ donors is originally continuous or circular and that Hfr strains arise by opening up the circle at a point characteristic for each Hfr type to yield a linear, transferable structure (3).

The circular nature of the *E. coli* chromosome has been confirmed by gentle liberation and dispersion of its DNA and visualization by **autoradiography** or by **electron microscopy**. Therefore there is indisputable, direct evidence that this chromosome is indeed a continuous, circular thread of double-stranded DNA (4,5).

REFERENCES

1. F. Hayes (1953) *Cold Spring Harbor. Symp. Quant. Biol.* **18**, 75–93.
2. E. L. Wollman and F. Jacob (1955) *C. R. Acad. Sc. Paris* **240**, 2449–2451.
3. E. L. Wollman and F. Jacob (1958) *Ann. Inst. Pasteur* **95**, 641–666.
4. J. Cairns (1962) *J. Mol. Biol.* **4**, 407–409.
5. J. Cairns (1963) *Cold Spring Harbor Symp. Quant. Biol.* **28**, 43–46.

Suggestion for Further Reading

F. Jacob and E. L. Wollman (1961) *Sexuality and the Genetics of Bacteria*, Academic Press, New York, pp. 164–165.

CIRCULAR DICHROISM

ROBERT W. WOODY

Circular dichroism (CD) is a form of **spectroscopy** that uses circularly polarized light and is highly sensitive to the **conformations** of molecules. Because of this conformational sensitivity, CD is one of the most widely used techniques for characterizing the conformations of **proteins**, **nucleic acids**, and carbohydrates, and for monitoring conformational transitions induced by temperature, solvent composition, **ligand binding**, etc.

The principal conformations of polypeptide chains—**alpha-helix**, **beta-sheet**, **β-turns**, and the so-called unordered conformation—all have distinctive CD spectra that permits these **secondary structures** to be readily recognized in their pure forms or in simple combinations. Even in globular proteins, which represent complex mixtures of these and other conformations, in most cases the CD spectrum can be analyzed to provide reasonably accurate estimates of the fraction of residues in the various secondary structures. CD signals due to aromatic side chains and nonprotein chromophores provide sensitive probes of protein **tertiary structures** and of protein–ligand interactions. In the field of nucleic acids, CD also provides valuable information about the nature of the secondary and higher order structures (see **Secondary structure, nucleic acids**) detecting the binding of small molecules and proteins. CD has more restricted application to carbohydrates and polysaccharides, but it has some important applications to studies of carbohydrate conformation.

In addition to its pronounced sensitivity to conformation, CD has other advantageous features. Only small amounts of material are required, comparable to those required for a UV-visible absorption spectrum, and it is relatively easy to survey a broad range of conditions of temperature, pH, ionic strength, and solvent composition. Moreover, the cost of the instrumentation is moderate, and it is easy to operate and maintain. The principal limitation of CD is that it provides lower resolution structural information than **X-ray crystallography** or **NMR** spectroscopy. A CD signal can rarely be assigned to a specific residue or group in a macromolecule. Whereas CD gives an estimate of the fraction of residues in a globular protein that are in α-helical segments, it cannot provide information on their location in the sequence. These limitations, however, are outweighed by the conformational sensitivity and other experimental advantages of CD.

FUNDAMENTALS OF CIRCULAR DICHROISM

Definition of Circular Dichroism

Circular dichroism is a type of spectroscopy in which the difference in absorption of right- and left-circularly polarized light is measured. In circularly polarized light (cpl), the vector **E** describing the electric field of the light, rotates about the direction of propagation once in each period or wavelength of the light. As seen by an observer looking toward the light source, the electric vector may rotate in either a clockwise sense, corresponding to right-circularly polarized light (rcpl), or in a counterclockwise sense, corresponding to left-circularly polarized light (lcpl). At a given instant, the tip of the electric vector of rcpl follows a right-handed helix in space, whereas the tip of the electric vector of lcpl follows a left-handed helix. Therefore, cpl is **chiral**, and rcpl and lcpl are related as mirror images, that is, they are **enantiomeric**. Because of their enantiomeric character, it is to be expected that rcpl and lcpl interact differently with a chiral molecule, giving rise to differences in optical properties for a chiral molecule measured with rcpl and lcpl.

CD is defined as the difference in absorbance of rcpl and lcpl:

$$\mathrm{CD} = \Delta A = A_\mathrm{l} - A_\mathrm{r} = \epsilon_\mathrm{l} c\ell - \epsilon_\mathrm{r} c\ell = (\epsilon_\mathrm{l} - \epsilon_\mathrm{r})c\ell = \Delta\epsilon_\mathrm{M} c\ell \tag{1}$$

Here A_l and A_r are, respectively, the absorbances for lcpl and rcpl; c is the molar concentration of the chiral species; ℓ is the pathlength of the sample cell (in cm); ϵ_l and ϵ_r are the molar extinction coefficients of the chiral species for lcpl and rcpl, respectively; and $\Delta\epsilon_\mathrm{M}$ is the molar CD, commonly called simply the CD.

To facilitate comparisons of the CD of proteins or nucleic acids of very different molecular masses, it is nearly universal practice to normalize the CD by dividing the molar CD by the number of amino acid or nucleotide residues, yielding the mean residue CD,

$$\Delta\epsilon = \Delta\epsilon_\mathrm{M}/n_\mathrm{r} = \Delta A/c_\mathrm{r}\ell \tag{2}$$

where n_r is the number of residues and c_r is the molar concentration of residues. Near-UV CD spectra of proteins are reported as either molar CD or mean residue CD, and visible CD spectra are usually reported as molar CD of the protein or the molar CD per visible chromophore, for example, per heme in vertebrate hemoglobins. The reader must always carefully check the basis for reporting CD spectra in the literature.

Although nearly all CD instrumentation measures ΔA and hence yields $\Delta\epsilon$ most directly, many instruments are calibrated in units of ellipticity, an angular unit based on an earlier method for measuring CD. The molar ellipticity is defined as

$$[\theta]_M = 100\theta/c\ell \quad (3)$$

where θ is the measured ellipticity in degrees and c and ℓ have the same meaning as before. Molar ellipticity is directly proportional to molar CD:

$$[\theta]_M = 3298\Delta\epsilon_M \approx 3300\Delta\epsilon_M \quad (4)$$

Mean residue ellipticity $[\theta]$ is defined by analogy to mean residue CD and is commonly used for reporting protein CD data in the far UV and for CD of DNA and RNA.

Instrumentation (1)

Measurement of CD requires high sensitivity because the difference between ϵ_l and ϵ_r is small in comparison with their average value $\epsilon = (\epsilon_l + \epsilon_r)/2$, which corresponds to the molar absorbance coefficient measured with unpolarized light. Typically, $\Delta\epsilon/\epsilon$ is of the order of $10^{-3}$to 10^{-4}. To achieve adequate signal-to-noise ratios, CD spectrophotometers use a modulation technique. The light incident on a sample is switched between left- and right-circular polarization at a specific frequency. The difference in the sample response to the two forms of circularly polarized light is detected directly by selecting the part of the response that has a frequency component matching that of the modulation of the incident light.

Figure 1 shows a highly schematic representation of a CD spectrophotometer. Light from the source L is dispersed in the monochromator MC, and then a narrow band of wavelengths passes through the linear polarizer lP. The polarizer splits the unpolarized monochromatic beam into two linearly polarized beams, one polarized in the x direction, the other in the y direction (z is the direction of propagation). Then one of the two linearly polarized beams passes through the photoelastic modulator (PEM), which consists of a plate made of a transparent, optically isotropic material bonded to a piezoelectric quartz crystal. When an alternating electric field is applied, the light emerging from the PEM switches from lcpl to rcpl and back with the frequency of the applied electric field, typically about 50 kHz. If the sample S exhibits CD, the amount of light absorbed varies periodically with the polarization of the incident light, and so the intensity of light that reaches the photomultiplier PM exhibits sinusoidal intensity variations at the frequency of the field applied to the PEM. Thus, the photomultiplier output consists of a signal with a small alternating current (ac) component superimposed on a direct current (dc) component. The ac component is filtered out and amplified. The ratio of the ac to the dc component is directly proportional to the circular dichroism of the sample, and this quantity is recorded as a function of wavelength to provide a CD spectrum.

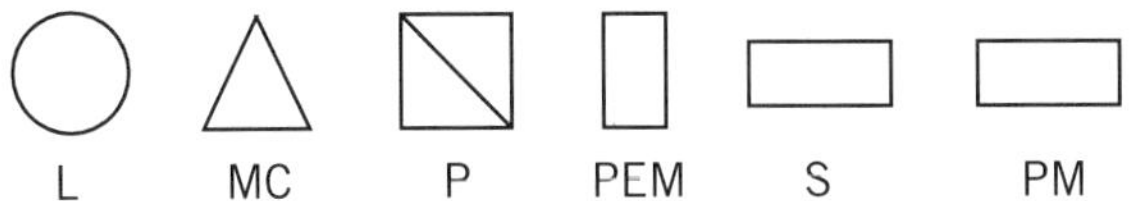

Figure 1. A block diagram of a CD spectrophotometer. L is the light source, MC the monochromator, P the polarizer, PEM the photoelasic modulator, S the sample, and PM the photomultiplier.

CD is most commonly measured in the visible and ultraviolet regions of the spectrum. Commercially available instruments are used in the 180 to 700-nm region. Some commercial instruments permit measurements to 170 nm, at the edge of the vacuum UV, or to 1000 nm, in the near infrared (IR). Xenon arc lamps, quartz optics, and photomultiplier tubes are used for the light source, the dispersing and polarizing elements, and the detectors, respectively.

CD is also observed in IR absorption bands, associated with vibrational transitions (see **Vibrational spectroscopy**). This type of CD is called vibrational circular dichroism (VCD) (2). Some VCD instruments use the same type of layout as that shown in Fig. 1, although the light sources, dispersing and polarizing optics, and detectors are necessarily different. In addition, VCD instruments have been constructed that utilize the Fourier transform principle.

In recent years, CD has been widely used to measure the kinetics (3) of conformational changes in proteins and nucleic acids, especially in investigating **protein folding in vitro** and unfolding. Reactions on the timescale of a minute or longer are readily measured by using manual mixing and conventional CD instrumentation. A conventional CD instrument coupled to a stopped-flow mixing device permits measurements of reactions with halftimes down to a few milliseconds. Still faster reactions, triggered by temperature or pressure jumps, or by laser or electron pulses, require specialized CD instrumentation (4) that does not utilize the modulation technique. Such instruments have been constructed for nanosecond and picosecond timescales.

Theoretical Background (5)

The connection between theory and experiment in CD spectroscopy is provided by a quantity called the rotational strength, R. Experimentally, the rotational strength is proportional to the area under a CD band, that is, the integral

$$R \propto \int (\Delta\epsilon/\lambda) d\lambda \approx (\lambda_{\max})^{-1} \int \Delta\epsilon d\lambda \quad (5)$$

The integral is taken over a CD band attributable to a single electronic or vibrational transition. CD bands are generally rather sharp, so it is a satisfactory approximation to replace the λ^{-1} factor in the integrand of Eq. 5 by $\lambda_{\max}^{-1}$, where $\lambda_{\max}$ is the wavelength at which the CD band has maximal intensity.

The rotational strength is related to molecular properties by the equation

$$R = \mathrm{Im}\{\boldsymbol{\mu}_{0a} \cdot \mathbf{m}_{a0}\}, \quad (6)$$

where Im indicates the imaginary part of a complex quantity. The first factor in the curly brackets is the electric dipole transition moment for the transition from the ground state 0 to the excited state a. The second factor is the magnetic dipole transition moment of the transition. Qualitatively, $\boldsymbol{\mu}_{0a}$ can be interpreted as the linear displacement of charge associated with the transition, and $\mathbf{m}_{a0}$ as the circular displacement of charge. Therefore a non zero rotational strength requires that the transition have both a linear and a circular displacement of charge and that the axis of the circulation not be perpendicular to the linear displacement. This corresponds to a helical motion of electronic charge, where the sense of the helix is determined by the relative orientation of $\boldsymbol{\mu}_{0a}$ and $\mathbf{m}_{a0}$ and this determines the sign of the CD band.

Related Phenomena

Phenomena that depend upon differences in the interaction of a sample with left- and right-circularly polarized light are called chiroptical phenomena. CD is the most widely used chiroptical phenomenon in molecular biology, but others are important for historical reasons or have specialized applications. The first chiroptical phenomenon discovered was optical rotation (OR), the rotation of the plane of polarization of plane-polarized light as it passes through a chiral medium. The angle through which the plane is rotated, commonly measured at 589 nm (the Na D line), is still widely used in organic chemistry to characterize chiral molecules. OR was also used in protein chemistry, but was supplanted in the 1950s by optical rotatory dispersion (6) (ORD), the wavelength dependence of OR, and in the 1960s by CD. The specific rotation α of a sample is defined as

$$[\alpha] = 10\alpha/c'\ell \qquad (7)$$

where α is the angle through which the plane of polarization is rotated by the sample, ℓ is defined as in Eq. (1), and c' is the concentration of the chiral substance in g/mL. Molar rotation and mean residue rotation are defined by analogy to the corresponding CD parameters. It should be noted that refraction is a scattering phenomenon and occurs at all wavelengths, not just in absorbance bands. For this reason, OR and ORD can be measured at visible wavelengths for substances that absorb only in the ultraviolet.

PROTEINS

General Aspects

In the far UV, the peptide groups of the backbone generally dominate the CD spectrum (see **Absorbance spectroscopy**). The peptide group has two electronic transitions in the normally accessible far UV. These are the $n\pi^*$ transition near 220 nm and the $\pi\pi^*$ transition near 190 nm in secondary amides and 200 nm in tertiary amides (X-Pro peptide groups). The $n\pi^*$ transition is weak in absorbance, but it gives rise to strong CD bands. The $\pi\pi^*$ transition is associated with strong absorbance and CD. Because of the strong electric dipole transition moment, $\pi\pi^*$ transitions in neighboring peptide groups interact with each other, giving rise to two or more absorbance and CD bands. This phenomenon, called exciton splitting, is most clearly seen in the α-helix CD spectrum, described later.

The aromatic side chains of **Phe, Tyr,** and **Trp** residues have strong absorbance bands in the far UV that contribute to the CD spectrum (7). In most cases, their contribution is small compared to those of the much more numerous peptide groups. For some proteins, however, aromatic CD bands are clearly discernible.

In the near UV, the CD spectrum of proteins is dominated by the aromatic and disulfide transitions. The near-UV CD bands of the aromatic side chains are generally relatively sharp and have a characteristic fine structure due to vibrational effects. In proteins with a small number of aromatic side chains, the near-UV CD bands can frequently be assigned to one of the three types of aromatic side chains and in some cases, through **site-directed mutagenesis**, to specific residues in the sequence. The CD bands due to the **disulfide bond** are generally distinguishable from aromatic CD bands by their much greater width.

For proteins containing only the normal amino acids, there are no CD bands at wavelengths above 300 nm. Many **prosthetic groups**, **coenzymes**, transition-metal ions, and other ligands have absorbance bands in this wavelength region, and these are associated with CD bands in complexes with proteins.

Secondary Structure Analysis

The various types of secondary structure (see **Secondary structure, protein**) in proteins have characteristic CD spectra, as established by studies of model oligo- and polypeptides. Figure 2 shows CD spectra of α-helix, β-sheet, unordered polypeptides, and β-turns. The α-helix has the most distinctive and strongest CD spectrum with two negative bands of comparable magnitude at ca. 222 and 208 nm and a stronger positive band near 190 nm. The 222 nm band results from the $n\pi^*$ transition of the amide group, whereas the 208 and 190 nm bands both arise from the amide $\pi\pi^*$ transition. The latter two bands are from exciton splitting of the $\pi\pi^*$ transitions in the amide groups that are held in a well-defined helical geometry. Interactions between the transition dipole moments in a very long helical array give rise to three absorbance bands, one at 208 nm polarized parallel to the helix axis, and two bands at 190 nm, polarized in the two independent directions perpendicular to the helix axis. For a right-handed α-helix, the parallel band is associated with the negative CD band at 208 nm, and the perpendicular bands with the positive CD band at 190 nm. The CD of the α-helix is largely, but not completely, independent of the solvent and of the sequence of amino acids. Aromatic residues (Phe, Tyr, Trp) modify the α-helix CD spectrum, especially if they constitute a sizeable fraction of the residues. In homopolymers of aromatic amino acids, the α-helix CD spectrum is modified beyond recognition.

The CD of β-sheets is more variable than that of α-helices. A negative band near 217 nm and a positive band in the 195 to 200-nm region are characteristic of β-sheets (Fig. 2). Theory (8) predicts that the absolute value of the ratio of the ellipticity at the positive maximum near 197 nm to that at the negative maximum near 217 nm increases with increasing twisting of the sheet, and is larger for parallel than for antiparallel twisted sheets. The relative amplitude of the 217 nm and 197 nm bands varies considerably among model systems, ranging from ~1.7 for the antiparallel β-sheet formed in poly(Lys) by heating to 52°C at pH 11.1 to 8.3 for the parallel β-sheet of Boc(Val)$_7$OMe in trifluoroethanol.

Several types of β-turn can be distinguished, but only two fundamental types are common in proteins: the type I turn (and its variant, type III), which can accommodate any L-amino acid at either of the positions in the turn, and the type II turn, which usually has a Gly at the second residue of the turn. The CD characteristics of β-turns vary in accordance with the range of conformations. However, studies of cyclic peptides with well-defined β-turns indicate that type I turns have α-helix-like CD spectra, with negative maxima at ~220 and 210 nm and a positive band near 190 nm, whereas type II turns have a CD spectrum like that of a β-sheet, but with the maxima shifted 5 to 10 nm to longer wavelengths, that is, the negative band in the 220 to 225-nm region, and the positive band between 200 and 210 nm (Fig. 2).

All models for unordered **random coil** polypeptides have a strong negative band near 200 nm (Fig. 2), but some have a long-wavelength positive band and others a negative shoulder at longer wavelengths. The resemblance of the CD spectra of charged poly(Lys) and poly(Glu) to that of poly(Pro) II led to the proposal that these polypeptides are not truly unordered but contain short segments of the poly(Pro)II helix. The current view (9) of unordered polypeptides is that those peptides with a negative shoulder in the 210- to 220-nm region have amino acid residues predominantly in the α-helix and β-sheet regions of conformational space (see **Ramachandran plot**) and those with a positive long-wavelength band have a substantial fraction of residues in the poly(Pro)II conformation. The latter systems undergo a noncooperative transition to the former as the temperature increases.

Protein Secondary Structure. Early applications of OR and ORD to the analysis of helix content in proteins have been reviewed (10). Initially, protein CD data were analyzed by fitting the CD spectra of the protein to a linear combination of data for model polypeptides, such as those shown in Fig. 2. However, such methods generally gave poor results because the CD spectra of globular proteins are too complex to be adequately represented by a simple linear combination of homopolymer spectra. Many methods for analyzing protein CD data to obtain secondary structure have been proposed (11,12). Two features are essential for satisfactory results. First, a basis set of proteins of known secondary structure is needed to calibrate the method by providing information on the CD contributions of the various types of secondary structure existing in real protein structures. Second, it is necessary to allow flexibility in weighting the proteins in this basis set when analyzing each protein. Several methods incorporating these features provide useful estimates of the fraction of α-helix, β-sheet, β-turn, and unordered conformations (11–14). VCD and IR absorption spectra (See **Vibrational spectroscopy**) in the amide I region also provide estimates of the secondary structure content of globular proteins by using analytical methods similar to those used for electronic CD (15). Combinations of electronic CD, VCD and IR absorption can also be used.

The methods used for analyzing the secondary structure of proteins should be applied with caution to **peptides**, because globular proteins are used to calibrate these methods. The CD measured at 220 or 222 nm has frequently been used to determine the helix content of peptides. The difference in CD between the 217 nm negative maximum and the positive maximum near 195 nm, characteristic of β-sheets, has been utilized to quantify the β-strand conformation of peptides (16).

Tertiary Structure of Proteins. Far-UV CD spectra of proteins have been used to assign proteins to the broad classes of all-α, all-β, $\alpha + \beta$, and α/β by cluster analysis (17). Aromatic side-chain bands (7) in both the near and far UV have been used as markers to identify **tertiary structures**. The **T-state** (deoxy) conformation of **hemoglobins** has a strong negative CD band at 287 nm, whereas the **R-state** (liganded) conformation has weak negative or positive CD at this wavelength. The magnitude of the negative CD band at 250 to 255 nm correlates with the T → R transition in **insulin**. Tertiary structural changes in the **chymotrypsinogen** → chymotrypsin conversion are associated with a reorientation of the Trp 175 - Trp 215 pair, leading to a large change in the 225 nm Trp CD band.

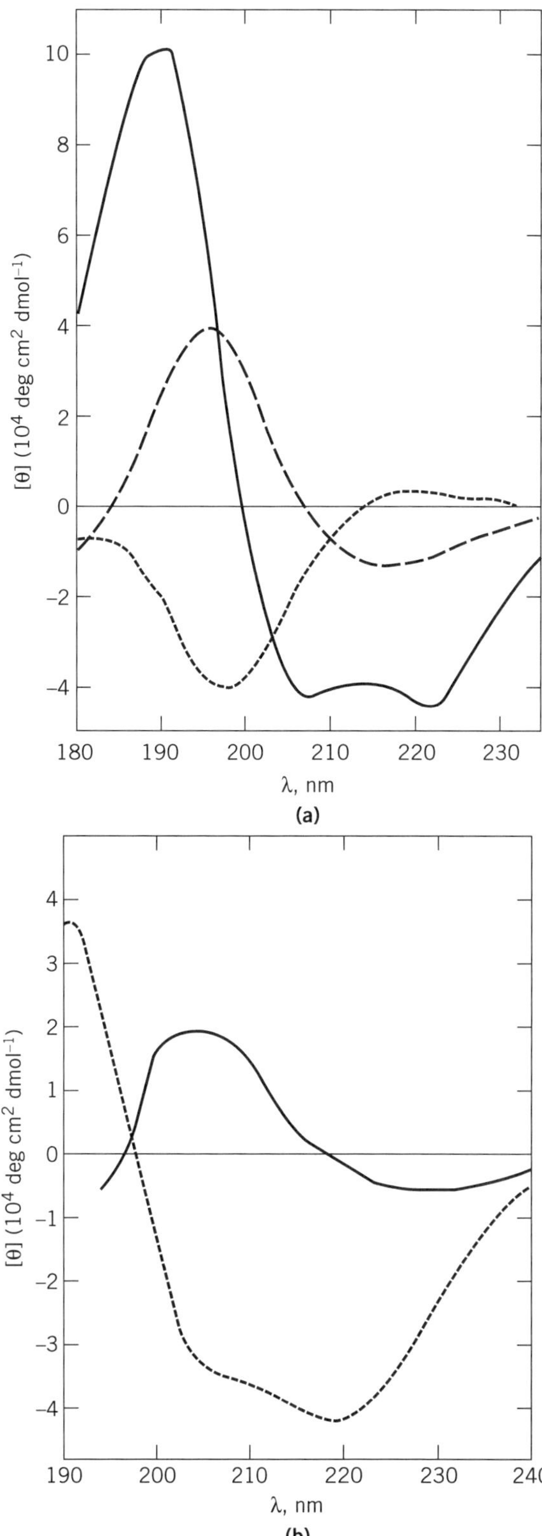

Figure 2. (**a**) CD spectra of model polypeptides in three conformations: α-helix (———), poly(Glu), pH 4.5 (33); β-sheet (– – – –), poly(Lys-Leu), 0.5 M NaF, pII 7.4 (34); and unordered (· · · · · ·), poly(Lys-Leu) in salt-free aqueous solution (34). (**b**) CD spectra of model peptides in two β-turn conformations: Type I β-turn (– – – –), cyclo(L-Ala-L-Ala-Aca), where Aca is ϵ-aminocaproyl; Type II β-turn (———), cyclo(L-Ala-D-Ala-Aca). Spectra acquired in water at 22°C (35).

Protein Folding. One of the major aims in equilibrium studies of protein folding (see **Protein folding, *in vitro***) is detecting intermediates. If equilibrium intermediates are not detectable, the folding/unfolding process is said to be two-state because only the fully folded and fully unfolded forms are present at significant levels. If only these forms are present, the extent of unfolding is the same regardless of the physical property used to monitor it. CD is especially useful in this regard because the far-UV and near-UV CD measure primarily the secondary and tertiary structures, respectively. If the extent of unfolding measured by these two probes is identical within experimental error, the process is likely to be two-state. On the other hand, lack of coincidence of the unfolding curves measured in these two spectral regions is evidence that one or more intermediates exist.

A type of folding/unfolding intermediate observed in a number of proteins is called the **molten globule**. This intermediate is nearly as compact as the native protein and has a secondary structure content comparable to that of the native protein. However, the side chains are quite mobile and are not locked into a well-defined tertiary structure. CD is one of the most useful techniques for identifying the molten globule form of proteins. The CD spectrum of a molten globule is similar to that of the native protein in the far UV and reflects a substantial amount of α-helix and/or β-sheet but has only weak features in the near UV because the aromatic side chains do not have well-defined positions and conformations.

Another aspect of equilibrium protein folding studies in which CD has played a major role is investigating the effect of sequence on the stability of intact proteins and peptide segments. Here, the far UV is used to follow the **denaturation** profiles of the wild-type protein and of a series of mutant forms in which specific residues have been altered. From reversible thermal denaturation studies, changes in the free energy of unfolding, $\Delta\Delta G_{unf}$, can be obtained that measure the effect of the amino acid substitution upon the net stability of the protein and thus shed light on the effects of steric bulk, **hydrophobic** character, and charge effects. There have also been numerous CD studies of model peptides that adopt the helix conformation to examine systematically the effects of amino acid substitutions on α-helix stability (18). Similar studies have recently been reported for β-sheet-forming peptides and protein fragments (19).

Kinetic studies of protein folding by CD have been limited to stopped-flow methods and thus to the millisecond timescale. During the so-called dead time of the experiment, many proteins studied thus far (20) acquire a substantial fraction of α-helix and β-sheet, as evidenced by $[\theta]_{220}$ values that are generally closer to those of the native form than to the fully unfolded form. By contrast, the CD in the near-UV detected immediately after the dead time is essentially that of the unfolded form. These results generally have been interpreted as resulting from a so-called burst phase in which a molten globule-like intermediate is formed on the timescale of milliseconds or less. Then subsequent slower processes lead to additional formation or remodeling of the secondary structure and formation of the tertiary structure.

Membrane Proteins. The study of **membrane proteins** in their native environment poses problems for CD. Two kinds of potential artifacts must be avoided. Strongly scattering suspensions of membrane fragments exhibit preferential scattering of rcpl or lcpl, which manifests itself as an apparent CD signal that distorts the true CD of the membrane protein. The particulate character of membrane fragments also distort the CD signals because of the CD analog of Duysens' flattening (21), which occurs in samples with a highly inhomogeneous distribution of absorbing chromophores that is associated with clustering of the chromophores. Methods have been devised that correct for or avoid each of these difficulties, but in the flattening effect, they require that the membrane protein be transferred from the native membrane to small unilamellar vesicles that have, on average, no more than one protein molecule per vesicle. Alternatively, these artifacts are circumvented by solubilizing the membrane protein in a nonionic **detergent**. This poses some risk of inducing conformational changes in the protein, although nonionic detergents are usually relatively benign.

Protein–Ligand Interactions. CD is widely used to characterize ligand binding to proteins (see **Ligand binding**) and yield binding constants, stoichiometry, and information about conformational changes in the ligand and/or protein, in favorable cases. If the ligand absorbs at wavelengths above 300 nm, it is convenient to monitor the corresponding CD bands. This is especially so if, as in many cases, the ligand is achiral or a rapidly interconverting racemate. In such cases, the free ligand contributes no background CD, and only the bound ligand contributes to the CD signal. This is effectively so even in chiral ligands, such as NADH, FMN, FAD, in which the chromophore is conformationally mobile in the free form and therefore has only a weak CD signal. Ligand binding is also studied at shorter wavelengths, where protein CD bands are normally studied, either as induced ligand CD bands or as a perturbation of the aromatic and peptide CD bands. These perturbations arise either through direct interactions with the ligand or through changes in the tertiary or secondary structure induced upon ligand binding.

The use of CD to determine **dissociation constants** (K_d) is straightforward, but there are limitations to the range of dissociation constants that can be determined accurately. To provide an accurate analysis, the concentration of the species being monitored must be comparable to the K_d. Given the typical magnitudes of CD, this requires $K_d >$ ca. 1 μM for successful measurements by conventional CD. Fluorescence-detected CD (22) may extend this range to ca. 10 nM.

If $K_d <$ ca. 1 nM, CD cannot be used to measure reversible dissociation, but the protein–ligand complex dissociates negligibly at μM concentrations and above. Studies of such complexes are very fruitful because the ligand serves as a tightly but noncovalently bound spectroscopic probe, capable of reporting conformational changes in the protein. Heme proteins (23) and flavoproteins are important examples. The CD signals from such proteins are highly sensitive to oxidation state, spin state (in heme proteins), and to substrate, inhibitor, and allosteric effector binding.

CD OF NUCLEIC ACIDS

General Aspects (24)

The **purine** and **pyrimidine** bases of **DNA** and **RNA** are largely responsible for the CD spectra of nucleic acids in the wavelength range normally studied (180 to 300 nm). The sugar and phosphate groups do not absorb significantly above 200 and 180 nm, respectively. From the standpoint of CD, their main function is maintaining the relative geometries of the bases. Each base has a characteristic set of $\pi\pi^*$ transitions in the 180- to 300-nm region. The corresponding absorbance and CD bands are relatively broad. All five natural bases have one or two moderately intense $\pi\pi^*$ bands near 260 nm and several more intense bands in the 180- to 200-nm region. In addition, each base is expected to have several $n\pi^*$ transitions in the 180- to 300-nm region, but these bands absorb weakly. Though potentially strong in CD, few $n\pi^*$ bands have been identified, and the CD spectra of nucleosides, nucleotides, and polynucleotides are dominated by $\pi\pi^*$ contributions.

The CD spectra of nucleosides and nucleotides are relatively weak compared to those of polynucleotides. In the monomer, only base–sugar and base–phosphate interactions contribute, and the CD is averaged over a broad range of conformations. By contrast, oligo- and polynucleotide CD spectra are dominated by base–base interactions, and the range of conformations is usually narrower.

The CD spectra of oligonucleotides differ markedly from the sum of the constituent monomer spectra. This is illustrated by the spectra of ApA and pA (Fig. 3). The large difference is attributed to coupling of the $\pi\pi^*$ transitions in the two bases, another example of exciton coupling. This led to a model in which the adenine rings are stacked in a right-handed helix, like two successive bases in A-RNA or A-DNA. In fact, the CD of ApA closely resembles that of poly(A), except that it is only about half as large. This model is supported by theoretical calculations of the CD and more recently by NMR. Interestingly,

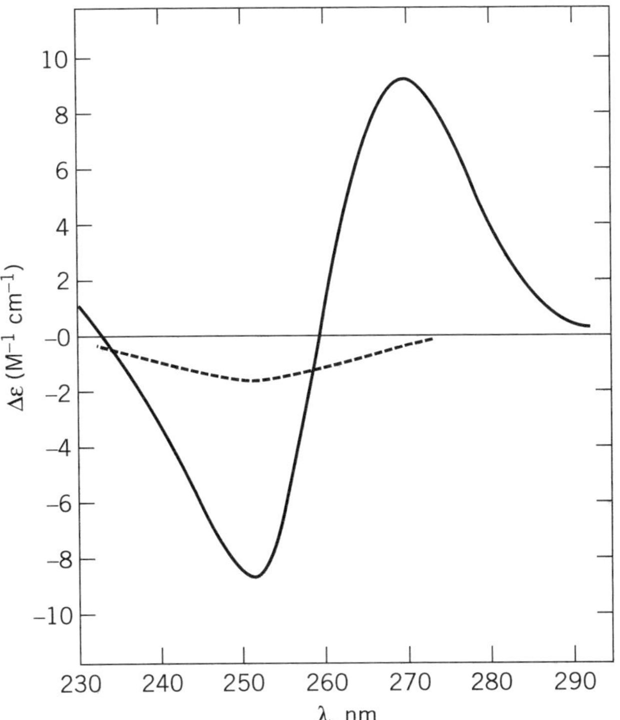

Figure 3. Near-UV CD spectra of the dinucleoside monophosphate ApA (———) and the mononucleotide pA (– – – –). Spectra obtained in 0.1 M Tris, 0.1 M NaCl, pH 7.4 at 5.5°C for ApA and at 20°C for pA (36).

although the CD spectra of dA, dAp, and dpA are similar to one another and to those of the riboA monomers, the CD spectrum of d(ApA) is quite different from that of ApA and reflect a difference in the geometry of base stacking.

The CD of nucleotide trimers can be described in most cases by the equation

$$[\theta]_{\text{ApBpc}} = (2[\theta]_{\text{ApB}} + 2[\theta]_{\text{BpC}} - [\theta]_{\text{Bp}})/3 \qquad (8)$$

This model assumes that (1) nearest neighbor bases stack similarly in dimers and in higher oligomers and (2) only nearest neighbor interactions are significant for CD. The model works reasonably well for higher oligomers and for trimers and has been extended to double-stranded oligonucleotides of DNA and RNA. The same model has been used to derive nearest neighbor frequencies for DNA and RNA from base-composition and CD data.

Secondary Structure. The CD of a DNA is highly diagnostic of its secondary structure (See **Secondary structure, nucleic acids**). The **B-DNA** conformation normally found in aqueous solution has a positive CD band near 275 nm and a negative band of similar magnitude near 245 nm, as shown in Fig. 4. A pair of closely spaced bands of opposite sign is called a couplet, and it is characterized by the sign of its long-wavelength component as a positive or negative couplet. A couplet in which the two lobes are of very similar amplitude is called a conservative couplet. In the far UV, the B-DNA spectrum has a positive couplet with peaks near 190 and 175 nm that is nearly an order of magnitude more intense than the near-UV bands.

The A-form of DNA (see **A-DNA**) is favored by low water activity and is induced by adding alcohols. In the near UV, the B → A transition is marked by a large increase in the positive band and a decrease in the magnitude of the 245 nm band. Therefore, the A-DNA CD spectrum is distinctly nonconservative. A strong negative band near 210 nm is also a characteristic feature of the A-DNA spectrum in contrast to the weak negative band in the B-DNA spectrum. Below 200 nm, a positive couplet is observed as in B-DNA, but the couplet is asymmetrical.

The B- and A-forms of DNA are both right-handed double helices. As expected, the left-handed Z form of DNA (see **Z-DNA**) has a very different CD spectrum, as shown for Z-form poly[d(G-C)] in Fig. 5. The near-UV bands for the Z-form are opposite in sign to those for the B-form and are shifted to longer wavelengths. The strong bands in the far UV are also reversed in sign and are red-shifted relative to those for the B-form. (Note that the CD spectrum of poly[d(G-C)] in the B-form differs from that of natural or "random-sequence" DNA. This is generally true for DNAs with simple di- and trinucleotide repeating sequences, in which specific features of geometry and electronic interactions are not averaged.) The reversed signs of the near-UV CD in the Z-form of poly[d(G-C)] were the first indications of a dramatically different conformation for this copolymer at high salt concentrations, but subsequent studies have shown that the signs of the near-UV CD bands do not reliably indicate helix sense for the double helix. However, the signs of the far-UV CD bands correlate with the helix sense in all known cases: a positive band between 180 and 192 nm for right-handed A-and B-forms and a negative band between 185 and 200 nm for the left-handed Z-form (25). Near-UV CD is widely used to monitor B → Z transitions in DNA.

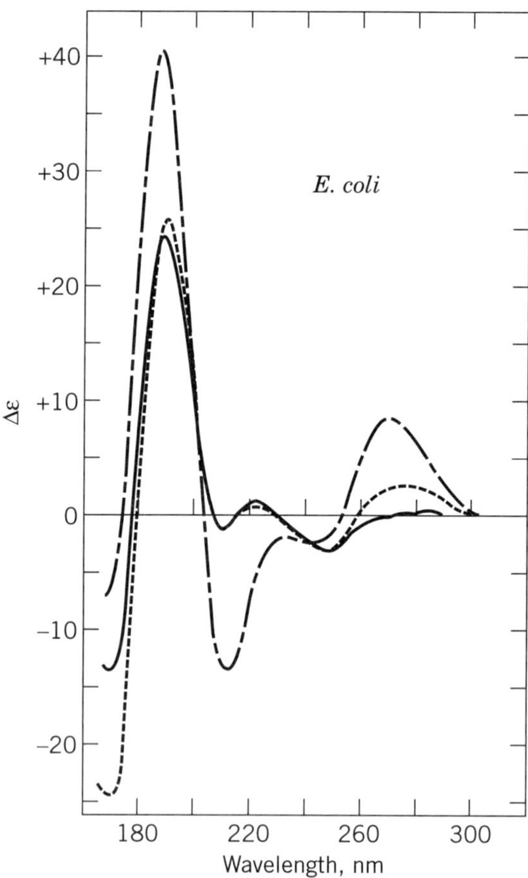

Figure 4. The CD spectra of *E. coli* DNA in the B form at low ionic strength (10 mM sodium phosphate buffer, ----), the A form (0.67 mM sodium phosphate buffer, 80% trifluoroethanol, —·—), and the B form at high ionic strengths (6 M NH_4F, ——) (37). (Reprinted with permission from Ref. 37, © 1985, Wiley-Liss.)

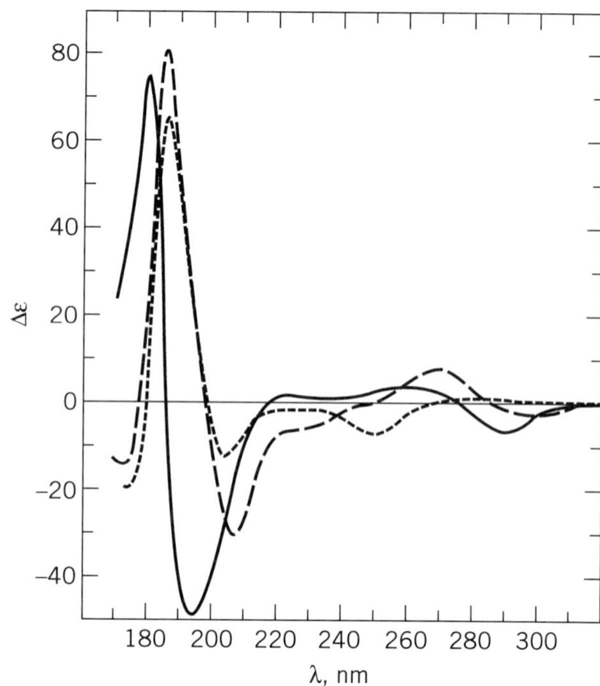

Figure 5. CD spectra of poly[d(GC)] in three conformations. B form (10mM sodium phosphate buffer, ······); A form (0.67 M sodium phosphate buffer, 80% trifluoroethanol, ----); Z form (10 mM sodium phosphate buffer, 2 M $NaClO_4$, ——) (38). (Reprinted with permission from Ref. 38, © 1985, IRL Press.)

The near-UV CD of B-DNA is highly sensitive to ionic strength. At ionic strengths comparable to physiological, the conservative CD pattern shown in Fig. 4 is observed. As ionic strength increases, the 275 nm band is selectively diminished, and at 6 M NH_4F, it nearly vanishes (Fig. 4). By contrast, the 245 nm band undergoes little change. Studies with closed circular DNA show that changes in ionic strength from ca. 0.1 M to 6 M change the number of base pairs per turn from ca. 10.4 to 10.2. The DNA in **chromatin** has a CD spectrum similar to that in 6 M NH_4F, indicating that it is a somewhat underwound form of B-DNA.

Double-stranded RNA is nearly always found in the A-form, and never in the B-form. Z-form RNA has been observed for poly[r(G-C)], but the conditions required to drive it into this form are more stringent than for the deoxy copolymer. A-form RNA has a CD spectrum similar to that of A-DNA, except that a weak negative band is commonly observed near 290 nm on the long-wavelength side of the 260 nm band. As with DNAs, the far-UV CD is diagnostic of the helix sense in double-stranded RNA, and the correlation between the sign of the far-UV couplet and the helix sense is the same.

In both RNA and DNA, the single-stranded forms have CD spectra that qualitatively resemble the double-stranded forms because base stacking has a much greater influence on the CD than does base pairing. Stand separation eliminates the latter but does not affect stacking significantly. Thermal denaturation of DNA leads to relatively small CD changes in the near UV but substantially larger far-UV CD changes. CD is also useful in characterizing **triplex** (26) forms of DNA and RNA.

Tertiary Structure. **Supercoiling** of DNA leads to readily detectable changes in the CD spectrum (27). The positive 275-nm band of B-DNA becomes more positive for negative superhelical supercoiling and conversely for decreased negative superhelical densities. For superhelical densities in the normal range, the CD changes are directly proportional to the superhelical density.

Although there is no evidence that the CD of RNA is directly sensitive to tertiary structure, it is sensitive to sequence and base pairing. Thus, CD has been used to test alternative models of RNA tertiary structure that differ in secondary structure distribution (28).

Ligand Binding to Nucleic Acids. Complexes of DNA and RNA with dyes, antibiotics, and other small ligands (See **Intercalation** and **Minor groove**) give rise to induced CD in the ligand absorption bands and to changes in the nucleic acid CD bands (29). These changes permit the determination of dissociation constants and stoichiometries for complexes of such small ligands with DNA and RNA. In some cases, additional structural information about the ligand or nucleic acid conformation is obtained.

CD is widely used to study DNA-protein and RNA-protein interactions (30) (see **DNA–protein interactions**). A very useful feature of CD in such studies is that the nucleic acid generally dominates the CD spectrum in the 260-nm region, whereas the protein dominates in the 200- to 240-nm region. Because each of these regions responds to the conformation of the dominant component, CD is of great value in monitoring conformational changes upon complex formation. Nonspecific protein–nucleic acid complexes often have dissociation constants in the micromolar range where CD is well suited for measuring K_d values. Specific protein–nucleic acid complexes usually have K_d s that are smaller by several orders of magnitude at physiological ionic strengths. In some cases, the K_d is shifted into the measurable range by increasing the ionic strength. Even if the binding is too tight to permit measuring the K_d by CD, conformational changes in the protein and/or nucleic acid are detected and characterized by CD. CD is also well suited for measuring the kinetics of protein–nucleic acid complex formation.

CARBOHYDRATES

Circular dichroism has important applications in studying carbohydrates (31), although these are more limited than for proteins and nucleic acids. Of the chromophores common in carbohydrates, only the amide (N-acetyl sugars) and **carboxyl groups** (uronic acids) have CD bands above 200 nm. The hydroxyl, ether, acetal, and ketal chromophores that are most typical have their first longest wavelength CD bands near the short-wavelength limit of conventional CD instruments, near 180 nm. Higher energy transitions are studied only with vacuum UV instruments, but they are usually obscured by solvent absorption, so studies are limited to thin solid films. For these reasons, optical rotation and ORD measurements in the near UV and visible regions continue to play an important role in studying carbohydrates.

Monomeric sugars have been investigated extensively, and sector rules relating the signs of CD bands near 170 nm to the conformation of the sugar have been derived. A useful model relating the optical rotation at the Na D line to the conformation of a sugar has been developed. This model is useful in testing features of the conformational energy surface in disaccharides. In the study of polysaccharides, CD and optical rotation have been used to characterize order-disorder, single- to multistranded, and sol to gel transitions (32).

BIBLIOGRAPHY

1. W. C. Johnson Jr. (1996) In *Circular Dichroism and the Conformational Analysis of Biomolecules* (G. D. Fasman, ed.), Plenum Press, New York, pp. 635–652.
2. T. A. Keiderling (1996) In *Circular Dichroism and the Conformational Analysis of Biomolecules* (G. D. Fasman, ed.), Plenum Press, New York, pp. 555–598.
3. K. Kuwajima (1996), in *Circular Dichroism and the Conformational Analysis of Biomolecules* (G. D. Fasman, ed.), Plenum Press, New York, pp. 159–182.
4. J. W. Lewis, R. A. Goldbeck, D. S. Kliger, X. Xie, R. C. Dunn, and J. D. Simon (1992) *J. Phys. Chem.* **96**, 5243–5254.
5. R. W. Woody (1996) In *Circular Dichroism and the Conformational Analysis of Biomolecules* (G. D. Fasman, ed.), Plenum Press, New York, pp. 25–67.
6. K. Imahori and N. A. Nicola (1973) In *Physical Principles and Techniques of Protein Chemistry* (S. J. Leach, ed.), Academic Press, New York, Part C, pp. 357–444.
7. R. W. Woody and A. K. Dunker (1996) In *Circular Dichroism and the Conformational Analysis of Biomolecules* (G. D. Fasman, ed.), Plenum Press, New York, pp. 109–157.
8. M. C. Manning, M. Illangasekare, and R. W. Woody (1988) *Biophys. Chem.* **31**, 77–86.
9. R. W. Woody (1992) *Adv. Biophys. Chem.* **2**, 37–79.
10. P. Urnes and P. Doty (1961) *Adv. Protein Chem.* **16**, 401–544.
11. W. C. Johnson Jr. (1990) *Proteins: Struct. Funct. Genet.* **7**, 205–214.
12. N. Sreerama and R. W. Woody (1994) *J. Mol. Biol.* **242**, 497–507.
13. S. Yu. Venyaminov and J. T. Yang (1996) In *Circular Dichroism and the Conformational Analysis of Biomolecules* (G. D. Fasman, ed.), Plenum Press, New York, pp. 69–107.
14. N. J. Greenfield (1996) *Anal. Biochem.* **235**, 1–10.
15. V. Baumruk, P. Pancoska, and T. A. Keiderling (1996) *J. Mol. Biol.* **259**, 744–791.
16. L. Zhong and W. C. Johnson Jr. (1992) *Proc. Natl. Acad. Sci. USA* **89**, 4462–4465.
17. S. Yu. Venyaminov and K. S. Vassilenko (1994) *Anal. Biochem.* **222**, 176–184.
18. J. M. Scholz and R. L. Baldwin (1992) *Ann. Rev. Biophys. Biomol. Struct.* **21**, 95–118.
19. D. L. Minor Jr. and P. S. Kim (1994) *Nature* **367**, 660–663.
20. H. Roder and G. A. Elöve (1994) In *Mechanisms of Protein Folding* (R. H. Pain, ed.), IRL Press, Oxford, pp. 26–54.
21. L. M. N. Duysens (1956) *Biochim. Biophys. Acta* **19**, 1–12.
22. D. H. Turner (1978) *Meth. Enzymol.* **49**, 199–214.
23. Y. P. Myer and A. J. Pande (1978) In *The Porphyrins* (D. Dolphin, ed.), Academic Press, New York, Vol. 3, pp. 271–322.
24. W. C. Johnson Jr. (1996) In *Circular Dichroism and the Conformational Analysis of Biomolecules* (G. D. Fasman, ed.), Plenum Press, New York, pp. 433–468.
25. J. H. Riazance, W. C. Johnson Jr., L. P. McIntosh, and T. M. Jovin (1985) *Nucleic Acids Res.* **15**, 7627–7636.
26. D. M. Gray, S.-H. Hung, and K. H. Johnson (1995) *Meth. Enzymol.* **246**, 19–34.
27. S. Brahms, S. Nakasu, A. Kikuchi, and J. G. Brahms (1989) *Eur. J. Biochem.* **184**, 297–303.
28. K. H. Johnson and D. M. Gray (1991) *Biopolymers* **31**, 385–395.
29. C. Zimmer and G. Luck (1992) *Adv. DNA Sequence-Specific Agents* **1**, 51–88.
30. D. M. Gray (1996) In *Circular Dichroism and the Conformational Analysis of Biomolecules* (G. D. Fasman, ed.), Plenum Press, New York, pp. 469–500.
31. E. S. Stevens (1996) In *Circular Dichroism and the Conformational Analysis of Biomolecules* (G. D. Fasman, ed.), Plenum Press, New York, pp. 501–530.
32. E. R. Morris, D. A. Rees, D. Thom, and E.J. Welsh (1977) *J. Supramol. Struct.* **6**, 259–274.
33. W. C. Johnson Jr. and I. Tinoco Jr. (1972) *J. Am. Chem. Soc.* **94**, 4389–4390.
34. S. Brahms and J. Brahms (1980) *J. Mol. Biol.* **138**, 149–178.
35. J. Bandekar, D. J. Evans, S. Krimm, S. J. Leach, S. Lee, J. R. McQuie, E. Minasian, G. Nemethy, M. S. Pottle, H. A. Scheraga, E. R. Stimson, and R. W. Woody (1982) *Int. J. Peptide Protein Res.* **19**, 187–205.
36. K. E. van Holde, J. Brahms, and A. M. Michelson (1965) *J. Mol. Biol.* **12**, 726–739.
37. W. C. Johnson Jr. (1985) *Meth. Biochem. Anal.* **31**, 62–125.

38. J. H. Riazance, W. A. Baase, W. C. Johnson Jr., K. Hall, P. Cruz, and I. Tinoco Jr. (1985) *Nucleic Acids Res.* **13**, 4983–4989.

Suggestions for Further Reading

G.D. Fasman (1996) *Circular Dichroism and the Conformational Analysis of Biomolecules*, Plenum Press, New York.

W.C. Johnson Jr. (1985) Circular dichroism and its empirical application to biopolymers, *Meth. Biochem. Anal.* **31**, 62–125.

K. Nakanishi, N. Berova, and R.W. Woody (1994) *Circular Dichroism: Principles and Applications*, VCH, New York.

D.W. Sears and S. Beychok (1973) Circular dichroism, In *Physical Principles and Techniques of Protein Chemistry* (S. J. Leach, ed.), Academic Press, New York, Part C, pp. 445–593.

R.W. Woody (1995) Circular dichroism, *Meth. Enzymol.* **246**, 34–71.

CIS CONFIGURATION

VERNON ANDERSON

The prefix *cis* has been used in chemistry to indicate "on the same side" (1). The original use came from describing **stereoisomers** of molecules with a ring or containing carbon–carbon double bonds. When similar substituents are on the same side of the ring or double bond, the **configuration** is referred to as *cis*, and when they are on opposite sides, they are referred to as ***trans***. The C2 and C3 hydroxyl groups of the furanose form of **ribose** are *cis*. The **conformation** about a single bond may also be noted as *cis* or *trans*, particularly when all the substituents lie in a plane. The *cis* conformation of a **peptide bond** has both α-carbons on the same side of the *N—C* amide bond (see ***Cis/trans* isomerization**). The notation **s-*cis*** (2) is used to emphasize that the conformation about a single bond is being described, as in *s*-1-*cis* retinal. These different uses of the chemical prefix *cis* are shown in Figure 1.

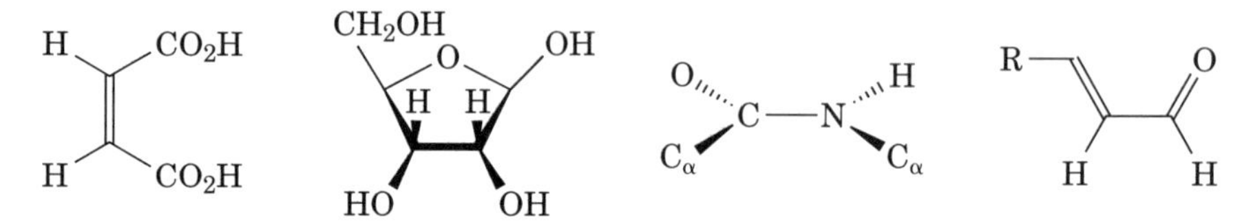

Figure 1. The various uses of the prefix *cis* are shown. The common feature is that the prefix denotes "on the same side."

BIBLIOGRAPHY

1. A. Baeyer (1888) *Liebig's Annalen der Chemie* **CCXLV**, 137.
2. R. Mullikan (1942) *Rev. Mod. Phys.* **14**, 265–267.

Suggestions for Further Reading

E. L. Eliel (1962) *Stereochemistry of Carbon Compounds*, McGraw-Hill, New York, pp. 318–371.

J. March (1985) *Advanced Organic Chemistry*, 3rd ed., Wiley-Interscience, New York, pp. 109–115.

CIS-ACTING

J. R. S. FINCHAM

This term is used in the context of regulation of gene **transcription** to denote DNA sequences that have to be physically joined to the **genes** in question to influence their activity. ***Trans*-acting** sequences, on the other hand, exert their regulatory effects regardless of whether or not they are present on the same chromosome as the regulated gene. *Trans*-acting genes generally encode proteins that bind to *cis*-acting DNA sequences.

Cis-acting sequences were classically identified as a result of their **mutation**, either silencing or activating the controlled gene, depending on whether the *cis*-acting sequence provided a binding site for **RNA polymerase** or other necessary **transcriptional factor** or for a **repressor** protein. These two kinds of *cis*-acting sequences, respectively positive and negative in their effect, were first identified in the *Escherichia coli* ***Lac operon***, where they were called the **promoter** and **operator** sequences. With the development of **recombinant DNA** technology, it became possible to detect protein-binding sites in DNA without mutation (e.g., by **footprinting** or **gel retardation assays**), and then to determine their effects on transcription by "engineered" deletions.

Cis-acting sequences that act otherwise than by providing binding sites for regulatory proteins are comparatively uncommon, but two quite different examples should be mentioned. In bacterial **operons**, a mutation that causes premature translational termination of an **upstream** gene can cause a drastic reduction in the expression of genes in the same **operon** further **downstream**. This is a strictly *cis*-acting effect, acting within a single unit of transcription. The mechanism is explained under **Lac Operon**.

The second example is from female mammals where one of the two **X-chromosomes** in each cell lineage is largely inactivated with the notable exception of one gene, active only in the otherwise "silent" X, that is transcribed into a long RNA molecule called Xist. This transcript, which does not encode a protein, plays an essential role in silencing the rest of the chromosome. It acts only in *cis* and remains associated with the chromosome from which it is transcribed without effect on the other, active, X-chromosome (see **X-chromosome inactivation**).

X-chromosome inactivation involves propagating a certain type of condensed and transcriptionally inactivating chromatin structure, generally called **heterochromatin**, which, in most chromosomes, is restricted to certain segments. In

Drosophila melanogaster, heterochromatin exerts a *cis*-effect in suppressing gene activity when, as a result of segmental chromosomal rearrangement, it is brought close to genes not normally associated with it. The multiprotein heterochromatin complex spreads along the chromosome, inactivating the genes in its path with a probability that decreases with distance (see **Position Effect**). Genes that are normally close to heterochromatin are apparently insulated in some way against this effect.

CIS-DOMINANCE

J. R. S. FINCHAM

Dominance, in the genetic sense, is a property of one form (**allele**) of a **gene** relative to another allele of the same gene, said to be recessive. Where dominant and **recessive alleles** are present together, the dominant, not the recessive, registers its effect on the organism. Most mutant alleles have lost part or all of their activity and then are usually recessive. When mutant alleles are dominant it is usually because they are hyperactive or have been freed from some normal constraint. The term *cis*-dominant means that the mutation is in an element that has to be physically joined to the gene to exert its effect (ie, in *cis*), not separated from it on another chromosome.

Insofar as a gene is, by definition, an integrated (*cis*-acting) unit of genetic function, any part of a dominant allele, including its coding sequence, is said to be *cis*-dominant. But, in practice, the term (if it is used at all) is usually reserved for DNA sequences that govern gene activity in *cis* from outside the transcribed or translated sequence.

The concept of *cis*-dominance arose from the classical work of Francois Jacob and Jacques Monod at the Pasteur Institute in Paris on utilizing sugar lactose by the bacterium *E. coli* (see ***lac* operon**). Lactose utilization by the bacterium depends on two proteins, the enzyme **β-galactosidase**, which cleaves lactose to glucose and galactose, and a **membrane protein**, β-galactoside permease, needed for uptake of lactose into the cell. These proteins are normally synthesized by *E. coli* only in the presence of lactose in the growth medium, or a lactose analogue, that acts as an **inducer**.

Two classes of mutations result in failure to grow on lactose as a carbon source: *lacZ* mutants lack βgalactosidase and *lacY* mutants lack the permease. *LacZ* and *lacY* are two genes that are closely linked (together with a third, seemingly inessential gene *LacA*) in what was later shown to be a single unit of **transcription**, or **operon**, transcribed in the sequence Z - Y - A.

Two other kinds of mutants differ from wild type in producing β-galactosidase and permease **constitutively**, that is, whether inducer is present or not. They were mapped respectively in a separate gene, *LacI*, closely linked to *LacZ*, and in a so-called **operator** segment (o) at the end of *LacZ* that overlaps the LacZ transcription start point. The *LacI* constitutive mutants (*LacI*−) are recessive to wild type (*LacI*+), which is interpreted as meaning that *LacI*+ encodes a transcriptional **repressor** and that the repressor function is nullified by inducer. The wild-type operator segment ($o+$) was given the hypothetical role of binding to the *LacI* repressor, blocking *lac* operon transcription. The operator-constitutive mutants (o^c) are interpreted as changes in the DNA sequence of the operator so that it no longer bound the repressor. The inducer, was thought to act by binding to the repressor protein and changing its conformation, so that it is no longer bound to the operator (see Fig. 1). All of these elements of the model have since been confirmed by molecular analysis.

The tests for dominance were made in partial diploids (**merodiploids**) that have a second copy of the *Lac* operon brought into the cell as part of of an **F′ plasmid**. Some of the key results are shown in the table of Fig. 1. The formal demonstration of *cis*-dominance comes from the comparison of items 5 and 6. The *trans* merodiploid, $o^+LacZ^-/o^c\ LacZ^+$, is constitutive, but the *cis* merodiploid, $o^+LacZ^+/o^c\ LacZ^-$, in inducible.

The *cis*-dominance of o^c is readily understandable. There is no way in which the binding of a protein to one piece of DNA could affect the transcription of an unconnected piece. The operator segment is an integral part of the *lac* operon and cannot function when separated from it. On the other hand,

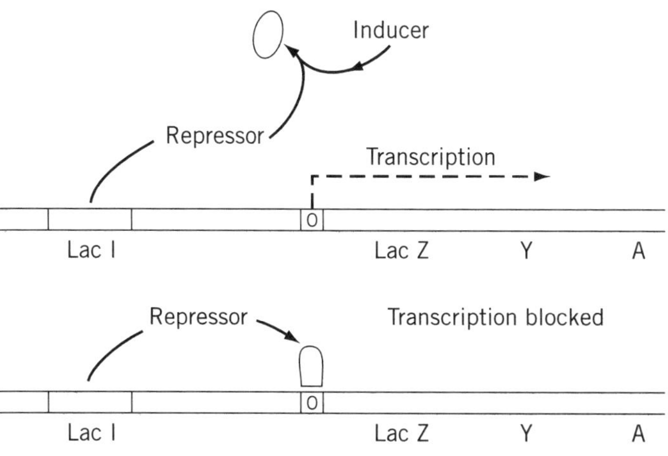

		Lac Z (β-galactosidase) activity	
		No inducer	With inducer
1	$I^+\ O^+\ Z^+$	−	+
2	$I^+\ O^c\ Z^+$	+	+
3	$I^s\ O^+\ Z^+$	−	−
4	$I^+\ O^+\ Z^-$	−	−
5	$I^+\ O^+\ Z^-$ / F′ $I^+\ O^c\ Z^+$	+	+
6	$I^+\ O^+\ Z^+$ / F′ $I^+\ O^c\ Z^-$	−	+
7	$I^s\ O^+\ Z^+$ / F′ $I^+\ O^+\ Z^-$	−	−
8	$I^s\ O^+\ Z^-$ / F′ $I^+\ O^c\ Z^+$	+	+

Figure 1. The regulation of the *Escherichia coli lac* operon and two kinds of mutations that have dominant effects: $LacI^s$ (non-inducible), dominant both in *cis* and in *trans*; and o^c (constitutive), dominant only in *cis*. The partial diploids for testing dominance (items 5–8) were made by introducing F′ plasmids that carry a second copy of the *lac* segment of the bacterial genome. o^c is resistant to the wild type ($LacI^+$) repressor and also to the $LacI^s$ superrepressor. $LacI^-$ mutants, not shown here, are constitutive and recessive. Information from Refs. 1 and 2.

LacI is a separate and functionally autonomous gene that acts on *LacZ* in *trans* just as well as in *cis*.

Mutations of the rather uncommon *LacI*^{s} type change the repressor into a superrepressor that is insensitive to inducer and confers noninducibility. Its dominance over *LacI*$^{+}$*LacI*$^{+}$ (both in *cis* and in *trans* - see item 7 in Fig. 1) is due to the *LacI* repressor functioning as a **dimer**. In mixed *LacI*$^{+}$/*LacI*^{s} protein dimers, the noninducibility of the *LacI*^{s} component is imposed on the whole.

The concept of *cis*-dominance was important at a time when the idea that gene sequences had regulatory and structurally determining functions was just emerging. Monod and Jacob found that *LacI* acts in *trans* on *LacZ* (and *LacY*) activity and the *o* segment only in *cis*. Today it is recognized that the activities of most genes are regulated by multiple *trans*-acting proteins, encoded by other genes, that bind to DNA sequences (**promoters, enhancers, silencers**) that act in *cis*. But it is questionable whether *cis*-dominance is an appropriate term in reference to all these *cis*-acting transcriptionally controlling sites. Dominance implies that there are recessive alleles over which the dominance can be demonstrated, which is seldom the case, and the *cis-trans* comparison, which could provide the evidence that the effect applies only in *cis*, is hardly ever available.

One dominant mutation, in ***Drosophila***, for which the *cis-trans* comparison has been made in Contrabithorax (*Cbx*), which maps close to the ***Ultrabithorax*** (*Ubx*) gene and extends its activity in the **embryo** in an anterior direction. Here the effect of *Cbx* is strong in flies of constitution *Cbx* + / + *Ubx* (*Ubx* being a virtually null allele) but weak, though still significant, in *Cbx Ubx*/ ++. In the latter genotype, the *Cbx* regulator is separated from the gene that it regulates. *Cbx* can be called *cis*-dominant, but the observation that it also has a small effect in *trans* is interesting. E. B. Lewis (3) showed that this and other *trans* effects to do with *Ubx* depend on the close pairing of homologous chromosomes that is a special feature of several *Drosophila* tissues (see ***Transvection*** and **Allelic complementation**).

The concept of *cis*-dominance is an important part of molecular biological history, even though the current usefulness of the term is rather limited. The principle of *cis*-acting controls of gene activity, to which the concept led, is now a commonplace and essential part of molecular biology.

BIBLIOGRAPHY

1. J. R. Beckwith (1970) In *The Lactose Operon* (J. R. Beckwith & D. Zipser, eds.), Cold Spring Harbor Laboratory Press, Cold Spring Harbor, New York, pp. 1–26.
2. F. Jacob and J. Monod (1961) On the regulation of gene activity. *Cold Spring Harbor Symp. Quant. Biol.* **26**, 193–209.
3. E. B. Lewis (1955) *Am. Nat.* **89**, 73–89.

CIS/TRANS ISOMERIZATION

G. FISCHER

The distinction between the *cis* and *trans* isomers of a molecule originates from a geometry-based classification of molecular structures. When the two substituents X and Y are on the same side of the structural unit [Fig. 1(**a**)] the isomer is *cis*. It is designated *trans* in the opposite arrangement. Ambiguities in the designation of the isomers are avoided by using the more sophisticated *E/Z* nomenclature, which often has *E* for the *cis* form and *Z* for the *trans* form. The interconversion between the isomers, the *cis/trans* isomerization, occurs by rotation about the central linkage [Fig. 1(**a**)]. There is a barrier to rotation quantified by the **free energy** of activation $\Delta G^{\ddagger}$ that is proportional to a first-order rate constant k_{obs}, where $k_{obs} = (k_{cis\ to\ trans} + k_{trans\ to\ cis})$ for the reversible isomerization. The bond order of the central linkage correlates with the magnitude of the energy barrier $\Delta G^{\ddagger}$. It ranges from high values of $\Delta G^{\ddagger} > 100$ kJ/mol for double bonds, via an intermediate range for linkages having a partial double bond character, down to low values in the range of $k_B T$ (where k_B is the Boltzmann constant and T the absolute temperature) for C–C single bonds. Under ambient conditions, a rotational barrier >90 kJ/mol indicates that the individual isomers are not readily interconverted, and they may exhibit quite distinct chemical properties. A corresponding $\Delta G^{\ddagger}$ value has been used to discriminate semantically between different **conformations** (barrier to rotation <90 kJ/mole) and **configurations** of a molecule.

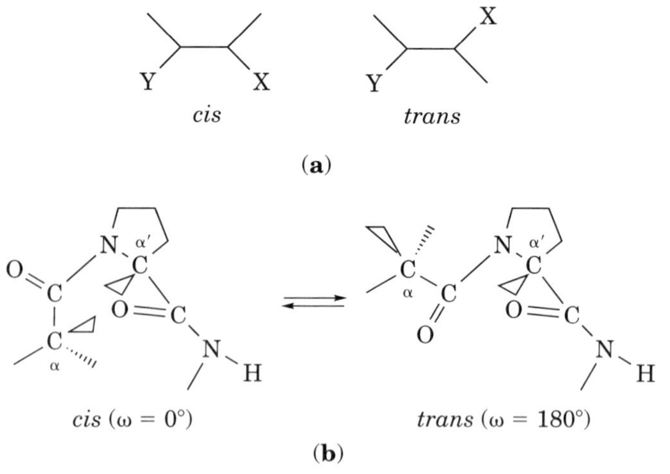

cis (ω = 0°) trans (ω = 180°)

Figure 1. (**a**) General representation of *cis/trans* isomerism (**b**) *Cis/trans* isomerism of a prolyl peptide bond.

In a polypeptide chain, both the amide -CONH- and the **imide** –CON < **peptide** groups are intrinsically competent to undergo *cis/trans* isomerization by rotation about the torsion angle ω of the peptide bond (see **Ramachandran plot**). Among the gene-encoded amino acids, the imide peptide bond is exclusively formed by **proline** residues. Because the lone electron pair of the nitrogen atom is delocalized over the peptide bond, the C–N linkage has a partial double-bond character. Typically, this destabilizes twisted conformations but stabilizes planar arrangements (*trans*, $\omega = 180°$ *cis*, $\omega = 0°$) of the two α-C atoms adjacent to the peptide bond [Fig. 1(**b**)]. The *cis/trans* isomerization of the -CONH- moiety is relatively fast (half-time <1 sec at room temperature) and leads to a very small percentage of *cis* isomer, ≫1%, at equilibrium. In contrast, the peptidyl-proline moiety (in the following referred to as a prolyl bond) often has *cis/trans* isomers in comparable amounts. Steric constraints, which favor a *trans* arrangement in the case of -CONH-, are similar in both isomers for prolyl bonds, and the *cis* isomer generally occurs in about 20% of the prolyl peptide bonds. There is a relatively high rotational barrier $\Delta G^{\ddagger}$ of about 80 kJ/mol for the prolyl bond. The combination of a substantial population of both isomers and their slow interconversion implies that *cis* prolyl bonds have considerable influence on biochemical reactions of the polypeptide backbone.

In the absence of a folded conformation, polypeptide chains can theoretically form 2^n *cis/trans* isomers, where *n* is the number of prolyl bonds in the molecule. In simple cases, as with a four-proline octapeptide derived from the prolactin **receptor**, all of the possible isomers have been detected and quantified in solution (1). Structural formation in the peptide chain, however, reduces the number of isomers substantially. A particular prolyl bond in native proteins is usually either *cis* or *trans* in all the molecules, although *cis/trans* isomerization in the native structure has been observed for a few proteins by **NMR** methods in solution. In the latter case, structural alterations propagate through the backbone around the isomeric proline residue, which can be accompanied by distinct biological activities of the isomeric proteins (2,3). Depending on the polypeptide structure, the **half-time** for prolyl bond isomerization ranges from seconds to hours.

From stereochemical considerations, the peptidyl transferase center on the **ribosome** is thought to be constructed for synthesizing all peptide bonds in the *trans* conformation (see **Protein biosynthesis**). However, in native, globular proteins of known three-dimensional structure, about 5 to 6% of prolyl peptide bonds are *cis* (4,5). When proline-containing proteins are unfolded in the presence of high concentrations of **denaturants**, such as **urea** or **guanidinium chloride** (GdmCl), the **random coil** polypeptides generally equilibrate slowly to a mixture of numerous *cis/trans* isomers. For the fraction of unfolded molecules that have one or more nonnative isomers of a prolyl bond, subsequent **refolding** of the protein has to start from different conformational states. If the *cis/trans* isomerization is slower than refolding, slow kinetic phases of folding may be apparent when the time course of refolding is monitored. For that reason, *cis/trans* isomerization is the rate-limiting step in folding for some proteins. It is generally most easily detectable when there is a *cis* prolyl bond in the native state, because then a large fraction of unfolded molecules has the incorrect isomer (6,7).

Chemical catalysis of prolyl bond isomerization is rare. Most organic solvents and micelles or phospholipid **vesicles**, cause a moderate decrease in the rotational barrier. The rate constants are independent of the pH value in the physiological range, unless dissociable groups are located adjacent to proline. An increased rotational barrier results from the O-protonation of the peptide bond in acidic solution, characterized by a $pK_a = -1$. In strong acids, N-protonated species become populated ($pK_a = -7$) that accelerate *cis/trans* isomerization (8). Many relationships between the peptide structure and prolyl bond isomerization have been elucidated. For example, when measured in oligopeptides, the increased barrier to rotation caused by aromatic amino acids that precede the proline residue is accompanied by an increase in the *cis* population of up to 40%, whereas small aliphatic side chains in the same position lead to lower $\Delta G^{\ddagger}$ values in conjunction and lower *cis* contents of 5 to 10% (9).

The conformational constraints on the polypeptide backbone induced by the prolyl *cis/trans* isomerization restrict bimolecular recognition processes. Conformational selectivity was demonstrated by the **hydrogen bond** directed preferential binding of proline peptides in the *cis* conformation to **β-cyclodextrins** (10) and to a synthetic, multidentate terephthaloyl amide (11). For the recognition of opioid peptides of the dermorphin type by μ- and ∂-receptors, specificity for the *cis* conformers was suggested (12). In a biological context, it may be important that endoproteinases, such as **chymotrypsin**, **trypsin**, **thrombin** and clostripain cannot readily cleave a peptide bond adjacent to a *cis* prolyl moiety, even if the isomeric bond occupies a position remote from the scissile bond (13,14). Due to this conformational specificity, the rate of the *cis* to *trans* isomerization of the prolyl bond limits the rate of proteolysis for good substrates in the presence of high protease concentrations, and this is a useful assay for isomerization. Even *in vivo*, the time course of bradykinin (Arg-Pro-Pro-Gly-Phe-Ser-Pro-Phe-Arg) degradation by pulmonary endothelial **peptidases** is controlled by the conformational specificity of the proteinases (15).

BIBLIOGRAPHY

1. K. D. Oneal et al. (1996) *Biochem. J.* **315**, 833–844.
2. A. P. Hinck, E. S. Eberhardt, and J. L. Markley (1993) *Biochemistry* **32**, 11810–11818.
3. J. Kordel, S. Forsen, T. Drakenberg, and W. J. Chazin (1990) *Biochemistry* **29,** 4400–4409.
4. M. W. MacArthur and J. M. Thornton (1991) *J. Mol. Biol.* **218**, 397–412.
5. D. E. Stewart, A. Sarkar, and J. E. Wampler (1990) *J. Mol. Biol.* **214**, 253–260.
6. J. F. Brandts, H. R. Halvorson, and M. Brennan (1975) *Biochemistry* **14**, 4953–4963.
7. F. X. Schmid (1986) *Methods Enzymol.* **131**, 70–82.
8. H. Sigel and B. R. Martin (1982) *Chem. Rev.* **82**, 385–426.
9. R. K. Harrison and R. L. Stein (1992) *J. Amer. Chem. Soc.* **114**, 3464–3471.
10. M. Lin et al. (1995) *Anal. Chim. Acta* **307**, 449–457.
11. C. Vicent, S. C. Hirst, F. Garciatellado, and A. D. Hamilton (1991) *J. Amer. Chem. Soc.* **113**, 5466–5467.
12. R. Schmidt et al. (1995) *Int. J. Peptide Protein Res.* **46**, 47–55.
13. G. Fischer, H. Bang, E. Berger, and A. Schellenberger (1984) *Biochim. Biophys. Acta* **791**, 87–97.
14. S. Meyer, M. Drewello, and G. Fischer (1996) *Biol. Chem.* **377**, 489–495.
15. M. P. Merker and C. A. Dawson (1995) *Biochem. Pharmacol.* **50**, 2085–2091.

Suggestions for Further Reading

T.E. Creighton, ed. (1992) *Protein Folding*, W.H. Freeman, New York; A book written by specialists that summarizes proline-limited protein folding in the context of other folding events.

B. Testa (1982) The geometry of molecules: basic principles and nomenclature, In *Stereochemistry* (C. Tamm, ed.) Elsevier Biomedical Press, Amsterdam, New York, Oxford, pp. 1–47.

CLAMP LOADERS, PROCESSIVITY COMPLEX

ZVI KELMAN

Chromosomal **replicases**, the **DNA polymerases** that replicate **chromosomes**, are multiprotein complexes characterized by the high processivity of their DNA synthesis. These replicating machines polymerize thousands of nucleotides without dissociating from the **template**. The replicases of the three well-characterized systems (***Escherichia** coli*, **eukaryotes**, and **bacteriophage** T4), which span the evolutionary spectrum, are similar in function, structure, and overall organization. Their remarkable processivity is achieved by a ring-shaped processivity factor (or "**sliding clamp**") that binds the polymerase catalytic unit and tethers it to the DNA. Another complex of accessory proteins is required to load the clamp onto the DNA. The sliding clamp does not have any affinity for DNA and, therefore, other proteins are needed to assemble the clamp around the DNA. The accessory protein complex ("clamp loader"; also called a "molecular matchmaker") binds to the primer terminus and **couples** ATP hydrolysis to the assembly of the ring around the DNA primer. Therefore, the replicase can be thought of as having three components: (1) a polymerase, (2) a clamp loader complex that assembles the clamp around DNA and (3) a DNA sliding clamp (Table 1).

THE CLAMP LOADERS

The three best-studied replicases are the *E. coli* DNA polymerase III holoenzyme (polIII), the eukaryotic polymerase δ (polδ), and the replicase of phage T4. These three multisubunit complexes share structural and functional similarities (reviewed in Refs. 1–3). The clamp loaders for these systems are the γ-complex of prokaryotes, the replication factor-C (RF-C) complex (also called Activator-1) of eukaryotes, and a complex of the products of gene 44 and gene 62 (gp44/62) of phage T4. The γ-complex is composed of five subunits called γ, δ, δ', χ and ψ. RF-C is also a five-subunit complex, composed of one large and four small subunits (p140, p40, p38, p37, and p36). gp44/62 contains only two polypeptides gp44 and gp62. In the complex, a tetramer of gp44 is tightly associated with one gp62. Interestingly, the three clamp loaders share amino acid similarities among the subunits (4,5).

A detailed mechanism by which these clamp loaders operate in assembling the sliding clamp around DNA is not yet known. The basic steps of the loading process, however, have been elucidated (Fig. 1). In general, the clamp loader recognizes the 3′ end of the junction between the single strand and duplex DNA (primer/template) and uses ATP hydrolysis to assemble the clamp around the primer. The status of the clamp loader after assembling the clamp around DNA is not yet clear. It has not yet been determined whether the clamp loader leaves the clamp on the DNA, where it interacts with the polymerase to initiate processive DNA synthesis, or if the clamp loader is also needed to assist binding the polymerase to the clamp (the different features of each clamp loader are discussed later).

Table 1. Three-Component Structures of Chromosomal Replicases

	DNA Polymerase	Clamp loader	Sliding Clamp
Eukaryotes	polδ (3 subunits)	RF-C complex (5 subunits)	PCNA
E. coli	Core polymerase (3 subunits)	γ-complex (5 subunits)	β-subunit
Phage-T4	gp43	gp44/62	gp45

In prokaryotes and eukaryotes, the clamp loader also functions as a clamp unloader (6). Upon completion of an **Okazaki fragment**, the polymerase rapidly dissociates from the clamp (7–9), leaving the clamp on the DNA (10). The clamps of polIII (the β subunit) and polδ [proliferating cell nuclear antigen (PCNA)] are relatively stable on DNA (6, and references therein). Therefore, it was postulated that the sliding clamp remains assembled around the DNA when the synthesis of an Okazaki fragment is complete. During lagging strand synthesis, a new sliding clamp is needed for the synthesis of each Okazaki fragment. In *E. coli* about 10 times more Okazaki fragments are formed during replication than there are β-subunits within the cell. In human cells, the number of Okazaki fragments formed during one round of chromosomal replication has been estimated to be 100 times greater than the amount of PCNA. Therefore, recycling of the clamp (β and PCNA) is needed to fulfill the requirement for a constant supply of clamps for further DNA synthesis. In phage T4, the clamp readily dissociates from DNA upon completion of an Okazaki fragment. Thus, the clamp loader has a dual function during DNA replication: (1) loading of the sliding clamp onto DNA to initiate processive DNA synthesis and (2) recycling the clamps on the lagging strand.

γ-Complex

The γ-complex is composed of five subunits (γ, δ, δ', χ, ψ) (reviewed in Ref. 11). The genes encoding them have been identified, and the purified proteins have been used to study the function of the individual subunits and subassemblies of the γ-complex (discussed later). Four of the subunits (δ, δ', χ and ψ) are encoded by unique genes. The γ-subunit is formed from the same gene (*dnaX*) that encodes τ (another subunit of polIII) by an efficient translational **frameshift** mechanism that produces γ in an amount equal to τ (reviewed in Ref. 12). As a result, the γ-subunit consists of the N-terminal 430 residues of τ followed by a unique C-terminal Glu residue. In addition, the γ-and δ'-subunits show amino acid sequence similarity.

No one subunit alone can assemble the β sliding clamp around DNA (11 and references therein). At low ionic strength, a combination of γ and δ assemble β onto DNA, but the reaction is feeble. The δ'-, χ- and ψ-subunits are needed for an efficient loading reaction under physiological conditions (13).

The γ-complex has only weak affinity for single-stranded and duplex DNA, and it exhibits weak DNA-dependent ATPase activity, but the ATPase activity is stimulated by the sliding clamp (β-subunit) (14). The best DNA effector for the ATPase activity is a singly primed template, indicating an interaction between the γ-complex and the primer/template junction. The γ-subunit is the only subunit of the γ-complex with an **ATP-binding motif**, and it binds ATP (15). Furthermore, mutation of the ATP-binding site of γ within the γ-complex destroys the ATPase activity and the ability to assemble the β-ring onto DNA. The γ-subunit, however, lacks significant ATP hydrolysis activity even in the presence of DNA (14,15), and the δ- and δ'-subunits are required for ATPase activity, implying that the $\gamma\delta\delta'$ complex recognizes DNA. The γ-complex binds β in the presence of ATP. The δ subunit interacts with β, with a strength similar to that of the entire γ-complex (16), indicating that β binds the γ-complex mainly through the δ-subunit. Interestingly, however, the interaction between the β-subunit and δ does not depend

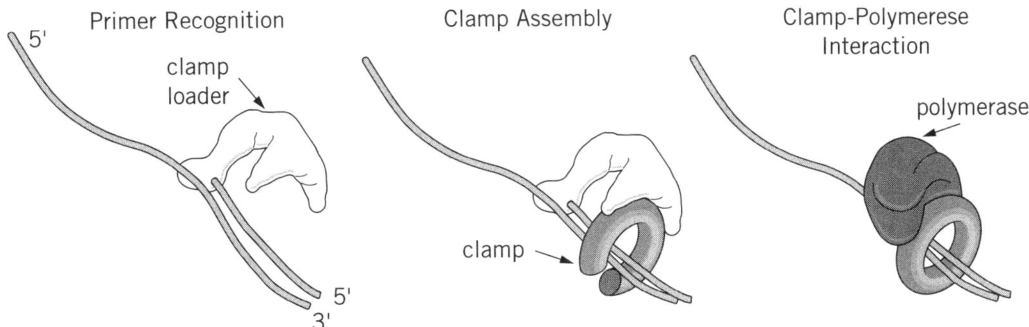

Figure 1. A model for clamp loader activity. The clamp loader recognizes the 3′ terminus of a primer and assembles the sliding clamp around the DNA. Following assembly, the clamp interacts with the polymerase and tethers it to the DNA for processive DNA synthesis.

on ATP (whereas the γ-complex requires ATP to bind the clamp). These apparently contradictory observations can be explained if the δ subunit is buried within the γ-complex and ATP induces conformational changes that lead to the interaction of δ with β (16). The γ-complex also interacts with **single-stranded DNA binding protein** (SSB) via the χ subunit (11). These interactions are important for efficient clamp assembly at physiological ionic strength.

The results obtained with the individual subunits and their subassemblies demonstrate the basic features of the γ-complex action during loading of the clamp onto DNA (Fig. 1). Upon binding of ATP, the γ-complex undergoes conformational changes leading to interaction of the δ-subunit with β. The γ-complex recognizes a primed template, aided by the interaction between χ and SSB. Assembly of the ring around the DNA primer stimulates the ATPase activity of the γ-complex. ATP hydrolysis results in conformational changes such that δ is again buried within the γ-complex leaving the clamp around the DNA primer. Now the clamp interacts with the polymerase to initiate processive DNA synthesis (Fig. 1).

Replication Factor-C (RF-C)

RF-C was first isolated from human 293 cells as an essential replication factor for the *in vitro* replication of simian virus-40 (**SV-40**) (17). The factor was purified later from different sources, including yeast and human cell lines (18 and references therein), based on its ability to stimulate processive DNA synthesis by polδ in the presence of PCNA and ATP.

Purified RF-C is a five-subunit complex (p140, p40, p38, p37, and p36). The genes that encode all RF-C subunits from yeast and humans have been identified (4, 5 and references therein), and all are essential in yeast. Although encoded by different genes, they show extensive amino acid sequence similarities in the central region of the protein (reviewed in Ref. 5).

Early biochemical studies determined the role of RF-C as a clamp loader of PCNA and suggested a mode of action during the assembly reaction (Fig. 1). RF-C binds preferentially to single-strand/duplex DNA (template/primer) junction in the presence of ATP (19,20). PCNA interacts with the RF-C/DNA complex. The DNA-dependent ATPase activity of RF-C is stimulated by the binding of PCNA. The hydrolysis of ATP leads to conformational changes within the RF-C, which locks the PCNA sliding clamp around the DNA. Then the clamp interacts with the polymerase to initiate processive DNA synthesis (Fig. 1).

Biochemical analysis of RF-C individual subunits and the protein–protein interactions within the RF-C complex is underway (eg, 21, 22). Purified proteins, subcomplexes and mutational analysis have shed light on the function of the individual subunits and subassemblies of the RF-C complex. Although each subunit contains an ATP-binding motif, only p40 binds ATP (20). Yeast p36 (23) and a subcomplex of human RF-C (p36-p37-p40) exhibit DNA-dependent ATPase activity. The DNA-binding activity resides within the large subunit (p140) (20), although the p37 subunit also binds DNA weakly. All of the subunits, except p37, bind PCNA (24–26). The intact complex, however, is needed for the assembly of PCNA around the DNA and, to date, no subunit combination analyzed performs this task. Unloading is probably a simpler mechanism, as the p40 subunit is sufficient to unload PCNA from DNA and also interacts with polδ (26). The role of this interaction in DNA synthesis, however, is not yet clear. RF-C also interacts with SSB. The interaction between RF-C and SSB might be needed for the assembly of the clamp, as demonstrated for the *E. coli* γ-complex.

gp44/62

Similar to the γ-complex and RF-C, the clamp loader of phage T4, the gp44/62 complex, is also composed of five subunits (Table 1). In this case, however, only two polypeptides form the pentamer, where a tetramer of gp44 binds a single subunit of gp62 (27). Like the clamp loader of eukaryotes and prokaryotes, the gp44/62 complex exhibits ATPase activity that is stimulated by the sliding clamp (gp45) and by DNA (28, reviewed in Ref. 29). A DNA structure that resembles a template/primer junction is the best effector for stimulating ATPase activity (27). Cross-linking and protein-DNA **footprinting** assays demonstrate that the clamp loader interacts with DNA in the absence of gp45, but these interactions are relatively weak, and the clamp is needed for binding to DNA (reviewed in Refs. 29,30). Studies with individual subunits of the gp44/62 complex revealed that the gp44 exhibits the ATPase activity, whereas gp62 interacts with the clamp (31).

The exact mechanism of action of the gp44/62 complex is not yet fully understood. The overall features of its activity, however, have been determined and are similar to those of the

γ-complex and RF-C (Fig. 1). The gp44/62 complex interacts with gp45 in an ATP-dependent manner. The complex recognizes the primer terminus, which stimulates the ATPase activity of the clamp loader, bringing about conformational changes that lock the clamp around the primer and allow it to interact with the polymerase to initiate processive DNA synthesis.

CONCLUDING REMARKS

The polymerases responsible for replicating chromosomal DNA during cell division are multiprotein complexes. The processivity of these enzymes relies on a ring-shaped protein that encircles DNA and tethers the catalytic unit to DNA for processive DNA synthesis. The mechanisms by which the clamp loaders assemble the ring around DNA and disassemble them from DNA are only beginning to be understood. The complexity of this mechanism is inferred from the observation that it requires the coordinate activity of several polypeptides in eukaryotes, prokaryotes, and phage T4. Future studies are needed to elucidate the mechanisms by which the clamp loaders operate in the loading and unloading reactions. Purification of the clamp loader complexes and the individual subunits will enable further analysis of these mechanisms.

BIBLIOGRAPHY

1. B. Stillman (1994) *Cell* **78**, 725–728.
2. Z. Kelman and M. O'Donnell (1994) *Curr. Opinion Genet. Dev.* **4**, 185–195.
3. B. Stillman (1996) *DNA Replication in Eukaryotic Cells.* M. L. DePamphilis, ed., CSH Laboratory Press, Cold Spring Harbor, NY, pp. 435–460.
4. M. O'Donnell, R. Onrust, F. B. Dean, M. Chen, and J. Hurwitz (1993) *Nucleic Acids Res.* **21**, 1–3.
5. G. Cullmann, K. Fien, R. Kobayashi, and B. Stillman (1995) *Mol. Cell. Biol.* **15**, 4661–4671.
6. N. Yao, J. Turner, Z. Kelman, P. T. Stukenberg, F. Dean, D. Shechter, Z.-Q. Pan, J. Hurwitz, and M. O'Donnell (1996) *Genes to Cell* **1**, 101–113.
7. C. A. Wu, E. L. Zechner, and K. J. Marians (1992) *J. Biol. Chem.* **267**, 4030–4044.
8. P. T. Stukenberg, J. Turner, and M. O'Donnell (1994) *Cell* **78**, 877–887.
9. K. J. Hacker and B. M. Alberts (1994) *J. Biol. Chem.* **269**, 24221–24228.
10. A. Yuzhakov, J. Turner, and M. O'Donnell (1996) Cell **86**, 877–886.
11. Z. Kelman and M. O'Donnell (1995) *Ann. Rev. Biochem.* **64**, 171–200.
12. C. S. McHenry (1988) *Ann. Rev. Biochem.* **57**, 519–550.
13. M. O'Donnell and P. S. Studwell (1990) *J. Biol. Chem.* **265**, 1179–1187.
14. R. Onrust, P. T. Stukenberg, and M. O'Donnell (1991) *J. Biol. Chem.* **266**, 21681–21686.
15. Z. Tsuchihashi and A. Kornberg (1989) *J. Biol. Chem.* **264**, 17790–17795.
16. V. Naktinis, R. Onrust, L. Fang, and M. O'Donnell (1995) *J. Biol. Chem.* **270**, 13358–13365.
17. T. Tsurimoto and B. Stillman (1989) *Mol. Cell. Biol.* **9**, 609–619.
18. U. Hubscher, G. Maga, and V. N. Podust (1996) *DNA Replication in Eukaryotic Cells*, M. L. DePamphilis (ed.), CSH Laboratory Press, Cold Spring Harbor, NY, pp. 525–543.
19. S.-H. Lee S-H, A. D. Kwong, Z.-Q. Pan, and J. Hurwitz (1991) *J. Biol. Chem.* **266**, 594–602.
20. T. Tsurimoto and B. Stillman (1991) *J. Biol. Chem.* **266**, 1950–1960.
21. F. Uhlmann, J. Chi, H. Flores-Rozas, F. B. Dean, J. Finkelstein, M. O'Donnell, and J. Hurwitz (1996) *Proc. Natl. Acad. Sci. USA* **93**, 6521–6526.
22. V. N. Podust and E. Fanning (1997) *J. Biol. Chem.* **272**, 6303–6310.
23. X. Li and P. M. Burgers (1994) *Proc. Natl. Acad. Sci. USA* **91**, 868–872.
24. M. A. McAlear, E. A. Howell, K. K. Espenshade, and C. Holm (1994) *Mol. Cell. Biol.* **14**, 4390–4397.
25. M. Chen, Z.-Q. Pan, and J. Hurwitz (1992) *Proc. Natl. Acad. Sci. USA* **89**, 2516–2520.
26. Z.-Q. Pan, M. Chen, and J. Hurwitz (1993) *Proc. Natl. Acad. Sci. USA* **90**, 6–10.
27. T. C. Jarvis, L. S. Paul, and P. H. von Hipple (1989) *J. Biol. Chem.* **264**, 12709–12716.
28. J. R. Piperno and B. M. Alberts (1978) *J. Biol. Chem.* **253**, 5174–5179.
29. M. C. Young, M. K. Reddy, T. C. Jarvis, E. P. Gogol, M. K. Dolejsi, and P. H. von Hippel (1994) *Molecular Biology of Bacteriophage T4.* (J. D. Karam, ed.), American Society for Microbiology, Washington, DC, pp. 313–317.
30. N. G. Nossel (1994) *Molecular Biology of Bacteriophage T4.* (J. D. Karam, ed.), American Society for Microbiology, Washington, DC, pp. 43–53.
31. J. Rush, T.-C. Lin, M. Quinones, E. K. Spicer, I. Douglas, K. R. Williams, and W. H. Konigsberg (1989) *J. Biol. Chem.* **264**, 10943–10953.

Suggestions for Further Reading

U. Hubscher, G. Maga, and V. N. Podust (1996) DNA replication accessory proteins, In *DNA Replication in Eukaryotic Cells.* (M. L. DePamphilis, ed.), CSH Laboratory Press, Cold Spring Harbor, NY, pp. 525–543.

Z. Kelman and M. O'Donnell (1994) DNA replication enzymology and mechanisms, *Curr. Opinion Genet. Dev.* **4**, 185–195.

M. O'Donnell (1992) Accessory protein function in the DNA polymerase III holoenzyme from *E. coli*. *BioEssays* **14**, 105–111.

B. Stillman (1996) Comparison of DNA replication in cells from prokarya and eukarya, In *DNA Replication in Eukaryotic Cells* (M. L. DePamphilis, ed.), CSH Laboratory Press, Cold Spring Harbor, NY, pp. 435–460.

M. C. Young, M. K. Reddy, and P. H. von Hippel (1992) Structure and function of the bacteriophage T4 DNA polymerase holoenzyme, *Biochemistry* **31**, 8675–8690.

CLASS SWITCHING

MICHEL FOUGEREAU

Upon stimulation with an **immunogen**, the circulating **antibodies** belong first to the **IgM** isotype, before being replaced by **immunoglobulins** of another class (most frequently an **IgG** in the bloodstream). This phenomenon is known as **isotype** switching (or switch) and is the consequence of a gene rearrangement that takes place in the centrocytes, located in the light zone of the germinal center. As a result, the V–D–J rearranged section of the mature B cell, which was initially associated with a Cμ **constant region** to make a complete μ chain, will now become associated with the constant region of another isotype. This gene rearrangement involves recognition sequences termed "**switch regions**" that are located at 5′ of each constant CH gene (with the exception of the δ chain gene). These regions are very similar in sequence and are presumably recognized by an **enzyme** system, which has not been identified to date, but does not involve **recombinase**. As a result of this recombination event, the portion of DNA localized between the two switch regions concerned is deleted. The light-chain genes are not affected by the isotype switching, so the antibody combining site is unaffected. In other words, isotype switching has changed the Ig isotype without modifying the specificity of the cell. It has simply conferred on the antibody a distinct biological function, which will amplify the physiological action of the immune system in a number of ways, such as to fight more appropriately against pathogens, ensure fetal protection through transplacental transfer, be expressed in secretions, and so on, depending upon the selected isotype. Isotype switching is another example of a mechanism that necessitates interaction between **T cells** and **B cells** and makes use of two sets of molecules. One involves CD40 ligand (CD40-L) and CD40, expressed at the cell surfaces of T_H and B lymphocytes, respectively, and ensuring direct contact between the cells. The second signal is provided by the T_H cell and is a **cytokine** that is released as a soluble factor in the immediate proximity of the B cell. Depending upon the nature of the cytokine, and therefore on the type of interacting T cell, the switch mechanism will generate a discrete isotype. For example, secretion of **interleukin**-4 (IL-4) will favor a switch toward the IgE class, whereas **transforming growth factor** $\beta 1$ will induce switching to IgA.

See also entries **Antibody, Immune Response, Immunoglobulin, Isotype, Switch Region**.

Suggestions for Further Reading

J. Stavnezer (1996) Immunoglobulin class switching. **Curr. Opin. Immunol.** **8**, 199–205.

T. Honjo and T. Kataoka (1978) Organisation of immunoglobulin heavy-chain genes and allelic deletion model. *Proc. Natl Acad. Sci. USA* **75**, 2140–2144.

CLATHRIN

B. M. F. PEARSE

Clathrin is the major component of purified **coated vesicles** (1), intracellular structures that derive from coated membrane involved in **endocytosis** and **receptor** recycling.

Clathrin-mediated endocytosis (2,3) is the major route for the uptake of specific macromolecules into cells for example, **low density lipoprotein** carrying cholesterol and **transferrin** bearing iron. Such macromolecules are concentrated from the surrounding cellular fluid by binding to the receptors exposed on the surface of cells. The receptors, in turn, have features in their cytoplasmic tails that enable them to cluster into clathrin-coated pits and thus be endocytosed efficiently in small vesicles. (Fig. 1). The vesicles, in addition, carry other molecules that target them to an internal compartment, the **endosome**, and allow them to fuse with the endosomal membrane, thus delivering the cargo of nutrients into the lumen for use by the cell. Meanwhile, the coat proteins and receptors are recycled through the endosomal compartment and back to the plasma **membrane**. The coat proteins diffuse through the cytoplasm to reform coated pits there for a further round of endocytosis.

This system of budding coated vesicles is used again and again throughout cellular biology. Thus clathrin-coated vesicles are involved in **antigen presentation** in immunology (4) in delivering receptors and signaling molecules at the right place and time during **development** (5) and in recycling synaptic vesicle components in nerve synapses (6). In essence, clathrin and its associated proteins provide a mechanism for efficient recycling of specific receptors that are controlled in various ways. Other non-clathrin coated vesicle systems that contain a spectrum of related molecules and certain quite distinct coat components mediate sorting and recycling steps throughout the secretory pathway (7,8). Clathrin-coated vesicles are more particularly associated with sorting specific molecules from the trans **Golgi** network into the pre-lysosomal/endosome network in addition to their endocytic function.

Unfortunately, **viruses** [eg, influenza virus (9)] exploit the endocytic system to infect cells, and various other pathogens and foreign substances do damage to cells by this route. Defects in the system are also the root cause of certain medical conditions, for example, familial hypercholesterolaemia (2), and I-cell disease (10).

CLATHRIN

The function of clathrin is to form the strong outer, or cytoplasmic, surface of the coat, a remarkable honeycomb of hexagons and pentagons (Fig. 2). This flexible, structural network accommodates extensive areas of flattish membrane or encloses a range of vesicles, the smallest of which (250 Å diameter) is contained within a truncated icosahedron composed of 20 hexagons and 12 pentagons (for review, see Ref. 11).

The modular design of the clathrin molecule is extraordinary and unexpected [Fig. 3(**a**)]. The overall shape is that of a triskelion (12,13). Three heavy chains (180 kDa) and three light chains (one LCa and 2 LCb) form the three-legged structure. The C-termini of the heavy chains come together to form the hub of the structure, and the N-termini are in the terminal domains of the extended legs. The light chains are located along the proximal leg regions, possibly contributing in some way to the geometry at the vertex. Beyond the light chains, the heavy chains exhibit a kink, and the distal ends bend yet again to form a more globular terminal domain.

The packing arrangement of the triskelions (lacking terminal domains for simplicity) to form a hexagonal lattice is shown in Fig. 3(**b**). The profiles of two complete triskelions are shown at adjacent vertices of a clathrin cage in the form of a hexagonal barrel in Fig. 3(**c**).

In vitro in artificial conditions, clathrin exhibits its versatility as a construction molecule (Fig. 4). Among other small particles, clathrin makes cubes [Fig. 4(**c**)] a variant of the normal cage (14). In turn the cubes pack together to form remarkable arrays that resemble the foundations of buildings [Fig. 4(**a**) and (**b**)].

cDNAs coding for clathrin heavy chains and light chains have been **cloned** and sequenced (15,16). The sequences suggest that part of the light chains form a **coiled-coil** structure

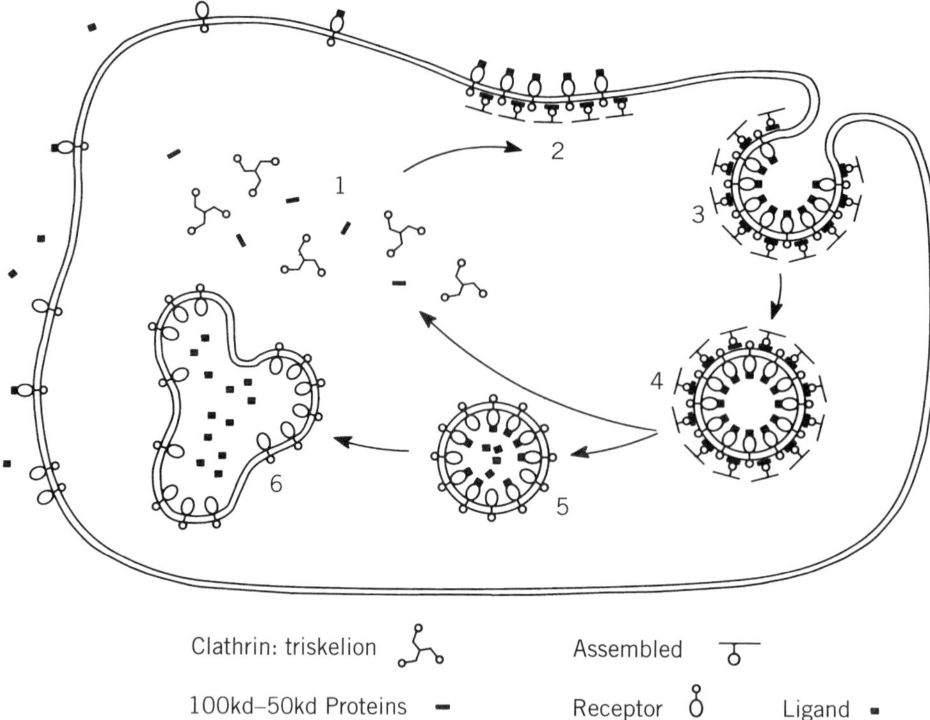

Figure 1. Endocytosis promoted by a round of clathrin assembly and recycling. [Taken from B.M.F. Pearse and R.A. Crowther (1987) *Ann. Rev. Biophys. Biophys. Chem.* **16**, 49–68.] (1) Cytoplasmic clathrin triskelions and adaptors (containing specific 100 kDa adaptins and associated subunits) are triggered to assemble on the membrane in a reaction involving GTP-binding proteins with a selection of receptors to form a coated pit(2). The coated pit invaginates as further receptors and coat proteins assemble(3). Pinching-off the completed coated vesicle(4) requires dynamin to promote fusion in the neck region. Uncoating follows, releasing the vesicle (5), which carries molecules primed for fusion with the endosome(6), and the soluble coat components, which recycle(1).

A similar coated vesicle cycle takes place in many nerve terminals. In this case, the cycle replenishes the synaptic vesicle population, whose efficient fusion with the presynaptic membrane, on stimulation, depends on, among other molecules, synaptobrevin, syntaxin, and munc18 (29–32).

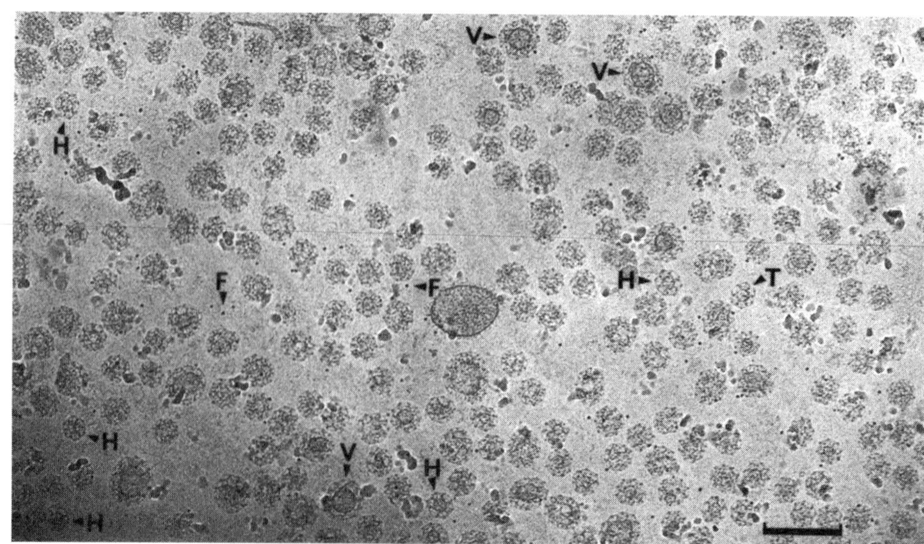

Figure 2. Field of unstained placental coated vesicles in ice. Hexagonal barrels (H), tennis ball structures (T), larger coats containing vesicle (V), and ferritin (F) are indicated. Scale bar 200 nm. [Taken from G.P.A. Vigers, R.A. Crowther, and B.M.F. Pearse (1986) *EMBO Journal* **5**, 529–534.]

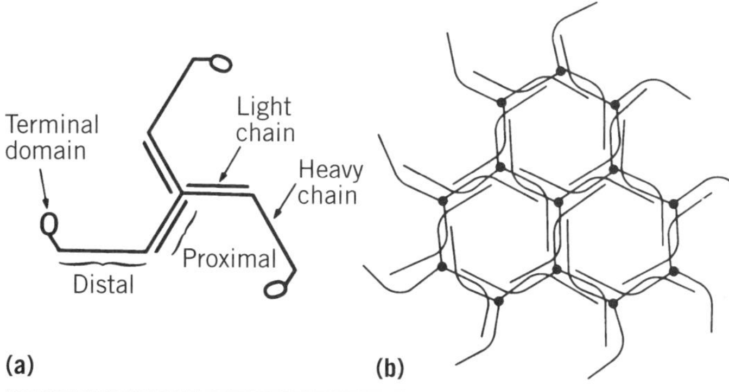

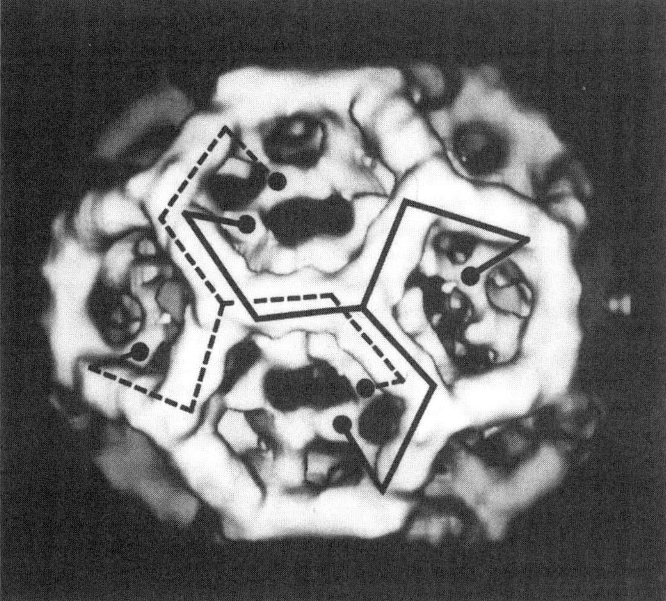

Figure 3. (**a**) Schematic drawing showing the modular structure of the triskelion. (**b**) Packing diagram showing how triskelions (lacking terminal domains for simplicity) form a hexagonal lattice. (**c**) Three-dimensional map of a clathrin cage containing 12 pentagons and 8 hexagons, computed from electron micrographs of unstained specimens embedded in vitreous ice. Each triskelion leg runs from one vertex along two neighboring polygonal edges and then turns inward. Its terminal domain forms the inner shell of density. [Taken from B.M.F. Pearse and R.A. Crowther (1987) *Ann. Rev. Biophys. Biophys. Chem.* **16**, 49–68.]

with a corresponding part of the heavy chain to form the proximal legs. Extensive studies of the interaction between light chains and heavy chains, using **antibodies** and **mutational** analysis, have provided further evidence for the positioning of the light chains along the proximal leg, studies also indicate that the C-termini of the light chains occur at the vertex, where they might influence the structural angles involved in cage assembly (see, eg, Ref. 17). Now the prospects are good for obtaining crystals suitable for determining the high-resolution structure by **X-ray crystallography** from material produced by expressing parts of the triskelion (e.g., the hub region) in E. coli (17).

An interesting problem is how spurious clathrin cage formation is prevented in the cytoplasm or indeed how clathrin triskelions are disassembled from the coat structure after vesicle budding but are not prevented from forming coated pits. Recently, other attendant "**chaperone**" proteins, hsp70c and cofactor auxilin, have been found, which modulate cage assembly in the cytoplasm (18).

Cloning of clathrin genes has also led to further exploration of the role of clathrin by gene knockout and mutation. Deletion of the clathrin gene in yeast makes various strains very sick, slow growing, or dead. In those that survive, membrane organization is affected. In particular, the processing of prepro-α-factor by the kex2 endoproteinase is disrupted, leading to secretion of the immature α-factor and failure in sexual reproduction, and endocytosis of specific molecules, for example, kex2, is reduced, and the cells fill with abnormal vacuoles (19). In a temperature-sensitive mutant of clathrin, transient defects have been observed in sorting to the vacuole. The picture emerging (20) is not dissimilar to that observed in mammalian cells, that is, clathrin is involved in specific sorting steps both in the trans Golgi region and during endocytosis at the plasma membrane. These are the sites where the characteristic coated structures, now confirmed as clathrin-coated pits, were seen in abundance in early **electron microscope** pictures of fixed cell sections (21,22). In a more complex creature, the **slime mold**, *Dictyostelium discoideum* (23), failure of clathrin heavy chain expression in cells impairs endocytosis and causes a lack of endosomes and contractile vacuoles, leading to defects in osmoregulation. Such cells also cannot follow the developmental program.

In *Drosophila melanogaster*, the clathrin heavy chain gene is essential (24). However, in flies, a dramatic effect is caused by the shibire mutation. This is a temperature-sensitive mutation in the molecule, now known to be **dynamin**, that is required for budding off a clathrin-coated vesicle (25,26). At the nonpermissive temperature, the flies drop down as if dead. They cannot recycle their synaptic vesicle components in synapses and therefore cease to fly. However, when cooled down again, they start to fly as usual. These results confirm and extend the original electron microscope observations of abundant clathrin-coated vesicles in nerve synapses (27) and particularly the neuromuscular junction (28). Recent studies have identified many more components in the synaptic vesicle cycle (6) and explored their role by genetic manipulation in *Drosophila* and *Caenorhabditis elegans* [see Fig. 1; (29–32)].

In the worm, *C. elegans*, clathrin-mediated sorting has been implicated in the development of the vulval region (5). This study suggests that even apparently quite subtle perturbations in sorting by the coated vesicle system have profound effects in the development of a complex organism.

Recently, a second clathrin heavy-chain gene (CLTD) has been identified in humans, which has its maximal level of expression in skeletal muscle (33). This gene was found in the region commonly deleted in velo-cardio-facial syndrome (VCFS). Based on the location and expression pattern of CLTD, the suggestion is that **hemizygosity** at this locus plays a role in the etiology of one of the VCFS-associated **phenotypes**.

In summary, the clathrin molecule is an extraordinary building unit that has an intricate packing arrangement and forms coated structures of striking beauty. It carries out an important function.

The assembly properties of clathrin with the coated structures and vesicles it encloses have allowed purifying and identifying many of the other functional components involved. Chief

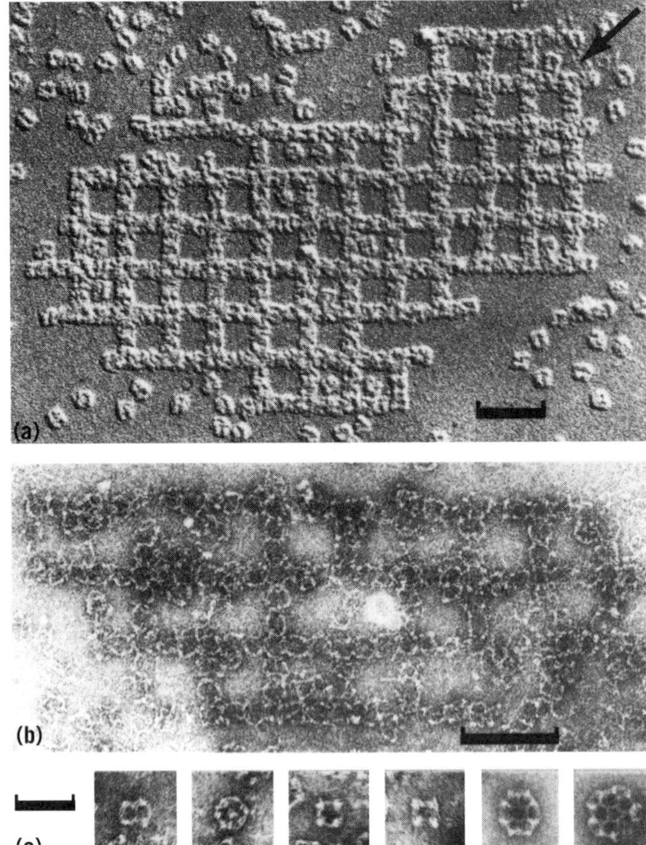

Figure 4. The most astonishing type of clathrin aggregate produced *in vitro* is an open square packing of cubes in a pattern reminiscent of the foundations of an ancient building visualized by (**a**) unidirectional shadowing; (**b**) negative staining in uranyl acetate (bar 0.2μm); (**c**) comparison of cubes with cages, showing from left to right, two-fold, three-fold, and two four-fold views of the cube plus cages of the hexagonal barrel and truncated icosahedron type (football). The edge of the cube is more than twice as long as the vertex to vertex distance in the cages. Bar, 0.1μm. [Taken from P.K. Sorger, R.A. Crowther, J.T. Finch, and B.M.F. Pearse (1986) *J. Biol.* **103**, 1213–1219.]

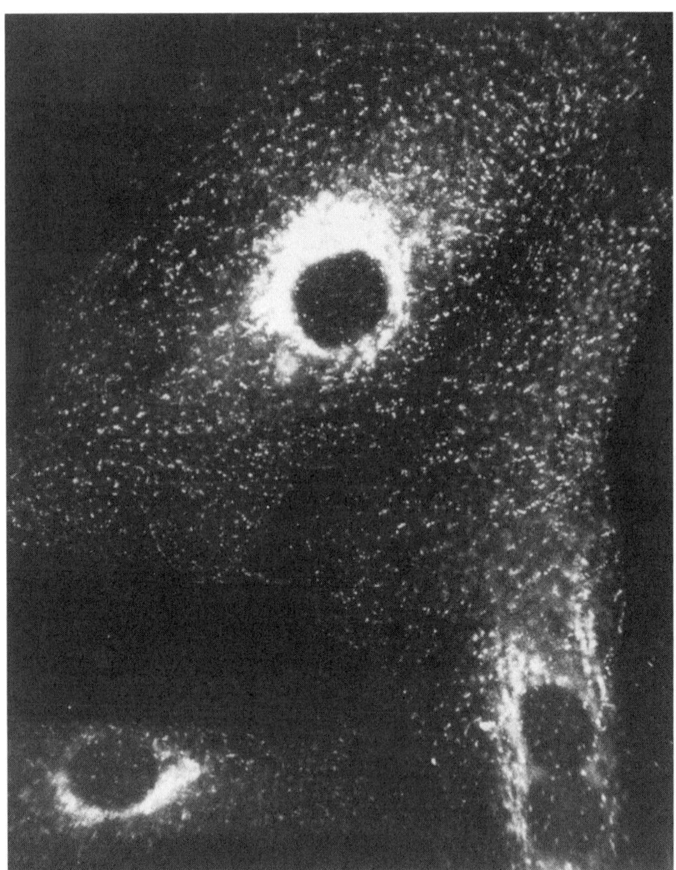

Figure 5. Immunofluorescence localization of clathrin in fibroblasts showing coated pits on the plasma membrane and in the Golgi region. [Taken from M.S. Robinson and B.M.F. Pearse (1986) *J. Cell Biol.* **102**, 48–54.]

among these are the clathrin adaptors, heterotetrameric complexes that coassemble with clathrin to form the vesicle coat.

Clathrin Adaptors

Two distinct types of clathrin adaptor complexes have been identified (3,34). One of these (the PM-adaptor) consists of an α-adaptin (~100 kDa) combined with β-adaptin (~100 kDa) and two smaller subunits, AP50 and AP17. This adaptor is found by **immunofluorescence** with a **monoclonal antibody** against α-adaptin mainly in plasma-membrane-coated pits. In contrast, the second, the Golgi adaptor, consists of γ-adaptin (~95 kDa) combined with β'-adaptin and two other subunits, AP47 and AP20, and is found by immunofluorescence using a monoclonal antibody against γ-adaptin in coated pits in the trans Golgi network (Fig. 5).

Then the problem immediately arises of how these different adaptor coat complexes assemble on the appropriate membrane and not on others. In fact, the problem increases as new related adaptors (35), coats and coatomer complexes (7,8,36) are identified throughout the intracellular membrane system. The problem is still incompletely understood, although numerous new components have been identified in the cytoplasm and on membranes by a combination of **yeast** genetics and biochemistry. These are implicated in various ways in the assembly of particular coated vesicles, their budding, and the control of membrane traffic (7,8). Intriguingly, one of the earlier control proteins identified in post-Golgi membrane transport is Sec4, a small **GTP-binding** protein (37). Since then, a whole range of GTP-binding proteins have been implicated throughout the recycling system, including Rab, ARF, dynamin, and in some sense, **Rho** (38) and heterotrimeric G-proteins (39). For instance, distinct Rab proteins take up a characteristic distribution on particular membrane compartments in the cell and are found in coated pits on those membranes (40). Their function is concerned with the kinetics of fusion of vesicles from one compartment with another (41), although precisely how this is achieved in molecular terms is unclear. ARF, on the other hand, has been implicated in coat binding to membranes notably in the Golgi region (42,43), and dynamins are concerned with actual budding of coated vesicles, as previously mentioned (see Fig. 1).

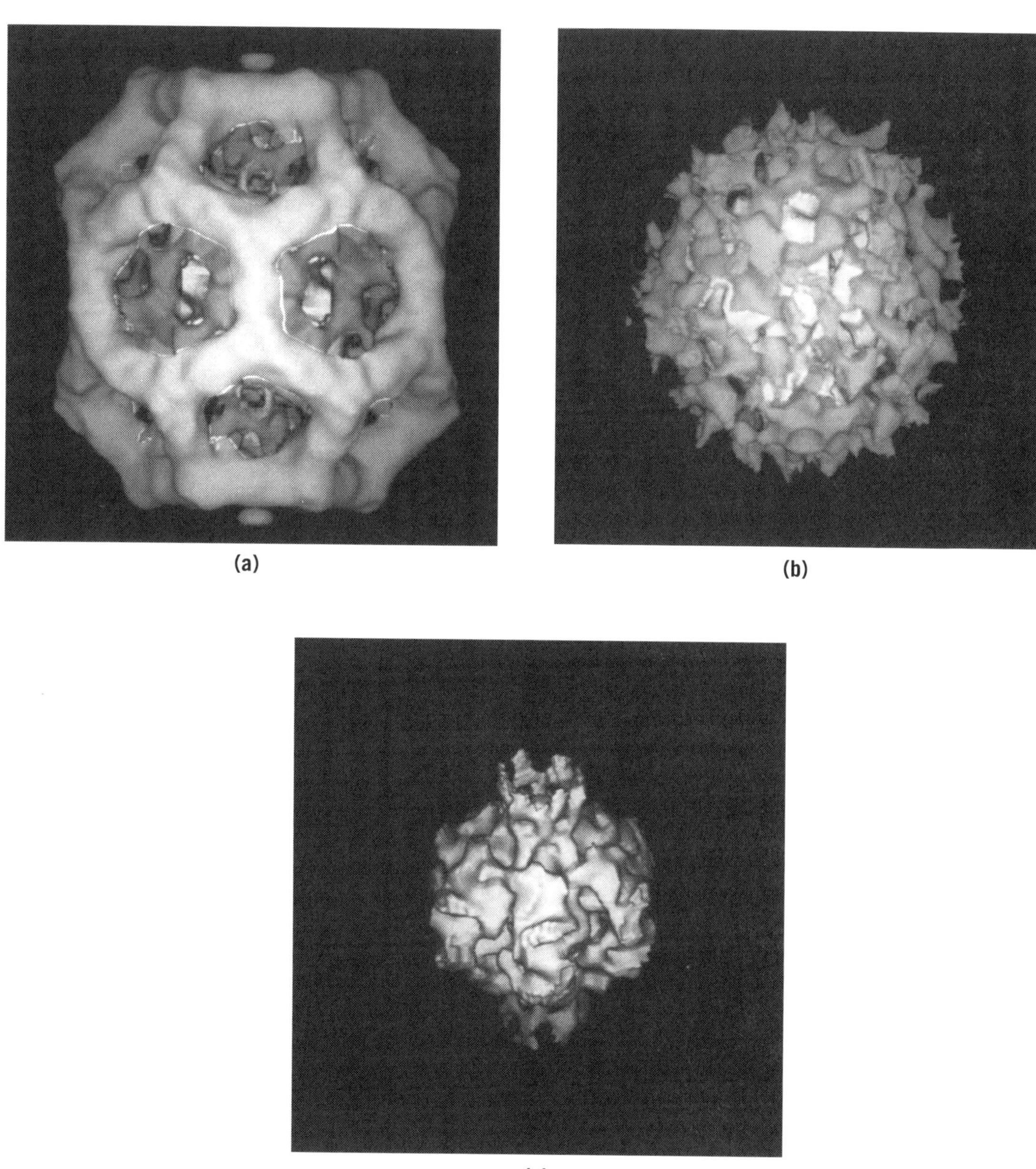

Figure 6. See color insert section. Clathrin encapsulates coated vesicles, the organelles responsible for uptake of essential nutrients into living cells (eg, especially across the placenta in mammals and into oocytes in chickens). (**a**) A three-dimensional map of a clathrin coat generated from electron micrographs of specimens reassembled from their constituent protein complexes. Colors highlight the outer polyhedral clathrin lattice (red), the shell of clathrin terminal domains (green) revealed in (**b**) and an inner shell of adaptors (blue) exposed in (**c**). These three maps led to the description of this structure as a cell biologist's LEGO model. (Courtesy of G.P.A. Vigers, R.A. Crowther, and B.M.F. Pearse from data presented in *EMBO J.* (1986) **5**, 2079–2085.)

When clathrin-coated vesicles assemble on a membrane, they concentrate certain specific receptors in the budding membrane, leaving other membrane proteins behind. These specific receptors are recognized by features in their cytoplasmic tails. Each receptor seems to find its own characteristic **steady-state** distribution throughout the sorting system because each receptor has an individual amino acid sequence. There is no single, defined, clear-cut sequence indicating that a receptor will assemble efficiently into a clathrin-coated pit. Nevertheless, at least in endocytosis, mutational analysis has pinpointed certain small groups of residues in the cytoplasmic tails of receptors that constitute primary sequence 'features' which allow the normal accumulation of their receptors in clathrin-coated pits. Typically, such a 'feature' contains four residues, the first and probably most important of which is **tyrosine** (2), followed by two residues (one or both of which is often charged) followed by a bulky aliphatic residue. Such tetrapeptides are capable of interaction with the AP50 subunit of the plasma membrane adaptor (44). However, as more examples are studied (45), it is clear that other possible signals exist and that other factors, such as phosphorylation of receptor tails, play a role in promoting entry into coated pits (46).

One of the most investigated examples of a receptor that is routed via clathrin-coated vesicles containing γ-adaptin in the trans Golgi network (TGN) is the cation- independent mannose 6-phosphate/IGF-II receptor (MPR). The MPR recognizes the first identified lumenal sorting signals, **mannose 6-phosphate**-containing sugar chains on **lysosomal** enzymes, and thus concentrates these enzymes as cargo inside coated vesicles on their way to lysosomes (47,48). The MPR has a 163-residue cytoplasmic tail that contains multiple features involved in directing the receptor along its complex, intracellular, recycling pathway, including a typical endocytosis sequence Tyr-Lys-Tyr-Ser-Lys-Val (49,50). The tail region containing these residues, apparently, also acts as a signal for efficiently sorting lysosomal enzymes, separately or in combination with the C-terminal region (51).

Both types of coated vesicle adaptors bind to the MPR tail. Recognition by the plasma membrane adaptor is abolished by mutation of the tyrosine residues important for the endocytosis signal, but these same mutations do not abolish sorting (51) or binding of the Golgi adaptor (52). There is also evidence that the Golgi adaptor preferentially binds to the MPR tail when it is **phosphorylated** (53). In addition to these cytoplasmic features that determine the routing of the MPR, the extracellular domain and **ligand binding** perhaps play some role in the precise steady-state distribution of the receptor, at least in CV1 and COS cells (54).

Clathrin Coats

A low (~50 Å) resolution three-dimensional map of a clathrin coat has been generated from electron micrographs of reconstituted coats in vitreous ice (Fig. 6) (see **Electron imaging**). The major components of the coat, clathrin, and adaptor complexes were solubilized from purified coated vesicles and separated. Three types of particles were reassembled from these preparations. The first type was the clathrin cage itself. The second was derived from the first by partial trypsin digestion to remove the terminal domains from the clathrin triskelions, and the third was the complete coat containing both clathrin and adaptors. Maps of each structure were generated by computer imaging tilted specimens of several examples of individual particles of the three different types. By subtracting the maps of the incomplete structures from those of the complete coat, three different layers of the overall coat were distinguished, shown color coded in the image presented (Fig. 6). Thus the outer polyhedral clathrin lattice is highlighted in red, the shell of clathrin terminal **domains** extends into the structure (green), and the adaptors form an inner shell (blue). These particles are hexagonal barrels, made in the absence of a vesicle, and in fact are actually too small to accommodate a vesicle. Probably the smallest coat structure able to contain a vesicle is a truncated icosahedron, which can also be reconstituted as a coat. If purified coated vesicles are observed in projection they exhibit the same coat thickness composed of three shells of density, whereas the vesicle itself appears as a fourth shell of high density contained within the coat. Now a higher (~20 Å) resolution version of the coat structure can be obtained from these types of hexagonal barrel preparations and the equivalent icosahedral specimens by using improved electron microscopes and the more highly developed imaging techniques currently available. This will allow further identification of smaller domains within the structure, especially if combined with specific antibodies to decorate those domains. It is also possible to generate populations of defined coated vesicles from their cytoplasmic constituents and membranes enriched in relevant proteins in physiological conditions. This is an exciting prospect as it may lead to a greater understanding of the specificity underlying the assembly system, and it may be possible to locate some of the controlling elements in the structure.

BIBLIOGRAPHY

1. B. M. F. Pearse (1975) *J. Mol. Biol.* **97**, 93–98.
2. J. L. Goldstein, M. S. Brown, R. G. W. Anderson, D. W. Russell, and W. J. Schneider (1985) *Ann. Rev. Cell Biol.* **1**, 1–39.
3. B. M. F. Pearse and M. S. Robinson (1990) *Ann. Rev. Cell Biol.* **6**, 151–171.
4. P. R. Wolf and H. L. Ploegh (1995) *Ann. Rev. Cell Dev. Biol.* **11**, 267–306.
5. J. Lee, G. D. Jongeward, and P. W. Sternberg (1994) *Genes Dev.* **8**, 60–73.
6. P. R. Maycox, E. Link, A. Reetz, S. A. Morris, and R. Jahn (1992) *J. Cell Biol.* **118**, 1379–1388.
7. R. Schekman and L. Orci (1996) *Science* **271**, 1526–1533.
8. J. E. Rothman and F. T. Wieland (1996) *Science* **272**, 227–234.
9. T. Stegmann, R. W. Doms, and A. Helenius (1989) *Ann. Rev. Biophys. Biophys. Chem.* **18**, 187–211.
10. S. Kornfeld (1986) *J. Clin. Invest.* **77**, 1–6.
11. B. M. F. Pearse and R. A. Crowther (1987) *Ann. Rev. Biophys. Biophys. Chem.* **16**, 49–68.
12. E. Ungewickell and D. Branton (1981) *Nature* **289**, 420–422.
13. T. Kirchhausen and S. C. Harrison (1981) *Cell* **23**, 755–761.
14. P. K. Sorger, R. A. Crowther, J. T. Finch, and B. M. F. Pearse (1986) *J. Cell Biol.* **103**, 1213–1219.
15. A. P. Jackson, H. Seow, N. Holmes, K. Drickamer, and P. Parham (1987) *Nature* **326**, 154–159.
16. T. Kirchhausen, S. C. Harrison, E. Ping Chow, R. J. Mattaliano, K. L. Ramachandran, J. Smart, and J. Brosius (1987) *Proc. Natl. Acad. Sci. USA* **84**, 8805–8809.
17. S. Liu, M. Wong, C. S. Craik, and F. M. Brodsky (1995) *Cell* **83**, 257–267.
18. E. Ungewickell, H. Ungewickell, S. E. H. Holstein, R. Lindner, K. Prasad, W. Barouch, B. Martin, L. E. Greene, and E. Eisenberg (1995) *Nature* **378**, 632–635.
19. G. S. Payne and R. Schekman (1989) *Science* **245**, 1358–1365.
20. E. Conibear and T. H. Stevens (1995) *Cell* **83**, 513–516.
21. T. F. Roth and K. R. Porter (1964) *J. Cell Biol.* **20**, 313–332.
22. D. S. Friend and M.G. Farquhar (1967) *J. Cell Biol.* **35**, 357–376.
23. T. J. O'Halloran and R. G. W. Anderson (1992) *J. Cell Biol.* **118**, 1371–1377.
24. C. Bazinet, A. L. Katzen, M. Morgan, A. P. Mahowald, and S. K. Lemmon (1993) *Genetics* **134**, 1119–1134.
25. T. Kosaka and K. Ikeda (1983) *J. Cell Biol.* **97**, 499–507.
26. H. Damke, T. Baba, A. M. van der Bliek, and S. L. Schmid (1995) *J. Cell Biol.* **131**, 69–80.
27. E. G. Gray and R. A. Willis (1970) *Brain Res.* **24**, 149–168.

28. J. E. Heuser and T. S. Reese (1973) *J. Cell Biol.* **57**, 315–344.
29. A. DiAntonio and T. L. Schwarz (1994) *Neuron* **12**, 909–920.
30. K. Broadie, A. Prokop, H. J. Bellen, C. J. O'Kane, K. L. Schulze, and S. T. Sweeney (1995) *Neuron* **15**, 663–673.
31. Y. Fujita, T. Sasaki, K. Fukui, H. Kotani, T. Kimura, Y. Hata, T. C. Sudhof, R. H. Scheller, and Y. Takai (1996) *J. Biol. Chem.* **271**, 7265–7268.
32. E. M. Jorgensen, E. Hartwieg, K. Schuske, M. L. Nonet, Y. Jin, and H. R. Horvitz (1995) *Nature* **378**, 196–199.
33. H. Sirotkin, B. Morrow, R. DasGupta, R. Goldberg, S. R. Patanjali, G. Shi, L. Cannizzaro, R. Shprintzen, S. M. Weissman, and R. Kucherlapati (1996) *Human Molecular Genetics* **5**, 617–624.
34. S. Ahle, A. Mann, U. Eichelsbacher, and E. Ungewickell (1988) *EMBO J.* **7**, 919–929.
35. F. Simpson, N. A. Bright, M. A. West, L.S. Newman, R. B. Darnell, and M. S. Robinson (1996) *J. Cell Biol.* **133**, 749–760.
36. P. Cosson, C. Démolliére, S. Hennecke, R. Duden, and F. Letourneur (1996) *EMBO J.* **15**, 1792–1798.
37. A. Salminen and P. J. Novick (1987) *Cell* **49**, 527–538.
38. C. Lamaze, T. Chuang, L. J. Terlecky, G.M. Bokoch, and S. L. Schmid (1996) *Nature* **382**, 177–179.
39. R. H. Kehlenbach, J. Matthey, and W. B. Huttner (1994) *Nature* **372**, 804–809.
40. P. Chavrier, R.G. Parton, H. P. Hauri, K. Simons, and M. Zerial (1990) *Cell* **62**, 317–329.
41. V. Rybin, O. Ullrich, M. Rubino, K. Alexandrov, I. Simon, M. C. Seabra, R. Goody, and M. Zerial (1996) *Nature* **383**, 266–626.
42. A. Peyroche, S. Paris, and C. L. Jackson (1996) *Nature* **384**, 479–481.
43. P. Chardin, S. Paris, B. Antonny, S. Robineau, S. Béraud-Dufour, C. L. Jackson, and M. Chabre (1996) *Nature* **384**, 481–484.
44. W. Boll, H. Ohno, Z. Songyang, I. Rapoport, L. C. Cantley, J. S. Bonifacino, and T. Kirchhausen (1996) *EMBO J.* **15**, 5789–5795.
45. I. V. Sandoval and O. Bakke (1994) *Trends Cell Biol.* **4**, 292–297.
46. A. Pelchen-Matthews, I. J. Parsons, and M. Marsh (1993) *J. Exp. Med.* **178**, 1209–1222.
47. S. Kornfeld (1992) *Ann. Rev. Biochem.* **61**, 307–330.
48. J. Klumperman, A. Hille, T. Veenendaal, V. Oorschot, W. Stoorvogel, K. von Figura, and H. J. Geuze (1993) *J. Cell Biol.* **121**, 997–1010.
49. P. Lobel, K. Fujimoto, R. D. Ye, G. Griffiths, and S. Kornfeld (1989) *Cell* **57**, 787–796.
50. M. Jadot, W. M. Canfield, W. Gregory, and S. Kornfeld (1992) *J. Biol. Chem.* **267**, 11069–11077.
51. K. F. Johnson and S. Kornfeld (1992) *J. Cell Biol.* **119**, 249–257.
52. J. N. Glickman, E. Conibear, and B. M. F. Pearse (1989) *EMBO J.* **8**, 1041–1047.
53. R. Le Borgne, A. Schmidt, F. Mauxion, G. Griffiths, and B. Hoflack (1993) *J. Biol. Chem.* **268**, 22552–22556.
54. E. Conibear and B. M. F. Pearse (1994) *J. Cell Science* **107**, 923–932.

CLEVELAND MAP

T. BERGMAN
H. JÖRNVALL

This method of **peptide mapping** involves generating of **protein** fragments via limited **proteolysis**, usually in the presence of **SDS**, followed by electrophoretic separation in **SDS-PAGE**. The resulting one-dimensional peptide patterns are characteristic of the protein substrate and are used to evaluate structural relationships between proteins and to aid identification of individual polypeptide chains. The technique is named after the first author of a 1977 publication describing the procedure (1), and many variants of this method have subsequently been presented (2).

After initial separation and excision from a primary gel, the protein substrate is digested in solution or, typically, in the stacking zone of the SDS-PAGE "mapping gel." The solution approach relies on purifying the protein by conventional chromatographic procedures or preparative SDS-PAGE followed by excision and **electroelution** from the gel (3). Protein preparation in SDS-containing gels is feasible because the subsequent digestion is performed in the presence of SDS. Frequently, SDS-PAGE isolation of the protein is possible at an early stage of purification, which is important when only a small amount of the protein is available. For Cleveland mapping, the efficient approach is to place the gel slice that contains the protein band of interest directly into a well of a second SDS/polyacrylamide gel and to add a suitable proteolytic enzyme. After electrophoresis just long enough for the protein substrate to migrate out of the gel piece and enter the stacking zone with the proteinase, the power is temporarily switched off and the gel left for digestion to take place. When the power is turned on again, the fragments of the protein substrate are separated from the proteinase and are mapped in the resolving gel. This procedure is preferred when handling small amounts of protein because of the direct transfer between primary and secondary gels.

The original Cleveland protocol (1) involved brief staining of the primary gel with **Coomassie Brilliant Blue**, followed by rapid destaining and excision of the band corresponding to the protein substrate. Because fixation of the protein is undesirable due to the electrotransfer, negative staining with 1 M KCl may be an alternative and has been used for subsequent electroblotting and sequence analysis (4). The excised gel piece is equilibrated in buffer containing 0.1% (w/v) SDS to unfold the protein substrate and to obtain a suitable pH of 6.8 for stacking gel electrophoresis and digestion (1). Often, 5 to 10 μg of protein substrate is required for application onto the primary gel to generate a proteolytic pattern detectable by Coomassie Blue staining, but the minimum amount can be substantially lowered if **silver staining fluorescence** techniques or **radiolabeling** are used to visualize the peptide map.

After equilibration, the gel piece that contains the protein is placed in a well of the "mapping gel," and, in the original protocol, overlaid with a solution of the appropriate proteinase in the same buffer. In later protocols, loading of the proteinase below the gel piece is suggested to avoid a variable depth of enzyme solution around the gel piece (2). Once a narrow stack has formed after the initial period of current, usually when the **tracking dye** (e.g., bromphenol blue) is approaching the bottom of the stacking gel, the power is turned off for 30 to 60 min. This is normally sufficient for the proteinase to cleave susceptible peptide bonds, which typically generates 10 to 20 fragments of 50- to 70-kDa proteins (1). It is crucial that the proteolytic enzyme is active in 0.1% SDS (up to 0.5% for in-solution digestions). Suitable enzymes are **trypsin**, **α-chymotrypsin**, and **papain**, in addition to the popular V8 proteinase from *Staphylococcus aureus*. All are added in proportions corresponding to an enzyme to substrate ratio in the range of 1:10 to 1:50 (w:w).

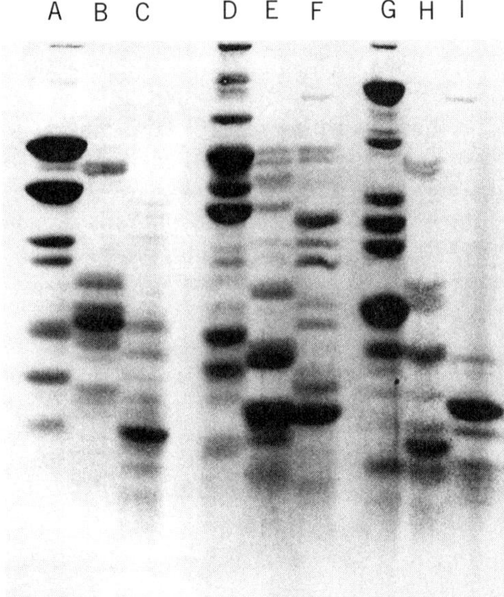

Figure 1. One-dimensional Cleveland maps of **albumin** (A,D,G), **tubulin** (B,E,H), and **alkaline phosphatase** (C,F,I), generated by cleavage with papain (A–C), *Staphylococcus aureus* V8 proteinase (D–F), and chymotrypsin (G–I), followed by SDS-PAGE in a 20% (w/v) acrylamide gel. Each combination of protein substrate and proteinase gives a unique pattern. From Ref. 1 with permission.

After digestion, electrophoresis is resumed, and the resulting fragments are resolved, typically in a 15 or 20% acrylamide gel of the discontinuous type (Fig. 1). A unique proteolytic pattern is obtained for each protein substrate and for each individual proteinase. This discriminating feature makes Cleveland mapping useful as a tool for visually judging the similarity between proteins.

In addition to enzymes, specific chemical cleavage is employed where the protein substrate within a gel is treated in a test tube. Suitable reagents are **cyanogen bromide** for cleavage after **methionine** residues, *N*-chlorosuccinimide for cleavage at **tryptophan**, **hydroxylamine** for cleavage at Asn-Gly peptide bonds, and formic acid for cleavage at Asp-Pro bonds. As for the size of the protein substrates, the method is generally applicable to polypeptide chains in the range 10 to 100 kDa. Peptide fragments smaller than 10 kDa are difficult to resolve into informative peptide maps, although separating such small peptides is possible by using different SDS-PAGE systems. An alternative is to recover small fragments separately by **size exclusion chromatography** for subsequent peptide mapping by reverse-phase **HPLC**. This can be useful also for proteins large than 10 kDa in order not to waste small fragments (5).

BIBLIOGRAPHY

1. D. W. Cleveland et al. (1977) *J. Biol. Chem.* **252**, 1102–1106.
2. W. J. Gullick (1986) In *Practical Protein Chemistry: A Handbook* (A. Darbre, ed.), Wiley, Chichester, U.K., pp. 207–225.
3. M. W. Hunkapiller et al. (1983) *Methods Enzymol.* **91**, 227–236.
4. T. Bergman and H. Jörnvall (1987) *Eur. J. Biochem.* **169**, 9–12.
5. T. Bergman et al. (1991) *J. Protein Chem.* **10**, 25–29.

CLONAL SELECTION THEORY

MICHEL FOUGEREAU

At the beginning of the twentieth century, Ehrlich was the first to propose that **antibodies** did exist prior to the introduction of **antigen** or **immunogen** and that antigenic stimulation only promoted amplification of molecules of the appropriate specificity. This proposal was quite pertinent, because it contained several key ideas, proven correct some 70 years later. One is the selective process of a preexisting **repertoire**, the other is that **immunoglobulins** (Igs) may exist as surface or secreted molecules. The only aspect that was not verified is that Ehrlich imagined a cell to express many different antibodies. After a long period during which **hapten** analysis indicated the exquisite precision of **antibody–antigen interactions**, it seemed impossible that antibody structures might preexist to recognize laboratory artifacts, like the haptens being studied. This resulted in the template theory of antibody formation, which required the presence of antigen prior to the synthesis of the corresponding antibodies. Revival of the selection theory was proposed in 1955 by Jerne, but it took another 4 years before the final form of the "clonal selection theory" was put forward by Burnet. The essence of the theory was that antibody specificities were preformed, but in a clonally distributed manner, so that one cell makes only one antibody of one given specificity. This was crucial to provide a mechanism accounting for acquisition of discrimination between self and nonself, after the observation on induction of immunological **tolerance** by the group of Medawar in 1953,when it was demonstrated that tolerance was an acquired phenomenon, resulting from a contact between the immature immune system and the self antigens during gestation or the perinatal period. Burnet most simply explained tolerance by a deletion of those clones that could interact with self components, leading to the concept of "forbidden clones." Once the basic anti-nonself repertoire established, an antigen was seen as merely selecting and amplifying the cell population of clones expressing the corresponding specific antibodies.

Many investigators worked to define models that would prove the clonal theory, and the literature is amply documented on this point, with many very elegant approaches, such as those based on limiting dilutions, either *in vitro* or *in vivo* after transfer in irradiated mice. The ultimate proof came from molecular analysis, first at the **protein** level on myeloma proteins and then when the mechanism of **gene rearrangement** of Ig **genes** was elucidated, under the control of **allelic exclusion**. Some exceptions to the clonal expression of Ig molecules have been periodically reported. Some were not exceptions, as in the case of lymphocytes coexpressing multiple **isotypes** that, in fact, share identical V_H and V_L regions, and therefore having the same specificity. Upon reactivation of the **recombinase** genes upon **immunization**, for example, a transient coexpression of different V genes may occur, leading to secondary gene rearrangements. But this rarely results in long-term production of antibodies of different specificities. The clonal theory is widely accepted and considered proven, and the fantastic use of **monoclonal antibodies** is (for most biologists) daily living proof of its pertinence.

See also entries **Allelic exclusion**, **Antibody**, **Gene rearrangement**, and **Immunoglobulin**.

Suggestions for Further Reading

M. F. Burnet (1959) *The Clonal Selection Theory of Acquired Immunity*, Vanderbilt University Press, Nashville, TN.

K. Rajewsky (1996) Clonal selection and learning in the antibody system. *Nature* **381**, 751–758.

M. Nussenzweig et al. (1987) Allelic exclusion in transgenic mice that express the membrane form of immunoglobulin. *Science* **236**, 816–819.

CLONING

DAVID B. WILSON

The word "cloning" has several different meanings. For example, it is possible to clone cells, that is, to cause cells to reproduce themselves so as to make a population of identical cells. In molecular biology, any piece of **DNA** can be cloned—inserted into a **vector** that replicates in a host to produce many copies of the same recombinant vector. It is also possible to clone **genes**. Gene cloning is the process of identifying and isolating a specific gene of interest. Cloning genes is a major focus of molecular biology, and the ability to clone genes has revolutionized biology. Once a gene is identified and cloned into an appropriate vector, it can be manipulated in many different ways. Cloning dramatically amplifies the DNA so that it can be **sequenced**. A cloned gene inserted into an **expression system** creates an organism that produces up to 25% of its total protein or more as the gene product, allowing large-scale production of the protein (1). A cloned gene can be mutated, and the mutated form inserted back into an organism that lacks a functional copy of the gene (2). This allows structure-function analysis of the gene product. A cloned gene or set of genes can be introduced into a new host to create a new metabolic pathway or to modify an existing pathway (3).

In many cases, gene cloning is carried out by fragmenting the **genomic** DNA from the organism containing the gene of interest and inserting the fragments into self-replicating DNA molecules (vectors) using **recombinant DNA** techniques; **transforming** the resultant population of recombinant molecules, which comprise a **DNA library**, into a host organism, screening or selecting the transformants to identify cells that contain the desired DNA, and isolating the vector DNA (see **Shotgun cloning**).

PREPARATION OF DNA FOR CLONING

Genomic DNA must be fragmented before cloning into a suitable vector. It is important to have fragments that contain the gene intact. One way to ensure this is to use a **restriction enzyme** that makes frequent cuts in the DNA under conditions of partial digestion (so that only some of the sites are cleaved) so as to obtain fragments of the desired size. The restriction enzyme *Sau*3A is often used because it has a four-base recognition sequence and therefore cuts DNA at many sites in the genome. Additionally, it produces **cohesive, sticky ends** that ligate with BamHI, a site often present in **polylinkers**. Mechanical shearing also fragments the DNA, but sheared DNA cannot be cloned directly. First, the ragged ends must be filled in and polylinkers added.

Because small fragments of DNA **ligate** to a vector more readily than large fragments, the DNA to be cloned is often fractionated by sucrose density gradient centrifugation (see **Sedimentation velocity centrifugation**) to obtain DNA of the desired size (pp. 2.85–2.86 of Ref. 4). It is important to set up the conditions for ligation so that most of the recombinant molecules produced contain only a single DNA fragment. This means that nearly equal numbers of vector molecules and fragment molecules should be present and that the total DNA concentration should not be too high (~40 μg of vector DNA per mL).

It is necessary to produce a large enough number of transformants or recombinant **bacteriophage** to have a high probability that a fragment carrying the intact gene is present in the population. The exact number depends on the size of the genome from which the gene is being cloned (the larger the genome, the more transformants need to be screened) and the size of the inserts (the larger the insert size, the fewer transformants need to be screened). A major problem in cloning is restriction enzymes in the host, which can degrade foreign DNA. Mutant strains of *Escherichia coli* are available that lack restriction enzymes, and these are best for constructing large **libraries** for screening or selection. Once a gene is cloned, restriction enzymes in a host are not usually a problem because large amounts of DNA overcome the restriction barrier (5).

Prokaryotic genes are usually cloned from genomic DNA, whereas **eukaryotic** genes are often cloned from **complementary DNA** (cDNA), which is prepared by copying **messenger RNA** with **reverse transcriptase.** Most eukaryotic genes contain **introns**, which prevent them from being expressed in prokaryotes, but the introns are not in the cDNA. The cDNA clone is often used as a probe to isolate the gene from a **genomic library**. The genomic clone can be sequenced to identify the nature and location of the introns and the **upstream** and **downstream** sequences of the gene.

CLONING VECTORS

There are four general types of vectors that are used in *E. coli*: **plasmids**, phage, **cosmids** or bacterial artificial **chromosomes** (BACs):

1. Plasmids have the advantage that they are smaller than the other vectors and usually give larger amounts of vector DNA from a given volume of culture because of their high **copy number**.
2. Phage vectors allow larger sized inserts of foreign DNA, so that a smaller number of **plaques** are needed to give a high probability that the entire genome of the organism is present in the library that is being screened. Furthermore, phage systems give more plaques per weight of DNA than plasmids give colonies, and plaques are easier to screen than colonies because they are smaller and more uniform. Phage T4 vectors allow cloning inserts as large as 120 kbp (6), whereas phage P1 vectors allow 80-kbp inserts (7), and **lambda phage** vectors allow up to 40-kbp inserts (pp. 2.2 to 2.125 of Ref. 4). Once a positive plaque is identified, the insert DNA is subcloned into a plasmid vector to take advantage of the

greater amplification and ease of DNA preparation associated with plasmid vectors.

3. Cosmid vectors combine features of the preceding two classes because they can be packaged into phage but also replicate as plasmids. They also allow inserts up to 40 kbp (8).
4. Bacterial artificial chromosomes are based on the single-copy **F factor** plasmid and accept >300-kbp DNA fragments (9). Another important system for cloning very large DNA pieces is **yeast artificial chromosomes** (YAC). YACs accept fragments over 200 kbp and are used in genome sequencing and positional cloning of genes (10). Although **yeast** is not as easy to work with as *E. coli*, it is useful for studying eukaryotic genes that cannot be expressed in *E. coli*.

IDENTIFYING THE GENE OF INTEREST

Screening is usually the most difficult step, and there are many different approaches available. The simplest method is to complement a mutant gene in a host organism (see **Complementation**), which allows direct selection of the desired clone (11). A powerful example of direct selection is cloning a **transposon-tagged** gene where selection is for **antibiotic resistance** encoded in the **transposon**. Another approach is to screen transformant colonies for the gene product, either with specific **antibodies** (pp. 12.11–12.29 of Ref. 4) or by an enzymatic assay (12). This requires a sensitive, specific assay because foreign genes are often expressed at a low level, and enzymes from the host organism must not interfere with the assay. In addition, the gene to be cloned needs to be expressed in the host organism.

The most general screening approach is to transfer the colonies to a filter, lyse the cells, **denature** the DNA on the filter and **hybridize** a labeled nucleic acid probe to the filter (see **Southern blots**). The probe pairs with complementary sequences in the DNA on the filter, and the colony to which the probe hybridizes is visualized by the label. There are various methods to design or obtain an oligonucleotide probe that specifically hybridizes to the desired gene.

If one wants to clone the gene coding for a known protein, it is possible to sequence the protein, usually at its N-terminus (see **Edman degradation**), and to use **reverse translation** to determine the probe sequence. However, sequences from internal peptides produced by specific cleavage of the polypeptide chain (eg, by an **arginine**-specific **proteinase** or by **cyanogen bromide** cleavage at **methionine residues**) are also used if the N-terminus is blocked or if the N-terminal sequence is unsuitable for reverse translation (pp. 11.44–11.57 of Ref. 4).

Another method is using a segment of the DNA from the gene of a closely related organism as the hybridization probe (13). A modification of this approach is looking at **homologous** sequences from a set of organisms to identify highly conserved regions that are used to design an oligonucleotide probe specific for the desired gene. It is important to choose organisms with DNA base compositions close to that of the organism from which the gene is to be cloned.

With the large amount of genomic and cDNA sequence information (see **Expressed sequence tag**) becoming available in computer **databases**, genes of interest are frequently identified during database searches. Hybridization probes are designed from the sequence information and used to screen a library or to amplify the gene directly.

An ingenious way of selecting for a full-length clone in a phage vector, when a partial clone is available, is inserting the sequence next to the *sup*F gene in a plasmid, infect *rec*A$^+$ cells carrying the plasmid with the phage library, and plate the resulting phage on an *E. coli* strain, where only phage containing *sup*F grows (14). Recombination between the complementary sequences introduces the *sup*F only into the desired phage.

POSITIONAL CLONING OF A GENE IDENTIFIED BY ITS MUTANT PHENOTYPE

Often genes are identified because they have an interesting **phenotype** when mutated. The mutation is mapped by using DNA-based markers. The most closely linked marker is used as a hybridization probe to screen a YAC, BAC, or cosmid library, thus initiating a **chromosome walk** to the gene (15). Once a library clone is identified that is deemed to contain the gene of interest, it is subcloned into a vector used to transform the mutant. The subclone containing the gene is identified by its ability to complement the mutant phenotype.

Potential Problems with Library-Based Cloning

One problem with direct selection is potential contamination of the library with host DNA, which leads to cloning a host gene, rather than the gene from the original organism. Therefore it is necessary to show that genes isolated by complementation are from the desired organism, which can be done by a Southern blot. Other potential cloning problems are that some DNA fragments cannot be cloned because they contain a **poison sequence**. DNA fragments containing **repeated DNA** sequences are often unstable, because they recombine easily. So *rec*A mutant strains of *E. coli* are often used as a host for cloning to minimize **recombination** (16).

A major problem in cloning a DNA fragment that has identical ends (cohesive or blunt) is recircularization of vector molecules that do not contain an insert because the two ends of the cut vector are also identical. This gives a high background of transformants or plaques lacking an insert. One method of minimizing this problem is treating the cut vector with calf **alkaline phosphatase** to remove its 5′ phosphate groups (pp. 1.60–1.62 of Ref. 4). Dephosphorylated vector molecules cannot circularize, but they ligate to the 5′-phosphate groups on the DNA fragments to be cloned. The resulting recombinant molecule circularizes, producing a circular molecule with a single-stranded nick in each strand. These molecules give transformants. It is necessary to completely inactivate the alkaline phosphatase before the DNA to be cloned is added to the vector. The dephosphorylation reaction must be run under conditions that remove most 5′-phosphate groups but do not inactivate the plasmid, so the reaction has to be monitored carefully.

Another method of dealing with the problem of vector molecules lacking inserts is blue-white screening. Many *E. coli* vectors contain an N-terminal portion of the **beta-galactosidase** gene that codes for the α-fragment that complements the inactive β-fragment. A polylinker is introduced into the β-galactosidase gene so that the α-fragment is inactivated when a DNA molecule is ligated into the polylinker.

The ligation mixture is transformed into an *E. coli* host that produces the β-fragment, and the transformants are plated on selective plates. The colonies that contain an insert are white, whereas colonies without an insert are blue (pp. 1.85–1.86 of Ref. 4).

When a probe that hybridizes to the gene to be cloned is available, the cloning process is simplified by using the probe to identify an appropriately sized restriction fragment that contains the gene by using Southern blots run on digests of a number of restriction enzymes. The appropriately sized fragments are eluted from a preparative agarose **gel electrophoresis** run on genomic DNA digested with the chosen restriction enzyme. The fragments are ligated into a vector cut with a compatible restriction enzyme, and the resulting transformants are screened with the probe. This process significantly reduces the number of colonies or plaques that need to be screened (1 of 40 positive vs. 1 of 1,000 for a *Thermomonospora fusca* gene) (17).

PCR TO CLONE GENES

Whenever sequence information is available for a gene, it is possible to design **PCR** primers that amplify all or part of a gene directly from DNA isolated from the desired organism. The PCR product is used as a probe to screen for the clone, or the PCR product is cloned.

INSERTIONAL MUTAGENESIS

Transposons (see **Transposon tagging**) and **retroviruses** are used as insertional mutagens to "tag" genes. If a previously identified transposon or retrovirus inserts into a gene, the sequence of the transposon or retrovirus is used as a tag to identify the host DNA flanking the insertion. Techniques, such as IPCR (see **PCR**), plasmid rescue, and library screening (using the element as a probe) are all successful in identifying flanking DNA from a "tagged" individual.

CLONING TISSUE- OR TREATMENT-SPECIFIC GENES

In some cases, researchers do not want to identify a specific gene, but instead they are interested in a certain class of gene, for example, liver-specific genes or genes induced by a **hormone**. There are numerous methods for identifying such specific types of gene, including **subtractive hybridization**, differential display, database analysis, and analysis of microarrays (see **PCR** and **Expressed sequence tag**).

IMMUNOCHEMICAL METHODS

Another method for cloning a gene, where the protein has been identified previously, is using an **antibody** that recognizes the protein to immunoprecipitate **polysomes** making the protein and then to isolate its mRNA. The mRNA produces a cDNA copy that is cloned. The cDNA clone is also used as a probe to isolate a genomic clone.

CONCLUSION

There are a large number of ways to clone a gene or set of genes of interest. The method used depends on what materials are available, such as protein sequence, antibodies to the protein coded for by the gene, gene sequence information, useful libraries for screening, or a "tagged" or "untagged" mutation in the gene of interest. The purpose of cloning a gene is to generate large amounts of DNA for further analysis, such as sequencing or mutational analysis. A cloned gene is also useful for producing the protein or for **antisense** studies, and in some cases it is used to disrupt the endogenous gene by homologous recombination. A cloned gene is also useful as a probe in **gene expression** studies. Cloning a gene is one of the first steps to understanding its function.

BIBLIOGRAPHY

1. C. Gellissen and L. P. Hollenberg (1997) *Gene* **190,** 87–97.
2. S. N. Ho, H. D. Hunt, R. M. Horton, J. K. Pullen, and L. R. Pease (1989) *Gene* **77**, 51–59.
3. R. A. Dixon, C. J. Lamb, S. Masoud, V. J. H. Sewalt, and N. L. Paiva (1996) *Gene* **179**, 61–71.
4. J. Sambrook, E. F. Fritsch, and T. Maniatis (1989) *Molecular Cloning; a Laboratory Manual*, Cold Spring Harbor Laboratory Press, Cold Spring Harbor, NY.
5. G. G. Wilson and N. E. Murray (1991) *Annu. Rev. Genet.* **25**, 585–562.
6. V. B. Rao, V. Thaker, and L. W. Black (1992) *Gene* **113**, 25–33.
7. J. C. Pierce and N. L. Sternberg (1992) *Methods Enzymology* **216**, 549–574.
8. N. Fairweather (1997) *Methods Mol. Biol.* **68**, 137–148.
9. B. A. C. Shizuya, B. Birren, U.-J. Kim, V. Mancino, T. Slepak, Y. Tarhiri, and M. Simon (1992) *Proc. Natl. Acad. Sci. USA* **89**, 8794–8797.
10. D. T. Burke, G. F. Carle, and M. V. Olson (1987) *Science* **236**, 806–812.
11. D. Mazel, E. Loic, S. Blanchard, W. Savrin, and P. Marliere (1997) *J. Mol. Biol.* **266**, 939–949.
12. R. M. Teather and P. J. Wood (1982) *Appl. Environ. Microbiol.* **43**, 777–780.
13. J. Agnan, C. Korch, and C. Selitrennikoff (1997) *Fungal Genet. Biol.* **21**, 292–301.
14. A. J. Hanzlik, M. M. Osemlak-Hanzlik, and D. M. Kurnit (1992) *Gene* **122**, 171–174.
15. S. D. Tanksley, M. W. Ganal, and G. B. Martin (1995) *Trends Genet.* **11**, 63–68.
16. A. C. M. Radding (1982) *Annu. Rev. Genet.* **16**, 405–437.
17. S. Zhang, G. Lao, and D. B. Wilson (1995) *Biochemistry* **34**, 3386–3395.

Suggestions for Further Reading

D. M. Glover (1985) *DNA Cloning; a Practical Approach,* Vol. 2, IRL Press, Washington, D.C., describes methods for cloning into microorganisms other than *E. coli.*

D. M. Glover and B. D. Hanes (1995) *DNA Cloning: a Practical Approach*, IRL Press, New York, describes methods for introducing genes into mammalian cells and mammals.

D. W. S. Wong (1997) *The ABCs of Gene Cloning*, Chapman and Hall, New York; an overview of cloning.

ClpAP AND ClpXP PROTEINASES

S. H. Lecker
A. L. Goldberg

These large ATP-dependent **proteinases** are involved in **protein degradation** in prokaryotes. Each enzyme is composed of two subcomplexes, both of which are essential for

ATP-dependent **proteolysis**. (A similar enzyme system is also found in chloroplasts of plant.)

ClpP

The ClpP proteolytic component is a **serine proteinase** composed of 14 identical subunits, organized in two seven-membered rings. The rings enclose a central chamber where the **active sites** are localized. To be degraded, proteins must enter this chamber through openings in the rings. By itself, ClpP can hydrolyze small **peptides** but not polypeptides. To degrade proteins, this proteolytic component must associate with either of two homologous ATPase complexes (ClpA or ClpX), which determine the substrate specificity.

ClpA AND ClpX

ClpA and ClpX complexes are hexameric rings that associate with each end of the ClpP to form a four-ring active enzyme. These ATPases bind different types of polypeptides, probably unfold them, and facilitate their entry into ClpP for degradation.

Suggestion for Further Reading

S. Gottesman, M. R. Maurizi, and S. Wickner (1997) Regulatory subunits of energy-dependent proteases. *Cell* **91**, 435–438.

CODING STRAND

J. R. S. FINCHAM

This is an ambiguous term, and perhaps best avoided. Because **DNA** is double-stranded and only one strand of the DNA of a **gene** is usually **transcribed** to make **RNA**, there is a distinction between the transcribed and the nontranscribed strand at any point in the **genome**. Using the useful terminology "upstream"/"downstream" to refer to the direction in which the gene is transcribed and translated, the nontranscribed DNA strand runs upstream to downstream in the chemical direction 5′ to 3′, whereas the complementary transcribed strand has the opposite 3′ to 5′ polarity. The transcript RNA runs 5′ to 3′, like the nontranscribed DNA strand. Thus, the nontranscribed DNA strand matches the mRNA in its base sequence (with thymine replacing uracil), not the transcribed strand which has the back-to-front complements of the codons, as explained in Figure 1. Consequently, in predicting encoded **amino acid** sequence of the encoded **protein** from a gene DNA sequence, one uses the sequence of the nontranscribed DNA strand because this is the strand from which the codons can be read directly. For this reason, the nontranscribed strand is sometimes called the coding strand.

Some workers, however, adopt the opposite convention because, after all, the transcribed strand provides the information for the codon sequence of the mRNA. Therefore, it is best to eschew the coding versus noncoding terminology and to distinguish the two DNA strands as transcribed versus nontranscribed.

It should be remembered that both strands of a particular segment of DNA can be transcribed, though probably never in the same cell type and at the same time. This may be the case, for example, when genes are "nested" within the introns of other genes (see **Gene structure**).

CODON USAGE AND BIAS

J. R. S. FINCHAM

The **genetic code** is degenerate, in the sense that (except for methionine and tryptophan) a given **amino acid** can be specified by either two, three, four, or six alternative **codons**. When the same amino acid is specified by three, four, or six codons, they cannot all be recognized by the same **transfer RNA** anticodon, and so in these cases at least two or three distinct transfer RNAs must carry the same amino acid (isoaccepting) but with different anticodons (see **Transfer RNA** and **Aminoacyl tRNA synthetases**).

Not all codons for a particular amino acid are used equally frequently. Generally, strong biases favor some codons and are against others, and different codons are favored in different organisms (Table 1). Some genes within an organism show much greater codon usage bias than others, and there is a strong correlation between the degree of bias shown by a gene and its level of expression in terms of the amount of encoded protein produced. For example, in all organisms where the matter has been examined, the genes that encode ribosomal proteins, which are always relatively abundant, are among the most highly biased. The excellent correlation in *Escherichia coli* between codon bias and level of protein product is shown in Fig. 1. This suggests that the preferred codons are more efficiently translated than the relatively rarely used codons.

Different translation efficiencies of different codons is largely accounted for by a number of factors, pointed out by Ikemura ("Ikemura's rules"; 1). First, when different isoaccepting tRNAs are present at very different concentrations, the favored codons will be those that are serviced by the most abundant tRNAs. There is evidence for this generalization from *E. coli, Saccharomyces* (1,2) and *Drosophila* (3). Where a particular amino acid has both codon bias and unequal concentrations of isoaccepting tRNAs, the rule holds virtually without exception. For example, the favored arginine codons in *E. coli* are CGU and CGC, whereas AGA is strongly favored in *Saccharomyces*. In the former organism, the tRNA that has anticodon GCI (I for inosine) recognizing CUCU is present at several-fold higher concentration than that recognizing AGA, with anticodon UCU* (U* is a modified uridine), whereas in yeast the tRNA inequality is very much the opposite.

The different relative abundances of tRNAs do not, however, explain codon usage bias when it operates between codons served by the same isoaccepting tRNA. Here the bias can be explained in terms of Ikemura's other rules, which account for the fact that an anticodon may pair more comfortably with one codon than with another, even though it recognizes both. Thus an anticodon that has a thiolated uridine or 5-carboxymethyl uridine in the 5′ position (the "**wobble**" **pairing** position, see **Anticodon**) prefers a 3′ A in the codon over 3′ G (rule 2); 5′ inosine (I) prefers U or C to A, even though it pairs with all three (rule 3); and codon-anticodon interactions dependent on the relatively weak A–U or T–A pairings in both of the first two positions work best with a stronger G–C or C–G pairing in the third (rule 4). Examples from *E. coli* consistent with these

Table 1. Codon Bias in Different Organisms[a]

Amino Acid	Possible Codons	Favored codons: *Drosophila*	Favored codons: *Saccharomyces*	Favored codons: *Escherichia coli*
Leucine	CUU CUC CUA CUG UUA UUG	CUG	UUG	**CUG**
Arginine	CGU CGC CGA CGG AGA AGG	CGU CGC	AGA	CGU CGC
Proline	CCU CCC CCA CCG	CCC	CCA	CCG
Glutamine	CAA CAG	**CAG**	**CAA**	CAG
Lysine	AAA AAG	**AAG**		
Alanine	GCU GCC GCA GCG	GCC	GCU,GCC	
Valine	GUU GUC GUA GUG	[**GUA**]	GUU GUC	[GUC]
Glycine	GGU GGC GGA GGG	[**GGG**]	**GGU**	GGU GGC
Serine	UCU UCC UCA UCG AGU AGC	[**AGU**]	UCU UCC	UCU UCC AGC
Threonine	ACU ACC ACA ACG	ACC	ACU ACC	ACC,ACU
Isoleucine	AUU AUC AUA	AUC	AUU AUC	AUC
Asparagine	AAU AAC	**AAC**	AAC	**AAC**
Phenylalanine	UUU UUC	**UUC**	UUC	UUC
Tyrosine	UAU UAC	UAC	UAC	UAC
Glutamic acid	GAA GAG	**GAG**	GAA	GAA
Cysteine	UGU UGC	**UGC**	UGU	
Histidine	CAU CAC	CAC		CAC
Aspartic acid	GAU GAC			

[a] *Drosophila* data from a pool of 15 high-expression genes (3); *E. coli* and *Saccharomyces* from pools of highly expressed genes (1,2). The codons indicated show strong majority usage (heavy type indicates 90%) except those italicized in brackets where the bias is strongly against. Blanks indicate no strong bias or insufficient data.

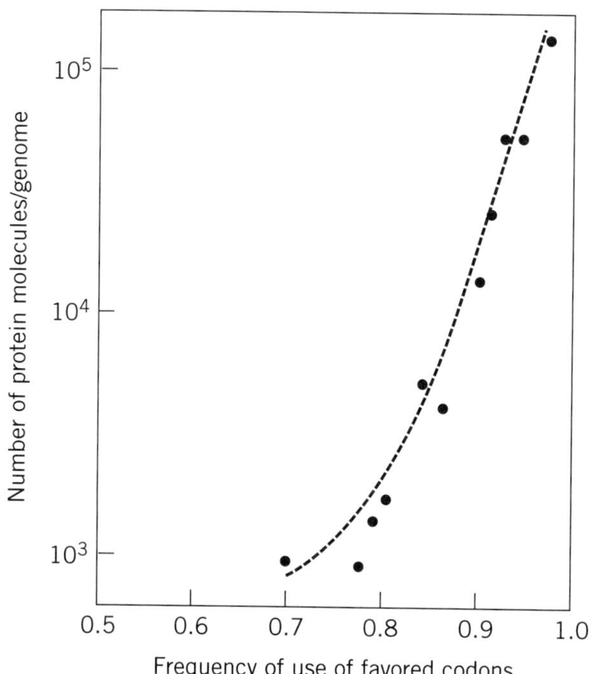

Figure 1. Frequency of using favored ("optimal") codons in *E. coli* genes plotted against the quantity of the gene protein products. From Ref. 1 by permission.

rules are the biases in favor of GAA over GAG (rule 2), CGU or CGC over CGA (rule 3) and AUC over AUU (rule 4) (1).

EFFECTS OF ALTERING CODONS

If the more rarely used codons really are translated relatively slowly, one would expect that their introduction into normally highly expressed genes would decrease protein yields. This prediction has been tested experimentally, and substantial effects have indeed been found, especially when multiple and contiguous rare codons have been introduced into the upstream ends of open reading frames. Hoekema et al. (4) carried out a total replacement of the normal codons by synonymous infrequently used codons in the upstream third of the normally highly expressed *Saccharomyces PGK1* (phosphoglycerate kinase) gene and found a tenfold reduction in enzyme yield. There was also a threefold reduction in the level of **messenger mRNA**. In the gene encoding glutamate dehydrogenase, in *Neurospora crassa*, a double frameshift mutant that generated three successive rare codons, numbers 54 to 56 of a 453-codon sequence, brought about a threefold reduction in enzyme without appreciable effect on mRNA level (5). And in *E. coli*, the insertion of an additional mRNA segment containing five contiguous rare AGG arginine codons 24 bases downstream of the initiation codon of *lacZ* (**β-galactosidase**) reduced the enzyme yield by 90% in the early exponential phase and virtually to zero thereafter. Another insertion, identical except that AGG was replaced by the favoured CGU codon (6), had little effect. However, when the distance between the inserted AGG-containing sequence and the initiation codon was increased by additional insertions, the yield of β-galactosidase increased almost linearly with distance, as if the rare codons have maximum effect when translation is just starting and progressively less effect as it proceeds. The mechanism of such distance-dependence is unclear. Chen and Inouye (6), reviewing the positions of rare codons in *E. coli* genes, noted that they fall within the first 25 codons after the initiating AUG and suggested that their function is to modulate the rate of translation. This would only be a regulating mechanism if there were some means of increasing or decreasing the effect, for instance, by adjusting the levels of the least abundant tRNAs. But there is as yet no evidence for this.

If rare codons have strong negative effects on the protein yields of genes, they may be an obstacle to the expression of **transgenes** in alien host cells. A codon that is abundant in the donor organism may be rare in the recipient. Therefore, to obtain a good yield of a foreign protein, it may be necessary to adjust the coding sequence of the introduced gene by *in vitro* DNA manipulation. Perhaps the best example of successful application of this idea is provided by the work of Cormack *et al.* (7), who wished to confer on the fungus *Candida albicans* the ability to produce the **green fluorescent protein** (GFP) native to the jellyfish *Aequoria victoria*. Introduction of the natural GFP gene into *Candida* on a replicating plasmid **vector** resulted in the formation of some GFP mRNA in the fungus, but no detectable GFP. Replacing the single CUG codon with CUU, in case the *Candida* coding idiosyncracy of translating CUG as serine rather than leucine (8) was responsible for the failure, brought no improvement. But when all the *Aequoria* codons that are comparatively rare in *Candida* were replaced by codons that are common in the fungus, introduction of the transgene produced an abundance of GF protein. The level of GFP mRNA was also considerably increased, which suggests, as does the yeast *PGK1* example mentioned before (3), that blocked translation, and hence failure to cover the mRNA with ribosomes, may expose the mRNA to enzymic degradation.

BIBLIOGRAPHY

1. T. Ikemura (1985) *Mol. Biol. Evol.* **2**, 13–34.
2. P. M. Sharp, T. M. F. Tuohy, and K. R. Mosurski (1986) *Nucleic Acids Res.* **14**, 5125–5143.
3. D. C. Shields, P. M. Sharp, D. G. Higgins, and F. Wright (1988) *Mol. Biol. Evol.* **5**, 704–716.
4. A. Hoekema, R. A. Kastelein, M. Vasser, and H. A. de Boer (1987) *Mol. Cell. Biol.* **7**, 2914–2924.
5. J. A. Kinnaird and J. R. S. Fincham (1991) *J. Mol. Biol.* **221**, 733–736.
6. G.-F. T. Chen and M. Inouye (1990) *Nucleic Acids Res.* **18**, 1465–1473.
7. B. P. Cormack et al. (1997). *Microbiology* **143**, 303–311.
8. M. A. S. Santos and M. F. Tuite (1995) *Nucleic Acids Res.* **23**, 1481–1486.

COENZYME, COFACTOR

JOHN F. MORRISON

The term cofactor has been used as a general term to indicate that a compound was required, in addition to the **enzyme**, for a reaction to proceed and that the compound remained unchanged at the end of the reaction. Cofactors were considered to be either activators, such as metal ions, or *coenzymes*, organic molecules that participated in enzymic reactions. **NAD** and NADP were named originally as Coenzyme I (CoI) and Coenzyme II (CoII), respectively, because in studies on metabolic reactions they appeared to function as electron carriers. They certainly perform that role in coupled reactions, such as those catalyzed by glyceraldehyde-3-phosphate dehydrogenase and **lactate dehydrogenase**:

$$\text{Glyceraldehyde-3-phosphate} + \text{NAD} + P_i \rightleftharpoons 1,3-\text{bis glycerophosphate} + \text{NADH} + \text{H}^+ \quad (1)$$

$$\text{Pyruvate} + \text{NADH} + \text{H}^+ \rightleftharpoons \text{lactate} + \text{NAD} \quad (2)$$

with the sum of reactions (1) and (2) being

$$\text{Glyceraldehyde-3-phosphate} + P_i + \text{pyruvate} \rightleftharpoons \text{1,3-bis glycerophosphate} + \text{lactate} \quad (3)$$

However, this is not the role of NAD/NADH when functioning as substrates for a dehydrogenase reaction. Similar comments can be made about Coenzyme A (CoA), which functions as an acyl carrier, and Coenzyme Q (CoQ, ubiquinone), which acts as an electron carrier. In this connection, it is of interest that nucleoside phosphates, such as ATP, are not usually classified as carriers of phosphoryl groups.

Other compounds that have been considered as coenzymes are **biotin**, flavin adenine di- and mono-nucleotide (**FAD**, **FMN**), **lipoic acid**, **pyridoxal phosphate**, and thiamin pyrophosphate. Biotin, lipoic acid, and pyridoxal phosphate are bound covalently to carboxylases, acyl transferases, and aminotransferases, respectively. FAD may, or may not, be bound covalently to oxidases. The interaction of thiamin pyrophosphate with decarboxylases is noncovalent, although the binding is tight.

It would seem preferable to consider a compound as a coenzyme only under conditions where it is functioning as a carrier and not when it is simply a substrate for a single enzyme. Compounds that are covalently or tightly bound to an enzyme might better be regarded as **prosthetic groups**. The features that these compounds have in common are the helping groups that they provide to enzymes. These groups possess chemical attributes that the enzymes do not have and that are essential for catalysis.

COHESIVE, STICKY ENDS

DAVID B. WILSON

Cohesive ends are short single-stranded sequences of **DNA**, usually 1 to 3 bases long, that are produced at the end of a double-stranded DNA molecule by the action of a type II **restriction enzyme** that makes **staggered cuts** in a symmetrical, **palindromic** sequence (1). For example, *Pst*1 cuts the sequence 5′ ..CTGCAG.. to give the ends ..C and TGCAG.. . The two ends 3′ ..GACGTC.. ..GACGT.. C.. produced in this way are complementary, so they bind together by base pairing and allow **DNA ligase** to repair the break. Furthermore, any fragment cut by an enzyme of this type binds to any other DNA fragment cut by the same enzyme or to one that creates the same cohesive ends. When the cohesive ends ligated together are cut by enzymes that have different recognition sequences, the new sequence is not cut by either enzyme. In the case of *Bam*HI, 5′ ..GGATCC.. and *Sau*3a, ..GATC.., which produce the same sticky ends ..CCTAGG.. ..CTAG..but have different length recognition sites, the sequence created can always be cut by *Sau*3a but can be cut only by *Bam*HI if the sequence cut by *Sau*3a is a *Bam*HI site. Some restriction enzymes recognize degenerate sequences, so that only when both DNA fragments cut with this type of enzyme have the same sequence do they pair and allow ligation. Depending on the position of the cuts in the two strands, either a 5′-strand or the 3′-strand, creating a 5′-overhang or a 3′-overhang is possible, but each enzyme produces only one type.

Cohesive ends are also present on some **bacteriophages**. For example, on **lambda phage** they are 12 bases long and are produced by staggered cutting of a symmetrical sequence site (cos site) by a phage **enzyme** during DNA packaging (2).

BIBLIOGRAPHY

1. G. G. Wilson and N. E. Murray (1991) *Ann. Rev. Genet.* **25**, 585–562.
2. O. Yang, A. Hanagan, and C. E. Catalano (1997) *Biochem.* **36**, 2744–2752.

COILED-COILS

DAVID PARRY

A coiled-coil is a generic name for any helical structure that has an axis that is itself helical. A general feature of all such coiled-coil conformations is that the handedness of coiling alternates at successive levels of structure, in line with the well-established practice employed by ropemakers over the centuries. This maximizes the interactions between strands and minimizes relative slippage. There are two particularly well-known classes of coiled-coil structures in biological structures: (i) the α-fibrous proteins and (ii) the **collagens**. The structure of each will be described.

α-Fibrous proteins display a characteristic **heptad** quasi-**repeat** of the form $(a–b–c–d–e–f–g)_n$, where about 75% of the a and d are positions are occupied by **nonpolar** residues such as **leucine**, **isoleucine**, and **valine** (1,2). These sequences adopt a right-handed α**-helical** conformation with about 3.6 residues per turn. Because the apolar residues in the heptad repeat are 3.5 residues apart, on average, it follows that they will form an apolar stripe on the surface of the α-helix, and this will wind around the helix in a left-handed manner (3). In an aqueous environment, apolar residues tend to pack as closely together as possible to shield each other from the **water** and, in doing so, provide the **hydrophobic** driving force for assembly. This can be facilitated in this case by two or more α-helices coming together, optimizing the packing of the apolar residues along their interface (so-called knobs-into-holes packing) and winding around one another to generate a left-handed coiled-coil structure (Fig. 1). Favorable **electrostatic interactions** can also be made between the chains, which help to specify both the relative chain direction and the axial displacement between the chains (1). The interchain ionic interactions occur predominantly between oppositely charged residues in positions e and g of different chains. Calculations and experimental observations on two-stranded α-helical coiled-coils have shown that the strands are parallel (rather than antiparallel) and in axial alignment. The types of apolar residues in positions a and d are important in specifying the number of chains in the coiled-coil molecule (see **Leucine zipper**). Two-stranded coiled-coils occur in muscle **myosin**, **intermediate filaments**, plectin, streptococcal M proteins, centrophilin, **kinesin**, β-giardin, and a host of other proteins. Three-stranded structures are found in the **laminins**, **fibrinogen**, the **spectrin** superfamily of proteins, bacteriophage leg proteins such as gp17, cartilage matrix protein, mannose-binding protein, **macrophage** scavenger receptor protein, and many others. Four-stranded ropes are found in the silks of the bees, wasps, and ants (*Hymenoptera aculeata*), as well as in globular proteins (such as the **four-helix motif**).

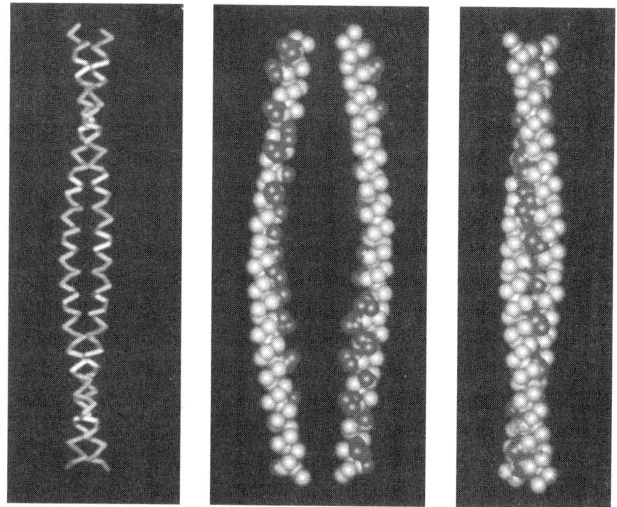

Figure 1. The structures of two-stranded coiled-coils. (Left) The backbone model of the two-stranded coiled-coil from a portion of the N-terminal end of **tropomyosin**. Two right-handed α-helices coil around each other in a left-handed manner. (Center) A space-filling model of the same structure with the strands shown separated. The apolar residues are shown in black. (Right) The two strands brought together. The apolar residues are interlocked in a systematic way along the axis of the coiled-coil and are shielded from water as a result. (From Ref. 9 with permission.)

Five-stranded coiled-coils have recently been described in **HIV** capsid protein and in collagen oligomeric matrix protein (COMP) (Fig. 2). As Cohen and Parry (3) have pointed out, it is not the bending of the axis of the α-helix in the supercoiled conformation that is crucial but rather the way in which it permits systematic apolar side-chain interactions to be made. It must also be pointed out that many sequences show discontinuities in their heptad substructures. A recent study by Brown et al. (4) has shown that all such (short) discontinuities can be classified as either "stutters" or "stammers"; these correspond to deletions of three and four residues, respectively, from an otherwise continuous repeat. Physically, a stutter results in a region in which the coiled-coil undergoes a degree of local underwinding to generate a longer supercoil pitch length, whereas the stammer causes a degree of local overwinding, thus giving rise to a shorter supercoil pitch length. The latter is likely to be stereochemically more difficult to achieve. The former has been observed experimentally. A coiled-coil is not confined to fibrous proteins with long rod-like domains as seen in myosin, intermediate filaments, and desmoplakin (for example), but occurs commonly in globular proteins in the form of α-helical bundles. These are generally short in length (say 3 to 10 heptads) and contain anything from two or three α-helices to sizable bundles containing six or even more. The same underlying heptad repeat is present, but it becomes less easy to recognize as the helix length decreases.

The α-chains in collagen contain a triplet repeat of the form $(Gly–X–Y)_n$, where X and Y can be almost any amino acid residue but are commonly proline and hydroxyproline, respectively. The chain folds up into a left-handed helical structure that is very similar to that seen in polyglycine II and

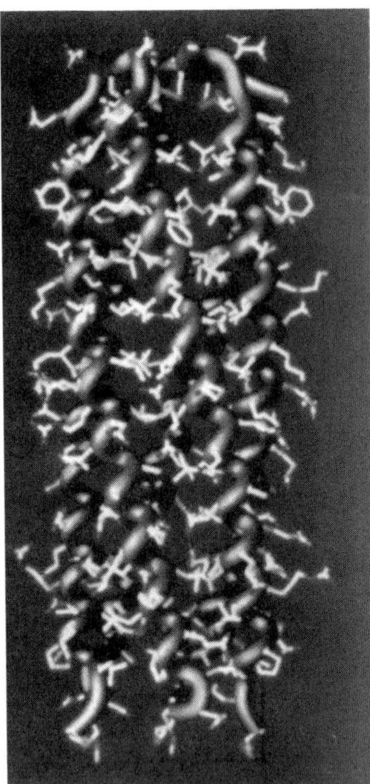

Figure 2. The five-stranded coiled-coil oligomerization domain of cartilage oligomeric matrix protein (COMP). Each of the α-helices forms about one-third of a complete turn over the length of the structure. The N-terminal end is at the bottom of the page. (Courtesy of R. A. Kammerer.) See color insert.

polyproline II. Three α-chains then aggregate in parallel with a one-residue relative axial displacement between chains to generate a right-handed, triple-helical structure in which the glycine residues lie along the axis of the molecule (see **Collagen**). This class of conformation was originally formulated by Ramachandran and Kartha (5) and Rich and Crick (6) and has subsequently been refined by Fraser et al. (7,8). The structure has 10 repeating units in three turns, a unit rise of 0.2894 nm, and a pitch length of 0.9647 nm. The individual α-chains have a pitch length of 8.68 nm. The conformation is stabilized by a single hydrogen bond per triplet between the peptide NH group of a glycyl residue and the peptide carbonyl group of an X residue in another chain. Furthermore, water molecules bridge the glycyl NH to the C═O of a prolyl residue in another chain and the OH group of a hydroxyprolyl residue in the same chain. The diameter of a collagen molecule is about 1.4 nm. Its length varies from one type of collagen to another, but for a Type I collagen molecule its length is close to 300 nm. The molecule thus has a high axial ratio in excess of 200.

BIBLIOGRAPHY

1. J. F. Conway and D. A. D. Parry (1990) Structural features in the heptad substructure and longer range repeats of two-stranded α-fibrous proteins. *Int. J. Biol. Macromol.* **12**, 328–334.
2. J. F. Conway and D. A. D. Parry, Three-stranded α-fibrous proteins: the heptad repeat and its implications for structure. *Int. J. Biol. Macromol.* **13**, 14–16.
3. C. Cohen and D. A. D. Parry (1986) α-Helical coiled-coils—a widespread motif in proteins. *Trends Biochem. Sci.* **11**, 245–248.
4. J. H. Brown, C. Cohen, and D. A. D. Parry (1996) Heptad breaks in α-helical coiled coils: stutters and stammers. *Proteins Struct. Funct. Genet.* **26**, 134–145.
5. G. N. Ramachandran and G. Kartha (1955) Structure of collagen. *Nature (London)* **176**, 593–595.
6. A. Rich and F. H. C. Crick (1961) The molecular structure of collagen. *J. Mol. Biol.* **3**, 483–506.
7. R. D. B. Fraser, T. P. MacRae, and E. Suzuki (1979) Chain conformation in the collagen molecule. *J. Mol. Biol.* **129**, 463–481.
8. R. D. B. Fraser, T. P. MacRae, A. Miller, and E. Suzuki, (1983) Molecular conformation and packing in collagen fibrils. *J. Mol. Biol.* **167**, 497–521.
9. C. Cohen and D. A. D. Parry (1990) α-Helical coiled-coils and bundles: how to design an α-helical bundle. *Proteins Struct. Funct. Genet.* **7**, 1–15.

Suggestions for Further Reading

C. Cohen and D. A. D. Parry, α-Helical coiled-coils and bundles: how to design an α-helical bundle. (1990) *Proteins Struct. Funct. Genet.* **7**, 1–15.

R. D. B. Fraser and T. P. MacRae (1973) *Conformation in Fibrous Proteins and Related Synthetic Polypeptides*, Academic Press, London.

A. Lupas (1996) Coiled-coils: new structures and new functions. *Trends Biochem. Sci.* **21**, 375–382.

COINTEGRATIVE VECTORS

RICHARD WALDEN

Cointegrative vectors were the first type of **vector** developed for transferring foreign **DNA** from the bacterium ***Agrobacterium*** to **plant** cells, for use in **plant genetic engineering** (1). Although cointegrative vectors may be relatively difficult to work with in practice, compared with the alternative binary vectors, they offer **plasmid** stability in the **bacterial** cell.

The general concept of using cointegrative vectors relies on homologous **recombination** within *Agrobacterium* to introduce into a modified T-DNA (see **T-complex**) the DNA to be transferred to the plant cell. First, the DNA to be transferred is introduced into an intermediate **cloning** vector based on an *Escherichia coli* plasmid. The intermediate vector that contains the foreign DNA is introduced by conjugation into *Agrobacterium* that contains the cointegrative vector. The cointegrative vector is a **Ti plasmid** from which the T-DNA genes that encode oncogenic function have been removed and/or replaced with a sequence that is also contained in the intermediate vector. The intermediate vector is not stable in *Agrobacterium*. Homologous recombination between the intermediate vector and the cointegrative vector results in transferring the foreign DNA to the cointegrative vector. These vectors are designed so that once the foreign DNA, is integrated into the cointegrative vector, it is located between its border sequences. The most generally used cointegrative vector system is the "split-end vector system" (2)

BIBLIOGRAPHY

1. P. Zambryski et al. (1983) *EMBO J.* **2**, 2143–2150.
2. R. Fraley et al. (1985) *Bio/Technology* **3**, 629–635.

Suggestions for Further Reading

R. Walden (1988) *Genetic Transformation in Plants*, Prentice-Hall, Englewood Cliffs, NJ.

F. F. White (1993) Vectors for gene transfer in higher plants. In *Transgenic Plants*, Vol. 1 (D. Kung and R. Wu, eds.), Academic Press, San Diego, pp. 15–48.

COLCHICINE

LESLIE WILSON
DULAL PANDA

Colchicine (Fig. 1), a compound obtained from the autumn crocus, *Colchicum autumnale*, is an ancient drug that has been used for centuries for the treatment of gout (1). Its use in molecular biology began in the mid-1930s when its potent ability to inhibit eukaryotic cell proliferation at mitosis was discovered. Since then, colchicine has had a remarkable history as an experimental tool for characterizing the biochemical properties of **tubulin**, the protein subunit of **microtubules**, for characterizing the diverse processes in eukaryotic cells that are dependent upon drug-sensitive microtubules, and for studying the polymerization and dynamics properties of microtubules. For example, colchicine has been used to determine the role of microtubules in mitosis, protein **secretion**, axonal transport, and the development and maintenance of asymmetric cell shape. It has also been used extensively as a cytogenetic tool to determine **chromosome** numbers in **karyotypes**. Also, radiolabeled colchicine was used as an affinity marker to effect the first purification of tubulin from brain (2).

COLCHICINE BINDING TO TUBULIN

The binding of colchicine to tubulin is not a simple process (3–6). The binding reaction is slow and highly temperature-dependent. Binding is extremely slow at 0°C, and 2 to 3 h is required to reach equilibrium at 37°C. The **activation energy** for the forward reaction is high, ~20 kcal/mol. The kinetics of binding are biphasic. The initial step is the formation of a reversible preequilibrium complex that is followed by a slow step in which conformational changes in tubulin lead to the formation of a final-state tubulin-colchicine (TC) complex. Although the binding of colchicine to tubulin is noncovalent, the final state TC complex is very poorly reversible.

Colchicine binds to tubulin at a single site with a dissociation constant in the range of 0.02 μM to 5 μM. Colchicine has very different affinities for tubulin isolated from different sources.

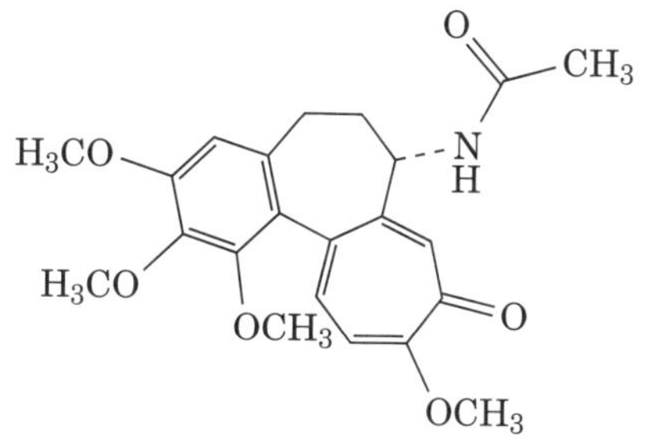

Figure 1. Structure of colchicine.

For example, colchicine binds strongly to vertebrate brain tubulin, but it binds very weakly to plant, fungal, and yeast tubulins. Also, there are different tubulin isotypes, and colchicine binds differently to the isotypes purified from same tubulin source (7). Despite intensive investigation, the precise location in tubulin of the colchicine-binding site is not clear. The binding site is thought to reside in the β-tubulin subunit, near residues Cys 354 and Cys 241, close to the intradimer interface.

The binding of colchicine to tubulin induces conformational changes in tubulin, as well as in colchicine (6,8,9). For example, the binding of colchicine to tubulin quenches the intrinsic tryptophan **fluorescence** of tubulin, indicating that it induces a small change in the **tertiary structure** of the protein. It also perturbs the far-ultraviolet **circular dichroism** spectrum of tubulin, indicating that it changes the **secondary structure** of the protein. In addition, colchicine binding to tubulin strongly increases the intrinsic **GTPase** activity of tubulin, increases the affinity of the $\alpha\beta$ dimer association by threefold, changes the exposure of **thiol groups** in tubulin, and, under certain conditions, induces tubulin to self-assemble into nonmicrotubule polymeric structures. Tubulin is also subject to a time- and temperature-dependent irreversible decay of its **protein structure**. Binding of colchicine to tubulin slows the rate of decay. The concept that colchicine undergoes conformational changes upon binding to tubulin is evident from the development of colchicine-tubulin fluorescence and the change of the colchicine circular dichroism spectrum upon binding to the protein.

INHIBITION OF MICROTUBULE POLYMERIZATION BY COLCHICINE

Colchicine inhibits microtubule polymerization at concentrations that are far below the total concentration of tubulin (10,11), indicating that colchicine inhibits microtubule polymerization by acting at the microtubule ends. In order to produce its potent actions on microtubule polymerization, colchicine must first form a TC complex. Substoichiometric concentrations of TC-complex only partially depolymerize microtubules, and relatively low concentrations of TC-complex can stabilize microtubules against dilution-induced disassembly (12). These studies support the hypothesis that the TC-complex forms a stabilizing "cap" at the end of the microtubule. Also, the TC-complex can form copolymers with unliganded tubulin when microtubules are assembled in the presence of TC complex (13).

KINETIC SUPPRESSION OF MICROTUBULE DYNAMICS

Microtubules exhibit two kinds of nonequilibrium dynamics: **treadmilling** and dynamic instability (see **Microtubules**). Recent studies have revealed that colchicine and other compounds that depolymerize microtubules (see **Vinblastine**) strongly suppress these dynamics at relatively low concentrations in the absence of appreciable microtubule depolymerization. Colchicine was found some years ago to suppress treadmilling *in vitro* and the rate of microtubule disassembly upon dilution of the microtubules [now called "kinetic capping" (11,12)]. However, its stabilizing effects on dynamics were only fully appreciated with the introduction of real-time differential-interference contrast video microscopy, which enabled one to visualize directly in real time the stabilizing action of the drug on the growing and shortening dynamics of individual microtubules (14). Small numbers of incorporated TC complexes strongly suppress the rates and extents of growing and shortening and greatly increase the percentage of time that the microtubules spend in an attenuated state. In addition, the TC complex strongly suppresses the catastrophe frequency and increases the rescue frequency. At low submicromolar concentrations, TC complex suppresses the dynamics without reducing the polymer mass. Significant reduction of polymer mass requires relatively high TC complex concentrations. However, the surviving microtubules are extremely stable. Colchicine appears to suppress microtubule dynamics by binding at the microtubule ends, most probably by inducing a conformational change and/or by steric hindrance at the ends.

BIBLIOGRAPHY

1. P. Dustin (1984). *Microtubels*, Springer-Verlag, Berlin, pp. 1–482.
2. R. C. Weisenberg, G. G. Borisy, and E. W. Taylor (1968) *Biochemistry* **7**, 4466–4478.
3. G. G. Borisy and E. W. Taylor (1967) *J. Cell. Biol.* **34**, 525–533.
4. L. Wilson and M. Friedkin (1967) *Biochemistry* **6**, 3126–3135.
5. B. Bhattacharyya and J. Wolff (1974) *Proc. Natl. Acad. Sci. USA* **71**, 2627–2631.
6. D. L. Garland (1978) *Biochemistry* **17**, 4266–4272.
7. R. F. Luduena (1983) *Mol. Biol. Cell* **4**, 445–457.
8. J. M. Andreu and S. N. Timasheff (1982) *Biochemistry* **21**, 6465–6476.
9. T. David-Pfeuty, C. Simon, and D. Pantaloni (1979) *J. Biol. Chem.* **254**, 11696–11702.
10. J. B. Olmsted and G. G. Borisy (1973) *Biochemistry* **12**, 4282–4289.
11. R. L. Margolis and L. Wilson (1977) *Proc. Natl. Acad. Sci. USA* **74**, 3466–3478.
12. R. L. Margolis, C. T. Rauch, and L. Wilson (1980) *Biochemistry* **19**, 5550–5557.
13. H. Sternlicht and I. Ringel (1979) *J. Biol. Chem.* **254**, 10540–10550.
14. D. Panda, J. E. Daijo, M. A. Jordan, and L. Wilson (1995) *Biochemistry* **34**, 9921–9929.

Suggestions for Further Reading

O. J. Eigsti and P. Dustin Jr. (1955) *Colchicine, in Agriculture, Medicine, Biology, and Chemistry*, The Iowa State College Press, Ames, IA.

L. Wilson and K. W. Farrell (1986) *Ann. NY Acad. Sci.* **466**, 690–708.

S. B. Hastie (1991) *Pharmacol. Ther.* **512**, 377–401.

L. Wilson and M. A. Jordan (1994) in *Microtubules*, J. S. Hyams and C. W. Lloyd, eds., Wiley-Liss, New York, pp. 59–83.

A. Vandecandelaere, S. Martin, M. Schilstra, and P. Bayley (1994) *Biochemistry* **33**, 2792–2801.

COLD-SENSITIVE MUTANTS

WILLIAM A. ROSCHE
PATRICIA L. FOSTER

Cold-sensitive **mutants** are a class of **conditional lethal mutants** that cannot grow at a temperature below the organism's normal optimum. The low temperature may be lethal because the mutant form of the **protein** loses its function or because it forms inappropriate or inhibitory interactions with other proteins. Finding and characterizing second-site revertant mutations that **suppress** the cold-sensitive phenotype identifies **protein–protein interactions** (see **Temperature-sensitive mutations**).

COLICINS

C. J. LAZDUNSKI
V. GELI

Colicins produced by and active against coliform **bacteria** constitute a subset of the **bacteriocins** generated by many groups of bacteria. They have been a subject of interest since very early in this century. In 1925, Gratia demonstrated that *Escherichia coli* strain V (virulent in experimental infection) in liquid media produces a heat-stable substance that in high dilution inhibits the growth of *E. coli* (1). This protein was designated as colicin V. It has now been shown that it fits better with the description of the so-called "microcins" (see **Bacteriocins**) (2). Then a whole series of colicins produced by *E. coli* and closely related members of Enterobacteriacae were discovered. Mainly as a result of the influence and efforts of Fredericq (3), knowledge of the colicins advanced at a great rate, and more than 20 different types were identified on the basis of their action against a set of specific resistant (generally receptor-deficient) mutants (Table 1). In contrast to the bacteriocins from **Gram-positive** bacteria that kill species other than those that are likely to have the same ecological niche, colicins are active only against *E. coli* and closely related bacteria (there is a similar relationship between the cloacins and *Enterobacter cloacae*, the klebicins and *Klebsiella* species, and

Table 1. Characteristics of Colicins

Colicin	Colicin Activity Group		Receptor	Translocation System
E2,E7,E8,E9	A	DNase	BtuB	TolA, B, Q, R
E3,E6,DF13	A	RNase	BtuB	TolA, B, Q, R
E1	A	Pore-forming	BtuB	TolC, TolA, TolQ
A	A	Pore-forming	BtuB	OmpF, TolA, B, Q, R
N	A	Pore-forming	OmpF	OmpF, TolA, Q
K	A	Pore-forming	Tsx	OmpF, OmpA, Tol A, B, Q, R
Col5	B	Pore-forming	Tsx	TolC, TonB, ExbB, D
Col10	B	Pore-forming	Tsx	TolC, TonB, ExbB, D
Ia,Ib	B	Pore-forming	Cir	TonB, ExbB, D
D	B	Inhibition of protein synthesis	FepA	TonB, ExbB, D
M	B	Inhibition of synthesis of murein and LPS	FhuA	TonB, ExbB, D

the pyocins and pseudomonacae). A characteristic feature of colicinogenic bacteria is to be specifically immune to the colicin they make but not to other colicins. A large number of colicins have now been identified, and each has been characterized by the corresponding specific immunity protein. Colicins are **plasmid**-encoded, and each plasmid also bears an immunity protein, thus ensuring that plasmid carriers are protected from the colicin they themselves produce. There is probably a selective advantage for Enterobacteriacae to produce a colicin because 30 to 40% of natural isolates of *E. coli* carry colicinogenic plasmids (4). The colicins exert their lethal effect through a single-hit mechanism (5). There was some controversy about the meaning of this terminology, but now there is a consensus that sensitive cells are killed by a single event, implying that a single molecule kills, although not every one does.

Colicins are soluble proteins of 29 to 70 kDa (for colicin V, see previous). Their amino acid sequence is known, and they share the property of being linearly organized in three **domains** that have specific functions. They are also highly asymmetrical molecules and have axial ratios of 8 to 10 (6,7). Another common feature is that their production is induced by exposure of colicinogenic bacteria to agents like UV light and genotoxic chemicals, such as **mitomycin C**, that elicit the **SOS response**. After induction, they are produced in large amounts, so they provide useful model systems to study fundamental biological problems, such as protein–protein interactions, polypeptide translocation and insertion across and into **membranes**, functioning of voltage-gated pores, etc.

The classification of colicins reflects bacterial rather than colicin properties. According to its activity spectrum against a variety of mutants, a particular colicin is unambiguously assigned to one of two groups, A and B (8,9). Group B colicins are inactive on strains that have a lesion in the *tonB* gene but are active against strains mutant in *tolA* and *tolB* genes. Group A colicins show the opposite specificity. Now we know that group A colicins (A, E1 to E9, K, L, N, and cloacin DF13) and filamentous **bacteriophage** (f1, fd, and M13) need the Tol proteins to penetrate into cells (9–11). Group B colicins (B, D, Ia, Ib, M, 5, and 10) and phages T1 and Φ80 need TonB and its associated proteins (8,9). In all cases studied so far, the determinants for colicinogeny are located on plasmids (Table 1) of two different types: small multicopy plasmids or large low-copy-number plasmids that generally correspond to the A and B groups of colicins. In general, group A colicins are encoded by small plasmids and are actively released to the extracellular medium, whereas group B colicins are encoded by large plasmids and are very poorly secreted.

At least three hypotheses have been proposed to explain the evolution of colicin plasmids: (1) positive selection of diversity, (2) recombinational shuffling, and (3) transposition (12). The different colicins may have evolved by DNA recombination of fragments encoding different colicin domains (see later). This is best exemplified by the common uptake route for colicins B and D, which have a highly homologous N-terminal and central polypeptide sequence (defining the translocation and receptor domains respectively; see later) but very different C-terminal domains with different types of activities (13).

GENES FOR COLICINS AND ASSOCIATED PROTEINS

Regardless of the plasmid type, the genes for the colicin, immunity, and lysis (when there is one) are always clustered (14). The difference between immunity to the enzymatic colicins and to the channel-forming colicins and colicin M (whose target is the cytoplasmic membrane) is reflected in the regulation of their synthesis and the arrangement of the various colicin **operons**. In enzymatic colicins, the immunity gene is transcribed in the same direction as the colicin and lysis genes (15–17), whereas for the channel-forming colicins and for colicin M, immunity is encoded on the opposite DNA strand and is thus transcribed in the direction opposite to the colicin and lysis genes (18–20). Enzymatic colicin and cognate immunity genes form an operon under the control of an SOS **promoter** (see later). Thus, induction of colicin synthesis results in increasing the amount of immunity protein synthesized, but the presence of a terminator causes partial transcription arrest upstream of the lysis gene. In contrast, inducing the synthesis of pore-forming colicins from their SOS promoter does not result in a concomitant increase in the immunity protein. The immunity is constitutively expressed from a weak promoter (19,21–23). The situation is similar for the murein-synthesis-inhibiting colicin M and its immunity gene (24).

Colicin production is inducible and suicidal. Under normal conditions, very few cells produce colicin because the expression is repressed by the **LexA repressor** and is switched on only during the SOS response normally associated with repairing damaged DNA. This means that, in culture, colicin production is inducible in a growing population of cells by **mutagens** (UV light, mitomycin C). This induction is very efficient. In the absence of mutagens, LexA exhibits a very tight repression (25) and ensures that only a small proportion of cells go down the dead-end road of colicin expression and export. Once they have done so, it is important to the plasmid clone as a whole that the sacrificial cells produce as much colicin as possible. This maximizes selection against non-plasmid-bearing cells. Meanwhile, the silent preservation of many identical copies of the same plasmid in immune cells ensures no significant loss of the plasmids in the producer cell.

EXTRACELLULAR RELEASE OF COLICINS

Most of the colicins (group A, but not group B) are actively released into the growth medium. Their release mechanism differs from that of proteins secreted by Gram-negative bacteria (see **Protein secretion**):

1. they do not contain an N-terminal or C-terminal signal sequence like **Sec**-dependent exported polypeptides or those depending on ABC transporters;
2. their release depends on the function of a lytic protein (variously called *kil*, *lys*, *brp*) whose gene is part of the colicin operon, downstream of the colicin gene;
3. they are released from the host cell some time after synthesis;
4. their extracellular release is not specific, and quasi-lysis proteins have various effect on cells in addition to causing the release of assorted proteins and small molecules (26).

The lysis genes encode short proteins produced in a precursor form with a **signal sequence** and a typical cysteine lipid-modification consensus box, implying that the precursor undergoes the following **posttranslational modifications**: (1) acylation of the cysteine residue, (2) cleavage of the signal sequence by signal peptidase II, and (3) fatty acylation of the **amino group** of the now N-terminal acylated cysteine. The maturation and processing of lysis proteins occurs slowly, so every intermediate form can be observed. Both the mature form and the signal sequence accumulate in *E. coli* after processing (27,28). The addition of globomycin to *E. coli* cells stops processing of lysis protein precursors and inhibits colicin release (29). Acylation is essential for quasi-lysis (29,30). The mature lysis proteins activate the normally dormant **phospholipase** A in the outer membrane of colicin-producing cells, thus causing production of lysophospholipids, quasi-lysis, and colicin release (31). However, lysis proteins must also affect the permeability of the inner membrane (32). The lysis protein must reach a critical concentration within the cells before quasi-lysis and colicin release, which also indicates that the lysis protein directly affects the inner membrane. Local modifications in the structure of the bilayer are induced by interactions with lipopeptide micelles, as previously suggested in the case of Iturin A and bacillomycin L (33).

MECHANISM OF ENTRY AND DOMAIN ORGANIZATION OF COLICINS

Colicins bind to specific receptors in the bacterial outer membrane, from where they are translocated to, and eventually through, the inner membrane to reach their targets. Consistent with these three steps, their polypeptide chains are linearly organized into three domains: the *N*-terminal domain is involved in the translocation step, the central domain is responsible for receptor binding, and the C-terminal domain carries the lethal activity (Fig. 1). The boundaries of these domains were defined first by limited **proteolytic** digestion (34–36), then by genetic engineering (37,38), and from sequence homologies between colicins with the same type of lethal activity and using the same receptors (E-type, for example) or the same translocation pathways. Even colicin M, which has a molecular mass of 29 kDa, has three functional domains (24).

To enter the cells, colicins have borrowed multiprotein systems used by sensitive cells for important biological functions. These proteins include **porins**, **vitamin B_{12}**, siderophore, nucleoside receptors, and multiprotein systems that cooperate with these proteins. The group A and group B colicins correspond to two different pathways of entry beyond receptor binding: the TonB and the Tol pathway. The receptors for the various colicins are given in Table 1. The protein most frequently used is BtuB, the vitamin B_{12} receptor, which defines the so-called E-type colicins (E2 to E9) (12). High-affinity receptors or iron siderophores are also used by many colicins. The nucleoside Tsx porin and the major porin OmpF are also receptors (Table 1). Depending on the colicin class, the colicin receptors function cooperatively with either the Ton B system or the Tol system.

The *tol* locus was originally defined as comprising four loci, *tol A*, *B*, *Q*, *R*, on the basis of group A colicin tolerance (39). These genes are organized in two operons that comprise three additional genes called *orf*1, *orf*2 and *pal* (40), whose gene products are not directly involved in colicin uptake. However, one of them, the Pal protein (peptidoglycan associated protein), a **lipoprotein** of the outer membrane, interacts with

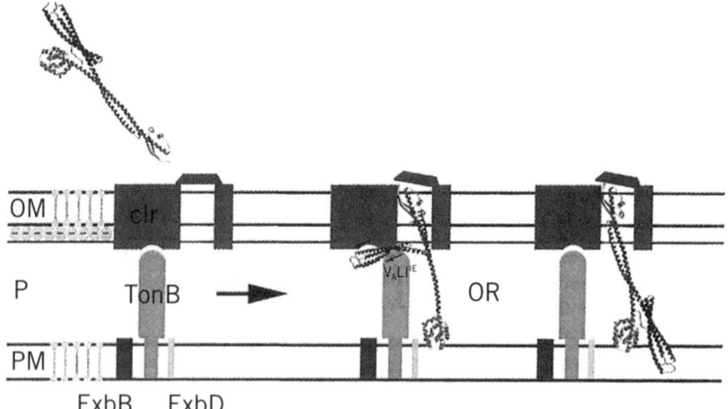

Figure 1. The mechanism of colicin Ia attachment and translocation. The outer membrane (OM) receptor for colicin Ia consists of Cir, TonB, and accessory plasma membrane (PM) proteins ExbB and ExbD. The translocation domain (blue), receptor binding domain (green), and channel-forming domain (red) separated by a pair of helices (gray), each 16 nm long, are indicated in colicin Ia. The TonB box of colicin Ia may compete with the TonB box of Cir for linkage to TonB. The channel-forming C domain reaches the plasma membrane, where it forms an ion-conducting channel by subsequent insertion and rearrangement of helices within the membrane (not shown). The translocation T domain may remain near the periplasmic surface of the outer membrane during channel activity. However, the presence of the 16 nm-long T3 helix indicates an alternative possibility in which the T domain crosses the periplasmic space (P) to participate in channel formation *in vivo*. The location of the TonB box on the upper surface of the T domain of sequence Glu(E)-Ile(I)-Met(M) Ala(A)-Val(V) is indicated, reading right to left, as EIMAV. The arrows indicate the locations of the TonB box in the receptor and colicin Ia. Reproduced from (68) with permission.

TolB (41). Both *tol* and *pal* mutants have an altered outer membrane (39). The only soluble Tol protein is TolB, which is principally located in the **periplasm** (42) but is linked to the outer membrane through interaction with Pal. The other Tol proteins are associated with the cytoplasmic membrane. TolQ contains three membrane-spanning segments, and TolA and TolR are anchored to this membrane by their *N*-terminal regions. All three proteins interact through membrane-spanning **α-helices** (43,44). There is yet no direct evidence that TolB interacts with membrane-associated Tol proteins, but each of the Tol proteins cofractionates with a membrane fraction accounting for the presumed contact sites between the inner and outer membranes of *E. coli*. In colicin A-treated cells, the toxin was also found in this fraction, and the relative amount of Tol proteins was doubled, suggesting that the colicin itself may recruit more Tol proteins at the contact sites (45,46).

Although both TolB and TolR are required to take up most group A colicins, neither is required for colicin E1, which in contrast requires the so-called TolC protein, a minor outer membrane protein that forms pores *in vitro* (47) and is also involved in secreting hemolysin (48) and colicin V (49). OmpF, another porin, is required for translocating many group A colicins (A, E-type, N) (50,51). The N-terminal domain of group A colicins contains all the information needed for the translocation step

(52) beyond receptor binding, including binding to OmpF (or TolC for colicin E1) and to Tol proteins like TolA (53) and TolB (54). There are about 1000 translocation sites for colicin A per bacterium (55). In addition, it has been shown that: (1) unfolding occurs upon receptor binding (55), (2) the N-terminal domain interacts with Tol proteins, (3) the polypeptide chain spans the entire cell envelope after inserting the pore-forming domain (56), and (4) a reduced constriction of the lumen of the OmpF pore prevents translocation of colicin A and N (57). A hypothetical model of colicin translocation based on these results has been presented (11,58).

Group B colicins are imported into sensitive cells through high-affinity siderophore receptors and the Ton B pathway. The Ton B system consists of a complex of proteins TonB-ExbB-ExbD, which facilitates the flow of energy from the cytoplasmic to the outer membrane for the energy-dependent transport of ferric siderophores and of vitamin B_{12}. ExbB and ExbD are physically and functionally homologous to TolQ and TolB (59). Ton B, like Tol A, has its N-terminal anchor in the cytoplasmic membrane and spans the periplasm (60). It has been proposed that an energized conformation of TonB, like FepA or FhuA, opens channels in the outer membrane, which act as receptors for the ferric siderophores. Until recently, each of the group B colicins studied uses a Ton B-dependent receptor, but it has now been reported that a new colicin (colicin 10) uses the Ton B system and binds to a Ton B-independent receptor Tsx. Colicin 10 also requires TolC for its uptake (61). Thus the interaction between the colicin receptor and Ton B is not obligatory if TolC is involved. This situation may be comparable to that of the group A colicin E1, which does not need TolB or TolR but does need TolC.

The high-affinity receptors for siderophores and vitamin B_{12} contain a pentapeptide motif close to the N-terminus, called the "Ton B box" (62). Mutations in the Ton B box reduce receptor activity dramatically. They are also **suppressed** by mutations in TonB (63), indicating physical and functional interactions between TonB and the receptors. Group B colicins, such as colicins B and D, which bind to the FepA receptor, and colicin M, which binds to the FhuA receptor, also contain a TonB box sequence close to the N-terminus. This indicates that these colicins interact directly with TonB during translocation. Mutations in the TonB box of colicin B affect its uptake and are suppressed by secondary substitutions clustered in TonB (63). A similar situation is observed with TonB, FhuA, and colicin M. These results suggest that group B colicins need to interact with TonB to be translocated, just as group A colicins need to interact with Tol proteins. Both the N- and C-terminal domains of the protein also interact with the inner membrane and therefore are protected from proteolytic degradation (64). This suggests that the Y-shaped structure (65) allows colicin Ia to span the entire cell envelope, as previously reported with colicin A (66). The receptor-binding domain would be the only part remaining exposed outside the cells.

A molecular model of colicin M uptake has been proposed (24). After first binding to FhuA in the outer membrane, colicin M is taken up in an energy-coupled process through the action of the TonB protein, which is anchored to the inner membrane and spans the periplasmic space. TonB binds to FhuA, inducing the release conformation of FhuA, which, in turn, results in the vectorial translocation of colicin M across the outer membrane. TonB itself has two conformations, energized and unenergized. Energization of TonB takes place in the cytoplasmic membrane. Induction of the receptor release conformation consumes energy, so TonB dissociates from the FhuA receptor and has to be reenergized to induce the next round of colicin M uptake. Because both FhuA and colicin M contain a TonB box, the TonB protein most likely interacts with both of them sequentially. The role of ExbB and D is still unclear. They are involved either in energy transduction (cycling of the TonB conformational change) from the cytoplasmic membrane to TonB or in stabilizing TonB, which is physically and functionally unstable (24,67,68).

The 626-residue colicin Ia is about 21 nm long and consists of three functional domains separated by a pair of α-helices 16 nm long. A central domain at the bend of the hairpinlike structure mediates binding to the outer membrane receptor. A second domain mediates translocation across the outer membrane via the TonB pathway. The TonB box recognition motif of colicin Ia is on one side of three 8 nm-long helices arranged as a helical sheet. The third domain, made up of 10 α-helices, is the pore-forming domain (see below). The two exceptionally long 16 nm α-helices enable colicin Ia to span the periplasmic space between the outer and inner membranes (Fig. 1) (68). This type of structure, which will very likely be found in each colicin, strongly suggests how these toxins exploit the machinery of the target cell to get across the periplasmic space (Fig. 1).

MODE OF ACTION OF COLICINS

The biochemical effects of colicins on sensitive bacteria fall into four classes: (1) the RNase colicins typified by colicin E3, (2) DNase colicins typified by colicins E2 and E9, (3) pore-forming colicins like colicin E1, and the still unique colicin M, which inhibits cell wall biosynthesis (24).

RNase Colicins

The two most extensively studied members of this class are colicin E3 and cloacin DF13 (derived from a strain of *Enterobacter cloacae* closely related to *E. coli*). The 11 kDa and 12 kDa C-terminal segments of these bacteriocins cleave the 16S ribosomal RNA, either in isolation or in intact 30S ribosomal subunits, which explains their inhibitory effect on protein synthesis (69). Colicins E5 and E6 have the same type of activity (70). Although they are highly homologous in their RNase regions (71), colicins E3, E5, E6, and cloacin DF13, each have a specific immunity protein. No structures are yet available for any of these RNase domains.

DNase Colicins

In 1970, it was shown that colicin E2 causes both single- and double-strand breaks in the DNA of sensitive cells (72), and three new types of DNase colicins were subsequently described (colicin E7, E8, and E9). The sequential identity between the various DNase domains is greater than 80% (71), yet each is inhibited by a specific immunity protein. This is similar to the situation in the RNase-type colicins. Structural and biophysical analysis of the DNase immunity protein complexes has been aided by the *E. coli* overexpression systems developed by Wallis et al. (73–75).

Pore-Forming Colicins

Pore-forming colicins constitute the largest group. Colicin V must be set apart in this class because it fits the definition of microcins better than that of colicins (2). Pore-forming colicins dissipate the membrane potential (76–80), which causes a series of metabolic effects, such as inhibition of **active transport** and of protein and nucleic acid synthesis, decrease of the internal ATP concentration (see **Adenylate charge**), and leakage of potassium ions (81–85). Cell death results from the depletion of the ATP pool following efflux of phosphate through the channel and ATP hydrolysis (86). Ten pore-forming colicins have been identified thus far. Sequential homologies indicate that they should be separated into two groups: (1) E1, 5, Ia, Ib and (2) A, B, and N (87), differing mainly in a segment containing the loop between helices 8 and 9, as defined in the X-ray crystal structure of the C-domain of colicin A (88). The structures of soluble forms of the channel domains of colicins A, Ia and E1, are known to varying extents (89–92). The pore-forming domain is composed of a bundle of 10 α-helices arranged in three layers in a novel protein fold (93). The N-terminal layer (helices 1 and 2) is connected to another layer consisting of two pairs of **amphipathic** helices (helices 3 + 4 and 6 + 7), and a single helix (helix 5) connects the two helix pairs. The middle layer is composed of helices 8 and 9, which form a **hydrophobic** hairpin buried in the core of the molecule (Fig. 1). The external sides of the peripheral helices are **hydrophilic**, which explains the paradox of how the same protein domain exists either in a water-soluble form or in a membrane-inserted state (see later). Because the C-terminal domain of colicins is easily purified by proteolytic digestion of the whole colicins or by genetic engineering , many biophysical studies have been carried out to improve our understanding of this step in the action of pore-forming colicins (for a review, see Suggestions for further reading).

Schein et al. (78) were the first to demonstrate that colicin A forms well-defined channels in planar lipid bilayers that permit ions to cross the membrane (94,95). Pore-forming colicins form channels of 10 to 30 picosiemens (pS), which correspond to 10^7 ions/sec/channel. These channels are characterized by their sensitivity to the electrical potential across the membrane. The gating voltage (corresponding to the activation or to the inactivation of 50% of the channels) varies for a given colicin. Values are +21, +50, and +70 mV for colicins N, A, and E1, respectively. There are multiple membranes states at the single-channel level. With colicin E1, channels of 10 pS are initially observed. With increased activation in 1 M KCl, channels of 20, 40, and 60 pS with substates appear, and flickering is observed. Similar observations with other colicins suggested the existence of many closed, open, and inactivated states. Thus, the concept of a single-membrane-inserted conformation appears to be incorrect, although a simplified model with a single or a few membrane-inserted forms has been very helpful.

The channel properties of the C-domain and of whole colicin are qualitatively similar for colicin E1, but not for colicins A (96) and Ia (97). Differences are ascribed to the influence of the other domains on the pore-forming domain in the intact molecule. The pores are permeable to cations and anions, but the rates at which they are transported are low (98). There is a relative preference for cations versus anions that is modulated by the pH and the lipid composition (99,100). Using the relative permeabilities of large organic cations and anions and taking into account their asymmetrical shapes, the size of the channel diameter was estimated at only 0.4–0.5 nm. This implies strong interactions between permeable ions and the side chains of residues exposed in the channel lumen (100).

The transition of pore-forming colicins from a water-soluble to a membrane-inserted state involves large structural changes, which is of interest in the context of **protein folding**, protein translocation, and protein insertion into membranes. From studies with model membrane systems, a likely sequence of events for membrane insertion and pore formation has been proposed for colicins A and E1 (see Suggestions for Further Reading). Briefly, the C-domain first binds to the outer face of the cytoplasmic membrane. The negative charge density plays a part in the binding and the kinetics of insertion (101,102). After binding, but before insertion, the pore-forming domain must have lost the tertiary structure of the water-soluble form and adopted primarily a **molten globular** conformation, which is still compact and has kept its native secondary structure (87,103,104). As mentioned later, the interaction of the N-terminal and central domains with the translocation machinery might trigger the appearance of a molten globular conformation of the channel domain *in vivo* (105). The topology of the membrane-inserted state has been extensively studied, mainly with the pore-forming domains of colicins E1, A, and Ia. The membrane-bound state involves insertion of the hydrophobic helical hairpin formed by H8 and H9 into the bilayer. In contrast to colicin A, which is more dependent on **electrostatic interactions**, the protein would lie flat on the membrane surface. The hydrophobic hairpin could not insert because the density of charged residues on the membrane surface is too high (103).

The polypeptide translocation involved in voltage-dependent gating has been studied through three main experimental approaches: (1) By using lipophilic radiolabeled probes, Merill and Cramer (106) demonstrated that a 42-residue region of colicin E1 corresponding to helices 5 and 6 was labeled in the presence, but not in the absence, of a membrane potential in a vesicle system with **valinomycin**-induced diffusion potential. Another segment corresponding to the hydrophobic hairpin was strongly labeled in the presence or absence of membrane potential. (2) The existence of a drastic change triggered by a transnegative membrane potential *in vivo* was also demonstrated by **disulfide-bond** engineering (107). The same technique further showed that α-helices 1, 2, 3, and 10 remain at the membrane surface after application of a membrane potential. (3) Colicin Ia with **biotin**, conjugated to specific cysteine residues introduced by **site-directed mutagenesis**, were inserted into a lipid bilayer, and the channels were opened or closed by varying the membrane potential. Then **streptavidin** was added on the *cis*- or *trans*-side of the membrane. The results demonstrated (108) that a region of the pore-forming domain of colicin Ia reversibly flips across the membrane, implying a large conformational change when potential is applied. A region of at least 68 residues is able to flip back and forth with channel opening and closing. Several open, channel structures probably exist (109).

Inhibition of Murein Biosynthesis

Colicin M is unique among the colicins in that it inhibits murein biosynthesis (110) by interfering with the dephosphorylation of C_{55}-polyisoprenyl pyrophosphate, which leads to cell lysis (111). It also has a much shorter polypeptide chain than other colicins (29 kDa as compared to 40 to 70 kDa). Yet, it has the same basic design of three functional domains (see previous). The largest part of the C-terminal domain of colicin M resides in the cytoplasmic membrane because the carrier lipid and pyrophosphatase activity were found in the membrane fraction (112). Colicin M does not penetrate deeply into the cytoplasmic membrane but acts mostly at the periplasmic side of this membrane (24). Furthermore, cytoplasmic colicin M does not kill the cells, and only external colicin M, after its import through the energy-dependent TonB system (see previous), has access to the target in the cytoplasmic membrane. A chimeric protein of colicin M linked to a signal peptide directing the C-domain to the periplasm (the outer face of inner membrane) causes cell lysis. The amount of colicin M at the target was so high that immunity broke down (24).

THE SYSTEMS IMMUNE TO COLICINS

The immunity of colicinogenic cells to the action of the colicin they produce was first noted by Fredericq in 1957 (113). The cell relies for its survival on the immunity protein. The individual cell is safe from its own colicin when that gene is carried on the same plasmid as the *imm* gene. Much progress with regard to the specificity determinants has been made during recent years on the basis of the immune activity, especially for nuclease-type and pore-forming colicins.

Immunity Against RNases

The determinants for specificity of colicin-immunity interactions with colicins E3, E6 and cloacin DF13 are likely to reside in a limited number of amino acid residues, eight in the nuclease domains and up to nine in the corresponding immunity proteins (114,115). These determinants are located in the N-terminal regions of both the RNase domain and the immunity protein. A single residue either in Im3 or in Im6 is critical for defining the specificity (116). Other residues are also likely to be involved in the interaction but have more peripheral roles in defining specificity. The 84-residue Im3 (117) is folded into a four-stranded antiparallel β-sheet connected by loops and a single short α-helix, in marked contrast to the structure of a DNase-specific immunity protein Im9 (118), which contains four α-helices. The specificity-determining residues of Im3 at the two main positions are exposed and line one face of the β-sheet (117). No structures are yet available for any RNase domains.

Immunity Against DNases

The sequences of the DNase domains of colicins E2, E7, E8, and E9 are more than 80% identical, yet they have specific immunity proteins. Site-directed mutagenetic studies identified six Col E9 residues in the C-terminal DNase domain as possible specificity determinants (119). The sequences of Im2, Im7, Im8, and Im9 are 58% identical but are not homologous to the RNase-type immunity proteins. Two amino acid residues were identified as specificity determinants of Im8 and Im9 by using homologous recombination (74) and site-directed mutagenesis (118,120).

Further structural and biophysical studies of DNase-immunity protein complexes have shown that the DNase E9-Im9 complex is extremely stable (like the Col E9-Im9 complex) with a K_d of 9.3×10^{-17} M, one of the highest affinities ever measured for a protein interaction (121). The association of the nuclease with the Im9 protein is essentially **diffusion-controlled**, involves electrostatic steering, and follows a two-step mechanism in which the proteins form an initial encounter complex before undergoing a conformational change to yield the final stable complex (121). Although there is no cross-reactivity between DNase-type colicins (E2, E7, E8, and E9) and noncognate immunity proteins under normal levels of expression, biophysical studies indicated that the latter bind to the DNase domain of colicin E9 and inhibit its activity (122). The K_d values range from 10^{-17} M (IM9) to 10^{-4} M. Consistent with this result, overexpressing each of the noncognate *Im* genes in bacteria results in significant levels of cross-reactivity toward the ColE9 toxin, and the order of *in vivo* cross-reactivity (Im9 > Im2 > Im8 > Im7) mirrors exactly the measured *in vivo* affinities (122). The specificity for the DNase-Im complexes is controlled through the dissociation constant, which is more than 10^6-fold faster for the noncognate Im proteins. The Im9 fold consists of a distorted antiparallel four-helix bundle in which the second helix (from the N-terminus) is the main specificity determinant (118,120). The surface regions of Im9 that interact with the DNase include the two central helices of the molecule, and the binding surface is heavily negatively charged, consistent with a positively charged partner DNase domain (118).

Immunity to Pore-Forming Colicins

The immunity proteins directed against nuclease-type colicins described previously are expressed at a level similar to that of their cognate colicins. In contrast, the proteins immune to pore-forming colicins are expressed constitutively at very low levels (10^2 to 10^3 molecules per cell). Another major difference is that these proteins protect the cells against external colicin because the membrane potential has the wrong orientation for colicin activity from the inside. Thus the immunity is directed against colicins produced by other cells.

Membrane vesicles from cells immune to colicin Ia are depolarized by colicin E1, but not Ia (123), which first indicated a cytoplasmic membrane localization for these types of immunity proteins. In addition, the construction of hybrids between colicins Ia and Ib, A and E1, or A and B, demonstrated that, independent of the translocation pathway, immunity is directed specifically against the C-terminal, pore-forming domain (19,52,124). ImmA has four transmembrane α-helices, and both the N- and C-termini are located in the cytoplasm (125). The shorter ImE1 has three transmembrane α-helices. The N- and C-termini are on the cytoplasmic and periplasmic sides of the membrane, respectively (127). These orientations are in agreement with the inside-positive prediction rule (128).

As with Im proteins directed against nucleases, there is no cross-reactivity between homologous immunity proteins directed against homologous A, B, or N colicins. Similarly, overproduction of the immunity protein leads to partial cross-reactivity (124). The main determinant for specific immunity recognition is located in the hydrophobic hairpin of the channel domain of the colicin (124,129). Either the whole bacteriocin or the C-terminal domain of pore-forming colicins produced in the cytoplasm of *E. coli* are devoid of cytotoxicity. However, when the C-terminal domain is fused to a signal sequence, the channel is inserted and functional, and cell death follows. This could be inhibited by coproduction of the cognate immunity protein (130,131). The cytotoxocity of the hybrid protein is independent of the uptake machinery normally used, demonstrating that the C-domain alone forms the channel *in vivo*, as in *in vitro*. The interaction of ImA with colicin A requires the immunity protein to assemble functionally but does not require the channel to be in the open state (131). In other words, the immunity protein interacts with the hydrophobic helical hairpin, as first suggested by the studies mentioned previously (124). This interaction was demonstrated directly using an epitope-tagged immunity protein (131). Site-directed mutagenetic studies of ImA indicate that a role for polar regions of ImA cannot be excluded, in addition to the intramembrane helix–helix interactions. Two roles are proposed for the hydrophilic loops in this protein: (1) they may stabilize its interactions with the channel on both sides of the membrane; (2) they may be required for the functional assembly of the transmembrane helices of ImA (132). The colicin E1 immunity protein tolerates a higher degree of substitution than ImA (127).

Immunity to Colicin M

Like the protein it inhibits, the immunity to colicin M is unique when compared with other Im proteins. It prevents colicin M from inhibiting murein synthesis. This 14 kDa protein has been localized in the cytoplasmic membrane, and a substantial portion is exposed to the periplasmic space. There is indirect evidence that the colicin M-immunity interaction occurs at the periplasmic side of the cytoplasmic membrane (133).

BIBLIOGRAPHY

1. A. Gratia (1925) *C.R. Soc. Biol.* **93**, 1041–1041.
2. R. Kolter and F. Moreno (1992) *Ann. Rev. Microbiol.* **46**, 141–163.
3. P. Fredericq (1963) *Ann. Rev. Microbiol.* **11**, 7–22.
4. M. Riley and D. Gordon (1992) *J. Gen. Microbiol.* **138**, 1345–1352.
5. F. Jacob, L. Siminovitch, and L. Wollman (1952) *Ann. Inst. Pasteur* **83**, 295–315.
6. J. Konisky (1982) *Ann. Rev. Microbiol.* **36**, 125–144.
7. D. Cavard, P. Sauve, F. Heitz, F. Pattus, C. Martinez, R. Dijkman, and C. Lazdunski (1988) *Eur. J. Biochem.* **172**, 507–512.
8. J. K. Davis and P. Reves (1975) *J. Bacteriol.* **123**, 96–101.
9. J. K. Davis and P. Reeves (1975) *J. Bacteriol.* **123**, 102–117.
10. R. Nagel del Zwaig and S. E. Luria(1967) *J. Bacteriol.* **94**, 1112–1123.
11. C. Lazdunski (1995) *Mol. Microbiol.* **16**, 1059–1066.
12. M. A. Riley (1993) *Mol. Biol. Evol.* **10**, 1048–1059.
13. U. Ross, E. E. Harkness, and V. Braun (1989) *Mol. Microbiol.* **3**, 891–902.
14. S. Luria and J. Suit (1987) In *Escherichia coli* and *Salmonella typhimurium* (F. E. Neidhardt, ed.), ASM, Washington, DC, pp. 1615–1624.
15. H. Masaki and T. Ohta (1985) *J. Mol. Biol.* **82**, 217–227.
16. S. T. Cole, B. Saint-Joanis, and A. P. Pugsley (1985) *Mol. Gen. Genetics* **198**, 465–472.
17. K. F. Chak and R. James (1985) *Nucleic Acids Res.* **13**, 2519–2530.
18. P. T. Chan, H. Ohmori, J. I. Tomizawa, and J. Lebowitz (1985) *J. Biol. Chem.* **260**, 8925–8935.
19. J. A. Mankovich, C. H. Hsu, and J. Konisky (1986) *J. Bacteriol.* **168**, 228–236.
20. J. Morlon, M. Chartier, M. Bidaud, and C. Lazdunski (1988) *Mol. Gen. Genetics* **211**, 231–243.
21. R. Lloubes, D. Baty, and C. Lazdunski (1986) *Nucleic Acids Res.* **14**, 2621–2636.
22. A. P. Pugsley (1988) *Mol. Gen. Genetics* **211**, 335–341.
23. S. Zhang, L. Yan, and G. Zubay (1988) *J. Bacteriol.* **170**, 5460–5467.
24. V. Braun, S. Gaisser, C. Glaser, R. Harkness, T. Ölschager, and J. Mende (1992) In *Bacteriocins, Microcins and Lantibiotics* (R. James, C. Lazdunski, and F. Pattus, eds.), Springer-Verlag, Berlin, Heidelberg, NATO ASI Series, Vol. 65, pp. 119–125.
25. R. Lloubes, M. Granger-Schnarr, C. Lazdunski, and M. Schnarr (1991) *J. Mol. Biol.* **217**, 421–428.
26. D. Cavard and B. Oudega (1992) In *Bacteriocins, Microcins and Lantibiotics* (R. James, C. Lazdunski, and F. Pattus, eds.), Springer-Verlag, Berlin, Heidelberg, NATO ASI Series, Vol. 65, pp. 297–305.
27. B. Oudega, A. Ykema, F. Stegehuis, and F. de Graaf (1984) *FEMS Microbiol. Lett.* **22**, 101–108.
28. D. Cavard, R. Lloubes, J. Morlon, M. Chartier, and C. Lazdunski (1985) *Mol. Gen. Genet.* **199**, 95–100.
29. D. Cavard, D. Baty, S. P. Howard, H. Verheij, and C. Lazdunski (1987) *J. Bacteriol.* **169**, 2187–2194.
30. A. P. Pugsley and S. T. Cole (1987) *J. Gen. Microbiol.* **133**, 2411–2420.
31. A. P. Pugsley and M. Schwartz (1984) *EMBO J.* **3**, 2393–2397.
32. S. P. Howard, D. Cavard, and C. Lazdunski (1991) *J. Gen. Microbiol.* **137**, 81–89.
33. R. Maget-Dana, F. Heitz, M. Ptak, F. Peypoux, and M. Guinaud (1985) *Biochem. Biophys. Res. Commun.* **129**, 965–971.
34. Y. Ohno-Iwashita and K. Imahori (1980) *Biochemistry* **19**, 652–659.
35. F. de Graaf and B. Oudega (1986) *Curr. Top. Microbiol. Immunol.* **125**, 183–205.
36. R. Dreher, V. Braun, and B. Wittman-Liebold (1985) *Arch. Microbiol.* **140**, 343–346.
37. D. Baty, M. Frenette, R. Lloubes, V. Géli, S. P. Howard, F. Pattus, and C. Lazdunski (1988) *Mol. Microbiol.* **2**, 807–811.
38. M. Frenette, H. Bénédetti, A. Bernadac, D. Baty, and C. Lazdunski (1991) *J. Mol. Biol.* **217**, 2509–2514.
39. R. Webster (1991) *Mol. Microbiol.* **5**, 1005–1011.
40. A. Vianney, M. Michelle Muller, T. Clavel, J. C. Lazzaroni, R. Portalier, and R. E. Webster (1996) *J. Bacteriol.* **178**, 4031–4038.
41. E. Bouveret, R. Derouiche, A. Rigal, R. Lloubes, C. Lazdunski, and H. Bénédetti (1995) *J. Biol. Chem.* **270**, 11071–11077.
42. M. Isnard, A. Rigal, J. C. Lazzaroni, C. Lazdunski, and R. Lloubes (1994) *J. Bacteriol.* **176**, 6392–6396.
43. R. Derouiche, H. Bénédetti, J. C. Lazzaroni, C. Lazdunski, and R. Lloubes (1995) *J. Biol. Chem.* **270**, 11078–11084.
44. J. C. Lazzaroni, A. Vianney, J. L. Popot, H. Bénédetti, S. Samatey, C. Lazdunski, R. Portalier, and V. Géli (1995) *J. Mol. Biol.* **246**, 1–7.
45. J. P. Bourdineaud, S. P. Howard, and C. Lazdunski (1989) *J. Bacteriol.* **171**, 2458–2465.
46. G. Guiard, P. Boulanger, H. Bénédetti, R. Lloubes, M. Besnard, and L. Letellier (1993) *J. Biol. Chem.* **269**, 5874–5880.
47. R. Benz, E. Maier, and I. Gentscher (1993) *Zbl Bakt* **278**, 187–196.
48. C. Wandersman and P. Deleplaire (1990) *Proc. Natl. Acad Sci. USA* **87**, 4776–4780.
49. L. Gilson, H. Mahanty, and R. Kolter (1990) *EMBO J.* **9**, 3875–3884.
50. H. Bénédetti, M. Frenette, D. Baty, R. Lloubes, V. Géli, and C. Lazdunski (1989) *J. Gen. Microbiol.* **135**, 3413–3420.
51. J. P. Bourdineaud, H. P. Fierobe, C. Lazdunski, and J. M. Pagès (1990) *Mol. Microbiol.* **4**, 1737–1743.

52. H. Bénédetti, M. Frénette, D. Baty, R. Lloubes, M. Knibiehler, F. Pattus, and C. Lazdunski (1991) *J. Mol. Biol.* **217**, 429–439.
53. H. Bénédetti, C. Lazdunski, and R. Lloubes (1991) *EMBO J.* **10**, 1989–1995.
54. E. Bouveret, A. Rigal, C. Lazdunski, and H. Bénédetti (1997) *Mol. Microbiol.*, **23**, 909–920.
55. D. Duché, D. Baty, M. Chartier, and L. Letellier (1994) *J. Biol. Chem.* **269**, 24820–24825.
56. H. Bénédetti, R. Lloubes, C. Lazdunski, and L. Letellier (1992) *EMBO J.* **11**, 441–447.
57. D. Jeanteur, T. Schirmer, D. Fourel, V. Simonet, G. Rummel, C. Widner, J. P. Rosenbusch, F. Pattus, and J. M. Pagès (1994) *Proc. Natl. Acad. Sci. USA* **91**, 10675–10679.
58. H. Bénédetti, L. Letellier, R. Lloubes, V. Géli, D. Baty, J. M. Pagès, and C. Lazdunski (1992) In *Dynamics of Membrane Assembly* (J.A.F. Op den Kamp, ed.), NATO ASI Series, Springer-Verlag, Berlin, Vol. 63, pp. 316–332.
59. K. Eick-Helmerich and V. Braun (1995) *J. Bacteriol.* **171**, 5117–5126.
60. K. Postle (1990) *Mol. Microbiol.* **4**, 2019–2925.
61. H. Pils and V. Braun (1995) *Mol. Microbiol.* **16**, 57–67.
62. E. Schramm, J. Mende, V. Braun, and R. M. Kemp (1987) *J. Bacteriol.* **169**, 3350–3357.
63. V. Braun (1995) *FEMS Microbiol. Rev.* **16**, 295–307.
64. S. F. Mel, A. M. Falick, A. L. Burlingame, and R. M. Stroud (1993) *Biochemistry* **312**, 9473–9479.
65. P. Ghosh, S. Mel, and R. M. Stroud (1994) *Nature Struct. Biol.* **1**, 597–604.
66. H. Bénédetti, R. Lloubes, C. Lazdunski, and L. Letellier (1992) *EMBO J.* **11**, 441–447.
67. K. Postle and J. Skare (1988) *J. Biol. Chem.* **263**, 11000–11007.
68. M. Wiener, D. Freymann, P. Ghosh, and R. Stroud (1997) *Nature* **385**, 461–464.
69. K. Jakes (1982) In *Molecular Action of Toxins and Viruses* (P. Cohen and S. von Heinegen, eds.), Elsevier, Amsterdam.
70. M. Mock and A. P. Pugsley (1982) *J. Bacteriol.* **150**, 1069–1076.
71. P. C. K. Lau, M. Parsons, and T. Uchimura (1992) In *Bacteriocins, Microcins and Lantibiotics* (R. James, C. Lazdunski, and F. Pattus, eds.), Springer-Verlag, Berlin, pp. 353–378.
72. P. S. Ringrose (1970) *Biochim. Biophys. Acta* **213**, 320–334.
73. R. Wallis, A. Reilly, A. Rowe, G. Moore, R. James, and C. Kleanthous (1922) *Eur. J. Biochem.* **207**, 687–695.
74. R. Wallis, G. R. Moore, C. Kleanthous, and R. James (1992) *Eur. J. Biochem.* **210**, 925–930.
75. R. Wallis, A. Reilly, K. Barnes, C. Abell, D. Campbell, G. Moore, R. James, and C. Kleanthous (1994) *Eur. J. Biochem.* **220**, 447–454.
76. J. Weiss and S. Luria (1978) *Proc. Natl. Acad. Sci. USA* **75**, 2483–2487.
77. H. Tokuda and J. Koninsky (1978) *Proc. Natl. Acad. Sci. USA* **76**, 6167–6171.
78. S. Schein, B. Kagan, and A. Finkelstein (1978) *Nature* **276**, 159–163.
79. W. Cramer, J. Dankert, and Y. Uratami (1983) *Biochim. Biophys. Acta* **737**, 173–193.
80. J. P. Bourdineaud, P. Boulanger, C. Lazdunski, and L. Letellier (1990) *Proc. Natl. Acad. Sci. USA* **87**, 1037–1041.
81. K. Fields and S. Luria (1969) *J. Bacteriol.* **97**, 57–63.
82. K. Fields and S. Luria (1969) *J. Bacteriol.* **97**, 64–77.
83. A. Kopecky, D. Copeland, and J. Lusk (1975) *Proc. Natl. Acad. Sci. USA* **72**, 4631–4634.
84. C. Plate, J. Suit, A. Jetten, and S. Luria (1974) *J. Biol. Chem.* **19**, 6138–6143.
85. J. Gould and W. Cramer (1977) *J. Biol. Chem.* **252**, 5491–5497.
86. G. Guihard, H. Bénédetti, M. Besnard, and L. Letellier (1994) *J. Biol. Chem.* **268**, 17775–17780.
87. C. Lazdunski, D. Baty, V. Géli, D. Cavard, J. Morlon, R. Lloubes, P. Howard, M. Knibiehler, M. Chartier, S. Varenne, M. Frenette, J. L. Dasseux, and F. Pattus (1988) *Biochim. Biophys. Acta* **947**, 445–464.
88. M. Parker, F. Pattus, A. Tucker, and D. Tsernoglou (1989) *Nature* **337**, 93–96.
89. M. Parker, J. Postna, F. Pattus, A. Tucker, and D. Tsernoglou (1992) *J. Mol. Biol.* **224**, 639–657.
90. P. Ghosh, S. Mel, and R. Stroud (1994) *Nat. Struct. Biol.* **1**, 597–504.
91. P. Elkins, H. Y. Song, W. Cramer, and C. Stauffacher (1994) *Proteins Struct. Funct. Genet.* **19**, 150–157.
92. M. Wormald, A. Merril, W. Cramer, and R. Williams (1990) *Eur. J. Biochem.* **191**, 155–161.
93. L. Holm and C. Sander (1993) *FEBS Lett.* **315**, 301–306.
94. F. Pattus, D. Cavard, R. Verger, C. Lazdunski, and H. Schindler (1983) in *Physical Chemistry of Transmembrane Ion Motions* (G. Spach, ed.), Elsevier, Amsterdam, pp. 407–413.
95. F. Pattus, D. Massote, H. Wilmsen, J. Lakey, D. Tsernoglou, A. Tucker, and M. Parker (1990) *Experientia* **96**, 180–192.
96. M. Collarini, G. Amblard, C. Lazdunski, and F. Pattus (1987) *Eur. Biophys. J.* **14**, 147–153.
97. P. Gosh, S. Mel, and R. Stroud (1993) *J. Membrane Biol.* **134**, 85–92.
98. L. Raymond, S. Slatin, and A. Finkelstein (1985) *J. Membrane Biol.* **84**, 173–181.
99. J. Bullock (1992) *J. Membrane Biol.* **125**, 255–257.
100. J. Bullock, E. Kolen, and J. L. Shear (1992) *J. Membrane Biol.* **128**, 1–16.
101. F. van der Goot, N. Didat, F. Pattus, W. Dowhan, and L. Letellier (1993) *Eur. J. Biochem.* **213**, 217–221.
102. G. van der Goot, J. Gonzalez-Manas, J. Lakey, and F. Pattus (1991) *Nature* **354**, 408–410.
103. J. Lakey, G. van der Goot, and F. Pattus (1994) *Toxicology* **87**, 85–108.
104. M. Parker and F. Pattus (1993) *Trends Biochem. Sci.* **18**, 391–395.
105. D. Duché, D. Baty, M. Chartier, and L. Letellier (1994) *J. Biol. Chem.* **269**, 24820–24825.
106. A. Merill and W. Cramer (1990) *Biochemistry* **29**, 85298534.
107. D. Duché, M. Parker, J. Gonzalez-Manas, F. Pattus, and D. Baty (1994) *J. Biol. Chem* **269**, 6332–6339.
108. S. Slatin, X-Q. Qiu, K. Jakes, and A. Finkelstein (1994) *Nature* **371**, 158–161.
109. X. Q. Qiu, K. Jakes, P. Kienker, A. Finkelstein, and S. Slatin (1996) *J. Gen. Physiol.* **107**, 313–328.
110. K. Schaller, J. Höltje, and V. Braun (1982) *J. Bacteriol.* **152**, 994–1000.
111. R. Harkness and V. Braun (1989) *J. Biol. Chem.* **264**, 6177–6182.
112. G. Siewert and J. Strominger (1967) *Proc. Natl. Acad. Sci. USA* **57**, 767–773.
113. P. Fredericq (1957) *Ann. Rev. Microbiol.* **11**, 7–21.
114. H. Masaki, S. Yajima, A. Akutsur-Koide, T. Ohta, and T. Uozumi (1992) In *"Bacteriocins, Microcins and Lantibiotics"* (R. James, C. Lazdunski, and F. Patus eds.), Springer-Verlag, Berlin, pp. 379–395.
115. A. Akutso, H. Masaki, and T. Ohta (1989) *J. Bacteriol.* **171**, 6430–6436.

116. H. Masaki, A. Akutsu, T. Uozumi and T. Ohta (1991) *Gene* 107, 133–138.
117. S. Yajima, Y. Muto, S. Yokoyama, H. Masaki, and T. Uozumi (1992) *Biochemistry* **31**, 5578–5586.
118. M. Osborne, A. L. Breez, L. Y. Lian, A. Reilly, R. James, C. Kleanthous, and G. Moore (1996) *Biochemistry* **35**, 9505–9512.
119. M. Curtis and R. James (1991) *Mol. Microbiol.* **5**, 2727–2733.
120. M. Osborne, L-Y. Lian, R. Wallis, A. Reilly, R. James, C. Kleanthous, and G. Moore (1994) *Biochemistry* **33**, 12347–12355.
121. R. Wallis, G. Moore, R. James, and C. Kleanthous (1995) *Biochemistry* **34**, 13743–13750.
122. R. Wallis, K. Y. Leung, A. Pomoner, H. Videler, G. Moore, R. James, and C. Kleanthous (1995) *Biochemistry* **34**, 13751–13759.
123. C. Weaver, A. Redborg and J. J. Konisky (1981) *J. Bacteriol.* **148**, 817–828.
124. V. Géli and C. Lazdunski (1992) *J. Bacteriol.* **174**, 6432–6437.
125. V. Géli, D. Baty, and C. Lazdunski (1988) *Proc. Natl. Acad. Sci. USA* **85**, 689–693.
126. V. Géli, D. Baty, F. Pattus, and C. Lazdunski (1989) *Mol. Microbiol.* **3**, 679–687.
127. H. Song and W. Cramer (1991) *J. Bacteriol.* **173**, 2935–2943.
128. G. von Heijne (1992) *J. Mol. Biol.* **225**, 487–494.
129. Y. Zhang and W. Cramer (1993) *J. Biol. Chem.* **268**, 1–8.
130. D. Espesset, Y. Corda, K. Cunningham, H. Bénédetti, R. Lloubes, C. Lazdunski, and V. Géli (1994) *Mol. Microbiol.* **13**, 1121–1131.
131. D. Espesset, D. Duché, D. Baty, and V. Géli (1996) *EMBO J.* **15**, 2356–2364.
132. D. Espesset, P. Piet, C. Lazdunski, and V. Géli (1994) *Mol. Microbiol.* **10**, 1111–1120.
133. T. Ölschläger, A. Turba, and V. Braun (1991) *Mol. Microbiol.* **5**, 1105–1111.

Suggestions for Further Reading

H. Bénédetti and V. Géli (1996) Colicin transport, channel formation and inhibition. In *Handbook of Biological Physics* (W. Konings, H. Kaback, and J. S. Lolkema, eds.), Elsevier, Amsterdam, Vol. 2, pp. 665–691.

W. Cramer, J. Heymann, S. Schendel, B. Deriy, F. Cohen, P. Elkins, and C. Stauffacher (1995) Structure-function of the channel-forming colicins. *Ann. Rev. Biophys. Biomol. Struct.* **24**, 611–641.

R. James, C. Kleanthous, and G. Moore (1996) The biology of E colicins: Paradigms and paradoxes. *Microbiology* **142**, 1569–1580.

A. P. Pugsley (1984) The ins and outs of colicins: Part 1. Production and translocation across membranes. Part 2. Lethal action, immunity and ecological implications, *Microbiol. Sci.* **1**, 168–175 and 203–205.

V. Braun, H. Pilsl, and P. Grob (1994) Colcins: Structures, modes of action, transfer through membranes and evolution. *Arch. Microbiol.* **161**, 199–206.

COLLAGEN

T. HAYASHI
K. MIZUNO

Proteins of the *collagen* family are the major components of the **extracellular matrix** and facilitate the formation and maintenance of a multicellular system. Collagens serve as solid-state regulators for cellular function and as scaffolding of the tissue architecture, particularly in large vertebrates. They contain many **proline** and **glycine** residues and a distinct **secondary structure**: the **polyproline** II-like helix, which is distinct from the **α-helix**, **β-sheet** and **turn** secondary structures in other **protein structures**. This regular secondary structure arises because the sequences the collagen polypeptide chain consist largely of the repeated sequence Gly–X–Y, with abundant proline residues at the X positions and rich in *hydroxyproline* (Hyp) residues at the Y positions. Three polyproline II-like helices constitute the **supersecondary structure** of the collagenous triple helix, which is stabilized through **hydrogen bonds** nearly perpendicular to triple helical axis (see **Triple-helical proteins**). The collagen protein family includes all the structural proteins of the extracellular matrix with triple-helical collagenous **domains** in their molecular architecture.

The collagen superfamily is classified into groups (Table 1) according to their molecular and/or supramolecular structures. The molecular structures of the collagenous proteins can be depicted with ball-and-stick models (Fig. 1). The balls represent noncollagenous or globular domains, without abundant glycine and proline residues, while the sticks show the collagenous triple helices. Some of the triple-helical domains, including that of the type IV collagen, have interruptions in the Gly–X–Y triplets, in that glycine residues do not always occupy every third position.

The structure, assembly, and supramolecular aggregation of type I collagen is the prototype from which has developed our understanding of collagenous structure, particularly the fibrillar collagens. Type I collagen is one of the major components of the fibrous collagens that occur in the greatest amounts. The structure and characteristic properties of triple-helical domains have been deduced from the study of type I collagen, together with comparative studies of other types of collagen. Unless otherwise mentioned, the description of collagen triple helices given below is based primarily on the information obtained through the studies type I collagen. These properties are generally shared with the triple-helical domains in other collagen types, especially regarding the characteristic features distinct from α-helix or β structure in noncollagenous proteins. However, most recent studies suggest that the triple-helical regions have structures and properties specific for each type, particularly in their intermolecular interactions. The

Table 1. Collagen Classification

Family	Types of Molecuels or Chains
Fibrillar collagen	Type I, type II, type III, type V, type XI
Meshwork-forming collagen	α1, α2, α3, α4, α5, and α6 chains of type IV
Fibril associated collagen with type interrupted triple helices (FACIT)	Type IX, type XII, type XIV, type XVI
Collagen with long triple helix	Type VII
Collagen with short triple helix	Type VIII, type X, type VI
Membrane associated collagen	Type XVII
Others	Type XIII, type XV, type XVIII, type XIX

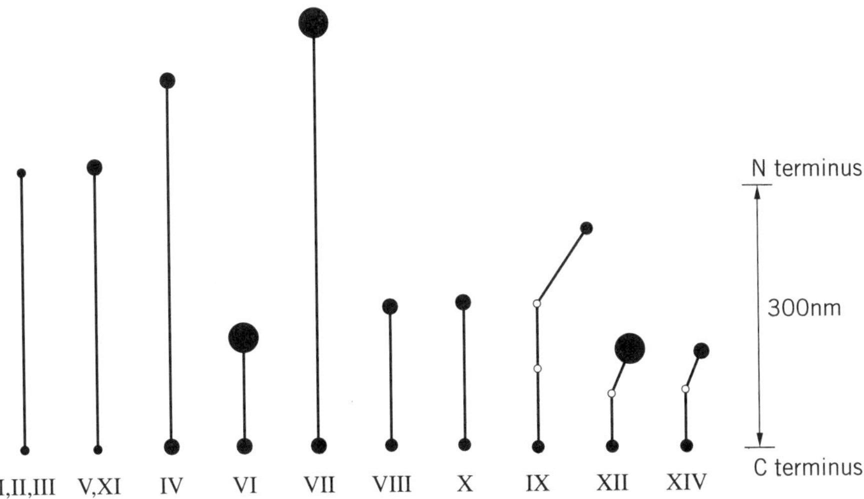

Figure 1. Molecular architecture of collagen superfamily depicted with ball and stick models (stick = collagenous, triple-helical domain; ball = noncollagenous domain).

characteristic features of the various collagen types are due to differences in the distribution of charged residues along the triple helix, content of glycosylated hydroxylysine and bulky **hydrophobic** residues, as well as imperfection of the Gly–X–Y repeat, which is a prerequisite for the triple-helical conformation.

COLLAGENOUS TRIPLE HELIX; COMMON AND SPECIFIC FEATURES AMONG DIFFERENT TYPES OF COLLAGEN

Shape and Structure of the Triple-Helical Domains

A variety of physicochemical studies, including direct visualization of individual molecules by rotary shadowing techniques in the **electron microscope**, have demonstrated the rod-like nature of the collagen triple-helical domain. The molecules of type I collagen have a length of about 300 nm (Fig. 1) and a diameter of about 1.4 nm. The rod is neither rigid nor randomly flexible, but appears to possess an intermediate level of semiflexibility.

Details of the three-dimensional structure of the collagen triple helix were established by model building to fit X-ray **fiber diffraction** data. Frequent occurrence of proline and hydroxyproline residues (which together account for approximately two-ninths (~22%) of the amino acid residues) favor a polyproline II-like conformation. The axial distance between one amino acid and the next in the polyproline II-like helical structure is 0.286 nm, close to twice that in the α-helix (0.15 nm). The triple-helical structure of the synthetic peptide $(Pro-Pro-Gly)_{10}$ was also determined by **X-ray crystallography**. The overall helical symmetry is left-handed, with 10 residues per three turns (108°/residue) or 7 residues per two turns (103°/residue), with a pitch of 2.9 nm. The three polyproline II-like helical chains are further coiled about a central axis, to form a right-handed helix (Fig. 2). The occurrence of glycine as every third residue gives rise to a polymer of repeating tripeptide units with the formula of (–Gly–X–Y–).

Thermal Stability of the Triple-Helical Conformation

Flexibility of the collagenous triple helix may vary along the chain. Collagen triple helices of different types have varying flexibility, depending on what residues occupy the X and Y positions. Yet the thermal stability in terms of the helix-to-coil transition temperature (see **Helix-coil theory**) is similar, regardless of the type of collagen, in the same animal or animal tissues. The collagenous domains are heat-stable up to the upper limit of animal body temperature. Heat denaturation of collagenous domains starts around 37°C in mammalian collagens (Fig. 3). The most recent study on the denaturation temperature of the bovine type IV collagen triple-helical domain indicated that the domain also has denaturation temperature above 37°C. This suggests that interruptions in the triplet repeats do not greatly decrease the thermal stability.

Primary Sequence Required for the Triple-Helical Conformation

The characteristic **primary structure** of polypeptides adopting collagenous triple-helical structures consists of repeats of the sequence Gly–X–Y, where Gly represents glycine and X or Y represents any other amino acid residue. Glycine is the only amino acid that can pack tightly at the center of the triple-stranded collagen monomer, where it provides HN groups for hydrogen bonding to O=C– groups in the **peptide bonds** of the other chains (Fig. 4). The stability of the triple-helical conformation is due in part to these hydrogen bonds, which are aligned nearly perpendicular to the helical axis. Interruptions of the Gly–X–Y repeats decrease the stability of the conformation. Substitution of a glycine residue occurring within the sequence Gly–X–Y of triple-helical domains destabilizes and disrupts the helical conformation. In the triple helix, the side chains of all the X and Y residues are exposed on the surface of the triple helix.

Proline Residues at Position X and Hydroxyproline Residues at Position Y

The helical conformation of individual α chains arises largely as a result of steric repulsion between the proline residues in the X position (approximately 120 residues per $\alpha1(I)$) and 4-hydroxyproline residues in the Y position (approximately 100 residues per $\alpha1(I)$ chain) and because the five-membered rings of the imino acids are rigid and limit rotation about the peptide N—C bond. The proline and hydroxyproline residues also stabilize the triple helix. The contribution to helix stability from the pyrrolidine rings of proline and hydroxyproline is thought to be entropic, in that these residues may not acquire as much freedom on **denaturation** as other residues. Another interpretation for contribution of the pyrrolidine rings to the

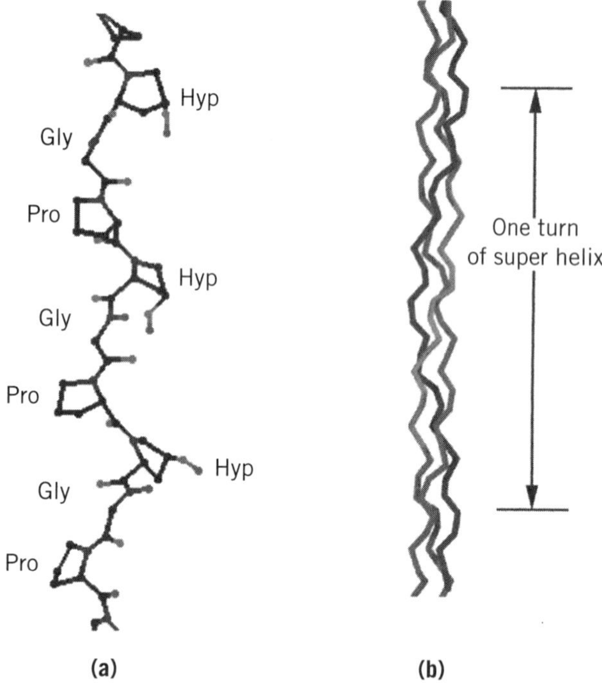

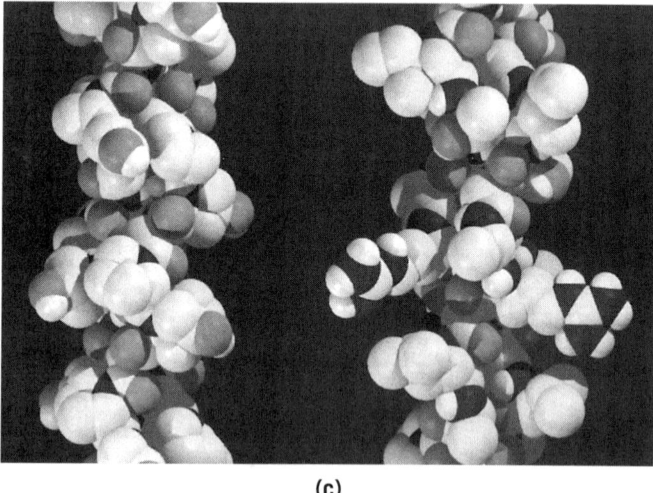

Figure 2. Collagenous triple helix. (**a**) Gly–Pro–Hyp repeated sequence of α chain in a model of part of one α chain in the collagenous triple helix. Both NH and CO groups project perpendicular to the fibrillar axis (C, black; N, blue; O, red; H, gray). (**b**) Backbone of the collagenous triple helix. (**c**) Gly–Pro–Hyp trimer (*left*). Note that there is a groove on the surface of the helix. A part of human type III collagen molecule, $[\alpha 1(\mathrm{III})]_3$ in the sequence of GITGARGLAGP (*right*). Note that all the residues except Gly project to the outer surface of the molecule (C, yellow; N, blue; O, red; H, white). See color insert.

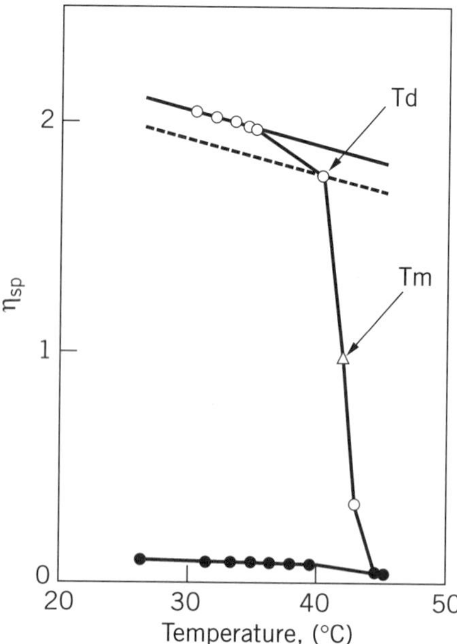

Figure 3. Denaturation temperature of collagen triple helix. The temperature was raised stepwise by 1.5°C at intervals of 20 min, and the specific viscosity (η_{sp}) was measured. Pepsin-treated acid-soluble collagen from calf skin (0.8 mg/mL) was dissolved in 0.15 *M* potassium phosphate buffer, pH 6.8, and 1 *M* glucose. Open circles indicate the specific viscosity of the native collagen solution with increasing temperature; filled circles correspond to the values of the denatured collagen solution when the temperature was lowered the same way as it was raised. The broken line is drawn through the values 5% less in the specific viscosity compared with that expected for the native collagen solution. The triangle (T_m) corresponds to the denaturation temperature obtained by the conventional method of taking the midpoint of the curve.

stability of triple-helical conformation is related to the fact that these side chains are located on the surface of the triple helix. Pyrrolidine rings are surprisingly favorable in contact with water. Furthermore, **hydroxylation** of the proline residues before Gly or Y positions increases the thermal stability greatly, although hydroxyproline residues at X positions or after Gly decrease the stability. Whether the hydroxyl group is at the 3 or 4 position of proline residues also influences greatly the thermal stability of the triple-helical conformation.

The amino acid sequence responsible for the formation and stability of the collagenous triple helix is also susceptible to **proteolysis** by collagenases. The sequence specifically recognized by bacterial collagenase is usually in sequences such as –GlyXYGlyProYGlyXHypGlyXY–. Cleavage occurs at the amino side of Gly residues.

Resistance of the Triple-Helical Domains to Pepsin or Other Proteinases

The intact triple-helical domain is generally resistant to most **proteinases**. However, when heated above physiological temperatures, it undergoes a helix-to-coil transition and, once melted, becomes susceptible to degradative enzymes. Protein consisting primarily of collagenous domains stays in solution at acidic pH, so the collagenous domains have generally been isolated by **pepsin** treatment of otherwise insoluble tissues. In a number of triple helices of recently discovered collagen family members, the occurrence of glycine at every third residue is occasionally interrupted. The interrupted sites are speculated to lower the stability of the triple helix and might form kinks in the rod-like triple helix. Thus, collagen

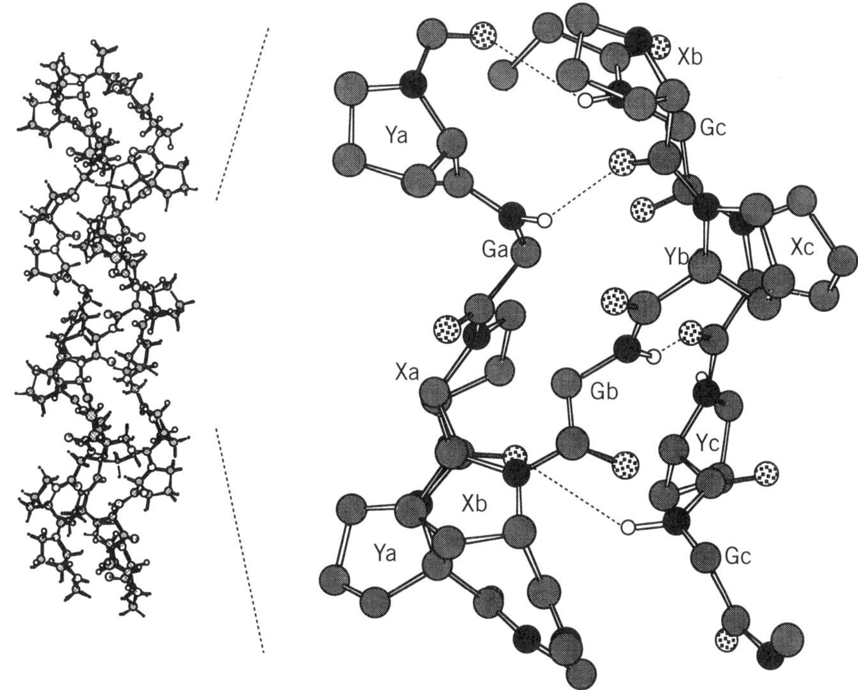

Figure 4. A schematic drawing illustrating direct interchain hydrogen bonding between (Gly)NH··CO(Pro in X position) found in the collagenous triple-helix. The three charms (a, b, and c) with repeated Gly-Pro(x)-Pro(y) sequence, eg, -Ga-Xa-Ya, -Gb-Xb-Yb-.

helix that is otherwise unsusceptible to many proteinases might become susceptible at the interrupted sites of Gly–X–Y repeats. Thus, type IV collagen may be degraded gradually at the interrupted sites of glycine–X–Y triplets, before being liberated from the aggregates with pepsin treatment.

Other Amino Acid Residues at Positions X and Y

All sides chains at positions X and Y protrude out along the surface of the triple helix. Consequently, they contribute to the **hydrophilicity**, **ionization**, hydrophobicity (see Fig. 5), and steric roughness of the protein **molecular surface**, including the size of the groove of the triple helix (Fig. 2). Charged groups, together with their neighboring sequences, may also affect the stability of the triple helix, presumably due to differential contributions from **water** of **hydration** on the surface.

The amino acid residues other than proline or hydroxyproline residues at positions X and Y can provide another classification of the collagen protein family. In general, the collagenous triple-helical domains contain a higher content of basic (arginine + lysine) than of acidic (aspartate + glutamate) residues, resulting in a basic **isoelectric point**. Two groups of collagens can be classified on the basis of their relative contents of arginine and lysine: the high arginine group (Arg/Lys >1) and the low arginine group (Arg/Lys <1). The greater content of Lys residues may result in a greater content of hydroxylysine residues, which further provides a possibility for additional **glycosylation**. A high content of glycosylated hydroxylysine should contribute greatly to the surface roughness of the triple-helical domains. Another general feature about the amino acid composition of collagens is the low content of hydrophobic amino acid residues. The interior of the triple helix is not stabilized by hydrophobic interactions, but they are strong between triple helices, because all the hydrophobic residues are exposed on the surface of the triple helix. The content of large hydrophobic residues in the collagenous triple helices also classifies the collagenous proteins into two groups. One group contains a high ratio of Ala/hydrophobic amino acids (Val, Leu, Ile, Phe, and Met), while the other group contains a low ratio.

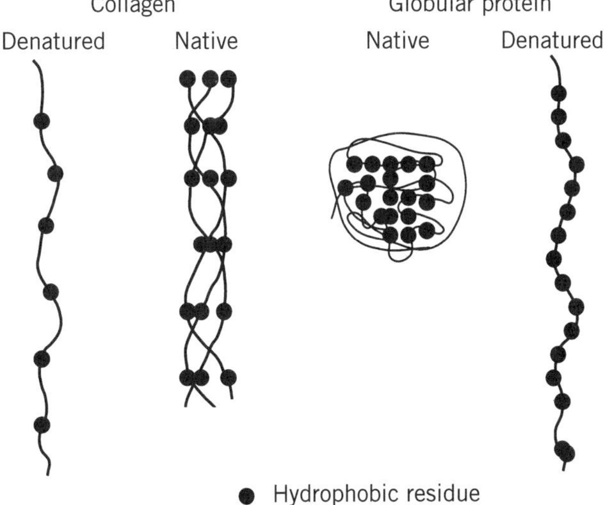

Figure 5. Hydrophobic residues in collagen polypeptide chains. The content of hydrophobic residues in collagenous domains (indicated as filled circles) is relatively low compared to the globular proteins. However, they projected to the surface of the triple helix, while the globular protein keeps most of the hydrophobic residues inside.

CLASSIFICATION OF COLLAGENS BASED ON THE PRIMARY SEQUENCE

In the fibrillar collagens, the triple-helical conformation occurs throughout 95% of the length of the rod-like monomer

(Fig. 3). Thus, of the 1057 residues in the α1(I) chain of human collagen, 1014 occur in repeating Gly–X–Y triplets. The *N*-terminal 17 residues and the *C*-terminal 26 residues (referred to as *telopeptides*) do not have glycine as every third residue. The type I collagen helical molecule is a heterotrimer comprising two identical α1(I) chains and one α2(I) chain. The α1(I) and α2(I) chains are very similar, but their primary structures are coded by separate genes and are sufficiently different for the chains to be separated by **ion-exchange chromatography** and by **SDS-PAGE**. The α-chains each contain over 1000 amino acid residues and have molecular weights of approximately 95,000.

The existence of a family of collagenous proteins in the connective tissues of vertebrates was first identified when cartilage collagen (type II) was found to be genetically distinct from the type I collagen of skin, bone, and tendon. A third collagen (designated type III) was detected in skin. More than 30 different collagenous polypeptides have been found in the extracellular matrix in the form of at least 19 different collagen types (designated from I to XIX; see Table 2).

The collagen numbering system (with Roman numerals for each collagen type and Arabic numerals for individual α-chains) to some extent reflects the relative abundance of the various collagens, in that generally the more abundant collagens were identified earliest. In addition to these collagens, there exists a number of secreted proteins that contain collagenous amino acid sequences and short triple-helical conformations, such as the **complement** component Clq, acetylcholine esterase, lung surfactant protein, conglutinin, serum mannose-binding protein, scavenger receptors (AR-I and AR-II), and MARCO. The collagenous sequences in these proteins contribute to their distinctive structures and functions. Since they have no known structural role in the extracellular matrix, however, they are not classified as collagens.

From the data derived from amino acid and gene sequencing, collagen molecules can be grouped into the groups shown in Figure 1 and Table 1. Fibrillar collagen molecules are characterized by an uninterrupted helical domain of approximately 300 nm. They are synthesized as procollagens comprised of three proα chains that undergo processing to α chains and subsequently assemble into collagen fibrils and fibers. Fibrillar collagen molecules (ie, types I, II, III, V, and XI) exhibit several common structural features that reflect the highly conserved exon–intron structure of the genes.

Polygonal meshwork-forming collagens (type IV collagen polypeptides) have large triple- helical domains (>160 kDa) with a length of >350 nm. Their primary structures are characterized by imperfections in the –Gly–X–Y– triplet sequence. These interruptions are a particular feature of type IV collagen, in which the helical domain contains more than 20 short stretches of non-helix-forming amino acids.

Short triple-helical collagen molecules and (types VI, VIII, X) contain interruptions in the helical domains (as in types IX, XII, and XIV). Collagen types VIII and X show remarkable homology and might have similar roles in tissues. Type XII and type XIV collagens have similarities to type IX collagen in the helical domain structures. A portion of these triple-helical domains have the potential to interact with fibrillar collagen. Thus, these three types of collagen, plus type XVI, comprises a group of fibril-associated collagens with interrupted triple helices (FACIT collagens).

An alternative approach for classifying the collagens depends on supramolecular structures that might be related to their physiological function. Individual collagen types may themselves represent a family or group of related collagenous structures in the extracellular matrix. Type IV collagen is a family of six homologous α chains (α1, α2, α3, α4, α5, and α6), and type V/XI is a family of 6 α chains: α1(V), α2(V), α3(V), α1(XI), α2(XI) and α3(XI).

SELF-ASSEMBLY INTO SUPRAMOLECULAR STRUCTURE (FIBRILLAR AGGREGATES OR POLYGONAL MESHWORKS)

Lateral Interactions Between Triple-Helical Domains

The collagen triple helix is essentially a one-dimensional molecule, with its near-constant axial separation between amino acid residues throughout the triple-helical domain. This has permitted a direct correlation of structural data obtained by electron microscopy with chemical sequence data. Lateral association of the collagenous triple-helical domains, to form fibrils, is a common feature of the collagen family. Fibrils will self-assemble spontaneously from solutions of extracted type I collagen when the pH, temperature, and ionic strength are adjusted to physiological values. This lateral association appears to be promoted by elevated temperature, suggesting that **hydrophobic interactions** are the principal driving force (Fig. 6). Fibril formation is controlled to a large extent by the amino acid sequence of the collagen and, in particular, by the distribution of **polar** and hydrophobic residues that are exposed on the surface of the triple-helical domain. Hydroxylysine and glycosylated hydroxylysine residues might be a most potent candidate involved in limiting the lateral growth of the collagenous domains. The residues on the surface of collagen triple helix may well be bulky enough to cause steric hindrance in the lateral association of the triple-helical domains (Fig. 15, see later). The situation would be particularly pronounced in the triple-helical domains of type V and type IV collagens, which respectively contain more than 20 and 50 glycosylated hydroxylysine residues in a single polypeptide.

There is stringent control over lateral growth of the collagen fibrils *in vivo*, in that the distributions of fibrillar diameter are sharp in a wide range of extracellular matrices of various tissues. That type I collagen can be organized in fibrils with different diameters may indicate the involvement of other regulatory elements. Heterotypic fibrils containing molecules of type III and type V, in addition to type I, collagen molecules may predominate *in vivo*. The presence of the **propeptide** of type III procollagen, for example, at the surface of collagen fibrils has led to the suggestion that this peptide may have a role in limiting the fibril diameter. Type V collagen was shown to form banding fibrils with finite diameter (Fig. 7). It has been postulated that the type V collagen can form hybrid fibrils with the type I collagen. The responsible domain in the type V collagen appears to be the triple-helical domain, since pepsin-treated type V collagen formed only fine fibrils. This finding suggests that type V collagen functions as one of the regulatory elements of fibril diameter, since the triple-helical domain of type V collagen forms D-period banded fibrils, with a potent ability to limit the lateral growth.

Table 2. Polypeptide Chain Composition of Genetically Distinct Collagen Types

Type	Known α chains	Known or Putative Chain Compositions of Molecules at Present	Length of Triple-Helical Domain (nm)	Main or Known Distribution	Aggregate Structure of Purified Protein or that Estimated with Immunohistochemical study, etc.
Fibrillar Collagen					
I	α1(I)	$[\alpha 1(I)]_2\alpha 2(I)$	300	Almost all connective tissues without hyaline cartilage	Fibril
	α2(I)	$[\alpha 1(I)]_3$?			
II	α1(II)	$[\alpha 1(II)]_3$	300	Cartilage	Fibril
III	α1(III)	$[\alpha 1(III)]_3$	300	Almost similar to type I	Fibril
V	α1(V)	$[\alpha 1(V)]_2\alpha 2(V)$	300	Almost similar to type I, adult cartilage	Fibril
	α2(V)	α1(V)α2(V)α3(V)			
	α3(V)	$[\alpha 1(V)]_3$?			
XI	α1(XI)	α1(XI)α2(XI)α3(XI)	300	Cartilage	Fibril
	α2(XI)				
	α3(XI)				
V/XI	α1(XI)	$[\alpha 1(XI)]_2\alpha 2(V)$	300	Vitreous body	Fibril
	α2(V)				
FACIT (Fibril-Associated Collagen with Interrupted Triple-Helices)					
IX	α1(IX)	α1(IX)α2(IX)α3(IX)		Surface of the cartilage fibril	Aggregated with fibrillar collagen periodically on the cartilage collagen fibrils
	α2(IX)				
	α3(IX)				
XII	α1(XII)	$[\alpha 1(XII)]_3$		Tissues rich in type I	
XIV	α1(XIV)	$[\alpha 1(XIV)]_3$		Tissues rich in type I	
XVI	α1(XVI)	$[\alpha 1(XVI)]_3$			
Others					
VI	α1(VI)	α1(VI)α2(VI)α3(VI)	100	Almost all connective tissues	Beaded microfibril
	α2(VI)	$[\alpha 1(VI)]_2\alpha 2(VI)$			
	α3(VI)				
VII	α1(VII)	$[\alpha 1(VII)]_3$?	420	Anchoring fibril	Short dimer
VIII	α1(VIII)	$[\alpha 1(VIII)]_3$	150	Basement membrane of endothelial cell	Hexagonal array
	α2(VIII)	$[\alpha 1(VIII)]_2\alpha 2(VIII)$, $[\alpha 2(VIII)]_3$			
X	α1(X)	$[\alpha 1(X)]_3$	130	Hypertrophic cartilage	Hexagonal array
XIII	α1(XIII)	$[\alpha 1(XIII)]_3$?			
XV	α1(XV)	$[\alpha 1(XV)]_3$?	150		
XVII	α1(XVII)	$[\alpha 1(XVII)]_3$			
XVIII	α1(XVIII)	$[\alpha 1(XVIII)]_3$?			
XIX	α1(XIX)	?			FACIT?
Meshwork-Forming Collagen					
IV	α1(IV)	$[\alpha 1(IV)]_2\alpha 2(IV)$	350	Basement membrane	Polygonal meshwork
	α2(IV)	α3(IV)α4(IV)α5(IV), $[\alpha 3(IV)]_2\alpha 4(IV)$		Sinusoid	
	α3(IV)	$[\alpha 5(IV)]_2\alpha 6(IV)$?			
	α4(IV)				
	α5(IV)				
	α6(IV)				

D-Staggered Lateral Association

The fibrillar structures with banded patterns are accounted for by the specific parallel and mutually staggered alignment of the triple-helical domains of the fibrillar collagens (Fig. 8). The sequence of hydrophobic and charged residues distributed along the triple helix provide maximum electrostatic and hydrophobic interactions between the neighboring molecules. The reconstituted fibrils exhibit a cross-striated banding pattern in the electron microscope, demonstrating conclusively that the aggregation of collagen molecules into axially ordered fibrillar structures is basically a **self-assembly** process, where the information for association is contained within the assembling molecules themselves.

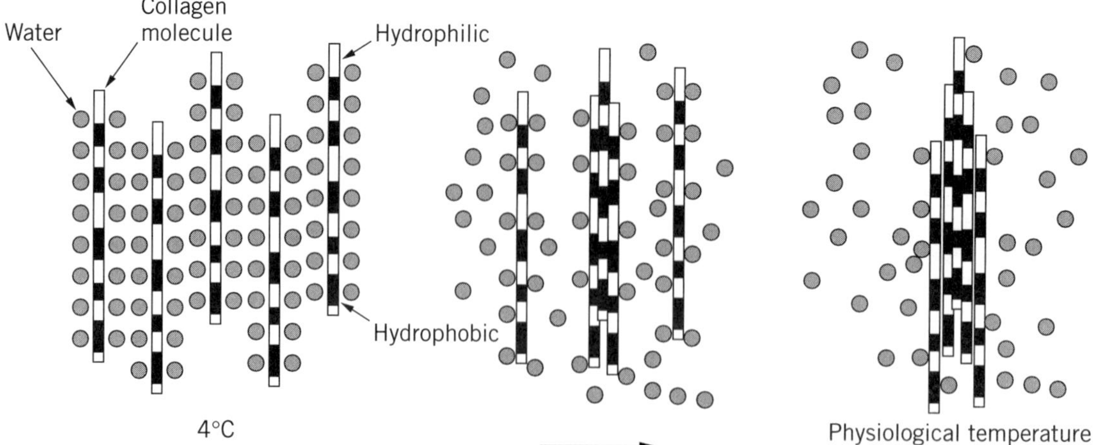

Figure 6. Lateral association of triple-helical domains. Triple-helical domain is highly hydrated at 4°C. At physiological temperature such as 37°C, water molecules on the hydrophobic domains will be dehydrated. Collagen molecules accompanied the lateral association of the triple-helical domains through the exposed hydrophobic regions.

Figure 7. Reconstituted type V collagen fibrils. Purified pepsin-treated type V collagen from human placenta with the chain composition of $[\alpha 1(\mathrm{V})]_2\alpha 2(\mathrm{V})$ forms the banded fibrils with finite thickness. The periodicity of the banding pattern is approximately 65 nm. Fibril thickness is 35 nm in average.

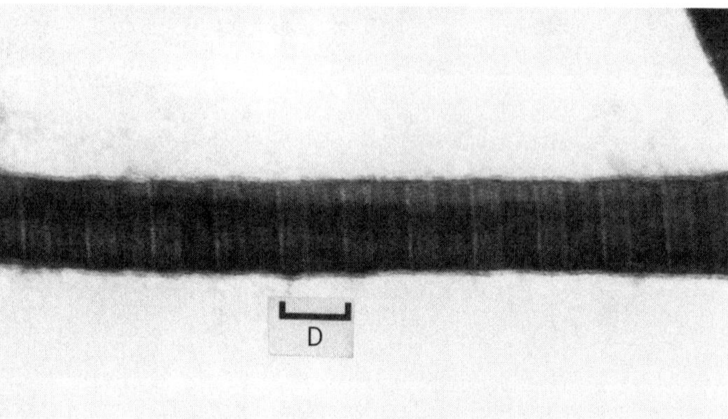

Figure 8. D-staggered array of type I collagen in banded fibrils.

It has been recognized for over 30 years that the cross-striated periodic structure of the native collagen fibril is a consequence of the assembly of molecules in a parallel array, but mutually staggered (ie, axially displaced with respect to one another) by approximately one-quarter of their length—often, which is referred to as the *quarter-staggered array*. The periodicity of the cross-striations in the fibril is explained by the fact that each collagen monomer has eight highly charged regions 67 nm apart that appear under appropriate conditions as stained bands.

The banded period (the D period) is confirmed by low-angle **X-ray scattering** of rat-tail tendon fibrils, and the overall length of a fibrillar collagen monomer is 4.4 D units (when D = 234 amino acids, the length of one cross-striation period is 67 nm). Within a collagen fibril, the molecules are staggered by integral multiples of the distance D (Fig. 8).

Collagen Supramolecular Aggregates in Tissues

The organization and arrangement of the collagen fibrils and polygonal meshwork in tissues are subject to considerable variation. Fibrillar collagen, type I collagen in particular, provides the major mechanical strength of skin, tendon, bone, dentine, cornea, sclera, and so on. The fibrillar collagen types form long, unbranched, banded fibrils with a characteristic periodicity of 60–70 nm. The fibrillar aggregate of collagen is classified as a quasi-crystal, because of its highly symmetrical insoluble structure built up of essentially identical subunits. Especially in tendon, type I collagen accounts for approximately 90–95% of the dry weight of the tissue and is present as large, highly orientated fibers. All the fibrillar units are arranged in large parallel bundles, with the average diameter of the fibrils varying between 50 and 500 nm. In dermis, where type I collagen accounts for 80–90% of the collagenous proteins, the fibrils form a coarse network partially oriented in the plane of the skin, with the average diameter of the fibrils between 40 and 100 nm. The orthogonally arranged and precisely packed fibrils in corneal stroma are of particular interest, since the arrangement may well be related to tissue transparency. In the cornea, which has a high content of type V and type I collagens, the collagen fibrils have a uniform diameter of approximately 25 nm. Concentric circles of collagen fibrils in cortical bone provides another type of organization and arrangement.

The skeletal structure of the basal lamina has a special three-dimensional meshwork with an average pore size of about 18 nm. Isolated type IV collagen can reconstitute into polygonal meshwork (Fig. 9) with an average pore size also of about 18 nm (Fig. 10). This suggests that the self-assembled supramolecular structure of the type IV collagen is formed primarily through the lateral interactions of the triple-helical domain, which has kinks or bending points due to interruptions of the Gly–X–Y repeat.

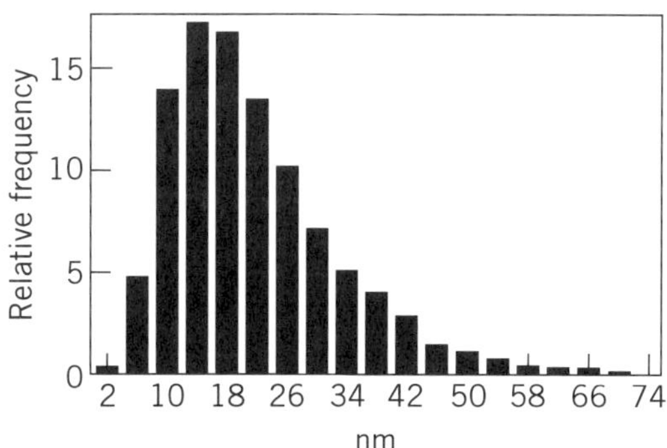

Figure 10. Histograms of the distances between two branching points of polygonal meshwork of type IV collagen aggregates.

BIOSYNTHESIS AND CHAIN ASSEMBLY OF COLLAGEN

Biosynthesis of the collagen molecule is a complex process that begins with **transcription** of the individual collagen **genes** and culminates in the chains assembled into collagen molecules (Fig. 11). This biosynthetic process involves a number of co- and **posttranslational modifications** unique to collagenous sequences. Intracellular modifications of the newly synthesized polypeptide chains result in the formation of triple-helical molecules. At least 10 different enzymes have been implicated in the posttranslational processing of the collagen molecule. The biosynthesis of type I collagen can be regarded as a useful model exemplifying many of the common features of collagen biosynthesis. The nonfibrillar collagens may, however, deviate from the general scheme.

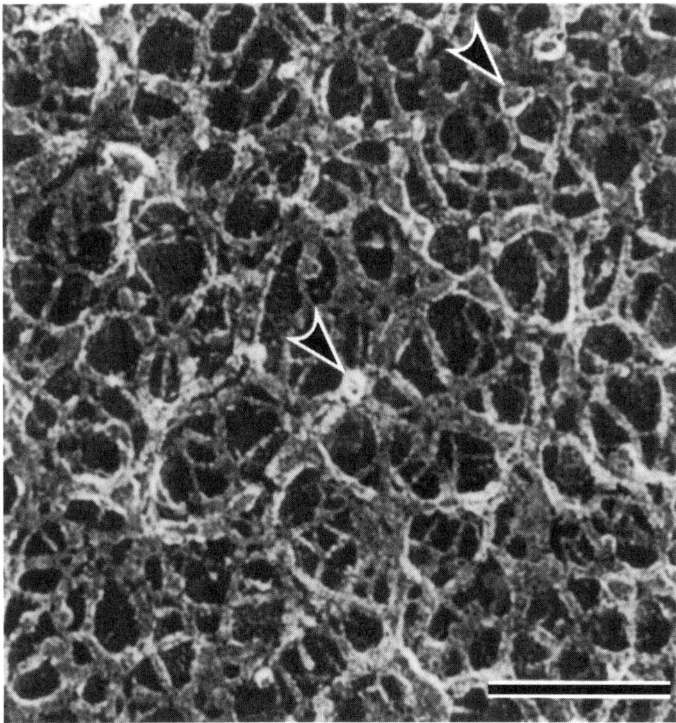

Figure 9. Polygonal meshwork structure of type IV collagen aggregates. Bar in lower right corner is equal to 100 nm. Arrowheads indicate globules presumably formed through NCl domains.

Pre-procollagen mRNA Translation

The pre-proα collagen chains that are the primary **translation** product contain hydrophobic *N*-terminal **signal sequences** of 22–26 amino acid residues, similar to those in most other secreted proteins and are essential for targeting of nascent polypeptides to the **endoplasmic reticulum** (see **Protein secretion**). During or shortly after translocation, the signal sequence is removed by the **signal peptidase** on the luminal side of the endoplasmic reticulum membrane.

Hydroxylation of proline and lysine residues of the newly synthesized collagen polypeptide chains results in the formation of procollagen molecules containing an array of hydroxyproline and hydroxylysine residues. This hydroxylation is produced by collagen-specific enzymes within the cell (see **Hydroxylation**).

Helix Formation and Secretion

The assembly of three polypeptides into a triple-helical collagen molecule is a complex process initiated by association of the three *C*-terminal **propeptides** (in the case of type I collagen), which involves chain alignment, nucleation, and propagation into the triple-helical conformation. Requiring the *C* terminus of the polypeptide chain, chain association and alignment (in the case of type I collagen) takes place only as the chains near full elongation, and folding may proceed in the *C*- to *N*-terminal direction. Association of the *C* termini is stabilized by disulfide bond formation catalyzed by the enzyme **protein disulfide isomerase** (PDI). Assembly of the triple helical molecule might be limited by ***cis–trans* isomerization** (Fig. 12) of the peptide bonds preceding proline residues, which is catalyzed by a cytosolic enzyme, **peptidylprolyl *cis–trans* isomerase** (PPI). Formation of the triple-helical conformation appears to preclude further enzymatic post-translational modifications. The procollagen in a triple-helical conformation is ready to be secreted into the extracellular space.

Stable triple helix formation requires 4-hydroxyproline in the Y position of a high proportion of –Gly–X–Y– triplets

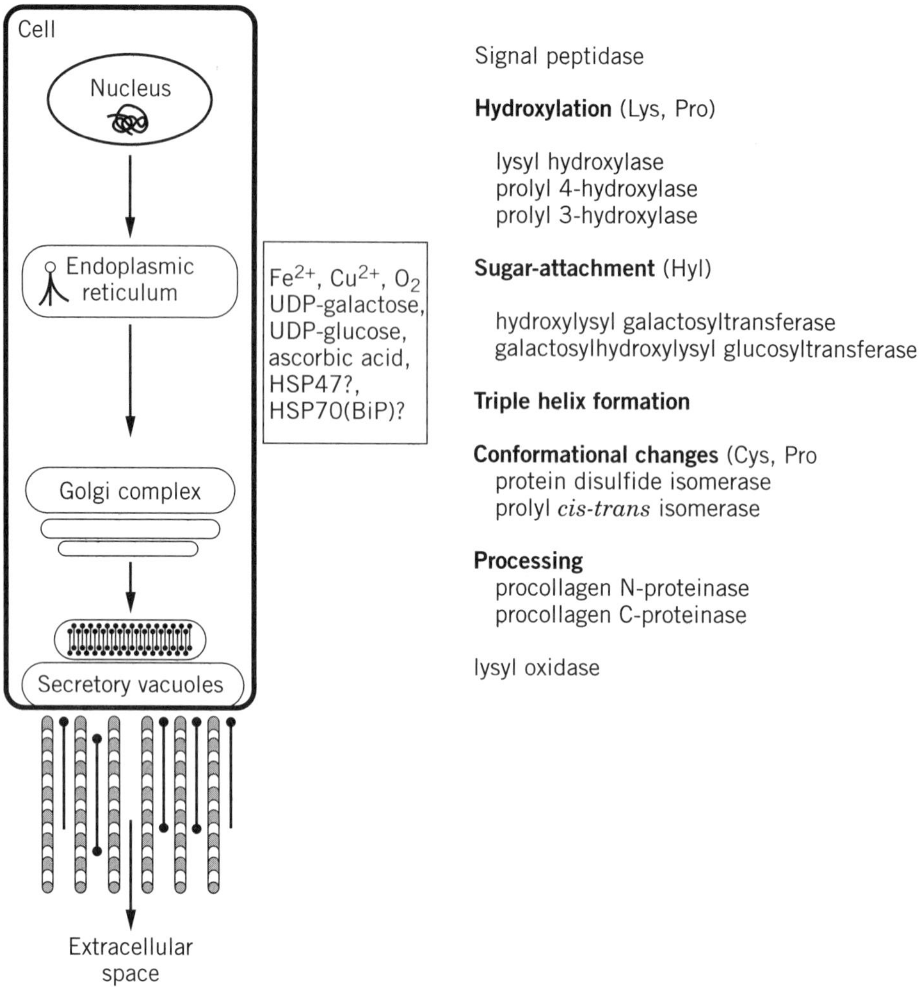

Cell

Nucleus

Endoplasmic reticulum

Golgi complex

Secretory vacuoles

Extracellular space

Fe^{2+}, Cu^{2+}, O_2
UDP-galactose,
UDP-glucose,
ascorbic acid,
HSP47?,
HSP70(BiP)?

Signal peptidase

Hydroxylation (Lys, Pro)

lysyl hydroxylase
prolyl 4-hydroxylase
prolyl 3-hydroxylase

Sugar-attachment (Hyl)

hydroxylysyl galactosyltransferase
galactosylhydroxylysyl glucosyltransferase

Triple helix formation

Conformational changes (Cys, Pro
protein disulfide isomerase
prolyl *cis-trans* isomerase

Processing
procollagen N-proteinase
procollagen C-proteinase

lysyl oxidase

Figure 11. Schematic drawing of the complex process of biosynthesis of collagen. Collagen translation products are directed into the endoplasmic reticulum by the *N*-terminal signal peptide that is the cleaved from the polypeptide chain by signal peptidase. Individual procollagen chains undergo complex enzyme-catalyzed post-translational modifications prior to the completion of chain assembly and the folding of the triple helix. Hydroxylation of Lys and Pro and attachment of sugars to hydroxylysine (Hyl) are characteristic post-translational modifications of collagenous polypeptides.

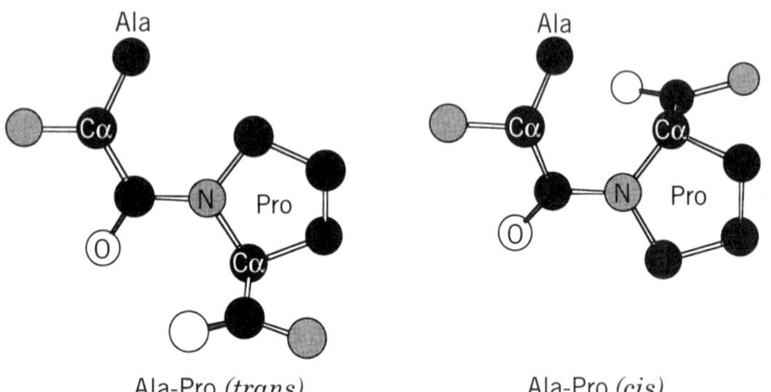

Figure 12. *Cis–trans* isomerization of Ala-Pro peptide bonds. H atoms are not shown.

(at least 100 of the 1000 amino acids of the helical domain of type I collagen) (see **Hydroxylation**). The normal rate of procollagen secretion is apparently dependent on triple-helical conformation. If triple helix formation is prevented by inhibiting prolyl hydroxylase (by the use of α,α'-dipyridyl or by limiting the availability of the cofactors Fe^{2+} or O_2), the nonhelical proα(I) chains first accumulate within endoplasmic reticulum.

Extracellular Polymerization

Fibril formation involves the specific enzymatic cleavage of the procollagen *N*- and *C*-terminal propeptides in the extracellular space (Fig. 13), and these enzymes must therefore play a key role in regulating the aggregation. Such processing is characteristic of the fibrillar collagens and may not be a universal prerequisite for the assembly of all collagen types. The cleavage of the *N*- and *C*-terminal propeptides from the procollagen I molecule at specific peptide bond cleavage sites is achieved by two specific neutral **metalloproteinases**—procollagen *N*-proteinase and procollagen *C*-proteinase—both of which require Ca^{2+} for activity and are inhibited by metal chelators. These enzymes act essentially only on molecules in the triple-helical conformation. Type I procollagen *N*-proteinase cleaves the *N*-terminal propeptides of types I and II procollagens between a proline and a glutamine residue. A separate enzyme is involved in the processing of the *N*-propeptide of type III procollagen. The type I procollagen *C*-terminal proteinase cleaves the *C*-terminal propeptides of both the proα1(I) and proα2(I) procollagen chains at an Ala–Asp bond. Most recently, procollagen *C*-proteinase was found to be BMP-1(bone morphogenetic protein-1), and recombinant BMP-1 was shown to act as procollagen *C*-proteinase. The enzyme is also able to cleave the *C*-terminal propeptides of type I procollagen

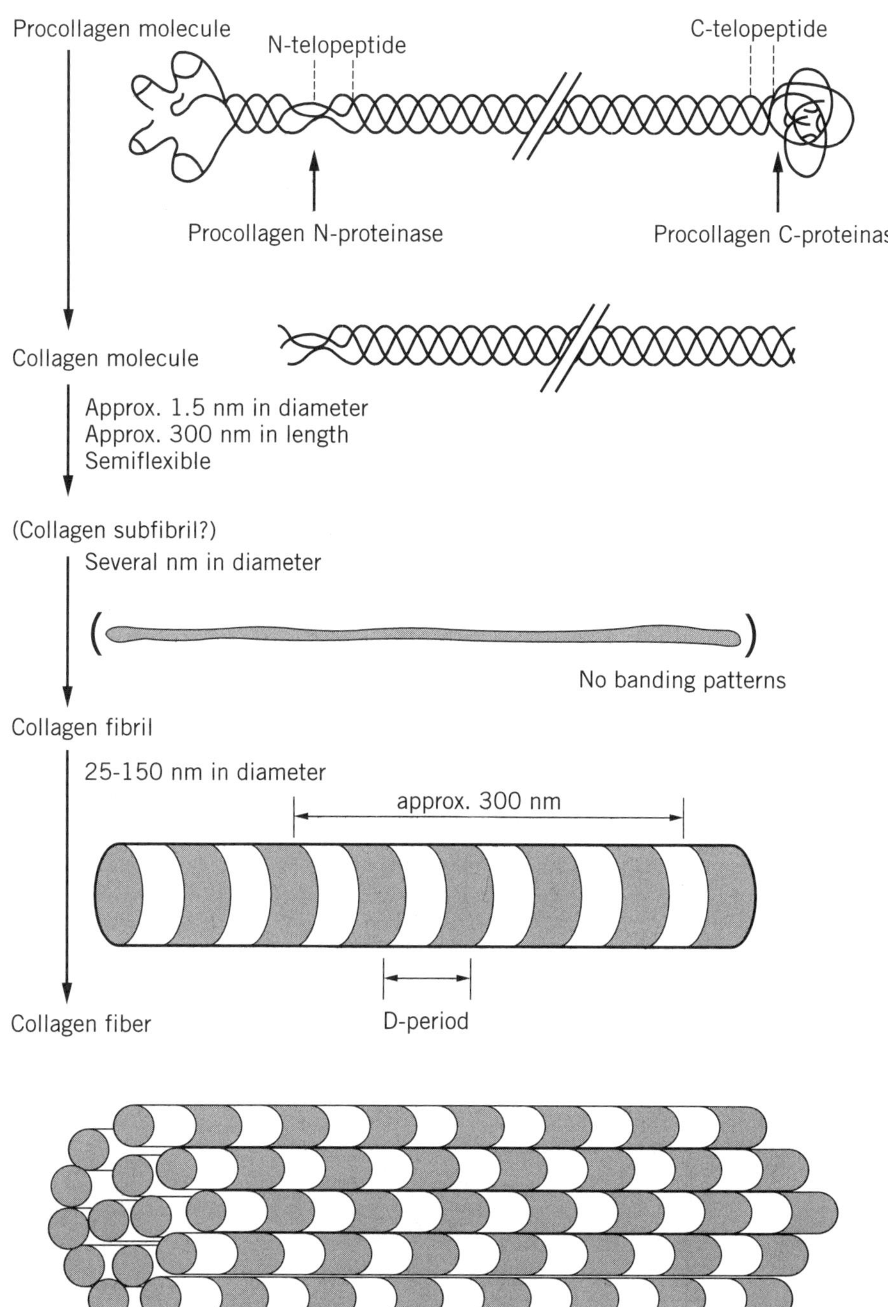

Figure 13. Processing of type I procollagen to fibril formation.

homotrimer, type II procollagen, type III procollagen, and other proteins, including laminin-5 and prolysyl oxidase.

Crosslinking of Type I Collagen in the Fibrils

Covalent crosslinking of type I collagen in the fibrils involves oxidative deamination of specific lysine or hydroxylysine residues in the nonhelical (telopeptide) regions at the ends of the triple helix. Thus, lysyl and hydroxylysyl aldehydes are generated by lysyl oxidase (protein lysine 6-oxidase) (Fig. 14). The catalytic activity of lysyl oxidase is dependent on strict steric requirements (in the case of type I collagen, the quarter-stagger arrangement of molecules in the fibril) and on the sequence of amino acids surrounding the "target" lysyl/hydroxylysyl residues. The recombinant enzyme was produced as a proenzyme, which could be processed for activation by BMP-1 or procollagen *C*-proteinase.

The majority of covalent crosslinks stabilizing the fibrillar collagens involve the telopeptides and, consequently, pepsin treatment of insoluble, crosslinked fibrils tends to release the triple-helical domain, which can be recovered in its native conformation.

POSTTRANSLATIONAL MODIFICATIONS OF PROLINE AND LYSINE RESIDUES

Hydroxylation of proline and lysine residues is a posttranslational modification that is specific to collagen (see **Hydroxylation**). The reason for hydroxylation of proline residues at the 3 position is not known, but the presence of

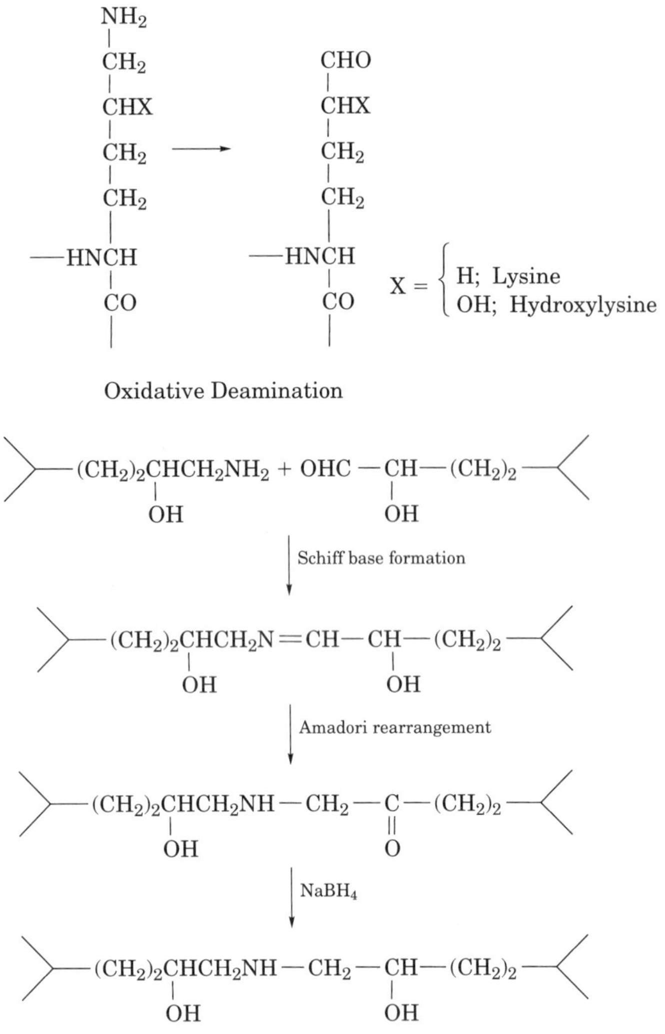

$$
\begin{array}{c}
NH_2 \\ | \\ CH_2 \\ | \\ CHX \\ | \\ CH_2 \\ | \\ CH_2 \\ | \\ —HNCH \\ | \\ CO \\ |
\end{array}
\longrightarrow
\begin{array}{c}
CHO \\ | \\ CHX \\ | \\ CH_2 \\ | \\ CH_2 \\ | \\ —HNCH \\ | \\ CO \\ |
\end{array}
\qquad X = \begin{cases} \text{H; Lysine} \\ \text{OH; Hydroxylysine} \end{cases}
$$

Oxidative Deamination

$$
\rangle\!—(CH_2)_2\underset{|\atop OH}{C}HCH_2NH_2 + OHC—\underset{|\atop OH}{C}H—(CH_2)_2—\!\langle
$$

↓ Schiff base formation

$$
\rangle\!—(CH_2)_2\underset{|\atop OH}{C}HCH_2N{=}CH—\underset{|\atop OH}{C}H—(CH_2)_2—\!\langle
$$

↓ Amadori rearrangement

$$
\rangle\!—(CH_2)_2\underset{|\atop OH}{C}HCH_2NH—CH_2—\underset{\|\atop O}{C}—(CH_2)_2—\!\langle
$$

↓ $NaBH_4$

$$
\rangle\!—(CH_2)_2\underset{|\atop OH}{C}HCH_2NH—CH_2—\underset{|\atop OH}{C}H—(CH_2)_2—\!\langle
$$

Figure 14. Oxidative deamination and crosslinking of lysine or hydroxylysine residues. The oxidative deamination, Schiff base formation, and Amadori rearrangements take place *in vivo*. The last step, treatment with sodium borohydride, is carried out in the laboratory to stabilize the crosslinks for chemical analysis.

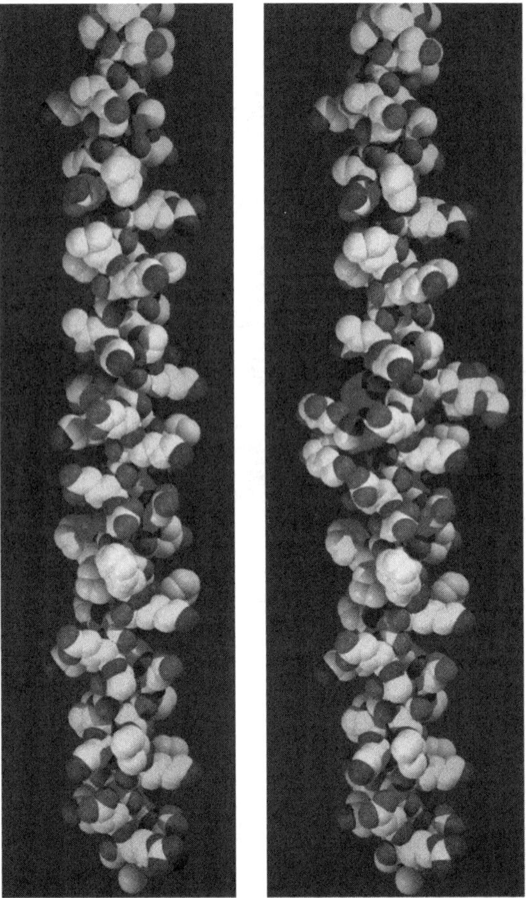

Figure 15. Glycosylated hydroxylysine residues on the surface of the collagen triple helix. The molecular model of a portion (36 residues/chain) of human type V collagen, $[\alpha1(V)]_2\alpha2(V)$, was built up by the McMolw (Molecular Images Software, San Diego, CA, USA) and Chem3D (CambridgeSoft Corporation, Cambridge, MA, USA) programs, using the 2CLG file of the Protein Data Bank, which is the theoretical model of the $(Gly–Pro–Hyp)_4$ trimer. The two lysine residues of the two $\alpha1(V)$ chains are either left unchanged (*left*) or changed to glulcosylgalactosylhydroxylysine (*right*). The glycosylated hydroxylysine residues are evidently bulkier than other residues.

an appropriate number of 4-hydroxyproline residues in the –Y– position of –Gly–X–Y– is a crucial determinant of the stability of the triple helix under physiological conditions. The hydroxyl group forms hydrogen bonds that contribute to the thermal stability. The transition temperature (thermal denaturation temperature; melting temperature, T_m) of triple helices formed *in vitro* by nonhydroxylated collagen is about 25°C, which is 15°C lower than the fully hydroxylated collagen triple helix (Fig. 3). The importance of hydroxylysine may be inferred from the inherited connective tissue disorder, Ehlers–Danlos syndrome type VI, in which lysyl hydroxylase deficiency results in impaired crosslink formation and consequent susceptibility to mechanical disruption of tissues.

Glycosylation of Hydroxylysine Residues

Vertebrate collagen molecules contain the monosaccharide galactose and the disaccharide glucosyl–galactose. The glycosides are covalently linked through the hydroxyl group to hydroxylysine residues within the triple-helical domains (see Figs. 15 and 16).

This glycosylation is unique to collagen, and its extent is variable both between collagen types and within the same collagen in different tissues and at different ages. A potential biological role of the specific carbohydrate units would be the regulation of the lateral assembly of the triple-helical domains. An inverse relationship exists between carbohydrate content and collagen fibril diameter, suggesting that the steric hindrance to the formation of highly ordered fibrils might be caused by the bulky glycosylated hydroxylysine residues. The glycosylation of hydroxylysine residues within the triple helix proceeds through the action of two isolated enzymes: hydroxylysyl galactosyl transferase (UDP galactose:

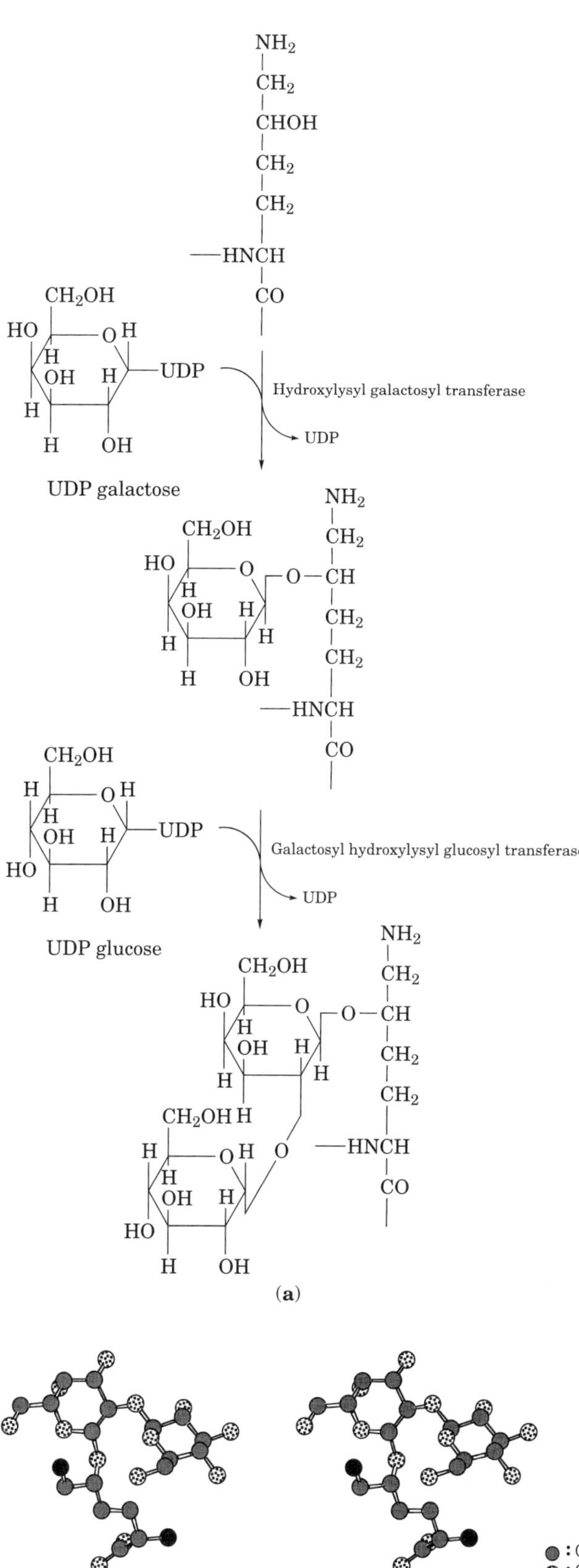

Figure 16. Attachment of galactose and glucose groups to hydroxylysine residues of collagen. (**a**) Enzymatic steps in coupling the two sugars; (**b**) stereo view (direct) of glucosylgalaclosylhydroxylysine. H atoms are not shown.

5-hydroxylysine–collagen galactosyltransferase) and galactosyl hydroxylysyl glucosyl transferase (UDP glucose: 5-hydroxylysine-collagen glucosyltransferase). Galactose is transferred to hydroxylysine residues by the former enzyme, while glucose is transferred to galactosyl-hydroxylysine residues by the latter enzyme. The carbohydrates are provided in both reactions by the appropriate UDP-glycosides. Both enzymes require a divalent cation (preferably Mn^{2+}) for the activity. A free ϵ-amino group in the substrate hydroxylysyl residue and a nonhelical polypeptide conformation are requirements for both transferases. The reactions are more favorable with longer peptides. Hydroxylysyl galactosyl transferase can bind at least two manganese ions per molecule of active enzyme.

Suggestions for Further Reading

B. Brodsky (1995) *FASEB J.* **9**, 1537–1546.

E. Adachi, I. Hopkinson, and T. Hayashi (1997) Basement-membrane stromal relationships: interactions between collagen fibrils and the lamina densa. *Intl. Rev. Cytolo.* **173**, 73–156.

P. Yurchenco, D. E. Birk, and R. P. Mecham, eds. (1994) *Extracellular Matrix Assembly and Structure*, Academic Press, San Diego.

J. F. Bateman, S. R. Lamandé, and J. A. M. Ramshaw (1996) Collagen superfamily, in *Extracellular Matrix*, Vol. 2, W. D. Comper, ed., Harwood Academic Publishers, Amsterdam.

C. M. Kielty, I. Hopkinson, and M. E. Grant (1992) Collagen, in *Connective Tissue and Its Heritable Disorders*, R. M. Royce and B. Steinmann, eds., Wiley, New York.

COLONY-STIMULATING FACTORS

JUDITH LAYTON

Colony-stimulating factors (CSFs) are **glycoproteins** that stimulate the proliferation, differentiation, and survival of hematopoietic (blood) cells and also activate mature myeloid cell functions. CSFs were first described in the 1960s as activities in various conditioned media that stimulated the growth of colonies of bone marrow cells in semisolid cultures. These colony-forming cells are the precursors of our red and white blood cells (see **Hematopoiesis**). Purification and characterization of these activities were difficult with the biochemical techniques available because they were present in conditioned media in very low concentrations and also because they were heterogeneous, largely due to variable glycosylation. Eventually, in the late 1970s and 1980s, four different murine CSFs were purified, followed by their human counterparts. These were called

1. Multi-CSF or interleukin-3 (IL-3)
2. Granulocyte-macrophage (GM)-CSF
3. Granulocyte (G)-CSF
4. Macrophage (M)-CSF, or CSF-1

Since then, the development of molecular biology has led to the **cloning** of CSF **complementary DNAs** and the production of purified **recombinant proteins**, facilitating *in vitro* and *in vivo* studies of their activities. All four CSFs have been tested in clinical trials, and both G-CSF and GM-CSF are used in clinical practice to elevate leukocyte levels in a variety of situations. In addition to these CSFs, many **interleukins** and **growth factors** have been described that stimulate or

Table 1. Biochemical Characteristics of the CSFs

Property	IL-3		GM-CSF		G-CSF		M-CSF	
	Hu	Mu	Hu	Mu	Hu	Mu	Hu	Mu
Amino acids	133	140	127	124	174(177)[a]	178	224/406/522[a]	519
M_r (kDa)								
Predicted	15.1	15.7	16.3	16	19	19	60	61
Expressed	14–30	28	18–33	18–33	22	25	45–90[b]	45–90[b]
Glycosylation								
N-linked	2	4	2	2	0	0	4	3
O-linked	ND	ND	2	ND	1	1	1	1
Disulfides	1	2	2	2	2	2	7–9	7–9
α-Helix type	Short		Short		Long		Short	
Cross-species activity[c]	–	–	–	–	+	+	+	–

[a] Different forms resulting from alternative splicing.
[b] Disulfide-bonded dimer of proteolytically cleaved transmembrane protein.
[c] Between human and murine. Data derived from Refs. 4,18, and 19.

Table 2. Characteristics of the Genes Encoding the CSFs

Property	IL-3		GM-CSF		G-CSF		M-CSF	
	Hu	Mu	Hu	Mu	Hu	Mu	Hu	Mu
Size (kbp)	2.2	2.7	2.5	2.5	2.5	2.5	20	20
Chromosomal location	5q23-31	11A5-B1	5q21-31	11A5-B1	17q11-22	11D-E2	1p13-21	3F3
Number of exons	5	5	4	4	5	5	10	10
Alternative splicing	No	No	No	No	Yes	No	Yes	Yes
mRNA species (kb)	0.9	0.9	0.78	0.78	1.5	1.5	4, 2.3, 2.0, 1.6	4, 2.6, 1.4
Promoter elements	AP-1, ELF-1, NIP[a], NF-IL-3, CRE, CK-1, CK-2, GC		CK-1, CK-2, IgκB, GC, AP-1, ELF-1		PU box, CK-1, NF-IL-6, octamer, GPE-3, CK-2		SP-1, AP-1, AP-2, NF-1, NF-IL-6, CK-1, CK-2	

[a] Repressor element. Data from Refs. 18 and 19.

co-stimulate bone marrow cell colony growth. The most important of these are the early-acting stem cell factor (SCF) and FLT3/FLK-2 ligand (FL) (1), the erythroid lineage stimulator erythropoietin, and the megakaryocyte and platelet stimulator thrombopoietin (TPO). Recently, TPO has been found to have multilineage activities (2).

BIOCHEMICAL CHARACTERISTICS

The CSFs belong to a family of **cytokines** that have little sequence identity but were predicted to be similar in structure (3). The **protein structures** of all four have now been determined, although for IL-3 a mutant molecule with enhanced stability was used, rather than the native molecule (4). They are all **four-helix bundles** with up–up–down–down connectivity. IL-3, G-CSF, and GM-CSF are monomeric secreted proteins, but the soluble form of M-CSF is a **disulfide-bonded** homodimer produced by **proteolytic** cleavage of cell-membrane molecules. Other characteristics are given in Table 1.

CSF GENES AND THEIR EXPRESSION

The cloning of **genomic** DNA encoding the CSFs and analysis of the regions 5′ of the genes has resulted in a better understanding of the control of CSF expression (see Table 2). A single **messenger RNA** species is transcribed for IL-3 and GM-CSF, whereas a minor **alternatively spliced** mRNA is produced for human G-CSF that results in a three-amino-acid-residue insertion in the protein after residue Leu35 and a 20-fold reduction in activity. Transcription of M-CSF mRNA is more complex, with four species resulting from alternative splicing of the 10 exons (see **Introns/exons**). The largest two differ only in the 3′ untranslated region and encode identical 522-residue **transmembrane** proteins. The smaller two result from splicing out of part of exon 6, which results in loss of the proteolytic cleavage site from the protein produced from the 1.6-kb **transcript** and expression of a stable transmembrane protein (5). The DNA regions upstream of the coding regions contain **promoter** and **enhancer** elements that have been partially characterized and shown to respond to CSF inducers such as **tumor necrosis factor** (TNF), IL-1, lipopolysaccharide (LPS), and **antigens**. The control of CSF gene expression is not understood completely, but it involves control of mRNA stability by elements in the 3′ untranslated region of the mRNAs, as well as control of transcription.

CSF PRODUCTION

Many cell types have been shown to produce CSFs *in vitro* in response to various stimuli, such as bacterial products (LPS) and inflammatory cytokines (TNF, IL-1), but the role of CSFs in maintaining white blood cell levels in normal healthy individuals is not entirely clear (see later). Whether unstimulated cells produce CSFs is still debated (6). GM-CSF, G-CSF, and M-CSF are produced by macrophages, endothelial cells, fibroblasts, and stromal cells. In addition, T lymphocytes produce GM-CSF, M-CSF (in humans), and IL-3. Mast cells have been shown to produce IL-3 and GM-CSF. Other cell types have been reported to produce GM-CSF and IL-3, but these studies have yet to be confirmed.

CSF RECEPTORS

The CSF receptors are transmembrane proteins belonging to the cytokine receptor class I family, except for the M-CSF receptor, which is a **tyrosine kinase receptor**. The class I receptors contain a structurally conserved module called the cytokine-binding **domain** (CBD), the hemopoietin domain, or the cytokine receptor homology domain. This module comprises two **fibronectin** type III domains with four conserved **cysteine** residues in the *N*-terminal domain and a conserved Trp–Ser–X–Trp–Ser sequence in the *C*-terminal domain. Ligand binding causes dimerization or oligomerization of the receptors, which initiates **signal transduction**. The IL-3 and GM-CSF receptors form heterodimers of a ligand-specific α-chain and a common β-chain that is also shared with the IL-5 receptor. In the mouse, there is an additional β-chain that is unique to the IL-3 receptor. The G-CSF and M-CSF receptors form homodimers. Receptor dimerization activates tyrosine kinases (intrinsic in the M-CSF receptor, cytoplasmic **JAK** kinases for the other receptors) that **phosphorylate** the receptors, providing binding sites for other signaling molecules that are in turn activated by phosphorylation of **tyrosine** residues and, in some cases, **serine** or **threonine**. Ultimately, these signaling cascades lead to alterations in gene transcription, but the components of the biochemical pathways are still being defined (reviewed in Refs. 7–9).

The CSF receptors are expressed at low levels on hematopoietic cells, usually a few hundred receptors per cell. IL-3 receptors are expressed on neutrophil, eosinophil, and macrophage lineages, on mast cells, and on undifferentiated blast cells. GM-CSF receptors are expressed on neutrophil, eosinophil, and macrophage lineages. G- and M-CSF receptor expression is more restricted; G-CSF receptors are present on neutrophils and at a low level on macrophages, and M-CSF receptors are found on macrophages, osteoclasts, and placental trophoblasts. Recent studies have detected expression of receptors on purified stem-cell populations. Human $CD34^+$ cells expressed SCF, FL, G-CSF, and, at a low level, GM-CSF receptors (10). Murine stem cell populations expressed IL-1α, IL-3, IL-6, and G-CSF receptors (11).

BIOLOGICAL ACTIVITY

In Vitro

All four of the CSFs stimulate the proliferation and differentiation of bone-marrow progenitors to form colonies in semisolid agar. The type of mature cells in the colonies varies with the stimulator (see Fig. 1). IL-3 was otherwise called multi-CSF because it stimulates the largest number of different cell lineages. Colonies of granulocytes, macrophages, eosinophils, megakaryocytes, undifferentiated blast cells, and mixed cell lineages are obtained. In liquid cultures, proliferation of mast cells is strongly stimulated by IL-3, and some types of B lymphocytes are reported to respond. GM-CSF stimulates growth of granulocytic, macrophage, and eosinophil colonies. Production of dendritic cells, which are important **antigen-presenting** cells in the immune response, is stimulated by GM-CSF. G-CSF and M-CSF are more lineage restricted, producing mainly granulocytic and macrophage colonies, respectively. The CSFs can give enhanced or synergistic responses when used in combination with each other and with other interleukins and growth factors. Some effects of CSFs on nonhematopoietic cells have been reported. G- and GM-CSF can stimulate proliferation of endothelial cells. Placenta-derived cells and oligodendrocytes proliferate in response to GM-CSF. M-CSF stimulates proliferation of trophoblast cells and may be important in pregnancy.

In Vivo

The effects of CSFs *in vivo* have been determined both by injection in mice and humans and by analysis of CSF-deficient mice created by gene targeting and, in the case of M-CSF, a naturally occurring mutation. Injection of IL-3 or GM-CSF elevates circulating levels of neutrophils, monocytes, and eosinophils. G-CSF causes a greater elevation of neutrophils than does IL-3 or GM-CSF, but only a weak increase in monocytes and no effect on other lineages. M-CSF causes a selective elevation of monocytes. The magnitude of these increases depends on the dose of CSF administered. It is interesting that in response to infection, circulating IL-3 or GM-CSF are not usually detected,

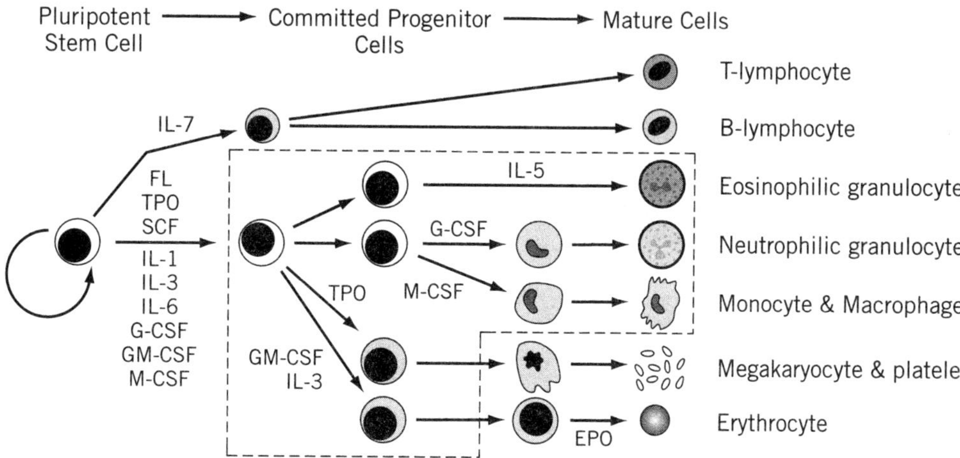

Figure 1. Schematic representation of hematopoiesis and the CSFs and other cytokines required for mature cell production. IL, interleukin; FL, FLT3/FLK-2 ligand; SCF, stem cell factor; TPO, thrombopoietin. GM-CSF and IL-3 stimulate cells inside the dotted line box.

whereas G-CSF and, to a lesser extent, M-CSF are frequently elevated in serum. These observations suggest that IL-3 and GM-CSF may be important in local inflammatory responses in the tissues rather than in a systemic response to infection.

The importance of the CSFs in maintaining basal hematopoiesis has been assessed by analysis of CSF-deficient mice. For IL-3, GM-CSF, and G-CSF, gene "knockout" mice were created by introduction of a defective version of the gene of interest into embryonic stem cells by homologous recombination. The embryonic stem cells were injected into blastocysts to create chimeric mice that were then bred to create heterozygous and homozygous deficient mice. For M-CSF, mice with a naturally occurring point mutation (*op/op*) that resulted in a frameshift in the sequence and a loss of M-CSF production were discovered. The IL-3- and GM-CSF-deficient mice were able to produce normal levels of circulating blood cells and bone marrow progenitors. The GM-CSF-deficient mice developed pulmonary alveolar proteinosis and increased pulmonary infections, presumably as a consequence of defective pulmonary macrophage function. Recent studies in IL-3-deficient mice have shown that IL-3 is important for elevation of basophils and mast cells in response to a parasitic infection (12) and for some delayed-type hypersensitivity (T-lymphocyte-dependent) responses (13). G-CSF defective mice were neutropenic, with neutrophil numbers reduced to 20–35% of wild-type control mice. Bone marrow cellularity was normal, but some granulocytic and macrophage progenitors were reduced in number. These mice were more susceptible to a bacterial infection (*Listeria monocytogenes*) than wild-type mice. Young M-CSF-deficient mice have osteopetrosis, a severe macrophage and osteoclast deficiency resulting in failure of teeth eruption and overgrowth of bone that occludes the marrow cavity and reduces marrow cellularity. Unlike the other CSF-deficient mice, the *op/op* mice overcome their defects with age, which suggests that other cytokines may be able to compensate for the lack of M-CSF. Thus, the CSF-defective mice have shown that each CSF has a unique function *in vivo*. GM-CSF and IL-3 are not required for steady-state hematopoiesis. G- and M-CSF are required for normal neutrophil and macrophage production, respectively. All four CSFs contribute in different ways to control of infection. They may have other nonunique functions that are not revealed in the deficient mice because of compensatory activities of other cytokines. These may be detected in future studies of multiply deficient mice.

MYELOID LEUKEMIA

Acute and chronic myeloid leukemias (AMLs and CMLs) are clonal diseases of bone marrow cells in which the predominant cell types are granulocytes and macrophages, although leukemic cells of other lineages are also found. There are potentially two ways in which CSFs could be involved in leukemia: They may be required for the proliferation of the leukemic cells, and they may cause differentiation and clonal extinction of the leukemia. In semisolid agar cultures, CML cells will produce colonies of normal appearance containing granulocytes, macrophages, and eosinophils. The colony growth is dependent on CSF stimulation, and all four CSFs are active. In contrast, AML cells produce small clones with little maturation, and growth is very variable from patient to patient. The cells are responsive to CSFs to a variable extent, and in a proportion of patients (<30%) there is colony growth in the absence of CSFs. The colonies do not contain clonogenic cells that are able to form further colonies in secondary cultures. The leukemic stem cells have not been identified, and their dependence on CSFs is unclear. Whether autocrine production of CSFs by leukemic cells is important is also unknown, with the exception of leukemic monocytes, which produce CSFs like their normal counterparts. Lopez and colleagues have produced a GM-CSF antagonist that induced remission in a murine model of juvenile myelomonocytic leukemia (14). Although there are some leukemic cell lines that differentiate and cease proliferation in response to CSFs *in vitro*, many cell lines are not suppressed by CSFs. There may be defects in the signal transduction pathways that prevent differentiation. In a small number of patients with congenital neutropenia who develop AML, a mutation in the G-CSF receptor expressed by the leukemic cells has been described that prevents differentiation in response to G-CSF (15). However, it is not known whether this mutation contributed to the development of AML.

CLINICAL USE OF CSFS

The use of chemotherapy in cancer treatment is limited by its toxicity, primarily neutropenia. Both G-CSF and GM-CSF have been tested extensively in clinical trials and shown to reduce neutropenia, allowing the intensification of chemotherapy doses. Clinical trials have not yet shown, however, whether dose intensification improves survival. G-CSF has fewer side effects and is more effective at elevating neutrophil levels than GM-CSF and thus is usually the CSF of choice, but GM-CSF may be more effective in reducing infectious complications (16). In high-dose chemotherapy followed by autologous bone marrow transplantation, both CSFs reduced the time taken for recovery to ≥500 cells/mm^3 by about one week and, in some trials, reduced the duration of hospital stay. G-CSF (and GM-CSF) treatment increased the levels of circulating progenitor and stem cells so that G-CSF-mobilized peripheral blood progenitor cells are now used instead of bone marrow cells for reconstitution after high-dose chemotherapy. This treatment results in more rapid recovery than bone marrow transplantation.

Patients with congenital neutropenia or idiopathic chronic neutropenia respond to G-CSF treatment with elevation of neutrophil levels, resulting in fewer infections and less time in hospital. A few of these patients (3% in Ref. 17) develop myelodysplasia or leukemia, but at present there is no evidence whether G-CSF increases the frequency of these cases. Despite initial concerns that treatment of leukemia patients with CSFs might exacerbate the disease, there is no evidence that this occurs. Neutrophil recovery has occurred more rapidly in leukemia patients receiving G-CSF or GM-CSF after chemotherapy, but reduction in duration of hospital stay was not usually observed.

There have been fewer clinical trials with M-CSF and IL-3. M-CSF appears to improve recovery from fungal infections and is approved in Japan for use in allogeneic transplantation, induction therapy of AML, and dose-intensive therapy of ovarian cancer. IL-3 given after chemotherapy reduced the recovery time from leukopenia and elevated the levels of neutrophils, monocytes, and eosinophils. In the future, it is likely that combinations of cytokines will be tested.

BIBLIOGRAPHY

1. Y. Yonemura, H. Ku, S. D. Lyman, and M. Ogawa (1997) *Blood* **89**, 1915–1921.
2. G. P. Solar, W. G. Kerr, F. C. Zeigler, D. Hess, C. Donahue, F. J. de Sauvage, and D. L. Eaton (1998) *Blood* **92**, 4–10.
3. J. F. Bazan (1990) *Immunol. Today* **11**, 350–354.
4. Y. Feng, B. K. Klein, and C. A. McWherter (1996) *J. Mol. Biol.* **259**, 524–541.
5. M. Baccarini and E. R. Stanley (1990) In *Growth Factors, Differentiation Factors, and Cytokines* (A. Habenicht, ed.), Springer-Verlag, Berlin, pp. 188–200.
6. F. H. M. Cluitmans, B. H. J. Esendam, J. E. Landegent, R. Willemze, and J. H. F. Falkenburg (1995) *Blood* **85**, 2038–2044.
7. T. Hara and A. Miyajima (1996) *Stem Cells* **14**, 605–618.
8. B. R. Avalos (1996) *Blood* **88**, 761–777.
9. L. R. Rohrschneider, R. P. Bourette, M. N. Lioubin, P. A. Algate, G. M. Myles, and K. Carlberg (1997) *Mol. Reprod. Dev.* **46**, 96–103 (Abstract).
10. A. W. Roberts, M. Zaiss, A. W. Boyd, and N. A. Nicola (1997) *Exp. Hematol.* **25**, 289–305.
11. W. J. McKinstry, C. Li, J. E. J. Rasko, N. A. Nicola, G. R. Johnson, and D. Metcalf (1997) *Blood* **89**, 65–71.
12. C. S. Lanz, J. Boesiger, C. H. Song, N. Mach, T. Kobayashi, R. C. Mulligan, Y. Nawa, G. Dranoff, and S. J. Galli (1998) *Nature* **392**, 90–93.
13. N. Mach, C. S. Lanz, S. J. Galli, G. Reznikoff, M. Mihm, C. Small, R. Granstein, S. Beissert, M. Sadelain, R. C. Mulligan, and G. Dranoff (1998) *Blood* **91**, 778–783.
14. P. O. Iversen, I. D. Lewis, S. Turczynowicz, H. Hasle, C. Niemeyer, K. Schmiegelow, S. Bastiras, A. Biondi, T. P. Hughes, and A. F. Lopez (1997) *Blood* **90**, 4910–4917.
15. F. Dong, R. K. Brynes, R. K. Tidow, K. Welte, B. Lowenberg, and I. P. Touw (1995) *N. Engl. J. Med.* **333**, 487–493.
16. J. Nemunaitis (1997) *Drugs* **54**, 709–729.
17. D. C. Dale, M. A. Bonilla, M. W. Davis, A. M. Nakanishi, W. P. Hammond, J. Kurtzberg, W. Wang, A. Jakubowski, E. Winton, P. Lalezari, W. Robinson, J. A. Glaspy, S. Emerson, J. Gabrilove, M. Vincent, and L. A. Boxer (1993) *Blood* **81**, 2496–2502.
18. D. Metcalf and N. A. Nicola (1995) *The Hemopoietic Colony-Stimulating Factors. From Biology to Clinical Applications*, Cambridge University Press, Cambridge, England.
19. R. Callard and A. Gearing (1994) *The Cytokine Facts Book*, Academic Press, London.

Suggestions for Further Reading

G. J. Lieschke (1997) CSF-deficient mice—what have they taught us? *Ciba Foundation Symposium* **204**, 60–77. A detailed analysis of CSF deficient mice.

D. Metcalf (1998) Lineage commitment and maturation in hematopoietic cells: the case for extrinsic regulation. *Blood* **92**, 345–348.

T. Enver, C. M. Heyworth, and T. M. Dexter (1998) Do stem cells play dice? *Blood* **92**, 348–351. Two sides of the debate about hematopoietic lineage commitment.

S. S. Watowitch, H. Wu, M. Socolovsky, U. Klingmuller, S. N. Constantinescu, and H. F. Lodish (1996) Cytokine receptor signal transduction and the control of hematopoietic cell development. *Annu. Rev. Cell Dev. Biol.* **12**, 91–128.

References 18 and 19 above. Reference 18 gives a detailed, well referenced description of the four CSFs. Reference 19 is less detailed, but covers other cytokines in addition to the original four.

COMBINATORIAL LIBRARIES

BARRY HENDERSON
JACK KEENE
DANIEL KENAN

Few fields have emerged as rapidly as the technologies collectively referred to as combinatorial chemistry. The origins of combinatorial "thinking" as applied to biological problems can be traced to our understanding of the immune system. Humoral immunity is based on the generation of billions of different **antibody** molecules through unique **gene recombination** events. Upon exposure to a foreign **immunogen**, the subset of the antibody **repertoire** with specificity for the invader is clonally expanded. During expansion, the active antibody subpool undergoes additional **recombination** and **mutation** events, and the most active variants are further selected. This process is reiterated until mature antibody molecules are produced that are highly optimized against the foreign **antigen**. In essence, antibody repertoires are combinatorial libraries—collections of related but distinct molecules representing different combinations of genetic elements. The recognition of these facts preceded the creation of the first combinatorial library, which was based upon the synthesis and screening of large numbers of synthetic peptides in order to identify selected antigenic determinants of a viral pathogen (1). Given the parallels between the immune system and combinatorial libraries, it is not surprising that the original application of this emerging technology was in the derivation of immunogens capable of stimulating immune protection in animals not yet exposed to certain pathogens.

The hallmarks of a combinatorial approach are the generation and use of complex compound mixtures to identify molecules with desirable properties. Four basic operations underpin combinatorial approaches: (i) generation of libraries containing related but diverse compounds, (ii) selection of molecules with desirable properties from the library, (iii) characterization of the selected molecules, and (iv) amplification of the identified products. Depending upon the chemical composition of the library, the selected molecules may require amplification before structure determination. For example, **DNA libraries** can be amplified using the polymerase chain reaction (**PCR**) before the final products are identified by sequencing. Synthetic peptides and small molecules cannot be amplified per se and must be subjected to structural elucidation, and then resynthesized for subsequent use. Regardless of the library composition, the four operations enumerated above can be applied in an iterative fashion to progressively refine the selected compounds.

Several entries in this volume deal with the application of combinatorial approaches to complex biological problems. Issues central to the implementation of combinatorial technologies in biological systems will be described below, including cognitive aspects of library design, practical aspects of library construction, basic issues in affinity selection and library screening, and discussions of different library types and their applications. The breadth of the topic Combinatorial Libraries is so expansive that it is impossible to cover any topic in great detail. The entries entitled **Libraries, Combinatorial Synthesis**, and **Affinity Selection** are intended to provide a generic introduction to the principles underlying combinatorial operations. The remaining entries, **DNA li-**

braries, **Genomic libraries**, **cDNA libraries**, **Expression libraries**, **Peptide libraries**, and **Phage display libraries**, provide more specific details regarding the major types of combinatorial libraries in widespread use. Small molecule synthesis or drug discovery applications are not dealt with in detail, although many basic concepts from these fields are presented. The reader is referred to the suggested reading list for additional discussion of these related topics.

See also **Libraries**, **Combinatorial synthesis**, **Affinity selection**, **DNA libraries**, **Genomic libraries**, **cDNA libraries**, **Expression libraries**, **Peptide libraries**, **Phage display libraries**.

BIBLIOGRAPHY

1. H. M. Geysen, R. H. Meleon, and S. J. Barteling (1984) *Proc. Natl. Acad. Sci. USA* **81**, 3998–4002.

Suggestions for Further Reading

J. D. Watson (1987) *Molecular Biology of the Gene*, 4th ed., Benjamin-Cummings, Menlo Park, CA.

J. Sambrook, E. F. Fritsch, and T. Maniatis (1989) *Molecular Cloning: A Laboratory Manual*, 2nd ed., Cold Spring Harbor Laboratory Press, Cold Spring Harbor, NY.

D. J. Kenan, D. E. Tsai, and J. D. Keene (1994) *Trends Biochem. Sci* **19**, 57–64.

J. N. Abelson, ed. (1996) *Combinatorial Chemistry*, Vol. 267, *Methods in Enzymology*, Academic Press, San Diego.

L. A. Thompson and J. A. Ellman (1996) *Chem. Rev.* **96**, 555–600.

B. K. Kay, J. Winter, and J. McCafferty (1996) *Phage Display of Peptides and Proteins*, Academic Press, San Diego.

COMBINATORIAL SYNTHESIS

BARRY HENDERSON
JACK KEENE
DANIEL KENAN

The diversity of a molecular **library** may be derived from natural sources, or it may be created through specialized techniques in synthetic chemistry. Traditionally, synthetic chemists focus on preparing individual compounds, paying careful attention to stereochemical control and to achieving the greatest possible yield and purity. This often requires the use of highly specialized reactions and can be a painstakingly slow process. In contrast, chemists engaged in combinatorial organic synthesis employ transformations that are more general in scope and can be applied to the simultaneous preparation of many chemically distinct compounds. In exchange for the ability to prepare many compounds in a short time frame, compromises are accepted that may result in sacrificing, within acceptable limits, yield and purity of individual reaction products. This entry deals with the basic issues faced in preparing and screening synthetic **combinatorial libraries** and some of the relevant technologies and methods. No attempt has been made to provide exhaustive coverage of the rapidly expanding range of chemistries now amenable to library synthesis, nor have we attempted to cover the application of combinatorial chemistry as it applies directly to the field of drug discovery. See **Combinatorial libraries**.

Combinatorial chemistry seeks to create all possible combinations of a set of building blocks and then extract the identity of members that exhibit a desired property. Thus, if a traditional synthesis involves incorporating building block A, followed by B, and then C to give the desired compound A–B–C, then the analogous combinatorial synthesis would employ a set of monomers, A_1–A_n, followed by a second set of monomers, B_1–B_n, and then a third set of monomers, C_1–C_n, giving all possible combinations of $(A_{1\text{ to }n})$–$(B_{1\text{ to }n})$–$(C_{1\text{ to }n})$. Such processes can be described as n^k sets, where n is the number of building blocks used in each cycle of synthesis and k is the total number of synthetic cycles. The number of individual species generated by these approaches become vast very quickly. For example, a relatively short DNA library of 100 random nucleotide positions has a theoretical complexity of 4^{100}, or 1.6×10^{60} different sequences. This number exceeds the number of protons in the sun. Complexities of more realistic peptide libraries using the 20 amino acids incorporated into proteins are as follows: A library of all possible tripeptides will contain 8000 (20^3) distinct peptide sequences; a library of all possible tetrapeptides will contain 160,000 (20^4) distinct peptide sequences; and a library of all possible pentapeptides will contain 3,200,000 (20^5) distinct peptide sequences. Specialized methods are required to prepare, assay, and elucidate the identity of active compounds from collections of this size. In order to facilitate product purification, compounds are typically synthesized while covalently attached to inert solid supports, including polymeric pins or resin beads, although other materials have been used (see **Solid phase synthesis**). Use of solid phase synthesis enables many reactions to be carried out simultaneously using parallel arrays of individual reaction chambers. If each reaction yields a single product, the result is a library of spatially discrete compounds. This approach has the advantage that the identity of compounds can be ascertained immediately, based on the position in the synthetic array and the associated synthetic history. The initial application of this approach was reported by Geysen et al. for synthesis of peptide libraries (1); however, many variations have been reported since then, involving a wide range of chemistries. Although convenient and straightforward to implement, the number of compounds that can be synthesized and screened using this approach is limited by the physical dimensions of the arrays to approximately 10,000 different species. Miniaturization and photolithographic masking techniques have enabled arrays containing more than 100,000 compounds to be synthesized on silica chips (2). Although this has been applied to a variety of chemistries, it is perhaps most widely recognized as the **DNA chip** technology of Affymetrix (3). Unfortunately, this approach requires highly specialized instrumentation for the photolithographic masking, as well as for assay of the resulting arrays.

The synthesis of larger libraries requires pooling methods that overcome the size limitations of spatial arrays (4). Pooling methods employ mixtures of reactants at each synthetic step (5). For example, in random DNA oligonucleotide synthesis, two or more deoxyribonucleotide phosphoramidites are introduced during each coupling cycle. One major problem with mixture synthesis is that variation in reactivity can result in some building blocks being under- or overrepresented in the final compound pool. At least in some cases, the initial concentration of reactants can be adjusted to correct for these variations in reactivity. A more attractive approach, first introduced by Furka, is the "split synthesis" strategy (6). Briefly, a portion of synthesis resin is "split" or divided into equal portions and placed into individual reaction vessels (Fig. 1). Each vessel receives a single building block in large excess, which drives the reaction to completion. The resin from all the vessels is then recombined, mixed thoroughly, and then redistributed. A second set of building blocks is then introduced. This process is repeated the desired number of cycles, to produce all possible combinations of the desired building blocks. Because each resin bead is exposed to a single building block at each step, there will be only one compound (but many copies) attached to each bead. This has led to the term one-bead, one-compound (OBOC) libraries (7,8).

Compound identification is straightforward for spatially discrete libraries; however, identifying active species in the large and complex mixtures that result from pooling strategies presents a formidable problem. Typically, each member of a pool is present in a quantity that, although sufficient for determining biological activity, is insufficient for isolation and structure determination by standard analytical methods. As a result, strategies to identify the active members must be built

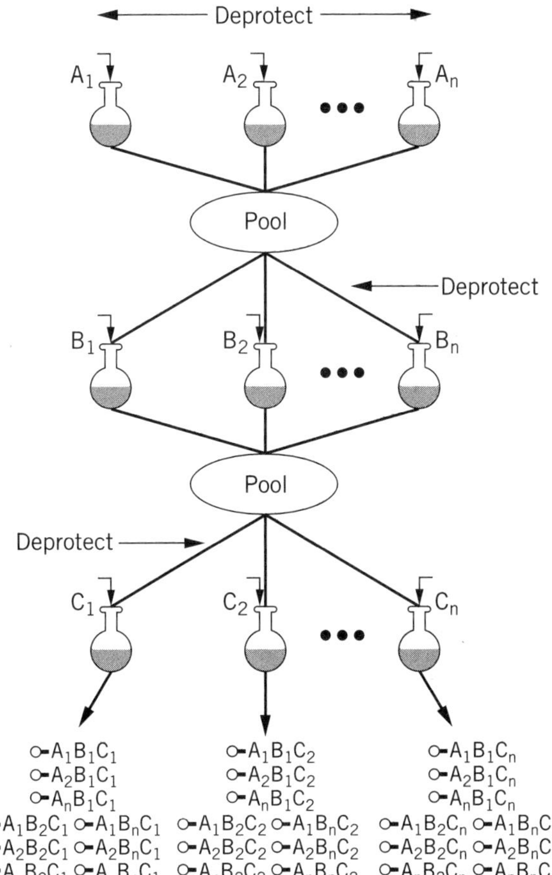

Figure 1. Schematic flow of a split synthesis. The resin is first divided into n equal portions. Following deprotection, each resin batch is reacted with a different monomer from set A. The resin is pooled, mixed, and divided into equal portions. After deprotection, individual monomers from set B are coupled. The process is repeated, and monomers from set C are added. Each final reaction batch is held separately for subsequent assay.

into the synthetic schemes and closely coupled to the types of assays that will be performed with a particular library.

Compounds can be identified from soluble libraries using deconvolution strategies such as those introduced by Houghten (9,10). This approach relies on cycles of synthesis and screening to focus progressively in on active species. An initial library is prepared in which one position of diversity is held constant. Each individual pool of compounds contains a different building block at the constant position but a mixture of building blocks at the remaining positions of diversity. Screening identifies the pool and therefore the optimal building block at the defined location. A second library is then synthesized in which all members have the optimal building block at this first position, but each pool contains a different building block at a second position of diversity. This process is repeated until the preferred building block for each position is determined. The chief disadvantage of this approach is that it is a laborious and time-consuming process. Moreover, this strategy does not guarantee that the most potent library member will be identified since selection of a "preferred" building block depends on the cumulative activity of a pool. Selection of a particular building block could, therefore, result from the presence in a pool of a relatively few number of potent compounds or a large number of relatively poor inhibitors.

Several strategies have been developed that allow the active members of a library to be determined without requiring cycles of synthesis and screening. The method of positional scanning (11) involves synthesizing a series of combinatorial libraries, each fixing a single position of diversity. Synthesis of each library is divided such that each component pool is prepared using a single building block at the defined position, while the remaining positions are synthesized using mixtures of building blocks. Although this method does not require resynthesis, it does require additional synthetic effort to prepare a complete set of scanning libraries, and the total number of syntheses is greater than for deconvolution approaches. Moreover, because each position is defined independently, there is an increased chance of missing the most active members in the library.

Orthogonal pooling strategies also avoid the resynthesis burden of deconvolution methods (12). In this approach, two separate libraries are synthesized such that any given pool of the first library has only one compound in common with any pool of the second library. Determining the common species between positive pools from the two libraries reveals the identity of active compounds. The chance for misidentifying the optimal building block at any given position is avoided by this approach, but there is a considerable synthetic burden, as well as compound tracking issues.

The second general assay format is designed to measure the interaction of a target (enzyme, receptor, cell, etc) with compounds while they remain attached to the solid support used for library synthesis. Geysen pioneered this approach for assaying libraries of peptides arrayed on pins (1); however, the full potential was realized only when it was coupled to screening of OBOC libraries (8). Because each resin bead in an OBOC library carries a single library compound and the beads are physically distinct, each bead behaves as an isolated assay system. Briefly, a target is incubated with a pool of beads. The excess target is removed by washing, leaving behind only target that associates specifically with a compound attached to a resin bead. The active compounds are identified following visualization of those beads to which the target specifically interacts.

Specialized techniques are required to identify the compounds attached to individual beads, because each bead can carry at most a few nanomoles of compound, which is insufficient for structure determination except in a few rare cases (eg, peptides). These methods seek to "tag" or "encode" the beads with the synthetic history in a manner that is easily deciphered (13,14). The code is introduced either prior to library synthesis or following each cycle of building block addition. In the latter case, the chemistries for library synthesis and code synthesis must be mutually compatible, which is one of the major obstacles in the development and utilization of this approach. Examples of encoding methods include the use of peptides read by **Edman degradation** (15), nucleic acids read by PCR amplification and sequencing (16), several small molecule chemical classes identified by various analytical methods (17,18), nonradioactive isotopic tags read by mass spectrometry (19), and radioisotopes encapsulated within the solid-phase support (20,21). The development of encoding methods is an active area of research because the split synthesis method is currently the only tractable approach to preparing large libraries in which all possible compounds are represented.

See also **Combinatorial libraries**, **Libraries**, **Affinity selection**, **DNA libraries**, **Genomic libraries**, **cDNA libraries**, **Expression libraries**, **Peptide libraries**, and **Phage display libraries**.

BIBLIOGRAPHY

1. H. M. Geysen, R. H. Meleon, and S. J. Barteling (1984) *Proc. Natl. Acad. Sci. USA* **81**, 3998–4002.
2. S. P. Fodor, J. L. Read, M. C. Pirrung, L. Stryer, A. T. Lu, and D. Solas (1991) *Science* **251**, 767–773.
3. A. C. Pease, D. Solas, E. J. Sullivan, M. T. Cronin, C. P. Holmes, and S. P. Fodor (1994) *Proc. Natl. Acad. Sci. USA* **91**, 5022–5026.
4. C. Pinilla, J. Appel, S. Blondelle, C. Dooley, B. Dörner, J. Eichler, J. Ostresh, and R. A. Houghten (1995) *Biopolymers (Peptide Sci.)* **37**, 221–240.
5. H. M. Geysen, S. J. Rodda, and T. J. Mason (1986) *Mol. Immunol.* **23**, 709–715.
6. A. Furka, F. Sebestyen, M. Asgedom, and G. Dibo (1991) *Int. J. Pept. Protein Res.* **37**, 487–493.
7. K. S. Lam, M. Lebl, and V. Krchnak (1997) *Chem. Rev.* **97**, 411–448.
8. K. S. Lam, S. E. Salmon, E. M. Hersh, V. J. Hruby, W. M. Kazmierski, and R. J. Knapp (1991) *Nature* **354**, 82–84.
9. C. T. Dooley, N. N. Chung, B. C. Wilkes, P. W. Schiller, J. M. Bidlack, G. W. Pasternak, and R. A. Houghten (1994) *Science* **266**, 2019–2022.
10. S. E. Blondelle, E. Takahashi, P. A. Weber, and R. A. Houghten (1994) *Antimicrob. Agents Chemother.* **38**, 2280–2286.
11. R. A. Houghten, C. Pinilla, S. E. Blondelle, J. R. Appel, C. T. Dooley, and J. H. Cuervo (1991) *Nature* **354**, 84–86.
12. B. Deprez, X. Williard, L. Bourel, H. Coste, F. Hyafil, and A. Tartar (1995) *J. Am. Chem. Soc.* **117**, 5405–5406.
13. S. Brenner and R. A. Lerner (1992) *Proc. Natl. Acad. Sci. USA* **89**, 5381–5383.
14. K. D. Janda (1994) *Proc. Natl. Acad. Sci. USA* **91**, 10779–10785.
15. J. M. Kerr, S. C. Banville, and R. N. Zuckermann (1993) *J. Am. Chem. Soc.* **115**, 2529–2531.
16. M. C. Needels, D. G. Jones, E. H. Tate, G. L. Heinkel, L. M. Kochersperger, W. J. Dower, R. W. Barrett, and M. A. Gallop (1993) *Proc. Natl. Acad. Sci. USA* **90**, 10700–10704.
17. M. H. J. Ohlmeyer, R. N. Swanson, L. W. Dillard, J. C. Reader, G. Asouline, R. Kobayashi, M. Wigler, and W. C. Still (1993) *Proc. Natl. Acad. Sci. USA* **90**, 10922–10926.
18. Z.-J. Ni, D. Maclean, C. P. Holmes, M. M. Murphy, B. Ruhland, J. W. Jacobs, E. M. Gordon, and M. A. Gallop (1996). *J. Med. Chem.* **39**, 1601–1608.
19. H. M. Geysen, C. D. Wagner, W. M. Bodnar, C. J. Markworth, G. J. Parke, F. J. Schoenen, D. S. Wagner, and D. S. Kinder (1996) *Chem. Biol.* **3**, 679–688.
20. K. C. Nicolaou, X.-Y. Xiao, Z. Parandoosh, A. Senyei, and M. P. Novz (1995) *Angew. Chem. Int. Ed. English* **34**, 2289–2291.
21. E. J. Moran, S. Sarshar, J. F. Cargill, M. M. Shahbaz, A. Lio, A. M. M. Mjalli, and R. W. Armstrong (1995) *J. Am. Chem. Soc.* **117**, 10787–10788.

COMPETENCE

D. LANE

Competence is the condition of bacterial cells that enables them to take up naked **DNA**. When this DNA can integrate into the **chromosome** by **recombination** or, in the case of **plasmids**, establish itself in the cytoplasm, the cells are said to be transformed (see **Transformation**). Competence is manifested by several species of bacteria—representatives of 29 different genera according to a recent count (1)—but far from all. Even in transformable strains, competence generally depends on the establishment of a particular physiological state. Four species have served as the major experimental models for studies of competence, and we shall therefore confine discussion largely to these: *Bacillus subtilis, Streptococcus pneumoniae* ("pneumococcus"), *Haemophilus influenzae*, and *Neisseria gonorrhoeae*. In view of the universal utility of *Escherichia coli*, which does not acquire competence naturally, we shall also review briefly the artificial induction of competence in this species.

The basic mechanism of DNA uptake appears to be the same in all naturally transformable organisms. Duplex DNA binds to the cell surface, and is then processed so that just one of the two strands passes through a specialized transmembrane channel into the cytoplasm, where it undergoes recombination with the resident chromosome. In **gram-negative** bacteria, a means of first passing through the outer membrane is also needed. The systems that regulate the uptake process, however, as well as the environmental stimuli to which these systems respond, appear to be more idiosyncratic, presumably reflecting the very different natural habitats in which the capacity to take up DNA is induced. Furthermore, the regulatory pathways of competence intersect with those of other metabolic activities. It is with the regulation of competence that this entry is chiefly concerned, the uptake process itself being deferred largely to the entry **Transformation**. We begin with the bacterium for which most is known at both the mechanistic and regulatory levels.

B. SUBTILIS

B. subtilis survives and grows in many environments, but its natural reservoir is the soil. It was early appreciated that competence in this bacterium is a phenomenon induced by decelerating growth (2), such as occurs at the transition from exponential to stationary phase or upon depleting the medium of amino acids. Dependence of competence on transition to the stationary phase appears to reflect inhibition of competence by added amino acids (3). Prototrophs in minimal medium can attain competence in exponential phase (4). We now know that it is just one of a number of late-growth responses, which also include excretion of degradative **enzymes**, antibiotic production, development of motility, and **sporulation** (5). Regardless of the many overlapping inputs from these other pathways, induction of competence is mediated by a single transcriptional activator, the ComK protein (6). This protein stimulates expression of the genes that encode the 12 components of the DNA-uptake machine identified to date (7–9). These genes are distributed among four polycistronic **transcription** units, *com G, C, E* and *F*, which specify the DNA-binding apparatus and its assembly, as well as the functions that translocate a single strand of DNA into the cytoplasm. The **nuclease** that degrades the other strand has not been identified. ComK also stimulates (a) the synthesis of **RecA**, which catalyzes the final step of transformation, (b) the recombination of the imported strand with the chromosome, and (c) the synthesis of AadAB, a nuclease/**DNA helicase** that assists the process.

Several intersecting metabolic currents determine the net level of active ComK at a given moment, and hence the degree of competence. Some of these are indicated in Figure 1. What is remarkable about this system is that positive input from all these factors is necessary for high levels of competence; the absence of any one of them diminishes activation of ComK or reduces transcription of its gene. Stochastic variability, leading to limits on one or more of these factors, may explain why in *B. subtilis* cultures, unlike those of other species, only about 10% of the cells become competent, even with optimum inducing treatments. The special condition of these cells is underscored by assorted observations of their relatively quiescent metabolic state and reduced buoyant density (4).

The finding that competence was attained earlier in cultures growing in medium already used for growth of cells to the competent state implied that extracellular signaling molecules promote development of competence (10), and this proved to be the case. Two such molecules are known, both generated by the cell and exported to the medium. The ComX pheromone interacts with the cell-surface protein ComP to initiate a cascade of interactions that culminate in competence (11,12). Competence stimulating factor (CSF) is probably imported to stimulate this pathway. These interactions are shown in Figure 1 and described in the legend. It is possible that the apparent redundancy of competence-inducing function reflects only the co-opting of the CSF sporulation factor to protect the primary product of the competence-specific pathway, the phosphorylated form of ComA, ComA $\sim$ P (13). Unlike ComX, which seems to be specific for competence, CSF production is intimately linked with the system that regulates sporulation; transcription of its gene requires the Spo0H **sigma factor** and is stimulated, indirectly, by the Spo0A $\sim$ P transcription regulator. Grossman (14) has suggested that CSF is also a signaling molecule for sporulation.

The external actions of ComX pheromone and of CSF have led to the suggestion that these factors communicate cell-density information to the system that regulates competence (14). ComA thus serves as a depot for these signals, and in its phosphorylated form it sends them on by activating transcription of *comS* (12,15–18).

This general induction pathway is also subject to nutritional signals: amino acids via CodY (3), and glucose via SinI and SinR (19). How these, as well as signals from the sporulation and enzyme secretion pathways (AbrB, DegU ...), are integrated to allow the cell to decide whether to commit itself to competence remains an intriguing issue.

S. PNEUMONIAE

This organism is an obligate parasite of the respiratory tract. In contrast to *B. subtilis, S. pneumoniae* readily attains competence during exponential growth in rich medium. Moreover, all the cells become competent, albeit briefly. Competence in *S. pneumoniae* does share with that of *B. subtilis*, however, an association with a distinct physiological state. This takes the form of altered surface properties, cessation of cell-wall synthesis, increased cell-wall fragility, and release of DNA into the medium. The onset of competence is very sensitive to culture conditions: temperature, the nature of the medium buffer, the concentrations of magnesium and calcium, and pH. In low-pH medium, there is only one peak of competence, at high cell density, whereas in high-pH medium there is a succession of peaks (20).

The competence induction pathway of *S. pneumoniae* shows notable similarities to that of *B. subtilis*, as depicted in Figure 1. Induction of competence is mediated by an extracellular signaling molecule, the competence stimulating peptide (CSP) (21–23), and this initiates the induction cascade via a sensor-regulator protein pair (24–26). Induction is also modulated by imported peptides (27,28), though in this case the effect is negative (29).

Several proteins other than those shown in Figure 1 make their appearance at competence, some of them associated with the membrane (30); but the only one demonstrated to have a clear role in transformation, indeed in competence itself, is the RecA protein, which mediates homologous recombination (31).

S. pneumoniae cultures exhibit a particular aspect of competence that, though detected long ago (32), has received little attention. Filtrates of cultures that have passed their peak of competence prevent development of competence when added to another culture, suggesting that a specific inhibitor of competence is secreted into the medium. Such a factor might be induced at competence just in order to turn it off again, accounting for the brevity of the competence peak.

H. INFLUENZAE

H. influenzae lives in the mucus of the respiratory tract, where it may be exposed to large amounts of naked DNA (33). Only its own DNA, however, is taken up with high efficiency. This selectivity results from the presence throughout the genome of numerous copies of a short, specific base sequence, the uptake-signal sequence (USS), which is recognized by the uptake apparatus (34,35). The outer membranes of competent cells extrude vesicles, termed *transformasomes* (or "blebs"), which encapsulate the DNA and then release it into the **periplasm** to allow transport into the cell (36–38). As in *B. subtilis*, growth-inhibiting conditions, such as transient anaerobiosis,

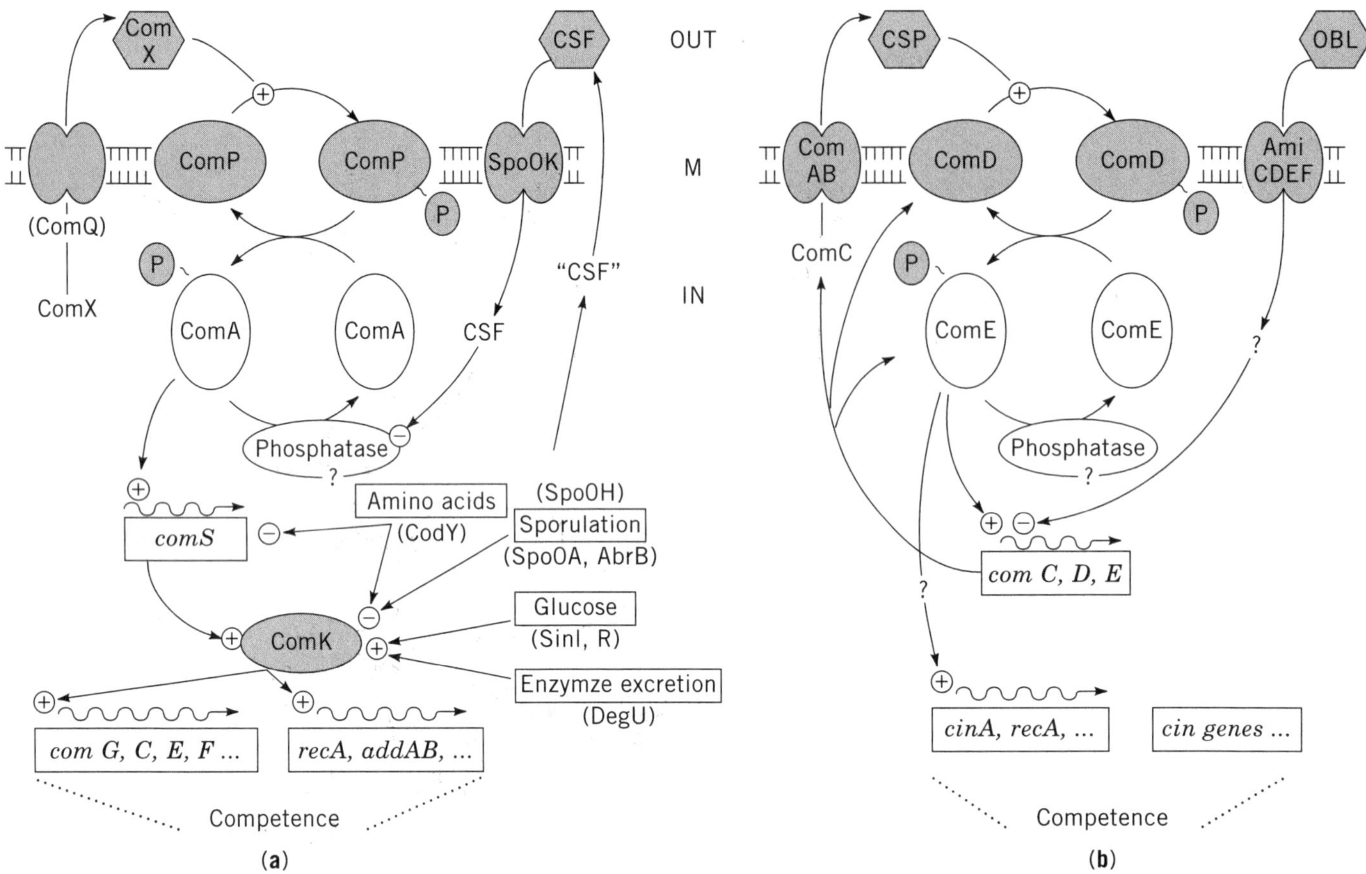

Figure 1. Competence induction in *B. subtilis* and *S. pneumoniae*. The pathways are drawn to emphasize similarities and have been simplified (!) to highlight what are thought to be the main participants. M indicates the membrane, OUT indicates the medium, and IN indicates the cytoplasm. Most proteins are defined below and in the text; OBL denotes oligopeptide-binding lipoprotein.; cin stands for competence induction; + indicates stimulation of either transcription or protein activity, − indicates repression or inhibition; question marks indicate factors not yet identified or presumed activities not yet verified. Wavy arrows denote transcripts. *B. subtilis*: The ComX pheromone is a peptide of 10 amino acid residues that is cleaved from the longer *comX* gene product (11). It is exported through a *comQ*-dependent pore to act at a surface domain of the membrane protein, ComP, causing the latter to phosphorylate itself at a site on the cytoplasmic side. ComP, the sensor of a two-component regulatory system, in turn phosphorylates ComA, converting it to a transcriptional activator of the *comS* gene (11,12). Another short (4–7 residue) peptide, competence stimulating factor [CSF (12)], is cleaved from a precursor during export: It is thought to enter the cytoplasm through an oligopeptide **permease** pore and to protect ComA ~ P from inactivation by inhibiting a **phosphatase** (12,13). The ComS peptide acts by displacing ComK from a complex in which it is normally sequestered by the chaperone-like MecA and MecB proteins, liberating it to stimulate expression of the late competence genes (17,18). *ComG* specifies the DNA-binding apparatus, the *comC* gene product is responsible for processing ComG peptides and assembly of the ComG complex in the membrane (9), *comE* encodes the DNA translocation functions, and *comF* appears to specify a mechanism for actively transporting the single stranded DNA into the cytoplasm (8). The *recA* and *addAB* genes encode recombination functions. ComK also stimulates its own synthesis. The inhibitory effect of amino acids in the medium is mediated by the CodY protein, which represses transcription of *comS* and *comK* (3). The stimulatory effect of glucose seems to involve the freeing of an activator of *comK* transcription, SinR (19). *S. pneumoniae*: Competence stimulating peptide (CSP) is an unmodified 17 amino acid residue peptide, probably generated by removal of the **signal sequence** from the ComC prepeptide during secretion by a specific ABC **transporter**, ComAB (21–23). CSP activates the sensor–regulator protein pair, ComD–ComE, and the resulting ComE ~ P protein stimulates transcription of the *comCDE* operon (24). This system transmits the CSP signal to the known late competence genes, as well as, presumably, others not yet identified (*cin* genes). There is evidence that in *S. gordonii*, the ComD homologue interacts directly with CSP (25). Because CSP ultimately stimulates transcription of its own gene, initial secretion by a few cells can lead to wholesale induction of the culture, effectively synchronizing competence development. The *comC, D*, and *E* genes constitute a single **operon** (26), thus allowing coregulation. The importation of other peptides appears to modulate competence induction (27–29). Strains mutant for the oligopeptide transport proteins AmiA and AliAB develop competence at much lower cell densities than the wild type, suggesting that importation of peptides serves as a cell density signal (29). It is possible that the peptides repress *comCDE* transcription until exhausted at higher cell densities.

nutritional downshift, and approach to stationary phase, induce competence. The level of competence attained varies and can be close to 100% of the population. Nevertheless, there appears to be no cell density sensing involved, each cell developing competence independently. *H. influenzae* cells implanted in the rat peritoneum rapidly develop the capacity to import DNA added with the inoculum and manifest other changes characteristic of the competent state (39).

Several factors important to the development of competence in *H. influenzae* have been identified, but the picture remains fragmented. Of crucial importance is the intracellular level of **cyclic AMP (cAMP)**. Competence was abolished by mutations in the **cyclic AMP receptor protein** (CRP) and **adenylate cyclase** (Cya) genes, but could be restored to a *cya* mutant by addition of cAMP (40,41). The cell senses sugar availability through a carbohydrate phosphotransferase system that increases cAMP synthesis in response to declining sugar availability: Mutations in two genes specifying components of this system, *ptsI* and *crr*, caused partial loss of competence, a deficiency that can be compensated for by added cAMP (42,43). Competence in *H. influenzae* thus responds directly to nutritional state.

Some clues as to how cAMP might act in competence induction have recently emerged. Notably, cAMP stimulates transcription from a divergent **promoter** region of two genes essential for transformation, *tfoX* and *rec-1*. *TfoX* was originally identified by a mutation, *sxy*-1, that increased transformation 100-fold (44). Null mutations in the gene eliminated DNA binding and transformation, whereas overexpression of *tfoX* caused constitutive competence (45,46), implying that TfoX is a positive regulator of competence. The finding that TfoX is needed for expression of other competence genes (*com101A*, *dprA*, and *rec-2*) (46,47) supports this suggestion and hints at a central role in competence similar to that played by the ComK regulator of *B. subtilis*. Whereas *tfoX* transcription normally increases only as cells approach stationary phase, addition of cAMP to a log-phase culture causes an immediate stimulation (46).

On the other hand, transcription of *rec-1*, which encodes the general recombination enzyme that is analogous to RecA of *E. coli*, is only weakly stimulated by cAMP-CRP (48). It is, however, induced by single-stranded DNA, as befits its function of integrating transforming DNA. This common property of Rec proteins has been proposed to endow Rec-1 with another quite distinct function, that of triggering competence. Induction of competence in *H. influenzae* may be associated with the conversion of up to 5% of the chromosomal DNA to single-stranded form. Observing that overproduction of Rec-1 led to early development of competence, Stuy (49) suggested that formation of Rec-1/single-stranded DNA complexes might induce competence gene expression, as the equivalent complexes induce the SOS regulon in *E. coli*. The idea remains untested.

Several new proteins appear in the outer membrane at competence. Some of them, rather than being synthesized *de novo*, are redistributed from the inner membrane. One mutant that fails to do this also fails to bind DNA. It carries a lesion in the gene *por*, which encodes a periplasmic disulfide oxidoreductase, suggesting the need to assemble or fold **disulfide bond**-containing proteins (50) (see **Protein disulfide isomerase**). Two proteins probably involved in conducting DNA into the cytoplasm are induced at competence: Rec-2, which shares **homology** with the presumed transmembrane channel protein ComEC of *B. subtilis*, and DprA (47).

There is as yet no indication that *H. influenzae* competence responds to a specific extracellular signal. If it does, it now seems unlikely to do so via a sensor-regulator protein pair, such as ComP-ComA in *B. subtilis*: Mutation of the genes encoding four such systems, as deduced from the genome sequence, failed to affect competence (43). What is clear is that the identity and function of most of the components of the competence regulatory system are still unknown, and that the true place of the known ones in the system likewise remains to be shown.

N. GONORRHOEAE

Unlike the bacteria already discussed, for which competence is a transitory response to growth conditions, transformable *Neisseria* strains are competent at all phases of growth (51). An early indication (52) that competence could be induced by a **proteinase**-sensitive factor present in conditioned medium has not been confirmed. The transformation mutants so far reported are defective in one or more of the individual functions known to be involved in transformation-**pili** and proteins for DNA binding, transport, and recombination (53–55), but none have suggested the existence of competence factors or regulators. The competent state may be so integrated into the general physiology of *Neisseria* that a specific regulatory system is not needed, or the regulatory elements are essential and therefore missed in mutant hunts. Transformation of cells in culture is influenced by pH and by the concentrations of mono- and divalent cations (53), but these factors probably affect DNA-cell envelope interactions directly, rather than by regulating competence gene output.

E. COLI

In 1970, Mandel and Higa (56) discovered that *E. coli* could be induced to take up DNA by treatment of the cells with Ca^{2+} ions. Although this artificial competence has been a key technique in molecular genetics, its basis is not understood. Suspension of *E. coli* cells in cold $CaCl_2$ solution is known to induce the formation of transmembrane pores consisting of an outer shell of polyhydroxybutyrate, connected by Ca^{2+} to an inner sheath of polyphosphate (57). The role of these pores is unclear, because their internal diameter appears to be too small to allow the passage of double-stranded DNA. They do, however, have marked effects on temperature-induced membrane fluidity, and it is possible that, by anchoring an otherwise mobile membrane, they create tension that opens holes elsewhere in the membrane to admit the DNA (57,58).

SUMMARY

One of the most striking aspects of research on competence to emerge in recent years is the interrelationships between the proteins and regulatory factors involved and those of other phenomena previously considered distinct, perhaps the clearest example being *B. subtilis* sporulation (14). Indeed, acquisition of competence is usually accompanied by other major physiological changes. Because of this, and because our knowledge of competence is based largely on laboratory observation, it is fair to question whether DNA uptake itself is relevant in natural bacterial habitats. In fact there are good reasons for thinking that it is, and this issue is taken up in the entry **Transformation**.

BIBLIOGRAPHY

1. M. G. Lorenz and W. Wackernagel (1994) *Microbiol. Rev.* **58**, 563–602.
2. C. Anagnostopoulos and J. Spizizen (1961) *J. Bacteriol.* **81**, 741–746.
3. P. Serror and A. L. Sonenshein (1996) *J. Bacteriol.* **178**, 5910–5915.
4. D. Dubnau (1993) In *Bacillus subtilis and Other Gram-Positive Bacteria* (A. L. Sonenshein, J. A. Hoch, and R. Losick, eds.), ASM, Washington, D.C., pp. 555–584.
5. D. C. Dooley, C. T. Hadden, and E. W. Nester (1971) *J. Bacteriol.* **108**, 668–679.
6. D. van Sinderen and G. Venema (1994) *J. Bacteriol.* **176**, 5762–5770.
7. D. van Sinderen, A. Luttinger, L. Kong, D. Dubnau, G. Venema, and L. Hamoen (1995) *Mol. Microbiol.* **15**, 455–462.
8. D. Dubnau (1997) *Gene* **192**, 191–198.
9. Y. S. Chung and D. Dubnau (1994) *Mol. Microbiol.* **15**, 543–551.
10. H. Joenje, M. Gruber, and G. Venema (1972) *Biochim. Biophys. Acta* **262**, 189–199.
11. R. Magnuson, J. Solomon, and A. D. Grossman (1994) *Cell* **77**, 207–216.
12. J. Solomon, R. Magnuson, A. Srivastava, and A. D. Grossman (1995) *Genes Dev.* **9**, 547–558.
13. J. Solomon, B. Lazazzera, and A. D. Grossman (1996) *Genes Dev.* **10**, 2014–2024.
14. A. D. Grossman (1995) *Annu. Rev. Genet.* **29**, 477–508.
15. N. M. Nakano and P. Zuber (1991) *J. Bacteriol.* **173**, 7269–7274.
16. J. Hahn and D. Dubnau (1991) *J. Bacteriol.* **173**, 7275–7282.
17. L. Kong and D. Dubnau (1994) *Proc. Natl. Acad. Sci. USA* **91**, 5793–5797.
18. T. Msadek, F. Kunst, and G. Rapoport (1994) *Proc. Natl. Acad. Sci. USA* **91**, 5788–5792.
19. U. Bai, I. Mandic-Mulic, and I. Smith (1993) *Genes Dev.* **7**, 139–148.
20. J. D. Chen and D. A. Morrison (1987) *J. Gen. Microbiol.* **133**, 1959–1967.
21. A. Tomasz and J. L. Mosser (1966) *Proc. Natl. Acad. Sci. USA* **55**, 58–66.
22. L. S. Havarstein, G. Coomaraswami, and D. A. Morrison (1995) *Proc. Natl. Acad. Sci. USA* **92**, 11140–11144.
23. L. S. Havarstein, D. B. Diep, and I. F. Nes (1995) *Mol. Microbiol.* **16**, 229–240.
24. E. V. Pestova, L. S. Havarstein, and D. A. Morrison (1996) *Mol. Microbiol.* **21**, 853–862.
25. L. S. Havarstein, P. Gaustad, I. F. Nes, and D. A. Morrison (1996) *Mol. Microbiol.* **21**, 965–971.
26. Q. Cheng, E. A. Campbell, A. M. Naughton, S. Johnson, and H. R. Masure (1997) *Mol. Microbiol.* **23**, 683–692.
27. G. Alloing, P. de Philip, and J.-P. Claverys (1996) *J. Mol. Biol.* **241**, 44–58.
28. B. J. Pearce, A. M. Naughton, and H. R. Masure (1994) *Mol. Micrbiol.* **12**, 881–892.
29. G. Alloing, B. Martin, C. Granadel, and J.-P. Claverys (1998) *Mol. Microbiol.* **29**, 75–83.
30. M. N. Vijayakumar and D. A. Morrison (1986) *J. Bacteriol.* **165**, 689–695.
31. B. Martin, P. Garcia, M.-P. CastaniÈ, and J.-P. Claverys (1995) *Mol. Microbiol.* **15**, 367–379.
32. A. Tomasz and R. D. Hotchkiss (1964) *Proc. Natl. Acad. Sci. USA* **51**, 480–487.
33. L. Matthews, S. Spector, L. Lemm, and J. Potter (1963) *Am. Rev. Respir. Dis.* **88**, 199–204.
34. K. L. Sisco and H. O. Smith (1979) *Proc. Natl. Acad. Sci. USA* **76**, 972–976.
35. D. B. Danner, R. A. Deich, K. L. Sisco, and H. O. Smith (1980) *Gene* **11**, 311–318.
36. M. Kahn, M. Concino, R. Gromkova, and S. H. Goodgal (1979) *Biochem. Biophys. Res. Commun.* **87**, 764–772.
37. R. A Deich and L. C. Hoyer (1982) *J. Bacteriol.* **152**, 855–864.
38. M. E. Kahn, G. Maul, and S. H. Goodgal (1982) *Proc. Natl. Acad. Sci. USA* **79**, 6370–6374.
39. M. Dargis, P. Gourde, D. Beauchamp, B. Foiry, M. Jaques, and F. Malouin (1992) *Infect. Immun.* **60**, 4024–4031.
40. M. S. Chandler (1992) *Proc. Natl. Acad. Sci. USA* **89**, 1626–1630.
41. I. R. Dorocicz, P. M. Williams, and R. J. Redfield (1993) *J. Bacteriol.* **175**, 7142–7149.
42. L. P. Macfadyen, I. R. Dorocicz, J. Reizer, M. H. Saier Jr., and R. J. Redfield (1996) *Mol. Microbiol.* **21**, 941–952.
43. M. L. Gwinn, D. Yi, H. O. Smith, and J. F. Tomb (1996) *J. Bacteriol.* **178**, 6366–6368.
44. R. J. Redfield (1991) *J. Bacteriol.* **173**, 5612–5618.
45. P. M. Williams, L. A. Bannister, and R. J. Redfield (1994) *J. Bacteriol.* **176**, 6789–6794.
46. J. J. Zulty and G. J. Barcak (1995) *Proc. Natl. Acad. Sci. USA* **92**, 3616–3620.
47. S. Karudapuram and G. J. Barcak (1997) *J. Bacteriol.* **179**, 4815–4820.
48. J. J. Zulty and G. J. Barcak (1993) *J. Bacteriol.* **175**, 7269–7281.
49. J. H. Stuy (1989) In *Genetic transformation and expression* (L. O. Butler et al., eds.), Intercept Inc., Andover, pp. 85–112.
50. J.-F. Tomb (1992) *Proc. Natl. Acad. Sci. USA* **89**, 10252–10256.
51. P. F. Sparling (1966) *J. Bacteriol.* **92**, 1364–1369.
52. A. Siddiqui and I. D. Goldberg (1975) *Biochem. Biophys. Res. Commun.* **64**, 34–42.
53. G. D. Biswas, T. Sox, E. Blackman, and P. F. Sparling (1977) *J. Bacteriol.* **129**, 983–992.
54. G. D. Biswas, S. A Lacks, and P. F. Sparling (1989) *J. Bacteriol.* **171**, 657–664.
55. I. J. Mehr and H. S. Seifert (1997) *Mol. Microbiol.* **23**, 1121–1131.
56. M. Mandel and A. Higa (1970) *J. Mol. Biol.* **53**, 159–162.
57. R. N. Reusch and H. L. Sadoff (1988) *Proc. Natl. Acad. Sci. USA* **85**, 4176–4180.
58. C. E. Castuma, R. Huang, A. Kornberg, and R. N. Reusch (1995) *J. Biol. Chem.* **270**, 12980–12983.

Suggestions for Further Reading

J. M. Solomon and A. D. Grossman (1996) Who's competent and when: regulation of natural genetic competence in bacteria. *Trends Genet.* **12**, 150–155. A brief but comprehensive review, providing access to the important literature.

A. D. Grossman (1995) Genetic networks controlling the initiation of sporulation and the development of genetic competence in *Bacillus subtilis*. *Annu. Rev. Genet.* **29**, 477–508. A review of competence development in *B. subtilis* and its interconnections with sporulation and nutritional state.

J.-P. Claverys, A. Dintilhac, I. Mortier-Barrière, B. Martin, and G. Alloing (1997) Regulation of competence for genetic transformation in *Streptococcus pneiumoniae*. *J. Appl. Microbiol. Symp. Suppl.* **83**, 32S–42S. A review of competence regulation in this pathogen, emphasizing the influence of peptide transport and other factors.

COMPETITIVE INHIBITION

JOHN F. MORRISON

Analogues of the substrate of an **enzyme** can often bind at the same place within the enzyme's **active site** as the substrate but are not acted on by the enzyme. They are inhibitors of the enzyme, giving rise to **dead-end inhibition**, and they usually compete with the substrate, giving rise to competitive inhibition.

LINEAR

The simplest type of competitive inhibition occurs when an inhibitor combines reversibly at the active site of the same form of the enzyme as the substrate. The inhibitor may be a structural analogue of the substrate or a product of that substrate. In the following scheme, the inhibitor I combines reversibly with the free enzyme to form a dead-end EI complex that can only

$$\begin{array}{ccccc} E & + A & \rightleftharpoons & EA & \longrightarrow E + P \\ I \big\updownarrow K_i & & & & \\ EI & & & & \end{array}$$

dissociate back to the components from which it was formed. Therefore, the interaction of E and I is at thermodynamic equilibrium, and the inhibition constant K_i is a **dissociation constant**. The general equation that describes this type of inhibition is shown in Equation 1:

$$v = \frac{VA}{K_a(1 + I/K_{is}) + A} \quad (1)$$

where v is the observed velocity of the reaction, V the maximum velocity, K_a the Michaelis constant for the substrate present at concentration A, I the concentration of inhibitor, and K_{is} the apparent inhibition constant. For a single-substrate reaction, K_{is} would be equal to K_i. The double-reciprocal form, **Lineweaver–Burk plot**, of the equation for linear competitive inhibition (Eq. 2):

$$\frac{1}{v} = \frac{K_a}{V}\left(1 + \frac{I}{K_{is}}\right)\frac{1}{A} + \frac{1}{V} \quad (2)$$

indicates that a primary plot of $1/v$ against $1/A$ at different concentrations of I would yield a family of straight lines that intersect at the same point on the vertical ordinate (Fig. 1). This is the signature of competitive inhibition; sufficiently high concentrations of substrate can compete successfully with the inhibitor. The slope of each curve is given by Equation 3:

$$\text{Slope} = \frac{K_a}{VK_{is}}I + \frac{K_a}{V} \quad (3)$$

A plot of slope against inhibitor concentration would yield a straight line that intersects the abscissa at the point where

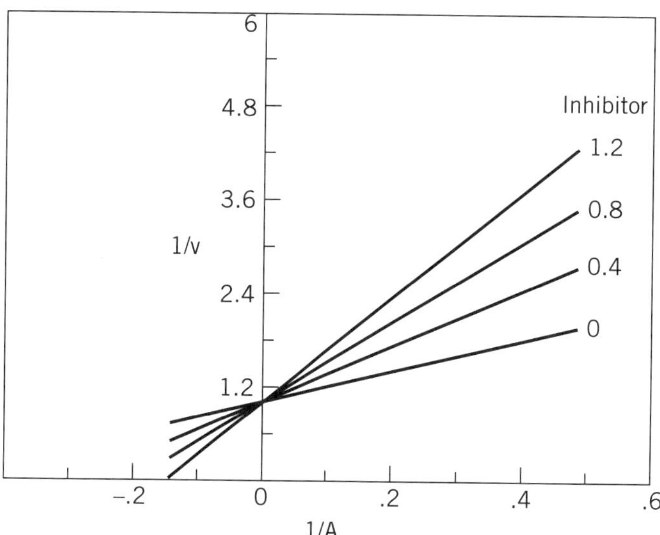

Figure 1. Primary double-reciprocal plot for linear competitive inhibition.

$I = -K_{is}$. Since this secondary plot should be linear, the inhibition would be classified as linear competitive. Although values for K_{is} can be obtained by such graphical procedures, it is preferable to obtain its value by an overall fit of the primary data to the general equation by use of a least-squares fitting method (1).

When two substrates are involved in a reaction, the value determined directly for K_{is} may be only an apparent constant, with its value dependent on the concentration of the nonvaried substrate (see **Lineweaver–Burk plot**). In this case, it would also have to be stated that the inhibitor was linear competitive with respect to a particular substrate.

Linear competitive inhibition will also be observed with a two-substrate reaction when the inhibitor is an analogue of the first substrate to add in an equilibrium ordered mechanism, A, and the second substrate to add, B, is varied (Fig. 2). The inhibition by I with respect to A will be linear competitive because both A and I compete directly for the same form of enzyme (E). The inhibition by I with respect to B will also be linear competitive, even though I and B combine with different enzyme forms, because the reaction between E and A in an equilibrium-ordered mechanism is at thermodynamic equilibrium. Thus, increasing concentrations of B will ultimately reduce the concentrations of both EA and E to zero, so that no free enzyme remains to interact with I.

HYPERBOLIC

This type of inhibition occurs when an inhibitor binds reversibly at a site on the enzyme, other than the active site, so as to make it more difficult for the substrate to combine at the active site. Thus, the inhibitor and substrate can be present on the enzyme at the same time. The reactions involved can be illustrated as follows:

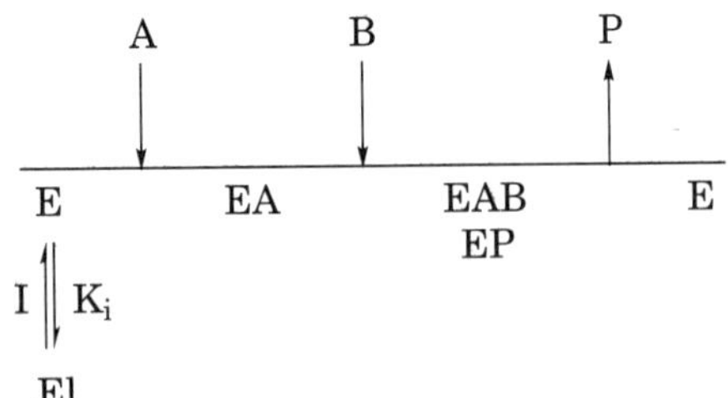

Figure 2. Equilibrium-ordered mechanism for which inhibition by I would be linear competitive with respect to both substrates.

$$\begin{array}{ccccc} E & \underset{K_a}{\overset{A}{\rightleftharpoons}} & EA & \xrightarrow{k} & P \\ I \Big\Updownarrow K_{is} & & I \Big\Updownarrow K_{id} & & \\ EI & \underset{K_A}{\overset{A}{\rightleftharpoons}} & EIA & \xrightarrow{k} & P \end{array}$$

where K_{is} and K_{id} denote dissociation constants for EI and EIA, respectively. It is a thermodynamic requirement that the affinity of the substrate A for the E and EI forms differs, so that

$$K_A = K_a \frac{K_{id}}{K_{is}}$$

For the inhibition to be competitive, the presence of the inhibitor on the enzyme must not affect the rate of product formation. The general equation that describes hyperbolic competitive inhibition is given in Equation 4:

$$v = \frac{VA\left[1 + \frac{I}{K_{id}}\right]}{K_a\left(1 + \frac{I}{K_{id}}\right) + A} \quad (4)$$

which can be arranged in double-reciprocal form as

$$\frac{1}{v} = \frac{K_a}{V}\left[\frac{1 + \frac{I}{K_{is}}}{1 + \frac{I}{K_{id}}}\right]\frac{1}{A} + \frac{1}{V} \quad (5)$$

This is an equation of a straight line for a plot of $1/v$ against $1/A$, but it is apparent that the slopes of the lines will vary as a hyperbolic function of the concentration of I and the intercept of the curves with the vertical ordinate is independent of I. For a plot of slope as a function of I, the curve would be a concave-down, nonrectangular hyperbola, with limiting values of K_a/V and $K_aK_{id}/(VK_{is})$ at zero and infinite concentrations of I, respectively. Values for K_{id} and K_{is} may be obtained from a plot of $1/(\text{slope}_o - \text{slope}_i)$ against $1/I$, where slope_o and slope_i are slopes of the primary plot in the absence and presence of inhibitor, respectively. Values for K_{is} and K_{id} are best obtained by an overall least-squares fit of the data to the equation that describes hyperbolic competitive inhibition (1).

It should be noted that the same kinetic mechanism would apply if I were an activator that enhances the combination of the substrate with the enzyme. In this case, the plot of slope against I would yield a concave-up, nonrectangular hyperbola (2).

BIBLIOGRAPHY

1. W. W. Cleland (1979) *Meth. Enzymol.* **63**, 103–138.
2. V. L. Schramm and J. F. Morrison (1970) *Biochemistry* **9**, 671–677.

COMPETITIVE LABELING

T. IMOTO

Competitive labeling is a method of **chemical modification** for determining the pK_a and the reactivity of functional groups in proteins. The extent of modification of a particular protein group at various pH values is compared to that of the same group in a standard compound. This gives the reactivity of the protein group relative to its intrinsic reactivity, and the pH-dependence gives the pK_a value. These data reflect the **accessibility** of the group in the folded protein, which provides information about the chemical reactivity of the group and its environment in the folded protein. This indirectly gives conformational information about the protein. Conformational transitions associated with function, association, and **ligand binding** can be monitored with this method.

Competitive labeling was originally employed in a study of the **amino groups** in **elastase** (1). A mixture of elastase and phenylalanine (the standard compound) was acetylated with a small amount of ^{3}H-labeled acetic anhydride (see **Anhydrides**). It is important that the amount of the labeling agent is less than that of the groups being modified. Under these conditions, the relative amounts of label in the standard and in the specific protein group will depend upon their relative ionization and reactivities. The fraction of the reactive, nonionized form of the amino group (α_i) and the ratio of reactivity ($r = k_i/k_{standard}$) of the ith group are derived from the known pK_a of the standard (α_{standard}) and the measured ratio of labeling:

$$\alpha_i r = \alpha_{\text{standard}}(\text{label on } i\text{th group/label on standard}) \quad (1)$$

The label on the ith group could be determined by the radioactivity of the appropriate peptide separated by **peptide mapping**. The values on the right side of the equation are obtained experimentally at each pH and are plotted against pH. The values of the pK_a and of r of the protein group are obtained by theoretical curve fitting.

Besides using acetic anhydride to react with amino groups, 1-fluoro-2,4-dinitrobenzene has been used to react with amino groups and with **histidine** and **tyrosine** residues, and **iodoacetate** with **thiol groups**. A small agent is most suitable for of modification. Double labeling with ^{3}H-label, followed by ^{13}C-labeling after **denaturation** gives a much more sensitive analysis. In this case, the same group in the denatured protein is the standard.

BIBLIOGRAPHY

1. H. Kaplan, K. J. Stevenson, and B. S. Hartley (1971) *Biochem. J.* **124**, 289–299.

Suggestion for Further Reading

N.M. Young and H. Kaplan (1989) Chemical characterization of functional groups in proteins by competitive labelling. In *Protein Structure: A Practical Approach* (T.E. Creighton, ed.), IRL Press, Oxford, U.K., pp. 225–245.

COMPLEMENT FIXATION

J. FOOTE

Complement is a set of serum proteins that undergo a sequential series of cleavage and association reactions in response to antibody–antigen complexes. If these complexes occur on a cell surface, complement reactions can cause cell lysis (1). The lysis of erythrocytes is particularly easy to detect visually, due to the release of hemoglobin, and this lysis forms the principle for a type of immunoassay. Antibody–antigen complexes in solution cause complement reactions to occur without cell lysis. Since the complement reactions are stoichiometric and the products short-lived, these non-cell-associated complexes deplete a portion of the complement proteins, a process termed "complement fixation." This depletion reduces the lysis observed when the complement is subsequently added to antibody-coated erythrocytes.

Key reagents for complement fixation assays, all available commercially, are complement, erythrocytes, and hemolytic antibody (2,3). Guinea pig serum is usually used as a source of complement, since guinea pig complement reacts with antibodies of most experimentally significant mammalian species. Sheep erythrocytes, washed free of serum components, are used for lysis. Hemolytic antibody is obtained by immunizing rabbits with sheep erythrocyte membranes. The erythrocytes are sensitized in preparation for complement lysis by incubation with the hemolytic antibody. Once amounts of complement and sensitized erythrocytes appropriate for easy detection of lysis are determined, the reagents can be used to measure the interaction of an antibody–antigen pair (4). The antibody and antigen to be tested are incubated for a set time in the presence of complement. Standards containing known amounts of antibody and antigen are incubated simultaneously. Sensitized erythrocytes are next added, and lysis is measured after further incubation, by spectrophotometric determination of hemoglobin. Diminished lysis relative to a control indicates that the antibody–antigen pair in question has formed an aggregate and induced complement reactions to occur in solution, reducing the amount of complement available to lyse cells. Results observed with the standards are used to calibrate the correspondence between antibody–antigen complex formation and lysis. Because antibody–antigen aggregates are the most active species in fixing complement, bell-shaped response curves result when percent complement fixed is plotted against added antibody or antigen. Complement fixation is most efficient at roughly equivalent amounts of antibody and antigen, whereas either component in excess will favor formation of binary and ternary complexes that do not fix complement well. Consequently, several concentrations of antigen or antibody must be tested to establish whether the observed response lies on the increasing or decreasing slope of the standard curve.

As an analytical technique, complement fixation is sensitive to nanogram quantities of antigen, but is subject to reagent variation and interference by chemical components of the sample. A notable value, however, is in study of cross-reactivity between structurally similar antigens. Proteins that differ slightly in sequence give easily measured differences in complement fixation ability when they are probed with a single antiserum. This effect is so regular that complement fixation by homologous antigens can be used to establish the evolutionary relationship between closely related species (5,6).

BIBLIOGRAPHY

1. J. Bordet and O. Gengou (1901) *Ann. Inst. Pasteur* **15**, 289–302.
2. M. M. Mayer (1961) Complement and complement fixation, in *Experimental Immunochemistry*, Thomas, Springfield, IL, pp. 133–240.
3. L. Levine and H. Van Vunakis (1967) *Meth. Enzymol.* **11**, 928–936.
4. E. Wasserman and L. Levine (1961) *J. Immunol.* **87**, 290–295.
5. E. M. Prager and A. C. Wilson (1971) *J. Biol. Chem.* **246**, 5978–5989.
6. A. B. Champion, E. M. Prager, D. Wachter, and A. C. Wilson (1974) Microcomplement fixation, in *Biochemical and Immunological Taxonomy of Animals*, C. A. Wright, ed., Academic Press, London, pp. 397–416.

COMPLEMENT SYSTEM

DAVID L. HAVILAND
RICK A. WETSEL

INTRODUCTION

After the discovery of humoral immunity, Hans Buchner found that if fresh serum possessing an antibacterial antibody was added to bacteria, the bacteria were quickly lysed. However, if the serum was heated to 56°C, the lytic capacity of the serum was lost. He also demonstrated that the loss of lytic capacity could be restored by using unheated nonimmune serum in conjunction with heated immune serum. These studies were extended by others, including Jules Bordet, who concluded that serum contained two components necessary for cellular lysis. These two components are heat-stable **antibodies** and a heat-labile component that "complements" the lytic function of antibodies. He reasoned that antibodies had two binding sites, one for **antigen** and the other for heat-labile substance that was given the name ***complement***.

Complement is now known not to be a single component, but it consists of at least 20 chemically and immunologically distinct plasma proteins capable of interacting with one another in a highly regulated manner to provide at least four main biological functions. First, ***cytolysis*** is mediated by the association, or polymerization, of specifically activated complement components on the surface of targets cells. These components form pores that disrupt the integrity of the lipid membrane, and the cell is killed by osmotic lysis. Second, antibody and antigen combine to form ***immune complexes*** that, unless removed, can result in damage of body tissues. The binding of complement proteins prevent the damage from immune complexes by mediating their solubilization and clearance. Third, the ***opsinization*** of foreign particles is mediated by the binding of complement proteins (opsonins). Phagocytic leukocytes bear receptors for these complement proteins, so that opsinized particles are cleared from the body by **phagocytosis**. Finally, through the activation of complement, **proteolytic** fragments are released whose function are to ***mediate inflammation***.

These fragments, or "anaphylatoxins," can act on several target cells such as neutrophils, smooth muscle, and vascular endothelium, as well as organ systems of the body.

The complement proteins are normally present in the circulation and are produced primarily by the liver and other extrahepatic sources, such as **macrophages** and fibroblasts. Some of the components are produced as functionally inactive **proteinases** that are activated only when proteolytically cleaved themselves by previously activated complement proteins. Complement activation occurs only at localized sites under specific conditions. The binding of specific antibody to antigen can initiate complement activation through what is called the ***classical pathway***. Additionally, in the absence of antibody, some complement components are directly activated by binding to the surfaces of infectious agents, such as bacteria, **fungi**, and **viruses**, through what is called the **alternative pathway**.

OVERVIEW OF THE COMPLEMENT SYSTEM

Individual proteins of the complement system are present in the serum as functionally inactive molecules. The individual precursor proteins are designated numerically, such as C1, C2, up to C9. Other proteins involved with the complement system have retained their common, or trivial, names such as properdin, factor B, and factor H. Individual components must be activated sequentially and specifically for the complement cascade to progress and mediate inflammation and lysis of foreign organisms. The activation of complement is a dynamic process that allows the proteins of the system to interact functionally and is not a static singular event.

An overview of the operation of the complement system is shown in Figure 1, and biochemical and physical information of the proteins involved in the complement system are listed in Table 1. A protein called C3 is the central component of the complement system. The alternative and classical pathways are two parallel, but entirely independent, mechanisms that lead to the formation of C3 ***convertases*** whose function is to cleave C3 into C3a and C3b. In the classical pathway, initial events involve binding of specific antibody to antigen. This antigen–antibody complex binds complement C1 and sequentially activates complement proteins C4 and C2, leading to the formation of the C4b2b complex that functions as the ***classical pathway C3 convertase***. In contrast, activation of complement via the alternative pathway occurs independent of antigen–antibody complexes. Instead, alternative pathway activation is dependent on activators that allow the formation and deposition of the C3bBb complex (**alternative pathway C3 convertase**) on their surfaces. Activators of the alternative pathway include many bacteria, fungi, and viruses. Both classical and alternative convertases can cleave additional C3, generating more C3b (see Fig. 1). The binding of additional C3b to these C3 convertases changes them conformationally to ***C5 convertases***, which specifically cleave C5. Once cleaved, both the alternative and classical pathways share the same terminal steps. These steps do not involve the proteolytic cleavage of additional components but instead involve the sequential binding of components C6, C7, C8, and C9. This terminal assembly leads to the formation of the ***membrane attack complex*** (**MAC**), resulting in the osmotic lysis or cytolysis of bacteria or other affected cells.

During activation, through either the classical or alternative pathway, various cleavage products mediate a variety of biological functions. As discussed, cytolysis is mediated through the membrane attack complex. Opsinization and subsequent phagocytosis by leukocytes is mediated primarily through C3b, a cleavage fragment of C3 that binds receptors for C3b. Cleavage products of C3, C4, and C5 (C3a, C4a, and C5a anaphylatoxins) mediate inflammatory events and the recruitment and activation of leukocytes.

ASSEMBLY AND ACTIVATION OF THE CLASSICAL PATHWAY

The classical pathway is activated by antigen–antibody complexes or aggregated immunoglobulin acting on complement

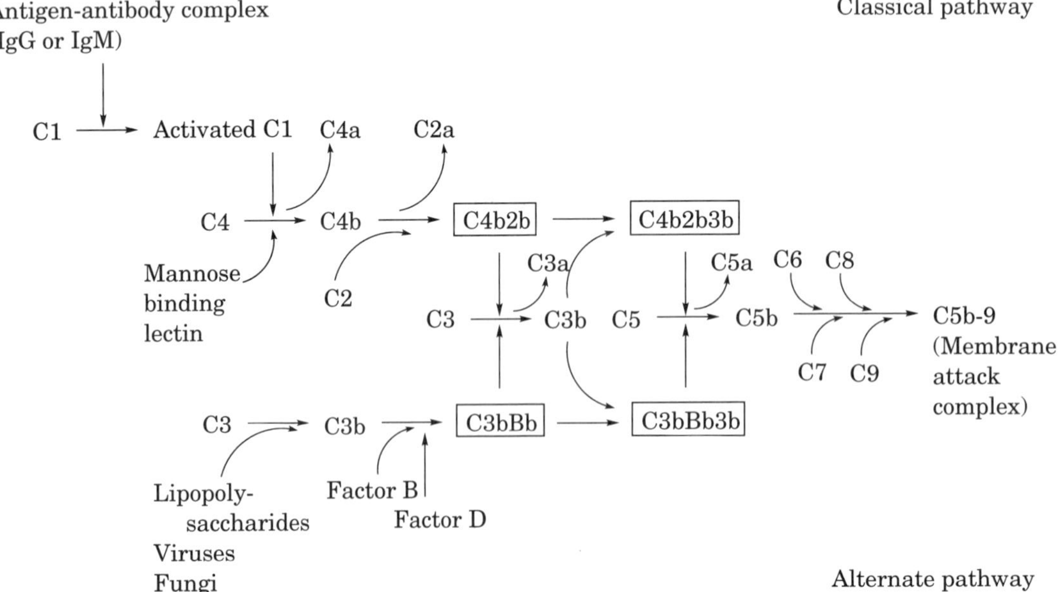

Figure 1. Schematic representation of the alternative and classical activation of the complement pathway leading to the assembly of the membrane attack complex.

Table 1. Protein Components of the Classical and Alternative Pathways[a]

Component	Serum Concentration, μg/mL	Molecular Weight	Number of Polypeptide Chains	mRNA[b] Size, kb	Chromosome Location	Gene[b] Size, kb
			Classical Pathway			
C1q	75	410,000	6 of A at 24,000	0.52	1	2.5
			6 of B at 23,000	1	1	2.6
			6 of C at 22,000	1.6	1	3.2
C1r	34	85,000	Single chain	2	12	10.5[c]
C1s	30	85,000	Single chain	2	12	10.5[c]
C4	450	210,000	α-Chain, 93,000; β-chain, 75,000; γ-chain, 33,000	5.3	6	16
C2	25	95,000	Single chain	2.9	6	18
C3	1500	195,000	α-Chain, 110,000; β-chain, 75,000	5.2	19	41
C5	75	180,000	α-Chain, 115,000; β-chain, 75,000	5.5	9	80
C6	60	128,000	Single chain		5	85
C7	60	121,000	Single chain	3.9	5	78
C8	80	150,000	α-Chain, 64,000;	2.5	1	70
			β-chain, 68,000;	2.6	1	40
			γ-chain, 22,000	1	9	1.8
C9	58	79,000	Single chain	2.4	5	80
Mannose binding Lectin	1	~ 600,000	Multimer of 32,000	3.5	10	7
			Alternative Pathway			
Properdin	25	220,000	4 at 56,000	1.6	X	6.4
Factor B	225	100,000	Single chain	2.9	6	6
Factor D	1	25,000	Single chain	1	4	2.5[d]
			Inhibitors—Soluble			
Factor I	34	105,000	46,000	2.4	4	63
Factor H	500	150,000	Single chain	4.4	1	7
C1 inhibitor	275	105,000	Single chain	1.8	11	17
C4 binding protein	150	560,000	6, α-Chain, 70,000; 1,	2.5	1	40
			β-chain, 45,000	1	1	10
S protein	500	83,000	Single chain	1.6	17	5.3
			Membrane—Inhibitors and Receptors			
DAF (CD55)		70,000	Single chain	3.1	1	40
MCP (CD46)		58,000–63,000	Single chain	4.2	1	43
CR1 (CD35)		190,000–280,000	Single chain	7.3–13	1	130–160
CR2 (CD21)		140,000	Single chain	5	1	20
CR3 (CD11b/CD18)		260,000	α-Chain, 165,000;	4.1	16	55
			β-chain, 95,000	3	21	32
CR4 (CD11c/CD18)		245,000	α-Chain, 150,000;	4.7	16	25
			β-chain, 95,000	3	21	32
MIRL (CD59)		18,000	Single chain	0.6–6.0	1	27
C5aR (CD88)		42,000	Single chain[e]	3.0	19	9
C3aR		50,000	Single chain[e]	3.0	12	8

[a] All data for proteins listed are for human complement components.
[b] Approximate size.
[c] C1r size estimated from proximity and homology to C1s.
[d] Complete factor D cDNA hybridizes within a 2.5-kb genomic fragment.
[e] Single chain with 7-transmembrane domains.

component C1 (Table 1). C1 consists of three distinct protein molecules, C1q, C1r, and C1s, which are held together by Ca^{2+}-dependent bonds. C1 is present in serum as a firm C1q-C1r-C1s complex, while individual C1 components are found only in pathologic conditions. C1 contains one molecule of C1q and two molecules of both C1r and C1s.

A molecule of C1q is comprised of 18 **polypeptide chains** of three distinct types (A, B, and C) which are twisted together comprising one subunit. Each C1q molecule is then made up of such six subunits connected together; once assembled, its structure is similar to that of **collagen**. The C1q molecule bears the sites that enable the entire C1 molecule to bind the Fc region of **IgM** and **IgG** immunoglobulin molecules. A single C1 molecule can bind about six IgG molecules. After binding to Ag-Ab complexes, C1q undergoes a conformational change that causes C1r to activate itself by a limited self-cleavage.

Activated C1r cleaves a single **peptide bond** in C1s, which then acquires enzymatic activity of its own. This newly generated enzyme is a serine esterase type and mediates the cleavage of component C4. With the activation of the C1s enzyme, the initial process is complete, and the earlier reactants, including antibody, antigen, C1q, and C1r, are no longer needed.

C4 is composed of three nonidentical polypeptide chains and is synthesized as a single-chain precursor in a beta–alpha–gamma orientation. C1 cleaves a single bond on the C4 alpha chain leading to the production of C4a and C4b. This cleavage leads to the formation of a labile binding site in the larger fragment of C4b, which enables it to bind to the activating surface (activator). The C4b alpha chain, like C3, contains an internal ***thioester*** bond formed between a **glutamic acid** and **cysteine** residue. Cleavage of the alpha chain of C4 is followed by stress-induced hydrolysis of the thioester bond. This permits the reactive acyl group of the glutamyl residue to form a covalent bond with a reactive hydroxyl or **amino group** on the surface of the activator.

C2 cleavage by C1s also generates a labile binding site of unknown chemical composition in the larger C2b fragment, which allows it to bind to C4b. Mg^{2+} ions are required for the formation of the C4b2b complex. Formation of the C4b2b complex is not very efficient, as the majority of C2 and C4 molecules entering into this reaction lose their labile binding sites before achieving union with membranes or with each other and diffuse away as inactive reaction products. The newly formed C4b2b is a proteolytic enzyme that assumes the role of continuing the complement reaction cascade, so earlier activating components are no longer required. The C4b2b complex is also called the **classical C3 convertase**, which acts to cleave C3.

The substrate for C4b2b is C3, which is synthesized as a single chain of beta–alpha orientation that is processed after translation. The larger, alpha chain, is cleaved at a single site located near the amino terminus. The smaller resulting fragment C3a (9000 M_r (relative molecular mass) is a biologically potent peptide that mediates inflammation and will be discussed later. A labile binding site is generated in the larger fragment C3b, which enables the molecule to attach to membranes at sites near, but distinct from, those utilized by antibody and C4b2b. Often described as the "lynchpin," C3 is the precursor of several biologically active fragments that function by association with the other proteins of the complement system leading to lysis of the target cell cell via either classical or alternative pathways. C3 has numerous distinct binding sites, one of which is the thioester domain that allows the molecule to bind covalently to target sites and particles, such as immune complexes and membrane surfaces.

The chemical site of the C3 thioester has the sequence Gly-Cys-Gly-Glu-Glu-Asn with the Cys and second Glu residues joined by a thioester bond, specifically a β-cysteinyl–γ–glutaminyl thioester bond. This thioester bond is present in the C3d domain of the alpha chain. With the cleavage of C3 into C3a and C3b, the thioester undergoes a stress-mediated hyrdolysis, and the reactive acyl group of the glutamyl residue forms a covalent bond with a reactive hydroxyl or amino group on the activator surface. A major amount of reactive C3 fails to achieve binding with activators as most of these reactive thioesters have reacted with water.

The attachment of C3b to membranes in the vicinity of C4b2b molecules leads to the generation of the last enzyme of the classical pathway, C4b2b3b, **or the classical C5 convertase**. To this end, the classical pathway has finished the initial steps of activation. From here, the C5 convertase acts on C5 to initiate steps in the forming of the MAC.

ACTIVATION OF THE CLASSICAL PATHWAY

The activation of the classical pathway can occur by one of two ways: activation through the use of immunoglobulin or through nonimmunoglobulin activation by cleavage of C1. Immunoglobulin activation of the classical pathway involves the use of antibodies belonging to subclasses IgG_1, IgG_2, and IgG_3, as well as IgM, are capable of initiating the classical pathway. Immunoglobulins IgG_4, IgA, IgD, and IgE are inactive in this regard. Among the three listed above, IgG_3 is the most active in interacting with C1, then IgG_1 and IgG_2. Activation occurs when the first component C1 binds to a site in the Fc region of the IgG or IgM. Only one molecule of bound IgM is capable of initiating the activation of the classical pathway; however, it is estimated that six molecules of IgG are required.

Nonimmunologic activation of the classical pathway can be accomplished by diverse substances such as DNA, certain viruses, and **trypsin**-like enzymes, by direct proteolytic attack on the C1 molecule. Of particular interest are **lectins** that also act as a nonimmunological activator of the classical pathway. Mannose binding lectin (MBL) is a serum protein found in all mammals and is regulated as an acute-phase protein. MBL is capable of binding carbohydrate moities present on many microorganisms and subsequently capable of activating the classical pathway through C4 without the need for specific antibody or C1q.

ALTERNATIVE COMPLEMENT PATHWAY OR PROPERDIN PATHWAY

The alternative pathway may be activated immunologically by aggregates of human IgA and by certain complex polysaccharides, fungi, viruses, bacterial lipopolysaccharides, and trypsin-like enzymes. This system was originally described as the properdin system, a group of proteins involved in resistance to infection, but distinct from complement. The proteins of the alternative pathway, C3, factor B, factor D, and properdin (factor P) perform the functions of activation, recognition, and amplification of the pathway, resulting in the formation of the activator bound C3/C5 convertase. The alternate pathway system was found to be involved in the destruction of certain bacteria, neutralization of some viruses, and red blood cell of (RBC) lysis from patients with paroxysmal nocturnal hemoglobinuria.

ACTIVATION OF THE ALTERNATE PATHWAY

Alternate pathway activation initially proceeds in a manner different from that for the classical pathway. The exact mechanism by which the first C3b molecule is produced is still controversial. However, it is clear that antibody is not required since mixtures of only highly purified alternative pathway proteins behave as well as serum. The ultimate aim of the proteins, C3, factor B, factor D, and properdin is the initiation, recognition, and amplification of the pathway that results in the formation of the activator bound **C3/C5 convertases**. The first requirement for activation is the presence of C3, which participates in the initiation and amplification of the pathway. It is generally accepted that activated C3 (but not-yet cleaved) [C3*] reacts with proenzyme B, and this complex is then cleaved by factor D, yielding two fragments, Ba and Bb. Bb then attaches to C3b forming the alternative pathway C3 convertase. The C3 convertase is able to cleave additional C3 into C3a and C3b. While most C3b remains in the fluid phase, some binds to various cell surfaces. Thus C3b is continuously generated and is deposited on the surface of the activator. The alternative C3 convertase is unstable and decays rapidly unless another alternative pathway member, **properdin**, binds to the C3 convertase and stabilizes it. Properdin binding of C3bBb extends its functional half-life eightfold. The function of the C3 convertase is to produce enough meta-stable C3 to deposit on the surface of surrounding particles. Surface-bound C3b then interacts with factor B and is activated by D to form more C3bBb. This enzyme is capable of cleaving large amounts of C3, and forms more C3bBb (C3bBb3b); thus, a positive feedback mechanism is formed that amplifies the initial stimulus. Many of the C3b molecules bind to the surface of the activator in close proximity to these enzymes, forming C3bBb3b or the alternative pathway C5 convertase. Like its counterpart in the classical pathway (C4b2b3b), the function of C3bBb3b is to proteolytically cleave C5 and initiate the assembly of the membrane attack complex.

THE MEMBRANE ATTACK COMPLEX: C5–C9 REACTION

The terminal portion of the complement sequence is known as the MAC (complex). C5–C9 must become membrane bound for damage to occur. This complex may attach to a cell bearing the activation enzymes of the classic or alternate pathways, or it may attach to a bystander cell. The MAC is initiated by the cleavage of C5 by the alternative and classical C5 convertases.

Similar to C3, pro-C5 is secreted in a beta–alpha orientation prior to **post translational modification**. C5 shares sequence **homology** with C3 and C4, including the domain corresponding to the thioester region; however, C5 lacks the essential cysteine and glutamic acid residues necessary for thioester formation. Assembly of the MAC initiates with the cleavage of C5. Cleavage at residue 74 of the alpha chain, results in the generation of C5a anaphylatoxin and C5b (Table 1), where C5b can then bind C6. C6 is a single chain peptide that, when bound to C5b, remains loosely attached to the membrane until the addition of C7. Once a single C7 has bound the C5b6, the resulting complex is highly lipophilic and inserts into the lipid bilayer. When inserted, the C5b67 complex serves as a high affinity integral membrane receptor for a single C8 molecule. C8 is composed of three nonidentical polypeptide chains and has an interesting structure in that the alpha and gamma chains are covalently linked by one or more disulfide bonds and the beta chain is joined to the alpha–gamma chains by noncovalent forces. The C5b-8 becomes stably attached to the membrane by insertion of the C8 gamma chain into the lipid bilayer. With this early "MAC" formed, cellular leakage can occur at this stage. The cytolytic process is greatly accelerated by the attachment of C9, the final complement component. The C9 monomer polymerizes at the site of the C5b-8 complex. As few as 4 C9 molecules can result in lysis for many microorganisms. However, when the MAC contains as many 12 to 15 C9 molecules, the "poly-C9" forms a pore in the membrane and permits passive exchange of cytoplasmic contents with the extracellular media with subsequent osmotic lysis. The pores have an internal diameter of about 110 Å and appear similar in structure to those formed by lytic T cells when the related protein **perforin** is deposited on target cells.

CONTROL MECHANISMS OF THE COMPLEMENT SYSTEM

Given the lytic potential of the complement system, uncontrolled activation can lead to the formation of the MAC on normal tissues and significant generation of C3a, C4a, and C5a anaphylatoxins. Multiple fluid phase and membrane bound proteins tightly regulate both the alternative and classical pathways by acting at specific steps during activation. In addition, the uncontrolled activation of the complement cascade is also prevented by the necessity of specific activators, such as antibody–antigen complexes, as well as the lability of the activated complement components such as C3b. Deficiencies in regulatory proteins result in unregulated complement activation and are apparent in various disease conditions. An overview of the regulation complement pathway activation by both membrane-bound and soluble inhibitors are shown in Figure 2. The biochemical and physical information of these inhibitors are listed in Table 1.

CLASSICAL PATHWAY INACTIVATORS

A protein celled ***C1 inhibitor*** (**C1INH**) acts to inhibit the initiation of the classical pathway. C1INH is a member of the serine protease inhibitor family and acts by blocking C1r and C1s of the C1 complex. As with other serpins, such as α_1-antitrypsin, C1INH uses a "bait" sequence that mimics the normal substrates of C1r and C1s. However, when C1INH is cleaved by C1r or C1s, it forms a covalent ester linkage that inhibits cleavage of C4 and C2, thus inhibiting the progression of the classical pathway. Most of the C1 in the serum is already bound to C1INH. On specific binding of C1 to antigen–antibody complexes, C1INH is released allowing classical pathway activation to proceed.

Other means to control the classical pathway involve interfering with the formation of the C3 convertase. Such inhibition can be mediated either through interfering with the assembly of the convertase or in promoting the dissociation of the convertase. Three proteins called ***C4-binding protein*** (***C4bp***), ***complement receptor type I*** (***CR1***) (also known as CD35), and

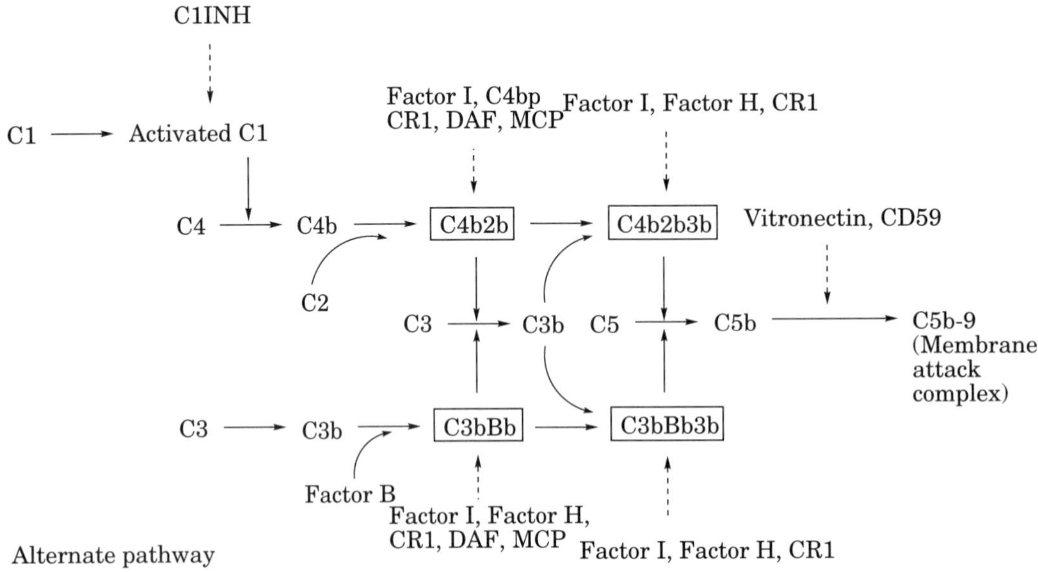

Figure 2. Regulation of classical and alternative complement pathway activation by membrane and soluble inhibitors. Principal components and C3/C5 convertases are retained from Figure 1. Regulators of complement activation are shown in italics.

decay accelerating factor (***DAF***) (also known as CD55) function as inhibitors of the classical pathway by binding to C4b. In doing so, they compete with C2 for C4b and thus inhibit the formation of the classical C3 convertase. In binding C4b, they also promote the dissociation of the C3 convertase. All three proteins are structurally similar and are homologous members of a family of proteins called the regulators of complement activity (RCAs).

Another way that the C3 convertase can be inhibited is by the cleavage of C4b by a protein called **factor I**. Factor I is a serine esterase that cleaves C4b into two additional fragments: C4c and C4d. C4d remains bound to the original activating surface; however, it is unable to contribute to the formation of the classical C3 convertase. Cleavage of C4b by factor I can only occur in the presence of a "cofactor." The proteins that can serve a cofactor for factor I cleavage include CR1, C4b, and membrane cofactor protein (MCP), or CD46. MCP is also a member of the RCA family but does not possess decay accelerating activity.

ALTERNATIVE PATHWAY INACTIVATORS

Regulation of the alternative pathway is also accomplished by several circulating and membrane bound proteins, many of which also act on the classical pathway. In analogous fashion, the inhibition of convertase activity through either decay acceleration or by providing cofactor activity for factor I are the major ways in which this is accomplished.

Factor H is a soluble serum protein that possesses decay-accelerating activity that is specific for the Bb fragment of the alternative pathway. By competitively binding Bb, the formation of the alternative C3 covertase is impaired. Factor H is also a member of the RCA family and in binding Bb can also promote the dissassembly of the alternative C3 convertase.

In addition to its role in inhibiting the classical pathway activation, factor I can also cleave C3b using factor H, MCP, and CR1 as cofactors. The resulting fragment of C3b (iC3b) is unable to participate in the formation of the alternative C3 convertase.

INACTIVATORS OF THE MEMBRANE ATTACK COMPLEX

With the alternative and classical pathways activated, the potential for normal or bystander host cells to be subjected to lysis is significant. Excessive cell lysis is normally prevented by a number of proteins that act to inhibit the formation of the membrane attack complex.

A membrane bound protein that is capable of inactivating the membrane attack complex is the ***membrane inhibitor of reactive lysis*** (***MIRL***), more commonly known as ***CD59***. CD59 is broadly expressed among different cell types and is probably the most important molecule in minimizing damage to bystander cells. It is thought that CD59 acts by binding C8 and C9 and thus inhibits the polymerization of C9 as well as the insertion of the MAC into the membrane.

As discussed above, when C7 binds C5b6, the resulting complex is lipophilic and it inserts into the membrane. This process of membrane insertion is inhibited by **S protein**, also known as **vitronectin**, a protein related to fibronectin and laminin. The S protein is thought to function by binding the C5b67 complex before it inserts into the membrane.

COMPLEMENT RECEPTORS AND BIOLOGICAL CONSEQUENCES OF COMPLEMENT ACTIVATION

Through the activation of the complement cascade, through either the classical or alternative pathways, numerous complement derived fragments are generated. Many of these fragments possess biological activities that are mediated through the use of specific receptors.

In addition to regulating the activation of complement, ***CR1 (CD35)*** serves as a high affinity receptor for C3b and C4b. CR1 functions as a receptor for C3b- and C4b-coated particles wherein macrophages ingest and remove these coated particles. Another receptor, ***CR2 (CD21)***, is expressed primarily on B cells, dendritic cells, and epithelial cells. CR2 is known to specifically bind iC3b and C3dg—both factor I–generated cleavage products. In addition, CR2 is also known as the Epstein–Barr virus receptor. ***Mac-1*** or ***CR3 (CD11b/CD18)*** is related to the integrin family and is a specific receptor for the factor I cleavage product of C3b, iC3b. CR3 is expressed on many different cells derived from the bone marrow including neutrophils, mononuclear phagocytes, mast cells and natural killer (NK) cells. The primary role of CR3 is the phagocytosis and clearance of iC3b-coated particles. **CR4 (CD11c/CD18)**, like CR3, is also a member of the integrin family and also binds iC3b as well as C3dg.

The cleavage of C3, C4, and C5 results in the production of **C3a**, **C4a**, and **C5a anaphylatoxins**. These hormone-like peptides mediate a variety of cellular and biochemical responses from leukocytes, including the generation of bacteriocidal superoxide radicals, proteolytic enzyme release from intracellular granules, cellular aggregation, smooth muscle contraction, and phagocytosis. In addition, C5a induces chemotaxis, or the migration, of leukocytes into areas of complement activation. The induction of chemotaxis was thought to be unique for C5a; however, recent findings suggest that C3a may be chemotactic for eosinophils, a cell whose function is mediating allergic reactions. These anaphylatoxins mediate their effects through specific receptors that are coupled to GTP-binding proteins and whose signal transduction can be abrogated by pertussis toxin. The C3a and C4a receptors have been found to be expressed on mast cells, basophils, lymphocytes, and smooth muscle cells. The C5a receptor has been found on mast cells, basophils, neutrophils, monocytes/macrophages, endothelial cells, smooth muscle, liver parenchyma, and epithelial lining of lung as well as astrocytes and microglia of the central nervous system. Of the three anaphylatoxins, the most potent is C5a in mediating biological effects. C3a is approximately 20-fold less potent, while C4a is about 2500-fold less potent than C5a. C5a has been shown to stimulate the release of **tumor necrosis factor** (TNF) from mast cells as well as stimulating P-selectin (CD62P) expression on vascular endothelium to promote neutrophil binding.

COMPLEMENT GENES AND RELATIONSHIPS

On the basis of sequence homologies, many of the complement proteins can be grouped together as members of families. Members of these families share structural and/or functional similarities by which relationships can be assessed.

One of the more defined families is that of the regulators of complement activity (RCAs) as previously discussed. Factor H, CR1, CR2, DAF, C4bp, and MCP are all members of the RCA, and all share the ability to bind both C3b and C4b. At the peptide level, all members of this family possess multiple, tandemly arranged repeated structures called short consensus repeats (SCRs). Each SCR is 65–70 amino acids long with 11–14 conserved amino acid residues. The composition and number of SCRs each of these members contain vary, but structurally and functionally they form the foundation by which this family is based. The genes encoding H, CR1, CR2, DAF, C4bp, and MCP are found within an 800-kb genomic segment on the long arm of chromosome 1.

Another group of complement genes that are linked together are C2, factor B, and C4 that all map within the major histocompatibility complex of both humans and mice. In the human, these complement genes, often called the Class III genes, map between the Class II HLA-DR and Class I B loci of chromosome 6. Similar to the Class I and Class II genes, the Class III genes are polymorphic as multiple alleles exist for each of these genes.

Complement components C3, C4, and C5 constitute a structurally homologous family of proteins that includes **α_2-macroglobulin** and pregnancy zonal protein. All except C5 are characterized by the presence of an interal thioester bond allowing these proteins to covalently bind with cell surfaces or other proteins. However, unlike the RCA family, C3, C4, and C5 reside on separate chromosomes (Table 1).

PATHOLOGIES RELATED TO THE COMPLEMENT SYSTEM

The complement cascade is a potent and powerful system whose function is the destruction of target cells and infectious agents. Unwanted deposition of complement as well as the production of significant quantities of inflammatory mediators can result in substantial damage to normal healthy tissues. The regulatory mechanism of the MAC and alternative and classical pathways establishes a fine balance between activation and inhibition so as to protect autologous cells but allow the destruction of foreign agents. Deficiencies in nearly any component of the complement system compromise the health of an individual by tipping the balance one way or another. In a general sense, complement deficiencies can be due to the absence of a component due to structural defect in the gene, or the production of a component that is incapable of functioning in its specific role in the cascade. Table 2 summarizes complement deficiencies, the resulting abnormalities, and associated pathologies.

Deficiencies in alternative and classical pathway components usually manifest clinically in recurrent bacterial infections. The most serious deficiencies are those in C3. Given its role in opsinization, phagocytosis, and lysis of bacteria, homozygous deficiencies in C3 often prove fatal. In contrast, C9 deficient individuals cannot effectively generate MAC formation, yet these patients have minimal or no associated pathology as the addition of C8 to the C5b-7 complex can result in osmotic lysis of target cells. Individuals deficient in the early classical pathway components (C1, C2, and C4) are impaired in their ability to solubilize and clear immune complexes, resulting in local inflammation associated with autoimmune diseases such as ***systemic lupus erythematosus (SLE)***.

Patients that are deficient in the terminal complement components (C5 through C9) have impaired ability to assemble the MAC. Of particular interest is that these patients seem to have a propensity for Neisseria bacterial infections suggesting that cytolysis may be the major defense mechanism against these bacteria.

Deficiencies in the soluble and membrane bound regulatory components of complement result in abnormal activation

Table 2. Complement Deficiencies and Associated Clinical Abnormalities

Component	Biological Defect	Associated Disease
	Classical Pathway	
C1q, C1r, C1s	Defective classical pathway	Systemic lupus erythematosis (SLE), bacterial infections
C4, C2	Defective classical pathway	SLE, bacterial infections, glomerulonephritis
C3	Defective classical and alternative pathways	Bacterial infections, glomerulonephritis
	Alternative Pathway	
Properdin, factor D	Defective alternative pathway	Bacterial infections
	Membrane Attack Complex	
C5, C6, C7, C8	Defective MAC assembly	Recurrent neisserial infections
	Inhibitors	
Factor I, factor H	Deregulation of complement activation Consumption of C3	Bacterial infections
CR3 (CD11b/CD18)	Impaired opsinization	Bacterial infections
C1 inhibitor	Deregulation of classical pathway	Hereditary angioneurotic edema (HANE)
DAF (CD55)	Deregulated C3 convertase	Paroxysmal nocturnal hemoglobinuria
MIRL (CD59)	Impaired MAC regulation	Paroxysmal nocturnal hemoglobinuria

and deposition of complement. Defects in C1INH results in a condition called ***hereditary angioneurotic edema*** (***HANE***), in which the intermittent accumulation of edema (swelling) in the skin and mucosa occurs. The exact mechanism of how the edema occurs is not known; however, without C1INH to inhibit non-specific C1 activation, C2 and C4 are readily cleaved and their by products are implicated. Deficiencies in the expression of phosphytidylinositol-linked membrane proteins result in a condition called ***paroxysmal nocturnal hemoglobinuria*** (*PNH*), where patients suffer from recurrent bouts of intravascular hemolysis. As DAF and CD59 are PI-linked, they also serve to inhibit C3 convertase formation. However, erythrocytes are especially sensitive to lysis as they do not possess membrane bound form of CD59 as other cell types do.

SUMMARY

The proteins of the complement system consists of at least 20 chemically and immunologically distinct plasma proteins that can dynamically interact with one another in a highly regulated manner. The complement system can be activated by using either of two converging initiation pathways: the classical pathway, which is activated by antibody–antigen complexes; and the alternative pathway, which is activated by the surface of invading organisms. This interaction of proteins results in (*1*) the lysis of target cells and bacteria, (*2*) the binding and clearance of immune complexes throughout the body, (*3*) the binding of complement based proteins to foreign particles and subsequently cleared by phagocytosis, and (*4*) the generation of inflammatory mediators to enhance the humoral and cellular immune response. The activation of complement is tightly regulated so as to protect host cells from nonspecific attack deposition while allowing the destruction of foreign agents. Deficiencies in complement components of either pathway or in regulatory components are associated with recurrent bacterial infections and autoimmune diseases.

Acknowledgments

This is publication 133-IMM from the Institute of Molecular Medicine for the Prevention of Human Diseases, University of Texas-Houston Health Science Center. This work was supported by GM56050 (DLH) and AI25011 (RAW). This article is dedicated to the memory of Dr. Hans Müller-Eberhard.

Suggestions for Further Reading

A. K. Abbas, A. H. Lichtman, and J. S. Pober (1994) *Cellular and Molecular Immunology*, 2nd ed., Saunders, Philadelphia.

H. Müller-Eberhard (1986) The membrane attack complex, *Annu. Rev. Immunol.* **4**, 503–528.

H. Müller-Eberhard (1988) Molecular organization and function of the complement system, *Annu. Rev. Biochem.* **57**, 321–347.

R. A. Wetsel and H. R. Colten (1990) in *Inheritance of Kidney and Urinary Tract Diseases*, A. Spitzer and E. D. Avner, eds., Kluwer Academic, Chapter 18, pp. 401–429.

G. D. Ross, ed. (1986) *Immunobiology of the Complement System*, Academic Press, San Diego, CA.

K. Rother and G. O. Till, eds. (1988) *The Complement System*, Springer-Verlag, New York.

K. Rother and U. Rother, eds. (1986) *Hereditary and Acquired Complement Deficiencies in Animals and Man, Progress in Allergy*, Vol. 39, Karger.

COMPLEMENTARY DNA (CDNA)

ANTHONY J. ROBERTSON
SAMUEL H. WILSON

The base-pair complementarity between a **gene** and its **transcript** leads to the formation of a DNA–RNA hybrid under experimental conditions that favor specific **hybridization**. Pioneering work by Gillespie and Spiegleman that made use of the complementarity between viral genomic DNA and RNA in infected cells (1) helped establish the beginnings of a new technology for the detection of specific RNA molecules hidden among an RNA background, in this case of viral RNA among host cell RNA. The potential advantages of *complementary DNA* (cDNA) as a probe for RNA detection were recognized before techniques existed to make specific cDNA conveniently for tagging genes. The great abundance in the genome of extra, nontranscribed DNA, which is a common feature of higher organisms, complicates the use of genomic DNA as a cDNA probe. Some time after the first references to cDNA, a

DNA polymerase activity that depends on the presence of an RNA **template** was observed in RNA **viruses** (2,3). This discovery, in addition to extending our knowledge of RNA virus replication intermediates, led to the identification and isolation of a previously unknown form of DNA polymerase called *reverse transcriptase* and to the identification of buffer conditions that allow some DNA-dependent DNA polymerases to use a DNA-primed RNA template to form a DNA product (4,5). Largely due to the discovery of reverse transcription and the technical advantages it provides for *in vitro* cDNA synthesis, cDNA has become a fundamental research tool for use with higher organisms.

CDNA COPIES OF MESSENGER RNA

Transcriptionally active genes represent only a relatively small fraction of the total host genomic DNA in higher organisms, only 3% in human cells, with the noncoding or nonexpressed DNA representing the majority of the genome. Reverse transcribed cDNA synthesized under carefully controlled conditions is a faithful and stable double-strand DNA copy of cellular RNA and is trimmed of excess genomic sequence. A full-length cDNA molecule made from **messenger RNA** (mRNA) contains only the protein-coding region of an expressed gene, together with the adjacent untranslated regulatory sequences; it lacks any unexpressed gene or genomic sequences not contained in the mature mRNA, such as **introns** (Fig. 1).

For research that involves mRNA, the preparation of cDNA not only provides a DNA strand complementary to RNA, but it also serves to minimize the technical problems of working directly with RNA, which is chemically unstable at alkaline pH and susceptible to notoriously stable and ubiquitous cellular ribonucleases. A further advantage of cDNA is that in many applications it is integrated into a **prokaryotic** DNA **vector to** form **recombinant DNA** capable of autonomous replication when **transformed** into a host organism with a favorable genetic background. Engineered vectors with a variety of features are available for many different methods of cDNA analysis. As recombinant DNA, cDNA may be amplified through culturing of the host cells. Because each transformed host cell receives a single recombinant vector, clonal lines of host cells differing only in the recombinant vector received are established. As a copy of total cellular mRNA, however, cDNA is a complex mixture representing the complexity of the template mRNA used to direct its synthesis. Thus, cDNA integrated into a vector, transformed into a host, and cultured represents a complex mixture or **library** of transcripts, a **cDNA library**. The use of cDNA frequently requires the selection of a single **clone** or cDNA copy from a library, since the analysis of expression from a single gene is often the desired aim. The efficient selection of a previously uncharacterized cDNA clone from an entire cDNA library poses a major technical challenge. Selection of a new cDNA is complicated by the lack of any previous information to assist in the selection process. Often, the selection might start with little information other than observations about the encoded function of the desired cDNA clone or atypical cellular characteristics that could be associated with it.

As a stable copy of RNA, a single selected cDNA clone may be transcribed through vector signals to regenerate the RNA, which may in turn be used for synthesis of the encoded protein, either within a host cell or through the use of **protein biosynthesis *in vitro*** system. The purity of RNA or protein prepared from a single cDNA clone is absolute, thus justifying the effort to prepare a cDNA library and select a desired clone. This is an alternative to the task of biochemical purification of a protein from a tissue or cell extract. A purified cDNA in a regulated protein expression system may be used to provide a continuous renewable protein source for analysis of the structure, activity, or any characteristic of the encoded protein. Also, the effect that a cDNA's expression has on a relevant **eukaryotic** cell line may provide information about the biological role of the encoded function. Thus, cDNA-directed protein expression not only provides a source of pure protein for either *in vitro* or *in vivo* analyses, but it may also provide a method of cloning (selection) of a cDNA from a cDNA library. Many selections of new cDNAs start without protein or gene sequence information, but instead rely on the detection of the function or biological effect. This process is known as direct expression or functional cloning. Any direct expression cloning method mimics the classic genetic methods of rescue or **complementation** and so could be considered genetics in the absence of mating. To describe functional cloning in a general way, an altered host cell's **phenotype** or measured activity is restored to the wild-type when the genetic material encoding the host's deficiency is supplied. Although experimental approaches that rely on classic genetic methods have produced much information about cellular processes, they depend on good genetic systems to which the genetic material can be delivered. Systems that lend themselves to an imposed manipulation and selection, generally single-cell organisms with a short generation time, are best for this purpose. A myriad of eukaryotic cell cultures that have been purposefully altered or carefully isolated from the affected organs of diseased patients and that show an atypical phenotype have been widely used as host cells for functional cloning. The applications for the functional cloning of new cDNAs from higher organisms also depend on a selectable phenotype or activity to isolate the host cell that harbors the relevant genetic material. As an example, the anchorage-dependent phenotype of nearly all primary tissue cells is not exhibited by many tumor cells that will grow in suspension (6). When the genetic material used is cDNA, the details of the selection method must be designed so that the cDNA library contains the cDNA encoding the deficient host activity, and so that the cDNA is incorporated into the host cells and expressed functionally. With expression of a cDNA library in eukaryotic host cells, the selection of a desired clone depends on acquisition of a new phenotype. In the example of the anchorage-independent phenotype exhibited by some tumor cells, the induction of the anchorage dependency allows selection for a cDNA-encoded activity related to the anchorage phenotype and possibly to the cause of tumor formation (7,8). For any selection where the desired cDNA is of an mRNA that is in low abundance, the selection will require a large-scale transformation of host cells, in order to produce even a single positive clone.

Acquiring a cDNA opens up many types of experimental approaches for the analysis of gene expression and studies of the protein encoded by the cDNA. As an example, protocols are available to target a single codon in a cDNA for **site-directed mutagenesis**. By using available methods to express the altered cDNA, the effects of the mutation on the encoded

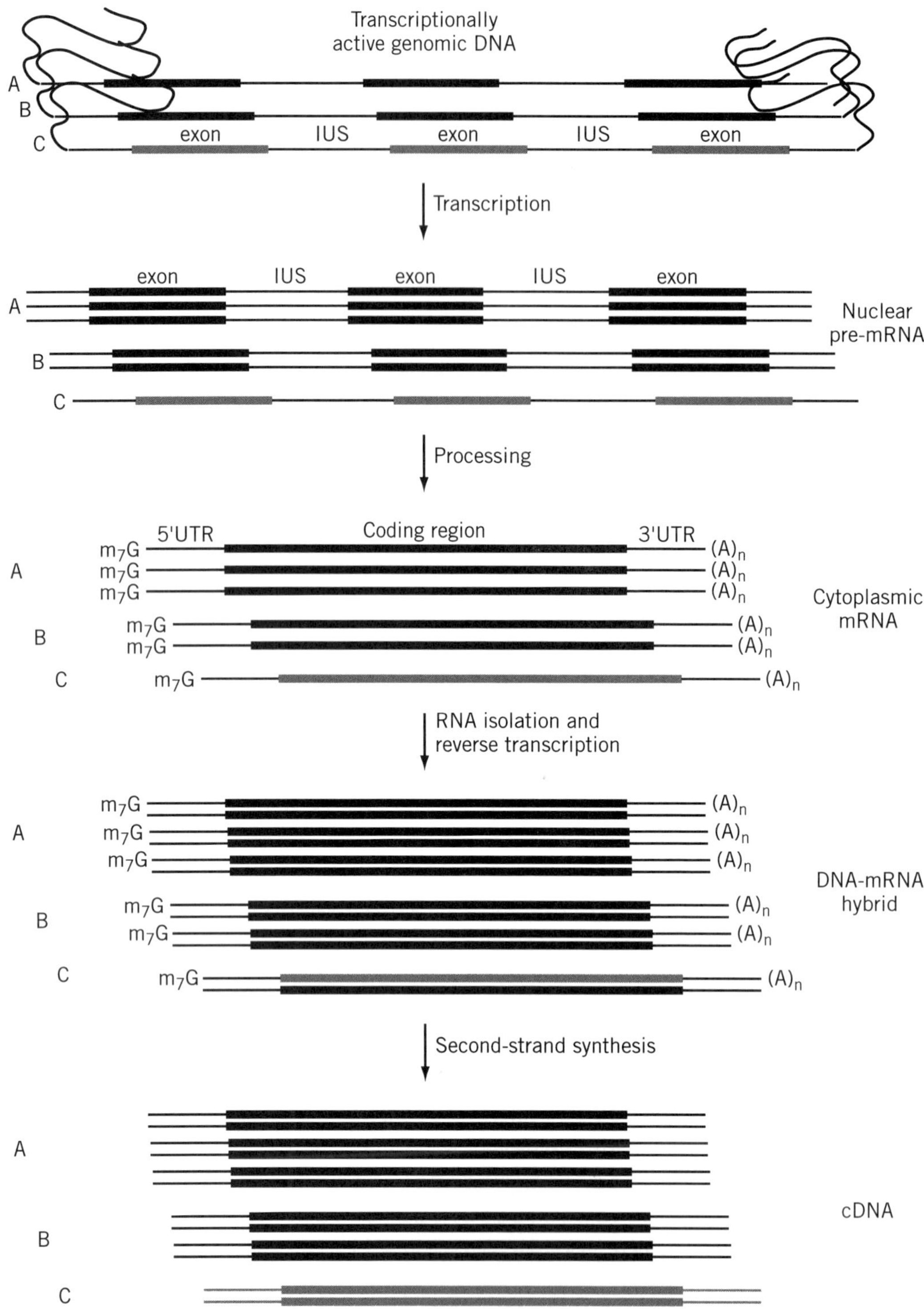

Figure 1. Schematic representation of cDNA synthesis from mRNA. Transcriptionally active genes are transcribed, resulting in pre-mRNA that is processed to the mature form. Nuclear processing includes the removal of intervening sequences (IVS, introns) and addition of a 5′ m_7G cap and 3′ poly dA tail. Different gene activities result in different mRNA abundance shown as A, B, or C. After mRNA isolation and reverse transcription, a population of DNA–mRNA hybrid molecules are formed where the DNA is anti-sense in relation to the mRNA. Second-strand synthesis produces a cDNA population in which the more active gene is represented more than the less active gene.

protein's activity may be measured, and in this way, a structure–function relation is probed.

Although the methods to prepare and make use of cDNA are advanced and commercially available, it remains a challenge to select a completely new cDNA clone from a cDNA library. This may limit the uses of cDNA technology to better characterize known cellular processes rather than uncharacterized cellular processes. The successful recovery of a novel cDNA depends on a specific selection strategy that ultimately will test each cDNA for an encoded activity. The task of individually testing each clone in a cDNA library is nearly overwhelming. If a screening method designed to test each cDNA in a collective manner cannot be designed, it forces the need for biochemical methods to purify the protein activity in order to provide enough sequence information to obtain the cDNA by more conventional methods, such as library screening with a sequence probe or selective **PCR** library amplification with degenerate oligonucleotide primers. Evolving methods of protein expression have demonstrated the potential of cDNA selection as a possible alternative to protein purification. However, functional cloning may not always be possible. An alternative approach, which can temporarily bypass the challenge presented by the need to design or apply a new isolation strategy for each new cDNA, is to randomly isolate and analyze cDNA clones first and then subsequently to identify the encoded function. The use of sequence information derived from randomly selected cDNA samples has become part of the Human Genome Project as a way of providing an aid for the physical mapping of expressed genes, as gene tags (9,10). The cDNA sequences obtained are collected into the **expressed sequence tag** (EST) database, which is a collection of nonrepetitious human cDNA sequences of randomly selected cDNAs from 250 cDNA libraries from RNA isolated from 37 distinct human organs and tissues (11). The EST information is providing genetic markers for physical gene mapping and also information about the expression pattern of the mapped genes as the cell lineage is known for each sample analyzed. Some hints about the encoded function of the mapped genes may be revealed by the expression pattern. Furthermore, the large volume of sequence information produced by random sequencing may be compared to the available sequence information of the growing number of fully sequenced genes, in order to identify the presence of conserved protein domains with established activities in a process that has come to be known as *functional genomics* (12,13). This process of extending available information about function to randomly acquired new sequence information, which necessarily requires a unified, large-scale approach, will undoubtedly contribute greatly to the identification of the encoded functions of previously unknown human genes.

PREPARATION OF CDNA

Isolation of RNA Template

The preparation of cDNA begins with the isolation of the RNA that will serve as a template to direct the cDNA synthesis. Since cDNA is representative of the RNA template, an inciteful choice of a cellular source of RNA, such as a tissue that is known or at least thought to contain the specific mRNA of interest, is important. In some cases, the RNA might be tested for the presence of an RNA of interest. In a simple case, the RNA sample could be analyzed prior to use for cDNA preparation by use of a sequence probe. This sort of analysis requires, of course, some definite sequence information about the mRNA of interest. In a more complex case, when no sequence information about the mRNA is available, but detection of the encoded protein is possible, the RNA source could be analyzed for the ability to direct synthesis of the protein. Detection of a protein may be accomplished by any known function, such as **enzymatic** activity or **ligand binding** activity. The presence of a protein in a cell does not insure the presence of the protein-encoding mRNA, as the protein could be present as a stable form from an earlier expression of an mRNA that is now degraded. On the other hand, a mRNA encoding a protein of interest could be present in a tissue, but remain untranslated. One technique that can be used to insure the presence of the mRNA of interest in an RNA preparation is to translate the total RNA in an *in vitro* translation reaction and then assay the total resulting protein product for the expected protein. The two high-activity *in vitro* translation systems in common use are those prepared from rabbit reticulocytes and wheat germ (14,15) (see **Protein biosynthesis *in vitro***).

Given that a tissue source can be identified that contains the RNA of interest, there may be a possibility of enriching for the desired mRNA. Only a small fraction of total cellular RNA is mRNA that is directed to **ribosomes** for **translation** into protein (Table 1). The majority of cellular RNAs function in other ways to aid translation, often with structural roles. Enrichment methods aim to remove the more abundant structural RNAs selectively from the mRNAs. Enrichment of **polyA** + mRNA by hybridization with immobilized oligo (dT), followed by stringent washing to remove the abundant, structural, poly A − RNAs, is a common method used to prepare the mRNA template for synthesis of cDNA representing protein-encoding mRNAs (16). To reduce contamination by mitochondrial poly A + RNA, the **mitochondria** may be removed by cell fractionation prior to total RNA extraction, although the extra time and handling of cells required could lead to some loss or degradation of RNA. Isolation of full-length mRNA of nuclear origin based on the unique m^7G **5′ cap** structure might also serve as an alternative to cell fractionation to reduce those RNA messages of mitochondrial origin often present in RNA preparations (17). Translational inhibitors have been used to stabilize ribosome-bound mRNA that is degraded during translation (18), and as selective RNA

Table 1. Cellular Distribution of RNA in HeLa Cells

Cellular Location	% of Total	Average $t_{1/2}$
Nucleus	**10**	
hnRNA	4.0	20 min
snRNA	2.5	100 hours
pre-rRNA	3.5	15 min
Cytoplasm	**85**	
mRNA	2.0	8–10 hours
rRNA	75.0	107 hours
tRNA	8.0	80 hours
Mitochondrion	**5**	
mit-mRNA	0.5	8–10 hours
mit-rRNA	3.5	107 hours
mit-tRNA	1.0	80 hours[a]

[a] *Note*: These values are approximate ones based on experimental observations made during repeated isolation of RNA from **HeLa cell** cultures.

degradation pathways are increasingly well characterized, there is an increasing promise of specific inhibitors of RNA degradation for use in preparing mRNA that normally has a relatively short half-life.

Among poly A + RNA, there are messages of low, middle, and high abundance. The process of normalizing cDNA is used to reduce the relative number of high- and middle-abundance mRNA represented in the cDNA product. In this way, relatively fewer cDNA clones need to be screened in order to isolate a low-abundance message copy from a cDNA library. Preferred removal of cDNAs representing high- or middle-copy abundance mRNA is based on the rate of strand reannealing in a denatured cDNA sample (19). The two strands of a cDNA representing a high-abundance message have a higher probability of finding each other in a mixture and anneal before those of low-abundance cDNA. When annealed or double-stranded, the cDNA is separated from unannealed or single-strand cDNA by the use of **hydroxyapatite chromatography**. After a short annealing time, the low-abundance cDNA will remain in the single-strand chromatography fractions. These fractions are collected and allowed to anneal fully for use as normalized cDNA. Alternatively, **subtractive hybridization** is a technique used for the enrichment of, eg, tissue-specific or developmental stage-specific mRNA in the preparation of subtraction cDNA libraries.

Synthesis of the cDNA First Strand

The two cDNA strands are prepared independently. In the first-strand synthesis, the use of an RNA-dependent DNA polymerase results in a mixture of cDNA–RNA hybrid molecules in which the DNA strand is complementary to the original RNA, but possibly of shorter length (Fig. 2). The reverse transcription systems in common use are based on the avian myeloblastosis reverse transcriptase (AMV-RT) prepared from purified avian myeloblastosis virus particles or on the Moloney murine leukemia virus reverse transcriptase (MMLV-RT) prepared from a recombinant *Escherichia coli* strain (20). The purified AMV-RT has a stronger endogenous RNAse H activity than the cloned MMLV-RT, thus increasing the possibility of unwanted side reactions when using AMV-RT. However, purified AMV-RT may be more stable than MMLV-RT at the elevated temperatures used to reduce premature termination caused by **secondary structure** in the RNA template. RNase H is an endoribonuclease that specifically hydrolyzes the phosphodiester bonds of RNA in a DNA–RNA hybrid to produce products with terminal groups 3′-OH and 5′-P. Both AMV-RT and MMLV-RT have a tendency to begin second-strand synthesis prematurely through a reaction in which reverse transcriptase uses DNA rather than RNA as a template by forming a hairpin structure that serves as a template-primer substrate (21). Hairpin formation may be inhibited by the addition of sodium pyrophosphate and **spermidine**, although the RNase H activity is not inhibited with these conditions. The RNase H activity present in AMV-RT may provide a better opportunity for favorable hairpin formation by leaving single-stranded regions of DNA available for self-hybridization; this side reaction is more of a problem when AMV-RT is used. Heat-stable DNA polymerase isolated from *Thermus thermophilus* is capable of reverse transcription in the presence of MnCl2, which is a common alternative to the use of reverse transcriptases (22). Each system of reverse transcription has some advantages, as will be discussed, and should be considered with a view of how the second-strand synthesis will be performed. In addition to reverse transcriptase, first-strand synthesis requires deoxyribonucleotide triphosphates, dATP, dGTP, dTTP, and dCTP (or 5-metyl dCTP which may be substituted for dCTP in order to protect the cDNA from 5-methyl dCTP-sensitive **restriction enzymes** that may be used in subsequent cloning steps). The first-strand synthesis reaction also requires primers.

Unlike **RNA polymerases**, which bind a recognition sequence in a double-strand DNA template to initiate polymerization, replicative-type DNA polymerases require primers with a free 3′-OH group and a short double-stranded sequence. Note that a consideration of the primers to be used for the first-strand synthesis may serve to direct a more representative cDNA or cDNA better suited to the research aim. Because all mRNA, except animal **histone** mRNA, terminate with a 3′ poly A tract, the use of a $p(dT)_{10-15}$ oligonucleotide as a first-strand primer ensures that the synthesis begins at the 3′ end of mRNA, even when total RNA is used as a template for reverse transcription. A disadvantage of this primer is that the 5′ end of longer mRNA may be underrepresented in the cDNA product, since the chance that processive synthesis will continue to the 5′ end decreases with product length. The mRNA 3′ terminal poly A tracts with undesirable average lengths of ≈200 nucleotides can be avoided by the use of an "anchored p(dT) oligo," which has an equal concentration of dA, dC, or dG at the 3′ end of $p(dT)_{12}$ dN. Such an oligo mixture initiates first-strand synthesis at the position immediately adjacent to the beginning (5′) of the poly A tract, with an expected reduction in the length of the poly A tract captured in the cDNA. A random primer of six oligodeoxynucleotides used in place of $p(dT)_{10-15}$ may result in a more representative first-strand synthesis, since at least some of the primers should find matching hybridization sequences near the middle and 5′ end of the RNA and thus are frequently extended to the 5′ end of the template. A random primer is a better choice for use with poly A(−)RNA, which is not complementary to $p(dT)_{10-15}$. A technical consideration for random primed first-strand synthesis is the amount or ratio of primer to template that should be used, since more than one primer for each template could lead to interference between polymerase initiating on different primers hybridized to the same RNA template molecule. This will result in premature termination and a smaller average cDNA product length in the sample. Mixtures of random-primed and oligo dT-primed first-strand synthesis products should equally represent the length of RNA.

Another choice of primer to initiate first-strand synthesis is a sequence-specific primer, a primer complementary to a known sequence within the mRNA of interest. The design of a sequence-specific primer depends, of course, on some available sequence information. A very conserved sequence common to a group of possibly related mRNAs used for the preparation of a limited cDNA library is an example of the application of a sequence-specific primer. Whichever type of primer is used for the initiation of first-strand synthesis, extra sequence at the 5′ end of a primer, such as a restriction site sequence or a second primer-binding site, should not interfere with polymerase-directed extension of the 3′ primer end. Thus, the 5′ end of the primer may serve as a "sticky end" to incorporate the cDNA into a vector or as a primer target for subsequent amplification of the cDNA by PCR.

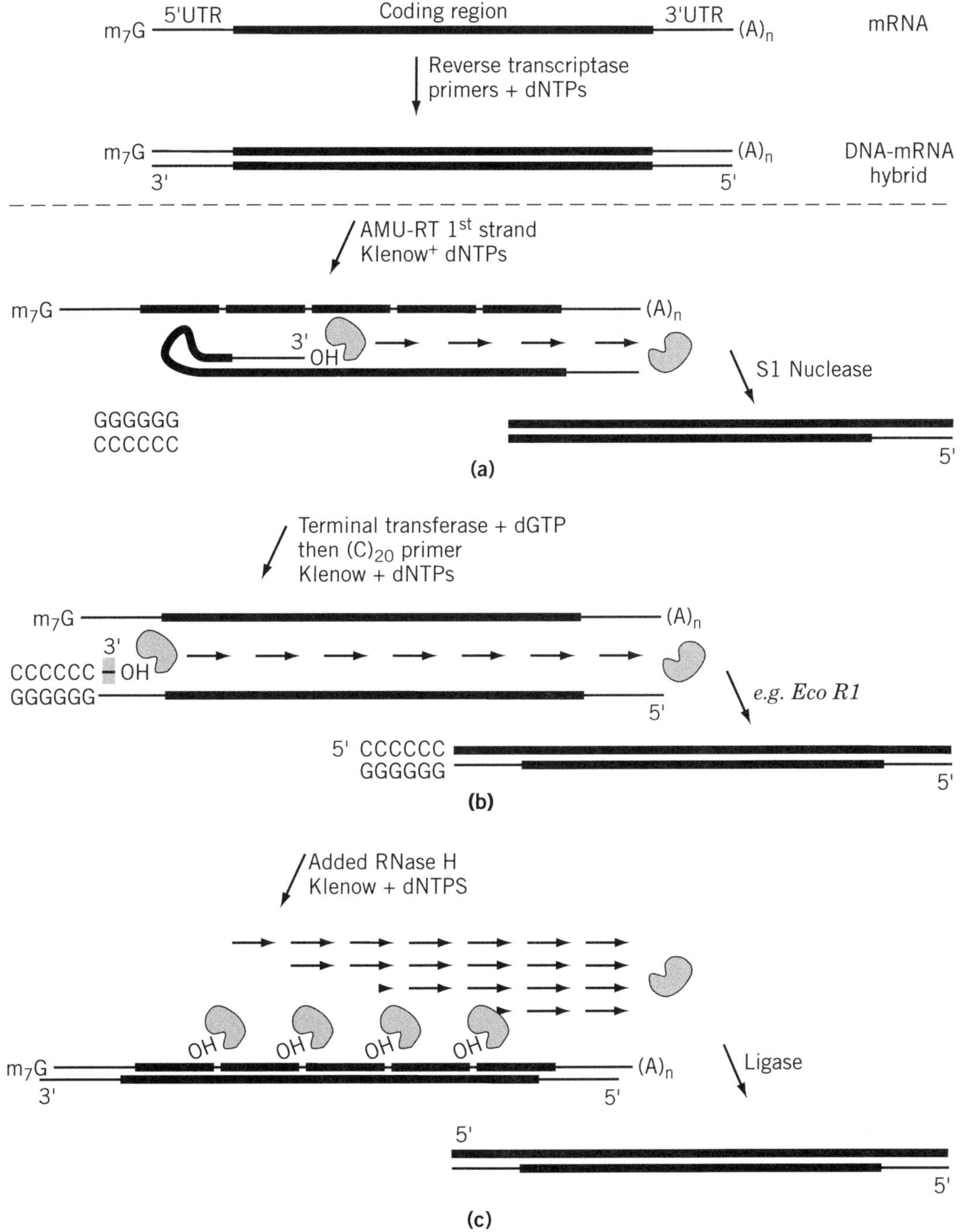

Figure 2. Three separate schemes of second-strand cDNA synthesis from the DNA–mRNA hybrid. (**a**) A classic method makes advantage of a looped back first strand as a primer for second-strand synthesis. After second-strand synthesis, S1 nuclease digestion of the single-strand loop structure results in an available free end for vector incorporation. (**b**) Terminal transferase-directed extension of the 3′ end of the first strand results in a dG stretch that provides a binding sight for an oligo dC primer to initiate second-strand synthesis. The undesired dG stretch on the first-strand oligo dT primer is removed by restriction endonuclease digestion targeted to a site present in the oligo dT primer or in the recipient vector if vector priming is used. (**c**) Added RNase H degrades the RNA in the DNA–RNA hybrid product of first-strand synthesis, resulting in a series of RNA primers that are extended during second-strand synthesis. Subsequent ligation may aid the recovery of full-length cDNAs.

Synthesis of the Second Strand

As for first-strand synthesis, second-strand synthesis requires a polymerase, dNTPs, and a primer added to a template. In this case, the template is the product of the first-strand synthesis reaction. Since the template is DNA, a DNA-dependent DNA polymerase, most commonly *E. coli* DNA polymerase I or the **Klenow fragment** of DNA polymerase I, is used for second-strand synthesis. Additionally, reverse transcriptase added as a "chaser" reportedly facilitates second-strand synthesis through obstructions caused by secondary structure in the template (23). The product of the second-strand synthesis reaction is double-strand cDNA ready for steps leading to recombinant vector integration. The choice of primers that might serve to direct the synthesis of the second strand is limited, since the 3′ terminal sequence of the first strand varies greatly among the first-strand synthesis product population; there is no common sequence at the 3′ end of the first-strand product available for primer hybridization. Primer design for second-strand synthesis has conventionally been approached in one of three ways: (1) the classic method of hairpin-loop priming followed by **S1 nuclease** digestion, (2) RNAse H treatment to generate RNA primers, or (3) homopolymeric tailing, also known as vector priming (see Fig. 2). RNAse H is commonly available in a pure form for the second-strand synthesis reaction (Fig. 2**c**). Added RNAse H is used to generate short RNA primers that are extended to form the second strand (24,25). The relative technical simplicity, high yield, and large average product length achieved with the RNAse H method have proven it to be the superior approach generally, although other methods may continue to have specific applications and certainly have had a central role in characterization of the potential side reactions that affect the yield and average size of the cDNA product.

The classic method of second-strand synthesis (26,27) takes advantage of the tendency for the 3′ end of the first strand to fold back and form a DNA duplex with sequences located further 5′ or upstream on the same DNA strand (Fig. 2**a**). A consequence of the duplex or hairpin formation is availability of the self-primed template for second-strand synthesis. After removal of the RNA, reagents for second-strand synthesis are added, and so the second strand is generated. The product is double-strand cDNA covalently closed at the end corresponding to the 5′ end of the original mRNA. The addition of S_1 nuclease, a single-strand-specific nuclease, results in cleavage of the single-strand loop to produce free 5′ and 3′ ends, which are necessary for subsequent integration into a cloning vector. Although the use of S_1 nuclease, which has a potential of hydrolyze double-stranded DNA when used in excess, is often cited as the major disadvantage of the classical method, it is the necessary loss of sequence information that is the true technical fault in the method. Since at least some sequence is involved in the hairpin formation, the sequence of the 5′ end of the original mRNA is not represented in the cDNA product when this classical method of second-strand synthesis is used.

The method of homopolymeric tailing or vector priming relies on the use of *terminal deoxynucleotidyltransferase*, an enzyme that catalyzes the sequential addition of available nucleotide triphosphates to the free 3′ end of a DNA strand. Thus, a homopolymer tail made by extension of the free 3′ end of the cDNA strand resulting from the first-strand synthesis reaction serves as the primer site for second-strand synthesis (Fig. 2**b**). The application of homopolymer tailing to second-strand synthesis preserves the cDNA sequence representing the original 5′ end of the mRNA template, since priming is from sequence added by terminal transferase (28,25). After first-strand synthesis initiated from a T-tailed vector, the 3′ end of the newly formed cDNA strand is the target for homopolymer tailing. The use of dGTP for the tailing reaction consistently results in an average of 20 G nucleotides on both the 3′ end of the cDNA first strand and on the free 3′ end of the vector. The consistent length of the tail with use of dGTP is apparently due to structural constraints, which cause decreased tailing after 20 Gs (29,30). A restriction site in the vector is used to remove the G-tail from the vector, without affecting the G-tail on the cDNA, which as a DNA–RNA hybrid is a poor substrate for restriction endonucleolytic cleavage (31). Finally, an oligo $p(dC)_{20}$ primer is added, along with other reagents required for second-strand synthesis. The $p(dC)_{20}$ primer may be designed to have a 5′ end complementary to the overhang of the vector restricted end to aid in subsequent circularization of the recombinant DNA; otherwise, both the cDNA and vector ends may be made flush or blunt prior to circularization. Many variations of the vector priming method briefly described here have been described in detail elsewhere, but all make use of terminal transferase (32). The disadvantages of being more technically difficult and having more steps, as well as having a stretch of 20 G nucleotides at the vector-cDNA boundary (which could require special techniques for sequencing and subcloning), might be balanced by the important potential of increasing the number of full-length clones when using this method of second-strand synthesis.

Isolation of a cDNA corresponding to the intact full-length mRNA molecule often presents a challenge. Methods that increase the frequency of obtaining full-length clones in a cDNA synthesis reaction are often preferred. Losses of cDNA may occur at each step of the synthesis, from RNA isolation to vector incorporation. Losses of full-length cDNA may also occur due to incomplete replication of template by polymerase. For example, the presence of secondary structure in the RNA template may impede the progress of reverse transcriptase during first-strand synthesis, resulting in the formation of a partial cDNA beginning at the primer site and ending at the site where the secondary structure is encountered by the polymerase. Technical manuals that contain detailed protocols for cDNA synthesis often describe ways to monitor the progress of cDNA synthesis to give an indication of the reaction yields, or the incurred losses, and also the average size of the cDNA strand. A method designed to extend a partial cDNA isolated by a selection procedure called RACE (rapid amplification of cDNA ends) may be used to obtain the corresponding full-length clone (33). The actual 5′ end of an mRNA should, however, be mapped against the gene and/or correlated with the N-terminal sequence of the protein where possible. **Primer extension** is frequently used to determine the precise 5′ end of an mRNA, and this method only requires enough sequence information for primer design.

USES OF CDNA

The final steps in cDNA synthesis are often those leading to vector integration. The use or application of cDNA is largely determined by the type of vector that is chosen to contain the cDNA. Vector integration is, therefore, the first step toward fitting the cDNA to the type of application that will be made or toward developing a selection strategy for the isolation of a cDNA of interest from a library. Recent progress in the development of cloning vectors for specialized applications and of methods of efficiently adapting the cDNA ends to fit the vector ends have resulted in a lot of flexibility in the combination of available vector features when making a choice from the long list of commercially available cloning vectors. The immediate aim of retaining all the newly synthesized cDNA through the process of vector integration and host transformation, while keeping the desired vector features, is frequently satisfied by the use of a **phagemid** (34). A phagemid vector combines features of **lambda phage**, filamentous phage, and plasmid vectors into one cloning vector (35,36). The efficiency of host cell transformation through packaging of cDNA into viral particles, and subsequent viral infection as performed with the use of a λ phage cloning vector, cannot be matched by even the optimal conditions for direct host-cell transformation with plasmid DNA. There is consequently less potential for cDNA loss during host cell transformation with a λ phage system when compared to a plasmid system. The lower background that results from the screening of plaques rather than colonies with sequence hybridization probes or **antibody** probes offers another significant advantage to the use of viral vectors. Plasmid vectors offer advantages over λ phage vectors for the characterization of the cDNA inserts, due to the necessarily large size of λ phage vectors. Phagemid vectors contain a plasmid vector within a λ phage vector and additional sequences derived from filamentous phage that contribute to the *in vivo* excision of the plasmid from the λ phage vector. Thus, for characterization of individual selected cDNA clones, the original phage clones are converted to plasmid clones. Although it is technically straightforward to subclone a single cDNA clone from one vector to another, it is technically difficult to subclone an entire cDNA library from one to another vector without some loss of cDNA. Whichever type of vector is used to receive newly made cDNA, consideration of the vector features required for selection of a cDNA of interest from a library and the efficiency of the process of integration and host-cell transformation should be given prior to vector integration of cDNA as a library.

Possibly the greatest benefit of aquiring a new cDNA of interest from a cDNA library is the sequence information that can be derived from it. In addition to the encoded protein sequence, comparison of a cDNA sequence with the corresponding gene sequence reveals or confirms the location of any introns present in the gene. Analysis of a set of positives identified by rescreening of a library with the now available hybridization probe may provide some indication of the variability of the gene transcripts, eg, the presence of alternate splice sites or multicopy genes with minor sequence variations. The presence of viral RNA polymerase recognition sequences, most commonly SP6 and T7 **promoters**, in many cloning vectors are useful for the *in vitro* production of anti-sense RNA probes that may be used for the high-sensitivity detection of gene expression. Likewise, PCR primers may be designed from the sequence information to provide an alternate method of sensitive or quantitative gene expression analysis. The presence of potential post-transcriptional regulatory elements in the untranslated regions represented in the cDNA may be identified through computer-assisted **sequence analysis** and **database** comparisons. Although any number of extremely sophisticated uses for cDNA sequence information may be envisioned for a particular application, a generally useful benefit is the possibility to generate peptide antibodies to encoded peptide **epitopes** predicted to be on the protein surface.

Vectors are available with many combinations of features. A particularly useful vector feature for many types of applications is the inclusion of sequences that support cDNA expression. As the sequences required for correct initiation of translation differ among organisms, expression vectors, or an expression cassette within a cloning vector, are designed for use as a set with a matching host cell. The promoter sequences included to drive the transcription of the expression cassette and efficient transcriptional termination signals are likewise active in the same host. As there are no introns to be removed, expression of many encoded eukaryotic proteins is supported in strains of *E. coli*, provided that these proteins are not toxic to the *E. coli* host strain used. In order to avoid the possible toxic effects of some eukaryotic proteins, expression vectors with inducible promoters are available to postpone the cDNA expression to a point of vigorous growth so that significant protein production is accomplished before the host culture dies. The activity of eukaryotic proteins produced in prokaryotic host cells may be deficient due to the absence of factors required for proper **protein folding *in vivo***, for **post-translational modifications** essential for the activity, or even for correct intracellular localization (see **Protein targeting**). Eukaryotic expression systems may better support the production of eukaryotic proteins with natural levels of activity and are designed for maximum protein production. Such expression systems may support continued replication of the vector containing a cDNA insert, in addition to containing an **expression cassette** with an active promoter and signals for transcription termination and translation initiation. Both prokaryotic and eukaryotic expression systems with matched vector and host are designed to produce proteins in quantities required for their structural characterization from manageable volumes of cell culture. Accurate physical measurements that require purified proteins may take advantage of "built-in" purification aids available in some expression vectors. Expression of a cDNA as a **fusion protein** aids in the isolation of protein from cell cultures. Provided a cDNA is cloned in-frame in relation to a vector sequence that encodes a **polypeptide chain** with a known ligand affinity, the resulting fusion protein may be purified from cell lysates by passage through an **affinity chromatography** column containing the immobilized ligand. The presence of a **proteinase** peptidase recognition sequence facilitates proteolytic removal of the vector-encoded polypeptide and elution of the cDNA-encoded protein from the affinity column.

An additional type of expression vector that is adapted to cDNA selection is one designed for use with a yeast host strain. The yeast 1, 2, and 3-hybrid systems have been designed to isolate a cDNA clone from a library based on the encoded protein's recognition or specific binding to a DNA, protein, or RNA sequence, respectively (37–39) (see **Two-hybrid systems**). In this way, cDNA selection is facilitated by its expression inside a yeast host cell that has been genetically altered for the selection. In order to perform the selection, an entire cDNA library is cloned into a shuttle vector that will support replication in *E. coli*, as well as replication and expression (as a fusion protein) in the yeast host. The library is amplified for transformation into the recipient yeast host strain, which is altered in such a way that specific binding of a target sequence or "bait sequence" results in a positive selection. The yeast hybrid systems typify advances in the use of traditional genetic strains adapted for functional cloning of cDNA from higher organisms, and yeast has consequently become a tool for functional cloning through molecular recognition, extending the use of cDNA as an aid to the selection of novel factors. Thus, a factor that binds to a DNA sequence, such as a **transcription factor** that binds to a gene promoter element, may be difficult to isolate by biochemical means due to its low intracellular abundance, but it might be cloned with the yeast 1-hybrid system. Likewise, a factor that binds to a protein might be cloned with the 2-hybrid system, or an RNA-binding protein with the yeast 3-hybrid system.

SUMMARY

cDNA is a faithful double-stranded DNA copy of RNA and, as such, can represent those genes that are expressed as RNA in a source tissue. When integrated into an appropriate cloning vector to form recombinant DNA, cDNA may provide a continuous renewable source of genetic material for hybridization probes, designed reporters of biological activity, or pure encoded protein. The benefits of protein production and a myriad of specific research applications have made cDNA a universal tool for basic and applied cell and molecular biology research. The potential for using cDNA in molecular methods designed to select for activities that contribute detectable cellular phenotypes has provided possible alternatives or improvements to genetic and biochemical approaches.

BIBLIOGRAPHY

1. D. Gillespie and S. Spiegleman (1965) *J. Mol. Biol.* **12**, 829–842.
2. H. M. Temin and S. Mizutani (1970) *Nature* **226**, 1211–1213.
3. D. Baltimore (1970) *Nature* **226**, 1209–1211.
4. M. G. Sarngadharan et al. (1972) *Nature New Biol.* **240**, 67–72.
5. E. M. Scolnick (1971) *Develop. Biol.* **26**, 175–176.
6. J. C. Barrett et al. (1979) *Cancer Res.* **39**, 1504–1510.
7. M. Noda (1990) *Molec. Carcinogenesis* **3**, 251–253.
8. C. P. Carstens et al. (1995) *Gene* **164**, 195–202.
9. L. Rowen et al. (1997) *Science* **278**, 605–607.
10. M. Boguski and G. D. Schuler (1995) *Nature Genet.* **10**, 36971.
11. M. D. Adams et al. (1995) *Nature* **377**, Suppl., 3–174.
12. S. Henikoff et al. (1997) *Science* **278**, 609–614.
13. R. L. Tatusov et al. (1997) *Science* **278**, 631–637.
14. B. E. Roberts and B. M. Paterson (1973) *Proc. Natl. Acad. Sci. USA* **70**, 2330.
15. C. W. Anderson et al. (1983) *Methods Enzymol.* **101**, 635.
16. H. Aviv and P. Leder (1972) *Proc. Natl. Acad. Sci. USA* **69**, 1408–1412.
17. I. Edery et al. (1995) *Mol. Cell Biol.* **15**, 3363–3371.
18. J. Ross (1997) *Bioessays* **19**, 527–529.
19. T. G. Coche (1997) *Methods Mol. Biol.* **67**, 359–369.
20. M. Roth et al. (1985) *J. Biol. Chem.* **260**, 9326–9335.
21. M. S. Krug and S. L. Berger (1987) *Methods Enzymol.* **152**, 318.
22. C. Rüttimann et al. (1985) *Eur. J. Biochem.* **149**, 41–46.
23. U. Gubler (1987) *Methods Enzymol.* **152**, 325.
24. H. Okayama and P. Berg (1982) *Mol. Cell Biol.* **2**, 161–170.
25. U. Gubler and B. J. Hoffman (1983) *Gene* **25**, 263–269.
26. A. Efstratiadis et al. (1976) *Cell* **7**, 279–288.
27. F. Rougeon and B. Mach (1976) *Proc. Natl. Acad. Sci. USA* **73**, 3418–3422.
28. H. Land et al. (1981) *Nucl. Acids Res.* **9**, 2251–2266.
29. A. Dugaiczyk et al. (1980) *Biochemistry* **19**, 5869–5873.
30. A. Otsuka (1981) *Gene* **13**, 339–346.
31. W. H. Eschenfeldt et al. (1987) *Methods Enzymol.* **152**, 339.
32. P. L. Deininger (1987) *Methods Enzymol.* **152**, 376.
33. M. A. Frohman et al. (1988) *Proc. Natl. Acad. Sci. USA* **85**, 8998–9002.
34. M. A. Alting-Mees et al. (1992) *Methods Enzymol.* **216**, 483.
35. D. Hanahan (1983) *J. Mol. Biol.* **166**, 557–580.
36. J. M. Short et al. (1988) *Nucl. Acids Res.* **16**, 7583–7600.
37. S. Fields and O. Song (1989) *Nature* **340**, 245–247.
38. D. J. SenGupta et al. (1996) *Proc. Natl. Acad. Sci. USA* **93**, 8496–8501.
39. Z. F. Wang et al. (1996) *Genes Develop.* **10**, 3028–3040.

Suggestions for Further Reading

V. B. Chanda, ed. (1997) *Current Protocols in Molecular Biology*, John Wiley & Sons, New York.

I. G. Cowell and C. A. Austin, eds. (1997) *Methods in Molecular Biology—cDNA Library Protocols*, Humana Press, Totowa, NJ.

Suppliers of the reagents used for cDNA synthesis often supply excellent technical manuals.

COMPLEMENTATION

J. R. S. Fincham

The term complementation usually refers to the situation where two defective genomes in the same cell together support a normal or nearly normal **phenotype**, each of which supplies the function that the other lacks. Another meaning of the term, in the context of DNA manipulation, is the repair of a mutational deficiency in an organism or cell culture by artificially introducing a **gene** that supplies the missing function. This is a potent way of **cloning** genes selected according to function. In either sense, complementation provides a way of discriminating between gene functions and of defining genes as functional units.

COMPLEMENTATION AND THE DEFINITION OF THE FUNCTIONAL GENE

Classical genetics was based upon clear-cut heritable variants, either induced by mutagenic treatments or, in earlier days, just turning up as "sports" in wild or cultivated populations. The geneticists' favorite organisms, notably **maize** (*Zea mays*) and the fruit-fly (***Drosophila** melanogaster*), were **diploid**. Most mutations were recessive to **wild type**, which is what would be expected if they mostly represented losses of function that could be more or less adequately supplied if one gene out of two was normally active.

As mutants accumulated, they were attributed to the same or different genes on the basis of two criteria. First, it was supposed that mutations in the same gene were inseparable by **recombination**. A diploid that carries two different mutant **alleles** (ie, alternative forms) of a particular gene should never produce nonmutant or double-mutant germ cells by free reassortment or **crossing-over** at **meiosis**. According to the second criterion, mutations in the same gene should affect the organism in similar or at least related ways. Similarity or overlap of functional effects was judged by individual appearance (phenotype) and also and more rigorously by failure of the mutant chromosomes to complement each other's defects in the hybrid diploid. (In fact, the term complementation did not come into use until the late 1950s, and I use the word with hindsight.)

A classical example of using the complementation criterion, and one of the first in which it was thrown into doubt, involves the sex-linked *white* (w) gene of *Drosophila*, one of several genes governing eye color. A considerable number of recessive mutant w alleles had been identified. They diluted the normal red-brown eye pigment to various degrees when the same allele was present on both **X-chromosomes** (***homozygous***) in females, or on the single X in males. Females that had two different mutant w alleles (***heterozygous***) had dilute eye colors generally intermediate between the colors of the two homozygotes. Thus females of constitution *white/apricot* (w/w^a) have pale apricot eyes, not wild-type red as would be expected if each mutant X-chromosome could provide the pigment-forming function lacking in the other.

In 1952, E. B. Lewis (1) found that, at very low frequency (about 1 in 10,000), w/w^a females produce eggs of recombinant types, with neither mutation or both. By the attached-X technique, which need not be described here, he showed that these recombinants arise by crossing-over between the X-chromosomes, just as if *white* and *apricot* were mutations at distinct, though very closely linked loci. Accordingly he gave them different symbols, w and a, and rewrote the constitution of the female parent wa^+/w^+a, where the + superscripts denote the respective wild-type alleles.

The remaining anomaly was what Lewis called the *cis/trans* position effect (Fig. 1). Two kinds of doubly heterozygous constitution could be compared, the original so-called *trans* arrangement wa^+/w^+a, that has the two mutations on opposite chromosomes, and the *cis* arrangement, w^+a^+/wa. They had identical overall gene content, but the former had pale apricot eyes and the latter was fully wild type. w^+ and a^+ could act cooperatively to promote normal levels of eye pigment only when they were together on the same chromosome. Lewis called their relationship ***pseudoallelic***: allelic according to their failure to complement each other but nonallelic according to their ability to recombine. A formally identical situation at the *Drosophila lozenge* locus had already been reported by M. M. Green (2).

For Lewis and Green at that time, the recombination criterion for allelism took priority, and the reason for the position effect was a matter for speculation. Before long, however, high-resolution recombination analysis of series of mutants in the microbial world, first in the fungus ***Aspergillus*** *nidulans* (3) and then in the bacterial virus (**bacteriophage**) T4 (4), revealed that pseudoallelism, in Lewis's sense, was much more the rule than the exception. It was rather rare for independently occurring mutations to be "truly" allelic. Almost all pairwise combinations of noncomplementing mutants showed some low frequency of recombination when crossed together. Soon afterward, the same was shown to be true in two bacterial species, *Escherichia coli* and *Salmonella typhimurium* (5), which have no regular diploidy or meiosis but other means exist for obtaining partial diploids and tests for complementation

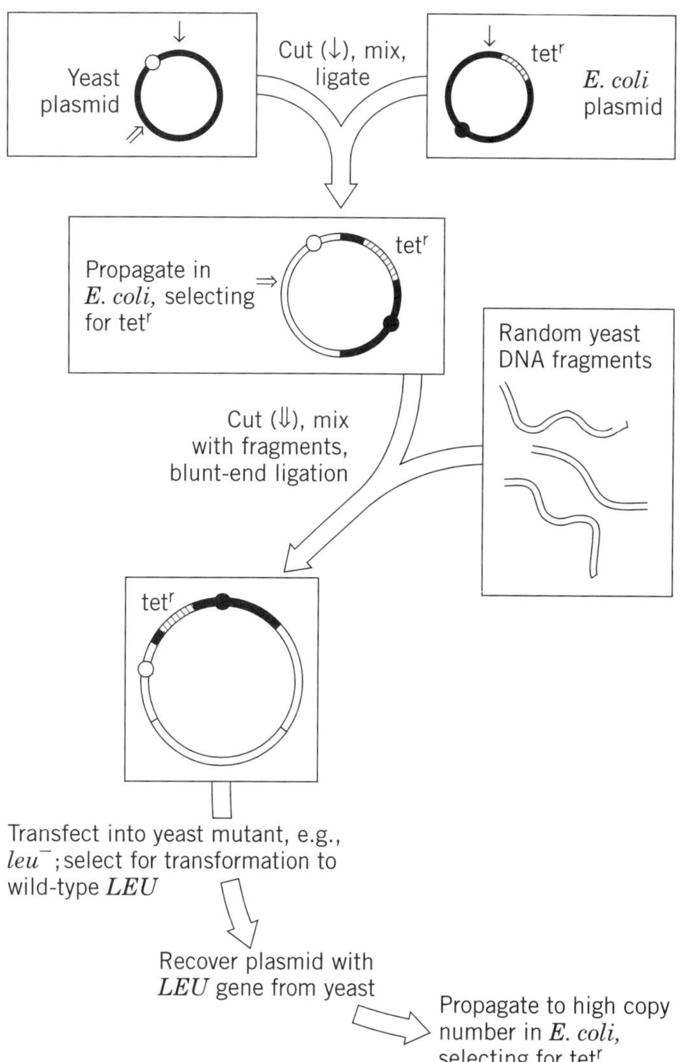

Figure 1. Cloning and amplifying a yeast gene in a shuttle vector selected by its complementation of a yeast auxotrophic (leucine-requiring) mutant. Open and filled circles indicate replication origins for propagation in yeast and *E. coli*, respectively. Based on Beggs' first demonstration of the method (7).

and recombination (see **Complementation tests**). In one example after another, it was shown that most mutations within functional genes are at different mutually recombinable sites which, by different methods in different organisms, could be mapped in a closely spaced linear sequence that was continuous with the much longer sequence of the whole chromosome.

Following his very extensive analysis of the *rII* series of mutants in phage T4 (4, see **Complementation Tests**), S. Benzer proposed replacing the word gene with a new term, **cistron**, meaning the unit of function as defined by Lewis's *cis/trans* comparison. When the *cis* and *trans* phenotypes are identical (wild type, if the mutants involved were recessive), the mutations are in different cistrons, affecting different functions. If *trans* is mutant and *cis* normal (or relatively so) the mutants are in the same cistron. The concept has been universally accepted, but the term cistron has dropped out of use. Now "gene" is generally used in the same sense, and the full *cis-trans* test is rarely applied. The theoretical reason for the comparing trans ($m1 + / + m2$) with *cis* ($m1m2/ + +$), rather than just *trans* with the wild type, is that it controls the possibility that the cumulative effect of two heterozygous mutations in different genes could be a sub-wild phenotype, even though each is individually recessive to its wild-type allele. In practice, however, recessive alleles usually remain recessive even when several of them are present together, and so the *trans*/wild comparison, the simple complementation test, is usually deemed sufficient.

Although diploidy provides the means for complementation testing in higher plants and animals and also in budding **yeast**, alternative methods, described under **Complementation tests**, have to be used for habitually haploid organisms, such as bacteria, viruses, and most fungi.

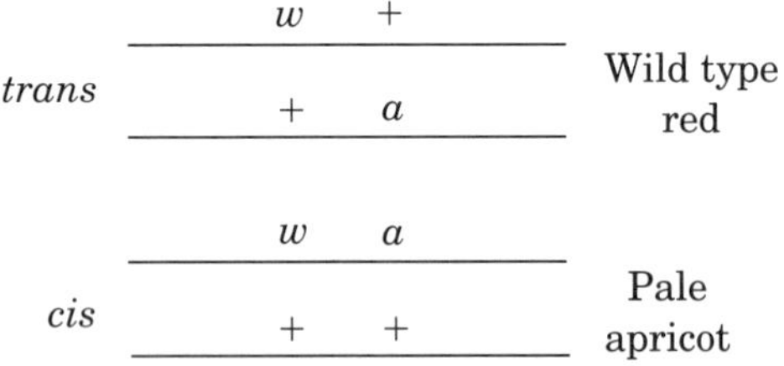

Figure 2. Lewis's (1) *cis-trans* comparison using two mutations in the *Drosophila melanogaster* X-linked *white* (w) gene. The mutations *white* (w) and *apricot* (w^a, shown here as a) are each recessive to wild type in females but do not complement one another in *trans*. By this criterion they are allelic in spite of being separable at low frequency by recombination.

Complementation and gene cloning *In yeasts*. Cloning wild-type alleles of yeast genes with metabolic functions and mutating to give **auxotrophic** phenotypes, became straightforward following the development of the Saccharomyces two-micron (2-μm) **plasmid** as a cloning vehicle (7). This plasmid also replicates in the other major experimental yeast species, the fission yeast *Schizosaccharomyces pombe*. It has been most useful as a component of a series of hybrid **shuttle vectors**, capable of replication either in yeast or in *Escherichia coli*. These usually incorporate the replicative origin of the *E. coli* plasmid ColE1 and a gene for **antibiotic resistance** (e.g., to **tetracycline**) that can be selected for in the bacterial host.

To select and clone a yeast gene, fragments of total yeast DNA, produced by digestion with a **restriction endonuclease** (or sometimes by sonication), are ligated into the closed-loop, double-stranded DNA of a shuttle vector and introduced en masse by one of the standard **transformation** procedures into mutant yeast cells, which then are plated on a medium on which they grow to form colonies only if their mutational deficiency has be repaired. (see **Cloning**) The first yeast genes to be cloned were selected because they conferred on auxotrophic (nutritionally exacting) mutants the ability to form colonies on a minimal (unsupplemented) growth medium. But any gene that mutates to give a conditional no-growth phenotype can be cloned in essentially the same way. Temperature sensitive mutants haved been a particularly rich source of cloned genes. Almost any essential protein can be mutated to a temperature-sensitive form, functional at, say, 30°C but not at 36°C. Mutant cells grown at the permissive temperature are transformed with a gene "**library**" and selected for repair of function at the restrictive temperature.

The shuttle vector bearing the selected gene replicates autonomously either in yeast or after transfer to *E. coli*. The bacterial host is more efficient for mass propagation of the clone. When sufficiently amplified, the cloned gene is cut out of the vector, and its structure and function are analyzed in various ways, most obviously by DNA sequencing. The general method is outlined in Fig. 2.

The complementation of yeast mutants is used for cloning yeast genes and also genes from such distant organisms as *Drosophila* and humans. One good example is cloning the human equivalent of the *Schizosaccharomyces pombe* (fission yeast) *CDC2* gene (8). *S. pombe* temperature-sensitive *cdc2* mutants cannot complete their cell cycle at elevated temperature (36°C) because the mutant *cdc2* gene product, a protein **kinase** that provides a signal essential for initiating cell division, is inactivated at this temperature (see ***cdc* genes**). Transformation of mutant cells with a human **cDNA library** cloned in a replicating plasmid led to the isolation of a few colonies that harbor hybrid plasmids carrying the desired cDNA sequence. The cloning plasmid used in this instance differs from that illustrated in Fig. 2 in that its replication in fission yeast depends on a DNA sequence from the monkey **SV40 virus**, which replicates well in *S. pombe*, not on the yeast 2-μ plasmid. To clone mammalian gene sequences in yeasts, it is generally necessary to use cDNA rather than genomic DNA. Mammalian **introns** are not effectively spliced out in yeast cells and, in any case, often make the gene too long for cloning in one piece.

In Neurospora. Although yeast species that have a single-cell, colony-forming habit and are accessible to replicating plasmids are especially convenient for cloning genes, the complementation principle can also be applied to filamentous fungi and has been particularly successful in *Neurospora crassa*. This species has the disadvantage that it has no convenient plasmid that stably replicates without integration into the chromosomes. Consequently, although auxotrophic mutants are "rescued" by transformation with fragments of total *Neurospora* DNA, it is not so easy to recover the transforming sequences from the transformed cultures. This difficulty was overcome by sib selection (9).

The method is making a library of *Neurospora* DNA sequences in some suitable vector (sufficient in number so that it is probable that the great majority of genes are represented), dividing the clones into a number of pools, and testing DNA isolated from each pool for its ability to repair the mutant of interest. Then a pool that works (and there is usually at least one) is subdivided into smaller pools for further tests, and then to still smaller pools, until the hunt is narrowed down to a small number of clones that can be tested individually. Though obviously laborious, the method has been very successful. It has been used for cloning genes of metabolic function and also for the loci (actually gene complexes) that govern fungal **mating type** by complementation of mating-deficient mutants (10,11).

In Drosophila. Many gene functions involved with basic cellular processes are common to yeast and mammals. Then it is not surprising to find that others, more to do with tissue differentiation, are common to most animals. Because the genetic control of **development** is better understood in *Drosophila* than in any other animal, it is of interest to ask whether some of the relatively well-characterized *Drosophila* genes have mammalian equivalents. Flies cannot be used for the kind of mass screening of gene libraries that is possible with yeast, but it is not difficult to find out whether some particular mouse gene or cDNA clone complements a *Drosophila* mutant.

It is relatively easy to get exotic DNA into the *Drosophila* genome by inserting it into the transposable **P element** and injecting the construct into early embryos. Thus the mouse gene M33, cloned on the basis of its sequence similarity to the *Drosophila polycomb* (*Pc*) gene, was successfully introduced into *Pc* mutant flies (12). The mutants, which usually have sex combs on all legs of the male instead of just on the front pair as in the wild type, were partially normalized by inserting one copy of M33 into their genome and were restored almost to wild type by two copies. *Pc* protein is one of a group of chromosomal proteins that act together to prevent certain developmentally important genes from being expressed in the wrong places or at the wrong times. This experiment strongly indicates that there is a mammalian gene that has a similar function.

BIBLIOGRAPHY

1. E. B. Lewis (1952) *Proc. Natl. Acad. Sci. USA* **3**, 953–961.
2. M. M. Green and K. C. Green (1949) *Proc. Natl. Acad. Sci. USA* **35**, 586–591.
3. R. H. Pritchard (1953) *Heredity* **9**, 343–371.
4. S. Benzer (1955) *Proc. Natl. Acad. Sci. USA* **41**, 344–354.
5. P. E. Hartman, J. C. Loper, and D. Serman (1960) *J. Gen. Microbiol.* **22**, 323–368.
6. S. Benzer (1958) In *The Chemical Basis of Heredity* (W. D. McElroy and B. Glass, eds.), Johns Hopkins Press, Baltimore, pp. 70–93.
7. J. D. Beggs (1978) *Nature* **275**, 104–109.
8. M. G. Lee and P. M. Nurse (1987) *Nature* **327**, 31–33.
9. S. J. Vollmer and C. Yanofsky (1986) *Proc. Natl. Acad. Sci. USA* **83**, 4867–4873.
10. N. L. Glass, J. Grotelueschen, and R. L. Metz (1990) *Proc. Natl. Acad. Sci. USA* **87**, 4912–4916.
11. C. Staben and C. Yanofsky (1990) *Proc. Natl. Acad. Sci. USA* **87**, 4917–4921.
12. J. Muller, S. Gaunt, and P. A. Lawrence (1995) *Development* **121**, 2847–2852.

COMPLEMENTATION TESTS

J. R. S. FINCHAM

The **gene** is best defined as an integral unit of genetic function, in the sense that none of its parts function normally in isolation from the others. If so, two different forms (**alleles**) of the same gene, both more or less inactivated by **mutation**, should not make good each other's defects when together in the same cell. They should not, in other words, show **complementation**. On the other hand, two **genomes** that have mutations in different genes should complement each other to produce a wild-type phenotype, provided that each mutant allele is recessive to its **wild-type** counterpart. This complementation criterion is of great importance because it permits assigning mutations to the same or to different genes without a detailed analysis of gene functions. When mutations are closely linked and have fairly similar phenotypic effects, it is always likely that they are in the same gene. But it is not uncommon to find close linkages of separate genes whose function is related. A complementation test provides much stronger evidence.

The interpretation of complementation tests is itself subject to certain difficulties, which are discussed separately under the heading **Allelic complementation**. Here we make the simplifying assumption that complementation between recessive mutations means that they are in different genes.

For a complementation test, it is necessary to bring two mutant genomes, or at least relevant parts of genomes, together in the same cell. In organisms including animals, higher plants and budding **yeasts**, that have a stable **diploid** phase in their life cycles, this can be simply achieved by making a sexual cross between mutant strains. In microorganisms that have no stable diploid phase, various other means have to be employed. In **haploid** filamentous **fungi**, **heterokaryons** that have two genetically different kinds of nuclei in a common cytoplasm provide a satisfactory substitute for diploidy. In **bacteria**, where neither diploidy nor heterokaryosis is usually available, there are various ways of introducing fragments the genome from one cell to another to make partial diploids (**merodiploids**). This makes it possible to test for complementation of genes located within the duplicated fragment (Fig. 1). In **viruses**, especially bacterial viruses (**bacteriophages**), complementation tests can be carried out simply by mixed infections.

COMPLEMENTATION TESTS IN DIPLOIDS

The most extensive complementation testing using diploids has been carried out in the budding yeast *Saccharomyces cerevisiae* and the fruit fly ***Drosophila*** *melanogaster*. These are the diploid organisms that have most experimentally induced mutants sorted into genes.

Saccharomyces has the great advantage as an experimental organism of propagating itself equally well as a haploid or a diploid. Haploid cells are of two different mating types, called *a* and α. In wild strains, the cells are constantly interconverted by a genetic switching mechanism but are stabilized in nonswitching laboratory strains. Most yeast genetics are based on the use of **auxotrophic** mutants that grow only if given some specific nutritional supplement which they can no longer synthesize—most commonly an **amino acid**, a **purine**, or **pyrimidine** base, or a vitamin. Such mutants can be readily obtained in haploid strains. For comprehensive complementation testing, it is necessary to isolate each mutant in both mating types from a cross to wild type. Then haploid cells of pairs of mutants of opposite mating types are mixed at marked positions on plates of unsupplemented agar growth medium. Sexual fusion to form diploid cells occurs almost immediately, and the diploids grow if the two mutations complement each other's nutritional deficiencies. The result is quite clear after overnight incubation. Auxotrophs that have different nutritional requirements always complement, but mutants that have the same requirement may or may not do so because several different genes acting in sequence are usually necessary to synthesize a single end product. The results split the mutants that have a common requirement, say, for the amino acid **histidine**, into clear-cut complementation groups -

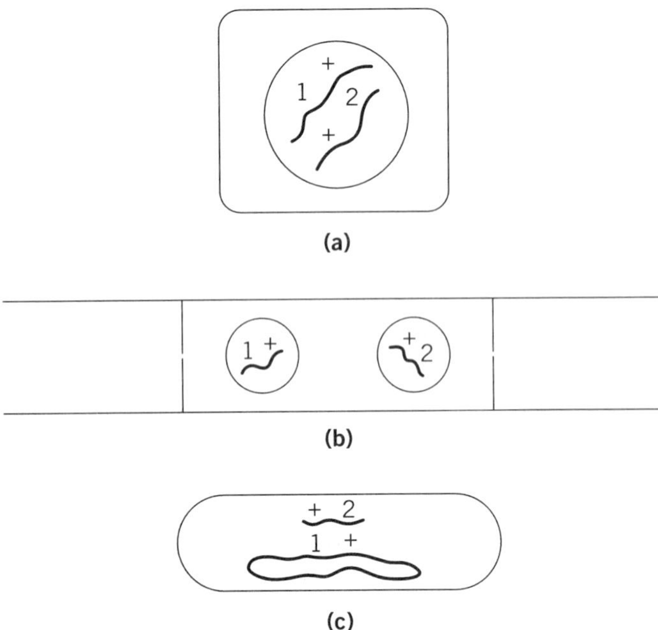

Figure 1. Three general ways of making complementation tests: (**a**) by diploidy (higher plants, animals and the budding yeast *Saccharomyces cerevisiae*); (**b**) by heterokaryosis (fungi, such as *Neurospora crassa* and sometimes cultured mammalian cells); (**c**) by addition of a genomic fragment to a haploid genome (bacteria, such as *Escherichia coli* and *Salmonella typhimurium*). In each diagram, 1 and 2 are different mutations, and + stands for the corresponding nonmutated sites.

his1, *his2*, *his3*, etc. Every member of one group complements every member of any other group, but only sporadic and fairly infrequent complementation occurs within groups (see Fig. 3 and **Allelic complementation**). In general, complementation groups equate to genes, and each one is deficient in one enzyme of the biosynthetic pathway.

In the fruit-fly, *Drosophila melanogaster*, which is certainly the most intensively investigated multicellular organism, many mutants have been assigned to allelic series based on the mutant phenotypes of their heteroallelic combinations. The famous example of multiple alleles of the *w* (white-eye) gene is referred to under **Complementation**. Most of the allelic series have consisted of viable mutants, all within the same short **chromosome** segment, and obviously related phenotypes that made their allelic relationship are very likely even without complementation testing. However, nonvisible mutations, **recessive lethals**, have been much used in attempts to establish the total number of genes that have essential functions residing within a particular chromosome segment. Complementation is the only guide for classification of lethals.

A broadly applicable experimental scheme is shown in Fig. 2. The method is selecting, after heavy mutagenization, a large number of **recessive lethal** chromosomes based on their failure to complement a deletion of a short, defined chromosome segment, and maintaining them in breeding stocks combined with a "balancer" chromosome. A *Drosophila* balancer chromosome carries a dominant viable mutation to show that it is present and a recessive lethal mutation to ensure that any fly that carries it also has the chromosome against which the balancer is supposed to be balanced. The balancer chromosome also contains an inverted segment, or complex of inverted segments, to inhibit its recombination with the mutagenized chromosome, or else has genetic markers to ensure that recombination within the mutagenized region can be avoided. Crosses between flies that have different recessive lethals, each in combination with the same balancer, produce a predicted one-third of their progeny free of the balancer, if the two are in different genes and therefore complement one another, but no progeny without the balancer if they are in the same gene.

In the study on which Fig. 2 is based (1), a total of 268 recessive lethals that fall within a segment of chromosome 3 comprising 26 **polytene** chromosomal bands could be assigned to 25 complementation groups. The numbers were sufficiently great to make it likely that most, if not all, of the genes that have essential functions and therefore can mutate to recessive lethality had been identified. The gene number was probably underestimated because there are certainly genes that can be deleted without lethality. Nevertheless, the agreement between chromosome band number and the apparent number of essential genes (now found in several different experiments) is too close to be without significance.

THE USE OF HETEROKARYONS

Filamentous fungi, or at least those (eg, ***Neurospora*** *crassa*, ***Aspergillus*** *nidulans*) that have been well studied genetically, are naturally haploid except for the immediate product of sexual fusion, which immediately undergoes meiosis to provide nuclei for haploid **spores**. In *Aspergillus*, however, rare vegetative diploids are obtained by selection (2). In *Neurospora*, where this is not possible, complementation tests are easily carried out with haploid heterokaryons, mixed cultures in which the filamentous hyphae have undergone fusions to bring the two different kinds of haploid nuclei together in the same cytoplasm (3).

For heterokaryon formation, *Neurospora crassa* strains have to be of the same mating type, and identical with respect to a number of other genes governing vegetative compatibility. These requirements are automatically met if the mutants under test have all been isolated in the same wild-type strain, as is usually the case. Given vegetative compatibility, fusions occur readily as soon as mixed mutant inocula of conidia (asexual spores) begin to germinate. *Neurospora* filaments (hyphae) are coenocytic. Numerous nuclei are present together in each cytoplasmic compartment, and the nuclear ratio in a heterokaryon is quite variable but is determined at least approximately by the ratio of the two kinds of conidia in the initial mixed inoculum. The nuclear ratio is, in fact, not usually critical. If, as is usually the case, the mutants under test are auxotrophic (ie, have special nutritional requirements) and the mixed inoculation is on a minimal medium on which neither mutant grows by itself, the heterokaryon grows vigorously if the mutants complement one another.

At least in *Neurospora*, different genes can complement each other perfectly efficiently in heterocaryons even though they are separated in different nuclei. This is not at all surprising. Genes act, after all, by exporting their **messenger RNA** to

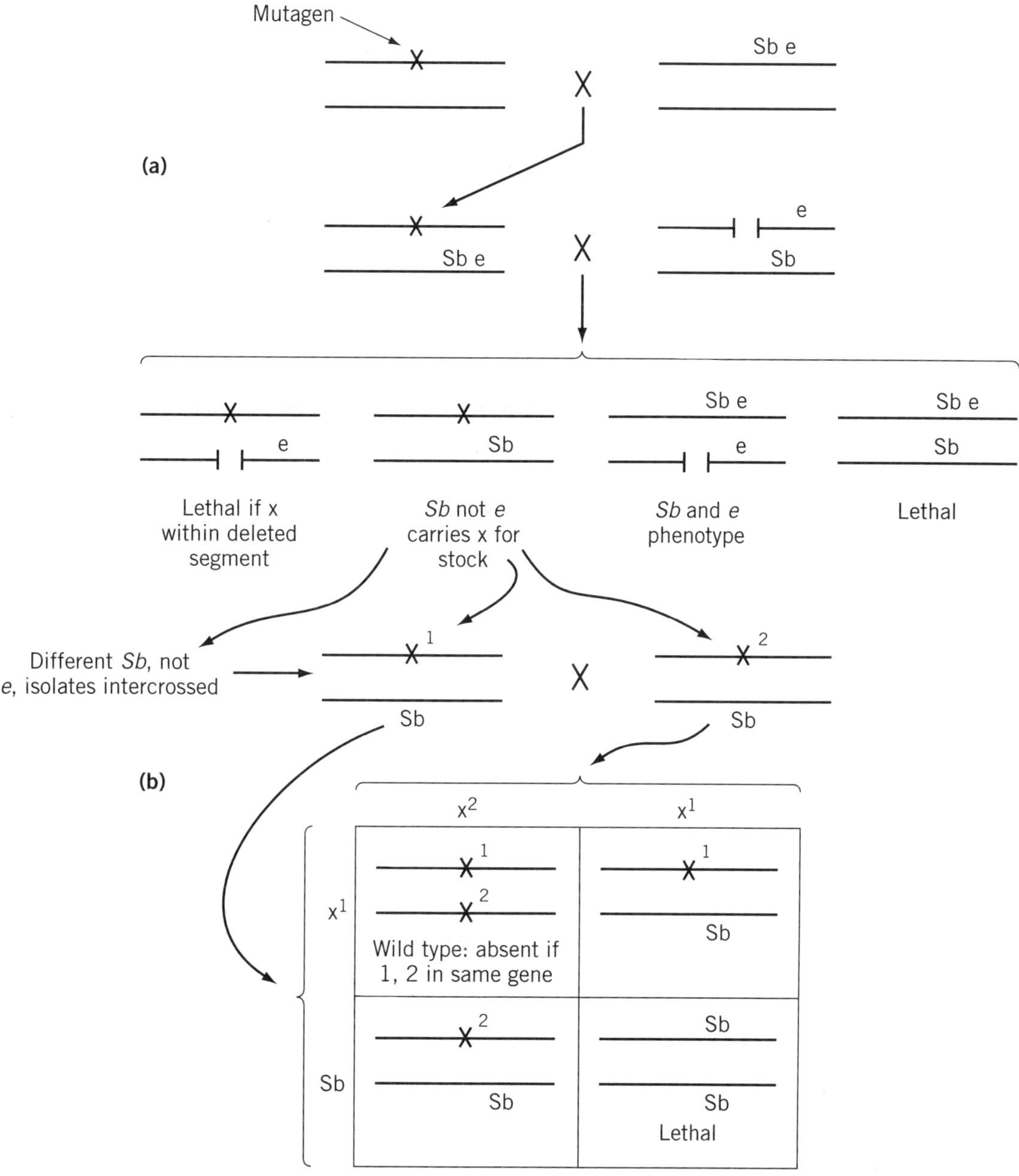

Figure 2. The isolation of recessive lethal mutations in *Drosophila* and testing them for complementation. (**a**) Male flies are treated with a mutagen and mated to females carrying a balancer third chromosome that has the marker mutations *Sb* (dominant bristle phenotype, recessive lethal) and *e* (recessive ebony body colour). Flies in the next generation that have mutagenized chromosome 3 covered by the balancer, are crossed to flies that have a third-chromosome deletion and the *e* marker, balanced against *Sb*. If the mutagenized third chromosome has a recessive lethal mutation (x) at a site within the deletion, there are no viable progeny that do not carry *Sb*. Then, the recessive lethal is present, balanced against the *Sb* chromosome in all viable progeny that are not ebony (*e/e*). (**b**) When *1/Sb* and *2/Sb* flies are crossed (*1* and *2* are two different lethals), one-third of the viable progeny should be wild type (without *Sb*) if *1* and *2* are in different complementation groups (equivalent to genes), but all exhibit *Sb* if *1* and *2* are in the same complementation group. Based on Ref. 2.

the **cytoplasm**. Reports indicate that complementation in *Aspergillus* is more efficient in diploids than in heterokaryons, but one suspects that this is because, in this fungus, the different nuclear types are not sufficiently closely intermingled. A large number of *Neurospora* auxotrophic mutants have been efficiently classified into complementation groups, equivalent to genes, by heterokaryon testing. An example is shown in Fig. 3 (4).

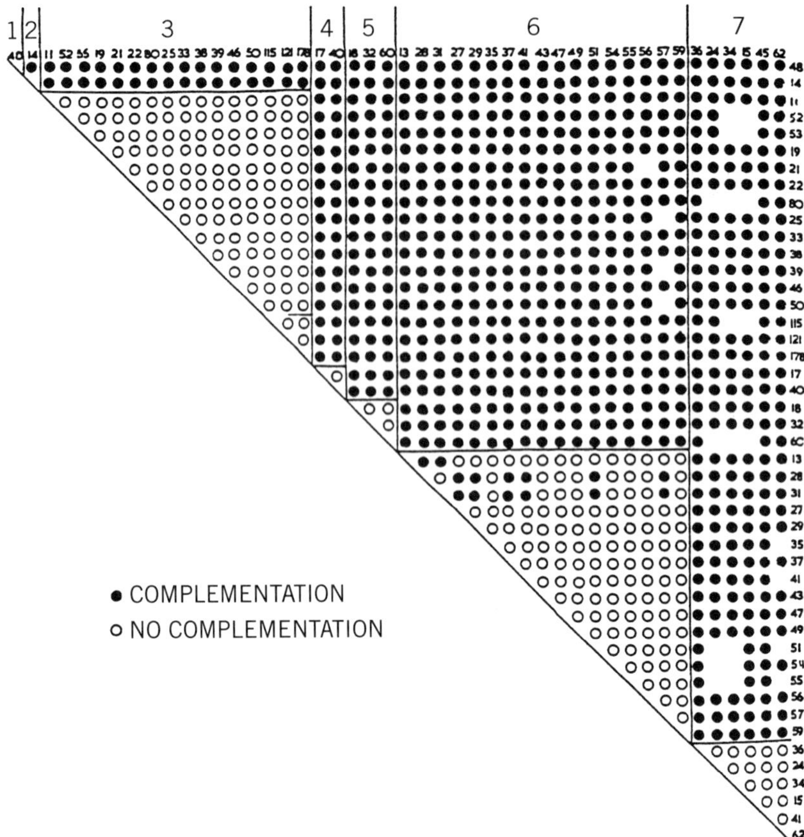

Figure 3. The result of complementation tests by heterokaryon formation on 46 *Neurospora* mutants that cannot utilize acetate. Nearly all possible pairwise mixtures of mutant conidia were inoculated into a medium containing acetate as the sole carbon source. Growth occurred only if the mutants showed complementation. In the matrix shown, the mutants, designated by their original isolation numbers, are divided into seven different complementation groups. All intergroup combinations showed complementation and grew like wild type. Most or all combinations within a group did not grow. The growth shown by a few pairs of mutants within group 6 is evidence for *inter***allelic complementation**. From Ref. 4.

Humans, of course, are diploid, but complementation testing by controlled sexual crossing is obviously out of the question. Here again, heterokaryosis may provide an alternative, and cells in culture may serve as surrogates for people. One good example is an analysis of the human *xeroderma pigmentosum* syndrome, caused by genetic deficiency in **DNA repair** (5). Cultured cells isolated from different patients were induced to fuse by treating them with heat-inactivated **Sendai virus**. Complementation in the resulting heterokaryotic cells was assessed by their ability or inability to incorporate a radioactive precursor into their DNA following damage by UV light. The results permitted assigning the different mutations to four different complementation groups, putatively different genes (Fig. 4).

In Viruses by Mixed Infection

S. Benzer's very extensive complementation testing of bacteriophage T4 mutants (see also **Complementation**) depended on infection of bacterial cells with mixtures of phage particles (6). Mixed infection is analogous to heterokaryosis because two different whole genomes are present in a somewhat indeterminate ratio, but the difference is that the simple viral genomes are not enclosed in **nuclei**. The mutants under study (class *rII*) were distinguished by growing on one *Escherichia coli* strain (B) but not on another (K). Complementation between pairs of mutants was checked by superimposing *rII* phage inocula on lawns of K cells, usually with a small admixture of B cells to allow at least some viral propagation. Three results were clearly distinguished: (1) complementation of the two mutants—total clearing of the mixedly infected patch due to lysis of all the cells in that area; (2) no complementation, but some formation of wild-type phage by recombination during mixed growth, shown by scattered spots of lysis; and (3) much less commonly, neither complementation nor recombination—no lysis at all (Fig. 5**a**). These results were interpreted to mean that (1) mutant sites are in different cistrons (ie, different functional genes); (2) mutants are at different sites in the same cistron; and (3) mutants are at the same site, or one is a deletion overlapping the other. The complementation results split the

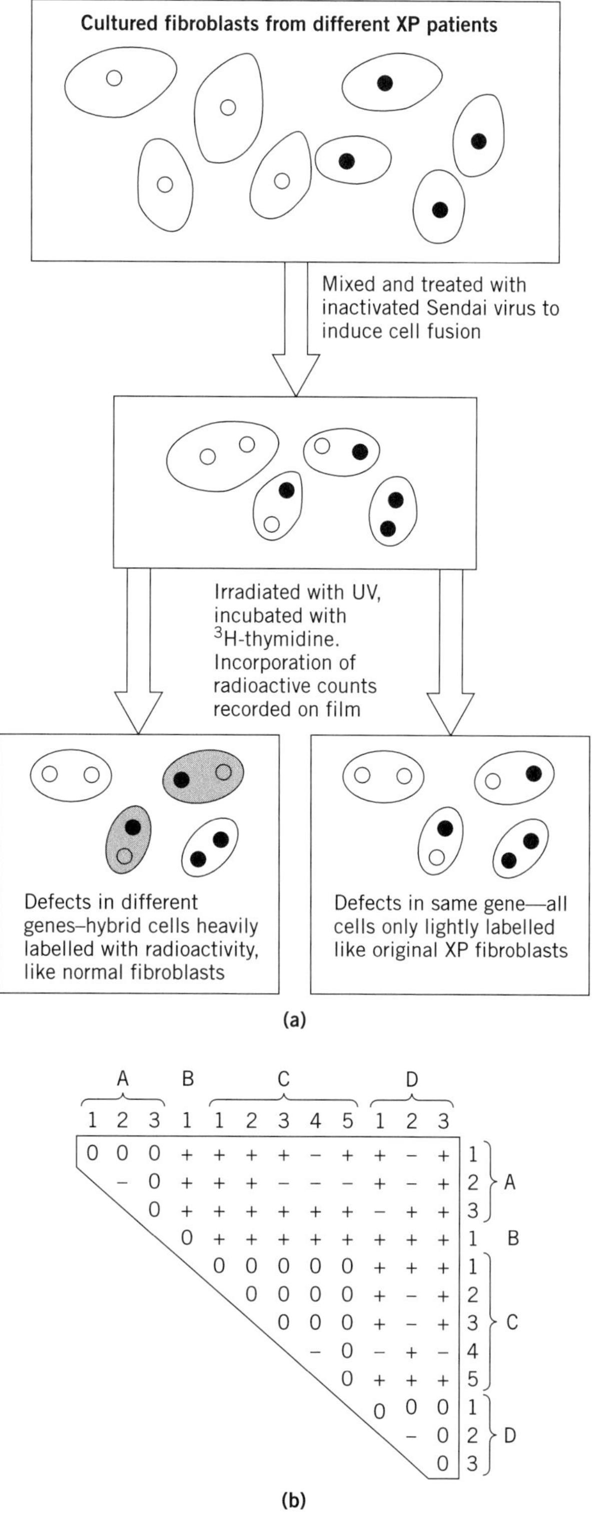

Figure 4. The use of heterokaryosis for complementation testing of cultured cells from human Xeroderma Pigmentosum (XP) patients. The syndrome that is excessive sensitivity to uv light is a consequence of failure to repair damaged DNA by excision followed by new synthesis. (**a**) Cultured fibroblasts from different patients were mixed and induced to undergo fusions. The resulting cells, some now heterokaryotic, were irradiated and cultured in the presence of tritiated **thymidine** to see whether they responded to the DNA damage by incorporating this radioactive DNA precursor into their nuclei. The original XP mutant cells, or noncomplemented heterokaryons, incorporated no tritium, but certain pairs of mutant cells showed complementation. (**b**) The complementation matrix with the division of 13 XP mutations into four complentation groups A-D. Here the cells are diploid and homozygous for the different mutations instead of haploid as in the *Neurospora* example (Fig. 3), but the principle is the same.

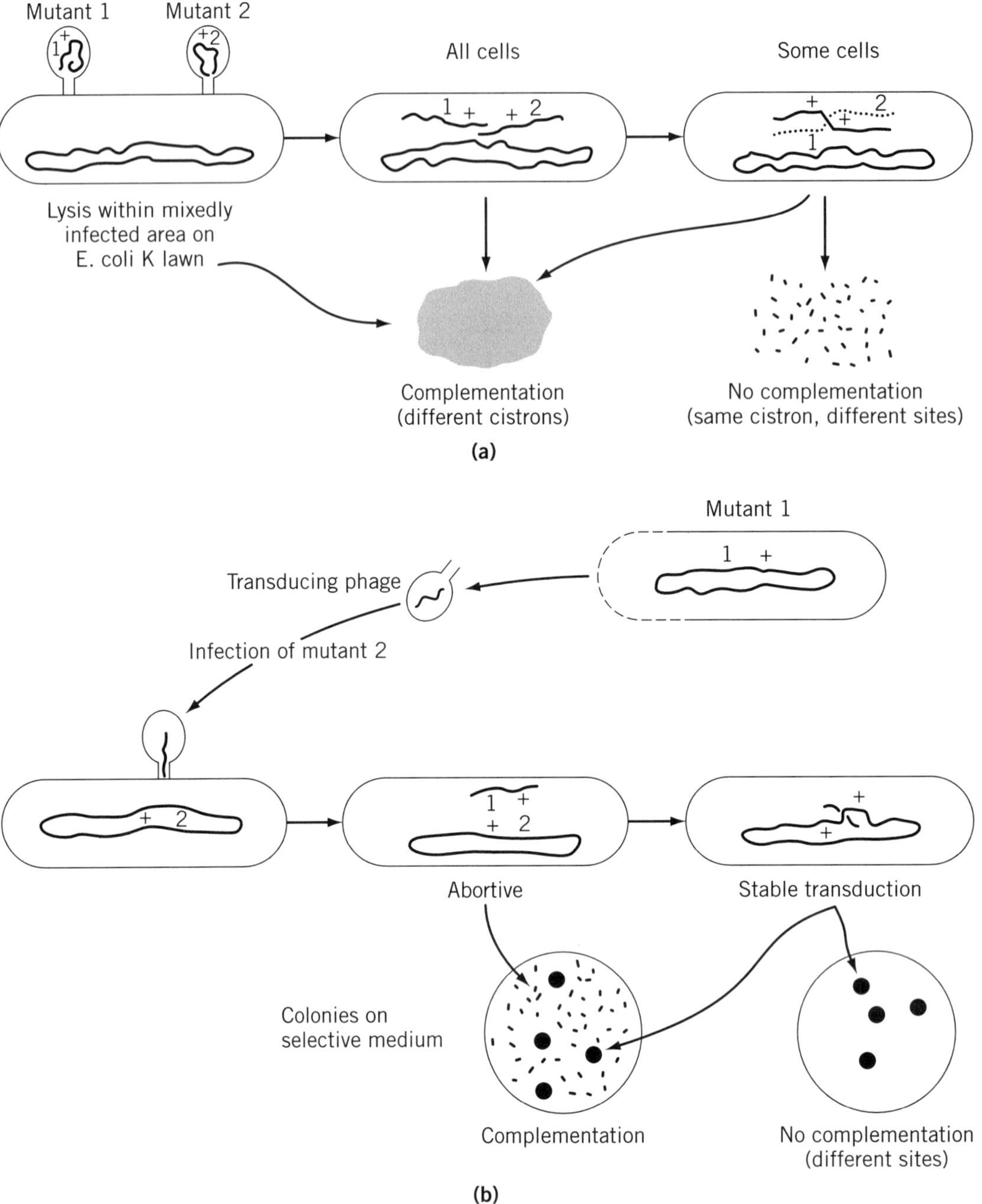

Figure 5. Two examples from bacteria of complementation tests that distinguish between complementation and recombination. (**a**) Testing for complementation between *rII* mutants of bacteriophage T4. These mutants grow well in *Escherichia coli* strain B but not at all in strain K12(λ). Mixed infections on a lawn of K cells that have a minor admixture of B cells to permit some joint growth give total clearing (lysis) if mutants *rII-1* and *rII-2* are in different genes (cistrons), and lysis only in spots because of recombination, if they are at different sites in the same gene. Based on Ref. 6. (**b**) Abortive transduction as a criterion for complementation in *Salmonella typhimurium*. Phage P22 grown on histidine-requiring (*his-*) mutant 1 was used to infect *his-* mutant 2. Surviving bacteria were tested for their ability to grow on a histidine-free medium. Rather numerous tiny (abortive) colonies showed that a nonreplicating fragment of DNA from mutant 1 complements mutant 2. The absence of such tiny colonies signifies no complementation. In either case, vigorously growing colones arise by recombination between the bacterial chromosome and the incoming fragment, provided that 1 and 2 are at different sites. Based on Ref. 7.

rII mutants cleanly into two mutually complementing groups or cistrons. Recombination analysis using a set of overlapping deletion mutants showed that these two cistrons correspond to two adjacent, nonoverlapping segments of the phage "chromosome", which is not actually a chromosome in the cytological sense but a single molecule of **double-stranded DNA**.

By Partial Diploids (Merodiploids) in Bacteria

It is not possible to test in bacteria for complementation between whole genomes because, at least in those species that are well-analyzed genetically, there is neither a stable diploid phase nor any sexual process that transfers a whole genome from one cell to another. However, bacteria acquire extra genomic fragments in three general ways: (1) by infection with defective bacteriophages which, "by mistake", have picked up bacterial DNA (**transduction**); (2) by joining a fragment of bacterial DNA to the DNA of a plasmid that brings about its own intracellular transfer (**conjugation**); and (3) by uptake of free DNA (**transformation**). The first two methods have been important in complementation analysis.

Transduction has been the principal means of classification and mapping of auxotrophic mutants of the bacterium *Salmonella typhimurium*. Bacteriophage **P22** grown on one mutant strain (the donor) is used to infect a second mutant strain, perhaps requiring supplementation with the same nutrient, such as an amino acid. Most of the infected bacteria are lysed, but some survive, are now lysogenic, and harbor the phage in latent form (see **Lysogen**, **Lysogeny**). Some survivors have acquired some fragment or other of the genome of the donor bacterium by infection from defective bacteriophage. The donor DNA is replicated in the recipient cell only if it recombines into the recipient chromosome and replaces the equivalent recipient segment. More commonly the donor DNA persists without replication, and its genes are expressed only in the primary recipient cell and in a few surrounding cells by diffusion of gene products. The appearance of tiny colonies of cells (abortive transductants) that grow transiently without the nutritional supplement, demonstrates complementation between the recipient genome and the donor genomic fragment. Usually a much smaller number of large strongly growing colonies results from chromosomal integration of a donor fragment and recombination between donor and recipient mutational sites to reconstitute a wild-type gene. The presence of large colonies without abortive ones shows that the mutations are at different sites within the same gene or cistron (Fig. 5**b**). Failure to form any colonies at all means (apart from possible experimental error) that the two mutations are at the same site or that one is a deletion overlapping the other (7).

The plasmid that is used to bring about partial diploidy in *E. coli* is the transmissible "sex factor" F (see **F, F′ factor**), a closed-loop DNA molecule that replicates autonomously in the bacterial cell and is occasionally integrated by **crossing-over** into the much larger loop of the bacterial chromosome. The integrated plasmid DNA undergoes occasional excision, usually precisely, but sometimes carries with it an adjacent segment of bacterial DNA, the nature of which depends on where the plasmid happened to have been inserted. This modified F is called F′ and, like F itself, is transmitted from cell to cell. F′-*lac* plasmids that carry segments of DNA from the **lac operon** of the bacterial chromosome, were of great importance in elucidating the complementation relationships of *E. coli lac* mutants (a topic dealt with separately under **cis-dominance**).

BIBLIOGRAPHY

1. J. Gausz, H. Gyurovics, G. Beneze et al. (1981) *Genetics* **98**, 775–789.
2. J. A. Roper (1952). *Experientia* **8**, 14–15.
3. G. W. Beadle and V. L. Coonradt (1944) *Genetics* **29**, 291–308.
4. R. B. Flavell and J. R. S. Fincham (1968) *J. Bacteriol.* **95**, 1056–1062.
5. K. H. Kraemer, H. G. Coon, R. A. Petinga et al. (1975) *Proc. Natl. Acad. Sci. USA* **72**, 59–63.
6. S. Benzer (1959) *Proc. Natl. Acad. Sci. USA* **45**, 1607–1620.
7. P. E. Hartman, Z. Hartman, and D. Serman (1960) *J. Gen. Microbiol.* **22**, 354–358.

COMPLEX LOCI

J. R. S. Fincham

In classical **genetics**, the word locus referred to a position on a **chromosome** marked by a **gene** difference that has a **phenotypic** effect. Neither the locus nor the gene were thought to be subdivisible by **recombination**, and so locus and gene were treated almost as synonyms. The idea of complex loci arose from examples, especially from *Drosophila* and **maize**, of different **mutations**, apparently at the same chromosome locus, that give a range of different phenotypic effects, suggesting that the locus harbors some rather complex kind of gene.

With our modern knowledge, the word locus can take either of two very different meanings. It may mean just the genetically mapped site of a mutation, which, at the molecular level, is quite likely to be no more than a single base-pair change within the long DNA sequence of a gene. But, in the context of complex loci, it can mean a relatively short chromosomal segment (not frequently split up by genetic recombination) that contains a cluster of genes among which there is some functional connection. So complex loci, though of somewhat confused etymology, is a useful heading under which to review the evidence for chromosomal units of function at a level higher than that of the individual gene.

This article does not cover examples of loci that mutate to give various phenotype effects when the varied effects are due to mutations in a single gene that has multiple functions. Such genes that encode "multiheaded" enzymes (see **Multifunctional proteins**) are sometimes called "cluster genes" in fungal genetics to distinguish them from the **gene clusters** with which they may initially be confused. Some examples are reviewed under **Allelic Complementation**. Nor are we concerned with single genes that have multiple modes of **intron** splicing, which yield alternative **messenger RNAs** (see examples under **Gene Structure, Allelic Complementation** and **Introns**). And we obviously cannot include groups of functionally diverse genes that just happen to be very closely linked, for that would make virtually all genes parts of one complex locus or another. To qualify as a complex locus, in our present sense, a gene cluster must have some kind of unity in structure, in function, or in both.

Gene clusters that fulfil this criterion have been found in **fungi**, flowering **plants**, *Drosophila* and mammals, and this article reviews examples from each of these kinds of organisms. We do not deal here with **bacteria**. **Operons**, which might be viewed as complex loci but are not usually so termed, are dealt with under their own heading. Also excluded from this article are highly repetitious tandemly arrayed genes required by the cell in very high dosage. The preeminent example is **ribosomal** RNA genes in the **nucleolus organizer** chromosomal regions of eukaryotes. **Histone** genes are also highly repetitive and are concentrated in clusters in some organisms, including *Saccharomyces* yeast. This is arguably more repetition than complexity.

FUNGI

In fungi, the self-incompatibility and cross-compatibility in sexual mating in many species is determined by **mating-type** loci, which virtually always consist of more than one gene. A relatively simple example is the maize smut fungus, *Ustilago maydis*, whose two loci, *a* and *b*, control the specificity of sexual fusion and the formation of an invasive dikaryotic mycelium after fusion, respectively. The *a* locus has two "**alleles**," *a1* and *a2*, each of which consists of two genes, one that encodes a polypeptide pheromone and the other a pheromone **receptor**. The *a1* pheromone is recognized by the *a2* receptor. **Homology** is not close between *a1* and *a2*, and they do not recombine with each other. The *b* locus exists in numerous alternative forms. Each consists of two closely spaced, divergently transcribed genes called *bE* and *bW* that encode the two components of a dimeric transcriptional activator. But dimerization occurs only between *bE* and *bW* products of different mating types (1).

The ink-cap mushroom fungus, *Coprinus cinereus* (like *Ustilago*, a member of the Basidiomycete class), also has two mating-type loci, *A* and *B*. The *A* locus, which controls the ability to form a fruiting dikaryon, is the most thoroughly investigated. Here the situation resembles that at the *Ustilago b* locus, except that genes that encode dimerizing monomers, thought to be transcriptional activators on the basis of their **homeobox**-like sequence motifs, are distributed in multiple copies between two subloci, $A\alpha$ and $A\beta$, between which rare recombination occurs (2) (Fig. 1). Again, dimerization occurs only between the products of different mating types.

Ascomycete fungi, such as *Neurospora crassa* and *Saccharomyces cerevisiae*, simpler than Basidiomycetes, have only two mating types (*A* and *a* in *Neurospora, a* and α in *Saccharomyces*), but the genetic loci that determine them are still complex. *Neurospora A* contains three genes (Fig. 2) and *Saccharomyces* α contains two, though each of the *a* mating-type loci has only a single gene. The two alternative mating-type loci in both organisms are completely nonhomologous, and the same is true for the *Ustilago a* locus. It has been customary to call the alternative occupants of these mating-type loci "alleles," but in the light of present knowledge this terminology is inappropriate because it implies different forms of a single gene. The new word idiomorphic has been suggested for the relationship between quite different genes or gene clusters that are alternative occupants of the same locus (3). Much the same applies to the yeast mating-type system that has the additional complication of mating-type switching.

Fungi also provide examples of another kind of locus complexity, the tight clustering of separate genes that co-operate in a particular area of metabolism. The *Neurospora crassa Qa* cluster is a particularly good example (Fig. 3). Here a DNA segment of only 18 kbp harbors seven different genes, the transcription of which is increased 50- to 1000-fold when the fungus is forced to grow using quinate as a sole carbon source. Three of these genes encode the three enzymes that, acting in sequence, oxidize quinic acid to protocatechuic acid. Two genes encode regulatory proteins that control the **transcription** of the whole cluster. One gene is a transcriptional activator and the other a repressor that cancels the function of the activator but whose own effect is nullified by binding to quinate. The remaining two genes have no identified function and are recognized solely as **open reading frames** (4). The significance of the gene clustering is not fully understood, but it seems

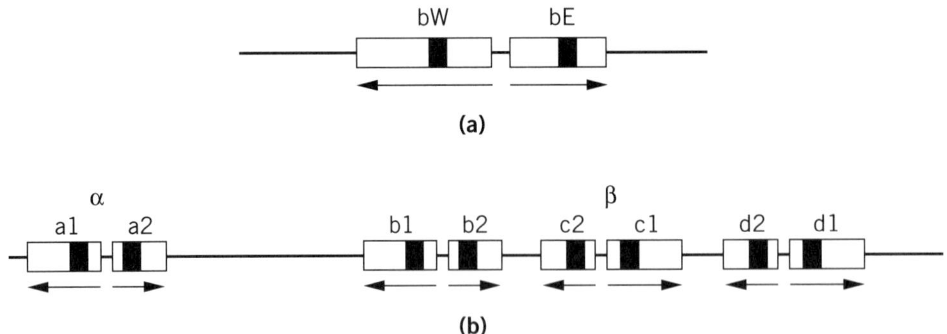

Figure 1. The structures of mating type loci in Basidiomycete fungi. (**a**) The *b* locus of *Ustilago maydis*. The *bW* and *bE* gene products form W-E protein dimers to activate transcription of genes necessary for sexual development. Each exists in multiple allelic forms. Effective dimers are not formed by W-E combinations from the same strain. Gene segments that encode homeodomain-like protein sequences are filled. (**b**) The archetypal *A* mating type locus of *Coprinus cinereus*. The system works in principle as in the *Ustilago b* locus, except that here there are four pairs of genes, *a*, *b*, *c*, and *d*, any one of which can function in promoting fertility. The archetype is a composite of known mating-types, all of which are missing one or another part of it. From Refs. 1 and 2 by permission.

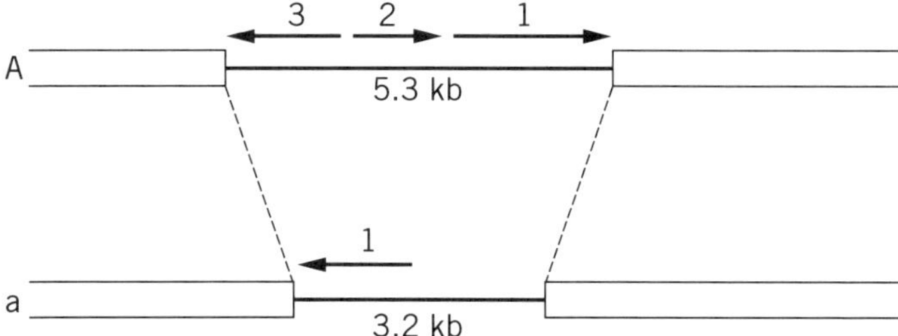

Figure 2. The mating type loci of the Ascomycete fungus *Neurospora crassa*. *A1* and *a1*, quite unrelated in sequence, act in a complementary way to permit sexual reproduction. *A2* and *A3* function in postfertilization fruiting-body development. From Ref. 3 by permission.

plausible to suggest that derepression of transcription is facilitated by an induced unfolding of **nucleosomal** structure that extends across the entire complex.

FLOWERING PLANTS

A long-standing example from what used to be, pre-*Arabidopsis*, the geneticists' favorite plant, is the *R* locus of *Zea mays* that is necessary for pigmentation and encodes transcriptional activators for genes that encode pigment-synthesizing enzymes. It was an early example of locus complexity because it has two independent functions. Some recessive loss-of-pigment alleles affect only the red color in the plant whereas others affected only the seed. Analysis at the DNA level has revealed the situation shown in Fig. 4 (5). The *R* locus contains three genes. One gene (*P*) is specific for plant pigment, and the other two (*S1*, *S2*) apparently duplicate the function in seeds. Presumably their distinct tissue specificities are due to different promoter/**enhancer** sequences. There is also a partial and presumably functionless duplication of *P*. An origin by **duplication** and subsequent partial **divergence** of function is the most obvious explanation for the clustering of these functionally related genes, although the opposite orientations of *S1* and *S2* are a puzzle.

A long-standing problem in plant biology has been the various systems of self-incompatibility, often controlled by what are multiple alleles at a single locus, usually called *S*. The general principle is that pollination is prohibited if the pollen grain and stigma (or, in the *Brassica* family, the pollen mother cell and stigma) have an allele in common. The *Brassica S* locus, one of the best analyzed, contains two genes, each present in the population in multiple allelic forms, that control the incompatibility response of the stigma. It is conjectured that it may also harbor one or more genes that are expressed in the pollen (6). Here, as in the analogous self-incompatibility in fungi, the functional reason for the multiple genes all in the same locus is obvious. For the system to work, the different specificities have to be mutually exclusive in the **haploid** genome.

DROSOPHILA

In the fruit fly, *Drosophila melanogaster*, a classical example of locus complexity, perhaps originally generated by tandem duplications, is provided by the scute (*sc*) and achaete (*ac*) families of mutations (7,8). These affect the bristles in the wings and mid-thorax of the fly, really parts of the fly's neural system, the origin of which can be traced back to the transcription of scute and achaete, usually both together but sometimes one more than the other, in rather precisely positioned bristle mother cells in the wing **imaginal disc** of the larva. It seems that *sc* and *ac* largely duplicate each other's functions and, before molecular analysis, the basis for recognizing two different genes was not clear. Different mutations in the *sc*/*ac* locus eliminate different bristles or different combinations of bristles, and, to the extent that their effects are nonoverlap-

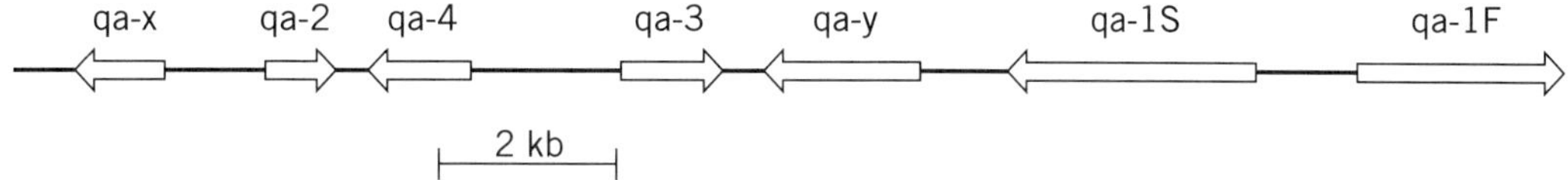

Figure 3. The *Qa* (quinate-utilization) gene cluster of *Neurospora crassa*. The genes *qa-3, qa-2,* and *qa-4,* respectively, encode quinate dehydrogenase, catabolic dehydroquinase, and dehydroshikimate dehydratase, which act sequentially to degrade quinate. *qa-1S* and *qa-1F* encode regulatory proteins, respectively, that repress and activate transcription of the whole cluster. *qa-Y*, it is thought, encodes a quinate transporter protein, and *qa-X* has no known function (4).

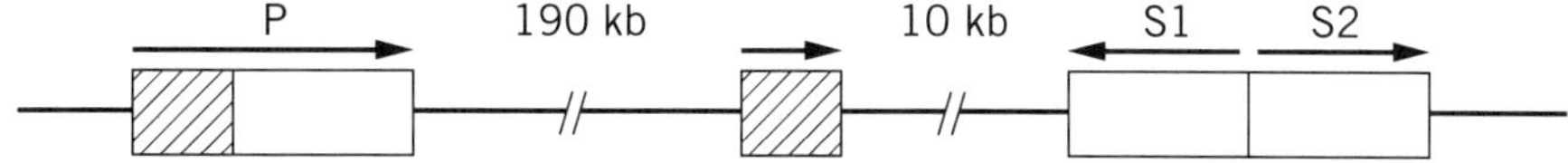

Figure 4. The complex *R* locus of *Zea mays*. *P* controls red pigment synthesis in the plant generally and the duplicate *S* genes in the seed. *S1* and *S2*, transcribed divergently, are each similar to *P* except that they lack the segment shown hatched. This segment is present as a tandemly duplicated fragment, presumably nonfunctional. Sequence information suggests that this scrambled arrangement is due to the activity of a transposable element (5). "Alleles" of *R*, R^g (red seed, green plant), and r^r (colorless seed, red plant) are, respectively, due to loss of *P* and *S* function. The *P* and *S* components can be separated by meiotic crossing over at frequencies on the order of 10^{-4} to 10^{-3}.

ping, complement one another in *trans* (see **Allelic Complementation**). Now that the *sc/ac* locus has been analyzed at the DNA level, it is seen that there are six genes within the approximately 100 kbp that comprise the *sc/ac* complex (7,8) (Fig. 5), of which at least four are involved with the neural/bristle system. *sc* and *ac* apparently have the major roles. Most of the mutations that affect the positioning of bristles actually fall between the genes. Several are chromosomal rearrangements that have breakpoints within the *ac/sc* complex, and it is thought they exert their effects by separating *ac* and/or *sc* from enhancer sequences, several of which have been identified in the *sc/ac* region. The patterns of transcriptional activation of *sc* and *ac* and hence the positioning of bristles, it is thought, reflect the distribution of a number of enhancer-binding proteins, the identity of which remain to be established. The genes and the enhancers together may be seen as an integrated system for responding to a pattern of positional information set by the products of other genes.

MAMMALS

Many clusters of related genes have originated through tandem duplication and subsequent functional divergence. Obvious examples are the globin gene clusters in animals and humans (see under **Globins, Gene Structure, Locus Control Region**). The β-globin gene cluster has a common ancestral origin and is also functionally integrated through the locus control region.

The clustered **HOX genes**, which are of profound importance in animal development and encode homeobox-containing proteins, were discovered in mammals and now in other animal groups through their significant similarities to the genes of the linked *bithorax* and *Antennapedia* complexes of *Drosophila*. Five unlinked clusters of HOX genes in mouse and humans each consist of up to 13 tandemly arranged genes within a region on the order of 100 kbp of DNA. Through their homeobox-encoding segments, the genes of one cluster have only a very limited degree of sequence similarity to each other but have a greater resemblance to genes in corresponding positions in other clusters in the same species or even in different species. Because they were not identified through the classical genetic route of mutation mapping, the HOX clusters of mammals are not usually called loci though, confined as they are within chromosomal segments that correspond to a small fraction of a **centimorgan**, they are sufficiently tightly linked to qualify for that title, and they also fulfil the criterion of functional integration. The almost uncanny correspondence between the linear sequence of genes within each cluster and the anterior-posterior order of the body segments that the genes affect clearly indicates some function for the clustering and ordering of the genes, though this is still mysterious.

The mammalian **major histocompatibility complex** (MHC), intensively investigated in mouse, and its human

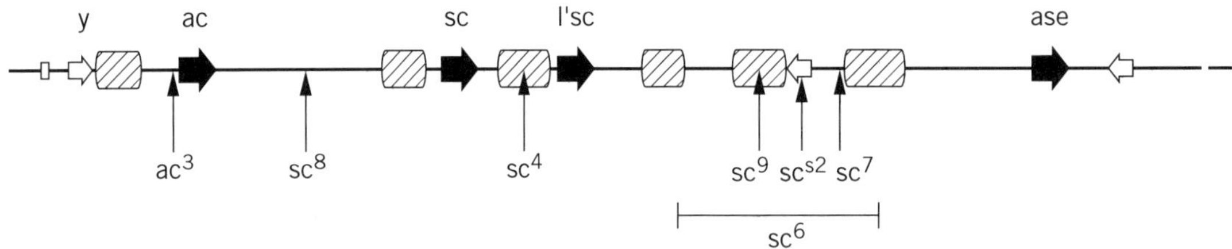

Figure 5. The *scute-achaete* (*sc/ac*) complex at the tip of the *Drosophila melanogaster* X-chromosome. Genes that have functions in neural/bristle development are shown as filled arrows, and other genes, apparently unrelated in function, as open arrows (*y* = *yellow*). Thin vertical arrows indicate the breakpoints of segmental rearrangements that affect the bristle phenotype (sc^6 is a deletion). Regions that include enhancer sequences are shown hatched. The different bristle phenotypes are believed to be due to separation of *sc* and/or *ac* from enhancers. Redrawn and simplified from Ref. 8 by permission.

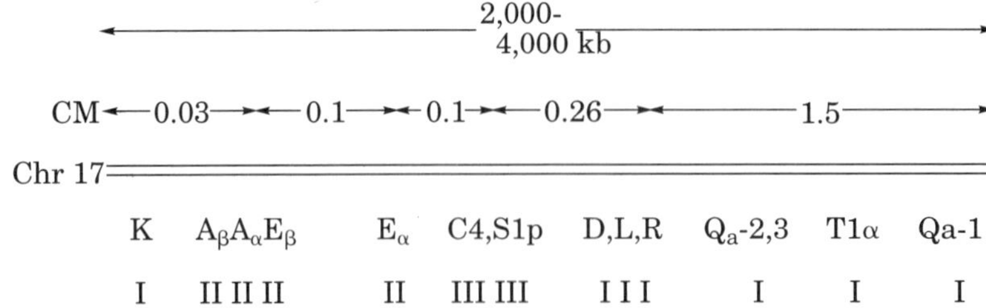

Figure 6. The arrangement and spacing of genes within the major histocompatibility complex (MHC) of mouse. Class I genes encode transplantation antigens mediating graft rejection (K, D, L), or other cell-surface antigens (Qa, Ti). Class II genes encode antigens involved in the immune response. Class II genes encode components of the complement system. Meiotic recombination within the complex is infrequent but not very rare (genetic map distances are shown in centimorgans, cM). After Ref. 9 by permission.

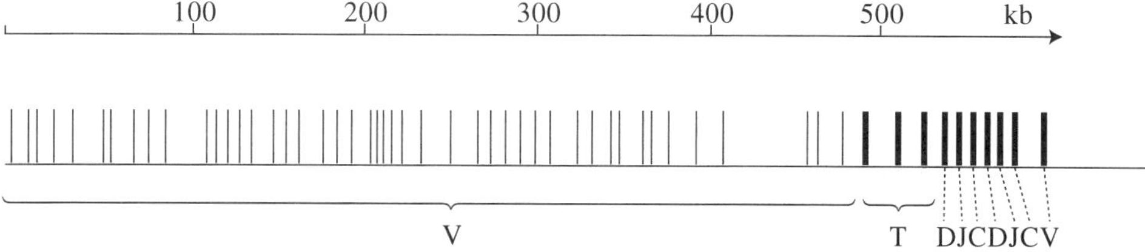

Figure 7. The arrangement and spacing of the part-genes which, when appropriately spliced together, encode β-chains of the human T-cell receptor. Splicing of DNA during T-cell differentiation brings together one C "constant", one J ("junction"), one D ("diversity"), and one V ("variable") sequence to encode a large number of alternative β-chains. J on the map stands in each case for six or seven short alternatively used junction sequences. The other marked elements are all single units. All are transcribed/translated from left to right, except for the rightmost V which is inverted and doubtfully functional. The splicing occurs in stages, first D to J, then DJ to V, and finally VDJ to C. The diagram has been simplified by omitting the **pseudogenes** and incomplete V sequences, which approximately equal the functional genes in number and are interspersed among them. T = trypsinogen gene, the presence of which, in multiple copies, within the β-chain complex, has uncertain functional significance. After Ref. 11 by permission.

equivalent the human leucocyte antigen (HLA) complex, is encoded by genes spread over rather too large a section of chromosome to be called a locus. Indeed, the positions of individual antigen-encoding genes within it are themselves often called loci. In humans it encompasses about four megabases of DNA and 3 to 4 centimorgans. But it may be regarded as at the upper end of the size range of clusters of functionally related genes, and so it is relevant to mention it here. A diagram showing of the structure of the complex, and its scale in relative to the whole chromosome is shown in Fig. 6. Because there is significant homology among the different genes, at least within each of the two main subclusters, the close linkage is likely to be due to duplication by short-range **transposition** and subsequent diversification. But it may also have been maintained by **natural selection**, because their continued coupling is advantageous if certain combinations of alleles of different genes confer disease resistance (9).

An entirely different kind of complex locus is involved in determining hypervariable proteins of the mammalian immune system, **immunoglobulins** and **T-cell receptor** proteins. These are not really gene clusters but rather banks of gene components that are assembled by DNA splicing reactions in a large number of different ways to provide coding sequences for proteins of almost any required specificity. In mammals, as exemplified by mouse, there are six known rearranging complexes of the mouse immune system, three (heavy-chain and two light-chain) for immunoglobulin components and three (α, β and γ polypeptides) for T-cell receptor components. A diagram to scale of the completely sequenced human β-chain locus that extends over 600 kb is shown in Fig. 7 (11). Here the functional reason for clustering is obvious. It is only surprising that the sequences destined to be spliced together are as widely separated as they are. Presumably they are brought close during the splicing process by the formation of some very extensive DNA loops.

BIBLIOGRAPHY

1. F. Banuett (1992) *Trends in Genetics* **8**, 174–180.
2. E. H. Pardoe, S. F. O'Shea, and L. S. Casselton (1996) *Genetics* **144**, 87–94.
3. N. L. Glass and M. A. Nelson (1994) In *The Mycota* I. (J. G. H. Wessels and F. Meinhardt, eds.), Springer-Verlag, Berlin, pp. 295–306.
4. R. F. Geever, L. Huiet, J. A. Baum, B. M. Tyler, V. B. Patel, B. J. Rutledge, M. E. Case, and N. H. Giles (1989) *J. Mol. Biol.* **207**, 15–34.
5. E. L. Walker, T. P. Robbins, T. E. Bureau, J. Kermicle, and S. L. Dellaporte (1995) *EMBO J.* **14**, 2350–2363.
6. J. B. Nasrallah (1997) *Proc. Nat. Acad. Sci. USA* **94**, 9516–9519.
7. S. Campuzano and J. Mondolell (1992) *Trends in Genetics* **8**, 202–207.
8. J. L. Gomez-Skarmeta, I. Rodriguez, C. Martinez, J. Culi, D. Ferres-Marco, D. Beamonte, and J. Mondolell (1995) *Genes Dev.* **9**, 1869–1882.
9. L. Hood, M. Steinmetz, and B. Malissen (1983) *Annu. Rev. Immunol.* **1**, 529–568.
10. I. P. M. Tomlinson and W. F. Bodmer (1995) *Trends Genet.* **11**, 493–498.
11. L. Rowen, B. F. Koop, and L. Hood (1996) *Science*, **272**, 1755–1762.

COMPRESSIBILITY

K. GEKKO

Compressibility is an important measure of the atomic packing and flexibility of protein molecules. Measurement of this quantity is possible, in principle, only for proteins in solution. The partial specific compressibility of a protein in solution $\overline{\beta}$ is defined as the change in the **partial specific volume** $\overline{v}_2{}^0$ with increasing pressure. Since the atomic **van der Waals volume**, v_c, may be assumed to be incompressible, the experimentally determined $\overline{\beta}$ of a protein can be mainly attributed to its cavities v_{cav}, and to changes in **hydration** Δv_{sol}:

$$\overline{\beta} = -(1/\overline{v}_2^0)(\partial \overline{v}_2^0/\partial P) = -(1/\overline{v}_2^0)[(\partial v_{\text{cav}}/\partial P) + (\partial \Delta v_{\text{sol}}/\partial P)] \quad (1)$$

Table 1. Partial specific volume ($\bar{v}_2^0$), compressibility ($\bar{\beta}_s$ and $\bar{\beta}_T$), and volume fluctuation (δV_{rms}) of proteins in water at 25°C [1,4)

Protein	$\bar{v}_2^0$ (cm^3g^{-1})	$\bar{\beta}_s$ (10^{-12} cm^2dyn^{-1})	$\bar{\beta}_T$[a] (10^{-12} cm^2dyn^{-1})	δV_{rms}[b] (cm^3mol^{-1})
		Globular proteins		
Cytochrome c	0.725	0.07	4.27	30.8 (0.34)
Soybean trypsin inhibitor	0.713	0.17	4.44	41.1 (0.20)
Trypsin	0.717	0.92	5.16	46.0 (0.28)
Ribonuclease A	0.704	1.12	5.48 (5)	36.2 (0.38)
Peroxidase	0.702	2.36	6.70	68.3 (0.24)
α-Chymotrypsin	0.717	4.15	8.32	62.2 (0.33)
Lysozyme	0.712	4.67	7.73 (12.3)	44.2 (0.43)
Carbonic anhydrase	0.742	6.37	10.5	76.0 (0.34)
α-Lactalbumin	0.736	8.27	12.4	56.9 (0.54)
β-Lactoglobulin	0.751	8.45	11.8	63.6 (0.46)
Myoglobin	0.742	8.98	13.1	64.1 (0.51)
Ovalbumin	0.746	9.18	12.1	101 (0.30)
BSA	0.735	10.5	14.6 (13.4)	135 (0.27)
		Fibrous proteins		
Gelatin	0.689	−2.5		
F-actin	0.720	−6.3 (20°C)		
Myosin	0.724	−18 (20°C)		
Tropomyosin	0.733	−41 (20°C)		

[a] The values in parentheses represent the experimental data; the other values were calculated with Equation 2 of the text.
[b] The values in parentheses represent the ratio (%) of δV_{rms} to the total protein volume.

(see **Partial specific volume**). The first term on the right-hand side contributes positively, and the second term negatively, to $\bar{\beta}$. Thus, a positive $\bar{\beta}$ value can be ascribed to a large cavity effect overcoming the hydration effect. There exist two types of compressibilities, depending on the experimental conditions: adiabatic and isothermal. Most compressibility studies of proteins have been of adiabatic compressibility because of its accuracy and technical convenience.

ADIABATIC COMPRESSIBILITY

The partial specific adiabatic compressibility $\bar{\beta}_s$ is calculated with the Laplace equation, $\beta_s = 1/u^2d$, where u is the sound velocity and d the density of the protein solution. The sound velocity is usually measured with an accuracy of 1 cm sec^{-1} by using a resonance or sing-around pulse method at 3 to 6 MHz. All amino acids show large negative $\bar{\beta}_s$, with values between $-62.5 \times 10^{-12} cm^2dyn^{-1}$ (for **glycine**) and -21×10^{-12} cm^2dyn^{-1} (for **tryptophan**) at 25°C, because they have a negligibly small amount of cavity. Data for $\bar{\beta}_s$ have been reported for about 40 different proteins; some values in water (or in dilute buffer at neutral pH) are listed in Table 1. Fibrous proteins show negative $\bar{\beta}_s$, indicating the dominant effects of the surface on hydration relative to internal cavity volume. On the contrary, most globular proteins have positive $\bar{\beta}_s$, due to the large contribution of the internal cavities overcoming the hydration effect. This is consistent with the fact that more than 50% of the total surface area is buried in the interior of a folded protein molecule.

Some assumptions are necessary for estimating separately the contributions of cavities and hydration to $\bar{\beta}_s$. A rough estimation suggests that the intrinsic compressibility of globular proteins free from hydration is of the order of (10 to 20) $\times 10^{-12}$ cm^2dyn^{-1}. This value is comparable to the adiabatic compressibility of normal ice, suggesting that an internal protein structure is as rigid as ice.

As can be seen in Table 1, the value of $\bar{\beta}_s$ varies over a wide range and is sensitive to the structural characteristics of individual proteins. For example, **lysozyme** has a considerably smaller $\bar{\beta}_s$ than does α-lactalbumin, in spite of their great similarities in primary and **tertiary structures**. However, the compressibility-structure relationships of proteins have been scarcely discussed on a molecular level, because of the complicated contribution of **hydration**. The contributions of some structural factors have been deduced by the statistical analyses of $\bar{\beta}_s$ data on globular proteins (Ref. 1). From the ratio of accessible surface area to volume, $\bar{\beta}_s$ would be expected to increase with increasing molecular weight, but there appears no definite correlation between $\bar{\beta}_s$ and molecular weight. Instead, a positive correlation is found between $\bar{\beta}_s$ and the partial specific volume. Hydrophobic proteins show large positive $\bar{\beta}_s$, probably due to an enhanced imperfect packing of nonpolar residues localized in the interior of the molecule. Typical α-helical proteins, such as **myoglobin** and bovine **serum albumin** have very large $\bar{\beta}_s$, even though the α-helix itself is rigid, predominantly nonhelical proteins, such as **trypsin** and **soybean trypsin inhibitor**, have small $\bar{\beta}_s$. These results suggest that the α-helix could be a dynamic domain for the thermal fluctuation of proteins. Four amino acids (Leu, Glu, Phe, and His) show statistically a strong ability to increase $\bar{\beta}_s$, whereas another four (Asn, Gly, Ser, and Thr) decrease it, although the meaning of this is obscure. A single amino acid substitution can bring about a noticeable change in $\bar{\beta}_s$, probably due to modified atomic packing, even though it is rare to observe any visible changes in the tertiary structures of mutants by **X-ray crystallography**.

Adiabatic compressibility data for nonnative conformations and for unfolding processes, in combination with partial specific volume, also present useful information on the principles of protein structure that cannot be obtained by spectroscopic techniques. Unfolding of a protein with strong denaturants such as **guanidinium** chloride decreases the compressibility and specific volume, but it is known that the compressibility increases and the partial specific volume decreases on thermal and pressure denaturation.

ISOTHERMAL COMPRESSIBILITY

The volume fluctuations and the pressure-dependent properties of proteins are theoretically related to the partial specific isothermal compressibility $\overline{\beta}_T$, rather than the adiabatic one $\overline{\beta}_s$. To determine $\overline{\beta}_T$, the solution density or partial specific volume must be measured as a function of pressure, under hydrostatic pressure or centrifugal force. Such an experiment is very difficult, however, and few $\overline{\beta}_T$ data are available for proteins, because high pressure may cause protein denaturation and modify preferential solvent interactions. Approximate values of $\overline{\beta}_T$ may be estimated from β_s by utilizing the relationship

$$\beta_T = \beta_s + \alpha^2 T / C_p d \tag{2}$$

if the thermal expansion coefficient α and the **heat capacity** at constant pressure C_p, are known or can be reasonably inferred. Such assumed $\overline{\beta}_T$ values are listed in Table 1, along with some experimental data. Despite many assumptions, the calculated values of $\overline{\beta}_T$ are not so very different from the experimentally observed ones. The value of $\overline{\beta}_T$ is greater than $\overline{\beta}_s$ by (3 to 4) $\times 10^{-12}$ cm^2dyn^{-1}; these differences are comparable to those found for the amino acids.

There are some other methods to evaluate the isothermal compressibility of proteins, although the value obtained is not a partial quantity in solution. Kundrot and Richard (2) reported 4.7×10^{-12} cm^2dyn^{-1} for β_T of lysozyme by comparing its X-ray crystallographic volume at atmospheric pressure and 1000 atmospheres. Contraction of the molecule was distributed nonuniformly; one domain was essentially incompressible, but another had a large compressibility. Computer simulations, such as **molecular dynamics**, **Monte Carlo**, and normal mode analyses, are also available. Normal mode analysis of myoglobin yielded $\beta_T = 9.37 \times 10^{-12} cm^2dyn^{-1}$.

According to statistical thermodynamics, the volume fluctuation δV_p of a protein with volume V_p is related to its isothermal compressibility $\overline{\beta}_T$ (Ref. 3):

$$\overline{\delta V_p^2} = kTV_p\beta_T \tag{3}$$

where k is the Boltzmann constant and T the absolute temperature. The root-mean-square fluctuation of the partial molar volume $\delta V_{\rm rms}$, which was estimated by using $\overline{\beta}_T$ instead of β_T, is listed in the last column of Table 1. The volume fluctuation is only about 0.3% of the overall dimensions of protein, but the $\delta V_{\rm rms}$ values estimated by this method are the lower limit of the probable fluctuations, since the solvation factor is still included in the value of $\overline{\beta}_T$ that was used. If the intrinsic isothermal compressibility of a protein itself, which is not easy to determine, is used instead of $\overline{\beta}_T$, a greater volume fluctuation (up to 50% greater) might be expected for a protein without hydration. Although many assumptions are used in estimating $\overline{\beta}_T$ and the fluctuation of volume, the values obtained are considered to be reasonable; if concentrated in one area at one particular moment, the fluctuation in volume could produce sufficient cavities or channels to allow the entry of solvent or probe molecules to account for phenomena such as **hydrogen exchange** and fluorescence quenching observed with folded proteins.

BIBLIOGRAPHY

1. K. Gekko and Y. Hasegawa (1986) *Biochemistry* **25**, 6563–6571.
2. E. Kundrot and F. M. Richards (1987) *J. Mol. Biol.* **193**, 157–170.
3. A. Cooper (1976) *Proc. Natl. Acad. Sci. U.S.A.* **73**, 2740–2741.
4. K. Gekko (1991) In *Water Relationship in Food* (H. Levine and L. Slade, eds.), Plenum Press, New York, pp. 753–771.

Suggestions for Further Reading

A. P. Sarvazyan (1991) Ultrasonic velocity of biological compounds. *Annu. Rev. Biophys. Chem.* **20**, 321–342.

K. Gekko (1996) Hard proteins and soft proteins: Structural flexibility as revealed by compressibility. *Protein, Nucleic Acid and Enzyme* (*Tanpakusitu, Kakusan, Koso*, in Japanese) **41**, 2025–2036.

COMPUTER SIMULATION OF BIOLOGICAL MOLECULES

A. WARSHEL

Biological molecules are systems of enormous complexity. This makes it exceedingly difficult to elucidate the details of biological activities using only experimental techniques. Despite enormous progress in structural studies of **proteins**, it is hard to determine uniquely how the particular function is determined by the given structure. A mathematical description of the energy of a macromolecule in terms of its atomic coordinates is, in principle, sufficient to predict relevant biological properties. However, such a description can only be handled by very powerful computers. Having the proper model of a macromolecule should allow one to simulate its properties as if one were dealing with the actual molecule. Such an approach, called *computer simulation*, offers what is probably the best way of describing the functions of biological molecules.

The relationship between the energy of a molecule and the position of its atoms is called a **potential function** or a *force field*. A reasonable potential function provides the first step for modeling a molecule. In principle, one can obtain potential functions by using quantum mechanical calculations describing the entire molecule as a supermolecule. However, such approaches are not yet practical for describing macromolecules. Some progress can be made using hybrid QM/MM approaches, but in general it is necessary to use approximated potential functions to obtain parameters from empirical data. (1–5). Potential functions can be based on all-atom molecular models that describe the interactions between all the atoms of the given molecule or on simplified models, where the forces due to several atoms are described by a single interaction center. Examination of molecular properties using the corresponding potential functions is known as **molecular mechanics**. This term is usually reserved for the use in energy minimization and normal mode analysis methods. The use of potential functions to solve the equation of motion of atoms is known as **molecular dynamics** (MD). This method gives the atomic position as a function of time and provides a powerful way of obtaining dynamical and time-average properties. In some cases one can obtain useful insight by using Brownian dynamics. Frequently it is sufficient to know the properties of macromolecules at their energy minima. Generating starting configurations in the search for such minima can be done by **Monte Carlo calculations** and other sampling approaches. Apparently, the most important biological properties are reflected in most molecules by their thermodynamic properties, not their dynamical properties. In particular, the **free energy** provides the most important connection between structure and function, while the knowledge of the individual contributions to free energy from **enthalpy** and **entropy** is less critical for reproducing the functional properties of macromolecules. The free energies of macromolecules can be evaluated by several approaches and in particular by the free energy perturbation (FEP) and linear response approximation (LRA) methods. **Free energy calculations** are frequently based on the use of thermodynamic cycles (3,4,6).

The potential functions that describe the energetics of macromolecules reflect different types of interactions. These include covalent bonding and **van der Waals interactions**, which determine the steric properties of the given molecule, and **electrostatic interactions**, which include **hydrogen bonds** and other dipolar and ionic interactions. The free energy of a macromolecule is largely dependent on electrostatic energies, which reflect the effects of **polar** and ionized groups and their compensation by the polarization of the solvent around the protein (6,7) These energies involve long-range interactions, whose accurate evaluation present a major challenge. Electrostatic energies can be modeled by using implicit models where the solvent molecules (and sometimes the protein dipoles) are represented implicitly. The use of a dielectric constant in an electrostatic model is a form of implicit model, and in fact the dielectric constant in many electrostatic models is not related at all to the true protein dielectric constant but to the properties that are not simulated explicitly (6).

The most effective uses of computer simulations have been in the following subjects.

ENERGETICS (THERMODYNAMICS AND EQUILIBRIUM PROPERTIES)

One of the most effective tests of free energy calculations and of the nature of biomolecular energetics is provided by calculations of absolute pK_a values of ionizable groups in proteins. The basis for calculating this quantity is the relationship (6)

$$2.3RT(pK_a^p - pK_a^w) = \Delta\Delta G_{sol}^p(AH \rightarrow A^-) - \Delta\Delta G_{sol}^w(AH \rightarrow A^-) \qquad (1)$$

where the superscripts p and w denote protein and water, respectively, and $\Delta\Delta G_{sol}(AH \rightarrow A^-)$ is the difference in solvation energy between the protonated and deprotonated states of the acid. Although early models only considered the change of pK_a values due to the effect of protein ionized groups (8), it is now recognized that the most important factor is the self-energy (6) (or "solvation energy"), which is the energy of forming the charge in its protein environment when the other charges are turned off. The modulation of this factor is largely determined by the protein permanent dipoles (6). Obtaining reliable estimates of the self-energy by FEP is very challenging, but significant progress has been made (3,9). Apparently, it is essential to treat long-range effects in a consistent way in order to obtain meaningful results in pK_a calculations (9). It is frequently simpler to obtain better results by simplified solvent models (6) and by models that treat the solvent as a dielectric continuum (10). However, one must bear in mind that models that treat the protein as a uniform dielectric without considering the microscopic nature of the protein permanent dipoles cannot reproduce the correct physics of ionized groups in proteins (6).

Redox energies of biological cofactors play a major role in **electron transfer** processes. Such energies can be evaluated by simulation methods using a similar approach to that used in equation 1 and evaluating the change in solvation energy upon moving the charge from water to the protein. FEP calculations of redox energies were reported by several workers, and related studies in **photosynthetic** reaction centers were also presented. Here the proper treatment of the solvent around the protein is quite crucial (11). As in the case of pK_a calculations, one finds that simplified models (12) and continuum models (13,14) are quite effective.

Reliable calculations of free energies of **ligand binding** are crucial for quantitative progress in rational drug design. FEP has been used extensively in studies of binding free energies (3). Most studies involve simple cases of small changes in the ligand (15). Calculations of absolute binding free energies are much more challenging (3,16) and involve major convergence problems. Here the use of the LRA has been found to be quite effective (16,17). Extensive progress has also been made with simplified models (16) and with more phenomenological approaches (18).

The permeation of ions through membrane channels (see **Membrane transport**) plays a major role in biophysics. Here the challenge is in evaluating the energy of an ion in the channel relative to its energy in solution and in using the energetics to determine the conductance of the channel. Reliable simulation of such systems requires one to represent properly the channel, plus its surrounding membrane and water environment. The energy contribution of the membrane-induced dipoles must be included in consistent calculations. Consistent FEP studies of the energetics of moving a Na^+ ion to the center of the **gramicidin** A channel have been reported with a reasonable result (19). Incomplete modeling of the boundaries of the simulated system and long-range interactions can lead to significant overestimates of the penetration barrier (eg, the results of Ref. 20). Semimacroscopic studies of ion channels are quite useful and much less challenging than full microscopic calculations (9,21).

STRUCTURAL PROPERTIES

Molecular simulations can be used to refine and sometimes to predict structural properties. Energy minimization approaches were the first to be used successfully in relaxing bad steric contacts in structures of proteins determined by **X-ray crystallography** (22). Such approaches were also used in more systematic refinements of protein structures (23). The emergence of powerful computers allows one to use more rigorous statistical mechanical approaches, coupled with MD simulations, in exploring structural properties (24).

Simulations are also very useful in evaluating and analyzing **temperature factors** and interpreting the meaning of the corresponding experimental results.

All-atom simulation approaches are still quite ineffective in predicting protein structures, and one has to resort to simplified models. Even when starting from the X-ray structure, it is not trivial to find the simulated structure identical to the observed one. The problems involved include the need for reliable potential functions and, perhaps more importantly, a realistic description of the solvent (25) and proper treatment of long-range forces.

DIELECTRIC PROPERTIES

Although proteins cannot and should not be described as a medium of a uniform dielectric constant, it is important to understand their dielectric properties. Experimental attempts to determine local dielectric constants are far from being unique, and simulation studies can provide great insight. In relating the simulated dielectric constant to experimental observations, one should take into account the effect of the solvent around the protein, which is also referred to as the *reaction field* (26). Consistent simulations of the overall dielectric constant of solvated proteins have been reported by several workers (27–29). It has been found that the local dielectric at protein **active sites** is quite large and quite different from the low value usually deduced from dielectric measurements. Simulations are also very effective in analyzing the fast time components of dielectric **relaxation times** (27,29).

DYNAMICAL PROPERTIES

MD simulations and related approaches provide powerful ways of evaluating dynamical properties. Many of these properties can be cast in terms of the corresponding time correlation function (30):

$$C_A(t) = \frac{1}{\tau}\int_0^{\tau} A(t')A(t + t')dt' \qquad (2)$$

where A is the property of interest and τ is a sufficiently long simulation time. For example, the **diffusion** constant, D, can be obtained from the velocity autocorrelation, $C_v(t)$, by

$$D = \frac{1}{3}\int_0^{\infty} C_v(t)dt \qquad (3)$$

This relationship can be used in both comparing simulations to experiments and in analyzing the microscopic meaning of experimentally-deduced parameters. The dynamical range accessible to simulation studies can be probed by several approaches, and significant progress has been made in comparing simulations to the results of inelastic **neutron scattering** (31).

Molecular fluctuations are characterized by correlation times that are closely related to the underlying relaxation processes. Relevant relaxation times can be evaluated by simulation studies using

$$\tau_A^{-1} = C_A(0)^{-1}\int C_A(t)dt \qquad (4)$$

The fluctuations of the macroenvironments of proteins can play an important role in fast reactions. This is true in particular with regards to the effect of electrostatic fluctuations on charge-transfer processes, such as electron transfer reactions. The rate of such a reaction is determined by the effect of the protein fluctuation on the difference between the energies of the reactant and product state. Such an approach has been used in studies of the dynamics of the primary event in **photosynthesis** (32–34). The same approach can be used in studies of photoisomerization processes, although in such cases one needs to account for a large coupling between the electronic states. Regarding the rate constants of regular chemical reactions, the most important factors are **activation energies** (see below) and not dynamical effects. Nevertheless, the probability that a trajectory that reaches the **transition state** will lead to reaction, the transmission factor, is a dynamical effect. This factor can be evaluated using MD simulation by running trajectories downhill from the transition state (35). However, the transition factor is usually close to unity in reactions with significant activation barriers (Fig. 1) (4).

RATES OF BIOLOGICAL PROCESSES

The rate constant of a process in a condensed phase can be expressed as

$$k = F(k_BT/h)\exp(-\Delta g^{\ddagger}/k_BT) \qquad (5)$$

where k_B is the Boltzmann constant, T is the absolute temperature, h is the Planck constant, F is the transmission factor (see above), and $\Delta g^{\ddagger}$ is the **activation free energy**. For most reactions with large $\Delta g^{\ddagger}$, one finds that F is close to unity and therefore the most important factor is $\Delta g^{\ddagger}$. **Enzymes** catalyze their reactions by reducing activation free energies (36), and one of the most important questions in structure function correlation is how this reduction of $\Delta g^{\ddagger}$ is accomplished (4). Molecular simulations should allow one to evaluate activation free energies and transmission factors and to probe the molecular origin of enzymatic reactions. However, such reactions involve bond-making and bond-breaking processes that cannot be described by simple force fields. It is also impractical to describe the complete enzyme/substrate complex by high-level quantum mechanical approaches. A possible alternative is provided by hybrid quantum mechanics/molecular mechanics (hybrid QM/MM) methods (4,37–40). Such methods describe only parts of the reacting system quantum mechanically, while treating the rest of the protein classically. QM/MM approaches are very promising, but the corresponding results are not yet quantitative. An alternative that is at present the most reliable way of simulating enzymatic reaction is provided by the empirical valence bond (EVB) approach (4,41,42). The EVB method describes the reaction potential surface by a valence bond method that considers the different resonance structures of the reactant and product fragments and their interaction with the surrounding protein and solvent. This method is calibrated by considering the reference reaction in water, so it focuses on the difference between the reaction in protein and in solution, which is much easier than predicting the reaction potential surface in the protein from first principles.

The EVB approach can be used to calculate, conveniently and efficiently, activation free energies of enzymatic reactions using a combination of FEP and umbrella sampling approaches (4,41). The activation energies can be used in obtaining rate constants of enzymatic reactions using equation 5.

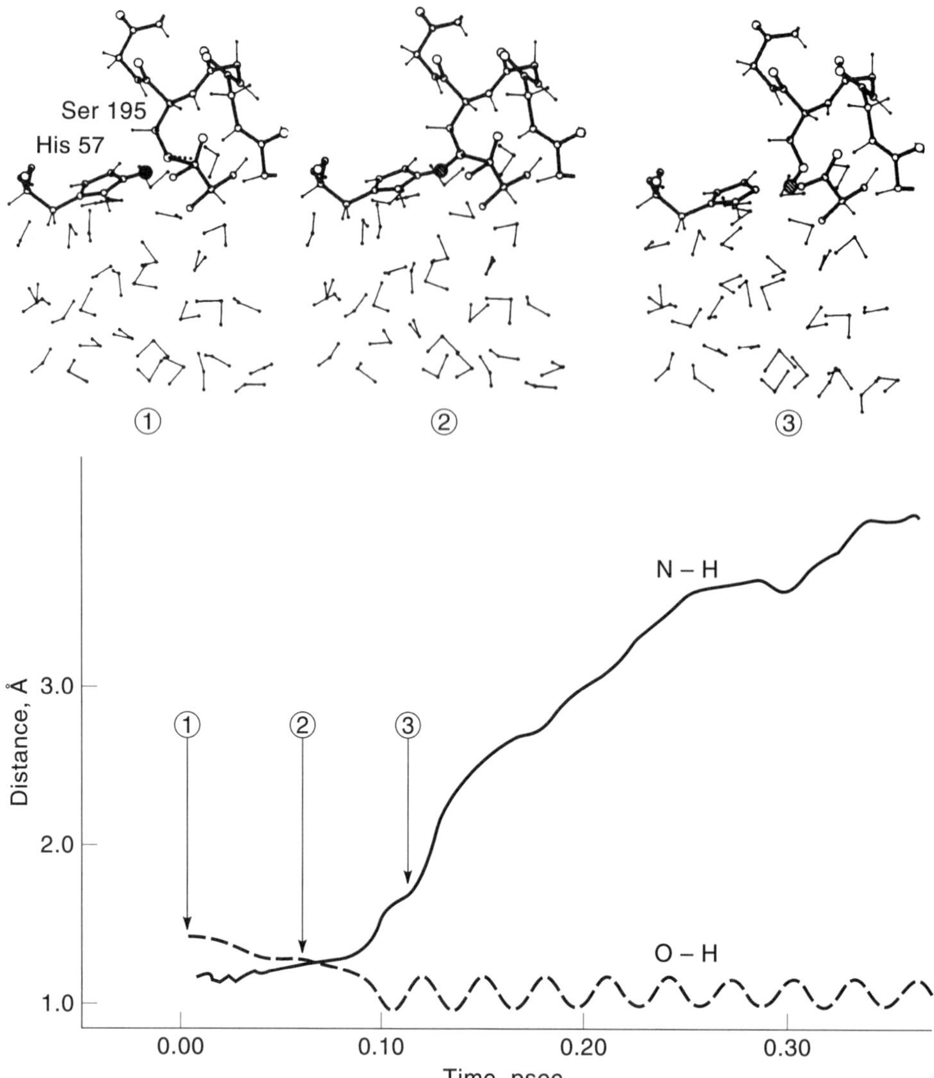

Figure 1. A downhill trajectory for the proton transfer step in the catalytic reaction of trypsin. The trajectory moves on the actual ground state potential, from the top of the barrier to the relaxed enzyme–substrate complex. 1, 2, and 3 designate different points along the trajectory, whose respective configurations are depicted in the upper part of the figure. The time reversal of this trajectory corresponds to a very rare fluctuation that leads to a proton transfer from Ser 195 to His 57. (Adapted from Ref. 4.)

PROTEIN FOLDING

Models that use simplified representations of the protein amino acid residues (43) have provided a powerful understanding of the protein folding process (44–47). Such models are sometimes classified as *on-lattice* or *off-lattice* models. On-lattice models consider the simplified representation along a regular cubic lattice (thus allowing exact enumeration, but at the expense of somewhat unrealistic structure), while the off-lattice models try to provide more realistic representation (43). Simplified models are used extensively in understanding the relationship between sequence to folding probability and to the specific folded structure (43–47) and in deducing the sequence of events in the folding process (44).

BIBLIOGRAPHY

1. S. Lifson and A. Warshel (1968) *J. Chem. Phys.* **49**, 5116.
2. U. Burkert and N. L. Allinger (1982) *Molecular Mechanics*, American Chemical Society, Washington, D.C.
3. P. Kollman (1993) *Chem. Rev.* **93**, 2395.
4. A. Warshel (1991) *Computer Modeling of Chemical Reactions in Enzymes and Solutions*, Wiley, New York.
5. A. T. Hagler, E. Huler, and S. Lifson (1977) *J. Am. Chem. Soc.* **96**, 5319.
6. A. Warshel and J. Åqvist (1991) *Annu. Rev. Biophys. Biophys. Chem.* **20**, 267.
7. P. E. Smith and B. M. Pettitt (1994) *J. Phys. Chem.* **98**, 9700–9711.

8. C. Tanford and J. G. Kirkwood (1957) *J. Am. Chem. Soc.* **79**, 5333.
9. F. S. Lee, Z. T. Chu, and A. Warshel (1993) *J. Comp. Chem.* **14**, 161.
10. A. S. Yang, M. R. Gunner, R. Sampogna, K. Sharp, and B. Honig (1993) *Proteins Struct. Funct. Gen.* **15**, 252–265.
11. R. G. Alden, W. W. Parson, Z.-T. Chu, and A. Warshel (1995) *J. Am. Chem. Soc.* **117**, 12284.
12. P. J. Stephens, D. R. Jollie, and A. Warshel (1996) *Chem. Rev.* **96**, 2491.
13. M. R. Gunner and B. Honig (1991) *Proc. Natl. Acad. Sci. USA* **88**, 9151–9155.
14. H.-X. Zhou (1994) *J. Am. Chem. Soc.* **116**, 10362–10375.
15. T. P. Straatsma and J. A. McCammon (1992) *Annu. Rev. Phys. Chem.* **43**, 407.
16. F. S. Lee, Z. T. Chu, M. B. Bolger, and A. Warshel (1992) *Protein Eng.* **5**, 215.
17. J. Åqvist, C. Medina, and J.-E. Samuelsson (1994) *Protein Eng.* **7**, 385.
18. G. Verkhivker, K. Appelt, S. T. Freer, and J. E. Villarfranca (1995) *Prot. Eng.* **8**, 677–691.
19. J. Åqvist and A. Warshel (1989) *Biophysical. J.* **6**, 91.
20. B. Roux and M. Karplus (1993) *J. Am. Chem. Soc.* **115**, 3250–3260.
21. V. Dorman, M. B. Partenskii, and P. C. Jordan (1996) *Biophys. J.* **70**, 121–134.
22. M. Levitt and S. Lifson (1969) *J. Mol. Biol.* **46**, 269–279.
23. A. Jack and M. Levitt (1978) *Acta Cryst.* **A34**, 931.
24. A. T. Brunger (1988) *XPLORE Manual*, Yale University, New Haven, CT.
25. M. Levitt and R. Sharon (1988) *Proc. Natl. Acad. Sci. USA* **85**, 7557–7561.
26. L. Onsager (1936) *J. Am. Chem. Soc.* **58**, 1486.
27. G. King, F. S. Lee, and A. Warshel (1991) *J. Chem. Phys.* **95**, 4366.
28. T. Simonson and D. Perahia (1995) *J. Am. Chem. Soc.* **117**, 7987.
29. W. F. vanGunsteren and A. E. Mark (1992) *Eur. J. Biochem.* **204**, 947.
30. D. A. McQuarrie (1976) *Statistical Mechanics*, Harper and Row, New York.
31. S. Cusack (1989) *Chem. Scripta* **29A**, 103–107.
32. A. Warshel and W. W. Parson (1991) *Annu. Rev. Phys. Chem.* **42**, 279.
33. S. Creighton, J.-K. Hwang, A. Warshel, W. W. Parson, and J. Norris (1988) *Biochemistry* **27**, 774.
34. K. Schulten and M. Tesch (1991) *Chem. Phys.* **158**, 421–446.
35. C. H. Bennet (1977) *Algorithms for Chemical Computations*, ACS, Washington, D.C.
36. L. Pauling (1946) *Chem. Eng. News* **263**, 294.
37. A. Warshel and M. Levitt (1976) *J. Mol. Biol.* **103**, 227.
38. V. Thery, D. Rinaldi, J.-L. Rivail, B. Maigret, and G. G. Ferenczy (1994) *J. Comp. Chem.* **15**, 269–282.
39. J. Gao (1995) *Reviews in Computational Chemistry*, VCH, New York.
40. A. J. Mulholland, G. H. Grant, and W. G. Richards (1993) *Protein Eng.* **6**, 133–147.
41. J. Åqvist and A. Warshel (1993) *Chem. Rev.* **93**, 2523.
42. J. Lobaugh and G. A. Voth (1996) *J. Chem. Phys.* **104**, 2056–2069.
43. M. Levitt and A. Warshel (1975) *Nature* **253**, 694.
44. J. N. Onuchic, P. G. Wolynes, Z. Luthey-Schulten, and N. D. Socci (1995) *Proc. Natl. Acad. Sci. USA* **92**, 3626–3630.
45. A. Godzik, A. Kolinski, and J. Skolnick (1993) *J. Comput. Aided Mol. Des.* **7**, 397–438.
46. K. Yue and K. A. Dill (1992) *Proc. Natl. Acad. Sci., USA* **89**, 4163–4167.
47. E. Shakhnovich, G. Farztdinov, A. M. Gutin, and M. Karplus (1991) *Phys. Rev. Lett.* **67**, 1665–1668.

Suggestions for Further Reading

M. P. Allen and D. J. Tildesley (1987) *Computer Simulation of Liquids*, Oxford University Press, Oxford.

A. Warshel (1991) *Computer Modeling of Chemical Reactions in Enzymes and Solutions*, Wiley, New York.

D. Chandler (1987) *Introduction to Modern Statistical Mechanics*, Oxford University Press, New York.

CONCATEMERS

GEORGES N. COHEN

A concatemer is a multimeric molecule of **DNA** formed by identical monomers arranged linearly in the same head-to-tail orientation. Closure of a **bacteriophage** DNA molecule upon entry into the bacterial host cell is a relatively simple process. The complementary single-stranded ends anneal in a reaction that is thermodynamically favored and by conditions that restrict **diffusion** and keep the ends close to each other. **Annealing** is followed by covalent joining, catalyzed by the host **DNA ligase**, which forms phosphodiester bonds. Opening of the circular DNA molecule, coupled to its encapsidation to form mature phage particles, however, is rather complex.

Capsid precursors interact with the newly replicated DNA molecules in the form of concatemers (1,2), which are the product of DNA replication proceeding in two stages. First, the DNA injected into the host is circularized and covalently closed, as previously described. In the case of the **lambda phage**, the DNA undergoes several stages of DNA replication in which the circular DNA rings generate daughter rings. Then, during phage vegetative growth, **rolling-circle** replication gives rise to *linear* concatemers of λ DNA. These concatemers, which in the absence of packaging may be up to 10 chromosomal units long, are the normal substrates for packaging (see **Bacteriophage infection**).

RNA primers for DNA synthesis are excised during the process. This raises the question of how the extreme ends of linear DNA can be completed. Not only must a small RNA fragment initiate at the 3′ terminal nucleotide of the template strand, but the sequence must be filled in after ribonucleotide removal. This dilemma was resolved using phage T7 DNA, which has the property of having redundant ends. The sequence of about 160 bp found at the left end is repeated exactly at the right end of the molecule. When the linear T7 molecules replicate, they do not generate unit-length progeny molecules, but very long concatemers (units linked end to end) containing the complete genome sequence repeated over and over.

BIBLIOGRAPHY

1. D. Kaiser and T. Masuda (1973) *Proc. Natl. Acad. Sci. USA* **70**, 260–264.
2. D. Kaiser, M. Syvanen, and T. Masuda (1974) *J. Supramol. Struct.* **2**, 318–328.

CONDITIONAL LETHAL MUTATIONS

WILLIAM A. ROSCHE
PATRICIA L. FOSTER

Mutations that cause lethality under one condition (the restrictive or **nonpermissive condition**) but not another (the **permissive condition**) are called conditional lethal mutations, or loosely, conditional lethals. Conditional lethal mutations have an honored position in molecular biology. The recognition that a **gene** is not indivisible, the determination of the triplet nature of the **genetic code**, and the discovery of **nonsense codons**, all resulted from studies of a particular class of conditional lethal mutants of **bacteriophage** T4. T4 *rII* mutants grow on *Escherichia coli* strain B (the permissive host) but not on *E. coli* strain K-12 when it is **lysogenic** for bacteriophage **lambda** (the nonpermissive host). Interestingly, the function of the proteins encoded by the two *rII* genes remains unknown. The conditional lethality of the *rII* mutants allowed geneticists to grow the phage, to perform genetic crosses on the permissive host, and then to analyze the results on the nonpermissive host.

Conditional lethal mutations are the only genetic way to identify essential genes in **haploid** organisms or cells. By definition, a mutant with a **null mutation** in an essential gene cannot be isolated in the haploid state (in **diploid** organisms, **recessive lethal mutations** on **autosomes** are **complemented** by the other functional **allele**). Conditional lethal mutations are usually identified by screening mutant clones for growth under permissive conditions but not under restrictive conditions. Lethal in this case does not always mean conveying death. For example, **auxotrophies** are considered a class of conditional lethals because the auxotrophic cells grow only in the presence of the required growth factor, although they do not necessarily die in its absence. The more typical classes of conditional lethals are host-range mutants (described above), **temperature-sensitive mutants**, and **cold-sensitive mutants**. **Nonsense mutations** are considered conditionally lethal because the mutation is suppressed when a suppressor **transfer RNA** is present (see **Nonsense suppressors**).

Conditional lethal mutations are a powerful way to identify **protein–protein interactions**. A mutation that renders a protein inactive at high or low temperature, for example, may be suppressed by a compensating mutation in the gene encoding a protein that interacts with it. To conclude that the two proteins interact, the suppression must be allele-specific, ie, not caused by general suppression (as in the case of tRNA nonsense suppressors). In addition, other events, such as **gene duplication**, suppress conditional lethals. Interacting proteins are also identified by synthetic lethal mutations, which are mutations in each of two genes, neither of which is lethal alone but which are lethal when together in the same cell.

Suggestions for Further Reading

A. Adams, D. E. Gottschling, C. Kaiser, and T. Sterns (1998) *Methods in Yeast Genetics: A Laboratory Course Manual*, Cold Spring Harbor Laboratory Press, Cold Spring Harbor, NY.

E. M. Phizicky and S. Fields (1995) Protein-protein interactions: methods for detection and analysis, *Microbiol. Rev.* **59**, 94–123.

A. Griffiths, J. H. Miller, D. Suzuki, R. Lewontin, and W. Gelbart (1996) *An Introduction to Genetic Analysis*, 6th ed., W. H. Freeman, New York.

J. Beckwith and T. Silhavy (1992) *The Power of Bacterial Genetics: A Literature-Based Course*, Cold Spring Harbor Laboratory Press, Cold Spring Harbor, NY.

T. D. Brock (1990) *The Emergence of Bacterial Genetics*, Cold Spring Harbor Laboratory Press, Cold Spring Harbor, NY.

CONFIGURATION

VERNON ANDERSON

Although configuration and **conformation** are listed as synonyms in the *Oxford English Dictionary* (1), in chemistry and biochemistry they have specific and separate meanings differentiated by time or energy. Both terms refer to the organization of atoms in space, but configuration refers to the element that remains invariant in time in the absence of any covalent bonds being altered, whereas conformation refers to the relative orientation in space of atoms that can vary by rotation about single bonds and consequently varies in real time for biological molecules. The absolute configuration identifies the arrangement of atoms that generates **chirality**in a molecule. Following Pasteurs identification of the two chiral forms of tartaric acid, it was only known that they differed by being nonsuperimposable mirror images. This ambiguity was resolved when the crystal structure of (+) tartaric acid was determined by Bijvoet and co-workers (2). By this means, the absolute configuration of many related molecules became identifiable.

The designation of the configuration of a chiral center was problematic for nearly a century. Fischer introduced a general procedure that identified enantiomers as either D- or L- (3) based on whether the nonhydrogen substituent was on the right or left when the molecule was drawn as a "Fischer projection" (Fig. 1). This nomenclature permeates biochemistry through the now colloquial names of **amino acids**, carbohydrates, and **lipids**.

D-ribose

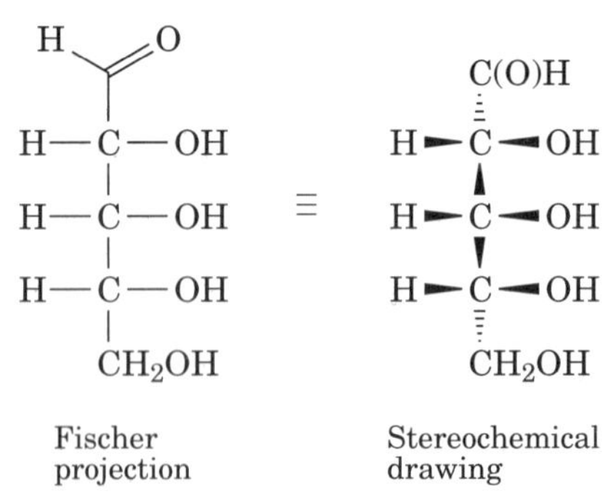

H O / C(O)H
H—C—OH ≡ H—C—OH
H—C—OH H—C—OH
H—C—OH H—C—OH
CH_2OH CH_2OH

Fischer projection — Stereochemical drawing

Figure 1. A Fischer projection. By convention, it has the vertical bonds directed away from the viewer and horizontal bonds directed out toward the viewer. The configuration of each chiral carbon in D-ribose is D because the nonhydrogen substituent is drawn to the right. For sugars, the enantiomer is defined by the bottom chiral carbon when the carbon chain is oriented vertically with the carbonyl carbon at the top; thus, D-ribose is pictured.

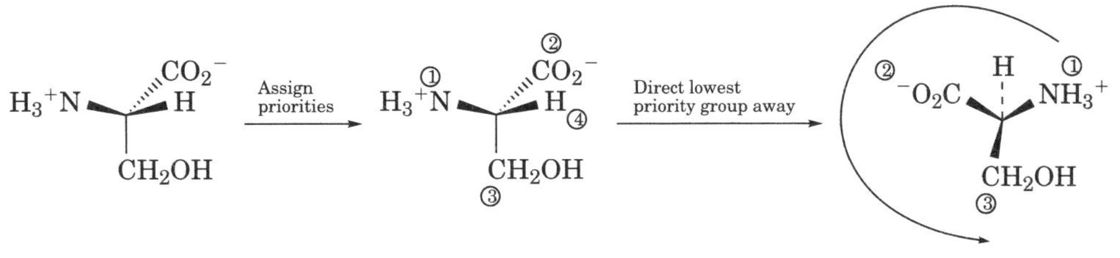

Figure 2. Assignment of the relative priorities of the four substituents on the chiral α-carbon of serine. The carboxylate carbon takes precedence over the hydroxymethyl carbon because it has three bonds to oxygen. The assignment of the (S) configuration results from the counterclockwise arc that connects the substituents in order of precedence.

The Fischer nomenclature leads to ambiguities (4). A rigorous and unambiguous method of identifying configuration proposed by Cahn et al. has been adopted for specifying absolute configuration as either (R) or (S) for chiral tetrahedral centers (5). The procedure requires the assignment of priority to the four substituents that generate a chiral center, followed by a procedure to identify the arrangement as either (R) or (S). The four rules for assigning priority are:

1. Substituents are assigned their priority in order of decreasing atomic number of the atom directly bonded to the chiral center.
2. If two or more atoms receive the same priority in step 1, the atoms bonded to each of the equal priority atoms are examined one at a time. If the two groups are not differentiated by the atom of greatest priority, the second and third atoms are compared successively.
3. Heavier isotopes take precedence over lighter isotopes, eg, ^2H over ^1H and ^{14}C over ^{12}C.
4. Double bonds are counted as two bonds to the same atom.

Once the relative priorities of the four substituents are assigned, the bond to the lowest priority group is oriented directly away from the observer, and the remaining three groups are viewed in a plane. If the arc connecting the three groups in order of highest to lowest priority is clockwise, the chiral center is assigned the (R) configuration (from the Latin rectus), whereas if the arc is counterclockwise, the center is assigned the (S) configuration (from the Latin sinister). These assignments are shown in Figure 2.

BIBLIOGRAPHY

1. *Oxford English Dictionary*, 2nd ed. (1989) Clarendon Press, Oxford.
2. J. M. Bijvoet, A. F. Peerdeman, and A. J. van Bommel (1951) *Nature* **168**, 271.
3. E. Fischer (1891) *Ber. Dtsch. Chem. Ges.* **24**, 2683.
4. E. L. Eliel (1962) *Stereochemistry of Carbon Compounds*, McGraw-Hill, New York, Chap. 5.
5. R. S. Cahn, C. K. Ingold, and V. Prelog (1966) *Angew. Chem. Int. Ed.* **5**, 385–415.

Suggestions for Further Reading

J. March (1985) *Advanced Organic Chemistry*, Wiley-Interscience, New York, pp. 93–99.

K. Mislow (1966) *Introduction to Stereochemistry*, W. A. Benjamin, New York, pp. 86–97.

CONFOCAL MICROSCOPY

GUY A. PERKINS
TERRENCE G. FREY

Confocal microscopy is now a well-established tool for the examination of subcellular structure and function and complements light and electron microscopy. Because of its high temporal resolution, confocal microscopy allows for the visualization of living as well as fixed tissues and cells, and therefore dynamic processes can be examined and analyzed quantitatively as they actually occur. Confocal microscopy offers several advantages over conventional light microscopy, among which are an increase in contrast, resolution, and clarity (1). Conventional light microscopy provides a two-dimensional (2-D) image of the specimen in the focal plane of the objective lens, but this image is contaminated by out-of-focus images of the specimen above and below the focal plane. Confocal microscopy provides a 2-D image in the focal plane without the out-of-focus information. Furthermore, the resolution of images from a confocal microscope is improved by a factor of 1.4–1.75 (2). Through computer control of the focus and acquisition of images, modern confocal light microscopes can collect a series of 2-D images (or "optical sections") through the specimen producing a three-dimensional (3-D) image. Four-dimensional imaging (4-D), defined as 3-D imaging over time, is a recently developed extension in which 3-D images are recorded at periodic time intervals (3).

The basic principle of the confocal microscope is to eliminate the scattered, reflected, or fluorescent light from out-of-focus planes by making the illumination, specimen, and detector all have the same focus, ie, they are confocal. In effect, this microscope will image only the very thin optical section on which the beam is focused. Matched pinholes are used, one at the light source which is imaged onto the specimen to function as a probe that is scanned over the specimen, and one at the detector to capture only a narrow plane of focus. Thus, the out-of-focus blur from areas above and below the focal plane are eliminated. The matched pinhole apertures improve the lateral resolution over conventional light microscopes by a factor of 1.4 with the use of circular apertures and 1.75 with annular apertures (2). Confocal microscopes are often designed to scan in a raster pattern over the sample in which the microscope illuminates one spot at a time, scanning the spot along parallel lines in the focal plane. Lasers are an ideal illumination source for raster scanning because they provide an intense beam of monochromatic radiation that can be condensed onto a small spot. Hence many confocal microscopes are laser scanning (LSCM).

Applications of LSCM include (i) determining the location of **proteins**, **lipids** and nucleic acids, cytoskeletal structures and organelles within cells (4,5), using fluorescent dyes, antibodies, **phalloidin**, and **lectins**, (ii) observing ionic fluctuations, such as calcium, magnesium, and pH, in cells and organelles (6), and, (iii) measuring **membrane potential** using fluorescent dyes (7). The power of 4-D imaging in molecular biology was illustrated by the noninvasive monitoring by LSCM of mitotic events, and cleavage and migration patterns of fertilized sea urchin eggs labeled with $DiOC_6$ (8). **Macromolecules** and subunits can be characterized by immunocytochemical fluorescence probes, which involves the use of antibodies labeled with fluorophores (9). When more than one fluorophore is used, 3-D multilabel (multicolor) imaging can be used to map the 3-D relationship of the labeled structures. For example, z-series collected from two different channels, fluorescein and rhodamine, can be merged into a single reconstruction. By rotating the rendered volume, particular nuances of the structural relationships highlighted by the bound fluorophores may be revealed (10,11). It is not only important to know that a particular compound is present in a specific cell type or subcellular component, but it is also important to detect and quantify changes in local concentrations of such compounds. Quantitative immunocytochemistry utilizes the principles of stereology and statistical analyses (12) and can be coupled to LSCM to detect differences in immunoreactivity. This method has been used to measure the distribution of integral **membrane proteins** in the vertebrate retina and to determine the distribution of transport vesicles (13).

Many of the advantages of a confocal light microscope can be achieved with a standard light microscope equipped with a digital camera interfaced to a computer, which controls the microscope stage and focus control as well as run rapid deconvolution algorithms (14). With knowledge of the 3-D point spread function of the microscope, the deconvolution algorithms remove out-of-focus signal to produce images comparable in quality to LSCM images. The advantages of the deconvolution confocal microscope are that (i) the system expense is a fraction of the cost for a LSCM system, and (ii) lower illumination levels can be used, since all of the light emitted from the specimen is used in forming the 3-D image. Lower illumination minimizes the problem of photobleaching of fluorophores used to label the specimen. A disadvantage of this approach is that the full 3-D deconvolution process can take significantly more time to produce a 3-D image than does a confocal microscope.

BIBLIOGRAPHY

1. A. Boyde (1990) Confocal optical microscopy, In *Modern Microscopies* (P. J. Duke and A. G. Michette, eds.), Plenum Press, New York, pp. 185–204.
2. E. M. Slayter and H. S. Slayter (1992) *Light and Electron Microscopy*, Cambridge University Press, Cambridge, UK.
3. S. A. Stricker, S. Paddock, and G. Schatten (1990) *J. Cell Biol.* **111**, 113a.
4. A. H. Cornell-Bell et al. (1993) Membrane glycolipid trafficking in living polarized pancreatic acinar cells: Assessment by confocal microscopy, In *Methods in Cell Biology*, Vol. 38: *Cell Biological Applications of Confocal Microscopy* (B. Matsumoto, ed.), Academic Press, San Diego, pp. 222–241.
5. I. L. Hale and B. Matsumoto (1993) Resolution of subcellular detail in thick tissue sections: Immunohistochemical preparation and fluorescence confocal microscopy, In *Methods in Cell Biology*, Vol. 38: *Cell Biological Applications of Confocal Microscopy* (B. Matsumoto, ed.), Academic Press, San Diego, pp. 290–325.
6. A. Boyde (1995) Confocal optical microscopy, In *Image Analysis in Histology: Conventional and Confocal Microscopy* (R. Wootton et al., eds.), Cambridge University Press, Cambridge, UK, pp. 151–196.
7. L. M. Loew (1993) Confocal microscopy of potentiometric fluorescent dyes, In *Methods in Cell Biology*, Vol. 38: *Cell Biological Applications of Confocal Microscopy* (B. Matsumoto, ed.), Academic Press, San Diego, pp. 195–210.
8. S. A. Stricker et al. (1992) *Dev. Biol.* **149**, 370–380.
9. T. C. Brelje, M. W. Wessendorf, and R. L. Sorenson (1993) multicolor laser scanning confocal immunofluorescence microscopy: Practical application and limitations, In *Methods in Cell Biology*, Vol. 38: *Cell Biological Applications of Confocal Microscopy* (B. Matsumoto, ed.), Academic Press, San Diego, pp. 98–182.
10. W. Galbraith et al. (1989) *Soc. Photo-Opt. Instrum. Eng.* **1063**, 19–20.
11. M. W. Wessendorf (1990) In *Handbook of Chemical Neuroanatomy*, Vol. 8 (A. Bjorklund et al., eds.), Elsevier, Amsterdam, pp. 1–45.
12. J. T. McBride (1995) Quantitative immunocytochemistry, In *Image Analysis in Histology: Conventional and Confocal Miscroscopy* (R. Wootton et al., eds.), Cambridge University Press, Cambridge, UK, pp. 339–354.
13. B. Matsumoto and I. L. Hale (1993) In *Methods in Neuroscience*, Vol. 15 (P. C. Hargrave, ed.), Academic Press, San Diego, pp. 54–71.
14. D. Agard (1984) *Ann. Rev. Biophys. Bioeng.* **13**, 191–219.

CONFORMATION

VERNON ANDERSON

Nonidentical spatial arrangements of the atoms of a molecule achieved by rotation about single bonds are referred to as different conformations (1). Two molecules differing in their conformation may be referred to as conformers. Because rotations about single bonds are usually rapid, conformers are rapidly interconverted, making it difficult to separate individual conformers. Note one exception to this rule is rotation about the C—N single bond of the **peptide bond** of polypeptides, which has partial double-bond character, so it is planar and the *cis* and *trans* conformations are only slowly interconverted (see **Cis/trans isomerization**). The conformation of a molecule may then be largely specified by characterizing the rotations about its single bonds. These rotations are quantified by determining the **torsion angle** or **dihedral angle**. Individual torsion angles may be qualitatively described as eclipsed, **gauche**, or **anti** (Fig. 1), or more

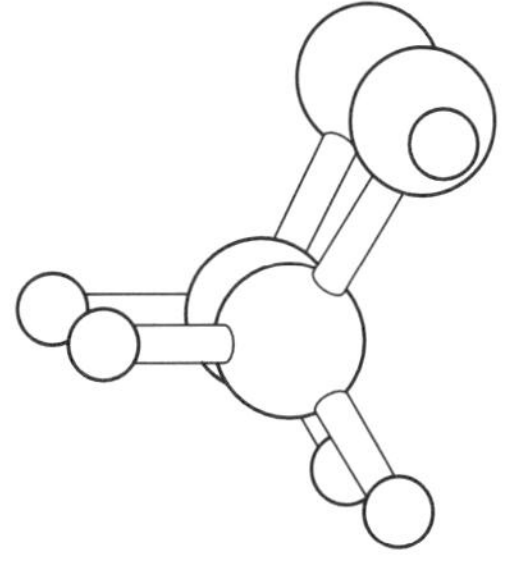

eclipsed or *syn*periplanar

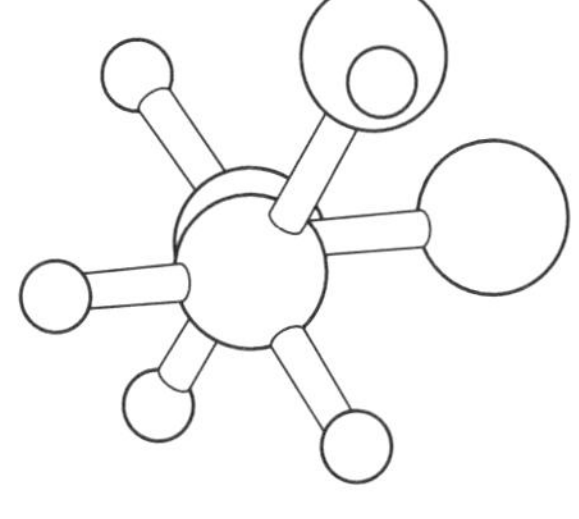

gauche or *syn*clinal

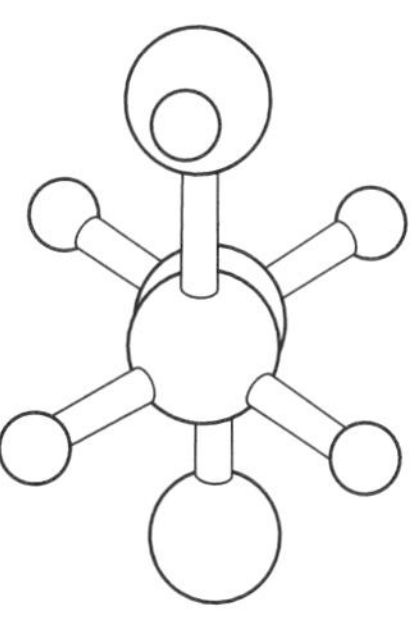

anti or *anti*periplanar

Figure 1. The common description of a conformation, illustrated by ethylene glycol. When the substituents are superimposed, the conformation is described as *eclipsed*; when they are staggered, the conformation is *gauche* if the large substituents are roughly separated by 60° and *trans* or *anti* if they are opposite.

formally named *syn* periplanar, clinal, and *anti* periplanar; they are quantified, as described in **Torsion angle**.

Other common combinations of torsion angles lead to descriptive names for common groups of atoms, eg, the 2′-endo conformation of **ribose** describes all five of the torsion angles of the furanose ring. This practice extends to larger biomacromolecules. The conformation of a small repeating segment of a biomacromolecule is characterized by its torsion angles along the backbone. Thus specification of the phi (ϕ) and psi (ψ) angles for an oligopeptide (see **Ramachandran plot**) describes its conformation as alpha-**helical** or beta-**sheet** (2). The specification of the torsion angles along the phosphodiester backbone of a polynucleotide also will determine whether the conformation is of the A, B, or Z type (3) (see **DNA structure**).

A stable three-dimensional structure of a biological macromolecule is referred to as its conformation. Because rotation about single bonds is energetically easy, a unique conformation does not exist, as at least one single bond will be changing rapidly. A cooperative change of several torsion angles, such as occurs in the protein unfolding or melting of double-stranded DNA (see **DNA melting**), may have a significant energetic barrier. The interconversion of the two conformations is then identified as a *conformational change*, where the implication is that the macromolecule has been converted from one family of conformations to an experimentally distinguishable family of conformations.

BIBLIOGRAPHY

1. E. L. Eliel et al. (1967) *Conformational Analysis*, Wiley-Interscience, New York.
2. G. E. Schulz and R. H. Schirmer (1979) *Principles of Protein Structure*, Springer Advanced Texts in Chemistry (C. R. ed.), Springer-Verlag, New York.
3. W. Saenger (1984) *Principles of Nucleic Acid Structure*, Springer Advanced Texts in Chemistry (C. R. Cantor, ed.), Springer-Verlag, New York.

Suggestions for Further Reading

C. R. Cantor and P. R. Schimmel (1980) *Biophysical Chemistry, Part I: The Conformation of Biological Macromolecules*, W. H. Freeman, San Francisco, CA.

E. L. Eliel (1962) *Stereochemistry of Carbon Compounds*, McGraw-Hill, New York, pp. 124–137.

CONOTOXINS

C. MONTECUCCO

Perhaps the largest array of **channel**-binding **toxins** is produced by marine gastropods of the genus *Conus*, which includes more than 500 species (1,2). These animals appear to have developed the ability of generating a very large number of toxin variants starting from a basic peptide structure consisting of relatively few residues (Fig. 1). In such a way, they produce venoms containing complex mixtures of toxins, capable of binding and blocking the activity of several ion channels. Moreover, some conotoxins are capable of binding differentially to the various **isoforms** of an ion-specific channel, thus providing precious tools to pharmacologists and neuroscientists (1).

The μ-conotoxins are 22 residue-long basic peptides and include three hydroxyproline residues and six **cysteines** that form three **disulfide bonds** (Fig. 1). These conotoxins have an ellipsoid shape with the basic residues, which are essential for the binding activity, clustered on one side of the molecule (2,3). μ-conotoxins bind to the tetrodotoxin binding site of the **sodium channel**, thereby inhibiting the propagation of the action potential and causing flaccid paralysis. Another family of conotoxins specific for sodium channels is that of the δ-conotoxins, whose binding causes an increased conductance of such voltage-gated channels (4).

ω-conotoxins act on voltage-gated **calcium channels** and inhibit calcium entry through the presynaptic membrane, thus preventing the release of neurotransmitters. Their peptide chain consists of 24 to 30 residues and three disulfide bonds (Fig. 1). A frequently used ω-conotoxin is G_{VIA}, a 27 residue-long peptide that is very specific for calcium channels containing the α1B subunit, such as the N-type calcium channel present at the neuromuscular junction of vertebrates. The specificity of different ω-conotoxins for different vertebrates is exploited in studying the function and anatomical distribution of calcium channels (1).

The largest family of conotoxins is that of the α-conotoxins, 13 to 18 residue-long peptides, whose folding is dominated by two disulfide bonds (Fig. 1). These paralytic neurotoxins bind specifically to the nicotinic **acetylcholine receptor** and different α-conotoxins are able to distinguish between different neuronal isoforms of the receptor. *Conantokins* are *Conus* peptides with yet another channel specificity: They bind to the NMDA-sensitive glutamate channels. They are peptides of 17 to 21 residues with no cysteine but four **gamma-carboxyglutamate** residues (1) (Fig. 1). Very short *Conus*

μ-Conotoxins R D C C T X X ⊕ K C K D R X C K O X ⊕ C C A α

δ-Conotoxins E A C X X X G X X C X X X X X X C C X X X C

Conantokins G E γγ X X X X X γ X X ⊕ γ X X X <

Figure 1. Peptide toxins contained in *Conus* venoms. *Conus* spp. produce a highly complex mixture of toxins. Shown here are the consensus sequences and disulfide connections of toxins specific for the acetylcholine receptor (α-conotoxins), sodium channels (μ- and δ-contoxins), calcium channels (ω-conotoxins), and the NMDA glutamate channel (conantokins). O indicates *trans*-4-hydroxyproline, plus a positively charged residue, γ a γ-carboxyglutamate, and α an amidated α-carboxyl. The disulfide connectivity of δ-conotoxins is not established.

peptides are conopressins, which consist of nine amino acid residues with a single disulfide bond and are specific for the vasopressin receptor.

BIBLIOGRAPHY

1. R. Rappuoli and C. Montecucco (1997) *Guidebook to Protein Toxins and Their Use in Cell Biology*, Sambrook and Tooze, Oxford University Press, Oxford, UK.
2. B. M. Olivera, G. Miljanich, J. Ramachandran, and M. E. Adams (1994) *Ann. Rev. Biochem.* **63**, 823–867.
3. J.-M. Lancelin et al. (1991) *Biochemistry* **30**, 6908–6916.
4. K. Shon (1995) *Biochemistry* **34**, 4913.

CONSERVATIVE SUBSTITUTIONS

WILLIAM A. ROSCHE
PATRICIA L. FOSTER

Missense mutations cause the substitution of one **amino acid** for another in the **protein** encoded by the mutant **gene**. The effect of such a change on the protein's function will often depend on how similar the new amino acid is to the original one. Amino acids are classified by their acidity, polarity, **hydrophobicity**, and structure. Substitutions are considered conservative if the change is between amino acids of the same class. However, a better criterion is whether the substitution has occurred during the **evolution** of related proteins. The mutations that give rise to conservative amino acid substitutions are considered evolutionarily **neutral mutations**, or nearly so. Lists of likely conservative substitutions are found in many laboratory manuals, for example (1).

BIBLIOGRAPHY

1. A. Ellington and J. M. Cherry (1991) In *Current Protocols in Molecular Biology* (F. M. Ausubel, R. Brent, R. E. Kingston, D. D. Moore, J. G. Seidman, J. A. Smith, and K. Struhl, eds.), Wiley, New York, pp. A.1C.1–A.1C.10.

CONSTANT (C) REGION

MICHEL FOUGEREAU

The heavy and light chains of **immunoglobulins** contain (a) **variable (V) regions** of approximately equal size (ie, about 110 amino acid residues), and (b) constant (C) regions that contain a multiple of 110 residues, one for the L chain and three or four for the H chains, depending upon the **isotype**. Whereas the V regions are involved in **antigen** recognition, C regions are devoted to effector functions, such as **complement** fixation, transplacental passage, or attachment to various cell types. C regions are organized on a basic **domain** structure, corresponding to the 110-residue subunit of Ig with an autonomous three-dimensional **protein structure**, as shown by **X-ray crystallography** (see **Immunoglobulin structure**). This structure, known as the immunoglobulin constant domain fold, provides the basis for the general organization of a large number of proteins, known as the Ig **superfamily**.

The existence of different C regions was first shown by immunochemistry, because specific antisera could distinguish two types of light chains, known as **kappa** (κ) **and lambda** (λ) **light chains**, and five discrete isotypes of H chains, which dictate the organization into the five classes of immunoglobulins (**IgA**, **IgD**, **IgE**, **IgG**, **IgM**). Definitive understanding of the basis of isotype structure was obtained by **protein sequencing** and finally by **DNA sequencing**. Within each Ig class, one may distinguish subclasses, as in the case of the γ chain or the α chains in humans.

Most of the effector functions of immunoglobulins are supported by the last two or three COOH-terminal domains of the heavy chains, depending upon the class. The most interesting feature of constant regions is their ability to bind to a receptor family, termed **Fc receptors** (FcRs), because binding involves the Fc region of Ig. FcRs represent adaptors that can confer on a variety of cell types any antibody specificity, depending on which Ig will bind to the receptor. There is an FcR specific of each soluble isotype, IgG, IgA, and IgE. One distinguishes FcRs of high affinity (FcR type I) that fix Ig monomers and FcRs of low affinity (FcR type II) that bind Ig only when aggregated or as multivalent complexes. Depending on which cell type FcR is anchored, they express an cytoplasmic tail of variable size, which may result from alternative splicing and that contains various signaling motifs such as an immunoreceptor tyrosine-based activation motif and an immunoreceptor tyrosine-based inhibition motif. FcεRI, present on mast cells, basophils, and eosinophils, plays a major role in anaphylaxis, because it will fix monomeric IgE. Upon cross-linking by antigen, this FcR is triggered, leading to the cascade of events that are responsible for degranulation of the corresponding cells and thus to the liberation of pharmacologically active molecules that account for immediate hypersensitivity. FcεRII (CD23) binds aggregated IgE, or IgE already cross-linked by antigen. It is present at the surface of many cell types, including **B cells**. It may also be cleaved and release several soluble products that have a **cytokine**-like activity. FcγRI is found on monocytes, eosinophils, and neutrophils and binds to monomeric IgG, whereas FcRαI is present on macrophages, neutrophils, and T and B cells.

This demonstrates the extraordinary plasticity and amplification of molecules of the immune system and illustrates the huge diversity of functions that are linked to the constant regions of immunoglobulins.

See also entries **Antibody**, **Immunoglobulin**, and **Isotype**.

Suggestions for Further Reading

F. Shakib (1986) *Basic and Clinical Aspects of IgG Subclasses, Monographs in Allergy*, Vol. 19, Karger, Basel.

J. V. Ravetch and J. Kinet (1993) Fc receptors. *Annu. Rev. Immunol.* **9**, 457–492.

CONTACT INHIBITION

R. I. FRESHNEY

Cultured cells, particularly when grown as a monolayer attached to a substrate, will go through a reproducible growth cycle following subculture (1) (see **Cell lines**). The phases normally defined are (1) the lag phase, before the culture starts to proliferate; (2) the exponential, or log, phase, when the cells are dividing rapidly and the population doubles over a fixed time; and (3) the plateau phase, when the cell number remains stable with further time in culture. As a culture reaches the end of the exponential phase and enters plateau, it responds to cell contact, reduced substrate availability, a reduction in nutrients, and a number of other stimuli that inhibit a further increase in cell numbers. These have been divided into two major components: contact inhibition and density limitation of cell proliferation. Contact inhibition is the cessation of cell motility that occurs when a cell culture reaches confluence. It was first described from observations of randomly migrating fibroblasts made in time-lapse cinemicrography (2). Normal fibroblasts cultured at low cell density show a spindle-shaped morphology, with a reversible polarity along their long axis, which is also the direction of cell migration. When the cell encounters another similar cell, contact induces a reversal of the polarity; the ruffling in the leading tip of the cell ceases and transfers to the opposite tip, which then becomes the anterior end, and migration resumes in the opposite direction. When contact is made with another cell, the process of reversal of polarity is repeated, unless the cell is surrounded by other cells, when membrane ruffling ceases and cell migration is inhibited. In the cases of normal human or chick diploid fibroblasts, the cell assumes a parallel array that gives the appearance of whorls in the cell monolayer under low power **microscopy** or naked eye observation.

At the time that motility ceases, the cells are still capable of proliferation, but the culture becomes crowded following one or two population doublings, the cells assume a narrower spindle shape, and cell spreading is reduced beyond the point where the cell is able to re-enter the **cell cycle** (3). At this point, proliferation ceases in a normal cell culture, and the culture remains as a monolayer. Further growth is inhibited by density limitation of cell proliferation, and the culture enters the plateau phase of the growth cycle (see **Cell lines**). Density limitation of cell proliferation is distinct from contact inhibition, which affects primarily membrane ruffling, polarity, and cell migration. The mechanisms are, however, similar, although the detail remains obscure. Subconfluent cells have many more adhesions to the substrate than to other cells, favoring motility and proliferation. When the culture becomes confluent, the number of cell–cell adhesions (cadherins) increases, but they are still exceeded by cell–substrate adhesions (integrins), which inhibit cell motility but not proliferation. When the culture becomes crowded, the number of cell–cell adhesions increases, and cell–substrate adhesions are reduced, maintaining inhibition of motility but now inhibiting re-entry into the cell cycle. As contact with adjacent cells and contact with the substrate are mediated via different classes of **receptors** (4), the signaling that results from activation of these receptors is potentially different, although exactly how remains unclear.

In addition to alterations in cell spreading, adhesion receptors, and the **actin** cytoskeleton when cells reach a high density, the cells are also exposed to reduced nutrient levels, due to a higher rate of utilization, and increased levels of catabolites, both of which will tend to inhibit proliferation. In a static culture, this may generate a depleted layer above the cells, across which nutrients and **growth factors** must diffuse to reach the cells. The **diffusion** rate will be significantly slower with higher molecular weight components of the medium, including **growth factors** and peptide hormones such as **insulin**. Support for this diffusion boundary layer hypothesis was provided by experiments where local irrigation was increased, by means of a micropump, and resulted in a localized increase in the frequency of labeling with [^{3}H]-thymidine (5). As the range of the diffusion boundary was short, it was proposed that the limitation was principally in growth factors, depleted by binding to cell-surface receptors and **endocytosis**, which was confirmed by the observation that addition of growth factors or serum to the medium will induce further proliferation (6).

On the other hand, one of the main observations that confirms that contact inhibition and density limitation of cell proliferation are not simply evidence of environmental deterioration is the result of the so-called "wounded monolayer" experiment (7). If a confluent, growth-arrested monolayer is scored with a sharp instrument, creating a bare patch devoid of cells, the surrounding cells respond as follows. The edge cells start to show ruffling of the cell membrane bordering the wound, spread out, and migrate into the space. When they have spread beyond a critical point, they enter the cell cycle and divide, and they continue to do so until the space of the wound is filled. They then stop migrating, but continue to proliferate until the cell density in the wound matches the crowded state of the rest of the monolayer; at this point they stop dividing. Meanwhile, no other cells distant from the wound show any sign of entering the cell cycle. This confirms that a major component of density limitation of cell proliferation is due to geometry and cell shape, rather than simply nutritional or growth factor limitation.

If a culture of normal fibroblasts that is growth arrested at high density is fed with medium containing serum or growth factors, many of the cells will reenter the cell cycle, resulting in the formation of a second cell layer over the first; if this is repeated, multilayered cultures can be produced. It has been proposed that the first layer of cells secretes a **collagen** overlay (which may produce a further diffusion barrier) and the second cell layer grows on top of this, technically not disobeying the rules of contact inhibition and density limitation of cell proliferation. This second layer, however, is still limited by contact and density, and when it reaches confluence, it will form a second layer of parallel arrays of cells, not colinear with the first. If perfused, these cultures may continue to increase up to 20 or 30 cell layers deep (8).

When cultures of normal epithelium reach confluence, they also become contact-inhibited. In less dense cultures, however, they do not show the random migration seen in cultures of fibroblasts, but tend to grow in patches with ruffling of the membrane, cell spreading, and proliferation restricted to the outer edge of the patch (9). Their motility is limited, but when it occurs, it tends to involve the whole patch, which moves as a unit. Multilayering in normal epithelial cultures also occurs, but tends to imply maturation perpendicular to the substrate, rather than overgrowth. Epidermal cells, for example, will become stratified, with the upper layers of cells becoming progressively more keratinized. This is accentuated if the cells are grown on collagen, particularly in the presence of dermal fibroblasts and due to integrin down-regulation. Epithelial cells, which are derived from simple epithelium that is only one cell thick *in vivo*, tend to be obligate monolayers in culture. Likewise, normal endothelium from the lining of blood vessels will not pile up in culture, although after some time at high density it will tend to curl up and form secondary structures resembling capillaries (10), particularly if grown on an extracellular matrix, such as Matrigel.

Neoplastic transformation allows cells to escape from the restrictions of contact inhibition and density limitation of cell proliferation. On one hand, they often lack, or have modified, the appropriate receptors or **cell-adhesion molecules** to recognize each other on contact; on the other hand, they are no longer dependent on cell spreading to allow entry into the cell cycle. Many transformed cell lines will propagate in suspension, without any substrate attachment. In addition, transformed cells often produce autocrine growth factors or have permanently active steps in the **signal transduction** cascades that promote cell proliferation (11–13). Hence, transformed cells will grow to a higher density and form multilayered cultures quite readily. The limitations imposed by density now tend to be diffusion-related, principally nutrient, catabolite, and gaseous (dissolved O_2 in and CO_2 out). One major limitation is the release of lactic acid by the cells, as they are generally more anaerobic in their metabolism than normal cells, and this depresses the pH, initially inhibiting proliferation and ultimately killing the cells. Although transformed cell cultures enter a plateau phase of culture when they reach a high density, the growth fraction in transformed cultures in plateau can be quite high (10–20%), unlike the plateau in normal cell cultures, where the growth fraction is very low (<5%). A steady state is reached when cell proliferation is matched by cell **necrosis** or **apoptosis**, and the cells deteriorate rapidly and irreversibly if the medium is not replenished.

The differences in response to high cell density between normal and transformed cells has been exploited in the isolation of transformed foci. If a monolayer of 3T3 cells is **transfected** with an **oncogene**, especially in the exponential phase of growth, when transfection is more efficient, it will show foci of transformed cells when the culture reaches confluence, because of the lack of contact inhibition of cell motility and density limitation of cell proliferation in successfully transformed cells (14). Likewise, transformed cells plated on a confluent monolayer will continue to grow and may form foci, or they may infiltrate the monolayer. Normal cells plated on a confluent monolayer will give a varying response, depending on the cell lineage of the confluent monolayer. If the plated normal cells are of the same lineage as the confluent monolayer, they will not grow; for instance, fibroblasts will not grow on a confluent monolayer of fibroblasts, and normal glial cells will not grow on a confluent monolayer of normal glial cells (15), whereas their transformed counterparts will. On the other hand, normal glial cells will grow on a confluent monolayer of fibroblasts, and normal keratinocytes will grow on a monolayer of normal 3T3 cells (16).

Cultures that are propagated in suspension will, of course, not encounter contact inhibition. Nevertheless, they are subject to density limitation of cell proliferation. Part of this is undoubtedly due to nutrient depletion, catabolite buildup, and pH depression, but even if limiting nutrients are replaced and the pH is stabilized, high density suspension cultures do not increase significantly above $1{-}2 \times 10^6$ cells/ml. Quite often suspension cultures, such as **hybridomas**, will remain in the plateau phase for only a short time and then deteriorate rapidly, leaving few viable cells in the culture. The reason for this appears to be that the cells tend to enter apoptosis when they reach a high density. Some moderate success has been achieved overexpressing *bcl* in these cells to inhibit apoptosis (17).

BIBLIOGRAPHY

1. R. I. Freshney (1994) *Culture of Animal Cells, a Manual of Basic Technique*, Wiley-Liss, New York, pp. 153–157.
2. M. Abercrombie and J. E. M. Heaysman (1954) *Exp. Cell Res.* **6**, 293–306.
3. I. Folkman and A. Moscona, (1978) *Nature* **273**, 345–349.
4. S. Levenberg, B.-Z. Katz, K. M. Yamada, and B. Geiger (1998) *J. Cell Sci.* **111** 347–357.
5. M. G. P. Stoker (1973) *Nature* **246**, 200–203.
6. G. A. Dunn and G. W. Ireland (1984) *Nature* **312**, 63–65.
7. B. Alberts, D. Bray, J. Lewis, M. Raff, K. Roberts, and J. D. Watson (1989) *Molecular Biology of the Cell*, 3rd ed., Garland Publishing, New York, p. 898.
8. P. F. Kruse Jr. and E. Miedema (1965) *J. Cell Biol.* **27**, 273.
9. I. McKay and J. Taylor-Papadimitriou (1981) *Exp. Cell Res.* **134**, 465–470.
10. J. Folkman and C. Haudenschild (1980) *Nature* **288**, 551–556.
11. P. Kahn and T. Graf (1986) *Oncogenes and Growth Control*, Springer, Berlin.
12. K. Siegfried, Y. H. Han, M. A. A. DeMichele, J. D. Hunt, A. L. Gaither, and F. Cuttitta (1994) *J. Biol. Chem.* **269**, 8596–8603.
13. A. Balmain and K. Brown (1988) *Adv. Cancer Res.* **51**, 147–182.
14. R. I. Freshney (1994) *Culture of Animal Cells, a Manual of Basic Technique*, Wiley-Liss, New York, p. 233.
15. C. M. MacDonald, R. I. Freshney, E. Hart, and D. I. Graham (1985) *Exp. Cell Biol.* **53**, 130–137.
16. J. G. Rheinwald and H. Green (1975) *Cell* **6**, 331–344.

17. S. Terada, Y. Itoh, H. Ueda, and E. Suzuki (1997) *Cytotechnology* **24**, 135–141.

CONTACT MAPS

ANDRZEJ KOLINSKI
ADAM GODZIK
JEFFREY SKOLNICK

Contact maps are two-dimensional representations of three-dimensional **protein structures**. A three-dimensional description of a protein structure composed of N structural units could be expressed as an $N \times N$ array of the pairwise distances (see **distance geometry**). This could be done for all pairs of atoms, for selected types of atoms (eg, Cα atoms), for groups of atoms (eg, **side-chain** centers of mass), or for entire **amino-acid** residues. Contact maps are generated from such matrices by taking a certain cutoff value for the pairwise distances. For example, the $N \times N$ matrix of the distances between protein Cα atoms (1,2) can be transformed into a Cα-based contact map (3). Those Cα atoms that are closer to each other in the protein structure than the chosen cutoff distance are considered to be "in contact." This produces a binary $N \times N$ matrix—a so-called black and white contact map. Alternatively, one may assume a set of several critical values for the distances between Cα atoms (or for other atoms or groups of atoms) and generate an integer matrix—the equivalent of a contact map with a colored or gray scale. A map of main-chain **hydrogen bonds** could also be considered as a variant of a protein contact map. The choice of structural units being mapped and the choice of cutoff distances determine the quality and range of structural information being stored in a contact map (see Fig. 1).

CONTACT MAPS AS A FINGERPRINT OF PROTEIN THREE-DIMENSIONAL STRUCTURE

A contact map constitutes a structural "fingerprint" of a protein (4). Each protein can be identified based on its contact map. The **secondary structure**, fold topology, and side-chain packing patterns (for side-chain contact maps) can be visualized conveniently and read from the contact map.

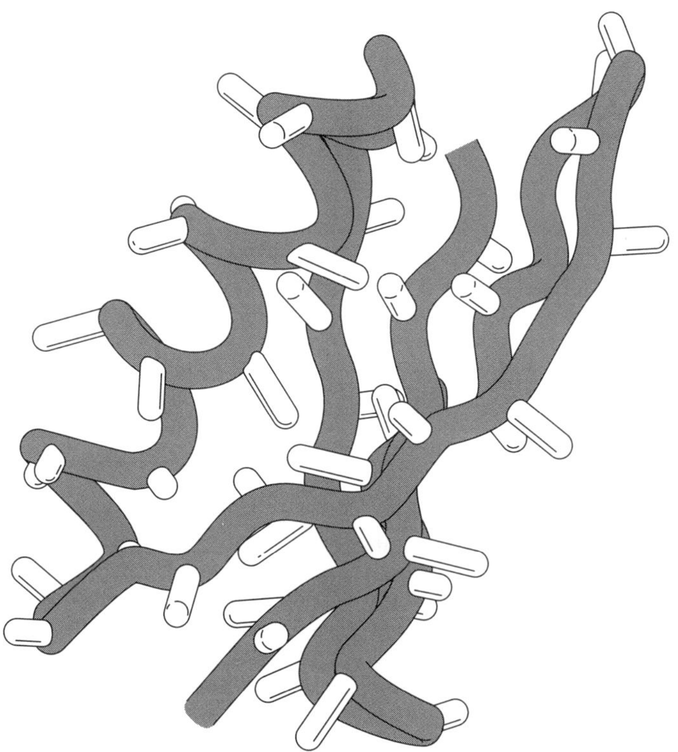

Figure 1. Schematic drawing of the structure of the B domain of protein G and its contact maps. (a) Three-dimensional structure of the B domain. (b) Cα-based contact maps: above the diagonal, the black and white map with a cutoff distance of 8 Å; below the diagonal is the gray-scale map where various shades of gray correspond to three values of the cutoff distance, 8 (darkest), 10, and 12 Å. (c) Side-chain-based contact map. Above the diagonal, the dark squares correspond to the pairs of side chains for which the distance between at least one pair of heavy atoms is less than 5 Å. Below the diagonal, a 6.25-Å cutoff criterion has been applied to the centers of mass of the side chains. (d) The main chain hydrogen bond map for the same structure. The amino acid sequence of the protein is given along each axis in one-letter code, and the secondary structure is indicated as follows: E is extended β-strand, H is helix, and S and T are two types of turns.

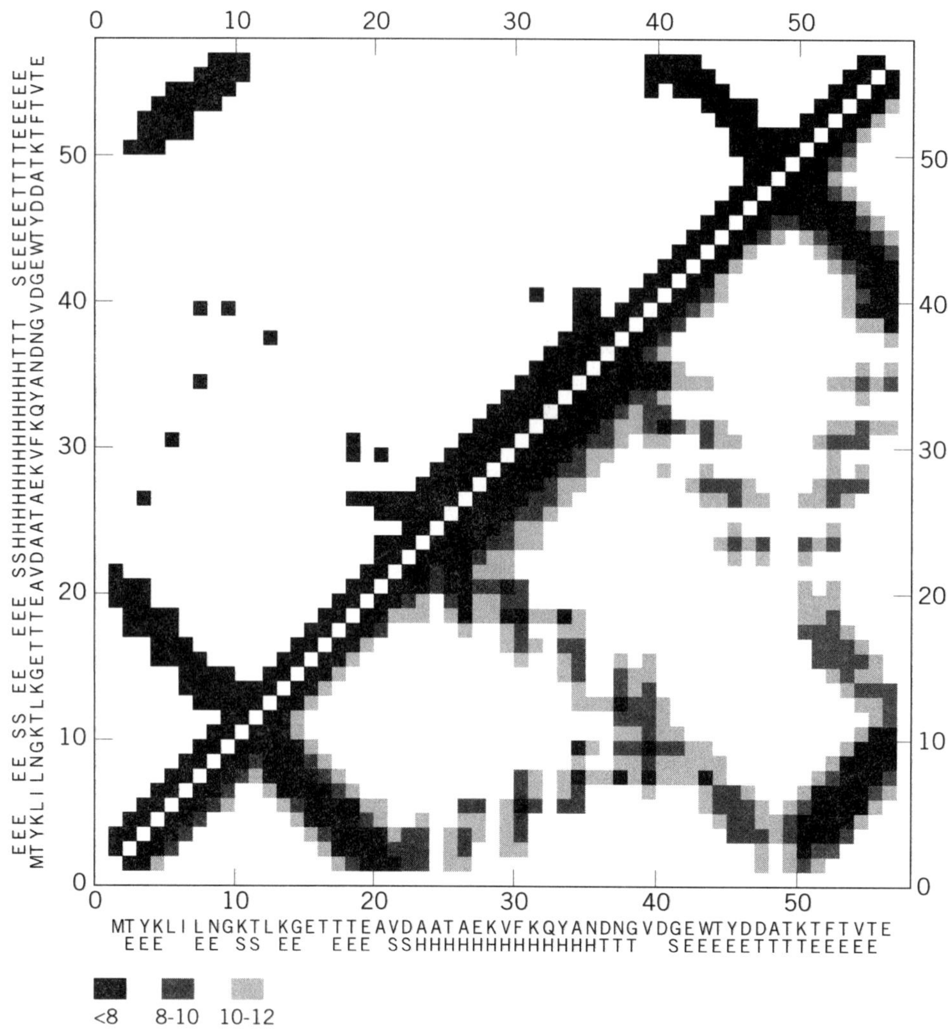

Figure 1. *(Continued)*

Furthermore, structural similarity between a pair of proteins is immediately apparent by a very pronounced similarity of their contact maps; in comparing two protein structures, there is no need to search all their possible relative orientations. The reconstruction of a protein structure from its contact map is more complex, although low-to-moderate resolution three-dimensional models can be easily built, even from a fragmentary contact map (5). The accuracy of the model depends on the type of contact map and the computational tools employed. A combination of protein nuclear magnetic resonance **NMR** spectra (see **NOESY spectra**; **COSY spectra**) constitutes a hybrid contact map of a protein, and model building from these data is an example of a map-to-structure modeling procedure (6,7).

Cα-BASED CONTACT MAPS

Cα-based contact maps and distance matrices were perhaps the first commonly used maps for visualization of protein structures (3,8,9). An example is given in Figure 1b. These contact maps reflect well the overall topology of the protein fold, but only rather coarse structural details can be read from them. This is due to the fact that the Cα−Cα distance distributions extracted from protein structures have several convoluted peaks. These peaks correspond to various distances between pairs of various secondary structure elements (α-helices, **beta-strands**, etc.). Thus, a single cutoff distance is always inadequate: Too small a value would miss some helix-to-helix contacts, while too large a value may create some problems with identification of the secondary structure patterns. Gray scale maps (several cutoff ranges) communicate much more detailed structural information.

SIDE-CHAIN CONTACT MAPS

Side-chain contact maps contain much richer information, not only about the topology of a protein fold and its secondary structure, but also many fine details about the packing patterns of the protein side-chains. Various conventions can be used to build side-chain contact maps; two are illustrated in

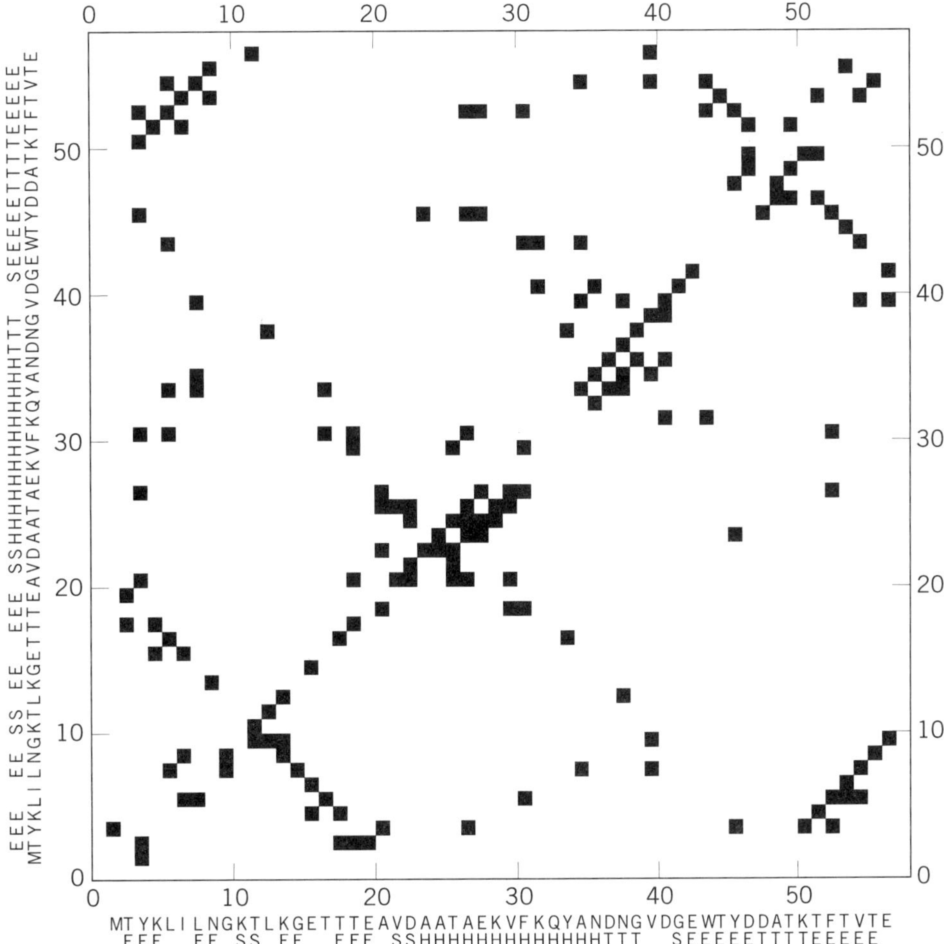

Figure 1. *(Continued)*

Figure 1. In one case, two residues are assumed to be in contact when any two heavy atoms (ie, all except hydrogen) are a shorter distance from each other than some assumed cutoff. Due to the comparable size of all the united atom types constituting the side chains (eg, CH_2, CH, NH_2, etc.), a good choice of cutoff distance is between 4.5 and 5.0 Å (4,10). In this range, the number of detected contacts is not sensitive to the particular choice of cutoff, and the packing pattern of the side chains is always described with high fidelity. Characteristic patterns of contacts between elements of secondary structure are an important and useful feature of these contact maps (11). Alternatively, one may build a side-chain contact map using the side-chain centers of mass as a reference. In this case, a larger value of the cutoff distance needs to be used. These cutoff values for the distance between side-chain centers of mass could be made specific for certain amino acid pairs, on the basis of the different sizes of the side chains, which produce different average contact distances for various pairs. As seen in Figure 1, the two approaches (atom based and center of mass based) lead to very similar protein representations. The patterns of the atom-based contact maps are slightly better defined.

REGULARITIES OF THE CONTACT MAPS REFLECT REGULARITIES OF PROTEIN STRUCTURES

Different types of contact maps reflect different aspects of the regularities seen in protein structures. The side-chain-based contact maps are a very good example (10). Near the diagonal of the map, the distinct features of the protein secondary structure can be easily read. Indeed, for extended fragments of the polypeptide chain, only residues i and i + 2 can be in contact. For α-helices, the i, i + 3 and i, i + 4 patterns of contacts are well pronounced. Furthermore, characteristic clusters of contacts further away from the diagonal reflect the packing between particular pairs of β-strands within the β-sheets. Parallel and antiparallel structures have very different features on the contact patterns. Very characteristic patterns could also be observed for other pairs of secondary-structure elements. It is even very easy to distinguish between the patterns for two helices in a helical or α/β protein and in the **coiled-coil** structural motifs. Figure 2 shows some typical side-chain contact map patterns.

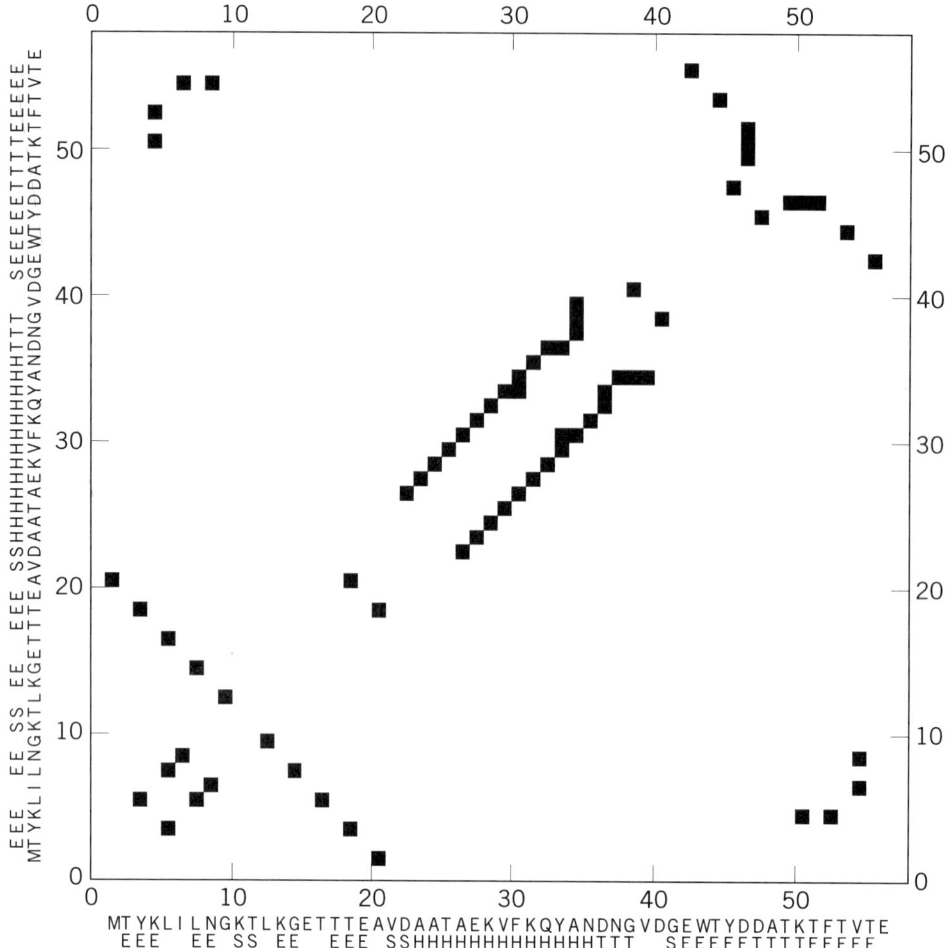

Figure 1. *(Continued)*

APPLICATION OF CONTACT MAPS IN MODELING PROTEIN STRUCTURE

Contact maps provide a very convenient way of identifying pairwise interactions within protein structures. This has been applied in various algorithms for **threading protein sequences**, where contact maps are actually used as ersatz template structures (4). In **computer simulations of biological molecules**, contact maps provide a convenient way of displaying structural changes. Also, regularities of the patterns seen in all classes of proteins can provide a guideline for designing knowledge-based multibody potentials (12,13) and for protein modeling in a reduced (and perhaps also in all-atom) representation (13,14). Such potentials may be necessary to reproduce the all-or-none character of protein folding transitions in model simulations (15).

One can easily recognize the distinct features of well-defined contact maps, with their characteristic patterns, after inspecting several maps of various proteins. This is an excellent example of a pattern recognition problem that could be learned by **neural network calculations**. Then such a trained network can be used for the automated recognition of good versus poor models of protein structure (16).

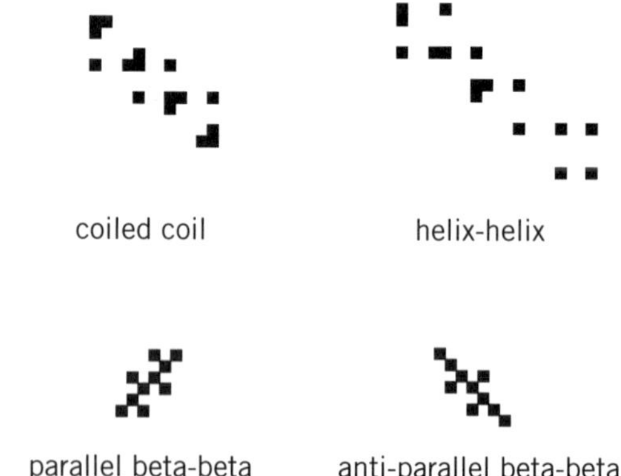

Figure 2. Representative patterns of side-chain contact maps describing interactions between α-helices (top) and between β-strands in parallel and antiparallel **beta-sheets** (bottom).

BIBLIOGRAPHY

1. I. D. Kuntz (1975) *J. Am. Chem. Soc.* **97**, 4362–4366.
2. F. M. Richards and C. E. Kundrot (1988) *Proteins* **3**, 71–84.
3. M. Levitt (1976) *J. Mol. Biol.* **104**, 59–107.
4. A. Godzik, J. Skolnick, and A. Kolinski (1992) *J. Mol. Biol.* **227**, 227–238.
5. J. Skolnick, A. Kolinski, and A. R. Ortiz (1997) *J. Mol. Biol.* **265**, 217–241.
6. W. Braun and N. Go (1985) *J. Mol. Biol.* **186**, 611–626.
7. R. Kaptein, R. Boelens, R. M. Scheek, and W. F. van Gunsteren (1988) *Biochemistry* **27**, 5389–5395.
8. M. N. Liebman, C. A. Venanzi, and H. Weinstein (1985) *Biopolymers* **24**, 1722–1758.
9. D. C. Philips (1970) *Biochem. Soc. Symp.* **30**, 11–28.
10. A. Godzik and C. Sander (1989) *Protein Eng.* **2**, 589–596.
11. A. Godzik, J. Skolnick, and A. Kolinski (1993) *Protein Eng.* **6**, 801–810.
12. A. Kolinski, A. Godzik, and J. Skolnick (1993) *J. Chem. Phys.* **98**, 7420–7433.
13. A. Kolinski and J. Skolnick (1996) *Lattice Models of Protein Folding, Dynamics and Thermodynamics*, R. G. Landes, Austin, TX.
14. K. A. Olszewski, A. Kolinski, and J. Skolnick (1996) *Proteins* **25**, 286–299.
15. A. Kolinski, W. Galazka, and J. Skolnick (1996) *Proteins* **26**, 271–287.
16. M. Milik, A. Kolinski, and J. Skolnick (1995) *Protein Eng.* **8**, 225–236.

CONTIGUOUS GENES

ALAN P. WOLFFE

The order of genes on a **chromosome** and the existence of directly contiguous genes can be useful in understanding their individual functions and in mapping the genome to understand the function of chromosomal **domains**. In **prokaryotes** there is often a close linkage of genes involved in a common metabolic pathway. This type of linkage is uncommon in **eukaryotes**, although examples exist, including five genes in *Neurospora crassa* involved in synthesis of chorismic acid (1) and two genes in *Drosophila* melanogaster involved in purine metabolism (2).

There are several families of closely linked genes in humans. These include genes for **alcohol dehydrogenase**, α-amylase, the **major histocompatibility complex**, α-**globins**, β-globins, chorionic gonadotropin, luteinizing hormone, and **growth hormone**. Every chromosome has at least one family cluster of undispersed genes. These clusters vary in size from several thousand base pairs to millions of base pairs with a large cluster like the major histocompatibility complex. The members of a single **gene family** are derived from the **gene duplication** of single ancestral genes. Whereas gene families are rare in prokaryotes, the more complex differentiative states in eukaryotes require similar gene products with subtle differences in function. An example of a well-studied locus, where genes with closely related functions are linked together, occurs at the α- and β-globin loci. At these chromosomal sites, globin genes specific for embryos, neonatal and adult animals are linked together and expressed in a sequential fashion during **development** (Fig. 1). The globin loci also present interesting examples of where gene linkage helps understanding the chromosomal function. All of the globin genes depend on a super **enhancer** known as a **locus control region** (LCR), which functions over an entire chromosomal **domain**.

The relative order of genes is often conserved within a chromosome.This is termed linkage conservation. Linkage conservation often occurs in chromosomes isolated from distinct species. For example, in the **X-chromosome** there are many genes that have the same order along the chromosome in both **mouse** and humans. This is a useful aid in mapping new genes. There are several potential evolutionary advantages to linking certain genes together, which include common regulatory mechanisms, at the level of establishing a functional chromosomal domain, as described for the globin genes (3), or in terms of coordinate expression, as described for two genes involved in a particular metabolic pathway. There may also be

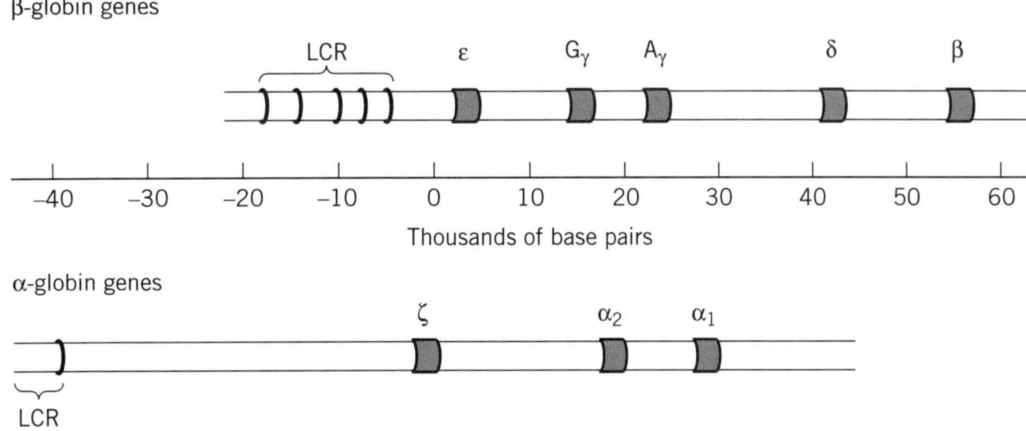

Figure 1. The β-globin gene cluster, which is on the short arm of chromosome 11 in humans and includes five genes that are sequentially expressed during development. Its locus control region (LCR) consists of five short stretches of DNA extending over a span of 12 bp. Similarly, there are three α-globin genes on the short arm of chromosome 16, but that LCR comprises only one DNA sequence far upstream.

regulatory mechanisms where genes compete with each other for a common regulatory element, as occurs in the globin genes at certain times in development (4).

BIBLIOGRAPHY

1. F. H. Gaertner and K. W. Cole (1977) *Biochem. Biophys. Res. Comm.* **75**, 259–270.
2. M. E. Johnstone (1985) *Biochem. Genet.* **23**, 539–546.
3. F. Grosveld, G. B. van Assendelft, D. R. Greaves, and G. Kollias (1987) *Cell* **51**, 975–985.
4. M. Wijgerde, F. Grosveld, and P. Fraser (1995) *Nature* **377**, 209–213.

CONTRAST VARIATION

JILL TREWHELLA

"Contrast" in scattering and diffraction experiments is defined as the difference between the mean scattering density of a component and its background, which can be a solvent or other components of an assembly of biological molecules. The greater the contrast, the more readily a component can be distinguished from its surroundings. "Contrast variation" involves the manipulation of the contrasts of specific components in a system in order to extract structural information on individual components. Contrast variation used in combination with **small-angle scattering** is a powerful method for examining the shapes and interactions of biological molecules in solution. If one has ordered samples, then contrast variation used in combination with **neutron diffraction** experiments can give important information on the location of disordered components that cannot be seen in X-ray diffraction experiments.

CONTRAST VARIATION WITH NEUTRONS

Either contrast variation experiments take advantage of inherent differences in scattering density between components of a complex or assembly, or the experimenter introduces contrast into the system by manipulating the scattering density of a specific component. Scattering densities are calculated by summing the scattering amplitudes of each atom within a volume and dividing by that volume. Because X-rays are scattered by electrons and the X-ray scattering amplitudes of atoms increase monotonically with the number of electrons, it is difficult to change the scattering density of a biological molecule in a benign way. On the other hand, neutrons provide extremely elegant and practical means for contrast variation. Neutrons are scattered by atomic nuclei in a sample. Hence neutron scattering amplitudes depend upon the complex properties of the neutron–nucleus interaction, and they show no systematic dependence on atomic number. Furthermore, isotopes of the same element can have very different neutron scattering properties. For neutrons, one of the largest differences in neutron scattering amplitude is between the isotopes of hydrogen (^1H and ^2H). Table 1 lists the coherent, elastic neutron scattering amplitudes for the atoms commonly found in biological systems. Note that the scattering amplitudes for most nuclei are positive and approximately equal. The exception is the scattering amplitude for ^1H which is negative, resulting from a 180° phase shift between neutrons scattered by ^1H compared to the other nuclei. As a consequence, the neutron scattering density of a particle depends strongly on its mean hydrogen content. The basic biological constituents, proteins, polynucleotides, and lipids, each have quite different mean neutron scattering densities (Fig. 1). Furthermore, the scattering densities for pure ^1H_2O and ^2H_2O bracket the values for biological constituents. Thus by simply varying the ^2H_2O:^1H_2O ratio in the solvent, one can do a contrast variation experiment on a protein–DNA complex, or a membrane protein in a bilayer. For a complex of proteins, selective deuteration of individual proteins provides a way of altering their mean neutron scattering densities for contrast variation studies. Recently, it has been demonstrated that one can obtain dramatically increased contrast effects using a novel method of "spin contrast variation." This method uses polarized neutron scattering from samples labeled with dynamically polarized nuclei and has been demonstrated to be effective for studying biological systems by locating RNA (1) and proteins (2,3) in ribosomes.

Table 1. Coherent Neutron Scattering Lengths, b_{coh}, and Corresponding X-Ray Scattering Factors, $f_{X\text{-ray}}$, for Biologically Relevant Nuclei

Atom	Nucleus	(10^{-12} cm)	f_X for $\theta = 0(10^{-12}$ cm)[a]
Hydrogen	^1H	−0.3742	0.28
Deuterium	^2H	0.6671	0.28
Carbon	^{12}C	0.6651	1.69
Nitrogen	^{14}N	0.940	1.97
Oxygen	^{16}O	0.5804	2.25
Phosphorous	^{31}P	0.517	4.23
Sulfur	mostly ^{32}S	0.2847	4.5

[a] X-ray scattering amplitudes are normally given in units of electrons, but have been converted to cm here for comparison with neutron scattering amplitudes. At very short wavelengths and low-Q, the X-ray coherent scattering cross-section of an atom with Z electrons is $4\pi(Zr_0)^2$, where $r_0 = e^2/m_ec^2 = 0.28 \times 10^{-12}$ cm.

SMALL-ANGLE SCATTERING WITH CONTRAST VARIATION

The **scattering intensity distribution** from a homogeneous solution of monodisperse particles in a solution can then be expressed as

$$I(Q) = \int\int \Delta\rho(\mathbf{r}_1)\Delta\rho(\mathbf{r}_2)\frac{\sin Q|\mathbf{r}_1 - \mathbf{r}_2|}{Q|\mathbf{r}_1 - \mathbf{r}_2|}d\mathbf{r}_1 d\mathbf{r}_2 \qquad (1)$$

The integration is taken over the volume of the particle. Q is the momentum transfer or scattering vector amplitude and is equal to $4\pi(\sin\theta)/\lambda$, where θ is half the scattering angle, and λ is the wavelength of the incident neutron radiation. $\Delta\rho(\mathbf{r}) = \rho(\mathbf{r}) - \rho_s$ is the "contrast," or neutron scattering density difference between the particle and the solvent. For uniform scattering density particles, if one can "match," or make equal, the scattering density of a particle and its solvent, then the small-angle scattering from that particle will be zero; that is, the particle will become "invisible" in the scattering experiment. For a multicomponent assembly, one can use this technique to study the shape of an individual component within the assembly. Rigorous solvent matching can be difficult, however, and is made more difficult when there are significant internal density fluctuations within the component being matched.

For a two-component complex in solution, one can conduct a series of measurements at different contrast points ("contrast

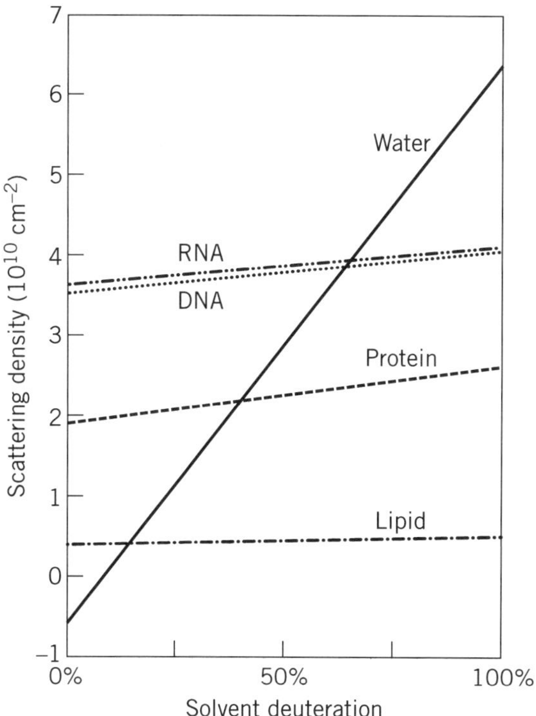

Figure 1. Average scattering length densities for biological molecules as a function of the fraction of D_2O in the solvent. The slope of the lines arises from the exchange of labile hydrogens.

series") to extract the basic scattering functions of individual components. Ignoring internal scattering density fluctuations, the scattering from a two-component complex in solution can be written as

$$I(Q, \Delta\rho_1, \Delta\rho_2) = \Delta\rho_1^2 I_1(Q) + \Delta\rho_1 \Delta\rho_2 I_{12}(Q) + \Delta\rho_2^2 I_2(Q) \quad (2)$$

The subscripts 1 and 2 refer to each component; and $\Delta\rho_{1(2)} = \rho_{1(2)} - \rho_s$, where $\rho_{1(2)}$ is that mean scattering density for component 1 (or 2). $I_1(Q)$ and $I_2(Q)$ represent the scattering of components 1 and 2, respectively, while $I_{12}(Q)$ is the cross term. A set of scattering measurements with different solvent scattering densities, or contrasts, gives a set of equations in the form of equation 2, which can be solved by linear least-squares regression to give the three basic scattering functions $I_1(Q)$, $I_2(Q)$, and $I_{12}(Q)$. Using these functions, the individual structural parameters for each component such as molecular weight, radius of gyration, R_g, and the pair distribution function, $P(r)$, can be calculated, as well as information on the relative dispositions of the individual components (see **Small-angle scattering** and **Radius of gyration** for details). $P(r)$ is the probable frequency distribution of all vector lengths between pairs of volume elements within the scattering component, weighted by the product of the scattering densities of each pair of volume elements. $P(r)$ is calculated as the indirect Fourier transform of the basic scattering function for each component, and its second moment gives R_g for that component. The indirect Fourier transform of the cross term $I_{12}(Q)$ gives the pair distribution function containing intercomponent vector lengths, which can give information on the distance between the two components.

Ibel and Stuhrmann (4) proposed an alternative approach to determining the relative dispositions of components in a complex. They showed that the R_g^2 dependence on the scattering contrast can be written as

$$R_g^2 = R_m^2 + \frac{\alpha}{\Delta\rho} - \frac{\beta}{\Delta\rho^2} \quad (3)$$

where R_m is R_g at infinite contrast; and $\Delta\rho = \overline{\rho} - \rho_s$, where $\overline{\rho}$ is the mean scattering density for the complex and ρ_s is the scattering density of the solvent. The coefficient α is related to the second moment of the scattering density fluctuations about the mean value for the complex, while β is related to the square of the first moment of the density fluctuations about the mean. If the sign of α is positive, the lower scattering density component is more toward the inside of the complex than the higher scattering density component. A negative α indicates the reverse. β is proportional to the square of the separation of the centers of mass of the two components. If β is zero, then the centers of mass are coincident.

DEUTERATION OF BIOLOGICAL MOLECULES

Neutron scattering and/or diffraction experiments using **contrast variation** frequently depend upon the ability to reconstitute complexes with specific components labeled uniformly with deuterium. Deuteration of biological molecules is achieved by growing microorganisms on deuterated media. Relatively high average levels of deuteration (50% for nucleic acids, 70 to 80% for proteins) can be achieved by growing organisms on ^2H_2O with normal protiated carbon sources (5). Typically, however, the aromatic groups are not highly deuterated using this approach. Higher levels of enrichment (> 99%) are achieved using deuterated nutrients and ^2H_2O (5). Deuterated algal hydrolysate provides an excellent rich growth medium for *E. coli* containing an expression system with a specific protein of interest. If the expression is sufficiently robust, deuteration can be achieved in minimal growth media with deuterated glucose, glycerol, or succinate in ^2H_2O. Intermediate levels of uniform deuterium labeling can be achieved by diluting ^2H with ^1H in the growth medium.

EXAMPLES OF THE APPLICATION OF CONTRAST VARIATION

Small-Angle Scattering and Contrast Variation

Examples of significant advances based on small-angle neutron scattering with contrast variation include: the solution structure of the **nucleosome** core particle showing the DNA winding around the outside of the **histone** core (6, 7); mapping the radial distribution of protein and nucleic acid components in the **influenza virus** (8); determining the subunit structure of DNA-dependent **RNA polymerase** (9) and its relationship to a 130-bp DNA fragment (10); the subunit structure of a ribosome particle, showing the dispositions its components (11,12) (see **Neutron diffraction and scattering**); and the determination of the conformations and dispositions calmodulin complexed with its target enzymes (13) and of the evolutionarily related muscle protein troponin C (TnC) complexed with its regulatory target troponin I (TnI) (13) which acts as a calcium-sensitive "switch" to regulate

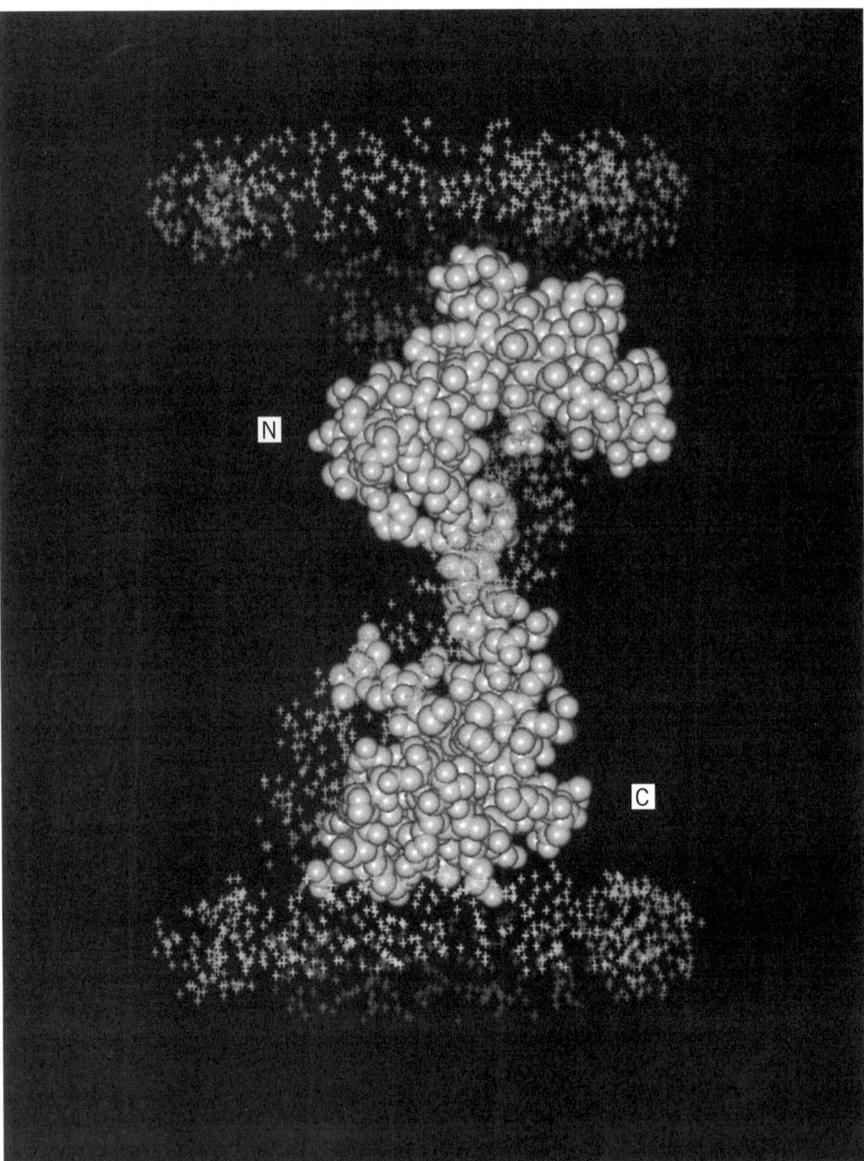

Figure 2. Model structure for the muscle protein complex troponin C/troponin I derived from small-angle neutron scattering and contrast variation (14). Troponin C is represented as a space filling model based on its crystal structure, and the green crosses represent the volume occupied by the troponin I. This figure is based on work done at Los Alamos National Laboratory. See color insert.

the contractile apparatus. Figure 2 illustrates this latter experiment, showing the model derived from a neutron contrast series on complexes of TnI with deuterated TnC in its calcium-saturated form (14). The neutron data show TnC in the complex has the unusual dumbbell shape evident in the TnC crystal structure (15,16), while TnI forms a spiral structure that passes through two hydrophobic clefts in each globular domain of TnC. The diameter of the TnI central spiral is close to that expected for an α-helix. The model suggests a mechanism for the action of the calcium-sensitive TnC/TnI molecular switch. The hydrophobic cleft of the *N*-terminal regulatory domain of TnC alternately opens and closes (in response to calcium binding and release) and thus binds and releases TnI, which in turn switches “off” and “on” TnI's inhibition of the interactions between the muscle fibers required for contraction.

Diffraction Experiments with Contrast Variation

In crystals of large biomolecular complexes, there are frequently disordered regions, even when the diffraction data extend to relatively high resolution. The scattering from the disordered components is seen predominantly at low-Q—that is, in the “small-angle” region where the contrast effects can be strong. Deuteration strategies similar to those used in small-angle neutron scattering allow for contrast manipulation. Phase information can be derived for the low-Q data using contrast-variation procedures (17), beginning with starting phases from a **X-ray crystallography** or a model structure and using the contrast-variation relationships to produce the best scattering-density maps (18). Examples of experiments using contrast variation and low-resolution neutron crystallography include the determination of the location of 25% of the protein and the RNA in crystals of the **Tomato Bushy Stunt Virus** (19) that were “missing” in the 2.8 Å resolution X-ray structure. The most important conclusion of this study is that at the virus radius where the protein and RNA interact, the quasi-equivalence of the 180-protein subunits breaks down, and the RNA interacts preferentially with 60 of the subunits. This observation favors a model for self-assembly in which the RNA interacts with trimers of protein to form an icosahedral scaffolding in which the gaps are filled by the other 120 subunits which have nonspecific contacts with the RNA (see **Capsid, viral**). Another example of the application of low-resolution neutron diffraction and contrast variation is in **membrane protein** crystallography, where it has been used to locate and orient **detergent** molecules with respect to the protein (20,21). These experiments take advantage of the inherent high contrast between detergent and ^2H_2O; and they provide information on the role of detergent in membrane protein crystallization, as well as the potential lipid–protein interactions.

BIBLIOGRAPHY

1. J. Wadzack et al. (1997) *J. Mol. Biol.* **266**, 343–356.
2. R. Willumeit et al. (1996) *J. Mol. Struct.* **383**, 201–211.
3. J. Zhao and H. B. Stuhrmann (1993) *J. Phys. IV* **3**, 233–236.
4. K. Ibel and H. B. Stuhrmann (1975) *J. Mol. Biol.* **93**, 255–265.
5. H. L. Crespi (1977) *Stable Isot. Life. Sci., Proc. Tech. Comm. Meet. Mod. Trends Biol. Appl. Stable Isot.*, IAEA, Vienna, pp. 111–121.
6. J. F. Pardon et al. (1975) *Nucleic Acids Res.* **2**, 2163–2176.
7. P. Suau et al. (1977) *Nucleic Acids Res.* **4**, 3769–3786.
8. S. Cusack, R. W. H. Ruigrok, P. C. J. Krygsman, and J. E. Mellema (1985) *J. Mol. Biol.* **186**, 565–582.
9. P. Stöckel et al. (1980) *Eur. J. Biochem.* **112**, 411–417 and 419–423.
10. H. Lederer, K. Mortensen, R. P. May, G. Baer, H. L. Crespi, and H. Heumann (1991) *J. Mol. Biol.* **219**, 747–755.
11. M. S. Capel, D. M. Engelman, B. R. Freeborn, M. Kjeldgaard, J. A. Langer, V. Ramakrishnan, D. G. Schindler, D. K. Scheider, B. P. Schoenborn, I. Y. Sillers, S. Yabuki, and P. B. Moore (1987) *Science* **238**, 1403–1406.
12. R. P. May, V. Nowotny, P. Nowotny, H. Voß, and K. H. Nierhaus (1992) *EMBO J.* **11**, 373–378.
13. J. K. Krueger, G. Zhi, J. T. Stull, and J. Trewhella (1998) *Biochemistry* **37**, 13997–14004.
14. G. A. Olah and J. Trewhella (1994) *Biochemistry* **33**, 12800–12806.
15. O. Herzberg and M. N. G. James (1985) *Nature (London)* **313**, 653–659.
16. M. Sundaralingham et al. (1985) *Science* **227**, 945–948.
17. M. Roth, A. Lewitt-Bentley, and G. A. Bentley (1984) *J. Appl. Crystallogr.* **17**, 77–84.
18. M. Roth (1991) In *Int. Union Crystallogr., Crystallogr. Symp.* **5** (*Crystallogr. Comput.* **5**) (D. M. Moras, and A. D. Podjarny, eds.), Oxford University Press, Oxford, U.K., pp. 229–248.
19. P. Timmins, D. Wild, and J. Witz (1994) *Structure* **2**, 1191.
20. M. Roth et al. (1989) *Nature* **340**, 659–662.
21. P. Timmins, E. Pebay-Peyroula, and W. Welte (1994) *Biophys. Chem.* **53**, 27–36.

Suggestions for Further Reading

L. A. Feigen and D. I. Svergun (1987) *Structure Analysis by Small-Angle X-Ray and Neutron Scattering,* Plenum Press, New York.

B. Jacrot (1976) *Rep. Prog. Phys.* **39**, 911–953.

P. B. Moore (1982) Small-Angle Scattering Techniques for the Study of Biological Macromolecules and Macromolecular Aggregates, In *Methods of Experimental Physics*, Vol. 20 (G. Ehrenstein and H. Lecar, eds.), Academic Press, New York, pp. 337–390.

B. P. Schoenborn and R. B. Knott (eds.) (1996) *Basic Life Sciences*, Vol. 64: *Neutrons in Biology*, Plenum Press, New York.

H. B. Stuhrmann (1987) Molecular Biology. In *Methods of Experimental Physics*, 23, Part C, Academic Press, New York.

CONTROLLING ELEMENT

N. Craig

Controlling element is the term used by Barbara McClintock (1) to describe the new genetic elements she found in the late 1940s that "can modify and control the action of the genes themselves" and "may move from location to location within the chromosome complement without losing its identity." We now know such elements as **transposable elements**—that is, discrete pieces of DNA that can move from place to place within a **genome**. Such elements are extremely widespread and have been found in virtually all organisms that have been examined. It is interesting that she discovered these elements that we now know to be mobile DNA before the structure of DNA was established in the early 1950s. The first molecular characterization of transposable elements came nearly 20 years later, after their discovery in bacteria, where DNAs containing them could be isolated and characterized (2).

McClintock was a maize geneticist and cytogeneticist. Her discovery of controlling elements resulted from the study of irregular patterns of pigment in maize kernels (ie, variegations of pigment patterns) that arose in some ears. Such variegation results in changes in pigment gene expression during kernel development. We now know that such variegation reflects the insertion of a transposable element to inactivate a gene and the subsequent excision of the element at a later time, restoring gene function. Having earlier studied chromosome breaks that resulted from unusual chromosome fusions, she was also quick to appreciate that some variegation resulted from chromosomal breakage at a particular site. She proposed that there was an element *dissociation* (Ds) at the breakage site and suggested that variegation results from the loss of genes downstream of this site following chromosome breakage.

Identification of an element that could cause chromosome breaks was novel, but even more dramatic was her finding that the element could move to another chromosomal site and again cause instability via chromosome breakage. McClintock realized that another element in addition to Ds was required to cause breaks at Ds and also to promote the translocation of Ds to another chromosomal position—that is, that another element encodes a product that promotes the movement of Ds. She named this other element *activator* (Ac) and also established that Ac could move from place to place.

We now know that Ac is an intact (autonomous) element (about 2.9 kbp) encoding a **recombinase**, a **transposase**, that acts on special sequences at the tips of the element to promote its translocation; Ds is a deleted version of Ac that lacks the transposase but still has the special terminal recombination sequences, so that Ac transposase can promote movement of Ds. Ds is called a **nonautonomous element** because it requires the presence of transposase provided by another element.

BIBLIOGRAPHY

1. B. McClintock (1956) *Cold Spring Harbor Symp. Quant. Biol.* **21**, 197–216.
2. P. Starlinger (1977) In *DNA Insertion Elements, Plasmids, and Episomes* (A. I. Bukhari, J. A. Shapiro, and S. L. Adhya, eds.), Cold Spring Harbor Laboratory, Cold Spring Harbor, NY, pp. 25–30.

CONVERGENT EVOLUTION

T. GOJOBORI

Convergence is defined as an evolutionary event in which two morphological or molecular traits become similar during **evolution** due to a similarity in environment or selection pressure, even though these traits have independent ancestors or origins. In the case of amino acid or **nucleotide sequences**, there is little known evidence for convergence; the probability of two sequences becoming similar by convergence seems vanishingly small, and any similarities in sequence are attributed to **divergent evolution**. Yet even the absence of sequence similarity cannot be taken as evidence of convergence, as this will also result from extensive divergence between two sequences that originated from a common ancestor. In many such cases where proteins related functionally have no detectable sequence **homology**, their three-dimensional **protein structures** are very similar; protein structure appears to change more slowly than does the **primary structure**, and such cases seem to be further examples of divergent evolution.

There are, however, some examples of convergence that are apparent in the protein structure and the **active sites** of **enzymes**. For example, there are several kinds of **serine proteinases**. One family of these proteinases, represented by **chymotrypsin**, has very similar **tertiary structures** and active sites, containing a **catalytic triad** of serine, histidine, and aspartic acid residues. The serine proteinase **subtilisin** also contains an extremely similar triad structure, even though the rest of its **tertiary structure** and sequence is unrelated to the chymotrypsin family (Fig. 1) (1). Moreover, the three residues constituting the catalytic triad occur in different orders in the primary structures, making it virtually inconceivable that the two families of proteins arose from a common ancestor. Furthermore, serine carboxypeptidase II also contains a catalytic triad at its active site, even though its catalytic mechanism is entirely different. It is considered that the very similar triad structures of these different protein families emerged from their respective and independent ancestors by positive **natural selection**. This is one of the best characterized examples of convergent evolution.

There are other examples of enzymes with the same functions but sufficient differences to make it clear that they have not arisen from the same ancestor, but that the same function has arisen independently by a type of convergent evolution (Table 1) (2).

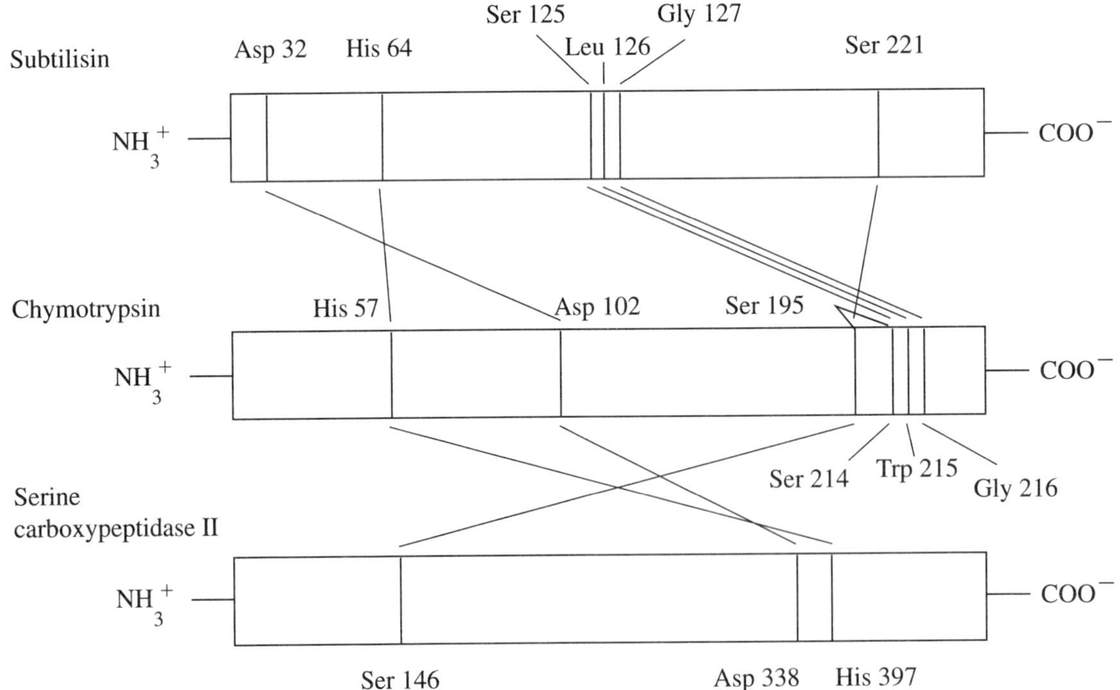

Figure 1. Convergent evolution of the active site residues in subtilisin, chymotrypsin, and serine carboxypeptidase II. Although the relative positions of active-site residues differ in the amino acid sequences among subtilisin, chymotrypsin, and serine carboxypeptidase II, they have a common catalytic triad consisting of Ser 221, His 64, and Asp 32 in subtilisin; of Ser 195, His57, and Asp102 in chymotrypsin; and of Ser 146, His 397, and Asp 338 in serine carboxypeptidase II. Thus, they are considered as examples of convergence or convergent evolution (2).

Table 1. Some Enzymes that Have Evolved Independently on More Than One Occasion, by Convergent Evolution

Superoxide dismutases
Adolases
Sugar kinases
Serine proteinases
Alcohol dehydrogenases
Aminoacyl tRNA synthetases
Ribonucleotide reductases
Topoisomerases
Phosphoenoylpyruvate carboxykinases
Malate dehydrogenases

BIBLIOGRAPHY

1. D. Voet and J. G. Voet (1995) *Biochemistry*, 2nd ed., Wiley, New York.
2. A. Doolittle (1994) *Trends. Biochem. Sci.* **19**, 15–18.

COOMASSIE BRILLIANT BLUE

T. E. CREIGHTON

Proteins are most frequently detected after **gel electrophoresis** by fixing them in the gel, so that they do not diffuse further, and then staining them to produce a colored zone. The most sensitive stain is **silver stain**, but the stain used most frequently is Coomassie Brilliant Blue. It has two frequently encountered forms, R250 and G250, whose structures are depicted in Figure 1. Although the two dyes are structurally very similar, they require different physical and staining procedures and are not interchangeable in any particular protocol. These two dyes do not react chemically with proteins but merely form noncovalent complexes. The interaction with proteins is believed to be primarily ionic and involves the acidic sulfonate groups on the dye and basic groups on the protein, but nonpolar **van der Waals** forces are probably also involved. Consequently, the dyes do not bind equally to all proteins, so two electrophoretic bands with the same blue color need not contain the same amount of protein (1).

Electrophoresis gels are readily stained by placing the gel in a liquid containing the Coomassie dye plus an agent to fix the protein bands, usually acetic acid plus methanol; a solution of 10% (w/v) **trichloroacetic acid** plus 10%(w/v) sulfosalicylic acid is very effective for fixing and staining. Fixed and insoluble proteins bind the dye tightly. It is usually necessary to remove the excess dye from the staining mixture before the bands on the gel can be seen. This is usually accomplished simply by washing the gel in the fixative solution, but it can also be accomplished by adding an agent, such as polyurethane foam, that binds the excess dye tightly. It is important not to remove all of the excess dye from the solution because then the dye bound to the protein bands dissociates and the gel becomes bleached. The procedure is simplified if the Coomassie dye is only slightly soluble in the original staining mixture; the protein takes up the dye from the solution, but the background color remains weak.

A number of other dyes, such as **Ponceau S** and Amido black, can be used in the same way, but they are not as sensitive as the Coomassie dyes. Coomassie Brilliant Blue detects approximately 0.1 μg of protein in a band on a **polyacrylamide** gel.

Coomassie Brilliant Blue, particularly G250, is also used to quantify the amount of protein in solution, which is known as the Bradford assay (2). The dye complexed to protein has an altered **absorbance** spectrum. Under certain acidic conditions, the absorbance maximum shifts from 465 to 595 nm upon binding to a protein. Even though different proteins give somewhat different responses in this assay, its simplicity makes it widely used.

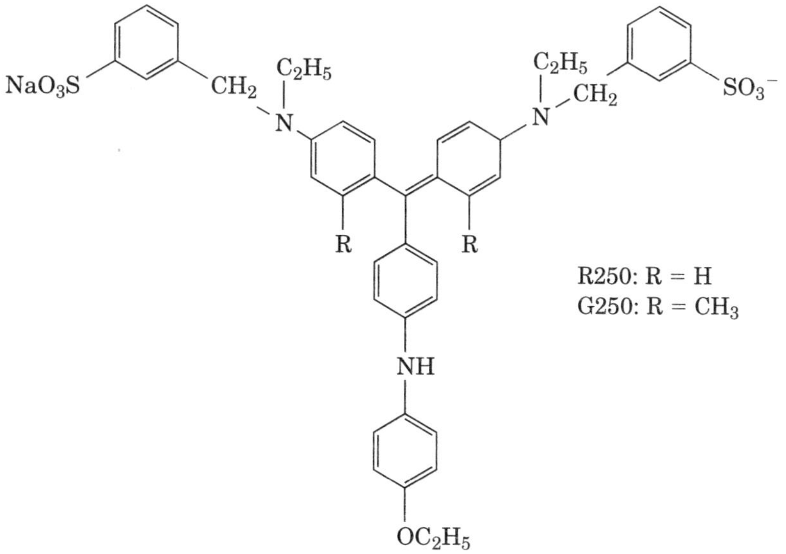

Figure 1. The chemical structures of the R250 and G250 forms of Coomassie Brilliant Blue. They differ only in the nature of the group R.

BIBLIOGRAPHY

1. M. Tal, A. Silberstein, and E. Nusser (1980) *J. Biol. Chem.* **260**, 9976–9980.
2. M. M. Bradford (1976) *Anal. Biochem.* **72**, 248–254.

CORDYCEPIN

J. D. LEWIS

Cordycepin is 3′-deoxyadenosine, but is normally used in molecular biology as the triphosphate form. 3′-Deoxyadenosine triphosphate is a chain-terminating nucleotide analogue of ATP that can be incorporated into an **RNA** by **polyadenylate polymerase**. This inhibits further elongation of the **poly A** tail due to the absence of the 3′-hydroxyl group. The cordycepin triphosphate has been used primarily to study the **polyadenylation** reaction *in vitro*, where it has allowed this reaction to be resolved into two distinct steps: cleavage of the **messenger RNA** precursor and the addition of the poly A tail (see **Polyadenylation**) (1,2).

BIBLIOGRAPHY

1. M. Sheets, P. Stephenson, and M. Wickens (1987) *Mol. Cell Biol.* **7**, 1518–1529.
2. C. Moore, H. Skolnik-David, and P. Sharp (1986) *EMBO J.* **5**, 1929–1938.

Suggestion for Further Reading

W. Keller and L. Minvielle-Sebastia (1997) A comparison of mammalian and yeast pre-mRNA 3′-end processing, *Curr. Opin. Cell Biol.*, **9**, 329–336.

COREPRESSOR

ANDREW TRAVERS

A corepressor is a molecule that binds to a sequence-specific DNA binding protein and enables that protein to direct transcriptional repression. The term is used to apply to two types of mechanism. In the first case, a small molecule binds to and effects a conformational change in a DNA-binding protein, thus, enabling binding to a specific DNA sequence. One example of this mode of action is the binding of tryptophan to the tryptophan repressor. The second type of corepressor is a protein that forms a specific complex with a sequence-specific DNA binding protein and then itself inhibits transcription either directly or indirectly. One example of such an activity is the complex formed in eukaryotes between histone deacetylases and DNA-targeted regulators via an adaptor protein.

COS CELLS

R. I. FRESHNEY

COS cells are monkey kidney fibroblasts (Fig. 1); they have become popular in recent years because they support certain strains of replication-deficient animal **viruses** (1,2) and can be **transfected** easily (3,4). Two lines are in common use, COS-7 (ATCC CRL-1651) and COS-1 (ATCC CRL-1650).

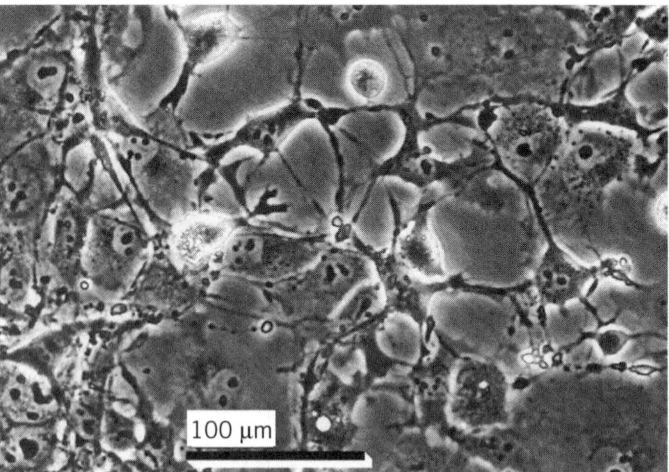

Figure 1. Confluent culture of COS-7 cells. Phase contrast, Olympus CK microscope, 20× objective.

ORIGIN

Both cell lines were derived from CV-1 cells (ATCC CCL-70), an established **cell line** of simian fibroblasts, by transformation with an origin-defective mutant of **SV40 virus** (1). Both are **T-antigen**-positive, permist lytic growth of SV40, and support the replication of temperature-sensitive mutant SV40 tsA209 and of SV40 mutants with deletions in the early region. COS-1 cells contain a single integrated copy of the complete early region of the SV40 **genome**.

PROPERTIES

COS cells are propagated as a monolayer in the alpha modification of Eagle's minimal essential medium (α-MEM) supplemented with 10% fetal bovine serum (see **Serum dependence**) or in Dulbecco's modification of Eagle's medium. They grow rapidly, with an approximate population doubling time of 18 h.

USAGE

COS cells have been used extensively in DNA transfer studies (5–7) and have been transfected by a number of different methods, including calcium phosphate (8), DEAE (4), and lipid transfection (9).

BIBLIOGRAPHY

1. Y. Gluzman (1981) SV40-transformed simian cells support the replication of early SV40 mutants. *Cell* **23**, 175–182.
2. F. Unckell, R. E. Streeck, and M. Sapp (1997) *J. Virol.* **71**, 2934–2939.
3. Y. Cao, D. M. Stafforini, G. A. Zimmerman, T. M. McIntyre, and S. M. Prescott (1997) *J. Biol. Chem.* **273**, 4012–4020.
4. J. J. Schwartz and R. Rosenberg (1998) in *DNA Transfer to Cultured Cells*, K. Ravid and R. I. Freshney, eds., Wiley-Liss, New York, pp. 179–192.
5. V. Misra, H. J. Klamut, and A. M. Rauth (1998) Transfection of COS-1 cells with DT-diaphorase cDNA: Role of base change at position 609. *Br. J. Cancer* **77**, 1236–1240.

6. S. G. Plonk, S. K. Park, and J. H. Exton (1998) *J. Biol. Chem.* **273**, 4823–4826.

7. B. Corsi, F. Perrone, M. Bourgeois, C. Beaumont, M. C. Panzeri, A. Cozzi, R. Sangregorio, P. Santambrogio, A. Albertini, P. Arosio, and S. Levi (1998) *Biochem. J.* **330**, 315–320.

8. J. V. O'Mahoney and T. E. Adams (1998) in *DNA Transfer to Cultured Cells*, K. Ravid and R. I. Freshney, eds., Wiley-Liss, New York, pp. 125–145.

9. V. V. Bickko (1998) In *DNA Transfer to Cultured Cells* (K. Ravid and R. I. Freshney, eds., Wiley-Liss, New York, pp. 193–212.

COSY SPECTRUM

J. T. GERIG

Correlation spectroscopy (COSY) was among the first two-dimensional (2D) nuclear magnetic resonance (NMR) experiments to be developed. In COSY–type experiments, cross peaks arise in a 2D map at the intersections of two **chemical shifts** if a resolved spin-coupling interaction exists between the nuclei characterized by the two shifts. The cross peaks have considerable fine structure, consisting of positive and negative signals if the lines of the spectrum are sufficiently narrow.

The appearance of the cross peaks in a COSY–type experiment relates to a quantum-mechanical phenomenon known as coherence transfer. In NMR, a coherence is characterized by a specific Larmor precessional frequency that is essentially defined by the magnetic field of the spectrometer and the chemical shifts of the spins involved in the creation of the coherence. A cross peak in COSY–type experiments develops because magnetization that was initially characterized by one precessional frequency is converted at some point to a coherence that precesses at a different frequency, defined by another chemical shift. The COSY 2D experiment in effect measures the precessional frequency of the coherence during both parts of the experiment. Coherence transfer occurs only if spins are J-coupled to each other. In peptide and protein systems, the appearance of a cross peak in proton-proton COSY experiments generally identifies sets of protons that are adjacent to each other in the covalent structure. With sufficiently narrow lines, it is possible to analyze the structure of COSY cross peaks to obtain values for the spin-spin coupling constant involved in the coupling interaction signaled by the cross peak.

Many variations on the basic ideas of the COSY experiment are now available, including DQFCOSY (double quantum filtered COSY). The advantages of the DQFCOSY experiment include a much narrower set of diagonal peaks (so that cross peaks close to the diagonal can be detected more readily) and drastic reduction of the intensities of singlets. Such singlets tend to be very intense and can cause noise and various artifacts in a 2D spectrum.

Note the requirement that a resolved spin-coupling interaction be present for optimum detection of a COSY–type cross peak. With greater molecular weights of samples, proton NMR spectral lines increase in width. Under these conditions, the positive and negative components of COSY cross peaks can begin to cancel one another, with the detection of the cross peak relative to noise becoming more difficult. Thus, although the presence of a cross peak in a COSY–type spectrum indicates mutual spin coupling between (probably) adjacent groups of protons, the absence of a cross peak in a COSY–type spectrum cannot be taken as evidence that two nuclei are not coupled to each other. See also (**NMR, Scalar coupling**.)

Suggestions for Further Reading

R. J. Abraham, J. Fisher and P. Loftus (1988) *Introduction to NMR Spectroscopy*, Wiley, New York.

Two-dimensional NMR Spectroscopy: applications for chemists and biochemists, 2nd ed. (1994) (W. R. Croasmun and R. M. K. Carlson, eds.), V.C.H., New York.

A. E. Derome (1987) *Modern NMR Techniques for Chemistry Research*, Pergamon, Oxford

S. W. Homans (1992) *A Dictionary of Concepts in NMR*, Clarendon, Oxford.

R. S. Macomber (1998) *A Complete Introduction to Modern NMR Spectroscopy*, Wiley, New York.

D. L. Turner (1985) *Prog. NMR Spectrosc.* **17**, 281–358.

D. E. Wemmer (1988) In *Recent Advances in Organic NMR Spectroscopy* (J. B. Lambert and R. Rittner, eds.), Norell Press, Landisville, New Jersey.

COUNTING RESIDUES

T. E. CREIGHTON

Natural **proteins** have integral numbers of each of the twenty **amino acids** in a **polypeptide chain**, but most currently accepted methods of determining this number by **amino acid analysis** yield only a nonintegral ratio of moles of amino acid per mole of protein. This value is rarely close to an integer because of experimental error and uncertainty about the **molecular weight** of the protein. The method described here determines the integral number of residues of certain amino acids, independently of any other property of the polypeptide chain, including its molecular weight. Then this information is combined with amino acid analysis to determine a more accurate value for the molecular weight and the number of other amino acids. The method also provides information about the net charge on a protein.

The general procedure is to modify the N residues of a given amino acid gradually and specifically, so as to generate a complete spectrum of molecules with 0, 1, 2, 3, ..., N of the groups modified. A single modification reaction can be used, varying the average extent of the reaction. Alternatively, one uses the competition with a related, but distinguishable reagent and carry out the modification to completion. Then the modified species generated are counted after separating them by a technique sensitive only to the number of groups modified by the particular reagent. The most useful modifications are those that alter the electric charge of a residue. Then the number of modifications introduced is determined by separating the various species by **electrophoresis**, **isoelectric focusing**, or **ion exchange chromatography**. The separation must be sensitive only to the number of groups modified in charge, not which particular ones in the polypeptide chain. This is usually accomplished by carrying out the modification and the

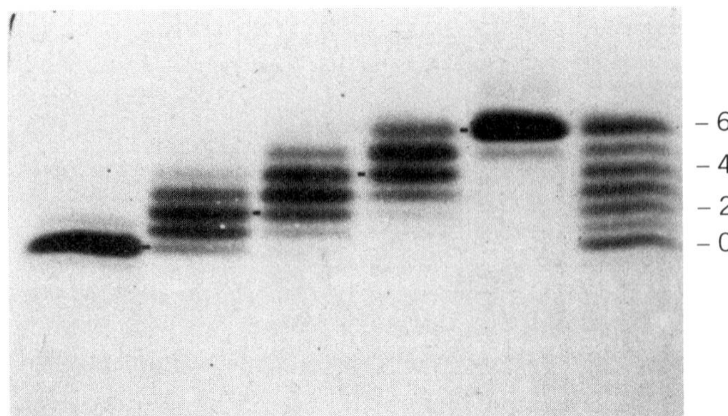

Figure 1. Counting the six cysteine residues of reduced **BPTI**. The reduced protein was reacted with the neutral iodoacetamide (1st lane), acidic iodoacetic acid (5th lane), and mixtures of the two in the ratios 1:1, 1:3, and 1:9 in lanes 2, 3, and 4, respectively. Lane 6 contains a mixture of equal portions of the samples applied to lanes 1 to 5. Electrophoresis of the basic protein was in 8 *M* **urea** from top to bottom. The electrophoretic mobility is decreased in proportion to the number of acidic carboxymethyl groups on each molecule; the number for each of the electrophoretic bands is indicated at the right. The competition between the two reagents indicates that iodoacetamide reacts three times more rapidly than iodoacetic acid under the conditions used here. From Ref. 1.

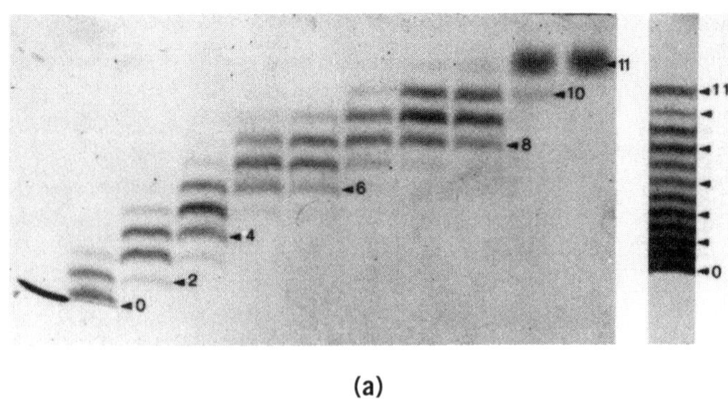

(a)

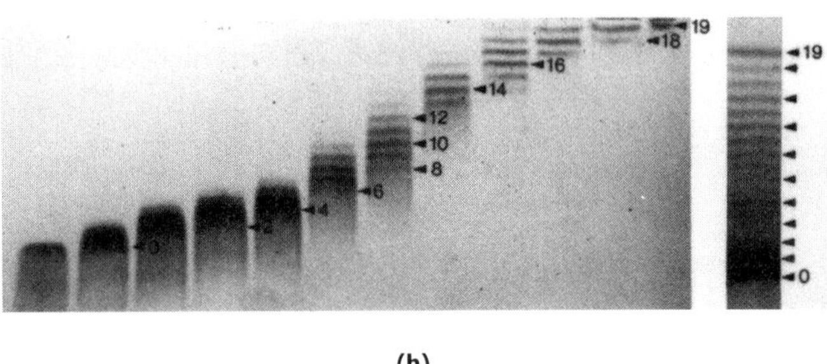

(b)

Figure 2. Counting amino groups of bovine **ribonuclease A** (**a**) and horse ferri**cytochrome c** (**b**) by electrophoretically separating the mixtures produced by progressive succinylation. The original unmodified protein is on the left, and the degree of succinylation increases to the right. Electrophoresis was in 8 M urea at pH 3.6 from top to bottom. The separate lane on the far right is of a mixture obtained by combining all of the individual samples. In this case, electrophoresis was at the slightly lower pH of 3.45. Alternating bands are marked by arrows, and the number of succinyl groups is indicated for a few of the bands. The results confirm the presence of 11 and 19 amino groups in ribonuclease A and cytochrome *c*, respectively. From Ref. 2.

separation on unfolded molecules, where differences between the *N* residues are minimized. There should be $(N + 1)$ species present.

The **thiol groups** of **cysteine** residues are reacted with a reagent like **iodoacetic acid**, which introduces an acidic group. This reaction competes with that with the closely related iodoacetamide, which is neutral and does not introduce a charged group. The reaction of the thiol groups is taken to completion, and the number of acidic groups introduced is varied by varying the ratio of iodoacetic acid to iodoacetamide (Fig. 1). The number of acidic groups introduced is determined by electrophoresis under denaturing conditions. In addition to the number of total cysteine residues or thiol groups, the number of **disulfide bonds** can be determined by counting the thiol groups present before and after reduction of the disulfide bonds.

The amino groups of **lysine residues** and at the N-terminus can be modified specifically by reacting them with a reagent like succinic anhydride, which replaces a basic amino group with an acidic group. Consequently, the net charge of the protein can change by up two units for each amino group reacted. The reaction is varied in its extent by varying the amount of reagent added to the unfolded protein (Fig. 2).

Other amino acids for which specific modifications are possible may be counted in the same way, in theory, but the specific techniques have not yet been developed.

In each of the previous procedures, the number of groups modified is counted by the number of new bands apparent by electrophoresis. Electrophoretic mobility is proportional to the net charge of the protein molecule, which is changed gradually by modifiying specific residues. Therefore, the net charge on the original protein molecule can be determined by measuring the degree of modification necessary to give the protein zero electrophoretic mobility or by using stepwise modification to establish a scale for electrophoretic mobility as a function of changes in net charge.

BIBLIOGRAPHY

1. T. E. Creighton (1980) *Nature* **284**, 487–489.
2. M. Hollecker and T. E. Creighton (1980) *FEBS Lett.* **119**, 187–189.

Suggestion for Further Reading

M. Hollecker (1997) In *Protein Structure: A Practical Approach*, 2nd ed. (T. E. Creighton, ed.) IRL Press, Oxford, pp. 151–164.

COVALENT CATALYSIS

JOHN F. MORRISON

Covalent **catalysis** by an **enzyme** involves the formation, along the reaction pathway, of a covalent enzyme intermediate that can be formed and broken down at a rapid rate. Covalent bonds are formed most commonly as a result of the attack by an enzyme nucleophilic (electron-rich) group on an electrophilic (electron-deficient) moiety of the substrate that is bound at the **active site**. The nucleophilic groups that are present on the side-chains of **amino acid** residues are $RCOO^-$, RNH_2, R—OH, RS^-, and the nitrogen atoms of the imidazole ring of **histidine** residues. The electrophilic moieties of substrates may be acyl, phosphoryl, or glycosyl groups, so the covalent intermediates would be acyl-, phosphoryl-, and glycosyl-enzyme complexes, respectively. The reactions involved in the formation of an acyl-enzyme are

$$\text{RC(=O)—Y} + \text{Enz—}\ddot{\text{X}} \rightleftharpoons \text{R—C(O}^-\text{)(Y)—X—Enz} \rightleftharpoons \underset{\text{Acyl enzyme}}{\text{R—C(=O)—X—Enz}} + \text{Y}^-$$

The second step in covalent catalysis is the attack by a low-molecular-weight nucleophile on the covalent intermediate to release the second reaction product. If the attacking nucleophile is water, the enzyme would catalyze a hydrolytic reaction that conforms to a **kinetic mechanism** with the ordered release of products. If the attacking nucleophile were a second substrate, the reaction would have a ping-pong kinetic mechanism.

Enzyme molecules are poor in electrophilic groups. But electrophilic catalysis does occur with those enzymes that contain metals (see **Metal-requiring enzymes**) or prosthetic groups (see **Coenzyme, cofactor**) that act as electron sinks during catalysis.

As a large number of enzymes catalyze reactions through the formation of covalent intermediates, it appears that the this type of catalysis offers some general advantages (1). It has been considered that the immobilization of a covalent intermediate should provide a significant **entropic** driving force and catalytic efficiency is increased by having multiple bond-making and bond-breaking steps occurring within a single active site.

BIBLIOGRAPHY

1. C. Walsh (1979) *Enzymatic Reaction Mechanisms*, W. H. Freeman and Company, San Francisco, Calif., pp. 39–41.

CPG ISLANDS

GEORGES N. COHEN

In some animals, cytosine bases occurring in the sequence CG are frequently methylated to **5-methylcytosine** (see **Methylation, DNA**). Yet even in species where most such residues are methylated, there are genome segments, known as *CpG islands*, which are conspicuously not methylated. These are short, dispersed regions of DNA with a high frequency of CpG dinucleotides relative to the bulk genome, but they are not methylated. In the haploid human genome, it is estimated that there are some 30 to 45 $\times 10^3$ CpG islands. Such regions of DNA are at least 200 bp in length and have a $G + C$ content greater than 50%. Outside these islands, the frequency of CpG sequences is depleted to about only 20% of that expected on a random basis. This bias against CpG sequences that would be methylated is believed to be caused by methylcytosine bases that deaminate spontaneously 10 to 20 times more readily than normal cytosine bases. Furthermore, the product of deamination of 5-methylcytosine is thymine, a normal base, so this mutagenic event is more difficult to repair than usual (see **DNA repair**).

CpG islands are associated with the 5′ ends of **housekeeping genes** and of a few tissue-specific genes (highly tissue-specific genes usually lack islands). They have an open **chromatin** structure, and it has been postulated that they are sites of interaction between **transcription factors** and **promoters**. All known widely expressed genes are associated with more than one CpG island, which usually includes the **transcription** start site, and a few of these genes have an additional island in the 3′ direction. The average length of CpG islands is about 240 bp in widely expressed genes, but there is no typical size. Most are between 200 and 1400 bp, and a majority of islands are 200 to 400 bp. The average size of islands associated with genes with limited expression is slightly smaller.

Less than half of tissue-specific expressed genes and of those with limited expression are associated with CpG islands that occur in one or more exons. This implies that CpG islands may be used to identify transcripts, since they occur in exons. Furthermore, CpG islands cover the whole or part of the promoter regions containing canonical **TATA box** and **CAAT box**, and they include GC-rich promoters lacking a typical TATA box. Such promoters are considered typical of housekeeping genes.

Suggestion for Further Reading

S. H. Cross and A. P. Bird (1995) CpG islands and genes, *Curr. Opin. Genet. Dev.* **5**, 309–314.

CRISTAE

L. FRONTALI

Electron microscopy of **mitochondria** demonstrates that the inner mitochondrial membrane (IM) folds towards the inner part of the mitochondrion and then folds back (Figure 1), thereby forming structures that are called cristae (see **Mitochondria**). The same morphological evidence has shown that a relationship exists between the development of cristae and the functional activity of mitochondria, which is, in turn, related to the energy requirement of the cell. In fact, folding of the inner mitochondrial membrane is a prerequisite for the structural organization of respiratory enzyme complexes (see **ATP synthase**). The IM defines the intermembrane space (IS) and the mitochondrial matrix (M), where most soluble enzymes and the **protein biosynthesis** apparatus are localized, while the respiratory enzyme complexes are localized

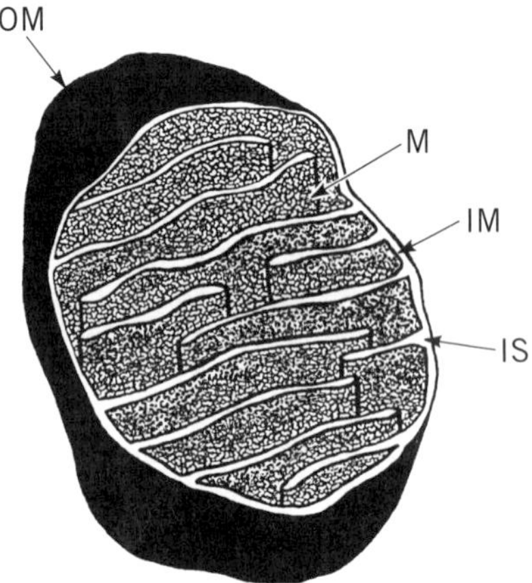

Figure 1. Morphology of a mitochondrion derived from electron microscopy (OM = outer membrane; IM = inner membrane; IS = intermembrane space, M = matrix).

in the IM. Recently, detailed analysis of mitochondrial inner and outer membranes (OM), of their contacts, and of the localization of proteins in the two membranes, in the intermembrane space, and in the matrix, has led to very important developments in our understanding of the essential aspects of mitochondrial transport. Since the IM and OM separate two distinct spaces in mitochondria, proteins encoded by the cell **nucleus** and targeted to the mitochondria can be destined to either one of these spaces, or to be integral components of the IM or OM.

Sorting of proteins to mitochondria is generally mediated by presequences, which are by signals composed of 15—35 amino acid residues at the amino terminus (see **Signal peptides** and **Protein targeting**). Most proteins to be targeted to the mitochondrion are therefore synthesized on cytoplasmic **ribosomes** as precursors and then transferred to **receptor** proteins on the mitochondrial surface. The pre-sequences are then inserted across both membranes at contact points, through distinct outer and inner protein-conducting channels. Many components of this complex machinery for importing proteins have recently been identified (1).

Transport requires unfolding and refolding of proteins by **chaperonins** and is followed by removal of the presequence by a **proteinase** in the mitochondrial matrix. **Signal peptides** are needed not only for routing proteins to mitochondria but also to reach the correct mitochondrial location. Proteins may actually remain in the matrix, assemble with other proteins in the IM, or be transported to the IS through the IM. An inner membrane–associated proteinase is then used to cleave a second signal peptide from these proteins. However, this very interesting scheme is not general, since some proteins lacking a signal peptide can also be imported into mitochondria.

BIBLIOGRAPHY

1. T. Lithgow, B. S. Glick, and G. Schatz (1995) *Trends Biochem. Sci.* **20**, 98–101.

CRITICAL MICELLE CONCENTRATION

PROBAL BANERJEE

Many properties of a **detergent** in aqueous solution depend on its critical micelle concentration (cmc), which in turn is determined by the chemical structure of the detergent. In aqueous solution, detergents remain either in aggregated (micellar) or monomeric states, and the cmc is the maximum concentration at which the detergent molecules remain as monomers. An increase in detergent concentration above the cmc leads to the formation of micelles or mixed micelles with the solubilized lipid and protein molecules (see **Detergents**). Although the formation of mixed micelles of detergents with lipid and protein molecules is an important requirement in detergent solubilization of plasma **membrane proteins** and lipids, a low cmc may not be essential for efficient membrane solubilization because the cmc value alone does not always reflect a detergent's ability to dissociate membranes. For example, sodium dodecyl sulfate (**SDS**) is strongly ionic and has a cmc of ~0.23% at 25°C, whereas MEGA-10 is **polar** but nonionic and has virtually the same cmc of ~0.22%. Yet SDS is probably the strongest **denaturing** detergent known, whereas MEGA-10, just like MEGA-8, is much less disruptive to biomembranes (1,2). A very similar picture emerges from comparing sodium cholate, which is strongly ionic and denaturing, with CHAPS, which is moderately zwitterionic and much less denaturing, even though both have a cmc of ~0.5% (2). Because the cmc values are also the result of certain more fundamental factors, such as structural features and ionic properties of the detergents (ie, nonionic, moderately ionic, or strongly ionic, and the membrane-dissociating property of a detergent is directly related to such factors), one effective way of classifying detergents is based on their structure and ionizability, rather than their cmc values (see **Detergents**).

BIBLIOGRAPHY

1. M. Hanatani, K. Nishifuji, M. Futai, and T. Tsuchiya (1984) *J. Biochem.* **95**, 1349–1353.
2. A. C. Newby, A. Chrambach, and E. M. Bailyes (1982) *Techniques in Lipid and Membrane Biochemistry*, B409, Elsevier/North-Holland, pp. 1–22.

CROSSLINKING

G. M. CARLSON
O. W. NADEAU

Crosslinking refers to linking separate chemical functionalities covalently. Crosslinking is usually achieved through use of a **bifunctional crosslinking reagent**, specifically, a compound having two reactive groups that react chemically with the two functional groups that become crosslinked. In this case, only two functional groups are usually crosslinked, but, with other types of reagents, higher order crosslinking is possible. Crosslinking can be either intramolecular (between separate functional groups on the same molecule) or intermolecular (between functional groups located on separate molecules). In the latter case, it is usually between two macromolecules or between a macromolecule and a relatively small molecule. Crosslinking is usually used to detect groups that are in much greater proximity than would be expected from just their bulk concentrations, due to the structures and associations of the molecules of which they are part. In biochemical applications, the chemical functions that can be crosslinked may be components of all classes of **macromolecules** (**proteins**, **nucleic acids**, carbohydrates, and **lipids**). For simplicity, this article focuses on crosslinking involving the functional groups of proteins, predominantly **amino**, **thiol**, imidazole and **carboxyl groups** (ie, the nucleophilic side chains of certain amino acids); however, the concepts are the same for the other classes of macromolecules. It should be noted that crosslinking also occurs naturally, resulting from oxidation, radiation, or the action of certain **enzymes**.

A distinction is sometimes made between the terms *crosslinking* and *conjugation* (or bioconjugation). In this case, crosslinking is used to denote the covalent linkage of only molecules that are associated naturally, such as the subunits of an **oligomeric protein** or a **hormone** and its **receptor**. Conjugation, in contrast, refers to the covalent linkage of molecules that lack affinity for one another. The resulting covalent complexes, referred to as conjugates, have a multitude of uses, especially in biotechnology (see **Enzyme immobilization and conjugation**). The distinction between crosslinking and conjugation is maintained in this article.

USES OF CROSSLINKING

On the basis of the variety of the crosslinking reagents that are available and of the molecules that can be crosslinked, plus the types of experimental questions that can be answered, crosslinking is among the most versatile of biochemical techniques. In the case of proteins, for example, some information about their three-dimensional structure can be gained through identifying specific amino acid residues that are readily crosslinked intramolecularly (1). Crosslinking can also be used as a general conformational probe to detect structural changes in a protein, such as might be induced by **ligand binding** (2). Inasmuch as crosslinking can hinder conformational transitions and the dissociation of oligomers, it can also be used to stabilize the **tertiary** or **quaternary structures** of proteins. Thus, it can "lock" a protein into a particular functional state (3) or stabilize an oligomeric protein for subsequent physical studies, such as **electron microscopy**. A careful analysis of the mass and composition of crosslinked complexes allows a minimal subunit stoichiometry to be determined; this is especially useful for insoluble oligomers, such as those found in **membranes** (4). Information concerning the symmetry of oligomeric proteins can also be obtained from crosslinking (5,6). One of the most common uses of crosslinking is to identify neighboring proteins in a complex (7), and this can be extended to include determination of the specific regions of each that are crosslinked (8). By varying the length of the crosslinker, one can also obtain a limit for the maximum distance that can separate neighboring proteins (6). In addition to giving both relative and absolute structural information, crosslinking can be used for general screening. For example, one can identify the receptor for a particular ligand, such as a hormone (9).

It must be emphasized that for all the above applications, interpretations can be made only for those crosslinked complexes that are generated. The absence of crosslinking does not mean that two molecules do not interact, only that they are not crosslinked under the specific set of experimental conditions chosen, which includes the crosslinking reagent used. Because crosslinking requires the proper positioning of those functional groups having the appropriate chemical reactivities, one cannot expect all interacting molecules to be crosslinked by any given reagent.

CROSSLINKING PROCEDURES

Crosslinking, like chemical modification of proteins, is an empirical procedure, and both processes are influenced by the same variables, namely temperature, pH, concentrations of reactants, and time. In crosslinking, however, the influence of the latter two variables is more complex. Because nucleophilic side chains of amino acids are considerably more reactive when deprotonated, the pH of the reaction is important in controlling the extent of crosslinking and the selectivity for particular side chains. Regarding concentrations, the amounts of both the crosslinker and the proteins can influence crosslinking, with low concentrations of protein and crosslinker generally favoring intramolecular crosslinking. Adding the crosslinker over time, instead of as a single addition, favors a greater extent of crosslinking. Differences in selectivity of the reactive groups of bifunctional reagents, especially heterobifunctional crosslinkers, for protein functional groups can be maximized using a two-step reaction. In this procedure, one protein is first derivatized with the crosslinking reagent and then purified to remove excess free reagent before being exposed to the second protein. This two-step procedure, which allows for each of the two reactions to be performed under different conditions, is particularly useful for **enzyme immobilization and conjugation**, when the two molecules do not normally associate with each other.

Following crosslinking, the covalent complexes formed are commonly identified as new **electrophoresis** bands in sodium dodecyl sulfate–polyacrylamide gel electrophoresis (**SDS-PAGE**), with altered molecular weights, although care must be taken, because intramolecular crosslinking can alter the migration of individual polypeptides with this technique. Alternatively, in simple systems, covalently linked complexes can be identified as components with increased **Stokes radii** in **size exclusion chromatography**. If either the crosslinker or the crosslinked protein has identifiable physical or chemical characteristics, these can aid in the identification of complexes. Once crosslinked complexes have been identified, complete analysis of their composition and stoichiometry can still be challenging, especially when multiple proteins are present. One useful method for identifying the proteins composing a complex is to use a cleavable crosslinker so that the original components can be regenerated from the fractionated complex (7). Another useful method is to perform **Western blots** on the fractionated complex using antibodies against potential components (2). Once the components have been identified, their stoichiometry in the complex is deduced from its overall mass. The ultimate step in analysis is to determine directly the actual amino acid residues that are crosslinked. Using conventional **protein sequencing** by **Edman degradation**, the sequences of the two crosslinked peptides are obtained simultaneously, with the crosslinked residues represented by gaps in the sequences; however, identification of crosslinked peptides by **mass spectrometry** is becoming increasingly common.

BIBLIOGRAPHY

1. F. C. Hartman and F. Wold (1967) *Biochemistry* **6**, 2439–2448.
2. O. W. Nadeau, D. B. Sacks, and G. M. Carlson (1997) *J. Biol. Chem.* **272**, 26296–26201.
3. R. E. Benesch and S. Kwong (1991) *J. Prot. Chem.* **10**, 503–510.
4. E. Heymann and R. Montlein (1980) *Biochem. Biophys. Res. Commun.* **95**, 577–582.
5. F. Hucho, H. Mullner, and H. Sund (1975) *Eur. J. Biochem.* **59**, 79–87.
6. J. Hajdu, S. R. Wyss, and H. Aebi (1977) *Eur. J. Biochem.* **80**, 199–207.
7. D. Dey, D. E. Bochkariov, G. G. Jokhadze, and R. R. Traut (1998) *J. Biol. Chem.* **273**, 1670–1676.
8. V. Rossi, C. Gaboriaud, M. Lacroix, J. Ulrich, J. C. Fontecilla-Camps, J. Gagnon, and G. J. Arlaud (1995) *Biochemistry* **34**, 7311–7321.
9. T. H. Ji (1977) *J. Biol. Chem.* **252**, 1566–1570.

Suggestions for Further Reading

G. Mattson, E. Conklin, S. Desai, G. Nielander, M. D. Savage, and S. Morgenson (1993) A practical approach to crosslinking, *Mol. Biol. Rep.* **17**, 167–183.

S. S. Wong (1993) *Chemistry of Protein Conjugation and Cross-Linking*, CRC Press, Boca Raton, FL. (An outstanding book describing all aspects of the chemistry and uses of bifunctional reagents in crosslinking and conjugation, with an extensive bibliography for each chapter.)

CRUCIFORM (HOLLIDAY JUNCTION)

A. H.-J. WANG
H. ROBINSON

A key intermediate in DNA **recombination** is the **Holliday junction** that involves the joining of four homologous DNA strands into a cruciform arrangement, as visualized by **electron microscopy** some years ago. **Supercoiling** also can induce the formation of the cruciform for **inverted repeat** sequences. Detailed structural analysis of the cruciform **DNA structure** became possible when specific sequences were designed to form a so-called immobile junction. Studies on those molecules using **gel electrophoresis** and **fluorescence** indicated that the structure has a stacked X shape (ie, not tetrahedral) in the presence of metal ions (eg, Mg^{+} or $Co(NH_3)_6^{3+}$). Models of the detailed structure at the junctions have been proposed. One such model involves a right-handed, antiparallel X structure, with two stacked helices crossing over each other with a crossover angle of ~60° (or 120°). Within each helix of the X structure, the two arms are collinear and the junction site is folded. A number of **B-DNA** structures determined by **X-ray crystallography** showed helix–helix packing contacts in the crystal lattice that are remarkably similar to the X structure. Recently the crystal structures of two proteins that are involved in the recombination process have been solved in the presence of their DNA substrate. The crystal structure at 2.4 Å resolution of Cre recombinase bound to a loxP DNA substrate showed that four recombinases and two loxP sites form a synapsed structure in which the bound DNA adopts a conformation similar to the four-way Holliday junction (1). The synapse assembly has a pseudofourfold symmetry, and the junction core is less compact than that proposed for the "free" Holliday junction.

BIBLIOGRAPHY

1. F. Guo, D. N. Gopaul, and G. D. Van Duyyne (1997) *Nature* **389**, 40–46.

CRYOELECTRON MICROSCOPY

GUY A. PERKINS
TERRENCE G. FREY

Cryoelectron microscopy can be divided into two classes, depending on whether the specimen is *hydrated* or *dry*. Table 1 provides a brief description of the possible methods of specimen preparation. In each method, the specimen is initially frozen and can subsequently be observed in a cryoelectron microscope, or can be further preserved by drying or making a replica for observation in a conventional electron microscope.

The development of techniques for visualizing frozen hydrated tissues and **macromolecules** has been a boon for studying structures in a state that closely approximates their native condition (1). Because freezing takes place very rapidly, these techniques offer the ability to examine dynamic events with high time resolution. For example, the structure of the open-channel form of the **acetylcholine receptor** was elucidated from membrane crystals that were briefly activated (<5 ms) with droplets containing acetylcholine (2). Small conformational changes triggered by acetylcholine were detected in the frozen hydrated crystals. The spatial resolution of periodic structures, eg, protein crystals, can also be high, extending to 3.3 Å (3; Perkins, unpublished results). Moreover, these structures can be examined unstained by utilizing the intrinsic contrast difference between the biological specimen and vitreous ice. Thus, the effects of solvent composition on the structure and aspects of interacting components can be examined (2).

Fundamental to examining frozen hydrated specimens is the preservation of the environment by rapid freezing (4). Rapid freezing causes the water of the specimen and its environment to be trapped in a vitreous (amorphous or glass-like) state. This vitreous state prevents damage to the specimen caused by the formation of ice crystals when the specimen is frozen more slowly. The specimens preserved in thin films of vitreous ice generally retain a high degree of structural integrity. However, it is crucial to maintain the specimen at below ~−140°C because vitreous ice tends to crystallize above this temperature (4).

A wide range of frozen hydrated specimens has been examined in cryoelectron microscopes. Most studies have concentrated on macromolecules and their assemblies hydrated and frozen in thin films. To a lesser extent cryosections of biological tissues have also been studied (5–9), but the difficulty of freezing of bulk material without formation of ice crystals has hampered progress. The first published results obtained of macromolecules were from viruses, fairly robust particles (10).

Table 1. Methods of Specimen Preparation in Cryoelectron Microscopy

Frozen Hydrated Specimens	Dry Specimen Previously Frozen
1. *Vitrified film of ice* A thin layer of macromolecules in aqueous suspension is rapidly frozen by plunging in a cryogen, eg, ethane, and transferred to a cryoelectron microscope. 2. *Frozen section* Cells or tissues are rapidly frozen by contact with a cold metal block, sometimes under high pressure, then sliced into thin sections at ~<−160°C. Subsequently, sections are transferred to a cryoelectron microscope.	1. *Freeze drying* Cells or tissues are frozen, then lyophilized. The remaining material is dried and stabilized for transfer to a conventional electron microscope. Alternatively, a replica is made. 2. *Freeze fracture* Cells or tissues are frozen, then fractured. A replica of the fractured surface is made. 3. *Freeze etching* Freeze fracture followed by freeze drying. 4. *Freeze substitution* Tissues are frozen, then the ice is replaced at low temperature by an organic solvent. Conventional plastic embedment follows. 5. *Cryoimmunohistochemistry* Tissues are first chemically fixed, then suffused with a cryoprotectant and frozen for sectioning. Immunolabelling of the thawed sections follows.

Early work produced 3-D reconstructions at relatively low resolution, ~2–3 nm (10–13), although certain viruses produced data to ~1 nm resolution (14). Important new information about viral structures was obtained even at lower resolution. Recent work has extended the resolution of complex viruses beyond 1 nm (15,16). By combining cryoelectron microscopy with **X-ray crystallography**, near atomic resolution detail of the overall architecture of large complexes and of the interactions between the components can be obtained. An instructive example is the 3-D structure of rhinovirus 14 complexed with Fab fragments determined at 0.4 nm resolution (17).

Although the contrast in frozen hydrated specimens is low compared to negatively stained specimens, the signal can be significantly boosted by averaging many identical particles together using Fourier averaging for crystals (2,17) and correlational averaging for single particles as discussed below (15,16,18). Vitreous ice embedment is particularly useful for visualizing the specimen interior compared to negative stain, which accumulates around areas accessible to the aqueous phase, thus usually showing only the molecular envelope. Hence, in general only those portions of integral **membrane proteins** extending beyond the lipid bilayer are observed in negative stain. However, exceptions to this limitation of negatively stained membrane proteins have been found (19–22). In certain instances, frozen hydrated crystals of membrane proteins do not provide resolution as high as the same crystals in negative stain (23). Vitreous ice embedment of delicate specimens, such as **chromosomes** or DNA fragments (9,24,25) has provided impressive results, since this type of specimen is commonly poorly preserved in other embedment media. Another example is the direct visualization by cryoelectron microscopy of A-, P-, and E-site transfer RNAs in the *E. coli* **ribosome** (26).

Cryofixation of tissues is an alternative to conventional chemical fixation of subcellular ultrastructure. Vitrification of ice can frequently be achieved to a depth of 10–20 μm when tissues are frozen by slamming them against a polished copper block cooled to the temperature of liquid nitrogen or liquid helium (27). High-pressure freezing can freeze samples in vitreous ice to depths as great as 100 μm, since the increase in volume going from water to ice is impaired by a high-pressure environment inhibiting the growth of ice crystals. Cryofixation is usually the choice for immunocytochemistry because it retains the antigenic properties of the specimen better than chemical fixation. With good vitrification, freeze-etching or freeze-substitution (Table 1) can provide information about structures stabilized with a mechanism different from chemical fixation.

BIBLIOGRAPHY

1. J. Jaffe and R. M. Glaeser (1984) *Ultramicroscopy* **13**, 373–378.
2. P. N. T. Unwin (1995) *Nature* **373**, 37–43.
3. J. Brink, W. Chiu, and M. Dougherty (1992) *Ultramicroscopy* **46**, 229–240.
4. J. Dubochet et al. (1982) *J. Microscopy* **128**, 219–237.
5. J.-J. Chang et al. (1983) *J. Microscopy* **132**, 109–123.
6. J. Dubochet et al. (1983) *J. Bacteriol.* **155**, 381–390.
7. A. W. McDowall et al. (1983) *J. Microscopy* **131**, 1–9.
8. A. W. McDowall et al. (1984) *J. Mol. Biol.* **178**, 105–11.
9. A. W. McDowall, J. M. Smith, and J. Dubochet (1986) *EMBO J.* **5**, 1395–1402.
10. J. Adrian et al. (1984) *Nature* **308**, 32–36.
11. J. Lepault and K. Leonard (1985) *J. Mol. Biol.* **182**, 431–441.
12. F. P. Booy et al. (1985) *J. Mol. Biol.* **184**, 667–676.
13. R. H. Vogel et al. (1986) *Nature* **320**, 533–535.
14. J. Lepault (1985) *J. Microscopy* **140**, 73–80.
15. B. Bottcher, S. A. Wynne, and R. A. Crowther (1997) *Nature* **386**, 88–91.
16. J. F. Conway et al. (1997) *Nature* **386**, 91–94.
17. T. J. Smith (1996) *Nature* **383**, 350–354.
18. M. van Heel, G. Harauz, and E. V. Orlova (1992) *J. Struct. Biol.* **116**, 17–24.
19. B. Karlsson et al. (1983) *J. Mol. Biol.* **165**, 287–302.
20. B. Bottcher, P. Graber, and E. J. Boekema (1992) *Biochem. Biophys. Acta* **1100**, 125–136.
21. S. Karrasch et al. (1996) *J. Mol. Biol.* **262**, 336–348.
22. G. A. Perkins et al. (1997) *Biophys. J.* **72**, 533–544.
23. J. M. Valpuesta, R. Henderson, and T. G. Frey (1990) *J. Mol. Biol.* **214**, 237–251.
24. J. Dubochet et al. (1994) *Nature Struct. Biol.* **1**, 361–363.
25. J. Bednar et al. (1995) *J. Mol. Biol.* **254**, 579–594.
26. R. K. Agrawal et al. (1996) *Science* **271**, 1000–1002.
27. J. Dubochet (1995) *Trends Cell Biol.* **5**, 366–369.

CRYSTALLIZATION

SERGE N. TIMASHEFF

The determination of the three-dimensional structures of **proteins** by **X-ray crystallography** requires the availability of large, well-ordered crystals. These are difficult to obtain. The crystallization of any given protein requires long painstaking work, including much trial and error in the choice of environmental conditions (pH, temperature) and crystallizing agents.

The basic requirement is the availability of the protein in *very* pure form, free of all other molecules, such as traces of nucleic acids, in a single form (absence of **isoforms**), in a solution that is free of any dust particles. If liganded to any **coenzymes**, cofactors, or effectors, all molecules must be equally liganded; all molecules must be in the native state, and there must be no partial oligomer formation.

The basic approach is to prepare a protein solution at high concentration (10 to 100 mg/ml) in a crystallizing solvent at conditions at which the solution is close to saturation. Then the solution is gradually brought to supersaturation by, eg, slow variation in pH or temperature. All changes must be brought about slowly and the solution not subjected to shocks, such as mixing. This allows the formation of crystal nuclei. Amorphous aggregates must be totally avoided. Once nuclei are formed, the system is induced to crystallize further by the addition of protein to these nuclei, avoiding the formation of new ones, since the aim is to form large crystals. See the specialized literature for further details.

Typical crystallizing agents are (1) salts $(NH_4)_2SO_4$, Na_2SO_4, Na citrate, $MgSO_4$; (2) organic molecules [2-methyl-2,4-pentanediol (MPD), ethanol, isoproponal, acetone, dioxane, 1,3-propanediol]; (3) polyethylene glycols (molecular weight between 2,000 and 20,000). See also **Vapor phase crystallization**.

Suggestions for Further Reading

T. L. Blundell and L. N. Johnson (1976) *Protein Crystallography*, Academic Press, New York.

H. W. Wyckoff, C. H. W. Hirs, and S. N. Timasheff, eds., (1985) *Methods in Enzymology*, Chaps. 2–15, Academic Press, New York.

CRYSTALLOGRAPHY

JAN DRENTH

Crystals are very important for molecular biology because they are required for detailed structure determination of macromolecules using **X-ray crystallography**. The scientific study of crystals is very old. Its origin can be traced back to 1669 when Nicolaus Steno noticed that the angles between the faces in growing crystals remain constant. Individual crystals may differ in shape appreciably, but the angles between corresponding faces are always the same. A century later, René Just Hauy proposed that crystals are built from small units whose shape is a parallelepiped. An extensive review of the history of crystallography is given in Ref. 1. The physical properties of crystals are, generally not equal in all directions. Their optical, magnetic, and electric properties vary with crystal direction, and this is the reason that crystals are widely used for physical measurements and in industrial processes, for instance, a quartz crystal in a clock or watch. The wealth of variable characteristics is a consequence of the regular packing of the ions, atoms, or molecules in the crystal. This internal regularity had been suggested since Hauy, but it could only be proven with the introduction of X-ray crystallography. In this technique, the crystal acts as a grating and diffracts X-rays in directions specific for the repeating distances in the crystal. In addition, it became possible to elucidate the internal structure of crystals, that is, the arrangement and structure of the molecular composition.

In the regular packing inside the crystal, three repeating vectors can be recognized: **a**, **b**, and **c**, with angles α, β, and γ between them. These three vectors define a **unit cell** in the crystal lattice. From morphological crystal symmetry considerations, that is, the rotational axes, mirror planes, and centers of symmetry observed in the shape of crystals, 32 combinations of these symmetry elements (point groups) are possible. They can be assigned to seven, and not more than seven, crystal systems (2): triclinic, monoclinic, orthorhombic, tetragonal, trigonal, hexagonal, and cubic. Internally, a crystal has more symmetry operators. Translational symmetry is added to the symmetry axes and the mirror planes. These operators can be combined in 230 different ways, leading to 230 **space groups**, introduced by Schoenflies and Fedorow at the end of the nineteenth century (Ref. 1, p. 114; 2).

Because biological macromolecules have an asymmetrical structure, they can only form crystals without mirror planes or centers of symmetry. This restricts the number of space groups appreciably for these molecules.

BIBLIOGRAPHY

1. P. Groth (1926) *Entwickelungsgeschichte der Mineralogischen Wissenschaften*, Julius Springer, Berlin.
2. International Union of Crystallography (1992) *International Tables for Crystallography*, Vol. A (T. Hahn, ed.) Kluwer Academic Dordrecht, Boston, London.

Suggestion for Further Reading

J. P. Glusker, M. Lewis, and M. Rossi (1994) *Crystal Structure Analysis for Chemists and Biologists*, VCH, New York, Weinheim, Cambridge.

C_0T CURVE

GEORGES N. COHEN

A C_0t curve describes the sequential complexity of a DNA sample. It describes the **kinetics** of reassociation and re**annealing** of a sample of double-stranded DNA that has been (*1*) fragmented to small pieces, approximately 400 nucleotides, (*2*) denatured to single strands, and (*3*) then permitted to reanneal. The extent to which two complementary DNA strands reassociate is proportional to C_0t. C_0 is the original molar concentration of the complementary strands of DNA, and t is the incubation time. $C_0t_{1/2}$ is the value of C_0t required for completing half of the reannealing reaction. If the sequences of all initial DNA fragments were identical, complementary fragments encounter each other readily and the mixture reanneals readily at a low value of C_0t. At the other extreme, reannealing is very slow and occurs only at high C_0t values if the original DNA contains many different sequences, so that the concentrations of complementary strands are very low and they encounter each other only infrequently. The value of $C_0t_{1/2}$ is a measure of the sequential complexity of the original DNA.

Such studies on genomic DNA were one of the first indications that **genomes** often contain **repetitive DNA**. Such DNA segments are present in higher concentrations than those segments present in only one copy per genome, and this fraction of the DNA reanneals at a correspondingly lower C_0t value. The results indicate that different sequences are present in widely varying frequencies throughout the genome, and only a fraction occurs only once per genome.

CUBITUS INTERRUPTUS GENES

Y. Chen
S. M. Smolik

In Drosophila **development**, the **phenotype** associated with a gain-of-function **mutation** of the *cubitus interruptus* (*ci*) gene was first described in 1971 by Hochman (1). Later, this gene was categorized as a segment polarity gene because the loss of *ci* function results in the loss of naked regions in the segments of larval cuticle (2). The *ci* gene encodes a **zinc-finger** protein, CI, that stimulates **transcription** in response to the ***hedgehog*** (*hh*) signal transduction cascade, which helps to establish the anterior-posterior positional information in the embryonic segments and **imaginal discs**. The **morphogen** in vertebrates, ***Sonic hedgehog*** (*Shh*), induces and patterns many developmental processes, at least in part through regulating the expression of the vertebrate *ci* **homologues** *Gli1*, *Gli2*, and *Gli3*, which then transcribe a second level of *Shh* target genes. The importance of these genes in growth and development is illustrated by the fact that the Gli1 protein is amplified in glioblastoma cells.

PROTEIN STRUCTURE OF CI AND GLI

Both the *ci* and *Gli* genes encode proteins with five highly homologous, tandemly repeated zinc fingers that act as **DNA-binding** domains and are distinguishable from other zinc-finger domains (3–6). The protein products of these genes are required for the transcription of *hh* target genes during **development**. Because the *Gli* and *ci* genes respond to the same signaling pathway and share a common DNA-binding domain, they are considered to be homologues, even though there is very little conservation on the amino acid sequence level between CI and Gli proteins outside of the zinc-finger region. Three members of the *Gli* gene family (*Gli1*, *Gli2*, and *Gli3*) have been identified in mouse, chicken, and **zebrafish** (7,8). Two *Gli* genes have been identified in ***Caenorhabditis elegans*** (9), four in **Xenopus** (10). They share various degrees of similarity with the zinc finger region, which represents the greatest degree of homology among the family. The zinc fingers are responsible for the DNA-binding activity of both CI and Gli proteins as demonstrated by both in vitro binding assays and in vivo experiments, in which expression of a transgene is driven by Gli binding sites (4,5,11,12). The consensus DNA-binding site for CI and Gli proteins has also been characterized by isolating and comparing **genomic** DNA fragments that can bind to Gli1. The core sequences were further confirmed by DNase **footprinting** (13). The **X-ray crystallography** structure of the Gli1 zinc fingers bound to the DNA binding site has been solved (14). Zinc finger 1 does not contact the DNA, whereas the rest of the four fingers wrap around the DNA, with fingers 4 and 5 making most of the base contacts with the DNA. This structure might explain the higher homology shared by zinc fingers 3–5 among the members of the *Gli* gene family. **Bandshift assays** demonstrate that full-length Gli1, full-length Gli3, or a GST protein fused to the CI zinc finger domain translated in vitro can bind to the Gli DNA binding site **consensus sequence** (5,11,13). When the CI zinc-finger **domain** is fused to a heterologous activation domain, it functions as a transcriptional activator in fly imaginal discs and embryos (15). Furthermore, CI can transactivate a **reporter gene** driven by Gli binding sites in yeast (12) and in fly cell lines (16). All of these experiments suggest that *ci* functions as a **transcription factor** that binds DNA through its zinc-finger domain. The amino acid residues of CI that are N-terminal to the zinc-finger domain are rich in alanine residues, a characteristic of transcriptional repressor domains (17). When a CI protein that is truncated C-terminal to the zinc-finger region is expressed in the anterior compartment of the fly embryonic segments, it causes downregulation of *ci* target genes (15). This result supports the idea that the N-terminus of CI has a **repressor** function. The C-terminus of the CI protein is highly acidic, a characteristic of transcriptional activators (18). When introduced into Drosophila embryos, a CI protein in which the N-terminus is deleted can function as a transcriptional activator to enhance *ci* target gene expression, indicating that the activation domain resides in the C-terminus (15).

EXPRESSION AND FUNCTION OF *CI*

The repeating segments of the fly embryo are established during blastoderm stage through a series of hierarchical interactions among the maternal gap and **pair-rule genes**. The segment polarity genes specify the anterior-posterior polarity of each segment during gastrulation. Consequently, it is no surprise that most of the segment polarity genes describe interdependent signal transduction cascades. *Ci* is a member of the *hh* signal transduction pathway and encodes a zygotic gene that is not detected until early cellular blastoderm stage (stage 5). Before stage 11, expression of CI protein closely follows the pattern of *ci* transcripts. Both protein and message are detected initially on the dorsal surface of the embryo. As development progresses, the domain of *ci* transcription and protein expression extends ventrally and, by stage 10, encompasses the entire embryo. Coincident with the onset of ectodermal segmentation, both the message and the protein are restricted to the anterior compartment of each segment by the ***engrailed*** (*en*) gene, resulting in 15 broad, metamerically repeating stripes. At stage 11, the protein expression pattern no longer coincides with the transcript pattern. Although transcripts are still uniformly distributed throughout the anterior compartment of each segment, the protein stripe within each segment is graded so that higher levels of protein are detected in the rows of cells that define the posterior edge of the anterior-posterior (A/P) boundary and lower levels are seen in the middle of the CI-expressing stripe. This pattern of expression also exists in imaginal discs, where *ci* is uniformly transcribed in the entire anterior compartment and higher levels of CI are detected along the A/P boundary (3,19). It should be noted that the protein staining patterns were obtained with **antibody** made against the C-terminus of CI and detected only the full-length protein. The discrepancy found between transcript and protein levels in the segments and discs indicates that CI is subject to **post-transcriptional regulation**.

The *ci* transcription pattern is regulated by various unidentified factors and is mediated through the regulatory sequences in the *ci* promoter. Using transgenic flies that carry the *LacZ* reporter gene (see **Beta-galactosidase**) under the control of different *ci* regulatory sequences, it has been shown that various elements in the *ci* promoter differentiate embryonic from disc expression (20). Although the sequences required for the repression of *ci* transcription in the posterior compartment of the embryonic segments and imaginal discs are different, they both contain binding sites for the *engrailed* protein, EN, consistent with the genetic data showing that *en* inhibits *ci* transcription in the posterior compartment. EN protein expressed in bacteria protects specific DNA sequences in the *ci* regulatory region that have been shown to mediate *ci* repression in the posterior compartment (20). In addition, EN binds the *ci* gene in **polytene chromosomes**, and misexpression of *en* by the **heat-shock** promoter greatly reduces *ci* mRNA levels (20). Furthermore, Eaton and Kornberg (21) have shown that *ci* transcripts expand into the posterior compartment of both embryonic segments and imaginal discs in *en* mutant flies (21).

Genetic **epistasis** experiments place *ci* within the *hh* signaling cascade. It acts downstream from the *hh* receptor *patched* (*ptc*) and *smoothened* (*smo*), a transmembrane protein that is coupled to *ptc* and transduces the *hh* signal. *Ci* also acts upstream of the *hh* target genes (see ***hedgehog* signaling**). CI is an activator of transcription in the cells that receive an *hh* signal. These cells abut the posterior edge of the A/P boundary where high levels of CI are expressed. The cells expressing lower levels of CI are outside the range of *hh* activity and, in these cells, CI is a repressor of the *hh* target genes (19,22). Recently, great progress has been made toward understanding the cellular mechanism involved in determining whether *ci* is an activator or a repressor. The first insight comes from the observation that CI exists in two forms: the 155-kDa full-length activator form and a 75-kDa proteolytic fragment (22). In the absence of an *hh* signal, a proteolytic event generates the 75-kDa form by cleaving the sequences C-terminal to the zinc-finger domain from the full-length protein. Because the resulting 75-kDa proteolytic product has both the zinc-finger DNA-binding domain and the N-terminal repressor domain, it functions as a repressor, preventing the activation of *ci* target genes. Antibodies made against the N-terminus of CI (AbN), which detect both full-length CI and the proteolytic fragment, stain the entire anterior compartment of the embryonic segments and imaginal discs. On the cellular level, the antibody staining is distributed evenly in both the **nucleus** and the cytoplasm (15,22). On the other hand, full-length CI resides primarily in the cytoplasm. The levels are very low in the cells of the anterior compartment that do not receive an *hh* signal and are very high along the A/P boundary that contacts the *hh*-producing cells (19). This expression profile suggests that *hh* signaling regulates full-length CI levels post-transcriptionally. Genetic and cell culture experiments support this hypothesis. The hh^{Mrt} mutation results in a dominant misexpression of *hh* in the wing disc. In these animals, high levels of CI protein are detected along the anterior wing margin, where ectopic *hh* is expressed (23). Suppression of *hh* expression by shifting a temperature-sensitive *hh* mutant (hh^{9k94}) to nonpermissive temperatures causes decreased levels of CI along the A/P boundary in imaginal discs (15). The transmembrane protein PTC, the product of the *patched* gene, is the receptor for HH and, in the absence of HH binding, inhibits the *hh* signaling pathway. The binding of HH to PTC relieves the inhibition of *ptc* on *smo*, the positive transducer of *hh* signaling (see ***hedgehog* signaling**). In *ptc* mutants, *smo* is constitutively active, and *hh* target gene expression becomes independent of a *hh* signal (24–28). As expected, high levels of full-length CI are distributed uniformly in the anterior compartment of *ptc* mutants (15). Overexpression of *ptc* in wing discs results in the downregulation of full length CI along the A/P boundary without changing the *ci* **messenger RNA** levels, providing further evidence that *hh* regulates CI protein levels post-transcriptionally (23).

The mechanism involved in the *hh* regulation of full-length CI levels is currently under intense investigation. It has been shown that a domain C-terminal to the zinc-finger domain and N-terminal to the sites **phosphorylated** by protein kinase A is responsible for the retention of the 155-kDa full-length CI in the cytoplasm (22). Because the 75-kDa repressor form of CI does not contain this domain, its nuclear localization is not regulated. The 155-kDa CI protein forms a complex with *costal2* (*cos2*), a protein related to **kinesin**, *fused* (*fu*), a **serine/threonine protein kinase**, *suppressor of fused* (*su*(*fu*)), and other unidentified proteins. This complex is tethered to microtubules, thus anchoring CI in the cytoplasm. When cells receive an *hh* signal, the complex dissociates from microtubules, releasing the full-length CI from the cytoskeletal structure and presumably making it more accessible to nuclear translocation. *Hh* signaling also inhibits CI proteolysis, which accounts for the increased levels of full-length CI detected along the A/P boundary (22,29,30).

Protein kinase A (PKA) antagonizes *hh* and negatively regulates *ci* activity (see ***hedgehog* signaling**). Recent studies have helped to elucidate the mechanism involved in this regulation. Loss of PKA function or inhibition of PKA activity increases the levels of full-length CI in both embryos and discs (23). Cell culture experiments have provided details about the mechanism involved in the PKA regulation of CI (31). CI has consensus PKA phosphorylation sites at four serine residues in its C-terminus. Substitution of the serine residue with alanine in any of the first three PKA sites inhibits CI proteolysis, thus increasing the levels of full-length CI and stimulating CI-mediated transcription, suggesting that CI is the direct target of PKA regulation. Recently an F-box/WD40-repeat protein encoded by *slimb* has been identified in flies (32). Loss of *slimb* function causes the formation of supernumerary limbs and the accumulation of high levels of full-length CI. Because *slimb* is related to a yeast protein, cdc4p, that is involved in targeting cell-cycle regulators to the **ubiquitin**-mediated **protein degradation** pathway, *slimb* has been hypothesized to mediate CI proteolysis.

Evidence obtained both *in vitro* and *in vivo* suggests that *ci* directly mediates transcription of the *hh* target genes. Consensus CI binding sites have been identified in the promoter regions of *ptc* and *wg*, and CI has been shown to bind to these sites both in vitro and in vivo (11,12,22). In cell culture, CI transactivates a reporter gene driven by the *wg* (***wingless***) promoter region that contains the CI binding sites (11). Both endogenous and ectopic *hh* signals transactivate a *lacZ* gene driven by the 758-bp minimal *ptc* promoter in a CI binding site-dependent manner in transgenic flies (12). These experiments support the idea that CI is a transcription factor that mediates the *hh* signal in the nucleus. *Ci* also functions as a suppressor of *hh* gene expression. *Ci* mutant clones generated in the anterior compartment of the imaginal discs ectopically express HH and induce the ectopic expression of *ci* and *hh* target genes in the surrounding wild type cells, presumably through the inductive function of the ectopic *hh* (33).

GLI EXPRESSION AND FUNCTION IN VERTEBRATES

A number of different *hh* homologues exist in vertebrates, and they also play important roles in the specification of cell fate. The transcription factors that respond to these vertebrate signaling cascades are considered homologues of *ci* because the zinc-finger motif that defines their DNA binding domains is more similar to that of CI than any other zinc-finger proteins. To date, three *ci* homologues, *Gli1*, *Gli2* and *Gli3*, have been studied in detail.

In mouse, all three *Gli* genes are detected initially during gastrulation at 7.5 days postcoitum (dpc) in broad domains in both ectoderm and mesoderm. As development progresses, the expression patterns of the three genes become more restricted and, by the completion of organogenesis, their expression is no longer detected. *Gli1* expression overlaps with *Sonic hedgehog*(*Shh*) expression initially in the ventral midline; however, by 9.5 dpc *Gli1* is excluded from *Shh*-expressing cells and is restricted to a narrow stripe of cells that lies along the caudal-rostral axis of the ventral neural tube and lateral to the *Shh*-expressing floor plate cells. The *Gli2* and *Gli3* expression patterns are widespread within the dorsal neural tube in domains that do not overlap with regions of *Gli1* expression. While *Gli2* expression is relatively homogenous and broad, *Gli3* is expressed as a gradient, with highest concentrations found in the dorso-lateral cells of the neural tube. This dorsal-ventral pattern of expression is maintained throughout the development of the brain and spinal cord (7,34,35). In the somites, all three *Gli* genes are expressed in the dorsal-lateral mesochymal cells (dermomyotome), suggesting that the *Gli* genes are involved in myogenesis. The *Gli1* message is also detected in the ventral medial somite (sclerotome), whose development is patterned by *Shh*. *Gli3* is expressed throughout the newly formed somite, and its expression becomes restricted to the dorsal medial myotomal cells as development progresses (7,36). As in the developing neural tube, the initial *Gli1* expression pattern in the limb bud overlaps that of *Shh* in the posterior mesochyme and then becomes restricted to the posterior cells immediately adjacent to the ZPA (zone of polarization activity) that expresses *Shh*. Except in the posterior margin, *Gli2* and *Gli3* are expressed throughout the limb bud, with higher levels of *Gli3* expression detected in the autopod. During bone formation, *Gli1* is expressed strongly in the perichondrium surrounding the cartilage elements where *Indian hedgehog*(*Ihh*) is expressed, and *Gli2* and *Gli3* are expressed in the tissues surrounding the perichondrium (37). *Gli1* transcripts are also detected in developing lung, gut, gonad, and eye (7,38).

That *Gli1* is expressed in cells adjacent to those that express *Shh* or *Ihh* is reminiscent of the relationship between *hh* and *ci* and suggests that *Gli1* is one of the *hh* target genes and may transduce the *hh* inductive pathways. Transgenic or Strong's luxoid mutant mice that misexpress *Shh* induce ectopic *Gli1* expression in the neighboring cells (37,38). Removal of the notochord from wild-type quail embryos abolishes *Gli1* expression, and supplementing the same embryos with SHH restores *Gli1* expression (36), supporting the notion that *Gli1* is a target of *Shh*. Ectopic expression of *Gli1* induces ectopic expression of the ventral neural tube markers *HNF-3β* (hepatocyte nuclear factor-3β), *patched* (*ptch*), and *Shh*. In addition, ectopic *Gli1* expression in mouse and Xenopus suppresses the expression of the dorsal neural tube marker *Pax-3*. This phenotype is reminiscent of the ectopic floor plate differentiation obtained with a gain of *Shh* function. *Gli1* also induces ectopic ventral neuronal differentiation, as shown by its ability to induce the expression of the ventral neuron markers, serotonin and dopamine. This inductive activity may be a secondary effect of *Shh*, because *Gli1* also induces *Shh* expression (34,35,39). The promoter region of *HNF-3β* contains Gli binding sites that can bind to *Gli1* *in vitro*. Tissue culture experiments demonstrate that *Shh* can transactivate a reporter gene that is controlled by the Gli binding sites found in the *HNF-3β* promoter. Mutating the Gli consensus binding site in the *HNF-3β* promoter abrogates the *Shh* response, showing that this site is required for *Shh*-dependent transactivation (40). The best evidence that *Gli1* is a mediator of *Shh* signaling comes from experiments in transgenic mice that express a *LacZ* reporter gene driven by the *HNF-3β* minimal promoter. In these animals, β-galactosidase expression is detected in the floor plate, and this expression depends on the presence of an intact Gli binding site (40). Taken together, these studies imply that the transcription of the *Gli1* gene is induced by *Shh*, whereupon the *Gli1* protein activates an array of genes needed to differentiate the floor plate and the ventral neurons. This pattern of regulation also exists in limb development. Marigo et al (8) have shown that misexpression of *Shh* in the anterior limb bud causes ectopic *Gli1* expression, and expression of a Gli-VP16 fusion protein activates the *Shh* target gene *ptch*. *Gli* is also a target for *Ihh* and mediates *Ihh* function in bone morphogenesis. In a developing bone, the prehypertrophic cells of cartilage express *Ihh* and the receptor for parathyroid/parathyroid related protein (PTH/PTHrP). On secretion, *Ihh* stimulates *Gli1* expression in the perichondrial cells, leading to the synthesis of PTHrP, which is then secreted and binds to its receptor on the *Ihh*-producing cells. Stimulation of the PTHrP signaling pathway ensures the proliferative state of the prehypertrophic cells and prevents their premature ossification (41,42).

Our understanding of *Gli2* function comes mainly from experiments with *Gli2* knock-out mice. *Gli2* loss-of-function mutations result in the absence of floor plate differentiation, severe axial skeletal abnormalities, and downregulation of the *Shh* target genes, such as *ptch* and *Gli1*. These results suggest that *Gli2* functions as a transcriptional activator of *Shh* target gene expression in certain structures during development (43). Injecting frog embryos with *Shh* or *Gli1* causes ectopic expression of *Gli2*, implying that *Gli2* transcription is activated by *Gli1* in response to *Shh* signaling (35).

Gli3 has been proposed for several reasons to function as a repressor of *Shh*. First, the expression patterns of *Gli3* and *Shh* are mutually exclusive (8,34,35,44). Second, *Gli3* can suppress the activation of the *HNF-3β* promoter by *Gli1* in cotransfection assays (40). Third, *Gli3* knock-out mice (*Xt*) misexpress *Shh* in the anterior limb bud and dorsal neural tube, domains that normally express *Gli3* (35). Genes that are involved in limb outgrowth and patterning, such as *ptch* and *Hox*, and those for **fibroblast growth factors** (FGFs) and bone morphogenic proteins (*Bmp*) are also ectopically expressed in *Gli3* mutant mice due to the misexpression of *Shh* (45). This cascade of misexpression might explain the extra autopods associated with the heterozygous *Xt* mouse. Whereas *Gli3* is a repressor of *Shh*, *Shh* acts to repress *Gli3*. The misexpression of *Shh* by viral infection downregulates *Gli3* in the chick limb, and the ectopic expression of *Gli1* in the dorsal tube of transgenic mice suppresses the normal dorsal *Gli3* expression (8,39). These results support the notion that CI function is accomplished by two Gli proteins in vertebrates, with *Gli1* acting as an activator and *Gli3* as a repressor, respectively, of *Shh* target gene expression.

Although Gli proteins have distinct expression patterns and cause different phenotypes when disrupted individually, they share some functional redundancies. Both *Gli1* and *Gli2* can induce motor neuron differentiation, and this redundancy may explain the fact that motor neuron induction is unaffected in homozygous *Gli2* mutant mice (43).

From an evolutionary perspective, many interesting parallels exist between the Drosophila and vertebrate *hh* signaling pathways. In Drosophila, *hh* suppresses the activity of the CI repressor by inhibiting the proteolysis of the 155-kDa CI protein. *Shh* signaling suppresses the activity of the *Gli3* repressor by inhibiting its transcription. CI suppresses *hh* expression, and *Gli3* suppresses the transcription of *Shh* in the cells where CI or Gli is active. In the presence of an *hh* signal, the 155-kDa CI protein is an activator of *hh* target genes, and this activity has evolved in vertebrates into the functions of *Gli1* and *Gli2*. Clearly, continued studies of CI and the Gli family members will elucidate the evolution of the signaling systems that drive the developmental process.

BIBLIOGRAPHY

1. B. Hochman (1971) *Genetics* **67**, 235–252.
2. C. Nusslein-Volhard and E. Wieschaus (1980) *Nature* **287**, 795–801.
3. T. V. Orenic, D. C. Slusarski, K. L. Kroll, and R. A. Holmgren (1990) *Genes Dev.* **4**, 1053–1067.
4. K. W. Kinzler, J. M. Ruppert, S. H. Bigner, and B. Vogelstein (1988) *Nature* **332**, 371–374.
5. J. M. Ruppert, B. Vogelstein, K. Arheden, and K. W. Kinzler (1990) *Mol. Cell. Biol.* **10**, 5408–5415.
6. D. C. Hughes, J. Allen, G. Morley, K. Sutherland, W. Ahmed, J. Prosser, l. Lettice, G. Allan, M.-G. Mattei, M. Farrall, and R. E. Hill (1997) *Genomics* **39**, 205–215.
7. C.-C. Hui, D. Slusarski, K. A. Platt, R. Holmgren, and A. L. Joyner (1994) *Dev. Biol.* **162**, 402–413.
8. V. Marigo, R. L. Johnson, A. Vortkamp, and C. J. Tabin (1996) *Dev. Biol.* **180**, 273–283.
9. D. Zarkower and J. Hodgkin (1992) *Cell* **70**, 237–249.
10. J.-C. Marine, E. J. Bellefroid, H. Pendeville, J. A. Martial, and T. Pieler (1997) *Mech. Dev.* **63**, 211–225.
11. T. v. Ohlen, D. Lessing, R. Nusse, and J. E. Hooper (1997) *Proc. Natl. Acad. Sci. USA* **94**, 2404–2409.
12. C. Alexandre, A. Jacinto, and P. W. Ingham (1996) *Genes Dev.* **10**, 2003–2013.
13. K. W. Kinzler and B. Vogelstein (1990) *Mol. Cell. Biol.* **10**(2), 634–642.
14. N. P. Pavletich and C. O. Pabo (1993) *Science* **261**, 1701–1707.
15. J. Hepker, Q.-T. Wang, C. K. Motzny, R. Holmgren, and T. v. Orenic (1997) *Development* **124**, 549–558.
16. H. Akimaru, Y. Chen, P. Dai, D.-X. Hou, M. Nonaka, S. M. Smolik, S. Armstrong, R. H. Goodman, and S. Ishii (1997) *Nature* **386**, 735–738.
17. K. Han and J. L. Manley (1993) *EMBO J.* **12**, 2723–2733.
18. P. J. Mitchell and R. Tjian (1989) *Science* **245**, 371–378.
19. C. K. Motzny and R. Holmgren (1995) *Mech. Dev.* **52**, 137–150.
20. C. Schwartz, J. Locke, C. Nishida, and T. B. Kornberg (1995) *Development* **121**, 1625–1635.
21. S. Eaton and T. B. Kornberg (1990) *Genes Dev.* **4**, 1068–1077.
22. P. Aza-Blanc, F.-A. Ramirez-Weber, and T. B. Kornberg (1997) *Cell* **89**, 1043–1053.
23. R. L. Johnson, J. K. Grenier, and M. P. Scott (1995) *Development* **121**, 4161–4170.
24. A. Martinez-Arias, N. E. Baker, and P. W. Ingham (1988) *Development* **103**, 157–170.
25. T. Tabata and T. B. Kornberg (1994) *Cell* **76**, 89–102.
26. J. Capdevila, M. P. Estrada, E. Sanchez-Herrero, and I. Guerrero (1994) *EMBO J.* **13**, 71–82.
27. J. Capdevila and I. Guerrero (1994) *EMBO J.* **13**, 4459–4468.
28. W. Li, J. T. Ohlmeyer, M. E. Lane, and D. Kalderon (1995) *Cell* **80**, 553–562.
29. D. J. Robbins, K. E. Nybakken, R. Kobayashi, J. C. Sisson, J. M. Bishop, and P. P. Therond (1997) *Cell* **90**, 225–234.
30. J. C. Sisson, K. S. Ho, K. Suyama, and M. P. Scott (1997) *Cell* **90**, 235–245.
31. Y. Chen, N. Gallaher, R. H. Goodman, and S. M. Smolik (1998) *Proc. Natl. Acad. Sci. USA* **95**, 2349–2354.
32. J. Jiang and G. Struhl (1998) *Nature* **391**, 493–496.
33. M. Dominguez, M. Brunner, E. Hafen, and K. Basler (1996) *Science* **272**, 1621–1625.
34. J. Lee, K. A. Platt, P. Censullo, and A. R. i. Altaba (1997) *Development* **124**, 2537–2552.
35. A. R. i. Altaba (1998) *Development* **125**, 2203–2212.
36. A.-G. Borycki, L. Mendham, and C. P. Emerson Jr. (1998) *Development* **125**, 777–790.
37. K. A. Platt, J. Michaud, and A. L. Joyner (1997) *Mech. Dev.* **62**, 121–135.
38. J. C. Grindley, S. Bellusci, D. Perkins, and B. L. M. Hogan (1997) *Dev. Biol.* **188**, 337–348.
39. M. Hynes, D. M. Stone, M. Dowd, S. Pitts-Meek, A. Goddard, A. Gurney, and A. Rosenthal (1997) *Neuron* **19**, 15–26.
40. H. Sasaki, C.-c. Hui, M. Nakafuku, and H. Kondoh (1997) *Development* **124**, 1313–1322.
41. A. Vortkamp, K. Lee, B. Lanske, G. V. Segre, H. M. Kronenberg, and C. J. Tabin (1996) *Science* **273**, 613–621.
42. B. Lanske, A. C. Karaplis, K. Lee, A. Luz, A. Vortkamp, A. Pirro, M. Karperien, L. H. K. Defize, C. Ho, R. C. Mulligan, A.-B. Abou-Samra, H. Juppner, G. V. Segre, and H. M. Kronenberg (1996) *Science* **273**, 663–666.

43. Q. Ding, J. Motoyama, S. Gasca, R. Mo, H. Sasaki, J. Rossant, and C.-c. Hui (1998) *Development* **125**, 2533–2543.
44. R. Burke and K. Basler (1997) *Curr. Opinion Neurobiol.* **7**, 55–61.
45. D. Buscher, B. Bosse, J. Heymer, and U. Ruther (1997) *Mech. Dev.* **62**, 175–182.

Suggestions for Further Reading

A. R. Altaba (1997) Catching a Gli-mpse of hedgehog. *Cell* **90**, 193–196.

L. G. Biesecker (1997) Strike three for *GLI3*. *Nature Genet.* **17**, 259–260.

U. Radhakrishna, A. Wild, K.-H. Grzeschik, and S. E. Antonarakis (1997) Mutation in *Gli3* in postaxial polydactyly type A. *Nature Genet.* **17**, 269–271.

CYCLIC AMP (3′,5′-CYCLIC AMP, CAMP)

ANTOINE DANCHIN

Nucleotides play a universal role in life as components of **nucleic acids**, as forms in which chemical free energy is stored (see **Adenylate charge**), and as regulators of **genetic expression** or **enzymatic** activity. Cyclic adenosine 3′,5′-monophosphate (cAMP) (Fig. 1) and **cyclic GMP** play a universal role in controlling gene expression and in integrating metabolic functions. It is present in both **eukaryotes** and **prokaryotes** [see (1) for early recognition of the presence of cAMP in **bacteria**]. cAMP is absent only from **archaebacteria**, although a possible gene for its synthesis has been noticed (see **Adenylate cyclase**). Its presence in **plants** is controversial. Since the writing of this entry, it has been shown that the paper demonstrating the existence of an adenylate cyclase in plants is a fake. It has been reported in **cyanobacteria** and in **algae**. This ubiquity explains the major interest displayed in its mode of synthesis and the vast amount of literature devoted to the enzymes, the adenylate cyclases (E. C. 4.6.1.1), that produce cAMP from ATP. Because cAMP is a regulatory molecule, it must be excreted into the environment or inactivated so that it does not accumulate in ever increasing amounts. It is inactivated by 3′,5′-cyclic-nucleotide **phosphodiesterases** (EC 3.1.4.17). These enzymes are generally specific for cyclic nucleotides (namely, cAMP and cGMP), and sometimes specific for cAMP or cGMP alone. A variety of natural inhibitors (nucleoside triphosphates, pyrophosphate, and especially methylated xanthines, such as theophylline) modulate their activity by a variety of processes involving protein **phosphorylation** and/or by **calcium signaling**.

Adenylate cyclases comprise four independent classes of enzymes, and this raises the questions of the origin of cyclic nucleotides as regulatory molecules and their universal implication in regulatory networks. Because it is very **polar** and negatively charged, cAMP does not cross membranes and permeate easily into cells, and specific **transporters** are generally unknown at present. Therefore, more lipophilic analogs, such as N_6,O_2′-dibutyryladenosine-3′,5′monophosphate, are used to modulate the concentration inside cells and to mimic the effect in cell cultures *ex vivo*. Where the cAMP concentration must be altered *in vivo*, however, inhibitors of the phosphodiesterases or specific mediators, neuromediators in particular, are used for therapeutic purposes.

Cyclic AMP was discovered in 1958 by E. Sutherland, who received the Nobel prize in 1971 for this and other discoveries of hormone action. As he has himself written, it was within the scope of molecular biology that cAMP was discovered:

> *When I first entered the study of hormone action, some 25 years ago, there was a widespread feeling among biologists that hormone action could not be studied meaningfully in the absence of organized cell structure. However, as I reflected upon the history of biochemistry, it seemed to me there was a real possibility that hormones might act at the molecular level (2).*

Sutherland built up a cell-free system where well-known hormones could control glycolysis *in vitro*. Using this system, he isolated a small thermostable molecule that activates **glycogen phosphorylase**. Chemical analysis of the molecule identified it as adenosine 3′-5′ cyclic monophosphate. Synthesis of cAMP resulted from the action of an enzyme, adenylate cyclase, that when activated by adrenaline generated cAMP and PP_i from ATP. Since this pioneering work, the study of cAMP-mediated effects has required identifying the structure, function, and regulation of adenylate cyclases, the cAMP-synthesizing enzymes. Contrary to expectation, this did not yield a unifying picture of the role of cAMP but, instead, demonstrated that this molecule has been used over and over again by living organisms for very different functions.

When cAMP was discovered, the aphorism of Jacques Monod, *What is true for Escherichia coli is true for the elephant*, induced biochemists to try bacterial systems to unravel cAMP function. In 1963, Mackman and Sutherland demonstrated that glucose-starved *E. coli* cells accumulate cAMP (3). Ullmann and Monod later established that the phenomenon of **catabolite repression** is controlled in part by cAMP (4). This discovery raised hopes that the study of this mediator in bacteria would help understanding what happens in eukaryotes (perhaps even in higher eukaryotes). It soon became clear, however, that cAMP in eukaryotes was generally, as found by Sutherland, a **"second messenger"** used as an intracellular relay molecule to modulate the action of extracellular hormones, whereas in *E. coli* it acted directly on gene **transcription** via its receptor, the **cyclic AMP receptor protein (CRP)/catabolite gene activator protein (CAP)** (5). Study of the **slime mold** *Dictyostelium discoïdeum* revealed another function of cAMP, phylogenetically linked to its hormone-mediated action in higher eukaryotes, namely, a pulsatile synthesis and degradation used by bacteria as a signal to control their aggregative properties as differentiating multicellular organisms (6).

The universal role of cAMP in controlling such diverse metabolic processes simultaneously is puzzling because the enzymes needed for its synthesis, adenylate cyclases, are extremely varied and subject to a wide variety of regulation (see **Adenylate cyclases**). Why do all of these different signals result in the synthesis of the same molecular species, cAMP? Is not all the specificity of the regulatory process lost in this way? How can the cAMP signal generated by one type of enzyme be distinguished from that generated by another? Compartmentalization is often invoked to explain this, but although this is relatively easily accounted for in the case of macromolecules, it is difficult to imagine in the case of small diffusing molecules, such as cAMP. Another usual answer is to say that it is the combination of hormonal receptors of

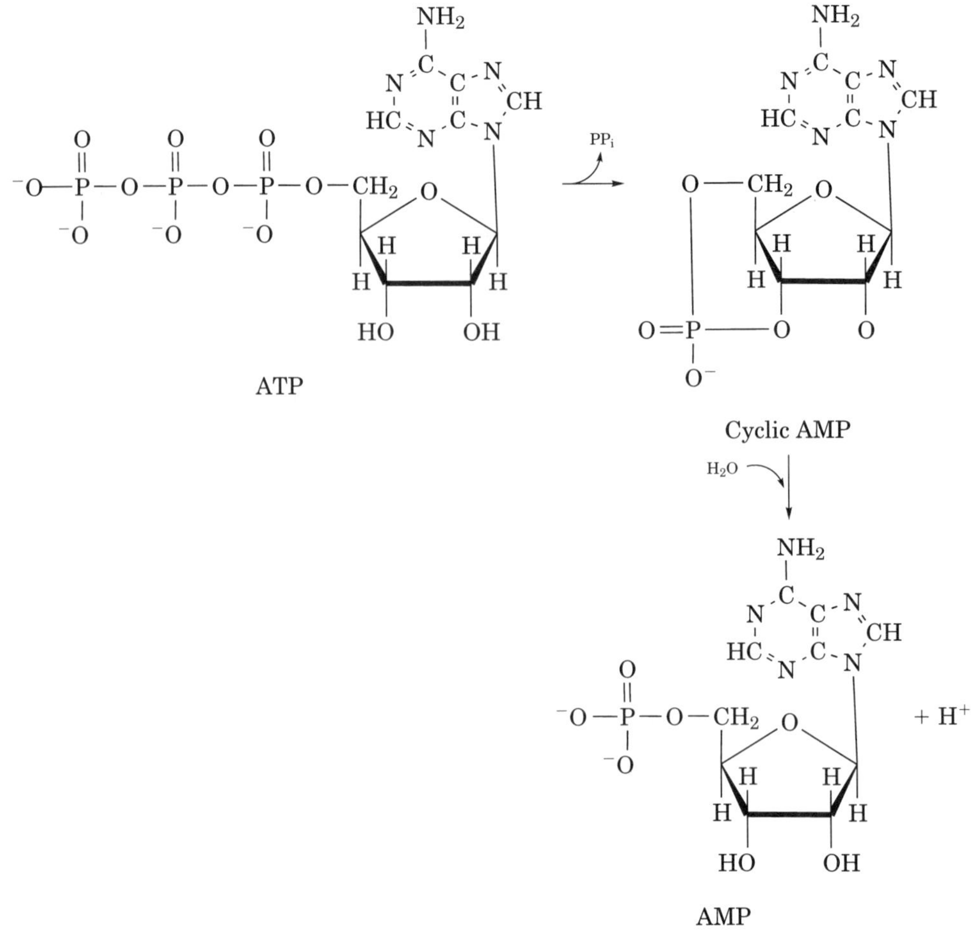

Figure 1. Synthesis of 3′,5′-cyclic AMP from ATP and its degradation to AMP. The first reaction is catalyzed by adenylate cyclase, and the second by cAMP phosphodiesterase.

differing types plus cAMP—and not cAMP alone—that is required for specificity. But would not cAMP synthesized from different sources also be recognized? Another answer is that cAMP is known to be only one among many second messengers. **Cyclic GMP** has been added to the list, as well as **inositol phosphates**, **diacylglycerol**, **calcium**, etc. This certainly permits generating a combinatorial control of activities, but it would certainly be very sensitive to accidental synthesis of cAMP.

Cyclic AMP is not synthesized continuously (even in bacteria). Therefore it is important not to consider cAMP as such, at some average concentration, but to consider its time-dependent variation in concentration. In fact observations are accumulating strongly suggesting that cAMP does not have the same effect when it is delivered in a steady-state fashion or in a pulse (or a series of pulses) (6).

The motile and aggregating amoebae *D. discoïdeum* has been used as a paradigm for cell **differentiation** because undifferentiated cells start to differentiate into specific tissues some time after starvation. Secretion of cAMP into the external medium is necessary for aggregation. The genetics, biochemistry, cellular biology, and physiology of phenomena involving cAMP have been investigated in detail in this organism, where it controls, as in higher eukaryotes, a protein phosphorylation cascade, initiated after a regulatory cAMP-binding subunit of a protein kinase detaches from its target enzyme (see **Signal transduction**). This cascade is necessary for **chemotaxis** and cell aggregation and also for triggering genes involved in differentiation. Regulation of the pulsed cAMP concentration is mediated by two sets of enzymes, adenylate cyclases and phosphodiesterases. In contrast to the situation in higher eukaryotes, however, control by cAMP and phosphodiesterase does not operates from the interior of the cell but from the external medium. This requires specific membrane receptors for cAMP, and a process of signal transduction. In *D. discoïdeum* the pulses of cAMP are generated by an appropriate coupling between adenylate cyclase activity, phosphodiesterase activity, and diffusion. The main observation is that variation in the cAMP pulse frequency changes the response of the cell. Many biochemical models can account for such cAMP pulses. These models require simple enzyme properties, in particular, standard nonlinear features, such as self-activation and desensitization after saturating activity. The models do not require the existence of many gene products but only the appropriate behavior, maximum velocity and K_m (Michaelis constant) of the biosynthetic and degradative enzymes. Therefore, assuming that variation of the cAMP concentration with time is crucial to control does not present a biochemical paradox.

In bacteria, Utsumi et al. investigated cyclic AMP synthesis during the cell cycle of *E. coli.* using synchronized cells. They unambiguously demonstrated that there is a strong correlation between cAMP synthesis and replication or cell division, suggesting that the molecule may play some role in the cell cycle

(7). This is also correlated with the position of the adenylate cyclase gene on the *E. coli* **chromosome** near the **origin of replication** and with its very low level of expression, suggesting that expression is strongly coupled to DNA replication. This observation has long been overlooked because cells deficient for adenylate cyclase or cyclic AMP receptor protein are viable, suggesting that cAMP is dispensable. Specific time-dependent variation of the concentration of cAMP for finely coordinating replication and division in *E. coli* is achieved by excretion of the nucleotide, rather than coupling to the activity of a phosphodiesterase (5). In this respect, it is interesting that high concentrations of cAMP produced by foreign genes in *E. coli* are not toxic until they reach a very high level (at least 10 times the normal concentration), whereas lowered concentrations of cAMP produced by the endogenous adenylate cyclase are toxic (8). Cyclic AMP has been formally linked to catabolite repression, but there are many catabolite-sensitive operons that do not respond to cAMP. In addition, cAMP synthesis is very strong in *E. coli* when cells enter the stationary phase of growth, suggesting that it could be a signal between cells, as in *D. discoïdeum*. This may be one of its functions in other bacteria (such as *Rhizobium* species), where it is clearly not linked to catabolite repression.

In the same way, intracellular and extracellular levels of cAMP vary during the cell cycle of the **yeast** *Saccharomyces cerevisiae*. Using centrifugal elution, Smith *et al.* showed that the intracellular cAMP concentration follows the stages of the cell cycle. The concentration is highest during the division cycle and lowest immediately before or just after cell separation. At the same time, the external cAMP concentration do not vary (9). Therefore, in yeast as in *E. coli*, the role of the external medium is to behave as a sink. These observations substantiate the demonstration that, under normal conditions, appropriate enzyme systems generate a specific time-dependent pattern of cAMP concentration. As in the case of *E. coli*, it is known that adenylate cyclase is dispensable in *S. cerevisiae* mutants of the cAMP receptor, and adenylate cyclase is dispensable in *S. pombe* during vegetative growth (10). But, as in the former case, cells that carry a mutation and are deficient in adenylate cyclase have several growth defects. In this respect, the function of the time-dependent cAMP pattern could be optimization of transient processes, in particular cell division and chromosome segregation.

In all of these instances cAMP is recognized on the cell surface by a specific receptor. Therefore, it is interesting to identify instances, apart from *D. discoïdeum*, where membrane targets of cAMP have been demonstrated. Nerve cells typically generate and are sensitive to transient signals, and they also have very complex patterns regulating adenylate and guanylate cyclases (see **Adenylate cyclase** and **Guanylate cyclase**). In this respect, it is important to observe that cGMP (and also cAMP) are involved as central molecules in vision, taste, and olfaction. In particular, in addition to their role as second messengers in protein phosphorylation cascades, cyclic nucleotides are involved at the membrane surface, but intracellularly, in gating **ion channels** in olfactory and taste neurons. This certainly permits generating a variety of time-dependent patterns for cAMP regulation as a function of environmental inputs, the fine molecular structure of the enzyme or its subunits (11,12). Because ion channels are involved in the main functions of neurons (firing patterns), this makes cyclic nucleotides important in learning processes.

Indeed, many experiments have demonstrated that cAMP is a mediator of learning and memory in invertebrates, for example, *Aplysia* (13) and *Drosophila melanogaster* (14), and in vertebrates (15,16). The study of mutants of *D. melanogaster* that are defective in learning and/or memory has been of major importance in our understanding of the physiology, biochemistry, and anatomy underlying conditioned behavior. *D. melanogaster* learning mutants have been separated into two general classes, those with structural defects in the brain and those without obvious defects. From studies of mutants affected in the brain structure, two areas are involved in conditioned behavior, the mushroom bodies and the central complex. Analysis of the mutants has shown that many types of molecules are involved in learning, but the cAMP-mediated phosphorylation cascade has emerged as especially important. During learning, time-dependent processes are involved in stabilizing synapses. A general view is that they are created during growth as transient entities that either regress or are stabilized. In this process, evolution of the synaptic pattern depends on the pattern of neurotransmitter delivery. Analysis of the minimal requirements for synapse stabilization suggests that neurotransmitter release must be retrogradely coupled to some other transient metabolic process to yield a stable geometry (17). When cAMP is involved, one can therefore speculate that the role of this mediator is to trigger an appropriate biochemical process with the proper time-dependent control of its synthesis (18). Accordingly, once again, it is not the cAMP concentration that is important but instead the time-variation of its concentration. Therefore in the process of learning, the regulation of adenylate cyclase activity would be exquisitely tuned to permit delivery of the molecule in the proper time-dependent manner.

In *D. melanogaster*, five different genes have proven important for normal learning: dunce (a cAMP phosphodiesterase), rutabaga (an adenylyl cyclase), amnesiac (a product similar to adenylate cyclase-activating peptides), DCO (protein kinase A), and dCREB2 (a cAMP-**response element** binding protein). The products of many of these learning mutants are enriched in mushroom bodies. A process involving control of transcription by the cAMP response element-binding protein (CREB) plays a central role in forming long-term memory in *D. melanogaster*, *Aplysia*, and mammals. This is one of the examples where cAMP is involved in the control of transcription in eukaryotes, as it is in **eubacteria**, although through a different chain of events. Agents that prevent CREB activity interfere with the formation of long-term memory, whereas agents that increase the amount or activity of the **transcription factor** accelerate the process, thus indicating that CREB is essential for the switch from short-term memory to long-term memory, which depends on **protein biosynthesis** (19).

BIBLIOGRAPHY

1. M. Ide (1971) *Arch. Bioch. Biophys.* **144**, 262–268.
2. E. W. Sutherland *Science* (1972) **177**, 401–408.
3. R. S. Mackman and E. W. Sutherland (1963) *Fed. Proc.* **22**, 470.
4. A. Ullmann and J. Monod (1968) *FEBS Lett.* **2**, 714–717.
5. A. Ullmann and A. Danchin (1983) *Adv. Cyclic Nucleic Acid Res.* **15**, 1–52.
6. J. Dallon and H. Othmer (1997) *Philos. Trans. R. Soc. Lond. B Biol. Sci.* **352**, 391–417.

7. R. Utsumi, M. Kawamukai, H. Aiba, M. Himeno, and T. Komano (1986) *J. Bacteriol.* **168**, 1408–1414.
8. A. Danchin (1993) *Adv. Second Messenger Phosphoprotein Res.* **27**, 109–162.
9. M. E. Smith, J. R. Dickinson, and A. E. Wheals (1990) *Yeast* **6**, 53–60.
10. T. Maeda, N. Mochizuki, and M. Yamamoto (1990) *Proc. Natl Acad. Sci. USA* **87**, 7814–7818.
11. T. Leinders-Zufall, M. Rand, G. Shepherd, C. Greer, and F. Zufall (1997) *J. Neurosci.* **17**, 4136–4148.
12. W. Zagotta and S. Siegelbaum (1996) *Annu. Rev. Neurosci.* **19**, 235–263.
13. C. Bailey, C. Alberini, M. Ghirardi, and E. Kandel (1994) *Adv. Second Messenger Phosphoprotein Res.* **29**, 529–544.
14. R. Davis (1996) *Physiol. Rev.* **76**, 299–317.
15. R. Bourtchuladze, B. Frenguelli, J. Blendy, D. Cioffi, G. Schutz, and A. Silva (1994) *Cell* **79**, 59–68.
16. Z. Xia and D. Storm (1997) *Curr. Opinion Neurobiol.* **7**, 391–396.
17. J. P. Changeux and A. Danchin (1976) *Nature* **264**, 705–712.
18. Y. Zhong and C. F. Wu (1991) *Science* **251**, 198–201.
19. J. Yin and T. Tully (1996) *Curr. Opinion Neurobiol.* **6**, 264–268.

CYCLIC AMP RECEPTOR PROTEIN (CRP)/ CATABOLITE GENE ACTIVATOR PROTEIN (CAP)

H. AIBA

The cyclic AMP receptor protein (CRP), or catabolite gene activator protein (CAP), of *Escherichia coli* is one of the best characterized **transcription factors**. It was discovered more than 25 years ago as a protein that binds **cyclic AMP** (cAMP) and stimulates **gene expression** of the ***lac* operon** (1,2). Since then, CRP has attracted attention as the paradigm of gene activator proteins. It is now known that CRP, in conjunction with cAMP, participates as a global transcription factor in a wide regulatory network, both activating and repressing a large set of **operons**. One mechanism of **catabolite repression** is due to the reduction in the intracellular concentrations of both cAMP and CRP.

The CRP protein is composed of two identical subunits of 209 amino acid residues each. It is active when complexed with cAMP, which behaves as an **allosteric** effector. It is a prototype of sequence-specific **DNA-binding proteins** containing a **helix–turn–helix motif**. The cAMP–CRP complex binds to specific DNA sites in various operons. On binding DNA, it bends the DNA and interacts with **RNA polymerase** and/or other regulatory proteins to regulate **transcription** of the target operons.

Following early studies on the properties of purified CRP, the *crp* gene was **cloned** and sequenced in 1982 (3,4). The **X-ray crystallography** structure of CRP complexed with cAMP was solved in 1981 (5). This confirmed the dimeric nature and **domain** structure of CRP deduced from biochemical studies. The subunit has two domains (Fig. 1**a**). The larger *N*-terminal domain is responsible for cAMP binding and for dimerization. The smaller *C*-terminal domain contains a helix–turn–helix DNA-binding motif. CRP is one of the first regulatory proteins where the helix–turn–helix motif was identified. CRP alone is able to bind to DNA nonspecifically. The binding of cAMP induces a conformational state that binds to specific sequences with a dyad symmetry. The detailed nature of the structural changes in CRP caused by cAMP remains unknown.

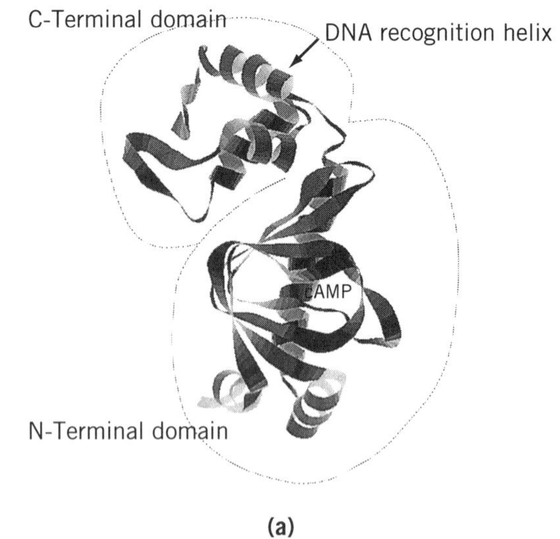

(a)

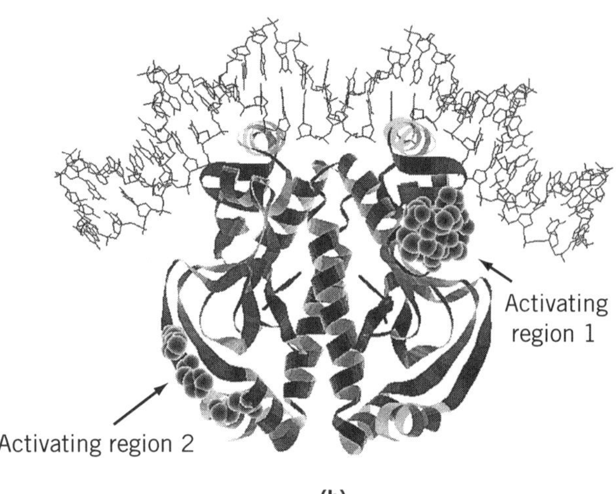

(b)

Figure 1. Structures of the CRP monomer and the CRP-DNA complex. The crystallographic coordinates were obtained from the Brookhaven Protein databank (accession code 1CGP). The images were generated by Protein Adviser (FQS). (**a**) The two domains of a CRP monomer. The larger *N*-terminal domain is responsible for cAMP binding and for dimerization. The smaller *C*-terminal domain contains a helix–turn–helix motif that is involved in DNA binding. (**b**) The CRP dimer bound to a consensus DNA site. The positions that are responsible for transcription activation (activating regions) are highlighted by space-filling models. Activating region 1 contacts the *C*-terminal domain of α subunit of RNA polymerase (8), while activating region 2 contacts the *N*-terminal domain of α subunit of RNA polymerase (9).

The DNA sequences of many CRP binding sites, and a resulting **consensus sequence**, have been determined. They include variations of the 22-bp **palindromic** sequence, 5′-AAATGTGATCTAGATCACATTT-3′, in which the two

TGTGA motifs are relatively well conserved in different promoters. The binding of cAMP–CRP to target DNA induces bending of the DNA (6). The structure of cocrystals of cAMP-CRP bound to a 30-bp DNA target was determined in 1991 (7). It shows the helix-turn-helix motifs of the two subunits inserted into successive major groves of a DNA that is bent by 90°C (Fig. 1**b**).

In promoters where CRP alone is sufficient to activate transcription, the role of the cAMP–CRP complex is to enhance functional binding of RNA polymerase (8,9). Binding of cAMP–CRP or RNA polymerase to these promoters stimulates the other's binding. An example is the *lac* promoter, where the CRP-binding site is centered on base pair −61, which is one of the preferable positions for CRP action. Another position where CRP activates transcription efficiently is −41, as in the *gal* promoter (see ***gal*** **operon**). Changing these two standard distances reduces or eliminates CRP action. Only when this change is by an integral number of turns of DNA double helix is the ability of CRP to activate transcription retained to some extent, suggesting that both CRP and RNA polymerase must bind to the same face of the DNA (10,11). Genetic and biochemical analysis revealed that cAMP–CRP activates transcription by directly contacting RNA polymerase at these simple promoters. A surface-exposed loop in the *C*-terminal domain of CRP and the *C*-terminal part of the α subunit of RNA polymerase are primarily responsible for the contact between two proteins (8). An additional contact between the *N*-terminal domain of CRP and the *N*-terminal domain of α subunit is also involved in transcription activation at the promoters where the CRP site lies at −41 (9). The interaction between CRP and RNA polymerase is needed transiently to stimulate events leading to the formation of an open complex in these simple promoters (12). The contribution of CRP-induced DNA bending to transcription activation is still unclear. The CRP-binding site lies well upstream in several CRP-dependent promoters. In these cases, CRP acts as a coactivator of a second activator that is presumed to be responsible for the direct interaction with RNA polymerase. In the example of the *araBAD* promoter, CRP binds around −94 and the second activator AraC binds in the region between the CRP and RNA polymerase sites (13) (see ***ara*** **operon**).

CRP also acts as a repressor or a corepressor in several operons. A simple example is the *cya* promoter, where cAMP–CRP inhibits the action of RNA polymerase by binding a target site located within the promoter (14). A more complex mechanism of repression is found in the operons that are coordinately regulated by CRP and CytR. Most of these promoters possess tandem CRP-binding sites that flank the CytR operator. An example is the *deo* promoter, where the CRP-binding sites are located at −41 and −94. The binding of cAMP-CRP to these sites dramatically enhances the binding of CytR to its operator, resulting in the formation of a repression complex containing both cAMP–CRP and CytR (15).

Although many studies have been identifying the regions of both transcription factors and RNA polymerase that are responsible for protein–protein and protein–DNA interactions, how these interactions lead to transcription activation is largely unknown. CRP, along with its target promoters, will continue to be a useful system for further understanding of molecular mechanisms of transcriptional regulation, including this fundamental question.

BIBLIOGRAPHY

1. G. Zubay, D. Schwartz, and J. R. Beckwith (1970) *Proc. Natl. Acad. Sci. USA* **66**, 104–110.
2. M. Emmer, B. deCrombrugghe, I. Pastan, and R. L. Perlman (1970) *Proc. Natl. Acad. Sci. USA* **66**, 480–487.
3. H. Aiba, S. Fujimoto, and N. Ozaki (1982) *Nucleic Acids Res.* **10**, 1345–1361.
4. P. Cossart and B. Gicquel-Sanzey (1982) *Nucleic Acids Res.* **10**, 1363–1378.
5. D. B. McKay and T. A. Steitz (1981) *Nature* **290**, 744–749.
6. H.-M. Wu and D. M. Crothers (1984) *Nature* **308**, 509–513.
7. S. C. Schultz, G. C. Shields, and T. A. Steitz (1991) *Science* **253**, 1001–1007.
8. S. Busby and R. H. Ebright (1994) *Cell* **79**, 743–746.
9. S. Busby and R. H. Ebright (1997) *Mol. Microbiol.* **23**, 853–859.
10. K. Gaston, A. Bell, A. Kolb, H. Buc, and S. Busby (1990) *Cell* **62**, 733–743.
11. C. Ushida and H. Aiba (1990) *Nucleic Acids Res.* **18**, 6325–6330.
12. H. Tagami and H. Aiba (1998) *EMBO J* **17**, 1759–1767.
13. R. B. Lobell and R. F. Schleif (1991) *J. Mol. Biol.* **218**, 45–54.
14. H. Aiba (1985) *J. Biol. Chem.* **260**, 3063–3070.
15. L. Søgaard-Anderson and P. Valentin-Hansen (1993) *Cell* **75**, 557–566.

Suggestions for Further Reading

G. Zubay (1980) The isolation and properties of CAP, the catabolite gene activator. *Methods Enzymol.* **65**, 856–877.

J. L. Botsford and J. G. Harman (1992) Cyclic AMP in prokaryotes, *Microbiol. Rev.* **56**, 100–122.

D. M. Crothers and T. A. Steitz (1992) Transcriptional activation by *Escherichia coli* CAP protein, in *Transcriptional Regulation*, S. L. McKnight and K. R. Yamamoto, eds., Cold Spring Harbor Press, Cold Spring Harbor, New York, Vol. 1, pp. 501–534.

A. Kolb, S. Busby, H. Buc, S. Garges, and S. Adhya (1993) Transcriptional regulation by cAMP and its receptor protein. *Annu. Rev. Biochem.* **62**, 749–795.

CYCLIC GMP (CYCLIC GUANOSINE 3′,5′-MONOPHOSPHATE, CGMP)

ANTOINE DANCHIN

Cyclic GMP, a structural analog of **cyclic AMP**, occurs at similarly low concentrations (*i.e.*, in the micromolar range) in many animal tissues. It is synthesized by **guanylate cyclases**, which correspond to the **adenylate cyclases** that synthesize cyclic AMP, and is destroyed by specific **phosphodiesterases**.

A functional role for cyclic GMP in **bacteria** is controversial. It is present in *Escherichia coli* but at such a low intracellular concentration (nanomolar, corresponding to about one molecule per cell) that hardly makes it a significant molecule. In addition, the sequence of the *E. coli* **genome** does not reveal any gene product that could code for a guanylate cyclase. There are many reports suggesting the presence of cGMP in bacteria (1), but its presence is perhaps most likely in myxobacteria, where it could be involved in cell aggregation and **differentiation**.

In contrast, cGMP is universally in **eukaryotes** (except in **plants**). It is generally involved in processes leading to activation of specific regulation cascades, which are differentially controlled by appropriate mediators, or leading to control specific processes for neuronal activation of sense organs, such as the sensitivity to light of retina receptor cells and the triggering of olfaction and taste (2–4). The organization of guanylate cyclase and the control elements is sometimes similar to but distinct from the organization of hormonally regulated adenylate cyclase. In particular, **G-protein**-mediated regulation of vision operates on the phosphodiesterase rather than on the cyclase.

Phototransduction systems in vertebrates and invertebrates share a great deal of overall strategic similarity but differ significantly in the underlying molecular machinery. In the dark, vertebrate retinal rod cells synthesize a high level of cGMP that keeps gated **sodium channels** open in the plasma membrane of the outer segment (5). Light closes these channels by activating an enzymatic cascade that leads to rapid hydrolysis of cGMP by cGMP-specific phosphodiesterase (6). This hyperpolarizes the cell and modulates transmitter release at the synaptic buttons. Photoexcited **rhodopsin** triggers a transducer protein (**transducin**), which is related to G-proteins by catalyzing the exchange of GTP for bound GDP. Subsequently, the activated GTP-form of transducin switches on the phosphodiesterase. The cascade that has an overall gain of 10^5, is turned off by the **GTPase** activity of transducin and by the action of two proteins, rhodopsin kinase and arrestin. The **kinetics** of the reactions in the cGMP cascade limit the temporal resolution of the visual system, and statistical fluctuations in the reactions limit the reliability of detecting dim light. Together with **calcium** ions and **inositol phosphates**, cGMP controls visual excitation and adaptation. A light-induced fall in the internal free Ca^{2+} concentration subsequently stimulates resynthesis of cGMP, antagonizes the catalytic activity of rhodopsin, and restores the high affinity of the light-regulated sodium channel for cGMP, allowing the cell to adapt to background light (7).

The initial events in mollusks and arthropods are probably similar to those of vertebrates. However, whereas light activation of vertebrate photoreceptors leads to activation of cGMP-phosphodiesterase and generation of a hyperpolarizing response, activation of photoreceptors of invertebrates like Drosophila leads to stimulation of **phospholipase C** and generation of a depolarizing response (8). Cyclic GMP has also been implicated in modulating behavior in insects (9).

Cyclic GMP is also a **second messenger** in regulation mediated by natriuretic peptide hormones, but it is probably in the recently discovered **nitric oxide** (NO) regulation cascade that cGMP has the most unexpected role. Nitric oxide and atrial natriuretic peptide hormones play key roles in a number of neuronal functions, including learning, memory, and in blood circulation. Most experiments suggest that they exert converging actions by elevating of intracellular cGMP levels by activating soluble and membrane-bound guanylate cyclases. Cyclic GMP is the starting point for multiple **signal transduction** cascades, which are now beginning to be unraveled (10). In more than a quarter of a century since the discoveries of atrial granules and volume receptors in the heart atria, the search for natriuretic hormones has led to the isolation and identification of many atrial natriuretic factors (ANF) [(11), as a specific example see (12)]. In the heart, for example, ANF peptides are synthesized and stored in the **Golgi apparatus** of cardiac myocytes and are released in response to atrial wall stretch following acute plasma volume expansion and increased central blood volume. The mechanisms of the renal action of these potent natriuretic hormones are not yet completely unraveled. The renal hemodynamic, tubular, and adrenal, and systemic vascular effects are related to enhanced cGMP synthesis in specific medium-sized arteria, in the glomeruli and specific tubular segments, and in adrenal tissue. Specific ANF-binding sites have been detected in these target organs. A primary action of elevated cGMP levels is stimulating cGMP-dependent protein kinase (PKG), the major intracellular receptor protein for cGMP, which phosphorylates substrate proteins to trigger a regulation cascade (10).

Cyclic GMP-dependent protein kinases (PKG) also mediate some of the neuronal effects of cGMP (13), but unfortunately few PKG substrates are known in the brain. In striatonigral nerve terminals, for example, NO mediates phosphorylation of the protein phosphatase regulator, dopamine- and cyclic AMP-regulated phosphoprotein by PKG. PKG substrates are critically placed in the protein phosphorylation network and regulate protein phosphatases, intracellular calcium levels, and the function of many ion channels and neurotransmitter receptors. Nitric oxide is a signaling molecule in the nervous system of both mammals and insects. In contrast to classical transmitters, NO permeates membranes and acts on neighboring targets normally limited by diffusion barriers. This diffuse signaling is evolutionarily highly conserved. The NO forming enzyme, NO synthase, is present mostly in the nervous system, especially the brain. A soluble form of guanylate cyclase is the major target of NO action. Usually there is cellular separation of the release site and target site of NO, although exceptions to this rule exist.

As with cAMP, cGMP is important for memory in insects. In the honeybee, for example, the NO/cGMP system in the antennal lobes is implicated in the processing of adaptive mechanisms during chemosensory processing, and experimental data support a specific role of the NO system in memory formation (14).

Signal transduction in gastric and intestinal smooth muscle is mediated by receptors coupled via distinct G-proteins to various effector enzymes. Calcium is implicated in signal transduction in different ways according to the cell type (e.g., circular and longitudinal muscle cells). The initial steps involve Ca^{2+}/calmodulin-dependent activation of myosin light-chain kinase and the interaction of **actin** and **myosin**. Relaxation is mediated by cAMP-and/or cGMP-dependent protein kinases. A specific cascade involves G-protein-dependent stimulation of Ca^{2+} influx, leading to Ca^{2+}/calmodulin-dependent activation of a constitutive NO synthase in muscle cells that activates soluble guanylyl cyclase. The resulting activation of protein kinase A and PKG is jointly responsible for muscle relaxation (15).

Therefore, cyclic GMP is a secondary messenger that acts on targets that are sometimes similar to those of cAMP but proceed through a completely different cascade, in which the diffusible NO (and sometimes carbon monoxide) play a major role.

BIBLIOGRAPHY

1. N. B. Bhatnagar, R. Bhatnagar, and T. A. Venkitasubramanian (1984) *Biochem. Biophys. Res. Commun.*, **121**, 634–640.

2. L. Stryer (1986) *Ann. Rev. Neurosci.* **9**, 87–119.
3. T. Misaka, Y. Kusakabe, Y. Emori, T. Gonoi, S. Arai, and K. Abe (1997) *J. Biol. Chem.* **272**, 22623–22629.
4. T. Nakamura and G. H. Gold (1987) *Nature* **325**, 442–444.
5. J. T. Finn, M. E. Grunwald, and K. W. Yau (1996) *Ann. Rev. Physiol.* **58**, 395–426.
6. M. Chabre, J. Bigay, F. Bruckert, F. Bornancin, P. Deterre, C. Pfister, T. Vuong, and D. Baylor (1988) *Cold Spring Harbor Symp. Quant. Biol.* **53** (*Pt 1*), 313–324.
7. D. Baylor (1996) *Proc. Natl. Acad. Sci. USA* **93**, 560–565.
8. C. Zuker (1996) *Proc. Natl. Acad. Sci. USA* **93**, 571–576.
9. K. Osborne, A. Robichon, E. Burgess, S. Butland, R. Shaw, A. Coulthard, H. Pereira, G. RJ, and M. Sokolowski (1997) *Science* **277**, 834–836.
10. S. M. Lohmann, A. B. Vaandrager, A. Smolenski, U. Walter, and H. R. DeJonge (1997) *Trends Biochem. Sci.*, **22**, 307–312.
11. H. J. Kramer and B. Lichardus (1986) *Klinische Wochenschrift* **64**, 719–731.
12. L. R. Forte, X. Fan, and F. K. Hamra (1996) *Am. J. Kidney Dis.* **28**, 296–304.
13. X. Wang and P. J. Robinson (1997) *J. Neurochem.* **68**, 443–456.
14. U. Muller (1997) *Prog. Neurobiol.* **51**, 363–381.
15. G. Makhlouf and K. Murthy (1997) *Cell Signal* **9**, 269–276.

CYCLINS

MARK SOLOMON

From a simple start as a family of proteins with an interesting pattern of accumulation during the **cell cycle**, the cyclins have grown to become key regulators of diverse cellular processes, in particular the cell cycle. The first cyclins were found during studies of **translational** control before and after fertilization of sea urchin eggs conducted as part of the Physiology course at the Marine Biological Laboratory in Woods Hole (1). These proteins were synthesized continuously, and they accumulated until their abrupt **protein degradation** during mitosis. This sawtooth pattern of accumulation hinted that cyclins might play an important role during the cell cycle, either as inducers of cell cycle transitions or, perhaps less interestingly, as proteins that responded to cell cycle states to perform functions important for that stage. Later work showed that the injection of cyclin **messenger RNA** caused frog oocytes to mature (2)—that is, to progress through meiosis and to arrest in second meiotic metaphase. This result suggested that cyclins were actually inducers of the transition into meiosis or mitosis. Subsequent work revealed that the mitotic cyclins are the regulatory subunits of **maturation promoting factor** (MPF), the key inducer of entry into M phase, and whose catalytic subunit is Cdc2, a protein **kinase** identified by genetic studies in *Schizosaccharomyces pombe* as the key regulator of this transition. (For further discussion of this and other aspects of cyclin function, see **Maturation promoting factor (MPF)** and **Cell cycle**.)

One of the irreversible ratchet steps in the cell cycle is the degradation of cyclins by the **ubiquitin** system (3). Ubiquitin is a 76-residue protein whose covalent attachment to proteins can target them for **proteolysis** by the **proteasome**, a huge multiprotease unwinding and degrading machine. Ubiquitin is activated by the ATP-dependent covalent attachment to an enzyme called E1. E1 then transfers ubiquitin to one of many E2 enzymes. The E2s can ubiquitinate substrates, often with the help of an E3. The E3s may be the most diverse and interesting components of this system. Some E3s receive the covalently bound ubiquitin; others serve as matchmakers that bring together a substrate and the appropriate E2. For the mitotic cyclins, the E3 was first termed the *cyclosome*, though it is now generally termed the *anaphase promoting complex* (APC), an 8- to 12-subunit complex. The APC is the regulated component of the ubiquitin system for the degradation of the mitotic cyclins and serves as the target for checkpoint signals that can block cyclin degradation. The ubiquitin-dependent degradation of cyclins that act earlier in the cell cycle generally does not require the APC and has been best studied in *S. cerevisiae*.

The cyclins now comprise a large family of proteins with diverse functions, each bound to a cyclin-dependent kinase (Cdk) catalytic partner. All cyclins resemble the first mitotic cyclins (cyclin A and B), but not all behave in a cyclic manner. Cyclins A, B, D, and E play major roles in regulating cell cycle transitions. When each kinase functions is largely determined by when each cyclin partner accumulates during the cell cycle. Quite a few cyclins (and their associated Cdks) function in **transcription**. For example, cyclin H is one of the subunits of TFIIH, a general **transcription factor** for **RNA polymerase II** involved in the **phosphorylation** of the *C*-terminal domain (CTD) of the large subunit of the polymerase. Cyclin C is a subunit of the RNA polymerase II holoenzyme, which can also phosphorylate the CTD. As more cyclins are discovered (and the alphabet is exhausted!), the majority of all cyclins will probably be noncycling regulatory subunits of protein kinases functioning in processes other than the cell cycle. The observation that the first cyclins had cyclic patterns of accumulation and were involved in cell cycle control may represent more a historical footnote owing to their relative ease of discovery than a reflection of fundamental properties of this family of proteins.

BIBLIOGRAPHY

1. T. Evans, E. T. Rosenthal, J. Youngblom, D. Distel, and T. Hunt (1983) *Cell* **33**, 389–396.
2. K. I. Swenson, K. M. Farrell, and J. V. Ruderman (1986) *Cell* **47**, 861–870.
3. R. W. King, R. J. Deshaies, J.-M. Peters, and M. W. Kirschner (1996) *Science* **274**, 1652–1659.

Suggestions for Further Reading

D. O. Morgan (1997) Cyclin-dependent kinases: engines, clocks, and microprocessors. *Annu. Rev. Cell Dev. Biol.* **13**, 261–291.

A. Murray and T. Hunt (1993) *The Cell Cycle*, W. H. Freeman, New York.

CYCLODEXTRINS

A. GUTTMAN
N. ROOS

Cyclodextrins are cyclic oligosaccharides with a truncated cone shape and an axial void cavity (Fig. 1). The diameter and the volume of the cavity vary with the number of glucose

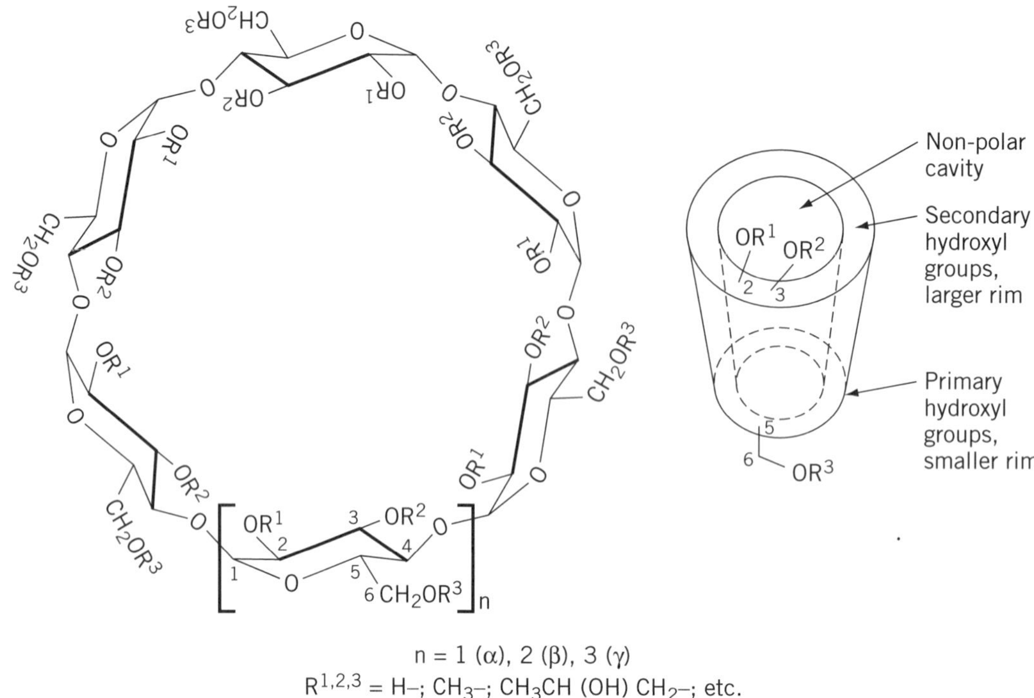

Figure 1. Cyclodextrin structure.

units in the cyclodextrin ring (1). The most commonly used cyclodextrin is β-cyclodextrin, which has seven glucose units and a cavity with a diameter of 0.78 nm and a volume of approximately 35 nm^3. Other natural cyclodextrins, such as α- and γ-cyclodextrin, with six and eight glucose units and 0.57- and 0.95-nm cavity diameters, respectively, are also frequently used. The outer surface of the cyclodextrin molecule is **hydrophilic**, because the majority of the hydroxyl groups project outward, resulting in good water solubility. The internal cavity is relatively **nonpolar**, and it can encapsulate nonpolar solutes of appropriate dimensions, with binding occurring through various nonpolar interactions. Such binding is known as *inclusion complexation*.

The conformation of cyclodextrins in aqueous solution is believed to be that of the truncated cone of Figure 1. Molecules of **hydrophobic** compounds of appropriate size and shape penetrate into the cavity and are bound mainly through hydrophobic interactions, whose strengths depend on the efficiency of the contact. The edge of the torus of the larger circumference consists of secondary hydroxyl groups that are attached to **chiral** carbons (C2 and C3 of the glucose units). This structure results in variable binding affinities for different **enantiomers**, probably due to interactions of the chiral solute with the chiral entrance to the cavity (2). The primary hydroxyl groups of the glucose monomers make up the smaller edge of the cone. Chemical derivatization of natural cyclodextrins via modification of their hydroxyl groups is currently an area of very active research (3–6). Such modifications are yielding materials of varying complexation selectivities and physicochemical properties, such as improved solubility (7).

Hoffmann and Bock (8) have examined complex formation between different cyclodextrins and **nucleotides**. They found that adenosine, cytidine, guanosine, uridine, inositol, and deoxythymidine monophosphates (AMP, CMP, GMP, UMP, IMP, and dTMP) did not form a complex with α-cyclodextrin, but β-cyclodextrin readily bound AMP and IMP. It was concluded that that these six nucleotides are too bulky to fit into the cavity of α-cyclodextrin. When a complex is formed with β-cyclodextrin, the ribose and phosphate groups of the nucleotides exert a stabilizing effect by establishing hydrogen bonds with the outer rim of the cyclodextrin molecules. The position of the phosphate group is not important; increasing distance of the phosphate group from the base increased the stability of the complex. Larger oligonucleotides exhibited decreasing tendencies for complex formation, but the extent of complexation depended significantly on their base composition. Interestingly, polynucleotides with double- or triple-helical structures did not show any complexation with β-cyclodextrin (9). **Transfer RNA**, which comprises both helical and nonhelical structures, can interact with cyclodextrins, and these complexes have been used extensively for tRNA studies.

Cyclodextrins have also been used in labeling nucleic acid molecules in various biochemical analysis applications, especially in **DNA sequencing**. Cyclodextrin labels provide potentially high signal efficiency and versatility in label colors, while maintaining uniform chemical and physical properties. Cyclodextrin tracers are prepared by coupling the cyclodextrin to specific binding substances, such as nucleic acids, and forming inclusion complexes with the fluorophores. Then the DNA molecules are sequenced using cyclodextrin-labeled chain terminators. This method allows DNA sequencing with high sensitivity and high throughput (10).

Ikeda et al. (11) reported the use of anthryl(alkylamino)–cyclodextrin complexes as chemically switched DNA intercalators that were **allosteric**. On adding a ligand that is tightly bound in the cyclodextrin cavity, such as 1-adamantol, the host molecule releases the anthryl unit, which then leads to strong intercalation with the double-stranded DNA molecule. This principle could be extremely useful in nucleic acid reactions of medicinal and biotechnological importance, particularly in view of the established methods to modify cyclodextrins for specific interactions with a wide range of substances, for new drug delivery systems.

Cyclodextrins and their analogs can also be used as carriers to increase cellular uptake of phosphorothioate **antisense oligonucleotides**. Cellular uptake of phosphorothioate oligodeoxynucleotides in the presence of various cyclodextrin analogs was found to depend on the concentration and the time. In particular, 2-hydroxypropyl-β-cyclodextrin (HPβCD), 2-hydroxyethyl-β-cyclodextrin (HEβCD), and a mixture of various HPβCDs having different degree of substitution were observed to increase the uptake of phosphorothioate oligodeoxynucleotides two- to threefold in 48 h (12).

Cyclodextrins and derivatives are used to enhance the bioavailability of many water-insoluble pharmaceuticals (13). The solubilization results in fast and quantitative *in vivo* delivery for intravenous and intramuscular dosing. A decrease in irritation at the administered site can be observed (14). The solubility of **hormones** such as hydrocortisone was enhanced 72-fold using randomly methylated β-cyclodextrin (15). Sublingual administration of γ-cyclodextrin complex of testosterone avoided rapid first-pass loss of the hormone and directed it effectively into the circulation; administration of the complex into the stomach resulted in much lower circulatory hormone levels (16).

The relatively good water solubility of the cyclodextrins makes these materials useful in **chromatography** and **electrophoresis** separation methods, such as in **HPLC** (high-performance liquid chromatography) and **capillary zone electrophoresis** (CZE) (17,18). Buffer systems, even with organic modifiers, can be used to control the pH or modify other secondary equilibrium. Temperature also has a significant effect on selectivity in cyclodextrin-mediated separation systems.

ACKNOWLEDGMENT

The authors gratefully acknowledge Professor József Szejtli for his stimulating discussions.

BIBLIOGRAPHY

1. J. Szejtli (1988) In *Cyclodextrins and Their Inclusion Complexes*, Cyclodextrin Technology, Kluywer, Dordrecht, The Netherlands.
2. A. Guttman et al. (1988) *J. Chromatogr.* **488**, 41–53.
3. B. Sébille (1987) In *Cyclodextrin Derivatives, Cyclodextrins and their Industrial Uses*, D. Duchêne, ed., Editions de Santé, Paris, pp. 353–385.
4. K. Uekama (1985) *Pharm. Intl.* **6**, 61–65.
5. J. Pitha et al. (1986) *Intl. J. Pharm.* **29**, 73–82.
6. B. W. Müller and U. Brauns (1985) *Intl. J. Pharm.* **26**, 77–88.
7. Y. Y. Rawjee, D. E. Stark, and Gy. Vigh (1993) *J. Chromatogr.* **635**, 291–306.
8. L. Hoffmann and R. M. Bock (1970) *Biochemistry* **9**, 3542–3550.
9. J. Szejtli, ed. (1982) *Proceedings of the 1st International Symposium on Cyclodextrins*, Reidel, Dordrecht, The Netherlands.
10. K. M. Kosak, *PCT Intl. Appl.* (WO 9102040, 2/21/1991), 69 pp.
11. T. Ikeda, K. Yoshida, and H. J. Schneider (1995) *J. Am. Chem. Soc.* **117**, 1453–1454.
12. Q. Zhao, J. Temsamani, and S. Agrawal (1995) *Antisense Res. Devel.* **5**, 185–192.
13. R. A. Rajewski and V. Stella (1996) *J. Pharm. Sci.* **85**, 1142–1169.
14. T. Loftsson and M. E. Brewster (1996) *J. Pharm. Sci.* **85**, 1017–1025.
15. J. Szejtli (1994) *Med. Res. Rev.* **14**, 353–386.
16. J. Pitha, E. J. Anaissie, and K. Uekama (1987) *J. Pharm. Sci.* **76**, 788–790.
17. S. Ahuja (1991) In *Chiral Separations by Liquid Chromatography*, American Chemical Society, Washington, DC.
18. A. Guttman (1996) In *Handbook of Capillary Electrophoresis*, 2nd ed., J. P. Landers, ed., CRC Press, Boca Raton, pp. 75–100.

CYCLOHEXIMIDE

YOSHIKAZU NAKAMURA
EISHUN MUTO

$C_{15}H_{23}NO_4$ (Fig. 1; molecular mass, 281.35). Cycloheximide is an antibiotic produced by *Streptomyces griseus*, and it inhibits protein biosynthesis in eukaryotes. Cycloheximide does not affect protein synthesis in prokaryotes or in **mitochondria**. It binds to the 60S subunit of eukaryotic **ribosomes** and inhibits the **peptidyl transferase** activity. The inhibition of peptidyl transfer causes a stack of ribosomes on the **messenger RNA**, generating polysome accumulation in cells. Cycloheximide is frequently used for the study of gene expression in eukaryotes as an inhibitor of protein synthesis *in vivo*, as well as *in vitro*.

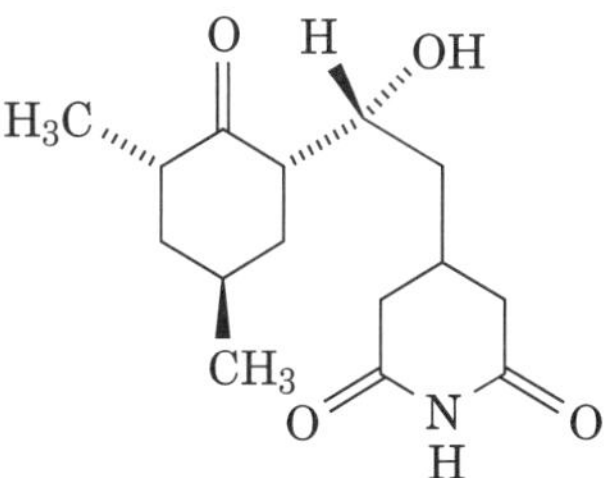

Figure 1. Cycloheximide.

CYCLOPHILIN

G. FISCHER

Historically, an 18-kDa cytosolic protein from pig kidney was the first **peptidyl prolyl *cis/trans* isomerase** (abbreviated PPIase) discovered and characterized enzymatically (1). These enzymes catalyze the rotation of a peptidyl prolyl moiety in **peptides** and **proteins**. In 1989 it was found that this protein is nearly identical in its **primary structure** to the already known cytosolic receptor of the immunosuppressive undecapeptide **cyclosporin** A in human T lymphocytes, previously named cyclophilin [Cyp18; for nomenclature see (2)] (3–5). This enzyme represents the archetype of cyclophilins in that it constitutes the catalytic **domain** of larger cyclophilin-like proteins. Now two additional families of such isomerases are known, those that bind **FK506** (FKBP) and the parvulin family.

Alignment of amino acid sequences of the cyclophilin family defines residues that are highly conserved among vertebrates, plants, fungi, and bacteria. Nearly perfect conservation is located primarily in the central segment of the protein, including the sequences -Phe-His-Arg-Ile/Val-Ile-$(Xaa)_5$-Gln-Gly-Gly- at positions 53 to 65 and -Met-Ala-$(Xaa)_{9-10}$-Gln-Phe-Phe/Tyr-Ile/Val- at positions 100 to 114, where Xaa is any amino acid and / separates alternatives. These sequences are used as typical **motifs** to search for other cyclophilins. The three-dimensional structure of the archetypal cyclophilin consists of an eight-stranded antiparallel **β-sheet** barrel capped by **α-helices** (6,7).

The characterization of 11 distinct cyclophilin genes in the genome of the **nematode** *Caenorhabditis elegans* demonstrates that organisms use numerous cyclophilins (8). As many as 12 human homologues have been identified thus far from the initial assessment of gene diversity by **expressed sequence tag** analysis. The C-terminal and/or N-terminal amino acid extensions of larger cyclophilins consist of additional **domains**, directing the proteins into specific cellular compartments, mediating protein/protein and DNA/protein interactions, or generating other biochemical functions.

The cytosolic Cyp18 often represents the major cyclophilin of cells. Kidney tubules and endothelial cells, for example, contain 10 μg of Cyp18 per mg total protein (9). Despite the numerous studies reporting the three-dimensional structures of Cyp18-bound oligopeptide substrates and inhibitors (10–12), a definite catalytic mechanism for cyclophilins and other PPIases is still lacking. It is currently believed that catalysis arises from a combination of desolvation of the substrate and substrate-assisted catalysis. Electrophilic assistance, for example, by the Arg55 residue of Cyp18, can also be envisaged to account for the more effective catalysis found for cyclophilins, compared to FKBPs. The positively charged side-chain of this amino acid residue is located within **hydrogen-bonding** distance and perpendicular to the plane of the substrate proline ring in Cyp18/substrate complexes. This environment may cause additional weakening of the reactive C–N linkage by immobilizing the lone pair of electrons of the nitrogen atom. Indeed, a Cyp18 variant in which Arg55 has been replaced retains only 0.1% of the wild-type enzymatic activity (13).

Because of the putative functional redundancy and overlapping of the many PPIases present in most organisms, gene deletion experiments often do not lead to a recognizable **phenotype** under normal growth conditions. Among the seven cyclophilins found in *S. cerevisiae*, deletions of only two are associated with any phenotype. Disruption of *Cpr3*, the gene encoding the **mitochondrial** isoform of yeast cyclophilin, affects growth on L-lactate medium (14), whereas *cpr7Δ* cells are defective in normal cell growth (15). The expression of many PPIase genes seems to be sensitive to **heat shock** and to the chemical **stress response**.

Recently, it was shown that host cell cytosolic Cyp18 is required for **HIV**-1 infection before **reverse transcription** but subsequent to receptor binding and membrane fusion in T cells (16). A proline-rich segment of the capsid domain of $Pr55^{gag}$ mediates incorporation of Cyp18 into HIV-1 virions (17). This conserved segment that contains four proline residues occurs in the sequence -Pro-$(Xaa)_4$-Pro222-$(Xaa)_2$-Pro-$(Xaa)_5$-Pro-. Mutant proteins of HIV-1_{HXB2} that have **site-directed mutations** in which Pro222 is replaced by Ala or Gly221 by Ala, fail to bind to the **fusion protein** glutathione-S-transferase/Cyp18. Virions that contain these altered proteins cannot sequester Cyp18 into the released virions, emphasizing the importance of the Gly-Pro^{222} bond for $Pr55^{gag}$ /Cyp18 complex formation. The ability of many cyclosporin derivatives to dissociate the $Pr55^{gag}$/Cyp18 complex reveals a quantitative relationship between Cyp18 inhibition and complex decomposition. Obviously, the antiviral effect of cyclosporin A must be caused by a pathway distinctive from immunosuppression because cyclosporin A derivatives with negligible immunosuppressive activity, but high affinities for the active site of Cyp18, retain potent anti-HIV activity (18–20).

A genetic **cDNA** screen was used to identify the bovine homologue of the retina-specific cyclophilin NinaA of *Drosophila*. The membrane-localized PPIase of the secretory pathway of the fly is required for proper folding and trafficking of Rh1-6 opsin in photoreceptor cells (21). The bovine RanBP2 cyclophilin has a domain that binds to the **GTPase** Ran, as well as to red/green opsin. A still unknown modification of opsin, possibly a prolyl bond isomerization catalyzed by the Cyp-domain of RanBP2, augments and stabilizes the interaction between the Ran-binding domain and opsin. This modification is important in membrane trafficking of long-wavelength opsin in photoreceptors (22).

BIBLIOGRAPHY

1. G. Fischer, H. Bang, and C. Mech (1984) *Biomed. Biochim. Acta* **43**, 1101–1111.
2. G. Fischer (1994) *Angew. Chem., Int. Ed. Engl.* **33**, 1415–1436.
3. R.E. Handschumacher et al. (1984) *Science* **226**, 544–547.
4. G. Fischer et al. (1989) *Nature* **337**, 476–478.
5. N. Takahashi, T. Hayano, and M. Suzuki (1989) *Nature* **337**, 473–475.
6. R.T. Clubb, S.B. Ferguson, C.T. Walsh, and G. Wagner (1994) *Biochemistry* **33**, 2761–2772.
7. H. Ke (1992) *J. Mol. Biol.* **228**, 539–550.
8. A.P. Page, K. Macniven, and M.O. Hengartner (1996) *Biochem. J.* **317**, 179–185.
9. B. Ryffel et al. (1991) *Immunology* **72**, 399–404.
10. G. Pflügl et al. (1993) *Nature* **361**, 91–94.
11. H.M. Ke et al. (1994) *Structure* **2**, 33–44.
12. Y.D. Zhao and H.M. Ke (1996) *Biochemistry* **35**, 7362–7368.
13. L.D. Zydowsky et al. (1992) *Protein Sci.* **1**, 1092–1099.
14. E.S. Davis et al. (1992) *Proc. Natl Acad. Sci. U.S.A.* **89**, 11169–11173.
15. A.A. Duina, J.A. Marsh, and R.F. Gaber (1996) *Yeast* **12**, 943–952.
16. D. Braaten, E.A. Franke, and J. Luban, (1996) *J. Virol.* **70**, 3551–3560.
17. E.T. Franke, H.E.H. Yuan, and J. Luban (1994) *Nature* **372**, 359–362.
18. B. Rosenwirth et al. (1994) *Antimicrob. Agents Chemother.* **38**, 1763–1772.
19. S.R. Bartz et al. (1995) *Proc. Natl Acad. Sci. USA* **92**, 5381–5385.
20. C. Aberham, S. Weber, and W. Phares (1996) *J. Virol.* **70**, 3536–3544.
21. E.K. Baker, N.J. Colley, and C.S. Zuker (1994) *EMBO J.* **13**, 4886–4895.
22. P.A. Ferreira, T.A. Nakayama, W.L. Pak, and G.H. Travis (1996) *Nature* **383**, 637–640.

Suggestions for Further Reading

A. Galat and S.M Metcalfe (1995) Peptidylproline *cis/transisomerases*, *Prog. Biophys. Molec. Biol.* **63**, 67–118; contains an exhaustive bibliography of many aspects of PPIases.

C.C. Trandinh, G.M. Pao, and M.H. Saier (1992) Structural and evolutionary relationships among the immunophilins—two ubiquitous families of peptidyl-prolyl *cis/trans* isomerases, *FASEB J.* **6**, 3410–3420; contains a detailed analysis of the amino acid sequences of PPIases.

G. Wiederrecht and F. Etzkorn (1994) The immunophilins, *Persp. Drug Disc. Design* **2**, 57–84.

D.M. Armistead and M.W. Harding, (1993) Immunophilins and immunosuppressive drug action, *Ann. Rep. Medicinal Chem.* **28**, 207–213.

CYCLOSPORIN

G. FISCHER

The cyclosporin family of secondary metabolites from the **fungus** imperfectus *Beauveria nivea* are cyclic undecapeptides characterized by an enormous spectrum of biological effects. The most widely used derivative, cyclosporin A (CsA), is immunosuppressive, antiinflammatory, and antichemotactic, and combats multidrug-resistance and viral, fungal, and parasitic infections, without being cytotoxic to mammalian cells (1,2). Among a number of other *N*-methyl amino acids, CsA contains an uncommon amino acid at position 1, the (4*R*)-4([(*E*)-2-butenyl]-4-*N*-dimethyl-L-threonine (Fig. 1). Several minor congeners of CsA, designated CsB, CsC ... CsZ, have been found as by-products of the fermentation process. The [D-MeVal^{11}]-CsA derivative (CsH) is devoid of most biological effects, thereby serving as a control for nonspecific interactions. Other derivatives have been developed that exhibit a more restricted spectrum of biological activities compared to CsA. For example, [L-MeLeu(3-OH)^{1},MeAla^{4,6}]-CsA is nonimmunosuppressive but has enhanced antiviral efficacy (3). In the fungus, non-ribosomal synthesis of cyclosporins occurs on a huge, single-chain enzyme with a molecular mass of 1,689 kDa, termed cyclosporin synthetase. It utilizes unmethylated natural amino acids, ATP/Mg^{2+} and **S-adenosyl-L**-methionine as starting materials in biosynthesis (4). Beside its use as a powerful tool for unraveling the molecular basis of the cellular **immune response**, the therapeutic application of CsA has revolutionized the likelihood of success for organ transplantation since its introduction in 1983 (5). Other disorders with immune components, like psoriasis and juvenile diabetes, are relieved with the aid of cyclosporins.

The dramatic effects of CsA on the cellular immune response, in which the T cells remain in a nonproliferative state, results primarily from the impaired expression of a number of **lymphokine** genes. This could include a variety of **interleukins** (IL-2, IL-3, IL-4), factors, such as granulocyte macrophage **colony stimulating factor** (GM-CSF), **tumor necrosis factor** α (TNF-α) and γ-**interferon**, and the **oncogene** c-myc, all synthesized subsequent to antigen stimulation of T cells. Initial investigations identified cytosolic **cyclophilin** Cyp18 as the specific receptor for CsA in T cells, but minor **isoenzymes** cannot be completely ruled out yet.

The three-dimensional structure of the inhibitory CsA/Cyp18 complex was determined by **X-ray crystallographic** and **NMR** methods (6–8). The structure of Cyp18 is hardly affected by complex formation, whereas that of CsA bound to Cyp18 is dramatically different when in nonpolar organic solvents or in the solid state. In response to the altered conformation of CsA in the complex, the inhibition of the PPIase activity of Cyp18 by CsA is time-dependent, including a ***cis/trans* isomerization** of the peptide bond between MeLeu^{9}-MeLeu^{10}. **Site-directed mutagenesis** of Cyp18 indicates the involvement of a **hydrophobic** pocket of side-chains, with

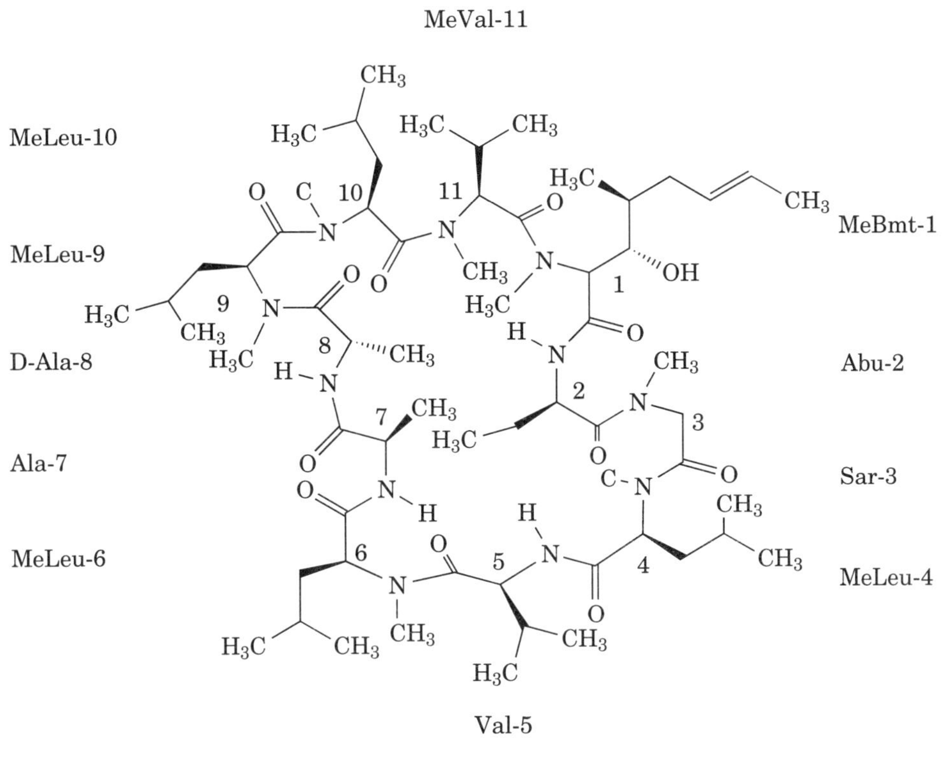

Figure 1. Structure of cyclosporin A.

Trp121 as a major determinant, in the tight binding of CsA to Cyp18, with an inhibition constant, K_i of 2.3 nM. Comparison of Cyp18/substrate complexes with Cyp18/CsA indicates that the peptide chain of CsA runs in the opposite direction. The MeVal11 side-chain of the inhibitor occupies the place of the substrate proline ring.

In some cases, binding of cyclosporins to the targeted cyclophilin and inhibition of its PPIase activity may explain the biological effect of cyclosporins. In contrast, binding of CsA to Cyp18 is necessary, but not sufficient, to arrest signals for the cellular immune response. For example, [MeAla^6]-CsA is an excellent inhibitor of the PPIase activity of Cyp18 but has only 1% of the immunosuppressive effect of CsA, whereas the weak inhibitor [MeBm$_2$t^1]-CsA ($K_i > 1 \mu M$) still shows a considerable fraction of the CsA effect (9).

Thus, for immunosuppression CsA (like **FK506** and **rapamycin**) is best viewed as a pro-drug that is activated functionally when bound to Cyp18 (10). Once formed from the components, the complexes CsA/Cyp18 or FK506/FKBP12 bind to and inactivate reversibly, in a noncompetitive mechanism, the Ca^{2+}- and calmodulin-dependent, heterodimeric protein **phosphatase** 2B (calcineurin). This enzyme catalyzes the dephosphorylation of Ser/Thr residues in phosphorylated proteins (11,12) (see **Phosphorylation**). As a result of dephosphorylation, this enzyme induces the phosphorylated cytoplasmic subunit of NF-AT to translocate into the **nucleus** for association with the newly synthesized nuclear subunit of NF-AT to form a functional **transcription factor** (13).

BIBLIOGRAPHY

1. J. F. Borel, C. Feurer, H. Gubler, and H. Stahelin (1976) *Agents Actions* **6**, 468–475.
2. J. F. Borel (1989) *Pharmacol. Rev.* 41, 259–371.
3. S. R. Bartz et al. (1995) *Proc. Natl Acad. Sci. U.S.A.* **92**, 5381–5385.
4. A. Lawen and R. Zocher (1990) *J. Biol. Chem.* **265**, 11355–11360.
5. C. R. Stiller (1996) *Transplant. Proc.* **28**, 2005–2012.
6. J. Kallen et al. (1991) *Nature* **353**, 276–279.
7. H. M. Ke et al. (1994) *Structure* **2**, 33–44.
8. S. W. Fesik et al. (1991) *Biochemistry* **30**, 6574–6583.
9. N. H. Sigal et al. (1991) *J. Exp. Med.* **173**, 619–628.
10. S. L. Schreiber (1991) *Science* **251**, 283–287.
11. J. Liu et al. (1991) *Cell* **66**, 807–815.
12. J. Liu et al. (1992) *Biochemistry* **31**, 3896–3901.
13. N. A. Clipstone, D. F. Fiorentino, and G. R. Crabtree (1994) *J. Biol. Chem.* **269**, 26431–26437.

Suggestions for Further Reading

A. Lawen (1996) Biosynthesis and mechanism of action of cyclosporins, *Prog. Med. Chem.* **33**, 53–97. This nice review about cyclosporin A-mediated immunosuppression includes a summary of the biosynthesis of cyclosporin derivatives.

P. F. Halloran and J. Madrenas (1991) The mechanism of action of cyclosporine—a perspective for the 90s, *Clin. Biochem.* 24, 3–7.

M. Thali (1995) Cyclosporins: Immunosuppressive drugs with anti-HIV-1 activity, *Mol. Med. Today*, **6**, 287–291.

J. Clardy (1995) The chemistry of signal transduction, *Proc. Natl Acad. Sci. U.S.A.* 92, 56–61.

H. Fliri, G. Baumann, A. Enz, J. Kallen, M. Luyten, V. Mikol, R. Movva, V. Quesniaux, M. Schreier, M. Walkinshaw, R. Wenger, G. Zenke, and M. Zurini (1993) Cyclosporins—structure-activity relationships, *Immunosuppressive Antiinflammatory Drugs* **696**, 47–53.

G. Fischer (1994) Peptidyl-prolyl *cis/trans* isomerases and their effectors, *Angew. Chem., Int. Ed. Engl.* **33**, 1415–1436.

CYSTATIN

M. LASKOWSKI JR.

Cystatin is a protein inhibitor of cysteine proteinases from chicken egg white that gave its name to a family and superfamily of such inhibitors [see **Cysteine**; **Proteinase inhibitors, protein**].

CYSTEINE (CYS, C)

T. E. CREIGHTON

The **amino acid** cysteine is incorporated into the nascent **polypeptide chain** during **protein biosynthesis** in response to two **codons**—UGU and UGC—and represents only 1.7% of the residues of the proteins that have been characterized. The cysteinyl residue incorporated has a mass of 103.15 Da, a **van der Waals volume** of 86 Å^3, and an **accessible surface** area of 140 Å^2. Cys residues are one of the most conserved type of residue during **divergent evolution**; they are interchanged in **homologous** proteins most frequently with **serine** residues.

The **thiol group** of the Cys side-chain

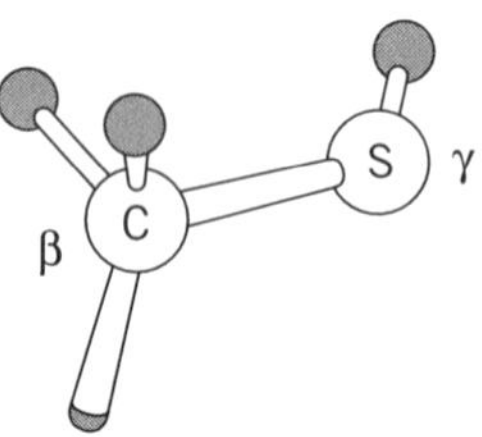

is the most reactive of any of the normal amino acids. It undergoes numerous chemical reactions with a variety of reagents (see **THIOL GROUPS**). These reactions provide means for counting and measuring the number of Cys residues present in a protein (see **COUNTING RESIDUES** and **ELLMAN'S REAGENT**). It is the ionized thiolate anion that is reactive, and only it has **absorbance** in the UV region; the pK_a value of a typical exposed Cys thiol group is 8.7, but deviations as low as 3.5 and greater than 11 are known, depending on the environment (1).

In the nonionized form, the thiol group is not very **polar** or reactive, and thiol groups may be buried in folded proteins without the need for any **hydrogen bonding**. Approximately 40–50% of Cys residues are fully buried in folded **protein structures**. They occur predominantly in *beta*-**sheet** type of **secondary structure**. Cys residues can serve as efficient "capping" residues at the *N*-termini of α-helices in which their thiol group interacts with the hydrogen bonding groups of the backbone; the pK_a can then be decreased and the thiol groups made more reactive; probably for that reason, Cys residues occur at such positions in folded proteins only when important for function (2).

A notable tendency is for two Cys residues of the same or different polypeptide chains to form **disulfide bonds**. This occurs only if the protein conformation permits, and disulfide formation is a useful probe of **protein folding** *in vitro*. *In vivo*, disulfide bonds are usually inserted primarily into proteins to be secreted by the catalyst **protein disulfide isomerase**.

If a protein is hydrolyzed to amino acids under conditions in which the disulfide bond persists, two cysteine amino acids linked by a disulfide bond result, which historically was known as the amino acid cystine. Peptide bonds preceding Cys residues can be cleaved by cyanylation with 2-nitro-5-thiocyanobenzoic acid, followed by incubation at alkaline pH (3). The reaction is specific for Cys residues but has the disadvantage that the new *N*-terminus generated is blocked and not amenable to further **sequencing** by the **Edman degradation**.

Thiol groups have intrinsic affinities for many metal ions, and they are used in proteins to ligate ions of zinc, iron, and copper. In an extreme example, **metallothioneins** about a third of the residues are Cys. The thiol groups of Cys residues are also used as nucleophiles in catalysis in the **thiol proteinases**.

Cys residues are the sites of several types of **post-translational modifications**, in addition to disulfide formation. **Farnesyl groups** and geranylgeranyl groups can be added to Cys residues at the *C*-termini of certain proteins, after removal of a few *C*-terminal residues. **Palmitoyl groups** can be attached in thioester linkages to the side chains of Cys residues.

BIBLIOGRAPHY

1. J. W. Nelson and T. E. Creighton (1994) *Biochemistry* **33**, 5974–5983.
2. T. Kortemme and T. E. Creighton (1995) *J. Mol. Biol.* **253**, 799–812.
3. G. R. Stark (1977) *Meth. Enzymol.* **47**, 129–132.

CYSTEINE PROTEINASE INHIBITORS

M. LASKOWSKI JR.

Cysteine proteinase inhibitors attracted wide attention and, after serine proteinase inhibitors [see **Serine proteinase inhibitors, protein**], have the most frequently described members. Most but not all of these belong to the **cystatin** superfamily, which is frequently divided into three families: the stefin, the cystatin, and the kininogen families. The stefins are low molecular weight (about 100 residues) molecules without attached carbohydrate and without any intramolecular disulfide bridges. They are extremely stable intracellular proteins. The word *stefin* comes from the name of the J. Stefin Institute in Ljubljana (Slovenia), where the first stefin was discovered. Cystatins proper are only slightly bigger than stefins (120–130 residues). They contain two disulfide bridges near their COOH terminals and typically are secretory proteins. The family and superfamily are named after chicken egg white cystatin, which was discovered long before the other cystatins became popular. The third family of cystatins is the kininogen family. Proteins are assigned to it on the basis of the high molecular weights of its members, predominantly HMW (high molecular weight) (120 kD) and LMW (low molecular weight) (68 kD) kininogens. Kininogens are glycosylated and contain numerous disulfide bridges. Each contains three cystatin like repeats.

All cystatin family members inhibit only cysteine proteinases and only those related to papain but, within this broad group, they exhibit little discrimination between individual enzymes. In strong contrast to serine proteinase inhibitors, cystatins react with enzymes whose catalytic sulfhydryl group is significantly blocked. The mode of association was elucidated in detail from the three-dimensional structure of a complex of stefin A with papain.

In avian and mammalian cells, there are present cysteine proteinases called calcium-activated neutral proteinases (CANP), or calpains. These complicated multidomain enzymes are of great importance in the metabolism of muscles. They are controlled by a closely related family of molecularly complex endogenous protein inhibitors—calpastatins.

Suggestions for Further Reading

A. J. Barrett (1981) Cystatin, the egg white inhibitor of cysteine proteases. *Methods Enymol.* **80**, 771–778.

W. M. Brown and K. M. Dziegielewska (1997) Friends and relations of the cystatin superfamily—new members and their evolution. *Protein Sci.* **6**, 5–12.

H.-H. Otto and T. Schimeister (1997) Cysteine proteases and their inhibitors. *Chem. Rev.* **97**, 133–171.

CYSTINE KNOT

JENNIFER MARTIN

The cystine knot is a common **protein motif** that occurs in some **protein structures**. It incorporates an antiparallel **β-sheet** and three **disulfide bonds**, one of which passes through a ring formed by the other two disulfides (Fig. 1). Two families of cystine knot motifs have been identified. One is the **growth factor** cystine knot family (1–2) that includes **nerve growth factor**, **transforming growth factor β2**, **platelet derived growth factor**, and human chorionic gonadotrophin. In this topology, the six cysteine residues (designated CI through CVI) form three disulfide bridges (CI-IV, CII-V, CIII-VI), with CI-IV passing through the ring formed by the disulfides of CII-V and CIII-VI. The size of the ring varies from 8 to 14 residues. The β-sheet is formed from four antiparallel β-strands. These cystine knot growth factors are all dimeric, but their dimer interfaces differ.

The second group is the inhibitor cystine knot family (3) and includes the neurotoxin **ω-conotoxin** GVIA and the uterotonic peptide kalata B1. The inhibitor cystine knot motif has the same disulfide bond pattern as the growth factor cystine knot, but in this case the knot is formed by CIII-VI passing through the ring formed by disulfides CI-IV and CII-V. The ring of the inhibitor cystine knot varies from 8 to 12 atoms, and the β-sheet is triple stranded and antiparallel.

[See also **Disulfide bond** and **Beta-sheet**.]

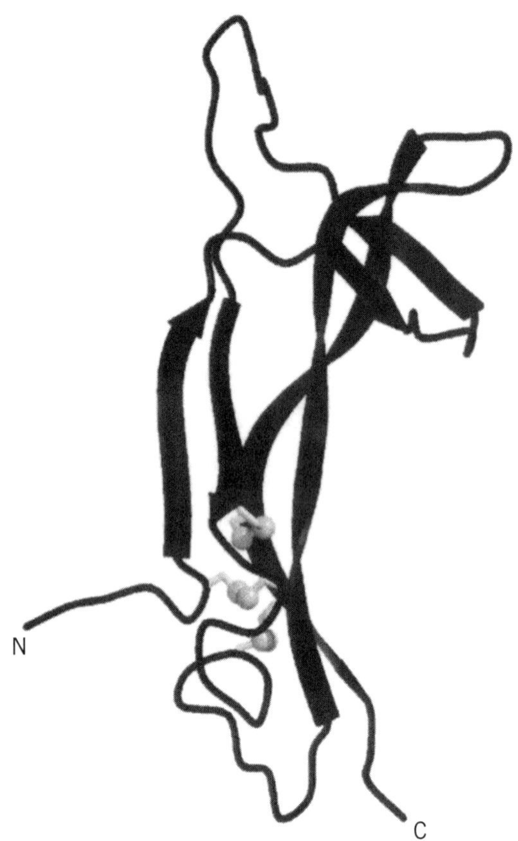

Figure 1. Schematic representation of the backbone structure of the growth factor cystine knot protein, nerve growth factor (4). β-Strands are shown as arrows, and the three disulfide bonds of the cystine knot are shown in gray, with the sulfur atoms depicted as gray spheres. The *N*- and *C*-termini are labeled. This figure was generated using Molscript (5) and Raster3D (6,7).

BIBLIOGRAPHY

1. N. Q. McDonald and W. A. Hendrickson (1993) *Cell* **73**, 421–424.
2. N. W. Isaacs (1995) *Curr. Opin. Struct. Biol.* **5**, 391–395.
3. P. K. Pallaghy, K. J. Nielsen, D. J. Craik, and R. S. Norton (1994) *Protein Sci.* **3**, 1833–1839.
4. N. Q. McDonald et al. (1991) *Nature* **354**, 411–414.
5. P. J. Kraulis (1991) *J. Appl. Crystallogr.* **24**, 946–950.
6. E. A. Merritt and M. E. P. Murphy (1994) *Acta Crystallogr.* **D50**, 869–873.
7. D. J. Bacon and W. F. Anderson (1988) *J. Mol. Graphics* **6**, 219–222.

CYTOCHALASIN

EDWARD H. EGELMAN

Cytochalasins are a group of fungal metabolites that can permeate cell membranes and exert very strong effects on the **cytoskeleton**: ruffling and **translocation** are stopped, and cells round up (1,2). In addition, cytochalasins A and B also inhibit glucose transport across the plasma membrane. Although it was originally believed that the main action of the cytochalsins on the cytoskeleton was binding with high affinity to the barbed, or fast-growing end of the **actin** filament, and inhibiting subunit addition and loss, it has become clear that the effects are more complex. The concentrations of cytochalasins used in cell biological experiments will have strong effects on actin monomers and may convert much of a cell's monomeric G-ATP actin to G-ADP actin (3). In addition, even at saturating concentrations cytochalasins do not completely block the fast-growing ends with respect to subunit addition or loss (3). Thus, the action of cytochalsin on a living cell is likely to be complicated and may affect the interactions of actin with many **actin-binding proteins**. Nevertheless, as a result of the large effect cytochalsins have on actin, they have been widely used to study the role of the actin cytoskeleton in cellular processes.

BIBLIOGRAPHY

1. I. Yahara, F. Harada, S. Sekita, K. Yoshihira, and S. Natori (1982) *J. Cell Biol.* **92**, 69–78.
2. M. Schliwa (1982) *J. Cell Biol.* **92**, 79–91.
3. P. Sampath and T. D. Pollard (1991) *Biochemistry* **30**, 1973–1980.

CYTOCHROME P450

B. KEMPER

Cytochromes P450 (designated here as P450) arise from a **superfamily** of **genes** that encode hemoproteins with molecular weights of 40 to 55 kDa and extraordinarily diverse properties and functions. P450 is nearly ubiquitous in the biotic world, with over 500 genes characterized in **plants**, animals, **fungi**, and **bacteria**, although it is not present in all organisms. Of the substrates of P450, 1000 are known, and there are perhaps as many as 10^6 in total; more than 40 different reactions are catalyzed (1). These enzymes play a critical role in:

1. Detoxification of bioactive molecules in vertebrates, insects, and plants
2. Metabolic activation of compounds that are both beneficial (drug precursors, **hormones**) and detrimental (**carcinogens**)
3. Biogenesis and catabolism of many endogenous compounds, including **steroid hormones**, vitamin D, biological insecticides, **fatty acids**, insect pheromones, and plant lignins and pigments

The P450 that metabolize xenobiotics appear to have no critical physiological functions, since mice with disrupted genes for these proteins had no obvious phenotypic abnormalities (2,3). Most P450 proteins are membrane-bound in the **endoplasmic reticulum (microsomes)** or **mitochondria**, but soluble forms are present in bacteria. In mammals, the greatest concentrations of P450, particularly the detoxification forms, are present in the liver, but they are probably present in most tissues of the body, with relatively high concentrations in the intestines, lungs, nasal epithelia, and steroidogenic tissues.

The genes for these enzymes differ in size, **intron** number, and **chromosomal** location. **Gene expression** of subsets of the P450 is regulated by a variety of substances, including xenobiotics such as aromatic hydrocarbons, barbiturates, and **peroxisomal** proliferators, and by endogenous steroid and polypeptide hormones.

The name cytochrome P450 is based on the characteristic increase in the absorbance of the protein heme group at 450 nm upon binding of CO and reduction of its iron atom. P450 differ structurally from most **cytochromes** because the fifth coordination group of the iron atom of the heme is occupied by a thiolate group from a cysteine residue and the sixth is occupied by water. They differ functionally because P450 are monooxygenases and not simply electron carriers (see **Cytochromes**).

CLASSIFICATION

On the basis of amino acid sequence similarity, the P450 are divided into more than 100 families and subfamilies (4). The various P450 within a **gene family** generally have >40% sequence identity, and members within a subfamily have identities >55% within mammalian species and >46% overall. P450 with >97% identity are arbitrarily considered **allelic** variants unless there is independent evidence for separate gene loci. Genes from animals, fungi, plants, or bacteria are assigned family numbers 1–49, 51–69, 71–99, and 101 and above, respectively. Family 51 is represented in plants and mammals, as well as fungi.

The recommended nomenclature for P450 genes is an italicized *CYP* (*Cyp* for mouse and *Drosophila*) followed by an Arabic number designating the family, a letter for the subfamily, and a second number for the member of the subfamily, eg, *CYP1A1* (4). Identical terminology is used for gene products, except that the name is not italicized. Trivial names of P450 and other designations of the proteins, such as cytochrome P450 1A1, P450 1A1 or 1A1, are also considered acceptable in publications so long as the official name is specified.

CATALYTIC ACTIVITY

The overall reaction catalyzed by P450 is

$$RH + O_2 + NADPH + H^+ \rightarrow ROH + H_2O + NADP^+$$

where R is an organic substrate. One oxygen atom of the O_2 is incorporated into the substrate and the other is reduced to water, so P450 is both a monooxygenase and mixed-function oxidase. Further reactions and rearrangements lead to more than 40 different types of chemical reactions, including aliphatic and aromatic oxidations, N-, S-, and O-dealkylations, oxidative deamination, and peroxidation.

The P450 reaction cycle starts with the binding of the substrate to the ferric form of P450, which is followed by reduction of the heme iron atom by acceptance of an electron, binding of O_2, activation of the oxygen by acceptance of a second electron, and scission of the oxygen molecule, followed by formation of water, a hydroxylated product, and ferric P450. The two electrons from NADPH are donated either to microsomal P450, through a flavoprotein, P450 reductase, or to mitochondrial and bacterial P450, sequentially through a flavoprotein (a ferredoxin reductase) and an iron-sulfur **ferredoxin**. In rare cases, the P450 and P450 reductase are fused into a single protein (5). Cytochrome b_5 increases the activity of some P450 *in vitro*, either as a source of the second electron or **allosterically** (6).

STRUCTURE

The catalytic domain of membrane-bound P450 is oriented toward the cytoplasm for the microsomal forms and toward the matrix for the mitochondrial forms. The microsomal P450 are inserted into the membrane in a cotranslational, signal recognition particle-dependent manner (7) (see **Protein targeting**). The N-terminal 20 to 25 amino acid residues function as a membrane-insertion and halt-transfer signal and are not cleaved, so they anchor the protein to the membrane. The mitochondrial P450 contain standard cleavable mitochondrial targeting signals at their N-terminus, but the mode of insertion into the inner membrane of the mitochondria is not known.

The three-dimensional structures of soluble bacterial P450, but not any membrane-bound P450, have been determined. Although the primary sequence is dissimilar among these bacterial proteins, a common "P450 fold" is present in each (see **Protein structure**), which suggests that a similar core structure will also be present in eukaryotic membrane-bound P450 (8). The P450 structure contains an **alpha-helical** region and a **beta-sheet** region, which constitute about 70 and 20%, respectively, of the polypeptide chain. The sequence FXXGXXXCXG, which includes the heme-binding **cysteine** residue, is conserved in nearly all P450. Structures of eukaryotic P450 have been predicted from amino acid sequence alignments and homology molecular modeling with bacteria P450 (see **Protein structure prediction**), and predictions of residues that interact with substrates have been generally consistent with the observed effects of mutations (9,10).

GENE STRUCTURE

The sizes of the eukaryotic P450 genes, from the **transcription** initiation site to the **polyadenylation** site, vary dramatically, from about 3 kb for the *CYP21* genes to >70 kb for *CYP19* genes (11). In higher eukaryotes, the number of **introns** varies from 6 to 13, and the sites of insertion of introns are almost always conserved within a family, but only rarely between families. **Polymorphism** is common in P450 genes and underlies genetic differences in drug metabolism and congenital diseases in humans. The families of P450 genes are dispersed among at least nine chromosomes in humans and mice, but occasionally families cluster together, such as families 1, 11, and 19. Subfamilies within a family also may be dispersed to separate chromosomes or may cluster together, and members of the *CYP2A*, *CYP2B*, and *CYP2F* subfamilies are intermingled (4,12). The **gene duplication** leading to the *CYP21A* locus included the C4 **histocompatibility** gene, so the two P450 genes alternate with the C4 gene. In spite of the diversity of P450 genes, their similarity in **primary structures**, the common three-dimensional fold, and some conserved intron sites among P450 genes support the idea that the P450 superfamily evolved by **divergence** from a common ancestral gene.

GENE EXPRESSION

P450 genes are expressed in tissue-, developmental-, and sex-specific patterns. The mechanisms regulating these patterns of expression are best understood in the liver and adrenal gland. An **orphan receptor** containing a **zinc finger**, termed Ad4BP or steroid factor-1 (SF-1), has been shown to be critical for the developmental and tissue-specific expression of P450 genes that are required for the biogenesis of adrenal and sex steroid hormones (13). The SF-1 gene has been disrupted in mice, with major effects on differentiation of the adrenal gland and gonads, pituitary gonadotropin function, and the ventral medial hypothalamus (14), so this gene has important developmental functions as well as effects on the expression of P450 genes. *CYP19*, aromatase, contains SF-1 sites, but its tissue-specific expression is also dependent on alternate **promoters** and transcription initiation sites (15).

All the known liver-enriched regulatory factors have been implicated in the expression of one or more P450 genes. Examples in which functional binding sites for liver-enriched factors have been demonstrated include HNF-1α for *CYP2E1*, HNF-4 for *CYP2C1/2/3*, DBP for *CYP2C6* and *CYP7*, HNF-3 for *CYP2C6*, C/EBPβ for *CYP2D5*, and C/EBPα and C/EBPβ for *CYP2B1/2* (16,17). A variety of ubiquitously expressed factors also contribute to basal expression of the *CYP* genes. The developmental expression of *CYP2C6* correlates with the developmental activation of DBP (16), and the diurnal expression of DBP underlies the diurnal regulation of *CYP7* (18). Activation of *CYP2D5* by C/EBPβ also requires the constitutive factor Sp1, which has a binding site near that of C/EBPβ (16).

Sex-dependent Expression

The sex-dependent expression of several P450 genes in rodents may be mediated by more than one mechanism. In rats, the sex-specific expression of *CYP* genes is dependent on different patterns of **growth hormone** secretion in the two sexes. STAT **transcription factors** (19) and **phospholipase$_{A2}$** (20) may be involved in the growth hormone-regulated male-specific expression of specific *CYP* genes, but additional studies are required to establish the exact mechanism. In mice, a sex difference information (SDI) element is present in male-specific *Cpy2d9* and female-specific *Cyp2a4* (21). The SDI contains a CpG sequence that is **methylated** in a sex-dependent way, and a factor has been identified that binds in a methylation-dependent manner (22). The exact role of this factor is not yet resolved, as its tissue and developmental expression differs from that of the *CYP* genes.

Aromatic Hydrocarbons

The expression of subsets of the xenobiotic-metabolizing P450 genes is regulated by aromatic hydrocarbons, peroxisomal proliferators, and barbiturates. The induction of *CYP1A1* by aromatic hydrocarbons, including the halogenated aromatic hydrocarbon 2,3,7,8-tetrachlorodibenzo-*p*-dioxin, is best understood. Genetic differences in the induction of *Cyp1a1* in mice and in variants of cultured mouse hepatoma Hepa 1 cells contributed critically to the identification of a regulatory locus *AhR*, which encodes the aromatic hydrocarbon receptor, and a second gene, the Ah receptor nuclear translocator *Arnt*, which encodes a protein required for the **nuclear import** of AhR (23,24). The AhR is complexed with the **molecular chaperone** hsp90 in the cytoplasm in uninduced cells, which is essential to maintain a functional AhR. Binding of the ligand results in the dissociation of hsp90 and localization of AhR in the nucleus. Arnt is a nuclear protein that binds to liganded AhR and is required for the binding of AhR to the DNA binding site (25). AhR/Arnt bind to multiple sites within an **enhancer** about 1 kb from the start site of transcription and disrupt the **nucleosomal** structure of the enhancer (23). In turn, the nucleosomal structure of the promoter region is also disrupted in a manner dependent on the **transactivation** domain of AhR, which then allows the binding of regulatory proteins to the proximal promoter. The core sequence of the binding sites, which have been designated xenobiotic-, dioxin-, or Ah-**response elements** by different investigators, is 5′-TNGCGTG-3′. Arnt binds to the core sequence, and AhR binds to the flanking sequence. Both AhR and Arnt contain (1) basic loop-helix-loop (bLHL) motifs near their N-termini; (2) PAS **domains**, which exhibit homology with the *Drosophila* proteins, Per and Sim; and (3) glutamine-rich, putative transcriptional activation domains in the C-terminal region (24). The ligand and hsp90 interact with sequences near the PAS domain of AhR, and the binding of AhR to Arnt involves the bHLH and PAS domains of each protein. Arnt is capable of transactivation independently, but AhR transactivation requires the association of AhR with Arnt.

Peroxisomal Proliferators

Induction of expression of the *CYP4A* genes by peroxisomal proliferators and fatty acids requires peroxisomal proliferator-activated receptor-α (PPARα), a member of the steroid **hormone receptor** family (26). CYP4A P450 catalyze the ω-hydroxylation of fatty acids, including arachidonic acid, which may ultimately result in the degradation of the fatty acids by peroxisomes. The induction of *CYP4A* genes is defective in mice with disrupted PPARα genes (27). PPARα forms a heterodimer with retinoid-X-receptor (RXR) to activate the gene. The predominant sequence in *Cyp4A6* that confers the response to peroxisomal proliferators contains an imperfect repeat of the sequence AGGTCA, with a single base pair between the repeats and additional sequence to the 5′ side of the repeat. The 5′ sequence appears to be important for the specificity of binding of PPARα/RXRα to the nonconsensus, imperfect repeat of the peroxisomal proliferator response element present in *CYP4A6* (26).

Barbiturates

Induction of P450 genes by barbiturates is not well understood. Neither a barbiturate receptor nor specific barbiturate response elements in mammalian genes have been defined, and different mechanisms may be involved for different P450 (28). The system best understood is the induction by barbiturates of Bm-1, Bm-2, and Bm-3 P450 in ***Bacillus*** *megaterium* (29). In untreated cells, a repressor, Bm3R1, binds to a palindromic 20-bp sequence upstream of a promoter that drives the expression of both the repressor and Bm-3. In the presence of barbiturate, the binding is reversed, which may result from direct effects on the repressor, but positively acting factors are also induced that compete for the binding of the repressor at the palindromic site and at sites termed "barbie boxes" present in the 5′ regions of the Bm-3 and Bm-1 genes. In the mammalian *CYP2B1/2* genes, evidence has been presented for phenobarbital responsive sequences in both the proximal promoter region, including barbie-box-like sequences (30), and in distal phenobarbital-responsive enhancers (31,32). Identification of specific responsive elements and characterization of the cognate binding proteins will be required to establish the role of these regions in phenobarbital induction.

Glucocorticoids

A number of *CYP* genes are also modulated by **glucocorticoids**. In some cases, the effect is mediated through the classical glucocorticoid receptors, such as modulation by glucocorticoids of aromatic hydrocarbon induction of *CYP1A1* and phenobarbital induction of *CYP2B* genes (33). In other cases, such as *CYP3A* genes, the response to glucocorticoids occurs by nonclassical mechanisms that remain to be elucidated (34).

Cyclic AMP

Genes for P450 involved in steroid metabolism are induced by polypeptide hormones via **cyclic AMP**-mediated (cAMP) mechanisms (35). In the adrenal gland, *CYP11A*, *CYP11B*, *CYP17*, and *CYP21* are regulated in this way. Interestingly, even though each of these genes probably evolved from a common ancestor, different regulatory elements in the promoter of each gene mediate the cAMP response. A classical CREB binding site is present in *CYP11B*, but binding sites for proteins other than CREB are present in other genes, and different sites are present in ***orthologous*** *genes* in different species. In bovine *CYP11A*, a binding site for Sp1, a constitutive factor, mediates the cAMP response (36), and members of the **homeodomain** PBX family of **helix-turn-helix motif** proteins bind to the cAMP-responsive element in *CYP17* (37). Multiple proteins bind to many of these elements, and those that predominate *in vivo* remain to be identified.

BIBLIOGRAPHY

1. M. J. Coon, A. D. N. Vaz, and L. L. Bestervelt (1996) *FASEB J.* **10**, 428–434.
2. H. C. L. Liang et al. (1996) *Proc. Natl. Acad. Sci. USA* **93**, 1671–1676.
3. S. S. T. Lee, J. T. M. Buters, T. Pineau, P. Fernandez-Salguero, and F. J. Gonzalez (1996) *J. Biol. Chem.* **271**, 12063–12067.
4. D. R. Nelson et al. (1996) *Pharmacogenetics* **6**, 1–42.
5. L. O. Narhi and A. J. Fulco (1987) *J. Biol. Chem.* **262**, 6683–6690.
6. H. Yamazaki, W. W. Johnson, Y.-F. Ueng, T. Shimada, and F. P. Guengerich (1996) *J. Biol. Chem.* **271**, 27438–27445.
7. B. Kemper and E. Szczesna-Skorupa (1989) *Drug Metabol. Rev.* **20**, 811–820.
8. C. A. Hasemann, R. G. Kurumbail, S. S. Boddupalli, J. A. Peterson, and J. Deisenhofer (1995) *Structure* **2**, 41–62.
9. O. Gotoh (1992) *J. Biol. Chem.* **267**, 83–90.
10. C. von Wachenfeldt and E. F. Johnson (1995) In *Cytochrome P450—Structure, Mechanism, and Biochemistry* (P. Ortiz de Montellano, ed.), Plenum Press, New York, pp. 183–223.
11. B. Kemper (1993) In *Frontiers in Biotransformation* (K. Ruckpaul and H. Rein, eds.), Vol. 8, Akkademie Verlag, Berlin, pp. 1–58.
12. S. M. G. Hoffman, P. Fernandez-Salguero, F. J. Gonzalez, and H. W. Mohrenweiser (1995) *J. Mol. Evol.* **41**, 894–900.
13. D. S. Lala, D. A. Rice, and K. L. Parker (1992) *Mol. Endocrinol.* **6**, 1249–1258.
14. K. L. Parker and B. P. Schimmer (1996) *Trends Endo. Metab.* **7**, 203–207.
15. M. S. Mahendroo, C. R. Mendelson, and E. R. Simpson (1993) *J. Biol. Chem.* **268**, 19463–19471.
16. F. J. Gonzalez and Y.-H. Lee (1996) *FASEB J.* **10**, 1112–1118.
17. Y. Park and B. Kemper (1996) *DNA Cell Biol.* **15**, 693–701.
18. D. J. Lavery and U. Schibler (1993) *Genes Develop.* **7**, 1871–1884.
19. D. J. Waxman, P. A. Ram, S. H. Park, and H. K. Choi (1995) *J. Biol. Chem.* **270**, 13262–13270.
20. P. Tollet, M. Hamberg, J. Å. Gustafsson, and A. Mode (1995) *J. Biol. Chem.* **270**, 12569–12577.
21. H. Yoshioka, M. Lang, G. Wong, and M. Negishi (1990) *J. Biol. Chem.* **265**, 14612–14617.
22. N. Yokomori, R. Kobayashi, R. Moore, T. Sueyoshi, and M. Negishi (1995) *Mol. Cell Biol.* **15**, 5355–5362.
23. J. P. Whitlock Jr. et al. (1996) *FASEB J.* **10**, 809–818.
24. O. Hankinson (1995) *Ann. Rev. Pharmacol. Toxicol.* **35**, 307–340.
25. R. S. Pollenz, C. A. Sattler, and A. Poland (1993) *Mol. Pharmacol.* **45**, 428–438.
26. E. F. Johnson, C. N. A. Palmer, K. J. Griffin, and M.-H. Hsu (1996) *FASEB J.* **10**, 1241–1249.
27. S. S.-T. Lee et al. (1995) *Mol. Cell Biol.* **15**, 3012–3022.
28. D. J. Waxman and L. Azaroff (1992) *Biochem. J.* **281**, 577–592.
29. Q. Liang and A. J. Fulco (1995) *J. Biol. Chem.* **270**, 18606–18615.
30. L. Prabhu et al. (1995) *Proc. Natl. Acad. Sci. USA* **92**, 9628–9632.
31. E. Trottier, A. Belzil, C. Stoltz, and A. Anderson (1995) *Gene* **158**, 263–268.
32. Y. Park, H. Li, and B. Kemper (1996) *J. Biol. Chem.* **271**, 23725–23728.
33. R. A. Prough, M. W. Linder, J. A. Pinaire, G.-H. Xiao, and K. C. Falkner (1996) *FASEB J.* **10**, 1369–1377.
34. L. C. Quattrochi, A. S. Mills, J. L. Barwick, C. B. Yockey, and P. S. Guzelian (1995) *J. Biol. Chem.* **270**, 28917–28924.
35. M. Waterman (1994) *J. Biol. Chem.* **269**, 27783–27786.
36. P. Venepally and M. Waterman (1995) *J. Biol. Chem.* **270**, 25402–25411.
37. N. Kagawa, A. Ogo, Y. Takahashi, A. Iwamatsu, and M. R. Waterman (1994) *J. Biol. Chem.* **269**, 18716–18719.

Suggestions for Further Reading

FASEB Journal series of 12 reviews, *"The cytochrome P450: structure, function, regulation, and genetics"* (1996/97) Initial review in *FASEB J.* (1996) **10**, 205. This series presents four reviews each on the structures of P450, regulation of xenobiotic-metabolizing P450, and regulation of P450 that metabolize endogenous compounds.

C. Ioannides, ed. (1996) *Cytochromes P450—Metabolic and Toxicological Aspects*, CRC Press, New York. Comprehensive reviews of the drug metabolizing P450, primarily families 1–4.

T. Omura, Y. Ishimura, and Y. Fujii-Kuriyama, eds. (1993) *Cytochrome P450*, VCH, New York. A concise introduction to P450.

P. R. Ortiz de Mantellano, ed. (1995) *Cytochrome P450—Structure Mechanism, and Biochemistry*, 2nd ed., Plenum Press, New York. Excellent comprehensive reviews of P450, including chapters on the regulation of gene expression by xenobiotics, hormones, and cAMP-mediated mechanisms.

M. A. Schuler (1996) Plant cytochrome P450 monooxygenases. *Crit. Rev. Plant Sci.* **15**, 235–284. A comprehensive review of plant P450.

CYTOCHROMES

MICHAEL A. CUSANOVICH

Cytochromes are iron-containing proteins that facilitate the movement of electrons in a wide variety of metabolic processes (see **Electron transfer proteins**). Cytochromes are widely distributed throughout nature. They are found in all **eukaryotes** and in most, but not all, **prokaryotes**. The prosthetic group or chromophore of cytochromes is heme, an iron-containing porphyrin. The prototypic form of heme in cytochromes is protoheme IX, which consists of iron, two vinyl side chains, four methyl groups and two propionic acid side chains on a conjugated tetrapyrole ring (Fig. 1). Protoheme IX is the prosthetic group of *b*-type cytochromes and the family of proteins known as **cytochrome P450**. The *c*-type cytochromes have heme *c* as the prosthetic group, in which protoheme IX is covalently bound to the protein through one or, more commonly, two thioether bonds to **cysteine** side-chains. Heme *a*, found in *a*-type cytochromes, has a long isoprenoid tail substituted on one of the vinyl groups and a formyl group replacing a methyl. Further variations in heme structure are found in isolated cases, for example, cytochrome *d* and cytochrome *o* (1) but will not be discussed here. Cytochrome P450 and **cytochrome *c*** are discussed elsewhere in this volume.

The iron in cytochromes, is either penta- or hexa-coordinate. Porphyrin nitrogen atoms fill four coordination positions (in-plane coordination, see Fig. 1) and appropriate ligands are positioned below the heme plane (5-coordinate) and, in most cases also above the heme plane (6-coordinate). The out-of-plane ligands in typical *c*-type cytochromes, are the N_3 atoms of **histidine** residues and the sulfur atom of **methionine** (Fig. 2). Ligation in the *b*-type cytochromes is generally by two His residues (Fig. 2) and by one or two His in the cytochromes *a*. Variations exist, however. For example, the out-of-plane ligation is by two His residues in cytochrome c_3, and by one

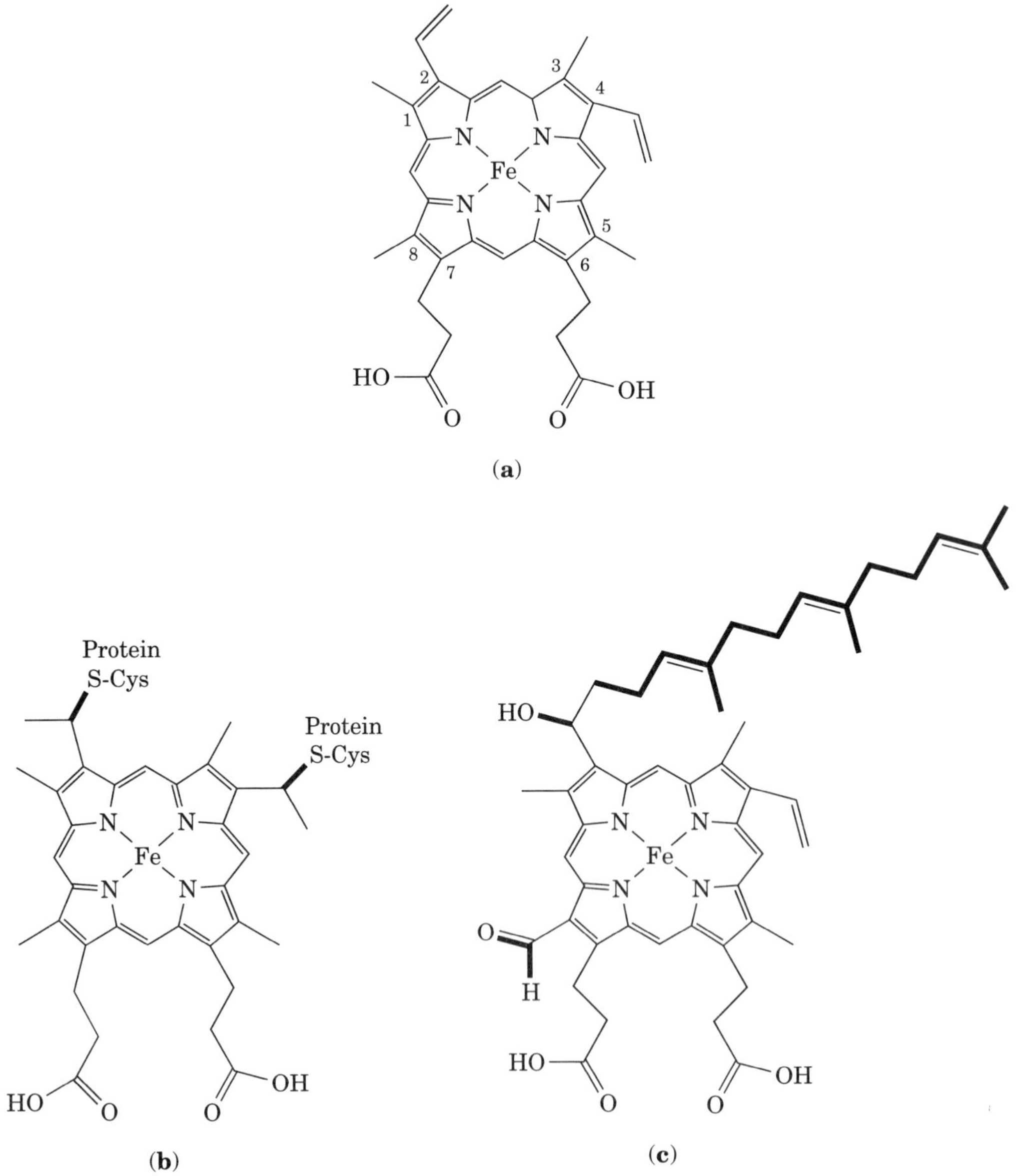

Figure 1. The structures of (**a**) protoheme IX, (**b**) heme *c*, and (**c**) heme *a*.

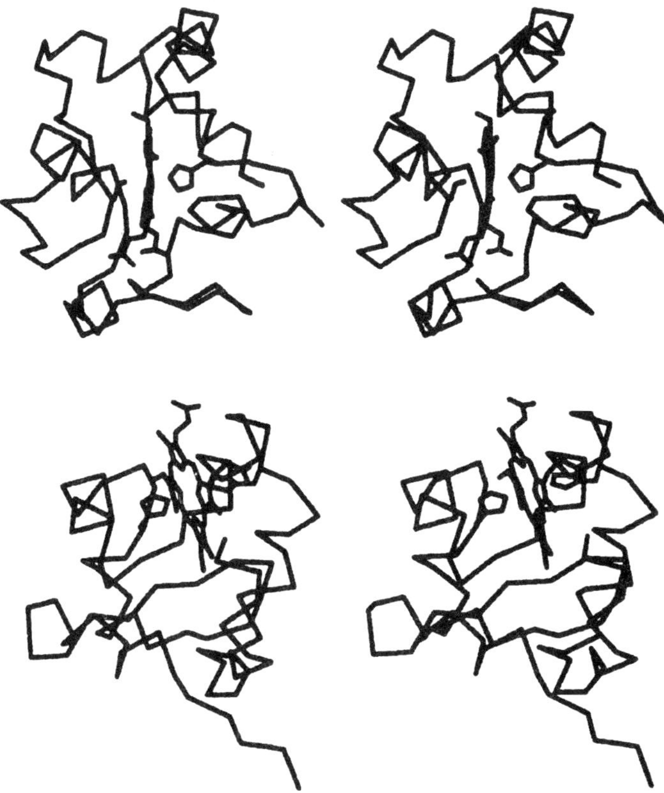

Figure 2. Stereo representations of the structure of tuna cytochrome *c* (upper stereo pair) and microsomal cytochrome b_5 (lower stereo pair). In cytochrome *c*, the N-terminal α-helix is at the *top*, the Met80 ligand is on the *left*, and the His18 ligand on the *right*. In cytochrome b_5, the N-terminus is at the lower *right* the His61 ligand is on the *left*, and the His37 ligand is on the right. The heme propionates are oriented toward the solvent.

His and an open position in cytochrome *c′*. Cytochromes are associated with the inner membrane of **mitochondria**, that of **microsomes**, the **thylakoid** membrane in **chloroplasts**, and the plasma membrane in **bacteria** (see **Membranes**). However, soluble forms (typically *c*-type cytochromes) are found in the space between the inner and outer mitochondrial membranes and in the **periplasmic** space in bacteria.

The physiological activity in all cytochromes is the reversible oxidation and reduction of the iron atom, which cycles between the ferric (Fe^{3+}) and ferrous (Fe^{2+}) states. From a physiological standpoint, an important property of cytochromes is their oxidation-reduction **(redox)** potential, which is E_0' at pH 7. The redox potential measures the relative stabilities of the oxidized (ferric) and reduced (ferrous) forms and therefore their tendencies to donate and accept electrons, respectively. At pH 7 in aqueous solution, the redox potential of H_2 is −414 mV and that of O_2 is +816 mV. With relatively minor exceptions, these define the biological range. Cytochromes have redox potentials somewhere between these limits. In biological electron transfer, the direction of electron flow is from lower to higher potentials in the limit from hydrogen to oxygen. Thus the redox potential of a particular cytochrome determines, to a first approximation, where it is positioned relative to other redox compounds in electron transfer pathways (see **Electron transfer proteins**). The *c*-type cytochromes have redox potentials ranging from −400 to +450 mV, depending on the structural family and species, although they typically vary between only +200 and +300 mV in eukaryotic *c*-type cytochromes. The *b*-type cytochromes have redox potentials that range from −200 to +150 mV, and the *a*-type cytochromes have potentials in the +220 to +400 mV range.

The role of the protein moiety in cytochromes is threefold. First, the protein provides the out-of-plane ligands and, in the case of *c*-type cytochromes, the covalent attachment sites for the heme. Secondly, the protein provides the proximal environment for the heme, a nonpolar molecule that is not water-soluble, controls the exposure of the heme to the solvent, and interacts with the heme to modulate the redox potential. Thirdly, the protein surface provides the sites for the interaction with electron donors and acceptors because in all known cases the heme iron is buried in the protein's interior. Thus, the redox potential is modulated by a combination of the nature of the out-of-plane ligands, the immediate protein environment, and the solvent exposure of the heme.

The redox potentials dictate the direction of electron flow. For example, mitochondrial cytochrome *c* ($E_0' = +260$ mV) cannot significantly reduce oxidized pyridine nucleotide (**NAD**$^+$, $E_0' = -320$ mV). They do not, however, explain the specificity and efficiency of biological electron transfer. Within a biological milieu, a large number of redox compounds exist close to each other. These include small organic compounds, such as flavins and pyridine nucleotides (for example, NAD^+); oxygen (in aerobes); inorganics, such as nitrates, nitrites, sulfates, and sulfhydryls; and a variety of redox proteins, including non-heme-iron sulfur proteins, flavoproteins, cytochromes, and copper proteins (see **Electron transfer proteins**). Thus, from an energetic standpoint, substantial opportunities exist for electrons to flow in a great variety of directions. This does not happen, and electron flow occurs with very high efficiency (approaching 100%) only within specific pathways (for example, in respiration or **photosynthesis**). This incredible efficiency, the essence of life, is mediated by the surface of the protein, which provides recognition sites that determine which redox compounds interact on a physiological timescale, that is, it determines how fast electrons are transferred, and the direction of electron flow (2,3). This kinetic control means that energetically equivalent reactions can take place on timescales that differ by many orders of magnitude. For example, reduced pyridine nucleotide reduces mitochondrial cytochrome *c* slowly (in seconds), whereas cytochrome c_1 ($E_0' = +220$ mV) reduces the same cytochrome in milliseconds, even though reduction by pyridine nucleotide is more strongly favored energetically in the first case (65.7 kJ/mol versus 3.8 kJ/mol).

The interactive domains of the cytochromes are defined by the amino acid side chains positioned on their molecular surface that dictate their sites of interaction. The structure of the complex between two electron transfer proteins determines the proximity and orientation of the relevant prosthetic groups (the two hemes in the case of electron transfer between two cytochromes). Three factors play a role in defining the interactive domains. First, the positioning of charged amino acid side chains on the surface of interacting molecules result in electrostatic interactions, either attractive or repulsive, that define the interaction domain (4). Secondly, the distance between redox centers (heme irons in the case of cytochromes) influences the rate of electron transfer. This distance is controlled by the size and nature of the amino acid side chains in the interactive domain (2). Third, two interactive domains that interact must be complementary physically and structurally, even if defined by nonpolar **van der Waals interactions** (2,3). Apparently, during the course of evolution, mutations have occurred that optimize the rate of electron flow through relevant metabolic pathways by modulating the amino acid side chains defining the interactive domains, thus allowing competing redox reactions to occur in close proximity and channeling the flow of electrons through the appropriate pathway.

From the standpoint of solubility, cytochromes exist in three types of environments. Soluble forms are typically *c*-type cytochromes that are free to diffuse between electron donors and acceptors. Integral **membrane proteins**, that is, transmembrane proteins, have only part of the molecule exposed to solvent. The majority of the molecule is integrated into the membrane. Examples are *a*-type cytochromes and some *b*-type cytochromes. Peripheral membrane proteins are strongly associated with a membrane or with integral membrane proteins and have a substantial portion of the molecule in solution. Examples are cytochrome b_5 and cytochromes c_1 and *f*.

A-TYPE CYTOCHROMES

Cytochromes *a* and a_3 are components of cytochrome oxidase, or complex IV, of the mitochondrial electron transfer chain and are found in all eukaryotic organisms and many aerobic prokaryotes. Cytochrome oxidase catalyzes the four-electron reduction of O_2 to H_2O, utilizing cytochrome *c* as the electron donor. Cytochrome *a*, which is hexa-coordinate, mediates the transfer of electrons from cytochrome *c* to the penta-coordinate cytochrome a_3 (with the intermediate participation of a copper atom; see **Copper-binding proteins**), which provides the site for oxygen reduction. Variants of cytochrome oxidase are found in some aerobic bacteria that do not contain hemes other than heme *a*. Examples are cytochrome *bo*, which uses quinols as the electron donor in *Escherichia coli* and cytochrome cbb_3 in *Rhodobacter sphaeroides*, which has cytochrome *c* as an electron donor (1). Thus, the heme type can vary, although the function, the reduction of molecular oxygen to water, remains constant. Cytochrome oxidase is an integral membrane protein and couples the transfer of electrons and the corresponding redox energy with the movement of protons across the inner mitochondrial or periplasmic membrane, driving the synthesis of ATP (see **Proton motive force**). From an evolutionary standpoint, there is structural and amino acid sequence homology between eukaryotic and bacterial cytochrome oxidase (1). In addition, a variety of other enzymes exist that are structurally related to cytochrome oxidase and generally use cytochrome *c* or an electron donor, such as nitrous oxide reductase, which reduces N_2O to N_2 and utilizes copper as the prosthetic group, but has sequence homology with cytochrome oxidase and utilizes cytochrome *c* as its electron donor (1). Another example is nitric oxide reductase, which catalyzes the reduction of **nitric oxide** to nitrous oxide. Although it is a cytochrome *bc* complex, it has a transmembrane component and homology to portions of cytochrome oxidase (1).

B-TYPE CYTOCHROMES

The *b*-type cytochromes fall into two broad categories: (1) a family of relatively large cytochromes containing protoheme IX that are part of integral membrane protein complexes like cytochrome bc_1 and cytochrome b_6f; (2) a family of cytochrome b_5-like proteins that are peripheral proteins, typically anchored to a membrane, or are components of protein complexes like flavocytochrome b_2 and sulfite oxidase. The cytochrome bc_1 complex and its homologue in plants, cytochrome b_6f, catalyze the transfer of electrons from quinones to cytochrome *c* (in animals) or to plastocyanin (in plants) coupled with the movement of protons across the inner mitochondrial or the thylakoid membranes, which drives the synthesis of ATP (see **ATP synthase**). The mitochondrial cytochrome bc_1 complex, also known as complex III, has the cytochrome *b* component integrated into the membrane through a number of transmembrane **α-helices** (5). The cytochrome c_1 component is anchored to the membrane and is the electron donor to cytochrome *c*. Cytochrome *b* is also a component of succinate dehydrogenase (also known as complex II), a complex integral membrane protein that participates in the mitochondrial electron transfer pathway (6). Succinate dehydrogenase oxidizes succinate to fumarate. The electrons reduce quinone and feed into the mitochondrial electron transfer chain.

Because of their large size and strong association with membranes, cytochromes *b* are not as well understood as the generally small and water-soluble *c*-type cytochromes (cytochromes c_1 and *f* are exceptions in terms of size and solubility). Thus the extent of structural and functional diversity among the cytochromes *b*, particularly in prokaryotes, remains an open question.

Proteins of the cytochrome b_5 family contain a cytochrome b_5 domain (Fig. 2) and, in the case of the mitochondrial and microsomal members, a hydrophobic tail or anchor that associates the cytochrome with the relevant membrane (7). The structure of cytochrome b_5 is shown in Fig. 2. In nitrate reductase, sulfite oxidase, and flavocytochromes b_2, the cytochrome b_5 domain is tightly associated with a flavoprotein that specifies the particular enzymatic activity. The cytochromes b_5 domain acts as an electron acceptor from the molybdenum (Mo)-pterin cofactor in sulfite oxidase or from protein-bound flavin in cytochrome b_5 reductase, nitrate reductase, P450 reductase, and flavocytochrome b_2. The cytochrome b_5 electron acceptor is cytochrome *c* in the flavocytochrome b_2 and sulfite oxidase systems, and a Mo-pterin cofactor in nitrate reductase. In microsomal systems, cytochrome P450 and a variety of acyl-CoA desaturases are the electron acceptors. In red blood cells, cytochrome b_5 reduces methemoglobin (see **Hemoglobin**). Thus, cytochrome b_5 has evolved the ability to function in a diverse group of electron transfer pathways and with a variety of electron donors and acceptors (7).

C-TYPE CYTOCHROMES

These cytochromes are discussed in more detail elsewhere in this encyclopedia (see **Cytochrome c**), but it is important to note that this is a diverse group of cytochromes consisting of a variety of structural motifs and functions (8). Although restricted to mitochondrial cytochrome *c* and cytochrome c_1 (discussed previously) in eukaryotes and key components in the respiratory electron transfer chain, a variety of bacterial *c*-type cytochromes have evolved into a diverse and complex array of structures and function. In terms of structural families, the cytochromes *c* include the mitochondrial cytochrome *c* family and their bacterial homologies, cytochrome c_1 and *f*, and the cytochromes c_3, cytochromes c', flavocytochromes *c*, cytochrome c_4, and cytochromes c_5. Now, only cytochrome c_3 and *c*-type cytochromes that are related to mitochondrial cytochrome *c* are well understood in both structure and function.

The basic structural motif of the mitochondrial cytochrome *c* family has been maintained throughout evolution. Figure 2 presents the structure of tuna cytochrome *c*. In contrast to cytochrome b_5, the heme propionates are buried and not solvent accessible. Cytochromes utilizing the cytochrome *c* structural motif are found in almost all bacteria but differ primarily in their redox potentials, which vary from approximately +50 to +450 mV, and in the type and distribution of amino acid side chains on their molecular surfaces.

SUMMARY

Cytochromes evolved early in the development of life on this planet, and they have retained the basic functional group protoheme IX (with relatively minor variations) throughout evolution. A number of basic structural motifs have evolved that are found in both prokaryotes and eukaryotes, including the mitochondrial cytochromes *c* and their bacterial precursors, the cytochrome b_5 family, the cytochrome oxidases, the membrane-spanning cytochromes *b*, and the cytochrome c_1/b_6 family. It is striking that a relatively small number of structural families have evolved to take advantage of diverse metabolic opportunities, capturing redox energy to produce ATP and adapting to available ecological niches.

BIBLIOGRAPHY

1. M. Saraste, J. Castresana, D. Higgins, M. Lübben, and M. Wilmanns (1996) In *Origin and Evolution of Biological Energy Conversion* (H. Baltscheffsky, ed.) VCH, New York, pp. 255–289.
2. G. Tollin, T. E. Meyer, and M. A. Cusanovich (1986) *Biochim. Biophys. Acta* **853**, 29–41.
3. T. E. Meyer, G. Tollin, and M. A. Cusanovich (1994) *Biochimie* **76**, 480–488.
4. J. A. Watkins, T. E. Meyer, G. Tollin, and M. A. Cusanovich (1994) *Protein Sci.* **3**, 2104–2114.
5. P. N. Furbacker, G.-S. Tae, and W. A. Cramer (1996) In *Origin and Evolution of Biological Energy Conversion* (H. Baltscheffsky, ed.), VCH, New York, pp. 221–253.
6. B. A. C. Ackrell, M. K. Johnson, R. P. Gunsalus, and G. Cecchini (1992) In *Chemistry and Biochemistry of Flavoenzymes*, Vol. III (F. Müller, ed.), CRC Press, Boca Raton, pp. 229–297.
7. F. Lederer (1994) *Biochimie* **76**, 674–692.
8. T. E. Meyer, J. J. van Beeumen, R. P. Ambler, and M. A. Cusanovich (1996) In *Origin and Evolution of Biological Energy Conversion* (H. Baltscheffsky, ed.), VCH, New York, pp. 71–108.

Suggestions for Further Reading

T. E. Meyer and M. A. Cusanovich (1989) Structure, function and distribution of soluble bacterial redox proteins, *Biochim. Biophys. Acta* **975**, 1–28.

R. H. Scott and A. G. Mauk, eds. (1996) *Cytochrome* c. *A. Multidisciplinary Approach*, University Science Books, Mill Valley.

G. R. Moore and G. W. Pettigrew (1990) *Cytochromes* c, Springer-Verlag, New York.

CYTOGENETICS

ALAN P. WOLFFE

Cytogenetics is the study of **genetics** by visualization of **chromosomes** and chromosomal aberrations, usually through light **microscopy**. There is a distinguished history of chromosomal visualization for analytical purposes. Longley defined the morphology of **maize** chromosomes (reviewed in Ref. 1), which opened the door for subsequent work on maize cytogenetics by McClintock (2). Staining of **prophase** and early **metaphase** maize chromosomes with acetocarmine allowed McClintock to identify 10 distinct linkage groups. In turn this laid the foundation for other maize cytogeneticists to identify chromosomal **translocations** and **inversions** visually. For example, it was possible to demonstrate that there is a correspondence between marker genes and cytologically defined chromosomal domains in crosses between heterozygous strains (3). Cytogenetics in *Drosophila melanogaster* received an enormous stimulus from the systematic study of the **polytene chromosomes** in salivary glands (4). These chromosomes have visibly distinct banding patterns under a phase microscope and after staining. Each of the four *Drosophila melanogaster* chromosomes have easily differentiated band patterns. The thousands of bands served as markers that allowed cytogeneticists to discern small deletions, duplications, translocations, and transpositions. Cytological experiments in *Drosophila* also allowed the discovery of the **position effect**, the phenomenon in which the position of a gene relative to heterochromatin influences the efficiency with which it is expressed (5).

Improvements in cytogenetics eventually allowed visualizing metaphase chromosomes in somatic cells in culture. Only in 1956 was it established by these methodologies that human cells contain 46 chromosomes (6). Likewise cytogenetic approaches allowed the discovery that children with Down's syndrome are **trisomic** for chromosome 21 (see **Autosome**). Medical cytogenetics has since been advanced by chromosome stains, such as Giemsa, that enable separating individual chromosomes in sufficient numbers for highly focused analysis of their biochemistry and genetics (see **G-banding**, **Giemsa banding**).

BIBLIOGRAPHY

1. A. E. Longley (1952) *Bot. Rev.* **18**, 399–412.
2. B. McClintock (1929) *Science* **69**, 629–632.
3. H. B. Creighton and B. McClintock (1931) *Proc. Natl. Acad. Sci. USA* **17**, 485–491.
4. E. Heitz and H. Bauer (1933) *Z. Zellforsch Mikrosc. Anat.* **17**, 67–81.
5. A. H. Sturtevant (1925) *Genetics* **10**, 117–147.
6. J. H. Tijo and A. Lean (1956) *Hereditas* **42**, 1–6.

CYTOKERATINS

DAVID PARRY

Cytokeratins form long filaments (they are one of the family of **intermediate filament** proteins; see also **Keratins**) that extend in a sinuous manner between the cytoplasmic surface of the cell **nucleus** and the plasma membrane of epithelial cells. The cytokeratin filaments appear to attach to the cytoplasmic side of the **nuclear envelope**, often close to the **nuclear pore complexes**. From there they radiate outward and link to or loop through the portions of the **desmosomes** or adhesion plaques that lie on the inner side of the cell membrane.

Cytokeratin chains can be divided into two groups: "acidic" Type I and "neutral-basic" Type II. In general, the Type II chains have higher **isoelectric points** and higher molecular weights than their Type I counterparts. Each chain comprises nonhelical end **domains** separated by a flexible rodlike structure. The *N*- and *C*-terminal domains in the epidermal keratins (or cytokeratins) have a characteristic substructure arranged about the rod domain with bilateral symmetry: E1, V1, H1 in the *N*-terminal domain and H2, V2, E2 in the *C*-terminal domain (1). The H1 and H2 subdomains, located respectively immediately *N*- and *C*-terminal to the rod domain, have sequences that are highly **homologous** regions within each chain type. The V1 and V2 subdomains are variable in length and sequence but always have a high content of **glycine** and **serine** residues. The E1 and E2 subdomains lie at the extreme ends of the molecule and are short, generally basic regions.

Between the terminal domains, the amino acid sequence has a **heptad repeat** substructure characteristic of a **coiled-coil** conformation (2). The latter thus consists essentially of coiled-coil rod domains (segments 1A, 1B, 2A, and 2B), joined by short linking domains (L1, L12, and L2), and results in a rod of total length 46.2 nm (Fig. 1). Each molecule contains a pair of different chains (Type I and Type II) aligned parallel to one another and in axial register (3–5). Segments 1B and 2

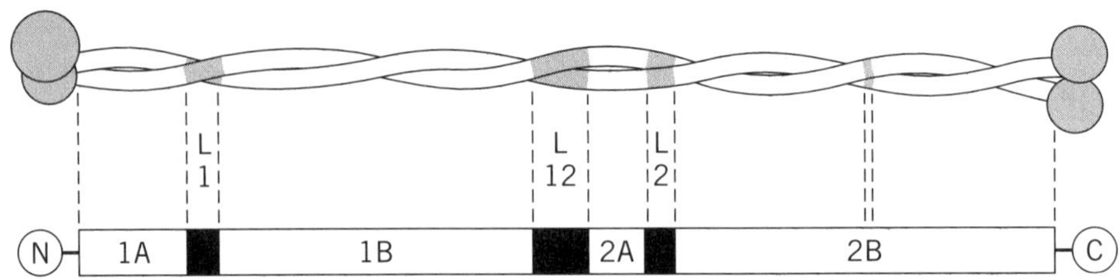

Figure 1. Schematic cytokeratin intermediate filament molecule. The rod domain is of approximate length 46 nm, with coiled-coil segments 1A, 1B, 2A, and 2B and non-coiled-coil links L1, L12, and L2. A break in the phasing of the heptad repeat, known as a stutter, occurs at the center of segment 2B. The *N*- and *C*-terminal domains contain E (end), V (variable but rich in glycine and serine residues), and H (homologous) subdomains arranged about the rod domain with bilateral symmetry. (From Ref. 8, with permission.)

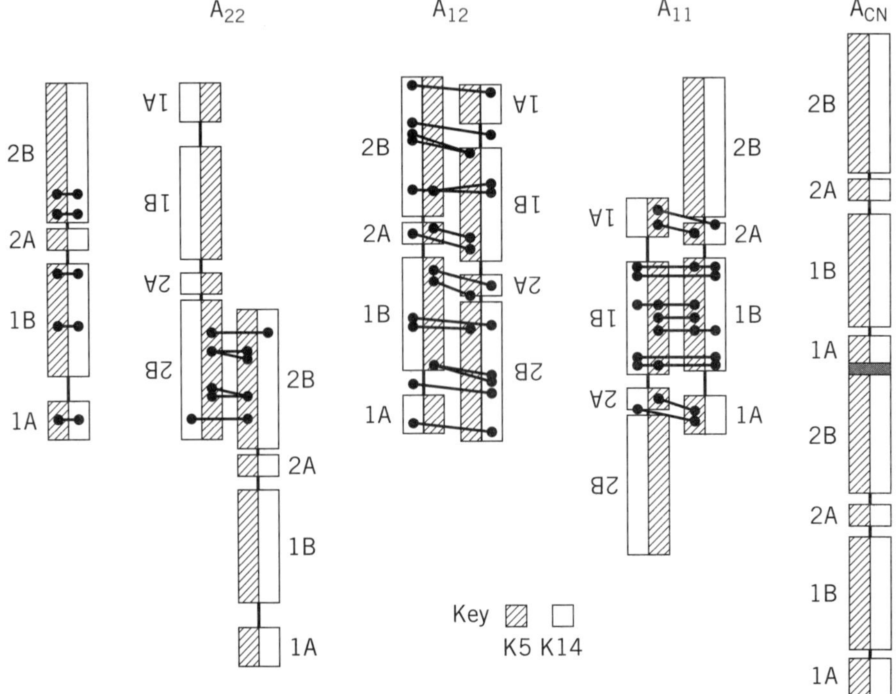

Figure 2. The cross-links induced chemically between lysine residues are shown for human K5/K14 epidermal keratin IF. The cross-links between the two polypeptide chains within a molecule (*left*) show that the chains are parallel and in axial register. Cross-links between molecules indicate that the molecules aggregate antiparallel through one of three modes, with (i) their 2B segments largely overlapped (A_{22}), (ii) the entire molecules largely overlapped (A_{12}), and/or (iii) their 1B segments largely overlapped (A_{11}). Combination of A_{22} and A_{11} results in a small head-to-tail overlap of about 1 nm between parallel molecules. (From Ref. 7, with permission.)

(= 2A + L2 + 2B) each contain highly regular distributions of both acidic and basic residues (2). Thus each rod segment can be modeled as alternating bands of positive and negative charge, with a period of about 1.4 to 1.5 nm. Pairs of molecules aggregate through their rod domains in three antiparallel modes: (i) with their 1B segments largely overlapped, (ii) with their 2B segments largely overlapped, and (iii) with their entire lengths largely overlapped. In cytokeratins there is also a short head-to-tail overlap of about eight residues between parallel molecules (6). The antiparallel interactions originate in large part by the maximization of **electrostatic interactions**, which are possible as a consequence of the regular charge distribution in neighboring rod domains. They are also both specified and stabilized by interactions between the H1 (and presumably H2) subdomains, the head-to-tail overlap region, and the link L2. Cross-links formed between **lysine** residues in cytokeratins have been characterized (Fig. 2), thus confirming the modes of aggregation originally proposed on theoretical grounds (see earlier). This work also allowed the projected (axial) lengths of the links L1, L12, and L2 to be determined, together with the relative axial staggers between various coiled-coil rod segments (6,7). These data allow a surface lattice structure to be constructed that predicts a near-axial period in the cytokeratin of 22.6 nm (the experimentally observed value is 22.7 nm). **Scanning transmission electron microscope** (STEM) data indicate that *in vitro* assemblies may contain about 16, 24, 32, or 40 chains, but it seems that only the 32-chain variant is present *in vivo*. The STEM data, nonetheless, would indicate that the cytokeratin may have a substructure in which an eight-chain entity, a protofibril, is present. *In vitro* two, three, four, or five of these protofibrils may aggregate in a nonspecific manner.

Characterization of a large number of gene mutations leading to keratinopathies has been achieved in recent years. The resulting sequence changes lie very largely in or close to the head-to-tail overlap region comprising the conserved sequences at the *N*-terminal end of segment 1A and the *C*-terminal end of segment 2B. A reasonable conclusion is that this interaction is subtly modified by the mutated residue and becomes less able to stabilize the cytokeratin *in vivo*.

BIBLIOGRAPHY

1. P. M. Steinert, D. A. D. Parry, W. W. Idler, L. D. Johnson, A. C. Steven, and D. R. Roop (1985) Amino acid sequences of mouse and human epidermal type II keratins of M_r 67000 provide a systematic basis for the structural and functional diversity of the end domains of keratin intermediate filament subunits. *J. Biol. Chem.* **260**, 7142–7149.
2. W. G. Crewther, L. M. Dowling, P. M. Steinert, and D. A. D. Parry (1983) Structure of intermediate filaments. *Int. J. Biol. Macromol.* **5**, 267–274.
3. P. M. Steinert (1990) The two-chain coiled-coil molecule of native epidermal keratin intermediate filaments is a type I–type II heterodimer. *J. Biol. Chem.* **265**, 8766–8774.
4. M. Hatzfeld and K. Weber (1990) The coiled-coil of *in vitro* assembled keratin filaments is a heterodimer of type I and type II keratin: use of site-specific mutagenesis and recombinant protein expression. *J. Cell Biol.* **110**, 1199–1210.
5. P. A. Coulombe and E. Fuchs (1990) Elucidating the early stages of keratin filament assembly. *J. Cell Biol.* **111**, 153–169.
6. P. M. Steinert, L. N. Marekov, R. D. B. Fraser, and D. A. D. Parry (1993) Keratin intermediate filament structure: crosslinking studies yield quantitative information on molecular dimensions and mechanism of assembly. *J. Mol. Biol.* **230**, 436–452.
7. P. M. Steinert, L. N. Marekov, and D. A. D. Parry (1993) Conservation of the structure of keratin intermediate filaments: molecular mechanism by which different keratin molecules integrate into pre-existing keratin intermediate filaments during differentiation. *Biochemistry* **32**, 10046–10056.
8. D. A. D. Parry and R. D. B. Fraser (1985) Intermediate filament structure: 1, Analysis of IF protein sequence data. *Int. J. Biol. Macromol.* **7**, 203–213.

Suggestions for Further Reading

D. A. D. Parry and P. M. Steinert (1995) *Intermediate Filament Structure*, Springer- Verlag, Heidelberg.

R. D. Goldman and P. M. Steinert (1990) *Cellular and Molecular Biology of Intermediate Filaments* (eds.), Plenum Press, New York.

E. Fuchs and K. Weber (1994) Intermediate filaments: structure, dynamics, function and diseases. *Annu. Rev. Biochem.* **63**, 345–382.

CYTOKININS

DOMINIQUE VAN DER STRAETEN
MARC VAN MONTAGU

HISTORY

The first discovery that plants make compounds that stimulate cell division dates back to 1913, when Haberlandt (1) demonstrated that an unknown substance present in vascular tissues of various plants caused cork cambium formation and wound healing in cut potato tubers. These molecules were later called *cytokinins*, from the Greek "cytokinesis" (cell division). In the 1940s, van Overbeek et al. (2) showed that similar compounds were also present in the milky endosperm of immature coconuts. In 1954, Miller et al. (3) isolated a substance, formed by partial breakdown of herring sperm **DNA**, that, in synergy with **auxin**, stimulated cell division in tobacco pith cells very actively. The molecule was called *kinetin* (6-furfurylamino purine). *Trans*-zeatin (6-(4-hydroxy-3-methylbut-*trans*-2-enylamino) purine) was the first cytokinin to be identified from maize endosperm (4) (Fig. 1).

BIOSYNTHESIS AND METABOLISM

Natural cytokinins (CKs) are primarily purine derivatives with an N^6-substituted isopentenylated side chain (iP-type CKs) (5). However, a series of CKs with an aromatic side chain (eg, benzyl-adenine) have been demonstrated in certain species (6,7). The most abundant naturally occurring CKs are CK nucleosides and CK nucleotides (5). Certain phenolic compounds also possess cell-division-promoting activity, but are not classified as true cytokinins (8).

The origin of CKs in plants remains a subject of debate. Roots are the primary sites of cytokinin biosynthesis, although limited synthesis in the shoot has also been demonstrated (9,10). CKs have been proposed to be transported from root to shoot by the transpiration stream in xylem tissues, but conclusive evidence is lacking (11). Even though present throughout the plant, CKs have never been proven to be synthesized by endogenous, plant-borne **enzymes**. Because

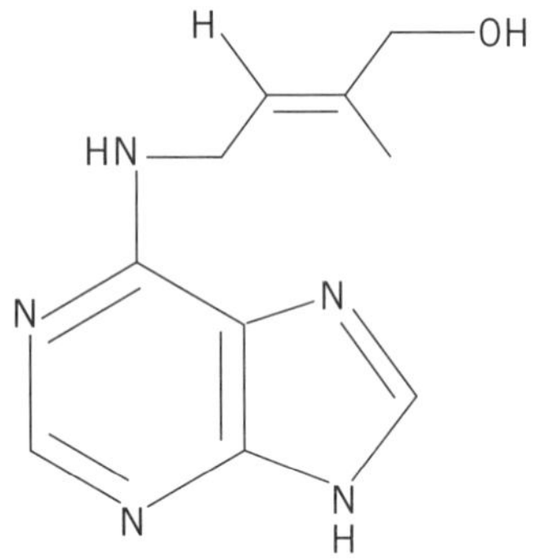

Figure 1. Structural formula of *trans*-zeatin.

many root-associated bacteria are cytokinin producers, it has been hypothesized that CKs are made exclusively by these microbial symbionts (12).

Cytokinins can be synthesized via a direct (*de novo*) or an indirect (**transfer RNA**) pathway. The main source of free CKs is the *de novo* pathway. The key step is the formation of N^6-(Δ^2-isopentenyl) adenosine-5′-monophosphate from Δ^2-isopentenyl pyrophosphate and AMP catalyzed by isopentenyltransferase. The isopentenyl side chain originates from mevalonate, which also serves as a precursor to **abscisic acid**, **gibberellins**, and **brassinosteroids**. Isopentenyltransferase was first characterized from bacteria and slime molds (5,7). Different constructs using the ***Agrobacterium*** isopentenyltransferase *IPT* (*tmr*) **gene** were used to generate constitutive or controlled cytokinin overproduction in **transgenic** plants (13–15). These studies mainly confirmed the classical view of cytokinin action in plants. Specific manipulation of **senescence** was achieved by placing the *IPT* gene under the control of a senescence-specific **promoter** (16). Recently, isopentenyltransferase activity was also demonstrated in plants (17), but the enzyme was never purified to homogeneity, nor have plant genes encoding isopentenyltransferase been **cloned**. This is the major reason why CK production by the plant itself is still questioned. An **Arabidopsis** mutant with sixfold elevated CK levels was isolated from a screen for seedlings with multiple cotyledons (18,19). The **phenotype** of this *amp1* (*altered meristem program1*) mutant is reminiscent of several CK effects, including enhanced lateral branching, elevated rate of leaf formation, induction of flowering and floral morphological abnormalities, and delayed senescence. In addition, *amp1* displays a de-etiolated phenotype in the dark and is an **allele** of *cop2* (*constitutive photomorphogenic 2*). It was proposed that CK induces light-regulated genes in a light-independent fashion (19). The *AMP1* gene product may be required for CK degradation, or it could be a negative regulator of CK biosynthesis. A second mutant *cri* (*cristal*) that accumulates sixfold more CK than the wild type was recently identified (20). *Cri1* mutants are phenotypically different from *amp1* and show symptoms of vitrification.

The second biosynthetic route for CKs is via tRNA. Modified bases in tRNA, including isopentenylated bases, have been found in virtually all organisms investigated (21). CK modification in tRNA is always at the adenosine residue immediately 3′ to the **anticodon** of tRNAs that read codons starting with U (5). These CK residues in tRNA may serve as a direct, albeit not major, precursor of free CK, as a result of tRNA turnover. The rate of CK release in this way is dependent on the metabolic activity of the cell. The major CK in tRNA is *cis*-zeatin, with very low CK activity, but it can be converted to the biologically active *trans* isomer (Fig. 1) by a *cis–trans* isomerase (22) (see ***Cis–trans isomerization***).

Several modifications can occur to the N^6-substituted side chain and purine ring. Formation of zeatin and its nucleoside and nucleotide derivatives, most probably by hydroxylation of the isopentenyl adenine moiety, is common in most plant tissues. The hydroxylase that mediates this conversion was isolated from cauliflower (23). ***O*-Glycosylation** of hydroxylated CKs is often observed, but leads to a considerable decrease in activity (5). CK-*O*-glucosides can be converted to free active CK by the action of β-glucosidases. In contrast, modification of the purine ring by *N*-glucosylation of the 3, 7, or 9 position yields inactive, stable derivatives that cannot generate free CKs. *N*-Glucosylation is thus expected to play an important role in regulating the levels of active CKs. An additional control is exerted by CK oxidase (24,25). This enzyme catalyzes degradation of the N^6-(Δ^2-isopentenyl) side chain and irreversibly leads to loss of CK activity. CK nucleotides are not accepted as a substrate. A general role for CK oxidase is to contribute to CK homeostasis in plants (26).

SIGNAL PERCEPTION AND TRANSDUCTION

A number of CK-binding proteins have been described, one of which is a potential CK **receptor** (27,28). The latter is a cytosolic protein that reversibly binds zeatin and is involved in transcriptional activation in the presence of zeatin, but not in its absence. Only the *trans* isomer elicited the observed effect.

Two candidate plasma membrane CK receptors were identified by mutational analysis (29,30). Mutants of *Arabidopsis* that exhibit shoot formation in tissue culture in the absence of CK were obtained by an activation-tagging approach. The *CKI1* (*cytokinin-independent 1*) gene was isolated and encodes a protein similar to two-component regulators of bacteria (29). As in the case of the ethylene receptor ETR1 (see **Ethylene**), CKI1 is a hybrid **kinase** with fused sensor and receiver domains. Based on this analogy and on the fact that constitutive overexpression of *CKI1* results in typical effects of CK action, CKI1 is hypothesized to encode a CK receptor. Overexpression would enable cells to sense endogenous concentrations below the threshold, which normally leads to CK response. Intriguingly, phytochromes also display similarities with two-component systems. It has been speculated that they might regulate other two-component systems, thus enabling cross-talk between the ethylene and CK pathways (31). A second candidate CK receptor is a seven-transmembrane domain receptor (7TM), characteristic in **G-protein** signaling and involved in transduction of extracellular signals (30). The *Arabidopsis GCR1* (*G-coupled receptor 1*) gene was isolated by polymerase chain reaction (**PCR**), and **antisense** mutagenesis with a 35S-a*GCR1* construct resulted in reduced cotyledon and leaf expansion and a single flowering stem, a phenotype reminiscent of the *cyr1* (*cytokinin resistant 1*) mutant (see text below). However, *GCR1* and *CYR1* do not represent the same gene (30).

The above findings confirm the involvement of both phosphorylation and G proteins in CK signaling (32). Sense and antisense expression in tobacco of a gene encoding a small GTP-binding protein from rice resulted in a dwarf phenotype and abnormal flower development, correlating with a sixfold elevation of CK levels compared to wild type (33). **Phosphorylation** plays an important role in **cell cycle** control by CK (32). In addition, evidence was given for involvement of intracellular Ca^{+} in CK signaling (see **Calcium signaling**) mostly from studies in mosses (32). Finally, the similarity of a CK-binding protein to *S*-adenosyl-homocysteine hydrolase raised the possibility that some CK effects might be mediated through control of **methylation** of DNA and/or proteins (34).

Further progress in understanding CK **signal transduction** is expected from mutant studies. Several CK-resistant mutants have been identified, mainly in *Arabidopsis* (35). Five loci have been studied in detail. The majority were isolated using a root elongation response in the presence of CK. In certain cases, resistance to other hormones was found. For instance, the *ckr1* (*cytokinin resistant 1*) mutant is allelic to *ein2* (*ethylene-insensitive 2*), indicating cross-talk with the ethylene pathway. It was demonstrated that the inhibition of root elongation by CK results from an induction of ethylene biosynthesis (36). This was exploited by screening for the absence of triple response in the dark, in the presence of kinetin (37). The *cin5* (*cytokinin-insensitive 5*) mutant affects the 1-aminocyclopropane-1-carboxylate synthase gene 5 in *Arabidopsis*. Disruption of the carboxyl terminus of the same isoform leads to ethylene overproduction, previously identified as an *ethylene overproducer* (*eto2*) mutant. A third class of CK–resistant mutants is *cyr1* (38). In contrast to *ckr1* and *cin5*, *cyr1* displays an abnormal shoot phenotype: cotyledons and leaves fail to expand, a limited number of leaves are formed, and a single infertile flower is made. No resistance to other hormones was observed; however, *cyr1* is hypersensitive to abscisic acid (38). *Cyr1* is probably allelic to *emf2* (*embryonic flower 2*). *EMF* gene products might function in the maintenance of vegetative growth (39). The *stp1* (*stunted plant 1*) mutant has a reduced sensitivity to CKs, both in assays for elongation and radial swelling of the root (40). The general morphology was not affected, but a reduction in growth rate was observed. The function of the *STP1* gene product remains to be elucidated. Finally, a recent study describes identification of three loci involved in control of cell division and plant development (41). The corresponding mutants were called *pasticcino* (*pas*) and showed hypertrophy of their apical parts when grown on CK-containing medium, a phenotype reminiscent of abnormal shoots regenerated on media with unbalanced auxin/CK ratio and strikingly similar to the fasciation disease caused by *Rhodococcus fascians* (41). The altered embryo, root, and leaf development result from uncoordinated cell divisions. *PAS* genes are hypothesized to play a role in CK signaling, because no differences in CK levels were found compared to wild type.

DOWNSTREAM TARGETS

Primary CK response genes have not yet been isolated, nor have CK responsive ***cis*-acting** elements or ***trans*-acting** factors been identified (42). A large number of genes that are either induced or repressed upon CK treatment have been described. Controls are exerted at a transcriptional or post-transcriptional level. Changes in gene expression were observed between 2 h and 48 h after treatment. Many of these secondary CK response genes are involved in light regulation and nutrition (32,42).

EFFECTS

Depending on temporal and spatial factors, a particular subset of downstream target genes is activated by CK treatment and results in one of many described effects. Cytokinins regulate many aspects of plant growth and development, including cell division, enlargement, and differentiation, as well as **chloroplast** development, release of apical dominance, nutrient mobilization, flowering, and delay of senescence (43,44).

BIBLIOGRAPHY

1. G. Haberlandt (1913) *Sitzungsber. K. Preuss. Akad. Wissensch.*, 318–345.
2. J. van Overbeek, M. E. Conklin, and A. F. Blakeslee (1941) *Science* **94**, 350–351.
3. C. O. Miller, F. Skoog, M. H. Von Saltza, and F. Strong (1955) *J. Am. Chem. Soc.* **77**, 1392–1393.
4. D. S. Letham (1963) *Life Sci.* **8**, 569–573.
5. C.-m. Chen (1997) *Physiol. Plant.* **101**, 665–673.
6. M. Strnad, W. Peters, E. Beck, and M. Kamínek (1992) *Plant Physiol.* **99**, 74–80.
7. E. Prinsen, M. Kamínek, and H. Van Onckelen (1997) *Plant Growth Regul.* **23**, 3–15.
8. R. A. Teutonico, M. W. Dudley, J. D. Orr, D. G. Lynn, and A. N. Binns (1991) *Plant Physiol.* **97**, 288–297.
9. A. Carmi, and J. Van Staden (1983) *Plant Physiol.* **73**, 76–78.
10. C.-M. Chen, J. R. Ertl, S. M. Leisner, and C.-C. Chang (1985) *Plant Physiol.* **78**, 510–513.
11. P. D. Hare, W. A. Cress, and J. van Staden (1997) *Plant Growth Regul.* **23**, 79–103.
12. M. A. Holland (1997) *Plant Physiol.* **115**, 865–868.
13. J. I. Medford, R. Horgan, Z. El-Sawi, and H. J. Klee (1989) *Plant Cell* **1**, 403–413.
14. T. Schmülling, S. Beinsberger, J. De Greef, J. Schell, H. A. Van Onckelen, and A. Spena (1989) *FEBS Lett.* **249**, 401–406.
15. J. J. Estruch, E. Prinsen, H. Van Onckelen, J. Schell, and A. Spena (1991) *Science* **254**, 1364–1367.
16. S. Gan and R. M. Amasino (1997) *Plant Physiol.* **113**, 313–319.
17. J. R. Blackwell and R. Horgan (1994) *Phytochemistry* **35**, 339–342.
18. A. M. Chaudhury, S. Letham, S. Craig, and E. S. Dennis (1993) *Plant J.* **4**, 907–916.
19. A. N. Chin-Atkins, S. Craig, C. H. Hocart, E. S. Dennis, and A. M. Chaudhury (1996) *Planta* **198**, 549–556.
20. V. Santoni, M. Delarue, M. Caboche, and C. Bellini (1997) *Planta* **202**, 62–69.
21. N. Murai (1981) In *Cytokinins: Chemistry, Activity and Function* (D. W. S. Mok and M. C. Mok, eds.), CRC Press, Boca Raton, FL, pp. 87–99.
22. N. V. Bassil, D. W. S. Mok, and M. C. Mok (1993) *Plant Physiol.* **102**, 867–872.
23. C.-m. Chen and S. M. Leisner (1984) *Plant Physiol.* **75**, 442–446.
24. D. J. Armstrong (1994) In *Cytokinins: Chemistry, Activity, and Function* (D. W. S. Mok and M. C. Mok, eds.), CRC Press, Boca Raton, FL, pp. 139–154.
25. R. J. Jones and B. M. N. Schreiber (1997) *Plant Growth Regul.* **23**, 123–134.
26. S. Eklöf, C. Astot, T. Moritz, J. Blackwell, O. Olsson, and G. Sandberg (1996) *Physiol. Plant.* **98**, 333–344.

27. C. Brinegar (1994) In *Cytokinins: Chemistry, Activity, and Function* (D. W. S. Mok and M. C. Mok, eds.), CRC Press, Boca Raton, FL, pp. 217–232.
28. O. N. Kulaeva, N. N. Karavaiko, S. Y. Selivankina, Y. V. Zemlyachenko, and S. V. Shipilova (1995) *FEBS Lett.* **366**, 26–28.
29. T. Kakimoto (1996) *Science* **274**, 982–985.
30. S. Plakidou-Dymock, D. Dymock, and R. Hooley (1998) *Curr. Biol.* **8**, 315–324.
31. J. W. Reed (1998) *Trends Plant Sci.* **3**, 43–44.
32. P. D. Hare, and J. van Staden (1997) *Plant Growth Regul.* **23**, 41–78.
33. H. Sano, S. Deo, E. Orudgev, S. Youssefian, K. Isizuka, and Y. Ohashi (1994) *Proc. Natl. Acad. Sci. USA* **91**, 10556–10560.
34. S. Mitsui, T. Wakasugi, and M. Sugiura (1996) *Plant Growth Regul.* **18**, 39–43.
35. J. Deikman (1997) *Plant Growth Regul.* **23**, 33–40.
36. A. J. Cary, W. Liu, and S. H. Howell (1995) *Plant Physiol.* **107**, 1075–1082.
37. J. P. Vogel, K. E. Woeste, A. Theologis, and J. J. Kieber (1998) *Proc. Natl. Acad. Sci. USA* **95**, 4766–4771.
38. J. Deikman and M. Ulrich (1995) *Planta* **195**, 440–449.
39. C.-H. Yang, L.-J. Chen, and Z. R. Sung (1995) *Dev. Biol.* **169**, 421–435.
40. T. I. Baskin, A. Cork, R. E. Williamson, and J. R. Gorst (1995) *Plant Physiol.* **107**, 233–242.
41. J.-D. Faure, P. Vittorioso, V. Santoni, V. Fraisier, E. Prinsen, I. Barlier, H. Van Onckelen, M. Caboche, and C. Bellini (1998) *Development* **125**, 909–918.
42. T. Schmülling, S. Schäfer, and G. Romanov (1997) *Physiol. Plant.* **100**, 505–519.
43. A. N. Binns (1994) *Annu. Rev. Plant Physiol. Plant Mol. Biol.* **45**, 173–196.
44. P. J. Davies (1995) *Plant Hormones: Physiology, Biochemistry and Molecular Biology*, Kluwer, Dordrecht, The Netherlands.

CYTOMEGALOVIRUS

AIKICHI IWAMOTO

Cytomegaloviruses (CMVs) are a distinct subgroup of **herpesvirus** and are classified as subfamily *Betaherpesvirinae* in the *Herpesviridae* family. CMV forms a characteristic nuclear inclusion, which gives an "owl's eye" appearance to infected cells. Many animal species have their own species-specific CMV.

Human CMV maintains a collinear genomic organization, but no significant nucleotide sequence **homology** with other animal CMVs. Human CMV is also designated as human herpes virus 5 (HHV-5). A great majority of the human population are infected with CMV, without any symptoms by adulthood. CMV is excreted in the urine of infants and intermittently in saliva of healthy seropositive adults. Human CMV may cause diseases such as congenital cytomegalic inclusion disease and infectious mononucleosis. The importance of human CMV as a pathogen has been increasingly correlated with the rising number of immunodeficient patients due to immunosuppressive therapy after transplantation, acquired immune deficiency syndrome (AIDS), and so on.

Mature virions of CMV are 150 to 200 nm in diameter and have an envelope composed of a lipid bilayer and viral glycoproteins. Inside the envelope, a tegment or matrix surrounds an icosahedral capsid, which contains a double-stranded, linear DNA **genome**. Two types of noninfectious particles, in addition to infectious virions, are produced in cells infected with human CMV. Dense bodies do not contain nucleocapsid or viral DNA and are essentially enveloped tegment protein. Noninfectious enveloped particles have a capsid but no electron-dense DNA core.

Equine CMV has a genome of 180 kbp, while the genomes of human, simian, and murine CMVs are 230 to 240 kbp. Human CMV is unique among animal CMVs in that it has a genome structure similar to herpes simplex virus 1 (HSV-1, class E structure). Two unique sequences (UL and US) are flanked by **inverted repeats**. Inversion of the UL and US segments mediated by inverted repeats can give rise to four different genomes. Other animal CMVs are less complicated in genome structure, and genome inversion is not known for them.

The genome of the AD169 strain of human CMV has been sequenced. Two hundred and eight open reading frames (ORFs) with a coding capacity of at least 100 amino acid residues were identified in the genome. Although only a few of these ORFs have been actually shown to encode proteins, at least 40 have been shown to be dispensable for replication. ORFs have been designated by their location with regard to the unique and repeated sequences (TRL, UL, IRL, IRS, and US) and sequential numbers. By analogy to HSV-1, roughly one-quarter of these ORFs are predicted to encode proteins required for **DNA replication** and metabolism functions, and three-quarters for virion maturation and virion structure.

Human CMV shows a highly restricted host range and takes 48 to 72 h to yield detectable amounts of progeny virions. Primary human differentiated cells, such as skin or lung fibroblasts, as well as chimpanzee cells, support human CMV replication, but undifferentiated, transformed, or **aneuploid** cells are usually nonpermissive for CMV replication. The genes of CMV have been classified into three sequential, kinetic classes: α(immediate early), β(delayed early) and γ(late) genes. Expression of the latter two relies on the α gene products. The β gene products mainly serve DNA replication and metabolism, while γ gene products encode structural proteins.

Suggestion for Further Reading

E. S. Mocarski Jr. (1996) Cytomegaloviruses and Their Replication. In *Fields Virology*, 3rd ed. (B. N. Fields et al., eds.), Lippincott-Raven, Philadelphia, pp. 2447–2492.

CYTOPLASMIC INHERITANCE

ENRIQUE CERDÁ-OLMEDO
JAVIER AVALOS

Most of the genetic information of a eukaryotic cell resides in its **nucleus**, and some is found in the cytoplasm. Each **mitochondrion** and **chloroplast** contains several copies of specific DNA molecules that carry some of the genetic information for the organelle; the rest is encoded in nuclear **genes**. Organelle genes are **transcribed** and **translated** in the organelles themselves through a **genetic code** that diverges in many cases from the standard one, while the **proteins** encoded in nuclear genes are imported into the organelle after their **protein biosynthesis** on cytoplasmic **ribosomes**. As a consequence of this shared genetic determination, there are many cases of nuclear and organelle mutants that exhibit similar phenotypes.

Most organelle **genomes** consist of circular double-stranded DNA molecules (Table 1). After considerable controversy, it seems clear that many mitochondrial DNA molecules are linear rather than circular.

Organelles may contain molecules of DNA other than their own **chromosomes**. Examples are the two linear plasmids of the mitochondria of *Zea mays* that result in the absence of functional pollen (androsterility); they are very useful for the commercial production of hybrid seed (see **Heterosis**). Another example is *kalilo*, a linear plasmid of *Neurospora crassa* that blocks mitochondrial functions when it integrates into the mitochondrial DNA; the result is **senescence**, the inability for indefinite vegetative growth.

Other cytoplasmic bodies with their own genetic information range from various self-replicating particles to **viruses** and complete endosymbiotic cells. Their genomes may consist of DNA or RNA. First to be discovered were the *kappa* particles of *Paramecium aurelia* (1), but many others are known, such as the *killer* particles in *Saccharomyces cerevisiae* and the σ particles that render *Drosophila melanogaster* specially sensitive to carbon dioxide.

HETEROPLASMONS AND MOSAICISM

A cell contains numerous copies of organelle DNA molecules, up to several thousand, and these need not be identical (heteroplasmon). Organelle genomes are notoriously unstable and suffer frequent point mutations and large rearrangements. Heteroplasmons can be produced by these spontaneous changes or by cell fusion. Contrary to the usual behavior of **heterozygotes**, the relative proportion of two variants in a heteroplasmon is not fixed, giving rise to mosaicism in cell clones.

Table 1. Some Genomes from Mitochondria and Chloroplasts, with Indication of Their Size, Number of Confirmed or Putative Translatable Genes (Open Reading Frames, ORFs), and number of genes for transfer RNA and ribosomal RNA[a]

Species	Number	Size (kb)	ORF	tRNA	rRNA
Mitochondrial Genomes					
Saccharomyces cerevisiae	L36885	78	13	24	4
Schizosaccharomyces pombe	X54421	19	10	25	2
Drosophila melanogaster	U37541	20	13	22	2
Caenorhabditis elegans	X54252	14	12	22	2
Homo sapiens	X93334	17	13	22	2
Chlamydomonas reinhardtii	U03843	16	8	3	14
Arabidopsis thaliana	Y08501, Y08502	367	81	19	3
Chloroplast Genomes					
Euglena gracilis	X70810	143	66	41	11
Nicotiana tabacum	Z00044	156	107	37	8
Oryza sativa	X15901	135	113	46	8

[a] Data taken from the indicated EMBL sequence accession numbers.

In some cases, one of the variants imposes itself because of its faster reproduction. Thus, most *petite* mutants of the yeast *Saccharomyces cerevisiae*, lacking aerobic metabolism, carry changes in their mitochondrial DNA. Some of these are called *suppressive*, because mutant mitochondria displace wild-type mitochondria in heteroplasmons. Wild-type mitochondria, on the other hand, impose their phenotype when mixed with mitochondria that carry petite mutations called *neutral*.

In other cases, the two variants may be found in various relative proportions, or even pure, in different cells, as would be expected from random **genetic drift**. Many variegated ornamental plants exhibit patches or streaks of different colors in their stalks and leaves, ranging from normal green to whitish. This phenotype may have many causes; in the plant *Mirabilis jalapa*, the colors depend on the presence of mutant chloroplasts, mixed with the normal ones in various proportions.

Recombination can occur between the various forms of chloroplast DNA or mitochondrial DNA in heteroplasmons, as shown for example for chloroplast markers of *Chlamydomonas reinhardtii* and mitochondrial markers in *Saccharomyces cerevisiae*.

HETEROKARYON TEST

The fusion of two cells that differ in both nuclear and cytoplasmic markers results in cells that are at the same time **heterokaryons** and heteroplasmons. Vegetative multiplication of the heterokaryon produces three kinds of cells: heterokaryons and two homokaryons, one for each of the kinds of nuclei. Cytoplasmic traits may segregate in various ways, all independent of the segregation of the nuclei. Homokaryotic segregants may show new combinations of cytoplasmic and nuclear markers. This test requires that the nuclei in the heterokaryon do not fuse, but remain separate, so that they can segregate to different daughter cells. The test has been extended to the yeast *Saccharomyces* by the use of dominant *kar* mutations that block nuclear fusion after mating, thus giving rise to heterokaryons.

RECIPROCAL CROSSES

The results of reciprocal crosses (see **Crosses**) offer a test for cytoplasmic inheritance when the two parents do not contribute equally to the cytoplasm of the zygote. The most frequent case in animals and plants is *matrilinear inheritance*, in which all individuals inherit from their mothers only, because of the larger size and contribution of female gametes. Examples are the variegation in *Mirabilis*, the first report of this heredity pattern (2), androsterility in *Zea*, and several human diseases due to mitochondrial mutations, such as Leber's hereditary optic neuropathy.

The matrilinear rule is not absolute. The σ particles of *Drosophila* are transmitted most often by the ova (**egg**), but sometimes by **sperm**. Chloroplasts may be provided by the male, as in the conifers, or by both parents, as in *Oenothera*. Mammals have powerful mechanisms to exclude the mitochondrial DNA of the sperm from the embryo, and the failure of these mechanisms might produce exceptions to the matrilinear rule.

Uniparental inheritance is observed with the *poky* mitochondrial mutants and the *kalilo* plasmid in the fungus *Neurospora crassa*. When two haploid strains of different **mating type**, *A* or *a*, are crossed, either can provide the large protoperithecia or the small conidia that fuse to produce the zygote. All offspring inherit the cytoplasmic markers from the protoperithecia donor, irrespective of its mating type.

Uniparental inheritance can occur even when the gametes are similar in size, as in *Chlamydomonas reinhardtii*. The zygote is formed by the fusion of seeming identical cells of the two mating types, called mt^+ and mt^-. Almost all the offspring inherit the chloroplast markers of the mt^+ parent only, because the DNA of the mt^- parent is usually destroyed (3).

There are cases of maternal inheritance that are not due to cytoplasmic genes (see **Maternal effects**).

BIBLIOGRAPHY

1. T. M. Sonneborn (1938) *Science* **88**, 503.
2. C. Correns (1909) *Z. Indukt. Abst. Vererbungsl.* **1**, 291–329.
3. R. Sanger (1954) *Proc. Natl. Acad. Sci. USA* **40**, 356–363.

Suggestions for Further Reading

G. J. Hermann and J. M. Shaw (1998) Mitochondrial dynamics in yeast. *Annu. Rev. Cell Dev. Biol.* **14**, 265–303.

R. N. Lightowlers, P. F. Chinnery, D. M. Turnbull, and N. Howell (1997) Mammalian mitochondrial genetics: heredity, heteroplasmy and disease. *Trends Genet.* **13**, 450–455.

N. G. Larsson and D. A. Clayton (1995) Molecular genetic aspects of human mitochondrial disorders. *Annu. Rev. Genet.* **29**, 151–178.

J. Nosek, L. Tomaska, H. Fukuhara, Y. Suyama, and L. Kovac (1998) Linear mitochondrial genomes: 30 years down the line. *Trends Genet.* **14**, 184–188.

M. Sugiura (1995) The chloroplast genome. *Essays Biochem.* **30**, 49–57.

CYTOSKELETON

DAVID PARRY

The cytoskeleton consists primarily of three types of filament: **microtubules** (about 25 nm in diameter), **actin**-containing **microfilaments** (initially thought to be 5 to 7 nm in diameter but now known to be about 10 nm in diameter), and **intermediate filaments** (about 10 nm in diameter). The latter were so named because they were intermediate in size between the long-established and well-characterized microtubules and microfilaments. Microtubules are filamentous structures often many microns in length that occur in the cytoplasm of eukaryotic cells. The actins are a highly conserved family of eukaryotic proteins, and actin filaments are perhaps best known for the highly specific interactions they make with **myosin** thick filaments that results in **muscle** contraction. Many actin-binding proteins play important roles in establishing the cytoskeleton *in vivo* and in allowing it to function appropriately. For example, fimbrin and α-**actinin** aggregate and cross-link microfilaments into bundles, severin severs actin filaments, β-actinin acts as an actin filament-capping protein, and ponticulin (a **transmembrane protein**) binds actin filaments and **spectrin** binds and links the microfilaments to the cell membrane. See **Actin**, **Microfilaments**, **Microtubules**, **Intermediate filaments**, **Cytokeratins**, and **Keratins** for further information on various components of the cytoskeleton.

CYTOTOXIC T LYMPHOCYTES

P. GOLSTEIN

The main interest to molecular biology of cytotoxic T lymphocytes is probably mechanistic. These cells are able to destroy other cells. How do they achieve this? The 40-odd year-old history of this field started with immunology and defense, and the cellular phenomenology that was possible at the time. It now culminates with the realization that the long-sought mechanisms of cytotoxicity convey in fact molecular signals of **cell death**. Thus, fundamental research on cytotoxic T cells now goes beyond immunology, becoming part of the booming research effort on **programmed cell death**.

It was shown in 1960 that white peripheral blood cells from alloimmune dogs were able to lyse *in vitro* target cells bearing the corresponding **alloantigens** (1). We now know that many different cell types have this ability to destroy other cells, in particular lymphoid cells, and among these T (i.e., *thymus-derived*) lymphocytes. There are two varieties: CD4+ and CD8+ lymphocytes. When activated, both varieties can be cytotoxic through a Fas-based mechanism, and in addition CD8+ cells can be cytotoxic via a perforin-granzyme-based mechanism. These two mechanisms will be considered in more detail below. A main role of cytotoxic T cells is to lyse syngeneic cells presenting surface determinants originating from **viruses** or other intracellular microorganisms. These, when within cells, are out of reach of other immune system effectors. Destruction of infected cells destroys the microorganism or exposes it to other effector mechanisms. Cytotoxic T cells thus fight intracellular pathogens. Interestingly, cytotoxic T cells can also destroy normal, uninfected cells, such as Fas-bearing activated lymphocytes, thus in part ensuring down-regulation of immune responses. Neither the perforin- nor the Fas-based mechanisms in isolation, nor both together, seem to account fully for graft rejection.

When an effector cell approaches a target cell, it "recognizes" it through an array of diverse molecules at the surface of the effector cell (TCR/CD3/peptide, CD4 or CD8, LFA-1, etc.) and of the target cell (Class I or Class II MHC, ICAM-1, etc.). How, then, does the cytotoxic cell destroy the target cell? First indications were obtained at a phenomenological level. Engagement of the above-mentioned molecules leads to effector cell activation, including an increase in the concentration of free cytoplasmic calcium (2) within a matter of minutes, immediately followed by reorientation within the effector cell of the secretory apparatus towards an area facing the target cell (3). This suggested the possible participation of secretory phenomena in at least one mechanism of cytotoxicity. Many cytotoxic T cells indeed contain peculiar cytoplasmic granules. Their content can be released by degranulation in the extracellular fluid on target cell recognition (see **Protein secretion**).

Among the phenomena then observed in the target cell is the early disintegration of nuclear DNA (4,5) into fragments of around 180 bp, and multiples thereof, corresponding to the size of **nucleosome**-shielded DNA, which is due to the action of one or several **endonucleases** activated early in the sequence of events leading to target cell death. Other events in this sequence include an influx of calcium, condensation of the cytoplasm, fragmentation of the cytoplasm and nucleus, and only very late secondary membrane disruption. This phenomenology of cell death is very similar, if not identical, for example in developmental circumstances. In all cases, the same program seems to govern the course of events in a cell that dies, including, as we now know, a similar cascade of molecular events and the same "apoptotic" morphological traits (6). What may be different is merely the signaling of this death program. The nature of the signals produced by cytotoxic cells has been largely unraveled by studies at the molecular level.

THE PERFORIN/GRANZYME B MECHANISM OF CYTOTOXICITY

A first approach to identify molecules involved in cytotoxicity has been to look for effector cell molecules themselves endowed with cytotoxic activity. This approach led to the detection (7,8), characterization, and **cloning** of *perforin*, a 60-kDa protein present in granules in the cytoplasm of many cytotoxic T and NK (natural killer) cells. At high concentrations, it can act on membranes as a calcium-dependent channel-former; most interestingly, it shows significant homology to the terminal components of the **complement system** cascade. A formal demonstration that perforin was involved in a mechanism of cytotoxicity was provided by several groups (9), all showing that cytotoxicity mediated by T cells was greatly impaired in mice made perforin-deficient through **gene targeting**.

Other molecules present in granules of cytotoxic T cells were identified by subtractive cloning. Prominent among molecules isolated this way were a number of **serine proteinases**, such as CTLA-1/CCP1/Granzyme B (10–12) and H Factor/CTLA-3/Granzyme A (10,12,13). Granzyme B was demonstrated through gene targeting to be required in a mechanism of cytotoxicity (14).

A current view on the cooperation between perforin and Granzyme B within the same mechanism of cytotoxicity is as follows. On recognition of the target cell, perforin- and granzyme B–containing granules in the effector cell reorient towards the target cell and are triggered to **exocytose**. Granzyme B may somehow enter the target cell and migrate into a target cell compartment where it would be innocuous, but from which it can be released by perforin at sublytic concentrations (15–17). Intracellularly released Granzyme B would then activate the standard programmed cell death cascade, through direct cleavage and therefore activation (18) of some of the ICE-family cysteine proteases called **caspases**. However, while target cell death mediated by Granzyme B (or at least by purified granules) may involve caspase activation for phenomena such as DNA fragmentation, it may, interestingly, use a noncaspase pathway for cell membrane disruption (19).

THE FAS-BASED MECHANISM OF CYTOTOXICITY

Through a **somatic cell** genetic approach it was found (20) that another mechanism of T cell–mediated cytotoxicity required a protein called Fas/APO-1/CD95 (21,22) at the target cell surface. This in turn led to the cloning of the Fas ligand that is expressed at the effector cell surface (23). Fas belongs to a family of **receptors** (including Fas, TNF-R1, DR3, DR4, DR5), most or all of which are very efficient at transducing a signal interpreted as a cell death signal, while Fas ligand belongs to a family of corresponding ligands (including the Fas ligand, the TNFs, TRAIL, etc.). Involvement of Fas in cytotoxicity may be a reflection *in vitro* of the main roles of the Fas system *in vivo*, such as down-regulation of the **immune response** (24), protection against the immune system (25,26), or potential major physiopathological effects (27).

Target cell Fas and effector cell Fas ligand define molecularly the Fas-based mechanism of cytotoxicity. In this mechanism, when an effector cell encounters a target cell, engagement of the **T cell receptor** of the effector cell by target cell **major histocompatibility complex** (MHC) leads to expression of the Fas ligand at the effector cell surface. Fas-ligand expression can be induced on CD8+ and on Th0 and Th1 CD4+ **T cells** (28–30) following specific recognition of **antigen** through the T cell receptor. This induction seems to involve the activation of **tyrosine kinases**, requires RNA and protein synthesis and the presence of calcium, and is inhibited by **cyclosporin** A (31–34). Other molecules may be involved, such as Myc and Max, 9-*cis* -**retinoic acid**, and its receptors, the **orphan receptor** for steroids, Nur77, and ALG-3.

Once expressed, effector cell Fas ligand engages target cell Fas, which leads to target cell death. In more detail, engagement of Fas leads in a matter of seconds to protein recruitment via the Fas "death domain," the cytoplasmic segment of Fas that is necessary and sufficient to transduce a death signal (35,36). First, FADD/MORT-1 (37,38) directly binds the death domain of Fas via its own *C*-terminal death domain. FADD/MORT-1 also includes an *N*-terminal death effector domain through which it associates with another molecule, FLICE/MACH (39,40). FLICE/MACH includes two death effector domains at its *N* terminus and, strikingly, a caspase-homology domain at its *C* terminus (39,40). Thus, the Fas-FADD-FLICE complex provides a remarkably direct link from a membrane signal to caspase activation, which is required for programmed cell death.

CONCLUDING REMARKS

The perforin-based and the Fas-based pathways account for most, and perhaps all, of T cell–mediated cytotoxicity (41–44), at least as assessed in a 4-hr assay *in vitro*. Other molecules, such as TNF-R1 or the TRAIL receptors, may play a role in longer assays. The Fas pathway is used by both $CD4^+$ (mostly Th1) and $CD8^+$ effector cells, while the latter also use the perforin/granzyme pathway. Thus, in $CD8^+$ T cells, two signals may stem from the T cell receptor/CD3 complex on antigen-specific recognition, one of them leading to granule exocytosis and the other one leading to **transcription** of the Fas ligand gene. Neither the exact nature of these distinct signals (45), nor whether they can be triggered simultaneously in a given cytotoxic cell is known as yet.

More generally, both mechanisms of T cell–mediated cytotoxicity seem to act by signaling the caspase activation step of the evolutionarily conserved programmed cell death cascade within the target cell. In this sense, the emergence in evolution of T cell–mediated cytotoxicity has not required the invention of new mechanisms of killing, but merely of new ways of signaling a preexisting programmed cell death cascade.

BIBLIOGRAPHY

1. A. Govaerts (1960) *J. Immunol.* **85**, 516–522.
2. M. Poenie, R. Y. Tsien, and A. -M. Schmitt-Verhulst (1987) *EMBO J.* **6**, 2223–2232.
3. A. Kupfer, G. Dennert, and S. J. Singer (1985) *J. Mol. Cell. Immunol.* **2**, 37–49.
4. J. H. Russell, V. Masakovski, T. Rucinsky, and G. Phillips (1982) *J. Immunol.* **128**, 2087–2094.
5. J. J. Cohen et al. (1985) *Adv. Exp. Med. Biol.* **184**, 493–508.
6. J. F. R. Kerr, A. H. Wyllie, and A. R. Currie (1972) *Br. J. Cancer* **26**, 239–257.
7. P. A. Henkart (1985) *Annu. Rev. Immunol.* **3**, 31–58.
8. E. R. Podack (1985) *Immunol. Today* **6**, 21–27.
9. D. Kägi et al. (1994) *Nature* **369**, 31–37.
10. J.-F. Brunet et al. (1986) *Nature* **322**, 268–271.
11. C. G. Lobe et al. (1986) *Science* **232**, 858–861.
12. D. Masson and J. Tschopp (1987) *Cell* **49**, 679–685.
13. H. K. Gershenfeld and I. L. Weissman (1986) *Science* **232**, 854–858.
14. J. W. Heusel et al. (1994) *Cell* **76**, 977–987.
15. C. J. Froelich et al. (1996) *J. Biol. Chem.* **271**, 29073–29079.
16. D. A. Jans et al. (1996) *J. Biol. Chem.* **271**, 30781–30789.
17. L. Shi et al. (1997) *J Exp. Med.* **185**, 855–866.
18. A. J. Darmon, D. W. Nicholson, and R. C. Bleackley (1995) *Nature* **377**, 446–448.
19. A. Sarin et al. (1997) *Immunity* **6**, 209–215.
20. E. Rouvier, M.-F. Luciani, and P. Golstein (1993) *J. exp. Med.* **177**, 195–200.
21. S. Yonehara, A. Ishii, and M. Yonehara (1989) *J. Exp. Med.* **169**, 1747–1756.
22. B. C. Trauth et al. (1989) *Science* **245**, 301–305.
23. T. Suda, T. Takahashi, P. Golstein, and S. Nagata (1993) *Cell* **75**, 1169–1178.
24. J. H. Russell, B. Rush, C. Weaver, and R. Wang (1993) *Proc. Nat. Acad. Sci. USA* **90**, 4409–4413.
25. D. Bellgrau et al. (1995) *Nature* **377**, 630–632.
26. T. S. Griffith et al. (1995) *Science* **270**, 1189–1192.
27. L. E. French and J. Tschopp (1997) *Nat. Med.* **3**, 387–388.
28. T. Suda et al. (1995) *J. Immunol.* **154**, 3806–3813.
29. F. Ramsdell et al. (1994) *Internat. Immunol.* **6**, 1545–1553.
30. L. L. Carter and R. W. Dutton (1995) *J. Immunol.* **155**, 1028–1031.
31. A. Anel et al. (1994) *Eur. J. Immunol.* **24**, 2469–2476.
32. A. Anel et al. (1995) *Eur. J. Immunol.* **25**, 3381–3387.
33. F. Vignaux et al. (1995) *J. Exp. Med.* **181**, 781–786.
34. M.-F. Luciani and P. Golstein (1994) *Roy. Soc. Phil. Trans. B* **345**, 303–309.
35. N. Itoh and S. Nagata (1993) *J. Biol. Chem.* **268**, 10932–10937.
36. M. P. Boldin et al. (1995) *J. Biol. Chem.* **270**, 387–391.
37. M. P. Boldin et al. (1995) *J. Biol. Chem.* **270**, 7795–7798.
38. A. M. Chinnaiyan, K. O'Rourke, M. Tewari, and V. M. Dixit (1995) *Cell* **81**, 505–512.
39. M. P. Boldin, T. M. Goncharov, Y. V. Goltsev, and D. Wallach (1996) *Cell* **85**, 803–815.
40. M. Muzio et al. (1996) *Cell* **85**, 817–827.
41. D. Kägi et al. (1994) *Science* **265**, 528–530.
42. B. Lowin, M. Hahne, C. Mattmann, and J. Tschopp (1994) *Nature* **370**, 650–652.
43. H. Kojima et al. (1994) *Immunity* **1**, 357–364.
44. C. M. Walsh et al. (1994) *Proc. Natl. Acad. Sci. USA* **91**, 10854–10858.
45. M. T. Esser, B. Krishnamurthy, and V. L. Braciale (1996) *J. Exp. Med.* **183**, 1697–1706.

Suggestions for Further Reading

G. Berke (1994). The binding and lysis of target cells by cytotoxic lymphocytes: Molecular and cellular aspects. *Annu. Rev. Immunol.* **12**, 735–773.

P. A. Henkart (1994). Lymphocyte-mediated cytotoxicity: two pathways and multiple effector molecules. *Immunity* **1**, 343–346.

S. Nagata and P. Golstein (1995). The Fas death factor. *Science* **267**, 1449–1456.

M. E. Peter, F. C. Kischkel, S. Hellbardt, A. M. Chinnaiyan, P. H. Krammer, and V. M. Dixit (1996). CD95(APO-1/Fas)-associating signalling proteins. *Cell Death Differ.* **3**, 161–170.